HUMAN PHYSIOLOGY

fourth edition

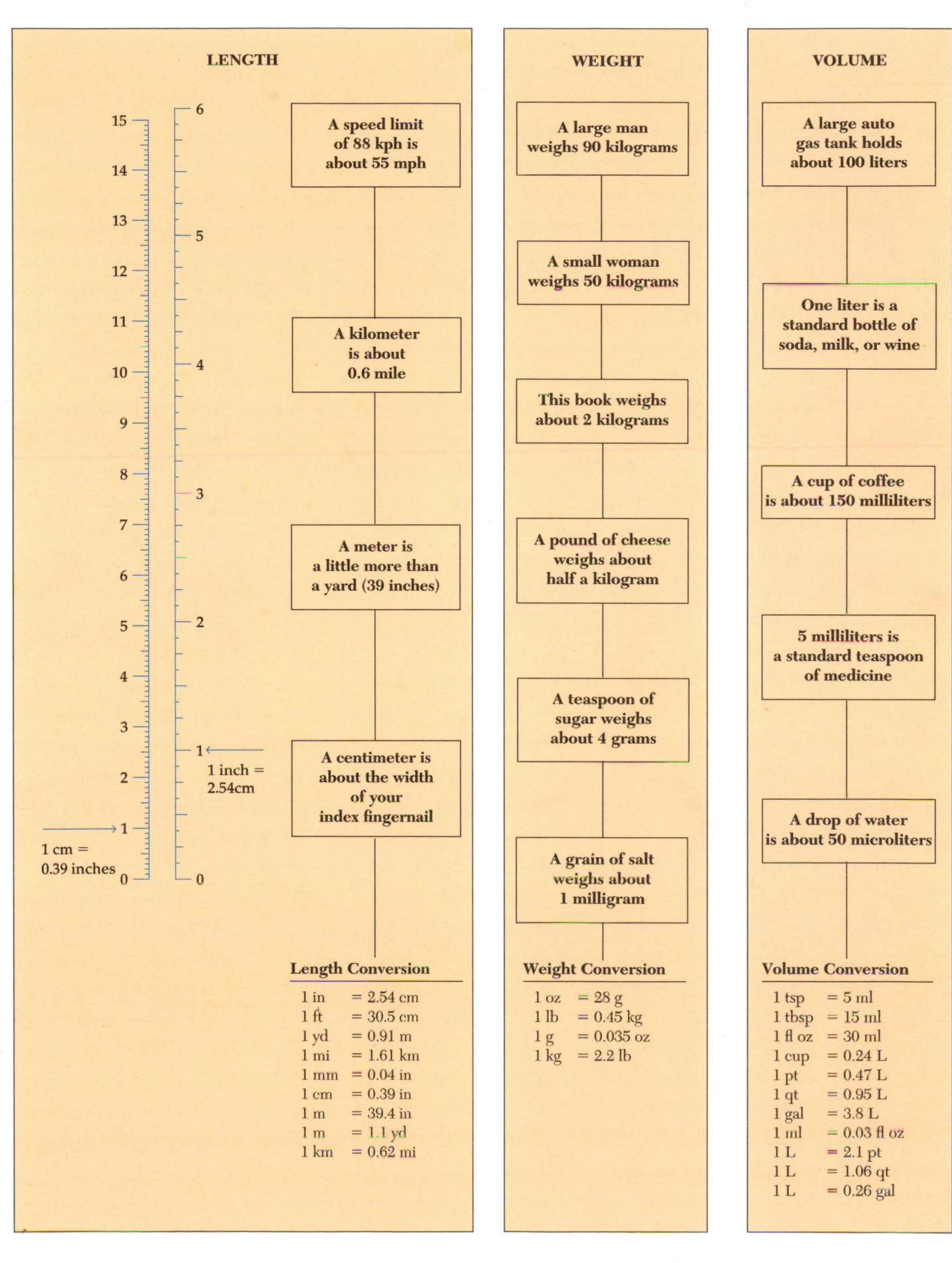

LENGTH

15
14
13
12
11
10
9
8
7
6
5
4
3
2
1
0

6
5
4
3
2
1
0

1 inch = 2.54cm

1 cm = 0.39 inches

A speed limit of 88 kph is about 55 mph

A kilometer is about 0.6 mile

A meter is a little more than a yard (39 inches)

A centimeter is about the width of your index fingernail

Length Conversion

1 in	= 2.54 cm
1 ft	= 30.5 cm
1 yd	= 0.91 m
1 mi	= 1.61 km
1 mm	= 0.04 in
1 cm	= 0.39 in
1 m	= 39.4 in
1 m	= 1.1 yd
1 km	= 0.62 mi

WEIGHT

A large man weighs 90 kilograms

A small woman weighs 50 kilograms

This book weighs about 2 kilograms

A pound of cheese weighs about half a kilogram

A teaspoon of sugar weighs about 4 grams

A grain of salt weighs about 1 milligram

Weight Conversion

1 oz	= 28 g
1 lb	= 0.45 kg
1 g	= 0.035 oz
1 kg	= 2.2 lb

VOLUME

A large auto gas tank holds about 100 liters

One liter is a standard bottle of soda, milk, or wine

A cup of coffee is about 150 milliliters

5 milliliters is a standard teaspoon of medicine

A drop of water is about 50 microliters

Volume Conversion

1 tsp	= 5 ml
1 tbsp	= 15 ml
1 fl oz	= 30 ml
1 cup	= 0.24 L
1 pt	= 0.47 L
1 qt	= 0.95 L
1 gal	= 3.8 L
1 ml	= 0.03 fl oz
1 L	= 2.1 pt
1 L	= 1.06 qt
1 L	= 0.26 gal

HUMAN PHYSIOLOGY

fourth edition

Rodney Rhoades, Ph.D.

Professor and Chair

Department of Cellular and Integrative Physiology
INDIANA UNIVERSITY SCHOOL OF MEDICINE

Richard Pflanzer, Ph.D.

Associate Professor

Departments of Cellular and Integrative Physiology and Biology
INDIANA UNIVERSITY SCHOOL OF MEDICINE
PURDUE UNIVERSITY SCHOOL OF SCIENCE

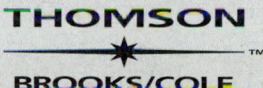

THOMSON

™

BROOKS/COLE

Australia Canada Mexico Singapore Spain United Kingdom United States

THOMSON
™
BROOKS/COLE

Biology Editor: Nedah Rose
Developmental Editor: Melanie Cann/Lee Marcott
Marketing Strategist: Ann Caven
Production Manager: Alicia Jackson
Project Editor: Dana Passek/UG / GGS Information Services, Inc.
Text Designer: Carol Bleistine/Kimberly Menning
Art Director: Carol Bleistine/Jacqueline LeFranc
Photo Researcher: Amy Dunleavy

Copy Editor: Carrie Bachman
Illustrator: Rolin Graphics
Cover Designer: Jacqueline LeFranc
Cover Printer: Lehigh Press
Compositor: UG / GGS Information Services, Inc.
Printer: Quebecor World — Versailles
Cover Image: Neuron/Synapse, Serotonin/SSRI
by Scott Thorn Barrows

For more information about our products, contact us at:
Thomson Learning Academic Resource Center
1-800-423-0563

For permission to use material from this text, contact us by:
Phone: 1-800-730-2214
Fax: 1-800-730-2215
Web: http://www.thomsonrights.com

Library of Congress Catalog Number: 2002102993
Human Physiology, fourth edition
ISBN: 0-03-032129-8

Asia
Thomson Learning
60 Albert Street, #15-01
Albert Complex
Singapore 189969

Australia
Nelson Thomson Learning
102 Dodds Street
South Melbourne, Victoria 3205
Australia

Canada
Nelson Thomson Learning
1120 Birchmount Road
Toronto, Ontario M1K 5G4
Canada

Europe/Middle East/Africa
Thomson Learning
Berkshire House
168-173 High Holborn
London WC1 V7AA
United Kingdom

Latin America
Thomson Learning
Seneca, 53
Colonia Polanco
11560 Mexico D.F.
Mexico

Spain
Paraninfo Thomson Learning
Calle/Magallanes, 25
28015 Madrid, Spain

Preface

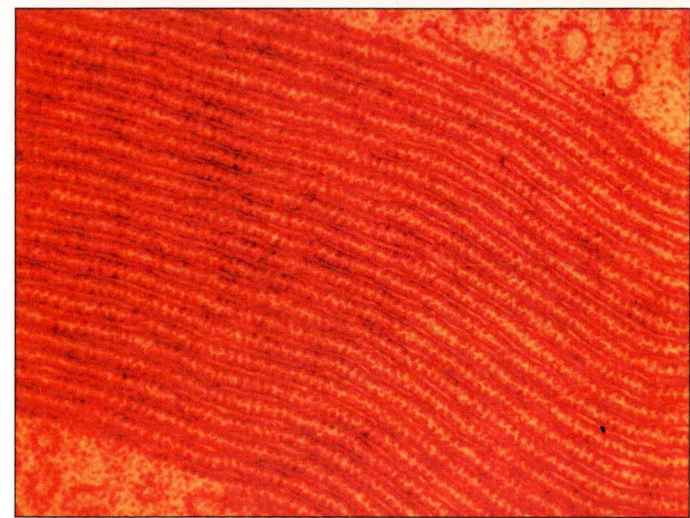

Physiology (from the Greek: *physio* = nature; *logy* = the study of) originally meant "an inquiry into nature" and included the systematic study of both living and nonliving matter. Today the focus of physiology is on the study of the integration of molecular and cellular mechanisms and how theses events are coordinated at the tissue and organ levels for normal function and survival of a living organism.

The human body is a complex organism comprised of several major systems — nervous, musculoskeletal, endocrine, circulatory, respiratory, gastrointestinal, and reproductive, to mention a few. The study of human physiology, therefore, is most logically presented by the "systems" approach. However, we believe it is important, especially for the beginner student of physiology, to recognize that the body does not function as a set of independent systems and that no single system is more important than another in maintaining life. Rather, it is the interactions and coordination of the multisystems to maintain normal functions that is essential to life.

Human physiology is a fascinating discipline. Most of us, professor and student alike, have a natural curiosity about how our bodies work. The recent explosion of scientific information has afforded the life and health sciences tremendous opportunities and tools for a better understanding of the world within us. As educators and authors, we face an important challenge in presenting to the students a vast amount of scientific information as a cohesive body of unifying principles and concepts, instead of as a catalogue of disconnected facts and mysteries. *Human Physiology*, Fourth Edition, develops and explains these underlying principles and concepts while acknowledging the excitement of unsolved problems.

The Fourth Edition, more than previous ones, presents important current questions in physiological and biomedical research and provides the background tools for approaching these problems with possible solutions. One of the biomedical research successes between the editions of *Human Physiology* was the approval of a genetically engineered drug, which is inhaled through an atomizer that cuts down the mucus development in patients with cystic fibrosis. Another is the discovery that mutations in a particular segment of DNA are associated with colon cancer as well as Huntington's disease, a finding that will improve detection and treatment of people with these diseases. Recently, thrombopoietin, which induces the body's production of platelets, was genetically engineered, an accomplishment especially important to patients who have low platelet counts and who are at a risk for uncontrolled bleeding, such as those with AIDS. The recent discovery of the gene for PKD (polycystic kidney disease) will lead to more informed decisions and an improved quality of life for inheritors of this disease.

The five years between editions have witnessed incredible changes and upheavals in the health care and biomedical research professions in movements promoting both streamlined budget consciousness and accelerated but improved quality. Never before has there been a greater need for well-trained, motivated, inquiring scientific thinkers.

STUDENT AUDIENCE

We have written this text for students who are planning careers in the life or health sciences. The Fourth Edition should be very useful to students of diverse backgrounds and levels of preparation. We have not presumed, for instance, that all students have had extensive experience with math, chemistry, physics, and biology. The background information on basic chemistry, physics, and molecular/cell biology in the first six chapters will help those students who have studied little or no college-level science. Students who have stronger backgrounds in the sciences will find the basic science coverage a useful review.

In addition, underlying principles of functional anatomy have been introduced where knowledge of basic structure is necessary for understanding human function. The emphasis of this text, however, is on architectural function rather than on basic anatomy. There are several reasons for writing this text, and our objectives include the following:

- To develop concepts and principles of physiology in a logical, clear, and concise manner
- To improve student ability to reason scientifically
- To promote understanding and application of basic concepts under normal and abnormal conditions
- To engender excitement about physiological research and other related career interests

The organization and approach of *Human Physiology*, Fourth Edition, have evolved through many years of teaching the subject to students of various backgrounds and abilities. Our research, along with that of many colleagues and other scientists, helped us a great deal in shaping the content of this text.

The field of physiology is burgeoning with new discovery and change. No single author, or even two authors, can possibly be the authority on the amount of basic information in a textbook on human physiology. An accurate and current presentation of the most important principles and research issues and details requires the collaborative enterprise of a team of research specialists. We are fortunate to have the skills and knowledge of nine other contributing authors. Please get to know them in the "About the Authors" section that follows the Preface.

We sincerely hope that this presentation of human physiology will capture the interest of student and professor alike and stimulate them to pursue the discipline beyond the limits of this book.

ORGANIZATION

The book is organized so that topics progress from the molecular level to integrated organ function. Chapter 1, designed to develop an appreciation of science, provides a historical perspective of human physiology as a science, and introduces basic concepts such as homeostasis. This chapter lays the foundation for Part I, which is basic cellular functions and includes the essential chemistry, biochemistry, physics, and mathematical principles required to enhance understanding of fundamental physiological concepts.

Part II focuses on the body's internal environment and analyzes the nature of biological control systems including the properties of specialized cells from the nervous and endocrine systems that regulate body function.

Part III discusses the coordinated functions of other body systems that all stem from the integration of specialized tissue function and includes chapters that present concepts dealing with integrated systems function, such as temperature regulation, exercise physiology, environmental physiology, and sexual physiology.

CONTENT FEATURES

- Chapter 1 provides an historical perspective of the science of physiology and a clear explanation of the scientific method. This helps to define the discipline and offers the student an exciting sense of direction for mastering the text.
- Early coverage of molecular biology and cell physiology provides a strong basis for understanding systemic functions. The concepts that are derived from cellular and molecular biology are applied throughout the text.
- Physiological control systems (Part II) are presented before organ functions (Part III) to establish the concept of homeostasis early.
- Metabolism and energy transformation are covered in Chapter 6 in order to present intermediary metabolism as a fundamental concept rather than as a process specific to nutrition.
- There are separate chapters on the autonomic nervous system, acid-base balance, exercise physiology, temperature regulation, calcium and phosphate metabolism, environmental physiology, body defense and the immune response, and sexual physiology. These chapters provide unique, in-depth coverage of these topics, which are often not found in competitive books.

NEW TO THE FOURTH EDITION

- Chapter 1 contains an expanded discussion of scientific method, control systems theory and homeostasis, and seven newly developed models of biological and physical concepts that recur in subsequent chapters. These models provide a framework for explaining different physiological mechanisms. They are designed to serve with homeostasis as a unifying theme for the Fourth Edition.
- Each chapter begins with an expanded list of key concepts that helps the student to focus attention on chapter content.
- Case histories have been added to each chapter as a means of providing an interesting, relevant, and thought-provoking introduction to chapter content. The case history concludes with questions that promote critical thinking skill development and relate chapter content to applied or clinical physiology. Answers to the case history questions appear at the end of the chapter.
- Focus questions have been incorporated into the chapter content to emphasize key concept areas and to direct attention to related narratives in the chapter section.
- Each chapter contains at least one new focus box adding important current information about *Current Concepts in Physiology* or *Applications of Physiology*. Each box concludes with a question that requires some critical thinking initiative. Answers to the questions are in an accompanying *Instructor's Manual with Test Bank*.
- Icons mark recurring concepts incorporated into the content of each chapter. For example, principles of Ohm's law applied to blood flow (Chapter 19) and to airflow (Chapter 20) are noted in each chapter by a *Mass and Heat Flow* icon.
- Chapter summaries have been revised and expanded so as to be more useful in reinforcing key concepts and bibliographies have been extensively revised to reflect updating of the Third Edition content and the addition of new material.
- The review questions at the end of each chapter have been condensed to 20 to 25 five-part multiple choice questions that serve as a short test regarding mastery of chapter subject material. A more extensive series of review questions, including a variety of question formats that develop critical thinking skills, are contained in an accompanying *Instructor's Manual with Test Bank*.
- A significantly revised art program, with 30% new art, presents concepts and processes in a clearer way.

LEARNING AIDS

Many pedagogical aids incorporated in the Second and Third editions have been retained and refined in the Fourth Edition to help students learn physiology. They are the following:

- Each chapter begins with a set of declarative key concept statements that help focus the student's attention on chapter content.
- Chapter content ends with a bulleted summary of key concepts that provide an abbreviated review of chapter content.
- Each chapter concludes with a set of review questions, a list of suggested readings, and key terms that are referenced to specific pages of first time usage in the text.
- The artwork and photographs emphasize and reinforce important physiological concepts. Numerous flow charts clarify complex physiological processes.
- Throughout the text, boldface type is used to emphasize important new terms.
- Most chapters contain focus boxes to encourage interest in applied physiology and physiological research. The focus boxes demonstrate relevance of the chapter content, help the student understand applied and/or abnormal physiology, and suggest directions future research may take.
- Principles of measurement, computation, and graphic analysis have been added to Appendix A and provide a quantitative approach to the study of physiology.
- Appendix A provides insight for the student in data and graphic interpretation. Also Appendix A provides background information for the student about current research techniques used in cell physiology.
- Appendix B lists normal physiological values for total body water, core temperature, and blood, cardiovascular, renal, respiratory, and acid-base measurements.
- Appendix C presents elements of medical terminology, which includes Latin and Greek prefixes and suffixes, their meaning and an example of usage, and a list of Latin and Greek terms for body parts for the purpose of building a medical vocabulary.
- A large glossary with phonetic pronunciations for many of the most difficult terms appears at the back of the text.
- The index has been greatly expanded and provides extensive cross-referencing.
- Front endpapers contain tables of Decimal Factors and Prefixes, Conversion Factors, and Equivalents for the Metric and English systems of measurement.
- The rear endpapers contain an alphabetized table of commonly used clinical abbreviations.

ANCILLARIES

A comprehensive teaching and learning package accompanies the text. These include:

- Study Guide: A tool for students including chapter summaries and key terms, practice quizzing, and suggestions for approach to the material.
- Instructor's Manual with Test Bank: A tool for instructors including lecture outlines and test questions/answers.
- ExamView®: Create, deliver, and customize tests and study guides—both print and online—in minutes with this easy-to-use assessment and tutorial system. ExamView® offers both a Quick Test Wizard and an Online Test Wizard. It will guide you step-by-step through the process of creating tests, while its "what you see is what you get" capability allows you to see the test you are creating on the screen exactly as it will print or display online. You can build tests of up to 250 questions using up to 12 question types. Using ExamView®'s complete word processing capabilities, you can enter an unlimited number of new questions or edit existing questions.
- Overhead Transparencies: Acetates of line art from the text for instructor use.
- Multimedia Manager: This digital library and presentation tool is available on one convenient multiplatform CD-ROM. With its easy-to-use interface, you can take advantage of Thomson·Brooks/Cole's already-created text-specific presentations, which consist of text art, photos, tables, and more in a variety of e-formats that are easily exported into presentation software or used on Web-based course support materials. You can even customize your own presentation by importing your personal lecture slides or other material you choose. The result is an interactive and fluid lecture that truly engages your students. Contact your Thomson·Brooks/Cole representative for ordering details.
- WebTutor on WebCT & Blackboard: A great study tool, a great course management tool, a great communication tool! For students, WebTutor offers real-time access to a full array of study tools, including flashcards with audio, practice quizzes, online tutorials, and Web links. Professors can use WebTutor to provide virtual office hours, post syllabi, set up threaded discussions, track student progress with quizzing material, and more. Contact your Thomson·Brooks/Cole representative for ordering information.
- Web site (student and instructor resources): Each site is chapter specific. The instructor resource site makes available much of the material to be found in the instructor manual (with the exception of testing) and other resources. The student resource site adds flashcards for studying key terms, practice quizzes and more.

For more information on these outstanding resources, please visit The Brooks/Cole Biology Resource Center at http://biology.brookscole.com

ACKNOWLEDGMENTS

We want to express our deepest thanks and appreciation to all of the contributing authors of each edition of this book. Without their expertise, talent, and effort this book would not have been possible. A very special thanks goes to Harold Modell for his development and application of the physiological models used throughout the book, and for his careful reviews and suggestions for pedagogical improvement of each chapter.

We would also like to give thanks for a job well done to the editorial staff and management of Saunders College Publishing for their help and guidance in significantly improving each edition of this book. Special thanks go to our Developmental Editor, Lee Marcott, whose patience and editorial talents were absolutely essential to the successful completion of the fourth edition of this book. We are indebted, as well, to Carol Bleistine, Art Director, Kimberlee Heldt and Rolin Graphics/Julie Martinez, Art Developmental Editors, Dana Passek, Project Editor, and Amy Ellis Dunleavy, Photo Researcher, for their advice and guidance during the production phase. A special thanks goes to Kimberlee Heldt and Julie Martinez for their dedication, patience, and artistic talent. Their work has greatly improved the book's art and facilitates the presentation as well as the understanding of important physiological concepts. Finally, we would like to thank Nedah Rose, our Executive Editor, for her support and commitment to this book and for sharing with us her administrative talent in managing the human and material resources of this very complex project.

We are indebted and offer a special thanks to Marlene Brown for her administrative assistance. Without her dedication and administrative talent, our work as authors and editors would have been much more difficult.

Last, we thank our families and friends for their patience and understanding of our need to devote a great deal of our personal time and energy to the development of this book. Our very warm gratitude goes to our wives, Diane Pflanzer and Judy Rhoades, without whose love, understanding, and support this project could not have been completed.

REVIEWERS

We express our sincere appreciation to the following faculty members and scientists for their conscientious reviews and assessments of the manuscript. Their suggestions, comments, and insights have been invaluable.

Reviewers of the First Edition:

Jerome Yochim, *University of Kansas*
Arnold J. Sillman, *University of California, Davis*
Rick Turnquist, *Augustana College*
Andy Anderson, *Utah State University*
Byron A. Schottelius, *University of Iowa*
John T. Fales, *University of Missouri, Kansas City*
John P. Harley, *Eastern Kentucky University*
L. Stephen Whitley, *Eastern Illinois University*
S. J. Coward, *University of Georgia*
David Noyes, *The Ohio State University*
Pegge Alciatore, *University of Southwest Louisiana*
Stephen Williams, *Glendale Community College*
Kenneth H. Bynum, *University of North Carolina, Chapel Hill*

Reviewers of the Second Edition:

Steve Simasko, *SUNY at Buffalo*
Jack Wood, *Western Michigan University*
Ralph Ferges, *Palomar College*
John Scheide, *Central Michigan University*
Mark Beidebach, *California State University, Long Beach*
Daniel Gibson, *Worcester Polytechnic Institute*
Lin Aanonsen, *MacAlester College*
Mary Lynne Stephanou, *Santa Monica College*
Sunny Boyd, *University of Notre Dame*
Pegge Alciatore, *University of Southwest Louisiana*
Audrey Mackey, *Austin Community College*
Carl Hammen, *University of Rhode Island*
John G. Moner, *University of Massachusetts, Amherst*
Katherine Mechlin, *Wright State University*
Fred D. Hinson, *Western Carolina University*
Sheldon Lustick, *Ohio State University*
Harry Bernheim, *Tufts University*
Lawrence Wit, *Auburn University*
Leon Goldstein, *Brown University*
Robert Slechta, *Boston University*
Barbara Howell, *SUNY at Buffalo*
Stuart Coward, *University of Georgia*
Pinapka Murthy, *Fayetteville State University*
Edmund Tong, *Wheaton College*
Leonard Lundquist, *Lander College*

Reviewers of the Third Edition:

Robert Hazelwood, *University of Houston*
John Moner, *University of Massachusetts, Amherst*
Sunny Boyd, *University of Notre Dame*
Margaret Gould, *Georgia State University*
Harold Falls, *Southwest Missouri State University*
Katherine Mechlin, *Wright State University*
Carl Hoegler, *Marymount College*
Jim Goldinger, *SUNY at Buffalo*
Frederick Hagerman, *Ohio University*
Steve Overmann, *Southeast Missouri State University*
George Veomett, *University of Nebraska at Lincoln*
Ron Wiley, *Miami University*
Kim Cooper, *Arizona State University*
Don Whitmore, *The Univeristy of Texas at Arlington*
John Hranitz, *University of Central Oklahoma*
Lewis Lutton, *Mercyhurst College*
James Junker, *Campbell University School of Pharmacy*
J.H.U. Brown, *University of Houston*
Thomas Adams, *Michigan State University*
Alan P. Brockway, *University of Colorado at Denver*
Jerry Stinner, *University of Akron*
Evelyn H. Schlenker, *University of South Dakota School of Medicine*
Andrew Pellett, *Louisiana State University Medical Center School of Allied Health*
Henry Kermott, *St. Olaf College*

Theresa Bacon-Bagley, *Grand Valley State University*
Stephen T. Bishoff, *University of South Carolina at Sumter*
Leonard Beuving, *Western Michigan University*
Lois Jane Heller, *University of Minnesota at Duluth School of Medicine*
Richard L. Hébert, *University of Ottawa*
Vicki Motz, *University of Lowell*
Kathleen M. Susman, *Vassar College*
Judy Van Liew, *SUNY at Buffalo*
Charles Costa, *Eastern Illinois University*
Hugo Lane, *Wake Forest University*
Bruce Grayson, *University of Miami*
James L. Blank, *Kent State University*
Daniel Lemons, *City College of CUNY*
Roger Theis, *University of Oklahoma Health Sciences Center*
Jay Farber, *University of Oklahoma Health Sciences Center*
S. S. Hull, *University of Oklahoma Health Sciences Center*
David Bruce, *Wheaton College*
Michael Lynes, *University of Connecticut at Storrs*
T. Earle Bowen, *University of Kentucky, A. B. Chandler Medical Center*
Ian Tizard, *The Texas Veterinary Medical Center of Texas A&M University*
Brenda Alston-Mills, *North Carolina State University*
Sandra D. Michael, *SUNY at Binghamton*
Michael Philips, *University of Florida College of Medicine*

Reviewers of the Fourth Edition:

Thomas Adams, *Michigan State University*
Brenda Alston-Mills, *North Carolina State University*
Roland Bagby, *University of Tennessee*
Delon Barfuss, *Georgia State University*
Marylynn Barkley, *University of California, Davis*
David Bell, *Fort Wayne Center for Medical Education*
Christina Benishin, *University of Alberta*
Mark Biedeback, *California State University*
Stephen T. Bishoff, *University of South Carolina*
William Blaker, *Furman University*
Sunny Boyd, *University of Notre Dame*
Michael J. Buono, *San Diego State University*
Robert G. Carroll, *East Carolina University*
Thomas T. Chen, *Santa Monica College*
Henry F. Cole, *Butler University*
John Cummings, *Waynesburg College*
Jean-Pierre Dujardin, *Ohio State University*
Richard Falvo, *Southern Illinois University*
Richard Friedman, *Indiana University School of Medicine*
Jack Goldberg, *University of California, Davis*
H. Maurice Goodman, *University of Massachusetts Medical School*
Joe R. Haeberle, *University of Vermont*
Richard Hahin, *Northern Illinois University*

Dr. Garth Hall, *University of Massachusetts, Lowell*
Dr. Carol Hayes, *St. Joseph's College*
Richard Heisey, *Michigan State University*
Lois Jane Heller, *University of Minnesota, Duluth*
Douglas N. Henry, *Michigan State University*
Richard L. Herbert, *University of Ottawa*
Patrick K. Hidy, *Central Texas College*
Christine N. Hinko, *University of Toledo*
Sarah Jerome, *University of Central Arkansas*
Robert Jinks, *Franklin & Marshall College*
J. Kelly Johnson, *University of Kansas*
Stephen A. Kempson, *Indiana University School of Medicine*
Thomas M. Linder, *University of Washington*
David Mailman, *University of Houston*
Michael Malachowski, *City College of San Francisco*
Bruce Martin, *Indiana University School of Medicine*
Kip L. McGillard, *Eastern Illinois University*
Katherine Mechlin, *Wright State University*
Richard A. Meiss, *Indiana University School of Medicine*
William C. Michel, *University of Utah*
Charles S. Nicoll, *University of California, Berkeley*
Karl Olson, *Michigan State University*
Mark Osadjan, *University of Colorado, Boulder*
Anthony T. Paganini, *Michigan State University*
Frederick M. Pavalko, *Indiana University School of Medicine*
Shawn D. Pearcy, *Wayne State College*
Daniel E. Peavy, *Indiana University School of Medicine*
Andrew Pellet, *Louisiana State University*
Linda N. Peterson, *University of Ottawa*
Michael Ian Phillips, *University of Florida*
David Quadagno, *Florida State University*
David S. Roane, *University of Louisiana, Monroe*
Sharon M. Russell, *University of California, Berkeley*
Evelyn Schlenker, *University of South Dakota*
Amanda Starnes, *Emory University*
Leeann Sticker, *Northwestern State University*
James Stockand, *University of Texas*
Robert B. Talliitsch, *Augustana College*
George A. Tanner, *Indiana University School of Medicine*
R. Brent Thomas, *University of South Carolina, Spartenburg*
Donna Van Wynsberghe, *University of Wisconsin, Milwaukee*
William Ventura, *Pace University*
Wiltz W. Wagner, *Indiana University School of Medicine*
Benjamin Walcott, *State University of New York, Stony Brook*
Cheryl Watson, *Central Connecticut State University*
Judy Williams, *Southeastern Oklahoma State University*

RODNEY A. RHOADES
RICHARD G. PFLANZER
March 6, 2002

About the Authors

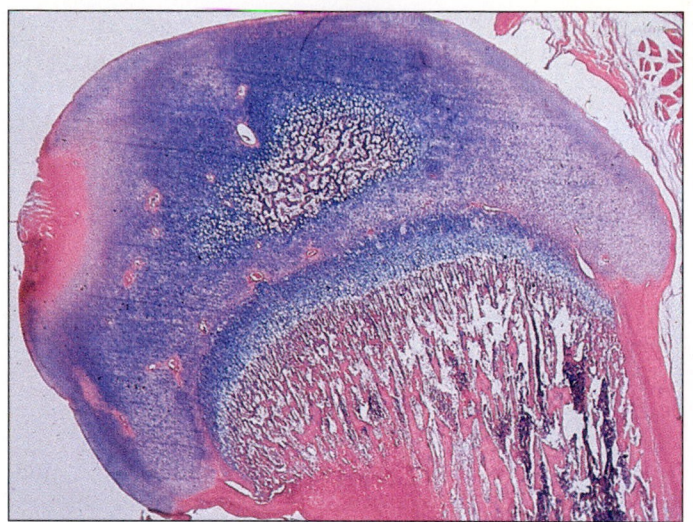

Rodney A. Rhoades

Professor and Chairman of Cellular and Integrative Physiology at Indiana University School of Medicine, Indianapolis. He received his Ph.D. in Physiology at The Ohio State University. Respiratory physiology is the focus of Dr. Rhoades' research. He is currently investigating hypoxia-induced signal transduction in smooth muscle growth and proliferation in the pulmonary arterial smooth muscle, and hypoxia-induced gene/protein expression in vascular smooth muscle.

Richard G. Pflanzer

Associate Professor of Biology at the Purdue University School of Science and Associate Professor of Cellular and Integrative Physiology at Indiana University School of Medicine. He received his Ph.D. in Physiology from Indiana University in 1969, with research in comparative cardiovascular physiology. Dr. Pflanzer has authored many articles and books related to teaching physiology at the undergraduate, graduate, and professional levels.

CONTRIBUTORS

David R. Bell

Associate Professor of Cellular and Integrative Physiology at Indiana University School of Medicine. He received his Ph.D. in Physiology and Biophysics from the University of Alabama–Birmingham in 1983. Dr. Bell's current research interests focus on the effects of estrogen and phytoestrogens on coronary vascular function.

Joe R. Haeberle

Professor of Molecular Physiology and Biophysics at the University of Vermont. He earned his Ph.D. in Physiology at Indiana University School of Medicine in 1981. Dr. Haeberle is currently investigating the molecular mechanism by which actin- and myosin-associated proteins regulate the contraction of vascular smooth muscle.

Stephen A. Kempson

Professor of Cellular and Integrative Physiology at Indiana University School of Medicine. He received his Ph.D. in Biochemistry from the University of London, U.K. His research work is focused on cellular regulation of membrane transport systems in the kidney.

Bruce J. Martin

Professor of Cellular and Integrative Physiology in the Medical Sciences Program at Indiana University in Bloomington. He earned his Ph.D. in Physiology at Indiana University in 1976. His research interests include delineating the pathophysiological factors that contribute to development of glaucoma.

Richard A. Meiss

Professor of OB/GYN and Cellular/Integrative Physiology at Indiana University School of Medicine. His Ph.D. degree was awarded by the University of Illinois of Urbana–Champaign in 1969. Dr. Meiss' current research focuses on the mechanical properties of smooth muscle and the interactions between muscle cells and tissues.

Fred M. Pavalko

Associate Professor of Cellular and Integrative Physiology at Indiana University School of Medicine, Indianapolis. He received his Ph.D. in Cell Biology at Florida State University in 1987. Dr. Pavalko's research is focused on the role of the cytoskeleton and adhesion molecules in the transduction of mechanical signals in cells.

Daniel E. Peavy

Associate Professor of Cellular and Integrative Physiology at Indiana University School of Medicine, Indianapolis. He received his Ph.D. in Physiology from the University of California at Davis in 1976. Dr. Peavy's research interests focus on insulin signal transduction and the biochemical effects in the pathway that lead to insulin resistance and diabetes.

George A. Tanner

Professor of Physiology and Biophysics at Indiana University. He earned his Ph.D. in Physiology from Harvard University and his current research interest is the pathophysiology of polycystic kidney disease.

Wiltz W. Wagner, Jr.

The V.K. Stoelting Professor of Anesthesiology and Pediatrics. He also holds a secondary appointment in Cellular and Integrative Physiology at Indiana University. He received his Ph.D. in physiology and biophysics from Colorado State University in 1974. Dr. Wagner's research is focused on pulmonary microcirculation, investigating how red and white cells cross the capillary bed and how the capillary bed responds to exercise and to exposure to high altitude.

Contents Overview

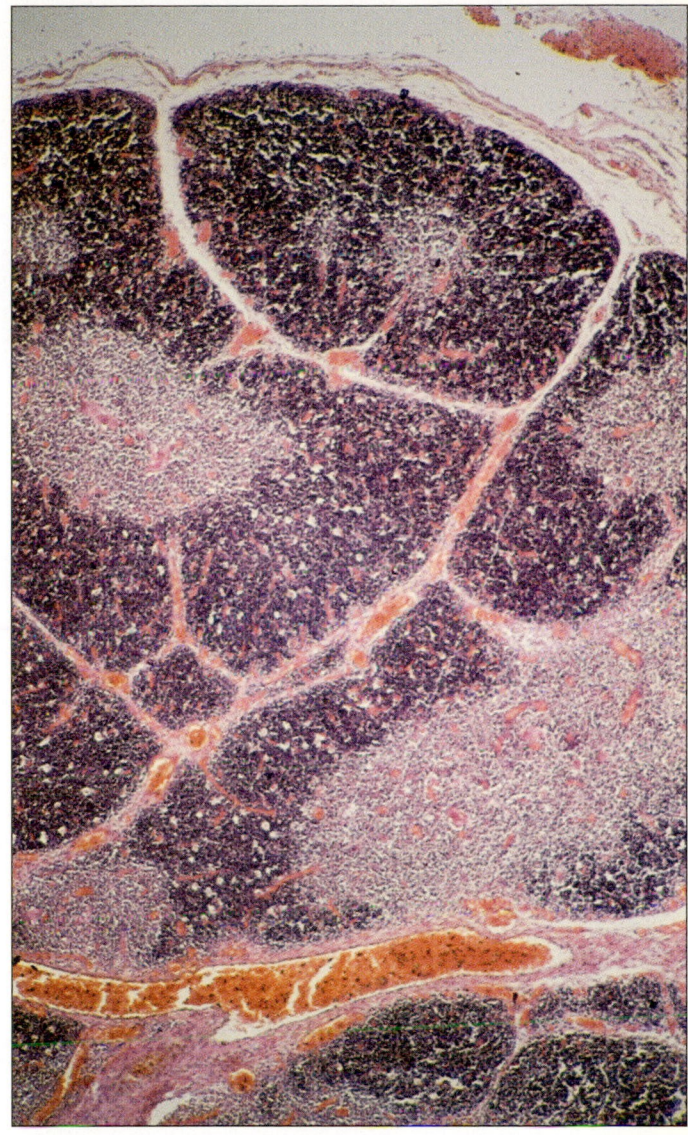

Contents

Applications of Physiology and Current Concepts in Physiology Boxes

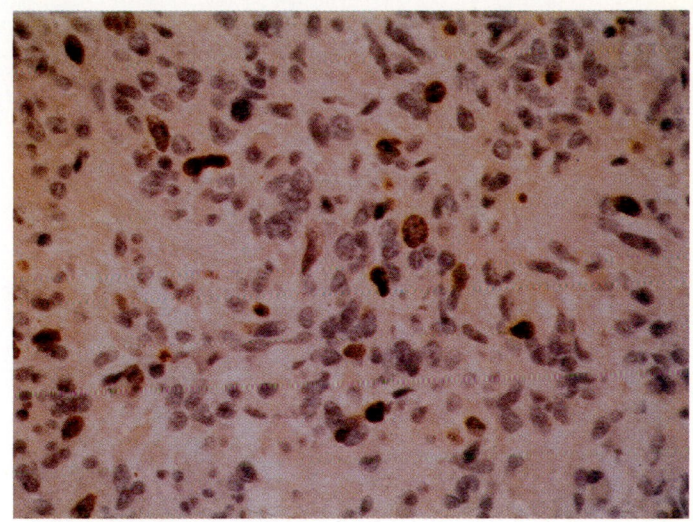

The Applications of Physiology and Current Concepts in Physiology Boxes are intended to apply basic concepts and to illustrate how diseases are characterized by altered functional changes.

HUMAN PHYSIOLOGY

fourth edition

Chapter 1

THE SCIENCE OF PHYSIOLOGY

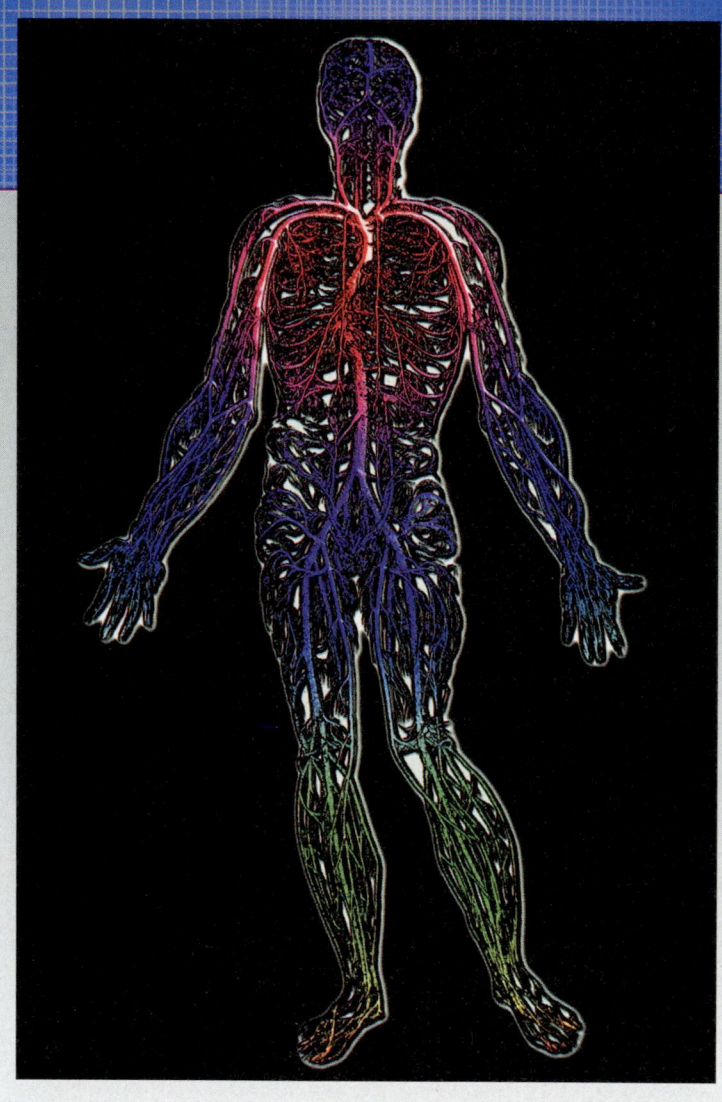

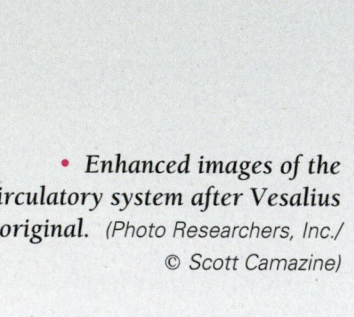

- *Enhanced images of the circulatory system after Vesalius original.* (Photo Researchers, Inc./ © Scott Camazine)

KEY CONCEPTS

- *Physiology is a branch of biology that deals with function and coordinated activities of cells, tissues, and organs.*

- *Discoveries made during the European Renaissance laid the foundation for modern advances in physiology.*

- *The scientific method involves formulation of a hypothesis, collection of data through observations or experiments, and testing of hypotheses.*

- *The body is organized into cells, tissues, organs, and coordinated systems.*

- *A constant internal environment is maintained for cellular and organ function.*

- *Models are used to understand physiology.*

CASE HISTORY

Two days after attending a post-game party, Larry, a 22-year-old college senior, went to the student health center to see if he could get medication to alleviate his flulike symptoms. He complained of nasal congestion and discharge, coughing and a sore throat, muscle aches and weakness, and chills. His skin was pale, and he said he felt cold, occasionally shivering despite having put on additional clothing. His sublingual (oral) temperature was taken and recorded as 40°C (normal sublingual temperature is 36–38°C). His recent history and the results of a preliminary physical examination suggested a diagnosis of upper respiratory tract infection, or common cold. Cultures of the nasal discharge and sputum, along with a complete blood count, ruled out more serious disease. He was advised to rest, drink plenty of fluids, use a vaporizer, and take over-the-counter medications, such as aspirin, for temporary relief.

Homeostasis is the maintenance of a relatively stable internal environment of the body so as to optimize function of body cells. Invasion of the body by pathogenic (disease-causing) bacteria or viruses, as in Larry's case, disturbs the body's internal environment and elicits defensive and protective measures that minimize the effects of the disturbance and help return stability to the internal environment. One such protective response is the regulated elevation of core body temperature, known as *fever*. A fever helps limit harm by killing bacteria or adversely affecting their growth, by increasing the release of enzymes that destroy infected cells, and by increasing the release of chemicals used for communication and defense by cells of the body's immune system.

Core body temperature is precisely regulated by a neuroendocrine control system containing a thermostat that balances heat gain and heat loss so as to maintain a relatively stable and desired internal temperature close to 37.0°C (98.6°F). The desired internal temperature is called the *set point* of the control system. If core body temperature rises above the set point, the control system increases the rate of heat loss compared with the rate of heat gain. Dilation of peripheral blood vessels and sweating help the body lose excess heat. If the core body temperature falls below the set point, the control system increases the rate of heat gain compared with the rate of heat loss. Constriction of peripheral blood vessels and shivering help the body gain heat. Fever occurs when an agent called a *pyrogen* elevates the set point of the body's thermostat, increasing heat gain until the new, higher core body temperature is reached and maintained. Near the end of the course of a fever, at a stage called the *crisis*, the fever is broken when the set point is lowered. The rate of heat loss is increased until a normal core body temperature is reached and maintained.

Concepts of homeostasis and physiological control systems that serve to maintain stability of the body's internal environment recur throughout the chapters of this book. They are central to both the understanding and the application of basic human physiology.

Questions

1. Antipyretic medicine, such as aspirin, opposes the influence of a pyrogen on the body's thermostat. How do antipyretics reduce fever?

2. What is the principal disadvantage of using an antipyretic to relieve the discomfort of a fever?

3. Why are chills associated with the onset of a fever?

INTRODUCTION

The ancient Greeks considered the universe to be composed of four elemental substances: air, water, earth, and fire. The Greeks further believed that each human being was a miniature universe in whom these substances appeared in the form of four "humors," or fluids: *blood* (corresponding to air); *phlegm* (representing water), *black bile* (corresponding to earth), and *choler* or *yellow bile* (representing fire; Fig. 1–1).

According to the Greek view, the fluids were not present in each person in equal amounts; one fluid usually predominated and characterized the individual's temperament. For example, one temperament was the "sanguine" type. We still use this word today to describe someone with a cheerful, confident personality. In ancient times, such a personality would have been attributed to a predominance of blood over the other three humors within that person's body, as evidenced perhaps by a ruddy complexion (*sanguine* comes from the Latin *sanguineus*, meaning "bloody"). A "melancholy" person was thought to have more black bile than any of the other humors, whereas a "choleric" person was considered to have a predominance of yellow bile (the humor associated with fire), which produced an excitable, easily angered temperament. A "phlegmatic" person generally was calm and unemotional, even sluggish, supposedly because of a pre-

dominance of phlegm, the cold, waterlike humor. The Greeks thought that the particular quantities of each humor and the resulting proportion, or balance, determined a person's physical and psychological makeup.

According to this classical scheme, while a normal predominance of one fluid within each person's unique humoral balance produced a characteristic personality, an abnormal lack or excess of one or more humors resulted in disease. Doctors from ancient times until well into the Renaissance shaped their therapeutic methods around attempts to restore the proper balance of humors within the sick person's body. Purging, bloodletting, and similar treatments were all intended to adjust humors.

In contrast to the ancient Greek philosophers and physiologists, today we organize the universe of matter into 105, rather than 4, elemental substances, and we attribute each person's innate uniqueness to genetic makeup rather than to relative quantities of bodily fluids. Yet despite the revolutions in our conceptions of human **physiology,** some classical principles influence modern science. One enduring principle is the idea that the human body, if not a world in miniature, is composed of elements from the universe that obey the laws of nature. Another "modern" idea cherished by the ancient philosophers is that of a harmonious balance within the body. While today we speak of homeostasis and recognize it as a self-regulatory process involving not only fluids but also dynamic processes, we share the classical view that health is the evidence of balance among body functions and substances. Finally, we have inherited the Renaissance attitude — which echoed the ancient Greek belief — that the human body is worthy of the most rigorous study and eloquent description.

THE HISTORY OF PHYSIOLOGY

The story of the development of physiological understanding in modern culture is fascinating. The following brief survey of the past 2000 years of physiological theory and research provides a context for the study of twentieth-century physiology in all its complexity and sophistication. Furthermore, the discussion communicates a sense of wonder at the vast advances physiologists have achieved since human beings first began to record observations and theories about the workings of the human body.

Physiological Knowledge in the Classical and Medieval Periods Was Steeped in Dogma

Because so many Western attitudes about science, philosophy, and the place of human beings in the universe have been influenced by the traditions of the classical Greeks, that period seems a logical starting point to examine the development of physiology as we understand it today. The studies of the Greek philosophers and the investigations of the Greek physiologists at the great museum in Alexandria, a famous center of Greek culture and learning in what is now Egypt,

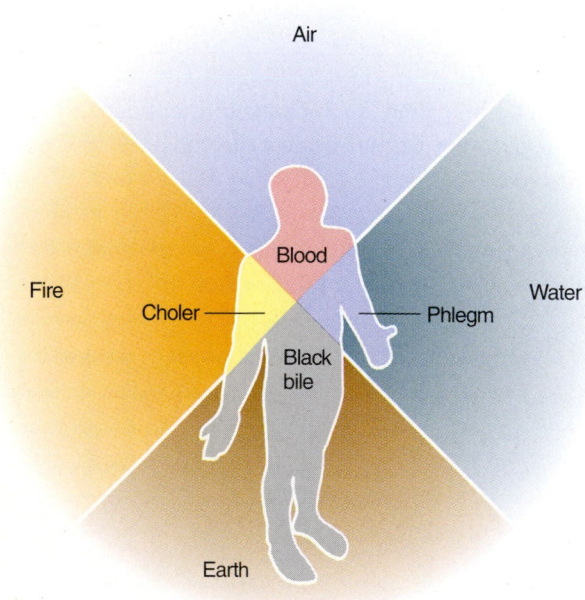

Figure 1–1

The four bodily "humors" of classical and medieval philosophy and physiology. The Greeks thought that the elements of the universe were represented in each human body by blood, phlegm, black bile, and choler. According to this scheme, a predominance of one humor produced each person's temperament.

influenced European medical and physiological practice until well into the Renaissance because of the mediating influence of Galen. Galen's writings preserved the ancient studies and, in turn, were preserved through the medieval period for examination by Renaissance scholars.

The Observations of Aristotle

Although he is most often thought of as a philosopher, Aristotle (384–322 BC) did important work in biology and was one of the first to describe the blood vessels as a system with the heart at the center. Unlike other investigators of his time, he used a bloodless method of killing animals for dissection, finding that the vessels were easier to see and trace when the blood remained.

Some of Aristotle's concepts were corrected through the discoveries of later investigators. Aristotle thought of the heart as both the seat of the intellect and a furnace that heated the blood to provide needed warmth. Because warmth disappeared from the body so soon after death, it was thought to be the cause or source of life. Just as the blood needed a source of heat, so too the furnace needed ventilation. According to Aristotle, this was the purpose of breathing; the lungs were a ventilation system and the air, a cooling agent.

Aristotle correctly observed that blood flowed between arteries and veins, but he could not see the microscopic capillaries that we now know provide the connection between them. He explained the flow by saying, in essence, that the vessels were themselves made of blood and that they disappeared when blood flow stopped.

> *... the channels of the blood vessels may be compared to the mud that a running stream deposits, they are as it were deposits left by the current of blood in the blood vessels. Thus, just as in the irrigation system the biggest channels persist whereas the smallest ones quickly get obliterated by the mud, though when the mud abates they reappear; so in the body the largest blood vessels persist while the smaller ones become flesh in actuality, though potentially they are blood vessels as ever before.*

The Theories of Herophilus and Erasistratus

At the museum at Alexandria, two men pursued a number of biological investigations in a continuation of Aristotle's search for an understanding of blood flow. Physiological research flourished at Alexandria partly because it was the only place that permitted dissection of human cadavers for the study of anatomy and physiology.

Herophilus (c. 335–280 BC), who is considered by some to be the first physiologist in the Western tradition, identified the brain as the seat of the intellect and demonstrated that the walls of the arteries are thicker than those of the veins. Using the newly invented water clock, he measured the pulse and showed that it varied when disease was present.

Erasistratus (310–240 BC) began his physiological studies as an assistant to Herophilus. Erasistratus believed that blood was made in the liver from food and was delivered to the or-

gans of the body by an "ebb and flow" in the veins. He thought of the arteries as air vessels, not blood vessels. The air, called *pneuma,* was thought of as a living force taken in through the lungs to the left side of the heart, where it was transformed into a "vital spirit" and then transported, in "airy" form, through the arteries to the rest of the body. (To explain why blood, not air, flowed from a cut artery, in apparent contradiction with his theory, Erasistratus postulated the existence of tiny connections between veins and arteries that sometimes opened to let blood flow into the arteries.)

The "Pneumatology" of Galen

In a simple experiment with a goose feather, Galen (c. 130–201 AD) demonstrated that blood, not air, flowed through the arteries. Galen created a hollow tube by cutting off the ends of the feather. Then he inserted the tube into a tied-off artery, from which blood immediately flowed into the tube. Because, by being tied off, the artery was supposedly separated from the veins, the blood that flowed from it had to have been inside the artery from the beginning.

Galen was a Greek who studied physiology both in Greece and at Alexandria. His practice as a surgeon to gladiators in his native city of Pergamum allowed him to observe the internal structure of the body at a time when dissections for the purpose of study were forbidden. Later, he was appointed physician to a Roman emperor. Galen is known for his voluminous writings about philosophy, medicine, and physiology, in which he commented and expanded on the work of the investigators who preceded him. Even though many of his ideas have been disproved, we still consider Galen an important figure because of the intellectual sophistication of his physiological schemes, the breadth and scope of his explanations, and the length of time and extent to which his teachings were accepted. His was the prevailing view of human physiology from his own time until the Renaissance.

Like Erasistratus, Galen believed that the blood was produced from food, in the liver (Fig. 1–2). He thought that the blood took on "natural spirits" and carried them through the veins to the bodily organs, which needed the spirits to perform different functions. After its supply of spirits was depleted, the blood returned along the same venous pathways to the liver to be resupplied. (This idea that the blood flowed both ways in the vessels became one of the most revered of Galen's teachings. When Renaissance anatomists correctly challenged it because of their knowledge of valves, they found it very difficult to overthrow.)

According to Galen, some of the blood containing the "natural spirits" went first to the right side of the heart and then to the left side, where it contacted *pneuma. Pneuma* was a substance produced from air in the lungs and carried into the left side of the heart. When the "natural spirits" contacted *pneuma,* they were transformed into "vital spirits," a higher form of *pneuma.* These "vital spirits" were carried on up to the brain, where they were further transformed into yet a higher form of *pneuma* called "animal spirits." These vari-

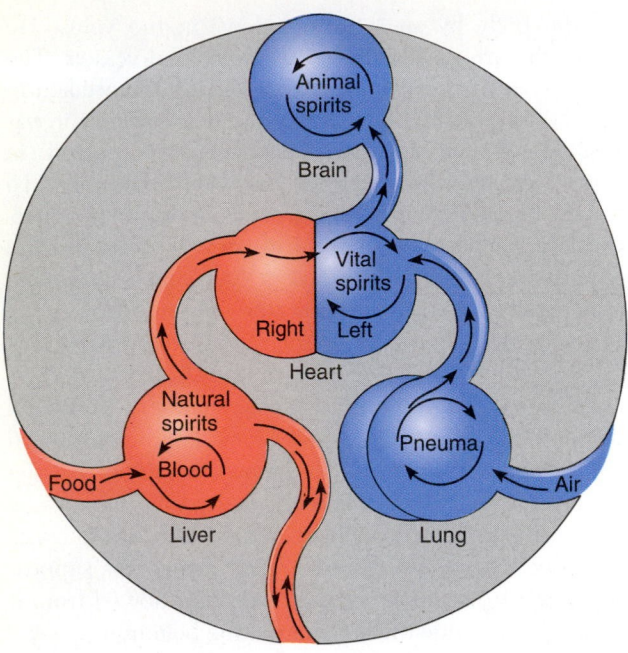

Figure 1–2

In the "pneumatology" of Galen, blood was made in the liver from food and carried "natural spirits" to the bodily organs, traveling both ways in blood vessels. *Pneuma* made from air in the lungs met blood in the heart and was transformed into "vital spirits" and then "animal spirits."

ous forms of *pneuma* drove the various processes of the body, according to Galen.

Despite his insistence on the importance of observation in the study of physiology, Galen often made incorrect assumptions about human anatomy based on his dissections of animals. As mentioned previously, however, the logic and elegance of his theories about bodily functions, which unified both the traditions he inherited and the knowledge available to his contemporaries, were convincing. His copious writings further established his authority, and throughout the late classical and medieval periods, Galen's physiology was accepted. It prevailed largely unquestioned until the technological advances and humanistic spirit of the Renaissance permitted a fresh look at the body.

The Discoveries of the European Renaissance Period Revolutionized Physiology

In the revival of classical learning that characterized the European Renaissance, Galen's writings about anatomy, physiology, and medicine first began to be widely translated from the Greek and Arabic into Latin. Although the writings were known to exist, and the general outlines of Galen's physiology had dominated medieval and early Renaissance medicine, only in the sixteenth century did the most important writings become available for close scrutiny by European physicians. By this time, the Galenic physiology had already taken hold of the imagination and had come to be regarded almost as religious doctrine. Thus, Renaissance professors of medicine taught the anatomy of Galen, and anyone who departed from Galen's concepts was considered a secular heretic. To explain the discrepancies between the anatomy described by Galen and that visible in a dissected corpse, a professor might claim that the body simply had changed since Galen studied it.

The task of the Renaissance physiologists was thus to study Galen's physiology, discover its errors, and replace it with a modern scheme. Many independent-minded investigators contributed to the revolution in physiological knowledge that occurred in the late Renaissance, although they too defaulted to many of Galen's tenets whenever they could not observe or explain for themselves certain structures and functions. In his teachings, writings, and drawings, Andreas Vesalius corrected many of the inaccuracies of Galen's anatomy while perpetuating others and retaining much of Galen's erroneous physiology. English physician William Harvey (1578–1657), in his discovery of the circulation, finally revised Galen's ancient notion that the air and the blood met in the heart, although he still viewed the lungs, as had Galen, merely as cooling organs.

The Drawings of Vesalius

Andreas Vesalius (1514–1564) was educated as a doctor and, at the age of 23, became a professor of surgery and anatomy at the great medical school in Padua, in what is now Italy. He had been taught the physiology of Galen and, in turn, began teaching it to his students. Unlike other professors of his time, however, who taught almost exclusively from Galen's writings, Vesalius centered his lectures not on the ancient texts but on human cadavers that he himself dissected as he lectured. (Previous instructors might not have used a human cadaver or might have lectured from afar while another person conducted the dissection.)

Another pedagogical device that Vesalius introduced into the study of physiology was the anatomical drawing. Galen's texts had been published with few or no illustrations; Vesalius created six anatomical "tables" that showed labeled parts of the body and allowed students to follow his lectures.

Despite his new approach to the teaching of anatomy, Vesalius considered himself a part of the tradition of Galen and initially intended to simply explicate Galen's teachings. Hence, in the beginning, he unwittingly reproduced some of Galen's erroneous assumptions about human anatomy. For example, Galen had thought that an organ called a "rete mirabile," similar to one he saw in hoofed animals, existed at the base of the human brain. This organ was supposedly where "vital spirits" were changed into "animal spirits." Vesalius showed this organ in his anatomical drawings, although, like Galen, he had never seen one in a human body.

Gradually, however, Vesalius began to realize that Galen's physiological dogma had been based not on an earlier version of the human body but on the bodies of apes, pigs, and other animals. He decided to write a new anatomy text that would reflect his own personal observations of a human body. The result was *De Humanis Corporis Fabrica* ("The Structure of the Human Body"), often simply called the *Fabrica* and now considered the first modern anatomy textbook. Published in 1543, the text contained plentiful illustrations and corrected many of Galen's errors.

Despite Vesalius's insistence on verification of anatomical details and his intention to correct Galen's theories of anatomy, he perpetuated some of Galen's mistaken ideas. One ancient notion that reappeared in Vesalius's text was that the blood passed directly from the right to the left ventricle of the heart. Vesalius did not recognize what we now know as the *pulmonary circulation* (the movement of blood between the lungs and the heart) despite evidence of this movement that was accumulating in his time. He claimed, as had Galen, that air traveled from the lungs to the left ventricle of the heart, where it cooled the heart "furnace" and endowed the blood in the heart with "vital spirits."

The Theories of William Harvey

The picture of the blood flow handed down to the early Renaissance physiologists showed the heart as both a furnace and a meeting place for blood and air. We know now that the heart is a pump, not a furnace, and that the lung, not the heart, is where blood and air meet. Renaissance anatomists thought that blood moved directly from the right ventricle of the heart to the left to be imbued with "vital spirits" brought from the lungs by the *pneuma*. Modern theory explains that before blood can get from the right ventricle to the left, it must first leave the heart and pass through the lungs. Before Harvey, Renaissance physiologists accepted Galen's view that blood returned along the same pathways by which it had come in its travels through the body. We now know that valves prevent this from happening and that blood flows only one way through each vessel. When it has gone as far as it can in one direction through the arteries, it crosses by way of the capillaries over into the veins to return to the heart.

Various pieces of this new picture were available to the Renaissance physiologists even as they taught the traditional scheme of Galen's *pneuma* and spirits. They knew, for example, that valves existed in the veins and that the large vessels that connect the lungs to the heart contained flowing blood, not air. Only William Harvey, however, was able to make sense of these details and deduce from them the notion of the heart as a pump that maintains a *circular* movement of blood.

Born in England, Harvey was educated at Cambridge University and then at Padua, where he attended the lectures of the famous anatomy professor Fabricius (Fig. 1–3). Fabricius was renowned for his construction of a special amphitheater that allowed large groups of students to observe dissections. He is noted for his observation of valves as well, which later played a crucial role in his famous student's conception of the circulation (Fig. 1–4). For his part, Fabricius dismissed the valves as merely slowing, not preventing, the backflow of blood through the veins. He held firmly to the physiology of Galen while patronizing the student whose physiological theories would soon turn Galen's views upside down.

Figure 1–3

William Harvey, the discoverer of circulation. *(Archive Photos)*

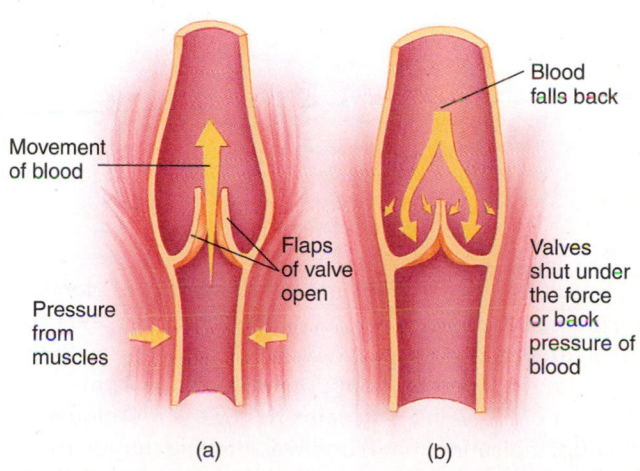

Figure 1–4

Valves in veins. The diagram shows how valves open to allow blood to flow in one direction but then close to prevent blood flow in the opposite direction.

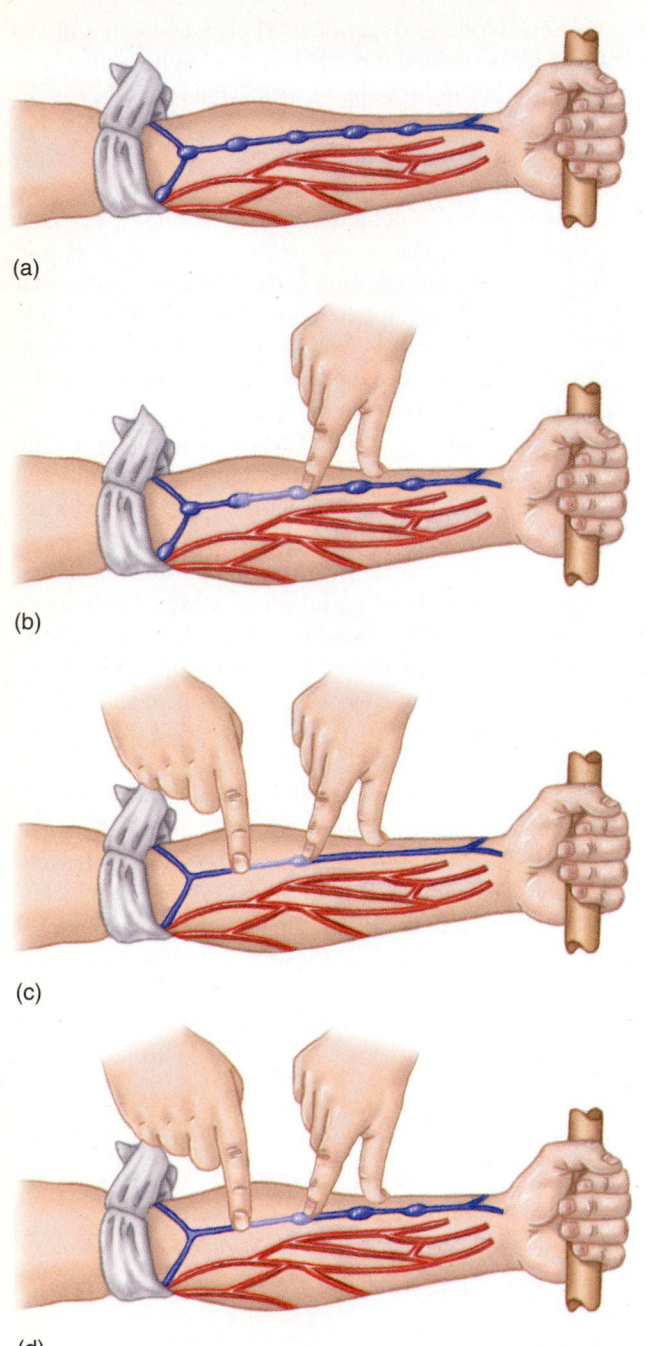

(a)

(b)

(c)

(d)

Figure 1–5

William Harvey's experiment showing direction of blood flow in veins and the connection between arteries and veins. *(a–d)* A tourniquet allowed blood to flow through the arteries into the lower arm but prevented blood from flowing out of the veins. The veins below the tourniquet swelled up, indicating that blood was flowing into them from the arteries. When the swellings were removed, they reappeared in the end of the vein farthest from the heart, showing that blood was flowing into the veins from the arteries instead of from the heart.

After completing his course at Padua, Harvey returned to England and began a career as a medical doctor (serving as royal physician) and lecturer. He also wrote a brief study of the circulation that was to become a classic in the history of science, revolutionizing the Western view of the movement of blood. *De Motu Cordis et Sanguinis in Animalibus* ("On the Motion of the Heart and of Blood in Animals"), published in 1628, described his search for the true role of the heart and the blood pathways.

The key features of Harvey's theories of blood flow that marked clear breaks from the physiology of Galen, Vesalius, and Fabricius were (1) the heart was a pump that worked by contracting to expel blood, rather than by expanding to fill with blood; (2) the blood went from the right ventricle of the heart first to the lungs, not directly to the left ventricle; (3) the air stayed in the lungs, where it met the blood, rather than coursing to the heart to meet the blood; and (4) the blood moved in a circular path through the body, taking different routes coming and going rather than traveling both directions in the same vessels.

Fifteen centuries earlier, Galen had used a goose feather to refute Erasistratus's claim that arteries were air vessels. Now Harvey used a similar experiment to refute Galen's claim that blood flowed both ways in a vessel (Fig. 1–5). He tied off the upper part of his arm to allow blood to flow into the arm through the arteries but to prevent blood already in the arm from flowing back out through the veins. As a result, blood coming into the arm through the arteries poured into the veins below the block but could not pass through them, causing them to swell. If, as Galen had claimed, blood could have flowed back through the arteries through which it had entered the arm, the veins would not have swelled.

Thus, Harvey's discovery of the circulation came about through a combination of creative speculation and simple experiments. He not only acknowledged the evidence he saw but also went on to conceive an overall scheme that made sense of that evidence (Fig. 1–6). While others of his time knew of valves and the pulmonary circulation, usually they tried to fit these details into Galen's traditional physiology. Here, Harvey describes how he combined direct observation with creative reasoning to break with the traditional view:

If I started from the root of these vessels and tried with all the skill that I could muster to pass a probe in the direction of the small vessels, I was unable to do so over any great distance because of the obstacles provided by the valves; on the other hand, it was very easy to pass a probe from without inwards, that is, from the small branches towards the root of the veins.

Since calculations and visual demonstrations have confirmed all my suppositions . . . to wit, that the blood is passed through the lungs and the heart by the pulsation of the ventricles, is forcibly ejected to all parts of the body, therein steals into the veins and the porosities of the flesh, flows back everywhere through those very veins from the circumference to the centre. . . .

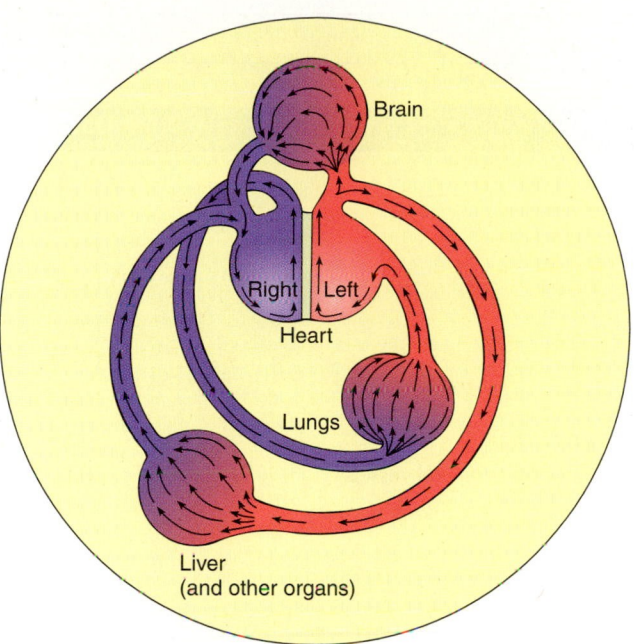

Figure 1–6

A simplified diagram of the circular movement of the blood as envisioned by Harvey.

I am obliged to conclude that in animals the blood is driven round a circuit with an unceasing, circular sort of movement, that this is an activity of function of the heart which it carries out by virtue of its pulsation, and that in sum it constitutes the sole reason for that heart's pulsatile movement.

Another unique aspect of Harvey's studies was his use of *quantitative methods*, which included the use of the metric system for his measurements. He calculated that the total amount of blood pumped over the course of an hour, for example, was much greater than what the body could hold or produce from food. He reasoned that, instead of each beat pumping a new quantity of blood, the same blood must be passing through the heart repeatedly. This idea reconciled the ceaseless beating of the heart with the small amount of blood that Harvey determined to be present in the body.

Harvey's discovery of the circulation was crucial for our modern conception of the workings of the body, but alone he could not supply all of the details of the picture. While recognizing the pulmonary circulation, Harvey did not know its true purpose, gas exchange. He still thought of the lungs as a cooling system and continued to view blood as an elemental, irreducible substance and an innately vital force that provided the impulse for the heartbeat. We now know that the autonomic nervous system governs the beating of the heart and that blood comprises various types of cells and other matter. Later, the Italian physiologist Malpighi discovered the existence of capillaries, which Harvey suspected but could not confirm.

The Microscopic Studies of Malpighi

Harvey's explanation of the flow of blood required the existence of some physical connection between the arteries and veins so that, when blood reached the end of its course at the "circumference" of the body, it could immediately enter the veins for its return to the "centre." Harvey was unable to see the capillaries that we now know provide this connection. The invention of the microscope and the brilliant experiments of Marcello Malpighi (1628–1694) provided our first glimpse of these structures, in the lungs of a frog.

The Study of What We Breathe

The Greeks thought that the air was an elemental, irreducible, and living substance drawn into the lungs to cool the heart and to become *pneuma*. Galen taught that this *pneuma* was transformed into a succession of "spirits" that drove the life processes.

We now recognize that air is made up of several gases. It is not transformed in the heart but contributes oxygen to the blood in the lungs. We can trace this modern understanding to the work of seventeenth- and eighteenth-century physicists and chemists. A series of experiments by European physiologists and physicists (including Torricelli, Borelli, Boyle, Hooke, and Lower) had shown, by the mid-seventeenth century, that (1) despite its intangibility, air had weight and therefore substance; (2) part, not all, of air was necessary for both breathing and combustion; (3) the purpose of breathing was not to cool the body's "furnace," wherever it might have been, but simply to fill the lungs with air; (4) air did not move to the heart but stayed in the lungs, where it met blood that had traveled through the pulmonary circulation from the heart; and (5) in the lungs, air transformed blood in some way, as evidenced by the color change of blood from dark to bright red when air contacted it.

An English chemist, Joseph Priestley (1733–1804), reaffirmed the earlier findings of John Mayow that the actions of both combustion and breathing used up or removed some part of air. Using an experiment similar to Mayow's, Priestley placed a burning candle in a sealed jar. The candle soon went out and could not be relit while still in the jar. Priestley then placed a green plant in the jar with the candle. After several days, he was able to relight the candle. He repeated the test with a mouse. After a period in the sealed jar, the mouse died. After a green plant was left in the jar for a few days, however, another mouse was able to live in the jar for a longer period. Priestley deduced that the processes of burning and breathing depleted some substance present in air and that a plant could resupply the substance. (This observation led to the discovery of photosynthesis.)

At the time, Priestley did not recognize the meaning of his findings. Although he succeeded in isolating the pure form of oxygen from a mercury compound, he called it "dephlogisticated air," according to a prevailing theory that an element called "phlogiston" was present in air and other

TABLE 1–1

Major Advances in Physiology and Medicine

Name	Discovery	Date°
Priestley	Discovered oxygen	1774
Laennec	Invented the stethoscope	1816
Hutchins	Invented the spirometer	1840
Morton, Wells, Jackson	Introduced general anesthesia	1846
Koch	Discovered the tuberculosis bacillus	1882
Roentgen†	Discovered X-rays	1895
Abel, Takamine	Discovered epinephrine	1897
Pavlov†	Discovered the physiology of digestion	1904
Krogh†	Discovered the mechanism controlling capillary blood vessels	1920
Hill†, Meyerhof†	Discovered the oxygen/lactic acid mechanism in working muscles	1922
Banting/Best†, MacLeod†	Discovered insulin	1923
Einthoven†	Discovered the electrocardiogram	1924
Fleming†	Discovered penicillin	1928
Landsteiner†	Discovered the blood groups	1930
Warburg†	Discovered tissue respiration	1931
Sherrington†, Adrian†	Discovered how nerves transmit their messages	1932
Dale†, Loewi†	Discovered that electrical nerve impulses are transmitted by chemical reactions	1936
Heymans†	Discovered the links between heart function and breathing regulation	1938
Muller†	Discovered the damaging effects on genes by X-rays	1946
Hess†	Discovered how distant organs can be controlled by the brain	1949
Krebs†	Discovered the citric acid cycle	1953
Clark	Invented platinum PO_2 electrode	1953
Enders†, Weller†, Robinson†, Salk†, Sabin†	Isolated polio virus and introduced polio vaccine	1955
Cournand†, Forssmann†, Richards†	Developed heart catheterization and showed circulatory pathophysiology	1956
Ochoa†, Kornberg†	Described the biological synthesis of RNA and DNA	1959
Crick†, Watson†	Discovered the double helix of DNA	1962
Eccles†, Hodgkin†, Huxley†	Discovered the ionic mechanisms by which nerves function	1963
Ikeda	Introduced flexible fiberoptic bronchoscope	1966
Granit†, Hartlin†, Wald†	Described the physiological processes of vision	1967
Swan, Ganz	Introduced right heart catheter (Swan-Ganz catheter)	1970
Axelrod†, Katz†, Euler†	Discovered neurotransmitters	1970
Sutherland†	Described mechanisms by hormones action	1971
Guillemin†, Schally†, Yakow†	Discovered the hormones of the hypothalamus	1977
Cohen†, Levi–Montalcini†	Discovered epidermal growth factor	1986
Bishop†, Varmas†	Discovered growth-regulating genes	1989
Gilman†, Rodbell†	Discovered G-proteins and the role that they play in signal transduction in the cell	1994
Furchgott†, Ignarro†, Murad†	Discovered nitric oxide as a signaling molecule in the cardiovascular system	1998

°Dates denote when inventions or discoveries were first announced or when the Nobel prize was awarded.

†A Nobel prize was awarded for their work. The first Nobel prize was awarded in 1901.

combustible matter and was destroyed by combustion. Priestley shared his findings, however, with the French scientist Antoine Lavoisier (1743–1794), who then identified the essential substance as oxygen.

Lavoisier thought of breathing as a physiological version of combustion because both processes required the same element. He claimed that the purpose of oxygen was to sustain the body's innate heat by allowing blood, the "combustible element," to burn. Lavoisier considered body heat the source of life and the lungs both a ventilation system and the place where combustion occurred. Later chemical advances would show that oxygen was involved in a complex trail of chemical reactions distributed throughout all the cells of the body.

The Advancement of Cell Theory

The Renaissance English scientist Robert Hooke (1635–1703), working with an early microscope, first detected the presence of small compartments he called "cells" in a piece of cork. Although he coined the term we now use to describe the basic unit of all life, it was not until two centuries after Hooke that the theory of the cell as we know it today arose in the writings of the German biologists Matthias Schleiden and Theodor Schwann. The **cell theory** is the broad basis of all modern biology and physiology and includes the following key concepts: (1) all organisms are made up of cells and their products; (2) new cells arise only from preexisting cells; (3) all cells have the same fundamental chemical makeup and metabolic processes; and (4) the activities and processes of the organism as a whole result from the interdependent and cooperative workings of groups of cells. The concept that cells originate only from other cells and not from nonliving matter, as had been thought previously, was stated first by Rudolf Virchow (1821–1902) and then confirmed by Louis Pasteur's (1822–1895) experiments with bacteria.

By the late nineteenth century, chemists knew that the breaking of chemical bonds that occurred when substances such as air and food were taken into the body produced chemical energy. Louis Pasteur's research into the process of fermentation helped increase chemists' understanding of the energy-releasing processes that go on in each cell. Researchers since have identified high-energy phosphate bonds (adenosine triphosphate, or ATP) as the repository for this energy in the cell. Twentieth-century scientists now recognize that oxygen is the substance that is linked to the utilization of ATP. The release of energy from ATP allows cell function to continue. Thus, ATP, not blood (as Lavoisier thought), is the substance whose "ignition" fuels the life of cells. Table 1–1 lists major advances in science, with emphasis on those advances that have profoundly influenced physiology and the practice of medicine.

The Idea of Homeostasis

Before the invention of the thermometer, physicians and physiologists thought that body temperature varied from one individual to the next, and that, within each individual, the temperature could vary with such factors as age, location, climate, and time of day. Early scientists thought that the temperature of the body was subject to external manipulation and thus was unstable. The advent of the thermometer allowed physiologists to measure each person's temperature against the same scale and observe that, apart from minute differences and variations due to abnormal (disease) states, the internal temperature of all human beings fell within a narrow range regardless of what climate they lived in.

This realization led the French physiologist Claude Bernard (1813–1878) to develop the idea of the *"milieu intérieur,"* a constant internal state in which stable conditions of temperature and chemical conditions prevail inside the cell despite changes in the external environment. The American physiologist Walter Cannon (1871–1945) termed this process **homeostasis.** Other twentieth-century scientists continue to reveal the mechanisms by which the body maintains the optimal conditions for its own functioning.

GENERAL ORGANIZATION OF THE BODY

In order to study functions of the human body, one must have a basic knowledge of anatomical terminology because function is often described in terms of bodily structures. Conventional lay terms, such as *top, bottom, above, below, behind, under,* and *on the side of,* can be ambiguous when used in complex descriptions of anatomy. To ensure precise understanding, anatomists have devised a special vocabulary that is a part of the language of anatomy and physiology. The following survey of anatomical terms is by no means exhaustive, but will introduce the correct meanings and uses of common anatomical terms that appear in subsequent chapters. When encountering new terms, you will find a medical dictionary an invaluable aid for mastering their meaning and usage.

Anatomical Planes and Positions Provide Spatial Orientation of the Body

Structures of the body are always described with reference to a standard static position called the **anatomical position** (Fig. 1–7). In this position, the body is erect, the face and feet are directed forward, and the arms are straight and directed toward the ground, with the hands rotated so that the palms face forward.

Planes

Imaginary **anatomical planes** are used to divide the body into parts (see Fig. 1–7). A **sagittal plane** divides the body into right and left portions. If such a plane passes directly along the midline of the body, it divides the body into right and left halves and is called a **median sagittal (midsagittal) plane.** A **transverse** or **horizontal plane** divides the body into upper and lower portions. A **frontal** or **coronal plane** divides the body into front and back portions.

APPLICATIONS OF PHYSIOLOGY

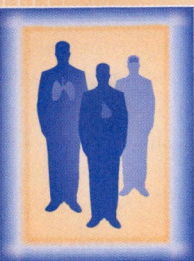

Radiographic Imaging Techniques

The spectrum of electromagnetic radiation contains wavelengths that are suitable for use in imaging techniques. Visible light, wavelength 10^{-7} to 10^{-6} m, is the most familiar form of electromagnetic radiation. It bounces off the surface of solid structures, allowing us to see them. In contrast, X-rays (wavelength 10^{-11} to 10^{-8} m), gamma rays (10^{-18} to 10^{-10} m), and radio waves (10^{-1} to 10^{-4} m) pass through most solid objects and can be used to study the internal structure of the human body. Some of the specific techniques are discussed briefly here.

Radiography is a technique for studying internal structures using X-rays. The rays are passed through the body and onto a photographic plate to produce a two-dimensional photograph. Some parts of the body, such as bone, absorb more of the X-rays than do other parts and can be distinguished in the photograph. The photograph, usually called an *X-ray* or *radiograph*, used to be known as a **roentgenogram** because X-rays were discovered by Wilhelm Roentgen, a German physicist, at the end of the nineteenth century. Roentgen used *X*, the symbol for an unknown quantity, because he did not understand completely the phenomenon he had found. In modern times, radiography has become a very useful diagnostic tool. **Contrast radiography** is useful for viewing certain soft tissues that are not seen easily on normal X-rays. For example, the upper gastrointestinal tract can be observed if the patient swallows a drink containing barium prior to an X-ray. Because X-rays cannot pass through barium, the position of the gastrointestinal tract, now lined with barium, can be located in the X-ray photograph. **Computed tomography (CT scanning)** is a system in which X-rays from many different angles are passed through a selected plane of the body in a fraction of a second. Detectors measure the absorption of X-rays by the various body structures; all the information is fed into a computer. The computer displays a reconstructed image of the cross-section of the body on a television screen. Hundreds of different cross-sections, like the slices in a loaf of bread, can be obtained and combined to form a three-dimensional image of the contents of large cavities in the body, such as the abdomen or thorax. **Positron emission tomography (PET scanning)** produces very clear images and can identify chemical and physiological changes. It uses gamma rays released by chemicals made especially for this purpose. These chemicals, which are similar to those normally present in the body, are injected into the patient. The PET scan detects the gamma rays released inside the patient and uses them to form an image. This image provides information about where the chemicals are used in the body. One use of the technique has been in studies of the brain to gain information about blood flow and utilization of glucose, an important source of energy.

Magnetic resonance imaging (MRI) is also known as nuclear magnetic resonance (NMR) imaging. It uses low-energy radio waves and a strong magnetic field and can produce excellent cross-sectional views without exposing the patient to high-energy gamma rays or X-rays or to a contrast medium, such as barium. MRI also can provide information about structure/function changes as well as the location of specific chemicals in vivo (in living cells).

Ultrasonography is based on a different principle. It uses high-frequency sound waves (**ultrasound**) to view internal anatomy and is better than X-rays for imaging soft tissues. The equipment consists of a transducer that produces and receives the silent, high-frequency sound waves. The transducer is placed against the body and slowly passed over the area to be examined. Sound waves pass through the skin into the body. As they strike various organs, they send echoes back to the transducer. Different kinds of tissue, bone, blood, and other fluids produce different echoes, which are separated and identified by the transducer. The transducer forms a visual representation of the varying echoes as it changes sound waves into electrical energy. This electrical signal is converted into an image on a video monitor. Ultrasound is particularly useful for visualizing the pregnant uterus and for detecting abnormalities of fetal organs. It can be repeated as often as required without harm. **Echocardiography** is a form of ultrasound used to examine the pumping action of the heart, especially the position and movement of valves and other structures within the heart.

All of these techniques are painless and are described as **noninvasive** because they allow the internal structures of the body to be studied without physical intervention, such as surgery.

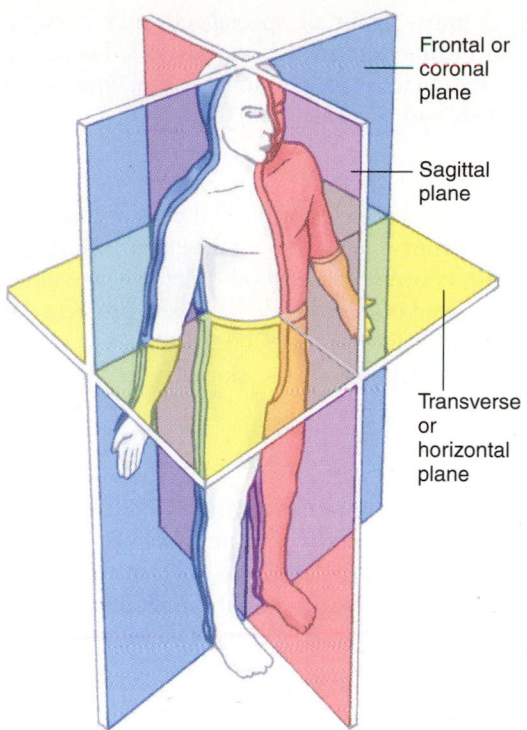

Frontal or coronal plane

Sagittal plane

Transverse or horizontal plane

Figure 1-7

Basic anatomical planes.

Positions

Terms of position are used to locate a structure relative to other structures. **Anterior** means "nearer to the front of the body," whereas **posterior** means "nearer to the back of the body." (Sometimes the comparative anatomical terms **ventral** and **dorsal** are used in place of *anterior* and *posterior,* respectively.)

Medial means "nearer to the midsagittal plane," whereas **lateral** means "farther from the midsagittal plane." **Superior** means "nearer to the head," whereas **inferior** means "nearer to the lower end of the body." (Sometimes the terms **cranial** and **caudal** are used instead of *superior* and *inferior,* respectively.)

Internal means "nearer the center of an organ, cavity, or part of the body"; **external** means "farther from the center." **Superficial** means "nearer to the surface of the body"; **deep** means "farther from the surface."

Two special terms are used in describing extremities or structures related to their long axes. **Proximal** means "nearer to the origin or point of attachment," whereas **distal** means "farther from the origin or point of attachment." For example, the humerus (the bone in the upper arm) is attached at its proximal end to the shoulder and at its distal end to the elbow.

Human Cells Can Differentiate and Exhibit Over 200 Modes of Specialization

Living matter is made up of **protoplasm,** a complex mixture of chemicals that displays the attributes of life. These attributes include (1) organization into specific kinds of structural units, (2) the ability to enter into chemical activities that include the transformation of energy and the maintenance or synthesis of protoplasm, (3) the ability to respond to changes in the environment, and (4) the ability to grow and reproduce.

Protoplasm is organized into cells, which are the basic structural units of the human body and of all life. A **cell** can be defined as a microscopic bit of organized protoplasm surrounded by a membrane called the **plasma membrane.** The adult human body comprises about 100 trillion cells, all of which function collectively to maintain an individual's life.

Each human life begins as a single cell, a fertilized ovum, which then divides to form two cells, four cells, eight cells, and so on. In addition to undergoing numerous cell divisions during development, cells also begin to exhibit specialized functions. There are about 200 different cell types in the body, as determined by differences in both structure and function.

The process by which a single cell type (the fertilized ovum) develops into many different cell types is known as **cell differentiation.** As you will see, these different cell types are arranged first into **tissues,** or groups of related cells. Tissues, in turn, are arranged in groups to form organs; organs, in turn, form systems. An analogy can be drawn between the cells of the body and a society of individuals. Just as mail carriers, police, doctors, and teachers have specific roles for the common good of the entire society, so do specialized cells such as muscle, nerve, connective tissue, and epithelial cells serve specific functions to promote the survival of the organism. The body can be thought of as a society or social order of cells.

In a multicellular organism, such as a human being, labor is divided among groups of specialized cells, with each group performing one principal function, such as movement, digestion of food, or reproduction. Specialization of cells is an essential factor in the development of the large size of multicellular animals. The different groups of cells work in concert to maintain the life of the organism; considerable interdependence exists between the different cell types. For example, most of the cells of vertebrates depend on the red blood cells to provide them with oxygen, while at the same time the red blood cells depend on the pumping action produced by the muscle cells of the heart to propel them through the body. Such a division and sharing of labor has allowed multicellular organisms to overcome the limitations of diffusion and surface area to grow to great size, whereas a unicellular organism must carry out for itself all the life processes and can exist independently only by remaining microscopically small.

Tissues Are Organized Colonies of Specialized Cells

Because multicellular organisms require a division of labor and specialization between cells, the cells in an individual's body differ in size, shape, internal architecture, and function. Some cells are spherical; others are shaped like cubes, discs, stars, or cylinders. Each cell manifests the form and structure best suited to its function (Fig. 1–8). The neuron (a nerve cell), for example, bears numerous cellular extensions that receive and transmit the chemical and electrical signals that form the basis for nervous function. A skeletal muscle cell has a long, cylindrical shape, which helps the cell move other parts of the body when it contracts. The "biconcave-disc" shape of a red blood cell helps it in its work of transporting oxygen through the body.

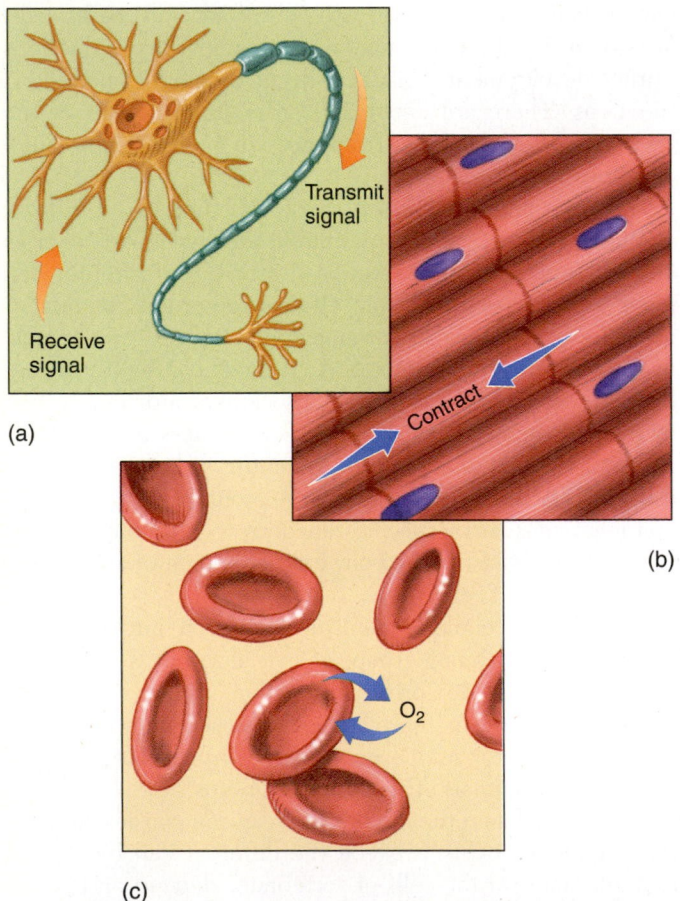

(a)

(b)

(c)

Figure 1–8

Cell shapes are designed to allow for optimal functioning of specialized cells. *(a)* The long extensions of the neuron enable it to receive and transmit nervous signals. *(b)* The fibers of a muscle cell allow it to change shape during contraction and relaxation. *(c)* The biconcave-disc shape of a red blood cell assists it in carrying oxygen to the tissues and cells of the body.

When one or more types of specialized cells become closely associated in function, they form a tissue. There are four classes of tissues in animals: epithelial tissue, connective tissue, muscle tissue, and nervous tissue (Fig. 1–9).

Epithelial Tissue

Epithelial tissue consists of closely packed cells arranged in flat sheets ranging from one to several layers in thickness. This type of tissue covers exposed body surfaces and lines body cavities, tubes, and organs. Epithelial tissue protects the body parts and absorbs, filters, and secretes substances. For example, the superficial layers of the skin are composed of epithelial tissue.

Connective Tissue

The cells that make up **connective tissue** are somewhat loosely arranged and separated by an intercellular **matrix** (ground substance). The matrix material, made by the connective tissue cells themselves, frequently contains cellular products, such as fibers, soluble proteins, and crystalline complexes. Connective tissue is the most widely distributed tissue in the body; the arrangement is largely what determines tissue growth. It supports, protects, binds, and partitions nearly all bodily components. Specific examples are cartilage, bone, ligaments, blood, and adipose (fat) tissue.

Muscle Tissue

The cells of **muscle tissue** are elongated and contain many contractile cells called **myofibrils.** There are three types of muscle tissue: skeletal muscle, cardiac muscle, and smooth muscle. **Skeletal muscle** is a specialized tissue for contraction and functions to accomplish movement. It is attached to the bones and is used for conscious, voluntary movements of the body, such as walking. **Cardiac muscle** is present only in the walls of the heart and produces the automatic pumping contractions of the organ. **Smooth muscle** is found in hollow internal organs, such as the blood vessels, and allows them to undergo automatic, "involuntary" contraction and expansion.

Nervous Tissue

The chief functional component of **nervous tissue** is **neurons,** cells that are specialized for receiving and conducting electrical impulses. The characteristic shape of a neuron is an enlarged main cell body with hairlike extensions termed **axons** that receive and transmit the nerve impulses. **Glia,** sometimes called **glial cells** or **neuroglia,** are connective tissue cells of the nervous system that provide structural support and protection. In addition, bundles of axons called simply **nerves** are found in all parts of the body.

Organs Are Integrated into Systems

Combinations of the primary tissues make up the organs of the body. An **organ** is defined as a group of tissues that have

CURRENT CONCEPTS IN PHYSIOLOGY

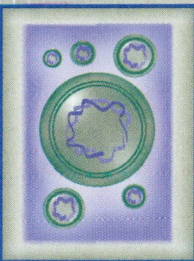

One Genome, Different Protomes

Genomics and proteomics have taken center stage in biology and medicine. **Genomics** involves the study of the structure, sequencing, and subsequent analysis of large numbers of genes of an organism simultaneously. **Proteomics** is the study of the proteins produced (expressed) by genes in an organism, tissue, or cell. The genome is composed of four chemically similar components arranged in the linear coding order of DNA, that is, the large, double-stranded, helical molecule that contains all of the genetic material of the organism. Determining the sequence of DNA involves fragmenting DNA strands through an automated repetitive process. DNA molecules are so alike that the sequencing process applies to DNA from any source, a technological universality that explains why genomics was undertaken first and why the exponential rate of progress has occurred worldwide. The complete genomes have already been worked out on simple organisms, such as the *Escherichia coli* bacteria, as well as multicellular organisms, such as the mouse. A survey (working draft) of the human genome, which consists of 3 billion base pairs, was completed in June, 2000.

DNA has no function except to store information. The complete genome of on organism gives only a relatively static overview of the functional potential of an organism and does not describe the immense dynamic process that occurs in a living organism. In sharp contrast, proteins are made of approximately 20 amino acids that exhibit a variety of different chemical properties. Amino acids are arranged differently in different proteins, and in addition proteins fold to give three-dimensional structures that give rise to thousands of different functions. In essence, while DNA stores information almost without modification for thousands of years, proteins do everything else. Furthermore, while DNA in all of the estimated 252 different somatic cell types in the human has basically the same sequence, the protein composition of these cells differs both qualitatively and quantitatively. For example, the somatic cells of the human liver and brain contain identical genetic information. However, the conversion of this genetic information (i.e., the expression of the different genes into proteins) takes place during the different developmental stages of an organism. This leads to an enormous individual phenotypic diversity and function within the organism. Moreover, the unique advantage of proteomics is that changes in protein-expression patterns can be studied during normal function as well as during cellular dysfunction, often seen with injury and diseases. Thus, the proteomics database will not only characterize the differences between the estimated 252 different cell types in the human but will also provide a basic foundation for systematic discovery of the molecular changes underlying altered cellular function, and markers for them. The latter will provide new insight into molecular medicine and the disease process. As a result, a new approach, called **functional genomics** or **physiological genomics**, is emerging and gaining prominence. This involves the analysis of the mRNA and protein products expressed by the genome. Their expression patterns can be correlated with specific effects of a disease and/or altered cell function. Thus, proteomics is an important field that will provide powerful tools for our understanding of protein profiling (i.e., protein signatures) of expressed genes in tissue and individual cells. Physiological genomics will be very promising in elucidating the molecular effects of xenobiotics and diseases.

What have we learned from the Human Genome Project? Without a doubt, the genome data have, for the first time, revealed the immense complexity of nature. In place of the dogma "one gene, one protein, one function," a new understanding of interwoven regulatory networks has developed. We have learned, for example, that humans have approximately 30,000 genes — about half of what was previously estimated. Of these 30,000 human genes, some 200 genes were somehow inserted into our early vertebrate ancestors by bacteria. We have learned that approximately 1.5% of the human DNA carries instruction for making proteins compared with the 3.0% to 5.0% previously thought. We know now that along the stretches of the DNA molecule, genes tend to occur in clusters, like cities with vast stretches of countryside. This clustering turns out to be much more pronounced than originally thought. We have learned that inherited genetic mutations arise about twice as often in males as in females. With proteomics, we have learned that the expression of different genes into proteins determines the many types and functions of different somatic cells in the body. The field of proteomics has provided a new understanding of cell function under normal and altered physiological conditions, which will provide new insight into the treatment of diseases in the future.

Organizational Level: Specialized Cells ⟶ Tissues ⟶ Organ ⟶ System

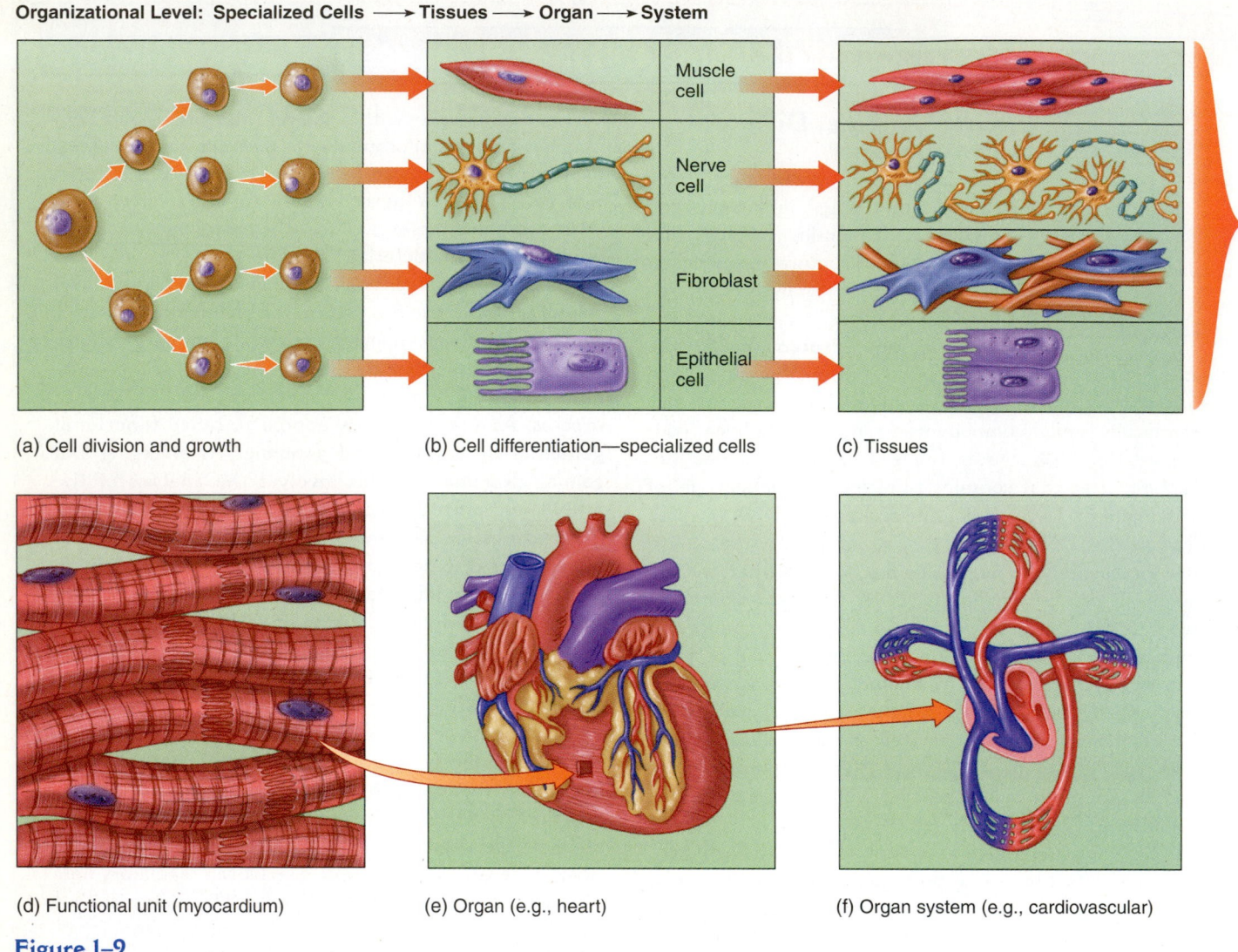

(a) Cell division and growth

(b) Cell differentiation—specialized cells

(c) Tissues

Muscle cell

Nerve cell

Fibroblast

Epithelial cell

(d) Functional unit (myocardium)

(e) Organ (e.g., heart)

(f) Organ system (e.g., cardiovascular)

Figure 1–9

(a–f) Organizational level: Specialized cells → tissues → organ → system.

been combined for the performance of a specific function or series of related functions. The stomach, for example, contains all four classes of tissues, organized for the purpose of receiving, storing, and digesting food and for moving partially digested food into the small intestine for further processing. No single tissue can perform as well as can several tissues working together.

In a similar manner, organs and other structures that share in the performance of related tasks are grouped into **systems** (see Fig. 1–9). The cardiovascular system, for example, consists of the heart, arteries, veins, and a capillary bed. These organs work toward the common purpose of providing nutrients and oxygen to cells and removing carbon dioxide and other waste products.

The grouping of organs and related structures into systems provides the approach for this textbook. Keep in mind as you proceed through the various chapters, however, that

although many levels of organization and function exist in the human body, all of the body's components function collectively to maintain an individual's integrity and life.

Bodily Fluids Are Divided into Two Compartments

The body can be thought of as mostly a watery solution. Approximately 60% of its weight is water. The body's fluids contain various mineral ions (e.g., sodium, potassium, and chloride) and organic substances (e.g., proteins and glucose) dissolved in the body water. Bodily fluids usually are grouped into two major divisions or compartments: intracellular fluid and extracellular fluid.

The two compartments are separated by the plasma membranes of each of the body's cells; the chemical compositions of the two compartments differ strikingly. The differences in chemical composition are maintained by the cells themselves and have important effects on the cell membrane potential, cell excitability, cellular metabolism, and the ability of the cell membrane to transport substances into and out of the cell. Much of cell physiology is concerned with the mechanisms for maintaining differences in the ionic composition of intra- and extracellular fluids and the consequences of these differences for cell function.

Intracellular Fluid

The **intracellular fluid** is the fluid within cells. It is not, therefore, one single continuous compartment, but rather a conglomeration of trillions of cellular subcompartments. Intracellular fluid has a relatively low concentration of sodium ions and a high concentration of potassium ions.

Extracellular Fluid

The **extracellular fluid** is the fluid outside of the cells. Extracellular fluid, in contrast to intracellular fluid, has a high concentration of sodium ions and a low concentration of potassium ions. The extracellular fluid that bathes our cells is a medium of exchange between one cell and another and between cells and the external environment.

Extracellular fluid consists of four types: interstitial fluid, plasma, lymph, and cerebrospinal fluid. **Interstitial fluid** is the fluid found between body cells; **plasma** is the fluid portion of the blood; **lymph** is the fluid in lymphatic vessels; **cerebrospinal fluid** is an ultrafiltrate of plasma but lacks many of the plasma proteins. Together, these extracellular fluids make up the **internal environment** of the body, the constant regulation of which is the purpose of the complex physiological processes. When certain concentrations of ions exist in its internal environment, the body will thrive; outside those ranges, it will be unable to function well or to live.

A Stable Internal Environment Is Essential for Normal Cell Function

In a single-celled organism (e.g., an amoeba living in a pond), the internal environment of the cell is difficult to control and is subjected to the vagaries of the external world. Changes in temperature, light, and concentrations of various chemicals in the external environment dictate its existence. It has little freedom because the internal environment of the cell is directly exposed to the external environment and is unable to manipulate or use a buffer or mediator between itself and the external world. If the water (its external environment) freezes, the amoeba freezes also. If it does not die, it is at least immobilized.

A more highly developed organism, such as a frog, has a greater degree of independence from external conditions. By virtue of its multicellularity, it has both an external *and* an internal environment. Its internal environment is the extracellular fluid that surrounds its cells; this fluid is *outside* the cells but still *inside* (and therefore part of) the frog. The frog, however, is unable to regulate its own body temperature.

When the weather turns cold, it may not freeze, but it will be forced into inactivity. Reptiles, such as snakes and alligators, must spend a great deal of time simply sitting in the sun after periods of exposure to cold air or water in order to raise the temperature of their internal environment to a level that permits life processes to continue at full speed.

In contrast to reptiles, mammals have developed a great deal of freedom from external conditions by evolving sophisticated mechanisms for maintaining a stable internal environment. Most noticeably, their internal body temperatures remain virtually constant under a wide range of external temperature conditions by virtue of physiological, behavioral, and adaptive mechanisms that counteract the effects of cold or heat in the external environment. Mammals are able to maintain a constant body temperature whether the air around them is at 0°C or 45°C. Furthermore, unlike frogs, many mammals are fully active under all climatic conditions; they do not require periods of physiological inactivity while waiting for their bodies to warm. For example, the arctic fox in the cold north and the fennec fox in the hot North African desert both have distinct morphological features and habits that adapt them to their habitat. The fennec, which avoids the midday heat in its burrow, possesses oversized ears to radiate heat. The arctic fox, which faces the need to conserve heat, has reduced ear surface and remarkable insulation that can make shivering unnecessary until ambient temperatures fall to −40°C.

A Constant Internal Environment Maintains Protein Conformation

Intuitively, one can appreciate that cells, as living things, can flourish and prosper in certain sets of conditions and not in others. By providing a constant environment for the cells of the body, the various homeostatic mechanisms cooperate to create optimal conditions for cell function.

Beyond this, however, one might ask, "What is the molecular basis of this need for a constant internal environment?" The answer to this question lies in the fact that the primary components of cells, macromolecules called **proteins,** change their shape at different environments. Proteins serve a variety of functions in cells. They form **enzymes,** which are responsible for the various metabolic processes by which cells obtain energy and synthesize other macromolecules in the cell. They also serve as structural components in the cell interior and the cell membrane. Every protein in the cell has its own distinctive **conformation,** or shape, that allows it to best carry out its own particular function. The shapes, and therefore the functions, of proteins are affected by such factors as ion concentrations, temperature, and pH. Changes in these factors can change **protein conformation.** With the right conditions of ion composition, temperature, and pH, proteins assume their proper conformations and carry out their tasks perfectly. Under abnormal chemical, temperature, and pH conditions, however, their shape is altered and their ability to function is diminished or lost. Protein malfunction, in turn, prevents the cells as a whole from carrying out their own specialized tasks, and the malfunctions accumulate until they become manifest as a disease condition in the organism.

Proteins function outside of cells as well as inside. The internal environment is *extra*cellular fluid. What is the connection? As we shall see in Chapter 4, the connection is everywhere — anywhere cells and their membranes are found. The plasma membranes of cells serve as the boundary between the two bodily fluid compartments — the extracellular fluid and the trillions of subcompartments of the intracellular fluid. At the same time, they serve as a transport system between the two compartments, constantly permitting or encouraging movement of substances between cells and the extracellular fluid. The composition of that fluid determines whether or not the cells can obtain from it what they need to make proteins and other molecules essential for their existence.

Thus, the molecular basis of the homeostatic process is, in essence, the maintenance of protein conformation. By constantly ensuring that proper conditions exist in the extracellular fluid, homeostatic mechanisms allow proper conditions inside cells, thereby permitting cell proteins to assume their proper shapes and carry out their special functions. The optimal functioning of specialized cells, in turn, ensures the coherent functioning of tissues, organs, and organ systems and the health of the body.

THE STUDY OF PHYSIOLOGY

This book is a starting point for your exploration of the human body. Many tools and approaches will be presented that will allow you to become fascinated with science and the way it is applied to understand how the body functions in health and disease. Physiologists work primarily in the laboratory but sometimes in the field. Their research ranges from the study of how molecules interact within the cell to how a whole organism adapts to high altitude. You may decide to become a research physiologist and help unravel how the brain works; discover a new drug that inhibits fat synthesis in cells and prevents obesity; or discover new treatment for diseases, such as diabetes, heart arrhythmia, hypertension, stroke, or cancer. Physiology is a science, and is derived from the Latin word meaning "natural science." In a broad sense, **science** is a method of thinking as well as a method of investigating how things work in a systematic manner. Science enables us to unlock the secrets of nature and uncover more information and insight about the world we live in. The new knowledge gained from science helps us expand our appreciation of the universe.

Regardless of the scientific discipline (e.g., chemistry, physics, biology, genetics, or physiology), all scientists use a similar approach in their research. This approach employs a rigorous thought process in a rational, systematic, error-minimizing way to make new discoveries. Science is systematic because scientists attempt to organize their ideas so that anyone who wishes to build upon their work can obtain the necessary information. By using a set of commonly accepted rules and procedures, scientists strive to provide precise and accurate knowledge about the universe.

The Scientific Method Is a Systematic Pursuit of Knowledge

All of the basic ideas in this textbook about how the body works have been gained by the use of the **scientific method** — a systematic, ordered approach of gaining information how the natural world, including ourselves, works. The scientific method provides a framework used by all scientists. However, many significant discoveries have depended on the creativity of the individual scientist. Often, they will approach the questions and **experimental design** (a planned procedure) that lead to new discoveries differently. Albert Einstein is a good example of an individual with a creative mind who approached problems differently than did his colleagues.

Reasoning Processes

Although many techniques are used, all are based on the idea of objective observations and reproducible data. Use of the **scientific method** requires a reasoning process to devise experiments.

The thought processes that scientists use can be grouped into two general categories: deductive reasoning and inductive reasoning. The two types of reasoning can be thought of as opposites. In **deductive reasoning,** one proceeds from a general rule to a specific conclusion, whereas in **inductive reasoning,** one goes from a specific observation to a general conclusion (Fig. 1–10).

In deductive reasoning, a scientist begins with supplied, or "given" information, called a **premise,** often stated in the form of an absolute rule using a word such as *all* or *always*. From that premise or general rule, the scientist draws a conclusion about a specific example that matches the premise. Frequently, the premise is actually a set of related premises, one general and one more specific. Following is an example of deductive reasoning:

Premise(s):	1. All tissues are composed of cells.
	2. Muscle is a type of tissue.
Conclusion:	Muscle is composed of cells.

In inductive reasoning, a scientist proceeds from one or more specific observations to a general conclusion. A conclusion drawn by inductive reasoning is true only until a specific observation is found that disproves it; one specific exception is enough to invalidate the conclusion. Following is an example of inductive reasoning:

Observation(s):	1. Muscle cells have nuclei.
	2. Epithelial cells have nuclei.
	3. Nerve cells have nuclei.
	4. Muscle, epithelial, and nerve cells are all human cells.
Conclusion:	All human cells have nuclei.

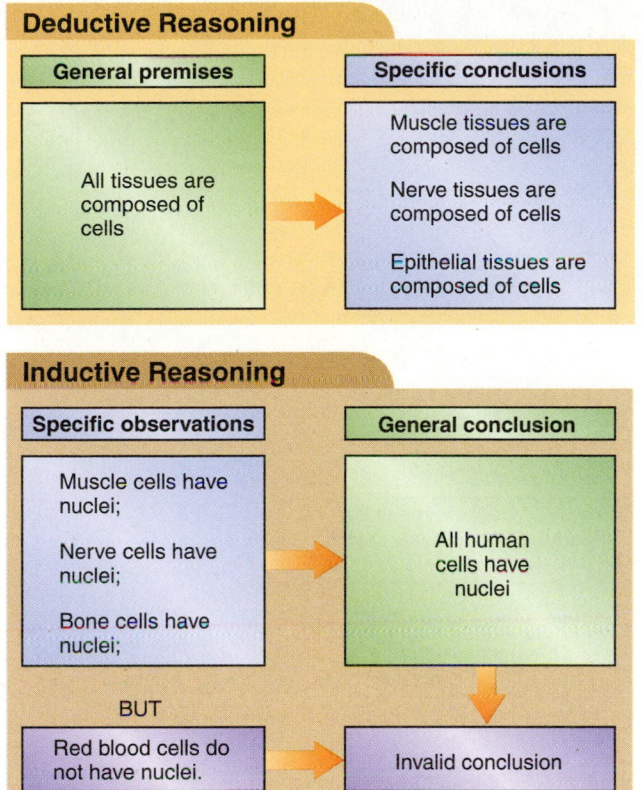

Deductive Reasoning

General premises	Specific conclusions
All tissues are composed of cells	Muscle tissues are composed of cells
	Nerve tissues are composed of cells
	Epithelial tissues are composed of cells

Inductive Reasoning

Specific observations	General conclusion
Muscle cells have nuclei;	All human cells have nuclei
Nerve cells have nuclei;	
Bone cells have nuclei;	
BUT	
Red blood cells do not have nuclei.	Invalid conclusion

Figure 1–10

A schematic view of the deductive and inductive reasoning processes. In deductive reasoning, one proceeds from a premise or general law to a specific conclusion. In inductive reasoning, one goes from specific observations to a general conclusion, which can be disproved by a single exception.

However, there is one exception that invalidates this conclusion. Mature human red blood cells lack nuclei; therefore, not all human cells have nuclei. This one example is enough to require that the conclusion be modified or discarded.

Steps Used in the Scientific Method

The scientific method can be thought of as a sequence of steps, broadly classified as (1) observation; (2) development of critical questions; (3) hypothesis; (4) prediction of an outcome; (5) experimentation; (6) conclusion; and, sometimes, (7) retesting and (8) publication.

OBSERVATION. Many significant discoveries are made by scientists who are in the habit of looking critically at nature. This means they make careful observations and ask critical questions derived from their data or from someone else's results. Often, chance and luck are involved in a new scientific discovery. A good example of this is the discovery that Alexander Fleming, a British bacteriologist, made in 1928.

Fleming was studying the *Staphylococcus* bacteria that caused boils and skin infection. He observed that one of his bacterial culture dishes had been contaminated with a blue mold. Before discarding the Petri dish, he noticed that the surrounding area where the mold invaded was void of bacteria. He reasoned that anything that could kill these *Staphylococcus* bacteria was worth saving. Fleming was not the only bacteriologist studying the *Staphylococcus* bacteria at this time. Many of the other bacteriologists had the same type of problem, in which mold had contaminated their experiments. However, they failed to make the critical observation and the connection and simply discarded the contaminated cultures. Fleming benefited from his critical observations, formulated critical questions, and published his results. The mold that Fleming saved turned out to be a variety of *Penicillium* (blue bread mold) and was subsequently discovered to be the mold that produced a substance that prevented the growth of bacteria. This substance was penicillin, one of the first antibiotics (see Table 1–1), which can kill bacteria but is harmless to animal and human cells. Although Fleming recognized the potential benefits of penicillin, he did not promote it; the development of practical applications was left up to other scientists. It took more than 10 years from Fleming's initial discovery before penicillin was put to significant use as an antibiotic.

HYPOTHESIS. Usually a scientific discovery begins with a set of observations that are not well understood. For example, a neurophysiologist who is doing vision research observes a high incidence of cataracts (opacity in the lens of the eye that causes cloudiness and light scattering) in individuals who have been exposed to sunlight for long periods of time. Next, he or she will most likely begin to mull over the observation and wonder *why* cataracts are much more prevalent in these individuals. Usually at this stage, the scientist also talks to colleagues about the finding and searches the scientific literature for reports of the same observation. Next, the neurophysiologist may speculate about various possible causes of the cataracts, wondering, for example, whether ultraviolet rays (part of sunlight) cause the higher incidence of cataracts. Following this speculation, he or she will form a **hypothesis,** which is an idea that tentatively explains the observation. In formulating the hypothesis, the scientist attempts to make a predication that is *testable*, realizing that it may not be true or that it can lead to new hypotheses (see Fig. 1–9). A good hypothesis usually addresses the following questions:

Is the hypothesis consistent with the established data?

Is the hypothesis capable of being tested?

Is the hypothesis capable of being proven false?

In the case of cataracts, the neurophysiologist proposed the following hypothesis: *Exposure to ultraviolet light causes cataracts in the lens of the eye*. Such a statement will lead to the next logical step: a prediction that can be tested by a set of controlled experiments (Fig. 1–11).

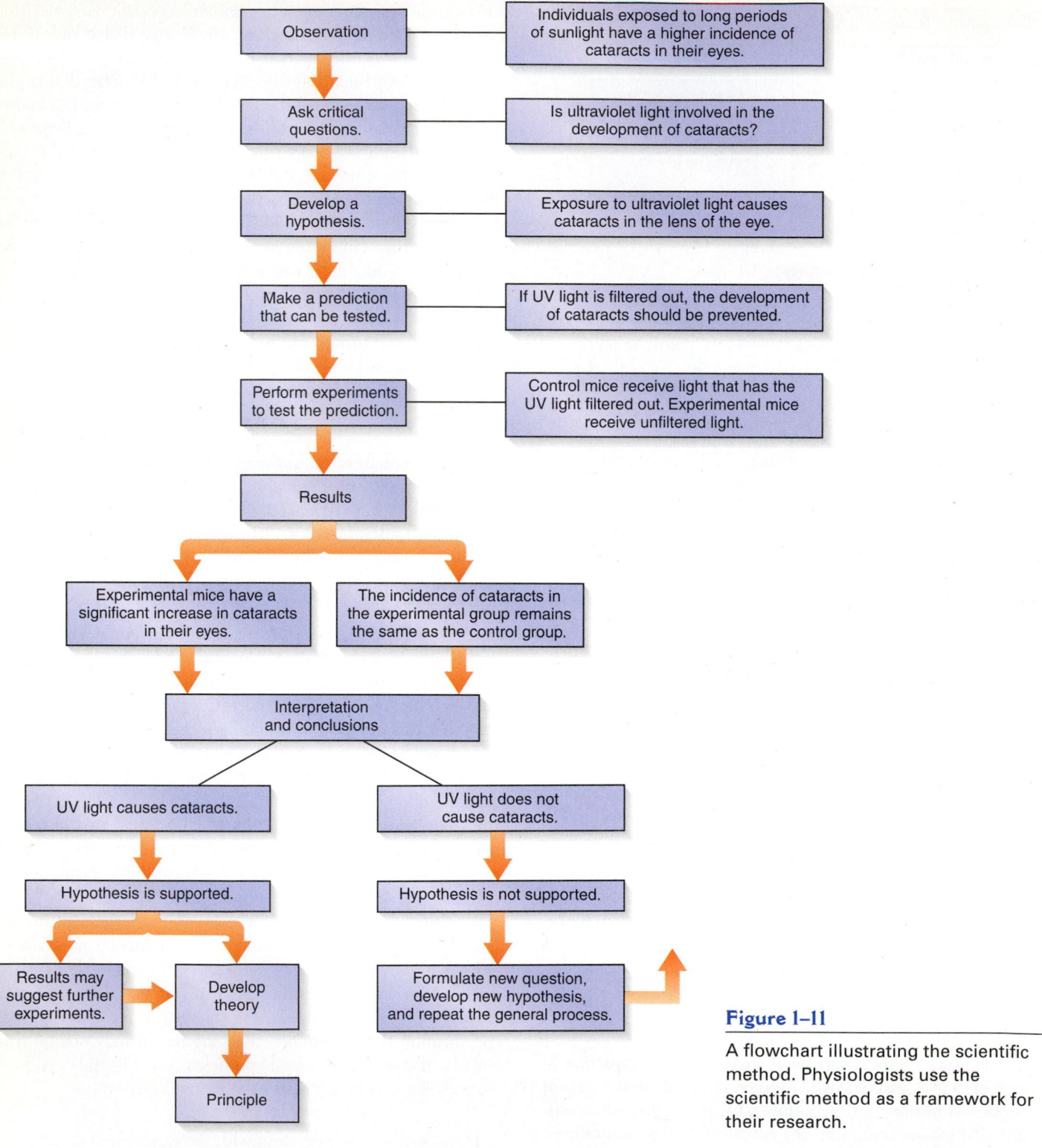

Figure 1–11

A flowchart illustrating the scientific method. Physiologists use the scientific method as a framework for their research.

EXPERIMENT. A hypothesis is tested in an **experiment,** in which carefully specified conditions are created deliberately to produce the observations in question. The purpose of an experiment is to determine whether a cause-and-effect relationship exists between an observation and a **variable** (the hypothesized cause of the observation).

Ideally, the experiment is designed in such a way that only one explanation exists for the observation. To accomplish this goal, most experimenters set up two groups of subjects with two different sets of conditions. One group is the **experimental group,** in which the variable is present. The other group is the **control group,** which is identical to the experimental group except that the variable is omitted. In essence, experiments are designed so that the control group provides a standard against which the effect of the variable in the experimental group can be compared (Fig. 1–12). The control group must contain conditions identical to the experimental group; any variations or differences that creep into the design are considered variables and must be considered in the results and conclusions of the experiment.

In scientific studies, care must be taken to avoid **bias** (a predetermined outcome).

Clinical studies are those that involve patients. An example of a clinical study is an experiment in which the effectiveness of a new drug is being tested in the treatment of a specific disease. Many clinical studies are carried out in a **double-blind** fashion. In double-blind studies, one group of patients is given the new drug; the other group is given a **placebo,** an inactive substance (usually a pill) that is similar in size, color, shape, taste, and texture to the drug being tested. This study design is called *double-blind* because neither the physician-scientist nor the patients know who is getting the experimental drug.

Recent studies have shown that approximately 25% of patients who receive placebo show an improvement in their condition. This phenomenon, known as the **placebo effect,** is caused by a subject's *expectations* concerning the treatment rather than by the treatment itself (i.e., simply because the subject *believes* that he or she is receiving the drug, his or her condition improves). If the experimental group also shows only a 25% improvement, then the drug is considered ineffective; the results could be attributed to the placebo effect. Forty years ago, the placebo effect was often ignored or discounted. In clinical studies today, however, much more than 25% of the experimental group must show an improvement; otherwise, the drug or treatment is considered ineffective.

Figure 1–9 illustrates how the scientific method can be used to test the hypothesis that ultraviolet rays cause cataracts. In this experiment, the ultraviolet rays are considered the variable. One group of animals, the experimental group, is exposed to light containing the variable, ultraviolet rays. Another group, the control group, is exposed to light from which the ultraviolet rays have been filtered or removed. Apart from this difference, all other conditions — time of day, temperature of the environment, length of exposure, and so on — are the same for the two groups. If one group were to be exposed inadvertently to the light for a longer period, the length of exposure would become a variable that would bias the results.

CONCLUSION. Scientists gather data from the experiments, interpret their results, and then formulate conclusions. For example, in the neurophysiology study, the results show that cataracts appeared in 20% of the animals in the experimental group but that none appeared in the control group. The neurophysiologist makes the following interpretation from these studies: *Mice exposed to ultraviolet light have a greater incidence of cataracts than do mice not exposed to ultraviolet light.* From this interpretation, the neurophysiologist can conclude that individuals exposed to sunlight for long periods of time, without protection, are more susceptible to cataracts than are those who use protection against sunlight. Statistical methods are used to determine biological significance of the data. His or her interpretation and conclusion, in turn, allows him or her to accept the original hypothesis that ultraviolet rays cause cataracts.

RETESTING. To be able to accept the interpretation of the data and conclusions more confidently, the results must be accurate and reliable. One reason for inaccurate conclusions is the **sample error.** Because all cases cannot be studied (we cannot test *all* mice), scientists often must rely on a small **sample size,** or a subset of them. When this approach is taken, we do not know whether the sample truly represents what we are investigating. This problem can be corrected by using large number of animals or subjects and applying statistical analysis to the data. Also, **replication** — the ability to produce the same results with repetition of the experiment — is important. For example, if the experiment involving ultraviolet rays were repeated six times, and only one of the six results showed a higher incidence of cataracts, then the study cannot be replicated, which will lead to an inaccurate conclusion.

Another approach to strengthen the hypothesis is **retesting** — to repeat the experiment, introducing another variable to determine whether any other conditions in addition to ultraviolet rays can increase or decrease the incidence of

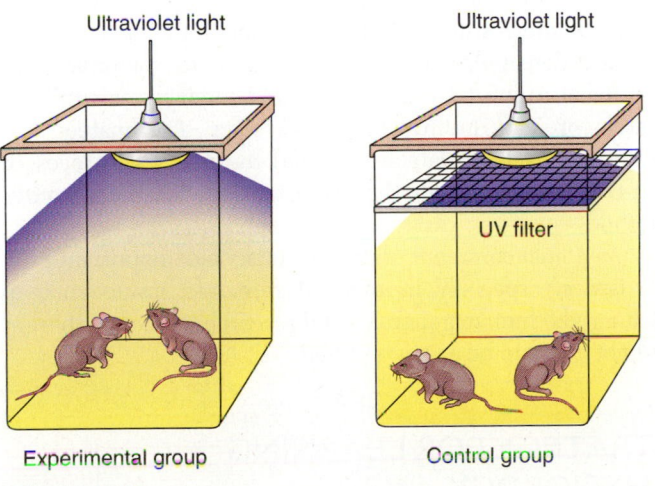

Figure 1–12

A controlled experiment to test the hypothesis that ultraviolet light causes cataracts. The experimental group of animals is exposed to ultraviolet light for a certain period of time (+ uv light). The control group is exposed to light from which ultraviolet rays have been filtered out (– uv light). All other conditions are the same for both the control and the experimental groups.

cataracts. For example, the neurophysiologist might dilate the pupils of the experimental mice, thereby increasing the area of the lens that is exposed to the ultraviolet rays. The rationale for this procedure would be that such an increase in exposure should cause a greater incidence of cataracts if ultraviolet light does in fact cause cataracts.

In such an experiment, four groups might be tested. This could include Group I, the control group, not exposed to ultraviolet rays; Group II, those exposed to ultraviolet light but without dilation of the pupils; Group III, those exposed to ultraviolet light with pupils dilated; and Group IV, those with pupils dilated but exposed to light that does not contain ultraviolet rays.

The results of such an experiment might show, for example, that Groups I and IV had no cataracts, Group II had a 20% incidence of cataracts, and Group III had a 30% incidence of cataracts. From these results, the neurophysiologist could conclude that greater exposure to ultraviolet light causes a higher incidence of cataracts. This conclusion strengthens the original hypothesis that ultraviolet rays cause cataracts.

PUBLICATION. Finally, the neurophysiologist compiles the data from these experiments and submits a manuscript describing the results to a scientific journal. The publication process is an important part of the scientific method. Sharing new information is consistent with the goals of modern science, which strive to maintain an ongoing dialogue among scientists in order to foster greater advancement of knowledge. Many experiments have important ramifications for biology as well as the health of human beings. For example, the finding that ultraviolet light causes cataracts has prompted government agencies to place warning labels on sunglasses that do not filter out ultraviolet rays because sunglasses can potentially cause the pupils to dilate, thereby increasing the wearer's exposure to the harmful rays.

Not all investigations produce such clear-cut and dramatic results. Many experiments fail because of uncontrolled variables, equipment malfunction, or poor design. In other cases, experimental results do not support the hypotheses, and new hypotheses and experiments must be formulated. When experiments fail, scientists often talk to colleagues and return to the literature, reviewing the work of other investigators for helpful clues. Often, a finding in a related field provides insights not available from a study of the scientist's own specialty. The history of science contains many examples of experiments that seemed at first to have failed, only to lead to a major breakthrough as the researcher pondered the reasons for the failure. Many of the discoveries that Thomas Edison made came from initial failures.

Modern Advances in Physiology Opened New Frontiers

In the twenty-first century, explorations of molecular/cellular biology, genomics, and bioenergetics remain the frontiers of physiological research. The pace of science has quickened. One example is the discovery of atrial natriuretic factor. In 1981, scientists showed that if a part of the heart (the atrium) was ground up, made into an extract, and injected into rats, a large amount of sodium was excreted by the kidneys. This was an important breakthrough because some mechanism of this type had long been hypothesized to explain certain aspects of water and salt balance in the body. Within 3 years, atrial natriuretic factor (a chemical messenger) had been purified, its amino acid sequence (basic units in proteins) defined, assay methods (defined procedures) developed, and synthesis by recombinant DNA technology (a procedure for determining structure and sequence) accomplished. Scientists learned where it was made, how it was made, how it was metabolized, and what it did. The astonishing rapidity of this work was possible because of strict adherence to the scientific method, the availability of sophisticated research methods, the large number of scientists working in this promising area, and the rapid communication facilitated by international meetings and publications.

Scientists Have an Ethical Responsibility

Scientists who publish their work in scientific journals must describe their experiments in sufficient detail in order that other researchers can independently repeat their findings. This fosters an objective approach to science and increases the reliability of the data. This also allows other investigators to detect possible errors or bias in the original experiments.

Scientists depend on a strong commitment to practical ideals, such as honesty, and a truthful obligation to communicate the results. Considerable damage can be done if an investigator knowingly disseminates false data: Not only is the image of the institution tarnished, but until the fabricated data are detected, many investigators in the scientific community may devote a lot of time and money to a futile research project due to erroneous data. Fortunately, the publishing of false data is rare and the scientific process is self-correcting through the consistent use of the scientific method.

As a final comment, many scientists face important ethical issues, especially in medical research. Issues such as cloning, the human genome, fetal research, and gene therapy are not going to be easily resolved.

STRATEGY FOR LEARNING PHYSIOLOGY

Many phenomena that occur in the body share basic underlying principles. To understand these basic principles and their applications, models are used that aid in the understanding of many physiological processes. These models provide ways to help recognize that many of the underlying principles reoccur in physiology. Consequently, these models can be applied to many different organ functions and help reinforce basic principles. There are seven basic models that will be used in this textbook. These models will provide a framework for explaining different physiological mechanisms. These models will be used to indicate recurring themes in subsequent chapters of the book. The seven basic models include **control systems, conservation of mass, elastic properties, membrane transport, cell-to-cell communication, chemical mass action,** and **mass and heat flow.** These seven general models, used alone or in combination, provide a framework in which physiological mechanisms can be analyzed and/or interpreted. These models, along with their basic components, are listed in Table 1–2.

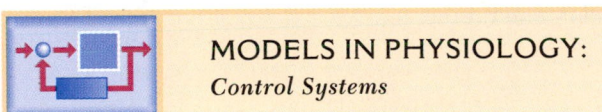

MODELS IN PHYSIOLOGY:
Control Systems

Coordinated Body Activity Requires a Control System

This model provides a way to understand how the body maintains its internal environment and is known collectively as **control systems** (Fig 1–13). Control systems share several features. They have a **regulated variable,** in which the control system tends to remain relatively constant. Examples of a regulated variable include blood glucose level, blood pressure, body temperature, and the amount of carbon dioxide in the blood. The control system uses a **sensing device,** or **sensor,** to detect changes in the regulated variable and sends a signal to a **comparator.** The comparator is the component that compares the signal coming from the sensing device and the **reference signal or set point.** The reference signal may be thought of as a set point that represents the desired value of the regulated variable. The comparator generates an error signal that will be sent to the **controller,** which inter-

---

TABLE 1–2

General Models Used in Physiology

General Model	Components	Functional Topic
Control systems	Sensor Controller Set point	Regulation Negative feedback Positive feedback
Conservation of mass	Input = output	Blood volume Alveolar gas composition Renal clearance Plasma hormone level
Mass and heat flow	Driving force Resistance Flow	Pressure–flow relationships Diffusion and osmosis Ion flux Heat flow
Tissue elastic properties	Elastic recoil Compliance Stiffness	Lung recoil Pressure–volume relationships Length–tension relationships
Membrane transport	Driving force Permeability Lipid bilayer	Cell diffusion Osmosis Cotransport Carrier-mediated transport
Cell-to-cell communication	Signal molecule Receptor Ligand binding	Hormone action Electrical synapses Motility
Chemical mass action	Reactants Products Equilibrium Dissociation constants	Oxygen–hemoglobin equilibrium

Control Systems

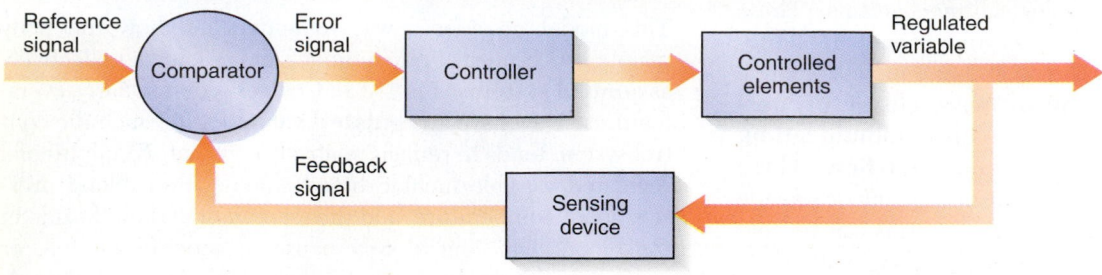

Figure 1–13

A schematic illustration of a control systems model. The basic components include a regulated variable, sensing device, reference signal, comparator, controller, and the error signal and feedback signals.

prets the error signal and responds by manipulating the controlled elements to minimize the error signal. It is important to distinguish between controlled elements and the regulated variable. For example, in the cardiovascular system, one of the regulated variables is mean arterial pressure. The controlled elements are those that can be manipulated by the system (e.g., heart rate, cardiac contractility, and blood vessel diameter).

Two general classes of control systems are active in the body: negative feedback systems and positive feedback systems. **Negative feedback systems** work to restore normal values of a variable and thus exert a stabilizing influence. **Positive feedback systems** destabilize and thus have limited value in maintaining the regulated variable. An example of negative feedback is a decrease in blood pressure, which leads to an increase in heart rate to return blood pressure back to normal. In positive feedback, the initial disturbance sets off a chain of events that actually increases the disturbance even more. One example of positive feedback is fever. With fever there is an increase in shivering, which acts as a positive feedback to further increasing body temperature.

Neurophysiology as a Control System

The **central nervous system** acts as the major control system in the body. For example, it monitors the temperature of the body and attempts to maintain it at 37°C. It has sensors that monitor the mean arterial blood pressure and tries to keep it at approximately 90 millimeters of mercury (mm Hg). It monitors the concentration of hydrogen ions in the extracellular fluid and attempts to maintain a pH of 7.4. The hydrogen ion concentration affects the ability of cells to carry out essential enzymatic reactions, and pH measures how acidic or basic the cellular fluids are.

The sensors that are part of the homeostatic function of the nervous system are associated with the autonomic nervous system (see Chapter 10). Sensors called **baroreceptors** are nerve endings that wrap themselves around blood vessels

to monitor blood pressure. The information they transmit to the brain is used by the brain to determine whether blood pressure must be adjusted. This adjustment is accomplished by the activation or inactivation of effector systems, such as the heart, causing it to change its rate and force of contraction, and the smooth muscles, causing them to dilate or constrict blood vessels.

Chemical sensors, also called **chemoreceptors,** monitor the concentrations of hydrogen ions and carbon dioxide in the extracellular fluid. Alterations in the levels of hydrogen ions and carbon dioxide are detected by chemoreceptors in the brain and in blood vessels located in the neck. The information these chemoreceptors transmit to the brain eventually affects the ways in which the lungs and kidneys work to maintain a constant pH in the body.

Sensors within the brain are also capable of monitoring body temperature. These **thermoreceptors,** located in the hypothalamus, detect the temperature of the blood flowing in the brain. When the temperature falls below 37°C, effector mechanisms are activated. These mechanisms reduce the amount of blood that flows to the skin, thus minimizing heat loss. Heat is retained in the body, thereby raising the temperature. In addition, hormones (chemical messengers) are released to increase the rate of cellular metabolism, which releases energy in the form of heat. Moreover, neural mechanisms go into action to produce shivering, which generates more heat. Together, these mechanisms generate and conserve heat in the body until the temperature is raised to 37°C. If too much heat is retained, different effector mechanisms take action to dissipate heat. Blood flow to the skin increases, and sweat glands are activated. Heat is lost from the body, and the temperature falls to 37°C.

Endocrine Function as a Control System

In addition to the nervous system, the other major control system in the body is the **endocrine system.** In the endocrine system, hormones serve as an important communication link in the effector component of several control systems. Many of the sensors involved are chemical sensors that serve to monitor changes in components in the extracellular fluid. Specialized cells in the pancreas and in the hypothalamus of the brain monitor the glucose concentration. Two hormones produced by the pancreas, insulin and glucagon, produce opposing effects in the body that lower and raise the blood glucose concentration, respectively. Similarly, much of the regulation of the concentration of calcium in the extracellular fluid involves hormones serving as the information link between sensors and effectors.

While there are many examples of homeostatic control mechanisms involving just the nervous system, or just the endocrine system, in a large number of cases, the two systems interact and work in concert. For example, when there is a loss of blood volume, a drop in blood pressure is detected by the baroreceptors. This information is relayed by the central nervous system, which in turn directly activates the heart and thirst mechanism. Specific hormones are also released as a result of nervous stimulation. These include epinephrine from the adrenal gland and antidiuretic hormone (ADH). Epinephrine acts on the heart and on blood vessels to produce effects that lead to the restoration of blood pressure to normal. ADH acts on the kidneys to retain body fluids.

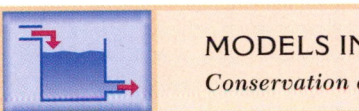

MODELS IN PHYSIOLOGY:
Conservation of Mass

Conservation of Mass Helps Maintain a Steady State

Conservation of mass, sometimes referred to as the "reservoir model," simply states that during a **steady-state condition,** the amount of mass that goes into a region in a period of time must equal the amount out of that region in a period of time. This means that in a steady-state condition, the reservoir stays level and the mass per unit of time that goes into the reservoir must equal the mass per unit of time that leaves the reservoir. The reservoir model can be applied to a cell, an organ, or a system. For example, during steady state, the cell volume remains constant and the amount of fluid that enters the cell per unit of time equals the amount leaving the cell per unit of time. The same applies for the amount of blood that enters an organ, such as the heart, kidney, or lung. During a steady-state condition, the amount of blood entering the heart per unit of time must equal the amount of blood leaving the heart per unit of time. Similarly, the amount of carbon dioxide entering the lungs during a steady state must equal the amount that leaves the lungs per unit of time. The reservoir model is also helpful to think about non–steady states, when what goes in per unit of time does not equal what goes out per unit of time. Consider the following two examples. In the first example, the aorta is considered the reservoir. However, when blood is being ejected from the heart, more blood enters than leaves the aorta per unit of time, and the volume of the aorta increases. When the ventricle relaxes, the aortic valve closes; more blood leaves the aorta per unit of time than enters it, and the volume in the reservoir decreases. In this example, a steady state is never reached.

In the second example, the reservoir is the lung. In the steady state, the amount of carbon dioxide entering the lungs per minute from the blood equals the amount of carbon dioxide that leaves the lungs through the airways per minute. If the subject decreases the volume of fresh air that enters the lung in a minute (i.e., a decrease in minute ventilation), the system will no longer be in a steady state for a period of time. The rate at which carbon dioxide enters the lung will be greater than the rate at which carbon dioxide leaves the lung. As a result, the level of carbon dioxide in the lungs (reservoir) will increase until a new steady state is achieved.

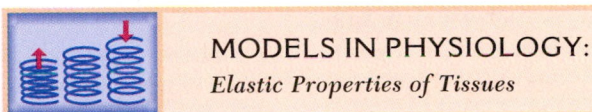

MODELS IN PHYSIOLOGY:
Elastic Properties of Tissues

Elastic Properties of Tissue Help Link Structure with Function

Elastic properties of tissue are important concepts to understand because much of the mechanical behavior of cells and tissue is linked to many physiological functions. A key example of this is the lung because most of the respiratory functions are tied to the elastic behavior of the lung. Consequently, many of the underlying mechanisms causing lung diseases are linked to some dysfunction of the elastic properties of the lung.

However, the word *elastic* often leads to confusion because the lay concept of *elastic* is different than that used in biophysics or physiology. As a result, many misconceptions arise about the elastic properties of tissues. The elastic model is based on principles that use a strict set of terminology. The term **elastic** refers to material that resists being stretched but, if stretched, recoils, or returns to its unstretched position. Thus, the key feature of elastic structures is **recoil,** a unique property of elastic material, as opposed to **plastic** material, which also is capable of being stretched but does not recoil. Examples of elastic and plastic material are a rubber band and putty, respectively. Both are capable of being stretched, but only the rubber band is elastic because of its ability to recoil. One of the misconceptions about elastic material (especially fabric such as a waistband elastic) is the notion that the "more elastic" it is, the more stretchable it will be, which is not the case. The more elastic the material is, the greater the recoil. Recoil is directly related to **stiffness,** or the ability to resist stretch. Thus, if two rubber bands have the same dimensions (length, width, and thickness) but one is twice as stiff as the other, then the stiffer one will also have greater recoil (i.e., greater ability to snap back). This means that elastic tissue that exhibits more recoil is more elastic but is stiffer and requires more force to stretch the material. This relationship holds true for a linear structure, such as a ligament; or for a volume structure, such as a blood vessel; or for an organ, such as the lung. For an elastic structure that has volume (e.g., cell, bladder, blood vessel, airways, heart, or lung), the recoil force equals zero at its resting volume (unstretched volume). At volumes greater than its resting volume, the recoil force is inward. At volumes less than its resting volume, the recoil force is outward. The direction of the recoil force is illustrated in the three springs used to illustrate elastic structure, seen in the previous icon. The center spring is at its resting position, and the recoil equals zero. The one on the left is compressed, which means the length is less than its resting position; the arrow indicates that the recoil is upward, toward the resting position. The spring on the right is stretched; the arrow indicates that the recoil is downward, toward the resting position.

Another term that is used to assess elastic properties of tissues is **compliance,** which is a measure of how stretchable or distensible the tissue is. Compliance is inversely related to recoil and stiffness. Thus, tissue that exhibits more recoil is less compliant. Conversely, tissue that is highly compliant is less elastic (i.e., easily stretched and has less recoil). The elastic model will be used extensively in the discussions of cardiovascular and respiratory physiology.

MODELS IN PHYSIOLOGY:
Transport Across Membranes

Membrane Transport Provides a Selective Mechanism for What Goes Into and Out of Cells

The **membrane transport model** provides a way of understanding the movement of solutes and water across cell membranes. Every cell is surrounded by a plasma membrane that separates the contents of the cell from its environment. The plasma membrane is a highly selective filter that permits certain material to enter and leave the cell. The selective nature allows the composition of the cell cytosol (fluid within the cell) to be very different than its surrounding environment. For example, the concentration of potassium ions inside the cell is approximately 25-fold higher than the concentration outside. The membrane transport model helps to illustrate how material is moved into and out of cells and different compartments of the body. There are two basic types of membrane transport that occur in the body: passive transport and active transport. **Passive transport** includes diffusion of

material and does require cellular energy. **Active transport** is linked directly or indirectly to cellular energy. Both transport systems involve two basic components, driving force and membrane permeability, which are described very well in Chapter 4.

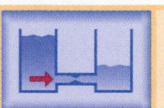

MODELS IN PHYSIOLOGY:
Mass and Heat Flow

Mass and Heat Flow Depend on Gradients and Resistance

The **mass and heat flow model** is based on **Ohm's law,** which states that flow results when an energy gradient exists, and the amount of flow resulting from the energy gradient depends on the degree of resistance. This model holds whenever flow exists in a biological system, and flow could be the movement of ions, water, solutes, blood, gas, or heat in the body. The model incorporates two very important principles: **gradient** and **resistance.** An energy gradient is the difference between two energy quantities separated by a physical distance. For solutes, this energy gradient is a **concentration gradient** that is separated by physical distance. For charged ions, the energy gradient may be an **electrical gradient.** For blood, gas, and heat, the energy gradient is a **pressure gradient** separated by a physical distance. For example, water will flow through a pipe only if there is a pressure gradient, that is, pressure is greater at one end than at the other end. Similarly, blood will flow through a blood vessel, or air will flow in an airway, only if there is a pressure gradient. Water moves by **osmosis** — diffusion of fluid through a semipermeable membrane until there is an equal concentration of fluid on both sides of the membrane — when an energy gradient between water molecules on either side of a semipermeable membrane (e.g., a cell membrane) exists.

Although gradients in biological systems have different terminology, it is important to remember that the energy gradient, that is, the difference between the two energy quantities separated by a physical distance, is basically the same.

Flow is met with resistance, and resistance can be in the form of physical distance, membrane permeability, diameter of a tube, and surface area. All of these cause resistance and impede flow. For example, if two cells have the same pressure gradient for oxygen, but one cell has a membrane thickness twice that of the other, then the flow of oxygen into the cell with a thicker membrane will be less, even though the pressure gradient for oxygen is the same for both cells. In this example, membrane thickness offers resistance to the movement of oxygen into the cell; the thicker the membrane, the more the flow of oxygen will be impeded. A similar example occurs with blood flow. If a pressure gradient is the same for two blood vessels with the same length, but one has a diameter that is half of that of the other, then flow in the smaller vessel will be much less because of the higher resistance.

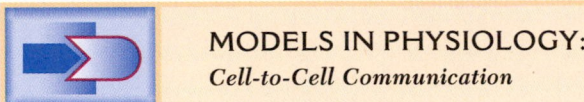

MODELS IN PHYSIOLOGY:
Cell-to-Cell Communication

Cell-to-Cell Communication Uses Chemical Signals

The coordination of different functions in tissues of the body is achieved by well-developed communication systems. The **cell-to-cell communication model** is used to illustrate how cells communicate with each other. Two broad types of communication occur in the body. Cells that are in direct physical contact with one another have structures known as **gap junctions** in their plasma membranes. These junctions allow cells to have direct communication by allowing small molecules to pass directly from the cytosol of one cell to the cytosol of an adjacent cell. Cells that are some distance apart communicate with each other by sending **chemical signals,** molecules secreted by one cell that bind to the plasma membrane of a distant cell, or the **target cell.** The model for chemical signaling involves two basic components: a signaling molecule and a ligand–receptor interaction. The **signaling molecules** can be a **local chemical mediator** (e.g., histamine), a **neurotransmitter,** which is produced by nerve cells (e.g., acetylcholine), or a **hormone** (e.g., epinephrine), which is released and carried by the blood to a target tissue. The second component, **the ligand–receptor interaction,** can occur on the plasma membrane or in the cytoplasm of the target cell. The cell-to-cell communication model is well described in Chapter 5 and will serve as a useful model to illustrate cell communication under different physiological conditions.

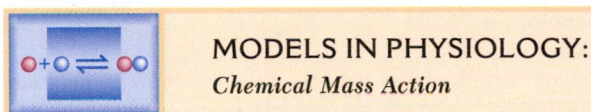

MODELS IN PHYSIOLOGY:
Chemical Mass Action

Chemical Mass Action Model Describes Molecular Interactions

In general, chemical reactions are described in terms of reactants and products. For example:

Reactants Products
$$A + B \leftrightarrow AB$$

In reversible reactions, the concentrations of reactants and products determine when equilibrium is reached. The rate of the reaction in either direction depends on a rate constant. If the concentrations of reactants decrease, the reaction reverses. This decreases the concentration of products and increases the concentrations of reactants until a new equilibrium state is reached.

This concept from chemistry is especially helpful when thinking about a certain class of reactions that occur in physiological systems. In this class of reactions, the "product" consists of one "reactant" molecule (e.g., oxygen, O_2) bound reversibly to another "reactant" molecule (e.g., hemoglobin, Hb):

$$O_2 + Hb \leftrightarrow HbO_2$$

The equilibrium between unbound reactants (O_2 and Hb) and bound products (HbO_2) depends on the relative concentrations of each. Consider, for example, the binding of oxygen to hemoglobin. The amount of binding depends on the amount of oxygen dissolved in the cytoplasm of red blood cells. If oxygen is bound to hemoglobin and the amount of dissolved oxygen in the cell decreases, oxygen will dissociate (unbind) from the hemoglobin and enter the dissolved state until a new equilibrium is reached.

Similar examples can be found in the endocrine system (hormone–receptor reactions), in the nervous system (neurotransmitter–receptor reactions), in membrane transport mechanisms (substrate–carrier reactions), in intracellular reactions (e.g., calcium binding to receptor sites), and in many other physiological mechanisms in which reversible chemical binding occurs.

CHAPTER REVIEW

Summary

- Early Greeks, such as Aristotle, searched for explanations of blood flow, temperature, and other aspects of how the body worked.
- Renaissance scientists corrected many errors of Galen's teachings. William Harvey's discovery of closed circulation and the use of quantitative methods marked an end of Galen's dominance and the beginning of modern physiology.
- The scientific method, which emphasizes experiment, quantitative results, retesting, and information exchange, forms the basis

of all modern scientific research. In conducting investigations, scientists use both inductive and deductive reasoning.
- Epithelial, connective, muscle, and nerve tissues are the four major tissue types in the body.
- Groups of tissues that work together form organs; organs, in turn, combine their functions to maintain systems.
- The body comprises two fluid compartments: the fluid within each plasma membrane, the intracellular fluid; and the extracel-

lular fluid (plasma, lymph, and interstitial fluid), which surrounds the cells.

- Homeostasis refers to control of the internal environment, and the extracellular fluid forms the "internal environment" of the body, allowing it to maintain constant conditions, such as temperature.

- Models provide a useful tool in the learning and understanding of physiology.

Review Questions

Choose the Correct Answer

1. The imaginary plane that divides the body into right and left portions is the:
   a. transverse plane.
   b. sagittal plane.
   c. frontal plane.
   d. horizontal plane.
   e. coronal plane.

2. The individual who discovered the circulation of the blood was:
   a. Galen.
   b. Harvey.
   c. Malpighi.
   d. Vesalius.
   e. Hippocrates.

3. The ancient Greeks considered the universe to be composed of four elemental substances:
   a. carbon, hydrogen, oxygen, and nitrogen.
   b. sodium, potassium, calcium, and magnesium.
   c. air, water, earth, and fire.
   d. gold, silver, platinum, and mercury.
   e. water, salt, pepper, and sugar.

4. The concept of the heart as a furnace for heating the blood and as the seat of intelligence originated with:
   a. Aristotle.
   b. Galen.
   c. Harvey.
   d. Erasistratus.
   e. Herophilus.

5. Which of the following is the first step in the scientific method?
   a. hypothesis
   b. experimentation
   c. observation
   d. conclusion
   e. retest

6. Which of the following properties distinguishes the difference between plastic and elastic material?
   a. stretch
   b. compliance
   c. stiffness
   d. recoil
   e. elastance

7. "To remain the same" is the literal definition of the word:
   a. homogenized.
   b. homeopathic.
   c. *Homo sapiens.*
   d. homosexual.
   e. homeostasis.

8. In a negative feedback control system:
   a. the variable is returned toward normal after a disturbance.
   b. the system operates to destabilize.
   c. the variable never changes.
   d. when the value of the variable increases, the system operates to increase it further.
   e. the system always operates to decrease the value of the variable.

9. Positive feedback systems:
   a. add stability to the body's internal environment.
   b. tend to restore the normal value of an abnormal controlled variable.
   c. are not common in the body because they act to destabilize or reinforce the direction in which a variable is moving.
   d. involve only the brain and processes of learning.
   e. raise the blood glucose level when it falls below normal.

10. According to its literal definition, the word *physiology* means the study of:
    a. animals.
    b. nature.
    c. plants.
    d. function.
    e. structure.

11. The smallest group of different types of cells that share common functional objectives with homeostasis is called a(n):
    a. membrane.
    b. tissue.
    c. organ.
    d. system.
    e. organelle.

12. Historically, processes of the body were believed to be driven by various forms of *pneuma*. The principal proponent of "pneumatology" was:
    a. Vesalius.
    b. Harvey.
    c. Herophilus.
    d. Erasistratus.
    e. Galen.

13. Which of the following is not one of the five steps in the scientific method?
    a. observation
    b. conclusion
    c. deduction
    d. hypothesis
    e. experiment

14. In physiology, the term *homeostasis* means:
    a. everything is the same inside the body.
    b. nothing ever changes inside the body.
    c. the body's internal environment is normally stable.
    d. to prefer one's own sex.
    e. to mix bodily fluids until they are uniform.

15. The concept of stability in the body's internal environment as a requisite for a free and independent existence was first pointed out by:
    a. Walter B. Cannon.
    b. William Harvey.
    c. Claude Bernard.

d. Vesalius.
e. Malpighi.

16. The imaginary plane that divides the body into anterior and posterior halves is the:
    a. sagittal plane.
    b. coronal plane.
    c. frontal plane.
    d. horizontal plane.
    e. vertical plane.

17. Two or more tissues that share common functional objectives relative to maintaining homeostasis form a(n):
    a. system.
    b. organ.
    c. membrane.
    d. organelle.
    e. receptor.

18. Which of the following is not an extracellular fluid?
    a. lymph

b. blood
c. interstitial fluid
d. cerebrospinal fluid
e. saliva

19. An anatomical term meaning "further from the midsagittal plane" is:
    a. medial.
    b. lateral.
    c. posterior.
    d. superior.
    e. inferior.

20. In an inverse relationship between X and Y, as X increases in value, the value of Y:
    a. does not change.
    b. increases.
    c. decreases.
    d. increases, then decreases.
    e. decreases, then increases.

Answers to Case History Questions

1. Antipyretics directly or indirectly reduce the set point of the body's thermostat (hypothalamus) from a febrile level to normal.
2. An antipyretic, by reducing fever, compromises the benefits of fever and may actually prolong illness.
3. At the onset of fever, heat gain and heat conservation mechanisms are brought into play to increase core body temperature.

Peripheral vasoconstriction conserves body heat but causes the skin to feel cold. Shivering generates body heat. "Goose pimples" reflect a vestigial mechanism of erecting body hair to trap an insulating layer of air near the skin to reduce heat loss. All of these responses contribute to chills.

Key Terms

anatomical planes (p.11)
cell differentiation (p. 13)
cell theory (p. 11)
connective tissue (p. 14)
control systems (p. 23)
epithelial tissue (p. 14)

experimental design (p. 18)
extracellular fluid (p. 17)
homeostasis (p. 11)
hypothesis (p. 19)
intracellular fluid (p. 17)
muscle tissue (p. 14)

negative feedback system (p. 24)
nervous tissue (p. 14)
organ (p. 14)
physiology (p. 14)
plasma membrane (p. 13)

positive feedback system (p. 24)
protein conformation (p. 17)
protoplasm (p. 13)
scientific method (p. 18)
sensor (p. 23)
variable (p. 20)

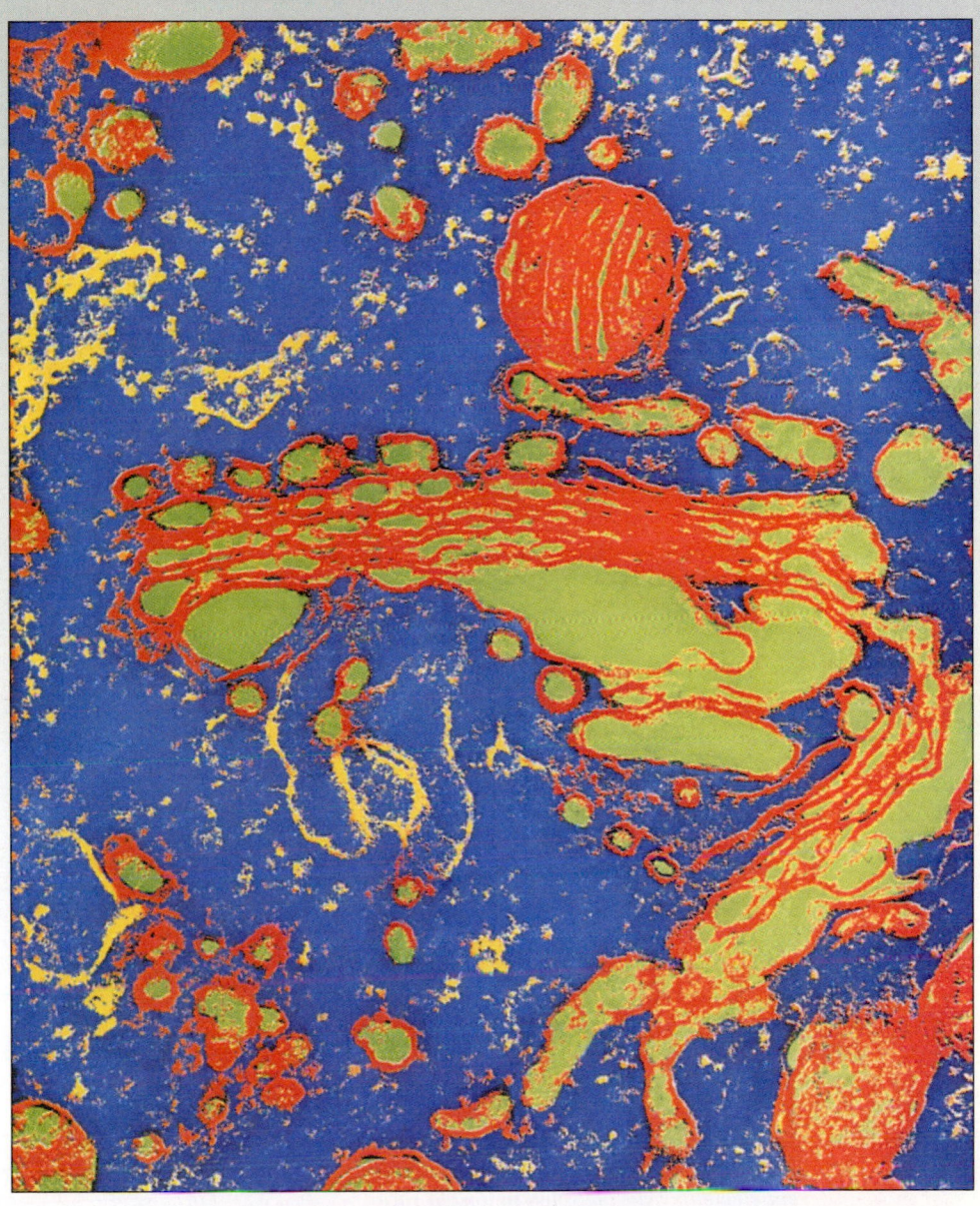

• *A false-color transmission electron micrograph of a section through one stack of the Golgi apparatus. The Golgi stack is the paired membranes (green outlined in red) in the middle of the image. The large red sphere near the top is a mitochondrion.* (× 31,800)

(© Dr. Gopal Murti/SPL/Photo Researchers, Inc.)

Chapter 2

CHEMICAL AND PHYSICAL PRINCIPLES

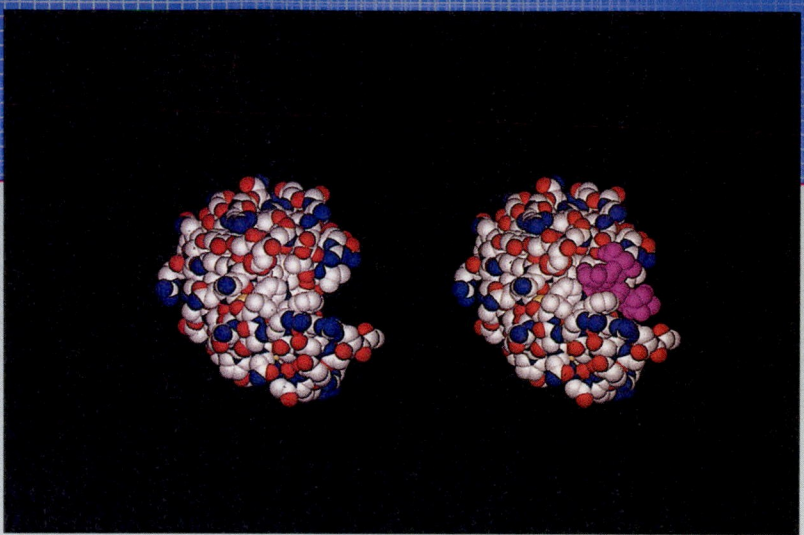

KEY CONCEPTS

- An element cannot be broken down into simpler substances by chemical reactions. An atom is the smallest unit of an element that retains all the chemical properties of the element. All elements are composed of atoms, and all atoms are composed of neutrons, protons, and electrons.

- Chemical bonds between atoms are formed by sharing, gaining, or losing electrons. A compound is formed when two or more different elements are combined in a fixed ratio.

- A chemical reaction occurs when the bonds between the atoms of a compound are broken and the atoms recombine in different ways to form new compounds. Cells contain enzymes that increase reaction rates by binding the reactants so that specific reactions can occur.

- Water is essential for living organisms and is the major chemical compound in most cells. Water is an excellent solvent; many biologically important reactions require an aqueous environment.

- Macromolecules play vital roles in cells and organisms. The most important are carbohydrates, lipids, proteins, and nucleic acids.

- Living organisms are highly organized systems; the organization is maintained by constant work. The chemical energy required to perform this work is held in the chemical bonds of molecules involved in metabolic reactions.

- Energy released from food molecules is trapped temporarily in the form of adenosine triphosphate (ATP) and can be used to drive synthetic reactions.

CASE HISTORY

A young couple take their only child, a 7-month-old boy, for a checkup. The pregnancy and delivery were uneventful; the child was pronounced normal and healthy at birth. The parents are now very concerned because the child's abdomen seems to be enlarged and they have noticed that the child has become less active and less interested in his surroundings. Upon physical examination, the pediatrician observes that the protruding abdomen is caused by an enlarged spleen and liver, a condition known as hepatosplenomegaly. The pediatrician also concludes that the reduced physical activity is caused by retardation of psychomotor functions. This refers to the psychological processes that are associated with the movement of muscles and that control voluntary movement. The baby's weight is on the low side of the normal range. The doctor is immediately concerned that this combination of symptoms, the enlarged internal organs and possible mental retardation, may be the result of an abnormality of lipid metabolism. The doctor schedules a liver biopsy, which is typically performed under local anesthesia by inserting a thin, hollow needle through the skin beneath the ribs and directly into the liver. Small amounts of liver tissue are trapped inside the needle and can be safely removed for measurement of enzymes and other indicators of liver function. In this case, the doctor is interested specifically in the activity of sphingomyelinase, which is the enzyme that catalyzes the breakdown of sphingomyelin, one of the phospholipids in cell membranes. Accumulation of sphingomyelin is one of the most common results of abnormal lipid metabolism. In spite of the doctor's recommendation, the biopsy is not performed because the parents decide that the procedure is too invasive. Three months later, they return to the doctor; by this time, the child shows significant retardation of physical and mental development. They finally consent to the liver biopsy; subsequent laboratory analysis of the liver tissue confirmed the doctor's initial suspicions. The analysis showed that the sphingomyelinase activity was decreased to less than 10% of normal. The liver sphingomyelin content was measured in the same sample and was found to be increased to 25-fold greater than normal.

Questions

1. Why is there a large increase in sphingomyelin content in the liver?

3. Why are the liver and spleen enlarged?

2. Why does sphingomyelin accumulation lead to mental retardation?

4. What disease is suggested by the baby's symptoms?

INTRODUCTION

What basic chemistry vocabulary should be reviewed before examining the chemical and physical principles that are relevant to physiology?

Living organisms are composed of **matter.** Matter is anything that occupies space, has mass, and can assume the form of either a solid (e.g., stone), a liquid (e.g., water), or a gas (e.g., oxygen). Matter exists as elements. An **element** is a substance that cannot be broken down into simpler substances by chemical reactions. There are 92 elements that occur naturally, although scientists have made more than a dozen new elements. In nature, the four most abundant elements are oxygen, silicon, aluminum, and iron. Together, they account for 87% of the atoms in the earth's crust. Elements are commonly referred to by their abbreviations, called **chemical symbols** (e.g., "O" for oxygen and "Fe" for iron).

Elements are composed of atoms. An **atom** is the smallest unit of an element that retains the properties of the element. Atoms of the same element combine in specified ways to form different varieties of the same element. Atoms of different elements combine in specified ways to form compounds. A **compound** is a substance formed by the union of the atoms of two or more elements. Many substances, whether made of one element or more than one element, have molecules (not atoms) as their smallest stable units. A **molecule** is formed by the union of two or more atoms, whether they are of the same element or not. Oxygen atoms, for example, in samples of pure oxygen, form pairs called oxygen molecules (O_2) rather than existing singly. Water, in contrast, is a compound substance; one water molecule consists of two hydrogen atoms combined with one oxygen atom (H_2O). A molecule or other group of atoms is designated by a **molecular formula,** such as H_2O (water), which shows the number of each type of atom in the group. The subscript "2" after the H, for example, means that two hydrogen atoms are present in a molecule of water. If there is no subscript number following a chemical symbol, the symbol represents a single atom.

In living cells, molecules combine further to form large molecules known as **macromolecules.** Macromolecules all contain carbon, the unique structure of which allows macromolecules to assume huge proportions and a great variety of shapes. Proteins, sugars, lipids, and **nucleic acids** are the chief macromolecules found in most living cells. The specific properties of these macromolecules, and the organization of their many interactions within the cell, form the subcellular structures and facilitate the chemical reactions that are the basis for all physiological processes and for life itself.

OVERVIEW: MATTER AND ENERGY IN THE LIVING CELL

As stated, the predominant elements in the earth's crust are oxygen (O), silicon (Si), aluminum (Al), and iron (Fe). In contrast, the matter of the human body is largely hydrogen

TABLE 2–1	
Essential Elements of Living Organisms*	
Element	**Chemical Symbol**
Hydrogen	H
Oxygen	O
Carbon	C
Nitrogen	N
Phosphorus	P
Sulfur	S
Sodium	Na
Potassium	K
Magnesium	Mg
Calcium	Ca
Chlorine	Cl

Trace elements (present in very small amounts)

Manganese (Mn), Iron (Fe), Cobalt (Co), Copper (Cu), Zinc (Zn), Boron (B), Aluminum (Al), Vanadium (V), Molybdenum (Mo), Iodine (I), Silicon (Si), Tin (Sn), Nickel (Ni), Chromium (Cr), Fluorine (F), and Selenium (Se).

*Not all of these elements are essential for every species.

(H), oxygen (O), carbon (C), and nitrogen (N). Together, they account for 99% of the atoms in the body. The marked difference between the chemical compositions of the human body and its external environment indicates that hydrogen, oxygen, carbon, and nitrogen must be uniquely suited for the processes that maintain the living state. In addition to these 4 most abundant elements, 23 others have been found to be essential components of living organisms. The 27 elements essential for life are listed in Table 2–1.

The matter of a living organism is highly organized. It also undergoes frequent transformation. The maintenance of this organized state and the processes of transformation require a constant input of energy. **Energy** is the capacity to produce change. It is measured by the amount of work performed to achieve a given change. In the universe, the energy used for work comes from various sources, such as heat, light, electricity, and chemical reactions. In the cell, the energy used for work is mostly **chemical energy** released by reactions that take place within cells. In the cell, the work to be done consists of moving matter from one place to another and changing it from one form to another — that is, breaking down or building up molecules from other molecules.

The different forms of energy also exist in two different states: **Potential energy** is stored energy; **kinetic energy** is energy that produces motion. Both states of energy enter into the metabolic processes of cells.

The chemical basis of the matter of widely diverse forms of life is strikingly similar. All living cells have fundamentally similar constituents and activities. Thus, both a complex, multicellular organism, such as a human being, and a simple, unicellular bacterium, such as *Escherichia coil* (*E. coli*), use the same simple molecules to synthesize, or make, larger molecules. Both human cells and bacterial cells store energy in the form of **adenosine triphosphate (ATP)**. Both types of cells store genetic information in **deoxyribonucleic acid (DNA)**.

The chemical content of an *E. coli* cell has been analyzed in detail. Its composition is 70% water, 15% proteins, 7% nucleic acids, 2% carbohydrates, 2% lipids, 1% inorganic ions, and 3% other small molecules. The cell is mostly water, and the major macromolecules present are proteins and nucleic acids. The chemical composition of the typical mammalian cell is markedly similar to that of this bacterial cell.

To determine why certain chemical structures dominate all forms of life so regularly, this chapter provides a survey of general chemical and physical principles. Then it focuses more closely on the specific principles that influence the life processes. The chapter begins with the simplest components of matter — atoms and small molecules — and the chemical and physical laws that govern their behavior. It then proceeds to the major classes of "giant" molecules, such as sugars, proteins, lipids, and nucleic acids. A look at their components and properties provides a sound basis for the description, in later chapters, of their unique roles in the lives of cells.

ATOMS AND ELEMENTS

> ***What conventions should be reviewed before examining chemical processes?***

In solid matter, atoms are closely packed. In liquid and gaseous matter, atoms are more loosely arranged. The same element can appear in different forms, depending on the temperature and pressure of its surroundings. For example, at room temperature and pressure, the element carbon is a solid because its atoms are packed closely together. At higher temperatures, carbon becomes liquid; at still higher temperatures, it takes the form of a gas as its atoms become separated from each other.

Atoms Are Composed of Subatomic Particles

Each atom, regardless of its element, is composed of certain **subatomic particles.** Three types of particles are important to an understanding of the behavior of atoms in life processes: (1) neutrons, (2) protons, and (3) electrons. **Neutrons** are electrically neutral, whereas **protons** are positively charged, and **electrons** are negatively charged. Protons and neutrons stay in the **nucleus,** or innermost area, of the atom, whereas electrons constantly orbit the nucleus (Fig. 2–1). The electrical attraction between the positively charged protons in the nucleus and the negatively charged electrons in orbit around the nucleus stabilizes the structure of the atom.

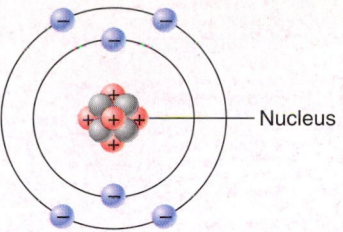

— Nucleus

Figure 2–1

The basic structure of an atom. Protons (positively charged particles) and neutrons (neutral particles) are found in the nucleus. The number of protons is the atomic number. Electrons (negatively charged particles) orbit the nucleus. A neutral atom contains equal numbers of protons and electrons.

In order to count large numbers of small atoms conveniently, scientists have come up with a counting method called the **mole.** Just as in everyday speech we talk of a dozen eggs, a dozen pencils, or a dozen pages — all the time meaning 12 eggs, 12 pencils, or 12 pages — scientists speak of a mole of atoms to mean 6.022045×10^{23} atoms. The number 6.022045×10^{23} is called **Avogadro's number.** A mole is a convenient and accurate way to express the same number of atoms of any substance. Thus, a mole of sodium atoms would be 6.022045×10^{23} sodium atoms; a mole of chlorine atoms would be 6.022045×10^{23} chlorine atoms.

Uniqueness of Elements: Protons and Atomic Number

Atoms of different elements are distinct from one another because of the different numbers of each subatomic particle they contain. In fact, the essential identity of an atom is found in the number of protons in its nucleus, called the **atomic number.** In the uncharged, or electrically neutral, atom, the number of electrons is the same as the number of protons. An atom of the element hydrogen has 1 proton in its nucleus and 1 electron in orbit around the nucleus. An oxygen atom typically has 8 protons and 8 electrons. A carbon atom has 6 protons and 6 electrons; a nitrogen atom has 7 of each particle; a uranium atom has 92 of each particle.

Variation Within Elements: Neutrons, Mass Number, and Isotopes

The **mass number** of an element is the total number of protons and neutrons in the nucleus. Because the number of protons is different for every element, the mass number is also different for every element. For example, the mass number of the most common form of hydrogen is 1 because the hydrogen atom has one proton (p) and no neutrons (n) in its nucleus (Fig. 2–2). Oxygen has 8 protons and 8 neutrons, so its mass number is 16. Nitrogen has 7 protons and 7 neutrons, so its mass number is 14.

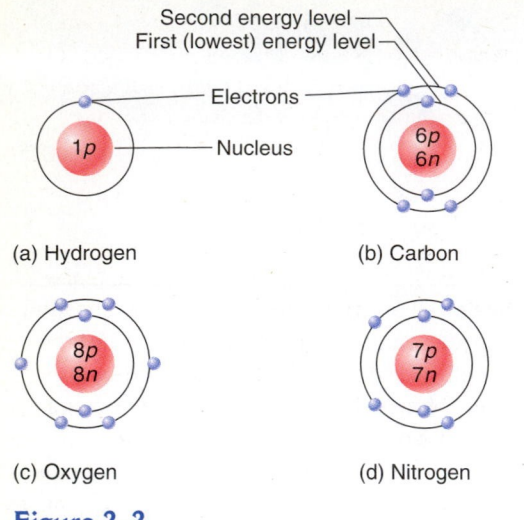

Figure 2–2

The electron configurations of four important elements. Hydrogen **(a)** needs one more electron to complete its only shell. Carbon **(b)** needs four electrons; oxygen **(c)**, two; and nitrogen **(d)**, three to complete their outer shells. The electron orbits are depicted as concentric circles for the sake of simplicity, but their true movements are more complex.

Moles and grams (the metric unit of mass) can be related in the following way: one mole of an element, or 6.022045×10^{23} atoms, contains x grams of the element, x being equal to the mass number of the element. For example, one mole of sodium atoms (6.022045×10^{23}) weighs 23 grams (g), 23 being the mass number of sodium. One mole of chlorine atoms weighs 35 g because the mass number of chlorine is 35.

While all atoms of the same element have the same atomic number (number of protons), atoms of the same element may have differing mass numbers because they may have differing numbers of neutrons. For example, hydrogen exists in three forms: one form contains no neutrons in its nucleus, another contains one neutron, and another contains two neutrons. These three forms of hydrogen are called **isotopes** of hydrogen. Because the mass number of an atom takes into account the number of neutrons, different isotopes of an element have different mass numbers. Generally, isotopes have the same chemical properties but their physical properties differ. Many isotopes are used widely in biomedical research for this reason.

An isotope is designated with both the atomic number (number of protons) and the mass number (number of protons plus neutrons) written to the left of the chemical symbol. For example, the isotope of hydrogen that has no neutrons has an atomic number of 1 (because all hydrogen atoms have 1 proton) and a mass number of 1, written as $_1^1\text{H}$. The isotope of hydrogen that has one neutron (deuterium) would be written as $_1^2\text{H}$; the isotope with two neutrons (tritium), as $_1^3\text{H}$.

Ions Are Atoms with Electrical Charges

A neutral, or uncharged, atom has the same number of protons (positive charges) and electrons (negative charges). When a neutral atom gains or loses an electron, it acquires an electric charge and becomes an **ion.** An atom that loses one or more electrons acquires a positive charge because it now has more protons (positively charged particles) than electrons (negatively charged particles). Such an atom is called a **cation.** An atom that gains one or more electrons acquires a negative charge and is called an **anion.**

For example, a potassium cation is formed when neutral potassium loses one electron. The single charge is denoted by a plus sign written to the upper right of the chemical symbol for potassium: K^+. The chloride anion, formed when chlorine gains an electron, is written as Cl^-. A calcium ion, formed by the loss of two electrons, would be denoted by a number 2 and a plus sign: Ca^{2+}. An oxide ion, formed by the gain of two electrons, is written as O^{2-}.

The potassium ion is a very important ion in the study of physiology because it is the major cation found inside most cells, while the sodium ion (Na^+) is the major cation present outside cells in fluids such as plasma. Chloride ions are the most abundant anions in plasma.

Atoms Prefer a Full Outermost Shell

The electrons, in their orbit of the nucleus, move in a specified **electron configuration,** consisting of a series of **shells** that can be thought of as concentric spheres (although this is a very simplified view of the way electrons actually move). The shells represent different energy levels and can hold different numbers of electrons. The shell with the lowest energy level, located closest to the nucleus, can accommodate two electrons. The shell with the next highest energy level can hold eight electrons. The electron configurations of hydrogen, oxygen, carbon, and nitrogen atoms are illustrated in Figure 2–2. Larger atoms have more electrons occupying several different energy levels. For example, the 92 electrons in the uranium atom are arranged in seven different shells, each with a different energy level.

The number of electrons in the outermost shell of an atom determines many of the chemical properties of the element. An atom may be stable and neutral even if its outermost shell is not filled to capacity. It can fill this outermost shell by interacting in certain ways with other atoms both of the same element and of different elements. An atom of carbon, for example, has four electrons in its outermost shell. Because this is the second shell from the nucleus, it can hold up to eight electrons. The carbon atom can fill that shell with four more electrons.

An atom can fill its outermost shell in one of three ways: (1) it can share electrons with one or more atoms that do not have complete outer shells; (2) it can obtain electrons outright from other atoms; or (3) it can donate all of the electrons in its currently incomplete outer shell to other atoms so that its next-lower shell, which is filled to capacity, becomes its outermost shell.

CURRENT CONCEPTS IN PHYSIOLOGY

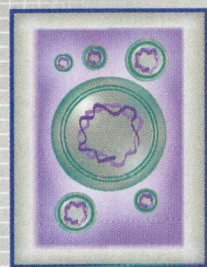

Fluorescence Microscopy of Living Cells and Tissues

The ability to distinguish objects invisible to the naked eye has revolutionized our approach to disease, public health, and agriculture. The Dutch cloth merchant Antony van Leeuwenhoek constructed the first crude microscope in the seventeenth century; the magnifying power of the conventional light microscope has not increased dramatically since that time. The limitation is due to the properties of light waves themselves, which interfere with each other. The result of this interference is that particles that are closer together than 200 nm cannot be **resolved** and appear to the eye as a single particle. The **electron microscope** uses a beam of high-energy electrons to dramatically improve magnification and resolution, but the tissue sample must be fixed and embedded in resin to withstand the electron bombardment. This severe treatment kills all cells, so it is not possible to monitor events within living cells and tissues using this technique.

The **confocal microscope** allows for the examination of live and unfixed tissues at the highest resolution available to the light microscope. When combined with **fluorescent** techniques, such as fluorescently labeled proteins, it allows scientists to examine interactions between specific molecules within a cell. In fluorescence microscopy, high-energy photons of light at a defined short wavelength (excitation) strike the fluorescent label, or fluorophore, and are absorbed. This excites the electrons of the fluorophore and moves them into an electron shell at a higher energy level. When the electrons resume their original energy state, a few nanoseconds later, the extra energy held by the fluorophore is emitted as new photons, but these have longer wavelengths (lower energy) than the incoming excitation light. Optical filters exclude all colors of the incoming light so that only the emission wavelength can be seen. The disadvantage of conventional fluorescent microscopes is that fluorophores above and below the plane of focus also become excited and the light emitted from these sources can blur the image and cause a loss of resolution. This problem is solved by the confocal microscope, which has the ability to eliminate the light emitted by fluorophores that are out of focus. In the confocal microscope, a tightly focused light source, usually a laser, is used for excitation so that the fluorophores at the plane of focus receive the most intense illumination. Emissions from sources above and below the plane of focus are blocked by a tiny pinhole that lets only the emission from the plane of focus reach the detector. Resultant images are striking; computer software packages allow for manipulations, such as progressive sectioning along multiple focal planes and construction of three-dimensional images.

A limitation of confocal microscopy is that the excitation light is of short wavelength and high energy, which can cause direct damage to the sample and photobleaching of the fluorophores, or loss of fluorescence, throughout the sample. The design of the **two-photon microscope** avoids these pitfalls. Instead of exciting the fluorophore with a single high-energy photon, the two-photon microscope uses low-energy light of very long wavelength that is provided by a laser emitting ultrashort pulses of very high power. Two-photon excitation occurs when the fluorophore absorbs two photons simultaneously; the additive effect causes emission of fluorescence at the usual wavelength. For example, an infrared laser at 1050-nm wavelength is used for two-photon excitation of a green-absorbing (525-nm) fluorophore. Two-photon excitation can only happen right at the focal plane because that is where the photon density is sufficiently high. The high-powered laser ensures that there are a sufficient number of two-photon excitation events while the ultrashort pulses, each lasting only 10^{-13} to 10^{-12} seconds, ensure that the average laser power to the sample is fairly low. There are several advantages of this new technology. The long-wavelength light directed at the sample is less damaging and, because it is only absorbed directly at the focal plane, fluorophores outside this spot are not excited and are spared from photobleaching. Because there is no out-of-focus absorption, more of the excitation light reaches the point of focus, allowing for greatly increased penetration of the sample. In fact, the two-photon microscope excels in dynamic imaging of thick, multicellular systems that often are encountered by physiologists. It can be used to examine the changes in distribution of Ca^{2+} ions in nerves of living mice and to perform time-lapse imaging of hamster embryo development over a period of 10 hours. Two-photon microscopes are now available from commercial suppliers; increasing numbers of researchers are able to use this state-of-the-art technology.

MOLECULES, GROUPS, AND COMPOUNDS

 How do atoms interact with other atoms?

The process of sharing, gaining, or losing electrons leads to the creation of a **chemical bond** between atoms. Atoms combine in definite proportions to form variations of elemental substances or new substances called **compounds.** Two atoms from the same element may join to form a **"diatomic" molecule.** Oxygen and hydrogen, for example, rarely exist as single atoms; instead, they are found as O_2 and H_2 molecules, pairs of atoms bonded together.

The difference between molecules and compounds is determined by the type of bond formed between the atoms from different elements as they come together. The result of bond formation may be a molecule (a discrete group of atoms, e.g., H_2O) or a larger structure known as a **crystal lattice,** which is made up of atoms arranged in precise patterns (e.g., sodium chloride). Both water and sodium chloride are compounds, but water forms molecules, whereas sodium chloride forms a crystal lattice. Whether or not separate molecules are formed, atoms bond in definite proportions, which are indicated by a molecular formula for the compound. For example, even though sodium chloride exists as a crystal lattice made up of many alternating sodium and chloride ions rather than discrete molecules, the compound has the formula NaCl, indicating that sodium and chloride are always present in a ratio of 1:1.

The molecular formula of a compound shows the relative quantities of atoms; the **structural formula** shows their general arrangement in two-dimensional space, for example:

	water	methane
Molecular formula	H_2O	CH_4
Structural formula	H—O—H	

$$H-\underset{\underset{H}{|}}{\overset{\overset{H}{|}}{C}}-H$$

A **chemical group** is a small cluster of bonded atoms that functions almost as a single atom. Chemical groups behave in distinctive ways and have special names. Examples of chemical groups are —OH (the "hydroxide group"), —COOH (the "carboxyl group"), and —PO_4 (the "phosphate group"). The dash written before the chemical symbols indicates that the group is not a free molecule but is attached to a larger molecule.

Just as atoms are counted in groups called *moles,* so are molecules. A mole of molecules, regardless of the substance, is 6.022045×10^{23} molecules. A mole of oxygen is 6.022045×10^{23} O_2 molecules. A mole of water is 6.022045×10^{23} H_2O molecules. The **molecular mass** of a compound is the sum of the atomic masses, or mass numbers, of its elements. Molecular mass is expressed in units known as **daltons** (Da), after John Dalton, who proposed the atomic theory of matter. The dalton is defined as 1/12 of the mass of carbon (mass number = 12). Based on this system, methane (CH_4) has a molecular mass of $12 + 4 = 16$ Da.

Regardless of whether a molecule is made of atoms from the same element or different elements, the type of bond formed depends on the manner in which the atoms interact. If they complete their outermost shells by sharing electrons, the bond formed is a **covalent bond.** If the atoms gain or lose electrons outright, the bond formed is an **ionic bond.**

 How do chemical bonds form?

A Covalent Bond Is Produced by Electron Sharing

Chemical bonds formed when two or more atoms share electrons are known as covalent bonds. These bonds are very stable and strong and are the type of bond most frequently found in biological compounds. Hydrogen, oxygen, carbon, and nitrogen, the major elements in living cells, readily form covalent bonds with each other. Atoms of these four elements all have an incomplete outer shell: hydrogen lacks one electron, oxygen, two; carbon, four; and nitrogen, three. They can form not only single bonds by sharing one pair of electrons but also double and (in the case of carbon) triple bonds. This ability increases their versatility in chemical reactions. A single carbon can form covalent bonds with up to four other carbon atoms to produce an enormous variety of linear, branched and cyclic structures that are the backbones of large molecules.

Covalent bonds and the molecules that they form may be either polar or nonpolar. In a **polar covalent bond,** a pair of electrons is shared unequally; that is, the electrons in the pair are attracted more strongly to the nucleus of one bonded atom than they are to the nucleus of the other bonded atom. The electrons can be thought of as spending more time near the nucleus of one atom, creating a partial negative charge, or pole, at one end of the molecule and a partial positive charge, or pole, at the other end. In a **nonpolar covalent bond,** the electrons in the pair are shared equally by each atom, so the molecule overall has no positive or negative poles. In general, compounds formed with carbon and hydrogen are nonpolar covalent compounds.

An Ionic Bond Is Formed by Gain or Loss of Electrons

When atoms complete their outer shells by either gaining or losing electrons instead of sharing them, they form ionic bonds to create **ionic compounds.** When two atoms form an ionic bond, one atom acquires one or more electrons and therefore has one or more extra negative charges. The other atom, by losing the electrons to the first atom, has an equal number of extra positive charges. The equal and opposite electrical charges of the two atoms attract them and hold them together; this attraction is the basis of the ionic bond. A specific example is the formation of sodium chloride. The neutral sodium atom (Na) has 11 protons and 11 electrons. After donating the lone electron in its outermost shell, it has only 10 electrons left — one less than the number of protons — and a net electrical charge of +1. The neutral chlorine atom (Cl), with 17 protons and 17 electrons, is able to fill its outermost electron shell by accepting the single electron that the sodium atom has given up. The chlorine atom then has 18 electrons — one more electron than protons — and a net electrical charge of −1. An ionic bond has been formed between the oppositely charged sodium and chlorine atoms, and the chemical compound sodium chloride (NaCl) results.

Ionic bonds between atoms do not produce molecules *per se;* rather, they make crystal lattices, or orderly patterns of cations and anions. As long as the cations and anions are bound together in the lattice, the compound overall is neutral. If the ions become dissociated, as in a solution, they exist as charged particles.

Hydrogen Bonds Occur Between Polar Molecules

Hydrogen bonds occur among polar covalent molecules, such as water, made up of hydrogen and certain elements, such as nitrogen, oxygen, and fluorine. In water, for example, the partial positive charge (δ^+) of the hydrogen atom of one water molecule is attracted to the partial negative charge (δ^-) of the oxygen atom of another water molecule (Fig. 2–3). Hydrogen bonds can occur between molecules or within the same molecule.

Hydrogen bonds are still relatively weak compared with covalent bonds. The input of energy needed to disrupt the hydrogen bonds between water molecules is only 4% of that needed to break the covalent bond between the hydrogen

and oxygen in the same molecule. Hydrogen bonds form from the hydrogen atoms in many biological compounds. They are important for maintaining the shapes of large molecules, such as proteins and nucleic acids.

CHEMICAL REACTIONS: BASIC PRINCIPLES

 What is a chemical reaction; how does it happen?

All atoms and molecules are moving continuously and therefore frequently collide with one another. These frequent collisions sometimes disrupt the electron configurations of the atoms involved, thereby breaking the covalent or ionic bonds that the atoms have formed with other atoms. When bonds are broken, compounds break down into their elements; when these elements recombine in new ways, new compounds are formed. This process is called a **chemical reaction.** According to the **law of conservation of mass,** the total number of atoms is the same before and after a chemical reaction; the atoms are simply rearranged into different groups, which represent new compounds with different chemical properties than the original compounds.

A chemical reaction is represented by a **chemical equation,** which shows the relative quantities of each compound in terms of its molecular formula. A simple example is the reaction of perchloric acid ($HClO_3$) and potassium carbonate (K_2CO_3) to form potassium perchlorate ($KClO_3$), carbon dioxide (CO_2), and water (H_2O). The two original compounds are called **reactants;** the resulting compounds are called **products.** Note that the same number of atoms of each element is present both before and after the reaction. This fact is ensured by providing two molecules (indicated by the number 2 to the left of the molecular formula) of the reactant perchloric acid for each molecule of potassium carbonate.

Reactants		Products		
$2HClO_3$ + K_2CO_3 $\rightarrow$		$2KClO_3$ +	CO_2 +	H_2O
perchloric acid	potassium carbonate	potassium perchlorate	carbon dioxide	water

total atoms	total atoms
2 H	2 H
2 Cl	2 Cl
9 O	9 O
2 K	2 K
1 C	1 C

Sometimes, the reaction proceeds in one direction, giving stable products. Some reactions, however, form unstable products. Either they break down and recombine into the

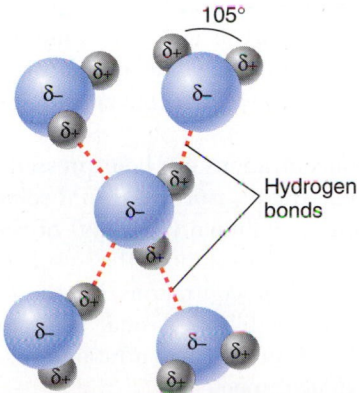

Figure 2–3

The polarity of the water molecule allows it to form hydrogen bonds with other water molecules.

original reactants or they break down into new products altogether. An example of this is the reaction of carbon dioxide (CO_2) and water (H_2O) to form carbonic acid (H_2CO_3). The product, H_2CO_3, is unstable, breaking down into the original reactants. Hence, the reaction is called *reversible* and is indicated by a set of opposite arrows. Many metabolic reactions that occur in the cell are reversible.

$$\boxed{H_2O} \; + \; \boxed{CO_2} \; \rightleftharpoons \; \boxed{H_2CO_3}$$

water   carbon   carbonic acid
          dioxide

original reactants   **unstable product**

When a reaction reaches a point at which the formation of products is exactly balanced by formation of the original reactants, the reaction is said to be in **equilibrium.**

The rate or speed at which a reaction occurs depends on a number of factors, two of which are temperature and chemical concentration.

Temperature. An increase in temperature generally accelerates the rate of a chemical reaction by increasing the velocity at which the reactant molecules and atoms are moving. The particles collide more frequently and more violently, increasing the possibility that electron configurations will be disrupted. The rate of many reactions doubles when the temperature increases by only 10°C. The metabolic reactions in human cells occur at approximately 37°C, a temperature higher than most environments. A further increase in temperature would be harmful to cells because proteins and nucleic acids would become unstable, breaking down and losing the ability to perform their functions.

Concentration. At least initially, an increase in the concentration of reactants usually increases the rate of reaction. The presence of more reactant molecules in the same space increases the frequency with which moving particles collide, thereby disrupting more bonds. Eventually, however, the rate decreases because the reactants are quickly used up in the formation of products. In the cell, the concentrations of molecules are carefully controlled so that metabolic reactions proceed at optimal rates.

WATER AND AQUEOUS SOLUTIONS

 What is "concentration?"

Water is essential for living organisms and is the major chemical compound in most cells, accounting for up to 70% of cell weight. Many biologically important compounds are soluble in water; in fact, the chemical reactions that they undergo require an **aqueous** (watery) medium or environment.

In chemistry, a **solution** is a type of mixture in which one substance, called the **solute,** is uniformly distributed throughout another substance, usually a liquid, called the **solvent.** A solvent dissolves, or breaks up, the solute without reacting chemically with it; that is, electrons are not shared or exchanged between the two substances.

Any substance that dissolves in water is said to be **hydrophilic** ("water-loving"). Ionic compounds and polar covalent compounds are generally hydrophilic. Nonpolar covalent compounds and groups are not easily soluble in water; they are said to be **hydrophobic** ("water-hating").

Concentrations of Solutions Are Described in Several Ways

We have said that some chemical reactions require an aqueous medium in order to occur and that the concentration of reactants can be an important factor in determining the rate of reaction. A solution may contain several different solutes, which may then react with one another. The concentration of reactants in these metabolic processes is determined by their concentration in solutions. Scientists have devised convenient ways of expressing these concentrations in order to understand the rates at which certain reactions occur both in the laboratory and in the living cell.

The concentration of a solution can be expressed in several ways. One way is to compare the number of moles of solute with the kilograms of solvent; this is known as **molality.** In aqueous solutions, this is the same as comparing moles of solute to liters of solution because 1 kilogram (kg) of water is 1 liter (L). For example, if 1 mole (58.5 g) of sodium chloride (the solute) is dissolved in 1 L of water, the concentration is expressed as 1 mole per kilogram (1 mol/kg). This is a "1 molal" solution, abbreviated as "1 m solution."

An alternative way to express concentration is **molarity.** Molarity is a unit of concentration, based on the number of moles of solute per liter of solution. A solution with a concentration of 1 mol/L is called a "1 molar" solution, abbreviated as "1 M solution." The solution is said to have a molarity of 1. A solution with a concentration of 2 mol/L is a 2 M solution.

Very low concentrations can be expressed with the same prefixes used in metric measures. A 1 mM solution of sodium chloride contains 1 millimole (58.5 mg) of sodium chloride per liter of solution. The solutions that occur in living organisms have a variety of concentrations, but most are in the millimolar range. Human blood plasma, for example, contains sodium and potassium at concentrations of approximately 140 mM and 4 mM, respectively.

Osmolarity is another expression of concentration that considers the total number of solute **particles** rather than the relative weights of specific solutes. Osmolarity is discussed in the section on osmosis in Chapter 4.

The pH Scale Describes the Concentration of H⁺ Ions

 What terms are used for talking about hydrogen ions?

Water (H_2O or HOH) has a slight tendency to dissociate into hydrogen ions (H^+) and hydroxide (OH^-) ions. This dissociation process can be represented as:

$$HOH \rightleftharpoons H^+ + OH^-$$

At the same time, the ions tend to reassociate, forming new water molecules. When the rate of dissociation equals the rate of reassociation, the reaction is said to be in equilibrium. At this point, the net concentrations of HOH molecules and of H^+ and OH^- ions remain constant, even while individual molecules and ions may be breaking down and reforming. The dissociation of water reaches equilibrium very rapidly, so that only very few of the water molecules are dissociated at any instant.

Ideally, the concentrations of H^+ and OH^- ions in water are exactly equal and very small, making the water a neutral compound. In reality, however, the concentrations of these ions are not equal because of the presence of solutes in the water that, by dissociating in it, increase or decrease the amount of H^+ ions relative to the number of OH^- ions. A solution in which the H^+ concentration is greater than the OH^- concentration is said to be **acidic.** A solution in which the number of OH^- ions is greater than the number of H^+ ions is a **basic** solution.

An **acid** is any substance whose dissociation in water releases H^+ ions. The hydrogen ion does not contain any neutrons and can be thought of as essentially equivalent to a proton. Therefore, an acid is often referred to as a **proton donor** because it gives off protons in the form of H^+ ions. A **base** is a **proton acceptor,** a substance that accepts H^+ ions and thereby decreases the H^+ concentration in a solution.

We can express the acidity or basicity of a solution by specifying the concentration of H^+ ions. Because this concentration is very small, scientists have devised a logarithmic method of expressing it, called **pH.** This expression is the negative logarithm (log) of the H^+ concentration:

$$pH = -\log[H^+]$$

where [ ] indicates the H^+ concentration in moles per liter.

In 1 L of neutral water, there are 1×10^{-7} moles of H^+ ions (and 1×10^{-7} moles of OH^- ions). The pH of neutral water then is:

$$
\begin{aligned}
pH &= -\log(1 \times 10^{-7}) \\
&= -[\log(1.0) + \log(10^{-7})] \\
&= -[(0.00) + (-7)] \\
&= 7
\end{aligned}
$$

Suppose the concentration of H^+ ions were to increase tenfold, to 1×10^{-6} mol/L. The pH of the solution would then be:

$$
\begin{aligned}
pH &= -\log(1 \times 10^{-6}) \\
&= -[\log(1.0) + \log(10^{-6})] \\
&= -[(0.00) + (-6)] \\
&= 6
\end{aligned}
$$

The solution is more acidic, and the pH has decreased. Conversely, if the H^+ concentration were to decrease, the solution would become more basic, and the pH would increase.

The pH scale goes from 0 to 14. A solution with a pH of 0 is highly acidic; one with a pH of 14 is highly basic. In terms of pure numbers, the scale is small, so that a 1000-fold increase in H^+ concentration produces a pH decrease of only 3 units. It is important to remember, therefore, that a *small decrease* in pH represents a *large increase* in the H^+ concentration, and vice versa.

Acids and Bases Change the pH of Solutions

The purpose of a **buffer system** is to minimize the change in pH that occurs when an acid or base is added to any solution. The pH of blood is controlled very closely and does not vary outside a small range, usually 7.38 to 7.42. This is a remarkable achievement because the chemical reactions in the cells of the body produce more than 10,000 nanomoles (nmol) of H^+ ions each day. An average blood pH of 7.4 represents an H^+ concentration of only 40 nmol/L, indicating that highly efficient mechanisms must exist to absorb the H^+ ions added to the blood. These mechanisms are addressed in detail in the chapter on acid-base physiology (Chapter 25).

MACROMOLECULES

 What is a macromolecule; what role do macromolecules play in biological systems?

The unique properties of the carbon atom allow it to form four bonds at once and lead to a huge variety of molecules of different shapes and enormous sizes. These molecules are called **macromolecules,** and they play important roles in the life processes of all cells and organisms. The most important macromolecules in cell physiology are carbohydrates, lipids, proteins, and nucleic acids. The following brief survey provides a glimpse of these compounds and a review of their chief functions in cells.

Carbohydrates Form Branched Chains

Carbohydrates are often referred to as **sugars.** All **carbohydrates** are composed of carbon, hydrogen, and oxygen atoms. Carbohydrates are classed in three broad groups:

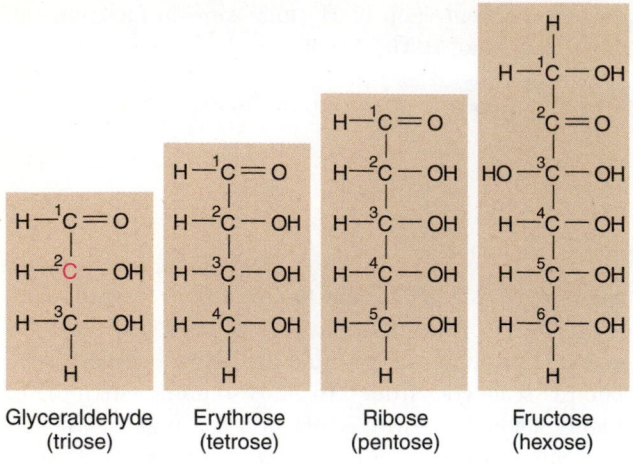

Figure 2–4

The structures of the monosaccharides glyceraldehyde, erythrose, ribose, and fructose.

(1) monosaccharides, (2) disaccharides, and (3) polysaccharides. In addition, they form compounds with proteins and lipids to form glycoproteins and glycolipids.

 How are the chemical structures of carbohydrates represented on paper?

Monosaccharides

The simplest type of carbohydrates are called **monosaccharides.** The general chemical formula for a monosaccharide is $(CH_2O)_n$, where n is a whole number indicating the number of carbons. A monosaccharide with three carbons, such as

glyceraldehyde, is a **triose** (Fig. 2–4). A monosaccharide with four carbons is a **tetrose;** one with five carbons, a **pentose;** one with six carbons, a **hexose.**

In structural formulas of sugars and other "organic" molecules, the carbon atoms often are numbered for identification. Specific carbons often have specific roles based on their position in the molecule and on the other atoms attached to them. A carbon atom, such as the second carbon atom ("carbon 2") in the glyceraldehyde molecule, is described as asymmetrical when it is linked covalently to four different chemical groups. Carbon 2 in glyceraldehyde (Fig. 2–4) is linked to a hydrogen (H), an aldehyde (—CHO) group, a hydroxide (—OH) group, and —CH$_2$OH.

Ring Structures. Monosaccharides can have different forms—either linear chains or rings. In solutions, monosaccharides most often form rings. One common hexose is **glucose,** $C_6H_{12}O_6$, or $(CH_2O)_6$. In glucose, ring formation occurs when the aldehyde (—CHO) group on carbon 1 reacts with the hydroxyl (—OH) group on carbon 5 to form a stable covalent bond, giving rise to a six-member ring structure (Fig. 2–5).

Although drawn as a flat structure, the plane of the ring is perpendicular to the plane of the page, and the hydrogen atoms and hydroxyl groups lie above and below the plane of the ring. The edge of the ring nearest the reader is indicated by a thicker line between the carbon atoms. In simplified drawings of ring structures, the letter C, indicating a carbon atom, is omitted, but the carbons are understood to be present wherever two "sides" of the ring meet. Furthermore, the letter H, indicating a hydrogen atom, also is omitted but understood to be present wherever needed to supply each carbon with a total of four bonds. In the most simplified drawings, only the —OH groups are identified by letters (Fig. 2–5).

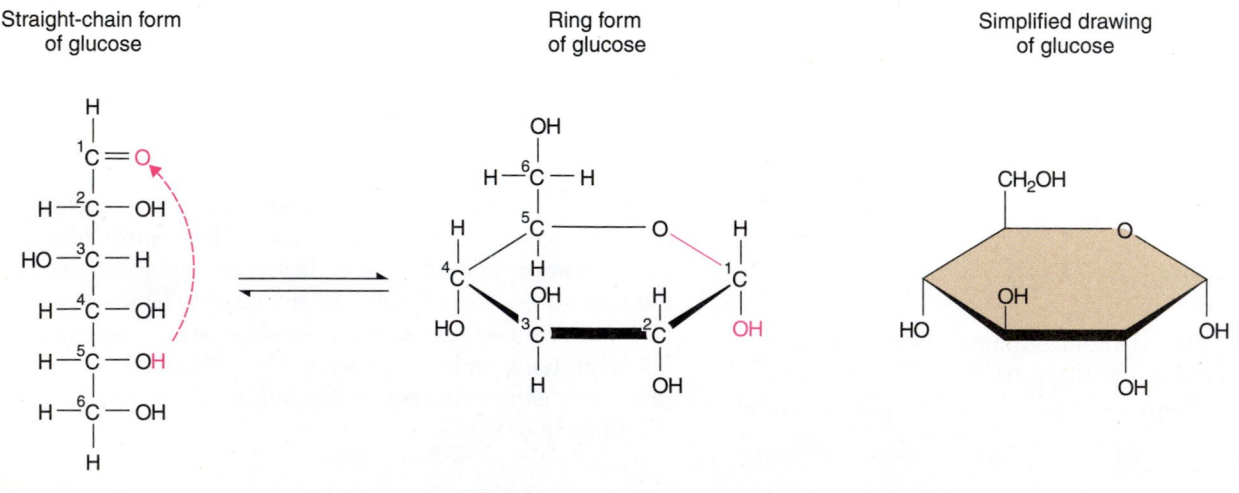

Figure 2–5

The linear and ring structures of the sugar glucose, a common hexose. When the —OH group on carbon 5 comes near the =O group on carbon 1, the two carbons are joined by the —O— atom, and a ring is formed.

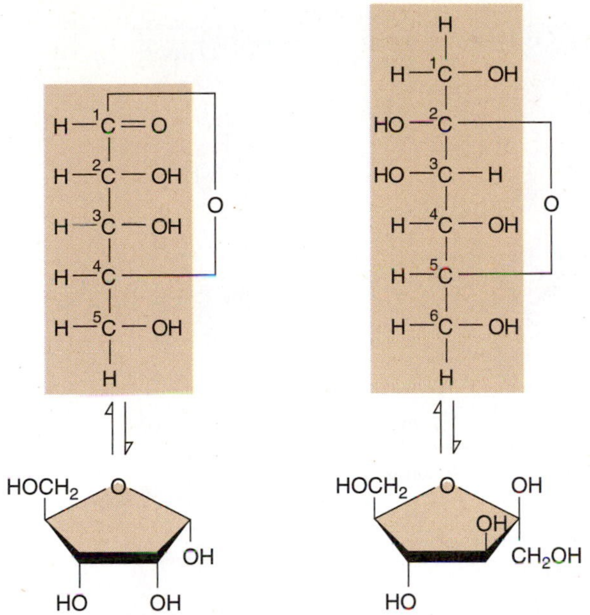

Figure 2–6

The cyclic forms of ribose (a pentose) and fructose (a hexose) both contain five-member rings.

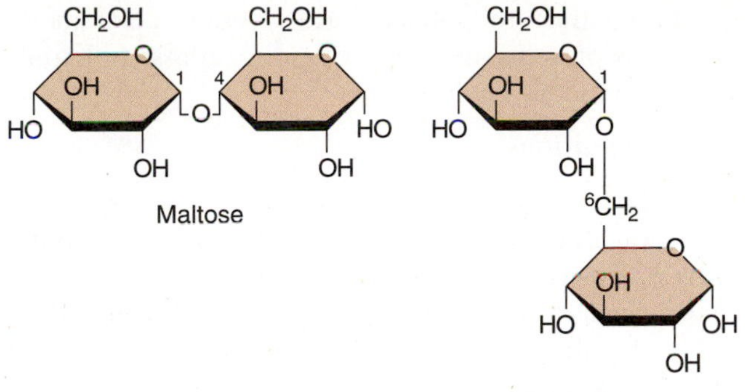

Ribose
(pentose)

Fructose
(hexose)

Pentoses form five-member ring structures, as do some hexoses, such as fructose (Fig. 2–6). Two important pentoses are ribose and deoxyribose, which are present in nucleic acids.

When a hexose forms a five-member ring, carbon 1 is left outside the ring because the ring forms when the carbonyl (C=O) group on carbon 2 reacts with the hydroxyl (—OH) group on carbon 5 (Fig. 2–6).

> *What are some common examples of molecules that form by combining monosaccharide units?*

Disaccharides

Two monosaccharides can combine to form a **disaccharide.** The most common disaccharides are **maltose** (two glucose molecules), **lactose** (one glucose molecule plus one galactose molecule), and **sucrose** (glucose plus fructose).

Isomaltose, like maltose, is formed from two glucose molecules; both isomaltose and maltose have the same chemical formula. However, they have different structures because their monosaccharide units join at different carbon atoms. In maltose, the glucose molecules join at carbons 1 and 4; in isomaltose, they join at carbons 1 and 6 (Fig. 2–7). Maltose and isomaltose are called **isomers,** molecules with the same chemical formula but different structural arrangements of the atoms.

Maltose

Isomaltose

Figure 2–7

Maltose and isomaltose are isomers. Both are formed from two molecules of glucose; both have the same molecular formula. They have different structural formulas, however, and therefore constitute two different disaccharides.

Polysaccharides

Most of the carbohydrates found in nature occur as **polysaccharides,** molecules composed of many monosaccharide units usually arranged in branched chains. Polysaccharides are distinguished from one another mainly by their monosaccharide units and the length and amount of branching of their chains. Glucose is the most common monosaccharide unit of polysaccharides. Glucose is the fuel that most animal cells require, and it can be stored in certain tissues in the form of the polysaccharide **glycogen.** This important polysaccharide is especially abundant in the liver, where it can account for up to 10% of the wet weight of the tissue. It consists of a chain of glucose units that are linked by chemical bonds between carbons 1 and 4 of adjacent glucose units. Branches occur about every 10 glucose units, arising from a bond between carbons 1 and 6 (Fig. 2–8). The molecular mass of glycogen has been estimated to be in the range of 3×10^5 to 3×10^6 Da, which means that the number of glucose units present in a glycogen molecule could be as high as

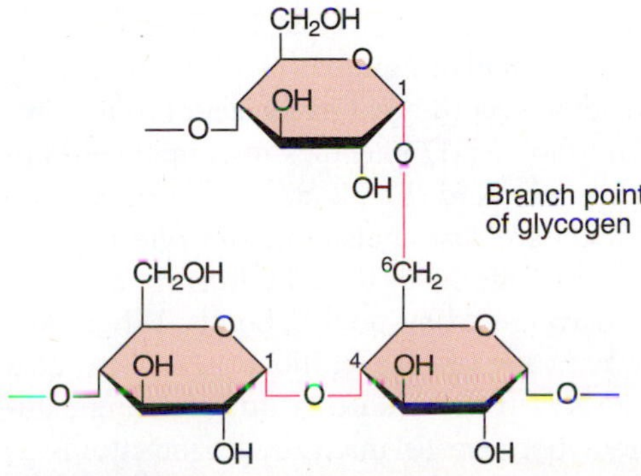

Branch point
of glycogen

Figure 2–8

A branch point in the polysaccharide glycogen.

16,667. In contrast to polysaccharides, other macromolecules, such as proteins and nucleic acids, can assemble only in linear, unbranched chains.

Sugar Combinations

Many of the polysaccharides present in living organisms occur in covalent combination with proteins or lipids. **Glycoproteins** and **glycolipids** are present in cell membranes. The polysaccharide chains in these compounds tend to be short but highly branched, containing several different monosaccharide units arranged in complex sequences. They are thought to be important in recognition processes, such as those involved in the action of hormones. A hormone released into the blood is distributed to all the cells in the body but affects only those cells that have an external site that the hormone "recognizes" and binds. The carbohydrate portions of membrane glycoproteins and glycolipids are thought to be important parts of these binding sites for hormones.

 How do lipids differ from carbohydrates?

Lipids Have Low Solubility in Water

Lipids are sometimes called **fats** and are distinguished from the other cellular macromolecules because they are poorly soluble in water and highly soluble in organic lipids, such as ether and chloroform. They are composed of carbon, hydrogen, and oxygen atoms; the same elements that make up carbohydrates; however, the proportions of these elements differ in lipids. The lipids contain only small amounts of oxygen relative to carbon and hydrogen.

Three major classes of lipids are considered here: (1) fatty acids; (2) lipids containing glycerol; and (3) steroids, or lipids that do not contain glycerol.

Fatty Acids

A **fatty acid** is a long chain of covalently linked carbon atoms to which only hydrogen atoms are attached, except for a carboxylic acid (—COOH) group at one end. The long hydrocarbon chain is hydrophobic, or poorly soluble, in water, whereas the —COOH group is highly water-soluble and reacts readily with other chemical groups, forming covalent bonds. Most of the fatty acids in animal cells are covalently linked to other molecules via the carboxylic acid group. Two of the most abundant fatty acid molecules in animal cells are **palmitic acid** and **stearic acid** (Fig. 2–9).

Specific fatty acids are distinguished from one another by the number of carbon atoms in their hydrocarbon chains and the number of carbon-carbon double bonds. When only single bonds exist between all the carbons in a fatty acid chain, the fatty acid is said to be **saturated**, meaning saturated with hydrogen, because as many hydrogen atoms as possible are bonded to each carbon. When one or more double bonds are present between carbon atoms, the fatty acid is **unsaturated.** The most abundant unsaturated fatty acids in

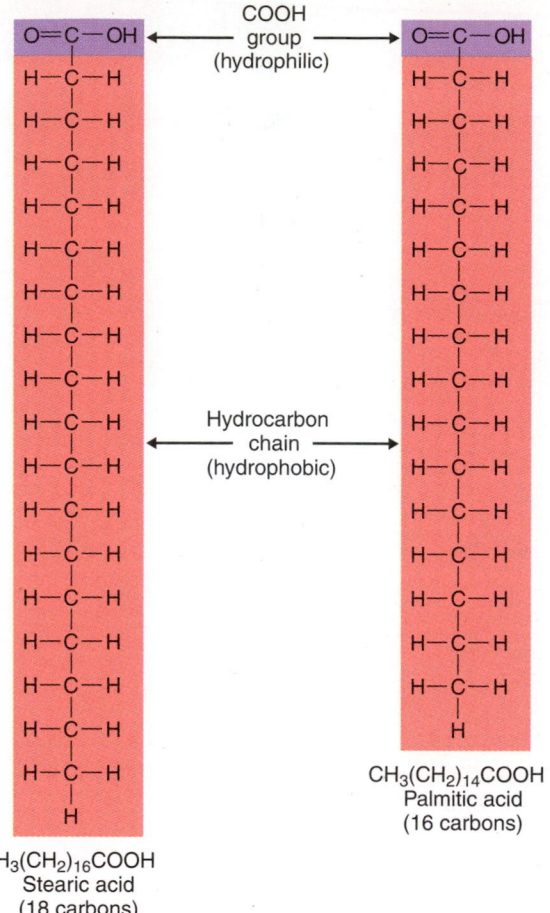

$CH_3(CH_2)_{16}COOH$
Stearic acid
(18 carbons)

$CH_3(CH_2)_{14}COOH$
Palmitic acid
(16 carbons)

Figure 2–9

The structures of stearic acid and palmitic acid, two common saturated fatty acids. The molecules consist mostly of hydrophobic hydrocarbon chains. The carboxylic acid (—COOH) group at one end, however, is water-soluble, or hydrophilic. Fatty acids are sometimes given the general symbol RCOOH, *R* representing the hydrocarbon chain.

animal cells are oleic acid, linoleic acid, and arachidonic acid, which contain one, two, and four double bonds, respectively:

oleic acid $CH_3(CH_2)_7CH{=}CH(CH_2)_7COOH$

linoleic acid
$CH_3(CH_2)_4CH{=}CHCH_2CH{=}CH(CH_2)_7COOH$

arachidonic acid $CH_3(CH_2)_4(CH{=}CHCH_2)_4(CH_2)_2COOH$

Fatty acids are a valuable source of fuel for cells. Considerably more energy can be obtained from the breakdown of 1 g of a fatty acid than from 1 g of carbohydrate. Arachidonic acid is an essential fatty acid that is a precursor of an important group of molecules called **prostaglandins,** which affect a wide variety of physiological processes.

Glycerol Compounds and Phospholipids

Glycerol is a molecule with three hydroxyl groups (Fig. 2–10). These hydroxyl groups can form bonds with the carboxylic acid groups of fatty acids. The attachment of one, two, or three fatty acids to glycerol produces a **monoacylglycerol,** a **diacylglycerol,** or a **triacylglycerol,** respectively. Both mono- and diacylglycerols are important natural compounds, but the triacylglycerols (Fig. 2–10) are the most abundant and are the form found in fat tissue. Cells can convert excess carbohydrate and protein to fatty acids and then store the fatty acids in triacylglycerols. The continued accumulation of triacylglycerol deposits in the body, however, can lead to obesity. Triacylglycerols often are referred to as **triglycerides.**

Another group of glycerol-containing lipids are the **phospholipids.** The parent compound of this group is **glycerol-3-phosphate,** a glycerol molecule to which a phosphate ($-PO_4$) group has been attached (Fig. 2–11a). A phospholipid is formed from this basic structure by (1) the addition of fatty acids to the free hydroxyl groups on carbons 1 and 2 to give **phosphatidic acid** and (2) the addition of another chemical group to the phosphate on carbon 3. In step 1, the fatty acid added to carbon 1 of the glycerol unit usually is saturated and contains 16 to 18 carbon atoms, whereas the fatty acid added to carbon 2 usually is unsaturated and contains 16 to 20 carbon atoms. The most common groups added in step 2 are the polar molecules *choline, ethanolamine, serine,* and *inositol.* The phospholipids that result are called *phosphatidylcholine, phosphatidylethanolamine, phosphatidylserine,* and *phosphatidylinositol,* respectively (Fig. 2–11b).

At **physiological pH** (7.40), the phospholipids are in the charged forms shown, and one end of the molecule is a polar (hydrophilic) region. In contrast, the long hydrocarbon chains of the two fatty acid groups constitute a highly hydrophobic region of the molecule (Fig. 2–11c). The dual nature of these phospholipids — hydrophilic at one end and hydrophobic at the opposite end — plays an important structural role in the plasma membrane surrounding all cells (see Chapter 4). The membrane is made up of a "lipid bilayer" — two layers of phospholipid molecules oriented so that their hydrophobic hydrocarbon chains point toward each other, away from the aqueous surroundings both inside and outside the cell (Fig. 2–11d). The hydrophilic parts of the phospholipids in each layer are oriented outward. This arrangement represents a very stable physical structure with unique properties, as is discussed in Chapter 4.

Sphingomyelin is the only phospholipid not derived from glycerol. The basic unit of sphingomyelin is **sphingosine:**

$$CH_3(CH_2)_{12}-CH=CH-CHOH-CHNH_3^+-CH_2OH$$

Sphingosine contains an unsaturated hydrocarbon chain and is converted to sphingomyelin by the addition of a long-chain fatty acid, a phosphate group, and a choline group. Overall, the structure of sphingomyelin is very similar to that of phosphatidylcholine (Fig. 2–11c) because it has two hydrophobic hydrocarbon chains and a hydrophilic phosphorylcholine group. Excessive accumulation of sphingomyelin occurs in individuals with a specific lipid storage disorder known as **Niemann-Pick disease.** In these individuals, sphingomyelin accumulation in brain cells inhibits their function and leads to mental retardation.

Steroids

Some of the most important lipids do not contain glycerol but are synthesized from cholesterol; these are the **steroids.** The basic structure of a steroid is four interlocking rings of carbon atoms (Fig. 2–12) and is quite different from that of other lipids. The male and female sex hormones and some hormones secreted by the adrenal glands, such as aldosterone, are steroids. The term **sterol** is often used when a hydroxyl group is present. **Cholesterol** is an example of a sterol that plays an important role in the structure of plasma membranes. Like the phospholipids, cholesterol has a polar hydrophilic region (the hydroxyl group) and a nonpolar hydrophobic region (the rest of the molecule).

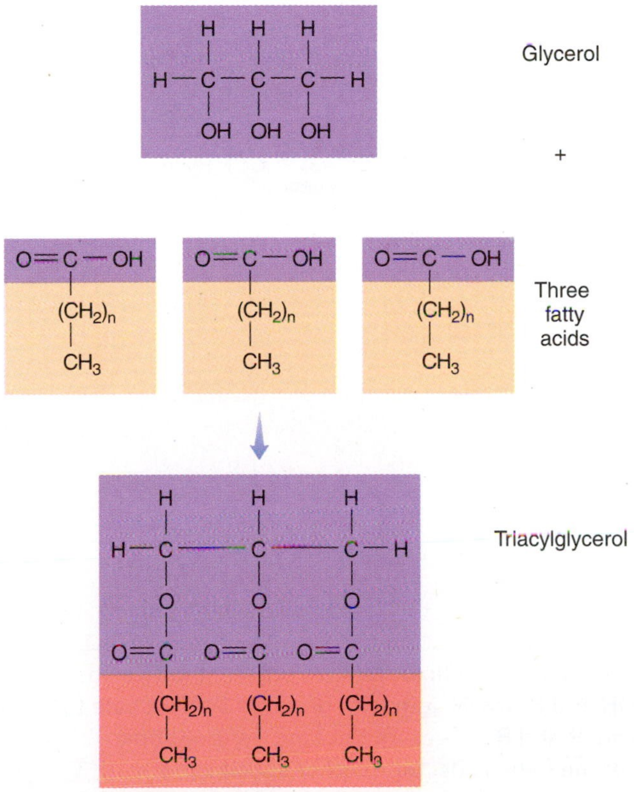

Figure 2–10

Steps in the formation of a triacylglycerol, also termed a *triglyceride,* from glycerol and three fatty acids. The fatty acids may be identical or different. *n* stands for a whole number, the number of carbon atoms in the parentheses representing all but the last carbon in the fatty acid hydrocarbon chain. For stearic acid, *n* is 16 (see Fig. 2–9).

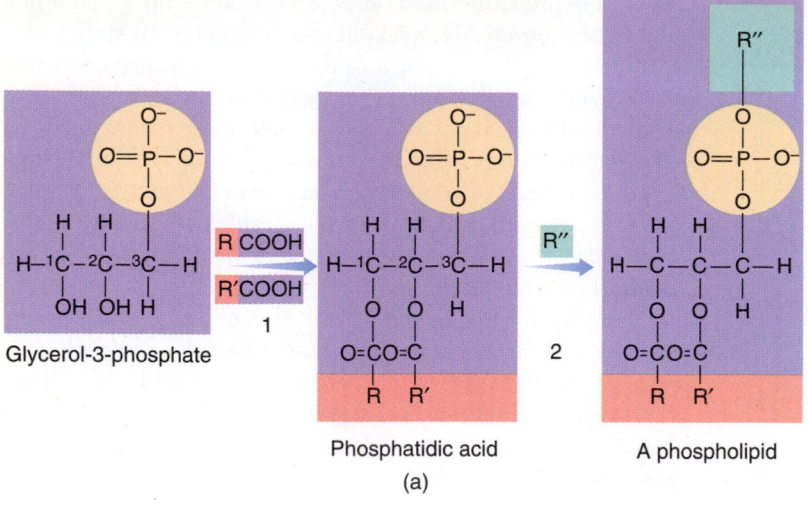

Glycerol-3-phosphate

Phosphatidic acid
(a)

A phospholipid

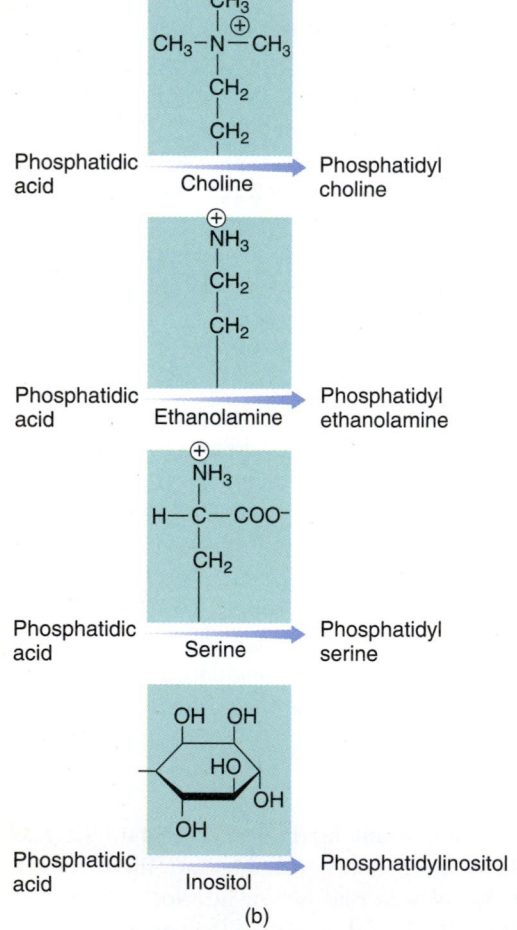

Phosphatidic acid → Choline → Phosphatidyl choline

Phosphatidic acid → Ethanolamine → Phosphatidyl ethanolamine

Phosphatidic acid → Serine → Phosphatidyl serine

Phosphatidic acid → Inositol → Phosphatidylinositol
(b)

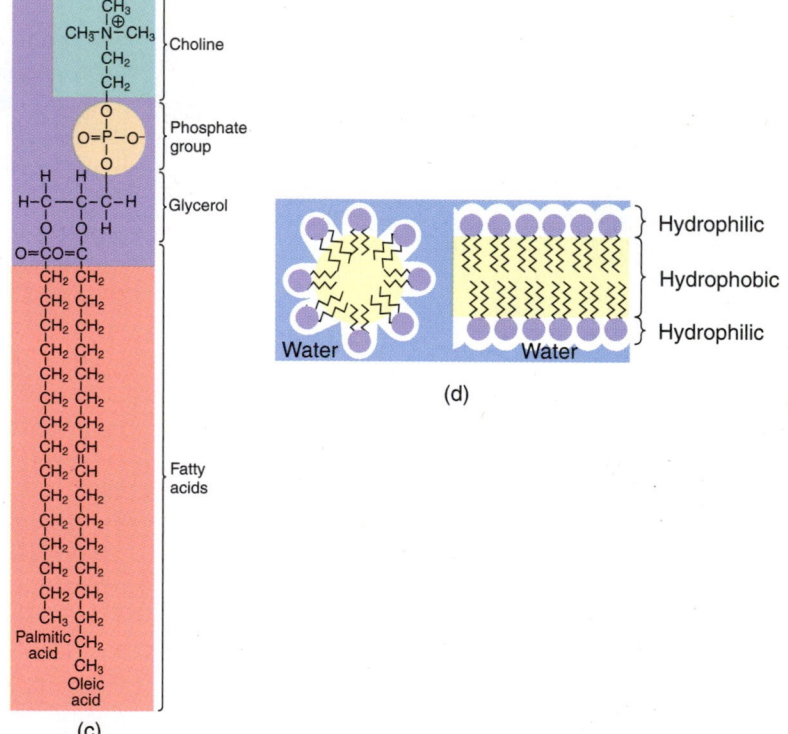

(c)

(d)

Figure 2–11

Steps in the formation of a phospholipid. *(a)* Two different fatty acids, symbolized by RCOOH and R'COOH, join with glycerol-3-phosphate to form phosphatidic acid. R and R' represent the long, hydrophobic hydrocarbon chains of the fatty acids. *(b)* Another specific group, (R"), either choline, ethanolamine, serine, or inositol, is added to phosphatidic acid to form a specific phospholipid. *(c)* An example of a specific phospholipid, phosphatidylcholine, formed by the addition of choline to phosphatidic acid. *(d)* In water, phospholipid molecules become oriented with their hydrophobic tails as far from the water as possible, forming either a sphere or a double layer.

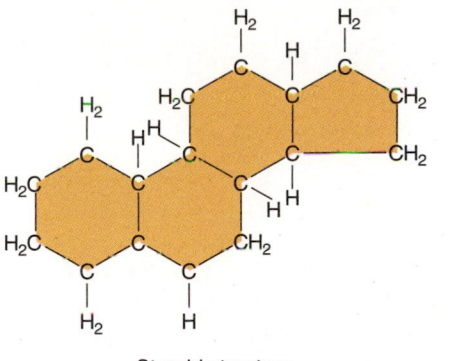

Steroid structure

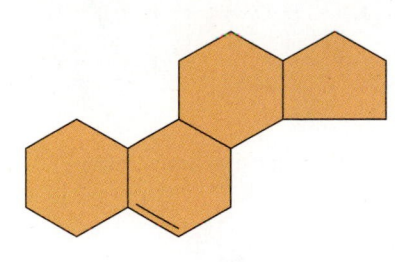

Simplified form

Figure 2–12

Basic structure of steroids, and examples: cholesterol and aldosterone are both formed from a basic structure of four interlocking rings.

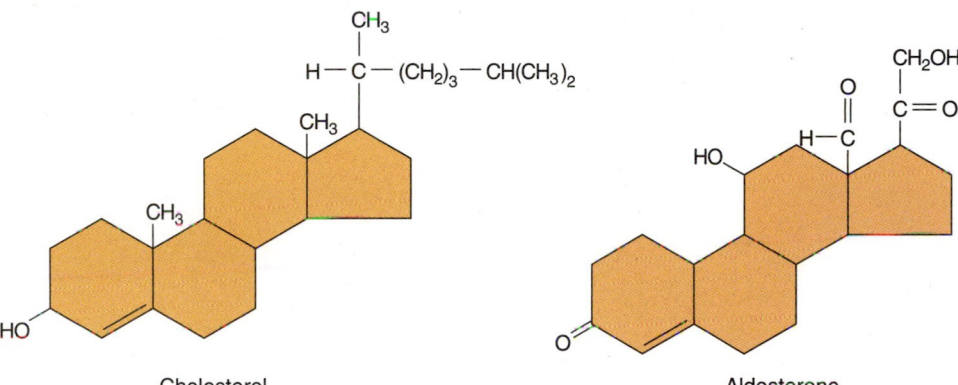

Cholesterol

Aldosterone

Glycolipids

Glycolipids are lipids, also derived from sphingosine, that contain monosaccharide units. They have two hydrocarbon chains, as sphingomyelin does, but between one and seven monosaccharide units usually are present instead of the phosphorylcholine group. The monosaccharides make up a hydrophilic region of the molecule. Glycolipids occur in cell membranes.

Proteins Are Unbranched Chains of Amino Acids

> *What distinguishes proteins from carbohydrates and lipids? What is the physiological significance of protein structure?*

Proteins are some of the largest macromolecules in the cell. The molecular mass of a protein is usually between 5000 and several million daltons. Proteins carry out remarkably diverse functions in the cell. For example, enzymes, the molecules that catalyze specific biochemical reactions, are proteins. Carrier molecules that transport small molecules and ions are proteins. Many hormones are proteins, as are antibodies that combine with foreign bacteria. In addition, proteins are a major component of the muscles, allowing for coordinated movement.

Two key aspects of protein molecules are important for an understanding of cell function: (1) the nature of the amino acids themselves, based on their side chain groups, and (2) the shape of the protein macromolecule, called its **conformation.**

Amino Acids

The basic units of protein molecules are **amino acids.** The general structure of an amino acid is illustrated in Figure 2–13. The "alpha" (α) carbon atom is asymmetrical, bonded to four different chemical groups. It is called the **alpha-carbon** to distinguish it from carbons that may be present in the **side chain (R) group.** Twenty different amino acids are available to make proteins, and the same 20 amino acids occur in all proteins whether they are made by bacterial, plant, or animal cells.

Amino acids are distinguished from one another by their side chain groups. These groups have important properties that determine the properties of the protein molecules they make up.

Proteins are formed from linear, unbranched chains of amino acids. The amino acids are covalently linked to one another via **peptide bonds,** formed between the alpha-carboxyl (—COOH) group of one amino acid and the alpha-amino (—NH$_2$) group of an adjacent amino acid (Fig. 2–13). A **dipeptide** is a chain of two amino acids; a **tripeptide,** a chain of three; and a **polypeptide,** a chain of eight or more amino acids. Protein molecules typically are polypeptide chains of between 50 and 2500 amino acids. The exact number of amino acids and the sequence in which they combine are characteristic features of a specific protein that determine its properties and functions.

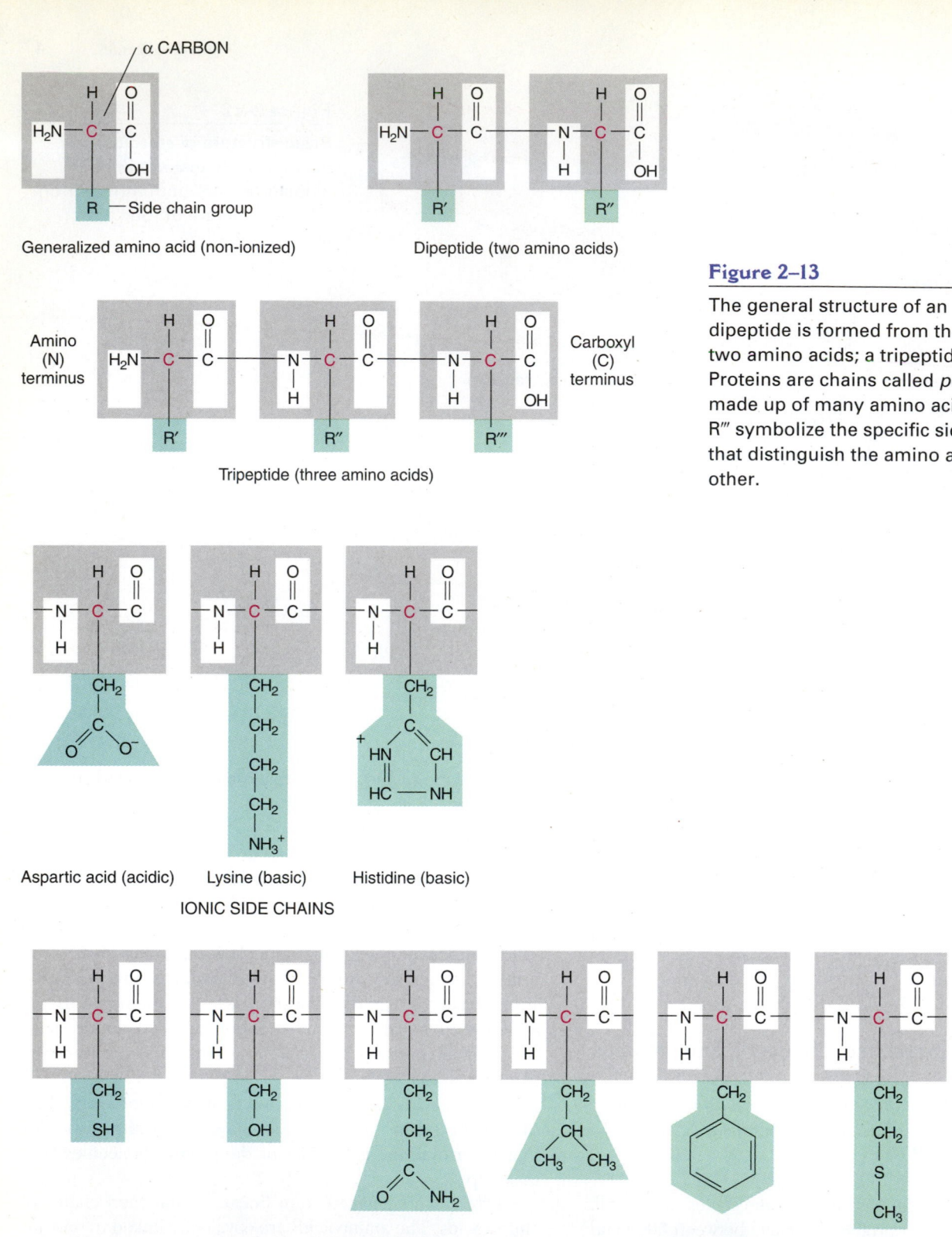

α CARBON

H₂N—C—C

Side chain group

Generalized amino acid (non-ionized)

Dipeptide (two amino acids)

Amino (N) terminus

Carboxyl (C) terminus

Tripeptide (three amino acids)

Figure 2–13

The general structure of an amino acid. A dipeptide is formed from the attachment of two amino acids; a tripeptide, from three. Proteins are chains called *polypeptides,* made up of many amino acids. R, R′, R″, and R‴ symbolize the specific side chain groups that distinguish the amino acids from each other.

Aspartic acid (acidic)   Lysine (basic)   Histidine (basic)

IONIC SIDE CHAINS

Cysteine   Serine   Glutamine   Leucine   Phenylalanine   Methionine

NONIONIC POLAR SIDE CHAINS   NONPOLAR SIDE CHAINS

Figure 2–14

Structures of some amino acid side chain groups. Ionic side chains, such as those that produce the amino acids aspartic acid, lysine, and histidine, are hydrophilic (water-soluble), as are nonionic polar side chains, such as those that form cysteine, serine, and glutamine. Nonpolar groups, such as those that form leucine, phenylalanine, and methionine, are not water-soluble. (The —NH₂ and —COOH groups of the alpha-carbon are shown as if they were involved in the formation of peptide bonds.)

The amino acids are divided into three broad groups, based on the properties of their side chains at physiological pH (Fig. 2–14): (1) amino acids with ionic polar side chains, (2) those with nonionic polar side chains, and (3) those with nonpolar side chains.

Ionic Polar Side Chains. Ionic side chains found in amino acids contain terminal —COOH or —NH₂ groups, which donate or accept, respectively, H⁺ ions in solution at pH 7.4 to become —COO⁻ or —NH₃⁺. Side chains that donate H⁺ ions are, of course, acids (e.g., aspartic acid). Side chains that accept H⁺ ions are bases (e.g., lysine and histidine). Note that the —NH₂ and —COOH groups on the alpha-carbon are involved in formation of peptide bonds and cannot donate or accept H⁺ ions. The ionic side chains are found on the external surface of protein molecules, where they can interact with the polar water molecules.

Nonionic Polar Side Chains. Some groups found in amino acids are polar but have no net charge at physiological pH. Examples are the —NH₂ group of glutamine, the —OH group of serine, and the —SH group of cysteine. These groups may be exposed on the protein surface, as are ionic side chains, or they may be buried within the protein molecule, where they are able to form hydrogen bonds. The side chain —SH groups of adjacent cysteines also interact to form a covalent bond known as a **disulfide bridge:**

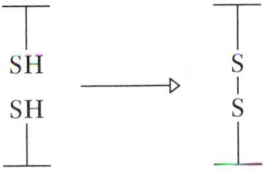

Nonpolar Side Chains. Some side chain groups are nonpolar and therefore hydrophobic. They are usually "hidden" from water by sequestration inside the protein structure where they cluster together via hydrophobic interactions. The side chains of leucine, methionine, and phenylalanine are good examples.

Protein Structure

There are four different levels of organization in the structure of a protein molecule. These levels are important to understand because protein shape affects protein functioning in cells. The sequence in which the amino acids are linked together determines the primary structure of a protein; the way in which the resulting chain bends or folds into sheets or helixes is the secondary structure. Further folding and stable associations between different regions create the tertiary structure. In some complex proteins, the physical relation of separate chains makes up the quaternary structure.

Primary Structure. The chemical bonds that are important for maintaining the **primary structure,** or the sequence of amino acids, in a protein molecule are the covalent peptide bonds between the adjacent amino acids. By convention, the amino-, or *N*-, terminus is considered to be the beginning of the chain, and the carboxyl-, or *C*-, terminus is the end (see Fig. 2–13).

The hormone **insulin** was the first protein for which the primary structure was discovered. It consists of two separate polypeptide chains, one containing 21 amino acids and the other containing 30 amino acids. Two disulfide bridges hold the chains together, and a third disulfide bond joins two cysteines in the shorter chain (Fig. 2–15). Eighteen different amino acids are present in the molecule.

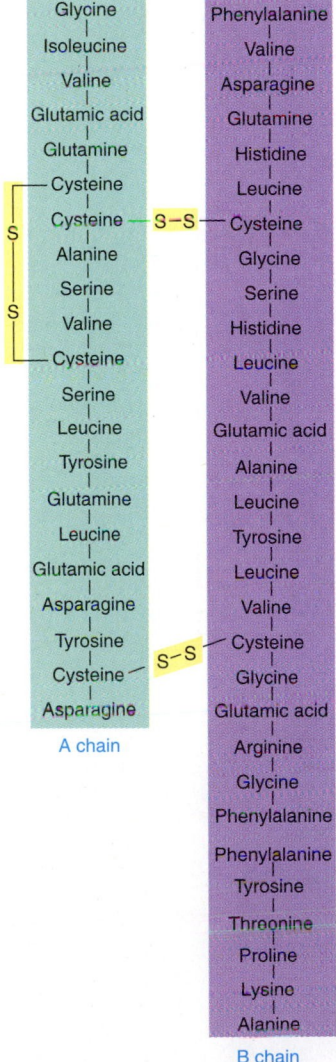

Figure 2–15

The primary structure of the hormone insulin, a protein (polypeptide) consisting of two chains of amino acids in specific sequence joined by two disulfide bridges.

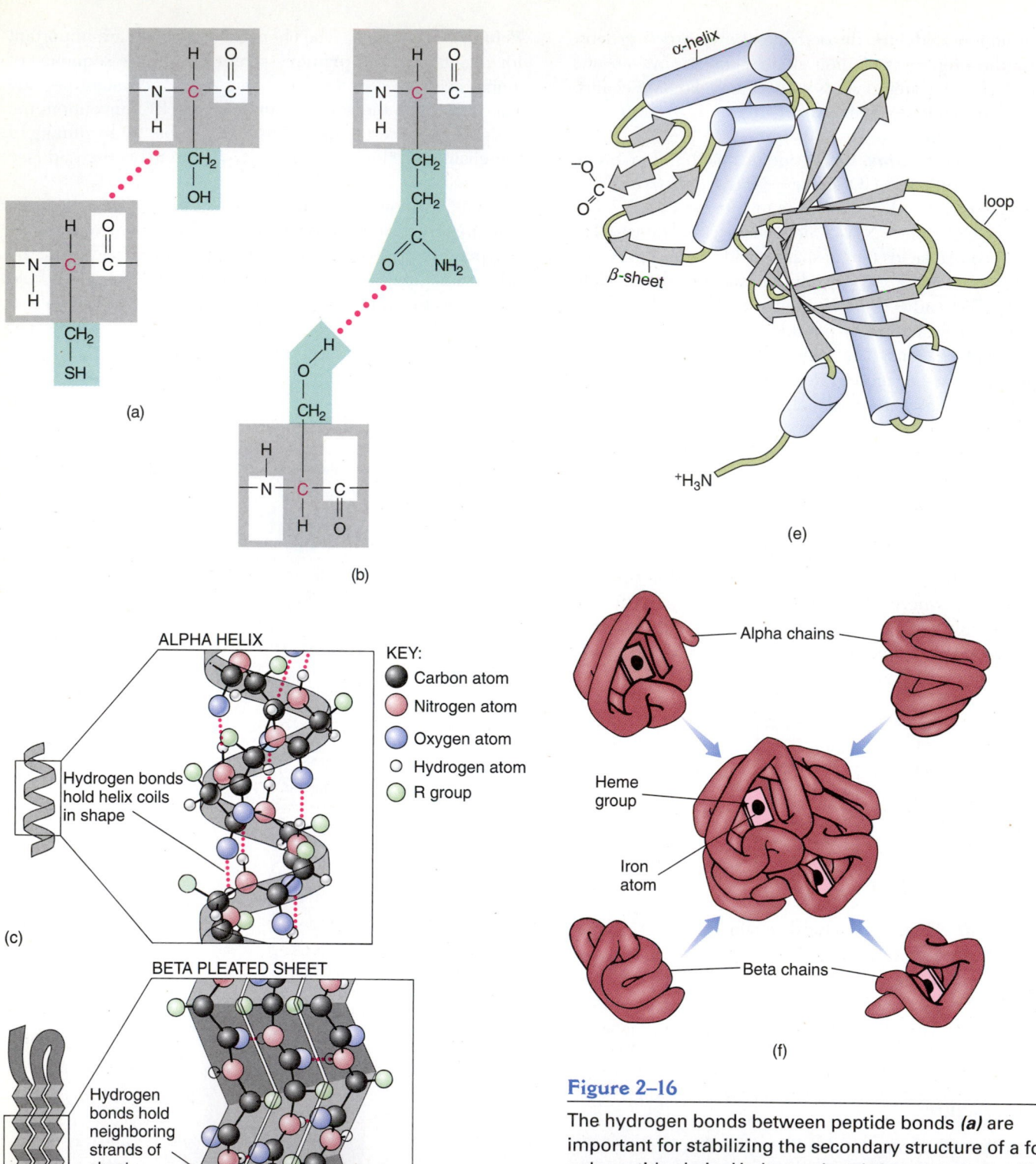

ALPHA HELIX

KEY:
- Carbon atom
- Nitrogen atom
- Oxygen atom
- Hydrogen atom
- R group

Hydrogen bonds hold helix coils in shape

(c)

BETA PLEATED SHEET

Hydrogen bonds hold neighboring strands of sheet together

(d)

α-helix

loop

β-sheet

^+H_3N

(e)

Alpha chains

Heme group

Iron atom

Beta chains

(f)

Figure 2–16

The hydrogen bonds between peptide bonds *(a)* are important for stabilizing the secondary structure of a folded polypeptide chain. Hydrogen bonds between polar side chains of amino acids *(b)* help to stabilize the tertiary structure. The secondary structure of a protein can be either an alpha-helix *(c)* or a beta-sheet *(d)*. The tertiary structure of a hypothetical protein is maintained by several different types of bonds. Tubes represent alpha-helices and flat arrows represent beta-sheets that are joined by unfolded loops *(e)*. Four separate polypeptides assemble together to form the quaternary structure of hemoglobin *(f)*.

Secondary Structure. A linear polypeptide chain is usually flexible and, under biological conditions, becomes folded into a specific shape or conformation. This conformation is called the **secondary structure** of the protein. The folding occurs spontaneously and produces a unique secondary structure for each protein. All of the specific conformations can be broadly classed as either alpha-helixes or beta-sheets (Fig. 2–16*c* and *d*). An **alpha-helix** is a type of conformation that results when a single polypeptide chain coils about itself to form a uniform cylindrical structure. Each complete turn of the helix is occupied by 3.6 amino acids. This structure is stabilized mainly by hydrogen bonds between the atoms of the peptide bonds (Fig. 2–16*a*). A **beta-sheet** results when the polypeptide chain runs back and forth upon itself. It is stabilized, like the alpha-helix, by hydrogen bonds between atoms of peptide bonds in neighboring segments of the chain. Alpha-helixes and beta-sheets are stable structures that are more rigid than extended polypeptide chains, and both types of conformation may occur within the same protein molecule.

Tertiary Structure. The next level of protein structure, the **tertiary structure,** is formed by the folding of the alpha-helix upon itself or by stable associations of beta-sheets and alpha-helixes. This additional level of folding is the result of interactions between the side chains of amino acids that are widely separated in the primary structure. The tertiary structure is stabilized by (1) hydrophobic interactions between the nonpolar groups that are buried inside the structure, (2) hydrogen bonds and ionic bonds between polar side chain groups on the protein surface, or (3) covalent disulfide bridges between cysteine–SH groups.

In most proteins, combinations of alpha-helixes and beta-sheets, connected by unfolded loop regions, fold into compact globular structures called **domains.** These are the basic units of tertiary structure (Fig. 2–16*e*). Large proteins may contain a number of different domains, which often have distinct functions.

Quaternary Structure. Many large, globular proteins consist of two or more separate polypeptide chains. The specific way in which the individual folded chains or subunits fit together to form the biologically active protein is known as the **quaternary structure. Hemoglobin,** the protein in red blood cells that binds and carries oxygen, is comprised of 574 amino acids arranged in four separate subunits (Fig. 2–16*f*). The subunits are held together by noncovalent interactions. One molecule of oxygen can bind reversibly to each of the four subunits.

The final conformation, or shape, of the intact functional protein depends on the interactions among the amino acid side chain groups. These in turn depend on the primary structure, or **amino acid sequences,** of the polypeptide chains. A change in just one amino acid in the polypeptide chain can alter the physiological function of

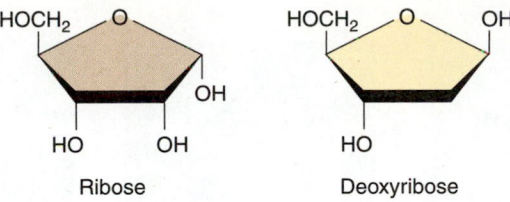

Figure 2–17

The structures of ribose (the pentose sugar that forms part of ribonucleic acid or RNA) and deoxyribose (the pentose sugar that forms part of deoxyribonucleic acid or DNA).

the protein. The disease **sickle cell anemia** results when a single amino acid is changed in two of the subunits of hemoglobin. The function of the altered hemoglobin is abnormal and seriously threatens the survival of the individual.

Nucleic Acids Are Long, Unbranched Chains of Nucleotides

What are the components of nucleotides; why are nucleotides important?

Nucleotides exist both as free molecules and as subunits of **nucleic acid** molecules. As free molecules, they are important in cell metabolism. As subunits of **ribonucleic acid (RNA)** and **deoxyribonucleic acid (DNA),** they are essential elements of genetic processes and heredity.

Components of Nucleotides

A nucleotide is composed of three smaller subunits: (1) a sugar (either ribose or deoxyribose), (2) a phosphate group or groups, and (3) a specific chemical group called a **base.** The subunits are linked by covalent bonds.

Sugars. The nucleotides of ribonucleic acids contain the pentose **ribose;** the nucleotides of deoxyribonucleic acids contain the pentose **deoxyribose** (Fig. 2–17).

Bases. The bases that form a part of nucleotides are nitrogen-containing ring compounds derived from either **purine** or **pyrimidine** (Fig. 2–18). They are called *bases* because they are proton acceptors. In DNA, the principal pyrimidine-derived bases are **thymine** and **cytosine;** in RNA, **uracil** and **cytosine.** The principal purine-derived bases in both DNA and RNA are **adenine** and **guanine.**

Phosphate Groups. A nucleotide molecule can have either one, two, or three phosphate groups.

PYRIMIDINE BASES

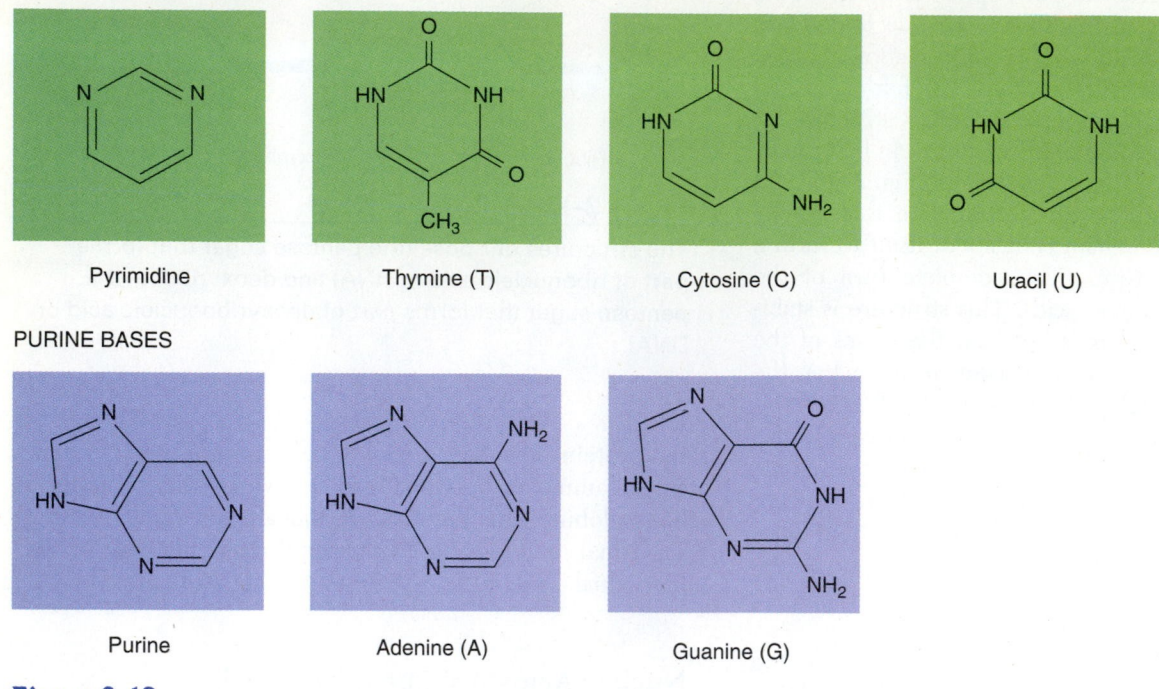

PURINE BASES

Figure 2–18

Structures of the pyrimidine- and purine-derived bases that are present in nucleotides. The single-letter abbreviations for the bases are shown in parentheses.

Formation of Nucleotides and Their Roles in Cell Metabolism

The basic structure of a nucleotide can be symbolized as follows:

Nucleotide = Base + Sugar + Phosphate

The base-plus-sugar group that makes up a part of a nucleotide is called a **nucleoside.** Thus the basic formula for a nucleotide also can be represented as:

Nucleotide = Nucleoside + Phosphate

For example, the base adenine becomes a nucleoside — **adenosine** — through the formation of a covalent bond between one of its ring nitrogen atoms and carbon 1 of ribose (Fig. 2–19). The nucleoside adenosine is converted to the nucleotide **adenosine monophosphate (AMP)** when a phosphate group is attached covalently to adenosine at carbon 5 of the ribose unit.

The addition of phosphate groups to adenosine monophosphate produces two new nucleotides: (1) **adenosine diphosphate,** or **ADP** (when one phosphate group is added to AMP), and (2) **adenosine triphosphate,** or **ATP** (when two phosphate groups are added to AMP; see Fig. 2–19).

Bases other than adenine can form nucleosides and nucleotides. Some examples are shown in Figure 2–20.

Nucleotides are important subunits for nucleic acids, the store of genetic information, but free nucleotides have important roles in the cell energy processes as well. ATP provides the chemical energy required for many of the biochemical reactions in the cell. **Guanosine triphosphate (GTP)** is required for protein synthesis; **uridine triphosphate (UTP),** for glycogen synthesis; and **cytidine triphosphate (CTP),** for phospholipid synthesis.

In addition, some compounds derived from nucleotides fulfill important functions in the cell. One example is **cyclic AMP** (cAMP; Fig. 2–20), which mediates the action of many neurotransmitters and hormones in cells. Adenine nucleotides also are components of many **coenzymes,** compounds required for many biochemical reactions. An important coenzyme is **nicotinamide adenine dinucleotide** (NAD; Fig. 2–20).

Nucleic Acids

Nucleic acids are composed of monophosphate nucleotide units covalently linked in long unbranched chains. The different nucleotides are linked by covalent bonds between the sugar of one nucleotide and the phosphate of the next (Fig. 2–21a). The phosphate group of the first nucleotide is attached to the hydroxyl group on carbon 3 of the pentose

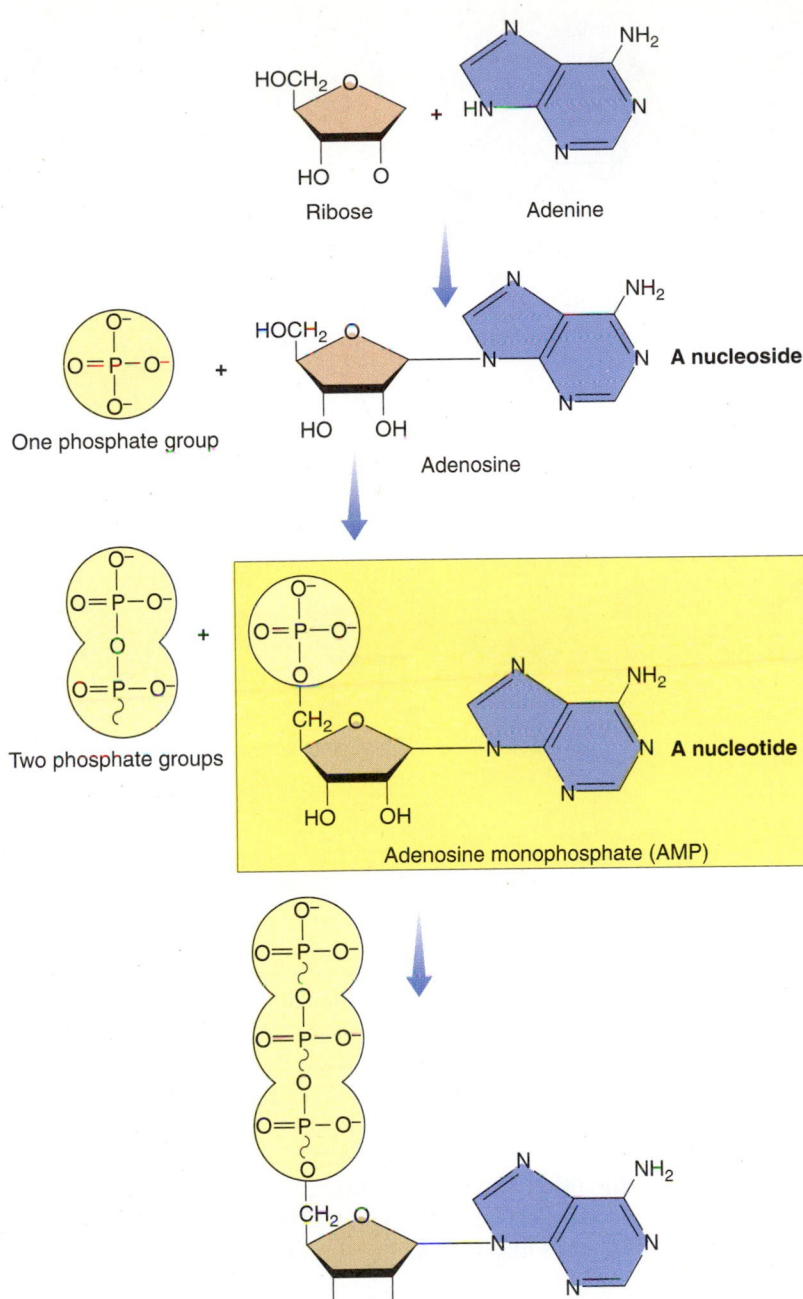

Ribose **Adenine**

One phosphate group Adenosine **A nucleoside**

Two phosphate groups **Adenosine monophosphate (AMP)** **A nucleotide**

Adenosine triphosphate (ATP)

Figure 2–19

Assembly of the units of a nucleotide. First, the pentose sugar ribose (or deoxyribose) attaches to a base, such as adenine, to form a nucleoside (in this case, adenosine). Next, a phosphate group (—PO$_4$) attaches to the nucleoside to form a nucleotide, adenosine monophosphate (AMP). The attachment of two more phosphate groups produces a different nucleotide, adenosine triphosphate (ATP).

sugar of the next nucleotide. Thus, each sugar in a nucleic acid has a base attached to carbon 1 and phosphate groups attached to carbons 3 and 5. The sugar–phosphate chain, analogous to the peptide bond in proteins, is a constant feature throughout any nucleic acid molecule and is known as the **sugar–phosphate backbone** of the nucleic acid (Fig. 2–21*b*). The variable part of the nucleic acid, analogous to the variable amino acid side chain groups in a protein, is the sequence of bases.

The precise sequences of the different bases in both DNA and RNA molecules represent the genetic information of the cell and of the organism as a whole. DNA and RNA interact in complex processes that transmit this genetic information to cellular systems responsible for the synthesis of all proteins involved in cell activities. The specific base sequences of DNA and RNA determine the specific amino acid sequences in the protein molecules, thereby dictating protein shape and function. These factors in turn affect the function-

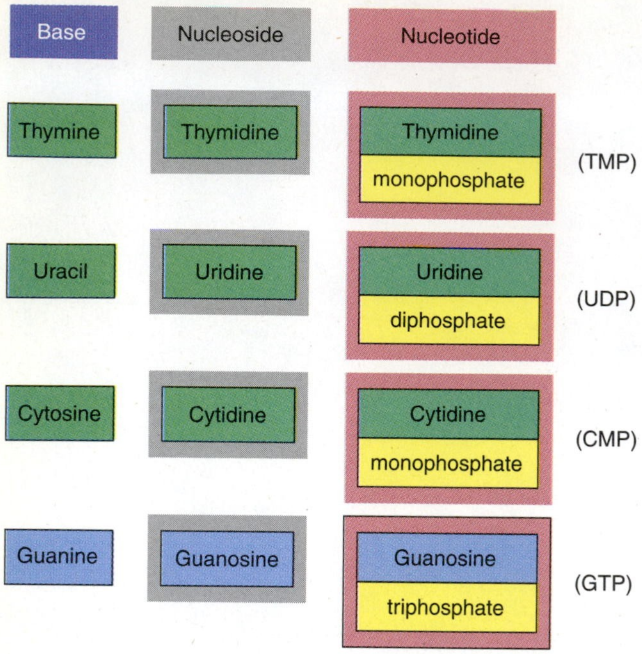

(a)

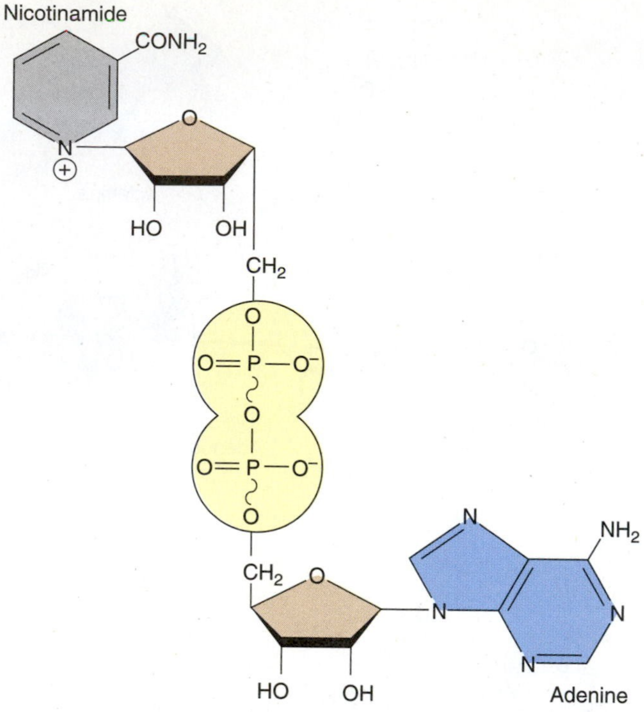

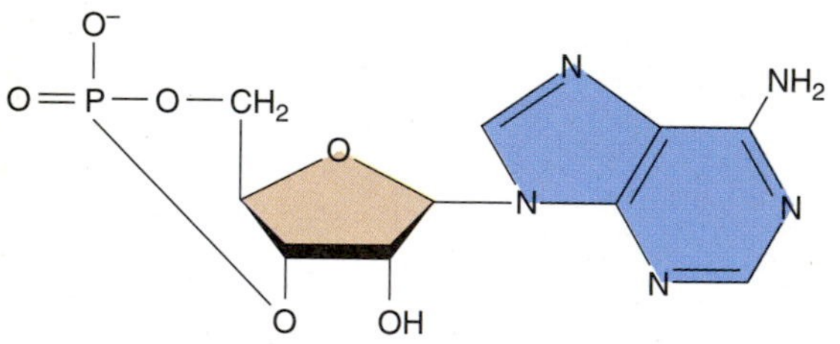

(b) Cyclic adenosine monophosphate (cyclic AMP)

(c) Nicotinamide adenine dinucleotide (NAD)

Figure 2–20

(a) Examples of some nucleosides and nucleotides and the structures of *(b)* cyclic AMP and *(c)* NAD, two important derivatives of adenine nucleotides.

ing of cells and the shape and characteristics of the organism as a whole.

The Structure of DNA

The DNA molecule differs from the RNA molecule in that it consists of two separate nucleotide chains, or "strands," instead of one. DNA also contains the sugar deoxyribose rather than ribose and contains thymine instead of uracil. Other aspects of the DNA molecule that distinguish it from RNA will become important in later chapters that discuss the replication of genetic material and cell reproduction.

The Double Helix. The two nucleotide strands in DNA are coiled around each other in a configuration called a **double helix** (Fig. 2–22). Each turn of the double helix contains ten bases.

The two nucleotide strands that form the double helix are oriented so that their bases are inside the column of the helix, with the bases from one strand situated close to the bases

from the other strand. As a result, hydrogen bonds form between a base in one strand and a base in the other. These hydrogen bonds help to hold the strands together in the helix.

Base-pairing. A hydrogen bond forms only between a purine-derived base and a pyrimidine-derived base. Hence, these interactions, called **base-pairings,** are highly specific and are the most important aspect of the DNA double helix. Adenine (a purine) always forms two hydrogen bonds only with thymine (a pyrimidine; Fig. 2–23*a*). Guanine (a purine) always forms three hydrogen bonds only with cytosine (a pyrimidine; Fig. 2–23*b*).

Complementary Strands. As a result of the base-pairing, the two DNA strands in the double helix are not identical; instead, they are **complementary** (Fig. 2–23*c* and *d*). Adenine in one strand always lies opposite thymine in the other strand. Cytosine in one strand always lies opposite guanine in the other strand.

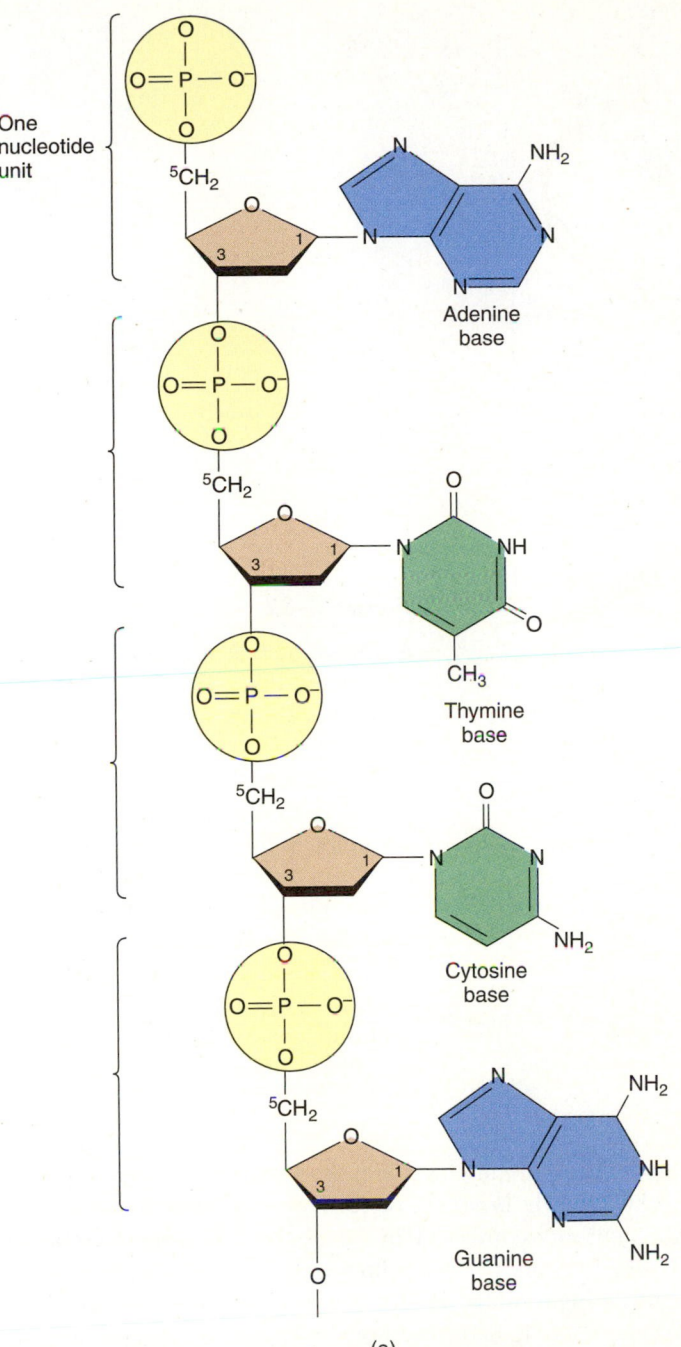

(a)

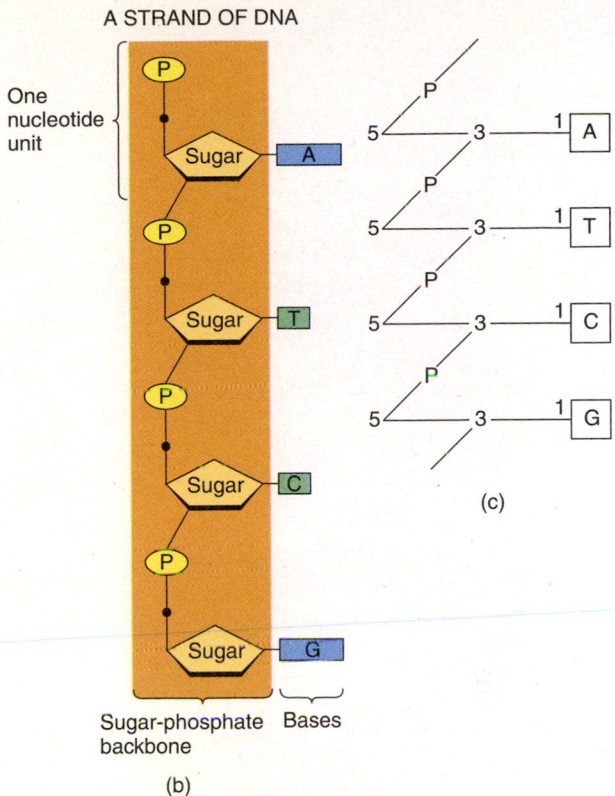

A STRAND OF DNA

One nucleotide unit

Sugar-phosphate backbone   Bases

(b)

(c)

Figure 2–21

(a) A single strand of DNA is formed by monophosphate nucleotides linked together via the sugar (deoxyribose) and phosphate groups. *(b)* The "sugar-phosphate backbone" is distinct from the bases, which are not directly linked to each other. *(c)* The shorthand notation for a DNA single strand.

Antiparallel Strands. The two nucleotide strands are **antiparallel** as well as complementary. This term refers to the arrangement of the sugar-phosphate backbone within each strand of DNA. An antiparallel arrangement is depicted in Figure 2–23*d*. Nucleoside T at the top of the left strand is linked to nucleoside C below it by a phosphate group attached to carbon 3 on the sugar of T and carbon 5 on the sugar of C. In contrast, nucleoside A at the top of the right strand is linked to nucleoside G below it by a phosphate group attached to carbon 5 on the sugar of A and carbon 3 on the sugar of G. Both the complementary and the antiparallel properties of the DNA strands play an essential role in the mechanism that passes on identical copies of DNA to offspring during cell division.

Length of DNA Molecules. A striking characteristic of DNA molecules is the length of the nucleotide chains. The bacterium *E. coli,* for example, contains a single molecule of DNA that is 1.2 millimeters (mm) long. It contains 3.4 million nucleotides and has a molecular mass of 23×10^8. In a

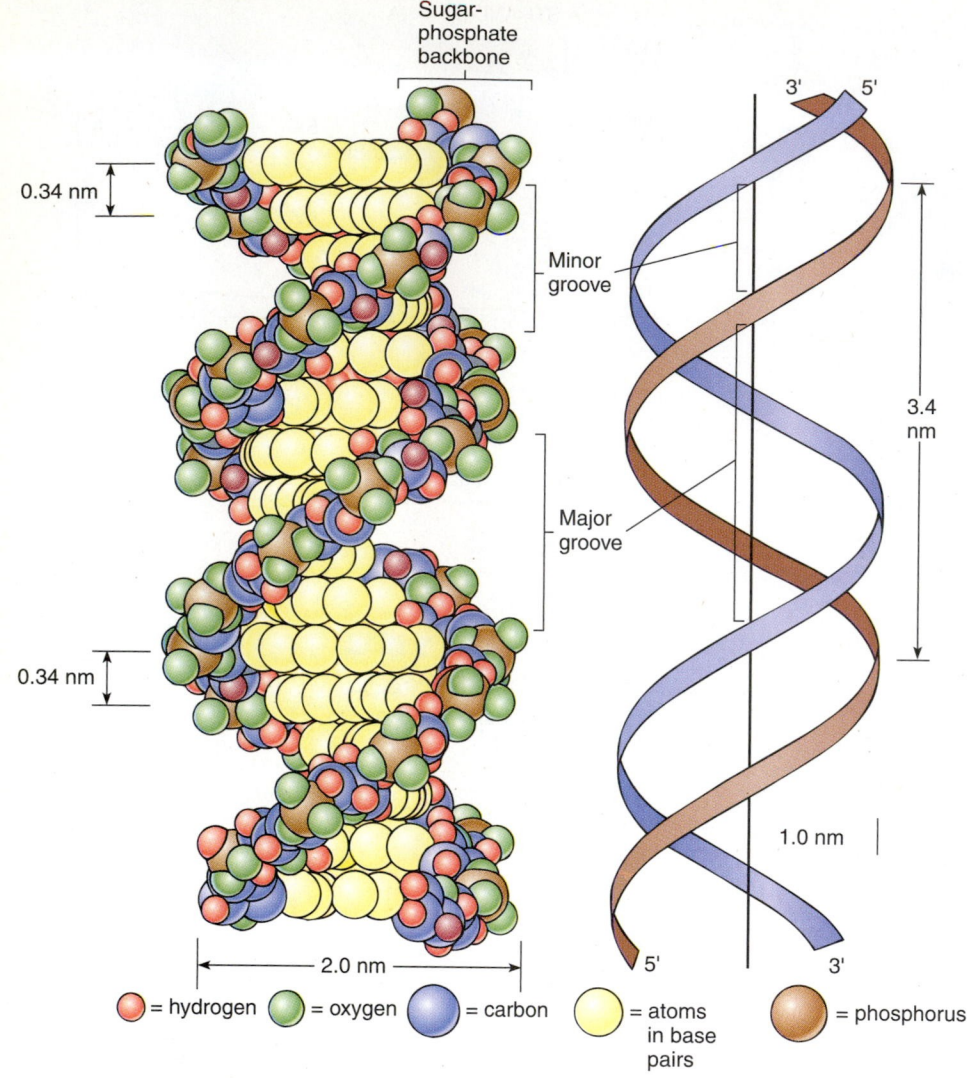

Sugar-phosphate backbone

0.34 nm

Minor groove

Major groove

0.34 nm

2.0 nm

3' 5'

3.4 nm

1.0 nm

5' 3'

 = hydrogen　　● = oxygen　　● = carbon　　○ = atoms in base pairs　　● = phosphorus

Figure 2–22

A double helix is formed when two complementary strands of DNA coil around each other. On the left is a space-filling model showing that the sugar–phosphate backbones of the two strands form the outside of the helix. In the model on the right, the ribbons represent the sugar–phosphate backbone of each strand. Some structural features and dimensions are shown. Exactly ten base pairs are present within each full turn of the helix, which occupies a distance of 3.4 nm.

typical animal cell, the DNA double helixes are themselves further coiled into tight bundles. A single animal cell might contain 1 meter (m) of DNA, with 3×10^9 nucleotides.

THERMODYNAMIC PRINCIPLES

> *What determines whether a chemical reaction will work?*

Living organisms are highly organized, or **ordered,** systems. Maintaining this ordered state requires constant work. The capacity to do work is provided by energy, which can exist in several different forms, such as radiant energy (e.g., light and heat), mechanical energy, chemical energy, and electrical energy. Various physiological processes use different forms of energy. For example, some of the energy released during metabolism is heat (**radiant energy**) that helps to maintain normal body temperature. **Mechanical energy** is involved in moving a part of the body, such as an arm, by the contrac-

tion of skeletal muscles. **Chemical energy** is released or absorbed during breakage or formation of chemical bonds in the molecules involved in metabolic reactions. **Electrical energy** results from the flow of charged particles such as ions and is essential for the transmission of nerve impulses. As elsewhere in nature, energy in the body is often converted from one form to another.

The unit used to measure energy is the **kilocalorie (kcal).** One kilocalorie is defined as the amount of heat that raises the temperature of 1 kg of water by 1°C. Most forms of energy are readily converted to heat and measured in kilocalories. An alternative unit of energy is the **joule (J);** 1 kcal is equivalent to 4.2 kilojoules (kJ).

The study of energy is known as **thermodynamics** (literally, "heat changes"). **The first law of thermodynamics** states that *energy can neither be created nor destroyed.* Any system can absorb or give up energy to its surroundings, but the total energy content of the system and its surroundings is always the same. **The second law of thermodynamics** states that the *disorder* (also called *randomness* or *entropy*)

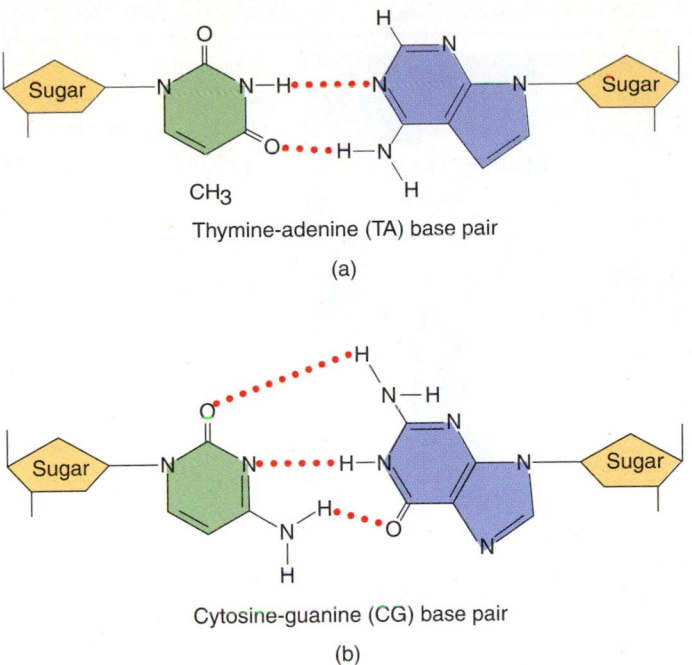

Thymine-adenine (TA) base pair

(a)

Cytosine-guanine (CG) base pair

(b)

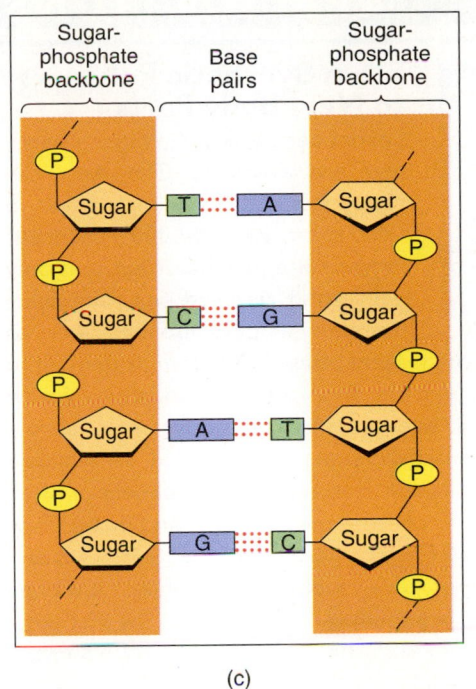

(c)

Figure 2–23

Base-pairing between bases of separated DNA strands.
(a) Thymine and adenine form one base pair. These two
bases always form two hydrogen bonds between them.
(b) Cytosine and guanine form the other base pair. These
two bases always form three hydrogen bonds between
them. **(c)** The two strands are complementary because
specific bases on one strand form hydrogen bonds only
with specific bases on the other strand. **(d)** The shorthand
notation indicating complementary and antiparallel strands.

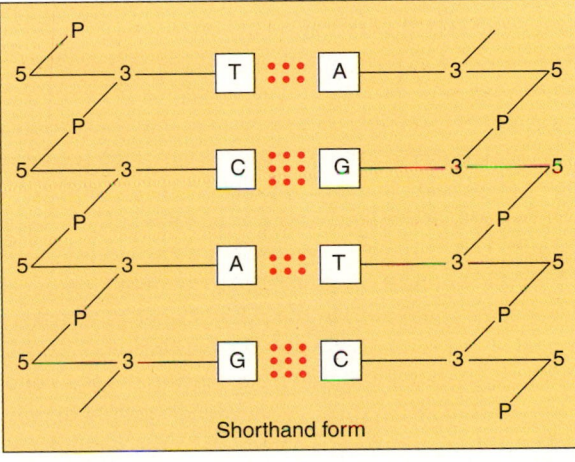

Shorthand form

(d)

in a system is always increasing over time. This means that as
energy is converted from one form to another, some usable
energy (available to do work) is degraded into a less usable
form (heat). Energy conversion is never 100% efficient; the
heat disperses to the surroundings.

Even relatively simple tasks, such as assembling a new
bicycle, require work to create the necessary order. The
parts of a bicycle, when shaken from the package, do not
fall to the ground in just the right position and sequence
needed for spontaneous assembly of the bicycle. Assembly
always requires some work. Similarly, a living organism,
which must impose order on its internal functions as well
as on its surroundings, must do work to resist the universal
tendency toward disorder. Much of this work takes place in
the individual cells of the organism, in the form of reac-

tions that build up macromolecules to support the structures and functions of the organism. This "biological" work
requires energy; however, energy does not simply exist in
the cell. The cell must transform the potential energy held
in the bonds of other molecules into a form it can use to do
work.

The Direction of Chemical Reactions Is Determined by the Free Energy Change

Some of the bonds that hold atoms together in a molecule or
other chemical group represent potential energy, or a source
of energy that can be released to do work. According to the
laws of thermodynamics, however, not all of the energy can
be used; some energy always is lost because of the tendency

APPLICATIONS OF PHYSIOLOGY

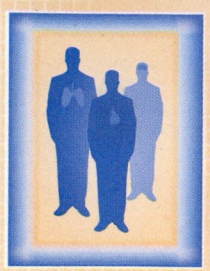

From Synthetic Polymers to New Body Parts

Spare body parts are in high demand; few are available. The American Heart Association estimated that in 1997 there were 40,000 Americans in need of a new heart but that only 2300 (6%) received one. Lifesaving livers and kidneys also are scarce, as is skin for burn victims and others with wounds that fail to heal. Tissue engineering has emerged as a thriving new field of medical science. Until recently, most people thought that damaged human tissue could be replaced only with direct tissue transplants from donors or with fully artificial parts constructed of plastic and metal. Few believed that bioartificial organs, hybrids of living cells and synthetic polymers, ever would be possible. However, through innovative and imaginative science, the first engineered tissue (living skin) became available in 1998 in the United States and Canada. Today, even though some important problems must be solved, biotechnology companies that are trying to develop tissue-engineered products have a market worth of nearly $4 billion.

The creation of new organs applies the recent gains in cell physiology and biology to the problems of tissue and organ reconstruction, just as advances in materials science facilitate new types of architectural design. Two scenarios are being explored. In one, specific peptides, known as **growth factors**, are placed at the site of a wound or damaged organ and stimulate the patient's own cells to migrate into the wound site, change into the right kind of cell and regenerate the tissue. In the other scenario, which is a major goal of the tissue engineer, the patient receives cells that were harvested previously from the patient or a donor and incorporated into a three-dimensional scaffold of polymers. The entire structure is attached to the wound site, where the cells reproduce, reorganize, and form new tissue. The polymers slowly break down, leaving only a new living organ in the body. The purpose of the scaffold is to maintain a space for formation of the tissue and to guide its structural development (see the ear in the INSERT figure). Circles of DNA known as **plasmids** can be embedded in the scaffold material for release over extended periods. The released DNA finds its way into migrating cells on the scaffold and provides the cells with the genetic information to make specific proteins. This technique may make it possible to promote more precise tissue formation in stages to promote the growth of blood vessels within the tissue. Scaffolds can influence the direction of cell growth, which is especially important for regeneration of nerve cells. Neurophysiologists hope to direct growing nerve cells to extend across gaps in nerves caused by damage or illness. Most of the materials now used as scaffolds are either synthetic, such as biodegradable suture material, or natural,

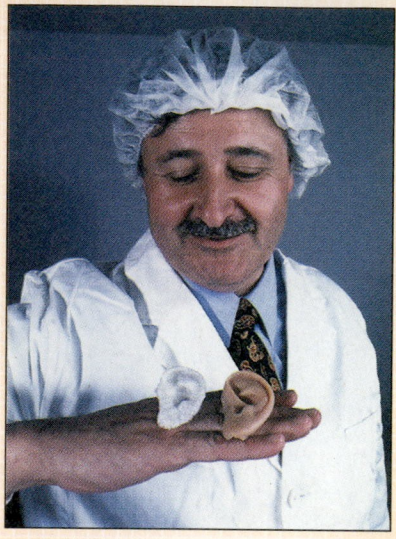

A polymer scaffold in the shape of an ear is "seeded" with cells that eventually replace the polymer with real tissue.

such as the protein known as **collagen** that is found in skin and bone. Cells tend to stick more easily to collagen, but the advantage of synthetics is that they can be mass-produced with close control of their strength, structure, and speed of degradation in order to optimize tissue formation. Researchers are trying to combine the best features of both by adding natural amino acids, peptides, or proteins to the surface of synthetic polymers. This approach allows the polymer surface to be "tailored" so that only specific types of cells are attracted and adhere, making rejection by the body unlikely. As this work proceeds, it may be possible to attract the right mix of cell types for a functioning organ, such as a liver. The liver, kidney, intestine, bladder, and even the heart are targets for regrowth. It may take 10 to 20 years to understand how to grow an entire heart, but heart valves and blood vessels may be available sooner.

Structural tissues, such as skin and bone, are relatively simple and probably will dominate the first wave of successes. The engineered skin product approved in 1998 was 60% better than conventional bandaging treatment for healing venous ulcers, skin lesions that affect the elderly population. Over a period of weeks, successive layers of laboratory-grown skin were placed over a patient's ulcerated skin. Blood vessels grew into the new tissue to sustain it, and no signs of rejection were present up to 3 months later. More than 1 million Americans each year require skin grafts. In addition to people who are burned severely, many others have skin removed because of skin cancer and many diabetic patients develop skin ulcers that are difficult to heal.

Although bioartificial tissues and organs are medical devices, they contain living cells that produce specific molecules that act like drugs. They present a problem for the US Food and Drug Administration, which must approve them. Clear-cut policies must be developed to determine that any implanted material and its by-products are safe for human use.

toward disorder. The term **free energy** is used to describe the energy that can be put to work.

Free energy is given the symbol G and, as with all energy, is measured in kilocalories. While it is difficult to determine the free energy content of a particular molecule, it is relatively easy to calculate the *change* in free energy that occurs in the course of a chemical reaction. The change in free energy is symbolized as ΔG (the Δ symbol is the capital Greek letter "delta").

The concept of free energy is useful for predicting the direction in which a reaction will tend to proceed; some specific examples are presented in Chapter 6. A reaction that releases free energy is said to be **exergonic** and occurs spontaneously. A reaction that requires input of free energy before proceeding does not occur spontaneously and is described as **endergonic.** Exergonic and endergonic reactions usually are linked in a cell, so that the free energy produced by an exergonic reaction can be used to drive an endergonic reaction.

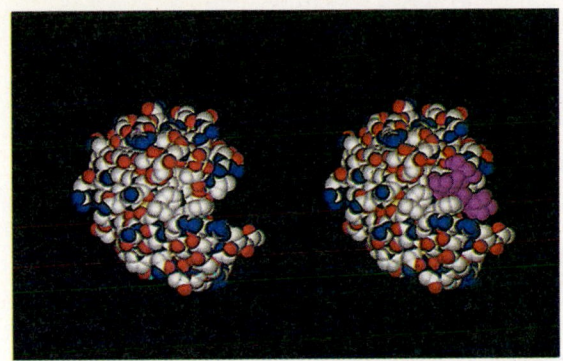

Figure 2–24

A computerized model of an enzyme protein before (*left*) and after (*right*) attachment to its substrate. (© *Visuals Unlimited*)

The Catalytic Role of Enzymes

A **catalyst** is a substance that increases the rate of a chemical reaction without actually participating in it and that is recovered unchanged when the reaction is completed. Catalysts do not influence the direction of a reaction, the final concentrations of the reactants and products, or the free energy change. They merely speed up a reaction that is already thermodynamically feasible. It is not unusual for a reaction to occur 1 million times faster than normal when an catalyst is present. Catalysts are widely used in industry to increase the rates of reactions and thereby the quantities of products that can be manufactured in a given amount of time.

Enzymes are natural catalysts in the cell that control the rates of many important reactions, speeding them up when necessary and slowing them down when necessary. The degree by which an enzyme can speed up a reaction can be very impressive. The increase is made possible in part because, in many reactions, a single enzyme molecule can catalyze the reaction of 10,000 reactant molecules every second. Such a number is called the **turnover number** for the enzyme. Typically, only a few enzyme molecules are needed to convert a relatively large number of reactant molecules into product.

The Mechanisms of Enzyme Action

> *How do enzymes work?*

Enzymes are proteins, and each enzyme has a characteristic three-dimensional shape and surface structure resulting from the folding of the polypeptide chain (Fig. 2–24). Enzymes facilitate the reactions of metabolism in cells by providing specific binding sites on their surfaces for the reactant molecules. These sites, which may be indentations on the surface of the enzyme, are known as **active sites.** The reactant molecules that bind to the active sites are known as the **substrates** of the enzyme.

We know of more than 1000 different enzymes. About 90% of the proteins present in a cell are enzymes. A key aspect of enzyme action is **specificity,** that is, a particular enzyme catalyzes only a very few reactions, and sometimes only one reaction, because its active sites accept and bind only certain reactants.

Although individual enzymes are "tailored" for specific substrate molecules, most enzymes can be classified according to the general types of reactions that they catalyze. A **dehydrogenase,** for example, catalyzes the removal of hydrogen from a molecule (oxidation), whereas an **oxidase** catalyzes the addition of oxygen. A **transferase** mediates the transfer of groups of atoms; an **isomerase** rearranges the atoms in a molecule. A **hydrolase** splits chemical bonds by adding the constituents of water (H and OH) to the separated atoms; a **ligase** joins ("ligates" or "ties") molecules together by forming new bonds in reactions coupled to breakdown of ATP. Note that the name for each class of enzymes always ends in *-ase.*

An enzyme (E) works by forming a temporary **enzyme–substrate complex** (E–S) with its substrates via hydrogen bonds at its active sites. This complex holds the substrate molecules in the proper orientation for the reaction to occur. When the reaction is complete, the E–S complex breaks apart to yield the reaction products (P) and the free enzyme:

$$E + S \rightarrow E\!-\!S \rightarrow E + P$$

This equation is represented in diagram form in Figure 2–25, which shows the "lock-and-key model" of enzyme action. Evidence suggests, however, that not all active sites

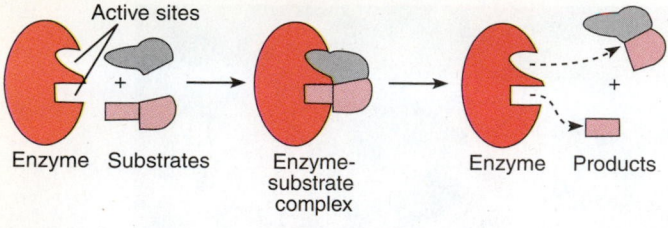

Figure 2–25

The "lock-and-key" model of enzyme action. The substrates fit tightly into the active sites on the enzyme molecule. When they have formed products, the enzyme releases them and is free to catalyze another reaction.

have the rigid conformation shown in this figure. In some cases, the active site may assume a shape complementary to the desired reaction only after the substrates are bound, a process known as **induced fit** (Fig. 2–26).

The particular reactant molecules that must come together for a specific reaction are not likely to meet spontaneously at a high rate. By binding only the reactant molecules specific to a desired reaction, an enzyme "favors" that reaction over the other possible reactions in which the molecules could participate. Formation of an $E-S$ complex increases the frequency with which the reactant molecules collide in the proper orientation for the reaction to occur.

To further encourage reactions, some enzymes bind a substrate molecule in a manner that distorts the structure of the substrate. This may place a strain on the existing chemical bond — helping to break it — or it may facilitate a chemical reaction between groups that are not normally close enough together to react. These effects contribute to a higher reaction rate.

From a thermodynamic standpoint, enzymes in effect decrease the **activation energy** necessary for the reaction to occur — for example, the energy needed to break old bonds and form new ones. More reactant molecules have the required energy to overcome this activation energy barrier and can therefore react to form products. Again, enzymes act specifically, decreasing activation energy only for the desired reaction. In the cell, products of reactions are often intermediate products in long chains of reaction. At each step, a product must be directed down one of several alternative pathways. Because an enzyme decreases the activation energy only for the specific reaction that it catalyzes, the reactant molecules are more likely to attain this level of activation energy and undergo the reaction catalyzed by the enzyme. They are less likely to undergo other possible reactions that require a higher activation energy.

Enzyme Kinetics

Researchers can study reaction rates and other characteristics of enzyme-catalyzed reactions using enzymes that have been isolated from the cell and purified. These in vitro (outside the body) studies of enzyme activity allow for comparisons of different enzymes under the same conditions.

The relationship between reaction rate and substrate concentration, at a constant enzyme concentration, is a particularly useful measure and is expressed graphically in Figure 2–27. Initially, the reaction rate increases rapidly as the concentration of substrate increases because, at low substrate levels, the enzyme is not working at full capacity.

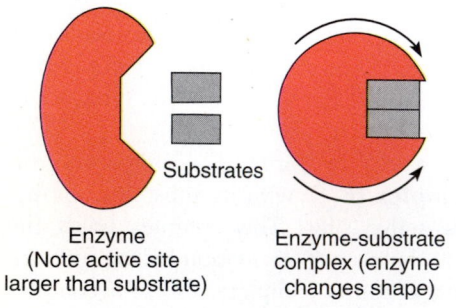

Figure 2–26

"Induced fit" is a more likely mechanism for some enzymes. According to this model, the active sites are somewhat larger than the substrates. When the substrates contact them, the active sites change shape to bind the substrates.

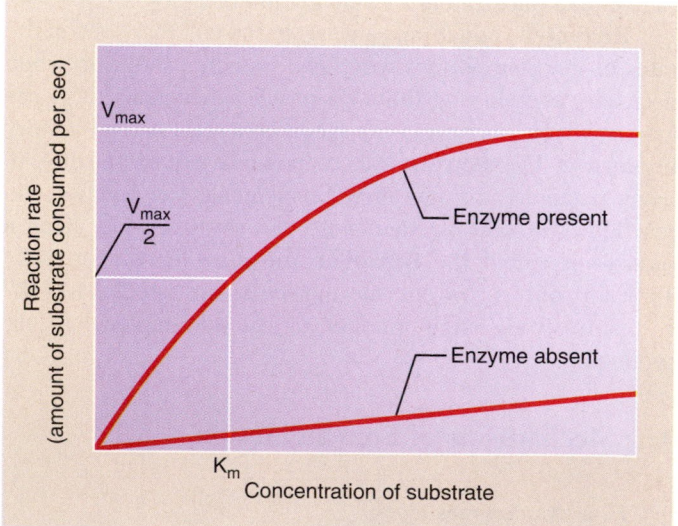

Figure 2–27

When an enzyme is present, a reaction proceeds initially at a very high rate, reaching a maximum rate (V_{max}) when the enzyme is saturated with substrate. The Michaelis constant (K_m) is the concentration of substrate that produces one half of the maximum reaction rate.

But eventually, a **maximum reaction rate (V_{max})** is reached. Further increases in substrate concentration produce no further increase in the reaction rate. A V_{max} is achieved because the enzyme molecule has a limited number of active sites. When substrate molecules occupy all of the sites on the enzyme, it is said to be "saturated" and is working at maximum capacity. The addition of more substrate cannot increase the rate of reaction. The reaction rate in the absence of the enzyme is shown for comparison. It is much slower and increases linearly without exhibiting a V_{max} over the same range of substrate concentration (Fig. 2–27).

Knowledge of the V_{max}, the maximum amount of substrate that can be utilized and turned into reaction product in a given time, allows calculation of the turnover number of the enzyme from the relationship:

$$V_{max} = k[E]$$

where k is the turnover number and $[E]$ is the enzyme concentration. For example, for a reaction catalyzed by the enzyme carbonic anhydrase:

$$CO_2 + H_2O \xrightarrow{\text{Carbonic anhydrase}} H_2CO_3$$

k can be calculated as follows: When carbonic anhydrase is fully saturated with substrate, just 10^{-6} mol $[E]$ of this enzyme catalyzes the formation of H_2CO_3 at a rate of 0.6 mol/sec (V_{max}). Hence, the turnover number k is 0.6 divided by 10^{-6}, which is 600,000 substrate molecules converted to products per second. This is one of the largest known turnover numbers for any enzyme. In contrast, the turnover number for DNA polymerase is only 15 molecules/sec, which is very small. Note that the V_{max} and the turnover number are determined by laboratory experiments and provide a measure of the maximum activity of an enzyme. However, most enzymes in the cell are not saturated by substrate and do not operate at their V_{max}. The advantage this provides to the cell is the capacity to deal with fluctuations in the concentrations of its molecules. Thus, when the demand for glucose catabolism is increased rapidly by physical exercise, the enzymes involved have the capacity to respond by processing more substrate molecules in the same amount of time.

Another useful term that can be obtained from the plot in Figure 2–27 is the **Michaelis constant, K_m,** which is defined as the substrate concentration that produces one half of the maximum reaction rate. In the appropriate circumstances, the K_m provides a measure of the affinity of the enzyme for a particular substrate. A high K_m indicates a low affinity; a low K_m indicates a high affinity. Thus, measurement of V_{max} and K_m values allows for quantitative comparisons of the activities and specificities of different enzymes for various substrates. This kind of analysis has greatly aided scientists' understanding of the reactions of cellular metabolism.

The Effects of pH and Temperature

In addition to concentration of substrate, enzyme activity depends on both the temperature and pH of the medium (surroundings). An increase in temperature will increase the rate of most enzyme-catalyzed reactions until temperatures in the range of 50°C to 60°C are reached. At this point, the temperature is high enough to **denature,** or break down the tertiary structure of, the enzyme protein. When this occurs, the enzyme rapidly loses its catalytic activity.

The pH range over which most enzymes can function is relatively narrow but varies from one enzyme to another. **Pepsin,** for example, which breaks down proteins in the stomach, works best at a pH of 2, the **pH optimum** for this enzyme. In contrast, **trypsin,** secreted by the pancreas, is most effective in breaking down proteins at a pH of 8; **alkaline phosphatase** removes phosphate groups from substrates most effectively when the pH is close to 10. Changes in pH result in the addition or removal of hydrogen ions (H^+) from the enzyme, thereby changing the number of positive and negative charges present at the active site and altering the catalytic efficiency of the enzyme.

Enzyme Cofactors: Coenzymes

A **cofactor** is a nonprotein component that some enzyme molecules need to function as catalysts. For example, several enzymes involved in glucose catabolism require a magnesium ion for their activity. Other enzymes use iron, copper, or zinc ions. In some instances, the metal ion actually carries out the catalytic reaction, although the reaction is facilitated by the presence of the enzyme protein. In other cases, the metal ion serves to maintain the enzyme molecule in the structure needed for biological activity. A change in pH can alter the binding of a cofactor and result in a change in the catalytic activity of the enzyme. It is difficult to remove metal cofactors that are tightly bound to an enzyme without disturbing the protein structure and denaturing the enzyme.

An enzyme cofactor that is an organic molecule is referred to as a **coenzyme** (Fig. 2–28). These types of cofactors can serve as donors or acceptors of the atoms or functional groups added to or removed from the substrate.

ATP Acts as a Free Energy Store

> *How is ATP involved in storing free energy in cells?*

The intricate processes by which cells transform chemical energy to do work are examined in detail in Chapter 6. A general overview at this point, however, will reveal the basic relationship — and an important molecule — linking the endergonic and exergonic reactions in the cell.

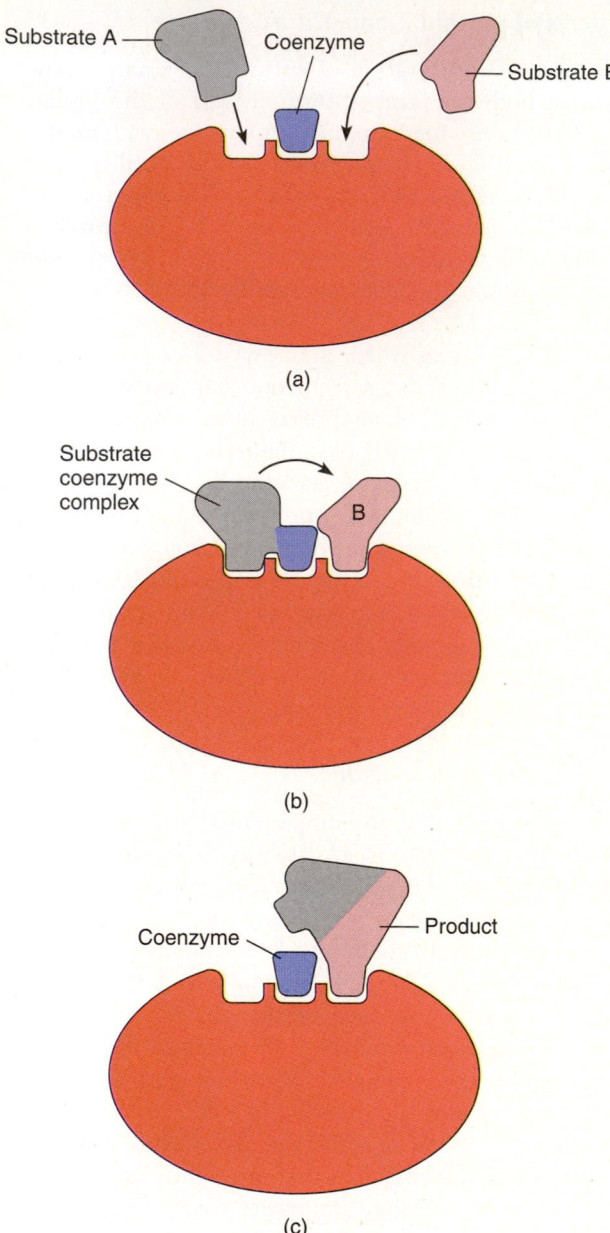

(a)

(b)

(c)

Figure 2–28

(a–c) Coenzymes are nonprotein molecules that aid enzymes in the attachment of substrates.

We have learned that the macromolecule ATP is a nucleotide consisting of three phosphate (P_i) groups, a ribose ring, and adenine. When a phosphate group is removed, ATP becomes ADP; when another is removed, ADP becomes AMP. The removal of one phosphate group from ATP releases a large amount of free energy; the attachment of one phosphate group requires a large amount of free energy. For these reasons, ATP acts as a source of free energy. The bonds among its phosphate groups are often represented by wavy lines in the ATP structural formula (see Fig. 2–19) and are referred to sometimes as "high-energy" bonds.

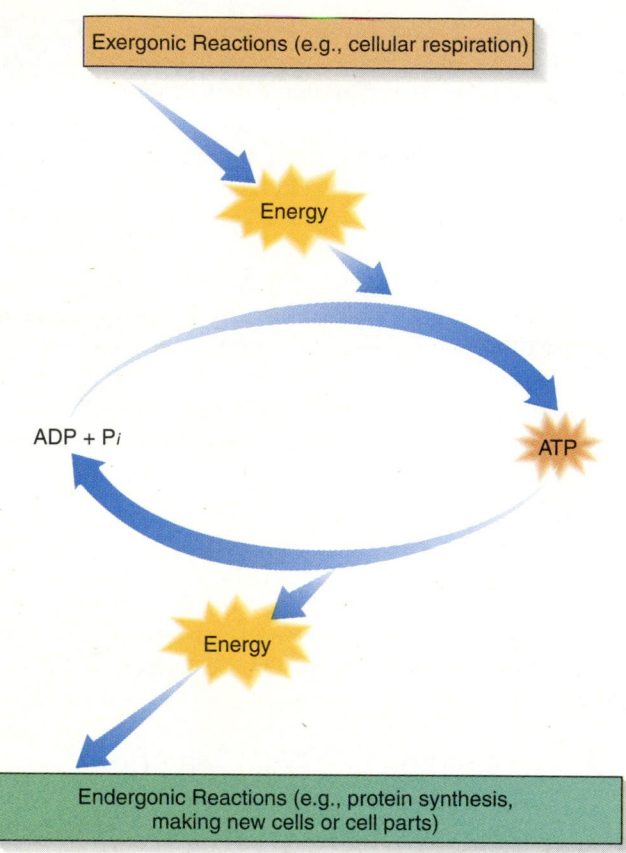

Figure 2–29

Both endergonic (require free energy) and exergonic (release free energy) reactions occur in the cell. Endergonic reactions obtain the free energy that they need to proceed from the free energy spontaneously released from exergonic reactions. This free energy is exchanged in the form of ATP.

Exergonic reactions that occur in the cell produce free energy that is used to **phosphorylate** (add a phosphate group to) AMP and ADP:

$$AMP + P_i + Energy \rightarrow ADP$$
$$ADP + P_i + Energy \rightarrow ATP$$

These reactions represent a process in which the cell temporarily deposits free energy released in exergonic reactions, such as the breakdown of nutrient molecules, in molecules of ATP (Fig. 2–29) through **phosphorylation.**

This energy is not deposited for long, however, because other reactions that need large amounts of free energy to begin also are occurring: these endergonic reactions are mainly those of biosynthesis. They obtain the free energy that they need from the ATP produced by the exergonic reactions:

$$ATP \rightarrow Energy + ADP + P_i$$

leaving ADP and a phosphate group to be used in the next round of energy release and storage.

CHAPTER REVIEW

Summary

- Matter is composed of elements, the smallest units of which are atoms.
- Atoms of the same element have the same number of protons. Uncharged atoms have equal numbers of protons and electrons. Charged atoms (ions) have unequal numbers of protons and electrons.
- Atoms share, gain, or lose electrons to form bonds with other atoms. Bond formation between atoms forms molecules, groups, and compounds.
- Chemical reactions occur when compounds react with other compounds to form new products. The rate of a chemical reaction depends on the temperature and the concentration of reactant molecules.

- The polarity of the water molecule makes water an excellent solvent for ionic and polar covalent compounds. Nonpolar compounds, in contrast, are insoluble in water.
- The unique covalent bonding capabilities of the carbon atom allow it to form molecules of huge proportions and intricately varied structures. These macromolecules include carbohydrates, lipids, proteins, and nucleic acids. They are essential for many cell processes.
- Enzymes increase the rates of specific reactions by bringing reactant molecules together in the proper orientation.

Review Questions

Choose the Correct Answer

1. Which of the following forms a base pair with adenine ?
   a. Thymine
   b. Guanine
   c. Cytosine
   d. Adenine
   e. Phosphate
2. Attachment of serine to phosphatidic acid forms a:
   a. dipeptide.
   b. glycolipid.
   c. phospholipid.
   d. nucleoside.
   e. fatty acid.
3. A dehydrogenase enzyme catalyzes:
   a. joining of molecules by forming new bonds.
   b. addition of oxygen to a molecule.
   c. rearrangement of atoms in a molecule.
   d. removal of hydrogen from a molecule.
   e. transfer of groups of atoms.
4. Nucleic acids are long chains of:
   a. amino acids.
   b. nucleosides.
   c. fatty acids.
   d. nucleotide monophosphates.
   e. ribose and fructose.
5. Adenosine triphosphate (ATP) is an important molecule that:
   a. contains guanine.
   b. is an example of a nucleoside.
   c. acts as an intracellular store of free energy.
   d. catalyzes the addition of phosphate groups to substrates.
   e. is an example of a coenzyme.
6. What is the molecular mass of water (in daltons)?
   a. 9
   b. 17

   c. 18
   d. 19
   e. 36
7. What is the concentration of hydrogen ions in a solution of HCl at pH 5.0?
   a. 5 M
   b. 5 mM
   c. 10^{-5} M
   d. 10^{-7} M
   e. 5^{-10} mM
8. Which of the following are present in a protein?
   a. Peptide bonds
   b. Hydrogen bonds
   c. Disulfide bonds
   d. Hydrophobic interactions
   e. All of the above
9. The two types of particles that make up the nucleus of an atom are:
   a. protons and neutrons.
   b. protons and electrons.
   c. anions and electrons.
   d. cations and protons.
   e. neutrons and electrons.
10. Which of the following bases is not present in DNA?
    a. Adenine
    b. Cytosine
    c. Guanine
    d. Thymine
    e. Uracil
11. Which of the following is true for disulfide bonds?
    a. They are known as *ionic bonds*.
    b. They are formed by hydrolase enzymes.
    c. They occur between two serines.
    d. They maintain the quaternary structure of proteins.
    e. They are found in the hormone insulin.

12. The K_m of an enzyme-catalyzed reaction is:
    a. the turnover number of the enzyme.
    b. a measure of the affinity of the enzyme for its substrate.
    c. dependent on the concentration of the enzyme.
    d. higher when the enzyme binds substrate more tightly.
    e. the substrate concentration when the reaction rate is maximal.

13. Which of the following acts as an enzyme?
    a. Glycogen synthase
    b. Cyclic AMP
    c. Cholesterol
    d. Phosphatidylcholine
    e. Glycogen

14. A cofactor is often:
    a. a protein.
    b. nonessential to the activity of enzymes that bind it.
    c. a metal ion.
    d. a monosaccharide.
    e. required to increase the activation energy of an enzymatic reaction.

15. The activity of an enzyme is dependent on:
    a. temperature.
    b. pH of the medium.
    c. substrate concentration.
    d. intact tertiary structure.
    e. all of the above.

16. Isotopes are atoms of the same element that have different numbers of:
    a. electrons.
    b. moles.
    c. neutrons.
    d. nuclei.
    e. protons.

17. How many grams of NaCl crystals must be dissolved in water to make 500 ml (final volume) of 0.3 M NaCl solution?
    a. 8.78
    b. 17.55
    c. 29.25
    d. 58.50
    e. 195.00

18. The nonvariable, repeating unit of DNA is a phosphate group linked by covalent bonds to:
    a. two bases.
    b. two sugars.
    c. a sugar and a base.
    d. two additional phosphates.
    e. a sugar, a base, and a phosphate.

19. Which of the following is true for DNA?
    a. It is a small molecule.
    b. It contains only eight different bases.
    c. The two DNA strands in the double helix are identical.
    d. It can catalyze chemical reactions involving small molecules.
    e. It contains the information that a cell needs to make proteins.

20. Which of the following is true for proteins?
    a. They are macromolecules.
    b. They are linear, unbranched polymers.
    c. They perform many diverse functions in the cell.
    d. Amino acids are the basic units of protein molecules.
    e. All the above are true.

Answers to Case History Questions

1. Sphingomyelin accumulates because it is produced at the normal rate, but its consumption is reduced as a result of the deficiency of sphingomyelinase.

2. Accumulation of sphingomyelin in specific cell types found in these organs causes the cells to enlarge and hence the entire organ becomes swollen in size. The spleen, for example, can be ten times its normal weight.

3. Sphingomyelin and sphingomyelinase are normal components of almost all cells. Mental retardation occurs because the progressive accumulation of sphingomyelin in brain cells gradually inhibits their function and causes cell death.

4. Niemann-Pick disease, specifically the infantile form (type A). This is the most common and is fatal. It is an inherited disease caused by an incorrect sequence of bases in the DNA molecule. This change in the genetic information means that the cells either cannot make the sphingomyelinase enzyme, a protein, or they make a protein that lacks normal enzyme activity. The disease is one example from the group of lipid storage disorders.

Key Terms

acid (p. 39)
activation energy (p. 58)
amino acid (p. 45)
atom (p. 32)
base (p. 39)
carbohydrate (p. 39)

chemical reaction (p. 37)
compound (p. 32)
covalent bond (p. 36)
domain (p. 49)
electron (p. 33)

element (p. 32)
enzyme (p. 57)
equilibrium (p. 38)
fatty acid (p. 42)
ionic bond (p. 36)

kinetic energy (p. 32)
lipid (p. 32)
macromolecule (p. 32)
nucleic acid (p. 32)
solution (p. 38)

Suggested Readings

Abbott, A. "Structures by numbers." *Nature* 408:130–132, 2000.

Branden, C., and Tooze, J. *Introduction to Protein Structure,* ed 2. New York, Garland, 1999.

Cooper, G. M. *The Cell: A Molecular Approach.* Sunderland, M. A., Sinauer, 1997.

Knowles, J. R. "Enzyme catalysis: Not different, just better." *Nature,* 350:121–124, 1991.
Kolodny, E. H. "Niemann-Pick disease." *Current Opinion Hematology,* 7:48–52, 2000.

McAdara, J. "The resolution solution: Confocal and multiphoton microscopy reign supreme for three-dimensional, live cell imaging." *The Scientist,* February 14, 2000.
Mooney, D. J., and Mikos, A. G. "Growing new organs." *Scientific American,* 280:60–65, 1999.

Piston, D. W. "Imaging living cells and tissues by two-photon excitation microscopy." *Trends Cell Biology,* 9:66–69, 1999 (April).

Service, R. F. "Designer tissues take hold." *Science,* 270:230–232, 1995.
Service, R. F. "Tissue engineers build new bone." *Science,* 289:1498–1500, 2000.
Simons, K., and Ilkonen, E. "How cells handle cholesterol." *Science* 290:1721–1726, 2000.

Vanier, M. T. "Biochemical studies in Niemann-Pick disease: Major sphingolipids of liver and spleen." *Biochimica et Biophysica Acta* 750:178–184, 1983.

Web sites

www.biomat.net
The Biomaterials Network is a free site that focuses on engineered skin, plastic joints, and artificial organs.

www.hhmi.org/index.html
The Howard Hughes Medical Institute supports biomedical research and science education.

www.sciam.com/
Scientific American publishes articles on a variety of scientific topics that are well written, with excellent illustrations, suitable for the nonspecialist. The Web site has a search feature for easy location of articles on specific issues.

www.thescientificworld.com
The Scientific World is a new Internet site that provides online resources, such as fun and facts about science, chat rooms, shopping for laboratory supplies, electronic publication of peer-reviewed scientific papers, and a database of articles. There is no charge to search the databases on this site.

www.the-scientist.com/
The Scientist is a free news journal (updated biweekly) for life scientists and covers general news and reviews exciting research and technological developments.

Answers to review questions

1. a	**2.** c	**3.** d	**4.** d	**5.** c	**6.** c	**7.** c	**8.** e
9. a	**10.** e	**11.** e	**12.** b	**13.** a	**14.** c	**15.** e	
16. c	**17.** a	**18.** b	**19.** e	**20.** e			

Chapter 3

THE STRUCTURE AND FUNCTIONS OF CELLS

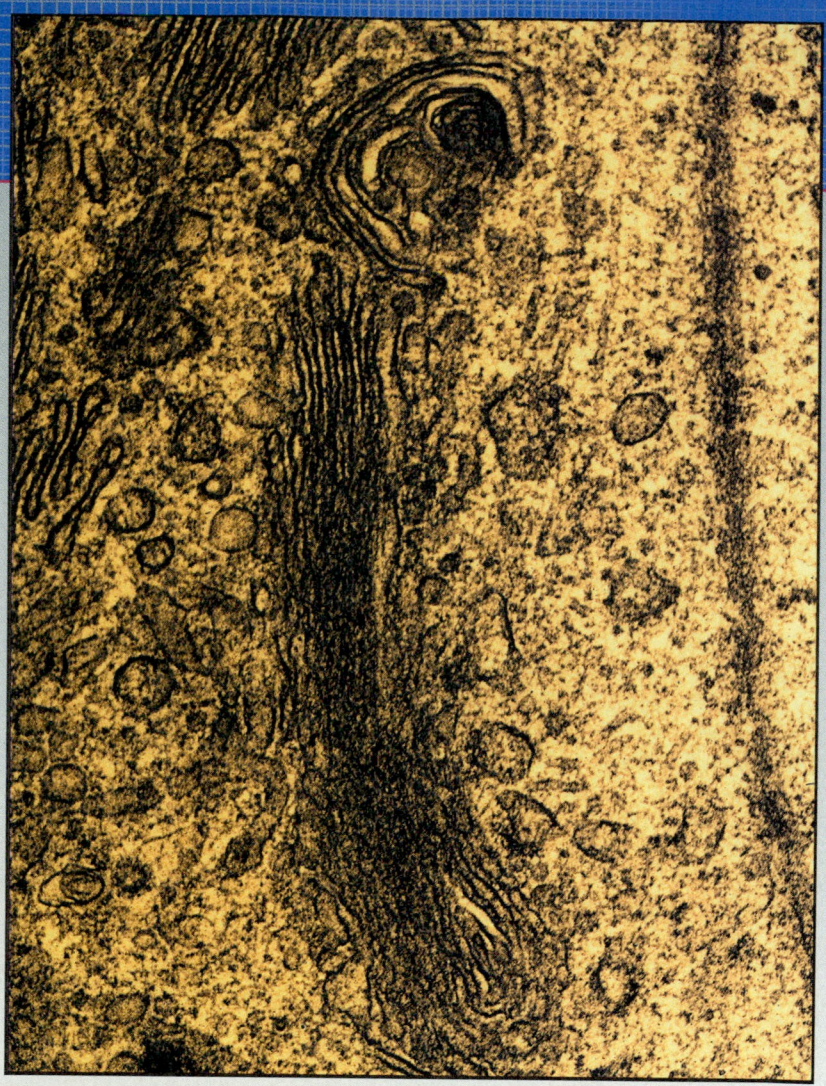

KEY CONCEPTS

- *The instructions that cells need to perform their required functions in the body and to reproduce themselves are contained in the DNA, the genetic material of the eukaryotic cell. Eukaryotic cells contain organelles that help them carry out their normal functions. For example, some organelles support a chain of chemical reactions that allow the cell to obtain energy from food and use it to support processes such as growth and reproduction.*

- *Cells have a plasma membrane consisting of a lipid bilayer and embedded proteins, which separates the cell interior from the extracellular fluid. Specialized proteins embedded in the membrane allow for communication between the extracellular environment and the cell interior. Intracellular membranes divide the interior of the eukaryotic cell into specific compartments to allow for specialized (and separated) activities. For example, in the endoplasmic reticulum (ER), lipid synthesis and protein synthesis occur, whereas in the Golgi complex, polysaccharide chains are modified and added to proteins in the rough ER.*

- *Communication between cells allows the specialized cells of higher organisms to coordinate their unique functions to benefit the organism. Consequently, not only does the plasma membrane separate the interior of the cell from the extracellular space but it also allows recognition and response to signals from the extracellular environment through a variety of proteins on the cell surface. Two important types of cell surface proteins are signal transduction receptors and adhesion molecules. Signal transduction receptors on the cell surface recognize and bind to signaling molecules, called ligands, present in the extracellular environment. Binding of cell adhesion molecules to extracellular matrix proteins can affect many cell processes, including cell growth and division, cell metabolism, and cell movement.*

- *The cell cycle is the time from the start of one cell division to the start of the next. Two main phases exist: (1) interphase and (2) cell division. Approximately 90% of the cycle is spent in interphase. Cell division consists of mitosis and cytokinesis. Entry into different phases of the cell cycle, such as mitosis, is under careful control by the cell. Cancer cells do not respond to normal controls of cell growth and division.*

CASE HISTORY

A two and a half-year-old boy was admitted to the hospital with fever, cough, an enlarged lymph node in the right side of his neck, and a 3-cm abscess on the left side of his face. The abscess felt cool to the touch and was not tender. The patient had been treated for two weeks prior to hospital admission with regular antibiotics and localized antibiotic treatment but had not improved. Significant medical history includes a late separation of the umbilical cord during an otherwise normal vaginal delivery at term and persistent infection of the umbilical cord stump. Subsequently, the patient has suffered from recurrent ear infections. These infections have not responded well to antibiotic therapy, and the child has been hospitalized on three previous occasions for *Staphylococcus pneumonia* infections. He had one brother who died at 26 months of age. During the current admission to the hospital, the child has maintained a persistent fever above 37.8°C (100°F), but his chest X-ray is normal. The white blood cell (WBC; also known as *leukocyte*) count was 55,000 cells/mm^3, with the majority (about 70%) being neutrophils (a type of white blood cell involved in inflammatory reactions). Normal WBC count is around 10,000 cells/mm^3. The neutrophils appeared normal under cell analysis.

Questions

1. What general medical problems are suggested by the patient's symptoms?

2. What is the significance of a high WBC count and the inability to fight infections?

3. What specific diagnosis is supported by these symptoms?

4. What types of treatment options are available for this patient?

INTRODUCTION

The **cell theory** forms the basis of all modern biology and physiology. The cell theory states that (1) all organisms are made up of cells and their products, (2) new cells arise only from preexisting cells, (3) all cells have the same fundamental chemical makeup and metabolic processes, and (4) the activities and processes of the organism as a whole result from the interdependent and cooperative workings of groups of cells.

The two broad classes of cells are (1) eukaryotic cells and (2) prokaryotic cells. The primary distinction between these classes is the presence or absence of a **nucleus,** a membrane-bound compartment surrounding the DNA. In **prokaryotic cells,** the DNA is not contained in a compartment that is separate from the rest of the cell, whereas **eukaryotic cells** have a well-defined nucleus contained within its own nuclear membrane.

Organisms, in turn, are broadly classified according to whether they have prokaryotic or eukaryotic cells. Bacteria are prokaryotic organisms, whereas mammals are eukaryotic organisms. Eukaryotic cells are generally larger (10–100 microns [μm] in diameter) than are prokaryotic cells (1–10 μm). In addition to a membrane-enclosed nucleus, all eukaryotic cells contain many other subcellular structures, most of which are also enclosed in a membrane. These structures are grouped under the general term **organelles.** They are distributed throughout the **cytosol,** the fluid surrounding the nucleus. Together, the organelles and the cytosol make up the **cytoplasm.** The cytoplasm and the nucleus make up the two major compartments of the cell and are enclosed by a cell membrane, called the **plasma membrane** (Fig. 3–1).

OVERVIEW: CELL PROCESSES AND PRODUCTS

> *What is the functional anatomy of the cell, and what advantages does this anatomy offer?*

Solitary cells are the simplest forms of life capable of an independent existence. The three features of cells that are recognized to be essential for independent life are (1) a set of genes that contain the information that the cell needs both to direct its own activities (primarily the synthesis of macromolecules) and to pass on to new cells; (2) the machinery for obtaining energy from foodstuffs to sustain growth, reproduction, and movement; and (3) a cell membrane that provides a physical boundary between the cell and its environment. Higher organisms are made of groups of cells that perform specialized functions. An important requirement of specialized cells in higher organisms is the ability to communicate with each other. This communication allows for the coordinated activity of complex cell processes that allows higher organisms to compete and survive in nature.

Cell processes can be thought of in two broad categories: (1) the **synthesis,** or creation from simpler molecules, or precursors, of macromolecules both for use by the cell itself and for export to other cells and (2) **degradation,** or breakdown, of other macromolecules for the purpose of obtaining energy. These processes together make up the **metabolism** of the cell.

Just as each organ of the body has a unique structure that permits it to perform a special function in a system of the body, each organelle of the cell has a unique structure and function that enables it to participate in the life

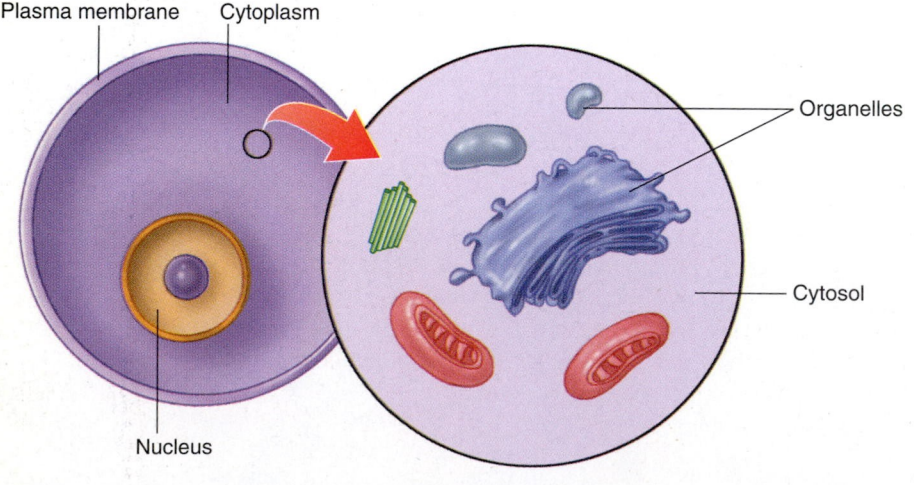

Plasma membrane Cytoplasm

Organelles

Cytosol

Nucleus

Figure 3–1

The cell has two major compartments: the nucleus and the cytoplasm. The cytoplasm contains the major cell organelles and a fluid called *cytosol.*

processes of the cell. This specialization allows cells to carry out unique functions in the body. One of the most important aspects of each cell's role in the body is the regulation of the types of proteins synthesized.

Cells have the potential to synthesize any protein encoded by DNA. However, different types of cells synthesize only certain proteins. For example, the chief function of some cells in the pancreas is the production of insulin for use by other cells. Insulin is a type of protein called a **hormone.** In order to produce insulin, the cell needs instructions. To ensure that future generations of pancreatic cells know how to produce insulin, the cell must preserve those instructions by copying them.

Even if they do not secrete proteins for use by other cells, all cells need certain proteins for their own internal functions. Many of these proteins are **enzymes.** As described in Chapter 2, they assist many important chemical reactions in the cell, controlling both synthetic and degradative processes.

The "instructions" that tell each cell how to make the specific proteins that it needs are contained in the DNA and RNA in the nucleus. The "machinery" for assembling the proteins is found outside the nucleus, however. When the cell needs to make a protein, the instructions are brought via RNA from the nucleus to the organelles that make the proteins. When the cell needs to divide, the instructions are duplicated in the nucleus so that each new nucleus will have a copy.

Following is a general summary of the major cell processes and products and the organelles that carry them out. A typical, generalized animal cell with all its organelles is depicted in Figure 3–2a.

Cellular Functions Are Controlled by Regulation of Gene Expression in the Nucleus

Genetic material, as mentioned previously, is necessary for the independent existence of most cells. Contained in the macromolecules of DNA, genetic material is essentially a set of instructions that the cell uses (1) to make the specific proteins and other macromolecules that it needs to carry out its own designated function in the body and (2) to reproduce itself. A set of related organelles in the cell work together to "express" the information contained in the DNA by synthesizing macromolecules according to its instructions.

The nucleus is the site where **DNA** is manufactured, stored, and duplicated. This DNA is used not only in **cell division,** the period during which the nucleus and then the cytoplasm divide to form two new cells, each containing one of the copies of the original DNA produced during the preceding interphase, but also throughout the life of the cell. Recall from Chapter 2 that the DNA molecule is a double strand of base-paired nucleotides. A **gene** is a portion of a DNA molecule containing a specific sequence of nucleotides that serve as a code dictating the sequence of amino acids to be followed when a particular polypeptide is assembled in

the cytoplasm. Gene expression involves transcription of a gene in the form of DNA into messenger RNA (mRNA), followed by translation of the mRNA into a polypeptide. Translation requires RNA-directed synthesis of a polypeptide on ribosomes. The function of a particular cell is largely a function of which genes are expressed by the cell.

Directing Synthesis

The nucleus is also the site of **RNA** manufacture. RNA carries genetic information (in the form of instructions about how to make proteins) from the DNA in the nucleus to the synthesizing organelles in the cytoplasm. RNA travels through **nuclear pores,** or openings, in the nuclear membrane (also called the *nuclear envelope*) to get from the nucleus to the cytoplasm.

Some RNA forms **ribosomes,** organelles that make proteins according to the instructions carried by other RNA molecules. The ribosomes can attach to the **endoplasmic reticulum** when they begin synthesis of proteins. Once made, many of the proteins are transported to the **Golgi complex.** There they are joined with other molecules and transported through the cytosol. Some of the proteins travel to the cell membrane for transport to the outside of the cell. Some proteins stay in the cell to participate in the internal cell processes. For example, some may be enzymes used to metabolize nutrients for the cell, while others become working parts of membranes in the cell.

Preserving Information

Many cells in the body are continually dividing. Unlike a human being, which can produce offspring and still continue to exist, a "parent" cell is completely replaced by its daughter cells. Therefore, any information to be passed on to the new cells must be prepared ahead of time, before cell division takes place.

The information about synthesis of macromolecules that must be passed on is contained within DNA. This molecule replicates, or copies, itself. Two copies of the same DNA then exist, each to become the nuclear DNA of one of the new cells. DNA replication is discussed in detail in Chapter 5.

Energy Obtained from Food Is Used to Support Life Processes

A second requirement for the independent existence of a cell is the ability to take energy from nutrients and convert it into a useful form to support life processes, including growth and reproduction. The cell has a set of organelles including **vacuoles, lysosomes,** and **mitochondria** that perform this function by supporting a complex chain of chemical reactions. A system of vacuoles conveys food matter in and waste matter out of the cell. Lysosomes are small bags of enzymes that surround food molecules that come into the cells in vacuoles. The enzymes in the lysosomes break down the mole-

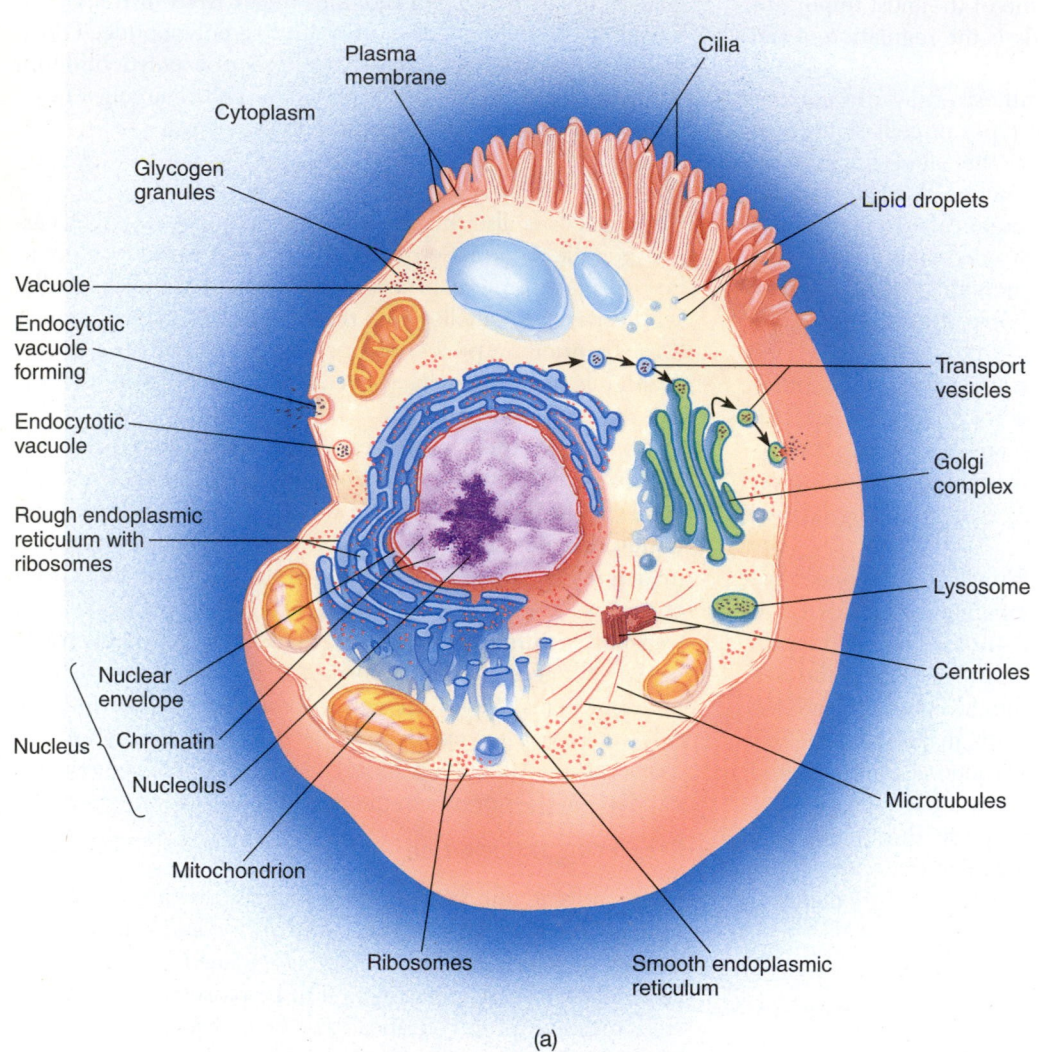

Plasma membrane

Cilia

Cytoplasm

Glycogen granules

Lipid droplets

Vacuole

Endocytotic vacuole forming

Endocytotic vacuole

Transport vesicles

Golgi complex

Rough endoplasmic reticulum with ribosomes

Lysosome

Nuclear envelope

Centrioles

Nucleus — Chromatin

Nucleolus

Microtubules

Mitochondrion

Ribosomes

Smooth endoplasmic reticulum

(a)

Figure 3–2

(a) A generalized view of a mammalian cell, showing organelles common to all cells (e.g., the Golgi complex) as well as specialized structures (e.g., cilia) found only in some cells. *(b)* An electron micrograph of a human cell. (× 15,000) *(© David M. Phillips/Visuals Unlimited)*

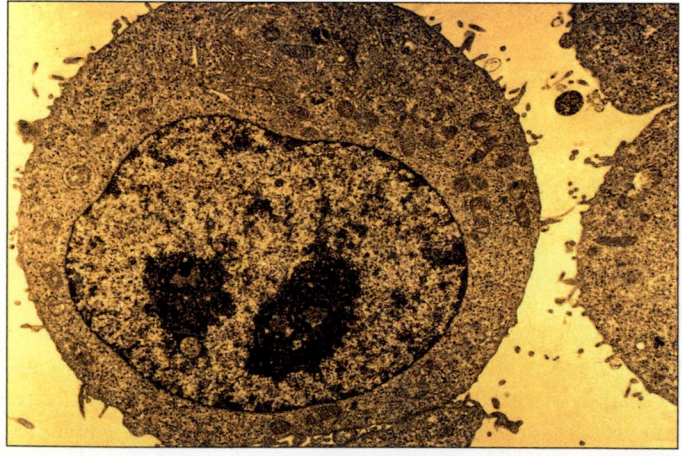

(b)

cules so that they can be used by mitochondria to produce ATP, the energy source for the cell. Mitochondria, organelles enclosed in a double membrane, are the sites where ATP is produced through the process of aerobic respiration.

Organelle Membranes Protect the Cell Interior and Divide It into Different Compartments

Recall from Chapter 1 that each of these two fluid compartments has a distinctive ionic composition that must be maintained. Without the plasma membrane that separates them, the two solutions would merge and their contents would mix, becoming indistinguishable.

Just as it is necessary for the cytoplasm to be segregated from the extracellular fluid, the contents of individual organelles within the cell must be protected from each other and from the cytosol by organelle membranes. Some cell products would destroy other cell products if contact were to occur. For example, secretory proteins would be destroyed by digestive enzymes if contact were to occur. Thus, the membrane system of the cell serves to segregate cytoplasm from the extracellular fluid and from various cell products

Protecting the Interior: The Plasma Membrane

The outer **cell membrane,** called the **plasma membrane,** is composed of two layers of phospholipid molecules. Recall from Chapter 2 that phospholipids consist of a long, hydrophobic (or water-insoluble) tail and a hydrophilic (or water-soluble) portion. The phospholipid molecules making up the cell membrane act according to the principles of hydrophobic interactions; their hydrophobic portions cluster together, leaving their hydrophilic portions exposed on the surface to the aqueous environment. The clustering of lipid molecules creates a sturdy barrier to the passage of water-soluble particles, keeping the watery cytosol from flowing out of the cell and the watery extracellular fluid from flowing in.

The plasma membrane not only separates the extracellular and intracellular environments but also regulates molecular traffic into and out of the cell. A variety of mechanisms exist for this purpose. **Endocytosis** involves portions of the membrane breaking off and enclosing material to be taken into the cell, with a reverse process (**exocytosis**) used to let material out of the cell. These processes are discussed later in this chapter.

Other mechanisms involve the passage of individual molecules through the membrane in strictly controlled ways via protein molecules called **channels** and **carriers.** Still other mechanisms involve simple diffusion, such as that occurring in any aqueous solution. These methods of transmembrane movement are covered in detail in Chapter 4. **Cell adhesion molecules** (CAMs), such as **integrins** and **cadherins,** interact with **extracellular matrix (ECM)** proteins and mediate signals that affect cellular functions depending on the composition of the ECM.

Separating Cell Processes: Membranes of Organelles

Because each organelle is bound by a distinct membrane, it follows that the outer cell membrane of a eukaryotic cell accounts for only a minor proportion (2–5%) of the total membrane present in the cell. The most extensive membrane-bound organelle present in most cells is the **endoplasmic reticulum** (ER; see Fig. 3–2). The membrane of the endoplasmic reticulum of a liver cell makes up more than 50% of the total membrane of the cell. Seven different subcompartments, or organelles, are formed by the intracellular membranes of most eukaryotic cells. A double membrane surrounds two organelles: the nucleus and mitochondria. These membranes are composed, as is the plasma membrane, primarily of two layers of phospholipid molecules.

Continuity exists to some extent between the subcompartments in that some of the materials present in one compartment may be passed on to another, but direct physical links between the compartments are very likely limited. Each of the compartments within the cell has a unique role, carried out at the molecular level by the enzymes that are either packed inside the organelle or components of the membrane of the organelle. Because each organelle has a different function, the set of enzymes within a specific organelle is not found in any of the other organelles. Catalase, for example, is an enzyme found only in organelles called *peroxisomes*. The enzyme that breaks down glycogen into glucose molecules is present only in the cytosol. The enzymes that catalyze the sequence of reactions in the citric acid cycle are present only within mitochondria.

Enzymes found in lysosomes, called **lysosomal enzymes,** are essential to normal cell function. The absence of just one of these enzymes, as occurs in some genetic diseases, has severe consequences for the individual. However, this group of **degradative enzymes** can break down all of the macromolecules responsible for the structure and function of the cell. Clearly, these enzymes cannot be allowed direct contact with other organelles or with the components of the cytosol because they would cause irreversible, fatal damage. The cell solves the problem very neatly by sequestering the enzymes within lysosomes so that they are always available within the cell. They are allowed access to structures and molecules targeted for degradation (breakdown) only when the lysosomes fuse with specialized vesicles, called endocytotic vesicles, that degrade proteins. This process ensures that the degradative process is confined locally within a specific compartment, the secondary lysosome (see Fig. 3–10).

Similarly, the presence of catalase within peroxisomes ensures that the toxic hydrogen peroxide produced within these organelles is destroyed locally before it has a chance to leak into the cytosol, where it would cause severe damage.

The presence of intracellular compartments also permits separation of the processes of synthesis and degradation of molecules. The advantage derived from physically separating these processes is that they can occur simultaneously within the cell and can be controlled independently. The metabolism of fatty acids is a good example. The cell continually synthesizes these molecules because they are components of the phospholipid molecules needed for the structure of cell membranes. At the same time, the cell is also degrading other fatty acids because they are an excellent source of chemical energy; in fact, they yield more energy than do equal weights of carbohydrates, such as glucose. The goal of synthesis is the opposite of the goal of degradation, and the two processes would not be compatible if both occurred openly in the cytosol. Eukaryotic cells solve the problem by localizing the reactions that degrade fatty acids in the mitochondria and those that synthesize fatty acids in the cytosol.

Finally, compartmentalization of cellular functions allows for intracellular processing and "sorting" of some of the synthesized molecules. We will discuss shortly some of the sorting and processing that occurs in the ER and the Golgi complex. For example, proteins destined to become components of membranes are trapped within the ER membrane during synthesis, whereas proteins destined for export from the cell are discharged into the lumen of the ER, where they

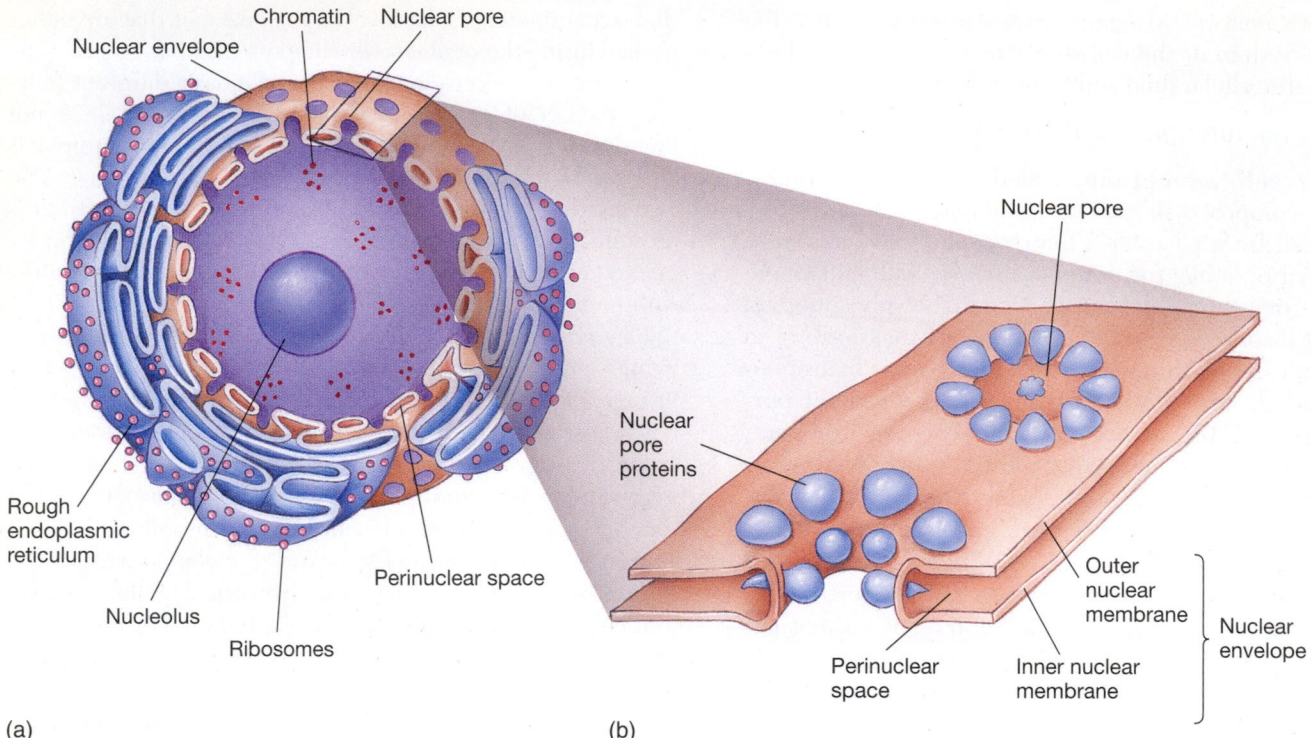

Chromatin Nuclear pore

Nuclear envelope

Rough endoplasmic reticulum

Nucleolus

Ribosomes

Perinuclear space

(a)

Nuclear pore

Nuclear pore proteins

Outer nuclear membrane

Inner nuclear membrane

Nuclear envelope

Perinuclear space

(b)

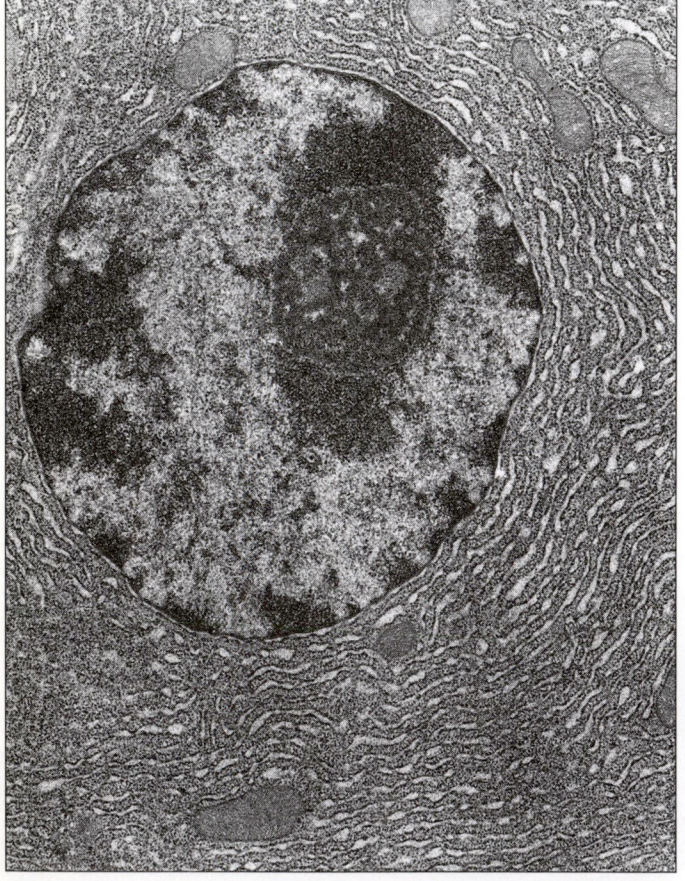

(c)

Figure 3–3

(a) The cell nucleus is enclosed in a double membrane called the *nuclear envelope*. Pores in the envelope permit the passage of molecules into and out of the nucleus. The outer layer of the nuclear envelope is continuous with the endoplasmic reticulum (ER), so that the lumen of the ER is continuous with the perinuclear space. In the nondividing nucleus, DNA is visible as chromatin. The nucleolus plays a role in the synthesis of ribosomes from RNA. *(b)* A nuclear pore is formed from the fusion of the two layers of the nuclear envelope. Proteins are thought to be located in the pores. *(c)* An electron micrograph of the nucleus and nucleolus of a glandular cell.   (© *Don W. Fawcett/Photo Researchers*)

are attached to carbohydrates to become glycoproteins. From there, they are transported to the Golgi complex, where the carbohydrate portion of the molecules is chemically modified. The Golgi complex also carries out another sorting step based on the specific structures present in the glycoprotein molecules: It distinguishes between the molecules that must be packaged inside specialized vesicles—called secretory vesicles, that move material from the Golgi complex to the cell membrane for export out of the cell. Other molecules are destined to become lysosomal enzymes and must be directed into primary lysosomes for use inside the cell.

CELL COMPARTMENTS AND THEIR FUNCTIONS

The Nucleus

 What are the important molecules in the nucleus?

The nucleus is the principal organelle that distinguishes eukaryotic from prokaryotic cells (Fig. 3–3). The nucleus is the site where the nucleic acids, DNA and RNA, are made and is also the site from which cellular functions ultimately are controlled. The information stored in DNA and RNA molecules directs the synthesis of proteins that determine the shape and function of a cell. By separating the nuclear contents from the cytoplasm, the nuclear membrane allows for regulation of gene expression that is not available to prokaryotic cells. (The red blood cell is the only common mammalian cell that lacks a nucleus; it survives for only a few months.) The

nucleus also contains many different types of proteins that are involved in controlling the organization and structure of nucleic acids. Other nuclear proteins are involved in regulating the expression of genes. The approximate composition of the nucleus in a liver cell, expressed as a percentage of the dry weight of the nucleus, is 80% protein, 15% DNA, and 5% RNA.

DNA

Recall from Chapter 2 that DNA (deoxyribonucleic acid) is an enormously long, unbranched, double strand of nucleotides arranged in an alpha-helix. A typical animal cell contains more than 1 m of DNA, all of which must fit inside a nucleus with a diameter of only 0.000006 m (6 μm). Each section of the chromosomal DNA that carries the information necessary for synthesis of a single polypeptide is known as a **gene.** The sum of all the chromosomal genes is termed the **genome** of the cell. Recently, nearly the entire genome of human cells was decoded in a massive collaborative project known as the Human Genome Project, funded by the National Institutes of Health and private industry. All cells in the body, except the reproductive cells, normally contain the same DNA, making it important to understand the complete DNA code. The DNA molecules typically remain unchanged throughout the life of the organism unless they are damaged in some way. Cells do not all look or function the same, however, because they differ in the specific proteins that are produced.

Specialized proteins called *histones,* which bind to the DNA and organize its structure in the nucleus, achieve the tight packing of DNA (Fig. 3–4). Histones are by far the most common protein found in the nucleus. They are rich in amino acids with basic (positively charged) side chains,

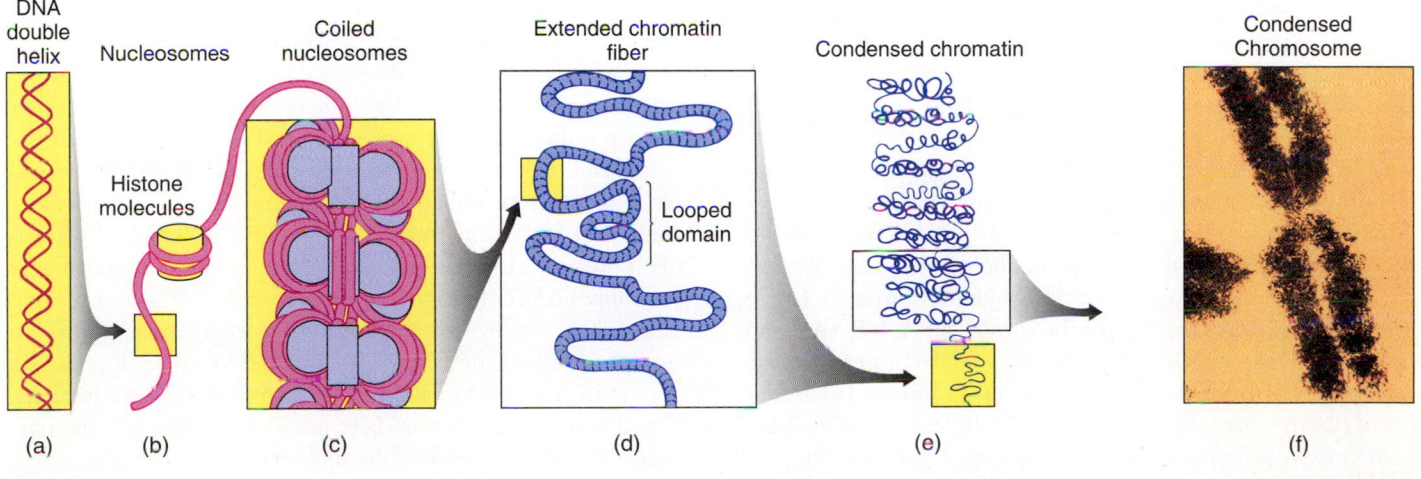

Figure 3–4

(a–f) The DNA is packed in a series of stages beginning with nucleosomes. In a nondividing cell, DNA appears as chromatin; in a dividing cell, it appears in the more tightly coiled form of chromosomes.

such as lysine and arginine (see Fig. 2–14). These positively charged side chain groups might be able to interact with and bind to the negatively charged phosphate groups in DNA (see Fig. 2–21), regardless of the specific nucleotide sequence.

CHROMATIN. Most of the DNA is associated with histones and other nuclear proteins, giving rise to a complex known as **chromatin.** The basic packing unit of chromatin is the **nucleosome** (Fig. 3–4*b*), in which parts of the DNA double helix are wrapped around histone molecules and linked by stretches of "free," or **linker,** DNA. The nucleosome is still a relatively extended form of chromatin; it is unlikely that much of the chromatin in the nucleus is present in this form.

A higher level of organization is produced by the packing together of nucleosomes in a regular array known as a **chromatin fiber.** Loops in these fibers give rise to **looped domains,** which reduce the initial length of the DNA double helix about 500-fold (Fig. 3–4*d*). A "scaffold" made of protein supports these loops.

CHROMOSOMES. Further levels of folding must occur, perhaps involving close packing of the loops, in order to condense the DNA into a single **chromosome** (Fig. 3–4*e* and *f*). Thus, each chromosome of a eukaryotic cell consists of a single, very large molecule of DNA arranged so that the double helix is folded in a compact and highly organized manner. Each species has a characteristic number of chromosomes; human cells, for example, have 23 pairs. When cells are not dividing, the chromosomes are in a relatively uncoiled state and are not readily identifiable as individual units. Only when cell division is about to occur do the chromosomes become apparent.

RNA

The RNA (ribonucleic acid) present in the nucleus is destined for export into the cytoplasm, where the machinery for protein synthesis is located. In cells such as rat liver cells, less than 10% of the total RNA in the cell is located in the nucleus at any one time.

An RNA molecule consists of a single strand of nucleotides. This molecule represents a copy of a nucleotide sequence from one of the strands of DNA. Three major functional types of RNA interact in the cytoplasm to synthesize proteins according to the information stored in DNA: **messenger RNA (mRNA); ribosomal RNA (rRNA);** and **transfer RNA (tRNA).** For now, the related roles of these three types of RNA can be summarized as follows (more detailed discussions follow in Chapter 5): Ribosomal RNA joins with certain proteins to form ribosomes. These are the protein-making structures that occur in the cytosol or are attached to the endoplasmic reticulum. Messenger RNA contains the information concerning which amino acids should be joined together, and in which specific sequence, on the ribosomes. Transfer RNA seeks out the required amino acids in the cytosol and brings them to the ribosome to be assembled into polypeptide chains.

One of the most obvious structures within the nucleus is the **nucleolus,** which is particularly prominent in cells that synthesize large amounts of protein. A membrane does not surround the nucleolus, which contains proteins and large loops of DNA from which rRNA is rapidly synthesized. Cells begin assembling rRNA and proteins into ribosomes in the nucleolus. However, ribosomes are not mature, or completely formed, until they reach the cytoplasm. This aspect of ribosome formation is discussed in Chapter 5.

The Nuclear Envelope

How do molecules move between the nucleus and cytoplasm? The interior of the nucleus is maintained as a separate biochemical compartment from the cytoplasm by the **nuclear envelope,** a set of two separate membranes that contain **nuclear pores** and provide a structural scaffold for the nucleus. Nuclear pores allow for selective movement of proteins and RNA between the nucleus and cytoplasm and are crucial for regulating gene expression. The two membranes (inner and outer) of the nuclear envelope are separated by a space about 10 to 50 nm wide, termed the **perinuclear space.** The outer membrane of the nuclear envelope is continuous with the ER and often has attached ribosomes that are engaged in protein synthesis. Thus, the perinuclear space is continuous with the lumen of the ER (Fig. 3–3*a*). The inner nuclear membrane contains nuclear-specific proteins that form the nuclear lamina, which provides an architectural framework for the chromosomes. The nuclear lamina is composed of numerous different proteins, several of which are called *nuclear lamins* and form a fibrous meshwork on the inner face of the nuclear envelope. The lamins, designated A, B1, B2, and C, are fibrous proteins that are structurally similar to the intermediate filaments of the cytoskeleton described later.

Much exchange occurs between the interior of the nucleus and the cytoplasm. Because DNA and RNA are synthesized only in the nucleus, the individual nucleotides and other parts required for their synthesis must be obtained from the cytoplasm. Conversely, RNA molecules produced in the nucleus must move out to the cytoplasm in order to carry out their roles in protein synthesis. The rapid exchange of materials between the nuclear and cytoplasmic compartments probably occurs through the nuclear pores. These are open channels in the nuclear envelope that form in areas where the inner and outer membranes are pinched together (Fig. 3–3*b*). A mammalian cell has about 3000 to 4000 pores per nucleus. The diameter of each pore is 9 nm. This size permits the passage of small, water-soluble molecules but excludes large, macromolecular structures. A number of proteins, including enzymes, are assembled in the cytosol, but their site of action is inside the nucleus. These proteins contain **nuclear localization signals (NLSs)** that cause the nuclear pores to open up just enough to allow the proteins to enter the nucleus.

Control of Nuclear Protein Import

By controlling which proteins are expressed from the genes, the nucleus can direct the ultimate function of the cell. The import of proteins into the nucleus is an important mechanism for regulating gene expression. Transcription factors, which are in proteins that regulate gene expression, cannot function in the cytoplasm; therefore, controlling their import into the nucleus is a powerful way to control gene expression. Proteins, such as histones, polymerases, and transcription factors, that control the structure and function of the genome must be imported into the nucleus from their site of synthesis in the cytoplasm. The first protein for which an NLS was described in detail was a virally encoded protein called *T antigen* from the simian virus 40 (SV40 T antigen). SV40 T antigen initiates viral DNA replication in infected eukaryotic cells. The NLS was defined as a seven–amino acid sequence that, when mutated, blocked the import of SV40 T antigen into the nucleus. Furthermore, the addition of this sequence to certain proteins normally found in the cytoplasm resulted in concentration of the proteins in the nucleus. Since then, many other nuclear localization signals have been identified, some of which are short, continuous stretches of amino acids rich in lysine and arginine. In other nuclear proteins, the NLS consists of nonadjacent amino acid residues called *bipartite motifs*, which are separated by stretches of amino acids not involved in nuclear import.

Movement through the nuclear pore complex can be separated into two phases. In the first phase, the nuclear target protein containing an NLS binds to a specific docking protein in the cytoplasm called *importin*, which is actually a complex of two proteins, importins α and β. This complex binds to and moves through the nuclear pore complex. This phase does not appear to require energy. The second phase involves dissociation of the nuclear target protein in the nucleus, dissociation of importins α and β, and translocation of importin β back into the cytoplasm. This last step requires the energy generated by hydrolysis of guanosine triphosphate (GTP). An important regulatory component of this type of nuclear trafficking is a protein called **Ran.** Ran is an enzyme that hydrolyzes a phosphate group from GTP (enzymes that do this are called *GTPases*). Ran is a member of a large family of GTPases that share homology with a GTPase called Ras. Like all GTPases, the Ras family GTPases catalyze the hydrolysis of GTP to guanosine diphosphate (GDP); however, they play many different roles in the regulation of cell function. Like other Ras family members, Ran's function in the cell is determined by whether it is bound to a molecule of GTP or GDP. A working model for Ran's role in the cells is illustrated in Figure 3–5. The nucleus contains primarily Ran-GTP and the cytoplasm contains primarily Ran-GDP. This gradient of GTP- and GDP-bound Ran helps drive trafficking of proteins from the cytoplasm into the nucleus. In the generalized model shown in Figure 3–5, a nuclear target protein in the cytoplasm containing an NLS forms a complex with importins α and β in the presence of primarily Ran-GDP. Once this complex moves through the nuclear pore into the nucleus, Ran-GTP binds to importin β, causing the dissociation of

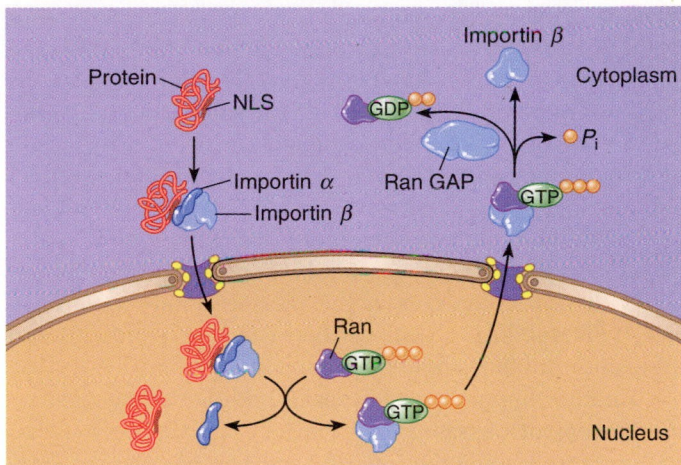

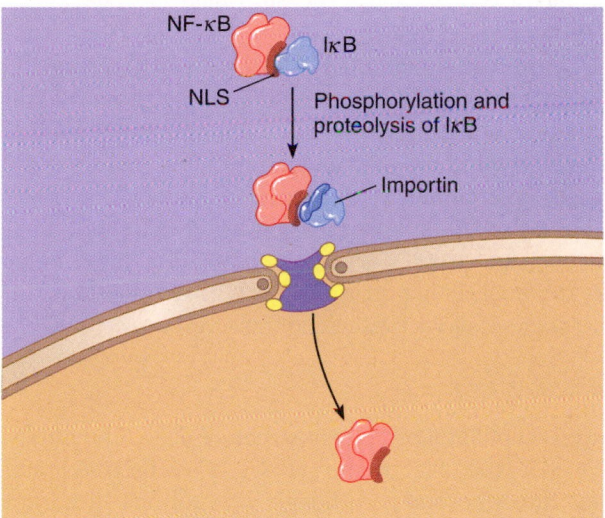

Figure 3–5

Import of proteins into the nucleus occurs through the nuclear pore complex and is mediated by several nuclear import proteins. *(a)* In one mechanism, target proteins that are destined for import into the nucleus contain a nuclear localization signal (NLS) that forms a complex with importins α and β in the cytoplasm. Once the target protein–importin complex moves through the nuclear pore, Ran-GTP binds to importin β, releasing the target protein and importin α in the nucleus. The Ran-GTP–importin β complex moves out into the cytoplasm, where hydrolysis of the Ran-GTP to Ran-GDP releases importin β into the cytoplasm and maintains a concentration gradient of high Ran-GTP in the nucleus. *(b)* In another mechanism, proteins, such as NF-κB, are maintained in the cytoplasm via binding to inhibitory proteins that mask their NLS. The NF-κB NLS is masked by binding to IκB, preventing import into the nucleus. When an activating signal is received by the cell, IκB is phosphorylated and degraded by proteolysis. This allows NF-κB to bind importin α/β and move into the nucleus through nuclear pore complexes.

the target protein and importin α. Ran-GTP–importin β moves back out into the cytoplasm through a nuclear pore, where the energy release by hydrolysis of GTP to GDP by Ran causes the release of importin β from Ran-GDP, making importin β available for another round of nuclear import.

Some nuclear proteins contain specific **nuclear export signals** that bind to proteins in the nucleus, called **exportins,** that allow them to shuttle back out into the cytoplasm. Significantly, while Ran-GTP dissociates importin–target protein complexes in the nucleus as just described, Ran-GTP stabilizes complexes between target proteins and exportins until they are in the cytoplasm.

Another important mechanism used by cells to regulate the movement of transcription factors into the nucleus has been intensely studied by researchers. This involves the transcription factor NF-κB and a protein called *IκB,* which inhibits the import of NF-κB into the nucleus (Fig. 3–5). When the **inhibitory protein** IκB associates with NF-κB in the cytoplasm, it masks the nuclear import signal on NF-κB so that NF-κB cannot move into the nucleus through nuclear pores. When the cell receives an appropriate signal from the extracellular environment, IκB becomes modified by the addition of a molecule of phosphate, a process called phosphorylation, and the protein is degraded by proteolysis. This unmasks the nuclear import signal on NF-κB, allowing it to enter the nucleus and activate gene transcription.

The Nucleolus

The nucleus contains a prominent structure called the *nucleolus.* This is the site where rRNA is transcribed and processed into ribosomes. Although a membrane does not surround the nucleolus, it is clearly a discrete structure in the nucleus that is located near the chromosomes that contain genes that encode rRNA. In general, the size of the nucleolus is proportional to the metabolic activity of the cell, with larger nucleoli found in cells that are more actively engaged in protein synthesis. The nucleolus also plays a role in regulating the cell cycle (described later). Many proteins that are important for driving the cell cycle appear to be sequestered in the nucleolus until they are needed.

The Cytosol and Organelles Are the Sites of Synthesis, Degradation, and Energy Release

> *Where do specific chemical processes occur in the cytoplasm?*

The normal structure, growth, and function of a cell require a constant supply of macromolecules, which are made from simpler molecules, or **precursors.** Synthesis of precursors, such as amino acids or nucleotides, is achieved by a set of chemical reactions that are part of the cell metabolism. Also important to intermediary metabolism are the reactions that break down small molecules and trap their chemical energy

as ATP, as discussed in the previous section. All the reactions of metabolism take place outside the nucleus (see Fig. 3–1).

The Cytosol

The cytosol of a mammalian cell, such as a liver cell, is a large compartment. It represents about 50% of the total volume of the cell and contains the small molecules that are precursors of macromolecules. It also contains thousands of enzymes, which account for its high protein content (about 20% by weight). Many of the enzymes are catalysts for the reactions of cell metabolism. Thus, the cytosol resembles a highly viscous fluid rather than a watery solution.

The cytosol contains some structures that are not organelles but are worthy of mention, including glycogen granules and lipid droplets. **Glycogen,** a polysaccharide, is the storage form of carbohydrate. The lipid droplets contain **triglycerides,** which are the storage form of fatty acids (see Chapter 2).

Ribosomes and the Endoplasmic Reticulum: Synthesis of Proteins and Lipids

The ER is like a flattened membrane sheet that is folded repeatedly to form a large, interconnected system (see Fig. 3–2), with two structurally and functionally different regions: (1) the **rough,** or **granular, ER** and (2) the **smooth,** or **agranular, ER** (Table 3–1). The rough ER is so called because ribosomes are bound to its outer surface, whereas the smooth ER has no ribosomes attached.

The ER makes new proteins and lipids, including molecules destined to be released outside the cell by a process known as **secretion.** Rough ER is particularly extensive in cells that synthesize and secrete proteins, such as certain hormones. Smooth ER is well developed in cells specializing in lipid synthesis or the synthesis and secretion of steroid hormones. Muscle cells have a highly specialized smooth ER known as the **sarcoplasmic reticulum.** An important function of this structure is to store calcium ions.

PROTEIN SYNTHESIS ON THE ROUGH ENDOPLASMIC RETICULUM. Individual amino acids are assembled into polypeptides on ribosomes. The ribosomes occur freely in the cytosol, but many become attached to the ER when

---

TABLE 3–1

Structural and Functional Differences Between Rough and Smooth Endoplasmic Reticulum (ER)

Characteristic	Rough ER	Smooth ER
Ribosomes present	Yes	No
Protein synthesis	Yes	No
Lipid synthesis	Yes	Yes
Steroid synthesis	No	Yes

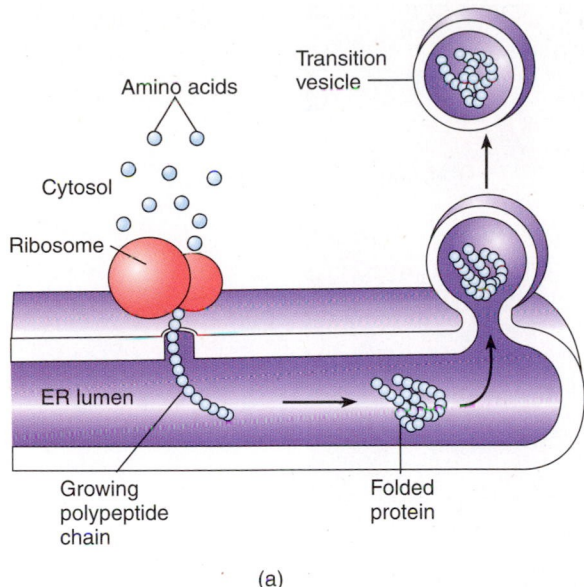

(a)

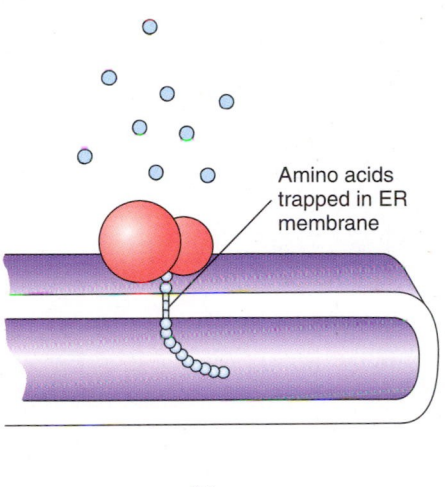

(b)

In the rough endoplasmic reticulum (ER), a ribosome synthesizes a polypeptide from amino acids found in the cytosol. *(a)* A protein destined to be a secretory product or an enzyme enters the ER lumen as it is assembled, folding into its various levels of structure. It is enclosed in a transition vesicle for transport to the Golgi complex. *(b)* A protein destined to be membrane protein is partially trapped in the ER membrane as it is synthesized. *(c)* The rough ER.   *(© Don W. Fawcett/Visuals Unlimited)*

they become involved in protein synthesis. mRNA dictates the sequence of the amino acids in each polypeptide. If an ER signal sequence appears in the emerging polypeptide, it is recognized and bound by a **signal recognition particle.** This particle directs the whole complex, including the ribosome, to a specific receptor on the ER. Polypeptide chains formed in the ribosomes are directed into the lumen of the ER and separated from the cytosol by the ER membrane (Fig. 3–6). Most of the proteins synthesized in the rough ER become modified by the addition of monosaccharide units to form glycoproteins. In contrast, the proteins synthesized on free ribosomes in the cytosol contain no ER signal sequence, are excluded from the ER lumen, and remain in the cytosol. They cannot be modified by addition of carbohydrate.

The proteins that are destined to become structural components of membranes (see Chapter 4) are trapped within the rough ER membrane during synthesis. The mechanism for trapping the polypeptide chain may be a specific sequence of amino acids that interacts strongly with the interior of the ER membrane and prevents further movement of the polypeptide into the ER lumen (see Fig. 3–6). The part of the polypeptide that is synthesized after the trapping sequence remains on the cytosolic side of the ER membrane. Movement of ER membrane to other membrane-bound organelles and to the plasma membrane is discussed shortly.

LIPID SYNTHESIS IN THE SMOOTH ENDOPLASMIC RETICULUM. The endoplasmic reticulum in many cells is the major site of the synthesis of lipids including triacylglycerols, phospholipids, and steroids. Lipids from each of these three general classes form complexes with proteins in the ER lumen. The resulting structures are the compounds known as **lipoproteins.** The lipoproteins that are secreted from the cell and that enter the blood play a crucial role in the transport and distribution of lipids, such as cholesterol, that are absorbed by the intestine.

Lipids, such as phospholipids and cholesterol, are also important structural components of membranes. The successive steps leading to the formation of a phospholipid occur on the ER membrane because the enzyme proteins that catalyze the synthesis are a part of the membrane. Thus, synthesis of phosphatidylcholine (see Fig. 2–11c) occurs on the cytosolic surface of the ER membrane; the molecules of phosphatidylcholine that are produced remain within the ER membrane

(c)

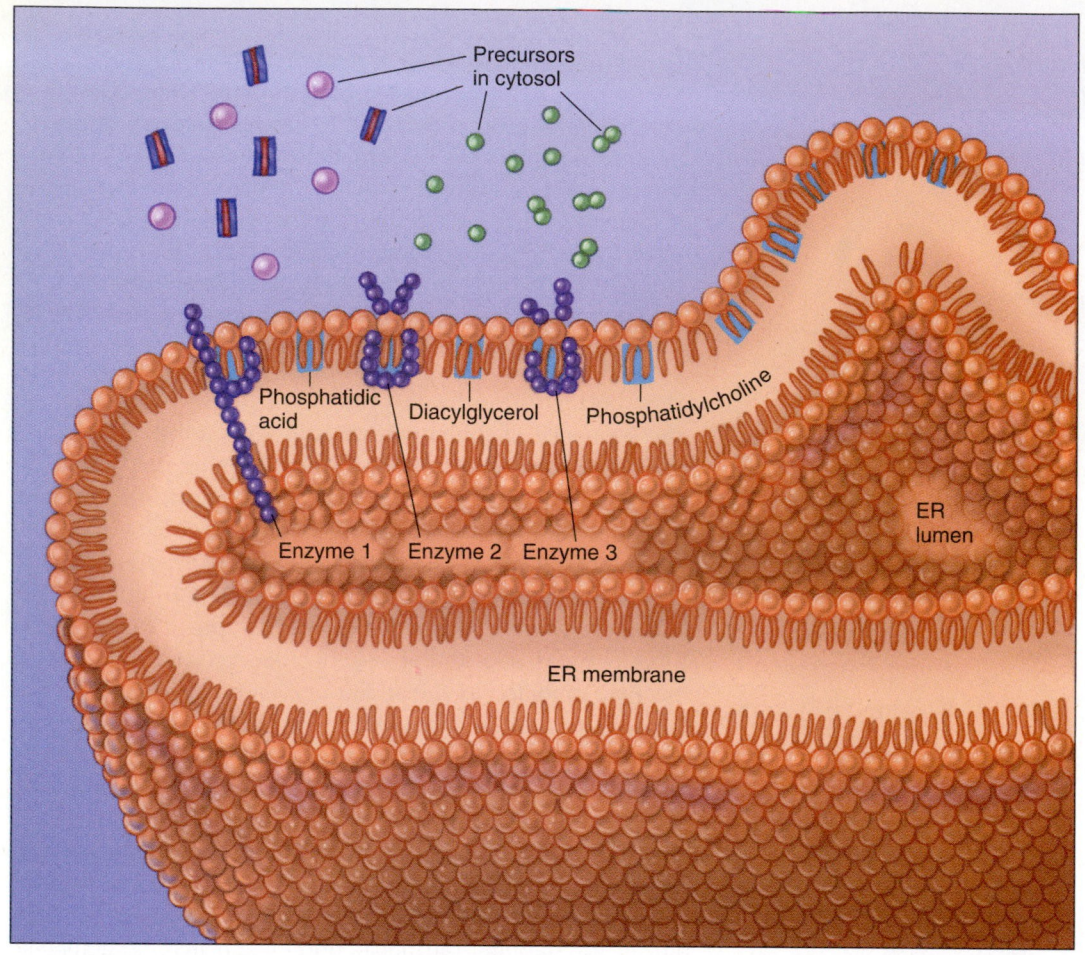

Precursors in cytosol

Phosphatidic acid   Diacylglycerol   Phosphatidylcholine

Enzyme 1   Enzyme 2   Enzyme 3

ER lumen

ER membrane

Figure 3–7

In the smooth ER, a phospholipid is assembled from precursors in the cytosol through a series of reactions catalyzed by enzymes located in the ER membrane. The developing phospholipid molecule moves through the layer of existing phospholipid molecules that make up one half of the ER membrane.

(Fig. 3–7). Clearly, synthesis of the proteins and lipids that are required for production of new cellular membrane structures is another important function of the ER.

A macromolecule that is to be secreted is first formed in the ER and then moves to the Golgi complex. This occurs via a specialized vesicle that moves between the ER and the Golgi complex, called a transition vesicle, which is formed from a portion of the ER membrane. In the Golgi complex, the macromolecule is processed and packaged for export from the cell or for storage or use within the cell.

The Golgi Complex: Processing and Packaging

The function of the Golgi complex is to modify the macromolecules synthesized in the ER. This is especially important in secretory cells. The Golgi complex processes and packages the macromolecules inside small membrane bags, or **vesicles,** ready for transport to the cell surface or the cytosol.

The Golgi complex is a collection of smooth membranes that look like flattened sacs. Each sac is called a **cisterna.** Commonly, about six cisternae are stacked together to form a structure, known as a **Golgi stack,** that is involved in processing proteins into their mature forms (Fig. 3–8). Small vesicles surround and radiate away from the Golgi stack.

The *cis,* or **forming, face** of the stack is oriented toward, and closely associated with, the rough ER. The two structures are thought to be connected by the small membrane-bound **transition vesicles,** which shuttle newly synthesized macromolecules from the rough ER to the Golgi complex (see Fig. 3–8).

The *trans,* or **mature, face** of the Golgi complex, on the other hand, is oriented toward the plasma membrane and away from the nucleus, especially in secretory cells. Membrane-bound secretory vesicles appear to arise from the *trans* face (see Fig. 3–8) to transport newly processed and packaged macromolecules from the Golgi complex to the plasma membrane for export.

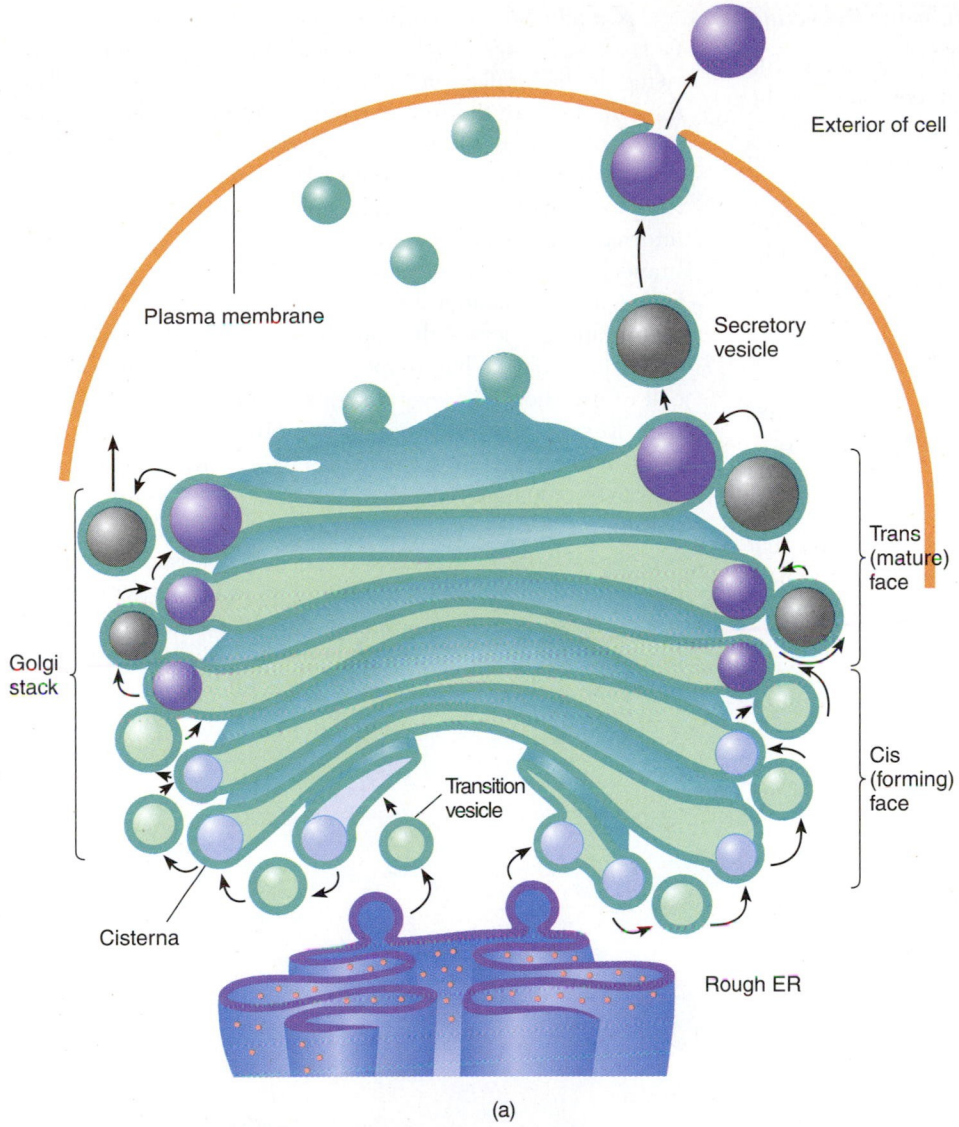

Plasma membrane

Exterior of cell

Secretory vesicle

Trans (mature) face

Golgi stack

Cis (forming) face

Cisterna

Transition vesicle

Rough ER

(a)

Figure 3–8

(a) One unit of the Golgi complex, called a *Golgi stack,* is composed typically of about six folded sacs, called *cisternae.* Macromolecules formed in the ER enter the *cis* face of the Golgi complex and pass through to the *trans* face via a series of vesicles. In the process, the lipids or proteins are modified. They leave the Golgi complex in secretory vesicles. *(b)* A false-color transmission electron micrograph of a section through one stack of the Golgi apparatus. The Golgi stack is the paired membranes (*green outlined in red*) in the middle of the image. The large red sphere near the top is a mitochondrion. (× 31,800)   (© *Dr. Gopal Murti/SPL/ Photo Researchers, Inc.*)

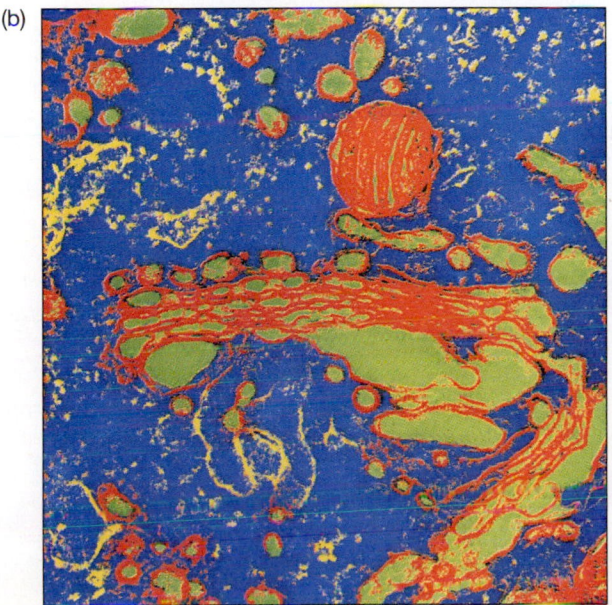

(b)

Substantial evidence shows that the Golgi complex modifies the macromolecules destined for secretion. One of its important activities involves the modification of the polysaccharide chains that were added covalently to the proteins in the rough ER. The processing of these glycoproteins within the Golgi complex involves a considerable amount of "trimming" as well as the addition of new monosaccharide units.

The form of the Golgi complex can vary considerably. In some cell types, it may be compact and limited, whereas in others, it may be more spread out, almost like a net. The position of the Golgi complex within the cytosol also may vary. In nerve cells, it often surrounds the nucleus, whereas in secretory cells, it usually is found between the nucleus and the side of the cell that is engaged in the secretory process. The number of Golgi stacks in a cell varies considerably with the type of cell. The Golgi complex in a liver cell, for example, accounts for about 7% of the total cell membrane.

Secretory Vesicles: Exocytosis and Membrane Recycling

A secretory vesicle containing a packaged macromolecule moves from the inner face of the Golgi complex to the plasma membrane. The vesicle membrane fuses with the plasma membrane in a process known as **exocytosis,** in which the secretory vesicle is allowed to open and expel its contents into the extracellular fluid. Some secretory vesicles are designed so that the exocytotic step occurs very rapidly when stimulation of the plasma membrane occurs from an outside signal (e.g., a specific hormone).

At the same time that the products of ER synthesis and Golgi complex packaging are conveyed to the exterior of the cell by exocytosis, the membrane of the secretory vesicle becomes a part of the plasma membrane. Recall that some proteins have been synthesized in the ER for use as membrane proteins. They also have been packaged in the Golgi complex and form the membranes of the same secretory vesicles that carry the proteins destined for export. In this way, the new membrane proteins and lipids synthesized and trapped within the ER membrane can be transported to the cell surface and inserted into the plasma membrane, by the fusion step, to replace damaged or "worn out" proteins and lipids.

Not all of the membrane lost from the *trans* face of the Golgi as secretory vesicles is replaced by new membrane arriving at the *cis* face in the form of transition vesicles from the ER. Some of the replacement membrane is derived from pieces of the plasma membrane that are internalized and delivered to the *trans* face of the Golgi by processes such as **endocytosis,** which is the reverse of exocytosis (Fig. 3–9). The internalized membrane may be returned to the plasma membrane by a subsequent exocytotic step. Exchange of membrane in this manner between the Golgi and the plasma membrane conserves the membranes of the transport vesicles, which may be highly specialized. It is economical for the cell to use them for several series of transport events rather than to synthesize new vesicle membranes each time. This **membrane recycling** mechanism can provide the cell with reusable raw materials because some of the internalized membrane is directed to **lysosomes,** where the component proteins and lipids are taken apart and deposited in the cytosolic pool of precursor molecules to await a new round of synthesis. Membrane recycling also helps to avoid an increase in the surface area of secretory cells when large numbers of secretory vesicles fuse with the plasma membrane.

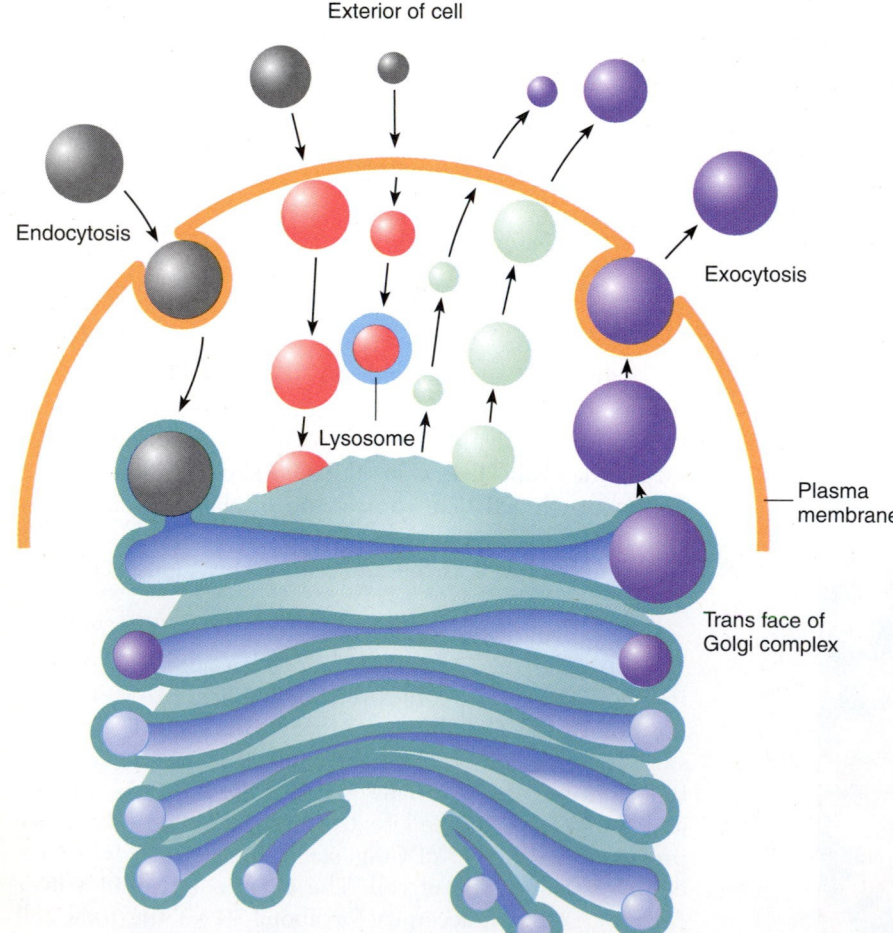

Exterior of cell

Endocytosis

Exocytosis

Lysosome

Plasma membrane

Trans face of Golgi complex

Figure 3–9

As portions of the Golgi complex fuse with the plasma membrane in the secretion process, portions of the plasma membrane are returned to the Golgi complex through various endocytotic events. Thus, the plasma membrane maintains the same size despite the continual addition of membrane from secretory vesicles.

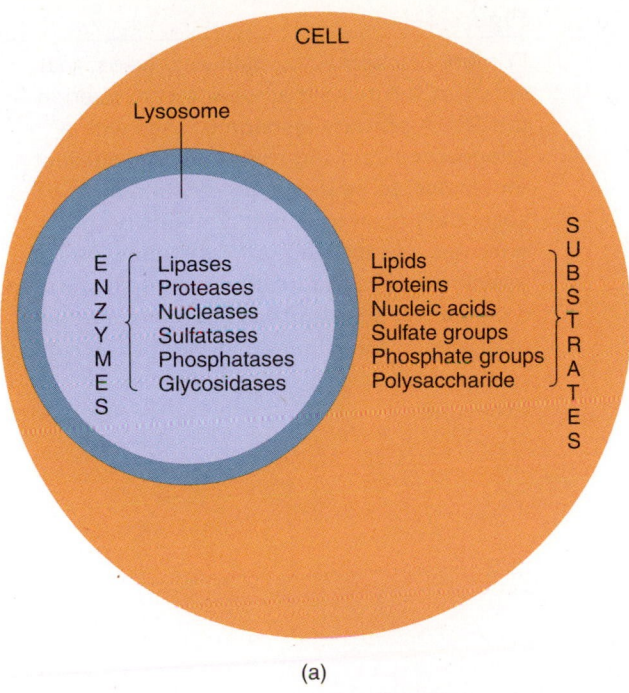

(a)

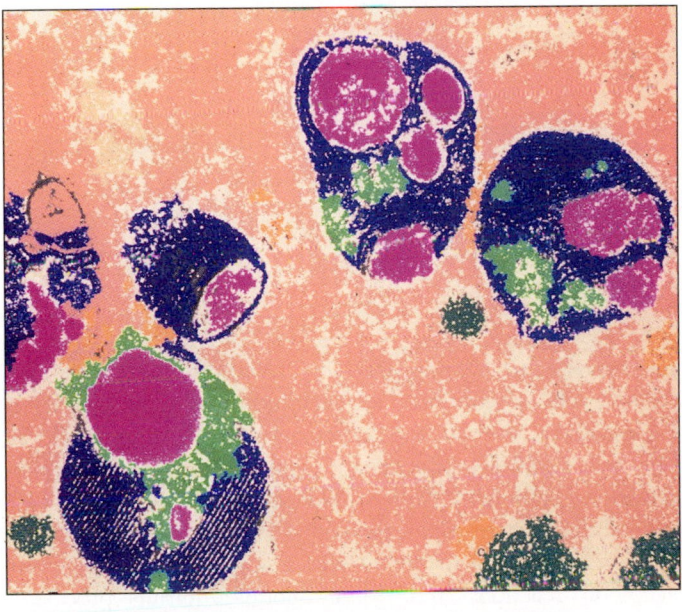

(b)

Figure 3–10

(a) Lysosomes protect the cell contents from degradation by keeping within their membranes enzymes that can break down all of the macromolecules in the cell. The only substrates allowed to contact the lysosomal enzymes are those within food vacuoles or other vacuoles containing material meant to be digested. **(b)** A transmission electron micrograph of lysosomes.   (© *Don W. Fawcett/Visuals Unlimited)*

Lysosomes and Peroxisomes: Digestion of Food and Waste

Lysosomes and peroxisomes are small, membrane-bound bags, about 0.5 μm in diameter. Together they account for less than 1% of the total membrane in a cell, but they are found in almost all cells in large numbers. One mammalian liver cell contains about 300 to 400 of each type of organelle.

LYSOSOMES. The **lysosome** is an organelle that contains about 50 different digestive enzymes. As a group, these enzymes allow cells to completely break down any of the macromolecules in the body (Fig. 3–10a). Because these enzymes would destroy the cells themselves, they are kept within a specific membrane-bound organelle. Two general classes of lysosomes have been distinguished: (1) newly formed **primary lysosomes** contain only the digestive enzymes; (2) **secondary lysosomes** contain both enzymes and molecules (substrates) that the cell must break down. (Recall that *substrate* is a general term for any material on which an enzyme is designed to work.)

Primary lysosomes are thought to arise by budding from the *trans* face of the Golgi complex (Fig. 3–11). Most of the lysosomal enzymes that they contain are glycoproteins that originate in the rough ER. When they enter the Golgi complex, the polysaccharide parts of the enzyme molecules present a unique signal to the Golgi complex, which responds by packaging these enzymes into the vesicles that become primary lysosomes.

Various types of secondary lysosomes exist. Among the largest are the **digestive vacuoles,** which arise after primary lysosomes fuse with an endocytotic vesicle containing particulate material, such as a bacterium. Internalization of particulate material is a special type of endocytosis known as **phagocytosis.** The resultant endocytotic vesicle is described sometimes as a **phagosome.** The process of fusion allows the digestive enzymes of the primary lysosome to gain access to the material to be broken down, whether it is bacteria or food.

After the lysosomal enzymes have performed their digestive task, the products of digestion can be either released outside the cell by exocytosis, for reuse by other cells, or absorbed from the digestive vacuole and used again within the same cell. Lysosomes are used not only to digest material originating from outside the cell but also to break down the cell's own organelles and other matter in a process known as **autophagy.** This process can be a response to cell injury, or it can occur during "remodeling" of the cell or during adverse nutritional conditions. The secondary lysosomes formed during autophagy are called **autophagic vacuoles.**

PEROXISOMES. The **peroxisome** also contains specialized enzymes. These enzymes are synthesized in the cytosol and transported into peroxisomes as they form, probably as buds from the smooth ER. The enzyme **catalase** is contained almost exclusively within the peroxisomes. Other enzymes within peroxisomes use oxygen (O_2) to carry out

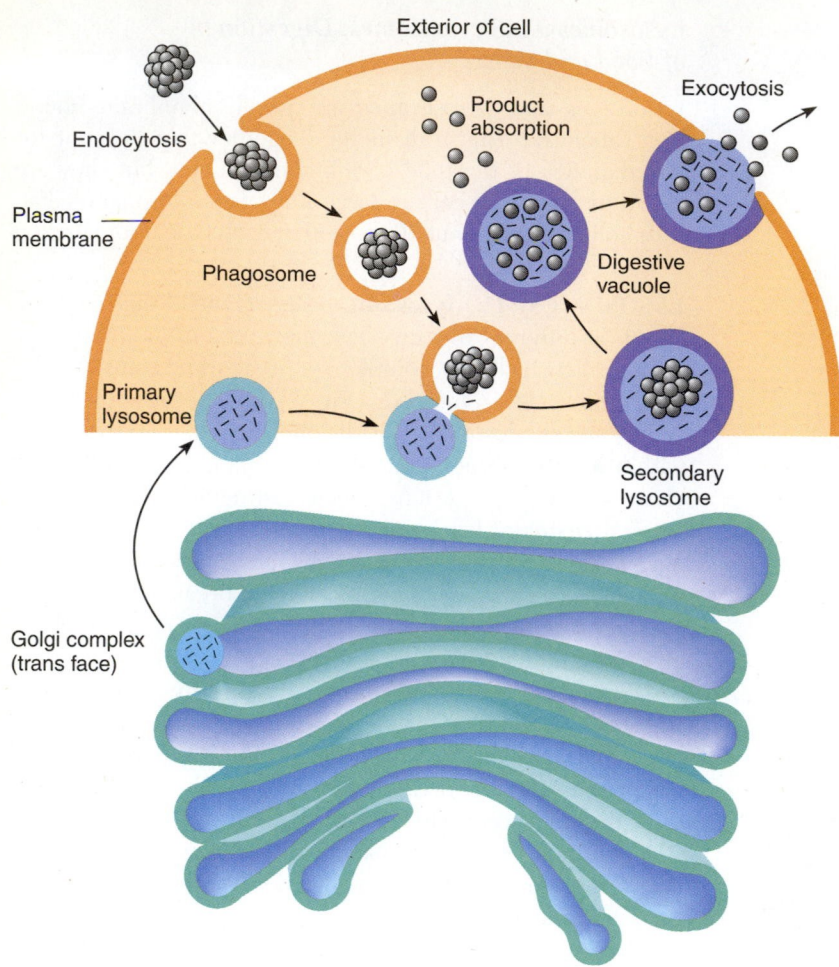

Figure 3–11

Primary lysosomes contain only lysosomal enzymes (proteins that were synthesized in the ER). A primary lysosome fuses with a phagosome carrying material (e.g., nutrient molecules) to be digested, creating a secondary lysosome that contains both enzymes and the nutrient substrates. The vesicle containing digested material is termed a *digestive vacuole;* from there, the products of digestion are either absorbed by the cytosol or expelled to the exterior of the cell by exocytosis.

oxidative reactions. One of the products of these reactions is hydrogen peroxide (H_2O_2), which is potentially harmful to the cell. Catalase helps to degrade the hydrogen peroxide to harmless water by the following reaction:

$$2\ H_2O_2 \xrightarrow{\text{catalase}} 2\ H_2O + O_2$$

The peroxisomes in many body cells are important in detoxifying a number of molecules.

Mitochondria: Energy Transformation

The **mitochondrion** is thought of as a "powerhouse" for the cell because its metabolic activities provide the energy needed to sustain the life of mammals and other **aerobic** (oxygen-breathing) organisms.

The principal fuels for most life forms are carbohydrates. After digestion and absorption in the digestive system, glucose is stored in liver and muscle cells in the form of glycogen. In the cytoplasm, glycogen is broken down into glucose. The primary mechanism for processing these molecules and extracting the energy stored in them is **aerobic respiration.** This metabolic pathway is completed within the mitochondrion by a process that uses most of the oxygen that mammals obtain from

the atmosphere. During aerobic respiration, about 36 molecules of ATP are produced for each molecule of glucose that is broken down. Without the participation of the mitochondrion, only two molecules of ATP are produced. Clearly, the mitochondrion is the major site of ATP production within the cell. ATP is the form in which the cell is able to trap chemical energy and transfer it to sites where it is needed to drive many synthetic reactions and other energy-requiring processes. These processes are described in detail in Chapter 6.

Mitochondria are much smaller than the nucleus. They are usually cylindrical and about 0.5 to 1.0 μm in diameter. The number of mitochondria per cell varies with the cell type, but animal cells usually contain large numbers. A liver cell may have as many as 1700 mitochondria, which would account for one fifth of the cell volume and 40% of the total cell membrane.

Like the nucleus, two separate membranes surround the mitochondrion. The **outer mitochondrial membrane** is smooth and unfolded, but the **inner mitochondrial membrane** is distinguished by numerous well-developed infoldings called **cristae,** which project deep into the interior of the organelle and increase its surface area considerably (Fig. 3–12). The two membranes divide the mitochondrion into

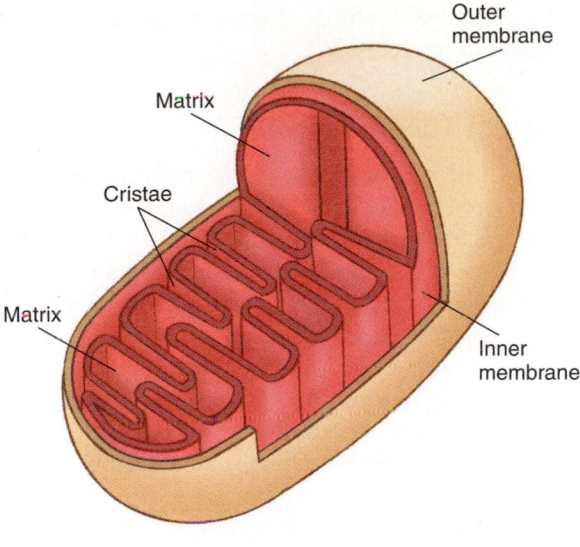

Outer
membrane

Matrix

Cristae

Matrix

Inner
membrane

(a)

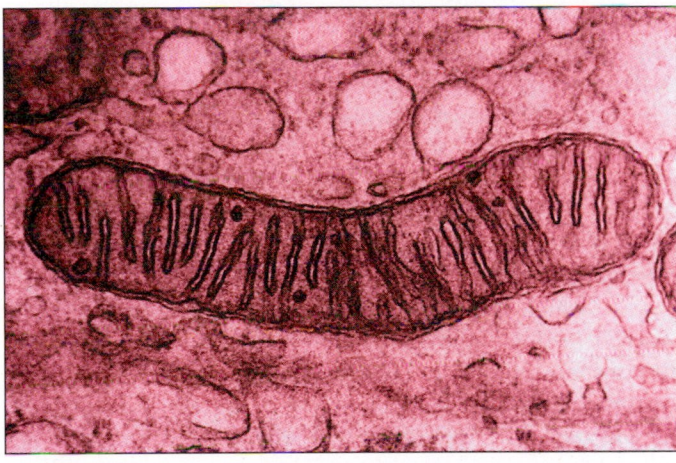

(b)

Figure 3–12

(a) The mitochondrion, the cell's "powerhouse," contains a double membrane. The numerous infoldings of the inner membrane, called *cristae,* create a large surface area on which the reactions of cellular respiration can take place. *(b)* An electron micrograph of a mitochondrion from an adrenal cortex cell.   (© *Don W. Fawcett/Visuals Unlimited*)

two compartments: (1) the narrow **intermembrane space** and (2) the much larger **matrix.**

The inner mitochondrial membrane is a highly impermeable structure, unlike the outer membrane. Most of the "work" of the mitochondrion is carried out by enzymes and other molecules located in the inner membrane and the matrix. The matrix compartment also contains ribosomes and 5 to 10 small double-helical DNA molecules that the mitochondria use to synthesize some of the mitochondrial proteins. Mitochondrial DNA has a circular structure. In

mammalian cells, it makes up less than 1% of the total cellular DNA.

The major component of mitochondria, besides water, is protein. Most (67%) of the protein of liver cell mitochondria is located in the matrix, and about 21% is in the inner membrane. Many of these protein molecules are specific enzymes; about 120 different enzymes are present. Some of the enzymes in the matrix participate in a series of reactions known as the **citric acid cycle,** also referred to as the **Krebs cycle** or the **tricarboxylic acid cycle.** This cycle is the final common pathway for the breakdown of all the fuel molecules in the cell — not only carbohydrates but also amino acids and fatty acids. These processes are discussed in detail in Chapter 6.

The Cytoskeleton

What is the cytoskeleton and how does it function?

Specialized proteins in the cytosol are organized into complex filamentous structures that form the **cytoskeleton** of the cell. The cytoskeleton plays an important role in coordinating a large variety of cellular activities, including maintaining cell shape, positioning and movement of the internal organelles, directing protein secretion, mediating movement of the cell, and relaying mechanical signals from the cell surface to the cytosol and into the nucleus. In many ways, the cytoskeleton represents a level of cytoplasmic organization that is functionally similar to the compartmentalization of biochemical functions into membrane-enclosed organelles. One of the earliest recognized functions of the cytoskeleton in eukaryotic cells was to provide a structural framework for maintaining the overall shape of the cell, hence the name *cyto* (= "cell") - "skeleton." It is important to recognize, however, that the cytoskeleton is not a rigid or static structure, as its name implies. Rather, the cytoskeleton is highly dynamic and undergoes constant change. One function of the cytoskeleton is to organize the cytoplasm. For example, the cytoskeleton serves to maintain and modify the positioning of organelles in the cytoplasm. Specific examples of organelle movement that have been discussed already in this chapter are (1) the movement of secretory vesicles from the Golgi complex to the plasma membrane and (2) the movement of endocytotic vesicles or phagosomes from the plasma membrane to the cell interior.

The cytoskeleton of eukaryotic cells consists of a complex array of three principal types of filamentous networks: (1) **microfilaments,** (2) **microtubules,** and (3) **intermediate filaments.** The structures and sizes of these three filament systems are compared in Table 3–2. These filament systems are organized by a variety of accessory proteins that control the size and shape of the filaments and link filaments to other structures in the cell, such as organelles, proteins in the plasma membrane, and even the chromosomes.

ACTIN MICROFILAMENTS. Actin is a globular protein that is the most abundant cytoskeletal protein in most eukaryotic cells. The cells of higher eukaryotic organisms con-

TABLE 3–2

Cytoskeletal Fibers

Fiber	Diameter (nm)	Protein Structure	Protein Subunit
Microtubules	25	Hollow	Tubulin
Intermediate filaments	10	Hollow	Several types
Microfilaments	7–8	Solid	Actin
Myosin filaments	15	Solid	Myosin

tain several variations, or **isoforms,** of actin that are encoded by a family of actin genes. In mammals, at least six types of actin exist that can be grouped into three classes: (1) muscle contains actins designated α-actins; nonmuscle actins are designated (2) β- and (3) γ-actins. In contrast, lower eukaryotes, such as yeast, contain only a single actin gene. Actin in nonmuscle cells is organized in the cytoplasm to form long strands called **microfilaments.** Individual actin molecules, or **monomers,** are called *globular actin* (G-actin). Monomers of G-actin polymerize to form filamentous actin (F-actin) microfilaments. Actin microfilaments are polarized so that a fast-growing (plus) end and a slow-growing (minus) end are seen. Actin monomers can bind a molecule of ATP; the binding and hydrolysis of ATP plays a crucial role in the polymerization of G-actin into filaments. Because the plus ends of microfilaments grow faster than do the minus ends, microfilaments undergo a dynamic process called *treadmilling,* which contributes to the motile activity of cells (Fig. 3–13). Although the different actin isoforms exhibit subtle differences in their biochemical properties, the amino acid sequence of actins has been conserved highly throughout evolution; their ability to assemble into filaments is indistin-guishable. The principal features of actin filament assembly are summarized in Figure 3–13.

Organization of Microfilaments

When actin was first isolated from muscle cells in the early 1940s, it was believed to have a unique function in muscle contraction. While it is true that actin is a key protein in mediating muscle contractility, actin is now known to be present in nearly all types of eukaryotic cells. Although the actin in muscle is very stable and highly organized, actin in nonmuscle cells is organized in many different ways. As mentioned earlier, nonmuscle actin generally is organized into either bundles or branching networks. These structures are often more **labile,** or subject to change, than are the actin filaments in muscle.

Actin microfilaments are very thin (approximately 7–8 nm in diameter) but can be up to several microns in length. Individual actin filaments are usually organized into more complex, three-dimensional structures, such as bundles made up of dozens or hundreds of microfilaments. These bundles often are referred to as **stress fibers** because they can generate tension or stress inside cells when combined with accessory proteins, such as **alpha-actinin** and **myosin.** The small GTPase, **rhoA,** plays a central role in organizing actin filaments into stress fibers. Like most GTPases, rhoA function is determined by the nucleotide to which it is bound. When bound to GDP, rhoA is inactive. However, when rhoA is bound to GTP (rho-GTP), it can activate other *effector* proteins, including a protein kinase called *rho-kinase,* which regulates myosin phosphorylation and stress fiber formation. Figure 3–14 shows stress fibers and focal adhesions in a fibroblast. **Fibroblasts** are the least specialized cells in the connective tissue family. Focal adhesions are sites on the ventral surface of the cell where stress fibers are anchored to the membrane. A model for how actin stress fibers are anchored to the cytoplasmic face of the plasma membrane also is shown in Figure 3–14.

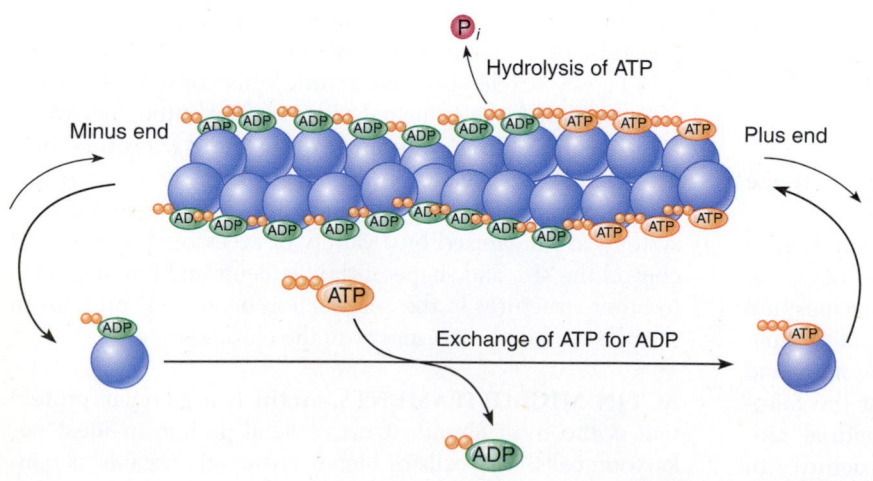

Figure 3–13

The dynamic process of treadmilling of actin filaments occurs for two key reasons: (1) at concentrations of actin monomer typically found in cells, actin monomers bound to ATP are added more easily to the fast-growing (plus) end of existing microfilaments than to the minus end. After incorporating into a filament, ATP-actin is hydrolyzed to ADP-actin. (2) Because ADP-actin dissociates from filaments more easily than does ATP-actin, monomers are more easily lost from the minus end of the filament. Consequently, there is a net loss of monomers from the minus end and a net addition of monomers to the plus end of microfilaments.

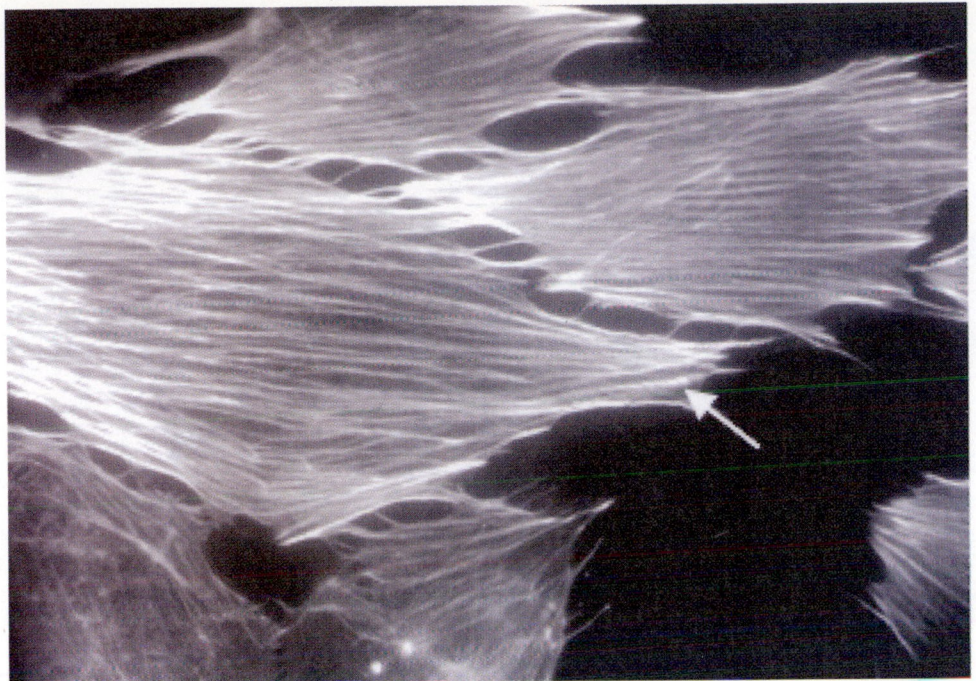

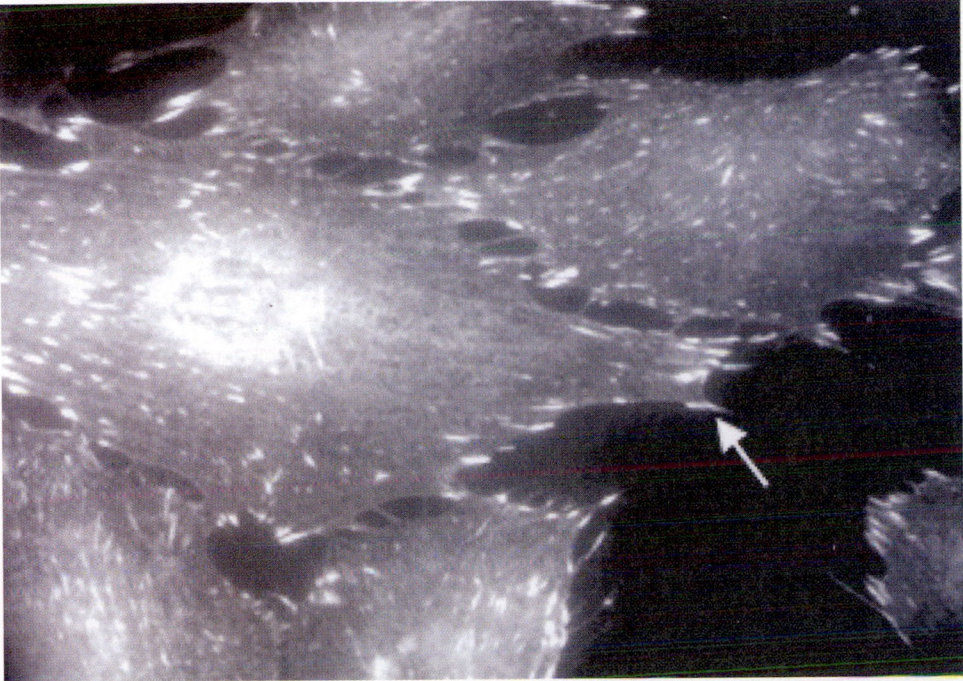

(a)

Figure 3–14

Actin in nonmuscle cells can be organized into bundles called *stress fibers,* which are anchored to the cytoplasmic face of the plasma membrane at sites of cell–extracellular matrix (ECM) contact called *focal adhesions.* *(a)* Fluorescence microscopy (× 7,500) of a mouse osteoblast stained with a fluorescent dye called *rhodamine–phalloidin,* which binds F-actin to reveal the location of stress fibers (*top panel*). When the same field of cells is stained with an antibody against the focal adhesion protein vinculin (followed by a second fluorescent antibody tagged with fluorescence against the vinculin antibody), the position of the focal adhesions is revealed (*bottom panel*). Note that the focal adhesions are present at the ends of the stress fibers, where the cell contacts the ECM. *(b)* A simplified model of the molecular organization of proteins in focal adhesions that anchor the actin filaments in stress fibers to integrins. Integrins also bind the ECM outside the cell, allowing for communication across the plasma membrane. Many other proteins are found in focal adhesions that are not shown here. *(Micrograph courtesy of Suzanne Ponik)*

Focal adhesion

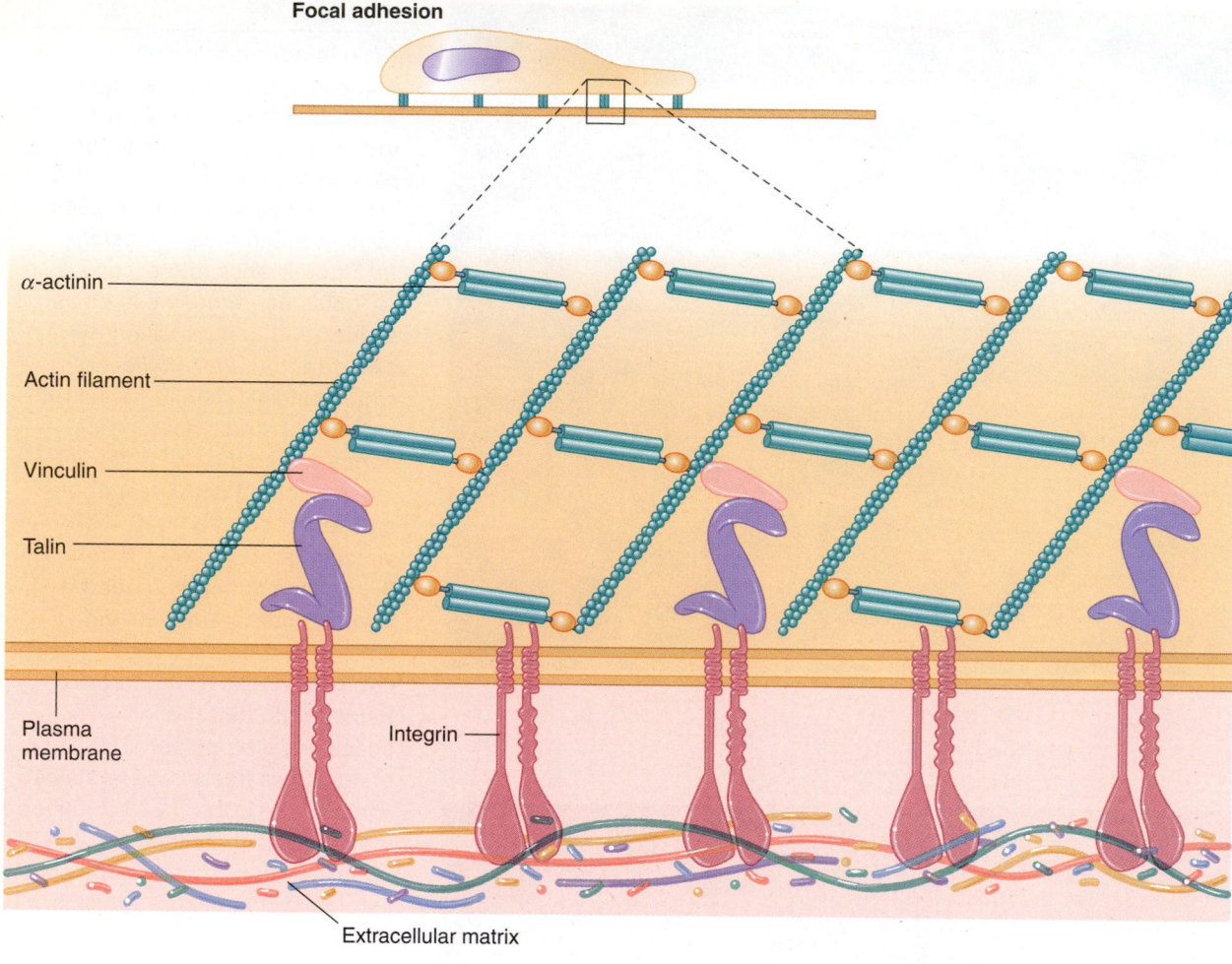

α-actinin

Actin filament

Vinculin

Talin

Plasma
membrane

Integrin

Extracellular matrix

(b)

Actin filaments can also be organized into complex, branching networks. Actin networks form through interactions with other accessory proteins, such as **filamin,** and become so viscous that they take on the properties of a gel (Fig. 3–15). Dozens of accessory proteins exist that bind actin (**actin-binding proteins**), regulate the polymerization of G-actin into F-actin microfilaments, and regulate the higher order three-dimensional organization of actin. Some of these accessory proteins are listed in Table 3–3.

In addition to stress fibers, a good example of actin bundles in nonmuscle cells is seen in the fingerlike projections called **microvilli,** which cover the inner surfaces of the intestine and the kidney proximal tubule. Figure 3–16 gives an indication of the huge size of the microfilaments relative to the accessory protein molecules present in the membrane of the microvillus.

While the actin microfilaments in microvilli are relatively stable (because microvilli persist more or less continuously), the actin bundles found in the contractile ring of

dividing cells during cytokinesis, the separation of the dividing cell into two daughter cells, are more labile. Together with accessory proteins, including alpha-actinin and myosin, the actin bundles in the contractile ring form to rapidly pinch off the cytoplasm, allowing the two daughter cells to separate during the final stage of cell division. After separation of the cells, the contractile ring falls apart and does not re-form until the next round of cell division and cytokinesis.

Large, actin-binding proteins, including filamin, regulate the organization of actin filaments into networks, which cross-link actin into gels. Another actin-binding protein, gelsolin, fragments actin filaments when activated by a localized increase in intracellular Ca^{2+} levels. Thus, gelsolin can not only reduce the viscosity of gels but also help increase network complexity by forming branching points for the growth of new microfilaments in the gel. The protein myosin is able to form **myosin filaments,** which are sometimes described as a fourth type of cy-

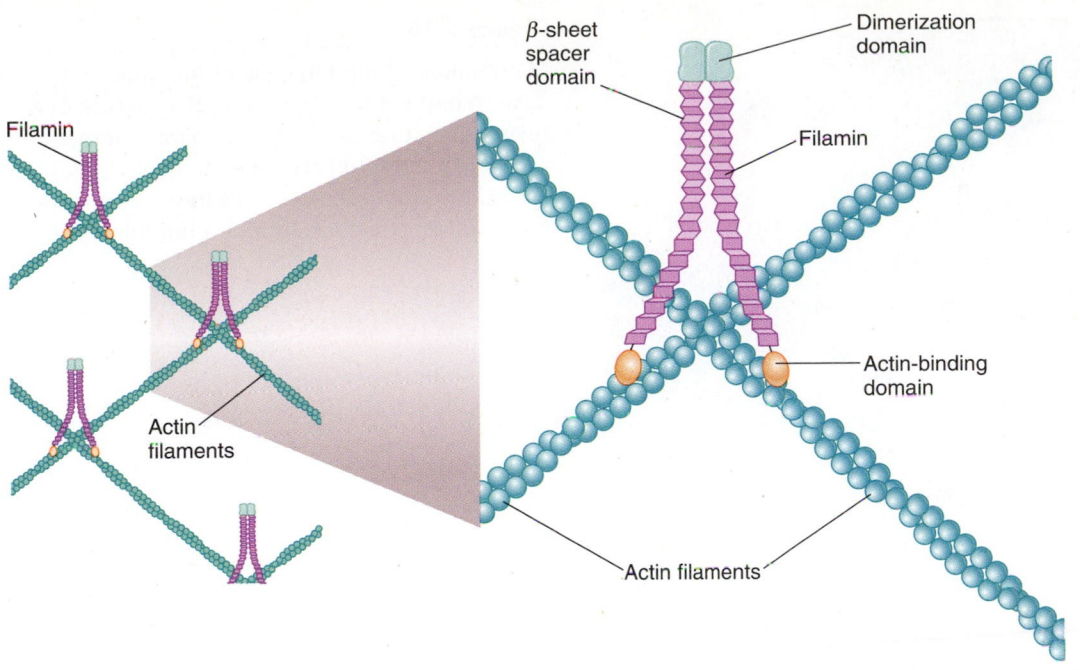

Figure 3–15

Actin filaments can be organized into networks by proteins, such as filamin, that can bind to and cross-link two actin filaments. Filamin is a dimer of two 280-kDa subunits that are attached by a dimerization domain at one end and that is separated from the actin-binding domains at the other end by an elongated beta-sheet spacer domain.

toskeletal filament in nearly all cell types. They are also known as **thick filaments** in muscle cells. They are most abundant in muscle cells in which, together with the thin actin filaments, form a part of the contractile machinery of these cells. The presence of **actin** and **myosin** molecules in nonmuscle cells indicates that nonmuscle cells possess a contractile mechanism. This contractile mechanism in nonmuscle cells is used for various cell functions, including cytokinesis, cell locomotion, and the generation of tension inside the cell.

TABLE 3–3

Some Actin Filament Accessory Proteins

Protein	Function
Alpha-actinin	Bundles actin filaments; links actin filaments to integrins
Fimbrin	Bundles actin filaments
Filamin	Cross-links actin filaments into gel
Spectrin	Attaches filaments to the plasma membrane
Gelsolin	Severs actin filaments, caps plus end of filament
Myosin I	Promotes actin filament sliding
Myosin II	Moves vesicles on actin filaments
Thymosin	Sequesters actin monomers
Villin	Bundles actin filaments; severs actin filaments
Arp 2/3	Nucleates actin filament polymerization; cross-links actin filaments
Ezrin-radixin-moesin	Attach actin filaments to plasma membrane proteins

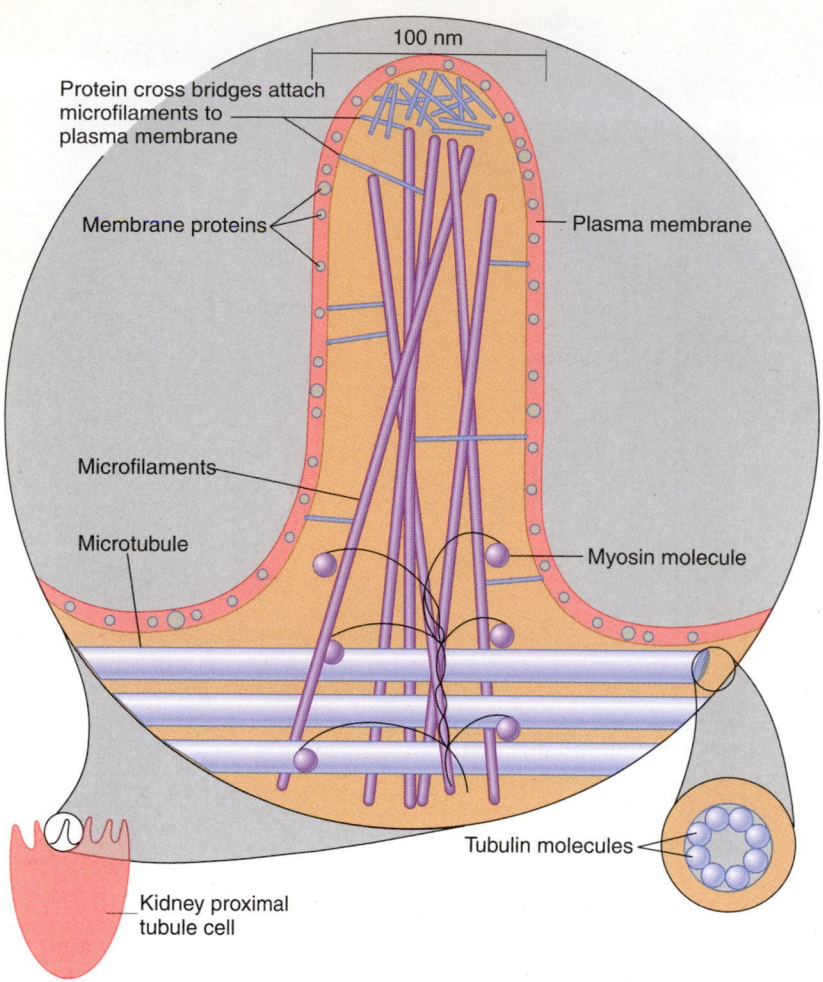

100 nm

Protein cross bridges attach microfilaments to plasma membrane

Membrane proteins

Plasma membrane

Microfilaments

Microtubule

Myosin molecule

Tubulin molecules

Kidney proximal tubule cell

Figure 3–16

Microvilli are found in cells of the digestive system and kidney. The internal structure of a microvillus, drawn to scale, shows that the microfilaments that create its fingerlike shape are relatively large structures in the cell. Microtubules also are found in the microvilli.

MICROTUBULES. Microtubules are composed of a single protein subunit, **tubulin,** which polymerizes into hollow rods approximately 25 nm in diameter. Like microfilaments, microtubules are a dynamic cytoskeletal filament system. Microtubules are constantly undergoing phases of net assembly or disassembly inside the cell. Figure 3–17 illustrates the principal features of microtubule assembly and structure. Microtubules are involved in various types of movement inside the cell, including organelle movement and separation of the chromosomes during mitosis. Unlike G-actin, which is a monomer, tubulin is a **dimer,** formed from one subunit of α-tubulin and one subunit of β-tubulin. A third type of tubulin, γ-tubulin, is found in the centrosome, where it plays a unique role in initiating microtubule growth from this organelle.

Tubulin dimers are arranged head to tail to form protofilaments. Microtubules usually are formed from 13 protofilaments arranged in parallel to form a filament with a hollow core. Like actin filaments, microtubules are polar, having a fast-growing (plus) end and a slow-growing (minus) end (Fig. 3–18). Both α- and β-tubulin bind a molecule of GTP, and the hydrolysis of GTP functions to regulate micro-

tubule polymerization in a manner that is analogous to ATP-regulated polymerization of actin microfilaments. When the concentration of GTP-tubulin is high, microtubule growth is favored; when GTP-tubulin concentrations are low, shortening of microtubules occurs. Microtubules usually are oriented so that their minus, or slow-growing, ends are embedded in the centrosome, while their plus, or fast-growing, ends radiate to the cell periphery. Microtubules undergo dramatic bouts of rapid shortening known as **dynamic instability.** When the addition of monomers to the fast-growing (plus) end of a microtubule slows to less than a crucial rate, the hydrolysis of GTP to GDP on the end of the microtubule leads to rapid depolymerization. This phenomenon, first recognized in the mid-1980s, led to the discovery of useful anticancer agents. Drugs were developed that alter the dynamics of microtubule polymerization and block cell division in rapidly growing cancer cells, such as vinblastine and taxol. These drugs are able to impede the growth of cancer cells because microtubules play an important role in the separation of chromosomes during mitosis.

Microtubule motor proteins, including kinesin and dynein, bind to microtubules and are able to use the energy

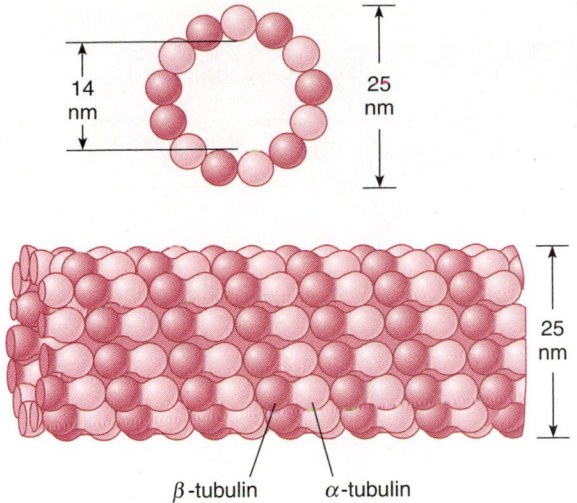

β-tubulin   α-tubulin

(a)

Figure 3–17

Microtubules are formed from dimers of α- and β-tubulin. *(a)* Tubulin dimers polymerize to form protofilaments, which are organized around a hollow center to form microtubules. Microtubules usually are made up of 13 protofilaments. *(b)* Microtubules polymerize when there is a high concentration of free tubulin bound to GTP. Because the GTP-tubulin that is incorporated into microtubules is rapidly hydrolyzed to GDP-tubulin, an excess of free GTP-tubulin is needed to maintain a GTP-tubulin cap at the fast-growing end of the microtubule. When the free GTP- tubulin concentration is low, the addition of GTP-tubulin cannot keep pace with hydrolysis of GTP-tubulin in the microtubule, and rapid shortening of the microtubule occurs. This process is described as *dynamic instability* and results in cycles of microtubule growth and rapid shortening of microtubules inside the cell.

High concentration of tubulin bound to GTP

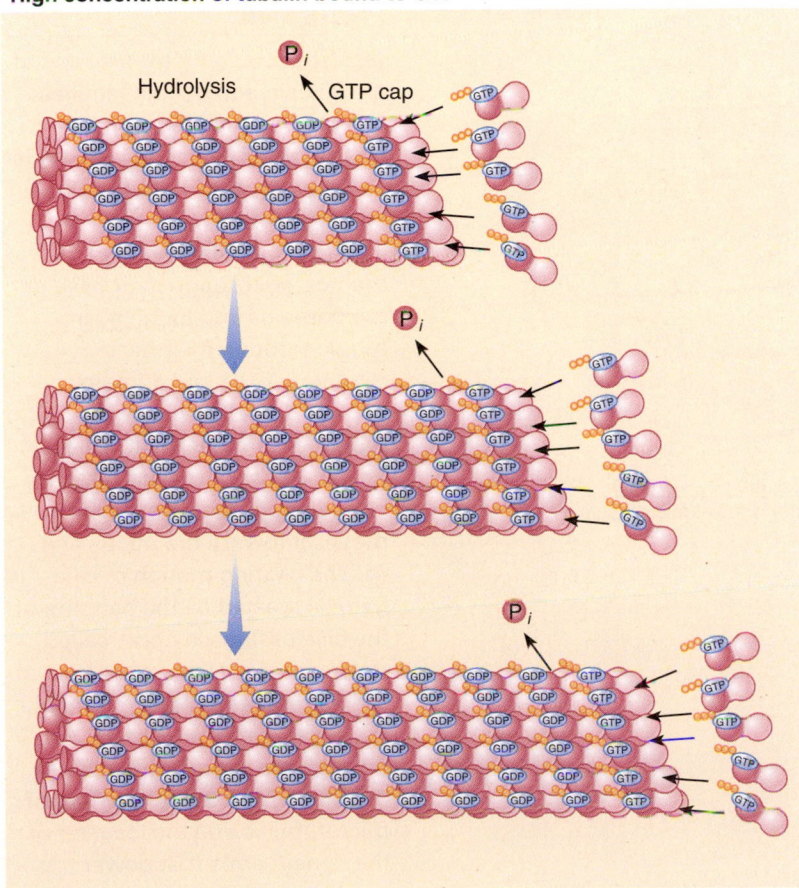

Low concentration of tubulin bound to GTP

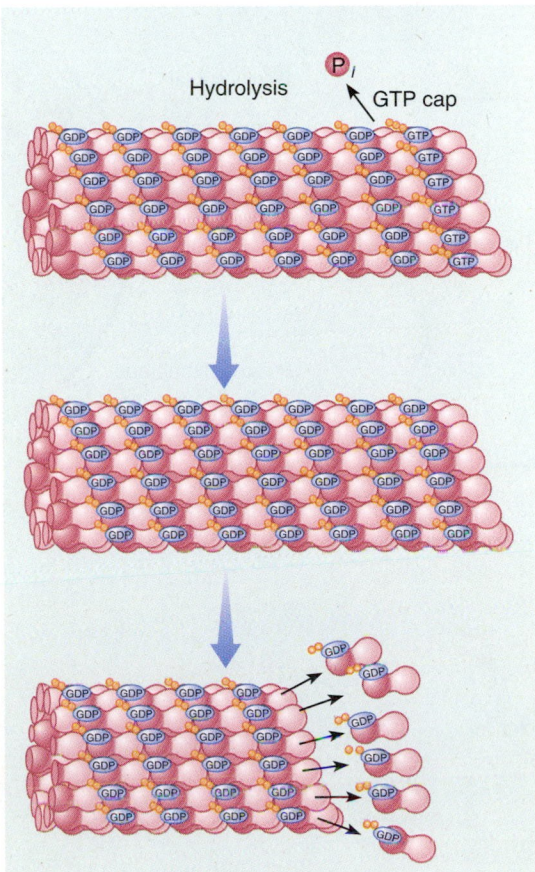

(b)

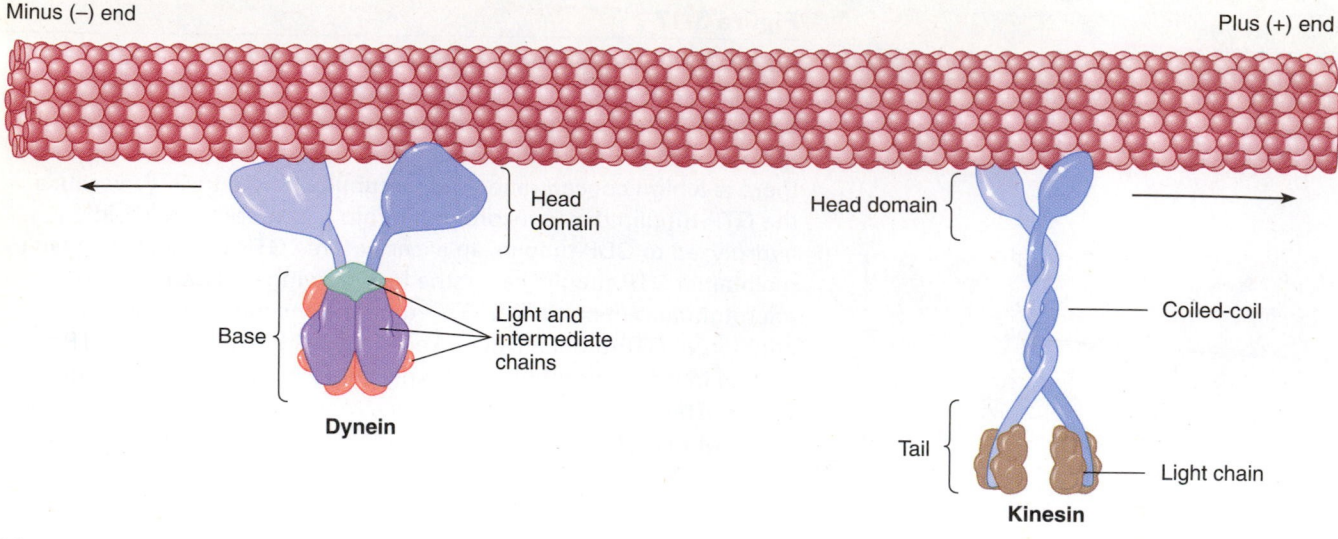

Minus (−) end

Plus (+) end

Head domain

Head domain

Base

Light and intermediate chains

Coiled-coil

Dynein

Tail

Light chain

Kinesin

(a)

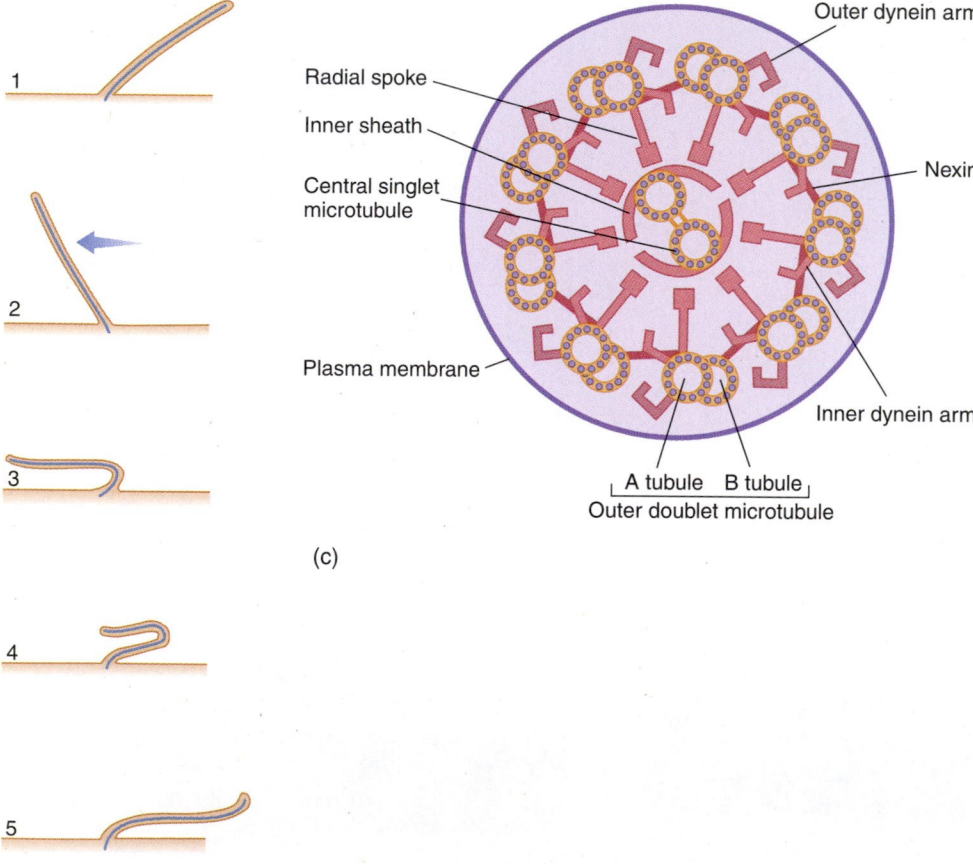

1

2

3

4

5

(b)

Radial spoke

Inner sheath

Central singlet microtubule

Plasma membrane

Outer dynein arm

Nexin

Inner dynein arm

A tubule   B tubule
Outer doublet microtubule

(c)

Figure 3–18

(a) Microtubule motor proteins of the kinesin and dynein families use energy obtained from the hydrolysis of ATP to move cellular cargo along microtubules. Dynein and kinesin are made up of multiple subunits and bind to their cargo through their base or tail domains and to the micro-tubule through their head domains. The head domains contain the ATPase activity required for movement. Kinesin moves its cargo toward the plus end of micro-tubules; dynein moves it toward the minus end. **(b)** The beating motion of cilia and flagella created by the bending of the cilia or flagella body, called the axoneme, moves fluid past cells or moves cells through fluid. **(c)** A diagram of a cross-section through the axoneme illustrates the unique 9 + 2 arrangement of microtubules and the location of the dynein arms that power axoneme movement.

APPLICATIONS OF PHYSIOLOGY

Gene Therapy

About 1% of all babies are born with an inherited genetic defect. We know of more than 4000 different types of inherited disorders, many of which lead to early death because there is no fully effective treatment. Due to the recent advances in recombinant DNA technology, scientists are on the verge of repairing certain genetic defects by introducing the normal, undamaged genes into patients. The first two human gene therapy trials in the United States were approved in September 1990 by the National Institutes of Health (NIH) Recombinant DNA Advisory Committee. This marked the start of a new era for biomedical science, in which genetic engineering will be used to alleviate the suffering caused by genetic diseases and perhaps even severe illnesses, such as cancer.

The best candidates for gene therapy are disorders caused by a single damaged gene, such as cystic fibrosis, rather than by multiple genes, such as cancer or heart disease. Ideally, the new gene will exactly replace the damaged one, and the disease will be cured for life with a single treatment and no side effects. (Random insertion of a new gene anywhere in one of the chromosomes is less ideal because of the risk of activating a cancer-inducing gene, or **oncogene.**) In reality, however, it is extremely difficult to control the fate of DNA introduced into a cell. For this reason, one of the approaches currently being studied is the introduction of a healthy gene into a cell without removal of the defective gene. In this way, the healthy gene supplies the missing product of the defective gene, and the risk for side effects is reduced. The disadvantage of this approach is that several treatments would be required throughout life, and so it is not a cure. The procedure, known as **gene augmentation**, involves removing cells from the patient, introducing the healthy gene into the cells, and returning the altered cells to the patient. Genetic flaws do not have to be corrected in all of the body's cells, provided that the altered cells can supply enough of the missing product. Bone marrow and skin and liver cells are the best candidates for gene therapy: They withstand removal from the body, can be returned without great difficulty, and survive for a long time after replacement within the body.

An important strategy for introducing a healthy or "therapeutic" gene into the isolated cells uses the native ability of an independently replicating genetic agent called a **virus** to enter cells. The most promising system uses a retrovirus, a type of virus that uses RNA as genetic material. Once inside a human cell, the **retrovirus** converts its RNA to DNA and inserts the DNA into a chromosome, causing the cell to begin making proteins needed for the virus to replicate. Many retroviruses have been engineered to serve primarily as gene delivery vehicles. Investigators substitute a therapeutic gene for viral genes and delete the instructions for making new viral proteins. The modified retrovirus still enters cells and inserts the therapeutic gene into cellular DNA, but now it has lost the ability to reproduce.

One of the NIH trials is focused on a rare condition in children known as **severe combined immunodeficiency**, which often arises because the function of white blood cells, a part of the body's defense against disease, is impaired by a damaged gene. The patients in this study are infused with their own white blood cells, which have been previously removed and modified using the retrovirus technique to contain the therapeutic gene. The other NIH trial recognizes that gene therapy also can be used to enhance the ability of cells to fight severe diseases, such as cancer. A class of white blood cells that penetrate tumors is being modified to contain a new gene that will increase their anticancer activity. Their therapeutic potential will be evaluated in patients with advanced cancer.

Gene therapy is an exciting development but is not without problems, both technical and ethical. There is dramatic success in some patients, while others show no improvement. At the University of Pennsylvania, the recent death caused by multiple organ failure of a patient undergoing a gene therapy trial for a non–life-threatening condition reminds scientists that application of gene therapy in humans is complicated. The technique requires extensive research in a high-technology environment for several more years. Intentional modification of human genes also is disconcerting for many people on religious or other ethical grounds. However, the consensus among scientists is that human gene therapy can be beneficial if carefully scrutinized and is worth pursuing if it proves successful in combating human disease.

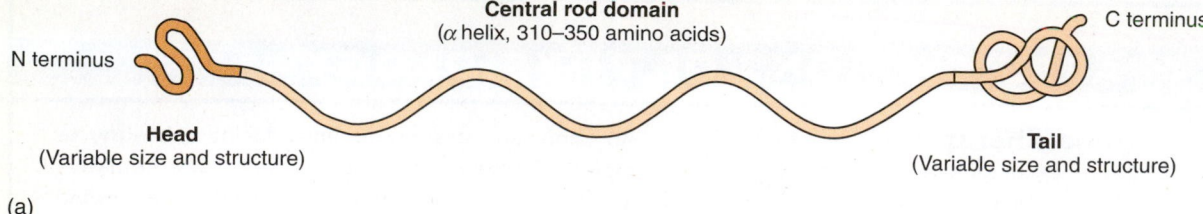

(a)

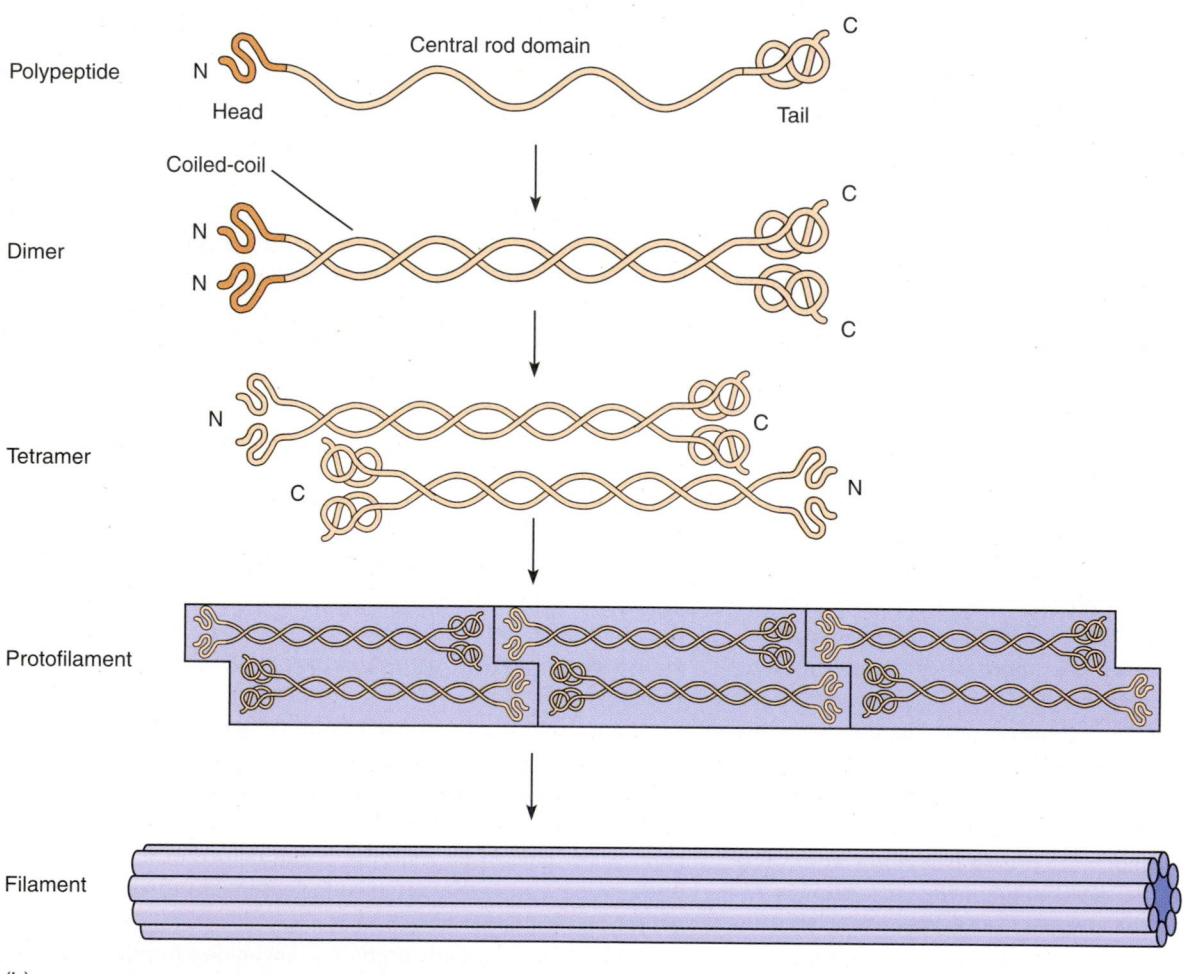

(b)

Figure 3–19

The structure of intermediate filament proteins controls their assembly into filaments. *(a)* Intermediate filament proteins have a central rod domain, which governs filament assembly, and head and tail domains, which are variable in size and determine the specificity of function. *(b)* Formation of polypeptide dimers is the first step in intermediate filament assembly and involves the association of the central rod domains. Dimers next form antiparallel tetramers, which are slightly staggered and join end to end with other tetramers to form protofilaments. The mature intermediate filament is made of multiple protofilaments that wind around each other to form a ropelike structure.

released from ATP hydrolysis to power a variety of movements inside cells. Organelles, chromosomes, and vesicles are among the intracellular cargo moved by microtubule-based motors. Kinesin moves its cargo toward the plus end of microtubules, while dynein moves toward the minus end of microtubules (see Fig. 3–18). The **cilia** and **flagella** of eukaryotic cells are closely related, microtubule-based, motile structures found on the surface of many types of cells. Cilia play an important role on the epithelial cells that line the respiratory tract, where they beat in a coordinated manner to move fluid over the surface of the cells. The beating of the single flagellum of most types of sperm cells is responsible for sperm motility. In these structures, microtubules are arranged in a characteristic "9 + 2" pattern, in which nine microtubule doublets surround a central pair of microtubules. It is the motor activity of molecules of dynein that extends from each of the outer nine microtubule doublets and interacts with the adjacent doublet to drive the beating movement of cilia and flagella.

INTERMEDIATE FILAMENTS. Microtubules and microfilaments are described as *dynamic structures* because their formation is readily (and often) reversed and because the distribution of these structures within a cell is subject to change as physiological conditions change. In contrast, the intermediate filaments are much more stable structures. Intermediate filaments are approximately 10 nm in diameter, intermediate in diameter between microfilaments (7–8 nm) and microtubules (25 nm). Intermediate filaments do not appear to be involved in cell movements, as are microfilaments and microtubules. Instead, intermediate filaments appear to play primarily a structural role in the cell. They increase the resistance of cells to mechanical forces and help hold together sheets of epithelial cells. Also in contrast to microfilaments and microtubules, which are composed of a single type of subunit protein, intermediate filaments are polymers of several types of proteins that differ by cell type. Although more than 50 different intermediate filament subunit proteins have been discovered, they can be grouped into six protein families based on similarities in their amino acid sequence: keratins (acidic, type I; neutral/basic, type II), which are found in epithelial cells; vimentins and desmins (type III), found in fibroblasts, leukocytes (white blood cells), muscle cells, and some cells of the peripheral nervous system; neurofilament proteins (type IV), found in neurons; nuclear lamins (type V), found in the nuclear lamina of all cell types; and nestin (type VI), found in the stem cells, which give rise to mature nerve cells of the central nervous system. Despite their differences, all intermediate filament subunit proteins share a common structural organization—they have a long central alpha-helical domain of 300 to 350 amino acids in length that drives their assembly into filaments. The principal features of intermediate filament assembly are illustrated in Figure 3–19. They also have an *N*-terminal head domain and a *C*-terminal tail domain, which determine the unique specific functions of the different types of intermediate filaments.

The organization of intermediate filaments is complex and differs greatly among cell types. The nuclear lamins form a ring around the nucleus and extend out to the plasma membrane. Vimentin filaments and keratin filaments help position the nucleus in the cell and can attach to the nuclear envelope and to the plasma membrane. In addition, intermediate filaments can interact with microfilaments and microtubules, help integrate all three filaments, and form a cytoskeletal "scaffold" that can detect and respond to changes in the mechanical forces experienced by cells.

In epithelial cells, keratin filaments help form unique plaquelike structures at the plasma membrane that help maintain tight attachments between cells that are necessary for maintaining the structural integrity of epithelial sheets. These structures, called **desmosomes** and **hemidesmosomes,** maintain cell–cell and cell–extracellular matrix contacts and are the major cell surface attachment sites for intermediate filaments at cell–cell and cell–substrate contacts, respectively. The transmembrane molecules in desmosomes are cadherin family members; those in the hemidesmosome are members of the integrin family. These adhesion molecules are discussed in more detail later. These sites not only play an important structural function in tissue organization but mediate cell signaling pathways. The organization of desmosomes and hemidesmosomes is illustrated in Figure 3–20. Specialized proteins at these sites called *plakins* bind to intermediate filaments and mediate attachment to cell adhesion molecules (CAMs) in the membrane. A number of skin diseases, such as epidermolysis bullosa, in which severe blisters form, are caused by genetic defects in proteins that maintain the organization and strength of these plaques. Such diseases are presumably caused by a loss of mechanical integrity in the epithelial cell layers. Abnormalities in members of the type IV neurofilament proteins also have been linked to diseases of motor neurons, including amyotrophic lateral sclerosis (ALS), a degenerative disease of the nervous system, better known as Lou Gehrig's disease.

The overall organization of the cytoskeleton is complex. It has been suggested that the major fibers are linked by another set of thin filaments, the **microtrabecular lattice.** These filaments appear to hold the fibers and the internal organelles in their places (Fig. 3–21). They even hold ribosomes in suspension. However, the microtubules may have the major role in the overall design of the cytoskeleton because they appear to influence the distribution of microfilaments and intermediate filaments. They may provide the framework on which the rest of the cytoskeleton is erected. The precise molecular mechanisms by which the cytoskeleton can mediate the movement of cells and the movement of organelles within cells remain the subject of intense biomedical research.

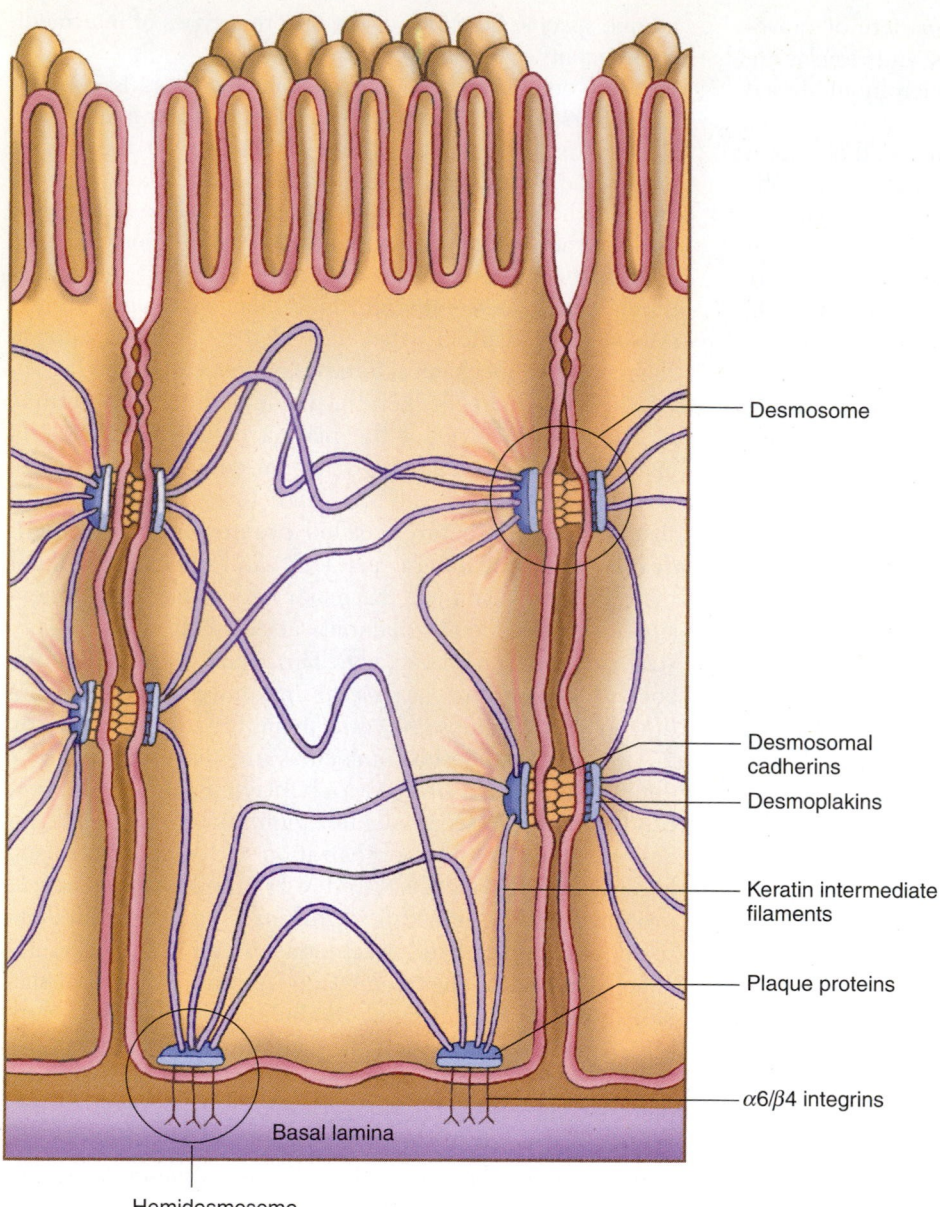

Figure 3–20

Desmosomes and hemidesmosomes anchor epithelial cells to each other and to the basal lamina, respectively. Intermediate filaments composed of keratin help form these unique plaquelike structures at the plasma membrane that maintain the structure of epithelial sheets. Desmosomal cadherins are found in desmosomes and link to the intermediate filaments via desmoplakins. The $\alpha6/\beta4$ integrin is found in the hemidesmosomes and links to intermediate filaments via specialized plaque proteins called *plakins*.

Desmosome

Desmosomal cadherins

Desmoplakins

Keratin intermediate filaments

Plaque proteins

$\alpha6/\beta4$ integrins

Basal lamina

Hemidesmosome

CELL–CELL AND CELL MATRIX INTERACTIONS

The ability of cells to interact directly and specifically with other cells and with the extracellular matrix (ECM) is crucial to the development and function of organisms. While some interactions between cells are transient, such as the interactions between leukocytes and endothelial cells at sites of inflammation, others are very stable, such as the interactions between cells in adult tissues. CAMs on the cell surface mediate interactions between cells and the ECM. Interactions between the cytoplasmic domains of CAMs and cytoskeletal proteins regulate the adhesive functions of CAMs.

Cell Surface Adhesion Molecules

What are cell surface adhesion molecules, and how do they function?

Adhesion molecules on the surface of eukaryotic cells allow cells to specifically interact with each other and with the extracellular matrix. Specialized structures formed by CAMs allow cells to regulate extracellular interactions and allow for communication between cells and the surrounding environment. Four families of CAMs mediate the majority of adhesive interactions: (1) integrins, (2) cadherins, (3) immunoglobulin super-family members, and (4) selectins. These molecules play crucial

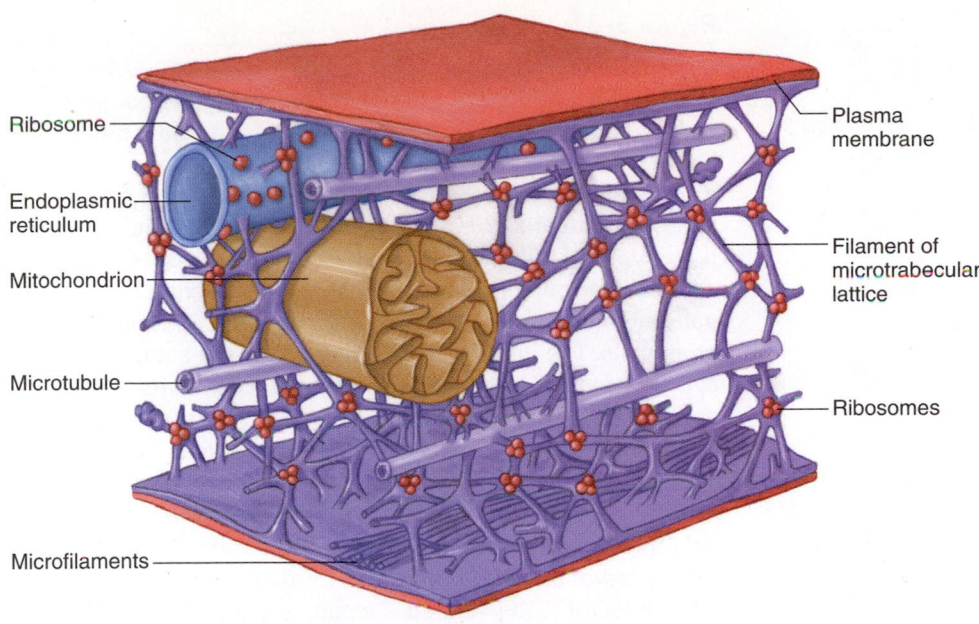

Ribosome

Endoplasmic reticulum

Mitochondrion

Microtubule

Microfilaments

Plasma membrane

Filament of microtrabecular lattice

Ribosomes

(a)

Figure 3–21

(a) Microtubules, intermediate filaments, and microfilaments are found throughout the cell. They may be linked together to form an interconnected structure called the *microtrabecular lattice.* *(b)* Fluorescent antibodies show the position of microtubules in a human cell ($\times$ 6,300).

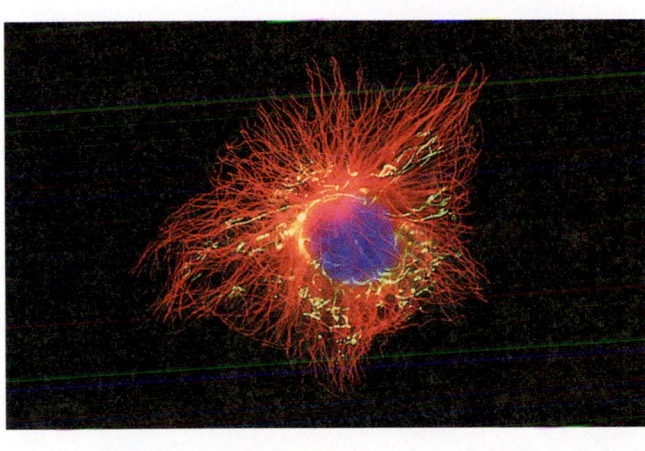

(b)

number of different matrix proteins, or **ligands,** to which integrins bind (Table 3–4). This combination of many different integrins and ligands provides cells with a mechanism to precisely control the sites to which they will adhere.

The extracellular domain of most integrins interacts with the extracellular matrix, while the intracellular or cytoplasmic tails of integrin β subunits interact indirectly with the actin cytoskeleton via accessory proteins including alpha-actinin and talin (see Fig. 3–14). Integrins form adhesive structures with the extracellular matrix known as **focal adhesions.** Focal adhesions can transmit force applied to the outside of the cell across the cell membrane. Focal adhesions are, therefore, sites where transduction of mechanical signals from the extracellular environment occurs. Integrins utilizing β subunits of the $\beta1$, $\beta3$, and $\beta5$ subtype are found in focal adhesions and can be stable structures in stationary cells. In locomoting cells, focal adhesions are very dynamic structures that form at the leading edge of a crawling cell and are disassembled at the rear as the cell moves past.

Integrins containing the $\beta2$ subunit are unique in that they are found exclusively on leukocytes. These integrins mediate interactions between leukocytes and other cells rather than with the ECM. The importance of the $\beta2$ integrins in the immune system is highlighted by leukocyte adhesion deficiency type I (LAD-I) syndrome. This inherited disease is characterized by the absence of the $\beta2$ subunit from the surface of leukocytes, rendering them unable to bind to endothelial cells lining blood vessels. Consequently, leukocytes are unable to emigrate from the bloodstream and enter tissues to fight infection. Patients with LAD-I suffer from recurring, severe bacterial and fungal infections that are eventually fatal if untreated. Another inherited disease, Glanzmann's thrombasthenia, is a familial bleeding disorder

roles in a variety of important cellular functions, including growth and differentiation, locomotion, wound repair, and embryonic development. Because most cells express many different types of adhesion molecules, the regulation of cell adhesion processes is complex. The structures of these four families of CAMs are compared in Figure 3–22.

Integrins

Integrins are a large family of CAMs composed of noncovalently linked α and β subunits (see Fig. 3–22). Eight different β subunits and at least 16 different α subunits pair to form a diverse array of α/β heterodimers. Integrins are found on essentially all cells in the human body, with the exception of erythrocytes, and are required for embryonic development, regulation of cell division and cell differentiation, blood clotting, and the immune response. There are a large

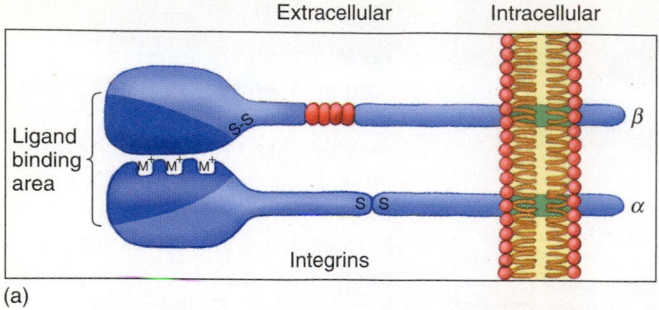

(a)

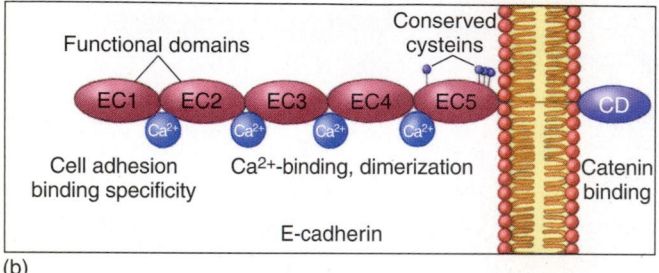

(b)

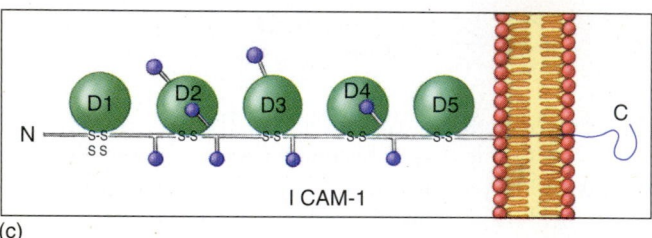

(c)

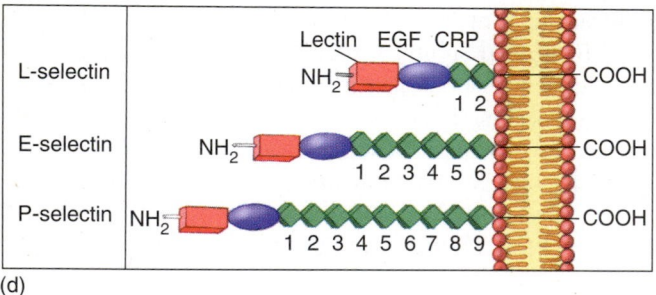

(d)

Figure 3–22

The structure of the four major families of cell adhesion molecules (CAMs). *(a)* Integrins are heterodimers composed of α and β subunits. The large extracellular domain binds ligands; the intracellular domain links to the actin cytoskeleton via α-actinin, talin, vinculin, and other focal adhesion proteins. *(b)* Cadherins are single-subunit homotypic CAMs containing variable numbers of repeating extracellular domains (e.g., EC1) that bind calcium ions and determine the binding specificity for similar cadherins on other cells. The intracellular domain binds the actin cytoskeleton via catenins. *(c)* Immunoglobulin family CAMs are typified by the ICAM-1 molecule, which contains five immunoglobulin-like repeats that are disulfide-linked and a short cytoplasmic tail that can interact with the actin cytoskeleton via alpha-actinin. *(d)* The selectin family contains only three members. Each contains an extracellular domain composed of an *N*-terminal lectin domain, an epidermal growth factor (EGF)–related domain, and a variable number of domains related to complement regulatory proteins (CRPs). Like integrins and ICAM-1, the short cytoplasmic domain of selectins also interacts with the actin cytoskeleton via alpha-actinin.

caused by the loss of the $\beta3$ integrin subunit. In animal experiments researchers have blocked the expression of the $\beta1$ integrin subunit and shown that this integrin subunit is required for embryos to form beyond an early stage of embryonic development.

Cadherins

Cadherins are a large family of homotypic CAMs. **Homotypic** means that one cadherin binds to a similar cadherin on another cell. Cadherins are Ca^{2+}-dependent (hence the name *cadherin*), which means they require Ca^{2+} ions to bind to each other. This fact makes cadherins well suited to their primary function of maintaining adhesion between similar cells within a tissue. In fact, all cells that are capable of forming solid tissues express cadherins. Cadherins are found in several membrane structures, including adherens junctions,

desmosomes, tight junctions, and gap junctions. The cadherin family can be subdivided into two large subfamilies: (1) classic cadherins and (2) desmosomal cadherins. The classic cadherins are found in adherens junctions that form between cells. These cadherin proteins contain a large *N*-terminal extracellular domain with Ca^{2+} binding sites, while the C-terminal cytoplasmic domain mediates interactions with the actin cytoskeleton (see Fig. 3–22). A family of proteins called **catenins** mediates binding of the cytoplasmic tail of cadherins to the cytoskeleton. β-catenin binds directly to the cytoplasmic tail of classic cadherins and to the protein α-catenin. F-actin microfilaments bind to α-catenin to complete the linkage. Classic cadherins can be further subdivided into types I and II. Type I cadherins include E-, P-, and N-cadherins and are expressed in *epithelial* cells, *placenta*, and *neural* cells, respectively. Although they were named based

TABLE 3–4

Some Integrins and Their Ligands

Integrin	Ligand
$\alpha 1/\beta 1$	Collagen, laminin
$\alpha 2/\beta 1$	Collagen, laminin, fibronectin
$\alpha 4/\beta 1$	Fibronectin, VCAM-1
$\alpha 5/\beta 1$	Fibronectin, collagen
$\alpha 6/\beta 1$	Laminin
$\alpha 9/\beta 1$	Tenascin
$\alpha L/\beta 2$	ICAM-1, -2, -3
$\alpha 6/\beta 4$	Laminin
$\alpha v/\beta 3$	Vitronectin, fibronectin, fibrinogen, von Willebrand factor, tenascin, collagen, thrombospondin
$\alpha v/\beta 1$	Vitronectin, fibronectin
$\alpha M/B2$	ICAM-1, iC3b, fibrinogen

on these tissues, in which they were first discovered, E-, P-, and N-cadherins also are found in other types of tissues. Type II classic cadherins include several recently discovered proteins that are slightly less similar to the classic cadherins and have been designated cadherins 5 to 12. Cadherin 5, also called VE-cadherin, is particularly interesting because it is restricted to endothelial cells and may play a role in the physiologically important process of new blood vessel formation, or **angiogenesis.**

The second major group of cadherins, the desmosomal cadherins, include desmoglein and desmocollin. These proteins are found in specialized cell–cell junctions called *desmosomes.* The cytoplasmic domains of these proteins interact with intermediate filaments through plaque proteins called *desmoplakin* and *plakoglobin.*

Immunoglobulin Superfamily

The **immunoglobulin (Ig) superfamily** of CAMs contains domains in their extracellular regions that are structurally similar to the 70 to 100 amino acid disulfide-bonded domains (D domains) in immunoglobulin (antibody) molecules (see Fig. 3–22). Over 100 molecules in this family have been identified, and several have been shown to play important roles in cell adhesion, particularly in cells of the immune system. One particularly important Ig superfamily member is intercellular adhesion molecule–1 (ICAM-1), which is a major CAM in inflammation. ICAM-1 is expressed at low levels on the surface of most endothelial cells. However, when endothelial cells are activated by cytokines, molecules that regulate immune responses—such as interleukin-1 (IL-1), interferon-γ (IFN-γ), or tumor necrosis factor-α (TNF-α)—expression of ICAM-1 on the surface increases significantly. This results in increased binding of leukocytes that express $\beta 2$ integrins.

These integrins bind to ICAM-1 and represent an important intercellular interaction necessary for a normal inflammatory response, which is critical to fighting off infection. Several other Ig superfamily members are also important in the immune system, including leukocyte function–associated molecule-2 (LFA-2) on thymocytes, T cells, and natural killer cells — cells that play important roles in the immune response to infection. Platelet–endothelial CAM-1 (PECAM-1) is an Ig family member present on platelets and in the cell–cell junctions between endothelial cells that is thought to play an important role in the migration of leukocytes between endothelial cells during inflammation. Vascular CAM-1 (VCAM-1), like ICAM-1, is expressed on endothelial cells; its expression is increased during inflammation and is recognized by the $\alpha 4/\beta 1$ integrin.

Selectins

A fourth family of CAMs, the **selectins**, also functions in the immune system. The selectins are a relatively small family of adhesion molecules, represented by only three members: L-, P-, and E-selectin (see Fig. 3–22). These were first discovered in leukocytes, platelets, and endothelial cells, respectively. We now know that the selectins are also expressed in other cell types. Each of the selectins has a similar structure consisting of a lectinlike domain at the N-terminus, a domain with homology to epidermal growth factor, and a variable number of repeated domains with homology to proteins involved in immune regulation. Selectins recognize carbohydrate-containing proteins on the surfaces of cells on leukocytes and endothelial cells.

Selectins mediate an early phase in the recruitment of leukocytes to sites of inflammation in a response known as *rolling.* During leukocyte rolling, leukocytes that normally flow rapidly in the bloodstream slow down and become loosely tethered to endothelial cells but do not adhere firmly. Thus, they appear to tumble or roll along the endothelial layer lining blood vessels. Subsequently, firmer adhesion and transendothelial migration, mediated by integrins, occur. A genetic defect known as *leukocyte adhesion deficiency type II* (LAD-II) is due to defects in the carbohydrate-containing ligands of E- and P-selectins. This results in patients having difficulty fighting infections because their leukocytes cannot slow down by rolling and therefore have difficulty beginning the integrin-mediated firm adhesion and transendothelial migration phases. Selectins also mediate targeting of leukocytes to lymph nodes in a process called *leukocyte homing.*

Cell Signaling

 How do cells communicate with other cells?

All cells detect and respond to signals from the surrounding environment. This is possible because of a variety of **signaling molecules** that are secreted by cells and by **re-**

ceptors, for those signaling molecules that are expressed on the surfaces of cells. Signaling molecules bind to cell surface receptors and regulate the activities of individual cells and allow complex organisms to function properly. The recognition and binding of signaling molecules to their specific receptors initiates a cascade of intracellular events that control nearly all aspects of cell function, including cell survival, growth, division, movement, differentiation, and metabolism. An exploding area of scientific research is aimed at understanding the molecular events that regulate signaling pathways in cells. It has become clear that many human diseases, including inflammatory diseases and cancers, result from defects in cell signaling pathways. Understanding how signaling pathways are integrated in cells will facilitate the development of treatments for many human diseases.

Cell Matrix and Cell–Cell Signaling

Many types of signaling molecules, including proteins, hormones, peptides, and gases. are able to transmit information to other cells. Signaling molecules that bind to specific receptors on other cells are referred to as **ligands.** While most ligands bind to receptors on the cell surface, others cross the cell membrane and bind to receptors inside the cell. Some cells are very close to the source of the signal, allowing signaling molecules to act over very short distances. In fact, cell signaling can occur through direct interaction of cells with the ECM or with other cells through CAMs. The integrins are an important molecule in cell–matrix signaling events. Binding of integrins to the ECM triggers an intracellular signaling cascade that can result in altered gene expression and changes in cell proliferation and differentiation. One or more members of the integrin family of adhesion molecules are present on nearly all cells in the human body.

The **cadherins** are also important in cell–cell signal transduction. Cell biologists have discovered that cadherins not only mediate adhesion of cells to each other but also transmit signals into the cell to regulate cell survival, cell division, and cell locomotion. The adhesive and signaling events mediated by cadherins are particularly important for coordinating cell functions during embryonic development as well as in adult tissues.

Secreted Signals

Signaling molecules that do not act by direct cell adhesion mechanisms must be secreted and diffuse over some distance to reach their target cell. This type of signaling often is divided into three broad categories: (1) **autocrine,** (2) **paracrine,** and (3) **endocrine.** Autocrine signals act on the same cells in which they are produced. This type of signaling is particularly important in the immune systems, in which T-lymphocytes respond to the binding of a foreign antigen by secreting molecules that promote cell proliferation, called growth factors, to stimulate their own replication.

Paracrine signals diffuse over a short distance and act on adjacent cells. The release of neurotransmitters at the synapse between nerve cells, discussed in Chapter 7, is an example of paracrine signaling. Another important paracrine signaling molecule is the gas nitric oxide (NO). Smooth muscle cells that surround blood vessels relax or contract based, in part, on the amount of NO released by endothelial cells that line the blood vessels. Control of vascular tone by NO released from endothelial cells acting on adjacent smooth muscle is a key regulatory mechanism in the circulatory system. Endocrine signals, usually hormones, are secreted by endocrine tissues and diffuse into the bloodstream. They are then carried through the body by the circulation and delivered to their target cells often very far away from where they were released. These types of signals are discussed further in Chapter 12.

A group of small molecules known as **growth factors** represents one of the most diverse and important types of signaling molecules. Examples of polypeptide growth factors include nerve growth factor (NGF), which binds to receptors on neurons and promotes their survival and development; epidermal growth factor (EGF), which binds to the receptors on epidermal and other cells to promote cell proliferation; and platelet-derived growth factor (PDGF), which is secreted by platelets during wound healing and promotes growth and differentiation of fibroblasts at the site of the wound.

THE LIFE CYCLE OF THE CELL

> *How does cell division take place?*

The many different cell types present in the human body have different patterns and rates of division. Cells such as red blood cells and nerve cells stop dividing when they reach maturity, whereas certain epithelial cells in the intestine and skin divide rapidly and continually.

The **cell cycle** is the period of time from the beginning of one cell division to the beginning of the next division (Fig. 3–23). Cell cycle times vary from just 8 hours for rapidly dividing cells to more than 100 days for rarely dividing cells.

The cell cycle consists of two main phases: (1) interphase and (2) cell division. **Interphase** is the period between the end of one cell division and the beginning of the next. During interphase, the cell carries out all the normal cell processes of growth and metabolism, including the replication of DNA.

The process of cell division, sometimes called the **M phase,** involves two steps: (1) mitosis and (2) cytokinesis. In **mitosis,** the nucleus divides into two new nuclei. In **cytokinesis,** the cytoplasm divides, each half taking with it one of the two new nuclei to form a new cell.

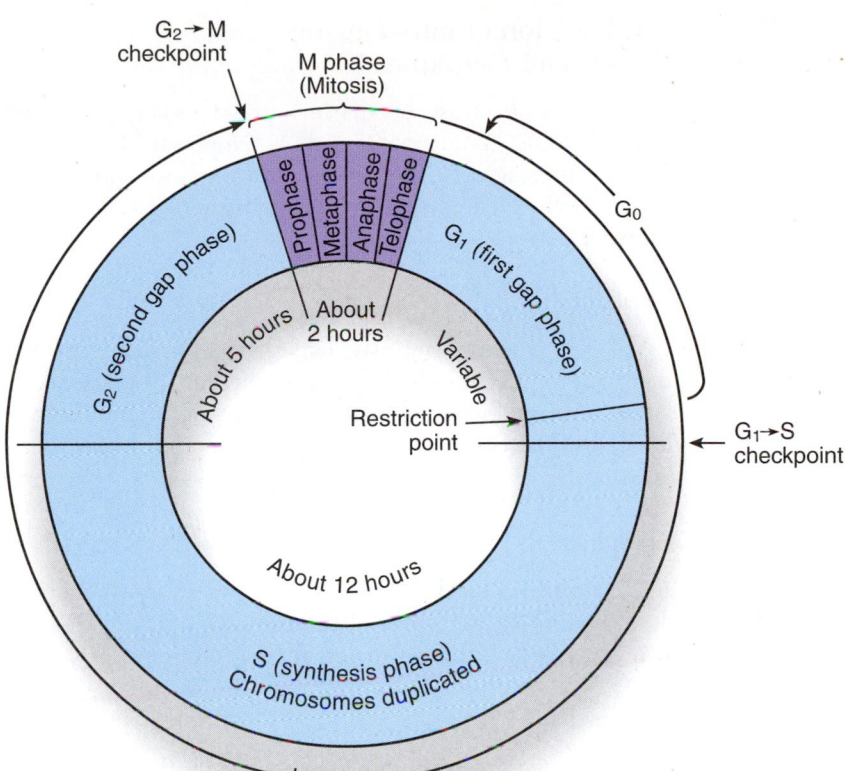

Figure 3–23

The life cycle of the cell contains two phases: (1) interphase and (2) mitosis. During interphase, the cell carries out normal metabolic activities and duplicates DNA. The cell must receive appropriate signals from the extracellular environment (adhesion and growth factors) in order to pass a restriction point and enter the cell cycle. Otherwise, cells remain out of the cell cycle in a phase called G_0. Once the cell passes the restriction point, two additional checkpoints must be passed to proceed through the cell cycle. One checkpoint is between the G_1 and S phases; another is between the G_2 and M phases. During M phase (mitosis), the cell divides into two new cells in four steps called (1) *prophase*, (2) *metaphase*, (3) *anaphase*, and (4) *telophase*.

Cell Growth and DNA Replication Occur During Interphase

Much of the preparation for cell division occurs during **interphase** (see Fig. 3–23), which occupies 90% or more of the cell cycle. Interphase itself is divided into three phases: the synthesis, or S, phase and two gap, or G, phases. The period of interphase that directly follows cell division is the **first gap**, or **G_1**, phase. During this period, the new cell begins its growth as the organelles it inherited from the parent cell resume their biosynthetic activities (which slowed during the M phase). Before progressing through each of these phases, however, the cell must receive specific signals indicating that it is permissible to proceed. These signals allow cells to move beyond the points in the cell cycle called **checkpoints** and prevent the cell from entering too soon into the next phase. Cells that do not receive the appropriate signals from the environment to enter the cell cycle are unable to pass a **restriction point** (also called **START** in yeast) and are referred to as being arrested, or in **G_0**. Other important checkpoints at the G_1 to S phase and the G_2 to M phase have been identified. One important function of these checkpoints is to help ensure that incompletely replicated or damaged DNA is not passed on to daughter cells. The failure of the G_1-to-S checkpoint to prevent DNA damage has been identified as

an important cause of many types of cancers. In mammalian cells, a 53-kilodalton (kDa) protein, called *p53*, mediates arrest of the cell cycle in G_1. The *p53* protein is rapidly induced in cells in response to DNA damage, such as that caused by radiation or other DNA mutagens. Normally, *p53* helps ensure that damaged DNA is repaired before proceeding through the cell cycle. However, when the gene encoding the *p53* protein is mutated, *p53* may not function properly, allowing for progression through the cell cycle before damaged DNA is repaired. Consequently, the damaged DNA is passed on to daughter cells. Analysis of many types of human cancers has, in fact, shown that the *p53* gene is frequently mutated. One way in which *p53* functions is to bind to DNA and activate transcription of a 21-kDa protein (*p21*) that can inactivate specific protein kinases that drive the cell cycle engine (cyclin–CDK complexes, described later) to arrest the cell cycle. Figure 3–24 illustrates the function of *p53* as a G_1 to S phase checkpoint control.

The **S, or synthesis, phase** follows the first gap phase. This is the part of interphase during which the cell synthesizes DNA in the nucleus in order to duplicate the DNA in preparation for cell division.

Directly following the S phase is the **second gap phase (G_2)**, during which the cell increases protein synthesis in final preparation for division. The transition be-

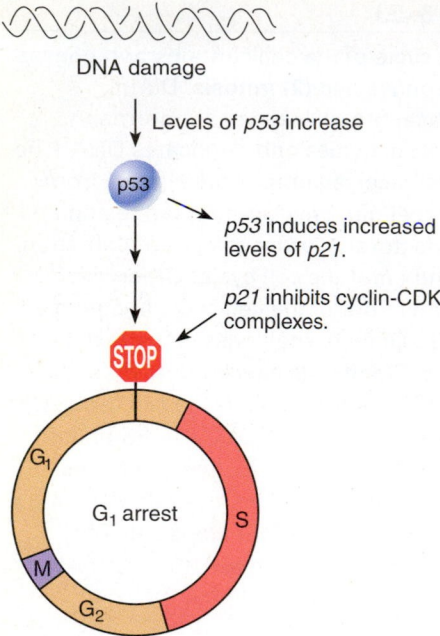

DNA damage

Levels of *p53* increase

p53

p53 induces increased levels of *p21*.

p21 inhibits cyclin-CDK complexes.

STOP

G₁ arrest

G₁ S M G₂

Figure 3–24

The *p53* protein functions to regulate progression through the cell cycle and prevent mutations in the DNA from being passed on to daughter cells. Following DNA damage, *p53* levels increase in the cell. The *p53* protein increases expression of another cell cycle regulatory protein, *p21*, which binds to and inhibits cyclin–CDK complexes to arrest cells at the G_1 to S checkpoint. Presumably, this allows the cell time to repair damaged DNA.

tween G_2 and the next phase, M (mitosis), is an important one. The cell must not enter M phase prior to completion of DNA replication but must also not initiate a second round of DNA replication. Restriction of DNA replication to once per cell cycle is mediated by a family of proteins that bind to replication origins on the DNA. These proteins, known as *MCM proteins,* can bind to replication origins only during G_1 to allow initiation of DNA replication when the cell enters S phase. Once replication of initiation occurs, MCM proteins no longer bind to replication origins, so that the cell must enter another G_1 phase of the next cell cycle before MCM proteins again bind to the DNA. The time taken to progress from the beginning of the S phase to the end of the M phase is markedly similar in all cell types. The principal difference between cells with short cell cycles and those with long cell cycles is the amount of time spent in the G_1 phase; slowly dividing cells stay in G_1 for days or years.

Cell Division Occurs During Mitosis and Cytokinesis

The process of mitosis occurs in four stages: (1) prophase, (2) metaphase, (3) anaphase, and (4) telophase. (Fig. 3–25). Each of the stages merges smoothly with those coming before and after. The stages are distinguished mainly by the different movements of the chromosomes.

Prophase

In the first stage of mitosis, called **prophase,** the DNA that was duplicated during the preceding S phase condenses into chromosomes. The nuclear membrane breaks down, and the contents of the nucleus become mixed with the cytoplasm.

Metaphase

In **metaphase,** the duplicated chromosomes, not yet separated, line up at the center of the cell and attach to the **mitotic spindle,** which is formed from microtubules.

Anaphase

The third stage, **anaphase,** begins when the duplicated chromosomes separate. The two sets, each representing a complete copy of the original DNA, are pulled to opposite ends of the cell.

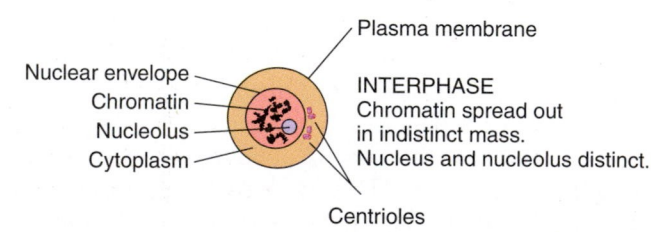

Nuclear envelope — Chromatin — Nucleolus — Cytoplasm

Plasma membrane

INTERPHASE
Chromatin spread out in indistinct mass.
Nucleus and nucleolus distinct.

Centrioles

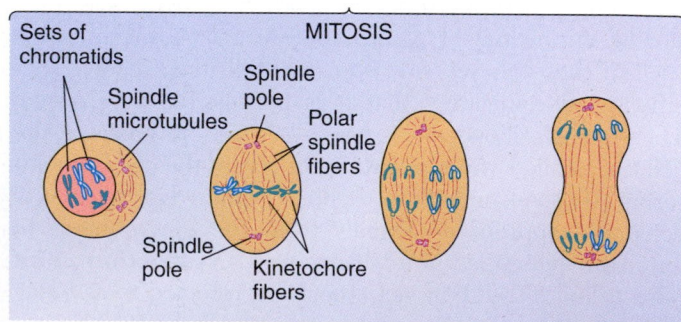

Sets of chromatids

MITOSIS

Spindle microtubules

Spindle pole

Polar spindle fibers

Spindle pole

Kinetochore fibers

(a)

Figure 3–25

(a) During the interphase portion of the life cycle of a cell, the chromatin form of the DNA in the nucleus replicates, or copies itself.

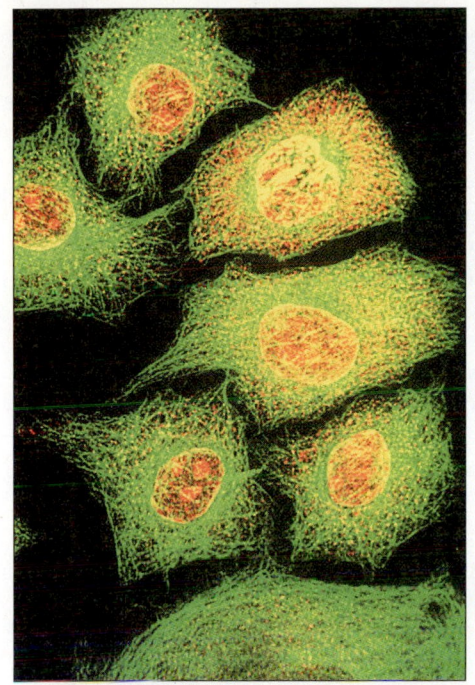

(b)

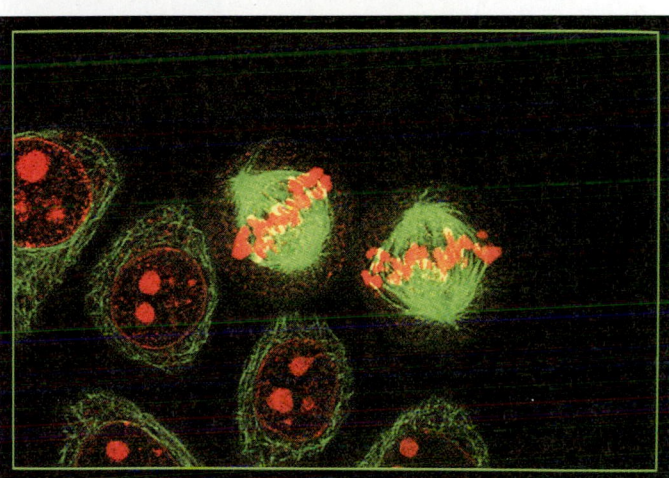

(c)

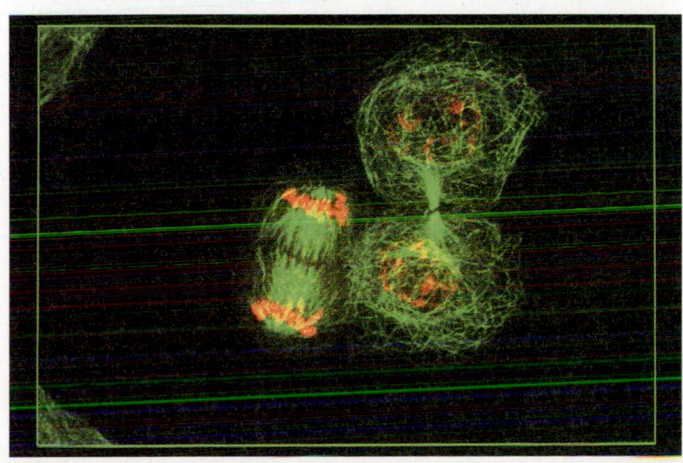

(d)

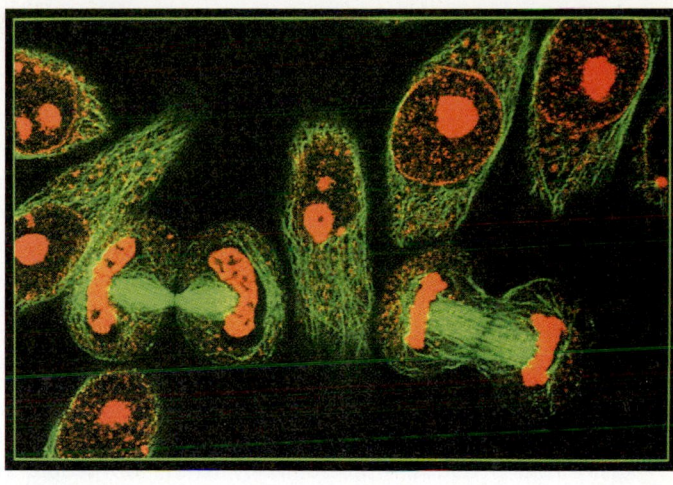

(e)

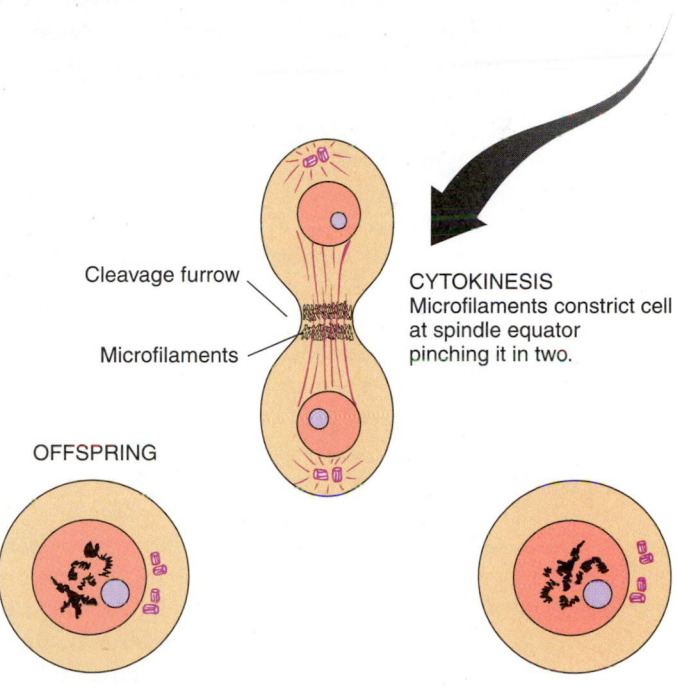

Cleavage furrow

Microfilaments

CYTOKINESIS
Microfilaments constrict cell at spindle equator pinching it in two.

OFFSPRING

(f)

Figure 3–25 (*cont.*)

(b) At the start of mitosis, the chromatin forms the tightly coiled bundles known as *chromosomes,* with the pairs of identical chromosomes linked together to form chromatids. (× 450) **(c–e)** The chromatids separate after lining up at the spindle equator, such that one copy of each chromosome becomes part of one of the nuclei of the two new cells. Thus, each new cell has the same genetic information contained in the parent cell. (× 315) **(f)** Cytokinesis, the physical separation into two cells, begins during telophase.   (b, © M. Abbey/Photo Researchers, Inc.; c–e, © John D. Cunningham/Visuals Unlimited)

CURRENT CONCEPTS IN PHYSIOLOGY

Osteoporosis

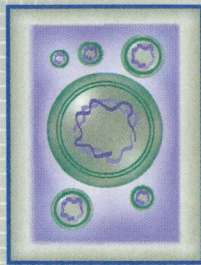

Osteoporosis is a disease that causes bone to become weak and porous. It is characterized by decreased bone mass and deterioration of the structure of bone, resulting in fragile bones that are more susceptible to fractures, especially in the hip, wrists, and spine. Osteoporosis is a serious threat to public health and affects people of all ethnic backgrounds. Approximately 10 million people in the United States have been diagnosed with the disease, and up to 20 million others are at increased risk for this condition due to dangerously low bone mass. Although this condition can affect both young and old people, it is estimated that half of the women and one-eighth of the men over age 50 will experience an osteoporosis-related bone fracture in their lifetime.

Although bone may appear to be a hard, lifeless structure, it is in fact a living, dynamic tissue. In addition to providing structural support for muscles, our bones perform many important functions, including protecting vital organs and storing the calcium essential for bone density and strength. Specialized cells in bone are responsible for maintaining the structure and integrity of this important tissue. One of the current goals of research in bone physiology is to understand how the activities of bone cells are regulated to help maintain bone strength throughout life.

Because bone is a tissue that must undergo constant remodeling to stay healthy, research is focused on understanding how the cells that erode bone, **osteoclasts**, and the cells that deposit new bone, **osteoblasts**, function to maintain strong bone. It is clear that maintaining bone strength requires a well-balanced diet that includes appropriate amounts of numerous minerals, including calcium, phosphorus, and magnesium, and other nutrients, such as protein. Regular exercise is also important in maintaining bone strength. Consequently, researchers are interested in understanding how the mechanical stimulation of bone cells during exercise stimulates bone remodeling. The term **mechanotransduction** has been used to describe the ability of cells, such as osteoblasts, to respond to mechanical stimulation with a change in their biochemical activity. Current research in mechanotransduction suggests that osteoblasts use cell adhesion molecules and the cytoskeleton to sense and respond to mechanical signals by increasing their anabolic activity — they deposit more new bone.

Until about age 30, most people have little trouble remodeling and maintaining healthy bones. As we age, our bones begin to deteriorate faster than new bone can be formed. In women, the rate of bone loss increases significantly after menopause, when the ovaries stop producing an important hormone, **estrogen**, which protects against bone loss. Therefore, physiologists also are studying the role of estrogen in bone formation.

Regular measurement of bone mass by your doctor is the only way to tell whether you have osteoporosis. **Bone density tests** measure bone density in different sites throughout the body. A bone density test can detect osteoporosis before a fracture occurs. This test can also help predict whether you might be at risk for fractures in the future and can monitor the effects of treatment if the test is conducted at intervals of one year or more.

Most scientists and physicians believe that osteoporosis is a preventable disease. Before the pain and costs of it can be reduced, a commitment to further osteoporosis research must be increased. The importance of this issue is highlighted by the World Health Organization in their declaration of the decade 2000 to 2010 as the "Bone and Joint Decade." With increased research, the future for definitive treatment and prevention of osteoporosis is very bright.

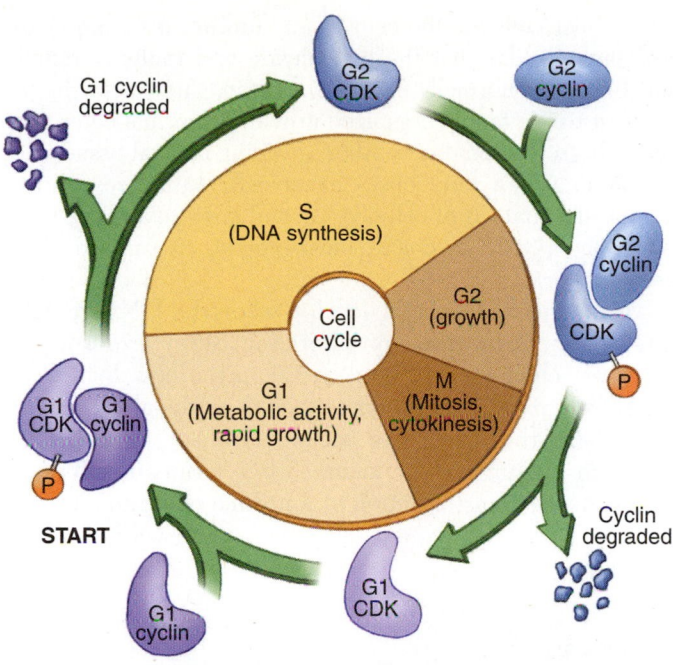

Figure 3–26

Progression through the cell cycle is regulated by cyclins and cyclin-dependent kinases (CDKs). These are evolutionarily conserved protein families that are found in all eukaryotic cells. CDKs phosphorylate various substrates to promote cell cycle progression. CDKs are inactive unless bound to cyclins, the levels of which vary through synthesis of new protein and proteolytic degradation. A cyclin–CDK complex that functions at the G_2 to M phase checkpoint was originally identified as mitosis-promoting factor (MPF). A G_1 to S cyclin–CDK complex was originally identified in yeast as START.

Telophase

During the final stage of mitosis, **telophase,** a new nuclear membrane forms around each set of chromosomes, and they begin to uncoil. **Cytokinesis,** or division of the cytoplasm into two daughter cells, also occurs during this stage. Following cytokinesis, the new cells begin interphase (see Fig. 3–24).

The Cell Cycle Is Driven by an "Engine"

 How is the cell cycle regulated?

Regulation of cell division must be coordinated carefully with growth of the cell and with replication of the genome. Important cues that tell cells when to progress through the cell cycle are received from the extracellular environment. One extracellular factor important to the cell cycle, cell adhesion

(including binding of extracellular matrix ligands to integrins and association of cadherins on adjacent cells), was mentioned earlier in this chapter. Another important factor is the presence of growth factors that bind to growth factor receptors and trigger intracellular signals that turn on the cell cycle engine.

An exciting area of research has focused on understanding how the cell cycle is regulated. One of the most important developments has been the discovery of a family of protein kinases, called **cyclin-dependent kinases (CDKs).** These proteins play a key role in the cell cycle and function as an "engine" that can be turned on and off to drive the cell through the cell cycle at appropriate times. The function of the engine is to phosphorylate substrate proteins inside the cell. The precise molecular details surrounding CDK substrates and how they regulate the cell cycle remain unclear. It is known that members of another protein family, called **cyclins,** regulate this engine. The general mechanism of cell cycle regulation by CDKs and cyclins is illustrated in Figure 3–26. The complex signals received by cells affect the kinase activity of CDKs. There are at least four molecular mechanisms through which CDK is regulated. First, CDK must associate with a cyclin to become active. Cyclins are made in the cells, associate with CDK, and then are rapidly degraded after moving into the next stage of the cell cycle. Second, CDK must be phosphorylated on a specific amino acid called threonine to become active. Third, phosphates must be removed from specific threonine and tyrosine residues on CDK in order to become active. This process is called dephosphorylation by enzymes called phosphatases. Fourth, CDK must be dissociated from any other proteins that are inhibitory to CDK, called CDK inhibitory proteins (CKIs).

The discovery of CDKs and cyclins began with the identification of an important factor that induced cells to enter mitosis from G_2. This factor became known as **maturation promoting factor (MPF).** When MPF was activated in oocytes, the nucleus responded with chromosome condensation, breakdown of the nuclear membrane, and formation of the mitotic spindle. Upon inactivation of MPF the cell exited from mitosis (via anaphase, telophase, and cytokinesis), the nucleus reassembled, and the new cells entered the next interphase. The cell cycle was driven through the mitotic phase by the periodic activation and inactivation of MPF. MPF and the factors that regulate its activity have been identified as CDKs and cyclins. One important engine component is **cyclin B,** a protein that increases in abundance during each interphase but is rapidly destroyed at the end of mitosis. In a simple model of the cell cycle engine, CDK, which is inactive during interphase, is activated by the accumulation of cyclin B. Active CDK drives the cell into mitosis and also induces degradation of cyclin B. Loss of cyclin B leads to inactivation of CDK, and the cell proceeds to interphase.

Cyclins are found in all eukaryotes, and differ widely, with different functions in the cell cycle. For example, cyclin A is important in the control of DNA replication during S phase, and cyclins C, D, E, and F are important in moving the cell cycle from G_1 into S phase.

Cancer Cells Have Abnormal Cell Cycles

The rate of cell division in normal tissues is controlled so that the cells divide only when new cells are needed. The removal of a part of the liver, for example, stimulates rapid division and growth of the remaining cells. This new growth ceases when the normal liver mass has been restored. The normal functions of tissues and organ systems would be destroyed rapidly if cell division were uncontrolled.

Cancer cells are the cells of any **malignant neoplasm.** *Malignant* means that the cells divide and multiply rapidly and behave abnormally. *Neoplasm* means "new growth." In contrast to the highly organized growth of normal cells, cancer cells grow upon one another, invading normal tissues and threatening their functions. Cancer cells do not respond to the normal controls of cell division and growth, and the body is simply a very favorable environment to support their growth.

Cancer cells probably develop when the DNA of a cell experiences **mutation,** or alteration, after exposure to radiation, chemicals, or viruses. When the cell divides, all the cells derived from it contain copies of the altered DNA and inherit the defect. An accumulation of several DNA mutations arising independently from various insults may complete the transformation of a normal cell into a cancer cell.

CHAPTER REVIEW

Summary

- Cell functions are controlled from the nucleus, where DNA stores the information that the cell uses to synthesize proteins. Many of these proteins are enzymes that catalyze specific reactions within the cell. Others are secreted from the cell to perform a specific function elsewhere in the body.
- A plasma membrane surrounds the cell to protect it from fluctuations in the external environment. Intracellular membranes surround specific organelles so that their activities can occur independently without endangering other organelles.
- The nucleus is the site where nucleic acids are synthesized. DNA remains in the nucleus. The information stored in DNA is copied in the form of RNA, which leaves the nucleus and is used by the protein-making organelles in the cytoplasm.
- The cytoplasm consists of various organelles suspended in cytosol. Cytosol is a viscous fluid that contains many enzymes and precursors for macromolecules.
- Extracellular signals direct many cellular functions by altering gene expression. Some signals are soluble and diffuse to the cell; others, such as the extracellular matrix, are insoluble and signal through cell adhesion.

- Regulation of protein import into the nucleus controls gene expression. Proteins move into and out of the nucleus through nuclear pores. Specific import proteins regulate movement between the nucleus and cytoplasm.
- The cytoskeleton mediates cellular movements and organizes the contents of the cell in the cytoplasm.
- Adhesion molecules mediate communication between the cell and the extracellular environment. Cytoskeletal filaments interact with adhesion molecules to facilitate this communication.
- The life cycle of a cell is divided into interphase (the growth phase) and mitosis (cell division). During interphase, which occupies most of the cell cycle, the cell conducts normal activities and replicates its DNA. During mitosis, the nucleus and cytoplasm divide into two, and each daughter cell receives an exact copy of the original DNA.
- Checkpoints in the cell cycle ensure that cell division occurs at appropriate times and that the daughter cells are genetically identical to the parent.

Review Questions

Choose the Correct Answer

1. The fluid surrounding the nucleus of a cell is called the:
   a. cytosol.
   b. plasma.
   c. extracellular matrix.
   d. membrane.
   e. lipid.

2. The complex formed between DNA and nuclear proteins such as histones is a:
   a. chromosome.
   b. nucleosome.
   c. fiber.
   d. gene.
   e. nucleotide.

3. The process by which secretory vesicles fuse with the plasma membrane and expel their contents outside the cell is called:
   a. endocytosis.
   b. vesicles.
   c. exocytosis.
   d. bilayer.
   e. joining.

4. Division of the cell cytoplasm to form two daughter cells after completion of mitosis is called:
   a. cytokinesis.
   b. condensation.
   c. replication.
   d. interphase.
   e. splitting.

5. The site of protein synthesis is the:
   a. nucleus.
   b. lysosome.
   c. nucleolus.
   d. rough endoplasmic reticulum.
   e. Golgi apparatus.

6. Which of the following organelles is surrounded by a double membrane?
   a. Golgi complex
   b. Nucleus
   c. Peroxisome
   d. Smooth endoplasmic reticulum
   e. Nucleolus

7. Intermediate filaments are structural components of the:
   a. cytoskeleton.
   b. mitochondrion.
   c. ribosome.
   d. chromosome.
   e. extracellular matrix.

8. The growth phase of the cell cycle is known as:
   a. mitosis.
   b. metaphase.
   c. interphase.
   d. M phase.
   e. telophase.

9. The endoplasmic reticulum of a liver cell accounts for what percentage of the total membrane of the cell?
   a. Less than 0.5
   b. 2–5
   c. 10–20
   d. 30–40
   e. more than 50

10. Which of the following proteins is involved in transport of proteins from the cytosol to the nucleus?
    a. α-Actinin
    b. Histone
    c. Ran
    d. DNA
    e. Both b and d

11. Which of the following are involved in energy production?
    a. Citric acid cycle
    b. Krebs cycle
    c. Tricarboxylic acid cycle
    d. Mitochondria
    e. All of the above

12. Which of the following is involved in organizing actin filaments into stress fibers?
    a. rhoA
    b. Filamin
    c. Importin
    d. Endoplasmic reticulum
    e. mRNA

13. The unique structure of microvilli is due primarily to:
    a. microtubules.
    b. intermediate filaments.
    c. actin microfilaments.
    d. lamins.
    e. myosin.

14. Microfilaments and microtubules have the following in common:
    a. Polarized
    b. Involved in cell motility
    c. Use ATP for polymerization
    d. Both a and b
    e. a, b, and c

15. Which of the following is/are true statements about intermediate filaments?
    a. They are static filaments.
    b. They are composed of a single type of subunit.
    c. They are involved in vesicle transport.
    d. They have primarily a structural function.
    e. They undergo dynamic instability.

16. Which of the following are involved in epithelial cell function?
    a. Cadherins
    b. Desmosomes
    c. Actin filaments
    d. Catenins
    e. All of the above

17. Which of the following are homotypic cell adhesion molecules that require Ca^{2+} to function?
    a. Integrins
    b. Cadherins
    c. Selectins
    d. Myosin
    e. Antibodies

18. Cell adhesion molecules are placed into families based on:
    a. common function.
    b. common shape.
    c. common amino acid sequence.
    d. common domain structure.
    e. both c and d.

19. Nitric oxide is:
    a. a paracrine signaling molecule.
    b. an anesthetic.
    c. an endocrine signaling molecule.
    d. not a signaling molecule.
    e. both a and b.

20. Points in the cell cycle that prevent improper advancement through the cell cycle are called:
    a. cyclins.
    b. kinases.
    c. checkpoints.
    d. centrioles.
    e. microtubules.

Answers to Case History Questions

1. This patient appears to be suffering from an inability to mount an effective immune response.

2. The presence of neutrophils in high numbers suggests that these cells are defective in their ability to fight infections.

3. This patient's symptoms suggest a primary immunodeficiency disease called *leukocyte adhesion deficiency type I* (LAD-I). This disease is characterized by reduced or absent expression of the leukocyte β integrin subunit called $\beta 2$. Leukocytes lacking this integrin subunit have difficulty with mobilization to inflammatory sites. Clinically, these patients have recurrent infections of soft tissues. Infections in early infancy that result in death are common with the severe form of LAD-I, characterized by CD_{18} expression of less than 0.5% normal. Patients with a more moderate form of the disease, in which CD_{18} levels are 1% to 10% of normal, have a moderate phenotype and may survive into young adulthood. The characteristic symptoms of LAD-I are persistent fever and an increase in white blood cells, often to more than 50,000 cells/mm^3, with histologically normal white blood cells called neutrophils. The diagnosis is confirmed by analyzing cells for the missing adhesion molecule. Defects in neutrophil function may cause "cold" abscesses because the warmth and tenderness are usually the result of the release of inflammatory mediators by the WBCs.

4. Usual therapy involves treatment with antibiotics to control infections and bone marrow transplantation, when possible. LAD-I also may be a candidate for gene therapy in the future by replacing the defective CD_{18} gene with a normal copy. Experimental studies suggest that in vivo administration of interferon-gamma (IFN-γ) increases CD_{18} expression in normal patients and alters leukocyte trafficking. It is hoped that modest increases in CD_{18} expression in LAD-I patients or alterations in CD_{18} independent trafficking could result in increased survival when combined with gene therapy.

Suggested Readings

Alberts, B., Bray, D., Lewis, J., Raff, M., Roberts, K., and Watson, J. D. *Molecular Biology of the Cell,* ed 3. New York, Garland, 1994.

Chien, S. *Molecular Biology in Physiology.* New York, Raven Press, 1989.

Ellis, R. E., Yuan, J., and Horvitz, H. R. "Mechanisms and functions of cell death." *Annual Review of Cell Biology,* 7:663–698, 1991.

Lodish, Berk, Zipursky, Matsudaira, Baltimore, and Darnell, J. *Molecular Cell Biology,* ed 4. New York, W. H. Freeman, 2000.

Murray, A., and Hunt, T. *The Cell Cycle.* New York, W. H. Freeman, 1993.

Rosenberg, S. A. "Adoptive immunotherapy for cancer." *Scientific American,* 262:62–69, 1990.

Ueda, N., and Shah, S. V. "Apoptosis." *Journal of Laboratory and Clinical Medicine,* 124:169–177, 1994.

Verma, I. M. "Gene therapy." *Scientific American,* 263:68–84, 1990.

Web sites

www.cellbio.com/
Cell and Molecular Biology online. An excellent cell biology resource.

vl.bwh.harvard.edu/
Harvard University's virtual cell biology library.

mindquest.net/biology/cell-biology/cell-biology.html
Provides local and distant links to resources for students studying cell structure and function.

www.ascb.org/ascb/
The American Society for Cell Biology home page.

Key Terms

actin (p. 83)
cell adhesion (p. 71)
cell cycle (p. 98)
cell signaling (p. 97)
chromatin (p. 74)
chromosome (p. 74)

cyclin (p. 103)
cyclin-dependent kinase (CDK) (p. 103)
cytokinesis (p. 98)
cytoplasm (p. 68)
cytoskeleton (p. 83)

cytosol (p. 68)
deoxyribonucleic acid (DNA) (p. 68)
endoplasmic reticulum (ER) (p. 69)
exocytosis (p. 80)

extracellular matrix (ECM) (p. 71)
gene (p. 69)
genome (p. 73)
Golgi complex (p. 69)
hormone (p. 69)

integrin (p. 95)
intermediate filament (p. 83)
lysosome (p. 81)
metabolism (p. 68)
microfilament (p. 83)

microtubule (p. 83)
mitochondrion (p. 69)
mitosis (p. 98)
nuclear pore (p. 69)
nucleolus (p. 76)

nucleus (p. 68)
nuclear import (p. 75)
organelle (p. 68)
phagocytosis (p. 81)

plasma membrane (p. 68)
prokaryotic cell (p. 68)
ribonucleic acid (RNA) (p. 69)
ribosome (p. 69)

Answers to Review Questions

1. a   **2.** b   **3.** c   **4.** a   **5.** d   **6.** b   **7.** a   **8.** c
9. e   **10.** c   **11.** e   **12.** a   **13.** c   **14.** d   **15.** d
16. e   **17.** b   **18.** e   **19.** a   **20.** c

Chapter 4

TRANSPORT THROUGH THE CELL MEMBRANE

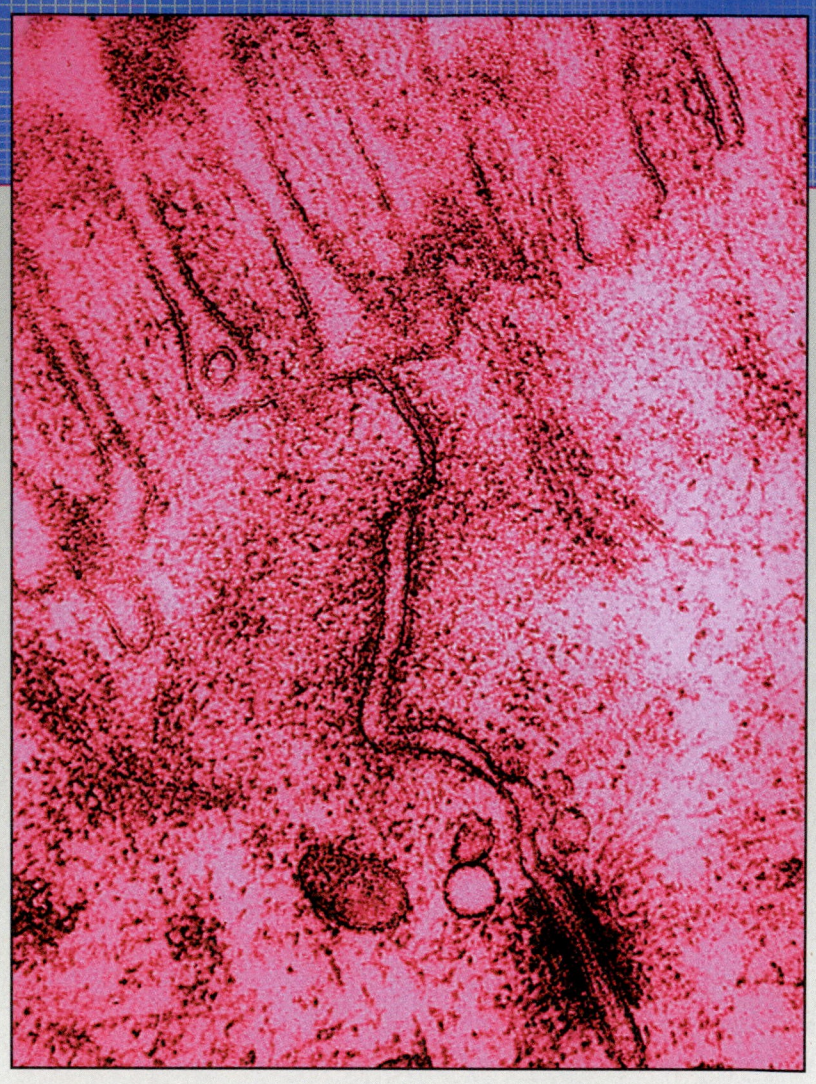

KEY CONCEPTS

- *The plasma membrane is a selective barrier that separates the cell cytosol from the external environment. It is also important for communication between cells because it contains binding sites for chemical signals.*

- *All cell membranes are composed mainly of phospholipids and proteins. The proteins and lipids can move rapidly within the plane of the bilayer, but movement from one half of the bilayer to the other half is very slow.*

- *Gradients are responsible for the diffusion of ions and molecules in solution.*

- *Many substances cross the plasma membrane by moving spontaneously down a gradient (passive transport). Other substances that cross the plasma membrane must move against a gradient (active transport). The passage of ions and large molecules is assisted by membrane transport proteins.*

- *Epithelial cells are organized structurally and functionally for transcellular transport of solutes and water.*

CASE HISTORY

During a routine visit to his doctor, a 19-year-old man complains that he frequently feels tired and, while exercising, he experiences a loss of stamina, shortness of breath, and a rapid heart rate. A physical examination reveals pale skin and a markedly enlarged spleen. Subsequent tests show that the number of red blood cells in his blood is below the normal range, a finding that is indicative of anemia. The normal range for a male is $4.6–6.2 \times 10^6$ cells/μL. One function of red blood cells is to carry oxygen from the lungs to the tissues to support normal metabolism. Upon review of his medical records over the previous 8 years, a history of severe anemic episodes is revealed, usually following systemic infections and fever. He also has experienced occasional attacks of jaundice. When a sample of his blood is viewed under the microscope a large number of spherical red blood cells, known as spherocytes, are observed. These spherocytes are somewhat smaller than the normal biconcave red blood cells, which are called erythrocytes. The osmotic fragility of the patient's red blood cells is much greater compared to erythrocytes from normal individuals. This refers to the finding that more of the patient's red cells burst (also called hemolysis) when they are suspended in dilute NaCl solution. The increased fragility is also observed when undiluted whole blood from the patient is included under sterile conditions at 37°C for 48 hrs, a test that measures spontaneous hemolysis. Under these conditions about 25% to 30% of the patient's red blood cells burst compared to fewer than 4% in a sample of blood from a healthy young male. When glucose was added to the blood at the start of the incubation period, the spontaneous hemolysis was largely prevented. The patient's spleen was surgically removed. The anemia was corrected by the surgery but the spherocytes remained abundant in his blood, indicating that the red blood cell defect persisted.

Questions

1. Why is the osmotic fragility of the patient's red blood cells increased compared with normal erythrocytes?

2. Why is spontaneous hemolysis increased in the patient's blood; why is this prevented by the addition of glucose?

3. Why is the patient anemic; why did he experience more severe anemia after illnesses involving fever?

4. Why was the anemia corrected by removing the patient's spleen?

5. What medical problem is suggested by these findings?

6. What is the defect at the molecular level that leads to spherical red blood cells?

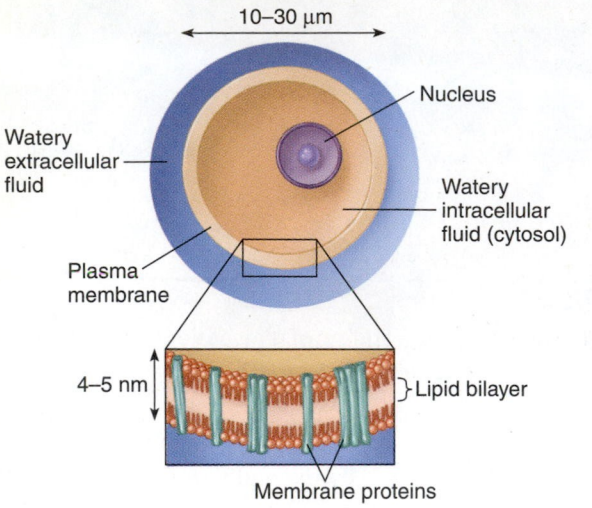

Figure 4–1

A simplified animal cell and its plasma membrane, which separates the intracellular fluid (the cytosol) from the extracellular fluid.

INTRODUCTION

In this chapter, we consider first the structure and organization of the main components of plasma membranes. Later sections discuss the mechanisms by which water, small ions, and solutes can move across membranes and epithelial cells. This chapter describes the elements of the membrane transport model, which was discussed in Chapter 1. Although the focus is primarily on the membrane that surrounds each cell, keep in mind that the membranes surrounding the various cell organelles are similar to the plasma membrane in composition and behavior.

OVERVIEW OF PLASMA MEMBRANE FUNCTIONS

Every cell is surrounded by a plasma membrane, which separates the contents of the cell from the environment (Fig. 4–1). The plasma membrane is a highly selective filter

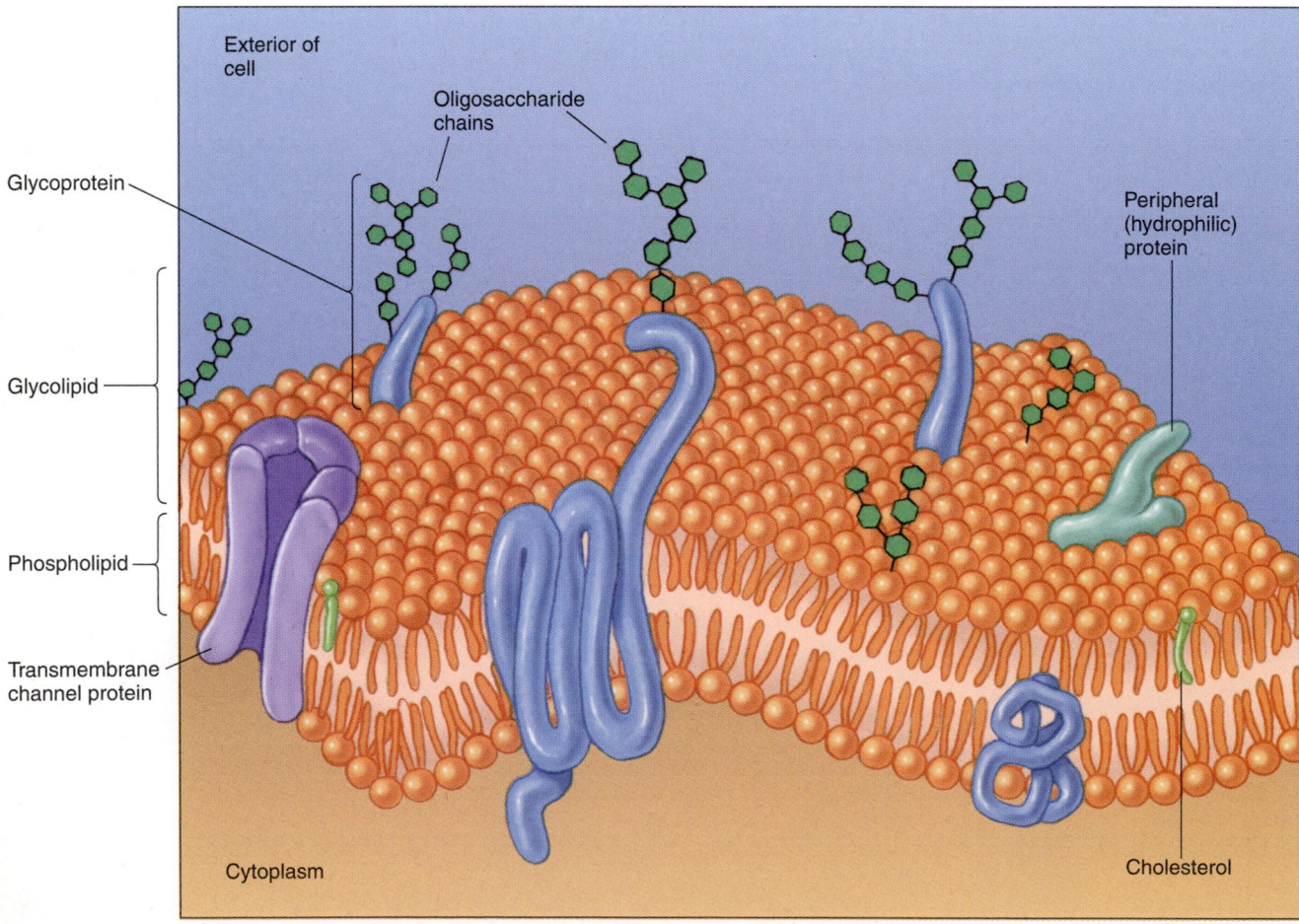

Figure 4–2

The current fluid mosaic model of the structure of plasma membranes.

that permits nutrients to enter, and waste products to leave, the cell. This selective nature allows the composition of the cell cytosol to be very different from its surrounding environment. For example, the concentration of potassium ions inside most cells is about 25- to 30-fold higher than the concentration of potassium outside, in the extracellular fluid. Other important differences must be maintained between the two fluid compartments. In addition to this function as a selective barrier, the plasma membrane also has many sites on its external surface that bind specific chemical signals produced by other cells. Thus, the plasma membrane has an important role in cell–cell communications.

THE COMPONENTS AND STRUCTURE OF THE MEMBRANE

> *What kinds of molecules are found in cell membranes; how do these determine the characteristic of the membrane?*

The following observations about the behavior of the major membrane components have been integrated into a generalized model of the structure of cell plasma membranes (Fig. 4–2). It is often referred to as the **fluid mosaic model** or the **Singer-Nicolson model,** Singer and Nicolson being the scientists who first put forward the concept. Although Figure 4–2 is designed to represent the structure of plasma membranes, the structure of intracellular membranes is very similar.

The plasma membrane consists primarily of lipids and proteins. The core structure of the plasma membrane is a double layer of phospholipid molecules known as the **lipid bilayer** (see Fig. 4–1). Lipids can account for as much as 50% of the mass of a plasma membrane.

Many of the functional properties of the membrane, however, are due to the presence of specific proteins that are closely associated with, and often embedded in, the lipid bilayer. Typically, proteins also account for about 50% of the mass of cell membranes, although this figure can vary from 18% in the membrane of the Schwann cell (a supporting cell in the nervous system) to 76% in the inner membrane of a mitochondrion (Fig. 4–3).

The relative quantities of lipid and protein in a plasma membrane usually reflect some aspect of the function of the cell or organelle that the membrane protects. Thus, the high lipid content of the Schwann cell membrane suits the role of this membrane as an insulator around certain nerve fibers. Conversely, the high protein content of the inner membrane of a mitochondrion reflects the rich complement of enzymes and other proteins vital for the processes of energy transformation that occur in these organelles.

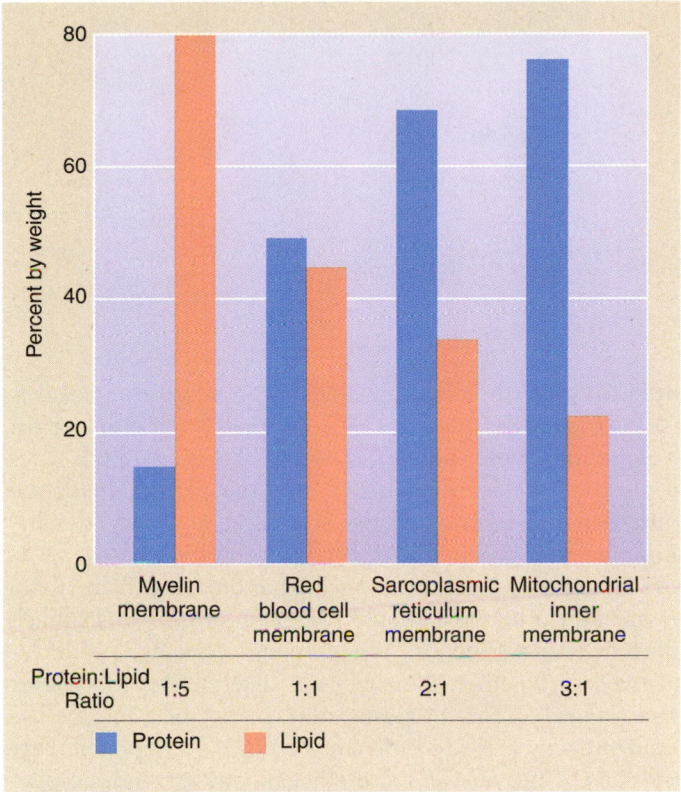

Figure 4–3

The proportions of membrane protein and lipid vary as the function of the membrane changes from primarily an insulator (myelin) to a complex variety of enzymatic and transport functions (mitochondrial inner membrane).

Membrane Lipids Are Amphipathic Molecules

> *How do lipids contribute to the properties of cell membranes?*

The specific types of lipids found in membranes are **phospholipids, cholesterol,** and **glycolipids.** All of these molecules are insoluble in water. However, they all have hydrophilic, or water-soluble, regions and hence are called **amphipathic** molecules (*amphi* = "both").

Look at the structure of **phosphatidylcholine** (see Fig. 2–11c) as an example of a typical amphipathic phospholipid. Phosphatidylcholine has a hydrophilic head group and two long hydrophobic fatty acid tails, each of which may contain 14 to 24 carbon atoms in an unbranched chain. (The 16- to 18-carbon fatty acids are the most common fatty acid chains found in phospholipids; typical examples are **palmitic acid** and **oleic acid.**)

Recall from Chapter 2 that nonpolar molecules and groups, when placed in aqueous solution, cluster together in hydrophobic pockets. Because they consist mainly of hydrophobic fatty acid chains, phospholipids form clusters in

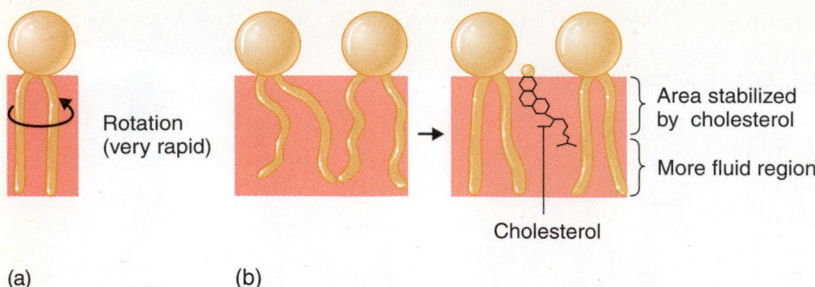

Rotation (very rapid)

Area stabilized by cholesterol

More fluid region

Cholesterol

(a)   (b)

Figure 4–4

Phospholipid molecules rotate on their axes **(a)** and undergo flexing motions. **(b)** The rigid steroid ring structures of cholesterol stabilize certain portions of the bilayer.

order to "hide" their hydrophobic tails from water molecules. They form either spheres, called **spherical micelles,** or **bimolecular sheets** (**bilayers;** see Fig. 2–11*d*). In both types of clusters, the hydrophobic tails of the phospholipids are buried in the interior of the structure, and only the hydrophilic head groups are exposed to the water.

The micelles are rather small structures, less than 20 nm in diameter, whereas a bimolecular sheet or bilayer is a much more extensive structure. When enough phospholipid molecules occur together, they are more likely to form a bilayer than a micelle. This is because the interior of a micelle has room for only a limited number of bulky hydrophobic fatty acid tails. A bilayer is a more efficient way for the phospholipids to protect their hydrophobic portions because it tends to close in on itself (forming a circle) to eliminate the free edges where the hydrophobic tails would be exposed to water. For the same reason, it reseals itself if punctured. Thus, a bilayer is the favored structure in an aqueous solution.

Movement of Phospholipids Within the Bilayer

Studies on the movement of individual phospholipid molecules within a lipid bilayer have revealed that the bilayer is a dynamic structure. Phospholipid molecules move laterally along one layer of the membrane, but crossing from one layer to the other is much more difficult.

MOVEMENT WITHIN MOLECULES: FLEXING AND ROTATION. Within each phospholipid molecule, there is a certain amount of movement. The phospholipid molecules are able to rotate about their longitudinal axes (Fig. 4–4*a*). In addition, at physiological temperatures (37°C), the hydrocarbon tails are mobile and undergo rapid flexing movements. The rate of this movement is higher if the hydrocarbon tail is short and contains double bonds.

STABILIZATION BY CHOLESTEROL. In eukaryotic cell membranes, as much as 20% of the total lipid may be cholesterol. Cholesterol stabilizes the lipid bilayer. Like the phospholipids, cholesterol is an amphipathic molecule; however, its steroid ring structure is more rigid than are the long fatty acid tails of the phospholipids. When present in the lipid bilayer, the ring structure of cholesterol can interact with and reduce the motion of the neighboring phospholipid hydrocarbon chains (Fig. 4–4*b*).

MOVEMENT WITHIN LAYERS. Phospholipid molecules exchange positions with their neighbors in the same half of the bilayer by very rapid lateral diffusion (Fig. 4–5*a*). By this process, a lipid molecule takes only 1 second to move about 2 μm, the entire length of a large bacterial cell.

MOVEMENT BETWEEN LAYERS. Lipids also move from one half of the lipid bilayer to the other in what is called **"flip-flop."** This motion requires that the polar head groups of the phospholipids in one half of the bilayer pass through the hydrophobic interior of the bilayer to reach the other half of the bilayer (Fig. 4–5*b*). This is a difficult task; hence, flip-flop movement is extremely slow.

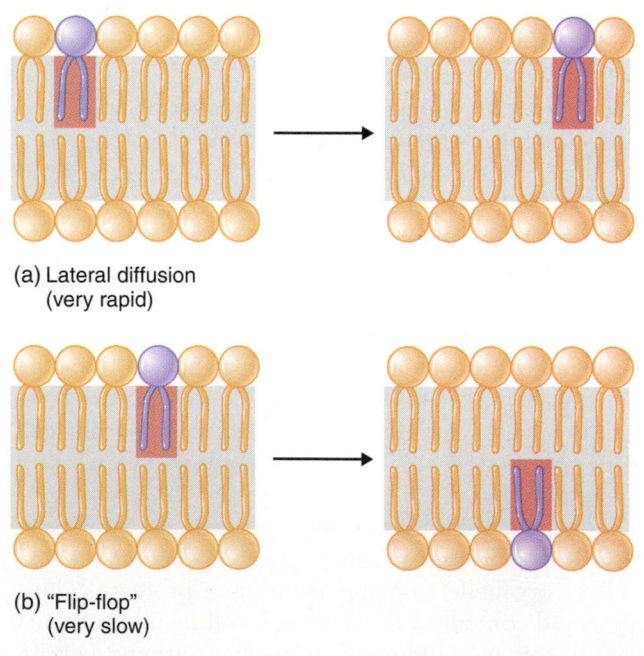

(a) Lateral diffusion (very rapid)

(b) "Flip-flop" (very slow)

Figure 4–5

(a) Phospholipid molecules diffuse laterally very rapidly within one layer of the plasma membrane. **(b)** Phospholipids move very slowly from one layer to the other.

Lipid Asymmetry in the Bilayer

Phosphatidylcholine is not the only phospholipid in biological membranes. Other important membrane phospholipids are **phosphatidylethanolamine, phosphatidylserine,** and **phosphatidylinositol** (all derived from glycerol) and **sphingomyelin** (derived from sphingosine). These different phospholipids are not distributed equally between the two halves of the lipid bilayer. A good example is the plasma membrane of the red blood cell, in which most of the phosphatidylcholine and sphingomyelin is present in the outer half of the bilayer, whereas most of the phosphatidylethanolamine and all of the phosphatidyl–serine are present in the inner half (Fig. 4–6).

The persistence of this **lipid asymmetry** in membranes is good evidence that the rate of phospholipid flip-flop is very slow. If flip-flop were rapid, the various phospholipids would probably mix randomly and thus be distributed equally between the bilayers. The phospholipid asymmetry originates during the initial assembly of the bilayer in the endoplasmic reticulum.

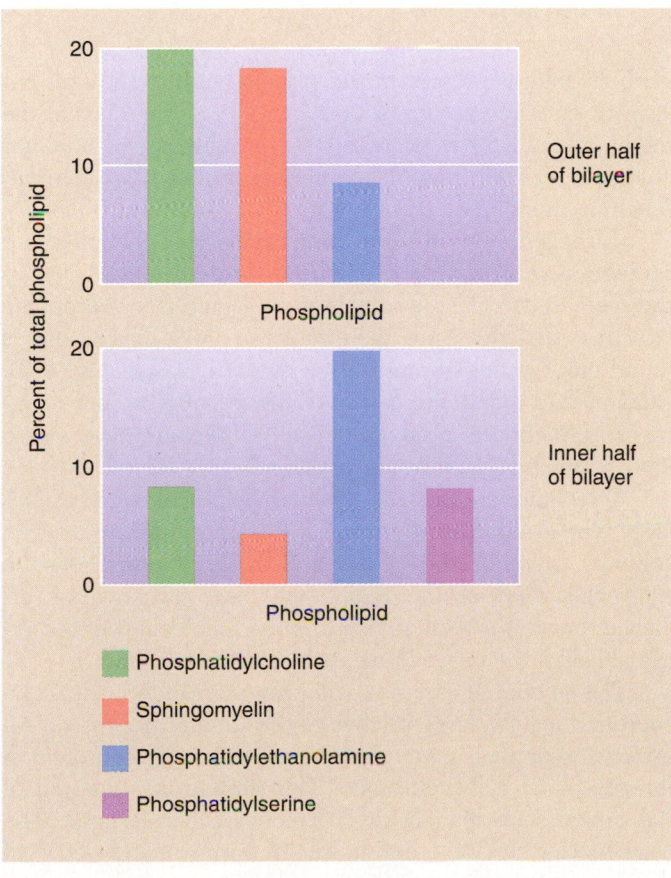

Figure 4–6

Asymmetrical distribution of lipids in the erythrocyte membrane. Note that each half of the bilayer contains 50% of the total lipids present in the membrane.

Texture of the Plasma Membrane

The net result of the movements of the phospholipids and their interactions with cholesterol is that, at 37°C, the interior of the lipid bilayer of biological membranes is more like a viscous fluid, such as olive oil, than a rigid crystalline matrix. This fluidity of the membrane structure is probably very important for the normal function of many membrane proteins.

Most Membrane Proteins Penetrate the Lipid Bilayer

 How do proteins contribute to the properties of cell membranes?

Proteins are an important component of most biological membranes. Some proteins are found only on the surface of the membranes, whereas others partially or completely penetrate the lipid bilayer.

Intrinsic Proteins: Amphipathic Molecules

The proteins that penetrate the bilayer, called **intrinsic (or integral) proteins,** account for about 70% of total membrane protein. Most of them probably span the bilayer completely, so that parts of the protein are exposed on both sides of the bilayer.

The three-dimensional structure or conformation (see Chapter 2) of much of the membrane-spanning region of an integral protein is usually an alpha-helix. Any polar (hydrophilic) amino acid side chains in this portion of the molecule are sequestered within the alpha-helix. Only the nonpolar (hydrophobic) side chains are exposed to the hydrophobic interior of the bilayer. In contrast, in regions of the protein that project outside the lipid bilayer, the nonpolar side chains are buried in the interior of the protein conformation, whereas the polar and ionic side chains are exposed to the hydrophilic environment (Fig. 4–7).

Many of the intrinsic membrane proteins display the same dynamic properties as do the membrane lipids, such as rapid lateral diffusion and rotation about an axis perpendicular to the plane of the membrane. The fluidity of the lipid bilayer makes these movements possible. However, flip-flop of intrinsic membrane proteins is even more difficult than it is for phospholipids; in fact, it probably does not occur.

The lateral mobility of these membrane proteins sometimes can be prevented by interactions with peripheral components of the membrane (e.g., other proteins) or with components of the cell cytoskeleton (e.g., microtubules and microfilaments).

Extrinsic Proteins: Hydrophilic Molecules

Membrane proteins that do not penetrate the lipid bilayer are termed **extrinsic (or peripheral) proteins** (see Fig. 4–7). They usually are not amphipathic molecules. They

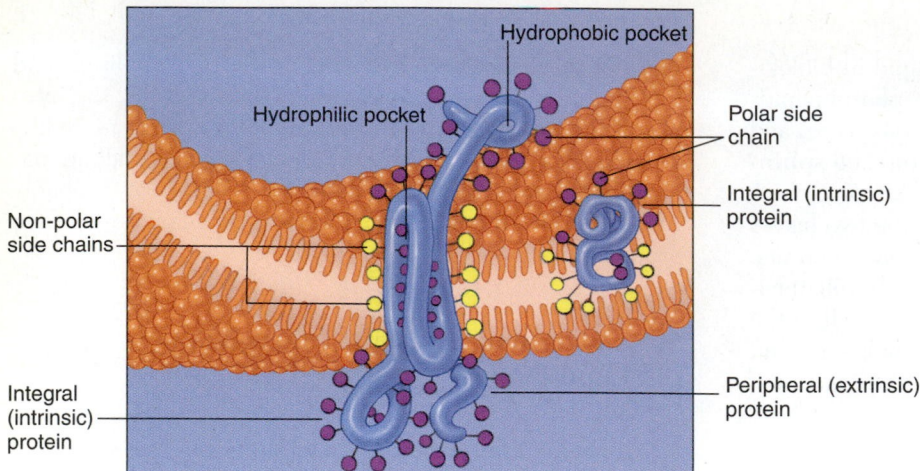

Figure 4–7

The arrangement of integral and peripheral proteins in a biological membrane.

occur on either the external or the cytoplasmic surface of the membrane and are attached primarily through interactions (electrostatic and hydrogen-bonding) with the polar portions of intrinsic membrane proteins and with the polar head groups of the phospholipids. These are relatively weak interactions.

Protein Asymmetry in the Membrane

Both extrinsic and intrinsic membrane proteins, like the phospholipids, often are distributed unequally between the two halves of the bilayer. That is, a specific protein may be associated with only one side of the bilayer. Specific advantages of protein asymmetry are better understood than are those of phospholipid asymmetry.

This asymmetry is related to cell function. In the membrane of the red blood cell, for example, the extrinsic protein called **spectrin** is found only on the cytoplasmic (inner) side of the membrane. It is a part of a filamentous network inside the cell that may be important in maintaining the biconcave shape while being flexible enough to allow the cell to change shape as necessary during its passage through narrow capillaries.

In contrast, many receptors for specific chemical signals, such as hormones, are located only on the external surface of the cell membrane and often represent a part of an intrinsic membrane protein. If such a protein spans the lipid bilayer, a potential mechanism exists for converting an extracellular signal into an intracellular response. For example, the binding of a hormone to its receptor could trigger a conformational change in the protein, especially if the protein is composed of several separate subunits. This conformational change could open a hydrophilic channel between the protein subunits and allow a specific ion to cross the plasma membrane.

Carbohydrate Is on the Surface of the Membrane

> *How do carbohydrates contribute to the function of the cell membrane?*

Carbohydrate is present in the plasma membrane of all eukaryotic cells and accounts for 2% to 10% of the mass of the membrane. In terms of quantity, carbohydrate is a minor component of the plasma membrane, but it has important functions in membrane physiology.

Carbohydrates, like the membrane phospholipids and proteins, are distributed unequally between the two layers of the membrane. The carbohydrate molecule is a hydrophilic structure that does not penetrate the lipid bilayer but is found only on the surface of the membrane (see Fig. 4–2). Most of the membrane carbohydrate is bound either to intrinsic membrane proteins, forming glycoproteins, or to lipids in the bilayer, forming glycolipids.

Glycolipids are present only in the external half of the bilayer. The glycoproteins are more abundant than are the glycolipids. Most of the proteins that are exposed on the external surface of the membrane have carbohydrate attached, but fewer than 10% of the lipid molecules in the external half of the bilayer have carbohydrate attached.

The amount of carbohydrate present in an individual glycoprotein may be large relative to the amount of protein. An extreme example is **glycophorin,** an intrinsic glycoprotein in the red blood cell membrane. This protein is composed of 131 amino acids and about 100 monosaccharide units. The monosaccharides together make up approximately 60% of the mass of the glycophorin molecule.

The location of membrane carbohydrates on the surface of the plasma membrane is probably an important factor in cell–cell recognition processes. These carbohydrates may also serve as specific membrane receptors for extracellular messengers, such as hormones. Unlike amino acids, which form only linear chains, monosaccharide units can be assembled in branching chain structures. This property provides almost limitless possibilities for construction of membrane receptor sites to which only one specific extracellular molecule can bind.

GRADIENTS AND MOVEMENT

What is a gradient; what role do gradients play in the movement of different substances?

A **gradient** is a difference between two quantities (measured in units of pressure or concentration) divided by the physical distance (measured in units of length) between them. Gradients are responsible for the movement of various substances throughout the body, including the movement of blood through the circulatory system, air into lungs, and ions and water through membranes of cells. All of these gradients represent energy gradients, but they take a variety of forms and manifest themselves as pressure gradients, electrical gradients, or concentration gradients.

A Concentration Gradient Drives Diffusion

How does diffusion occur?

Diffusion is a process that involves the movement of particles, such as ions and molecules, in solution. An understanding of these processes in general is essential for an understanding of the movement of particles into and out of cells.

Diffusion of Solutes in Liquid Solvents

In solutions as well as elsewhere in nature, the inherent kinetic energy of molecules, ions, and other particles causes them to move randomly from one location to another. The kinetic energy associated with a concentration gradient determines the magnitude and direction of the net movement of particles. A **concentration gradient** exists when two adjacent regions have different concentrations of particles. The larger the difference in concentrations, the greater the net movement between them. Particles tend to move from regions of higher concentration to regions of lower concentration, that is, "down" the concentration gradient. If there is no concentration gradient (a situation that occurs when particles have dispersed evenly throughout the solution), then the net movement of particles is zero, even though individual particles may move back and forth between the two regions. Such a solution is said to be in **dynamic equilibrium.**

The movement of sugar molecules in water illustrates the effects of a concentration gradient (Fig. 4–8). In Figure 4–8a, the sugar concentration in the lower region of the beaker is higher than elsewhere in the beaker. Although the sugar molecules are moving constantly, and some are moving toward the bottom of the beaker, most move up and away from the region of concentration until they are evenly dispersed throughout the water (Fig. 4–8d). Even at this point, individual particles still are moving in all directions, but there is no net movement and no net change in concentration. The concentration gradient that existed when the particles were

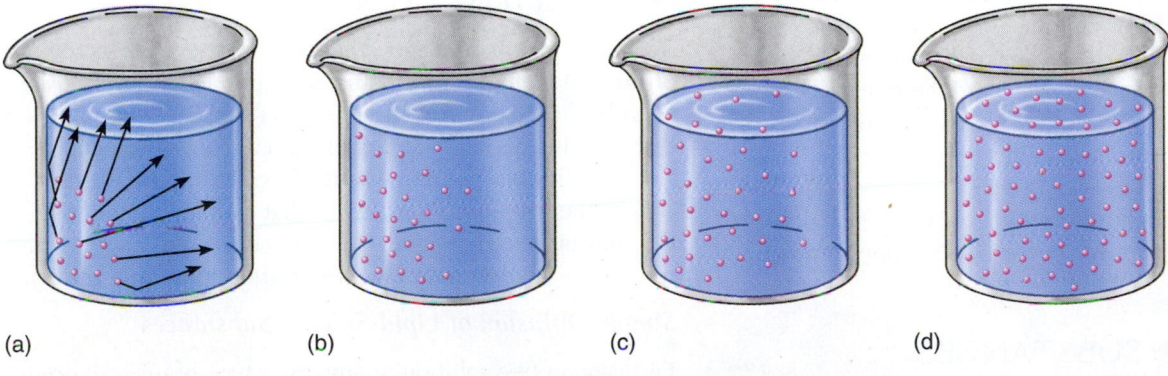

(a)　　　　(b)　　　　(c)　　　　(d)

Figure 4–8

Diffusion occurs when one substance moves randomly throughout another and becomes evenly mixed with it. Diffusion occurs only when a concentration gradient exists, such that the substance to be mixed is more concentrated in one region than another. Here, sugar molecules move from a region of high concentration to regions of low concentration in a beaker of water. When they are fully dispersed, a gradient no longer exists and diffusion ceases. Although the individual sugar particles are still moving in all directions, no net movement occurs in any one direction because no gradient exists.

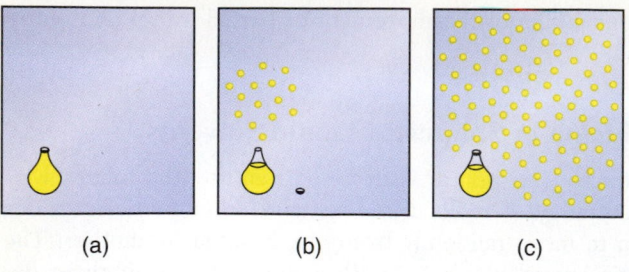

Figure 4–9

A model for the diffusion of gas molecules down a concentration gradient. *(a)* A capped bottle of perfume is placed in a closed space. *(b)* When the bottle is opened some of the liquid evaporates, and the gaseous perfume molecules pass into the surrounding space. A concentration gradient then exists because the molecules are more concentrated near the bottle than in the rest of the room. *(c)* When the bottle is recapped, the perfume molecules slowly diffuse throughout the room until they are evenly distributed, and the concentration gradient no longer exists.

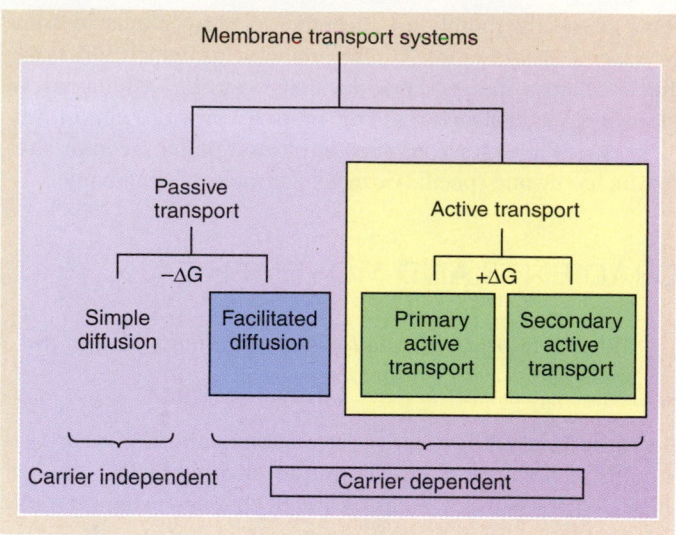

Figure 4–10

An overview of membrane transport systems.

more concentrated in one region has been abolished by diffusion, and there is no driving force for net movement. **Flux** is a general term used to describe the rate of movement of solute molecules. It is expressed in terms of amount of solute moving per unit of time (e.g., mmol/min).

Diffusion of Gases in Gaseous Solutions

Gas molecules diffuse in the same way as do solid or liquid solute particles. Imagine a room in which there is no movement of air molecules (Fig. 4–9). If a bottle of perfume is opened, some of the liquid perfume evaporates. The gas molecules come out of the bottle into the room. Slowly, the perfume molecules spread and eventually reach an equilibrium in which the distribution of the scent is even throughout the room. The perfume spreads by diffusion because of the concentration gradient that exists because initially the molecules are more concentrated near the bottle than in the rest of the room. The net movement of perfume molecules is from this region of high concentration in all directions to regions of low concentration.

MOVEMENT OF SUBSTANCES THROUGH THE MEMBRANE

> *How do substances move across cell membranes? What are the characteristics of our membrane transport general model?*

Two types of movement occur as substances pass through the plasma membrane of a cell: (1) **passive**, or **spontaneous**, processes, which occur in response to gradients and produce a *less* concentrated state overall and (2) **active**, or **energy-**

requiring, processes, which occur in opposition to gradients and produce a *more* concentrated state where a high degree of concentration already existed. **Passive transport** processes include both **simple diffusion** and **facilitated diffusion**. **Active transport** processes include **primary** and **secondary active transport** (Fig. 4–10). Note that the same substance — sodium ions, for example — can move one way through the membrane by a passive process and the other way by an active process.

Passive Transport Is Spontaneous Movement Down a Gradient

Spontaneous, random movements of substances into or out of the cell include simple diffusion and facilitated diffusion. **Simple diffusion** takes place when a substance, usually one that can dissolve in lipids, passes directly through the lipid interior of the plasma membrane. **Facilitated diffusion** occurs in the presence of a protein that enables the passage of the substance. However, both processes are passive and spontaneous movements that occur in response to a gradient.

Simple Diffusion of Lipid-Soluble Substances

Diffusion in free solution or any other type of mixture occurs because of the random movement that all particles exhibit. This motion is an expression of the kinetic energy of the molecules. As a result of diffusion, a substance, such as a solute, tends to become distributed in an equal concentration throughout the entire space available.

A SELECTIVELY PERMEABLE MEMBRANE. The previous section discussed the diffusion of solute particles in free solution, from a region of high solute concentration to an adjacent region of low concentration. There was no physical

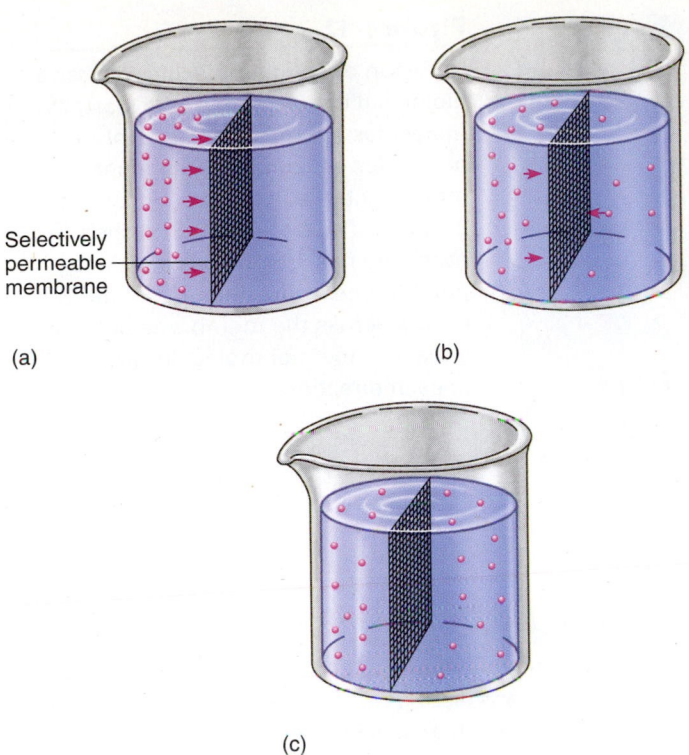

Selectively permeable membrane

(a) (b)

(c)

Figure 4-11

(a–c) Diffusion can occur in the presence of a selectively permeable membrane. As long as solute particles can pass through the membrane, they will diffuse from a region of high concentration to one of low concentration.

division between these two regions; they were more or less continuous.

Diffusion also can occur in the presence of a **selectively permeable membrane.** If a solute is highly concentrated on one side of a membrane but able to pass through the membrane, it will do so, diffusing into the adjoining region of low concentration by the same process that would occur if the membrane were not there (Fig. 4–11). If another solute were present to which the membrane were not permeable, that solute would remain highly concentrated on one side of the membrane.

The plasma membrane, as a mixture of lipids and proteins, represents a barrier to the diffusion of some solutes into or out of the cell. An inherent, important property of any lipid bilayer is that it is nearly impermeable to ions and most polar molecules. In contrast, a lipid bilayer is permeable to nonpolar molecules — that is, molecules that can dissolve in its nonpolar interior. The lipid bilayer of the plasma membrane is selectively permeable; it readily permits some substances to cross it but prevents or retards the free passage of others.

Consider the diffusion of glycerol, a polar molecule, from the plasma to the interior of a red blood cell. The **diffusion coefficient** *D* is a measure of the mobility of a specific solute molecule within a specific solute. *D* for glycerol in water is 1.0×10^{-5} cm²/sec. This is several orders of magnitude larger than the value of *D* for glycerol inside the lipid phase of the plasma membrane: 1.7×10^{-10} cm²/sec. This large difference reflects the fact that the mobility of glycerol is greatly diminished once it enters the membrane.

THE PERMEABILITY COEFFICIENT. The rate of diffusion of solute through a cell membrane is governed by the thickness of the membrane and by the lipid solubility of the solute. The combined effect of these two factors and the diffusion coefficient can be expressed as the **permeability coefficient *P*.** The value of *P* varies with the specific solute and the specific cell membrane.

Despite the barrier represented by the cell membrane, an uncharged solute still tends to move across it if the solute concentration inside the cell is different from the solute concentration outside — that is, if there is a concentration gradient. In simple diffusion, the solute always moves down the concentration gradient, from the region of high solute concentration to the region of low concentration, just as in free solution. The flux (*J*) of an uncharged solute across a membrane can be represented as follows:

$$J = PA[C_1 - C_2]$$

where C_1 and C_2 are the solute concentrations on either side of the membrane, and *A* is the area of membrane through which diffusion can occur. Thus, when C_1 is greater than C_2, there is net movement of solute from side 1 to side 2 (Fig. 4–12). As the solute concentration builds up on side 2, the rate of movement back to side 1 increases. Eventually, the solute concentration gradient is abolished, and these opposing fluxes become equal. At this point, the system is in a dynamic equilibrium — that is, individual solute particles continue to move from one side of the membrane to the other, but the net movement of solute ceases. This can also be seen from the equation: When $C_1 = C_2$, the value of *J* is zero.

THE ELECTROCHEMICAL GRADIENT. When the solute is electrically charged, as are Na⁺ and Cl⁻ ions, for example, the flux of solute is influenced not only by a concentration gradient but also by an electrical gradient. A typical animal cell has a small potential difference across the plasma membrane of about −70 millivolts (mV). By convention, this refers to the change in electrical potential at the cytosolic side relative to the extracellular side. The potential difference tends to cause positively charged solutes to move across the plasma membrane and into the cell while it opposes the entry of negatively charged solutes. This electrical gradient represents an additional factor in diffusion that must be considered when the solute bears a charge. The total gradient is now the sum of the chemical (concentration) and electrical components and is termed the **electrochemical gradient.**

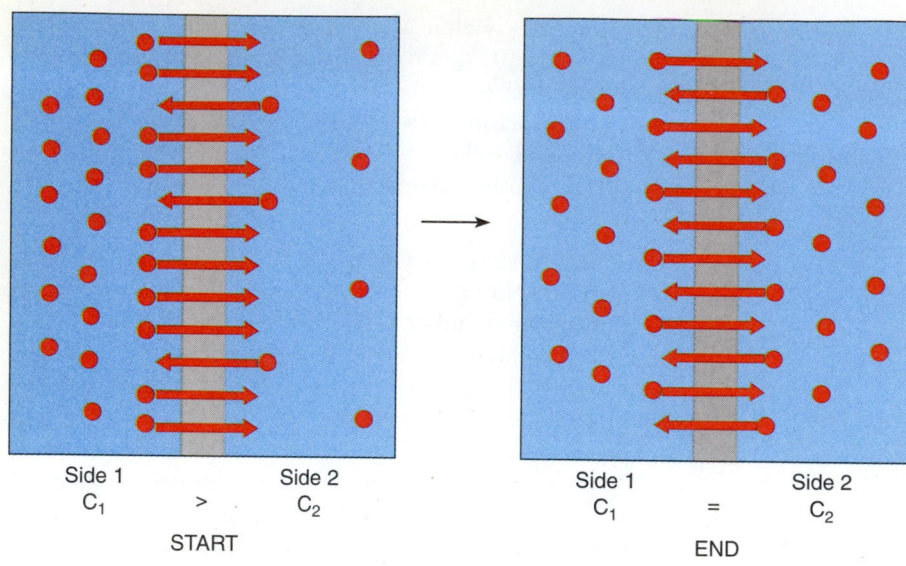

Side 1   Side 2
C_1 > C_2

START

Side 1   Side 2
C_1 = C_2

END

Figure 4–12

Diffusion of solute molecules across a biological membrane. At the start, 25 molecules are on side 1 and only 7 molecules are on side 2, so a net movement of solute from side 1 to side 2 occurs. The end point is reached when each side has 16 molecules. At this point, no net movement of solute occurs across the membrane because equal numbers of molecules are moving in each direction.

Considered separately, the chemical and electrical gradients can favor solute movement in the same direction. For example, the passive entry of Na^+ into cells is favored by both the concentration gradient (low Na^+ concentration inside the cell) and the electrical gradient (negative potential inside the cell).

On the other hand, electrical and chemical gradients can oppose each other. For example, the passive leak of K^+ out of cells goes against, or "up," the electrical gradient (negative potential inside the cell). However, the concentration gradient is so large — because of the high K^+ concentration inside the cell — that the electrochemical gradient force favors the exit of K^+ from the cell.

MOLECULAR SIZE. The principal barrier to solutes that cross cell membranes by simple diffusion is the lipid bilayer. The rate at which a solute diffuses through the membrane depends in large part on two factors: (1) its ability to dissolve in lipids and (2) its molecular size. In general, a nonpolar, and therefore lipid-soluble, solute diffuses across a cell membrane much more rapidly than does a polar (water-soluble) solute of the same size. The effect of the size of the solute molecule is particularly important for polar solutes. Membrane permeability for these molecules, already limited by their polarity, decreases further as their molecular size increases.

These relationships are summarized as follows: (1) the permeability of cell membranes to small nonpolar solutes is directly proportional to the lipid solubility of the solute and (2) the permeability of cell membranes to polar solutes is inversely proportional to the molecular size of the solute. Thus, a small, highly lipid-soluble molecule has the best likelihood of rapidly penetrating cell membranes by simple diffusion.

An exception to these general statements is water itself. Despite its small size, a water molecule is highly polar (lipid-insoluble) and yet it crosses cell membranes very rapidly.

The precise mechanism was not fully understood until recently, when it was discovered that the plasma membranes of many cell types contain specific protein channels that allow water molecules to cross the bilayer without entering the lipid phase. The diffusion of water is considered in the section on osmosis.

An electrical charge on the solute is another factor that decreases its lipid solubility. Thus, a lipid bilayer is almost impermeable to ions, even though an ion represents a solute of small size. Any ionic diffusion that occurs probably uses specific ion channels formed by proteins that span the lipid bilayer (see the section on facilitated diffusion later in this chapter).

WEAK ACIDS AND BASES. The very low permeability of cell membranes to solutes bearing an electrical charge has important consequences for the diffusion of **weak acids** and **weak bases.** These compounds, unlike strong acids and bases, do not dissociate completely when dissolved in water. Recall from Chapter 2 that an acid is a proton donor, whereas a base is a proton acceptor. We can write the following general equations to represent the behavior in solution of a weak acid (HA) and a weak base (B):

$$HA \rightleftharpoons H^+ + A^-$$
(weak acid)

$$B + H^+ \rightleftharpoons BH^+$$
(weak base)

A solution of a weak acid or weak base contains a mixture of the undissociated form and the dissociated form, the proportions of these two forms varying with the H^+ concentration of the solution. In the case of the weak base, for example, a high H^+ concentration (low pH) shifts the equilibrium to the right, resulting in more BH^+. Conversely, a low H^+ concentration (high pH) shifts the equilibrium to the left, resulting in more B and more H^+. The pH at which the concentration of the associated form is exactly equal to that of the dissociated form is defined as the *pK* and is a characteristic property of the weak acid or the weak base. Weak acids and bases in the body fluids act as buffers that maintain pH in the physiological range (7.38–7.42). The closer the *pK* of a buffer to the pH of body fluids, the more powerful the buffer system is.

DIFFUSION TRAPPING. The neutral form of a weak base diffuses across a cell membrane, whereas the charged form does not, which gives rise to the phenomenon of **diffusion trapping.** This is illustrated in Figure 4–13 using ammonia as an example of a weak base. Ammonia is produced in the cells of the kidney tubules as an end product of the metabolism of amino acids, especially glutamine. Ammonia is neutral and lipid-soluble and easily diffuses out of the cell, moving down its concentration gradient and into the fluid present in the lumen of the tubule. Typically, this fluid has a low pH; it is slightly acidic relative to the cell cytosol. When the ammonia contacts the excess H^+ in this acidic solution, it gains a proton to form the ammonium ion. The ammonium ion is positively charged and lipid-insoluble and cannot diffuse back into the tubule cell. It is effectively "trapped" in the lumen and eventually is excreted in the urine.

Protonation of available ammonia in the tubular fluid maintains a low concentration of ammonia outside the tubule cells; continued production of ammonia by cellular metabolism maintains a relatively high concentration of ammonia inside the tubule cells. The combination of these processes preserves the concentration gradient that favors the passive diffusion of ammonia out of the cells.

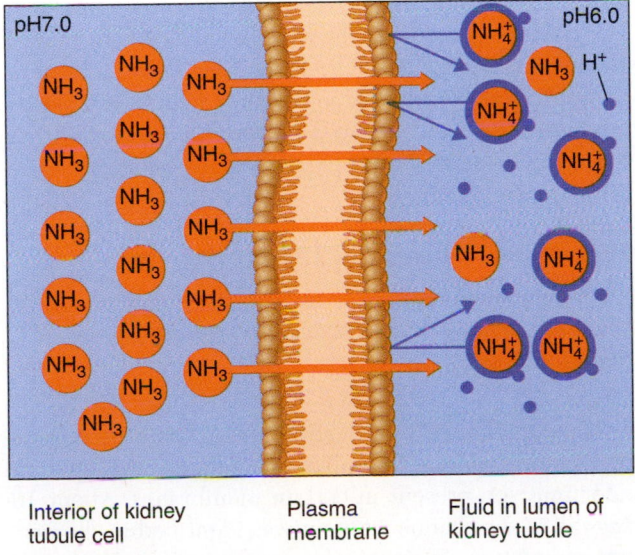

Figure 4–13

Diffusion trapping of a weak base. NH_3 can cross the plasma membrane because it is a neutral molecule. When it gains a proton in acidic solution, it becomes a charged ammonium ion (NH_4^+), which cannot cross the membrane.

Aspirin (or **acetylsalicylic acid**), one of the most widely used drugs for the control of pain and inflammation, is a weak acid. Following ingestion by mouth, it is absorbed rapidly by the stomach and the early part of the small intestine. Absorption depends on diffusion trapping. In the highly acidic fluid (pH 2–3) of the stomach, the negatively charged acetylsalicylate ion gains a proton and, in this uncharged form, becomes lipid-soluble. It easily crosses the plasma membrane of stomach epithelial cells by diffusing down its concentration gradient. Inside the cells, the pH of the cytosol is neutral (7.1). The proton dissociates, and the charged acetylsalicylate ion cannot diffuse back to the stomach lumen. Prolonged use of aspirin unfortunately interferes with the production of the mucus that lines the surface of the stomach. The role of the mucous layer is to protect the stomach from the strongly acidic contents. Loss of the mucus allows excess H^+ ions to enter and damage the epithelial cells, leading to ulceration and bleeding.

Simple Diffusion of Water: Osmosis

Osmosis is the diffusional flow of water across a membrane in response to a difference in solute concentration on either side of the membrane. Water flows from the side of the membrane with the lower concentration of solutes to the side of the membrane with the higher solute concentration. Water molecules move because of a difference in free energy states. The free energy state of water molecules is reduced by the presence of dissolved solute. Thus, the free energy of water molecules in the dilute solution (less solute) is relatively high compared with the free energy of water molecules in the concentrated solution (more solute). Water molecules move spontaneously from the dilute solution to the concentrated one because this is accompanied by a decrease in their free energy.

A practical application of osmosis is the preservation of meat by salting. When bacteria contact salt on the surface of meat, a concentration gradient exists because the salt is more concentrated outside the bacterial cell than inside. The plasma membrane of the cell is impermeable to the salt; salt cannot diffuse into the cell to abolish the gradient. However, the membrane is permeable to water and, because of the osmotic gradient that is created, water flows out of the bacterial cell toward the region of high salt concentration, where its free energy is lower. Water loss inhibits the normal functions of the bacterial cell. By a similar osmotic mechanism, the salt also removes water from the meat tissue. The combination of these two effects prevents the meat from spoiling and allows it to be preserved for long periods.

OSMOTIC CONCENTRATION. When discussing osmosis, we use a special measure of concentration in order to better understand the relationship of water and solute particles. The number of solute particles a solute releases in solution is not related to the mass of the solute. Chemical concentration, such as molality or molarity, is based essentially on weight and does not always reveal the extent to which solute particles are present in the solution. For example, in a solution of glucose in water, the glucose molecules ($C_6H_{12}O_6$) stay intact rather than dissociating into atoms. In a solution of sodium chloride in water, by contrast, the Na^+ and Cl^- ions dissociate completely, and this doubles the total number of particles that are present in the solution.

Osmotic concentration, also known as **osmolarity,** is expressed in units called **osmoles (Osm).** For physiological solutions, we use **milliosmoles (mOsm)** because the concentrations are generally low.

Just as molarity expresses moles of solute per liter of solution, osmolarity expresses osmoles of solute per liter of solution. The relationship between molarity and osmolarity can be summarized as:

$$\text{Osmoles} = n \times \text{moles}$$
$$(\text{milliosmoles} = n \times \text{millimoles})$$

where n is the number of particles derived from each molecule of solute when in solution. Note that this relationship applies only to dilute solutions that are assumed to behave ideally. Deviation from ideal behavior occurs as the concentration of the solution increases.

For example, the osmotic concentration of a glucose solution with a chemical concentration of 60 mmol/L is 60 mOsm/L because n equals 1 for glucose. Glucose exists as single, intact molecules in water, producing only one solute particle per molecule.

In contrast, the osmotic concentration of a solution of NaCl whose chemical concentration is 60 mmol/L is 120 mOsm/L because $n = 2$. Each molecule of NaCl produces two solute particles (Na^+ and Cl^-), and both become osmotically active solute particles. A $CaCl_2$ solution with a chemical concentration of 20 mmol/L will have an osmotic concentration of 60 mOsm/L because each $CaCl_2$ group dissociates completely in solution into three ions: one Ca^{2+} and two Cl^-. A 20 mmol/L solution of $CaCl_2$ is said to have an osmolarity of 60 mOsm/L.

OSMOTIC PRESSURE. Osmotic pressure is a property of a solution that is proportional to the solute concentration. A dilute solution has a lower osmotic pressure than does a concentrated solution. Thus, water moves by osmosis from a region of low osmotic pressure to a region of high osmotic pressure.

If we know the concentration of a solution, it is relatively simple to calculate the osmotic pressure. Osmotic pressure is related to the solute concentration by the **van't Hoff equation:**

$$\pi = RTC$$

in which π is the osmotic pressure, R is the ideal gas constant (0.082 [L · atm]/[degree · mole]), T is the absolute temperature in kelvins (°C + 273), and C is the total solute concen-

tration expressed as osmoles of solute per liter of solution (osmolarity). For example, the osmotic pressure of a solution of 1.0 Osm/L at 0°C is calculated as follows:

$$\pi = RTC$$
$$= 0.082 \times 273° \times 1.0 \text{ Osm/L}$$
$$= 22.4 \text{ atm}$$

Thus, a 1.0-Osm/L solution exerts an osmotic pressure of 22.4 atm at 0°C. A similar calculation reveals that human blood plasma, which is normally a 0.31 Osm/L solution, has an osmotic pressure of 7.9 atm at 37°C. Thus, the osmotic pressure of body fluids is large compared with normal hydrostatic pressures in physiological systems. The osmotic pressure of a solution depends on the total osmolarity of the solution and not on the nature of the specific solute or solutes. The value of C for pure water containing no dissolved solutes is zero; hence, the osmotic pressure of pure water is zero.

The osmotic pressure specifically due to the proteins present in blood plasma is termed the **colloid osmotic pressure,** or **oncotic pressure.** It is a relatively small fraction of the total osmotic pressure of plasma, but it plays a very important role in regulating the distribution of water between the blood plasma and the interstitial fluid. This is discussed in detail in the chapter on circulation.

Because R and T in the van't Hoff equation are constants, if no temperature change occurs, the osmotic pressure is directly related to osmolarity. In this case, there is no need to calculate osmotic pressure in order to determine which of several solutions has the highest osmotic pressure. A comparison of the osmolarity of the solutions suffices; the solution with the highest osmolarity has the highest osmotic pressure.

Two solutions that have the same osmolarity are said to be **iso-osmotic.** Thus, 300-mOsm/L NaCl is iso-osmotic with 300-mOsm/L sucrose. A solution with a higher osmolarity than another is said to be **hyperosmotic;** a solution with a lower osmolarity is said to be **hypo-osmotic** relative to one with a higher osmolarity.

Changes in Cell Volume: The Erythrocyte

> *What happens to cells that are exposed quickly to osmotic stress?*

Nearly all cell membranes are permeable to water; scientists had predicted for a long time that a protein channel specific for water must be present in certain membranes. About 10 years ago, a water channel protein was identified in the erythrocyte membrane. Its molecular mass was 28,000 Da, and it was termed **aquaporin-1** or **AQP1.** It allows for rapid movement of the polar water molecule across the hydrophobic lipid bilayer. AQP1 consists of six membrane-spanning alpha-helical segments, and it is believed that two of the polypeptide loops between these segments fold into the lipid

Outside

Lipid
bilayer

Inside

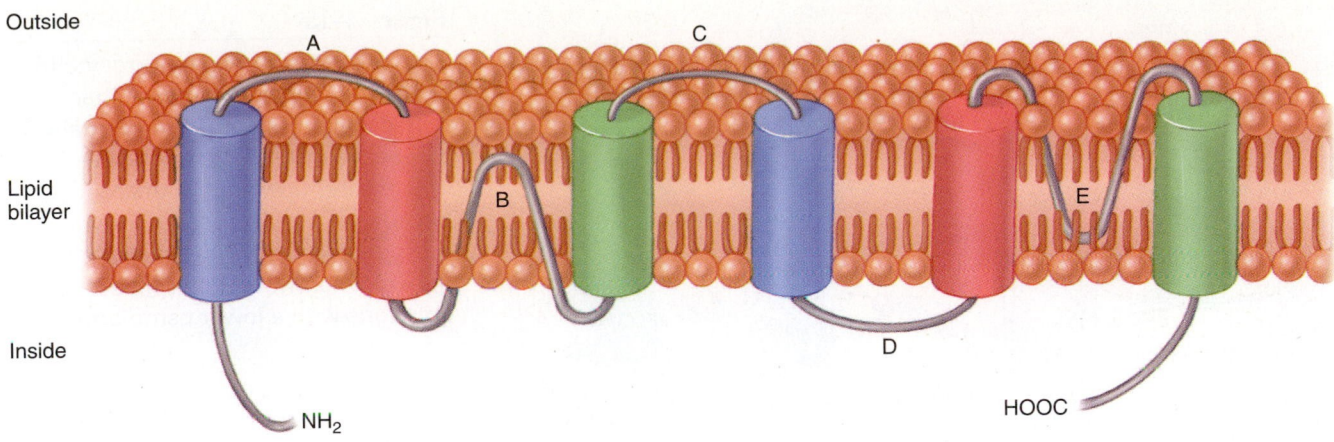

NH₂   HOOC

Figure 4–14

Secondary structure of aquaporin-1 in the erythrocyte membrane showing six transmembrane alpha-helical segments joined by three extracellular loops (A, C, and E) and two intracellular loops (B and D). A part of loops B and E folds into the lipid bilayer. In the functionally active protein, the different transmembrane segments probably form a cluster so that loops B and E can come together to form the aqueous channel across the lipid bilayer.

bilayer to form the aqueous channel across the bilayer (Fig. 4–14). A total of ten different mammalian aquaporins have been identified to date and numbered from AQP0 to AQP9. They occur in several tissues, including the lung, brain, eye, and kidney. **Antidiuretic hormone** increases reabsorption of water in the kidney by inserting AQP2 proteins into the plasma membrane of the collecting duct cells.

Cells must regulate the amount of water that enters. Plant cells and many bacteria have a rigid cell wall that surrounds the plasma membrane and prevents an increase in the volume of the cell due to an influx of water. Animal cells lack an exterior cell wall and have developed other mechanisms to deal with osmotic stress.

We will use the red blood cell, or **erythrocyte,** to illustrate the changes that occur in animal cells when the osmolarity of the external medium changes. The erythrocyte is a favored experimental tool of cell physiologists because it has a simple structure, has no intracellular membranes, and is easily obtainable in large numbers.

Previously in this chapter, we stated that many biological membranes are permeable to solutes as well as to water. This is certainly true of the erythrocyte plasma membrane, but the net movement of the major solutes, such as Na^+, K^+, and Cl^-, is relatively small during osmotic experiments of short duration.

The osmolarity of the erythrocyte cytosol is close to 300 mOsm/L. Suppose an erythrocyte is placed in an iso-osmotic solution of 300-mOsm/L NaCl. The total osmolarity (and hence the osmotic pressure) of the solution inside the cell is the same as that of the solution outside. Here, no concentration gradient favors osmotic water flow and no net movement of water into or out of the cell occurs; the intracellular volume does not change.

Next, suppose the erythrocyte is placed in hyperosmotic 400-mOsm/L NaCl. The erythrocyte is no longer in osmotic equilibrium because the external solution has a higher osmolarity than does the cell cytosol. In response to the higher solute concentration outside the cell, water flows out of the cell (Fig. 4–15). However, solutes remain in the cell, and therefore the osmolarity of the cell cytosol increases. Eventually, the osmolarity of the cytosol increases to the point at which it is equivalent to that of the external solution, and osmotic flow of water ceases. The cell returns to an osmotic equilibrium with the external medium, but this occurs at the expense of a part of the intracellular volume. The decrease in volume is not accompanied by a decrease in surface area, and the cell assumes a spiky, or **crenated,** shape (see Fig. 4–15).

When a normal erythrocyte is placed in a hypo-osmotic solution of 200-mOsm/L NaCl, the osmolarity of the cytosol is higher than that of the external solution. Water enters the cell, leading to an increase in intracellular volume (see Fig. 4–15). Osmotic flow of water into the cell continues until the osmolarity of the cytosol has been diluted to 200 mOsm/L. At this point, net movement of water into the cell ceases, and osmotic equilibrium is reestablished, but with an increase in cell volume (see Fig. 4–15).

The erythrocyte has a unique characteristic that enables it to accommodate a considerable increase in cell volume. The normal cell is biconcave, not spherical. When water enters from a hypo-osmotic solution, the cell simply fills out to become more like a sphere (see Fig. 4–15). When the cell has become a perfect sphere, the intracellular volume will have increased by 67% over the volume of the original biconcave cell. In this way, the erythrocyte can undergo a large increase in volume without experiencing a change in the

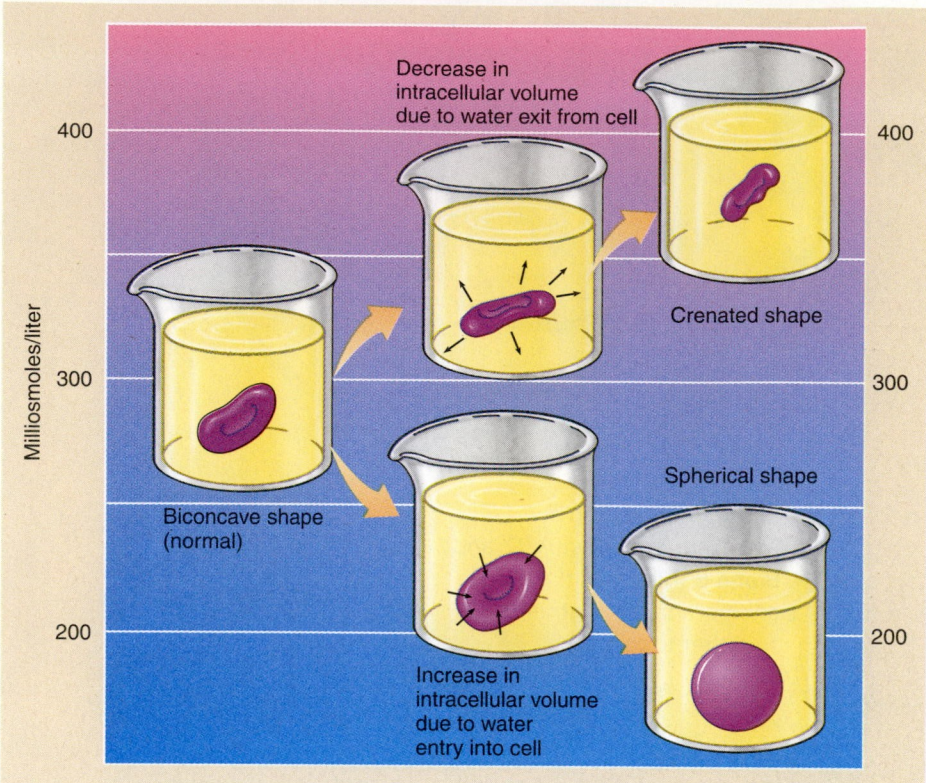

Figure 4–15

Response of a human erythrocyte to changes in the osmolarity of the extracellular medium. The normally biconcave disc becomes crenated when water flows out of the cell because of a greater osmolarity in the surrounding fluid. The cell becomes a sphere when water flows into it from solutions with a lower osmolarity.

surface area of the cell. Most other types of animal cells are not biconcave and cannot respond in this way to osmotic entry of water. The plasma membrane cannot stretch, and the cells can accommodate only a very small increase in intracellular volume. If osmotic entry of water continues, the cells undergo osmotic **lysis** — that is, holes develop in the plasma membrane, and the intracellular contents are lost. The cells literally burst open. The process is also called **hemolysis** to reflect the loss of **hemoglobin,** the principal intracellular protein.

OSMOTIC FRAGILITY. An erythrocyte bursts open, or **lyses,** if it is placed in a medium that is very dilute (e.g., 100-mOsm/L NaCl). The difference in osmolarity between the inside and the outside of the erythrocyte is now so large that the cell cannot achieve osmotic equilibrium. Water continues to enter, even when the cell has become a sphere, and so the cell lyses.

Osmotic fragility is a term used to characterize how a sample of erythrocytes responds to osmotic stress. It is expressed as the osmolarity of NaCl that causes 50% of the erythrocytes to lyse. In the case of normal erythrocytes, 50% of the cells lyse at an osmolarity of 150 mOsm/L (Fig. 4–16). Some cells, presumably the older ones, are more fragile and lyse at a slightly higher osmolarity than 150 mOsm/L. Other cells, presumably the younger ones, are less fragile and lyse at a slightly lower osmolarity. However, almost all of the population of cells lyse over a fairly narrow range of osmolarities, as shown by the steep slope of the curve in Figure 4–16.

Osmotic fragility can be an important test for diseases that involve a change in erythrocyte behavior. When the erythrocytes become less resistant to osmotic stress, they are said to be more fragile, and the curve (see Fig. 4–16) shifts to the left. A specific example of a more fragile population is

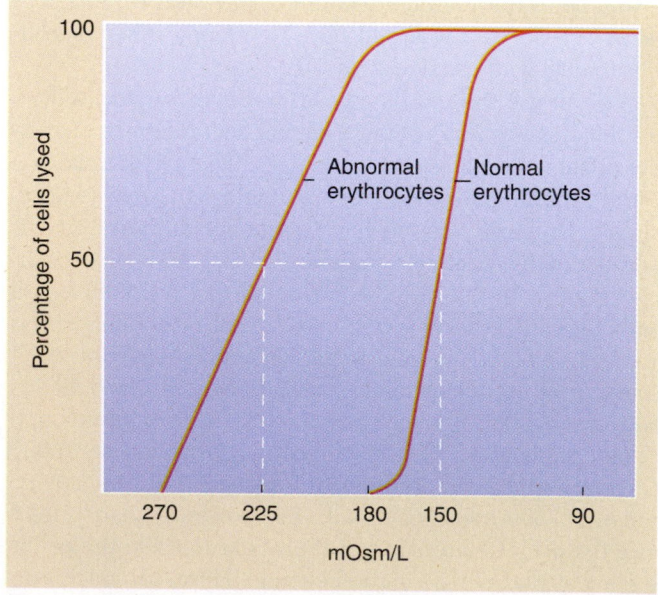

Figure 4–16

Abnormal erythrocytes are more fragile and lyse in solutions of higher osmolarity compared with normal erythrocytes.

the erythrocytes from mice with **hereditary spherocytosis** (see Case History). These cells tend to be spherical rather than biconcave. As a result, the cells can no longer accommodate the volume increase caused by osmotic flow of water into the cells. As seen in Figure 4–16, some of the cells lyse when the osmolarity of the NaCl medium decreases only slightly. Fifty percent of the cells lyse at an osmolarity of 225 mOsm/L, which represents a considerable change from the behavior of the normal cell population. The erythrocytes in outdated blood from a blood bank are another example of increased osmotic fragility. These cells are less resistant to osmotic stress than are erythrocytes in freshly drawn human blood.

After a discussion of some specialized transport mechanisms, we return to the concept of osmosis and the way in which the cell transports materials to maintain appropriate cell volume.

Facilitated Diffusion of Lipid-Insoluble Substances

> *How do solutes that are not lipid soluble pass through the membrane?*

A lipid bilayer is essentially impermeable to ions and large, water-soluble molecules, such as D-glucose, amino acids, and other metabolites. Still, these substances diffuse across the membrane in response to electrochemical gradients. They do so by a process of **facilitated diffusion,** so called because their passage is facilitated by intrinsic membrane proteins that act as **carriers** or **channels.** Ions and other lipid-insoluble substances normally would not be able to pass through the nonpolar interior of the membrane. The proteins enable them to get from one side of the lipid bilayer to the other, in part by shielding them from the lipid molecules. The molecular mechanisms of action of membrane carriers and channels are the subjects of intense investigation.

Simple diffusion and facilitated diffusion both occur spontaneously and passively in response to an electrochemical gradient; no additional input of energy is required. Similarly, both processes cease once the gradient has been abolished. Because facilitated diffusion is mediated by a carrier or channel protein, however, a number of characteristics distinguish it from simple diffusion: (1) facilitated diffusion allows for a very high rate of solute transport, (2) it is a saturable process — the transport rate is limited, (3) it is a highly specific process, and (4) solute transport can be blocked by competitive inhibitors.

CARRIER PROTEINS. A hypothesis to explain the operation of a carrier is depicted in Figure 4–17. This model proposes that (1) the carrier protein is composed of at least two subunits and (2) binding of solute to a site exposed to the cell exterior triggers an intramolecular rearrangement in the structure of the protein such that the solute binding site is now exposed to the cell interior. The solute is always surrounded by hydrophilic regions of the protein subunits and is never exposed to the hydrophobic lipid phase of the membrane. Once the solute is exposed to the interior of the cell, where the solute concentration is low, it dissociates from the binding site. This step allows the carrier protein to revert back to its original conformation, with the empty binding site again exposed to the cell exterior. The cycle then can be repeated to move another solute molecule into the cell.

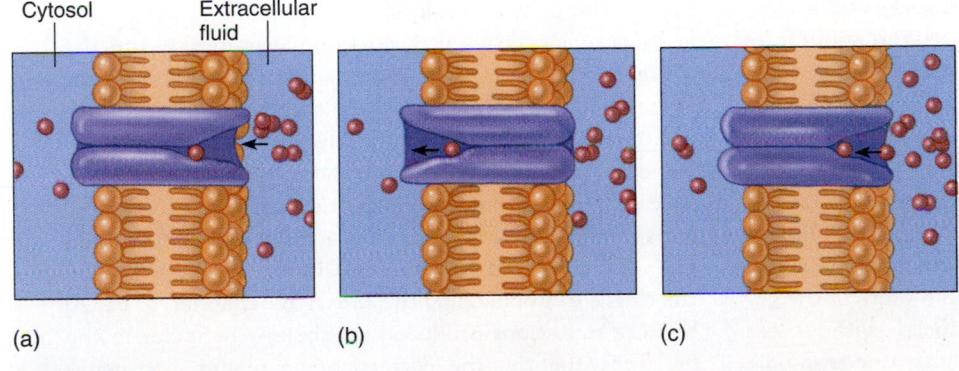

(a) (b) (c)

Figure 4–17

A possible mechanism for carrier-dependent transport. *(a)* Solute concentration is high outside and low inside the cell. A solute molecule binds to the carrier protein at a specific site that is exposed on the exterior surface of the cell membrane. *(b)* Binding of the solute to the carrier triggers a change in conformation, and the solute binding site becomes exposed to the cell interior. *(c)* The solute easily dissociates from the carrier because of the low concentration of solute inside the cell. This dissociation allows the carrier protein to revert to its original conformation. The solute binding site is again exposed to the exterior of the cell, and the cycle can be repeated. Probably, the changes in protein conformation are much more subtle than depicted here.

CURRENT CONCEPTS IN PHYSIOLOGY

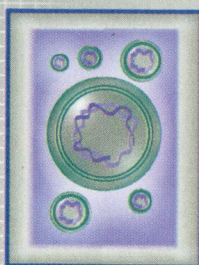

Mice, Flies, and Ion Channels

Animals carrying defective copies of genes are used increasingly in physiological and biomedical research laboratories as model systems for investigating the causes, and preventing the progression, of human disorders such as cancer, arthritis, AIDS, cystic fibrosis, and Alzheimer's disease. The use of mice has many advantages because they reproduce rapidly and can be genetically manipulated at the molecular level. The shift away from larger animals, such as cats, dogs, and primates, stems from the explosion in molecular biology that began in the 1970s. Just as foreign genes can be inserted into bacterial DNA, genes from other organisms can be inserted into mice, creating **transgenic** animals. A transgenic mouse could carry a DNA sequence from one of many different organisms, ranging from viruses to humans. Practically any gene that can be isolated by cloning techniques can be added to mouse DNA. Transgenic mice develop from eggs that have been directly injected with pieces of foreign DNA. The injected eggs are then transferred to a female mouse to develop normally. Some of the offspring that develop from these eggs have the foreign DNA (**transgene**) added to their normal complement of mouse genes. Researchers can then study the function of the transgene in specific mouse tissues. If a human transgene is used, it closely replicates a human disease in the mouse, so that the animal studies are much more applicable to human disease. Specific strains of transgenic mice, known as **knockouts**, are receiving a lot of attention. In these mice, the addition of the transgene has been designed to disrupt ("knock out") the function of a specific mouse gene. Researchers can then determine the effect of losing a normal gene and, in this way, determine its proper function in the intact animal. Knockout mice provide valuable models of human diseases because the genetic patterns of mice and humans are very similar. With careful breeding, the transgene can be passed on to future generations so that more of the genetically modified mice can be produced without experimental interventions. Consider just one example involving the epithelial sodium channel (**ENaC**) that is important for salt homeostasis and control of extracellular volume and blood pressure. Functional ENaC consists of three protein subunits, and it is located in the apical membrane of many salt-absorbing epithelial cells. ENaC is hyperactive in patients with **Liddle's syndrome**, a human genetic disease leading to severe hypertension. The use of transgenic mice has shown that each of the three ENaC subunits is essential for survival of the animal and for regulation of sodium transport in colon and kidney. A mouse model of Liddle's syndrome has been developed and is expected to show whether the hypertension is a direct result of hyperactive ENaC.

For almost a century, the *Drosophila* fruit fly has been a workhorse for genetic studies, and a wealth of techniques now permit genetic manipulations and creation of stable lines of transgenic flies. The entire sequence of the *Drosophila* genome (about 13,600 genes) was published in *Science* (March 24, 2000). This remarkable achievement, the result of technical innovation and unprecedented collaboration between academia and industry, has enabled researchers to compare the fly's genes with all known human genes. The importance of *Drosophila* for studying human biology and disease was enhanced considerably when it was discovered that the fly has counterparts for 177 of the 289 human genes that are known to be involved in human disease. Thus, if *Drosophila* has a gene that is the counterpart of an important but poorly understood human gene, the powerful arsenal of genetic techniques available in the *Drosophila* system can be applied to its characterization. These include genes encoding structural proteins, chromosomal proteins, signaling proteins, and ion channels. Ion channel proteins tend to be highly conserved throughout evolution, and the first systematic analysis of ion channel dysfunction was initiated in the *Drosophila* system. The important advantage of using *Drosophila* is that genetic tools can be combined with direct measurement of channel function and animal behavior. This approach provided the first molecular characterization of ion channel polypeptides and their functional roles in vivo. These include several types of **voltage-gated Na$^+$ channels** and **K$^+$ channels** in the nervous system, the classic example being the *shaker* K$^+$ channel, so called because mutations produced leg-shaking behavior in the fly. Subsequently, the corresponding mammalian genes for these ion channels were identified, including several that are associated with human diseases.

CHANNEL PROTEINS AND GATING. A channel mechanism is particularly important for the rapid transmembrane movement of ions, such as Na^+ and K^+. Several different channel structures and subunit arrangements have been characterized for channel proteins. K^+ channels consist of four identical subunits, each one containing six transmembrane segments. Na^+ channels consist of a single polypeptide chain that contains four repeated "domains" corresponding to the four subunits of the K^+ channel. In general, a hydrophilic channel or pore across the membrane is formed when the membrane-spanning segments or protein subunits cluster together, similar overall to the arrangement discussed earlier for the aquaporin water channel (see Fig. 4–14). The open channel permits a higher rate of transport than that provided by a carrier. For example, some channels enable ions to pass through them at the rate of 10^8 ions/sec, whereas the fastest carriers move solutes across membranes at no greater than 10^5 molecules/sec.

It is unlikely that an ion moves through a channel by simple diffusion down the electrochemical gradient. Most likely, the ion undergoes specific interactions with charged groups along the sides of the channel. These reactions help to move the ion through the channel. The types of charged groups present in the channel also may be important in determining the specificity of the channel for one ion type. For example, K^+ channels are 100-fold more permeable to K^+ than to Na^+.

Another difference between a channel and a carrier, in addition to the overall rate of transport, is that a channel can be kept closed by a **"gate"** (Fig. 4–18). A closed channel does not permit movement of ions, even though the electrochemical gradient may favor it. The mechanisms used to regulate the opening and closing (**"gating"**) of ion channels include changes in membrane potential (**voltage-gated channels**) and binding of specific molecules to the channel protein (**ligand-gated channels**).

THE HIGH RATE OF FACILITATED DIFFUSION. A solute that crosses a cell membrane by facilitated diffusion does so at a rate that is much more rapid than that of a solute crossing by simple diffusion, even though both solutes are of the same molecular size and have the same lipid solubility. Consider a hypothetical example in which 17 solute molecules are outside a cell and only 1 is inside; assume that the solute is electrically neutral. Net movement of solute inside the cell ceases when 8 more molecules have entered — that is, when 9 molecules are on each side of the membrane. It takes 8 seconds to achieve this balance if the solute moves by simple diffusion at a rate of 1 molecule entering the cell per second. In contrast, it takes only 1 second if the solute moves by facilitated diffusion, which allows for solute entry at a rate of 8 molecules/sec.

This example illustrates that the endpoint of both processes is the same but is achieved faster by facilitated diffusion. In this way, the carrier protein serves a function analogous to that of an enzyme in a biochemical reaction. The rates of transport used in this example are arbitrary and are not meant to represent experimentally determined values.

SATURATION IN FACILITATED DIFFUSION. A facilitated diffusion system can become saturated because the number of binding sites on the carrier for solute molecules is limited. The rate of transport is at a maximum when all the sites are occupied by solute (Fig. 4–19). A graph drawn to relate transport rate to solute concentration (Fig. 4–20) shows that, initially, the rate of solute transport increases in proportion to the solute concentration. However, once the solute concentration has reached a level at which the carrier binding sites are always full (saturation), a further increase in solute concentration produces no change in the solute transport rate. The rate of transport is now at its maximum (V_{max}). The solute concentration that gives rise to one half of the V_{max} is known as the **Michaelis constant (K_m).** These kinetic parameters are a very useful way of comparing the characteristics of different transport systems. It is worth adding that transport by simple diffusion is not saturable. The rate of transport increases in proportion to the solute concentration (Fig. 4–20). The limitation posed by carrier saturation is rarely problematic because the overall rate of transport by this mechanism is so much faster than that by simple diffusion. The V_{max} for transport by facilitated diffusion is analogous to an enzyme-catalyzed reaction, which proceeds at a maximum rate once the enzyme is fully saturated with substrate (see Chapter 2). Any further increase in substrate concentration has no effect on the reaction rate. The analogy with enzymes is limited, however, because the solute is not permanently altered as a consequence of binding to the carrier protein.

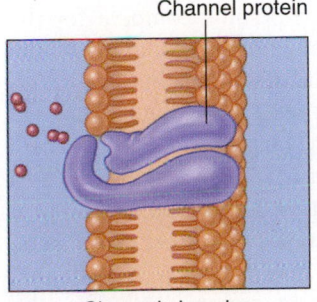

Channel protein

Channel closed   Channel open

Figure 4–18

A highly simplified view of the gating mechanism for regulating ion transport through a membrane channel protein.

THE SPECIFICITY OF FACILITATED DIFFUSION. Unlike simple diffusion, facilitated diffusion is a highly specific process. Because the transport of the solute involves binding to a carrier protein, it is likely that there are specific carriers in

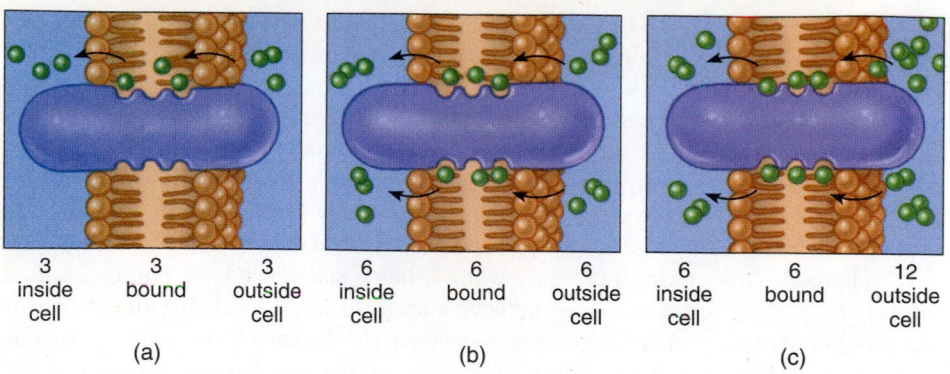

| 3 inside cell | 3 bound | 3 outside cell | 6 inside cell | 6 bound | 6 outside cell | 6 inside cell | 6 bound | 12 outside cell |

(a)   (b)   (c)

Figure 4–19

The saturation of a membrane carrier with solute. *(a)* Assume that a carrier protein in a membrane has specific binding sites for 6 solute molecules. If 3 solute molecules outside a cell arrive at the carrier, all 3 molecules are transported inside the cell in one cycle of the carrier. *(b)* If 6 molecules outside the cell arrive at the carrier, all 6 molecules are moved inside the cell in one cycle. That is, the rate of net transmembrane transport of solute doubles when the amount of solute doubles. *(c)* When 12 solute molecules arrive at the carrier, the rate of solute movement is not increased because the carrier can bind and transport only 6 molecules per cycle. Once the carrier is saturated with solute, any further increase in solute concentration outside the cell has no effect on the overall rate of solute transport. At this point, the rate of solute transport has reached a maximum.

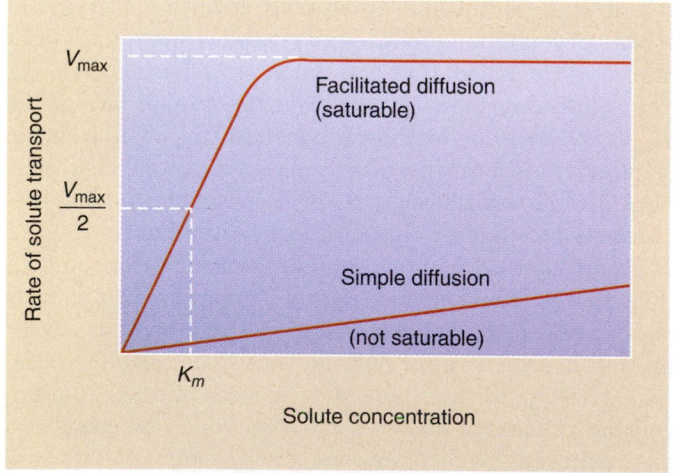

Figure 4–20

Graphical representation of the saturability of facilitated diffusion. The maximum rate of transport possible is termed the V_{max}, and the K_m is the concentration of solute necessary to reach one half of the V_{max}. Although simple diffusion is not saturable, the rate of transport is much slower at the physiological range of solute concentration.

the cell membrane for each of the solutes that enters or leaves the cell by facilitated diffusion. The specificity of the carrier is not absolute. For example, structurally related amino acids often are able to share the same carrier. The transport of sugars across the erythrocyte plasma membrane is another example. The transport system works best when glucose is the solute being transported, but related sugars, such as galactose, can utilize the same system to cross the membrane.

COMPETITIVE INHIBITION IN FACILITATED DIFFUSION. Structurally related compounds compete for the same binding site on a membrane carrier, giving rise to the phenomenon of **competitive inhibition** of solute transport, as characterized by an increase in the K_m without a change in the V_{max} of the transport system. Thus, amino acid *A*, which shares the same transport system as amino acid *B*, acts as a competitive inhibitor of the transport of *B* if both are present simultaneously (Fig. 4–21). The inhibitory effect of *A* can be overcome by an increase in the concentration of *B* so that more molecules of *B* are available to compete with *A* for the carrier binding sites. In this way, the rate of transport of *B* can be increased to the same V_{max} that was achieved in the complete absence of *A*, but a higher concentration of *B* (i.e., an increased K_m) is required (see Fig. 4–21).

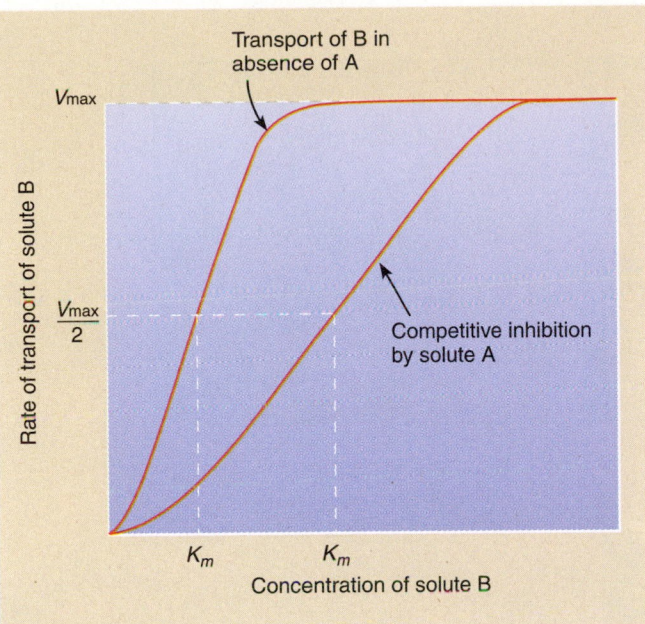

Figure 4–21

Competitive inhibition of facilitated diffusion of solute *B* by the structurally similar solute *A*. In the presence of *A*, the concentration of *B* must increase to reach the same V_{max} as that attained in the absence of *A*.

Active Transport Is Energized Movement Against a Gradient

> *How do substances move through the membrane against a concentration gradient?*

Thus far, we have discussed passive transport processes, in which molecules and ions move into or out of the cell from regions of high to regions of low concentration. These movements, while sometimes requiring the presence of a special protein molecule, do not require additional energy. They are driven by the inherent energy in the electrochemical gradient. The result is always a *decrease* in concentration of a once highly concentrated region.

Frequently, however, the cell requires an increase in concentration of an already highly concentrated solute. For example, the electrochemical gradients that drive so many important passive transport processes must be maintained by the cell. If they are not maintained, they will dissipate and all passive transport will cease. The maintenance of an electrochemical gradient, or an increase in its magnitude, requires the movement of solute from a less concentrated region outside the cell to a more concentrated region inside — that is, *against* the prevailing gradient. This process requires input of energy. The transport system continues to operate for as long

as the energy supply and the solute are available. Active transport, like facilitated diffusion, requires carrier proteins. The energy supply for active transport is derived from cellular metabolism; hence, active transport systems are susceptible to inhibition by compounds that act primarily as metabolic inhibitors and have no direct effect on the carrier proteins.

The mechanisms of active transport are of two general types, depending on whether the movement of solute is linked directly or indirectly to energy-yielding reactions (see Fig. 4–10). When the movement of solute is coupled directly to an energy-yielding reaction, typically the hydrolysis of high-energy phosphate bonds, the transport process is termed **primary active transport.** When active transport of a solute is not coupled directly to the energy-yielding reactions, the transport mechanism is described as **secondary active transport.**

Primary Active Transport

The best known example of a primary active transport system is probably the **Na^+/K^+-ATPase pump,** so called because Na^+ is pumped out of the cell at the same time as K^+ is pumped in and because the ion movements are coupled directly to breakdown of ATP. This pump is present in the plasma membrane of most animal cells and is particularly important for the normal function of nerves and muscles. One of the subunits of the pump has a binding site for ATP and the enzymatic activity necessary to remove one of the phosphate groups. The free energy released by ATP hydrolysis is used to drive the movement of both Na^+ and K^+ against their respective electrochemical gradients. It is chiefly through this active transport system that the cell is able to maintain a high K^+ concentration and a low Na^+ concentration in the cytosol.

The breakdown of ATP is tightly coupled to the transport of Na^+ and K^+; the one cannot occur without the other. For every molecule of ATP that is split into ADP and phosphate, 3 Na^+ are pumped out of the cell and 2 K^+ are pumped in. The molecular events that link ATP breakdown to ion transport are not fully understood, but it seems clear that the enzyme protein is temporarily phosphorylated by attachment of the phosphate group released from the ATP. The phosphorylation step is Na^+ dependent — that is, it occurs only when Na^+ is present and presumably leads to a change in protein conformation that moves bound Na^+ out of the cell. The subsequent dephosphorylation (removal of phosphate) is K^+ dependent and probably allows the protein to return to its original conformation, moving bound K^+ into the cell. These steps are depicted in the scheme in Figure 4–22. The Na^+/K^+-ATPase is represented as a protein with four subunits. This active transport system is so important that about one third of the energy supply of the cell is used to maintain its activity.

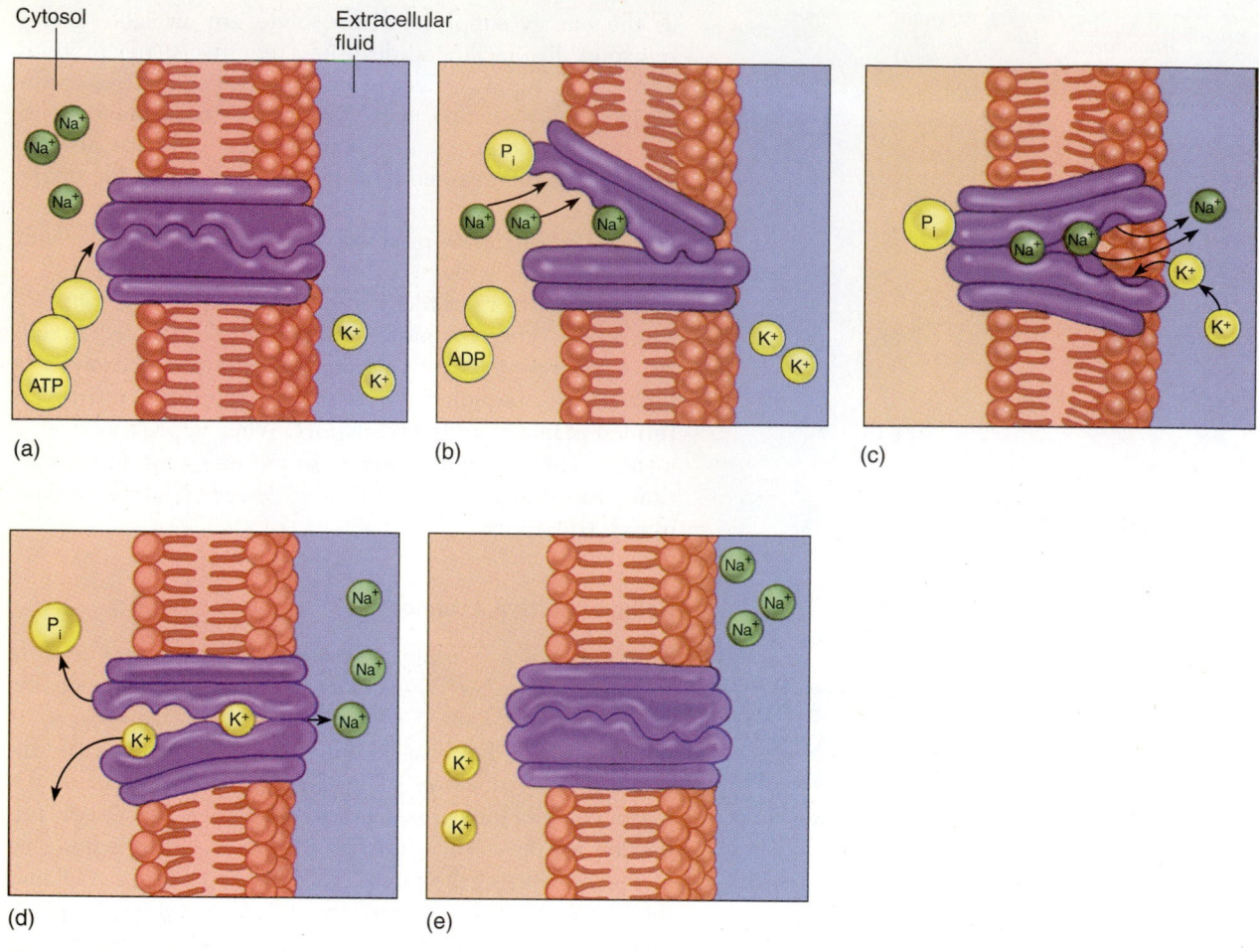

Cytosol

Extracellular fluid

(a)

(b)

(c)

(d)

(e)

Figure 4–22

Hypothetical scheme of action of the Na^+/K^+-ATPase pump. *(a and b)* In the presence of Na^+ inside the cell, ATP binds to and phosphorylates one of the protein subunits. Na^+ then binds to specific sites on the phosphorylated protein. *(c and d)* These events trigger a conformational change that exposes the bound Na^+ to the cell exterior, and Na^+ dissociates from the protein. The binding site for K^+ is now exposed, and external K^+ binds. *(e)* Binding of K^+ promotes dephosphorylation (removal of the phosphate) of the protein, which allows it to revert to the original conformation. The conformational change moves the bound K^+ inside the cell, where it dissociates from the protein. The cycle now can be repeated. The conformational changes are greatly simplified in the drawing.

Secondary Active Transport

When transport of a solute against a gradient is not coupled directly to energy-yielding reactions, it is described as *secondary active transport.* A common example of this kind of mechanism is a transport system that is driven by the energy stored in the electrochemical gradient for another solute. In animal cells, this solute is usually Na^+, which enters cells passively by moving down a very favorable electrochemical gradient.

SYMPORT AND ANTIPORT. Recall that a low sodium concentration and a negative potential exist inside the cell relative to the outside. Many different solutes are trans-

ported across cell membranes against their electrochemical gradients by coupling to the movement of Na^+. Movement of the solutes in the same direction as Na^+ is known as **symport,** or **cotransport;** movement in the opposite direction is known as **antiport,** or **exchange** (Fig. 4–23). Solute transport systems that are coupled to Na^+ movement have a specific requirement for Na^+; most other ions of similar size, such as K^+, cannot be used in place of Na^+.

The Na^+ gradient across the cell membrane is maintained by the Na^+/K^+-ATPase, which actively pumps Na^+ out of the cell against its electrochemical gradient. As we have just discussed, this process is driven directly by a metabolic energy supply in the form of ATP. Thus, secondary

APPLICATIONS OF PHYSIOLOGY

Glucose–Galactose Malabsorption

Glucose–galactose malabsorption is a genetic disorder that is quite rare (about 200 cases worldwide) but illustrates well the severe, often life-threatening problems that can arise when a membrane transport system does not function properly. Infants with glucose–galactose malabsorption develop a profuse, watery diarrhea when fed milk or foods containing glucose, galactose, sucrose, or lactose. Recall from Chapter 2 that sucrose and lactose are disaccharides. Sucrose contains glucose plus fructose, whereas lactose contains glucose plus galactose. The monosaccharides are liberated in the intestine by enzymatic hydrolysis of the disaccharides. Fructose and milk formulas that are carbohydrate-free are tolerated well. A specific defect in absorption of glucose and galactose in the intestine can be demonstrated by tolerance tests in which oral administration of glucose produces little or no increase in the levels of plasma glucose. The primary defect lies in the Na^+-coupled glucose transporter (termed **SGLT1**), located in the apical microvilli of the intestinal epithelial cells. Galactose also uses SGLT1, but fructose transport is not affected because fructose absorption involves facilitated diffusion via a specific fructose transporter named **GLUT5**. The diarrhea can be fatal within a few weeks but disappears if a diet free

of glucose, galactose, sucrose, and lactose is adopted. This regimen must be maintained carefully because the transport defect persists throughout life. The human SGLT1 transporter was cloned in 1989 and was shown to be mutated in glucose–galactose malabsorption. Almost 30 different mutations have now been discovered, many of which result in either premature cessation of SGLT1 protein synthesis or normal synthesis followed by disruption of trafficking to the apical membrane. In a few, the mutation produces a protein that is made, trafficked, and inserted normally in the apical membrane but no longer transports glucose. The end result is the same in all cases: Functional SGLT1 proteins are not present in the apical microvilli and so glucose and galactose cannot be absorbed from the intestinal lumen. Accumulation of these solutes in the lumen increases the osmolarity and retards the absorption of water, leading to diarrhea and severe intestinal water loss. Recent advances in molecular biology have led to a better understanding of the genetic defect at the cellular and molecular levels and how this leads to the clinical symptoms. Conversely, the identification of the specific changes in SGLT1 protein structure in affected individuals has provided clues about the specific amino acid residues that are essential for the normal function of SGLT1. Thus, glucose–galactose malabsorption is an example of the interface between physiology and disease in which information from one aspect reinforces and illuminates the other.

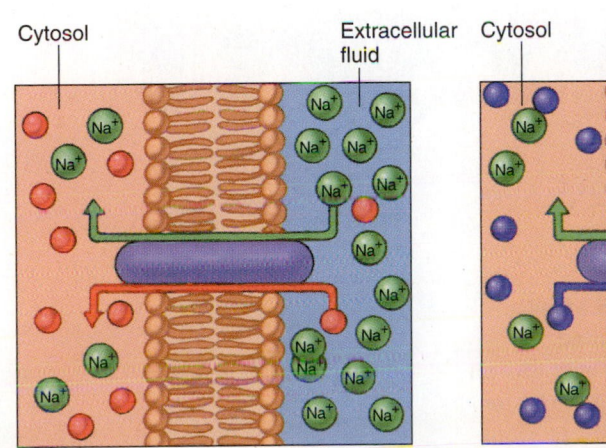

Symport

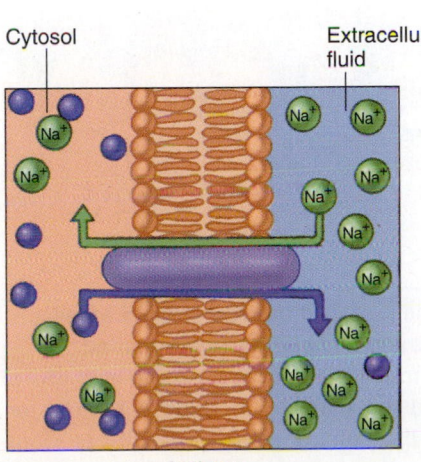

Antiport

Figure 4–23

Secondary active transport occurs when solute transport is driven by the electrochemical gradient for Na^+. Both solute and Na^+ bind to the membrane carrier. The Na^+ gradient is maintained by primary active transport of Na^+.

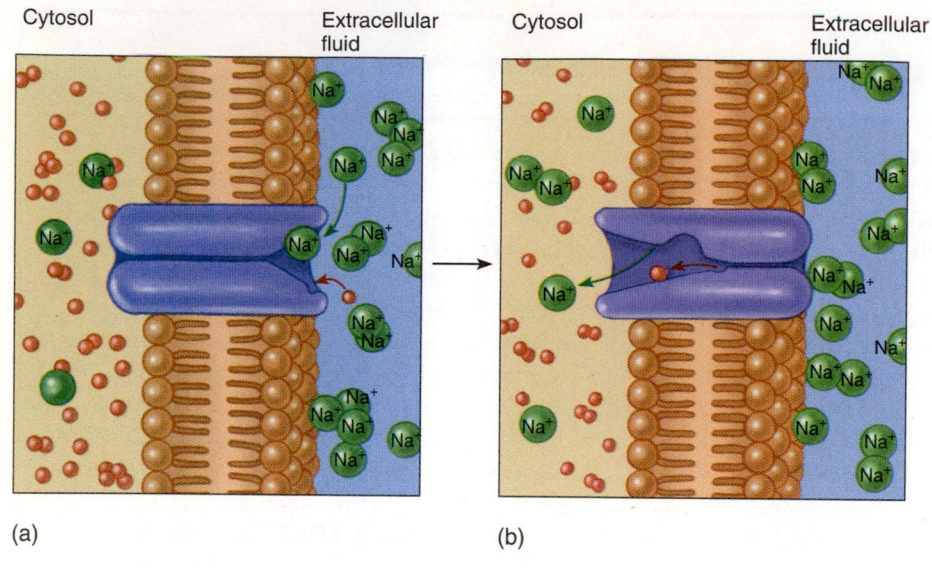

Cytosol   Extracellular fluid   Cytosol   Extracellular fluid

(a)   (b)

Figure 4–24

Hypothetical scheme of action of a two-site carrier for Na^+-coupled solute symport. *(a)* Binding of Na^+ outside the cell to the carrier increases the affinity for the solute, which also binds, even though it is present at a low concentration. *(b)* A conformational change in the carrier protein exposes the binding sites to the cell interior. Na^+ dissociates from the carrier because the Na^+ concentration is low inside the cell. This dissociation step leads to a decrease in carrier affinity for solute, so the latter also dissociates from the carrier, even though the solute concentration is high inside the cell.

active solute transport that is driven by the Na^+ electrochemical gradient depends ultimately on a supply of ATP but is not linked directly to the reactions that break down ATP.

The mechanism of Na^+-coupled transport systems is believed to involve a two-site carrier that binds both Na^+ and the solute. A possible mechanism is illustrated in Figure 4–24. Na^+ outside the cell binds readily to the carrier and encourages binding of solute to the carrier, even though the solute concentration is low outside the cell. A conformational change in the carrier protein now exposes the bound Na^+ and solute to the cell interior. Na^+ readily dissociates from its site because the intracellular Na^+ concentration is very low. Loss of Na^+ from the carrier may alter the binding of solute

to the carrier and cause the solute to dissociate from the carrier, even though there is already a high intracellular concentration of the same solute.

Molecular characterization of the Na^+-coupled glucose transporter (referred to as **SGLT1**) from intestinal epithelial cells suggests that 14 transmembrane segments are present and has confirmed that conformational changes are responsible for coupling Na^+ and glucose transport. The pathway for glucose movement is thought to be formed by segments 10 to 13. A mutation at amino acid 457 located on segment 11 blocks glucose transport and causes **glucose–galactose malabsorption** (see Applications box). The pathway for Na^+ movement has not been identified yet (Fig. 4–25).

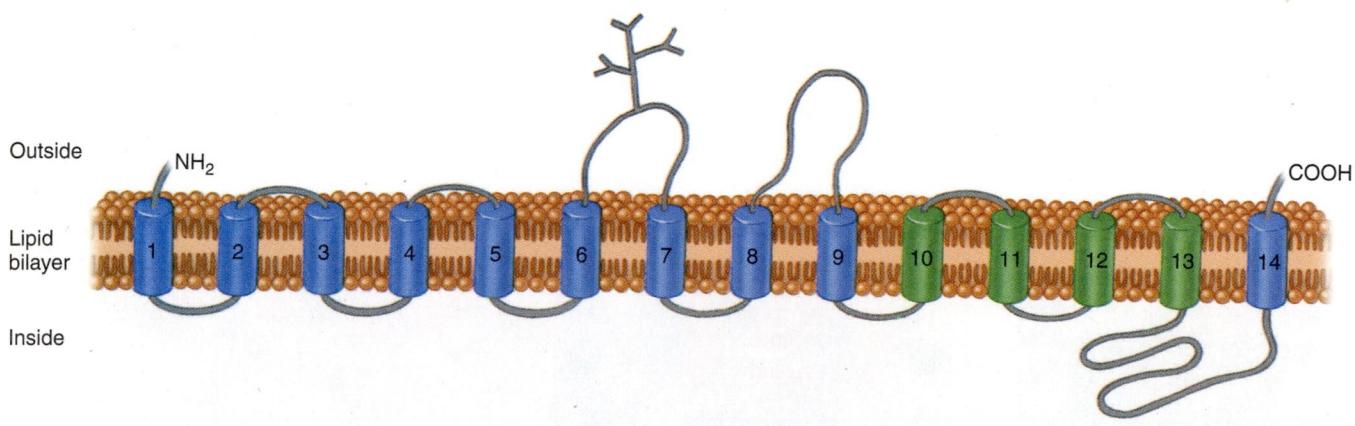

Figure 4–25

Secondary structure of the Na^+-coupled glucose transporter protein from small intestine showing 14 membrane-spanning segments. Carbohydrate attached between membrane spans 6 and 7 is on the extracellular surface. Amino acid 457 on segment 11 is crucial for glucose transport, and segments 10 to 13 probably cluster together in vivo to form a pathway that allows the polar glucose molecules to cross the lipid bilayer.

Transport Processes and Cell Volume Are Interdependent

> *How does the Na$^+$/K$^+$-ATPase pump affect water movement across the cell membrane?*

The cell cytosol normally contains macromolecules, such as proteins, nucleic acids, and organic phosphates, such as metabolic intermediates and ATP. These molecules are too large to cross the plasma membrane, but because they are negatively charged at the pH of cell cytosol, they tend to attract positively charged ions (Na$^+$ and K$^+$) in the extracellular fluid. The entry of cations, if unchecked, would increase intracellular osmolarity, leading to osmotic entry of water and an increase in cell volume. The Na$^+$/K$^+$-ATPase pump, which uses ATP to pump 3 Na$^+$ out of the cell in exchange for 2 K$^+$, is a key factor in the mechanism for controlling the osmolarity of the cytosol, and hence the intracellular volume, of animal cells. By returning Na$^+$ to the extracellular fluid and simultaneously keeping the intracellular Na$^+$ concentration low, the pump is able to counteract and balance the osmotic effects of nonpermeable intracellular molecules.

The importance of the Na$^+$/K$^+$-ATPase pump in the control of cell volume is evident in the fact that animal cells swell and sometimes burst when the pump is inhibited. Decreased activity of the pump probably contributes to the increased osmotic fragility of erythrocytes in stored blood. During storage, the ATP supply tends to decrease. Because the Na$^+$/K$^+$-ATPase pump cannot operate without ATP, any Na$^+$ that "leaks" (diffuses) back into the cell is not pumped out. This abnormal buildup in intracellular Na$^+$ leads to a gradual increase in the osmolarity of the cell cytosol, with osmotic entry of water causing the cell to swell. In this swollen state, the cells are less resistant to osmotic stress than are normal biconcave erythrocytes because they cannot accommodate the same increase in intracellular volume.

Typically, animal cells are exposed to a high concentration of external Na$^+$. There is no net increase in intracellular Na$^+$ because the rate at which Na$^+$ leaks into the cell is closely matched by the rate at which Na$^+$ leaves the cell via the Na$^+$/K$^+$-ATPase pump. This mechanism prevents the external and internal Na$^+$ concentrations from reaching equilibrium, even though Na$^+$ constantly enters and leaves the cells.

Thus, under normal conditions, Na$^+$ outside the cell can be considered to behave as if it were a non-permeable solute. No net movement into or out of the cell occurs even though the internal and external concentrations are markedly different. Some solutes, such as urea and glycerol, penetrate the plasma membranes of an animal cell very rapidly. The cell does not have a mechanism for pumping these solutes back out again. Such **permeant solutes** exert only a transient effect on cell volume, however. Unlike Na$^+$, their external and internal concentrations rapidly equilibrate.

Consider an erythrocyte placed in a large volume of a solution containing 300-mOsm/L NaCl and 60-mOsm/L

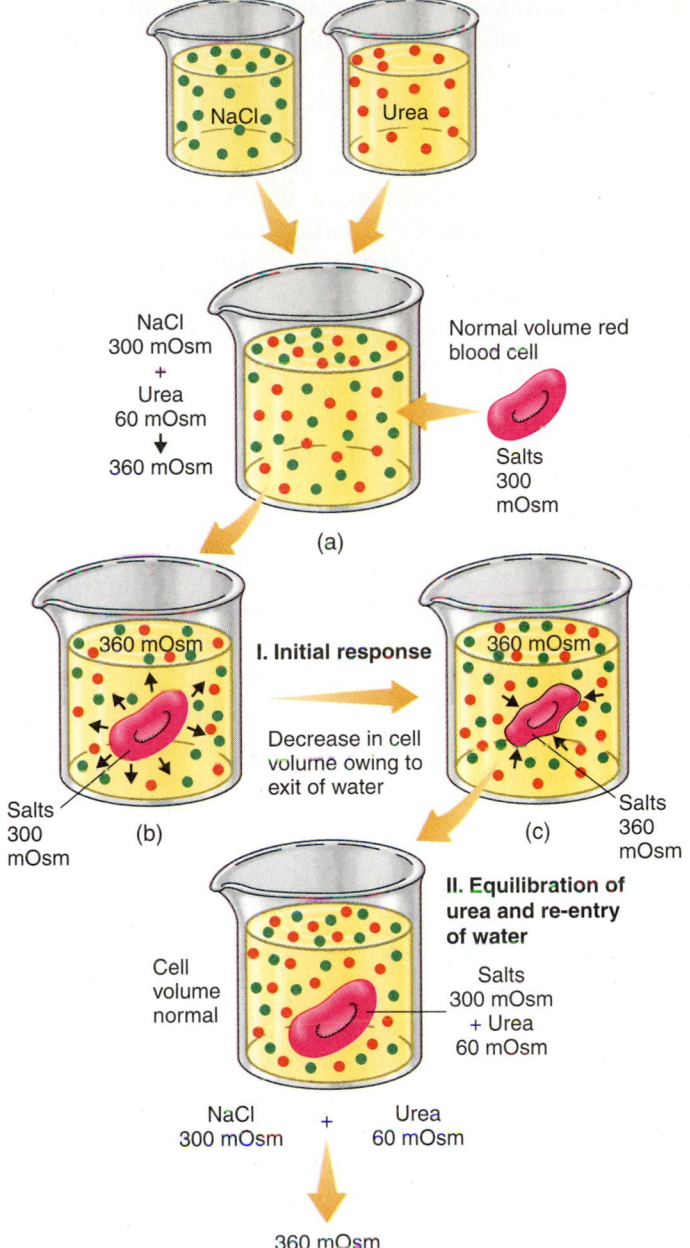

Figure 4–26

Permeant solutes, such as urea, exert only a transient effect on the volume of an erythrocyte because their intracellular and extracellular concentrations rapidly equilibrate. In reality, steps I and II probably overlap considerably.

urea (Fig. 4–26). Initially the cell experiences a hyperosmotic environment: Water flows out, and the cell volume decreases. However, as the urea rapidly enters the cell and the intracellular urea concentration increases toward 60 mOsm/L, water also enters and the cell begins to swell. The final volume of the cell at equilibrium is determined only by the osmolarity of the non-permeable solutes in the extracellular medium. In this case, the nonpermeable solutes (NaCl) have a total osmolarity of 300 mOsm/L, the same as the cell cytosol, and the cell assumes its original volume.

Regulation of Cell Volume

Na$^+$ is the most abundant solute present in blood plasma. As an effectively nonpermeable solute, it is the major determinant of intracellular volume in the animal, even though plasma contains many membrane permeable solutes. A decrease in plasma Na$^+$ concentration reduces plasma osmolarity, and osmotic flow of water into cells causes them to swell. Conversely, an increase in plasma Na$^+$ concentration leads to cell shrinkage because of a loss of intracellular water. This is one of the reasons that plasma Na$^+$ concentration is closely regulated. A change in plasma Na$^+$ concentration induces a series of rapid responses designed to return it to the normal range. For example, an increase in plasma Na$^+$ is detected indirectly as an increase in plasma osmolarity and, in response, more water is added to the plasma to decrease the osmolarity. This also reduces the plasma Na$^+$ concentration to the normal level.

Close control of the plasma Na$^+$ level means that most normal human cells are not subjected to large shifts in plasma osmolarity. However, challenges to the internal osmolarity can occur in cells designed to move water and specific solutes from one side of the cell to the other. This is true particularly for epithelial cells (see next section). Changes in internal osmolarity cause water to flow into or out of the cell and lead to changes in intracellular volume. Many cells have mechanisms for rapidly correcting the volume changes, and these mechanisms do not utilize the Na$^+$/K$^+$-ATPase pump. Most cells respond to an increase in cell volume by activating pathways that increase the efflux of K$^+$ and Cl$^-$ from the cell. The loss of solute decreases intracellular osmolarity which, in turn, causes efflux of water from the cell and leads to a return to normal cell volume. This response is termed a **regulatory volume decrease.** Some cells can respond to a decrease in normal cell volume by a complementary process known as a **regulatory volume increase.** In this case, shrinkage of the cell activates pathways that increase the influx of solutes, such as Na$^+$ and Cl$^-$. A gain of solutes increases intracellular osmolarity, causing an influx of water and a return to normal cell volume. These rapid volume regulatory responses may help epithelial cells maintain a near-constant volume, even with large changes in the amount of water and solutes passing through the cells.

Abnormal situations can impose chronic osmotic stress, and some types of cells have developed mechanisms to maintain normal volume under these conditions. For example, elderly people who are forgetful about drinking fluids for 2 to 3 days are subjected to prolonged dehydration. Their blood plasma is hyperosmotic during this period of dehydration. Cells in the brain respond to the chronic hyperosmotic stress by synthesizing organic solutes, such as **sorbitol,** in order to increase intracellular osmolarity and avoid osmotic loss of intracellular water. Sorbitol can be accumulated without adverse effects on normal brain cell functions. It should be noted that rehydration corrects the problem, but the water must be added very slowly to allow for time for the brain cells to discharge the intracellular sorbitol and other solutes. If rehydration is done too quickly, the brain cells accumulate water by osmosis and their volume increases. Brain swelling within the unyielding skull leads to increased pressure, which compresses blood vessels and destroys brain tissue.

EPITHELIAL TRANSPORT

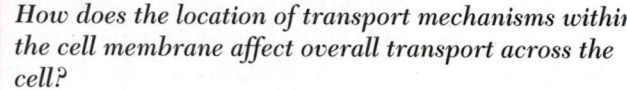

How does the location of transport mechanisms within the cell membrane affect overall transport across the cell?

The transport systems in the plasma membrane of the cell are important for supplying nutrients and removing waste products. In addition to these roles, the specific arrangement of these transport systems in epithelial cells permits the net movement of solutes and water from one side of the epithelium to the other.

Epithelial Cells Are Organized Asymmetrically

Asymmetrical organization is crucial for the function of epithelial cells, such as those lining the lumen of the small intestine and the renal proximal tubule — tissues that are specialized for reabsorption. These cells allow for **transepithelial transport** of solutes and water because the entry and exit pathways are on opposite sides of the cell (Fig. 4–27a).

In an absorptive process, the entry step is at the apical membrane. This is the membrane where the Na$^+$-coupled symport systems for amino acids, glucose, and phosphate are located. The exit step is across the basolateral membrane, where the Na$^+$/K$^+$-ATPase is located. The Na$^+$ gradient across the apical membrane is maintained because the ATPase in the basolateral membrane constantly pumps Na$^+$ out of the other side of the cell to keep intracellular Na$^+$ low.

The asymmetrical distribution of transport systems between the apical and basolateral plasma membranes persists because of the **tight junctions** between the cells. These junctional complexes occur at the point where the plasma membranes of neighboring cells come into close contact; they prevent lateral diffusion of the proteins in the cell membrane (Fig. 4–27b). Thus, the carrier proteins in the apical membrane are physically prevented from leaving the apical membrane and mixing with the carrier proteins in the basolateral membrane.

In some types of epithelia, the tight junctions are relatively permeable and allow water and ions to move between the epithelial cells, providing an additional or **paracellular** pathway across the epithelium (see Fig. 4–27a). This type of epithelium is termed a **"leaky"** epithelium because water and some solutes can leak between the cells in addition to moving through the transcellular pathway. A good example of this type of epithelium is the proximal tubule of the kidney. In other epithelia, such as the small intestine and the collect-

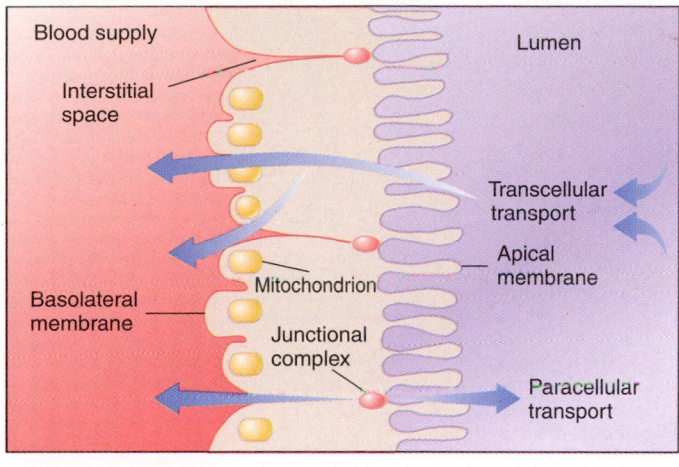

(a)

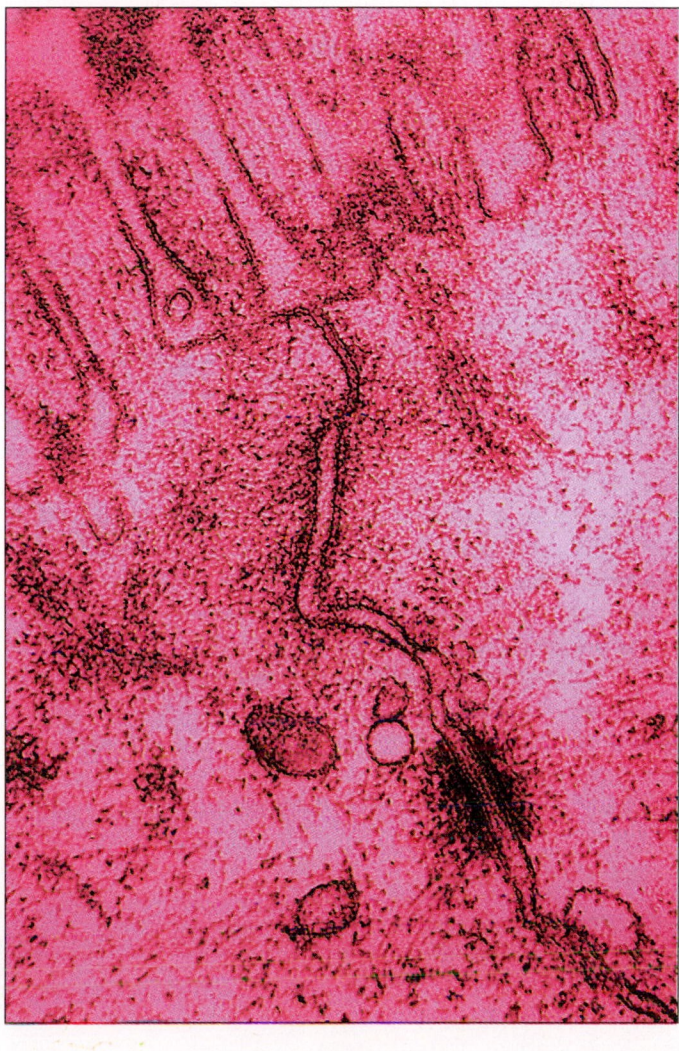

(b)

Figure 4–27

(a) Transport pathways across a layer of epithelial cells.
(b) An electron micrograph of two intestinal epithelial cells
and typical junctional complexes (*arrows*). (× 70,000)
(© *John Hansen/Photo Researchers, Inc.*)

ing tubule of the kidney, the tight junctions are much less
permeable; the transcellular pathway is the major route by
which water and solutes cross the epithelia. These types of
epithelia are termed **"tight"** epithelia because they do not
permit leakage between the cells.

Water movement across epithelia is always passive and is
driven by the osmotic gradient. This gradient is set up by the
active transport of solutes from the lumen and their addition
to the fluid in the interstitial space.

A problem that leaky epithelia face is **backflux** of solute
through the paracellular pathway. Thus, in the proximal
tubule, the Na^+ that is reabsorbed from the lumen by the
transcellular route is added to the fluid in the interstitial
space (see Fig. 4–27a), tending to increase the Na^+ concentration of the fluid. If the Na^+ concentration increases to the
point at which it exceeds the concentration in the fluid in the
tubular lumen, a Na^+ concentration gradient exists. In addition to favoring water reabsorption from the lumen, this Na^+
gradient drives the movement of some of the Na^+ back into
the lumen through the paracellular pathway. This backflux
limits the net reabsorption of Na^+ from the lumen. Movement of ions through the paracellular pathway is entirely passive and not under any direct control. Ions move in the
direction favored by their electrochemical gradient.

The transcellular route achieves net movement from the
lumen at one side of the cell to the blood supply at the other
side. It is the principal route across tight epithelia, and it can
be regulated by hormones. This regulation allows almost
complete control of the reabsorptive process by specific hormones that are able to regulate either the entry step at the
apical membrane or the exit step at the basolateral membrane. For example, the reabsorption of water in the collecting tubule of the kidney is regulated by antidiuretic
hormone, which increases the water permeability of the apical membrane of the tubular cells.

In addition to functional specialization of reabsorptive
epithelia is marked structural specialization. The surface area
of the apical membrane is often increased considerably by
numerous finger-like extensions known as **microvilli.** Similarly, the surface area of the basolateral membrane is increased by extensive **infoldings** that project into the cell.
Large numbers of mitochondria are present. They are frequently found between the basolateral infoldings (Fig. 4–28),
where presumably they provide a plentiful supply of ATP to
maintain the activity of the Na^+/K^+-ATPase pump in the basolateral membrane. In other words, the cells are adapted for
the rapid transport of large amounts of solutes and water.

The Functions of Different Transporters Are Coordinated

We can understand how the different transport systems interact to achieve net movement of a solute across an epithelial cell
by considering how inorganic phosphate crosses the proximal
tubule in the kidney (Fig. 4–29). The initial step in this process
is the entry of phosphate into the cell across the apical mem-

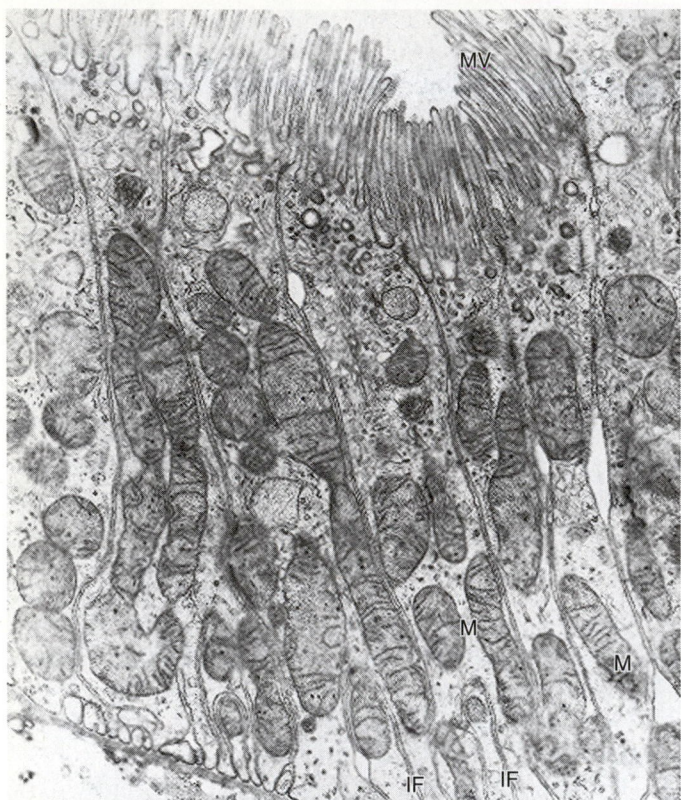

Figure 4–28

An electron micrograph of a kidney proximal tubule cell showing microvilli (MV) of the apical plasma membrane, extensive infoldings (IF) of basal and lateral membranes, and mitochondria (M) between the basolateral infoldings. (× 10,000)   (© *C. Craig Tischer, MD/University of Florida*)

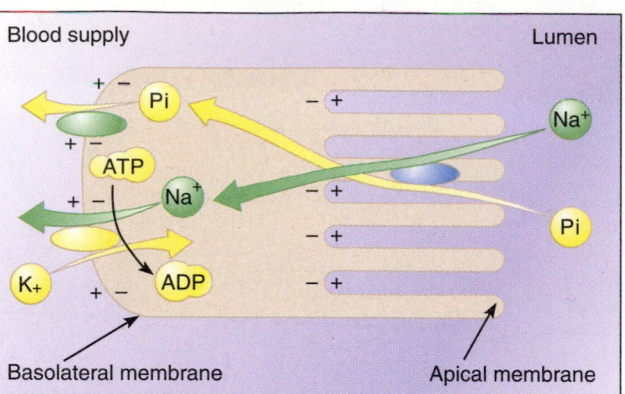

Figure 4–29

Transepithelial movement of inorganic phosphate (Pi) in the renal proximal tubule. Note that the Pi ion is negatively charged, Na^+ is positively charged, and there is a negative potential across the plasma membrane.

brane. This step requires active transport because the negatively charged phosphate ion must move against both an electrical gradient (a negative potential inside the cell relative to the lumen) and a chemical gradient (a higher phosphate concentration inside the cell relative to the lumen). Phosphate enters the cell via a symport mechanism in the apical membrane that couples the movement of phosphate to the movement of Na^+. There is a very favorable electrochemical gradient driving Na^+

entry. Thus, phosphate entry is an example of secondary active transport. Once inside the cell, the phosphate is able to leave by a passive mechanism (possibly facilitated diffusion) located in the basolateral membrane. The electrochemical gradient that opposed phosphate entry at the apical membrane now drives the exit step at the basolateral membrane. After diffusing across the interstitial space, the reabsorbed phosphate enters the blood in the capillary system and is returned to the general circulation. The Na^+ that enters the cell across the apical membrane is pumped out across the basolateral membrane by the Na^+/K^+-ATPase pump. The exit of Na^+ is an example of primary active transport because it is coupled directly to ATP hydrolysis. Active removal of Na^+ from the cell maintains a low intracellular Na^+ concentration and a favorable electrochemical gradient for continued Na^+ entry across the apical membrane.

An analogous arrangement of the appropriate transport systems is used to achieve the reabsorption of glucose and amino acids across the cells of the renal proximal tubule and the intestinal epithelium. The solutes enter the cell from the lumen via secondary active Na^+-coupled symport across the apical membrane. Then they exit from the cell via passive facilitated diffusion across the basolateral membrane.

CHAPTER REVIEW

Summary

- The core of a plasma membrane is the lipid bilayer, a double layer of lipid molecules. The bilayer is highly impermeable to charged or polar solutes but allows for the passage of small, lipid-soluble molecules.
- Most membrane proteins are embedded in the lipid bilayer, but a few are not. Those membrane proteins that are not are loosely associated with one side of the bilayer. The proteins carry out

many membrane functions by acting as specific receptors, enzymes, or transporters.
- Carbohydrates are a relatively minor component of the plasma membrane. They are attached to both proteins and lipids and are located exclusively on the extracellular surface of the membrane.
- Gradients of pressure, concentration, or voltage drive the movement of many substances in the body.

- Many solutes can cross the plasma membrane by diffusion down the electrochemical gradient, a process known as *passive transport*. Lipid-soluble, nonpolar molecules pass through the lipid bilayer, but the passage of charged or polar solutes is "facilitated" by specific membrane-spanning proteins.
- Osmosis is the diffusion of water across a plasma membrane in response to a difference in solute concentration on either side of the membrane. The rapid movement of polar water molecules occurs through specific water channel proteins termed *aquaporins*.

- Active transport produces movement of a solute across the plasma membrane, even though the solute is moved against its electrochemical gradient. This process requires a specific membrane carrier protein and an input of energy (usually in the form of ATP) derived from cellular metabolism. The energy-yielding reaction can be linked directly or indirectly to the transport step.
- The different types of transport systems act in concert in epithelial cells to achieve directional movement of solutes and water from one side of the cell to the other. This organization is aided by the structural and functional specialization of the cells.

Review Questions

Choose the Correct Answer

1. Which of the following processes requires a supply of metabolic energy?
   a. Simple diffusion
   b. Facilitated diffusion
   c. Primary active transport
   d. Osmosis
   e. Diffusion trapping
2. Simple diffusion differs from facilitated diffusion because it:
   a. is much slower.
   b. can be blocked by competitive inhibitors.
   c. reaches a maximum rate.
   d. requires expenditure of energy.
   e. moves solute against a gradient.
3. The plasma membrane of a cell separates:
   a. organelles from cytosol.
   b. cytosol from chromosomes.
   c. cytosol from mitochondrial matrix.
   d. extracellular fluid from cytosol.
   e. extracellular fluid from blood plasma.
4. In the phospholipid bilayer of a plasma membrane:
   a. phospholipids are distributed asymmetrically.
   b. proteins are distributed asymmetrically.
   c. both phospholipids and proteins move laterally.
   d. the interior is a viscous fluid at 37°C.
   e. all the above are true.
5. The plasma membrane Na^+ pump:
   a. moves K^+ out of the cell.
   b. exchanges 3 Na^+ for 2 K^+.
   c. uses energy released by ADP breakdown.
   d. moves Na^+ into the cell.
   e. is an ion channel.
6. A membrane that is permeable only to water separates two solutions of glucose dissolved in water. On one side of the membrane (side A), the glucose concentration is 0.1 g/ml. On the other side (side B), the glucose concentration is 0.5 g/ml. Initially, the rate of water flow is:
   a. most rapid from side *A* to side *B*.
   b. most rapid from side *B* to side *A*.
   c. the same in both directions.
   d. zero (no flow in either direction).
   e. none of the above.
7. The plasma membrane of a typical cell is least permeable to which of the following?
   a. Carbon dioxide
   b. Oxygen
   c. Sodium ions

   d. Ethanol
   e. Water
8. The rate of solute diffusion decreases in response to an increase in:
   a. the concentration gradient.
   b. membrane permeability.
   c. temperature.
   d. the molecular weight of the solute.
   e. the diffusion coefficient.
9. Which of the following uses a membrane carrier protein to cross the plasma membrane?
   a. Ethanol
   b. Glycerol
   c. Glucose
   d. Oxygen
   e. Water
10. The process that transports amino acids across the lumenal membrane of intestinal epithelial cells is:
    a. primary active transport.
    b. simple diffusion.
    c. the sodium pump.
    d. cotransport with chloride ions.
    e. cotransport with sodium ions.
11. Which of the following transport pathways across plasma membranes is not highly selective for specific molecules?
    a. Active transport via carrier proteins
    b. Simple diffusion through the lipid bilayer
    c. Facilitated diffusion through ion channels
    d. Facilitated diffusion via carrier proteins
    e. Diffusion of water through aquaporins
12. Which of the following is true for membrane proteins that perform secondary active transport?
    a. They contain multiple transmembrane alpha-helical segments.
    b. They are able to "flip" within the membrane to release the transported molecule on the other side.
    c. They provide a permanent open pore through the membrane.
    d. They possess ATPase activity for direct coupling of ATP hydrolysis to transport.
    e. Their activity is controlled by various gating mechanisms.
13. Which of the following statements is *false*?
    a. Osmosis can cause cells to burst.
    b. Osmosis involves water movement through specific protein channels.
    c. Osmosis may involve large-scale movement of water molecules.
    d. Osmosis is involved in meat preservation by salting.
    e. Osmosis is water flow from high solute to low solute concentration.

14. The osmolarity of 50 mM Na_2SO_4 is:
    a. 50 mOsm.
    b. 100 mOsm.
    c. 150 mOsm.
    d. 200 mOsm.
    e. 300 mOsm.
15. Inhibition of the Na^+/K^+-ATPase in the basolateral membrane of intestinal epithelial cells results in decreased absorption of glucose and amino acids because this:
    a. ATPase directly transports glucose and amino acids.
    b. disrupts the Na^+ gradient that drives transport of glucose and amino acids across the apical membrane.
    c. disrupts the K^+ gradient that drives transport of glucose and amino acids across the basolateral membrane.
    d. blocks the paracellular transport pathway.
    e. decreases backflux of glucose and amino acids through the paracellular pathway.
16. Transmembrane proteins are:
    a. peripheral proteins.
    b. excluded from the lipid bilayer.
    c. the major component of nerve myelin membranes.
    d. integral proteins that span the lipid bilayer.
    e. attached to carbohydrates that lie on the cytosolic side of the membrane.
17. What is the most abundant cation (highest concentration) in the cytosol of a mammalian cell?
    a. Chloride
    b. Sulfate
    c. Calcium
    d. Sodium
    e. Potassium
18. Which of the following statements is *false?* One function of the plasma membrane is to:
    a. synthesize proteins for the cell.
    b. transport water and solutes into and out of the cell.
    c. bind chemical signals from other cells.
    d. structurally link cells together.
    e. keep the composition of the cytosol different from extracellular fluid.
19. According to the fluid mosaic model of membrane structure, plasma membranes are:
    a. a fluid lipid bilayer with carbohydrates embedded in it.
    b. a lipid bilayer with outer surfaces covered by proteins.
    c. mainly phospholipids with a few nucleic acids.
    d. perfectly symmetrical structures.
    e. none of the above.
20. A mammalian plasma membrane composed of 25% lipids and 75% proteins is likely to:
    a. be found in cells that insulate nerves.
    b. have important transport functions.
    c. perform the same functions as a membrane with a lipid to protein ratio of 3:1.
    d. be heavily involved in energy transformation by the cell.
    e. none of the above.

Answers to Case History Questions

1. The patient's blood contains many spherical erythrocytes (spherocytes), whereas normal blood contains only biconcave erythrocytes. When placed in dilute NaCl solution, the normal biconcave erythrocytes first swell due to entry of water, but they do not burst until they have reached a spherical shape. In contrast, the spherocytes are already at a maximum volume and cannot swell because the plasma membrane does not stretch. Any entry of water causes them to burst. Thus, the spherocytes hemolyze more readily when placed in the same dilute NaCl solution.

2. Na^+ is prevented from accumulating inside the red blood cell by the membrane Na^+/K^+-ATPase, which pumps them out as fast as they enter. The pump mechanism requires a source of energy (ATP) to perform this active transport. When red blood cells are incubated at 37°C, the intracellular ATP supply decreases due to the continuous activity of the Na^+/K^+-ATPase pump. If the ATP supply is not replenished, the pump stops working and Na^+ accumulates. Accumulation of Na^+ inside the cells causes water to enter the cells, which swell and eventually hemolyze. Under these conditions, spherocytes hemolyze more readily because they have no room for the water that enters. The addition of glucose allows the cells to make more ATP and maintain the activity of the pump to drive Na^+ out. This prevents water entry and delays appreciable hemolysis in the patient's spherocytes.

3. The life span of a spherical red blood cell is significantly shorter than the 90- to 120-day life span of a biconcave red blood cell. If the premature destruction of the spherocytes is not matched by production of new ones, the overall population of red blood cells decreases, resulting in anemia. If the patient is otherwise healthy, the decreased life span is partially compensated by an increase in the production of new red blood cells. However, the rate of new cell production (erythropoiesis) is decreased during an illness involving fever; this process makes the anemia temporarily more severe until the fever subsides.

4. A major reason for the shortened life span of spherocytes is that they are destroyed much faster in the spleen. Removal of the spleen increases the life span of the red blood cells, which corrects the anemia. The biconcave shape of a normal erythrocyte allows it to be flexible, so it can squeeze through narrow spaces in the spleen (see Chapter 28, "Body Defense and the Immune Response). Spherocytes are much less flexible and are delayed as they pass through the spleen. The increased size of the spleen may be a response that is triggered in part by the trapped red blood cells. The trapped cells are destroyed either by splenic cells that engulf them or by hemolysis that results from the inability to replenish ATP while in the spleen.

5. Hereditary spherocytosis. It often escapes detection until adult life and is most prevalent in people from northern Europe, where the prevalence may be 1 case in 5000 people. Most cases are due to inheritance from parents, but 20% of patients have unaffected parents, suggesting that the mutation also can arise spontaneously.

6. The molecular abnormality involves the proteins of the cytoskeleton (see Chapter 3) that underlies the plasma membrane, primarily the proteins (e.g., spectrin) that tie the membrane to the cytoskeletal network. When these proteins are defective in a red blood cell, there are regions of the membrane that are not well anchored to the cytoskeleton. These regions tend to break off, and the membrane left behind reseals to keep the cell intact. As a result, the cell has less and less surface membrane to enclose the same volume and it gradually assumes an overall spherical shape.

Key Terms

active transport (p. 125)
amphipathic molecule (p. 109)
antiport (p. 126)
aquaporin (p. 118)
carrier protein (p. 121)
channel gating (p. 123)
cholesterol (p. 110)
diffusion coefficient (p. 115)
diffusion trapping (p. 117)
electrochemical gradient (p. 115)
epithelial transport (p. 130)
extrinsic protein (p. 111)
facilitated diffusion (p. 114)
fluid mosaic model (p. 109)
intrinsic protein (p. 111)
iso-osmotic (p. 118)
lipid (p. 109)
lysis (p. 120)
Na^+/K^+-ATPase pump (p. 125)
osmosis (p. 117)
osmotic concentration (p. 118)
osmotic fragility (p. 120)
passive transport (p. 114)
permeability coefficient (p. 115)
phospholipid (p. 109)
simple diffusion (p. 114)
symport (p. 126)
tight epithelium (p. 131)
tight junctions (p. 130)
transepithelial transport (p. 130)
volume regulation (p. 130)

Suggested Readings

Ashcroft, F. M. *Ion Channels and Disease.* San Diego, Academic Press, 1999.

Beutler, E., and Luzzatto, L. "Hemolytic anemia." *Seminars in Hematology,* 36:38–47, 1999.

Bonny, O., and Hummler, E. "Dysfunction of epithelial sodium transport. From human to mouse." *Kidney International,* 57:1313–1318, 2000.

Brown, D., and Breton, S. "Sorting proteins to their target membranes." *Kidney International,* 57:816–824, 2000.

Curran, M. E. "Potassium ion channels and human disease: Phenotypes to drug targets?" *Current Opinion in Biotechnology,* 9:656–572, 1998.

Ganetzky, B. "Genetic analysis of ion channel dysfunction in *Drosophila.*" *Kidney International,* 57:766–771, 2000.

King, L. S., Yasui, M., and Agre, P. "Aquaporins in health and disease." *Molecular Medicine Today,* 6:60–65, 2000.

Ko, K. S., and McCulloch, C. A. "Partners in protection: Interdependence of cytoskeleton and plasma membrane in adaptations to applied forces." *Journal of Membrane Biology,* 174:85–95, 2000.

Marples, D., Frokiaer, J., and Nielsen, S. "Long-term regulation of aquaporins in the kidney." *American Journal of Physiology,* 276:F331–F339, 1999. (January)

Sharon, N., and Lis, H. "Carbohydrates in cell recognition." *Scientific American,* 268:82–89, 1993.

Skach, W. R. "Defects in processing and trafficking of the cystic fibrosis transmembrane conductance regulator." *Kidney International,* 57:825–831, 2000.

Tse, W. T., and Luz, S. E. "Red blood cell membrane disorders." *British Journal of Haematology,* 104:2–13, 1999.

Wright, E. M. "Genetic disorders of membrane transport: I. Glucose galactose malabsorption." *American Journal of Physiology,* 275:G879–G882, 1998.

Wright, E. M. "Renal Na^+-glucose transporters." *American Journal of Physiology,* 280:F10–F18, 2001.

Web sites

www.ncbi.nlm.nih.gov/Omim
The Omim database is a catalog of human genes and genetic disorders developed by The Johns Hopkins University and the National Center for Biotechnology Information. It provides information, pictures, and references for physicians, genetics researchers, and advanced students in science and medicine. Links to additional resources are included.

nips.physiology.org
Web site of the journal *News in Physiological Sciences,* which specializes in publishing short reviews of important developments in all areas of physiology. Full use of the site and access to the articles require a subscription, but there is no charge for access to article abstracts and a search function to locate topics of interest.

www.the-aps.org
The American Physiological Society fosters scientific research and education and provides the physiological community with scientific information that is current and usable.

Chapter 4

1. c **2.** a **3.** d **4.** e **5.** b **6.** a **7.** c **8.** d
9. c **10.** e **11.** b **12.** a **13.** e **14.** c **15.** b
16. d **17.** e **18.** a **19.** e **20.** b

Chapter 5

CELLULAR CONTROL MECHANISMS

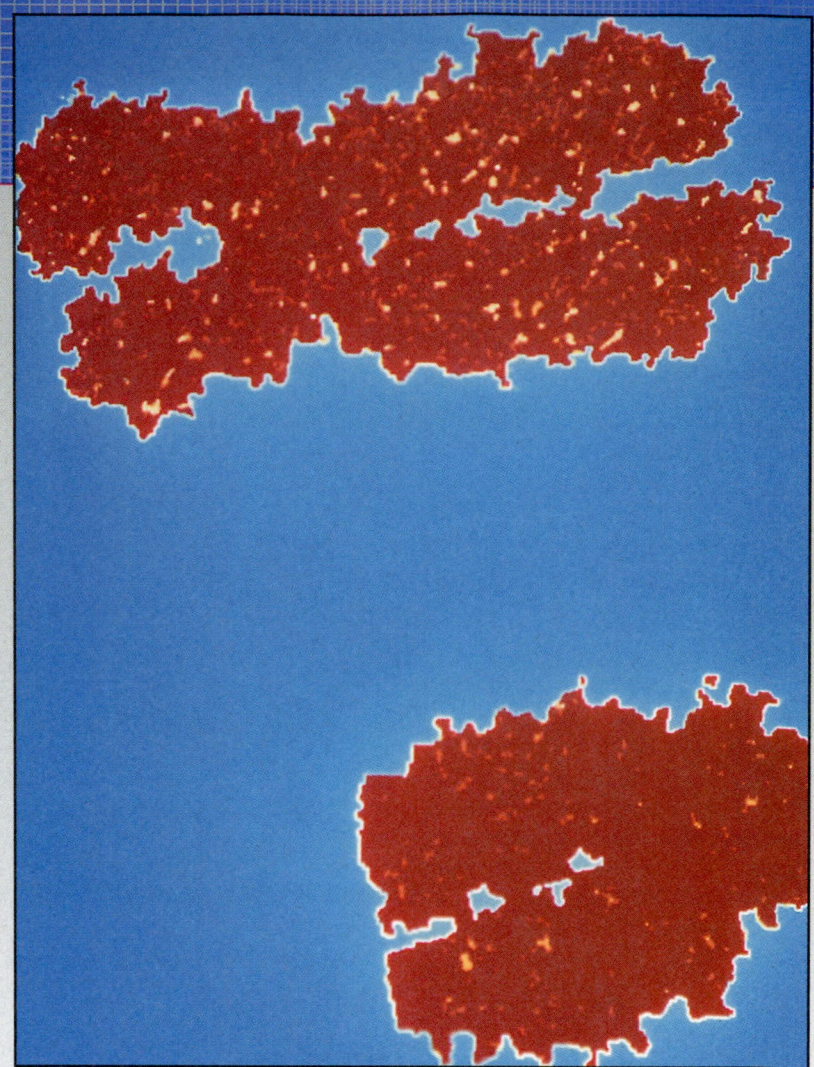

KEY CONCEPTS

- *Different genes are expressed in different cells. The end products of the expression of most genes are proteins, and the specific proteins made by a cell determine its structure and function.*

- *Communication between cells allows for coordination of the functions of the different organs and tissues in the body. Adjacent cells can communicate via gap junctions; cells that are far apart release chemical signals.*

CASE HISTORY

A physician-scientist in the pediatrics department of a major university medical center specializes in understanding the genetic basis of developmental diseases of the brain. The scientist was consulted by clinical staff about a 4-year-old girl who was recently admitted to the hospital because of seizures and breathing difficulties. The girl was discharged after the acute symptoms subsided. Her medical history showed an uneventful pregnancy and delivery, and her postnatal growth and development proceeded normally until she reached 12 months of age. However, after this point came gradual loss of expressive speech and of acquired motor skills, such as purposeful and coordinated use of the hands. Growth of the head had slowed; the head circumference at her current age was significantly smaller than the normal range. The child was now severely mentally retarded and unable to sit up or walk. No metabolic problems or physical abnormalities were detected. After obtaining the consent of the parents, the clinical staff asked the scientist to examine some of the patient's genes, using a sample of her blood drawn previously for routine testing. The scientist isolated DNA from the white blood cells and checked some known genes to determine whether they were normal or abnormal (mutated). Two weeks later, the scientist reported that only the **MECP-2** gene was mutated in this patient, all the other genes were normal. On reviewing the literature, the scientist discovered that the *MECP-2* gene encodes for **methyl-CpG-binding protein 2,** a protein that binds to methyl groups on DNA. Specific mutations in *MECP-2* have been found only in females.

Questions

1. Why does the problem appear to be specific to females?

2. What is a possible cause of the clinical symptoms?

3. How does methylation control gene expression?

4. What happens to the brain when the *MECP-2* gene is mutated?

5. What clinical problem is suggested by these findings?

INTRODUCTION

The success of multicellular organisms is based on the organization of groups of differentiated cells into specialized tissues and organs. The tissue functions are coordinated to carry out all the processes necessary for the organism's survival.

The specialization and coordination of cell functions require effective communication, both within each cell and among the different tissues. A good example is the process involved in avoiding a predator. When the tissues responsible for sight, sound, or smell detect the presence of a predator, the survival of the organism often depends on the almost instantaneous response of other tissues that can carry out rapid evasive movement. Thus, we must have systems that provide communication and coordination among the different organs. These systems are the **nervous system** and the **endocrine system,** and a large portion of this book is devoted to these sophisticated systems.

The emphases of this chapter are on the specific control systems that reside and operate within a cell and the ways in which these controls are influenced by signals from other cells. We begin by discussing the use of the genetic information stored in DNA, perhaps the fundamental **intracellular control system.** We will learn how intracellular factors regulate this system. The latter part of the chapter introduces the concept of **extracellular control systems,** or **cell-cell signaling,** the ways in which one cell communicates with and influences the function of another.

AN INTRACELLULAR CONTROL SYSTEM: DNA, RNA, AND PROTEIN SYNTHESIS

The cells of a complex eukaryotic organism, such as a human, have developed a variety of structural and functional specializations, giving rise to epithelial, connective, muscle, nervous, and other tissues. The molecular basis for the differentiation of structures and functions of cells is the structure of their proteins. In Chapter 4 we learned, for example, that different cells have different types and amounts of proteins in their membranes. These proteins are distributed in specific ways in the membrane in order to effectively perform their roles as carriers, channels, receptors, or shaping agents. An islet cell in the pancreas produces insulin, but a nerve cell makes none of this protein. All cells in the body make and use Na^+/K^+-ATPase, but different cells make different amounts of it. Most cells in the body produce myosin (a contractile protein), but only muscle cells make it in the abundance necessary to produce the type of motion that we associate with visible body movements.

Proteins Are the End Result of Expression of Most Genes

Markedly different types of cells contain the same **genome** in their nuclei—that is, each cell, regardless of tissue type, has the same DNA molecules, the same "permanent files" or

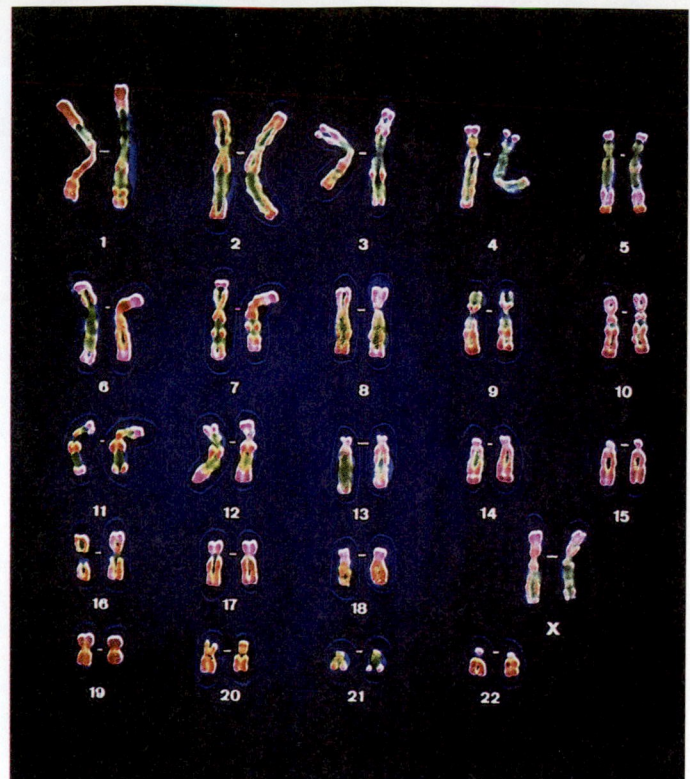

Figure 5–1

Human chromosomes. Each nonsex cell in a human body has 46 chromosomes identical to those in all other cells of the same body. The sex chromosomes are XX in the female (above) and XY in the male (not shown). Chromosomes are visible during cell division; at other times, some of the DNA in the nucleus is in a loosely coiled form. Each chromosome consists of a molecule of DNA wound around proteins called *histones*. Certain sequences of the nucleotides on each DNA molecule, called *genes*, contain the code for the order of amino acids in a protein molecule to be made by the cell.   *(CNRI/SPL/Photo Researchers, Inc.)*

master set of instructions for protein synthesis. These DNA molecules are in groups called **chromosomes.** Every cell in the human body, except for the reproductive cells and mature red blood cells, has the same number (46) of chromosomes containing the same DNA molecules (Fig. 5–1). These DNA molecules are made up of **genes,** and most genes carry the code for a specific protein. Each gene consists of a certain sequence of nucleotides that dictates the sequence of amino acids in a specific protein produced by the cell. Every cell in the body, except for the sex cells and mature red blood cells, can potentially make every protein found in the body.

It seems that if all cells in the body contained the same genetic information, they would all make the same proteins in the same amounts. How, then, do cells become **differentiated,** producing some proteins and not others? For example, how does a pancreatic islet cell know to produce an abundance of insulin, whereas an erythroblast (a developing red blood cell) knows *not* to produce insulin but to produce hemoglobin instead? How does a cell in the kidney know to produce myosin in the controlled amount for its own shape and internal movements, but not in the abundant amounts required for movement of the entire organ? How does a skeletal muscle cell know to produce myosin in abundance, enough to produce movement of other tissues?

Genetic Expression: Selective Use of the Permanent Files

The answer to the previous questions lies in the phenomenon of **genetic expression** or, more precisely, the **control of genetic expression.** An individual cell has much more genetic information than it ever needs. Most of the genetic information in the cell is unused, or **suppressed.** At any given time in a particular cell, only a small part of the DNA is being used as a source for information about protein synthesis. In other words, only certain genes are being tapped for information about how to make the specific protein they describe; only certain genes in each cell are **expressed** in the form of **protein products.**

Thus, gene expression is the principal factor that determines the structural, functional, and behavioral characteristics of a cell. In complex eukaryotes, different genes are being expressed as the characteristic protein products of each different cell type. Furthermore, in a given cell, certain genes may be expressed only at certain times, such as when the cell receives the appropriate stimulus, possibly a hormone.

The Genetic Code: From Nucleic Acids to Amino Acids

> *How does information stored in DNA determine which proteins will be produced by the cell?*

The term **genetic code** refers to the way in which information is stored in the sequence of bases in DNA. The base sequence specifies both the specific amino acids and the order in which they must be assembled to form a functional polypeptide molecule.

DNA and RNA are each composed of only four different nucleotides, whereas polypeptides can contain up to 20 different amino acids. Because there are only four nucleotides in each type of nucleic acid to specify the 20 different amino acids, it is **combinations** of nucleotides, rather than single nucleotides, that are used to describe the amino acids. The genetic code must specify, in the sequence in which the four nucleotides are arranged, the sequence in which the 20 common amino acids should be arranged. Think of DNA, RNA, and proteins as three different languages, each with a different alphabet and vocabulary. Protein synthesis, then, is a process of conversion, from the language of nucleic acids to the language of proteins.

THE FOUR LETTERS OF THE DNA AND RNA ALPHA-BETS. DNA molecules are formed from **deoxyribonu-cleotides,** four different molecules distinguished by four different **bases:** adenine, cytosine, guanine, and thymine

(Fig. 5–2*a*). The DNA "alphabet" consists of four letters—A, C, G, and T—representing the four DNA bases (Table 5–1). These four letters can be combined in a total of 64 different "words" of three letters, or **triplets,** such as CGA, CGT, CGG, and so on.

RNA molecules are formed from **ribonucleotides**—again, four different molecules distinguished by four different bases: adenine, cytosine, guanine, and uracil (see Fig. 5–2*a*). Note that uracil is used in RNA, instead of thymine. Thus, the RNA alphabet also contains four letters—A, C, G, and U (Table 5–1). These can also be combined in groups of three to make a total of 64 different "words," called **codons,** found on a special type of RNA called messenger RNA (mRNA). As we will learn, the codons on an mRNA molecule are **complementary** to the triplets on the DNA molecule from which they were transcribed. For example, the DNA triplet CGA has the complementary mRNA codon GCU (uracil is the RNA base complementary to adenine). Each codon represents a "word" in the language of mRNA, translated from the triplet "word" in the language of DNA.

THE TWENTY LETTERS OF THE PROTEIN ALPHA-BET. Each mRNA codon, in turn, refers to one of the 20 "letters" in the protein alphabet, one of the 20 common amino acids. Scientists have deduced the individual codons that represent the links between the nucleic acid strands and the polypeptide chains. Figure 5–2*b* gives examples of the way in which the relationship among the three languages results in an RNA molecule and a polypeptide.

Of the 64 different codons, 61 specify a particular amino acid. Several examples of specific codons are shown in Figure 5–2*b*. The codon for **methionine** (AUG) has a dual role because it serves as a **start codon,** which is involved in initiating **translation** of the nucleotide triplets in mRNA. The remaining 3 codons (UAA, UAG, and UGA) do not specify amino acids; instead, they are used as **stop codons** to terminate translation.

THE VAST LANGUAGE OF PROTEINS. In contrast to the language of nucleic acids, which is restricted to words of only three letters, there are no restrictions in the length of words used in the language of proteins. Consider the enormous number of words that can be generated from the English alphabet of 26 letters. Then consider the huge number of different polypeptide chains that could be constructed from nearly that many amino acids, without the constraints on "word" length found in verbal language. This will give you some sense of the variety of proteins that can be synthesized in the cell and the great possibilities for specialization of function that this variety allows. There may be as many as 10,000 different proteins in a typical cell of the human body.

Think of the difference in meaning one letter change makes in a word—for example, the change from *deride* to *decide*. This is similar to the difference that one amino acid can make in the nature of a protein. No other type of macromolecule allows for this great variety and specificity of structure

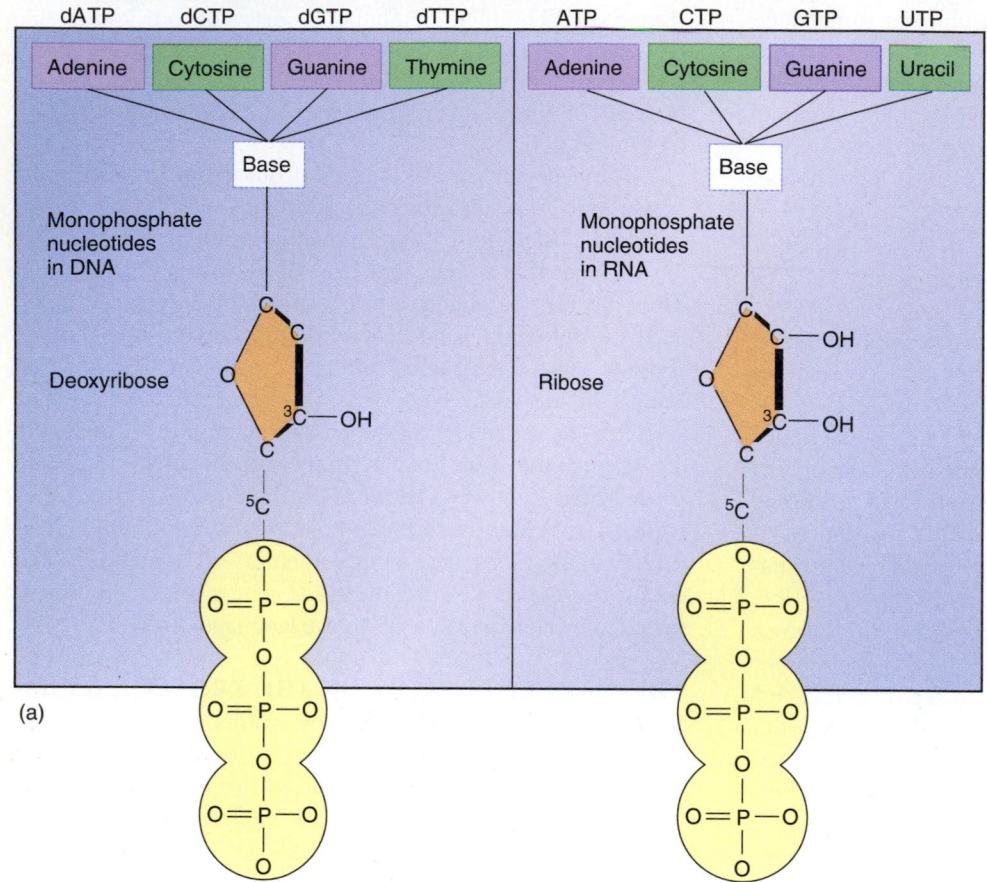

(a)

(b)

Figure 5–2

(a) The DNA "alphabet" consists of four nucleotides, derived from triphosphate nucleotides and distinguished by their bases: adenine, cytosine, guanine, and thymine. The RNA "alphabet" contains nucleotides formed from a different sugar molecule than that in DNA and distinguished by a slightly different set of bases: adenine, cytosine, guanine, and uracil. *(b)* The genetic code on DNA is copied from the template strand as a complementary strand of RNA, and then a matching chain of amino acids is formed. Each set of three bases on the DNA template strand, called a *triplet,* is transcribed into a complementary set of RNA bases, called a *codon.* Each codon refers to a specific amino acid.

TABLE 5–1					
DNA, RNA, and Protein "Alphabets"					
DNA Bases	**RNA Bases**		**Amino Acids**		
A	A	Alanine	Cysteine	Histidine	Methionine	Threonine
C	C	Arginine	Glutamic acid	Isoleucine	Phenylalanine	Tryptophan
G	G	Asparagine	Glutamine	Leucine	Proline	Tyrosine
T	U	Aspartic acid	Glycine	Lysine	Serine	Valine

and function. Compared with lipids, carbohydrates, and nucleic acids (which have only a few subunits to arrange), proteins (with 20 different subunits) have an enormous evolutionary advantage because of the variety of different functional macromolecules that can be assembled. Indeed, the molecular basis of the evolution of species lies in the ability of cells to make new proteins in response to changes in their environment.

A CONSISTENT AND UNIVERSAL CODE. The genetic code is said to be **consistent** because each codon specifies only a single amino acid. It is essentially **universal** because it is the same in diverse types of organisms, including plants and humans. Because there are 64 different codons for only 20 amino acids, some amino acids are specified by more than one codon. For example, there are two mRNA codons for phenylalanine (UUU and UUC), two for tyrosine (UAU and UAC), and three for isoleucine (AUU, AUC, and AUA). For this reason, the code is said to be **degenerate.**

A gene and its polypeptide product are usually regarded as **colinear** molecules. This term means that the codon representing the beginning of the gene determines the amino acid present at the beginning of the polypeptide, the codon representing the second triplet in the gene determines the second amino acid in the polypeptide, and so on. Many eukaryotic genes are interrupted with nucleotide sequences known as **introns;** in fact, introns are now known to account for much of the DNA in the genomes of eukaryotic cells. Most introns have no known cellular function; they generally are thought to be remnants of sequences that were important earlier in evolution.

Let us now look at how the nucleic acids themselves are assembled and translated into proteins. Figure 5–3 gives a simplified overview of the flow of information during these processes. We will discuss **DNA synthesis** (also known as **DNA replication**) and **RNA synthesis** (also known as **DNA transcription**). We will also look at methods of altering newly synthesized RNA, known as **RNA processing,** before it leaves the nucleus, methods that result in the production of functional RNA molecules. Finally, we will examine the process of **protein synthesis** (also known as **RNA translation**), in which the information in RNA is translated into a polypeptide chain.

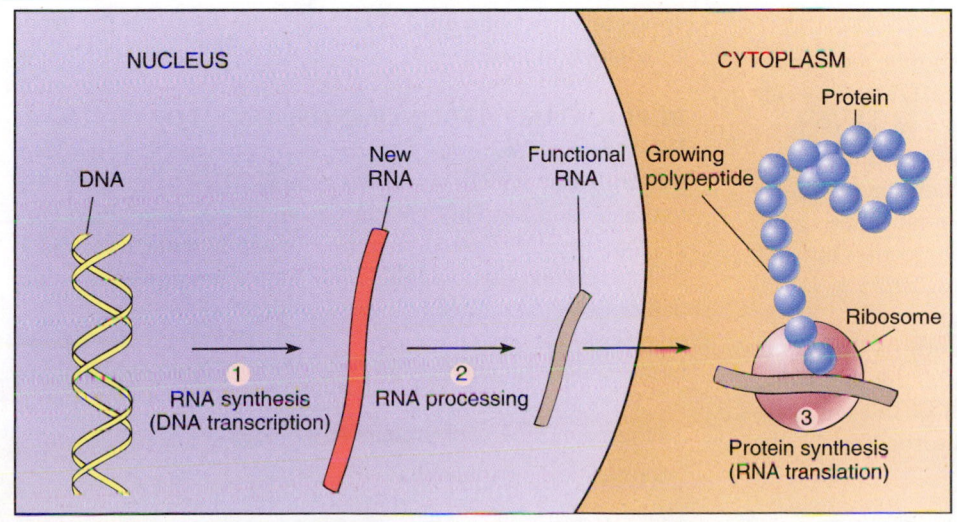

Figure 5–3

A simplified overview of the flow of information in the expression of the genetic code. The information on a DNA molecule is transcribed, or reproduced in coded form, on an RNA molecule. After the RNA molecule is processed, it leaves the nucleus and travels to a ribosome, where the coded information is translated into a sequence of amino acids that forms a polypeptide chain.

New Nucleic Acids Are Synthesized by Copying a Template

 How is the information stored in DNA passed on to subsequent cell generations?

As discussed in Chapter 2, the macromolecules deoxyribonucleic acid (DNA) and ribonucleic acid (RNA) are each composed of molecules called *nucleotides.* Nucleotides contain a ribose sugar and one to three phosphate groups, in addition to one of five bases. DNA and RNA are distinguished by the type of ribose sugar and bases their nucleotides contain. All nucleotides forming a part of a nucleic acid strand have one phosphate group.

Synthesis of DNA: Replication of the Permanent Files

When a cell divides into two, the contents of both the cytoplasm and the nucleus divide. Division of the nucleus, a process known as **mitosis,** ensures that both of the new cells have an identical copy of the DNA of the parent cell. Thus, the new cells will be identical genetically to the parent.

Before the nucleus can divide, the DNA molecules within the nucleus (in the form of **chromatin**) must be copied, or **replicated.** The replication process must be accurate so that the genetic information is not altered, and it must be fast so that it can occur in time for a cell to divide. DNA replication occurs during the **interphase** step that precedes mitosis. Each chromosome, a long DNA molecule, is copied to produce two identical DNA molecules called **chromatids.**

A DNA molecule consists of the two long strands of nucleotides arranged in a double helix (see Chapter 2). The two strands are held together by hydrogen bonds between the bases of the nucleotides. Adenine always forms bonds with thymine; guanine always forms bonds with cytosine. The specific base-pairing produces two strands that complement each other.

SEMICONSERVATIVE REPLICATION. During the process of DNA replication, the two strands of the DNA molecule first are separated. Each one then acts as a mold, or **template,** upon which a new complementary strand of nucleotides is assembled. In this way, two new molecules of DNA are produced. Each new molecule contains one strand that was the template from the parent molecule and one newly synthesized strand (Fig. 5–4). The process is described as **semiconservative** because the structure of the parent DNA molecule is disrupted, but both of the parent strands are conserved intact, one in each of the new DNA molecules. This process ensures that both of the new cells receive an exact copy of the genetic information in the parent cell.

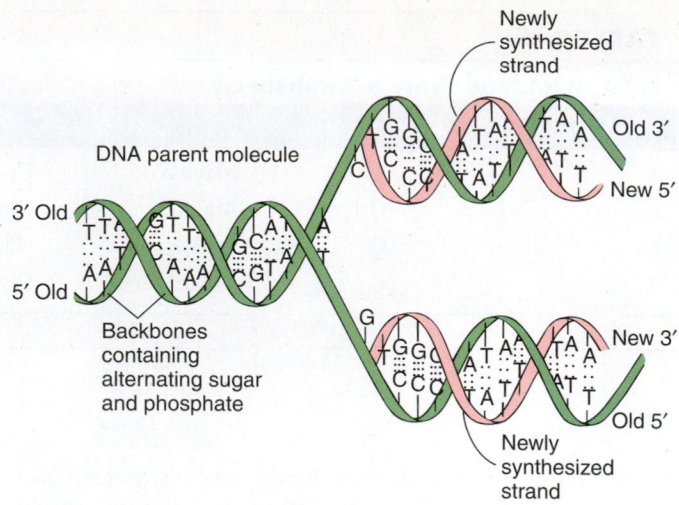

Figure 5–4

DNA replication is known as *semiconservative replication* because each new DNA molecule contains one new strand and one original strand. Thus, single strands of DNA are preserved and passed on for many generations.

SEPARATING THE STRANDS OF THE DOUBLE HELIX. Before replication of DNA can take place, the double helix must be unwound to separate the two strands. Unwinding is accomplished by an enzyme known as **DNA helicase. DNA topoisomerases** act downstream of the helicase to relieve the torsion that develops as the double helix unwinds. The separated strands are stabilized by proteins that bind to the single DNA strands and therefore assist the opening of the helix. They are known as **helix-destabilizing proteins.** Each strand then acts as a template for the assembly of a complementary chain. Separation of the two original DNA strands produces a structure called a **replication bubble** at the site of active synthesis of the new strands (Fig. 5–5). Many replication bubbles can exist on a DNA molecule at once, and the replication forks within each replication bubble move in opposite directions until they meet the fork coming from an adjacent bubble.

REPLICATING THE STRANDS: DNA POLYMERASE. DNA replication is catalyzed by an enzyme called **DNA polymerase,** which assembles a new DNA strand on the template strand. The specific reaction catalyzed by DNA polymerase is the joining together of the various nucleotides that are components of DNA. The individual nucleotides initially must contain deoxyribose and three phosphate groups; otherwise, the enzyme cannot use them. Thus, DNA polymerase takes the nucleotides:

deoxyadenosine triphosphate (dATP)

deoxycytidine triphosphate (dCTP)

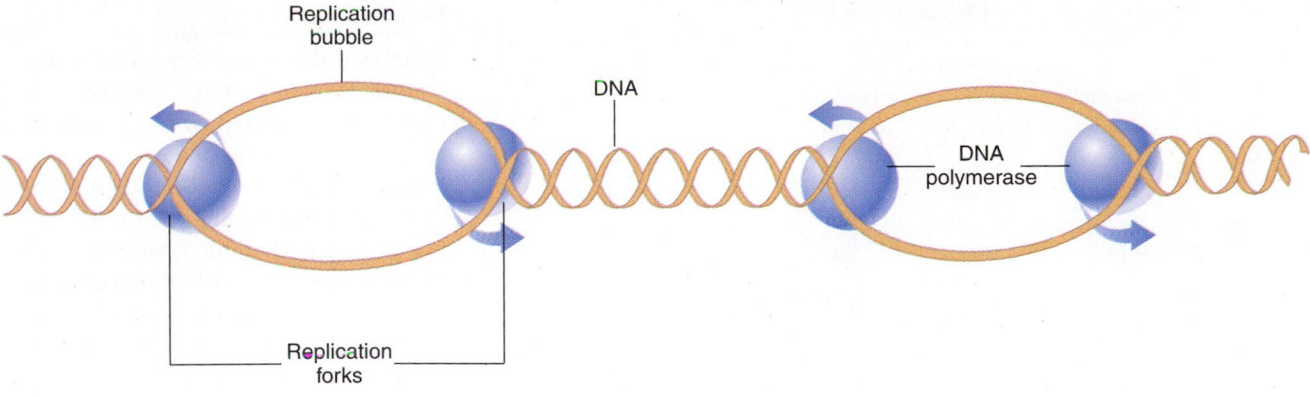

Figure 5–5

A replication bubble, containing two forks, is formed when double-stranded DNA is separated. DNA replication is bidirectional in mammalian cells because synthesis occurs at both forks in the bubble. Many bubbles are present because replication starts simultaneously at multiple sites. (Only the parent DNA strands are shown here.)

deoxyguanosine triphosphate (dGTP)

deoxythymidine triphosphate (dTTP)

and removes two of their phosphate groups to form the monophosphate nucleotides:

dAMP

dCMP

dGMP

dTMP

and joins them in the order specified by the complementary bases on the DNA template to form a new strand of DNA. Recall that the base C always forms a pair with the base G; and A, with T. Hence, a DNA strand of nucleotides with bases in the order ACAGTG serve as a template for a set of nucleotides with bases in the order TGTCAC. Base-pairing ensures that the order of bases in the new DNA strand is identical to the order in the original strand from which the template recently has separated.

Figure 5–6a depicts the end of a DNA strand as it is synthesized on a DNA template. The incoming nucleotide dCTP has been selected for attachment to the end of the strand because its base, cytosine (C), is the only base that can form hydrogen bonds with guanine (G), the base of the next nucleotide on the DNA template. The preliminary hydrogen-bonding step serves to line up the individual nucleotide molecules on the DNA template. DNA polymerase then catalyzes a reaction that covalently attaches dCTP to the end of the new strand. During the reaction, the two phosphate groups on the end of the dCTP molecule are removed, and the rest of the molecule is attached via the remaining phos-

phate to the hydroxyl group on carbon 3 of the deoxyribose ring of the nucleotide at the end of the new strand (see Fig. 5–6a). DNA polymerase can add nucleotides to the DNA chain at a rate of more than 20 per second; each one is always attached to carbon 3 of the sugar of the last nucleotide. Thus, the new DNA chain grows from the end where there is a free hydroxyl group on carbon 3.

SYNTHESIS IN THE 5′-TO-3′ DIRECTION. A special numbering system commonly is used for nucleotides in order to distinguish between the two ring structures that are present in the molecule—that is, the sugar and the base. By convention, the numbers of the carbon atoms of the sugar are given a prime (e.g., 3′ or 5′), whereas the carbon atoms of the base are numbered in the usual way. A DNA strand is formed by linking nucleotides together via phosphate groups between carbon 3′ of one nucleotide and 5′ of another. The nucleotide at the beginning of the DNA strand has a free phosphate group on the 5′ carbon. This end of the strand is usually called the **5′ ("five prime") end.** The nucleotide at the other end of the strand has a free hydroxyl on the 3′ carbon. This is termed the **3′ ("three prime") end** of the strand (see Fig. 5–6).

In theory, a new nucleotide could be attached to either the 5′ end or the 3′ end of the chain. In practice, however, DNA polymerase attaches nucleotides only to the 3′ end of a strand, so that the new strand always grows from the **5′-to-3′ direction.**

ANTIPARALLEL STRANDS: LEADING AND LAGGING STRANDS. In the DNA molecule, the 5′—3′ directions of the two complementary nucleic acid strands are opposite (see Fig. 5–4); they are said to be **antiparallel.** A DNA template

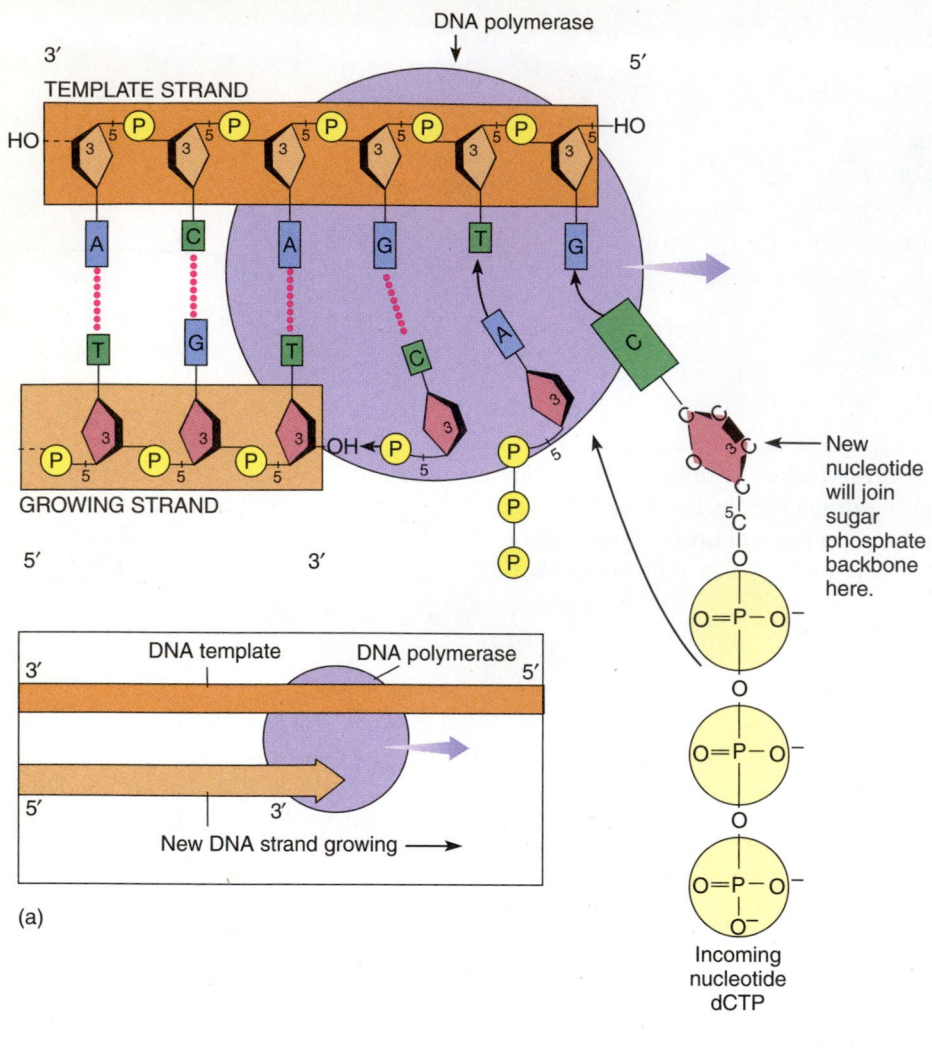

(a)

Figure 5–6

(a) DNA polymerase moves from the 3' end to the 5' end of a DNA template, joining each new nucleotide first to the complementary base on the template via hydrogen bonds and then to the previous nucleotide on the growing strand at the 3' carbon. DNA polymerase adds nucleotides only to the 3' end of the growing strand (so called because it has a free hydroxyl group attached to the 3' carbon of the sugar). *(b)* DNA polymerase works on both strands of the parent DNA molecule simultaneously. However, because the parent strands are antiparallel, and because DNA polymerase travels only in the 3'-to-5' direction on a DNA template, in effect it moves in opposite directions on the two parent strands. When DNA polymerase is moving in the same direction as DNA helicase, the growing strand is termed the *leading strand*. When DNA polymerase is moving away from the replication fork, the growing strand is called the *lagging strand*.

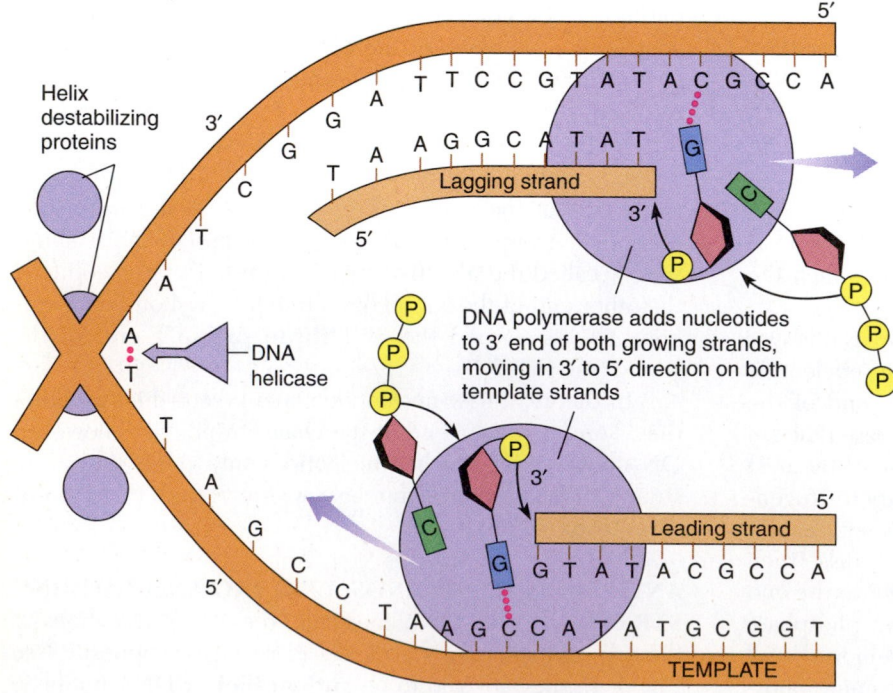

(b)

and its complementary growing strand are also antiparallel. Hence, a template with the nucleotide sequence 3′—TGGCGTATAC—5′ will direct the synthesis of the complementary and antiparallel strand 5′—ACCGCATATG—3′ (see Fig. 5–6*b*). This complicates the process of DNA replication because DNA polymerase moves along the *template* strands in the 3′-to-5′ direction only and synthesizes new DNA by adding nucleotides only to the 3′ end of the growing strand.

In the simultaneous replication of both strands of the DNA double helix, one of the new DNA strands, the so-called **leading strand,** is synthesized in a straightforward manner because it can grow in the 5′-to-3′ direction on the template strand (see Fig. 5–6*b*). It is synthesized continuously and "follows" the DNA polymerase as the polymerase moves from the 3′ end of the template strand.

The other new strand of DNA, the **lagging strand,** is formed by a process of **discontinuous synthesis,** in a series of small sections (see Fig 5–6*b*). Each section of the lagging strand is assembled in exactly the same way as a section of the leading strand, in the 5′-to-3′ direction by a DNA polymerase moving from 3′ to 5′ on the template. By the time one section is finished, the replication fork has opened up to expose more of the template, allowing another DNA polymerase to synthesize a second fragment of new DNA behind the first. However, DNA polymerase does not join the segments of the lagging strand. This task is carried out by another enzyme, called **DNA ligase.**

In summary, the replication fork depicted in Figure 5–6*b* is always moving to the left, and the overall growth of both of the new DNA strands is from right to left following the movement of the fork. However, the synthesis of both of the new strands is exclusively in the 5′-to-3′ direction, which means that the lagging strand actually is synthesized in the direction opposite of the movement of the replication fork. As already pointed out (see Fig. 5–4), this mechanism of DNA replication produces two double-stranded DNA molecules, each consisting of one nucleic acid chain from the "parent" and one newly synthesized chain.

PROOFREADING IN DNA REPLICATION. Replication of DNA is a remarkably accurate process. There is probably only one error per 10^9 base pairs. This efficiency is well within the limits required to maintain the genome of a mammalian cell (about 3×10^9 base pairs). One of the reasons for the high degree of accuracy is that the polymerase is able to check, or "proofread," the new chain as it is synthesized, meaning that the polymerase responds to the incorporation of an incorrect nucleotide by immediately removing it and inserting the correct one before moving on to the next. DNA replication is also remarkably rapid. The average human DNA chain contains more than 10^8 nucleotides linked together. Such a chain can be replicated completely within a few hours. This rapidity is possible because many replication forks are present on each DNA molecule that is undergoing replication, so that many different sections of the huge molecule can be replicated simultaneously.

Damaged DNA Can Be Repaired

The genetic information stored in DNA must remain essentially unchanged as it is passed on from cell to cell. The replication process is designed to faithfully reproduce molecules that are identical to the parent molecule, but it cannot correct alterations to DNA that occur in the intervals between the replication steps. DNA is sensitive to damage by factors commonly present in the environment. Ultraviolet light is an example of an environmental factor that damages DNA. Other factors include a number of common chemicals and the radiation produced by the decay of radioactive isotopes. An estimated several thousand bases in the genome of a mammalian cell are damaged every day. If permanent, changes on this scale in the structure of DNA could have drastic consequences, leading to mistakes in genetic information, malignant cells, and a threat to the long-term continuance of the species. Fortunately, cells have developed repair systems to correct almost any kind of damage, and very few stable changes are allowed to accumulate in the genome of each cell.

One example of a repair mechanism is the selective cutting out (excision) of the section of damaged DNA. The remaining gap can be filled relatively easily because the complementary strand has all of the information for the correct sequence of nucleotides needed to fill the gap. The process is known as **excision repair** and is initiated by a specific **endonuclease** enzyme that breaks the damaged DNA segment. Next, DNA polymerase replaces it with a new segment that is complementary to the undamaged strand. Finally, the 3′ end of the new segment is joined to the rest of the chain by **DNA ligase.**

Synthesis of RNA: Selective Transcription of the Permanent Files

> *How does a nucleotide sequence in DNA become transformed into synthesis of a specific protein in the cytoplasm of the cell?*

DNA does not provide protein-synthesizing information directly to the **ribosomes,** the sites where amino acids are assembled into proteins; the information in DNA is first transcribed into RNA. In fact, RNA is synthesized in a process known as **DNA transcription.**

Three kinds of RNA are synthesized: **messenger RNA (mRNA), ribosomal RNA (rRNA),** and **transfer RNA (tRNA).** All three kinds of RNA molecules interact closely in the mechanism whereby a nucleotide sequence in DNA is translated into a specific protein. All three types of RNA are transcribed from the DNA template strand; all three types of RNA consist of ribonucleotides complementary to DNA nucleotides. The nucleotides on mRNA molecules are translated directly into amino acids, whereas the nucleotides on tRNA and rRNA molecules provide other information crucial to protein synthesis. The interactions of the three RNA molecules will be discussed in more detail soon. Strictly speaking,

TABLE 5–2

Principal Differences Between DNA Synthesis and RNA Synthesis

	DNA Synthesis	RNA Synthesis
Sugar required	Deoxyribose	Ribose
Bases required	Adenine	Adenine
	Cytosine	Cytosine
	Guanine	Guanine
	Thymine	Uracil
Fate of DNA template	Incorporated into reaction product	Not incorporated into reaction product
Type of product	Double strand of nucleotides	Single strand of nucleotides

genes code for functional RNA molecules. Only when the functional RNA is mRNA can it be said that the gene codes for a protein.

The base sequence in the RNA is complementary to the sequence in the DNA template strand, and the base pairs formed ensure that the correct base sequence is copied (see Fig. 5–2). Synthesis of RNA differs from synthesis of DNA in four principal ways, which are summarized in Table 5–2.

TRANSCRIBING THE DNA TEMPLATE STRAND: RNA POLYMERASE. During transcription of DNA, the new chain of RNA is synthesized by the enzyme **RNA polymerase.** There are three distinct types of RNA polymerase enzymes in eukaryotic cells, and all three are located in the nucleus. **RNA polymerase I** synthesizes the major rRNA molecules in the nucleolus. The other polymerases are active in the portion of the nucleus that is outside the nucleolus. **RNA polymerase II** synthesizes mRNA; **RNA polymerase III** synthesizes tRNA and small rRNA molecules.

PROMOTER AND TERMINATOR SITES. Transcription of a given gene begins when the RNA polymerase binds to a **promoter** site on the DNA template strand, a discrete sequence of nucleotides immediately preceding the 3' end of the gene (Fig. 5–7). Polymerase action stops when the enzyme reaches the **terminator** site, which is a sequence of nucleotides immediately following the 5' end of the gene (see Fig. 5–7). At this point, the polymerase releases both the finished RNA strand and the DNA template. The promoter and terminator act as recognition sites only; they are not transcribed. The promoter sequence also determines which strand of the double-stranded DNA will be transcribed by the RNA polymerase.

SYNTHESIS IN THE 5'-TO-3' DIRECTION: ANTIPARALLEL STRANDS. During synthesis of RNA, the RNA polymerase moves along the DNA template strand in the 3'-to-5' direction, adding ribonucleotides to the 3' end of the growing RNA strand. Thus, the new RNA strand grows in the antiparallel 5'-to-3' direction (Fig. 5–8). A portion of the DNA template

with the base sequence 3'—CCAGTAAC—5' directs the incorporation of the sequence 5'—GGUCAUUG—3' into RNA (see Fig. 5–8). On a single DNA template strand, several RNA molecules may be synthesized at the same time (Fig. 5–9).

RNA Processing: Post-Transcriptional Modification

Before they leave the nucleus to direct protein synthesis, newly synthesized RNA molecules are covalently modified in a series of reactions known as **RNA processing** or **post-transcriptional modification.** Several examples are discussed later.

The RNA that is synthesized by the action of RNA polymerase II is known as **premessenger RNA (pre-mRNA).**

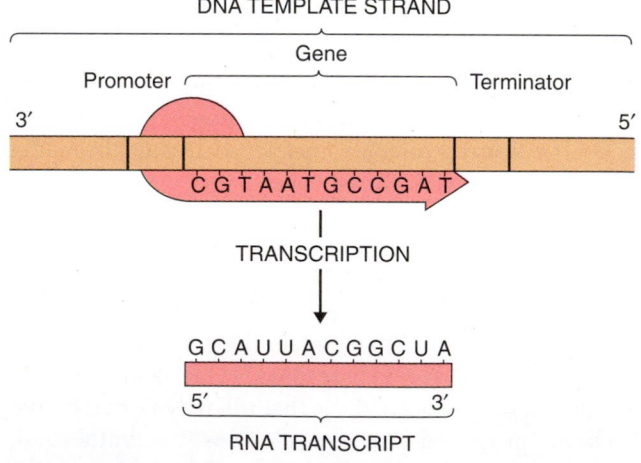

Figure 5–7

The synthesis of a molecule of RNA begins when the enzyme RNA polymerase encounters a sequence of DNA nucleotides called a *promoter.* Ribonucleotides are assembled in an order complementary to the order of nucleotides on the DNA template strand in a process known as *transcription,* producing an "RNA transcript." Transcription stops when the RNA polymerase encounters a terminator sequence on the DNA template strand.

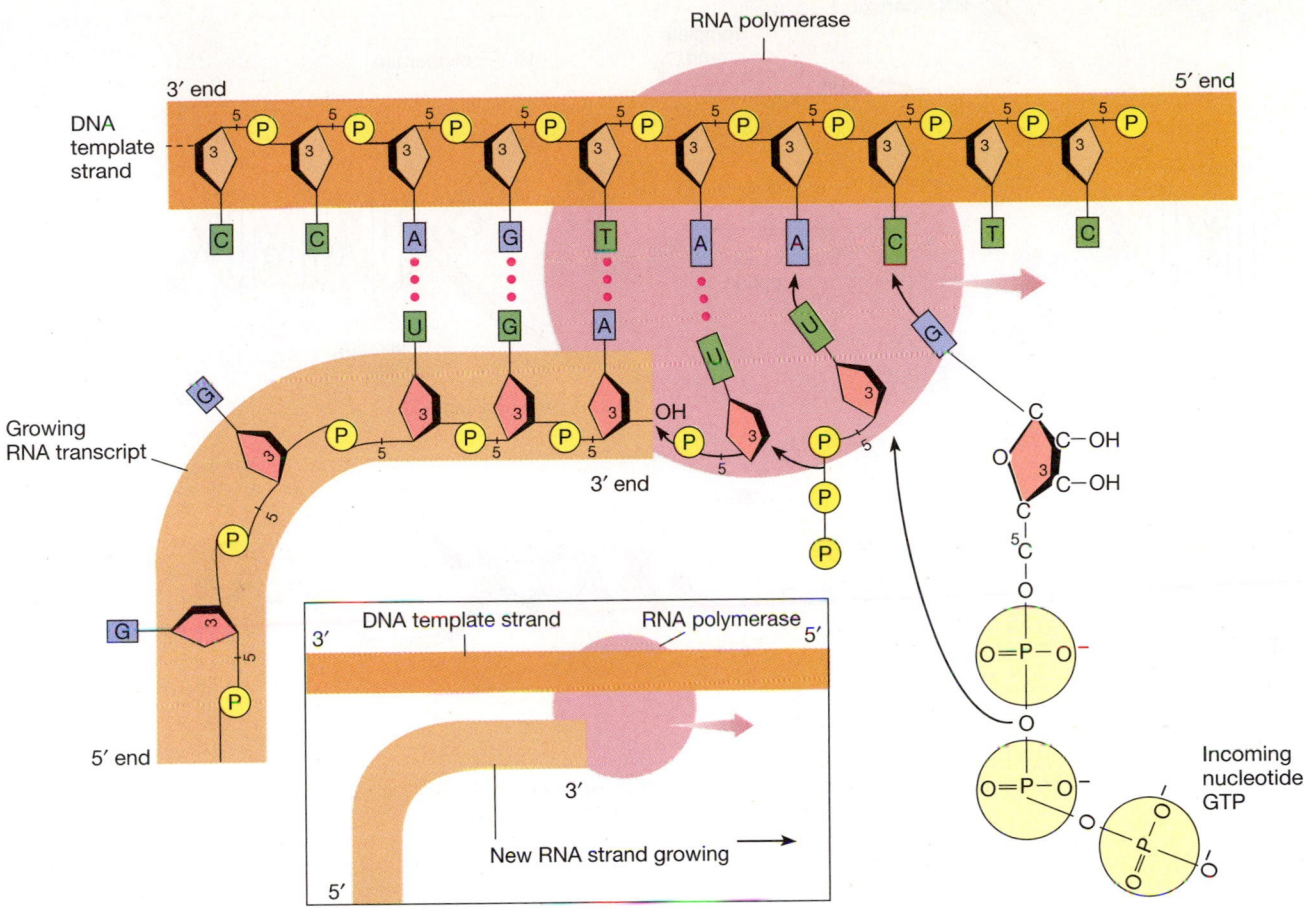

Figure 5–8

RNA polymerase works in the 3'-to-5' direction on a DNA template strand,
adding ribonucleotides to the 3' end of a growing RNA strand. The RNA is made
in the 5'-to-3' direction. Ribonucleotides initially form hydrogen bonds with
complementary bases on the DNA template, but when transcription is completed,
the RNA transcript is released.

The synthesis of pre-mRNA accounts for more than 50% of
the total RNA synthesis in a typical mammalian cell. Many of
these pre-mRNA molecules are destined to leave the nucleus
in the form of much smaller mRNA molecules.

THE 5' CAP AND THE POLY(A) TAIL. Processing of pre-
mRNA helps to provide the resultant mRNA with characteristics that distinguish it from rRNA and tRNA, the products
of the action of the other RNA polymerases. These characteristics are a **5' cap** and a **poly(A) tail.** The 5' cap is formed
by addition of **7-methylguanosine** and three phosphate
groups, represented in shorthand notation as **Gppp** (Fig.
5–10), to the nucleotide at the 5' end of the RNA transcript.
This is the end of RNA that is synthesized first, and it is
capped immediately—before the rest of the molecule is finished. The cap may serve to protect mRNA from enzymatic
degradation and to help align it on the ribosome during
translation. The poly(A) tail consists of 100 to 200 adenine
residues that are added to the 3' end of the RNA after tran-

scription is completed (see Fig. 5–10). The poly(A) tail may
be important for the stability and life span of the mRNA.

RNA SPLICING: INTRONS OUT. Newly synthesized RNA
molecules in eukaryotic cells contain intervening nucleotide
sequences known as *introns*, nucleotides transcribed from
DNA nucleotide sequences that contain no useful information (noncoding segments). These sequences are reproduced
in the pre-mRNA during transcription and must be removed
before the RNA leaves the nucleus. After the addition of the
5' cap and the poly(A) tail, the introns are cut out in order to
convert the RNA to mature mRNA that can be used to direct
the synthesis of a complete protein. This is another example
of RNA processing, and it is known specifically as **RNA splicing** (see Fig. 5–10) because the remaining segments, called
exons, are joined together to form functional mRNA.

It is estimated that at least 100 proteins are involved in
RNA splicing, making the process as complex as protein
synthesis. Splicing usually takes place in large complexes of

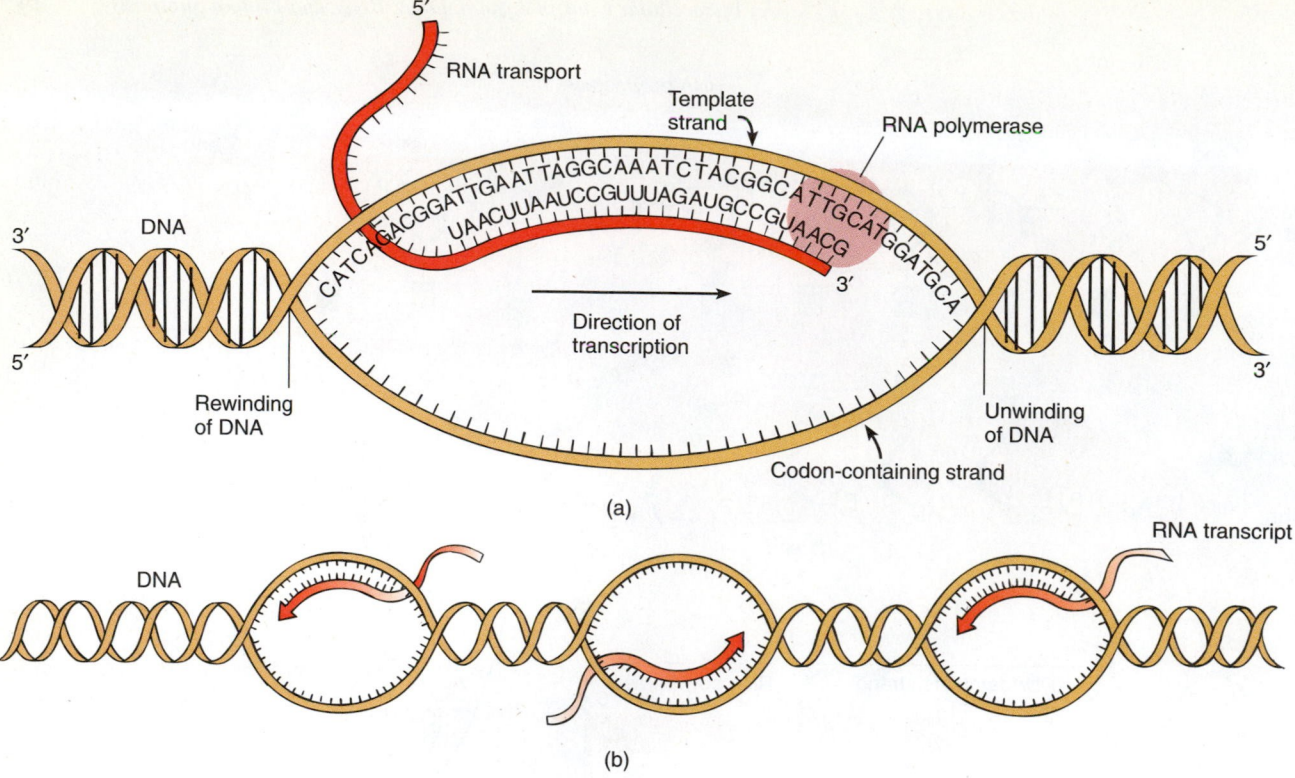

Figure 5–9

(*a* and *b*) Several different RNA transcripts may be synthesized simultaneously on a single DNA template strand.

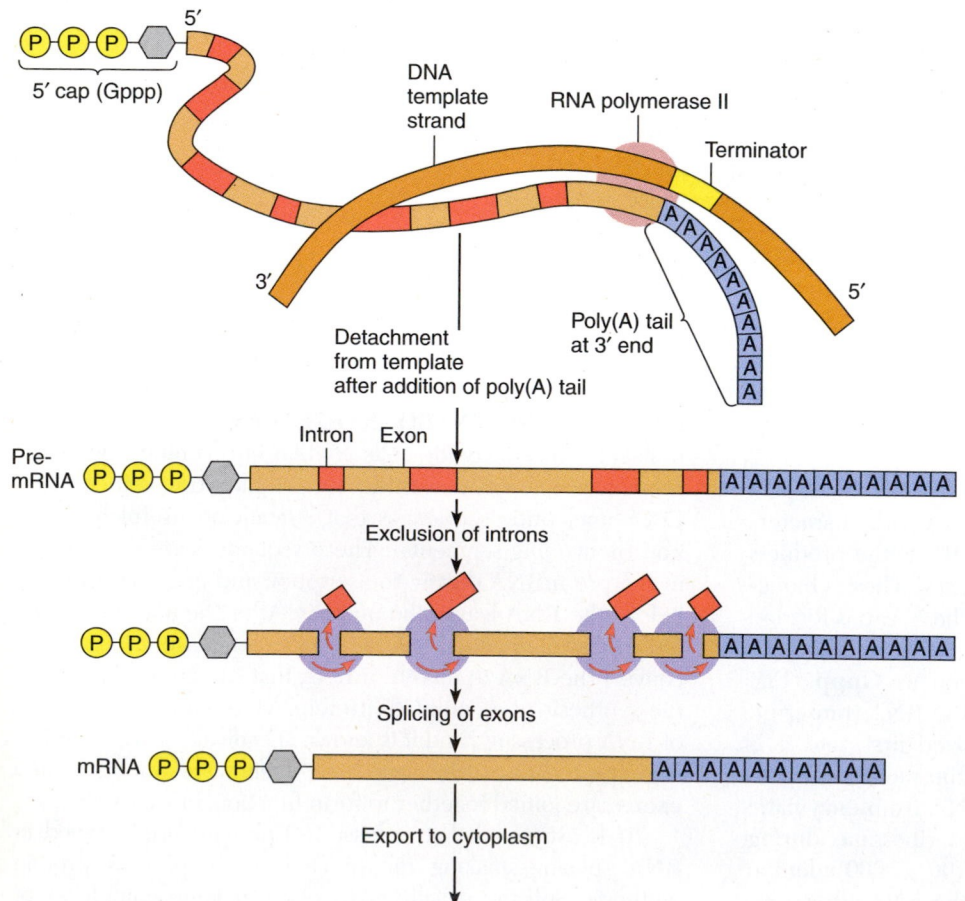

Figure 5–10

As part of the processing of newly synthesized RNA molecules, a "5' cap" and a "poly(A) tail" are added to the 5' and 3' ends, respectively. In addition, introns (transcripts of meaningless segments of the DNA template strand) are excised, with the remaining RNA segments, known as *exons*, spliced together.

proteins and RNAs, known as **spliceosomes,** which are similar in size to ribosomes. The RNA components of spliceosomes are **small nuclear RNAs (snRNAs),** which are between 50 and 200 nucleotides in length. Studies on the splicing mechanism led to the important discovery that the reactions were catalyzed by the snRNAs rather than the proteins of the spliceosome. These were some of the first findings that revealed the unexpected enzymatic activity of some forms of RNA.

RNA processing removes most of the mass of the pre-mRNA. The mature mRNA that is produced by these reactions accounts for only 3% of the total RNA present in a cell.

FORMATION OF FUNCTIONAL TRNA. A molecule of tRNA is relatively small, usually 70 to 80 nucleotides long, compared with the other types of RNA. The structure adopted by a tRNA molecule is shown in Figure 5–11*a*; the three-dimensional L shape is depicted in Figure 5–11*b*. Transfer RNA molecules have two important features. One, called the **amino acid accepting end** or the **aminoacyl attachment site,** is where an amino acid destined to join a

growing polypeptide chain attaches. The other, called the **anticodon,** is a set of three nucleotides complementary to a codon on an mRNA molecule. Each amino acid has a specific tRNA molecule designed to convey it to a ribosome and link it in the correct order to another amino acid in a process to be described later.

The presence of specific bases, known as *identity elements,* on a tRNA molecule ensures that it links up with the correct amino acid. In some tRNA molecules, these elements include the bases that form the anticodon as well as other bases in the molecule. In others, the anticodon is not included. Experiments have shown that if the identity elements are removed from a tRNA molecule, the molecule no longer recognizes its amino acid. It may link up with another amino acid or simply fail to function altogether.

FORMATION OF RIBOSOMAL SUBUNITS FROM rRNA. Ribosomal RNA is formed in the nucleolus, an area of the nucleus. There, rRNA combines with special proteins to form the large and small subunits that later combine to produce

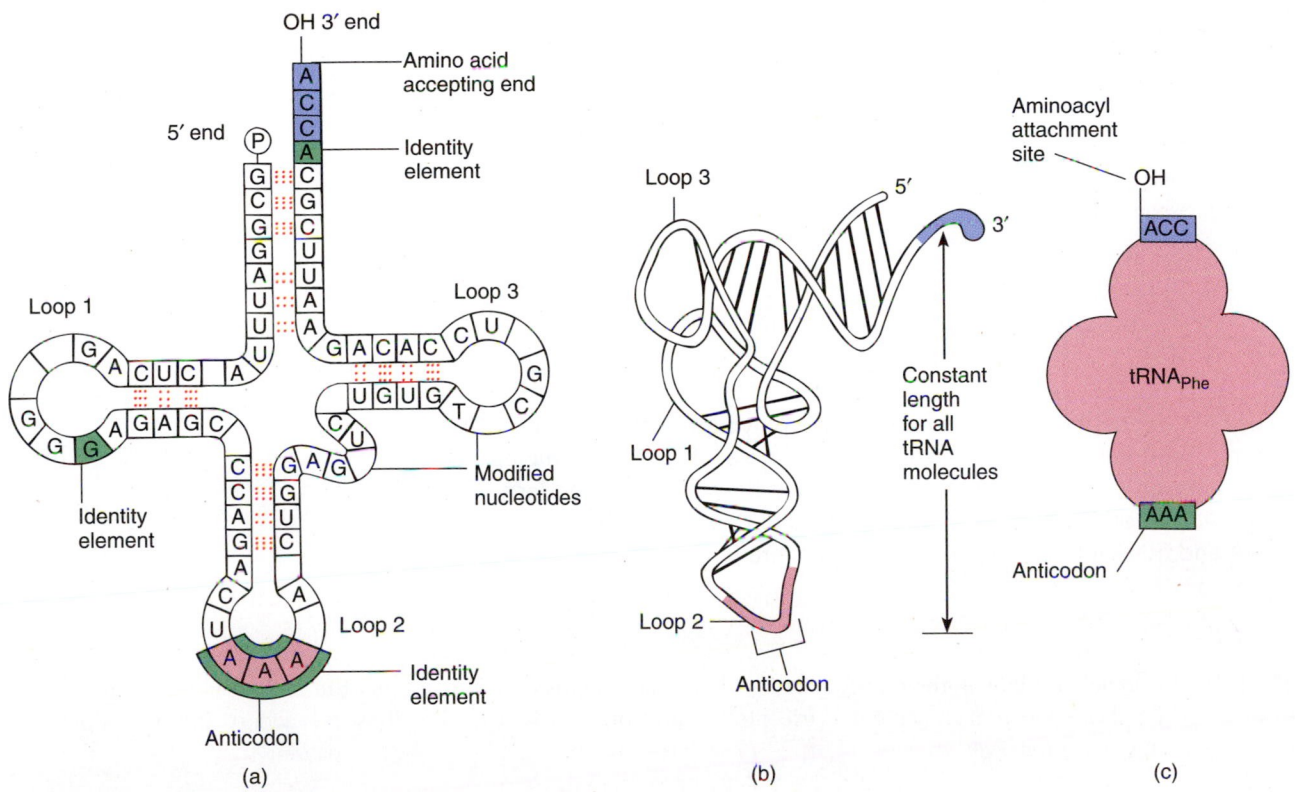

Figure 5–11

A molecule of tRNA contains two important sites: (1) the aminoacyl attachment site, where an amino acid attaches to the molecule, and (2) the anticodon, a triplet of bases complementary to a specific codon on a molecule of mRNA. In addition, equally important "identity sites," sometimes including the anticodon, ensure that the molecule attaches to the correct amino acid. *(a)* structure of tRNA with base pairing. *(b)* three-dimensional, L-shaped arrangement of tRNA. *(c)* Simplified view used in subsequent figures.

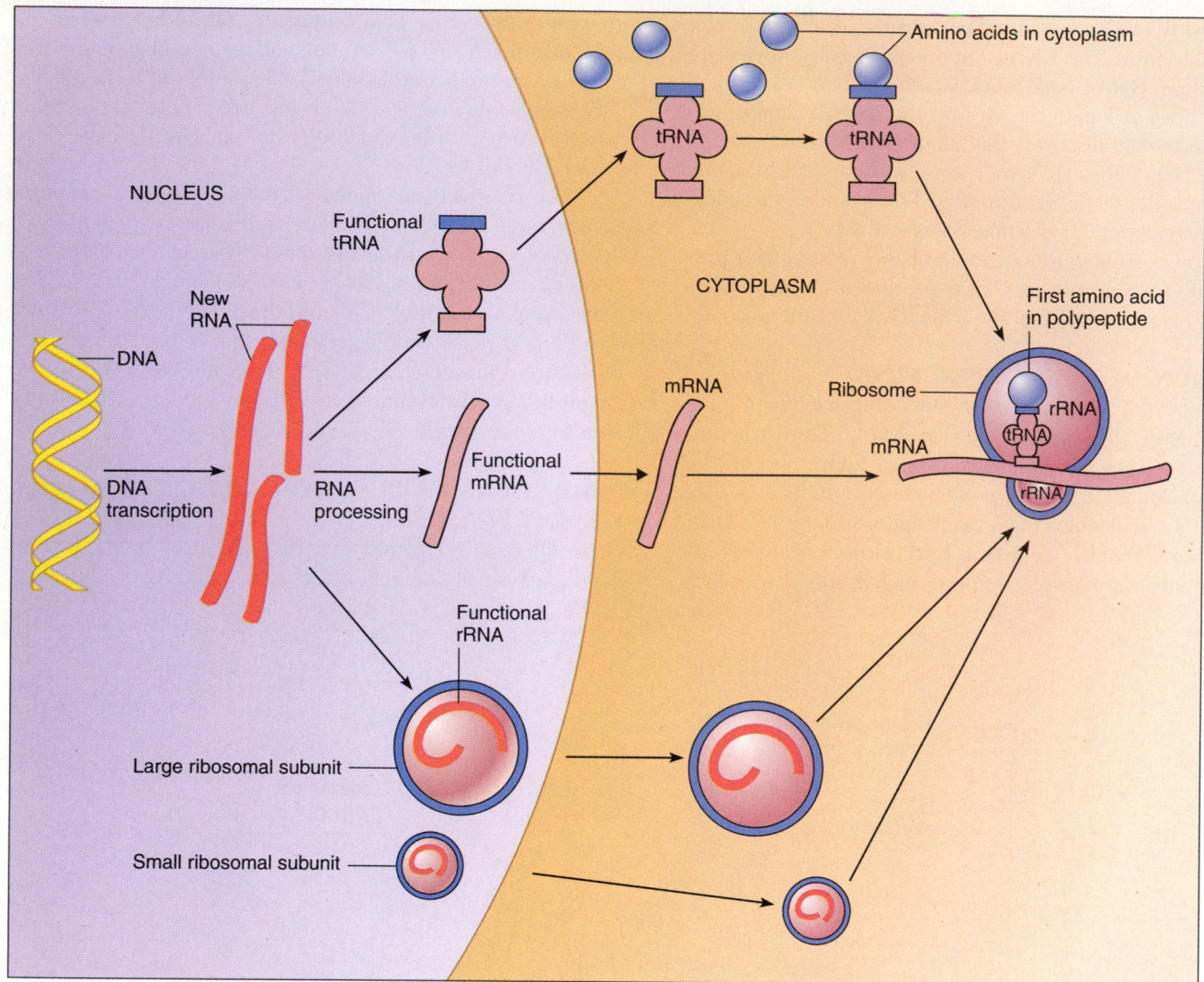

Figure 5–12

The three types of RNA cooperate in the synthesis of a polypeptide. Messenger RNA molecules carry the codons that specify the order of amino acids to be assembled. Transfer RNA molecules obtain individual amino acids in the cytoplasm and ferry them to the mRNA. Ribosomal RNA forms the subunits of ribosomes, the structures on which mRNA and tRNA interact to join amino acids into polypeptide chains.

ribosomes. The subunits do not combine in the nucleolus; they are exported into the cytoplasm and remain separated except when they are participating in protein synthesis.

New Proteins Are Synthesized by Translating mRNA

After export from the nucleus, all three kinds of RNA cooperate in protein synthesis in the cytoplasm. Ribosomal RNA makes up a part of the ribosome, the organelle responsible for protein synthesis. Messenger RNA contains the codons specifying the amino acids to be used in the polypeptide chain. Transfer RNA obtains the amino acids in the cytosol and brings them to the ribosome, where the polypeptide is assembled.

RNA in the Cytoplasm: A Three-Way System

Although all three types of RNA are complementary transcripts of a DNA template, only mRNA nucleotides are directly translated into amino acids. The nucleotides on rRNA and tRNA serve other purposes in the protein-synthesizing process, interacting with mRNA to produce a polypeptide. Figure 5–12 gives an overview of the role each type of RNA plays in protein synthesis.

RIBOSOMAL RNA: THE PROTEIN-SYNTHESIZING ORGANELLE. The several steps involved in translation of mRNA and synthesis of a polypeptide chain take place on, and are coordinated by, ribosomes. Eukaryotic ribosomes are composed of approximately equal weights of rRNA and protein. The molecular weight of these ribosomes is about 4.5×10^6. Each ribosome is composed of two subunits. Some ribosomes are found free in the cytosol or in clusters called polysomes (see Fig. 5–16); other ribosomes are attached to the endoplasmic reticulum (ER).

TRANSFER RNA: PROCURING AMINO ACIDS. A preliminary step in protein synthesis involves coupling amino acids to the tRNA molecules, a process driven by the energy derived from ATP and requiring the catalytic action of an enzyme called aminoacyl-tRNA synthetase (AAS):

$$\text{amino acid} + \text{tRNA} + \text{ATP} \rightarrow \text{aminoacyl-tRNA} + \text{AMP}$$

The amino acid is attached to the ribose ring of the nucleotide at the 3' end (the amino acid-accepting end) of the tRNA.

Each of the 20 amino acids has one specific synthetase enzyme and one specific tRNA. The synthetase must be able to "recognize" both the amino acid and the tRNA that is specific for that amino acid (Fig. 5–13). Free amino acids cannot participate in protein synthesis because they are unable to form peptide bonds (—CONH—) spontaneously. Participation is mediated by the tRNA molecules to which the amino acids are covalently linked.

The tRNA molecules help to maintain the accuracy of translation by making sure that the correct amino acid is inserted at the correct location in the growing polypeptide chain. The basis of this mechanism is complementary base-pairing between the anticodons in the tRNA molecules (see Fig. 5–13) and the codons in mRNA. Thus, it is the tRNA molecule, not the attached amino acid, that recognizes the codon in mRNA. Only one of the many kinds of tRNA molecules in the cell can form base pairs with a specific codon in mRNA, which is how the amino acid specified by the codon can be selected accurately and added to the polypeptide chain.

MESSENGER RNA: CONVEYING THE AMINO ACID SEQUENCE. Messenger RNA is the focal point of protein synthesis, a meeting place for rRNA and tRNA, and the mediator between the information stored in DNA and the final polypeptide product.

Initiation of Protein Synthesis at the Ribosome

In eukaryotic cells, the synthesis of a polypeptide is initiated at the AUG start codon that is nearest to the 5' cap of the mRNA. The other AUG codons in the mRNA do not act as start codons. They specify incorporation of

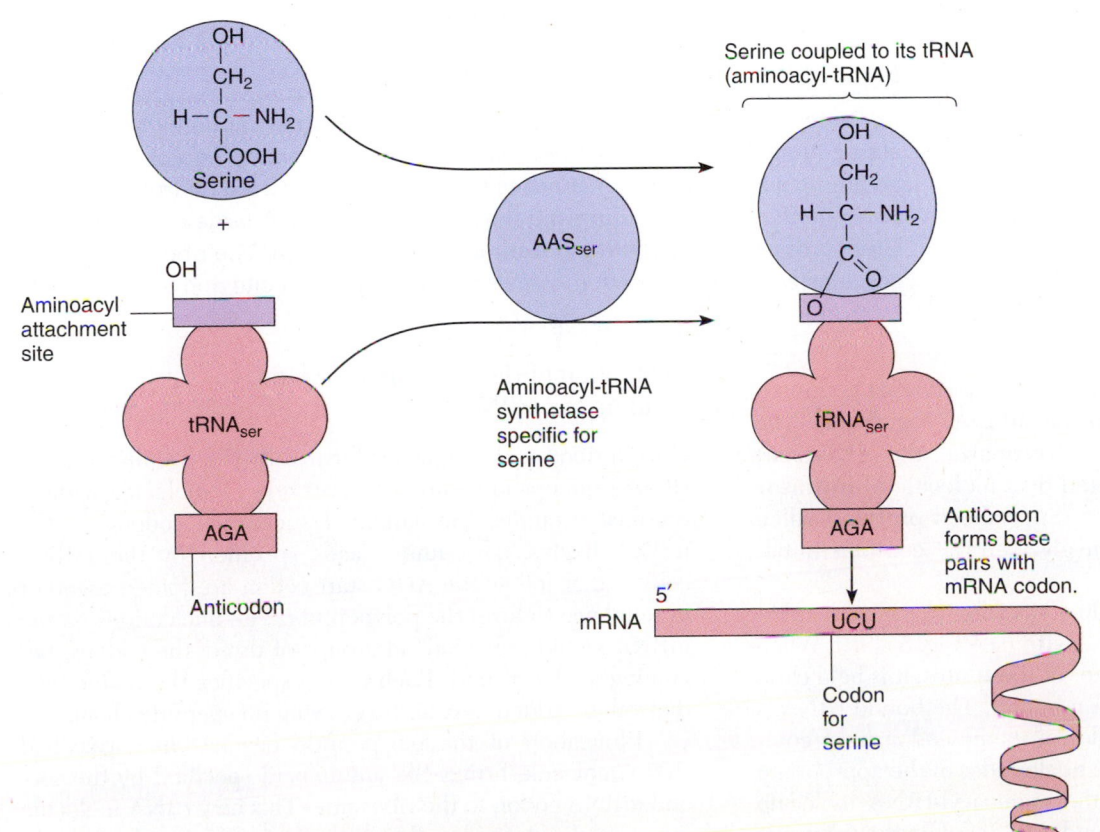

Figure 5–13

Each amino acid has a specific tRNA molecule and a specific enzyme designed to join it to the tRNA. For example, the amino acid serine has a specific tRNA molecule designated as tRNA$_{ser}$. A specific aminoacyl tRNA synthetase (AAS) for serine couples the amino acid to the tRNA molecule via a bond at the aminoacyl attachment site. Later, the amino acid–tRNA complex will join the mRNA molecule via base pairs formed between the tRNA$_{ser}$ anticodon (AGA) and the mRNA codon for serine (UCU).

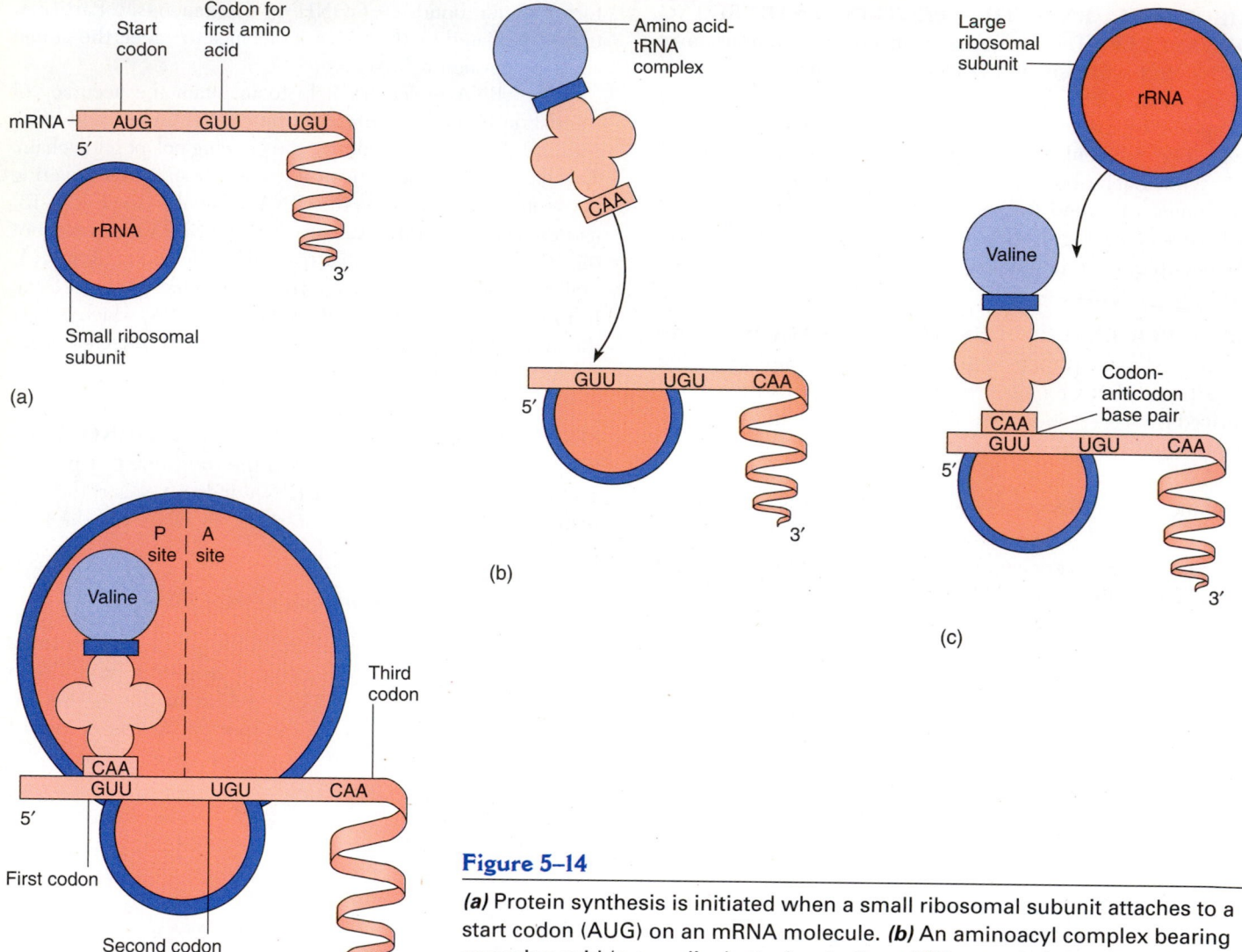

Figure 5–14

(a) Protein synthesis is initiated when a small ribosomal subunit attaches to a start codon (AUG) on an mRNA molecule. *(b)* An aminoacyl complex bearing an amino acid (e.g., valine) attaches to the mRNA strand via hydrogen bonds between the tRNA anticodon and the mRNA codon. *(c)* A large ribosomal subunit then attaches, forming a complete ribosome. *(d)* The ribosome contains two sites for tRNA molecules to occupy, the A-site and the P-site.

methionine into the polypeptide. The AUG start codon allows the ribosome to bind to the mRNA (Fig. 5–14). The ribosome itself is not able to "recognize" the AUG start codon. This process is mediated by a molecule of **initiator tRNA** bound to the ribosome. Specialized proteins called **initiation factors** are also involved in the complex initiation process.

A ribosome has two binding sites for tRNA molecules, denoted as the **P-site** and the **A-site** (see Fig. 5–14*d*). When a tRNA molecule binds to either of these sites, it is held close to the mRNA strand that is also bound. The bound tRNA is oriented so that the nucleotides in its anticodon form complementary base pairs with the nucleotides of the appropriate codon in the mRNA. Thus, the aminoacyl-tRNA molecule cannot fit into the ribosomal binding site unless the appropriate mRNA codon also is present.

Elongation of the Polypeptide Chain: Reading the mRNA Strand

Once a ribosomal complex is formed by the conjunction of a tRNA–amino acid complex and an mRNA molecule with ribosomal subunits, translation of successive codons on the mRNA begins. The amino acids specified by the mRNA codons that follow the AUG start codon are joined together in sequence to form the polypeptide. The nucleotides on the mRNA strand are "read" in groups of three, the codons, beginning at the 5' end. Each codon specifies the amino acid that will be added next to the growing polypeptide chain.

Elongation of the polypeptide begins when a second tRNA molecule brings the amino acid specified by the second mRNA codon to the ribosome. This new tRNA molecule occupies the A-site on the ribosome, right beside the first tRNA molecule, which occupies the P-site (Fig. 5–15*b*). The

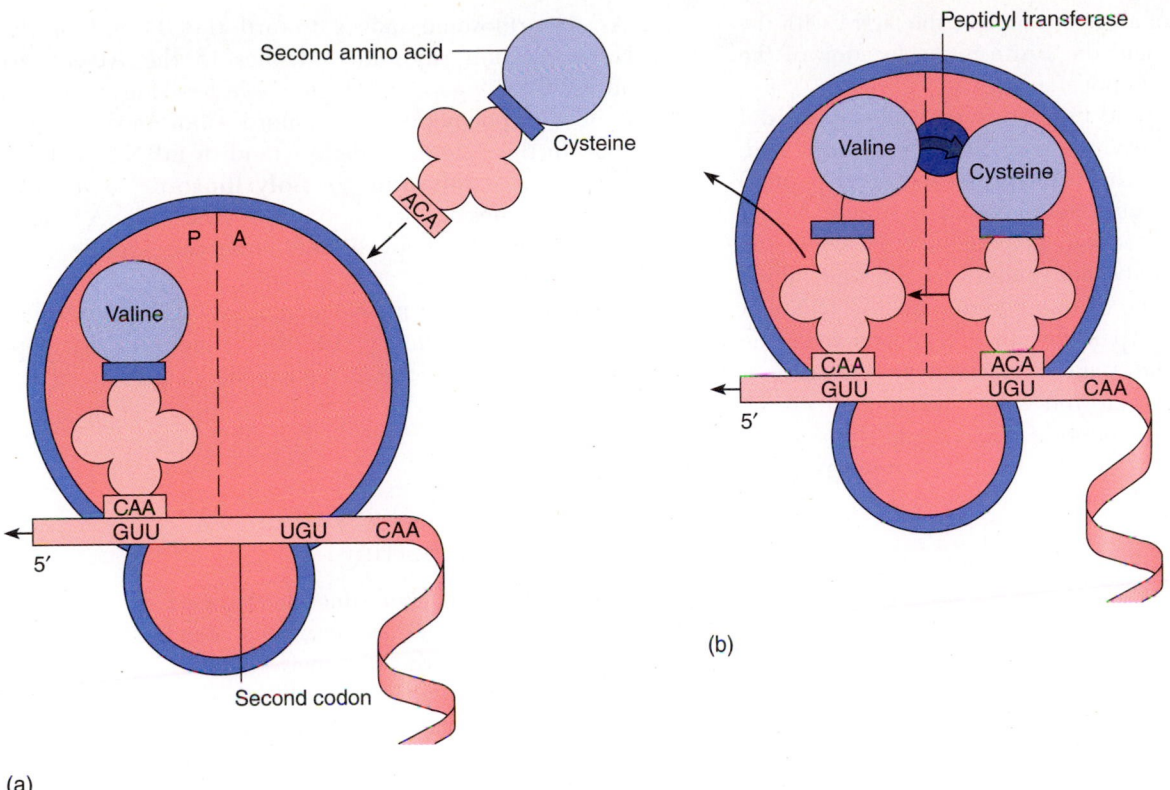

(a)

(b)

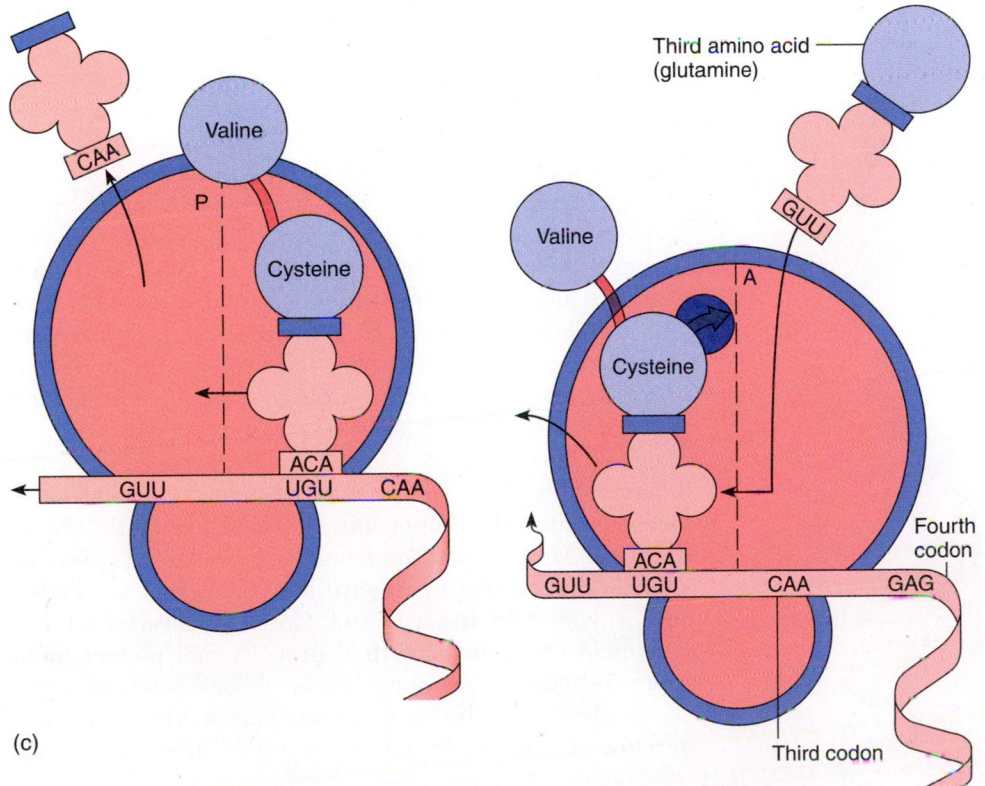

(c)

(d)

Figure 5–15

(a) Elongation of a polypeptide begins when a tRNA molecule bearing another amino acid enters the A-site of the ribosome. The amino acid is specific to the second codon on the mRNA strand. Aided by the enzyme peptidyl transferase and energy from GTP, a peptide bond forms between the two amino acids that now sit side by side on the ribosome. *(b)* At the same time, the first amino acid dissociates from its tRNA molecule. *(c)* The free tRNA molecule leaves the P-site, and the ribosome moves to the right along the mRNA strand, moving the second codon and attached tRNA–amino acid complex from the A-site to the P-site. *(d)* A third amino acid–tRNA complex, corresponding to the third codon, now enters the vacant A-site.

A-site can be thought of as the location of the tRNA with the incoming **A**mino acid, and the P-site is the location of the tRNA with the growing **P**eptide chain.

In a reaction catalyzed by an enzyme called **peptidyl transferase,** the carboxyl end of the first amino acid, bound to the tRNA on the P-site, is joined via a peptide bond to the second amino acid, bound to the tRNA at the A-site. After this reaction, the now-free tRNA at the P-site is ejected. The ribosome moves to the right along the mRNA by a distance of one codon, and the tRNA at the new end of the polypeptide chain is moved from the A-site to the P-site. The A-site is now vacant and ready to accept the tRNA bearing the next amino acid to be incorporated into the polypeptide. The various steps of this cycle are illustrated in Figure 5–15. It is a rapid process; for example, it occurs 20 times each second on bacterial ribosomes.

As one ribosome moves toward the 3′ end of the mRNA, a second ribosome attaches to the AUG start codon at the 5′ cap and begins synthesizing a second copy of the polypeptide. As more ribosomes become attached in this way to a single strand of mRNA, a structure known as a **polysome** (or **polyribosome**) is formed. Several ribosomes are attached to the same mRNA strand, each synthesizing a copy of the same polypeptide (Fig. 5–16).

Translation of the mRNA ceases when a **stop codon** is encountered. At this point, the completed polypeptide is released, and the ribosome dissociates from the mRNA.

Destinations of Newly Synthesized Proteins: Protein Sorting

It is crucial for normal cell function that each of its numerous proteins be localized to the proper membrane or cellular compartment. Proteins that function as hormone receptors must be delivered to the plasma membrane, as must specific ion channels and other transporter proteins. Enzymes, such as RNA and DNA polymerases, must be sent to the nucleus; proteolytic enzymes must go to the lysosomes. Proteins, such as hormones, must be directed to the cell surface and secreted. Directing each newly synthesized protein to a specific destination is referred to as **protein sorting** or **targeting.**

The very first sorting event takes place while the new polypeptide chain is still growing on ribosomes in the cytosol. Some of these polypeptides contain an amino acid signal sequence that directs the ribosomes synthesizing them to attach to the ER. When completed, the polypeptides move to the Golgi complex (see Chapter 3). However, proteins that are folded abnormally or lack important subunits are retained in the lumen of the ER and often are transported directly to the cytosol for destruction. After processing in the Golgi, the proteins are sorted to various destinations. Many proteins that are synthesized by this route, known as the **secretory pathway,** are secreted from the cell. Others remain in the cell in the lumen of the ER, Golgi, and lysosomes; some proteins become integral components of various membranes in the cell. Most plasma membrane and secretory proteins contain one or more carbohydrate chains that are added and modified in the ER and Golgi. The carbohydrates may help the proteins to fold properly and protect them from degradation. Enzymes destined for lysosomes arrive at this destination because they contain multiple mannose-6-phosphate residues that are recognized by the sorting machinery.

Synthesis of most other proteins is completed on ribosomes that remain "free" in the cytosol, and the completed proteins are released into the cytosol. These proteins remain in the cytosol unless they contain a targeting sequence of

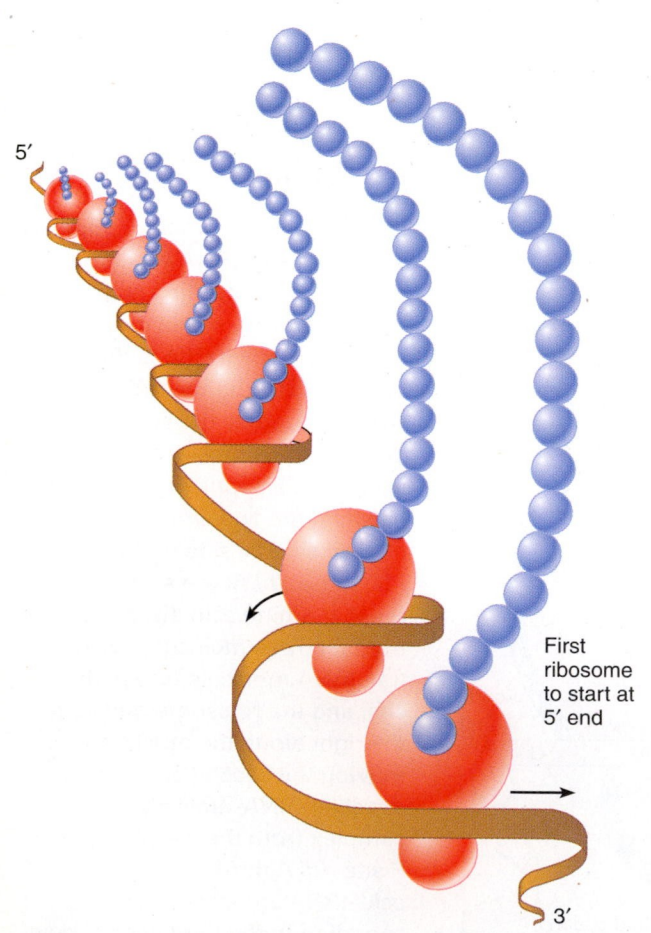

5′

First ribosome to start at 5′ end

3′

Figure 5–16

Several ribosomes may be at work at the same time on the same mRNA strand, creating a structure known as a *polysome.*

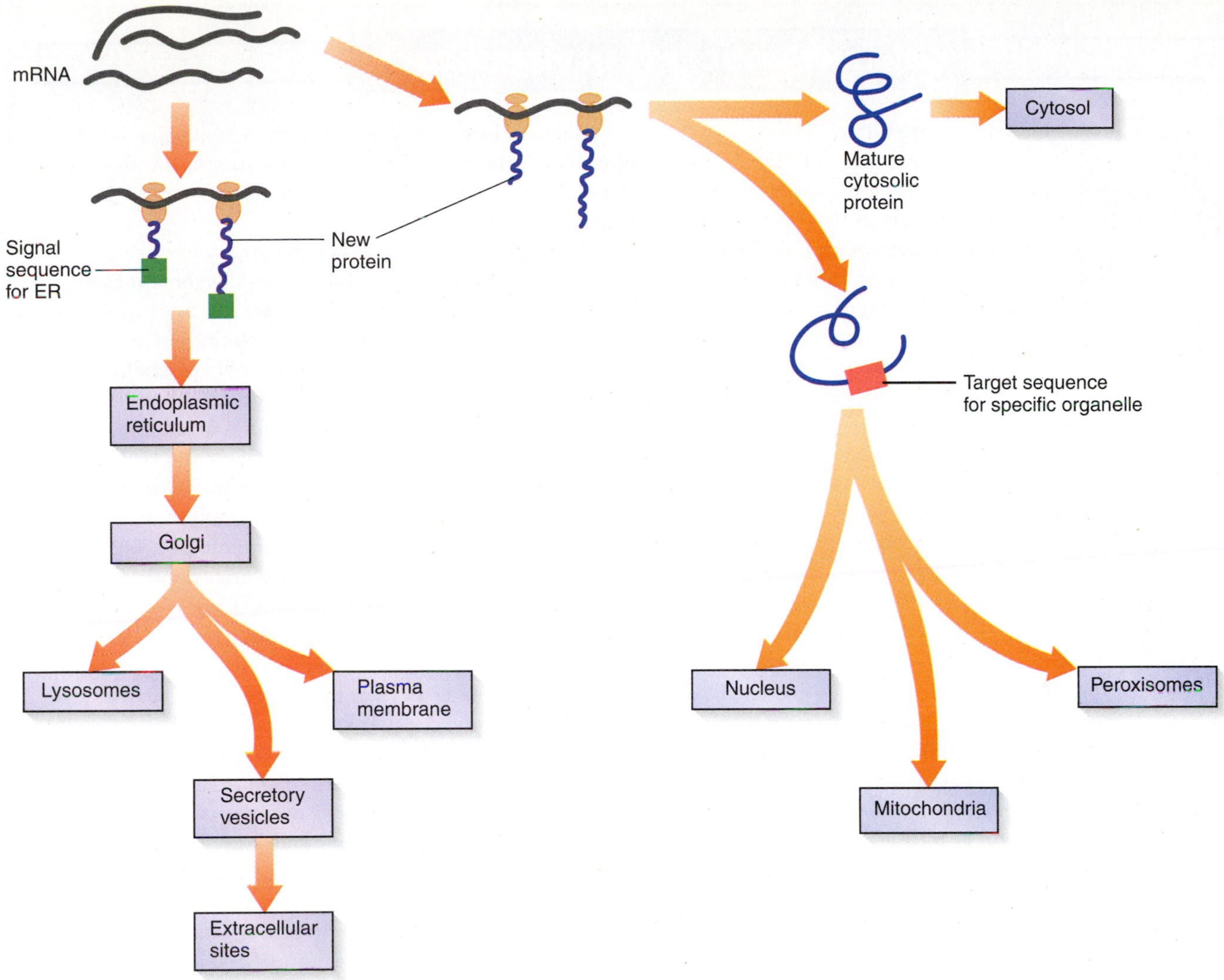

mRNA

Signal
sequence
for ER

New
protein

Mature
cytosolic
protein

Cytosol

Endoplasmic
reticulum

Golgi

Target sequence
for specific organelle

Lysosomes

Plasma
membrane

Nucleus

Peroxisomes

Secretory
vesicles

Mitochondria

Extracellular
sites

Figure 5–17

Overview of sorting of newly synthesized proteins to their final destinations. All mRNA from the nucleus is translated on ribosomes in the cytosol. Ribosomes synthesizing proteins for the secretory pathway are directed to the endoplasmic reticulum (ER) by an ER signal sequence. Synthesis of proteins that lack an ER signal is completed on free ribosomes, and the proteins are released into the cytosol. Those proteins with a targeting sequence for a specific organelle are bound and enter the organelle by a specific transport mechanism.

amino acids that directs them to organelles, such as mitochondria, peroxisomes, or the nucleus. On arriving at a specific organelle, additional signal sequences direct the protein to the appropriate location within it.

Each sorting step involves binding of a specific amino acid sequence to one or more receptor proteins on the surface or interior of an organelle. The pathway of the principal events is summarized in Figure 5–17. Pioneering work on protein-sorting mechanisms led to the award of the 1999 Nobel Prize in Physiology or Medicine to Dr. Gunter Blobel and colleagues at Rockefeller University, New York.

Gene Expression Is Regulated at Several Different Steps

Through what mechanisms is gene expression regulated?

Transcription

The expression of eukaryotic genes is regulated primarily at the level of initiation of transcription, rather than translation, but the mechanisms are complex, and many are still

CURRENT CONCEPTS IN PHYSIOLOGY

Bioinformatics

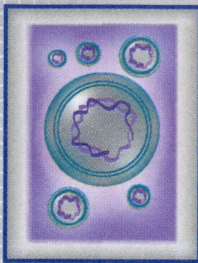

Enormous amounts of DNA sequence have been obtained already, and new sequences are being determined at an increasingly faster pace with the advent of robotics and automated DNA-sequencing machines that run almost continuously. Some biotech companies claim to have the ability to sequence 20 million DNA base pairs in just one day. This enormous volume of information must be stored by computers. The human genome contains 50,000 to 100,000 genes; storing the sequence of 3 billion base pairs that comprises the genome will occupy about three gigabytes of disk space, equivalent to more than 2100 standard diskettes. This is just the tip of the iceberg compared with the information that is expected to be generated in the near future. This will include databases that describe when and in which tissues of the body various genes will be turned on, the shapes of the proteins they encode, how the proteins interact with each other, and the role these interactions play in disease. **Bioinformatics** is the term for the rapidly growing area of computer sciences devoted to collecting and organizing DNA and protein sequences. These sequences are stored in two principal data banks: (1) the **GenBank** at the National Institutes of Health (NIH) in Bethesda, Maryland, and (2) the **EMBL Sequence Data Base** at the European Molecular Biology Laboratory in Heidelberg, Germany. These databases continuously exchange newly reported sequences and make them available via the Internet to cell physiologists and molecular biologists all over the world. Determination of the complete genome sequences of several organisms has allowed for comparison and analysis of entire genomes and has developed into a branch of bioinformatics known as **genomics**. Bioinformatics allows for access to, and evaluation of, the importance of the information provided by genomics.

In-house bioinformatics offers large pharmaceutical companies the prospect of finding better drug targets earlier in the drug development process by facilitating database access between multiple departments, such as product development, toxicology, and clinical testing. This access drives down costs, decreases research and development time, and allows the drug to become commercially available sooner. Smaller pharmaceutical companies are willing to hire one of the more than 50 companies that offer bioinformatics products and services. It has been estimated that bioinformatics could become a $2 billion business within five years.

A popular set of software programs for comparing DNA sequences is **BLAST** (Basic Local Alignment Search Tool), which is a part of a suite of search tools available from many database providers. The National Center for Biotechnology Information (NCBI) at NIH offers **Entrez**, a search tool that covers most of the NCBI databases. This includes three-dimensional protein structures, complete genomes of organisms such as yeast, and references to scientific journals that back up the database entries. One of the most basic procedures in bioinformatics involves searching for similarities, or **homologies**, between a newly sequenced piece of DNA and previously sequenced DNA from various organisms. Protein-coding regions often are translated into amino acid sequences before making comparisons because related proteins may exhibit more homology than their genes as a result of the degenerate genetic code. If a protein encoded by a newly sequenced gene exhibits homology with proteins of known function, this can reveal much information about the function of the new gene. This approach has proved to be invaluable to gain a better understanding of diseases that arise as a result of inheriting a gene that is mutated, such as the *BRCA-1* gene. Women who inherit a mutated *BRCA-1* gene have a high risk for breast cancer before age 50. After the normal gene was sequenced and the amino acid sequence of the normal BRCA-1 protein was deduced, careful and complex comparisons with the sequences in databases revealed that the BRCA-1 protein is distantly but significantly related to a protein that controls cell division in yeast. Investigations are now underway to determine whether the BRCA-1 protein serves a function similar to the yeast protein. If it does, the next step will be to determine whether a mutation in *BRCA-1* results in loss of control of cell division, as occurs in cancer cells. The yeast protein may be a good model for this study because it is relatively easy to introduce mutations in specific genes of the yeast cell.

As researchers move into the postgenomic era, beyond assembling genomes and finding genes, the broad interdisciplinary science of bioinformatics will play a crucial role in understanding the complexities of cell development, physiology, and populations.

unclear (see Fig. 5–12). In general, transcription is controlled by proteins known as **transcription factors,** which bind to specific regulatory sequences and modulate the activity of RNA polymerase. The intricate task of regulating gene expression in the many differentiated cell types of humans and other multicellular organisms is accomplished by the combined actions of many transcription factors. Further levels of complexity are added by **gene-silencing** mechanisms, such as packaging DNA into chromatin, where it is inaccessible for transcription. Another mechanism is the modification of certain genes by **methylation.** The subsequent binding of specific proteins to the methyl groups blocks transcription of the gene (see Case History section).

DNA Binding Proteins

Many of the steps by which steroid hormones control gene expression are well known. Steroid hormones include estrogens (the female sex hormones produced by the ovary and the placenta); testosterone (the male sex hormone from the testis); and mineralocorticoids, such as aldosterone from the adrenal glands. The general action of these hormones is as follows. On arriving at the cell, the steroid hormone diffuses through the plasma membrane and binds to a specific receptor molecule inside the cell. The hormone-receptor complex is able to enter the nucleus and bind to the DNA at a specific site. The receptor protein belongs to a family of DNA-binding proteins that contain **zinc finger domains,** which are regions of the protein that bind zinc ions and fold into looped structures ("fingers") that can bind to DNA. The binding of this hormone receptor to DNA increases the transcription of specific genes, which changes the amount of mRNA and leads, in turn, to increased synthesis of specific proteins. It is these few, specific proteins that produce the changes in cellular function that represent the characteristic response of the cell to the hormone.

Alternative Splicing

In some eukaryotic cells, gene expression can be regulated at steps following transcription. For example, RNA molecules can be modified by changing the order in which exons are spliced together (see Fig. 5–10). This process, known as **alternative splicing,** generates multiple mRNAs from the same pre-mRNA and is an important mechanism for tissue-specific gene expression and for changing gene expression during development.

Stability of mRNA

A control system at the level of translation provides a means for synthesizing large amounts of a single protein. This occurs when transcription of the gene produces a very stable mRNA molecule that remains in the cytoplasm for days and can be used repeatedly for translation and synthesis of the specific protein product. In other words, the mRNA molecules that are involved in the production of the major proteins of a cell tend to be long-lived. A specific example is the synthesis of the protein hemoglobin in the immature red blood cell. As red blood cells mature, their nuclei are lost and new RNA cannot be synthesized, but hemoglobin synthesis continues for weeks. This continuation is possible because the mRNA molecules that code for this protein have a much longer life span than do the other mRNA molecules in the cell. These observations suggest that controlling the life span or stability of mRNA molecules in the cytoplasm may be a mechanism for regulating the expression of certain genes.

Post-Translation

The final step at which gene expression may be regulated is that following translation. Many newly synthesized polypeptides undergo one or more types of covalent alterations, known collectively as **post-translational modification** (not to be confused with the post-transcriptional modification of RNA, discussed earlier; see Fig. 5–10). Some of the covalent modifications of polypeptides were discussed in Chapters 2 and 3—for example, the attachment of carbohydrate to produce a glycoprotein and the regulation of the biological activity of a protein by alternate cycles of attachment and removal of a phosphate group. Other examples include the permanent covalent attachment of a coenzyme molecule essential for the activity of an enzyme protein, and the production of a biologically active polypeptide from a much larger inactive polypeptide molecule (known as a **precursor**) through the activity of enzymes that selectively remove small sections. The latter mechanism is involved in the production of the hormone **insulin,** a polypeptide of 51 amino acids, from an inactive precursor called **proinsulin,** a polypeptide of 84 amino acids.

CELL-TO-CELL SIGNALING: HORMONES, NEUROTRANSMITTERS, AND OTHER MOLECULES

 How do cells communicate?

The coordination of the different functions of the many tissues in a multicellular animal is achieved because cells have developed communication systems. Cells that are in direct physical contact with one another have structures known as **gap junctions** in their plasma membranes (Fig. 5–18a). Gap junctions allow small molecules to pass directly from the cytosol of one cell to the cytosol of the adjacent cell. Cells some distance apart are able to communicate by sending **chemical signals,** molecules secreted by one cell that

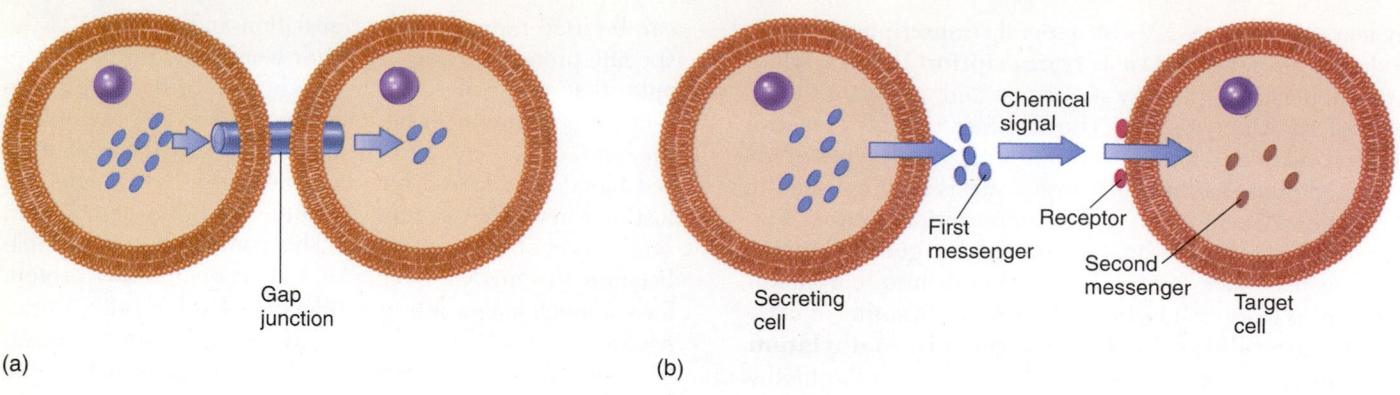

(a)   (b)

Figure 5–18

The two broad types of communication between cells: *(a)* gap junctions (between adjacent cells) and *(b)* chemical signals (between distant cells). Chemical signals usually involve an extracellular signal produced by one cell, a receptor in or on the target cell, and an intracellular signal in the target cell.

bind to the plasma membrane or enter the cytoplasm of the distant cell (see Fig. 5–18*b*). Chemical signaling systems usually involve three separate components: (1) extracellular signals, called *first messengers;* (2) receptors; and (3) intracellular signals, called *second messengers.* The following discussion includes a description of the features of the general cell-to-cell communication model developed in Chapter 1.

Adjacent Cells Communicate Directly via Gap Junctions

 What are gap junctions; why are they important?

Gap junctions are common structures in tissues of all animals. They consist of a group of small pores or channels, each about 1.5 nm in diameter, that allow small, water-soluble molecules with molecular weights of less than 1000 to move directly from the cytosol of one cell to the cytosol of an adjacent cell (Fig. 5–19). This includes ions, sugars, amino acids, and nucleotides but excludes the macromolecules. Note that gap junctions are different structurally and functionally from the tight junctions discussed in Chapter 4. The role of tight junctions is to restrict the diffusion of small molecules through the spaces between cells.

The major advantages of this direct communication system are its speed and reliability. The "message" must travel only a short distance, with little chance of intervention by outside forces because it does not leave the cytosol. In principle, this system could also allow for **metabolic coupling** between different cells present in a tissue, which means that molecules made by only one group of cells within the tissue could be shared with the other cell types present by being passed through the gap junctions.

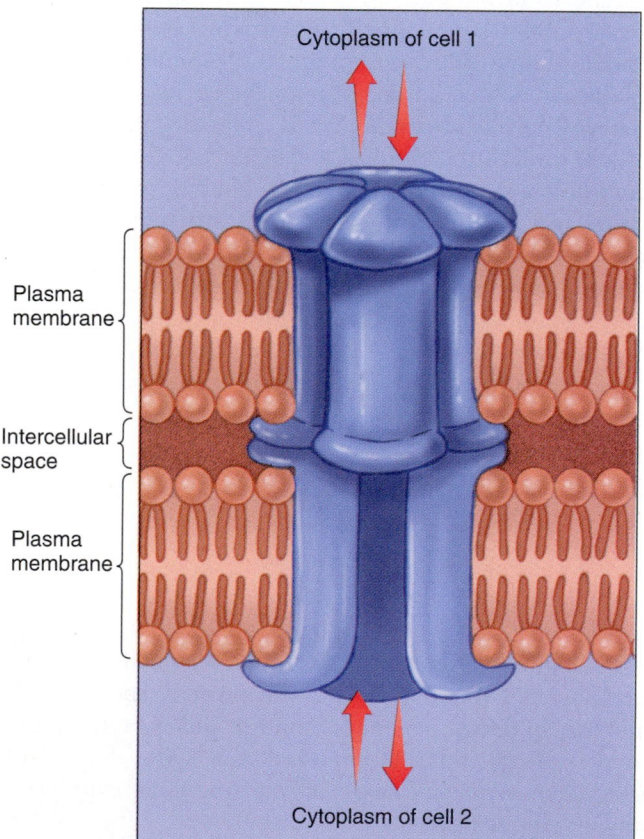

Figure 5–19

A gap junction forms direct connections between the cytoplasm of adjacent cells. Proteins create a tunnel-like structure in the plasma membrane of each cell, and two tunnels of adjacent cells connect to allow materials to pass between them.

Gap junctions also allow for electrical communication between cells because they permit the rapid flow of ions. Electrical transmission is nearly instantaneous—much faster than chemical transmission—and sometimes this difference is very important. For example, **electrical coupling** is essential in mammalian tissues, such as the heart, because all the muscle cells must be synchronized to contract almost simultaneously in order for the heart to pump blood efficiently. Another example is the electric eel, which can generate a powerful electric shock only because all of the cells in its electrical organ are electrically coupled and can be synchronized to "fire" at the same time. Chemical transmission would be too slow to achieve the necessary degree of precision. Electrical coupling via gap junctions also occurs in the early embryos of vertebrates. As specific groups of cells develop their identities, they uncouple from the surrounding tissue but remain coupled to one another. It is thought that electrical coupling allows their behavior and further development to proceed in a coordinated way.

Distant Cells Use Chemical Signals for Communication

> *How do cells communicate with distant cells; what determines how the target cell will respond to the "message"?*

Chemical signals take the form of specific molecules released by one cell and sent to produce a change in a distant **target cell.** Chemical signaling systems generally involve three components: (1) the signal molecules, also called *extracellular signals* or **first messengers;** (2) **receptors,** molecules in either the membrane or the cytoplasm of the target cell; and (3) intracellular signals, or **second messengers.** The cumulative effect of these components is to elicit the desired physiological response in the target cell. The remainder of this chapter is concerned with the operation of chemical signaling systems.

Extracellular Signals: First Messengers

Cells use various types of chemical signals to communicate with other cells. **Local chemical mediators** affect only nearby cells because they are rapidly destroyed before they can diffuse very far. **Neurotransmitters** are produced by nerve cells and also affect cells that are nearby. Finally, **hormones** are sent into the blood by secretory cells to affect target cells, which may be nearby or far away from the hormone-producing cells. Figure 5–20 summarizes these three types of signaling systems.

With the exception of the prostaglandins and related compounds, most of these chemical signals are synthesized by specialized cells, where they are stored until the appropriate stimulus triggers their release. Specific examples of chemical signals are listed in Table 5–3, which summarizes the wide variety of molecules used.

These groups of chemical signals can be classified into two large groups: (1) **water-soluble signals** and (2) **lipid-soluble signals.** Most hormones, most local mediators, and all of the neurotransmitters are water-soluble. These molecules are removed or destroyed very rapidly (within seconds, or a few milliseconds for neurotransmitters) after being released. Steroid hormones and thyroid hormones (e.g., thyroxine), in contrast, are lipid-soluble molecules that can persist in the bloodstream for several hours, or even days in the case of thyroid hormones. They are made soluble for transport in the blood by binding to a specific carrier protein. In general, water-soluble signal molecules are involved in responses that are needed rapidly but only for a short time, whereas lipid-soluble signal molecules are involved in responses that may develop more slowly but last for a long time. Prostaglandins and related compounds are an exception: They are lipid-soluble but mediate rapid, short-lasting responses.

A common feature of the chemical communication systems is the presence of specific receptors that allow the signaling molecule to recognize and bind to the target cell. Receptor function is discussed in more detail in a later section of this chapter.

LOCAL CHEMICAL MEDIATORS. Some signal molecules released by a cell affect only the cells in the immediate vicinity because the molecules are destroyed rapidly once released. If present in high enough concentrations, the chemical signal can bind to specific receptors on the nearby target cell (see Fig. 5–20a and b). If the signal molecule dissociates from the receptor, because binding is reversible, the signal molecule is likely to be degraded by enzymes present in the interstitial fluid or on the surface of the target cell. These chemical signals are called **local chemical mediators;** specific examples include histamine, the prostaglandins, and nitric oxide.

Histamine is derived from the amino acid histidine and is stored in the specialized **mast cells** present in connective tissues. It is released rapidly in response to injury, infection, and allergic reactions. One of its principal actions is to cause dilation and leakiness of local blood vessels. This effect allows antibodies and white blood cells access to the site of injury, where they perform their tasks of destroying foreign substances.

The signaling molecules synthesized from essential fatty acids containing 20 carbon atoms, such as arachidonic acid (see Chapter 2), are known collectively as *arachidonic acid derivatives,* or **eicosanoids.** They include the **prostaglandins,** which were named after the prostate gland, once thought to be their only source. It is now recognized that prostaglandins are formed by most tissues, rather than by specialized cells, and that they are released continuously instead of being stored. Prostaglandins are a specific set of different molecules that fall into nine classes, commonly abbreviated as PGA to PGI. They bind to cell surface receptors and have diverse effects on different tissues. Smooth muscles are major target tissues that either contract or relax, depending on the specific type of prostaglandin. Prostaglandins also mediate inflammatory

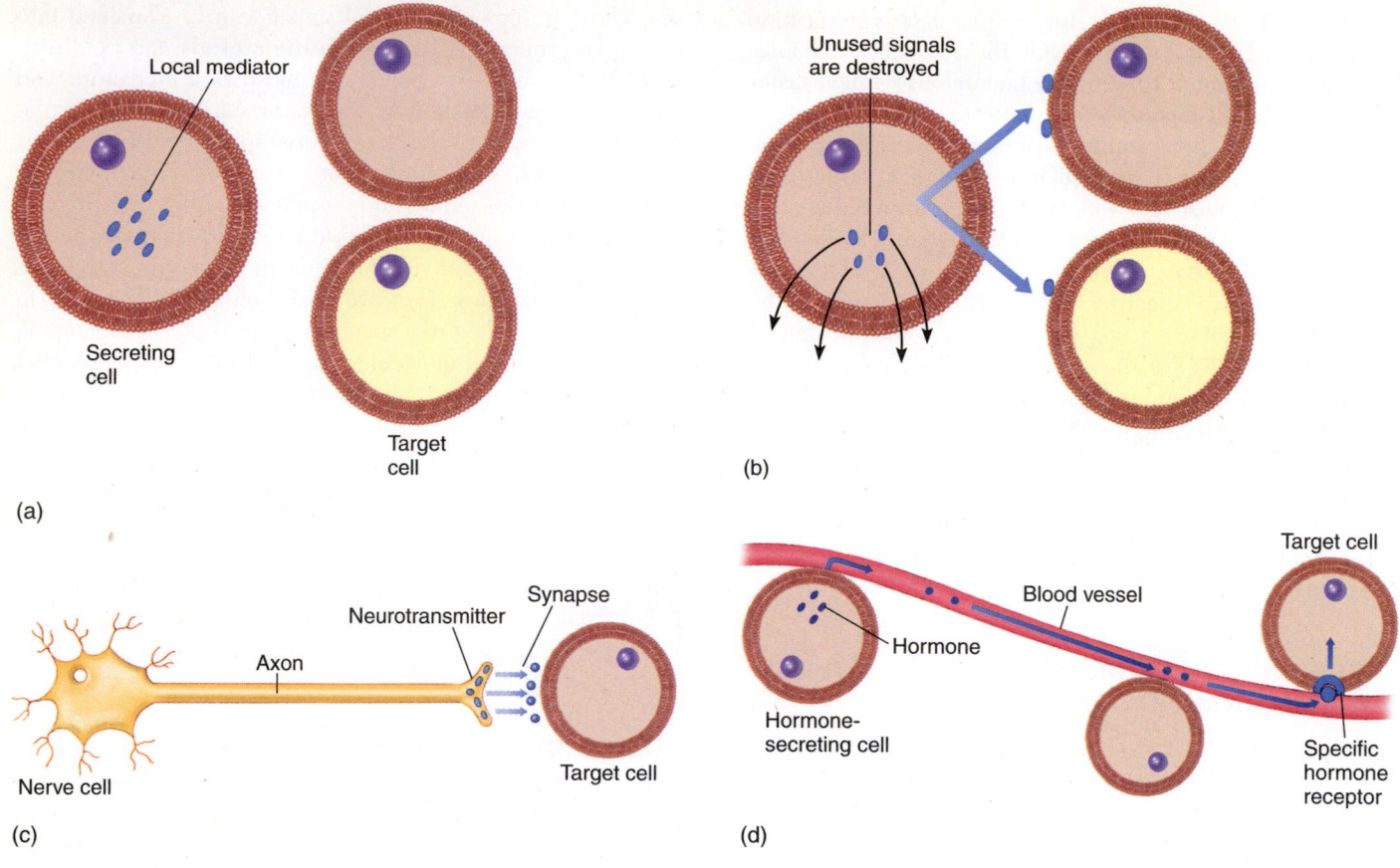

Figure 5–20

The three basic types of chemical signaling systems. (**a** and **b**) Local chemical mediators act only on cells near them because they are quickly destroyed once released. (**c**) Neurotransmitters are produced by nerve cells and are delivered via a synapse at a target cell, which may be a long distance away from the nerve cell body. (**d**) Hormones travel via the blood from hormone-secreting cells to target cells that have specific receptors for them.

responses. The effectiveness of aspirin in reducing fever and inflammation is largely due to its ability to inhibit prostaglandin synthesis. Many of the actions attributed originally to the prostaglandins may be due to their highly unstable but potent relatives known as **prostacyclins** and **thromboxanes.**

The **leukotrienes,** another group of eicosanoids, act locally to mediate allergic and inflammatory reactions.

Nitric oxide gas is synthesized from arginine in endothelial cells lining the walls of blood vessels. Its action is unusual because it is a local mediator that does not bind to surface re-

TABLE 5–3

Examples of Molecules That Serve as Chemical Signals

Type of Molecule	Local Mediator	Neurotransmitter	Hormone
Peptides	Chemotactic factors	Substance P	Vasopressin
Small proteins	Interleukin-1	—	Insulin
Amino acids and derivatives	Histamine	Glycine	Epinephrine
Steroids	—	—	Testosterone
Fatty acid derivatives	Prostaglandins	—	—
Other small molecules	Nitric oxide	Acetylcholine	—

ceptors. Instead, because it is lipid-soluble, it readily diffuses out of the endothelial cells and into the adjacent vascular smooth muscle cells, where it triggers relaxation by generating a second messenger known as **cyclic GMP.** Relaxation of the muscle results in dilation of the blood vessel. Nitroglycerin has long been used to treat angina, pain caused by inadequate blood flow to heart muscle. Nitroglycerin is effective because it is converted to nitric oxide, which dilates blood vessels in the heart, thereby increasing the blood flow to heart muscle.

NEUROTRANSMITTERS. A second group of chemical signals, known as **neurotransmitters,** is used by nerve cells to influence their target cells. A typical nerve cell is represented in Figure 5–20*c*. Notice that the neurotransmitter is delivered directly to the target. The neurotransmitter is released by the nerve axon at the **synapse** (the gap between the end of the axon and the target cell) and diffuses across to the target. Diffusion occurs in less than 1 millisecond because the gap is very small (only a few nanometers). Receptor molecules on the surface of the target cell in the region of the synapse bind the neurotransmitter. Delivery of the neurotransmitter directly to the target cell, and the presence of specific receptors there, ensure that the effects of the chemical signal are restricted to a very small area, even though the target cell may be many centimeters from the main body of the nerve cell.

More than 40 different molecules serve as neurotransmitters at chemical synapses; only a few examples are noted here. Synapses where the target is a muscle cell use **acetylcholine** as the neurotransmitter. The amino acids **glycine, glutamate,** and gamma aminobutyric acid (**GABA**) and small peptides, such as **substance P** and **endorphins,** are important in the central nervous system. **Norepinephrine, dopamine,** and **serotonin,** often called *biogenic amines,* or **catecholamines,** also are found in the central nervous system, where they affect mood. Their imbalance has been linked to several mental disorders, such as depression and schizophrenia. The nervous system is a unique communication system because, in addition to transmitting information, it can store a vast amount of information in the cells of the brain.

HORMONES. Only a few specialized tissues synthesize and secrete chemicals known as **hormones** into the blood, which then carries them to distant target cells (see Fig. 5–20*d*). This type of communication system relies on the flow of blood and requires the chemical signal (the hormone) to travel distances that are much greater than the distance a neurotransmitter must travel in crossing a synapse. Consequently, a typical hormone-secreting cell transmits information much more slowly than does a nerve cell.

Once released, many hormones affect more than one type of cell or tissue in the body. For example, a decreased level of Ca^{2+} in plasma increases the production of **vitamin D** by the kidney. The major role of vitamin D is to stimulate the absorption of Ca^{2+} in the small intestine. It also increases reabsorption of Ca^{2+} by the kidney, although this response is weaker. The action of vitamin D on these two tissues helps to return plasma Ca^{2+} to the normal range. Vitamin D is able to "recognize" the correct target cells and does not affect the other cells that it encounters. Accurate recognition is achieved because the target cells have receptor molecules to which the hormone can bind. Nontarget cells lack the specific receptor and thus are not affected by the hormone.

The various hormone-secreting cells of the body make up the **endocrine system.** This system is influenced in part by the nervous system. For example, the secretion of several specific hormones is subject to overall regulation by the nervous system.

Hormones are as varied a group of molecules as are the other types of chemical signals. Although the vast majority are polypeptides, they also include amino acid derivatives and steroids (see Table 5–3). Insulin and parathyroid hormone are small proteins. Vasopressin (also called *antidiuretic hormone*) is a small peptide of nine amino acids. Epinephrine is derived from the amino acid tyrosine. The female and male sex hormones (estrogens and testosterone, respectively) are steroids.

Receptors: Mediating Between First and Second Messengers

When a chemical signal arrives at a target cell, it ultimately changes processes, such as enzyme-catalyzed reactions, the transport of ions, and transcription of genes. The initial event, however, is the interaction of the signal molecules with specific receptors located either on the surface of the target cell or in its cytoplasm. The receptors determine whether and how the target cell responds to the chemical signal.

THE NATURE OF RECEPTORS. A receptor is a protein to which the signal molecule can bind with high specificity. This means that other molecules do not bind easily to the receptor. This specificity may arise because the specific shape of the signal molecule may facilitate its binding to the receptor. Other molecules with slightly different shapes may be unable to fit into the binding sites on the receptor. This point is illustrated in Figure 5–21. In addition to specificity, the signal molecule binds to the receptor with **high affinity,** which means that the binding is strong.

The combined purpose of these receptor properties is to ensure that only the signal molecule binds and remains attached to the receptor. These properties are particularly important for hormone receptors because a hormone that is delivered to a cell via the bloodstream is usually present in a low concentration (10 nmol/L or less). In addition, the hormone is only one of the many different types of molecules in the blood that are "seen" by the receptor.

Little is known about the interactions that stabilize the binding of signal molecules to receptors. The binding is noncovalent and reversible and probably involves contributions by hydrophobic interactions, hydrogen bonds, and ionic bonds. The binding of a hormone to its receptor in a target cell triggers a sequence of events that ultimately leads to the changes that represent the cellular response to the hormone. The cellular location and action of hormone receptors vary with the type of hormone.

Chemical signal A   Chemical signal B   Chemical signal C

Figure 5–21

Receptor molecules have specific shapes designed to discriminate among various chemical signals. Although chemical signals A, B, and C are very similar in shape, only B fits the shape of the receptor exactly, so only B binds with the receptor and affects the target cell.

RECEPTORS IN THE CYTOPLASM. The receptors for all steroid hormones, vitamin D (a sterol), and **thyroid hormones** are soluble proteins located inside the target cell (Fig. 5–22). The hormones must diffuse through the cell membrane in order to gain access to the receptors. This process is facilitated by the lipid-soluble nature of the hormones. Binding of steroids and vitamin D to their specific receptors in the cytosol is followed by the movement of the hormone-receptor complex into the nucleus, where it binds to specific acceptor sites on the DNA and regulates the transcription of specific genes. Thyroid hormones exert their effects through a similar mechanism, but the receptors for these hormones are concentrated in the nucleus even in the absence of hormone.

RECEPTORS IN THE PLASMA MEMBRANE. Circulating catecholamine hormones, such as **epinephrine,** and polypeptide hormones initiate a completely different series of events upon binding to their receptors. The receptors for these hormones are located on the plasma membrane of the target cell. The hormones bind to the receptors, producing a response within the cell without gaining entry to it (see Fig. 5–22). The effects of the hormones are mediated by second messengers, which are generated within the cell when the hormones bind to their receptors.

Plasma membrane receptors are intrinsic membrane proteins that usually account for less than 1% of the total mass of protein present in the membrane. The receptor is more than just a binding site; it acts more like a **transducer.**

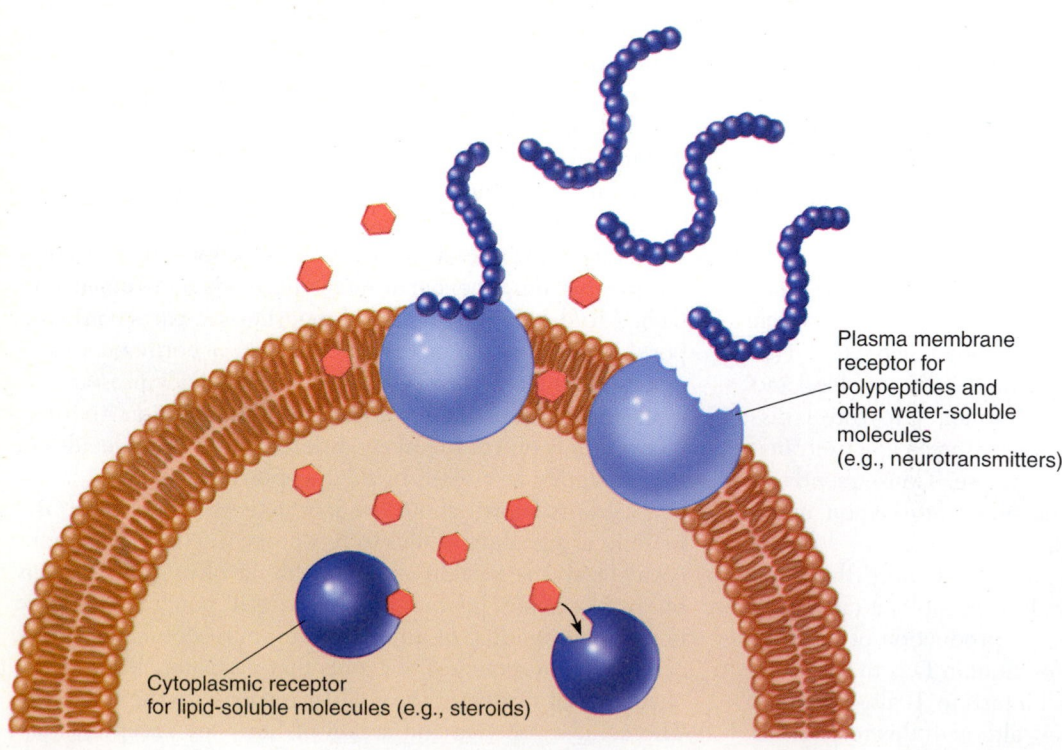

Plasma membrane receptor for polypeptides and other water-soluble molecules (e.g., neurotransmitters)

Cytoplasmic receptor for lipid-soluble molecules (e.g., steroids)

Figure 5–22

Receptors can be located in either the plasma membrane or the cytoplasm of a target cell. Receptors for water-soluble signals are often intrinsic membrane proteins, whereas receptors for lipid-soluble signals, which can diffuse through the plasma membrane, often are found inside the cell.

It can convert an extracellular event—the binding of a signal molecule—into an intracellular response that modifies the behavior of the target cell.

The largest family of plasma membrane receptors are the receptors that are linked to **G-proteins,** which are proteins bound to a guanine nucleotide, either GTP or GDP. G-proteins are found on the inner surface of the plasma membrane and provide a link between the receptor and its **effector protein.** An unactivated G-protein is a trimeric complex: the three subunits are called α, β, and γ, and

a GDP molecule is bound to the α subunit. Different G-proteins are activated by binding of different first messengers. Binding of a first messenger to the receptor changes the shape of the receptor so that it can interact with a specific G-protein. This interaction results in the loss of GDP from the α subunit and the attachment of GTP. The binding of GTP activates the α subunit so that it breaks away from the βγ complex and moves along the inner surface of the plasma membrane until it reaches the effector protein (Fig. 5–23). The effector is either an enzyme or an ion channel.

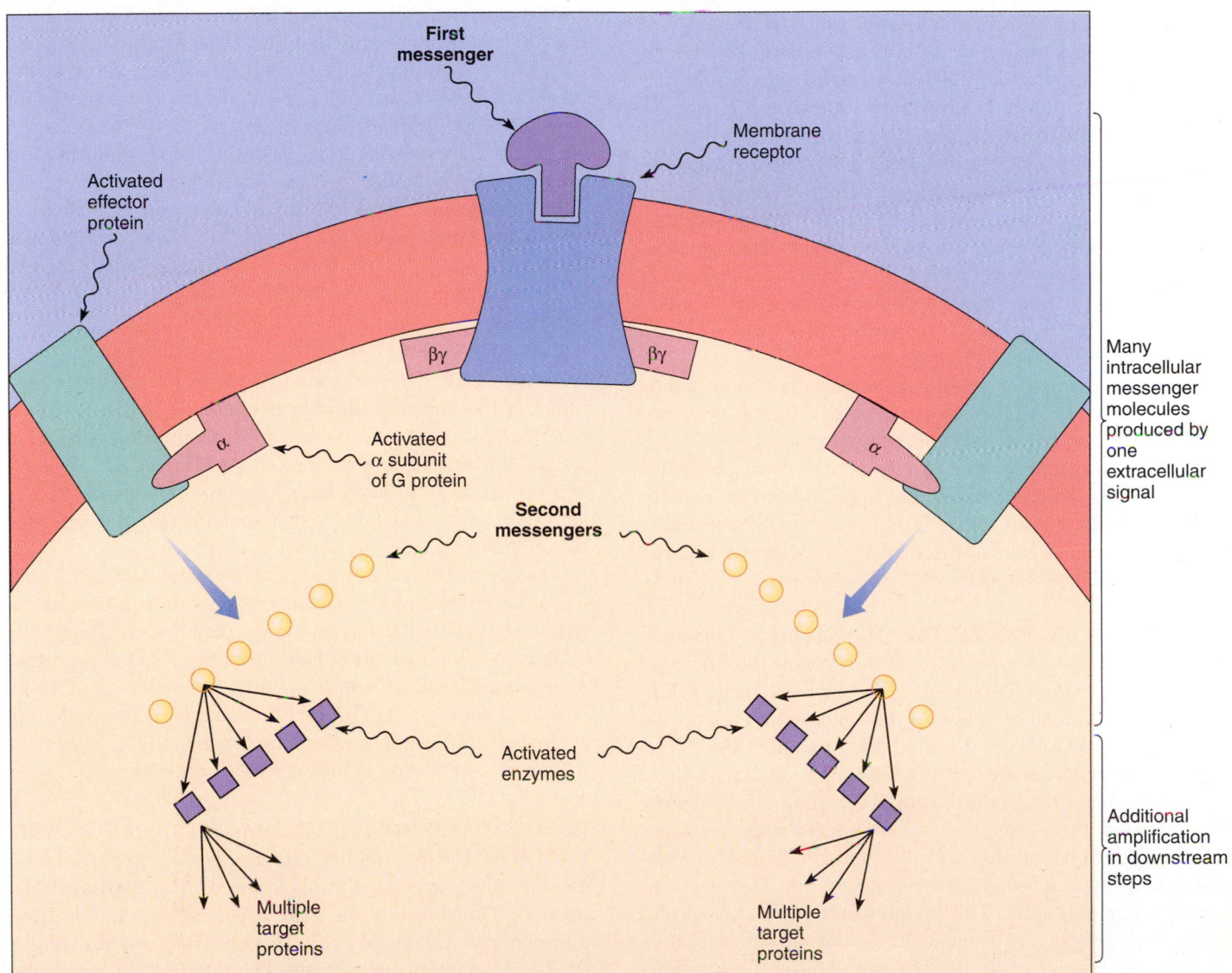

Figure 5–23

An example of amplification at almost every step in a signaling pathway. When a single extracellular signal molecule (first messenger) binds to a G-protein–coupled receptor, the receptor changes its conformation and activates several G-protein α subunits. Each α subunit breaks away from the βγ complex, and activates a single effector protein which, in turn, generates many intracellular second-messenger molecules. One second messenger activates many enzymes, and each activated enzyme can regulate many target proteins.

Interaction with the α subunit either activates or inhibits a specific enzyme, or opens or closes a specific ion channel. The precise outcome depends on the first messenger. More than one thousand **G-protein–coupled receptors** have been identified; these include receptors for many peptide hormones, neurotransmitters, and neuropeptides. The large number of different receptors emphasizes the point that the receptor at the target cell, and the associated mechanisms, determine the specific response of the cell to the hormone.

LIGANDS: AGONISTS AND ANTAGONISTS. Any compound, such as a signal molecule, that binds with high specificity to a receptor is termed a **ligand.** Any ligand capable of both binding to a plasma membrane receptor and evoking a physiological response is known as an **agonist.** A ligand that binds with high affinity to a receptor but evokes no response is called an **antagonist** because, by occupying receptors, it interferes with (or "antagonizes") the action of an agonist.

Antagonists of hormone receptors can be of great therapeutic importance in medicine. Circulating catecholamines act on the heart to increase both the heart rate and the amount of blood pumped with each contraction. Too much catecholamine action can cause high blood pressure. A catecholamine antagonist called **propranolol** is used to block these unwanted effects.

Another example is **spironolactone,** which is an aldosterone receptor antagonist. Aldosterone is a steroid hormone that increases the retention of Na^+ by the kidneys; hence, water is also retained (see Chapter 25). Treatment with spironolactone blocks the action of aldosterone and produces an increase in the excretion of water, which is important in cases of severe fluid retention and edema. Spironolactone is known as a **diuretic,** a substance that increases water excretion.

VARIATIONS IN RECEPTOR QUANTITIES. Quantitative studies of receptor numbers have revealed that the concentration of many receptors in cells can be regulated. For example, a sustained elevation of the concentration of an agonist circulating in the blood often leads to a decrease in the number of receptors present in target cells. This phenomenon is known as **down-regulation** of receptor number and is accompanied by a decrease in the physiological response **(desensitization)** of the cells to the circulating agonist. When the agonist concentration decreases, the receptor number increases again. This phenomenon makes many cells particularly sensitive to a change in the normal concentration of a chemical signal, as opposed to its absolute concentration.

A decrease in the number of plasma membrane receptors frequently occurs because binding of the agonist molecules to their receptors stimulates internalization of the receptor-agonist complexes into the cell by an endocytotic process, which is followed by transfer to lysosomes and degradation. The mechanism of down-regulation is less clear for cytoplasmic receptors, but it may involve any one of a number of processes, including inactivation of receptors and changes in the rate of synthesis or rate of degradation of the receptor proteins.

Intracellular Signals: Second Messengers

> *What role do second messengers play in determining the response of the cell to an extracellular signal (ligand-receptor interaction)?*

It is generally accepted that the conformation (three-dimensional shape) of a receptor molecule is altered when an extracellular signal molecule binds to it. The change in receptor conformation not only generates a second messenger but also increases, or **amplifies,** the signal. Amplification occurs because many G-proteins (up to 100) are activated for each extracellular signal molecule that binds to a receptor. Although each active G-protein probably activates only a single effector protein, each effector generates many second-messenger molecules (see Fig. 5–23). Additional amplification occurs at each of the subsequent steps in the signaling pathway. For example, each second-messenger molecule can activate many enzyme molecules, each activated enzyme can regulate multiple target proteins, and the targets function together to mediate a specific cellular response. This intracellular **cascade** of events serves to greatly amplify the original extracellular signal.

CYCLIC AMP. One effector protein in the plasma membrane is the enzyme **adenylate cyclase.** This catalyzes a reaction that uses ATP to generate many molecules of a second messenger, known as **cyclic AMP (cAMP).** Adenylate cyclase is stimulated by a G-protein α subunit that is activated by binding of a signal molecule to the receptor (see Fig. 5–23).

Cyclic AMP modifies the behavior of a cell by activating a protein kinase enzyme referred to as **protein kinase A,** or **A-kinase.** Once activated, this kinase catalyzes the phosphorylation (addition of phosphate groups) of specific protein molecules. Phosphorylation of a protein usually modifies its biological activity (Fig. 5–24). For example, if the protein is an enzyme, the result of phosphorylation could be either activation or inactivation of its enzymatic capacity.

INOSITOL TRIPHOSPHATE, DIACYLGLYCEROL, AND CALCIUM IONS. Another effector protein in the plasma membrane is a specific phospholipase, known as **phospholipase C.** The substrate for this enzyme is a phosphorylated phospholipid called **phosphatidylinositol bisphosphate (PIP_2)** in the plasma membrane. This compound is produced by the covalent attachment of two phosphate groups to phosphatidylinositol (see Chapter 2). When activated by a G-protein α subunit, phospholipase C catalyzes the breakdown of PIP_2 into inositol triphosphate and diacylglycerol, a pair of second messengers.

Inositol triphosphate releases Ca^{2+} from intracellular stores, such as the ER. In many cells, this effect is transient but it triggers a more sustained increase due to an influx of extracellular Ca^{2+} through plasma membrane Ca^{2+} channels.

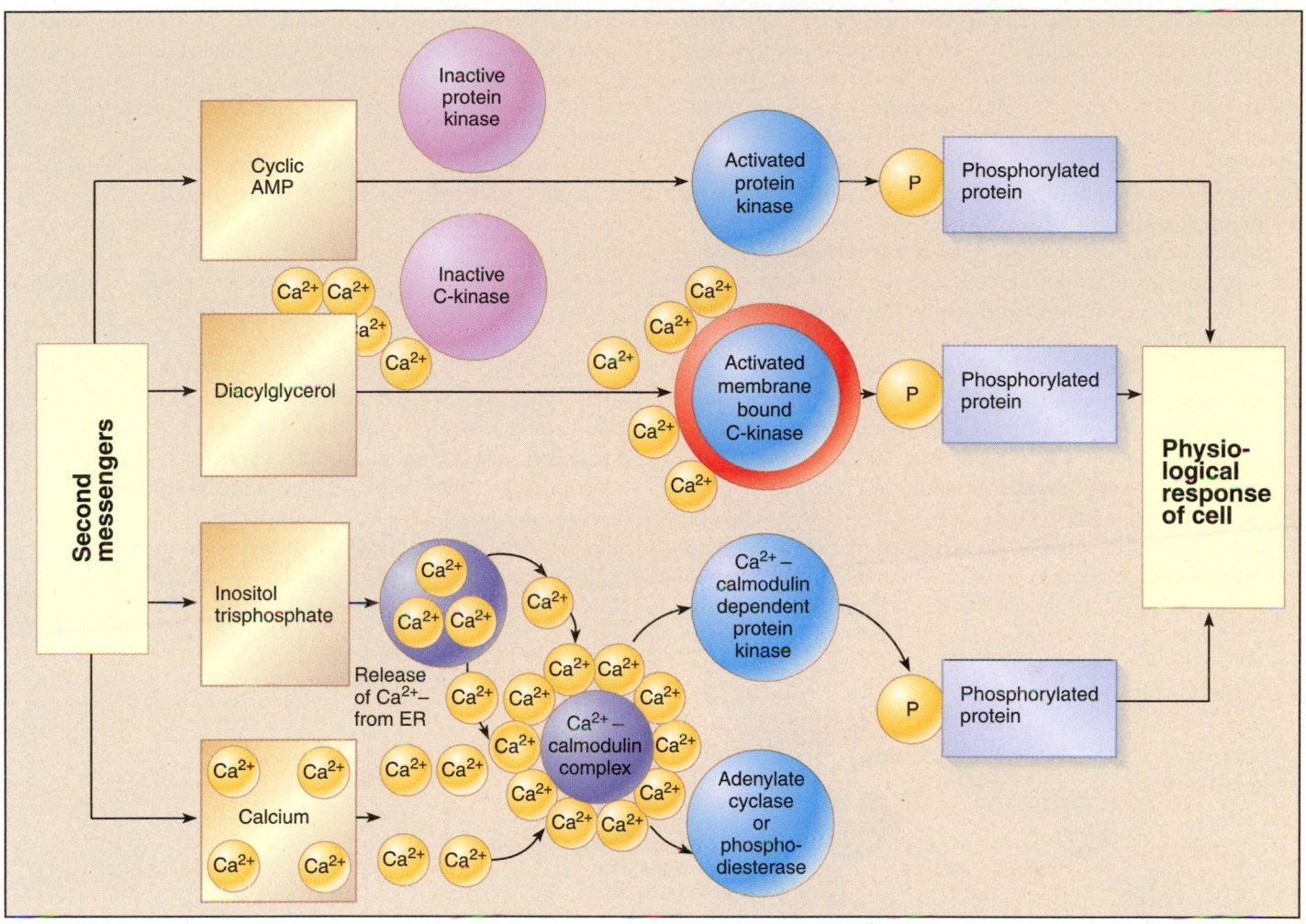

Figure 5–24

Second messengers may activate certain enzymes that catalyze the phosphorylation of certain proteins, which in turn produce the physiological response of the cell to the extracellular signal.

There is a highly favorable electrochemical gradient for passive entry of extracellular Ca^{2+} when the channels are opened. The overall result is to increase the cytosolic concentration of free Ca^{2+} from the resting level of 0.1 μmol/L to somewhere in the range of 1 to 10 μmol/L. The increased Ca^{2+} becomes bound with high affinity to **calcium-binding proteins** and produces a change in the conformation of these proteins and, hence, a change in their function. Some Ca^{2+}-binding proteins have no enzymatic activity but, upon binding Ca^{2+}, alter other proteins. Specific examples of this type of Ca^{2+}-binding protein are **troponin C**, found in skeletal and cardiac muscle cells, and **calmodulin**, which is found in almost all cells. Binding of a Ca^{2+}–calmodulin complex by a Ca^{2+}/calmodulin-dependent protein kinase (**CaM kinase**) activates the kinase and leads to protein phosphorylation (see Fig. 5–24), a mechanism similar to that resulting from activation of protein kinase A. The Ca^{2+}–calmodulin system is more versatile than the cAMP system because, in addition to

producing physiological changes via protein phosphorylation, the Ca^{2+}–calmodulin complex can directly activate enzymes, such as adenylate cyclase and **cAMP phosphodiesterase,** the enzymes that make and break down cAMP (see Fig. 5–24). Clearly, this allows interaction between the cAMP and the Ca^{2+} intracellular signaling pathways.

Diacylglycerol, together with Ca^{2+}, activates another important class of protein kinases, the phospholipid-dependent kinases known as **protein kinase C** or **C-kinase.** Protein kinase C in the cytosol is not affected by Ca^{2+}. In the presence of diacylglycerol, however, it becomes bound to the membrane, where it is activated by the phospholipids. In its activated state, protein kinase C is extremely sensitive to stimulation by Ca^{2+}. When the concentration of cytosolic Ca^{2+} is increased, the activated protein kinase C phosphorylates specific proteins (see Fig. 5–24). Among the substrates of protein kinase C are proteins involved in the control of cell division.

MEMBRANE EXCITABILITY. In some cells, G-protein–coupled receptors regulate an effector protein that is a plasma membrane ion channel. This changes the ion permeability and hence the electrical **excitability** of the membrane. An example is the action of acetylcholine to reduce both the rate and strength of heart muscle contractions. Binding of acetylcholine to its receptor activates a G-protein α subunit that directly opens K$^+$ channels in the muscle cell plasma membrane. Exit of K$^+$ makes the membrane potential more negative inside the cell, and the cell becomes less excitable. This contributes to the inhibitory action of acetylcholine on the heart.

Some plasma membrane receptors are also ion channels, and the conformational change that occurs in the receptor upon ligand binding causes the ion channel to open. Many of the neurotransmitters work in this way; the acetylcholine receptor at the nerve–muscle junction (synapse) is a good example. Acetylcholine is released by the nerve cell and crosses the synapse to bind to its receptor on the skeletal muscle cell. The receptor forms an ion channel, which opens transiently when acetylcholine binds. Na$^+$ enters the muscle

cell through this channel, making the membrane potential less negative and more excitable in a localized area. This triggers an **action potential,** which spreads rapidly throughout the membrane of the skeletal muscle cell and signals the muscle to contract.

Note the different responses to acetylcholine in these two examples. The receptor in the heart muscle is G-protein–linked, whereas the receptor in the skeletal muscle is a ligand-gated ion channel (see Chapter 4). Different receptors mediate different responses to the same extracellular ligand.

CYCLIC GMP. Another important second messenger in animal cells is **cyclic GMP (cGMP),** although its roles are not understood as well as are those of cAMP. Cyclic GMP is formed from GTP by **guanylate cyclase.** Its role in blood vessel dilation, mentioned earlier, appears to involve activation of specific cGMP-dependent protein kinases. The best-characterized role of cGMP is in the vertebrate eye, where it is the second messenger responsible for converting visual signals received as light into nerve impulses.

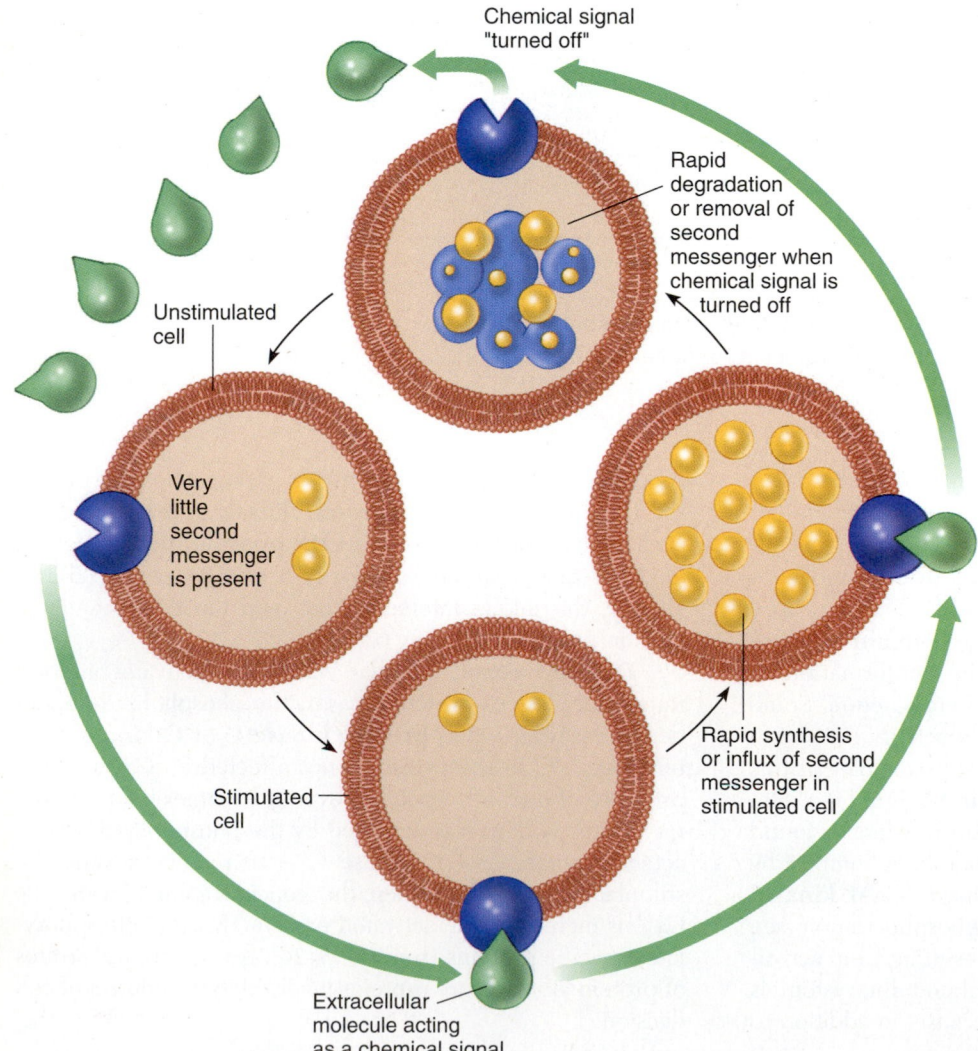

Chemical signal "turned off"

Rapid degradation or removal of second messenger when chemical signal is turned off

Unstimulated cell

Very little second messenger is present

Rapid synthesis or influx of second messenger in stimulated cell

Stimulated cell

Extracellular molecule acting as a chemical signal

Figure 5–25

A second messenger is synthesized rapidly when a first-messenger molecule (an extracellular signal) binds to a receptor. When the extracellular signal is removed, however, the second messenger is rapidly destroyed or removed from the cell. Thus, the next time a signal binds to the receptor, the second messenger must be synthesized anew.

ENZYME-LINKED RECEPTORS. In contrast to G-protein–linked receptors, which indirectly regulate the activity of a separate plasma membrane target protein, enzyme-linked receptors function directly as enzymes or are directly linked to an enzyme. The ligand binding domain is on the extracellular surface of the plasma membrane. The cytosolic domain either has an intrinsic enzyme activity or associates directly with an enzyme protein rather than a trimeric G-protein. The majority of these receptors have protein kinase activity and phosphorylate specific intracellular proteins. The largest group is the **receptor tyrosine kinases,** which are key signaling pathways in the response of cells to stimulation by growth factors. Ligand binding to the receptor activates its kinase activity, and the receptor phosphorylates tyrosine side chains on specific cellular proteins. Growth factors that act through receptor tyrosine kinases include insulin, epidermal growth factor, and nerve growth factor.

Removal of Second Messengers

The concentrations of second messengers in the cytosol of an unstimulated cell are usually very low because these molecules are continuously and rapidly destroyed or removed from the cytosol. The cytosolic concentration of a second messenger is altered principally by changes in its rate of synthesis or influx into a cell, which ensures that an increased rate of synthesis or influx will rapidly increase the cytosolic concentration of the messenger, allowing the target cell to respond quickly to the extracellular signal (Fig. 5–25). The concentration of cAMP is usually only 1 μmol/L, but this can be increased fivefold within seconds after a hormone binds to the plasma membrane receptor and stimulates adenylate cyclase. Synthesis or influx of the second messenger decreases to zero when the extracellular signal is "turned off." Because the rapid rate of destruction or removal of the messenger continues, the cytosolic concentration of the messenger molecules decreases rapidly to a point at which the cell ceases to respond. In other words, the response of the cell to the second messenger is reversed rapidly once the second messenger is no longer present (see Fig. 5–25).

Cyclic AMP, for example, is degraded very rapidly to adenosine 5′-monophosphate (AMP) by a reaction catalyzed by the enzyme cAMP phosphodiesterase. Cyclic GMP is degraded to GMP by a cGMP phosphodiesterase. Likewise, inositol triphosphate and diacylglycerol are rapidly degraded by specific enzymes.

To give another example, Ca^{2+} is removed from the cytosol by several active transport systems working together (Fig. 5–26). One mechanism moves Ca^{2+} out of the cell against its electrochemical gradient by primary active transport. This is achieved by a specific protein, known as Ca^{2+}-ATPase, which utilizes ATP to pump Ca^{2+} out of the cell. Another mechanism involves secondary active transport using the Na^+/Ca^{2+} antiport system. Entry of Na^+ into the

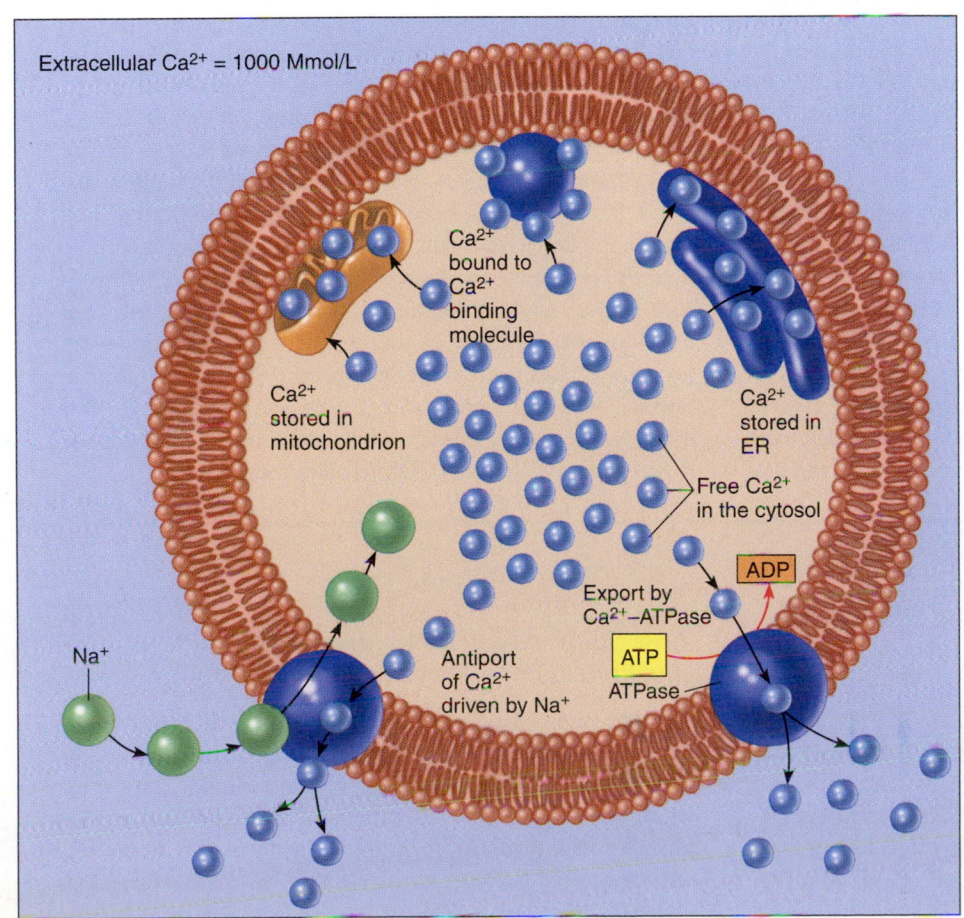

Figure 5–26

Examples of the ways in which a second messenger is removed from a cell. The second-messenger calcium ions in the cytosol may be removed by active transport, by antiport, by sequestration inside certain cytoplasmic organelles, or by binding to special molecules. These mechanisms reduce intracellular Ca^{2+} in nonstimulated cells to 0.1 μmol/L.

APPLICATIONS OF PHYSIOLOGY

Therapeutic Use of Phosphodiesterase Inhibitors

The era of second messengers and intracellular signaling was born with the discovery more than 35 years ago that some hormones and neuro-transmitters exerted their regulatory effect on cell functions by increasing the intracellular content of an adenine nucleotide known as **cyclic AMP (cAMP)**. Shortly after cAMP was identified, an analogous guanine nucleotide known as **cyclic GMP (cGMP)** was found, which had a similar, but not identical, role in intracellular signaling. The cellular content and, by extension, the biological effects of cAMP or cGMP are defined by the balance between the activities of synthesizing enzymes (adenylate cyclase and guanylate cyclase, respectively) and catabolizing enzymes. The latter are a group of enzymes termed cyclic-3′,5′-nucleotide phosphodiesterases, abbreviated as **PDE**. As discussed in this chapter, the PDE enzymes hydrolyze cAMP and cGMP to the biologically inactive noncyclic nucleotides AMP and GMP, respectively. Most of the interest in cAMP and cGMP has focused on regulation of synthesis, and, until recently, little attention was paid to regulation of hydrolysis. In retrospect, this is surprising because it has long been known that in most cells and tissues the cat-alytic capacity for hydrolysis of cAMP and cGMP by PDE enzymes is tenfold greater than the capacity for synthesis. It follows that PDE activity in intact cells must be tightly regulated so that it is normally operating well below full capacity. PDE activity influences not only the amplitude of the increase in cyclic nucleotide content but also the duration of this signal. The explosion of research interest in PDE enzymes has now established their importance as components of cellular signaling systems.

Vertebrate PDEs are a large superfamily of related enzymes that are diverse in structure, intracellular loca-tion, and sensitivity to inhibitors. Eleven distinct families have been identified so far and numbered from PDE1 to PDE11. Structural features that are shared by the differ-ent PDE enzyme proteins are a "regulatory domain" near the **N**-terminus, a conserved catalytic domain (about 300 amino acids), and a small **C**-terminal domain of unknown function. The regulatory domain contains amino acid se-quences that are targets for ligand binding, phosphoryla-tion, and protein-protein interactions. When this domain is removed experimentally there is a many-fold increase in PDE catalytic activity. This is consistent with the notion that PDE enzymes in cells are normally in an inhibited state and the various interactions confined to the regula-tory domain enhance the catalytic activity by relieving this "autoinhibition."

Many PDE inhibitors are recognized pharmacological agents and some were employed as drugs in medical prac-tice long before it was appreciated that they also inhib-ited PDE activity. Inhibitors of specific PDE families now are being explored vigorously, both experimentally and clinically, for effectiveness as therapeutic agents. Some are already approved for clinical use, such as the PDE3 inhibitors **milrinone** for treatment of congestive heart failure and **cilostazol** as an antithrombotic agent.

The drug **sildenafil**, better known as **Viagra**, that is used to treat penile erectile dysfunction in men is an in-hibitor of the PDE5 family. Experimental work suggests that some inhibitors of PDE3 and PDE4 may be useful to prevent or diminish renal transplant rejection. Other in-hibitors of the PDE4 family were found to reduce airway constriction and to suppress inflammation in the lungs of an-imals. PDE4 has now become an important molecular tar-get for the design of novel therapies for asthma and chronic obstructive pulmonary disease. The three-dimensional struc-ture of the catalytic domain of one of the PDE4 enzymes has now been resolved at the atomic level. The active site for catalysis has been identified and this structural informa-tion will help in understanding the catalytic mechanism and improve the rational design of clinically useful inhibitors of PDE4 and possibly other PDE families too.

---

cell, which diffuses down a favorable electrochemical gradi-ent, is used to drive Ca^{2+} out of the cell.

Free Ca^{2+} also is removed by binding to cytosolic cal-cium-binding proteins and other molecules and by sequestra-tion inside intracellular organelles, such as mitochondria and the ER. The large number of different mechanisms that can regulate cytosolic Ca^{2+} levels may be, at least in part, a de-fense system to guard against possible toxic effects of uncon-trolled concentrations of this ion.

Case Study: Glycogen Metabolism and Protein Phosphorylation

Protein phosphorylation is one example of a way in which the production of second messengers can mediate a change in cell function that represents the cellular response to a chemical signal. The metabolism of **glycogen** in skeletal muscle cells is a good example. Recall from Chapter 2 that glycogen is a polysaccharide, the form in which glucose is stored until its chemical energy is needed by the muscle cell. The catecholamine hormone **epinephrine** stimulates glycogen breakdown and shuts off glycogen synthesis——the net effect being to provide the cell with an adequate supply of glucose for strenuous activity.

Epinephrine works by binding to a plasma membrane receptor and elevating intracellular cAMP. The cAMP activates a protein kinase A, initiating a cascade of phosphorylation reactions (Fig. 5–27a). Phosphorylation by the protein kinase of enzyme 1 activates this enzyme; once activated, enzyme 1 phosphorylates enzyme 2, which begins to remove individual glucose molecules from glycogen. The same protein kinase phosphorylates a third enzyme (enzyme 3) that is involved in synthesis of glycogen from glucose. In contrast to enzymes 1 and 2, phosphorylation of enzyme 3 causes inactivation.

The overall effect of this signal cascade is to make available a maximum supply of glucose by simultaneously stimu-

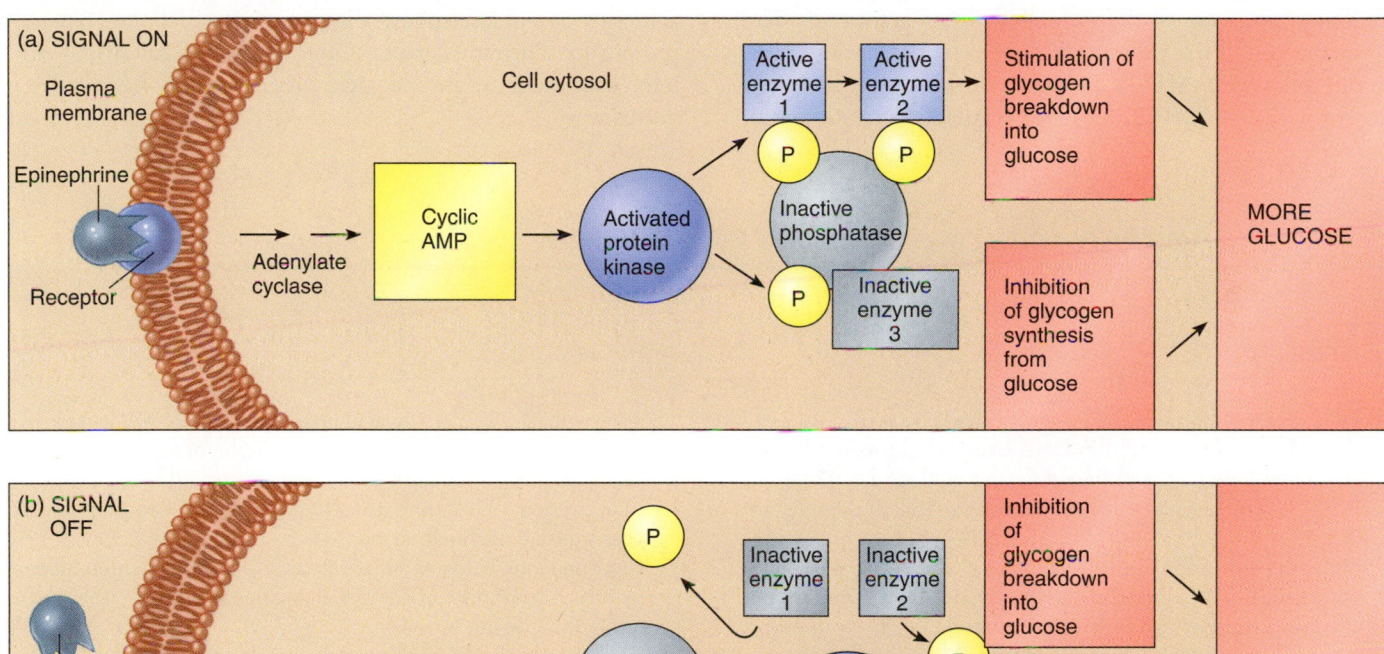

Figure 5–27

An example of the response of a cell to an extracellular signal and the reversal of that response once the signal is turned off. In this example, the desired effect of the chemical signal epinephrine on the target cell is an increase in available glucose as opposed to its storage in the form of glycogen. *(a)* When epinephrine binds to the G-protein–coupled receptor, adenylate cyclase is activated by the G-protein α subunit (not shown) and produces the second-messenger cyclic AMP. This, in turn, activates protein kinase A, which causes the phosphorylation of three enzymes. Phosphorylated enzymes 1 and 2 stimulate glycogen breakdown, while enzyme 3, which normally stimulates glycogen synthesis from glucose, is inactivated by phosphorylation. The net result of the signal cascade is the desired response of the target cell, an increase in the availability of glucose. *(b)* When the concentration of epinephrine decreases, the receptors are no longer occupied, the level of cyclic AMP decreases rapidly, and there is a decrease in activity of the signal cascade. As a result, all three enzymes are dephosphorylated by a protein phosphatase, glycogen synthesis is resumed, and less glucose is available.

lating breakdown of glycogen into glucose molecules while inhibiting the use of glucose for synthesis of glycogen.

When the epinephrine signal is turned off, the extracellular concentration of epinephrine decreases and the receptors no longer are occupied. This causes a decrease in the activity of the signal cascade, and the phosphorylation steps just described are reversed rapidly by reactions catalyzed by a **protein phosphatase** enzyme (see Fig. 5–27b). In other words, the phosphate groups on each of the three key enzymes are removed by the phosphatase, with the net effect of inhibiting glycogen breakdown (enzymes 1 and 2 are inactivated) and stimulating glycogen synthesis (enzyme 3 is activated). The opposing action of the protein phosphatase does not interfere with the action of cAMP because the phosphatase is inhibited while the level of cAMP is elevated. Regulation of both break-

down and synthesis of glycogen is an important feature of this cAMP control system. It increases the response to an elevated level of cAMP in the cell, the specific response being a readily available supply of glucose molecules.

We now know of more than 25 different enzymes regulated by a reversible phosphorylation process. All cells contain protein phosphatases that reverse the effects of protein phosphorylation. As shown in this example, the phosphatases are subject to regulation by agonists and second messengers, just like the kinases. Thus, the extent of phosphorylation of a regulated protein depends on the balance of the relative activities of the kinase and the phosphatase. These types of protein modifications are so abundant that many of the proteins that mediate physiological processes may be regulated in this way.

CHAPTER REVIEW

Summary

- Genetic information, the instructions a cell uses to synthesize RNA and proteins, is stored in DNA by a code formed from the specific linear sequence of four different nucleotides. Different cell types arise because the genes expressed in these cells are different.
- DNA is copied before cell division by a replication procedure catalyzed by DNA polymerase. RNA is transcribed from DNA in a process catalyzed by RNA polymerase. Newly synthesized RNA is converted to mature functional RNA by processing in the nucleus.
- Translation of the information in mRNA is coordinated by the ribosomes. Amino acids are linked to specific tRNA molecules that are bound to ribosomes, along with mRNA, during polypep-

tide synthesis. The amino acid attached to the tRNA is added to the growing polypeptide chain by formation of a peptide bond.
- Expression of genes can be regulated at several levels including transcription, RNA processing, RNA stability, translation, and post-translational modifications of the protein product.
- Gap junctions between adjacent cells allow for communication by direct movement of certain molecules from the cytoplasm of one cell to the other.
- Cells that are far apart release chemical signals that bind to specific receptors in the target cell. When the receptor is at the cell surface, binding of the chemical signal generates an intracellular signal known as a second messenger. The second messenger mediates the cellular response to the extracellular signal.

Review Questions

Choose the Correct Answer

1. Which of the following describes the function of a nucleolus?
   a. Produces ATP
   b. Synthesizes DNA polymerase
   c. Produces ribosomal subunits
   d. Stores genetic information for protein synthesis
   e. Proofreads in DNA replication

2. Amino acids are carried to the ribosomes in the cytoplasm by:
   a. mRNA.
   b. mitochondrial DNA.
   c. rRNA.
   d. tRNA.
   e. nuclear DNA.

3. Unwinding the DNA double helix prior to replication is aided by:
   a. DNA polymerase.
   b. DNA topoisomerase.
   c. DNA endonuclease.

   d. DNA ligase.
   e. RNA polymerase.

4. Chemical signaling systems generally involve:
   a. an extracellular signal molecule.
   b. a receptor molecule.
   c. a G-protein.
   d. an intracellular signal molecule.
   e. all of the above.

5. The ion that acts as a second messenger is:
   a. calcium.
   b. magnesium.
   c. potassium.
   d. sodium.
   e. zinc.

6. Inositol triphosphate is an example of a:
   a. second messenger.
   b. membrane receptor.

  c. first messenger.
  d. calcium channel.
  e. calcium-binding protein.

 7. The enzyme responsible for making cAMP is:
  a cAMP-dependent protein kinase.
  b. cAMP phosphodiesterase.
  c. adenylate cyclase.
  d. DNA glycosylase.
  e. C-kinase.

 8. The DNA that encodes RNA or protein accounts for approximately what percentage of the total DNA present in a mammalian cell?
  a. 90%
  b. 50%
  c. 5%
  d. 0.1%
  e. 0.01%

 9. Which of the following is correct? A ribosome:
  a. consists of one large and two small subunits.
  b. is composed of protein and carbohydrate.
  c. is the site of DNA replication.
  d. contains double-stranded rRNA.
  e. has two sites to which tRNA can bind.

10. In which of the following polymers are the monomers added one at a time during synthesis?
  a. DNA
  b. RNA
  c. tRNA
  d. Protein
  e. All of the above

11. Which of the following is correct? RNA:
  a. contains uracil.
  b. is single-stranded.
  c. is found in all cells.
  d. has three primary roles in protein synthesis.
  e. all the above are correct.

12. Primary RNA transcripts (pre-mRNA) of protein-encoding genes
  a. contain only introns.
  b. undergo capping and polyadenylation.
  c. encode the product of several genes.
  d. are translated immediately.
  e. are synthesized by RNA polymerase I.

13. Which of the following base pairs is found in double-stranded DNA?
  a. C-T
  b. U-C
  c. A-U
  d. T-A
  e. G-A

14. The various molecules that function in cells as second messengers:
  a. have related structures.
  b. are usually rapidly degraded or recycled after release.
  c. bind to receptors on the cell surface.
  d. always activate the enzyme C-kinase.
  e. include peptides, amino acids, and fatty acid derivatives.

15. An example of a signaling molecule that is derived from a fatty acid and acts locally is:
  a. a prostaglandin.
  b. a G-protein.
  c. a steroid.
  d. epinephrine.
  e. histamine.

16. The genetic code is defined as a series of:
  a. codons in DNA.
  b. anticodons in mRNA.
  c. codons in rRNA.
  d. codons in mRNA.
  e. anticodons in tRNA.

17. Inositol triphosphate initially increases Ca^{2+} in cytosol by causing Ca^{2+} release from
  a. mitochondria.
  b. Ca^{2+}-calmodulin complexes.
  c. lysosomes.
  d. the Golgi complex.
  e. the endoplasmic reticulum.

18. What is the term for the specialized structure between closely apposed plasma membranes of adjacent cells that directly connects the cytoplasm?
  a. Synapse
  b. Gap junction
  c. Tight junction
  d. Ion channel
  e. Calcium-binding protein

19. Steroid hormones and thyroid hormones diffuse across the plasma membrane of target cells and bind to:
  a. DNA.
  b. RNA.
  c. intracellular receptors.
  d. protein kinases.
  e. transcription factors.

20. Most of the effects of Ca^{2+} in cells are mediated by protein phosphorylations catalyzed by a family of:
  a. Ca^{2+}/calmodulin-dependent protein kinases.
  b. cAMP-dependent protein kinases.
  c. phosphoprotein phosphatases.
  d. cGMP phosphodiesterases.
  e. phospholipases.

Answers to Case History Questions

1. The *MEPC2* gene is located on the X chromosome. In humans (and many animals), females have two X chromosomes and males have only one. Thus, females have two *MEPC2* genes. If one is normal and the other is mutated, a female embryo can develop normally at first but eventually acquires Rett syndrome, a progressive neurodevelopmental disorder that is one of the most common causes of mental retardation in females, in early childhood. It is a fatal problem for males. A male embryo has a single *MEPC2* gene; if this is mutated, the embryo experiences the full effect of this defect and dies before birth or shortly afterward.

2. The probable result of the genetic defect is that the brain is unable to develop and function normally during postnatal growth.

Abnormal levels of neurotransmitters have been implicated, but not proven, in both the behavioral changes and the reduced brain size.

3. A common means to turn off, or silence, a gene so that it is not transcribed into protein is to attach a methyl group to it. The methyl group is attached to the cytosine nucleotide of a cytosine-guanine dinucleotide sequence (CpG) within the DNA of the gene. A methyl-binding protein then attaches itself to the methyl group and the final complex that is formed prevents the transcription machinery from binding to the DNA. The encoded proteins cannot be synthesized and the gene is said to be silenced.

4. A mutated *MECP-2* gene produces an abnormal methyl-binding protein that may be unable to turn off methylated genes because it cannot bind to them. This process disrupts the precisely regulated pattern of development in the brain, which, as in other cells, depends on turning genes on or off in the correct sequence and at the proper time. The genes that bind the normal MECP-2 protein are not known, but some scientists have speculated that they may be the genes that control neurotransmitter levels during brain development.

5. Rett syndrome, a progressive neurodevelopmental disorder that is one of the most common causes of mental retardation in females. The incidence ranges from 1 in 10,000 to 1 in 23,000 population; mutations in the *MECP-2* gene have been discovered in 76% of Rett syndrome patients. Only the brain seems to be affected.

Key Terms

alternative splicing (p. 159)
anticodon (p. 151)
bioinformatics (p. 158)
calcium ions (p. 166)
chromosome (p. 140)
cyclic AMP (p. 166)
diacylglycerol (p. 167)
DNA polymerase (p. 144)
DNA replication (p. 142)
DNA transcription (p. 142)
exon (p. 149)

gap junction (p. 159)
gene silencing (p. 159)
genetic code (p. 141)
genetic expression (p. 141)
genome (p. 140)
hormone (p. 160)
inositol triphosphate (p. 166)
intron (p. 143)
ligand (p. 166)
messenger RNA (p. 147)
methylation (p. 159)

neurotransmitter (p. 161)
phosphodiesterase (p. 167)
phospholipase C (p. 166)
phosphoprotein phosphatase (p. 172)
post-translational modification (p. 159)
protein kinase (p. 166)
protein sorting (p. 156)
protein synthesis (p. 143)

ribosomal RNA (p. 147)
ribosome (p. 147)
RNA polymerase (p. 148)
RNA processing (p. 143)
second messenger (p. 161)
spliceosome (p. 151)
template (p. 144)
transfer RNA (p. 147)
transcription factor (p. 159)
zinc finger (p. 159)

Suggested Readings

Branden, C., and Tooze, T. *Introduction to Protein Structure*, ed 2. New York, Garland, 1999.

Brown, K. "The human genome business today." *Scientific American*, 283:50–55, 2000. (July).

Cech, T. R. "The efficiency and versatility of catalytic RNA: Implications for an RNA world." *Gene*, 135:33–36, 1993.

Dousa, T. P. "Cyclic-3′,5′-nucleotide phosphodiesterase isozymes in cell biology and pathophysiology of the kidney." *Kidney International*, 55:29–62, 1999.

Ezzell, C. "Beyond the human genome." *Scientific American*, 283:64–69, 2000. (July).

Halim, N.S. "Methylation: Gene expression at the right place and right time." *The Scientist*, 13:21, 1999.

Hendrich, B. "Methylation moves into medicine." *Current Biology*, 10:R60-R63, 2000.

Howard, K. " The bioinformatics gold rush." *Scientific American*, 283:58–63, 2000. (July).

Lander, E. S., and Weinberg, R.A. "Genomics: Journey to the center of biology." *Science*, 287:1777–1782, 2000.

Lockhart, D. J., and Winzeler, E. A. "Genomics, gene expression and DNA arrays." *Nature*, 405:827–836, 2000.

Lodish, H., Berk, A., Zipursky, S. L., Matsudaira, P., Baltimore, D., and Darnell, J. *Molecular Cell Biology*, ed 4. New York, Freeman, 2000.

McLaren, A. "Cloning: Pathways to a pluripotent future." *Science*, 288:1775–1780, 2000.

Nebl, T., Oh, S. W., and Luna, E. J. "Membrane cytoskeleton: PIP$_2$ pulls the strings." *Current Biology*, 10:R351-R354, 2000.

Neer, E.J. "Heterotrimeric G proteins: organizers of transmembrane signals." *Cell*, 80:249–257, 1995.

Pandey, A., and Mann, M. "Proteomics to study genes and genomes." *Nature*, 405:837–846, 2000.

Pawson, T., and Scott, J. D. "Signaling through scaffold, anchoring and adaptor proteins." *Science*, 278:2075–2080, 1997.

Scott, J. D., and Pawson, T. "Cell communication: The inside story." *Scientific American*, 282:72–79, 2000. (June).

Service, R.F. "Creation's seventh day." *Science*, 289:232–235, 2000.

Soderling, S. H., and Beavo, J. E. "Regulation of cAMP and cGMP signaling: New phosphodiesterases and new functions." *Current Opinion in Cell Biology*, 12:174–179, 2000.

Spengler, S. J. "Bioinformatics in the information age." *Science*, 287:1221–1223, 2000.

Van den Veyver, I. B., and Zoghbi, H. Y. "Methyl-CpG-binding protein 2 mutations in Rett syndrome." *Current Opinion in Genetics and Development,* 10:275–279, 2000.

Venter, J. C. et al. "The sequence of the human genome." *Science* 291:1304–1351, 2001 (Part of a special issue on the human genome).

Web Sites

http://smart.embl-heidelberg.de/
Web site for the European Molecular Biology SMART database. Scientists can search the database to determine if the amino acid sequence of any protein they are interested in contains specific domains (motifs) that are known to be involved in protein signaling.

http://www.nhgri.nih.gov/index.html
Web site for the National Human Genome Research Institute. Includes information on the Human Genome Project and also the ethical, legal, and social implications of research on human genetics.

http://www.tigr.org
The Institute for Genomic Research is a not-for-profit institute that studies genomes of viruses, bacteria, plants, animals, and humans. The Web site contains a listing of genome sequences and related links.

http://www.ncbi.nlm.nih.gov
A group of bioinformatics databases maintained by the National Center for Biotechnology Information. It includes GenBank, a collection of all publicly available DNA sequences. There were more than 6 million sequence records as of April 2000. Many scientific journals require sequence information to be submitted to GenBank so that an accession number can be included in the published paper. This number allows easy retrieval of the sequence by other scientists. The database also can be searched for specific DNA sequences and there is no charge for access.

http://www.nsf.gov
The National Science Foundation was established in 1950 as an independent agency of the U.S. Government. The Foundation provides grants and contracts to support research and education in science and engineering. The home page includes a "News" feature that highlights recent important discoveries.

http://workbench.sdsc.edu
The San Diego Supercomputer Center hosts the Biology Workbench, a web-based tool for biologists. The Workbench provides a point and click interface for rapid access, so that anyone with a web browser can use bioinformatics for research, teaching, or learning. Visitors can search a number of worldwide biology databases, such as protein and nucleic acid sequences. Searching is integrated with application programs, such as analytical and modeling tools, that perform operations on the databases. Users can compare molecular sequences and visualize and manipulate molecular structures using high-performance computing facilities. Links to tutorials and a student version are provided. Registration is required, but there is no charge for access and use.

Answers to Review Questions

1. c	**2.** d	**3.** b	**4.** e	**5.** a	**6.** a	**7.** c	**8.** c
9. e	**10.** e	**11.** e	**12.** b	**13.** d	**14.** b	**15.** a
16. d	**17.** e	**18.** b	**19.** c	**20.** a

Chapter 6

ENERGY AND CELLULAR METABOLISM

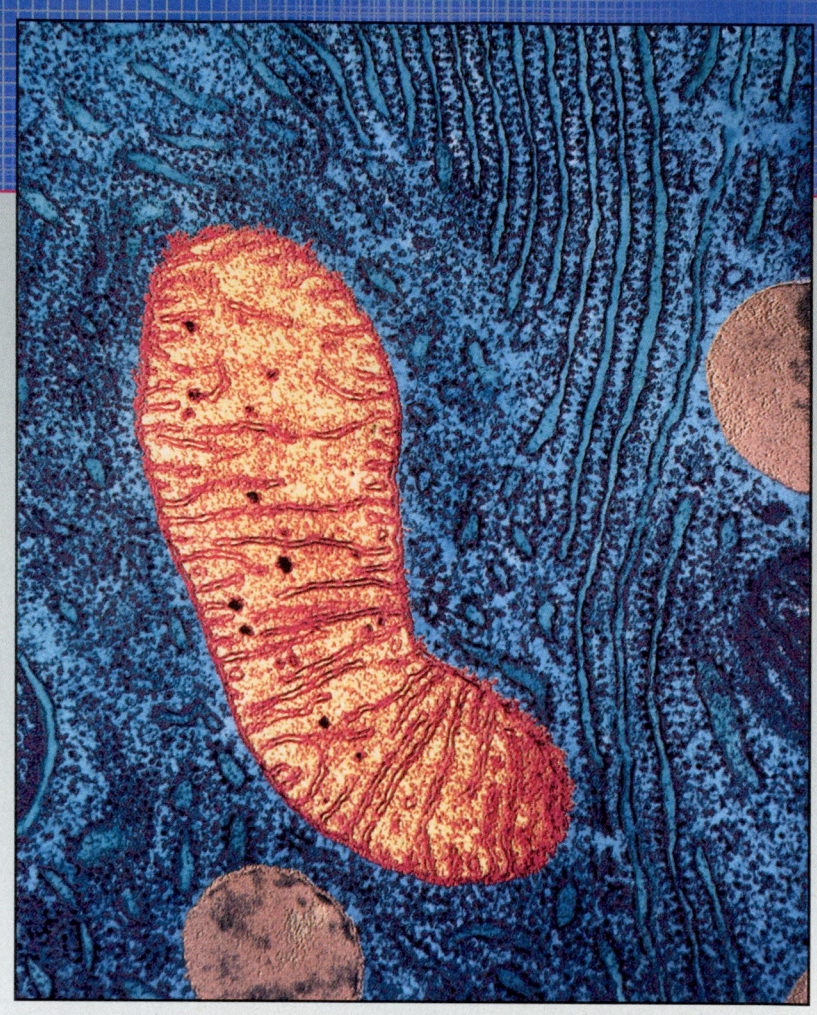

KEY CONCEPTS

- Living cells transform energy from one form to another. Plants use radiant energy from the sun to photosynthesize high-energy organic molecules. Animals transform energy by eating plants; breaking down these molecules; and converting their energy into ATP, which they use to drive biological processes.

- Digestion of the proteins, carbohydrates, and fats present in food produces amino acids, glucose, and fatty acids. Cells degrade these macromolecules by pathways that culminate in the citric acid cycle in the mitochondrial matrix. The high-energy electrons released by this cycle are trapped by coenzymes and fed into the electron transport chain to generate ATP.

- Glucose molecules are converted by glycolysis into pyruvate, and the acetyl group derived from each pyruvate is fed into the citric acid cycle in the form of acetyl coenzyme A.

- Long-chain fatty acids are a rich source of energy. The chain of carbon atoms is cut into 2-carbon fragments, and these fragments (acetyl groups) are fed into the citric acid cycle. The glycerol enters the glycolytic pathway.

- Many amino acids are degraded by the initial removal of the amino groups, followed by entry of the carbon atoms at various points in the pathway of glucose catabolism.

- The 2-carbon fragments produced by catabolism are transferred as acetyl groups to coenzyme A. These acetyl groups are the precursors required for synthesis of fatty acids and cholesterol. Depending on metabolic demands, they are either shunted into the citric acid cycle or used in biosynthetic pathways.

CASE HISTORY

ick, a 28-year-old man, visited his family physician, complaining that he had experienced progressive loss of vision in one eye during the past three weeks. He described the problem as dimming or darkening of vision rather than blurring. No pain or preexisting symptoms occurred. Neither of his parents nor his two older sisters has experienced any visual losses. The physician found that the eye was free of cataract and that no other clinical symptoms were present. Rick was a casual smoker but used alcohol rarely and had no nutritional deficiencies. Rick was immediately referred to an ophthalmologist, who discovered that his vision was depressed in a central area corresponding to and interfering with the point of fixation of the eye, a condition known as **central scotoma.** Further examination revealed thickening and swelling of the optic nerves in both eyes. Both optic discs appeared mildly pink, with dilated surface capillaries, but angiography showed that no vascular leakage was occurring. In order to confirm a preliminary diagnosis, the ophthalmologist scheduled Rick to have a blood sample drawn for analysis of his mitochondrial DNA. The ophthalmologist's suspicions were confirmed when the clinical laboratory reported that there was a single mutation at base pair 11778 in the mitochondrial DNA. Rick was informed that he had a disease known as **Leber's hereditary optic neuropathy (LHON).** With no effective treatment available, the long-term prognosis was poor. Rick was warned to expect a similar problem with the other eye during the next several months, and it was likely that he would be effectively blind in both eyes within two years.

Questions

1. Why does LHON cause blindness?

2. What proteins are encoded by mitochondrial DNA; why does the mutation at base pair 11778 produce a clinical problem?

3. Why is the optic nerve the only tissue that becomes impaired when the mutant mitochondrial DNA is present in all cells of the body of a patient with LHON?

4. What might account for the long and variable period of normal vision that precedes the onset of LHON?

INTRODUCTION

Metabolism is the term used to describe the highly organized and integrated set of chemical reactions and energy transformation that sustain a living organism. These reactions are myriad; at least 1000 different reactions can take place within the unicellular bacterium *Escherichia coli.*

In a series of steps known as **catabolism,** the cell extracts energy from its environment, the extracellular fluid. Catabolism is usually achieved in animals by the breakdown of the molecules present in food. This process has the additional purpose of converting complex macromolecules into simple molecules that can be used as "building blocks" to make new macromolecules. In a series known as **anabolism,** or, more descriptively, **biosynthesis,** the energy extracted in catabolism is used to assemble the building-block molecules into needed proteins, nucleic acids, lipids, and other new macromolecules for the cell.

Every cell is always engaged in both catabolism and biosynthesis, and normally these processes are carefully balanced. The amount of a specific macromolecule present in the cell at any given time depends on the balance between biosynthesis and catabolism; ideally, there is no shortage or excess. The fine control that this balance requires is remarkable, both at the cellular level and, on a larger scale, at the level of the whole animal. Consider, for example, that the average adult human can eat about 6 tons of food over a span of 40 years without losing or gaining weight and with little change in the composition of the body.

Although cellular metabolism is a highly complex process, the central pathways are not difficult to understand. Moreover, they are essentially similar in most animal cells. This chapter examines some of the basic principles that drive all metabolic processes and their specific involvement in the metabolism of carbohydrates, fats, and proteins.

BASIC PRINCIPLES OF ENERGY TRANSFORMATION IN THE CELL

 How do animals obtain energy from the environment?

Cells obtain from their environment the energy that they need to carry out work and, ultimately, release some of that energy back to the environment in the form of heat. Some of the ways by which living cells obtain and transform energy are depicted in Figure 6–1.

Plants Use Radiant Energy to Synthesize Organic Molecules

Plants use light from the sun, a form of radiant energy, for the **photosynthesis** of ATP, the most versatile form of chemical energy (see Chapter 2). The **ATP** drives biosynthe-

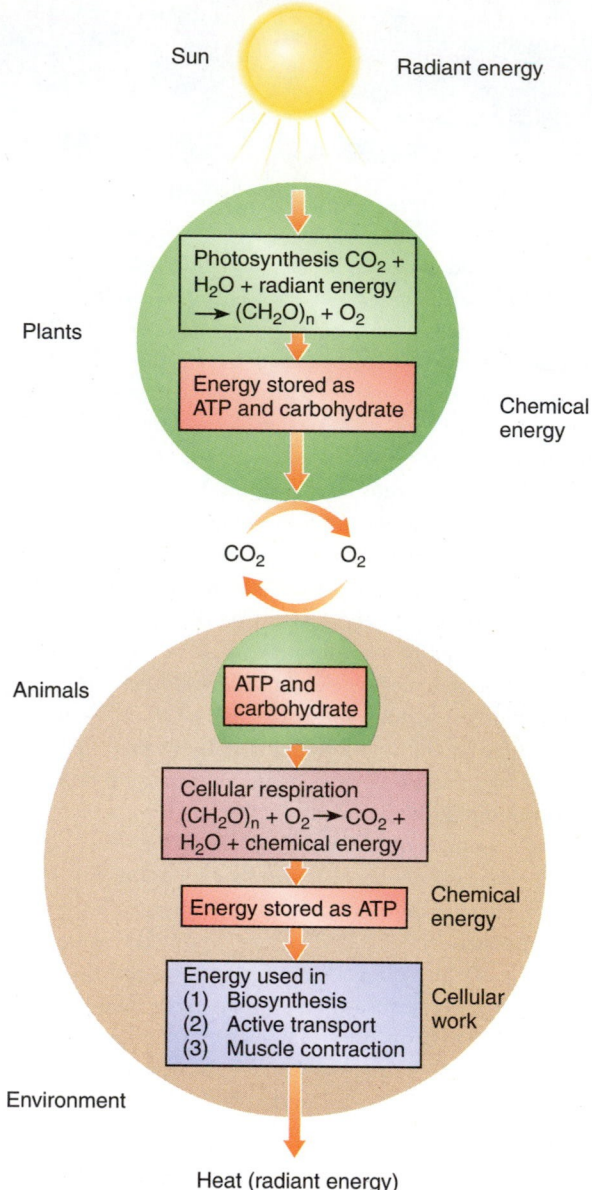

Figure 6–1

The flow of energy. Plants use radiant energy directly from the sun to photosynthesize high-energy organic molecules. Animals obtain these molecules by eating plants and then transform their energy into ATP, which they use to do biological work.

sis within plant cells, including the series of reactions known as **CO_2 fixation,** in which CO_2 from the atmosphere is used in the creation of carbohydrates and other high-energy organic compounds. They are considered high-energy compounds because their many covalent bonds represent solar energy stored as chemical energy. Photosynthesis requires

CO_2 and produces O_2 as a by-product. The process can be summarized as:

$$CO_2 + H_2O + \text{radiant (solar) energy} \rightarrow (CH_2O)_n + O_2$$

low-energy
compounds

organic
molecule
(high-energy
compound)

Plants convert some newly synthesized carbohydrates, by a series of metabolic reactions, to the other types of high-energy organic molecules that they need.

Animals Obtain Chemical Energy from Plant Molecules

The organic molecules made by the plants are a source of chemical energy and building-block molecules for the animals that eat the plants. Unlike plants, animals cannot use the energy from sunlight directly. They can obtain this energy only after its transformation by plants.

The organic "fuel" molecules that animals ingest when they consume plants are catabolized by reactions that require atmospheric O_2, and are termed **cellular respiration.** They produce CO_2 and water as by-products and can be thought of as the reverse of photosynthesis:

$$(CH_2O)_n + O_2 \rightarrow CO_2 + H_2O + \text{chemical energy}$$

high-energy
organic
compounds

low-energy
compounds

(ATP)

These complex reactions are directed by **enzymes** and designed to release the chemical energy of the organic molecules in a controlled fashion so that it can be "trapped" and stored as ATP. This is a form of chemical energy that cells can use for biosynthesis, active transport, and the generation of force and movement required for muscle contraction.

There Is a Constant Demand for ATP in Living Cells

The transformation of energy by animal cells is not a perfect process. Not all of the energy released from a fuel molecule can be used to form ATP. The untrapped energy is lost to the environment in the form of heat. Because they constantly need ATP and are continually losing heat energy, living cells require constant input of energy in order to survive. ATP is not a long-term storage form of chemical energy. It is used almost as soon as it is formed; the "turnover" of an ATP molecule in a cell is very rapid. At any moment, less than 1 g of ATP is present in the human body. However, even in a person who is resting with minimal activity, about 45,000 g of ATP is formed and used every 24 hours. If the person undergoes strenuous physical activity during the day, the rate of ATP utilization can be as high as 500 g/min.

Many Factors Influence the Rate and Direction of Chemical Reactions in Cells

What types of chemical reactions that require or produce energy are found in cells?

Free Energy Change in Cell Reactions

Recall from Chapter 2 that some reactions, called **exergonic reactions,** are highly likely to proceed spontaneously. The driving force for such reactions is the difference in total free energy between reactants and products. If a large difference exists—one in which the reactants contain much more free energy than the products—a large amount of useful energy is released when the reaction occurs.

Other reactions, called **endergonic reactions,** cannot proceed without input of free energy. Both types of reactions occur in the cell. Endergonic reactions (in which ΔG is positive) obtain the free energy they need by being **coupled** to exergonic reactions (in which ΔG is negative).

One of the first steps in glucose catabolism, for example, is to add a phosphate group (P_i) to carbon atom 6 in the glucose molecule, an endergonic reaction called a **phosphorylation:**

$$\text{Glucose} + P_i \rightarrow \text{Glucose-6-}P_i$$
$$\Delta G = +3 \text{ kcal/mol (or } +13 \text{ kJ/mol)}$$

This reaction takes place because the P_i that is added comes from a molecule of ATP. The P_i group has been removed from ATP by a strongly exergonic reaction:

$$\text{ATP} \rightarrow \text{ADP} + P_i$$
$$\Delta G = -7 \text{ kcal/mol (or } -29 \text{ kJ/mol)}$$

In the coupling of these two reactions, the 7 kcal/mol of free energy that ATP gives up by losing a P_i is used to drive the endergonic phosphorylation of glucose. A summary equation shows the overall reaction and the net change in free energy:

$$\text{Glucose} + \text{ATP} \rightarrow \text{Glucose-6-}P_i + \text{ADP}$$
$$\Delta G = -4 \text{ kcal/mol (or } -17 \text{ kJ/mol)}$$

The ΔG for the overall reaction is negative, indicating that the phosphorylation of glucose by this mechanism is thermodynamically feasible.

Reversible Reactions in the Cell

Reactions that are highly exergonic proceed in one direction, toward the formation of products. However, many reactions are **reversible;** that is, the products may immediately recombine by the reverse reaction. The following equation describes a reversible reaction:

$$AB \rightleftharpoons A + B \qquad \Delta G = 0$$

Such a reaction is said to be in **equilibrium** when the rate of formation of the products (A + B) is exactly balanced by the rate at which the products recombine to yield the reactant (AB). In this situation, there is no net change in free energy. In an equilibrium reaction, it might be difficult to obtain enough of the product B because it tends to recombine with A as soon as it is formed. One way to ensure that such a reaction proceeds "to the right," toward "completion" or the formation of A and B, is to remove any B as soon as it is formed. Then AB continues to break down because there is no B to recombine with A.

Similarly, a metabolic reaction can be driven to completion even though the formation of products is almost exactly balanced by the reformation of reactants. In the cell, metabolic reactions are driven to completion because the product of one reaction (e.g., B) is constantly being "removed" by participating in another reaction:

$$(1)\ AB \rightarrow A + B$$
$$\downarrow$$
$$(2)\ B + C \rightarrow D$$

Many reactions of metabolism, while not at equilibrium, take place with a very small change in free energy and can be driven in either direction relatively easily. This fact allows a cell to reverse a pathway of metabolism according to the specific demands of any given moment. For example, the synthesis of glucose from precursors takes place by reversal of many of the reactions that are used to catabolize it.

These last points emphasize one of the characteristic features of cellular metabolism: No single reaction is taking place independently or in isolation. Hundreds of reactions are occurring in the cell at the same time, and most of them are interrelated; the products of one reaction are being used for the next reaction in a metabolic pathway. Even the so-called "final products" of one pathway are likely to be used in another pathway.

Oxidation-Reduction Reactions in the Cell

Oxidation-reduction (redox) reactions are combinations of **oxidation reactions** (involving loss of electrons) with **reduction reactions** (involving gain of electrons). A compound becomes "reduced" by gaining electrons.

$$Ae^- + B \rightarrow A + Be^-$$

| electron | electron | has been | has been |
| donor | acceptor | oxidized | reduced |

> **What role does hydrogen play in oxidation-reduction reactions in the cell?**

HYDROGEN: THE "ELECTRON" IN BIOLOGICAL REDOX REACTIONS. Organic compounds do not give up electrons easily. Removal of an electron during oxidation is achieved only by loss of an entire atom. The specific atom given up is almost always hydrogen, and so the oxidation process is actually a **dehydrogenation** reaction; the reduction process is a **hydrogenation** process. In general, compounds that are highly reduced—that is, contain a high proportion of hydrogen atoms—are high-energy compounds. Oxidized compounds contain a low proportion of hydrogen atoms and are low-energy compounds.

$$AH + B \rightarrow A + BH$$

electron	electron	has been	has been
donor	acceptor	oxidized	reduced
high	low	low	high
energy	energy	energy	energy

CELLULAR RESPIRATION: AN OXIDATION PROCESS. Conversely, the breakdown of an organic molecule in cellular respiration can be thought of as an oxidation process (the removal of electrons in the form of hydrogen atoms) that produces the highly oxidized, low-energy compound CO_2:

$$(CH_2O)_n + O_2 \rightarrow CO_2 + H_2O + energy$$

electron donor	electron	carbon	has been
(high-energy	acceptor	has been	reduced
organic		oxidized	
compound)			

Note that the electrons (H atoms) removed from the organic molecule are accepted by the O_2 molecule, forming water. Oxygen is therefore called the *electron acceptor*. (Because oxygen is the most common electron acceptor, the loss of electrons is called *oxidation*.) As we have said, organic fuel molecules, such as glucose ($C_6H_{12}O_6$), are highly *reduced* compounds because they contain many hydrogen atoms. Thus, the catabolism, or breakdown, of these molecules is essentially an *oxidative* process, involving the removal of electrons in the form of those many hydrogen atoms. The hydrogen atoms often combine with oxygen to form water. For example, the complete catabolism of one glucose molecule uses up six molecules of oxygen and can be summarized as:

$$C_6H_{12}O_6 + 6\ O_2 \rightarrow 6\ CO_2 + 6\ H_2O$$

Before combining with water, however, the hydrogen atoms that glucose gives up during oxidation are transferred temporarily to molecules called **coenzymes,** chiefly **NAD** (see Chapter 2), which thereby becomes reduced. The reduced forms of NAD and other coenzymes in turn donate electrons to the **electron transport chain,** known also as the **respiratory chain,** which is embedded in the inner mitochondrial membrane. The flow of electrons along the respiratory chain occurs with the release of free energy that is used to drive endergonic synthesis of ATP (Fig. 6–2). This process is discussed later in this chapter.

> **What role does oxygen play in cellular respiration?**

ELECTRON ACCEPTORS: KEY MOLECULES IN CELLULAR RESPIRATION. Electrons cannot simply be released but must be accepted by another molecule. Oxygen and coenzymes play crucial roles in cellular respiration

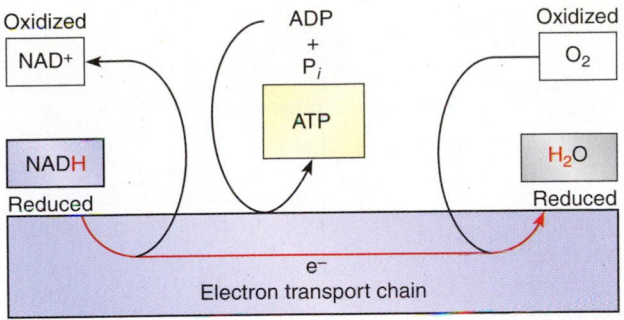

Figure 6–2

Energy transformation occurs in the mitochondria of animal cells through a series of redox reactions, in which electrons in the form of hydrogen atoms are transferred from highly reduced compounds, such as NADH, to highly oxidized compounds, such as O_2. The energy released in these reactions is used to phosphorylate ADP to produce ATP, the energy-storage molecule.

by accepting the electrons that organic molecules release when oxidized. Coenzymes accept electrons in intermediate steps in the process. Oxygen accepts electrons in the final step.

COENZYMES: INTERMEDIATE ELECTRON ACCEPTORS. An example of a biological redox reaction that involves a coenzyme is the conversion of lactic acid into pyruvic acid:

$$
\begin{array}{ccc}
\text{COOH} & & \text{COOH} \\
| & & | \\
\text{CHOH} & \xrightarrow{\text{oxidation}} & \text{C}=\text{O} \quad + \quad 2\,\text{H} \\
| & & | \qquad\qquad \text{H atoms}\\
\text{CH}_3 & & \text{CH}_3 \\
\text{lactic acid} & & \text{pyruvic acid} \\
\text{electron} & & \\
\text{donor} & &
\end{array}
$$

The freed hydrogen atoms usually are transferred to a coenzyme. The coenzyme most frequently used as a hydrogen acceptor in metabolic reactions is NAD (see Fig. 6–7).

The formation of pyruvic acid by this reaction is a reversible process in the cell. Pyruvic acid can be converted back to lactic acid by the addition of two hydrogen atoms; by accepting the hydrogen atoms, the pyruvic acid becomes reduced to lactic acid:

$$
\text{Pyruvic acid} \quad + \quad 2\,\text{H} \xrightarrow{\text{reduction}} \text{Lactic acid}
$$

electron
acceptor   H atoms

The hydrogen atoms for this reaction are provided by the same coenzyme (NAD) involved in the oxidation of lactic

acid. Note that during the interconversion of lactate and pyruvate, the NAD is donating or accepting hydrogen atoms from the reactants, thereby becoming either oxidized or reduced. The process can be summarized as:

$$
\begin{array}{cccc}
\text{Lactic} + \text{Oxidized} & \rightleftharpoons & \text{Pyruvic} + \text{Reduced} \\
\text{acid} \qquad \text{NAD} & & \text{acid} \qquad \text{NAD}
\end{array}
$$

This illustrates the point that the oxidation and reduction of the organic molecules involved in cellular metabolism are true redox reactions. The oxidation of one molecule, such as lactic acid, is coupled to the reduction of another, such as NAD.

In some oxidative reactions involving the removal of hydrogen atoms from an organic compound, the protons and electrons of the hydrogen atoms depart separately or become separated after removal of the atoms. The hydrogen atom contains one proton bearing a single positive charge and one electron with a single negative charge. Thus, a hydrogen atom that has given up its electron has a single positive charge, due to the remaining proton, and can be represented as H^+. A single electron can be represented as e^-. By way of example, the oxidation of hydroquinone can be represented in two stages, as follows:

$$
\text{OH} \quad \xrightarrow{2\,H^+} \quad O^- \quad \xrightarrow{2\,e^-} \quad O
$$

reduced   two protons   two electrons   oxidized
               removed   removed

OXYGEN: THE ULTIMATE ELECTRON ACCEPTOR. Organic molecules vary in their tendencies to give up or accept electrons. Their affinity for electrons can be measured in terms of their oxidation-reduction potential, better known as the **redox potential.** The redox potential of molecules is expressed on a scale in which the redox potential of hydrogen gas (H_2) is 0 millivolts (mV). A molecule that has a negative redox potential on this scale has a lower affinity than H_2 for electrons. A specific example is NADH, the reduced form of NAD. This molecule, with a redox potential of -320 mV, has a low affinity for electrons and donates them to another electron carrier, thereby reverting to its oxidized form, NAD^+.

At the other end of the scale of redox potentials is molecular oxygen, which, with a redox potential of $+820$ mV, has a high affinity for electrons, and readily accepts them, becoming reduced to water in the process. Molecular oxygen is the ultimate acceptor of the electrons donated by NADH to the mitochondrial respiratory chain (see Fig. 6–2).

High-Energy Phosphate Compounds

> **What determines whether a molecule will serve as a phosphate donor to ATP or as a phosphate acceptor from ATP?**

Earlier in this chapter, we discussed the use of the chemical energy stored in a molecule of ATP to drive an endergonic reaction, such as the phosphorylation of a glucose molecule. ATP is an example of a special group of compounds known as **high-energy phosphate compounds**, which have a strong tendency to transfer their phosphate to an acceptor molecule. This transfer is accompanied by the release of a large amount of free energy.

The ΔG for a given reaction is influenced by the concentrations of the reactants and products; it is not a very useful term for comparing the energetics of different reactions. For comparisons of reactions, the free energy difference between the reactants and the products is considered under a set of "standard conditions": temperature of 25°C, pressure of 1 atmosphere (atm), and a concentration of 1.0 M for all reactants and products present. Under these conditions, the free energy change for a reaction is called the *standard free energy change* and is given the symbol $\Delta G°$. This term is a constant for a specific reaction. For example:

$$ATP \rightarrow ADP + P_i$$
$$\Delta G° = -7 \text{ kcal/mol (or } -29 \text{ kJ/mol)}$$

In the cell, however, the "standard conditions" are not present. The temperature is 37°C, and the concentration of ATP is very high relative to ADP and P_i. Under these conditions, the free energy change (ΔG) for ATP breakdown by the above reaction is about -12 kcal/mol (-50 kJ/mol). The large difference between $\Delta G°$ and ΔG in this example indicates that it is the ΔG that must be used to predict the direction of reactions within the cell.

The **standard free energy change** ($\Delta G°$) can be used to compare the tendency of different high-energy phosphate compounds to transfer their phosphate group to an acceptor molecule. The $\Delta G°$ values for the phosphorylated compounds found in cells range from about -2 to -13 kcal/mol (-8 to -54 kJ/mol). ATP is unique in that it occupies a central position in this scale (Fig. 6–3). All of the high-energy phosphate compounds for which $\Delta G°$ is more negative than ATP have a lower affinity for the phosphate group and can form ATP by donating their phosphate group to ADP:

$$ADP + P_i \rightarrow ATP$$

Specific examples are given in Figure 6–3. Phosphoenolpyruvate and 1,3-diphosphoglycerate are intermediates formed during glucose catabolism and donate phosphate groups to ADP to form ATP. Phosphocreatine is

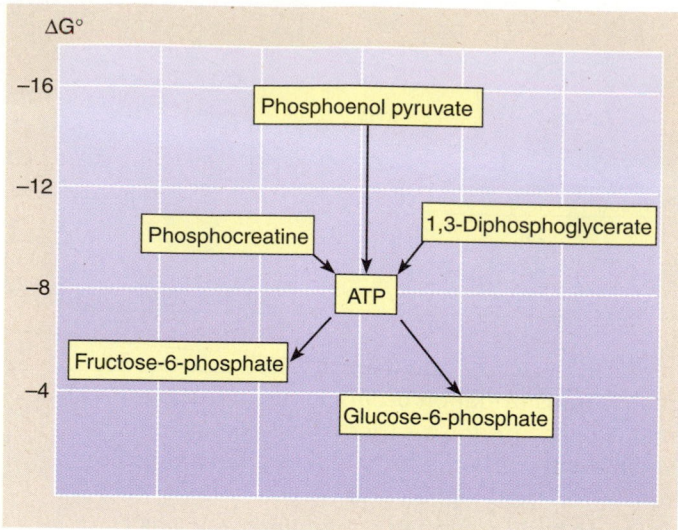

Figure 6–3

The standard free energy of hydrolysis of some phosphate compounds important in metabolism. Arrows indicate that phosphate groups can be transferred from high-energy donors to low-energy acceptors.

a phosphate donor that replenishes the ATP supply in skeletal muscle during periods of intense activity when the metabolic production of ATP cannot meet the demand. In fact, phosphocreatine is the major phosphorylated compound present in the resting muscle cell. The amount of phosphocreatine is threefold to eightfold greater than the amount of ATP. Compounds for which $\Delta G°$ is less negative than ATP have a greater affinity for the phosphate group and accept phosphate groups from ATP. For example, glucose accepts a phosphate group to form glucose-6-phosphate.

The direct formation of a high-energy phosphate compound, such as ATP, by a chemical reaction that incorporates a phosphate group is a process known as **substrate-level phosphorylation.** We shall see that this reaction does not require oxygen, in contrast to the mechanism for ATP formation by the electron transport chain in the mitochondrion (see Fig. 6–2). Phosphorylation of ADP in the mitochondrion is coupled to loss of electrons (oxidation) by the components of the electron transport chain and is termed **oxidative phosphorylation.** Most of the ATP supply for the cell is generated by oxidative phosphorylation. While ADP is the only molecule that can be phosphorylated by mitochondrial oxidative phosphorylation, substrate-level phosphorylation can generate other high-energy phosphate compounds. An example is the phosphorylation of GDP to form GTP during the citric acid cycle.

The large release of free energy that occurs upon removal of a phosphate group from a high-energy phosphate compound is a property of the whole molecule; the energy does not reside solely in the covalent bond by which the

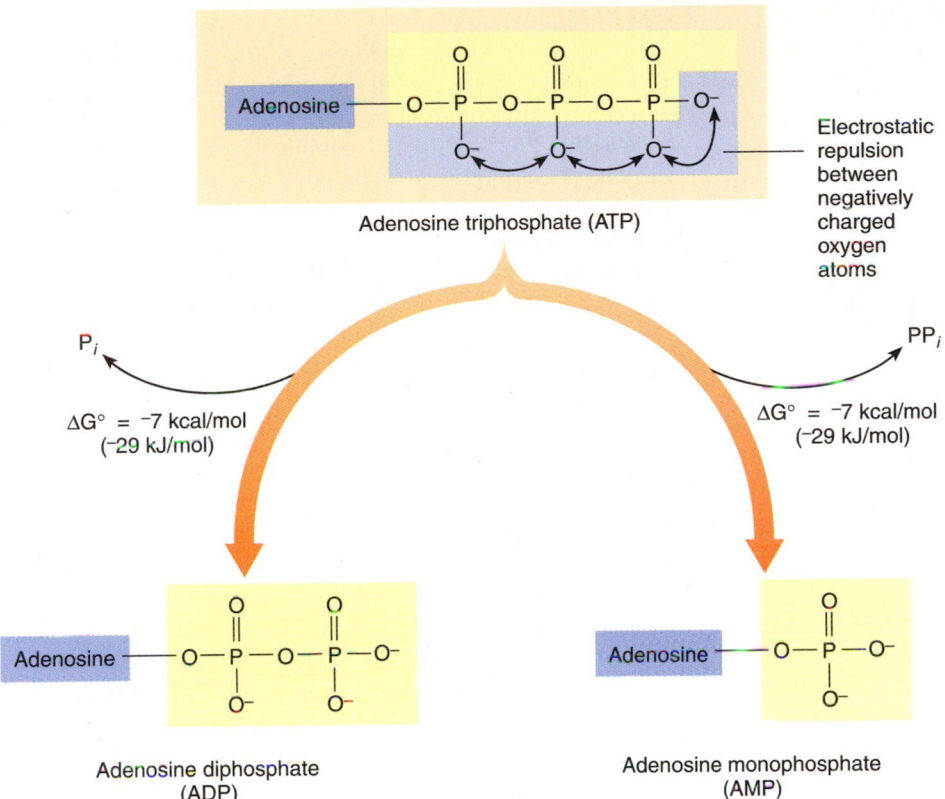

Figure 6–4

Release of phosphate (P$_i$) or pyrophosphate (PP$_i$) groups reduces the electrostatic repulsion between negatively charged phosphate groups in adenosine triphosphate (ATP).

phosphate group is attached. There is nothing special about this bond, even though it may be referred to as a "high-energy bond." Several factors account for the free energy that is released. In the specific case of ATP, one important factor is the reduction of **electrostatic repulsion** that is achieved by loss of a phosphate group (Fig. 6–4). Removal of the terminal phosphate group yields ADP, while the removal of two phosphate groups (which occurs in some reactions) yields AMP. In either case, the free energy released is the same.

Enzymes in Cellular Metabolism

> *What roles do other molecules play in determining which cellular reactions will occur and how fast they occur?*

By binding reactants to their active sites, enzymes ensure that reactants meet in the most favorable orientation for reactions to occur. For example, ATP and glucose are both bound by the enzyme **hexokinase,** which facilitates the phosphorylation reaction in which a phosphate group is added to glucose. The enzyme in effect couples an exergonic reaction (the removal of a phosphate group from the ATP) to an endergonic reaction (the addition of a phosphate group to the glucose). Glucose-6-phosphate and ADP are the reaction products.

Enzymes are also important in directing various products of cell reactions down the correct pathways. In Chapter 2 we discussed how enzymes lower the activation energy for specific reactions. Enzymes lower the activation energy only for specific desired reactions, so that, at each step, reactants are more likely to participate in those reactions than in others that might be possible.

Most enzymes in the cell are not saturated by substrate and do not operate at their V$_{max}$. This provides the cell with the capacity to deal with fluctuations in the concentrations of its metabolites. Thus, when physical exercise rapidly increases the demand for glucose catabolism, for example, the enzymes involved have the capacity to respond by processing more substrate molecules in the same time.

The cell must maintain all of the various metabolic substrates in the required concentrations, ultimately by controlling the rates of the chemical reactions that produce and consume them. All of these reactions are believed to require catalysis by enzymes, and one of the great advantages of an enzyme catalyst is that its activity in the cell can be controlled. One control mechanism, discussed already in Chapter 5, is covalent modification of the enzyme protein by attachment of a phosphate group, a reaction that is itself catalyzed by a specific **kinase** enzyme. Phosphorylation of an enzyme either activates or inactivates it, and this change can be reversed by removal of the phosphate group with a **protein phosphatase** enzyme (Fig. 6–5).

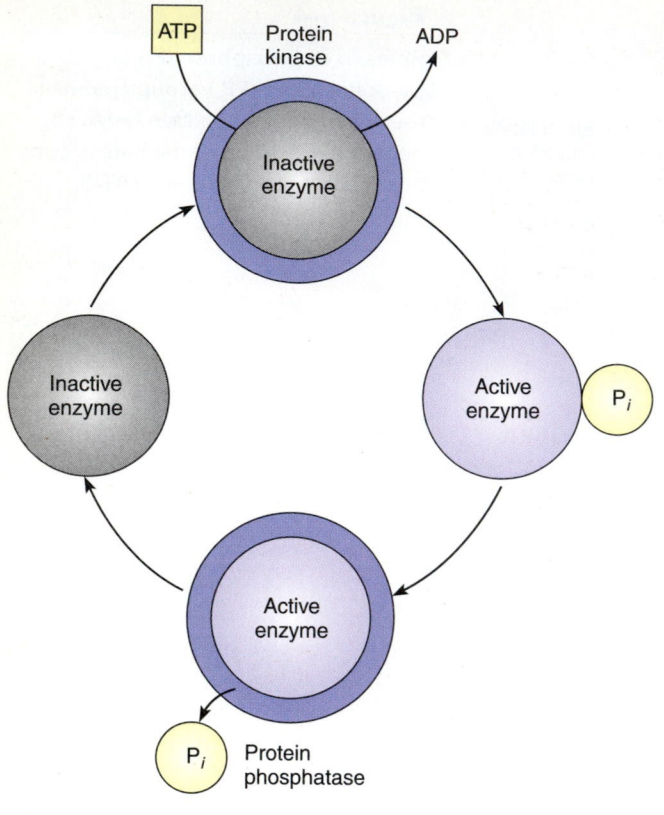

Figure 6–5

The phosphorylation of certain enzymes can activate them, whereas the removal of a phosphate group inactivates them.

ALLOSTERIC INHIBITION. A specific enzyme usually catalyzes each of the distinct reactions of a pathway of metabolism. A control mechanism that regulates the output of an entire pathway is known as **allosteric inhibition.** An allosteric inhibitor inhibits an enzyme by binding to it at a site that is separate from the active site, causing a conformational change in the protein that impairs the function of the active site (Fig. 6–6, *right panel*). **End-product inhibition** occurs when the allosteric inhibitor is the final product of a metabolic pathway that "feeds back" to inhibit the enzyme catalyzing the *first step* of that pathway (Fig. 6–6, *left panel*). If the first enzyme (e.g., enzyme *A*) in the pathway is inhibited, the first reaction does not occur. Hence, no substrate is available for the next reaction, catalyzed by enzyme *B*, and so on. In this way, the rate of formation of the final product *X* rapidly decreases when the first step is inhibited.

Allosteric inhibition by the end product of a metabolic pathway is a reversible process. As formation of product *X* decreases, owing to inhibition of enzyme *A* (see Fig. 6–6), the concentration of *X* in the cell also decreases because now it is used and consumed faster than it is produced. The concentration of *X* decreases until there is no longer enough of *X* to inhibit enzyme *A*. When the inhibitor of enzyme *A* is removed, the pathway leading to production of *X* is "turned on" again.

Cells also can control enzyme-catalyzed reactions by increasing or decreasing synthesis of the enzyme proteins, a process that changes the amount of enzyme present in the cell. If a cell requires more of product *X* produced by the pathway depicted in Figure 6–6, for example, it can synthesize more of enzyme *A*, enzyme *B*, and the other enzymes involved in this pathway. This type of control is relatively slow to take effect; it may take several hours or days for the enzyme levels to change significantly. It does not provide the rapid and fine control possible with covalent modification or allosteric inhibition.

REDOX REACTIONS AND COENZYME ACTION. Coenzymes, which already have been mentioned as intermediate electron acceptors in cell metabolism, play an important role in enzyme catalysis. Some coenzymes, such as NAD, are separated easily from the enzyme. NAD and its phosphorylated version NADP are derivatives of the vitamin **niacin,** while the coenzyme flavin adenine dinucleotide (FAD) is derived from another vitamin, called **riboflavin.**

The oxidized forms of the coenzymes NAD and NADP have a single positive charge (Fig. 6–7); when reduced, they accept electrons and protons from two hydrogen atoms. Because a hydrogen atom consists of one electron and one proton, two hydrogen atoms are equivalent to two electrons and two protons. NAD^+ and $NADP^+$ accept both electrons but only one of the protons:

$$\underset{\text{oxidized}}{NAD^+} + 2\,H^+ + 2\,e^- \rightarrow \underset{\text{reduced}}{NADH} + H^+$$

FAD, like NAD^+, can accept two electrons, but it also accepts two protons. In other words, FAD can accept two hydrogen atoms when it is being reduced (Fig. 6–7):

$$\underset{\text{oxidized}}{FAD} + 2\,H^+ + 2\,e^- \rightarrow \underset{\text{reduced}}{FADH_2}$$

Recall that the reduction of these coenzymes is coupled to oxidation of a molecule, such as glucose. NAD^+ and FAD are important coenzymes used in catabolic pathways, such as the oxidation of glucose and other fuel molecules. $NADP^+$ is important in biosynthetic pathways of metabolism, such as the synthesis of fatty acids.

AN OVERVIEW OF CATABOLISM: DIETARY PROTEINS, CARBOHYDRATES, AND FATS

How do cells obtain energy from carbohydrates, fats, and proteins?

The simple act of consuming food is only preliminary to the complex process by which the body obtains the energy and matter it needs to live. Even the complex process of digestion

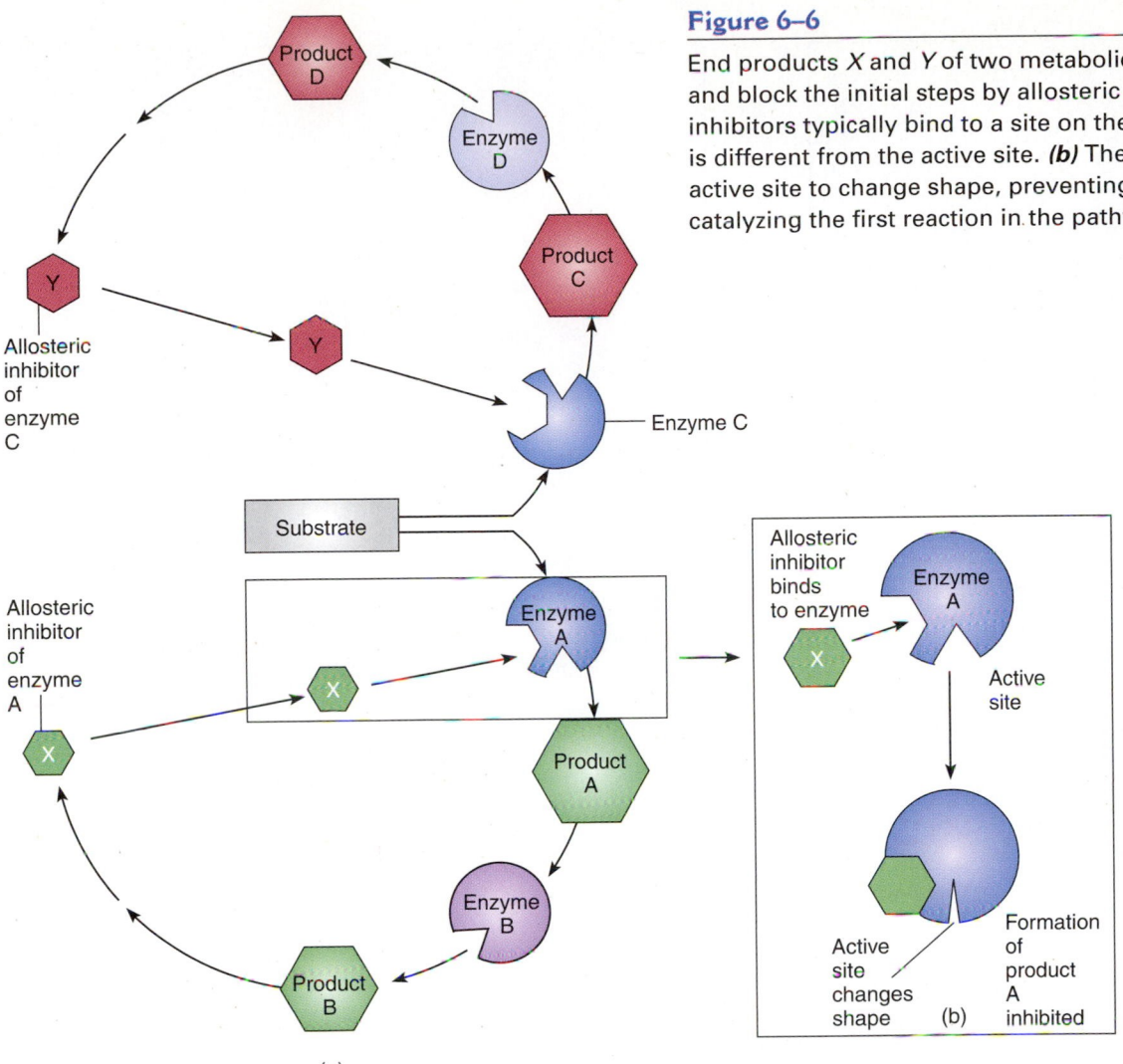

Figure 6–6

End products *X* and *Y* of two metabolic pathways feed back and block the initial steps by allosteric inhibition. **(a)** Allosteric inhibitors typically bind to a site on the enzyme molecule that is different from the active site. **(b)** The binding causes the active site to change shape, preventing the enzyme from catalyzing the first reaction in the pathway.

is not the ultimate destiny of food substances. Only the individual cells can enact the specialized reactions required to transform the energy and matter contained in food into a form that the body can use.

Before examining in detail the complex processes by which cells transform matter and energy, let us look at all of them in general and discuss the ways in which they relate. We first review the way in which cells obtain food molecules and what they do with those molecules before delivering them to the mitochondria, the organelles finally responsible for the release of energy.

The cells of the human body obtain energy from food that has been ingested in the form of dietary proteins, carbohydrates, and fats. These food substances are broken down in the digestive tract by the action of certain enzymes. Proteins consumed in the form of meat, for example, are broken down into amino acids. Carbohydrates eaten in the form of fruits, vegetables, and grains are broken down into simple sugars (monosaccharides), such as glucose. Dietary fats, such as margarine and oils, are degraded into fatty acids and glycerol.

After digestion, these macromolecules, the products of digestion, circulate in the blood to all cells in the body (Fig. 6–8).

The chemical bonds that hold together these various macromolecules represent a store of energy the cells need to tap. They do so by obtaining the macromolecules from the blood and breaking them down. Each type of food molecule is broken down by a specific process. Amino acids obtained by the cell are broken down by **deamination,** which separates nitrogen atoms and carbon atoms. Monosaccharides (most commonly glucose) are broken down by **glycolysis.** These reactions occur in the cytosol of the cell. Fatty acids are broken down by a process known as **fatty acid oxidation** in the mitochondrial matrix (see Fig. 6–8). All of these processes are regulated and facilitated by enzymes.

Most of the molecules that result from the breakdown of amino acids, glucose, and fatty acids are converted to **acetyl** (CH_3CO—) **groups** in the matrix of the **mitochondrion,** the "powerhouse" of the cell. The acetyl groups are fed into a series of reactions called the citric acid cycle, also known as the **tricarboxylic acid (TCA) cycle.** This cycle, occurring in

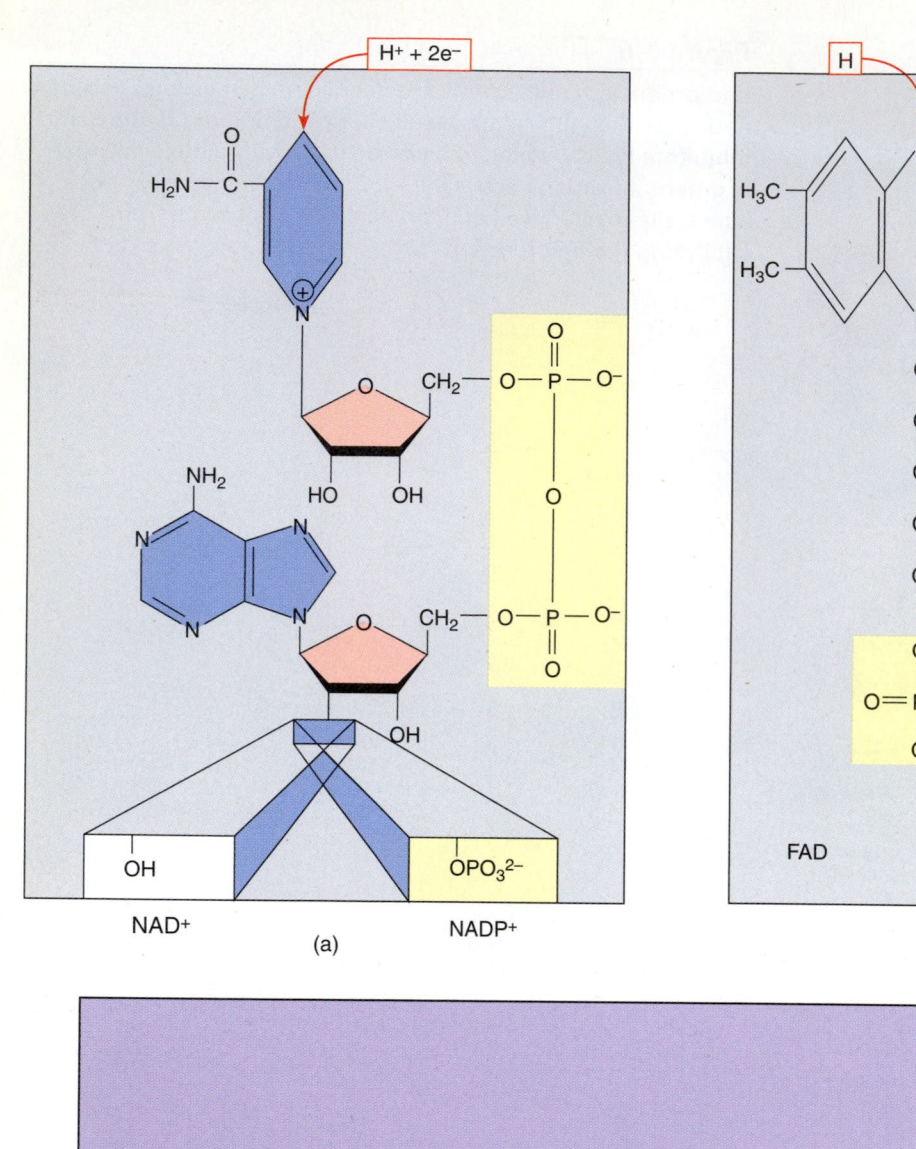

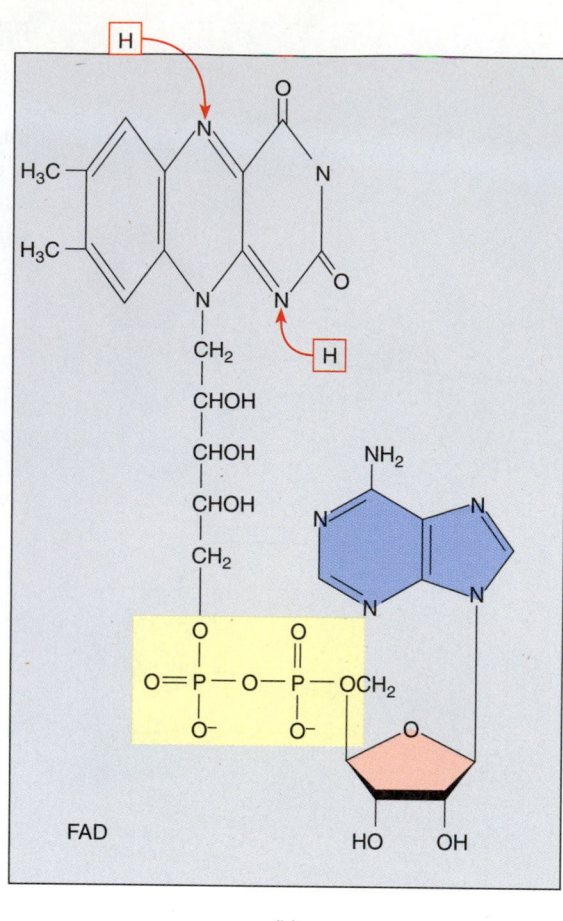

Figure 6–7

(a-c) The structures of common coenzymes. Arrows indicate the points that accept hydrogen atoms (NAD⁺, NADP⁺, and FAD) or acetyl groups (coenzyme A).

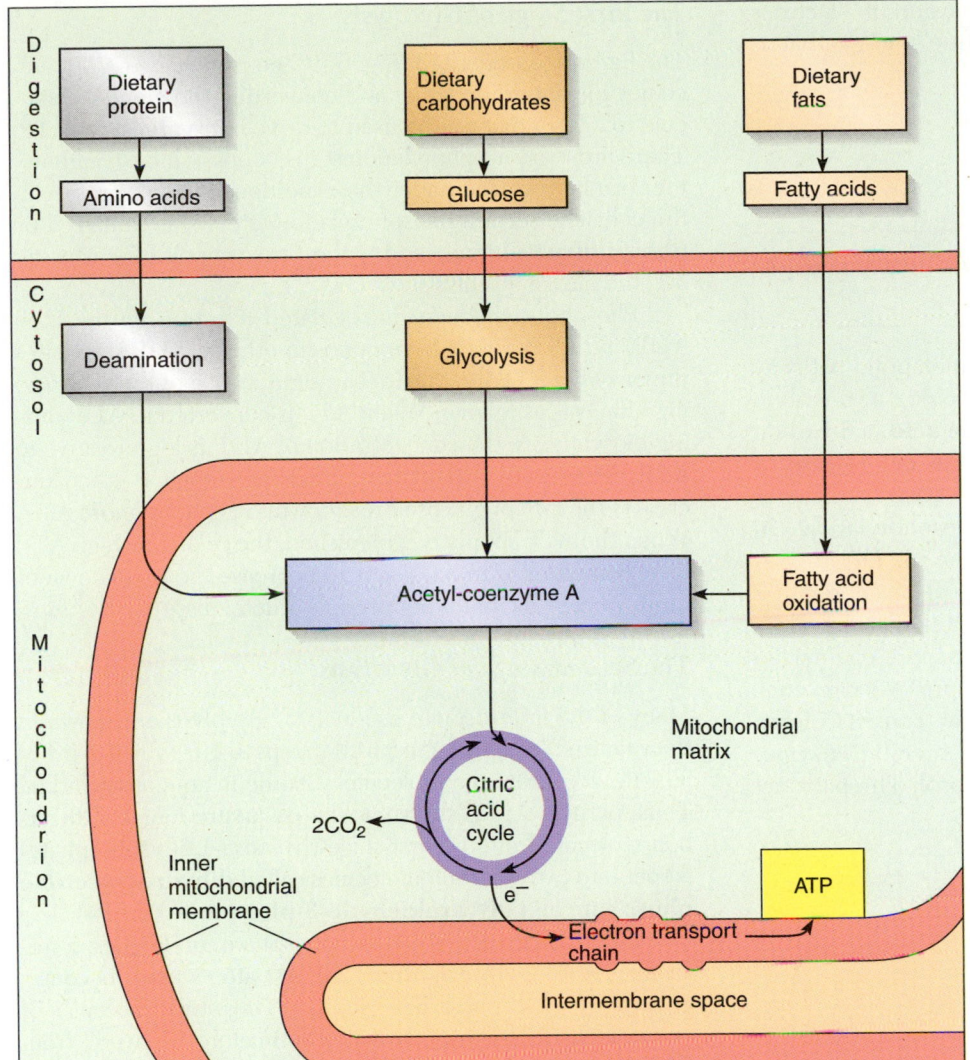

Figure 6–8

An overview of catabolism, showing the locations of important processes. Note that all three pathways feed carbon fragments into acetyl coenzyme A. (The outer mitochondrial membrane has been omitted.)

the mitochondrion, is the final common pathway for the breakdown of all the fuel molecules in the cell (see Fig. 6–8).

In the mitochondrial matrix, all of the acetyl groups produced from food molecules are coupled to the terminal—SH group of coenzyme A (see Fig. 6–7) to produce **acetyl coenzyme A.** In this form, the acetyl groups are able to enter the citric acid cycle, which converts each acetyl group to two molecules of CO_2 (see Fig. 6–8). This conversion releases high-energy electrons that are temporarily trapped by the coenzymes NAD^+ and FAD. The coenzymes later feed the electrons into the electron transport chain, which is a series of electron carriers, many of them proteins, embedded in the inner mitochondrial membrane. The redox reactions of the electron carriers release the energy used to synthesize ATP.

The final step in this process requires oxygen, making cellular respiration an **aerobic (oxygen-requiring) process.** The electron transport chain uses oxygen as the final acceptor of the electrons, a step that converts the oxygen to water. As

the electrons pass from one electron carrier to the next, their energy is released in several small bursts that are used to drive the energy-dependent synthesis of ATP from ADP and inorganic phosphate (P_i). Thus, the ADP becomes phosphorylated; this step is tightly coupled to the flow of electrons. The electron flow does not occur unless the phosphorylation step can proceed. The loss of electrons, as we have said, is defined as *oxidation.* Because electron transport consists of successive oxidations and is coupled to ADP phosphorylation, the entire process is known as **oxidative phosphorylation.**

METABOLISM OF CARBOHYDRATES: THE OXIDATION OF GLUCOSE

The metabolism of carbohydrate involves three main processes: (1) **glycolysis,** the conversion of glucose to pyruvic acid; (2) **the citric acid cycle,** the oxidation of pyruvic acid, releasing electrons that are trapped by coenzymes; and

(3) **oxidative phosphorylation,** the transfer of electrons from coenzymes to oxygen, releasing free energy that is stored in ATP.

Glycolysis Is the Conversion of Glucose to Pyruvate

> *How is energy derived from the breakdown of glucose when oxygen is not available?*

The overall reaction that occurs in this metabolic pathway is the breakdown of a **glucose** molecule, a 6-carbon compound, into two molecules of **pyruvic acid,** a 3-carbon compound. This is accomplished with the net release of enough free energy to phosphorylate two molecules of ADP to ATP by substrate-level phosphorylation. In addition, two molecules of the coenzyme NAD$^+$ are reduced to NADH, providing the potential for more ATP synthesis when the NADH molecules later give up electrons to the mitochondrial electron transport chain (see Fig. 6–2). The pathway of glycolysis is represented in condensed form in Figure 6–9. This pathway consists of ten reaction steps, each one catalyzed by a specific enzyme, and each one taking place in the cell cytosol. The pathway has two stages.

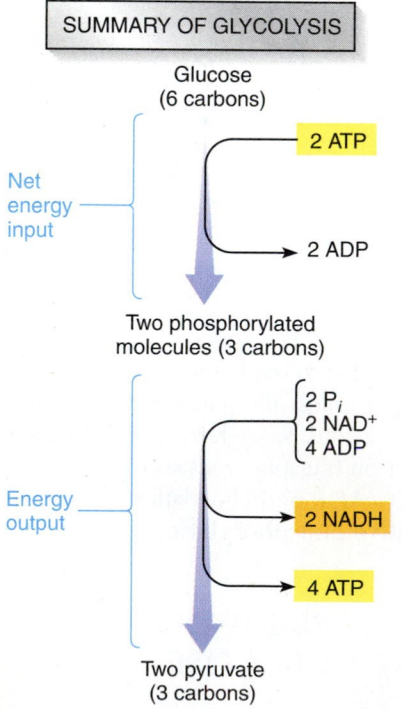

Figure 6–9

A summary of the pathway of glycolysis. The net yield is two molecules of ATP.

The First Stage of Glycolysis

The first stage (steps 1–3, Fig. 6–10) consumes two ATP molecules for phosphorylating and converting the glucose molecule to a form that can be used to drive net synthesis of ATP. There are two phosphorylation steps (steps 1 and 3), and neither can be reversed under the conditions that are present in the cell. The second phosphorylation (step 3) is catalyzed by **phosphofructokinase** and is the primary point for controlling the rate of glycolysis.

Phosphofructokinase is regulated allosterically by both ADP and ATP, which have opposite effects: ADP stimulates the enzyme; ATP inhibits it. The significance of these opposing effects is as follows. When ATP is converted to ADP during biosynthesis, the concentration of ADP is high relative to ATP. The ADP stimulates phosphofructokinase and increases the rate of glucose breakdown to generate more ATP. When the ATP supply is replenished, the relatively high ATP level compared with ADP acts to decrease the breakdown of more glucose by inhibiting phosphofructokinase.

The Second Stage of Glycolysis

Many of the intermediate products of glycolysis are shown in their ionized form in Figure 6–10 (steps 4–10), which is probably the form that the molecules assume in aqueous solution. Thus, pyruvic acid is shown as the **pyruvate** ion. The 6-carbon molecule from the first stage (fructose-1,6-bisphosphate) is split into two 3-carbon molecules called **dihydroxyacetone phosphate** and **glyceraldehyde-3-phosphate** (step 4). The molecule of dihydroxyacetone phosphate undergoes a rearrangement (**isomerization**) of its atoms and becomes glyceraldehyde-3-phosphate (step 5). Thus, two molecules of glyceraldehyde-3-phosphate are ultimately derived from fructose-1,6-bisphosphate. Each molecule of glyceraldehyde-3-phosphate is converted first to *1,3-diphosphoglycerate* (step 6) and then to *phosphoenolpyruvate* (steps 7–9). These are high-energy phosphate compounds that can donate a phosphate group to ADP to generate ATP (see Fig. 6–3). Formation of 1,3-diphosphoglycerate in step 6 is not only a substrate-level phosphorylation but also an oxidation reaction. The protons (H$^+$) and electrons that are removed from glyceraldehyde-3-phosphate during oxidation are accepted by NAD$^+$, which is reduced to NADH. The phosphorylation of ADP by phosphoenolpyruvate in step 10 is essentially irreversible under the conditions present in the cell.

The second stage of glycolysis produces a total of two NADH molecules and four ATP molecules from each glucose molecule that is metabolized. Because two ATP molecules are consumed in the first stage, there is net gain by the cell of two ATP molecules as a result of glycolysis.

The yield of free energy from glucose as it is catabolized by the glycolytic pathway to pyruvic acid is low. The standard free energy change ($\Delta G°$) is about −20 kcal/mol (or −84 kJ/mol), and only two molecules of ATP are gained from the process. Compare this with the $\Delta G°$ of −686 kcal/mol (or −2870 kJ/mol) when glucose is fully oxidized to CO$_2$ and

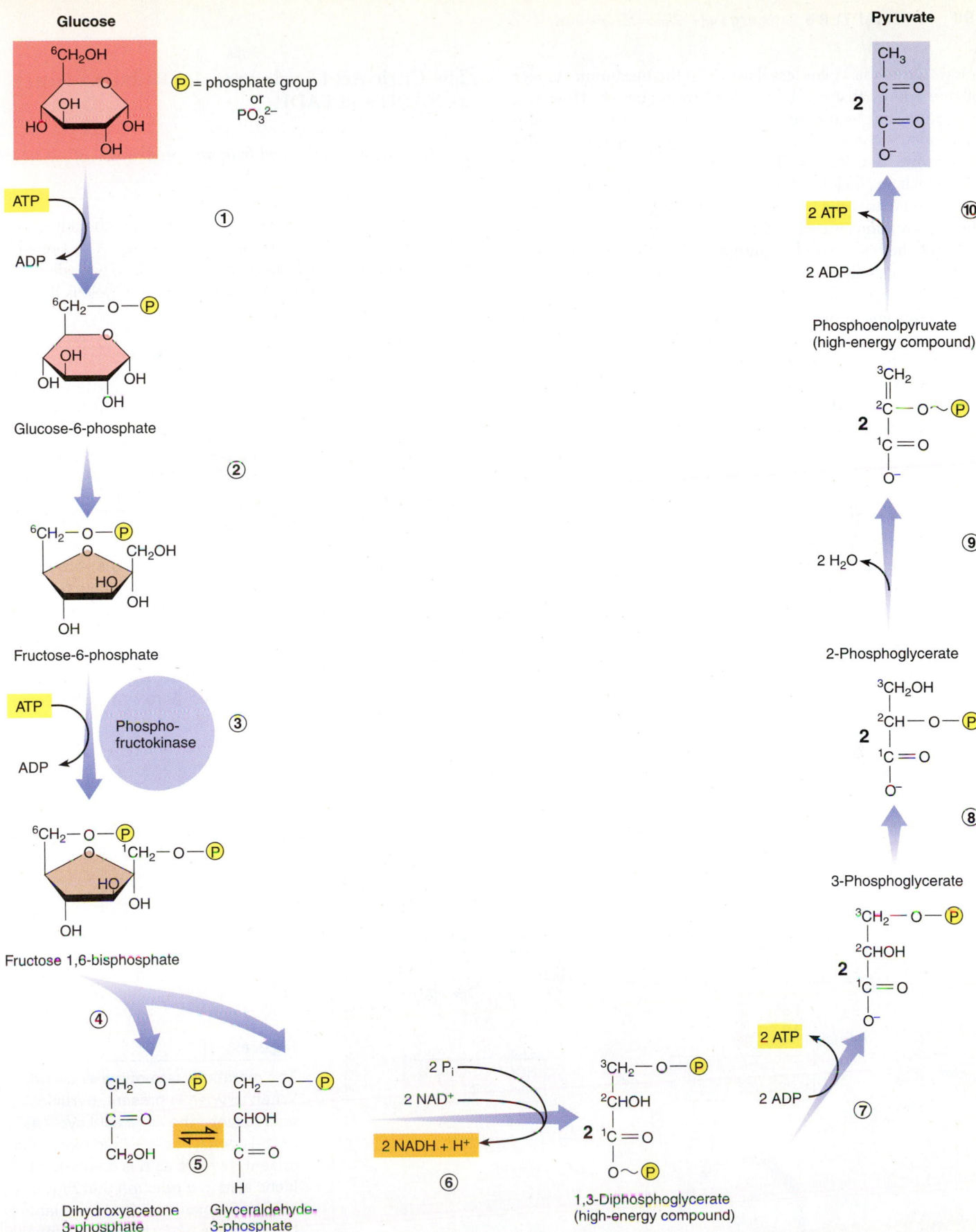

Figure 6-10

A detailed look at glycolysis. Note that step 4 converts a single 6-carbon molecule into two 3-carbon molecules that, after step 6, become two molecules of 1,3-diphosphoglycerate.

water. Glycolysis yields less than 3% of the maximum amount of free energy that can be obtained from glucose. However, it is important to realize that this pathway does not require molecular oxygen—that is, glycolysis can proceed under **anaerobic conditions.** This capability is important in tissues, such as skeletal muscle, when the demand for ATP rapidly increases at a time when oxygen supply is low, such as during strenuous exercise. The skeletal muscle cells are able to meet their increased requirement for ATP by increasing the rate of glycolysis, producing ATP despite the fact that little or no oxygen is available.

Under anaerobic conditions, the pyruvic acid produced by glycolysis is converted to lactic acid (Fig. 6–11). This step is coupled to oxidation of NADH to NAD$^+$, which ensures a continued supply of NAD$^+$ for continuation of anaerobic glycolysis. However, the lactic acid accumulates because it cannot be metabolized further. When exercise stops, the lactic acid in the muscle cells is transferred to the liver, where it is converted back to pyruvic acid and broken down by aerobic metabolism to CO_2 and water. This process requires additional oxygen, which is obtained by continued excessive breathing after muscular activity has ceased, a phenomenon known as **oxygen debt.**

Remember that two molecules of pyruvic acid are generated from every molecule of glucose that enters the glycolytic pathway (see Fig. 6–9). If the oxygen supply is plentiful—as is usually the case in most cells—the pyruvic acid produced by glycolysis is fed rapidly into the citric acid cycle, a pathway that takes place in the **mitochondrial matrix.**

The Citric Acid Cycle Provides Reducing Power as NADH and FADH$_2$

 How is energy derived from the breakdown of pyruvic acid?

Pyruvic acid produced in glycolysis next enters the mitochondrial matrix, where it is coupled to coenzyme A in preparation for processing via the citric acid cycle (also called the *tricarboxylic acid cycle*). The first step in the cycle is the formation of acetyl coenzyme A.

The Formation of Acetyl Coenzyme A

In a reaction catalyzed by pyruvate dehydrogenase (see Fig. 6–11), an acetyl group (CH_3CO—) is transferred from pyruvic acid to the —SH group of coenzyme A to form acetyl coenzyme A. The acetyl group contains two of the three carbons present in pyruvic acid. (The third carbon of pyruvic acid is lost as CO_2, which is ultimately exhaled via the lungs.) The reaction also removes protons and electrons from pyruvic acid and coenzyme A; these are accepted by NAD$^+$, which is reduced to NADH. The net result of the pathway up to this point is that glucose, a compound containing 6 carbon atoms, is fed into the citric acid cycle as two 2-carbon units in the form of acetyl groups (Fig. 6–12).

The 2-carbon acetyl groups enter the citric acid cycle in a reaction catalyzed by citrate synthetase, which transfers the acetyl group from acetyl coenzyme A to a 4-carbon com-

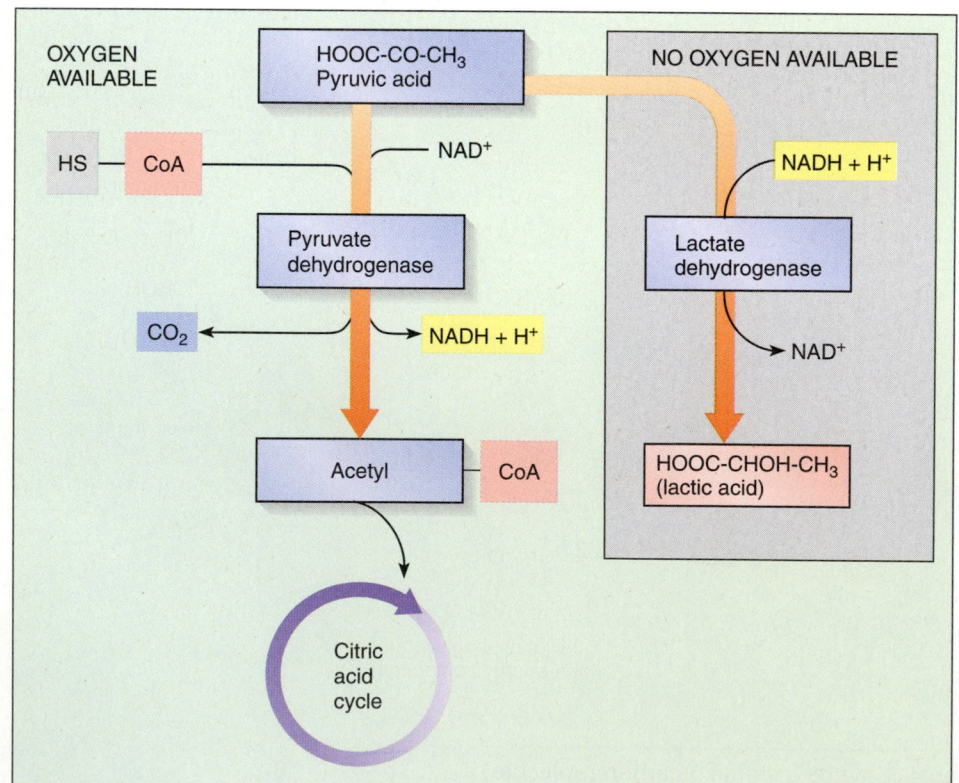

Figure 6–11

The metabolic fate of pyruvic acid. When oxygen is present, pyruvic acid reaches the citric acid cycle as acetyl coenzyme A. When oxygen is absent, pyruvic acid is converted to lactic acid in a reaction that requires NADH. Cytosolic NADH is available because it is generated by glycolysis but cannot be utilized by mitochondria under anaerobic conditions. (CoA = coenzyme A.)

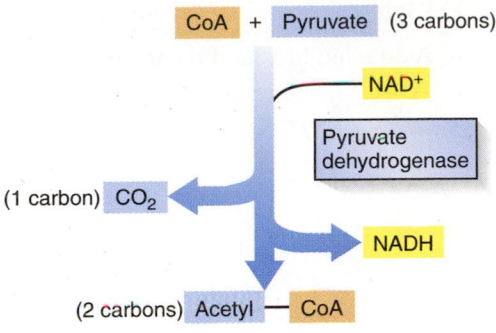

Figure 6–12

The formation of acetyl coenzyme A from pyruvic acid and coenzyme A (CoA). Pyruvate is produced in the cytosol and enters the mitochondrial matrix, where it is converted to acetyl coenzyme A and CO_2 in a reaction catalyzed by pyruvate dehydrogenase.

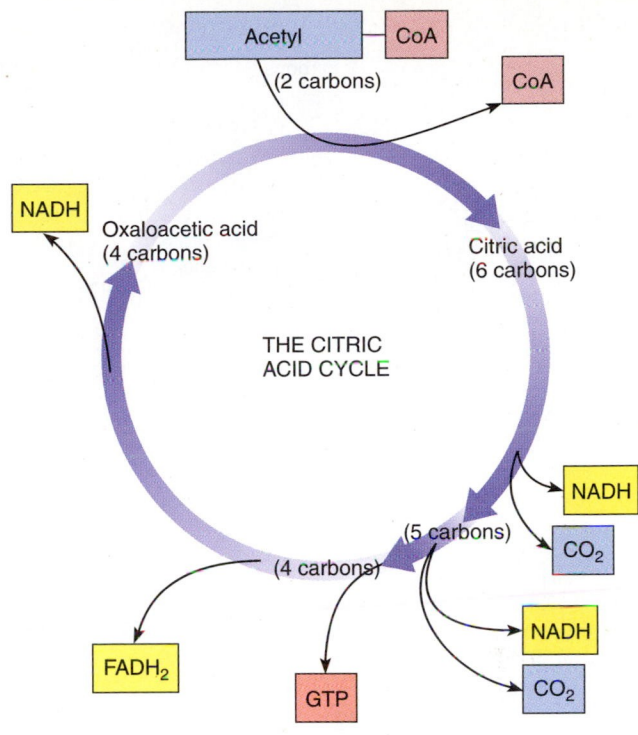

Figure 6–13

A summary of the citric acid cycle.

pound called *oxaloacetic acid*. One product of this reaction is the 6-carbon compound citric acid (hence the name for the cycle). The other reaction product is free coenzyme A, which becomes available to accept another acetyl group from another pyruvic acid:

oxaloacetic acid + acetyl coenzyme A →
   4 C atoms   2 C atoms
               citric acid + coenzyme A
                6 C atoms

The pathway of the citric acid cycle is a true cyclic process because the last reaction regenerates the 4-carbon oxaloacetic acid, which is free to accept an incoming acetyl group from acetyl coenzyme A. These steps are summarized in a simplified view of the citric acid cycle in Figure 6–13.

The Reduction of NAD⁺ and FAD

The reactions of the citric acid cycle regenerate free coenzyme A and produce two molecules of CO_2 for every 2-carbon acetyl group that enters the cycle. The process can be summarized as:

$$CH_3CO—\text{coenzyme A} → 2\ CO_2 + \text{coenzyme A}$$

The carbon atoms lost in one round of the cycle, in the form of CO_2 exhaled by the lungs, are those that entered as an acetyl group in a previous round. The steps involved in oxidation of the acetyl group are coupled to the reduction of the electron-carrying coenzymes NAD⁺ and FAD. There are four such reactions that generate, from each acetyl group entering the cycle, a total of three molecules of NADH and one molecule of $FADH_2$. In addition, one high-energy phosphate molecule (GTP) is formed by substrate-level phosphorylation (see Fig. 6–13).

Each reaction of the citric acid cycle is shown in detail in Figure 6–14. There are nine consecutive steps, beginning with the entry of an acetyl group (step 1). In steps 2 and 3, H and OH are interchanged to form **isocitrate.** Steps 4 and 5 are redox reactions. The net result is that the 6-carbon isocitrate is oxidized to the 4-carbon **succinyl coenzyme A** with the release of two CO_2 molecules. At the same time, two molecules of NAD⁺ are reduced to NADH. Step 4 is catalyzed by **isocitrate dehydrogenase.** Step 6 is a substrate-level phosphorylation reaction that utilizes the free energy released by breakdown of succinyl coenzyme A to drive phosphorylation of GDP to GTP:

$$\text{succinyl coenzyme A} + GDP + P_i →$$
$$\text{succinate} + GTP + \text{coenzyme A}$$

This is the only reaction in the cycle that produces a high-energy phosphate compound directly. The GTP formed in this reaction is used to phosphorylate ADP to produce ATP:

$$ADP + GTP → ATP + GDP$$

The next three reactions involve the 4-carbon compounds of the cycle and culminate in the regeneration of oxaloacetic acid. There are two oxidation reactions (steps 7 and 9) coupled to the reduction of one molecule of FAD to $FADH_2$ and one molecule of NAD⁺ to NADH.

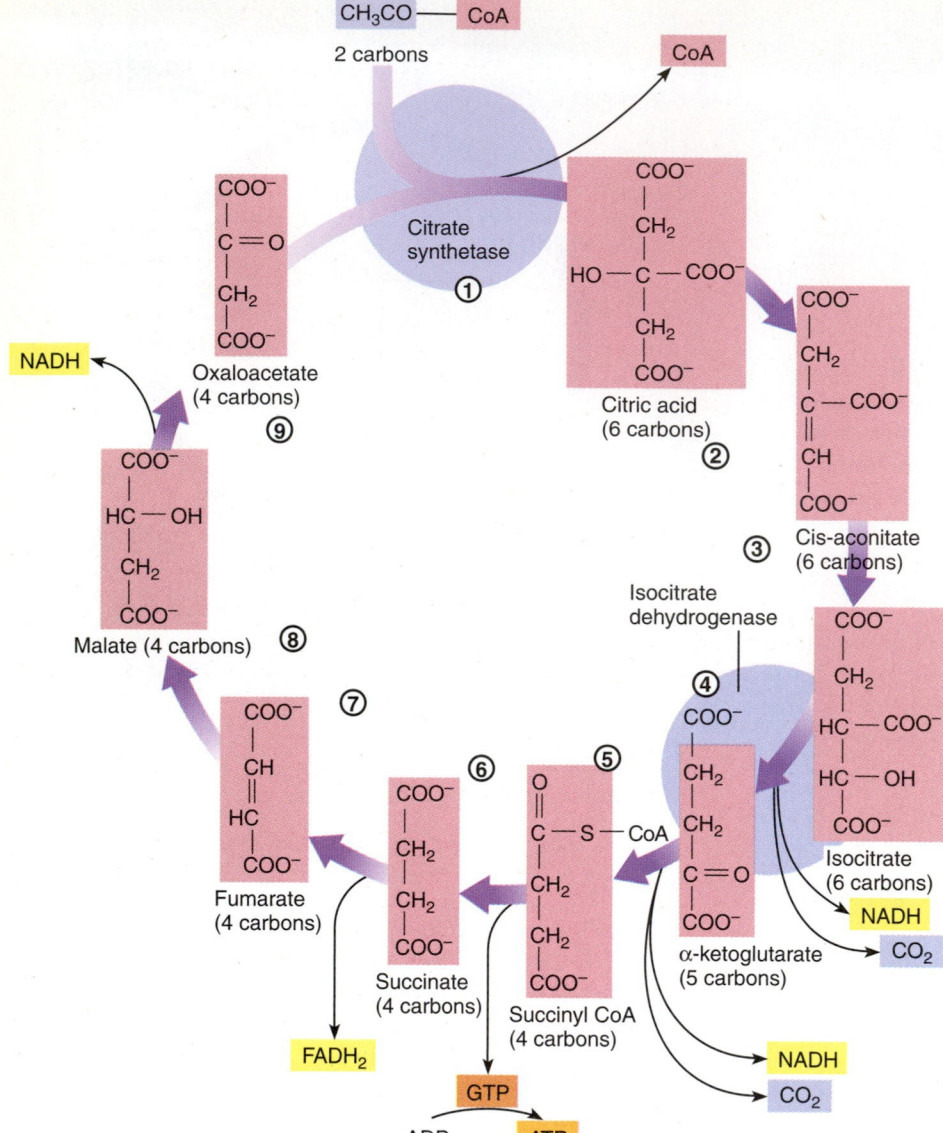

Figure 6–14

A detailed look at the citric acid cycle.

Molecular oxygen is not required for the reactions of the citric acid cycle, but it is essential for oxidation of the reduced coenzymes by the mitochondrial electron transport chain (see Fig. 6–2). A supply of oxidized coenzymes is needed for continued operation of the cycle. Hence the cycle cannot operate under anaerobic conditions because, under such conditions, all of the coenzyme molecules remain in a reduced form. In contrast, a supply of NAD^+ is always available for anaerobic glycolysis in the cytosol because the reduction of pyruvate to lactate is coupled to oxidation of NADH to NAD^+ (see Fig. 6–11).

The reactions of the citric acid cycle generate two molecules of ATP (from GTP) when the two acetyl groups from one glucose molecule are processed. There is also a net gain of two molecules of ATP during glycolysis (see Fig. 6–9),

which gives a total of four ATP molecules synthesized directly during the complete oxidation of glucose. Most of the chemical energy of the glucose molecule is conserved, by transfer of electrons, in the reduced coenzymes. Two NADH are produced by aerobic glycolysis of glucose to two pyruvates (see Fig. 6–9). Two more NADH are produced during formation of two acetyl coenzyme A molecules (see Fig. 6–12), and six NADH are produced when two acetyl groups are fed into the citric acid cycle (see Fig. 6–13). Thus, the complete oxidation of a single molecule of glucose generates ten molecules of NADH; in addition, two molecules of $FADH_2$ are formed. As we shall see, the subsequent transfer of electrons from these reduced coenzyme molecules to O_2 drives the synthesis of 32 more molecules of ATP.

Allosteric Inhibition in the Citric Acid Cycle

 How is the citric acid cycle controlled?

The citric acid cycle contains several control points. The cycle is slowed when the ATP supply is adequate and accelerated when the ATP supply is low. When the supply of ATP is high, ATP itself acts as an allosteric inhibitor of citrate synthetase to reduce the rate of formation of citrate (see step 1, Fig. 6–14) and slow down the cycle. When the supply of ATP is low and the concentration of ADP is relatively high, ADP acts as an allosteric activator of isocitrate dehydrogenase to increase the rate of production of α-ketoglutarate (step 4) and speed up the cycle. These and other mechanisms cooperate to match precisely the rate of the reactions of the cycle to the requirement of the cell for ATP.

Electron Transfer Is Coupled to Synthesis of ATP

How is ATP produced by the electron transport chain in the mitochondria?

The transfer of electrons from reduced coenzymes to O_2 takes place on the electron transport chain, which is an integral part of the inner membrane of mitochondria. These oxidation steps (loss of electrons) are coupled to phosphorylation of ADP to form ATP. The process produces most of the ATP supply for the cell and, because molecular oxygen is an essential reactant, it is an aerobic pathway called *oxidative phosphorylation.*

Cytochromes: Electron Carriers

The driving forces for oxidative phosphorylation are the negative redox potentials of NADH and $FADH_2$, which allow these molecules to readily donate electrons to proteins serving as **electron carriers** in the electron transport chain. Many of these electron carriers are proteins called **cytochromes,** and their redox potentials increase sequentially from the beginning to the end of the chain.

The importance of this increase in redox potentials is that each cytochrome has a greater affinity for electrons than the previous one in the chain. This increase causes the passage of electrons from one cytochrome to the next and, ultimately, to oxygen. Each cytochrome becomes reduced when it accepts electrons and oxidized when it donates the electrons to the next cytochrome. The components of the electron transport chain are organized into three major complexes (Fig. 6–15).

The flow of electrons from NADH to molecular oxygen represents a redox reaction that can be summarized as:

$$1/2\ O_2 + NADH + H^+ \rightarrow H_2O + NAD^+$$

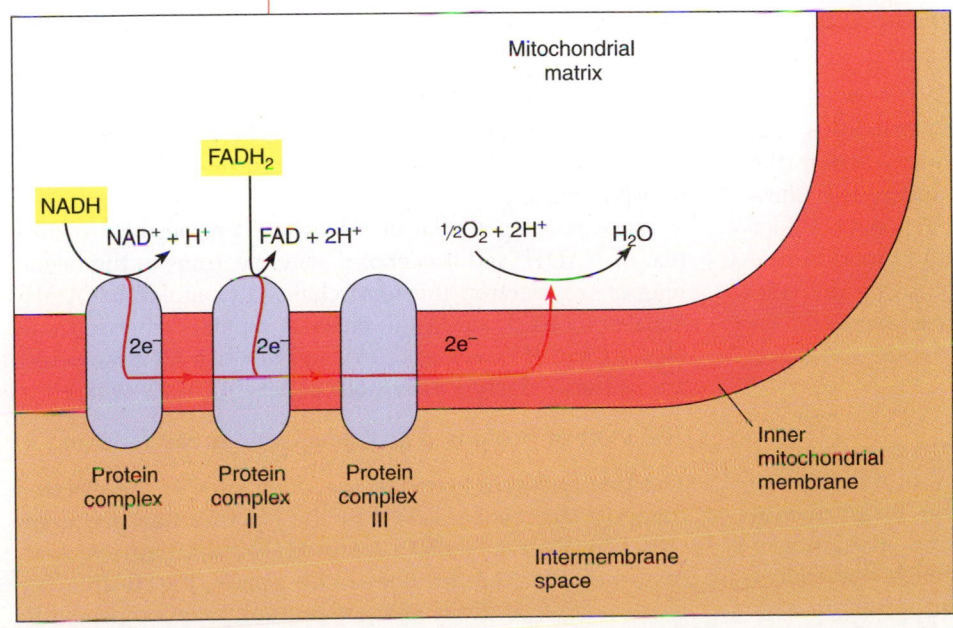

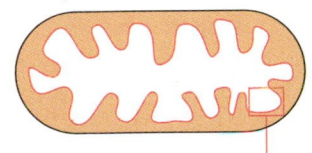

Figure 6–15

The oxidation of NADH and $FADH_2$ by the electron transport chain. Each of three electron transport complexes has one site where enough free energy is released to synthesize one molecule of ATP. The ATP is not synthesized directly. The free energy released at each site is used to pump protons (H^+) across the inner mitochondrial membrane.

CURRENT CONCEPTS IN PHYSIOLOGY

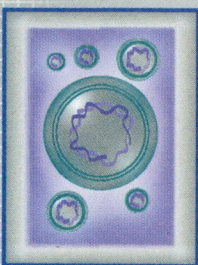

Heat Generation by Brown Fat

Some of the energy released by oxidation of food molecules cannot be used to drive synthesis of ATP in the mitochondria. Although it is released as heat, it is not wasted. It is a crucial source of heat for humans and other warm-blooded animals that maintain their internal temperature higher than that of their environment. In some animals, including humans, the ability of newborns to generate heat **(thermogenesis)** is vital for survival. Thermogenesis is also important for mammals during arousal from hibernation and for adapting to cold climates. Mammals are equipped for thermogenesis by the presence of brown fat **(adipose)** tissue in which fatty acids are the primary source of fuel for oxidative metabolism. Fat cells are termed **adipocytes**, and brown adipocytes are characterized by the presence of high numbers of mitochondria. Furthermore, the inner membrane of these mitochondria contains very high amounts of a specific protein known as the **uncoupling protein (UCP).** The UCP facilitates the return of H^+ ions to the mitochondrial matrix by facilitated diffusion across the inner membrane, thus bypassing the ATP synthetase. The diffusion is rapid enough to reduce the H^+ gradient without dissipating it completely. The leak of H^+ removes the controls on oxidative phosphorylation that are dependent on the relative amounts of ADP and ATP, and electron transport runs at a high rate whether ATP is needed or not. Consequently, the free energy produced by oxidation of fatty acids in brown adipocytes is "uncoupled" from synthesis of ATP and is released as heat. This is carried from the brown adipose tissue by the circulatory system and distributed to other parts of the body. Brown adipose tissue may represent only a small proportion of the body weight, but the thermogenesis is so efficient that body temperature can be maintained without shivering.

The level of UCP in vivo can be modulated by various physiological stimuli. A prominent example is the exposure of rats to cold, which triggers a marked increase in the UCP level and converts the mitochondria of brown adipocytes to heat-producing machines. It was found that exposure of rats to 4°C for just 15 minutes produced a tenfold increase in the rate of transcription of the *UCP* gene. The mechanisms that control expression of the *UCP* gene in large mammals, including humans, change soon after birth. The UCP level and the amount of brown fat decrease during the first few days of life. The gene remains responsive, however, because the amount of brown adipose tissue and its activity in adult humans can be changed by environmental conditions. Exposure to a hot climate, for example, switches off the thermogenic response and promotes degeneration of brown adipose tissue. In contrast, overeating stimulates the thermogenic response, which may be a device to rapidly oxidize excess fuel molecules so that they are not stored as fat. Although there is insignificant brown fat in most adult humans, UCP exists in most tissues, and many pharmaceutical companies are actively seeking drugs to control body weight by safely stimulating UCP action.

The standard free energy change $\Delta G°$ for this reaction is -53 kcal/mol (or -221 kJ/mol). It is an exergonic reaction; the free energy released is used to drive the endergonic phosphorylation of ADP to produce ATP (see Fig. 6–4). Under standard conditions, about 7 kcal/mol (or 29 kJ/mol) of free energy can be stored in one ATP molecule, so the free energy released by the earlier-mentioned reaction could be stored in seven molecules of ATP.

Sites of Free Energy Release

The electron transport chain ensures that the free energy is not released all at once, but even so, not all of this free energy can be used to synthesize ATP because, under physiological conditions, there are only three sites in the chain where the transfer of electrons releases enough free energy to synthesize a molecule of ATP. One site is present in each of the three electron transport complexes (see Fig. 6–15).

Thus, in the cell, three molecules of ATP are synthesized for each molecule of NADH that is oxidized in the electron transport chain.

The redox potential of $FADH_2$ is less negative than is that of NADH, and it cannot donate electrons to the beginning of the electron transport chain in Complex I. $FADH_2$ donates its electrons to the chain at a point between Complexes I and II (see Fig. 6–15), and just two ATP molecules are synthesized when each molecule of $FADH_2$ is oxidized.

The Flow of Protons: The Chemiosmotic Mechanism

The free energy released at the three sites in the electron transport chain is harnessed by the electron transport complexes in order to pump protons (H^+) from the mitochondrial matrix to the intermembrane space (Fig. 6–16). The movement of positively charged protons, unaccompanied by negatively charged anions, sets up an electrical gradient

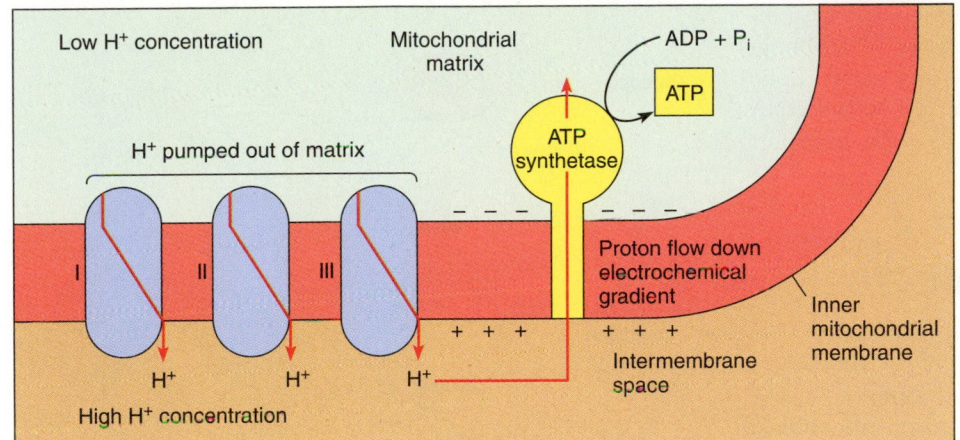

Figure 6–16

The chemiosmotic mechanism of ATP synthesis. Protons (H$^+$) are pumped from the matrix to the intermembrane space, then returned to the matrix by passive flow down the electrochemical gradient. ATP synthetase uses the passive proton flow to drive the synthesis of ATP.

across the inner mitochondrial membrane. In addition, the accumulation of protons in the intermembrane space builds a chemical gradient. The result of the net electrochemical gradient is a passive flow of protons from the intermembrane space back to the matrix. This flow can occur only through a specific protein in the inner membrane that bears the enzyme **ATP synthetase.** Just as the flow of sodium ions down an electrochemical gradient can be used to do "work," such as moving another solute against its electrochemical gradient (secondary active transport), the flow of protons is coupled by ATP synthetase to drive the endergonic synthesis of ATP from ADP and a phosphate group (see Fig. 6–16). In other words, the ATP synthetase transforms the energy of the proton electrochemical gradient into chemical energy in the form of ATP.

The proton pumps use the free energy released by electron transport to maintain the proton gradient, which, in turn, maintains the flow of protons through ATP synthetase. This is known as the **chemiosmotic mechanism** of ATP synthesis because it links chemical reactions to transport processes. The newly formed ATP is transferred out of the mitochondrion and used to drive metabolic reactions in the rest of the cell.

A "balance sheet" summarizing the pathways involved and the yield of 36 ATP molecules from complete oxidation of one glucose molecule is presented in Table 6–1. Only one additional comment is needed: The inner mitochondrial membrane is impermeable to NADH (and NAD$^+$), so the cytosolic "pool" of this coenzyme never mixes with the mitochondrial pool. Cytosolic NADH produced by glycolysis is oxidized by a shuttle mechanism that, in many cell types, transfers electrons and protons to FAD inside the mitochondrion. As a result of this mechanism, each cytosolic NADH molecule ultimately produces only 2 ATP molecules.

Mitochondrial Diseases

Mitochondria can be regarded as the "power plants" of the cell because they are the principal suppliers of ATP. Most mitochondrial proteins are encoded by nuclear DNA, but a few are encoded by a circular DNA molecule located in the mitochondrial matrix. The matrix also contains rRNA and tRNA molecules. The mitochondrial DNA molecule encodes 13 protein subunits for the oxidative phosphorylation system. The central nervous system has particularly high energy requirements, making it very susceptible to defects in mitochondrial function that impair the production of ATP. There is good evidence for mitochondrial dysfunction in several neurodegenerative diseases, including Parkinson's disease and Huntington's disease. Some problems arise from mutations in nuclear DNA that affect the functions of the electron transport complexes in mitochondria. Other prob-

TABLE 6–1	
Production of 36 ATP Molecules by Complete Oxidation of One Glucose Molecule (2 molecules of pyruvic acid are produced from each glucose molecule and both enter the citric acid cycle)	
Glycolysis 2 **ATP** and 2 NADH	Cytosol

Shuttle (inner mitochondrial membrane):
    2 mitochondrial FADH$_2$ produced from 2 cytosolic NADH
 	
Conversion of pyruvate to acetyl coenzyme A (mitochondrial matrix):
    2 NADH
 	
Citric acid cycle (mitochondrial matrix):
    2 **ATP** (from 2 GTP), 6 NADH and 2 FADH$_2$
 	
Oxidative phosphorylation (inner mitochondrial membrane):
    24 **ATP** from 8 mitochondrial NADH
    8 **ATP** from 4 mitochondrial FADH$_2$
	Mitochondria

Total: 36 **ATP**	

lems are due to mutations in mitochondrial DNA that directly impair oxidative phosphorylation. Loss of vision in both eyes, a condition known as Leber's hereditary optic neuropathy, is caused by a mutation in mitochondrial DNA that inhibits the function of one of the electron transport complexes.

The Pentose Phosphate Pathway Is an Alternative for Glucose Oxidation

The result of glucose oxidation, via glycolysis and the citric acid cycle, is production of ATP as an energy source for biosynthesis. Another pathway for the oxidation of glucose exists in the cytosol, called the **pentose phosphate pathway** or the **hexose monophosphate shunt.** The main result of the pentose phosphate pathway is generation of NADPH, the reduced form of the coenzyme $NADP^+$ (see Fig. 6–7), which drives the oxidation-reduction reactions in various biosynthetic pathways, such as the synthesis of fatty acids and steroids. It also provides the pentose sugars that are essential components of nucleotides and nucleic acids. This process uses enzymes different from those used in glycolysis.

The pathway starts with glucose-6-phosphate, a hexose. After several reactions, the final result is the reduction of two molecules of $NADP^+$ to NADPH and the formation of one molecule of ribose-5-phosphate, a pentose (Fig. 6–17). The carbon atom removed from glucose-6-phosphate is lost as CO_2. Excess ribose-5-phosphate not needed for biosynthetic reactions is recycled by conversion to fructose-6-phosphate and glyceraldehyde-3-phosphate, two of the intermediates of the glycolytic pathway. These intermediates can be used to synthesize glucose-6-phosphate, and the cycle begins again. Six rounds of the cycle result in the complete oxidation of one molecule of glucose-6-phosphate to six molecules of CO_2 with the generation of NADPH:

$$\text{glucose-6-phosphate} + 12\ NADP^+ \rightarrow 6\ CO_2 + 12\ NADPH$$

In contrast to aerobic oxidation of glucose, the pentose phosphate pathway does not use ATP, oxygen, or the citric acid cycle. It does not produce ATP; the only electron acceptor involved is $NADP^+$. Which of these pathways is used for glucose catabolism seems to depend on whether or not the cell is heavily engaged in biosynthesis. Skeletal muscle, for example, catabolizes glucose solely via glycolysis and the citric acid cycle. Adipose and liver tissue, on the other hand, catabolize a large fraction of glucose via the pentose phosphate pathway. These tissues are important sites of fatty acid synthesis, a reaction pathway that consumes a large amount of NADPH and is located in the cytosol where the pentose phosphate pathway also occurs.

Note the markedly different roles of NADH and NADPH. NADH is oxidized by the electron transport chain to generate ATP, whereas NADPH is oxidized directly in biosynthetic reactions.

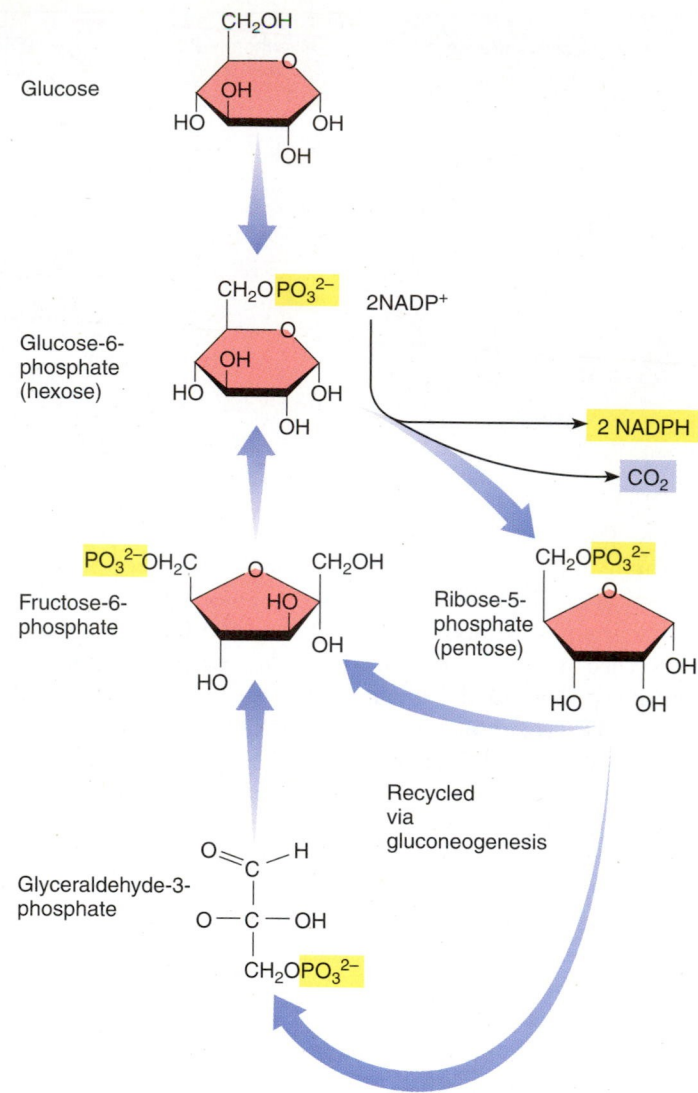

Figure 6–17

A summary of the pentose phosphate pathway.

METABOLISM OF FATS

> *How does the cell obtain energy from the breakdown of fats?*

Fats stored in adipose tissue provide a rich energy source for the body because they can be oxidized to produce ATP. The body can store more fat than carbohydrate, and, as we shall see, much more energy can be obtained from fats than from carbohydrates. For example, 1 g of lipid contains 9 kcal (or 38 kJ) energy compared with only 4 kcal (or 17 kJ) in 1 g of glucose. The reason is that fats contain many hydrogen atoms (and few oxygen atoms) and are more highly reduced than are carbohydrates. As a result, the amount of ATP produced by a given weight of fat exceeds that produced by the same weight of carbohydrates. Because a mobile animal, unlike a

Obesity

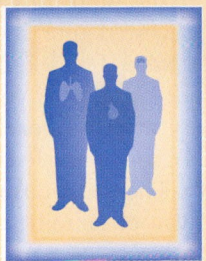

Obesity is a severe health problem in the Western world and in developing countries. Approximately one-third of US adults (about 60 million people) are overweight. The prevalence of childhood obesity has doubled during the past decade and continues to increase. In addition to suffering poor health and living with an increased risk for severe illnesses, such as hypertension, diabetes, and heart disease, obese people frequently are subjected to social discrimination. There are two essentially opposing views on the causes of obesity. One view suggests that obesity results from a fundamental lack of self-discipline by affected individuals. This view is supported by the diet industry, which has a huge financial interest (>$50 billion/year) at stake even though their remedies are likely to fail over the long term in most cases. More than 90% of people who lose weight by dieting alone eventually return to their original weight. The other view suggests that body weight (or the amount of body fat) is controlled physiologically and that an increase or decrease in weight elicits a counter-response that resists the change. In other words, an individual's body mass is determined by a physiological homeostatic mechanism. Within an advanced society where the food supply is adequate and the environment is much the same for everyone, the central question for both viewpoints is: Why are some people lean and others obese? The answer has been supplied by genetic studies that have identified a number of genes that are responsible for obesity in both humans and animals. These studies indicate that obesity is not simply a personal failing; it is largely determined by genetic factors that are inherited similar to the genes that determine other physical characteristics, such as height. For example, 80% of children with two obese parents become obese themselves, whereas only 14% of children of normal-weight parents become obese.

A key part of the physiological control system is the hormone **leptin**, which is produced by fat tissue and acts on the brain to control food intake. Increased body fat is associated with an increased level of leptin, which acts to reduce the intake of food. This is one mechanism that helps to maintain weight within a relatively narrow range. Obese people in general tend either to produce too little leptin or to be more resistant to its biological action. For example, a 3-year-old girl recently appeared in newspaper reports because she weighed 54 kg (120 lb), nearly three times the normal weight for this age. She was resistant to leptin because of a deficiency in leptin receptors. An altered response to leptin could be the result of genetic, environmental, or psychological influences. Thus, the range of body weight that is set by the internal physiological control system can be altered by external factors, such as diet, environment, age, and perhaps exercise. The mechanisms are poorly understood, but the evidence is compelling. For instance, adoption of a typical high-fat Western diet by several native populations is associated with an astonishing increase in body weight. Efforts to prevent obesity are long overdue to protect individuals from the potentially devastating consequences of the associated diseases. While the genetic factors involved in development of obesity must continue to be explored and understood, the fact remains that the worldwide epidemic of obesity during the past 20 years has emerged from a relatively constant genetic pool. Furthermore, the rate of weight gain in the population, an average of 0.45 kg (1 lb) per person per year, is too fast to be accounted for by genetic changes. This highlights the influence and importance of external factors. Effective programs that successfully encourage healthy eating and increased physical activity throughout entire populations are the best approach for long-term prevention of obesity and its related problems. People who have successfully maintained a decrease in body weight eat low-fat diets and consciously look for opportunities for physical activity, such as using stairs instead of elevators. They also exercise regularly at a level that is equivalent to walking 6.4 km (4 mi) every day. The diet and exercise become a natural part of their normal, healthy lifestyle rather than an "obsession."

When prevention fails, treatment with therapeutic drugs may be necessary in conjunction with a structured diet and exercise program. The drugs available at this time are effective only while being actively used. Weight gain is an inevitable consequence when the drugs are discontinued. An effective drug must be safe for long-term use and must reduce energy intake (food consumption), or increase energy expenditure, or both. Currently approved drugs include **sibutramine**, which inhibits food intake, and **orlistat**, which blocks fat digestion. Neither drug is free of side effects. Enzymes involved in fat metabolism may be important targets for new drugs. One enzyme, known as **fatty acid synthase (FAS)**, makes the long-chain fatty acids that are the building blocks of fats. A synthetic inhibitor of FAS was shown recently to produce a dramatic but reversible decrease in appetite and body weight when tested in mice. A decrease in food intake is normally compensated by a decrease in metabolic activity, which explains why losing weight can be so difficult. It is thought that the FAS inhibitor is so effective because it not only suppresses appetite but also tricks the internal control system by preventing the brain from sensing the reduced food intake. Metabolic activity continues at the normal rate, ensuring that energy expenditure exceeds energy intake. Leptin is not involved in the action of the FAS inhibitor, and its precise mechanism of action on the brain is being determined. Although the current research is a long way from human testing, the hope is that one day this approach will aid the treatment of human obesity.

stationary plant, must carry its energy store with it, the amount of ATP available per unit weight of the energy store is an important factor. This is a major reason that animals use fat as an energy store.

Catabolism of a neutral fat, or **triacylglycerol,** molecule begins with the release of the glycerol backbone and the three long-chain fatty acids. Glycerol is readily converted to glyceraldehyde-3-phosphate, which can be used for glycolysis. The fate of the fatty acids is illustrated in Figure 6–18, using palmitic acid as an example. Coenzyme A is attached to the carboxyl (—COOH) group and, in a series of complex reactions within the matrix of a mitochondrion, a fragment of two carbon atoms is removed. This fragment, which also contains coenzyme A, is actually a molecule of acetyl coenzyme A (see Fig. 6–18). The 16-carbon chain of palmitic acid is reduced to 14 carbons by the removal of acetyl coenzyme A. The 14-carbon fatty acid is recycled

through the same series of reactions, and another two carbon atoms are removed as acetyl coenzyme A. In this way, the original long-chain fatty acid is converted into several molecules of acetyl coenzyme A. While NADH and $FADH_2$ are produced by this series of reactions, no ATP is generated. All of the ATP derived from fatty acid catabolism is produced when the acetyl coenzyme A groups are fed into the citric acid cycle (see Fig. 6–18). The catabolism of palmitoyl-coenzyme A, the activated form of palmitic acid, can be summarized as follows:

palmitoyl-coenzyme A + 7 FAD + 7 NAD^+ + 7 coenzyme A →
8 acetyl coenzyme A + 7 $FADH_2$ + 7 NADH

When this yield of acetyl coenzyme A and reduced coenzymes is processed via the citric acid cycle and electron transport, it can be calculated that 131 ATP molecules are generated. Because two high-energy phosphate bonds are used to activate the fatty acid (attachment of coenzyme A), the net gain from oxidation of 1 molecule of palmitic acid is 129 molecules of ATP. This exceeds by far the 36 ATP molecules generated by oxidation of one molecule of glucose, reinforcing the point that fats are a much richer energy source than are carbohydrates because they are more reduced.

Fatty acids are synthesized in a different subcellular compartment (the cytosol) and by a different pathway from that used for this catabolism. The pathway begins with acetyl coenzyme A, and the carbon chain is elongated by a succession of reactions that add two carbon atoms at a time. The energy source for these reactions is ATP, with NADPH from the pentose phosphate pathway providing reducing "power" for oxidation-reduction reactions. Synthesis of palmitic acid can be summarized by these equations:

8 acetyl coenzyme A + 7 ATP + 14 NADPH →
palmitic acid + 7 ADP + 14 $NADP^+$ + 8 coenzyme A

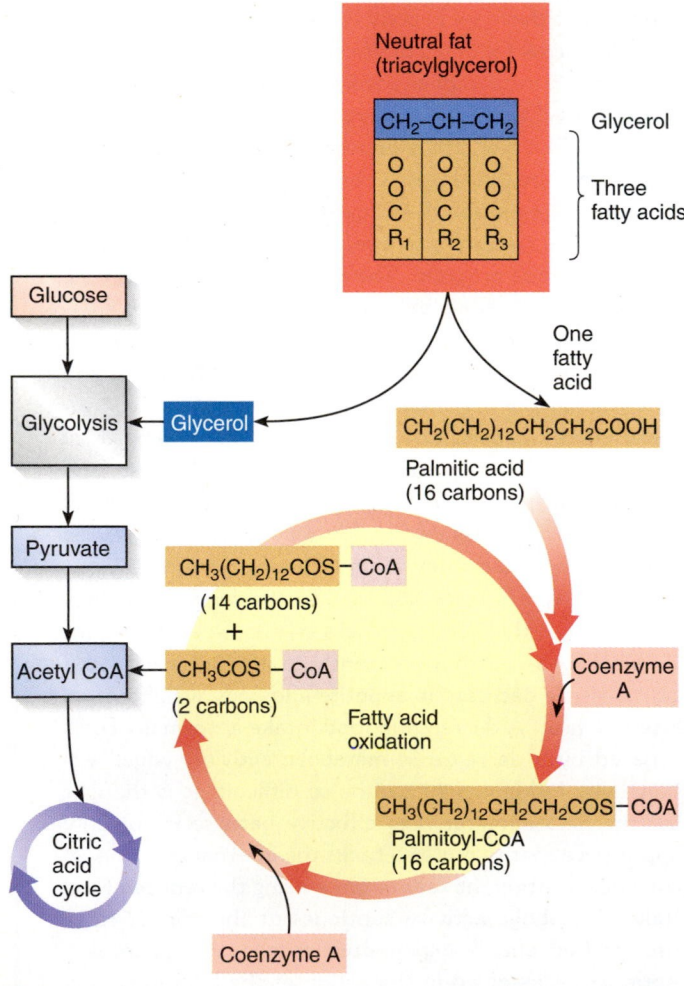

Figure 6–18

The catabolism of fatty acids to acetyl coenzyme A (acetyl CoA). The long hydrocarbon chains of fatty acids are represented by the symbols R_1, R_2, and R_3.

METABOLISM OF PROTEINS

How does the cell obtain energy from the breakdown of protein?

The building blocks of proteins are amino acids. In contrast to glucose and fatty acids, which can be stored until needed, excess amino acids cannot be stored and instead are used as a metabolic fuel. Many of the amino acids are catabolized by removal of the amino group (—NH_2) in the form of ammonia, a process known as **deamination.** Ammonia is potentially toxic and so is converted to urea and excreted from the body in the urine (Fig. 6–19).

The carbon atoms of the amino acids can enter the pathway of glucose oxidation, where they are converted to CO_2 and water with the generation of ATP. Some amino acids enter the

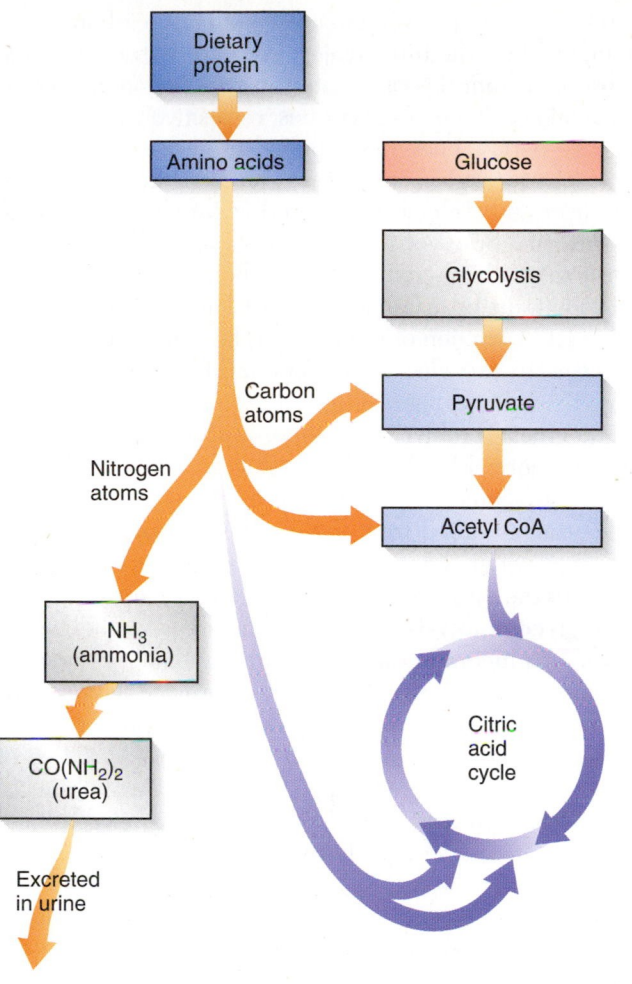

Figure 6–19

During catabolism, carbon atoms from amino acids are used to form pyruvate, acetyl groups, and various products of the citric acid cycle. Nitrogen atoms from amino acids are eventually excreted from the body in the urine.

pathway as pyruvate, some as acetyl coenzyme A, and others as intermediates in the citric acid cycle (see Fig. 6–19).

About half of the basic set of 20 amino acids required by humans are described as **essential amino acids** because they cannot be synthesized; they must be ingested as food. The other needed amino acids are termed **nonessential amino acids** because they can be synthesized by relatively simple pathways.

In addition to their use in protein synthesis, certain amino acids are the precursors for a number of important molecules. These molecules include the **purines** and **pyrimidines** used for synthesis of DNA and RNA. **Histamine** is a powerful, locally acting vasodilator synthesized from the amino acid histidine. The hormones **epinephrine** and **thyroxine** are formed from the amino acid tyrosine, and the nicotinamide ring of NAD$^+$ is derived from the amino acid tryptophan.

ACETYL COENZYME A: MEDIATING BETWEEN CATABOLISM AND BIOSYNTHESIS

 What is the significance of acetyl coenzyme A in the cellular metabolism story?

The citric acid cycle, together with mitochondrial electron transport, is the principal pathway for supplying the cell with ATP. Very little of the chemical energy present in the macromolecules of the cell is made available until small units of these macromolecules are fed into the citric acid cycle (Fig. 6–20).

Although the carbon fragments of several amino acids can enter the citric acid cycle directly, almost all of the other intermediate compounds produced by catabolism are fed

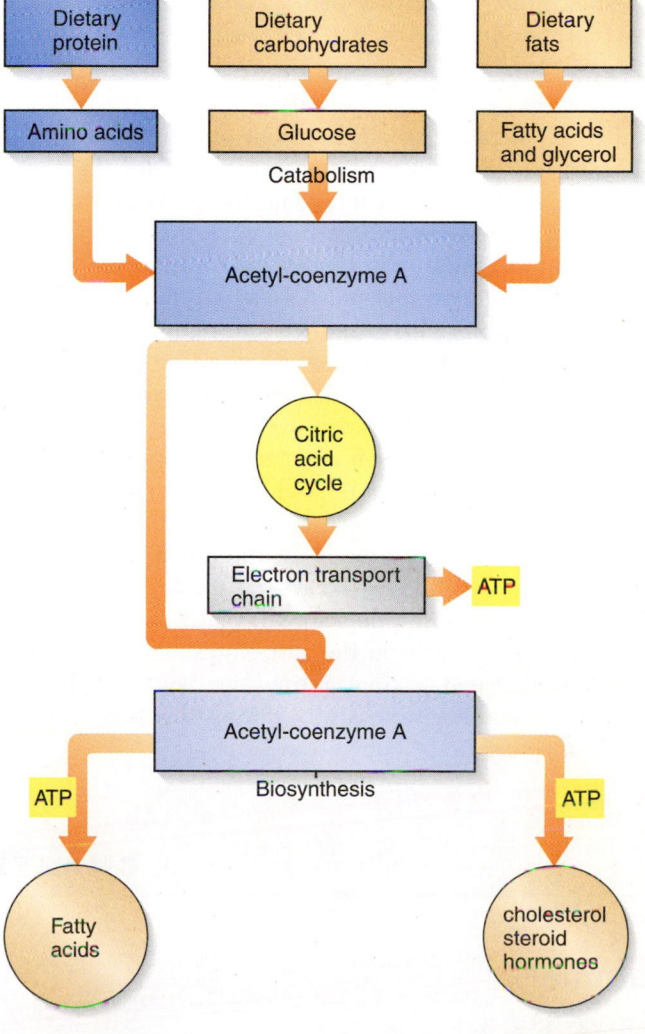

Figure 6–20

An illustration of the central role of acetyl coenzyme A in cellular metabolism in mammals.

into the cycle as acetyl groups coupled to coenzyme A. The acetyl groups are also the 2-carbon precursors used in the synthesis of fatty acids and cholesterol. Steroid hormones are derived from cholesterol. Thus, coenzyme A is required not only to derive energy from fuel molecules but also to provide the many carbon atoms required for biosynthesis of a large variety of organic molecules. Carbon atoms derived from catabolism of fats and proteins can be shunted through acetyl coenzyme A and used in biosynthesis (see Fig. 6–20), depending on the metabolic demands at any given time. In other words, acetyl coenzyme A provides a place for the different metabolic pathways of the major macromolecules of the cell to intersect. Note that in mammals, acetyl coenzyme A cannot be converted to pyruvate; hence, fats cannot be converted to carbohydrates.

INTEGRATION OF METABOLISM

> *What is the "big picture" with respect to metabolism of carbohydrates, fats, and proteins?*

The connections between the different metabolic pathways mean that not only carbohydrate but also protein and fat molecules can be used as sources of energy for the body. These sources are not used at random. They are accessed in a carefully integrated way in order to provide an uninterrupted energy supply to the various tissues and organs, even though the supply of external nutrients is likely to be intermittent. The absence or presence of ingested nutrients places different demands on internal sources of energy. Most people eat only at set times during the day and not at all during the night. The body is able to adapt to these periods of plenty and fasting. The period when ingested nutrients are entering the blood from the gastrointestinal tract is termed the **absorptive state.** Some of these nutrients are used immediately to supply energy for the body; the remainder is added to the energy stores. Because an average meal requires only 4 hours for complete absorption from the gastrointestinal tract, no nutrients are present in the gastrointestinal tract during the night and at certain times during the day. This phase is called the **postabsorptive state;** during this time, energy must be supplied by the body's stores.

Assume that an average meal contains carbohydrate, protein, and fat—the three major groups of nutrients. These are absorbed from the gastrointestinal tract as monosaccharides, amino acids, and triglycerides, respectively. During the absorptive period, energy is supplied primarily by absorbed monosaccharides (glucose). Excess glucose is stored as glycogen in liver and skeletal muscle and as fat (triglycerides) in adipose tissue. Some of the absorbed amino acids are used for protein synthesis and other metabolic pathways; in the typical adult, the extra amino acids are converted to carbohydrate or fat. A fraction of the ingested fat is oxidized by various organs to provide energy, but most is transported to adipose tissue for storage.

The crucial problem of the postabsorptive period is to maintain a normal level of plasma glucose, which is the primary energy source for the brain, even though glucose is no longer being absorbed from the gastrointestinal tract. Thus, net synthesis of glycogen, protein, and fat ceases and net catabolism of these stores begins. Glycogen is degraded to glucose by **glycogenolysis.** Additional glucose is synthesized in the liver by **gluconeogenesis,** a pathway that uses noncarbohydrate precursors, such as lactate, pyruvate, glycerol, or amino acids. The amino acids are provided by protein catabolism which, through this mechanism, is a major source of blood glucose during prolonged fasting. Catabolism of triglycerides, a process termed **lipolysis,** yields glycerol and fatty acids. Oxidation of the fatty acids provides much of the energy supply for the body. In the liver, the acetyl coenzyme A produced by fatty acid breakdown does not enter the citric acid cycle; instead, the acetyl groups are processed into compounds known as **ketones.** One of the ketones is acetone, some of which is exhaled and accounts for the distinctive odor of a person fasting for many days. As ketones begin to build up in the blood, the brain starts to use them as an energy source in addition to glucose. Use of ketones by the brain and other tissues as a energy supply is an important adaptation to severe starvation because it reduces the dependence on glucose. Hence, protein stores last longer because less protein is needed to provide amino acids for glucose synthesis. The combined effects of glucose production and fat utilization are so effective that, after one month of complete fasting, the plasma glucose level is decreased by only 25%. The total store of energy in the body allows the average person to survive for many weeks without food.

CHAPTER REVIEW

Summary

- Plants transform radiant energy from the sun into carbohydrates, a form of chemical energy that animals can use.
- Animals degrade the carbohydrates and transform the chemical energy into ATP. The ATP is used for processes such as biosynthesis and movement.

- ATP turnover in an animal cell is very rapid. Some of the chemical energy of a fuel molecule cannot be trapped as ATP and is released as heat.
- Endergonic reactions in the cell are driven by free energy released from exergonic reactions. All chemical reactions in the

- cell are catalyzed by enzymes, and many reactions are reversible.
- The citric acid cycle and electron transport chain in the mitochondrion are the final common pathways in the catabolism of all fuel molecules, whether amino acids, glucose, or fatty acids.
- The reactions of glycolysis split one glucose molecule into two molecules of pyruvic acid and generate two molecules of ATP.
- When one pyruvate is fed into the citric acid cycle as acetyl coenzyme A, the cycle generates three NADH, one $FADH_2$, and one GTP.
- The oxidation of reduced coenzymes by the electron transport chain drives phosphorylation of ADP to form ATP, a process known as *oxidative phosphorylation*. Complete oxidation of one glucose molecule generates 36 ATP molecules.

- Glucose oxidation by the pentose phosphate pathway generates NADPH, which is needed to drive redox reactions in biosynthetic pathways.
- Fats are a much richer energy source than are carbohydrates. Oxidation of one molecule of palmitic acid generates 129 molecules of ATP.
- The carbon atoms of amino acids are fed into various points of the glucose oxidation pathway. The nitrogen atoms eventually are excreted in the urine in the form of urea.
- The mitochondria, via the citric acid cycle and electron transport chain, are the principal source of ATP in the cell. Acetyl coenzyme A in the mitochondrial matrix is the central molecule that connects the different metabolic pathways.

Review Questions

Choose the Correct Answer

1. In a skeletal muscle cell undergoing intense activity, the normal metabolic production of ATP cannot meet the demand. Extra ATP is provided by another mechanism using:
   a. phosphocreatine.
   b. GTP.
   c. lactic acid.
   d. lactate dehydrogenase.
   e. ribose-5-phosphate.

2. Complete oxidation of carbohydrates, such as glucose, requires:
   a. oxidative phosphorylation.
   b. glycolysis.
   c. citric acid cycle.
   d. coenzyme A.
   e. all of the above.

3. An organic molecule that functions in metabolism as an acceptor of acetyl groups is:
   a. NADP.
   b. FAD.
   c. coenzyme A.
   d. pyruvate.
   e. ADP.

4. How many molecules of ATP are synthesized for each molecule of NADH oxidized in the electron transport chain?
   a. 1
   b. 2
   c. 3
   d. 4
   e. 5

5. How many molecules of ATP are gained from the complete metabolism of one molecule of pyruvic acid under anaerobic conditions?
   a. 0
   b. 4
   c. 8
   d. 12
   e. 16

6. How many molecules of $FADH_2$ are gained from the complete oxidation of one molecule of palmitic acid, a 16-carbon fatty acid?
   a. 0
   b. 5
   c. 10

   d. 15
   e. 20

7. In glycolysis, ATP synthesis is catalyzed by:
   a. hexokinase.
   b. diphosphoglycerate kinase.
   c. enolase.
   d. aldolase.
   e. phosphofructokinase.

8. Phosphofructokinase is stimulated by:
   a. ATP.
   b. ADP.
   c. AMP.
   d. NAD.
   e. NADH.

9. Triacylglycerols:
   a. are water-soluble.
   b. yield about the same amount of ATP as does glucose.
   c. are hydrated when stored.
   d. are a rich store of energy in the body.
   e are negatively charged molecules at neutral pH.

10. In humans, fatty acids:
    a. cannot be synthesized.
    b. are synthesized in mitochondria.
    c require NADH for synthesis.
    d. can be synthesized from excess carbohydrate or protein.
    e. none of the above.

11. The principal metabolic pathway that generates NADPH is:
    a. glycolysis.
    b. oxidative phosphorylation.
    c. deamination.
    d. fatty acid oxidation.
    e. the pentose-phosphate pathway.

12. Nitrogen produced by deamination of amino acids is excreted in the form of:
    a. urea.
    b. protein.
    c. NADH.
    d. lactic acid.
    e. palmitic acid.

13. Citrate synthetase is inhibited by:
    a. AMP.
    b. ADP.
    c. ATP.

d. NADH.

e. NADPH.

14. Much of the free energy released during metabolism of carbohydrates is trapped and stored as:
    a. NAD$^+$.
    b. ADP.
    c. GTP.
    d. FAD.
    e. ATP.

15. ATP synthetase:
    a. is located in the endoplasmic reticulum.
    b. uses proton flow to drive synthesis of ATP.
    c. pumps protons out of the mitochondrial matrix.
    d. hydrolyes ATP.
    e. converts AMP to ATP.

16. Allosteric inhibition of an enzyme is due to inhibition by:
    a. an end product of the reaction pathway.
    b. a substrate for the reaction pathway.
    c. a change in pH of the reaction mixture.
    d. a change in temperature.
    e. the absence of a coenzyme.

17. An enzyme that catalyzes phosphorylation of glucose is:
    a. phosphofructokinase.
    b. pyruvate kinase.
    c. hexokinase.
    d. pyruvate dehydrogenase.
    e. ATP synthetase.

18. Molecules that serve as important precursors for synthesis of hormones, such as thyroxine and epinephrine, are certain:
    a. carbohydrates.
    b. amino acids.
    c. cytochromes.
    d. nucleic acids.
    e. fats.

19. The reactions of the citric acid cycle take place in the:
    a. nucleus.
    b. cytosol.
    c. endoplasmic reticulum.
    d. mitochondrial matrix.
    e. inner mitochondrial membrane.

20. How many sites for release of free energy are present in the electron transport chain of the mitochondrion?
    a. 1
    b. 2
    c. 3
    d. 4
    e. 5

Answers to Case History Questions

1. LHON is a rare, inherited disease that leads to blindness because of degeneration of the optic nerve. Loss of vision between ages 15 and 35 years is usually the only clinical sign of the disease. More than 80% of cases occur in males; the basis for this male predominance is unknown.

2. The human mitochondrial genome is a small, circular DNA molecule that contains about 16,000 base pairs and encodes 13 proteins involved in electron transport and oxidative phosphorylation. The mutation at base pair 11778 involves only a single base pair ("point mutation"), but it lies within the gene that encodes one of the subunits of Complex I of the electron transport chain. It results in substitution of a histidine for an arginine. This change in Complex I reduces the capacity of mitochondria to carry out oxidative phosphorylation and generate ATP. In some tissues, the overall rate of mitochondrial respiration is reduced by 50% to 70%.

3. It is likely that impairment of oxidative phosphorylation and ATP generation in cells has the greatest effect on tissues that are most dependent on a copious supply of oxygen and ATP. The central nervous system, including the brain and optic nerves, is highly dependent on oxidative metabolism. Thus, the clinical signs of LHON tend to show up only in these tissues and not throughout the body.

4. This is one aspect of LHON that is not understood at the present time. Environmental factors, such as smoking, alcohol, unusual diets, and exposure to toxins, may play a role in initiating the expression of the LHON mutation. LHON patients and unaffected carriers of the mutation usually are advised to avoid tobacco smoking and alcohol consumption and to adopt a healthy lifestyle.

Key Terms

absorptive state (p. 200)
acetyl coenzyme A (p. 187)
adenosine triphosphate (ATP) (p. 178)
aerobic (p. 187)
anabolism (biosynthesis) (p. 178)
anaerobic (p. 190)
catabolism (p. 178)
chemiosmotic mechanism (p. 195)

citric acid cycle (Krebs cycle, tricarboxylic acid cycle) (p. 187)
coenzyme (p. 180)
cytochrome (p. 193)
deamination (p. 185)
electron carrier (p. 193)
electron transport chain (p. 180)
energy transformation (p. 178)

essential amino acids (p. 199)
fat (p. 196)
glycolysis (p. 185)
high-energy phosphate compound (p. 182)
nonessential amino acids (p. 199)
oxidative phosphorylation (p. 182)

pentose phosphate pathway (p. 196)
photosynthesis (p. 178)
postabsorptive state (p. 200)
redox reactions (p. 180)
standard free energy change (p. 182)
substrate-level phosphorylation (p. 182)

Suggested Readings

Block, S. "Real engines of creation." *Nature,* 386:217–218, 1997.

Boss, O., Hagen, T., and Lowell, B. B. "Uncoupling proteins 2 and 3: Potential regulators of mitochondrial energy metabolism." *Diabetes,* 49:143–156, 2000.

Boyer, P. D. "The ATP synthase: A splendid molecular machine." *Annual Review of Biochemistry,* 66:717–749, 1997.

Dickerson, L. M., and Carek, P. J. "Drug therapy for obesity." *American Family Physician,* 61:2131–2138, 2000.

Gura, T. "Enzyme blocker prompts mice to shed weight." *Science,* 288:2299–2300, 2000.

Hoganson, C. W., and Babcock, G. T. "A metalloradical mechanism for the generation of oxygen from water in photosynthesis. *Science,* 277:1953–1956, 1997.

Kopelman, P. G. "Obesity as a medical problem." *Nature,* 404:635–643, 2000.

Lowell, B. B., and Spiegelman, B. M. "Towards a molecular understanding of adaptive thermogenesis." *Nature,* 404:652–660, 2000.

Purvin, V. A. "Optic neuropathies for the neurologist." *Seminars in Neurology,* 20:97–110, 2000.

Rothman, S. M. "Mutations of the mitochondrial genome: Clinical overview and possible pathophysiology of cell damage." *Biochemical Society Symposia,* 66:111–122, 1999.

Saraste, M. "Oxidative phosphorylation at the *fin de siecle.*" *Science,* 283:1488–1492, 1999.

Stryer, L. *Biochemistry,* ed 4. New York, Freeman, 1995.

Went, L. N. "Leber hereditary optic neuropathy (LHON): A mitochondrial disease with unresolved complexities." *Cytogenetics and Cell Genetics,* 86:153–156, 1999.

Answers to Review Questions

1. a	**2.** e	**3.** c	**4.** c	**5.** a	**6.** d	**7.** b	**8.** b
9. d	**10.** d	**11.** e	**12.** a	**13.** c	**14.** e	**15.** b	
16. a	**17.** c	**18.** b	**19.** d	**20.** c			

Part II

PHYSIOLOGICAL CONTROL SYSTEMS

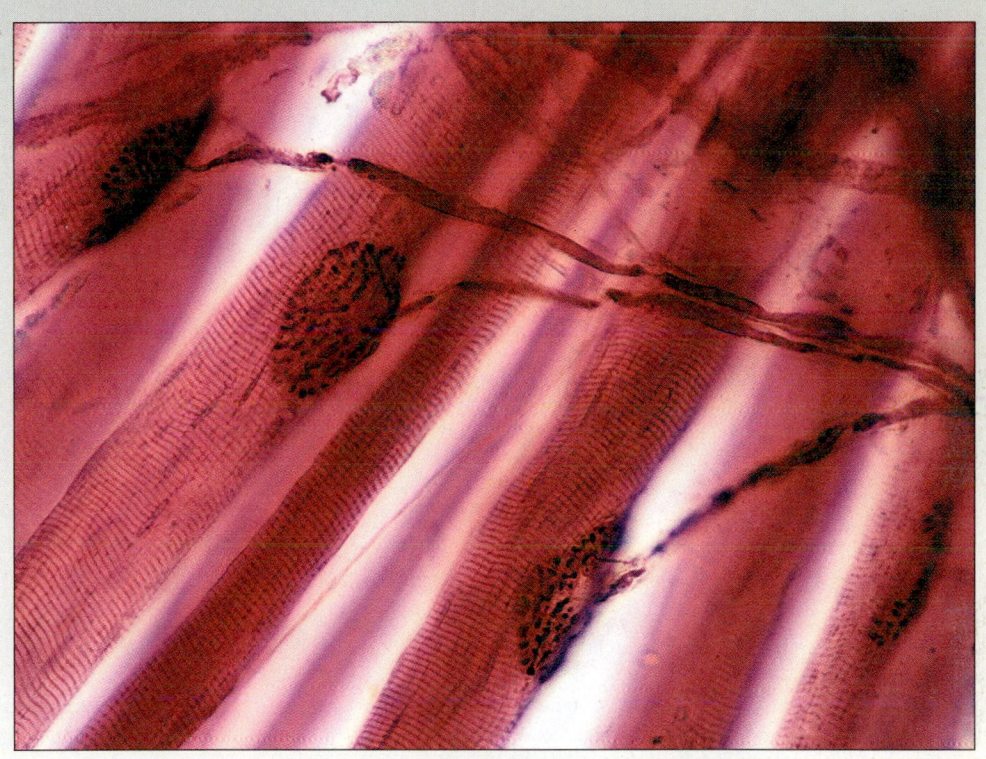

• *Photomicrograph of a nerve fiber innervating a skeletal muscle.*
(© J. & L. Weber/Peter Arnold, Inc.)

Chapter 7

FUNCTIONAL ORGANIZATION OF THE NERVOUS SYSTEM

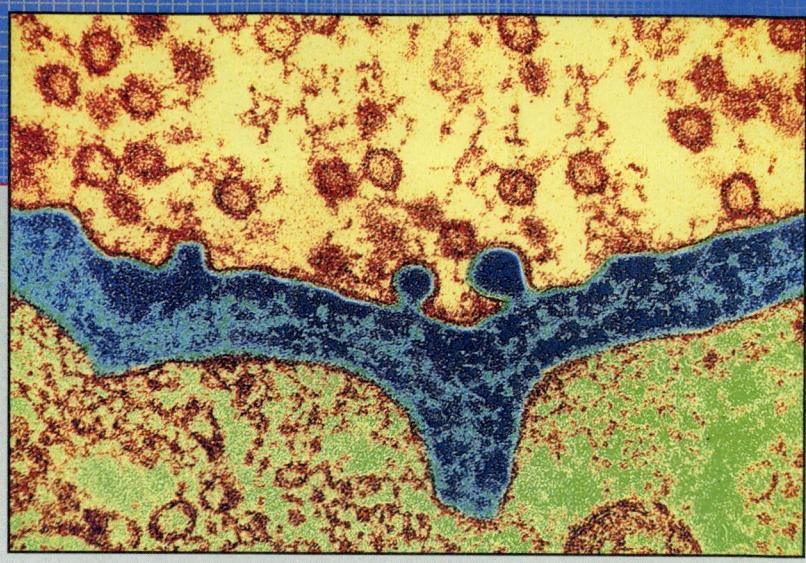

KEY CONCEPTS

- *The body's nervous system is organized into a central nervous system (CNS) and a peripheral nervous system (PNS). Sensory elements of the PNS detect changes in the body's internal and external environments and send sensory information to the CNS. The central nervous system receives the information, processes it, and, in response to it, may generate an output that is sent to effectors by way of motor elements of the PNS.*

- *Neurons and neuroglia are the principal types of cells found in the nervous system. Neurons, commonly called nerve cells, receive and integrate information coming from sensors or other neurons and transmit information to effector cells by electrochemical processes of nerve impulse conduction and synaptic transmission. Neuroglia provide structural and metabolic support for neurons and help to maintain an extracellular environment that is optimum for neuron function.*

- *The nerve impulse is a wave of depolarization immediately followed by a wave of repolarization, collectively called an action potential, propagated without decrement along the plasma membrane of a nerve fiber. Changes in ion conductances across the nerve membrane are responsible for the initiation and propagation of the action potential.*

- *The functional connection between a neuron and its effector cell is called a synapse. Neurons communicate with one another by the process of synaptic transmission. This process involves the depolarization of the nerve fiber terminal, which causes the activation of voltage-gated calcium channels, the fusion of neurotransmitter-containing vesicles with the terminal pre-synaptic membrane, and the release of neurotransmitter into the synaptic cleft. The neurotransmitter engages receptors on the post-synaptic membrane, changing ion conductance and post-synaptic membrane potential.*

- *Neurotransmitters can be grouped according to their molecular weight. The synthesis of low-molecular-weight transmitters typically takes place in the nerve fiber terminal, where they can be rapidly recycled for release. High-molecular-weight transmitters are synthesized in the region of the cell soma and are transported to the nerve fiber terminal for release.*

- *A neuron processes incoming information by summating post-synaptic membrane potential changes occurring at its synaptic inputs. Spatial summation involves the addition of post-synaptic potentials occurring simultaneously at two or more inputs. Temporal summation involves addition of post-synaptic potentials that occur very close together in time at the same synapse, one potential change occurring before a preceding one has decayed.*

CASE HISTORY

James, a 35-year-old computer software developer, suffers from relapsing-remitting multiple sclerosis (MS). His first attack occurred after a herpes viral infection, acquired when he was 25 years old. The attack lasted for more than 6 weeks before almost completely remitting. After 9 months in remission, he suffered his first relapse.

Multiple sclerosis is a chronic, progressive, disabling disease that damages and destroys the myelin sheath surrounding nerve cells (neurons) in the brain and spinal cord, at first slowing, and later blocking, transmission of electrical signals. Although the exact cause remains unknown, MS is believed to be an autoimmune disease in which the immune system abnormally recognizes the body's own myelin as a foreign substance and mounts a destructive attack.

Myelin is a cream-colored fat that helps form a lipid-protein sheath of insulation around a nerve fiber, somewhat like the vinyl or rubber insulation that covers an electrical wire. Normally, the myelin sheath allows nerve impulses to be rapidly conducted along a nerve fiber, thereby providing for quick and efficient communication and control by the central nervous system. Destruction and removal of the myelin sheath, a process called *demyelination*, may occur in other diseases, such as diabetes mellitus (sugar diabetes), but its most common occurrence is in MS.

Symptoms of an MS attack are unpredictable and variable. They depend on the location and extent of myelin damage. Common symptoms resulting from motor fiber damage include muscle weakness, loss of muscle coordination, muscle spasms, tremors, and slurred speech. Damage to sensory fibers can cause double and blurred vision, numbness and tingling sensations, and poor balance. Memory loss, depression, and extreme fatigue are also common.

MS symptoms occur during an attack on myelin. When the attack subsides, the severity of the symptoms is reduced and the myelin can be repaired or regenerated during this period of remission, often resulting in significant improvement in motor and sensory function. However, repeated attacks eventually cause scar tissue, called *sclerotic plaques*, to form in place of normal myelin. This results in permanent motor and sensory deficits.

Questions

1. Two persons, each suffering from MS, may experience very different symptoms of the disease. Explain.

2. How does the destruction of myelin block conduction of the nerve impulse in an affected nerve fiber?

INTRODUCTION

The primary function of the nervous system is to maintain homeostasis through the control of the other body systems. It coordinates the activities of other organ systems so that they operate in concert with one another to maintain a relatively stable internal environment for optimum cell function. During exercise, for example, the nervous system activates selective muscle groups for movement. It stimulates the heart to pump faster, resulting in an increased blood flow in skeletal muscle and it stimulates the respiratory system, increasing the rate and depth of breathing. This helps to increase oxygen delivery to working muscles and to remove carbon dioxide. These aspects of neural control involve motor commands to skeletal, cardiac, and smooth muscles of target organs.

Motor commands issued by the nervous system are commonly generated in response to sensory inputs, as when withdrawing the hand after touching a hot object. Sensors of the nervous system detect changes in the body's internal and external environments and transmit sensory information to the brain for storage and use in generating a response. We respond to our external environment based on what we hear, see, touch, smell, and taste. Initially, these sensations are processed in separate pathways within the nervous system. At higher levels, however, there is a convergence of various sensory inputs for more complex processing. Such is the case, for example, with information from the visual and auditory systems, which converge in areas of the brain involved in the interpretation of language.

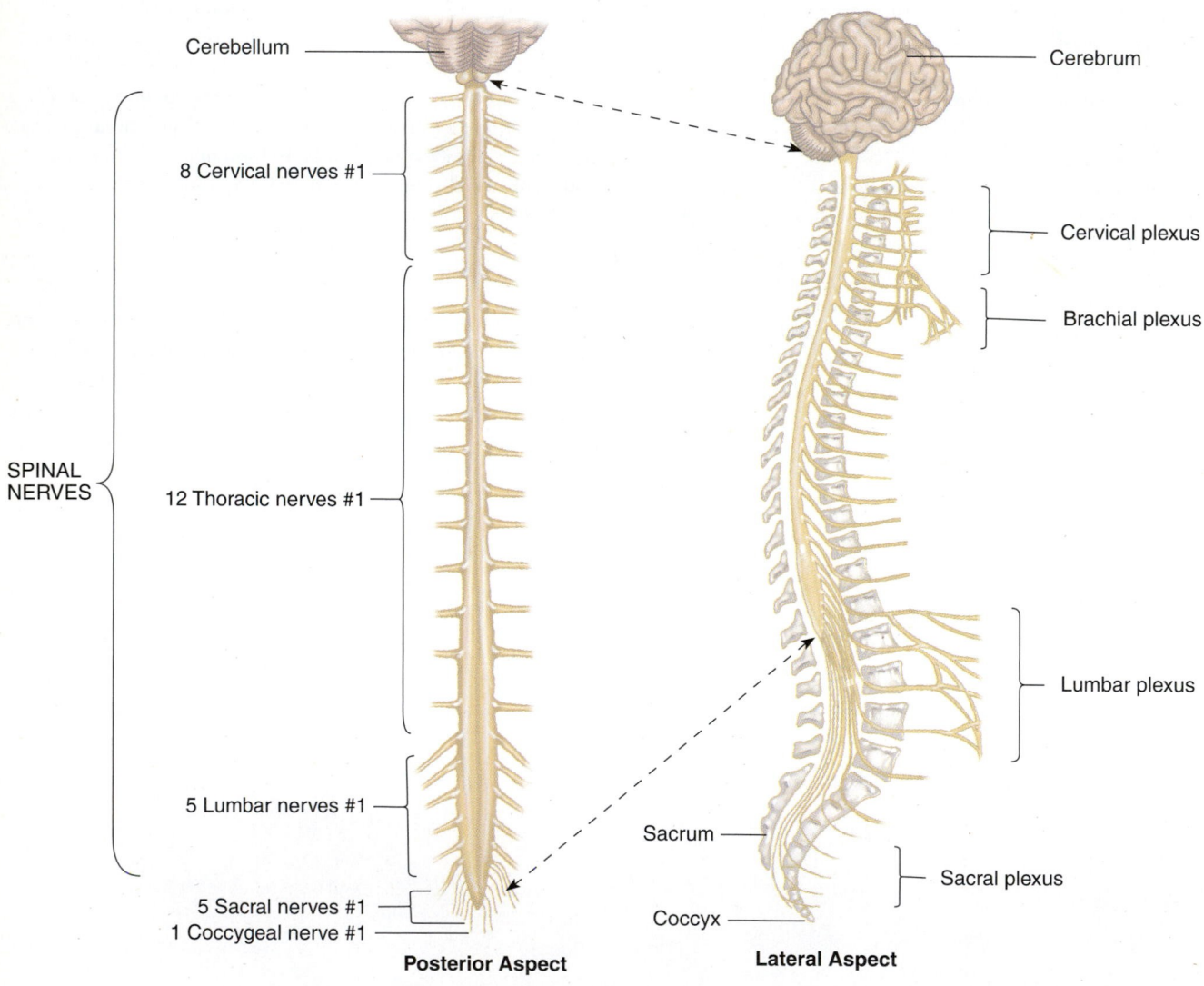

(a) Spinal cord and spinal nerves

Figure 7–1

(a) The spinal cord and 31 pairs of spinal nerves. *(b)* The anterior surface of the brain and 12 pairs of cranial nerves.

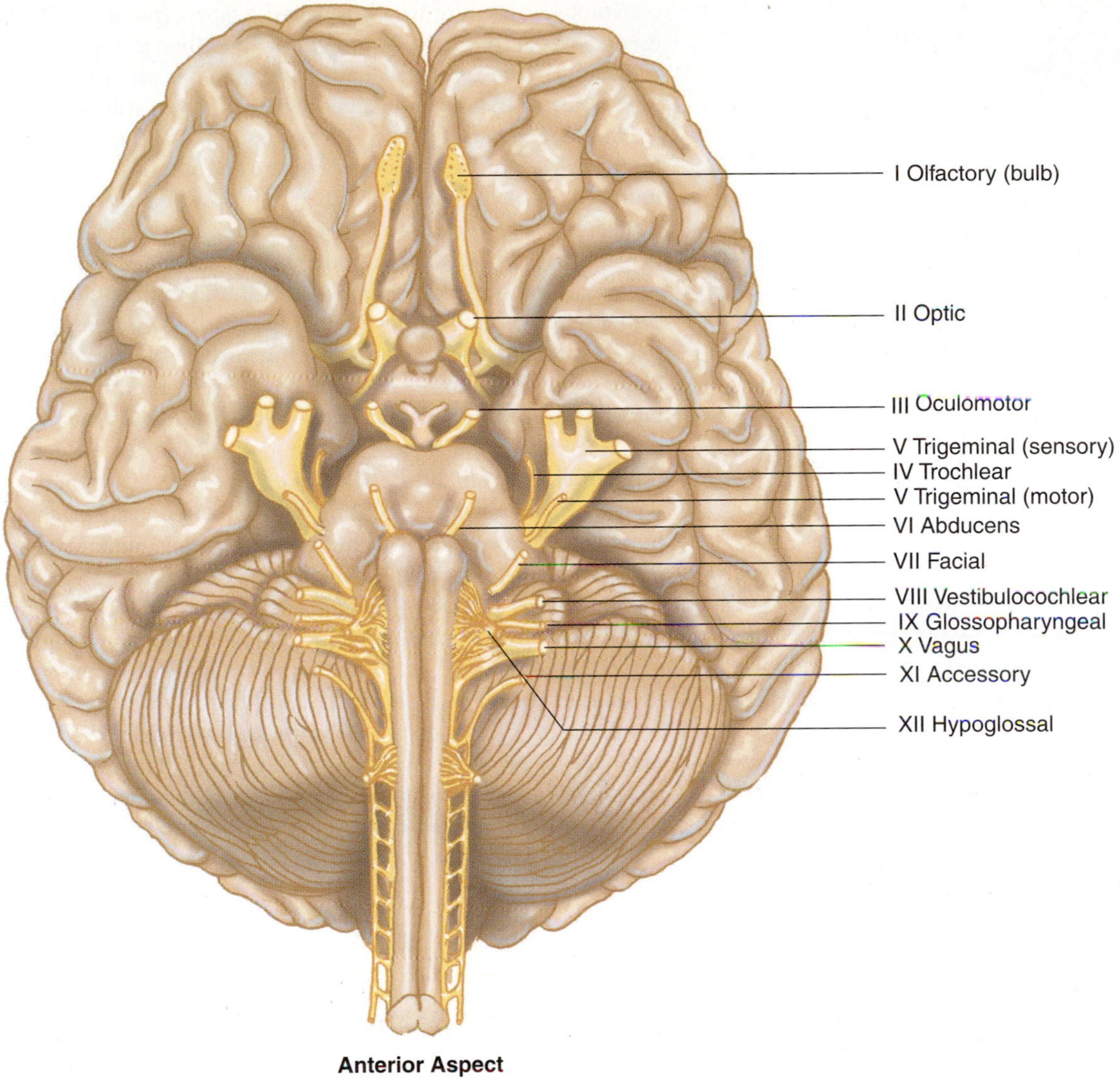

I Olfactory (bulb)

II Optic

III Oculomotor

V Trigeminal (sensory)
IV Trochlear
V Trigeminal (motor)
VI Abducens
VII Facial
VIII Vestibulocochlear
IX Glossopharyngeal
X Vagus
XI Accessory

XII Hypoglossal

Anterior Aspect

(b) Brain and cranial nerves

The basic structural and functional unit of the nervous system is the neuron. This chapter focuses on how neurons are organized in the nervous system and how neurons receive, integrate, and transmit information from one area of the body to another.

ORGANIZATION OF THE NERVOUS SYSTEM

> *What is the functional organization of the nervous system?*

The nervous system develops from **ectoderm**, an embryonic tissue that forms a **neural plate** in the developing embryo. The neural plate differentiates into a **neural crest** and a **neural tube.** The neural tube differentiates into the **central** **nervous system (CNS),** which consists of the **brain** and the **spinal cord**. The neural crest gives rise to most of the **peripheral nervous system (PNS),** which consists of 12 pairs of **cranial nerves** and 31 pairs of **spinal nerves** (Fig. 7–1).

The Central Nervous System

The brain develops at the cranial end of the embryonic neural tube. By the end of the first month of development, three primary vesicles have formed: the **forebrain** (prosencephalon), the **midbrain** (mesencephalon), and the **hindbrain** (rhombencephalon). A week later, the forebrain gives rise to the **telencephalon** and the **diencephalon,** and the hindbrain gives rise to the **metencephalon** and the **myelencephalon,** resulting in a total of five secondary vesicles (Fig. 7–2). In adults, the telencephalon develops into **cerebral hemispheres.** The diencephalon gives rise to the **thalamus** and the **hypothalamus** and other structures. The

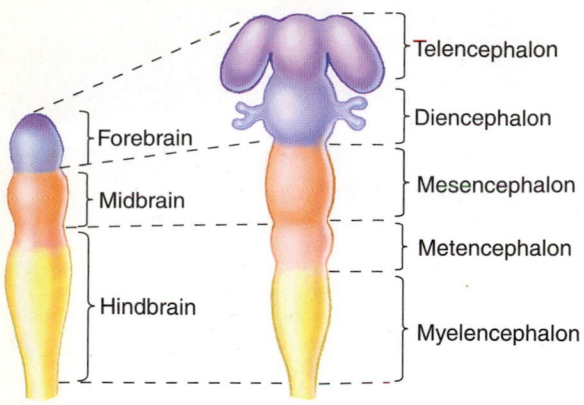

Figure 7–2

Early development of the human brain into primary and secondary vesicles.

mesencephalon becomes the **midbrain.** The metencephalon gives rise to the **pons** and **cerebellum,** and the myelencephalon becomes the **medulla oblongata** (Fig. 7–3). The medulla oblongata, pons, and midbrain form the adult **brain stem** (Fig. 7–4a).

The outermost layers of cells in each cerebral hemisphere form gray matter called the **cerebral cortex** or **neocortex.** The surface of the cerebral cortex is marked by convolutions, each of which is called a *gyrus* (pl., gyri). Each gyrus is separated from its neighbor by either a shallow furrow called a *sulcus* (pl., sulci) or by a deeper crevice known as a *fissure*. Fissures and sulci divide each hemisphere into a frontal lobe, a parietal lobe, a temporal lobe, and an occipital lobe, each named after overlying protective bones of the cranium (braincase). The neocortex is proportionally larger in humans than in most other species. It is functionally divided (Fig. 7–4b) into areas that control movement (**primary somatic motor cortex**), areas involved in the initiation and coordination of complex movements (**premotor cortex**), areas that respond to sensations felt on the surface of the skin (**somatosensory cortex**), visual images seen by the eye (**visual cortex**), and sounds detected by the ear (**auditory cortex**). Other areas of the neocortex are responsible for language (**Wernicke's area** and **Broca's area**), personality, planning, decision-making, differentiating between trivial and important sensory information (**frontal cortex**), and memory and emotion (**limbic cortex**). Several areas of the cortex are also involved with processes of reasoning and abstract thought. The core of the cerebral hemispheres is made up of bundles of white nerve fibers that connect the lobes to each other and to areas of the brain beneath the cerebrum.

Beneath the neocortex are subcortical areas of the brain involved in a variety of different functions. The **thalamus** (see Fig. 7–4a), for example, processes and relays information from the sensory organs, such as the eyes and ears, and

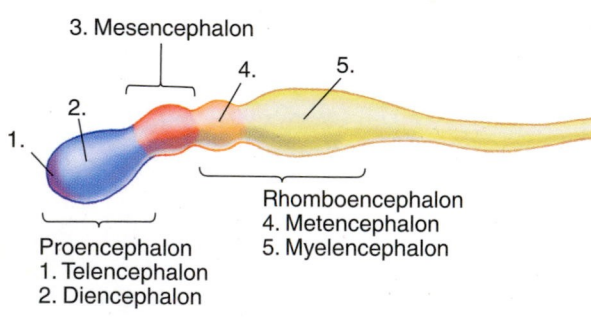

Figure 7–3

Embryonic development of major parts of the brain from five secondary vesicles.

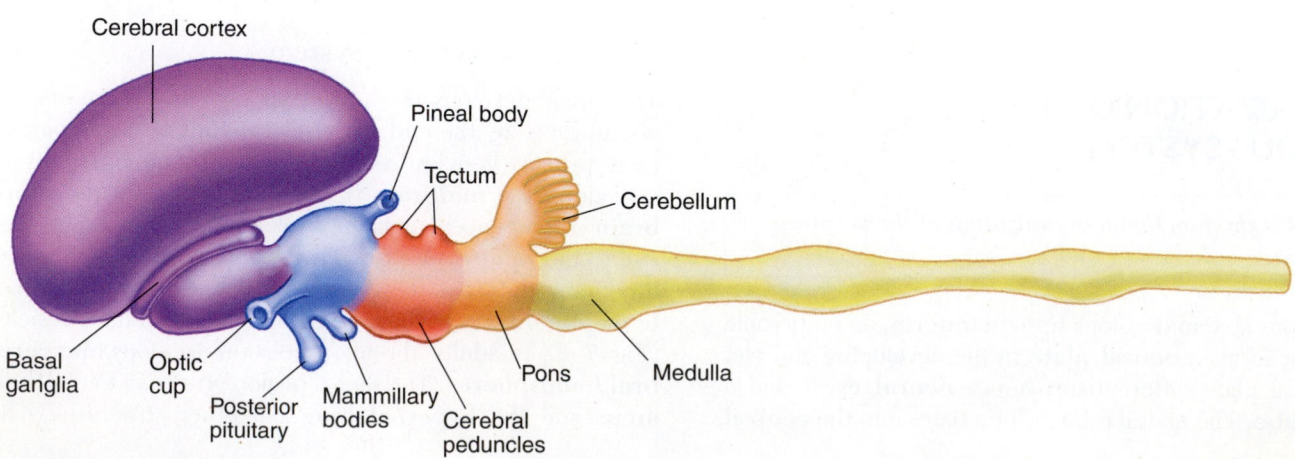

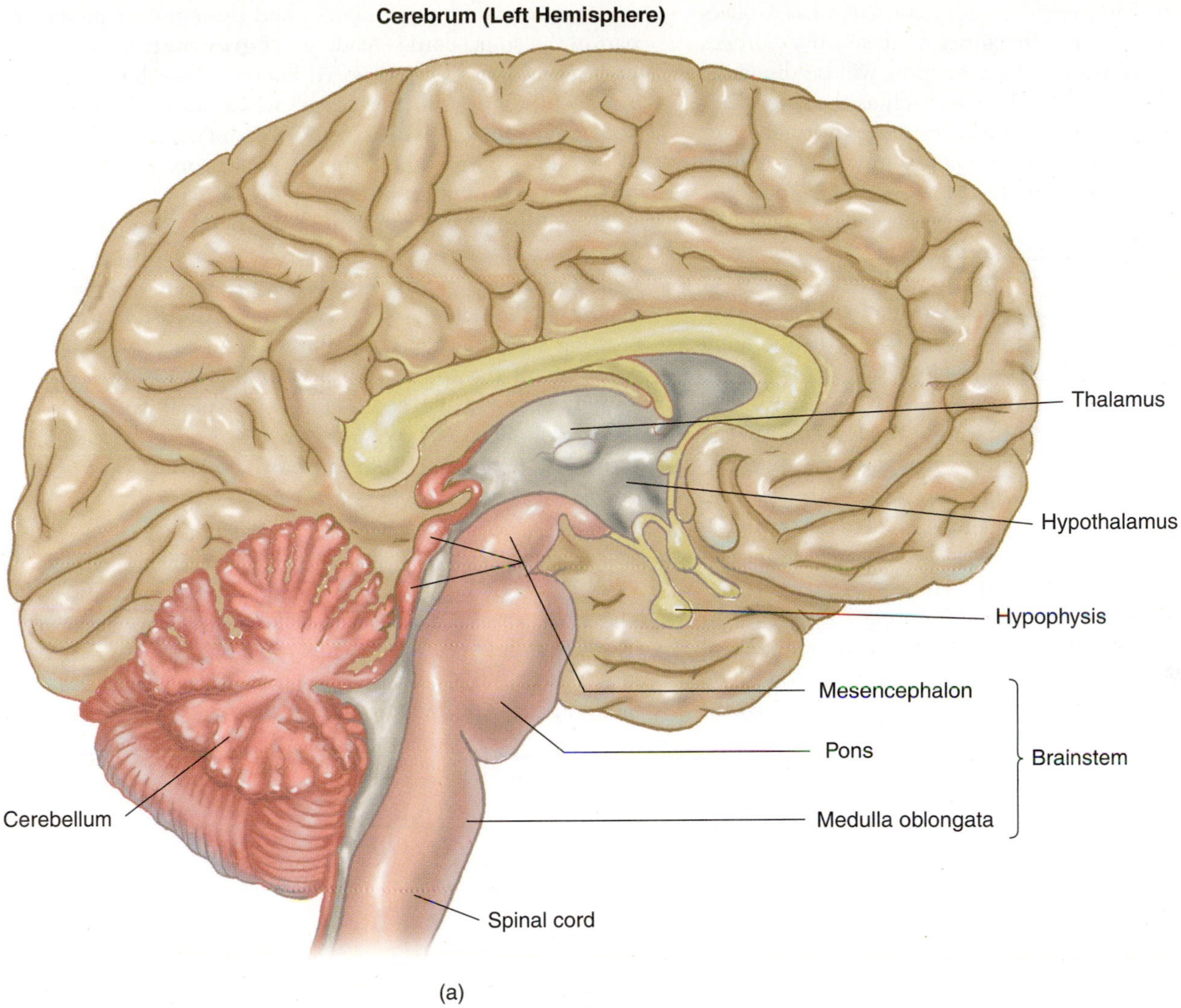

Cerebrum (Left Hemisphere)

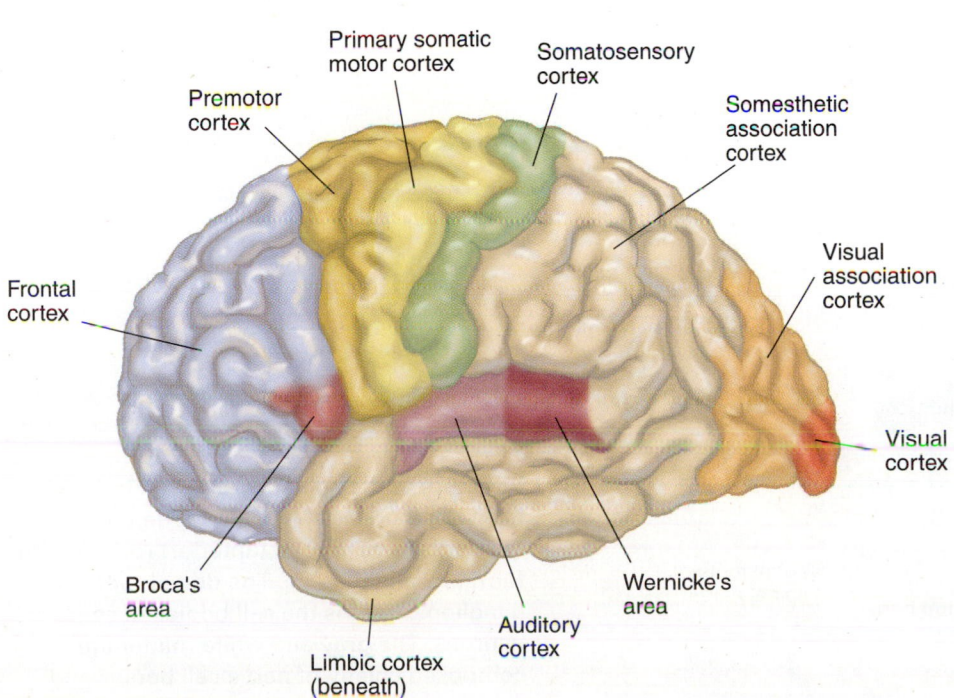

Thalamus

Hypothalamus

Hypophysis

Mesencephalon ⎤
Pons   ⎬ Brainstem
Medulla oblongata ⎦

Thalamus

Cerebellum

Spinal cord

(a)

Primary somatic motor cortex

Somatosensory cortex

Premotor cortex

Somesthetic association cortex

Frontal cortex

Visual association cortex

Visual cortex

Broca's area

Wernicke's area

Limbic cortex (beneath)

Auditory cortex

(b)

Figure 7–4

(a) Midsagittal section of an adult brain. *(b)* Functional areas of the cerebral cortex.

cutaneous receptors to specific sensory cortical areas. Collectively, the sensory organs, thalamus, and sensory cortices form the **sensory systems.** These systems will be discussed in Chapter 8. Other subcortical areas, such as the **hypothalamus** (see Fig. 7–4a), control the secretion of hormones for growth, reproduction, metabolism, and temperature regulation. Areas of the **brain stem** (see Fig. 7–4a) play important roles in several functions. By receiving information from the motor cortex and transmitting it to the spinal cord, the brain stem aids in maintaining body posture and body balance, as in riding a bicycle. It also transmits information to and from the **cerebellum,** another area involved in coordination and involuntary control of skeletal muscle. Collectively, these areas form the **motor systems,** which will be discussed in Chapter 9. Areas within the brain stem also control the breathing and heart rates, blood pressure and flow, digestive system secretion and movement, and other functions in which changes occur automatically and are not typically under conscious control. These areas of the brain stem are a part of the **autonomic nervous system (ANS).**

The **spinal cord** is connected to the brain at the base of the skull and extends within the vertebral canal to the level between the first and the second lumbar vertebrae. The outer or cortical part of the spinal cord is made up of **white matter** (insulated nerve fibers), and the inner or medullary part of the spinal cord is made up of **gray matter** (nerve cell bodies and noninsulated nerve fibers). The white matter is arranged into **dorsal (posterior) white, lateral white,** and **ventral (anterior) white** (Fig. 7–5) and consists of bundles of ascending nerve fibers that relay sensory information to the brain and bundles of descending nerve fibers that carry motor information originating in the brain. The gray matter appears in the form of the letter *H,* with a **dorsal (posterior) horn** and a **ventral (anterior) horn** in each half connected at the center of the cord by commissural gray (the crossbar in the letter *H*). In the thoracic and upper lumbar segments of the spinal cord a **lateral horn** is present. Although anatomically continuous, the spinal cord may be thought of as consisting of 31 segments: 8 **cervical** (neck), 12 **thoracic** (upper back), 5 **lumbar** (lower back), 5 **sacral** (lower back), and 1 **coccygeal.**

The Peripheral Nervous System

The PNS consists of bundles of nerve fibers, called *nerves,* bundles of nerve cell bodies, called *ganglia,* and sensory receptors. The PNS transmits sensory information to the CNS, and motor information away from the CNS. Anatomically, the PNS

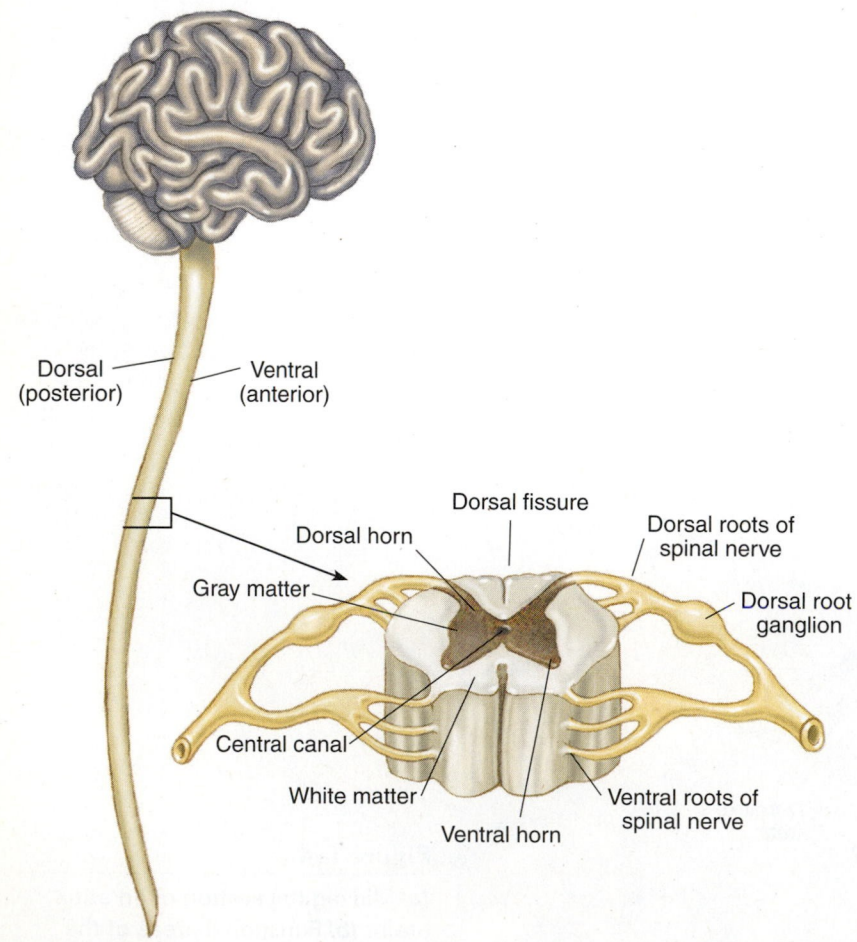

Dorsal (posterior)

Ventral (anterior)

Dorsal fissure

Dorsal horn

Dorsal roots of spinal nerve

Gray matter

Dorsal root ganglion

Central canal

White matter

Ventral horn

Ventral roots of spinal nerve

Figure 7–5

Structure of the spinal cord. A section taken from the thoracic spinal cord. The dorsal (posterior) and ventral (anterior) roots join to form the spinal nerve. The dorsal root ganglion contains the cell bodies of sensory neurons. The gray and white matter are composed chiefly of nerve cell bodies and myelinated axons, respectively.

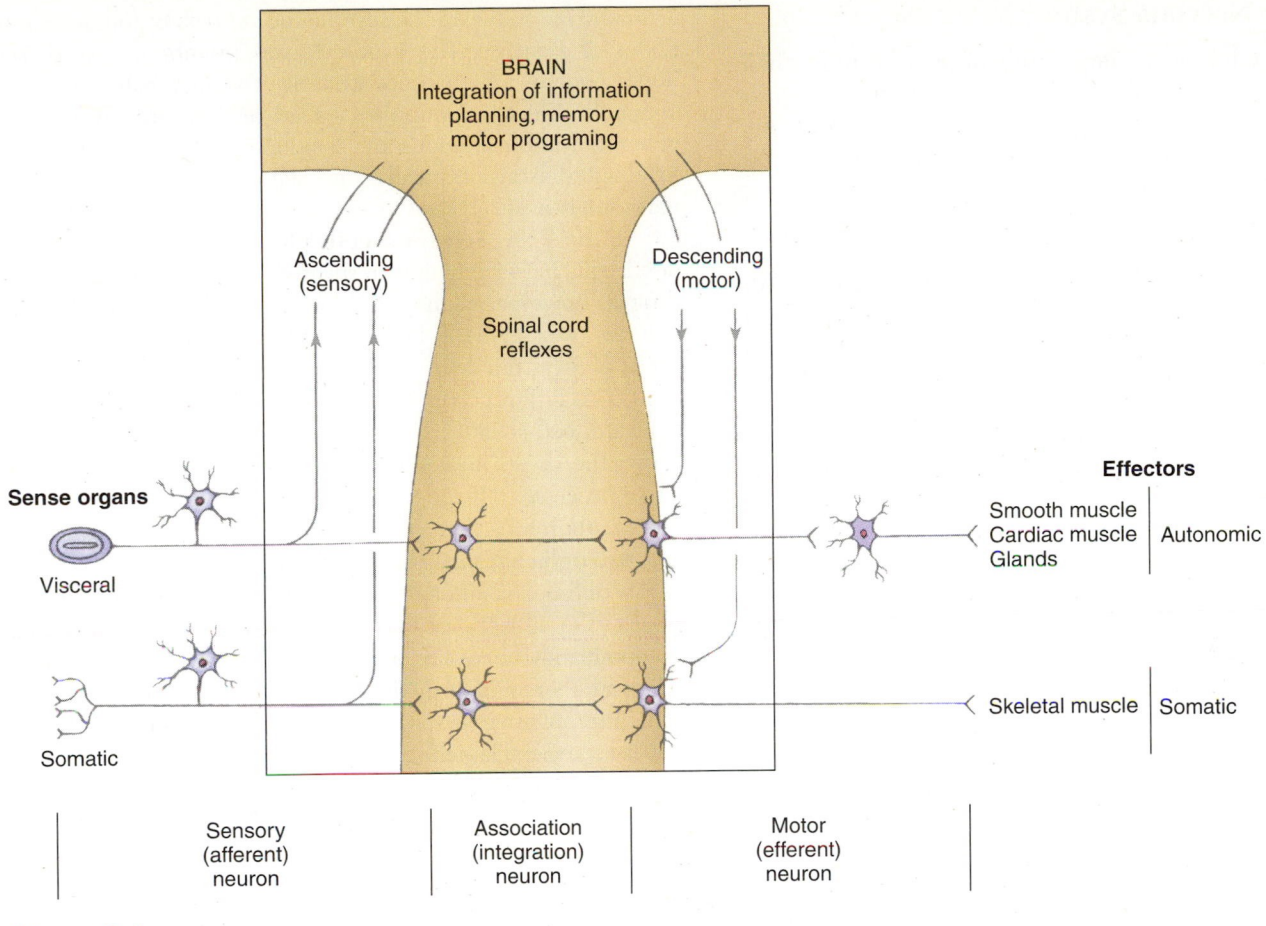

BRAIN
Integration of information
planning, memory
motor programing

Ascending
(sensory)

Descending
(motor)

Spinal cord
reflexes

Effectors

Sense organs

Visceral

Smooth muscle
Cardiac muscle | Autonomic
Glands

Somatic

Skeletal muscle | Somatic

Sensory
(afferent)
neuron

Association
(integration)
neuron

Motor
(efferent)
neuron

Figure 7–6

Functions and flow of information through various components of the nervous system.

contains 12 pairs of cranial nerves connected to the brain, and 31 pairs of spinal nerves connected to the spinal cord.

A pair of spinal nerves, one right and one left, are attached to each segment of the spinal cord by ventral and dorsal roots (see Fig. 7–5). The **dorsal root** contains sensory nerve fibers and is easily identified because of the **dorsal root ganglion,** a collection of sensory neuron cell bodies. The **ventral root** contains motor fibers attached to neuron cell bodies located in the spinal gray matter. Each spinal nerve is a mix of sensory and motor fibers that supply structures in the same side of the body as the origin of the spinal nerve.

Functionally, the PNS is comprised of a somatic portion and an autonomic portion (Fig. 7–6). Within the **somatic nervous system,** sensory nerves transmit information from somatosensory organs, such as pain receptors in the skin, to the CNS. In the CNS, the sensory information is processed and an appropriate motor response may be generated. Somatic motor nerves transmit instructions about appropriate responses to skeletal muscles.

Within the ANS, sensory nerves transmit information about the condition of internal organs to the CNS, and motor nerves instruct glands and involuntary muscles about appropriate responses. In the ANS motor nerves are assigned to one of two divisions: (1) the sympathetic division or (2) the parasympathetic division. **Sympathetic nerves** cause effector responses that generally prepare the body for stressful situations and help to maintain homeostasis as the body responds to an acute stress. For example, an increase in sympathetic activity elevates heart rate, dilates the pupils of the eye, increases airway diameters, increases blood flow to skeletal muscles, and causes numerous other visceral changes. **Parasympathetic nerves** help to maintain homeostasis on a daily routine basis, such as preparing the body for processes of food consumption and digestion. An increase in parasympathetic activity increases blood flow to the stomach and intestines, increases gastrointestinal motility and secretion, increases the formation of watery saliva, and also alters function in many other internal organs. Together, the sympathetic and parasympathetic divisions of the ANS control nearly all of the body's internal organs in a manner more complex than inferred here. Consequently, all of Chapter 10 is devoted to a discussion of autonomic function.

Cells of the Nervous System

The nervous system is a complex organization of more than 1 trillion cells. Many cells rapidly communicate with each other by producing electrical impulses that are sent from one cell to another. The remainder of this chapter focuses on the basic mechanisms by which cells in the nervous system operate and communicate with each other.

The nervous system is composed of two types of cells: (1) neurons (or nerve cells) and (2) neuroglia (or glial cells). Neurons are designed to transmit information rapidly from one cell to another. They are responsible for the functions normally associated with the nervous system, such as sensory perception and movement. Neuroglia, on the other hand, help to maintain the environment surrounding neurons and aid in their ability to transmit information rapidly.

Neuroglia

The five basic types of **glial cells** in the nervous system are (1) oligodendroglia, (2) Schwann cells, (3) astroglia, (4) microglia, and (5) ependymal cells (Fig. 7–7).

Oligodendroglia are located in the brain and the spinal cord. They wrap themselves around axons to form **myelin**, layers of lipid membrane, which insulate the axon to prevent the passage of ions through the axonal membrane (see Fig.

7–7a and b). Between the myelinated regions of the axon are nodes of Ranvier. The exposed axon membranes at these nodes contain a high concentration of voltage-gated sodium channels. Electrical impulses called **action potentials** are generated in these nodes and conducted along the length of myelinated axons. Typically, one oligodendroglial cell myelinates many axons in the CNS.

In the PNS, **Schwann cells** play a similar role to oligodendroglia in forming myelin sheaths. In contrast to oligodendroglia, however, in the PNS, one Schwann cell myelinates only an axon segment, between two nodes of Ranvier. Hundreds of Schwann cells may be required to myelinate a single axon. This difference in geometry is not trivial; the one-to-one Schwann cell sheath has an intact basement membrane that allows the sheath to act as a regeneration tube for damaged nerve processes. Oligodendroglia sheaths cannot guide regeneration; therefore, damage in the CNS is usually irreparable.

Astroglia are star-shaped cells that exhibit the greatest diversity of function among glial cells in the CNS. One type of astroglia, the fibrous astrocyte, is found in areas containing predominantly nerve fibers. It is called *fibrous* because of the large number of intermediate filaments found in these glial cells. Protoplasmic astrocytes are similar to fibrous astrocytes but have fewer filaments. They are found in areas containing predominantly nerve cell bodies, dendrites, and synapses. Both

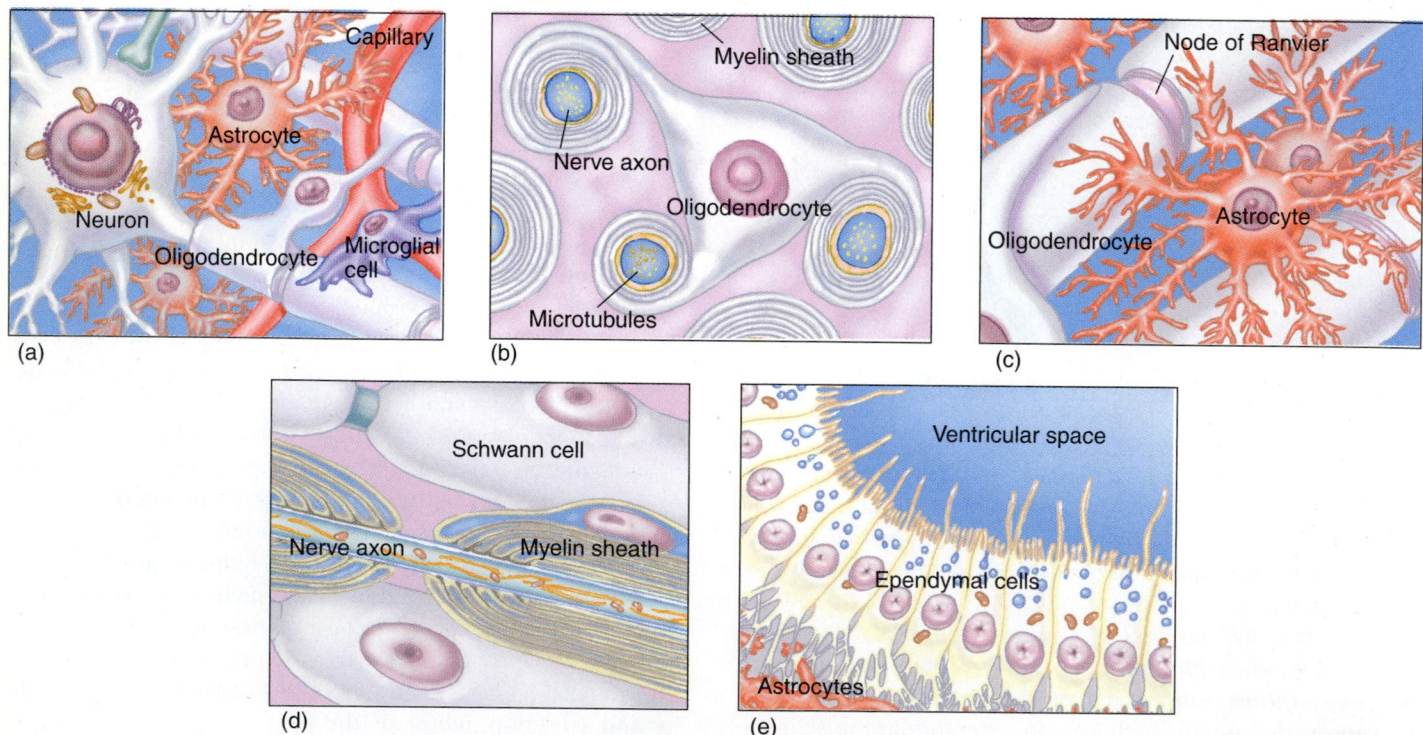

Figure 7–7

Types of glial cells and their functions in protecting neurons. *(a)* An astrocyte, an oligodendrocyte, and a microglial cell. *(b, c)* Oligodendroglia form myelin sheaths around axons. *(c)* Astroglia remove cellular debris from the central nervous system. *(d)* Schwann cells form myelin sheaths around peripheral axons. *(e)* Ependymal cells help to form and circulate cerebrospinal fluid.

CURRENT CONCEPTS IN PHYSIOLOGY

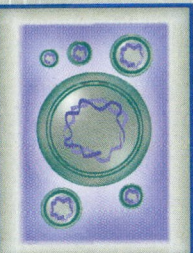

Gene Therapy for Brain Tumors

Oncogenes and tumor-suppressor genes are involved in the process of cell division during development. Under normal conditions, these genes are highly regulated to control cell proliferation. Recently, researchers have shown that the oncogene *c-myb* encodes a nuclear phosphoprotein involved in the regulation of cell proliferation and differentiation. The expression of the *c-myb* gene is greatest in cells as they prepare to enter and traverse the GI/S transition phase of the cell cycle. Its expression subsequently decreases as cells terminally differentiate. C-MYB protein binds to specific DNA sequences and transactivates transcription of *DNA polymerase-α* and *cdc2*, two genes that are crucial for

Human glioblastoma cells in tissue culture. (From Tabuchi, K., and Nishimot, A. Atlas of Brain Tumors. *New York, Springer-Verlag, 1988.*)

DNA synthesis. *C-MYB* mRNA levels are exceptionally high during the development of the nervous system and in cancerous glial cells called *glioblastomas*. One approach that researchers are now using in attempts to suppress the proliferation of glioblastoma cells is gene therapy with antisense oligonucleotides.

Antisense oligonucleotides are specific nucleic acid sequences that can interfere with the transcription of DNA into mRNA, and the translation of mRNA into protein. It has been estimated that antisense oligonucleotides 15 to 18 bases in length are sufficient to bind to specific genes and disrupt their function. Researchers have now developed an 18-base antisense oligonucleotide to the initiation region of the *c-myb* genes for codons 2 to 7. The administration of this antisense to human glioblastoma cells has suppressed their ability to proliferate in tissue culture and in mice that received injections of the tumor cells. Ultimately, this type of gene therapy may be attempted in treating patients with brain tumors.

---

types form "glial end-feet" on blood vessels (see Fig. 7–7*a*) and help to provide structural support for neurons and other elements of nervous tissue. During fetal brain development, astrocytes help to guide neurons to make proper connections.

During injury to the brain, another type of astrocyte, called a *reactive astrocyte*, appears. It removes degenerating debris by the process of phagocytosis. Reactive astrocytes proliferate after the degenerated neurons have been removed and, in conjunction with fibroblasts, form glial scars that are typically seen after brain injury.

Astrocytes regulate the concentration of potassium ions (K^+) in the extracellular space of the brain by absorbing and redistributing them to other astrocytes. This is accomplished by the transport of K^+ ions through the gap junctions that link networks of glial cells. This type of spatial buffering is needed to maintain the constant extracellular environment of potassium ion that is crucial to the function of neuron-generated electrical impulses.

Astrocytes also control the concentration of chemicals called **neurotransmitters,** after their release by nerve cells during the process of **synaptic transmission.** This uptake by glial cells is needed to terminate synaptic transmission and prevent the flooding of neurotransmitters between nerve cells.

Astrocytes are also thought to be involved in the maintenance of interstitial hydrogen ion concentration, which is required for the optimal function of many cellular processes. As in other organ systems, the enzyme **carbonic anhydrase** plays a crucial part in acid-base balance. In the CNS, this enzyme is found mainly in astrocytes and oligodendroglia.

Microglia are defensive cells found in the nervous system near blood vessels. Typically, their occurrence is somewhat rare in healthy brain tissue. After injury, however, they are found to migrate to the site of damage, where their main functions are the destruction of bacteria and the removal of cellular debris by phagocytosis. Unlike other glial cells, microglia originate from monocytes in the bone marrow and migrate to the CNS.

Ependymal cells are epithelial cells that line the cerebrospinal fluid–filled ventricles of the brain and central canal of the spinal cord. Many of the ependymal cells are ciliated. They contribute to the formation of cerebrospinal fluid and help to circulate cerebrospinal fluid. Ependymal cells also serve as neural stem cells with the potential of developing into other types of glial cells or even neurons.

Neurons

Neurons have four functionally distinct regions: (1) the soma, (2) dendrites, (3) axon, and (4) axon terminals (Fig. 7–8). The **soma** is the cell body of a neuron and its metabolic center. It contains the components necessary to fabricate and package proteins used in other parts of the cell to maintain a variety of cellular functions. **Dendrites** are usually short, branchlike structures that extend out from the cell body. The surfaces of the dendrites and soma are where most neurons receive information from other nerve cells.

The **axon** is a long fiberlike structure that extends from the soma to make contact with other **nerve** cells, or effectors,

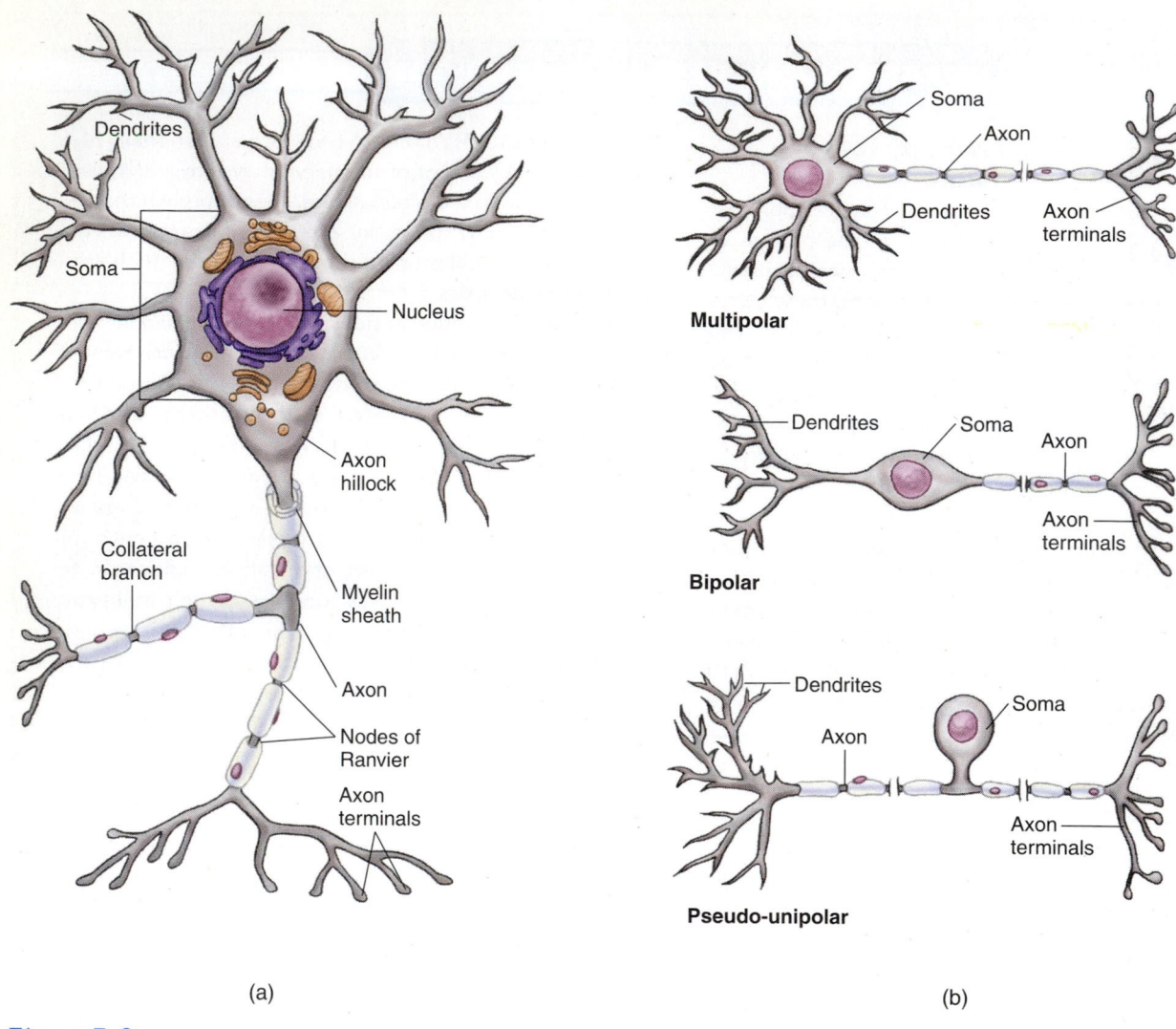

(a)

(b)

Figure 7–8

(a) The general structure of a neuron. *(b)* Types of neurons. Multipolar neurons have several short dendrites and one long axon, whereas bipolar neurons have one axon and one dendrite. Pseudounipolar neurons have one short branch connected to an axon having a dendritic end and an axon terminus.   *(Petit Format/Science Source.)*

such as muscle fibers. The axon hillock or **initial segment** of the axon is where action potentials are generated and transmitted down the axon as nerve impulses. **Myelinated** nerve fibers are electrically insulated from extracellular fluid by segments of myelin. **Unmyelinated** nerve fibers lack the myelin insulation. Near its distal end, the axon branches many times, forming **axon terminals,** which make contact with and transmit information to other nerve cells at junctions called **synapses.**

Types of Neurons. Neurons can be classified in two ways: (1) by their structure, as multipolar neurons, bipolar neurons, or pseudounipolar neurons, and (2) by function, as sensory neurons, motor neurons, or interneurons.

A **sensory neuron,** also called an **afferent** ("going toward") **neuron,** carries information about external or internal stimuli from sensory receptors to the CNS. A **motor neuron,** also called an **efferent** ("leading away") **neuron,** carries motor instructions from the CNS to muscle cells or secretory cells. An **interneuron,** found in the intermediate zone of the spinal cord, or gray matter of the brain, for example, often connects sensory and motor neurons and integrates their functions, as in spinal cord or brain stem reflexes. Interneurons are also found between neurons within nuclei and centers in the CNS (see Fig. 7–6).

A **multipolar neuron** has several short dendrites and one long axon (see Fig. 7–8). Multipolar neurons are the most common type of neuron. A **bipolar neuron** has one axon and one dendrite, on opposite sides of the soma. Examples can be found in the retina and in the olfactory cortex, the part of the brain that receives information regarding the sense of smell. A **pseudounipolar neuron** has two fibers that fuse a short distance from the cell body. Dorsal root ganglion cells are of this type. One fiber comes from a peripheral sensory receptor, and the other fiber enters the CNS. Motor neurons are typically multipolar neurons, whereas sensory neurons are often pseudounipolar.

Specialized Neuronal Structures. As with other types of cells, the nucleus of nerve cells contains the genetic information needed for the synthesis of specific proteins. Some proteins, for example, are involved in the formation of new functional connections with other cells as a result of signals from afferent inputs or during the process of regeneration.

In neurons, the nucleus is quite large. A substantial portion of the genetic information contained within the genome is continually transcribed by neurons. Based on hybridization studies, it is estimated that one third of the genome is actively transcribed, producing more mRNA than any other organ in the body. Because of the high level of transcriptional activity, the nuclear chromatin is in the form of **dispersed chromatin** (Fig. 7–9). In contrast, the chromatin in non-neuronal cells in the brain, such as glia, is found in clusters on the internal face of the nuclear membrane. The neuronal nucleus also contains one or two nucleoli, many of which have attached electron-dense bodies. One of the dense bodies associated with the nucleolus in females is the Barr body, which is the condensed chromatin of the inactive X chromosome.

Most of the proteins formed by free ribosomes and polyribosomes remain within the cell, whereas proteins formed by rough endoplasmic reticulum (ER) are exported. Polyribosomes and rough ER (RER) are found predominantly in the soma of neurons. Axons contain no RER and are unable to synthesize proteins. Dendrites, on the other hand, contain ribosomes and RNA and are thought to play an important part in the modulation of dendritic function.

The smooth ER is involved in the intracellular storage of calcium. Smooth ER within neurons binds calcium and

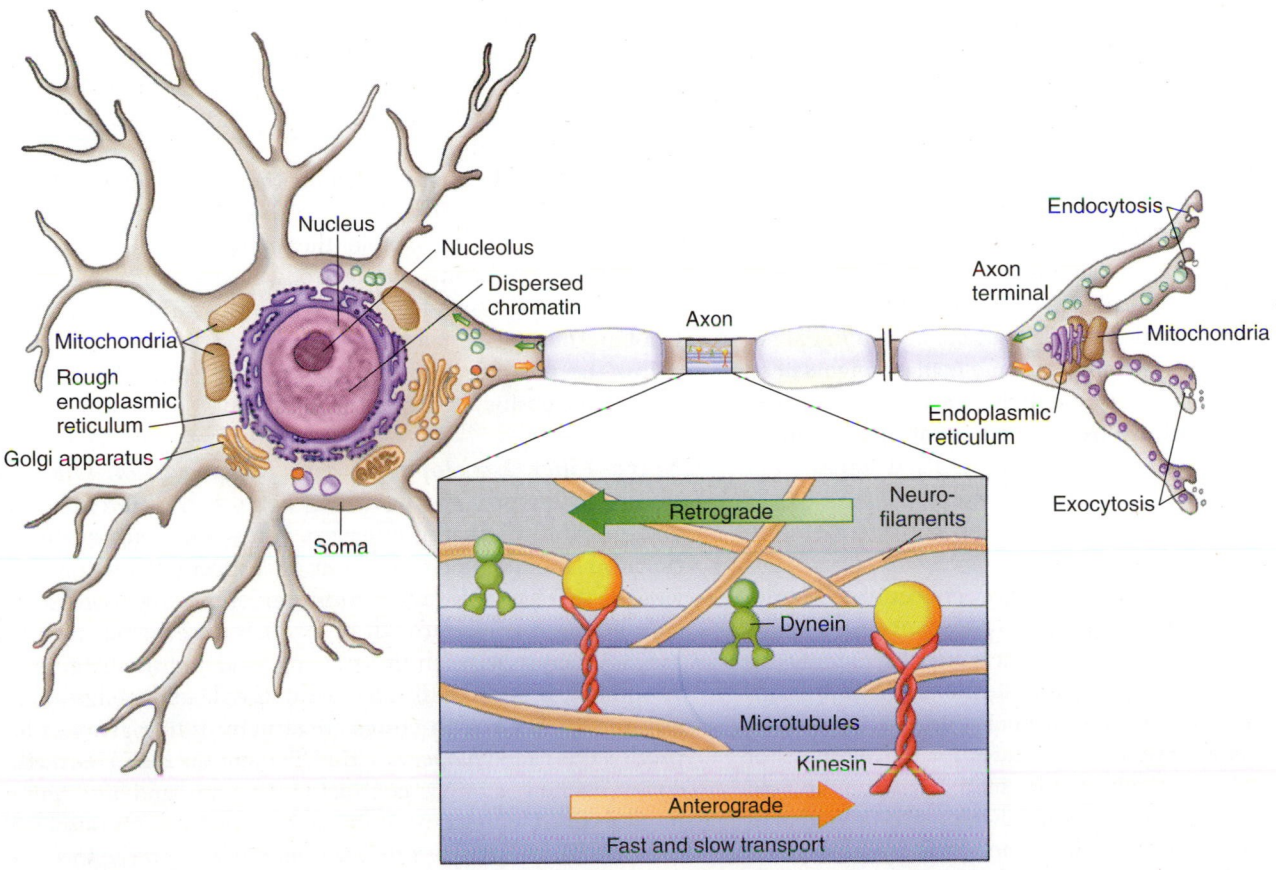

Figure 7–9

The cytoskeleton structure of an axon, consisting of microtubules, microtubule-associated proteins dynein and kinesin, and neurofilaments that aid in the transport of organelles and secretory products to and from the axon terminal. Materials move by anterograde fast transport from the soma to the axon terminal and by retrograde transport from the axon terminal to the soma.

maintains the intracellular cytoplasmic concentration at a low level of $0.1\mu M$. Prolonged elevation of intracellular calcium has been shown to lead to disruption of cytoskeletal elements, such as microtubules, and to cause neuronal degeneration and death.

In neurons, the Golgi apparatus is found in the soma. As in other types of cells, this structure is engaged in the terminal glycosylation of proteins synthesized in the RER. The Golgi apparatus takes proteins produced for exportation in the RER and forms protein-containing vesicles. These vesicles are released into the cytoplasm, where some are carried by axoplasmic transport to the axon terminals (see Fig. 7–9).

The Neuronal Cytoskeleton. The irregular, functionally related shape of the neuron is maintained by the internal framework of the cytoskeleton. The neuronal cytoskeleton is made of microfilaments, neurofilaments, and microtubules (see Fig. 7–9). **Microfilaments** are composed of actin, a contractile protein commonly found in muscle. They are 4 to 5 nm in diameter, and, in neurons, they are found in developing extensions of the neuron, such as spines of dendrites and the growth cone of an axon. **Neurofilaments,** on the other hand, are found in established axons and dendrites and are thought to provide structural rigidity. They are not found in the growing tips of axons and dendritic spines, which are more dynamic structures. Neurofilaments are approximately 10 nm in diameter and are composed of three proteins (70 kDa, 140 kDa, and 220 kDa in size). The core of the filament consists of the 70-kDa protein, similar to intermediate filaments in other cells. The two other neurofilament proteins are thought to be side arms that interact with microtubules.

Microtubules are responsible for the rapid movement of material in axons and dendrites. They are 23 nm in diameter and are composed of tubulin with molecular weights of 52 and 56 kDa. In neurons, microtubules have accessory proteins called **microtubule-associated proteins (MAPs).** Dendrites have high-molecular-weight MAPs, whereas axons have low-molecular-weight MAPs; these two types of MAPs are thought to determine whether material is distributed to dendrites or to axons.

In neurons, mitochondria are highly concentrated in the region of axon terminals. They produce adenosine triphosphate (ATP), which is required as a source of energy for many cellular processes. In the axon terminal, the mitochondria provide not only a source of energy for the process of synaptic transmission but also substrates for the synthesis of certain neurotransmitter chemicals, such as the amino acid glutamate. In addition, enzymes involved in the degradation of other types of neurotransmitter chemicals are embedded in the outer membrane of the mitochondria. Thus, the role of mitochondria in the neuron is multifunctional and varied.

Axonal and Dendritic Transport Mechanisms. The shapes of most cells in the body are relatively simple compared with the complexity of neurons, with their elaborate axonal and dendritic processes. Because of the length of nerve cell processes, however, neurons have developed specialized mechanisms to transport proteins, organelles, and other cellular material along the length of axons and dendrites needed for the maintenance of the cell. These transport mechanisms are capable of moving cellular components along fiber processes in an **anterograde** direction (away from the soma), or in a **retrograde** direction (toward the soma; see Fig. 7–9). **Kinesin,** a microtubule-associated protein, is involved in anterograde transport of organelles and vesicles via the hydrolysis of ATP. Retrograde transport, on the other hand, is mediated by another microtubule-associated protein called **dynein.**

In the axon, anterograde transport occurs at both a slow and a fast rate. The rate of **slow axonal transport** is 1 to 2 mm/day. Structural proteins, such as actin, as well as components of neurofilaments and microtubules, are transported at this speed. The rate of **fast axonal transport** is 400 mm/day. Fast transport mechanisms are utilized by organelles, vesicles, and membrane glycoproteins needed at the synaptic terminal. Another feature that distinguishes fast from slow transport mechanisms is that fast transport is dependent on oxidative metabolism, Ca^{2+}, glucose, and ATP.

In dendrites, anterograde transport occurs at a rate of 0.4 mm/day, and, like fast transport, it also requires ATP. Dendritic transport appears to be involved in the movement of ribosomes and RNA, suggesting that protein synthesis occurs within dendrites.

In retrograde axonal transport, material is moved from terminal endings to the cell body. This provides a mechanism for the cell body to sample the environment around its synaptic terminals. In some neurons, maintenance of synaptic connections depends on the **transneuronal** transport of trophic substances, such as nerve growth factor (NGF), across the synapse. After transport to the soma, NGF activates mechanisms for protein synthesis.

Nerve Fiber Development and Regeneration. One of the major features distinguishing nerve cell differentiation and growth from that of other cell types is the outgrowth of the axon from the nerve cell body, in a specific direction and along a specific pathway, to form synaptic connections with specific targets. The growth of axons is determined largely by interactions between the growing axon and the tissue environment. The growth cone is at the leading edge of a growing axon. **Growth cones** are structures that give rise to protrusions called *filopodia* ("tiny filamentous feet"). Growth cones contain actin, a contractile protein, and are quite motile, with filopodia extending and retracting at a rate of 6 to 10 μm/min. Newly synthesized membranes in the form of vesicles are also found in the growth cone and fuse with it as it extends. As the growth cone elongates, microtubules and neurofilaments are added to the distal end of the fiber and partially extend into the growth cone. They are transported to the growth cone via the process of slow axonal transport.

The direction of axonal growth is directed in part by **cell adhesion molecules (CAMs).** CAMs are glycoproteins that are expressed on cell surfaces to promote cell adhesion (see Chapter 3). Neuron-glia-CAM (Ng-CAM) is expressed in post-mitotic neurons and is particularly prominent in growing neurites (axons and dendrites); these neurites migrate along certain types of glial cells that provide a guiding path to target sites. The secretion of tropic factors by target cells also influences the direction of axon growth. Once the proper target site is reached and synaptic connections are formed, the processes of growth cone elongation and migration are terminated.

During development, axon terminals are found on target cells in excess of the numbers seen after the nervous system has matured, indicating a loss of synaptic contacts after development. This loss of synaptic contacts is a result of a selection process whereby the most active inputs predominate and survive at the expense of less active synaptic contacts, thereby optimizing neural development and function.

Growth cones are found during not only development but also the regeneration of nerve fibers. When a nerve fiber is cut, the distal end degenerates while the fiber segment proximal to the soma develops growth cones for elongation and extension. Regeneration occurs at a rate of 1 mm/day, the rate of slow axonal transport.

Ion Channels of Nerve Cell Membranes. Generation and conduction of the nerve impulse requires an electrical potential difference, or voltage, called the **membrane potential**, between the cytosol and the extracellular fluid. The plasma membrane of the neuron is polarized; that is, the electrical behavior of the inner and outer surfaces of the plasma membrane are opposites. The inner or cytosol surface of the membrane behaves as though it were lined with negative charges and the outer (extracellular fluid) surface as though it were covered with positive charges. Although the *number* of positive and negative charges in the intracellular fluid and the extracellular fluid are equal, their *distribution* is not equal or uniform. It is the unequal distribution of charged particles (ions) near inner and outer surfaces of the plasma membrane that gives rise to the membrane potential. The unequal distribution of ions is a consequence of the behavior of ion channels in the membrane; ion concentration gradients across the membrane; and membrane carriers, which transport ions and thereby maintain the charge separation reflected by the value of the membrane potential.

Certain ions (Na^+, K^+, Cl^-, and Ca^{2+}) can cross the membrane only through protein pores in the membrane that form **ion channels**. These channels typically allow only specific ions to pass while blocking others because of their size, charge, and so on. The three basic types of ion channels are (1) passive, (2) chemically activated, and (3) voltage-activated channels (Fig. 7–10).

Passive ion channels are found in membranes throughout all areas of the nerve cell. Each passive channel is identified according to the specific ion that it allows through (e.g., Na^+ channel, K^+ channel, Cl^- channel, and Ca^{2+} channels).

Chemically activated ion channels are located predominantly on dendrites and the soma. These channels are

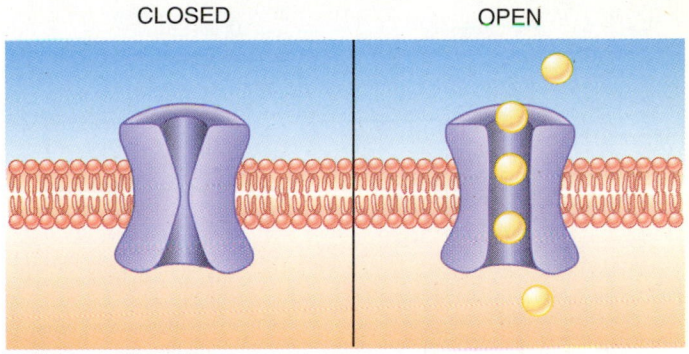

(a) Ion channel

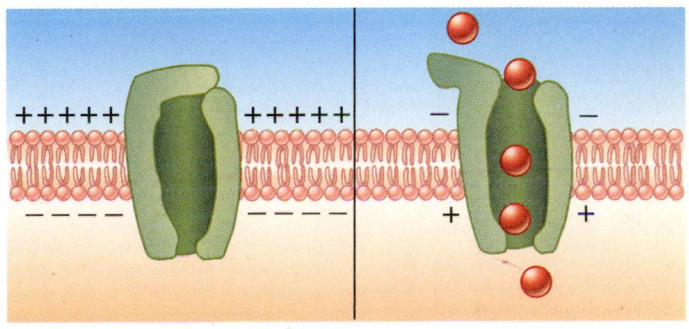

(b) Voltage-gated channel

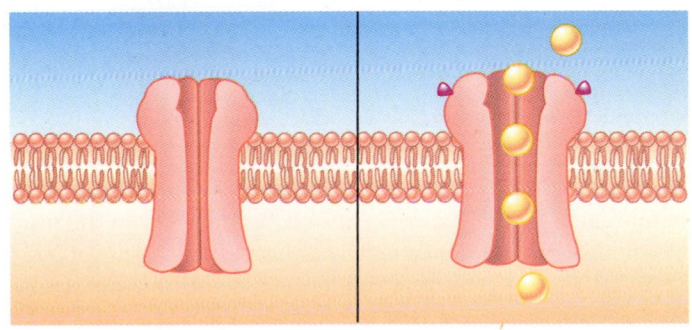

(c) Ligand-gated channel

Figure 7–10

Types of ion channels. *(a)* Passive channels are specific to individual ions. *(b)* Voltage-activated channels open or close with specific changes in the membrane potential. *(c)* Chemically activated channels depend on activation by a chemical transmitter.

generally closed by "gates" that prevent the flow of ions through the membrane. Chemicals called **neurotransmitters** and **neuromodulators** bind to receptor sites on these protein channels and open the channel gate, permitting the flow of ions through the channel.

Voltage-activated ion channels, primarily found in the membrane of the axon, are opened when the membrane voltage reaches a crucial or activating value. They play a key role in generating and conducting the nerve impulse.

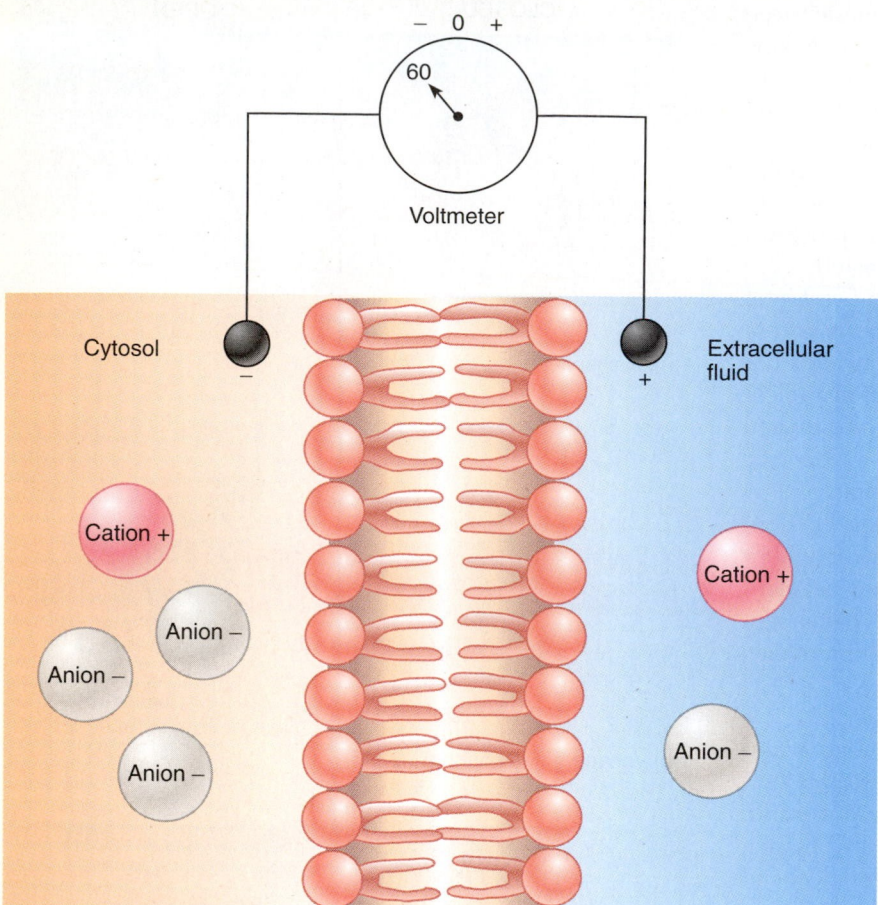

Figure 7–11

The membrane potential (E_m) of -60 mV indicates that the inside of the cell is more negative than is the extracellular fluid.

Electrical Properties of Nerve Cell Membranes. The membrane potential exists because of a difference in the distribution of positive and negative charges across the membrane. These charges are attributed to the **cations** (positively charged ions) and **anions** (negatively charged ions) that occur on each side of the membrane. When there is a difference in the net charge between the inside and outside of a cell, an **electrical potential (E),** or voltage, is established. The electrical potential of a plasma membrane is measured in units called **volts (V)** and is determined by the charge difference between two points. The charge at two points in the extracellular fluid is the same, for example, so the charge difference and electrical potential are therefore zero. In contrast, because a charge difference exists between a point within the cell and a point in the extracellular fluid, an electrical potential is established. In many nerve cells, the resting membrane potential (E_r) is approximately 0.060 V, or 60 millivolts (mV; Fig. 7–11). The convention that has been adopted for measuring electrical potential differences across the membrane is the use of the extracellular voltage as a reference. A minus sign thus indicates that the inside of the cell is more negative than is the extracellular fluid.

The electrical potential can be viewed as the electrical equivalent to pressure. A difference in pressure between two points causes water to flow from the point of higher pressure to

the point of lower pressure. In a similar fashion, an electrical potential results in the movement of ions or electrons. In physical systems, the movement of electrons in a copper wire is the basis of **current flow.** In biological systems, current flow is due to the movement of ions. This movement occurs when anions are attracted to positively charged regions and cations to negatively charged regions. The flow of ions is called **ionic current (I)** and is measured in units called **amperes (A).** The ionic sodium current, I_{Na^+}, that moves through a single channel, for example, is approximately 12×10^{-12}A, or 12 picoamperes.

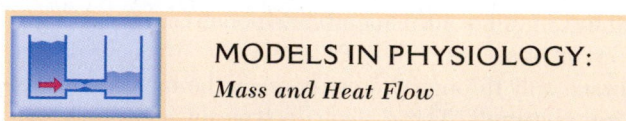

MODELS IN PHYSIOLOGY:
Mass and Heat Flow

The ease with which ions can flow through an ion channel is called **conductance (g).** Units of conductance are called **siemens (S).** Membrane conductance is inversely related to membrane **resistance (R).** In other words, conductance is the reciprocal of resistance: g = 1/R. As resistance decreases, conductance increases, and vice versa. Several factors affect the conductance of an ion through a channel. If the channel is large in comparison to the size of the ion, the ion passes with greater ease and the channel therefore has a lower resistance or higher conductance. A smaller channel increases the probability of the ion colliding with the walls of the channel, making it more difficult for the ion to pass. Smaller channels generally have a lower conductance. Other factors, such as the shape of the channel, the distribution of charges in the channel, and the charge of the ion, also affect conductance through the channel. Each type of channel has a specific conductance for its associated ion.

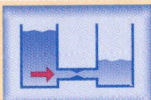

MODELS IN PHYSIOLOGY:
Mass and Heat Flow

The relationship between electrical potential, ionic current, and conductance is given by **Ohm's law.** This law states that the rate of ion flow through a channel is directly proportional to the conductance of the channel and the magnitude of the electrical potential, $I = g \times E$. Larger electrical potentials and greater channel conductances produce a faster flow of ions across the membrane. This is similar to the flow of water through a tube. Greater pressures and tubes with bigger diameters allow more water to flow.

The overall effect of all of the channels for a particular ion is called the **membrane conductance** for that ion. Membrane conductance of an ion, therefore, depends on the number or density of a particular type of channel within a region of the membrane and how many of those channels are open (Fig. 7–12). Membrane conductance can vary from one region of a nerve cell to another. The membrane conductance for sodium ions, for instance, is greater in the region of the initial segment of the axon and nodes of Ranvier than in other areas of the cell. These regional differences in Na^+ conductance determine where and how the nerve impulse is generated and conducted by the neuron.

THE NEURON: EXCITATION AND CONDUCTION

 How are membrane potentials generated?

Origin of the Plasma Membrane Electrical Potential Difference

Nearly all living cells in the human body are characterized by the presence of an electrical potential difference between the inside and outside of the cell. Using the neuron, this may be demonstrated by placing one electrode (−) within the cell, a second electrode (+) near the outer surface of the cell, and connecting the wires from the electrodes to a voltage-measuring device, such as a voltmeter (Fig. 7–13). While the cell is at rest, that is, not conducting an impulse, a stable voltage of approximately −75 mV, inside negative with respect to the outside, is recorded. This potential difference, measured across the plasma membrane of a resting cell, is called the *resting membrane potential* (E_r). The magnitude of the resting potential varies from one cell type to another but it is constant for a given cell. If the cell is lethally poisoned, the potential at first diminishes and then disappears while the cell is structurally intact, suggesting that the maintenance of the resting potential is dependent on cellular metabolism and biological properties of the membrane.

The basis of the resting potential is the separation and segregation of various small ions and charged macromolecules by the plasma membrane, resulting in an ionic imbalance between the intracellular and extracellular fluids. Table 7–1 indicates the imbalance that exists between the intracellular and extracellular concentrations of the major ions potassium, sodium, and chloride. The values in Table 7–1 are representative values that differ in various cell types;

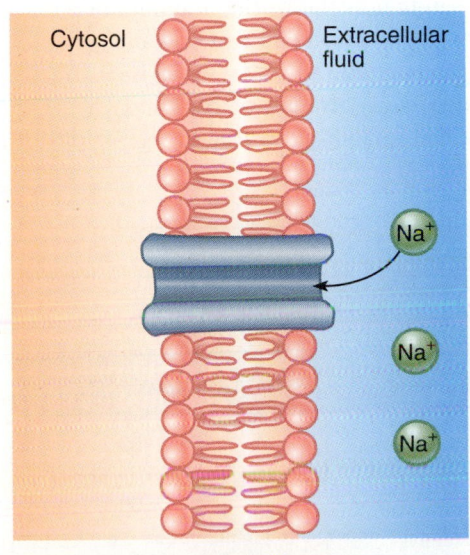

Low conductance
(a)

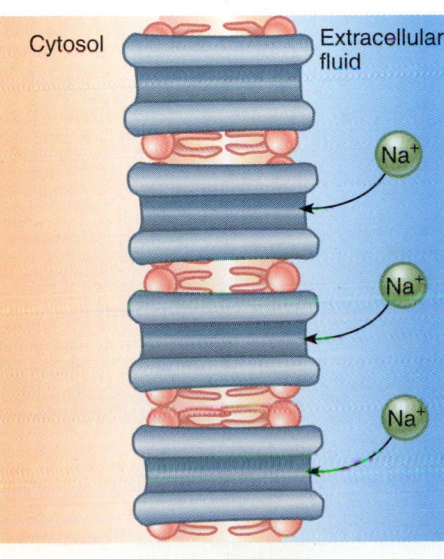

High conductance
(b)

Figure 7–12

Membrane conductance for a particular ion depends on the number of open ion channels present in the membrane. *(a)* Low conductance. *(b)* High conductance.

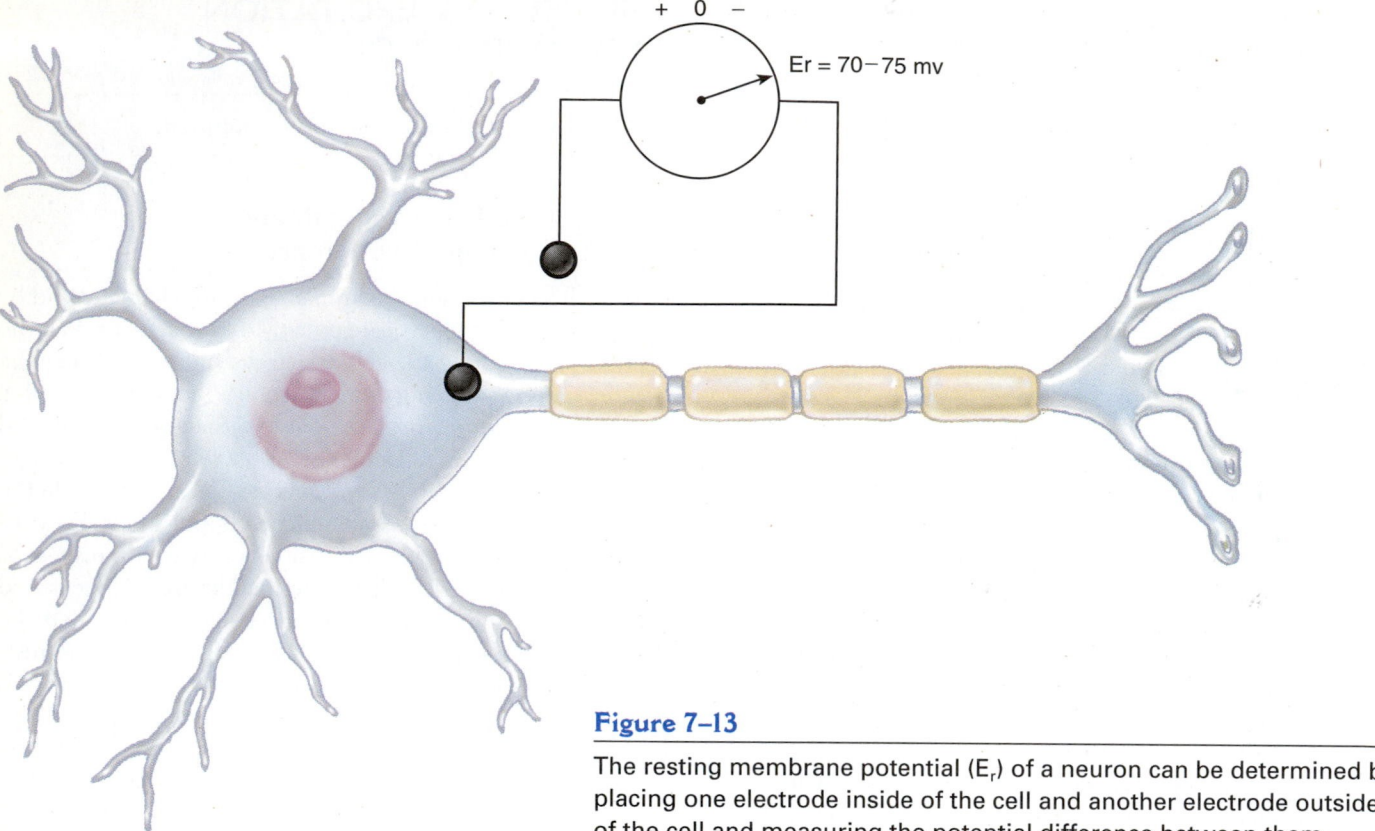

Figure 7–13

The resting membrane potential (E$_r$) of a neuron can be determined by placing one electrode inside of the cell and another electrode outside of the cell and measuring the potential difference between them.

however, in most cells, there is much more potassium inside than outside, and much more sodium outside than inside of the cell. Organic anions, which are primarily proteins are confined to the interior of the cell. Many other ions also are pres-ent in both intracellular and extracellular fluids, and imbalances exist for these as well; however, the sum of all negative charges, as determined by chemical techniques, equals the sum of all positive charges, outside of the cell as well as within the cell, so that each compartment (intracellular and extracellular) appears to be electrically neutral. Nevertheless, as mentioned earlier, in the immediate vicinity of the membrane there exists an actual difference, too small to be detected chemically, in the distribution of positive and

negative ions. More intracellular negative ions are found near the inner surface of the plasma membrane, and more extracellular positive ions are found near the outer surface. The positive charges near the outer membrane surface are attracted to the negative charges near the inner surface but remain separated by the plasma membrane. A membrane that maintains a separation of positive from negative charges is said to be polarized.

To see approximately how the resting membrane potential arises, consider a hypothetical cell that contains a higher concentration of K$^+$ and organic anion An$^-$ than the extracellular fluid and a lower concentration of Na$^+$ and Cl$^-$ (Fig. 7–14). Suppose that the membrane will allow K$^+$ to diffuse

TABLE 7–1

Representative Intra- and Extracellular Ion Concentrations and Equilibrium Potentials Calculated by the Nernst Equation

Ion	Extracellular Concentration (mM)	Intracellular Concentration (mM)	Equilibrium Potential (mV)
K$^+$	5.5	150	−90
Na$^+$	150	15	+60
Cl$^-$	125	9	−70

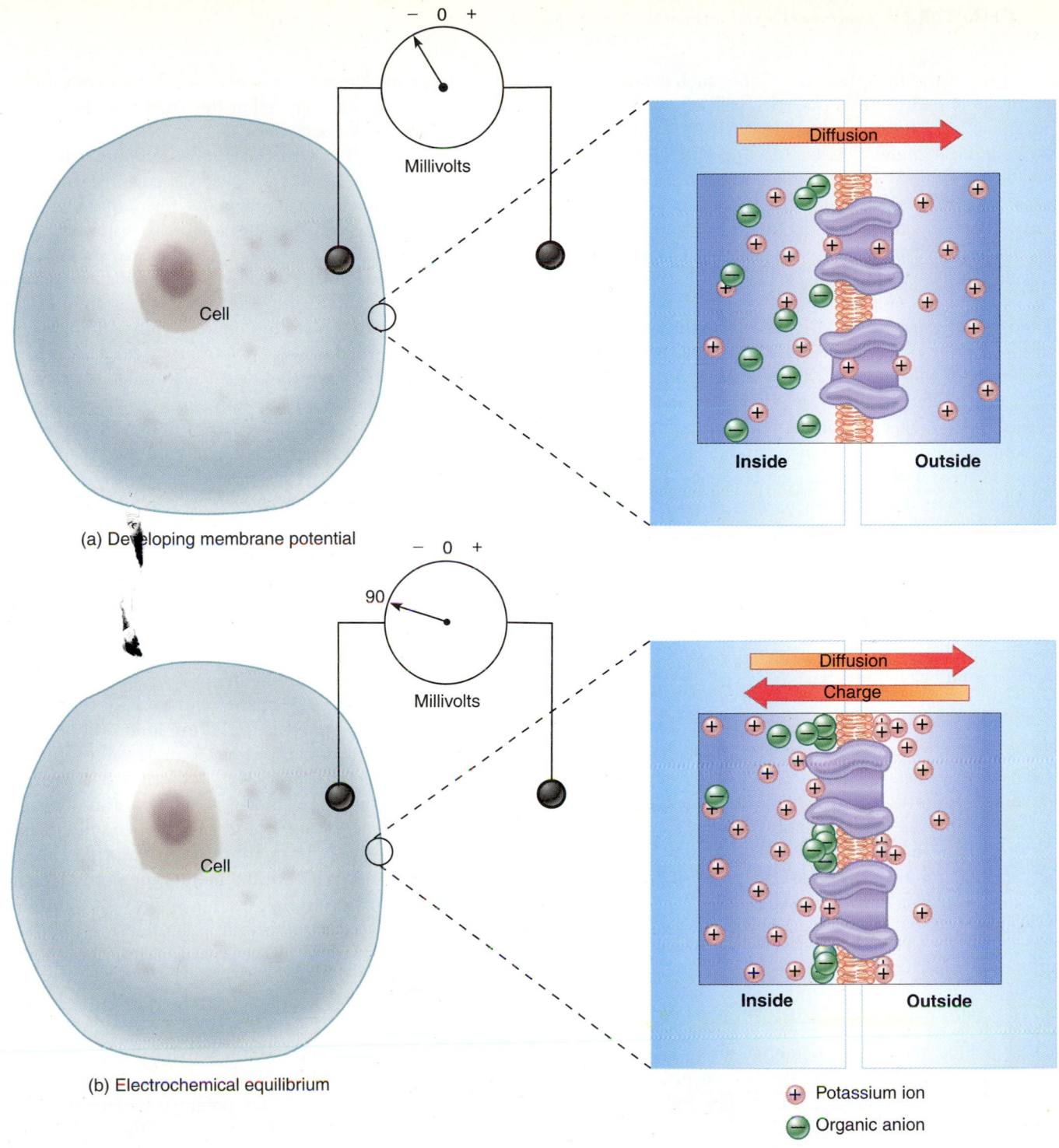

Inside   Outside

Diffusion

Diffusion
Charge

Inside   Outside

Millivolts

Cell

(a) Developing membrane potential

90

Millivolts

Cell

(b) Electrochemical equilibrium

+ Potassium ion

− Organic anion

Figure 7–14

A membrane potential develops when a difference in electrical charge between the cytosol and the extracellular fluid arises owing to the movement of ions down a concentration gradient. *(a)* K^+ ions flow out of the cell, down their concentration gradient, whereas anions must remain inside because the membrane is impermeable to them. Negative charges begin to accumulate near the cytosol surface of the membrane. K^+ ions that have diffused out of the cell are electrically attracted to the negative charges on the inner membrane surface and begin to accumulate near the outer membrane surface, and some flow into the cell because they are attracted to the negative charge of the cytosol, although other K^+ ions are still flowing out owing to the concentration gradient. *(b)* At electrochemical equilibrium, the chemical force (concentration gradient) favoring K^+ efflux is equal to the electrical force (membrane potential) favoring K^+ influx. Because K^+ is the only ion flowing through the membrane in this example, the membrane potential at electrochemical equilibrium equals the potassium equilibrium potential (E_K).

through but not any of the other ions. Under such conditions, the membrane is said to be permeable (will let pass) to K^+ but impermeable (will not let pass) to Na^+, Cl^-, and An^-. Because a chemical gradient exists for K^+ and the membrane is permeable to the ion, K^+ has a tendency to diffuse out of the cell. As K^+ diffuses down its concentration gradient, the anions An^- migrate with K^+ toward the membrane but cannot pass through, leaving an excess of negative charges near the inner membrane surface. The excess negative charge on the inside of the cell tends to pull positive ions back in. The membrane is impermeable to Na^+; therefore, the only positive ion to be pulled back in is K^+. The chemical gradient forcing K^+ out of the cell quickly becomes opposed by an electrical force pulling K^+ back in. When the magnitude of the electrical force (voltage) becomes equal to the magnitude of the chemical force (concentration gradient), further net diffusion of K^+ ceases, the membrane potential becomes stable, and the system is in electrochemical equilibrium (see Fig. 7–14).

The transmembrane potential difference necessary to balance a given concentration gradient for a single diffusible ion is called the *equilibrium potential of the ion* (E_{ion}). If the intracellular (I) and extracellular (O) concentrations of the diffusible ion are known, the Nernst equation can be used to calculate the equilibrium potential:

$$E_{ion} = (-RT/Fz) \ln ([O]/[I])$$

where R is the universal gas constant, T is the absolute temperature, F is the Faraday constant, and z is the ion valence. If we assume a temperature of 37°C and assume that the cell is freely permeable to the diffusible ions Na^+, K^+, and Cl^-, application of the Nernst equation using concentration values from Table 7–1 gives the following predicted equilibrium potentials: $E_{Na} = +60$ mV, $E_K = -90$ mV, $E_{Cl} = -70$ mV.

The Neuron Resting Membrane Potential

The actual resting membrane potential for a mammalian neuron is approximately −75 mV. It does not exactly agree with any of the previously predicted equilibrium potentials because the resting membrane is neither freely nor equally permeable to sodium, potassium, and chloride, and the actual concentrations of these ions are slightly different than the values in Table 7–1.

The resting neuron is much more permeable to potassium than to sodium and a chemical gradient favors potassium diffusing out of the cell. Potassium ions therefore tend to diffuse out of the neuron along their concentration gradient and give rise to an electrical potential that is negative inside. The magnitude of the resting membrane potential (≈ -75 mV) is less than the potassium equilibrium potential ($E_K = -90$ mV) that is predicted by the Nernst equation. In other words, there appears to be more potassium inside the neuron than is predicted by the resting potential. This sug-

gests a carrier-mediated transport system that moves some of the potassium ions into the cell at the same time that other potassium ions are diffusing out.

Because there is twelve times more sodium outside of the neuron than inside, and the inside is negative, both the concentration and electrical gradients for sodium are directed inward. However, the resting neuron is not very permeable to sodium. If it were more permeable to sodium, the resting potential would be closer to the predicted sodium equilibrium potential ($E_{Na} = +60$ mV). Small amounts of sodium do leak into the resting neuron, with a tendency to make the inside of the neuron less negative. The resting membrane potential nevertheless remains stable, suggesting that sodium is pumped out of the neuron at the same rate that it leaks in.

Many lines of evidence have established the existence of a carrier-mediated antiport system that moves two potassium ions into the neuron and three sodium ions out, that is, an electrogenic Na^+/K^+ pump. The Na^+/K^+ pumps develop and maintain the resting membrane potential. Na^+/K^+ pumps require energy from ATP to operate and help to maintain the resting membrane potential by returning potassium to the cell, thereby keeping the potassium concentration gradient high, and by preventing the intracellular accumulation of sodium, which would abolish the internal negativity.

The plasma membrane of the neuron is more permeable to chloride ions than it is to sodium ions, and because chloride ions are negative, they tend to move out of the cell down an electrical gradient. More chloride is therefore found outside of the cell than inside. The neuron does not actively transport or "pump" chloride as it does sodium and potassium. Therefore, the distribution of chloride (inside vs. outside) is more a result of the membrane potential and membrane conductance rather than the cause of it, and the previously calculated equilibrium potential for chloride ($E_{Cl} = -70$ mV) is close to the actual resting membrane potential.

The neuron plasma membrane is very permeable to K^+, the resting membrane potential is close to the equilibrium potential for potassium ion, and the Na^+/K^+ pumps maintain a high concentration gradient for potassium ion diffusion. Therefore, the resting membrane potential is sometimes referred to as a *potassium diffusion potential*. However, the actual value of the resting membrane potential is determined by the intracellular and extracellular concentrations of all ions in the immediate vicinity of the plasma membrane and the membrane's permeability to each kind of ion. For sodium, potassium, and chloride ions, these relationships are expressed in the Goldman equation:

$$E_r = \frac{RT}{F} \ln \frac{P_K[K^+]_0 + P_{Na}[Na^+]_0 + P_{Cl}[Cl^-]_i}{P_K[K^+]_i + P_{Na}[Na^+]_i + P_{Cl}[Cl^-]_0}$$

where P_{ion} is a permeability coefficient and the brackets denote millimolar concentration.

The Action Potential

> *What is an action potential and how is it generated?*

When an adequate positive stimulus is applied to an axon, a rapid and marked change occurs in the membrane potential at the point where the stimulus was applied (Fig. 7–15). The membrane potential is reduced, moving from a resting value of −75 mV toward 0 mV. The process of reducing the membrane potential is called **depolarization.** When the membrane potential is reduced approximately 20 mV to a crucial value, called the **threshold potential** (−55 mV), the rate of depolarization rapidly increases and the membrane potential is quickly reduced to 0 mV and then reverses polarity, the inner surface of the membrane becoming +30 mV with respect to the outer surface where the stimulus was applied. The change in membrane conductance associated with the reversal of membrane polarity stops the depolarization process, and the membrane potential returns toward resting value. The process of returning to the resting membrane potential is called **repolarization.** In some neurons, for a short period of time at the end of repolarization, the inner surface of the membrane may become more negative than is normal at rest. This increased internal negativity is termed **hyperpolarization.** Following hyperpolarization, the resting membrane potential is restored. The graphic representation (voltage vs. time) of membrane depolarization and repolarization is termed the **action potential** (see Fig. 7–15). The duration of the action potential for mammalian neurons is approximately 0.5 msec to 1.5 msec. The amplitude of the action potential is constant for a given axon but can vary from one axon to another.

What causes such a drastic change in the membrane potential? The voltage changes seen in the action potential are a result of the opening and closing of voltage-sensitive ion channels that control the influx of Na$^+$ and efflux of K$^+$. These channels are sensitive to the voltage across the membrane. Changes in membrane voltage cause channel proteins to change their conformation (shape or form), opening gates that control passage of ions through the channel. The Na$^+$ channel has two gates, (1) an **activation gate** and (2) an **inactivation gate,** that can open or close, determining three states of Na$^+$ channel operation: (1) a **resting state,** (2) an **activation state,** and (3) an **inactivation state** (Fig. 7–16). The K$^+$ channel has one gate, and thus two operational states: (1) **open** and (2) **closed.**

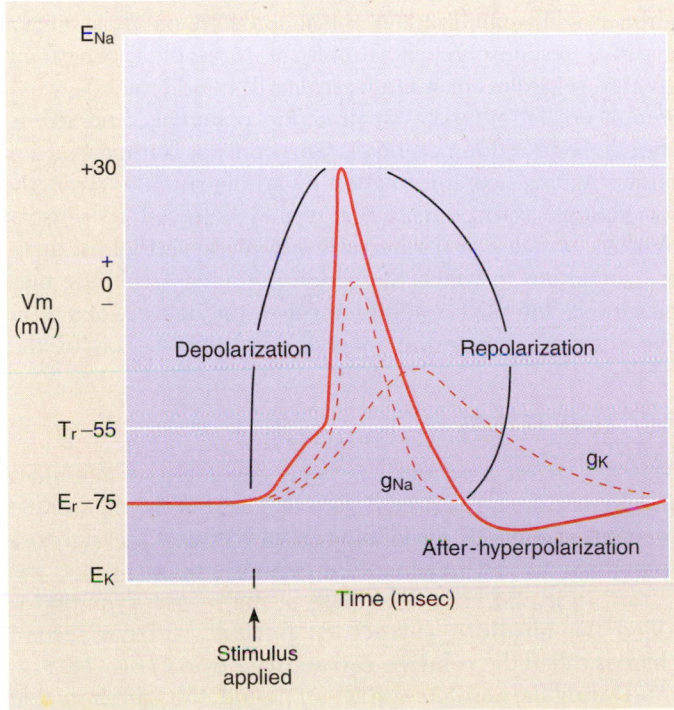

Figure 7–15

A graph showing depolarization, repolarization, and hyperpolarization phases of the action potential and associated changes in membrane conductances of Na$^+$ and K$^+$.

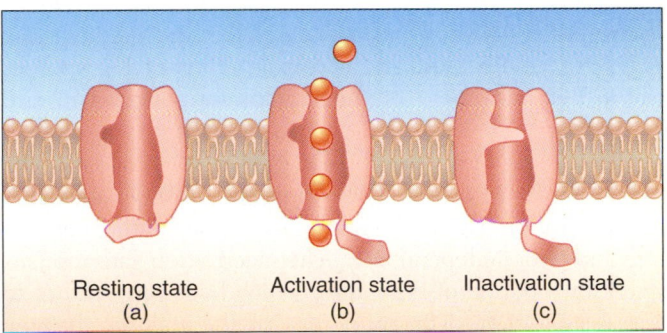

Voltage-gated sodium ion channel

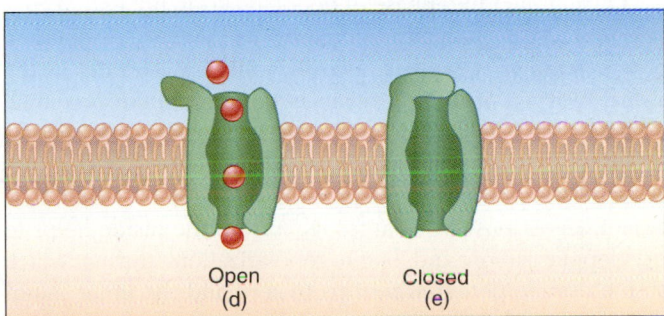

Voltage-gated potassium channel

Figure 7–16

States of operation of voltage-gated Na$^+$ (a,b,c) and K$^+$ (d,e) channels. **(a)** Resting state. **(b)** Activation state. **(c)** Inactivation state. **(d)** Open. **(e)** Closed.

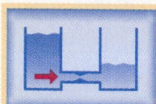

MODELS IN PHYSIOLOGY:
Mass and Heat Flow

When the nerve cell is at rest, the voltage-gated Na^+ channels of the axon are in a resting state; their inactivation gates are open, but their activation gates are closed preventing the passage of Na^+ ions (Fig. 7–17). When a positive stimulus is applied to the axon, many of the Na^+ channel activation gates open, and these Na^+ channels switch from a resting state to an activation state. Sodium ions pass through the open channels and into the cell, moving down favorable chemical and electrical gradients, gradually depolarizing the membrane toward its threshold potential. Once the threshold value of the membrane potential has been reached, many more voltage-sensitive Na^+ channels are activated, resulting in a greater influx of Na^+ ions. This influx of positive charges depolarizes the membrane still further, opening more $Na+$ channel activation gates, in turn leading to a further influx of Na^+ and consequently a further depolarization in a positive feedback manner. This regenerative process is characterized by a rapid depolarization of the membrane, recorded as the rising phase of the action potential. The primary cause of the depolarization is a local increase in membrane sodium conductance (g_{Na}), evidenced by the shift in membrane potential toward the sodium ion equilibrium potential ($E_{Na} = +60$ mV). However, E_{Na} is not reached by the end of depolarization because inactivation gates close the Na^+ channels, limiting the influx of Na^+. The opposing electrical gradient also brakes Na^+ influx, and the increasing efflux of K^+ opposes any further movement of the membrane potential toward E_{Na}.

The closing of the inactivation gate is independent of the opening of the activation gate. Thus, Na^+ channel activation and inactivation can be thought of as two separate gates that open and close independently. The inactivation gate remains closed for a period of time, after which both gates return to their resting state. The properties of the sodium channel, therefore, are dependent on voltage and time.

During the rapid phase of depolarization, potassium conductance (g_K) gradually increases as slower responding voltage-gated K^+ channels open. K^+ efflux accelerates as the inward movement of Na^+ reverses the membrane potential. K^+ leaves the cell for the same reasons that Na^+ entered; favorable electrical and chemical gradients coupled with an increase in membrane permeability. As K^+ efflux accelerates, the net loss of positive charges (K^+) from the inside helps to end depolarization and begin repolarization, during which the membrane potential returns to zero and the inner membrane surface becomes negative once again, reestablishing the resting potential. The inactivation gates of the Na^+ channels open and the activation gates close, restoring the resting state of the Na^+ channels. After the membrane potential returns toward the resting value, g_K decreases because voltage-gated K^+ channels close. In some neurons, more K^+ leaves

the cell than is actually required to restore the resting potential, and for a short time, the inner surface of the membrane is more negative than is normal at rest. This increased internal negativity is termed **after-hyperpolarization.** Following the hyperpolarization phase, the resting membrane potential is restored as ion conductances return to resting values and the Na^+/K^+ pumps maintain the normal internal/external ratios of these ions. The preceding activities that restore the resting membrane potential after depolarization of the membrane collectively characterize the phenomenon of repolarization. The primary cause of repolarization is an increase in potassium conductance (g_K) out of the neuron, evidenced by the shift in the membrane potential toward the potassium ion equilibrium potential ($E_K = -90$ mV).

The Action Potential as a Maximal Response

The application of a positive stimulus to an axon must reduce the membrane potential to threshold value before an action potential can be generated. If the depolarization fails to reach threshold voltage, too few voltage-gated sodium channels are activated to perpetuate the regenerative process, and no action potential is produced even though a subthreshold depolarization response to the stimulus can be recorded. If the stimulus strength is adequate to lower the membrane voltage to threshold, an action potential develops. If a stronger than threshold strength stimulus is applied, the action potential amplitude is the same as with the application of the weaker, but adequate strength, stimulus. Thus, for a given axon, the amplitude of the action potential represents a maximal response to all stimuli, of threshold strength or greater, regardless of stimulus strength. In other words, weak stimuli do not generate small action potentials, and strong stimuli, large action potentials. This concept is important for understanding how information regarding stimulus strength, for example warm versus hot stimuli detected by sensory neurons, is coded in the nervous system. Nevertheless, there are conditions in which the action potential amplitude may be varied. One such condition involves the delivery of a positive stimulus during the relative refractory period of the axon.

Refractory Periods

After an action potential has been generated, a minimum amount of time is required before that area of the membrane becomes capable of responding in an identical manner to a second stimulus. This minimum period of time is called the *refractory period.* The first phase of the refractory period is called the **absolute refractory period,** and the second phase is called the **relative refractory period** (Fig. 7–18). During the absolute refractory period, the axon is unable to generate a second action potential regardless of the strength of the applied stimulus. Many of the gated Na^+ channels have been inactivated and for a period are incapable of being opened. Consequently, g_{Na} is too low to trigger the regenerative response required to initiate an action potential.

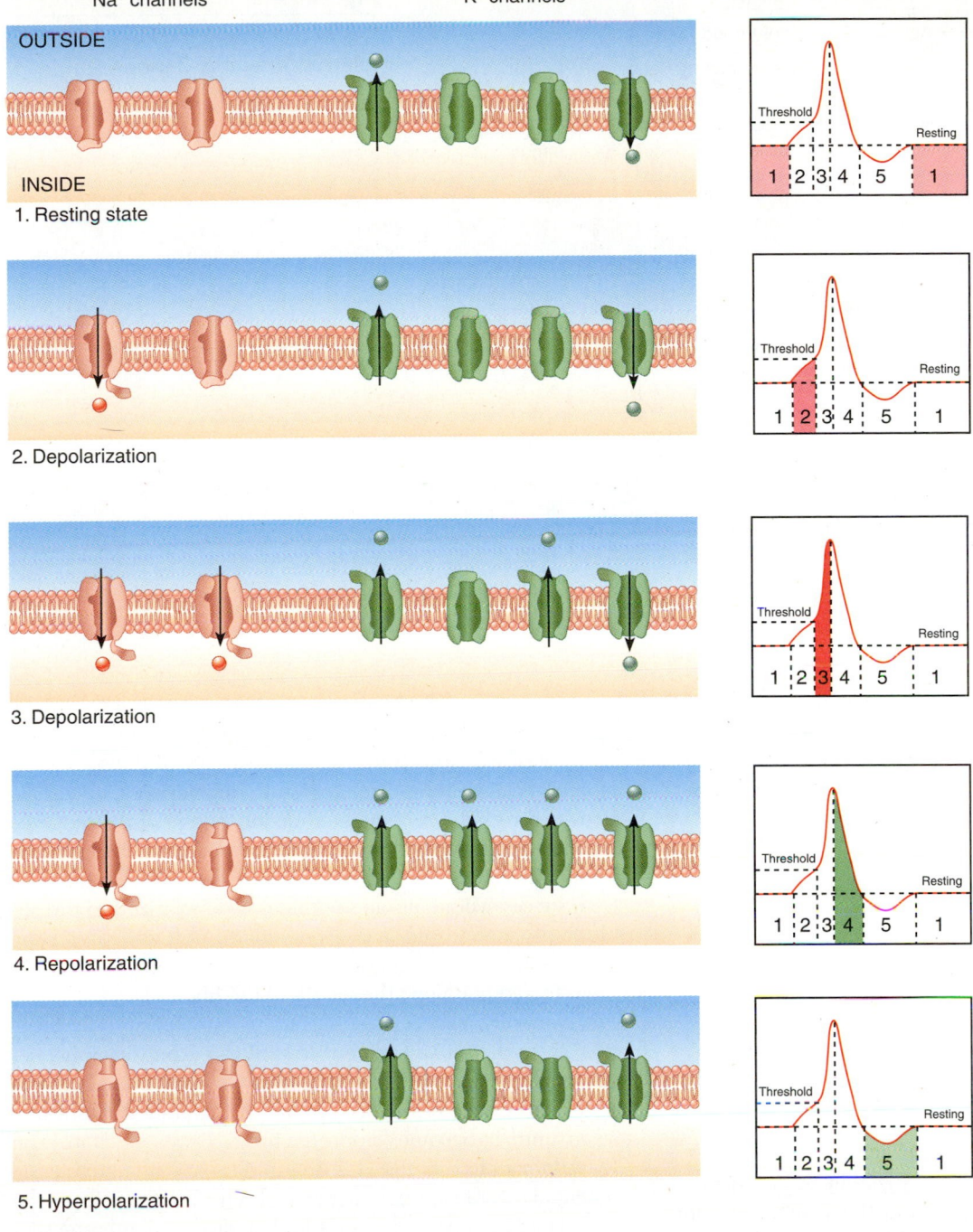

Na$^+$ channels K$^+$ channels

OUTSIDE

INSIDE

1. Resting state

2. Depolarization

3. Depolarization

4. Repolarization

5. Hyperpolarization

Figure 7–17

Phases of the action potential as they correspond to altered states of operation of the voltage-gated Na$^+$ and K$^+$ channels.

During the relative refractory period, some (but not all) of the Na$^+$ channels again become responsive to an applied stimulus, but the stimulus strength must be much greater than before in order to reach a higher threshold and elicit another action potential (see Fig. 7–18). The amplitude of the second action potential is lower than normal because g$_{Na}$ is still low and g$_K$ remains high for a short time following the delivery of the first stimulus. These action potentials are not propagated.

The significance of the refractory period is that its time interval determines the fastest frequency at which an axon can generate normal action potentials. Axons with short refractory periods can conduct impulses at a higher frequency than can fibers with long refractory periods.

The action potential represents the change in membrane potential only in the region of the membrane where an adequate positive stimulus has been applied. The entire plasma

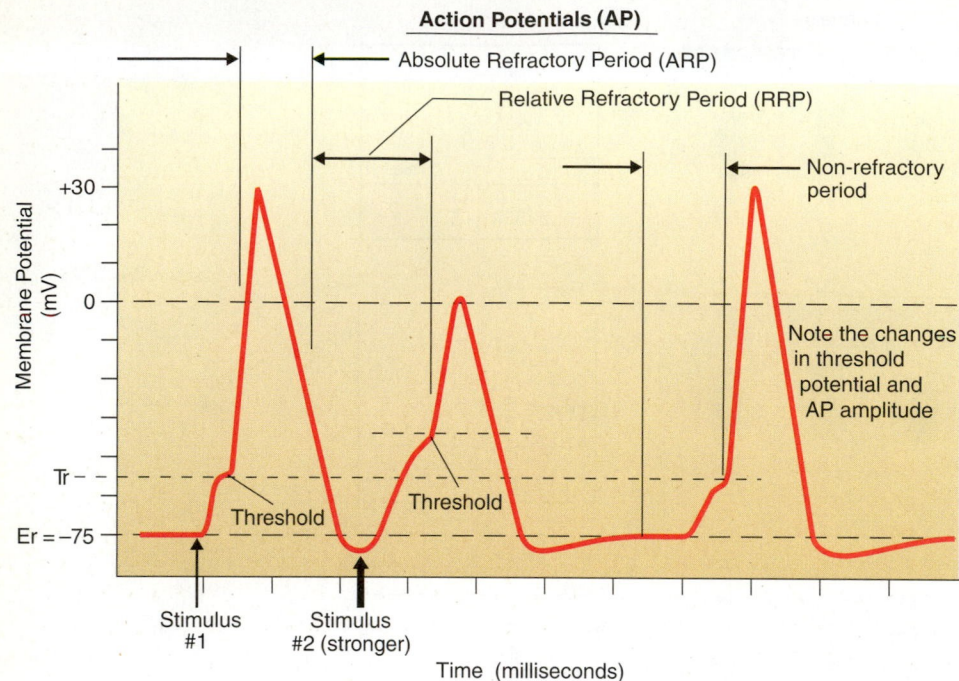

Action Potentials (AP)

Figure 7–18

A graph showing the absolute and relative refractory periods that follow the action potential.

membrane does not simultaneously depolarize and then repolarize in response to an adequate stimulus. However, once an action potential is generated, it will spread from one area of the axon membrane to another (i.e., propagate), resulting in the conduction of a nerve impulse.

MODELS IN PHYSIOLOGY:
Mass and Heat Flow

Propagation of the Action Potential

> **How is the action potential propagated along the axon?**

A nerve impulse is a wave of depolarization followed by a wave of repolarization that travels along the axon away from the point of stimulation. When a stimulus of strength or duration sufficient to reduce the membrane potential to threshold is applied to the axon, the membrane depolarizes, with the inside becoming positive with respect to the outside. This area of the membrane is called the *sink* because Na^+ ions flow inward across the membrane at that point (Fig. 7–19).

Because the Na^+ ions are drawn from the adjacent membrane region, the adjacent region is called the *source*. Current flow follows the path of a closed loop; therefore, the influx of Na^+ ions into the sink results in the movement of positive charges from the sink along the interior of the axon and onto the inside surface of the membrane of the source

region. This causes an accumulation of positive charge on the inside of the membrane, depolarizing the source region. The removal of Na^+ ions from the outer surface of the source region also causes the source region to depolarize. Depolarization to threshold generates an action potential, turning the source region into a sink. The next, more distal, membrane segment becomes the source for this sink. In this manner, each successive membrane segment is depolarized to threshold and generates an action potential, in effect, conducting the action potential sequentially from one membrane segment to another along the length of the fiber.

Normally, action potentials are conducted in one direction, from the initial segment toward the axon terminus, because the sink where the earlier action potential was generated is in a refractory period, so it cannot act as a sink again until it becomes nonrefractory. However, if an axon was adequately stimulated in the middle of its length, action potentials would be conducted in both directions from the stimulated area (sink) because both adjacent membrane segments could act as sources, become depolarized, and act as sinks.

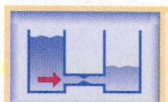

MODELS IN PHYSIOLOGY:
Mass and Heat Flow

The Velocity of the Action Potential

The propagation of an action potential along the axon occurs at a speed directly proportional to the diameter of the axon. The larger the diameter of the axon, the greater the speed of

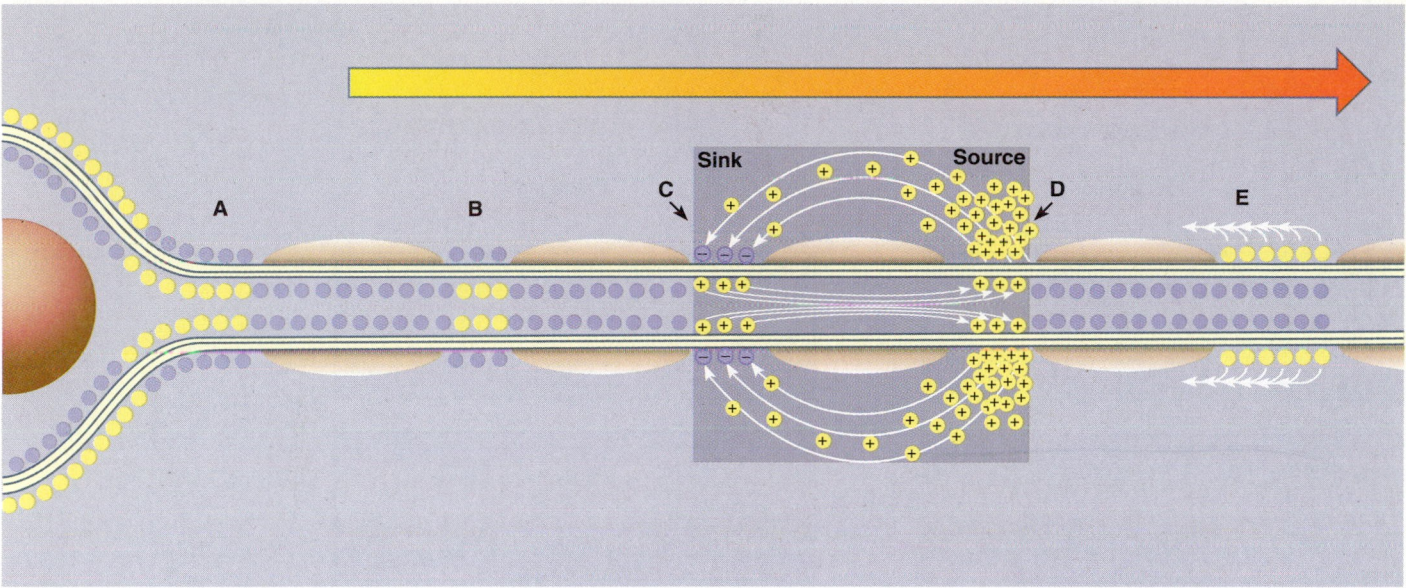

Figure 7–19

Propagation of the action potential. The action potential was generated at the axon hillock (A) which is now completing repolarization. The first node (B) is in early repolarization, and the second node (C) has depolarized. The depolarized node (C) acts as a sink, attracting (+) charges from the outer membrane surface at the next distal node (D), which acts as a source. Positive charges flow along the inner membrane surface from sink to source, completing the circuit for current flow and beginning the depolarization process at the source. As the source (D) begins to depolarize, it becomes a sink, attracting (+) charges from the outer membrane surface at the next more distal node (E). As the more distal node is depolarizing, the preceding node is repolarizing and therefore refractory, resulting in unidirectional propagation.

propagation. This is because axons with large diameters do not offer as much resistance to the flow of ions along the length of the axon. In an unmyelinated fiber the action potential spreads from one membrane segment to the next (Fig. 7–20*a*). Because each membrane segment is involved in action potential propagation, conduction velocity is generally slow, especially in small-diameter fibers. Conduction in an unmyelinated fiber can be likened to walking by repetitively and alternately placing the heel of one foot directly in front of the toes of the other foot.

Myelin is a lipid material that is a poor conductor of ion current. Therefore, the segments of the nerve fiber that are covered with myelin are electrically insulated from the extracellular fluid and do not allow the penetration of ions needed for the conduction of the action potential. Depolarization of the myelin-covered segments of the fiber does not occur. However, the exposed areas of the fiber, as at the axon hillock and the nodes of Ranvier, contain large numbers of Na^+ channels and can depolarize. Thus, the generation of an action potential at an exposed area of the fiber results in the spread of current from one node to another, and the nerve impulse is conducted along the fiber by traveling from one

node to another node or nodes. This form of transmission is referred to as **saltatory conduction** (Fig. 7–20*b*). The velocity of propagation is directly proportional to the distance between the nodes of Ranvier, as well as fiber diameter.

NEURON COMMUNICATION: THE SYNAPSE

A **synapse,** from the Greek word *synapsis* ("connection"), is a cellular junction where information is transmitted from one cell to another. In the nervous system, a synapse is usually a junction where signals are transmitted between two neurons. At a synapse, the membrane of the transmitting neuron is called the **pre-synaptic membrane** and the membrane of the receiving cell is called the **post-synaptic membrane.** The mechanism of information transfer between pre-synaptic and post-synaptic membranes is called **synaptic transmission.** In the nervous system, the two types of synapses are (1) the electrical synapse and (2) the chemical synapse.

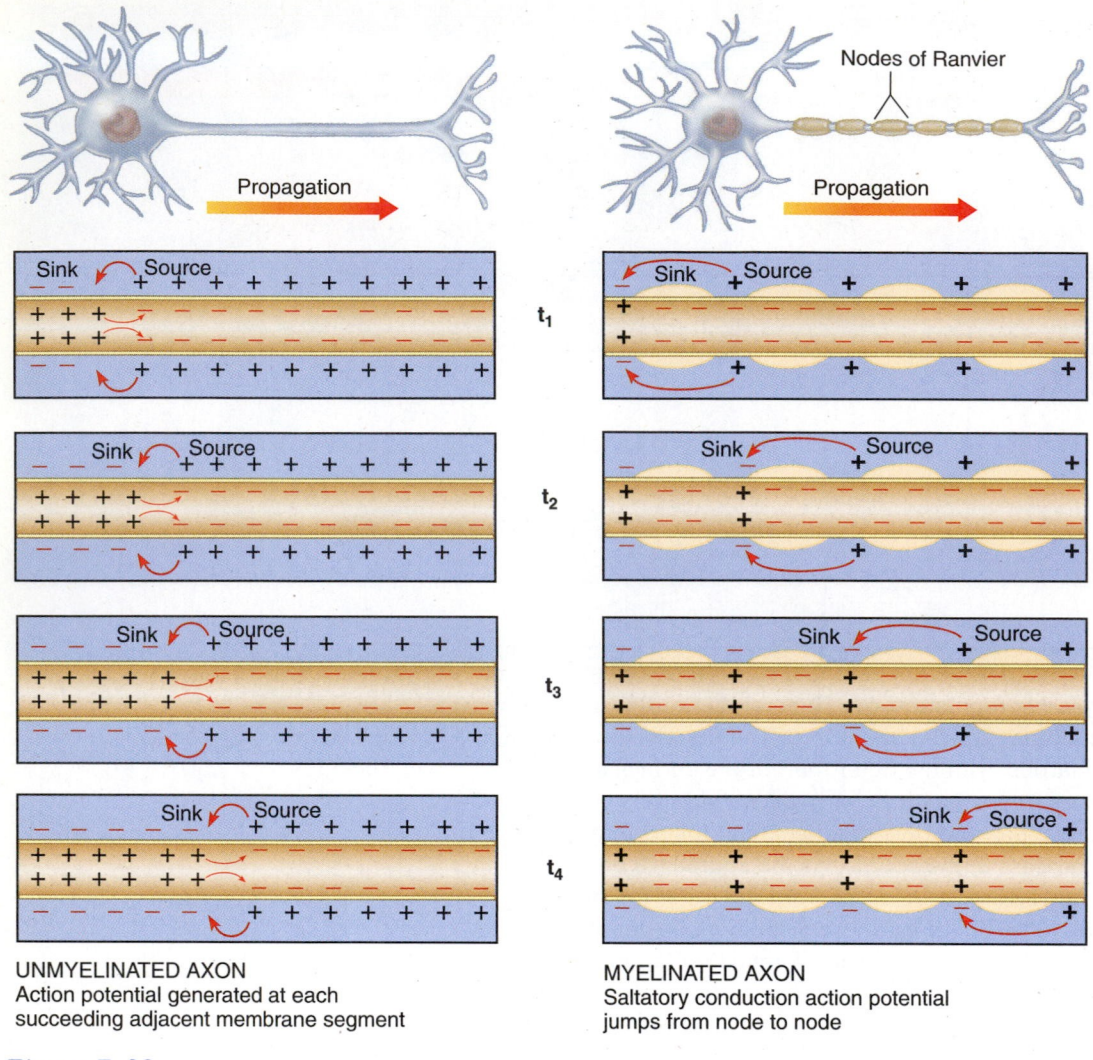

Figure 7–20

Conduction of the nerve impulse in unmyelinated and myelinated fibers. In an unmyelinated nerve fiber, the action potential is propagated through each segment of the axon membrane. In a myelinated nerve fiber, the impulse spreads from one node of Ranvier to succeeding nodes. This form of nerve impulse conduction is called *saltatory conduction.*

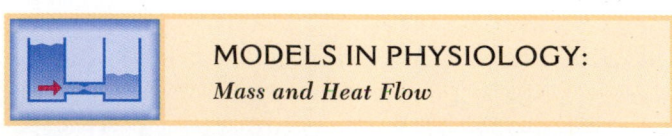

MODELS IN PHYSIOLOGY:
Mass and Heat Flow

The Electrical Synapse

At an **electrical synapse,** information is transmitted from one cell to another by the direct passage of ions between the cells. Electrical (ionic) current can flow from the pre-synaptic cell directly into the post-synaptic cell through connected membrane channels. Under the electron microscope, the membranes of two adjoining cells appear to be fused together. Studies have shown that pre-synaptic and post-synaptic cells are connected by protein channels or **gap junctions** that span

the gap between them (Fig. 7–21). Half of the channel is in the membrane of each of the opposing cells. Each half-channel is called a **connexon** and is made of six protein subunits called **connexins** that are arranged into a hexagonal assembly.

Although common in the nervous systems of invertebrates, electrical synapses are not as common in vertebrates. Electrical synapses are found between axons and soma, axons and dendrites, dendrites and dendrites, and soma and soma. Transmission at an electrical synapse may be bidirectional. The channels are large enough to allow the passage of molecules such as cAMP, sucrose, and small peptides. Thus, these "electrical" synapses may serve as channels for both electrical and metabolic communication.

In terms of electrical communication, these synapses synchronize the activity of many adjoining cells as well as

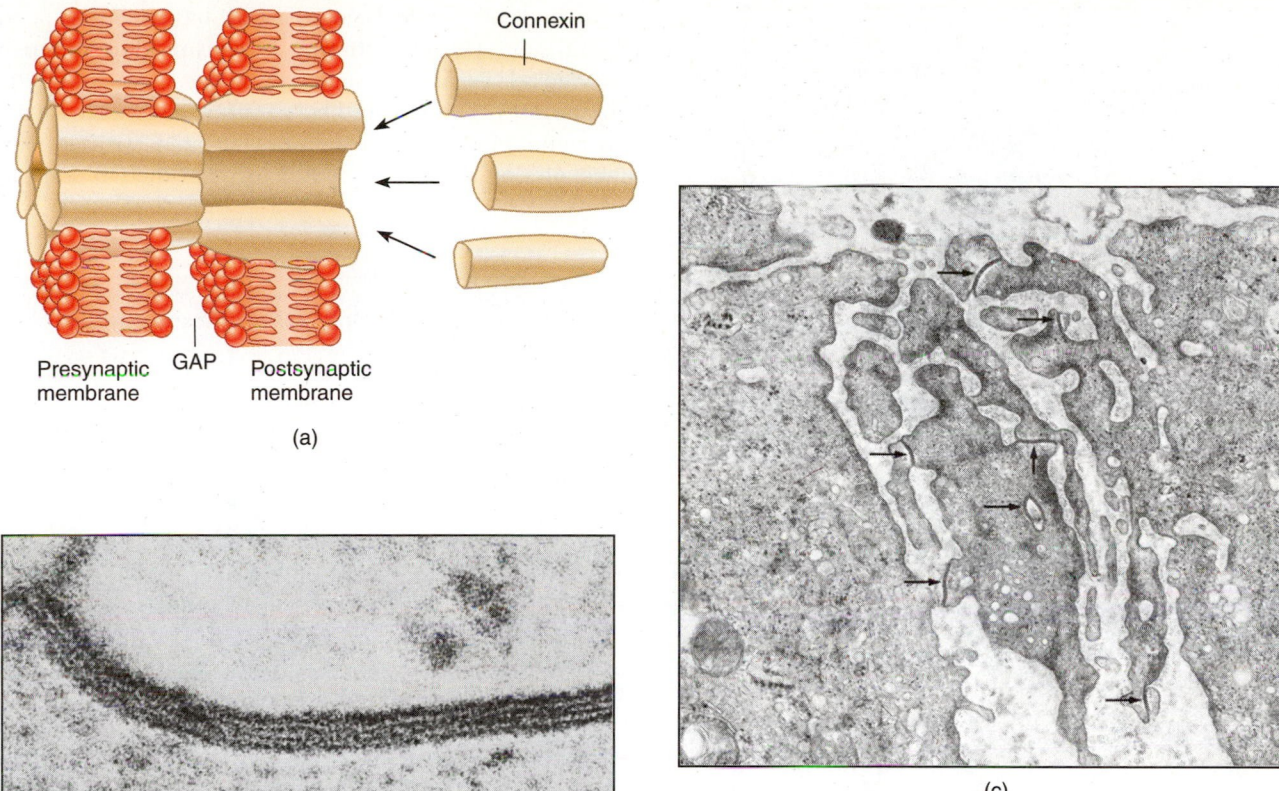

Figure 7–21

Electrical synapses: gap junctions. *(a)* Gap junctions are channels of communication between cells formed by proteins called *connexins* in adjacent cell membranes. Six connexins form a half-channel called a *connexon.* Two connexons, one connexon from each cell, join to form the connecting channel. *(b)* SEM of a gap junction between smooth muscle cells in the wall of the uterus. *(c)* Several gap junctions shown at a lower magnification (approximately $\times$ 36,000). *(Electron micrographs courtesy of Puri and Garfield, from* Biology of Reproduction, *27:967–975, 1982.)*

provide a pathway for a rapid communication between cells. As we will see in a later chapter, electrical synapses and also gap junctions are also found in cardiac muscle cells, many smooth muscle cells, and other cells that display a synchronization of activity.

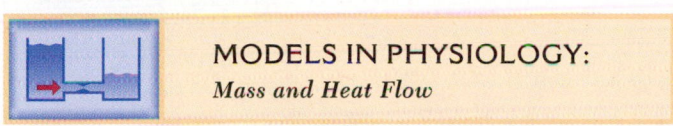

MODELS IN PHYSIOLOGY:
Mass and Heat Flow

The Chemical Synapse

At a **chemical synapse,** the pre- and post-synaptic membranes are separated by a space, called the **synaptic cleft.** A chemical called a **neurotransmitter,** released by exocytosis from the pre-synaptic neuron, relays the signal to the post-synaptic cell. Once released, the transmitter diffuses across the synaptic cleft and interacts with receptors on the post-synaptic membrane to accomplish transmission. In contrast to the electrical synapse, transmission at a chemical synapse is unidirectional from pre-synaptic to post-synaptic membranes.

Several types of chemical synapses exist based on structure (Fig. 7–22). The name of the synapse is determined by regions of the cells or the type of cells involved. For example, an **axodendritic** synapse is a junction between the axon of the pre-synaptic neuron and a dendrite of the post-synaptic neuron. Other common neural synapses include **axosomatic** and **axoaxonic** synapses. The synapse between an axon and a skeletal muscle cell is called the **neuromuscular junction,** and in smooth muscle, the junction is called **"synapse en passant"** (synapse in passing).

Two kinds of chemical synapses exist based on the speed of transmission. The **fast chemical transmission synapse** involves the release of a transmitter from specific pre-synaptic membrane sites. The released transmitter quickly engages a post-synaptic membrane receptor associated with an ion channel, altering membrane conductance of the ion and the

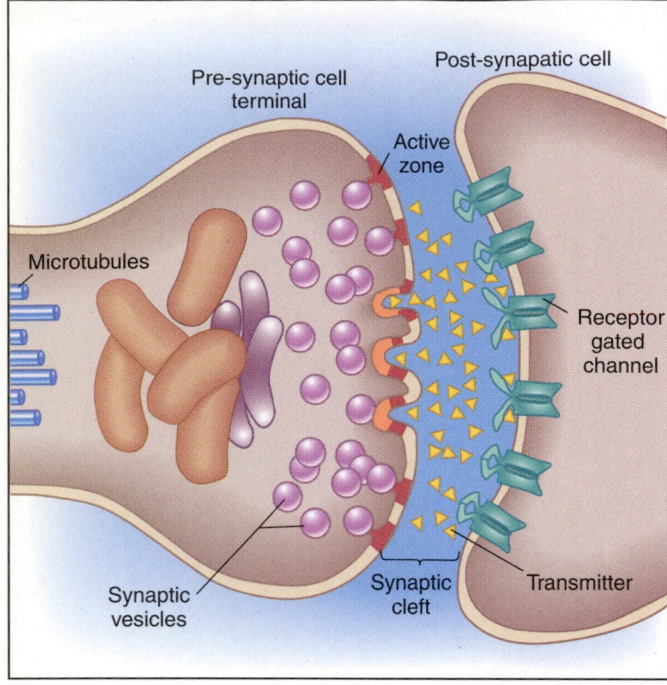

(a)

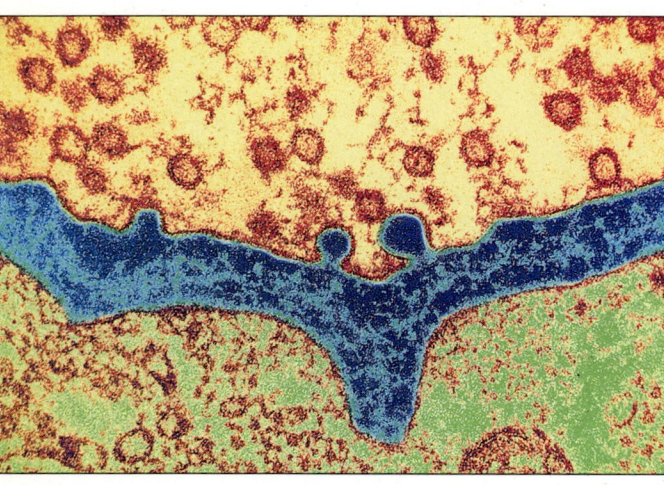

(b)

Figure 7–22

Chemical synapse. *(a)* An axodendritic synapse. *(b)* Synaptic vesicles and pre-synaptic exocytosis.

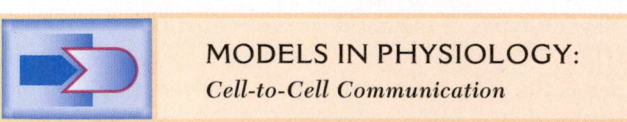

MODELS IN PHYSIOLOGY:
Cell-to-Cell Communication

A synapse may also be named based on the kind of neurotransmitter present. A **cholinergic synapse,** for example, is one where **acetylcholine (ACh)** is used for transmission. On the other hand, an **adrenergic synapse** requires the pre-synaptic release of norepinephrine (noradrenaline) to accomplish transmission.

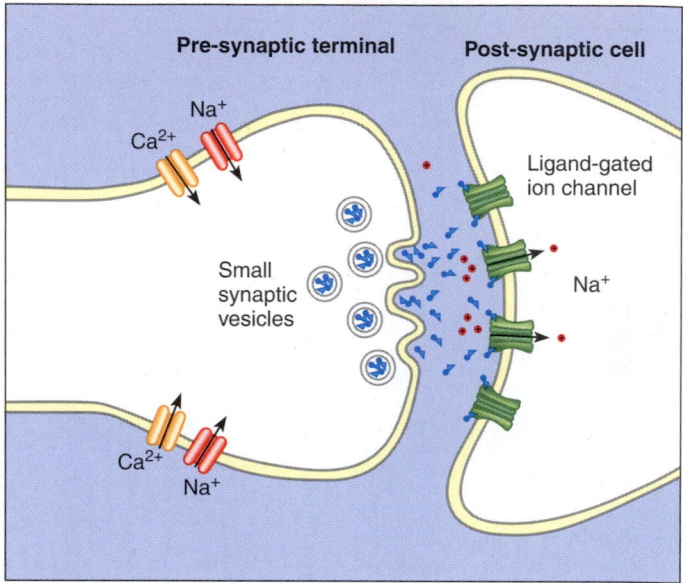

(a) Fast chemical transmission

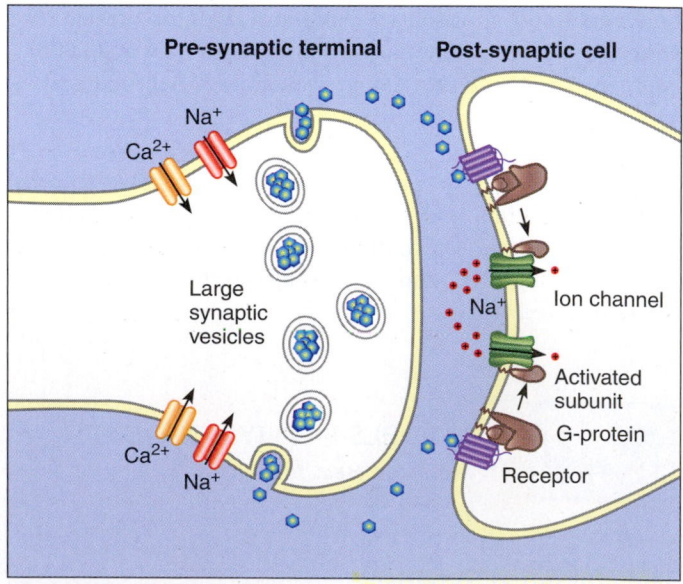

(b) Slow chemical transmission

Figure 7–23

(a) Fast chemical transmission. *(b)* Slow chemical transmission.

post-synaptic membrane potential. Typically, the transmitter is a small, low-molecular-weight molecule packaged in small vesicles located near pre-synaptic release sites (Fig. 7–23).

The **slow chemical transmission synapse** involves the release of a transmitter from larger vesicles more distant from the synaptic cleft. The transmitter is usually a large, high-molecular-weight molecule, such as a peptide, that diffuses more slowly and indirectly exerts its effect on the post-synaptic cell by engaging a receptor linked to a G-protein (see Fig. 7–23) in the post-synaptic membrane. In turn, the activated G-protein–second-messenger system may alter post-synaptic membrane ion conductance or intracellular processes.

Pre-synaptic Mechanisms of Chemical Transmission

The terminal ends of an axon are often enlarged to form knoblike structures, called *terminal boutons*, which make synaptic connections with an effector cell (Fig. 7–24a). Numerous synaptic vesicles are located in the terminal bouton, and each vesicle contains a fixed or **quantum** number of transmitter molecules. At a cholinergic synapse, for example, each vesicle contains approximately 10,000 molecules of ACh.

Vesicles are sequestered in the vicinity of release zones by their association with the actin cytoskeleton. An external vesicle protein, called **synapsin I,** binds the vesicle to an actin filament. This association is controlled by Ca^{2+}/calmodulin-dependent phosphorylation. Phosphorylation of synapsin I frees vesicles from association with actin filaments, making them available to release their transmitters in response to pre-synaptic membrane depolarization.

When a nerve impulse travels the length of the axon and arrives at the terminal bouton, depolarization of the pre-synaptic membrane activates voltage-sensitive Ca^{2+} channels triggering a rapid but transient influx of calcium ions into the terminal bouton (Fig. 7–24b). Normally, the concentration of free calcium in the terminal bouton is very low because it is sequestered by reticula in the terminus. The brief increase in free Ca^{2+} concentration leads to the fusion of transmitter-filled vesicles at **active zones** on the pre-synaptic membrane and subsequent release of transmitter molecules into the synaptic cleft by a process of **exocytosis.** After a vesicle releases its contents, the vesicle is reclaimed from the pre-synaptic membrane, reformed, and refilled with transmitter molecules.

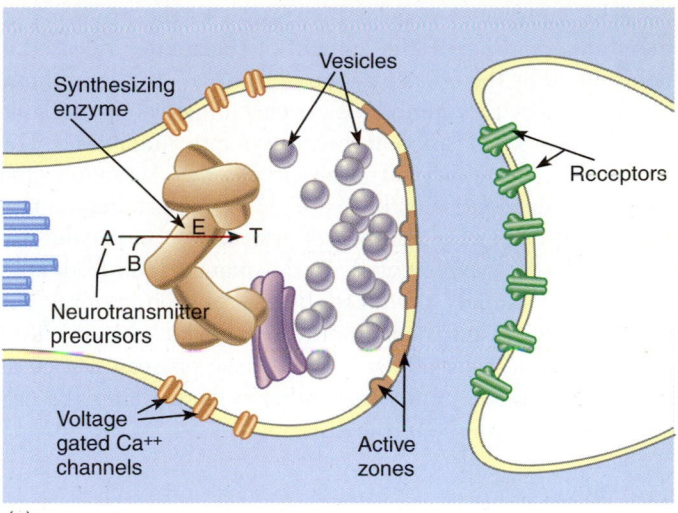

(a)

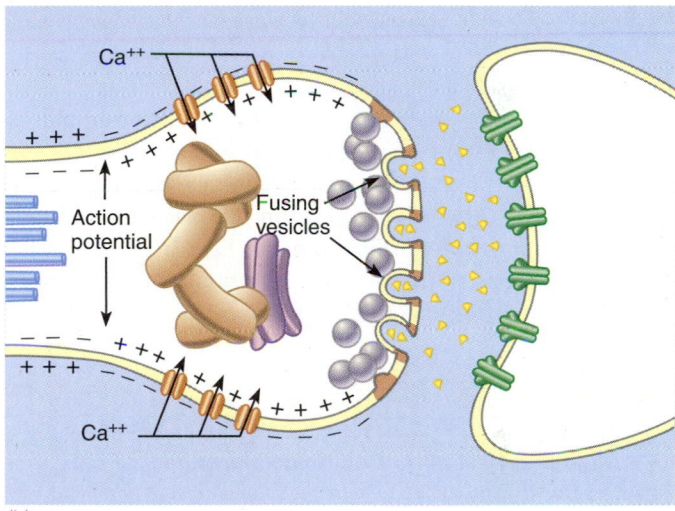

(b)

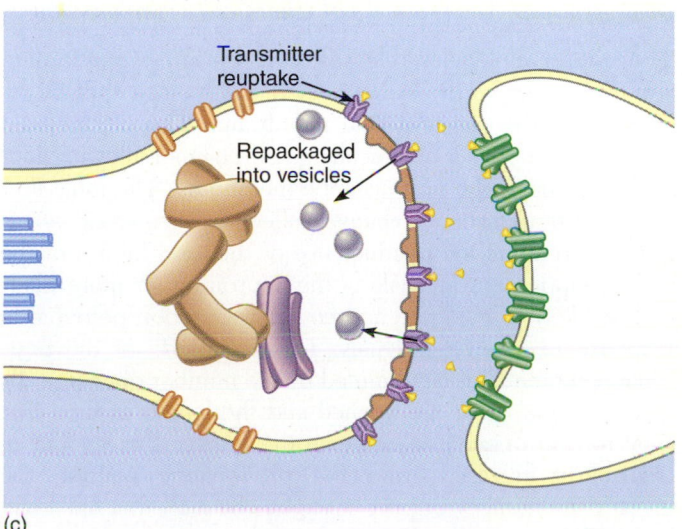

(c)

Figure 7–24

Transmission at a chemical synapse. *(a)* Synthesis of transmitter molecules from precursors and packaging into vesicles. *(b)* Depolarization of the pre-synaptic membrane causes calcium ions to flow into the cell, the secretory vesicles fuse with active sites on the pre-synaptic cell membrane, and the transmitters are released into the synaptic cleft. *(c)* Some transmitter molecules bind to receptor proteins on the post-synaptic membrane, whereas others are pumped back into the pre-synaptic terminal by a reuptake mechanism.

The number of vesicles that fuse with the pre-synaptic membrane and hence the amount of transmitter released is determined by the concentration of free Ca^{2+} in the terminal bouton and by the number of vesicles immediately available for Ca^{2+}-triggered fusion at the pre-synaptic membrane. The higher the frequency of pre-synaptic depolarization, the higher the concentration of free Ca^{2+} in the terminal bouton and the greater the number of transmitter molecules released into the synaptic cleft. In this manner, the frequency of action potentials produced by the neuron governs the amount of released transmitter. The amount of transmitter released can also be affected by extrinsic processes, such as axoaxonic synaptic inputs from other cells that alter the calcium levels within the terminals.

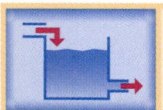

MODELS IN PHYSIOLOGY:
Conservation of Mass

When the neuron ceases conduction of nerve impulses, ending depolarization of the terminal bouton, free Ca^{2+} is rapidly removed from the terminal by uptake into organelles, such as the ER. Also, Ca^{2+} pumps in the plasma membrane move Ca^{2+} out of the terminal bouton. Low free Ca^{2+} in the terminal bouton stops vesicle migration and exocytosis, thereby ending pre-synaptic transmission.

MODELS IN PHYSIOLOGY:
Cell-to-Cell Communication

Post-Synaptic Mechanisms of Chemical Transmission

Transmitter molecules released from the pre-synaptic neuron diffuse across the synaptic cleft and engage specific receptors on the post-synaptic membrane. The engagement temporarily opens a chemically gated channel allowing ions to flow through the post-synaptic membrane. The ionic flow creates a graded voltage change, called a *post-synaptic potential,* which alters ion conductance of the axon hillock of the post-synaptic neuron, making the neuron either more likely or less likely to generate and conduct an action potential in response to incoming signals. The magnitude of the post-synaptic potential is determined by the number of chemically gated channels that are opened and the length of time that they remain open. For example, the greater the amount of transmitter released from the pre-synaptic neuron, the greater the number of post-synaptic channels that open and hence the greater the magnitude of the post-synaptic potential change. Post-synaptic potentials are graded, nonpropagated, localized changes in the post-synaptic membrane's resting potential. They cause ion current to flow toward the axon hillock, indicating changes in ion conductance at the axon hillock membrane. These changes may either increase or decrease the likelihood that an action potential will be generated at the axon hillock.

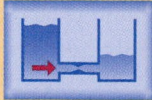

MODELS IN PHYSIOLOGY:
Mass and Heat Flow

At an excitatory synapse, the action of a transmitter leads to an increase in the inward flow of positive ions (e.g., sodium) thereby tending to depolarize the post-synaptic membrane (Fig. 7–25*a*). A depolarizing graded post-synaptic voltage change increases the probability that the post-synaptic neuron will generate an action potential and therefore is called an **excitatory post-synaptic potential (EPSP).**

Most EPSPs are caused by an increase in g_{Na}. Sodium ions diffuse into the post-synaptic neuron through open channels in the post-synaptic membrane because of favorable chemical and electrical gradients. In some post-synaptic cells, excitatory post-synaptic potentials can also be produced by the opening of channels that are somewhat nonselective and result in the simultaneous flow of both Na^+ into the cell and K^+ out of the cell. An excitatory or depolarizing potential develops because the increase in g_{Na} is greater than the increase in g_K; in other words, Na^+ influx is greater than K^+ efflux. This is characteristic of a type of ACh receptor found at the neuromuscular junction.

At an inhibitory synapse, the action of a transmitter leads to an increase in the outward flow of positive ions (e.g., potassium) and/or the inward flow of negative ions (e.g., chloride), thereby hyperpolarizing the post-synaptic membrane, making the post-synaptic neuron less likely to generate an action potential (Fig. 7–25*b*). A graded post-synaptic voltage change that results in a hyperpolarization of the post-synaptic membrane, making the neuron less excitable, is called an **inhibitory post-synaptic potential (IPSP).**

The activation of a single receptor results in an increase in the influx or efflux of one or more types of ion. The number of ions that flow through the channel per unit time is referred to as the **single channel current.** The **synaptic current** is made up of all of the single channel currents through the membrane at the synapse.

The flow of synaptic current resulting from the release of transmitters from one vesicle produces a change in the membrane potential referred to as a **unitary post-synaptic potential** (see Fig. 7–25). These post-synaptic potentials can summate, or act together at an excitatory synapse to increase depolarization of the post-synaptic membrane, or at an inhibitory synapse to increase hyperpolarization of the membrane. At any single synapse, the transmitter molecule is either excitatory or inhibitory, but never both. However, some transmitters, such as ACh, may be excitatory at one synapse and inhibitory at another.

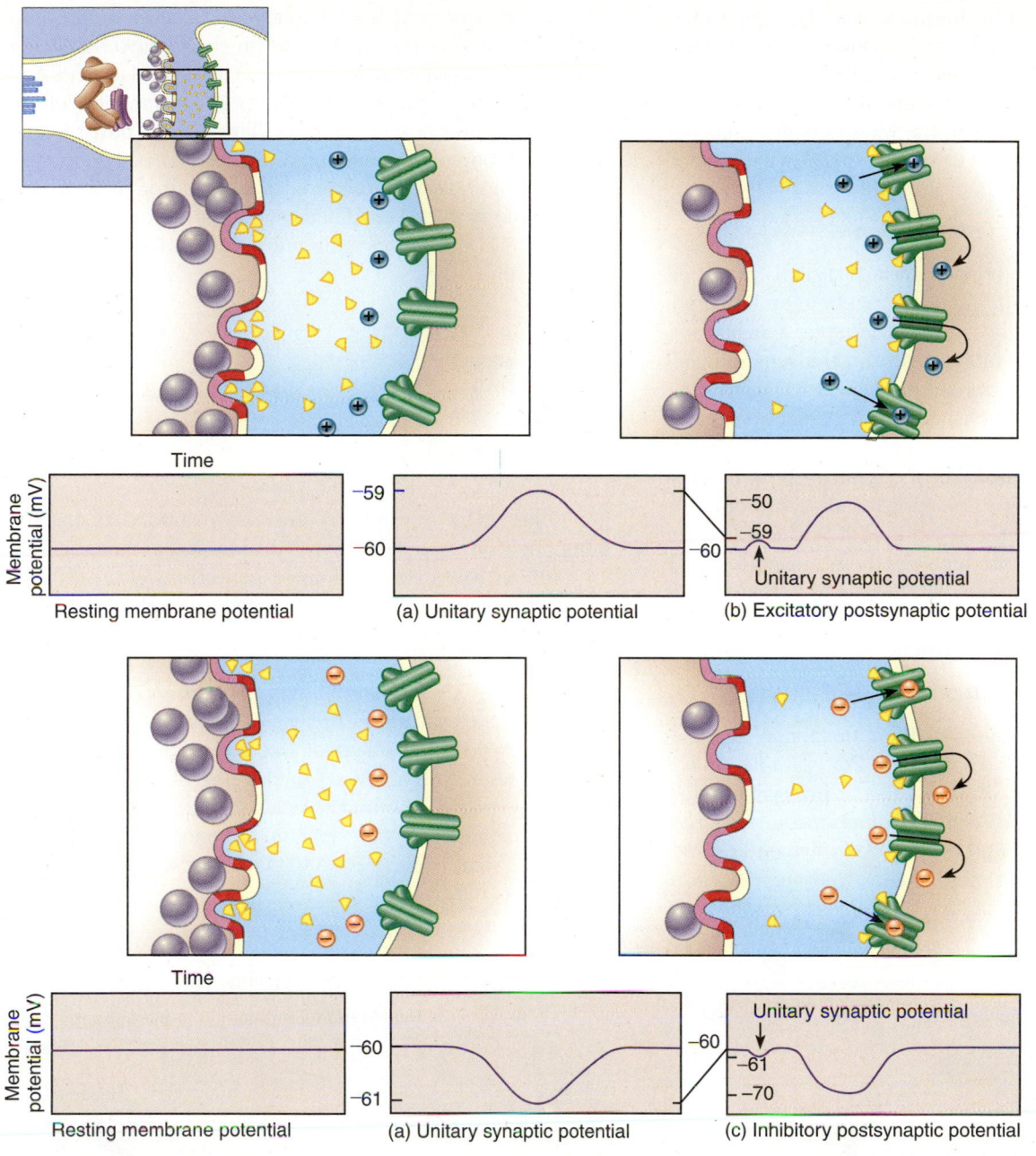

Figure 7–25

Post-synaptic potentials. *(a)* Unitary synaptic potentials. *(b)* Excitatory post-synaptic potential (EPSP). *(c)* Inhibitory post-synaptic potential (IPSP).

The properties of the chemically activated ion channels involved in post-synaptic potential changes are quite different from those of the voltage-activated ion channels responsible for the action potential. First, the chemically activated channels remain open as long as the transmitter is bound to the receptor. Second, the chemically activated channels are generally not sensitive to changes in the membrane potential. Our discussion of transmitter-receptor interactions has thus far focused only on the opening of ion channels. Another class of receptors, however, closes their ion channels when coupled to a transmitter. The closing of these chemically activated ion channels can also produce depolarizing or hyperpolarizing changes of the post-synaptic membrane potential. For example, the closing of Na^+ channels that are normally open prevents the influx of Na^+. As a consequence, the efflux of K^+ is the predominant effect, and the membrane hyperpolarizes. In a similar fashion, transmitter-receptor interactions that close K^+ channels prevent the efflux of K^+. In this situation, the influx of Na^+ predominates, and the membrane depolarizes.

The mechanisms of transmitter-activated channel closings are typically carried out by second-messenger systems from within the post-synaptic cell and last longer than transmitter mechanisms that involve the opening and/or closing of ion channel gates. The details of the ways in which channels close vary with different types of receptors and are discussed in relation to the associated neurotransmitters.

Termination of Synaptic Transmission

Inhibitory and excitatory post-synaptic potentials are transient, short-lived changes in the post-synaptic membrane potential because pre-synaptic and post-synaptic mechanisms act to remove transmitter molecules from the synaptic cleft or to inactivate the transmitter shortly after it has exerted its effect. This process is accomplished in most neurotransmitter systems by the transport of the transmitter back into the presynaptic terminal. A **reuptake** mechanism pumps most transmitters back into the pre-synaptic terminal. Other transmitters are removed from the synaptic cleft by degrading enzymes, and the metabolic products are then transported back into the pre-synaptic terminal. The pumping mechanism is specific for each type of transmitter substance or metabolite and can be affected by selective drugs.

Recycling of Neurotransmitters

Neurotransmitters or products of their metabolic degradation are recycled after they are released. For example, ACh is composed of **acetate** and **choline.** In the pre-synaptic terminus, a synthetic enzyme called *choline acetyltransferase* combines acetate and choline to form ACh, which is stored within the synaptic vesicles (Fig. 7–26). After the release of ACh into the synaptic cleft and its engagement with the post-synaptic receptor, ACh is degraded back into acetate and choline by an enzyme, **acetylcholinesterase (AchE),** present on the post-synaptic membrane. The degradation of ACh into its inactive components results in termination of the neurotransmitter action, thereby allowing the post-synaptic membrane potential to return toward resting level. Choline is taken up and reused by the pre-synaptic neuron to synthesize ACh. Some synthesis of ACh occurs in the soma, but most of the synthesis and repackaging into vesicles takes place in the axon terminus.

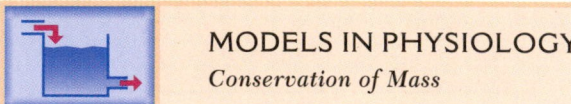

MODELS IN PHYSIOLOGY:
Conservation of Mass

Other neurotransmitters are also degraded to inactive components by specific enzymes present in the synaptic cleft. In addition, the activity of most neurotransmitters is controlled through the uptake of transmitter molecules by the pre-synaptic neuron. For example, norepinephrine, after being released from the pre-synaptic neuron in response to a nerve impulse, is rapidly pumped back inside. Thus, as the number of the transmitter molecules in the synaptic cleft decreases, so does the effect of the transmitter on the post-synaptic neuron. The intact molecules of norepinephrine may be recycled back into vesicles for later release or they may be degraded by the enzymes monoamine oxidase or catechol-*O*-methyltransferase (COMT) present in the terminal bouton (Fig. 7–27).

Figure 7–26

Acetylcholine synthesis, release, reuptake, and repackaging in a cholinergic neuron.

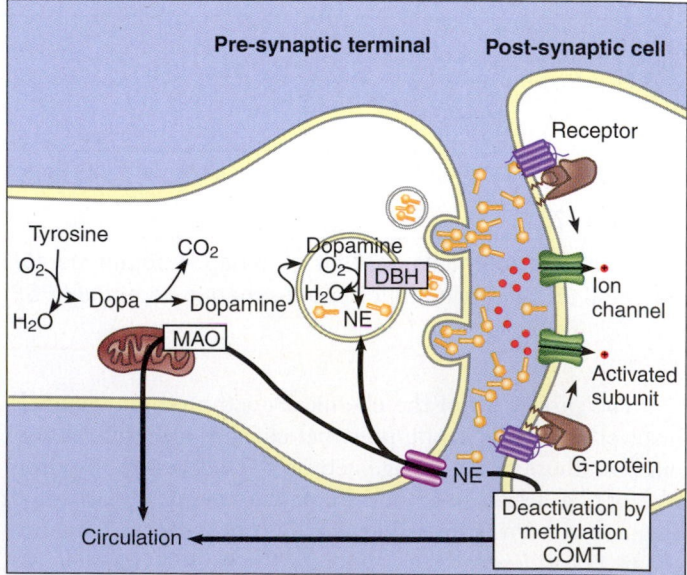

Figure 7–27

Norepinephrine synthesis, release, reuptake, and repackaging in an adrenergic neuron.

Diffuse and Discrete Chemical Synapses

Chemical synapses have been shown to have either discrete or diffuse actions. In **discrete synapses,** the chemical neurotransmitter is released from restricted areas of the pre-synaptic terminal, called **active zones,** into a small synaptic cleft; only 30 nm separate the pre- and post-synaptic membranes (Fig. 7–28). An example of a discrete synapse is the junction between nerve and striated muscle cells, which tend to activate a small region on the muscle fiber.

In **diffuse synapses,** on the other hand, transmitter release is not limited to specific active zones, and the distance between the pre-synaptic and post-synaptic membrane can be as much as 150 nm. These synapses have the form of beads on a chain, called **varicosities.** The varicosities are an extension of the axon and form an overlapping network of synapses **en passant,** or synapses in passage. As the action potential invades each varicosity, vesicles fuse with the pre-synaptic membrane, and the action potential continues on to the next varicosity. The effect of this type of synapse is to activate a large surface area of one cell or a large number of cells in a diffuse manner. The diffuse synapse is typical of nerve terminals of the sympathetic ANS and of nerve cells containing catecholamines within the CNS.

Neurotransmitters and Associated Receptors

How do neurotransmitters differ and how do they work?

Chemical neurotransmitters can be broadly classified into two groups: (1) **low-molecular-weight transmitters** and (2) **peptide transmitters** (Tables 7–2 and 7–3). The low-molecular-weight transmitters are synthesized within the pre-synaptic terminal. The necessary synthesizing enzymes are produced within the soma and transported to the terminal region. Neuropeptides, on the other hand, are fabricated in the soma and carried via axonal transport mechanisms to the synaptic terminal.

Although the processes of transmitter synthesis, release, degradation, uptake, and physiological effect are not completely understood for the peptides, they are fairly well characterized for many of the low-molecular-weight transmitters.

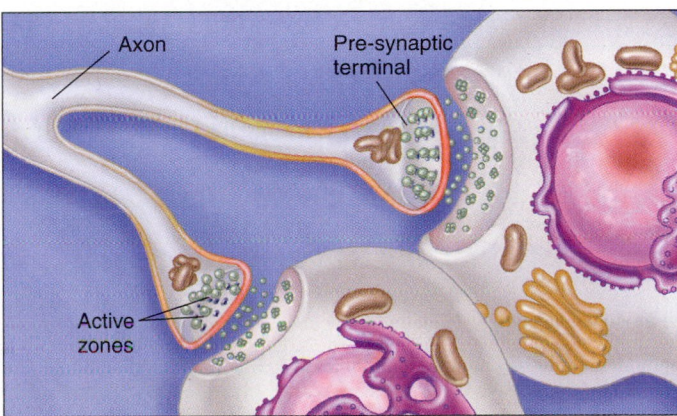

(a) Discrete terminals

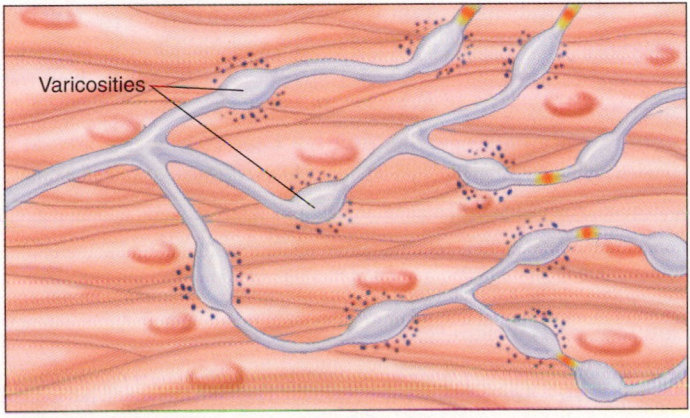

(b) Diffuse terminals

Figure 7–28

Discrete *(a)* versus diffuse *(b)* chemical synapses.

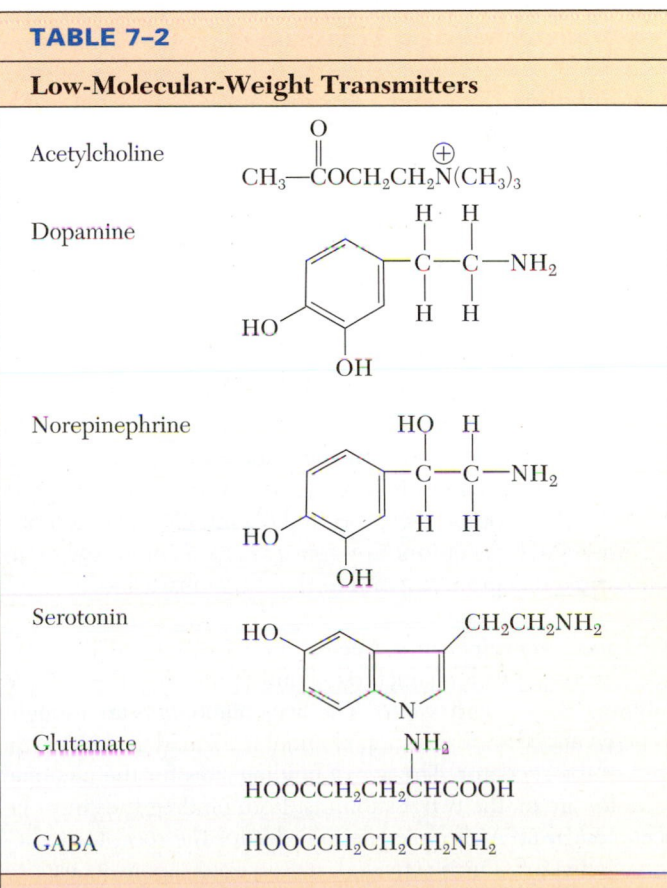

TABLE 7–2

Low-Molecular-Weight Transmitters

Acetylcholine	$CH_3-\overset{O}{\overset{\|}{C}}OCH_2CH_2\overset{\oplus}{N}(CH_3)_3$
Dopamine	
Norepinephrine	
Serotonin	
Glutamate	$HOOCCH_2CH_2CHCOOH$
GABA	$HOOCCH_2CH_2CH_2NH_2$

TABLE 7–3

Neuroactive Peptides Categorized According to Principal Tissue Localization

Hypothalamic-releasing hormones	Gastrointestinal peptides
Thyrotropin-releasing hormone	Vasoactive intestinal polypeptides
Gonadotropin-releasing hormone	Cholecystokinin
Somatostatin	Gastrin
Corticotropin-releasing hormone	Substance P
Growth hormone–releasing hormone	Neurotensin
	Methionine-enkephalin
Neurohypophyseal hormones	Leucine-enkephalin
Vasopressin	Insulin
Oxytocin	Glucagon
Neurophysin(s)	Bombesin
	Secretin
Pituitary peptides	Somatostatin
Adrenocorticotropin	Motilin
β-Endorphin	
α-Melanocyte-stimulating hormone	Others
Prolactin	Angiotensin II
Luteinizing hormone	Bradykinin
Growth hormone	Sleep peptide(s)
Thyrotropin	Calcitonin
	CGRP (calcitonin gene–related peptide)
	Neuropeptide Y

Low-Molecular-Weight Transmitters

Acetylcholine. One of the first low-molecular-weight transmitters to be studied was acetylcholine (ACh). It is found not only in the CNS but also in the PNS, where it acts as the chemical transmitter between nerve and muscle. ACh is synthesized from acetyl coenzyme A (CoA) and choline with the catalytic enzyme, choline acetyltransferase (see Fig. 7–26). ACh is packaged and stored in vesicles within the presynaptic terminal, where it is positioned for release near the active zones. The fusion of the synaptic vesicle with the presynaptic membrane causes the release of ACh into the synaptic cleft. It then diffuses across the cleft and binds with receptors on the post-synaptic membrane.

In the central and PNSs, there are two types of receptors for ACh: (1) nicotinic receptors and (2) muscarinic receptors. Nicotinic ACh receptors are sensitive to nicotine, whereas muscarinic receptors respond to the drug muscarine.

Nicotinic Acetylcholine Receptors. Nicotinic ACh receptors have been well characterized and found to consist of five subunits: β, γ, δ, and two αs. The five components are thought to be arranged so that an ion channel is formed at the central core of the receptor. The active binding sites for the nicotinic receptor are on the two α subunits. Both binding sites must be occupied by an ACh molecule in order for the receptor to become activated. Once activated, the receptor opens its gate to permit the simultaneous influx of Na^+ ions and the efflux of

K^+ ions (Table 7–4). As we have seen, the driving forces for Na^+ influx are much greater than for K^+ efflux.

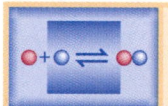

MODELS IN PHYSIOLOGY:
Chemical Mass Action

Consequently, the excess positive charge movement into the cell causes the membrane to depolarize. The ion channel of the nicotinic receptor remains in the open state until ACh uncouples from its receptor. After ACh dissociates from its receptor, the receptor channel closes and Na^+ and K^+ are no longer able to pass through the channel. ACh is then free to diffuse within the synaptic cleft, where it binds with the membrane-bound enzyme AChE. The AChE enzyme degrades ACh by hydrolysis to produce choline and acetate. Choline is then taken up into the pre-synaptic terminal by a high-affinity uptake mechanism and recycled to be used again to synthesize ACh. This process of cholinergic nicotinic transmission is found at the neuromuscular junction in addition to various locations in the CNS.

Muscarinic Acetylcholine Receptors. The structure and function of the second type of ACh receptor, the muscarinic receptor, are quite different from those of the nicotinic ACh receptor. Two types of muscarinic receptors, M_1 and M_2, have been identified. Both are composed of seven membrane-

TABLE 7–4

Types of Neurotransmitters and Receptors

Neurotransmitter	Receptor	Membrane Conductance	Membrane Potential	Second Messenger
Glutamate	Kainate	Increase g_{Na}, g_K	EPSP	
	Quisqualate	Increase g_{Na}, g_K	EPSP	
	NMDA	Increase g_{Ca}	EPSP	
Acetylcholine	Nicotinic	Increase g_{Na}, g_K	EPSP	
	Muscarinic M_1	Decrease g_K	EPSP	IP_3 and DAG
	Muscarinic M_2	Increase g_K	IPSP	cAMP
Serotonin	$5\text{-}HT_{1A}$	Increase g_K	IPSP	cAMP
	$5\text{-}HT_{1B}$			
	$5\text{-}HT_{1C}$	Increase g_{Cl}	IPSP	IP_3
	$5\text{-}HT_{1D}$			
	$5\text{-}HT_2$	Decrease g_K	EPSP	IP_3
	$5\text{-}HT_3$	Increase g_{Na}, g_K	EPSP	
GABA	GABA-A	Increase g_{Cl}	IPSP	
	GABA-B	Increase g_K	IPSP	cAMP?
Dopamine	D_1		EPSP?	cAMP
	D_2		IPSP	cAMP
Norepinephrine	α_1	Increase g_K	IPSP (CNS) [contraction (PNS)]	IP_3, DAG
	α_2	Decrease g_{Ca}		cAMP
	β_1		[heart acceleration]	cAMP
	β_2		[dilation (PNS)]	cAMP

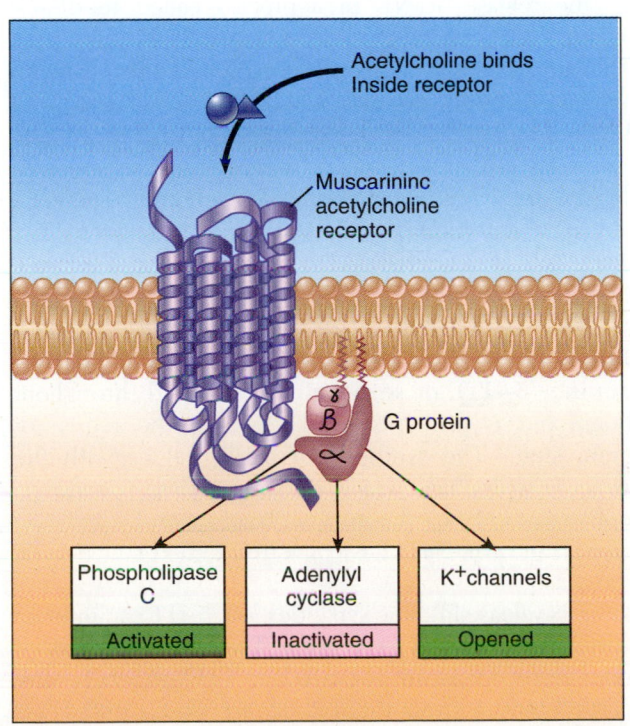

Figure 7–29

Muscarinic acetylcholine receptor, type M_2.

spanning domains and both exert their actions through a G-protein (Fig. 7–29). The activation of M_1 receptors, however, results in a decrease of K^+ conductance via phospholipase C, whereas M_2 receptor activation causes an increase in K^+ conductance via the inhibition of adenylate cyclase. As a consequence, when ACh binds to an M_1 receptor, it depolarizes the membrane, and when it binds to an M_2 receptor, it causes a hyperpolarization.

Biogenic Amines. Biogenic amines are a class of low-molecular-weight transmitters characterized by the presence of an amine group (NH_2). A subgroup of the biogenic amines, the catecholamines, have a catechol ring (see Table 7–2). This subgroup includes the transmitters **dopamine (DA)**, **norepinephrine (NE)**, and **epinephrine**. The synthesis of each catecholamine follows a similar biochemical pathway that starts with the synthesis of DA.

Dopamine. DA is synthesized from the amino acid tyrosine, which is converted to DOPA (Dihydroxyphenylalanine) by the enzyme tyrosine hydroxylase (TH). TH is the controlling enzyme that regulates the overall synthesis of DA. DOPA, in turn, is converted to DA by the enzyme DOPA decarboxylase. Based on their pharmacological properties, three subtypes of DA receptors have been identified: (1) D_1, (2) D_2, and (3) D_3. **D_1 receptors** are coupled to G_s-proteins, leading to the activation of adenylate cyclase, while **D_2 and D_3 receptors** are coupled to G_i-proteins, which cause a decrease in adenylate cyclase activity. The activation of D_2 receptors has been shown to hyperpolarize the post-synaptic membrane by increasing potassium conductance. After uncoupling with the receptor, DA is transported by a reuptake pumping mechanism back into the pre-synaptic terminal, where it is repackaged into synaptic vesicles. Eighty percent of the DA within the synaptic cleft is transported back to the pre-synaptic terminal. Thus, the reuptake process is the main mechanism by which DA transmission is terminated. The remaining 20% of DA is degraded within the cleft by COMT.

The synaptic transmission of DA is greatly affected by commonly used illegal drugs. Cocaine, for example, inhibits the reuptake of DA into the pre-synaptic terminal, and amphetamine increases the release of DA into the synaptic cleft. Both drugs effectively increase the levels of DA within the cleft to activate DA receptors.

Norepinephrine. Norepinephrine (NE) is another member of the catecholamine family that is found throughout the CNS and also at the junction between nerves and smooth muscles in the ANS. NE is synthesized from DA by the enzyme DA β-hydroxylase (see Fig. 7–27).

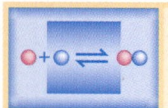

MODELS IN PHYSIOLOGY:
Chemical Mass Action

The formation of NE is quite labile and is controlled by mechanisms that can change the enzymatic activity associated with NE synthesis over either a short or long period of time. The short-term changes involve the modulation of TH. Typically, the activity of TH is controlled by both DA and norepinephrine through the process of end-product inhibition (see Chapter 5). When the enzyme is phosphorylated, however, the process of end-product inhibition is less effective. In addition, the phosphorylation of TH increases the affinity of the enzyme to the tyrosine substrate. Thus, the phosphorylation of the TH can increase its enzymatic activity. The phosphorylation of TH has been shown to occur by cAMP-dependent protein kinase and Ca^{2+}/calmodulin mechanisms (see Chapter 5). The short-term increases in the synthesis of NE occur within minutes and are easily reversible.

Long-term increases in NE can be caused by factors such as stress because stress can lead to increases in TH and DA β-hydroxylase enzymes. Environmental factors can therefore play a significant role in modifying nerve cell function on a long-term basis.

The binding of NE to α-receptors in the CNS opens K^+ channels and causes a hyperpolarization of the post-synaptic cell. In the PNS, two types of norepinephrine receptors have been identified: α-receptors and β-receptors. The α_1-receptors are found on the smooth muscles of blood vessels. When they are activated, an increase in Ca^{2+} ion influx occurs, which in turn causes the contraction of smooth muscles. The α_1-receptors also stimulate the hydrolysis of phosphatidylinositol, a lipid in the post-synaptic membrane that activates the intracellular messenger diacylglycerol. Diacylglycerol in turn activates protein kinase C, which initiates a variety of cellular functions (see Chapter 5).

β_2-receptors are also found on smooth muscles. Their activation, however, results in a relaxation of smooth muscles by mechanisms that remain to be determined. In addition to activating ionic channels, the binding of NE to β_2-receptors activates metabolic pathways by converting ATP to cAMP. As previously discussed, this second messenger in turn activates a cAMP-dependent protein kinase, which regulates a number of important cellular functions (see Chapter 5). The β_1-receptors are found in the heart, kidney, and adipose tissue, where they are responsible for the acceleration of heart rate, renin secretion, and lipolysis, respectively.

The α_2- and β_2-receptors are located on the membrane of the pre-synaptic terminal and appear to control the amount of NE that is released (Fig. 7–30). These receptors are often referred to as **autoreceptors.** As the amount of NE within the synaptic cleft is increased, more α_2-autoreceptors become activated. These autoreceptors then inhibit the release of NE from the pre-synaptic terminal in a process of **feedback inhibition.** The activation of β_2-receptors, on the other hand, increases the release of NE in a process called **feedback excitation.**

After uncoupling from its receptor, NE is taken back up into the pre-synaptic terminal, where it is repackaged into vesicles in preparation for release again into the synaptic cleft. As with DA, this reuptake process removes approximately 80% of the NE from the synaptic cleft and is the main mechanism that terminates NE synaptic transmission. The remaining NE is degraded within the cleft by the enzyme COMT.

Serotonin. Another common biogenic amine is **5-hydroxytryptamine (5-HT),** or **serotonin.** It is found throughout the brain, but is primarily synthesized in the region of the brain stem. The synthesis of 5-HT begins with the amino acid tryptophan, which is converted to 5-hydroxytryptophan (5-HTP) by the enzyme tryptophan hydroxylase (Fig. 7–31). In turn, 5-HTP is converted to 5-HT by 5-HTP decarboxylase.

After its release into the synaptic cleft, 5-HT can interact with several types of 5-HT receptors. The best characterized serotonergic receptors are the 5-HT_{1A}, 5-HT_{1C}, 5-HT_2, and 5-HT_3 receptors. Activation of either the 5-HT_{1A} or 5-HT_{1C} receptors results in IPSPs. The inhibitory potential of 5-HT_{1A} receptor activation, however, is mediated by an increase in K^+ conductance via cAMP as the second messenger, whereas

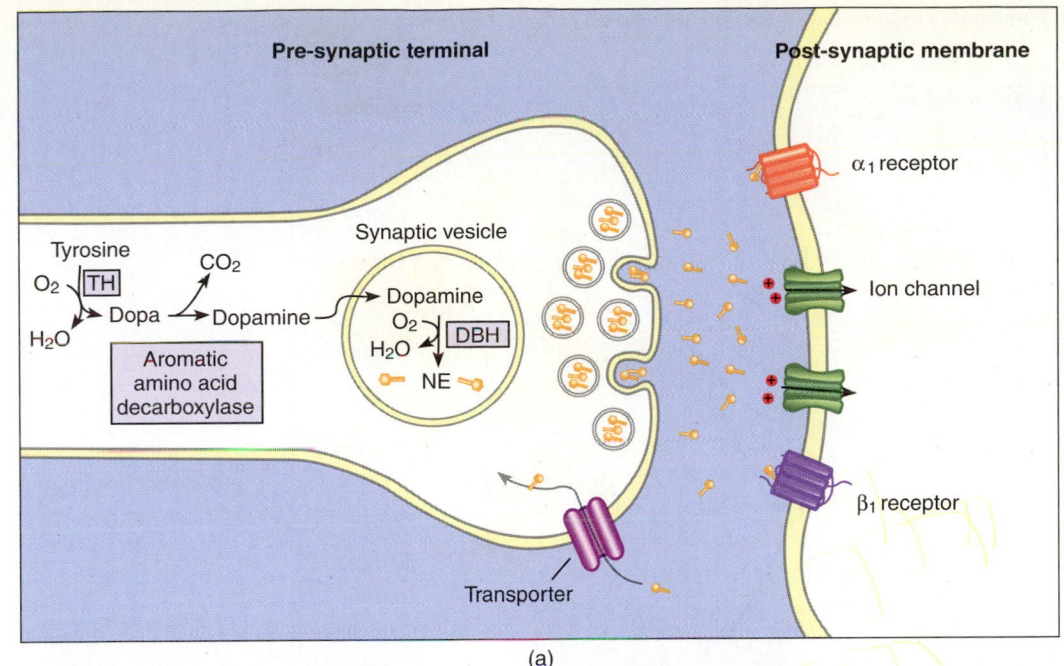

(a)

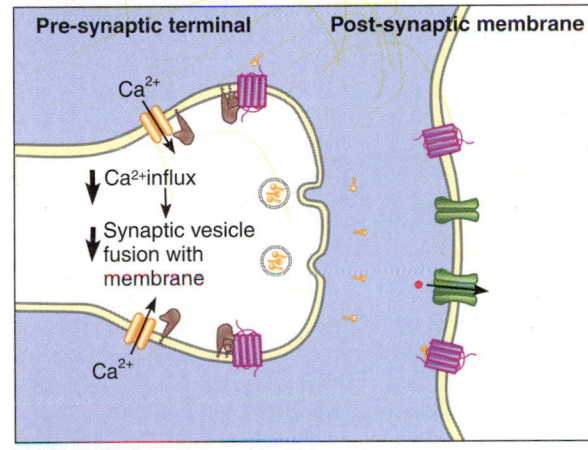

(b) FEEDBACK INHIBITION

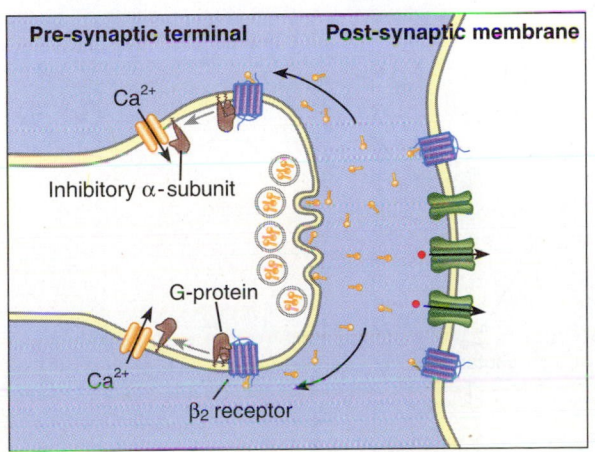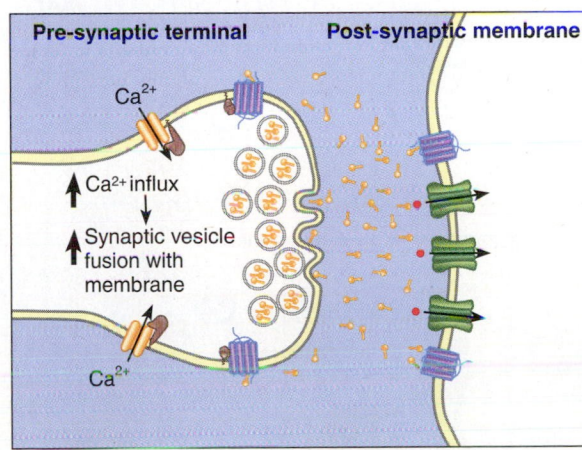

(c) FEEDBACK EXCITATION

Figure 7–30

(a) Norepinephrine (NE) binds to two types of receptors on the post-synaptic membrane: α_1 and β_1. *(b)* On the pre-synaptic membrane, binding of NE to α_2-receptors causes NE release to be inhibited, whereas *(c)* binding of NE to β_2-receptors causes more NE to be released.

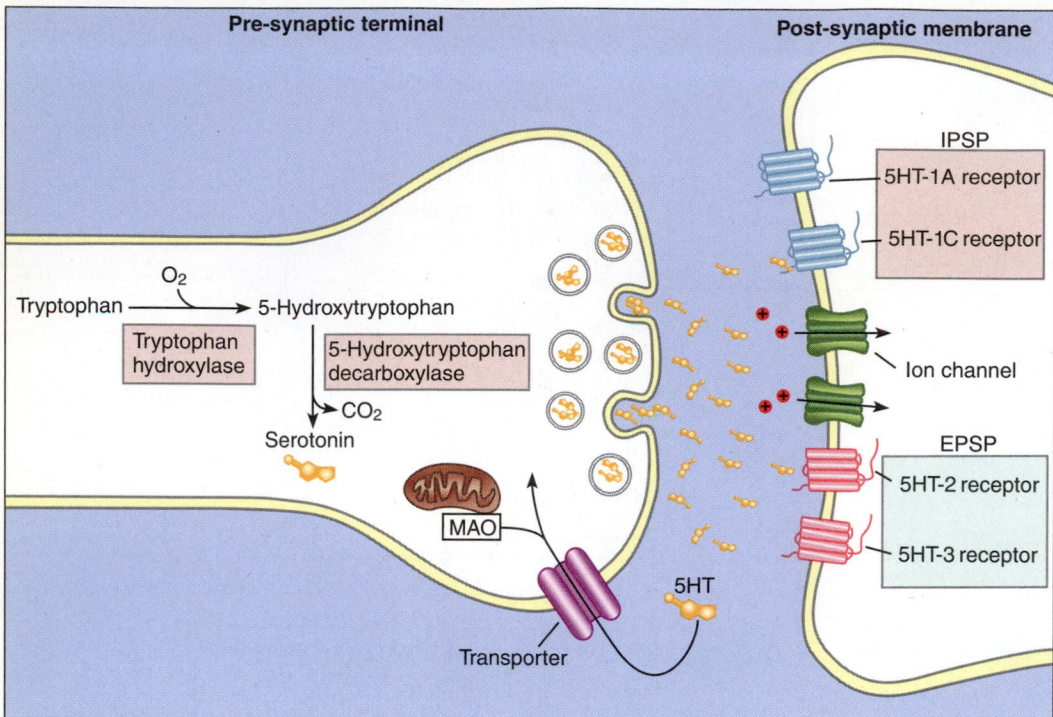

Figure 7–31

Serotonin neurotransmission.

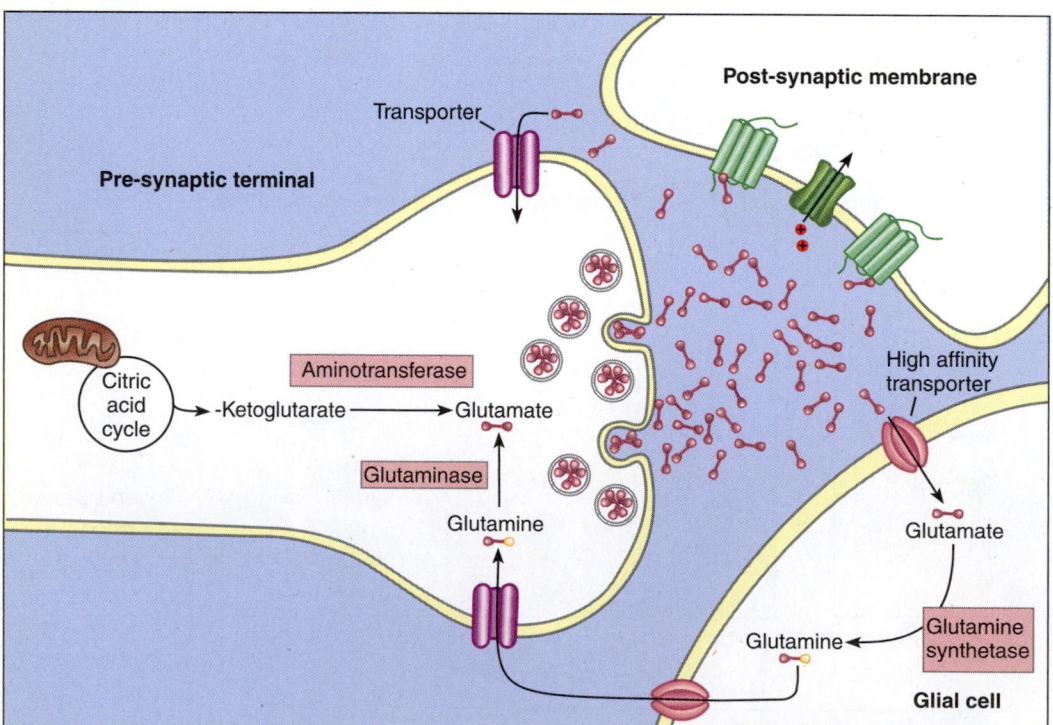

Figure 7–32

Glutamate is synthesized from a product of the citric acid cycle and is converted to glutamine in a glial cell before reuptake.

that produced by 5-HT$_{1C}$ activation is carried out by an increase in Cl$^-$ conductance via IP$_3$. Activation of either the 5-HT$_2$ or 5-HT$_3$ receptors results in the generation of EPSPs. The excitatory potential produced by the activation of the 5-HT$_2$ receptor is due to a decrease in K$^+$ conductance via IP$_3$, whereas the 5-HT$_3$ receptor is directly coupled to an ion channel that allows the influx of Na$^+$ ions and the efflux of K$^+$ ions.

After uncoupling from its receptor, 5-HT is transported back into the pre-synaptic terminal. As with NE and DA, approximately 80% of the 5-HT within the synaptic cleft is removed and recycled by this reuptake process. The remainder of the 5-HT is traken up and degraded by an enzyme called monoamine oxidase.

Amino Acids. A variety of amino acids also satisfy the conditions necessary to be classified as neurotransmitters. We will consider only glutamate and gamma aminobutyric acid because they are found in great abundance throughout the nervous system.

Glutamate. **Glutamate** is synthesized from α-ketoglutarate by way of the citric acid cycle (Fig. 7–32). It is one of the most potent excitatory neurotransmitters in the nervous system. The three subtypes of glutamate receptors are (1) kainate, (2) quisqualate, and (3) *N*-methyl-D-aspartate (NMDA). Activation of the **kainate** and **quisqualate receptors** produces excitatory post-synaptic potentials by

opening ion channels that increase Na$^+$ and K$^+$ conductance. **NMDA receptor** activation results in an increase in Ca^{2+} conductance. This receptor, however, is blocked by Mg^{2+} when the membrane is in the resting state and becomes unblocked when the membrane is depolarized. Thus the NMDA receptor can be thought of as both a ligand and a voltage-gated channel.

Synaptic transmission by glutamate is terminated in part by its reuptake into the pre-synaptic terminal. Much of the glutamate within the synaptic cleft, however, is transported into glial cells. There, it is converted to glutamine by the enzyme glutamine synthetase. The glutamine in turn is transported back into the pre-synaptic terminal, where it is then reconverted to free glutamate and repackaged into vesicles. Glial cells, therefore, play a crucial role in regulating glutamate neurotransmission.

Gamma Aminobutyric Acid. Another amino acid neurotransmitter found throughout the CNS is **gamma amino-butyric acid (GABA)**. GABA is a potent inhibitory neurotransmitter synthesized from glutamate by the enzyme glutamic acid decarboxylase (Fig. 7–33). The two types of GABA receptors are (1) GABA$_A$ and (2) GABA$_B$. The **GABA$_A$** receptor is a ligand-gated Cl$^-$ channel, and its activation produces inhibitory post-synaptic potentials by increasing the influx of Cl$^-$ ions. This receptor is composed of five subunits, each designated as α, β, or γ. All of the subunits bind GABA. The increase in Cl$^-$ conductance is facilitated by drugs called

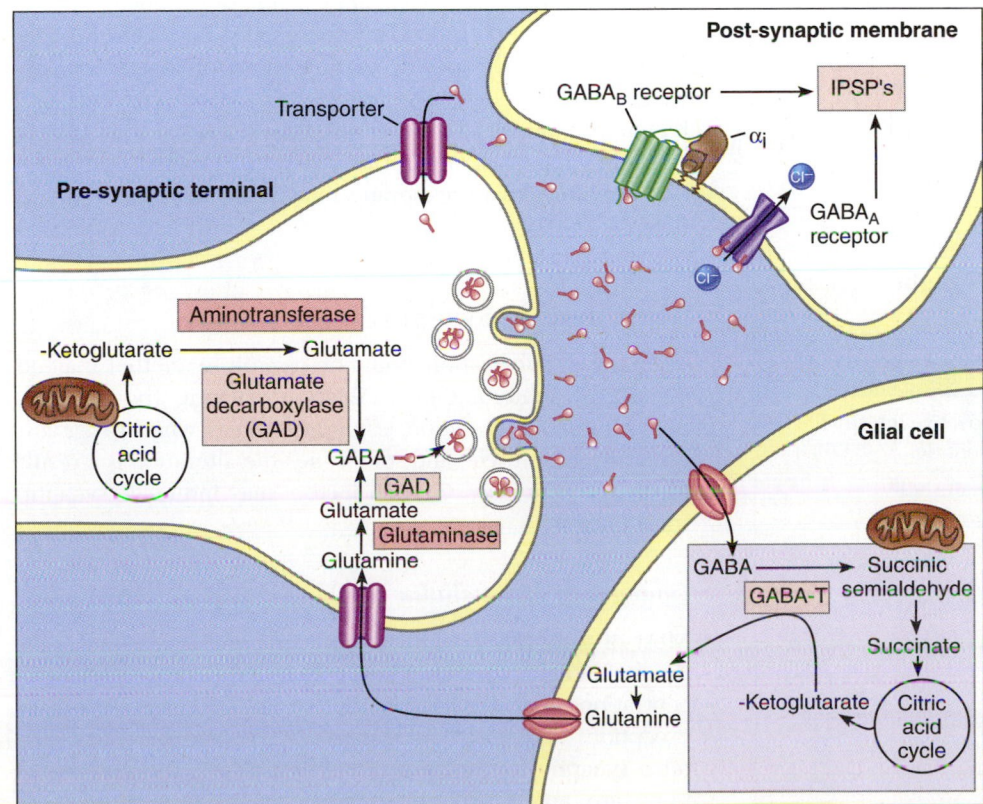

Figure 7–33

The GABA synapse. The GABAA receptor is made up of five subunits arranged to form a central pore which, when opened, allows for the passage of chloride ions into the post synaptic cell. GABA binds to a receptor subunit opening the channel and allowing chloride ions to pass into the post synaptic neuron, initiating an IPSP. After recognition, GABA is released from the receptor and taken up by surrounding glial cells, which recycle the neurotransmitter.

benzodiazepines, which bind to the α-subunits. Benzodiazepines, such as chlordiazepoxide, are used as anticonvulsants and sedatives.

The activation of the **GABA_B** receptor also produces inhibitory post-synaptic potentials. With this receptor, however, the IPSP results from an increase in K^+ conductance via the activation of a G-protein. The synaptic transmission of GABA is terminated by its reuptake into the pre-synaptic terminal and by its transport into glial cells. The mitochondria within glial cells convert GABA into succinic semialdehyde by the enzyme GABA-T. At the same time, this enzyme is coupled to the conversion of α-ketoglutarate to glutamate. Glutamate in turn is converted to glutamine by glutamine synthetase and is then transported to the pre-synaptic terminal. Within the pre-synaptic terminal, glutamine is converted into glutamate and subsequently into GABA to be packaged into synaptic vesicles.

High-Molecular-Weight Neuropeptides

Neuropeptides are chemical transmitters that consist of chains of amino acids. The processing of neuropeptides differs considerably from that of low-molecular-weight transmitters. Neuropeptides are synthesized in the soma of neurons rather than in the synaptic terminal. In addition, neuropeptides are created when large proteins, or **polyproteins,** are broken down. The various peptides are packaged within secretory vesicles and carried to the terminal area by mechanisms of fast axonal transport. Within the synaptic terminal, vesicles containing peptides are found to coexist with vesicles containing low-molecular-weight transmitters. Table 7–5 lists the low-molecular-weight and neuropeptide transmitters that have been found within the same synaptic terminals. It is thought that the released neuropeptide modulates the actions of the low-molecular-weight neurotransmitter.

TABLE 7–5	
Colocalization of Low-Molecular-Weight Transmitters and Neuropeptides	
Low-Molecular-Weight Transmitter	**Neuropeptide**
Acetylcholine	Vasoactive intestinal peptide
Norepinephrine	Somatostatin Enkephalin Neurotensin
Dopamine	Cholecystokinin Enkephalin
Adrenalin	Enkephalin
Serotonin	Substance P Thyrotropin-releasing hormone

At the terminal ending, the process of synaptic transmission of peptides is different from that of low-molecular-weight transmitters. Once they are released into the synaptic cleft, no reuptake mechanisms are available to recycle the neuropeptides. Therefore, the process of peptide transmission cannot be sustained as it is for the low-molecular-weight transmitters.

Several neuropeptides are listed in Table 7–5, but there are many others. **Opiates** are peptides that bind to opioid receptors. They appear to be involved in the regulation of pain information. Opioid peptides include met-enkephalin, leu-enkephalin, dynorphin, and β-endorphin. Structurally, they share homologous regions consisting of the amino acid sequence Tyr-Gly-Gly-Phe. The opiates are derived from three propeptides: **proenkephalin, proopiomelanocortin,** and **prodynorphin.** Proenkephalin gives rise to met- and leu-enkephalin; proopiomelanocortin gives rise to β-endorphin; and prodynorphin is the precursor of dynorphin. Several opioid receptor subtypes exist. β-endorphin binds preferentially to **mu (μ) receptors;** enkephalins bind preferentially to **delta (δ) receptors,** and dynorphin binds preferentially to **kappa (κ) receptors.** The enkephalins are metabolized by two enzymes: (1) aminopeptidase, which hydrolyzes the Tyr-Gly bond, and (2) enkephalinase, which hydrolyzes the Gly-Gly bond.

Nitric oxide (NO) is a gaseous neurotransmitter synthesized from arginine by NO synthase. It also acts as a second messenger, causing relaxation of smooth muscle in the walls of arterioles. It has numerous other regulatory roles and is involved in many signaling pathways.

From our discussion of chemical neurotransmitters, it can be seen that nerve cells are capable of producing a variety of transmitters. Individual nerve cells, however, are able to synthesize only specific combinations of low-molecular-weight and peptide transmitters based on the enzymes that they have available. This defined set of chemical transmitters is used by a neuron at all of its synapses to transduce the action potentials that it generates into chemical signals that are detected by target cells.

Alterations in Synaptic Transmission

The flow of information from one neuron to another may be altered by inhibiting or facilitating synaptic transmission. Sometimes the alteration is intentional, as when drugs are used to relax muscle, and sometimes the alteration is a result of disease processes. Several factors may influence synaptic transmission.

Changes in extracellular (H^+)

Most of the body's chemical reactions are dependent upon the maintenance of stable interstitial fluid hydrogen ion concentrations, and the chemistry of synaptic transmission is no exception. In general, a decrease of (H^+) (more basic) facilitates synaptic transmission. In severe cases, the manifestations are apathy, mental confusion, tetany (twitching of

APPLICATIONS OF PHYSIOLOGY

Neurotoxins

A **toxin** is a noxious or poisonous substance elaborated within (an **endotoxin**) a plant cell, a microorganism, or a higher animal cell as a normal intracellular component, or produced and released (an **exotoxin**) as an extracellular product. Toxins exert their poisonous effects by disrupting normal cell function in the target organism. One way to classify toxins is by the type of cell function that they disrupt. Hemolytic toxins, such as Crotalus toxin from rattlesnakes, destroy blood cells, compromising functions of the blood. Neurotoxins interfere with nerve impulse conduction and synaptic transmission.

Neurotoxins can alter normal nerve cell function by blocking voltage-activated Na^+ channels, voltage-activated K^+ channels, pre-synaptic neurotransmitter release, or post-synaptic receptor sites for neurotransmitters. **Tetrodotoxin** is found in the skin and some internal organs of puffer fish, considered delicacies in Japan, where they are carefully and expertly prepared and served in a dish called *fugu*. Tetrodotoxin is a highly specific, high-affinity Na^+ channel blocker that acts by preventing the activation gate from opening. Consequently, the nerve fiber is unable to generate and propagate an action potential. **Batrachotoxin**, a poison in the skin secretions of the Columbia frog, prevents Na^+ channel gates from closing. As a result, Na^+ gradients essential for depolarization cannot be restored or maintained, and subsequent stimuli fail to elicit an action potential.

Dendrotoxin (DTX) is a high-affinity blocker of voltage-activated K^+ channels. DTX is a peptide component in the venom of the black mamba, a poisonous snake in central and southern Africa. K^+ channel blockers prevent repolarization of the axon, thereby stopping conduction of the nerve impulse.

Botulinus toxin is a potent neurotoxin produced by the anaerobic bacterium *Clostridium botulinum*. It causes food poisoning. After ingested, usually from improperly prepared canned or preserved food, it inhibits the release of acetylcholine from pre-synaptic nerve terminals, disrupting communication in cholinergic pathways. Effects include weakness and paralysis of skeletal muscles, respiratory failure, and even death.

Cobratoxin, a component of cobra venom, binds nicotinic acetylcholine receptors on the post-synaptic membrane of autonomic neurons, nerve cells that control visceral functions such as blood pressure, heart rate, gastrointestinal secretion and motility, and many others. Cobratoxin occupies the receptor and prevents acetylcholine from engaging the receptor and opening the ligand-gated ion channel, effectively blocking synaptic transmission.

---

skeletal muscles), and spastic muscles. An increase in the (H^+) (more acidic) tends to depress synaptic transmission and in extreme cases may lead to disorientation, depression of the CNS, coma, and finally death.

Chemical Agents

Some chemical agents block synaptic transmission by inertly occupying the post-synaptic receptor, thereby preventing the neurotransmitter from acting. A classic example is **D-tubocurarine**, the active agent in **curare**, an arrow poison used by South American natives for hunting small mammals. In small doses, D-tubocurarine blocks ACh receptors in muscle, reducing the ability of the motor nerve to stimulate muscle contraction. Large doses can kill by paralyzing respiratory muscles.

Other drugs act to block synaptic transmission by inhibiting transmitter synthesis, blocking transmitter release, or accelerating destruction of released transmitter.

Some chemical agents facilitate synaptic transmission. For example, **amphetamine** stimulates the release of DA, the neurotransmitter associated with arousal and pleasure pathways in the brain. Other psychoactive drugs act by mimicking natural transmitters at their post-synaptic receptors (e.g., mescaline) or enhance the effects of natural transmitter by blocking its degradation at the synapse. As mentioned earlier, **cocaine,** a powerful stimulant, appears to act by blocking pre-synaptic uptake of NE and 5-HT, thereby potentiating the effects of these transmitters in the brain.

Disease

Many disease processes also affect synaptic transmission. For example, evidence suggests that **schizophrenia** is associated with the overproduction of DA or an increased sensitivity to the transmitter in regions of the brain involved in emotional behavior.

Parkinson's disease, a chronic disorder characterized by involuntary shaking of the arms, legs, and head, is related to a deficiency of the transmitter DA at synapses in brain structures that control and adjust movement.

Tetanus, commonly called "lockjaw," is caused by a toxin produced by a microorganism, the tetanus bacillus. The toxin blocks receptors at inhibitory synapses in nerve

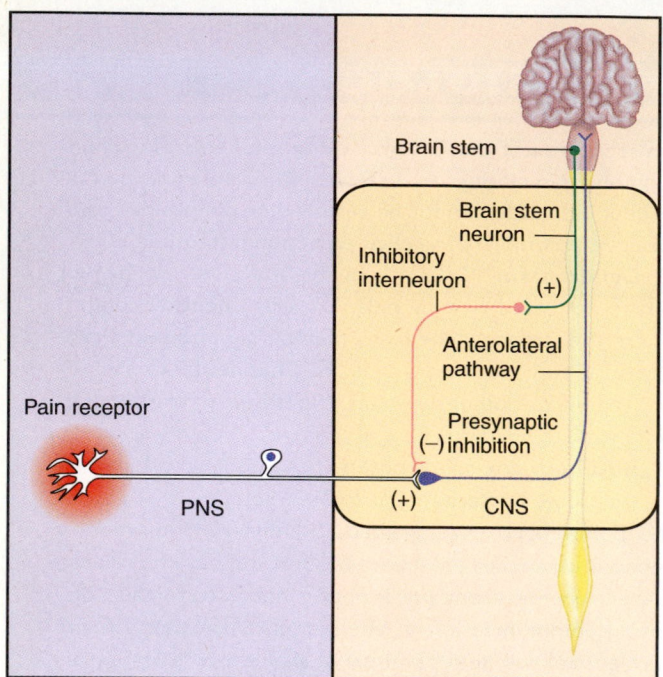

Figure 7–34

Pre-synaptic inhibition of the first-order neuron in the pain pathway.

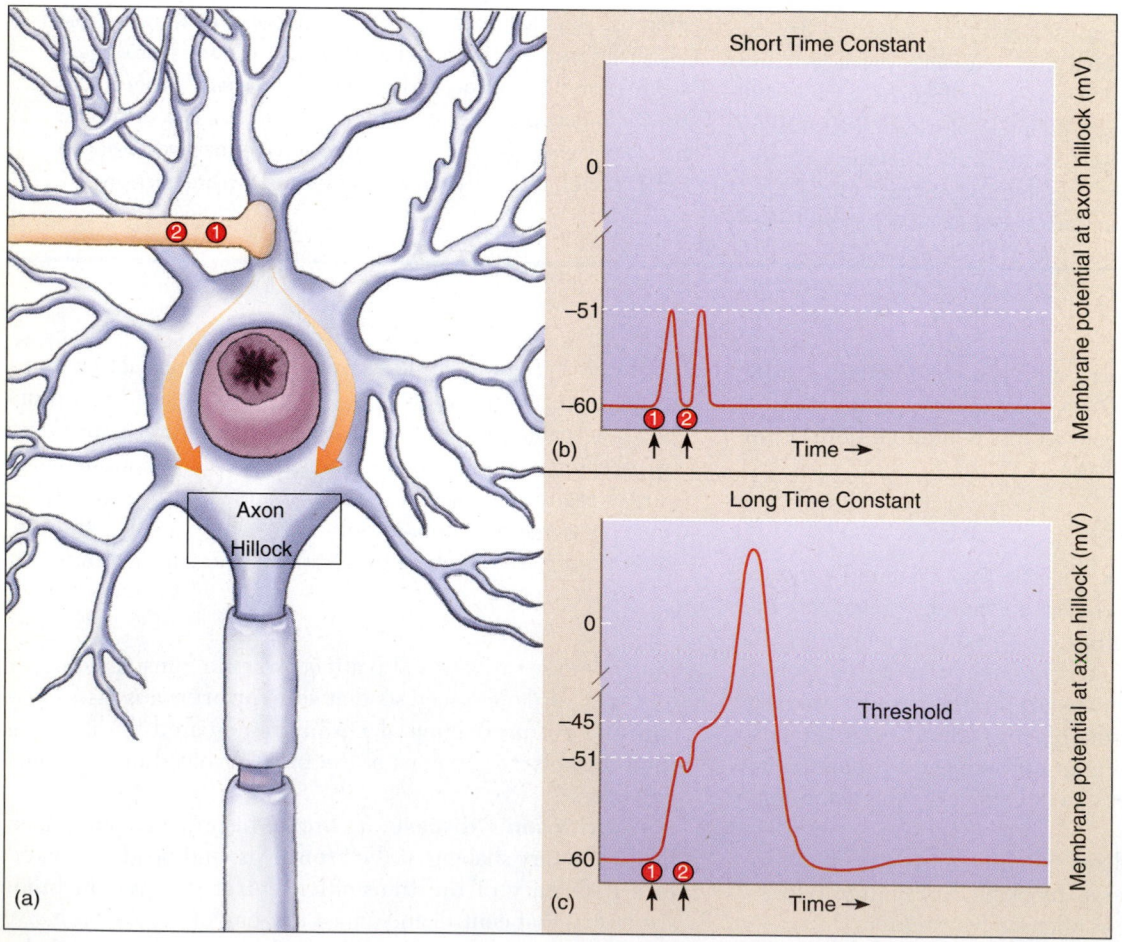

Figure 7–35

(a–c) Temporal summation of EPSPs.

pathways supplying skeletal muscle, resulting in muscle seizures and spasms, such as in jaw muscles.

Pre-synaptic Inhibition

Synaptic transmission at some excitatory synapses can be inhibited by decreasing the release of excitatory transmitter by means of an additional neuron that synapses with the pre-synaptic terminal (Fig. 7–34). Pre-synaptic inhibition provides a means whereby certain excitatory inputs to the brain and spinal cord may be depressed. For example, pre-synaptic inhibition is thought to regulate the transmission of pain information from peripheral pain receptors to the brain.

NEURONAL INTEGRATION

A principal function of the neuron is to relay information by transmitting nerve impulses. In performing this task, the neuron must necessarily receive input (excitatory, inhibitory, or both), integrate the input, and express the result in terms of its own rate of action potential generation. The process of integrating incoming signals is called *information decoding*. The process of expressing the result in terms of the frequency of action potential generation is called *information encoding*.

Information Decoding

Most neurons receive inputs from many terminals, sometimes hundreds or thousands. Some of these synapses are excitatory, and some are inhibitory. Furthermore, at any given moment, some synapses are active, and others, inactive. It is the sum of the activity at inhibitory and excitatory synapses that determines whether or not the neuron will generate an action potential and, if so, the frequency of generation.

A single EPSP is usually not sufficient to bring the axon hillock membrane of the post-synaptic neuron to threshold level and cause an action potential to be generated. However, two or more EPSPs generated in rapid succession at the same synapse may be added together to generate current sufficient for bringing the axon hillock membrane of the post-synaptic neuron to threshold, causing the neuron to generate an action potential (Fig. 7–35). Because this phenomenon involves two or more inputs close in time at the same synapse, it is called **temporal summation.**

Similarly, the effects of several EPSPs occurring simultaneously at different synapses on the same neuron, when added together, may be of sufficient strength to bring the axon hillock to threshold and cause the neuron to generate an action potential (Fig. 7–36). This phenomenon is called **spatial summation.**

Inhibitory post-synaptic potentials also are temporally and spatially summated. Generally, both EPSPs and IPSPs are summated simultaneously, their algebraic sum determining whether the current flow to the axon hillock is of sufficient strength to cause the neuron to generate an action potential.

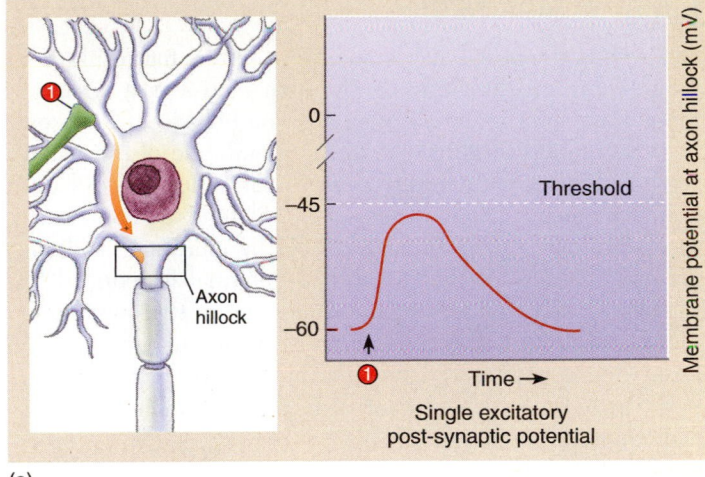

(a)

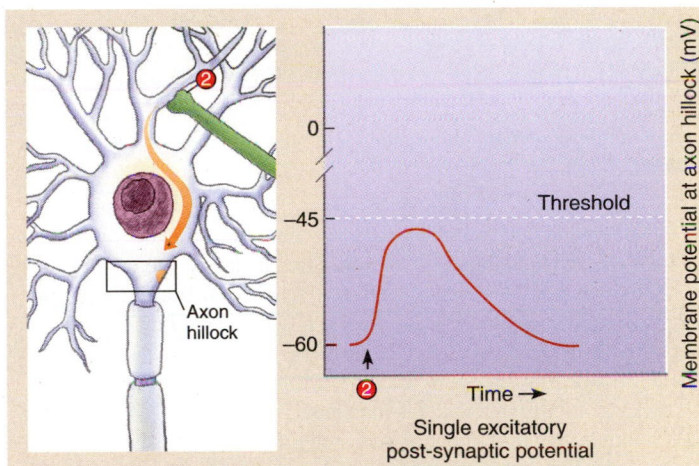

(b)

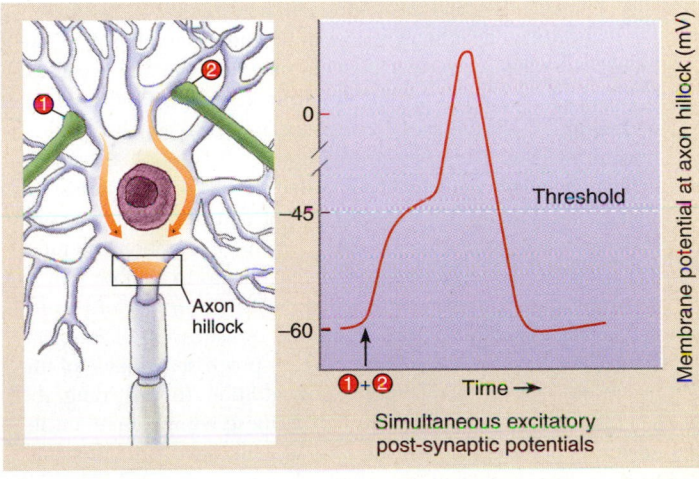

(c)

Figure 7–36

(a–c) Spatial summation of EPSPs.

Information Encoding

Changes in the post-synaptic membrane potential result in a flow of ion current between the area of the post-synaptic membrane and the initial segment (axon hillock) of the axon because the ion conductance of the membrane of the initial segment can be more easily altered than other parts of the axon. The current flow caused by EPSPs tends to depolarize the membrane of the initial segment by increasing sodium ion conductance. If the membrane potential is reduced to threshold, an action potential is generated at the initial segment, resulting in the conduction of an impulse down the axon. Similarly, IPSPs create a current flow that tends to hyperpolarize the initial segment, reducing the likelihood of the action potential generation.

Neurons usually respond to summed post-synaptic depolarization by generating a train of impulses proportional to the intensity of incoming stimuli. Frequencies vary from less than one per second to several hundred per second. The information that they carry is represented by the number of impulses generated per unit of time. This is known as **frequency coding.** The greater the intensity of synaptic stimuli, the higher the frequency of firing.

Many neurons relay the same kind of information (e.g., pain or temperature change), and in subsequent chapters, we shall see that information can also be coded in the nervous system by varying the number of nerve fibers that are conducting impulses to or from the same area of the body.

CHAPTER REVIEW

Summary

- The nervous system is composed of the brain, spinal cord, cranial nerves, and spinal nerves. The first two make up the central nervous system (CNS), whereas the latter two constitute the peripheral nervous system (PNS). The brain is subdivided into functional regions involved in sensory input from the skin, eye, ears, or other sensory organs. Other areas are involved with locomotor activity as well as other functions. This functional partitioning of brain areas is found throughout the nervous system. A general organizational theme of the nervous system is that the brain receives sensory input and provides the appropriate output for such functions as speech, movement, and the control of internal organs.

- Glia, or glial cells, aid in maintaining the environment of the brain. They also act as scavengers in the brain to remove cellular debris. Another important function of glial cells is the formation of myelin around the axon of nerve cells to increase the conduction and transmission of electrical impulses. Neurons are the basic cellular elements in the nervous system involved in the transmission of information. The membranes of neurons have structural and functional properties specialized for the integration and rapid transmission of information from one nerve cell to another.

- Neurons have extensive processes called *dendrites* and *axons* that are involved in the reception and transmission of information. The cytoskeletal structure of these processes is made of microtubules and neurofilaments. In addition to providing the structural framework for the neuron, these components are involved in the transport of cellular material to and from the soma to the outer reaches of axonal and dendritic processes. The fast anterograde axonal transport process carries organelles and vesicles at a rate of 400 mm/day, whereas slow anterograde axonal transport carries material at a rate of 1.0 to 2.0 mm/day. These processes play important roles in the maintenance and regeneration of nerve fibers.

- The membranes of nerve cells are composed of passive, voltage-activated, and chemically activated ion channels. These channels are made of proteins embedded within the membrane and permit specific ions to move through the membrane. Chemically activated ion channels are opened (or closed) when specific chemicals bind to receptor sites on the channel. Voltage-activated ion channels open (or close) when the membrane is depolarized (or repolarized). Passive ion channels remain in the open state and play a role in establishing the membrane potential.

- A difference in the number of positive and negative ions between the inside and outside of a cell produces the membrane potential. In general, the inside of a cell is more negatively charged than is the extracellular space and thus establishes an electrical driving force for the movement of ions across the membrane. Differences in the concentration of ions between the inside and outside of cells establish concentration gradients, which act as chemical driving forces for the movement of ions across the membrane. The summated effects of electrical and chemical driving forces determine whether an ion will move through the membrane in an inward or outward direction.

- The axon of nerve cells transmits information by generating action potentials. Action potentials are initiated at the axon hillock and propagated along the axon as a result of the influx and efflux of Na^+ and K^+ ions, respectively. These ionic currents that flow across the axonal membrane are a result of the opening and closing of voltage-sensitive ion channels for Na^+ and K^+, which are embedded within the membrane.

- Action potentials reach the axon terminal and initiate the process of synaptic transmission. At electrical synapses, the cytoplasm between two cells is in direct contact. Thus ions move directly from one cell to another to effect transmission. At chemical synapses, synaptic transmission occurs when an action potential depolarizes the membrane within the synaptic terminal to activate voltage-sensitive Ca^{2+} channels. The opening of these channels causes the influx of Ca^{2+} into the terminal, which in turn leads to the fusion of synaptic vesicles with the pre-synaptic membrane. As a consequence, chemical transmitters contained within the vesicles are released into the synaptic cleft, where they eventually bind to receptors located on the post-synaptic membrane.

- The binding of neurotransmitters with their receptors affects the ionic conductances of the target cells. These receptors are chemically activated ion channels that permit the passage of specific ions. Well-characterized neurotransmitters include acetylcholine, dopamine, norepinephrine, serotonin, glutamate, and gamma aminobutyric acid. Other transmitters exert their effect on the post-synaptic cell through activation of a G-protein–second-messenger system.

- The influx and efflux of ions through the chemically activated channels produce excitatory or inhibitory post-synaptic poten-

tials. These responses are integrated by the post-synaptic cell in a process of summation. If the summation of these potential changes in the post-synaptic membrane causes a depolarization that reaches threshold levels at the axon hillock, an action potential will be generated and conducted by the axon of the post-synaptic nerve cell.

Review Questions

Choose the Correct Answer

1. The time interval during which no stimulus can elicit an action potential in a nerve fiber is called the:
   a. latent period.
   b. relative refractory period.
   c. absolute refractory period.
   d. repolarization period.
   e. threshold.
2. In which of the following would the velocity of nerve impulse conduction be the greatest?
   a. Large-diameter unmyelinated fibers
   b. Small-diameter unmyelinated fibers
   c. Large-diameter myelinated fibers
   d. Small-diameter myelinated fibers
   e. None of the preceding because they conduct impulses at the same velocity
3. The principal cause of early repolarization of a nerve fiber after an adequate stimulus has been applied is:
   a. an increase in the diffusion of K^+ into the neuron.
   b. an increase in the diffusion of Na^+ out of the neuron.
   c. an increase in the diffusion of Na^+ into the neuron.
   d. an increase in the diffusion of K^+ out of the neuron.
   e. a decrease in the diffusion of Na^+ into the neuron.
4. The principal cause of depolarization of a nerve fiber when an adequate stimulus is applied is:
   a. an increase in the diffusion of K^+ into the neuron.
   b. an increase in the diffusion of Na^+ out of the neuron.
   c. an increase in the diffusion of Na^+ into the neuron.
   d. an increase in the diffusion of K^+ out of the neuron.
   e. a decrease in the diffusion of Na^+ into the neuron.
5. *Saltatory conduction* refers to conduction:
   a. between pre-synaptic and post-synaptic membranes.
   b. of a nerve impulse by a myelinated axon.
   c. of a signal between nerve and muscle.
   d. of electrical current through a body of water.
   e. of an impulse by muscle t-tubules.
6. What is the direction of the driving force(s) for the movement of potassium ions when a nerve cell is at rest?
   a. Inward electrical gradient
   b. Inward chemical gradient
   c. Outward electrical gradient
   d. Outward chemical gradient
   e. Both a and d
7. If energy input (in the form of ATP) to the Na^+/K^+ pump ceases, then:
   a. the resting membrane potential becomes larger.
   b. the resting membrane potential decreases toward zero.
   c. an EPSP develops.
   d. the cell loses sodium.
   e. Two of the preceding

8. Information is coded in the nervous system by:
   a. altering the amplitude of the action potential.
   b. varying the frequency of the action potentials.
   c. increasing the velocity of impulse conduction.
   d. changing the type of neurotransmitter used by a neuron.
   e. Two of the preceding
9. The period of excitability during which a greater than normal strength of stimulus is required to elicit a cell response is the:
   a. interphase period.
   b. relative refractory period.
   c. absolute refractory period.
   d. hyperexcitable period.
   e. None of the preceding
10. The establishment and maintenance of a cell's resting membrane potential is dependent on:
    a. selective permeability of the membrane to ions.
    b. active transport of ions.
    c. the presence of nondiffusible anions inside of the cell.
    d. energy expenditure.
    e. All of the above
11. The hyperpolarization phase of the action potential is caused by too much _____ leaving the cell.
    a. sodium ion
    b. potassium ion
    c. calcium ion
    d. phosphate ion
    e. water
12. When an adequate stimulus is applied to an axon:
    a. the amplitude of the action potential is directly proportional to the strength of the applied stimulus.
    b. the amplitude of the action potential is inversely proportional to the strength of the applied stimulus.
    c. the speed of nerve impulse conduction is inversely proportional to the diameter of the nerve fiber.
    d. for a given nerve fiber, the amplitude of the action potential does not vary with the strength of the stimulus.
    e. the first gate to open is the Na^+ channel inactivation gate.
13. Synaptic vesicles are induced to migrate toward, and blend with, the pre-synaptic membrane and then to release neurotransmitter into the synaptic cleft as a result of an increase in the concentration of free _____ in the pre-synaptic terminal.
    a. K^+
    b. Na^+
    c. Cl^-
    d. Ca^{2+}
    e. ACh
14. Somatic integration involving the additive effect of two or more EPSPs produced in rapid succession at one synapse defines:
    a. temporal summation.
    b. spatial summation.

c. motor unit summation.

d. mechanical summation.

e. None of the above

15. Excitatory post-synaptic potentials occur when a neurotransmitter engages a post-synaptic membrane receptor, resulting in the post-synaptic membrane becoming:

a. more permeable to Cl^-.

b. more permeable to K^+.

c. less permeable to Na^+.

d. less permeable to K^+.

e. more permeable to Na^+.

16. When the effects of two EPSPs occurring at different synapses combine to generate an action potential, the process is called:

a. synaptic summation.

b. temporal summation.

c. spatial summation.

d. somatic summation.

e. None of the above

17. The process whereby a neuron releases transmitter molecules into the synaptic cleft is called:

a. net diffusion.

b. osmosis.

c. endocytosis.

d. exocytosis.

e. phagocytosis.

18. Which of the following is a graded, nonpropagated, hyperpolarization of a neuron plasma membrane?

a. EPSP

b. IPSP

c. EPP

d. AP

e. TP

19. The most probable site on a neuron where incoming information is received is the:

a. terminal bouton

b. axon

c. perikaryon

d. dendrite

e. myelin sheath

20. Depolarization of a dendrite results in current flow between the dendrite and the _____, increasing the probability of action potential generation.

a. adjacent dendrites

b. terminal boutons

c. perikaryon

d. axon hillock

e. myelin sheath

21. Acetylcholinesterase:

a. is the neurotransmitter always found at the neuromuscular junction.

b. is a very common neurotransmitter and is found in central and peripheral nervous systems.

c. is an enzyme that degrades acetylcholine.

d. is an enzyme essential in the synthesis of acetylcholine.

e. has the same effect as does curare.

22. A nerve fiber has a resting potential of -70 mV and a threshold potential of -55 mV. Which of the following is most likely to generate an action potential?

a. $+10$ mV EPSP

b. -15 mV IPSP

c. $+5$ mV EPSP

d. $+20$ mV EPSP

e. $+8$ mV EPSP

23. Relative to the transmission of a nerve impulse, the membrane potential at which the membrane suddenly becomes highly permeable to Na^+ is the:

a. action potential.

b. resting potential.

c. threshold potential.

d. EPSP.

e. IPSP.

Answers to Case History Questions

1. The demyelinization that occurs in an attack of MS is nonspecific. It may involve nerve fibers in motor tracts or nerve fibers in sensory tracts. Symptoms of an MS attack depend on the extent and location of motor fiber demyelinization and sensory fiber demyelinization.

2. Nerve impulse conduction in a myelinated fiber depends on the rapid spread of ionic current from node to node(s). Voltage-gated ion channels necessary for the propagation of the nerve impulse are located only at the nodes of Ranvier. As myelin is destroyed, parts of the nerve fiber lose insulation and allow current to leak through, reducing the amount of current that would normally spread to the next node and depolarize it. Without sufficient current to reduce the membrane potential to threshold, the node fails to generate an action potential, and conduction of the impulse stops.

Key Terms

acetylcholine (p. 232)

action potential (p. 214)

autonomic nervous system (p. 212)

axon (p. 215)

axonal transport (p. 218)

brain (p. 209)

brain stem (p. 212)

central nervous system (p. 209)

cerebellum (p. 210)

dendrite (p. 215)

depolarization (p. 225)

electrical potential (p. 220)

glial cells (p. 214)

ion channel (p. 219)

membrane potential (p. 219)

microfilament (p. 218)

microtubule (p. 218)

myelin (p. 214)

neocortex (p. 210)

nerve (p. 215)

neuron (p. 215)

neurotransmitter (p. 215)

norepinephrine (p. 239)

peripheral nervous system (p. 209)

Schwann cell (p. 214)

spinal cord (p. 209)

synapse (p. 216)

synapsin I (p. 233)

thalamus (p. 209)

visual cortex (p. 210)

Suggested Readings

Brickley, S., Revilla, V., Cull-Candy, S., Wisden, M., and Farrant, M. "Adaptive regulation of neuronal excitability by a voltage-independent potassium conductance." *Nature,* 409:88–92, 2001

Cajal, S. R. "A new concept of the histology of the central nervous system," 1892. D. A. Rottenberg (trans.). *In* Rottenberg, D. A., and Hochberg, F. H. (eds): *Neurological Classics in Modern Translation.* New York, Hafner, 1977.

Dale, H. H., Feldberg, W., and Vougt, M. "Release of acetylcholine at voluntary motor nerve endings." *Journal of Physiology,* 86:353–380, 1936.

Eccles, J. C., Fatt, P., and Koketsu, K. "Cholinergic and inhibitory synapses in a pathway from motor-axon collaterals to motoneurons." *Journal of Physiology,* 126:524–562, 1954.

Eccles, J. C. *The Physiology of Synapses.* New York, Academic Press, 1964.

Ek, M., Engblom, D., Saha, S., Blomqvist, A., Jakobsson, P., and Ericsson-Dahlstrand, A. "Inflammatory response: Pathway across the blood-brain barrier." *Nature,* 410:430–431, 2001

Fischbach, G. D., and McKhann, G. M. "Cell therapy for Parkinson's disease." *New England Journal of Medicine,* 344:763–765, 2001

Hodgkin, A. L., and Huxley, A. F. "A quantitative description of membrane current and its application to conduction and excitation in the nerve." *Journal of Physiology,* 117:500–544, 1952.

Kandel, E. R., Schwartz, J., and Jessel, T. M. *Principles of Neural Science.* Amsterdam, Elsevier Press, 1991.

Katz, B., and Miledi, R. "The timing of calcium action during neuromuscular transmission." *Journal of Physiology,* 189:535–544, 1967.

Neher, F., and Sakmann, B. "Single-channel currents recorded from membrane of denervated frog muscle fibres." *Nature,* 260:799–802, 1976.

Nitta, A., Itoh, A., Hasegawa, T., and Nabeshima, T. "Beta-amyloid protein-induced Alzheimer's disease animal model. *Neuroscience Letters,* 170:63–66, 1994.

O'Dell, T. J., Huang, P. L., Dawson, T. M., Dinerman, J. L., Snyder, S. H., Kandel, E. R., and Fishman, M. C. "Endothelial NOS and the blockade of LTP by NOS inhibitors in mice lacking neuronal NOS." *Science,* 265:542–546, 1994.

Olson, L. "Combating Parkinson's disease-Step three." *Science,* 290:721–724, 2000.

Richmond, J. E., Weimer, R. M., and Jorgensen, E. M. "An open form of syntaxin bypasses the requirement for UNC-13 in vesicle priming." *Nature,* 412:338–341, 2001.

Sato, C., Yutaka, M., Ueno, Y., Asai, K., Takahashi, K., Sato, M., Engel, A., and Fujiyoshi, Y. "The voltage-sensitive sodium channel is a bell-shaped molecule with several cavities." *Nature,* 409:1047–1051, 2001.

Sherrington, C. S. *The Integrative Action of the Nervous System,* 1906. Reprint, New Haven, Yale University Press, 1947.

Answers to Review Questions

1. c	**2.** c	**3.** d	**4.** c	**5.** b	**6.** e	**7.** b	**8.** b
9. b	**10.** e	**11.** b	**12.** d	**13.** d	**14.** a	**15.** e	
16. c	**17.** d	**18.** b	**19.** d	**20.** d	**21.** c	**22.** d	
23. c							

Chapter 8

SENSORY SYSTEMS

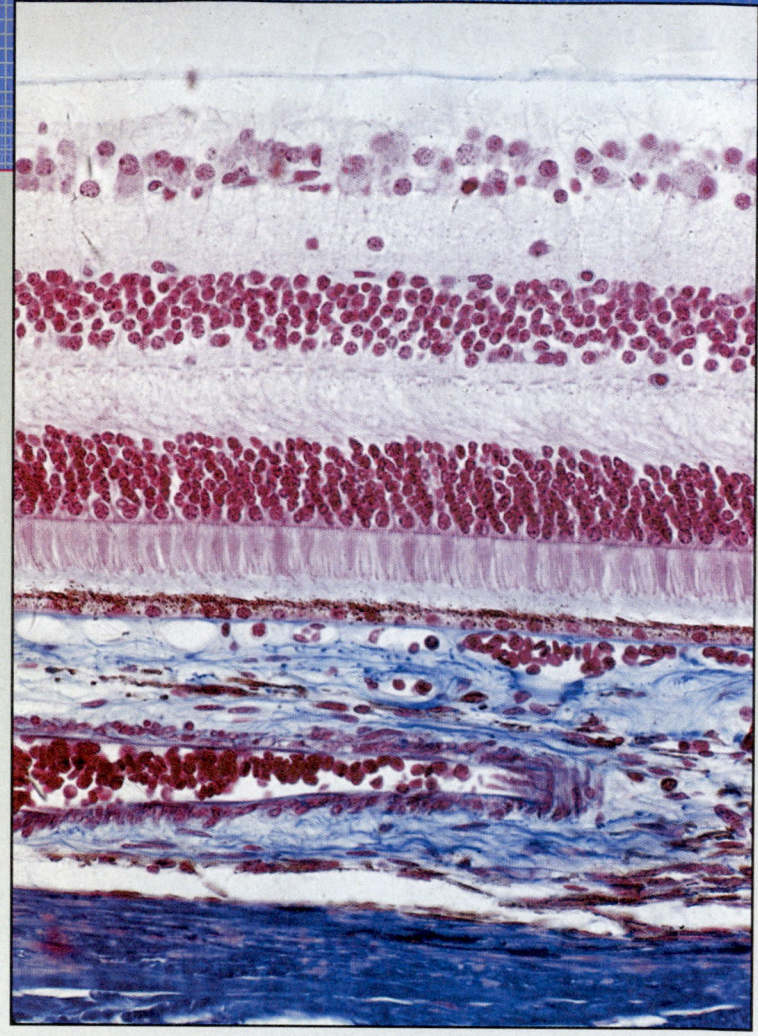

KEY CONCEPTS

- A sensation is what is perceived in the brain when a sensory receptor or, more commonly, a group of sensory receptors are stimulated. Sensations differ in various ways such as in modality, quality, and quantity.

- The basic function of a sensory receptor is to detect a change in the environment, called a stimulus, and convert the energy (e.g., heat, light, pressure, and so on) of the stimulus into electrical signals in the nervous system. This conversion process is known as transduction.

- Somatosensory pathways carry information about pain, temperature, touch, pressure, vibration, position of body parts and their movement. These pathways are organized into three neuron sets. The neurons in each set are designated in sequence, beginning with the receptor, first order, second order, and third order.

- The visual system detects the shape and color of objects and the movement of objects in the external environment. The features of shape, color, and movement are processed by different cell groups within the visual system, beginning with those found in the eye, and ending with those found in the visual cortices of the cerebrum.

- The auditory system detects complex sounds, such as spoken words, and breaks them into their basic sound frequencies. The fundamental frequency components are transduced into action potentials and transmitted through auditory pathways to the brain for interpretation.

- The vestibular system aids in maintaining the body's balance by detecting the position and motion of the head in space. The sensory cells of the vestibular system are in the ampullae of the semicircular canals and in the utricle and saccule.

- Taste receptor cells in the tongue detect chemicals associated with five basic taste qualities: umami (meaty flavor), sweet, salty, sour, and bitter. However, their areas of distribution overlap, and sensations of taste result from combinations of taste-receptor activation.

- Our ability to smell odors is a result of the activation of olfactory receptor cells in the nose. Human beings can detect seven general types of odors: camphoraceous, musk, floral, peppermint, ethereal, pungent, and putrid.

CASE HISTORY

A 37-year-old man went to his personal physician for his company-mandated annual physical examination. Previous annual physical examinations performed by the company's physicians had revealed no extraordinary finding. However, within the past month, he experienced two attacks of unexplained dizziness, and he was concerned about his ability to safely drive and operate machinery. Consequently, he scheduled his annual physical examination earlier than usual in the hope that his family physician would discover the cause and cure his dizzy spells.

The first attack of dizziness, he related to his physician, occurred midweek approximately an hour after lunch as he was walking from the restroom to his workstation. The dizzy spell came without warning, and soon he felt as though the world was spinning around him. He began to stagger from side to side, lost his balance, and fell down. A nearby worker came to his assistance but he had trouble hearing her voice and understanding her speech because of hissing or roaring in his ears. The coworker helped him to a safe spot on the floor, where he sat for approximately 15 minutes before his dizziness and roaring in the ears subsided. Although suffering from a mild headache, he was permitted and able to return to work approximately an hour later. The next day he felt fine and experienced no residual symptoms from the episode of the previous day. He attributed the dizzy spell to something he had eaten for lunch in the company cafeteria.

The second attack of dizziness also occurred without warning, but this time, approximately a half-hour before lunch. Again, in addition to vertigo (a dizzy spinning sensation), he experienced tinnitus (a roaring or ringing in the ears), and trouble hearing voices. His skin became cold and clammy, and he felt nauseous. The second attack lasted approximately an hour, although his hearing did not return to normal for more than a week. Worried about recurrent spells of unexplained dizziness, he sought the advice of his family physician.

The doctor reviewed his medical history and his symptoms, performed a physical examination, and referred him to a neurologist for a more complete neurological examination, including testing of inner ear functions. The diagnosis, suspected by the family physician and confirmed by the neurologist, was Meniere's disease.

Meniere's disease is caused by an excessive amount of fluid, called *endolymph*, in the membranous canals of the inner ear. The canals contain receptors for balance and hearing. Increases in fluid pressure within the canals stimulate the receptors, giving the brain false sensory information about balance and sound. Attacks may come and go over a person's lifetime. In between bouts, affected people can live perfectly normal lives. However, repeated attacks eventually destroy auditory receptors, resulting in progressive hearing loss. In this chapter, we will examine basic concepts of sensory physiology, including the normal roles the ears play in hearing and in balance.

Questions

1. Why does the person with Meniere's disease have difficulty in understanding speech during an episode?

2. What is vertigo and why does it occur in Meniere's disease?

INTRODUCTION

> **What is the vocabulary that pertains to the sensory system?**

In the early part of the 19th century, physiologists classified senses as either "special" or "general." General senses, such as pain, temperature, pressure, muscle stretch, joint movement, and others could be elicited by stimulating appropriate receptors that were widespread throughout the body. General senses were designated *somatic* if the receptors were located in the body wall or periphery, and *visceral* if the receptors were located within the organs of the body cavities. On the other hand, special senses, such as vision or hearing, could be elicited only by stimulating appropriate receptors confined to a specific area of the body. General senses were thought to be more diffuse and less distinct; special senses were believed to be more localized and specific. Although some of the early rationale is archaic and no longer valid, the classification of sensibilities into general and specific categories persists.

The five special senses are (1) *olfaction* (smell), (2) *gustation* (taste), (3) *audition* (hearing), (4) *equilibrium* (balance), and (5) *vision* (sight). The receptors mediating special sensations are, with the exceptions of taste and equilibrium, distance receptors through which we become cognizant of the external environment without the necessity of initiating immediate contact. Each type of special sensory receptor is limited in distribution and confined to specific parts of the head. Each is associated with one or more of the cranial nerves.

Visceral and somatic senses include *pain, temperature change* (warm and cold), *touch* and *pressure,* and *proprioception* (muscle, tendon, and joint capsule stretch). The receptors mediating general sensations are located in the skin and the connective tissues of muscles, tendons, joint capsules, and viscera of the thorax and abdomen. A basic understanding of general and special sensory physiology begins with defining sensation and its attributes. These are followed by consideration of the structure of receptors, the nature of the receptor discharge in response to a stimulus, processes of information coding, peripheral and central sensory pathways, and processing of sensory information within the central nervous system (CNS).

ATTRIBUTES OF SENSATION

A *sensation* is what is perceived in the brain when a sensory receptor or, more commonly, a group of sensory receptors are stimulated. Many sensations, such as pain or heat, are of a conscious nature, whereas others, such as blood pressure or muscle stretch, may not reach the conscious sphere. Sensations also differ in various other ways, such as in **modality, quality,** and **quantity.**

Modality is the characteristic that distinguishes one type of sensation from all other types. Sensory modalities include sight, smell, hearing, touch, pain, and others. Sensations of the same modality may differ qualitatively. For example, we can see red, blue, green, and other wavelengths of light. Sensations of the same modality and quality also may differ quantitatively; thus, we can distinguish between shades of color, such as red and pink. The *perception* of sensory stimuli is a complex process involving various sensory receptors; peripheral and central neural pathways; interactions between components of the pathways, such as in binocular vision or binaural hearing; and cerebral functions of learning and memory.

SENSORY RECEPTORS

For the body to react in a purposeful manner to changes in the external and internal environments, the CNS needs information concerning the nature of the environmental change. Such information is generated by specialized structures that receive stimuli (changes in the environment) and therefore are called **sensory receptors.** In some cases, the sensory receptor is a bare or free nerve ending; in other cases, the sensory nerve ending is encapsulated (a sense organ). In more complex sensory structures, such as the eye or the ear, stimuli are detected by specialized sensory cells that are not neurons but can relay the stimulus information to sensory neurons.

Classification of Sensory Receptors

> **Sensory receptors may be grouped into two major categories: exteroceptors and interoceptors.**

Exteroceptors

Exteroceptors detect changes in the body's external environment. *Integumentary (skin) receptors,* which detect touch, pressure, temperature change, hair movement, and painful stimuli applied to the body's surfaces, are examples of exteroceptors. Others include *taste receptors, visual receptors,* and *auditory receptors.*

Interoceptors

Interoceptors detect changes in the body's internal environment. Examples include *stretch receptors* in muscles, tendons, and ligaments; *chemoreceptors* in blood vessel walls; *baroreceptors* (pressure) in the vascular system; *labyrinthinoreceptors* (balance) in the inner ear; *osmoreceptors* in the brain, and others.

Receptors also may be differentiated on the basis of structure, location, or modality of sensation. A partial list of receptors correlated with modality of sensation and sense organs that contain the receptor is presented in Table 8–1. Some of the somatosensory receptors found in the skin are shown in Figure 8–1.

TABLE 8–1

Sensory Modalities and Receptor Cells

Sensory Mode	Receptor	Sense Organ
Vision	Rods and cones	Eye
Hearing	Hair cells	Ear (organ of Corti)
Rotational acceleration	Hair cells	Ear (semicircular canals)
Linear acceleration	Hair cells	Ear (utricle and saccule)
Smell	Olfactory neurons	Olfactory mucous membrane
Taste	Taste receptor cells	Taste buds
Touch—pressure	Nerve endings	Skin
Warmth	Nerve endings	Skin
Cold	Nerve endings	Skin
Pain	Naked nerve endings	Skin
Joint movement and position	Nerve endings	Various
Muscle length	Nerve endings	Muscle spindle
Muscle tension	Nerve endings	Golgi tendon organ

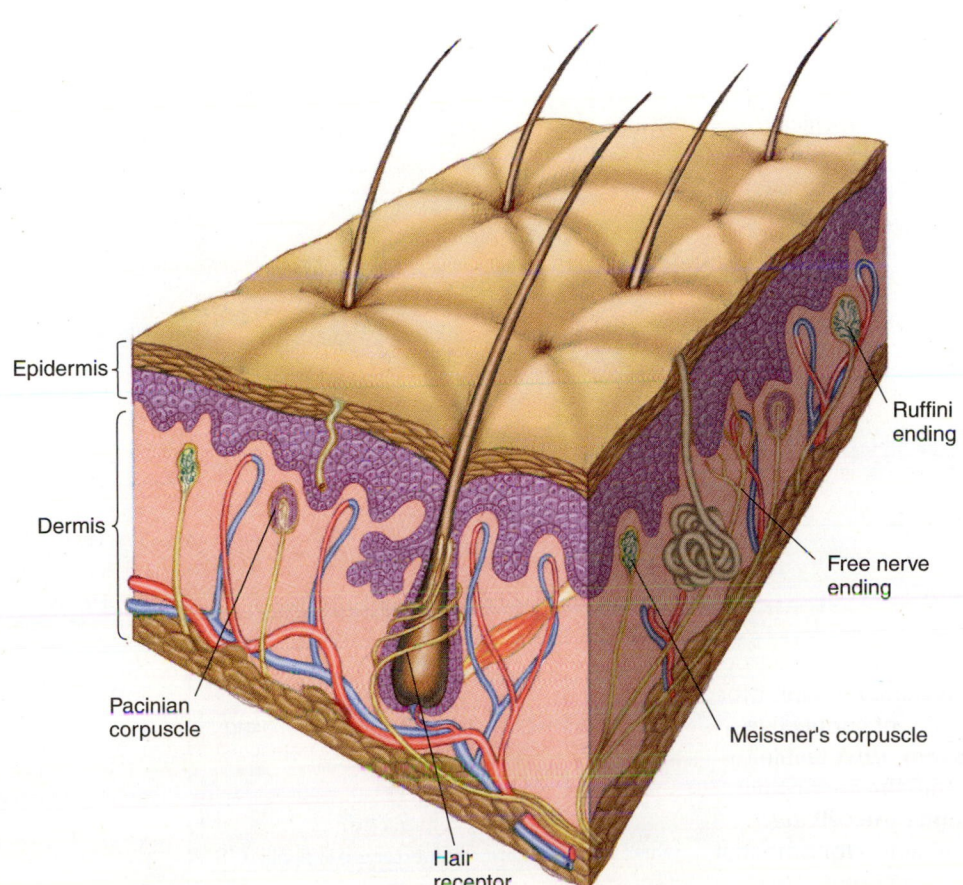

Epidermis

Dermis

Ruffini ending

Free nerve ending

Pacinian corpuscle

Hair receptor

Meissner's corpuscle

Figure 8–1

Somatosensory receptors in the skin.

Receptor Function

 How do receptors detect environmental change?

Transduction

The basic function of a sensory receptor is to detect a change in the environment, called a *stimulus,* and convert the energy (e.g., heat, light, pressure, and so on) of the stimulus into electrical signals in the nervous system. This conversion process is known as **transduction.** All sensory receptors are transducers; they respond to an adequate stimulus by generating a series of graded potentials or nerve impulses in their associated afferent nerve fiber.

Generator and Receptor Potentials

Some sensory receptors (see Fig. 8–1), such as the Pacinian corpuscle, are encapsulated nerve fiber terminals. When an adequate stimulus (pressure) is applied to the Pacinian corpuscle, the ion permeability of the nerve ending to Na^+ is increased, allowing Na^+ to enter the fiber terminal and decrease the membrane potential (Fig. 8–2). The resultant small, temporary change in the membrane potential is called a **generator potential.** It resembles an EPSP (see Chapter 7), and like an EPSP, it is graded and nonpropagated and can be summed. The generator potential in turn induces, via current flow, depolarization of the first node of Ranvier. If the generator potential is of sufficient amplitude to bring the first node to threshold, an action potential is generated and a

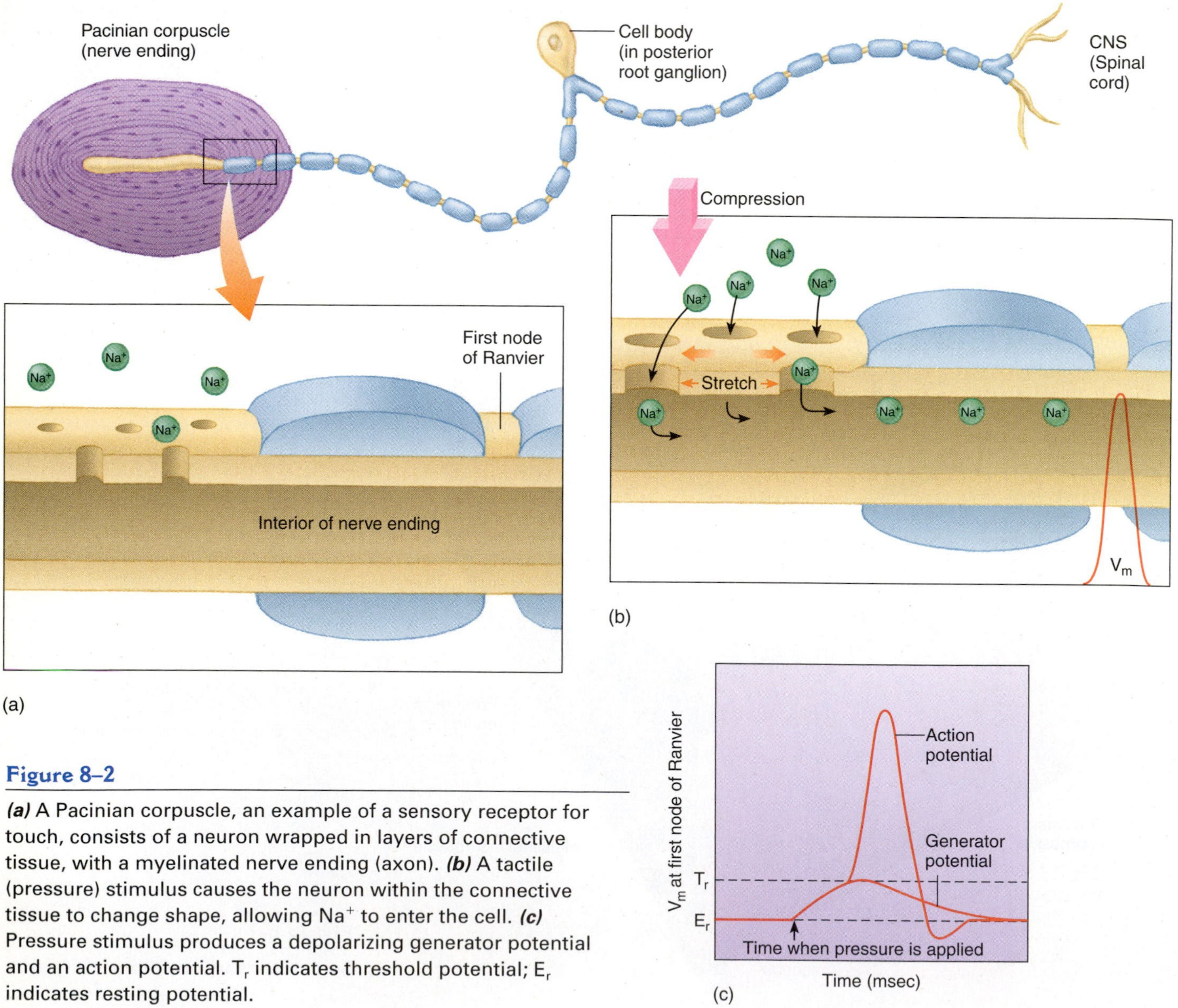

Figure 8–2

(a) A Pacinian corpuscle, an example of a sensory receptor for touch, consists of a neuron wrapped in layers of connective tissue, with a myelinated nerve ending (axon). *(b)* A tactile (pressure) stimulus causes the neuron within the connective tissue to change shape, allowing Na^+ to enter the cell. *(c)* Pressure stimulus produces a depolarizing generator potential and an action potential. T_r indicates threshold potential; E_r indicates resting potential.

nerve impulse is propagated along the fiber. The amplitude and duration of the generator potential is directly proportional to the intensity and duration of the applied stimulus and determines the frequency of firing (number of impulses conducted per second) of the afferent fiber.

Some receptors, such as gustatory (taste) receptors in the taste buds of the tongue, consist of a receptor cell that synapses with an afferent neuron (Fig. 8–3). The receptor cell responds to an adequate stimulus by producing a **receptor potential,** which may be a depolarization or a hyperpolarization of its plasma membrane. The receptor potential

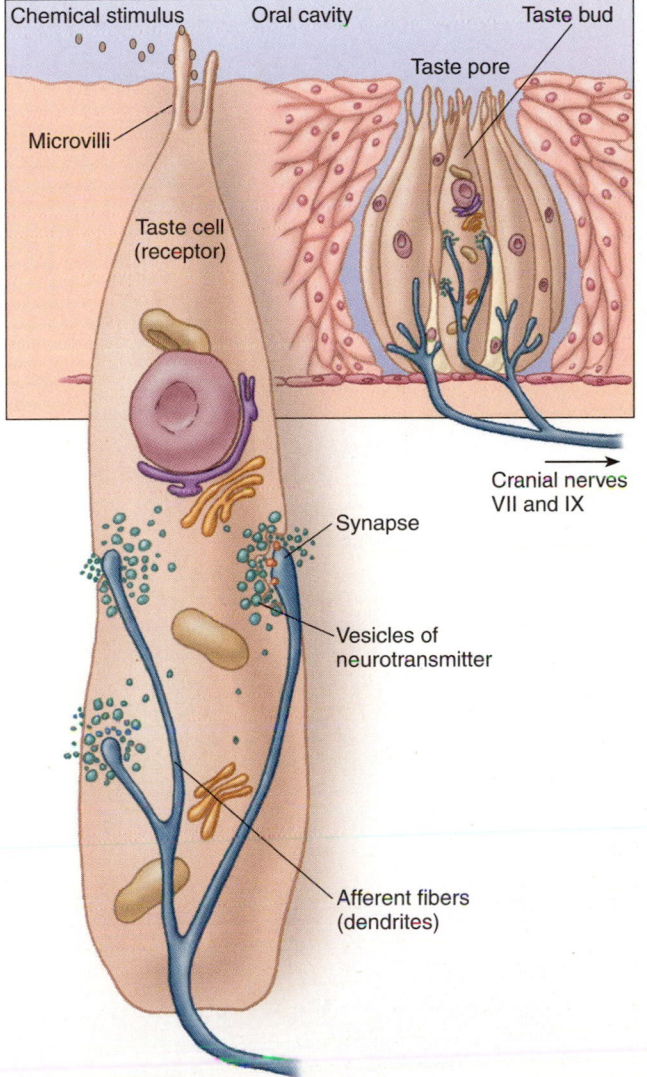

Figure 8–3

The taste cell responds to an adequate chemical stimulus by producing a receptor potential, which may be a depolarization or a hyperpolarization of its plasma membrane. The receptor potential controls the presynaptic release of neurotransmitter molecules into the synaptic cleft. The released neurotransmitter causes depolarization of the sensory neuron dendrites.

controls the presynaptic release of neurotransmitter molecules into the synaptic cleft. The amount of transmitter release per unit of time, as determined by the amplitude and duration of the receptor potential, controls the frequency of firing of the afferent fiber.

Adaptation/Accommodation

Many receptors, with the notable exception of free (bare) nerve endings that mediate pain, respond to the continued application of a constant-intensity stimulus with a decrease in the magnitude of the generator potential or the receptor potential, in other words, a return toward the resting membrane potential. The decline in the generator potential or receptor potential in response to maintained receptor stimulation is called **adaptation** or **accommodation.** When adaptation of the receptor is complete, sensation is no longer perceived from that receptor until the intensity of the stimulus is changed. With no adaptation, such as displayed by pain receptors, a maintained stimulus produces an undiminished sensation.

Some receptors adapt very rapidly and are therefore more useful in signaling a change in frequency of stimulus application rather than a change in stimulus magnitude. These receptors are called **phasic receptors** or **velocity receptors.** Some pressure receptors that detect vibration and the hair-ending plexus in the skin are examples of phasic receptors. A simple experiment demonstrating rapid adaptation of a receptor involves placing a small coin on the motionless forearm. The sensation perceived (pressure and hair movement) quickly disappears even though the coin remains. When the coin is removed, the same sensation as when the coin was placed is perceived, even though the coin is no longer on the forearm. This second sensation is called an *after-image* and is again the result of a receptor signaling a change.

Receptors that adapt very slowly or not at all are called **tonic receptors** or **intensity receptors.** Pain receptors and some stretch receptors of muscles, tendons, and ligaments are examples of tonic receptors. In general, tonic receptors provide continual information as long as the stimulus is being applied or until the receptor eventually adapts. Various receptors display different degrees of adaptation in response to an adequate stimulus of constant intensity (Fig. 8–4).

Stimulus Intensity Versus Receptor Response

The magnitude of the generator potential or receptor potential is directly proportional to the intensity of the applied stimulus, but the relationship between the two is more nearly logarithmic rather than linear (Fig. 8–5). However, the relationship between the magnitude (amplitude) of the generator potential and the frequency of firing of the afferent fiber is linear. As the amplitude of the generator potential increases, the frequency of firing of the afferent fiber increases in direct proportion. Therefore, the frequency of firing of the afferent fiber is proportional to the logarithm of the intensity of the applied stimulus. The latter statement, known as the

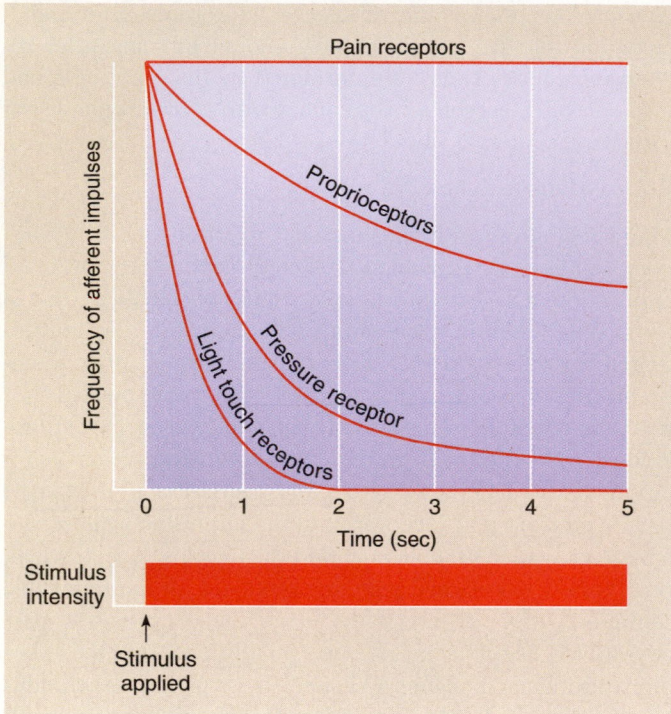

Figure 8–4

Rates of adaptation for selected receptors.

Weber-Fechner law, is an approximation and suggests the basis whereby receptors can provide for discrimination between a great range of stimulus intensities. In other words, a broad range of stimulus intensities may be compressed during transduction into a narrow range of nerve impulse frequencies. Such compression is essential if we are to sense stimuli intensities, which may vary 1000-fold or more, via afferent fibers that have an average maximum frequency, as determined by the duration of the action potential, of less than 200 impulses per second.

The Adequate Stimulus and Specific Nerve Energies

Specific types of sensory receptors are more sensitive to certain types of stimuli. For example, photoreceptors in the eye are most sensitive to light but also respond to pressure stimuli; Pacinian corpuscles are most sensitive to pressure but also respond to other stimuli. The type of stimulus (modality) to which the receptor is most sensitive, that is, for which the receptor has the lowest threshold, is called the *adequate stimulus*. Nearly all receptors respond to a stimulus other than an adequate stimulus if the stimulus intensity is great enough. For a receptor to respond, the stimulus strength must exceed a certain minimal (threshold) value. The threshold strength of a stimulus is lowest for the adequate stimulus, and higher for other types of stimuli.

Regardless of the type and strength of a stimulus applied to a given receptor, the sensation perceived when the receptor responds is always the same. This phenomenon is known

as the **law of specific nerve energies (Muller's law).** For example, in a darkened room, with eyes closed, pressure applied to the side of the eyeball by the finger elicits a visual sensation of light; a dark circle surrounded by a bright white ring is "seen" near the bridge of the nose. From Muller's law, it follows that mechanical stimulation of auditory receptors or nerve fibers (e.g., by a sharp blow to the head) produces a sensation of sound; electrical stimulation of taste receptors produces a sensation of taste, and so on.

The interpretation of Muller's law must be general rather than strict because there are sensations that we per-

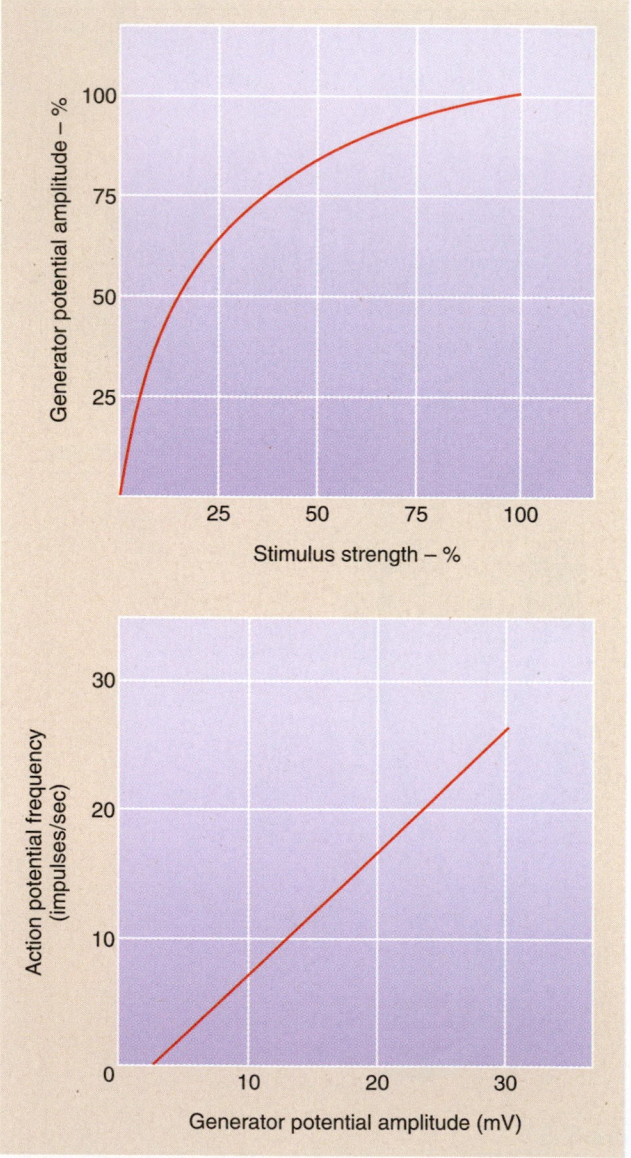

Figure 8–5

(a) The relationship between the generator potential and the stimulus strength is approximately logarithmic.
(b) The relationship between the strength of the generator potential and the frequency of the sensory fiber action potentials is approximately linear.

ceive because of the simultaneous stimulation of two or more receptors sensitive to different modalities of stimulus. For example, if a thin-rubber–gloved hand is placed in a container of cold water, a sensation of wetness is perceived, even though the hand remains dry, because the two components of wetness—temperature and pressure—are sensed.

Information Coding

Sensory information comes to the CNS in the form of nerve impulses. The CNS decodes and interprets or reacts to the incoming information. Decoding of sensory information is based on the location of the receptor, the adequate stimulus for the receptor, the frequency of nerve impulses being transmitted by the afferent fiber associated with the receptor, the number of receptors activated by a stimulus, and the central pathway.

The size of the action potential generated in the afferent fiber associated with a sensory receptor does not vary with the intensity or duration of the stimulus applied to the receptor. Therefore, the amplitude of the action potential cannot be used to code information about the nature of the stimulus. However, the frequency (number per second) with which action potentials are generated does vary according to the magnitude of the stimulus intensity, the rate of change of the stimulus intensity, summation of successive generator potentials, and adaptation of the receptor. Recall that the amplitude of the generator (receptor) potential determines the frequency of action potential generation in the afferent fiber. Information about stimulus intensity and its rate of change therefore may be coded in the form of nerve impulse frequency. This is known as the **frequency code of stimulus intensity.** For example, as increasing pressure is applied to a Pacinian corpuscle in the skin, the frequency of nerve impulse conduction from the receptor increases, and a sensation of increasing cutaneous pressure is perceived (Fig. 8–6). If pressure on the Pacinian corpuscle is gradually reduced (faster than the rate of receptor adaptation), the frequency of nerve impulse conduction from the receptor decreases, and a sensation of decreasing cutaneous pressure is perceived. If the receptor is allowed to adapt to a constant intensity stimulus, the frequency of impulses coming from the receptor decreases along with sensation until the stimulus intensity changes, at which time the receptor again responds and signals a change.

Many afferent neurons possess several dendrites, each with the same kind of receptor at its ending. Although each receptor has the same adequate stimulus, not all of the receptors have the same level of threshold for the adequate stimulus; some respond to weak stimuli, others respond to stronger stimuli. Impulses generated by these receptors are conducted along the dendrites toward the soma where they determine the frequency of impulse conduction by the axon of the sensory neuron. The spatial distribution of sensory receptors associated with the dendrites of a single sensory neuron forms that neuron's sensory field and al-

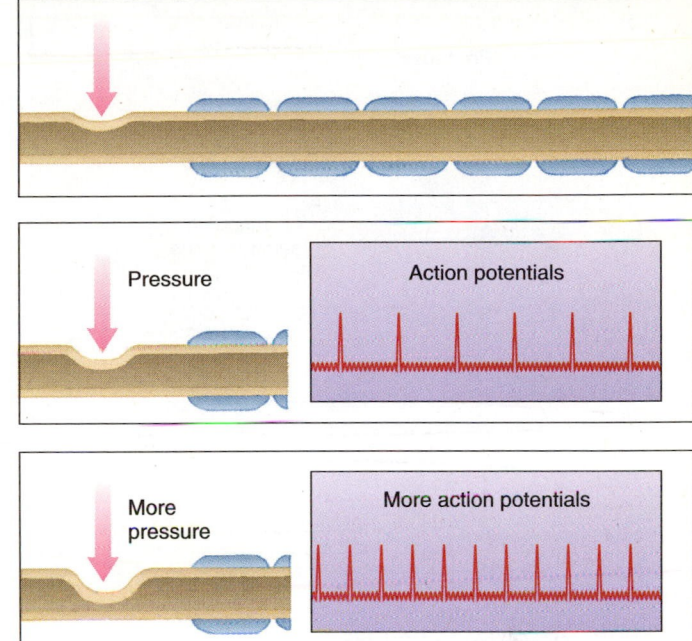

Figure 8–6

According to the frequency code of stimulus intensity, more pressure produces more action potentials from the same receptor.

lows for the coding of sensory information from a larger area without the need for maintaining larger numbers of sensory neurons.

The sensory fields of neurons relaying information from a specific area of the body (e.g., skin) generally overlap (Fig. 8–7). When a strong stimulus is applied to the sensory field of one neuron, the same type of receptors in adjacent and overlapping fields may respond because strong stimuli usually affect a greater area. The phenomenon of increasing the number of receptors responding to a stimulus of increasing strength is called *recruitment*. A second method of informing the brain about the stimulus is a **population code of stimulus intensity:** the greater the stimulus intensity, the greater the number of responding receptors and the greater the number of active afferent fibers (Fig. 8–8).

The type of information sent to the brain also is coded in the way that nerve fiber pathways leading to the brain are arranged physically. The skin, for example, contains sensory receptors for temperature in addition to pressure. The information about the temperature of the skin reaches the brain by a different nerve fiber pathway than does information about pressure. In this way, information about the quality, or type, of stimulus is maintained within each pathway without becoming mixed with other types of stimuli. This aspect of the nervous system, the mechanisms of coding for the **type** of stimulus detected by a sensory receptor, is called the **labeled-line code of stimulus quality** (Fig. 8–9).

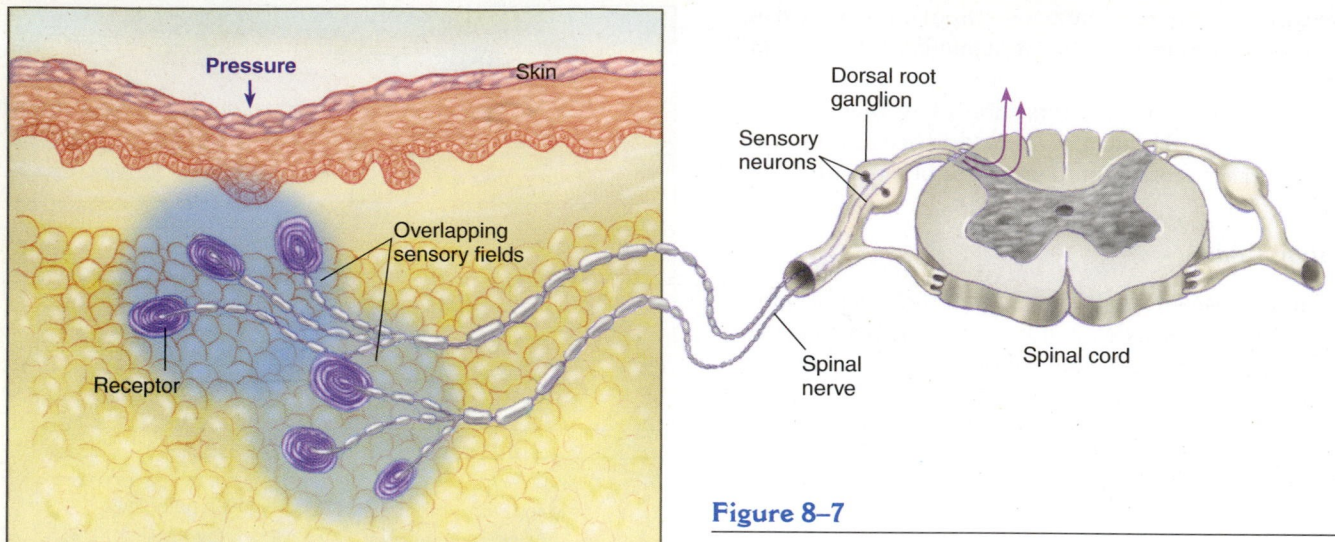

Figure 8–7

Overlapping sensory fields of cutaneous Pacinian corpuscles.

Receptor Distribution and Density

Although many types of sensory receptors are present near the body surface and within the body, receptors are not all equally distributed. Obviously, special sensory receptors, such as those involved with vision, hearing, equilibrium, taste, and smell are confined to relatively small areas or parts of the body, but even within the confines of the receptor's location, distribution is not uniform. Within the eye, for example, photoreceptors that function in daytime vision are more concentrated near the center of the retina, where light rays are focused; photoreceptors for vision in dim light are more concentrated near the periphery of the retina. Receptors found in the skin (integumentary or cutaneous receptors) are not distributed uniformly. Pain receptors (free nerve endings) are the most numerous and widely distributed receptor

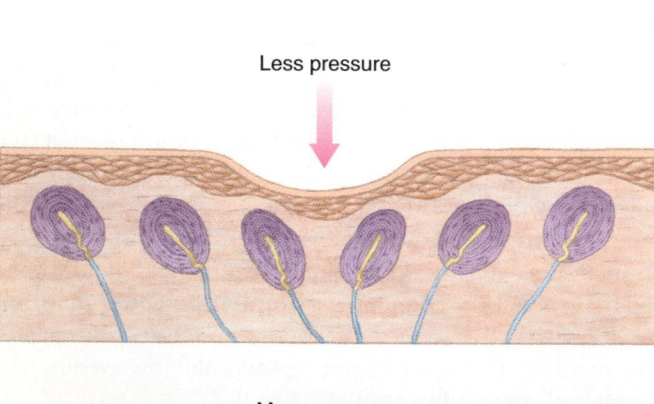

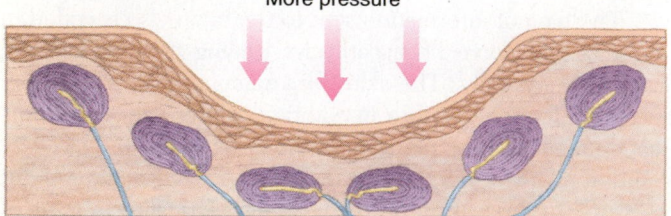

Figure 8–8

According to the population code of stimulus intensity, more pressure produces more action potentials because more receptors are affected.

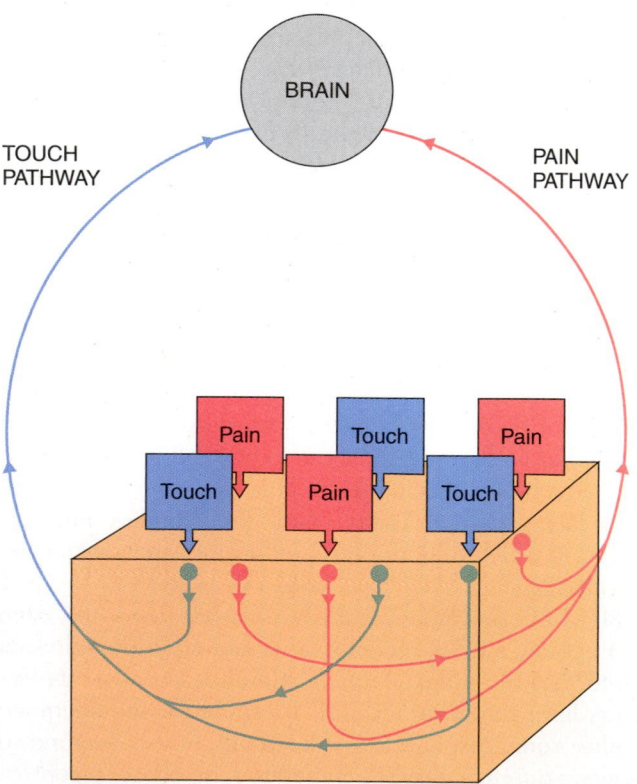

Figure 8–9

The labeled-line code of stimulus quality. Although different stimuli affect the skin, each type of stimulus has its own pathway to the brain, ensuring clarity of sensory messages.

type, but touch or pressure receptors, such as Meissner's corpuscle, are found in greater density in the skin of the extremities (particularly the digits), the lips, and tip of the tongue. The type, location, number, and density of receptors in the skin and other parts of the body depend on the probability of the receptor's coming in contact with an adequate stimulus. In general, skeletal muscles that can be more precisely controlled and areas of the skin that have a greater density of receptors are represented in larger areas of the primary somatosensory cortex of the cerebrum.

Types of Somatosensory Receptors

The types of somatosensory receptors include (1) tactile receptors, activated by the mechanical stimulation of the body's surface; (2) thermal receptors, activated by changes in temperature on the surface of the body; (3) nociceptors, or pain receptors activated by noxious (harmful) stimuli; and (4) proprioceptive receptors, activated by the movement of the limbs. Each type of sensation is detected by a variety of sensory receptors.

Tactile Receptors: Touch, Pressure, and Vibration. Tactile receptors are called **mechanoreceptors.** These receptors are responsible for detecting touch, pressure, and vibrations applied to the skin (see Fig. 8–1). Five types of cutaneous receptors that provide for a sense of touch, pressure, stretch, or vibration, are the *Merkel's discs, Meissner's corpuscle, Pacinian corpuscle, Ruffini corpuscle,* and *hair-follicle plexus.*

The hair-follicle receptor is a hair follicle on the surface of the skin that has a bare nerve fiber wrapped around its base. The bending of the hair shaft, such as that caused by a slight breeze, produces a mechanical displacement of the nerve fiber at the base of the follicle. This displacement, in turn, opens Na^+ channels, resulting in a generator potential and the initiation of action potentials. This type of receptor is very sensitive to a gentle touch across the surface of the skin. The hair-follicle plexus is the most sensitive, responding to very small deformations of the protruding hair, and is also the most rapidly adapting touch receptor.

Meissner's corpuscle is an elongated, encapsulated receptor found in the dermis of the skin close to the epidermis. It is found mainly in hairless portions of the skin, especially that of the finger, toes, hands, feet, lip, eyelids, and external genitalia. The adequate stimulus for Meissner's corpuscle is light touch or pressure, such as that experienced when stroking the skin with a feather.

Merkel's discs also respond to light touch, are more sensitive than Meissner's corpuscle, and are the only touch receptors in the epidermis of the skin. A single nerve fiber supplies many discs, and each disc is located below an epithelial cell. Pressure on the epithelial cell stimulates the disc below, resulting in a sensation of very light touch.

The Pacinian corpuscle is the largest, most widely distributed encapsulated receptor, detecting heavy pressure. It is located deep in the dermis of the skin and is prominent in the skin of the hands and feet, as well as in joints, ligaments, the walls of many viscera, membranes of the mesentery and peritoneum, connective tissue of the muscles, and elsewhere. Pacinian corpuscles have a much higher threshold for pressure stimuli than do other touch receptors. In addition to pressure, however, the Pacinian corpuscle is capable of detecting vibrations. The mechanisms that allow it to detect vibrations are extensions of its ability to rapidly adapt to applied pressures.

When pressure is applied to the Pacinian corpuscle for a length of time, it quickly stops generating action potentials. This occurs when the Pacinian nerve ending reverts back to its original shape even though the outer lamellae of the corpuscle remain distorted. As the applied pressure is released, the layers of connective tissue surrounding the nerve ending spring back to their original ellipsoidal form and then become compressed again in the opposite direction because of the elastic nature of the connective tissue. This recompression deforms the nerve ending and again causes Na^+ channels to open and produce a generator potential and action potentials. In this way, both the application and release of pressure result in the discharge of action potentials. This oscillation of pressure is the basis of a vibratory stimulus.

The Ruffini corpuscle is a slowly adapting, tactile mechanoreceptor that continues to generate action potentials as long as the stimulus is applied. This receptor is sensitive to stretching of the skin, such as that produced during a massage.

Thermal Receptors: Heat and Cold. Sensory receptors that respond to temperature are called **thermoreceptors.** Of the two types of thermoreceptors, one type responds to temperature decreases below 30°C, whereas the other type responds to temperature increases above 30°C. These thermoreceptors adapt at a moderate rate to a new temperature in the external environment. Neither type of receptor senses a constant value of skin temperature as would, for example, a thermometer; rather, the thermoreceptors sense the rate of heat change in the skin. If skin temperature is held constant at 30°C, neither receptor would sense the value. However, if skin temperature is lowered, cold receptors signal a decreasing skin temperature, resulting in a cold sensation. If skin temperature is raised, warm receptors signal an increasing skin temperature, resulting in a warm sensation. Neither of these thermoreceptors responds to an absolute value of temperature but rather to the appropriate directional change in temperature.

A simple experiment illustrating thermoreceptor function is to place the right hand in a beaker of cold (10°C) water and the left hand in a beaker of warm water (40°C) for approximately 5 minutes; then place both hands in a beaker of water at 30°C. The right hand will feel warm as skin temperature is raised from near 10°C toward 30°C, and the left hand will feel cool as skin temperature is lowered from near

40°C toward 30°C. When skin temperature becomes equal to water temperature, the receptors adapt and temperature sensation ceases.

The structural identity of thermoreceptors remains unclear. Certain types of free nerve endings are believed to be warm receptors, and the end bulbs of Krause (see Fig. 8–1) are believed to be cold receptors.

Nociceptors: Painful Touch and Temperatures. *Nociceptors* are free nerve endings that detect painful stimuli. In general, the two types of nociceptors are **mechanical nociceptors,** activated by intense mechanical stimulation, such as a knife cut on the arm or a slap to the head, and **heat nociceptors,** which respond to temperatures above 45°C. Pain receptors are the simplest, least specialized of receptors consisting of bare or free nerve endings (dendrites) of nonmyelinated sensory neurons. They are most numerous in skin but also are found within the walls of the viscera, skeletal muscles, joints, tendons, cornea of the eye, and mucous membranes. Compared with other types of cutaneous receptors, pain receptors have a much higher threshold and respond to many modalities of stimulus (e.g., chemical, thermal, mechanical, and so on). Sensations of pain may be sharp and localized (epicritic) or dull and diffuse (protopathic), as in aching. Visceral pain (e.g., heart, stomach, uterus, and so on) may be referred to a somatic area (e.g., arm, chest, back, and so on) because sensory impulses from the affected organ enter the spinal cord and stimulate neighboring neurons receiving sensory input from the skin. The brain then interprets the sensory information as coming from the skin even though the cutaneous receptors have not been stimulated (Table 8–2).

Proprioceptors: Position and Movement. The position of the body's limbs is detected by **proprioceptors.** A *static proprioceptor* detects the stationary position of the limbs in space with respect to the other parts of the body. A *dynamic proprioceptor* transmits information about ongoing limb movement to convey the sense of movement, or **kinesthesia.** The brain needs this information to determine where the arms and legs are located and how quickly they are moving in order to calculate how much further they need to go to complete a certain movement.

The operation of these receptors can be tested by trying to bring the tips of your left and right index fingers together with your eyes closed. If your proprioceptors are working properly, your fingers should touch without your having to watch and visually guide them together.

The sense of stationary or static position is transmitted to the brain by mechanoreceptors located in joint capsules, cutaneous mechanoreceptors, and mechanoreceptors in muscles that are specialized to transduce the stretching of the muscle. Receptor types include the annulospiral and flowerspray endings of the neuromuscular spindle and the Golgi tendon organ.

The static proprioceptors produce a continuous frequency of action potentials in response to different joint positions. If the joint is left in one particular position, the receptor generates action potentials at one specific frequency (Fig. 8–10*a*). This type of action potential response is called a **tonic discharge.** The dynamic proprioceptor generates action potentials only with a change in direction of movement. The burst of action potentials produced is very brief (Fig. 8–10*b*). This type of action potential response is called a **phasic discharge.**

TABLE 8–2

Referred Pain

True Point of Origin of Pain	Common Cause	Area to Which Pain Is Referred
Heart	Ischemia secondary to infarction	Base of neck, shoulders, pectoral area, arms
Esophagus	Spasm, dilation, chemical (acid) irritation	Pharynx, lower neck, arms, over heart ("heartburn")
Stomach	Inflammation, ulceration, chemical	Below xiphoid (epigastric area)
Gallbladder	Spasm, distention by stones	Epigastric area
Pancreas	Enzymatic destruction, inflammation	Back
Small intestine	Spasm, distention, chemical irritation, inflammation	Around umbilicus
Large intestine	As above	Between umbilicus and pubis
Kidneys and ureters	Stones, spasm of muscles	Directly behind organ or groin and testicles
Bladder	Stones, inflammation, spasm, distention	Directly over organ
Uterus, uterine tubes	Cramps (spasm)	Low abdomen or lower back

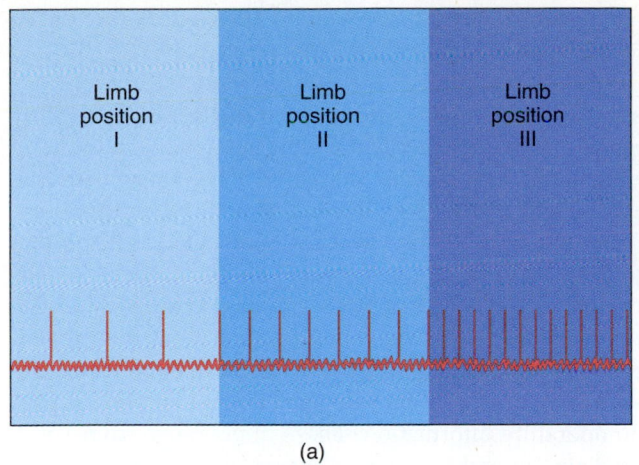

Figure 8–10

Action potential patterns produced by *(a)* static and *(b)* dynamic proprioceptors.

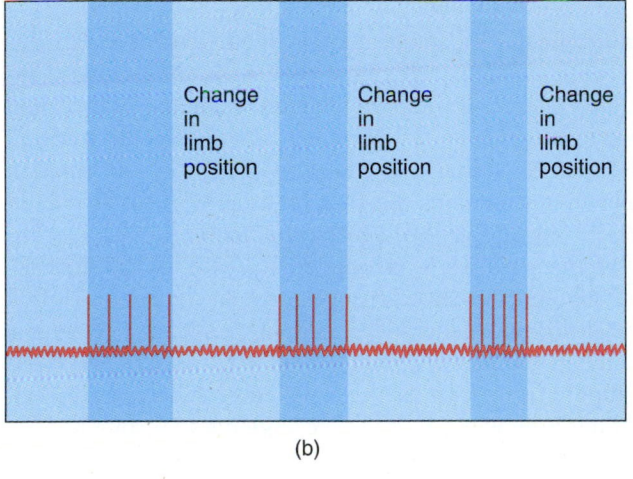

The information from somatosensory receptors is transmitted to the spinal cord and brain by several different pathways organized according to the following general principles: (1) somatosensory pathways from the receptor to the cerebral cortex are made up of three neurons arranged in sequence; (2) each type of somatic sensation has a specific, defined pathway; (3) most pathways cross over from one side of the spinal cord or brain to the other; (4) all spinal nerves and cranial nerves are paired, and each member of the pair, designated right or left, is distributed unilaterally, and somatosensory pathways are paired and are designated right or left, based on the position of the ascending axons in the spinal cord (for pathways originating with spinal nerves) or brain stem (for pathways originating with cranial nerves); (5) the nerve cells within each nucleus are topographically organized according to the location of their sensory receptors on the surface of the body. As we shall see, some of these principles also apply to the organization of other sensory systems.

SOMATOSENSORY PATHWAYS

> *How are the sensory pathways organized?*

General Organization

Somatosensory pathways carry information about pain, temperature, touch, pressure, vibration, position of body parts, and their movement. These pathways are organized into three neuron sets. The neurons in each set are designated in sequence, beginning with the receptor, first order, second order, and third order.

First-Order Neuron

The cell body of the first-order neuron is located in the **dorsal root ganglion** of a spinal nerve or the trigeminal ganglion of the fifth cranial nerve. The axon carries signals from the somatic receptor into the spinal cord by way of the dorsal

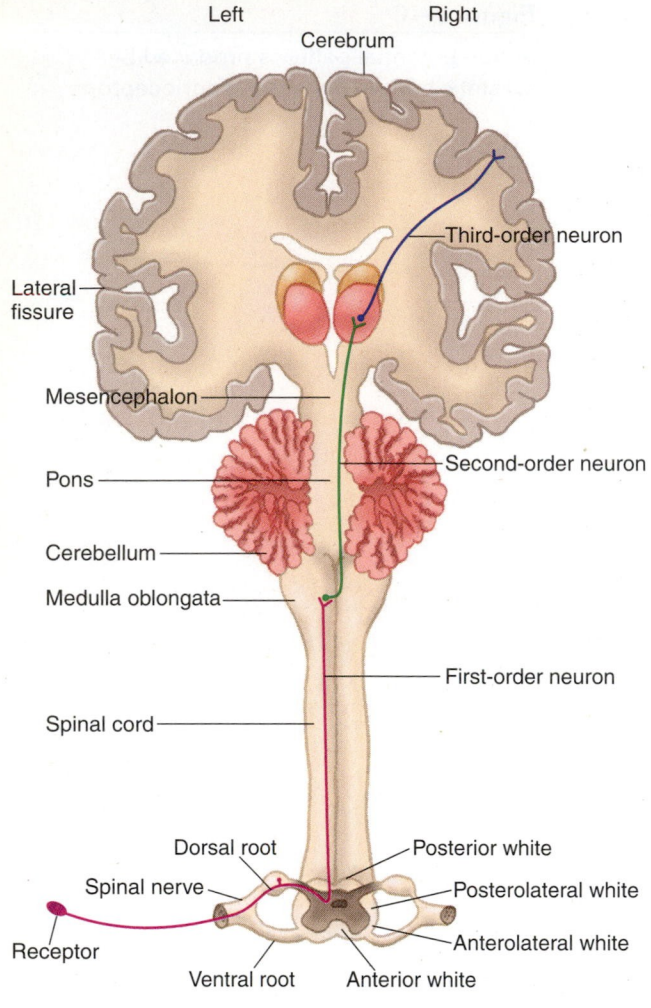

Figure 8–11

First-order, second-order, and third-order neurons in a typical ascending sensory pathway.

root of the spinal nerve (Fig. 8–11) or the sensory root of the trigeminal nerve.

Sensory Fiber Classification. Each of the somatosensory receptors previously described is associated with a peripheral nerve axon of a particular diameter. The fibers with the largest diameters are those associated with touch, pressure, and vibration sense. These are heavily myelinated fibers that range from 13 to 22 μm in diameter and have action potential conduction velocities of 70 to 120 m/s (Table 8–3). These fibers release the neurotransmitter glutamate at their terminal endings. The fibers with the smallest diameters are the lightly myelinated and unmyelinated fibers that convey pain and temperature information. These fibers range from 1 to 5 μm in diameter and have action potential conduction velocities of 2 to 15 m/s. These fibers release the neuropeptide substance P at their terminal endings.

Dermatomes. The spinal cord is subdivided into 31 segments (Fig. 8–12). Each spinal cord segment receives sensory information from particular areas of the skin called **dermatomes.** These dermatomes contain receptors for all of the somatic sensations. The loss of sensory information by certain dermatomes can thus be used as a clue to the location of damage to the spinal cord. In the uppermost region of the cord are eight cervical segments (C1 to C8). These segments receive sensory information from the back of the head, neck, shoulders, part of the arms, and hands. Next are 12 thoracic segments (T1-T12), which receive sensory inputs from parts of the arms and hands and trunk of the body. Following the thoracic segments are five lumbar segments (L1-L5), which receive sensory inputs from the waist, thighs, upper and lower legs, and parts of the feet. The remaining segments of the spinal cord are five sacral segments (S1-S5) and one sacrococcygeal segment. These

TABLE 8–3

Sensory Fibers

Type	Diameter (μm)	Conduction Velocity (m/sec)	Sensory Information
I$_a$ (A$_\alpha$)	22	120	Primary muscle spindle
I$_b$ (A$_\alpha$)	22	120	Primary touch and pressure
II (A$_\beta$)	13	70	Secondary muscle spindle, touch, and pressure
III (A$_\delta$)	5	15	Lightly myelinated fibers for touch, pressure, and pain
IV (C)	1	2	Unmyelinated fibers for pain and temperature

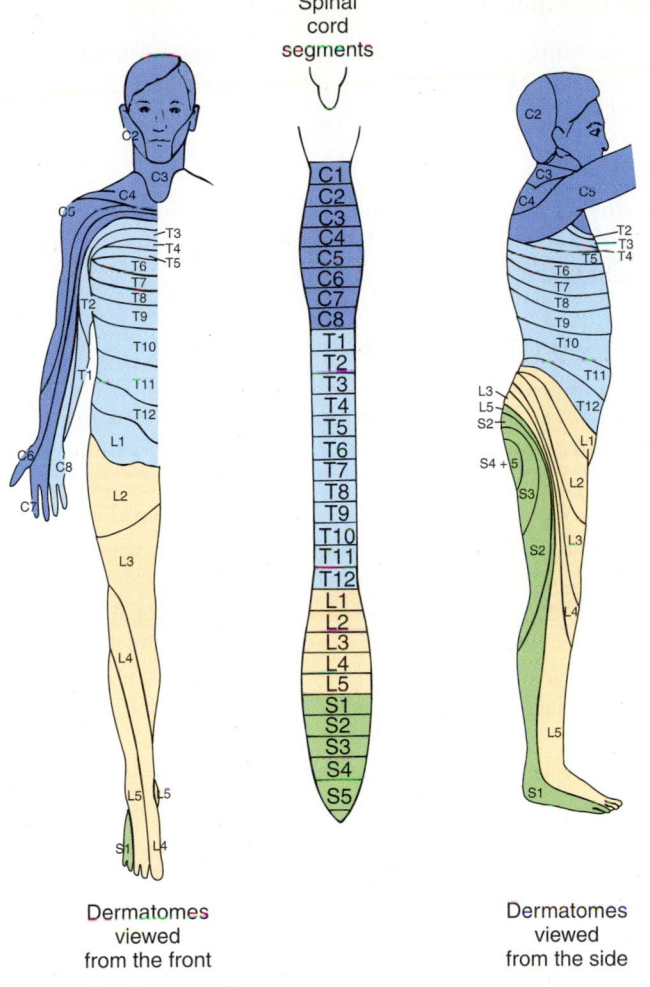

Figure 8–12

Spinal cord segments and dermatomes. Dermatomes represent the cutaneous sensory distribution of peripheral nerve fibers.

receive sensory input from the back of the legs, buttocks, and anus.

Somatosensory information from the face is carried to the brain stem by way of three sensory divisions of the trigeminal nerve: the ophthalmic division (upper eyelids, scalp), the maxillary division (lower eyelids upper jaw, cheeks), and the mandibular division (lower jaw) .

Second-Order Neuron

Axons of the second-order neurons carry signals from the spinal cord or brain stem to the thalamus. The axons cross the midline (decussate) in the spinal cord or in the brain stem and ascend to the contralateral thalamus, where they synapse with third-order neurons located in the **ventral posterolateral nucleus** (VPL) (body) or **ventral posteromedial nucleus** (VPM) (face).

Third-Order Neuron

Axons of the third-order neuron transmit somatosensory information from the right or left thalamus to the primary somatosensory area (the postcentral gyrus) of the ipsilateral parietal cortex. Conscious perception of somatic sensation occurs in the cerebral cortex and includes identification of stimulus modality, quality, and quantity and an accurate localization of the applied stimulus.

Dorsal Column-Medial Lemniscus Pathways

> *How does the general pattern of the somatosensory pathway apply to touch and proprioception sensations?*

Sensory fibers of the dorsal column system enter the spinal cord and ascend toward the brain in the dorsal white matter of the spinal cord (Fig. 8–13). Medially located fibers, primarily from the lower extremities and abdomen, form the *fasciculus gracilis,* and laterally located fibers from remaining body parts, except the face, form the *fasciculus cuneatus.* Together, the fasciculus gracilis and the fasciculus cuneatus are called the **dorsal columns.** These fibers continue to the medulla oblongata, where they form synapses with neurons in the dorsal column nuclei, called the **nucleus gracilis** and the **nucleus cuneatus,** respectively. Fibers from the dorsal column nuclei, in turn, cross the midline and become part of the **medial lemniscal** pathway. These fibers reach the **thalamus** on the opposite side of the brain and synapse with third-order neurons located in the VPL nucleus. Neurons in the thalamus, in turn, extend their fibers to the cortex.

The dorsal column-medial lemniscal pathways carry information regarding the following localized and well-differentiated (epicritic) somatic sensations.

Pressure and Stretch

Pressure and stretch are related to a person's ability to accurately assess the weight of an object placed in the hand.

Discriminative Touch

Discriminative touch allows a person to localize and differentiate two simultaneous stimuli applied to the skin with the points of a divider.

Vibration

The sense of cutaneous vibration begins with the ability to detect rapid fluctuations in skin pressure, such as that applied with a tuning fork.

Stereognosis

Stereognosis is the ability to identify an object exclusively on the basis of how it feels regarding shape, size, texture, and memory of previous tactile sensation. Blindfolded, a

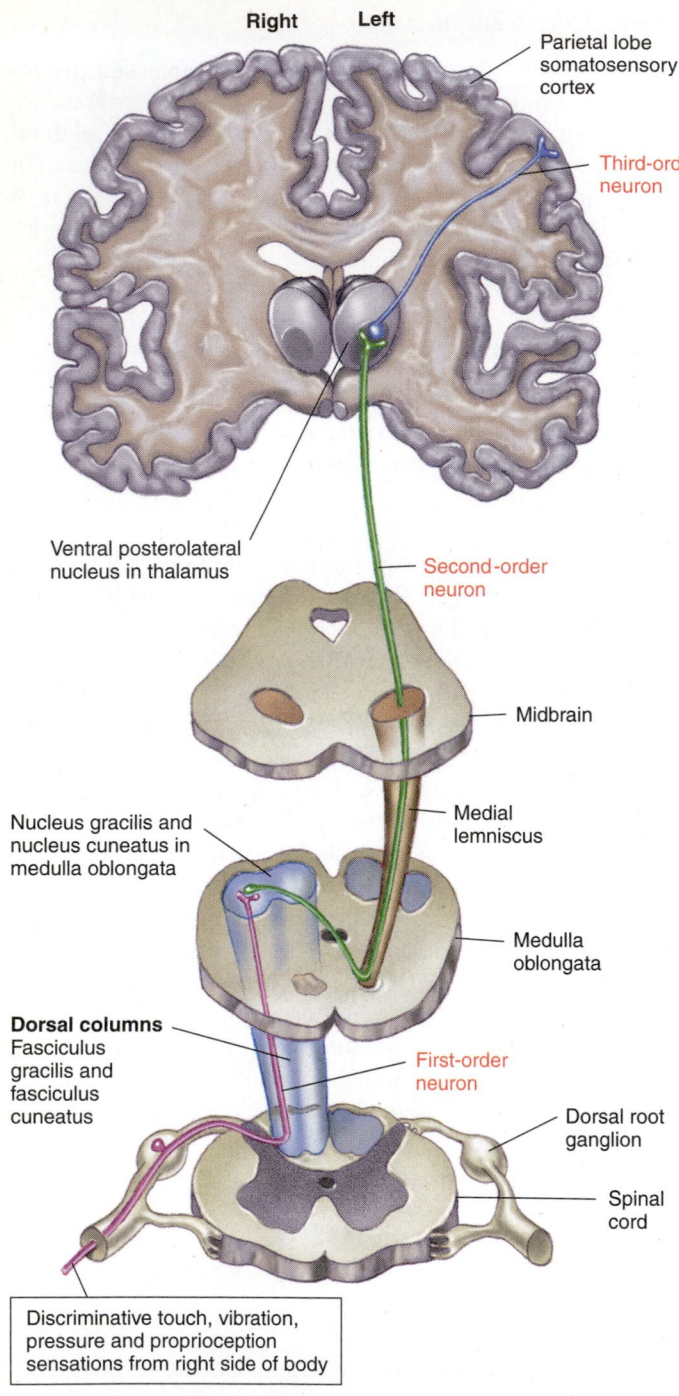

Dorsal column pathway

Figure 8–13

The right dorsal column pathway. The left dorsal column pathway, a mirror image of the right, has been omitted for clarity.

person usually can identify or at least describe the differences between a table-tennis ball placed in one hand and a small wooden cube simultaneously placed in the other hand.

Proprioception

Proprioception is an awareness of the position of body parts with respect to one another as well as with respect to gravitational force.

Kinesthesia

Kinesthesia is an awareness of movement resulting from activation of receptors in skeletal muscles, ligaments, tendons, and joint capsules.

Anterolateral (Spinothalamic) Pathways

> *How does the general pattern of the somatosensory pathway apply to pain and temperature sensations?*

The anterolateral pathways carry information regarding pain, temperature, touch, and pressure. First-order neurons enter the lateral part of the dorsal horn (Fig. 8–14). After entering the spinal cord, however, the majority of them form synaptic connections with second-order neurons located in the dorsal horn. This group of nerve cells collectively is called the **substantia gelatinosa.** These cells, in turn, cross the midline and ascend toward the brain as a bundle of fibers in the lateral white matter called the **lateral spinothalamic tract,** or in the anterior white matter of the spinal cord, forming the **anterior spinothalamic tract.** The anterior and lateral spinothalamic tracts collectively make up the **anterolateral system** (ALS). Ascending fibers of the ALS have inputs to the reticular formation of the brain stem and the thalamus. The reticular formation serves as an alerting system and activates other areas of the brain.

Inputs to the thalamus terminate in the **ventrobasal** (VB) nuclei. Third-order neurons in the ventrobasal nuclei project to the primary somatosensory cortex of the parietal lobe and to the frontal cortex.

Fibers of the anterior spinothalamic tract convey information about crude, poorly defined (protopathic) **touch, pressure, itch,** and **tickle.** The lateral spinothalamic tract carries information about **pain** and **temperature.** This information in turn is sent to the primary somatosensory cortex in a somatotopic fashion and to the frontal cortex. The pain information that reaches the primary somatosensory cortex is associated with acute pain, whereas the information that reaches the frontal cortex is associated with chronic pain.

The perception of pain can be altered by other sensory inputs from the periphery or controlled by descending inputs from the brain stem and other higher brain centers. This gating of pain information is carried out by an inhibitory interneuron located in the dorsal horn of the spinal cord (Fig. 8–15). As previously discussed, the pain fibers from the periphery release the peptide called *substance P* as the neurotransmitter. In the absence of substance P release, pain cannot be detected by the brain. The interneuron acts to inhibit the release of substance P by presynaptic inhibition. The axon terminal of the interneuron forms a synapse on the

sociated with touch, pressure, and vibration or by fibers from higher brain centers.

The influence of other tactile inputs is seen in the reduction of pain caused by rubbing the skin surrounding a region of injury; this causes the pain to subside more quickly. The influence of higher brain centers on pain perception is seen in cases in which individuals involved in emergency situations do not realize that they have been injured until much later. The mechanisms that prevent the perception of pain under these highly stressful conditions are part of an involuntary autonomic "fight-or-flight" response.

Another aspect of pain sensation is the perception that pain actually caused by the distress of internal organs is coming from the surface of the body. For example, in cases of heart attacks, the area of the chest and left arm may become painful. Because the pain is attributed to a location other than its source, this type of pain is called **referred pain.** The phenomenon of referred pain is the result of primary visceral and somatic pain fibers converging on the same group of secondary neurons within the pain pathway of the same spinal segment.

Trigeminal Pathways

> *How does the general pattern of the somatosensory pathway pertain to sensations from the face?*

Somatic sensation from the face is carried by nerve fibers in the ophthalmic, maxillary, and mandibular nerves to the bodies of first-order neurons located in the trigeminal ganglion, and from there into the pons. Axons carrying information about pain and temperature descend to the medulla and

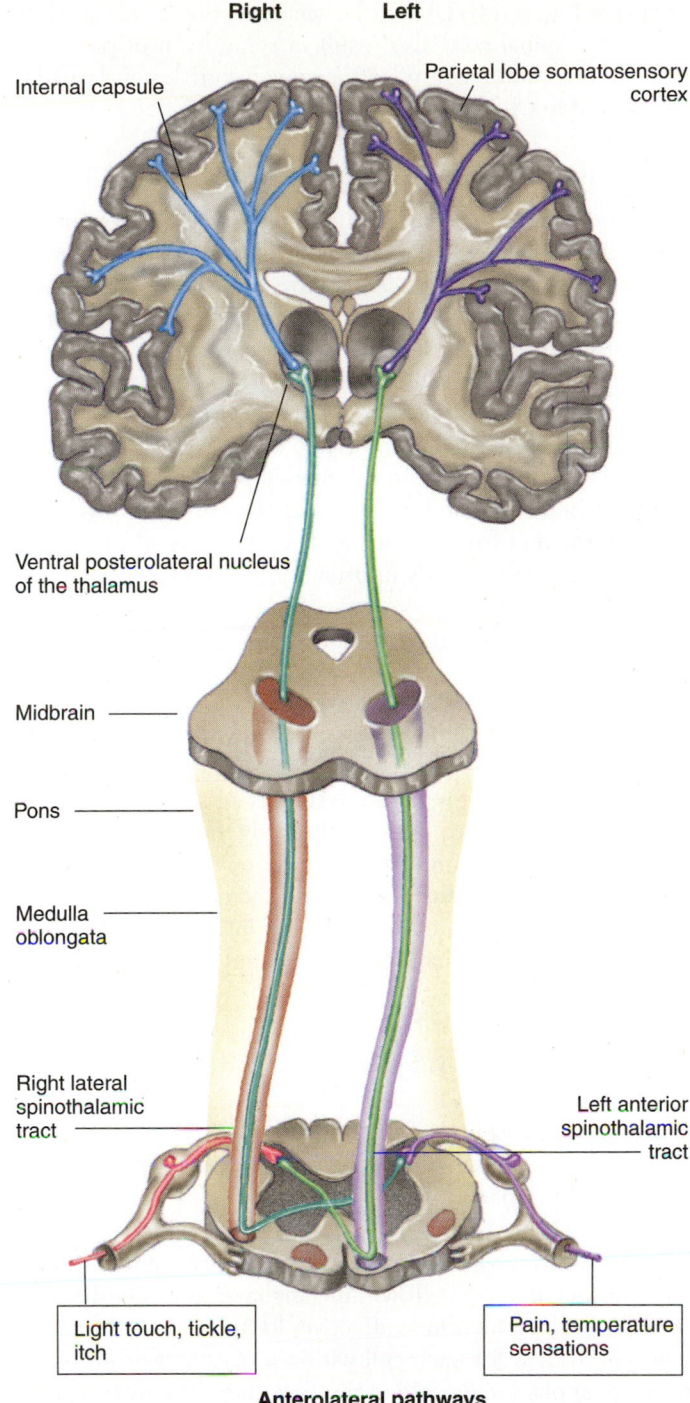

Right **Left**

Internal capsule

Parietal lobe somatosensory cortex

Ventral posterolateral nucleus of the thalamus

Midbrain

Pons

Medulla oblongata

Right lateral spinothalamic tract

Left anterior spinothalamic tract

Light touch, tickle, itch

Pain, temperature sensations

Anterolateral pathways

Figure 8–14

Anterolateral pathways consist of the anterior spinothalamic tracts and the lateral spinothalamic tracts. Only one member of each pair is shown for clarity.

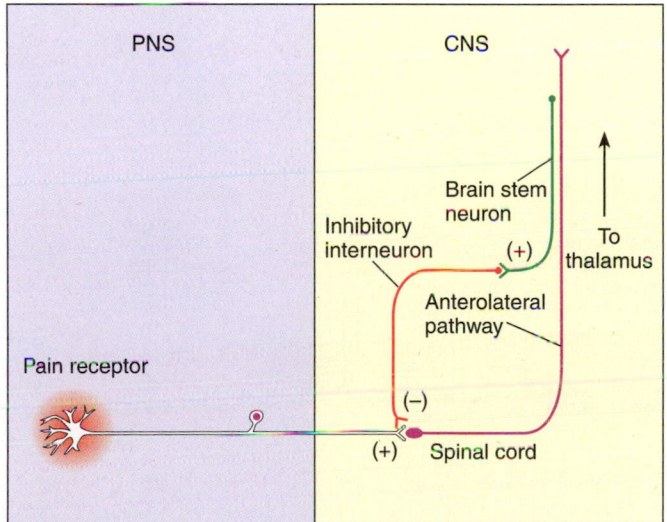

PNS   CNS

Brain stem neuron

Inhibitory interneuron

To thalamus

(+)

Anterolateral pathway

Pain receptor

(−)

(+)   Spinal cord

Figure 8–15

Presynaptic inhibition at the synapse between first-order and second-order neurons in the anterolateral pain pathway reduces the intensity of pain sensation. PNS = peripheral nervous system; CNS = central nervous system.

terminal of pain fibers. The terminals of the interneurons release the peptide called **enkephalin,** which in turn inhibits the release of substance P from the terminals of the pain fibers. Thus, the interneuron is the key element in altering the sensation of pain. It can be activated by sensory fibers as-

cervical spinal cord, where they synapse with second-order neurons in the **nucleus of the spinal tract of cranial nerve V**. Axons carrying information about touch and pressure synapse with second-order neurons in the **main sensory nucleus** of the trigeminal nerve. Some second-order axons from the trigeminal nuclei cross the midline and ascend contralaterally to the VPM nucleus in the thalamus. The other second-order axons ascend ipsilaterally to the VPM nucleus. Third-order neurons in the thalamus project to the primary somatosensory cortex.

Lesions of the Spinal Cord

In accidents that damage the spinal cord, the loss of cutaneous sensation appears in patterns that are characteristic of the level of the lesion. For example, an accident that severs half of the spinal cord on the right side of the body, at the tenth thoracic vertebra, causes a loss of pain and temperature sensation on the left leg and a loss of discriminative touch, vibration sense, and kinesthetic sense on the right leg (Fig. 8–16).

This pattern occurs because discriminative touch, vibration, and kinesthetic information is transmitted in the spinal cord along the same side of the body where the information enters the spinal cord. Pain and temperature information, on the other hand, crosses the midline once it enters the spinal

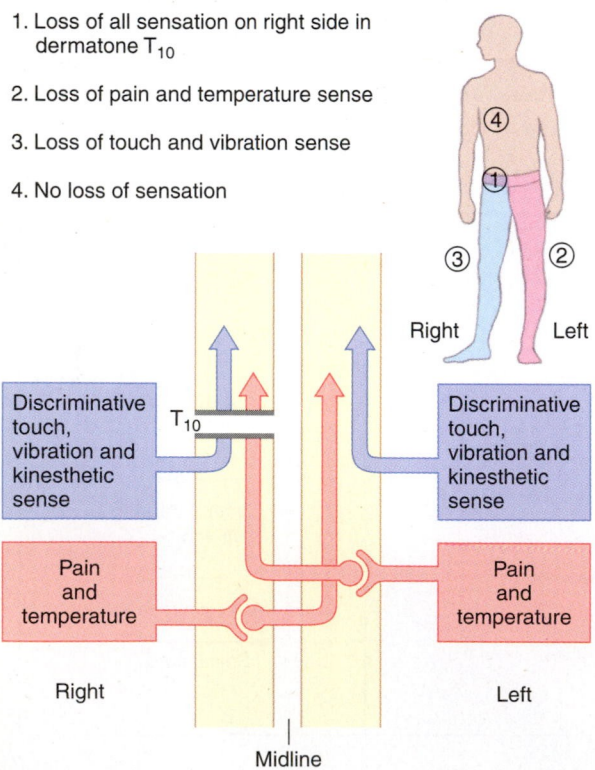

1. Loss of all sensation on right side in dermatone T₁₀

2. Loss of pain and temperature sense

3. Loss of touch and vibration sense

4. No loss of sensation

Figure 8–16

Some effects of spinal cord lesions. Damage to the spinal cord on the right side of the body at T10 causes the left leg to lose pain and temperature sensation, whereas the right leg loses the sense of discriminative touch and vibration.

cord (see Fig. 8–14). Obviously, accidents that transect all or half of the spinal cord also result in some form of paralysis. This loss of motor control after spinal cord lesions will be considered in Chapter 9.

The Somatosensory Cortex

The **neocortex** is the part of the brain that evolved most recently (*neo-* meaning "new"). It is divided into discrete areas that receive somatic, visual, auditory, and gustatory sensations as well as areas for the control of movement and other functions (Fig. 8–17). The area of the neocortex that receives somatic sensations is called the **somatosensory cortex.** A cross-section of the somatosensory cortex shows that it is organized like a map of the body (Fig. 8–18). Sensory information from the legs is sent to the medial portion of the cortex. Adjacent but more lateral regions of the cortex receive sensory information from the torso. The most lateral regions of the cortex receive sensory information from the arms, hands, and face.

This type of topographical organization of sensory information from various regions of the body on to the sensory cortex is called a **somatotopic organization.** The map that is created in the somatosensory cortex is a distortion of the true proportions of the body. The distorted figure, called the *sensory homunculus,* represents the density of receptors located in various body regions. Parts of the body that have the greatest density of sensory receptors are represented by a greater proportion of the cortex. The areas in the cortex representing the thumb and fingers of the hand, for instance, occupy substantially more space than that represented by the torso.

Receptive Fields of Neurons in the Somatosensory System

> *What is a receptive field; how do the receptive fields of the three cells in the general somatosensory pathway differ?*

Stimuli detected by a sensory receptor are transmitted to various clusters of nerve cells within ganglia or nuclei in the sensory pathway. The most effective location of a stimulus, however, differs for each cell within a cluster. For instance, pressure applied to an area of the skin directly above a Pacinian corpuscle activates the associated dorsal root ganglion (DRG) cell (Fig. 8–19).

If pressure is applied to another location on the surface of the skin, another DRG cell is activated; however, the first DRG cell is not affected by pressure applied to this second location on the skin. Thus, only a small area on the receptive surface of the skin can activate a nerve cell in the dorsal root ganglion. This small area of skin represents the **receptive field** for that cell. In the case of this receptor, the receptive field can be described as a circular area on the skin where pressure is applied to excite the cell. We will denote the ability to excite the cell with a "+," and the receptive field area of the skin, by a circle.

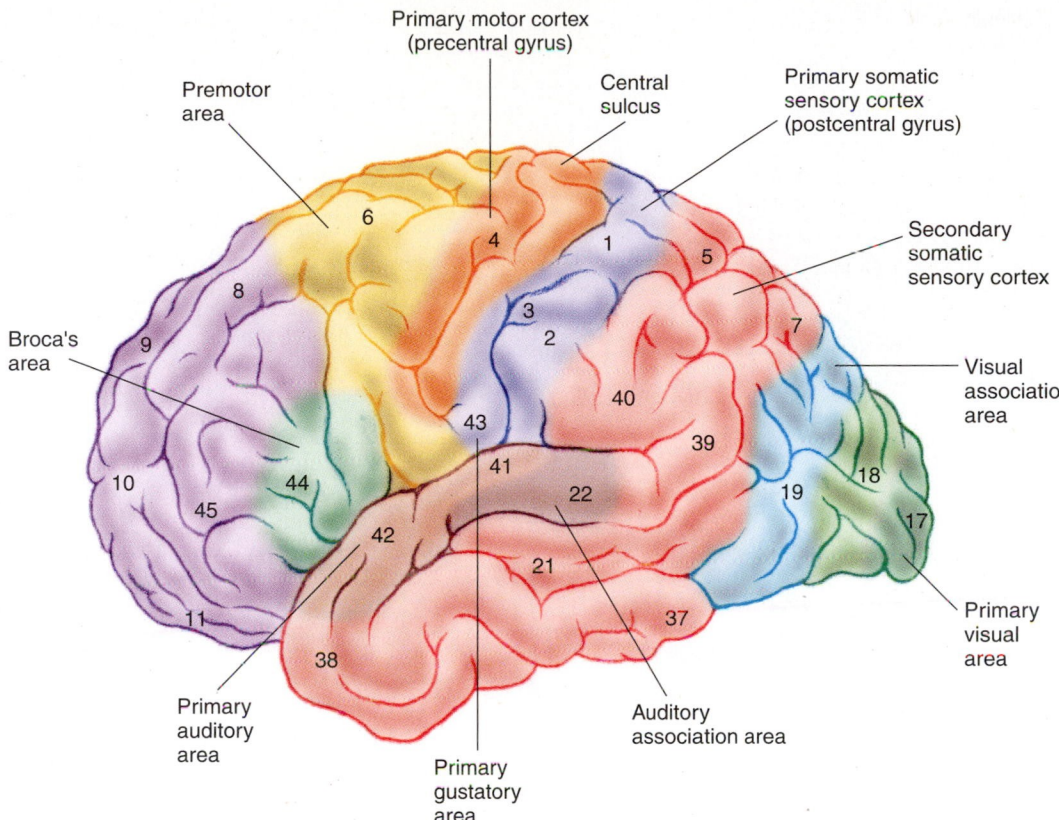

Figure 8–17

Functional areas of the neocortex shown on the left cerebral hemisphere. The numbers represent Brodmann areas.

Nerve cells in the dorsal column nuclei, the next cluster of cells along the dorsal column pathway, have different receptive fields than those in the dorsal root ganglia. Dorsal column nerve cells generate a low rate of spontaneous action potentials. This is their tonic baseline activity. When pressure is applied to an area of the skin, a dorsal column nerve cell (cell 1 in Fig. 8–20a) is excited to produce a higher frequency of action potential discharge than its baseline rate of activity. However, when the pressure is applied to a different location, the frequency of the action potentials is decreased to less than the baseline rate.

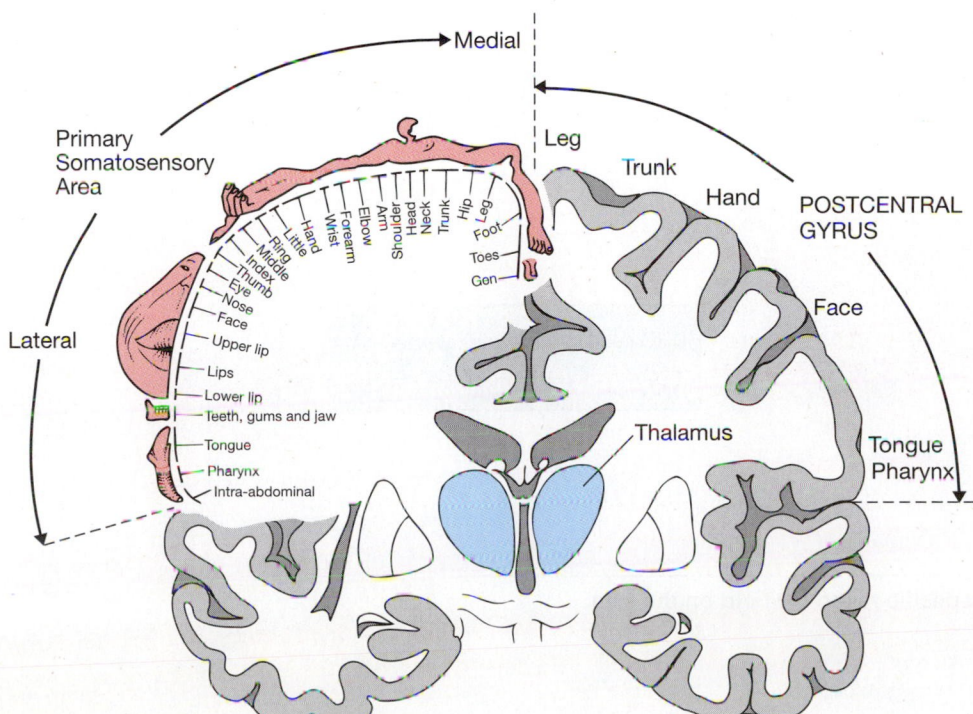

Figure 8–18

The somatotopic organization of the primary somatosensory cortex.

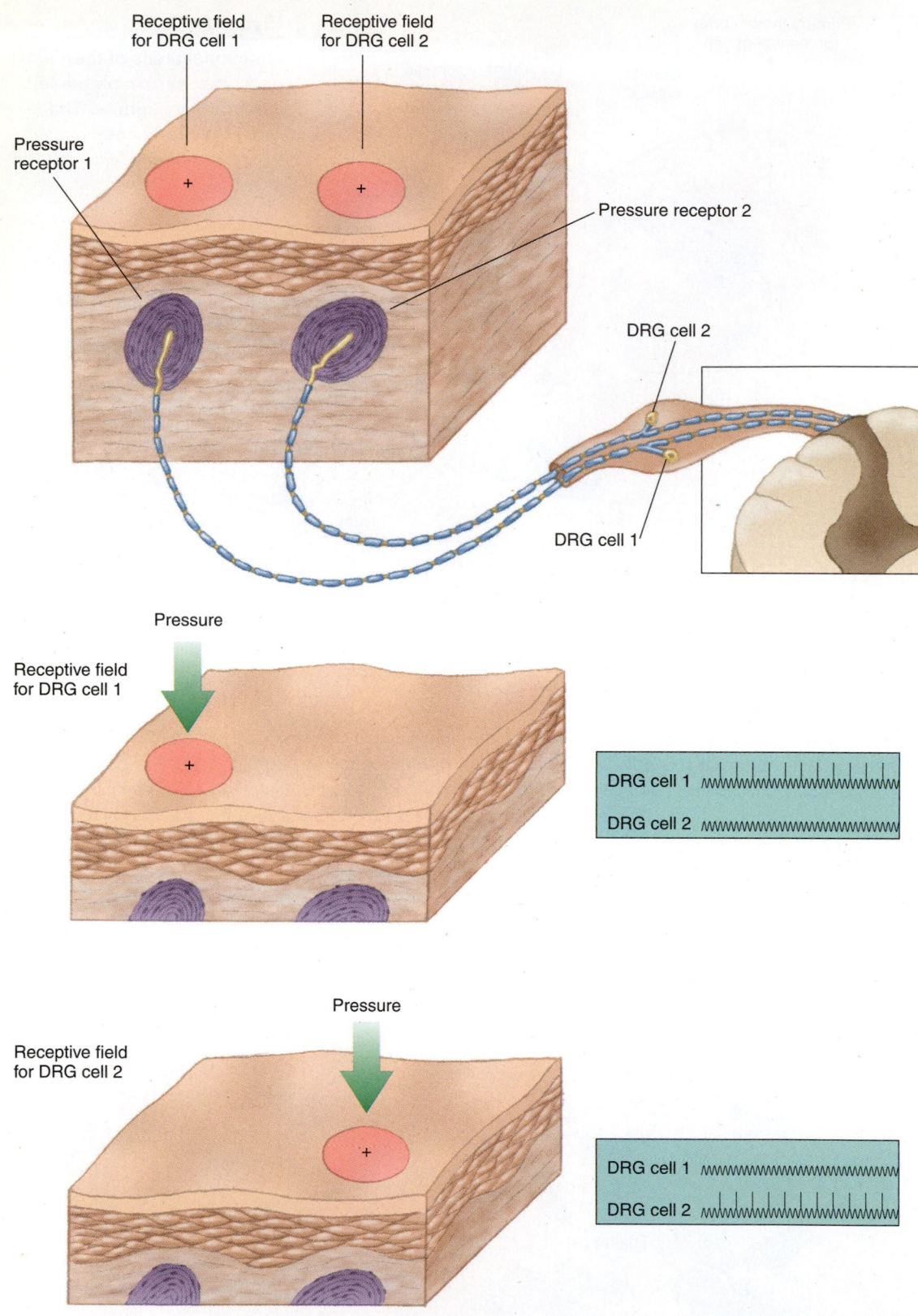

Figure 8–19

A cell in the dorsal root ganglion has a specific receptive field on the skin.

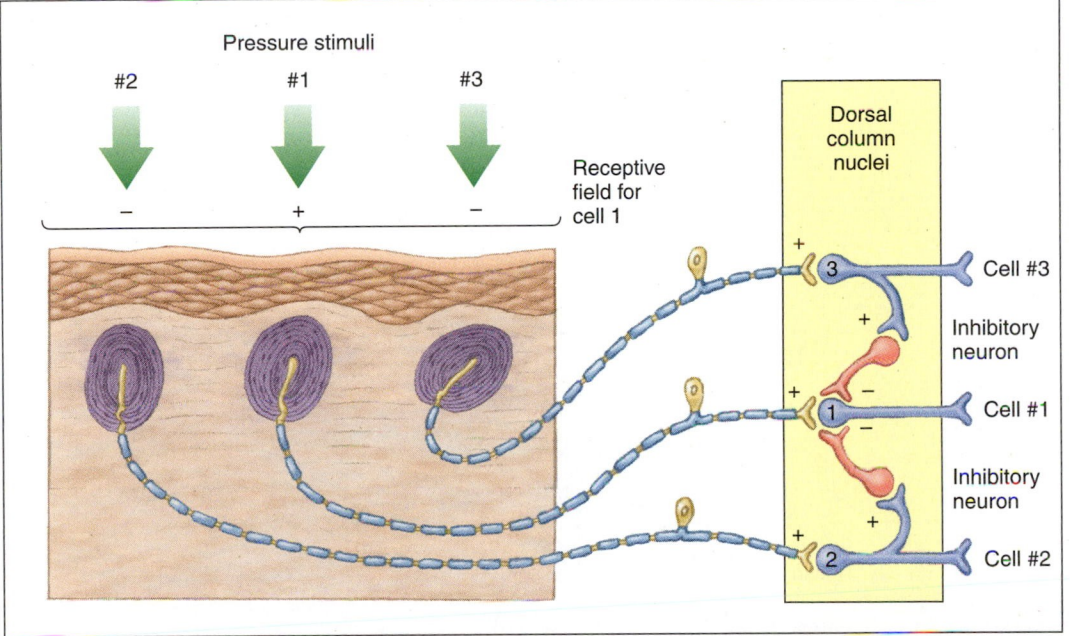

(a)

(b)

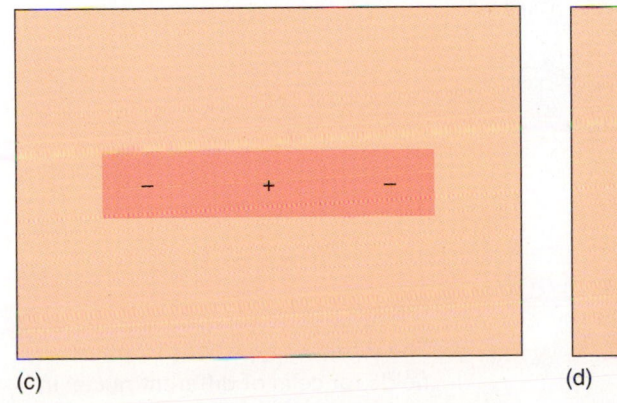

(c)

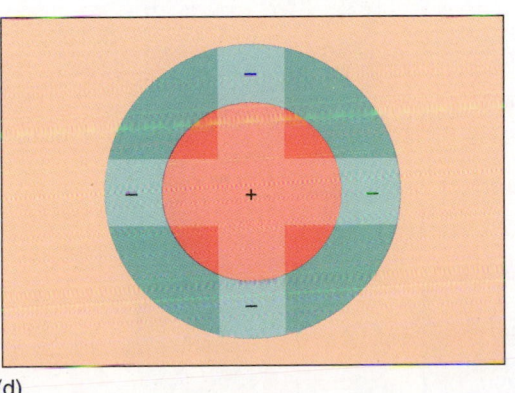

(d)

Figure 8–20

Annular receptive fields of neurons in the dorsal column nuclei are formed by inhibitory interneurons. **(a)** Activation of Pacinian corpuscle 1 excites dorsal column cell 1, whereas activation of Pacinian corpuscle 2 inhibits dorsal column cell 1 via an inhibitory interneuron (*red*). Bars under recordings indicate duration of stimulus application. **(b)** Mapping of receptive field on the surface of the skin for dorsal column cell 1. Pressure stimulus 1 activates this cell, whereas pressure stimuli 2 and 3 inhibit dorsal column cell 1. **(c)** One-dimensional top view of receptive field for dorsal column cell 1. **(d)** Two-dimensional top view of receptive field for dorsal column cell 1.

This inhibition, "−" of the dorsal column nerve cell, is a result of activating a second Pacinian corpuscle, which in turn activates a different nerve cell (cell 2) within the dorsal column. The activation of this second dorsal column nerve cell leads to the activation of an inhibitory interneuron. This inhibitory neuron, in turn, prevents the excitation of the first dorsal column neuron. Other areas of the skin, when pressed, also inhibit the response of dorsal column nerve cell 1 (Fig. 8–20*b*). When the excitatory and inhibitory areas on the skin are mapped for nerve cell 1, an annulus or doughnut-shaped receptive field is revealed with an excitatory center and an inhibitory ring (Fig. 8–20*d*).

This annular receptive field also is found for nerve cells in the thalamus and certain areas of the somatosensory cortex. The cells in the somatosensory cortex that exhibit the annular receptive field are all located in area 3 of the primary somatosensory cortex (see Fig. 8–17). This is a narrow strip of cortical tissue that is also called Brodmann

area 3 (named for the neuroanatomical studies by K. Brodmann in 1909).

In Brodmann areas 1 and 2, the receptive fields are no longer annular. The stimulus that most effectively activates nerve cells in areas 1 and 2 is pressure applied to the skin in a rectangular receptive field (Fig. 8–21). The effectiveness of a rectangular stimulus is the result of nerve cells in area 3 that have overlapping receptive fields. The output fibers of these cells converge on a nerve cell in area 1 (Fig. 8–22). Consequently, the simultaneous activation of all three input cells is needed in order to excite the one cell in area 1. This excitation occurs by the process of spatial summation. This process is described in Chapter 7.

Although sensory information of a particular type of stimulus is transmitted from area 3 to 1, and area 1 to 2, in any one Brodmann area of the somatosensory cortex, information from one type of somatic sensation tends to predominate. The nerve cells in Brodmann area 1, for example,

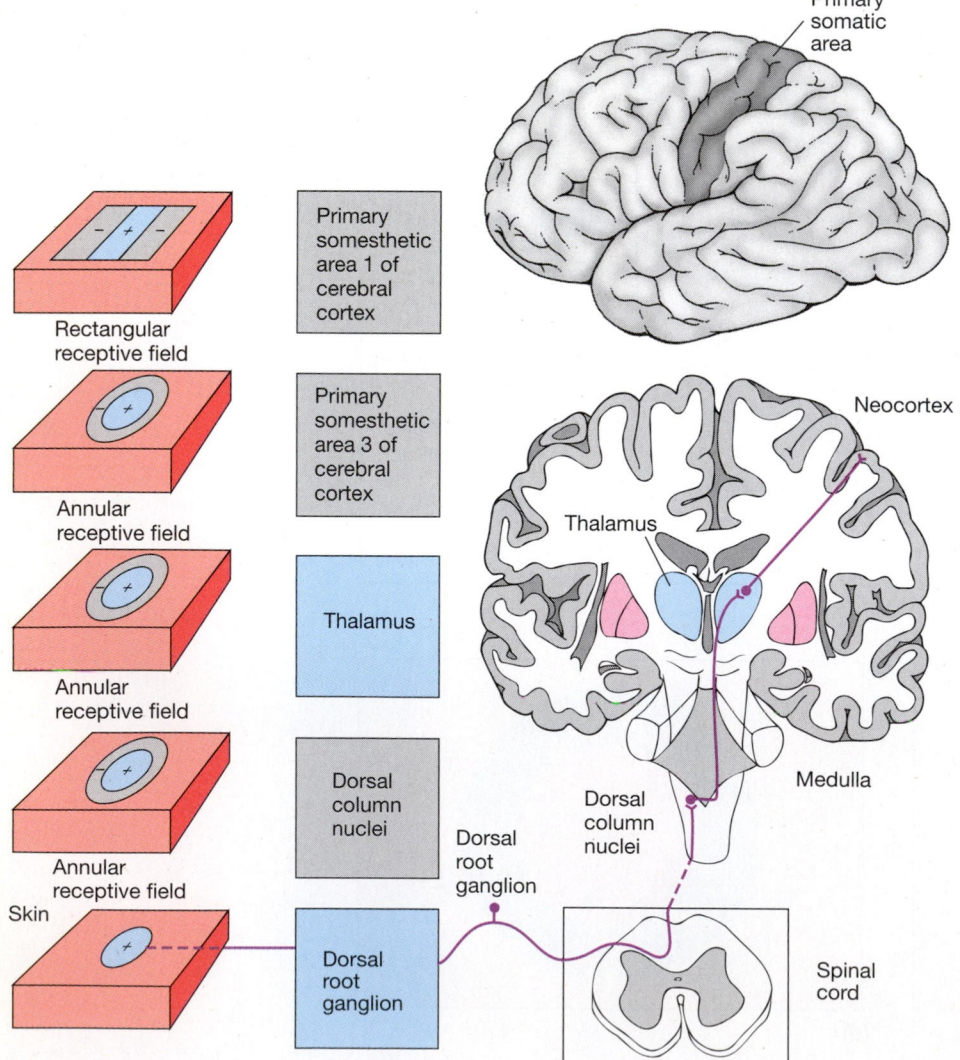

Figure 8–21

The different types of receptive fields for cells of different nuclei in a somatosensory pathway.

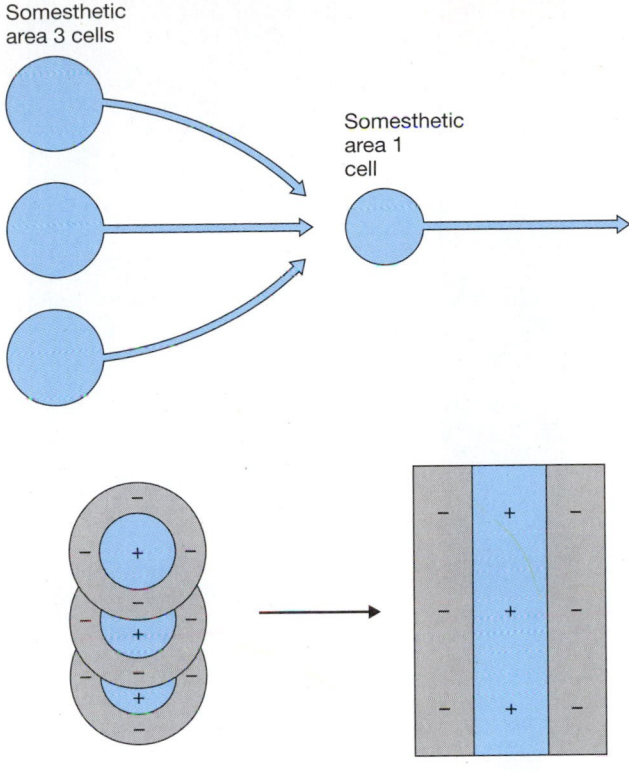

Somesthetic
area 3 cells

Somesthetic
area 1
cell

Figure 8–22

The receptive fields in the secondary somatosensory areas
are rectangular.

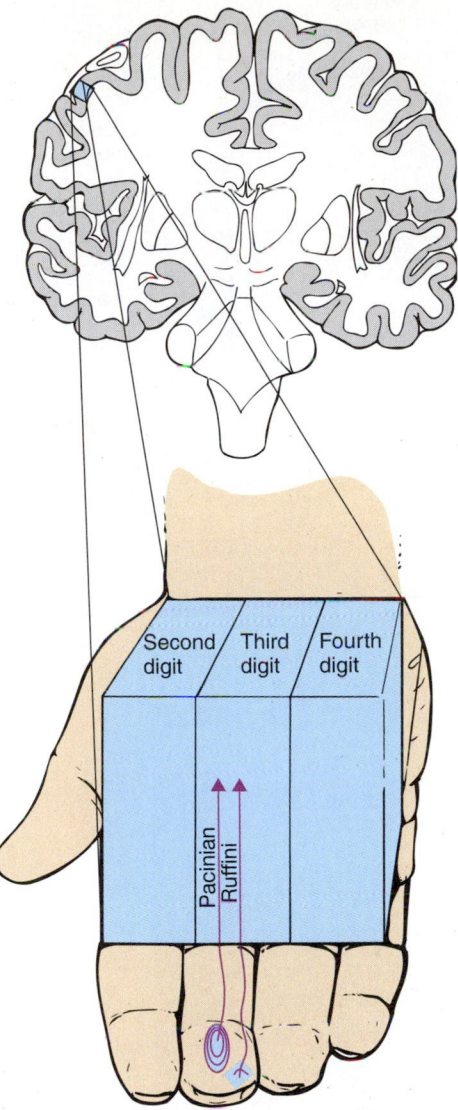

Second
digit

Third
digit

Fourth
digit

Pacinian
Ruffini

Figure 8–23

The somatosensory cortex is organized so that all of the
cells specialized for one sensation are grouped in a vertical
column.

respond to rapidly adapting cutaneous receptors; those in
area 2 respond to deep pressure; those in area 3a respond
to stretch receptors in skeletal muscle; those in area 3b respond to rapidly and slowly adapting cutaneous receptors.
In this way, the somatosensory cortex has several somatotopic maps of the body.

The most effective shapes of stimuli needed to activate
nerve cells in area 2 can become quite complex. In essence,
the physical features of an object in the environment are first
broken down into its fundamental components by the peripheral sensory receptors. This information then is assembled to reconstruct the holistic features of the object that are
capable of activating nerve cells in higher centers of the
brain. In our discussion of the visual system, we will see that
a similar synthesis and reconstruction of the external environment takes place for visual information.

The somatosensory cortex also is organized so that all of
the cells that respond to one type of sensation are located together within vertical columns. Each nerve cell within a vertical column of the somatosensory cortex, for example, is
activated by Pacinian corpuscles in one fingertip (Fig. 8–23).
The location of the receptive fields for cells within a vertical
column is also similar. Thus, the vertical columns in the neocortex represent basic units of sensation-specific and location-
specific function.

THE VISUAL SYSTEM

*How does the visual system receive and process
information from the environment?*

The visual system allows us to determine the shape and color
of objects, such as books, and the form of the letters printed
within their pages. Parts of the visual system also allow us to
recognize the grouping of letters in the form of words, and
the grouping of words to represent ideas. These latter functions will be discussed in Chapter 11, where we consider the
localization of language centers in the brain. In this section,
we consider how the visual system detects the shape and
color of objects and the movement of objects in the external

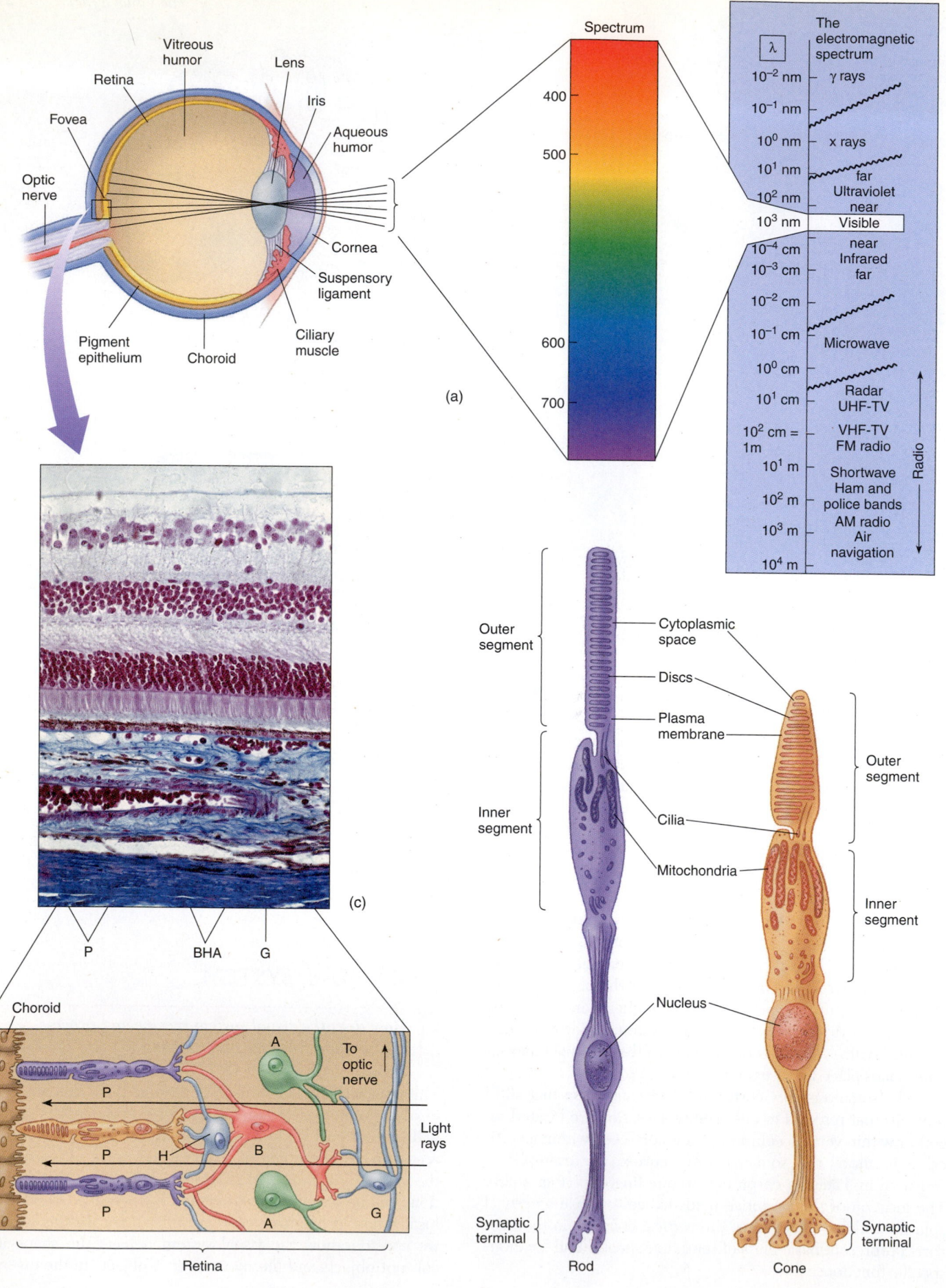

(a)

Spectrum

400

500

600

700

The electromagnetic spectrum

λ

10^{-2} nm — γ rays

10^{-1} nm — x rays

10^{0} nm

10^{1} nm — far

10^{2} nm — Ultraviolet near

10^{3} nm — Visible

10^{-4} cm — near

10^{-3} cm — Infrared far

10^{-2} cm

10^{-1} cm — Microwave

10^{0} cm

10^{1} cm — Radar UHF-TV

10^{2} cm = 1m — VHF-TV FM radio

10^{1} m — Shortwave Ham and police bands

10^{2} m — AM radio Air navigation

10^{3} m

10^{4} m

Radio

Vitreous humor

Retina

Lens

Iris

Fovea

Aqueous humor

Optic nerve

Cornea

Suspensory ligament

Pigment epithelium

Choroid

Ciliary muscle

(c)

P   BHA   G

Choroid

A

To optic nerve

P

P   H   B

Light rays

P

A   G

(b)   Retina

Outer segment

Cytoplasmic space

Discs

Plasma membrane

Inner segment

Cilia

Mitochondria

Nucleus

Synaptic terminal

Rod

Outer segment

Inner segment

Synaptic terminal

Cone

(d)

◄ **Figure 8–24**

(a) General structure of the eye and the visible portion of the electromagnetic spectrum. **(b)** Organization of photoreceptors (P, rods and cones), bipolar (B), horizontal (H), amacrine (A), and ganglion (G) cells in the retina. **(c)** Photomicrograph of the different cell types of the retina. **(d)** The general structure of a rod cell and a cone cell. Note the membranous discs within the outer segments of both cells. The membranes of these discs contain the pigment that, when activated by photons, produces visual sensation.   *(© Ed Reshke/Peter Arnold, Inc.)*

environment. The features of shape, color, and movement are processed by different cell groups within the visual system, beginning with those found in the eye.

The eye is the sensory organ of the visual system that transduces light into electrical impulses, which the nervous system then uses to transmit information to the brain. The light perceived is part of the continuum of **electromagnetic radiation** that ranges from radio waves to X-rays (Fig. 8–24). Each type of electromagnetic radiation is characterized by a **wavelength.** Radio waves have long wavelengths, whereas X-rays have very short wavelengths. The colors of light that we see have wavelengths in between radio waves and X-rays and range from 400 to 700 nm. These wavelengths correspond to the range of colors seen in a rainbow.

In addition to its wavelike nature, light also can be thought of as a composite of particles called **photons.** These photons have wavelengths and energies associated with different colors. All objects have the capability of selectively absorbing certain photons of light, and those photons not absorbed are reflected back into the environment. The reflected photons are those that give an object its color and that interact with the eye.

Ocular Accommodation

Figure 8–24 contains an illustration of a midsagittal section of the eye. It shows the cornea, iris, lens, vitreous humor, and retina. Together, the **cornea** and **lens** act like the lens of a camera. They operate to bend the light rays and focus them on to the retina. Focusing a camera changes the distance between the lens and the film. Our eyes accomplish this feat by changing the shape of the lens. *Ocular accommodation,* or focusing of the lens, is a function of the **ciliary muscle,** the **suspensory ligament,** and the **lens capsule.** Contraction of the ciliary muscle, a circular smooth muscle surrounding the lens, reduces tension on the suspensory ligament and lens capsule. The lens, by virtue of its own elastic recoil, tends to assume a more spherical shape, thus increasing its refractive power to focus on an object closer to the eye. Conversely, relaxation of the ciliary muscle increases tension on the suspensory ligament and the lens capsule, flattening the lens and causing the eye to become focused on a more distant object.

As an object is brought closer to the eye, a point is reached at which accommodation of the lens becomes maximum. This is known as the *near point of accommodation.* Objects brought closer to the eye than the distance of the near point cannot be properly focused on the retina and therefore appear blurred.

The **retina** is the part of the eye containing receptor cells called **photoreceptors** that absorb photons. As the illustration in Figure 8–24 shows, photons must first pass through the cornea and lens before reaching the retina. The most sensitive part of the retina is a region called the **fovea;** consequently, the lens focuses the image of an object onto the fovea. The image projected onto the retina, however, is upside down. The **inversion** of the image is similar to that produced by a camera lens onto a roll of film. Even though the retina detects the form of objects in an upside-down fashion, the brain reorganizes this information so that we interpret the outside world as right-side up.

The Nerve Cells of the Retina

We can best understand how the eye transduces the form of an object by first examining the functional organization of the nerve cells within the retina. In addition to photoreceptors, the retina contains other types of nerve cells called **horizontal cells, bipolar cells, amacrine cells,** and **ganglion cells** (see Fig. 8–24). Photoreceptors, either **rods** or **cones,** are responsible for absorbing the photons and are situated at the posterior portion of the retina, behind these other types of cells. Photons must therefore pass through several layers of cells before reaching the layer of photoreceptors (Fig. 8–24*b*).

Rods are cylindrically shaped photoreceptors (Fig. 8–24*d*) that contain one type of pigment capable of absorbing photons representing a broad range of wavelengths. Rods display a high degree of convergence onto other cells and thus as a group are very sensitive to low levels of illumination. They therefore operate well during the night.

Cones are conically shaped photoreceptors and absorb photons associated with a more narrow range of wavelengths for color. Cones display a low degree of convergence onto other cells and therefore as a group are not as sensitive as rods but have better spatial resolution. There are three types of cones, each responsive to photons of a different range of wavelengths. These cones form the basis of color perception and are used for day vision.

Photoreceptors and Phototransduction

> *How is light energy converted to electrical energy in the retina?*

The photoreceptor pigment that absorbs photons is called **visual pigment.** This pigment is embedded within the membranes of **discs,** lamellar structures in the **outer segment** of the photoreceptor. Also present in the disc membrane is a

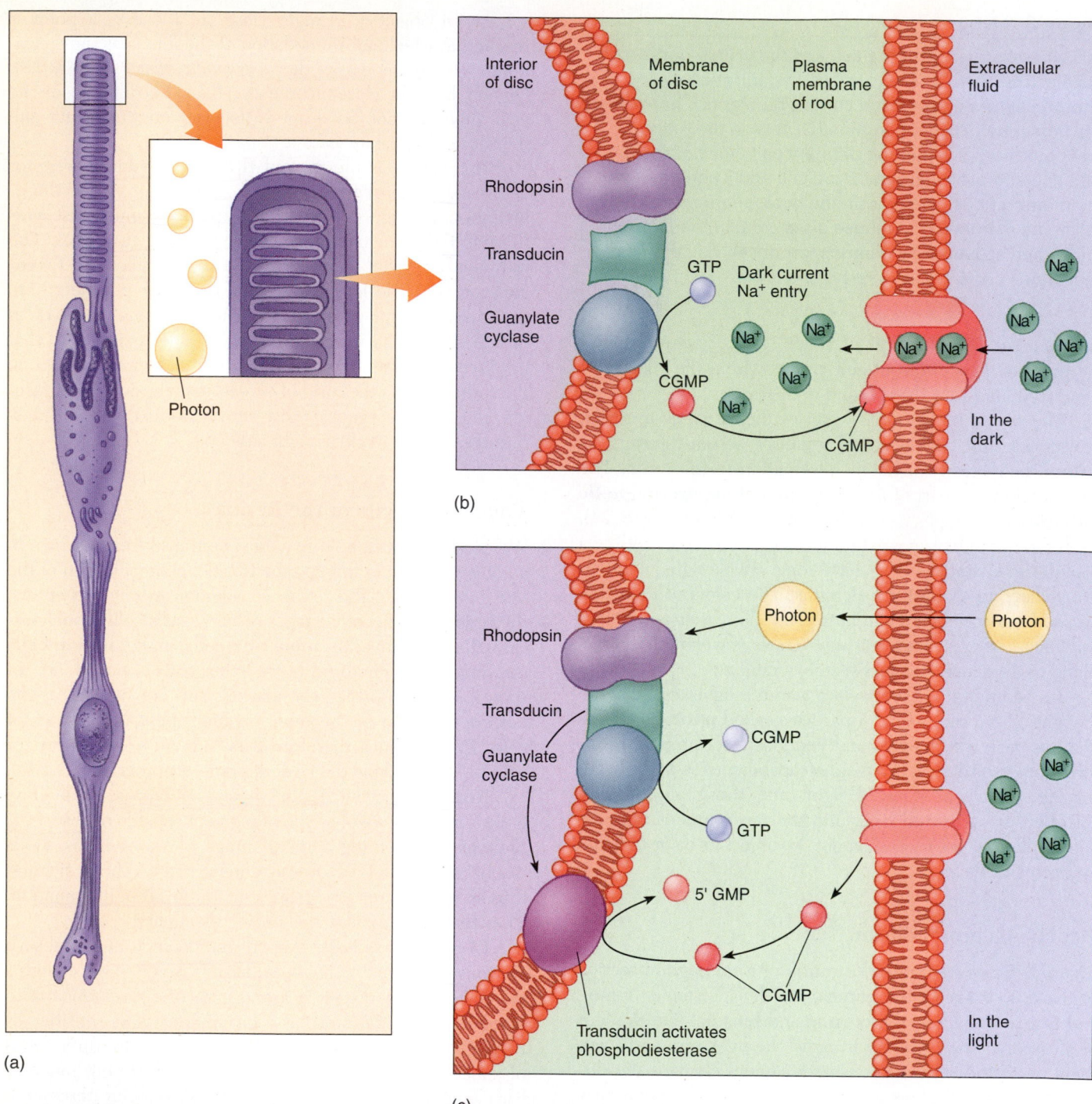

Figure 8–25

(a) Discs (*purple*) in a rod outer segment. *(b)* In the absence of light, sodium channels in the plasma membrane of rod cells are kept open by the presence of cGMP, and sodium ions enter the cell. *(c)* When a photon of light enters a rod cell, it causes a rhodopsin (pigment) molecule to activate transducin, which in turn activates phosphodiesterase. The enzyme hydrolyzes cytosolic cGMP of the rod to 5′-GMP. The reduction in the amount of cGMP causes Na^+ channels in the rod plasma membrane to close, reducing the amount of Na^+ entering the cell and producing a hyperpolarization.

protein called **transducin** and an enzyme called **phosphodiesterase** (Fig. 8–25a). The pigment in a rod cell disc is **rhodopsin.** When the rhodopsin molecule absorbs a photon, its three-dimensional form changes. In its new form, rhodopsin is now capable of activating transducin. Transducin, in turn, is now able to activate phosphodiesterase, which hydrolyzes cyclic GMP (cGMP) to 5'-GMP. In this process, the net effect is the reduction in the amount of intracellular cGMP.

What is the significance of decreasing the cGMP concentration? Normally, cGMP opens Na^+ channels in the photoreceptor plasma membrane. When cGMP is decreased, Na^+ channels begin to close, thus reducing the influx of positive ions into the cell. This results in a hyperpolarization of the membrane. Thus, the photoreceptor transduces the photon of light into a more negative membrane potential (Fig. 8–26). In order to understand how this hyperpolarization is transformed into a series of action potentials, we must first consider the organization and response of nerve cells in the retina.

Interaction of Cells in the Retina

The primary sequence of synaptic transmission in the optic pathway of the retina is photoreceptor–bipolar cell–ganglion cell. Axons of the **ganglion cells** form the optic nerve, which carries visual information from the eye to the brain. In the dark, ganglion cells generate continuous, low-frequency series of action potentials. In the light, some ganglion cell action potential frequencies increase, whereas others decrease. We can best understand the way in which visual information is processed in the retina by considering the interactions between the various types of retinal cells.

Photoreceptors always respond to light by hyperpolarizing and reducing their release of neurotransmitter. Photoreceptors form synaptic connections with **bipolar cells** (Fig. 8–27). There are two kinds of bipolar cells, each characterized by their response to transmitter release from the photoreceptor. One kind, found in an "on" center pathway, responds to reduced photoreceptor cell transmitter release by depolarizing. The other kind of bipolar cell, found in an "off" center pathway, responds to reduced photoreceptor cell transmitter release by hyperpolarizing.

Ganglion cells generate action potentials even in the absence of input from bipolar cells. The bipolar cells synapse with ganglion cells and alter their frequency of action potential generation. The on-center or **B1 bipolar cell,** when depolarized, releases more neurotransmitter, in turn increasing the frequency of ganglionic cell action potentials. The off-center or **B2 bipolar cell,** when hyperpolarized, releases less neurotransmitter, in turn reducing the frequency of ganglionic action potentials.

Ganglion cell activity also is influenced by **horizontal cells** and **amacrine cells.** The horizontal cell receives its inputs from photoreceptors located in the retina lateral to the

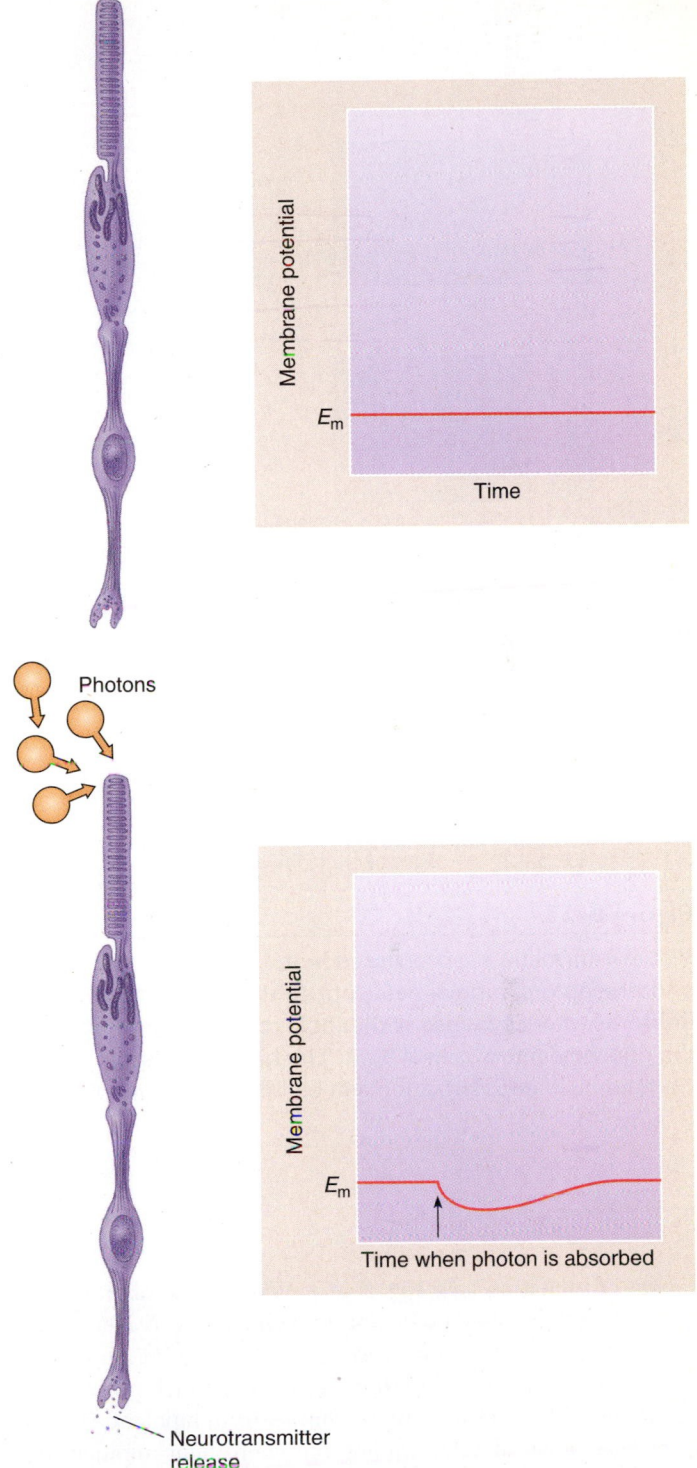

Photons

Neurotransmitter release

Figure 8–26

The hyperpolarization of a rod cell membrane caused by a photon results in neurotransmitter release and eventually action potentials that travel to the optic nerve.

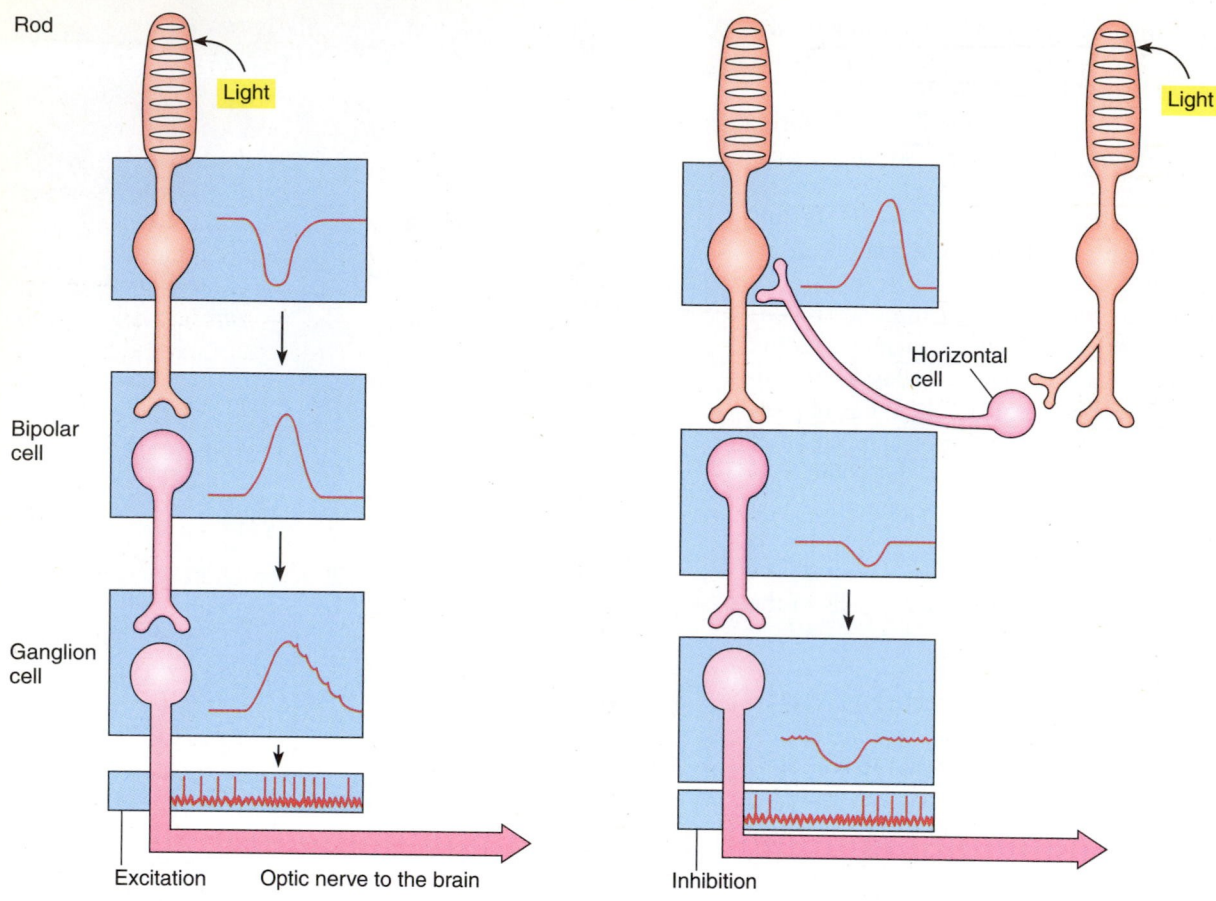

Figure 8–27

Visual information is processed in the retina by interactions between photoreceptors, bipolar cells, horizontal cells, amacrine cells, and ganglion cells. Photoreceptors synapse with bipolar cells, causing bipolar cells to depolarize when the photoreceptors detect light. The bipolar cells can depolarize ganglion cells. Ganglion cell depolarization can be inhibited by a horizontal cell.

region of excitation (see Fig. 8–27). When these photoreceptors detect light, they excite the horizontal cell. Excitation of the horizontal cell, in turn, results in lateral inhibition of off-center bipolar cells and excitation of on-center bipolar cells. Amacrine cells receive excitatory inputs from bipolar cells and excite ganglion cells. Depending on where light impinges on the retina, the frequency of ganglion cell action potentials thus can be increased or decreased.

The areas of the retina that excite or inhibit a ganglion cell make up the receptive field for that ganglion cell. Therefore—as in the case of the dorsal column nuclei cells of the somatosensory pathway—the retinal ganglion cells have receptive fields that are shaped like an annulus. For some ganglion cells, the center of the annulus is excitatory, whereas the periphery is inhibitory (Fig. 8–28). These neurons are

called **"on" center** and **"off" surround** cells. Other ganglion cells have **off centers** and **on surrounds.**

The spatial patterns of various types of activated ganglion cells in the two-dimensional plane of the retina code information about the object whose image is projected onto the surface of the retina. The image of a pencil, for example, would activate a selective array of ganglion cells (Fig. 8–29). Information about the image of the pencil thus is encoded by the spatial location of the types of activated ganglion cells within the retina. Additional information about the projected image, such as shape, texture, reflectivity, movement, color, shadows, edges, and so forth, results from processes such as lateral inhibition of ganglion cells by the horizontal cells and excitation of ganglion cells by amacrine cells.

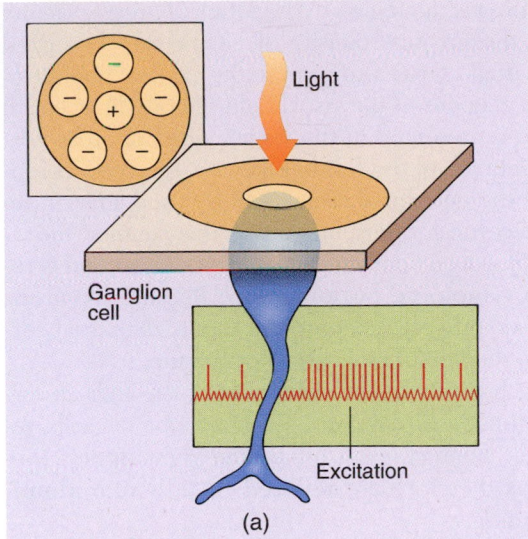

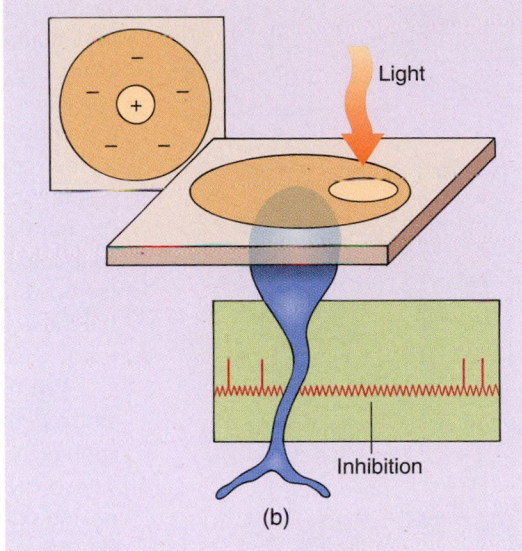

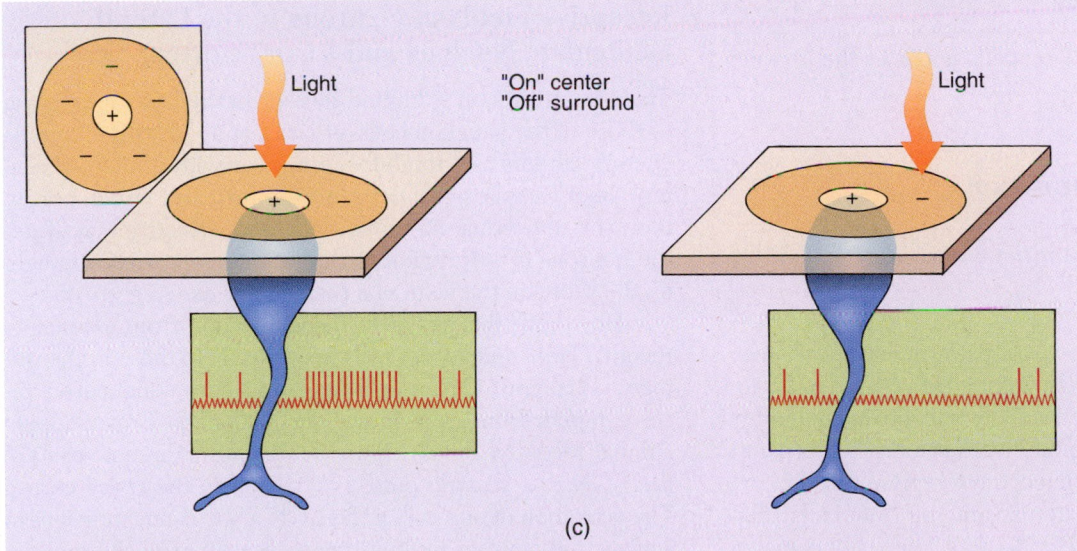

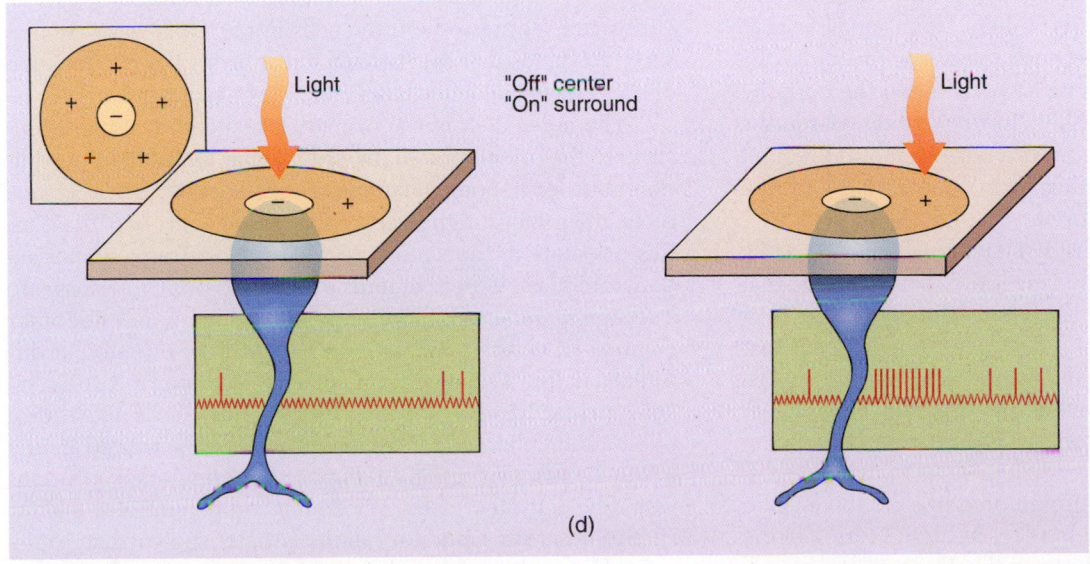

Figure 8–28

The effects of light on ganglion cell activity. Light on the center of the receptive field of a ganglion cell produces depolarization and action potentials *(a)*, whereas light in regions farther from the center may inhibit the formation of action potentials *(b)*. *(c)* A ganglion cell with a receptive field whose center is excitatory and whose surrounding ring is inhibitory is an "on" center and "off" surround cell. *(d)* Other ganglion cells have "off" centers and "on" surrounds.

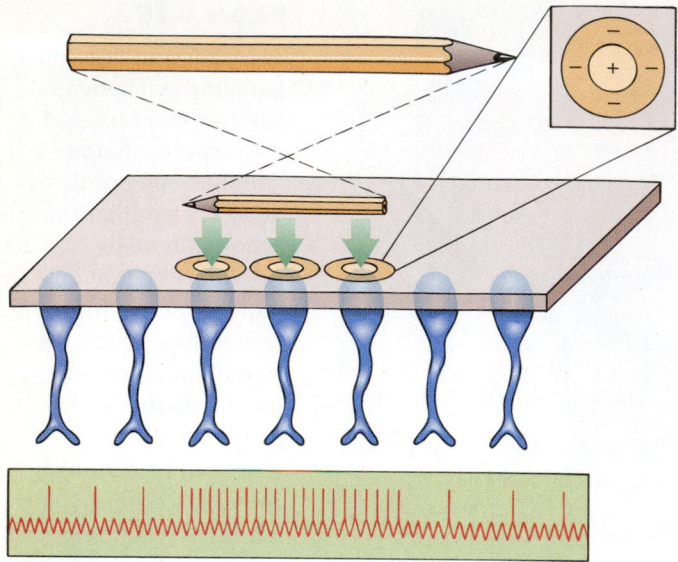

Figure 8–29

The location of activated ganglion cells encodes the image of an object.

Pathways for Visual Information

> *How does information about the visual environment reach the cortex?*

Other groups of neurons in the visual pathway have receptive fields that differ from those of retinal ganglion cells. In order to fully appreciate the visual receptive fields that are exhibited by neurons in other areas of the brain, we first need to consider the anatomical pathways that connect these regions together.

The ganglion cells of the retina are the only cells that send axons outside of the eye (Fig. 8–30*a*). These fibers make synaptic connections with neurons in the lateral geniculate nucleus (LGN) of the thalamus. Fibers from ganglion cells situated in the retina near the nose (nasal region) cross the midline and form connections with the LGN on the opposite side of the brain. In contrast, fibers from ganglion cells in the retina located farthest away from the nose (temporal region) do not cross the midline and thus form connections with the LGN on the same side of the brain.

The LGN contains six distinct layers of neurons (Fig. 8–30*b*). The first, fourth, and sixth layers contain cells that respond to light impinging on the eye on the opposite side of the brain. The second, third, and fifth layers contain cells that respond to light impinging on the eye on the same side of the brain. Nerve cells in these layers, in turn, have axons that synapse on neurons in the primary visual cortex (Brodmann area 17).

The cells of the visual cortex are organized in the form of a map of the external visual field (Fig. 8–30*c*). An area in the center of the visual field projects to the fovea of the retina.

Because the fovea is the region in the retina of greatest acuity and contains the greatest density of cones, the associated areas in the visual cortex also occupy the largest amount of cortical space. Regions of the visual field that are adjacent to the center are represented in the visual cortex in areas next to the representation of the fovea. Likewise, the cortical representation of the peripheral visual field is found adjacent to the parafoveal cortical representation. In this manner, the visual field is precisely mapped onto the visual cortex and is referred to as a **visuotopic organization.** Because the visual field also is mapped precisely onto the retina, the visual cortex also can be thought of as a **retinotopic map.**

Within each region of the visual cortex, the cells are organized into functional columns. Each column of cells responds to light impinging on either the right or left eye. These columns therefore are referred to as **ocular dominance columns.**

Receptive Fields of Neurons in the Lateral Geniculate Nucleus and Visual Cortex

The receptive fields exhibited by cells in the LGN are similar to those of the ganglion cells—that is, annular with either excitatory centers and inhibitory rings (or with inhibitory centers and excitatory rings). In the primary visual cortex, however, the visual stimulus that most effectively activates the neurons in this region of the brain is light that impinges on the retina in the form of a rectangular bar (Fig. 8–31).

How have the receptive fields changed from annular to rectangular in going from cells in the LGN to those in the primary visual cortex? The cells in the primary visual cortex receive inputs from cells in the LGN that have overlapping annular receptive fields. Figure 8–31*a* shows three neurons in the LGN that synapse onto a nerve cell in the visual cortex. The activation of one cell in the LGN alone is not sufficient to activate the neurons in the primary visual cortex; the simultaneous activation of all three is required to turn on the visual cortex cell. Because the three cells in the LGN have annular receptive fields that overlap in a linear array, the combination of these receptive fields has the form of a rectangular bar.

The nerve cells in the primary visual cortex also are sensitive to the orientation of the rectangular bar of light in addition to its form. Some columns of cells are most sensitive to a rectangular bar of light oriented vertically (Fig. 8–31*b*). With this orientation, they generate the greatest frequency of action potentials. Other columns of cells are most responsive to a rectangular bar of light oriented horizontally, and still other columns of cells are sensitive to bars of light oriented at different angles. These columns of cells are referred to as **visual orientation columns** (Fig. 8–31*c*). Cells in vertical orientation columns, for example, would be activated by the letter *I*, whereas cells in horizontal orientation columns would be activated by the horizontal component of the letter *H*. Studies with laboratory animals have shown that young animals raised in visual environments with vertical bars but

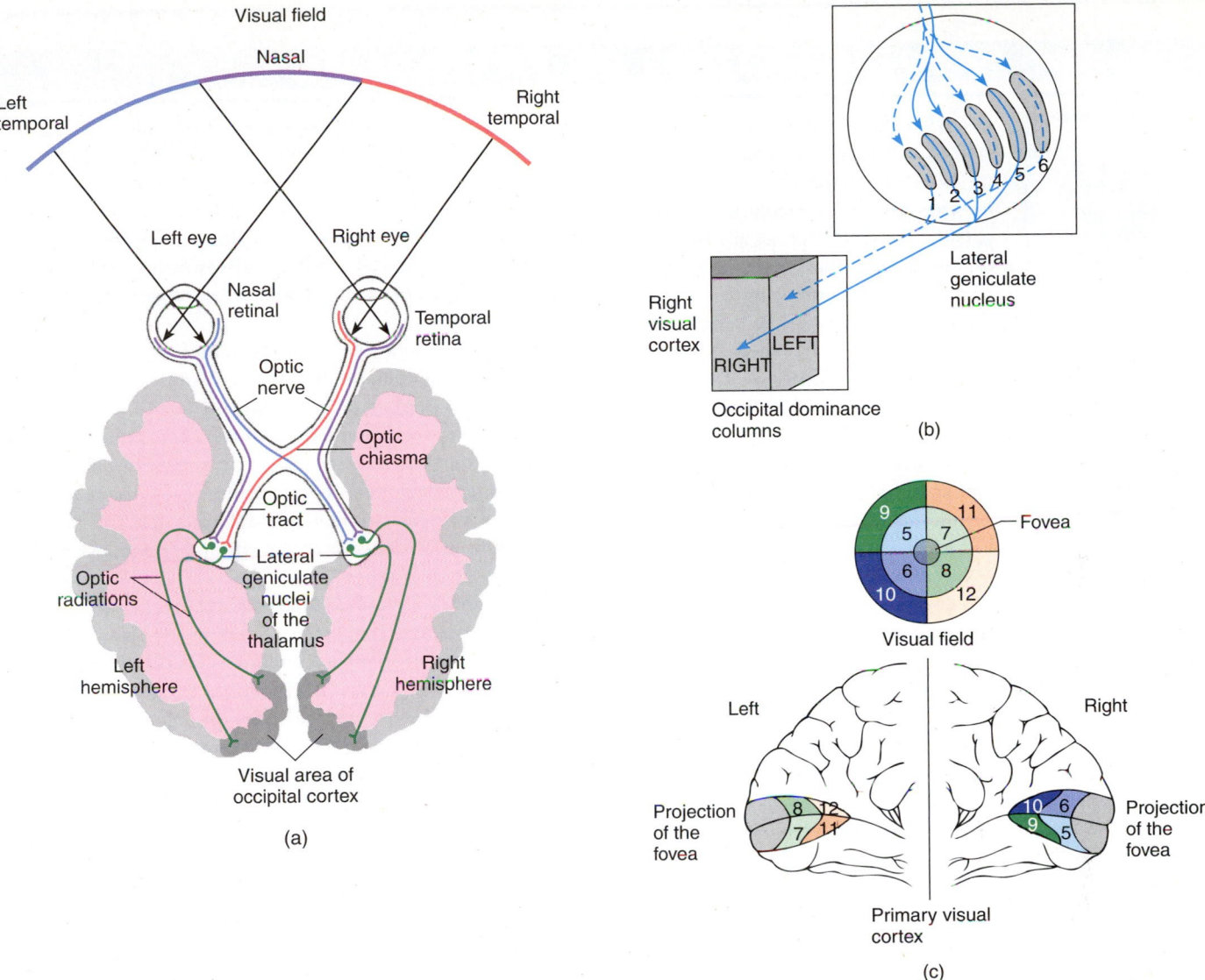

Figure 8–30

Ganglion cells extend their axons outside of the eye through the optic nerve. *(a)* Ganglion cell fibers in the nasal retina cross the midline at the optic chiasma and form connections with cells in the lateral geniculate nucleus on the opposite side of the brain. *(b)* In the lateral geniculate nucleus, visual information is sorted according to which eye it originated in and then sent to the visual cortex, where cells are arranged in ocular dominance columns, as well as in the form of a map of the external visual field *(c)*.

without horizontal bars do not perceive horizontal bars when tested later in life. These results indicate that the horizontal orientation columns do not exist in the visual cortex of these animals and demonstrate the importance of the visual environment in the development of visual perception.

Stimuli more complex than rectangular bars are required to activate neurons in the secondary visual cortices (areas 18 and 19). The most effective forms of visual stimuli capable of activating neurons in visual areas 18 and 19 have complex

shapes that are combinations of simple rectangular bars (Fig. 8–31*d*). Cells in these areas of the visual cortex combine the features of the receptive fields from the cells in the primary visual cortex. In this way, information about the form of the object being viewed is synthesized. Thus, like the somatosensory system, the photoreceptors of the visual system detect the basic features of a visual object. As this information is transmitted to the higher visual areas of the brain, the image is pieced together by the brain to reconstruct the basic features of the image.

Ocular Dominance and Critical Periods of Development

In the lateral geniculate nucleus (LGN) and visual cortex are clusters of nerve cells that respond to the stimuli of light impinging on either the left or right eye. The cells in the LGN are arranged in several layers, with each layer specific for inputs from one eye. The number of layers in the LGN is species dependent. Rodents, for example, have two layers, whereas monkeys have six. The segregation of left and right eye inputs emerges gradually during development from an initial pattern in which inputs from both eyes are intermingled. In cats in which this phenomenon has been studied in detail, ganglion cell fibers from the contralateral eye enter the LGN at embryonic day 32 (approximately half of a cat's gestational period). The ipsilateral inputs arrive a few days later. During the next 4.5 weeks, the inputs from the two eyes gradually segregate; by birth, the eye-specific layers are present as a result of a retraction of fibers from the nondominant input.

Laboratory studies have shown that competition between the ipsilateral and contralateral fibers is involved in the segregated pattern of eye inputs. The loss of one eye, for example, results in the expansion of fibers from the preserved eye into those layers typically occupied by inputs from the lost eye. Other studies have demonstrated that the segregated patterns are a result of activity-dependent processes. This has been shown by blocking, with the drug tetrodotoxin, the initiation and propagation of action potentials by ganglion axons. The administration of this drug in prenatal cats results in a loss of eye-specific layers within the LGN.

In the visual cortex the pattern of left-eye and right-eye inputs takes the forms of ocular dominance columns. These columns exhibit alternative left-eye and right-eye inputs. The width of each column is approximately 0.5 mm in monkeys.

The closure of one eye at various times after birth alters the pattern of ocular dominance columns. Closure of one eye at birth in monkeys greatly reduces the width of the associated ocular column in the visual cortex; closure at 4 weeks after birth or later causes no alterations in the ocular dominance columns. These results reveal the modifiable nature of ocular dominance columns and the fact that their formation depends on a postnatal critical period of development.

Visual stimuli of various types also have been shown to alter the functional properties of cells in the visual cortex. Kittens raised in an environment of vertical black and white stripes develop cells in the visual cortex that respond preferentially to vertical bars of light. Animals raised in visual environments that minimize movement perception have fewer movement selective cells.

Figure A   Eye-specific layers of the lateral geniculate nucleus. (*a*) Normal pattern of eye-specific layers in the adult cat. (*b*) Overlap of eye-specific layers in the prenatal cat. (*c*) Adult pattern of eye-specific layer in the cat after prenatal removal of one eye. (*d*) Adult pattern of eye-specific layers after prenatal administration of tetrodotoxin to both eyes.

Color Vision: A Three-Way Process

Our ability to discriminate between different colors is due to the three types of photoreceptor cones in the retina that form the basis of the **trichromatic theory** of color vision. Each cone has a different type of visual pigment capable of absorbing different wavelengths of light. The blue cone, for instance, absorbs photons with a wavelength of 443 nm. It also absorbs photons with wavelengths slightly greater than 443 nm and slightly less than 443 nm, but it does so less efficiently. Thus, the blue cones are most sensitive to blue light (Fig. 8–32). The other two types of cones are the **green cones** and **red cones.** These cones are most sensitive to photons with wavelengths of 535 and 570 nm, respectively; they absorb photons with other wavelengths less efficiently. The three photoreceptor cones, therefore, have overlapping probabilities of absorbing photons of various wavelengths.

If the retina contains only three cones, how can we perceive more than three colors? Each cone absorbs photons of a particular wavelength to a different degree. Consequently, each photoreceptor will be hyperpolarized to a different level in response to photons of a particular wavelength. The differential hyperpolarization of only one type of photoreceptor would give the ability to perceive colors, but in a somewhat limited way. Assume, for example, that the retina contained only red cones. If we were to view a picture painted with colors of 540 and 605 nm, we would not be able to tell that two different colors were used. This is because both colors emit photons that are absorbed by the red cone with equal probability (Fig. 8–33*a*). The addition of a green cone, however, results in a greater absorption of the 540-nm photons and

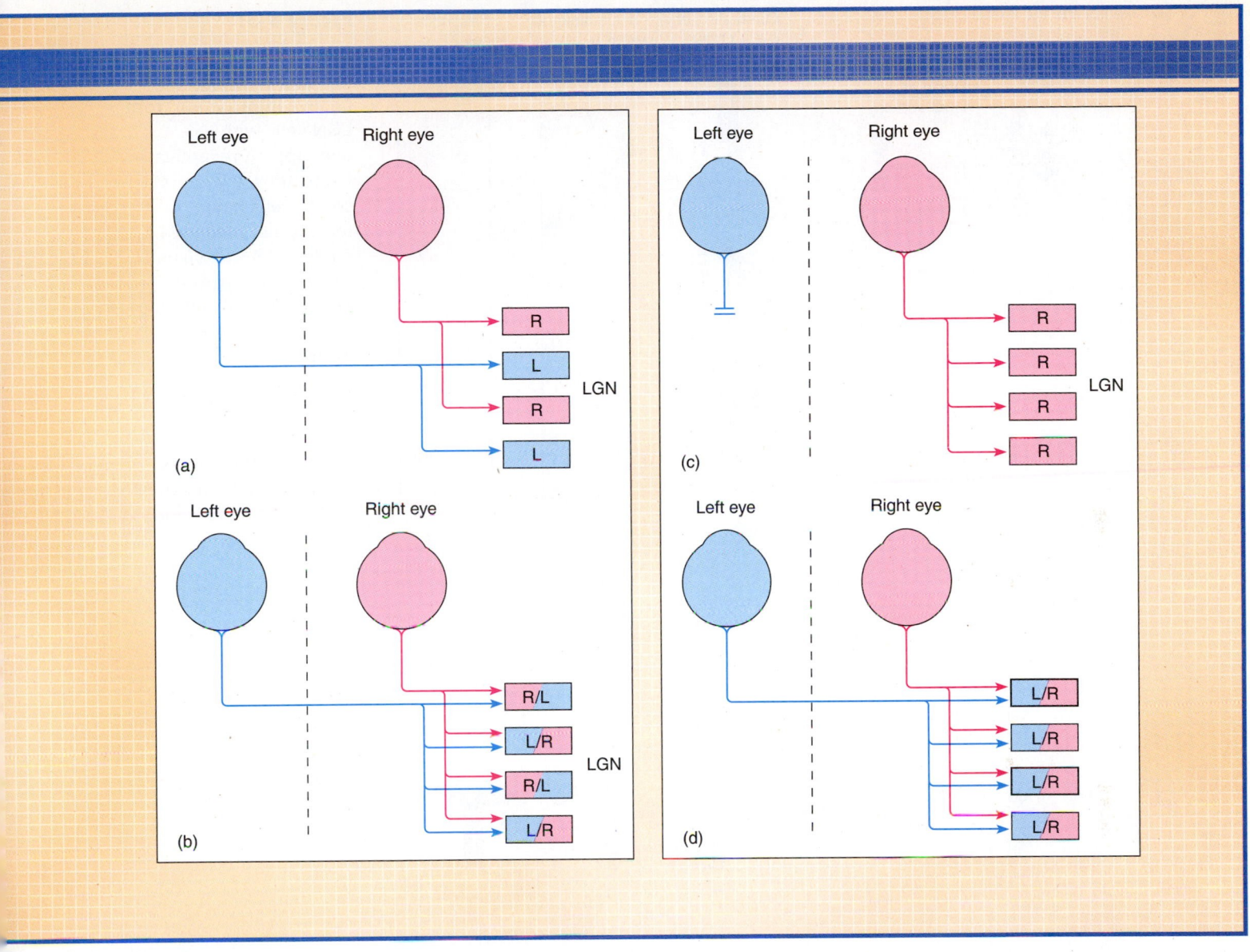

thus allows us to distinguish between the two colors. With three photoreceptor cones, our ability to distinguish between different colors becomes even better.

Color vision that relies on one type of cone is called **monochromatic color vision;** color vision that relies on two types of cones is called **dichromatic color vision.** Most people exhibit **trichromatic color vision.** Some people, however, are **color blind** and are not able to tell the difference between certain colors. Color blindness is a sex-linked trait that affects males predominantly. These males are usually **dichromats,** possessing only two types of cones, and represent approximately 2% of the population. Half of these dichromats do not have green cones, and the other half do not have red cones. These individuals cannot see the colored numbers embedded within the design of Figure 8–34.

The trichromatic theory of color vision alone is not sufficient to explain completely how we perceive color. After the selective absorption of photons by cones, cellular responses that are also responsible for the perception of color come into play. Ganglion cells, for example, differ in the way they process information from cones. Small ganglion cells are able to distinguish information about different colors. They in turn transmit this color information to small cells in the LGN (Fig. 8–35). These so-called **parvocellular neurons** in the LGN transmit information to neurons in the visual cortex that respond selectively to color information. Large ganglion cells, on the other hand, process information from cones in such a way that they derive information regarding shape and movement. They in turn transmit information to large cells in the LGN. These so-called **magnocellular neurons** transmit information to the primary and secondary visual cortices,

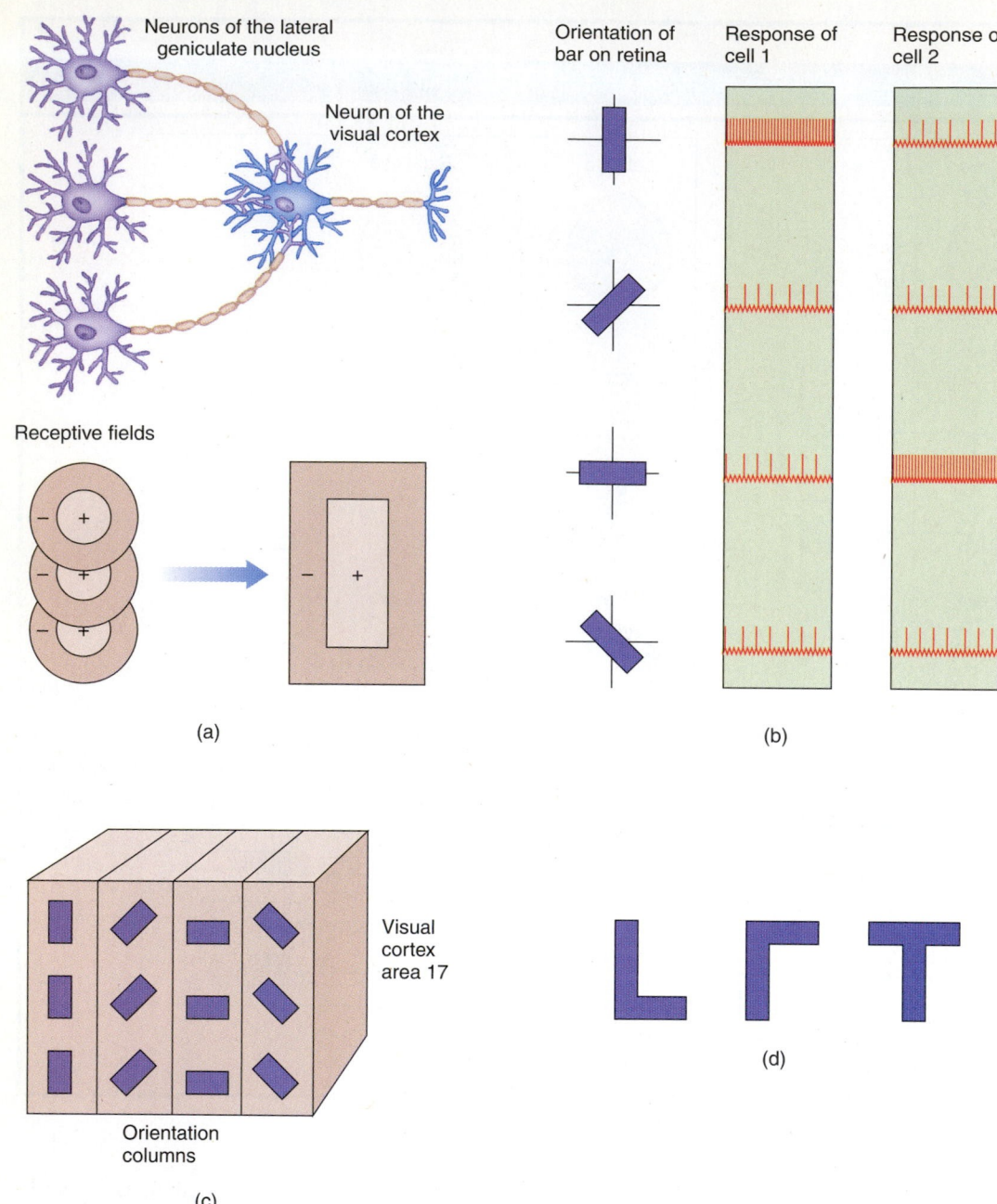

Neurons of the lateral geniculate nucleus

Neuron of the visual cortex

Receptive fields

Orientation of bar on retina

Response of cell 1

Response of cell 2

(a)

(b)

Visual cortex area 17

Orientation columns

(c)

(d)

Figure 8–31

(a) The cells in the primary visual cortex receive information from LGN cells with overlapping receptive fields; thus, the visual cortex cells have rectangular receptive fields. **(b)** Some visual cortex cells respond only to stimuli with the correct orientation. **(c)** Orientation-sensitive cells are arranged in columns in the visual cortex. **(d)** Some examples of complex stimuli needed to activate neurons in the secondary visual cortices.

where it is analyzed to extract features about an object's shape and movement.

Neurons that selectively respond to color are divided into three different groups: (1) **broad-band cells,** (2) **single-opponent cells,** and (3) **double-opponent cells** (Fig. 8–36). Broad-band cells are excited by a single color presented to the center of the receptive field and inhibited when that color is in the periphery. Single-opponent cells also are excited when one color of light is in the center of the receptive field; however, they are inhibited when another color is in the periphery. For example, one type of cell excited by red in the center of the receptive field is inhibited by green in the periphery. The opposing colors are usually paired in combinations of red and green, blue and yellow, or black and white, and oppose one another across the compartmental boundaries of the receptive field.

Double-opponent cells differ from single-opponent cells in that the color pairs oppose each other within the boundaries of the receptive field. In this type of cell, for example, red in the center of the field might excite the cell, whereas green within the center would inhibit the cell. In addition, green in the peripheral boundary of the field might excite the cell, whereas red in this region would be inhibitory.

The double-opponent cell also can be thought of as transmitting information about the contrast between two colors. In the previous example, the cell would become maximally

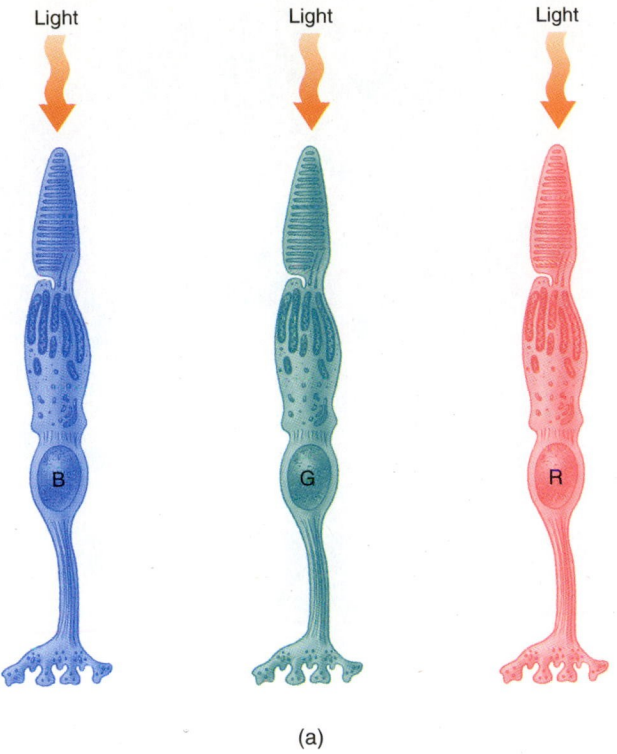

(a)

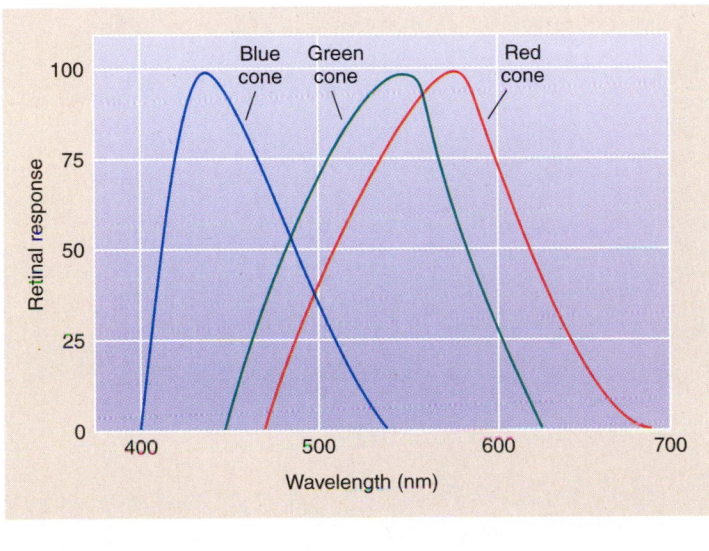

(b)

Figure 8–32

(a) The three types of cones—red, blue, and green—allow us to see and discriminate among colors. Each type of cone absorbs photons over a particular range of the visible spectrum, though the three ranges overlap *(b)*.

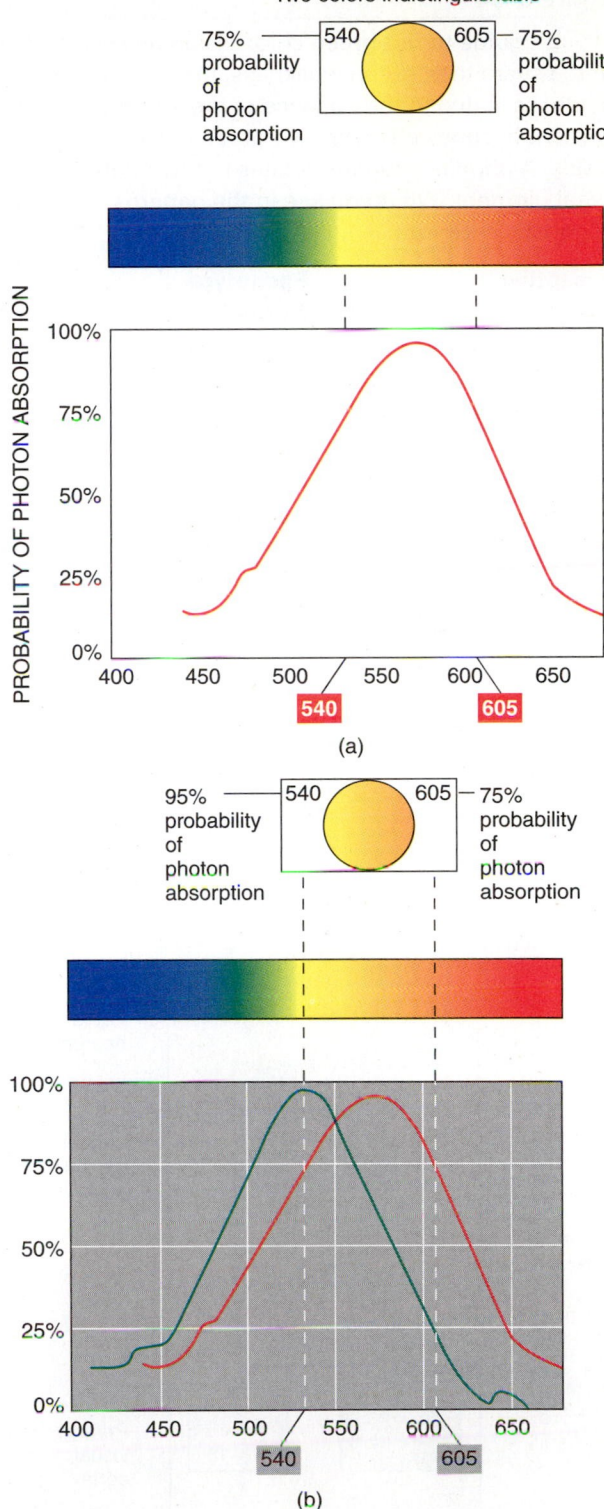

Figure 8–33

(a) A person with only red cones may not be able to distinguish two colors for which the rod has the same probability of absorbing photons. *(b)* The presence of a second type of cone allows the two colors to be distinguished by increasing the probability that the cone will absorb photons at one of the wavelengths.

Figure 8–34

A rough guide to red-green color-vision defects (Ishihara, 1968). More severe than green weakness, the phenotype green blindness is due to the absence of the green color-vision pigment in cone cells. The two green defects are controlled by alleles. A parallel situation relates the two defects of the red visual pigment. What you see in the patterns of dots is the diagnostic criterion:

Perception	Phenotype
42	Normal
4 (and less clearly a 2)	green-weak
4 only	green-blind
2 (and less clearly a 4)	red-weak
2 only	red-blind

This color plate is not intended for use in the diagnosis of color blindness. A diagnosis of color blindness must be based on the original plates, which appear in Ishihara, S. *Tests of Colour-Blindness*, 38 plate edition. Tokyo, Kanehara & Co., LTD, 1968.

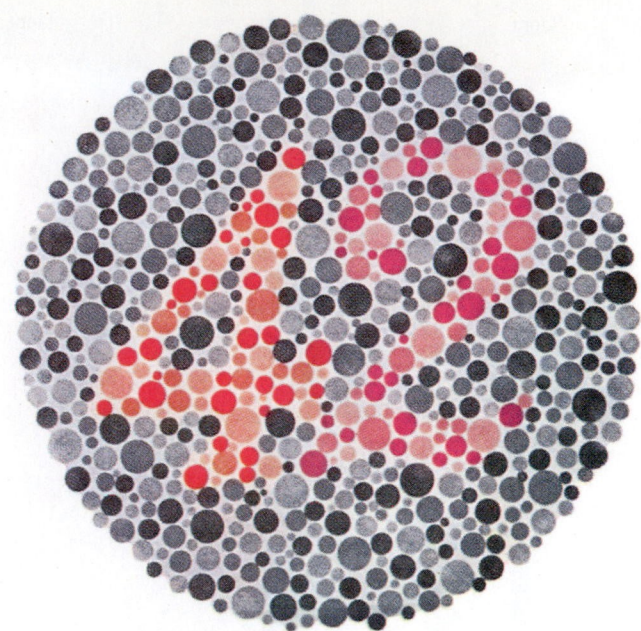

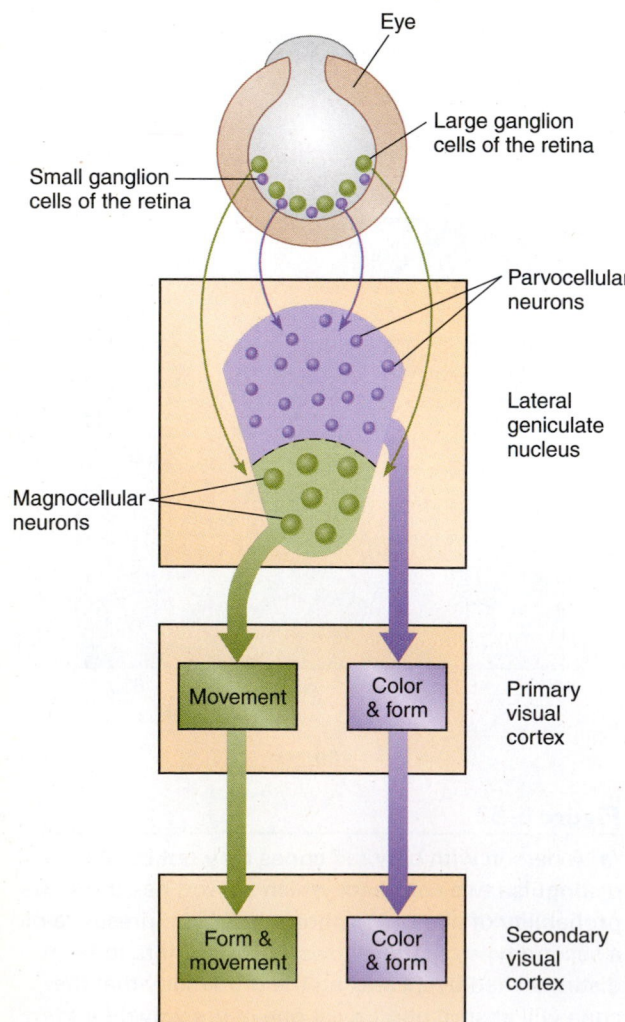

Figure 8–35

Small ganglion cells distinguish among different colors and transmit the information to small cells, or parvocellular neurons, in the lateral geniculate nucleus. Large ganglion cells transmit information about an object's movement to large, or magnocellular, neurons in the lateral geniculate.

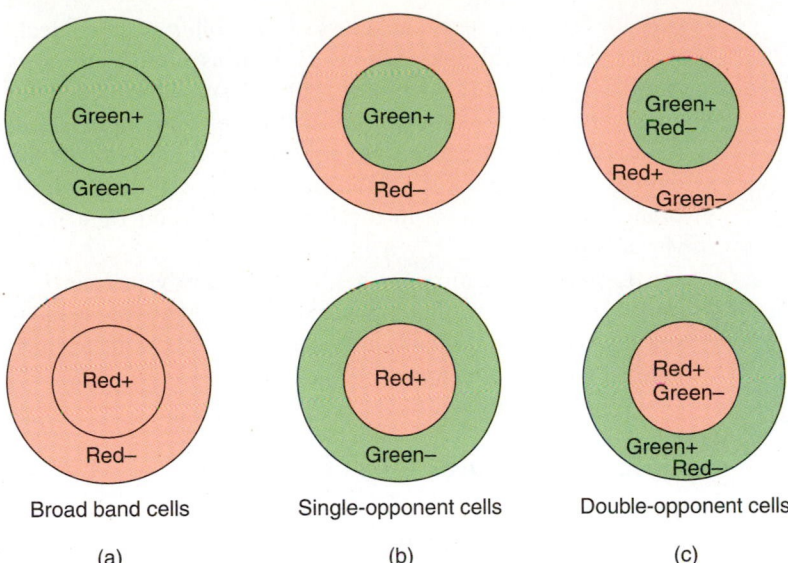

Figure 8–36

Neurons that respond selectively to color are classified as **(a)** broad-band, **(b)** single-opponent, or **(c)** double-opponent cells, dependent on the effects of different colors.

excited when red is present in the center of the field and green is in the periphery. The activation of this type of cell would therefore communicate to the brain that two contrasting colors are adjacent to one another in the receptive field.

Some cells in the visual cortex display the characteristics of double-opponent cells with annular receptive fields. Other cells have rectangular receptive fields. The response properties of these cells are a summation of the annular properties displayed by converging neurons (Fig. 8–37). The annular patterns of these converging neurons overlap and thus give rise to a bar-shaped receptive field. These cells, however, do not appear to be associated with orientation columns and re-

spond to rectangular forms of color in any orientation. The color-sensitive cells are clustered in functional columns that are parallel to orientation columns (see Fig. 8–35). This arrangement suggests that the color and form of an object are analyzed independently. In support of this theory are observations of stroke patients who are unable to determine the color of an object even though they recognize its form.

Disorders of the Visual System

With advancing age, the lens of the eye gradually stiffens, losing its capacity for accommodation. As a result, the near point of accommodation recedes (near point distance becomes larger), and it becomes increasingly difficult to focus properly on near objects, as when reading. The loss of near vision due to the gradual decline in the power of accommodation is known as **presbyopia** (old age vision).

In addition to presbyopia, other common visual defects of the eye that involve focusing include **myopia** (nearsightedness) and **hypermetropia** (farsightedness). Myopia usually results either from an imperfection in which the eyeball is longer than normal or from imperfections in the lens or cornea whereby the image is focused in front of the retina rather than on it. Myopia can be compensated for by placing in front of the eye a properly ground, biconcave lens that diverges the light rays just enough for the cornea and lens to focus them accurately on the retina (Fig. 8–38).

Hypermetropia, like myopia, is also the result of an improperly shaped eyeball, but in this case the eyeball is shorter than normal. This results in the images being focused at a point behind the retina instead of on it. If accommodation of the lens is normal, a person can compensate partially for hypermetropia by continuous partial contraction of the ciliary muscle; however, this generally leads to further complications (eyestrain, headaches). Hypermetropia can be corrected by placing in front of the eye an appropriate biconvex

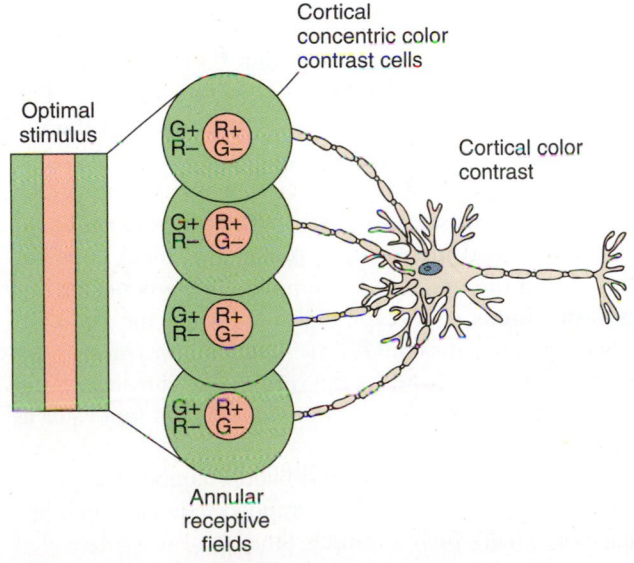

Figure 8–37

The annual receptive fields of some cells in the visual cortex overlap to give bar-shaped receptive fields for color.

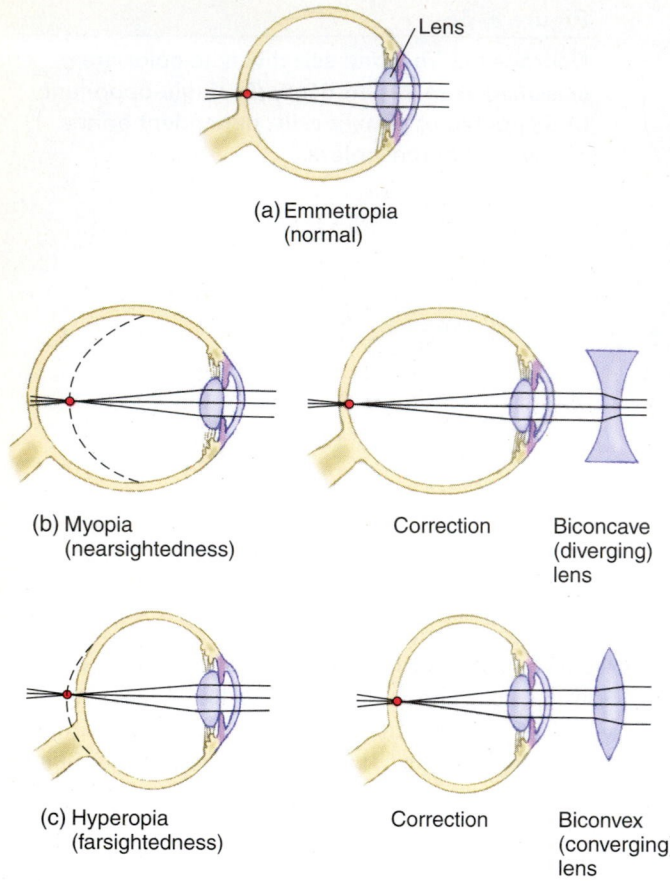

(a) Emmetropia
(normal)

(b) Myopia
(nearsightedness)

Correction

Biconcave
(diverging)
lens

(c) Hyperopia
(farsightedness)

Correction

Biconvex
(converging)
lens

Figure 8–38

(a) In the normal eye, an image is focused on the retina.
(b) In the nearsighted (myopic) eye, the image is focused in front of the retina and the light rays continuing past the focal point project an out-of-focus image on the retina.
(c) In the farsighted (hyperopic) eye, the focal point of the image is behind the retina. Light rays impinge on the retina before they are in focus, resulting in a blurred image. Biconcave lenses *(b)* diverge light rays and biconvex lenses *(c)* converge light rays. The appropriate lens can shift the focal point to fall on the retina.

lens that converges the light rays enough to permit the image to be sharply focused on the retina.

Visual disturbances also can occur when accidents or disease damages parts of the visual pathway. These disturbances are characterized by a loss of visual perception in some part of the **visual field** (Fig. 8–39). Each eye has a visual field that includes a nasal portion and a temporal portion. The nasal visual fields of each eye overlap. Processing of visual information from the area of overlap provides for binocular vision. Images in the nasal portion of the visual field for each eye are projected to the temporal portion of the retina in the respective eye. Images in the temporal portion of the visual field project to the nasal portion of the retina of the respective eye.

Trauma, such as a blow to the eye, sometimes can damage the optic nerve. Damage to the optic nerve of the right

eye, for example, results in a loss of the ability to see objects in the right visual field. Disorders such as pituitary tumors can damage nerve fibers in the optic chiasm, which results in an inability to perceive objects in the left and right visual fields. The optic tract on one side of the brain also can be damaged. Damage of the optic tract on the right side of the brain results in defects in the left temporal and right nasal visual field.

Lesions also can occur in the secondary visual cortices (areas 18 and 19), resulting in visual agnosia. Patients with visual agnosia have difficulty recognizing the faces of friends, familiar objects, or colors (Table 8–4).

THE AUDITORY SYSTEM

> *How does the auditory system receive and process information from the environment?*

The auditory system detects complex sounds, such as spoken words, and breaks them into their basic sound frequencies. The fundamental frequency components are transduced into action potentials by receptor cells in the ear and transmitted to the brain. Auditory areas of the brain subsequently recombine the basic sound features of the words, which then are interpreted in areas of the brain responsible for language (see Chapter 11).

Sound Transduction

> *How is sound converted to action potentials?*

Sound is the adequate stimulus of the auditory system and is transmitted through the air by the compression and expansion of air molecules in the form of air pressure waves. The sense organ for the transduction of sound is the ear. The ear consists of the *outer ear, middle ear,* and *inner ear* (Fig. 8–40*a*). The outer ear channels the sound waves to the ear drum, located at the interface between the outer and middle ear. The sound waves cause the ear drum to vibrate much like the vibration of the speakers of a stereo system. The vibration of the ear drum, in turn, causes the small bones or **ossicles** within the middle ear to move. These bones, called the **malleus, incus,** and **stapes** (also known as the hammer, anvil, and stirrup), are connected in series and transfer the vibrations of the ear drum to the inner ear. The stapes, the last in the series of ossicles, is connected to the inner ear at the oval window.

The inner ear contains a fluid-filled chamber called the **cochlea,** a cone-shaped structure wrapped in a coil much like the shape of a snail's shell. Internally, the cochlea is subdivided into three fluid-filled chambers by membranes: the **scala vestibuli,** the **scala media** (cochlear duct), and the **scala tympani.** The **vestibular membrane** separates the scala vestibuli and scala media. The **basilar membrane** separates the scala

VISUAL FIELD

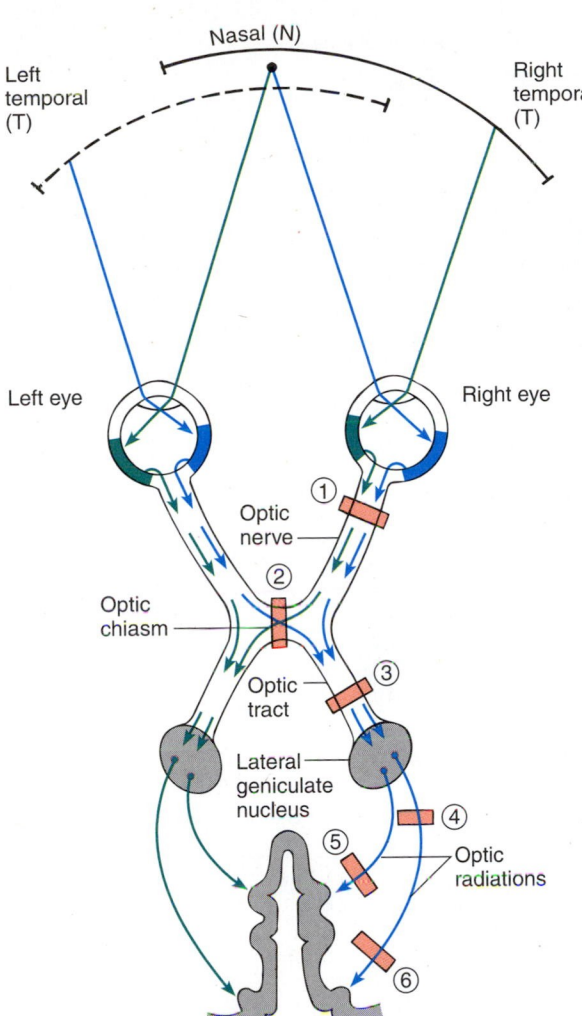

Visual area of occipital cortex

Figure 8–39

Damage to some parts of the visual pathway may result in defects in the visual fields of the eyes. The visual field defects shown to the far right of 5 and 6 depict the result of a lesion that affects both of these locations.

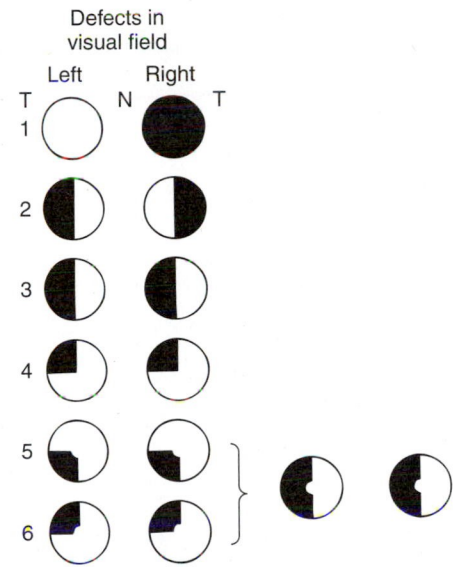

Defects in visual field

TABLE 8–4	
Visual Agnosia	
Type of Agnosia	**Visual Defects**
Agnosia for color	
Color agnosia	Associating colors with objects
Color anomia	Naming colors
Agnosia for form and pattern	
Object agnosia	Naming objects
Agnosia for drawings	Recognizing drawn objects
Prosopagnosia	Recognizing faces
Agnosia for depth and movement	
Visual spatial agnosia	Stereoscopic vision
Movement agnosia	Discerning movement of objects

media and the scala tympani. The **oval window,** a membrane-covered opening is located at the beginning of the scala vestibuli and the **round window** is located at the end of the scala tympani. The scala vestibuli and the scala tympani, confluent at the helicotrema, are filled with a fluid called **perilymph.** The scala media ends at the helicotrema and is filled with **endolymph.**

The **organ of Corti,** containing auditory receptor cells called *hair cells,* supporting cells, and the tectorial membrane, is attached to the basilar membrane and extends as a continuous sheet of cells from the base to the apex of the cochlea. The organ of Corti contains two morphologically and physiologically distinct populations of hair cells. The **inner hair cell** functions as the primary auditory receptor cell. The **outer hair cell** serves as an effector to modulate and fine-tune auditory stimulation of the inner hair cell. Both inner and outer hair cells are innervated by afferent and efferent fibers connected to the brain stem by way of the eighth cranial nerve.

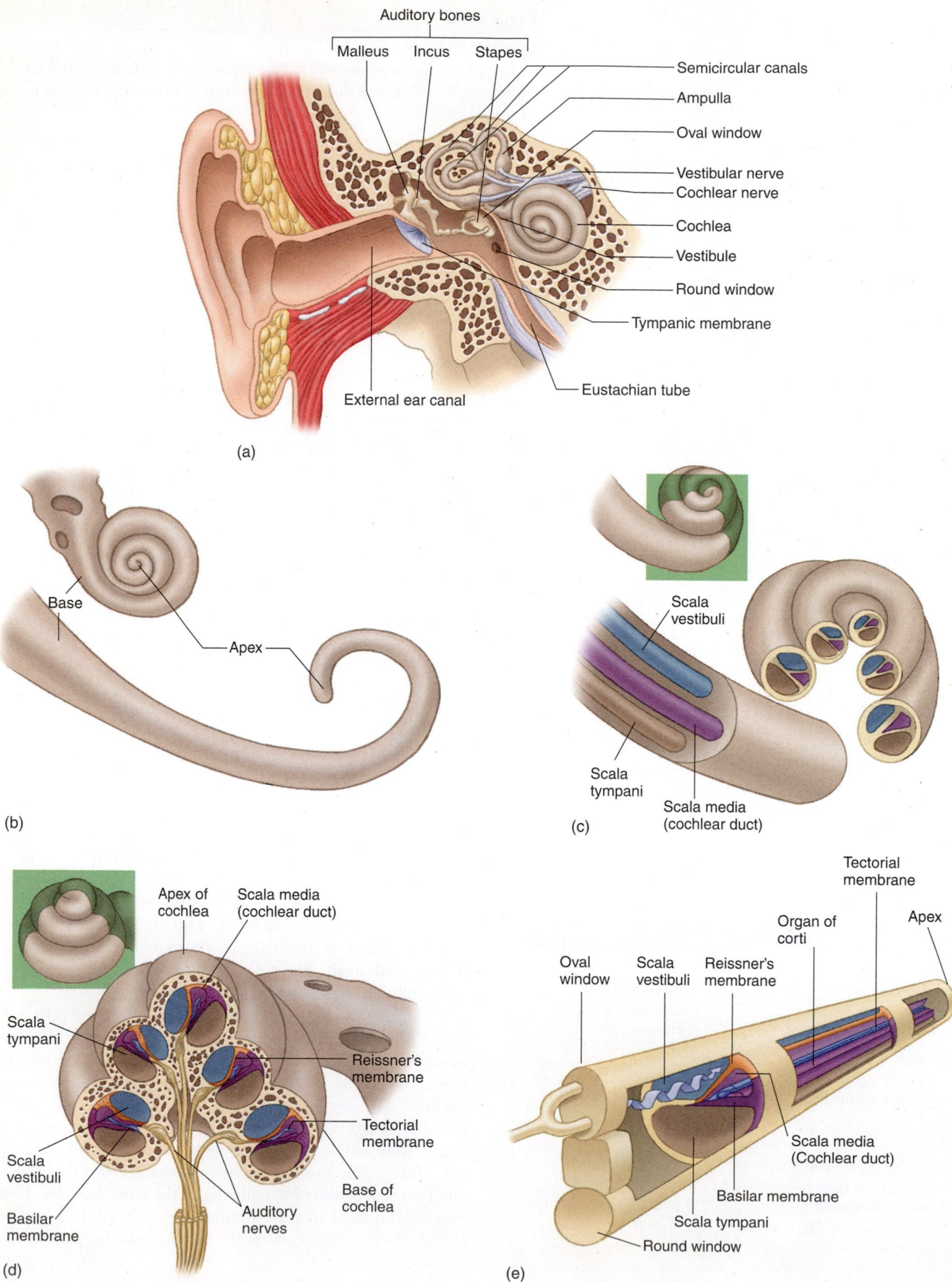

(a)

Auditory bones
Malleus   Incus   Stapes

Semicircular canals
Ampulla
Oval window
Vestibular nerve
Cochlear nerve
Cochlea
Vestibule
Round window
Tympanic membrane
Eustachian tube
External ear canal

(b)

Base
Apex

(c)

Scala vestibuli
Scala tympani
Scala media (cochlear duct)

(d)

Apex of cochlea
Scala media (cochlear duct)
Scala tympani
Reissner's membrane
Tectorial membrane
Scala vestibuli
Base of cochlea
Basilar membrane
Auditory nerves

(e)

Tectorial membrane
Organ of corti
Apex
Oval window
Scala vestibuli
Reissner's membrane
Scala media (Cochlear duct)
Basilar membrane
Scala tympani
Round window

(a) The general anatomy of the ear. *(b, c)* The cochlea is a coiled structure containing three fluid-filled chambers formed by a system of membranes. *(d)* A cross-section of the cochlea showing the fluid chambers and membranes. *(e)* The cochlea as if "unwound," showing the path of a sound pressure wave from the oval window through the scala vestibuli, around the apex of the cochlea, into the scala tympani, and back to the round window at the base of the cochlea. (Some pressure waves do not go all the way to the apex.)

The footplate of the stapes, attached to the membrane of the oval window, transfers vibrations of the tympanic membrane to the perilymph in the scala vestibuli and in turn to the perilymph of the scala tympani, causing the flexible basilar membrane to vibrate. The vibrations of the basilar membrane are in the form of traveling waves (Fig. 8–41) that deflect or deform the shape of the membrane. As the traveling wave progresses down the length of the membrane, the amplitude of the deformation becomes greater and greater until it reaches a peak. As the traveling wave progresses further, its amplitude diminishes. The location of the peak amplitude of the traveling wave along the basilar membrane depends on the frequency of the sound. The membrane's structure is such that it is narrower and stiffer at the base in comparison to the apex. Consequently, high-frequency sounds produce traveling waves that reach their peak amplitudes near the base of the cochlea. Low-frequency sounds cause the traveling waves to reach their peak near the apex of the cochlea. Intermediate frequencies of sound cause peak amplitudes to occur at locations along the membrane between the base and apex of the cochlea. Thus, the basilar membrane mechanically responds to different frequencies of sound by changing its degree of deflection at different points along its length.

The deformations of the basilar membrane are transduced into action potentials by **inner hair cells.** The hair cells sit on the basilar membrane and have fiberlike elements called **stereocilia** that protrude from the cell and move collectively as a unit (Fig. 8–42*a*). Many of these cilia are embedded in an overlapping membranous structure called the **tectorial membrane.** When the basilar membrane vibrates and reaches a peak amplitude, the stereocilia are bent in relation to the hair cell (Fig. 8–42*b*). This bending of the

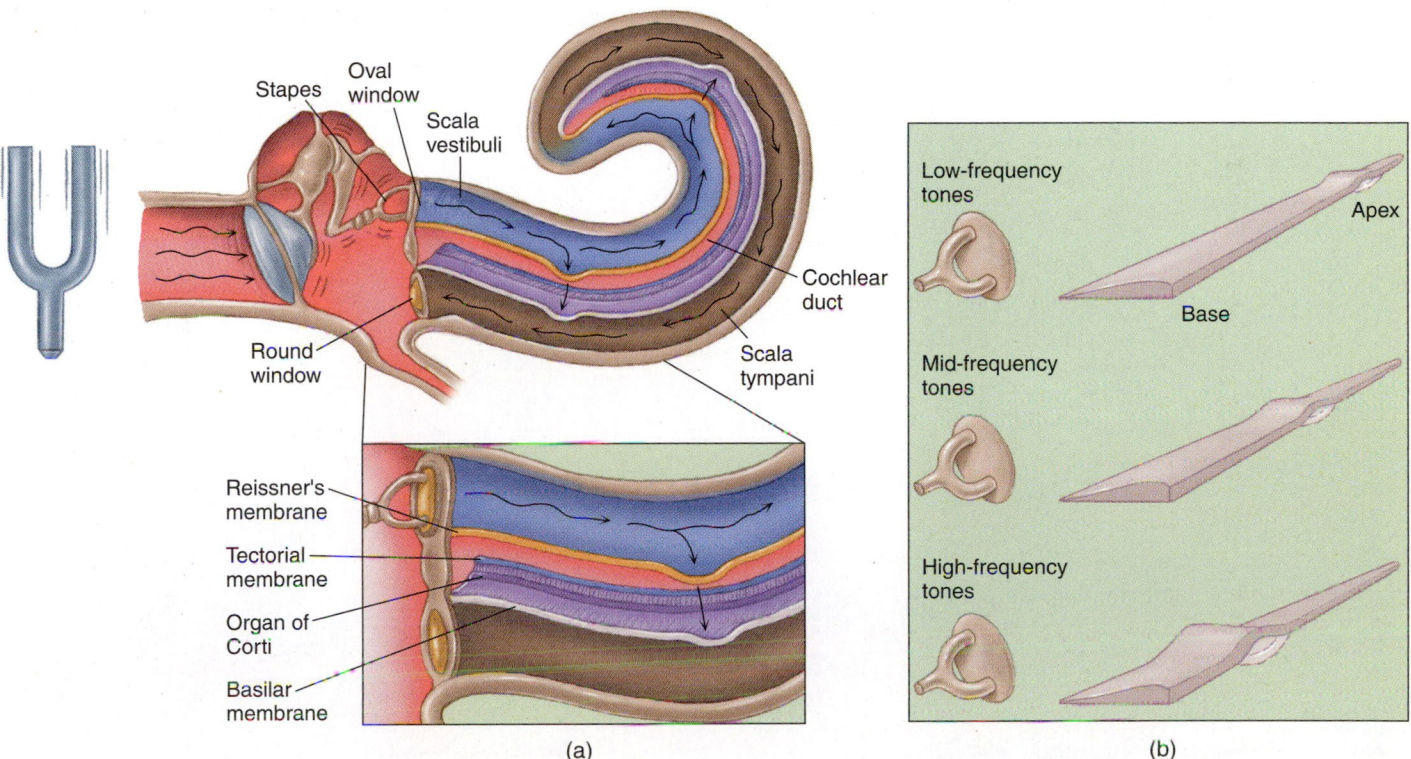

(a)

(b)

Figure 8–41

(a) The cochlear membranes shown in a different cross-sectional view. Sound causes the fluid in the scala vestibuli to exert pressure on Reissner's membrane, which in turn causes the round window to bulge. This movement sets up vibrations in the form of waves in the basilar membrane *(b)*. The waves peak at different locations on the basilar membrane, depending on the frequency of the sound.

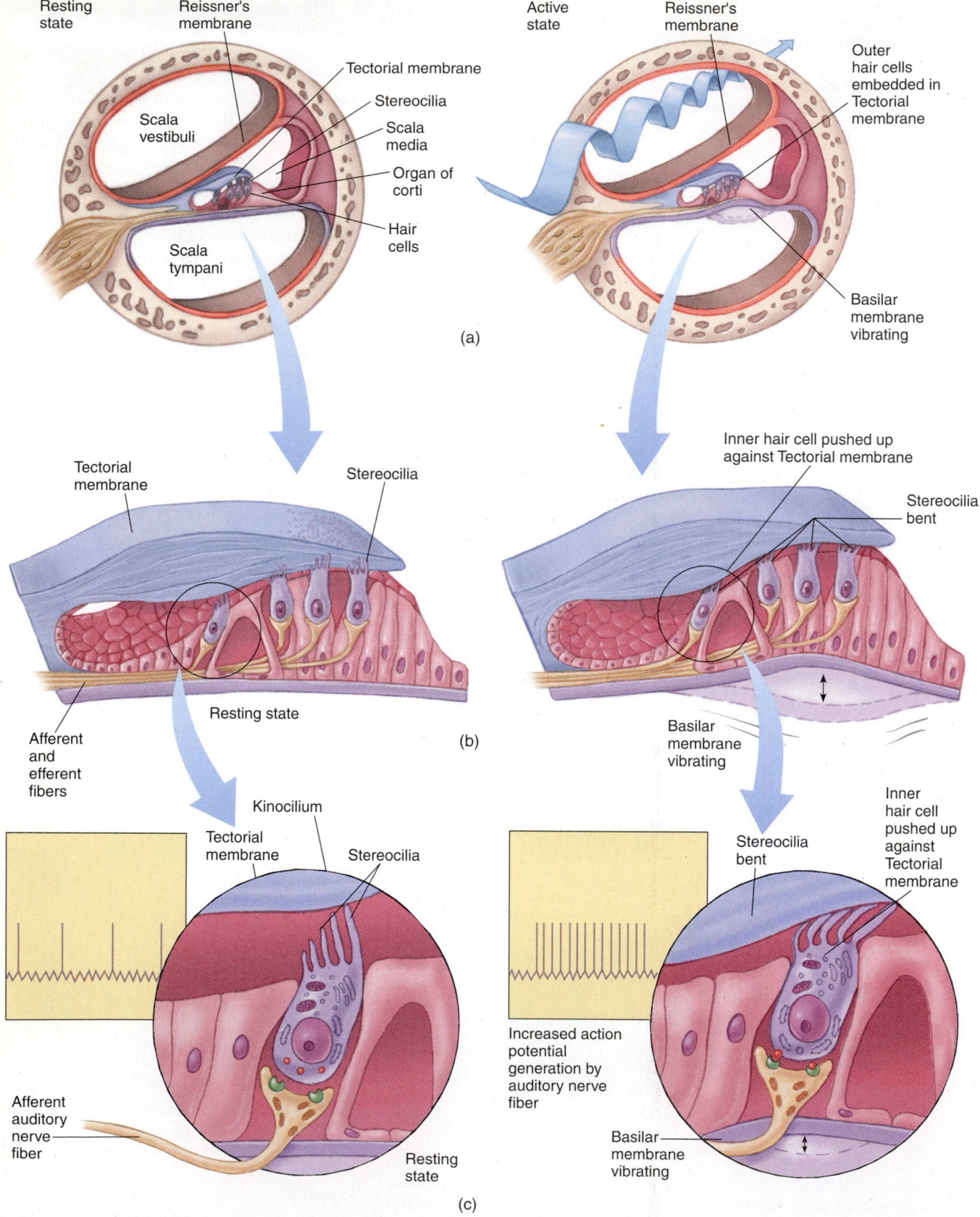

Resting state

Reissner's membrane

Scala vestibuli

Scala tympani

Tectorial membrane

Stereocilia

Scala media

Organ of corti

Hair cells

(a)

Active state

Reissner's membrane

Outer hair cells embedded in Tectorial membrane

Basilar membrane vibrating

Tectorial membrane

Stereocilia

Resting state

Afferent and efferent fibers

(b)

Inner hair cell pushed up against Tectorial membrane

Stereocilia bent

Basilar membrane vibrating

Kinocilium

Tectorial membrane

Stereocilia

Afferent auditory nerve fiber

Resting state

(c)

Increased action potential generation by auditory nerve fiber

Stereocilia bent

Inner hair cell pushed up against Tectorial membrane

Basilar membrane vibrating

(d)

Figure 8–42

(a) When the basilar membrane changes shape as waves travel along it, inner hair cells of the organ of Corti are distorted near the tectorial membrane so that *(b)* stereocilia that protrude from the hair cells bend *(c)*. The bending causes hair cell ion channels to open, leading to membrane depolarization, inflow of Ca^{2+}, and release of a neurotransmitter that causes action potentials to arise in auditory nerve fibers. *(d)* SEM of outer hair cells of the organ of corti. (×1705). *(© Dr. G. Bredberg/SPL/ Photo Researchers, Inc.)*

stereocilia causes ion channels of the hair cell's plasma membranes to open, leading to a depolarization of the plasma membrane. This depolarization is analogous to the generator potential described for the Pacinian corpuscle of the somatosensory system. Unlike the Pacinian corpuscle, however, the hair cell is not capable of producing its own action potential. When the hair cell is depolarized, Ca^{2+} ions flow into the cell and cause the release of transmitters that in turn depolarize afferent auditory nerve fibers (Fig. 8–42c). The auditory nerve fiber then generates action potentials that are transmitted to the brain to convey information relating to the frequency (pitch) and intensity (loudness) of sound.

Auditory Pathways

The first-order auditory neuron cell bodies are located in the spiral ganglia of the cochlea. Their axons form the auditory nerve, which enters the brain stem and synapses with second-order neurons in the cochlear nucleus (Fig. 8–43). Information from the cochlear nucleus is transmitted to both sides of the brain and eventually reaches the medial geniculate nu-

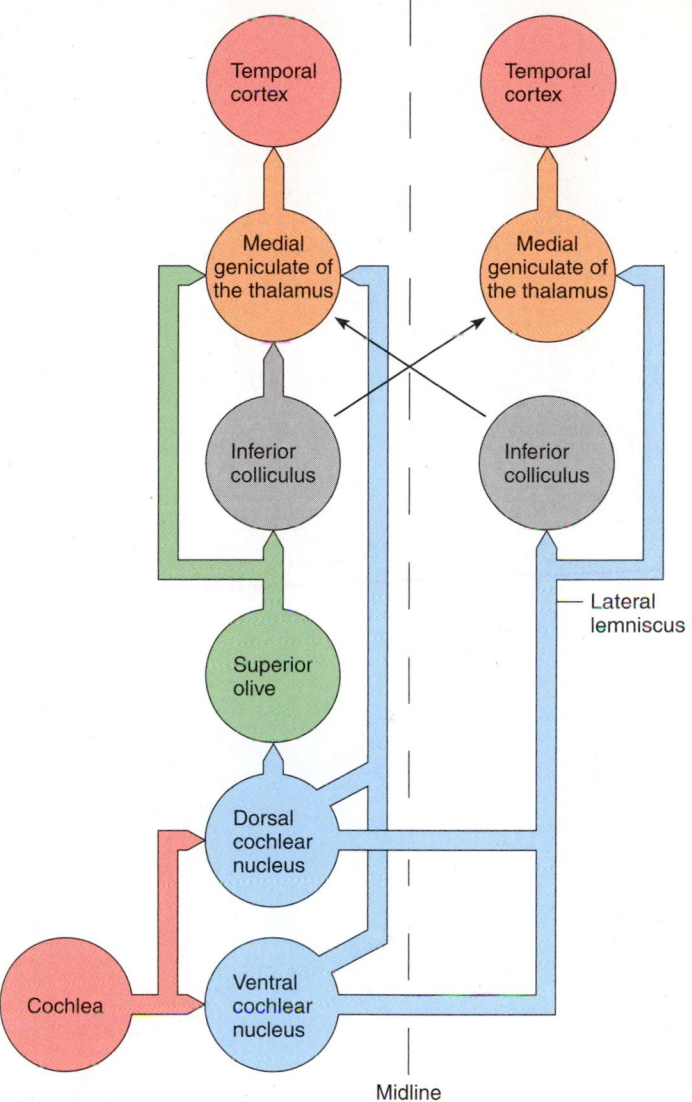

Figure 8–43

Auditory pathways, originating in the cochlea on one side of the brain and terminating bilaterally in the temporal cortices.

cleus of the thalamus. From the medial geniculate nucleus, information is transmitted to the primary auditory cortex in the temporal lobe.

The receptive field of neurons along the auditory pathways is characterized by the use of tones with different frequencies and intensities. A tone is a sound wave of one frequency. The sound produced by playing a single note on the piano, for example, produces nearly a pure tone. This tone, in turn, produces a sound pressure wave that causes the compression and expansion of air molecules at one frequency (Fig. 8–44). If many notes are played simultaneously, the sound produced is a combination of all of the individual frequencies. Tone can be represented graphically by an oscillating waveform; the peak of the waveform represents the

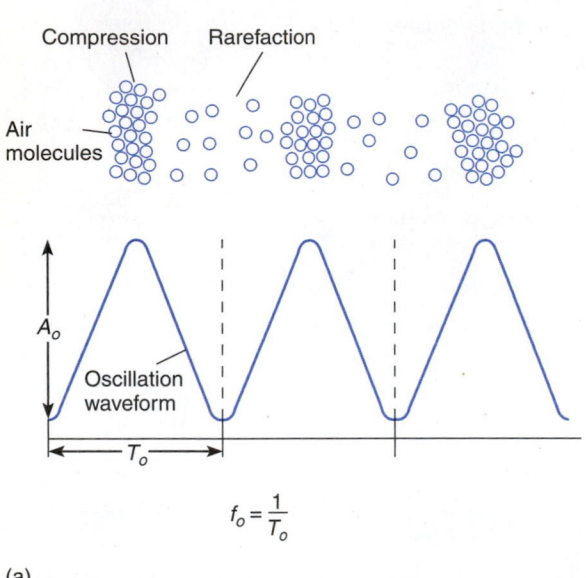

(a)

SOUND FREQUENCY   NERVE ACTION POTENTIAL

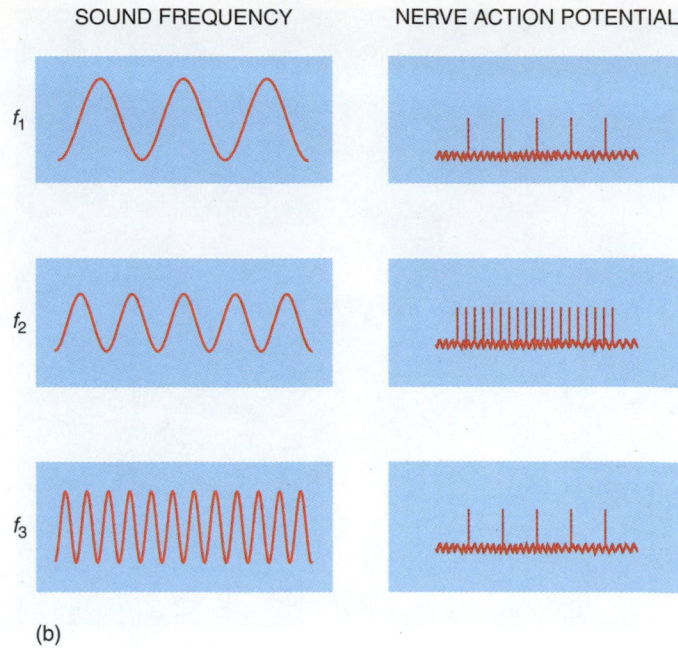

(b)

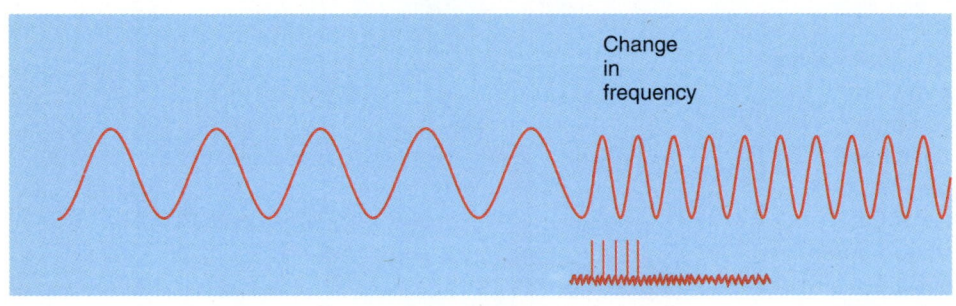

Change
in
frequency

(c)

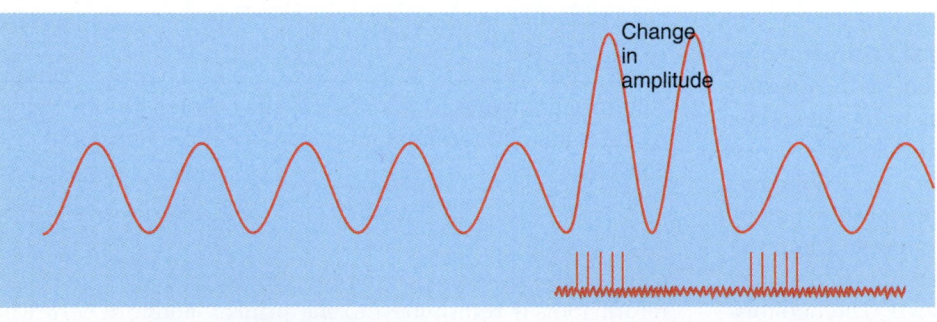

Change
in
amplitude

(d)

Figure 8–44

(a) A single tone produces a sound pressure wave that alternately compresses and rarefies air molecules at a particular frequency and amplitude. The characteristic frequency of a nerve fiber is the frequency that most effectively activates the fiber *(b)*. In this example, f_2 would be the characteristic frequency. *(c)* Some nerve fibers generate action potentials only in response to a change in the frequency or amplitude of a tone. Upper traces show frequency and amplitude of sound. Lower traces show nerve action potentials at changes in *(c)* frequency and *(d)* amplitude.

compression of the air molecules, and the valley of the waveform represents the expansion of the molecules. The difference between the peak and the valley is the **amplitude** (A_O), or intensity of the sound wave. The time between the peaks is the **time period** of the tone (T) and is measured in units of seconds. Its inverse is the **frequency** (f) of the tone and is measured in units of **Hertz** (Hz) (1 Hz = 1 cycle per second). The amplitude and frequency of a tone can be used to

characterize the receptive properties of nerve cells in the auditory pathway. If the frequency of the tones is varied, one frequency is found that most effectively activates a particular auditory nerve fiber. This frequency is known as the **characteristic frequency** (f_O) for the nerve fiber. Different auditory nerves have different characteristic frequencies. These auditory nerve fibers have a **tonotopic organization** such that those with a high characteristic frequency innervate hair cells near the base of the cochlea, whereas those with a low characteristic frequency innervate hair cells near the apex of the cochlea.

Neurons of the auditory system also respond in a variety of different ways to a particular tone. Some neurons generate action potentials only when the tone is on; other neurons are prevented from generating action potentials with the onset of the tone. Some neurons generate action potentials when the frequency of the tone is suddenly changed, whereas others generate action potentials only when the amplitude changes. The variety of responses suggests that nerve cells of the auditory system extract the basic features of a sound from its complex waveform. The fundamental features of a sound are reassembled by the convergence of neurons. These neurons converge onto other neurons that respond only to the reassembled complex auditory stimuli. For example, certain neurons in the auditory cortex of monkeys respond only to specific sounds used by that species of animal. Animal studies of auditory responses suggest that there are neurons in the human auditory cortex that respond only to the word *go*, and others that respond to the word *stop*.

Sound Localization

In addition to decoding the frequency and intensity of sound information, the auditory system also determines where sounds originate. This function of the auditory system allows us to orient toward a sound stimulus and respond accordingly. In crossing a street at a busy intersection, for example, the noise of a speeding car coming around the corner would cause us to stop, look in the direction of the oncoming car, and avoid a collision. The localization of sounds in space is accomplished by comparing differences in the timing and intensity of sounds as they reach each ear. Acoustic waves that originate on the right side of the head would first reach the right ear before traveling on to the left ear. The time delay between the two ears is called the **interaural time difference.** Sounds that originate along the midline would produce no interaural time difference. Sounds that originate off of the midline have been shown to produce interaural differences of 42 μsec for every 20° shift in space. These interaural time differences form the basis of an auditory field for sound localization as shown in Figure 8–45a. Each line represents an interaural time associated with a specific location in space.

In studies that have examined the mechanisms of sound localization in barn owls, researchers have found neurons that discharge in response to interaural timing differences. These neurons are found in regions of the medulla comparable to the human medial superior olive and anterior lemniscal nucleus (Fig. 8–45b). These neurons receive inputs from the cochlear nucleus on both sides of the brain and respond to specific frequencies and timing differences in the arrival of those frequencies to each ear. It is thought, therefore, that the mechanism underlying these responses is the coin-

cident arrival of information from each ear to a specific neuron. The summation of incoming activity at these so-called "coincidence cells" (Fig. 8–45c) in turn causes them to discharge. The concurrent arrival of information to specific cells despite interaural differences is thought to be due to propagation delays from the cochlear nucleus. For sound sources originating in the midline, for example, the propagation delays from the left and right ear are the same. In contrast, sound sources located closer to the right ear would result in the transmission of sound information through pathways with longer propagation delays than information coming from the left ear in order to reach a "coincidence cell" at the same time (see Fig. 8–45c). Likewise sound sources located closer to the left ear would result in the transmission of information with longer propagation delays than pathways from the right ear.

The concept of topographic maps of neurons in sensory systems now can be extended to include a mapping system in the auditory system for interaural time differences. Because these neurons overlap in their responses to specific frequencies of sound, the maps for tonotopic organization and interaural time differences thus can be thought of as a mapping of frequency and time information along orthogonal axes in the auditory system (see Fig. 8–45a).

Along with timing differences, sound information reaching the left and right ears from the same source is perceived as differences in intensity. In part, this is due to the attenuation of the acoustic wave as it propagates through air and as it is absorbed by the side of the head. In animals such as the barn owl, the ears are slightly offset to aid in the detection of intensity differences. A source of sound equidistant to both ears would produce a sound-intensity difference. However, for every 20° shift in space, there is a 4-decibel (dB) change in intensity difference (see Fig. 8–45a). A decibel is a unit of measure for sound (based on the logarithm of the ratio of a sound's intensity with respect to a reference value). A sound with an intensity of 10 dB is about equivalent to a soft whisper. Together, interaural intensity and time differences form the basis of an "auditory field" for the localization of sound. Each location is represented by a unique combination of interaural intensity and time differences. This auditory field is similar to the concept of the visual field for the location of objects within visual space.

Cells that respond to intensity differences are found in the lateral superior olivary nucleus and posterior lemniscal nucleus of the brain stem (see Fig. 8–45b). Information about interaural intensity and time differences is transmitted through different pathways in the medulla and finally converges in the inferior colliculus of the mesencephalon to form a neural map of the auditory field. Cells in the inferior colliculus thus respond to unique combinations of sound intensity and time differences to allow us to locate their place of origin. This information, in turn, ultimately is transmitted to areas of brain involved in moving our head and torso toward the direction in which the sound originates.

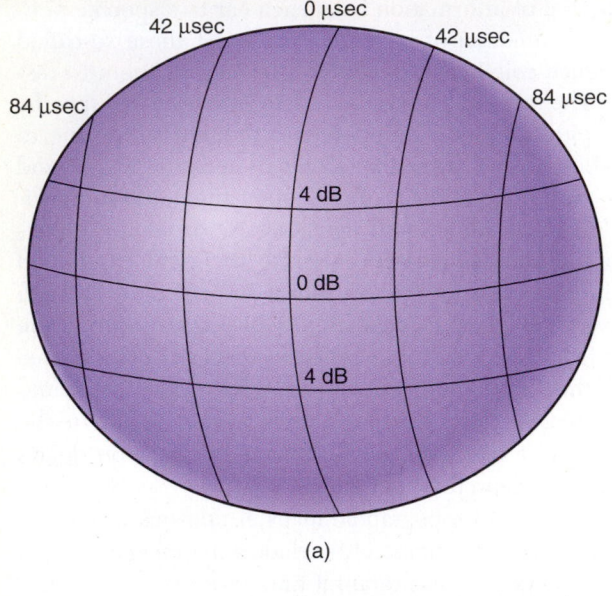

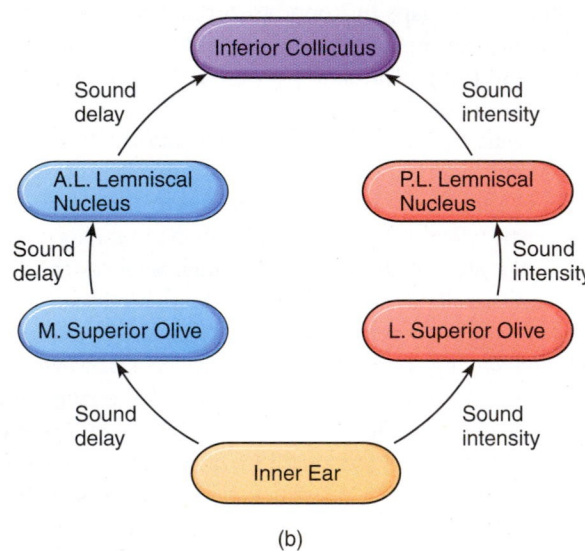

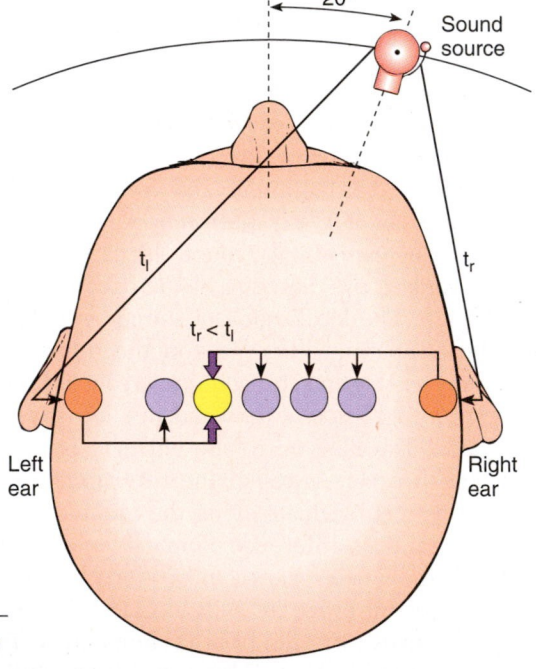

Figure 8–45

(a) The auditory field for sound localization is characterized by time and intensity differences for sound to reach the right and left ears. Vertical isocontours represent time differences. Sound sources originating in the midline reach both ears at the same time and thus have a 0-μsec time difference. Moving the sound source by 20° to the right or left produces a time delay of 42 μsec. Differences in sound intensity reaching the left and right ear are represented by the horizontal isocontours. Moving the sound source by 20° in either direction produces a 4-dB difference in sound intensity. *(b)* Auditory pathway for the transmission of sound delays and intensity differences. *(c)* Response of coincidence cells to sound sources originating at different locations in the auditory field. t_l and t_r are the times of arrival of the sound at the left and right ear, respectively.

Disorders of Hearing

Impairments of hearing are classified into three categories, according to the part of the auditory apparatus that fails to function properly: (1) conductive hearing loss, (2) sensorineural hearing loss, and (3) central hearing loss.

Conductive Hearing Loss

Conductive hearing loss involves inadequate transmission of sound through the outer and middle ear to the inner ear. It may be caused by blockage of the external auditory canal; perforation, inflammation, or scarring of the tympanic membrane; inflammation, pus, and fluid in the middle ear (otitis media); and sclerosis of the ossicles (otosclerosis). Conductive hearing loss produces a general reduction in hearing ability. Hearing by conduction through the skull, however, is still functional and can be tested by placing a tuning fork over the bony structure behind the ear to elicit the sensation of sound.

Sensorineural Hearing Loss

Sensorineural hearing loss is caused by damage to the cochlea, the organ of Corti, or the cochlear nerve fibers of cranial nerve VIII. Other causes include infectious organisms, degenerative bone disease, and functional derangement of the organ of Corti due to traumatic sound. Hair cells can be destroyed when the eardrum suddenly is exposed to a loud noise, such as an explosion, which is thought to crush hair cells between the basilar and tectorial membranes. Chronic exposure to high intensities of sound, such as associated with hard rock music, also may cause selective sensorineural hearing loss. The loss of hair cells is associated with frequencies of sound that make up the noise. Inflammation of the cochlear nerve also may result in sensorineural hearing loss.

Central Hearing Loss

Defects in the auditory tracts of the brain stem, thalamus, or the auditory cortex of the temporal cerebrum result in central hearing loss. Causes include malignancies of the brain, cerebrovascular disease, infections of the CNS, cerebral concussion, and hypoxia (lack of oxygen). Lesions in the primary auditory cortex (area 41) on one side of the brain have little effect because of the bilateral projection of sound information. However, lesions in the auditory association cortex (area 22) or Wernicke's area produce auditory sensory aphasia, a condition in which voices can be heard but the words appear to have no meaning.

Hearing disorders in infants can lead to lifelong language disorders. Ear infections, for example, that chronically impair the ability of infants to hear spoken words hinder the development of auditory nerve cell connections needed for hearing and speech. If these auditory connections are not formed within the first few years after birth, they will never be made. This developmental disorder illustrates the importance of the auditory environment and the formation of crucial nerve cell connections underlying language skills.

THE VESTIBULAR SYSTEM

How does the vestibular system sense the position and motion of the head in a gravitational field?

The vestibular system aids in maintaining the body's balance by detecting the position and motion of the head in space. The sensory cells of the vestibular system are in the **ampullae** of the **semicircular canals** and in the **utricle** and **saccule** (Fig. 8–46). Three semicircular canals are part of the vestibular apparatus: **the anterior (superior), posterior, and horizontal (lateral) semicircular canals.** Each canal is situated perpendicular to the other two. Thus, each semicircular canal is positioned in one of the three planes of space. When the horizontal (lateral) canal is oriented in the horizontal plane, the superior canal is in the anterior position. When the head is in its normal, resting position, the "horizontal" canal is tilted, anterior end upward, approximately 30°. With this arrangement, each semicircular canal detects the angular acceleration in one of three planes in space. The vestibular apparatus also consists of the utricle and saccule (see Fig. 8–46). These so-called **otolithic organs** detect linear acceleration in the horizontal plane by the utricle, and in the vertical plane by the saccule.

The semicircular canals and otolithic organs are filled with the same endolymph fluid that surrounds the hair cells of the cochlea. As does the cochlea, the semicircular canals and otolithic organs have hair cells responsible for the transduction of sensory stimuli. The appropriate stimulus for these vestibular hair cells is the acceleration of the head. As the head begins to move, the inertia of the endolymph fluid causes the stereocilia to bend (Fig. 8–47). This bending of the stereocilia causes the hair cell either to depolarize or to hyperpolarize.

The type of membrane potential change in the hair cell depends on the direction in which the stereocilia bend. The stereocilia of hair cells in the vestibular apparatus are aligned according to size, with the smallest cilium placed at one end of the group and the largest cilium, called the *kinocilium,* at the opposite end (see Fig. 8–47). If the stereocilia are bent toward the kinocilium, the hair cell depolarizes. If the stereocilia are bent toward the smallest cilium, the hair cell hyperpolarizes.

These hair cells, like those in the organ of Corti, are not capable of generating action potentials. They do, however, excite or inhibit the **vestibular nerve fibers** that innervate them. As the hair cell depolarizes, it releases a neurotransmitter that excites the innervating vestibular nerve. The hyperpolarization of a hair cell causes the vestibular nerve to produce fewer action potentials. This may be due to a

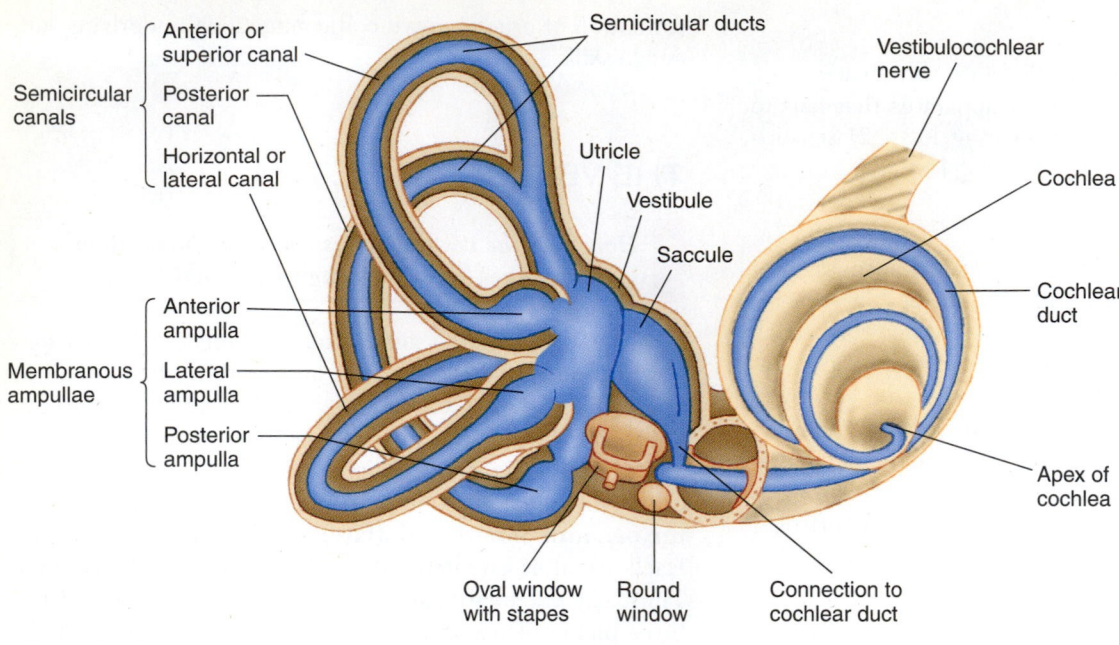

Figure 8–46

The vestibular apparatus consists of the semicircular canals, including the posterior, anterior (or superior), and horizontal (or lateral) canals, and the utricle and saccule.

decrease in the amount of an excitatory transmitter that is being released tonically from the hair cell. The hair cells are embedded in a gelatinous material composing the **cupula** (see Fig. 8–47) and arranged in a polarized manner on both sides of the head. Rotation of the head, termed *angular acceleration,* causes the inertia of the endolymph that fills the semicircular canals and ampullae to exert a force on the cupula in a direction opposite the rotation of the head. In the ampullae of the horizontal semicircular canals, for example, the stereocilia on each hair cell are positioned so that the kinocilium is nearest to the front of the head, whereas the smallest cilium is nearest the back of the head (Fig. 8–48). As the head rotates from right to left (counterclockwise), the inertia of the endolymph applies a force on the cupula from left to right (clockwise). The fluid's relative motion causes the stereocilia on the left side of the head to depolarize the hair cells, whereas the stereocilia on the right side of the head hyperpolarize the hair cells. Correspondingly, the vestibular nerves on the left side of the head increase their rate of action potential generation, whereas the nerves on the right side of the head decrease their rate of action potential generation. This information thus is transmitted to inform the brain that the head is rotating counterclockwise.

The information from the vestibular nerves is transmitted to the **vestibular nucleus,** located within the brain stem. From there, it is sent down to the spinal cord to act on nerve cells that regulate movement, and up to the cerebellum and somatosensory cortex. The cerebellum and somatosensory cortex use this information to coordinate muscular activity in order to maintain balance.

The vestibular system also helps the eyes to fixate on a spot in the visual field to provide a stable reference point during acceleration. This process is part of the vestibulo-ocular reflex and is achieved by the transmission of vestibular information for the control of eye movements. Outputs from the vestibular nucleus project to the motor nucleus of cranial nerve III (oculomotor nerve) ipsilaterally and cranial nerve VI (abducens nerve) contralaterally. These brain stem nuclei control muscles that cause the eyes to turn inward or outward, respectively. The movement of the head in a clockwise direction causes a slow, counterclockwise movement of both eyes to allow them to fixate on a point of reference (see Fig. 8–48). As the head continues to turn and the slow movement brings the eyes to an extreme position, there is a fast movement of the eyes in the direction of rotation. This allows the eyes to fixate on a new point of reference during rotation. The slow and fast movements of the eyes during accelerated angular rotation are called **nystagmus.** The fast-phase movement in the direction of rotation conventionally is chosen to describe the direction of the nystagmus.

The **maculae** are sense organs within the utricle and saccule that respond to change in the position of the head with respect to gravity. They also respond to linear acceleration of the body, as opposed to the angular or rotational acceleration detected in the semicircular canals. The stereocilia of the hair cells in the utricle and saccule are embedded in an **otolithic membrane** that contains calcium carbonate crystals called **statoconia** (also termed **otoconia** and **otoliths**) (Fig. 8–49). The statoconia give the otolithic membrane mass, so that when the head is tilted, the gravita-

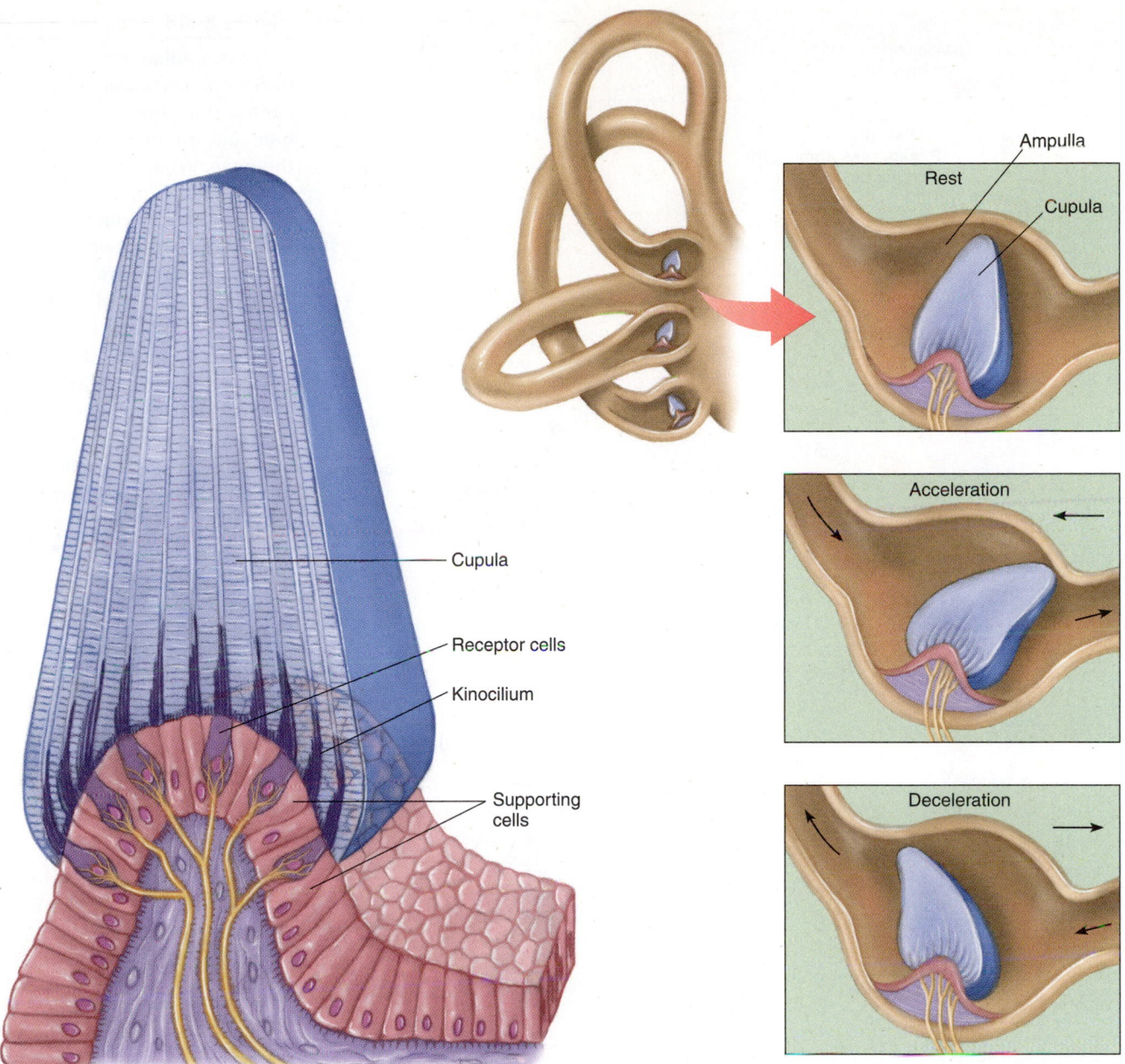

Figure 8–47

In the ampullae of the semicircular canals, the cilia of the hair cells are embedded in the cupula. When the head moves (acceleration), the inertia of the endolymph in the semicircular canals and ampullae exert force on the cupula in the direction opposite the movement, bending the stereocilia and kinocilia. *(Redrawn from Wersall, J. Acta Otolaryngol. (Stock.) Suppl. 126:1, 1956. [From: Berne and Levy,* Physiology, *The C. B. Mosby Co., 1983.])*

tional force causes the weighted otolithic membrane to bend the stereocilia. The stereocilia are arranged in the utricle in a complex pattern of orientation. When the head is tilted in any direction, part of the hair cell population depolarizes, part hyperpolarizes, and part is unaffected. Like the function of the stereocilia in the semicircular canals, bending the stereocilia toward the kinocilium depolarizes the hair cell, whereas bending away from the kinocilium produces a hyperpolarization. The complex signal generated by the total population of hair cells in the utricles provides the brain with information about the orientation of the head with respect to gravity.

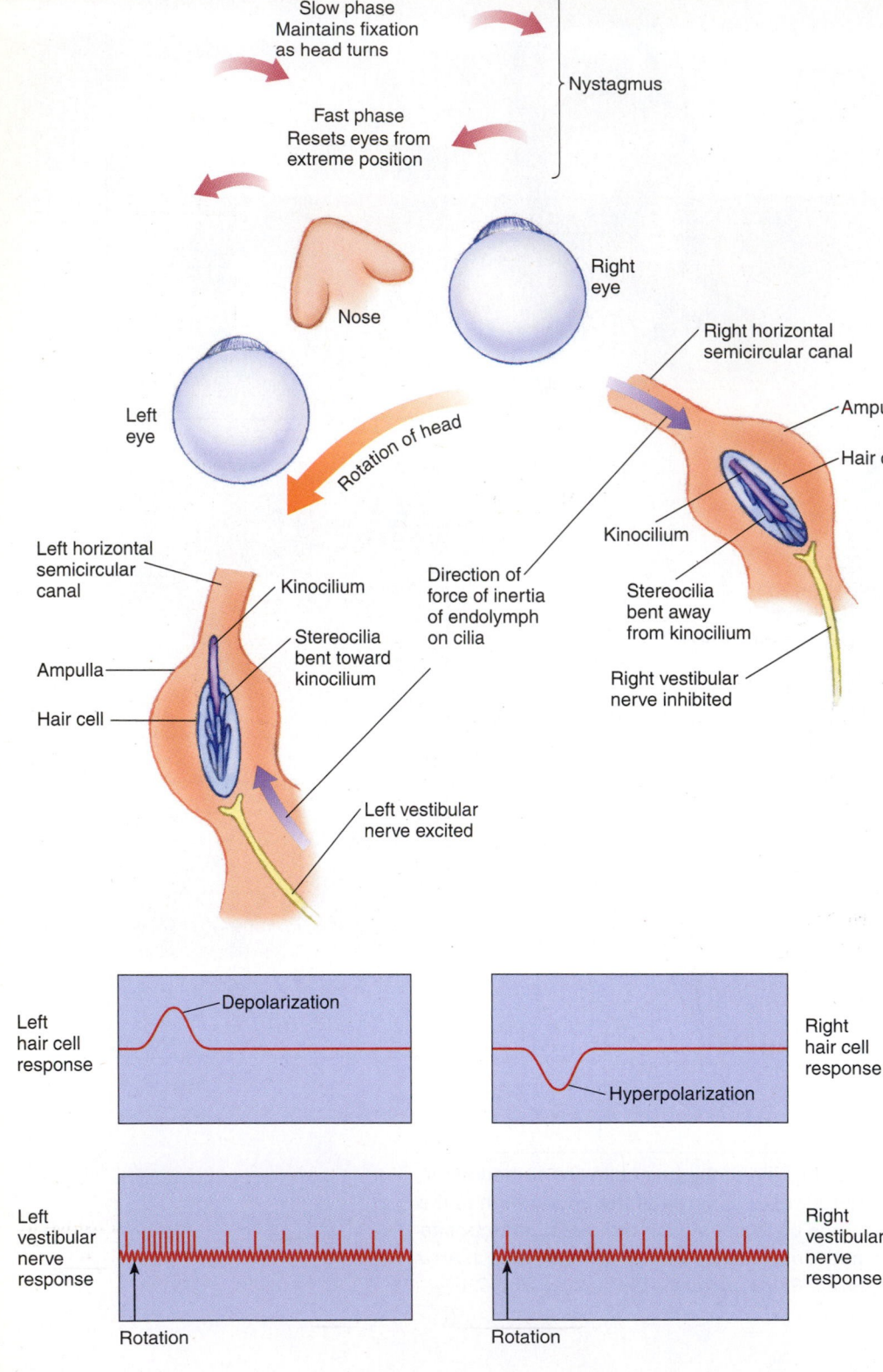

Slow phase
Maintains fixation
as head turns

Nystagmus

Fast phase
Resets eyes from
extreme position

Right eye

Nose

Left eye

Rotation of head

Right horizontal semicircular canal

Ampulla

Hair cell

Kinocilium

Stereocilia bent away from kinocilium

Right vestibular nerve inhibited

Left horizontal semicircular canal

Kinocilium

Direction of force of inertia of endolymph on cilia

Stereocilia bent toward kinocilium

Ampulla

Hair cell

Left vestibular nerve excited

Left hair cell response

Depolarization

Right hair cell response

Hyperpolarization

Left vestibular nerve response

Rotation

Right vestibular nerve response

Rotation

Figure 8–48

In the ampullae of the horizontal semicircular canals, the kinocilia of the hair cells are oriented toward the nose. When the head rotates in a counterclockwise direction, the inertia of the endolymph exerts pressure on the cilia embedded in the cupula in the opposite direction of the head movement. On the left side of the head, the hair cells depolarize because the fluid motion causes the cilia to bend toward the kinocilium, whereas the opposite effect takes place on the right side. The vestibular nerves detect the depolarization and hyperpolarization and thereby inform the brain about the direction of rotation. The vestibular system automatically regulates nystagmus, eye movements that allow stabilization of images on the retina as the head moves. Slow-phase movements maintain fixation of the image on the retina. When the slow phase moves the eyes to an extreme position, the fast phase resets the eye so that a new fixation point can be acquired. The cupula embedding the hair cells is omitted for clarity.

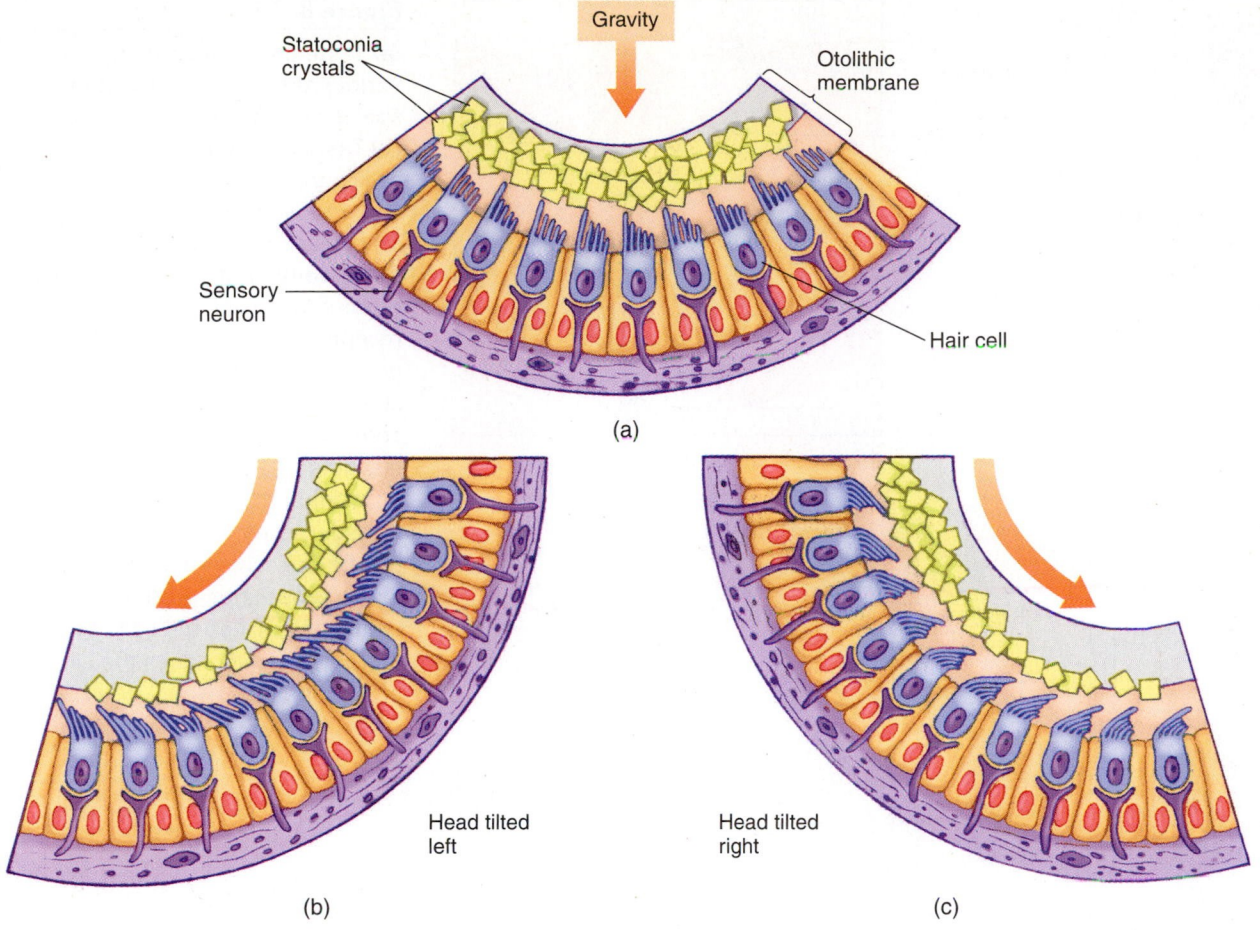

(a)

(b) Head tilted left

(c) Head tilted right

Figure 8–49

The stereocilia of the hair cells of the utricle bend under the weight of otolith crystals when the head tilts. For the hair cell population illustrated in **(a)**, when the head is tilted to the left, the force of the otolithic membrane bends the stereocilia toward the kinocilium, resulting in depolarization of the hair cells **(b)**. For the same hair cell population, a tilt to the right causes bending of the stereocilia away from the kinocilium and results in hair cell hyperpolarization **(c)**. Arrows indicate direction of otolith movement. Hair cells of the utricle are oriented in a complex pattern that provides information about the tilt of the head in any direction.

THE GUSTATORY SYSTEM

 How are different tastes detected; how is this information processed?

Sensations

Our ability to taste food is the result of the activation of taste receptor cells in the tongue. These receptors detect chemicals associated with five basic taste qualities: sweet, salty, sour, bitter, and umami (meaty flavor). However, their areas of distribution overlap. The tip of the tongue contains receptors that are sensitive to sweet substances (Fig. 8–50a).

The lateral region near the tip of the tongue is sensitive to salty substances; the lateral region from the tip to near the back of the tongue is sensitive to sour substances. The entire back of the tongue is sensitive to bitter substances. Taste receptors also are located in the palate and pharynx.

Taste receptors are called *gustatory receptors* and are a type of epithelial cell clustered with other cells called *supporting* and *basal cells* (Fig. 8–50b). Together, these cells compose the taste bud. Taste receptors degenerate every 10 days and are replaced by new receptor cells derived from the differentiation of basal and supporting cells.

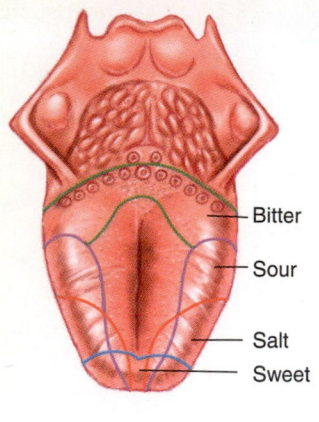

(a)

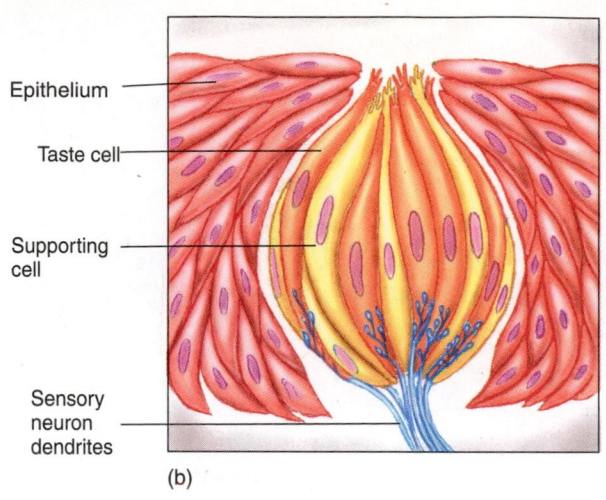

Epithelium

Taste cell

Supporting cell

Sensory neuron dendrites

(b)

Figure 8–50

(a) Taste buds sensitive to the various tastes are grouped in specific regions of the tongue. *(b)* Taste buds consist of taste cells surrounded by supporting and epithelial cells. *(c)* Receptors on taste cells respond to sweet, bitter, salty, and sour compounds. Binding of sweet compounds to receptor (R_s) elevates cAMP levels, which leads to closure of K^+ channels. Binding of bitter compounds to receptor (R_B) elevates IP_3 levels and increases intracellular Ca^{2+}. Salty taste is transduced by entry of Na^+ through passive ion channels. Sour compounds produce H^+ which penetrates the cell membrane and causes the closure of K^+ channels.

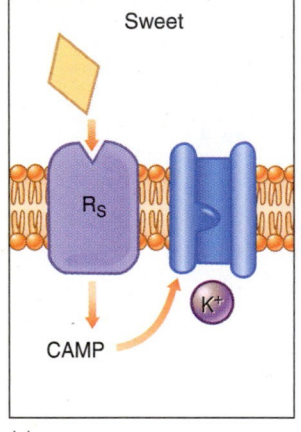
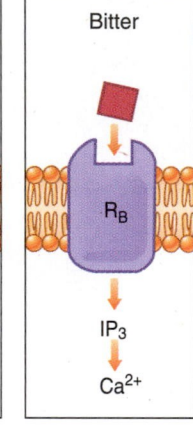
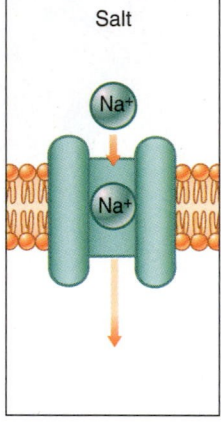
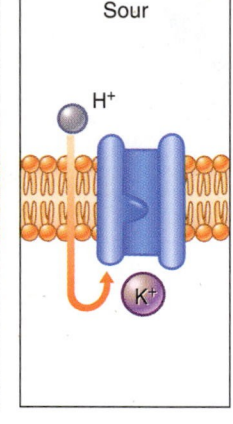

(c)

Transduction of Taste Stimuli

Various mechanisms appear to underlie the transduction of different taste stimuli. Sweet-tasting compounds exert their effect by binding to receptors that utilize cAMP as a second messenger (Fig. 8–50c). Elevations in cAMP, in turn, cause the closure of K^+ channels, thus depolarizing the membrane. The binding of bitter compounds to receptors in the base of the tongue results in an increase in intracellular Ca^{2+} via the second messenger IP_3. The transduction of saltiness is mediated by the direct passage of sodium ions through passive cation channels, thus altering the membrane potential of the taste receptor cell (see Fig. 8–50c). Acids responsible for sour tastes appear to penetrate the cell membrane and block voltage-gated K^+ channels, and thus sustain the depolarization of the membrane.

Increases in intracellular Ca^{2+} by IP_3 or by membrane depolarization-induced activation of voltage-gated Ca^{2+} channels ultimately result in the release of transmitters from taste bud cells. The binding of transmitters to receptors located on postsynaptic fibers would then cause the initiation of action potentials in the afferent fiber. Afferent fibers branch many times before they innervate a taste bud; consequently, one fiber often innervates several taste buds, and the action potentials from a single fiber may therefore represent the activation of many taste buds.

A single fiber can respond to stimuli from different categories of chemicals, but it responds best to one type of chemical group. This property of taste fibers is similar to the responses of fibers in the auditory pathway for sound frequencies and visual pathways for colors of light. As in the other sensory pathways, the higher brain centers of the taste pathway use this type of information to make comparisons and contrast the taste qualities of consumed substances.

Gustatory Pathways

Taste receptor cells are innervated at their base by nerve fibers that form part of the VII (facial), IX (glossopharyngeal), or X (vagus) cranial nerve (Fig. 8–51). Taste buds in the front of the tongue are innervated by the VII cranial nerve; those in the back of the tongue are innervated by the IX cranial nerve; and those in the pharynx are innervated by the X cranial nerve. Thus, information about bitter tastes is transmitted by way of a different pathway than is information

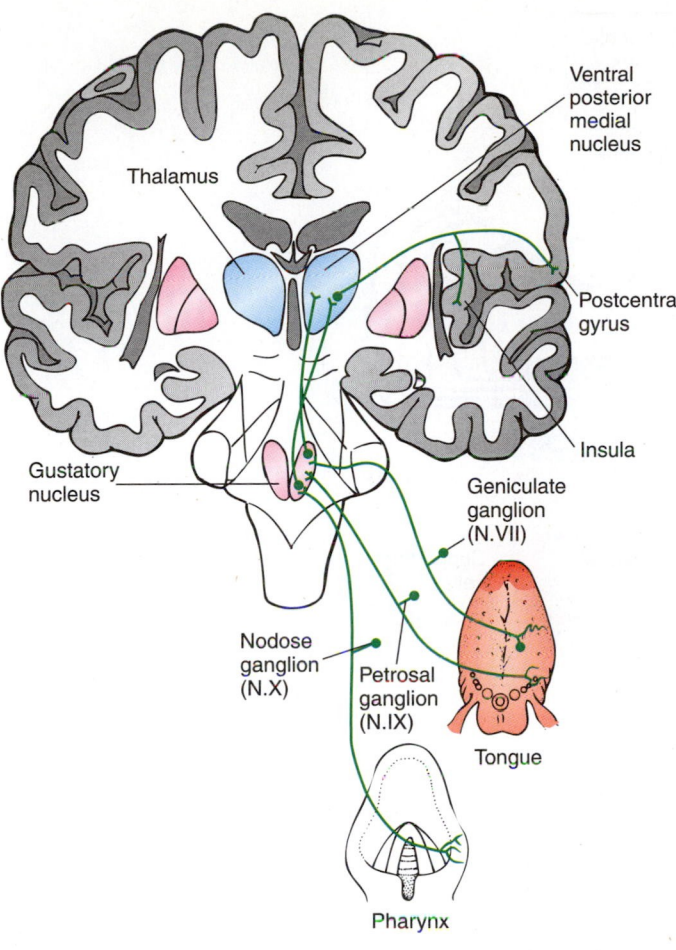

(d)

Figure 8–51

Taste pathways from the right side of the tongue and pharynx. First-order sensory neurons of cranial nerves VII, IX, and X relay taste information from taste cells to the gustatory nucleus in the brain stem. Second-order neurons in the gustatory nucleus relay taste information to the ipsilateral thalamus VPM and from there, third-order neurons carry the information to the cerebral cortex in the postcentral gyrus and insula.

about sweet, sour, and salty tastes. Fibers innervating the taste buds in the palate are part of the X cranial nerve.

All afferent fibers synapse on cells in the brain stem in the gustatory nucleus. Neurons in the gustatory nucleus, in turn, send their axons to the thalamus on the same side of the brain and terminate in the ventral-posterior-medial (VPM) nucleus of the thalamus. From the VPM, taste information is transmitted to the gustatory area of the neocortex just in front of the somatosensory region of the tongue.

Alternate pathways for taste sensation transmit information to the limbic system and hypothalamus. These pathways are responsible for the affective and emotional aspects of taste. For instance, some individuals become nauseated by the taste of certain foods. This can happen even if the food

was at one time a favorite dish. The consumption of spoiled food, for example, can lead to food poisoning. In some cases, a single exposure to the spoiled food can condition an individual to avoid it in the future.

THE OLFACTORY SYSTEM

> *How are smells detected; how is this information processed?*

Our ability to smell odors is a result of the activation of olfactory receptor cells in the nose. Olfactory receptor cells are contained within the olfactory mucosal membrane of the nasal cavity of the nose and surrounded by supporting and basal cells (Fig. 8–52a). As do the receptor cells in taste buds, the olfactory receptor cells degenerate and are replaced by basal cells that become new receptor cells. The regenerative cycle for olfactory cells, however, is 60 days.

Humans can detect seven general types of odors: camphoraceous, musk, floral, peppermint, ethereal, pungent, and putrid. These odor molecules bind to olfactory receptor cells along ciliated fiber processes that extend from the cell body and into the mucosa of the nose (see Fig. 8–52a). The coupling of odor molecules to receptors within the cilia activates olfactory-specific G-proteins, G_o, to regulate levels of intracellular cAMP (Fig. 8–52b). Increases in cAMP levels, in turn, result in the activation of Na^+ channels leading to the influx of Na^+ ions and the depolarization of the membrane. When threshold is reached, an action potential is initiated and is propagated along axons that pass through the cribiform plate and terminate on mitral cells located in the olfactory bulb. The bundle of fibers from mitral cells forms the olfactory tract. These fibers directly innervate areas of cortex known as the limbic system, and the orbitofrontal cortex, indirectly via the thalamus (see Fig. 8–52a). The olfactory innervation of the limbic system is exceptional in that it represents the only sensory system that directly innervates areas of cortex.

The olfactory system plays an obvious role in stimulating appetite and eating behavior. The activation of eating behavior is carried out by a motivational area of the brain called the *hypothalamus* and will be considered in Chapter 11, in the discussion about central integrative systems.

The olfactory system also plays a role in mating behavior. Perfumes and colognes are examples of the use of artificial odors to attract members of the opposite sex. In animals such as mice, the odor of a female stimulates the release of testosterone in males and enhances the growth rate of the male sexual organ in adolescents. The odor of a female also causes the release of hormones in the adult male that in turn activate the sebaceous glands to give off a male odor to signal its readiness for mating. Male odors also increase the rate of sexual maturation in adolescent female mice and accelerate the estrous cycle of adult females.

In humans, the olfactory system is poorly developed; thus, visual, tactile, and auditory stimuli play more important roles in human mating behavior.

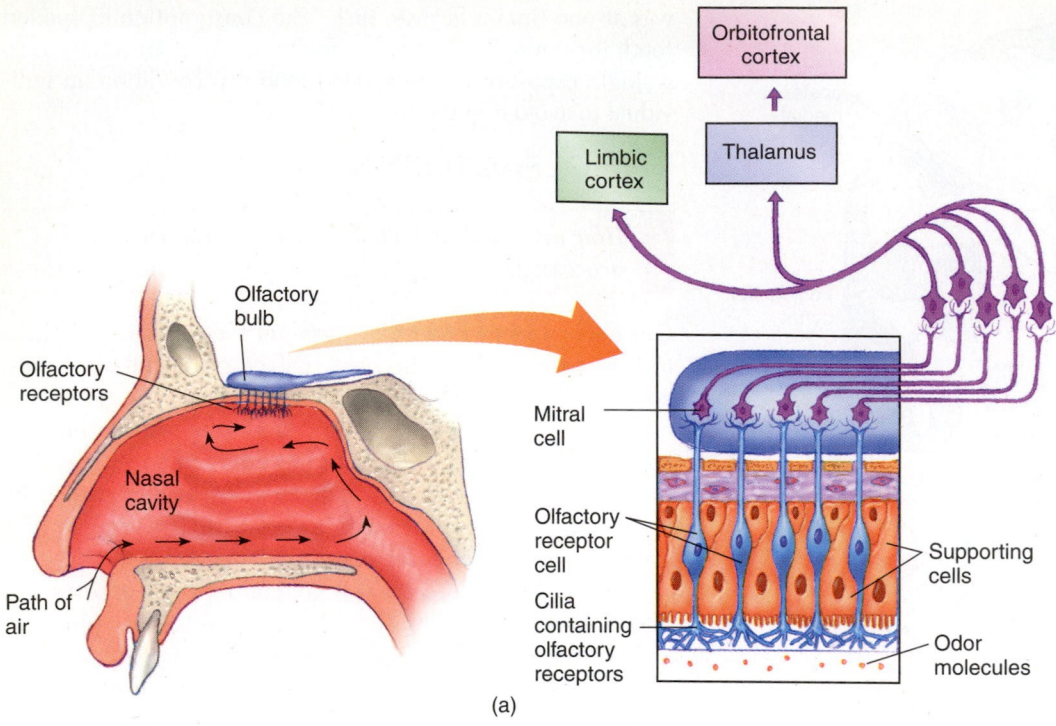

(a)

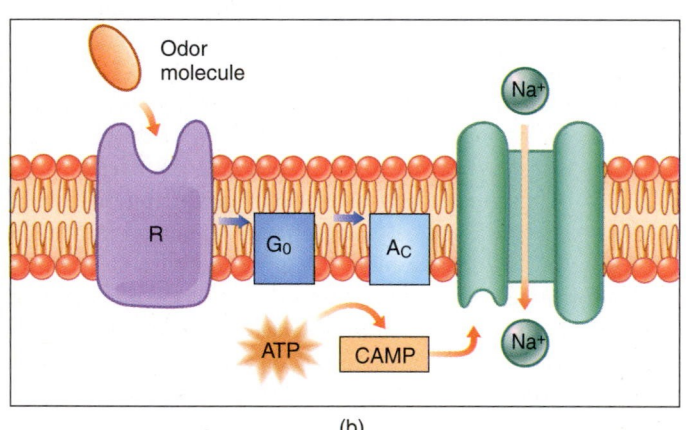

(b)

Figure 8–52

(a) Olfactory receptor cells in the nasal cavity contact mitral cells in the olfactory bulb. Mitral cell fibers form the olfactory tract, which innervates the limbic system and the orbitofrontal cortex. **(b)** Receptor in olfactory cilia. Binding of odor molecule to receptor (R) leads to activation of stimulatory olfactory specific G-protein (Go) and adenylate cyclase (Ac). Elevation of cAMP causes opening of cation channel and influx of Na^+ ions to depolarize the olfactory receptor cell.

CHAPTER REVIEW

Summary

- We detect information from the external environment with receptors that transduce sensory stimuli into action potentials, which are then conducted to the CNS. Sensory systems detect stimuli from the outside world in the form of light, sound, heat, pressure, chemical substances, and odors. Each type of stimulus is capable of activating a specific type of sensory receptor.

- The intensity of a stimulus is coded by the number of action potentials generated in a frequency code and by the number of receptors activated in a population code. The quality of a sensory stimulus is coded in the labeled-line code of stimulus quality. In sensory adaptation, receptors stop responding to a stimulus after a time.

- Somatic receptors on the surface of the skin detect pressures applied to the skin as well as temperature changes and painful stimuli. Pressure and touch sensation is transmitted ipsilaterally in the spinal cord along the dorsal column pathway.

- Pain and temperature sensation are transmitted contralaterally along the anterolateral pathway of the spinal cord. Relay stations that process pressure and touch sensation are located in the dorsal column nuclei of the medulla and the thalamus. In contrast, pain and temperature information are transmitted directly to the thalamus.

- From the thalamus, somatic information is sent to the somatosensory cortex. This cortex is organized in the form of a somatotopic map of the body, with the lower limbs located in the medial region of the cortex, and the upper extremities and face, in the lateral region.

- Nerve cells that process somatic information from the periphery to the somatosensory cortex have receptive fields. These receptive fields are areas of the skin that, when stimulated, can either excite or inhibit a particular nerve cell. The pattern of excitation or inhibition forms unique shapes that are characteristic of neurons at different levels along the somatosensory pathway.

- Sensory receptor cells respond to the basic features of a complex stimulus. In this way, they decompose the stimulus into its fundamental components. Contrasts and comparisons are made of the basic features by groups of nerve cells in each sensory pathway. These features eventually are reassembled by nerve cells in the sensory areas of the neocortex.

- Sensory receptors for vision are called *photoreceptors* and are located in the retina of the eye. Rod photoreceptors are used for night vision, whereas cones are involved in color vision.

- The three types of cones used in the process of color are red, green, and blue cones, so called because of their sensitivity to specific wavelengths or colors of light. Light in the form of photons is absorbed by photoreceptors, causing a hyperpolarization of the photoreceptors.

- Information then is transmitted to ganglion cells via bipolar cells. Visual information then is transmitted from ganglion cells in the retina to the LGN of the thalamus. Information from the lateral geniculate then is transmitted to the visual cortex. The receptive fields of ganglion cells are annular with "on" centers and "off" surrounds or with "off" centers and "on" surrounds. This type of receptive field also is exhibited by neurons in the lateral geniculate. In the primary visual cortex, neurons are more sensitive to bar-shaped stimuli. Ganglion cells detect the basic features of a visual image. These features are assembled to reconstruct the total image at the higher levels of the visual system.

- The auditory system transforms sounds into action potentials. The external ear channels sound to the ear drum, causing the ear drum to vibrate. The ear drum is connected to a series of small bones called the *malleus, incus,* and *stapes,* which also oscillate in response to sound pressures. The mechanical displacement of the stapes is transferred to a fluid-filled structure called the *cochlea.* The cochlea contains sensory receptor cells called *hair cells.* The movement of fluid within the cochlea also displaces the basilar membrane. The displacement of the basilar membrane activates the hair cell, which in turn excites fibers of the auditory nerve.

- Each auditory nerve responds to a narrow range of sound frequencies. In addition, the nerves are tonotopically organized so that those innervating hair cells near the base of the cochlea respond best to high frequencies, whereas those innervating the apex respond best to low frequencies. At the higher levels of the auditory system, neurons respond best to more complex sound frequencies.

- Taste information is detected by the gustatory system using taste buds located in the tongue. Taste buds near the back of the tongue respond best to bitter tastes, those in the middle respond best to sour and salty tastes, and those on the tip respond best to sweet tastes. The chemical activation of a taste receptor in turn excites afferent fibers that innervate the receptor. This information eventually is transmitted to the gustatory area of the neocortex near the somatosensory region associated with the tongue.

- The sense of smell is conveyed by olfactory receptors in the nose. These receptors are innervated by mitral cells whose axons constitute the olfactory tract. This pathway provides inputs to the limbic system and thalamus. Information reaching the thalamus then is transmitted to the frontal cortex.

- The sense of smell is evoked when odor-producing molecules interact with the cilia of the olfactory receptors. Humans can detect camphoraceous, musk, floral, peppermint, ethereal, pungent, and putrid odors. In humans, the sense of smell is poorly developed in comparison to other animals.

Review Questions

Choose the Correct Answer

1. Which one of the following factors applies to myopia, or nearsightedness?
   a. Light focuses behind the fovea.
   b. Myopia is most often caused by a deficient lens.
   c. The myopic eye has more cone cells in the fovea.
   d. The problem can be corrected with the proper biconcave lens.
   e. The eyeball of the myopic patient is identical in shape to that of an emmetropic (normal-vision) eye.

2. A complete lesion of the right optic nerve produces which one of the following visual field defects?
   a. Loss of the left temporal field only
   b. Loss of the right temporal field only
   c. Loss of the left and right nasal fields
   d. Loss of the left and right temporal fields
   e. Loss of both the right nasal and right temporal fields

3. In which component of the cochlea do sounds cause waves having the peak amplitude of a specific frequency (tone) at a specific location?
   a. Oval window
   b. Organ of Corti

   c. Basilar membrane
   d. Reissner's membrane
   e. Scala media (cochlear duct)

4. Which structure in the auditory pathway receives information about both sound delay and sound intensity?
   a. Inferior colliculus
   b. Superior colliculus
   c. Lateral superior olive
   d. Medial superior olive
   e. Anterior lateral lemniscal nucleus

5. Which one of the following events results when a photon of light interacts with rhodopsin in a rod cell?
   a. Phosphodiesterase is inhibited.
   b. 5'-GMP is converted to cyclic GMP.
   c. Sodium channels open and the cell depolarizes.
   d. Transducin is activated and activates phosphodiesterase.
   e. Potassium channels are blocked, prolonging depolarization of the cell.

6. Which one of the following mechanisms is an example of frequency coding in the nervous system?
   a. The location of displacement of the vestibular membrane in the cochlea represents a specific sound tone.

b. Overlapping annular receptive fields in the secondary somatosensory cortex respond maximally to rectangular tactile stimuli.

c. An increase in pressure on a Pacinian corpuscle results in an increased rate of action potential generation in its sensory nerve.

d. The release of neurotransmitter from a greater number of vesicles in a nerve terminal results in a greater postsynaptic potential amplitude.

e. The activation of an inhibitory interneuron in the spinal gray matter prevents a motor neuron from being brought to its threshold level.

7. Which one of the following patterns describes the behavior of a single opponent cell in the lateral geniculate body of the thalamus?
   a. Excited by green in center and periphery
   b. Inhibited by green in center and periphery
   c. Excited by green in center; inhibited by red in periphery
   d. Excited by green in center; inhibited by green in periphery
   e. Excited by green in center; inhibited by red in center; excited by green in periphery; inhibited by red in periphery

8. Constant exposure to loud, high-frequency noise should result in the loss of function of hair cells in and/or near which one of the following regions?
   a. Sacculus
   b. Utriculus
   c. Base of the cochlea
   d. Apex of the cochlea
   e. Ampulla of the semicircular canal

9. What type of receptor senses the position of the joints?
   a. Nociceptor
   b. Proprioceptor
   c. Ruffini ending
   d. Mechanoreceptor
   e. Pacinian corpuscle

10. What is the location of the cell bodies of sensory neurons that innervate somatosensory receptors in most body regions below the neck?
    a. Dorsal root ganglion
    b. Sympathetic ganglion
    c. Dorsal column white matter
    d. Dorsal horn of the spinal cord
    e. Ventral horn of the spinal cord

11. Light entering the eye encounters elements of the visual system in what order, beginning from outside the eye?
    a. Cornea, lens, ganglion cells, photoreceptors
    b. Cornea, lens, photoreceptors, ganglion cells
    c. Lens, photoreceptors, cornea, ganglion cells
    d. Lens, cornea, ganglion cells, photoreceptors
    e. Lens, cornea, photoreceptors, ganglion cells

12. Sensory information from which one of the following regions is represented on the most medial surface of the primary somatosensory cortex?
    a. Face
    b. Lips
    c. Foot
    d. Tongue
    e. Fingers

13. The optic nerve is composed of axons from which one of the following retinal cell types?
    a. Bipolar
    b. Ganglion
    c. Amacrine
    d. Horizontal
    e. Photoreceptor

14. Which of the following pairs of structures are both considered otolithic organs?
    a. Utricle and saccule
    b. Utricle and ampulla
    c. Ampulla and semicircular canals
    d. Cochlear duct and vestibular membrane
    e. Organ of Corti and Reissner's membrane

15. Which one of the following structures is not considered part of the olfactory tract?
    a. Thalamus
    b. Mitral cell
    c. Limbic cortex
    d. Gustatory nucleus
    e. Orbitofrontal cortex

16. Which one of the following taste sensations is mediated by Na^+ passing through cation channels?
    a. Salt
    b. Sour
    c. Sweet
    d. Bitter
    e. Umami

17. Which sensation(s) is/are carried to the brain via the dorsal column pathway of the spinal cord?
    a. Pain only
    b. Touch only
    c. Heat and cold only
    d. Pain, heat, and cold
    e. Touch and pressure

18. In the olfactory system, the mitral cells are located within which one of the following structures?
    a. Thalamus
    b. Olfactory bulb
    c. Gustatory nucleus
    d. Orbitofrontal cortex
    e. Olfactory mucosal membrane

19. When the head rotates clockwise (viewed from above with the nose starting at the 12-o'clock position), which one of the following mechanisms informs the brain about the movement?
    a. Statoliths (otoconia) apply pressure to hair cells.
    b. Hair cells are pushed against Reissner's membrane.
    c. Cilia in the ampullae bend in the direction opposite the head rotation.
    d. Fluid moving in the semicircular canals directly stimulates the vestibular nerve.
    e. Stereocilia in the saccule bend toward the kinocilium, hyperpolarizing the hair cell.

20. Which one of the following taste sensations results from a compound releasing hydrogen ions (H^+)?
    a. Sour
    b. Sweet
    c. Bitter
    d. Salty
    e. None of the above

Answers to Case History Questions

1. An attack of Meniere's disease often is accompanied by a loud ringing or roaring sensation from the ears, making it difficult for the ears to respond normally to speech.
2. Vertigo is a sensation of spinning or dizziness resulting in an inability to maintain normal balance in a standing or seated position. The vertigo of Meniere's disease is produced by abnormal stimulation of balance receptors in the vestibular apparatus subsequent to an increase in fluid pressure within the inner ear.

Key Terms

auditory system (p. 288)
bipolar cell (p. 275)
cochlea (p. 288)
cupula (p. 298)
dermatome (p. 264)
dorsal root ganglion (p. 263)
gustatory system (p. 301)

hair receptor (p. 261)
inner ear (p. 288)
macula (p. 298)
mechanoreceptor (p. 261)
middle ear (p. 288)
nociceptor (p. 262)
olfactory system (p. 303)

outer ear (p. 288)
photoreceptor (p. 275)
proprioceptor (p. 262)
retina (p. 275)
rhodopsin (p. 277)
saccule (p. 297)
semicircular canals (p. 297)

sensory system (p. 254)
somatosensory pathways (p. 263)
thermal receptor (p. 261)
utricle (p. 297)
vestibular system (p. 297)
visual field (p. 288)
visual system (p. 273)

Suggested Readings

Hubel, D. H., and Wiesel, T. N. "Receptive fields, binocular interaction and functional architecture in the cat's visual cortex." *Journal of Physiology*, 160:106–154, 1962.

Hubel, D. H., Wiesel, T. N., and LeVay, S. "Plasticity of ocular dominance columns in monkey striate cortex." *Philosophical Transactions of the Royal Society of London*, 278:377–409, 1977.

Hudspeth, A. J. "Transduction and tuning by vertebrate hair cells." *Trends in Neuroscience*, 6:366–369, 1983.

Mogdans, J., and Knudsen, E. I. "Representation of interaural level difference in the VLVp, the first site of binaural comparison in the barn owl's auditory system." *Hearing Research*, 74:148–164, 1994.

Moulton, D. G. "Spatial patterning of response to odors in the peripheral olfactory system. *Physiological Reviews*, 56:578–593, 1976.

Mountcastle, V. B. "Modality and topographic properties of single neurons of cat's somatic sensory cortex." *Journal of Neurophysiology*, 20:408–434, 1957.

Penfield, W., and Rasmussen, T. *The Cerebral Cortex of Man: A Clinical Study of Localization of Function*. New York, Macmillan, 1950.

Shatz, C. J., "Impulse activity and the patterning of connections during CNS development." *Neuron*, 5:745–756, 1990.

Stewart, W. B., Knauer, J. S., and Shepherd, G. M. "Functional organization of rat olfactory bulb analyzed by the 2-deoxyglucose method." *Journal of Comparative Neurology*, 185:715–734, 1979.

Takahashi, T. T., and Keller, C. H. "Representation of multiple sound sources in the owl's auditory space map." *Journal of Neuroscience*, 14:4780–4793, 1994.

von Bekesy, G. *Experiments in Hearing*. New York, McGraw-Hill, 1960.

Answers to Review Questions

1. d	**2.** e	**3.** c	**4.** a	**5.** d	**6.** c	**7.** c	**8.** c
9. b	**10.** a	**11.** a	**12.** c	**13.** b	**14.** a	**15.** d	
16. a	**17.** e	**18.** b	**19.** c	**20.** a			

Chapter 9

MOTOR SYSTEMS

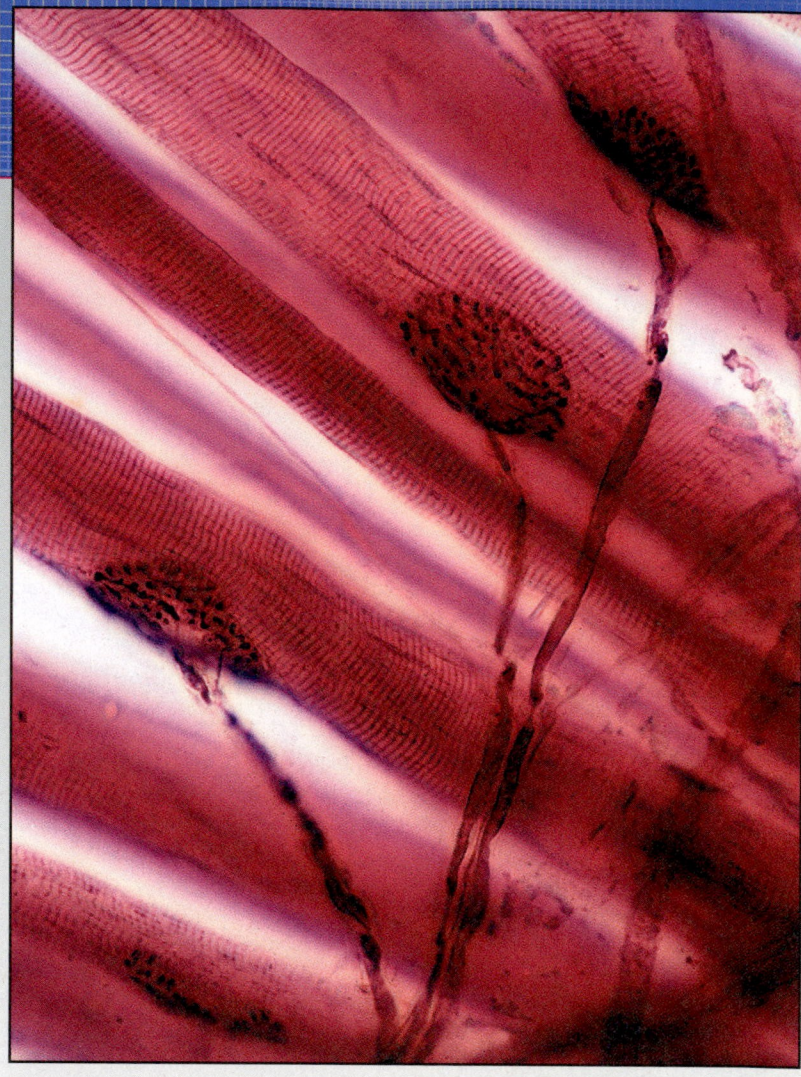

• *Photomicrograph of a nerve fiber innervating a skeletal muscle.*
(© Ed Reschke/Peter Arnold, Inc.)

KEY CONCEPTS

- The motor system conveys information to skeletal muscles that allows us to interact with the external environment. The functions are carried out by the activation of selective muscle groups.

- The spinal cord contains the neural circuitry needed for reflex activity and locomotion. The interaction of neurons within the spinal cord coordinates the activation of synergistic and antagonistic muscle groups to carry out intended movements, as well as reflexive movements initiated by sensory receptors.

- Conscious initiation of movement results from the flow of information from the primary motor cortex to the spinal cord. Inputs from the brain stem and cerebellum to the spinal cord

help us to maintain our balance and posture. Inputs from the motor cortex are responsible for the control of more complex movements.

- Specific sensory systems provide input to the motor system. Sensory receptors in muscles monitor and regulate muscle force, length, and velocity of shortening. These sensorimotor systems are needed for making appropriate movements and necessary adjustments in movement.

- Damage to the motor system produces movement disorders. These deficits can be manifested in the form of paralysis, uncontrolled movements, and a lack of coordination.

CASE HISTORY

The spectacular, inverting-can-opener-with-a-twist dive was designed to soak the maximum possible number of backyard, poolside guests at Jim's girlfriend's 16th birthday party. The dive was executed with the perfection attainable only by a totally motivated, strong, lanky teenager exploding off a cantilevered, spring diving board. Everyone thought he was joking when he floated limply to the surface and remained face down and motionless. Luckily, two guests who had had water safety and lifesaving training instantly recognized the problem and jumped to his aid, shouting for others to call 911. Jim was conscious and able to breathe but could not feel or move his arms or legs. The rescuers kept him in the warm pool, using head splint and hip- and shoulder-support techniques to keep his body in alignment to prevent further damage and awaited the emergency medical service. Thus began Jim's new life of struggles, first for life itself, and later to regain some ability to interact with the external world. His head had impacted the wall of the pool at a velocity sufficient to both fracture cervical vertebrae and rupture intervertebral discs. Movement of bone fragments into the spinal canal damaged the spine, eliminating sensation and voluntary movement of parts of his body below the injury. In this case, control of the legs and most muscles of the arms were lost, a condition called *quadriplegia*.

In the recent past, few victims of devastating spinal cord injuries survived. Now, because of advances in life-saving technologies, including life support, communication, transportation, and early medical intervention, many victims survive their injuries. Also, innovative developments in the rehabilitation sciences, including motorized wheelchair design and evolution of computer-based assistive devices, are helping victims of spinal cord injuries to achieve independence and become productive members of society. There is tremendous effort in the research community toward finding means of repairing the damaged spinal cord. Inducing nerve regeneration, fusing disrupted axons, and implanting stem cells that might replace lost neural elements are among the projects. Nevertheless, the motor nervous system that gives us control of much of our environment is complex and remains vulnerable to injuries that are not yet repairable.

Question

Why did the rescuers decide to keep Jim in the pool until professional help arrived?

BASIC PRINCIPLES OF THE MOTOR SYSTEM

 What general concepts apply to the motor system?

All of the information gathered from the external environment by sensory systems is of little use to humans if it cannot be used to organize and create behavior. The sensory systems take physical energy and transform it into neural signals. Conversely, motor systems transform neural signals into muscle contractions that produce movements. Motor responses range from automatic, involuntary **reflex actions,** such as a knee jerk, to complex, **voluntary** activities, such as those needed to play a piano. The **final common pathway** of all these actions is the **alpha motor neuron,** which originates in the ventral horn of the spinal cord. These neurons exit through the ventral root and innervate skeletal muscle. Activation of these neurons cause contraction of skeletal muscles, which results in either movement or tension development, stabilizing portions of the body.

Many reflex responses are controlled in the spinal cord, whereas more complex motor behaviors are organized in other higher motor areas of the central nervous system (CNS). The range of complexity of motor behaviors reflects the **hierarchical** (or **serial**) **organization** of the motor system (Fig. 9–1). The spinal cord is the lowest level of the hierarchy; it is followed in order by the **brain stem, motor cortex (Brodmann area 4),** and the **premotor and sup-**

plemental areas of the cortex (Brodmann area 6). Other noncortical areas that play important roles in motor function are the **cerebellum** and **basal ganglia** (Fig. 9–2). Along with the premotor and supplemental areas, the cerebellum and basal ganglia are used in planning, refining, and coordinating the signals for movement before they are sent to the skeletal muscles.

Most movement results from either sensory input resulting in **reflex actions** or from **volitional action** of the cerebral cortex. Signals to the spinal cord may directly stimulate alpha motor neurons or may contact interneurons that can excite or inhibit the alpha motor neuron (see Fig. 9–2). However, once an alpha motor neuron is stimulated and the signal leaves the CNS, there is no further opportunity to modify the signal, and a skeletal muscle is excited.

In the hierarchical organization of the motor system, the highest motor levels influence the levels below them. This arrangement connects different areas of the CNS together serially in a descending fashion. For instance, the premotor area has input to the motor cortex; the motor cortex, in turn, provides input to the brain stem; and the brain stem provides input to the spinal cord. However, some reciprocal connections exist between levels. Furthermore, motor divisions of cranial nerves that innervate skeletal muscles of the head, neck, and eyes do not enter the spinal cord. The cell bodies of these motor neurons are located in the cranial nerve motor nuclei in the brain stem. The motor system also has a **parallel organization.** For example, many fibers from the primary motor cortex and the brain stem directly synapse on spinal nerves. In this

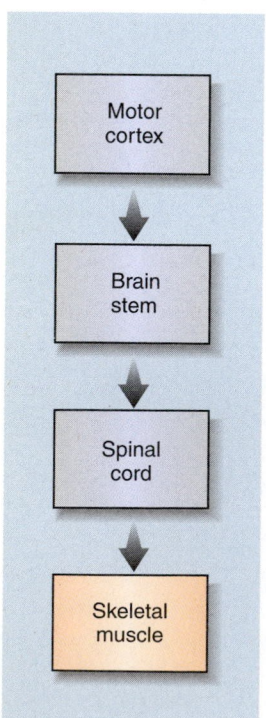

(a)

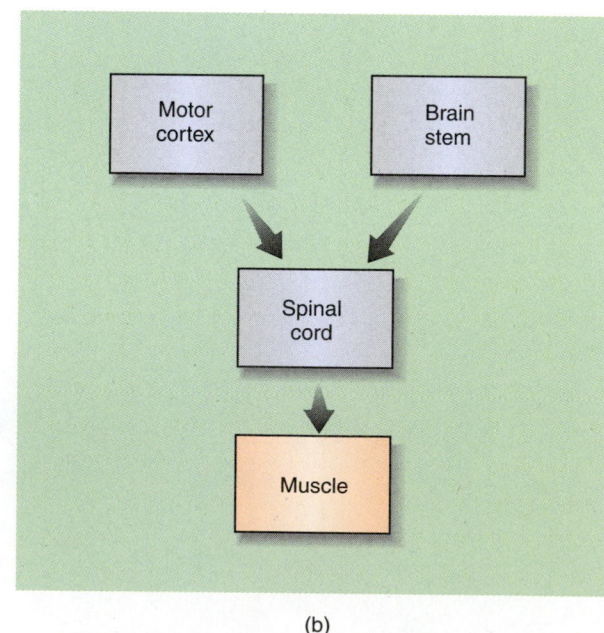

(b)

Figure 9–1

The motor system is organized in both *(a)* hierarchical and *(b)* parallel fashions.

Figure 9–2

Organization of the motor system showing flow of information among major components.   (From *Brown and Stubbs,* Medical Physiology, *John Wiley & Sons, 1983,* adapted from *Eyzaguirre and Fidone,* Physiology of the Nervous System, *Year Book Medical, 1975)*

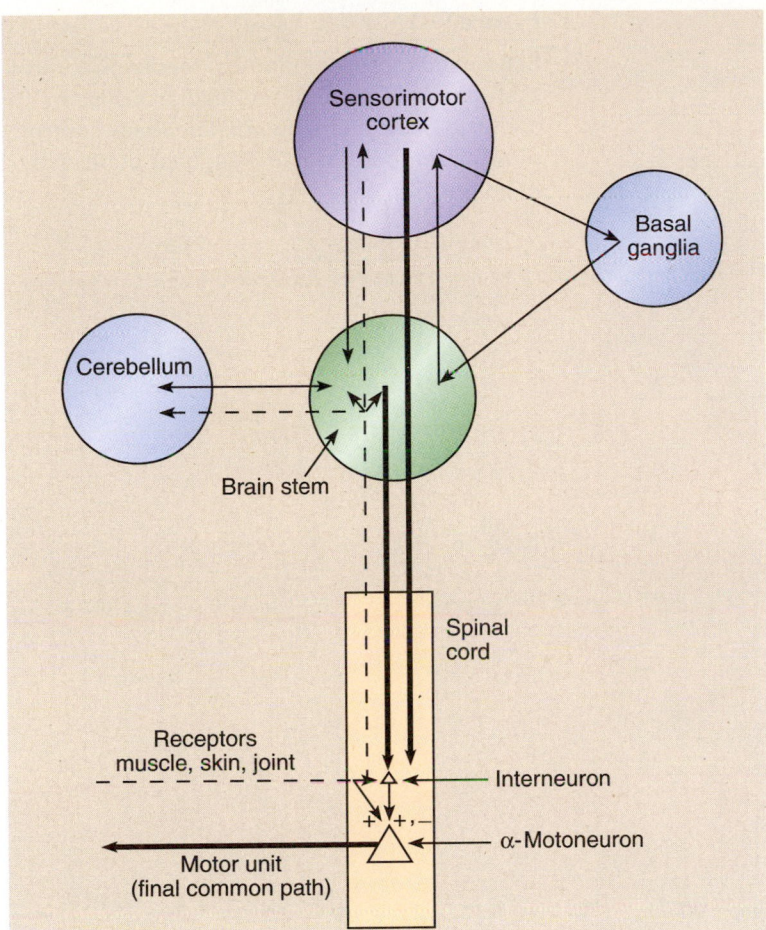

arrangement, information from these areas can simultaneously and independently affect the spinal cord (see Fig. 9–1).

In addition to the anatomical hierarchical and parallel nature of the motor nervous system, there is a recurrent theme in motor system control, as illustrated in Figure 9–3. Note that in this example, a **controlled system** (room temperature or hand position), is examined by a **sensor** (thermometer or proprioceptor) that sends information *back* to a **comparator** (thermostat or sensorimotor cortex). The comparator is a device that compares the sensor's information with a desired **reference** or **set point** (desired room temperature set on a thermostat or desired position of the hand). If there is a difference between the reference point and the sensor, a **difference signal,** sometimes termed an *error signal,* is sent to an **effector** (furnace or skeletal muscles) that can alter the controlled system toward the desired position. The cycle repeats until the controlled system is at the desired level, and no difference exists between the sensor reading and the reference. This type of control is called **negative feedback.**

Negative feedback control works well for control of relatively slow activities, such as maintenance of posture or slow, deliberate placements of the hand. However, such control systems are not rapid enough for rapid and complex movements of the hands, fingers, and arm, such as those required to play a fast, complicated violin piece; throw a baseball; or block a hockey puck approaching at 100 miles per hour. Furthermore, for certain desired outcomes, such as shooting a basketball goal, all of the coordinated muscle effort must be applied at the initial stage; there is no opportunity for control after the ball leaves the hands. These situations require the release of bursts of preprogrammed patterns of signals to appropriate muscle groups. These "packaged" messages are based on prior experience and early sensory input, such as visual information. Information about the present state of the body — such as its stance, balance, limb positions, and state of health — are also important factors in fast action control. Rehearsal and training refine the response programs. This type of activity is called **feedforward** control.

The effectors of both the feedback and feedforward control systems for moving the body are the skeletal muscles. The body contains three major types of muscle: (1) skeletal muscle; (2) **cardiac muscle** (the muscle of the heart); and (3) **smooth muscle,** which lines the blood vessels, airways, digestive tract, bladder, and reproductive organs. This chapter focuses on the areas of brain that control the **skeletal muscles** (also called **striated muscles** or **voluntary muscles**). These muscles are responsible for locomotion, fine movement, and the maintenance of balance and posture.

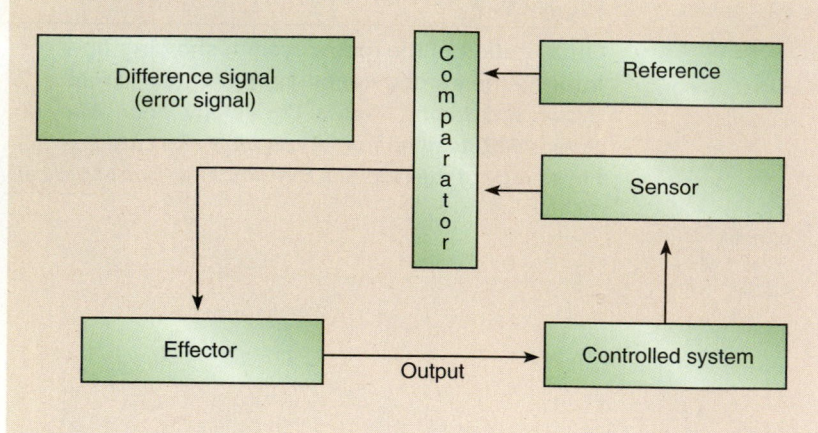

(a)

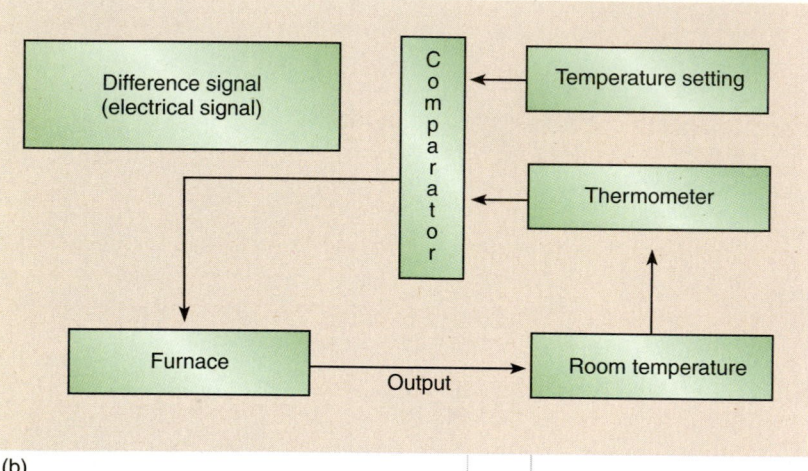

(b)

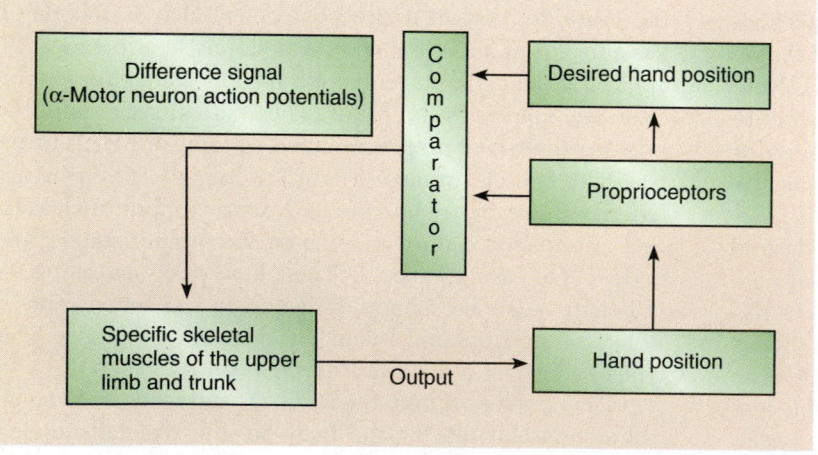

(c)

Figure 9–3

Schematic representations of feedback control systems. *(a)* A generalized feedback system showing basic elements. *(b)* Feedback control of room temperature. *(c)* Feedback control of hand position.

APPLICATIONS OF PHYSIOLOGY

"Brain Attack" and Stroke

Stroke can be defined as a local disruption of blood flow and, hence, oxygen delivery to the brain. This causes malfunction and eventual death of brain cells. This situation is analogous to the causes of heart attacks. Consequently, the term "brain attack" has come to be used as a term to describe stroke. Stroke results when brain blood vessels rupture or when they become clogged by a thrombus (i.e., blood clot). In addition, loss of blood flow to the brain during heart attacks can cause stroke. Approximately 500,000 new cases of stroke occur each year in the United States. It is the third leading cause of death in the United States, behind heart disease and cancer.

Patients who survive a stroke are often left with severe disabilities. The type of disability depends on which areas of the brain are affected. Often patients are paralyzed and unable to speak and communicate. Currently, the American Heart Association estimates that there are approximately 2.9 million survivors of stroke in the United States, many of whom are severely disabled.

Research has focused on developing ways to prevent the death of neurons within the brain following a stroke. Among the neural functions that are disrupted during stroke is the removal of excitatory amino acid neurotransmitters, such as glutamate. Abnormally high levels of extracellular glutamate, in turn, cause a prolonged excitation of postsynaptic neurons and an increase in the influx of calcium ions. The excessive accumulation of intracellular calcium has been shown to be a crucial step in the process of stroke-induced neuronal death.

A better understanding of stroke mechanisms has led researchers to develop new strategies to prevent stroke-induced cell death. These include the use of neuroprotective agents and hypothermia to reduce the metabolic demands of the brain. These approaches are undergoing clinical trials and may prove promising in reducing the extent of brain damage during a stroke.

The time in which these strategies are effective in preventing cell death, however, is soon after the onset of the stroke. Unfortunately, the warning signs of stroke often are not recognized by family and friends, resulting in considerable delay between when the stroke occurs and when treatment is initiated. Programs stressing public education and awareness of these warning signs is therefore crucial in reducing the morbidity and mortality rates of stroke and are currently underway in many areas of the United States.

A "brain attack" (i.e., stroke) is a medical emergency; the administration of thrombolytic agents within the first 3 hours of an ischemic stroke can reduce the damage and improve the outcome. If you have the following signs and symptoms, seek medical help immediately:

- Sudden weakness or numbness of the face, arm, or leg on one side of the body
- Sudden dimness or loss of vision, particularly in one eye
- Loss of speech, or trouble talking or understanding speech
- Sudden, severe headaches with no apparent cause
- Unexplained dizziness, unsteadiness or sudden falls, especially along with any of the previous symptoms

SKELETAL MUSCLE AND ITS CONTROL BY NERVES

General Structure of Skeletal Muscle

 How are form and function of skeletal muscle related?

A skeletal muscle bundle is made up of fascicles of **muscle cells,** which, because of their elongated shape, are called **muscle fibers.** Muscle fibers are **multinucleate;** one muscle fiber contains many nuclei. A **muscle** is composed of many muscle fibers arranged in parallel. Two types of muscle fibers make up a skeletal muscle: (1) **extrafusal fibers** and (2) **intrafusal fibers** (Fig. 9–4). Extrafusal fibers are the strongest and most common fibers and form the bulk of the muscle. Inside of the muscle are the tension-sensing, intrafusal fibers, arranged parallel to the extrafusal fibers. Extrafusal fibers are attached to tendons at either end of the muscle bundle. Muscle fibers shorten when activated, which results in contraction of the entire muscle and either movement or increased tension and thus stability in regions of the body.

Types and Actions of Skeletal Muscle

Most skeletal muscles are attached to bones at two points — (1) an **origin** and (2) an **insertion** — by strong connective tissue structures called **tendons** (see Fig. 9–4). Often, as skeletal muscles contract they move limbs by providing leverage on

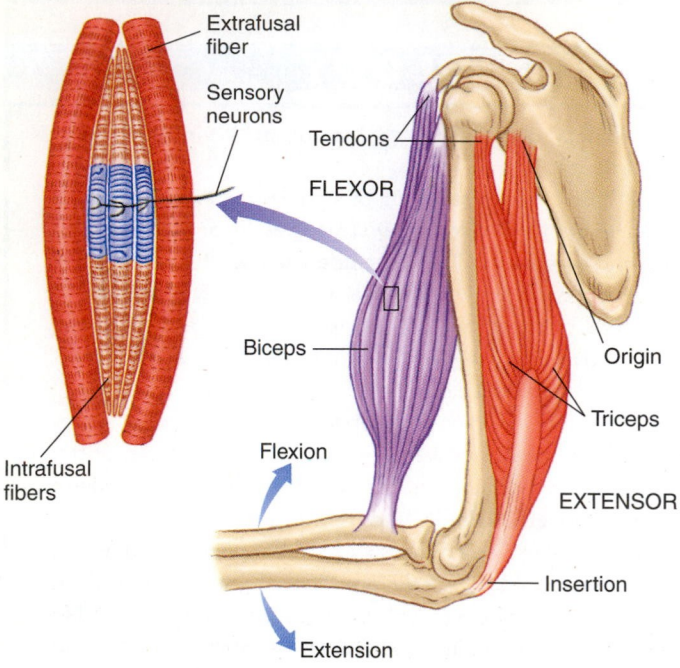

Figure 9–4

A skeletal muscle consists of extrafusal and intrafusal muscle fibers attached to tendons, which connect the muscle to a bone. Many skeletal muscles work in antagonistic pairs of flexors and extensors.

Antagonistic muscles are often arranged in parallel fashion across a joint where two bones meet. The muscle that, by contracting, causes the joint to open is called an **extensor,** whereas the muscle that, by contracting, causes the joint to close up is called a **flexor.** A flexor of a joint is said to be **synergistic** to another flexor for the same joint, whereas a flexor and an extensor for the same joint are said to be **antagonistic.** An example of an antagonistic pair producing flexion and extension are the biceps (flexor) and triceps (extensor) muscles, respectively. Skeletal muscles produce movements of the bones of the **axial skeleton,** consisting of the skull, backbone, and ribs. These muscles are essential in maintaining posture.

Skeletal muscles can be subcategorized according to their type of metabolism and their speeds of contraction. "Red" muscle produces slow, sustained contractions (called **slow twitch**). The red color of these muscles is due to their high myoglobin content, which is bright red when exposed to oxygen. Red muscle depends on aerobic (oxygen using) metabolism. These muscles are slow to fatigue (i.e., weaken with sustained stimulation) and are best suited for long-duration activities, such as maintaining posture. "White" muscle fibers have little myoglobin and produce fast, short-duration contractions (**fast twitch**) that fatigue rapidly. Such contractions are needed for rapid, finely controlled movements, such as those involved with movement of the eyes. Most muscles contain a mixture of units having these different characteristics and are classified by the predominant unit type.

bones that then cause joints to open or close. Skeletal muscles are usually arranged in **antagonistic** pairs or groups such that contraction of one muscle or muscle group causes lengthening of an opposing group of muscles. In fact, all movements are controlled by antagonistic muscles working in opposite ways.

The Motor Unit: The Myoneural Junctions of a Single Motor Neuron

The contraction of skeletal muscles necessary for movement is caused by firing of alpha motor neurons in the spinal cord (see Fig. 9–2). These neurons form synaptic connections with

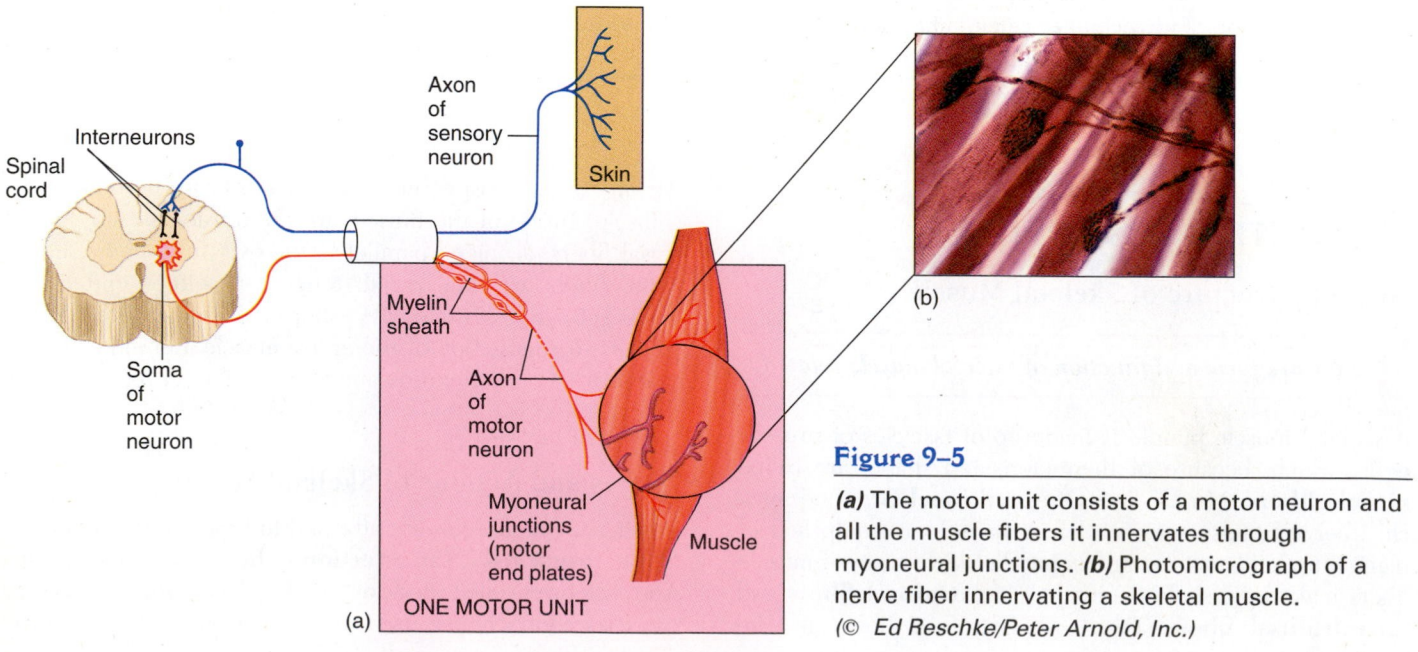

Figure 9–5

(a) The motor unit consists of a motor neuron and all the muscle fibers it innervates through myoneural junctions. *(b)* Photomicrograph of a nerve fiber innervating a skeletal muscle.
(© Ed Reschke/Peter Arnold, Inc.)

skeletal muscles at **neuromuscular junctions,** where the neurotransmitter **acetylcholine** is released from the nerve fiber terminal. Within the neuromuscular junction, acetylcholine binds to cholinergic **nicotinic receptors** that are located on the surface of the skeletal muscle membrane. As in the CNS, activation of the nicotinic receptor causes a depolarization of the membrane. In the case of skeletal muscles, this depolarization leads to a sequence of events resulting in the contraction of the muscle. Motor neurons that innervate fast skeletal muscles discharge action potentials with high frequencies and have rapid action potential conduction velocities. Motor neurons that innervate slow skeletal muscles have low rates of action potential discharge and exhibit slow conduction velocities.

One motor neuron can form synaptic connections with many muscle fibers. An individual motor neuron and all the muscle fibers that it innervates form a **motor unit** (Fig. 9–5). A small motor unit consists of a motor neuron that innervates few muscle fibers and is responsible for the precise control of very fine movements, such as contraction of a finger or darting movements of the eyes. A large motor unit consists of a motor neuron that innervates many muscle fibers. Such a unit is responsible for movements of a larger scale, such as contraction of the legs or the maintenance of posture. In the brain, the concept of motor unit size parallels that of the size of sensory receptive fields. Just as small receptive fields are associated with greater resolution of stimuli, small motor units are associated with greater precision of fine movement.

The Coding of Contractile Force

The amount of force that the muscle generates during contraction is altered by the nervous system in two ways. First, a motor neuron can control the tension developed by a single muscle fiber with the frequency of action potentials it generates (Fig. 9–6); the magnitude of sustained force increases with an increase in action potential frequency. Second, more motor neurons and, therefore, more motor units can be activated at one time. This activity increases the force generated by the muscle by recruiting more muscle fibers into the contraction and is called *motor unit recruitment.*

The Spinal Cord in the Motor Response System

 How is the anatomy of the spinal cord related to motor function?

The spinal cord plays a crucial role in the neuromuscular system. Automatic, or reflex, body movements are controlled solely by the spinal cord. Moreover, locomotion, or walking, can be produced by spinal cord neurons alone. Motor neurons are organized in the spinal cord in two ways: (1) the motor neurons in the dorsal area of the ven-

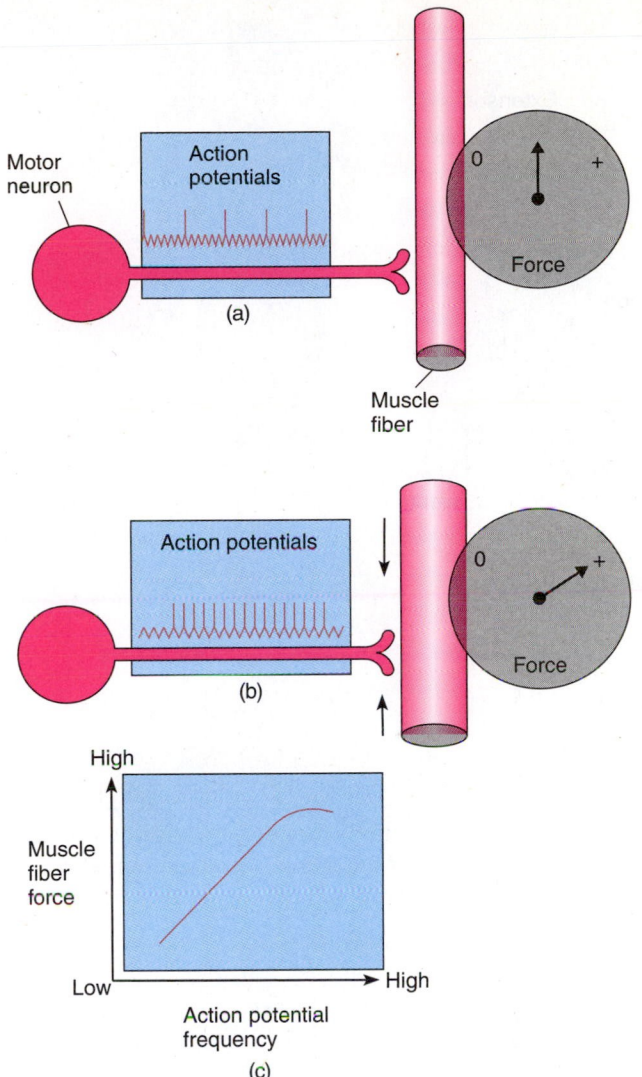

Figure 9–6

A motor neuron generates action potentials that cause a muscle fiber to contract and exert force. *(a)* Resting tension. *(b)* Increased motor neuron activation. *(c)* Plot showing change of muscle force with increasing action potential frequency.

tral horn are responsible for flexor movements, whereas cells in the ventral region are responsible for extensor movements (Fig. 9–7); and (2) those motor neurons in the dorsolateral region of the ventral horn innervate muscles in the extremities while those in the ventromedial region of the ventral horn innervate the axial muscles of the body to maintain posture.

Motor neurons that innervate a single muscle are functionally grouped in the spinal cord in **motor neuron pools.** Motor neurons within a single pool are located in several adjacent segments of the spinal cord (Fig. 9–8). The activation of a motor neuron pool thus coordinates the contractile action of the muscle.

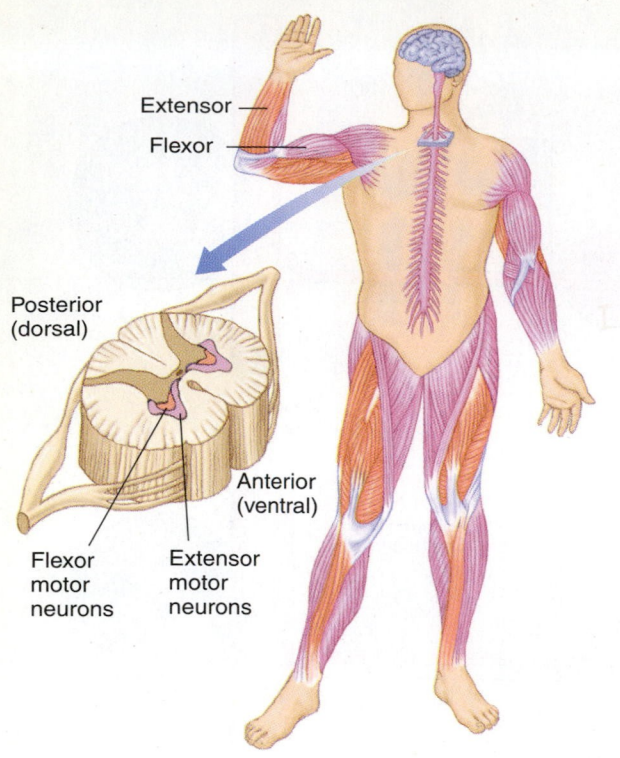

Extensor

Flexor

Posterior
(dorsal)

Anterior
(ventral)

Flexor
motor
neurons

Extensor
motor
neurons

Figure 9–7

Motor neurons that activate flexors originate in the dorsal area of the ventral horn of the spinal cord, whereas those that activate extensors originate in the ventral region of the ventral horn.

Interneurons of the Spinal Cord Coordinate the Response of Antagonistic Muscles

The interneurons of the spinal cord also play an important role in movement. These cells are located in the intermediate zone of the spinal cord. Those located in the lateral part of this zone have axons that synapse **ipsilaterally** (i.e., on the same side of the body) with motor neurons that innervate distal limb muscles. Interneurons lying closer to the midline have axons that synapse with motor neurons on both sides of the spinal cord that control muscles for posture. Interneurons form synapses with alpha motor neurons and mediate excitatory or inhibitory actions. The excitatory effects are associated with the activation of synergistic muscles, whereas the inhibitory effects are associated with suppressing the activation of antagonistic muscles.

Interneurons also send their axons up and down the spinal cord to form synaptic connections in regions several cord segments away from their cell bodies. They connect with motor neurons that control the contraction of several muscle groups. In this way, the activities of muscle groups are coordinated for performing functions such as maintaining posture and reaching movements.

Locomotor Generators in the Spinal Cord Control Walking Movements

The act of walking requires the coordinated contraction and relaxation of flexor and extensor muscles of the legs. This rhythmic alteration between flexion and extension is pro-

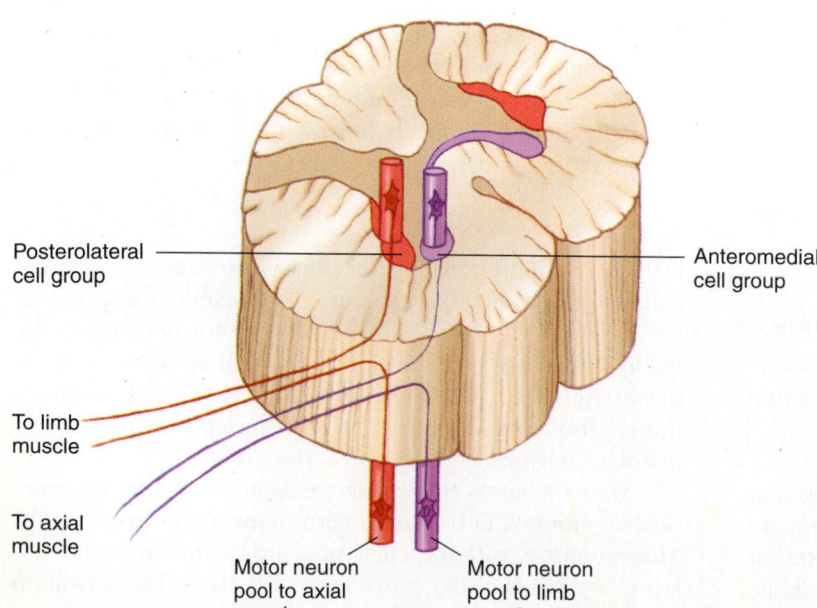

Posterolateral
cell group

Anteromedial
cell group

To limb
muscle

To axial
muscle

Motor neuron
pool to axial
muscle

Motor neuron
pool to limb
muscle

Figure 9–8

A motor neuron pool is a group of motor neurons that innervate a single muscle.

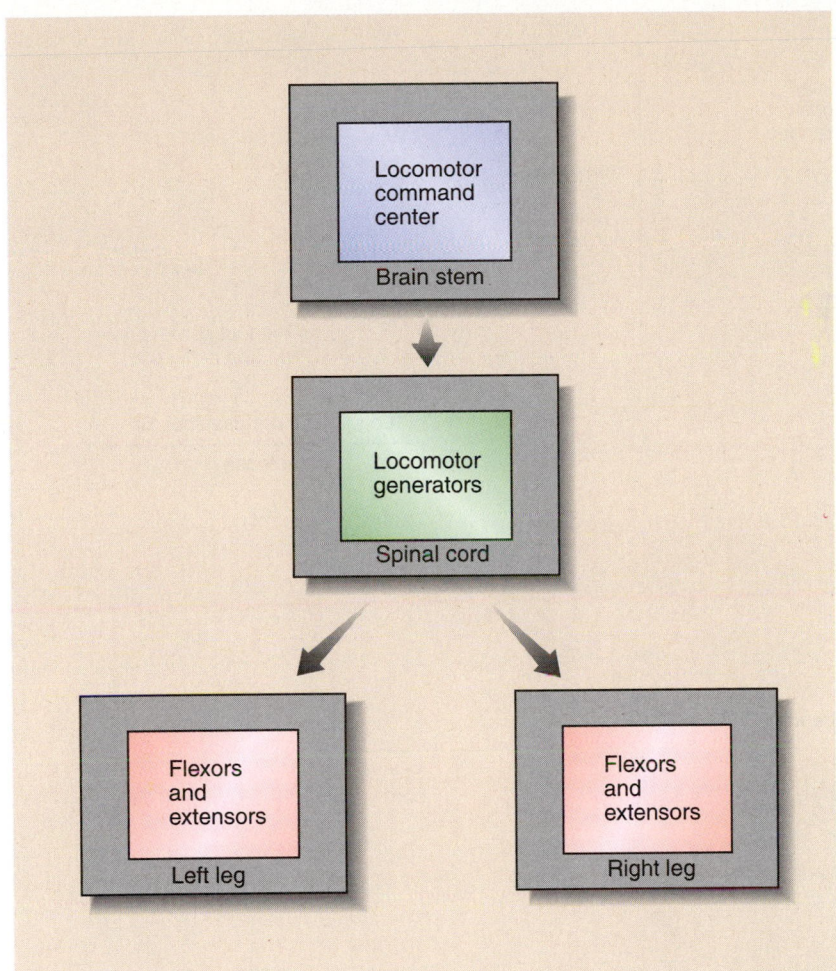

Figure 9–9

The locomotor command center and locomotor generators.

duced entirely by the interconnections of neurons in the spinal cord. These neurons and their interconnections compose the **locomotion generators** for walking. Animals whose spinal cords have been severed can walk without inputs from higher centers. This phenomenon occurs in chickens, for example, which remain capable of running for a short period after decapitation. Experimental animals with spinal cord transections are also capable of walking on treadmills. The speeds at which they walk are dictated by the speed of the treadmill. Furthermore, if one limb is prevented from walking, the other limb continues to walk. The locomotor generator for each limb, therefore, does not require activity from the other limb. However, when all limbs are active, the generators for limb movement are coupled to produce a coordinated response (Fig. 9–9).

The locomotor generators in the spinal cord are under the control of a "locomotor command center" in the brain stem. During the walking process in normal animals, the locomotor generators coordinate the flexor and extensor leg muscles to carry out the **stance phase** and the **swing phase**

of walking. In the stance phase, the leg is fully extended, with the foot on the ground, supporting the weight of the body. In the swing phase, the leg is flexed and the foot is off of the ground and swinging forward. Electrical stimulation of the command center causes animals with spinal cord lesions to walk on a treadmill. Weak stimulation causes walking, whereas stronger stimulation can cause the animal to run.

Sensory Fibers in Skeletal Muscle Convey Information About Muscle Length, Force, and Velocity of Shortening

 How are muscle position and force sensed?

Sensory neurons also have fiber processes within skeletal muscles. They transmit information about the state of contraction and tension in muscle fibers to the CNS so that it can revise its instructions as necessary to the motor

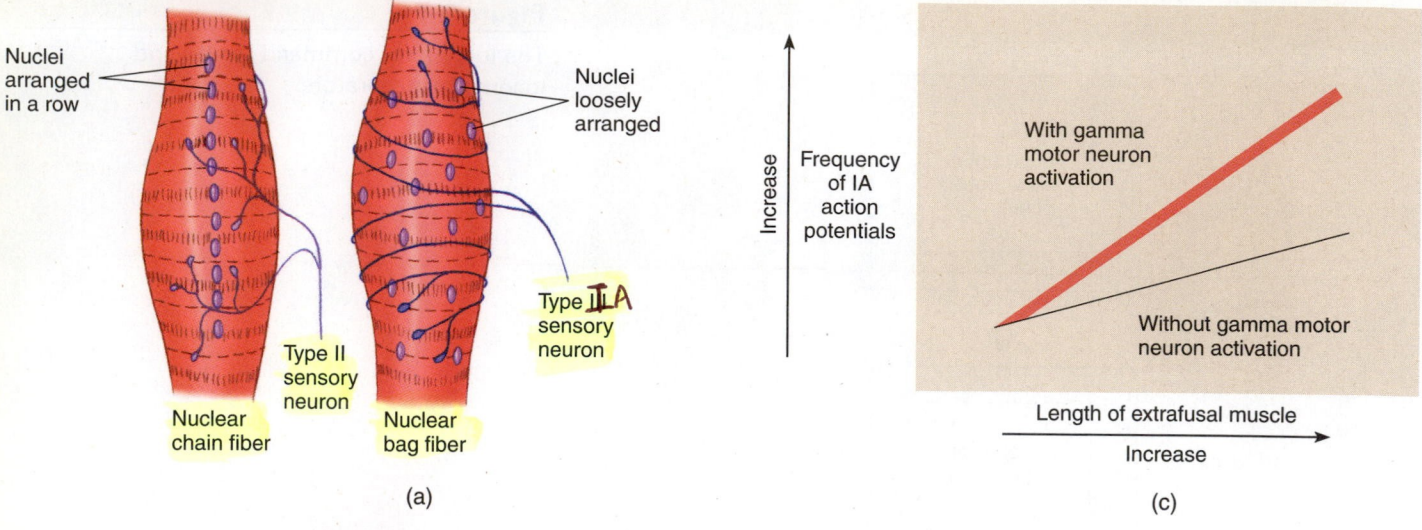

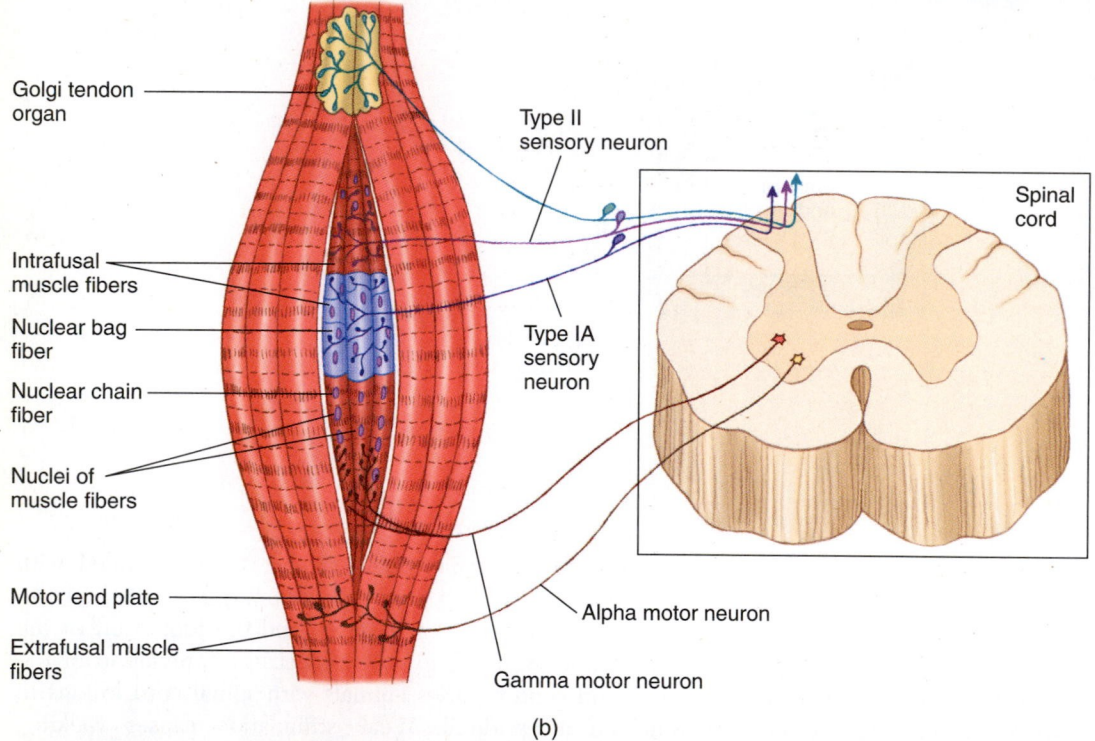

(a) Intrafusal muscle fibers consist of nuclear bag fibers and nuclear chain fibers. Labels: Nuclei arranged in a row, Nuclei loosely arranged, Type II sensory neuron, Type IA sensory neuron, Nuclear chain fiber, Nuclear bag fiber.

(b) Labels: Golgi tendon organ, Intrafusal muscle fibers, Nuclear bag fiber, Nuclear chain fiber, Nuclei of muscle fibers, Motor end plate, Extrafusal muscle fibers, Type II sensory neuron, Type IA sensory neuron, Gamma motor neuron, Alpha motor neuron, Spinal cord.

(c) Frequency of IA action potentials (Increase) vs Length of extrafusal muscle (Increase). With gamma motor neuron activation; Without gamma motor neuron activation.

Figure 9–10

(a) Intrafusal muscle fibers consist of nuclear bag fibers and nuclear chain fibers. *(b)* These fibers and the sensory neurons that innervate them make up a muscle spindle, a receptor for muscle length and state of contraction. Golgi tendon receptors innervate tendons and transmit information about the force exerted by the muscle. Gamma motor neurons activate intrafusal fibers, whereas alpha motor neurons activate extrafusal muscle fibers. *(c)* When activated by gamma motor neurons, Type IA sensory neurons generate action potentials with greater frequency than in the absence of gamma motor neuron activation in response to changes in extrafusal muscle length.

neurons producing the contraction or tension. Sensory neurons that innervate intrafusal muscle fibers form what is known as a *muscle spindle,* whereas sensory neurons that innervate tendons form *Golgi tendon organs.* If sensory feedback from muscle spindles and Golgi tendon organs is prevented from reaching the spinal cord during the stance phase, the step cycle can be stopped during the extension of the leg. In other phases of the walk cycle, however, the absence of sensory information does not inhibit the process of locomotion.

Sensory information can also change the activation of one motor response to another. The activation of tactile receptors on the top of the foot during the swing phase of walking, for example, results in a flexion response that is appropriate for stepping over an object. Sensory information to the spinal cord is therefore capable of changing ongoing motor programs and reflexes.

Length and Tension Receptors: The Muscle Spindle and Golgi Tendon Organs

The sensory receptors that detect the length of the muscle and its velocity of contraction are called **muscle spindles** (Fig. 9–10). These structures are formed by sensory neurons that entwine intrafusal muscle fibers. The muscle spindle is composed of two types of intrafusal fibers: (1) **nuclear bag fibers** and (2) **nuclear chain fibers** (see Fig. 9–10a). The nuclear bag fiber is innervated by **type Ia nerve fibers,** which transmit information about muscle length and velocity of contraction to the CNS. Nuclear chain fibers are innervated by **type II nerve fibers,** which transmit information about muscle length.

The muscle spindle acts as a **stretch receptor;** it increases its discharge of action potentials when the intrafusal fibers stretch and decreases its discharge when they shorten. Gamma motor neurons innervate contractile elements at the poles of the muscle spindles (see Fig. 9–10b). The activation of these contractile elements causes the central region of the muscle spindle to stretch as a rubber band would. Figure 9–10c shows the action potential activity of type Ia muscle spindle sensory nerves as a function of the length of the extrafusal muscle fiber. In the absence of gamma motor neuron activation, a small increase in the length of the extrafusal muscle fibers results in a small change in the frequency of action potentials generated by the sensory neurons. With gamma motor neuron activation, however, a small increase in the length of the extrafusal muscle fibers causes a great change in the frequency of generated action potentials. In this way, the activation of gamma motor neurons maintains the sensitivity of the muscle spindles even as the extrafusal muscle fibers shorten during contraction.

A second type of muscle receptor is the **Golgi tendon organ.** This type of receptor is located in **series** with the extrafusal muscle fiber (see Fig. 9–10b). In contrast to the muscle spindles, the Golgi tendon organ generates more action potentials as extrafusal muscle tension develops with contraction. These action potentials are carried to the spinal cord via type Ib nerves. Because the Golgi tendon organ and muscle spindle are somatosensory receptors, they transmit their information about muscle force, velocity, and length to the somatosensory cortex or to the spinal cord for reflex action.

Spinal Cord Reflexes Are Activated by Sensory Receptors

 How do spinal cord reflexes work?

Spinal cord reflexes represent the most basic of motor responses. These reflexes are carried out entirely within the spinal cord and are modified by inputs from higher brain centers to generate complex movements. Reflexes are also used to help diagnose disorders of the motor system.

Stretch Reflex

The **knee-jerk reflex** is an example of a **stretch reflex.** This reflex is activated by tapping the **patellar tendon** below the knee. This tendon then stretches the muscle spindles. Action potentials conducted along the muscle spindle sensory neurons enter the spinal cord via the dorsal roots (Fig. 9–11). The fiber from the muscle spindle branches after entering the spinal cord, with the ascending branch joining the dorsal column pathway. The other branch forms synaptic connections with motor neurons in the ventral horn. These motor neurons activate muscle fibers, causing the leg to extend quickly, or kick out. At the same time, antagonistic muscles that cause the leg to flex are inhibited by the activation of inhibitory interneurons within the ventral horn. By inhibiting flexors and activating extensors, the interconnections of nerve cells within the spinal cord produce well-coordinated motor responses.

Inverse Myotatic Reflex

The **inverse myotatic reflex** occurs when a person attempts to lift more weight than he or she can actually carry. Weightlifters, for example, often add increasing amounts of weight to barbells when lifting the weights in a maneuver called the "curl." In this maneuver, a person lifts the barbells by contracting the biceps muscles. If the weight is too great for the biceps to lift, however, they suddenly relax, and the barbells are dropped.

The inhibition of the biceps occurs because the Golgi tendon organ detects excessive force, which might tear or pull the tendon out of the bone insertion. With increasing force, the Golgi tendon organ begins to discharge action potentials (Fig. 9–12). These action potentials are transmitted to the dorsal column nuclei and subsequently to the somatosensory cortex. In the spinal cord, however, information

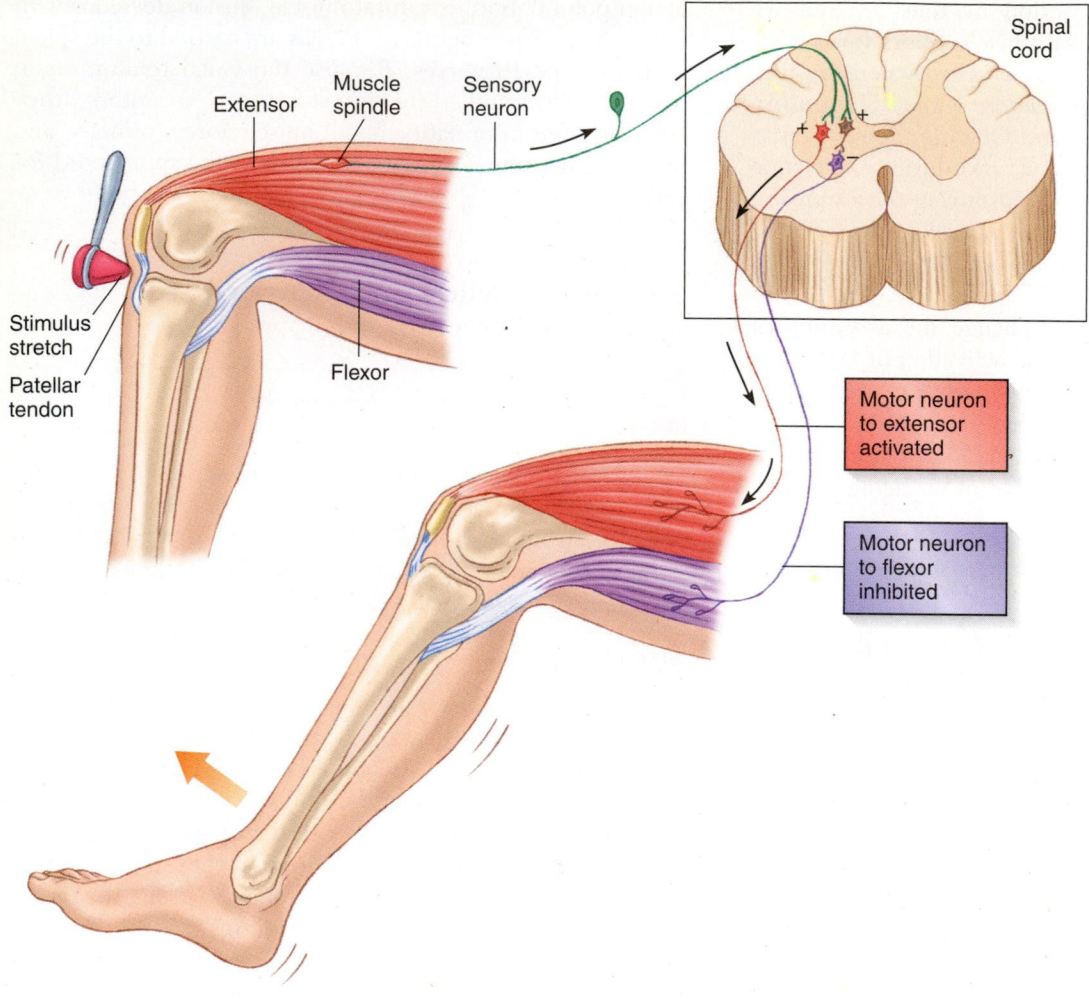

Extensor Muscle Sensory
spindle neuron

Stimulus
stretch

Patellar
tendon

Flexor

Spinal
cord

Motor neuron
to extensor
activated

Motor neuron
to flexor
inhibited

Figure 9–11

The stretch reflex occurs when, following a tap on the patellar tendon, muscle
spindles detect a stretch of the extensor muscle and inform the spinal cord, which
activates the leg extensor and inhibits the flexor, causing the knee joint to open up.

about excessive muscle tension is transmitted to inhibitory
interneurons that turn off the motor neuron innervating the
biceps. At the same time, interneurons turn off any muscles
(flexors) synergistic to the biceps and turn on any muscles
(extensors) that are antagonistic to the biceps, so that the arm
extends and the barbell drops.

Flexor Withdrawal Reflex

When you step on a tack, cutaneous pain receptors send
sensory information into the spinal cord, which results in
contraction of flexors needed to remove the foot from the
tack. This is called the **flexor withdrawal reflex** (Fig.
9–13). At the same time, extensors of the leg that would
normally keep the foot on the tack are inhibited. In order
to support the rest of the body, however, the extensor muscles in the other leg are activated, whereas its flexors are
inhibited.

THE BRAIN IN THE MOTOR RESPONSE SYSTEM

*What regions of the brain are involved in motor
function?*

Several important structures in the brain represent the highest control centers for the motor system (see Fig. 9–2). Various areas of the cerebral cortex organize and interpret
complex sensory input and program the responding movements. The brain stem controls spinal cord output in maintaining posture and balance of the body. The cerebellum and
the basal ganglia also plan, program, and coordinate specific
types of motor responses.

These components of the motor system interact with
each other to carry out specific functions. The desire to move

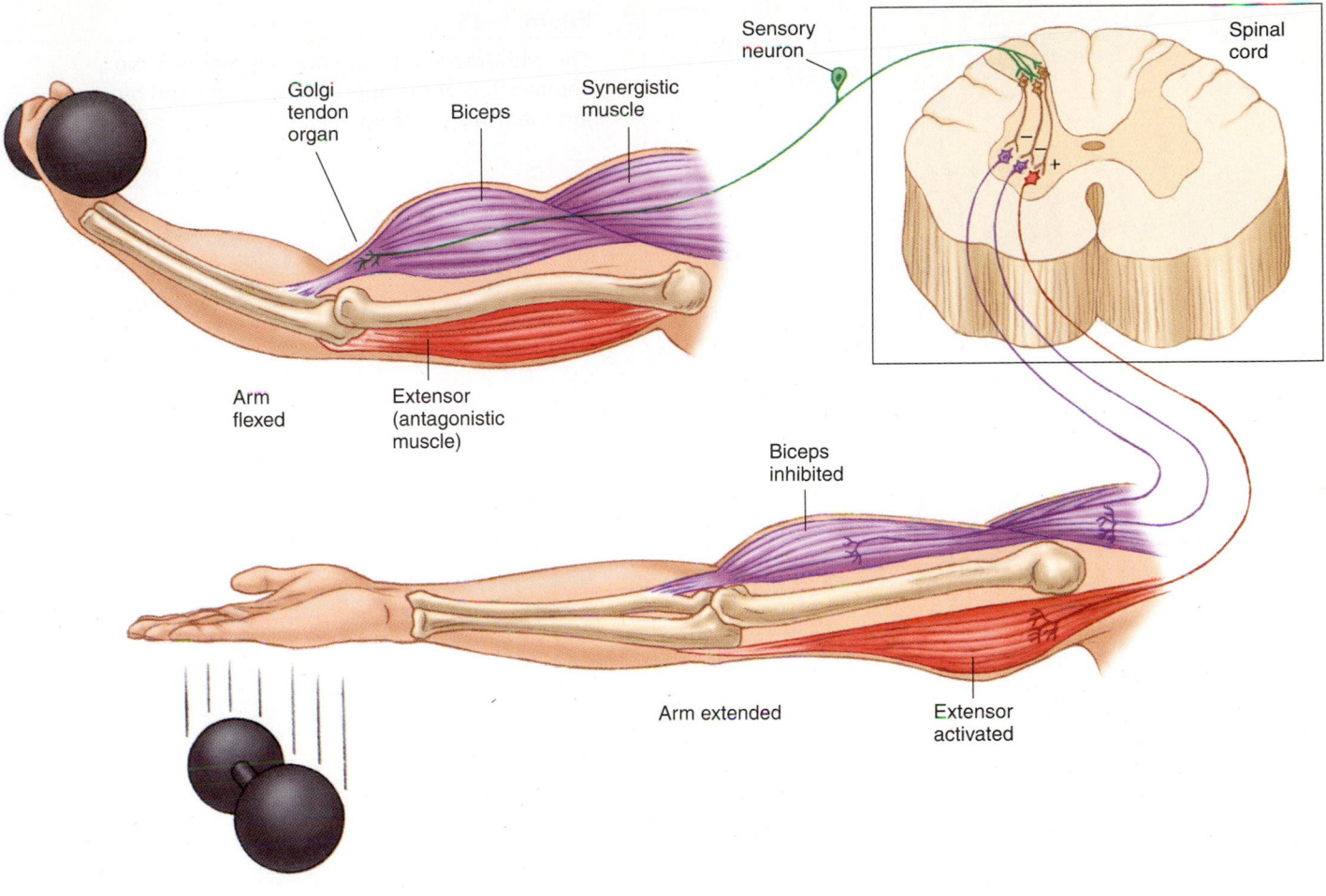

Figure 9–12

Detection of excessive weight by the Golgi tendon organ initiates the inverse myotatic reflex causing a flexed biceps muscle to relax and an extensor in the arm to contract so that the arm extends and drops a weight that might be harmful.

activates the supplemental motor cortex, basal ganglia, thalamus, and cerebellar components of the motor system. Information from these areas is transmitted to the premotor cortex, where movements are planned and programmed. From the premotor cortex, information about the planned movement is transmitted to the motor cortex. The motor program is then carried out when neurons in the motor cortex instruct those in the spinal cord to activate the skeletal muscles. The force and velocity of contraction and the length of muscle shortening are monitored by Golgi tendon organs and muscle spindles. This sensory information is fed back to the motor neurons at the spinal level and transmitted back to the motor cortex and cerebellum. Comparisons are made between the intended movement, as designed by the motor program, and the actual movement. Deviations from the intended movement are then corrected with the transmission of another set of instructions to the motor neurons in the spinal cord.

The Brain Stem Controls Posture

One of the primary roles of the brain stem is maintaining the body's posture and balance. Nerve cells within different clusters in the brain stem send axons that terminate in the spinal cord. Three pathways from the brain stem provide input to the motor neurons of the spinal cord: (1) the **ventromedial pathway,** (2) the **lateral reticulospinal pathway,** and (3) the **rubrospinal pathway.**

Medial Pathways

Three ventromedial pathways descend along the ventral and medial region of the spinal cord and impinge on motor neurons that control axial muscles (Fig. 9–14*a*). The first is the **vestibulospinal tract,** which originates in the vestibular nucleus and carries information for the reflex control of equilibrium (discussed in a later section). The second pathway is the **tectospinal tract,** which originates in the **tectum,** a structure

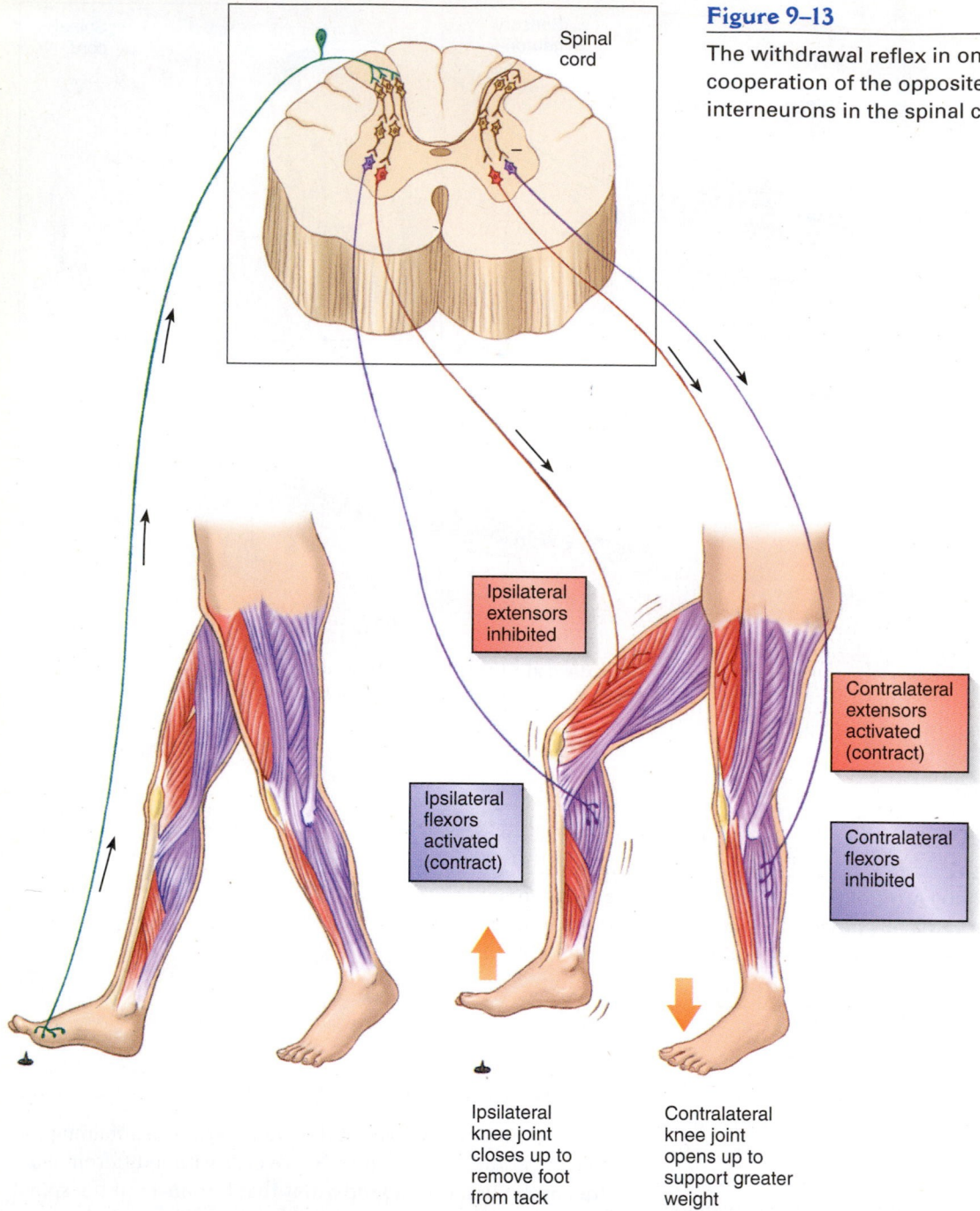

Spinal cord

Figure 9–13

The withdrawal reflex in one leg requires the cooperation of the opposite leg, arranged by interneurons in the spinal cord.

Ipsilateral extensors inhibited

Contralateral extensors activated (contract)

Ipsilateral flexors activated (contract)

Contralateral flexors inhibited

Ipsilateral knee joint closes up to remove foot from tack

Contralateral knee joint opens up to support greater weight

involved in the coordinated control of head and eye movements. The third pathway is the **medial reticulospinal tract,** which originates in the **reticular formation,** a structure involved in maintaining posture by the activation of extensor muscles.

Lateral Reticulospinal Tract

Nerve fibers in the **lateral reticulospinal tract** are derived from the lateral reticular nucleus and descend in the lateral

region of the spinal cord (Fig. 9–14*b*). These fibers innervate flexors in the control of posture.

Rubrospinal Tract

The rubrospinal tract has fibers that originate in the red nucleus of the brain stem. These fibers descend along the dorsal and lateral border of the cord to innervate motor neurons that control distal flexor muscles (see Fig. 9–14*c*). The effects of the major descending brain stem

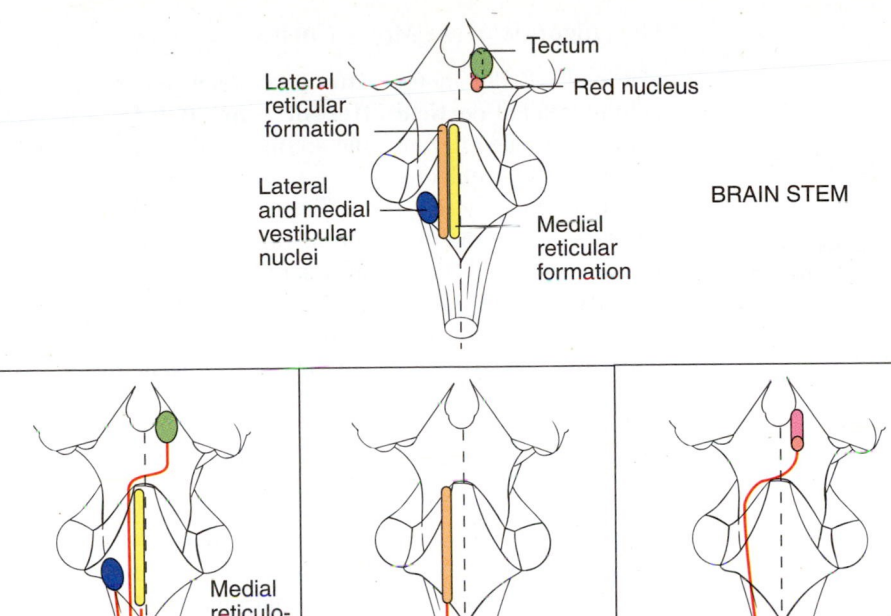

Figure 9–14

Pathways from the brain stem to the spinal cord: **(a)** the medial pathways, **(b)** the lateral reticulospinal tract, and **(c)** the rubrospinal tract.

pathways on flexor and extensor muscles in the maintenance of posture and equilibrium are summarized in Figure 9–15.

The Motor Cortex Controls Reaching and Fine Voluntary Movement

The control of reaching and fine voluntary movement is a result of instructions transmitted via descending pathways to the spinal cord from the **motor cortex.** The motor cortex occupies a cortical region rostral to the somatosensory cortex (Fig. 9–16). The cells in the motor cortex are organized in a somatotopic manner similar to that of the somatosensory cortex. The primary motor cortex region is the **precentral gyrus.** The **motor homunculus** illustrates the parts of the body receiving motor output from this region (Fig. 9–17). Cells in the medial region of the motor cortex cause the contraction of muscles in the leg. Nerve cells located in more lateral regions of the motor cortex activate the muscles of the torso, arm, hand, and face.

The parts of the body involved with fine movement, such as the fingers, occupy more space in the motor cortex

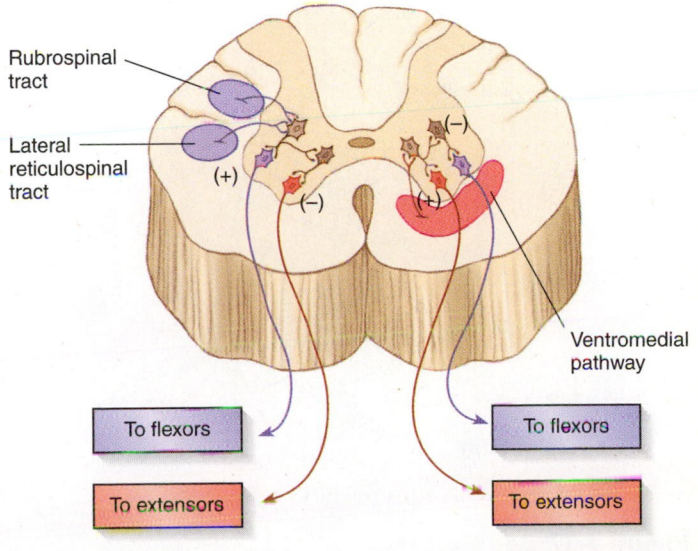

Figure 9–15

A summary of the effects of the brain stem on spinal motor neurons.

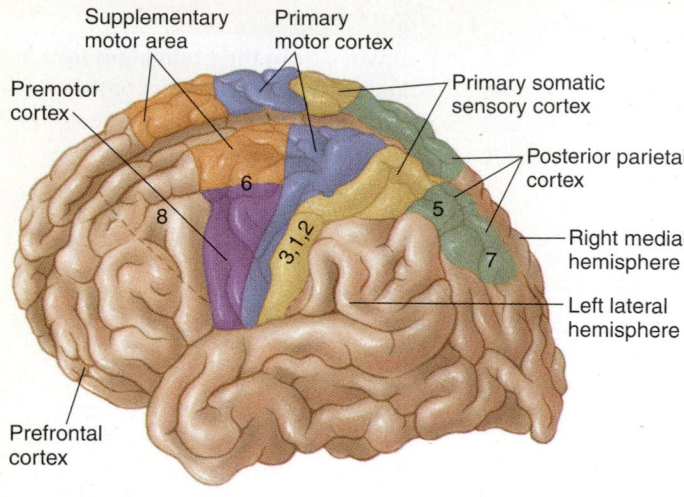

Figure 9–16

The functional areas of the cerebral cortex involved in motor control. Numbers indicate Brodmann's areas.

than parts of the body involved in gross movement, such as the torso. This relationship is in keeping with the general principle that the amount of cortical space devoted to different parts of the body is in proportion to each part's sensory sensitivity or degree of fine motor control.

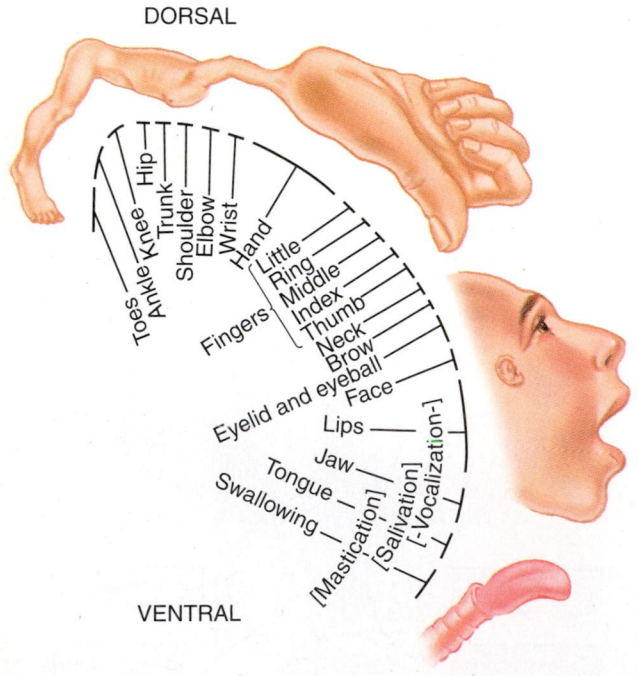

Figure 9–17

The somatotopic organization of the motor cortex. Areas innervated by efferents from the left motor cortex are illustrated and listed on a frontal (coronal) section of the left precentral gyrus.

Organization of the Motor Cortex

The cells of the motor cortex are organized in functional columns called **cortical efferent zones (CEZs).** All of the cells within one efferent zone are involved in the contraction of a given muscle. Neurons in the motor cortex are also organized horizontally into six different layers. Those located in layer V provide the output from the motor cortex. Pyramid-shaped nerve cells in this layer send axons to the spinal cord to synapse with spinal motor neurons and interneurons. Corticobulbar fibers arising in this area synapse on lower motor neurons not in the spinal cord. These cortical neurons are involved primarily in controlling distal muscles.

Descending Pathways from the Motor Cortex

Information from the motor cortex is transmitted to the spinal cord and brain stem by the **corticospinal pathway** and **corticobulbar pathway,** respectively. Inputs to the brain stem ultimately influence axial muscles near the midline for the maintenance of posture, whereas inputs to the spinal cord control the distal limb muscles. Approximately 30% of the corticospinal and corticobulbar fibers are from neurons in area 4; an additional 30% originate from the premotor cortex (area 6). The remaining fibers are derived from the somatosensory cortex.

As the corticospinal axons descend from the motor cortex, they form the **pyramidal tract.** When the pyramidal tract reaches the level of the brain stem, the majority of the fibers that continue to the spinal cord cross the midline of the body and continue their descent along the **lateral corticospinal tract** (Fig. 9–18). Nerve fibers exit from the lateral corticospinal tract at various levels of the spinal cord to innervate motor neuron pools that control distal limb muscles. The direct connections from motor cortex to spinal cord thus permit the independent control of individual muscles.

In addition to the lateral corticospinal tract, a minority of the fibers from the motor cortex descend to the spinal cord without crossing the midline of the body. These fibers form part of the **ventral corticospinal tract** and primarily innervate motor neurons in the medial region of the ventral horn associated with axial muscles of the body (see Fig. 9–18). Corticobulbar fibers travel to the motor nuclei of cranial nerves, providing motor innervation to muscles of the face and head.

Sensory Feedback to the Motor Cortex

Neurons in the motor cortex are informed of the consequences of movement through sensory feedback pathways. They receive input from either the muscles they project to or from areas of skin surrounding the muscle. This long loop of sensory feedback results in the alteration of information from the motor cortex to the spinal cord to correct any deviations from the intended movement. Sensory feedback to the motor cortex occurs by way of the somatosensory cortex (Fig. 9–19).

The nerve cells in the sensory cortex are connected to those in the motor cortex in a topographic manner. Cells in the sensory cortex receiving proprioceptive input from muscles in the thumb, for example, are connected with cells in

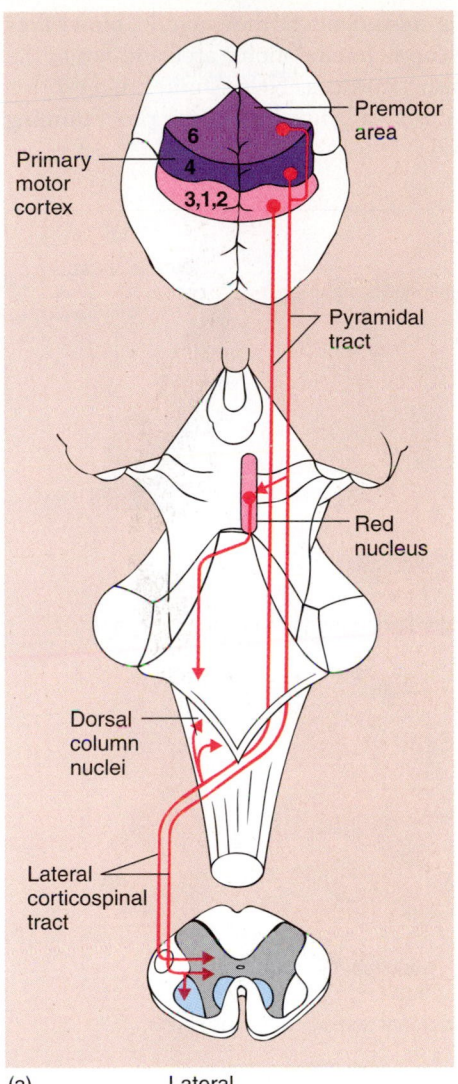

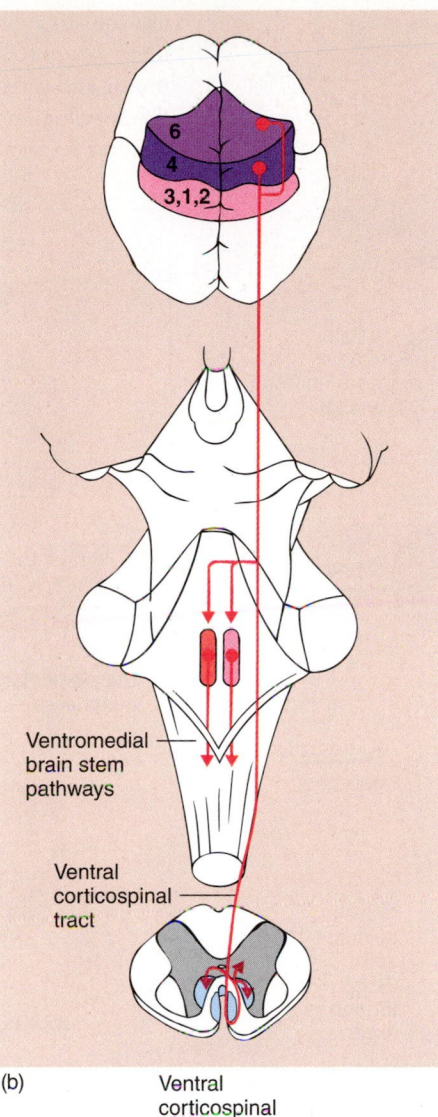

Figure 9–18

Descending pathways form the motor cortex: the *(a)* lateral and *(b)* ventral corticospinal tracts.

the cortical efferent zone responsible for the contraction of these same muscles. In this way, feedback is provided by the sensory system to inform the cells in the motor cortex whether the instructions they have transmitted for the muscular contraction have been faithfully executed. If not, then the sensory information that is sent back to the motor cortex causes the cells in the cortical efferent zone to modify their activity and thus modify the activity of the motor neurons.

The Supplemental and Premotor Areas Program Movement

The pyramidal cells in the motor cortex begin generating action potentials approximately 50 msec before a movement. Nerve cells in other areas of the brain involved in motor control become active much earlier than the motor cortex. These areas of the brain, called the **supplemental and premotor cortex, posterior parietal cortex, cerebellum,** and **basal ganglia,** are involved in the planning and programming of movement.

The motivational factors that cause specific movements most likely originate in subcortical areas of the brain, such as the **hypothalamus.** The motivational areas of the brain that respond to the need to accomplish a particular task transmit this information to the **supplemental and premotor cortex** (Brodmann area 6), which designs the necessary movements to accomplish the desired task. In the act of reaching for a glass of water, for example, the nervous system must determine which motor program will activate certain muscles, and when and how much they need to be contracted. The components of the motor program are thought to be developed by supplemental and premotor cortices (area 6).

The supplemental area has direct inputs to the motor cortex for the control of distal limb muscles and to motor neurons in the spinal cord for control of axial muscles involved with the maintenance of posture. Stimulation of the supplemental and premotor areas produces movements more complex than those produced by stimulation of the motor cortex. Activation of the premotor area causes

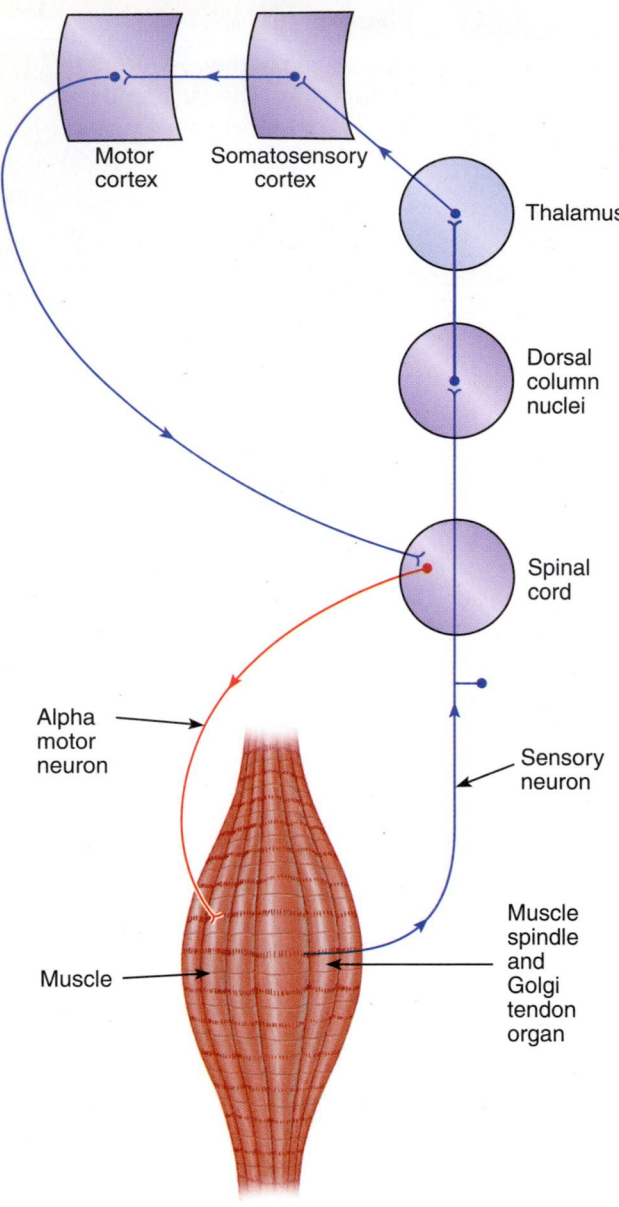

Figure 9–19

Sensory feedback pathways to the motor cortex.

are told to rehearse a movement in their minds, blood flow increases only in the supplemental motor area and not in the motor and somatosensory cortices. These studies suggest that the supplemental motor area is involved in the programming of complex movements.

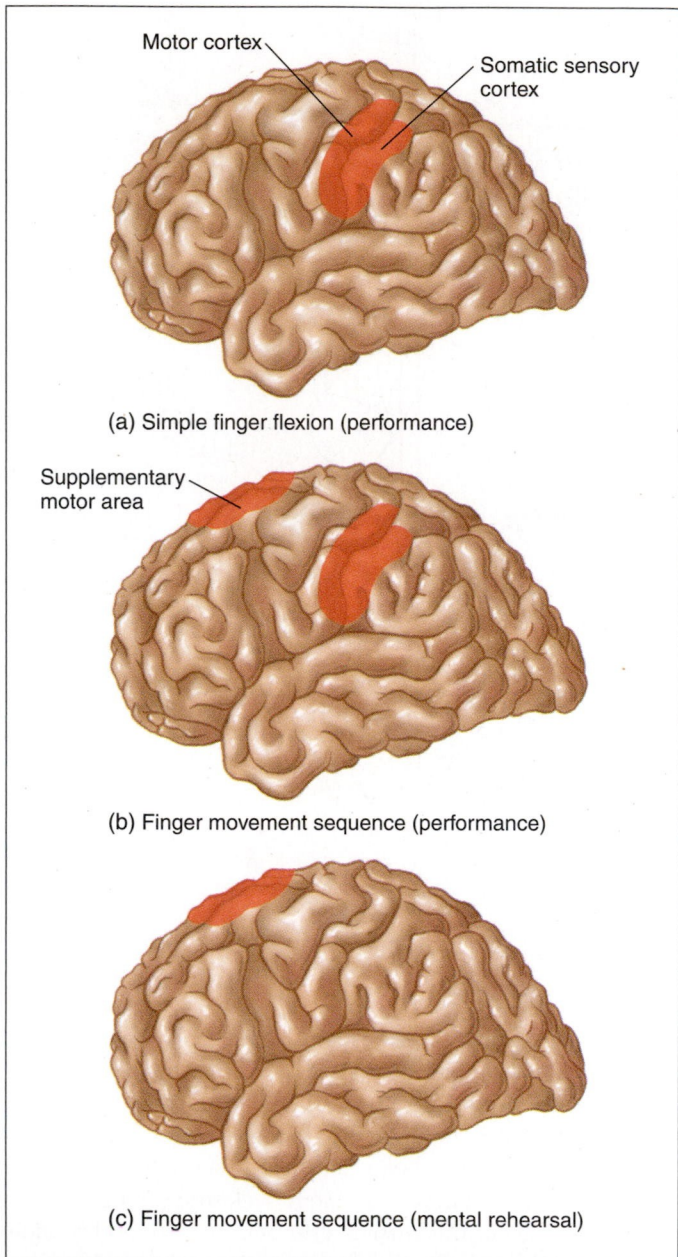

(a) Simple finger flexion (performance)

(b) Finger movement sequence (performance)

(c) Finger movement sequence (mental rehearsal)

Figure 9–20

The supplemental cortex is involved in the planning of complex movements. **(a)** During simple finger flexion, blood flow (indicated by red shading) increases in the motor and somatosensory cortices, whereas **(b)** during a complex sequence of finger movements, blood flow is also increased in the supplemental cortex. **(c)** When the same sequence is mentally rehearsed, blood flow increases in the supplementary cortex but not in the primary motor or somatosensory cortices.

movements of the torso or the opening and closing of the hand. In contrast, stimulating the motor cortex produces small, twitching movements.

An example of the role of the supplemental area in the programming of a motor behavior is seen in studies of simple and complex movements of the hand. In these studies, blood flow to certain brain regions is monitored. The amount of blood flow is an index of the activity of nerve cells in each area of the brain. With simple movements, blood flow increases in the motor and somatosensory cortices, but no increase is found in the supplemental motor area (Fig. 9–20). With more complex movements, blood flow also increases in the supplemental motor area. Interestingly, when subjects

The Posterior Parietal Cortex Integrates Sensory Stimuli for Purposeful Movement

The **posterior parietal cortex** is necessary for the processing of sensory stimuli leading to purposeful movement. It is located immediately behind the somatosensory cortex (see Fig. 9–16). This cortex receives both somatic and visual sensory information and transmits it to the supplemental and premotor areas.

Three main types of neurons have been identified in the posterior parietal cortex and become active only with very specific behavioral motor responses. **Arm projection neurons** generate action potentials when the arm reaches for an object; **hand-eye coordination neurons** are most active when the eye fixates on an object that is being touched; **hand manipulation neurons** increase their firing rate when the hand explores an object.

Patients with damage to this area of the cortex are unable to perform previously learned movements in the appropriate sequence. They appear to synthesize the spatial coordinates of objects in abnormal ways and behave as if their movements are not in accord with the coordinates of the objects in space. For example, when drawing a clock, a patient with a posterior parietal lesion places all of the numbers on one half of its face.

The Cerebellum and the Basal Ganglia Coordinate Movements

The basal ganglia and cerebellum (Fig. 9–21) are the major subcortical components of the motor system. They both receive inputs from the neocortex and transmit information back to the cortex by way of the thalamus. The inputs to the basal ganglia, however, are from the entire cortex, whereas those of the cerebellum are primarily from sensory and motor areas. In addition, the basal ganglia do not receive direct sensory information from somatic receptors, nor do they transmit descending information directly back to the spinal cord as does the cerebellum. These differences suggest that the basal ganglia are involved in more complex motor functions, whereas the cerebellum is more involved with the control of movement that requires constant monitoring by sensory feedback.

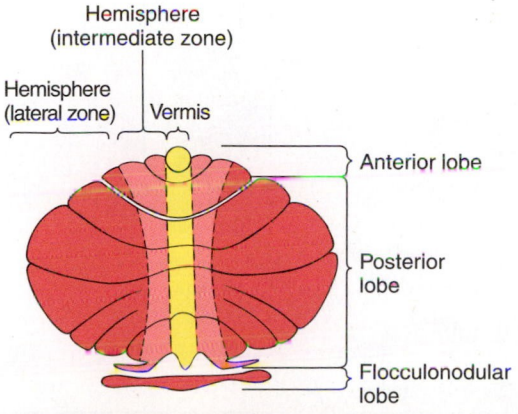

Hemisphere
(intermediate zone)

Hemisphere
(lateral zone)   Vermis

Anterior lobe

Posterior lobe

Flocculonodular lobe

Figure 9–21

A dorsal view of the cerebellum.

The Cerebellum: Planning, Coordination, and Posture

The cerebellum ("little brain" in Latin) has three basic functions: (1) planning movement, (2) controlling posture and equilibrium, and (3) controlling smooth limb movement. The cerebellum accomplishes the latter two functions by comparing information concerning an intended movement with sensory feedback about the actual movement and adjusting its output to compensate for differences between the two. It participates in the planning of a movement by receiving information from motor (supplemental and premotor) and parietal cortices and then uses these inputs to initiate a planned movement.

The cerebellum has three major lobes: (1) the **anterior lobe,** the (2) **posterior lobe,** and (3) the **flocculonodular lobe** (see Fig. 9–21). The flocculonodular lobe is involved in the maintenance of equilibrium and posture. The midline and intermediate regions of the anterior and posterior lobes are involved in limb movement, whereas the lateral regions of these lobes are involved in the planning and initiating of motor programs.

The most prominent nerve cell in the cerebellum is the **Purkinje cell,** found throughout the cerebellum and packed tightly into a single cellular layer (Fig. 9–22). The axons of Purkinje cells form synaptic connections with neurons in one of four deep cerebellar nuclei, the **dentate nucleus,** the **interpositus nuclei** (globase and emboliform), or the **fastigial nucleus.** These nuclei constitute the output of the cerebellum. Purkinje cells release the transmitter gamma-aminobutyric acid (GABA) onto these nuclei, causing hyperpolarization and thus inhibition.

The fastigial nucleus receives information from Purkinje cells in the flocculonodular lobe and transmits its information to the vestibular nucleus, which, in turn, controls motor neurons that innervate axial muscles. Therefore, these nuclei are involved in the maintenance of balance and posture. The interpositus nuclei are involved in the control of limb muscles. It receives information from Purkinje cells in the medial and intermediate regions of the anterior and posterior lobes and sends its information to the thalamus and red nucleus of the brain stem of the cerebellum. The red nucleus, in turn, controls motor neurons innervating distal limb muscles. The dentate nucleus receives information from Purkinje cells in the lateral regions of the anterior and posterior lobes of the cerebellum and is therefore involved in the planning and initiation of movement. The dentate transmits its information to the thalamus, which, in turn, forwards information to the motor and premotor cortex. The dentate also provides inputs to the red nucleus.

Research indicates that the input of information to the cerebellum is simultaneously transmitted to the deep cerebellar nuclei (Fig. 9–23). After processing by the cerebellum, it is transmitted by way of the Purkinje cells to the deep nuclei. The outputs from the deep nuclei are thus modulated by the inhibitory actions of the Purkinje cells in the comparison of intended and actual movements. Neurons of the deep nuclei then transmit information to correct for any deviation from the intended motion.

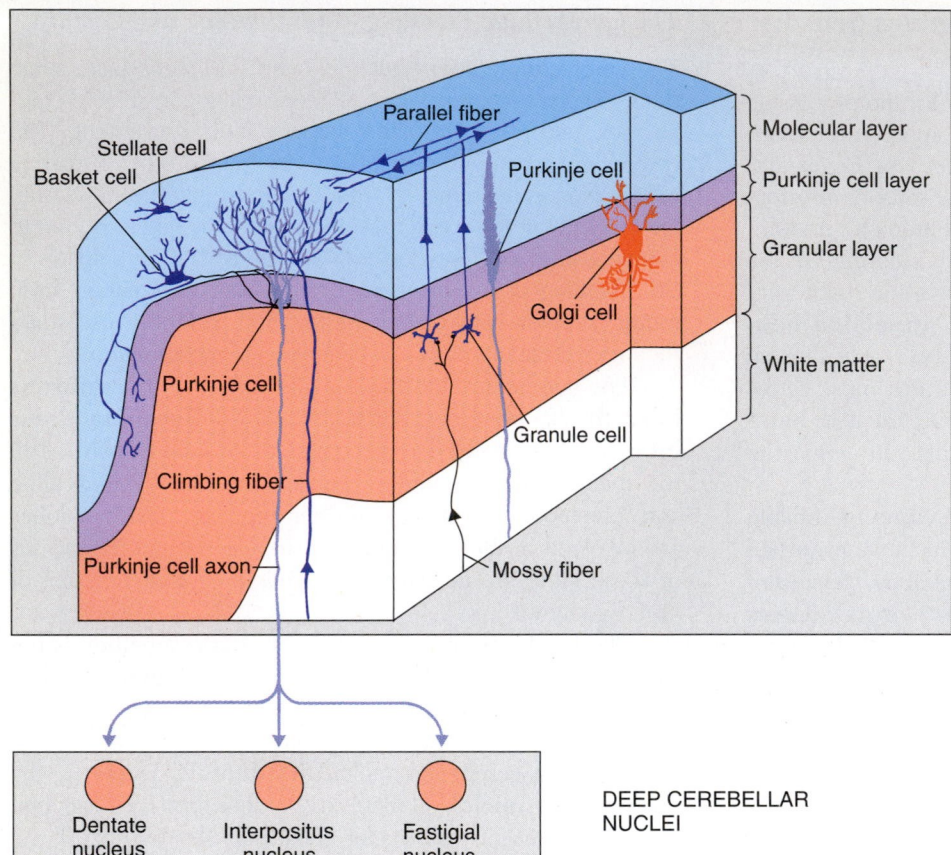

Figure 9–22

Cell types and circuits in the cerebellum.

The inputs to the Purkinje cells of the cerebellum traverse two routes: (1) the climbing fibers, which originate in the olivary nucleus on the lateral medulla (Fig. 9–24a), and (2) parallel fiber pathways. Each Purkinje cell receives synaptic connections from only one climbing fiber. The activation of this input causes the Purkinje cell to generate complex patterns of action potentials (Fig. 9–25). The parallel fiber input to the Purkinje cell originates from **granule cells.** Each Purkinje cell receives inputs from many granule cells, and the activation of the parallel fibers causes the Purkinje cells to generate a simple pattern of action potentials.

Inputs to the inferior olivary nucleus originate in supplemental and premotor areas of the cortex as well as from the spinal cord, the red nucleus, and the mesencephalon, whereas inputs to granule cells are derived from so-called **mossy fibers.** These mossy fibers also originate in the cerebral cortex as well as in the spinal cord. The mossy fibers and olivary climbing fibers respond quite differently during movement. Sensory stimuli and voluntary movements enhance the activity of mossy fibers but have little effect on climbing fiber activity. Climbing fibers are therefore thought to modulate the responsiveness of Purkinje cells to mossy fiber and hence to parallel fiber inputs, as is seen during the learning of motor behaviors.

When monkeys are trained to maintain a lever in one position with their hands, Purkinje cells generate a simple pattern

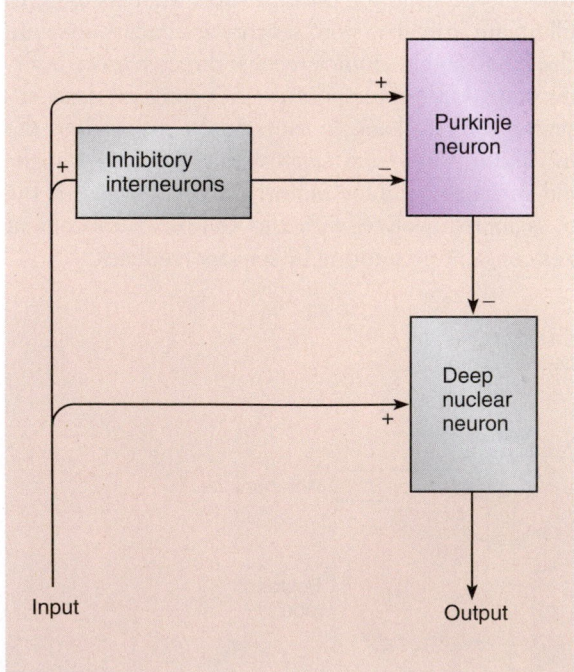

Figure 9–23

Information input to the cerebellum is simultaneously transmitted to the deep cerebellar nuclei.

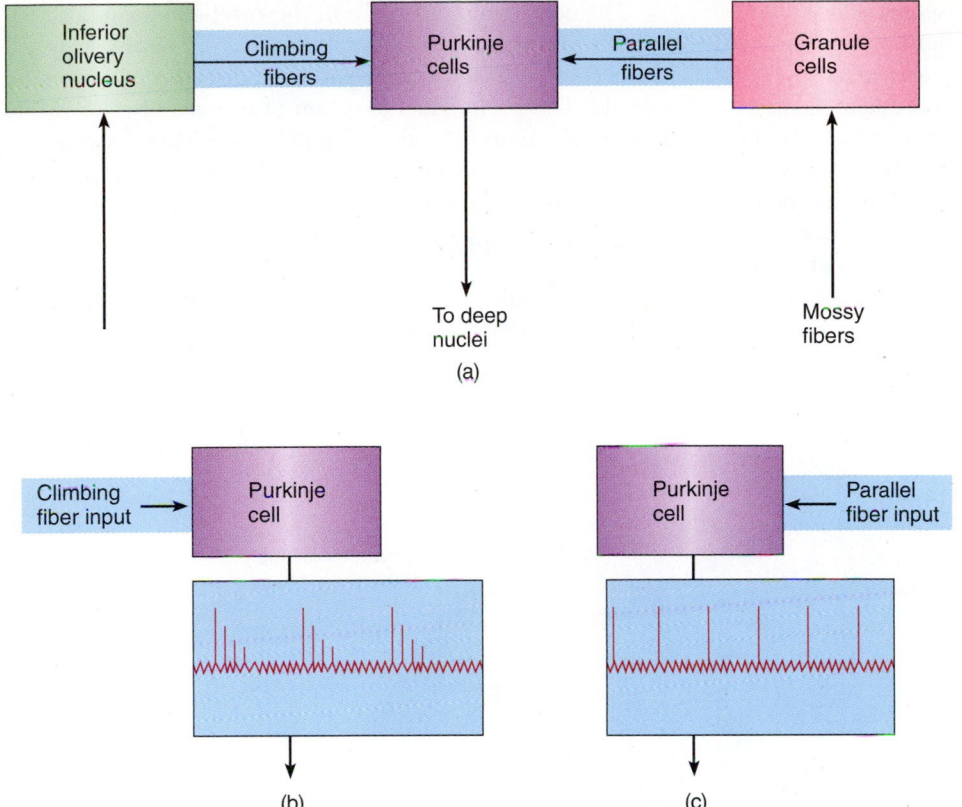

Figure 9–24

Climbing fibers to the cerebellum originate in the inferior olivary nucleus, whereas parallel fibers originate in the granule cells *(a)*. Activation of climbing fibers generates complex patterns of action potentials *(b)*, whereas activation of parallel fibers generates simple action potential patterns *(c)*.

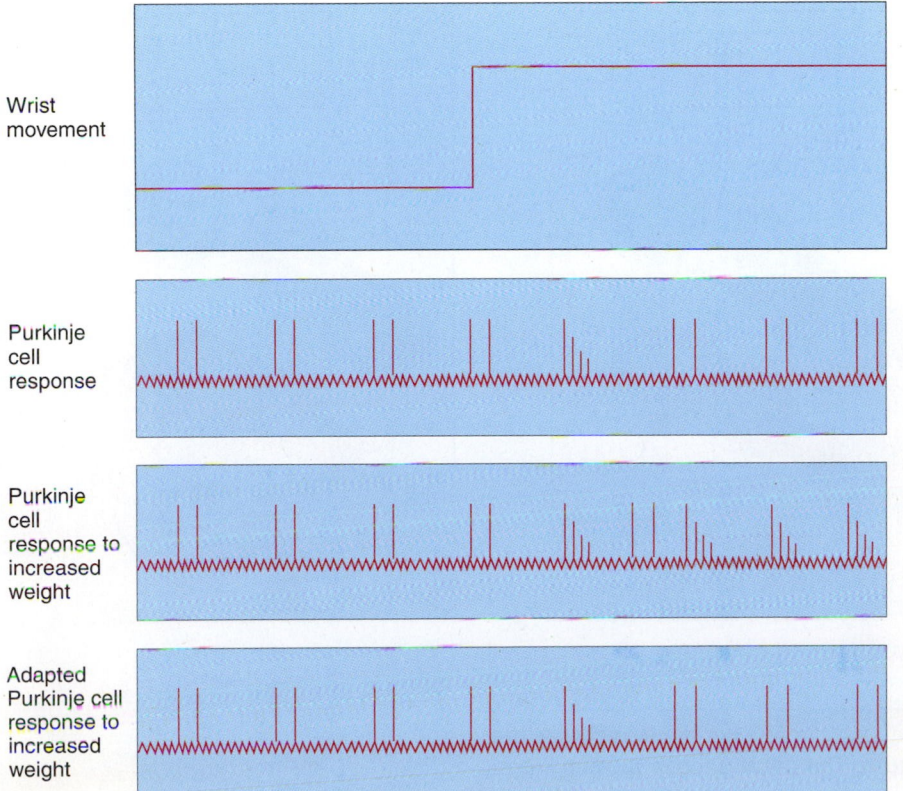

Figure 9–25

Purkinje cells can change their pattern of activity during adaptation to new motor responses. Complex action potential patterns show the involvement of climbing fibers in the learning of a new motor skill. As an animal adapts to the skill, simpler action potential patterns emerge.

of action potentials characteristic of that caused by activation by parallel fibers (see Fig. 9–25). When different weights are placed on the lever to change its position, the animal must compensate for this with added force in order to maintain its original position. When more weight is added, Purkinje cells begin to produce complex (rather than simple) patterns of action potentials. These complex patterns of Purkinje cell action potentials are due to their activation by climbing fibers. As the animal learns to adapt to the new weight, the complex pattern of action potentials diminishes and the simple pattern reemerges. Thus, the Purkinje cells are capable of changing their pattern of activity during the adaptation of new motor responses.

The activity of nerve cells in the cerebellum can also encode the direction of reaching. Individual cells in the cerebellum become activated when a particular direction of movement is made. This preferential direction of movement gives rise to directional vectors for cells and population vectors for cell clusters within the cerebellum. These cerebellar vectors, therefore, represent the neural code for the direction of limb movement.

Basal Ganglia: Planning of Movements

The basal ganglia are also involved in motor control and include the **caudate nucleus, putamen,** and **globus pallidus** (Fig. 9–26). The caudate nucleus and putamen receive inputs

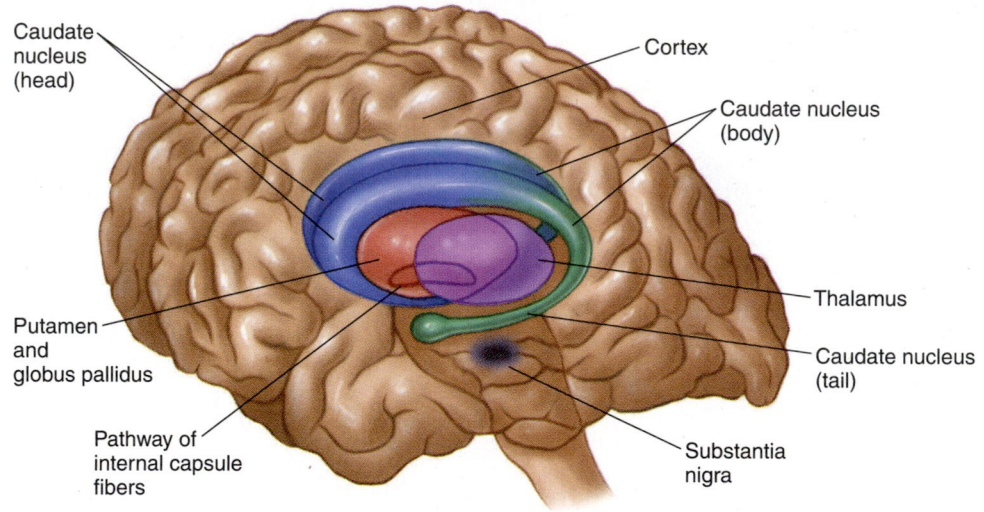

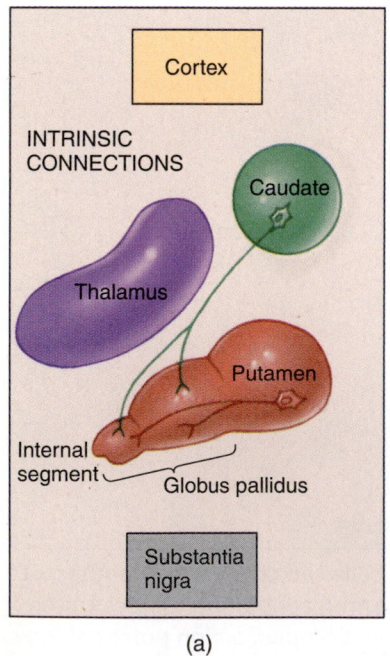
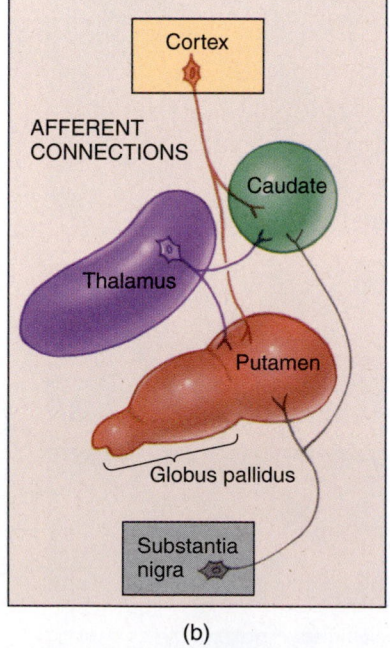
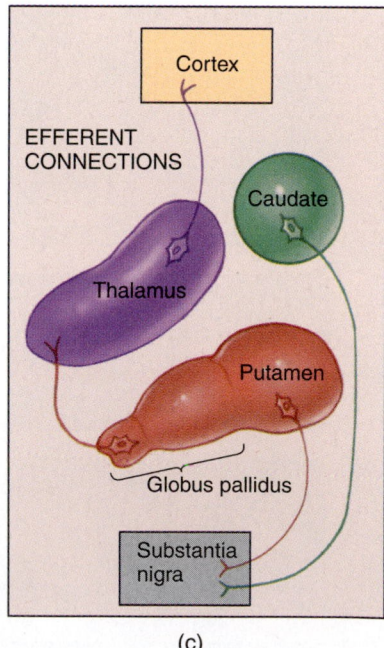

(a)   (b)   (c)

Figure 9–26

The caudate nucleus and putamen of the basal ganglia receive input, whereas the globus pallidus and thalamus provide the output. The basal ganglia receive inputs from the neocortex, thalamus, and substantia nigra. *(a)* Intrinsic, *(b)* afferent, and *(c)* efferent connections are shown.

CURRENT CONCEPTS IN PHYSIOLOGY

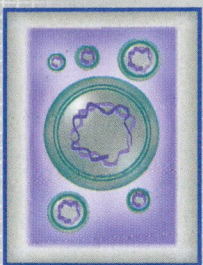

Huntington's Disease

Huntington's disease (HD) is a neurological disorder characterized by depression; dementia; psychiatric problems; and a progressive loss of motor control, producing an unsteady gait and quick, sudden jerking movements of the face, trunk, and limbs. In 1872, George Huntington described the motor components of this disorder as an "irregular and spasmodic action of certain muscles, as of the face, arms, etc. These movements gradually increase, when muscles hitherto unaffected take on the spasmotic action, until every muscle of the body becomes affected." Ultimately, the "patient presents a spectacle which is anything but pleasing to witness." The complications associated with this disease include loss of ability to care for ones self, inability to interact with others, injuries to self and others, increased risk for infection and depression. Symptoms of HD can begin at any time; the average age of onset, however, is approximately 40 years. At present there is no treatment that can cure the disease or slow its course of action. Dopamine blockers, such as haloperidol, provide some relief from abnormal movements and behaviors. The dementia associated with the disease can be treated symptomatically with reminders and memory aids. However, symptoms gradually worsen, and death occurs an average of 16 years after onset. The leading cause of death is infection, although suicide is not uncommon.

The cause of the abnormalities seen in HD is the progressive loss of neurons in the caudate and putamen of the neostriatum. These neurons receive nerve fiber inputs from the neocortex. They, in turn, innervate the globus pallidus and use GABA as their neurotransmitter. In HD, therefore, a major inhibitory input to the globus pallidus is missing. Various radiological scans of the head are now used to aid the clinician in identifying HD, including magnetic resonance imaging (MRI) and computed tomography (CT). CT may detect loss of brain tissue in the caudate region.

The reason for the degeneration of neurons in the caudate and putamen is still unknown. One theory speculates that fiber inputs from the neocortex, which release the transmitter glutamate onto neurons in the caudate and putamen, may be the source of the problem. Glutamate is an excitatory transmitter that when released in excess can damage postsynaptic neurons. Exactly how the *HD* gene may be involved in this process remains to be determined.

In addition to noting the motor abnormalities, Huntington made key observations regarding its hereditary nature. HD is passed from generation to generation by an autosomal dominant mode of genetic transmission. Both males and females are equally susceptible, and each offspring of an affected parent has a 50% risk for inheriting the gene.

The recent application of molecular techniques has determined the chromosomal location of the gene. Using genetic linkage analysis and DNA markers, James Gusella, Michael Conneally, and their collaborators have localized the *HD* gene to the short arm of chromosome 4. The exact site of the *HD* gene has been located and its nucleotide sequence identified. A striking feature of the *HD* gene is a repeating sequence consisting of the nucleotides CAG. Each Huntington patient has been found to exhibit a varying number of this sequence. Moreover, the greater the number of sequence repeats is correlated with an earlier onset of the disease process. DNA marker studies may be used clinically to aid physicians in identifying patients with a tendency to develop HD.

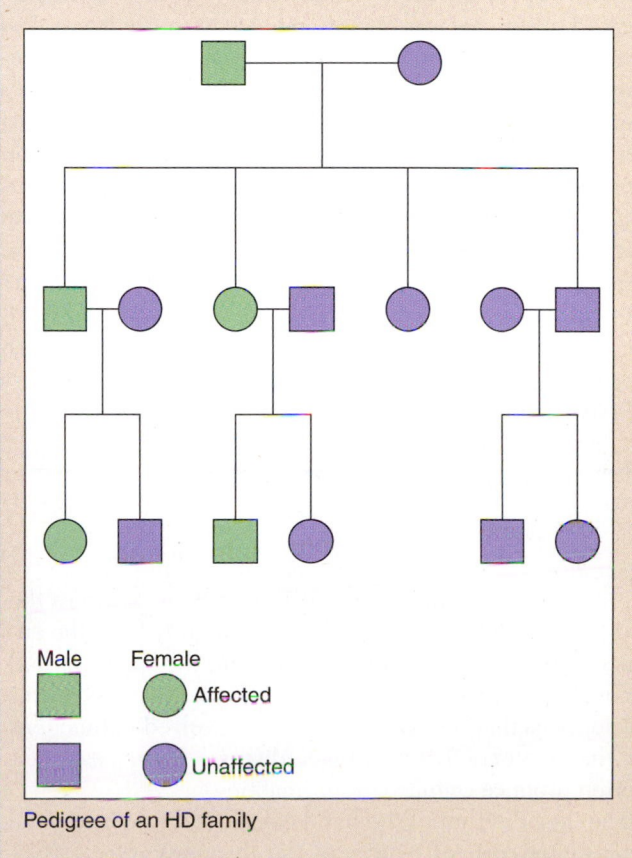

Pedigree of an HD family

APPLICATIONS OF PHYSIOLOGY

Parkinson's Disease: Understanding Physiological and Pathological Mechanisms Leads to Effective Treatment Strategies

Parkinson's disease (PD) is a disorder of the motor system characterized by slowness of movement (bradykinesia), poverty of movement (akinesia), tremor of the fingers and hands (most prominent at rest), absence of facial expression, slowness of speech, stooped body posture, and a shuffling gait and in extreme cases great difficulty in initiating movement. In these cases, patients are bedridden and incapable of feeding and caring for themselves. PD commonly appears in persons between the ages of 50 and 65 years. Currently in the United States, approximately 1 million patients are afflicted with PD, with 50,000 new cases reported each year.

Examination of brain tissue from patients with PD has revealed a loss of dopamine-producing neurons in an area called the substantia nigra pars compacta (SNc), located in the mesencephalon. As shown in the figure, loss of dopaminergic input to D_1 and D_2 receptors in the putamen results in severe alterations of the basal ganglia's normal role of aiding the cortex in planning and executing movements. The resulting combination of decreased inhibition and overstimulation of the internal globus pallidus (GPi) causes great inhibition of thalamic nuclei input to the cortex. This reduces cortical motor signals to the spinal cord. In addition, the overstimulated GPi's direct inhibition of the pedunculopontine nuclei produces akinesia. Control of PD has been achieved with drugs, such as L-dopa, that compensate for the loss of dopamine from SNc. However, long-term use of these drugs results in loss of effectiveness and can produce debilitating side effects, including uncontrolled movements. Recent advances in stereotactic neurological surgery allow very accurate three-dimensional positioning of electrode probes into deep brain structures, such as the basal ganglia. The electrode probe can be used to produce a very discrete lesion. If the lesion is placed in the sensorimotor region of the GPi, many of the PD symptoms can be alleviated.

To avoid potentially harmful effects of a permanent brain lesion, a new procedure was developed that implants a stimulating electrode into regions where it can disrupt the aberrant output of the GPi and its disruptive effects on movement. The figure shows stimulating electrode implant sites in the subthalamic nucleus (STN), the GPi, and the thalamus (Thal). The stimulation can be turned on or off and adjusted by an external control device. Results of clinical studies of these deep brain stimulator implant procedures are encouraging. It appears that deep brain stimulation can disrupt the harmful effects of the GPi output, freeing other basal ganglia regions to perform their normal function in motor activity.

In animal models of PD, scientists have taken a new approach, in which nerve cells that produce dopamine are transplanted directly into the neostriatum. These transplanted neurons form synaptic connections with those of the host. In addition, they are capable of correcting the locomotor abnormalities associated with PD. The sources of these dopamine neurons in laboratory animal experiments are fetuses of the same species. Because of social issues concerning the use of human fetal tissue, however, medical researchers are also exploring the possible use of alternative sources of dopamine cells for transplantation.

---

to the basal ganglia, whereas the globus pallidus provides the output. Inputs to the basal ganglia are primarily from the entire neocortex but also include those from the thalamus and substantia nigra of the brain stem. The extensiveness of this input suggests that the basal ganglia are involved in functions other than motor activities. In fact, diseases of the basal ganglia often produce cognitive abnormalities.

The major output of the basal ganglia is to the prefrontal and premotor cortices by way of the thalamus (Fig. 9–26c). Through this pathway, the basal ganglia can modulate the descending components of the motor system. The basal ganglia also have outputs to the substantia nigra. Nerve fibers from the substantia nigra that terminate in the basal ganglia release dopamine as the neurotransmitter. The degeneration of these dopamine fibers is responsible for the motor disorder called *Parkinson's disease*. Patients with Parkinson's disease exhibit stiff posture, a rhythmic tremor at rest, and difficulty in initiating movement. These symptoms can be alleviated, in part, with drugs that act as precursors to increase the synthesis of dopamine, such as levodopa (L-dopa).

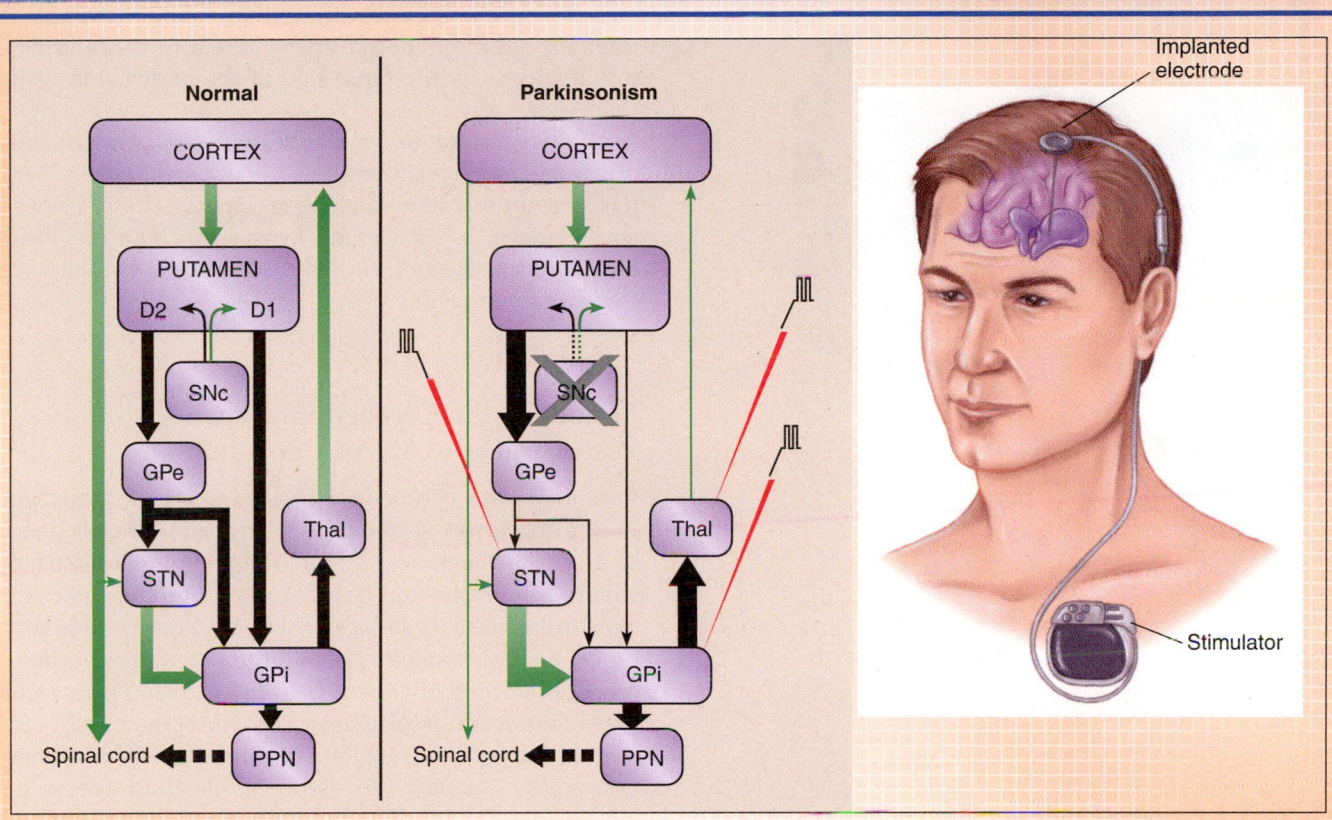

Left, Diagram of the basal ganglia-thalamocortical circuitry under normal conditions and in Parkinson's disease (PD). Inhibitory connections are depicted as black arrows, and excitatory connections, as green arrows. PD is characterized by a reduction in dopamine projections (depicted as the gray X) and differential changes in the striatopallidal projections, indicated by the thickness of the connecting arrows. In patients with PD, basal ganglia output from the internal segment of the globus pallidus to the thalamus (that is, GPi to Thal) is increased. This excess output can be reduced by surgical ablation or deep brain electrical stimulation (depicted as red electrodes) to GPi, the subthalamic nucleus (STN), or Thal. GPe, external segment of the globus pallidus; PPN, pedunculopontine nuclei; SNc, substantia nigra pars compacta. *Right,* Deep brain stimulating electrode implanted in the ventral intermediate (Vim) nucleus of the thalamus with the pulse generator control device and connecting electrical leads implanted under the skin.   *(Left panel, from Grafton, S. T., and DeLong, M. "Tracing the brain's circuitry with functional imaging." Nature Medicine, 3:602–603, 1997; with permission; right panel, courtesy of Medtronic, Inc.)*

DISORDERS OF THE MOTOR SYSTEM

 What kinds of deficits are associated with pathology in the motor system?

Because the different parts of the motor system have such distinctive roles, lesions in these areas result in characteristic movement disorders. These lesions often occur because of trauma, stroke, or tumors.

Lesions of the Corticospinal Tract Produce Paralysis

Upper motor neuron lesions result from damage to descending pyramidal cell fibers (Fig. 9–27). These lesions are characterized by paralysis on the side of the body opposite the site of the lesion, little muscle atrophy (decrease in size), increase in muscle tone, hyperactive reflexes, and the extension of the big toe and fanning of the other toes in response to stroking the bottom of the foot. This latter sign is called the **Babinski sign.**

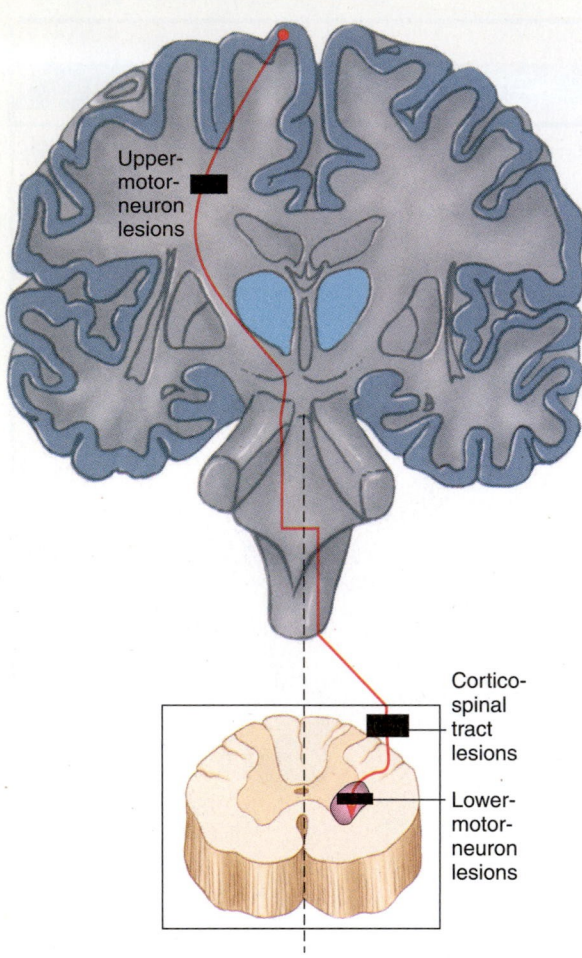

Corticospinal tract lesions occuring within the spinal cord disrupt the inputs to motor neurons. These lesions are characterized by loss of strength and movement in muscle groups, loss of strength in voluntary muscle contraction, and the Babinski sign. These effects are ipsilateral to (i.e., occur on the same side of the body as) the side of the lesion.

Lower motor neuron (alpha motor neuron) lesions can occur with spinal cord injuries that directly affect the motor neurons rather than their inputs. These lesions are characterized by ipsilateral hypoactive reflexes, paralysis limited to specific groups of muscles, and flaccid (limp) muscles with prominent atrophy.

Lesions of the Cerebellum Cause Disturbances in Motor Coordination

Lesions of the cerebellum cause ipsilateral disturbances. Lesions occurring in the lateral cerebellum result in a lack of limb coordination. Lesions occurring in the central region of the cerebellum (see Fig. 9–21) produce ataxia (loss of coordination); those localized in the flocculonodular lobe produce a disturbance in equilibrium as well as ataxia. Diseases of the cerebellum are also characterized by inaccurate range and direction of movement, diminished resistance to passive movement, inability to perform rapid alternating movements, inability to stop sharply, and tremor with movements. Interestingly, chronic alcoholism leads to the degeneration of cerebellar Purkinje cells. As a consequence, chronic alcoholics often exhibit a wide-base uncoordinated gait.

Figure 9–27

Lesions of the motor system result in characteristic motor disorders.

CHAPTER REVIEW

Summary

- The motor system controls locomotion, fine movement, body posture, and equilibrium by acting on motor neurons in the spinal cord that innervate skeletal muscles. The components of the motor system (the spinal cord, brain stem, and motor cortex) interact in both hierarchical and parallel fashion.
- A motor neuron and the muscle fibers that it innervates constitute a motor unit.
- Motor neurons have cell bodies located in the gray matter of the ventral horn of the spinal cord. The spinal cord contains interneurons, which play a role in coordinating the responses of antagonistic and synergistic muscles.
- Walking movements are created by locomotor generators in the spinal cord. These locomotor programs are driven by the loco-

motor command center in the brain stem and can be modified by sensory feedback.
- Sensory fibers within skeletal muscles detect muscle length, force, and velocity of muscle shortening. Reflex actions are initiated when sensory information is sent from muscle, tendon, or skin receptors to the spinal cord, which causes a reflexive movement in the appropriate muscles.
- Complex motor behaviors require inputs from the motor cortex, supplemental and premotor cortices, cerebellum, and basal ganglia. The maintenance of posture and balance is carried out by brain stem inputs to the spinal cord. Three pathways provide motor input to the spinal cord: (1) ventromedial, (2) lateral reticulospinal, and (3) rubrospinal.

- The motor cortex is a central motor command center responsible for fine movements. The motor cortex is organized in a somatotopic fashion. Regions of the body containing small motor units represent the largest portion of the motor cortex. The cortex contains neuronal populations responsible for the direction of limb movement.
- The supplemental and premotor cortices are involved in the planning of movements; they transmit this information to the motor cortex. The posterior parietal cortex is involved in the processing of sensory information leading to purposeful movement. This cortex receives somatic and visual sensory information and transmits this information to the supplemental and premotor areas.
- The cerebellum has three motor functions: (1) initiating and planning movement, (2) coordinating limb movement, and (3) maintaining posture and equilibrium. The cerebellum carries out these functions in concert with the basal ganglia. The basal ganglia are also involved in the planning of movements. They have inputs from all neocortical areas in addition to the motor cortex and are involved in functions such as cognition and movement.
- Lesions in different areas of the motor system produce unique movement disorders. Upper motor neuron lesions of the corticospinal tract result in bodily paralysis and increased muscle tone, with hyperactive reflexes and dorsiflexion of the big toe in response to stroking of the bottom of the foot on the side opposite to the lesion. Corticospinal lesions in the spinal cord produce loss of strength and movement in specific muscle groups, loss of strength in voluntary contraction, and dorsiflexion of the big toe in response to stroking of the bottom of the foot on the same side as the lesion. Lower motor neuron lesions of the spinal cord result in paralysis restricted to specific groups of muscles, hypoactive reflexes, and flaccid muscles. Cerebellar lesions are characterized by a lack of coordination and equilibrium.

Review Questions

Choose the Correct Answer

1. Identify the structure that is *not* directly involved in the stretch reflex (knee jerk).
   a. Muscle spindle
   b. Alpha motor neuron
   c. Gamma motor neuron
   d. Inhibitory interneuron
   e. Sensory nerve

2. In gamma motor neuron control of muscle length, what function most directly causes the muscle to shorten?
   a. Activation of the sensory nerve
   b. Activation of the alpha motor neuron
   c. Activation of the gamma motor neuron
   d. Stretching of the intrafusal muscle fibers
   e. Stretching of the extrafusal muscle fibers

3. What type of cell provides all of the output from the cerebellar cortex?
   a. Golgi
   b. Basket
   c. Granule
   d. Stellate
   e. Purkinje

4. What pathological process is thought to cause several symptoms of Parkinson's disease?
   a. Loss of GABA receptors in the temporal cortex
   b. Damage to the subthalamic nucleus due to a vascular accident
   c. Degeneration of dopaminergic neurons in the substantia nigra
   d. A lesion of the posterior parietal cortex due to trauma or stroke
   e. Autoimmune blocking of acetylcholine receptors at neuromuscular junctions

5. Which one of the following muscle groups is an example of a large motor unit?
   a. Muscles of the thumb
   b. Muscles mediating speech
   c. Axial muscles controlling posture
   d. Muscles mediating eye movement
   e. Muscles mediating facial expression

6. Which one of the following characteristics is a feature of Golgi tendon organs but not of muscle spindles?
   a. Sensory signals are carried by Ia afferent fibers.
   b. The receptors are nuclear chain and bag muscle fibers.
   c. Stretch of the muscle increases action potential generation.
   d. The receptors are located in parallel with the extrafusal fibers.
   e. Contraction of extrafusal fibers increases action potential generation.

7. Which one of the following actions is *not* a component of the flexor withdrawal reflex of the foot? (The ipsilateral side is the side receiving a painful stimulus.)
   a. Ipsilateral flexors contract.
   b. Ipsilateral flexors are inhibited.
   c. Contralateral extensors contract.
   d. Ipsilateral extensors are inhibited.
   e. Contralateral flexors are inhibited.

8. Which one of the following pathways carries motor nerve fibers that innervate the head and face?
   a. Tectospinal
   b. Rubrospinal
   c. Corticospinal
   d. Corticobulbar
   e. Vestibulospinal

9. Which region of the cerebellum is involved in the maintenance of equilibrium and posture?
   a. Vermis
   b. Flocculonodular lobe
   c. Lateral region, anterior lobe
   d. Lateral region, posterior lobe
   e. Intermediate region, anterior lobe

10. Alpha motor neurons, the final common path output of the motor nervous system, are located in which one of the following structures or regions?
    a. Anterior (ventral) horn of the spinal cord
    b. Substantia gelatinosa of the spinal cord
    c. Deep cerebellar nuclei
    d. Primary motor cortex
    e. Basal ganglia

11. Which one of the following structures is *not* considered part of the basal ganglia?
    a. Putamen
    b. Caudate nucleus
    c. Substantia nigra
    d. Internal globus pallidus
    e. External globus pallidus
12. Which structure provides dopaminergic input to receptors in the putamen?
    a. Thalamus
    b. Motor cortex
    c. Caudate nucleus
    d. Substantia nigra
    e. Internal globus pallidus
13. When a new motor task is encountered, such as weight being added to a lever that must be manually kept in a constant position, which one of the following statements correctly describes the effect on Purkinje cells in the cerebellum?
    a. Purkinje cell action potential firing patterns are not altered.
    b. Purkinje cells produce less and less complex action potential patterns.
    c. Purkinje cell action potential patterns become simpler, then increase in complexity with adaptation to the new motor situation.
    d. Action potential patterns generated by Purkinje cells become more and more complex with adaptation to the new motor situation.
    e. Purkinje cells produce more complex patterns of action potentials, which become simpler with adaptation to the new motor situation.
14. Action potentials in alpha motor neurons represent which component of a feedback control system?
    a. Effector
    b. Comparator
    c. Controlled system
    d. Difference signal
    e. Sensor
15. Fibers of the rubrospinal tract originate in what brain region?
    a. Red nucleus
    b. Inferior colliculus
    c. Superior colliculus
    d. Lateral vestibular nucleus
    e. Medial reticular formation
16. Ipsilateral hypoactive reflexes, paralysis limited to specific muscle groups and flaccid (limp) muscles, are characteristics of lesions of which one of the following structures?
    a. Cerebellum
    b. Basal ganglia
    c. Brain stem nuclei
    d. Lower motor neuron
    e. Upper motor neuron
17. What is the location of locomotor generators involved in walking?
    a. Putamen
    b. Thalamus
    c. Cerebellum
    d Spinal cord
    e. Motor cortex
18. During the lifting of a heavy object, which one of the following mechanisms is used to recruit additional motor units to increase a muscle's force of contraction?
    a. Population coding
    b. Frequency coding
    c. Feedforward control
    d. Inverse myotatic reflex
    e. Golgi tendon organ reflex
19. Output from the internal globus pallidus goes to what region before it is relayed to the cortex?
    a. Putamen
    b. Thalamus
    c. Spinal cord
    d. Substantia nigra
    e. Subthalamic nucleus
20. Which one of the following statements is *not* supported by the reason that follows it?
    a. The swing and stance phases of walking occur because locomotor generators coordinate the flexor and extensor muscles of the legs.
    b. When extrafusal fibers contract, sensitivity of the spindle is maintained because gamma motor neurons cause the intrafusal fibers to shorten.
    c. The effectiveness of the stretch reflex (i.e., knee-jerk reflex) is increased because the cerebellum coordinates the activation sequence of flexor and extensor muscles.
    d. Upper motor neuron lesions produce paralysis on the side of the body opposite the side of the lesion because the corticospinal tract crosses to the opposite side in the brain stem.
    e. Muscle fiber excitation occurs when alpha motor neurons generate action potentials because there is no means of modifying the motor signal once it leaves the CNS.

Answer to Case History Question

1. The rescuers knew that emergency medical help would soon arrive. The victim would be placed on a spine board. A neck support collar would be applied and rolled towels would be taped in place to prevent any movement of the head with respect to the neck and body. He could then be safely lifted from the pool. Had the pool not been warm, or if the accident had happened in a cooler body of water, such as a northern lake, removal from the water would have been crucial because, when spinal cord reflex mechanisms lose modulating control from the brain, they can become hyperactive. The resulting restrictive effects on blood vessel diameter below the level of the injury could produce wild variations of blood pressure and heart rate, a condition known as *autonomic dysreflexia*. This condition, especially in a maintained cold water environment, could quickly lead to seizures, stroke, and even death.

Key Terms

Babinski sign (p. 333)
basal ganglia (p. 310)
caudate (p. 330)
cerebellum (p. 310)
extensor (p. 314)

flexor (p. 314)
hierarchical motor organization
   (p. 310)
motor cortex (p. 323)
motor system (p. 310)

muscle (p. 313)
muscle fiber (p. 313)
muscle spindle (p. 319)
myoneural junction (p. 314)
reflex (p. 319)

skeletal muscle (p. 311)
smooth muscle (p. 311)
striated muscle (p. 311)
vestibulospinal tract (p. 321)

Suggested Readings

Camarata, P. J., Latchaw, R. E., and Heros, R. C. "'Brain attack': The rationale for treating stroke as a medical emergency." *Neurosurgery,* 34:144–158, 1994.

Fredericks, C. M., and Saladin, L. K. *Pathophysiology of the Motor Systems, Principles and Clinical Presentations.* Philadephia, F.A. Davis, 1996.

Grafton, S. T., and DeLong, M. "Tracing the brain's circuitry with functional imaging." *Nature Medicine,* 3:602–603, 1997.

Graybiel, A. M., Aosake, T., Flaherty, A. W., and Kimura, M. "The basal ganglia and adaptive motor control." *Science,* 265:1826–1831, 1994.

Kandel, E. R, Swartz, J. H., and Jessel, T. M. *Essentials of Neural Science and Behavior.* Norwalk, CT, Appleton & Lange, 1995.

Latash, M. L. *Neurophysiological Basis of Movement.* Champaign, Human Kinetics, 1998.

Liddell, E. G. T., and Sherrington, C. "Recruitment and some other features of reflex inhibition." *Proceedings of the Royal Society of London,* 97:488–518, 1925.

Purves, D., Augustine, G. J., Fitzpatrick, D., Katz, L. C., LaMantia, A. S., McNamara, J. O., and Williams, S. M. *Neuroscience,* ed 2. Sunderland, Sinauer Associates, 2001.

Stuart, D. G. "The segmental motor system." *Progress in Brain Research,* 123:3–28, 1999.

Wichmann, T., and DeLong, M. R. "Functional and pathophysiological models of the basal ganglia." *Current Opinions in Neurobiology,* 6:751–758, 1996.

Web sites

http://www.humankinetics.com
Human Kinetics. The information leader in physical activity books, journals, software, fitness, and medicine.

http://www.spinalvictory.org
The Kent Waldrop National Paralysis Foundation. Spinal injury resources.

http://my.webmd.com/content/article/1680.51745
An article on Parkinson's Disease.

http://my.webmd.com/content/article/1833.50158
An article on the genetics of Parkinson's Disease.

**http://mywebmd.com/content/asset/
   adam_disease_huntington_chorea**
An article on Huntington Disease.

**http://onhealth.webmed.com/conditions/resource/conditions/
   item%2c706.asp**

http://www.neurologychannel.com/huntingtons/

Answers to Review Questions

1. c **2.** b **3.** e **4.** c **5.** c **6.** e **7.** b **8.** d
9. b **10.** a **11.** c **12.** d **13.** e **14.** a **15.** a
16. d **17.** d **18.** a **19.** b **20.** c

Chapter 10

THE AUTONOMIC NERVOUS SYSTEM

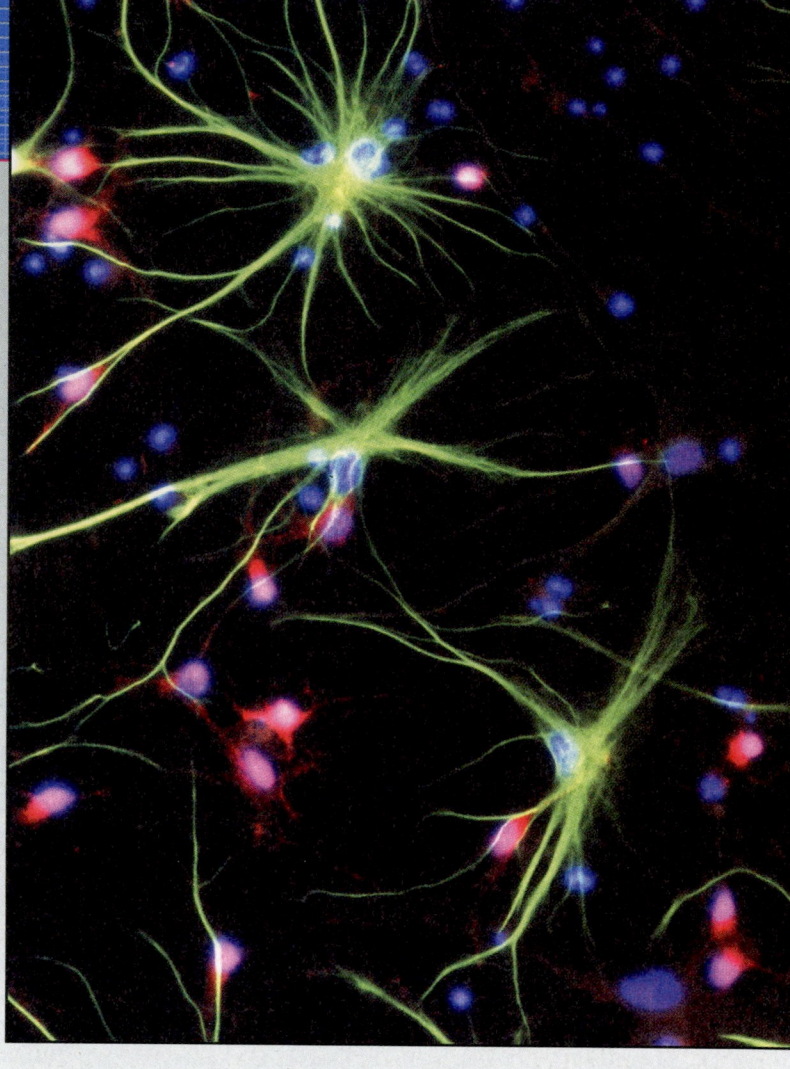

- *Flourescent labeled neurons from superior cervical ganglion.*
(© Nancy Kedersha/UCLA/Science Photo Library/Photo Researchers, Inc.)

KEY CONCEPTS

- *The autonomic nervous system (ANS) is organized into three divisions: (1) sympathetic (thoracolumbar), (2) parasympathetic (craniosacral), and (3) enteric. Central and peripheral components of the ANS are coordinated to constantly regulate vital functions of the body without conscious attention or effort.*

- *The sympathetic division coordinates the body's responses to stress, whereas the parasympathetic division is responsible for the regulation of several homeostatic functions. The enteric nervous system of the gastrointestinal tract can regulate gastrointestinal activity without input from other sources, although its activity is modulated through innervation from the other ANS divisions.*

- *The status of internal organs is monitored by sensory receptors that transmit information to regulate cardiovascular, respiratory, renal, digestive, excretory, and metabolic functions.*

- *One of the primary functions of the ANS is to regulate the body's internal environment by controlling parameters such as water volume, pH, oxygen (O_2), carbon dioxide (CO_2), blood pressure, and body temperature — and thus maintain conditions that are optimal for cells to function.*

- *The hypothalamus participates along with the ANS in the control of activity of different organ systems within the body so that they act in a coordinated fashion.*

CASE HISTORY

Shark! Erin was snorkeling in the warm waters of the cove in the late afternoon sun when she saw the unmistakable profile of the dorsal fin approximately 50 yards from her position. The profile shifted, nearly disappearing as it turned slowly toward her. Instantly, she felt her heart accelerate. Her breathing quickened. She felt a tremendous surge of energy; it was all she could do control the urge to turn and sprint for the beach. Remembering from her training that some sharks can swim faster than 30 miles per hour (48.3 km/h) and that panicky, thrashing movements invite instant attack, she calmed herself with deep, regular breaths. Keeping her eyes on the shark, she swam backward toward the beach using a strong, rhythmic kick, making sure that her fins did not break the surface and splash. She felt tireless. She remembered that sharks would usually circle, then move in and nudge their potential prey before attacking. It would be important to respond strongly to the nudge, striking or kicking the shark hard. She concentrated on her deep breathing and rhythmic kicking. Before the shark broke its approach to circle, she felt the sand bottom beneath her. She strode forcefully backward to the beach as the shark turned back toward deeper water. She collapsed on the beach, shivering and sweating profusely and her heart racing as a wave of nausea swept over her.

The autonomic nervous system that readied Erin to react to the perceived danger was still very much in control of her body. Without any conscious effort, her body was prepared for the "flight or fight" situation. Her adrenal glands released adrenaline (epinephrine) into her circulatory system. Her heart rate was increased and the resistance of her respiratory system was decreased, making a greater amount of blood and oxygen available to her skeletal muscles. Furthermore, glycogen was converted to glucose, increasing the circulating energy source for muscle activity.

All of these changes, and more, occurred in response to a visual stimulus. The sight of an approaching dorsal fin was combined with a wealth of memories recorded in the more primitive regions of her cortex, collectively called the *limbic system*. Circuits of this system coordinate factors such as emotion and memory with autonomic nervous system response mechanisms to optimize the likelihood of survival. Erin sat on the beach trying to calm herself. She told herself, "The shark wasn't really that big. And, besides, in these waters, with abundant marine life, the shark was probably well-fed and would not have bothered anything as large as a person." Sitting on the beach, all of this made great rational sense. However, Erin's body wasn't buying any of it, at least not for a while.

Question

Why was Erin's body still in a "fight or flight" state after the perceived-threat situation was clearly over?

THE AUTONOMIC NERVOUS SYSTEM

General Features

The term "autonomic" implies independent, self-controlling function. If we had to maintain our vital functions, such as heart rate, respiration, and blood glucose level, every second through conscious effort, we would accomplish little else. Sleep, of course, would be impossible. The **autonomic nervous system (ANS)** helps regulate our internal environment. It controls **visceral** functions, as opposed to the somatic nervous system, which controls exclusively skeletal muscles. This system is divided into **sympathetic** and **parasympathetic** divisions. A third, somewhat independent, nervous system, the **enteric nervous system (ENS),** can also be considered a part of the ANS. The enteric system controls many functions in the gastrointestinal tract. Sensory information about

several body functions are transmitted through the ANS, which then sends neural signals out to the body to adjust internal functions by altering the activity of smooth muscle, cardiac muscle, and glands. Many of the systems regulated by the ANS receive both excitatory and inhibitory signals from the sympathetic and parasympathetic divisions. This provides fine control of the regulated system.

In addition to regulation of the internal environment, the ANS adjusts the responses of our bodies and behavior to changes in the external environment, such as changes in temperature. The ANS also enables us to cope with extreme physical and psychological environments, such as those encountered in the arctic, space, or in deep-sea exploration. These coping mechanisms extend to emergency situations such as those encountered in accidents, crime, and warfare. Furthermore, the ANS helps us to survive internal challenges, ranging from the simple act of standing up from a re-

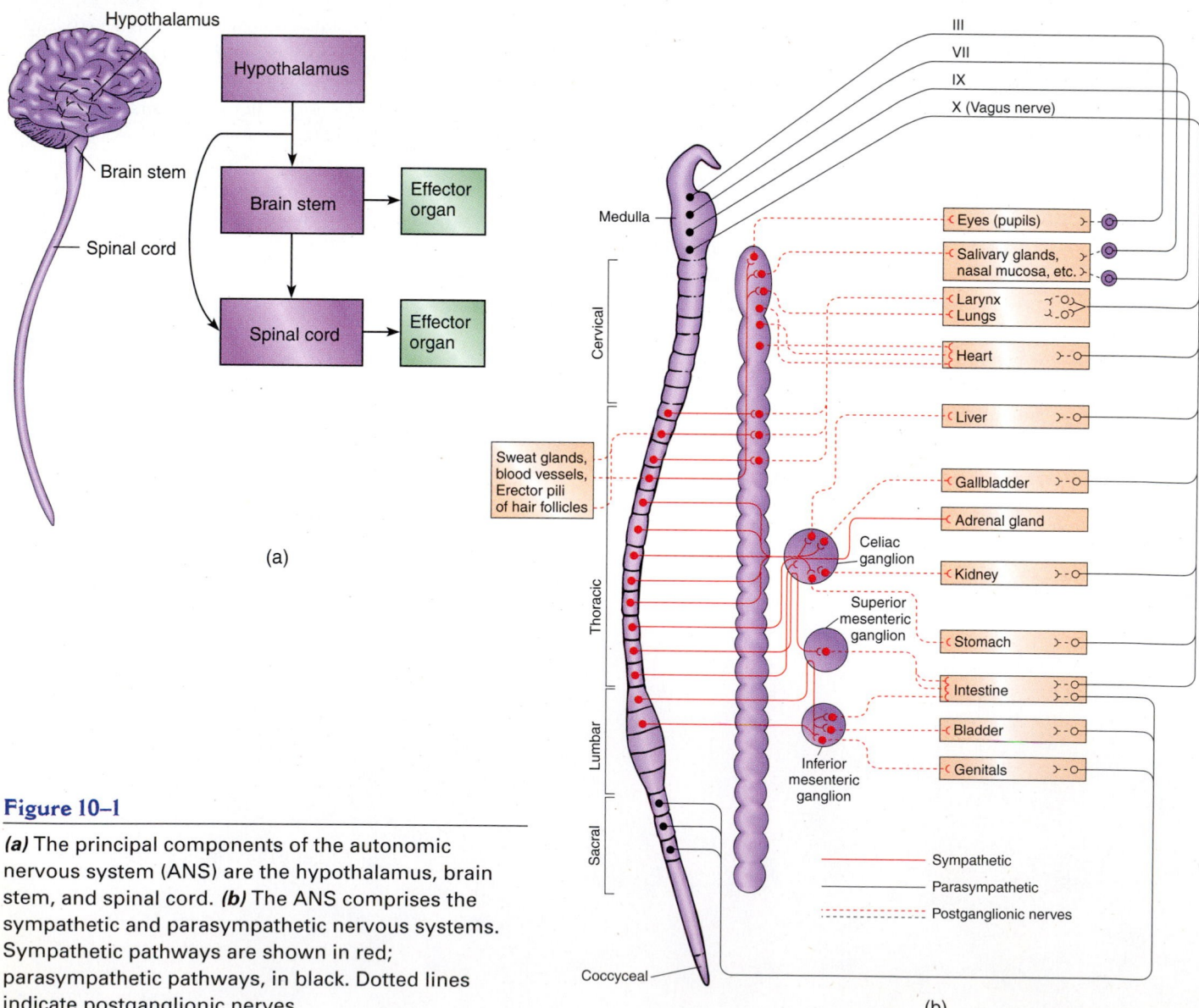

Figure 10–1

(a) The principal components of the autonomic nervous system (ANS) are the hypothalamus, brain stem, and spinal cord. *(b)* The ANS comprises the sympathetic and parasympathetic nervous systems. Sympathetic pathways are shown in red; parasympathetic pathways, in black. Dotted lines indicate postganglionic nerves.

clining position to maintaining blood flow to the brain in response to severe blood loss.

Conscious control of the ANS seems to be minimal. However, we have some conscious control over some ANS functions (e.g., bowel and bladder excretory functions). Also, there are striking demonstrations of control of respiration and heart rate by practitioners of certain Eastern religions and philosophies. Conscious control of the ANS is being practiced successfully for stress reduction. More recently, investigative efforts are being increased to examine the potential of conscious control of the ANS for combating disease processes.

This chapter examines the functional organization of the ANS. Mechanisms by which the ANS controls systems to supply the nutrients to the body, to remove waste products of metabolism, to coordinate the response to stress, and to regulate the internal environment are explored.

Organization of the Autonomic Nervous System

The ANS has both central and peripheral components. The central components of the ANS are the **limbic system,** including the **hypothalamus,** certain **brain stem** regions and nuclei, and the **spinal cord** (Fig. 10–1). The peripheral components consist of **ganglia,** where further processing of signals can occur after the information leaves the central nervous system (CNS), and the **nerves** that innervate the organs of the body, classified as either parasympathetic or sympathetic nerves. Autonomic nerve fibers include both sensory and efferent neurons.

The Autonomic Nervous System Controls the Activity of Organs and Systems in the Body

 What is the functional anatomy of the autonomic nervous system?

The sympathetic division of the ANS coordinates the body's response to stress, whereas the parasympathetic division coordinates the body's basic homeostatic functions, such as digestion, respiration, heart rate, and blood pressure. The sympathetic and parasympathetic divisions have distinct anatomical differences and release different neurotransmitters at their target sites. The sympathetic division simultaneously activates more organ systems than does its parasympathetic counterpart (called **divergence**). Divergence plays an important role during stress responses, when there is a need to coordinate changes in a broad range of systems. By contrast, in parasympathetic responses, different organ systems are activated more specifically and independently.

Nerve fibers of the sympathetic nervous system emerge from the spinal cord at the first thoracic (T1) through the second lumbar (L2) levels (Fig. 10–1b). The cell bodies of these fibers are located in the **intermediolateral cell columns (lateral horn)** of the spinal cord (Fig. 10–2). These cells are called **preganglionic nerve cells** and have short axons that innervate cells in **ganglia** located near the

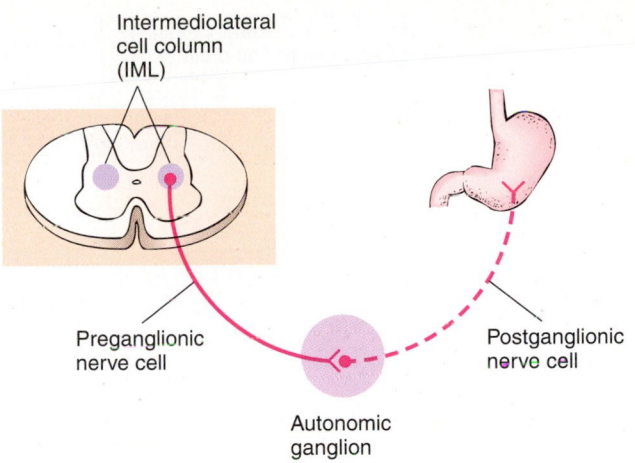

Figure 10–2

Nerve fibers of the sympathetic nervous system originate in the intermediolateral (IML) cell columns of the spinal cord. These preganglionic nerve cells innervate postganglionic nerve cells located within ganglia near the spinal cord.

spinal cord. The **paravertebral chain ganglia** lie laterally and slightly anteriorly on each side of the vertebral column. In addition, three prevertebral ganglia — (1) the **celiac** (solar plexus), (2) **superior mesenteric,** and (3) **inferior mesenteric** — lie anterior to the vertebral column, separated from it by the aorta. The cells within ganglia are **postganglionic nerve cells.** They have relatively long axons and synapse with target cells in other organ systems.

The parasympathetic nervous system has myelinated and unmyelinated fibers that exit from the brain stem via cranial nerves III (oculomotor), VII (facial), IX (glossopharyngeal), and X (vagus), and also from the spinal cord at sacral levels

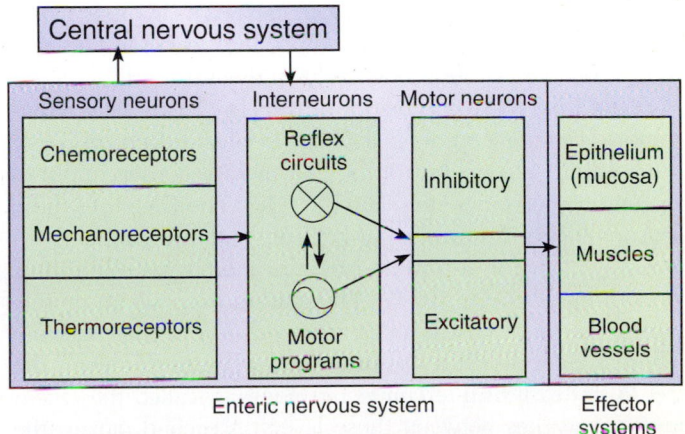

Figure 10–3

Conceptual model of the enteric nervous system.
(Rhoades, R. A., and Tanner, G. A. Medical Physiology. Boston, Little, Brown and Co., 1995)

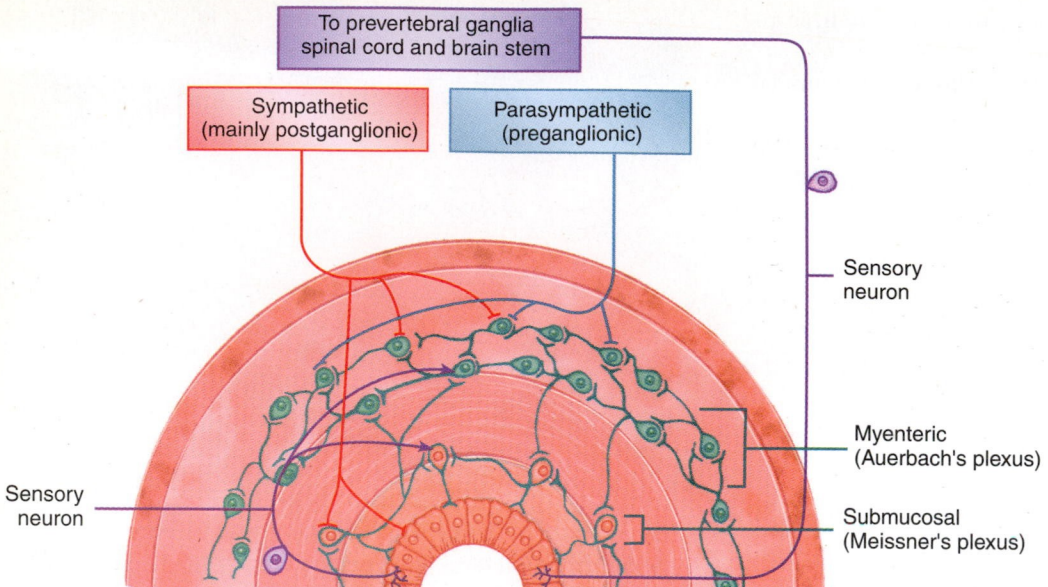

Figure 10–4

Neural control of the gut wall, showing (1) the myenteric and submucosal plexuses; (2) extrinsic control of these plexuses by the sympathetic and parasympathetic nervous systems; and (3) sensory fibers passing from the luminal epithelium and gut wall to the enteric plexuses and from there to the prevertebral ganglia, spinal cord, and brain stem. *(Guyton, A. C., and Hall, J. E. Textbook of Medical Physiology, ed 9. Philadelphia, W.B. Saunders, 1996.)*

S2, S3, S4, and occasionally S1 (see Fig. 10–1*b*). These fibers innervate postganglionic nerve cells clustered near or in the target organ which have very short axons.

Most organs are innervated by both sympathetic and parasympathetic nerve fibers (see Fig. 10–1*b*). As a general rule, one system increases the activity of the target organ, whereas the opposing system decreases its activity. For example, heart rate is increased by sympathetic nerve stimulation and decreased by parasympathetic stimulation. In addition, changes in end organ function are often brought about by simultaneously activating one of the ANS divisions while inhibiting the other.

The ENS is a complex and extensive system, having approximately 10^8 neurons — the same number of neurons as in the spinal cord. The independence of function of the ENS is demonstrated by peristalsis in intestinal preparations completely isolated from the CNS. Peristalsis is a rhythmic, coordinated contraction and relaxation of intestinal muscles that moves food and waste products down the intestine.

The ENS has receptors for sensing hydrogen ion concentration (chemoreceptors), stretch (mechanoreceptors), and temperature (thermoreceptors) that provide sensory input to the enteric nervous system and the CNS. The ENS uses these sensory inputs for activating reflexes and programmed sequences of motor activity to excite or inhibit gastrointestinal smooth muscles (Fig. 10–3). The gut has two smooth muscle layers; the innermost layer is arranged in a circular fashion, whereas the outer layer is oriented longitudinally along the gut. A network of interconnected neurons, called the *myenteric plexus*, lies between these layers. A second plexus, the submucosal, or Meissner's, plexus, lies in the submucosal layer. These plexuses are connected with each other and receive input from sensory receptors and from the sympathetic and parasympathetic divisions of the ANS (Fig. 10–4). The plexuses contain excitatory and inhibitory neurons that use a wide variety of different neurotransmitters. While the ENS

can function independently, the input from the other divisions of the ANS coordinate the responses of the gastrointestinal system to conform with the needs of the body.

The Central Nervous System Coordinates Organized Activity of the Organs of the Body

Organization of the ANS in the spinal cord is somewhat similar to that of the somatic nervous system. The preganglionic nerve cells are located in the interomediolateral cell column and have fibers that leave the spinal cord through the ventral roots (see Fig. 10–2), whereas sensory fibers that carry information from the target organ enter the spinal cord through the dorsal roots. The cell bodies of these sensory fibers are located in the dorsal root ganglia. This spinal organization is identical for the sympathetic and the parasympathetic components of the ANS.

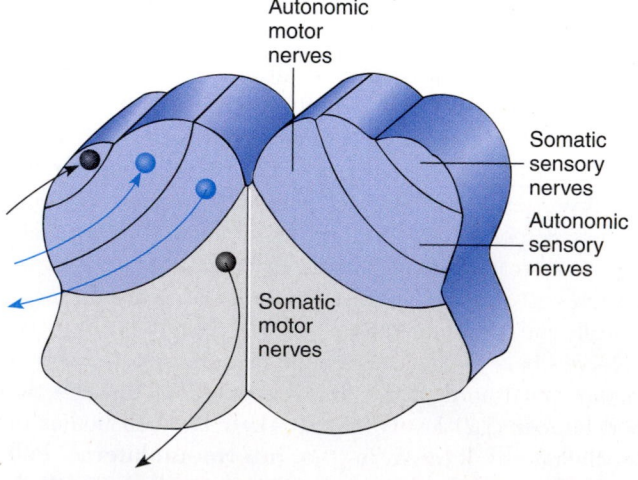

Figure 10–5

Autonomic sensory and motor nerves enter and leave the brain stem from different regions than do somatic nerves.

Sympathetic fibers, however, emerge from the thoracic and lumbar segments of the spinal cord, whereas spinal parasympathetic fibers emerge from sacral segments of the cord.

The organization of the ANS in the brain stem is different from that in the spinal cord. The autonomic sensory fibers coming into the brain stem are segregated from the somatosensory fibers (Fig. 10–5). Autonomic efferent nerves leave the brain stem in more lateral regions than do somatic efferents. The hy-

pothalamus and other structures of the limbic system also are involved in autonomic control of organ function.

Neurotransmitters in the Autonomic Nervous System Exhibit a Unique Pattern of Distribution

Figure 10–6 compares the structure and neurotransmitter of the somatic nervous system with those of the sympathetic and parasympathetic divisions of the ANS. The motor neurons of

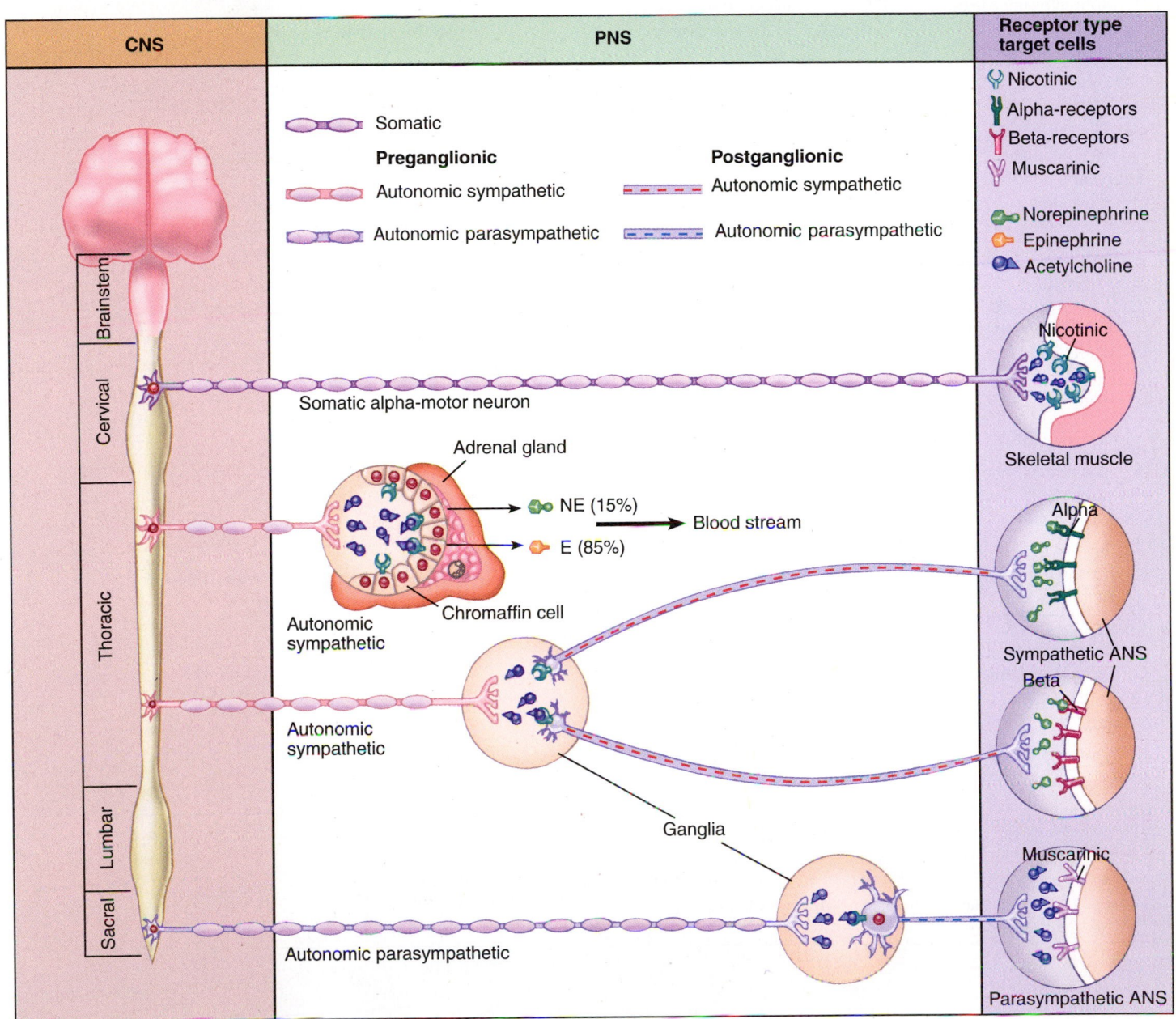

Figure 10–6

Structures and neurotransmitters of the somatic, parasympathetic, and sympathetic nervous systems. All preganglionic neurons release acetylcholine (ACh). Postganglionic parasympathetic nerve cells also release ACh, whereas postganglionic sympathetic cells release norepinephrine (NE). Exceptions to this, however, are fibers from sympathetic nerve cells that innervate sweat glands, and release ACh as well. Receptors for ACh are either muscarinic (M) or nicotinic (N); receptors for NE are either α or β receptors. Note similarities of the adrenal medulla and sympathetic ganglion.

TABLE 10–1

Summary of Cholinergic and Adrenergic Influences on Effector Organs

Effector Organs	Cholinergic Impulses	Adrenergic Impulses	
	Response	*Type**	*Response*
Eye			
Radial muscle of iris	—	α	Contraction (mydriasis)
Sphincter muscle of iris	Contraction (miosis)		—
Ciliary muscle	Contraction for near vision	β_2	Relaxation for far vision
Heart			
S–A node	Decrease in heart rate	β_1	Increase in heart rate
Atria	Decrease in contractility conduction velocity	β_1	Increase in contractility
A–V node and conduction system	Decrease in conduction velocity	β_1	Increase in conduction velocity
Ventricles	Slight decrease in contractility	β_1	Increase in contractility, conduction velocity, automaticity, and rate of idiopathic pacemakers
Arterioles			
Coronary	†	α	Constriction
		β_2	Dilation
Skin and mucosa	†	α	Constriction
Skeletal muscle	†	α	Constriction
		β_2	Dilation
Cerebral	†	α	Constriction (slight)
Pulmonary	†	α	Constriction
		β_2	Dilation
Abdominal organs	†	α	Constriction
		β_2	Dilation
Salivary glands	Dilation	α	Constriction
Kidney			
Proximal tubule	—	α	Stimulation of sodium reabsorption
Renin secretion	—	β_2	Increase renin secretion
Lung			
Bronchial muscle	Contraction	β_2	Relaxation
Bronchial glands	Stimulation		Uncertain
Stomach			
Motility and tone	Increase	α_2, β_2	Decrease (usually)
Sphincters	Relaxation (usually)	α	Contraction (usually)
Secretion	Stimulation		Inhibition (?)

*Where adrenergic receptor type has been established, α receptors are all α_1 and β receptors are all β_1 unless indicated otherwise.

†Parasympathetic influences are not present or are of minimal or uncertain physiological importance.

Modified from Goodman & Gilman, *The Pharmacological Basis of Therapeutics*, ed 9. New York, McGraw-Hill, 1996.

the somatic nervous system release acetylcholine onto nicotinic receptors to cause excitation at skeletal muscle neuromuscular junctions. The two major neurotransmitters in the ANS are **acetylcholine** and **norepinephrine (NE).** Acetylcholine is released at the synaptic terminals of all preganglionic neurons. In postganglionic nerve cells, NE is found in the terminals of sympathetic fibers, whereas acetylcholine is found in the terminals of parasympathetic fibers. Exceptions to this rule are that some sympathetic fibers to the arteries of skeletal muscle may release acetylcholine as do some postganglionic sympathetic fibers that innervate sweat glands. Nerve fibers that transmit acetylcholine are called **cholinergic** fibers. Fibers that release NE are referred to as **adrenergic** fibers.

Cholinergic receptors are classified as either **nicotinic** or **muscarinic** acetylcholine receptors, based on whether they are activated by the drug nicotine or muscarine. The acetyl-

TABLE 10–1

Summary of Cholinergic and Adrenergic Influences on Effector Organs

Effector Organs	Cholinergic Impulses	Adrenergic Impulses	
	Response	*Type**	*Response*
Intestine			
Motility and tone	Increase	α_2, β_2	Decrease
Sphincters	Relaxation (usually)	α	Contraction (usually)
Secretion	Stimulation		Uncertain
Gallbladder and ducts	Contraction	β	Relaxation
Urinary bladder			
Detrusor	Contraction	β_2	Relaxation (usually)
Trigone and sphincter	Relaxation	α	Contraction
Ureter			
Motility and tone	Uncertain	α	Increase
Uterus	Variable‡	α	Pregnant: contraction
		β_2	Nonpregnant: relaxation
Male sex organs	Erection	α	Ejaculation
Skin			
Pilomotor muscles	—	α	Contraction
Sweat glands	Generalized secretion	α	Localized secretion§
Spleen capsule	—	α	Contraction
		β_2	Relaxation
Adrenal medulla	Secretion of epinephrine and norepinephrine		—
Skeletal muscle	—	β_2	Glycogenolysis, K^+ uptake
Liver	—	α, β_2	Glycogenolysis and gluconeogenesis
Pancreas			
Acini	Secretion	α	Decreased secretion
Islets (β cells)	—	α	Decreased secretion
		β_2	Increased secretion
Fat cells	—	α, β_1	Lipolysis (thermogenesis)
Salivary glands	K^+ and water secretion	α	Potassium and water secretion
		β	Amylase secretion
Lacrimal glands	Secretion		Unclear
Nasopharyngeal glands	Secretion		—

‡Depends on stage of menstrual cycle, amount of circulating estrogen and progesterone, and other factors. Responses of pregnant uterus different from those of nonpregnant.

§On palms of hands and in some other locations ("adrenergic sweating").

choline receptor on the postganglionic neuron is a nicotinic receptor, whereas the acetylcholine receptor on target cells, such as smooth muscle, is muscarinic. NE activates two general types of receptors: (1) **α-adrenergic** and (2) **β-adrenergic** receptors. β-Receptors are distinguished from α-receptors by their greater sensitivity to isoproterenol, a drug similar to NE. α-Adrenergic receptors are subdivided into α_1 and α_2 subtypes based on their anatomical location. The α_1 receptor subtype is located on postsynaptic target cells of sympathetically innervated organs, whereas α_2 receptors are located primarily on the presynaptic terminals of adrenergic nerve fibers.

The β-receptor is also subdivided into subtypes β_1 and β_2. At the β_1 receptor, epinephrine and NE are more or less equipotent. In contrast, β_2 receptors respond better to epinephrine than to NE. Table 10–1 lists the effects of acetylcholine and norepinephrine on different organs.

APPLICATIONS OF PHYSIOLOGY

The Relaxation Response

The relaxation response is characterized by a decrease in the activity of the sympathetic nervous system, that results from conditioning and training. During this response, there is a decrease in oxygen consumption, heart rate, blood pressure, and respiration rate, and an increase in the alpha waves of the electroencephalogram. The relaxation response is not just simple relaxation. In simple relaxation, changes in the rate of respiration, oxygen consumption, and alpha wave activity do not occur.

Learning to induce the relaxation response is a common practice associated with various Eastern religions, such as Zen Buddhism. One can induce the relaxation response by sitting in a comfortable position, focusing on an object, and blocking out all other distractions while repeating a word or phrase.

The relaxation response is thought to modify the way in which stressful stimuli affect the sympathetic nervous system. A temporary activation of the sympathetic system is important in preparing the body to respond appropriately to stressful situations by the release of substances such as norepinephrine. The prolonged release of these substances, however, can have deleterious effects that include abnormally high blood pressure and a decrease in immune system function. Research indicates that the relaxation response reduces the body's

Man meditating, Shwedagon Pagoda, Yangon, Mayanmar. *(John Elk/Stone)*

response to norepinephrine secreted into the circulatory system.

The relaxation response produces changes in the sympathetic nervous system that extend outside of the training period. Clinical studies have shown that the relaxation response can alleviate heartbeat disorders, reduce pain, and lower blood pressure in hypertensive patients. If practiced daily, it may alleviate physiological conditions arising from undesirable increases in sympathetic nervous system activity and thus improve and maintain overall health.

Control of Organ Systems by the Sympathetic and Parasympathetic Systems

Sympathetic and parasympathetic inputs to an organ often have opposite effects. For instance, parasympathetic nerves release acetylcholine onto muscarinic receptors in the SA node of the heart to slow heart rate, whereas sympathetic nerves that release NE onto primarlily β-adrenergic receptors on the node have the opposite effect. Also parasympathetic stimulation to cardiac muscle reduces the strength of contraction, whereas sympathetic stimulation, by releasing NE onto β_1-adrenergic receptors, increases it. When the blood pressure becomes too low, activity of parasympathetic inputs to the heart is decreased while sympathetic inputs are increased. This increases the heart rate (the number of heart beats per minute) and the strength of each beat. The result is the heart pumps more often and ejects

more blood with each beat. These factors help to increase blood pressure.

Most arteries contain α_1- and β_2-adrenergic receptors. The α_1 receptors mediate vasoconstriction to NE and epinephrine, whereas β_2 receptors mediate vasorelaxation. In most organs, the α_1 receptors far outnumber the β_2 receptors on arterial smooth muscle. Thus, in most organs arteries and veins constrict in response to NE released from sympathetic nerve terminals and from epinephrine and NE released from cells in the adrenal medulla. Pressure inside veins increases when veins constrict which then helps to fill the heart with blood, that, along with the increased resistance to flow caused by constricting arteries, elevates blood pressure (Fig. 10–7). Little evidence shows significant parasympathetic influence on arteries or veins. Neurally mediated relaxation of blood vessels results from a reduc-

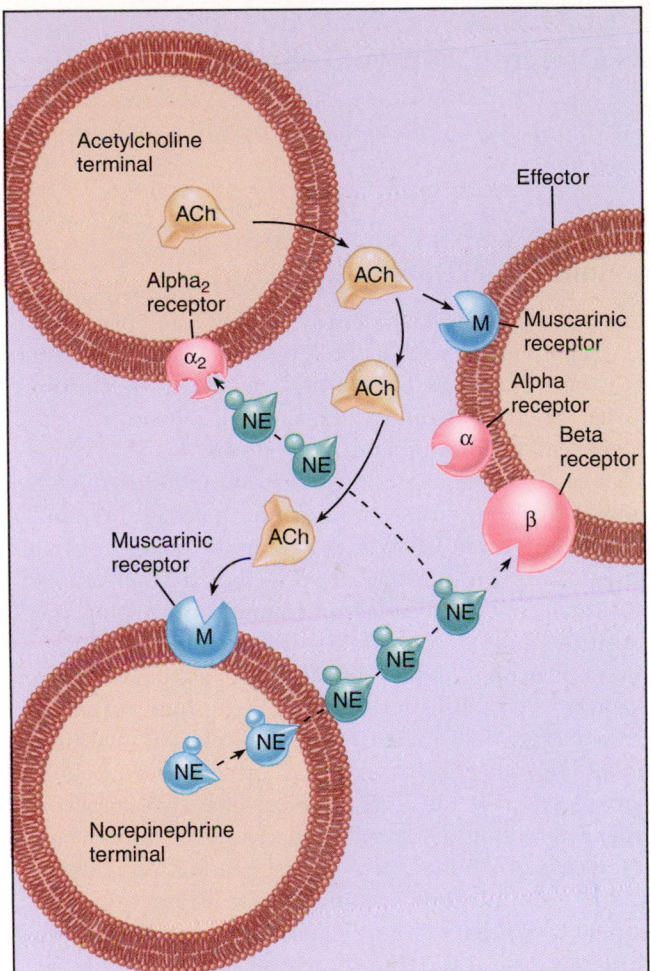

tion of sympathetic nerve stimulation and helps to reduce arterial pressure.

Receptors for NE and acetylcholine also exist in the membranes of nerve terminals. α_2-Receptors exist in the presynaptic terminals of preganglionic sympathetic and parasympathetic nerve fibers and inhibit the release of transmitter from those nerves. In addition, NE release from sympathetic neurons is inhibited by α_2 and muscarinic receptors on presynaptic terminals of sympathetic neurons (Fig. 10–8). These receptors, therefore, allow the ANS to enhance the effect of either the sympathetic or

Figure 10–8

Receptors for norepinephrine and acetylcholine located in the presynaptic terminal membranes can inhibit the release of these neurotransmitters, allowing the ANS to modulate sympathetic or parasympathetic activity.

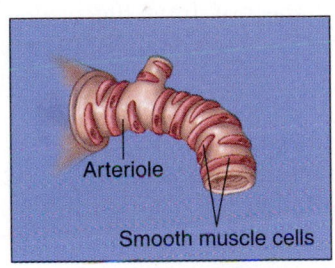

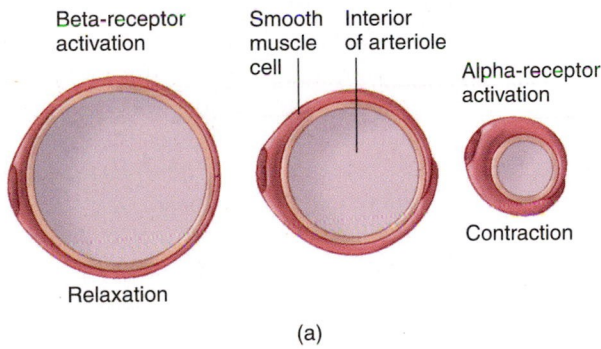

Figure 10–7

(a) α_1-Adrenergic receptors cause smooth muscle cells lining blood vessels to contract, reducing the inner space in the vessel and thereby increasing blood pressure. β_2-Adrenergic receptors cause smooth muscle cells to relax, thus widening vessels and decreasing blood pressure. *(b)* The net response, however, is a contraction of smooth muscle cells.

parasympathetic actions while inhibiting the opposite action at the level of the target organs. For example, when sympathetic inputs to the heart release NE, causing the heart to beat faster, the NE also binds to presynaptic α_2 receptors located on acetylcholine nerve terminals. The activation of the α_2 receptors inhibits the release of acetylcholine, which would normally decrease the heart rate. Likewise, when parasympathetic inputs to the heart release acetylcholine and reduce the heart rate, acetylcholine also binds to muscarinic receptors located on NE terminals. The activation of these muscarinic receptors decreases the amount of NE released, which would otherwise increase the heart rate.

REFLEXES GOVERNED BY THE AUTONOMIC NERVOUS SYSTEM

What are some of the actions of the autonomic nervous system?

Spinal Reflex Arcs Can Activate Autonomic Functions

A more complex coordination of sympathetic and parasympathetic responses is found at the level of the spinal cord and brain stem. These responses are viewed as **autonomic reflexes,** whereas at the hypothalamic level the responses are viewed as **autonomic control systems.**

The status of the target organ is transmitted to the CNS by sensory pathways using transduction mechanisms similar to those described in Chapter 8. Within the spinal cord, this information is then transmitted up to the brain and to neurons within the intermediolateral column of the spinal cord. Because the cells in the intermediolateral column provide the output from the ANS to the target organs, a loop of nerve cell activity is established; this forms the **spinal autonomic reflex arc** (Fig. 10–9). The formation of this neuronal circuit starts with the detection of visceral information by autonomic sensory receptors in the organ. Sensory fibers in turn activate interneurons within the spinal cord. These neurons synapse with the cells of the intermediolateral column. The intermediolateral cells integrate incoming information from many sensory neurons that enter the spinal cord within one spinal cord segment and discharge to activate postganglionic neurons. Finally, the postganglionic cells convey information to the target organ, causing it to make the appropriate responses. This reflex arc is quite similar to other reflex circuits, such as the knee-jerk reflex of the somatic nervous system (see Chapter 9).

The particular response in the autonomic reflex system varies with the target organ. In the bladder, this reflex is responsible for urination; in the colon and rectum, it is responsible for defecation.

Reflexes in the Bladder Aid in Urination

The bladder stores urine produced by the kidneys. It can store about 150 to 200 ml before the urge to urinate occurs. The ability to keep urine in the bladder is called **continence,** and the emptying of the bladder is called **micturition.** The bladder (Fig. 10–10a) contains a wall of smooth (involuntary) muscle (detrusor) that is innervated by sensory nerve endings. These nerves function as stretch receptors, sensing the enlargement of the bladder. Two groups of muscles close off the exit from the bladder and prevent urine from leaving. The first group is made of smooth muscle and is called the **internal sphincter.** These muscles are innervated by parasympathetic nerve fibers in the pelvic nerves and by sympathetic nerve fibers in the hypogastric nerves. The second group of muscles are striated (voluntary) muscles and are called the **external sphincter.** They are innervated by motor neurons that form the pudendal nerve. These motor neurons are under voluntary control of higher brain centers.

Sympathetic signals facilitate bladder filling by relaxing the muscles of the bladder wall and inhibiting parasympathetic ganglion activity to decrease bladder wall contraction. As urine begins to fill and distend the bladder, the stretch receptors become activated (Fig. 10–10b) and discharge action potentials. The action potentials from sensory fibers activate preganglionic parasympathetic nerve cells that result in the contraction of the smooth muscles surrounding the wall of the bladder. These actions constitute the urinary bladder reflex in the spinal cord.

In order for the bladder to empty, however, the internal and external sphincter muscles must be placed in a relaxed state. This involves the activation of a "micturition center" in the brain stem by inputs from the sensory stretch receptors that sense the enlargement of the bladder. The output from the micturition center inhibits the preganglionic sympathetic neurons and alpha motor neurons that normally cause the internal and external sphincters to contract. In addition, the micturition center further excites the preganglionic parasympathetic cells that cause the smooth muscles surrounding the bladder to contract and the internal sphincter to relax. Thus, the spinal bladder reflex is aided by the micturition center to coordinate the emptying of the bladder.

Individuals who have been in accidents that crushed their spinal cords can still control the bladder reflex. If the damage is above the level of the sacrum, the spinal reflex is intact. The reflex is initially lost, however, as a result of "spinal shock" but recovers over the next one to five weeks. Urination after spinal cord damage occurs with great frequency and is initiated when the bladder is slightly filled. This occurs because of the absence of descending inputs that cause the internal and external sphincters to contract. The

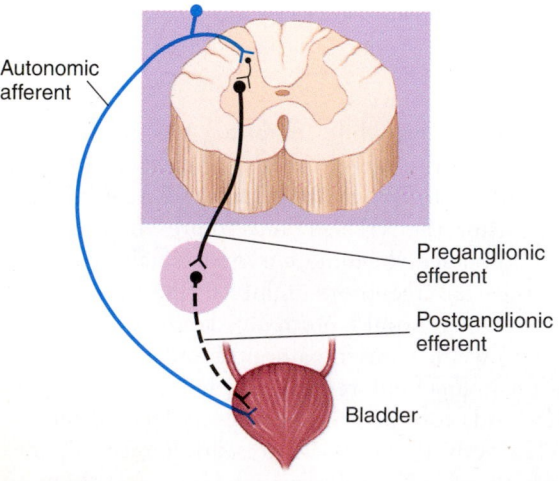

Autonomic afferent

Preganglionic efferent

Postganglionic efferent

Bladder

Figure 10–9

A spinal autonomic reflex arc.

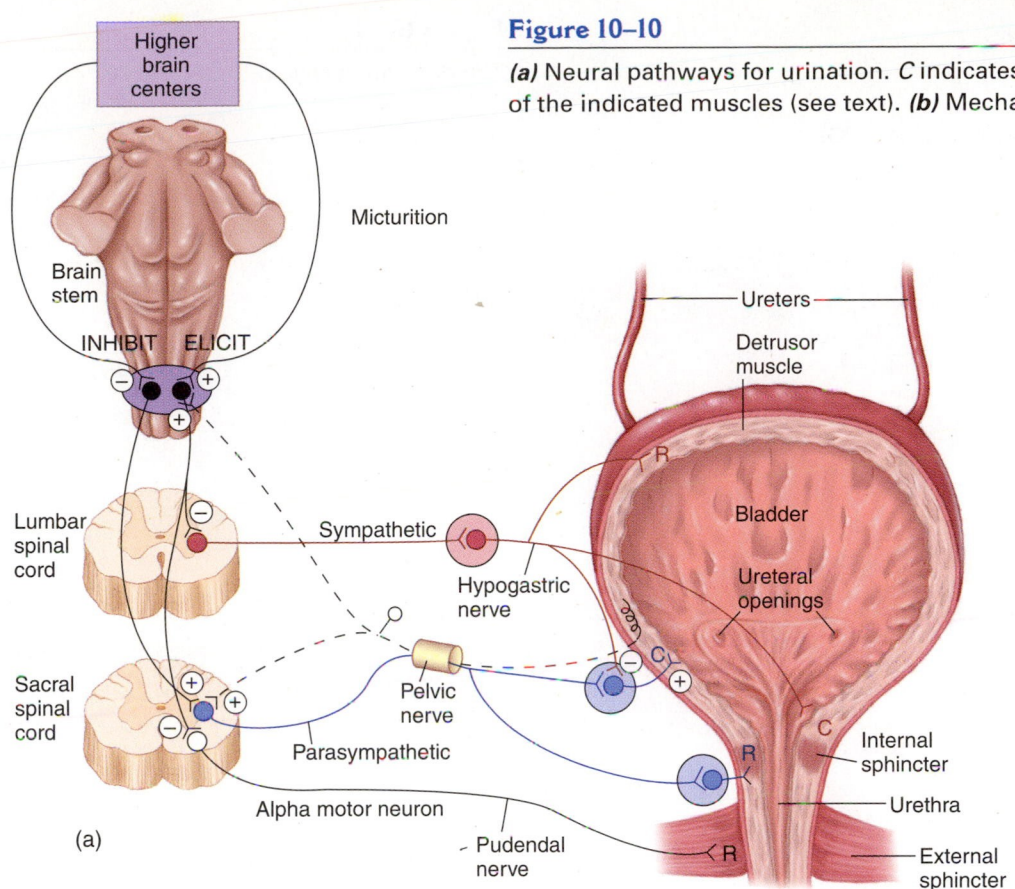

Figure 10–10

(a) Neural pathways for urination. *C* indicates contraction; *R* indicates relaxation of the indicated muscles (see text). *(b)* Mechanisms of urination.

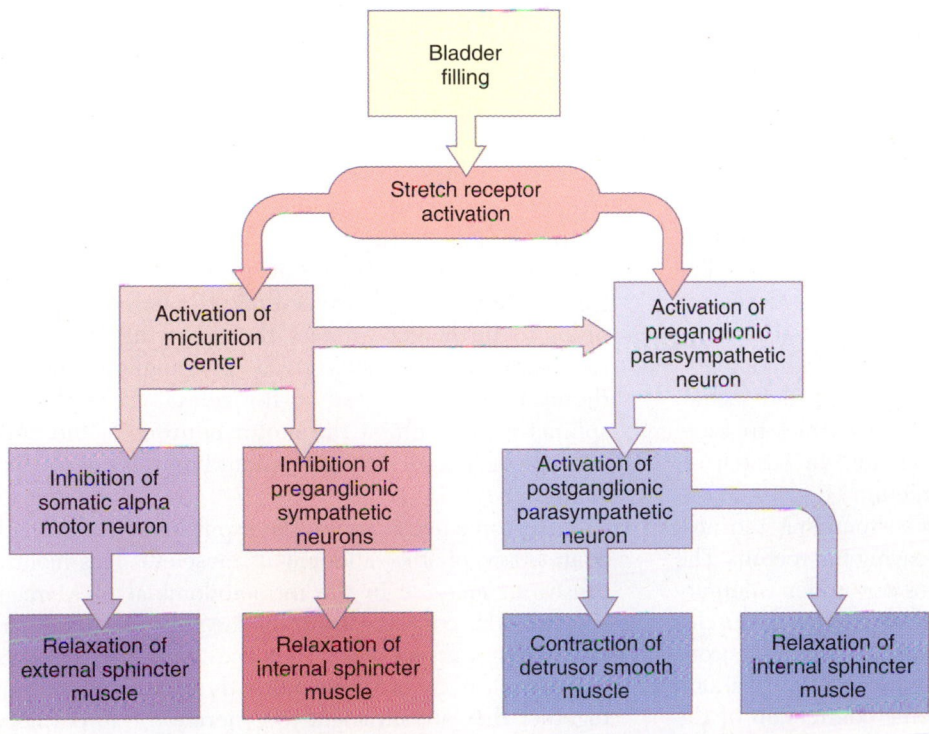

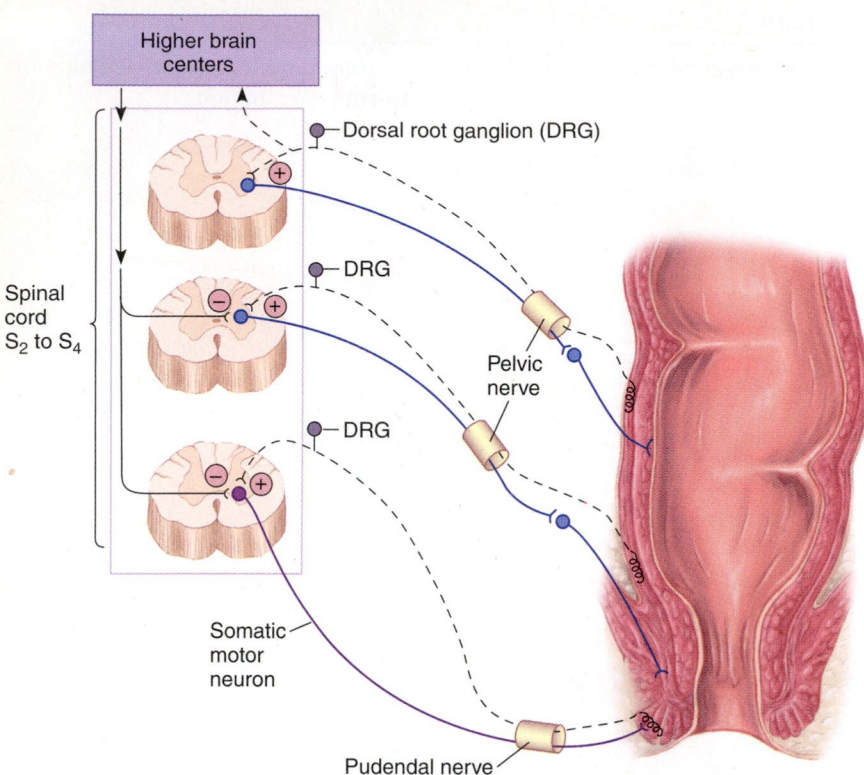

Figure 10–11

The defecation reflex.

bladder reflex can also be initiated after spinal cord damage by tapping the lower abdomen, which activates the segmental cutaneous-visceral reflex.

Reflexes in the Rectum Aid in Defecation

The colon and rectum store unabsorbed food products or fecal (waste) material from the small intestine. In an analogous manner to the urinary system, the ability to retain fecal material within the colon and rectum is also called **continence,** and the removal of fecal material is called **defecation.** Continence is controlled by spinal reflexes and descending inputs from higher brain centers, and is maintained until approximately 2 L of feces have accumulated.

As with the bladder, the walls of the colon and rectum are surrounded by smooth muscles that contract to expel fecal material. These muscles are innervated by stretch receptors that detect the filling of the rectum (Fig. 10–11). A set of internal and external muscles surrounding the anal opening prevents fecal material from leaving the rectum. The internal sphincters are smooth muscles not under voluntary control, whereas the external sphincters are striated muscles under voluntary control and innervated by motor neurons from spinal segments S2 to S4. Typically, both internal and external sphincters are closed. The tonic contraction of the external sphincter is maintained by the activation of the motor neurons from sensory nerves from the skin and from the external sphincter muscle itself.

As the rectum fills with fecal material, stretch receptors become activated and transmit this information to the spinal cord via the pelvic nerve. Within the spinal cord, these sensory inputs excite preganglionic parasympathetic neurons that in turn excite postganglionic nerve cells innervating the smooth muscles of the colon and rectum. This causes the smooth muscles to contract, forcing fecal material to move toward the anal opening. The sensory stretch receptors also transmit information to higher brain centers to initiate the urge to defecate. These higher centers activate descending inputs to the parasympathetic preganglionic nerve cells to facilitate the defecation reflex. In order to further coordinate the emptying of the rectum, the descending inputs activate preganglionic parasympathetic neurons that lead to the relaxation of the internal sphincters and inhibit the motor neurons of the pudendal nerve, which innervates the striated muscles of the external sphincters.

The force necessary for the expulsion of feces is aided by contraction of the abdominal muscles. This contraction causes an increase in the intra-abdominal pressure, which squeezes the contents inside of the rectum and colon. This increase in pressure works in concert with the contraction of the smooth muscles surrounding the colon and rectum, and together they are sufficient to generate the forces necessary for fecal expulsion. Thus, descending inputs from sensorimotor areas of cortex orchestrate defecation by coordinating both autonomic and somatic actions.

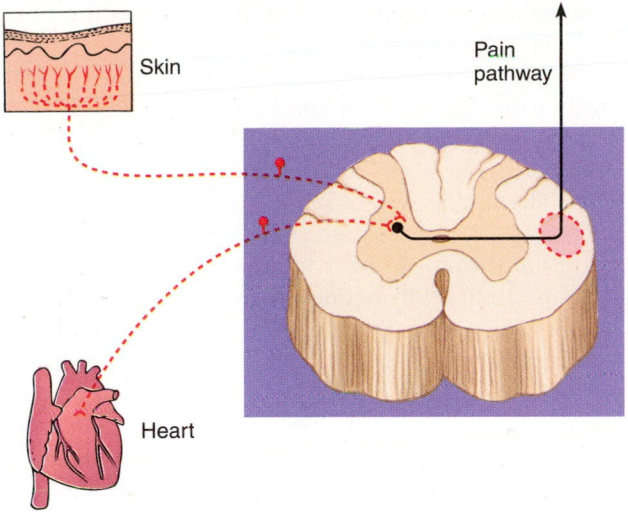

Figure 10–12

An example of a referred pain pathway.

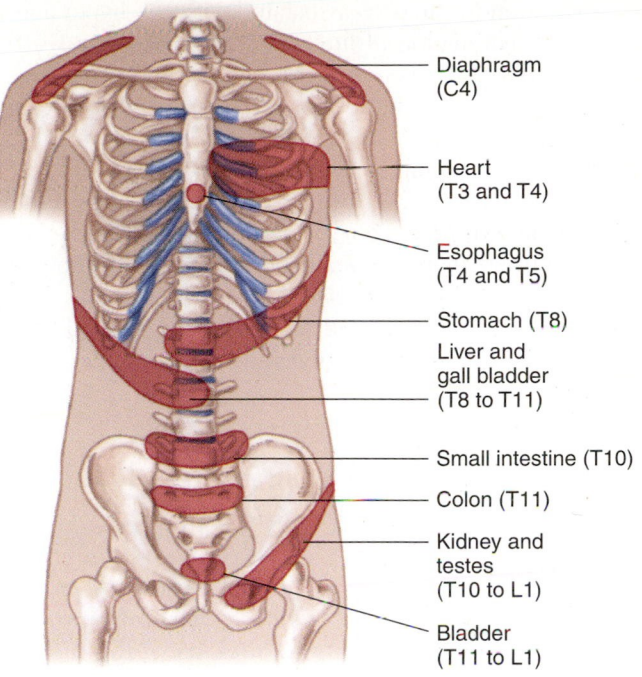

Figure 10–13

Hyperalgesic zones for pain originating in the major internal organs.

The defecation reflex is lost in individuals who have been in accidents that destroyed the sacral segments of their spinal cord. However, transections of the spinal cord above the sacrum preserve the reflex, and defecation can occur with the manual expansion of the external and internal sphincters.

Pain from Internal Organs Is Localized to Other Parts of the Body

People who experience stomach pains and heart attacks commonly perceive that their pain comes from the surface of their bodies. Because the pain is attributed to a location other than its source, it is called **referred pain.** This displaced perception of pain is a result of nerve fibers from pain receptors from internal organs and pain receptor fibers from the surface of the body converging onto the same neurons of the pain pathway.

This convergence of information occurs with visceral and somatic pain fibers that enter the spinal cord within the same spinal segment (Fig. 10–12). Pain fibers from the heart, for example, enter the spinal cord at thoracic segments T1 to T3 and synapse with neurons in the **substantia gelatinosa** of the pain pathway (see Chapter 8). Pain information from the left chest and the upper portion of the left arm also synapse with these same neurons. The brain associates pain information from spinal segments T1 to T3 with the left chest and arm; thus, heart attacks are often perceived as pain in the left arm and chest.

The painful areas on the surface of the body that are associated with inflammations of internal organs are called **hyperalgesic zones.** Figure 10–13 shows the hyperalgesic zones for pain originating in the heart, stomach, kidney, liver, gallbladder, and other internal organs. The hyperalgesic zones are useful in diagnosing disorders of visceral organs.

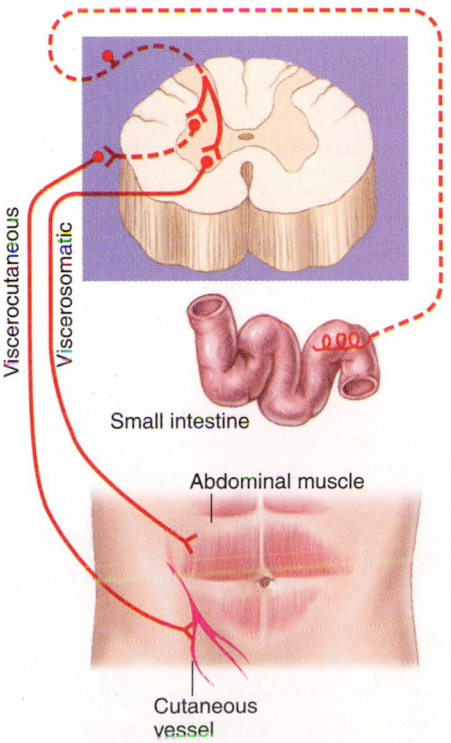

Figure 10–14

An example of a segmental spinal autonomic reflex, in which reddening of the skin signals internal inflammation.

In addition to the pain associated with the hyperalgesic zone, the inflammation of internal organs often causes the skeletal muscles in this zone to tighten and the skin to redden (Fig. 10–14). These signs are often used to help determine which organ is in distress. The reddening of the skin and the tightening of the muscles are results of reflex connections. The inputs from sensory fibers also inhibit preganglionic sympathetic neurons and thus decrease vasoconstriction and increase blood flow to the skin. The segmental spinal reflex connections, therefore, occur within the appropriate spinal cord segments where visceral sensation is received and transmitted outward through somatic and autonomic pathways.

REGULATORY SYSTEMS OF THE AUTONOMIC NERVOUS SYSTEM

Autonomic reflexes represent the simplest level of ANS control. Actions of the ANS involving the **limbic system,** the **hypothalamus,** the **solitary nucleus** of the medulla, and other **brain stem nuclei** are part of **autonomic regulatory systems,** which are more complex than spinal cord reflexes. The **limbic system** has been termed the "cerebral cortex of the ANS." Cortically stored past experiences, which may or may not be recalled at a conscious level, can be evoked by external stimuli, such as smells, sounds, or sights, to trigger emotional reactions leading to strong visceral responses coordinated by the ANS. The limbic system represents one of the highest levels of the hierarchy of normal control of the ANS. The exact position of conscious, cortical control of the ANS is unclear, although it has been shown to be capable of "overruling" other levels of control.

The hypothalamus coordinates ANS responses at the next, simpler level.

The limbic system is a network of structures of the more primitive regions of the cerebral cortex (Fig. 10–15). These structures are interconnected by intricate circuits that interact with both higher and lower regions of the brain. The limbic system seems to be the storage area for memories of experiences that shape subconscious feelings. Stimulation of areas of the limbic system can evoke a broad range of feelings and behaviors, including rage, anger, fear, and aggression. In addition, stimulation of the limbic system, either directly or by input from the senses, can evoke ANS-mediated physiological changes, such as increased heart rate, sexual arousal, and nausea. There is tremendous variation in the stimuli that affect individuals, as well as the magnitude of each individual's response. Religious and patriotic symbols, poetry and music may affect certain individuals. The arts are generally intended to evoke some degree of autonomic response. The range of magnitude of response can be from a remote feeling to, at the extreme, cardiac arrest. The setting in which a stimulus is received is another important factor in determining the quality and degree of the response. A shark, as experienced in the introductory vignette as a life-threatening predator, could be viewed with abject terror, but the same individual viewing the same shark in a controlled situation, such as a marine aquarium, may be enthralled with the creature's streamlined form and grace of motion. The participation of a range of conscious and unconscious stored information in the cortex has great influence on autonomic responses.

The important role of the hypothalamus in regulation of the ANS is discussed later. However, several brain stem sys-

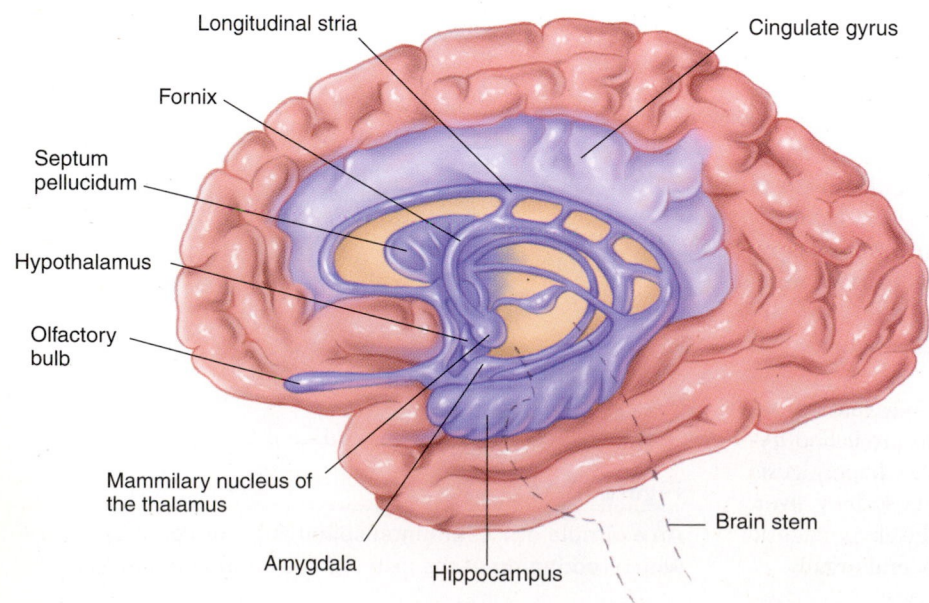

Figure 10–15

Anatomy of the limbic system shown by the colored areas of the figure. *(Guyton, A. C., and Hall, J. C.* Textbook of Medical Physiology, *ed 9. Philadelphia, W.B. Saunders, 1996; from Warwick and Williams.* Gray's Anatomy, *ed 35 Br. London, Longman Group, Ltd., 1973.)*

Longitudinal stria
Fornix
Septum pellucidum
Hypothalamus
Olfactory bulb
Mammilary nucleus of the thalamus
Amygdala
Hippocampus
Brain stem
Cingulate gyrus

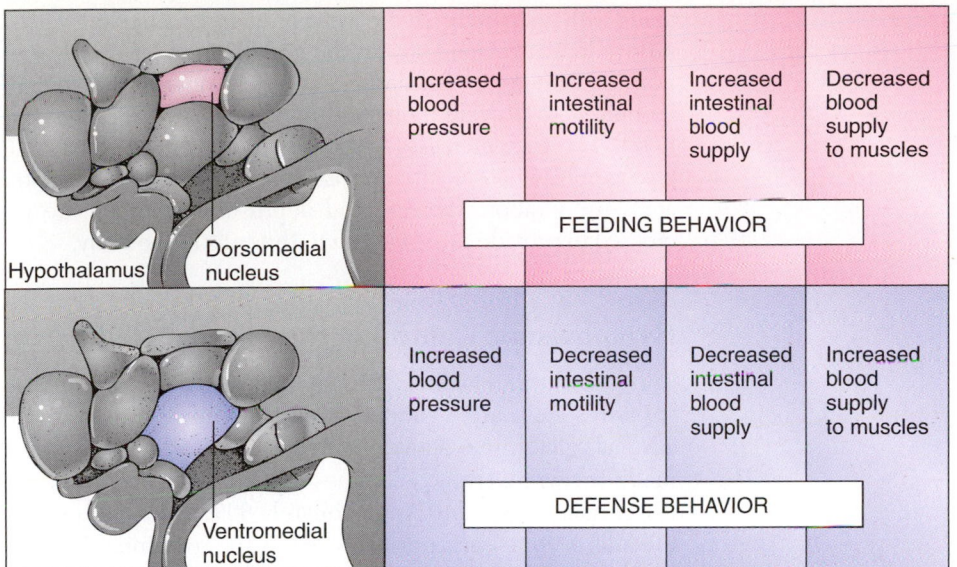

Figure 10–16

The hypothalamus coordinates the functions of different organs. Activation of the dorsal regions of the hypothalamus results in "feeding" behaviors, whereas activation of ventral regions produces the "fight or flight" response to a threat.

tem nuclei, including those mediating cardiac and respiratory activity, can function without hypothalamic regulation. The nucleus of the solitary tract in the medulla is the center for coordinating these activities. Nevertheless, this nucleus has extensive connections with structures at higher and lower levels, including the hypothalamus, limbic system, and other brain stem nuclei.

These regulatory systems are feedback controllers that maintain a constant internal environment for the optimal functioning of the various organs and their cells. The important regulatory role of the ANS will be emphasized by discussions in subsequent chapters on endocrine function, respiration, blood pressure regulation, gastrointestinal control, and renal function.

Autonomic Functions of the Hypothalamus

Because of its important and extensive regulatory functions, the hypothalamus is known as the main or head ganglion of the ANS. The activation of different regions of the hypothalamus produces a variety of coordinated autonomic responses. The activation of the **dorsal hypothalamus,** for example, increases blood pressure, intestinal motility, and intestinal blood supply but decreases the blood supply to the skeletal muscles (Fig. 10–16). These responses are associated with feeding behaviors.

Activation of the **ventral hypothalamus** increases blood pressure and the blood supply to the skeletal muscles but decreases intestinal motility and blood flow to the intestines. These responses are associated with defense behaviors or the so-called **"fight or flight" responses** (Table 10–2). For example, at the onset of intense pain or sudden fright, blood vessels to the skeletal muscles dilate, and heart rate increases

TABLE 10–2
Physiological Changes in the Fight or Flight Response
Fight or Flight Response

1. Increased blood pressure
2. Increased heart rate
3. Increased force of heart conraction
4. Increased heart conduction velocity
5. Increased depth and rate of respiration
6. Shift of blood flow distribution away from the skin and splanchnic regions and more to skeletal muscles and heart
7. Mobilization of liver glycogen to glucose (glycogenolysis)
8. Mobilization of free fatty acids from adipose tissue (lipolysis)
9. Contraction of spleen capsule (increased hematocrit)
10. Mydriasis (widening of pupil)
11. Accommodation for far vision (relaxation of ciliary muscle)
12. Widening of palpebral fissure (eyelids open wide)
13. Piloerection
14. Inhibition of gastrointestinal motility and secretion, contraction of sphincters
15. Sweating ("cold" sweats as skin blood vessels are constricted)

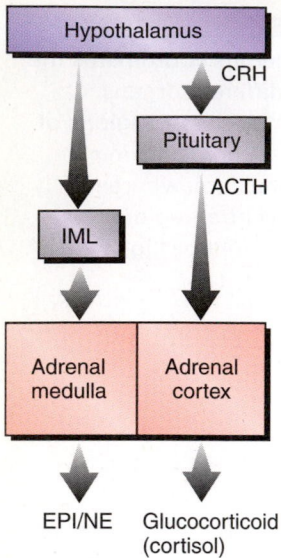

Figure 10–17

Hypothalamic control of the adrenal gland during the stress response. IML = interomediolateral nuclei; CRH = corticotropic-releasing hormone; ACTH = adrenocorticotropic hormone; EPI = epinephrine; NE = norepinephrine.

along with an increase in the force of heart contraction. These actions increase blood flow to the skeletal muscles in preparation for running or fighting. In addition, energy sources are mobilized to provide more fuel for the skeletal muscles and heart. The liver increases blood glucose levels by converting glycogen to glucose. Adipose tissue is converted to free fatty acids. In addition, the rate and depth of respiration are increased to provide more oxygen to the cells in the body.

Hypothalamic Control of the Adrenal Medulla

The coordinated actions of the hypothalamus during the stress response are mediated in part by the activation of the adrenal gland. As shown in Figure 10–17, the adrenal medulla secretes catecholamines (epinephrine and NE) into the circulatory system. Approximately 80% of the cells in the medulla secrete epinephrine, whereas the remaining 20% secrete NE. The release of catecholamines from the adrenal medulla is carried out by the direct connection of nerve fibers from the hypothalamus to the intermediolateral, and by nerve fibers from the intermediolateral to cells in the medulla. The catecholamine-secreting cells of the adrenal medulla, therefore, function as postganglionic sympathetic cells. The cells of the adrenal medulla are homologous to

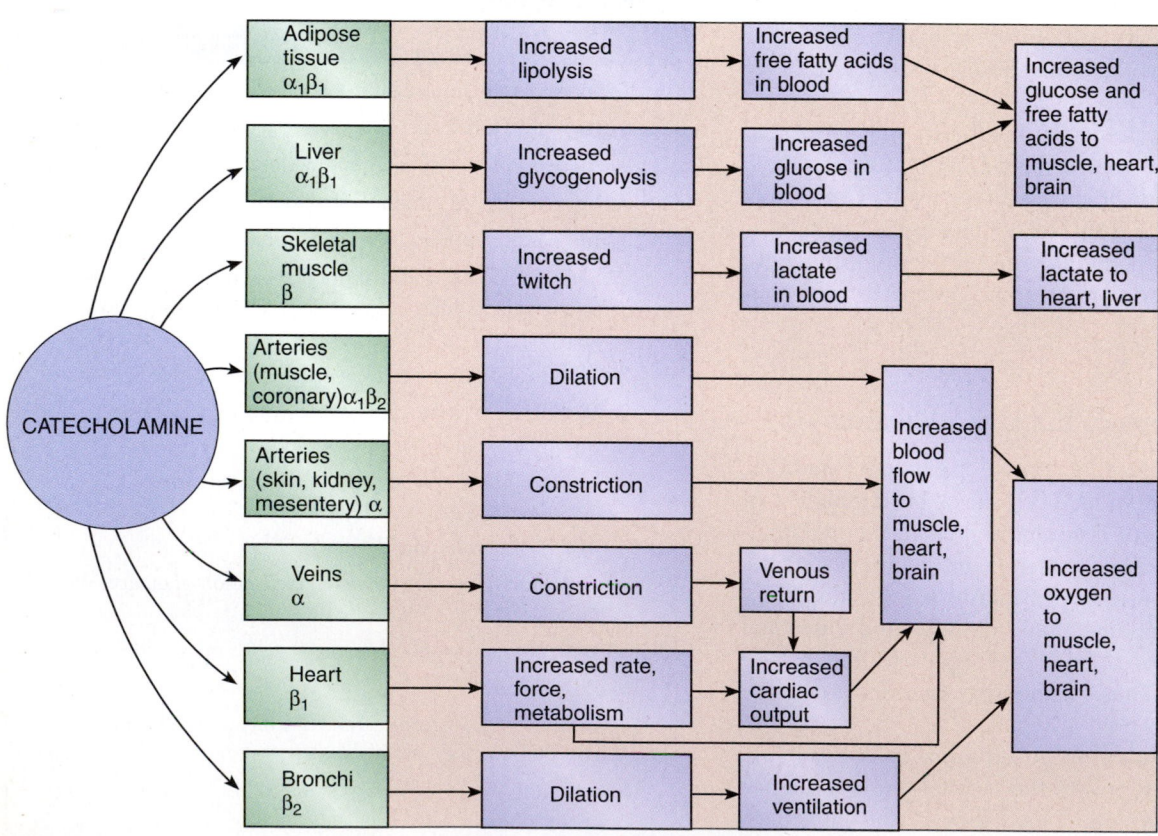

Figure 10–18

Circulating catecholamines produced by the adrenal medulla affect the same receptors and target organs that postganglionic sympathetic neurons do.

CURRENT CONCEPTS IN PHYSIOLOGY

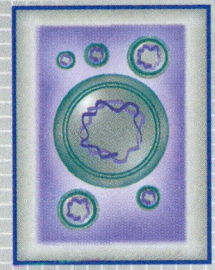

Chronic Stress and the Loss of Brain Cells

Glucocorticoids are steroids released by the adrenal cortex during stress. The work of Robert Sapolsky and his colleagues at Stanford University has shown that prolonged elevation of glucocorticoids can lead to the degeneration of nerve cells in rodents and monkeys. The area of the brain most sensitive to glucocorticoid damage is the hippocampal formation, a region long known to be involved in learning and memory. Earlier studies by Bruce McEwen and coworkers at the Rockefeller University revealed that the hippocampus is a major target site in the brain for circulating glucocorticoids.

In studies of vervet monkeys housed in primate colonies in Kenya, Sapolsky and his coworkers found that socially subordinate monkeys exhibited gastric ulcers, high incidences of bite wounds, and hyperplastic adrenal cortices suggestive of sustained glucocorticoid release. More-over, these characteristic signs of social stress were associated with a severe degeneration of nerve cells within the hippocampal formation.

In experiments to determine whether glucocorticoid secretion alone could lead to neuronal loss, glucocorticoid-containing pellets were implanted directly into the hippocampi of non–socially stressed monkeys. These animals ultimately displayed the same type of nerve cell loss as did socially stressed monkeys.

Other studies have shown that chronic stress causes premature aging of hippocampal brain activity. Conversely, reducing the exposure of glucocorticoids can protect against the age-related loss of hippocampal neurons. Furthermore, glucocorticoids in concentrations that are not in themselves toxic can impair the ability of hippocampal nerve cells to survive neurological insults, such as reductions in blood flow and oxygen to the brain, and seizures. Together, these studies suggest that prolonged exposure to glucocorticoids or stress can accelerate the loss of hippocampal neurons as well as increase their susceptibility to neurological insults.

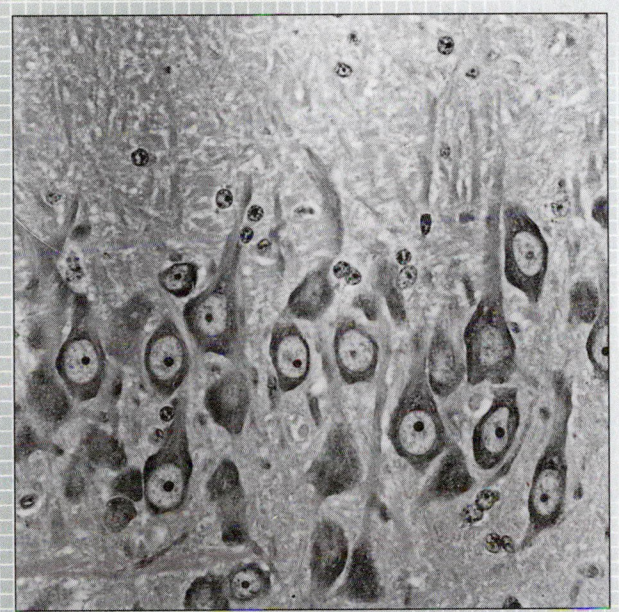 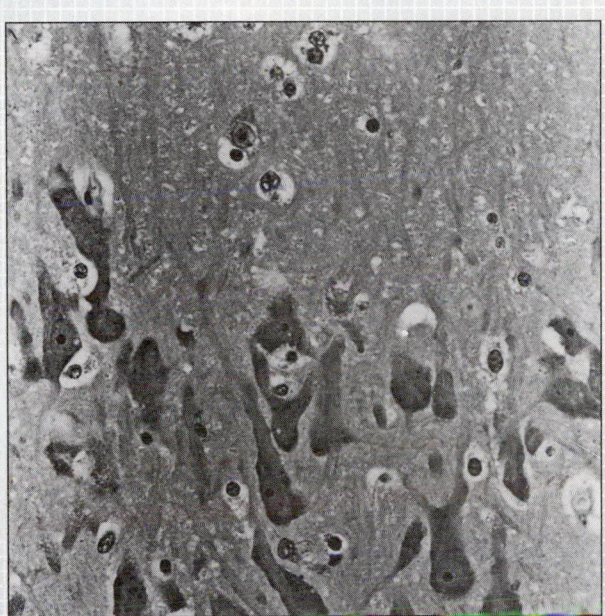

Effects of chronic stress on brain cells. Left panel shows the normal nerve cells of the hippocampal formation. Right panel shows the degenerating hippocampal nerve cells from the brain of a chronically stressed monkey. *(Courtesy of Hideo Uno, MD, PhD)*

postganglionic neurons because they are directly innervated by preganglionic sympathetic nerve fibers.

The catecholamines that circulate in the blood affect the same receptors and target organs as do postganglionic sympathetic nerve cells (Fig. 10–18). Because they enter the blood, however, their actions are not as discrete as those generated by sympathetic nerve cells. Circulating catecholamines cause the air passages in the lungs to dilate and increase the supply of oxygen to the skeletal muscles and heart. In the cardiovascular system, circulating catecholamines cause the heart to beat faster and with greater force; they also cause the constriction of arterioles in the skin, mesentery, and kidney. In contrast, the coronary arteries, which contain primarily β_2-adrenergic receptors, dilate, whereas blood vessels in the brain, which contain virtually no adrenergic receptors, are unaffected by the circulating catecholamines. Also, β_2 receptors on skeletal muscle arteries are activated by low levels of epinephrine that are not high enough to activate the α_1 constricting receptors. Thus, early after the release of catecholamines from the medulla, skeletal muscle arteries can dilate and skeletal muscle blood flow (and oxygen delivery) can increase. The net effect of these actions on the cardiovascular system is to provide more blood to skeletal muscles and heart and to reduce blood flow to organs in the body not essential to "fight or flight."

Circulating catecholamines also increase the supply of fuel for cells in the form of glucose and free fatty acids. Catecholamines stimulate **glycogenolysis** in the liver. Glycogenolysis produces glucose molecules from glycogen, which increases the amount of glucose in the blood. Circulating catecholamines cause adipose or fat tissue to release free fatty acids. The net effect of these metabolic actions is to increase the substrates for cells in the skeletal muscles, heart, and brain to support their activity.

Several factors influence the function of the adrenal medulla. One is an individual's emotional state. Stress-induced increases in nerve cell activity in the hypothalamus activate preganglionic sympathetic neurons. These preganglionic neurons in turn innervate and activate the adrenal medulla, causing the secretion of epinephrine and NE into the bloodstream. Other stress factors, such as the extreme loss of blood, hypothermia, hypoglycemia, and hypoxia, also activate the adrenal medulla in attempts to correct functional disorders and maintain a constant internal environment.

CHAPTER REVIEW

Summary

- The autonomic nervous system (ANS), a regulatory system that attempts to maintain a constant internal environment, has both central and peripheral components.
- The central components of the ANS consist of the limbic system, including the hypothalamus, certain brain stem nuclei, and the spinal cord. ANS peripheral components include ganglia and postganglionic neurons. The ANS can be divided into the sympathetic, parasympathetic and enteric nervous systems. The sympathetic nervous system prepares the body to respond to stressful conditions, whereas the parasympathetic nervous system helps regulate basic functions, such as digestion. The enteric nervous system is an independent division of the ANS that regulates gastrointestinal function.
- Both preganglionic sympathetic nerve fibers and preganglionic parasympathetic fibers release acetylcholine from their nerve terminals. Postganglionic sympathetic nerve fibers typically release NE from their nerve terminals, whereas postganglionic parasympathetic nerve fibers release acetylcholine.
- In many systems, the parasympathetic and sympathetic nervous divisions mediate opposite effects on their target organs.

- Autonomic reflexes are carried out primarily in the spinal cord. Reflexes, such as the urinary bladder reflex and the defecation reflex, are activated by stretch receptors and play important roles in coordinating the actions of smooth muscles and sphincters.
- The convergence of pain information from internal organs and regions on the surface of the body is the basis of referred pain.
- The hypothalamus is the center for the "fight or flight" response. It coordinates the variety of bodily changes needed for survival during life-threatening situations. It increases blood flow and oxygen to the heart and skeletal muscles while reducing blood flow to the organs involved with the digestion of food.
- The adrenal medulla participates in the body's response to stress by releasing epinephrine and NE. Epinephrine and NE in the bloodstream have effects similar to the activation of sympathetic nerve fibers; the onset of their actions, however, is delayed because they must course through the circulatory system.

Review Questions

Choose the Correct Answer

1. Which one of the following receptor types is found on dendrites and cell bodies of postganglionic parasympathetic neurons?
   a. Nicotinic cholinergic
   b. Muscarinic cholinergic
   c. α_1-Adrenergic
   d. β_1-Adrenergic
   e. α_2-Adrenergic

2. Which nervous system element produces an excitation of the target organ that cannot be modified after the signal leaves the central nervous system?
   a. Enteric
   b. Somatic
   c. Sympathetic
   d. Parasympathetic
   e. Intermediolateral cell column

3. Which one of the following types of structures secretes epinephrine and norepinephrine in response to stimulation by preganglionic sympathetic neurons?
   a. Pituitary
   b. Hypothalamus
   c. Adrenal cortex
   d. Adrenal medulla
   e. Myenteric plexus

4. A drug that activates alpha receptors would have which one of the following effects?
   a. Dilate coronary blood vessels
   b. Decrease glycogenolysis in the liver
   c. Cause renin secretion from the kidney
   d. Relax sphincters in the urinary bladder
   e. Constrict blood vessels in skeletal muscles

5. Which one of the following elements is *not* found in the enteric nervous system?
   a. Myenteric plexus
   b. Submucosal plexus
   c. Sensory nerve fibers
   d. Preganglionic sympathetic fibers
   e. Postganglionic parasympathetic fibers

6. Which one of the following actions of bladder function is *not* regulated by the autonomic nervous system?
   a. Relaxation of the internal sphincter
   b. Contraction of the internal sphincter
   c. Contraction of the external sphincter
   d. Contraction of detrusor muscles in the bladder wall
   e. Provision of the efferent component of the bladder reflex

7. Which structure is considered the main or head ganglion of the autonomic nervous system?
   a. Hypothalamus
   b. Dorsal root ganglion
   c. Intermediolateral cell column
   d. Paravertebral chain ganglion
   e. Celiac ganglion (solar plexus)

8. Which of the following features are characteristic of the parasympathetic nervous system?
   a. Preganglionic neurons near spinal cord in thoracic and lumbar regions
   b. Short preganglionic fibers that release acetylcholine
   c. Ganglia in or near the target organ
   d. Adrenergic postganglionic fibers
   e. Nicotinic receptors in the target organ

9. Which one of the following events occurs as a result of activation of the micturition center?
   a. Stretch receptor activation
   b. Contraction of internal sphincter
   c. Activation of somatic alpha motoneurons
   d. Relaxation of internal and external sphincters
   e. Excitation of preganglionic sympathetic neurons

10. Which one of the following functions is *not* considered part of the "fight or flight" response?
    a. Widening of the pupil
    b. Accommodation for far vision
    c. Inhibition of gastrointestinal functions
    d. Increased depth and rate of respiration
    e. Conversion of glucose to glycogen in the liver

11. An **increase** in which (one) of the following function(s) is the same for both the dorsomedial and ventromedial nuclei of the hypothalamus?
    a. Intestinal motility
    b. Blood pressure only
    c. Intestinal blood supply
    d. Blood supply to muscles only
    e. Blood supply to the muscles and blood pressure

12. When stimulated by preganglionic fibers of the sympathetic nervous system, which neurotransmitter(s) is (are) released from the adrenal medulla?
    a. Substance P
    b. Acetylcholine
    c. Epinephrine only
    d. Norepinephrine only
    e. Epinephrine and norepinephrine

13. What is the location of sympathetic preganglionic nerve cell bodies?
    a. Nuclei of cranial nerves only
    b. Intermediolateral cell columns only
    c. Thoracic and sacral spinal cord segments
    d. Brain stem nuclei and intermediolateral cell columns
    e. Nuclei of cranial nerves and sacral spinal cord segments S2, S3, and S4

14. Which one of the following central nervous system actions is *not* required for producing defecation?
    a. Relaxation of abdominal muscles
    b. Inhibition of certain alpha motor neurons
    c. Inhibition of preganglionic sympathetic nerve activity
    d. Coordination of somatic and autonomic nerve activity
    e. Enhancement of preganglionic parasympathetic nerve activity

15. Which one of the following characteristics is not representative of the sympathetic division of the autonomic nervous system?
    a. Upon activation, usually targets single organs to produce specific responses
    b. Always has at least two synapses after leaving the central nervous system
    c. Releases norepinephrine at all of its synapses

    d. Can dilate as well as constrict blood vessels
    e. Constricts the pupil of the eye

16. What is the function of an alpha$_2$ norepinephrine receptor on an acetylcholine-releasing nerve terminal?
    a. Remove norepinephrine from the synaptic cleft
    b. Enhance acetylcholine release into the synaptic cleft
    c. Decrease the release of acetylcholine into the synaptic cleft
    d. Stimulate the release of norepinephrine from postganglionic sympathetic terminals
    e. Increase the probability of acetylcholine binding to the post-synaptic effector organ

17. Stimulation of which one of the following structures would most likely evoke responses such as anger, rage, or aggressive behavior?
    a. Hypothalamus
    b. Limbic system
    c. Celiac ganglion
    d. Solitary nucleus
    e. Intermediolateral cell column

18. The programmed sequences of smooth muscle contraction in peristalsis are mediated chiefly by which one of the following components of the nervous system?
    a. Cranial nerve nuclei
    b. Alpha motor neurons
    c. Enteric nervous system ganglia
    d. Sympathetic nervous system ganglia
    e. Parasympathetic nervous system ganglia

19. In autonomic nervous system regulatory feedback mechanisms, which one of the following components would be analogous to an air conditioner or heater in a feedback temperature control system in a modern home?
    a. Hypothalamus
    b. Postganglionic neuron
    c. Vascular smooth muscle
    d. Visceral sensory neuron
    e. Myenteric or submucosal plexus

Answer to Case History Question

1. Even if the mind is completely convinced that a threatening crisis is over, some physiological consequences of activation of the "fight or flight" mechanism may persist. Epinephrine released from the adrenal medulla into the circulation is not immediately metabolized. Blood glucose levels also remain elevated following crisis resolution.

Key Terms

α-receptor (p. 345)
β-receptor (p. 345)
epinephrine (p. 354)
enteric nervous system (p. 340)

"fight or flight" response (p. 353)
hypothalamus (p. 341)
intermediolateral cell columns (p. 341)

limbic system (p. 341)
parasympathetic nervous system (p. 340)

referred pain (p. 351)
sympathetic nervous system (p. 340)

Suggested Readings

Benson, H. *The Relaxation Response.* New York, Morrow, 1975.

Ciriello, J., Calaresu, F. R., Renaud, L. P., and Polosa, C. *Organization of the Autonomic Nervous System: Central and Peripheral Mechanisms.* New York, Alan R. Liss, Inc., 1987.

Gebber, G. L., and Barman, S. M. "Lateral tegmental field neurons of cat medulla: A potential source of basal sympathetic nerve discharge." *Journal of Neurophysiology,* 54:1498–1512, 1985.

Gershon, M. D. *The Second Brain: The Scientific Basis of Gut Instinct and a Groundbreaking New Understanding of Nervous Disorders of the Stomach and Intestines.* New York, HarperCollins Publishers, 1998.

Irwin, M. "Stress-induced immunosuppression: Role of brain corticotropin releasing hormone and autonomic nervous system mechanisms." *Advances in Neuroimmunology,* 4:29–47, 1994.

Janig, W. "The Autonomic Nervous System." *In* Schmidt, R. F., and Thews, G: *Human Physiology,* ed 2. Berlin, Springer-Verlag, 1989, pp. 333–370.

Kregel, K. C., Stauss, H., and Unger, T. "Modulation of autonomic nervous system adjustments to heat stress by central ANG II receptor antagonism." *American Journal of Physiology,* 266: R1985–R1991, 1994.

Reis, D. J., Ruggiero, D. A., and Morrison, S. F. "The C1 area of rostral ventrolateral medulla: A critical brainstem region for control of resting and reflex integration of arterial pressure." *American Journal of Hypertension,* 2:363S–374S, 1989.

Rust, G., Burgunder, J. M., Lauterburg, T. E., and Cachelin, A. B. "Expression of neuronal nicotinic acetylcholine receptor subunit genes in the rat autonomic nervous system." *European Journal of Neuroscience,* 6:478–485, 1994.

Sapolsky, R. M., Uno, H., Rebert, C. S., and Finch, C. E. "Hippocampal damage associated with prolonged glucocorticoid exposure in primates." *Journal of Neuroscience,* 10:2897–2902, 1990.

Schramm, L. P. "Spinal factors in sympathetic regulation." *In* Magro, A. (ed): *The Molecular Basis for the Central and Peripheral Regulation of Vascular Resistance.* New York, Plenum Press, 1986, pp. 303–352.

Uno, H., Tarara, R., Else, J. G., Suleman, M. A., and Sapolsky, R. M. "Hippocampal damage associated with prolonged and fatal stress in primates." *Journal of Neuroscience,* 9:1705–1711, 1989.

Answers to Review Questions

1. b **2.** b **3.** d **4.** e **5.** d **6.** c **7.** a **8.** c
9. d **10.** e **11.** b **12.** e **13.** a **14.** b **15.** a
16. c **17.** b **18.** c **19.** c

Chapter 11

CENTRAL INTEGRATIVE SYSTEMS

• *Human brain waves during REM sleep (dreaming state).*
(© SIU/Visuals Unlimited)

KEY CONCEPTS

• *Many bodily functions are driven by central integrative systems within the brain that exhibit circadian rhythms. These integrative systems have their own periodicities but can be entrained by exposure to specific light-dark cycles.*

• *Motivational systems are activated in response to, for example, the body's need to eat to supply energy to fuel cellular processes or for procreation or the activation of reward pathways in the brain.*

• *The development of mechanisms for learning and memory is crucial to the adaptation to new situations and environments. Various areas of the brain are responsible for storing and retrieving different types of information.*

• *Language is localized in the dominant hemisphere, typically the left hemisphere for most people.*

• *Many neurological functions in humans are located in either one side of the brain or the other. Spatial and musical abilities are predominantly located on the right side of the brain.*

CASE HISTORY

Jim and Steve are brothers and close friends traveling together by car on their spring break from college. As they drive south at the start of their trip, both talk excitedly about the past week's midterm exams, their semester experiences, and their anticipated fun in the sun in Florida. It has been an emotionally trying semester for both of them following the death of their mother just two months ago. Jim, however, is buoyed by the progress of his academics and is enthused in particular about his course in Japanese. He explains to his younger brother Steve the fascinating double alphabet of the Japanese, composed of a system of letters used to form phonetic words, much like English, and a separate set of over 40,000 "symbols" or pictures that express concepts, ideas, and more complex items. Steve too has had a good semester despite a very stressful calculus exam taken that day. Steve finished his exam literally minutes before he hopped into the car with his brother and headed south.

Two hours into the trip, the brothers' car approaches a highway intersection with a green light. Suddenly, seemingly out of nowhere, a speeding car enters the intersection from the right, smashing into the brothers' car and slamming the left side of Jim's head into his door window, fracturing the temporal/parietal area. By the time the emergency medical service arrives, Jim is unconscious. Steve does not show signs of physical injury and is conscious but stares fixated at his bleeding, unconscious brother. When questioned about what happened, he is unable to speak. Upon arrival at the local hospital, Steve can finally speak reasonably well when asked questions but cannot identify himself. He recalls taking a calculus exam and traveling in the car, as well as arriving at the hospital, but cannot recall the automobile accident in any detail. He remembers that his mom passed away recently but cannot remember the funeral. He can remember the time he finished his calculus exam and what he did on his last birthday. He responds to some questions just fine, but then asks the questioner to repeat other questions minutes after he gave an answer. He does not seem to have any severe deficits in his motor skills.

Two days after the accident, Steve seems to have no central nervous system (CNS) deficits; his speech and short-term and long-term memory are normal. At this time, Jim has regained consciousness and is immediately self-aware. Jim can recall details of the accident, the death of his mother, and minutia relating to the emotional funeral that followed. He can also recall the state, county, and town where he lives. However, he is uncertain as to the day he and his brother left on their spring break and has to ask repeatedly what time of day it is, even though he was told repeatedly over just a 15-minute span. When trying to speak on his own or in answer to a question, his speech is not fluent and halting. He struggles to find even the most basic words in response to simple questions. When asked, he cannot spell *world* backward. He cannot identify Japanese words spelled out with letters, even though he knew them well before the accident, but can easily identify the Japanese "symbol" words. When asked, Jim can execute a simple, three-stage command, such as "pick up the pencil, touch it to the paper, and place the pencil in a box," but when asked to repeat the task 2 minutes later without being told, he cannot execute any of the stages. Jim is right-handed, but at this time he cannot move his right arm. Two months after his accident, all of Jim's CNS deficits remain.

Questions

1. Deficits in integrated brain functions, such as memory, learning, speech, and learned motor skills, are classified as *organic,* relating to actual physical trauma of the brain, or *psychogenic,* which refer to the presence of deficits in brain function in the absence of actual physical damage. To which category do the deficits experienced by Jim and Steve belong?

2. What is the likely cause of the brain dysfunction in Jim and Steve?

3. What are the differences in the brain dysfunction between Jim and Steve? Could these be used to clinically distinguish between patients with organic versus psychogenic brain dysfunction?

CENTRAL INTEGRATIVE SYSTEMS

 What are the central integrative systems?

The nervous system detects and processes sensory information from the external and internal environment; the motor system puts out information to skeletal muscles for locomotion and fine movement. However, these sensory and motor functions do not occur in isolation. In fact, human behavior is highly coordinated in order to carry out purposeful actions. The brain areas involved in these coordinated behaviors are a part of the **central integrative systems.** Central integrative systems can be subclassified into several broad categories. Motivational systems, for example, are those responsible for our drive to satisfy basic needs, such as hunger and thirst. Other systems carry out functions such as learning and memory, which endow us with the ability to acquire and store new information. Still other systems provide us with the ability to communicate in the form of written and spoken language. This chapter considers the brain areas responsible for these integrative functions. This chapter also examines how the brain brings together and integrates information from its two hemispheres and how some functions are localized on one side of the brain. We begin with how the brain controls bodily rhythms by acting as a master clock.

BODILY RHYTHMS, CYCLIC SYSTEMS, AND THE HYPOTHALAMUS

Some of the functions of the hypothalamus involve control of biological rhythms and motivational systems. The hypothalamus is composed of many different cell clusters that are involved in the regulation of body rhythms (Fig. 11–1). Rhythms of the body vary on daily or monthly cycles. **Monthly cycles** are those seen, for example, in the process of ovulation, in which the female body prepares itself for reproduction by the release of gonadotropic hormones from the hypothalamus. **Diurnal** (daily) or **circadian** (*circa* = "approximately"; *dies* = "day") **rhythms** are those that vary on a daily basis.

Biological Clocks in the Brain Regulate Circadian Rhythms

The "master clock" thought to be responsible for coordinating various biological rhythms is the **suprachiasmatic nucleus** of the hypothalamus (see Fig. 11–1). This nucleus receives information about light and darkness directly from the retina of the eyes. By this pathway, the light-dark cycle entrains the body's biological clock. The destruction of this nucleus in experimental animals disrupts their sleep-wake cycle and many other bodily functions.

Well over 100 bodily functions and biochemical processes have been found to vary in accordance with a 24-hour sched-

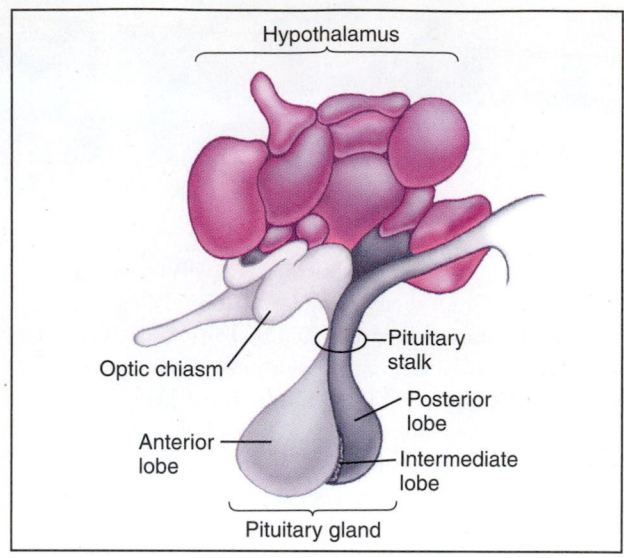

Figure 11–1

The hypothalamus (*purple*) comprises several different cell clusters or nuclei, some of which are involved in central integrative systems of the body.

ule. Body temperature, for example, decreases during the night and increases during the day (Fig. 11–2). This rhythmic variation is controlled by the hypothalamus. The hypothalamus also controls certain hormones, such as **adrenocorticotropic hormone (ACTH)** and **cortisol** in a similar circadian rhythm. Many of the body's rhythms have periods that are longer or shorter than 24 hours. The sleep-wake cycle, for instance, has a 25-hour period. However, the body's rhythms can be entrained and synchronized to external factors, such as the 24-hour light-dark cycle or to our work schedules.

The uncoupling of the body's rhythms can result in fatigue and poor mental and physical performance. Most of us have experienced the fatigue of "jet lag." When we cross different time zones, our bodily functions are no longer synchronized with the day-night cycle; hormonal fluctuations, body temperature, and the sleep-wake cycle are desynchronized. Eventually, the synchrony of these cycles is reestablished and the body's rhythms again function in harmony.

Sleep Has Different Stages

 How is sleep classified?

The need for a sleep-wake cycle remains a mystery. Sleep has various stages in which the activity of brain nerve cells varies from relatively quiescent to very active. According to some theories, the increased activity of certain brain neurons during sleep creates the internal stimuli necessary for the maturation of the brain during development. Other theories suggest that sleep and dreams strengthen the nerve cell con-

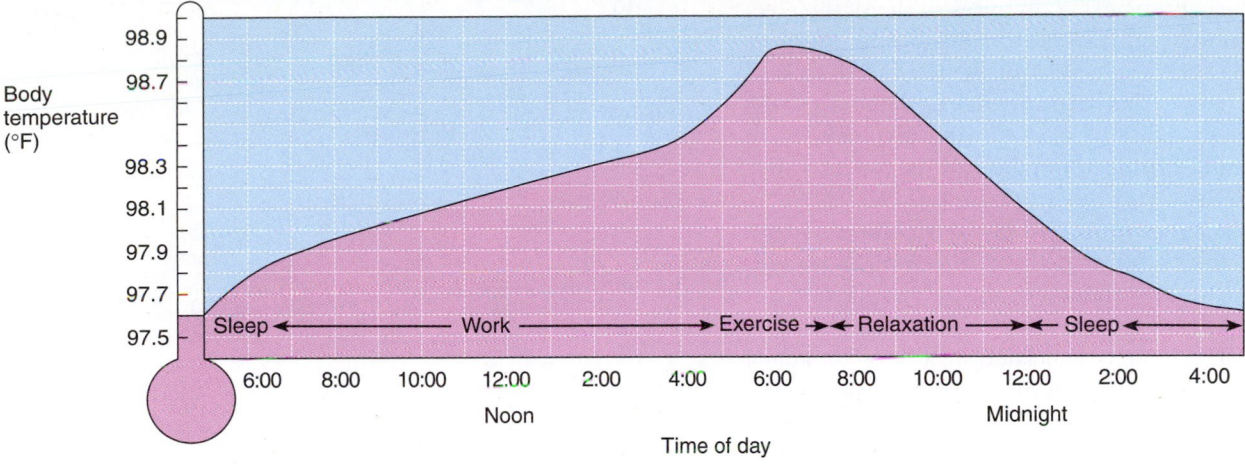

Figure 11–2

Body temperature shows a rhythmic variation throughout the day and night.

nections associated with the memories that were formed during the day or they reintegrate memory patterns to remove unnecessary activity.

Stages of Sleep

During one night of sleep, the brain displays different stages of electrical activity (Fig. 11–3). This activity can be monitored by electrodes placed on the scalp that produce a record, called an **electroencephalogram,** or **EEG.** The EEG of a person who is awake exhibits electrical activity that is low in amplitude and high in frequency. During the onset and progression of sleep, the frequency of the electrical activity decreases while the amplitude increases.

Stages 1 to 4. Four stages of sleep have been identified. Stage 1 of sleep has the lowest amplitude and highest frequency of electrical activity, whereas stage 4 has the highest amplitude and lowest frequency. An individual passes through each stage, in sequence, an average of three to five times during the night. In approximately the first 45 minutes of sleep, the brain progresses from stage 1 to stage 4. During this progression, the skeletal muscles are relaxed, but a sleeper frequently tosses and turns to readjust the sleeping position. In addition, heart rate and blood pressure decrease during stage 4, whereas gastrointestinal motility increases as a result of activation of the parasympathetic division of the autonomic nervous system.

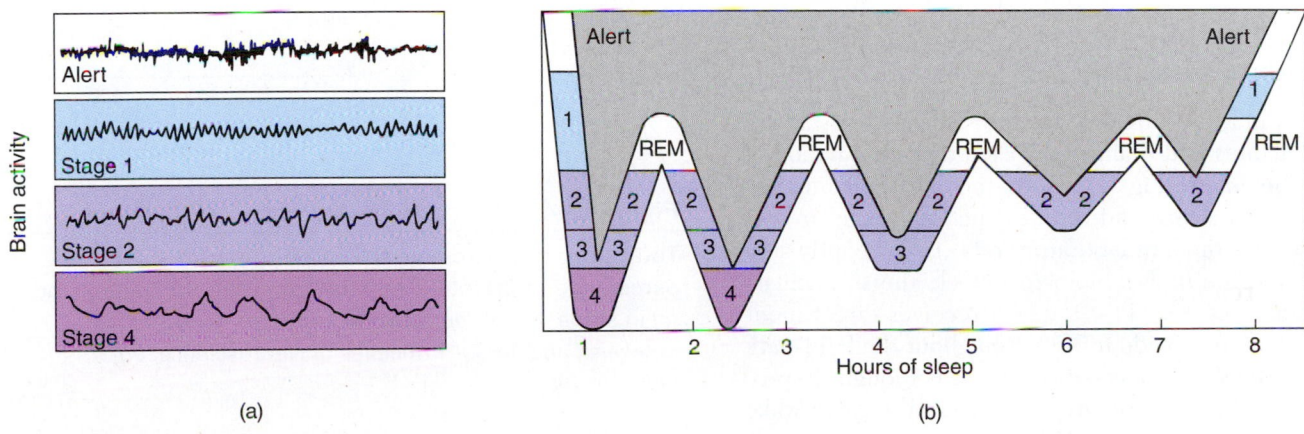

Figure 11–3

(a) The electrical activity of the brain during the various stages of sleep can be recorded by means of electroencephalography. *(b)* During one night's sleep, a person goes through three to five cycles, each of which includes several sequential sleep stages. REM indicates the rapid eye movement stage of sleep having an electroencephalographic pattern similar to that of stage 1 sleep.

REM Sleep. During the next 45 minutes of sleep, the electrical activity of the brain reverts from stage 4 to a pattern similar to that of stage 1. In this phase, the distal skeletal muscles, except those that control the movement of the eyes, are completely inhibited from moving. Eye movement becomes very rapid during this phase of sleep; thus, this phase is referred to as **rapid eye movement (REM)** sleep. During REM sleep, heart rate and blood pressure increase while gastrointestinal motility decreases as a result of activation of the sympathetic division of the autonomic nervous system.

In general, REM sleep is associated with visual dreaming although dreaming also occurs during slow wave sleep. One night's sleep contains approximately three to five REM episodes. Each successive episode is longer and is characterized by increases in the electrical activity of the brain and the number of visual images. Dreams are primarily visual; however, people who are born blind have auditory dreams. Individuals who lose their sight eventually lose their ability to dream visually.

The deprivation of REM sleep in humans has no apparent long-term effects on behavior. When permitted to sleep undisturbed after REM deprivation, subjects initially exhibit more REM sleep. Normal sleeping patterns, however, are soon reestablished.

Variations with Age

The amount of sleep and the proportion of REM and non-REM sleep vary with age (Fig. 11–4). During the last few weeks of gestation, for example, REM sleep occupies approximately 80% of the total sleep time. This amount decreases to 50% for full-term newborns and levels off at 25% by 10 years of age. In general, the amount of total sleep required decreases with age, with the greatest changes occurring for REM and stage 4 sleep, and the latter decreasing rapidly with age. Stage 4, but not REM, sleep often disappears after 60 years of age.

Neural Mechanisms of Sleep

The areas of brain responsible for the sleep-wake cycle are the hypothalamus and brain stem. Electrical stimulation of the **preoptic area** (Fig. 11–5) of the hypothalamus induces non-REM sleep, and the destruction of this area of brain causes insomnia in laboratory rats. The preoptic area is located near the body's biological clock, the suprachiasmatic nucleus (see Fig. 11–5), which receives direct input from the retina to provide information about the light-dark cycle. Thus, the suprachiasmatic nucleus is thought to provide inputs to the preoptic area to entrain the sleep-wake cycle.

Sleep also can be induced by injections of serotonin into the preoptic area. A possible internal source of serotonin may come from serotoninergic neurons from the **raphe nuclei** (see Fig. 11–5) of the brain stem. The destruction of the raphe causes insomnia in laboratory animals. It is thought that the raphe is responsible for the generation of REM

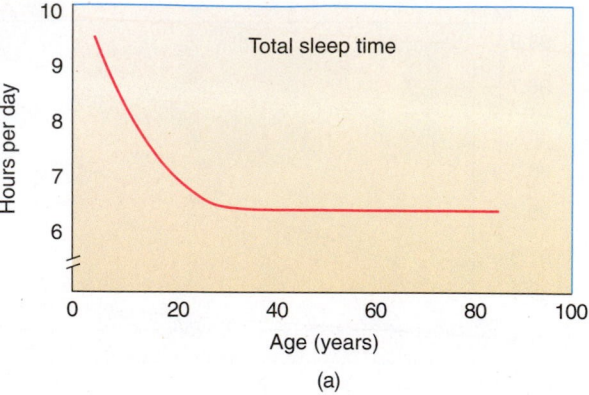

(a)

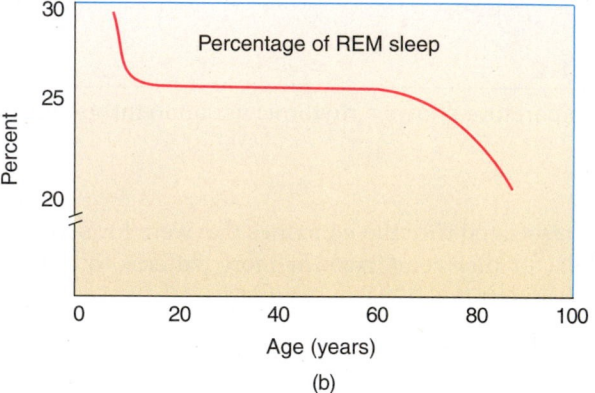

(b)

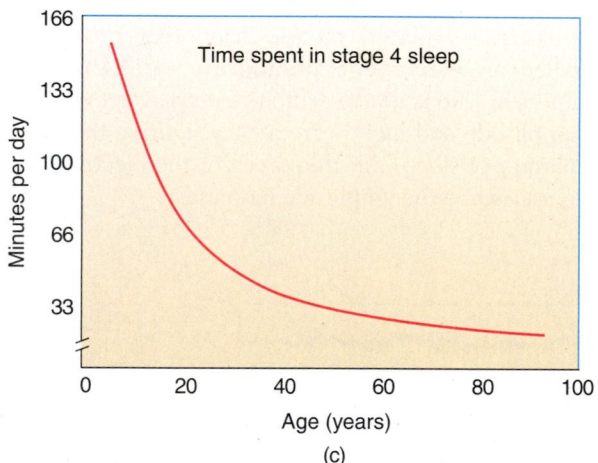

(c)

Figure 11–4

Total sleep time decreases *(a)* with age, as does the percentage of *(b)* total sleep time spent in REM sleep and *(c)* in stage 4. *(Adapted from Feinberg, 1969; In: Kandel, Schwartz, and Jessel,* Principles of Neuroscience, *ed 3. Elsevier, 1991.)*

sleep. Descending fibers from the raphe nucleus innervate the spinal cord and may be responsible for inhibiting the activity of motor neurons of skeletal muscles. In this way, the motor programs that are planned and programmed during dreaming and REM sleep are normally prevented from being carried out.

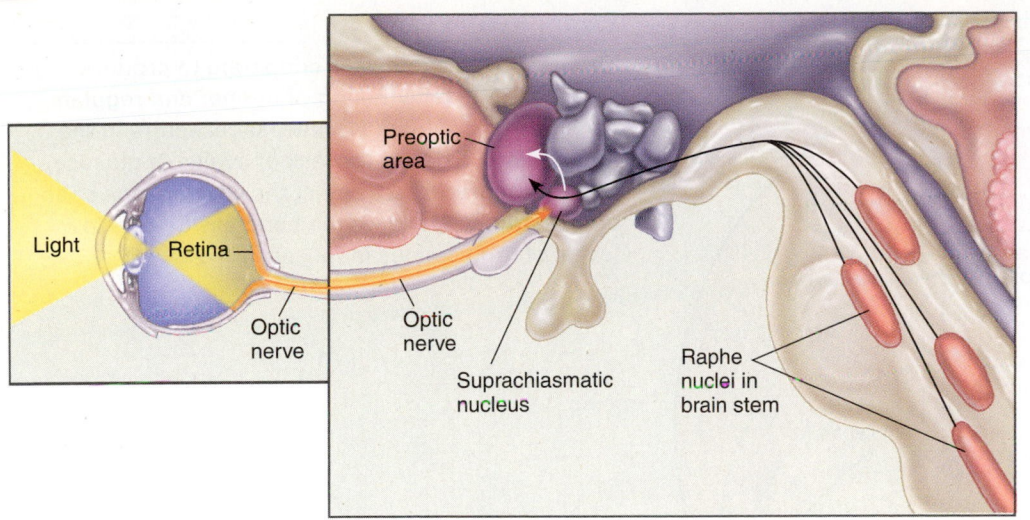

Figure 11–5

The suprachiasmatic nucleus, the body's "biological clock," receives information about light directly from the retina and is thought to regulate the sleep-wake cycle with the cooperation of the preoptic area. The raphe nuclei of the brain stem is thought to be involved in the generation of REM sleep.

Disorders of Sleep

Insomnia is a general term for the inability to sleep. Normal causes of insomnia include disruptions of the body's circadian rhythms (e.g., those due to the travel across time zones), fatigue, and various medications. **Narcolepsy** is a severe, inherited disorder in which an individual can suddenly lapse into sleep at any time during the middle of the day. Narcolepsy is thought to result from the activation of the REM-sleep generator. Narcoleptic attacks that occur during the waking state have all of the characteristics of REM sleep. They last approximately as long as REM sleep (90–100 minutes); are accompanied by vivid hallucinations similar to REM dreams; and are associated with a lack of muscle tone, except for eye movement. It is thought that an inability to inhibit the REM generator is responsible for narcolepsy. The primary inhibitory inputs to the REM generator are noradrenergic nerve fibers. Consequently, drugs that enhance norepinephrine release, such as amphetamines, are used in treating this disorder.

MOTIVATIONAL SYSTEMS: MAINTAINING AN INTERNAL BALANCE AND ENSURING SURVIVAL

 What functions are considered part of motivational systems?

Motivational systems initiate behaviors that satisfy bodily needs. These systems typically operate to ensure the survival of an individual or species. In some cases, motivational systems are driven by signals that represent the imbalance of certain chemicals within the body. Hunger, for example, is activated by internal signals that communicate the need for food in order to provide energy for the body. Other motivational systems, such as those responsible for sexual drive, are activated by external signals typically from members of the opposite sex, as well as from internal signals within the body.

Hunger Is Regulated by Levels of Glucose and Triglycerides

The consumption of food is required for the production of energy required by all cells in the body and for the growth, maintenance, and repair of all organ systems. The hypothalamus plays a central role in regulating food intake by acting as a sensor of glucose availability (Fig. 11–6). This process is part of the motivational system involved in the short-term regulation of food intake. **Satiety,** or the feeling of fullness after a meal, appears to be controlled in the ventromedial hypothalamus.

The long-term regulation of food intake appears to involve the maintenance of body weight. Research has shown that laboratory animals forced to gain weight by eating an excessive amount of food reduce their food consumption and revert back to their original body weight when left to eat at their own pace. Animals that are forced to lose weight by starvation also eventually gain weight back to their original level when allowed free access to food. These studies suggest that the body may have a "set point" for weight, which it attempts to maintain.

The maintenance of body weight is now thought to operate in an attempt to maintain a constant amount of **adipose tissue,** or **fat cells.** Fat cells store **triglycerides,** a major source of energy for the body. In adults, the amount of adipose tissue remains relatively constant, and there is a good correlation between body weight and the amount of adipose tissue. The variable that is thought to be regulated in the maintenance of adipose tissue is the volume of individual fat cells. These cells are thought to release a chemical signal in amounts proportional to their triglyceride content. This adipose signal affects the level of **insulin,** a hormone produced by the pancreas, in the circulatory system. Research has shown that basal insulin levels are proportional to body

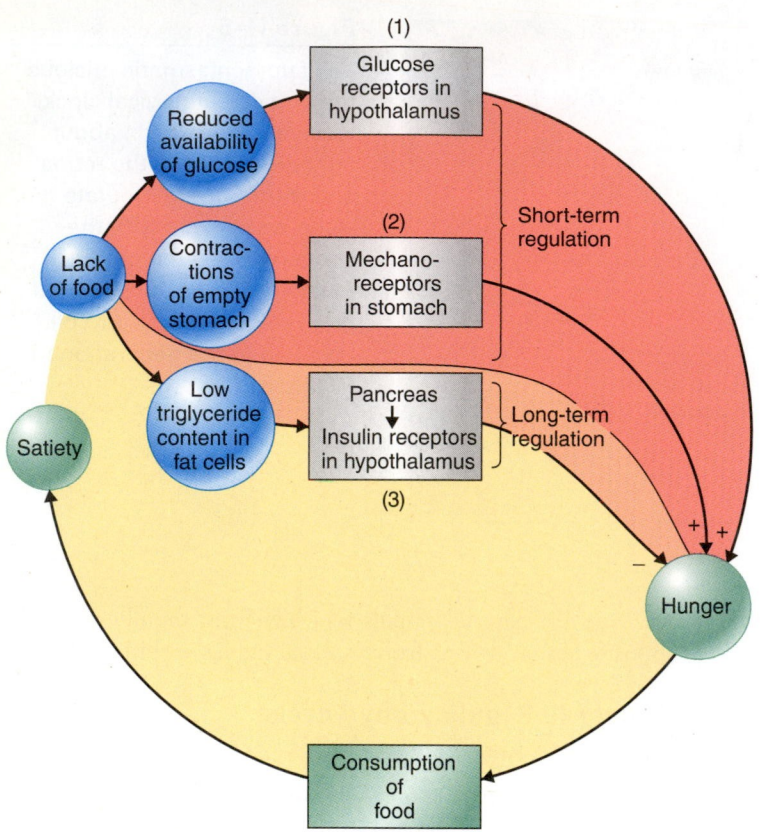

Figure 11–6

Three monitoring devices cooperate to produce (+) or decrease (−) the feeling of hunger and regulate consumption of food: (1) glucose receptors in the hypothalamus detect the lack of available glucose; (2) mechanoreceptors detect the contractions of an empty stomach; and (3) the pancreas detects low levels of triglycerides in fat cells.

adipose tissue. Circulating insulin, in turn, penetrates the blood-brain barrier and binds to receptors in the ventromedial hypothalamus. The activation of these hypothalamic insulin receptors lowers food intake and, subsequently, body weight. Thus, insulin is involved not only in the control of circulating glucose after a meal but also the long-term regulation of body fat and hence of body weight (see Fig. 11–6). It is now believed that adipose tissue releases a hormone called **leptin** into the blood; leptin decreases appetite, food intake, body weight, blood glucose, and blood insulin levels. Mice who have been genetically altered to be deficient in leptins become markedly obese but then lose weight when injected with leptins. This has led to speculation that genetic variations in leptin levels in individuals might be linked to obesity.

Sexual Drive and Sexual Behavior Can Be Altered by Exposure to Hormones During Development

Sexual behavior is controlled by both the nervous and endocrine systems. One method of study of sexual drive and behavior in animals is to observe their mating postures. Among different species is a great deal of similarity in the mating posture. During mating, males of many species display a **mounting** posture, and the female adopts a **receptive** posture called **lordosis.** In studies of laboratory animals, the frequency of these two postures is used as a measure of sexual drive. In both males and females, the **anterior hypothalamus** (see Fig. 11–1) appears to be one of the brain centers

for sexual drive. Lesions of this area of brain in experimental animals reduce their sexual activity. Under certain circumstances, both the male and female animals can adopt the sexual behavior of the opposite sex.

The stimulus for activating sexual behavior in many species is the release of chemicals called **pheromones.** Ovulating female monkeys (i.e., those that are prepared to conceive), for example, emit pheromones along with other vaginal secretions. These chemical odors are detected by the olfactory system of male monkeys and sent to the **limbic system** and indirectly to the hypothalamus. In lower animals, pheromones play a prominent role in sexual behavior. In humans, the use of perfume or cologne as a sexual attractant is related to this olfactory influence on sexual behavior, although other social cues play a significant role.

An important factor in the development of a particular pattern of sexual behavior is the presence or absence of particular steroid hormones called **androgens** during **critical periods** of development. In laboratory studies of mice, the presence of androgens during the fetal and early postnatal periods causes the brain to develop male sexual-behavioral characteristics, whereas the absence of these steroids results in the development of female characteristics (Fig. 11–7). In laboratory mice, for example, the removal of the testis (the primary source of androgens) from males soon after birth causes these animals to develop lordotic behavioral responses as adults. If given ovaries, these males ovulate as would a normal female. In analogous fashion, the sexual behavior of a female can also be altered. In-

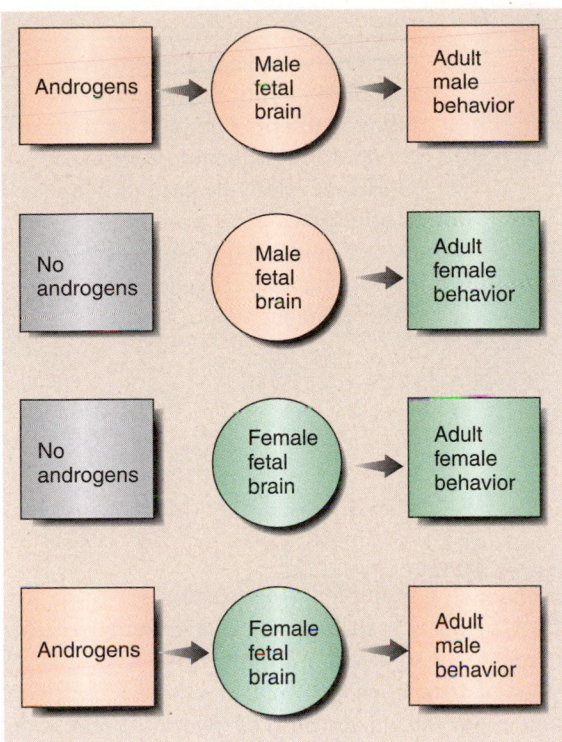

Figure 11–7

In laboratory experiments involving mice, exposure (or a lack thereof) to androgens during fetal development has been shown to affect adult sexual behavior. If a male fetal brain is exposed to androgens during critical periods of development, the adult will exhibit male behavior. If androgen exposure does not occur, the adult male will exhibit female behavior. If a female fetal brain is not exposed to androgens during critical periods, the adult will exhibit female behavior, but if the female fetal brain is exposed to androgens, the adult will exhibit male behavior.

jections of androgens into newborn female mice cause them to develop male mounting behaviors as adults. Research has shown that the critical time period for the exposure (or lack of exposure) to androgens in order to affect brain-sex alterations in laboratory animals is the first few days after birth. In humans, androgen activity is elevated during the 12th and 22nd weeks of gestation and during the first six weeks after birth.

The fetal environment of mice also influences the sexual behavior that they will display as adults. The positioning of a fetal mouse in the uterus is a random process. A female fetus, for instance, could be placed in between two male fetuses, two females, or in between a male and a female. Research has shown that female fetuses placed between two male fetuses are exposed to higher androgen levels than if placed between two female fetuses (Fig. 11–8). As adults, these "androgenized" females are more aggressive; they exhibit male mounting behaviors, display irregular estrus (ovulation) cycles, mate later, and cease ovulating earlier. A male fetus

situated between two females in the uterus has smaller testicles than does a male fetus surrounded by other male fetuses. In addition, these animals, when grown, require higher doses of testosterone in order to induce aggressive behavior.

How do androgens affect the genetic programming of nerve cells, leading to differences in "male" and "female" behavior? It is thought that when nerve cells in the preoptic area of the hypothalamus are exposed to androgens, they release **gonadotropin-releasing hormone (GnRH)** at relatively constant levels as adults (Fig. 11–9). Nonandrogenized brains, on the other hand, develop a cyclical pattern of GnRH release. In addition, nerve cells in the preoptic area of nonandrogenized female brains are sensitive to estrogen secreted by ovarian follicles. Androgenized brains are not affected by estrogen. Thus, the function of nonandrogenized preoptic neurons has been altered in not only their release of GnRH but also their sensitivity to estrogen.

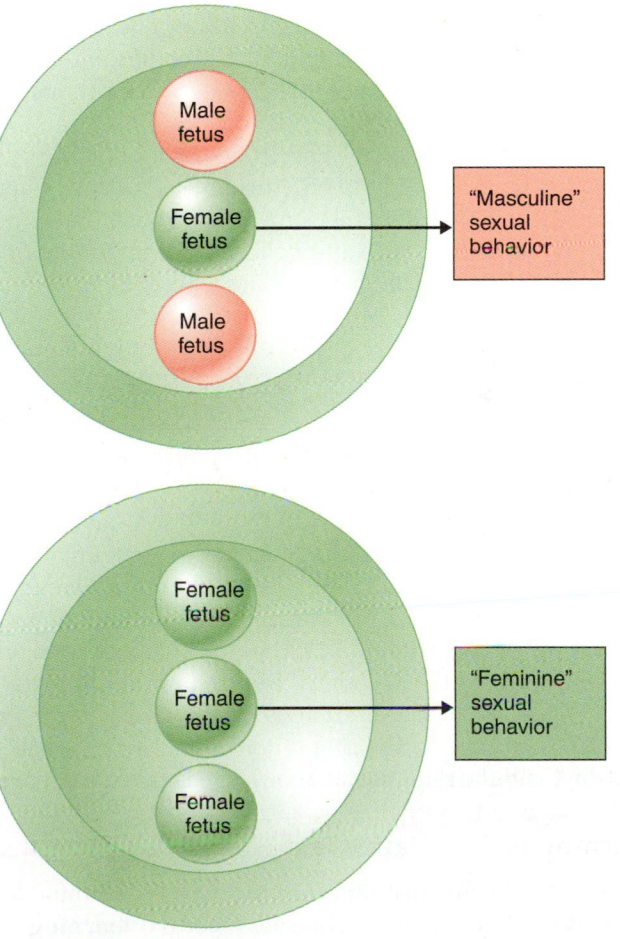

Figure 11–8

Fetal environment can affect the adult sexual behavior of laboratory animals. A female mouse fetus that develops between two males will show male behavior as an adult, whereas a female surrounded by other females in the fetal stage will develop as a female.

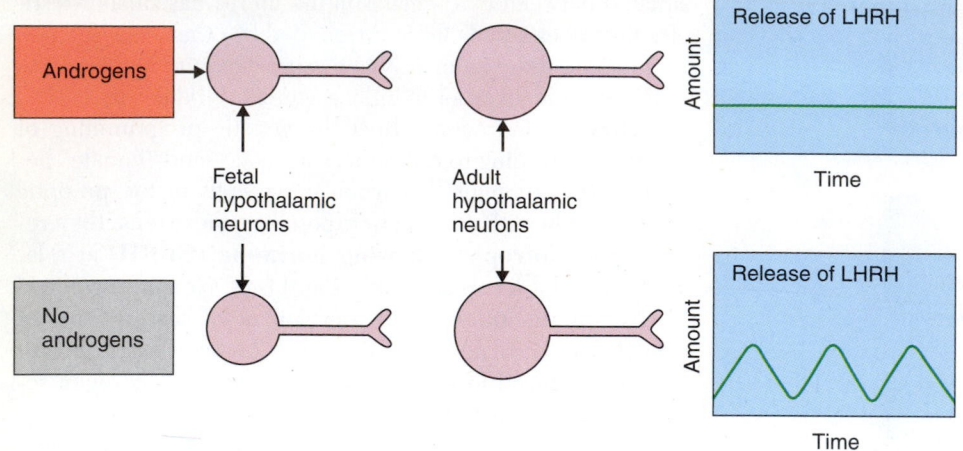

Figure 11–9

When nerve cells in the preoptic area of the hypothalamus are exposed to androgens, they release gonadotopin-releasing hormone (GnRH) at relatively constant levels as adults, whereas preoptic nerve cells of animals not exposed to androgen show a cyclic pattern of GnRH release.

Alterations in sexual behavior are also observed in humans. In a condition known as **androgen insensitivity,** males are incapable of responding to androgens. This disorder is a genetic defect transmitted by the mother. Males with this defect develop as females in their external appearance as well as in their psychosexual preference.

Drugs can also affect the androgenic environment of the developing brain. Barbiturates; pesticides, such as DDT; and hormones, such as diethylstilbestrol (DES; at one time given as a "morning after" birth control pill and also to prevent premature births) have been shown to masculinize the brain.

LEARNING AND MEMORY SYSTEMS

What do we know about the neurophysiology of memory and learning?

Prior to our birth, and from the moment we are born, we acquire information and knowledge about our environment. In the womb, babies learn to distinguish the sounds of their mothers' voices. Soon after birth, they are conditioned to orient their heads to their mothers' breasts in order to receive milk. Eventually, babies are trained to use a toilet in the excretion of body waste. The acquisition of knowledge is called **learning,** and the retention of knowledge is called **memory.**

Learning Is Associative and Nonassociative

Research has shown that there are two types of learning: (1) **associative learning** and (2) **nonassociative learning.** In nonassociative learning, we learn about a single type of stimulus and adapt to it according to its relevance to our desires or survival. The sound of a ticking clock, for example, may initially draw our attention, but after a period of time we ignore it. This type of behavior, in which we learn to decrease our response to a repeated stimulus, is called **habituation.** In contrast, we can become more sensitized to certain other sounds. In school, for example, we learn to recognize the sound of a fire alarm and respond accordingly. **Sensitization** is thus an increase in response to a stimulus.

In one type of associative learning, an animal acquires an understanding about the relationship or association between one stimulus and another. This learning process is called **classical conditioning.** An example of classical conditioning is the salivation of a conditioned dog in response to a stimulus of light. Normally, the dog salivates only in response to a piece of food. After a light stimulus (the **conditioned stimulus,** or **CS**) has been paired with the presence of a piece of meat (the **unconditioned stimulus,** or **UCS**) in a sufficient number of conditioning trials, the light alone is sufficient to cause the dog to salivate (salivation is the **conditioned response,** or **CR**). The dog has been **conditioned** to salivate in response to light.

The other type of associative learning is called **operant conditioning.** In this type of learning, animals acquire an understanding of the relationship between their own behavior and subsequent reward (or punishment). With this type of learning, animals tend to repeat behaviors that are rewarded and avoid behaviors that are punished.

In both classical and operant conditioning, a crucial factor is the time interval between the appearances of the two stimuli, or of a behavior and its reward or punishment. With classical conditioning, if the time between the appearances of a conditioned response and unconditioned stimulus is long, the conditioned response is poor. Likewise, with operant conditioning, if the reward or punishment is delayed, the conditioned behavior is weak.

Only certain types of stimuli are capable of producing a conditioned response. The most effective stimuli are those relevant to the survival of the animal. Food poisoning, for example, is a case of conditioning in which nausea acts as a form of punishment. Subsequent exposure to the taste of the food causes an animal to avoid it. The conditioning of food aversion is most effective when taste is used as the stimulus. If other stimuli, such as sounds or lights instead of taste, are

APPLICATIONS OF PHYSIOLOGY

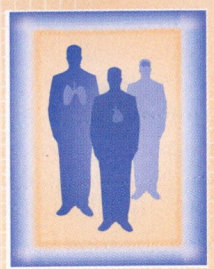

Drugs of Abuse and the Reward System of the Brain

The reward system of the brain was first discovered when stimulating electrodes were placed into the brains of laboratory animals, which were then allowed to control the amount of electrical current passing through the electrodes. By pressing a bar, animals were able to deliver current to the electrodes. When the electrodes were placed in specific areas of the brain, these animals would press the bar well over 100 times per minute (Figure (a)). When given a choice of receiving food instead of the electrical current, animals would choose the electrical stimulus, starving as a result. The reward areas of the brain thus constitute a powerful motivational system that overrides drives, such as

hunger; sex; and, in some cases, the avoidance of pain and punishment.

The areas of the brain that are associated with the reward system are shown in Figure (b). They include the **substantia nigra, hypothalamus, nucleus accumbens, caudate nucleus,** and the **frontal cortex.** The common thread linking these areas together is the **dopamine nerve fiber pathway,** which originates in the substantia nigra. This pathway contains axons that form synaptic connections with each of the reward areas of the brain.

The release of dopamine by these synaptic connections is crucial to the reward process. Drugs that enhance the release of dopamine, such as amphetamine and cocaine, also augment the rate of self-stimulation. Drugs that block dopamine release or its ability to bind to its receptor decrease the rate of self-stimulation. The reward system may therefore play a role in the process of addiction to certain abused drugs.

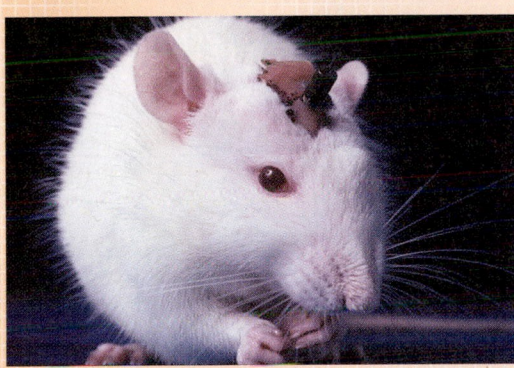

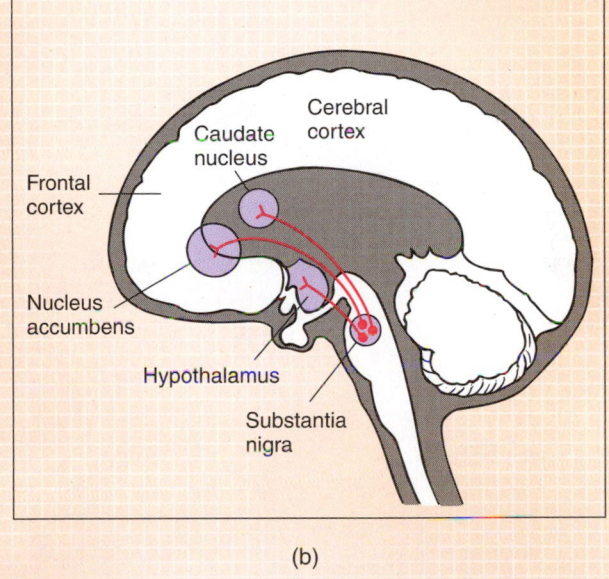

(a)

(a) Rat with a brain-stimulating electrode. *(© Yoav/Phototake.)* *(b)* The reward system of the brain involves the substantia nigra, the hypothalamus, the nucleus accumbens, the caudate nucleus, and the frontal cortex.

(b)

---

paired with nausea, the animal does not develop an aversion to these stimuli. Consequently, the suitability of a stimulus depends largely on the nature of the response to be learned.

Short-Term Memory Is Transformed into Long-Term Storage

Memory has different components and progresses through different stages. Information is first stored briefly in **short-term memory.** Short-term memory is used, for example, in the action of dialing a telephone number after looking it up

in the telephone book. After the phone call, we typically forget the number if we no longer need it. In contrast, if the telephone number happens to be one we use often, we will most likely retain it in **long-term memory.**

Short-term and long-term memory have different characteristics. Short-term memory is easily disrupted and short-lived, whereas long-term memory is stable, less volatile, and long-lasting. To be stored in long-term memory, information must first be transferred from short-term memory. In addition, protein synthesis is required for the formation of long-term memory. If the information is disrupted while it is in short-term memory, it is not permanently registered in long-term memory. The process of forming long-term memory from short-term memory is called **consolidation.**

Different Neuronal Pathways Are Used for Different Types of Learning and Memory

Learning and memory systems are found in several areas of the brain. These systems are associated with the interconnection of nerve cells in the form of neuronal **memory circuits** that are necessary for learning and memory to occur. Memory circuits have been identified for activities such as the learning of motor skills, visual discrimination, and several other types of learning.

The learning of skilled movements, such as playing the piano and typing, are examples of motor learning. These types of behaviors, however, are very complex and therefore difficult to study in terms of memory circuits. A simpler para-

digm that can be used to study motor learning is the conditioning of the eye-blink reflex. In this paradigm, which has been used in the rabbit, a puff of air blown on the cornea of the eye causes the eyelid to blink (an unconditioned stimulus producing an unconditioned response). If the puff of air is paired with a tone, the sound of the tone alone will ultimately cause the eye to blink (a conditioned stimulus producing a conditioned response). The neural pathways used in this learning paradigm are (1) somatosensory pathways that transmit information about the puff of air and (2) motor pathways that cause the eyelid to blink (Fig. 11–10). In addition, during the conditioning process, the sound of the tone becomes a predictor of the puff of air and eventually conditions the same eye-blink response. Consequently, information from the auditory pathway must ultimately affect the motor pathways to result in the eye blink. The crucial memory circuits involved in this conditioning process are the connections between the inputs to, and the nerve cells within, the cerebel-

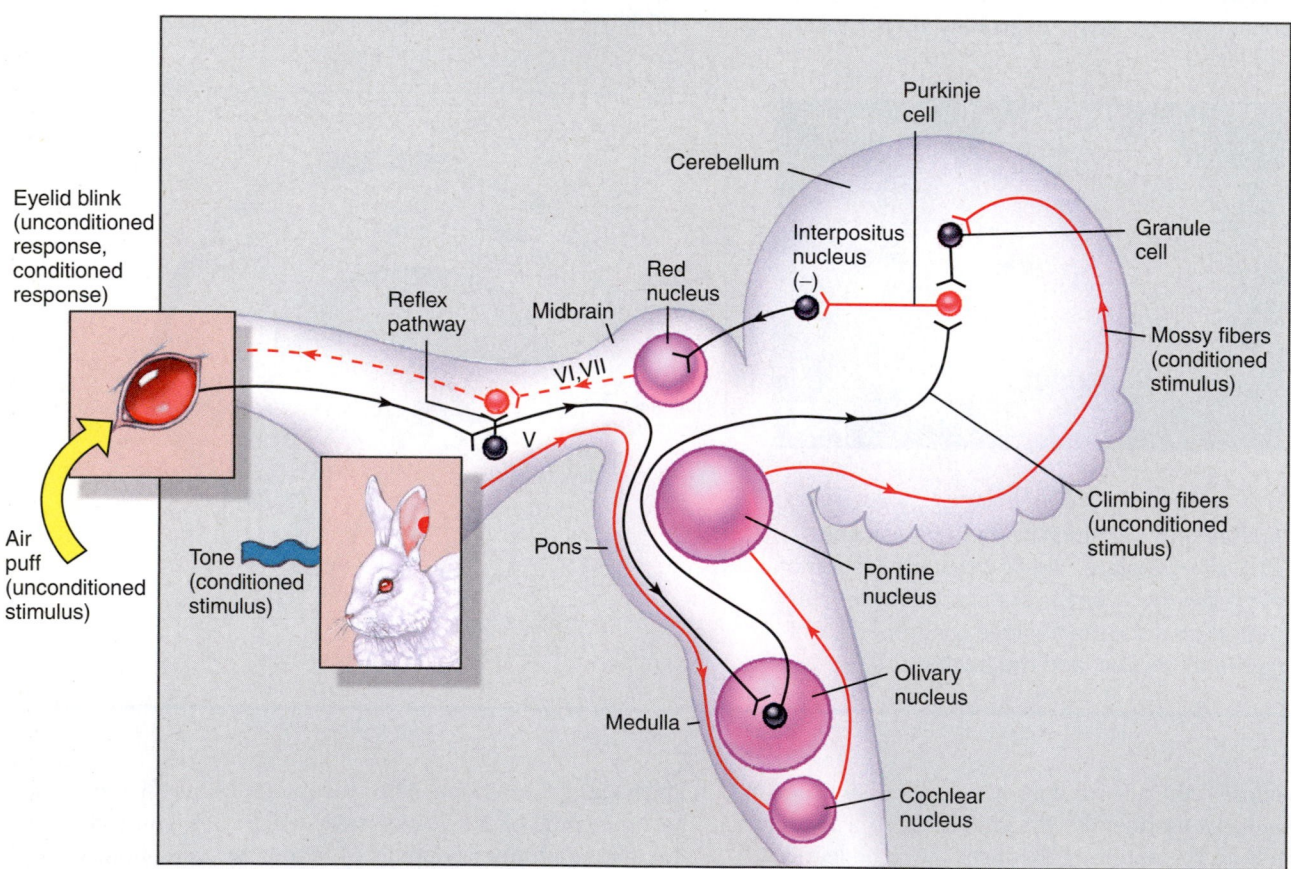

Figure 11–10

A learning paradigm for the eye-blink reflex in a rabbit. Somatic-sensory pathways transmit information about a puff of air (an unconditioned stimulus), which causes the eyelid to blink through the reflex pathway. This response eventually becomes linked to the sound of a tone (a conditioned stimulus). The auditory-motor pathways then cause the eyelid to blink. V, VI, and VII are the trigeminal, abducens, and facial cranial nerves, respectively.

lum (see Fig. 11–10). In the cerebellum, the pathway used by an unconditioned stimulus is the climbing fiber inputs to the Purkinje cells. The conditioned stimulus, the tone, is transmitted through conditioned stimulus pathways. In the cerebellum, this pathway is the mossy fiber input. The output of the conditioned responses from the cerebellum to motor nuclei that control eye blinking is by way of the interpositus nucleus and red nucleus (see Fig. 11–10).

What changes occur in the memory circuit to account for learning? The crucial dynamic changes that occur during learning take place in the Purkinje cells or within inputs that control the activity of Purkinje cells. During the conditioning process, activation of the climbing fiber pathway (the unconditional stimulus pathway) causes the Purkinje cells to generate a complex spiking pattern of action potentials (Fig. 11–11). As conditioning progresses, fewer complex spikes are produced in response to activation of the unconditional stimulus inputs. Well-trained animals display a decrease in the number of simple spikes generated in response to activation

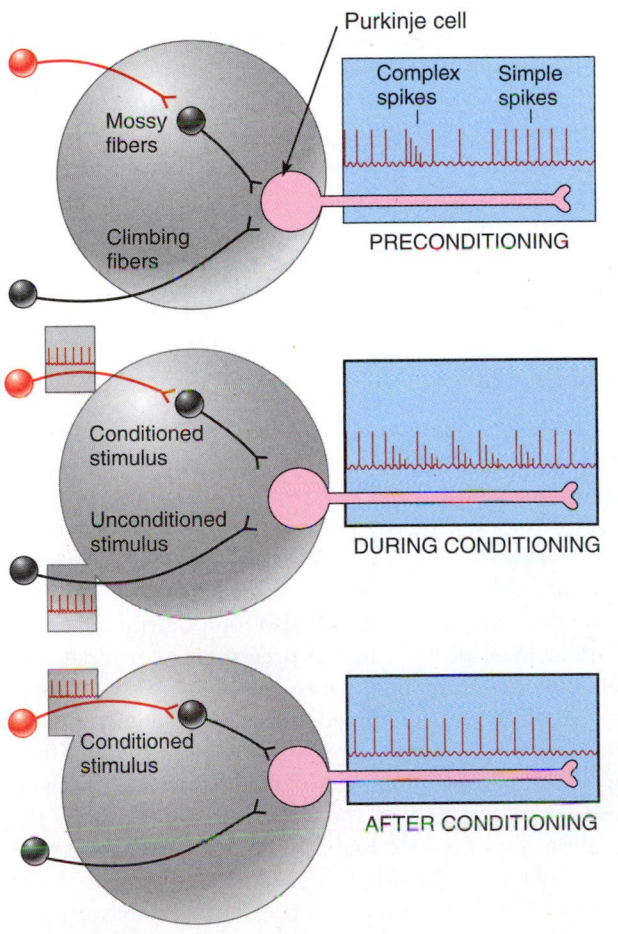

Figure 11–11

During the conditioning process, the Purkinje cells of the cerebellum show a complex spiking pattern of action potentials, which decreases in complexity as conditioning progresses.

of the mossy fiber inputs. The net effect is a reduction in Purkinje cell activity. As we saw in Chapter 9, the Purkinje cell terminals release gamma-aminobutyric acid (GABA), an inhibitory neurotransmitter. Consequently, a reduction in Purkinje cell activity in essence causes the nerve cells in the interpositus nucleus to become more active. The decrease in Purkinje cell activity is a form of conditioned habituation.

Other memory circuits that have been identified in the brain include the connections of the hypothalamus and amygdala in learning a conditioned cardiovascular response, and of the primary and secondary visual cortices in visual memory. Memory circuits in other areas of the brain are currently being studied in association with other memory functions, and a diversity of these circuits will most likely be identified throughout the brain.

Other areas of the brain, such as the temporal lobe and the hippocampus, also play a prominent role in learning and memory. Patients undergoing neurosurgical operations under local anesthesia report that electrical stimulation of the temporal lobes causes them to "hear" melodies that they had heard or learned in the past.

The hippocampus plays a role in the transformation of short- to long-term memory. Lesions of the hippocampus were at one time made in attempts to prevent seizures in epileptic patients. After the surgery, long- and short-term memory in these patients remained intact. However, their ability to convert short-term to long-term storage was lost. Patients who were taught a new task, for example, could perform the task as long as they were not interrupted. If they were temporarily distracted, however, and then asked to perform the task again, they were not able to do so. In addition, although they were able to recognize people they had known before the operation, these patients were unable to remember faces of new individuals to whom they were introduced after the operation. As a consequence, these patients continued to "live in the past" because they formed no new memories.

"Split Brains" Reveal Unique Learning and Memory Mechanisms

"Split-brain" studies of experimental animals reveal that information reaching the cerebral hemispheres is processed and stored differently for different species. Studies of monkeys demonstrate that learned responses stored in one hemisphere are not accessible by the other when the fibers connecting the two hemispheres are disconnected (Fig. 11–12). Thus, it is suggested that the hemispheres must normally work together in order to exchange stored information. In studies with cats, on the other hand, learned responses stored in one hemisphere are also transferred and stored in the other. Consequently, when the hemispheres are disconnected after storing learned information, each hemisphere is still capable of producing the correct response. Different species of mammals therefore approach the storage of information in different ways.

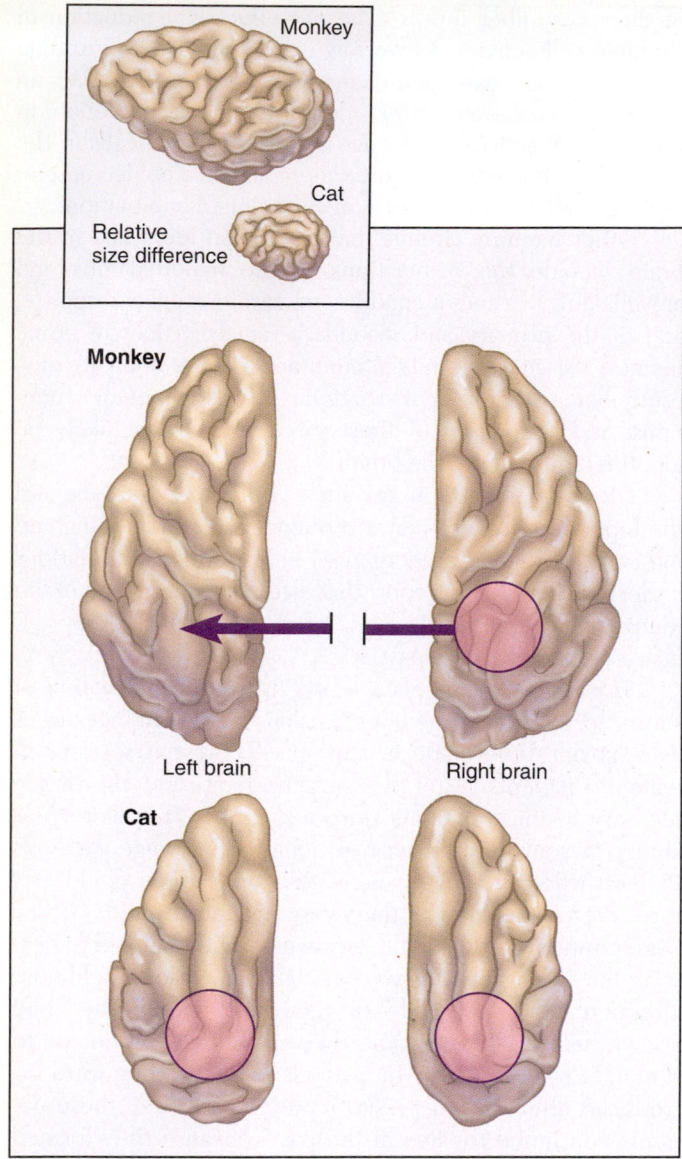

Figure 11–12

In monkeys, learning stored in one hemisphere of the brain is not accessible to the other if the two hemispheres have been disconnected, whereas in the brain of a cat, information is stored in both hemispheres and so is accessible to both even when disconnected.

Learning Causes Changes in Cell Function

Studies of many different species are beginning to reveal common cellular mechanisms of learning and memory that may also operate in humans. These cellular mechanisms involve alterations of synaptic function. These alterations can come about by changes in presynaptic or postsynaptic mechanisms of transmission or by the formation of new synaptic connections.

Habituation: A Decrease in Synaptic Transmission

Habituation, or the decrease in a behavioral response to a repeated stimulus, can be explained at the cellular level in terms of a decrease in synaptic transmission as a result of the inactivation of Ca^{2+} channels in the presynaptic terminal (Fig. 11–13a). The reduction of Ca^{2+} influx in turn decreases the amount of neurotransmitter released into the synaptic cleft. In addition, the number of synapses that contain active zones for the release of vesicles decreases, as does the area of the active zones. Together, these mechanisms diminish the functional capacity of the synapse.

Sensitization: An Increase in Synaptic Transmission

In contrast to habituation, cellular mechanisms of **sensitization** enhance an animal's response to a particular stimulus. The underlying mechanism is called **presynaptic facilitation;** it involves interneurons that form synapses with axons of other neurons (see Figs. 11–13b and 11–14a-c). These facilitating interneurons enhance the release of a transmitter by raising the levels of cAMP, which in turn phosphorylates and closes K^+ channels. The reduction in K^+ efflux effectively prolongs the action potential and, as a consequence, allows the voltage-sensitive Ca^{2+} channels to stay open longer, increasing Ca^{2+} influx and resulting in more transmitter released into the synaptic cleft. The sensitization process is also associated with an increase in the number of synapses with active zones and a doubling in the area of the active zones. These changes suggest that an overall increase in transmitter release is the basis of the sensitization process.

Classical Conditioning: A Sensitization Process

Although there may be a variety of different cellular mechanisms of classical conditioning for different memory circuits, one mechanism has been found to be similar to that of the sensitization process. This mechanism leads to an enhanced response to a conditioned stimulus. The activation of a conditioned stimulus by this mechanism not only excites the postsynaptic neuron but also temporarily raises the intracellular level of Ca^{2+} in the presynaptic terminal (see Fig. 11–14b). An increase in intracellular Ca^{2+} levels can activate a calmodulin-mediated increase in adenylate cyclase activity. As a consequence, cAMP levels are amplified, leading to the phosphorylation and inactivation of K^+ channels and a prolongation of the action potential. As in the sensitization response (see Fig. 11–13c), the prolongation of the action potential results in a greater Ca^{2+} influx at the terminals and an enhancement in transmitter release. Thus, the pairing of the unconditioned stimulus and the conditioned stimulus activates cellular processes that alter the excitability of the postsynaptic cell. As a consequence, the conditional response input alone is ultimately capable of activating the cell.

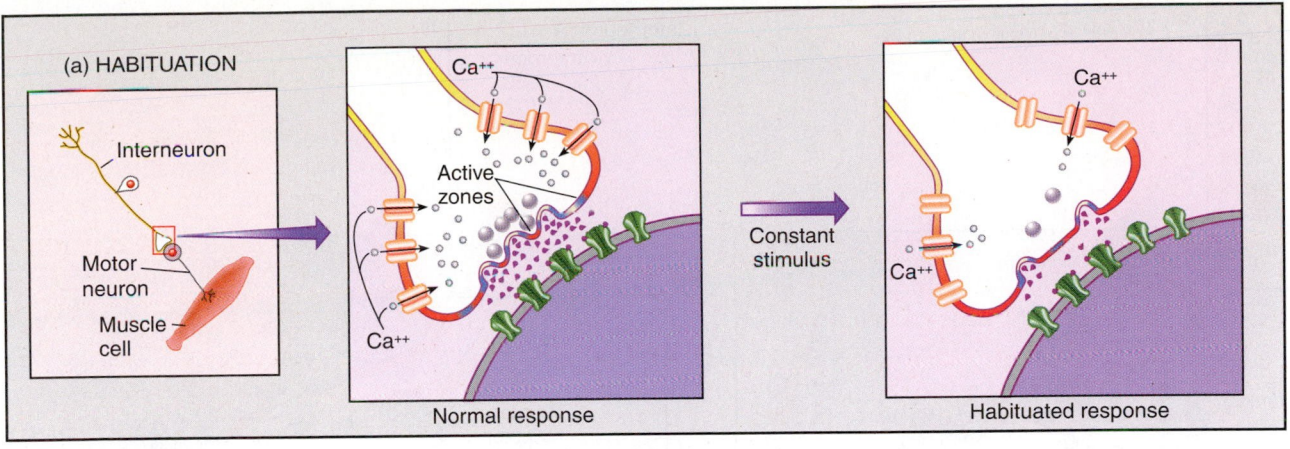

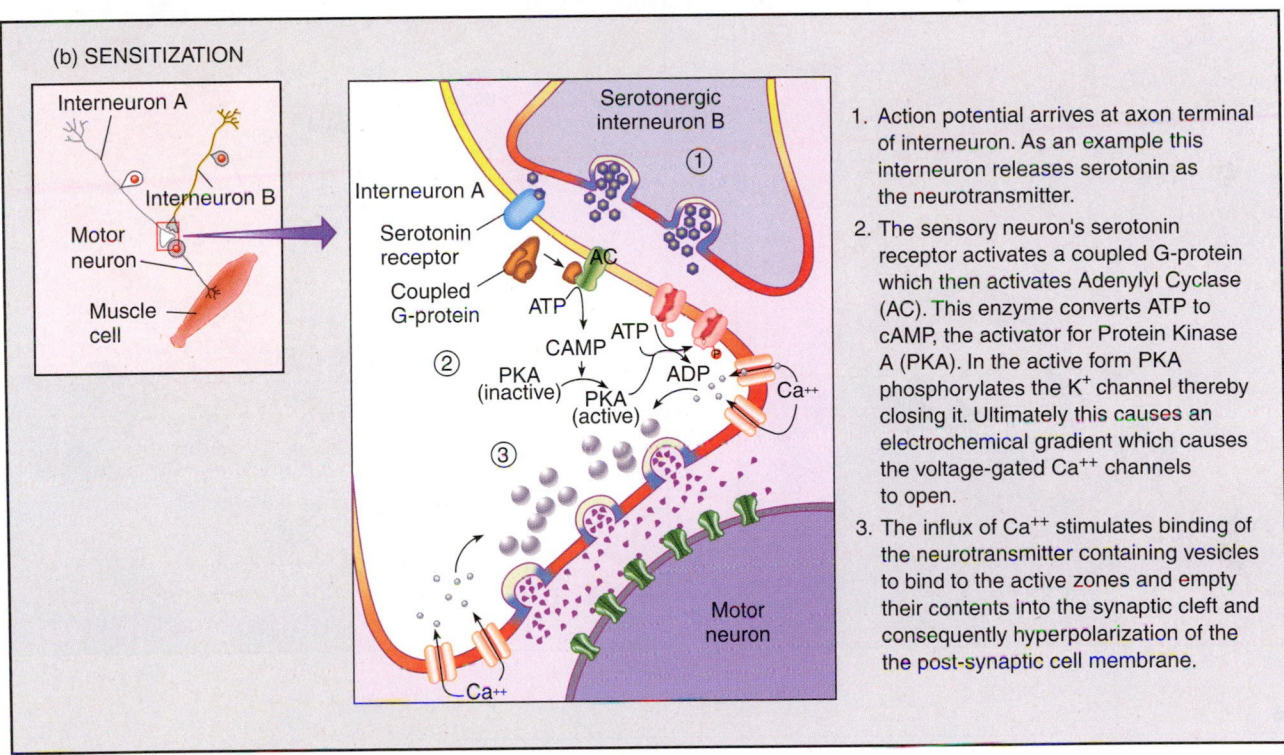

1. Action potential arrives at axon terminal of interneuron. As an example this interneuron releases serotonin as the neurotransmitter.

2. The sensory neuron's serotonin receptor activates a coupled G-protein which then activates Adenylyl Cyclase (AC). This enzyme converts ATP to cAMP, the activator for Protein Kinase A (PKA). In the active form PKA phosphorylates the K$^+$ channel thereby closing it. Ultimately this causes an electrochemical gradient which causes the voltage-gated Ca^{++} channels to open.

3. The influx of Ca^{++} stimulates binding of the neurotransmitter containing vesicles to bind to the active zones and empty their contents into the synaptic cleft and consequently hyperpolarization of the the post-synaptic cell membrane.

Figure 11-13

(a) At the cellular level, habituation is thought to occur because of a decrease in synaptic transmission due to a decrease in the influx of Ca^{2+}. **(b)** Sensitization at the cellular level is thought to involve an increase in synaptic transmission due to a prolongation of the influx of Ca^{2+}. This prolonged influx occurs as a result of action by a facilitating interneuron, which causes cAMP to phosphorylate a potassium channel, thereby closing it and preventing K$^+$ efflux. This prolongs depolarization of the terminal, prolonging the influx of Ca^{2+} through voltage-sensitive channels.

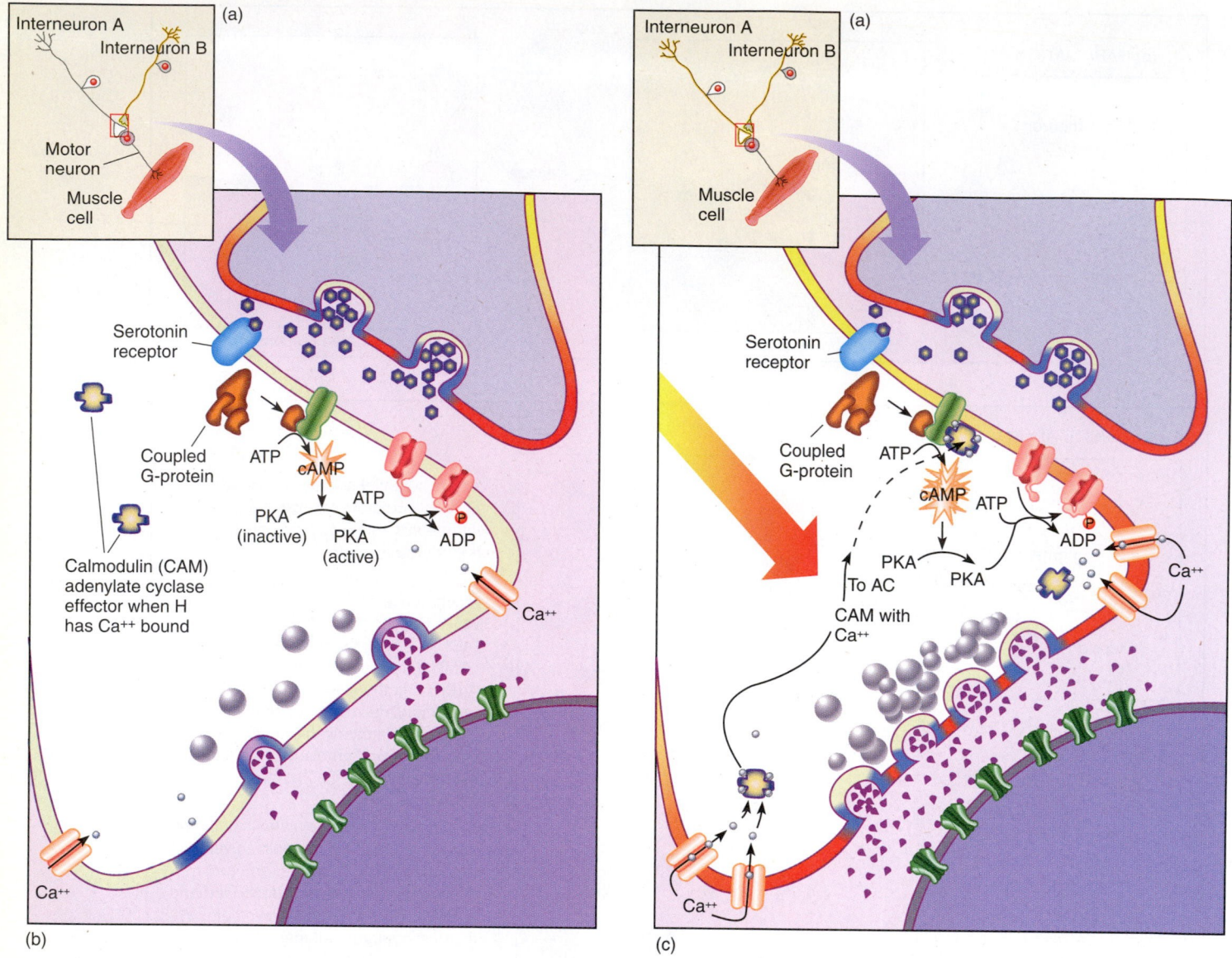

Figure 11–14

A mechanism for classical conditioning. *(a)* Interaction between a neuron along the conditional stimulus-response pathway (interneuron A) and a facilitating neuron (interneuron B). *T* indicates packets of neurotransmitter stored within synaptic vesicles. *(b)* Enlargement of boxed area shown in *(a)*, illustrating the effect of no activity along the conditioned stimulus-response pathway. *(c)* Activation of conditioned stimulus-response pathway temporarily raises intracellular levels of Ca^{2+}, activating a calmodulin-activated increase in adenylate cyclase production of cAMP. This activity closes K^+ channels and prolongs the action potential-induced depolarization of the terminal. The effect of prolonged terminal depolarization on transmitter release is shown in *(c)*.

Use and Nonuse of Neural Pathways Result in Maintenance or Atrophy of Connections

During muscle exercise, there is an eventual change in muscle tone and muscle mass. In a similar way, the use and nonuse of neural pathways can strengthen or weaken the connections between nerve cells. Animals raised in "enriched" environments, that is, those providing an abundance of sensory stimulation, have nerve cells that form more synaptic connections with other nerve cells than would normally appear. In contrast, animals raised in "impoverished" environments develop fewer synapses than normal.

The cerebral cortex exhibits **plasticity,** that is, the ability to change by forming new connections and circuits. The use and nonuse of pathways can alter topographical maps in the cortex. The map of the somatosensory cortex and the motor cortex, for example, can be modified when an animal is trained to use one finger more than the others. The amount of cortex representing the function of the exercised finger expands beyond that which is normally found. In contrast, if the nerve fibers innervating the finger are severed, the cortical space devoted to that finger is gradually eliminated and eventually incorporated into the maps of other fingers.

The results of these studies indicate that the brain is a dynamic organ system, capable of changing the strength of existing synaptic connections and, in some cases, forming new synapses as part of the process of learning and memory. Many of these alterations in synaptic function can take place only during certain critical periods of development, whereas other modifications can occur at all ages. These observations argue for continued learning and exposure to new challenges, regardless of age.

LANGUAGE SYSTEMS

> *What regions of the brain are associated with language?*

Language is a tool that allows for the communication of thoughts, knowledge, feelings, and emotions in both written and spoken form. The deciphering of words from complex sounds or visual symbols and the extraction of concepts from these words represent activities at the highest level of sensory integration. Written language allows us to transfer knowledge from generation to generation so that each generation can build upon the achievements of the past.

The development of language has thus played a crucial role in the progress of humankind. It is difficult to determine when language was first used; however, fossil evidence suggests that the areas of the brain responsible for language existed over 500,000 years ago. Spoken language obviously developed before written language. Consequently, another leap in human evolution may have been the formation of connections between the visual and language areas of the brain, allowing people to develop symbolic writing as a method of communication.

Comprehension and the Motor Aspects of Language Are Located in the Dominant Hemisphere

Language centers are located in the **dominant hemisphere.** Typically, the speech area in the dominant hemisphere is larger than the same region in the nondominant hemisphere. This difference in size is apparent in the human fetus by the 31st week of gestation and suggests that the de-

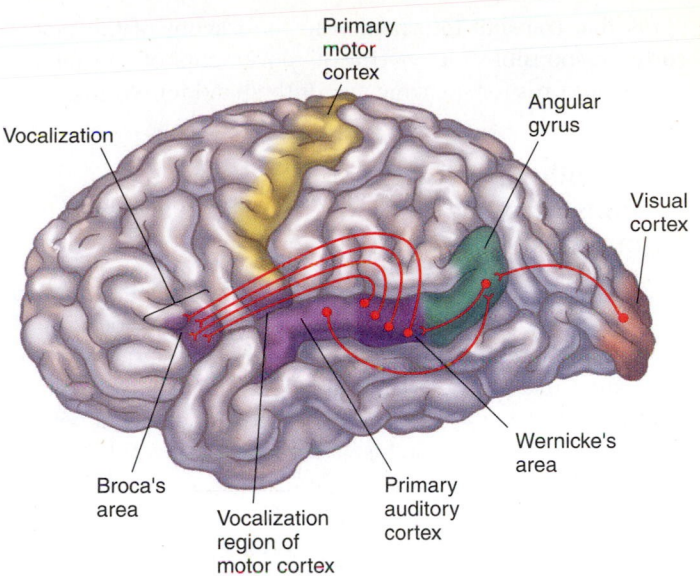

Figure 11–15

Language functions are localized in Wernicke's area and Broca's area. Other areas of the brain involved in language function are the primary auditory cortex, the visual cortex, the vocalization region of the motor cortex, and the angular gyrus.

velopment of language and speech may be an innate process. Indeed, infants are born with the ability to learn all human languages. This ability to learn other languages fluently after acquiring a first language, however, appears to diminish with age.

Language functions are located in regions of the brain called **Wernicke's area** and **Broca's area** (Fig. 11–15). Wernicke's area is responsible for the comprehension of language, while Broca's area is responsible for motor aspects of speech and language. In right-handed individuals, these language centers are found in the left hemisphere. The majority of left-handed people also have language centers in the left hemisphere; however, many left-handed individuals have language centers in both hemispheres or only in the right hemisphere.

For our understanding of spoken and written language, both auditory and visual information must reach the language centers. The reading of a word on a page, for example, requires that sensory information about the shape and form of the letters be transmitted from the retina to the primary and secondary visual cortices. This information is then transmitted to an association cortex called the **angular gyrus.** This area of brain is thought to be involved with the association of visual, auditory, and tactile sensations (see Fig. 11–15). From the angular gyrus, information about the word is transmitted to Wernicke's area, where it is thought to be recognized or comprehended. From Wernicke's area, information about the word is transmitted to Broca's area, where the motor programs for speech reside. Broca's area in turn contains nerve fibers that transmit information to the regions of the motor cortex responsible for control of movements of the mouth and vocal cords for speaking and of the hand for writing.

The Nondominant Hemisphere Is Responsible for Intonation and the Emotional Aspects of Language

The regions of the nondominant hemisphere of the brain that reside in the location corresponding to the location of Wernicke's and Broca's areas in the dominant hemisphere are responsible for the **affective** aspects of language (aspects related to mood or emotion). The manner in which a person speaks depends largely on his or her mood, and a listener can interpret simple statements in different ways depending on the speaker's tone of voice. The region in the nondominant hemisphere comparable to Wernicke's area is responsible for the comprehension of the tone of voice in speech, whereas the nondominant region comparable to Broca's area is responsible for the expression of intonation. Damage to these areas of brain causes **aprosodias,** disorders producing the inability to understand or express intonation.

Disorders of Speech and Language Have Anatomical Origins

Disorders of speech and language can also be explained in terms of the connections between the language centers in the dominant hemisphere. Disorders of language resulting from damage to the brain are called **aphasias.** A congenital disorder affecting reading is **dyslexia.**

Aphasia: Damage to the Language Centers

Several types of aphasia occur. In **Broca's aphasia,** damage to Broca's area disrupts the motor programs that control speech and writing. Patients with this disorder lose their ability to speak fluently and grammatically and are unable to express their ideas in writing. Damage to Wernicke's area produces an aphasia characterized by loss of language comprehension. Patients with this disorder can still speak and write but have difficulty understanding the spoken and written language.

Dyslexia: A Congenital Reading Disability

Dyslexia is a condition in which an individual has difficulty associating letters with their corresponding sounds and distinguishing letters that are similar in form, such as *p* and *b*. These mistakes occur in reading and writing. Dyslexics may also read words backward (e.g., confusing *saw* and *was*). Estimates suggest that between 10% to 30% of the population in the United States is affected by this disorder. However, dyslexia is more common among boys than among girls and among left-handers than among right-handers. Association of handedness and letter-configuration confusion with dyslexia suggest that this condition might involve a deficit in the development of dominance by the left hemisphere. Research

has revealed that the **planum temporale,** a region that includes Wernicke's area, is reduced in size in dyslexics. In addition, the layers of nerve cells in people with dyslexia have not properly separated in this region, suggesting that the migration of neurons has been altered during development.

LATERALITY OF BRAIN FUNCTION

 Do both sides of the brain perform the same function?

The localization of language to one hemisphere is only one example of the laterality of brain function. Other functions, such as musical, artistic, and spatial abilities, also exhibit a lateralization to one hemisphere. Tests show that, in right-handed subjects, the left ear is better at recognizing music, whereas the right ear is better at recognizing verbal material.

Spatial Abilities Are Localized to One Side of the Brain

Earlier, we saw how studies of split-brain animals provided information on the mechanisms of learning and memory. Isolation of one cerebral hemisphere from the other also has provided information about spatial abilities. Studies of so-called "split-brain patients" who have undergone surgery to disconnect the two hemispheres have shown that spatial abilities appear to reside in the right hemisphere. This surgical procedure cuts through the corpus callosum and is used as a last resort to control the spread of epileptic seizures. When lesions of the corpus callosum are made, the two hemispheres are no longer able to send information directly from one side to the other. Patients with lesions of the corpus callosum who are asked to reassemble a jigsaw puzzle are unable to do so with their right hands but are able to piece the puzzle together with their left hands. This is because the left hand is controlled by the right hemisphere, which houses the spatial abilities.

The spatial abilities of the right hemisphere are also seen in the processing of the written language. The Japanese, for example, use two writing systems: one consists of 71 letters that form the basis of phonetic (sound-related) symbols, and the second consists of over 40,000 **characters,** or ideographic symbols, each one representing a separate idea or object. The first system uses the letters to aid in the pronunciation of a word, whereas the second uses "pictures" to represent the word. Although both the phonetic and ideographic systems of writing use the language centers of the left hemisphere, the ideographic system also uses the right side of the brain.

Lesions of the angular gyrus near the language centers in the left hemisphere impair reading in the phonetic system of writing but generally not in the ideographic system. Even in cases in which ideographic reading is impaired, patients are able to explain the meaning of the word. For the same word written in the phonetic system, they lose this ability. Thus, ideographic writing, which uses the language centers in the

APPLICATIONS OF PHYSIOLOGY

Dyslexia: A Reading Disorder with Neuroanatomical Abnormalities

Dyslexia is a reading disorder of developmental origin. Children with this disorder cannot learn to read with normal proficiency, despite specialized instruction. Although these children appear to have normal intellectual ability and motivation, they have difficulty associating letters with their corresponding sounds and distinguishing letters that are similar in form, such as *p* and *b*.

Dyslexics represent approximately 2% to 16% of school-aged children. Male dyslexics outnumber female dyslexics by 3:1, and the disorder is more common among left-handed children. Research indicates that inheritance plays an important role in the transmission of dyslexia. Neurological signs often related to dyslexia include stuttering, abnormal electroencephalography (i.e., measures of brain electrical activity), and problems involving manual dexterity of one or both hands. Anatomical abnormalities are also apparent. The brain region required for the interpretation of language, Wernicke's area, is notably smaller in dyslexics, and the cell layers are disorganized. In addition, blood vessels are found in regions where they are not normally present. An abnormally high percentage of dyslexics have dominant right hemispheres.

Current theories argue that testosterone activity in dyslexics during the prenatal period is abnormally high. This alteration in testosterone preferentially affects the development and function of the left hemisphere. As a consequence, handedness and speech (stuttering) are affected, along with the manifestation of dyslexia.

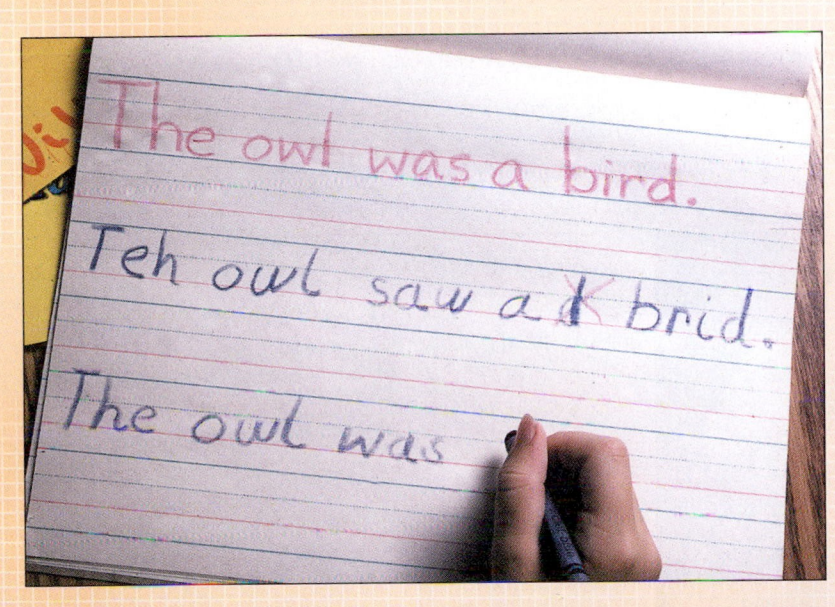

A person with dyslexia learning to write.
(© Will & Deni McIntyre/Science Source/
Photo Researchers, Inc.)

---

dominant hemisphere, also relies on the spatial abilities of the nondominant hemisphere.

Male and Female Brains Exhibit Differences in Laterality

Differences in the development of the lateralization of brain function exist between human males and females. In tests that evaluate the development of spatial function, males exhibit a lateralization of spatial tasks to the right hemisphere by 6 years of age. With females, spatial function is equally developed in both hemispheres until the age of 13 years.

The absence of early hemispherical specialization imparts certain advantages to females. Damage to the left hemisphere in childhood, for example, impairs language development in males more than in females. It appears that the delay in hemispherical specialization allows the transfer of language function from the left to right hemisphere to occur more readily in females.

To some extent, differences in the laterality in brain function between males and females are also seen in adults.

CURRENT CONCEPTS IN PHYSIOLOGY

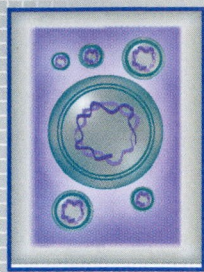

Magnetic Resonance Spectroscopy and the Brain

Magnetic resonance imaging (MRI) is a method used to visualize the internal organs of the body in a noninvasive manner (see Focus Box on Imaging of Internal Structures, Chapter 1). This approach uses a powerful magnet to align the molecules within the tissue of an organ. The most abundant element in the body is hydrogen. The nucleus of a hydrogen atom possesses a magnetic moment that causes it to behave like a small bar magnet. When an external magnetic field is applied to the hydrogen atom, it aligns itself with the magnetic field. When the magnetic field is turned off, the hydrogen molecules then return to their normal orientation. The rate at which hydrogen atoms align with the magnetic field and return to their original orientation gives rise to so-called "relaxation times." These relaxation times are different for hydrogen atoms embedded within a water environment, such as the cerebrospinal fluid (CSF), and in lipid and protein environments. These differences in relaxation times result in differences in magnetic signal intensity, which in turn can be detected to produce the types of brain images that we see with MRI.

The advancement of MRI technology has provided us with previously unattainable images of the living brain with tremendous detail. Recent developments in MRI now allow us to combine images of the brain with its chemistry. By using different pulsing sequences, the applied magnetic field can be manipulated to produce a spectrum of biochemical constituents within the brain. This form of "magnetic resonance spectroscopy" can be used to detect compounds, such as lactic acid in the brain, which forms during ischemic brain injury (i.e., brain injury caused by a loss of blood circulation). The combination of MRI with spectroscopy therefore allows us to map the distribution of a variety of neurochemical compounds in the brain, and their alterations during pathophysiological conditions. This approach might also be used to monitor the effectiveness of therapies to prevent nerve cell death and to destroy brain tumors.

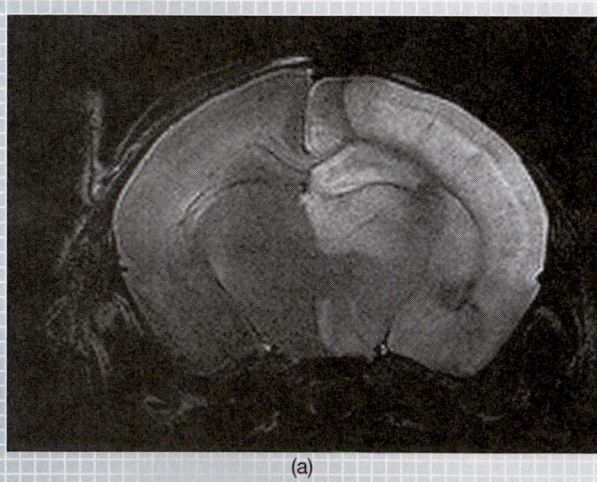

(a)

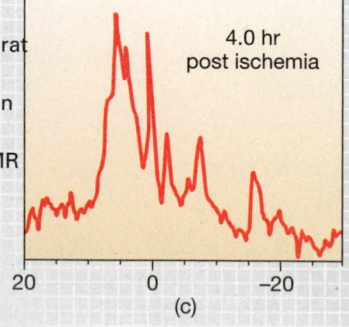

(c)

Magnetic resonance image (MRI) and spectroscopy of neonatal rat brain after ischemic brain injury. This approach permits the noninvasive monitoring of neurochemical alterations in the brain under pathophysiological conditions. (*a*) MRI of rat brain after ischemic injury. (*b*) Normal chemical spectra before injury. (*c*) MR spectra after injury.
PM = phosphomonoesters, PI = inorganic phosphate, PD = phosphodiesters, PCr = phosphocreatine, ATP = adenosine triphosphate. (*Photo courtesy of Elizabeth M. Jansen and Michael Garwood*)

Damage to the left hemisphere is more strongly associated with language disorders in adult males than in adult females. Women appear to be less lateralized for language functions. Likewise, damage to the right hemisphere is more strongly associated with nonlanguage disorders in males than in females. These observations suggest that the female brain is more symmetrical in function than is that of the male.

CHAPTER REVIEW

Summary

- Many bodily functions fluctuate on a monthly or daily basis. Those that vary daily are called *circadian rhythms*. Many of these rhythms are entrained to the 24-hour light-dark cycle by the suprachiasmatic nucleus of the hypothalamus.
- The cycle of sleep and wakefulness varies on a 25-hour schedule but is entrained to a 24-hour period. Sleep involves repeated cycles of brain activity, each having similar stages.
- Hunger is the desire for food, a motivational system that operates to satisfy the energy needs of the body and the need for growth, development, and maintenance of the body's organs. Hunger is regulated by both short- and long-term mechanisms that affect food consumption.
- Sexual drives are important for reproduction and the perpetuation of the species. They are activated by the interaction of social environment, hormones, and genetics. The development of male or female sexual behavior is influenced by androgens during development.
- In nonassociative learning, an understanding of the properties concerning a single type of stimulus leads to either a habituation or sensitization to the stimulus.
- In associative learning, we acquire an understanding of the relationship either between two stimuli (called *classical conditioning*) or between a stimulus and a behavior (called *operant learning*).

- The storage of information progresses from short-term memory to long-term memory. If information is disrupted while in short-term memory, it does not become consolidated into long-term storage. Memory is stored in a variety of memory circuits found in different areas of the brain. Memory circuits have been identified for the learning of motor responses and visual perceptions.
- Habituation is a decrease in response to a repeated stimulus due to a decrease in neurotransmitter release. Sensitization is an increase in response to a repeated stimulus and is due to an increase in the release of neurotransmitters.
- The connections between nerve cells can be altered by the amount of activity between the cells. The use of neural pathways can strengthen the connections between groups of nerve cells, whereas the nonuse of pathways weakens their connections.
- Language centers are located in one or the other side of the cerebral cortex, depending on the individual. Language is located in an individual's *dominant hemisphere*. Within the dominant hemisphere, Wernicke's area is responsible for the interpretation of language; Broca's area, for the motor aspects of speech.
- Spatial, artistic, and musical abilities are located predominantly in the right hemisphere. Laterality of brain function develops earlier in males than in females.

Review Questions

Choose the Correct Answer

1. Exposure of fetal hypothalamic neurons to androgens has what effect on the release of gonadotropin-releasing hormone (GnRH) from adult hypothalamic neurons?
   a. Release is at relatively constant levels.
   b. Release is in response to estrogen stimulation.
   c. Release is in a cyclic pattern of changing levels.
   d. Release is in an irregular pattern of changing levels.
   e. Release is inhibited by the failure of neurons to produce LHRH.
2. Which one of the following statements about speech and language is *not correct?*
   a. Dyslexia is more common among boys than among girls.
   b. Patients with damage to Wernicke's area lose the ability to write.
   c. Damage to Wernicke's area produces loss of comprehension of language.
   d. Patients with damage to Wernicke's area usually have the ability to speak.
   e. Damage to Wernicke's area in the nondominant hemisphere produces a loss in comprehending the intonation of speech.

3. What area(s) of the brain is (are) responsible for REM sleep?
   a. Raphe nucleus in the brain stem
   b. Suprachiasmatic nucleus of the hypothalamus
   c. Prepotic area of the hypothalamus
   d. Lateral geniculate body of the thalamus
   e. Supraoptic area of the hypothalamus
4. Which one of the following cellular mechanisms is *not* a factor in the sensitization of a response to stimulation?
   a. Increase in cAMP levels
   b. Activation of voltage-gated potassium channels
   c. Increase in presynaptic intracellular calcium levels
   d. Prolonged depolarization of the presynaptic terminals
   e. Calmodulin-mediated increase in adenylate cyclase activity
5. Which one of the following disorders indicates a condition in which the patient falls asleep at unusual or inappropriate times?
   a. Aphasia
   b. Dyslexia
   c. Insomnia
   d. Narcolepsy
   e. Somnambulism

6. What region of the central nervous system is currently thought to be the "master clock" for coordinating various biological rhythms?
   a. Pineal body
   b. Preoptic area
   c. Raphe nucleus
   d. Reticular formation
   e. Suprachiasmatic nucleus

7. Which one of the following factors is important in the *long-term* mechanism for maintaining body weight?
   a. Triglyceride content in fat cells
   b. Glucose receptors in the pancreas
   c. Levels of glucose in the circulation
   d. Contractions in the gastrointestinal tract
   e. Mechanoreceptors in the gastrointestinal tract

8. What stage of sleep is characterized by decreased heart rate and blood pressure and an electroencephalogram pattern of relatively high amplitude slow waves?
   a. Stage 1
   b. Stage 2
   c. Stage 3
   d. Stage 4
   e. REM

9. During which stage of development does REM sleep occupy approximately 80% of the total sleep time?
   a. The last few weeks before birth
   b. Full-term newborn
   c. Preadolescent (10 years of age)
   d. Young adult (20–30 years of age)
   e. Older adult (70+ years of age)

10. Pheromones are important chemicals in regulating which one of the following functions?
    a. Sleep
    b. Thirst
    c. Hunger
    d. Temperature
    e. Sexual behavior

11. The hypothalamus regulates the release of what substance to increase the rates of cellular oxygen consumption, glucose metabolism, and heat production?
    a. Thyroxin
    b. Cortisone
    c. Serotonin
    d. Cortisol
    e. Adrenocorticotropic hormone (ACTH)

12. Habituation is which one of the following processes?
    a. A decrease in response to a repeated stimulus
    b. An increase in response to a stimulus as it becomes more familiar
    c. An increase in attention to a stimulus as its frequency of presentation decreases
    d. A decrease in response to two different stimuli when the time separating their presentations increases
    e. An increase in understanding the relationship between two stimuli with their frequent, simultaneous presentation

13. Damage to brain areas regulating the ability to understand or express intonations of voice result in what type of disorder?
    a. Amnesia
    b. Aphasias
    c. Dyslexia
    d. Aprosodias
    e. Split-brain syndrome

14. Which one of the following processes occurs during the REM stage of sleep?
    a. Suspension of dreaming
    b. Increase in gastrointestinal activity
    c. Increase in skeletal muscle movement
    d. Decrease in eye movements while the eyes remain closed
    e. Activation of the sympathetic division of the autonomic nervous system

15. Which one of the following functions is *not* predominantly located in the right (nondominant) cerebral hemisphere?
    a. Artistic skills
    b. Spatial abilities
    c. Musical abilities
    d. Comprehension of language
    e. Comprehension of tone of voice

16. Abnormal development of which one of the following brain regions most likely underlies the disorder termed *dyslexia*?
    a. Broca's area
    b. Planum temporale
    c. Primary visual cortex
    d. Region of the nondominant cerebral hemisphere corresponding to Broca's area
    e. Region of the nondominant cerebral hemisphere corresponding to Wernicke's area

17. Damage to the left cerebral hemisphere in childhood impairs language development in males more than in females. Which one of the following statements is the most likely explanation for this fact?
    a. Female language abilities develop earlier than do those of males.
    b. Males exhibit right-to-left, but not left-to-right, transfer at a young age.
    c. Male lateralization of language function is not fully developed until the late teen years.
    d. Delayed lateralization of function in females helps the transfer of function between hemispheres occur more readily.
    e. Males exhibit lateralization of spatial tasks to the right hemisphere at 6 years of age, whereas female spatial function does not fully develop until the age of 13 years.

18. Which one of the following processes does *not* represent plasticity in the nervous system?
    a. An enriched environment increases the number of synapses formed between cortical neurons.
    b. Animals raised in impoverished environments develop fewer than normal numbers of synapses.
    c. When one sensory area of the brain loses input from its sensory receptors, the area is taken over by adjacent sensory areas.
    d. If animals do not receive adequate visual stimulation at a critical period during development, they will be blind.
    e. Certain regions of the brain form new synapses during learning and memory processes.

19. Which one of the following fetal environmental situations would cause an adult mouse to exhibit mounting as a sexual posture instead of lordosis, which is typical for its sex?
    a. Male develops between male pups.
    b. Male develops between female pups.
    c. Female develops between male pups.
    d. Female develops between female pups.
    e. Male develops between one male and one female pup.

Answers to Case History Questions

1. Jim has an organic brain deficit, whereas Steve had a psychogenic brain deficit.
2. Jim, physical trauma to the left side of the brain. Steve, momentary hysteria due to the emotional trauma of the accident and witnessing his brother's severe injury.
3. When conscious, Jim was self-aware and able to form at least some speech, whereas initially Steve was mute and not aware of who he was. Mutism and lack of self-awareness accompany hysteria and psychogenic brain disturbances but are not generally characteristic of organic brain disorders. Steve displays recollection of a mixture of recent and remote memories, especially those that circumscribed the "traumatic event" of the car accident but has blocked out the accident itself. This pattern is also characteristic of psychogenic memory loss. In contrast, Jim can recall remote memories better than recent ones, especially those reinforced by emotional events, such as the automobile accident and his mother's funeral. In addition, Jim's memory loss is more severe for time than it is for persons and places. These types of memory losses are typical of organic brain disturbances. Finally, Jim has difficulty finding words, recognizing letters, speaking, and moving the right side of his body. All of these are characteristic of organic brain damage in his left, dominant hemisphere.

   Overall, comparisons of patterns of memory loss, language deficits, and motor-skill deficits can be used clinically to distinguish between organic and psychogenic brain disturbances.

Key Terms

adipose tissue (p. 365)
androgens (p. 366)
angular gyrus (p. 375)
aphasia (p. 376)
associative learning (p. 368)

Broca's area (p. 375)
circadian rhythm (p. 362)
classical conditioning (p. 368)
dominant hemisphere (p. 375)
dyslexia (p. 376)

electroencephalogram (EEG) (p. 363)
habituation (p. 368)
long-term memory (p. 369)
nonassociative learning (p. 368)

operant conditioning (p. 368)
REM sleep (p. 364)
sensitization (p. 368)
short-term memory (p. 369)
Wernicke's area (p. 375)

Suggested Readings

Bjork, E. L., and Bjork, R. A. *Memory.* San Diego, Academic Press, 1996.

Crary, M. A. *Developmental Motor Speech Disorders.* San Diego, Singular Publishing Group, 1993.

Gillam, R. B. *Memory and Language Impairment in Children and Adults.* Gaithersburg, MD, Aspen Publishers, 1998.

Kamhi, A. G., and Catts, H. W. *Reading Disabilities.* Boston, Allyn and Bacon, 1991.

Kryger M. H., Roth, T., and Dement, W. C. *Principles and Practice of Sleep Medicine,* ed 3. Philadelphia, W.B. Saunders, 2000.

Lister, R. G., and Weingartner, H. J. *Perspectives on Cognitive Neuroscience.* New York, Oxford University Press, 1991.

Merzenich, M. M., Kaas, J. H., Wall, J. T., Sur, M., Melson, R. J., and Feller, D. J. "Progression of change following median nerve section in the cortical representation of the hand in areas 3b and 1 in adult owl and squirrel monkeys. *Neuroscience,* 10:639–665, 1983.

Monk, T. H. *Sleep, Sleepiness and Performance.* Chicester, John Wiley & Sons, 1991.

Pain, J. C. *Adult Neurogenic Language Disorders: Assessment and Treatment,* San Diego, Singular Publishing Group, 1997.

Parkin, A. *Memory: Phenomena, Experiment and Theory.* Oxford, UK, Blackwell, 1995.

Raichle, M. E. "Visualizing the mind." *Scientific American,* 270:58–64, 1994.

Spafford, C. S., and Grosser, G. S. *Dyslexia: Research and Resource Guide,* Boston, Allyn and Bacon, 1996.

Weisskopf, M. G., Castillo, P. E., Zalutsky, R. A., and Nicoll, R. A. "Mediation of hippocampal mossy fiber long-term potentiation by cyclic AMP." *Science,* 265:1878–1882, 1994.

Zola-Morgan, S. M., and Squire, L. R. "The primate hippocampal formation: Evidence for a time-limited role in memory storage." *Science,* 250:288–290, 1990.

Web sites

http://cancer.med.upenn.edu/causeprevent/screening/mrs_dx .html
Liu, L. Use of Magnetic Resonance Spectroscopy (MRS) in Cancer Diagnosis.

http://www.nida.nih.gov/TXManuals/IDCA/IDCA3.html
Mercer, D. E., and Woody, G. E. National Institute on Drug Abuse. Therapy Manuals for Drug Addiction. Manual 3: An Individual Counseling Approach to Treat Cocaine Addiction: The Collaborative Cocaine Treatment Study Model.

Answers to Review Questions

1. a **2.** b **3.** a **4.** c **5.** d **6.** e **7.** a **8.** d
9. a **10.** e **11.** e **12.** a **13.** b **14.** e **15.** d
16. b **17.** d **18.** d **19.** c

Chapter 12

ENDOCRINE CONTROL MECHANISMS

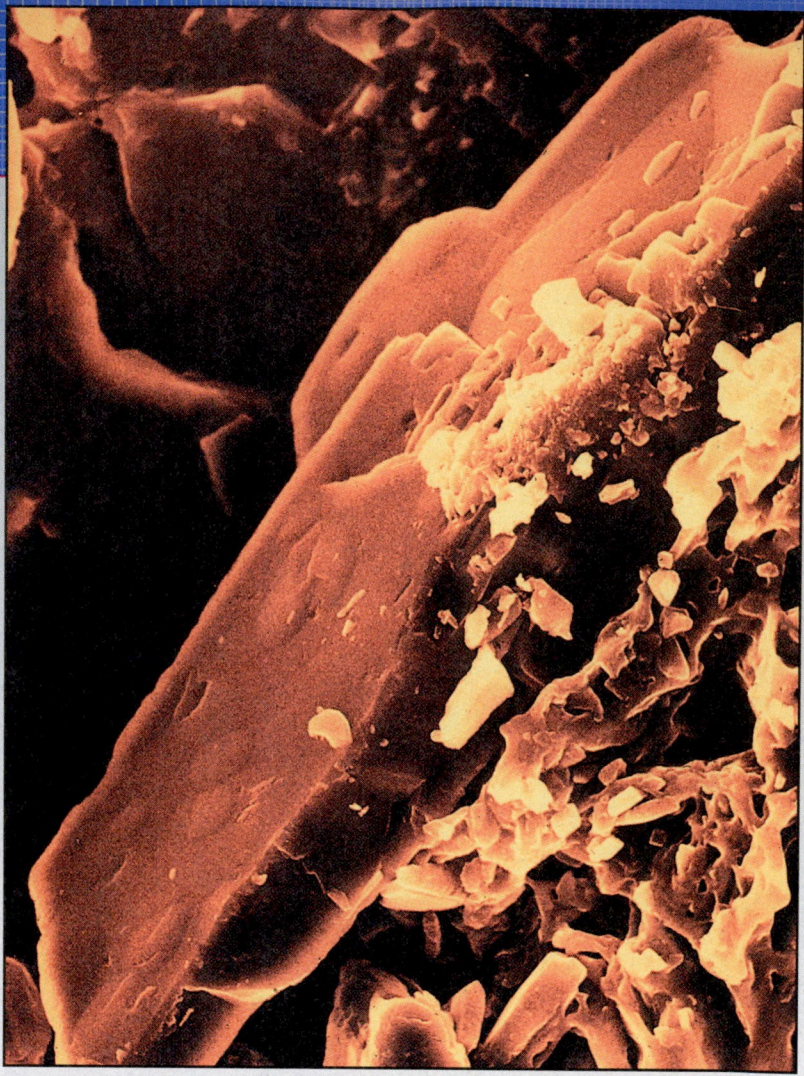

KEY CONCEPTS

- Hormones are chemical signals that are used for communication between cells.

- Chemically, hormones are a rather diverse group of substances. Some are derived from metabolism of amino acids, some are polypeptide in nature, and some are derived from cholesterol.

- Binding of a hormone to its receptor and activation of that receptor is the initial step in producing a hormone effect.

- Some hormones act via second messengers to alter activity of preexisting proteins in target cells.

- Other hormones act by altering gene expression in target cells to change the amount of a few key proteins.

CASE HISTORY

John is a 55-year-old man who was diagnosed with type II diabetes mellitus (maturity-onset or non-insulin-dependent diabetes mellitus) about 10 years ago. He has been able to control his blood glucose levels relatively well by controlling his diet and level of exercise and by taking a daily oral antidiabetic agent. John has been married to his wife, Helen, for 30 years, and they have raised three wonderful children. They still have an active sex life, but in recent months John has suffered from intermittent episodes of impotence (failure to achieve an erection). John consulted with his family physician who prescribed Viagra, which cured John's problem and allowed him to resume a normal relationship with his wife.

Viagra enhances the normal signal-transduction processes in the penis that lead to an erection. Activation of the parasympathetic division of the autonomic nervous system leads to release of acetylcholine. Acetylcholine acts on vascular endothelial cells to cause release of nitrous oxide (NO), which diffuses to vascular smooth muscle cells where guanylate cyclase is activated, resulting in increased formation of cyclic GMP (cGMP). Increased cGMP leads to relaxation of smooth muscle cells in the blood vessels and results in increased blood flow, followed by engorgement and subsequent erection. Viagra is an inhibitor of a specific form of cGMP phosphodiesterase found largely in vascular smooth muscle cells of the penis. Phosphodiesterases are enzymes responsible for the breakdown and therefore termination of the cGMP (and cAMP) intracellular signal in cells. By inhibiting cGMP phosphodiesterase in the smooth muscle cells of the penis, Viagra causes increased levels of cGMP in the cells, which leads to the vascular changes that cause an erection.

Question

Might other Viagra-like compounds have different pharmaceutical uses?

INTRODUCTION

Endocrinology is the branch of biology concerned with the actions of hormones and the organs or glands in which the hormones are produced. Endocrinology often overlaps with areas such as anatomy, biochemistry, and neurophysiology. Included within the field of endocrinology are such topics as the anatomy and physiological function of the endocrine organs, the chemical nature of the hormones these glands produce, the cellular mechanisms by which the hormones produce their biological effect, and clinical manifestations of abnormal endocrine function. This brief list is by no means complete; numerous other exciting areas of study exist within the field of endocrinology. However, the majority of current research efforts in endocrinology can probably be placed into one of these four categories.

GENERAL CONCEPTS OF ENDOCRINE CONTROL

> *What should we use as our working definition of a hormone?*

The word *hormone* is derived from the Greek *hormaein,* which means "to excite" or "to arouse." But what exactly is a hormone, and, conversely, what is not a hormone? Is the glucose that circulates in the blood a hormone? Is the lactic acid produced by a vigorously exercising muscle, or the carbon dioxide produced by virtually all cells in the body, a hormone? Why do we consider insulin, produced by the pancreas, and epinephrine (also known as *adrenaline*), secreted by the adrenal medulla, as hormones? Over the years, scientists have made many attempts to define *hormone,* and much discussion has ensued over the propriety of any particular definition. As new hormones were discovered, definitions of hormones were rewritten to include the newest members of the "hormone family."

Despite the complexity involved in describing hormones, at this point in the discussion, a broad, working definition is probably useful to help gain a general appreciation of what hormones are and how they function in the body. Hormones are generally considered to be substances secreted into the blood in very small amounts by specialized cells or glands and carried by the bloodstream to other parts of the body, where they interact with specific receptors in target tissue cells to produce a particular biological response (Fig. 12–1a).

This is a generalized definition of *hormone,* not an absolute one. Several bona fide hormones would probably not fit this definition in its entirety, and, likewise, several nonhormonal substances could be viewed as filling most of the criteria of this definition. (Some of the terms used in this definition, such as *receptor* and *target tissue,* will be described in greater detail later in this chapter.)

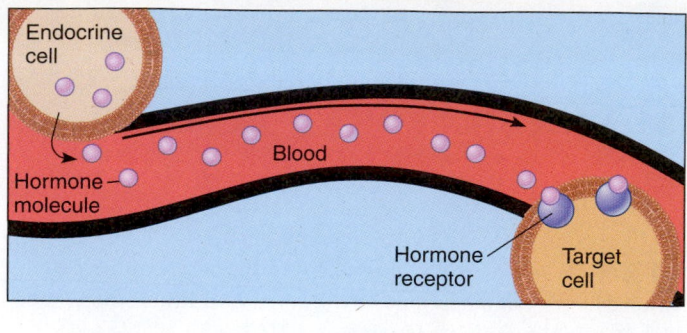

(a)

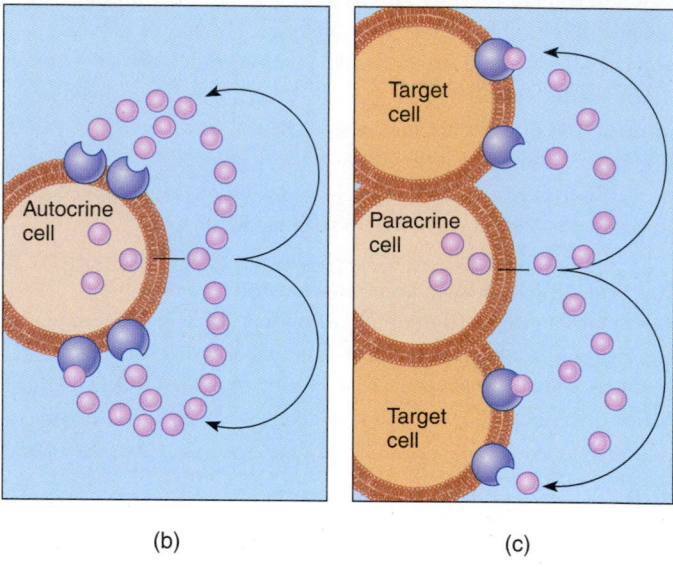

(b)   (c)

Figure 12–1

Comparison of endocrine *(a)*, autocrine *(b)*, and paracrine *(c)* mechanisms of cellular communication.

Hormones Are Chemical Signals Used in Cell-to-Cell Communication

In recent years, knowledge of the degree of the complexities involved in cell-to-cell communication mechanisms has increased dramatically. Virtually every cell in the body receives information in the form of chemical signals, from either nearby cells or distant cells or organs. Understanding the details of these very powerful control systems is still in a very basic state. A fuller understanding of these mechanisms may lead to important breakthroughs in the treatment of a wide variety of diseases.

Not all hormonelike substances are carried to their action sites by the blood. Some reach their target cells by **diffusion** through the interstitial fluid and exert their effects by what are known as **autocrine** or **paracrine** mechanisms. Autocrine regulation involves the secretion of a hormone product that exerts its effects on the cell in which it was produced (Fig. 12–1b). Paracrine regulation involves the secretion of a hormone that diffuses to and acts on adjacent cells (Fig. 12–1c). Although our knowledge of classic endocrine control is extensive, information about autocrine and paracrine function is rather limited and remains to be more completely characterized. However, these substances likely play a major role in the establishment and maintenance of a normal, healthy state.

The Nervous System and the Endocrine System Function Together to Promote the Maintenance of a Steady State

> *How do the nervous system and the endocrine system contribute to the overall control of physiological systems?*

As ancient single-cell organisms evolved into today's complex animals, their survival was enhanced by specialized function of cells. Specialization allowed the various cell types to perform unique functions in a cooperative manner and thus contribute to the overall survival of the organism. Along with this specialization came an increasing need for communication between cells or among groups of cells as they became more interdependent. Such communication ensures that the activities and functions of various tissues and organs are carried out in a coordinated, purposeful manner.

In the human body, cells communicate through two major systems: the **nervous system,** discussed in previous chapters, and the **endocrine system.** A considerable degree of interaction occurs between the two systems. This overlap between the nervous and endocrine systems constitutes an area of study known as **neuroendocrinology.**

In Figure 12–2, the different mechanisms whereby neural, endocrine, and neuroendocrine cells receive and transfer information is illustrated. Neurons, for example, receive information from other nerve cells in the form of chemical signals (neurotransmitters) that reach the cell by diffusion across synaptic clefts. The neuron in turn releases neurotransmitters at its synapses, which then activate subsequent cells in the neuronal chain. Thus, neurotransmission involves the transfer of information in the form of electrochemical signals from neuron to neuron (Fig. 12–2a).

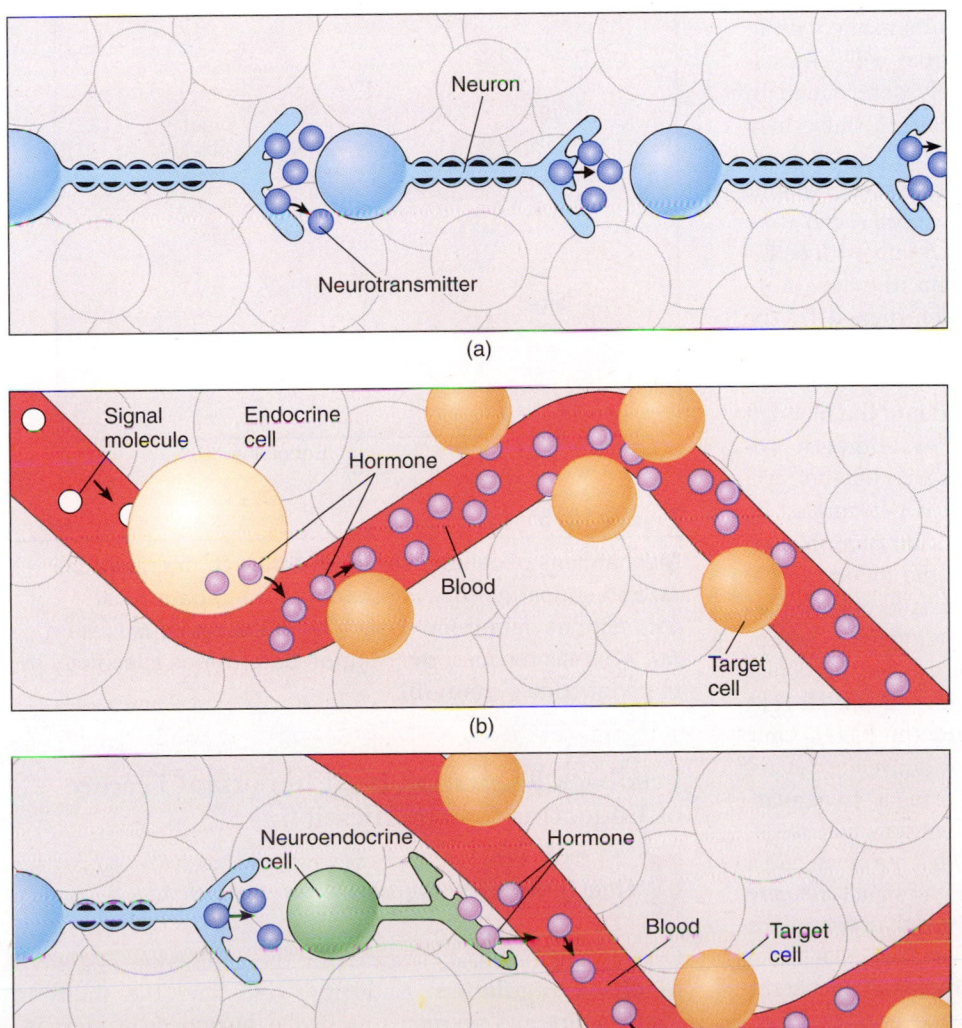

Figure 12–2

Basic mechanisms of neural **(a)**, endocrine **(b)**, and neuroendocrine **(c)** communication methods.

A typical endocrine cell receives information in the form of chemical messages ("signal" molecules) from the blood (Fig. 12–2b). In response to the appropriate input signal, the endocrine cell then secretes its own specific hormonal signal into the blood. The endocrine cell may sense the concentration of another hormone present in the blood, or it may perceive the concentration of some particular ion or nutrient, such as sodium or glucose. When the endocrine cell detects a significant change in the concentration of that particular factor in the blood, it appropriately alters the rate of release of its own secreted hormone product.

Neuroendocrine cells function as an interface between the nervous and endocrine systems, providing a crucial link between these two communication systems. As shown in Figure 12–2c, a neuroendocrine cell is directly innervated by nerve cells, but, unlike other neurons, in response to stimulation it secretes its product, a hormone, directly into the blood. Thus, neuroendocrine cells act as one-way translators. They convert **electrochemical signals** from the nervous system into **hormonal signals** in the blood.

Figure 12–3 illustrates a basic difference in the functional organization of the nervous and endocrine systems. The ability of one particular neuron to communicate with its target cells — whether they are other nerve cells, muscle cells, or other excitable cells — is generally determined by the anatomical connections within the system. A single neuron can transmit information — in this case, an electrochemical signal — only to those cells with which it makes synaptic contact. For example, in Figure 12–3a, nerve cell A can communicate only with cells d and e, nerve cell B only with cells e and f, and nerve cell C only with cell g. Thus, the flow of information within the nervous system is determined by the point-to-point anatomical connections made between different cells (i.e., the "hardwiring").

By contrast, once a hormone is secreted into the blood, it is carried to nearly every cell in the body. This difference is illustrated in Figure 12–3b. Whether or not a certain cell will respond to a particular hormone molecule is determined by the presence or absence in the cell of a specific **receptor** for that hormone. A receptor is a molecule in the membrane of or inside the cell that specifically recognizes and binds a particular hormone. As a result of hormone binding to the receptor, the biological responses characteristic of that particular hormone are initiated. As depicted in Figure 12–3b, all cells are exposed to the hormones distributed by the blood. Only those cells having a specific receptor for a hormone can respond to it. For example, only those cells having a "triangular" receptor can respond to the "triangular" hormone, and so on. In this case, we can draw an analogy between a hormone and a radio signal: A radio transmitter broadcasts its signal in every direction for all who care to listen; however, only those radios tuned to the appropriate station detect that particular radio signal. Similarly, a hormone circulates throughout the body, but only those cells "tuned" to that particular hormone (i.e., that possess specific receptors for that hormone) are able to sense its presence in the blood. The concept of receptor specificity will be discussed in greater detail later in this chapter.

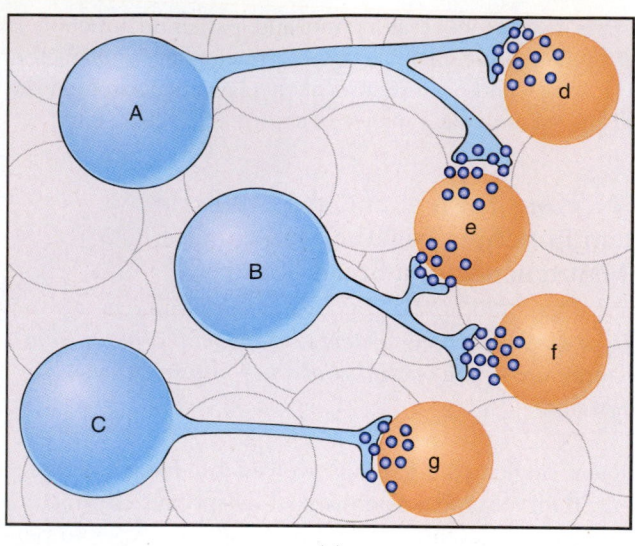

(a)

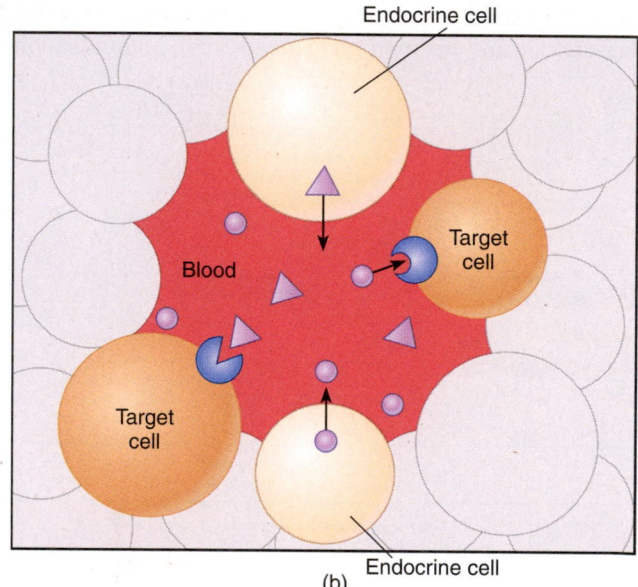

(b)

Figure 12–3

Mechanisms of cellular communication in the nervous *(a)* and endocrine *(b)* systems. Note that anatomical connections determine specificity in the nervous system *(a)*, whereas receptor distribution determines specificity in the endocrine system *(b)*.

Feedback Regulation Is an Important Feature of Endocrine Communication

 How is the release of hormones controlled?

One of the hallmark features of the endocrine system is **feedback regulation.** Endocrine cells have the ability to manufacture and secrete a particular hormone; in most instances, they are also equipped to detect or monitor the magnitude of the biological effect of that hormone. Feedback regulation permits the endocrine cell to adjust the rate of

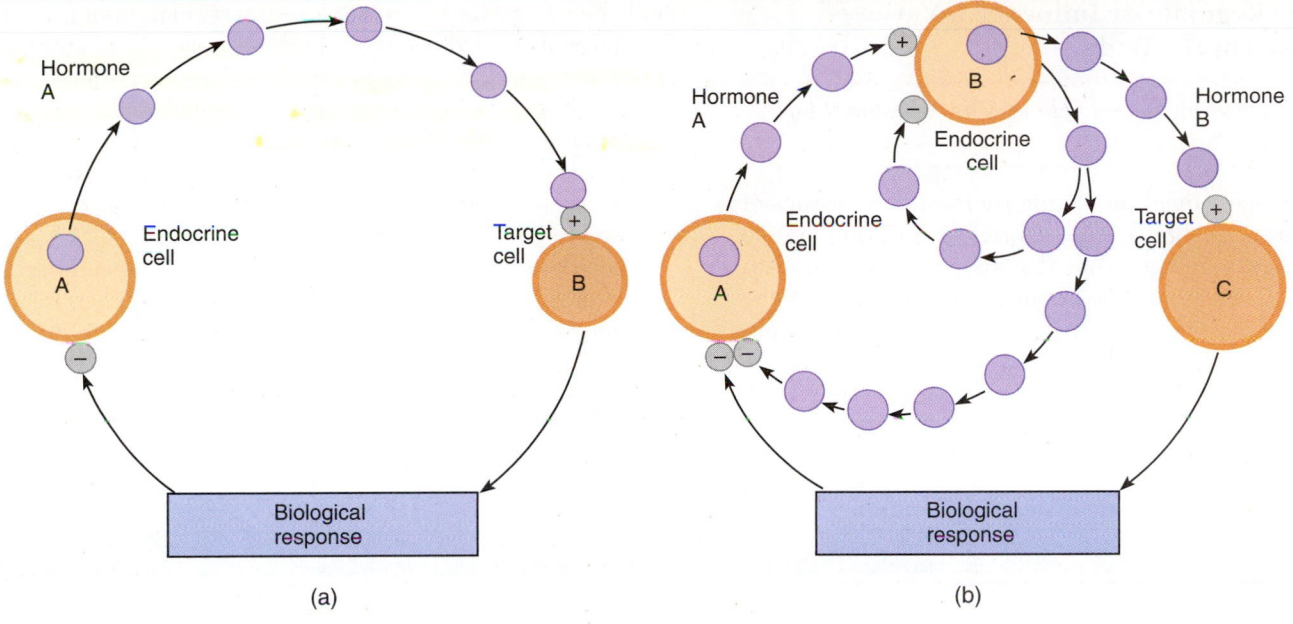

Figure 12–4

(a) A simple negative feedback loop. *(b)* A more complex negative feedback system involving two different endocrine cells and two different hormones.

hormone secretion in an appropriate manner to achieve the desired effect, which is the maintenance of a steady state.

In most instances, feedback regulation involves **negative feedback,** although some cases of **positive feedback** are also known. A very simple negative feedback system is illustrated in Figure 12–4*a.* In this case, the endocrine cell (*A*) secretes a hormone, which has an effect on one of its target cells (*B*). The hormone triggers a specific biological response in the target cell, and the magnitude of the response is in turn sensed or monitored by the endocrine cell that first secreted the hormone. If the response is too small, the endocrine cell produces and secretes more hormone in an attempt to elicit a larger, more normal response. Likewise, if the response of the target cell is too great, the endocrine cell reduces the amount of hormone it secretes until the response decreases into the normal range. In this case, the particular response being produced can be said to have a negative feedback effect on the amount of hormone secreted by the endocrine cell.

An example of this simple type of negative feedback regulation is the control of blood glucose concentration by the hormone **insulin,** which is produced by specialized endocrine cells within the pancreas. These cells are triggered to secrete insulin when the amount of glucose in the blood exceeds the normal concentration of about 100 mg of glucose per 100 ml of blood. Therefore, when blood glucose levels are increased above normal, such as might occur soon after a meal, the insulin-secreting cells sense this increase and respond by increasing the secretion of insulin. Insulin in turn stimulates the uptake of glucose from the blood by such tissues as muscle and adipose (fat). The overall response to insulin therefore is a reduction in the amount of glucose

circulating in the blood. The stimulus for insulin secretion is therefore lessened, and the secretion rate decreases. (These effects of insulin will be discussed in greater detail in Chapter 15.)

More complex types of negative feedback systems are also prevalent within the endocrine system. An example of one such system is illustrated in Figure 12–4*b.* In this case, endocrine cell *A* secretes a hormone (*A*), which has as its target endocrine cell *B.* In response to hormone *A,* endocrine cell *B* increases its production of hormone *B.* This hormone in turn acts on target cell *C* to produce some particular biological response. In similar fashion to the example shown in Figure 12–4*a,* the response produced by the final target cell — in this case, cell *C* — inhibits the production of hormone *A.* In addition, hormone *B* itself has negative feedback effects on cell *A,* and it also has negative feedback effects to limit its own production by cell *B.* Several examples of this type of multilevel feedback can be found within the endocrine system, especially in the regulation of secretion of pituitary hormones.

The more complex type of feedback control shown in Figure 12–4*b* offers certain advantages over the simple type shown in Figure 12–4*a.* It permits fine-tuning of the endocrine cells with regard to their hormone-secreting activity. This in turn leads to biological responses that are closely matched to the original stimulus. A nearly constant internal environment results. Additionally, these multilevel control systems provide some degree of backup in the event that one segment of the overall system functions abnormally. In such a situation, a multilevel control system is better able to achieve the desired biological response than is a simple system, such as the one shown in Figure 12–4*a.*

Hormones Regulate or Influence a Variety of Processes in the Body

What kinds of biological activities are regulated by hormones?

Table 12–1 lists primary human hormones, their tissues of origin, their target cells, and a brief description of the processes they regulate. Note that tissues or organs with endocrine functions can be found in many parts of the body. The specific effect of individual hormones can gen-

erally be categorized as relating to the regulation of one of four physiological processes: (1) the digestion and storage of nutrients and their metabolism and use for metabolic energy, (2) salt and water balance, (3) growth and development, and (4) reproductive function. Later sections of this chapter describe details of the biochemical mechanisms involved in hormonal regulation of a variety of cellular processes.

The descriptions of target tissues and hormonal effects presented in Table 12–1 are considerably abbreviated and simplified. Most hormones have several different target tis-

TABLE 12–1

Summary of the Primary Human Hormones

Hormone	Gland or Source	Target Cells	Primary Biological Effect
Release or release-inhibiting hormones	Hypothalamus	Anterior pituitary	Regulate hormone secretion
Oxytocin	Synthesized in hypothalamus, secreted from posterior pituitary	Breast Uterus	Triggers milk "let down" Stimulates uterine contraction
Vasopressin (antidiuretic hormone; ADH)	Synthesized in hypothalamus, secreted from posterior pituitary	Kidney Blood vessels	Increases water reabsorption Causes constriction
Growth hormone (GH)	Anterior pituitary	Many; especially bone, fat, and liver	Stimulates growth of skeleton and muscle
Prolactin (PRL)	Anterior pituitary	Breast	Stimulates milk production
Adrenocorticotropic hormone (ACTH)	Anterior pituitary	Adrenal cortex	Promotes adrenal steroid production
Thyroid-stimulating hormone (thyrotropin; TSH)	Anterior pituitary	Thyroid gland	Promotes thyroid hormone production
Follicle-stimulating hormone (FSH)	Anterior pituitary	Gonads (in females, ovarian follicle cells; in males, Sertoli cells of testes)	Stimulates growth and development
Luteinizing hormone	Anterior pituitary	Gonads, ovarian follicle cells Leydig cells of testes	Triggers ovulation Stimulates testosterone production
Insulin	Pancreas—islets of Langerhans	Primarily liver, muscle, and fat	Regulates metabolism and blood glucose
Glucagon	Pancreas—islets of Langerhans	Primarily liver	Regulates metabolism and blood glucose
Somatostatin	Hypothalamus	Anterior pituitary	Inhibits growth hormone secretion
	Pancreas—islets of Langerhans	Other cells of islets	Regulates insulin and glucagon secretion
Glucocorticoids	Adrenal cortex	Primarily liver, muscle, and fat	Regulate metabolism
Aldosterone	Adrenal cortex	Kidney	Regulates sodium excretion
Epinephrine	Adrenal medulla	Cardiovascular system	Stimulates cardiovascular function
Angiotensin II	Formed by conversion steps involving kidney, blood, and lung	Adrenal cortex	Stimulates aldosterone production

sues. In addition, many hormones produce multiple individual effects within their specific target cells. These multiple effects, when added together, produce the overall response characteristic of that particular hormone. For example, the male sex hormone, **testosterone**, has several different effects in different tissues. It is required for normal sperm formation to occur in the testes and also promotes the growth and development of most parts of the male reproductive tract, such as the prostate gland and seminal vesicles. Testosterone is also responsible for the development of secondary male sexual characteristics at puberty, including

beard growth and a deepening of the voice. Thus, this one single hormone has many different effects in a variety of different tissues.

Just as one hormone may influence many different processes, a single process may in turn be regulated by several different hormones. For example, the process of converting blood sugar into glycogen in the liver is regulated by insulin, glucagon, epinephrine, adrenal glucocorticoids, and thyroid hormone. Several other hormones have minor influences as well. The overall effect is an integrated response reflecting the circulating levels of all of these hormones. Thus, endocrine

TABLE 12–1

Summary of the Primary Human Hormones

Hormone	Gland or Source	Target Cells	Primary Biological Effect
Atrial natriuretic factor (ANF)	Atrial wall of heart	Primarily kidney	Regulates sodium excretion
Thyroid hormones (T_3, T_4)	Thyroid gland	Many cell types	Regulate energy metabolism
Somatomedins Insulin-like growth factor I (IGF-I)	Primarily liver	Bone and many tissues	Promotes bone growth and tissue growth and repair
Parathyroid hormone (PTH)	Parathyroid glands	Bone, kidney	Regulates plasma calcium and phosphate
1,25 Dihydroxyvitamin D_3	Conversion in kidney from precursor formed in liver	GI tract; bone	Regulates plasma calcium and phosphate
Calcitonin (CT)	Parafollicular cells of thyroid gland	Bone	Regulates plasma calcium and phosphate
Gastrointestinal (GI) hormones Gastrin, secretin, cholecystokinin, gastric inhibitory peptide, and somatostatin	Various cells of the GI tract	GI tract; gallbladder and pancreas	Regulate digestive processes, including secretion and motility
Androgens	Primarily testes; also adrenal cortex	Reproductive tract	Stimulate growth and development
Estrogens	Ovaries and placenta	Reproductive tract; breasts	Stimulate growth and development
Progesterone	Ovaries and placenta	Uterus; breasts	Promotes proper development and function
		CNS	Inhibits ovulation
Placental hormones Chorionic gonadotropin, placental lactogen, estrogen, and progesterone	Placenta	Various reproductive tissues	Maintain pregnancy
Mullerian inhibitory hormone (MIH)	Testes	Müllerian ducts	Causes regression of ducts in fetus
Inhibin	Testes	Hypothalamus and pituitary	Inhibits FSH secretion
Melatonin	Pineal gland	Reproductive system	Influences onset of sexual maturity
Erythropoietin	Kidney	Bone marrow	Stimulates red cell formation

regulation is characterized by single hormones with multiple effects and by processes regulated by several hormones.

THE CHEMISTRY OF HORMONES

Up to this point, the discussion of hormones has been rather general in nature. Let us now begin to examine the hormones in more detail, beginning first with the chemistry of hormones.

Most Hormones Can Be Characterized as Belonging to One of Three Chemical Classifications

 What characteristics do various hormones have in common?

Hormones generally fall into one of three different chemical categories: (1) amino acid derivatives, (2) polypeptides, and (3) steroids. Figure 12–5 shows an example of each type of hormone.

Hormones Derived from Amino Acids

Some hormones are simple derivatives of amino acids either obtained from the diet or synthesized in the body. With just a few, relatively simple chemical changes in the molecule, an ordinary amino acid can be converted into the powerful regulatory substance that is a hormone. Examples of hormones that are amino acid derivatives are **thyroxine, epinephrine,** and **melatonin;** their structures are shown in Figures 12–5 and 12–6.

Thyroxine consists of four iodide atoms joined to a pair of tyrosines that have been coupled end to end. Epinephrine is derived from tyrosine, and melatonin is formed from the amino acid tryptophan.

Thyroxine: An Iodide-Containing Compound

Although it is an amino acid derivative, thyroxine is synthesized in a somewhat different manner from amines. Details of thyroxine formation will be presented in Chapter 13. Briefly, thyroxine is first formed as part of a much larger precursor protein, known as **thyroglobulin,** which is stored in relatively large amounts within the **thyroid gland.** Because iodide is not always available in the diet, thyroxine is the one hormone in the body that is not made solely from readily available dietary constituents. A system has therefore evolved that allows for the storage of large amounts of thyroxine in its precursor form, thus providing a reservoir of hormone in the event that iodine is not available in the diet for an extended time period (weeks or months). Stimuli for thyroxine secretion result in the rapid breakdown of the thyroglobulin molecule to liberate thyroxine and stimulate the synthesis of new thyroglobulin protein.

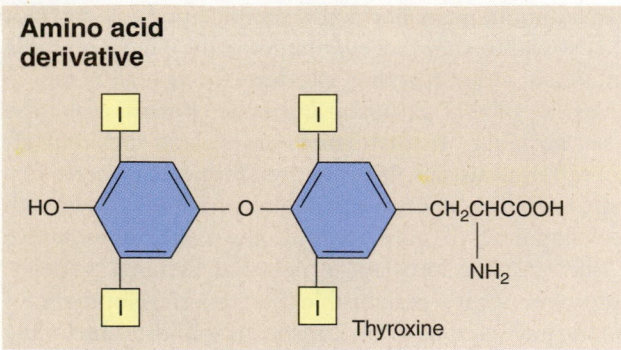

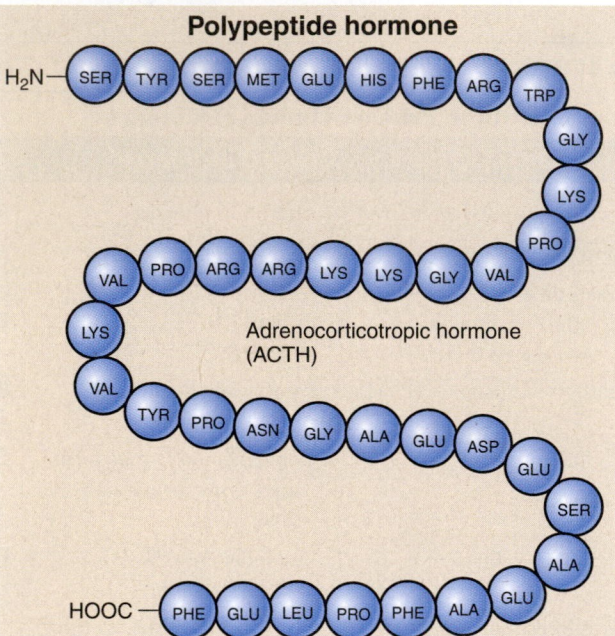

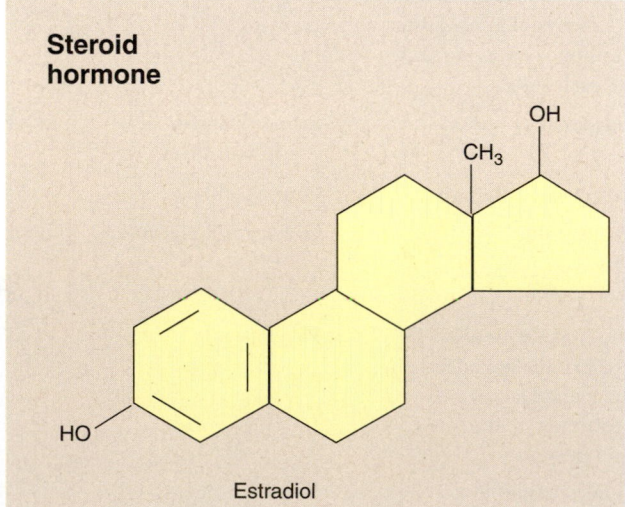

Figure 12–5

Examples of the three types of hormones.

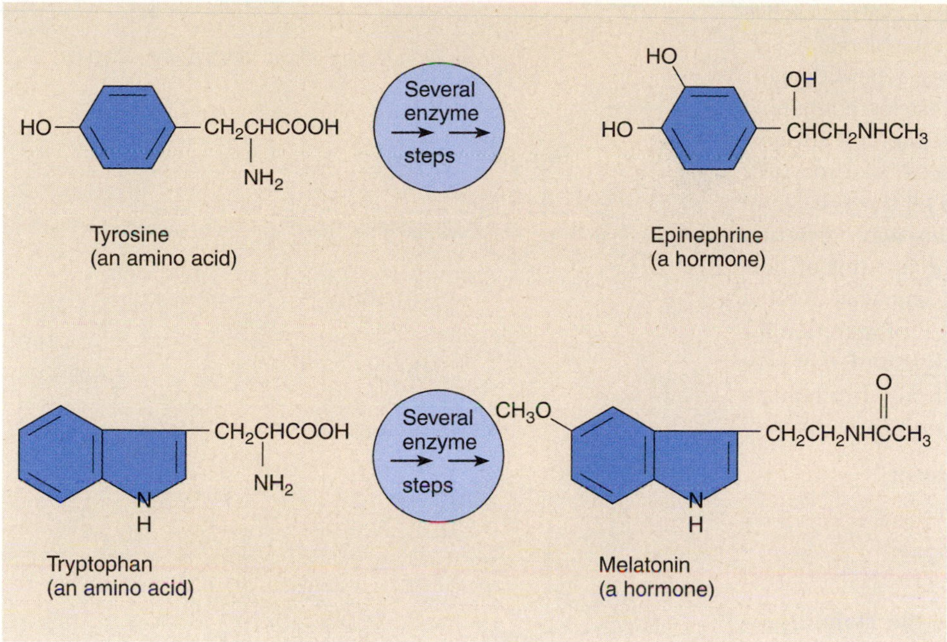

Figure 12–6

Examples of two hormones (epinephrine and melatonin) that are derivatives of amino acids.

Tyrosine (an amino acid) → Several enzyme steps → Epinephrine (a hormone)

Tryptophan (an amino acid) → Several enzyme steps → Melatonin (a hormone)

Epinephrine and Melatonin: Amines

Epinephrine and melatonin also represent a subset of hormones known as **amines,** which have a free amino group. Most amines are synthesized, stored, and secreted in a similar way. Within the endocrine cell, an amino acid is converted to an amine by a sequence of specific enzyme-catalyzed steps. The amine is packaged into secretory granules in the cell, where it is held until it is discharged into the blood in response to the appropriate stimulus to the endocrine cell. Stimuli that trigger amine secretion not only cause discharge of preformed granules from the cell but also tend to increase the activity of key regulatory enzymes involved in formation of the hormone. Thus, most stimuli not only increase the rate of secretion of previously formed hormone but also accelerate the rate of formation of new hormone molecules.

Hormones That Are Polypeptides

Many hormones are either peptides or proteins. They can be very small peptides, such as **thyrotropin releasing hormone (TRH),** which is a tripeptide produced by neuroendocrine cells in the hypothalamus. It plays a role in the regulation of thyroid hormone production. Polypeptide hormones can be intermediate in size, such as **adrenocorticotropic hormone (ACTH),** which contains 39 amino acids, or **parathyroid hormone (PTH),** which contains 84 amino acids. There can even be much larger polypeptide hormones, such as **growth hormone (GH),** which in humans is a chain of 191 amino acids. Some are simple, single-chain polypeptides, such as the ACTH and PTH molecules, whereas others, such as GH, contain disulfide bonds (S—S) that connect cysteine residues within different portions of the peptide chain. Still others, such as **follicle-stimulating hormone (FSH),** are composed of different peptide subunits joined together to form the active hormone molecule.

Some hormones also contain a significant amount of carbohydrate. One family of **glycoprotein hormones,** which includes **follicle-stimulating hormone (FSH), thyroid-stimulating hormone (TSH), luteinizing hormone (LH),** and **human chorionic gonadotropin (hCG),** in addition to having similar basic protein structures, average about 20% carbohydrate by weight. Thus, hormones in this category are quite diverse in terms of their chemical structure, ranging from small, single-chain peptides to very large, multiunit glycoproteins.

Precursors of Peptide Hormones

Similar to other polypeptides destined for secretion, polypeptide hormones are synthesized on ribosomes of the rough endoplasmic reticulum (ER; see Chapter 3). From the rough ER, they are transported to the Golgi complex, where they are packaged into vesicles prior to secretion from the cell. During these transport and packaging steps in the cell, some very important events take place. Many, if not most, polypeptide hormones are first synthesized in the form of much larger precursor molecules. In some cases, the precursor contains the amino acid sequence for several hormones or hormonelike substances. The precursor for ACTH, for example, contains the ACTH sequence as well as the sequences for at least four other bioactive peptides. The precursor for TRH contains several repetitions of the TRH sequence, so that a number of TRH molecules may be formed from a single precursor molecule. During the intracellular transport phase of hormone secretion, the precursor molecule is "clipped" by specific proteolytic ("protein-cleaving") enzymes to produce the mature, active hormone molecules that will be secreted into the blood.

When an endocrine cell that produces a peptide hormone receives an appropriate stimulus, vesicles containing the mature hormone rapidly move toward the cell surface, and the contents are released by the process of exocytosis (see Chapter 3). Much as they do in other cell types, calcium ions play a key role in coupling a stimulus with secretion from the endocrine cell. In most cases, the amount of hormone available in preformed vesicles is somewhat limited and usually only enough to support high rates of secretion for a matter of minutes. To compensate for this, stimuli that promote polypeptide hormone secretion usually also promote an increased rate of synthesis of the hormone or its precursor.

Hormones Derived from Cholesterol: Steroids

 What features do steroid hormones share?

The third major category of hormones is the steroid hormones. The structure of cholesterol, which is the precursor for all of the steroid hormones, is shown in Figure 12–7a. Differences in biological activity between the various steroid hormones are the result of sometimes very subtle alterations or substitutions in the cholesterol backbone. Figure 12–7a also shows the designation of each of the four rings (*A* to *D*) and the numbering convention used to designate each of the 27 carbon atoms in the cholesterol molecule. These are used when the various steroids are described according to strict chemical nomenclature. For simplicity, however, in most instances, the steroids are referred to by their common names.

In the body, the major sites of steroid hormone production are the adrenal cortex, ovaries, testes, and the placenta. Hormonally active metabolites of vitamin D, which are also derived from cholesterol and have structures similar to the steroid hormones, are produced by sequential modifications of their structure first in the liver and then in the kidney.

Types of Steroid Hormones

The steroid hormones can be grouped into five categories according to the type of biological process they regulate: (1) glucocorticoids, (2) mineralocorticoids, (3) androgens, (4) estrogens, and (5) progestins. Glucocorticoids, produced by cells in the cortex of the adrenal gland, are involved in regulating glucose and energy metabolism (see Table 12–1). Mineralocorticoids are the principal steroids produced by the outer portion of the adrenal cortex and are involved in regulating sodium ion and potassium ion balance by the kidneys. Androgens, the male sex hormones, are produced primarily in the testes of males, although the adrenal cortex produces and secretes physiologically significant amounts of androgens as well and can be an important source of androgens in females. Estrogens, the female sex hormones, are secreted by the ovaries and the placenta. Progestins are involved in the maintenance of pregnancy and are produced by the ovaries and the placenta in females.

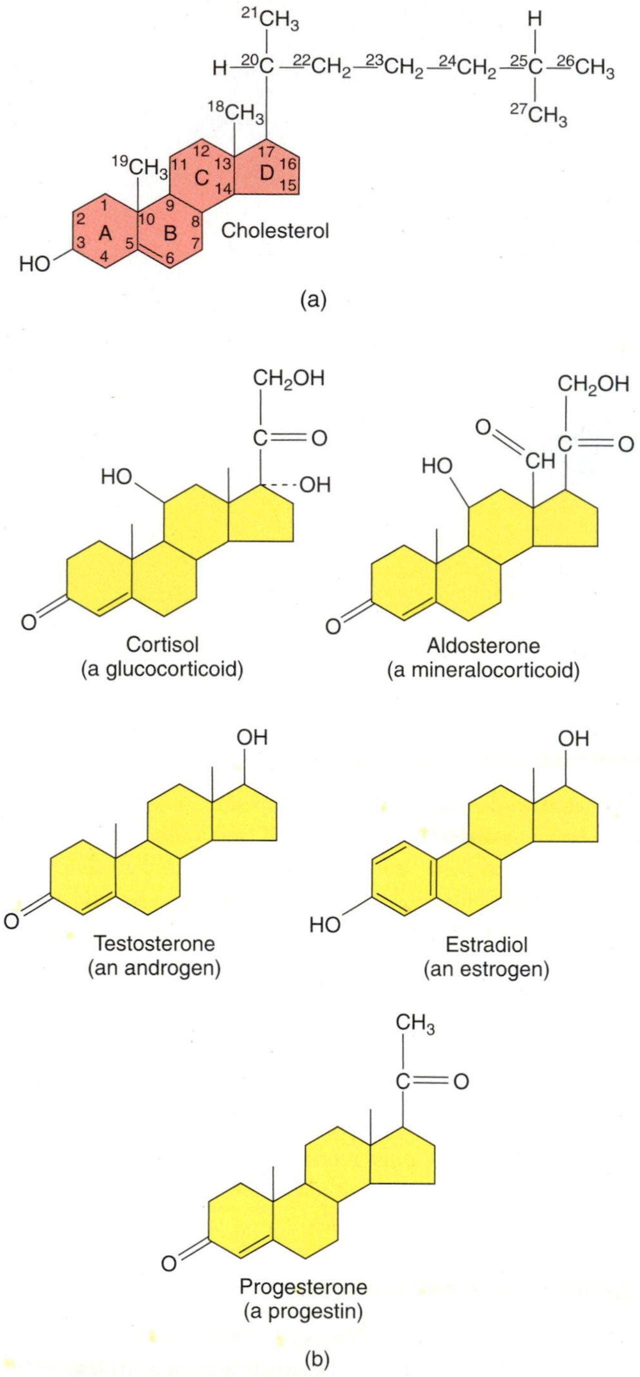

Figure 12–7

(a) Chemical structure of cholesterol. The letters *A* through *D* designate each of the four rings in the molecule. Numbers designate each of the 27 carbon atoms in cholesterol. *(b)* Examples of steroid hormone structure. Each steroid shown represents one of the five classes of steroid hormones.

Figure 12–7b shows examples from each category of steroid hormone. If you examine this figure closely, you will notice that differences in the structure of steroids are relatively small. In the body, however, these small changes in structure are translated into markedly different biological activities. This phenomenon is a result of the high degree of receptor specificity, causing the receptor to recognize and respond primarily to one particular hormone and only weakly or not at all to another.

Pathways in Steroid Hormone Synthesis

Conversion of the cholesterol precursor into each of the steroid hormones occurs by a number of well-defined steps, each catalyzed by a specific enzyme within the steroid-producing cell. Figure 12–8 is a simplified diagram of the primary biosynthetic pathway involved in the formation of each of the major steroids. It should be evident from this figure that several common steps are involved in the formation of each of the five classes of steroids. For example, both cortisol and aldosterone are formed from progesterone, and thus the conversion of pregnenolone to progesterone is common to the synthesis of both steroids. Similarly, testosterone, an androgen, also serves as an intermediate in the formation of estradiol, an estrogen.

Given these common steps, what factors determine which hormone a particular steroid-producing endocrine gland cell will produce? As indicated previously, each step in steroid hormone biosynthesis is catalyzed by a specific enzyme. The relative amount of each of these enzymes in the endocrine cell determines the particular steroids produced. For example, cells in the outer portion of the adrenal cortex are the site of most aldosterone production in the body. These cells have relatively high amounts of the enzymes needed to convert pregnenolone to aldosterone but have few or none of the enzymes needed to convert pregnenolone or progesterone to cortisol, testosterone, or estradiol. Similarly, the steroid-producing cells of the testes possess the enzymes needed to form testosterone but not those needed to form significant amounts of cortisol or aldosterone. Likewise, they have relatively low amounts of the enzyme needed to convert testosterone into estradiol. Estrogen-producing cells of the ovary, however, have a set of steroid-metabolizing enzymes similar to that of the testes, but they also have relatively high amounts of the enzyme needed to transform testosterone into estradiol. Thus, each of these tissues produces and secretes different steroid hormones because of the different enzymes they contain.

Factors Affecting the Rate of Steroid Synthesis

Unlike peptide or protein hormones, which are stored in secretory vesicles, steroid hormones are synthesized and secreted only on demand. They are not stored in the endocrine cell prior to secretion but exit the cell immediately after formation. The rate of steroid hormone synthesis therefore determines the rate of hormone secretion. As with other enzyme-catalyzed reactions, steroid hormone synthesis can be influenced by alterations in the activity of the "rate-limiting enzyme" (the enzyme that catalyzes the slowest step) in the pathway. The conversion of cholesterol to pregnenolone is the rate-limiting step in the synthesis of all steroid hormones (see Fig. 12–8). Factors that increase the rate of steroid hormone secretion do so primarily by increasing the activity of the enzyme that converts cholesterol to pregnenolone. In addition, to ensure that the amount of cholesterol inside endocrine cells is adequate to meet the needs for steroid hormone biosynthesis, factors that increase steroid synthesis also stimulate uptake of cholesterol from the blood. Concurrently, these factors also cause a release of free cholesterol from storage sites within the cell. Steroid hormone secretion is therefore increased by two mechanisms. One results in an increase in the uptake and availability of cholesterol inside the cell, and the other stimulates the conversion of cholesterol to pregnenolone.

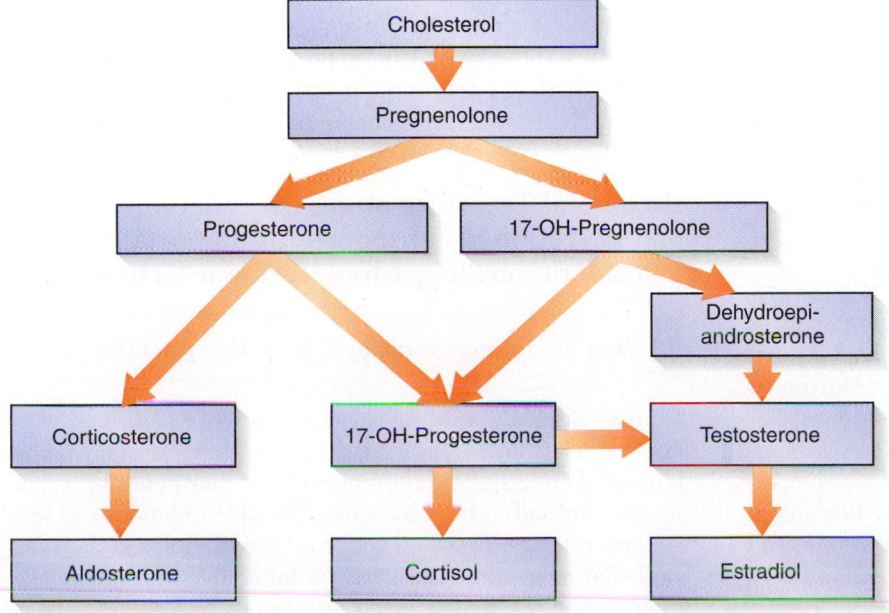

Figure 12–8

Simplified diagram of the pathways of steroid hormone synthesis. Note that the first step in synthesis of all steroids is the conversion of cholesterol to pregnenolone.

Some Hormones Are Carried in the Blood via Transport Proteins

 How does the chemical nature of a hormone affect its transport in the blood?

Once a hormone is released from the endocrine cell, it travels via the blood, in most cases, to reach its site of action in the body. But two problems may occur. The first and most important is that several hormones are only slightly soluble in the blood. For example, because steroid hormones are lipid in nature, they have limited solubility in aqueous solutions. Thyroxine, the principal hormone secreted by the thyroid gland, is also relatively insoluble in the plasma. Therefore, some mechanism is needed to increase the amount of these hormones that can be carried in the blood. The second problem is that, because of the small size of several of the hormones, they would be readily filtered and lost from the kidneys into the urine if allowed to exist free in solution in the plasma. Large peptide or protein hormones are generally freely soluble in water and large enough not to pass through the kidneys into the urine and thus do not present either of these problems.

Specific Transport Proteins

To solve the problems of solubility and excessive urinary loss, thyroxine and the steroids are bound to large carrier proteins for transport in the blood. (See Chapter 17 for a further discussion of plasma proteins.) Generally, these hormones bind to two types of carrier proteins: (1) specific transport proteins that bind one particular hormone with a relatively high degree of specificity and a relatively high affinity and (2) albumin and transthyretin, which bind hormones without high specificity or affinity.

An example of the first type of carrier protein is **cortisol-binding globulin,** or **CBG,** which primarily binds cortisol and relatively little, if any, of other steroids. Similarly, **thyroxine-binding globulin,** or **TBG,** binds and carries mainly thyroxine in the blood. These and other specific carrier proteins are listed in Table 12–2. These carrier proteins are synthesized and secreted by the liver; both nutritional and endocrine factors are known to influence the rate of synthesis of these specific carrier proteins.

Albumin and Transthyretin

Steroid hormones and thyroxine also bind to other plasma proteins for transport, primarily **albumin** and **transthyretin,** which are also produced in the liver. Hormones generally bind to albumin and transthyretin with low specificity and with low affinity. However, because these two proteins are abundant in the plasma, they have a relatively important role in the transport of some steroid hormones. For example, about 50% of the circulating aldosterone, and about 10% of the circulating cortisol is carried by binding to albumin.

TABLE 12–2

Circulating Carrier Proteins

Name	Principal Hormone Carried
Specific carrier proteins	
Cortisol-binding globulin (CBG)	Cortisol, some aldosterone
Thyroxine-binding globulin (TBG)	Thyroxine (T_4), some T_3
Sex steroid-binding globulin (SSBG)	Testosterone and estradiol
General carrier proteins	
Albumin	Many steroids, thyroxine
Transthyretin	Thyroxine, some steroids

Free and Bound Hormone: A Dynamic Equilibrium

The binding of steroid or other hormones to these plasma proteins occurs according to the principles of a simple chemical equilibrium. Thus, the free and the bound forms of the hormone exist in the plasma in a dynamic equilibrium with one another. In most instances, the equilibrium favors the bound form. However, a finite amount also is present in a free, or unbound, state. Only the hormone present in the free state is able to move from the blood across the walls of capillaries (the small blood vessels that bring blood into contact with most cells) to reach cells and interact with hormone receptors. The free hormone is therefore the physiologically active form of the hormone. As free hormone is lost from plasma, it is replaced by hormone newly released from carrier proteins. Similarly, if a sudden large burst of hormone secretion occurs, most of the newly secreted hormone molecules become bound to the carrier proteins, and the free-hormone concentration increases only slightly. Thus, in addition to facilitating hormone solubility and transport, carrier proteins also serve as a reservoir and buffer system to help maintain a relatively constant concentration of free hormone in the blood.

Peripheral Transformation, Degradation, and Excretion of Hormones Are Important Considerations Regarding Hormone Action

 How are hormones removed from the system?

In most cases, the form of the hormone secreted by the endocrine cell is the form directly responsible for producing the biological effects of the hormone; that is, hormones need not be transformed to be fully active. Notable exceptions to this general rule are thyroxine and, to some extent, testosterone. Both hormones are converted peripherally (within some of their target tissues) to forms displaying biological potency

APPLICATIONS OF PHYSIOLOGY

The Enzyme-Linked Immunosorbent Assay

As described in the text, the radioimmunoassay (RIA) allowed for significant advancements in clinical medicine and in research. Before the RIA, assays of hormone concentrations relied on the use of whole animals or tissues/cells derived from animals. In general, these assays lacked sensitivity and specificity and were somewhat costly to perform. The development of the RIA overcame a number of these problems, but not all.

The classic RIA, as the name implies, utilizes radioisotopes. In recent years, the disposal of even low-level radioisotopes, such as used in a RIA, has become an environmental concern. As the number of sites for radioactive waste disposal has dwindled to near zero, the economic cost to the producers of these wastes has skyrocketed. Thus the ultimate cost of these assays to the patient and the researcher has increased as well. In addition, a typical RIA is somewhat labor-intensive in that it may take one technician several days to process 100 or so samples, which contributes to the final cost per sample to be evaluated. Moreover, by nature of the assay, the time cannot be reduced. Thus, the time to obtain results may also be an issue in some instances.

The **enzyme-linked immunosorbent assay (ELISA)** provides an answer to many of the problems currently posed by use of the RIA. Results from an ELISA ultimately depend on development of a colored or fluorescent product that can be detected using optical devices, and thus radioisotopes are not involved. An ELISA is typically automated, which allows for analysis of a large number of samples in any one period of time. Robotic instruments can perform the various pipetting, addition, and withdrawal steps involved, and the wastes that are generated can be disposed of in the regular trash. Furthermore, the time between starting an ELISA and obtaining a final result is typically reduced compared with a RIA.

A typical ELISA is specific, sensitive, and economical. Thus, whenever possible and practical, the ELISA is replacing the RIA.

---

many times greater than the original hormone. Details of the conversion of thyroxine and testosterone to their more active forms will be presented in Chapters 13 and 30, respectively. Also included on this list of peripherally transformed hormones are **angiotensin II** and **1,25-dihydroxyvitamin D$_3$**, both of which are formed by sequences of conversion steps involving several different tissues.

The concentration of any one hormone circulating in the blood is determined by both its rate of secretion and the rate at which it is removed or metabolized to an inactive form. For most hormones, the major pathways for degradation and removal occur in the liver and kidneys, respectively.

Steroid hormones are primarily degraded by the liver. The liver first converts the steroid into a relatively less active form and then, to make the molecule more water-soluble, couples the steroid to a polar sulfate or glucuronide group in a process known as **conjugation.** The conjugated steroids are then primarily excreted in the urine with smaller amounts being secreted by the liver into the bile. Other hormones, such as epinephrine, are inactivated by specific degrading enzymes circulating in the blood and are then rapidly excreted in the urine. Determination of the rates of urinary excretion of a variety of steroid hormones or epinephrine metabolites has proven to be a useful, noninvasive, indirect index of the rate of secretion of the active hormone and has provided physicians with an important diagnostic tool.

Many of the larger peptide hormones, such as insulin and prolactin, are taken into their target tissue cells by a process known as **receptor-mediated endocytosis.** After the hormone binds to its receptor on the surface of target cells, the hormone-receptor complex is internalized, or taken into, the cell. Once internalized, the hormone is separated from its receptor. In most cases, the hormone molecule is then proteolytically degraded, and the majority of the internalized receptors are recycled back to the cell surface. Many peptide hormones are taken up and degraded by the kidneys, although the exact contribution made by the kidneys can vary considerably depending on the hormone.

Measurement of Hormone Concentrations in the Blood Provides Important Information to the Clinician and the Researcher

 How are plasma hormone concentrations measured clinically?

As indicated earlier, hormones are present in extremely low concentrations in the blood, usually in the range of 10^{-9} to 10^{-12} M. To put these low concentrations into perspective, consider that a hormone concentration of 10^{-9} M corresponds roughly to one hormone molecule in the blood per 50

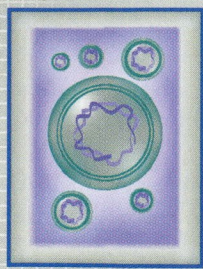

Patterns of Hormone Secretion

As described in the text, development of the RIA in the 1950s proved to be a pivotal turning point in the field of endocrinology. Development of this assay provided clinicians and researchers with a means to measure concentrations of hormones circulating in the bloodstream, which had not previously been possible.

Early investigations sought to determine the daily pattern of secretion for individual hormones as well as defining the factors or conditions that might trigger or inhibit their secretion. For example, it was determined that cortisol, a steroid produced by the adrenal cortex, was at its highest concentration in the blood around the time of awakening and lowest around the time of onset of sleep. This information was crucial to clinicians because it told them that, in order to assess function of the adrenal cortex, it was important to draw blood samples at times of these well-defined "peaks" and "valleys." Clearly, a random blood sample drawn at a variable time would be of little value in reaching a diagnosis. Thus, knowledge of the general pattern of secretion of any particular hormone became crucial for the practice of medicine.

As investigations into the patterns of hormone secretion became more refined, it became clear that many, if not most, hormones are secreted in a **pulsatile** fashion in addition to the overall general increases and decreases in hormone concentration. So, for example, cortisol concentrations may increase and decrease in the general pattern described earlier, but in addition there may be "spikes" of circulating cortisol at various times during the day.

The finding of pulsatile secretion of many, if not most, hormones has major implications for the practice of medicine. First is that these spikes may occur in an unpredictable fashion even in healthy patients. Depending on when a blood sample is drawn, a patient may be at the peak of a spike or at one of the valleys, but which of these would be the case could not be known.

One way to overcome this problem is to employ a **provocative test** of endocrine function. In this type of test, a specific amount of a known stimulus for secretion of a hormone is given to the patient. At prescribed times later, blood samples are taken and the hormone concentration is determined. In the medical literature there are very well-defined limits for a "normal" response in each of these provocative tests.

A second implication of pulsatile secretion that is now well recognized is that, quite often, knowledge of the normal pulsatile secretion pattern becomes important if the clinician wishes to provide hormone-replacement therapy that mimics the natural pattern. A pattern of replacement that does not generally match the pattern of normal hormone secretion is often not effective. With advances in technology, such as mini-pumps that allow the delivery of exogenous hormone in a pattern that closely matches the normal situation, physicians are able to provide more effective treatment for their patients.

---

billion water molecules. Thus, exceedingly sensitive techniques are needed to allow the measurement of circulating hormone concentrations. The measurement of circulating hormone concentrations, either in the basal (resting) state or after an appropriate stimulus has been given, is an important step in the diagnosis of many different endocrine disorders. Because of the similarity in structure and activity of many of the hormones, hormone assays (measurements) must be highly specific if meaningful results are to be obtained.

In the late 1950s, two American scientists, Solomon Berson and Rosalyn Yalow, developed an assay technique that is both extremely sensitive and highly specific. The technique can be performed quite rapidly and inexpensively. The assay that Berson and Yalow developed is known as the **radioimmunoassay (RIA).** The basic technique, described in Chapter 3, has been modified and adapted in a number of different ways, but these newer assays are still based on the original principle of **competition binding,** in which the amount of radioactive hormone that binds to an antibody reveals how much nonradioactive hormone is present. This assay technique can be used to measure the concentrations of hormones in the plasma and in various other bodily fluids, such as amniotic fluid or cerebrospinal fluid. In addition, the concentration of a variety of drugs and vitamins can also be measured by this technique. As a result, the radioimmunoassay has revolutionized the field of endocrinology, both clinically and in the area of basic research. For their efforts in the development of the technique, Berson and Yalow were awarded the Nobel Prize in Physiology and Medicine in 1977.

MECHANISMS OF HORMONE ACTION

 How do hormones produce their effects?

For convenience, hormones can be considered to use one of two different general mechanisms of action to produce their effects. One mechanism, used by amine and peptide hormones, involves membrane receptors and the generation of an intracellular signal, or "second messenger" (see Chapter 5). These hormones take effect by altering the activity of proteins that already exist in the cell. The other mechanism, used by steroid hormones and thyroxine, uses intracellular or nuclear receptors and involves alterations in the expression of particular genes within the nucleus of the cell. In the latter case, hormones work by initiating production of new proteins in the cell. The two general models of hormone action will be considered in more detail shortly. First, let us look at the nature of hormone receptors.

Receptors Are a Key Element in the Initiation of Hormone Action

As was described earlier, a **receptor** is a molecule inside the cell or in the plasma membrane that specifically recognizes and binds one particular hormone. Binding of the hormone to the receptor initiates a sequence of events that ultimately results in the biological effects typical of that hormone. Therefore, specificity within the endocrine system is determined at the level of the hormone receptor.

Receptors are themselves proteins or, in some cases, glycoproteins, and thus have definite three-dimensional structures. As illustrated in Figure 12–3b, only those hormones with a structure complementary to the structure of a specific receptor can bind to and activate the receptor. A very simple key and lock analogy works here: The key functions like a hormone, and the lock, like a receptor.

The ability of receptors to discriminate between hormones is not absolute, however. Similarities in structure between hormones can result in an overlap of hormonal activities. For example, cortisol primarily produces effects typical of a glucocorticoid, but it also interacts weakly with aldosterone receptors and therefore is considered a weak mineralocorticoid. Because of the similarities in structure among the steroid hormones, this type of overlapping activity is fairly common among these hormones. Similar examples of overlapping activities can be found among the peptide hormones. Growth hormone and prolactin have similar structures, and each displays a weak degree of the other's activity. Thus, while receptors are responsible for the specificity that exists within the endocrine system, the inability of the receptor to discriminate **absolutely** between hormones having similar structures leads to some overlap of activities. Normally, these weak secondary activities are of little or no physiological consequence. However, in an abnormal situation in which the circulating concentration of a particular hormone is markedly elevated, these secondary activities of the hormone may be strong enough to produce noticeable biological effects.

The binding of a hormone to its receptor involves weak chemical forces and is usually readily reversible. When a hormone binds to its receptor, the three-dimensional conformation of the receptor is altered. In this regard, a hormone functions in a manner similar to that of an allosteric modifier of an enzyme. The alteration in structure of the receptor is the first step in the sequence of events leading to the biological effects of the hormone. The magnitude of the biological response is therefore proportional to the number of receptors occupied by hormone molecules. However, receptors for any particular hormone are present only in a finite or limited number in their target tissue cells. It follows, therefore, that once a certain hormone concentration is reached, no greater response can be produced by further increases in the hormone concentration. Thus, for each biological effect of a hormone, there is a maximal response that can be produced by that hormone.

Current evidence indicates that all the biological effects of hormones are the result of interactions with their respective receptors. The hormones themselves do not directly affect other cellular constituents, such as enzymes, but instead use their receptors as intermediaries to produce the desired effects. The receptor, therefore, serves as a signal transducer, converting a specific extracellular hormonal signal into an intracellular signal that redirect or alter cell function.

Given the role of the receptor in signal transduction, what is the nature of the intracellular signal generated by the receptor, and how does a single message or signal within the cell lead to a variety of different effects of the hormone?

The "Second Messenger" Model of Hormone Action Describes How Hydrophilic Hormones Produce Their Effect

> *What is a "second messenger," and how do second-messenger systems work?*

Peptide hormones are generally hydrophilic ("water loving") and therefore not soluble in the lipid layer of the plasma membrane. As a result, they are unable to readily penetrate the cell membrane and gain access to the interior of the cell. The mechanism by which these hormones influence cell action involve intracellular signals, or second messengers. For this class of hormone, the receptor is an intrinsic membrane protein with its hormone-binding component exposed to the extracellular fluid. As illustrated in Figure 12–9, binding of the hormone to that portion of its receptor exposed to the outside of the cell results in generation of a second messenger. Many of the membrane receptors are glycoproteins and are oriented such that the carbohydrate portion of the molecule is exposed to the outside of the cell. As discussed in Chapter 5, several second messengers are known. Details of the mechanisms involved in the formation of each and the role they play in hormone action will be discussed shortly.

These surface receptors are not rigidly fixed in place in the membrane, as was once thought, but instead show considerable lateral mobility. In addition, as discussed in the earlier section about hormone degradation, a number of peptide hormones have been shown to undergo receptor-mediated endocytosis, in which both the receptor and the hormone are taken into the cell in an endocytotic vesicle. Binding of several peptide hormones to their receptors has also been shown to induce a clustering of the occupied receptors into patches in the plasma membrane. This clustering is believed to facilitate the endocytotic uptake of the hormone-receptor complexes.

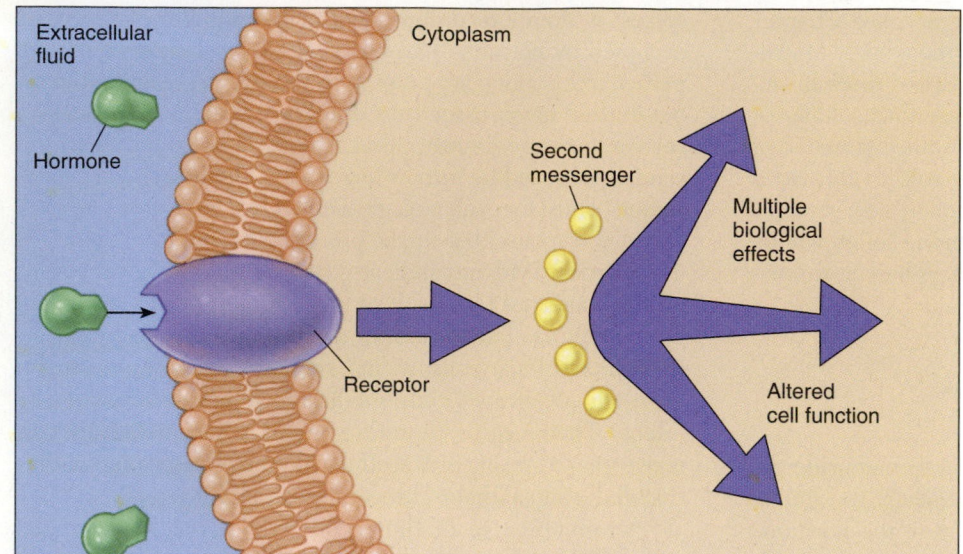

Figure 12–9

The general mechanism of action of hormones with membrane receptors.

As was indicated in Chapter 5, three general intracellular second-messenger systems have been identified and characterized: (1) the adenylate cyclase/cAMP system, (2) the phosphatidylinositol and diacylglycerol system, and (3) receptor-linked ion channels. For some peptide hormones, such as insulin, the mechanism of action has yet to be fully determined, so there may be additional, as yet unrecognized, second messengers.

Cyclic Adenosine Monophosphate as a Second Messenger

Cyclic 3,5-adenosine monophosphate, or, more simply, **cyclic AMP (cAMP),** was one of the first second messengers to be recognized and has been extensively studied and char-

acterized. As shown in Figure 12–10, this cyclic nucleotide is generated from cellular ATP by the action of the enzyme **adenylate cyclase,** which is located on the inner surface of the cell membrane. cAMP is rapidly inactivated by conversion to AMP, a process catalyzed by one or more enzymes known as **phosphodiesterases.** As a result of its rapid inactivation, any increase in cytoplasmic cAMP is short-lived, allowing for the rapid termination of any effects it mediates.

The activity of adenylate cyclase is coupled to membrane receptors by another group of integral membrane proteins known as **G-proteins,** so named because they bind to and require **guanosine triphosphate (GTP)** to function. One type of G-protein stimulates adenylate cyclase, whereas a second type of G-protein inhibits it. These

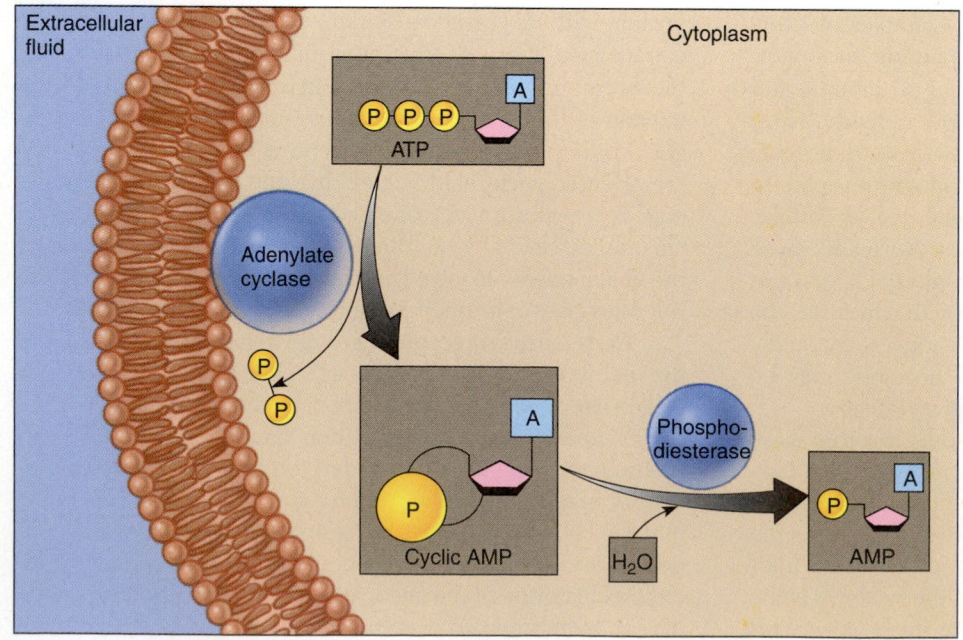

Figure 12–10

Conversion of ATP to cAMP by the enzyme adenylate cyclase and the subsequent conversion of cAMP to AMP by phosphodiesterase.

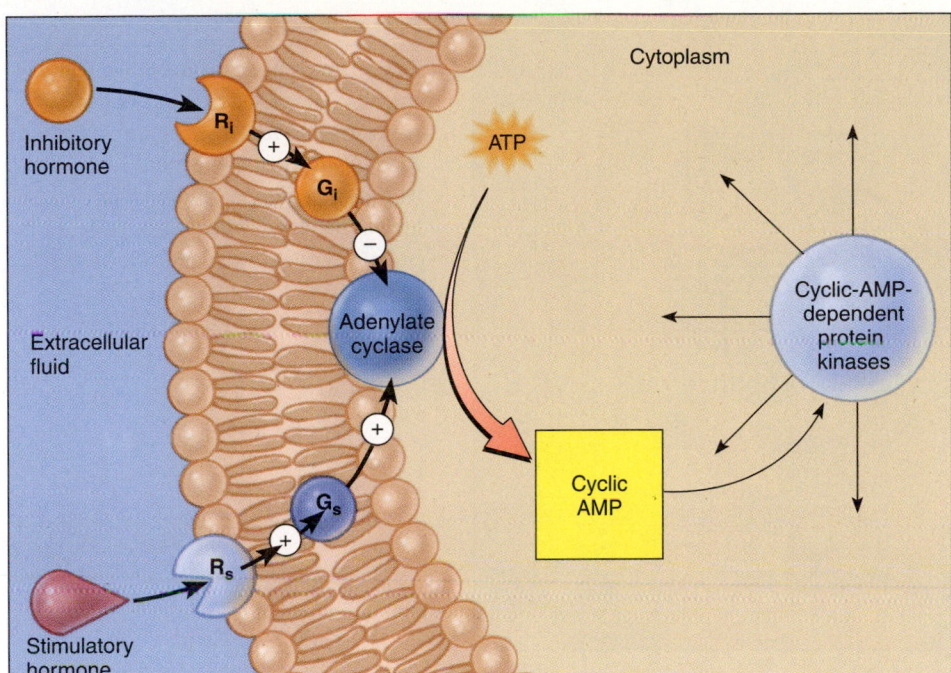

Figure 12–11

Illustration of the coupling of membrane receptors to adenylate cyclase by the G_s and G_i proteins.

are referred to as the **G_s** and **G_i proteins,** respectively (Fig. 12–11). Binding of hormones to membrane receptors (designated as **R_s** in the figure) coupled to adenylate cyclase through **G_s** proteins results in stimulation of adenylate cyclase and an increase in intracellular cAMP concentrations. Similarly, binding of a hormone to receptors (**R_i** receptors) linked to the **G_i** proteins results in inhibition of adenylate cyclase and a reduction in cAMP. Table 12–3 shows an abbreviated list of hormones that produce their effects at least in part as a result of changes in the intracellular concentration of cAMP. Most hormones act

through a stimulatory receptor and therefore produce their effects by increasing intracellular cAMP. These changes in cellular cAMP occur within a matter of just a few seconds after the hormone binds to its receptor. Likewise, the intracellular effects of cAMP occur very rapidly; hence, the overall system is well suited to producing very rapid, acute effects within the cell.

Given that certain hormones act by altering cellular cAMP concentrations, how can a change in the concentration of this single compound then lead to the many different effects that some hormones are known to produce within one cell type? As was described in Chapter 5, cAMP is known to produce its effects by activating enzymes known as **cAMP-dependent protein kinases.** Each particular protein kinase catalyzes the **phosphorylation** of (addition of a phosphate group to) a limited number of specific cellular proteins. Phosphorylation of these proteins results in alterations in their activities, which in turn leads to specific changes in cell function. As indicated in Figure 12–12, the activation of several different protein kinases within a cell by cAMP can result in a variety of different intracellular effects. In this way, binding of a hormone to its receptor on the surface of cells and generation of a single intracellular signal can regulate a variety of different processes within the cell. In addition, the particular set of kinases present in cells may vary from one tissue to the next, such that responses elicited as a result of increases in cAMP vary from tissue to tissue.

Although not as widespread as the cAMP second-messenger system, **cyclic GMP (cGMP)** serves as a second messenger in some tissues. Mechanisms for cGMP formation, second-messenger action, and removal from the cell are essentially analogous to that for cAMP.

TABLE 12–3	
Some Hormones That Use cAMP as Second Messenger*	
Hormone	**Target Tissue**
Adrenocorticotropic hormone (ACTH)	Adrenal cortex
Epinephrine	Heart skeletal muscle, and adipose
Glucagon	Liver
Luteinizing hormone (LH)	Testis and ovary
Parathyroid hormone (PTH)	Kidney and bone
Thyroid-stimulating hormone (TSH)	Thyroid gland
Follicle-stimulating hormone (FSH)	Testis and ovary

*This is only a partial list of hormones known to act via cAMP. Many other hormones and local mediators too numerous to list also stimulate adenylate cyclase and raise intracellular cAMP concentrations.

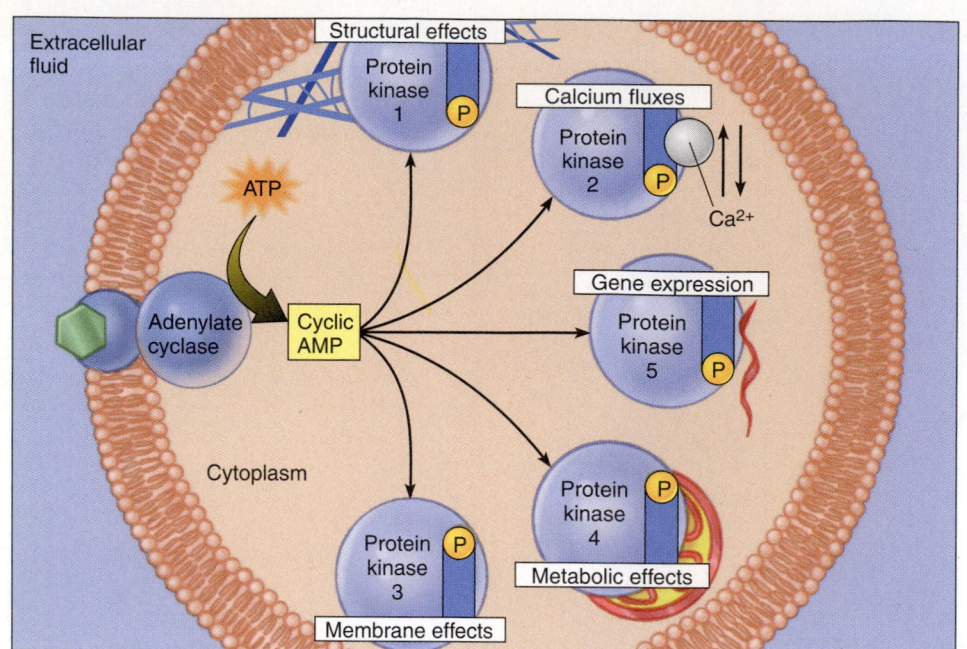

Figure 12–12

Example of how an increase in intracellular cAMP produced in response to a hormone can result in many different biological responses in the cell. The particular set of kinases shown in this figure is purely hypothetical. Some cells may have fewer kinases, some more, and some may have a different mixture altogether.

Inositol Trisphosphate and Diacyglycerol as Second Messengers

Although turnover of membrane lipids was first suggested as a possible mechanism for intracellular signaling in the 1950s, it is only within the past two decades that the importance of this mechanism has become widely appreciated. **Phosphatidylinositol** is a minor phospholipid constituent of the plasma membrane and is primarily located in the inner half of the bilayer. Upon the donation of two phosphate groups from ATP, phosphatidylinositol is converted to **phosphatidylinositol 4,5-bisphosphate (PIP$_2$),** which serves as the immediate precursor for generation of two intracellular second messengers. As shown in Figure 12–13, PIP$_2$ is split to form **diacylglycerol (DG)** and **inositol triphosphate (IP$_3$).** This cleavage is catalyzed by a membrane-bound enzyme known as **phospholipase C.** Several other enzymes function to replenish the supply of membrane PIP$_2$, therefore preventing its depletion.

As with adenylate cyclase, activity of phospholipase C is functionally coupled to membrane hormone receptors by a GTP-binding protein (G-protein). There appears to be only a stimulatory G-protein; as yet, no inhibitory G-protein has been identified for this system. Figure 12–14 illustrates the functional relationship between a receptor, G-protein, and phospholipase C. As indicated, activation of phospholipase C results in hydrolysis of PIP$_2$ to form IP$_3$ and DG. These two substances both play a role as intracellular second messengers, although they produce their effects by different mechanisms.

The IP$_3$ released from intracellular membranes acts inside cells by causing release of calcium ions from storage sites in the endoplasmic reticulum, thereby raising the concentration of free cytosolic calcium. As described in Chapter 5, calcium serves an important role as an intracellular signal or mediator in a variety of different processes. Thus, the cel-

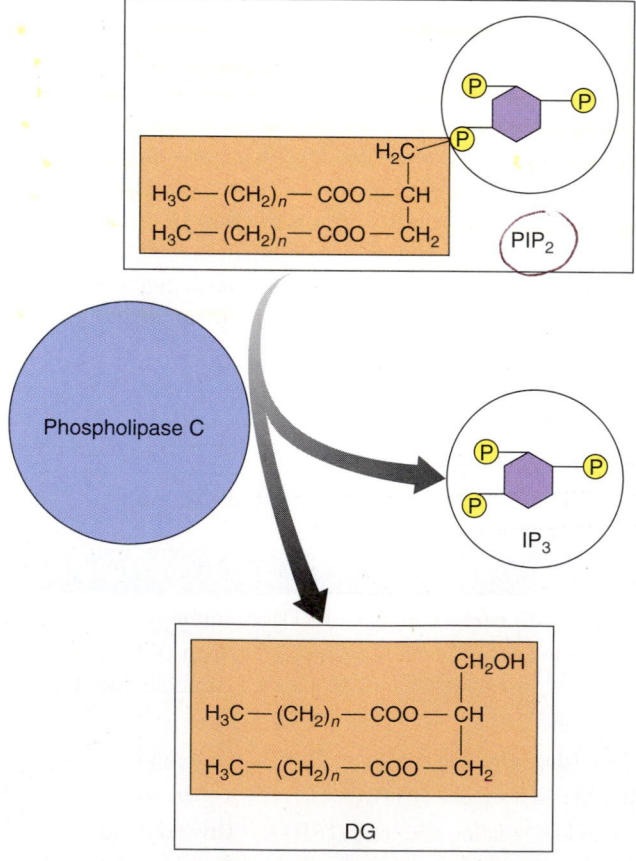

Figure 12–13

Pathway of conversion of phosphatidylinositol 4,5-bisphosphate (PIP$_2$) to inositol trisphosphate (IP$_3$) and diacylglycerol (DG) by phospholipase C.

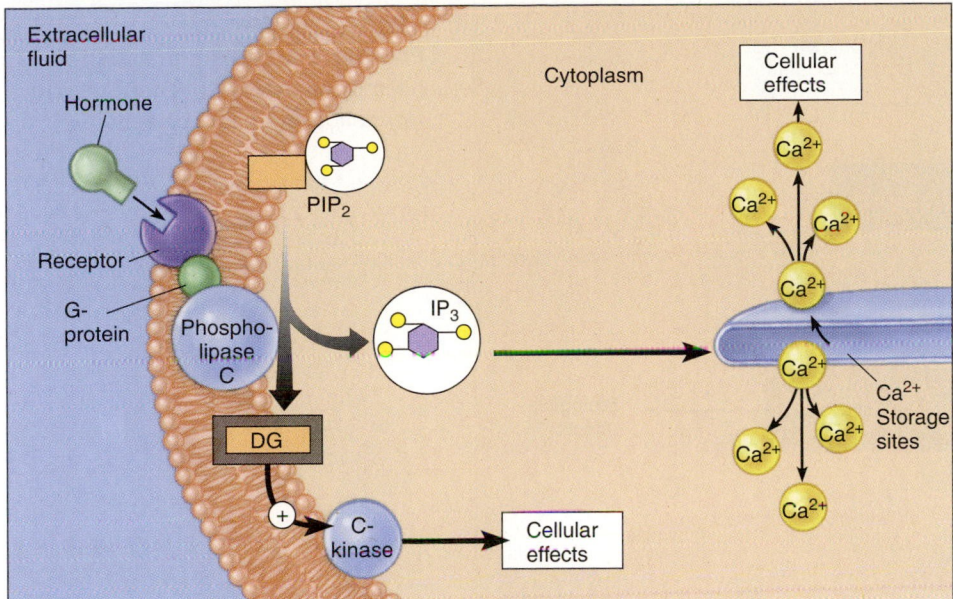

Figure 12–14

Diagram of the formation and the mechanisms of action of inositol trisphosphate (IP$_3$) and diacylglycerol (DG) as intracellular second messengers.

lular effects of one branch of this mechanism ultimately result from increased cytoplasmic calcium concentrations.

The other product formed from PIP$_2$ hydrolysis, DG, remains associated with the membrane. It activates a membrane-bound protein kinase known as **C-kinase.** In a manner analogous to that described for cAMP-dependent kinases, C-kinase catalyzes phosphorylation of a number of specific cellular proteins, thereby altering their activity and producing specific changes in cell function. Although many details of this two-branch IP$_3$/DG pathway have yet to be fully characterized and understood, the two mechanisms seem to act together to regulate a number of different processes.

The overall biochemistry of inositol phosphate metabolism is in fact much more complicated than described in the preceding paragraphs. A number of multiple-phosphorylated inositol phosphates, such as IP$_4$, IP$_5$, and IP$_6$, have now been identified in mammalian cells. However, details of the metabolic pathways leading to their formation and their postulated functions are not completely understood.

Ions as Second Messengers

A third signal-transduction mechanism used by membrane receptors is that of receptor-linked ion channels. In certain hormone/receptor systems, the change in configuration of a receptor caused by hormone binding results in opening of specific, gated, ion channels. Other membrane proteins, such as G-proteins, may also be involved in coupling the changes in receptor configuration to the functional changes in ion channels.

The opening of ion channels in the membrane generates a signal in one of two ways. The first, which operates mainly in electrically active cells, such as nerve cells and muscle cells, involves a small and transient flux of ions, which alters the membrane potential without any appreciable change in cytoplasmic concentrations of the ions. The second mechanism involves a major influx of ions into the cytoplasm, which in turn initiates a

cellular response. This mechanism operates primarily in cells that are not electrically active. Because of the important role of calcium ion as an intracellular signal, it is not surprising that a number of membrane receptors are linked to calcium channels, as illustrated in Figure 12–15. Hormone binding activates the receptors, causing calcium channels to open and allowing calcium ion to flow into the cell, down their electrochemical gradient. Once inside the cell, calcium functions as a second

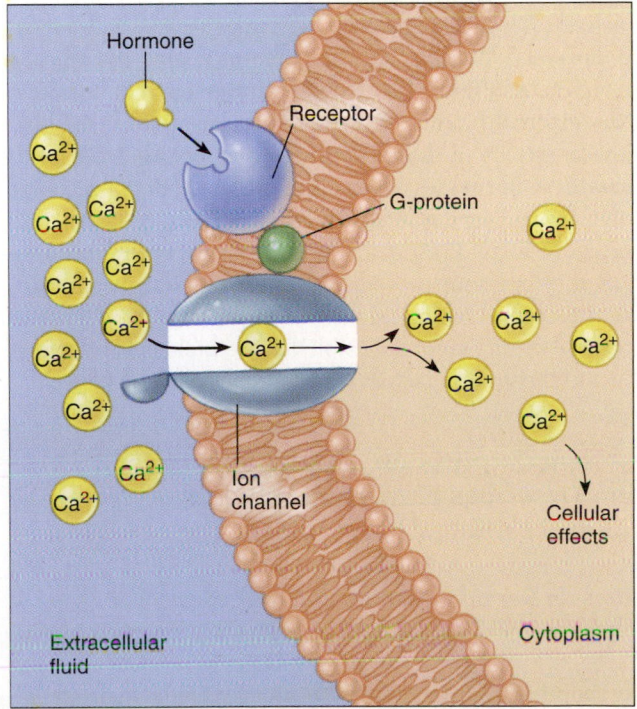

Figure 12–15

Coupling of membrane receptors to membrane calcium channels.

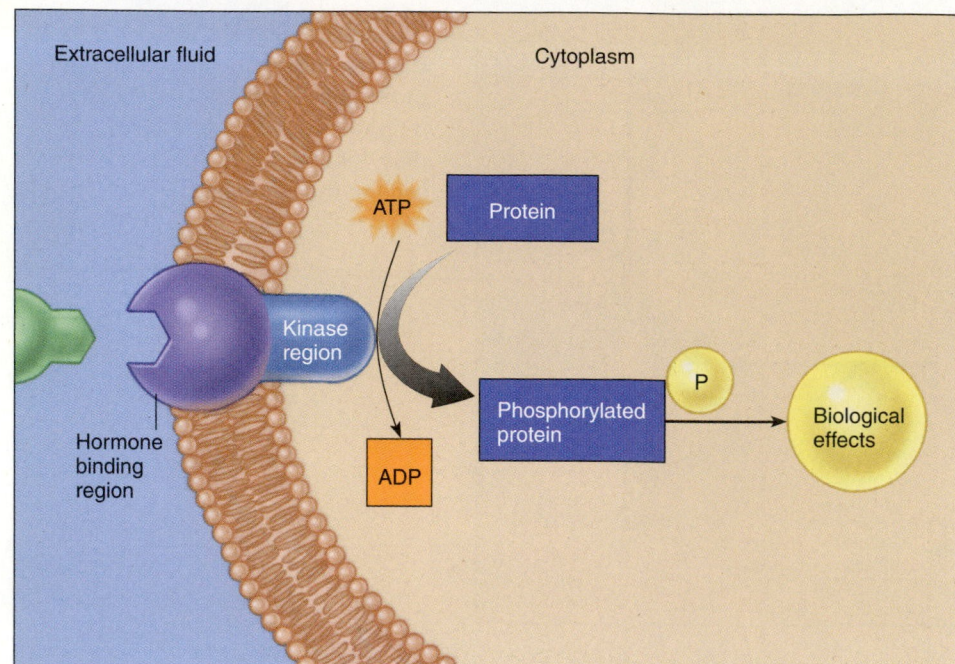

Figure 12–16

Possible mechanism of action of hormones that have receptors with intrinsic protein kinase activity.

messenger to produce a variety of effects by activating a number of calcium-dependent proteins (see Chapter 5).

As indicated earlier, the mechanism of action of some peptide hormones is currently unknown. However, insulin and certain peptide growth factors have unique receptor structures and activities that may be related to their mechanism of action. As illustrated in Figure 12–16, receptors for these agents are transmembrane proteins that exhibit an intrinsic protein kinase activity. Thus, a single receptor molecule has both an external hormone binding site and an internal enzymatically active region. As Figure 12–16 depicts, interaction of hormone with the external binding site results in activation of the protein kinase activity of the receptor. In this regard, it is easy to see how these hormones act as allosteric modifiers of receptor function, converting the relatively inactive receptor kinase into a more active form. Presumably, the kinase catalyzes phosphorylation of cytoplasmic proteins, altering their activity and producing the desired biological effects. At present, the degree to which these events are responsible for the biological effects of insulin and the growth factors is not completely understood.

The "Gene Expression" Model of Hormone Action Describes How Hydrophobic Hormones Produce Their Effects

 How do steroid hormones and thyroxine cause changes in cell function?

Unlike peptide hormones, which are generally hydrophilic, steroid hormones and thyroxine are hydrophobic, or lipid-soluble, and therefore can readily pass through plasma membranes and other membranes within cells. As a result, these hormones do not require a second-messenger system to produce their ef-

fects but instead can enter cells and bind to intracellular and nuclear receptors. In addition to this basic difference from peptide hormones, the manner in which these hormones regulate cellular function and the time course of changes they produce differ considerably from those of hormones that utilize membrane receptors and second messengers.

Figure 12–17 depicts the general mechanism of action for thyroid hormones and the steroids. After dissociating from carrier proteins in the plasma, these hormones freely move across the cell membrane into the cytoplasm. For many years, scientists believed that receptors for steroid hormones existed free in the cytoplasm of target cells. However, recent evidence suggests that many steroid hormone receptors exist either on the nuclear envelope or within the nucleus of the cell. Binding of a hormone molecule to the receptor causes a conformational change in the receptor so that the hormone-receptor complex then interacts with specific "acceptor sites" on the nuclear DNA. This interaction alters the rate of transcription of mRNA molecules that code for specific proteins. Because of the increased abundance of these particular mRNA molecules, synthesis of a few specific proteins is increased. Increased synthesis of these particular proteins leads to the altered cellular function characteristically produced by the particular hormone. As indicated in Figure 12–17, these newly synthesized proteins may be involved in a variety of different cellular functions.

The cellular response to a particular hormone acting via these mechanisms usually results from increased synthesis of several proteins. Although each of these proteins may serve a different role in the cell, each contributes in some way to producing the final, coordinated, biological effect. For example, aldosterone, a mineralocorticoid produced by the adrenal cortex, has the effect of increasing sodium (and potassium) transport across tubule cells of the kidney. As shown in

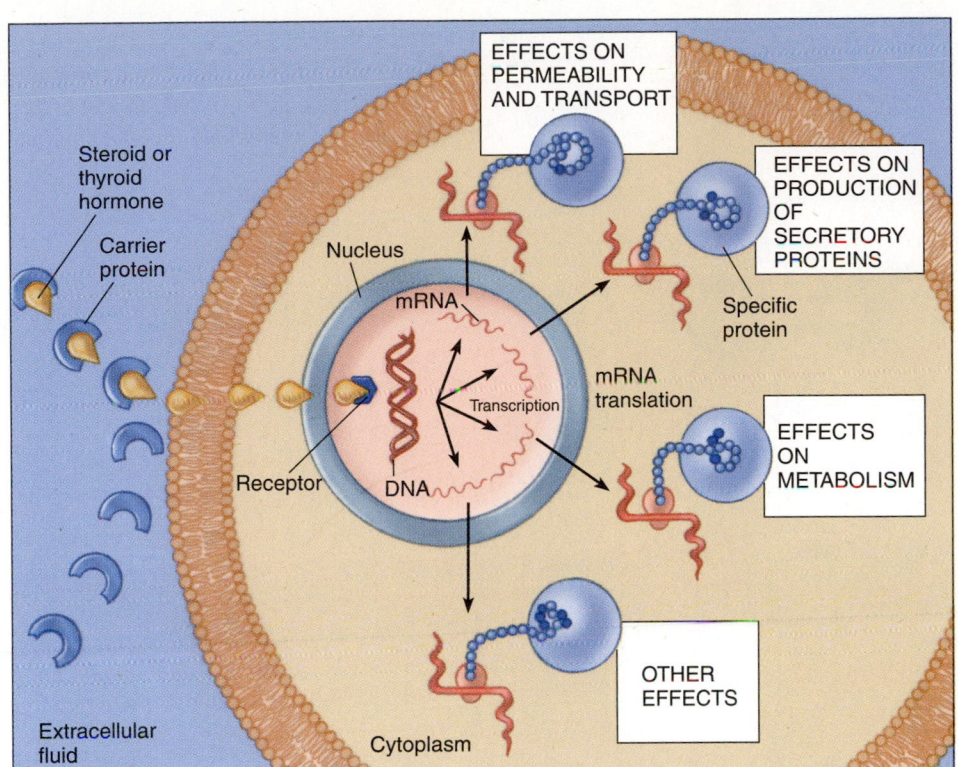

Figure 12–17

Mechanism of action of hormones with intracellular receptors, indicating the effects on gene expression and the possible ways in which cell function might be altered.

Figure 12–18, sodium first enters the cell from the lumen of the tubule (right side of Figure 12–18) by diffusion through specific sodium channels and is then pumped out on the opposite (left) side of the cell by an energy-requiring Na^+-K^+-ATPase pump. Aldosterone increases the synthesis of several mitochondrial citric-acid-cycle enzymes involved in ATP production, making more energy available to power the Na^+-K^+-ATPase pump; it stimulates synthesis of sodium channel proteins, making more sodium available to the Na^+-K^+-ATPase pump; and it also increases the synthesis of Na^+-K^+-ATPase proteins themselves. The overall effect is to increase the transfer of sodium across kidney tubule cells. Thus, as is typical of other steroid or thyroid hormones that take effect by regulating events within the nucleus, aldosterone alters the function of its target cells by changing the rate of synthesis and therefore the amount of several key proteins in the cell. Recall that, by contrast, hormones that act through membrane receptors using second messengers generally do so by modifying the activity of preexisting proteins, using such mechanisms as phosphorylation and dephosphorylation.

The Time of Action for the "Second-Messenger" Model and the "Gene Expression" Model Differs Considerably

How is the time that it takes to observe a biological response to a hormone signal related to its mechanism of action?

A basic difference between the cell surface and intracellular receptor systems involves the time course over which re-

sponses occur. In general, responses to hormones that use the second-messenger system can be observed in a matter of seconds or minutes. In addition, once the hormonal stimulus is withdrawn, cell function rapidly returns to its basal (resting) state. In contrast, responses to hormones that affect gene transcription are usually not evident for many minutes or hours, and in some cases, the response may even take days to develop. Likewise, after removal of the hormonal signal, it may take an equal period of time for the effect to diminish and the cells to return to their original state. With these two mechanisms, the endocrine system is able to closely regulate a number of cellular processes on both a short-term (acute) and a long-term (chronic) basis.

Amplification of Hormone Signals Is an Important Aspect of Hormone Action

How does the binding of just a few hormone molecules produce a large effect in its target cell?

One characteristic feature within the endocrine system is amplification, or increase in signal strength. As a result of this amplification, binding of just a few hormone molecules to their receptors can result in a dramatic change in cell function. Signal amplification occurs at nearly every step in hormone action, from binding of a hormone molecule to its receptor to the final biological effect produced by the hormone. Amplification occurs because each step in the sequence involves enzymes that are responsible for forming not just one but many "signal" molecules. For example, when glucagon binds to its receptor, it activates adenylate cyclase.

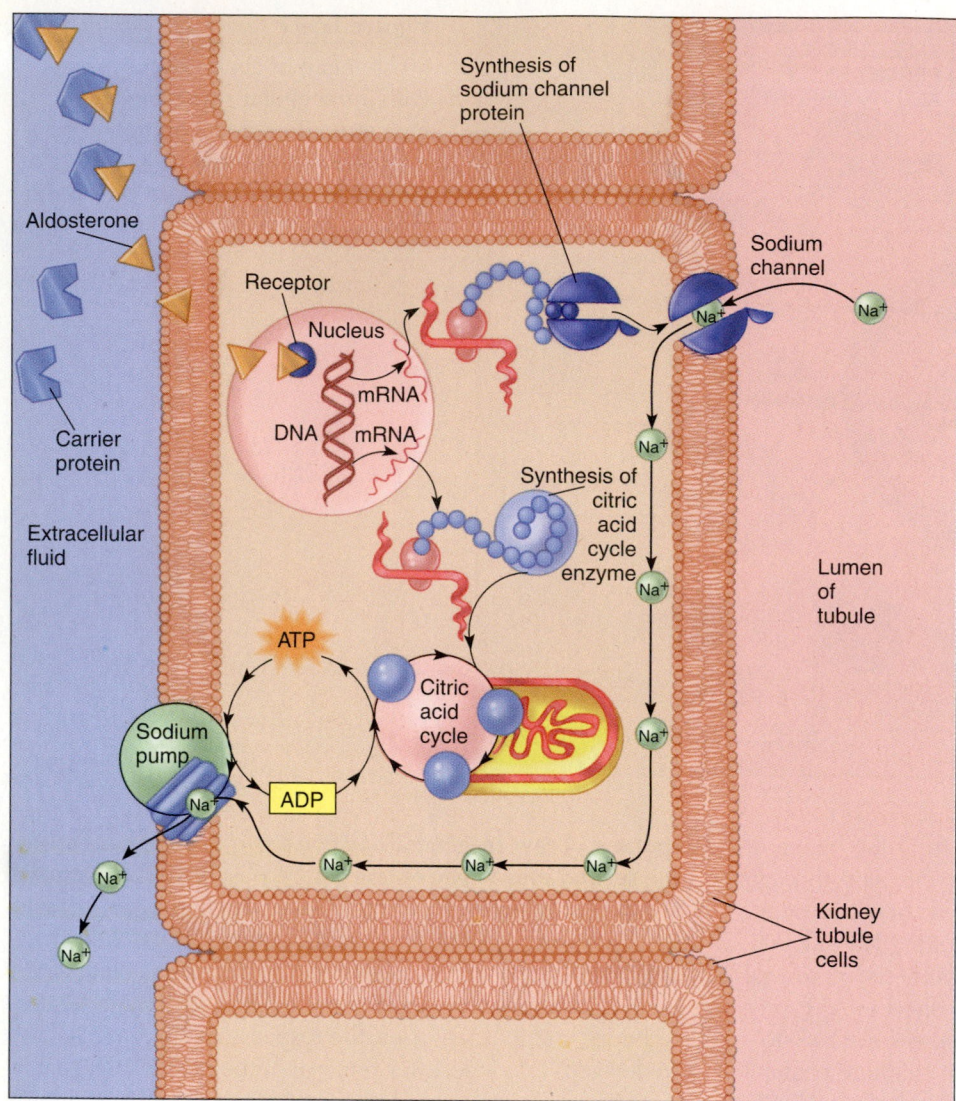

Figure 12–18

Mechanism of action of aldosterone to increase sodium transport across kidney tubule cells.

Adenylate cyclase catalyzes the formation of many cAMP molecules, each of which in turn activates a particular protein kinase. Each cAMP-activated protein kinase can now catalyze the phosphorylation of many copies of a particular enzyme, leading to an alteration in the activity of many enzyme molecules. Thus, just one glucagon molecule can result in the activation of many individual cell proteins. Amplification allows hormones to be effective regulators of cell function even though they are present in the blood in extremely small amounts.

CHAPTER REVIEW

Summary

• Hormones provide a chemical means for communication between cells or among groups of cells. The endocrine system and the nervous system are the two primary means for communication in the body; these two systems function in a coordinated manner to promote homeostasis. Neuroendocrine cells provide a means by which information in the nervous system can be transferred into endocrine signals.

• In general, hormones regulate processes related to either energy metabolism, salt and water balance, growth and development, or reproduction. Most hormones act on several tissues and produce several different specific effects.

• Chemically, hormones generally fit into one of three categories: amino acid derivatives, peptide or protein hormones, or steroid hormones. Steroids can be subdivided into glucocorticoids, mineralocorticoids, androgens, estrogens, and progestins.

- Steroid hormones and thyroid hormones travel in the bloodstream by binding to specific carrier proteins and to albumin and other plasma proteins. Peptide and protein hormones generally do not require carrier proteins for transport.
- The radioimmunoassay and ELISA allow for the measurement of hormone concentrations in the blood. Development of the radioimmunoassay was an extremely important advancement in the field of endocrinology.
- The receptor is the molecular entity of the cell that recognizes the hormone and initiates its biological effects. Receptors are largely responsible for the specificity that exists within the endocrine system, although the specificity is not absolute in every case.
- Peptide and protein hormones generally act via plasma membrane receptors present on the cell surface. Binding of a hormone to its receptor on the cell surface results in the generation of intracellular second messengers, which alter or regulate cell function. Currently known second messengers include cyclic AMP, inositol trisphosphate, diacylglycerol, and certain ions, most notably calcium.
- Steroid hormones exert their effects via intracellular or nuclear receptors. The hormone-receptor complex interacts with the DNA to alter transcription of messenger RNA molecules coding for specific proteins. Alterations in the rate of synthesis of these specific proteins lead to the particular cellular response characteristically produced by the hormone.
- Signal amplification occurs at nearly every step in hormone action, allowing hormones to be effective regulators of cell function even though they are present in the blood in extremely small amounts.

Review Questions

Choose the Correct Answer

1. Hormones that are secreted into the interstitial fluid (as opposed to being secreted into the bloodstream) are described as acting via a(an) _____ mechanism.
   a. endocrine
   b. autocrine
   c. paracrine
   d. isocrine
   e. b or c
2. Hormones are involved in regulating processes related to:
   a. production, use, and storage of metabolic energy.
   b. reproductive functions.
   c. growth and development.
   d. salt and water balance.
   e. all of the above.
3. Which of the following is a glycoprotein?
   a. Parathyroid hormone (PTH)
   b. Growth hormone (GH)
   c. Thyroxine (T_4)
   d. Follicle-stimulating hormone (FSH)
   e. Epinephrine
4. Cells in the body that convert information from the nervous system into endocrine signals in the blood are known as:
   a. paracrine.
   b. endocrine.
   c. neuroendocrine.
   d. pharmocrine.
   e. automatic.
5. By definition, a target tissue for a particular hormone contains:
   a. receptors for steroid hormones only.
   b. receptors for peptide hormones only.
   c. a cuboidal cell structure.
   d. enzymes that directly interact with the hormone.
   e. receptors for the hormone in question.
6. Feedback regulation is an important feature of endocrine communication. Which type of feedback predominates?
   a. Negative
   b. Positive
   c. Ergometric
   d. Neural
   e. Homeometric
7. Epinephrine and melatonin represent a subset of hormones known as *amines*. Epinephrine results from the transformation of:
   a. melatonin.
   b. glucagon.
   c. insulin.
   d. tryptophan.
   e. tyrosine.
8. Which of the following is not a steroid hormone?
   a. Aldosterone
   b. Cortisol
   c. Epinephrine
   d. Progesterone
   e. Estradiol
9. Which of the following is not one of the five categories of steroid hormones?
   a. Progestins
   b. Estrogens
   c. Prostaglandins
   d. Glucocorticoids
   e. Mineralocorticoids
10. Which of the following is not a general characteristic of peptide hormones?
   a. They are transported in the blood largely free in solution.
   b. They are synthesized and secreted on demand.
   c. They utilize cell membrane receptors and intracellular second messengers.
   d. They are usually first synthesized as a larger precursor protein.
   e. They are hydrophilic.
11. Which of the following hormones is not synthesized from readily available dietary constituents?
   a. Adrenocorticotropic hormone (ACTH)
   b. Thyroxine
   c. Progesterone
   d. Growth hormone
   e. Testosterone

12. Which of the following is thought to be transformed in its target tissues to a more active hormone form?
    a. Insulin
    b. Follicle-stimulating hormone (FSH)
    c. Estradiol
    d. Thyroxine
    e. Parathyroid hormone (PTH)
13. Receptors for many steroid hormones are located:
    a. within the nucleus.
    b. in the cytoplasm of nontarget cells.
    c. in the bloodstream.
    d. in the nuclear membrane.
    e. either a or d, not b or c.
14. Pregnenolone serves as a precursor in the formation of:
    a. progesterone.
    b. estradiol.
    c. aldosterone.
    d. cortisol.
    e. all of the above.
15. In the plasma, the majority of cortisol is:
    a. bound to albumin.
    b. bound to CBG.
    c. free in solution.
    d. bound to transthyretin.
    e. conjugated.
16. Steroid hormones are primarily degraded by:
    a. CBG.
    b. the adrenal cortex.
    c. muscle.
    d. testes and ovaries.
    e. liver.
17. Adenylate cyclase catalyzes the conversion of:
    a. ADP to ATP.
    b. ATP to AMP.
    c. ATP to cAMP.
    d. cAMP to AMP.
    e. cAMP to ATP.

18. Which of the following does not appear to use cAMP as a second messenger?
    a. Glucagon
    b. Epinephrine
    c. Thyroid-stimulating hormone (TSH)
    d. Insulin
    e. Adrenocorticotropic hormone (ACTH)
19. Which of the following exerts its effects primarily via the gene expression model of hormone action?
    a. Growth hormone
    b. Aldosterone
    c. Parathyroid hormone
    d. Luteinizing hormone (LH)
    e. Thyroid-stimulating hormone (TSH)
20. Hormonal activation of phospholipase C in the membrane results in the formation of:
    a. IP_3.
    b. cAMP.
    c. DAG.
    d. PIP_2.
    e. a and c.
21. Hormones that cause the opening of ion channels typically open the cell membrane to the movement of what ion?
    a. Lithium
    b. Potassium
    c. Calcium
    d. Chloride
    e. Inositol
22. The time of action for peptide hormones is typically:
    a. more rapid than steroid hormones.
    b. about the same as steroid hormones.
    c. slower than steroid hormones.
    d. related to the molecular size of the hormone.
    e. both a and d.

Answer to Case History Question

1. Yes, but it would depend on the specificity of the inhibitor and the particular distribution of the phosphodiesterase that is being affected. For example, if one form of cGMP phosphodiesterase were present in the vasculature of the heart, then an inhibitor of that enzyme would in theory be useful in the treatment of angina, which results from constriction of the coronary blood vessels.

Key Terms

adenylate cyclase (p. 398)
aldosterone (p. 388)
autocrine (p. 384)
cholesterol (p. 392)
cortisol (p. 392)
cortisol-binding globulin (p. 394)
cyclic AMP (cAMP) (p. 398)
cyclic AMP-dependent protein kinases (p. 399)
cyclic GMP (cGMP) (p. 399)

diacylglycerol (DG) (p. 400)
endocrine system (p. 385)
enzyme-linked immunosorbent assay (ELISA) (p. 395)
estrogen (p. 389)
G-proteins (p. 398)
glucocorticoid (p. 388)
glycoprotein hormones (p. 391)
hormone receptors (p. 397)

human chorionic gonadotropin (hCG) (p. 391)
inositol trisphosphate (IP_3) (p. 400)
neuroendocrine (p. 386)
paracrine (p. 384)
phospholipase C (p. 400)
progesterone (p. 389)
provocative test (p. 396)
radioimmunoassay (RIA) (p. 396)

receptor-mediated endocytosis (p. 395)
receptors (p. 397)
second messengers (p. 397)
steroid hormones (p. 397)
testosterone (p. 389)
thyroxine-binding globulin (p. 394)
transthyretin (p. 394)

Suggested Readings

Berne, R. M., and Levy, M. N. *Principles of Physiology, ed 3.* St. Louis, Mosby, 2000.

Berridge, M. J. "The molecular basis of communication within the cell." *Scientific American,* 253:142–152, 1986.

Berridge, M. J., and Irvine, R. F. "Inositol phosphates and cell signaling." *Nature,* 341:197–205, 1989.

Goodman, H. M. *Basic Medical Endocrinology, ed 2.* New York, Raven Press, 1994.

Griffin, J. E., and Ojeda, S. R. *Textbook of Endocrine Physiology, ed 4.* New York, Oxford University Press, 2000.

Kacsoh, B. *Endocrine Physiology.* New York, McGraw-Hill, 2000.

Linder, M. E., and Gilman, A. G. "G proteins." *Scientific American,* 267(1):56–65, 1992.

Norman, A. W., and Litwack, G. *Hormones, ed 2.* San Diego, Academic Press, 1997.

Web sites

http://www.viagra.com

http://www.endo-society.org/

Answers to Review Questions

1. e   **2.** e   **3.** d   **4.** c   **5.** e   **6.** a   **7.** e   **8.** c
9. c   **10.** b   **11.** b   **12.** d   **13.** e   **14.** e   **15.** b
16. e   **17.** c   **18.** d   **19.** b   **20.** e   **21.** c   **22.** a

Chapter 13

THE PITUITARY HORMONES

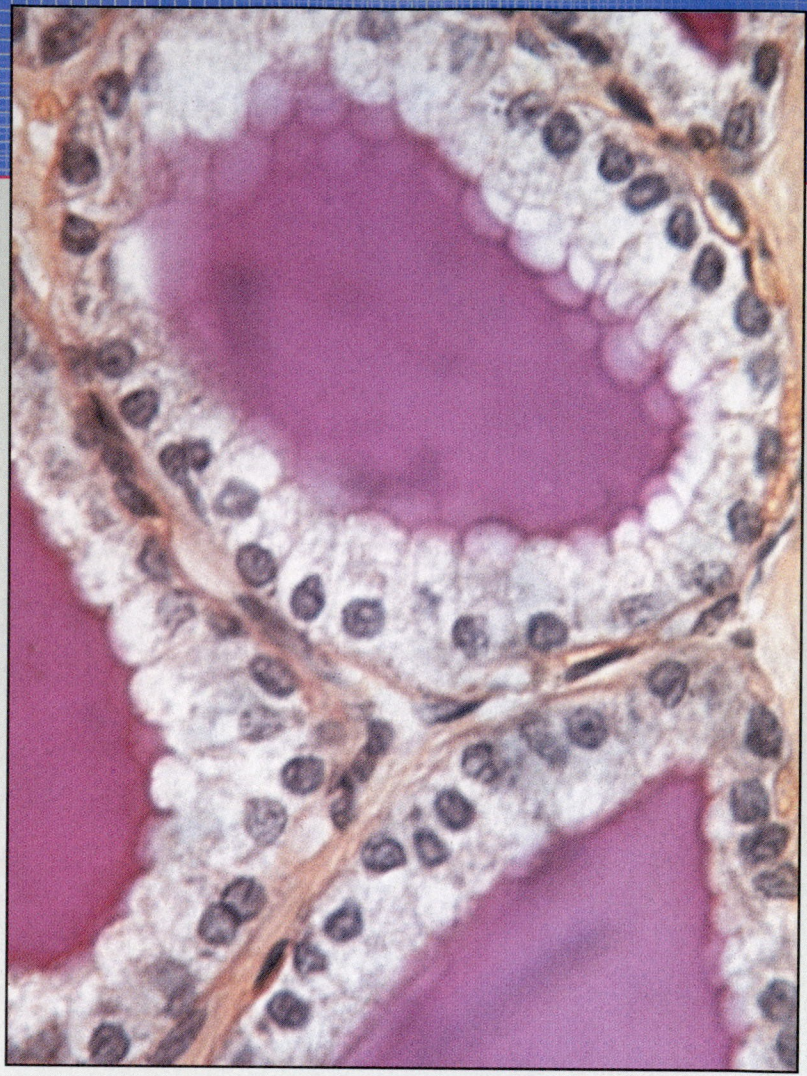

KEY CONCEPTS

- The pituitary consists of two main lobes (anterior and posterior) that have different embryonic origins and different regulatory mechanisms.

- Ultimately, the hypothalamus of the brain serves to regulate hormone secretion from both lobes of the pituitary.

- The anterior pituitary is regulated in a typical endocrine fashion and secretes six different hormones into the general circulation.

- The posterior pituitary is a typical neuroendocrine gland; it secretes two hormones and contains the terminals of neuroendocrine cells, the cell bodies of which are in the hypothalamus.

- Growth hormone is one of six hormones produced by the anterior pituitary. It affects skeletal growth as well as protein and carbohydrate metabolism.

- Thyroid-stimulating hormone (TSH) is also produced by the anterior pituitary and regulates thyroid hormone production by the thyroid glands. Thyroid hormones in turn regulate metabolism in most tissues in the body.

CASE HISTORY

Mary is a 32-year-old mother of two who recently sought medical attention. She complained of weakness and feeling rundown. She also complained of gaining 20 pounds over the past year even though her appetite was decreased. Further significant issues in her history included repeated bouts of constipation as well as dry, itchy skin. She reported that, on several occasions, her husband had commented that she seemed depressed and at times mentally sluggish. During the physical portion of the examination, her doctor noted a heart rate of 60 beats per minute (on the low side of normal); a sluggish knee-jerk reaction; puffiness about the face; and dry, brittle hair. During the examination, she became chilled and asked the doctor to please hurry so that she could change back from her gown into street clothes, even though the environmental temperature was quite warm.

The doctor suspected hypothyroidism (low thyroid hormone levels), but before she could make a firm diagnosis she needed additional information. She ordered some blood work on Mary that included a standard blood chemistry as well as measurement of thyroid hormone levels. The results revealed a high cholesterol level, which is often seen with hypothyroidism; extremely low thyroid hormone levels; and high levels of TSH, a hormone produced by the pituitary that normally stimulates thyroid hormone secretion. These results told Mary's doctor that her problem was in the thyroid gland, and not in the pituitary gland. Further tests revealed that Mary had a form of thyroid disorder known as Hashimoto's thyroiditis, which is an inherited autoimmune disorder in which follicle cells are attacked by the immune system and destroyed.

Thankfully, Mary was successfully treated; all of her original signs and symptoms were reversed to normal, and she has a much improved sense of well-being.

Questions

1. What would you think is the proper form of treatment for Mary; how long might she have to continue it?

2. Are any of the members of Mary's family likely to suffer from the same problem?

3. What would have been the indication if Mary's TSH levels had been low?

INTRODUCTION

The pituitary gland plays a central role in hormonal regulation of a wide variety of processes and has been one of the most extensively studied of all endocrine glands. As we will describe, pituitary function is largely controlled by the hypothalamus. The interaction of the pituitary and hypothalamus is therefore an excellent example of interaction between the nervous and endocrine systems.

We begin with a discussion of pituitary hormones and regulation of pituitary function by the hypothalamus. Later sections of this chapter describe actions of two pituitary hormones: growth hormone and thyroid-stimulating hormone.

ANATOMY OF THE PITUITARY GLAND AND ITS RELATION TO THE HYPOTHALAMUS

> *Based on the anatomical relationships between the hypothalamus and pituitary gland, what predictions can we make with respect to the interaction between these two structures?*

During Development the Pituitary Is Formed from Two Different Tissue Sources

In humans, the pituitary gland measures about 1 cm (0.5 in) in diameter and is located just below the hypothalamus at the base of the brain (Fig. 13–1). As described in Chapter 10, the hypothalamus serves as an integrative center for a variety of different bodily processes.

The hypothalamus also controls pituitary function and therefore serves as an important source of control for the endocrine system. The pituitary is not a single gland but consists of three separate glands, or lobes: the **anterior lobe,** the **intermediate lobe,** and the **posterior lobe.** The pituitary is connected to the hypothalamus by a thin segment of tissue known as the *pituitary stalk.* The hormones secreted by each of the three different lobes of the pituitary are listed in Table 13–1. As this table indicates, the pituitary gland secretes at least nine different hormones. However, as will be described later, the pituitary does not itself synthesize all nine.

Regulation of hormone secretion by each of the lobes of the pituitary involves different mechanisms. Cells of the anterior pituitary behave as true endocrine cells (see Fig. 12–1a) in that they receive signals or information via the blood and in turn release their own hormones directly into the blood. The posterior pituitary, in contrast, is an extension of the nervous system and functions like a group of neuroendocrine cells. In humans, the intermediate lobe is rudimentary and does not exist as a distinct lobe per se but consists of only a few cells interspersed along the borders of the anterior and posterior lobes. Although the intermediate lobe serves an important role in a number of lower vertebrates, it has little or no endocrine function in humans.

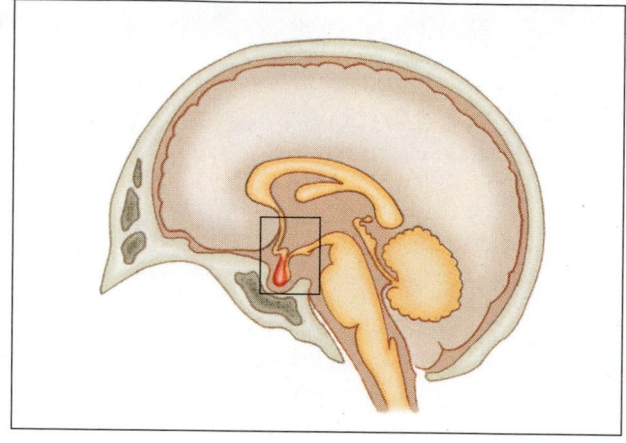

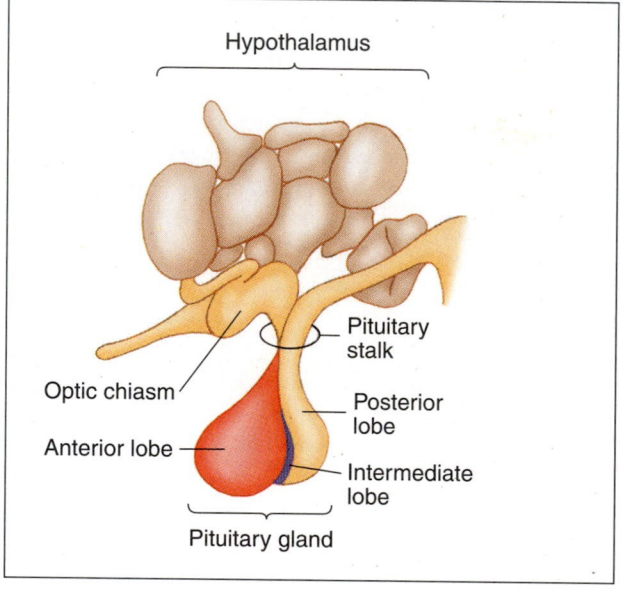

Figure 13–1

The location of the pituitary gland relative to the brain and hypothalamus.

TABLE 13–1
Hormones Secreted by the Pituitary Gland

Hormones secreted from the anterior lobe
  Luteinizing hormone (LH)
  Follicle-stimulating hormone (FSH)
  Prolactin (PRL)
  Adrenocorticotropic hormone (ACTH)
  Growth hormone (GH)
  Thryoid-stimulating hormone (TSH)

Hormones secreted form the intermediate lobe   Melanocyte-stimulating hormone (MSH)

Hormones secreted from the posterior lobe
  Oxytocin
  Antidiuretic hormone (ADH, also known as *vasopressin*)

Both Vascular and Neural Components Are Important for the Control of Pituitary Function

> *How does the functional anatomy of the posterior pituitary differ from that of the anterior pituitary?*

An examination of the embryologic development of the pituitary illustrates how both the endocrine and neuroendocrine portions of the gland are formed. As Figure 13–2 indicates, during early development, a pocket of cells known as **Rathke's pouch** pinches off from the roof of the primitive mouth cavity. At about the same time, a fingerlike projection of neural tissue extends downward from the base of the hypothalamus. Rathke's pouch meets this neural protrusion of the hypothalamus, and the two structures fuse to form the anterior and posterior lobes of the pituitary, respectively.

Neural connections between the hypothalamus and the posterior lobe of the pituitary are maintained during development. As Figure 13–3 shows, the posterior pituitary contains the axons and terminals of neurons that have their cell bodies within the hypothalamus. These neurons terminate close to numerous capillaries located throughout the posterior pituitary. As indicated in Figure 13–3, the cell bodies of these neurons are found within two distinct areas of the hypothalamus, the **supraoptic nucleus** and the **paraventricular nucleus.** The significance of these anatomical features will become evident when the regulation of posterior pituitary hormone secretion is discussed in the next section.

Hormones that are produced in the hypothalamus and are then carried to the anterior pituitary via blood vessels control the anterior pituitary. As shown in Figure 13–3, arterial blood reaching the hypothalamus enters a specialized region known as the **median eminence,** where vessels branch into a network of primary capillaries. From these primary capillaries, the blood enters the **long portal veins** and is carried down through the pituitary stalk to the anterior lobe of the pituitary, where a second capillary network is located. From this second capillary network in the anterior pituitary, blood leaves the gland through the veins and then mixes with the systemic venous blood.

This complex of blood vessels, which consists of the primary capillary network, the secondary capillary network, and the vessels that connect them, is known as the **hypothalamic-pituitary portal system.** As will be described later, the hypothalamic-pituitary portal system carries hormones from the hypothalamus to the anterior pituitary that serve to regulate the secretion of anterior pituitary hormones.

In addition to the long portal veins that connect the hypothalamus and anterior pituitary, a less prominent portal system connects the posterior and anterior lobes of the pituitary. This vascular system is referred to as the **short portal system.** Some evidence suggests that the short portal veins carry substances secreted by the posterior lobe to the anterior lobe, where they regulate hormone secretion from the anterior pituitary. The exact degree to which this system is involved in controlling anterior pituitary function is not well understood.

CONTROL OF PITUITARY HORMONE SECRETION BY THE HYPOTHALAMUS

The mechanisms by which the hypothalamus controls secretion of hormones from both the posterior and anterior lobes of the pituitary are shown in diagrammatic form in Figure 13–4.

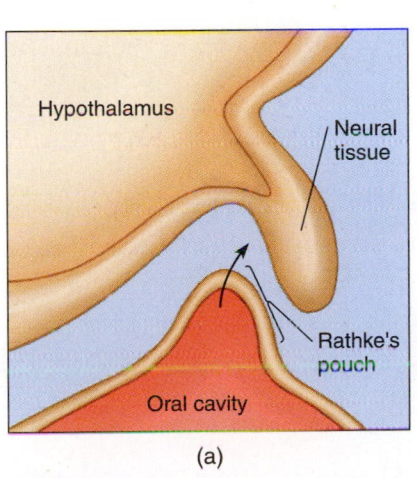
(a)

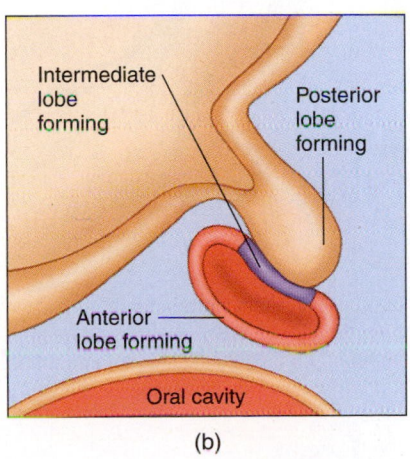
(b)

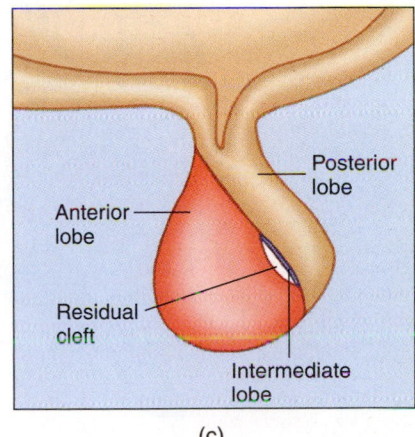
(c)

Figure 13–2

Formation of the anterior, posterior, and intermediate lobes of the pituitary during embryological development. The normal sequence of development is from *(a)* to *(c)*. Note that the posterior lobe arises from neural tissue, while the anterior lobe is formed from Rathke's pouch. The intermediate lobe forms from the portion of Rathke's pouch that makes contact with the neural tissue. The residual cleft *(c)* is the remnant of what was once the interior of Rathke's pouch.

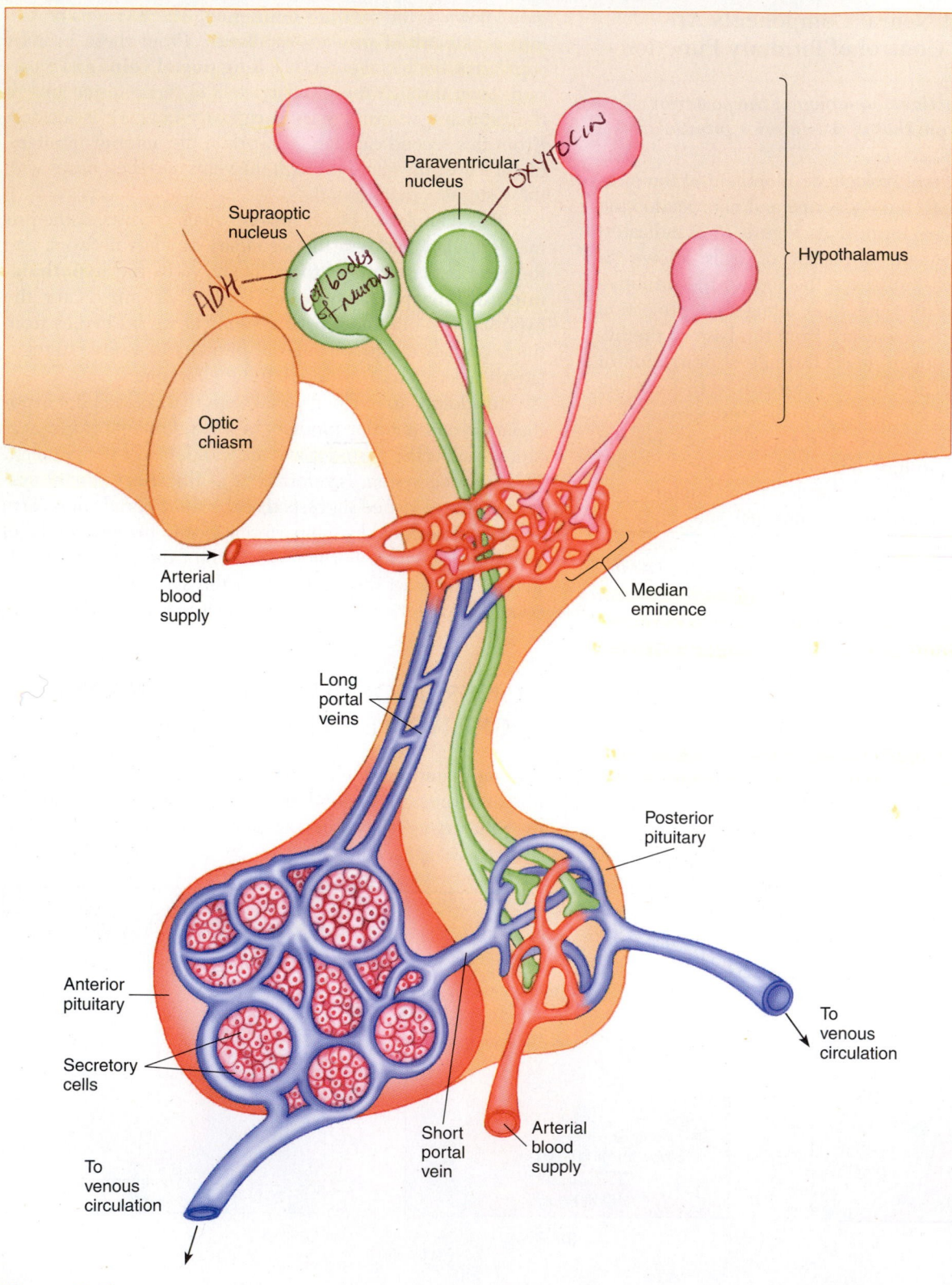

Figure 13–3

Neural and vascular connections between the hypothalamus and the anterior and posterior lobes of the pituitary.

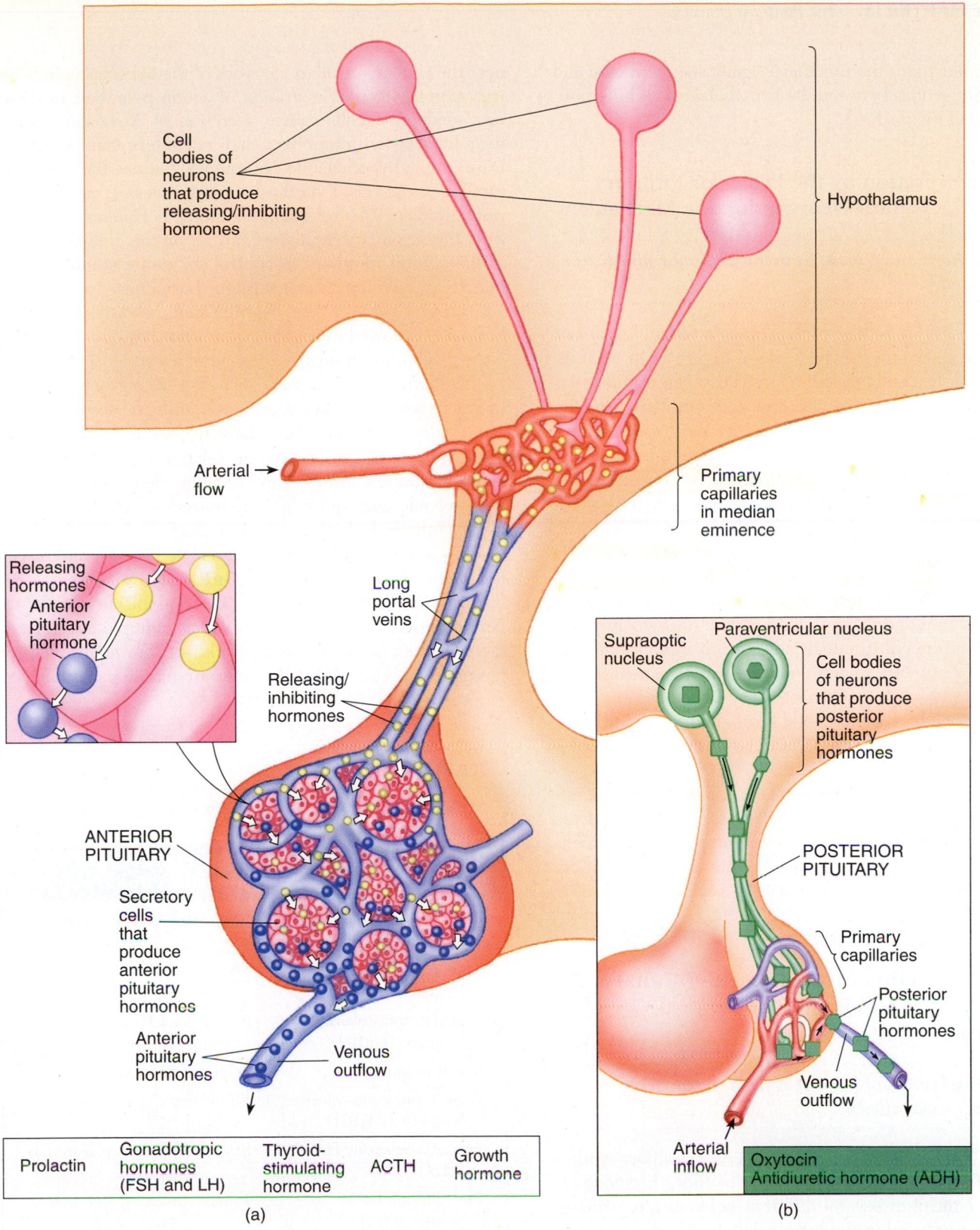

Cell bodies of neurons that produce releasing/inhibiting hormones

Hypothalamus

Arterial flow

Primary capillaries in median eminence

Releasing hormones
Anterior pituitary hormone

Long portal veins

Releasing/ inhibiting hormones

Supraoptic nucleus
Paraventricular nucleus

Cell bodies of neurons that produce posterior pituitary hormones

ANTERIOR PITUITARY

POSTERIOR PITUITARY

Secretory cells that produce anterior pituitary hormones

Primary capillaries

Posterior pituitary hormones

Anterior pituitary hormones

Venous outflow

Venous outflow

Arterial inflow

Prolactin	Gonadotropic hormones (FSH and LH)	Thyroid-stimulating hormone	ACTH	Growth hormone

Oxytocin
Antidiuretic hormone (ADH)

(a)

(b)

Figure 13–4

An illustration of the mechanisms involved in regulating anterior and posterior pituitary function. *(a)* Neuroendocrine cells of the hypothalamus transmit releasing/inhibiting hormones to the capillaries in the median eminence. These hormones travel via the long portal veins to the secretory cells of the anterior lobe, where they stimulate or inhibit the release of anterior pituitary hormones into the blood. *(b)* Neurons that originate in the supraoptic and paraventricular nuclei of the hypothalamus synthesize oxytocin and ADH and transmit them through axons that release the hormones into capillaries in the posterior pituitary. The hormones are then distributed to the general circulation.

This figure illustrates the functional significance of neural and vascular connections between the hypothalamus and the pituitary shown in Figure 13–3.

Hormone Secretion by the Posterior Pituitary Occurs by a Typical Neuroendocrine Mechanism

 How is hormone release from the posterior pituitary controlled?

Neural connections between the hypothalamus and posterior pituitary are shown on the right side of Figure 13–4. As indicated previously, cell bodies of neurons that send projections to the posterior pituitary are located primarily in the supraoptic and paraventricular nuclei of the hypothalamus. As indicated in Table 13–1, the posterior pituitary secretes two hormones: **oxytocin** and **antidiuretic hormone** (**ADH,** or **vasopressin**). These two hormones are actually synthesized within cell bodies of neurons in the supraoptic and paraventricular nuclei of the hypothalamus. Oxytocin and ADH are transported to the posterior pituitary through axons of the same neurons that produce them. Generation of action potentials in these nerves results in release of oxytocin or vasopressin from nerve terminals adjacent to capillary beds in the posterior pituitary, much the same way that neurotransmitters are released into a synapse. The hormones enter the blood vessels and are then carried throughout the body. Cells that produce oxytocin and ADH are therefore good examples of neuroendocrine cells.

Neural inputs to the supraoptic and paraventricular nuclei from other areas of the hypothalamus or from other areas of the brain modulate electrical activity of the neuroendocrine cells in these two nuclei. In this manner, a variety of stimuli are capable of regulating oxytocin and ADH secretion.

Hormone Secretion by the Anterior Pituitary Occurs by a Typical Endocrine Control Mechanism

 How is secretion of hormones from the anterior pituitary controlled?

As Table 13–1 indicates, cells of the anterior pituitary synthesize six different hormones. Secretion of these hormones is under the control of several different **releasing/inhibiting hormones** produced in the hypothalamus. All of the releasing/inhibiting hormones that have been chemically characterized thus far are either small peptides or amino acid derivatives. As illustrated in Figure 13–4a, these hormones are synthesized within the cell bodies of neuroendocrine cells located in a number of different sites in the hypothalamus. Axons of these neuroendocrine cells converge in the median eminence of the hypothalamus, where they terminate on or

near the primary capillary network of the hypothalamic-pituitary portal system. Generation of action potentials in these neuroendocrine cells causes secretion of particular releasing/inhibiting hormones from their nerve terminals. Upon release, the hypothalamic hormones enter the primary capillaries and travel via the long portal vessels to the anterior pituitary. The releasing/inhibiting hormones then exit from the secondary capillaries and act on the endocrine cells of the anterior pituitary to control (by either stimulating or inhibiting) secretion of anterior lobe hormones.

Table 13–2 lists the hypothalamic releasing/inhibiting hormones involved in regulating anterior pituitary secretion. Note that some stimulate secretion of a particular hormone, while others inhibit hormone secretion. As a group, they are therefore referred to as *releasing/inhibiting hormones.*

The system would be relatively easy to understand if there were a simple one-to-one relationship between each anterior pituitary hormone and a single releasing hormone or a pair of releasing and inhibiting hormones. However, a single hypothalamic hormone may influence the secretion of several anterior pituitary hormones. For example, gonadotropin-releasing hormone (**GnRH**) regulates the secretion of both luteinizing hormone (LH) and follicle-stimulating hormone (FSH). The degree to which multiple activities of these hypothalamic releasing/inhibiting hormones control hormone secretion from the anterior pituitary has not yet

TABLE 13–2

Releasing Hormones Involved in Regulating Anterior Pituitary Secretion

Hypothalamic-Releasing Hormone	Primary Effect on Anterior Pituitary*
Corticotropin-releasing hormone (CRH)	Stimulates ACTH secretion
Thyrotropin-releasing hormone (TRH)	Stimulates TSH secretion
Gonadotropin-releasing hormone (GnRH)	Stimulates LH and FSH secretion
Somatostatin	Inhibits GH secretion
Growth-hormone-releasing hormone (GHRH)	Stimulates GH secretion
Prolactin-releasing factor (PRF)+	Stimulates prolactin secretion
Prolactin-inhibitory hormone (PIH)	Inhibits prolactin secretion

*As described in the text, several of the releasing hormones influence the secretion of more than one anterior pituitary hormone. For simplicity, only the primary effects of the releasing hormones are presented here.

+The chemical structure of the substance which stimulates prolactin secretion has not been established. By convention, it is referred to as a "factor" until its structure is known.

been fully established. The general belief, however, is that secretion of any particular hormone from the anterior pituitary depends on the relative amounts of the various hypothalamic releasing/inhibiting hormones that reach the anterior pituitary.

Cell bodies of neurons that synthesize releasing/inhibiting hormones are not randomly distributed throughout the hypothalamus; instead, each releasing factor is synthesized by neurons located in different, discrete nuclei of the hypothalamus. For example, GnRH is produced primarily in one nucleus of the hypothalamus, while corticotropin-releasing hormone (CRH) is produced in another. Although they are synthesized in different regions of the hypothalamus, both hormones are ultimately released into the capillary network in the median eminence. Secretion of releasing/inhibiting hormones from these hypothalamic neurons depends on the input they receive from neurons located in other areas of the hypothalamus or in higher brain centers.

ACTIONS AND EFFECTS OF THE PITUITARY HORMONES

Now that we have examined the anatomical and physiological bases for the control of hormone secretion from the anterior and posterior lobes of the pituitary, a logical question might be, "What are the actions of the pituitary hormones?" The following sections describe specific biological effects of the pituitary hormones. Note the wide variety of physiological processes that the pituitary hormones serve to control.

Two Hormones Are Secreted from the Posterior Pituitary

Antidiuretic Hormone (ADH): An Increase in Water Retention

 What does antidiuretic mean; under what conditions is ADH released; how does ADH work?

Antidiuretic hormone, or **ADH,** is one of two small peptide hormones secreted by the posterior pituitary. It is nine amino acids in length and is synthesized primarily in the supraoptic nucleus of the hypothalamus, although smaller amounts are produced in the paraventricular nucleus as well.

The actions of ADH will be discussed in detail in Chapters 23 and 24, when the regulation of fluid balance by the kidney is described. Briefly, however, the primary effect of ADH is to increase water retention by the kidney. The net result of this effect is a decrease in urine volume and an increase in extracellular fluid volume. ADH can also act to constrict blood vessels (vasoconstriction). The other name for ADH, **vasopressin,** stems from this vasoconstrictor effect, but this response occurs only when the hormone is secreted in relatively large amounts.

The two primary physiological stimulators of ADH release are (1) an increase in osmolality of the blood or extracellular fluid and (2) a large decrease in blood volume, as in a hemorrhage (heavy bleeding). How do changes in osmolality or blood volume trigger ADH release? Specialized cells within the hypothalamus known as **osmoreceptors** are very sensitive to increases in osmolality of the extracellular fluid. Even a slight increase in osmolality stimulates osmoreceptor cells. Osmoreceptors in turn activate neuroendocrine cells of the supraoptic nucleus and trigger ADH release from the posterior pituitary. As a result of promoting water retention by the kidney, ADH causes a dilution of the extracellular fluid, which returns plasma osmolality to normal.

A decrease in blood volume also stimulates ADH release. However, stimulation of ADH release requires a 10% or greater blood volume decrease, which is a severe loss of blood. As a result of increasing fluid retention by the kidneys, ADH acts to restore blood volume to normal. In addition, a large loss of blood can result in sufficient ADH secretion to cause constriction of blood vessels. By constricting blood vessels, this effect serves to lessen the decrease in blood pressure that would occur with severe hemorrhage.

In addition to changes in osmolality and blood volume, several other factors are known to influence ADH release. For example, alcohol is a strong inhibitor of ADH release. When ADH secretion is inhibited, urine volume increases, and excess fluid is lost from the body. This accounts, in part, for the increase in urine volume and the thirst experienced following the consumption of alcoholic beverages. Nicotine, barbiturates, and certain anesthetics are known to stimulate ADH release. Overproduction of ADH caused by these agents results in excess fluid retention.

In the absence of ADH secretion, which can occur as a result of either a genetic abnormality or damage to the posterior pituitary, the kidneys are unable to conserve water, and large quantities of fluid are lost in the urine (a process called *diuresis*). This condition is known as **diabetes insipidus.** Diabetes insipidus is distinctly different from what is commonly termed *diabetes*, which is actually diabetes mellitus and will be discussed in Chapter 15.

Oxytocin: Control of Breast Milk Release

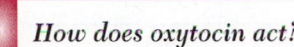 *How does oxytocin act?*

Oxytocin, the other hormone secreted by the posterior pituitary, is also a small peptide hormone (nine amino acids in length) and is very similar in structure to ADH. Like ADH, oxytocin is synthesized in the cell bodies of neurons in both the supraoptic and paraventricular nuclei of the hypothalamus, but its primary site of synthesis is in the paraventricular nucleus.

The principal action site of oxytocin is the female breast. In the breast, oxytocin stimulates contraction of specialized smooth muscle cells, which results in the transfer of milk from its site of synthesis in structures known as **alveoli** into the larger ducts of the breast. The net result of this effect is that milk is made available for a nursing infant. Actions of oxytocin will be described in Chapter 32, when the hormonal control of lactation is discussed. In addition to its effects on the breast, oxytocin can also stimulate contraction of smooth muscle in the uterus.

The primary stimulus for oxytocin secretion in a female is suckling by a nursing infant. When an infant nurses, touch sensors in the mother's nipple are activated and neural signals are transmitted to her brain. This information is processed in the brain, and the oxytocin-producing cells in the paraventricular nucleus are stimulated, resulting in oxytocin release from the posterior pituitary.

A variety of psychological stimuli can also influence oxytocin release. The mere sound of a baby crying can trigger oxytocin release in nursing mothers, whereas fear or apprehension can strongly inhibit its secretion. Thus, in addition to the neural mechanism that is triggered by suckling, input from other areas of the brain is also important in regulating oxytocin secretion.

Five Cell Types Are Responsible for the Secretion of Six Hormones from the Anterior Pituitary

In general, where are anterior pituitary hormones produced; what controls their secretion; what do they do?

Five different hormone-secreting cell types are found in the anterior pituitary. These are listed in Table 13–3 along with the hormone(s) produced by each cell type. Somatotropes, which produce growth hormone, are the most numerous cells of the anterior pituitary. Normally, about 50% of the cells present in the anterior pituitary are **somatotropes.** **Corticotropes,** the adrenocorticotropic hormone (ACTH)-secreting cells, compose about 20% of the cells, and **thyrotropes** the thyroid-stimulating hormone (TSH)-producing cells, about 5%. **Lactotropes** and **gonadotropes** are the other two cell types in the anterior pituitary.

TABLE 13–3

Hormone-Producing Cells of the Anterior Pituitary

Cell Type	Principal Hormone Produced
Somatotropes	GH
Corticotropes	ACTH
Thyrotropes	TSH
Lactotropes	Prolactin
Gonadotropes	LH and FSH

The size and activity of the latter two cell types vary under different physiological conditions. In lactating (nursing) females, the lactotropes, which secrete prolactin, are large and numerous, reflecting the high rate of prolactin secretion in these individuals. In males and nonlactating females, fewer lactotropes are present, and they are smaller as well. The gonadotropes, which secrete both LH and FSH, account for only about 5% of the cell population of the anterior pituitary of males. In females, their size varies in accordance with cyclical changes in LH and FSH secretion that occur during the monthly menstrual cycle.

As Table 13–3 indicates, these five cell types are responsible for production of six hormones in the anterior pituitary. Gonadotropes are responsible for the synthesis and secretion of both LH and FSH. As Chapter 31 will describe, concentrations of LH and FSH in the blood do not always change in parallel during the monthly ovarian cycle. Thus, although they are produced within the same cell, their secretion is controlled independently. At present, the mechanisms by which a single releasing hormone (GnRH) is able to differentially control the secretion of two hormones (LH and FSH) are not well understood.

The following sections describe the secretion and actions of each anterior pituitary hormone. The physiology of LH, FSH, and prolactin will be presented in detail in Chapters 31 and 32. Similarly, the physiology of ACTH will be presented primarily in Chapter 14. Therefore, these hormones will be discussed only briefly here. Instead, we will emphasize the remaining two hormones, growth hormone and TSH.

Luteinizing Hormone and Follicle-Stimulating Hormone: Control of Reproductive Function

LH and FSH are collectively referred to as the **gonadotropins.** This name reflects the fact that the target tissue of these hormones is the gonads — the ovaries and testes in females and males, respectively. In general, the gonadotropins have two primary effects: (1) to promote the development and maturation of sperm and egg and (2) to stimulate production of sex steroid hormones by the gonads. The principal sex steroids in males and females are testosterone and estradiol, respectively. As Table 13–2 indicates, secretion of gonadotropins is under the control of GnRH from the hypothalamus.

Prolactin: Control of Milk Synthesis

Prolactin has only minor function in males. In females, the primary effect of prolactin is to stimulate milk production by the breast. The target cells for prolactin are milk-producing alveolar cells of the breast. (Specific mechanisms by which prolactin stimulates milk formation will be presented in Chapter 32.) Prolactin secretion is under the dual control of a **prolactin-releasing factor (PRF)** and a **prolactin release-inhibiting hormone (PIH),** which has been shown to be dopamine. Similar to the situation for oxytocin described previously, prolactin release increases in response to stimulation of the mother's nipple by a suckling infant.

APPLICATIONS OF PHYSIOLOGY

Somatostatin in Medicine

As described in the text, the condition of acromegaly results from an overproduction of GH by somatotropes of the anterior pituitary. It is a disabling disease that is associated with reduced life expectancy. The goal of treatment is therefore to lower circulating growth hormone levels (and IGF-I levels), which almost universally improves patient well-being and decreases the risk for death.

Symptoms of acromegaly can be severe and include lethargy, headaches, sweating, painful joints, and a burning or tingling sensation in the extremities. Because of the antagonistic effects of GH on insulin action, the disease can also lead to diabetes mellitus. It can also cause hypertension and patients have a two- to threefold increase in the risk for death as a result of cardiovascular or respiratory disease.

For many years, the treatment of acromegaly has involved surgical removal of the pituitary gland, followed by radioablation therapy when needed. Surgical removal of the pituitary (hypophysectomy) is not an easy procedure and is best done by an experienced surgeon. However, with improvements in surgical techniques, the "cure" rate is in excess of 50%.

More recently, clinical studies exploring the use of long-acting somatostatin analogues in the treatment of acromegaly have shown promise. Octreotide is a long-acting synthetic somatostatin analogue that has a half-life of approximately 100 minutes (compared with a half-life of 1–3 minutes for natural somatostatin). When given as three injections per day, Octreotide was shown to be effective in lowering GH secretion in about half of the patients studied. More recently, Octreotide has been incorporated into microspheres of a biodegradable polymer. This microsphere form of somatostatin serves as a depot that is slowly released into the bloodstream over a prolonged period of time. Treatment of patients having acromegaly with a single injection of this preparation every 28 days proved effective in lowering GH levels in one study lasting a year.

Treating patients with these long-acting forms of somatostatin is now an accepted and fairly effective part of medical practice. As indicated earlier, surgery has typically been the first course of action in treating acromegaly. However, given the success in studies using somatostatin to treat the disease, there is optimism that this medical therapy will some day replace surgery as the first option in treating the disease.

Adrenocorticotropic Hormone: Control of Cortisol Synthesis and Secretion

The primary effect of **adrenocorticotropic hormone (ACTH)** is to promote the synthesis and secretion of the steroid hormone cortisol from the adrenal cortex (a portion of the adrenal gland, an endocrine gland located near the kidney). Cortisol produced by the adrenal gland in turn has numerous effects on metabolism in a variety of different tissues. As the name indicates, ACTH also has trophic or "growth-promoting" effects on cells of the adrenal cortex. As Table 13–2 indicates, the secretion of ACTH is controlled by corticotropin-releasing hormone (CRH), which is produced in the hypothalamus. The specific effects of ACTH on the adrenal cortex and the effects of adrenal steroids on metabolism will be covered in detail in Chapter 14.

ACTH is actually synthesized as part of a much larger precursor protein that contains not only ACTH but also several other biologically active peptides. This precursor has been named **proopiomelanocortin (POMC) peptide** to reflect the fact that it contains the sequences for endogenous (internally synthesized) opioid compounds (see Chapter 7), the sequence for **melanocyte-stimulating hormone (MSH),** and the sequence for ACTH.

Growth Hormone: Body Growth and Metabolism

 What factors influence the secretion of growth hormone and what does it do?

As was indicated in Chapter 12, **human growth hormone (hGH or GH)** is a polypeptide hormone consisting of 191 amino acids. The structure of GH is remarkably similar to that of a peptide hormone secreted by the placenta during pregnancy, **human placental lactogen (hPL)**. Growth hormone is also somewhat similar to prolactin, although to a much lesser extent. GH, hPL, and prolactin compose a family of related peptide hormones. The sections that follow will focus on the effects of GH on body growth and metabolism. The physiology of hPL and prolactin is presented in later chapters.

Control of Growth Hormone Secretion

As Tables 13–2 and 13–3 indicate, GH secretion from the somatotropes of the anterior pituitary is under the dual control of **growth-hormone-releasing hormone (GHRH) and somatostatin**, both of which are produced by the hypothalamus. As indicated previously in Table 12–1, somatostatin is also produced in the pancreas, where it is thought to play a

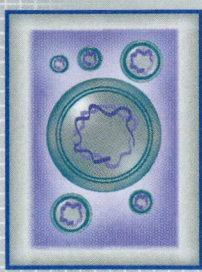

G-Proteins and Hormone Secretion

As described in Chapter 12, cell-to-cell communication mechanisms are vital for survival of higher organisms. Although these regulatory mechanisms open important avenues that allow for greater flexibility and survival potential, they also present the opportunity for error in the communication process. Perhaps it is not surprising that a variety of human diseases result from an abnormality in normal cell communication mechanisms.

GH from the pituitary is an important regulator of growth of the skeleton during adolescence. Overproduction of GH during adolescence leads to gigantism; during adulthood, overproduction leads to acromegaly. Most commonly, gigantism or acromegaly results from pituitary tumors involving GH-producing cells.

Regulation of GH secretion as well as growth of GH-producing cells involves changes in intracellular cAMP concentrations. These changes in intracellular cAMP are coupled to extracellular signals by G-proteins. In Chapter 12, the role of G-proteins in generating hormone second messengers was described. Mutations of G-proteins that result in either continuous activation or loss of activity have been identified in several human diseases. Recently, it has been demonstrated that in approximately 40% of GH-secreting pituitary tumors, the basal adenylate cyclase activity and cAMP levels are elevated. These changes result from a tissue-specific mutation in the G_s-protein gene that leads to a single amino acid substitution in the final protein product. This amino acid substitution occurs at a position in the molecule critical for normal G_s-protein function. The mutation results in continuous activation of the G_s-protein and increased cAMP production. Thus, in this situation, the disease of acromegaly results from a simple, single amino acid substitution in the somatotrope G_s-protein that leads to an abnormally elevated cAMP level.

Similarly, cases of overproduction of TSH have been found to be the result of constitutive activation of G_s-proteins in pituitary thyrotropes.

These are further examples of how an alteration in the normal process of cellular communication can lead to disease. One day, using gene therapy, it may be possible to correct this defect in cellular signaling.

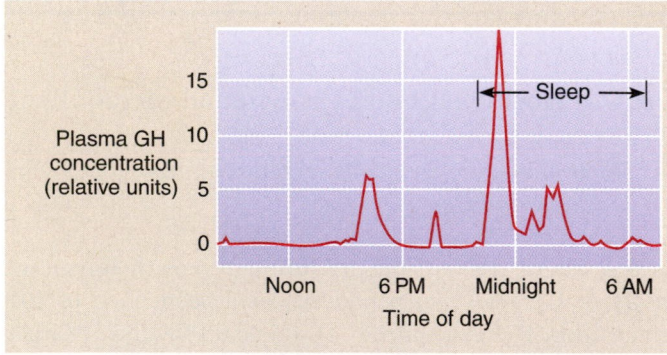

Figure 13–5

Pattern of GH secretion that might be seen in a normal individual.

TABLE 13–4
Partial Listing of Factors or Conditions Known to Influence Growth Hormone Secretion

Stimulators
  Deep sleep°
  Low blood glucose concentration
  Stress
    Physical trauma
    Infection
    Psychological stress
  Amino acids, especially arginine
Inhibitors
  REM sleep
  High blood glucose concentration

°See Chapter 11

local role in regulating pancreatic hormone secretion. This particular action of somatostatin will be described further in Chapter 15. At this point we should emphasize, however, that (1) the somatostatin produced by the hypothalamus and that produced by the pancreas serve different physiological functions and (2) the secretion of somatostatin from each site is independently controlled.

As Figure 13–5 shows, growth hormone is not secreted in a steady, continuous fashion throughout the day but instead is secreted in a pulsatile manner. The most consistent period of GH secretion occurs about 1 hour after the onset of deep sleep (Fig. 13–5). This can occur during the normal sleep period at night or during a daytime nap. The phase of sleep associated with dreams, known as REM sleep (see Chapter 11), initiates the return of GH secretion to basal presleep levels. The exact significance of this sleep-related surge in GH secretion is not known, but it has been suggested that it may be important in stimulating processes of tissue growth and repair during sleep.

In addition to the sleep-induced increase in GH secretion, a series of pulses often occurs 2 to 4 hours after a meal (see Fig. 13–5). The relative frequency and the size of these pulses tend to increase about the time of puberty. Some of the other agents or conditions known to influence GH secretion are listed in Table 13–4. Hypoglycemia (low blood glucose) stimulates GH secretion, whereas high blood glucose inhibits GH secretion. A useful test that physicians can perform to evaluate whether a patient's pituitary is capable of secreting GH involves the administration of a dose of insulin, followed by the monitoring of GH concentrations in the blood. As Chapter 15 will describe, insulin decreases the blood glucose concentration. In a normal individual in this test situation, the hypoglycemia caused by insulin triggers a pronounced increase in GH secretion. In contrast, the absence of increased GH secretion in response to insulin indicates to the physician that pituitary GH secretion is impaired. As Table 13–4 also shows, various types of stress, both physical and emotional, stimulate GH secretion. Several amino acids, especially arginine, also stimulate GH secretion.

How is such a wide variety of stimuli, such as those listed in Table 13–4, able to control GH secretion? As Figure 13–6 illustrates, GHRH and somatostatin serve as the final common pathway for determining GH secretion. Inputs from various areas of the brain and hypothalamus control GHRH and somatostatin release in the median eminence, thus controlling GH secretion from the anterior pituitary.

The Effects of Growth Hormone

The overall effect of GH is to promote tissue growth. In this regard, GH is considered to be an **anabolic hormone.** (Those hormones that tend to have the opposite effect, promoting breakdown of tissues, are known as **catabolic hormones.**)

Another important anabolic hormone is insulin, a hormone produced by the pancreas. While many of the effects of GH are similar to those of insulin, some are exactly opposite to those of insulin and actually antagonize or impair the actions of insulin. GH is therefore said to have both **insulin-like** effects and **anti-insulin,** or **"diabetogenic,"** effects.

The most striking and obvious effect of GH is that it stimulates linear growth of the skeleton. In addition, it also stimulates growth of a number of tissues in the body. Some effects are due to direct actions of GH on the tissue, whereas others result from the stimulation by GH of **somatomedin** production by the liver (Fig. 13–7). A later section of this chapter discusses the physiology and actions of the somatomedins.

As Figure 13–7 shows, GH directly stimulates the uptake of amino acids from the blood into muscle cells and also stimulates protein synthesis in muscle. In addition to these effects, which stimulate muscle growth, GH stimulates liver protein and RNA synthesis as well. These effects on muscle and liver are similar to those of insulin and are known as the *insulin-like effects of GH.*

However, as indicated previously, some effects of GH antagonize those of insulin. In adipose tissue, GH decreases glucose uptake and stimulates breakdown of fat stores (Fig. 13–7) — effects that are opposite to those of insulin. Other

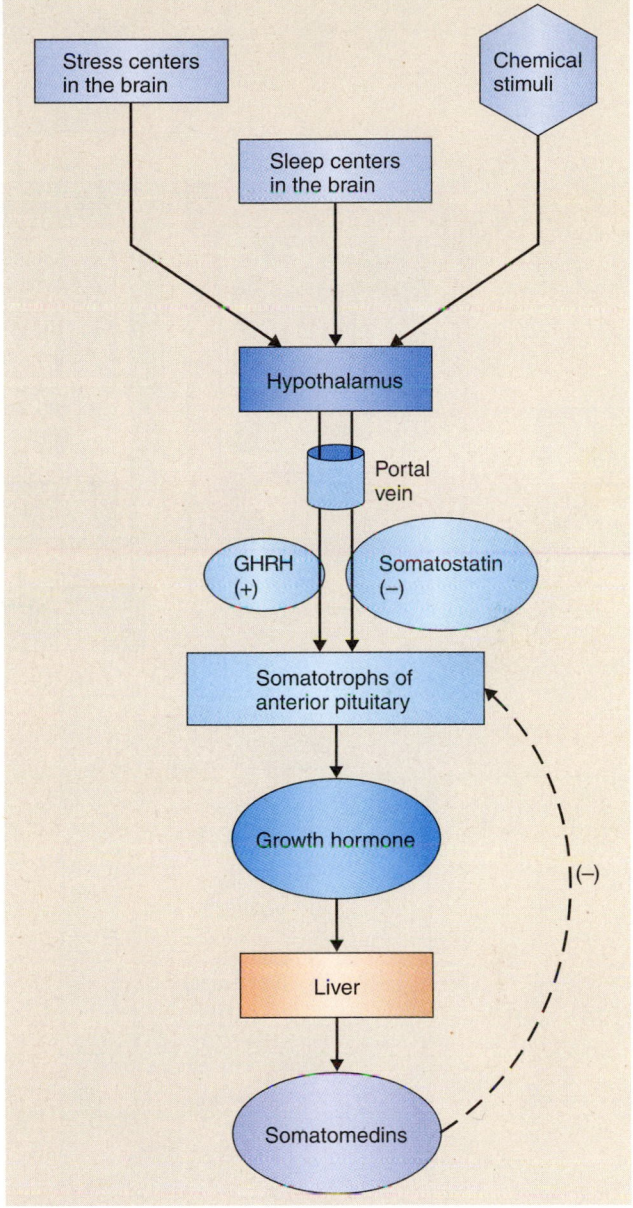

Figure 13–6

A variety of factors can influence growth hormone secretion by regulating the production of GHRH and somatostatin by the hypothalamus.

effects of growth hormone, which decrease glucose uptake into muscle and increase gluconeogenesis in the liver, tend to raise the blood glucose concentration. These effects also antagonize those of insulin, which normally causes a decrease in the blood glucose concentration.

As indicated in Figure 13–7, the net effect of GH on metabolism in adipose tissue and in muscle is to decrease fat storage and promote accumulation of muscle protein. The overall effect of the hormone is therefore to promote accumulation of lean body mass.

What are somatomedins; how do they affect growth?

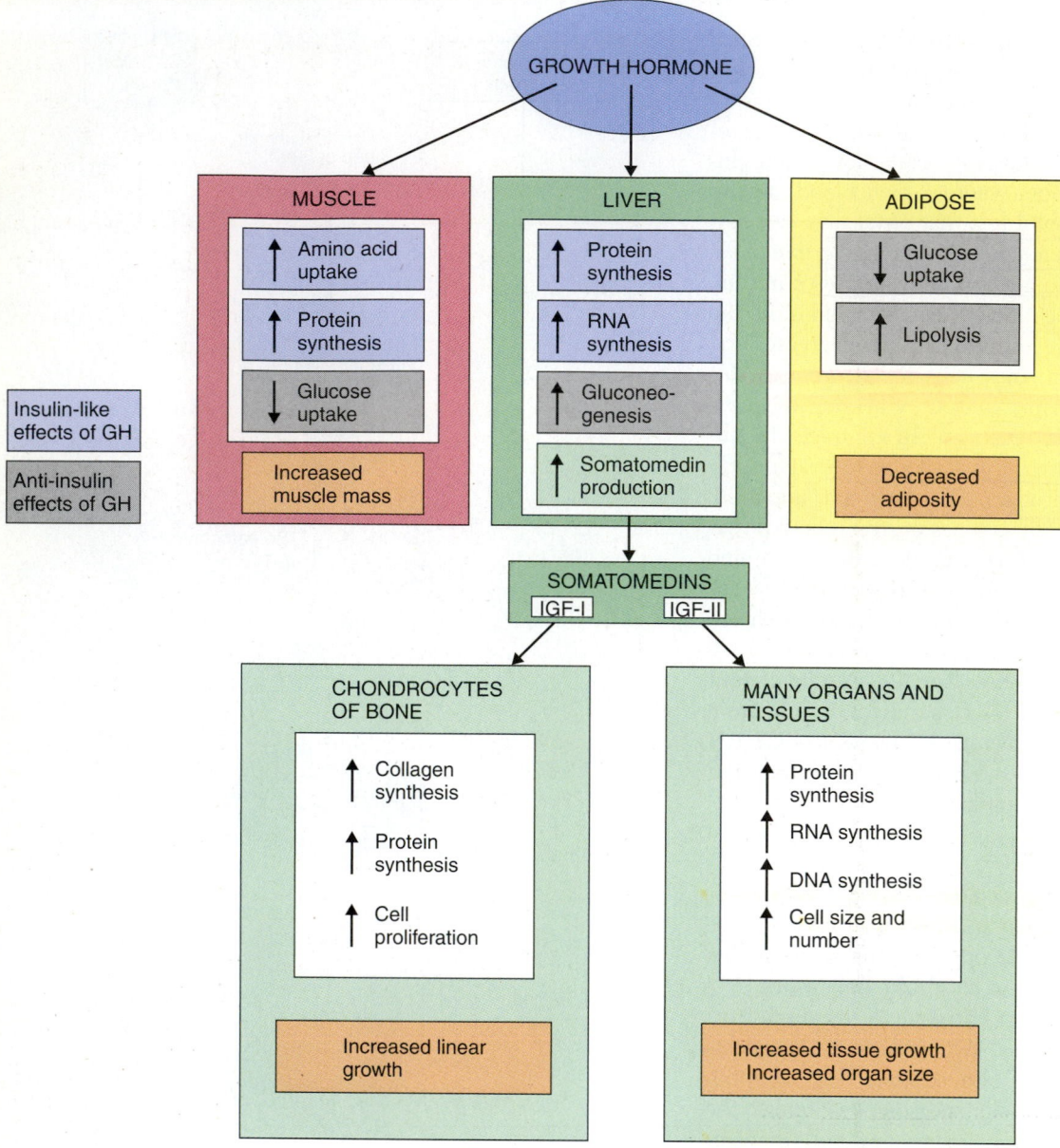

Figure 13-7

Growth hormone has direct effects on muscle, liver, and adipose tissue and indirect effects that occur as a result of somatomedin production. The overall effect of growth hormone is to promote skeletal growth and the accumulation of lean body mass.

Growth Hormone Stimulation of Somatomedin (IGF) Production

Early studies of GH indicated that not all effects of the hormone observed in the whole animal were the result of direct actions of GH on individual tissues. For example, although GH was known to promote bone growth when injected into young animals, it was without effect when directly added to pieces of bone incubated in a test tube. Later, scientists determined that GH stimulated the production of a group of peptide substances first called **somatomedins,** and that it was the somatomedins that were responsible for some of the effects of the hormone.

The liver is the predominant and probably most important production site of somatomedins, although many other tissues produce them as well. The two most abundant somatomedins produced by the liver are known as **insulin-like growth factor I (IGF-I)** and **insulin-like growth factor II (IGF-II).** This particular nomenclature for the somatomedins grew from the observation that, in addition to insulin itself, the plasma of most animals contains peptide substances exhibiting many of the growth-promoting activities of insulin. IGF-I and IGF-II are very similar to each other in structure and also bear a remarkable resemblance to **proinsulin,** which is the precursor for insulin found in the pancreas (see Chapter 15). This similarity in structure between the two IGFs and proinsulin accounts for much of the insulin-like activity of the two peptides.

IGF-I production by the liver appears to be primarily determined by the concentration of GH in the blood. As a result, IGF-I is markedly decreased in the blood of individuals deficient in GH and markedly increased in many patients who have higher than normal concentrations of GH in their blood. In addition to GH, insulin and thyroid hormones also tend to promote IGF-I production. In contrast to IGF-I, the production of IGF-II by the liver is much less dependent on the concentration of circulating GH. A deficiency in circulating GH usually results in a decrease in IGF-II to only about half of the normal value.

The full spectrum of activities of the IGFs has not yet been fully characterized. In general, however, the IGFs have two types of effects, as indicated in Figure 13–7. The first type of effect involves a stimulation of events leading to an increase in linear growth of the skeleton. Of the two growth factors, IGF-I is primarily responsible for producing this particular effect and has its actions in both prenatal and postnatal life. IGF-I stimulates skeletal growth by increasing the formation of cartilage in specialized regions near the ends of growing bone known as **epiphyseal plates.** The cartilage deposited in epiphyseal plates eventually is replaced by bone mineral, and the bone therefore increases in length. IGF-I promotes bone growth by stimulating those cells involved in cartilage formation, cells known as **chondrocytes.** Specifically, IGF-I stimulates the synthesis of collagen as well as total cell protein in chondrocytes and promotes proliferation of these cells as well (see Fig. 13–7). The process of bone formation will be described in greater detail in Chapter 26.

The second type of IGF effect is related to processes involved in stimulating tissue growth and tissue repair. IGF-II appears to be primarily responsible for producing these effects, especially during fetal development. By stimulating a number of cellular processes, such as protein synthesis and RNA synthesis, IGF-II is able to elicit a general anabolic response in a variety of different tissues and organs. This results in generalized tissue growth and an increase in organ size (see Fig. 13–7).

Conditions of Growth Hormone Deficiency or Excess

Given the role of GH in stimulating skeletal and soft tissue growth, one might imagine that the consequences of abnormalities that involve either under- or oversecretion of the hormone would be significant. The particular clinical symptoms that occur, however, depend on the age of the individual when the deficiency or excess occurs.

If a deficiency in GH secretion exists during childhood, the condition of **dwarfism** results. Because the deficiency in GH secretion usually occurs as the result of a defect in pituitary function, this condition is often referred to as **pituitary dwarfism.** Scientists now realize, however, that dwarfism may result not only from a pituitary deficiency of GH production but also from either an impairment in IGF-I production by the liver or an inability of target tissues to respond to the IGF-I. A tribe of Pygmies in Africa has been studied and found to have normal GH concentrations in their blood but low circulating IGF-I concentrations due to a genetic impairment in growth factor production by the liver. Other forms of dwarfism appear to result from an impairment in the responsiveness of target tissues. Thus, dwarfism can result from abnormal function in either the anterior pituitary, the liver, or the target tissues.

In contrast to the situation in children, however, if the deficiency in GH production occurs during adulthood after normal bone growth has occurred, generally no overt clinical symptoms are evident.

Overproduction of GH can result in one of two clinically significant conditions: (1) **gigantism** or (2) **acromegaly.** Gigantism results when increased GH secretion occurs during childhood. In these individuals, the skeleton is stimulated to grow excessively. A photograph of a person displaying an extreme case of gigantism is shown in Figure 13–8. Generally,

Figure 13–8

The world's tallest woman, Sandy Allen (7 ft, 7.25 in, 480 lb), pictured at home with her 12-year-old brother.
(© *Bettina Cirone/Photo Researchers*)

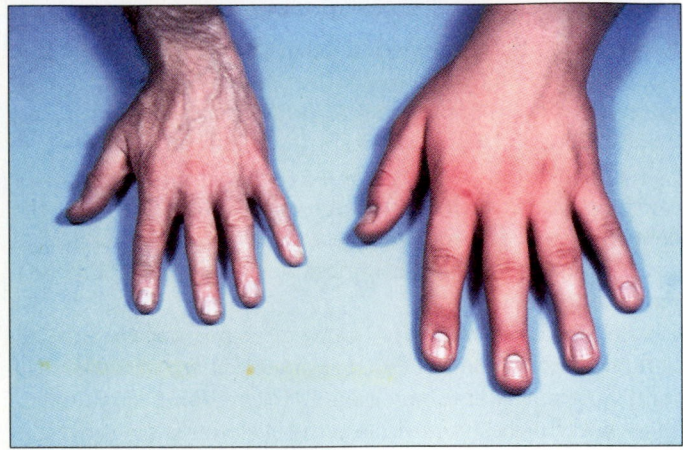

Figure 13–9

Hand of someone with acromegaly *(left)* placed next to a normal hand *(right).*   *(© Custom Medical Stock Photo)*

in cases of gigantism, the long bones of the body are stimulated to grow disproportionately, and as a result these persons have relatively long arms and long legs.

If overproduction of GH begins after the time of puberty, however, the condition of acromegaly results. Around the time of puberty, the growth centers, or epiphyseal plates, of the long bones of the body "close," meaning that they become unresponsive to hormonal stimulation. Thus, bones of the arms and legs cease to grow at this time. However, the bones of the hands, feet, skull, and lower jaw do not "close" and can still be stimulated to grow. As a result of the oversecretion of GH, these bones continue to grow and enlarge, leading to a distinctive physical appearance. Typically, individuals with acromegaly have an enlarged lower jaw, large hands and feet, and also generally large facial features due to the stimulation of soft tissue growth resulting from increased somatomedin production. The hand of a person with acromegaly is shown in Figure 13–9. In extreme situations, the diabetogenic effects of GH can also lead to overt cases of diabetes mellitus in individuals suffering from acromegaly.

Thyroid-Stimulating Hormone: Control of Thyroid Hormones

How is thyroid function controlled?

As its name implies, the primary target tissue of thyroid-stimulating hormone (TSH) is the thyroid gland. In the thyroid, TSH stimulates cell growth and secretion of thyroid hormones. Thyroid hormones in turn affect a variety of metabolic processes in the body. Therefore, this section provides a description of not only the physiology of TSH but also the thyroid gland and actions of thyroid hormones.

How is thyroid function reflected by the anatomy of the thyroid gland?

Anatomy of the Thyroid Gland

The gross anatomy of the thyroid gland is shown in Figure 13–10. The thyroid is one of the largest endocrine glands, weighing approximately 20 g in a normal adult. As Figure 13–10 indicates, the thyroid consists of two lobes that lie on either side of the trachea, just below the larynx. A thin band of tissue known as the *isthmus* connects the two lobes. The gland has an abundant blood supply and actually exhibits one of the highest blood flow rates of any tissue or organ in the body. The thyroid gland also has a tremendous capacity for growth. With an appropriate stimulus, the gland can become greatly enlarged, a condition known as *goiter.* An example of a goiter is shown in Figure 13–11. In this particular case, the goiter was caused by a dietary deficiency in iodide. A later section of this chapter describes the mechanism by which iodide deficiency causes goiter.

The thyroid gland consists of cells that are arranged in closely packed follicles or spheres of cells (Fig. 13–12); in a healthy adult, there are about 3 million of these thyroid follicles. Cells that form a follicle are usually cuboidal in shape; the interior of the follicle is filled with a protein-containing material termed **colloid.** The major component of colloid is a protein known as **thyroglobulin,** which serves as the precursor of thyroid hormones. A considerable amount of

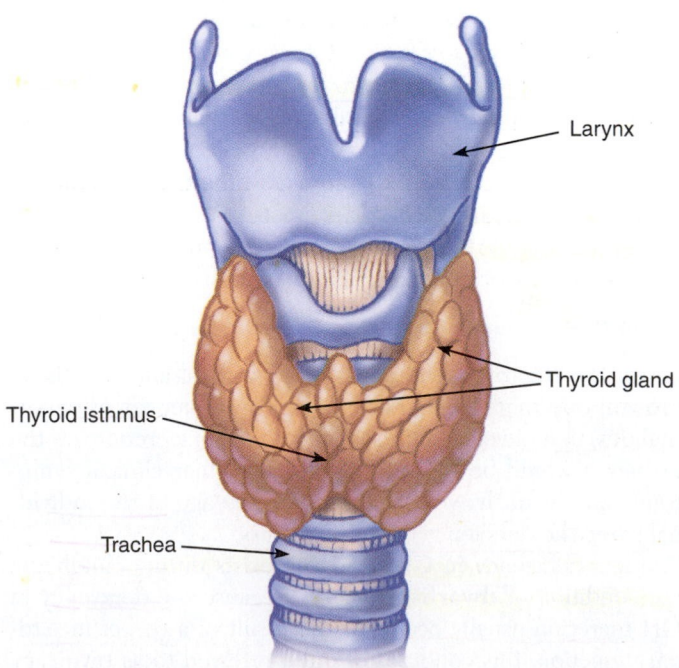

Larynx

Thyroid gland

Thyroid isthmus

Trachea

Figure 13–10

Gross anatomy of the thyroid gland showing its location relative to the trachea and the larynx.

Figure 13–11

An individual with a goiter.   (© *John Paul Kay/Peter Arnold, Inc.)*

colloid is usually present within each follicle, so that colloid is normally the major constituent of the total mass of the thyroid gland. An extensive capillary network surrounds each follicle.

Synthesis and Secretion of Thyroid Hormones

The major compounds relating to thyroid hormone production and metabolism are shown in Figure 13–13. **Thyroxine** (**tetraiodothyronine, or T$_4$**) is the primary hormone product of the thyroid gland. **Triiodothyronine (T$_3$)** is also produced by the thyroid gland but in lesser amounts. (In target tissues, T$_4$ is converted into T$_3$, as will be discussed later.) The other two iodinated compounds shown in Figure 13–13, **monoiodotyrosine (MIT)** and **diiodotyrosine (DIT),** are found primarily within the thyroid follicle cell, although small amounts are also secreted into the blood.

Each of the compounds shown in Figure 13–13 contains iodine as an integral part of the molecule. In comparison to other hormones, thyroid hormones are unique in that their synthesis requires a constituent (iodine) not always present in the diet. As a result, mechanisms have evolved that permit concentration of iodide in the gland and storage of large amounts of thyroglobulin, the thyroid hormone precursor. Normally, enough thyroglobulin may be stored in the gland to sustain thyroid hormone secretion for two months or more in the complete absence of any further thyroglobulin synthesis.

Thyroglobulin itself is a large glycoprotein synthesized by thyroid follicle cells (Fig. 13–14). As with other glycoproteins that are secreted, the protein component of thyroglobulin is synthesized in the rough endoplasmic reticulum of the cell, and the carbohydrate portion of the molecule is added in the Golgi complex. From the Golgi complex, thyroglobulin is then secreted from the apical side of the cell into the lumen of the follicle, where it is iodinated and stored as part of the colloid.

The first step in thyroid hormone formation involves the uptake of iodide (I⁻, the ionic form of the element iodine) into the thyroid follicle cell across its basal membrane from the plasma (Fig. 13–14). Uptake involves an active transport process that pumps iodide into the cells and can produce an intracellular iodide concentration 30-fold higher than that in the extracellular fluid. This transport system is so effective at pumping iodide into the follicle cell that the process is often referred to as the "**iodide trap.**"

Once inside the cell, free iodide diffuses down a concentration gradient toward the apical surface of the cell and the interior of the follicle. Soon after leaving the follicle cell, iodide undergoes **oxidation** and **iodination** reactions. Enzymes that rapidly oxidize iodide and convert it to iodine (I) are present on the surface of the follicle cell membrane that faces the colloid. Iodine formed by these enzymes then serves as a substrate for other enzymes that catalyze addition of iodine to tyrosine residues within the thyroglobulin protein (iodination). Tyrosine residues containing either one or two iodine atoms can be formed, and thus at this point thyroglobulin contains both MIT and DIT as part of its protein structure.

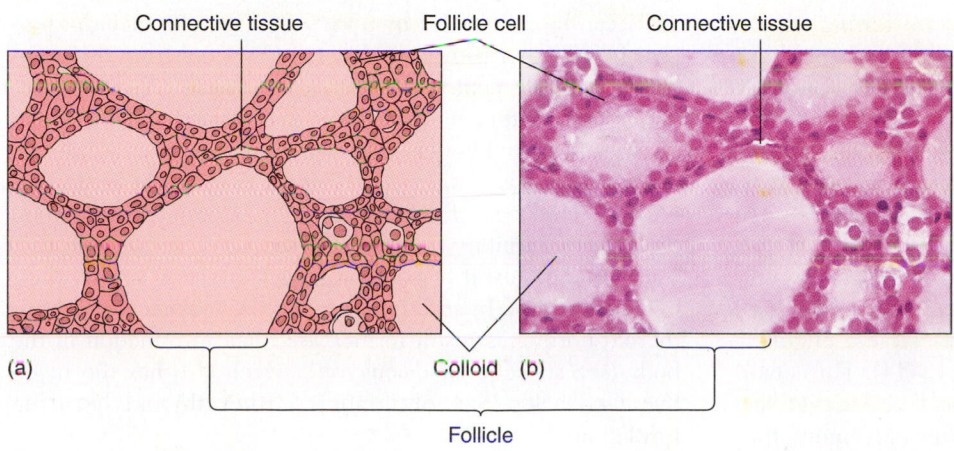

(a) Connective tissue   Follicle cell   Connective tissue

Colloid

Follicle

Figure 13–12

(a) Representation of a typical cross-section through a portion of the thyroid gland. Note the numerous blood vessels and the relative abundance of colloid in the gland. *(b)* Photomicrograph of a section through the thyroid gland. (× 425).   (© *Biophoto Associates/Science Source)*

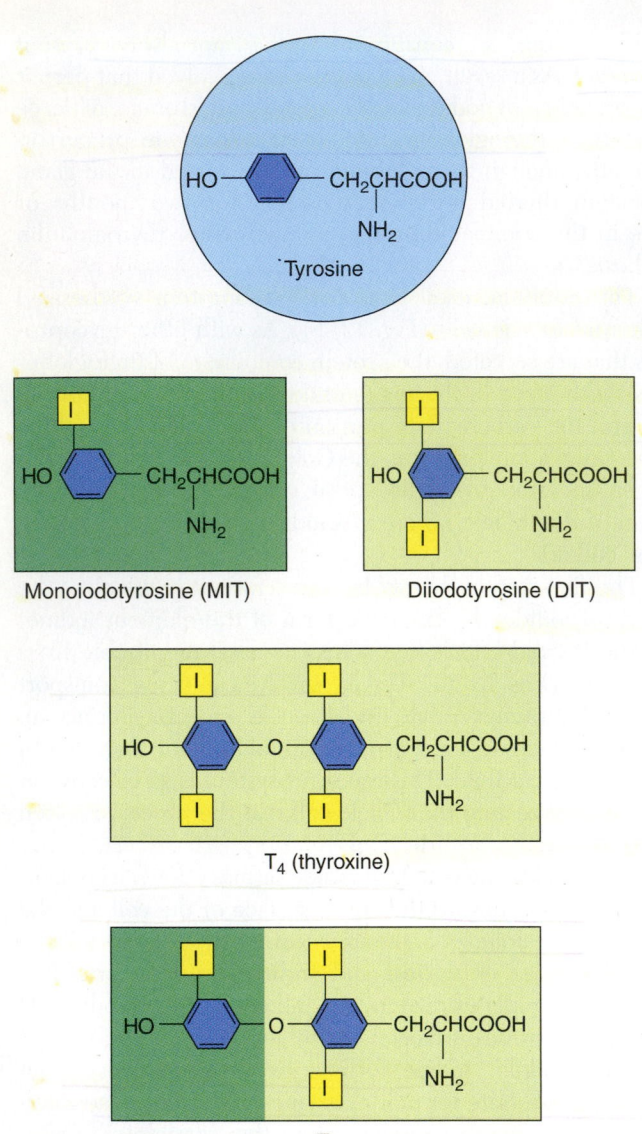

Figure 13–13

Structures of the primary iodinated compounds found in the thyroid gland and in the blood. Note the difference in number and location of the iodides between T_4 and T_3.

The next major step in thyroid hormone formation involves **coupling** two iodinated tyrosines within thyroglobulin to form either T_3 or T_4. If two DIT residues are coupled, then T_4 is formed. If, however, DIT and MIT residues of thyroglobulin are coupled, then T_3 is produced. T_3 and T_4 are not free but are still present as a part of the thyroglobulin peptide. In later steps to be described, thyroglobulin is broken down to release the free hormones.

When follicle cells are stimulated to produce thyroid hormones, numerous microvilli are formed on the apical surface. These microvilli extend out from the cell and engulf a portion of colloid by pinocytosis (see Fig. 13–14). This small droplet of colloid is taken into the follicle cell and moved toward the basal surface of the cell. During this movement, the colloid droplet meets with lysosomes, and the structures fuse to form a **phagolysosome** (see Fig. 13–14). Phagolysosomes move toward the basal end of the cell; as they do, thyroglobulin molecules are broken down by proteolytic enzymes to produce free T_4 and T_3. T_4 and T_3 are released into interstitial fluid, where they are picked up by nearby capillaries and carried by the circulation throughout the body.

Proteolytic breakdown of thyroglobulin also results in release of free MIT and DIT, which are present due to incomplete coupling reactions. MIT and DIT are subject to metabolism within the follicle cell by very active **deiodinases** present in the cytoplasm. These deiodinases remove iodide from MIT and DIT to produce iodide and tyrosine. The iodide can be recycled for use in further hormone synthesis, thereby conserving the iodide. Very little free MIT or DIT is actually released from the thyroid gland.

Control of Thyroid Hormone Synthesis and Secretion

The primary factor controlling thyroid follicle cell activity and thyroid hormone secretion is **thyroid-stimulating hormone, or TSH,** which is produced by thyrotropes of the anterior pituitary (see Table 13–3). Specific receptors for TSH are present on basal surfaces of follicle cells. These receptors are coupled to adenylate cyclase; hence, binding of TSH to its receptors leads to an increase in intracellular cAMP. The specific effects of TSH on thyroid hormone production are listed in Table 13–5. Note that TSH stimulates virtually every aspect of thyroid hormone formation. In addition to these specific effects on thyroid hormone formation, TSH also stimulates growth of the thyroid follicle cell. Thus, when TSH secretion is increased, follicle cells are stimulated to grow and become elongated and more columnar in shape. Likewise, when TSH secretion is low, the cells regress and become somewhat flattened.

A diagram outlining the control of thyroid hormone secretion is presented in Figure 13–15. The relationship between the hypothalamus, anterior pituitary gland, and thyroid gland indicated in this figure is known as the **hypothalamic-pituitary-thyroid axis.** As described previously, thyroid hormone secretion is under the control of TSH from the pituitary. Secretion of TSH in turn is controlled primarily by TRH produced by the hypothalamus. Secretion of TRH can be influenced by a variety of factors, including inputs from higher centers within the CNS and from temperature regulatory centers in the hypothalamus. Temperature regulatory centers in turn receive information concerning changes in body temperature and changes in environmental temperature. Also indicated in Figure 13–15 is the important negative feedback effect that thyroid hormones exert upon the anterior pituitary to limit TSH secretion. As a result, thyroid hormones exert negative feedback effects that limit their own production. In addition, some of the biological effects of thyroid hormones result in increased heat production in the body (see subsequent discussion), which also has the negative feedback effect of limiting further thyroid hormone formation.

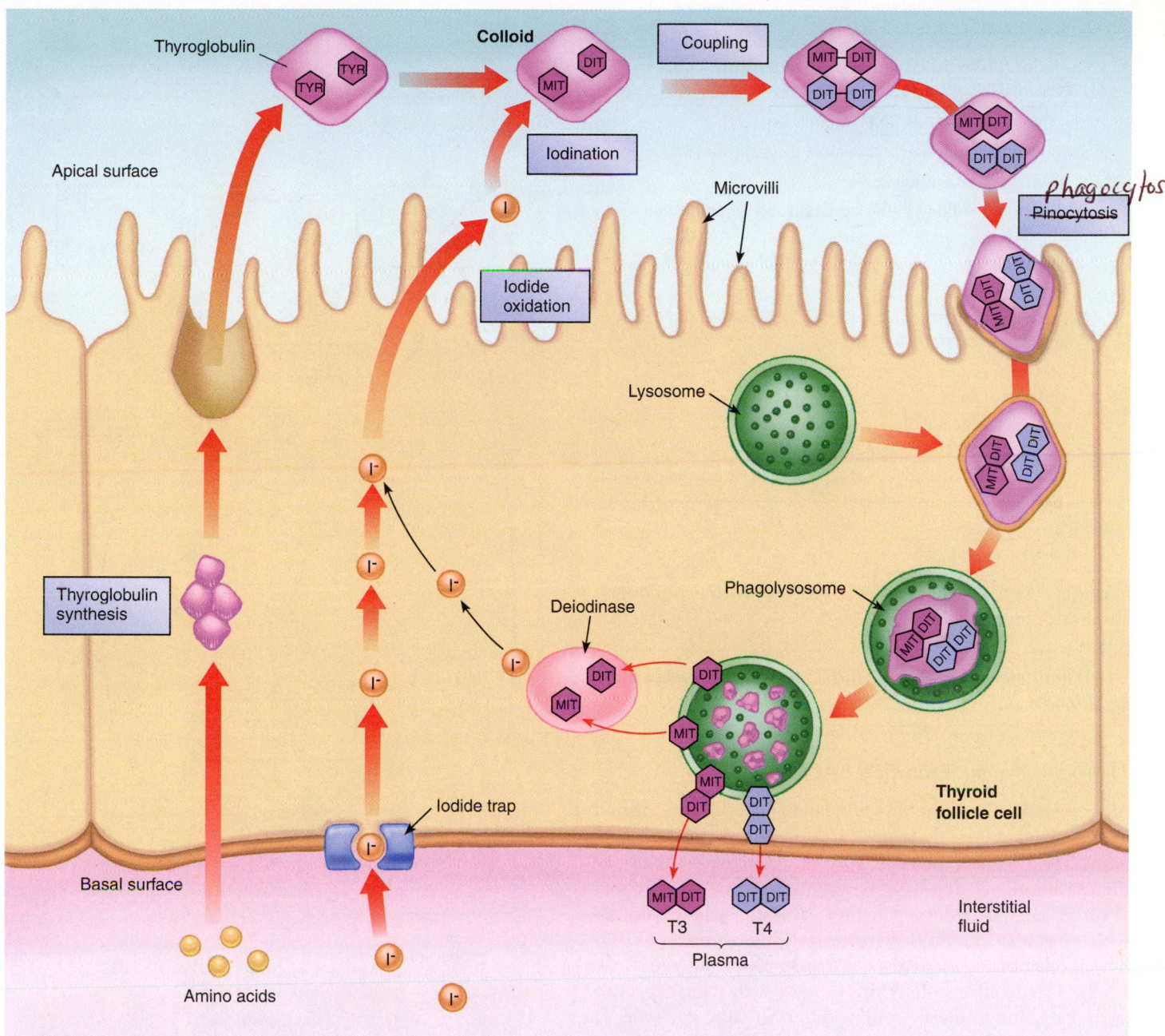

Figure 13–14

The major steps involved in thyroid hormone formation and release. For simplicity, the oxidation, iodination, and coupling reactions are shown separately. In actuality, an enzyme complex that is associated with the apical membrane of the follicle cell catalyzes these steps.

Prior to the introduction of iodized salt into the diet, the occurrence of goiter, or enlargement of the thyroid gland, was fairly common. Figure 13–15 provides the framework to understand how a deficiency of dietary iodide can lead to an enlargement of the thyroid gland. As described previously and indicated in Figure 13–15, iodide is a necessary component for thyroid hormone formation. In the absence of iodide, thyroid hormones cannot be produced, and in the absence of thyroid hormone secretion, negative feedback effects of T_4 and T_3 on TSH release are also absent (see Fig. 13–15). TRH and TSH secretion therefore increases, and the growth-promoting effects of TSH on thyroid follicle cells cause the gland to increase in size, despite the inability to actually synthesize active

TABLE 13–5
Effects of Thyroid-Stimulating Hormone (TSH) on Thyroid Hormone Production
In the thyroid follicle cell, TSH stimulates:
1. Iodide uptake by active transport mechanisms
2. Thyroglobulin synthesis
3. Reactions resulting in the oxidation and organification of iodide
4. Microvilli formation and engulfment of colloid at the apical cell surface
5. Movement of lysosomes from the basal toward the apical surface of the follicle cell
6. Movement of phagolysosomes from apical to basal surfaces of cells
7. Activity of the deiodinase enzymes
8. Growth of the follicle cell

hormone. As a result, goiter is produced with a prolonged absence of dietary iodine.

 What determines free T_3 and T_4 concentrations in the blood?

Thyroid Hormone Transport and Metabolism

After secretion from the thyroid gland, T_3 and T_4 are carried in the blood reversibly bound to a number of plasma proteins. These carrier proteins serve to increase solubility of thyroid hormones and to buffer acute changes in secretion, as was discussed earlier in Chapter 12. Both T_4 and T_3 are associated primarily with **thyroxine-binding globulin (TBG)** but are also bound to a lesser extent to **albumin** (see Table 12–2). The relative affinity or strength with which the two hormones bind to these proteins differs greatly, however. T_4 has a much higher affinity and therefore binds more tightly to the plasma proteins than does T_3. As a result, T_3 is more rapidly degraded and removed from the plasma. Thus, T_4 not only is secreted from the thyroid gland in much greater amounts than is T_3 but also is removed from the blood much more slowly than is T_3. However, in the case of both T_4 and T_3, it is the free hormone that interacts with the thyroid hormone receptor to produce biological effects.

Target tissues for thyroid hormones have the capacity to convert T_4 into T_3. Although T_3 and T_4 are capable of producing exactly the same biological effects, their relative strengths or potencies differ considerably. Within a target cell, T_3 is much more potent than T_4. It is currently thought that most, if not all, of the biological effects of T_4 are due to its conversion to T_3 within target tissue cells.

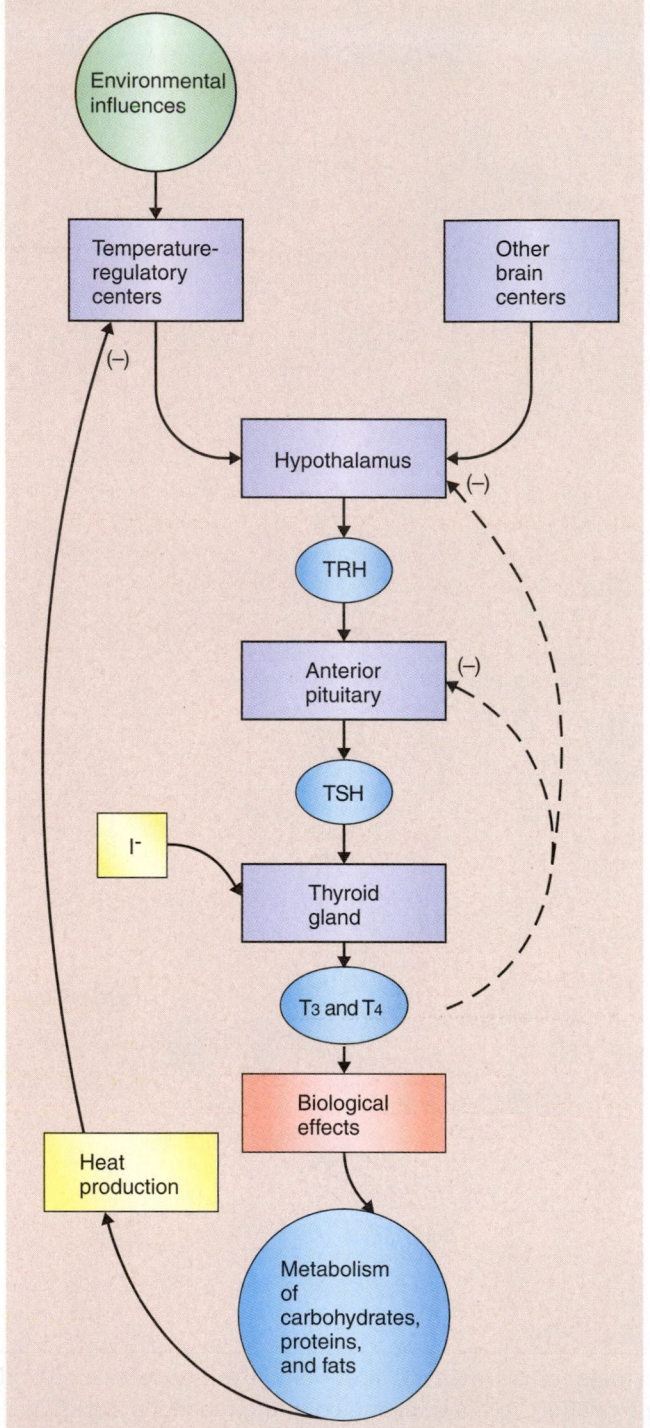

Figure 13–15

The primary steps involved in control of thyroid hormone production.

How do thyroid hormones affect tissue metabolism?

Effects of Thyroid Hormones on Metabolic Processes

Thyroid hormones exert numerous effects on metabolic processes in a wide variety of different tissues and cells; these are summarized in Table 13–6. The effects of thyroid hormones are so widespread that virtually no tissue or organ system escapes the adverse results of thyroid hormone deficiency or excess, which are especially evident during fetal development. In general, the effects of thyroid hormones on metabolic processes are slow in onset and long lasting compared with the effects of other hormones. Thus, thyroid hormones exert relatively long-term regulatory influences over metabolism.

One of the primary effects of thyroid hormones is the stimulation of **calorigenesis,** or heat production, in the body (see Table 13–6). This response occurs after a delay of several hours or days and is reflected by an increase in oxygen consumption. This particular effect is evident in most tissues, with brain, spleen, and testes being the most notable exceptions. Thyroid hormones also affect the cardiopulmonary system. In the heart, thyroid hormones increase the sensitivity of the heart to the sympathetic nervous system. As a result, the rate and strength of cardiac contractions increases and the cardiac output (the amount of blood being pumped) is in-

creased. The amount of oxygen carried by the blood is enhanced by thyroid hormones as a result of the number of red blood cells present in the blood.

Thyroid hormones influence several aspects of carbohydrate metabolism, although many of the effects that are produced depend on or are modified by other hormones. For example, normal concentrations of thyroid hormones increase glycogen formation in the liver (see Table 13–6), but this response occurs only when insulin is also present. Thyroid hormones also increase uptake of glucose from blood into adipose tissue and muscle and thereby act to potentiate the effect of insulin in this regard. In addition, many of the effects of epinephrine are enhanced by thyroid hormones as a result of enhancing responsiveness of the adenylate cyclase/cAMP system.

Thyroid hormones have a number of effects on lipid metabolism in both liver and adipose tissue (see Table 13–6). Thyroid hormones stimulate virtually all aspects of lipid metabolism, including the synthesis, mobilization, and oxidation of lipids. In general, however, oxidation of lipids is affected more than is their synthesis, such that in conditions of hormone excess the net effect is a decrease in size of most fat stores in the body.

Thyroid hormones play an important role in controlling protein metabolism in the body as well. When present in normal physiological concentrations, thyroid hormones increase protein synthesis and promote an overall accumulation of protein (see Table 13–6). However, when thyroid hormones are present in excess, they tend to cause a decrease in protein synthesis and an increase in protein breakdown. The result of these two effects is an overall loss of protein from the body. Effects of thyroid hormones on protein metabolism are therefore described as **biphasic.** Thyroid hormones have an additional effect on the pituitary, one that influences protein metabolism indirectly. In the anterior pituitary, thyroid hormones are known to stimulate somatotropes and increase GH secretion. As a result, owing to their influence on pituitary GH secretion, thyroid hormones also affect protein metabolism and growth.

In addition to specific effects on protein metabolism, thyroid hormones play a role in controlling the overall growth and development of several tissues. In humans, thyroid hormone is required for normal skeletal growth. This is due in part to stimulatory effects of thyroid hormones on GH production and in part to thyroid hormone stimulation of maturation of the epiphyseal growth centers in bone. In addition, as was noted in the earlier section describing somatomedins, thyroid hormones influence IGF-I production by the liver, which further contributes to promoting normal growth. Thyroid hormones are also required for normal development and maturation of teeth, hair follicles, and skin.

Development and maturation of the CNS are markedly affected by thyroid hormones. This effect of thyroid hormones is particularly evident in their absence. Branching of axons and dendrites that normally occurs during fetal and

---

TABLE 13–6

Summary of Effects of Thyroid Hormones

Stimulate calorigenesis in most cells

Increase cardiac output
    Increase rate of cardiac contractions
    Increase strength of cardiac contractions

Increase oxygenation of blood
    Increase rate of breathing
    Increase number of red blood cells in the circulation

Effects on carbohydrate metabolism
    Promote glycogen formation in liver
    Increase glucose uptake into adipose and muscle

Effects on lipid turnover
    Increase lipid synthesis
    Increase lipid mobilization
    Increase lipid oxidation

Effects on protein metabolism
    Stimulate protein synthesis
    Stimulate growth hormone secretion
    Promote bone growth
    Promote insulin-like growth factor I production by liver

Promote development and maturation of nervous system
    Promote neural branching
    Promote myelinization of nerves

early neonatal development occurs to only a limited degree in the absence of thyroid hormones. Normal myelinization of nerves is also impaired in the absence of thyroid hormones. As a result, severe mental retardation can occur if thyroid hormone is deficient during the period of fetal and early neonatal development. In adults, thyroid hormones influence mental alertness and responsiveness to external stimuli. The velocity of conduction of action potentials in peripheral nerves has also been shown to vary in response to an excess or deficiency of thyroid hormone secretion.

> *Based on the above discussion, how would too little (hyposecretion) or too much (hypersecretion) thyroid hormone be evident in a patient?*

Conditions of Abnormal Thyroid Hormone Secretion

Many of the specific effects of thyroid hormones that were discussed previously are readily evident in the condition of either thyroid hormone deficiency or thyroid hormone excess.

Hypothyroidism occurs when there is a deficiency in thyroid hormone production. In most instances, this occurs as a result of a defect in the thyroid gland itself. However, in some cases, the defect occurs in either the hypothalamus or pituitary, resulting in a deficiency in TSH production. Because of a slower metabolic rate and a reduced rate of heat production, hypothyroid individuals are intolerant of cold temperatures. Water tends to accumulate in the skin of these individuals, leading to a condition termed **myxedema,** in which the skin has a thickened, puffy appearance. This change is most evident in the facial features. The heart rate and the strength of cardiac contractions are reduced, contributing to an overall reduction in the cardiac output. There is also a general slowing of all intellectual functions, leading to a feeling of lethargy and some degree of speech impairment. The most common form of hypothyroidism is an autoimmune disorder known as **Hashimoto's disease.** For reasons that are not known, the body's immune system attacks and destroys the thyroid gland. The onset of this disease is typically between 30 and 40 years of age and shows a much higher incidence in women than in men.

Excess secretion of thyroid hormones results in **hyperthyroidism.** When tissues are presented with excessive quantities of thyroid hormones, a complex of biochemical and physiological events occurs. Hyperthyroid individuals exhibit such symptoms as a markedly increased heart rate, an intolerance of heat, and considerable weight loss owing to the lipid-mobilizing and protein-catabolic effects of excess thyroid hormones. In contrast to the lethargy that occurs in hypothyroidism, hyperthyroid individuals tend to be highly responsive to external stimuli. The most common form of hyperthyroidism is a condition known as **Graves' disease,** which occurs as a result of an interesting set of circumstances. In affected individuals, an abnormality in the immune system leads to the production of antibodies that recognize and bind to sites on the surface of thyroid follicle cells. When these antibodies bind to the follicle cell, the TSH receptor is activated, as though TSH itself were present. This results in a marked increase in thyroid hormone production, and a goiter forms as a result of the TSH-like effects of the antibodies on thyroid cell growth. The cause of the immune system abnormality and the mechanism by which the antibody activates the TSH receptor are not completely understood, however.

As indicated previously, a deficiency in thyroid hormone secretion during fetal development can have a dramatic impact, owing to the hormone's role in promoting normal development of the nervous system. A form of mental handicap known as **cretinism** can occur as a result of an untreated thyroid hormone deficiency. In addition to mental impairment, linear growth is also impaired in these individuals, so that dwarfism is also evident. If therapy with thyroid hormone is begun very soon after birth, severity of the impairment can be dramatically reduced. For this reason, most infants are checked very soon after delivery for the presence of adequate thyroid hormones in their blood.

CHAPTER REVIEW

Summary

- The pituitary is a small gland located just below the hypothalamus at the base of the brain and consists of two main lobes: the anterior lobe and the posterior lobe.
- The posterior pituitary is derived from neural tissue and contains the terminals of neuroendocrine cells that originate in the hypothalamus. Cells of the anterior pituitary are derived from non-neural tissue and are regulated by hypothalamic hormones carried from the hypothalamus to the anterior pituitary via the hypothalamic-pituitary portal blood vessels.

- The posterior pituitary secretes two hormones, ADH and oxytocin, which are synthesized in cell bodies of neuroendocrine cells in the hypothalamus and are then transported down their respective axons to the posterior pituitary, where they are released into capillaries when the hypothalamus receives an appropriate stimulus.
- Cells of the anterior pituitary synthesize and secrete at least six different hormones: LH, FSH, prolactin, ACTH, GH, and TSH. Secretion of each hormone is under the control of releasing/inhibiting hormones produced by the neuroendocrine cells in the

hypothalamus and released into blood vessels of the median eminence. The releasing/inhibiting hormones travel to the anterior lobe via the hypothalamic-pituitary portal system, where they act to regulate secretion of the six anterior pituitary hormones.

- The primary action of ADH is to increase water retention in the kidneys and, secondarily, to stimulate vasoconstriction. Oxytocin acts primarily on the breast and causes release of milk from the alveoli into the collecting ducts.
- LH and FSH, collectively referred to as the *gonadotropins,* regulate development of sperm and egg in males and females, respectively, and also regulate production of sex steroids by the gonads. GnRH produced in the hypothalamus regulates gonadotropin secretion.
- Prolactin stimulates milk production by the alveolar cells of the breast in females. Prolactin secretion increases in response to suckling stimuli.

- ACTH regulates the synthesis and secretion of cortisol from the adrenal cortex and also stimulates growth of the adrenal cortex. ACTH secretion is regulated by CRH.
- GH secretion from the anterior pituitary is regulated by GHRH and somatostatin. GH effects are generally anabolic, although specific effects can be either insulin-like or diabetogenic. GH affects metabolism in liver, muscle, and adipose tissue and stimulates production of somatomedins by the liver. Somatomedins stimulate both skeletal growth and tissue growth.
- Thyroid hormones promote calorigenesis in most cells of the body, with specific effects on carbohydrate, lipid, and protein metabolism. In addition, thyroid hormones have important influences on the maturation of the nervous system during fetal development.

Review Questions

Choose the Correct Answer

1. The region of the hypothalamus that contains the primary capillaries of the hypothalamic-pituitary portal system is the:
   a. optic chiasm.
   b. median eminence.
   c. supraoptic nucleus.
   d. arcuate nucleus.
   e. paraventricular nucleus.
2. The posterior pituitary secretes two hormones. They are:
   a. prolactin and ADH.
   b. prolactin and oxytocin.
   c. oxytocin and ADH.
   d. ACTH and oxytocin.
   e. ACTH and prolactin.
3. Which of the following are the most numerous hormone-producing cells of the anterior pituitary?
   a. Gonadotropes
   b. Corticotropes
   c. Lactotropes
   d. Somatotropes
   e. Thyrotropes
4. Which "hormone:site of synthesis" pair listed below is incorrect?
   a. Oxytocin : supraoptic nucleus
   b. Corticotropin-releasing hormone : hypothalamus
   c. Somatostatin : somatotropes
   d. FSH : gonadotropes
   e. T_4: follicle cells
5. Which of the following is not secreted by the anterior pituitary?
   a. Oxytocin
   b. GH
   c. ACTH
   d. FSH
   e. Prolactin
6. The net effect of ADH action on the kidney results in:
   a. an increase in urine volume.
   b. a decrease in blood pressure.

   c. increased fluid retention.
   d. decreased fluid retention.
   e. increased plasma osmolality.
7. Which of the following is not a stimulus for GH secretion?
   a. Deep sleep
   b. High blood glucose
   c. Infections
   d. Amino acids
   e. Psychological stress
8. GH promotes the accumulation of lean body mass. It generally exerts its metabolic effects on liver, muscle, and adipose tissue (fat) by:
   a. direct action on the tissue.
   b. insulin.
   c. IGF-I.
   d. IGF-II.
   e. T_4.
9. The insulin-like effects of GH include:
   a. increased protein synthesis in muscle.
   b. increased lipolysis in adipose tissue.
   c. increased gluconeogenesis in liver.
   d. increased glucose uptake in adipose tissue.
   e. all of the above.
10. Overproduction of GH can result in:
    a. cretinism.
    b. acromegaly.
    c. gigantism.
    d. dwarfism.
    e. either b or c.
11. Growth hormone (GH) promotes growth of the long bones via its effect to stimulate _____ production by the _____.
    a. hydroxyapatite, kidneys
    b. IGF-I, liver
    c. osteoid, parathyroid glands
    d. IGF-II, liver
    e. none of the above

12. The region of long bones that is responsible for growth of the bone is known as:
    a. osteoid zone.
    b. hypophyseal plate.
    c. zona fenestrae.
    d. epiphyseal plate.
    e. either a or d.
13. An abnormal enlargement of the thyroid gland is termed:
    a. myxedema.
    b. thyroid isthmus.
    c. goiter.
    d. colloid.
    e. Hashimoto's thyroiditis.
14. In the thyroid gland the "iodide trap":
    a. is responsible for the conversion of thyroglobulin into T_3 and T_4.
    b. removes iodide from MIT and DIT.
    c. is located on the apical surface of the follicle cell and is directly responsible for transporting iodide from the intracellular space into the colloid space.
    d. is located on the basal surface of the follicle cell and is responsible for transporting iodide from the extracellular space into the thyroid follicle cell.
    e. is responsible for the oxidation of iodide.
15. The primary hormone produced by the thyroid gland is:
    a. TSH.
    b. TBG.
    c. MIT.
    d. T_3.
    e. T_4.

16. One of the primary effects of thyroid hormones is:
    a. inhibition of GH secretion.
    b. slowing of the heart rate.
    c. inhibition of fatty acid oxidation.
    d. stimulation of calorigenesis.
    e. b and d.
17. Intolerance to cold, myxedema, and reduced cardiac output are typically seen in:
    a. hyperthyroidism.
    b. hypothyroidism.
    c. Grave's disease.
    d. acromegaly.
    e. none of the above.
18. The most common form of hyperthyroidism is a condition known as:
    a. Grave's disease.
    b. myxedema.
    c. acromegaly.
    d. goitrogen.
    e. cretinism.
19. A deficiency in thyroid hormones during fetal development and in the early neonatal period can result in what condition?
    a. Alzheimer's disease
    b. Cretinism
    c. Parkinson's disease
    d. Diabetes insipidus
    e. None of the above

Answers to Case History Questions

1. The standard treatment for hypothyroidism is straightforward. Thyroid hormone replacement is usually prescribed as pure synthetic thyroxine (T_4) in one dose (tablet) per day. Mary will have to take this for the rest of her life. It would be recommended to Mary that she return to her doctor once a year to have her T_4 and TSH levels measured. Based on these results, a change in dosage could be made.
2. It would be likely that other members of Mary's family are likely to suffer from the same problem because Hashimoto's disease is an inherited autoimmune disorder. Furthermore, the chances of other autoimmune disorders, such as lupus or type I diabetes, would also be increased.
3. If Mary's TSH level had been low, it would have indicated that the cause of her hypothyroidism rested with the pituitary. Normally, TSH stimulates the thyroid gland to produce and secrete thyroid hormones, so a deficiency in TSH secretion could lead to hypothyroidism.

Key Terms

acromegaly (p. 421)
adrenocorticotropic hormone (ACTH) (p. 417)
anabolic hormone (p. 419)
anterior pituitary gland (p. 410)
antidiuretic hormone (ADH) (p. 414)
calorigenesis (p. 427)
catabolic hormone (p. 419)
chondrocytes (p. 421)
cretinism (p. 428)
diabetes insipidus (p. 415)

diiodotyrosine (DIT) (p. 423)
dwarfism (p. 421)
follicle-stimulating hormone (FSH) (p. 416)
gigantism (p. 421)
goiter (p. 422)
growth hormone (GH) (p. 417)
human placental lactogen (hPL) (p. 417)
hypothalamic-pituitary portal system (p. 411)

insulin-like growth factor I (IGF-I) (p. 420)
insulin-like growth factor II (IGF-II) (p. 420)
luteinizing hormone (LH) (p. 416)
median eminence (p. 411)
monoiodotyrosine (MIT) (p. 423)
myxedema (p. 428)
osmoreceptors (p. 415)
oxytocin (p. 414)

paraventricular nucleus (p. 411)
posterior pituitary gland (p. 410)
prolactin (p. 416)
Rathke's pouch (p. 411)
somatomedin (p. 419)
supraoptic nucleus (p. 411)
tetraiodothyronine (T_4) (p. 423)
thyroglobulin (p. 422)
thyroid hormone (p. 422)
thyroid-stimulating hormone (TSH) (p. 422)
triiodothyronine (T_3) (p. 423)

Suggested Readings

Berne, R. M., and Levy, M. N. *Principles of Physiology,* ed 3. St. Louis, Mosby, 2000.

Ganong, W. F. *Review of Medical Physiology,* ed 18. Los Altos, Lange Medical Publishing, 1997.

Goodman, H. M. *Basic Medical Endocrinology,* ed 2. New York, Raven Press, 1994.

Griffin, J. E., and Ojeda, S. R. *Textbook of Endocrine Physiology,* ed 4. New York, Oxford University Press, 2000.

Guyton, A. C., and Hall, J. E. *Human Physiology and Mechanisms of Disease,* ed 6. Philadelphia, W.B. Saunders, 1997.

Koibuchi, N., and Chin, W. W. "Thyroid hormone action and brain development." *Trends in Endocrinology and Metabolism,* 11:123–128, 2000.

Norman, A. W., and Litwack, G. *Hormones,* ed 2. San Diego, Academic Press, 1997.

Stewart, P. M. "Current therapy for acromegaly." *Trends in Endocrinology and Metabolism,* 11:128–132, 2000.

Web sites

http://www.tsh.org
"A Patients Guide to Thyroid Disease" produced by the Thyroid Foundation of America. Contains information for the lay person with thyroid disease.

http://members.xoom.com/endocrine/faqframe.htm
"Common Endocrine Diseases." Contains useful information concerning thyroid diseases, adrenal diseases, and pituitary diseases.

http://www.pituitary.org
"Pituitary Network Association." Oriented toward patients with various forms of pituitary disease. Well-written sections on the function and regulation of the pituitary.

http://www.wfubmc.edu/surg-sci/ns/pituitary.html
Produced by the Wake Forest University School of Medicine. This clinically-oriented site contains additional links to other sites.

http://www.thyroid.org/patient/patient.htm

Answers to Review Questions

1. b	**2.** c	**3.** d	**4.** c	**5.** a	**6.** c	**7.** b	**8.** a
9. a	**10.** e	**11.** b	**12.** d	**13.** c	**14.** d	**15.** e	
16. d	**17.** b	**18.** a	**19.** b				

Chapter 14

THE ADRENAL GLANDS

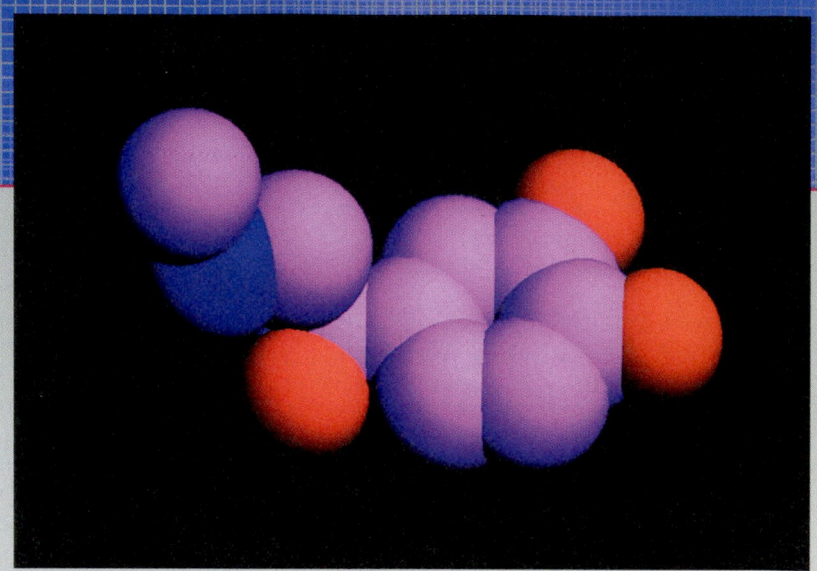

- *A computer-generated model of epinephrine (adrenaline).*
  (Pur=carbon, Blu=nitrogen, Red=oxygen)
  (© Ken Edward/Science Source/Photo Research)

KEY CONCEPTS

- The adrenal glands consist of two distinct endocrine glands, each under separate control and each producing different hormone products.

- Cells of the adrenal cortex produce steroid hormones, predominantly cortisol and aldosterone. Cortisol affects metabolism and generally results in the release of stored fuels. Aldosterone acts on the kidneys and influences sodium and potassium balance in the body.

- The adrenal medulla functions as part of the sympathetic nervous system and secretes primarily epinephrine. Epinephrine affects the cardiovascular and pulmonary systems and a variety of metabolic processes in the body.

- The net effect of the adrenal hormones cortisol and epinephrine is to increase the body's ability to effectively respond to stress.

CASE HISTORY

Helen is a 43-year-old woman who recently sought medical help because of an abrupt onset of lower back pain. A radiological exam showed a compression fracture of the second lumbar vertebra as well as some evidence of osteoporosis of the spine. During the history phase of her exam, Helen reported increasing muscle weakness and a tendency to bruise easily over the past several years. She also complained of increasing emotional lability, with rapid mood swings from euphoria to depression. She reported having more difficulty sleeping than in the past and had gained approximately 15 pounds over the past three years. Her menstrual periods, which had previously been regular, were now occurring every two or three months.

A physical exam showed an obese woman with excess adipose tissue in the trunk and face and above the clavicles. Her extremities were thin and appeared to be somewhat atrophied. There was excess hair growth on the upper lip and chin, her skin was thin, and there were multiple bruises that she could not account for by any known trauma. Her blood pressure was 172/110 mm Hg (normal, approximately 120/80) and pulse was 82 beats/min (normal, approximately 70). Results of laboratory studies showed a fasting plasma glucose level of 190 mg/dL (normal, < 124), and plasma electrolytes showed slightly elevated sodium and bicarbonate levels and a slightly decreased potassium.

Questions

1. What is the most likely cause of Helen's clinical picture?

2. How would you establish the presence of the suspected abnormal endocrine secretion?

3. What is the possible sequence of events that led to the suspected endocrine abnormality and what additional information would be helpful in discriminating between these possibilities?

INTRODUCTION

This chapter discusses the physiology of the adrenal glands and the adrenal hormones. Humans have two adrenal glands, each one composed of two distinct endocrine tissues that differ in cell type, hormone products, and control. The two endocrine components of the adrenal gland are the outer portion, or cortex, which constitutes about 80% of the gland, and the inner portion, or medulla, which forms the remaining 20%. The cortex produces and secretes several different steroid hormones, while the medulla produces and secretes catecholamines. The first section of this chapter focuses on the adrenal cortex, and a discussion of the adrenal medulla follows in the latter sections.

GROSS ANATOMY OF THE ADRENAL GLANDS

The shape of each adrenal gland roughly resembles that of a pyramid, and each of the two adrenals in humans rests like a cap just above the upper end of each kidney (Fig. 14–1).

There is no direct physical connection between the adrenals and kidneys, however. The arterial blood supply entering each adrenal gland is also separate from that entering the kidneys. In a normal adult, each adrenal gland weighs about 3 to 4 g and is about 5 cm (2 in) across at its widest point.

THE ADRENAL CORTEX

Cells of the Adrenal Cortex Show Three Specific Zones, and Each Zone Produces a Specific Class of Steroid Hormone

> *How is the function of the adrenal cortex related to its anatomy?*

Based on the arrangement of cells as seen under a microscope, the adrenal cortex can be divided into three different zones. Each zone secretes a different class of steroid hormone. The outermost region of the adrenal cortex is termed

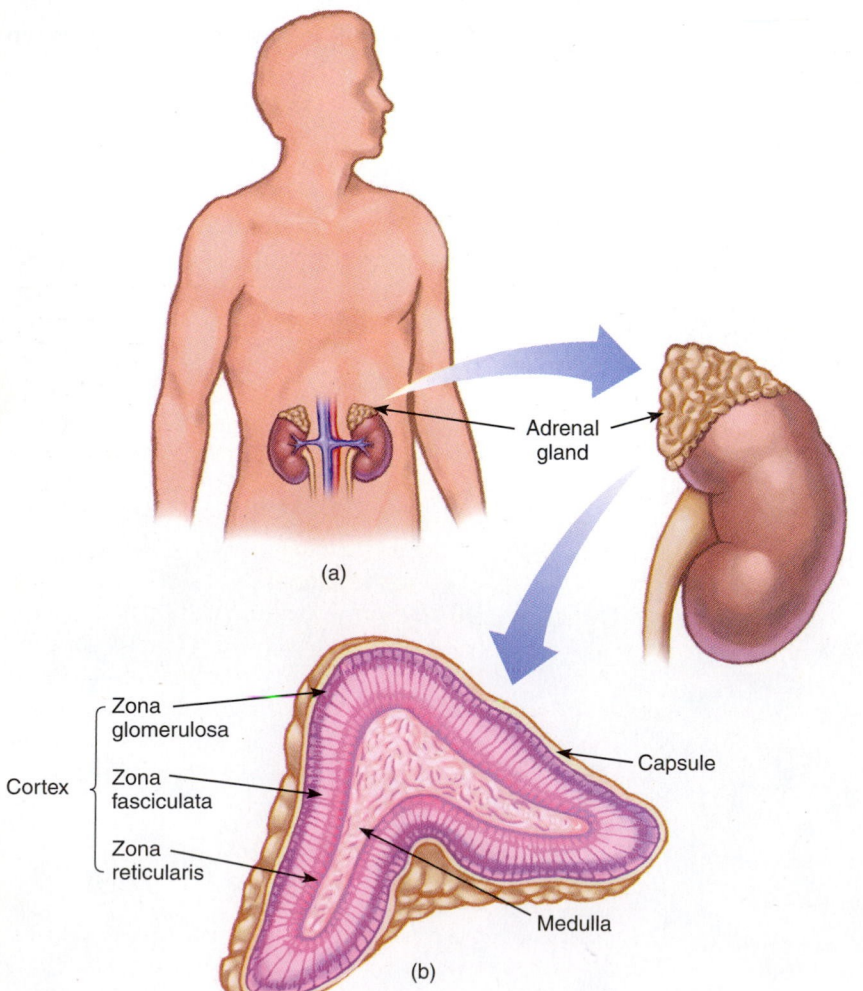

Cortex
- Zona glomerulosa
- Zona fasciculata
- Zona reticularis

Capsule

Adrenal gland

Medulla

(a)

(b)

Figure 14–1

(a) Location of the adrenal glands relative to the kidneys. *(b)* The adrenal gland shown in cross-section, with both the cortex and medulla visible.

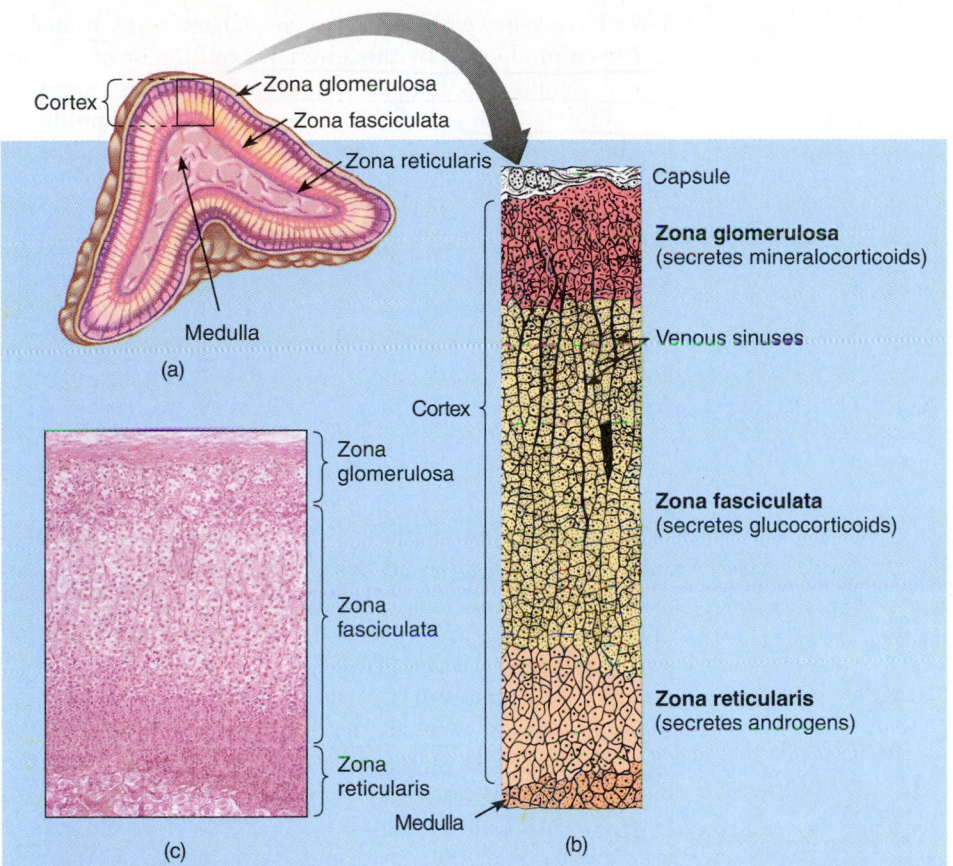

the **zona glomerulosa** (Fig. 14–2) and consists of cells arranged in "whorls" (**glomeruli**). Cells of the zona glomerulosa produce **mineralocorticoids.** The middle region of the cortex is widest and is termed the **zona fasciculata** because of the arrangement of cells in "**fascicles**," or cords, separated by venous sinuses (small channels for venous blood; see Fig. 14–2). Cells of the zona fasciculata produce **glucocorticoids.** The innermost region, which borders the medulla, is known as the **zona reticularis,** so named because cells are arranged in the form of a network, or **reticulum.** This zone of the cortex produces and secretes primarily **androgens.**

Numerous lipid droplets are present within cells of the adrenal cortex, especially in cells of the zona fasciculata. These lipid droplets represent stored cholesterol to be used to synthesize adrenal steroid hormones, as will be discussed later.

Activity of the Adrenal Cortex Is Under the Control of Adrenocorticotropic Hormone Produced by the Pituitary

Activity of the adrenal cortex is controlled primarily by **adrenocorticotropic hormone (ACTH),** which is produced by corticotropes of the anterior pituitary. As was described in Chapter 13, ACTH secretion is in turn controlled by **corticotropin-releasing hormone (CRH),** which is produced by neuroendocrine cells of the hypothalamus and carried to the anterior pituitary by the hypothalamic-pituitary portal vessels (refer to Figs. 13–3 and 13–4). This relationship between CRH, ACTH, and the adrenal cortex is illustrated in Figure 14–3 and is known as the **hypothalamic-pituitary-adrenal axis.**

ACTH has both **tropic** effects and **steroidogenic** effects on cells of the adrenal cortex. The tropic effects of ACTH include promoting the general growth and function of all three zones of the adrenal cortex. Steroidogenic effects of ACTH involve control of glucocorticoid secretion by cells of the zona fasciculata. In the absence of ACTH, adrenal cortex cells become smaller, the production of mineralocorticoids is reduced only by about 50%, while that of cortisol is dramatically reduced. As will be described later in this chapter, secretion of mineralocorticoids from the zona glomerulosa is primarily controlled by factors other than ACTH.

Three Steroids Are the Primary Products of the Adrenal Cortex

In all, some 30 to 50 different steroids have been isolated from the adrenal cortex. However, only three are produced and secreted in quantitatively significant amounts. These are

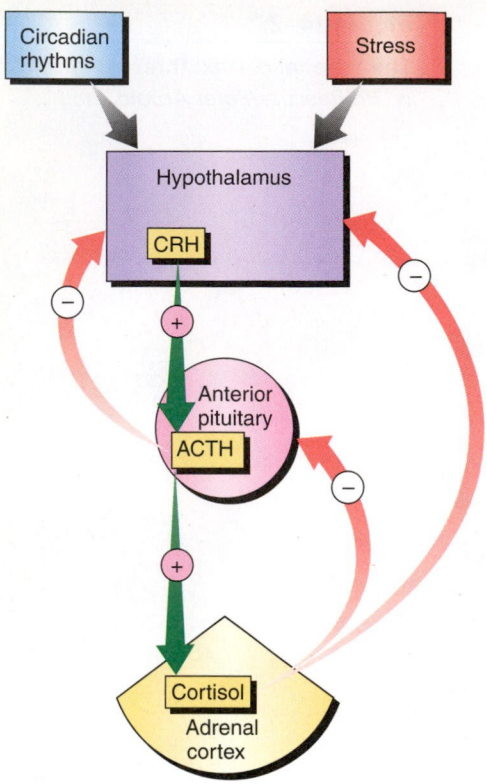

Figure 14–3

The hypothalamic-pituitary adrenal axis. Stimulatory effects are shown by the positive *(green)* arrows ⊕; inhibitory effects, by the negative *(red)* arrows ⊖.

a much more powerful androgen (see Chapter 31). In males, androgen production by the adrenal is of little or no physiological significance because far greater amounts are produced by the testes. In females, however, DHEA produced by the adrenal gland accounts for most of the androgens found circulating in the blood.

The general pathways involved in steroid hormone biosynthesis were previously described in Chapter 12 (see Fig. 12–8). For reference, a simplified diagram of the pathways is also presented here (Fig. 14–5). Recall that steroid hormones are synthesized and secreted on demand and are not stored. Therefore, control of steroid hormone secretion occurs at the level of hormone synthesis. As was described in Chapter 12, the first step in the synthesis of all steroid hormones begins with enzymatic conversion of cholesterol to pregnenolone. In most instances, this step in steroid synthesis is slowest, or rate-limiting. Once pregnenolone is formed, subsequent steps proceed fairly rapidly. Therefore, the enzyme that catalyzes conversion of cholesterol into pregnenolone represents the primary site of control of steroid synthesis (see step A, Fig. 14–6).

Increased supply of cholesterol inside cells also enhances steroid synthesis. Most of the cholesterol that cells of the adrenal cortex use to make steroids is taken up from the blood, where it is carried in the form of low-density lipoproteins (LDLs). Cells of the adrenal cortex can remove these lipoproteins from the blood and bring them into the cell, making cholesterol available for steroid synthesis (see step B, Fig. 14–6). These cells store any cholesterol not immediately used for steroid synthesis as lipid droplets. Later, stored cholesterol can be rapidly mobilized and used for steroid synthesis when needed (see step C, Fig. 14–6).

Cells within each of the three zones of the adrenal cortex contain enzymes specifically suited for producing the steroid hormone characteristic of that zone. For example, only cells of the zona glomerulosa contain enzymes required for conversion of corticosterone to aldosterone, and thus only these cells can produce aldosterone (see Fig. 14–5). These cells lack the enzyme required for the conversion of pregnenolone

aldosterone, cortisol, and **dehydroepiandrosterone (DHEA),** which are shown in Figure 14–4. Of the three, cortisol, a glucocorticoid, is produced in the greatest amount; aldosterone, a mineralocorticoid, is produced in the least amount. This chapter focuses on the physiology of these two steroids. The third, DHEA, is itself a rather weak androgen, but it can be converted in peripheral tissues to testosterone,

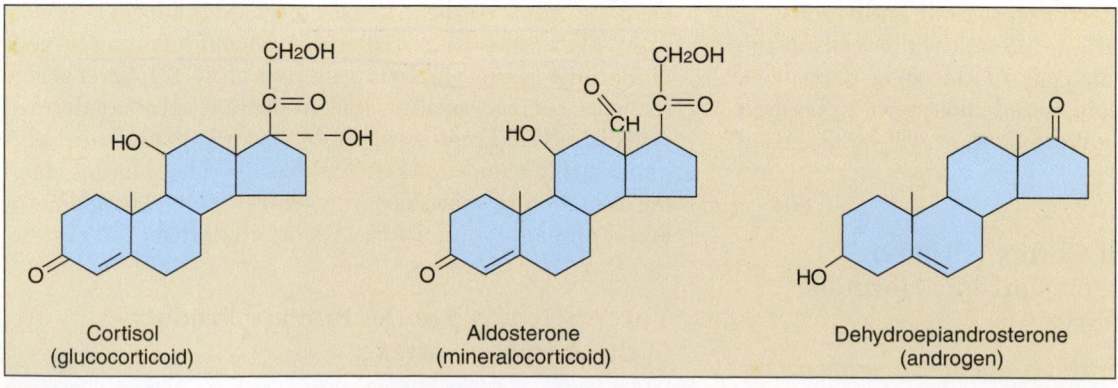

Figure 14–4

Chemical structures of the principal steroids of the adrenal cortex.

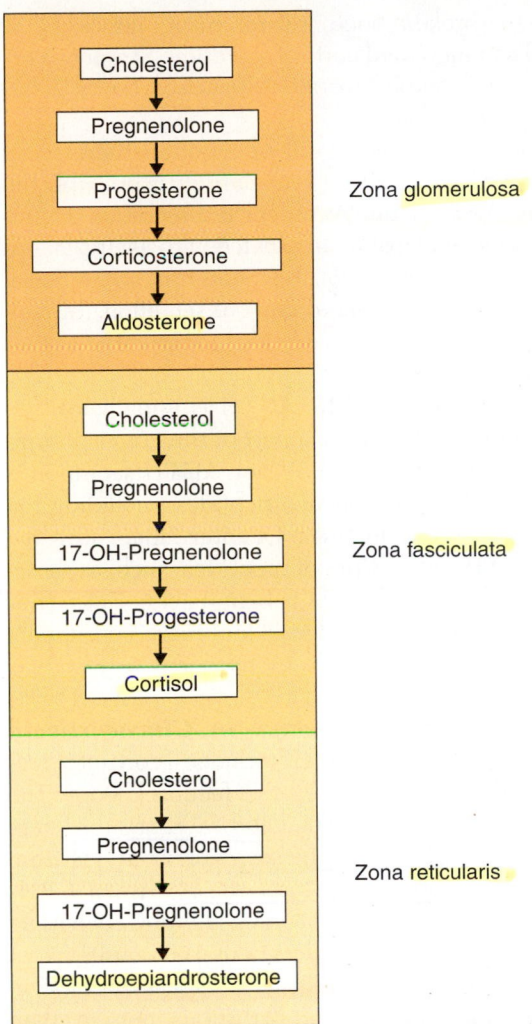

Figure 14–5

Pathways of steroid synthesis in the adrenal cortex.

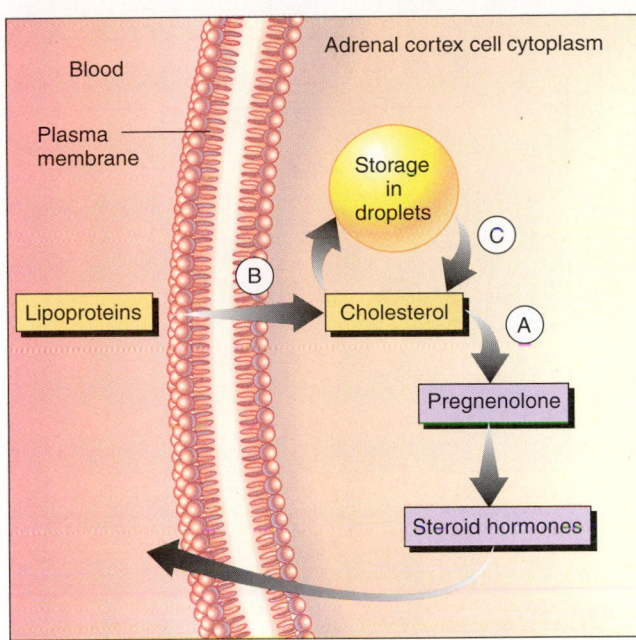

Figure 14–6

The regulatory steps in biosynthesis of adrenal cortex steroids. Step A, the conversion of cholesterol to pregnenolone, is the primary site of regulation of steroid synthesis. Steps B and C, the uptake of cholesterol from the bloodstream and its removal from intracellular storage sites, respectively, are secondary sites of regulation.

puts from a variety of higher brain centers. Therefore, much of the control of glucocorticoid secretion is due to these inputs to the hypothalamus. In general, glucocorticoid secretion from the adrenal glands is responsive to two types of stimuli. The first type involves a pattern of secretion that varies over the 24-hour period of a day; the second category involves increased secretion in response to a variety of specific stimuli.

The Circadian Rhythm of Glucocorticoid Secretion. Figure 14–7 shows typical changes in the concentration of plasma cortisol that might occur over a 24-hour period in a normal individual. As indicated, plasma cortisol concentration is highest around the time of awakening in the morning and lowest around midnight. Because this general pattern of secretion is repeated approximately every 24 hours, it is termed a **circadian rhythm.** These changes in cortisol secretion are due to similar circadian variations in CRH and ACTH secretion. This rhythm in cortisol secretion has its origin in higher brain centers that in turn influence the hypothalamic-pituitary system. These circadian rhythms are related primarily to the sleep-wake patterns of the individual rather than by environmental light-dark cycles. Changes in a person's sleep-wake cycle, such as working nights and sleeping days, result in a temporal shift in the daily rhythm of cortisol secretion.

to 17-OH-pregnenolone and so are unable to synthesize either cortisol or DHEA. Similarly, cells of the zona fasciculata contain the enzymes that favor the production of cortisol, and cells of the zona reticularis exhibit high levels of the enzymes favoring DHEA synthesis.

Glucocorticoids Play an Important Role in Metabolic Regulation

> *What determines when cortisol will be released by the adrenal cortex?*

Control of Glucocorticoid Secretion

As illustrated in Figure 14–3, secretion of glucocorticoids from cells of the adrenal cortex is controlled by a sequence of steps beginning with CRH secretion by the hypothalamus. CRH release by the hypothalamus is controlled by in-

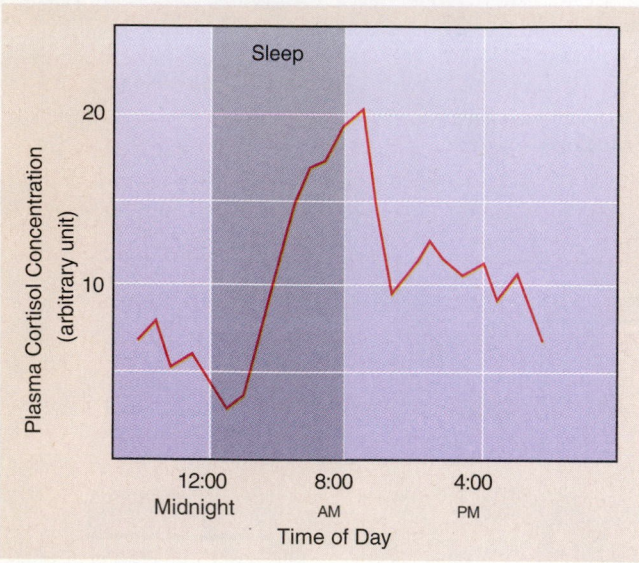

Figure 14–7

Graph of the daily rhythm of the plasma cortisol concentration.

Stress-Induced Glucocorticoid Secretion. A variety of different stress stimuli can also significantly increase cortisol synthesis and secretion (see Fig. 14–3). Increased secretion in response to stress occurs in addition to the normal circadian rhythm of secretion described previously. Table 14–1 lists several different types of stress known to cause increased glucocorticoid secretion. In general, these can be either physical or psychological in nature. Hypoglycemia (low blood glucose) is an important stimulator of cortisol secretion and provides for important adaptive changes that occur with fasting. Practically any type of physical trauma or in-

TABLE 14–1
Types of Stress Known to Increase Cortisol Secretion
Physical stress
Hypoglycemia
Trauma
Broken bones
Burns
Surgery
Cold exposure
Infection
Heavy exercise
Psychological stress
Acute anxiety
Anticipation of stressful situations; surgery, college exams, or an airplane flight, for example
Novel situations
Chronic anxiety

jury, such as a broken bone, severe burn, infection, or surgery, results in increased cortisol secretion. Strenuous exercise, such as in competitive athletics, also results in increased cortisol secretion.

Several types of psychological or emotional stress can also stimulate cortisol secretion. Acute anxiety is probably the strongest psychological stimulus for cortisol secretion. Examples might include anticipation associated with a surgical operation or final exams in school. Encountering a new social situation often triggers increased cortisol secretion. In contrast, chronic anxiety is considerably less important as a stimulus for cortisol secretion.

As described in Chapter 13, ACTH is initially synthesized in the anterior pituitary as a part of the larger proopiomelanocortin (POMC) peptide. When ACTH secretion is stimulated, several other bioactive peptides, including the morphinelike beta-endorphin (see Chapter 7), are secreted along with ACTH. Secretion of peptides such as beta-endorphin during times of stress is beneficial because they function as an endogenous analgesic and increase the pain threshold.

Negative Feedback Relationships in Glucocorticoid Secretion. As is true of most other hormones, cortisol secretion is controlled by a series of negative feedback loops. The red arrows in Figure 14–3 show that cortisol exerts dual negative feedback effects to inhibit the secretion of ACTH from the pituitary. First, cortisol decreases the activity of CRH-producing neurons in the hypothalamus, thereby decreasing CRH release. Second, cortisol decreases the sensitivity of corticotropes to CRH. As a result, CRH is less effective in stimulating ACTH release when cortisol is present than when it is absent. Thus, the combination of a decrease in CRH secretion as well as a decrease in sensitivity to CRH contributes to reduced ACTH secretion, leading ultimately to decreased cortisol secretion. In addition, as is indicated in Figure 14–3, ACTH itself exerts a negative feedback effect on the hypothalamus to reduce CRH secretion.

Transport of Cortisol in the Blood. Normally, about 80% of blood cortisol is bound to a specific carrier protein known as **corticosteroid-binding globulin (CBG).** An additional 15% of the cortisol is bound to albumin; the remaining 5% exists free in solution. As with other hormones, only free cortisol molecules are able to interact with receptors and produce physiological effects characteristic of the hormone. Recall from Chapter 12 that a dynamic equilibrium for the steroid hormones exists between free and bound hormone.

Physiological Effects of Glucocorticoids

> *How does cortisol help maintain blood glucose levels?*

Glucocorticoids play an important role in our day-to-day survival. Actions of glucocorticoids enable us to survive the potentially damaging effects of wide fluctuations in environmental

TABLE 14–2
Summary of Effects of Cortisol on Metabolism

Liver
    Increases gluconeogenesis
    Increases glycogen synthesis
Skeletal muscle
    Decreases protein synthesis
    Increases protein degradation
    Decreases glucose uptake
Adipose tissue
    Decreases glucose uptake
    Increases lipid mobilization

in cell function occur. Given these mechanisms, what are the specific changes in cell function that are produced by cortisol?

Cortisol exerts its primary effects on three different tissues in the body: liver, skeletal muscle, and adipose tissue. Specific physiological effects of the hormone in these three tissues are summarized in Table 14–2. The overall effect of cortisol under most conditions can be summarized as being **catabolic** owing to its promotion of protein breakdown in tissues.

Cortisol in the Liver: Increased Blood Glucose. In the liver, the net response to cortisol is increased **gluconeogenesis** (see Table 14–2), primarily as a result of increased conversion of amino acids into glucose (for a discussion of gluconeogenesis, refer to Chapter 6). Increased glucose formation results from specific effects of the hormone (Fig. 14–8). First, cortisol increases the activity of several enzymes that catalyze key steps in the gluconeogenic pathway. As a result, glucose formation from a variety of substrates is increased. Second, cortisol increases the activity of several enzymes involved in amino acid metabolism, facilitating the use of amino acids as substrates for gluconeogenesis. Third, glucocorticoids promote gluconeogenesis in the liver by stimulating the activity of enzymes of the **urea cycle**. This effect increases the ability of the liver to dispose of ammonia liberated during metabolism of amino acids. Together, these specific effects of cortisol permit increased conversion of amino acids into glucose (see Fig. 14–8). As a result, glucocorticoids tend to increase the blood glucose concentration. Glucocorticoids also tend to promote the formation of liver glycogen. This is due in part to increased glucose formation in the liver via gluconeogenesis and in part to a stimulation of enzymes involved in glycogen formation.

conditions, as well as a number of noxious or potentially noxious factors. As a result, glucocorticoids are considered essential for life. Even a modest period of fasting would not be possible without the actions of cortisol. Before we look at specific effects of glucocorticoids, let us briefly review the general mechanism of steroid action.

Glucocorticoids produce their effects via the general mechanism for steroid hormone action described in Chapter 12 (see Fig. 12–17). Recall that steroid hormones are lipid-soluble; they readily pass through plasma membranes and bind to receptors located in the nucleus. Hormone-receptor complexes then bind to specific sites on DNA, altering the transcription of mRNA molecules that code for specific proteins. As a result, the synthesis of a few specific proteins is appropriately increased or decreased, and the desired changes

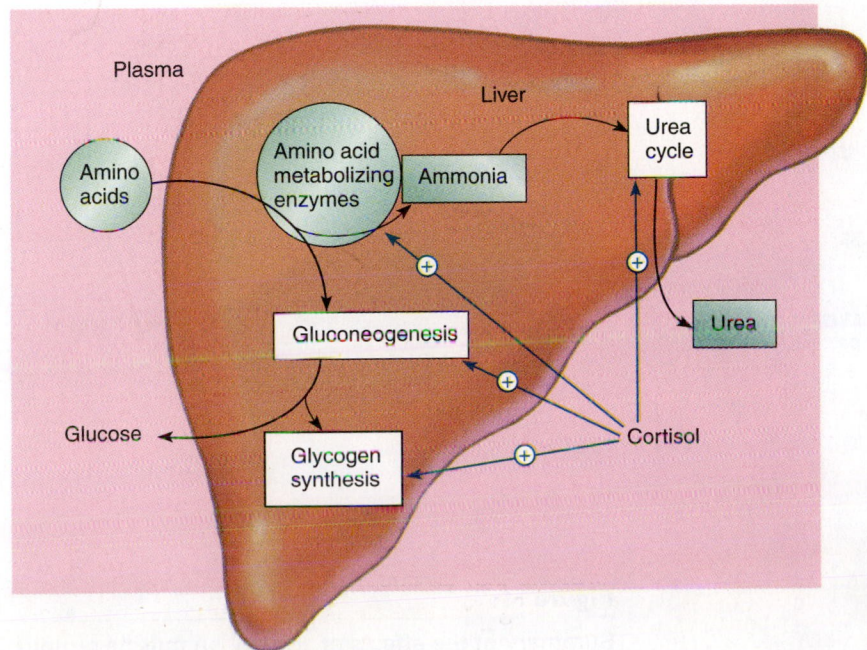

Figure 14–8

Summary of the effects of cortisol on liver metabolism. The net effect of cortisol is to stimulate the formation of glucose and glycogen at the expense of amino acids.

Cortisol in Skeletal Muscle: Decreased Protein. In skeletal muscle, cortisol has a catabolic action and tends to promote the net loss of protein. This occurs as a result of two different effects (Table 14–2 and Fig. 14–9). First, cortisol inhibits muscle protein synthesis. As a result, uptake of amino acids from blood and their subsequent incorporation into muscle protein are reduced. Second, cortisol increases degradation of existing muscle protein. Amino acids released during the degradation of muscle protein are then free to enter the blood. As a result of these two effects of cortisol, there is a net transfer of amino acids from muscle protein into the blood. Liver can then use these amino acids for the synthesis of glucose via gluconeogenesis. The combined effects of cortisol on liver and muscle help maintain blood glucose at the expense of muscle protein. These catabolic effects of cortisol on muscle occur in skeletal muscle only, not in cardiac muscle. Furthermore, the catabolic actions of cortisol and the anabolic actions of other steroids should not be confused. Anabolic steroids used to increase muscle mass consist primarily of androgens and not of glucocorticoids. Cortisol also decreases glucose uptake by skeletal muscle, an effect that opposes that of insulin and helps to increase blood glucose levels.

Cortisol in Adipose Tissue: Decreased Glucose Uptake. In adipose tissue, cortisol decreases glucose uptake (see Table 14–2). As in muscle, this effect is opposite to that of insulin and thus represents an anti-insulin effect of the steroid. Glucocorticoids also promote mobilization of lipid from some adipose tissue stores in the body. The effect of cortisol to promote lipid mobilization in adipose tissue is similar in concept to its effects in skeletal muscle; it promotes the mobilization of body fuel stores to help maintain blood glucose

levels. Not all adipose tissue is affected equally, however. This differential sensitivity of adipose tissue to glucocorticoids can best be seen when the glucocorticoids are secreted in excess. When this occurs, fat stores in the legs and arms decrease, and fat is redistributed to the trunk and shoulder-blade region. This leads to a very distinct physical appearance, which will be discussed in a later section describing the effects of cortisol hypersecretion.

 What other effects does cortisol have on target tissues?

The Permissive Actions of Glucocorticoids. In a number of instances, although cortisol itself does not directly control a particular metabolic process or initiate events inside the cell, it must be present for other hormones to produce their full effect. Cortisol therefore tends to have an amplifying effect on the action of other hormones. These are known as the permissive actions of glucocorticoids. One example of a permissive action of glucocorticoid occurs in adipose tissue, where stimulation of lipid breakdown by epinephrine is enhanced when cortisol is also present. Although numerous other permissive actions of glucocorticoids are known, the exact nature of the mechanisms for these effects is not well understood.

Effects of Glucocorticoids on Blood Vessels and Blood Cells. In addition to its effects on carbohydrate, lipid, and protein metabolism, cortisol has other important effects in the body. One of the more important effects of cortisol is to enhance responsiveness of blood vessels, or **vascular reactivity**. In the absence of cortisol, the diameter of arterioles (very small arteries) is not appropriately controlled during

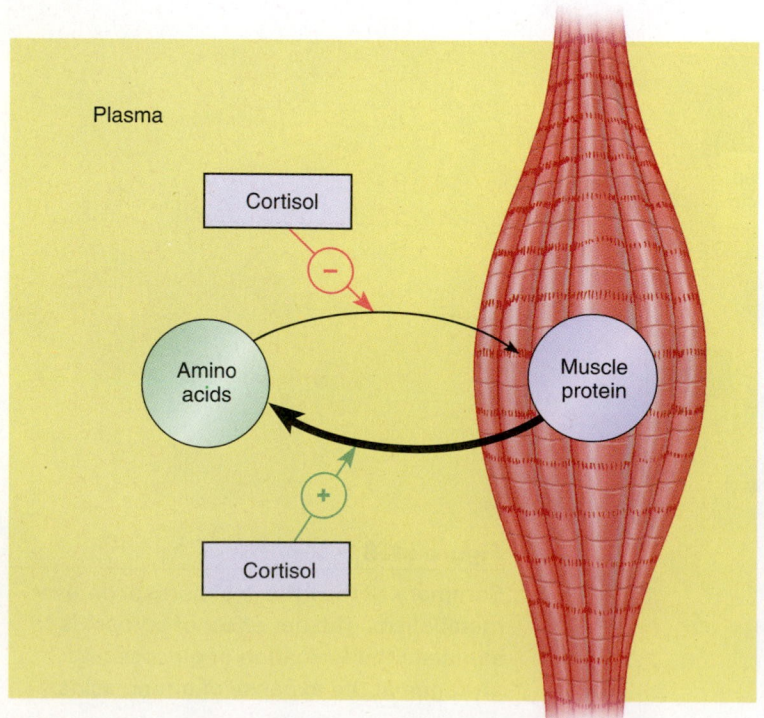

Figure 14–9

Summary of the effects of cortisol on muscle protein metabolism.

APPLICATIONS OF PHYSIOLOGY

Designer Steroids

Steroid hormones are widely used in medicine today, not only for purposes of replacement therapy when the body's own hormone production is lacking but also for anti-inflammatory and immunosuppressive effects that certain steroids produce. Because of similarities in structure, most naturally occurring steroids have some activity that overlaps with steroids of other classes. For example, cortisol has primarily glucocorticoidlike activity but can produce mineralocorticoidlike and androgenlike effects at high doses. Similarly, aldosterone has primarily mineralocorticoid activity but possesses a weak amount of glucocorticoid activity.

When treating patients with steroids or any other drug, it is beneficial if extraneous or undesirable activities are minimized. Recognizing this, the pharmaceutical industry has produced hundreds of synthetic steroids, each having its own spectrum of hormonal activity. For example, dexamethasone is a synthetic steroid widely used in medicine. On a molecule-per-molecule basis, it has much more glucocorticoidlike activity than does the natural steroid cortisol, but unlike cortisol, dexamethasone has essentially no mineralocorticoid activity. Prednisone is another steroid that is commonly prescribed. It has largely glucocorticoid activity but also exhibits a significant degree of mineralocorticoid action, making it useful in cases in which the adrenal glands have been removed or their func-

tion is impaired and the physician wants to provide some degree of replacement of both hormones. Furthermore, some steroids have been modified in such a way as to slow their rate of removal from the bloodstream and thus prolong the time course over which effects are produced. Others have been modified in such a way as to limit their solubility. This is particularly useful for steroids having anti-inflammatory activity because they can be formulated in a cream or ointment and be used topically to treat problems such as skin rashes, poison ivy, or small superficial wounds. Because of their low solubility, they are minimally absorbed and so do not produce whole-body effects but act primarily in the area to which they are applied.

Another form of steroid that has been modified to enhance certain biological effects are anabolic steroids. While glucocorticoids of the adrenal cortex generally tend to be catabolic in nature, anabolic steroids possess predominantly androgenlike activities and thus promote acquisition of muscle mass. Although they are occasionally used clinically, for the most part they are used (illegally) by athletes wishing to build up their muscles and therefore their strength. Although they obviously produce the desired short-term benefits, they also produce damage to other organs — particularly the liver — that accumulates over time. It is not uncommon to hear on the news of a prominent athlete who has died at a relatively early age as a result of the use of anabolic steroids. This is clearly a situation in which the use of designer steroids is not advantageous and is in fact highly detrimental to the well-being of the individual.

---

times of stress, and consequently blood pressure can decrease dramatically and may even result in death. Cortisol treatment restores responsiveness of the blood vessels and therefore is said to restore vascular reactivity.

Glucocorticoids also affect various blood cell types and lymphoid tissues. Cortisol decreases the number of eosinophils and basophils and increases the number of neutrophils, red blood cells, and platelets in the blood (see Chapter 17 for a further discussion of blood cell types). As will be discussed, high doses of glucocorticoids also influence inflammation and the immune response.

Pharmacological Effects of Glucocorticoids

 How do high levels of cortisol affect target tissues?

Effects of cortisol described in the foregoing paragraphs are those generally observed in response to physiological amounts of the hormone. At much higher concentrations, these effects are magnified, and other effects are produced as well, some of which have proven very useful for medical purposes. These are known collectively as **pharmacological** effects of glucocorticoids and other closely related steroids.

Perhaps the two most important pharmacological effects of glucocorticoids are their **anti-inflammatory** effects and their **immunosuppressive** effects. In large doses, glucocorticoids profoundly inhibit the inflammatory response to tissue injury or infection by inhibiting virtually every step in a normal inflammatory response, including vasodilation, increased capillary permeability, and increased phagocytosis. High doses of glucocorticoids also suppress the immune system, resulting in decreased antibody production. The number of circulating lymphocytes decreases owing to increased lymphocyte destruction in lymphoid tissues; the size of lymph nodes also decreases. As a result of these pharmacological effects, glucocorticoids are particularly useful in medicine for reducing the inflammatory response to injury or infection, in the treatment of asthma, or for suppressing the rejection response to transplanted organs. Although the immunosuppression produced by glucocorticoids is of great value following organ transplantation, it does leave the individual more susceptible to infections.

Mineralocorticoids Play an Important Role in Regulating Mineral Balance

Control of Mineralocorticoid Secretion

> *How is aldosterone secretion controlled?*

Aldosterone is the principal mineralocorticoid produced in the zona glomerulosa of the adrenal cortex. The two primary determinants of aldosterone secretion are a polypeptide hormone known as **angiotensin II** and **potassium ion.** The kidneys play a key role in determining plasma levels of both of these two factors and are also the major site of action for aldosterone in the body.

Regulation by Angiotensin II. Angiotensin II is formed by a sequence of steps that begin in the kidney and are collectively known as the **renin-angiotensin system.** These steps are outlined in Figure 14–10 and begin with the secretion of renin by specialized cells in the kidney known as **granular cells.** The location and function of the granular cells will be described in detail in Chapter 24. Renin is a proteolytic enzyme that acts upon a large protein, **angiotensinogen,**

which is synthesized by the liver and secreted into the blood (see Fig. 14–10). In the blood, renin splits a specific peptide bond in angiotensinogen, resulting in the formation of **angiotensin I,** which is a polypeptide of ten amino acids. Angiotensin I is carried via the blood to the lungs, where **angiotensin-converting enzyme (ACE)** removes two amino acids from the end of angiotensin I to form angiotensin II, a polypeptide of eight amino acids (see Fig. 14–10). From the lungs, angiotensin II travels via the blood to the tissues of the body.

In the adrenal glands, cells of the zona glomerulosa contain specific receptors for angiotensin II. Binding of angiotensin II to these receptors triggers increased synthesis and secretion of aldosterone. Angiotensin II utilizes the phosphatidylinositol pathway to provide the second messengers that couple hormone binding with the cellular effects of increased steroid synthesis (see Fig. 12–14).

As is indicated in Figure 14–10, the amount of aldosterone produced by the adrenal cortex is related to renin release from the kidney. Therefore, those factors that control renin secretion ultimately serve to control aldosterone secretion by the adrenal cortex. The primary stimulus for renin secretion is a decrease in blood pressure in the kidneys. Conversely, an increase in blood pressure inhibits renin release. In addition, sympathetic innervation of granular cells modulates the response in renin secretion to these and other stimuli.

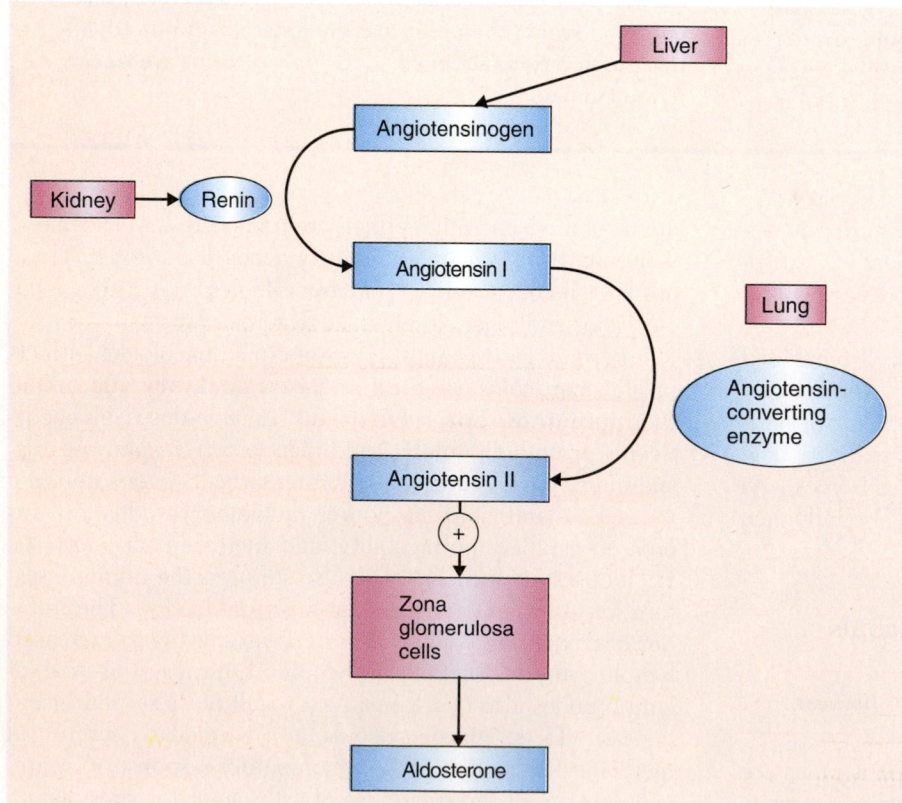

Figure 14–10

Summary of the renin-angiotensin system.

Regulation by Potassium Ion. The second major regulatory factor involved in controlling aldosterone secretion is the concentration of extracellular potassium ion. Aldosterone secretion is stimulated by an increase in the plasma potassium concentration. Potassium ions appear to directly affect cells of the zona glomerulosa. These cells are extremely sensitive to very small changes in the plasma concentration of potassium; an increase in potassium ion of as little as 5% above normal is sufficient to increase aldosterone secretion. As the following section will describe, one of the actions of aldosterone is to promote potassium secretion by the kidney and thus to lower the plasma potassium concentration. Thus, aldosterone and the kidneys make up two components of a feedback loop that regulate plasma potassium and sodium concentrations.

Effects of Mineralocorticoids

Owing to its effect to stimulate sodium reabsorption with water following, the net effect of aldosterone is to increase extracellular fluid volume. This increase in extracellular fluid volume results in increased blood volume, which tends to increase blood pressure and cardiac output. Thus, aldosterone indirectly affects blood pressure and blood flow in the cardiovascular system and is an important component of processes involved in regulating fluid balance.

The kidneys are the primary action site for aldosterone. In response to aldosterone, the kidneys retain more sodium in the body while excreting more potassium. Thus, in response to aldosterone, sodium reabsorption increases and potassium secretion increases. The specific mechanisms involved in these effects of aldosterone are presented in later chapters.

In addition to its effects on aldosterone secretion, angiotensin II has other important effects related to the maintenance of blood pressure and extracellular fluid volume. These include stimulation of smooth muscle contraction in blood vessels, activation of thirst centers in the brain, and stimulation of ADH release from the posterior pituitary. Each change that results (decreased blood vessel diameter, increased water intake, and increased fluid retention by the kidneys) contributes to maintenance of a normal blood pressure. These and other effects of angiotensin, as well as other characteristics of the renin-angiotensin system, will be described in Chapter 24, which also discusses in detail the regulation of fluid and electrolyte balance.

> *Based on the above discussion, how would too much (hypersecretion) or too little (hyposecretion) of adrenal cortical hormones be evident in a patient?*

Abnormalities of Adrenal Cortex Function Lead to Altered Physiological Regulation

Several common diseases result from either deficiencies or oversecretion of adrenal cortex steroids. As with other endocrine disorders, the symptoms expressed result from either absence or magnification of effects of the hormones involved.

A discussion of these conditions is therefore a useful way to review the actions of these hormones as well as to become familiar with endocrine disease states that may be fairly common in the general population.

Excess Secretion of Glucocorticoids

A common disorder of adrenal cortex function involves increased secretion of glucocorticoids. This can occur either as a result of a primary abnormality in steroid hormone production by the adrenal cortex or as a result of an overproduction of ACTH by the pituitary, causing excessive stimulation of the adrenal cortex. In either case, physiological changes that occur are due to the effects of excess cortisol. Excess cortisol secretion results in the condition known as **Cushing's syndrome.** Figure 14–11 illustrates the general appearance of a patient with Cushing's syndrome. Note the thin arms and legs, due in part to the loss of muscle mass as a result of the protein-catabolic effects of excess cortisol. Fat is redistributed from the extremities to the trunk, also contributing to a thinning of the arms and legs. Loss of fat in the extremities is accompanied by increased fat in the face and trunk, across the shoulder blades, and at the base of the neck. Accumulation of fat in the face gives it a characteristically rounded appearance. As a result of the protein-catabolic effects of cortisol, connective tissue is lost from the skin, causing it to become thinner. As a result, blood vessels are located closer to the surface, which is most often apparent in the red cheeks that characterize Cushing's patients. In addition, as the skin of the abdomen is stretched with increased fat deposition, it thins even further, and reddish-purple stretch marks (**striae**) form owing to the increased visibility of underlying blood vessels. Loss of connective tissue in small blood vessels makes them more fragile and easily broken than normal, resulting in an increased susceptibility to bruising.

Hypertension (high blood pressure) is also common among persons with Cushing's syndrome and, in many cases, is the primary contributing factor leading to death. These patients also suffer from increased susceptibility to infection and poor wound healing due to suppression of the inflammatory and immune systems caused by excess cortisol. Excess cortisol also impairs bone metabolism, and thus patients with Cushing's syndrome may also suffer from bone fractures and osteoporosis.

Many of the effects of cortisol antagonize the actions of insulin. Because a deficiency in insulin action results in diabetes, severe cases of Cushing's syndrome can actually produce diabetes if left untreated for a prolonged period of time.

Insufficient Secretion of Adrenal Cortex Steroids

A deficiency in steroid production by the adrenal cortex leads to the clinical condition termed **adrenal insufficiency.** If not treated, this particular disease state may lead to death. Adrenal insufficiency can be of two general causes. Primary adrenal insufficiency, or **Addison's disease,** results from disease of the adrenal cortex that impairs both glucocorticoid and mineralocorticoid production. Secondary

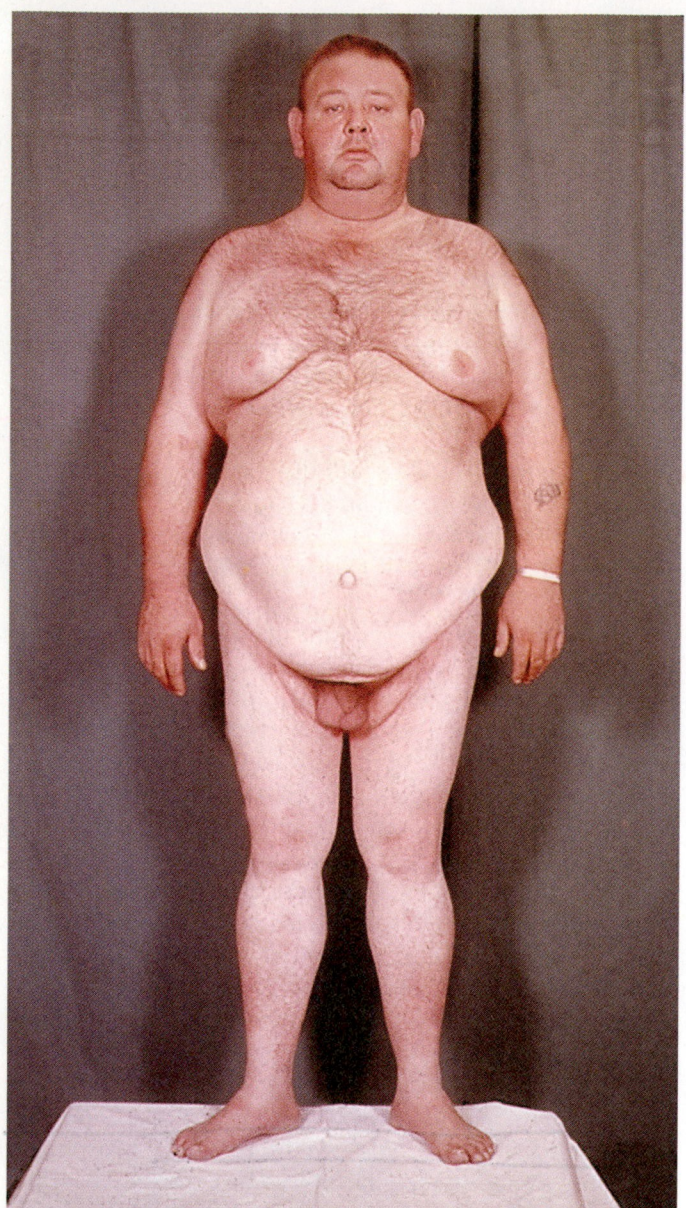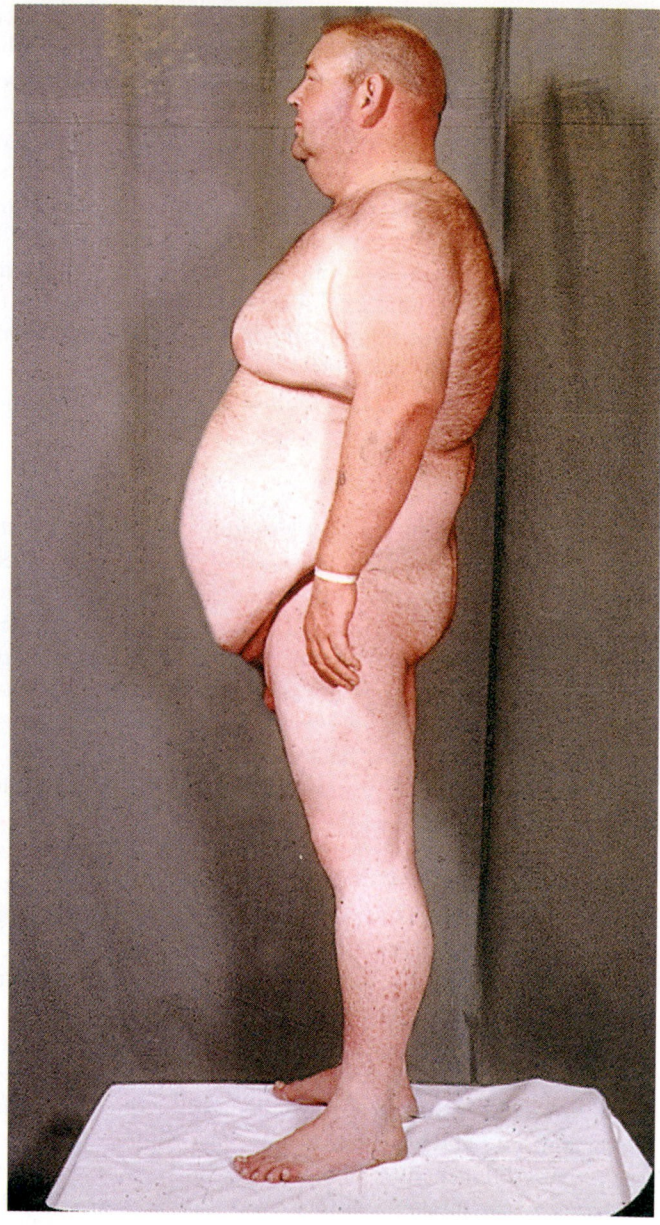

Figure 14–11

A man with Cushing's syndrome. Note the faint striae on the patient's abdomen.
(© VU/Visuals Unlimited.)

adrenal insufficiency results from inadequate ACTH production by the anterior pituitary. Because ACTH has only minor effects on mineralocorticoid production by the adrenal cortex, there is little or no change in aldosterone secretion in secondary adrenal insufficiency; it is primarily glucocorticoid secretion that is deficient in this situation.

Addison's disease is the more common form of adrenal insufficiency. Typical symptoms (Table 14–3) reflect a loss of both glucocorticoid and mineralocorticoid action. Due to loss of aldosterone, a person suffering from Addison's disease typically exhibits low plasma sodium and high plasma potassium levels. Low blood pressure is a common finding, in part due to electrolyte and water loss caused by the lack of aldosterone

(leading to reduced blood volume), and also due to the absence of the stimulatory effect of cortisol on vascular reactivity. Disturbances of electrolyte balance may lead to muscle weakness and easy fatigue as well as other symptoms, including vomiting and loss of appetite. Low blood glucose is also common in persons with Addison's disease, owing to an absence of the stimulatory effect of cortisol on gluconeogenesis in the liver.

In addition, hyperpigmentation of the skin and gums may occur in patients with Addison's disease (see Table 14–3). Recall from Figure 14–3 that in the absence of glucocorticoids, their negative feedback effect to inhibit ACTH secretion is lost, and plasma ACTH therefore increases dramatically. The structure of ACTH is similar to that of

TABLE 14–3

Typical Findings in Addison's Disease

Low plasma sodium concentration, high plasma potassium concentration

Low blood pressure

Muscle weakness, fatigue

Vomiting, loss of appetite, dehydration

Low blood sugar

Excess pigmentation of skin in some patients

melanocyte-stimulating hormone (MSH), which increases pigmentation in the skin. Although the MSH-like activity of ACTH is weak, greatly increased plasma concentrations of ACTH that occur in some cases of Addison's disease are sufficient to cause increased skin pigmentation.

THE ADRENAL MEDULLA

The adrenal medulla, which secretes **epinephrine** and **norepinephrine,** is an important component of the sympathetic nervous system. As a result, the sympathetic nervous system and adrenal medulla are often referred to together as the **sympathoadrenal system.**

The Medulla Is Centrally Located in Each Adrenal Gland

How is the anatomy of the adrenal medulla reflected by adrenal medullary function?

As is shown in Figure 14–1, the medulla is located in the interior portion of each adrenal gland. Within the adrenal gland, blood flows from the outer cortex inward to the medulla. As a result, blood that reaches the medulla contains high concentrations of glucocorticoids. This vascular orientation is important because it has been shown that glucocorticoids can influence hormone synthesis in the medulla.

The functional unit of the adrenal medulla is the **chromaffin** cell, which functions as a neuroendocrine cell. The name *chromaffin* originated from the fact that these cells are readily stained with solutions containing chromium. Functionally, chromaffin cells of the adrenal medulla are analogous to **postganglionic neurons** of the sympathetic nervous system (Fig. 14–12). Recall that, in the sympathetic nervous system, preganglionic fibers form synapses with postganglionic fibers in the sympathetic chain ganglia or in one of the outlying sympathetic ganglia. Postganglionic fibers then project to a particular target organ, such as the heart, as is illustrated in Figure 14–12. The arrangement is slightly different for the adrenal medulla, however. Preganglionic fibers to the medulla pass through the sympathetic chain without synapsing and directly innervate chromaffin cells of the adrenal

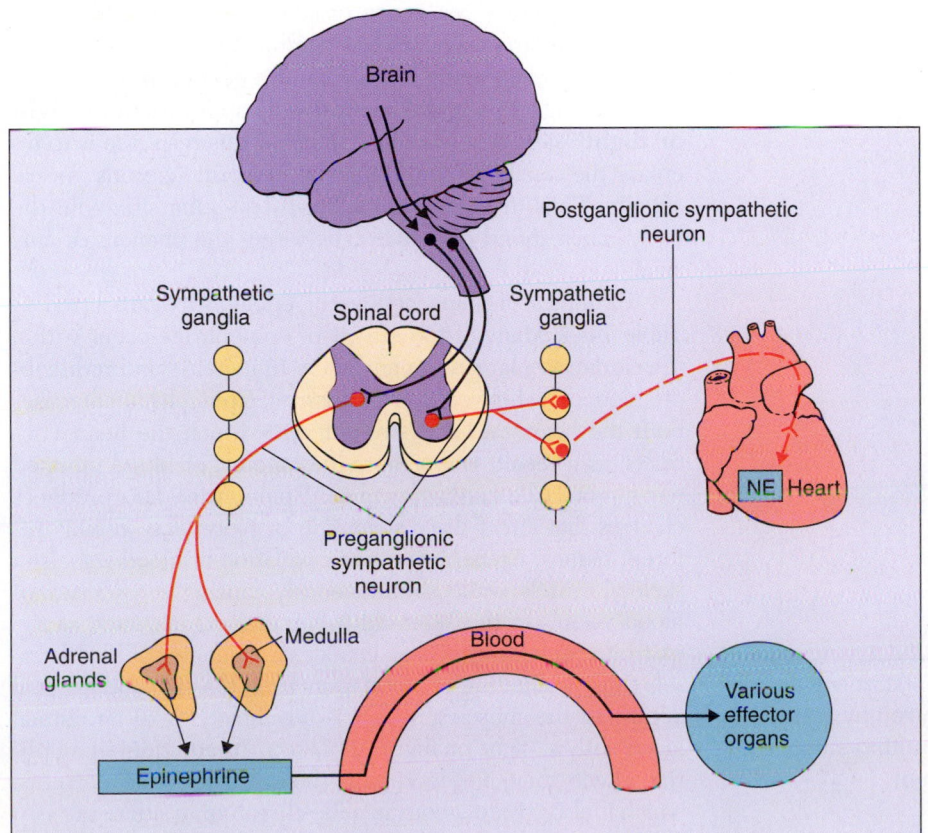

Figure 14–12

Figure illustrating the anatomical analogy between cells of the adrenal medulla and the sympathetic postganglionic neurons. Note that a particular postganglionic fiber *(dashed orange line)* has effects on one specific effector organ, such as the heart. In contrast, cells of the adrenal medulla, because they secrete epinephrine into the blood, are able to influence the activity of various effector organs throughout the body. NE = norepinephrine.

medulla. In response to stimulation, chromaffin cells secrete epinephrine directly into the blood, and epinephrine is carried to various effector organs in the body. Thus, although chromaffin cells of the medulla are analogous to sympathetic postganglionic neurons, they differ in that chromaffin cells influence the activity of a variety of effector organs throughout the body. In contrast, sympathetic postganglionic neurons generally influence the activity of single effector organs.

The Primary Hormone Secreted by the Adrenal Medulla Is Epinephrine

The synthesis of epinephrine in the adrenal medulla is outlined in Figure 14–13. Note that norepinephrine, another important catecholamine, is an intermediate in the synthesis of epinephrine. However, the adrenal medulla normally secretes only small amounts of norepinephrine.

Epinephrine synthesis begins with the amino acid tyrosine (see Fig. 14–13). Tyrosine is converted to **DOPA** (dihydroxyphenylalanine) by the enzyme **tyrosine hydroxylase.** This is the rate-limiting (slowest) step in epinephrine synthesis, and thus alterations in the activity of this enzyme directly influences the rate of epinephrine synthesis. Input of sympathetic stimulation to the medulla increases epinephrine synthesis by increasing activity of tyrosine hydroxylase. Subsequent steps resulting in conversion of DOPA to norepinephrine proceed rapidly. The final step in epinephrine synthesis is conversion of norepinephrine to epinephrine, which

is catalyzed by the enzyme **phenylethanolamine-N-methyl transferase (PNMT).** This enzyme is stimulated by cortisol, and thus a high concentration of this steroid in blood flowing from the cortex to the medulla enhances epinephrine synthesis by increasing the activity of PNMT.

Following synthesis, epinephrine is packaged and concentrated within chromaffin cells into dense, membrane-bound vesicles known as **chromaffin granules.** Stimulation of chromaffin cells as a result of sympathetic discharge causes the release of epinephrine from these granules into the blood.

Epinephrine Triggers a Sequence of Responses Appropriate to "Fight or Flight" Situations

 What are the actions of epinephrine on target tissues, and what are the consequences of these actions for the organism?

Stimulation of sympathetic nerves to the adrenal medulla results in the secretion of large amounts of epinephrine and smaller amounts of norepinephrine into the circulation. These hormones are carried in the blood to all tissues of the body, where they produce the same effects as those resulting from direct sympathetic stimulation of the tissue, except for one important difference. Because epinephrine and norepinephrine are removed from the blood more slowly than from the area of the sympathetic nerve terminal, effects of the circulating hormones are more prolonged than are effects resulting from direct sympathetic stimulation.

The adrenal medulla is a part of a general response system that triggers a sequence of events appropriate to **"fight or flight"** situations. The net result of this response is to increase the capability of the body to perform vigorous muscle activity. Thus, if faced with a life-threatening situation, the body can respond effectively, bettering the chances of survival.

Examples of some actions of epinephrine are listed in Table 14–4. Many of the effects of epinephrine occur within the cardiovascular system and serve to increase and redistribute blood flow between various tissues. Epinephrine increases both the heart rate and strength with which the heart contracts; as a result, it increases the amount of blood pumped per minute (the cardiac output). Epinephrine also produces changes that affect the distribution of blood flow among different tissues. Epinephrine causes dilation of blood vessels in skeletal muscle while simultaneously causing constriction of blood vessels in the skin and in visceral organs, such as the gastrointestinal tract. These changes in diameter of blood vessels cause a shunting of blood away from the skin and internal organs to the muscles, which is advantageous to an animal faced with a "fight or flight" situation. In addition, epinephrine affects the CNS, increasing the state of mental alertness, which is also advantageous in a life-threatening situation.

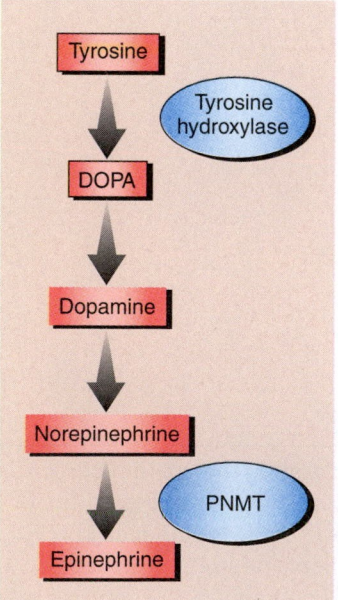

Figure 14–13

Pathways of epinephrine synthesis in adrenal chromaffin cells. Conversion of tyrosine to DOPA, catalyzed by tyrosine hydroxylase, is normally the rate-limiting step. The enzyme PNMT catalyzes the conversion of norepinephrine to epinephrine.

CURRENT CONCEPTS IN PHYSIOLOGY

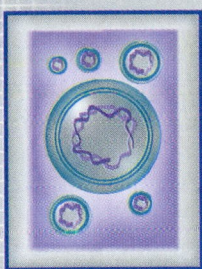

β-Adrenergic Receptors, Heart Failure, and Gene Therapy

When epinephrine interacts with one of its target cells, it does so through a group of receptors known as *adrenergic receptors*. The two principal adrenergic receptors are α-adrenergic receptors and β-adrenergic receptors. These two receptor types differ in their structure and in the type of hormone that they preferentially bind. Epinephrine has a higher affinity (strength of binding) to β-adrenergic receptors and lesser affinity for α-adrenergic receptors. Conversely, norepinephrine has a higher affinity for α-adrenergic receptors and a lesser affinity for β-adrenergic receptors.

The β-adrenergic receptor (β-AR) is one of the most intensively investigated hormone receptors, and information gained from studying this one receptor type has helped us understand many of the events in second-messenger cascades. Research has shown that β-AR is a member of a larger family of cell surface receptors that share common structures and common coupling mechanisms to downstream events in hormone-signaling pathways. They are known to be coupled to G-proteins that in turn activate adenylate cyclase and raise intracellular cAMP levels. Increased cAMP provides a signal that in some tissues re-

sults in elevated intracellular calcium. Calcium is the signal in the heart that leads to increased force of contractions.

When epinephrine binds to its receptor, it not only initiates the steps that lead to increased cAMP formation but also changes the shape of the receptor, making it a substrate for phosphorylation by a cytoplasmic enzyme, (β-adrenergic receptor kinase; β-ARK). The phosphorylated receptor is inactivated, resulting in the phenomenon of desensitization. It is known that β-ARs are desensitized to epinephrine in human heart failure and that there is increased expression of β-ARK.

Investigators have reasoned that, if they could inhibit β-ARK, they could alleviate the desensitization and restore normal responsiveness to epinephrine. Recently, using techniques of molecular biology, they were successful in manipulating the genes of mice to provide their cardiac cells with an inhibitor of β-ARK. Animals having this inhibitor showed much improved cardiac performance. Although considerably more research must be done before it becomes a reality, these results suggest the possibility of using gene therapy in certain forms of heart disease. One such situation might be congestive heart failure, in which the normal strength of cardiac contractions is considerably reduced. By giving the heart the inhibitor to block desensitization, the hope is that the heart would beat more forcefully and alleviate the need for treatment with drugs.

TABLE 14–4

Summary of the Actions of Epinephrine

Effects on the cardiovascular system
  Increased cardiac output
    Increased heart rate
    Increased strength of cardiac contractions
  Vasodilation in skeletal muscle
  Vasoconstriction in internal organs and skin

Effects on other tissues
  Relaxation of smooth muscle in gastrointestinal tract, urinary bladder, airways of lung
  Increased mental alertness

Effects on metabolism
  Increased glycogenolysis in liver and muscle
  Increased lipolysis in adipose tissue
  Decreased insulin secretion
  Increased glucagon secretion

Furthermore, epinephrine facilitates oxygen transport by the blood. By causing relaxation of smooth muscle in airways of the lungs, epinephrine reduces the work of moving air in and out of the lungs. In some species, epinephrine also causes the spleen to contract, which increases the number of red blood cells in the circulation. Both factors contribute to an increased oxygen-carrying capacity of the circulatory system.

Epinephrine affects metabolic processes in several different tissues, most notably the liver, skeletal muscle, and adipose tissue (see Table 14–4). The net result of epinephrine action is an increased breakdown of stored fuels into metabolizable substrates. The substrates that are released may serve as an energy source for local metabolism within the particular cell or tissue where they are formed, or they may be released into the blood for general distribution. In liver and muscle, epinephrine stimulates the breakdown of glycogen, while in adipose tissue, the hormone stimulates the breakdown of stored triacylglycerides into fatty acids and glycerol (see Table 14–4).

Epinephrine indirectly affects metabolism as a result of its effects on pancreatic hormone secretion. Insulin and glucagon are secreted by the pancreas and play key roles in regulating glucose metabolism. Insulin has metabolic effects that oppose those of epinephrine, and glucagon has effects similar to those of epinephrine. Epinephrine decreases insulin secretion while increasing glucagon secretion. Thus, this effect of epinephrine on the pancreas serves to indirectly promote its own metabolic actions. Discharge of the sympathetic nervous system also leads to an increased basal metabolic rate as a result of increased cellular metabolism throughout the body. This effect is primarily due to the release of norepinephrine from sympathetic nerve endings, rather than the release of epinephrine from the adrenal medulla.

A summary of the metabolic effects of epinephrine is presented in Figure 14–14. In liver, epinephrine stimulates glycogenolysis and thus increases glucose release from liver into the blood. Epinephrine also stimulates glycogenolysis in muscle, but muscle cannot release free glucose into the blood because it lacks the enzyme necessary to convert glucose-6-phosphate into glucose that the muscle cell could then release. Instead, glucose is metabolized to lactate, which is then released. Lactate can be taken up by liver and converted to glucose via gluconeogenesis. As glucose is metabolized to lactate in muscle, ATP, which the muscle can use for contraction, is produced.

In adipose tissue, epinephrine stimulates **lipolysis,** which is the process by which triacylglycerol units are broken down to fatty acids and glycerol. The fatty acids released can be utilized by other tissues to obtain energy. Heart and skeletal muscles in particular are able to meet a considerable amount of their energy needs by burning fatty acids; as a result, these tissues do not have to use glucose. This ability to use fatty acids as an alternative energy source is important because glucose is the primary fuel used by the nervous system. Provision of fatty acids is therefore said to have a "sparing effect" on blood glucose use. In addition, glycerol released from adipose tissue as a result of lipolysis can be taken up by liver and converted into glucose.

The net effect of epinephrine on metabolism is increased blood glucose concentration. The increase occurs as a direct result of increased glycogenolysis in liver and increased gluconeogenesis from lactate and glycerol in liver,

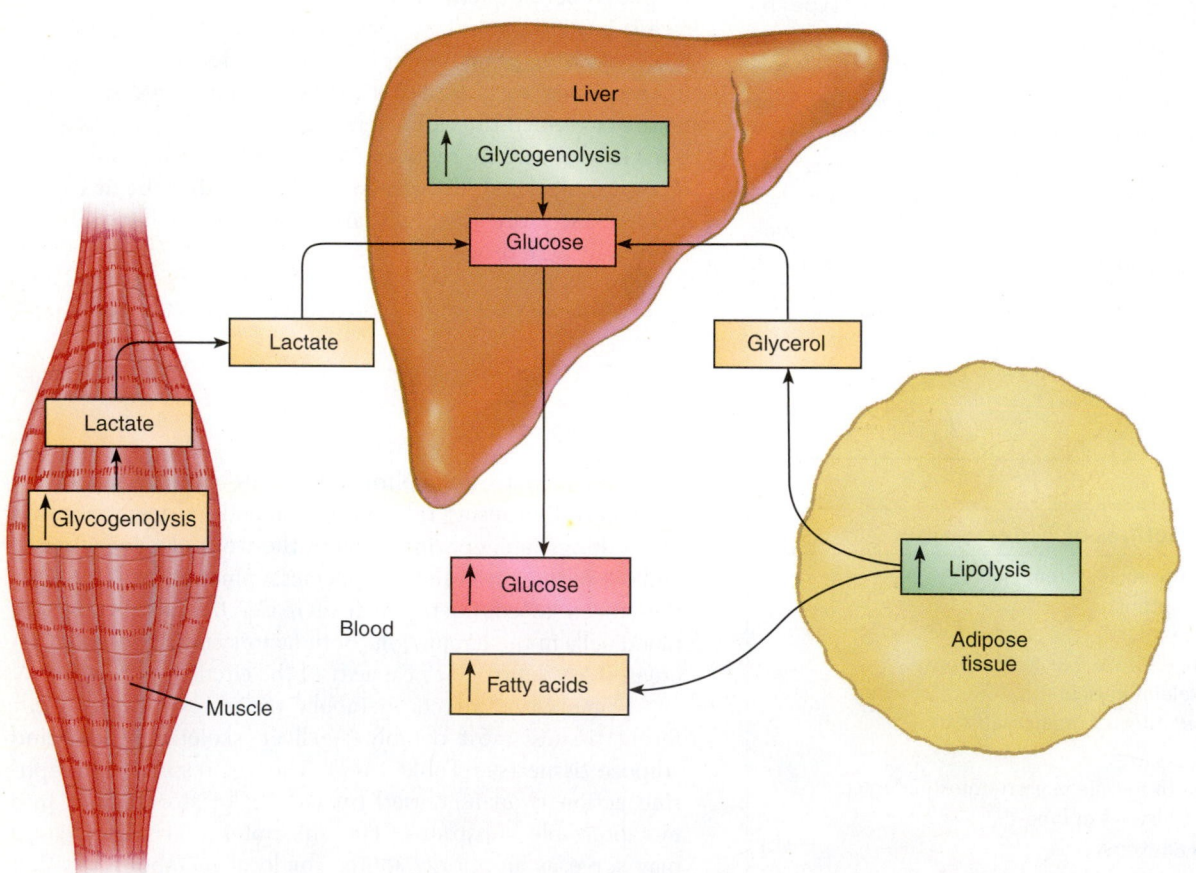

Figure 14–14

Summary of the principal effects of epinephrine on energy metabolism. The primary effects of epinephrine are to stimulate glycogenolysis in liver and muscle, and lipolysis in adipose tissue. The net result is an increase in blood glucose and blood fatty acid concentrations.

and as an indirect result of the sparing effects of fatty acids on blood glucose. The effect of epinephrine to inhibit insulin secretion and stimulate glucagon secretion from the pancreas also plays an important part in increasing blood glucose.

SUMMARY OF EFFECTS OF ADRENAL HORMONES

When we consider together the effects of hormones produced by the adrenal cortex and adrenal medulla, it is clear that the adrenal glands play an important part in the body's ability to respond effectively to stress. Epinephrine produces effects on metabolism and on the cardiovascular system and lungs that result in an increased supply of oxygen and metabolic fuels for the muscles and brain. Cortisol stimulates gluconeogenesis in the liver and thus helps to maintain blood glucose during prolonged periods of stress. As a result of either direct discharge of sympathetic nerves to the kidney or the release of epinephrine from the adrenal medulla, renin release can be increased during stress. An increase in renin results in increased aldosterone secretion. In response to increased aldosterone, the kidney retains more sodium and water, resulting in a greater blood volume. This is obviously beneficial when the body suffers blood loss. Thus, the hormones of the adrenal glands produce a number of effects that enable the body to adapt to a variety of situations and maintain a steady state.

CHAPTER REVIEW

Summary

- Humans have two adrenal glands, one located just above each kidney. Each adrenal consists of two distinct endocrine glands: the adrenal cortex and the adrenal medulla.
- The adrenal cortex primarily produces three steroids: aldosterone, cortisol, and dehydroepiandrosterone (DHEA). These steroids are produced in distinct zones of the adrenal cortex known as the *zona glomerulosa, zona fasciculata,* and *zona reticularis,* respectively.
- The zona fasciculata and zona reticularis of the adrenal cortex is controlled by ACTH from the anterior pituitary, which is in turn regulated by CRH secretion from the hypothalamus. Together, these organs form the hypothalamic-pituitary-adrenal axis. ACTH has both tropic and steroidogenic effects on the adrenal cortex.
- Cortisol is secreted in a circadian rhythm, with plasma levels being highest around 8 AM and lowest around midnight. Both physical and psychological stress also stimulate cortisol secretion. Cortisol has negative feedback effects on both the hypothalamus and the anterior pituitary.
- The overall effect of cortisol on metabolism is catabolic. In liver, cortisol produces effects that tend to increase blood glucose concentrations. In skeletal muscle, cortisol promotes protein loss, and in adipose tissue, it decreases glucose uptake and promotes the mobilization and redistribution of fat stores.

- In addition, cortisol enhances vascular reactivity; in large doses, glucocorticoids have anti-inflammatory and immunosuppressive effects that have proven useful in medical practice.
- Aldosterone is the principal mineralocorticoid produced by the adrenal cortex. Aldosterone secretion is regulated by the renin-angiotensin system and by changes in plasma potassium concentrations. Aldosterone acts on the kidneys to increase sodium retention and potassium excretion.
- When excess glucocorticoids are secreted, the result is Cushing's syndrome, in which there is a distinct physical appearance due to magnified effects of cortisol on protein and fat metabolism. A deficiency in steroid production by the adrenal cortex leads to the condition known as Addison's disease, in which there are abnormalities in electrolyte balance and blood pressure regulation.
- The adrenal medulla is coupled to the sympathetic nervous system and consists of neuroendocrine cells that produce and secrete catecholamines. Epinephrine is the primary product of chromaffin cells of the adrenal medulla.
- Epinephrine produces a variety of different effects that together prepare the individual for a fight or flight situation. Epinephrine affects the cardiovascular and pulmonary systems to increase delivery of oxygen to the muscles. In addition, epinephrine affects metabolism in liver, muscle, and adipose tissue in a manner to increase supplies of metabolic fuels, such as glucose and fatty acids, for the rest of the body.

Review Questions

Choose the Correct Answer

1. Dehydroepiandrosterone is primarily secreted from which region of the adrenal gland?
   a. Zona fasciculata
   b. Medulla
   c. Zona reticularis
   d. Zona fenestrae
   e. Zona glomerulosa

2. Which of the following "hormone class : hormone" pairs is incorrectly matched?
   a. Glucocorticoid : cortisol
   b. Catecholamine : epinephrine

    c. Androgen : aldosterone

    d. Mineralocorticoid : dehydroepiandrosterone

    e. c and d

3. Going from outermost to innermost, zones of the adrenal cortex are:

    a. zona fasciculata, zona reticularis, and zona glomerulosa.

    b. zona glomerulosa, zona reticularis, and zona fasciculata.

    c. zona glomerulosa, zona fasciculata, and zona reticularis.

    d. zona reticularis, zona fasciculata, and zona glomerulosa.

    e. zona fasciculata, zona glomerulosa, and zona reticularis.

4. Which steroid hormone is produced in the greatest amount in the adrenal cortex?

    a. Aldosterone

    b. Testosterone

    c. Cortisol

    d. Pregnenolone

    e. Progesterone

5. Cortisol has negative feedback effects at the level of the:

    a. posterior pituitary.

    b. anterior pituitary.

    c. hypothalamus.

    d. cerebellum.

    e. b and c

6. The predominant form of plasma cortisol is:

    a. bound to albumin.

    b. bound to TBG.

    c. bound to CBG.

    d. free in solution.

    e. bound to aldosterone.

7. Cortisol produces its effects via:

    a. IP_3 and DG.

    b. increased cellular sodium.

    c. decreased cellular cAMP.

    d. the gene expression model of hormone action.

    e. increased plasma cAMP.

8. Of the following conditions, when would plasma cortisol levels be highest?

    a. At the onset of deep sleep

    b. Just after awakening

    c. When ACTH levels are lowest

    d. Right after a meal

    e. b and c

9. Which of the following in the formation of cortisol is the step that is regulated by ACTH?

    a. 17-OH-progesterone → cortisol

    b. Progesterone → corticosterone

    c. Pregnenolone → 17-OH-pregnenolone

    d. Cholesterol → pregnenolone

    e. None of the above

10. Which of the following is not a stimulus for cortisol secretion?

    a. Hyperglycemia

    b. Trauma

    c. Heavy exercise

    d. Acute anxiety

11. As a part of its effect of increasing blood glucose, cortisol:

    a. stimulates glycogenolysis in liver.

    b. inhibits gluconeogenesis in liver.

    c. stimulates urea cycle in liver.

    d. stimulates protein synthesis in muscle.

    e. stimulates glucose uptake in adipose tissue.

12. Which of the following is least likely to be seen in a patient with Cushing's syndrome?

    a. High blood pressure

    b. Accumulation of fat in the trunk of the body

    c. Diabetes mellitus

    d. Osteoporosis

    e. Low blood sugar

13. The principal mineralocorticoid produced by the adrenal cortex is:

    a. cortisol.

    b. cholesterol.

    c. pregnenolone.

    d. aldosterone.

    e. androstenedione.

14. The net effect of aldosterone is to:

    a. increase blood glucose.

    b. decrease blood pressure.

    c. increase urine sodium.

    d. increase plasma sodium.

    e. increase plasma potassium.

15. Angiotensin-converting enzyme (ACE) is responsible for conversion of:

    a. angiotensinogen into angiotensin I.

    b. angiotensin II into renin.

    c. renin into angiotensin I.

    d. angiotensin II into aldosterone.

    e. angiotensin I into angiotensin II.

16. A typical finding in Addison's disease is:

    a. high blood glucose.

    b. high blood pressure.

    c. muscle weakness and fatigue.

    d. stimulated appetite.

    e. high plasma sodium.

17. Chromaffin cells of the adrenal medulla are analogous to:

    a. sympathetic preganglionic neurons.

    b. parasympathetic postganglionic neurons.

    c. parasympathetic preganglionic neurons.

    d. sympathetic postganglionic neurons.

    e. corticotropes of the anterior pituitary.

18. The rate-limiting step in epinephrine synthesis is catalyzed by:

    a. tyrosine hydroxylase.

    b. tyrosine deaminase.

    c. cholesterol desmolase.

    d. PNMT.

    e. DOPA dehydrogenase.

19. Effects of epinephrine on the cardiovascular system include:

    a. decreased strength of cardiac contractions.

    b. increased heart rate.

    c. redistribution of blood flow to internal organs.

    d. vasoconstriction in skeletal muscle.

    e. vasoconstriction in brain.

20. Which of the following is not an effect of epinephrine?

    a. Increased glycogenolysis in liver

    b. Decreased insulin secretion

    c. Increased lipolysis in adipose tissue

    d. Decreased urea synthesis

    e. Increased glucagon secretion

Answers to Case History Questions

1. Helen displays many of the classical signs and symptoms of Cushing's syndrome, which is characterized by excess secretion of cortisol by the adrenal cortex. The low back pain that initially sent her to the doctor was the result of a vertebral fracture, presumably caused by the osteoporosis often seen in Cushing's cases.

2. Obtain a late evening and early morning blood sample from Helen and measure her circulating cortisol levels. If she does have Cushing's syndrome, her cortisol levels will be elevated and there will not be a diurnal rhythm.

3. Cushing's is generally the result of one of two causes: either a tumor of the adrenal cortex or an ACTH-secreting tumor of the pituitary gland. In the former, high levels of cortisol secretion would feed back on the normal pituitary gland to inhibit ACTH secretion. So, in the case of an adrenal tumor, cortisol levels would be high; ACTH levels, low. In the case of the pituitary tumor, high levels of ACTH are being produced, which is in turn driving the adrenal cortex to produce high levels of cortisol. The key is to measure both cortisol and ACTH to discriminate between the two possibilities, as illustrated:

Tumor type	Cortisol	ACTH
Adrenal tumor	High	Low
Pituitary tumor	High	High

Key Terms

ACTH (p. 435)
Addison's disease (p. 443)
aldosterone (p. 436)
androgens (p. 435)
angiotensin I (p. 442)
angiotensin II (p. 442)
angiotensin-converting enzyme (ACE) (p. 442)
angiotensinogen (p. 442)
catabolic (p. 439)
chromaffin cell (p. 445)
circadian rhythm (p. 437)
corticosteroid-binding globulin (CBG) (p. 438)
corticotropin-releasing hormone (CRH) (p. 435)
cortisol (p. 436)
Cushing's syndrome (p. 443)
dehydroepiandrosterone (DHEA) (p. 436)
dihydroxyphenylalanine (DOPA) (p. 446)
fight or flight (p. 446)
glucocorticoid (p. 437)
hypothalamic-pituitary-adrenal axis (p. 435)
lipolysis (p. 448)
mineralocorticoid (p. 435)
permissive actions (p. 440)
postganglionic neurons (p. 445)
renin-angiotensin system (p. 442)
striae (p. 443)
sympathoadrenal system (p. 445)
tyrosine hydroxylase (p. 446)
urea cycle (p. 439)
vascular reactivity (p. 440)
zona fasciculata (p. 435)
zona glomerulosa (p. 435)
zona reticularis (p. 435)

Suggested Readings

Akhter, S., Eckhart, A. D., Rockman, H. A., Shotwell, K., Lefkowitz, R. J., and Koch, W. J. "In Vivo inhibition of elevated myocardial beta-adrenergic receptor kinase activity in hybrid transgenic mice restores normal beta-adrenergic signaling and function." *Circulation*, 100:648–653, 1999.

Berne, R. M., and Levy, M. N. *Principles of Physiology*, ed 3. St. Louis, Mosby, 2000.

Evans, R. M. "The steroid and thyroid hormone receptor superfamily." *Science*, 240:889–895, 1988.

Ganong, W. F. *Review of Medical Physiology*, ed 18. Los Altos, Lange Medical Publications, 1997.

Goodman, H. M. "Adrenal glands." *In Basic Medical Endocrinology*, ed 2. New York, Raven Press, 1994.

Griffin, J. E., and Ojeda, S. R. *Textbook of Endocrine Physiology*, ed 4. New York, Oxford University Press, 2000.

Guyton, A. C., and Hall, J. E., *Human Physiology and Mechanisms of Disease*, ed 6. Philadelphia, W.B. Saunders, 1997.

Kacsoh, B., *Endocrine Physiology*. New York, McGraw-Hill, 2000.

Norman, A. W., and Litwack, G. *Hormones*, ed 2. Orlando, Academic Press, 1997.

Web sites

http://www.pituitary.org
Home page of the Pituitary Network Association containing information for the layperson regarding pituitary disease.

http://medhelp.org/nadf/
Home page of the National Adrenal Diseases Foundation

http://members.xoom.com.endocrine/faqframe.htm
General information on a variety of topics related to endocrinology

Answers to Review Questions

1. c **2.** e **3.** c **4.** c **5.** e **6.** c **7.** d **8.** b
9. d **10.** a **11.** c **12.** e **13.** d **14.** d **15.** e
16. c **17.** d **18.** a **19.** b **20.** d

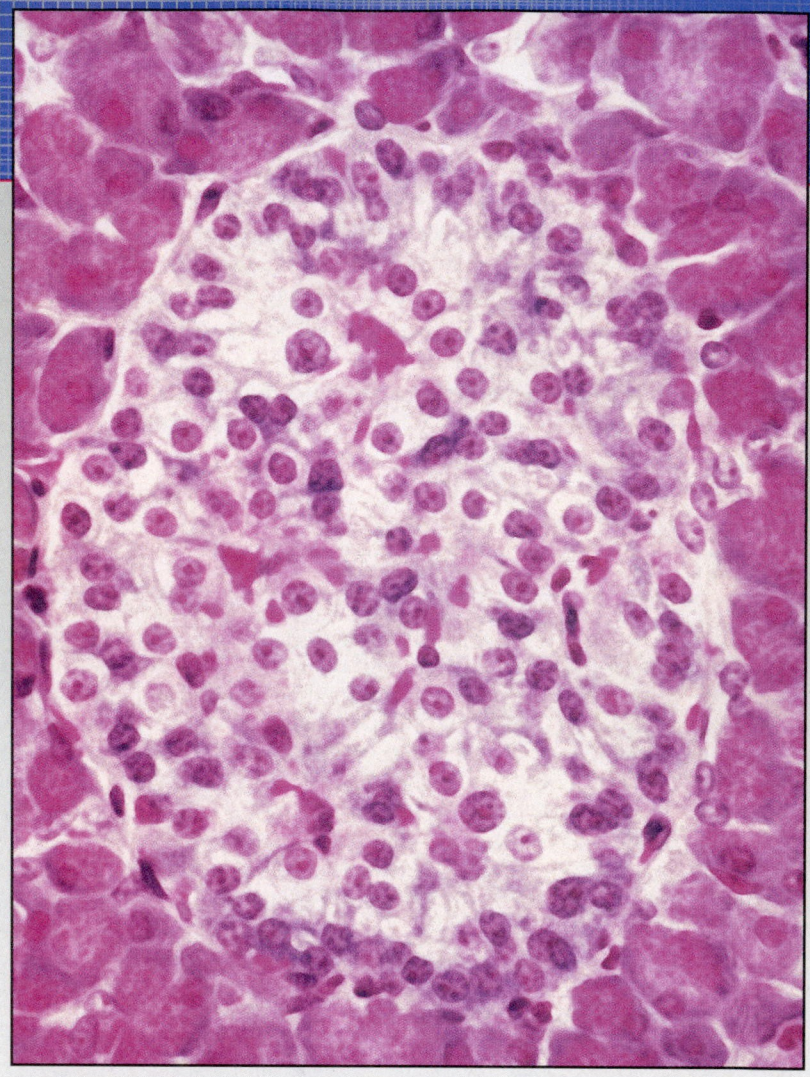

KEY CONCEPTS

- The endocrine pancreas produces the hormones insulin and glucagon, which play major roles in regulating fuel homeostasis in both the fed and fasted states.

- Insulin is secreted primarily in response to an increased blood glucose level. Glucagon is secreted in response to a decreased blood glucose level.

- Insulin directs the storage of excess nutrients in the form of glycogen, triacylglycerols, and protein. The major tissue targets of insulin are liver, muscle, and adipose tissue.

- Glucagon directs the movement of stored nutrients into the bloodstream. Liver is the primary physiological target of glucagon.

- In the fed state, the actions of insulin predominate in tissues, and nutrients are stored. In the fasted state, the actions of glucagon predominate, and stored nutrients are mobilized.

- Diabetes mellitus occurs when there is a deficiency in insulin action as a result of either an impairment in insulin secretion or an impairment in insulin action in its target tissues.

CASE HISTORY

enry is a 21-year-old man with a history of type 1 (insulin-dependent) diabetes mellitus who was recently found unconscious and rushed to the hospital. This was not the first time for Henry; this had happened three times in the past four years. It occurred twice because Henry had stopped taking his insulin injections, and once because Henry had mistakenly overdosed on his insulin. Henry was wearing a bracelet indicating that he was diabetic, which helped the emergency medical team diagnose his problem. When he arrived at the hospital, his respiration was exaggerated and his breath had an acetone odor. His blood pressure was 96/60 mm Hg (systolic/diastolic; normal, approximately 120/80) and his pulse was weak and rapid (120 beats/min; normal, approximately 70). Blood chemistry values were as follows:

Glucose:	830 mg/dL	[normal (fasting), 75–124]
Acetoacetate:	14.8 mg/dL	[normal, < 1.0 mg/dL]
β-hydroxybutyrate:	31.0 mg/dL	[normal, < 3.0 mg/dL]
bicarbonate:	12.2 mEq/L	[normal, 22–26 mEq/L]

Henry was quickly treated, and after a two-night stay in the hospital, he was released.

Questions

1. Was Henry experiencing insulin shock (too much insulin) or ketoacidosis? Why was his plasma bicarbonate low?

2. What is the cause of Henry's abnormal breathing, low blood pressure, and rapid heart rate?

3. How would you have treated Henry and what is the physiological basis for this treatment?

INTRODUCTION

The pancreas is located in the abdominal cavity adjacent to the upper part of the small intestine, as shown in Figure 15–1. The pancreas serves two functions, which are carried out by different groups of cells within the organ. These two groups of cells are designated as the **exocrine** and **endocrine** portions of the pancreas. The **exocrine pancreas** is responsible for secretion of fluid and various enzymes involved in food digestion. These secretions of the exocrine pancreas travel through a duct and empty into the upper part of the small intestine. The control and function of the exocrine pancreas will be described in Chapter 22.

The **endocrine pancreas** represents an anatomically small part of the pancreas. It secretes hormones that are important regulators of energy metabolism and fuel homeostasis in the body. This chapter describes the physiology of those hormones. Although the exocrine and endocrine portions of the pancreas are usually discussed in separate chapters in physiology textbooks, as they will be here, bear in mind that the pancreas as a whole serves functions related to the overall process of digestion, uptake, and use of metabolic fuels.

ISLETS OF LANGERHANS: FUNCTIONAL UNITS OF THE ENDOCRINE PANCREAS

 What clues does the anatomy of the endocrine pancreas provide with respect to its physiology?

The endocrine pancreas consists of groups of cells known as the **islets of Langerhans,** which are imbedded in the exocrine portion of the gland. Figure 15–2 shows a cross-section of the pancreas and the relationship of the islets to the exocrine portions of the gland. Although the average human pancreas contains approximately 1 million islets, they compose only approximately 1% to 2% of the total mass of the pancreas. Each islet is richly supplied with blood vessels, and the hormones that islet cells secrete enter these blood vessels. In contrast, cells of the exocrine pancreas are arranged to form blind-ended sacs and secrete their products into a duct system that eventually empties into the small intestine.

Islets are composed of four major cell types, listed in Table 15–1. Each cell type synthesizes and secretes a different hormone. As later sections will describe, **glucagon** and **insulin,** which are produced in the **alpha cells** and **beta cells,** respectively, are important hormones involved in the regulation of blood glucose concentrations and fuel homeostasis. **Somatostatin** produced by **delta cells** is identical to somatostatin produced in the hypothalamus. Recall that hypothalamic somatostatin functions as a release-inhibiting

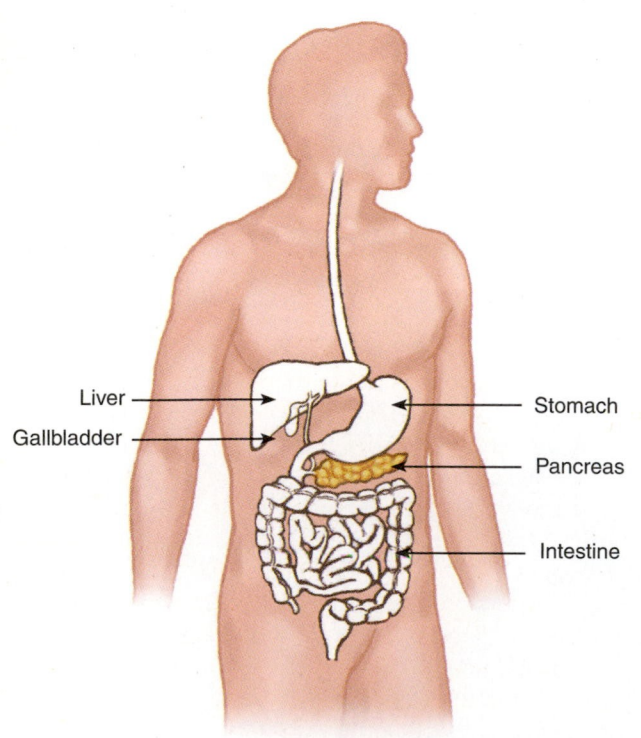

Figure 15–1

Location of the pancreas in the abdominal cavity.

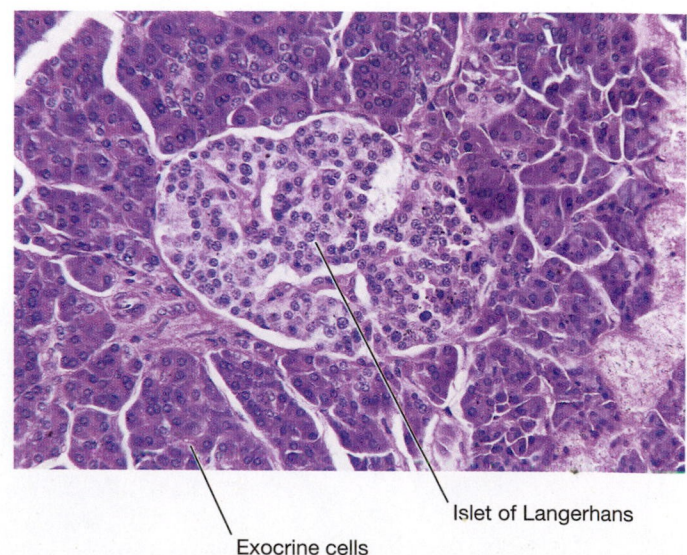

Islet of Langerhans

Exocrine cells

Figure 15–2

Photomicrograph of an islet of Langerhans and surrounding exocrine pancreas. (© *Peter Arnold, Inc./A.F. Michler*)

TABLE 15–1

Major Cell Types of the Islets of Langerhans and the Hormones They Produce

Name	Hormone Produced	Percentage of Total Islet°
Alpha cell	Glucagon	25
Beta cell	Insulin	60
Delta cell	Somatostatin	10
F cell	Pancreatic polypeptide	1

°The remaining 4% consists of connective tissue and blood vessels.

hormone, controlling the secretion of growth hormone from the anterior pituitary (Chapter 13). In the pancreas, somatostatin influences hormone secretion by the alpha and beta cells. The physiological function of **pancreatic polypeptide,** produced by the **F cells,** has not been fully characterized and is not discussed further.

Individual islets typically show an orderly arrangement of the different cell types. As Figure 15–3 illustrates, insulin-secreting beta cells tend to be located more toward the center of the islet and are generally the most numerous (see also Table 15–1). The less numerous glucagon-secreting alpha cells are located toward the edges of islets, forming a rim. The delta cells that produce somatostatin are scattered in between this rim of alpha cells and the core of beta cells (see Fig. 15–3). Somatostatin inhibits hormone secretion from both alpha cells and beta cells and, therefore, this arrangement of cells may be important with regard to paracrine actions of somatostatin. F cells show roughly the same distribution as the delta cells, although they are considerably fewer in number (see Table 15–1). Islets of Langerhans have a direct arterial blood supply. In human islets

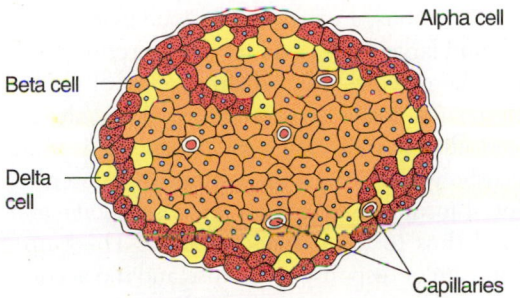

Figure 15–3

A typical islet of Langerhans, showing the anatomical relationship between the major hormone-producing cell types. Locations of the indicated cell types were determined by the technique of immunofluorescence microscopy.

vascular perfusion occurs in the direction from the core (beta cells) to the mantle (alpha cells). Thus the direction of blood flow is consistent with the known effect of insulin to stimulate glucagon secretion in certain circumstances (see subsequent discussion).

Islets also receive innervation from both the **sympathetic** and **parasympathetic** divisions of the autonomic nervous system (ANS). The importance of these neural inputs in regulating hormone secretion will be described in later sections.

SYNTHESIS AND SECRETION OF PANCREATIC HORMONES

As mentioned previously, the islets of Langerhans secrete three hormones, the actions of which have been well studied: **insulin, glucagon,** and **somatostatin.** Insulin is secreted in response to a number of factors, most importantly increased blood glucose. It promotes the uptake of glucose from the blood into cells, thereby decreasing blood glucose. Conversely, glucagon is secreted when blood glucose is low; it acts to increase the glucose concentration. Somatostatin is secreted in response to several factors as well, including increased blood glucose and glucagon levels.

Insulin Is Secreted in Response to Increased Glucose in the Blood

> *How is insulin synthesis and secretion controlled?*

As indicated in Figure 15–4, insulin is a polypeptide hormone that consists of an A-chain and a B-chain held together by two disulfide (S—S) bonds. In addition, there is a third disulfide bond within the A chain. These disulfide linkages are essential for the biological activity of the hormone.

Synthesis and Secretion of Insulin by Beta Cells of the Pancreas

Insulin is derived from a larger precursor molecule known as **proinsulin.** Proinsulin is first synthesized in the rough endoplasmic reticulum of beta cells. Proinsulin consists of a single peptide chain that contains both the A- and B-chains of insulin linked by a third connecting peptide segment. As the hormone is packaged into secretory vesicles within the beta cell, proinsulin is converted into insulin by proteolytic enzymes that clip the peptide chain in two places. The peptide segment that is removed is known as **C-peptide.** When insulin is secreted into the blood, a nearly equal amount of C-peptide is also secreted. C-peptide does not appear to have any biological function other than its role in insulin synthesis. However, C-peptide measurements are useful clinically because they provide an indirect index of insulin production.

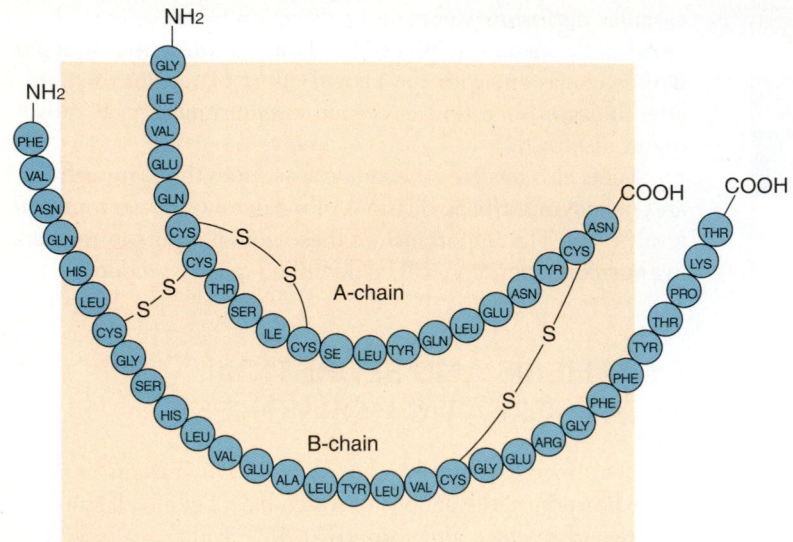

Figure 15–4

Insulin is a two-chain peptide consisting of 51 amino acids. There are 21 amino acids in the A-chain and 30 in the B-chain. The two chains are held together by a pair of disulfide (S—S) bonds, and a third disulfide bond is present in the A-chain.

Secretion of Insulin in Response to Various Stimuli

Several factors promote insulin secretion (Table 15–2), including increased blood glucose, amino acids, fatty acids, gastrointestinal hormones, neural and pharmacological stimuli, and other hormones secreted by the pancreas. When insulin secretion is stimulated, insulin-containing vesicles move toward the plasma membrane of the beta cell by a microtubule-microfilament system. The membrane of the vesicle fuses with the plasma membrane, and the contents of the vesicle (insulin and C-peptide) are released by exocytosis. In addition to stimulating release of preformed insulin, factors such as glucose that stimulate insulin secretion also increase proinsulin synthesis and the conversion of proinsulin to insulin.

After release from the beta cells into the surrounding capillary network, most of the insulin circulates free in the blood, although a small amount is loosely associated with plasma proteins, such as albumin.

TABLE 15–2

Factors Affecting Insulin Secretion from the Pancreas

Stimulators	Inhibitors
Increased blood glucose	Somatostatin
Amino acids	Epinephrine
Fatty acids	Norepinephrine
Gastrointestinal hormones (gastrin, secretin)	
Acetylcholine	
Sulfonylureas	
Glucagon	

Increased Blood Glucose. The most important physiological stimulus for insulin secretion is an increased concentration of glucose in the blood. Figure 15–5 shows average changes in plasma concentrations of glucose and insulin in a group of normal individuals who consumed 75 g of glucose at time zero following an overnight fast. Note that with a normal overnight fasting blood glucose concentration of 80 to 90 mg per 100 ml (prior to ingestion of the glucose), insulin secretion is low. As the plasma glucose concentration increases above approximately 100 mg per 100 ml, however, insulin secretion is stimulated. Notice in Figure 15–5 that after glucose ingestion, the concentration of glucose in plasma rises rapidly and is closely matched in time by an increase in the concentration of circulating insulin. In this case, the plasma insulin concentration increases to a peak value approximately tenfold that observed before glucose ingestion. With the administration of larger amounts of glucose, the blood glucose concentration can be driven even higher, and insulin secretion may reach values as high as 30-fold the basal value shown in Figure 15–5. Thus, an increase in blood glucose can elicit a very rapid and large increase in insulin secretion.

As a result of the glucose-induced stimulation of insulin secretion, plasma insulin concentrations are generally highest in the period immediately following a meal, especially if the meal is rich in carbohydrate. As later sections will describe, the overall effect of insulin is to increase glucose uptake and use by tissues and thus to lower blood glucose. Therefore, the secretion of insulin in response to glucose and the accompanying action of insulin to decrease blood glucose are a part of a feedback mechanism involved in the maintenance of a constant blood glucose concentration. Figure 15–6 shows the basic characteristics of this mechanism.

Amino Acids, Fatty Acids, and Gastrointestinal Hormones. Several other factors influence insulin secretion, although none is as physiologically important as glucose. A

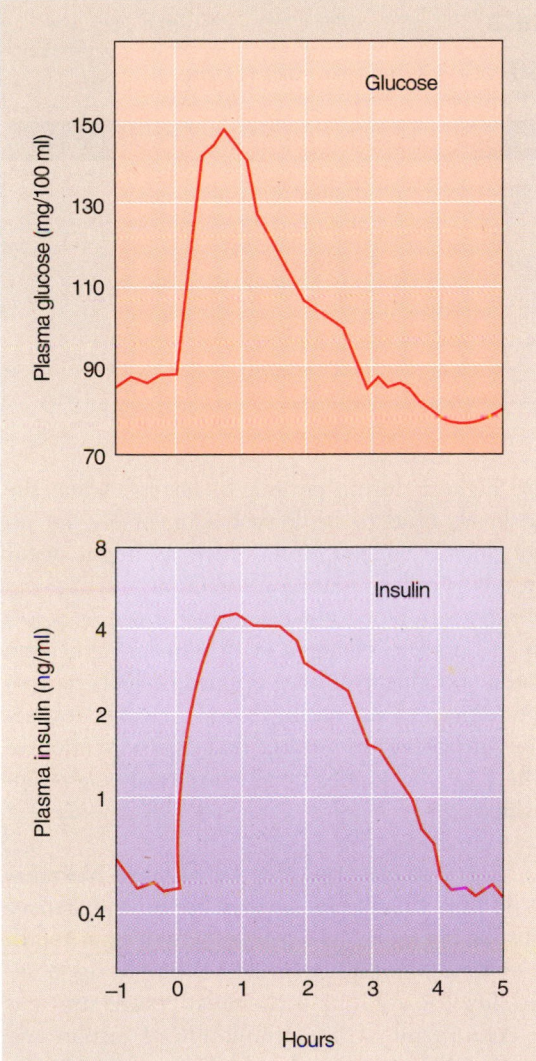

Figure 15–5

Time course of changes in plasma glucose and insulin after an oral glucose load. Values shown are the averages from a group of normal, healthy volunteers. After a 12-hour overnight fast, each person consumed 75 g of glucose at time zero. Blood samples were drawn at various times, and the concentrations of glucose and insulin in the plasma were determined. Values for glucose and insulin are given in terms of milligrams (mg) and nanograms (ng), respectively. Note that the values for insulin are expressed as a logarithmic scale. (Redrawn from Wilson, J. D., and Foster, D. W., eds. *Williams Textbook of Endocrinology*, ed 8. Philadelphia, W.B. Saunders, 1992, Figure 25–8, p. 996)

partial list of some of these other factors is given in Table 15–2. Several **amino acids,** especially arginine, are known to stimulate insulin secretion. Thus, ingestion of a high-protein meal results in increased insulin secretion. Similarly, fatty acids can also stimulate insulin secretion.

The gastrointestinal (GI) tract secretes a number of hormones that act to coordinate the various processes involved in digestion of food. A number of these GI hormones, including gastrin and secretin, also stimulate insulin secretion by the pancreas (see Table 15–2). It is well known that the administration route of glucose influences the amount of insulin secreted by the pancreas. If glucose is given orally, a greater increase in insulin secretion occurs compared with when the same amount of glucose is given intravenously. This difference is due to a stimulatory effect of GI hormones on insulin secretion. As a later chapter will describe, GI hormones are generally secreted in response to the presence of food in the GI tract. Soon after food enters the GI tract, the blood glucose concentration begins to increase. The effect that GI hormones have to stimulate insulin secretion therefore provides an anticipatory signal to beta cells that the blood glucose concentration is likely to increase in the near future.

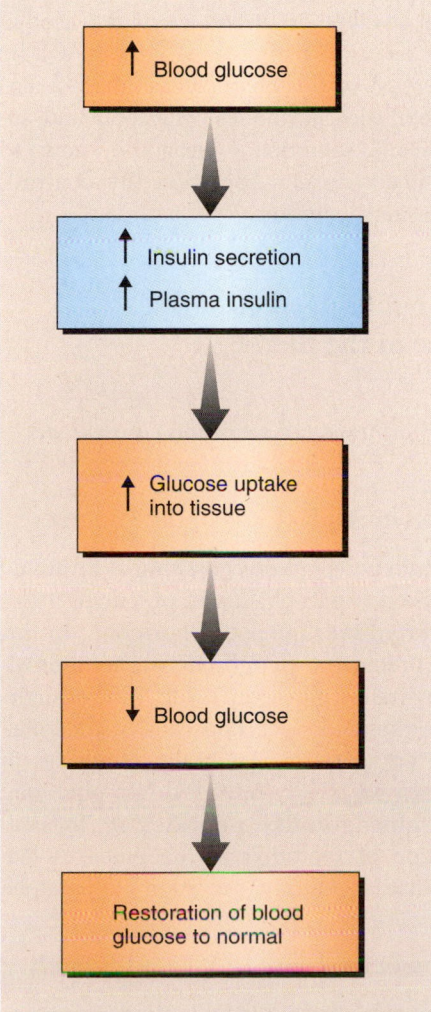

Figure 15–6

Role of insulin in the regulation of blood glucose concentration.

Autonomic Nervous System. Neural inputs to islets appear to play a role in regulating insulin secretion. Note in Table 15–2 that acetylcholine stimulates insulin secretion. Recall from Chapter 10 that acetylcholine serves as the neurotransmitter for postganglionic fibers in the parasympathetic division of the autonomic nervous system. Activation of parasympathetic neurons to the islets, as might be expected to occur during digestion, also results in increased insulin secretion. This activation also provides an anticipatory signal to the pancreas that an increase in blood glucose is likely to occur. Table 15–2 also indicates that epinephrine and norepinephrine inhibit insulin secretion. Activation of sympathetic fibers to the islets or release of epinephrine from the adrenal medulla, as might occur in response to stress, results in an inhibition of insulin secretion. The significance of stress-induced inhibition of insulin secretion was discussed in Chapter 14.

Drugs and Other Islet Hormones. A class of drugs known as **sulfonylureas** also promote insulin secretion (see Table 15–2). These drugs can be taken orally and are widely used in the treatment of some forms of diabetes mellitus, as will be discussed in a later section of this chapter. Table 15–2 also indicates that two other islet hormones, glucagon and somatostatin, influence insulin secretion. Glucagon is known to augment glucose-stimulated insulin secretion. By contrast, somatostatin inhibits insulin secretion.

Glucagon Is Secreted in Response to Decreased Glucose in the Blood

 How is glucagon synthesis and secretion controlled?

Synthesis of Glucagon by Alpha Cells of the Pancreas

Glucagon is a single-chain polypeptide consisting of 29 amino acids. It is first synthesized as part of a larger precursor molecule that is later converted into the active hormone. In this case, the immediate precursor, **proglucagon,** is approximately five times the size of glucagon itself. Interestingly, proglucagon is also synthesized in certain cells of the small intestine and in different parts of the brain. However, in these two areas proglucagon is not converted into glucagon. Instead, several other **glucagon-like peptides** are formed. However, the regulation of secretion of the glucagon-like peptides differs from that of glucagon secretion by the alpha cells of the pancreas.

Secretion of Glucagon in Response to Various Stimuli

Decreased Blood Glucose. As with insulin, the major physiological control factor of glucagon secretion is the blood glucose concentration. However, in direct contrast to insulin, a *decrease* in blood glucose concentration stimulates glucagon secretion (Table 15–3). Therefore, circulating glucagon con-

TABLE 15–3	
Factors Affecting Glucagon Secretion from the Pancreas	
Stimulators	**Inhibitors**
Low blood glucose	Fatty acids
Amino acids	Somatostatin
Acetylcholine	Insulin
Norepinephrine	
Epinephrine	

centrations are highest during periods of fasting, when the blood glucose levels tend to be lowest. Conversely, an increased blood glucose concentration, such as might occur after a meal, tends to reduce glucagon secretion.

Thus, alterations in blood glucose either above or below normal result in opposite changes in insulin and glucagon secretion. Insulin and glucagon have opposite effects on metabolism, decreasing and increasing blood glucose, respectively. It is not surprising, therefore, that the regulation of secretion of these two hormones by glucose involves a reciprocal relationship.

Amino Acids, Fatty Acids, and the Autonomic Nervous System. In addition to glucose, a number of substances that control insulin secretion also influence glucagon secretion (see Table 15–3). Amino acids stimulate glucagon secretion; as with insulin, arginine is the most potent stimulus among these. As a result of the stimulation of insulin and glucagon secretion by amino acids, the plasma concentration of both hormones increases after a high-protein meal. The parallel increases in both insulin and glucagon secretion in response to amino acids serve as a protective mechanism to ensure that blood glucose levels are maintained following the ingestion of meals rich in protein but low in carbohydrate, as will be discussed in later sections describing the actions of these two hormones. An increase in circulating fatty acids inhibits glucagon secretion, whereas a decrease in fatty acid concentrations increases glucagon secretion (see Table 15–3).

Insulin. Insulin also affects glucagon secretion from alpha cells (see Table 15–3). Normally, high blood glucose tends to inhibit glucagon secretion, but this effect depends on the presence of insulin. In the absence of insulin, alpha cells cannot detect the elevated blood glucose, and glucagon secretion continues at a high rate. This fact is important in those forms of diabetes mellitus in which insulin-secreting beta cells are destroyed. Not only is insulin secretion deficient; glucagon secretion is *inappropriately* high due to the absence of the inhibitory effect of insulin on glucagon secretion.

Somatostatin Is Secreted in Response to Increased Glucose and Glucagon in the Blood

> *What is the postulated role of pancreatic somatostatin secretion?*

As described in Chapter 13, somatostatin is a small polypeptide hormone. As was the case for insulin and glucagon, somatostatin is also first synthesized as part of a larger precursor molecule. The immediate precursor, **prosomatostatin,** is then converted into the smaller somatostatin peptide. Factors regulating somatostatin secretion from the delta cells are similar to those regulating insulin and glucagon secretion. The primary stimuli of somatostatin secretion are (1) increased blood glucose, (2) increased plasma glucagon, and (3) amino acids. The physiological significance of somatostatin produced by the islets of Langerhans is not well understood. It is known to inhibit insulin and glucagon secretion and to inhibit the secretion of various gastrointestinal hormones as well, influencing other aspects of GI function. However, a full picture of the physiology of pancreatic somatostatin remains to be determined.

METABOLIC EFFECTS OF PANCREATIC HORMONES

> *How do insulin and glucagon contribute to the regulation of blood glucose levels?*

Even while at rest or sleeping, the body is continually using energy to drive vital processes such as ion transport, synthesis of various cellular proteins, and the mechanical activity involved in respiration or cardiac contractions. Additional physical activity increases energy requirements above those in the basal (resting) state. However, although energy use is continuous, the intake of energy in the form of food is intermittent. Thus, excess fuels taken in with a meal must be stored for subsequent use between meals. As a result, mechanisms that coordinate and regulate fuel homeostasis have evolved. The following sections discuss the actions of insulin and glucagon, two primary hormones involved in regulating fuel homeostasis in the body.

As described earlier, alterations in the blood glucose concentration provide the primary signal controlling insulin and glucagon secretion. Glucagon and insulin have reciprocal effects that tend to increase and decrease the blood glucose concentration, respectively. Thus, these two hormones serve as key components of feedback loops whereby the blood glucose concentration is maintained within fairly narrow limits. As we have learned in previous chapters, other hormones, such as cortisol, growth hormone (GH), and epinephrine, also have effects that influence the blood glucose concentration. An appropriate question to ask at this point might be,

"What is the physiological basis for this tight regulation of blood glucose concentration?"

The answer to that question lies in the fact that the central nervous system (CNS) relies almost solely on glucose for its metabolic needs. Although other tissues, such as muscle and liver, can use alternate substrates such as fatty acids for fuel, the nervous system ordinarily must derive all of its energy from glucose. Thus, if the blood glucose concentration were to decrease to low levels, the brain could not obtain energy, and death would soon follow. Conversely, although a marked elevation of blood glucose would not produce such immediate deleterious effects, it can lead to wasteful loss of glucose in the urine as well as concurrent losses of large volumes of water. In a normal individual, the blood glucose concentration following an overnight fast is usually in the range of 75 to 115 mg of glucose per 100 ml of blood. Following ingestion of a large amount of glucose, such as with a high-carbohydrate meal, the concentration may reach as high as 200 mg glucose per 100 ml of blood. However, as shown in Figure 15–5, in a normal individual, the blood glucose concentration quickly returns toward the basal level owing to the secretion of insulin and its effects on glucose utilization.

Insulin Has Effects on Metabolism that Are Anabolic in Nature

> *How does insulin promote the storage of fuel?*

Insulin has been termed the **hormone of nutrient abundance.** When influx of nutrients is high, insulin provides the signal that directs storage of excess fuels while suppressing mobilization of preexisting fuel stores. Insulin therefore has **anabolic** effects.

The specific cellular mechanisms by which insulin produces its effects have yet to be determined, despite considerable research efforts in this area in the past several years. The insulin receptor is located in the plasma membrane, as might be expected, because insulin is a peptide hormone. Although the structure of the receptor has been well characterized biochemically, the exact details of how the binding of insulin to its membrane receptor is transduced into the intracellular events that the hormone produces is not fully known. Research has shown that the insulin receptor itself has tyrosine protein kinase activity, and thus insulin most likely produces some of its effects by stimulating its receptor to phosphorylate specific regulatory proteins within its target cells. Thus, insulin appears to act via a mechanism similar to the one depicted earlier in Figure 12–16.

The major target tissues of insulin are skeletal muscle, adipose tissue (fat), and liver.

Effects of Insulin on Carbohydrate Metabolism

The most obvious action of insulin is that it very rapidly and effectively lowers the blood glucose concentration. Most tissues in the body depend on insulin for the uptake of glucose

CURRENT CONCEPTS IN PHYSIOLOGY

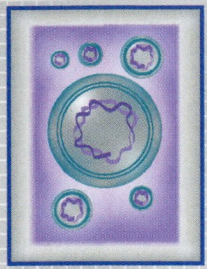

Insulin-Stimulated Glucose Uptake

As described in the text, insulin is a potent anabolic signal in humans and other mammals. It has multiple effects in a number of tissues, but the most central of its actions is the regulation of glucose homeostasis. Insulin decreases blood glucose concentrations by stimulating glucose uptake into skeletal muscle and adipose tissue and by inhibiting glucose release by the liver.

Glucose enters muscle and adipose tissue by facilitated diffusion utilizing specific carrier proteins in the plasma membrane. These carriers provide a means whereby glucose enters the cell faster than by simple diffusion alone, but the carriers do not require energy and cannot move glucose against a concentration gradient as occurs with active transport. A family of glucose carrier protein isoforms has been identified (GLUT-1, GLUT-2, GLUT-3, and GLUT-4). The one of greatest interest is GLUT-4, which is highly expressed in skeletal muscle and adipose tissue and is responsible for the insulin-stimulated increase in glucose uptake. In the basal state GLUT-4 is found both in the plasma membrane and in intracellular

vesicles. Under the influence of insulin, there can be a 10- to 20-fold increase in the rate of movement (exocytosis) of GLUT-4 intracellular vesicles to the plasma membrane and a somewhat less dramatic 2- to 3-fold decrease in the rate of GLUT-4 internalization (endocytosis). The net result is more GLUT-4 in the plasma membrane and therefore increased glucose uptake.

Of considerable interest to researchers at the present time is the biochemical basis for the intracellular trafficking of these GLUT-4-containing vesicles and their recognition, docking, and fusion with the plasma membrane. From what is currently known, it appears that the process is very similar to the movement of neurotransmitter-containing vesicles at the neuronal synapse.

Of equal interest is how insulin specifically causes the redistribution of the transporters. While the initial steps in insulin action are fairly well understood, the more downstream steps have yet to be clearly defined. Filling in the details of GLUT-4 trafficking and the final steps in insulin action will be a tremendous step forward and should allow for a much better understanding of dysregulation of glucose homeostasis and insulin resistance. It is hoped that improved therapeutic strategies for the treatment of type 2 diabetes will readily follow.

---

from the blood. In the absence of insulin, glucose does not readily enter cells and therefore cannot be metabolized and used for energy. Two primary exceptions to this rule are the brain and the liver, the cells of which are readily permeable to glucose even in the absence of insulin. In tissues such as muscle and adipose, glucose enters cells primarily by facilitated diffusion. Recall that facilitated diffusion differs from active transport in that it is not energy dependent and cannot move glucose against a concentration gradient. Instead, facilitated diffusion involves specific carrier proteins that "shuttle" glucose across the membrane at a faster rate than would occur by diffusion alone. Glucose transporters are located both in the plasma membrane and in intracellular sites. Insulin promotes movement of transporters from their intracellular location to the plasma membrane. By increasing the number of transporters present in the plasma membrane, insulin is able to increase the rate of glucose uptake into cells (see the "Current Concepts in Physiology" box regarding insulin-stimulated glucose uptake).

In addition to stimulating glucose uptake into cells, insulin also stimulates metabolism and use of glucose. In most cells, particularly those in liver and muscle, insulin stimulates glycogen synthesis while inhibiting glycogen breakdown (Fig.

15–7). The net result is an increase in the amount of glycogen stored. Insulin also stimulates a number of enzymes in the glycolytic pathway. Glucose metabolism via glycolysis therefore increases in most insulin-sensitive cells. In addition to stimulating glucose uptake and use by peripheral tissues, such as muscle and adipose tissue, insulin decreases glucose output by the liver in two different ways. First, insulin decreases the activity of several key enzymes in the gluconeogenic pathway. (Recall that gluconeogenesis is the process by which nonglucose substrates, such as amino acids, are converted into glucose.) Second, insulin promotes the use of amino acids in peripheral tissues (see the following section on protein metabolism) and so decreases the supply of amino acids, which are normally an important substrate for gluconeogenesis in liver.

Effects of Insulin on Lipid Metabolism

As a result of several specific effects of insulin on lipid metabolism, it promotes storage of excess fuel as triacylglycerols. In both liver and adipose tissue, insulin stimulates the synthesis of fatty acids. In adipose tissue, fatty acids combine with α-glycerol phosphate and are stored as triacylglycerols (Fig. 15–8). By contrast, liver stores only a small amount of

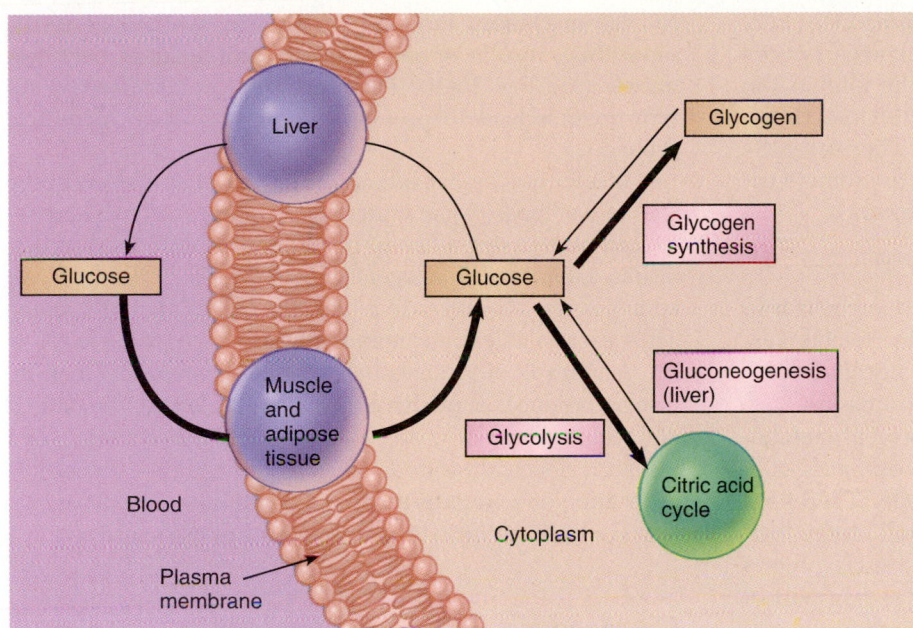

Figure 15–7

Summary of the effects of insulin on carbohydrate metabolism. Pathways or processes in which there is increased flow in the presence of insulin are shown with the thick arrows; those in which there is reduced flow are shown by the thin arrows.

the fatty acids it produces. Instead, the majority of fatty acids are packaged as lipoproteins and released into the blood. These lipoproteins are taken up by adipose tissue, and the fatty acids are then converted and stored as triacylglycerols (see Fig. 15–8). In adipose tissue, insulin stimulates the activity of **lipoprotein lipase,** an enzyme that promotes the uptake of lipoproteins from the blood (see Fig. 15–8). The effect of insulin to stimulate glucose uptake by adipose tissue also tends to promote triacylglycerol storage. Inside adipose

cells, glucose can be either metabolized and used for fatty acid synthesis or used to form α-glycerol phosphate. α-Glycerol phosphate serves as the backbone to which fatty acids are attached to form triacylglycerols (see Fig. 15–8). Thus, by increasing glucose entry into adipose cells, insulin promotes synthesis of both the fatty acid and α-glycerol phosphate components of triacylglycerols.

In addition to its stimulatory effect on fat synthesis, insulin also inhibits lipolysis (breakdown of triacylglycerols).

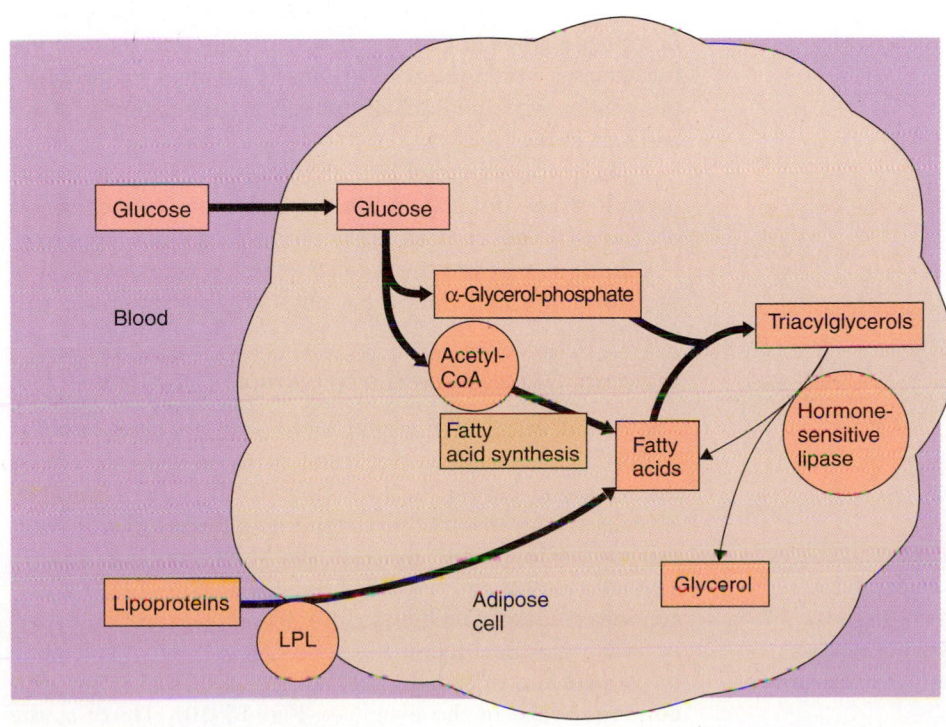

Figure 15–8

Summary of the effects of insulin on adipose cell metabolism. Pathways or processes in which there is increased flow in the presence of insulin are shown with the thick arrows; those in which there is reduced flow are shown by the thin arrows. The net effect of insulin is to promote triacylglycerol storage. For simplicity, lipoprotein lipase (LPL) is depicted as being present on the surface of the adipose cell; however, this enzyme actually is located on the surface of endothelial cells in the adipose tissue vasculature.

The breakdown of triacylglycerols into their component fatty acids and glycerol is catalyzed by an enzyme known as **hormone-sensitive lipase** (see Fig. 15–8). Insulin inhibits the activity of this enzyme both in liver and in adipose tissue, thus decreasing lipolysis in these two tissues. Therefore, the overall effect of insulin on lipid metabolism is to promote triacylglycerol storage while inhibiting its breakdown.

Effects of Insulin on Protein Metabolism

Insulin also has profound effects on protein metabolism in the body. These effects are most readily seen in skeletal muscle and liver, although the anabolic effect of insulin to promote the accumulation of protein occurs in most tissues.

The effects that insulin has on muscle protein metabolism are of particular significance because approximately 40% of the total body protein is present in muscle. Therefore, much of the influence that insulin has on overall protein balance in the body results from its effects on muscle protein metabolism. In muscle, insulin stimulates active transport of amino acids from the blood into individual muscle cells (see Fig. 15–9). As a result, more amino acids are available for muscle protein synthesis. Insulin also stimulates the process of protein synthesis (see Fig. 15–9), in part through increasing the number of ribosomes in each muscle cell and in part by increasing in the activity of individual ribosomes (the

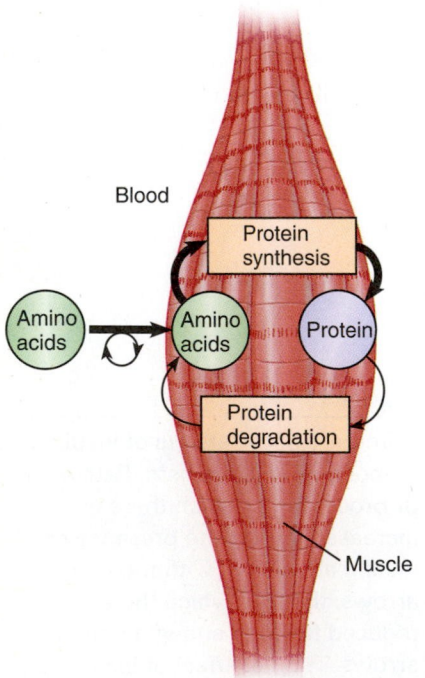

Figure 15–9

Summary of the effects of insulin on muscle protein metabolism. Pathways or processes in which there is increased flow in the presence of insulin are shown with the thick arrows; those in which there is reduced flow are shown by the thin arrows.

cellular machinery required for protein synthesis). At the same time, insulin strongly inhibits protein degradation in muscle (see Fig. 15–9). Therefore, the overall effect of insulin in muscle is to promote the accumulation of muscle protein.

Insulin has similar effects on protein metabolism in the liver and adipose tissue; it increases protein synthesis and decreases protein degradation. In addition, as described earlier, insulin inhibits gluconeogenesis in liver. As a result, fewer amino acids are converted into glucose, and more amino acids are available for protein synthesis.

As a result of its anabolic effects on protein metabolism, insulin produces a **positive nitrogen balance,** meaning a net accumulation of protein in the body. By contrast, when insulin is deficient, as in diabetes mellitus, there is a net loss of protein, or a **negative nitrogen balance.** Insulin therefore serves as an important determinant of tissue growth.

Glucagon Has Effects on Metabolism that Oppose Those of Insulin

> *How does glucagon promote mobilization of fuel?*

As will be evident in sections that follow, glucagon affects many of the same metabolic pathways as insulin. However, in each instance, the net effect of glucagon is opposite that of insulin.

The mechanism of glucagon action is well established. Because it is a peptide hormone, glucagon interacts with a receptor on the plasma membrane of its target cells. This receptor is coupled to adenylate cyclase and results in an increase in the concentration of cAMP inside the cell. As described earlier in Chapter 5, cAMP activates several protein kinases inside the cell, leading to changes in the phosphorylation state and activity of various regulatory proteins and enzymes.

The principal target site of glucagon is liver, where its overall effect on metabolism is catabolic. Some effects of glucagon on adipose tissue have also been reported, but these usually require a relatively high concentration of the hormone, and thus their physiological significance is uncertain.

Effects of Glucagon on Carbohydrate Metabolism

The overall effect of glucagon action is an increase in blood glucose. This occurs as the result of three separate effects that glucagon has on the liver (Fig. 15–10). First, glucagon stimulates conversion of glycogen into glucose (**glycogenolysis**) while inhibiting glycogen synthesis (see Fig. 15–10). The net result is an increase in the conversion of liver glycogen into glucose, which enters the blood. The second means by which glucagon increases blood glucose is by stimulating the conversion of nonglucose substrates into glucose (**gluconeogenesis**) in the liver (see Fig. 15–10). Third, as the

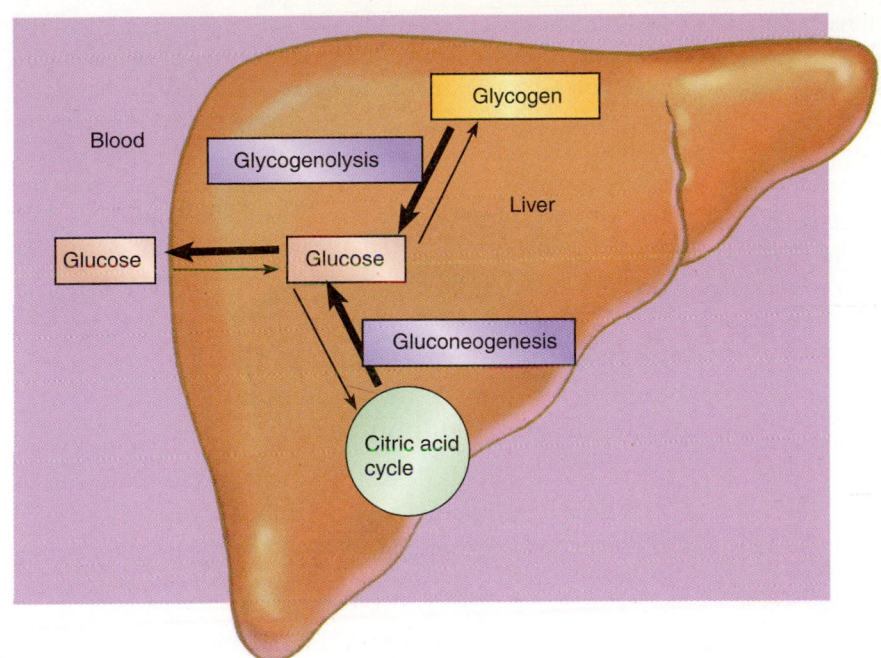

Figure 15–10

Summary of the effects of glucagon on liver carbohydrate metabolism. Pathways or processes in which there is increased flow in the presence of glucagon are shown with the thick arrows; those in which there is reduced flow are shown by the thin arrows.

following section will describe, glucagon affects lipid metabolism, resulting in glucose "sparing," which tends to increase blood glucose. To review the opposing effects of insulin and glucagon on carbohydrate metabolism, compare Figure 15–7 with Figure 15–10.

Effects of Glucagon on Lipid Metabolism

Effects of glucagon on lipid metabolism oppose those of insulin. In liver, glucagon stimulates hormone-sensitive lipase and thereby stimulates lipolysis (Fig. 15–11). In addition to

being oxidized and used directly for energy production in muscle, fatty acids can be partially metabolized and converted in liver into **ketone bodies.** The ketone bodies, which are four-carbon compounds, are released by the liver and can be taken up and used by peripheral tissues for energy. Glucagon stimulates the formation of ketones (**ketogenesis**) in liver. Skeletal muscle and heart both derive a considerable amount of their energy needs from the oxidation of ketones. In fact, ketones are a major source of energy for the heart. By stimulating ketogenesis in liver, glucagon

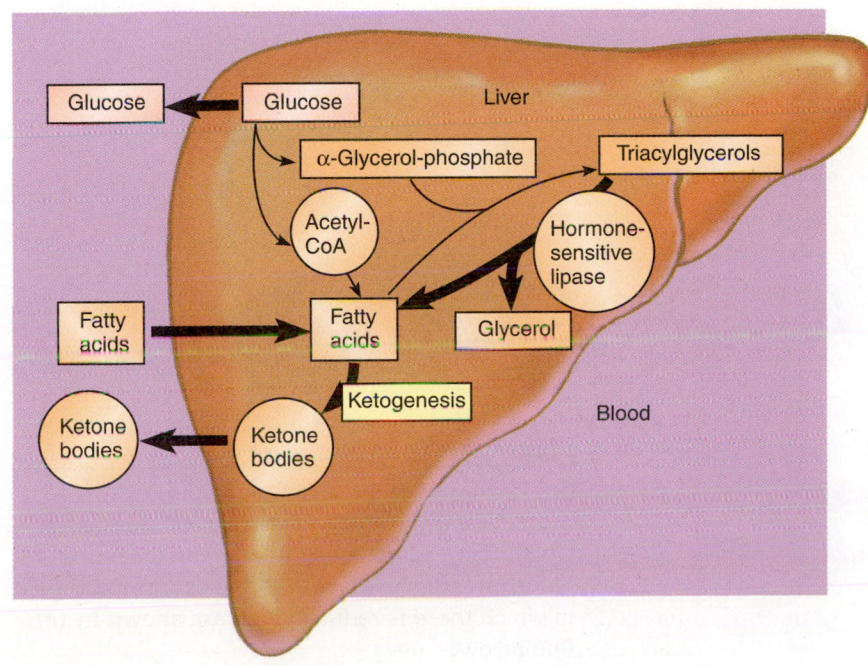

Figure 15–11

Summary of the effects of glucagon on liver fat metabolism. Pathways or processes in which there is increased flow in the presence of glucagon are shown with the thick arrows; those in which there is reduced flow are shown by the thin arrows.

provides increased amounts of ketones for use by muscle and heart. This has the effect of sparing glucose, which, when coupled with other effects of glucagon, also serves to increase blood glucose.

Effects of Glucagon on Protein Metabolism

In liver, glucagon is a potent stimulator of protein degradation (Fig. 15–12). As free amino acids are released when liver proteins are broken down, they can be used for glucose synthesis via gluconeogenesis. Glucagon has an additional effect of increasing the supply of amino acids for gluconeogenesis by stimulating transport of amino acids from blood into liver. As more amino acids are converted into glucose, more ammonia, a by-product of amino acid metabolism, must be excreted. To accomplish this, glucagon also stimulates **urea synthesis** (see Fig. 15–12).

As indicated earlier, amino acids stimulate secretion of *both* insulin and glucagon from the pancreas. In almost every case, actions of insulin are opposite to those of glucagon. Depending on the particular metabolic state, it is desirable to have either insulin or glucagon present in high concentrations in the blood, but usually not both, because their opposing actions would cancel out one another. However, the apparent paradoxical effect of amino acids to stimulate *both* insulin and glucagon secretion serves an important purpose. When one consumes a meal high in protein, it is desirable to use the amino acids in the meal for protein synthesis in tissues, and thus insulin secretion is beneficial under these conditions. However, when the meal is also low in carbohydrate,

very little glucose enters the blood from the GI tract, and the actions of insulin to decrease blood glucose would cause the blood glucose concentration to decrease to very low levels. It is therefore beneficial for glucagon to also be secreted in this situation. The effects of glucagon oppose the hypoglycemic (blood-glucose-lowering) actions of insulin, and therefore the blood glucose concentration is maintained near normal while insulin is able to promote the efficient use of amino acids for protein synthesis in various tissues.

The Insulin/Glucagon Ratio in the Blood Is an Important Determinant of the Flow of Metabolites

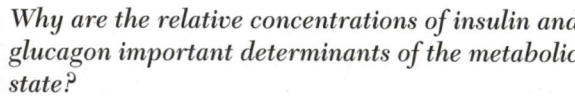

Why are the relative concentrations of insulin and glucagon important determinants of the metabolic state?

Because of opposing actions of insulin and glucagon, it has been suggested that the **insulin-to-glucagon (I/G) ratio,** rather than absolute concentrations of either hormone, is the primary factor determining metabolic status in the body. Thus, when the I/G ratio is high (insulin high, glucagon low), effects of insulin predominate, and an anabolic state exists. Excess fuels are stored as glycogen and triacylglycerols, and tissue protein synthesis increases. Conversely, when the I/G ratio is low (insulin low, glucagon high), effects of glucagon predominate, and tissues are in a catabolic state.

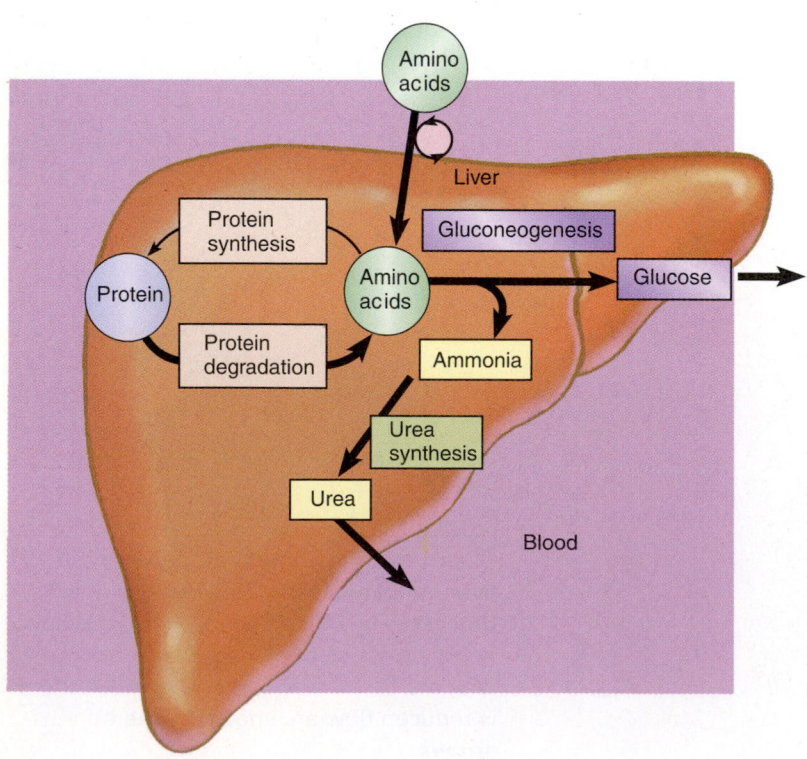

Figure 15–12

Summary of the effects of glucagon on liver protein metabolism. Pathways or processes in which there is increased flow in the presence of glucagon are shown with the thick arrows; those in which there is reduced flow are shown by the thin arrows.

Shortly after a high-carbohydrate meal, an individual's molar ratio of insulin to glucagon can be as high as 30. Conversely, when one awakens in the morning after an overnight fast, the I/G ratio is approximately 2. If the fast lasts for a day or two, the I/G ratio may decrease to 0.5 or less. Thus, in going from a fully fed state to a fasted state, the I/G ratio may change by a factor of 50 to 60.

As an example of the importance of the I/G ratio in determining the relative metabolic state, consider the situation that can exist in cases of diabetes mellitus in which insulin secretion is deficient. In the absence of insulin, blood glucose increases dramatically. Normally, a high blood glucose concentration inhibits glucagon secretion (refer to Table 15–3), but in the absence of insulin, this inhibitory effect is lost, and therefore plasma glucagon levels increase. In such cases, the I/G ratio is extremely low. The catabolic effects of glucagon greatly predominate, leading to increased glycogenolysis, lipolysis, ketogenesis, and gluconeogenesis. Blood glucose concentrations increase to extremely high levels, and a negative nitrogen balance exists. Thus, the I/G ratio is more important in determining the overall metabolic state than is the absolute concentration of either hormone.

OVERVIEW OF METABOLIC REGULATION BY PANCREATIC HORMONES: THE FED STATE VERSUS THE FASTED STATE

> *Based on the above discussions, what events would you expect to occur immediately following a meal and during a one-day fast?*

As described earlier, use of energy by the body is continuous, whereas the intake of energy in the form of food is intermittent. This situation requires that fuels be stored in the body for use during between-meal periods or for periods when food is not readily available.

Excess Fuel Is Stored in the Body in Several Forms

In Table 15–4 the chemical nature and tissue distribution of stored fuels in an average 70-kg (155-lb) person is shown. Note that most calories in the body are stored in the form of triacylglycerols in adipose tissue. By comparison, relatively little of the total stored fuel consists of glycogen in either liver or muscle. Because liver glycogen can be broken down directly into free glucose, which is then released into the blood, one might expect that it would be a better form in which to store excess calories. However, compared with glycogen, triacylglycerols are considered a much more efficient means of storing excess fuel for several reasons. First,

TABLE 15–4

Distribution of Stored Fuels in the Body

Storage Form	Site of Storage	Energy Stored (kcal)
Fat (triacylglycerols)	Adipose tissue	141,000
Glycogen	Muscle	480
Glycogen	Liver	280
Protein	Muscle	24,000

1 g of pure fat contains approximately twice the calories as does 1 g of carbohydrate. In other words, triacylglycerols have approximately twice the **caloric density** of glycogen. Secondly, glycogen is very hydrophilic, whereas fat is hydrophobic. Thus, glycogen attracts a considerable amount of water, which adds a significant amount of weight to the tissues in which glycogen is stored. If we stored all of our excess calories as glycogen and not as fat, the average person might weigh approximately 700 lb. Obviously, this would be a very inefficient way in which to store fuel because this excess fuel would represent a considerable burden to move about.

Note also in Table 15–4 that muscle protein represents approximately 17% of the total stored fuels in the body. During a prolonged fast, the body can and does call upon its muscle protein reserves as a source of stored energy.

Flow of Nutrients after a Meal Is Largely Directed by Insulin

By examining nutrient flow in both the fed and fasted states, we can review the effects of insulin and glucagon on metabolism and at the same time begin to appreciate nutrient flow between organs.

Figure 15–13 summarizes the metabolic state expected in liver, muscle, and adipose tissue of someone who has just eaten a meal rich in carbohydrate. Remember that in this situation the concentration of insulin in the blood is high, whereas glucagon is low. In liver, insulin promotes conversion of glucose into glycogen, so that liver glycogen stores are increased (see Fig. 15–13). Insulin also stimulates fatty acid synthesis from glucose in liver. Some of the fatty acids are stored as triacylglycerols, but most are packaged as lipoproteins and exported to adipose tissue (see Fig. 15–13). In adipose tissue, lipoprotein lipase (LPL) is stimulated by insulin, resulting in increased uptake of lipoproteins from the blood. Entry of glucose into adipose tissue is also stimulated under these conditions, as is synthesis of fatty acids and α-glycerol phosphate (see Fig. 15–13). Triacylglycerol formation from fatty acids and α-glycerol phosphate is also stimulated, resulting in expansion of fat stores in adipose tissue. In skeletal muscle, glucose uptake is stimulated, as is glycogen synthesis (see

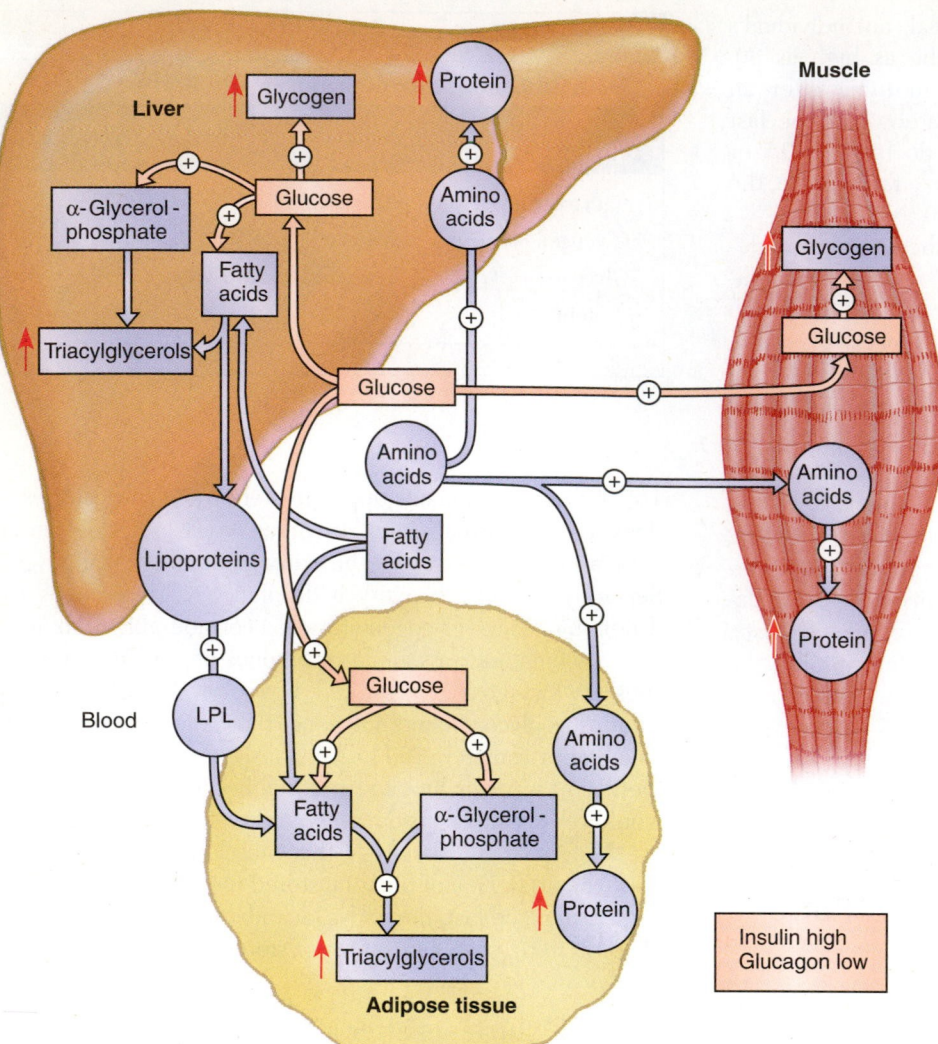

Figure 15–13

Summary of nutrient flow immediately after a meal. Those processes accelerated due to the prevailing high insulin/low glucagon concentrations are indicated by the ⊕.

Fig. 15–13). Thus, the amount of glycogen stored in muscle increases under these conditions. Uptake of amino acids into liver, muscle, and adipose tissue also increases in the fed state in response to insulin. In each tissue, insulin also stimulates the conversion of amino acids into protein while inhibiting protein degradation. Thus, protein stores in each of the three tissues increase (see Fig. 15–13). Any fatty acids taken in with a meal are either taken up directly by adipose tissue and stored as triacylglycerol or taken up by liver and then packaged as lipoproteins before export to adipose tissue.

Glucagon Influences the Flow of Nutrients During Fasting

Figure 15–14 summarizes the state of nutrient flow expected during a period of fasting. In this situation, the blood insulin levels are low while glucagon levels are high. In liver, glyco-

gen breakdown and gluconeogenesis are stimulated in an attempt to maintain blood glucose levels near normal (see Fig. 15–14). In the absence of insulin, breakdown of muscle protein is increased, and amino acids released by muscle are used in liver to synthesize glucose via gluconeogenesis. Glucagon also stimulates amino acid uptake and urea synthesis, thereby facilitating the use of amino acids by the liver (see Fig. 15–14). In the absence of insulin, glucose entry into muscle and adipose tissue is greatly reduced, which also contributes to maintenance of blood glucose levels. In adipose tissue, lipolysis predominates over lipid synthesis under these conditions. A result is increased release of fatty acids and glycerol from adipose tissue, which leads to increased fatty acids in the blood (see Fig. 15–14). Fatty acids can be used by muscle to obtain energy and therefore provide muscle with an alternate fuel to use instead of glucose. Glycogen breakdown in muscle also increases in the absence of insulin,

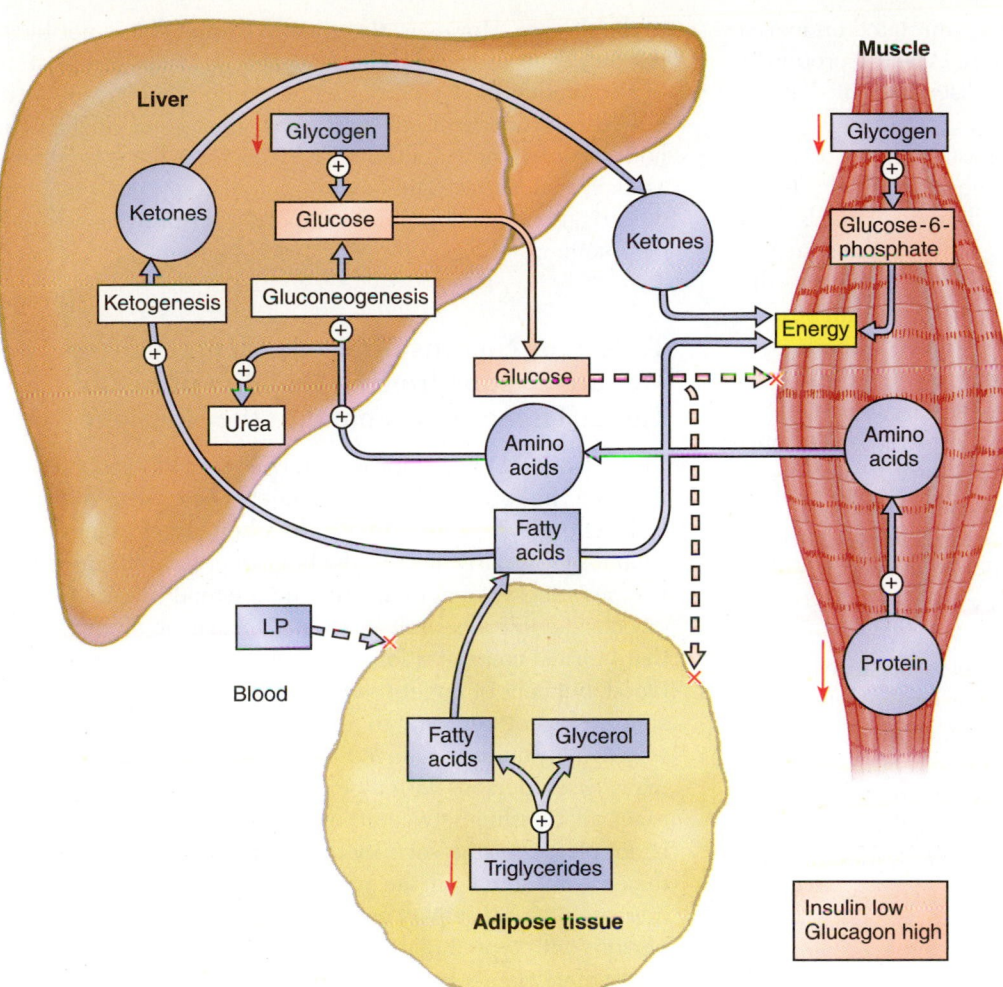

Figure 15–14

Summary of nutrient flow during fasting. Processes accelerated under the conditions of low insulin/high glucagon are indicated by the ⊕. (LP = lipoproteins.)

which provides another source from which muscle can derive energy.

In liver, glucagon stimulates ketogenesis, and, as a result of increased conversion of fatty acids into ketones, plasma ketone concentrations increase dramatically. Ketones can be used by muscle as an energy source, which again has a sparing effect on plasma glucose. If fasting continues for several days, an important adaptation involving ketones takes place. As stated earlier, the CNS normally requires glucose as its energy source. However, with fasting, the CNS adapts and begins to use ketone bodies for energy. This adaptation is important because it lessens the need to convert muscle protein into glucose. As Table 15–4 shows, considerably more calories are present in the body in the form of fat than are present as protein in muscle.

To summarize, with fasting, glycogen stores in both liver and muscle are depleted, and muscle protein is broken down; amino acids are used for glucose synthesis in liver, and triacylglycerol stores in adipose tissue are broken down,

As a result of these changes, plasma fatty acid and ketone concentrations increase, and the blood glucose concentration is prevented from decreasing to a dangerously low level.

DIABETES MELLITUS

Where in the mechanisms associated with insulin action can the system fail, and what are the consequences of such events?

Diabetes mellitus was recognized as early as the first century A.D., when the term *diabetes,* which stems from the Greek word for "siphon," was applied to a disease characterized by a marked increase in urine volume. Note that the term *diabetes* is commonly used (as it will be here) to refer to diabetes mellitus and should not be confused with diabetes insipidus, which is an entirely unrelated endocrine abnormality caused by a deficiency in ADH secretion or action (refer to other chapters for a discussion of diabetes insipidus).

Diabetes mellitus is by far the most common of all endocrine disorders and is a worldwide health problem. In the United States, approximately 10 million individuals have been diagnosed as having diabetes, and estimates suggest that an additional 5 to 6 million more may be borderline diabetics not yet diagnosed. Diabetes is the seventh leading medical cause of death and the leading cause of blindness in the United States. Therefore, research on the causes and cures of diabetes is an area of intense interest.

Diabetes mellitus is not a single disease as was once thought. Instead, it is now well known that diabetes comprises a heterogeneous group of disorders that differ in both cause and severity. The common characteristic of every form of diabetes is high blood glucose concentration, which in fact is the primary means by which the disease is diagnosed.

Elevation in blood glucose in diabetes is the result of a deficiency, either relative *or* absolute, in insulin action. The deficiency in insulin action is most commonly due to either (1) inadequate insulin secretion by beta cells of the pancreas or (2) a relative lack of response by target tissue cells to insulin. Inadequate insulin secretion results in the form of diabetes referred to as **insulin-dependent diabetes mellitus (IDDM), or type 1 diabetes.** The other common form is caused by a relative lack of insulin action in its target tissues and is known as **non-insulin-dependent diabetes mellitus (NIDDM), or type 2 diabetes.** Other classifications of diabetes are also well known but make up only approximately 2% of all cases. Often, these forms are secondary to or associated with other conditions. One example is **gestational diabetes,** a transient condition that occurs in approximately 3% of all pregnant women.

Type 1, or Insulin-Dependent, Diabetes Mellitus Results from an Inability of the Pancreas to Produce Adequate Amounts of Insulin

Type 1 diabetes (IDDM) is characterized either by very low levels of insulin or by an absolute lack of insulin secretion by the pancreas. To compensate for this deficiency and control their disease, individuals with IDDM must receive daily insulin injections. Approximately 10% of all diabetics suffer from this form of the disease.

The loss of beta cell function usually occurs during adolescence, although it can occur later in life. The loss of beta cell function is due to an autoimmune disorder that causes the immune system to attack and destroy the pancreatic beta cells. Causes of the autoimmune disorder are not well understood, but at least two major components appear to contribute to the disease state. The first is a genetic component in which certain individuals have an increased susceptibility

to the disease. However, the genetic component is not by itself sufficient to cause the autoimmunity. The effects of an as yet unidentified environmental component are also required to produce the disease. What that agent might be is not currently known, although viruses have been suggested as likely candidates. The environmental component in some way triggers the autoimmune response, resulting in beta cell destruction.

Type 2, or Non-Insulin-Dependent, Diabetes Mellitus Results from an Inability of Insulin Target Tissues to Respond to the Hormone

In the United States, approximately 90% of all diabetics suffer from the form of the disease known as **type 2 diabetes** (NIDDM). Type 2 diabetes results when target tissues of insulin lose responsiveness to the hormone. Reasons for loss of responsiveness are not currently understood but are an area of considerable research interest. Normal or often higher-than-normal concentrations of insulin may be present in the blood, but cells of target tissues, such as adipose and muscle, simply do not respond to the hormone as they normally would, a condition referred to as **insulin resistance.** In many cases, persons with type 2 diabetes are also obese. The exact relationship between the obesity and insulin resistance is not clear, but the severity of the disease can often be reduced considerably if the patients are placed on a diet and increase physical activity and lose weight. In addition to diet, persons with type 2 diabetes are often treated with oral hypoglycemic agents known as **sulfonylureas,** which, as indicated in Table 15–2, stimulate insulin secretion from beta cells. In addition to stimulating insulin secretion, sulfonylureas also appear to augment the actions of insulin in tissues and therefore help overcome the insulin resistance in target tissues.

The cause of insulin resistance in type 2 diabetes is poorly understood. As indicated previously, there does appear to be some link with obesity, but obesity per se is not sufficient to cause type 2 diabetes. There does, however, appear to be a strong genetic component to type 2 diabetes. Chances are nearly 100% that if one genetically identical twin develops type 2 diabetes, the other will also, even if they are raised separately under entirely different environmental, social, or economic conditions. In addition to genetics and obesity, there is also an association with type 2 diabetes and a sedentary lifestyle.

Diabetes Mellitus Can Result in both Short- and Long-Term Complications

Several immediate complications can arise if the glucose concentration in the blood is not checked by insulin; these problems include hyperglycemia, ketoacidosis, and electrolyte imbalance. In addition, long-term complications of the con-

APPLICATIONS OF PHYSIOLOGY

The Diabetic Foot

Despite efforts to control their disease and maintain a normal glycemic state, most diabetics eventually suffer from one or more secondary complications of the disease. These complications may be somewhat subtle in onset and slow in progression, but they nonetheless are responsible for most morbidity and mortality in diabetes. While the specific mechanisms involved remain areas of debate and research activity, it is clear that most secondary complications are vascular or neural in nature.

Vascular complications may involve atherosclerotic-like lesions in the large blood vessels or impaired function in the microcirculation. Damage to the basement membrane of capillaries in the eye (**diabetic retinopathy**) or kidney (**diabetic nephropathy**) is common. Although no satisfactory direct treatment exists for diabetic vascular disease, its progression is often monitored closely as an indirect indicator of the overall diabetic state.

Diabetic neuropathy typically involves symmetric sensory loss in the distal lower extremities, or autonomic neuropathy leading to impotence, gastrointestinal dysfunction, or anhidrosis (lack of sweating) in the lower extremities. The diabetic foot is an example of several complicating factors of diabetes exacerbating one another. From 50% to 70% of all nontraumatic amputations in the United States each year are due to diabetes. Breakdown of the foot in diabetics is commonly due to a combination of neuropathy, vascular impairment, and infection. In a typical scenario, small lesions or ulcers of the foot result from dryness of the skin due to a combination of neural and vascular complications. Impairments in sensory nerve function may result in these small lesions going unnoticed by the patient until a severe infection or gangrene has become well established.

Loss of the affected foot or limb often can be avoided with patient and physician education. The focus on management of diabetic patients often is aimed at maintaining a normal blood glucose level, avoiding primary complications, such as diabetic ketoacidosis or hyperosmolar coma, and the initial secondary complications, such as diabetic retinopathy. However, there is increasing awareness that the feet of the diabetic patient should be assessed on each patient visit. Results of one study show that the likelihood of amputation is reduced by half if diabetic patients simply remove their shoes for foot inspection during every outpatient clinic visit. Thus, while the underlying physiological mechanisms of the problem may be complex, the problem of the diabetic foot can be avoided or delayed by rigorous health care management.

---

dition include vascular problems, vision problems, kidney problems, and impairment of nerve function.

Acute Complications of Diabetes

As indicated previously, in the absence of either insulin secretion or insulin action, the blood glucose concentration increases markedly, a condition termed **hyperglycemia.** If hyperglycemia becomes severe, the concentration of glucose in the blood can exceed the capacity of the kidneys to recapture glucose by active transport, and glucose is excreted in the urine (a condition called **glucosuria**). Because of osmotic effects, glucose in the urine also attracts considerable amounts of water, which is excreted together with the glucose. As a result, urine volume and frequency of urination increase (a condition known as **polyuria**). As increased amounts of water are lost in the urine, **dehydration** of the individual can also occur.

In type 1 (IDDM) diabetics, the unopposed actions of glucagon result in increased ketone formation by the liver. The primary ketones that are formed are acids, and when they are produced in sufficiently large amounts, the normal acid-base balance in the body is considerably disturbed, resulting in the condition termed **ketoacidosis.** As with glucose, if ketones reach a high enough concentration in the blood, they will begin to "spill over" into the urine. As this spillover occurs, ketones also carry cations, such as sodium and potassium with them into the urine. Thus, accompanying severe ketoacidosis is a loss of these ions, resulting in an **electrolyte imbalance** in the body. If left untreated, severe ketoacidosis accompanied by dehydration can rapidly result in coma and death.

Chronic Secondary Complications of Diabetes

In addition to acute complications that may occur, several secondary complications usually accompany longstanding diabetes. These secondary complications often involve gradual changes that develop over a period of years and may shorten the life expectancy of these individuals. The most common secondary complications of diabetes are seen within the vascular system. Changes much like those seen with atherosclerosis lead to the narrowing of larger blood vessels in the brain, heart, and lower extremities. The resulting reduction in circulation to these areas may result in stroke, heart attack, or loss of limb, respectively. Lesions in the microvasculature (small blood vessels and capillaries) are also common in diabetics and are most detrimental in the kidney and in the retina of the eye. These changes can result in kidney disease (**diabetic nephropathy**) and in the condition termed **diabetic retinopathy.** Diabetic retinopathy is due to the deterioration of blood vessels that nourish the retina. As the blood vessels in the retina break

down, scar tissue is formed that eventually interferes with the light-sensing function of the retina. Each year, several thousand diabetics become legally blind as a result of retinopathy.

Another common secondary complication of diabetes involves impairment in nerve function (**diabetic neuropathy**). Fibers of the autonomic nervous system are often involved, frequently resulting in abnormalities in bladder or GI tract function or in impotence. Diabetic neuropathy also frequently involves peripheral sensory nerves, resulting in loss of feeling in the lower limbs in particular. As described in the "Applications of Physiology" box, there is a high risk of limb amputation in diabetes.

CHAPTER REVIEW

Summary

- Functional units of the endocrine pancreas are groups of cells known as the *islets of Langerhans.* The islets are composed of four major cell types: alpha cells, which secrete glucagon; beta cells, which secrete insulin; delta cells, which secrete somatostatin; and F cells, which secrete pancreatic polypeptide.
- Insulin is a peptide hormone. The most important physiological stimulus for insulin secretion is an increased blood glucose concentration. Glucagon is a polypeptide hormone whose secretion is stimulated by a decreased blood glucose concentration. Somatostatin is a polypeptide, the secretion of which is stimulated by increased blood glucose and glucagon.
- Insulin has metabolic effects that are anabolic in nature; its most obvious effect is to rapidly lower blood glucose as a result of increasing glucose uptake into muscle and adipose tissue by facilitated diffusion. Insulin also stimulates glycogen formation and glycolysis in its target cells and decreases glucose output by the liver. In liver and adipose tissue, insulin stimulates fatty acid synthesis and promotes the storage of fatty acids as triacylglycerols. In adipose tissue, insulin promotes the uptake of lipoproteins into cells and inhibits lipolysis. Insulin stimulates the accumulation of protein in most tissues of the body, thus promoting a positive nitrogen balance.
- Glucagon has catabolic effects on metabolism and acts primarily on liver. Glucagon increases blood glucose by stimulating glycogenolysis and gluconeogenesis in liver. Glucagon also stimulates lipolysis as well as ketogenesis in liver. In addition, glucagon provides increased substrate for gluconeogenesis in liver by stimulating protein degradation and amino acid transport.
- Because of the opposing actions of insulin and glucagon, the status of nutrient flow and metabolism in the body is determined by the relative amounts of the two hormones. A high I/G ratio exists in an anabolic state; a low I/G ratio is found in a catabolic state.
- The principal storage form of calories in the body is triacylglycerol (fat). Fat has a greater caloric density than glycogen and is hydrophobic. It is therefore a more efficient storage form than glycogen.
- Shortly after a meal, the actions of insulin on metabolism predominate over those of glucagon. As a result, glycogen stores are increased in liver and muscle, triacylglycerol stores are increased in liver and adipose tissue, and protein synthesis is promoted in all three tissues.
- In the fasting state, glucagon levels are high, insulin levels are low, and nutrient flow is exactly opposite to that in the fed state. Glycogen stores in liver and muscle are depleted; muscle protein is broken down, and the amino acids used for gluconeogenesis in liver and triacylglycerol stores in adipose tissue are broken down.
- Diabetes is a common endocrine disorder caused by either an absolute or relative deficiency in insulin action. Type 1 diabetes, or insulin-dependent diabetes mellitus (IDDM), is caused by destruction of insulin-secreting beta cells of the pancreas. Type 2 diabetes, or non-insulin-dependent diabetes mellitus (NIDDM), is caused by lack of responsiveness of insulin target tissues to the hormone, a condition termed *insulin resistance.*
- Hyperglycemia, glucosuria, polyuria, and dehydration are common acute complications of diabetes. Type 1 diabetics are also prone to ketoacidosis. Chronic secondary complications of diabetes include diabetic retinopathy, diabetic neuropathy, and vascular disease.

Review Questions

Choose the Correct Answer

1. Which cell type is most numerous in the islets of Langerhans?
   a. Alpha cell
   b. Beta cell
   c. Delta cell
   d. E cell
   e. F cell

2. Which of the following is *incorrectly* paired?
   a. Delta cell : somatostatin
   b. Alpha cell : pancreatic polypeptide
   c. Beta cell : insulin
   d. Exocrine cell : digestive enzymes
   e. All of the above pairs are correct

3. In a typical islet of Langerhans, the insulin-secreting cells are located:
   a. on the periphery of the islet.
   b. randomly distributed in the islet.
   c. in the core of the islet.
   d. in the exocrine tissue, just outside of the islet.
   e. none of the above.

4. The autonomic nervous system can:
   a. inhibit insulin secretion.
   b. stimulate insulin secretion.
   c. stimulate glucagon secretion.
   d. all of the above.
   e. b and c only.

5. Insulin secretion from the pancreas increases in response to:
   a. ingestion of a high protein meal.
   b. decreased blood glucose concentration.
   c. activation of the sympathetic nervous system.
   d. somatostatin.
   e. none of the above.

6. Of the following, which would be considered to be a normal fasting blood glucose value in a normal healthy adult?
   a. 100 mg/dL
   b. 800 mg/dL
   c. 100 mEq/dL
   d. 20 mg/dL
   e. 350 mg/dL

7. In general, the effects of insulin are considered to be:
   a. catastrophic.
   b. catabolic.
   c. anisotropic.
   d. enigmatic.
   e. anabolic.

8. Insulin stimulates glucose uptake into:
   a. the liver.
   b. the skeletal muscle.
   c. the brain.
   d. the adipose tissue.
   e. b and d.

9. The majority of stored fuel in the body is present as:
   a. triacylglycerol in muscle.
   b. glycogen in liver.
   c. blood glucose.
   d. protein in muscle.
   e. triacylglycerol in adipose tissue.

10. Insulin stimulates glucose transport into its target cells by:
    a. indirectly inhibiting glucagon secretion.
    b. increasing the number of ATP-dependent glucose transporters.
    c. promoting the insertion of glucose transporters into the plasma membrane.
    d. stimulating gluconeogenesis.
    e. inhibiting amino acid transport.

11. Among the effects of insulin on carbohydrate metabolism are:
    a. inhibition of glycogenolysis.
    b. stimulation of glycolysis.
    c. stimulation of glycogen synthesis.
    d. glucose uptake into muscle.
    e. all of the above.

12. In the liver, glucagon:
    a. stimulates ketogenesis.
    b. inhibits glycogenolysis.
    c. stimulates glucose uptake.
    d. inhibits gluconeogenesis.
    e. all of the above.

13. In adipose tissue, insulin:
    a. stimulates hormone-sensitive lipase and lipoprotein lipase.
    b. promotes lipolysis.
    c. stimulates urea synthesis.
    d. stimulates glucose uptake.
    e. stimulates glycogenolysis.

14. Which of the following is not usually associated with insulin-dependent diabetes mellitus (IDDM)?
    a. Autoimmune disease
    b. High blood glucose concentrations
    c. Ketoacidosis
    d. Low blood glucagon
    e. Retinopathy

15. The insulin/glucagon ratio (I/G ratio) would be expected to be lowest:
    a. immediately after a high-carbohydrate meal.
    b. immediately after a high-protein meal.
    c. after an overnight fast.
    d. after three days of fasting.
    e. shortly after taking an oral sulfonylurea drug.

16. NIDDM, or type 2 diabetes:
    a. bears a strong genetic component to the development of the disease.
    b. is typified by low or negligible circulating insulin.
    c. occurs only in obese individuals.
    d. is treated in the same manner as type 1 diabetes.
    e. is a relatively rare form of diabetes.

17. Which of the following would you least likely see in a person with long-standing NIDDM, or type 2 diabetes?
    a. Diabetic neuropathy
    b. Diabetic nephropathy
    c. Diabetic retinopathy
    d. Diabetic ketoacidosis
    e. c and d

18. Normally in the fed state the central nervous system relies on what substrate for its energy?
    a. Fatty acids
    b. Glucose

c. Endorphins
d. Amino acids
e. Prostaglandins

19. Insulin produces its effect in target tissues by:
    a. increasing intracellular cAMP.
    b. the IP_3 and DAG system.
    c. the gene expression model of hormone action.
    d. receptors having intrinsic phospholipase C (PLC) activity.
    e. receptors having intrinsic tyrosine kinase activity.

20. Of the total body content of protein, approximately what percentage is present in skeletal muscle?
    a. 40
    b. 10
    c. 2
    d. 0.1
    e. 80

Answers to Case History Questions

1. The extreme hyperglycemia (high blood glucose) exhibited by Henry argues against the idea of insulin shock, in which extreme hypoglycemia would be expected. Rather, all the findings in this case are consistent with the classic signs and symptoms of diabetic ketoacidosis. Two of the ketone bodies, acetoacetate and β-hydroxybutyrate are acids. In the insulin-deficient state, ketogenesis occurs unabated in the liver, leading to an increase in plasma ketone levels. Bicarbonate is the primary buffer defense against acidosis and so bicarbonate is consumed to combat the acidosis. The third ketone, acetone, is not an acid and does not contribute to the acidosis but may build up to the point that it can be detected on the breath.

2. The initial way in which the body compensates for a metabolic acidosis is to increase ventilation to the lungs. The deep and rapid breathing that Henry demonstrates is known as *Kussmaul breathing*. Henry's low blood pressure is the result of dehydration caused by the hyperglycemia and elevated blood ketones. As Henry spills glucose, ketones, and electrolytes into the urine, water follows passively by "osmotic drag." As blood pressure begins to decrease, the heart is stimulated to beat more rapidly in an attempt to maintain blood pressure near normal.

3. Henry's immediate need is to have his circulating blood volume, electrolytes, and acid-base status restored to normal. This would be accomplished by some fairly rapid intravenous drips of the appropriate solutions. Insulin treatment would also be initiated in order to bring Henry back into glycemic control.

Key Terms

alpha cell (p. 454)
beta cell (p. 454)
C-peptide (p. 455)
caloric density (p. 465)
diabetes mellitus (p. 467)
diabetic nephropathy (p. 470)
diabetic neuropathy (p. 470)
diabetic retinopathy (p. 470)
endocrine pancreas (p. 454)
exocrine pancreas (p. 454)

gestational diabetes (p. 468)
glucagon (p. 454)
glucosuria (p. 469)
hormone-sensitive lipase (p. 462)
hyperglycemia (p. 469)
insulin (p. 454)
insulin/glucagon (I/G) ratio (p. 464)
insulin resistance (p. 468)

islets of Langerhans (p. 454)
ketoacidosis (p. 469)
ketones (p. 463)
lipoprotein lipase (p. 461)
lipoproteins (p. 461)
negative nitrogen balance (p. 462)
pancreas (p. 454)
pancreatic polypeptide (p. 455)

polyuria (p. 469)
positive nitrogen balance (p. 462)
proinsulin (p. 455)
somatostatin (p. 454)
sulfonylureas (p. 458)
type 1 (IDDM) diabetes (p. 468)
type 2 (NIDDM) diabetes (p. 468)

Suggested Readings

Czech, M. P., "Molecular actions of insulin on glucose transport." *Annual Review of Nutrition,* 15:441–471, 1995.

Elmendorf, J. S., and Pessin, J. E., "Insulin signaling regulating the trafficking and plasma membrane fusion of GLUT4-containing intracellular vesicles." *Experimental Cell Research,* 253:55–62, 1999.

Ganong, W. F. *Review of Medical Physiology,* ed 18. Los Altos, Lange Medical Publications, 1997.

Rifkin, H., and Porte, D., Jr. *Ellenberg and Rifkin's Diabetes Mellitus: Theory and Practice,* ed 4. New York, Elsevier, 1990.

The Diabetes Control and Complications Research Group. "The effect of intensive treatment of diabetes on the development and progression of long term complications in insulin-dependent diabetes mellitus." *New England Journal of Medicine,* 329:977–985, 1993.

Wilson, J. D., and Foster, D. W., eds. *Williams Textbook of Endocrinology,* ed 8. Philadelphia, W.B. Saunders, 1992.

Web sites

http://www.nei.nih.gov/publications/retinopathy.htm
"Information for Patients: Diabetic Retinopathy." Produced by the National Eye Institute of the Federal government. Information for the lay person regarding diabetic retinopathy.

http://www.diabetes.org/default.asp
"Welcome to Your American Diabetes Association." Homepage of the ADA containing much useful information for the diabetic patient.

http://www.niddk.nih.gov/health/diabetes/diabetes.htm
"Diabetes." Produced by the NIDDK division of the National Institutes of Health. Contains links to a wealth of information about diabetes.

Answers to Review Questions

1. b   2. b   3. c   4. d   5. a   6. a   7. e   8. e
9. e   10. c   11. e   12. a   13. d   14. d   15. d
16. a   17. d   18. b   19. e   20. a

Part  III — INTEGRATIVE ORGAN FUNCTIONS

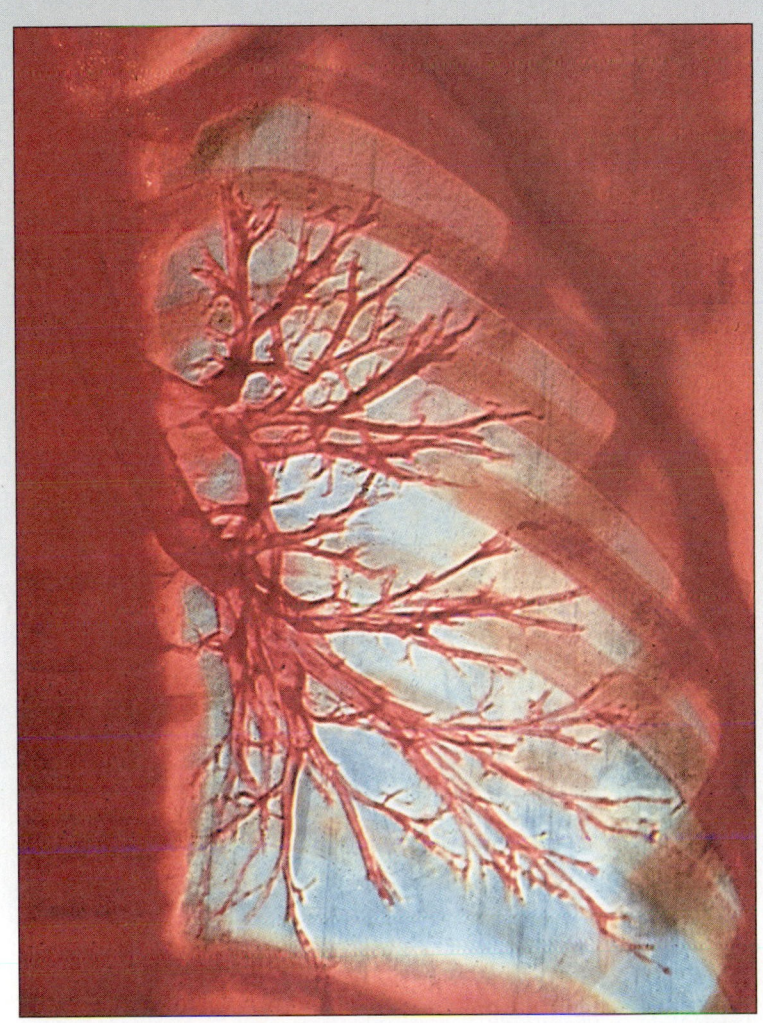

• *False-color arteriograph of a normal right lung showing the pattern of bronchial and bronchiolar branching.*

(© CNRI/SPL/Photo Researchers, Inc.)

Chapter 16

MUSCLE

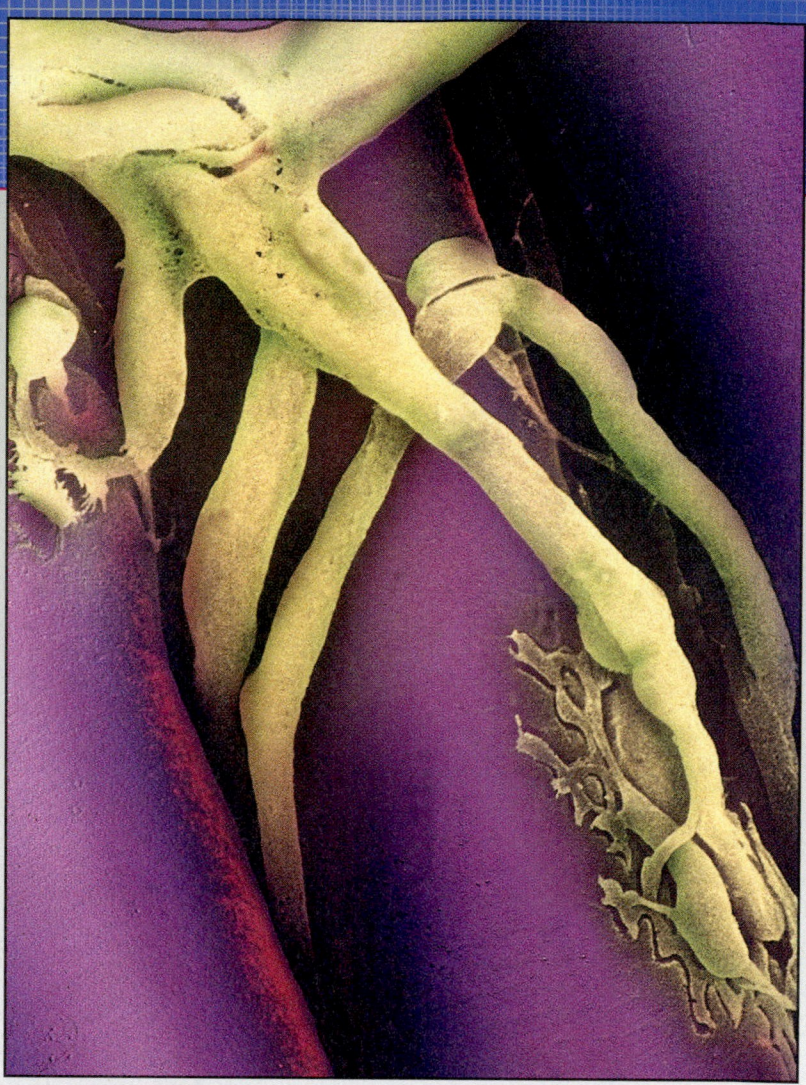

KEY CONCEPTS

- Muscles in the human body fall into three major categories. Skeletal muscle is a voluntary, striated, type used in functions such as locomotion and breathing. Smooth muscle is an unstriated type associated with blood vessels and visceral organs and involved with involuntary internal processes. Cardiac muscle is an involuntary, striated type that forms the muscle of the heart and powers the circulation of blood.

- Striated muscles are composed of two sets of overlapping protein myofilaments, called actin and myosin, the relative sliding motion of which, driven by the action of ATP on the myosin crossbridges, produces shortening and generates force. Smooth muscles, although they have a less regular internal structure, operate with a similar contractile protein system.

- Muscles can shorten at a constant force (isotonic contraction) or develop force without shortening (isometric contraction). The force-velocity curve characterizes isotonic contraction, and the length-tension curve summarizes the relationships involved in isometric contraction.

- The interface between the nervous system and skeletal muscle is the myoneural junction. This is an excitatory synapse that uses acetylcholine as a transmitter substance. Various chemicals and disease states can block transmission at the myoneural junction, and this action results in weakness or paralysis.

- The link between cell membrane action potentials and muscle contraction is provided by calcium ions. Skeletal and cardiac muscle have actin-linked control systems in which calcium acts on the thin filaments to release the inhibition of actin-myosin interaction. In smooth muscle, which has a myosin-linked control system, calcium ions cause the activation of the thick-filament myosin to promote contraction.

- The immediate fuel for muscle contraction is adenosine triphosphate, ATP. Its energy is derived from glucose and fatty acids by either aerobic or anaerobic metabolism. Muscles, according to their particular bodily role, are usually adapted to favor one or the other kind of metabolism. Cardiac muscle has a predominantly aerobic metabolism.

CASE HISTORY

ete is riding his bicycle through rush-hour traffic. He is aware, of course, that the skeletal muscles of his legs are providing the power that moves the bicycle forward. But many other muscles are at work to keep him moving safely on his way. The muscles of his arms, under the precise control of his central nervous system and his body's organs of balance, are performing hundreds of adjustments each minute to keep his cycle upright and on course. Other skeletal muscles are keeping his body erect and are making air move in and out of his lungs.

Deep within Pete's body, muscles of other types are also at work. The blood that carries oxygen and nutrients to his muscles and removes their waste products is pumped by the pulsing action of his heart muscle. The blood vessels supplying skeletal muscles and other vital organs are regulated by the smooth muscle in their walls, ensuring that an adequate supply of blood goes to the organs that need it most at the moment — the skeletal muscles and the brain.

Suddenly the horn of a truck sounds behind him. The reactions of all of Pete's muscles are rapid and — thanks to his nervous system — automatic. Tiny skeletal muscles in his middle ears quickly contract, protecting delicate hearing structures from the sudden loud noise. His heart speeds up and the contractions of its muscle become more powerful; his blood pressure rises, while the smooth muscles that control blood distribution allow even more of it to flow to the skeletal muscles. Even though the extra energy that his skeletal muscles must supply to move him out of harm's way outstrips the energy being brought to them by the bloodstream, specialized metabolic pathways and internal energy stores in the muscles allow for an emergency surge of power. All of these muscles working in concert allow him quickly to avoid the path of the speeding truck. As the truck roars past, Pete's body's systems gradually settle back to their nonemergency levels, and he continues on his way.

Questions

1. Compare the relative contraction speeds of the smooth and skeletal muscle responses when an emergency situation arises.

2. What energy sources were used by Pete's skeletal muscles during the emergency response?

3. In which direction (increase or decrease) did the contractility of Pete's heart muscle change?

4. Would smooth muscle or skeletal muscle be the first to react in this emergency situation? Why?

INTRODUCTION

For millennia prior to the Industrial Revolution, virtually all of human activity was powered by the muscles of human beings (slave and free) or by domesticated animals. Yet it is only within the past 100 years, and largely in the past 50 years especially, that we have developed our current understanding of how muscle is constructed and how it works. The development of research tools, such as the electron microscope; high-speed measuring and recording apparatuses; specialized biochemical techniques; and, most recently, the methods of molecular biology, has contributed to the extensive amount of knowledge that we now have about the function of muscle. While significant problems still remain to be solved, we have a solid understanding of muscle function based on studies in anatomy, physiology, biochemistry, and biophysics. This chapter attempts to outline the essentials of muscle physiology, especially as they relate to the overall function of the human body. The suggested readings at the end of the chapter provide the interested student with many opportunities for further study of this important subject.

 What do muscles do, and how can they be classified?

THE ROLES AND TYPES OF MUSCLE

The range of activities that muscles carry out in the body is extremely broad, so it is not surprising that muscles show a wide range of functional adaptations that specifically suit their many tasks in the overall function of the body.

Muscle Serves as the Output for the Central Nervous System

As described in Chapter 10, our central and peripheral nervous systems function by gathering information and instructing muscles to take useful action in response. This coordination between nerve and muscle enables us to walk, talk, digest our food, defend our bodies, propagate our species, and do almost everything else that we do. In addition to effecting changes prescribed by the nervous system, muscles also help to regulate what goes into that system by adjusting the sensitivity of our sensory organs. Some muscles are under neural control, whereas others are more independent. Very few muscles, however, are totally independent. The degree of neural control serves to adapt various muscles to their special roles.

Our other important control system, the endocrine system, also has muscle as one of its important effectors, or means of expression, although this function may not be very obvious when our bodies are functioning well.

Muscle Is a Biological Motor

Muscles can do what they do because they are a sort of biological motor. They enable our movements and bodily locomotion and help us to maintain posture. Like mechanical or electrical motors, they consume fuel while they do useful work, and waste some of the energy that they consume as heat. Muscles produce extra heat because they, like other motors, are not completely efficient. That is, they cannot turn all of the energy that they consume into mechanical work. The heat that they produce as a result of this inefficiency may at times be very important in overall bodily function. Under most conditions of activity, most of our body heat is produced as a result of the contraction of muscles. When the body temperature falls, brisk physical exercise (which may be relatively efficient) or shivering (which is very inefficient from a mechanical standpoint) provides the necessary warmth. During heavy exercise or under the stress of high environmental temperatures, our problem may involve getting rid of the excess heat produced by muscular contraction. Many of our conscious "cooling" mechanisms also involve the use of muscle, as may become apparent when the work of fanning to cool off produces more heating than cooling.

Muscle Aids in Regulating Physiological Processes

Muscles also help to regulate many important bodily functions. Muscles control the movement of substances through the tubular structures (e.g., blood vessels and intestines) of the body and expel those substances from the body at the proper time. For example, the regulation of blood pressure involves a complex interaction between the heart muscle, which pumps the blood, and smooth muscles that control the diameter of the blood vessels. Other regulatory mechanisms in which muscles play a central role include the maintenance of an upright posture and the control of the body temperature.

Muscle Can Be Classified in Several Ways

Several different sets of criteria are used to place muscles in separate categories relating to location and function, microscopic structure, or mode of control and action. These topics serve as convenient ways to describe and discuss particular aspects of muscle, but as with any attempt at classification, some exceptions have to be noted. The most usual sort of classification is shown in Figure 16–1.

Classification of Muscles by Location and Function

The most obvious muscles in the human body are called **skeletal muscles.** These are the large muscles used, among other things, for locomotion and for maintaining body posture. They usually attach to the skeleton and move the jointed bones with respect to one another. In some instances, the skeletal attachments may be indirect or may exist at only one end of the muscle (as is the case with the tongue, which is a skeletal muscle), or they may not exist at all (as is the case with the upper end of the esophagus). The bulk of skeletal muscles, however, are attached to bones at both ends.

Muscles that line the walls of the internal organs (the viscera) are called **visceral** muscles (called *smooth muscles* on the basis of their structure). Muscles of this type are also located in nonvisceral organs, such as blood vessels, and in sensory organs, such as the eye, where they aid in focusing and adapting to different levels of lighting.

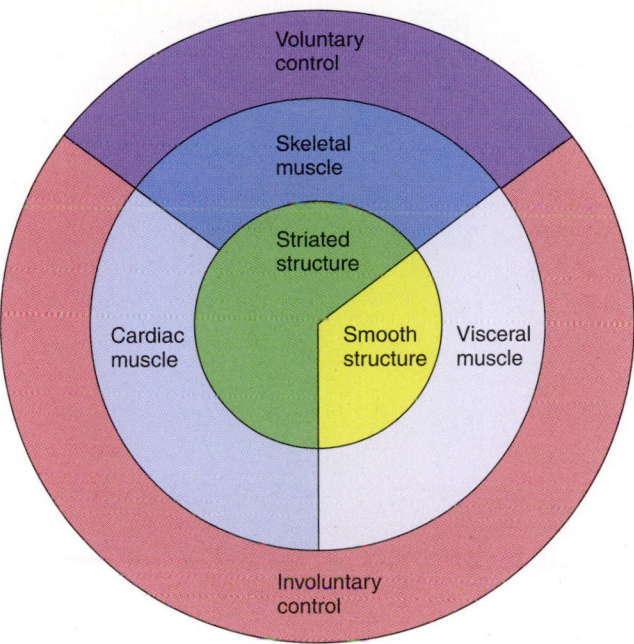

Figure 16–1

A simplified view of the classification of human muscle. Although there may be some difficulties with each type of categorization, muscle can be classified by three different means: (1) location (skeletal, cardiac, or visceral), (2) structure (smooth or striated), and (3) mode of control (voluntary or involuntary). Some of these categories and functions may overlap.

The location of **cardiac muscle** is obvious from its name; it makes up the pumping muscle mass of the heart. Cardiac muscle is found only in the heart (although it may extend a little way into the large vessels that enter and leave that organ). While confined to a relatively small anatomical region, its crucial role in body function in health and disease gives great importance to its understanding and study.

Classification of Muscles by Structure

When viewed at the level of microscopic structure, muscle falls into two general categories: **striated muscle** and **smooth muscle.** The term *striated* means "striped" and refers to the regular cross-striped appearance of the muscle cells under a microscope. This appearance is an important clue to the way in which these muscles function on a molecular level.

Skeletal muscle is a **striated muscle.** Cardiac muscle is also striated muscle. While its cells are considerably smaller than those of skeletal muscle, its molecular function is similar in many respects. Many of the details of the function of skeletal muscle have been found to apply to cardiac muscle also. In addition, cardiac muscle also shares some functional similarities with visceral muscle.

Visceral muscle cells lack the striations found in the other muscle types and are therefore referred to as **smooth muscle.** Their microscopic structure provides fewer clues to their function than in the case of skeletal muscle, although

many aspects of their internal processes are becoming better understood as research in this field progresses.

Classification of Muscles by Mode of Action and Control

Because most movements of skeletal (striated) muscle occur in response to a conscious (willed) effort, these muscles may be called **voluntary muscles.** However, many actions of skeletal muscle, such as the acts of breathing or walking, are almost automatic and are, thus, in a sense, involuntary. The strictest basis for this type of classification is that skeletal muscles, in order to function, must receive a specific signal or signals from the central nervous system (CNS). In the event of an accident or illness that disconnects a skeletal muscle from its nerve supply, paralysis results.

On the other hand, the actions of most smooth (visceral) muscles are involuntary. In many cases, these involuntary actions are initiated by the autonomic nervous system (ANS) in response to some internal reflex adjustment that may take place automatically. Smooth muscle is also capable of acting on its own, sometimes in response to hormonal or environmental factors. Voluntary control of some smooth muscle functions is also possible; for instance, specially trained persons using the techniques of "biofeedback" can cause blood vessels to dilate or constrict. In the majority of situations, however, the function of smooth muscle is completely involuntary.

Cardiac muscle is also an **involuntary muscle**, and no signals from the nervous system are required to start or maintain its contraction. The heart continues to beat indefinitely after all nerve connections to it are removed, although the rate and strength of the heartbeat are no longer under external nervous control. Because of this special autonomous nature of the cardiac muscle, it is not usually classified as being either voluntary or involuntary.

Smooth muscle is itself subject to further classification in the area of its mode of control. Some smooth muscle is called **multiunit smooth muscle;** this muscle is closely controlled by the ANS; as such, it is not subject to voluntary control. Other smooth muscle, that which makes up the bulk of our visceral organs, is called **unitary smooth muscle.** This sort of muscle is less strictly controlled by the nervous system, and large numbers of its cells function together as a single unit.

A summary of the three classifications is provided in Figure 16–1. When allowance is made for the exceptions noted previously, the classifications are useful for discussing and comparing the various muscle types. Within each type are important variations, which are taken up as the specific details of each muscle type are discussed.

> *How does the structure of skeletal muscle relate to its function?*

THE STRUCTURE OF MUSCLE

The study of muscle structure can be done at two different levels. At one level, large-scale structural aspects of muscle can be observed and described without microscopic aid. The position, size, shape, and attachment points of the origins and insertions of a skeletal muscle give important information about its function. At another level, light microscopes and electron microscopes reveal details of the fine structure responsible for the actual mechanism of contraction. Muscle is an excellent example of the way in which the structure and function of a biological tissue are intimately related. For this reason, we discuss in detail the structure of striated muscle before going on to other muscle types. Striated muscle can serve as a basis of comparison for the description of other muscle types.

Skeletal Muscle Structure and Function Are Closely Related

A major and obvious function of a skeletal muscle is its shortening and the resulting movement of parts of the skeleton. The large-scale structure of skeletal muscle reflects this function, and the organization of the cells and tissues reflects the structural and metabolic adaptations necessary for functions in which large forces are generated and large amounts of energy are expended.

Gross Structure of Skeletal Muscle

In mechanical terms, the skeleton works with the muscles as a **lever system.** Groups of muscles working together with jointed bones permit a larger range of movement than could muscles acting on their own. Muscles attach to bones by means of very strong connective tissue structures called **tendons** (Fig. 16–2). The more stationary (and proximal) attachment is called the **origin** of a muscle, and the other (distal) end is called the **insertion.** The wide or thick central portion of a muscle is called the **belly.** Some muscles, the **biceps** for example, have a double origin (*biceps* means "two heads") and a single insertion, whereas others have a narrow origin and a very wide insertion. The variable anatomy of skeletal muscles represents adaptations to the specific functions of the individual muscles. Muscles with many parallel (side-by-side) fibers can exert a large force but cannot shorten rapidly or by a very great amount. On the other hand, long muscles with many fibers in series (end-to-end) are able to shorten rapidly but can develop less force.

Because most skeletal muscles are attached to bones at either end, the range of motion of a muscle itself may be limited by the skeletal system. The needed range of motion must then be multiplied by the lever system provided by the skeleton. This arrangement has some important consequences. Large amounts of muscle shortening, as we shall see, can significantly affect the amount of force a muscle can exert, and limiting the range of muscle shortening helps to counteract this effect. However, this arrangement also means that the muscle must exert a much greater force at its insertion than is actually produced at the end of the limb. For instance, in the human forearm, the lever is hinged at the elbow. The hand is approximately seven times as far from the joint as is the insertion of the biceps, which means that if the hand is holding an object with a mass of 10 kg, the tendons of the biceps must exert seven times as much force, 70Kg (see Fig. 16–2). This sit-

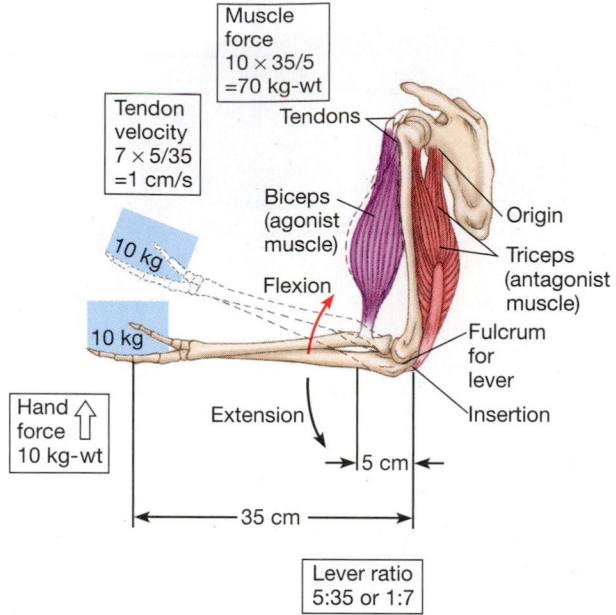

Figure 16–2

Antagonistic muscle arrangements and the skeletal lever system. The major muscles that move the upper arm work in opposition to each other. The biceps flexes (bends) the arm, whereas the triceps extends it. The tendon of the biceps attaches at a point approximately 5 cm away from the fulcrum of the lever system (the elbow joint), and the hand is 35 cm away. Therefore, the upward force exerted by the muscle must be seven times greater than the downward force supplied by a weight in the hand. However, any shortening of the muscle is multiplied by seven, meaning that the hand moves seven times as fast and far as the lower end of the biceps muscle.

uation can lead to serious muscle and tendon injury in athletes and in people who do heavy manual labor. On the other hand, it means that the hand can move seven times as fast as the biceps can shorten. Such an increase in speed is often valuable. For instance, it allows the leg muscles to propel the body at a high rate of speed or to jump up in the air much higher than the distance over which the muscles themselves can shorten.

As mentioned in Chapter 10, in most situations, (especially in the limbs) skeletal muscles are arranged in so-called **antagonistic pairs.** When a muscle moves a joint in one direction (e.g., **flexion,** in which a joint closes and a limb assumes a more withdrawn position), another muscle or muscles move the joint in the opposite direction, toward a more extended position (i.e., **extension**). This arrangement is necessary because muscles can only contract, pull; they cannot push.

Skeletal **muscle cells** (also called **muscle fibers**) are quite large compared with other cells. They are multinucleated, with the nuclei located at the periphery of the cell. Typical diameters are on the order of 100 micrometers (μm), whereas cells from some large muscles may be many centimeters in length. The way in which muscle cells are assembled into the complete muscle is diagrammed in Figure 16–3. In an intact muscle,

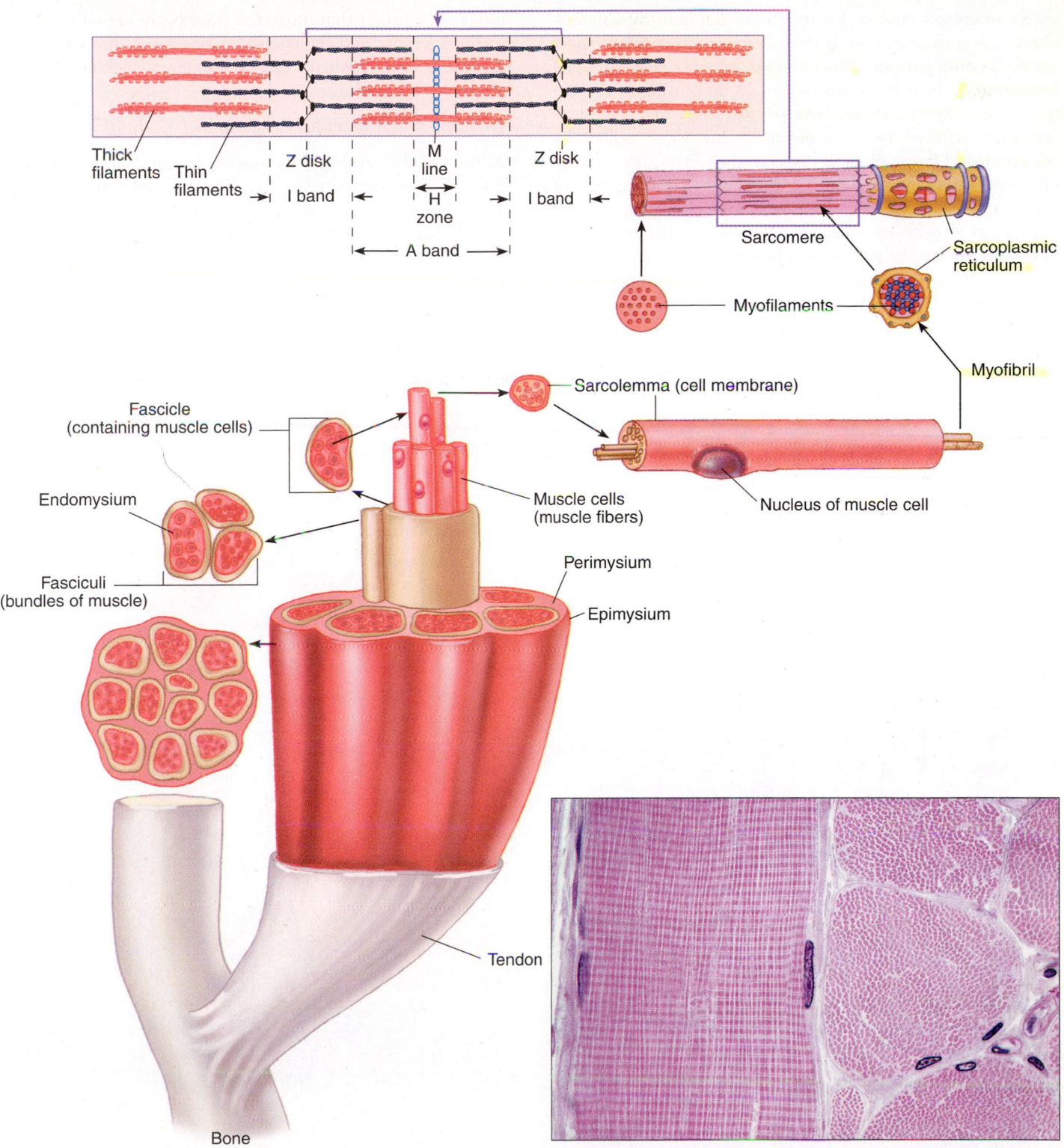

Thick filaments   Thin filaments   Z disk   I band   M line   H zone   Z disk   I band   A band

Sarcomere   Sarcoplasmic reticulum   Myofilaments   Myofibril

Fascicle (containing muscle cells)   Sarcolemma (cell membrane)   Muscle cells (muscle fibers)   Nucleus of muscle cell

Endomysium   Perimysium   Epimysium

Fasciculi (bundles of muscle)

Tendon

Bone

Figure 16–3

The organization of skeletal muscle. *(a)* Cross-section of the thigh, progressively magnified to show finer details. *(b)* Longitudinal section of the deltoid muscle, showing the bundles of fibers (fasciculi) that make up a whole muscle. *(c)* Further magnification shows that the individual muscle fibers, or cells, contain long structures called *myofibrils* that are made up of myofilaments and are organized into sarcomeres.

shown in cross-section in Figure 16–3*a*, the individual muscle fibers are surrounded by a delicate connective tissue sheet called the **endomysium.** Bundles of muscle fibers are grouped into **fasciculi** (shown in longitudinal section in Figure 16–3*b*) by a connective tissue called the **perimysium,** and the entire muscle is covered by a connective tissue sheet called the **epimysium.** During physical exercise, muscle requires a plentiful supply of oxygen and nutrients and must rapidly get rid of waste products. Accordingly, muscle is supplied with a plentiful network of blood vessels arranged among the fibers so that no muscle cell is very far away from a blood vessel.

Both the connective tissue sheets and the muscle fibers themselves are connected at either end into a very tough connective tissue extension of the muscle that comprise the tendon. Some tendons are thick and short, such as those that provide the insertion of biceps. Some tendons are very

long; the muscles that move the fingers are actually located in the forearm, and their tendons run across the wrist joint and the back and palm of the hand to connect the muscles to the bones of the fingers.

Cellular and Molecular Structure of Skeletal Muscle

At the cellular level of organization, a remarkably clear relationship exists between the structure and function of muscle. The molecules that make up the contractile substance of muscle have a dual role. They are both the major structural elements of the muscle cell; they are also the enzymes that convert the biochemical energy (derived ultimately from the food we eat) into mechanical energy.

When viewed through a light microscope, skeletal muscle cells show a repeating pattern of dark and light crossbanding. This striped, or striated, pattern gives skeletal

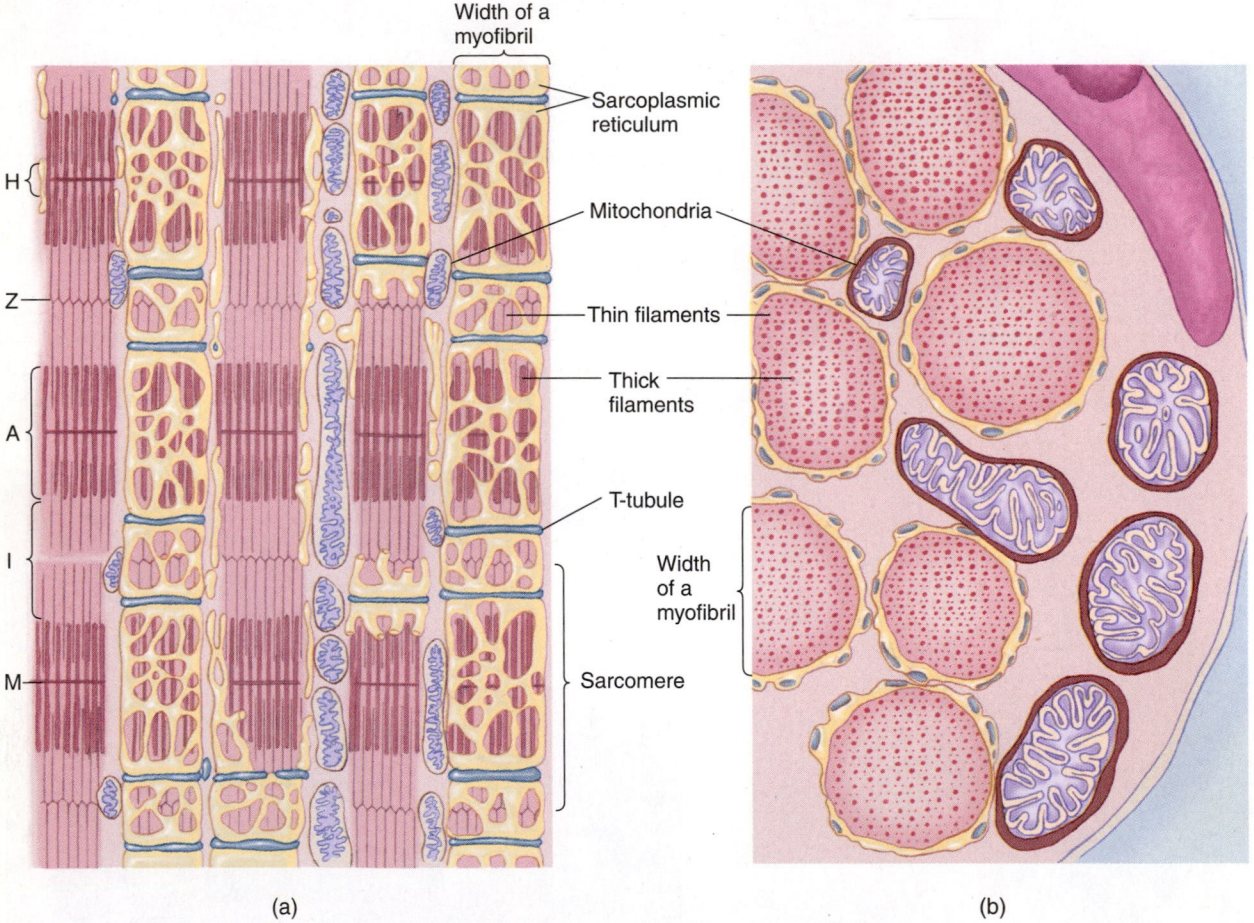

(a)   (b)

Figure 16–4

The microscopic structure of skeletal muscle. These drawings were made from electron micrographs of skeletal muscle. *(a)* A longitudinal section showing the arrangement of the sarcomeres. The letters *H, Z, A, I,* and *M* refer to the parts of the sarcomere diagrammed in Figure 16–5. *(b)* A cross-section of the same tissue. Several cells are shown in the inset at the upper left, and the area in the rectangle is magnified in the rest of the figure.   *(Modified from Krstic, R. V.,* Ultrastructur der Saugertierzelle. *Berlin: Springer-Verlag, 1976.)*

muscle one of its names; more important is that it provides a clue to how striated muscle actually works. Ultrostructure, analysis using electron microscopy and other methods, has revealed that the dark and light cross-bands, or striations, are composed of two types of protein filaments, with each type being arranged in sets side by side across the cell. These protein filaments are called **myofilaments,** and the two types differ from each other in their size and internal structure. The two sets overlap each other in a geometrically repetitive way so that the striation pattern changes as the muscle is lengthened or shortened.

The Structure of a Sarcomere. The two sets of myofilaments are organized into a structure called a **sarcomere;** this is the basic functional and anatomical unit of the contractile machinery of the skeletal muscle. Figure 16–4 shows both a cross section and longitudinal section view of the microscopic structure of skeletal muscle. The lateral regions of the sarcomere are made up of **thin filaments** (consisting largely of the protein *actin*), whereas the center portion of the sarcomere (the **A-band**) is made up of **thick filaments** (composed of the protein *myosin*), along with portions of the thin filaments that extend into the A-band. A structure called the **M-line** runs across the middle of the A-band and holds the thick filaments in side-by-side alignment. The thin filament array interdigitates (overlaps) with the thick filament array, and the depth of its penetration depends on the overall amount of shortening of the muscle. The region where only thin filaments are present is called the **I-band.**

Both sets of filaments are arranged in a hexagonal pattern; this is the most compact way in which the filaments could be packed (Fig. 16–5a). If a cut is made in a region in

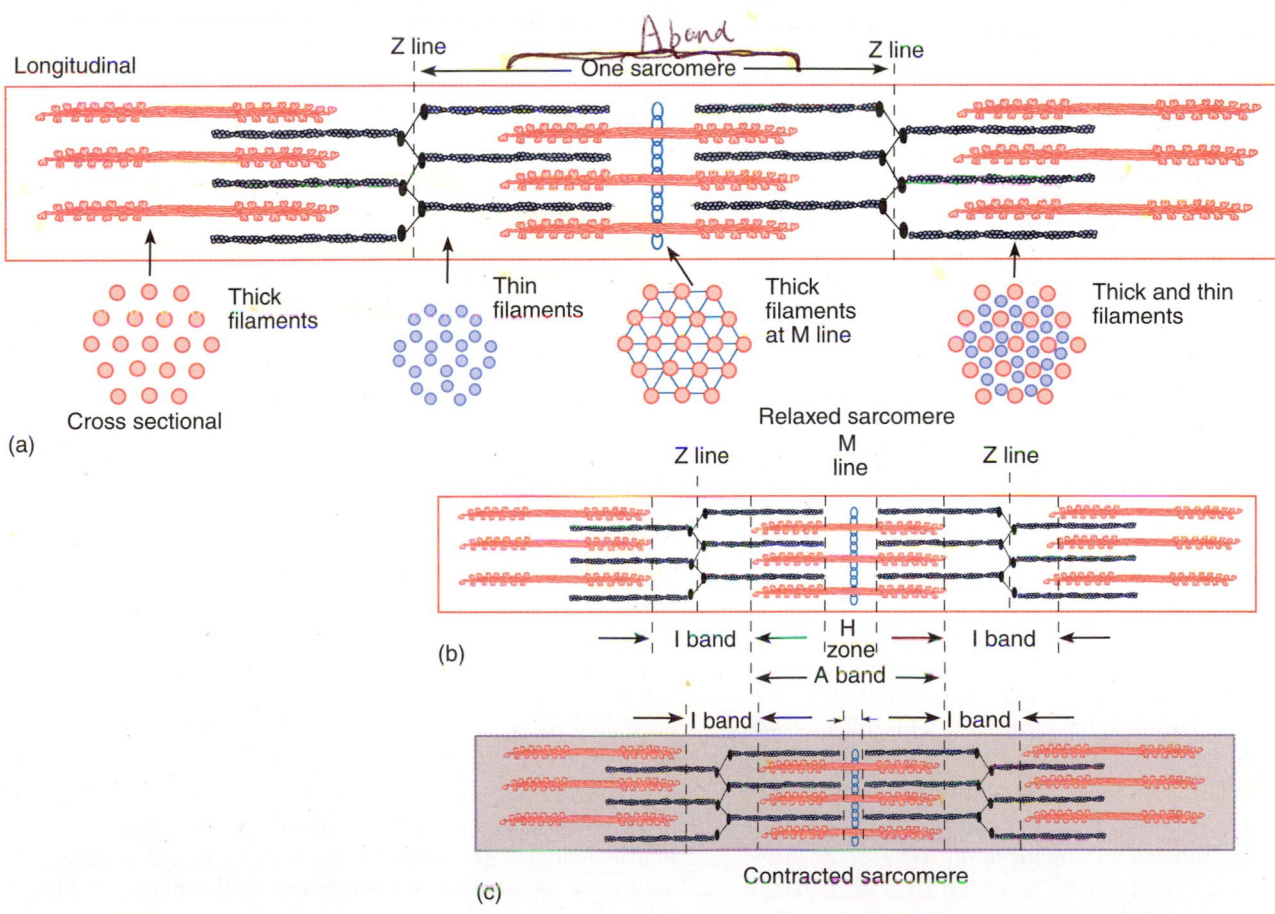

Figure 16–5

The structure of a sarcomere. **(a)** Diagram showing the relationship of the longitudinal and cross-sectional arrangement of the myofilaments. **(b)** The A-band is made up of thick filaments arranged side by side; the thin filaments extend from the Z-line into the A-band. The I-bands are made up of the portions of thin filaments where no overlap exists between thick and thin filaments, and the H-zone is the region of the A-band into which the thin filaments do not extend. The M-line runs down the center of the A-band and appears to help in maintaining the uniform arrangement of the thick filament array. **(c)** During contraction, the width of the A-band stays constant, while the H-zones and the I-bands become narrower and the Z-lines move closer together.

which sets of thick and thin myofilaments overlap, then the spaces provided by the hexagonal array of the thick filaments are occupied by an interlocking hexagonal array of the thin filaments. **Crossbridges,** which are projections from the thick filaments, reach out directly toward the thin filaments (Fig. 16–6*b* and *c*).

The usual names given to the parts of the sarcomere are shown in Figure 16–5*b*. The center of each I-band is marked by a structure called the **Z-line,** which serves to mark the boundary of the sarcomere. (Sometimes this structure is called the **Z-disc** in order to emphasize its three-dimensional nature.) The function of the Z-line is to connect the ends of the thin filaments from one sarcomere to those of the next one in line. The Z-lines also have cross-connections that keep the thin filaments in register with each other in a lateral direction. In the center region of the A-band, where the thin filaments of the I-band have not penetrated, is a somewhat lighter region called the **H-zone** (the M-line runs through the center of the H-zone because the filament overlap is symmetrical). Some other muscle proteins that are not directly involved in contraction are not shown in the figure. **Titin** is a very large, filamentous protein that extends from the Z-lines to the bare portions of the myosin filaments. It is highly elastic and may help to prevent overextension of the sarcomeres and maintain the A-bands in their proper place. **Nebulin** is a protein that extends along the thin filaments and may play a role in their formation in developing muscle. The protein **α-actinin** is involved in anchoring the thin filaments to the Z-lines. The important structural protein **dystrophin,** located just below the sarcolemma, aids in the transfer of force to the outside of the cells. A lack of this protein, and of some other related structural proteins, such as **laminin** and **integrins,** is associated with the severe disease called *muscular dystrophy.*

During the process of contraction, in which the muscle may shorten considerably, *neither the thick nor the thin filaments change their length.* A greater or lesser degree of overlap between the thick and thin filaments accounts for all of the change in the length of the muscle. This means that, when the muscle shortens, the length of the A-band stays the same, whereas that of the I-band decreases (see Fig. 16–5*c*). The thin filaments themselves do not change in length. Shortening of the muscle is brought about by a cyclic "rowing" motion of the crossbridges, in which they successively attach to the closest available thin filament site, change their shape and pull the actin filament along, and then let go of the actin and reattach further along.

Hundreds of sarcomeres connected in series (i.e., end-to-end) form a functional unit called a **myofibril,** which runs for the length of the cell. Most muscle cells contain a large number of parallel myofibrils, and the intracellular spaces between them are occupied by mitochondria and by molecules of glycogen, a source of chemical energy. Space between the myofibrils is also occupied by the internal membrane system.

Structure of the Myofilaments. A globular protein called *actin* makes up the thin filaments (see Fig. 16–6*a*). The slightly oblong actin molecules are each about 5.5 nanometers (nm) in diameter and are joined end to end to form a long structure that resembles a bead chain. Two such chains are wound about each other so that the pair makes a half-turn every seven actin monomers, which means that the thin filaments are helical structures, with a half-pitch of approximately 35 to 37 nm. Each half-turn of seven actin molecules is also associated with a long, fibrous protein molecule called **tropomyosin.** Each of these tropomyosin molecules bears a protein complex called **troponin.** Although it is the actin protein of the thin filaments that interacts with the myosin crossbridges of the thick filaments, the components of the troponin-tropomyosin system (known collectively as the **regulatory protein complex**) have an important function in controlling the contraction of skeletal muscle.

Thick filaments are made up of very large protein molecules called *myosin* (see Fig. 16–6*b*). A thick filament is composed of about 300 to 400 myosin molecules. Each myosin molecule has a long, rod-shaped section (a structural portion sometimes called the *tail region*) and a double globular head at one end. The head portion of the molecule, which has ATP and actin binding sites, contains the biochemical and enzymatic properties that participate in the actual contractile function. Also attached to each head portion are two protein molecules called **light-chains** because of their comparatively low molecular weights. These have some secondary regulatory functions in skeletal muscle.

When myosin molecules are packed together to form a thick filament, the tail regions of the molecules make up the structure of the filament itself, whereas the head portions project from the sides of the filament. The projections of the myosin heads are what form the structures called *crossbridges.* Because of the way the molecules are packed, the heads project from the filament in a radial fashion, and the projection makes a full turn for every six myosin molecules. The angle between the successive myosin heads is thus 60°, and they project out of the filament every 14.3 nm. In this way, they form a helical pattern (like a spiral staircase) that winds down the length of the thick filament.

The tail regions of molecules at either end of the filament point toward its center, making the thick filament symmetrical about its center, with myosin heads pointing in opposite directions in the two halves of the filament. This is important in creating the forces involved in contraction because, when both sets of filaments are assembled into a sarcomere (see Fig. 16–6*c*), the action of the myosin heads pulls actin filaments into the sarcomere from both directions. The construction of the thick filaments also means that there is a **bare zone** in the middle of each thick filament; this is an area containing only the tail portions of myosin molecules. The assembled myofilaments are held in place by the proteins of the Z-discs, the M-line, and by the accessory proteins nebulin and titin, which are associated with the thick and thin filament arrays.

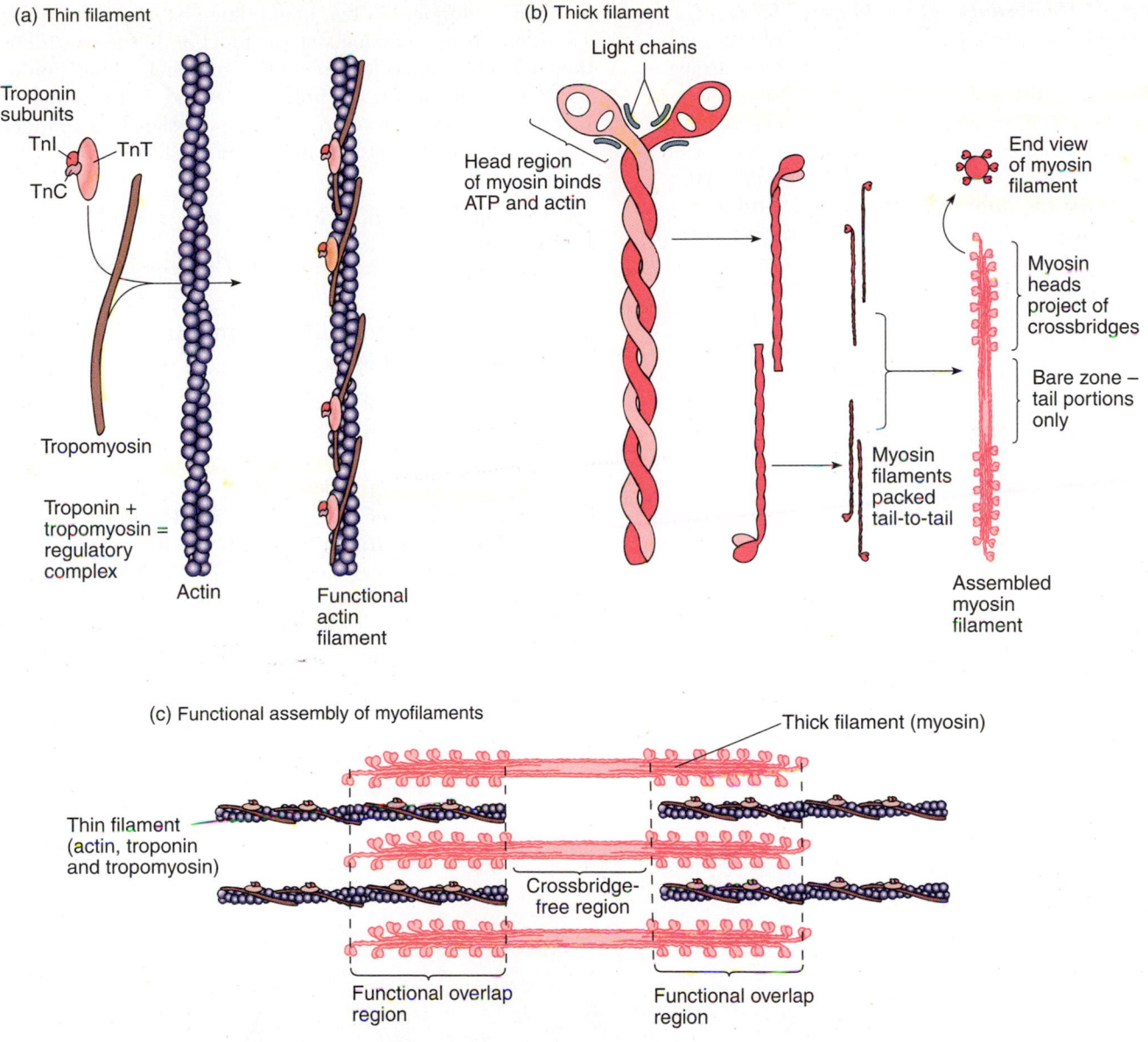

(a) Thin filament

Troponin
subunits

TnI — TnT

TnC

Tropomyosin

Troponin +
tropomyosin =
regulatory
complex

Actin

Functional
actin
filament

(b) Thick filament

Light chains

Head region
of myosin binds
ATP and actin

End view
of myosin
filament

Myosin
heads
project of
crossbridges

Bare zone –
tail portions
only

Myosin
filaments
packed
tail-to-tail

Assembled
myosin
filament

(c) Functional assembly of myofilaments

Thick filament (myosin)

Thin filament
(actin, troponin
and tropomyosin)

Crossbridge-
free region

Functional overlap
region

Functional overlap
region

Figure 16–6

The structure and assembly of myofilaments. **(a)** The thin filaments are made up of two chains of actin protein monomers that are twisted one half-turn for every seven monomers. The regulatory protein complex, made up of troponin (composed of its three subunits) and the long molecule tropomyosin, is associated with every seven actin monomers, with the tropomyosin extending along the groove formed by the twisting of the actin chains. **(b)** The individual myosin molecules that make up the thick filament are paired structures, with two adjacent head portions and long tails that are wound about each other. The head portions each contain an ATP binding site and an actin binding site, and associated with each of the heads are two protein light-chains that have regulatory functions. The thick filaments are composed of bundles of such individual myosin molecules packed together so that the tail portions form the "backbone" of the filaments and their heads project outward in a spiral fashion around the filaments. The center part of the filament contains only the tail regions because the molecules are oriented oppositely in either end of the filament, a bipolar arrangement. **(c)** When the filaments form a sarcomere, thin filaments extend into the thick filament array so that action of the crossbridges draws the thin filaments inward from both directions; this is because of the bipolar arrangement of the myosin molecules. Notice that the spacing of the crossbridges is different from the spiral period of the thin filaments.

The Internal Membranes of a Muscle Fiber. These membrane structures participate in the control of contraction and relaxation. (See Fig. 16–7 for their anatomical arrangements.) The plasma membrane and a thin layer of connective tissue fibers form the outer covering of the cell, called the **sarcolemma.** Extending inward from the sarcolemma into the depths of the cell is a set of membrane-lined passageways, the **transverse tubules** (often called **t-tubules** for short). Depending on the type of muscle being considered, the t-tubules may enter the fiber either at the region where the A- and I-bands overlap, or at the level of the Z-discs. The t-tubules are an inward extension of the plasma membrane, and the interior of the t-tubule system is continuous with the extracellular space. Closely associated with the t-tubular system is another set of membranes that run at right angles to the t-tubules and down the length of the sarcomere. This is the **sarcoplasmic reticulum (SR).** Unlike the t-tubules, the SR is not connected to the extracellular space; however, it does come into close contact with the t-tubules as they course across the fiber. This region of contact, which consists of the t-tubules and the enlarged end portions of the SR (called **lateral sacs** or **terminal cisternae**), is called a **triad** in skeletal muscle because it is made up of one t-tubule and the terminal cisternae from two adjacent sections of SR from adjacent sarcomeres. Periodic structures (sometimes called

feet) appear to join the two membrane systems at the triad. These internal membrane systems function in the control of the contraction and relaxation of the muscle. The function of the "feet" structures is not fully understood, but it is known that they play a role in communicating the signal for contraction from the t-system to the interior of the fiber.

> *How do the structures of cardiac and skeletal muscle differ?*

Cardiac Muscle Structure Is Similar to that of Skeletal Muscle

The cellular structure of heart muscle is similar in many ways to that of skeletal muscle, but some very important differences exist (Fig. 16–8). The cells of heart muscle are small compared with those of skeletal muscle. The cells have a single centrally located nucleus. They are 5 to 15 μm in diameter and may be 20 to 30 μm in length. The cells are connected end to end and, to a lesser extent, side by side. Some of the cells may branch so that one end connects to two other cells. At the regions of connection between the cells is a specialized structure called the **intercalated disc.** This is an area of extremely close contact in which intercellular space is essentially

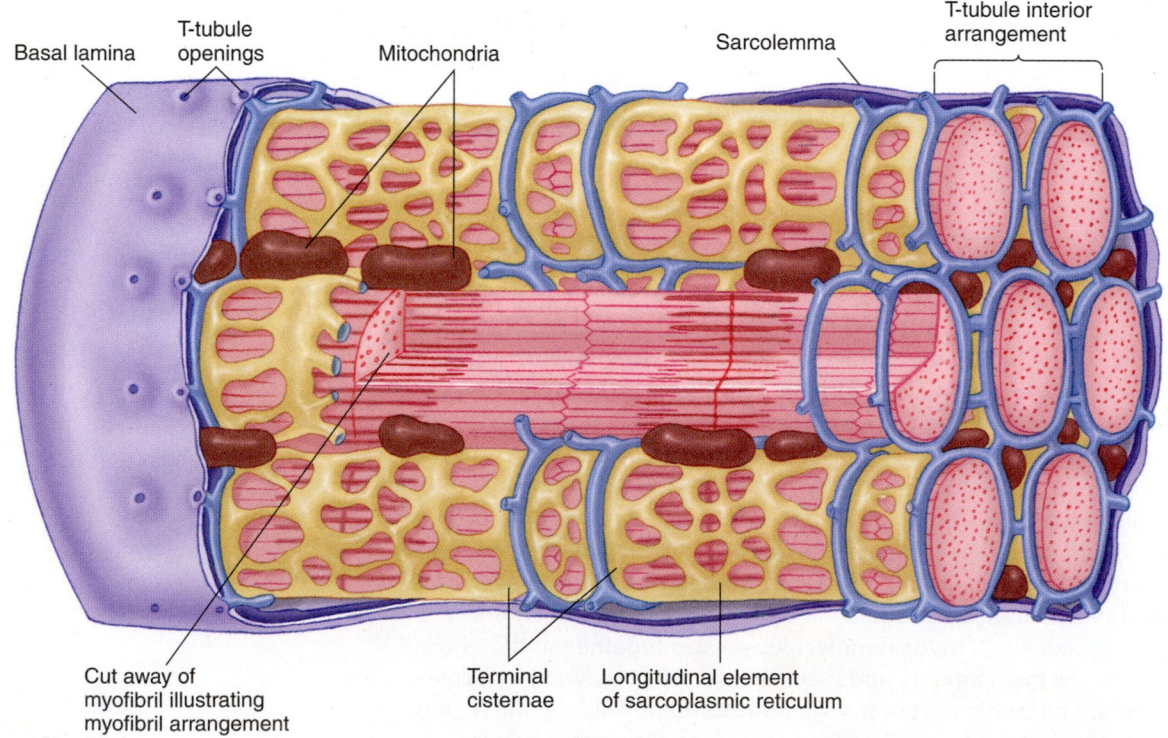

Basal lamina | T-tubule openings | Mitochondria | Sarcolemma | T-tubule interior arrangement

Cut away of myofibril illustrating myofibril arrangement | Terminal cisternae | Longitudinal element of sarcoplasmic reticulum

Figure 16–7

The internal membranes of skeletal muscle, showing the structure of the t-tubule system and the sarcoplasmic reticulum. *(Modified from Krstic, R. V.,* Ultrastructur der Saugertierzelle. *Berlin: Springer-Verlag, 1976.)*

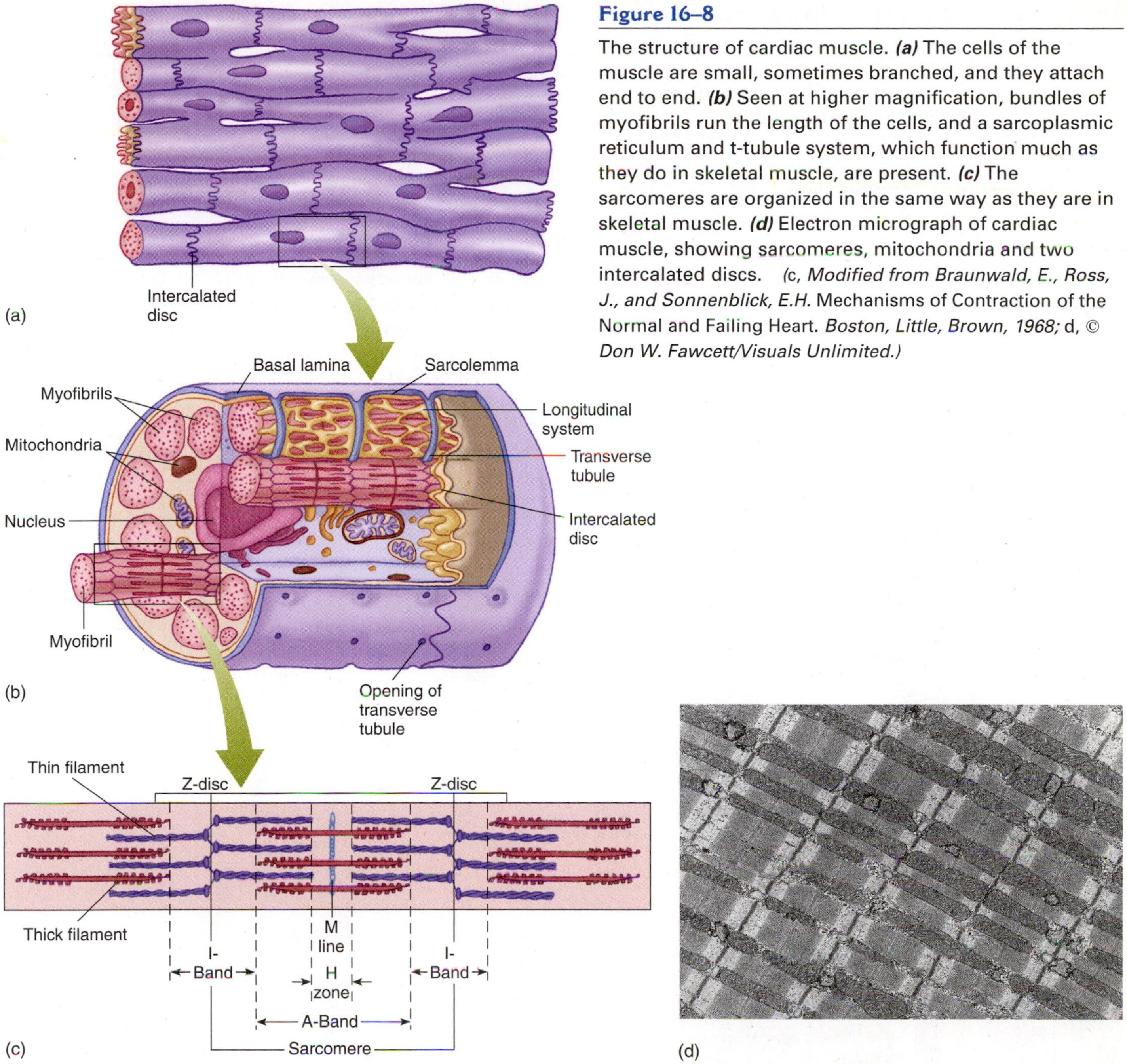

(a)

Intercalated disc

Figure 16–8

The structure of cardiac muscle. **(a)** The cells of the muscle are small, sometimes branched, and they attach end to end. **(b)** Seen at higher magnification, bundles of myofibrils run the length of the cells, and a sarcoplasmic reticulum and t-tubule system, which function much as they do in skeletal muscle, are present. **(c)** The sarcomeres are organized in the same way as they are in skeletal muscle. **(d)** Electron micrograph of cardiac muscle, showing sarcomeres, mitochondria and two intercalated discs.   (c, *Modified from Braunwald, E., Ross, J., and Sonnenblick, E.H.* Mechanisms of Contraction of the Normal and Failing Heart. *Boston, Little, Brown, 1968;* d, © *Don W. Fawcett/Visuals Unlimited.*)

(b)

Basal lamina   Sarcolemma

Myofibrils

Mitochondria

Nucleus

Longitudinal system

Transverse tubule

Intercalated disc

Myofibril

Opening of transverse tubule

(c)

Thin filament

Z-disc   Z-disc

Thick filament

M line

I- Band   H zone   I- Band

A-Band

Sarcomere

(d)

absent; many projections from one cell extend into another, and the other cell has the mirror image of this arrangement. There is a strong mechanical connection between cells in this region, as well as a close electrical connection. The mechanical connection is provided by a cellular structure called a **desmosome,** and the electrical connection is made through a **gap junction** (a structure found also between smooth muscle cells). The specialized contact structure of the intercalated disc allows a tissue made up of small cells to function electrically and mechanically in many ways as though it were a single, large cell, and this forms the cellular basis of the coordinated action of the muscle in the beating heart.

The intracellular organization of cardiac muscle is the aspect in which cardiac muscle is most like skeletal muscle. The structure and arrangement of skeletal and cardiac myofilaments are quite similar, as is their protein composition. The similar contractile structure gives rise to the same sort of striated appearance that skeletal muscle has, and many of the mechanical responses are similar. Because the muscle of the heart has a largely aerobic metabolism, it contains numerous mitochondria. A transverse tubular system is present in muscle from many areas of the heart, although the tubules are larger in diameter and usually make close contact with only one portion of longitudinal SR. This results in a structure called a **dyad,** rather than a triad as in skeletal muscle.

> *How does the structure of smooth muscle compare with that of skeletal and cardiac muscle?*

Smooth Muscle Is Made of Small Cells Without Striations

Smooth muscle is very similar to skeletal and cardiac muscle in the molecular structure of its contractile proteins, but it has a very different cellular and tissue structure, one that specially adapts it to its physiological roles. Because of its less regular structure, many details of the contractile system of smooth muscle are not yet well understood.

The Ultrastructure of Smooth Muscle

Drawings made from electron micrographs of smooth muscle are shown in Fig. 16–9. The cells are quite small, 5 to 15 μm in diameter and perhaps 200 to 500 μm in length, depending on which type of muscle is examined. The cells have a single, centrally located nucleus, and the ends taper off to a small diameter. In many smooth muscle tissues, numerous portions of cells are connected very closely together by a structure called the **nexus,** or gap junction. This structure has been described in some detail in Chapters 5 and 7. These close, cell-to-cell contacts in smooth muscle allow for coordination of smooth muscle activity (as in the heart, which also has gap junctions between cells). In the case of the smooth muscle of the uterus, gap junctions increase greatly in number and area at the end of pregnancy, just at the time at which labor is to

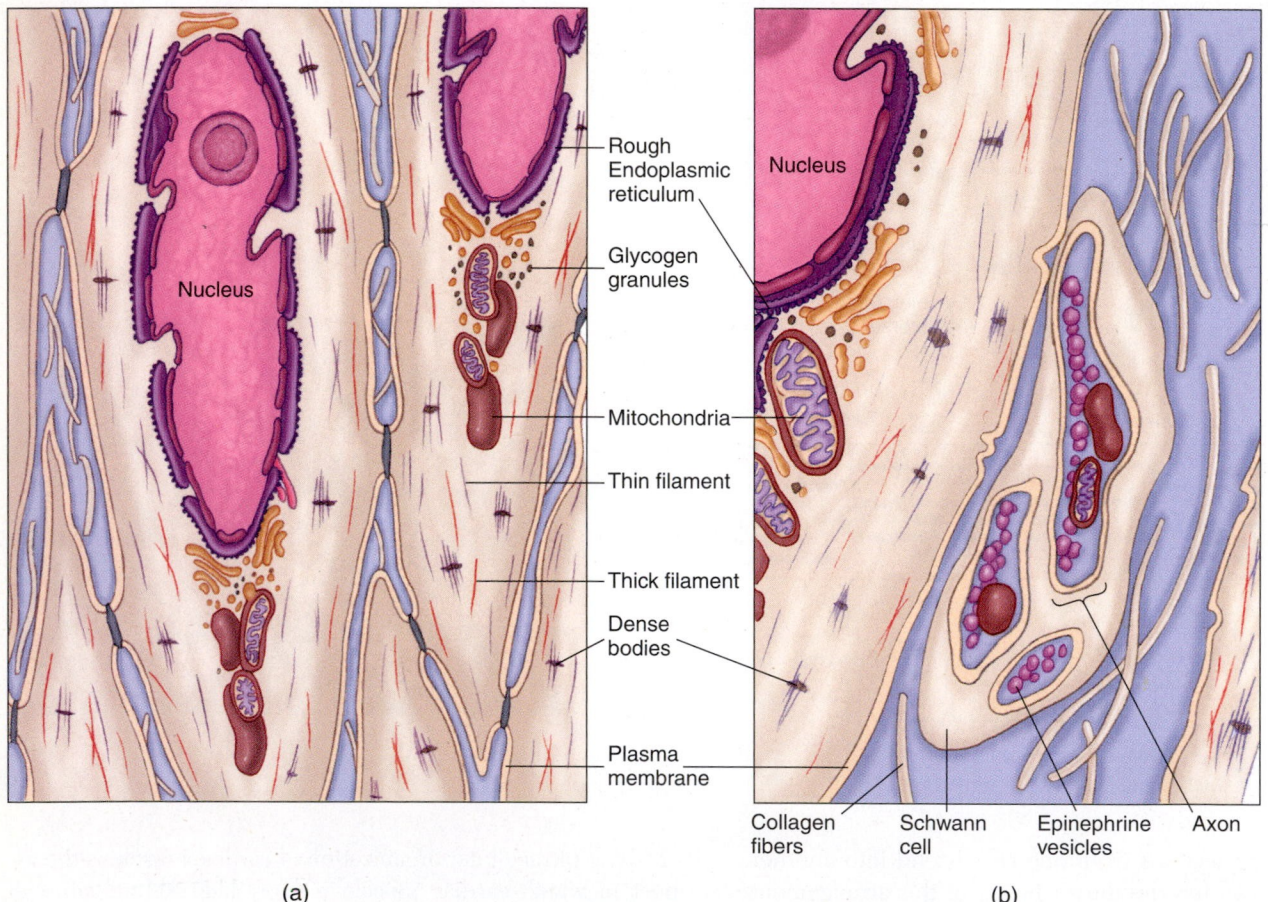

(a)   (b)

Figure 16–9

The structure of smooth muscle. **(a)** Drawing based on an electron micrograph of a longitudinal section of parts of several smooth muscle cells. Most of the cell organelles tend to cluster at the ends of the nucleus. The rectangular inset shows a lower-power view of cells in both longitudinal and cross-section. **(b)** A nerve muscle contact between a smooth muscle cell and a small bundle of autonomic nerve fibers. No specialized region of contact is present.   *(Modified from Krstic, R. V., Ultrastructur der Saugertierzelle. Berlin, Springer-Verlag, 1976.)*

begin. Their presence is reflected in the greatly increased coordination and effectiveness of the contractions that result in childbirth.

A network of connective tissue fibers (collagen and elastin) surrounds the cells and helps to link them together. It also gives strength to the whole tissue, particularly in those organs that may be subject to a high degree of stretch. The force of contraction of the individual cells is transmitted to the whole tissue by strong cell-to-cell connections and by the connective tissue network.

Mechanical Arrangements of Smooth Muscle

Smooth muscle makes up much of the walls of organs such as the stomach, intestines, and uterus. It is generally arranged in layers. In the small intestine, for example, one layer (the inner one) is arranged in a circular fashion around the lumen, and the outer layer runs lengthwise (longitudinally). When the circular layer contracts, the intestine is reduced in diameter, and when the longitudinal layer contracts, the intestine becomes shorter. Unlike skeletal muscles, which usually have antagonistic muscles to stretch them out after contraction, smooth muscles must return to their precontraction length in a different way. An increase in the volume of the contents of the organ restretches the muscles in the walls. This usually comes about by the entry of new fluid (e.g., by newly formed urine in the bladder) or by movement of contents from elsewhere along a tubular structure (as in the intestine). Coordinated movements of smooth muscle organs such as the intestine often involve the alternate contraction and relaxation of the two muscle layers. In some saclike organs, such as the uterus, muscle fibers are oriented in several directions, reducing the overall volume of the organ when the muscle contracts. The musculature of most blood vessels is simpler, usually consisting only of a circular layer of muscle. In very tiny blood vessels (e.g., an arteriole, which is pictured in Fig. 16–10), a single cell may wrap completely around the circumference of the vessel. Contraction of this muscle cell leads to a narrowing of the blood vessel and a reduction in the amount of blood flow. Such smooth muscle plays a vital role in the regulation of blood flow and pressure.

> *What clues do laboratory experiments with skeletal muscle provide us about how this muscle develops force?*

THE MECHANICAL FUNCTION OF MUSCLE

Most of what muscle does is obvious. We are aware that it shortens and develops force to lift an object, and some apparent relationship exists between how heavy a load is and how fast it can be lifted. These everyday impressions have been refined into a more precise set of concepts that are used to

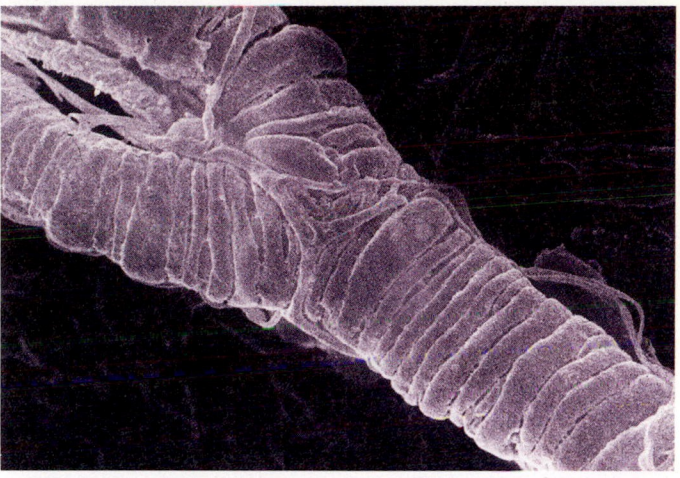

(a)

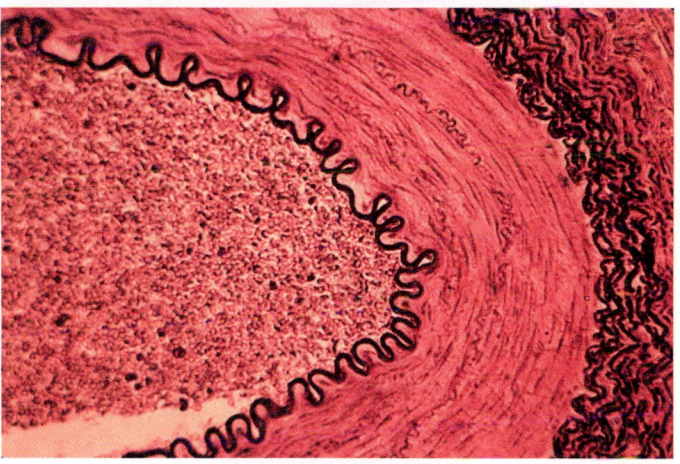

(b)

Figure 16–10

Special arrangements of smooth muscle. *(a)* An example of the smooth muscle cells in the wall of a blood vessel. This is a scanning electron micrograph (SEM) of a small blood vessel called an *arteriole.* At the cut end of this vessel, some of the smooth muscle cells have been spread apart so that they may be seen better, whereas further down the vessel, the cells are in their natural position. Contraction and shortening of the smooth muscle cells, which run in a circular direction, would cause the arteriole to become smaller in diameter. It is by this means that smooth muscle can regulate the flow of blood to the tissues. *(b)* Photomicrograph of a cross-section of an artery. Note again the circular arrangement of the smooth muscle. (a, *Courtesy of Dr. A. Evan;* b, © *Stan Elems/Visuals Unlimited.*)

describe, quantify, and compare the ways in which muscle functions. Before we approach this more exact analysis of muscle function, some basic definitions are necessary.

Skeletal Muscle Contraction Depends on External Conditions

In the terminology of muscle physiology, the term **contraction** has a special meaning. In everyday usage, it means "to shorten," but here it means any form of muscle activity in response to stimulation, whether it involves shortening or not. For instance, in an isometric contraction (*iso* = "same"; *metric* = "measure"), the muscle does not do any external shortening; instead, its *length remains constant,* and it develops **force,** or **tension,** while pulling against immovable attachments (Fig. 16–11a). In an **isotonic contraction** (*tonic* = "tension"), the muscle *force remains constant* while the muscle shortens. In an **auxotonic contraction,** the force

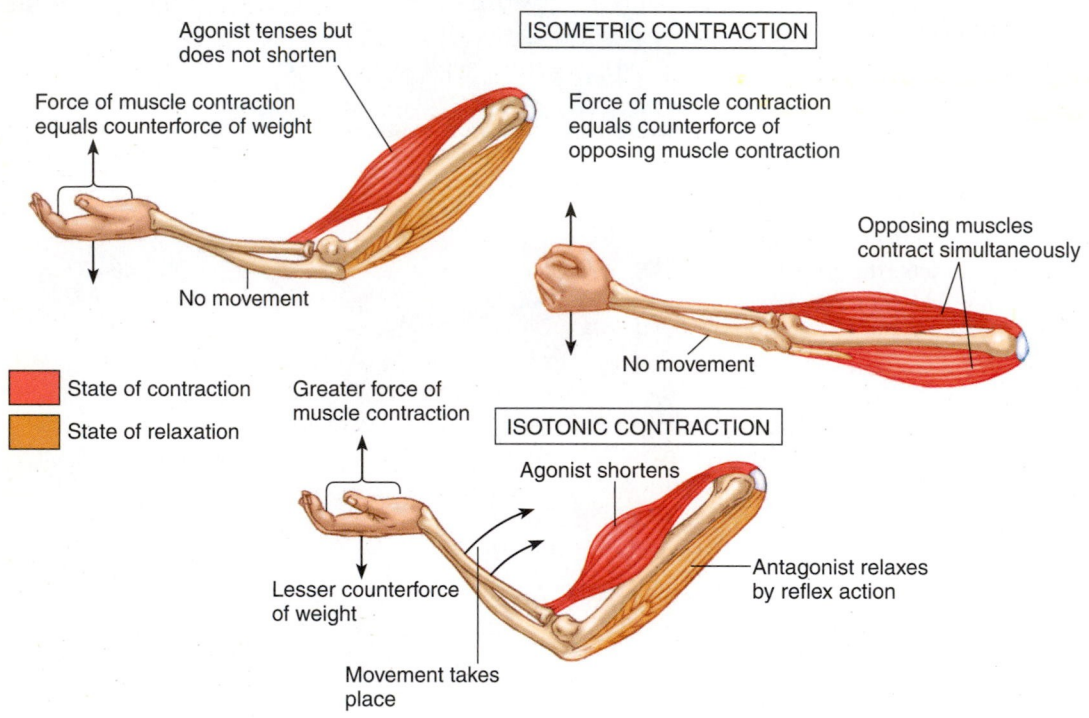

(a)

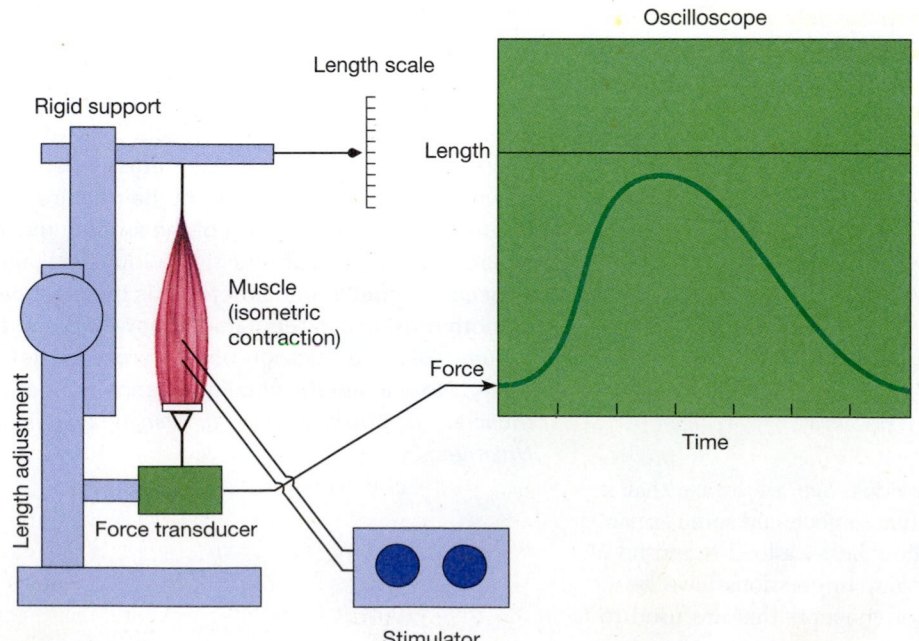

(b)

Figure 16–11

(a) Diagrams of isotonic and isometric conditions of contraction in the human arm. **(b)** Typical apparatus for measuring isometric contraction. The length of the muscle may be varied between contractions, and the record of the force of contraction is viewed on the face of an oscilloscope. The muscle is stimulated with an electric shock.

continually increases as the muscle shortens. Pulling back a bowstring or stretching a rubber band involves this sort of contraction. A **meiotonic contraction,** however, is one in which the force lessens as the muscle shortens. Many useful mechanical devices, such as gearshift levers and toggle switches, allow for this type of contraction in order to provide a clear indication that an action has been performed. These defined conditions represent special circumstances, and realistic contractions usually involve some combination of them. For purposes of analysis, however, contraction conditions can be set up to emphasize the specific properties desired.

Isometric Contractions in the Laboratory: Clues to the Sliding-Filament Hypothesis of Contraction

Under laboratory conditions, an isolated skeletal muscle (and other muscle types as well) can be made to produce contractions conforming to any of these definitions. The relationships between muscle and its external mechanical environment are usually studied by using some "standard" experimental conditions. The simplest type of setup, as diagrammed in Figure 16–11b, can record **isometric contractions.** Here, the muscle is securely attached to a device called a **force transducer,** which produces an electrical signal that varies in direct proportion to the force the muscle exerts. The other end of the muscle is attached rigidly to a fixed support, whose position can be adjusted. All connections, and the force transducer itself, are made very stiff so as to prevent any shortening of the muscle. The force signal from the transducer is fed to a device such as a chart recorder, which produces an ink-on-paper graph of the force as it changes with time. (An oscilloscope, which produces a similar, although temporary, record on a cathode ray tube, or computer may also be used.) The muscle is kept alive by being immersed in a physiological saline solution containing all of the necessary ionic and organic substances (along with oxygen and an energy source) that it requires.

The Muscle Twitch. A single electrical stimulus to a skeletal muscle sets off a series of electrical and chemical events that result in a single, brief contraction called a **twitch.** On the graph drawn by the chart recorder, the force rises rapidly to a peak and then declines a bit more slowly back to the resting value. For a given stimulus sufficient to activate all of the muscle fibers, and under a constant set of temperature and metabolic conditions, all twitches made at the same resting muscle length will produce similar contraction records.

The Isometric Length-Tension Curve. The mechanical factor with the most profound effect on an isometric contraction is the **resting length** at which the muscle is held. Figures 16–12 and 16–13 illustrate these length-dependent relationships. Making a series of isometric twitches, and changing the muscle length *between* (not *during*) the contractions, allows an important relationship to be observed. First the muscle is held at a very short length (shorter than it would ever be in the body) and then stimulated. If it is then progressively stretched out between contractions, the peak amount of contraction force in-

creases each time until some length is reached at which the twitch force is greatest. From this point on, further lengthening results in two effects. First, force begins to be measured in the muscle even before the stimulus is applied. This is called the **resting** (or **passive**) **force** and is due mostly to the connective tissue that holds the muscle together, acting much like a rubber band as it is stretched. Second, the amount of force (the so-called **active force**) that the muscle can develop over and above the resting force now diminishes as the muscle is lengthened further. At an extreme degree of stretch (provided that the muscle has not torn), there is a very high resting force and very little active force in response to stimulation.

This relationship, in which the force capability of the muscle is greatest at some intermediate length, is called the **isometric length-tension curve.** Usually the maximum force capability occurs at a length that is close to the natural length of the muscle in the body. This length is often termed the **optimal length** or L_0, whereas the **peak force** at this length is designated P_0 or F_0.

While this length-tension behavior is very important to an understanding of how muscle works, most skeletal muscles are limited in how far they can be stretched out by their attachments to the skeleton, and the length-tension curve is of relatively little importance in normal functioning. Heart muscle, however, does not have its length changes limited by skeletal attachments, and its resting length is set by the amount of blood that returns to fill the heart between beats. The length-tension curve thus allows the heart to make a more forceful contraction if it receives a larger filling of blood between contractions. See Starling Law of the Heart, Chapter 18.

Optimal Length and the Overlap of Myofilaments. Why does the resting length have such an important effect on the contraction of the muscle? The answer to this question has provided some important clues to the basic physiology of skeletal muscle, clues that relate the structure of the muscle to its function. Careful measurements with the electron microscope have shown that the A-bands do not change in width as the muscle is lengthened or shortened, but the width of the I-bands increases as the muscle is stretched (and vice-versa). The spacing between the Z-lines also changes in proportion to the muscle length. Other measurements have shown that the length of the myofilaments themselves also does not change. What does change, however, is the amount of overlap between thick and thin filaments. Stretching the muscle decreases this amount of overlap; in some muscles, the sarcomeres may be pulled completely apart. These observations (when compared with the length-tension curve of the living muscle) provided evidence that the overlap of the myofilament controls the amount of force that the muscle can develop (see Figs. 16–12 and 16–13). During delicate experiments with single fibers of skeletal muscle in specially controlled isometric contractions, the sarcomere length of the living muscle was measured as the length of the muscle was varied. These studies showed that the amount of force developed was indeed directly related to the amount of myofilament overlap at lengths greater than L_0.

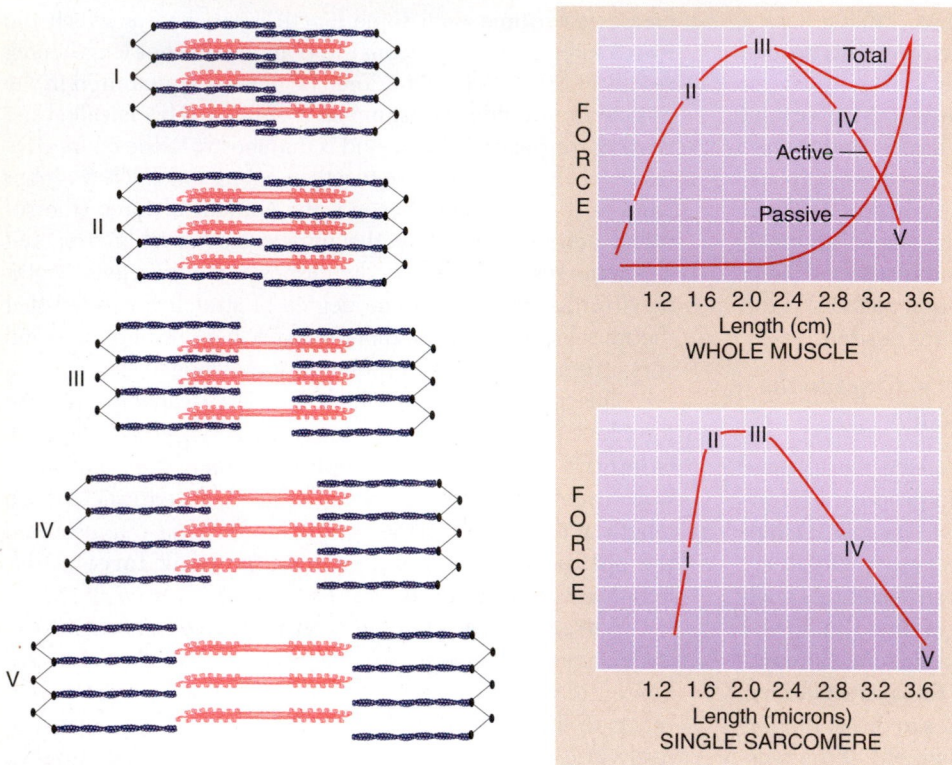

Figure 16–12

The length-tension curve. The upper graph shows how the resting length affects the tension (also called *force*) produced by a muscle during an isometric contraction. The curve labeled "Total" represents the "Active" force plus the "Passive" force. The key to understanding the basis for the length-tension relationship lies in the sarcomere diagrams to the left. They show the relative amounts of overlap in the sarcomeres of the whole muscle at various resting lengths (as shown by the roman numerals I to V). This may be seen better in the lower graph, which is the length-tension curve of a single sarcomere. This curve is made up of straight lines that correlate very well with the overlap of myofilaments; this correlation is one of the main pieces of evidence for the sliding-filament hypothesis. In the upper curve the collective effect of millions of sarcomeres obscures the simple relationship.

Over a small region of the length-tension curve near L_0, the force was independent of the length, but as the muscle was shortened to lengths less than L_0, the force decreased.

 How are these experimental findings interpreted?

Myofilament Overlap and the Amount of Force. These findings are interpreted in light of the structural evidence for the existence of crossbridges and the biochemical evidence that it is the head (crossbridge) portions of the myosin molecules that interact with actin and catalyze the release of energy from ATP. When the muscle is stretched out beyond L_0, there is incomplete overlap between the thick and thin filaments, and not all of the crossbridges on the thick filaments have the chance to attach to an actin filament. Therefore, when the muscle produces force, the amount of force is proportional to the myofilament overlap. In the region of the curve near L_0,

overlap is complete, and force is at its greatest. Because of the **bare zone** at the center of the thick filaments (where there are only myosin "tails" and no crossbridges), shortening the sarcomere does not allow any more crossbridges to attach to the thin filament as it slides further in. This is why the force does not change with length in this region. As the sarcomere shortens further, myofilaments from opposite sides of the sarcomere begin to interfere with each other, and less force can be produced. At the extreme shortening, in which force has virtually disappeared, the thick filaments have been pushed against the Z-lines, and there is no further opportunity for shortening or developing force. Under these conditions of extreme shortening, the normal activation processes are also disrupted, leading to further decrease in force.

It has also been shown that the processes that activate the myofilaments do not function as well at short lengths and that this factor also reduces the force. In experiments with whole muscles, the regions of the length-tension curve are

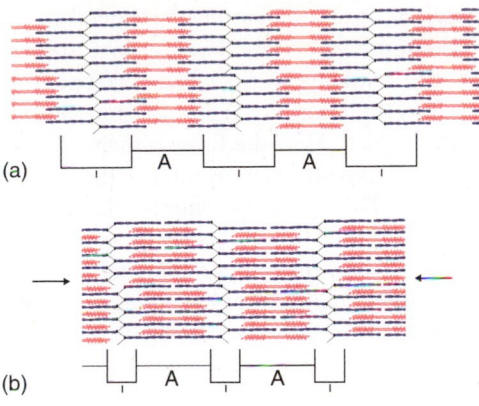

(a)

(b)

Figure 16–13

The shortening of sarcomeres in series. *(a)* Representation of a "snapshot" of two sarcomeres somewhere along a myofibril. An approximately 50% overlap of thick and thin filaments exists at this length. *(b)* Sarcomeres from a shortened muscle. Here, the overlap is complete, and three sarcomeres now fit in the space of two. The shortening of all of the sarcomeres in a muscle is added together, and the tiny amount of shortening of each sarcomere results in a very large shortening of the whole muscle.

not as clear-cut because the millions of sarcomeres making up the whole muscle are all at somewhat different places on this curve. For this reason, and because of the effects of the connective tissue, the length-tension curve for a whole muscle is a smooth curve without a flat region at its peak. Nevertheless, it is well established that the effects of length on muscle function do reflect a fundamental property of the molecular mechanisms responsible for muscle contraction.

The Sliding-Filament Model of Contraction in Other Muscle Types.

The account of muscle function just given is usually called the **sliding-filament hypothesis.** This is a very well-established body of knowledge that integrates much of what is known about skeletal muscle into a consistent framework. Because cardiac muscle is similar to skeletal muscle in structure and function, the same basic hypothesis appears to be valid there as well. Less is known about the relationship between structure and function in smooth muscle because of the different cell arrangement and lack of obvious structural regularity. However, what is known about the details of smooth muscle function and structure has not disproved the sliding-filament hypothesis, and as more information becomes available, such a mechanism will probably be experimentally supported.

> *What can we learn from experiments in which the muscle is allowed to shorten?*

Isotonic Contractions in the Laboratory: Clues to the Work of Muscle as a Motor

Analysis of isotonic contractions can reveal other aspects of muscle contraction. Because muscles shortening at a constant force are doing **work,** it is during isotonic contraction that the functions of *muscle as a motor* are best emphasized.

The equipment necessary to produce isotonic contraction from an isolated muscle is somewhat more complex than that used in isometric experiments. A force transducer is still used to measure the tension produced, but now the changes in muscle length must be measured, and a means of keeping the force constant during shortening must be provided. These requirements may be met by a lever system (Fig. 16–14). The muscle is attached to one end of a very light (but rigid) lever, and the desired load (e.g., a brass weight) is hung

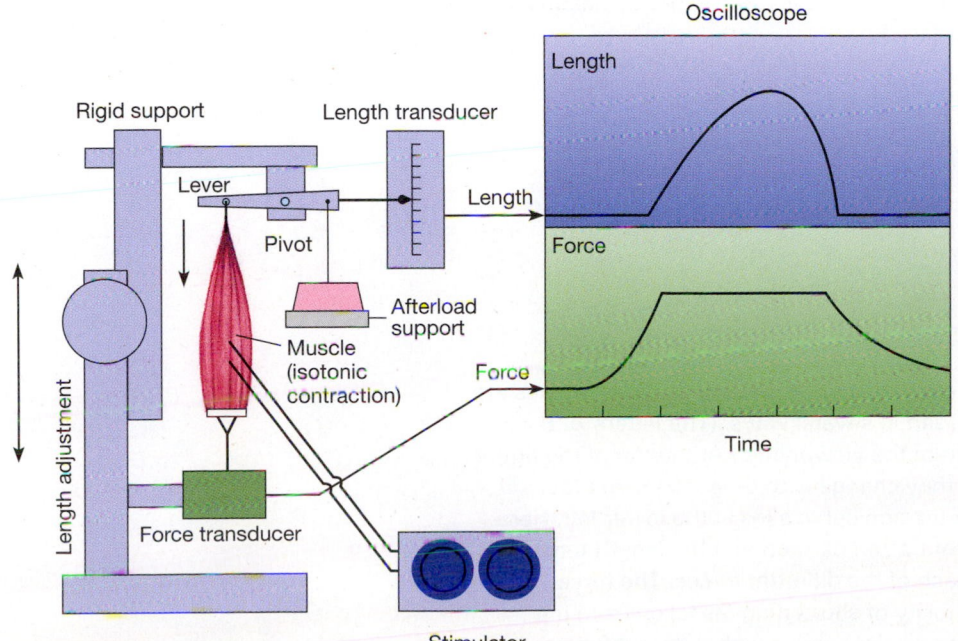

Figure 16–14

A typical apparatus for measuring isotonic contraction. The equipment is a bit more complicated than that shown in Figure 16–11*b* because the muscle is now free to shorten if it develops enough force, and the length change must be recorded along with the force. The afterload attached to the lever provides the force during isotonic shortening; early after the stimulus is applied, the muscle force is too little to lift the load, and that portion of the contraction is isometric. When force equal to the afterload is built up, the load is lifted, and the length change of the muscle is to be traced out by the pointer mounted on to the lever to which the muscle is attached.

from the other end. The muscle is anchored at its other end to the force transducer. A support is placed under the weight so that when the muscle is at rest it is not stretched out. The position of this support sets the resting length of the muscle. If the muscle length is appropriately set (see Fig. 16–14), there is no passive tension (recall this definition from the previous section). The force that the weight provides during contraction is called the **afterload** because the muscle bears this force only *after* beginning to contract. (If the muscle is stretched out by the attached weight so that some resting tension is present, this resting force is called the **preload.**)

The Mixed Contraction. When a stimulus is applied to an afterloaded muscle, the resulting contraction is at first isometric. It remains isometric until the muscle has developed an amount of force equal to the afterload. Then the muscle begins to shorten and lift the load. Now the force remains essentially constant and equal to the afterload as long as the load is lifted clear of its support. This is the isotonic portion of the contraction. As relaxation begins, the load is lowered (still isotonically) to the support. When the support prevents further lowering of the load, the muscle length no longer changes, and relaxation becomes isometric, just as it was in the first portion

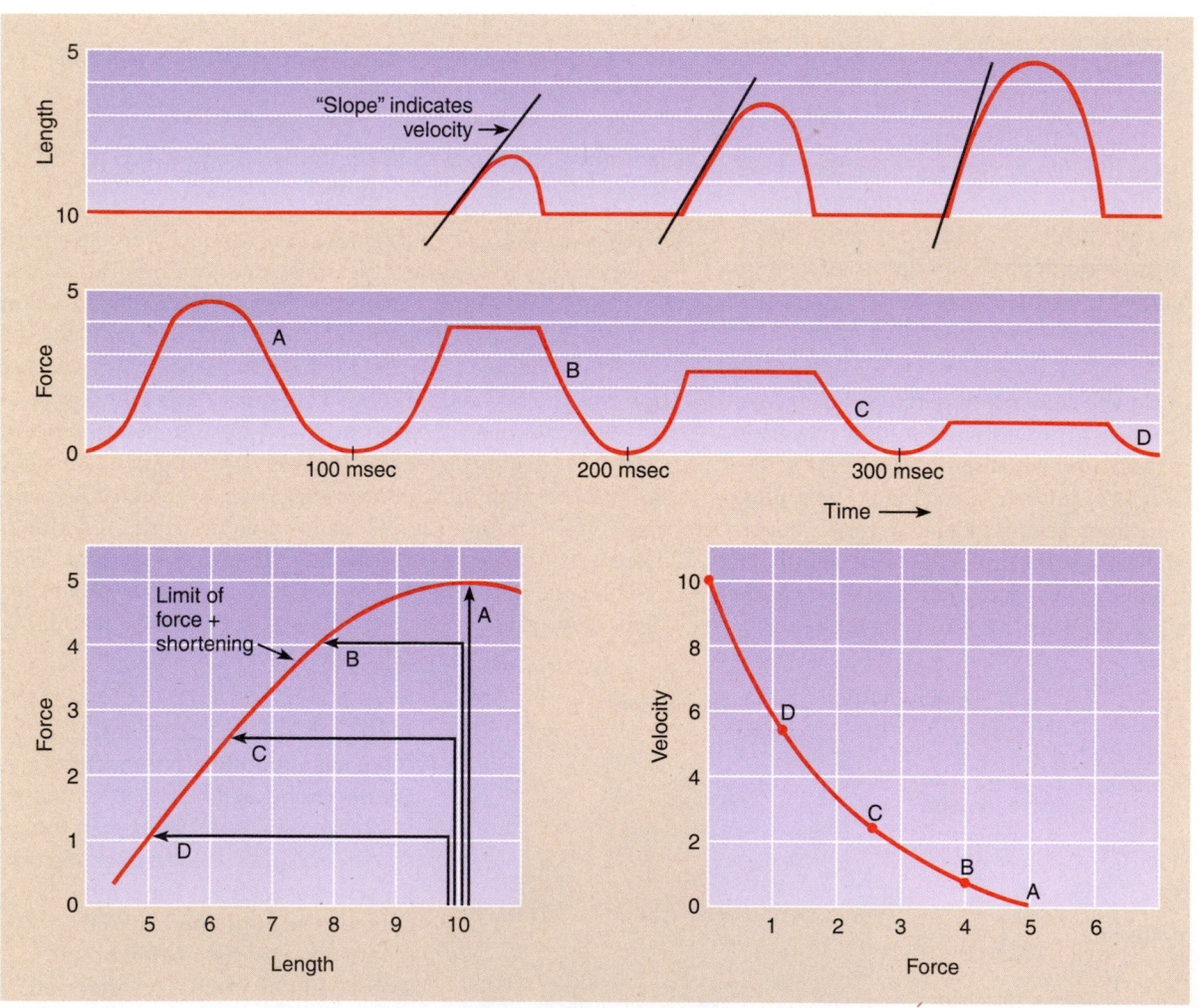

Figure 16–15

Force, velocity, length, and time in isotonic contraction. Four consecutive twitches are shown. Isotonic contractions can be visualized in several ways. (The letters *A, B, C,* and *D* refer to the same contraction in each of the viewpoints.) At the top of the figure is the usual display of force and length as they change with time. The same four contractions are also shown on the length-tension curve axes at the lower left. Here the dimension of time is not represented, but it can be seen that the length-tension curve provides the limit to shortening at each of the different forces. The force-velocity curve at the lower right shows how the velocity of shortening (as taken from the sloping tangent lines in the uppermost data display) varies with different forces.

of the contraction. The contraction would be entirely isotonic only if no load at all were applied. Under most circumstances, then, an isotonic contraction actually consists of an isometric portion followed by the isotonic shortening. Such a contraction is sometimes called a **mixed contraction.** With a large afterload, the isometric portion lasts relatively longer, because the muscle takes longer to build up enough force to lift the load.

Several important relationships can be expressed during isotonic (mixed) contractions (Fig. 16–15). First, the lighter the load, the sooner it can be lifted. (This information could actually also be obtained from the force record of an isometric contraction.) Second, the larger the load, the less the muscle is able to shorten. In fact, if the afterload is greater than the force the muscle can exert, the contraction is completely isometric.

The Force-Velocity Curve. Most importantly, the speed at which a muscle shortens is related to the amount of the afterload (see Fig. 16–15). This speed is measured at the very beginning of shortening because the muscle continually slows down during most of the shortening. The lighter the afterload, the greater the speed (velocity) of shortening. As the afterload is made larger and larger (in subsequent contractions), the speed with which the muscle begins its shortening becomes less and less. The relationship is described by the **force-velocity curve.** The highest velocity of shortening, called V_{max}, occurs with zero load. (Note: Not to be confused with the Umax associated with enzyme kinetics.) While useful contractions do not occur under conditions of no load, a measurement of V_{max} provides a good indication of the maximal rate at which the energy-converting processes of a muscle can operate. This information is often used for comparisons among different types of muscle. At the other extreme of loading, the lowest velocity of shortening is zero, and at this point the contraction is completely isometric.

The Power of a Muscle. The force-velocity curve emphasizes the function of muscle as a biological motor. It provides a basis for studying how our muscles interact with the loads

and forces encountered in our environment. We use muscles to do work, which is defined as *moving a force through a distance.* A concept related to this is the notion of power output, defined as the rate of doing work. Analysis of the information expressed in the force-velocity curve shows that the value obtained by multiplying the *force times the corresponding value of velocity* is the power output of the muscle. A graph of the way in which the power output of muscle varies with the afterload force (Fig. 16–16) shows that at the maximum force that the muscle can attain, *the power output is zero.* Because the isometric muscle does not shorten, it does no work and therefore produces no power. A related situation occurs at V_{max}, the velocity at no load. Here shortening is the most rapid, but no force is exerted. Again, this means that *the power output is zero.* Between these two extremes, the power output passes through a maximum when the force is at aproximately 30% of its maximal value.

The Efficiency of Muscular Contraction. For many skeletal muscles, it is at about the point of maximal power output that the **efficiency** is the greatest. (*Efficiency* is defined as the amount of work done for a certain input of energy.) This relationship has some practical consequences for the way in which we do work (or play). An athlete (e.g., a long-distance runner) who needs to maximize endurance runs for most of the race at a pace that is less than "all out." Only in the final stages of the race does the runner strive for the greatest speed. This high-speed running is done much less efficiently than is the running in the earlier part of the race and cannot be sustained as long. In human-powered machines, especially those designed to be operated for long periods of time, the force-velocity and power output properties of exercising muscle need to be considered. Machines such as bicycles (and now airplanes) that get their power from human muscle are designed so that the load that they present to the muscles is near to the optimum afterload on the force axis of the force-velocity curve. Many bicycles, in fact, are provided with a number of "gear ratios" so that the

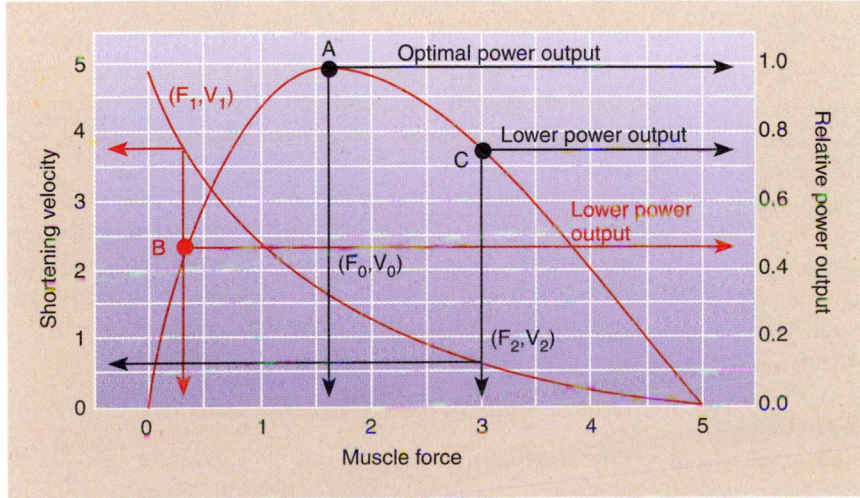

Figure 16–16

How the power output of muscle varies with the afterload. At the highest and lowest values of force, the power output is zero (see text). At an afterload of around one third the maximum (at *A*), the power output is maximal. Decreasing the force (at *B*) increases the shortening velocity but lowers the power output; increasing the force (at *C*) decreases the shortening and also decreases the power output. The efficiency of the muscle is also reduced at the force-velocity combinations at which the power output is reduced.

rider can match the muscle force to the leg-muscle force-velocity curves as the vehicle goes up and down hills, for example. While such considerations are not always important in everyday activities, all muscular activity is subject to the limitations described by the length-tension and force-velocity curves.

How do the concepts of length-tension and force-velocity apply to cardiac and smooth muscle?

Cardiac and Smooth Muscle Operate Under Specialized Mechanical Conditions

The contraction of heart muscle also can be described by length-tension and force-velocity curves, but its mechanical situation in the body is much different from that of skeletal muscle. Because it is not attached to the rigid bones of the skeleton, it does not have such a strictly limited range of lengths over which it must operate. In addition, unlike skeletal muscle, which is set by its attachments to operate at lengths near L_0, cardiac muscle generally operates at lengths somewhat less than L_0. This gives it some length "reserve" with which to produce more forceful contractions when they are needed.

Cardiac muscle, in normal body function, also relaxes differently from a skeletal muscle that has undergone isometric contraction. During isotonic relaxation, a skeletal muscle is stretched back to its original length by the load it has lifted or by antagonist muscles. The relaxation is thus mostly isotonic. Ventricular muscle (a type of cardiac muscle), however, at its maximum shortening, has the afterload removed by the action of the aortic and pulmonary valves (see Chapter 18), and the relaxation is mostly isometric. The muscle is relengthened by the refilling of the heart produced by returning blood. The mechanical consequences of such a cycle will become apparent when the function of the heart is discussed.

Because smooth muscle is usually found as a part of the makeup of the wall of a hollow organ or tube (rather than as a muscle stretched between two attachment points), it rarely can make a "typical" isotonic contraction. Most contractions in which smooth muscle shortens are done against a decreasing load as fluid or soft contents are propelled, or contents of an organ are expelled. Smooth muscle of the blood vessels is maintained in a partially shortened condition in which force is also constant for long periods. At the other extreme, smooth muscle sphincters contract isometrically and stay contracted most of the time. Even though the physiological situation of smooth muscle is very different from that of skeletal and cardiac muscle, its actions are also described by force-velocity curves similar to those of skeletal muscle. The differences are mostly in speed (smooth muscle is much slower than skeletal muscle) and in the shape of the length-tension curve (smooth muscle typically can function over a very wide range of lengths).

How does a "signal from a motor neuron initiate muscle contraction;" what is the mechanism of this muscle contraction?

THE CONTROL OF MUSCLE

Muscle is subject to a variety of bodily influences that control its function. Some of these controlling factors provide very precise and specific control of some types of muscle. Other control mechanisms may have a more subtle but still important control over the functions of various muscle types.

Cell Membrane Activity Is Important in Controlling Muscle Function

As with nerve cells, the cells of muscle are surrounded by an electrically excitable plasma membrane. The characteristics of this membrane vary widely among the muscle types. In the case of most skeletal muscles, the cell membrane functions in much the same way as neuronal membranes behave. We encounter additional complexity in the membrane properties of heart muscle, and a wide variety of membrane functions are found in smooth muscle. However, most of the basic features of membrane control can be illustrated with skeletal muscle as an example.

Action Potentials in the Plasma Membrane

The surface membrane of most skeletal muscle cells can be excited to produce action potentials. As in neurons, the action potential is caused by time- and voltage-dependent changes in the membrane permeability to sodium and potassium ions. These changes may be started by a brief electrical shock to the membrane, but stimulation usually occurs by way of the motor neurons that normally innervate voluntary muscle. The details of neural activation are discussed later on. At present, we will simply consider that stimulation has taken place and that an action potential is present in a muscle cell plasma membrane. This action potential spreads rapidly along the muscle fiber. As in a nerve fiber, the current from the active region causes depolarization of resting membrane areas nearby. Because muscle cells lack the myelin sheath found in many nerves, the propagation of the activity is somewhat slower than that found in myelinated neurons. The relative slowness with which action potentials travel along the muscle cell plasma membranes is not especially important from a practical standpoint because the muscle contraction itself is considerably slower than the speed of the action potential. A much more important problem involves the spread of activation inward from the surface plasma membrane to the depths of the fiber where the contractile material is located.

Action Potentials in the Transverse Tubular System

As cells go, those of skeletal muscle are quite large. The deepest parts of a muscle fiber may be as much as 50 μm away from the surface membrane. While this does not appear to be very far, it does pose an important problem. Suppose that muscle activation were due to the release of a chemical substance at the plasma membrane. This substance could then diffuse into the depths of the fiber. If this were actually the case, muscle contractions would have to be much slower than they are because diffusion over the distances involved would take a significant amount of time. If local electrical currents from the active plasma membrane flowed to the depths of the cell and caused its activation, they would not have sufficient strength to affect more than a few of the surface myofibrils.

In skeletal muscle these potential problems do not arise because of the special conduction that takes place in the **transverse tubular system (t-system).** The membrane-lined tubules of this system, the structure of which was outlined previously, penetrate the interior of the fiber at the level of each sarcomere. The insides of these t-tubules are open to the extracellular space, and their membrane linings are extensions of the surface membrane. At each sarcomere, they effectively bring the extracellular space and the muscle plasma membrane very close to the interior of the fiber. As with the surface membrane, the t-tubule membranes conduct action potentials. When an action potential sweeps along the surface membrane, it is also conducted into the depths of the fiber along the t-system and arrives at the innermost portions of the cell with very little delay. Because of the t-tubule system, this step in the activation of skeletal muscle becomes very rapid.

Calcium Ions in the Sarcoplasmic Reticulum

The t-tubule system does not actually open into the interior cytoplasm of the cell because the tubules are completely lined with cellular membrane. Instead, they come into close contact with the membranes of the SR, a closed membrane system. This means that it is a structure that maintains a separate region of the cell interior. This enclosed region is used to store and release the actual chemical substance that controls the contractile activity of the muscle proteins. This substance is ionized calcium (calcium ions), and in the resting muscle, it is highly concentrated within the SR. It is contained within the saclike areas called *terminal cisternae* (or *lateral sacs*). The longitudinal elements of the SR communicate with the terminal cisternae at each end of the sarcomere and form a highly branched network surrounding the myofilament bundles. The interior of the longitudinal elements is continuous with the terminal cisternae.

As this description suggests, each sarcomere has its own separate portion of SR. The junction between a t-tubule and the terminal cisternae of two adjacent sarcomeres (the **triad**) is the place where the electrical signal from the outside of the fiber actually is communicated with the interior of the fiber. As it travels along the t-tubule, the action potential reaches the region of a triad, where its presence is communicated to the terminal cisterna of the SR. The exact nature of this communication is not yet fully understood, but it appears that the action potential affects specific protein molecules termed **dihydropyridine receptors (DHPR),** which are located in the t-tubule membrane in groups of four. They serve as **voltage sensors,** responding to the t-tubule action potential. Each group is located in very close proximity to a specific channel protein called a **ryanodine receptor (RyR),** which is located in the membrane of the SR. The RyR serves as a controllable channel (termed a **calcium release channel**); when it is open, calcium ions can move readily through it and escape into the cytoplasm of the cell. When the muscle is at rest, the RyR is closed; as the t-tubule depolarization reaches the DHPR, some sort of linkage — perhaps a mechanical connection — causes the RyR to open and release calcium from the SR. In skeletal muscle, every other RyR is associated with a DHPR cluster; and the RyRs that lack this connection open in response to the presence of the calcium ions that were just released. This process takes only a few milliseconds, and the calcium ions just released can diffuse the short distance to the myofilaments and begin to initiate the contraction process.

Relaxation is associated with the movement of calcium ions back into the longitudinal elements of the SR. This ionic movement takes place up a steep concentration gradient via an ATP-dependent calcium ion pump present in this region of the SR membrane. While this mechanism (diagrammed in Fig. 16–17) may seem complicated, it quite effectively overcomes the diffusion and conduction obstacles posed by the large size of skeletal muscle fibers. It is capable, in some muscles, of allowing activation to be repeated as many as 60 to 100 times per second. Muscles without such a mechanism would be limited to very slow rates of contraction.

Calcium Ions Are a Vital Link in the Control of Muscle

The control of contraction in all types of muscle involves controlling the interaction of the actin and myosin myofilaments. This control is carried out through the action of calcium ions, although the actual process differs greatly between skeletal and cardiac muscle on one hand and cardiac and smooth muscle on the other.

The Molecular Mechanisms that Control Muscle Contraction

When an action potential, through the sequence of events described previously, causes the SR of skeletal muscle to release a quantity of stored calcium ions, these ions diffuse to the region of the myofilaments, where the resting calcium concentration is normally very low (Fig. 16–18). Here they bind to **troponin-C (TnC),** a subunit of the regulatory protein complex, which is attached to the tropomyosin molecule on the thin (actin) myofilaments. When calcium ions bind to TnC, they cause a conformational change in the other parts of the troponin molecule, which in turn causes the tropomyosin molecule to change its position along the "groove" of the actin filament. At rest, the tropomyosin

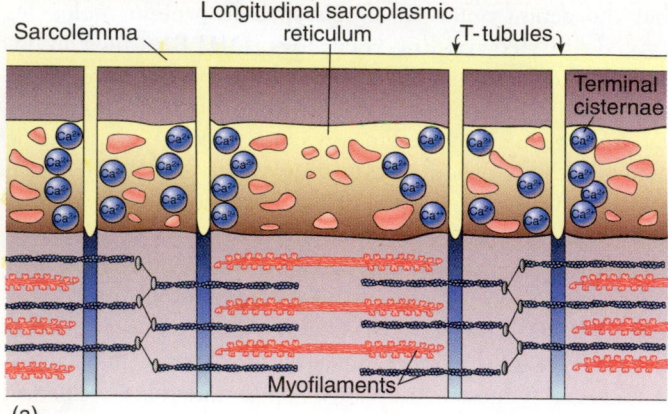

Sarcolemma, Longitudinal sarcoplasmic reticulum, T-tubules, Terminal cisternae, Myofilaments

(a)

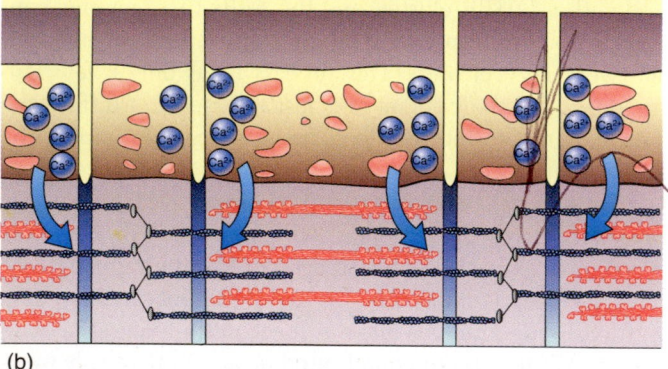

(b)

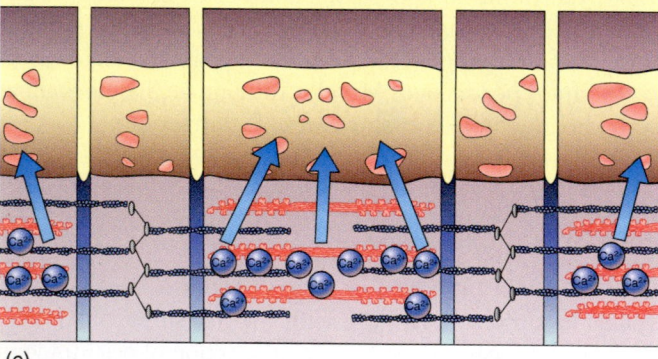

(c)

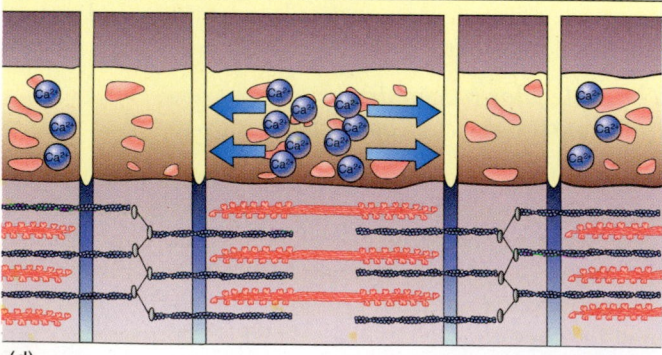

(d)

◄ **Figure 16–17**

How calcium moves during a contraction. **(a)** With the muscle at rest, the calcium ions are stored in the terminal cisternae of the sarcoplasmic reticulum. **(b)** As activation begins, calcium is released from the sarcoplasmic reticulum into the region of the myofilaments. **(c)** Activation ends as calcium is taken back up into the longitudinal elements of the sarcoplasmic reticulum. **(d)** Immediately after the contraction, the calcium moves back into the terminal cisternae.

interferes with the interaction of actin and myosin, but with the tropomyosin position slightly shifted, the actin filament becomes able to interact with the myosin crossbridge. Because the tropomyosin molecule covers seven adjacent actin monomers, it can exert a large degree of control over the actin-myosin interaction.

The contraction resulting from this interaction of actin and myosin in a series of steps is diagrammed in Figure 16–19. In resting skeletal muscle (with calcium ions not yet released), an ATP molecule is bound to each myosin molecule head that can potentially form a crossbridge with the actin filament. The ATP molecule has been "split," but its energy cannot be released until the actin is allowed to interact with the myosin. When the presence of calcium ions permits this interaction, the myosin head attaches to the actin filament at an angle of approximately 90° and the phosphate ion that was split from the ATP molecule is released. As the energy of the ATP molecule is released, the angle of attachment becomes closer to 45°. This change in attachment angle (the so-called **power stroke**) causes the actin filaments to be pulled in toward the center of the sarcomere, and the ADP molecule is released from the myosin. Only when a new ATP molecule binds to the myosin head can the crossbridge detach itself from the actin and become ready for a new cycle.

As long as enough ATP and calcium ions are present, the cyclic process of attachment, angle change, ATP binding and hydrolysis, and detachment can take place. Repeated crossbridge cycles result in the actin filaments being pulled further and further into the myosin filament array, and the whole muscle becomes shorter. The force that the muscle is working against is important in determining how fast the crossbridge cycle will run. If the ends of the muscle are held stationary (as in isometric conditions), the crossbridge cycles will still continue to some extent because some internal movement is allowed by the elastic nature of the muscle. When stimulation of the muscle is stopped, no additional calcium ions are released from the SR, and calcium ions that diffuse free from the troponin or that are free in the cytoplasm are taken back up by the SR, the calcium pump of which is continuously active. Under these conditions, tropomyosin resumes its blocking function. Crossbridges do not reattach, and relaxation takes place. If the supply of ATP becomes depleted, attached crossbridges are unable to detach, and the muscle becomes stiff and unable to relax, the

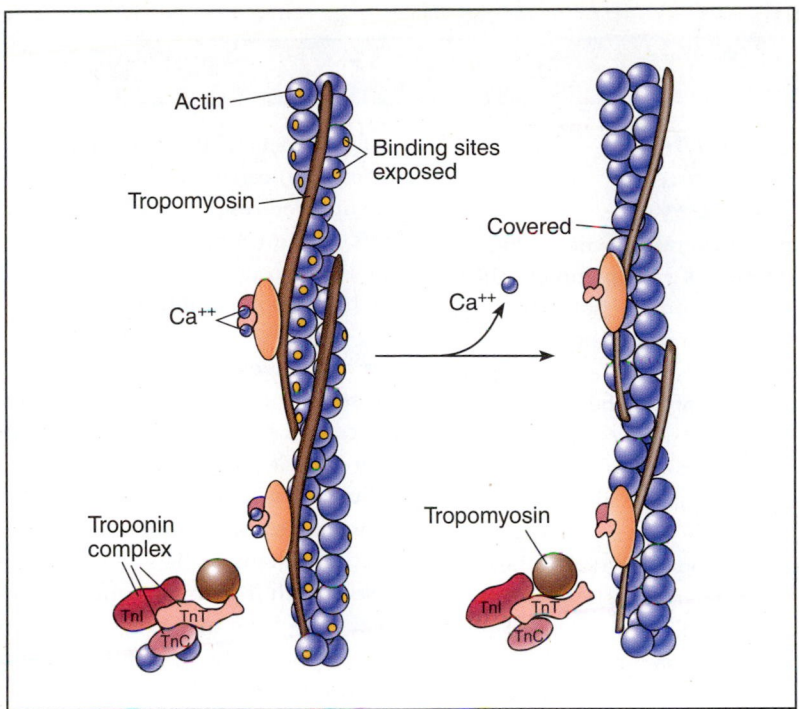

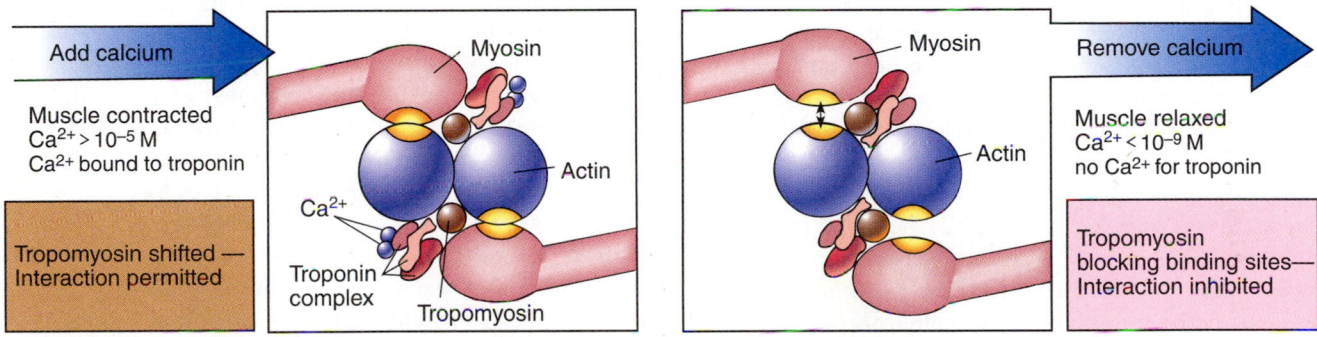

Figure 16–18

How calcium and the regulatory proteins control the interaction between actin and myosin. When internal calcium concentration is very low (the relaxed state, *lower right*), the position of tropomyosin along the thin filament (seen in cross-section) prevents interaction between the myosin heads and the active sites on the actin molecules. When calcium is released from the sarcoplasmic reticulum, the ions bind to the Tn-C subunit of troponin. This causes a shift in the position of tropomyosin, which allows for the interaction between the myosin heads and the actin monomers.

"rigor" complex. An extreme example of this situation is the condition known as **rigor mortis,** which occurs after death as the supplies of ATP in the muscle become exhausted.

Actin-Linked Regulation of Muscle Contraction. Muscle regulation that is accomplished through some action on the thin filaments is called *actin-linked regulation.* This type of regulation is also found in heart muscle, where the steps involved are quite similar to those in skeletal muscle. In cardiac muscle, however, the amount of calcium ions released in re-

sponse to an action potential is usually not sufficient to activate all of the possible actin molecules, and a maximal contraction is not obtained. This provides cardiac muscle with a "reserve capacity" for contraction, and there are several ways in which extra calcium ions can be released to meet additional requirements. In the specialized action potentials of cardiac muscle, one of the ions that flows into the cell is calcium. Because of the small size of cardiac muscle cells, it is possible for a significant amount of calcium ions to enter the cell through the plasma membrane during the course of an

CURRENT CONCEPTS IN PHYSIOLOGY

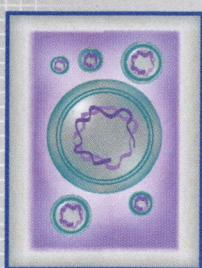

Measuring Muscle Contraction at a Molecular Level

Although the overall process of muscle contraction is very complex, the fundamental events that change chemical energy into mechanical motion are relatively simple, involving the interaction of the heads of the myosin molecules with actin filaments. In recent years, techniques have been worked out to allow actual mechanical measurements to be made on single filaments.

One of these methods, called an *in vitro motility assay*, uses myosin molecules and actin filaments that have been separately isolated in intact form from muscle tissue. In order to make them visible, the actin filaments are linked with modified antibody molecules (rhodamine-phalloidin) that give off light ("fluoresce") when they are illuminated with ultraviolet light under a powerful microscope. The single myosin molecules, separated out from the thick filaments, are chemically attached by their tail region to a specially coated glass surface so that the head portion is free to project into a bathing solution containing ATP as an energy source. When actin filaments are added to the bathing solution (which also contains necessary nonmuscle chemicals), the filaments begin moving over the myosin-coated surface, propelled by the actin-myosin crossbridges. A highly sensitive video system attached to the microscope observes and records the movement. The sliding speed is about 5 μm/s with myosin from fast skeletal muscle, and it is about 25 times slower if myosin from smooth muscle is used. The speed of sliding depends on the properties of the myosin and is correlated with the speed of the intact muscle that it is taken from. The actin and its source are much less important, although it is the orientation of the monomers of the actin filament that determine the direction of movement. The actin filaments can be taken from nonmuscle sources, with little effect on the rate of sliding.

The effects of the regulatory proteins (troponin and tropomyosin) can be studied by including them on the thin filaments, and researchers have found that their presence allows calcium ions to control the interaction between actin and myosin. When smooth muscle myosin is used, it must be phosphorylated in order for movement to take place. Mixtures of smooth and skeletal muscle myosin cause movement at intermediate speeds. Nonphosphorylated smooth muscle myosin molecules retard the movement caused by phosphorylated myosin molecules in the mixture coating the glass surface.

Further refinements of this technique allow the force of a single crossbridge to be measured. In one case, a single actin filament is chemically attached to a microscopic needle, which is then brought near to the myosin-coated surface. The force that a myosin molecule generates when it interacts with the actin filament is sufficient to cause a microscopic bending of the needle. This bending can be translated into a measurement of the tiny mechanical force. A newer technique uses two intensely focused laser beams that are able to trap and hold two tiny, plastic beads. A single actin filament is attached between the beads and brought near to a stationary myosin molecule. The microscopic movements of the beads that are caused by the actin interacting with the myosin molecule can be translated into values for the single-crossbridge force. This turns out to be approximately 2 to 4 picoNewtons (one trillionth of a Newton), depending on the experimental conditions. When the value for a single crossbridge is scaled upward to the dimensions of an intact muscle, rather good agreement occurs.

Use of the *in vitro* assay has answered many questions arising from experiments on intact muscle and has provided unique insight into biochemical processes that would be very difficult to interpret using strictly chemical means. A number of laboratories are conducting experiments that are probing individual steps in the crossbridge cycle in an effort to work out the details of actin-myosin interaction.

---

action potential. Factors that increase the duration of action potentials, or the number of action potentials per minute, or the amount of calcium ions entering with each action potential, can increase the force of contraction.

Myosin-Linked Regulation of Muscle Contraction. Regulation of contraction that takes place via thick filaments is called **myosin-linked regulation.** Smooth muscle is controlled in this way. Its thin filaments lack the troponin protein complex, although there is tropomyosin on the thin filaments. In smooth muscle, one of the two protein light-chains on the head of each myosin molecule has a crucial regulatory function. Interaction between myosin and actin can take place only when a phosphate group (from ATP) is bound to this light-chain. Because the thin filaments of smooth muscle lack the troponin regulatory complex, they are always ready for interaction with myosin, and it is the myosin, not the actin, that must be activated.

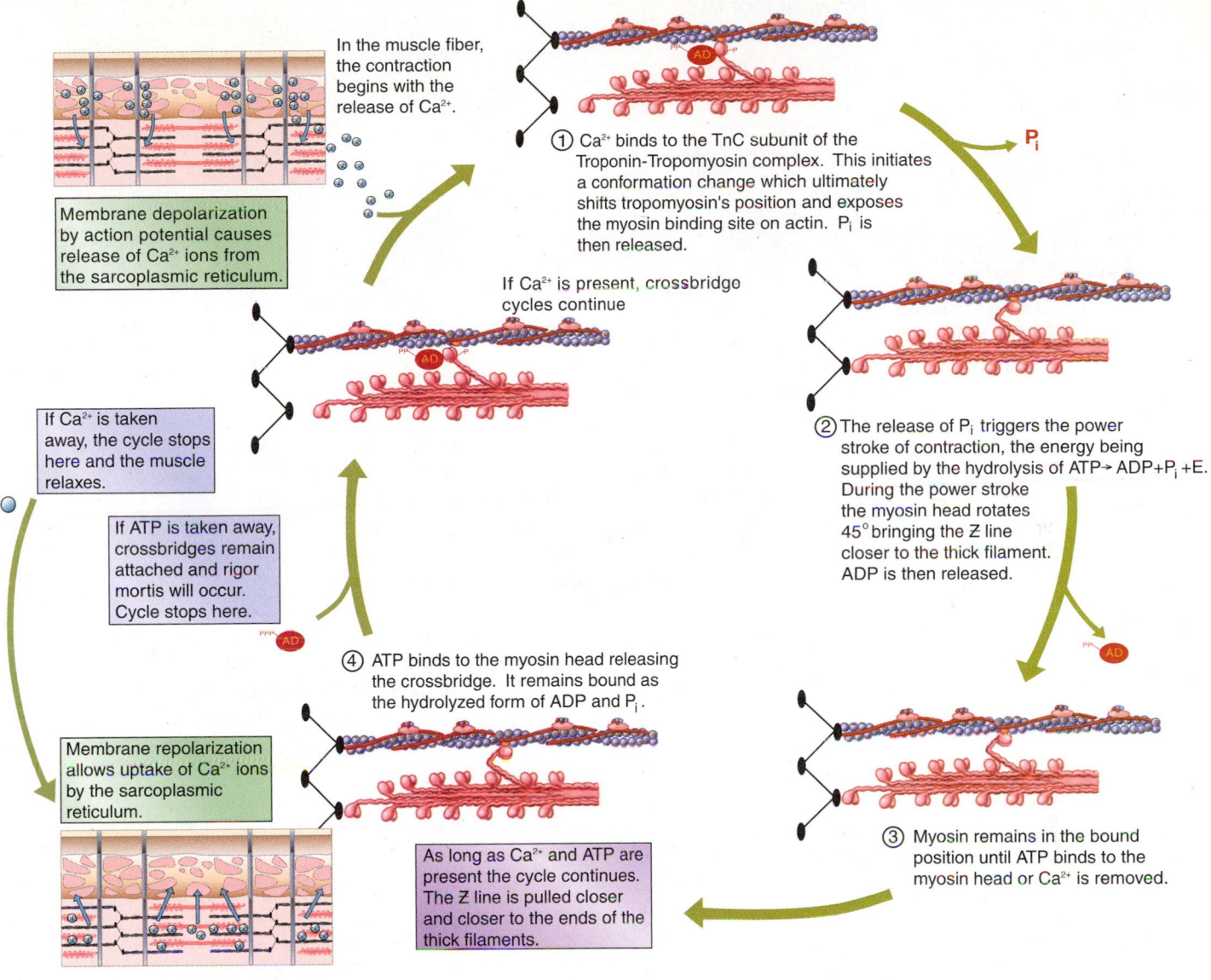

In the muscle fiber, the contraction begins with the release of Ca²⁺.

Membrane depolarization by action potential causes release of Ca²⁺ ions from the sarcoplasmic reticulum.

① Ca²⁺ binds to the TnC subunit of the Troponin-Tropomyosin complex. This initiates a conformation change which ultimately shifts tropomyosin's position and exposes the myosin binding site on actin. P$_i$ is then released.

If Ca²⁺ is present, crossbridge cycles continue

If Ca²⁺ is taken away, the cycle stops here and the muscle relaxes.

If ATP is taken away, crossbridges remain attached and rigor mortis will occur. Cycle stops here.

② The release of P$_i$ triggers the power stroke of contraction, the energy being supplied by the hydrolysis of ATP→ADP+P$_i$+E. During the power stroke the myosin head rotates 45° bringing the Z line closer to the thick filament. ADP is then released.

④ ATP binds to the myosin head releasing the crossbridge. It remains bound as the hydrolyzed form of ADP and P$_i$.

Membrane repolarization allows uptake of Ca²⁺ ions by the sarcoplasmic reticulum.

As long as Ca²⁺ and ATP are present the cycle continues. The Z line is pulled closer and closer to the ends of the thick filaments.

③ Myosin remains in the bound position until ATP binds to the myosin head or Ca²⁺ is removed.

Figure 16–19

The crossbridge cycle of skeletal muscle. The series of events goes on continuously as long as calcium and ATP are present.

The activation and regulation processes of smooth muscle can be divided into three groups of reactions that take place more or less simultaneously. These are shown in Figure 16–20. The first group (region 1) is the step where control of contraction begins. Calcium ions are released into the cytoplasm (*upper left*), either from the SR or from outside the cell via the cell membrane. These ions bind to a protein called **calmodulin (CAM)**, which is closely associated with an enzyme called **myosin light-chain kinase (MLCK)**. This binding activates the MLCK and allows it to catalyze one of the reactions shown in region 2. This reaction is the phosphorylation of the regulatory myosin light-chains; when the terminal phosphate group of an ATP molecule is transferred to a regulatory light-chain molecule, the catalytic function of myosin becomes activated and it is

able to interact with actin. A **phosphatase** enzyme simultaneously acts to remove other similar phosphate groups, but during activation, the phosphorylation reaction occurs at such a high rate that the dephosphorylation reaction is relatively unimportant.

The interaction between actin and activated myosin is shown in region 3. This is a crossbridge cycle similar to that of skeletal muscle, involving the cyclic attachment, rotation, and detachment of crossbridges to produce force and movement. This cycle also continues as long as there is a supply of ATP from the cellular energy stores and as long as myosin remains in its activated state. Notice that the ATP molecule that gives up its phosphate group to the light-chain to activate the myosin molecule is *not* the same ATP molecule whose energy is consumed in the contraction cycle.

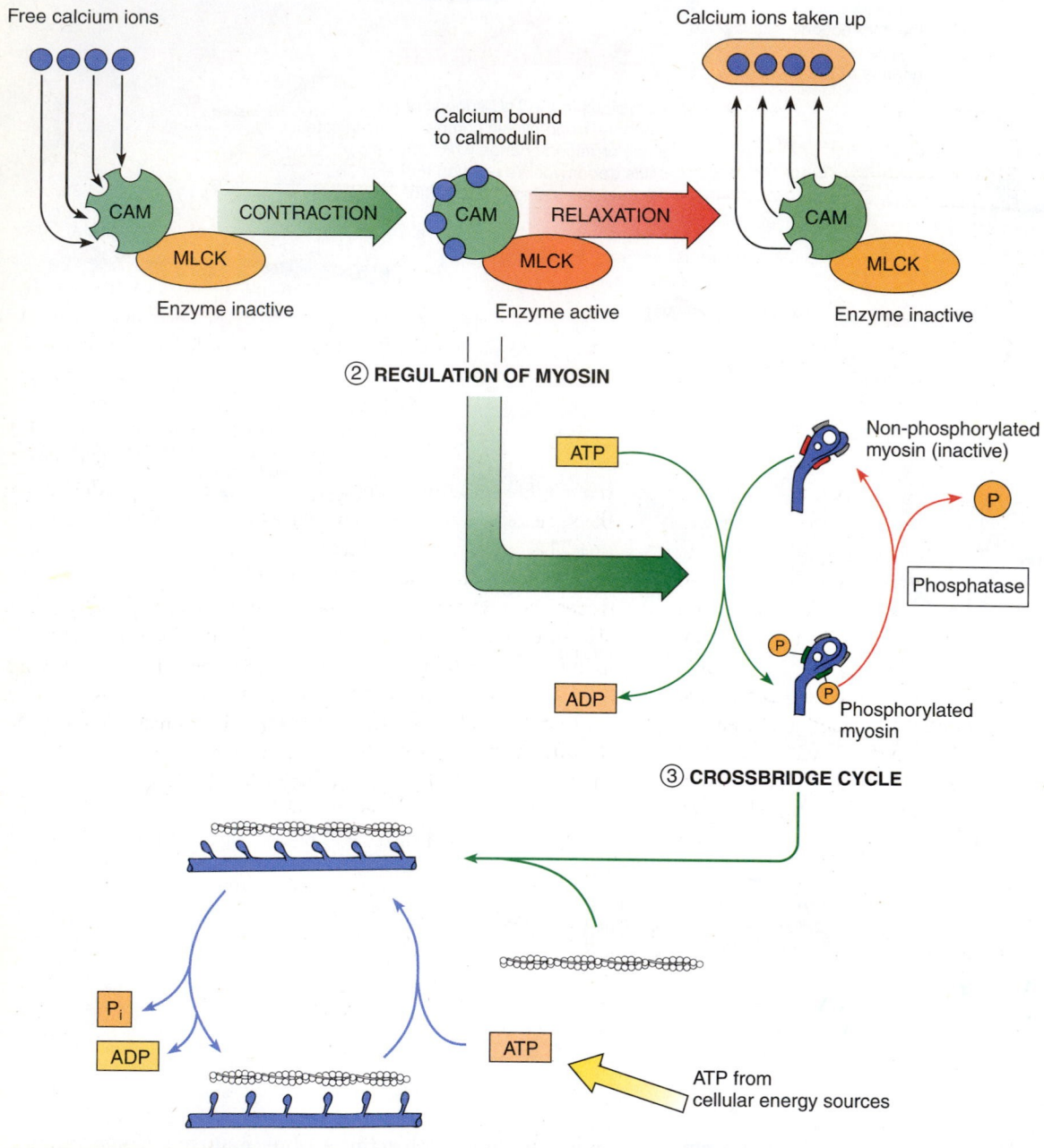

① ACTIVATION OF MLCK

Free calcium ions

Calcium ions taken up

Calcium bound
to calmodulin

CONTRACTION   RELAXATION

CAM

MLCK

Enzyme inactive   Enzyme active   Enzyme inactive

② REGULATION OF MYOSIN

ATP

Non-phosphorylated
myosin (inactive)

P

Phosphatase

ADP

Phosphorylated
myosin

③ CROSSBRIDGE CYCLE

P$_i$

ADP

ATP

ATP from
cellular energy sources

Figure 16–20

The major steps in the regulation of smooth muscle contraction. A contraction is
initiated by the release of calcium into the cytoplasm, and the myosin molecules are
activated by the steps in regions 1 and 2 so that the crossbridge cycle (shown in
region 3) begins to operate. Relaxation begins when the calcium is taken back up
and the phosphatase enzyme (in region 3) is no longer working against the MLCK
and can dephosphorylate (and thus deactivate) the myosin molecules.
CAM = calmodulin; MLCK = myosin light-chain kinase; P = phosphate.

Relaxation takes place when calcium ions are taken up
from the cytoplasm (*upper right*) by the action of calcium
pumps in the SR or plasma membrane. The MLCK becomes
inactive and can no longer catalyze the phosphorylation reaction. The phosphatase reaction is then able to dephosphorylate the myosin molecules, and so it can no longer interact
with actin molecules. The crossbridge cycle comes to a halt,
and the muscle relaxes. In addition to these basic processes,
smooth muscle also contains other regulatory processes responsible for its ability to maintain contraction for a long

period of time without consuming large amounts of energy. These processes appear to involve modifications in the properties of crossbridges; in some smooth muscles, crossbridges become "latch bridges" that detach very slowly and therefore spend most of their time in the attached, force-producing state. These latch bridges are thought to be normal crossbridges that have become dephosphorylated while still attached to actin filaments. In this condition, they can cycle, but at a rate much slower than normal. Thus, while the details of regulation of smooth muscle are much different from the regulatory processes in skeletal muscle, the control over contraction and relaxation is still exercised by calcium ions.

 How can the force developed by a muscle be increased?

The Mode of Stimulation Regulates Muscle Function

One of the important results of the way in which muscle is controlled is seen in the types of contractions produced in response to various types of stimuli. Many muscles (especially skeletal and cardiac) can respond to a single stimulus with a single action potential. This usually results in a *single* brief contraction. Such a response is called a *twitch*. The rapid contraction and relaxation of skeletal muscle during a twitch are the result of two processes that take place at the same time. As we have seen, the result of the action potential is the sudden release of stored calcium ions from the SR. While these calcium ions are diffusing to the myofilaments and activating the contraction, the processes that take the calcium ions back up are also operating. Even when the amount of calcium ions released is potentially enough for a maximal contraction (as is usually the case in skeletal muscle), their rapid uptake limits the amount of force that can develop. In most cases, the muscle does not have a chance to develop its full amount of force before relaxation processes have set in. Crossbridge cycling, necessary for the buildup of force, takes time, and this time is not available.

Summation of Action Potentials: Producing a Sustained Contraction in Skeletal Muscle

Rapid relaxation and low force mean that twitch contractions are not particularly useful for everyday tasks. However, a repetitive type of contraction readily overcomes this limitation. The action potential of twitch-type skeletal muscles is very brief compared with the duration of the twitch contraction. Even though the action potential has the usual absolute and relative refractory periods typical of excitable membranes in general, these also are brief compared with the twitch time. This means that the muscle can be restimulated before its relaxation is complete, and the force produced by the second stimulus can add to the force left from the first. This process is called *temporal summation,* and it results in a contraction, called a **tetanus,** that lasts longer and produces more force than a single stimulus could produce. Inside the muscle, the number of calcium ions available to activate the myofilaments has been increased by the second stimulus. Stimuli can be repeated at regular intervals in order to keep the internal calcium ions continuously available to the myofilaments. This continuous supply in turn keeps the muscle from relaxing completely, and the result is a sustained contraction that produces significantly more force than that developed in a twitch (Fig. 16–21). If the stimuli are spaced

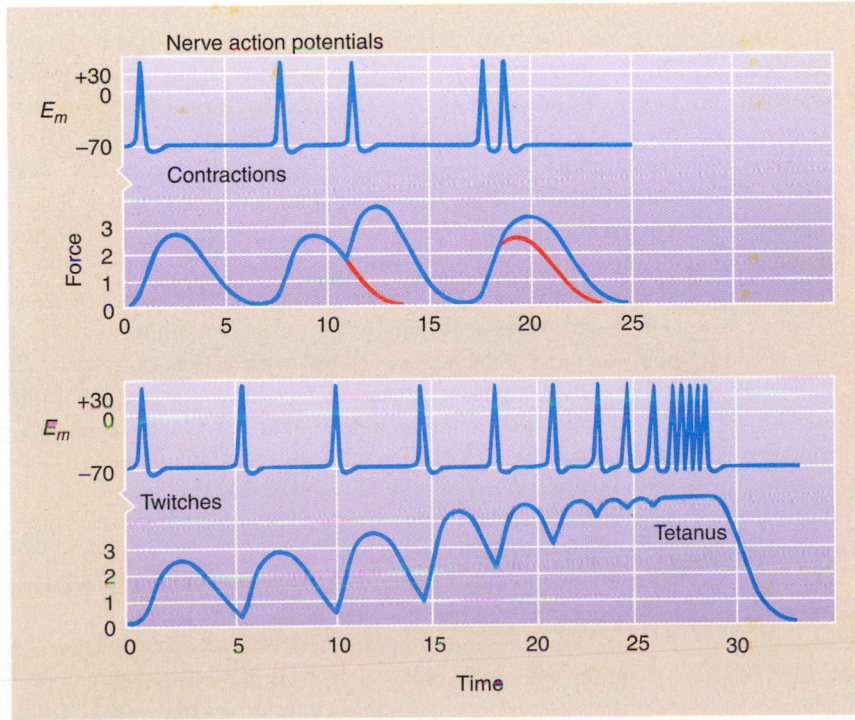

Figure 16–21

The relationship between twitches and a tetanus. Upper portion: A single motor nerve action potential produces a single twitch contraction (first event). Applying a stimulus during relaxation (second event) produces a contraction whose force adds to that of the first. Two stimuli delivered close together (third event) also produce an addition of force (this is made possible because of the short refractory period of the muscle cell membrane). *Lower portion:* When action potentials are made to occur closer and closer together, less and less time is available for relaxation, and the force produced by the individual twitches adds up. When the rate of stimulation is high enough, a smooth tetanus can be produced. A gradual buildup of twitches is not necessary for a tetanic contraction; a tetanus is produced immediately if the stimulus rate is high at the outset.

relatively far apart, the force increases and decreases somewhat between stimuli, but if the stimuli are close enough together, the developed force is steady. This is called a **complete, or fused, tetanus.** Just how many times per second a muscle must be stimulated in order to produce a fused tetanus depends on the relative rates of contraction and relaxation of the muscle. The lowest frequency that produces a fused tetanus is called the **tetanic fusion frequency.** It is approximately 20 to 60 times per second for most skeletal muscles.

Useful contractions — those that get us through daily life — are usually a mixture of twitches and partly fused tetanic contractions. These are distributed among the many individual fibers that make up a complete muscle, so that the contraction is "shared." For example, if only half of the fibers in a muscle were fully activated, the overall muscle could develop half of its maximal force. The CNS, in response to feedback from the muscle, visual feedback, and our prior experience, activates enough muscle fibers (usually with a partial tetanus) to allow us to make a contraction of the proper force and speed. Because different groups of fibers are activated at different times, their individual twitches or partial tetani are smoothed out, and the overall contraction is smooth and continuous.

   How does cardiac muscle differ from skeletal muscle with respect to repeated stimulation?

Spacing of Action Potentials: Producing Discrete Twitches in Cardiac Muscle

Cardiac muscle is a twitch-only type of muscle. A sustained and maximal tetanic contraction would be of little use in the task of pumping blood, and heart muscle has some special properties that guard against contractions of this sort. The spacing of the action potentials in cardiac muscle is controlled by the **pacemaker cells** in the heart, not by stimulation by motor nerves. These action potentials are quite long; they typically last almost as long as the contraction itself. This means that the membrane is refractory during most of the twitch, so that restimulation is not possible while the muscle has any significant tension. At higher rates of stimulation (high heart rates), the duration of both the action potential and the contraction are reduced. This allows for closely spaced twitches, with time for complete relaxation between them.

Even though cardiac muscle cannot develop a partial tetanus, the properties of its contractions can be adjusted by other means. In contrast to the situation in skeletal muscle, the amount of calcium released in response to a single action potential is generally not sufficient to provide for full activation. A number of factors can change the amount of calcium made available for each contraction. Making the heart contract more times per minute (up to a point, of course) results in the cellular accumulation of calcium and an increase in the strength of the contraction. Because calcium is one of the ions that enter the cell during each action potential, changes

in this calcium current can cause changes in the contraction. This is known as the **force-frequency relationship.** Some drugs, those of the digitalis family, for example, increase the overall internal supply of calcium, and more is released from the SR with each beat. Drugs called **calcium channel blockers** can reduce the strength of contractions, and the chemical epinephrine (both a drug and a hormone) can increase the strength of contraction by permitting the accumulation of internal calcium. All of the examples given here are agents that affect the **contractility** of the heart muscle. These influences are called **inotropic agents** (*ino* = "fiber"; *tropic* = "having an effect"). They permit changes in the properties of the contraction that do not depend on the possibility of a tetanic activation.

Along with the changes in contractility, changes in muscle length (as we saw previously) can have an important effect on the strength of the contraction of cardiac muscle. (Recall the relationships expressed in the length-tension curve.) When the heart is functioning properly in the body, these two types of influences work together to produce a very effective degree of control of its pumping function.

   How does the pattern of smooth muscle contraction differ from that of skeletal and cardiac muscle?

Phasic and Tonic Contractions: A Variety of Responses in Smooth Muscle

Smooth muscle can contract in a variety of different ways. Some smooth muscles can give twitchlike contractions in response to single stimuli or brief bursts of stimuli. The contractions are called **phasic** and are often found in smooth muscles that propel or move materials. Other smooth muscles give long and sustained contractions in response to single or continued stimulation by nerves, drugs, or hormones. These contractions are called **tonic** and are likely to be found in smooth muscle organs, such as sphincters or bladders, that must remain contracted for long periods of time. Many smooth muscles are almost never in a state of complete relaxation, and the steady force that they generate is called *tonus*. The level of tonus is often set by the concentration of specific hormone, such as norepinephrine.

In many cases, the contraction of smooth muscle is associated with electrical activity of the cell membranes. Most phasic contractions are associated with action potentials. In some tissues, such as intestinal muscle, which has rhythmic contraction patterns, there is a kind of electrical activity called the **slow wave.** The slow waves, which are many seconds in duration, periodically depolarize the membrane to the threshold for action potential generation; this mechanism acts to produce periodic contractions. Other smooth muscles show action potential activity that may be caused by a pacemaker cell and conducted to adjacent cells by means of the gap junctions previously mentioned.

Other smooth muscles may contract without any action potentials being present at all. For example, the walls of small blood vessels are lined with a layer of cells called *endothelial cells,* which lie just inside of the smooth muscle cells in the vessel walls. These endothelial cells produce an endothelium-derived relaxing factor (EDRF), which has been identified as nitric oxide (NO). This substance, as well as endothelium-derived contracting factors (EDCF), diffuses the short distance to the muscle cells and causes relaxation or contraction by setting in operation a train of responses known as a second-messenger system. These reactions result in the release of calcium from internal stores, which then goes on to control the activation of the contractile activity. This action takes place without the production of membrane action potentials. Numerous other natural chemical substances and drugs control the contraction and relaxation of smooth muscle without involving electrical activity at the cell membrane.

> *How does a motor neuron interact with muscle?*

The Central Nervous System Is Important in Controlling Muscle Function

Almost all of the muscle in the body is under the direct or indirect control of the CNS, although smooth muscle and cardiac muscle do have a degree of autonomous function that is suited to their specialized roles. Because neural control is so highly developed in the case of skeletal muscle, most of the general process can be understood by a study of how the action of a motor nerve controls the function of a skeletal muscle.

Nervous Control of Skeletal Muscle

Skeletal muscle depends on the somatic division of the CNS for the control of its contraction. As described in Chapter 10, under normal conditions, skeletal muscle does not contract unless it is stimulated by a motor neuron, which has its origin in the CNS. Connection between the motor nerve fibers, which are large myelinated axons, and the muscle fibers takes place at a structure called the **myoneural junction.** Because in some cases this structure spreads out somewhat on the muscle fiber, it is also called a **motor end-plate.**

The Myoneural Junction and the Motor Unit. The myoneural junction is a special case of a synapse. As in the vast majority of synapses in the CNS, the transmission of the impulse from the nerve to the muscle is by chemical means. The particular transmitter substance at the myoneural junction is acetylcholine (ACh). Unlike some CNS synapses, only excitation of the postsynaptic membrane is possible at the myoneural junction. Because many of the myoneural junctions in a muscle lie at its surface, they have been extensively studied, and the process of transmission is well understood. In fact, much of what we know about synapses in general has been learned by the study of this highly specialized synapse.

The structure of the myoneural junction provides keys to an understanding of its function. As a motor axon from the motor nerve bundle nears the muscle, it branches to innervate a number of individual muscle fibers. The assemblage of a *motor neuron,* together *with all of the muscle fibers that it innervates,* is called a **motor unit.** Because all of the muscle fibers served by a single branched axon are activated together when an impulse arrives from the axon, the fineness of control of a muscle depends on just how many muscle fibers a single axon controls. Muscles capable of fine and delicate movements, such as the *extraocular* muscles that move the eyeball, have a very few (or even only one) muscle fibers served by a single axon, whereas muscles used for rapid and relatively coarse movements, such as those of the posterior thigh, have many muscle fibers controlled by a single axon. Each muscle fiber normally receives innervation from only a single axon terminal, although some multiply innervated muscle fibers do exist.

Synaptic Transmission at the Myoneural Junction. As the motor axon branches, its myelin sheath becomes reduced to the cytoplasm and cell membranes of a single Schwann cell (the type of cell that forms the myelin sheaths around motor axons), and this covers the axon as it makes its final approach to the muscle (Fig. 16–22). The terminal region of the axon lies in a "groove" on the surface of the muscle. This arrangement increases the amount of nerve and muscle cell portions that make up the junctional region. The motor axon terminal contains small vesicles that carry the transmitter substance. Also present in this presynaptic region are a large number of mitochondria. These provide the metabolic energy for the synthesis of the transmitter substance and the ionic pumping mechanisms necessary for the recovery processes that follow the passage of an impulse. The muscle plasma membrane making up the postsynaptic side of the junction is folded into a series of deep grooves called *junctional folds,* which are lined with the receptors for the ACh transmitter substance. The receptors are complex protein molecules with dual function. Each receptor has two binding sites for ACh, and it also functions as an ion channel. Normally, the receptor is not permeable to ions, but when the transmitter substance has attached to the binding sites, sodium and potassium ions can pass through the channel down their electrochemical gradients. When the ACh is no longer bound to the receptor, the channel again becomes impermeable to ions. The postsynaptic membrane also contains an enzyme called **acetylcholinesterase**. This enzyme is essential for breaking down the transmitter to an inactive form once it has done its job of causing the channel to open.

Between the time when a motor nerve is stimulated and when the muscle actually contracts, a long series of events takes place (Fig. 16–23). The action potential travels down the motor axon by the mechanisms previously described. When the action potential spreads into the terminal regions of the axon branches, membrane calcium channels open in response to the depolarization, allowing for the entry of calcium ions into the neuron. This increase in the intracellular calcium ions causes the synaptic vesicles containing the ACh

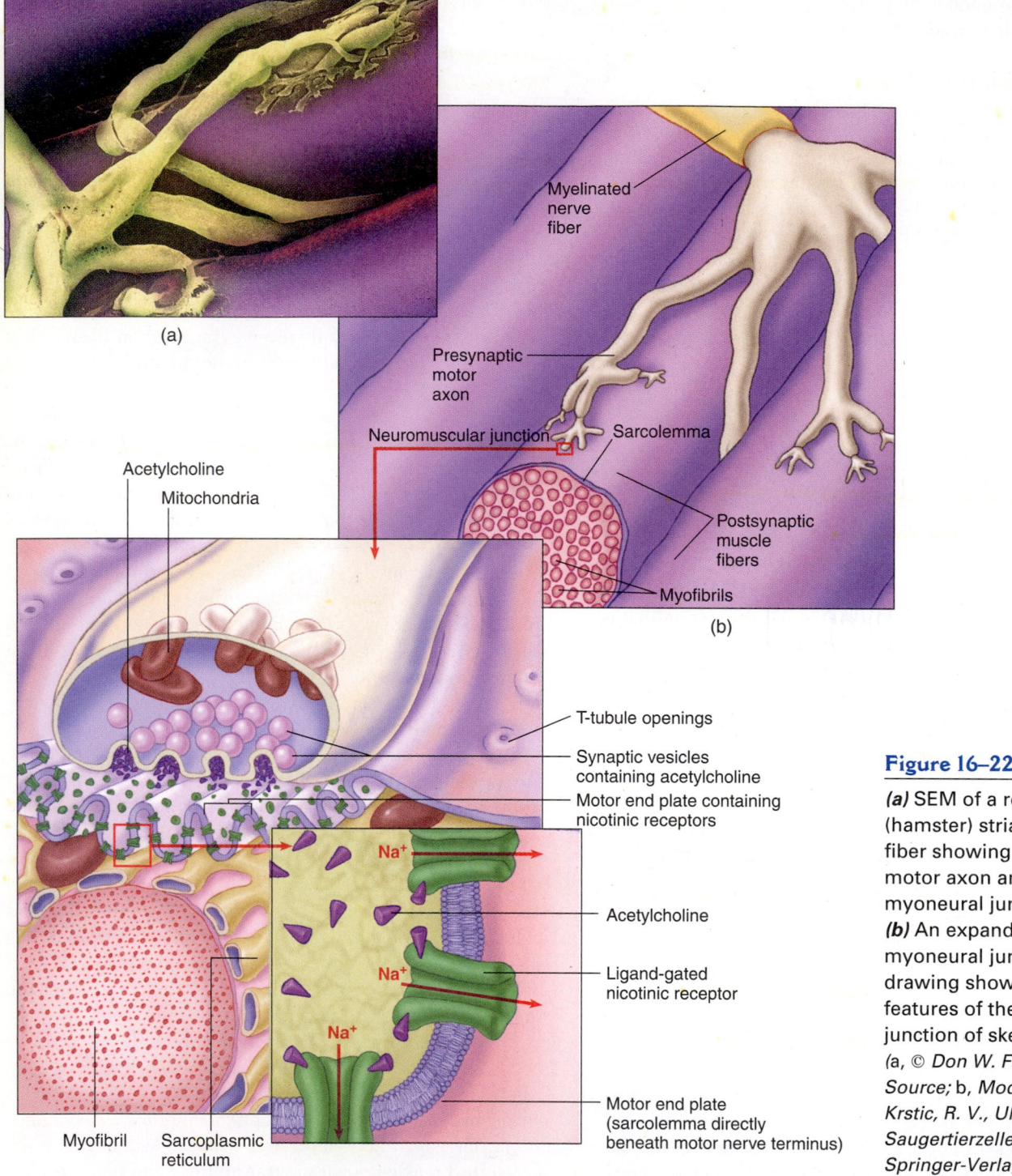

(a)

Myelinated nerve fiber

Presynaptic motor axon

Neuromuscular junction

Sarcolemma

Postsynaptic muscle fibers

Myofibrils

(b)

Acetylcholine

Mitochondria

T-tubule openings

Synaptic vesicles containing acetylcholine

Motor end plate containing nicotinic receptors

Acetylcholine

Ligand-gated nicotinic receptor

Na⁺

Na⁺

Na⁺

Motor end plate (sarcolemma directly beneath motor nerve terminus)

Myofibril   Sarcoplasmic reticulum

Figure 16–22

(a) SEM of a region of (hamster) striated muscle fiber showing the branching motor axon and two myoneural junctions. *(b)* An expanded view of the myoneural junction. This drawing shows the essential features of the myoneural junction of skeletal muscle. (a, © *Don W. Fawcett/Science Source;* b, *Modified from Krstic, R. V., Ultrastructur der Saugertierzelle. Berlin, Springer-Verlag, 1976.*)

to move to the region of the presynaptic membrane. Here, the vesicle membranes fuse with the plasma membrane. The fusion releases the contents of the vesicle into the synaptic cleft. These ACh molecules diffuse across the cleft to the postsynaptic region, where they attach to the specific binding sites on the receptor molecules. When this binding takes place, the membrane channels associated with the receptors become permeable to both sodium and potassium ions *simul-taneously*, which causes the membrane potential to move away from the resting potential (≈ -80 mV) and toward a new potential of approximately -15 mV, which is called the **end-plate potential.** (This is the potential expected from a membrane permeable to both sodium and potassium at the same time.) The postsynaptic membrane is unlike the rest of the plasma membrane in one important aspect. Because its channels are chemically activated and both permeability

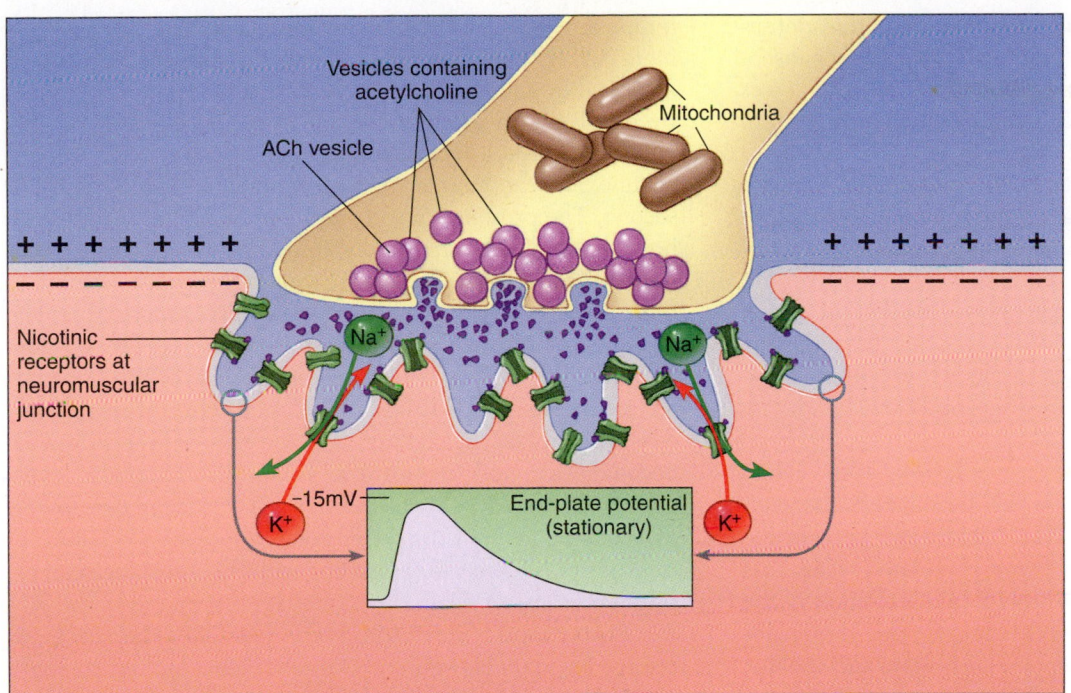

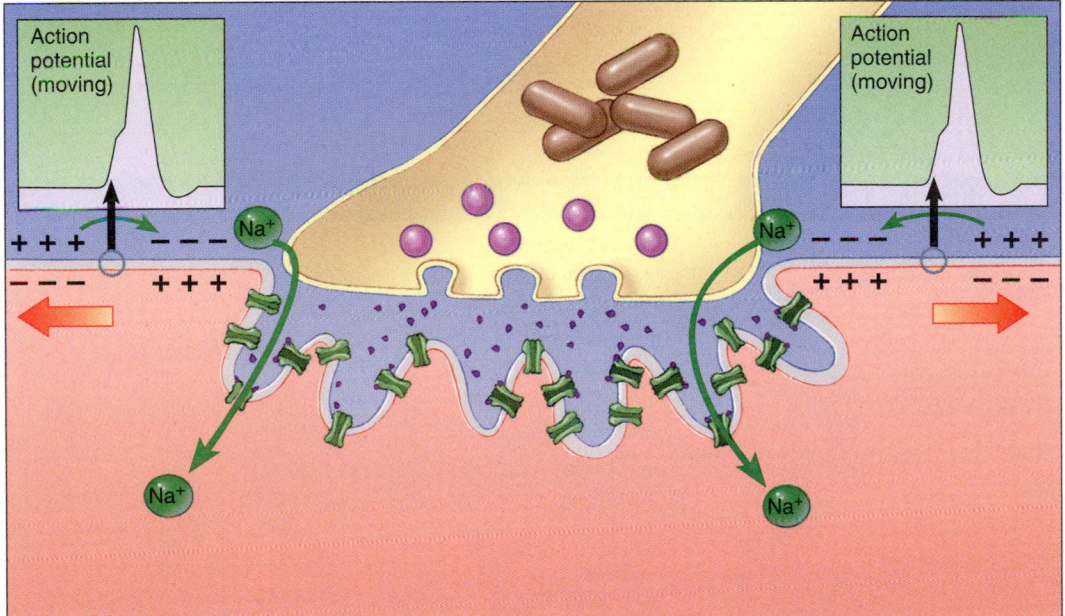

Figure 16–23

How the end-plate potential produces a muscle action potential. The region of membrane (the postsynaptic membrane) immediately opposite the axon terminal does not produce an action potential. Instead, its depolarization causes local currents to flow through the interior of the muscle cell and across the adjacent muscle cell membrane. This flow of current depolarizes the membrane to a threshold value and produces an action potential. This action potential then travels the length of the muscle cell membrane and enters via the t-tubule system to activate the contraction.

changes occur together, it cannot produce an action potential. What the end-plate potential does, however, is cause a small local current (the **end-plate current**) to flow across the membrane at this region. This inward current flows into the fiber and begins to spread down its length. The corresponding outward current is forced to flow across adjacent areas of the general plasma membrane, and it depolarizes these regions so that they reach their threshold for action potential generation. The action potentials produced by the end-plate potential then propagate down the length of the fiber and activate the contraction in ways that have already been described. After the passage of the action potential, the muscle membrane returns to its resting condition.

Factors that Block Myoneural Transmission. If the action of the transmitter substance were not terminated, the end-plate depolarization would continue. This would prevent the muscle plasma membrane from being stimulated again, and subsequent synaptic transmission would fail. Hydrolysis of the ACh by the postsynaptic acetylcholinesterase terminates ACh action and produces its components, acetate and choline. The choline is actively taken back up into the presynaptic terminal, where it is made back into ACh. The acetate fragment, which is a chemical common throughout the body, is not saved by any special mechanism. These events are summarized in Table 16–1.

All of these events take place in only a few milliseconds of time, and the muscle can be restimulated many times per second. The presence of so many steps crucial to the process, however, indicates that many things could go wrong. **Myoneural blockade** is the term given to this transmission failure. Some blockade can take place *presynaptically*. The presence of too many magnesium ions or too few calcium ions near the axon terminal prevents release of the transmitter substance. Some drugs, called **hemicholiniums,** interfere with uptake of choline, and the transmitter substance becomes depleted. **Botulinum toxin,** an extremely deadly poison produced by a bacterium (*Clostridium botulinum*) that may be present in spoiled food, also causes a failure of transmitter release, and poisoning with this substance can result in paralysis or death.

Postsynaptic blockade is also possible. The substance **curare** (used on blowgun darts by aboriginal hunters in South America) can bind to the ACh receptors, in preference to

TABLE 16–1

Sequence of Events During Neuromuscular Transmission

Event	Possible Blocking Agent	Event	Possible Blocking Agent
Action potential arrives at axon terminal.	Nerve-blocking agents such as procaine or tetrodotoxin	End-plate region (postsynaptic membrane) becomes depolarized; end-plate potential is set up.	
Extracellular calcium ions enter axon terminal.	Low extracellular calcium or high magnesium concentration	Local end-plate currents cause adjacent muscle membrane to depolarize.	
Transmitter vesicles migrate to axon membrane and fuse with it.		Muscle action potential is triggered.	
Acetylcholine is released from the transmitter vesicles into synaptic cleft.	Botulinum toxin	Acetylcholine diffuses away from receptors and is hydrolyzed by cholinesterase enzyme.	Eserine or other cholinesterase inhibitors
Acetylcholine molecules diffuse across synaptic cleft.		Choline is taken up by presynaptic terminal (now repolarized).	Hemicholinium drugs
Acetylcholine molecules bind to receptor proteins in postsynaptic membrane on muscle.	Curare, succinylcholine	Acetylcholine resynthesized and vesicles refilled.	
Ion channels in postsynaptic membrane open and allow the flow of sodium and potassium ions.			

APPLICATIONS OF PHYSIOLOGY

Myasthenia Gravis

Myasthenia gravis (meaning "muscle weakness associated with pregnancy") is an illness associated with muscular weakness and extreme fatigue following moderate exercise. Despite its name, it is not closely linked to childbearing, although it often makes its appearance in young women in their early childbearing years. In men, its onset is typically after middle age (in their 50s or later). The condition stems from a failure of neuromuscular transmission rather than from any defects in the muscle contraction mechanism itself. While no cure has been found for the disease, several kinds of therapy can give patients some symptomatic relief.

Scientists now know that myasthenia gravis is one of the so-called "autoimmune diseases." For unknown reasons, some people produce antibodies that react with the acetylcholinesterase receptors in the postsynaptic (motor end-plate) plasma membrane. This immune reaction destroys many of the receptors, and adequate end-plate potentials can no longer be produced. This means that a muscle action potential does not result from a nerve impulse, and so the muscle fails to contract. For a person afflicted with the condition, the result is that extreme effort may be required for even ordinary movements. In extreme cases, even breathing becomes very difficult.

Specific diagnosis of the condition involves study of the neuromuscular junction and immune system. Careful observation of the relationships between the nerve stimulation and muscle response can reveal delays in the transmission process even before it fails completely, and the presence of antibodies specific to the acetylcholine receptors may be detected in the blood of affected persons.

Several forms of treatment are possible. Drugs that inhibit the action of acetylcholinesterase allow the remaining postsynaptic receptors to function for a longer period of time, and adequate end-plate potentials can be produced. Unfortunately, the effect is only temporary, and the drug must be readministered repeatedly. This approach can lead to harmful side effects and, in any case, treats only a symptom of the underlying cause, with no effect on the general course of the disease.

Like other autoimmune diseases (e.g., rheumatoid arthritis), myasthenia gravis can undergo spontaneous remissions in its early stages, although the symptoms usually return. Removal of the thymus gland (which is known to play an important role in the immune system) can result in substantial improvement of the condition, especially if it is done early in the course of the disease.

Other therapies, also directed at the immune system, may be applied. Administration of gamma globulin may produce temporary improvement. Steroid drugs can be used to reduce the severity of the problem for a time, but they cause harmful side effects in many cases. Immunosuppressant drugs reduce the function of the entire immune system and can produce an improvement in the symptoms, although the patient will then be at risk from other infections and diseases. The technique of plasmapheresis, in which the proteins (including antibodies) are removed from the blood by a special transfusion method, can produce temporary improvement. But the procedure, though relatively safe, is time-consuming and expensive.

While an understanding of the effects of the disease can be obtained through study of the neuromuscular junction, its actual cause lies elsewhere. With immunological methods, myasthenia gravis can be produced in experimental animals, especially in rodents and monkeys. Because the animal condition very much resembles the human disease, studies involving animal immune responses offer hope for development of a more satisfactory treatment.

ACh. However, its binding does not produce a permeability change or depolarization. It is also resistant to breakdown by the acetylcholinesterase. When it is presenting sufficient quantity, therefore, myoneural transmission cannot take place.

A close chemical relative of ACh is the drug **succinylcholine,** which is used during surgical procedures to produce muscle relaxation. This drug is broken down only very slowly by acetylcholinesterase. Another possible site for blockade is the acetylcholinesterase enzyme. Drugs such as physostigmine and eserine inhibit this enzyme and are similar to the active (organophosphate) compounds in chemical warfare agents, the so-called "nerve gases." When the action of cholinesterase is inhibited, the end-plate membranes remain depolarized, and the muscle action potential mechanism cannot be reset.

In all of the situations mentioned, the myoneural blockade would result in death from respiratory failure. The diaphragm, a skeletal muscle, would fail to respond to nerve impulses, and along with other skeletal muscles, it would be paralyzed. Carefully controlled doses of blockers, such as succinylcholine, can be used as surgical muscle relaxants as long as respiration is maintained artificially.

Other factors also affect the myoneural junction. As is true of other synapses, it is subject to fatigue due to depletion of the transmitter substance. However, the transmission normally has a large safety factor. That is, more transmitter is released than is necessary for transmission, and the end-plate potential is normally larger than necessary to stimulate the muscle membrane. In addition, the process of contraction in the muscle is subject to fatigue far sooner than is the process of myoneural transmission. An abnormal situation, though, is present in the disease known as **myasthenia gravis.** In this condition, it appears that an immune reaction against the ACh receptors reduces their numbers, and the end-plate potentials may not be large enough to stimulate the muscle fibers. Weakness or paralysis results from this condition, but effective therapy, at least in the short term, is possible. Careful administration of **cholinesterase inhibitors** allows the ACh to remain active in the synaptic cleft for a longer time, and what receptors there are may be bound to and activated several times in succession. This allows sufficient end-plate current to flow to permit near-normal muscle function, at least over the short term.

> *How does nervous control of cardiac and smooth muscle compare to that of skeletal muscle?*

Nervous Control of Smooth and Cardiac Muscle

The two general types of smooth muscle, multiunit and unitary, differ considerably in the ways in which they are controlled by the nervous system. Multiunit smooth muscles are generally very densely innervated by autonomic nerve fibers and have rudimentary motor end-plates on each cell in the tissue. Myoneural transmission in these muscles is much like that found in skeletal muscles. The structures involving this muscle are quite small. They include some very small blood vessels; the arterioles; and the muscles that make our hair stand on end, the **piloerector muscles.** Because of the small size and the plentiful innervation of the tissues, the fact that the cells do not communicate with one another is of little practical consequence.

Most tissues composed of unitary smooth muscle are also innervated by postganglionic fibers of the ANS. There is nothing approaching a one-to-one connection between the nerve supply and the muscle cells. Instead of a nerve-muscle connection, such as a motor end-plate, the innervating fibers course among the smooth muscle cells. Periodically the nerves have swellings (called **varicosities**) that hold vesicles containing various transmitter substances (e.g., ACh or norepinephrine). Action potentials traveling down the nerves cause the release of the transmitter substance, which can then diffuse across the intercellular spaces to the vicinity of the muscle cells. Because there are so many cells in the tissue, only a few of them are directly affected by the transmitter substance. The cells that depolarize and/or produce action potentials, however, can affect neighboring cells through gap junctions. The gap junctions form low-resistance connections between adjacent cells; depolarizing one cell can cause an adjacent cell to depolarize by means of local current flow. This has the effect of spreading the activity throughout the tissue, and even though most individual cells are not directly innervated, the tissue tends to function as though they were. This is the source of the term *unitary,* and a tissue that behaves in this way is called a **functional syncytium** (*syn* = "together"; *cytium* = "cells"). Many unitary smooth muscle tissues are spontaneously active and can contract rhythmically on their own. In these cases, the effect of the autonomic innervation may be to modulate (increase or decrease) the ongoing function.

Heart muscle also receives an extensive autonomic innervation, but the function of the nerves in heart muscle is not to initiate contractions. The muscle of the heart does not contain motor end-plates, and its contractions begin directly in the muscle tissue, not in response to an external stimulus. As in some smooth muscles, the function of nerves in the heart is to modulate the ongoing timing and strength of contraction, not to conduct action potentials throughout the muscle. Cell-to-cell communication is also of very great importance in the function of heart muscle, because, as is true of smooth muscle, it is composed of cells that are very small in relation to the size of the tissue. The process of the control of heart muscle contraction is sufficiently specialized to be treated separately in Chapter 18.

> *How does muscle get its energy to perform work?*

THE METABOLISM OF MUSCLE CONTRACTION

Because muscle performs work, it must consume fuel in the form of energy-yielding biochemical compounds. The fuel most directly consumed by the actin-myosin contractile system is the universal high-energy compound, ATP. While muscle cells require energy for ion pumping, growth, and cell maintenance, as do other cells, their major function is that of contraction, and this function consumes more energy than does any other type of cell. Not surprisingly, the metabolic adaptations of muscle are specialized to provide an adequate and constant supply of ATP for the contractile process. The metabolic aspects of muscle contraction covered in this chapter are part of the broader subject of energy metabolism in general. Many essential details of these processes are found in Chapter 6, which provides a basis for the somewhat specialized processes in muscle.

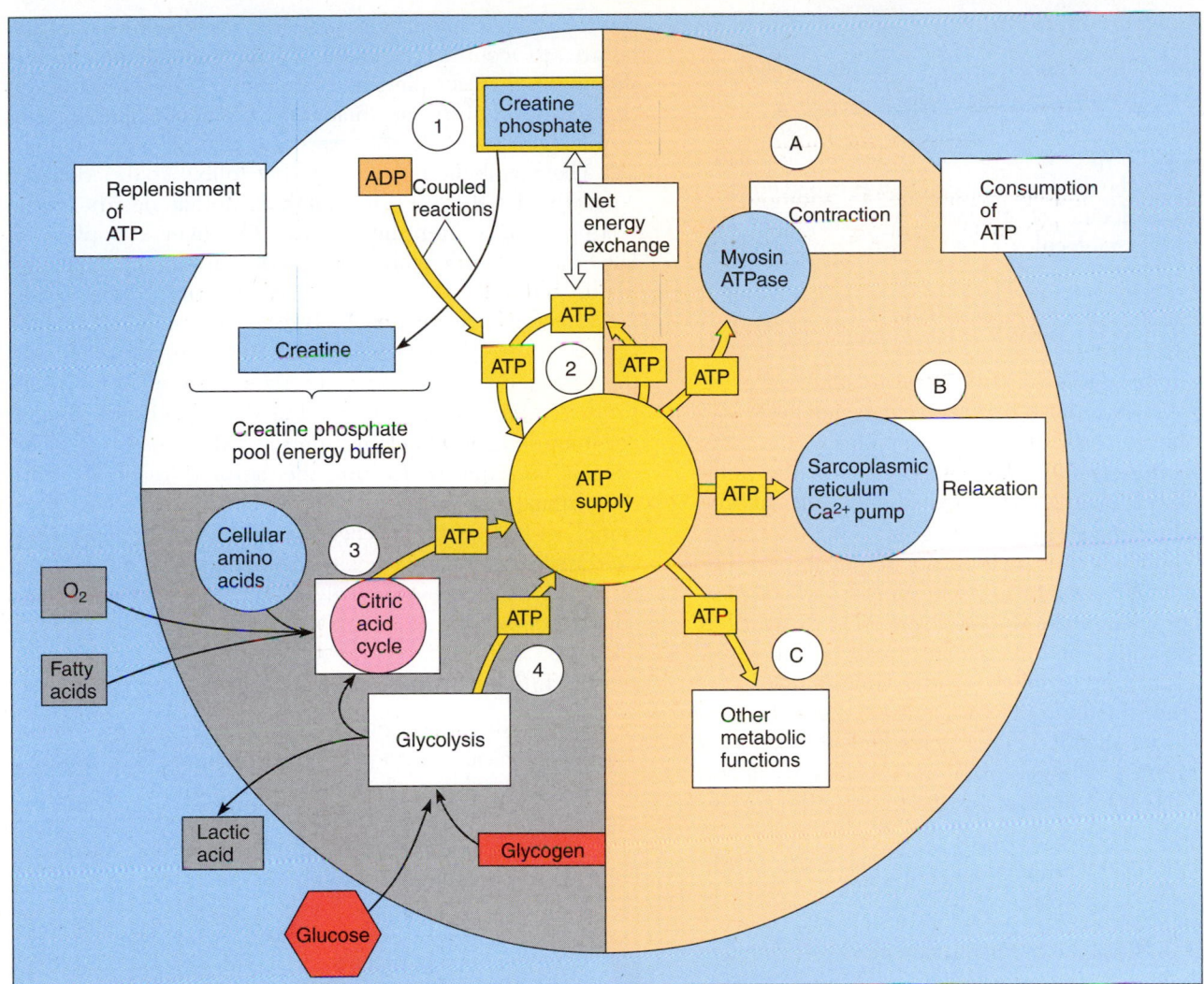

Figure 16–24

Summary of a cell's mechanisms for replenishing its supply of ATP, with the numbers indicating the order of replenishment. The cell uses ATP for a variety of energy-related functions. Processes that replenish ATP are on the left side, whereas processes that consume ATP are on the right side. The letters *A, B,* and *C* indicate the order in which cellular functions consume ATP.

Energy for Contraction Comes from Several Sources

As we have seen, ATP is broken down during the crossbridge cycle and provides the energy that is turned into mechanical work. If chemical inhibitors of the general cell energy metabolism are used to prevent replenishment of the ATP supply, a contracting muscle will shortly become exhausted and enter a state similar to rigor mortis because crossbridges are unable to detach at the end of their power stroke. However, a muscle maintains activity for a longer period of time than simple chemical measurements of the ATP content would indicate. This is because of a reserve "energy pool" in the cell in the form of a compound called *creatine*

phosphate. This high-energy compound is capable of transferring its energy to ATP very readily, so that as soon as the supply of ATP to the myofilaments begins to decrease, it is rapidly replenished. The ADP from the contraction process is *rephosphorylated* to ATP, and the creatine phosphate becomes creatine. This creatine is itself rapidly rephosphorylated by ATP derived from the general metabolism of the cell. The creatine phosphate pool provides a "buffer" for the rapid supply of ATP for the work of contraction, as well as a link to the cellular sources of ATP. Depending on the type of muscle fiber, this cellular ATP is produced by one or both of two common biochemical pathways, as diagrammed in Figure 16–24.

Glycolysis: The Anaerobic Pathway

Glycolysis, the first of these pathways, yields a relatively small amount of ATP from the metabolic fuel (glucose) in a series of reactions that take place in the cytoplasm of the cell. Recall from Chapter 6 that because this reaction pathway takes place in the absence of oxygen, it is called *anaerobic*. For each molecule of glucose metabolized by this pathway, only two molecules of ADP are "recharged" to ATP.

The Citric Acid Cycle: Aerobic Metabolism

When sufficient oxygen is present, however, the end products of glycolysis (and those of fatty acid metabolism) can enter an aerobic (or oxidative) reaction pathway, the citric acid cycle. These reactions, which take place in the mitochondria of the cell, yield an additional 36 molecules of ATP from a metabolized glucose molecule.

Given the much higher ATP production of aerobic metabolism, it is reasonable to ask why muscles would also have the less efficient anaerobic pathways. The answer lies in the rates at which energy is consumed and at which metabolic fuels can be supplied. Under conditions of normal activity, the blood supply to a muscle can bring the energy-supplying molecules (fatty acids and glucose) to a muscle at a rate that allows aerobic metabolism to supply almost all of the ATP needed by the contractile system. During moderate exercise as well, the blood supply and the oxygen it delivers can keep up with the needs of the muscle.

The Need for an Anaerobic Pathway

When the level of activity reaches approximately 70% of the maximum possible, however, aerobic metabolism is no longer able to supply sufficient ATP. At this point, glycolytic (anaerobic) metabolism begins to take over to provide the ATP. The reactions of the glycolytic pathway run more quickly than do the aerobic ones, and they can rapidly supply the needed amounts of ATP from glucose in the blood and from the glycogen (a polymer of glucose) stored in the muscle itself.

The Oxygen Debt

While they are quite rapid, the anaerobic reactions are not very efficient, and they produce large amounts of a temporary byproduct, **lactic acid.** When the bout of heavy exercise is over, the body is left with an overabundance of lactic acid and is deficient in its stores of creatine phosphate and glycogen. Muscle fatigue often results from prolonged exercise in which the energy supply fails to keep up with all of the energy demands. In a fatigued muscle, the maximum force that can be exerted decreases; after a period of rest, the capability of the muscle is restored. Paradoxically, chemical measurements made on fatigued muscle show that the total ATP content is not severely reduced; instead, the failure to contract appears to lie somewhere in the processes that couple the membrane depolarization to the activation of the myofilaments. Whatever the defect, it is quickly restored.

The metabolic imbalances that follow contraction are corrected by the aerobic metabolism that occurs in the resting muscle after the activity. The liver also plays an important role in metabolizing lactic acid that is circulating in the blood as a result of muscle activity. Some of this lactic acid serves as fuel for the heart muscle. The lactic acid is oxidized to compounds that can provide ATP, and the glycogen stores are resynthesized from glucose. These processes are responsible for the continued high oxygen consumption of a resting muscle after exercise. This oxygen is required to run the aerobic reactions until the original state of the muscle is restored. The burden of lactic acid and the glycogen deficit that occur during exercise make up the so-called **oxygen debt,** which must be "paid back" quantitatively by postcontraction oxygen consumption.

Muscle Efficiency and Heat Production

Another important by-product of muscle metabolism is heat. The biochemical reactions of contraction are only approximately 20% efficient; the other 80% of the energy is degraded into heat. This is the heat that the body must get rid of during heavy exercise; it is also the heat that warms the body by shivering or other exercise in the cold.

> *Are all skeletal muscles the same? If not, how are they different; what are the consequences of those differences?*

Skeletal Muscle Types Are Adapted for Special Tasks

In light of the many functions that muscles must perform, it is not surprising that specialized types of fibers are adapted to particular sorts of tasks. These adaptations involve structural and biochemical specializations that are interrelated and that fall into some fairly distinct categories. Table 16–2 gives the characteristics of these specialized muscle types.

Some muscle tasks, such as maintaining body posture, require that muscles maintain tension for long periods with a low expenditure of energy. Other tasks, such as sprinting in a race or running after a bus, require rapid contractions that may be made at a high energy cost. Between these extremes are muscles that require a mixture of metabolic and structural specializations. In general, muscle fibers fall into two broad categories, **white fibers** and **red fibers.** (For example, consider the light [white] and the dark [red] meat from poultry.)

TABLE 16–2

Classification of Skeletal Muscle

Feature	Fast		Slow
	White	Red	Red
Mechanical			
Contraction speed	Fast	Fast	Slow
Force capability	High	Medium	Low
SR Ca^{2+} pumping	High	High	Moderate
Motor axon velocity	100 M/s	100 M/s	85 M/s
Biochemical			
ATPase activity	High	High	Low
Source of ATP	Anaerobic glycolysis	Oxidative phosphorylation	Oxidative phosphorylation
Glycolytic enzymes	High	Moderate	Low
No. of mitochondria	Low	High	High
Myoglobin content	Low	High	High
Glycogen content	High	Moderate	Low
Rate of fatigue	Fast	Moderate	Slow
Structural			
Diffusion distance	Large	Small	Moderate
Fiber diameter	Large	Moderate	Small
Number of capillaries	Few	Many	Many
Sarcomere structure	Very regular	Less regular	Irregular Z-lines
Functional			
Role in body	Rapid, powerful movements	Medium endurance	Postural endurance
Example	Latissimus dorsi	Vastus lateralis (a mixed-fiber muscle)	Soleus

Red Muscle Fibers and Aerobic Metabolism

The color differences among muscle fibers arise because of the differing content of a protein called **myoglobin,** which imparts a red color to the tissue. The presence of myoglobin also gives a clue to the most important distinction between the muscle classes. Myoglobin is a protein (similar to the protein hemoglobin found in red blood corpuscles) that is capable of binding, storing, and releasing oxygen. It is found most abundantly in those muscle fibers that depend on oxidative (aerobic) metabolism, where it promotes the rapid diffusion of oxygen in times of heavy demand.

Red fibers can be further classified as **fast twitch** and **slow twitch** types. They both have plentiful mitochondria and a rich blood supply. Their speeds of contraction are correlated with their rates of ATPase activity — that is, with the rate at which the actin-myosin contractile proteins can break down ATP and release its energy. The slow fibers are very resistant to fatigue when fulfilling their usual function of maintaining posture or working at lower rates. Attempts to use such muscles for rapid, sustained tasks cause them to fatigue rather quickly as their ATP consumption outstrips the ability of the citric acid cycle to replenish the ATP supply. The fast fibers of the red group have a higher ATPase activity; they contract and relax more quickly and are suited to moderate endurance activities.

White Muscle Fibers and Anaerobic Metabolism

White muscle fibers, on the other hand, are adapted for fast and powerful contractions, but they fatigue quickly. They contain the enzymes necessary for glycolytic (anaerobic) metabolism. Glycolysis, a process that operates faster than oxidative metabolism, can use the stored glycogen at a high rate to produce ATP. Because their immediate contractile function does not depend heavily on their oxygen supply, these muscles are not as well supplied with a dense capillary network as are the red types. The white muscles show a high ATPase activity and very rapid contraction and relaxation. The rapid contraction appears to be due to specializations in the myosin molecules that permit them to convert the energy in ATP more rapidly. The rapid relaxation is due in part to a more aggressive calcium-pumping mechanism present in the SR. As might be inferred from the anaerobic character of their metabolism, the cells contain fewer mitochondria than do the red types.

These various characteristics are summarized in Table 16–2. A close look at the features compared there reveals many correlations among function and biochemical makeup. Note that the characteristics listed are for fibers, *not whole muscles.* Most whole muscles contain a mixture of these fiber types, so they are not strictly confined to a single type of activity. Studies on athletes have shown that, despite particular types of training programs, the fiber composition of a given muscle remains relatively constant. For an individual, the specific mixture of fiber types begins to be established before birth and becomes set in childhood. In some newborn mammals, the fibers are predominantly the slow type, a pattern that changes as the individual grows.

 How does energy use in smooth muscle compare to that in skeletal muscle?

Smooth Muscle Is Highly Economical in Its Energy Usage

While the basic chemical processes of skeletal muscle are shared by smooth muscle, some important adaptations allow smooth muscle to perform its specialized functions. The crossbridges of smooth muscle cycle much more slowly than those of skeletal muscle, and they spend much more of their cycle time in the attached state. This allows smooth muscle to maintain high levels of force with a much lower rate of ATP hydrolysis, although this property means that smooth muscle can shorten only very slowly compared with skeletal muscle. While some ATP must be used in skeletal muscle for noncontractile processes, such as ion pumping, this amounts to only a small portion of the total energy consumption. In smooth muscle, however, the ATP that must be used in the regulatory phosphorylation of myosin light-chains and in control of calcium concentration is a significant portion of the total ATP usage. This is more than offset, however, by the slower crossbridge cycle, and aerobic metabolism is sufficient to keep up with the demand.

 How do muscles change with use?

LONG-TERM PHYSIOLOGICAL CHANGES IN MUSCLE

While all muscle types are inherently adapted to their particular roles, special circumstances can cause modifications in existing muscles to further adapt them to changing physiological needs. These changes are usually beneficial, but sometimes they are the result (or cause) of a pathological condition.

Skeletal Muscle Can Increase or Decrease in Strength

Patterns of innervation and usage during early development play an important role in deciding the ultimate distribution of fiber types. Experiments have been done with animals in which the nerve supply to fast and slow muscles was surgically reversed. These studies have shown that the reinnervated muscles take on characteristics related to the type of nerve supplying them. That is, formerly slow muscles begin to take on the contraction characteristics of fast muscles. These changes are at least in part controlled by substances that are conveyed to the muscles via their motor axons over the course of time.

It is well known that the pattern of use is important in determining the size and strength of skeletal muscle. An increase in the mass of a muscle (without an increase in the actual number of cells) is called **hypertrophy.** Such hypertrophy is most readily brought about by exercises that emphasize *isometric* or *slow isotonic* contractions repeated many times. Muscles subjected to such routines increase in their cross-sectional area and in the amount of contractile protein they contain. The relative distribution of fiber types remain relatively constant. A negative aspect of this type of muscle adaptation (to a specific set of exercise conditions) is that the vascular supply to the muscle does not increase in proportion to the increase in muscle. This means that the endurance of the hypertrophied muscle is not greatly increased, and its improved function is largely limited to the sort of exercise that produced the hypertrophy.

Exercise that emphasizes *isotonic contractions* at moderate workloads produces a different type of modification in the muscle. The popular "aerobic" exercise regimens produce less marked increases in strength than do the isometric exercises mentioned previously but do result in an increased vascular supply to the muscle. This enhances the ability to produce increased performance for extended periods of time, such as for the duration of a cross-country race. Such exercises also provide conditioning for the heart muscle and can increase the efficiency of its pumping.

Adaptation in the opposite direction is also possible. Long-term immobilization of a muscle (such as might be produced by wearing a cast) causes **disuse atrophy,** in which the mass and strength of the muscle decrease. When a muscle is paralyzed by injury to its motor nerve and cannot contract on its own, it is also subject to atrophy. If there is a probability that nerve function may eventually be restored, periodic electrical stimulation of the muscle may prevent some of the atrophy.

Less severe atrophy occurs when a regular form of exercise is discontinued, and the muscle adapts to the lower level of use. This may occur during the "off-season" periods in the life of an athlete and is a factor in the "deconditioning" observed after long periods of weightlessness during space travel. Both of these situations can be at least partially reversed by a proper exercise program.

Cardiac Muscle Responds to Changes in Physiological Demand

Heart muscle is also subject to hypertrophy in response to increased demands. This occurs normally during athletic conditioning but may also be a result of some disease process. For example, if the improper functioning of a heart valve makes the heart do extra work to maintain the blood pressure, an enlarged heart results. Such enlargement can at least partially compensate for the condition that caused it, but it also increases the metabolic demands of the heart and may make it more prone to disturbances in the heartbeat rhythm. In addition, if the walls of the heart become too thick, the volume of the chambers is decreased, and less blood can be pumped with each beat. A failing heart may also enlarge, but this is due not to muscle hypertrophy but rather to stretching out of the weakened muscle due to the presence of extra blood that can no longer be pumped out.

Smooth Muscle Can Adapt to Physiological Needs

A number of factors can cause physiological changes in smooth muscle. For example, during the course of pregnancy, the muscle of the uterus undergoes a very significant hypertrophy that prepares it for the maintenance of pregnancy and the process of labor and childbirth. Shortly after the birth process is over, the uterus undergoes the process of **involution,** and it loses much of its contractile proteins as it reverts to the nonpregnant state. These changes are under the control of the hormonal changes associated with the reproductive processes.

Smooth muscle can also hypertrophy in response to increased mechanical demands. If the outflow of the bladder, for example, is restricted by an enlarged prostate gland, the extra effort required during urination produces a significant change in the mass and strength of the bladder muscle. In the vascular system, significant changes may occur that either cause high blood pressure or that occur as a result of the increased pressure. Under some conditions of physiological stress, smooth muscle cells can change from their so-called **contractile phenotype** to a **secretory phenotype.** These changed cells no longer have a contractile function; instead, they secrete connective tissue, such as collagen, that is added to the extracellular matrix of the tissue. Some of these changes in cell type are associated with the overall hypertrophy of an organ or tissue.

CHAPTER REVIEW

Summary

- Muscle functions as a motor for locomotion, for the pumping of blood, and for the regulation of internal body processes. It also regulates much of the input to the nervous system.
- Muscle may be classified in three ways: (1) on the basis of its anatomical location (as skeletal, visceral, or cardiac), (2) on the basis of its structure (as striated or smooth), and (3) on the basis of its mode of control (as voluntary or involuntary). There is some overlap in these categories.
- Skeletal muscle fibers are composed of sarcomeres, which are the fundamental units of contraction.
- The sarcomeres are composed of thick (myosin) and thin (actin) myofilaments joined by crossbridges. Contraction of the individual sarcomeres results in shortening or force development of the whole muscle.
- Cardiac muscle is striated, and the sarcomeres are also similar to those of skeletal muscle. The small cells are closely coupled together both mechanically and electrically.
- Smooth muscle is composed of small, unstriated cells. They contain contractile proteins, although no obvious sarcomere arrangement is present, and the internal membrane systems are not very extensive. Cells are coupled together mechanically and electrically.

- Muscles can shorten and exert a constant force (under isotonic conditions), or they can exert a force that does not result in movement (isometric conditions). Most practical contractions involve a mixture of these two conditions.
- The length of a muscle before contraction has an important effect on the amount of force that the muscle can exert when stimulated, and a particular length exists for optimal force. This relationship is described by the length-tension curve.
- The interaction between the actin filaments and the myosin crossbridges produces mechanical motion. The speed of this process and the overall amount of force generated are influenced by muscle length and the force during shortening.
- The crossbridge cycle is controlled by presence or absence of calcium ions and requires ATP.
- The speed at which muscle shortens is related to the amount of force exerted, a relationship described by the force-velocity curve. The power output of muscle also depends on the relationships presented in this curve.
- Cardiac muscle has a contractile mechanism similar to that of skeletal muscle, although the means of external control is different. Smooth muscle also appears to function with a sliding filament mechanism.

- Release of calcium in the region of the myofilaments allows the contractile proteins to interact in the crossbridge cycle. This action is exerted via control proteins on the actin (thin) filaments and is called *actin-linked regulation*. Cardiac muscle also has a similar actin-linked regulatory process.
- Smooth muscle is under the control of a myosin-linked regulatory process, in which calcium ions activate the myosin filaments by promoting the phosphorylation of the myosin crossbridges.
- Skeletal muscle contractions are controlled by the plasma membrane. A single, brief activation of a skeletal muscle results in a brief contraction called a *twitch,* and repeated activation results in a sustained contraction called a *tetanus.* Cardiac muscle is also controlled by its plasma membranes, but only twitch contractions are possible. A long-lasting contraction of smooth muscle is often called a *tonic contraction,* or *tonus.*
- The electrical activity of the skeletal muscle plasma membrane is controlled by the central nervous system (CNS). The link between a motor axon and a muscle fiber is called the *myoneural junction,* or *motor end-plate.* A nerve action potential causes the release of acetylcholine (ACh) from the nerve terminals to the muscle end-plate membrane, producing the end-plate potential. This in turn causes the muscle membrane to produce an action potential, which travels along the muscle fiber and activates contraction.

- Some smooth muscles are controlled by the autonomic nervous system (ANS), although they do not have well-developed myoneural junctions. Cardiac muscle functions under the control of specialized cells in the heart, and the nervous system acts to regulate both the rate and strength of the heartbeat.
- The energy for muscle contraction is supplied by glucose and fatty acids and is ultimately used by the muscles in the form of ATP. Glycolysis, an anaerobic process, produces relatively little ATP from each glucose molecule and produces lactic acid as a waste product. Such a pathway produces energy quickly but inefficiently and results in an oxygen debt in the form of lactic acid.
- Some skeletal muscles are specialized for powerful and rapid movements, but they become fatigued rapidly. These muscles have a primarily anaerobic metabolism, are composed of large fibers, and are light in color. They are termed *fast, white muscles*. Other muscles contract more slowly but can sustain activity for long periods of time. These muscles have a primarily aerobic metabolism, with many mitochondria and a rich blood supply. This group, which contains both slow and moderately fast muscle fiber types, is dark-colored; these are the red muscles.
- Muscle can undergo adaptive changes in response to physiological needs or environmental or pathological influences.

Review Questions

Choose the Correct Answer

1. According to the sliding-filament hypothesis for muscle contraction,
   a. movement occurs due to shortening of the myofilaments.
   b. cyclic interactions between myosin-filament crossbridges, and actin filaments cause the muscle to shorten.
   c. movement is caused by the action of calcium ions on the thick filaments.
   d. the length of the muscle does not affect the force of contraction.
   e. muscle cannot shorten if the myofilaments are overlapped.

2. From the standpoint of its function as a biological motor, which is the most crucial feature of a skeletal muscle sarcomere?
   a. Equal lengths of all of the thick filaments
   b. Equal lengths of all of the thin filaments
   c. Regular spacing of the Z-lines
   d. The overlap between the thick and thin filaments
   e. The presence of the M-line

3. Which statement about the relative speeds of muscle contraction is correct?
   a. Smooth muscle is faster than cardiac muscle but slower than skeletal muscle.
   b. Cardiac muscle is the fastest contracting of the three types.
   c. All three types contract at about the same speed.

   d. Skeletal muscle is faster than cardiac muscle, which is faster than smooth muscle.
   e. Cardiac and smooth muscle contract at about the same speed, which is slower than that of skeletal muscle.

4. During a tetanic contraction of skeletal muscle,
   a. the muscle cell membrane stays in a depolarized condition.
   b. the force produced a series of individual, repeated twitches that added up to produce a large, steady force.
   c. relaxation is essentially complete between each of the individual twitches, but the individual twitches are increased in size.
   d. shortening cannot occur.
   e. the force produced is less than in a twitch.

5. What determines the force produced in a single skeletal muscle twitch?
   a. The amount of calcium released from the sarcoplasmic reticulum
   b. The length of the muscle
   c. The temperature of the muscle
   d. The metabolic state of the muscle
   e. All of these factors have an effect on the twitch force

6. In an isotonic contraction of skeletal muscle,
   a. no movement occurs.
   b. the muscle shortens at a constant force.
   c. force increases as the muscle shortens.
   d. the force decreases as the muscle shortens.
   e. shortening is faster when the force is greater.

7. Increasing the length of a skeletal muscle prior to an isometric contraction:
   a. may either increase or decrease the force that can be developed, depending on the length at which muscle was originally held.
   b. has no effect on the force that the muscle can develop.
   c. can only decrease the force of contraction.
   d. can only increase the force of contraction.
   e. is not possible under isometric conditions.

8. When a weight is held in the hand, the lever action of the bones of the arm causes the flexor muscles of the upper arm to:
   a. exert a force greater than that provided by the weight.
   b. shorten more rapidly than the speed at which the weight itself is moved.
   c. move the weight even though they exert less force than the weight provides.
   d. generate a force that is equal to that provided by the weight.
   e. shorten farther than the distance by which the weight is moved.

9. The role of the transverse tubules (t-tubules) in the excitation of skeletal muscle contraction is to:
   a. provide an inward path for the spread of muscle action potentials.
   b. serve as the storage site for calcium ions.
   c. connect the sarcomeres end to end.
   d. conduct nutrients into the interior of the muscle fiber.
   e. allow the diffusion of oxygen to the interior regions of the fiber.

10. The primary role of calcium in the activation of skeletal and cardiac muscle is to:
    a. cause depolarization of the muscle cell plasma membrane.
    b. remove the inhibition of the reaction between the actin filaments and the myosin filaments.
    c. activate myosin molecules so that they can interact with actin filaments.
    d. provide the energy necessary for contraction.
    e. regulate the ionic composition of the interior of the cell.

11. The primary role of calcium in the activation of smooth muscle is to:
    a. cause depolarization of the muscle cell plasma membrane.
    b. remove the inhibition of the reaction between the actin filaments and the myosin filaments.
    c. activate myosin molecules so that they can interact with actin filaments.
    d. provide the energy necessary for contraction.
    e. regulate the ionic composition of the interior of the cell.

12. During the process of relaxation in all three types of muscle, the most important cellular event is:
    a. the consumption of all of the available fuels.
    b. a nerve signal that begins the relaxation process.
    c. a reduction in the amount of free intracellular calcium ions.
    d. the presence of a force that will re-extend the muscle.
    e. the entry of sodium ions into the muscle fiber.

13. The amount of power that a muscle produces during an isotonic contraction depends on:
    a. only the force of contraction.
    b. only the speed of shortening.
    c. a combination of force and speed of contraction.
    d. only the force of contraction when the speed is zero.
    e. only the speed of shortening when the force is zero.

14. During the process of transmission at the myoneural junction, the role of acetylcholine is to:
    a. cause the depolarization of the presynaptic (nerve) cell membrane.
    b. prevent depolarization of the muscle cell membrane.
    c. cause the depolarization of the postsynaptic (end-plate) membrane of the muscle cell.
    d. block the action of the enzyme acetylcholinesterase.
    e. provide energy for the transmission process.

15. Impulse transmission at the myoneural junction can be blocked by the drug curare. This drug works by:
    a. poisoning the enzyme that inactivates acetylcholine.
    b. preventing the presynaptic release of acetylcholine.
    c. keeping postsynaptic channels open and producing a steady depolarization.
    d. competing with acetylcholine for binding to the postsynaptic receptors.
    e. allowing sodium, but not potassium, to enter the postsynaptic membrane.

16. Which statement best describes the control of smooth muscle by the nervous system?
    a. Smooth muscle is completely inactive unless it is stimulated by a motor nerve.
    b. Nerve stimulation may either initiate mechanical (contractile) activity or it may modify ongoing activity.
    c. Nerve stimulation is usually inhibitory and serves mostly to prevent contraction.
    d. Most smooth muscles do not have any relationship to the nervous system.
    e. The nerves in smooth muscle are sensory, not motor, and thus have no effect on the contraction.

17. The chemical fuel that is most directly consumed by a contracting muscle is:
    a. glycogen.
    b. glucose.
    c. lactic acid.
    d. adenosine triphosphate.
    e. oxygen.

18. During the aerobic metabolism of exercising skeletal muscle:
    a. the primary source of energy is glucose, and relatively little ATP is formed.
    b. the production of ATP is associated with the consumption of oxygen.
    c. large amounts of lactic acid are produced.
    d. mitochondria do not play any important role.
    e. the blood supply to the muscle is restricted.

19. An oxygen debt in exercising skeletal muscle:
    a. is due to the buildup of lactic acid produced by anaerobic metabolism.
    b. can usually be "paid back" by anaerobic metabolism.
    c. occurs only when the muscle is at rest.
    d. must be "paid back" by further exercise.
    e. causes the muscle to stop contracting as soon as the nutrient supply from the blood becomes inadequate.

20. Which set of characteristics is found in those skeletal muscle fibers best adapted to produce fast and powerful movements?
    a. Resistance to fatigue, red color, mostly aerobic metabolism
    b. Small-diameter fibers, red color, irregular sarcomere structure

c. Little resistance to fatigue, large fiber diameter, white color, mainly anaerobic metabolism

d. Sarcomeres without Z-lines

e. Cell membranes that do not produce action potentials

21. How does shivering help to produce body heat?

    a. It causes blood vessels to constrict.

    b. It causes muscles to contract inefficiently and to waste energy.

c. It increases the insulating value of the muscles so that less heat is lost.

d. It tightens the outer surface of the body and lessens heat loss.

e. It allows the body to ignore the sensation of cold.

Answers to Case History Questions

1. Skeletal muscle, being much faster, would respond first and move Pete out of harm's way. Even during this time, the smooth muscle responses would be starting.

2. Because this is a brief and rapid reaction involving powerful muscle contractions, the primary energy source would be stored ATP, possibly aided in the longer term by energy derived from glycolysis.

3. Stimulation by circulating epinephrine and by sympathetic nerves to the heart would increase both the contractility of the cardiac rate and the rate of the heartbeat. The increased rate would further increase the contractility.

4. Both types of muscle would begin their responses instantly. Because of its greater speed, the skeletal muscle reactions would be apparent first.

Key Terms

actin (p. 483)

afterload (p. 494)

antagonistic pair (p. 480)

auxotonic contraction (p. 490)

contractility (p. 504)

crossbridge (p. 484)

end-plate potential (p. 506)

force-velocity curve (p. 495)

hypertrophy (p. 514)

involuntary muscle (p. 479)

isometric contraction (p. 491)

isotonic contraction (p. 490)

length-tension curve (p. 491)

meiotonic contraction (p. 491)

motor unit (p. 505)

muscle fiber (p. 480)

myofilament (p. 483)

myoglobin (p. 513)

myosin (p. 483)

sarcomere (p. 483)

sliding-filament hypothesis (p. 493)

tendon (p. 480)

tension: active, passive, resting (p. 491)

voluntary muscle (p. 479)

Suggested Readings

Andersen, J. L., Schjerling, P., and Saltin, B. "Muscle, genes, and athletic performance." *Scientific American,* 283:48–55, 2000.

Bagshaw, C. R. *Muscle Contraction,* ed 2. New York, Chapman and Hall, 1993.

Ford, L. E. *Muscle Physiology and Cardiac Function.* Carmel, IN, Biological Sciences Press, Cooper Group, 2000.

Gabella, G. "Structure of intestinal musculature." *In* Schultz, S. G., and Wood, J. D., eds. *Handbook of Physiology: The Gastrointestinal System I.* Bethesda, MD, The American Physiological Society, 1989.

Hall, Z. W., and Sanes, J. R. "Synaptic structure and development: The neuromuscular junction." *Cell,* 72(Neuron 10; suppl):99–122, 1993.

Jankovic, J., and Brin, M. F. "Therapeutic uses of botulinum toxin." *New England Journal of Medicine,* 324:1186–1194, 1991.

Junge, D. *Nerve and Muscle Excitation,* ed 3. Sunderland, MA, Sinauer Associates, Inc., 1992.

Katz, A. M. *Physiology of the Heart,* ed 2. New York, Raven Press, 1992.

Keynes, R. D., and Aidley, D. J. *Nerve and Muscle,* ed 2. New York, Cambridge University Press, 1991.

Kraemer, W. J., and Newton, R. U. "Training for muscular power." *Physical Medicine and Rehabilitation Clinics of North America,* 11:341–368, vii, 2000.

Matthews, G. G. *Cellular Physiology of Nerve and Muscle,* ed 2. Boston, Blackwell Scientific Publications, 1991.

Meiss, R. A. "Mechanical properties of gastrointestinal smooth muscle." *In* Schultz, S. G., and Wood, J. D., eds. *Handbook of Physiology: The Gastrointestinal System I.* Bethesda, MD, The American Physiological Society, 1989.

Meiss, R. A. "Mechanics of smooth muscle contraction." *In* Kao, C. Y., and Carsten, M. E., eds. *Cellular Aspects of Smooth Muscle Function.* New York, Cambridge University Press, 1997.

Meiss, R. A. "Mechanics of smooth muscles." *In* Barr L., ed. *A View of Smooth Muscle.* Greenwich, CT, JAI Press, Inc., 2000.

Rall, J. A. "Energetic aspects of skeletal muscle contraction: Implications of fiber types." *Exercise and Sport Science Reviews,* 13:33–74, 1985.

Ruegg, J. C. *Calcium in Muscle Contraction: Cellular and Molecular Physiology,* ed 2. New York, Springer-Verlag, 1992.

Shephard, R. J. *Physiology and Biochemistry of Exercise.* New York, Praeger, 1985.

Woledge, R. C., Curtin, N. A., and Homsher, E. *Energetic Aspects of Muscle Contraction.* New York, Academic Press, 1985.

Answers to Review Questions

1. b **2.** d **3.** d **4.** b **5.** e **6.** b **7.** a **8.** a
9. a **10.** b **11.** c **12.** c **13.** c **14.** c **15.** d
16. b **17.** d **18.** b **19.** a **20.** c **21.** b

Chapter 17

FUNCTIONS OF THE BLOOD

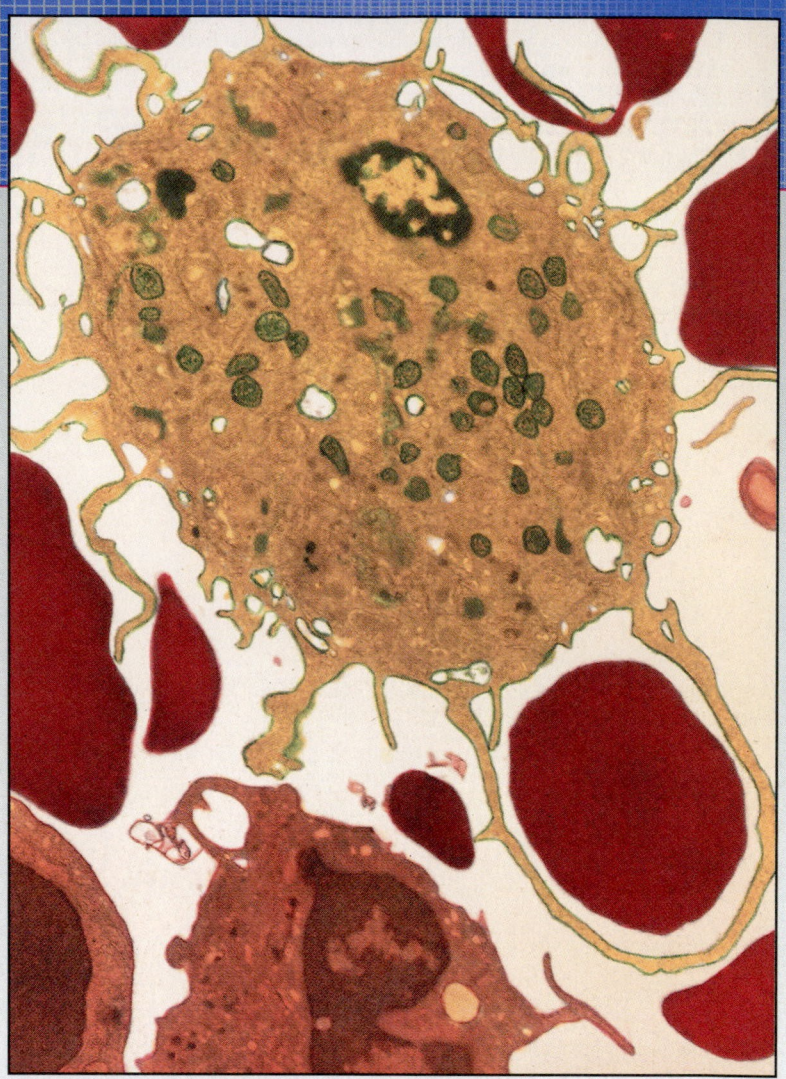

- Macrophage, *eating red blood cell. Coloured transmission electron micrograh (TEM) of a section through a macrophage engulfing a red blood cell. At top is the macrophage (green), a type of white blood cell which scavenges and plays a role in the immune response. Here it has a granular cytoplasm and pseudopodia "arms"; at lower right one of the "arms" has engulfed a red blood cell by a process known as phagocytosis. Red blood cells are unable to repair themselves, living on average 120 days, to then be consumed by macrophages. What signals the macrophage to engulf it is unknown. New red blood cells are constantly manufactured in the body. Magnification: x4,600 at 6x7cm size.*

KEY CONCEPTS

- *Blood helps maintain homeostasis by being a vehicle for transportation and communication among various cells, tissues, and organs of the body.*

- *Blood is a liquid connective tissue composed of approximately 5 million red blood cells per microliter, 7000 white blood cells per microliter, and 300,000 platelets per microliter, suspended in 3 L of intercellular fluid called plasma.*

- *Plasma is by weight 93% water and 7% solutes; 86% of the solutes are proteins.*

- *Red and white blood cells and platelets originate from pluripotential hemopoietic stem cells found mainly in the red bone marrow.*

- *Red blood cells lack internal organelles, contain a red pigment called hemoglobin, and transport oxygen from the lungs to body tissues and carbon dioxide from body tissues to the lungs for elimination.*

- *Each of the several types of human blood are characterized by the presence or absence of specific antigens on the surface of the red blood cell.*

- *All white blood cells are nucleated and play various important roles in defending the body against foreign chemicals and microorganisms.*

- *Thrombocytes (platelets) reduce blood loss by helping blood to clot and injured blood vessels to heal.*

- *Mechanisms of hemostasis that prevent or minimize blood loss include constriction of blood vessels, platelet aggregation at the site of blood vessel damage, coagulation (clotting) of blood, and clot retraction.*

A 35-year-old man complains of chronic physical fatigue, which began insidiously approximately 3 to 4 weeks ago. He said he felt tired all of the time even though his occupation as a software developer was mentally but not physically demanding. He breathed comfortably at rest but, when he exerted himself, he experienced difficulty in breathing and had a hard time catching his breath. He also complained of "more than the usual" mental fatigue, confessing an increasing inability to concentrate and focus his attention on tasks at hand. His mind often drifted, and he caught himself catnapping or dozing while working on his computer or attending seminars and meetings. Colleagues noticed his pallor and his inattentiveness at brainstorming sessions and suggested he reschedule his annual physical examination for an earlier date. He complained of vague abdominal pain and a sense of abdominal fullness. His appetite was depressed, and he thought perhaps his physical and mental symptoms were caused by poor diet. However, attempts to increase eating resulted in nausea. His stools, he said, were sometimes loose and tarry. Eventually, increased heart palpitations and chest pain made him seek medical advice. Physical examination and laboratory findings revealed the following:

Questions

1. What general medical condition is suggested by the person's symptoms?

2. What fundamental change in a function of blood related to the red blood cells could simultaneously affect the function of several systems (cardiovascular, respiratory, gastrointestinal, and others)?

3. What specific diagnosis is supported by the laboratory findings?

4. How could the stool be related to the laboratory findings?

Laboratory Test	Normal	Patient
RBC (erythrocyte count)	3.50×10^6 cells/μL	4.6–6.2×10^6 cells/μL
Hct (hematocrit ratio)	28.0%	42%–52%
Hb (hemoglobin content)	8.0 g/dl	14–18 g/dl
MCV (mean corpuscular volume)	80.0 pl	82–92 pl
MCH (mean corpuscular hemoglobin)	22.9 pg	27–31 pg
MCHC (mean corpuscular hemoglobin concentration)	28.6%	32%–36%

Peripheral smear: RBC: pale (hypochromic), 6.5-μm-diameter (microcytic), unequal in size (anisocytosis), and abnormally shaped (poikilocytosis).

INTRODUCTION

It has been known for centuries that abnormalities or deficiencies of blood are associated with illness and can even be the cause of death. The ancient Greeks considered blood to be one of four elemental substances (called *fluids* or *humors*) of the human body. Because of the life sustaining qualities of blood, many magical properties have been accorded to it by civilizations of the past, beliefs that continue in primitive societies and cults of the modern world. In this chapter, we will examine some physical and chemical properties of blood, formation and functions of each type of blood cell, normal and abnormal chemistry of the blood, and why one person's blood differs from another's.

GENERAL FUNCTIONS OF BLOOD

Circulation of blood through the vasculature provides a transportation and communications system between the body's cells, serving to maintain a relatively stable internal environment for optimum cellular activity. The role of the blood in homeostasis largely involves transport of molecules and heat, defense against foreign agents, and regulation of extracellular fluid pH and osmolarity.

Strictly speaking, blood is not a bodily fluid like tears, saliva, and urine; instead, it is a living connective tissue composed by volume of approximately 45% cellular elements and 55% intercellular fluid. Blood normally is confined to the blood vessels and chambers of the heart and is circulated throughout the body by the heart's pumping action. The lives of all cells depend on circulating blood because it helps maintain the relatively stable internal environment essential for normal cell function. When blood circulation ceases, so does the delivery of oxygen and nutrients, and cells of the brain, heart, liver, and other organs begin to die within a few minutes.

Blood Transports Molecules and Heat

Blood is the primary means by which substances are conveyed from one area of the body to another. Nutrients, such as glucose, amino acids, and fatty acids, electrolytes, and water, are absorbed from the gastrointestinal tract and carried by the blood to various body tissues. Oxygen is absorbed by blood as it passes through the lungs and is transported to respiring tissues. Carbon dioxide produced during cellular respiration is absorbed by the blood as it circulates through tissues and is transported to the lungs, where carbon dioxide is eliminated in expired air. In addition, other cellular waste, such as urea, uric acid, and excess water, are carried by the blood to the kidneys for excretion. Hormones and other chemical messengers produced within the body are transported by blood from sites of production to target cells. In essence, blood serves as a vehicle for moving many different substances to various areas within the body for a variety of purposes.

Body metabolism produces a considerable quantity of heat as a by-product, and the excess heat must be dissipated into the environment in order to prevent overheating. Blood conducts heat from the body's core to the upper respiratory passageways and the skin, where dissipation of heat occurs.

Blood Cells Defend Against Foreign Agents

Some of the blood cells are phagocytic; that is, they are capable of ingesting and rendering harmless substances foreign or toxic to the body and microorganisms that are pathogenic, or disease producing. Other blood cells produce and release chemicals (antibodies) that react with and nullify the harmful effects of foreign agents or chemicals that regulate blood flow and blood clotting during a defensive response to injury.

Blood Proteins Help Maintain Normal Extracellular Fluid pH and Osmolarity

Blood plays a major role in the regulation of extracellular fluid pH. Chemical buffers present in blood convert strong acids and bases into weak acids and bases, thereby minimizing large shifts in pH during the course of daily metabolism. (Mechanisms of buffering are discussed in Chapter 25.) The blood also transports acidic and basic substances to organs of excretion. Normal blood is slightly alkaline, with a pH between 7.35 and 7.45.

Most proteins in the blood do not readily pass through capillary membranes. As a result, plasma proteins create an osmotic pressure that influences the movement of water between blood and interstitial fluids and, in turn, the osmolarity of extracellular fluids. This role of the blood proteins is discussed in greater detail later in this chapter and in Chapters 19 and 24.

PROPERTIES OF WHOLE BLOOD

 What kind of measurements can be made on blood to detect pathophysiology?

Composition of Whole Blood

Whole blood is an opaque, red, liquid connective tissue consisting of microscopically visible, formed elements—the **red blood cells** (RBCs; erythrocytes), the **white blood cells** (WBCs; leukocytes), and the **platelets** (thrombocytes) suspended in an extracellular fluid called **plasma.** Blood may be transfused from one person to another as whole blood or as one of its components. The various components of whole blood can be separated using fractionation techniques. The components, called **blood fractions,** include red blood cells; white blood cells; platelets; plasma; and the protein fractions of the plasma, such as albumin, immunoglobulin, and clotting factors. Patients seldom require all of the components of whole blood. Therefore, a blood fraction, rather than whole blood, often is used clinically to alleviate symptoms of disease. This treatment, called **blood component therapy,** allows several patients to benefit from one unit of donated whole blood. For example, packed red blood cells are sometimes used to reduce the effects of severe anemia (lack of RBC or hemoglobin); immunoglobulin, to increase resistance to disease.

Blood accounts for approximately 6% to 8% of adult body weight. Normal adult **blood volume** ranges from 4.5 to 5.5 L for a female and 5.0 to 6.0 L for a male. Total blood volume varies, being altered by changes in body water balance, pregnancy, hemorrhage, and other factors. **Normovolemia** is a blood volume within normal range. An above-normal blood volume is called **hypervolemia,** and a below-normal volume is **hypovolemia.**

Specific Gravity and Viscosity of Blood

The **specific gravity** of a liquid is the ratio between the weight of a given volume of the liquid and the weight of an equal volume of pure water. The specific gravity of water is 1.000; the specific gravity of normal adult blood ranges from 1.050 to 1.060. The specific gravity of blood depends on the number of blood cells and the concentration of chemicals normally dissolved in the plasma. Diseases that alter cell numbers or chemical concentrations of the blood may change the specific gravity.

Viscosity is a measure of a fluid's resistance to flow. The greater the viscosity, the greater the fluid's resistance to flow. If we assume the viscosity of pure water to be 1.00, then whole blood viscosity by comparison is approximately 3.50 to 5.50 and that of blood plasma is 1.90 to 2.60; mean values for an adult are 4.5 and 2.2, respectively. Blood viscosity is determined by the number of cells present in whole blood, cells' resistance to being moved, and the mutual attraction of molecules dissolved in the plasma. When the total number of cells increases above normal or when the plasma concentration of large molecules, such as protein, is elevated above normal, the viscosity of the blood increases, forcing the heart to work more strenuously to maintain a normal rate of flow. A prolonged pathological increase in blood viscosity overloads the heart and may weaken its ability to pump. It also tends to elevate systemic blood pressure, which damages blood vessels and increases the possibility of hemorrhage, or it may reduce blood flow to the point at which circulation through certain organs (e.g., kidneys) becomes inadequate.

Erythrocyte Sedimentation Rate

If a small quantity of blood is removed from the body, mixed with a chemical that prevents clotting (an anticoagulant), and placed in a vertical graduated cylinder, the blood cells, having a higher specific gravity than the plasma, slowly sink to the bottom, leaving behind a transparent amber upper layer of plasma. The leukocytes are lighter than the erythro-

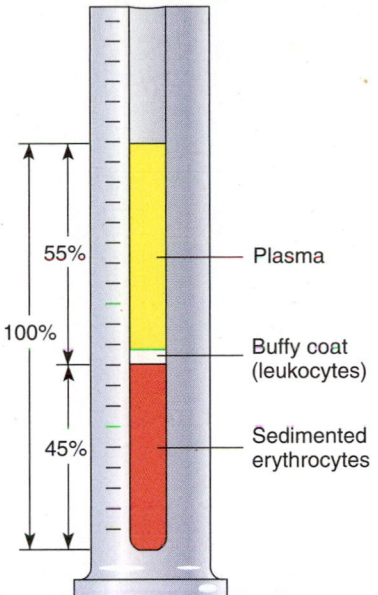

Figure 17–1

Sedimented blood.

cytes, and, as a result, the leukocytes form a thin, whitish layer (called the **buffy coat**) between the upper plasma and the lower sedimented erythrocytes (Fig. 17–1). The rate at which the erythrocytes settle is called the **erythrocyte sedimentation rate.** Normal erythrocyte sedimentation rates for adults are: men, 2 to 10 mm in 1 hour; women, 2 to 20 mm in 1 hour.

Sedimentation of erythrocytes proceeds in three phases: (1) formation of rouleaux (clumping of red cells together like a stack of coins), (2) rapid settling, and (3) final packing of the red cell mass. The formation of rouleaux does not occur with cells of abnormal shape, such as those in sickle cell anemia, and the sedimentation rate is therefore slower. Increases in the concentration of plasma proteins associated with inflammatory processes tend to increase the sedimentation rate of blood by facilitating rouleaux formation. Often, the erythrocyte sedimentation rate is a convenient and simple indicator of the level of inflammatory processes occurring in an individual. It tends to be elevated in acute and chronic infections, such as arthritis (and other inflammatory diseases), tuberculosis, rheumatic fever, and toxemia. It usually decreases in polycythemia (too many red blood cells), allergies, and hyperglycemia (elevated blood sugar).

The Hematocrit

In any sample of whole blood, the ratio between the volume occupied by all of the red cells and the volume of the sample is called the **hematocrit ratio.** The numerical value of the ratio is multiplied by 100 to express the ratio as a percentage (Fig. 17–2). Normal values for adults are: males, 47% plus or

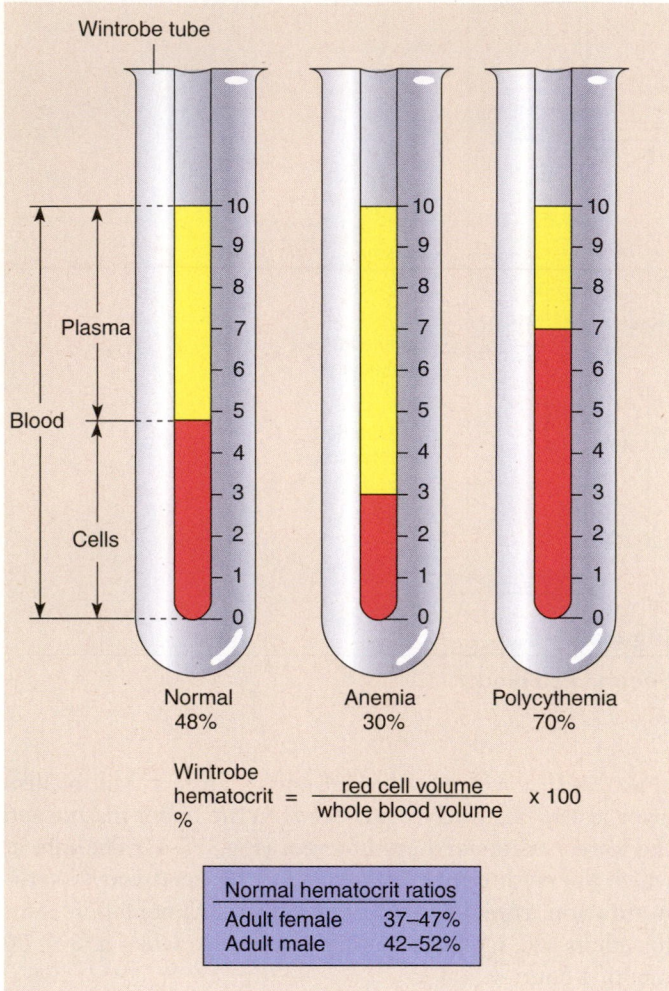

Figure 17–2

The hematocrit.

Abnormal hematocrits occur in diseases that alter the number of circulating erythrocytes, the volume of plasma, or both. In anemias characterized by a reduction in the number of circulating erythrocytes, the hematocrit may fall to 25, whereas in polycythemia (too many erythrocytes), the hematocrit may approach 70. The hematocrit is useful in diagnosing and assessing progress in treatment of diseases that alter the number of circulating normal erythrocytes or conditions that alter the normal ratio between erythrocyte and plasma volumes, such as dehydration. The hematocrit also is used to compute **red blood cell (RBC) indices,** which are indicators for abnormal erythrocyte size, shape, and color. RBC indices will be discussed later in the chapter.

PLASMA

What is plasma; what function does plasma perform?

The Composition of Plasma

Plasma is the complex fluid in which blood cells and platelets circulate. It is defined as whole blood minus the formed elements (cells and platelets). Plasma may be separated from whole blood by centrifugation, but, if left to stand, it coagulates within minutes, forming a gel. Serum is similar in composition to plasma, except that it lacks coagulation factors and therefore does not coagulate.

Normal plasma is transparent and light yellow in color and is approximately 93% water and 7% solutes. One liter of human plasma contains approximately 930 g of water; 60 g of protein; 8 g of inorganic solutes, such as Na^+, K^+, Cl^-, HCO_3^- and Ca^{2+}; and 2 g of nonprotein organic substances, such as glucose, glycerol, and fatty acids. Plasma contains dissolved gases (O_2, N_2, CO_2, hormones, enzymes, vitamins, pigments, and minerals); a variety of cell waste products (urea, uric acid, and so on); and cell nutrients, such as amino acids. Major constituents of human plasma are shown in Table 17–1.

The data in the table reveal a range of normal values for each constituent, suggesting variability in chemical composition of plasma. In fact, not only does the exact chemical composition of plasma vary from individual to individual, but a person's plasma composition varies slightly from minute to minute as substances are exchanged across capillary walls between plasma and interstitial fluids.

Figure 17–3 is a diagram of the three major fluid compartments in the body: (1) **the blood plasma,** (2) **the interstitial fluid,** and (3) **the intracellular fluid.** The interstitial fluid constitutes the immediate environment of all living cells of the body and is separated from the intracellular fluid by the plasma membrane. The capillary wall separates blood plasma from interstitial fluid. Normally, the three fluid compartments are in osmotic equilibrium. The exchange of water and small molecules, such as electrolytes, between plasma

minus 5%; females, 42% plus or minus 5%. Normal hematocrit values vary with age, sex, and the part of the body from which the blood sample was taken. The hematocrit value of arterial blood is slightly less than that of venous blood. The hematocrit varies from tissue to tissue, and the capillary hematocrit is generally lower than either the systemic arterial or systemic venous hematocrit.

Clinically, it is often convenient to use the **microhematocrit** method to determine hematocrit. A small-bore (1-mm) capillary tube coated internally with heparin (an anticoagulant) is filled with capillary blood taken from a fingertip puncture. The blood then is subjected to centrifugation and the hematocrit determined. After centrifugation, the hematocrit is calculated by dividing the measured red cell volume by the measured whole blood volume and multiplying the result by 100 to obtain percentage.

After centrifugation of the blood, approximately 4% of the plasma always remains trapped among the erythrocytes. To obtain a **corrected hematocrit,** the value of the **apparent hematocrit** is simply multiplied by 0.96.

TABLE 17–1

Normal Constituents of Blood

Compound[a]	Concentration and Fraction[b]
Acetoacetic acid + acetone (ketone bodies)	0.2–2.0 mg/100 ml (S)
Adenosine triphosphate	1 μM (E)
Aldosterone	3–10 ng/dl (P)
Amino acid, total	300 mg/dl (P)
Ascorbic acid	0.4–1.5 mg/dl (fasting B)
Bicarbonate	25 mM
Bilirubin	Direct: 0.1–0.3 mg/dl (S); indirect: 0.2–1.2 mg/dl (S)
Calcium	8.5–10.5 mg/dl; 4.3–5.3 mEq/L (S)
Carbon dioxide	26–28 mEq/L (S)
Chloride	100–106 mEq/L (S)
Cholesterol	150–240 mg/dl (S)
Copper, total	100–200 μg/dl (S)
Cortisol (17-hydroxycorticoids)	5–18 μg/dl (P)
Creatine	40 μM/L (P)
Creatinine	0.7–1.5 mg/100 ml (S); 90 μM/L (P)
2,3-Diphosphoglycerate (DPG)	4 μM (E)
Glucose (folin)	80–120 mg/dl (fasting, B)
Glucose (true)	70–100 mg/dl (B)
Iodine, protein-bound	3.5–8.0 μg/dl (S)
Iron	50–170 μg/dl (S)
Lactic acid	0.6–1.8 mEq/L (B)
Lipids, total	450–1000 mg/dl (S)
Magnesium	1.5–2.0 mEq/L (S)
Nitrogen, nonprotein	15–35 mg/dl (S)
Phosphatase, acid total[c]	Male: 0.13–0.63 sigma units/ml (S); female: 0.01–0.56 sigma units/ml (S)
Phosphatase, alkaline[c]	2.0–4.5 Bodansky units/ml (S)
Phospholipids	145–225 mg/dl (S)
Phosphorus, inorganic	3.0–4.5 mg/dl (adult,S)
Potassium	4 mM (P); 111 mM (E)
Protein	
Total	6.0–8.0 mg/dl (S)
Albumin	3.5–5.5 g/dl (S)
Globulin	1.5–3.0 g/dl (S)
Sodium	136–145 mEq/L; 310–340 mg/dl (S); 140 mM (P); 6 mM (E)
Sulfate	0.5–1.5 mEq/L (S)
Transaminase (serum glutamic oxaloacetic transaminase [SGOT][c])	10–40 units/ml (S)
Urea nitrogen (blood urea nitrogen [BUN])	8–25 mg/dl (B)
Uric acid	3.0–7.0 mg/dl (S)

[a]The substances and the values listed for them in this table give an indication of the biochemical complexity of blood and are not to be construed as absolute. Blood levels of specific components can vary among normal individuals or in the same individual at different times. Furthermore, the normal values for specific blood constituents can differ greatly among different laboratories, depending on the technique used for their determination.

[b]Abbreviations used for the several blood fractions are : B = whole blood; P = plasma; S = serum; E = erythrocytes.

[c]Activities of over 50 specific enzymes have been detected in various blood fractions. *Source:* Modified from Jensen, D. *The Principles of Physiology*, ed 2. New York, Appleton-Century-Crofts, 1980. Table 32–2.

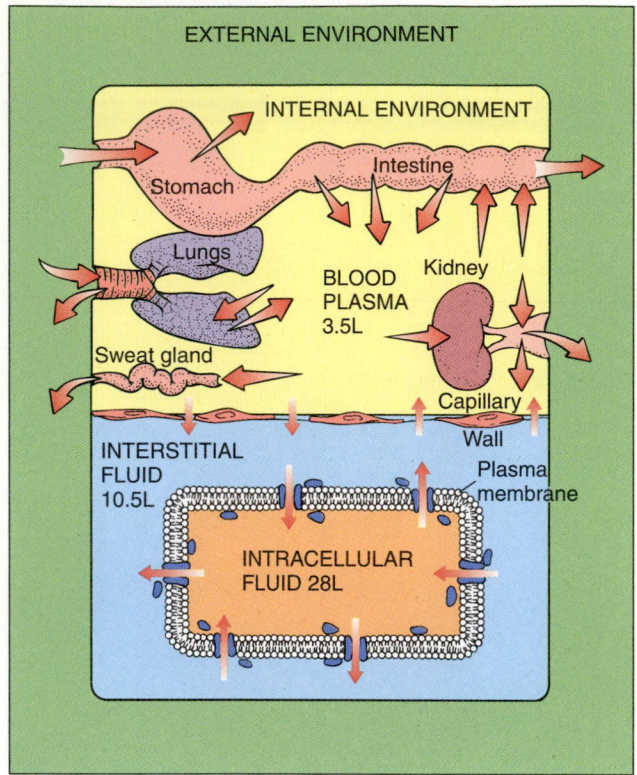

EXTERNAL ENVIRONMENT

INTERNAL ENVIRONMENT

Stomach
Intestine
Lungs
Kidney
BLOOD
PLASMA
3.5L
Sweat gland
Capillary
Wall
INTERSTITIAL
FLUID
10.5L
Plasma
membrane
INTRACELLULAR
FLUID 28L

Figure 17–3

The body's internal environment contains three major fluid compartments: blood plasma, interstitial fluid and lymph, and intracellular fluid. The chemical composition of plasma changes as various organs exchange solutes and water between the external and internal environments. The plasma compartment is separated from the interstitial fluid compartment by the capillary wall, which in most areas of the body allows for rapid exchange of water and small solutes between the two compartments. The interstitial and intracellular fluids are separated by the plasma membranes, most of which allow for the rapid passage of water through protein pores. Normally, the three fluid compartments are in osmotic equilibrium because water is able to shift rapidly from one compartment to another. Numerical values are for a 70-kg man.

and interstitial fluid is very rapid. Approximately 70% of the plasma fluid is exchanged with the interstitial fluid in 1 minute. Because of this rapid exchange, the water and electrolyte concentrations of plasma and interstitial fluid are nearly identical and vary little despite considerable variation in the uptake and release of substances by cells. Protein accounts for the major concentration difference between plasma and interstitial fluid because protein molecules are generally too large to readily pass through the capillary wall. The greater concentration of proteins in the plasma plays a role in determining the distribution of water between the blood and the interstitial fluid.

Plasma Protein Functions

Separation techniques can be used to divide the plasma proteins into the following fractions: **albumin, globulins,** (α_1, α_2, β_1, β_2, and γ), and **fibrinogen** (Fig. 17–4).

Plasma proteins are large molecules with high molecular weights (44,000 to 1.3 million) and configurations (arrangement of atoms) ranging from spherical (B-lipoprotein) to ellipsoid (albumin and globulins). Due to their large size, plasma proteins are classified as **colloids.** Plasma proteins collectively function in several different ways.

First, plasma proteins serve as an important reserve supply of amino acids for cell nutrition. If necessary, macrophages in liver, gut, spleen, lung, and lymphatic tissue can ingest plasma proteins, break them down, and release component amino acids into the blood so that other cells may use the amino acids for synthesis of new protein.

Second, plasma proteins serve as carriers for other molecules (see Fig. 17–4). Many types of small molecules bind to specific plasma proteins and are transported from absorptive organs (e.g., intestine) or storage organs (e.g., liver) to other tissues for utilization. Iron, for example, is carried in the blood combined with a transport protein called **transferrin.** Many of the ions in the blood bind to plasma proteins (e.g., Ca^{2+}), as do drugs, pigments, and hormones. Lipids (fats) absorbed into the lymph and blood by the intestine are water-insoluble, yet when combined with proteins in lymph and blood, they can be made soluble for transport.

Third, plasma proteins act as buffers, helping to maintain a stable blood pH. The normal range for the pH of blood is 7.35 to 7.45; that is, normal blood is slightly alkaline. Proteins are able to bind H^+ or OH^- depending on whether they exist as weak acids or weak bases, which in turn depends upon pH. In general, plasma proteins function as weak bases, binding excess H^+, and thus help to keep the blood slightly alkaline.

Fourth, several plasma proteins, some of which are zymogens (inactive precursors of enzymes) and others are not, interact in specific ways to cause blood to coagulate. Coagulation of blood is part of the body's response to vascular injury and helps protect against the loss of blood and invasion by microorganisms and viruses.

Fifth, plasma proteins produce a **colloid osmotic pressure,** also called *oncotic pressure,* which helps determine the distribution of water between the blood and the interstitial fluid.

The total osmotic pressure of normal plasma is approximately 7.3 atm (1 atm = 760 mm Hg), or 5550 mm Hg. The total osmotic pressure is the osmotic pressure plasma would exert if separated from pure water by a membrane permeable only to water. Solutions with the same osmotic pressure as plasma are called **isotonic.** An erythrocyte placed in an isotonic solution, such as 0.85% NaCl (clinically rounded off to 0.9% NaCl and called *physiological saline*) neither gains nor loses water (Chapter 4). Solutions with a higher osmotic pressure than that of plasma are called **hypertonic,** and

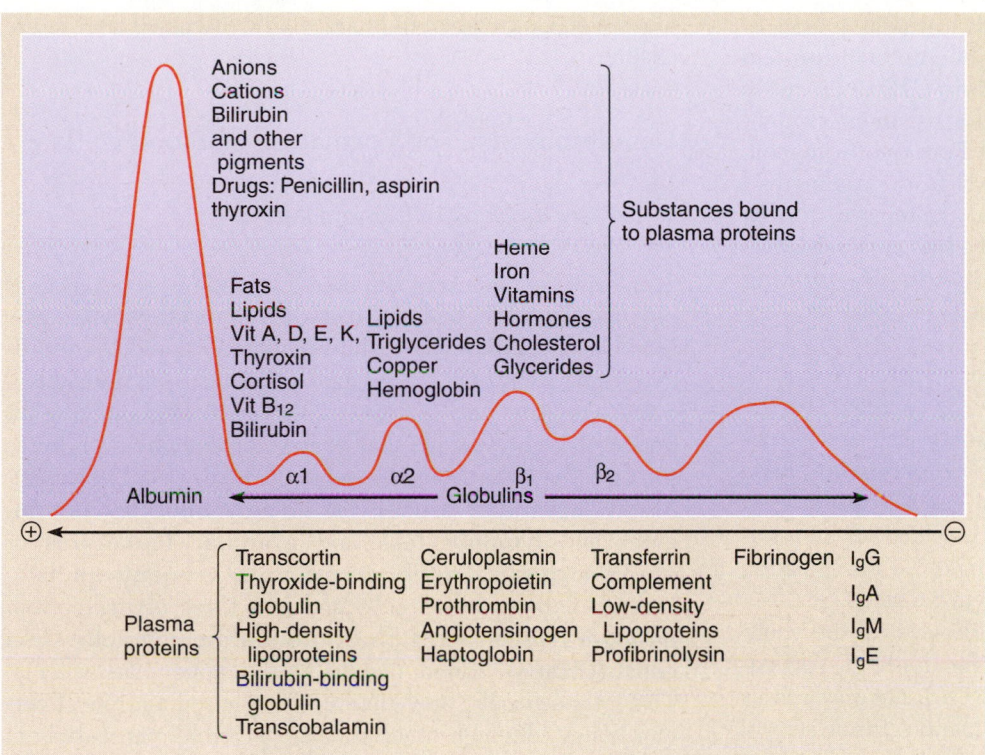

Figure 17–4

Electrophoretic survey of plasma proteins and substances bound to plasma proteins. Electrophoresis is a fractionation technique based on the movement of electrically charged particles, dissolved or suspended in a fluid, along a voltage gradient. Comparative mobilities of charged particles are a function of molecular size, shape, and electric charge, and the magnitude of the voltage gradient. In neutral or alkaline solutions, plasma proteins are electrically charged and can be fractionated by electrophoresis. The charged proteins migrate toward the anode (+ electrode) but at different velocities.

those with a lower osmotic pressure are called **hypotonic.** An erythrocyte will lose water and shrivel or crenate if surrounded by a hypertonic solution, whereas it will gain water, expand, and hemolyze (break open and lose hemoglobin) if surrounded by a hypotonic solution. In either case, red blood cells are distorted or destroyed and fail to function normally; thus, it is important when administering fluid intravenously to ensure that the tonicity of the fluid is correct for the circumstances.

Because plasma, interstitial fluid, and intracellular fluid are normally in osmotic equilibrium, homeostasis of all three depends critically on regulation of the osmotic pressure of the plasma. Any deviation from normal extracellular fluid osmotic pressure causes cells to gain or lose volume, and, like red blood cells, those cells may malfunction or die.

Approximately 99.5% of the total osmotic pressure of plasma is due to small molecules, such as electrolytes, urea, glucose, and others, that readily pass through capillary membranes along with water. Thus, the osmotic pressure due to these solutes is approximately the same in interstitial fluid as in plasma. The contribution of plasma proteins to the total osmotic pressure of plasma is approximately 0.5%. Normal colloid osmotic pressure (oncotic pressure) of the plasma is 28 mm Hg. Nevertheless, oncotic pressure plays an important role in determining the distribution of water between plasma and interstitial fluid because plasma proteins do not readily pass through capillary membranes. Small amounts of protein that do pass from blood to interstitial fluid do not collect in the interstitial fluid because of cell uptake and removal by the lymphatic system; therefore, a protein concentration gradient from blood to interstitial fluid exists, is maintained, and osmotically influences the movement of water and small molecules between plasma and interstitial fluid. Mechanisms of capillary exchange are presented in greater detail in Chapter 19.

In addition to functioning collectively as previously described, the individual types of plasma proteins have specific properties and functions.

Albumin

Approximately 60% of the total plasma protein is albumin, one of the smallest proteins (mol. wt. = 69,000) in the plasma. Being small and numerous, albumin accounts for approximately 80% of the colloid osmotic pressure; thus, the reductions in plasma albumin content that may occur in nutritional diseases (e.g., kwashiorkor), liver disease, and kidney disease may result in excess fluid moving out of the blood and into the interstitial space (peripheral edema).

In addition to its osmotic role, albumin helps many substances dissolve in the plasma by binding to them; thus, albumin plays an important role in plasma transport. Substances bound by albumin include drugs, such as barbiturates and penicillin; pigments, such as bilirubin and urobilin; hormones, such as thyroxin; and other substances, such as fatty acids and bile acid salts.

Globulins

Approximately 40% of the total plasma protein is globulin: 4% is α_1-globulin, 8% is α_2-globulin, 7% is β_1-globulin, 4% is β_2-globulin, and 17% is γ-globulin.

α_1-**Globulins** form **glycoproteins** (protein joined to carbohydrate) and small amounts of lipoproteins (protein joined to lipid). **High-density lipoprotein (HDL)** functions in lipid transport, carrying fats to cells for use in energy metabolism, membrane construction, and hormone formation. HDL also appears to prevent cholesterol from invading and settling in the walls of arteries. Other proteins in the α_1 group include **thyroxine-binding globulin, cortisol-binding globulin (transcortin),** and **vitamin B$_{12}$-binding globulin (transcobalamin),** all of which transport their respective substrates.

α_2-**Globulins** include **haptoglobin,** a protein that combines with free hemoglobin preventing its excretion by the kidney, and **ceruloplasmin,** a copper-containing oxidase enzyme. Other proteins in the α_2 group are **prothrombin,** a protein involved in blood coagulation; **erythropoietin,** a hormone that stimulates erythrocyte production; and **angiotensinogen,** a hormone precursor involved in regulating blood pressure, body fluid, and electrolyte balance.

The β-**globulins** (β_1 and β_2) include most of the apolipoproteins, which are carrier proteins for lipids. β_1-lipoprotein, called **low-density lipoprotein (LDL),** not only carries cholesterol and fats to tissues for use in manufacturing steroid hormones and building cell membranes but also favors the deposition of cholesterol in arterial walls and thus appears to play a role in disease of the blood vessels and heart. Other substances transported by β-globulins include phospholipids; glycerides; lipid-soluble vitamins (A, D, E, and K); and metals, such as copper and iron. **Transferrin,** a copper and iron transporting protein, is among the β-globulins.

The γ-**globulin** fraction contains the **immunoglobulins (Ig),** or antibodies. The five major classes of immunoglobulins are **IgG, IgA, IgM, IgD,** and **IgE.** More than 99% of all immunoglobulins in the plasma are either class G, A, or M. Class D and E immunoglobulins as free proteins are rare in plasma. The role of immunoglobulins in defense against foreign substances and disease-causing organisms will be discussed later, in Chapter 28.

The Albumin/Globulin Ratio

The total amount of plasma protein normally remains stable and contains approximately twice as much albumin as globulin. A typical adult normally consumes and replaces approximately 15 g of albumin and 5 g of globulin every 24 hours, but the amounts vary according to body needs for specific proteins. For example, inflammatory diseases cause an increase in the production of immunoglobulins, which is accompanied by a decrease of equal magnitude in the production of albumin. The albumin/globulin (A/G) ratio is reduced, but the total amount of protein in the plasma remains stable (A + G).

Fibrinogen

Fibrinogen is a soluble plasma protein manufactured in the liver. When blood clots, fibrinogen is converted into an insoluble protein called **fibrin,** which forms the foundation of the blood clot. Coagulation of blood will be discussed later in the chapter.

Hematopoiesis: The Formation of Blood Cells

 Where do blood cells originate?

The formed elements of the blood, sometimes referred to as **corpuscles** (little bodies), include the **erythrocytes** (red blood cells), **leukocytes** (white blood cells), and **thrombocytes** (platelets). Of the formed elements, only the leukocytes are true cells, possessing a nucleus and other organelles; however, we shall follow convention and refer to the erythrocyte as though it were still a true cell. Leukocytes include **agranular leukocytes** or **agranulocytes (monocytes** and **lymphocytes),** and **granular leukocytes** or **granulocytes (basophils, eosinophils,** and **neutrophils).**

All of these blood cells and structures originate from primitive cells known as **pluripotential hemopoietic stem cells (PHSCs),** found mainly in the bone marrow (Fig. 17–5). (Specifically, they are found in the red marrow of certain bones, although other sites are possible, particularly in the fetus.) These uncommitted stem cells are called "pluripotent" because they are capable of developing into any one of the various blood cells. In the process known as **hematopoiesis,** which is generally ongoing throughout life, these uncommitted stem cells (except those giving rise to lymphocytes) become committed progenitor cells called colony-forming units (CFUs) because they generate colonies of specific blood cell types. Some CFUs are multipotential (able to give rise to colonies containing more than two types of mature blood cells); some CFUs are bipotential (able to give rise to two types of mature blood cells); other CFUs are unipotential (committed to the development of a single kind of blood cell) (see Fig. 17–5).

Hematopoiesis is a general term for the development of all types of blood cells. Figure 17–5 gives an overview of the process. **Erythropoiesis** is a term that specifically describes the development of red blood cells from uncommitted stem cells; **leukopoiesis** is the term for the development of the white blood cells. The specific steps in each of these processes are described and illustrated in the relevant upcoming discussions of each cell type.

ERYTHROCYTES: RED BLOOD CELLS

Physical Characteristics and Red Blood Cell Count

The erythrocyte, or red blood cell, is a cell shaped like a biconcave disc with a diameter of approximately 8μm and a maximum thickness of approximately 2μm (Fig. 17–6). The shape of the cell results in a high ratio of surface area to volume, favoring more efficient diffusion into and out of the

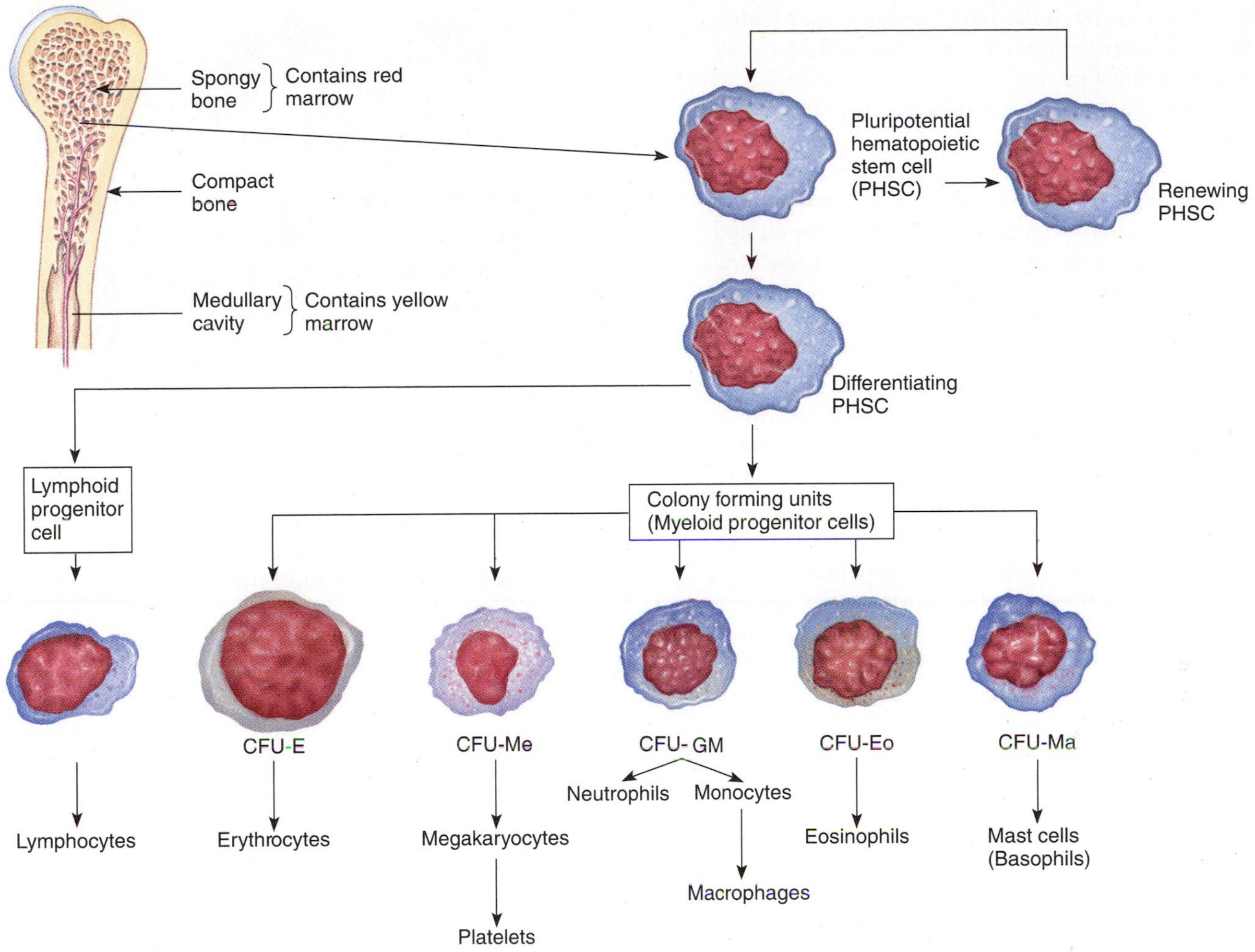

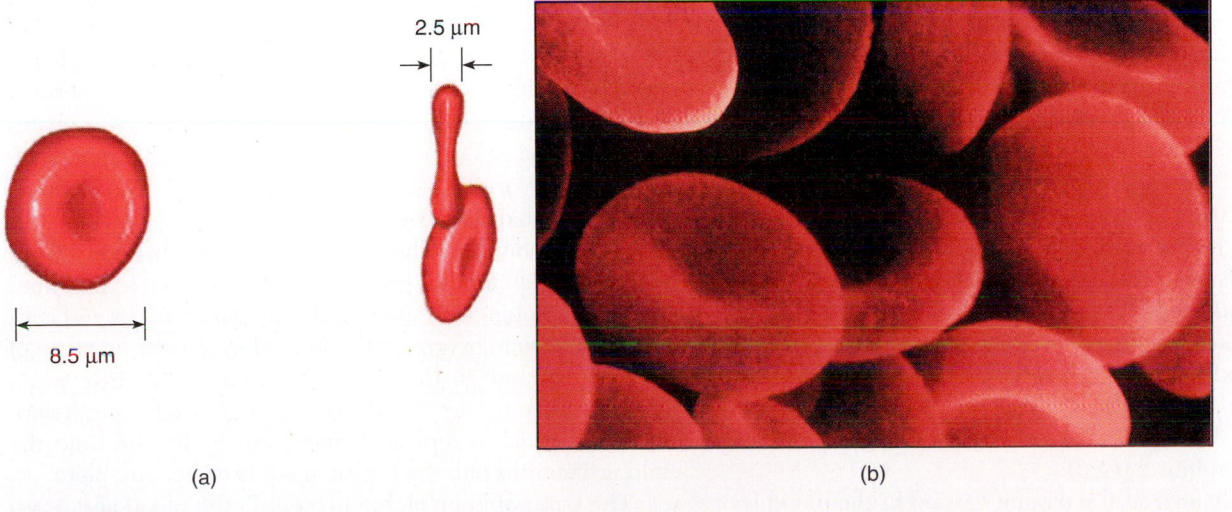

Figure 17–5

An overview of hematopoiesis.

Figure 17–6

(a) Typical dimensions of erythrocytes, or red blood cells. *(b)* A scanning electron micrograph of a red blood cell.   *(© Ken Elward/Science Source/Photo Researchers, Inc.)*

cell. The principal function of the erythrocyte is to transport oxygen and carbon dioxide in the blood and to buffer hydrogen ions from the extracellular fluid; as we will learn, its shape is optimal for this function.

The erythrocyte lacks a nucleus and the other organelles and therefore is not a true cell. The lack of internal organelles increases the erythrocyte's capacity for gas transport. The interior of the erythrocyte consists of a homogeneous matrix containing a loosely arranged meshwork of fibrous and globular proteins that form a cytoskeleton. The cytoskeleton is anchored to the inner surface of the plasma membrane so as to give the erythrocyte its discoid shape. Erythrocytes are pliable, elastic cells that bend, twist, and flex as they pass through tiny blood vessels; after such changes, they normally regain their discoid shape very quickly. Genetic abnormalities (e.g., **hereditary spherocytosis** or **hereditary elliptocytosis**) that affect the synthesis of cytoskeleton proteins may result in red blood cells that have abnormal spherical or elliptical shapes. Such cells do not function normally and are destroyed at a faster than normal rate, resulting in anemia.

Normal human blood contains approximately 5 million red blood cells per microliter of whole blood. Normal adult ranges are: men, 5.4 million cells/μL plus or minus 0.8 million cells/μL; women, 4.8 million cells/μL plus or minus 0.6 million cells/μL.

The Oxygen-Carrying Capacity of Blood

Erythrocytes contain **hemoglobin,** a pigment molecule that gives the cell a pale red color. Each erythrocyte contains 200 to 300 million molecules of hemoglobin, accounting for approximately one third of the mass of the red blood cell. Hemoglobin readily associates and dissociates with oxygen and carbon dioxide and is responsible for the erythrocyte's ability to transport these gases. Hemoglobin also plays an important role as a chemical buffer in the body's defense against changes in hydrogen ion concentration (pH).

The hemoglobin molecule consists of a **globin molecule** and four attached **heme groups** (Fig. 17–7). Each heme group contains an iron atom with which a molecule of oxygen may associate and dissociate; hence, each hemoglobin molecule can transport a maximum of four molecules of oxygen. When blood passes through the lungs, hemoglobin becomes saturated with oxygen, forming **oxyhemoglobin** and becoming bright red in color. Then, when this blood passes through body tissues, some of the oxygen dissociates from hemoglobin. **Reduced hemoglobin** is formed, and the blood turns a darker red.

Normal blood contains approximately 15 g of hemoglobin per deciliter (100 mL) of whole blood. Normal adult ranges are: men, 16.0 g/dL plus or minus 2.0 g/dL; women, 14.0 g/dL plus or minus 2.0 g/dL.

Approximately 98% of the oxygen carried in the blood is in the form of oxyhemoglobin. The **oxygen-carrying capacity** of blood is defined as the amount of oxygen that the hemoglobin in 100 mL of whole blood is capable of

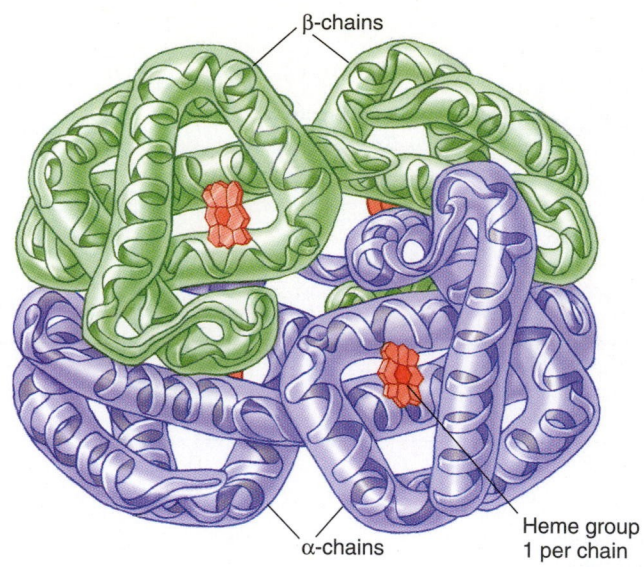

Figure 17–7

A model of the hemoglobin A molecule.

transporting. Each gram of hemoglobin can combine with 1.34 mL of oxygen. Because normal blood contains approximately 15 g of hemoglobin per deciliter, the oxygen-carrying capacity of normal adult blood is approximately 20 mL of oxygen per deciliter (15 g/dL $\times$ 1.34 mL O_2/g = 20.1 mL O_2/dL).

Hemoglobins

Several forms of hemoglobin have been found in human blood. **Type A** in adults and **type F** in the fetus are normal hemoglobins, whereas others are considered abnormal. Differences in the globin molecules give rise to the various types (the heme groups of all types are alike).

The *-globin* part of hemoglobin consists of two dissimilar pairs of peptide chains, each pair designated by the Greek letter alpha (α), beta (β), gamma (γ), or delta (δ). A heme group is bound to each of the four chains. Ninety-six percent of adult hemoglobin is **type A_1**, consisting of two α-chains and two β-chains. Two percent of adult hemoglobin is **type A_2**, consisting of two α-and two δ-chains, and 2% is type F, containing two α-and two γ-chains.

Hemoglobin F is the predominant type in fetal blood. Fetal hemoglobin has a greater affinity for oxygen than does adult hemoglobin. This greater affinity permits fetal red cells to carry adequate oxygen at the lower P_{O_2} (partial pressure of oxygen) prevalent in the fetal environment. At birth, when the lungs become the infant's organ of gas exchange, hemoglobin A begins to replace hemoglobin F. By the time the child is 6 months old, the replacement is nearly complete.

The types of hemoglobin present in the blood are determined genetically. Abnormalities in the types of hemoglobin present (**hemoglobinopathies**) usually result from altered DNA templates that produce minor variations in the amino

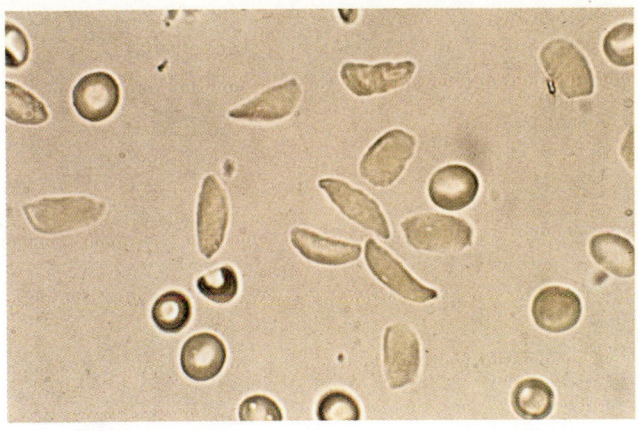

Figure 17–8

Blood from a patient with sickle cell anemia.
(© *Dr. Donald L. Rucknagel, Director of the Comprehensive Sickle Cell Center/Children's Hospital Medical Center/ Cincinnati, Ohio.*)

acid sequence or composition of either the β-chain (e.g., **sickle cell disease**) or the α-chain (e.g., **hemoglobin H disease**). **Sickle cell hemoglobin (type S),** for example, differs from hemoglobin A because of the substitution of a single amino acid in one of the β-chains. Oxyhemoglobin S, like normal hemoglobins A and F, is soluble in the intracellular fluid of the erythrocyte; however, deoxygenated hemoglobin S is insoluble, forming fibrous precipitates that change the red blood cell shape from a biconcave disc to a sickle shape (Fig. 17–8). Sickle cells are destroyed faster than are normal red blood cells, producing anemia. Sickle cells often become trapped in capillaries, blocking the flow of blood and thereby causing pain and tissue hypoxia (inadequate oxygen supply). These conditions, if severe, can result in tissue death.

Erythropoiesis: The Formation of Red Blood Cells

Erythropoiesis is the formation of erythrocytes from pluripotential hemopoietic stem cells. In the fetus, red blood cells are formed from stem cells in the yolk sac, liver, spleen, and red bone marrow. In the adult, red blood cell production is confined to the red marrow of the humerus (upper arm bone), femur (thigh bone), the ribs, sternum (breastbone), scapulae (shoulder blades), skull, vertebrae, and pelvis. In certain diseases during adult life, red blood cell formation occurring outside the bone marrow may be seen in the spleen and liver. Each day, the adult marrow produces approximately 230 billion red blood cells. If necessary, the marrow can increase its production of red blood cells sixfold.

An erythrocyte exists for approximately 4 months, after which it is removed from the blood and destroyed by macrophages in the spleen, liver, and marrow. Normally, the

rate of destruction equals the rate of production, so that the number of circulating erythrocytes is fairly constant.

Erythropoiesis begins with the differentiation of **pluripotential hemopoietic stem cells** (Fig. 17–9) into committed progenitor cells designated *colony-forming unit-burst* (CFU-B), which have a high rate of proliferation. CFU-Bs give rise to erythrocyte precursor cells and to another larger and more slowly proliferating progenitor cell called

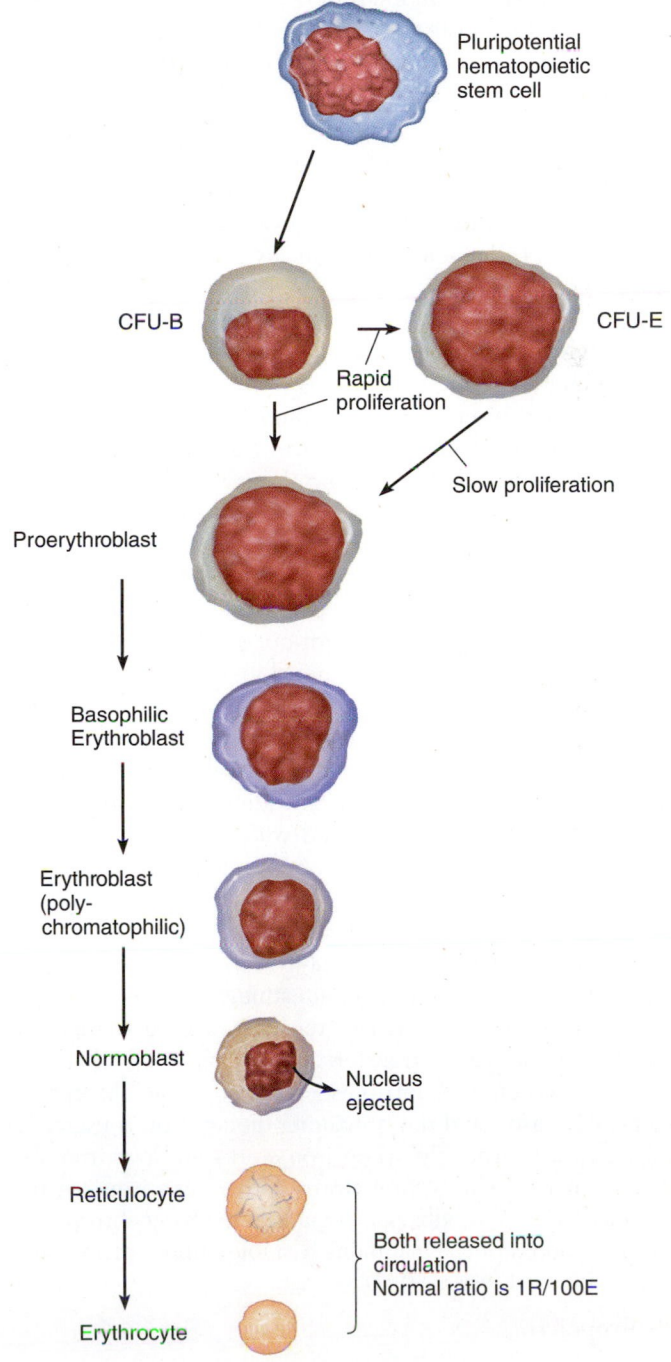

Figure 17–9

Erythropoiesis: The formation of red blood cells.

colony-forming unit-erythroid (CFU-E). Erythrocyte progenitor cells develop into precursor cells that proceed through a sequence of maturation stages during which they lose their nuclei and acquire hemoglobin.

Upon losing its nucleus, the erythrocyte precursor becomes a **reticulocyte.** Both mature erythrocytes and reticulocytes are released by the marrow into the blood. Released reticulocytes mature into erythrocytes approximately 1.5 days after entering the circulation. Normally, blood contains approximately one reticulocyte for every 100 erythrocytes, their production serving as an index of erythropoietic activity. The higher the reticulocyte proportion in the blood, the higher the rate of red blood cell production in the marrow.

Iron

Several substances are required for proper formation of red blood cells and hemoglobin: **amino acids, iron, copper, vitamins B₂ (riboflavin)** and **B₁₂ (cyanocobalamin), pyridoxine,** and **folic acid.** Deficiencies in any one of these substances may lead to anemia.

Iron is required for the production of heme; in fact, approximately two thirds of the body's iron content is found in hemoglobin. Most of the remaining iron is stored as **ferritin,** an iron-protein complex found in the liver and in the intestinal epithelium, spleen, and marrow. Stored iron is readily made available for hemoglobin synthesis via a protein called **transferrin,** found in the plasma. Transferrin combines with iron that has been absorbed from the intestinal tract or released from storage. It then transfers the iron to erythrocytes that are developing in the bone marrow. Transferrin also transports iron released from worn-out red blood cells to the bone marrow to be reused for hemoglobin synthesis or to the liver to be stored.

Nearly all body cells require iron because iron-containing enzymes are required in oxidative metabolism. The shedding of cells by the lining of the intestinal and reproductive tracts as well as the skin, coupled with smaller amounts of iron in sweat and urine, result in an average daily loss of approximately 4 mg of iron. Iron loss normally is replaced by dietary intake. The recommended daily allowance is 15 mg, although typically only approximately 4 mg of iron per day is absorbed, primarily in the small intestine.

Iron is absorbed by active transport, a process that occurs until all the plasma transferrin in the plasma is saturated with iron. When transferrin is saturated, iron absorption practically ceases, and the remaining dietary iron is excreted in the feces. Conversely, when iron stores are low, iron absorption increases until iron stores and plasma iron are replenished. Thus, a feedback mechanism involving absorption, transport, and storage maintains a stable supply of iron for hemoglobin synthesis.

Erythropoietin

Hemopoietin is a hormone that regulates the production of specific kinds of blood cells. **Erythropoietin** is a hormone that regulates erythropoiesis. A reduction in oxygen tension

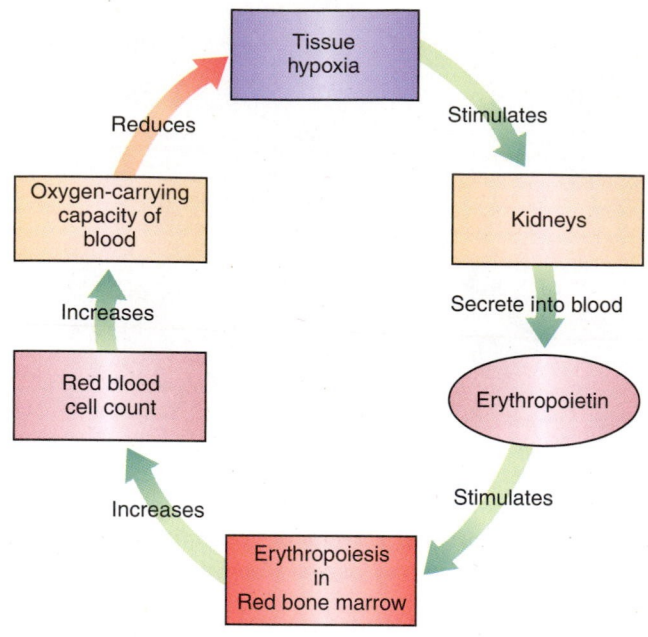

Figure 17–10

The mechanism of erythropoietin action.

in systemic arterial blood stimulates the kidneys to release erythropoietin, which travels via the blood to the marrow. In the marrow, erythropoietin stimulates the differentiation of erythrocyte progenitor cells and shortens the maturation time of precursor cells, thereby increasing the rate of red blood cell formation and release (Fig. 17–10). When the tissues severely lack oxygen, a condition termed severe **hypoxia,** the erythropoiesis can increase fivefold to correct the deprivation.

The synthesis of erythropoietin in turn is influenced by sex hormones. Male sex hormones, such as testosterone, stimulate production of erythropoietin; in part, this stimulation may account for the higher red blood cell count normally seen in men compared with women.

Erythrocyte Destruction and the Fate of Hemoglobin

The erythrocyte has no nucleus, mitochondria, endoplasmic reticulum, or other organelles required for cell maintenance and thus has a very limited life span. As erythrocytes age, changes in their plasma membranes make them susceptible to recognition and ingestion by **macrophages** in the spleen, liver, and marrow. Approximately 230 billion red cells are destroyed each day.

When the red cell is destroyed by a macrophage, hemoglobin is broken down, and important components of the molecule are recycled within the body (Fig. 17–11). The peptide components are reduced to individual amino acids, which may be reused by other cells for protein synthesis. The heme portions are broken down into iron (Fe^{3+}) and

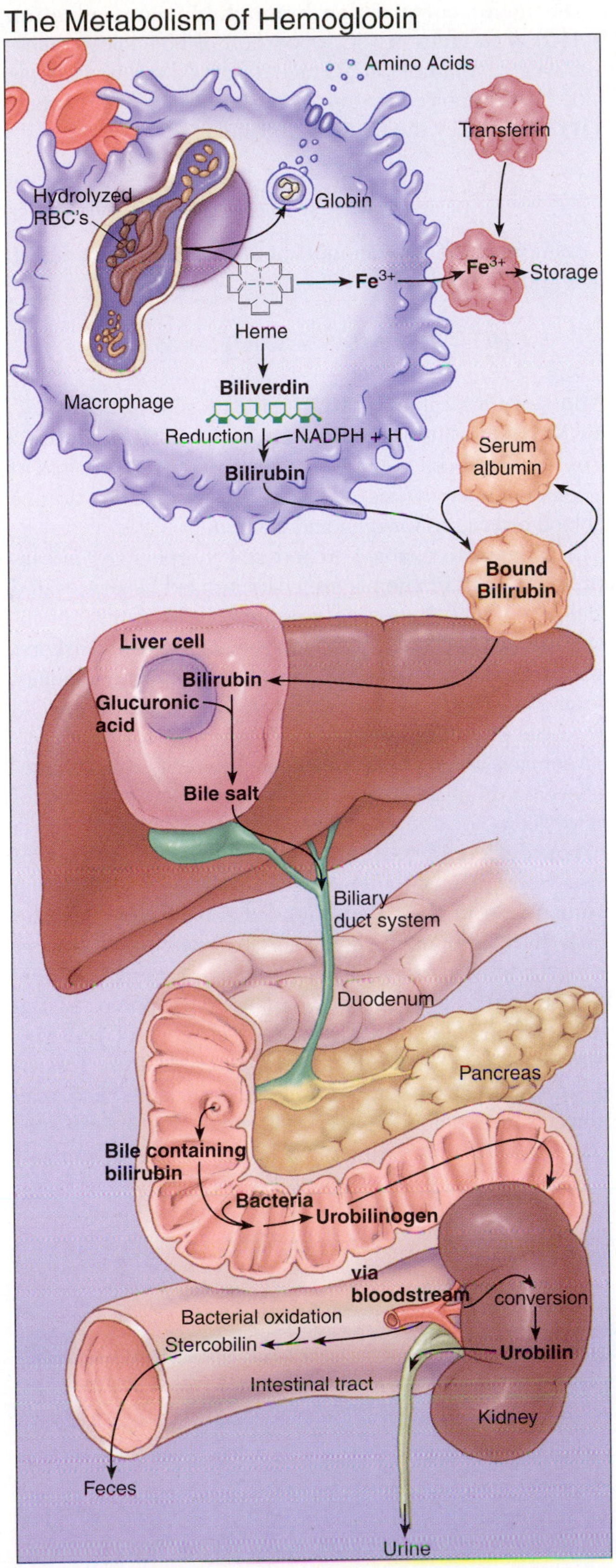

The Metabolism of Hemoglobin

Figure 17–11

The metabolism of hemoglobin.

biliverdin, a greenish pigment, which is reduced to **biliru-bin** and released into the plasma. In the plasma, bilirubin binds to **albumin,** a carrier protein, and is transported to the liver. The liver cells take up the bilirubin, join it to the **glu-curonic acid,** and excrete the complex as a bile salt into the biliary duct system, which empties into the small intestine. Intestinal bacteria convert bilirubin into **urobilinogen,** most of which is eliminated in the feces in the form of **sterco-bilin,** a brown pigment that colors stools. Some urobilinogen is absorbed from the intestine and excreted by the kidney into the urine, where it becomes oxidized to **urobilin,** the amber coloring matter in urine. The iron is released into the plasma, where it combines with transferrin and travels to other cells for storage or reuse.

Bilirubin is very toxic to the nervous system and, if al-lowed to accumulate in the body, can cause neural damage. An elevation of blood bilirubin above normally low levels is called **hyperbilirubinemia.** This condition may result from an excessively high rate of cell destruction, liver disease, or blockage of the liver biliary system. **Icterus,** or **yellow jaun-dice,** a symptom of hyperbilirubinemia, is caused by an ele-vation in plasma bilirubin to the point at which the eyes, skin, and mucous membranes appear yellow. Hyperbilirubinemia is common among newborn infants, who often become jaun-diced after the first day of life. The chief cause is an imma-ture liver unable to excrete bilirubin as fast as it is being formed. Fortunately, bilirubin is a photosensitive pigment that decomposes when exposed to near-ultraviolet light; thus, when the jaundiced infant's skin is exposed to the appropri-ate wavelength of light in an incubator, blood bilirubin can be kept at low levels until the infant's liver matures enough to handle the bilirubin load.

Although most old erythrocytes are destroyed by macrophages, a few erythrocytes actually disintegrate within the blood. When the red blood cell membrane is damaged, hemoglobin escapes into the blood, but it quickly becomes bound to a protein called **haptoglobin.** The resulting com-plex prevents the kidney from excreting the hemoglobin in the urine. Macrophages in the liver then remove and process the hemoglobin, as indicated in Figure 17–11.

Red Blood Cell Indices and Anemia

Approximately 98% of the oxygen transported by blood is carried within the red blood cells loosely bound to hemoglo-bin, a protein pigment that colors the blood red. A deficiency in the amount of normal hemoglobin present in each of the red blood cells, or a deficiency in the number of circulating red blood cells, reduces oxygen delivery to tissues and im-pairs systemic function.

Anemia is a condition characterized by an abnormally low oxygen-carrying capacity of the blood resulting from marked deficiencies in the number of circulating red cells, the amount of hemoglobin in the blood, or both. Anemia is considered to be present if the hemoglobin is less than 12 g/dL or the hematocrit is less than 37%. Because the oxygen-carrying capacity of their blood is reduced, anemic persons may experience physical weakness, shortness of breath, and difficulty in performing mental work. Chronic physical and mental fatigue are common symptoms.

Anemia may result from a failure of red blood cell production, an excessive rate of cell loss, or both. Anemias are classified on the basis of cell morphology (shape and size) and RBC indices (discussed later).

With respect to morphology, red blood cells are described in specific terms. A **normocytic** erythrocyte is normal in size and shape; a **normochromic** erythrocyte is normal in color (and thus in hemoglobin content). A **microcytic** erythrocyte is smaller than normal; a **macrocytic** erythrocyte, larger than normal. A **hypochromic** erythrocyte is paler in color than normal (indicating insufficient hemoglobin); a **hyperchromic** erythrocyte is deeper in color than normal. **Anisocytosis** is a variation in size; **poikilocytosis** is a variation in shape. The optimum situation is for erythrocytes to be normocytic and normochromic.

RBC indices are used to detect abnormalities in erythrocyte size, shape, and color. Three commonly used indices are the mean corpuscular volume (MCV) index, the mean corpuscular hemoglobin (MCH) index, and the mean corpuscular hemoglobin concentration (MCHC).

The **mean corpuscular volume (MCV) index** is a quantitative assessment of the red blood cell volume reported in picoliters (pL) per red blood cell. It is computed by the following formula:

$$MCV = (\text{hematocrit} \times 1000) \div \text{RBC count} \\ (\text{millions per } \mu L)$$

For example, if the hematocrit is 45% and the RBC count is 5.4 million per microliter:

$$MCV = (0.45 \times 1000) \div 5.4 = 83.3 \text{ pL per red blood cell}$$

The normal adult value of MCV is 87 pL plus or minus 5 pL. A low MCV indicates microcytosis, and a high MCV indicates macrocytosis.

The **mean corpuscular hemoglobin (MCH) index** is used to determine the concentration of hemoglobin within the erythrocytes. It is reported in picograms (pg) and is computed as follows:

$$MCH = (\text{g/dL hemoglobin} \times 10) \div \text{RBC count} \\ (\text{millions per } \mu L)$$

For example, if the hemoglobin content of whole blood is 16 g/dL and the RBC count is 5.4 million per μL:

$$MCH = (16 \times 10) \div 5.4 = 29.6 \text{ pg}$$

The normal adult value of MCH is 29 pg plus or minus 5 pg. A low MCH indicates microcytosis, hypochromia, or both. A high MCH suggests hyperchromia.

The **mean corpuscular hemoglobin concentration (MCHC)** is an index of the proportion of hemoglobin measured (weight/volume) per average red blood cell in a sample of blood. It is reported as grams per deciliter or a percentage and is calculated by the following formula:

$$MCHC = (\text{g/dl hemoglobin}) \div \text{hematocrit}$$

For example, if the hemoglobin content of whole blood is 16 g/dL and the hematocrit is 45%:

$$MCHC = (16) \div 0.45 = 35.6 \%, \text{ or } 35.6 \text{ g/dl}$$

The normal adult value for MCHC is 34% plus or minus 2%. A low MCHC means hypochromia; a high MCHC indicates a loss of red blood cell volume without a proportionate loss of hemoglobin (microcytosis). Table 17–2 summarizes normal red blood cell values for men and women.

In addition to changes in red cell morphology and indices, most forms of anemia are characterized by an elevated reticulocyte count. Increased numbers of circulating reticulocytes indicate increased bone marrow activity with early release of reticulocytes. This is common in hemolytic anemias, for example, when red blood cells are destroyed faster than marrow can produce them.

Anemia is a sign of disease. There are many types of anemia because there are many underlying causes. Only a few of the more common anemias will be discussed here.

Iron-deficiency anemia is rarely caused by lack of dietary iron. In fact, most Americans may be consuming too much iron in their diet. Good dietary sources of iron include liver, red meat, egg yolks, carrots, and spinach. The most common cause of iron-deficiency anemia is internal blood loss associated with hemorrhoids, polyps of the large intestine, bleeding ulcers, and cancers of the gastrointestinal tract. Inadequate absorption of iron, loss of iron stores, or abnormalities in iron utilization also may produce iron-deficiency anemia. Deficiencies in iron metabolism lead to a reduction in hemoglobin production and

TABLE 17–2

Normal Red Cell Values (Adult)

RBC Measurement	Male	Female
Red cell count (millions/μL)	5.4 ± 0.8	4.8 ± 0.6
Hemoglobin (g/dl)	16.0 ± 2.0	14.0 ± 2.0
Hematocrit (%)	47.0 ± 5.0	42.0 ± 5.0
MCV (pL)	87 ± 5	87 ± 5
MCH (pg)	29 ± 5	29 ± 5
MCHC (%)	34 ± 2	34 ± 2

pg = picogram
pL = picoliter

the release of microcytic, hypochromic red blood cells from the marrow.

Hemorrhagic anemias occur after acute or chronic blood loss, such as that associated with hemorrhoids, bleeding ulcers, excessive menopausal bleeding, and so on, and usually are characterized by normocytic, normochromic erythrocytes.

Megaloblastic anemias develop as a result of deficiencies of folic acid (a B vitamin), cobalamin (vitamin B$_{12}$), or both. These vitamins are essential for normal maturation of red blood cell precursors in the marrow. Lack of either vitamin impairs red cell maturation and allows the marrow to release large, nucleated red-cell precursors called **megaloblasts,** which contain more hemoglobin than normal red blood cells but that do not function normally. There are many causes of vitamin B$_{12}$ and folic acid deficiencies, ranging from dietary lack (least common) to impairments of absorption (most common). **Pernicious anemia** is a kind of megaloblastic anemia that occurs when autoimmune disease causes a lack of intrinsic factor and a resultant deficiency in the absorption of vitamin B$_{12}$. The lining of the stomach secretes a glycoprotein called the **intrinsic factor,** which is essential for the absorption of vitamin B$_{12}$ in the small intestine. Vitamin B$_{12}$ and the intrinsic factor combine to form the **erythrocyte maturation factor,** which is required for maturation of red blood cells in the marrow. In pernicious anemia, the person's immune system produces antibodies that attack the stomach cells producing the intrinsic factor, thereby creating a deficiency in erythrocyte maturation factor. Pernicious anemia and other megaloblastic anemias are characterized by the presence in the blood of macrocytic, hyperchromic, immature red blood cell precursors and a reduction in the numbers of normal erythrocytes.

Aplastic anemia results from an abnormal decrease in the production of all blood cells in the marrow. *Aplasia* means "faulty or incomplete development." Blood-forming cells do not mature at a normal rate in this disorder. The resultant anemia is characterized by reduced numbers of normal red blood cells. Aplastic anemia can be caused by radiation, cytotoxic drugs used for the treatment of malignant disease, exposure to chemicals (arsenic, DDT, benzene) that poison the marrow, or a genetic failure of bone marrow development.

Hemolytic anemia occurs when the rate of red blood cell destruction exceeds the capacity of the marrow to replace lost cells. Hemolytic anemias usually are accompanied by jaundice and elevated serum bilirubin. Causes of hemolytic anemia include damage to the red blood cell or red blood cell defects (e.g., hereditary spherocytosis and sickle cell disease) that cause early destruction by macrophages and various extracellular agents (toxins, antibodies) that cause hemolysis.

Thalassemias are a group of disorders caused by inheritance of one or more genes that limit the rate of synthesis of α- or β-chains of hemoglobin. The impairment leads to a hypochromic, microcytic anemia, the severity of which is determined by the number and kind of genes affected. **Thalassemia major** is usually homozygous and most often affects β-chain synthesis (**β-thalassemia major, Cooley's anemia).** β-Thalassemias occur most frequently in inhabitants of the Mediterranean area (*thalassa* = "sea"), and α-thalassemias are more common among Asians and central Africans.

Polycythemia

Polycythemia (*poly* = "many"; *cyte* = "cell") is a relative or absolute increase in the number of circulating red blood cells above 6.2 million cells/µL. It may result from an increase in the concentration of erythrocytes without an increase in the total red blood cell mass (relative polycythemia) due to dehydration and hemoconcentration, or it may result from an increase in the concentration of erythrocytes accompanied by an increase in total cell mass (polycythemia vera) due to the hyperactivity of the bone marrow.

Polycythemia vera (true polycythemia) is a chronic, slowly progressive disease involving excessive production of all blood cells produced in the marrow (**hyperplasia).** In most cases, the cause is unknown (**idiopathic),** whereas in others the polycythemia may be evidence of tumors in the bone marrow, kidney, or brain. Polycythemia vera is not associated with a lack of oxygen in the blood.

Secondary polycythemia (erythrocytosis) is a physiological polycythemia that occurs in response to arterial hypoxia and its stimulation of erythropoiesis. Exposure to chronic hypoxic (low oxygen) environments, such as that normally found at high altitudes, often is accompanied by secondary polycythemia. It also may occur in emphysema, pulmonary fibrosis, carbon monoxide poisoning, and other abnormal states in which oxygenation of the blood is reduced.

Polycythemias can be life-threatening because they increase the viscosity of the blood, placing an increased load on the heart and blood vessels. They also make an individual more susceptible to thrombosis, the intravascular formation of blood clots.

Blood Groups and Blood Types

 What are blood types; why are they important clinically?

All human blood contains erythrocytes, leukocytes, platelets, and plasma; however, not all blood is the same. When some types of blood are mixed, red blood cells aggregate into clumps in a process called **agglutination.** Agglutinated cell masses can become lodged in small blood vessels, blocking the flow of blood and severely impairing tissue and organ function (e.g., kidneys, heart, brain). If agglutination is widespread (major agglutination), death can occur. Because many circumstances require transfusion of whole blood from one person to another, the donor type must be selected carefully so as to avoid major agglutination in the recipient.

Incompatibilities of blood types are due to the presence of antigens embedded in the erythrocyte plasma membrane and specific antibodies in the blood plasma. An **antigen** is a molecule or a part of a molecule, recognized by the body as foreign. An antigen (*anti* = "against," *gene* = "to originate") stimulates specific defense mechanisms of the body (see Chapter 28), including the production and secretion of an antibody, a protein that reacts with the antigen normally to nullify its harmful effects. When antibodies react or bind to erythrocyte antigens, the affected red blood cells clump together, or agglutinate; therefore, erythrocyte antigens are called **agglutinogens** and the corresponding antibodies are called **agglutinins**. The presence or absence of erythrocyte antigens is determined genetically; that is, blood type is inherited.

Several hundred types of erythrocyte antigens have been identified. Nearly all of them are weakly antigenic; that is, they do not provoke a strong immune response when given to a person who does not normally possess it. The recipient would not have, nor would develop, large amounts of the corresponding antibody. However, a few erythrocyte antigens are strongly antigenic, causing the recipient to reject, by agglutination, the donated erythrocytes. Three strong antigens that may be present in erythrocytes are the **A, B,** and **D agglutinogens.** The corresponding antibodies, normally absent if the antigen is present, are designated as **agglutinin a (anti-A), agglutinin b (anti-B),** and **agglutinin d (anti-D).** The presence or absence of agglutinogens and agglutinins allows the blood to be classified into types or groups. We will discuss only the blood types of the ABO and D (Rh) blood groups.

The Blood Types of the ABO Blood Group

The presence or absence of the A and B antigens and antibodies determines blood type in the ABO blood group as follows (Fig. 17–12): **Type A** blood contains the A antigen on its red blood cell membranes. The plasma contains anti-B. **Type B** blood contains the B antigen on the plasma membranes of its red cells. The plasma contains anti-A. **Type AB** blood contains both the A and B antigens on its red blood cell membranes, but the plasma lacks both anti-A and anti-B. **Type O** blood contains neither the A nor the B antigens, but the plasma contains both anti-A and anti-B. The blood types of the ABO group and their constituents are listed in Table 17–3.

When type B blood is transfused into a type B recipient, no reaction occurs because the recipient's body does not recognize the blood as being foreign. When type B is transfused in a type A recipient, two reactions occur:

1. The anti-A in the donor's plasma reacts with the A antigen on the recipient's red cells (**minor agglutination**).
2. Anti-B in the recipient's plasma reacts with the B antigen on the donor's red cells (**major agglutination**).

The first reaction is of little consequence because the anti-A in the donor's plasma (serum) becomes greatly diluted on entering the recipient's blood and thus becomes inconsequential. Agglutination of recipient cells occurs, but the agglutination is minor. The second reaction, however, is of

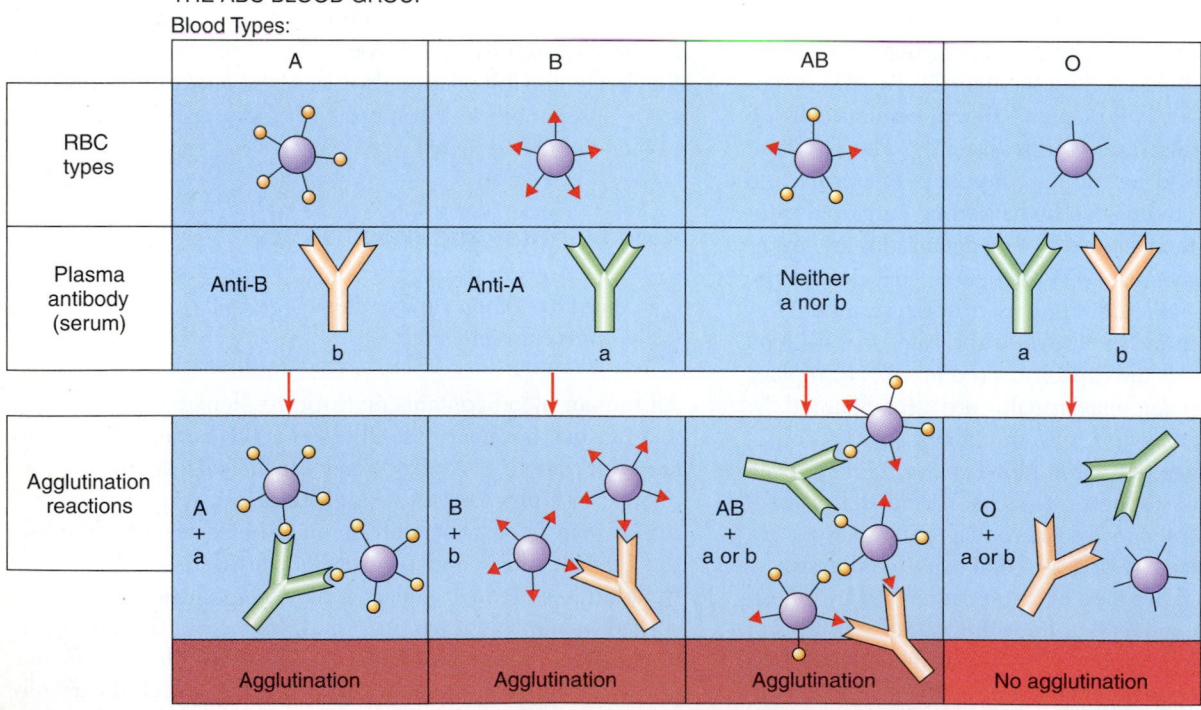

Figure 17–12

Antigens, antibodies, and antigen-antibody reactions of the ABO blood group.

TABLE 17–3

The ABO Blood Group and ABO-Rh Distribution

Blood Type	RBC Antigen (Agglutinogen)	Plasma Antibody (Agglutinin)	Population Distribution (Caucasian; %)	Rh+ (%)	Rh− (%)
A	A	b (Anti B)	40	34	6
B	B	a (Anti A)	11	9	2
AB	A and B	Neither a (Anti A) nor b (Anti B)	4	3	1
O	Neither A nor B	a and b (Anti A and Anti B)	45	38	7

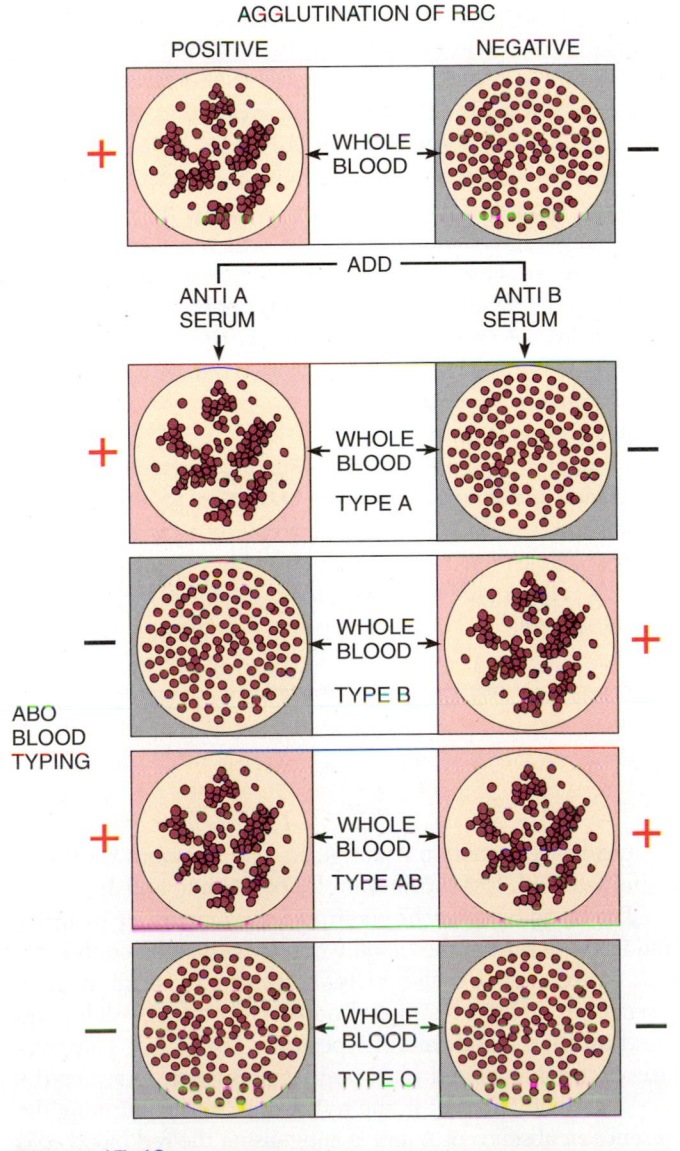

AGGLUTINATION OF RBC

Figure 17–13

Agglutination reactions in ABO blood typing.

major consequence because larger quantities of recipient anti-B (compared with much smaller quantities of donor anti-A) are available to react with the B antigens of the donor's cells. When large numbers (approximately 5 million per microliter of blood) of donor red cells enter the recipient's body, widespread agglutination occurs, and a **transfusion reaction** develops in the recipient as the donor's cells are agglutinated by the recipient's antibodies. A transfusion reaction is characterized by general discomfort, anxiety, difficulty in breathing, flushing of the face, pain in the chest and neck, and other variable symptoms leading to shock—as evidenced by rapid feeble pulse; cold, clammy skin; fall in blood pressure; nausea; and vomiting. These acute symptoms usually occur within the first 2 hours. Depending on the amount of blood administered, more severe complications involving kidney, heart, lung, liver, and brain damage may develop, leading to death several days later.

Blood may be typed and cross-matched to determine its compatibility with other blood. **Blood typing** for the ABO group involves placing serum containing a known antibody (anti-A or anti-B) on a microscope slide and mixing it with a small quantity of the blood to be typed (Fig. 17–13). The presence or absence of red blood cell agglutination determines the type of unknown blood (Table 17–4). For example,

TABLE 17–4

Hemagglutination Test for Blood Type

Whole Blood	Agglutination Observed with	
Blood Type	Anti-A serum	Anti-B serum
A	Yes	No
B	No	Yes
AB	Yes	Yes
O	No	No

CURRENT CONCEPTS IN PHYSIOLOGY

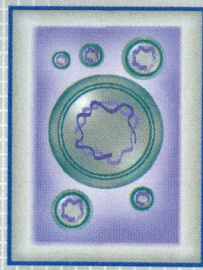

Universal Blood

According to the American Association of Blood Banks (AAAB), more than 12 million units of blood, including 630,000 autologous donations, are donated in the United States each year. *Autologous transfusions* refer to transfusions in which the blood donor and the transfusion recipient are the same. *Allogeneic transfusions* refer to transfusions in which blood is transfused to someone other than the donor. A unit of donated blood is equivalent to 500 mL of whole blood. These units are transfused to approximately 4 million patients per year.

A unit of donated blood, referred to as *whole blood,* usually is separated into blood components, such as red blood cells; plasma; cryoprecipitated AHF (a clotting protein complex); platelets; white blood cells; and plasma derivatives, including albumin, immunoglobulins, coagulation factors, and anticoagulation factors. Each component generally is transfused to a different patient, each with different needs. Blood-component therapy allows several recipients to benefit from a single unit of donated whole blood.

Red blood cells are the most recognizable as well as the most numerous component of whole blood. Red blood cells can be preserved and stored for 42 days or treated and frozen for extended storage up to 10 years. Allogeneic transfusions of red blood cells often are used to alleviate symptoms of chronic anemia resulting from malignancies, kidney failure, or gastrointestinal bleeding, and acute anemia resulting from surgery or trauma. The need for blood is great. On any given day, approximately 32,000 units of red blood cells are needed. Unfortunately, donated blood is often in short supply, in part because of an inadequate number of donations and in part because of red blood cell incompatibilities.

Most recipients are limited to specific types of red blood cells. For example, a type B recipient cannot receive type A or type AB red blood cells but could receive type B or type O red blood cells. Because type O, or "universal," red blood cells lack the A antigen and the B antigen, they can, assuming no other incompatibilities, be given to recipients of any ABO type (see text for a more detailed explanation). If type A, type B, and type AB red blood cells could be converted to type O red blood cells, then red blood cells could be transfused safely to anyone regardless of the recipient's blood type. Recent research has shown it possible, and commercially feasible, to enzymatically convert non-type O red cells to type O.

Different antigenic sugar chains on the surface of red blood cells distinguish the four main types (Fig. A). On type O cells, the chain ends in fructose. On type A cells and type B cells, *N*-acetylgalactosamine and galactose, respectively, branch off the end of the same chain found on type O cells. Recombinant enzymes and processing devices have been developed for converting type A and type B cells to type O by removing terminal portions of their sugar chains. A simplified version of the conversion process is shown in Figure B. Currently, the B-to-O and A-to-O conversion systems are in clinical trials and not available for commercial use.

Although Rh incompatibility is less of a problem than is ABO incompatibility, research is underway to alter the Rh factor so as to make the red cell effectively Rh-negative. The goal is to eventually be able to convert all other red cells to type O−, truly a more "universal" kind of red blood cell.

TABLE 17–5

Confirmation of Blood Type by Cross-Matching

Serum Antibody	Test Cells		
	A	*B*	*O*
a	+	−	−
b	−	+	−
a + b	+	+	−
Neither a nor b	−	−	−

+ = agglutination; − = no agglutination.

---

because its cells contain A antigens, type A blood would not agglutinate with test serum samples containing anti-B.

The blood type may be confirmed through testing of agglutinin in the serum of the blood whose type is to be confirmed. Such a procedure is called **cross-matching.** A small quantity of serum to be tested is placed on four microscope slides and mixed with red blood cells of type A, B, AB, and O. The presence or absence of red blood cell agglutination indicates the type of antibody present in the test serum, thus confirming the presence or absence of A and B antigens on the red blood cells in the blood from which the test serum was taken (Table 17–5). Cross-matching is clinically useful to confirm compatibility of a

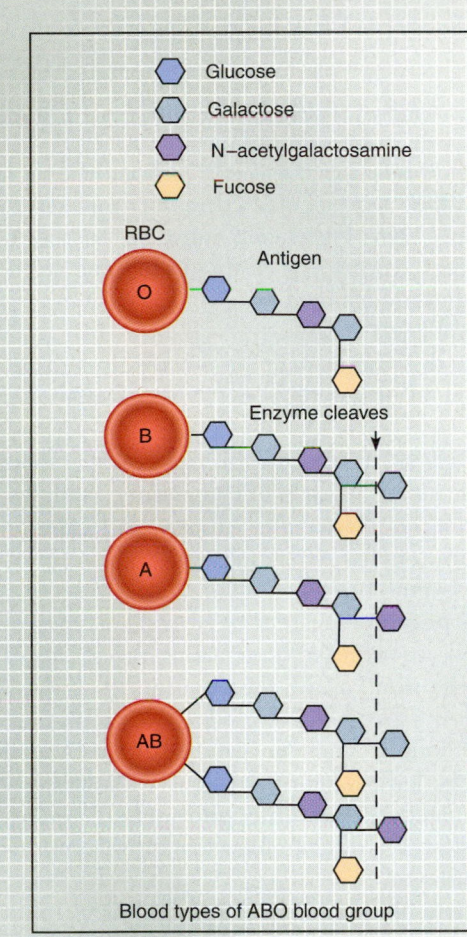

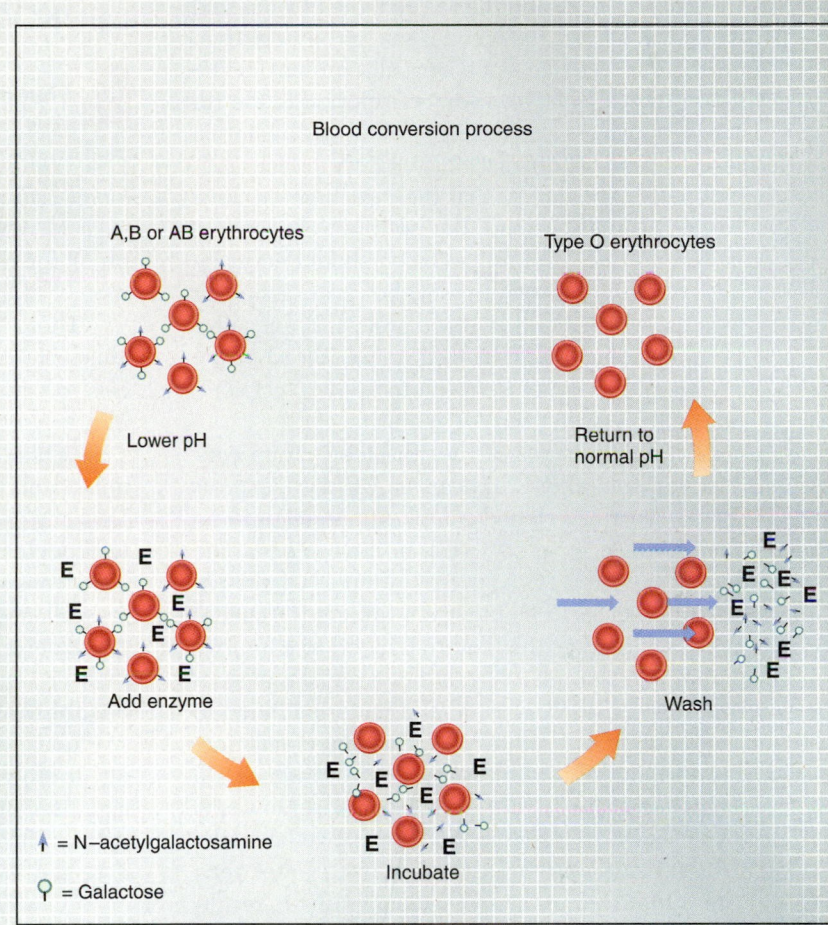

Figure A   Group ABO red blood cell types with attached antigen.

Figure B   An outline of the process converting type A, type B, and type AB erythrocytes to type O erythrocytes.

potential donor's red blood cells because the donor's erythrocytes may display an antigen, other than A or B, with which existing recipient antibodies would react.

Because type AB blood contains both the A and B antigens, neither anti-A nor anti-B persons with type AB blood may act as recipients for any of the other types (A, B, or O). Historically, such persons have been referred to as "universal recipients." By the same reasoning, persons with type O blood have been called "universal donors" because their erythrocytes contain neither the A nor the B antigen to oppose the recipient's anti-A and anti-B. In fact, there are no such things as a "universal donor" and a "universal recipient." The

presence of other antigens on the erythrocyte membrane or other erythrocyte antibodies in the plasma precludes this possibility. One such antigen, present in approximately 85% of the population, is the Rh factor. As we shall see, type O+ blood cannot be donated to an AB− recipient.

The Blood Types of the Rh Blood Group

The rhesus blood group, first described in 1940, is named after the monkey in which the antigens were discovered. As with the A and B antigen, the presence or absence of the Rh antigen is determined genetically. Most of the Rh antigens are weakly antigenic and thus of little clinical importance; however, one of the Rh antigens, called **antigen D,** is strongly antigenic. The D antigen is commonly called the **Rh factor.**

If the Rh factor is present on the red cell, the blood is said to be **positive;** if it is absent, the blood is said to be **negative.** Antibodies to the Rh factor are not normally present in the plasma of Rh-negative blood. However, on exposure to the Rh antigen (through transfusion or mixing of fetal and maternal blood during childbirth), Rh-negative persons produce antibodies against the Rh antigen, so that the next time negative blood is exposed to positive blood, the positive cells are destroyed. As a general rule, Rh-positive blood should never be transfused into an Rh-negative recipient; however, assuming compatibility in the ABO and other blood groups, Rh-negative blood may be given to an Rh-positive recipient. Blood of the type A+, for example, may not be given to an A− recipient, but A− blood could be given to an A+ recipient.

An important consequence of the Rh blood group concerns the Rh-negative mother who bears an Rh-positive fetus for the first time. Usually, no blood complications occur if the mother has never been exposed to Rh-positive blood. The mother would possess few, if any, Rh antibodies (anti-D), and the fetal erythrocytes containing the antigen D cannot cross the placenta and enter the maternal blood to stimulate the mother's immune system. At birth, however, the placental barrier is destroyed, allowing fetal and maternal blood to mix. When fetal erythrocytes enter the maternal circulation, they stimulate the mother's immune system to produce Rh antibodies, which remain in her blood for several years. Upon conceiving and carrying a second Rh-positive fetus, the maternal Rh antibodies can cross the placenta, enter fetal blood, and cause the fetal erythrocytes to agglutinate, resulting in a form of hemolytic anemia called **erythroblastosis fetalis.** Unless the agglutinated fetal erythrocytes are replaced, the fetus could die from hypoxia and other complications. Fortunately, the risk for erythroblastosis fetalis may be decreased significantly by the administration of Rh antibodies (RhoGAM) to the mother within 72 hours following the birth of an Rh-positive child. The antibodies react with any positive fetal erythrocytes that have entered the mother's immune system. As a result of this procedure, called *desensitization,* the circulating titer (concentration) of maternal Rh

antibodies usually does not increase to the point of interfering with the successful carrying of a second Rh-positive fetus.

LEUKOCYTES: WHITE BLOOD CELLS

 What are white blood cells; what do they do?

Leukocytes (*leuko* = "white"; *cyte* = "cells") are called *white blood cells* because they lack pigmented (color) molecules, such as hemoglobin. White blood cells must be stained in order for the different types to be distinguished and to be more visible when they are counted.

There are several types of leukocytes (Fig. 17–14). Each leukocyte is nucleated; some contain large cytoplasmic vesicles, or granules, while others do not. These granules are mostly lysosomes.

Leukocytes are classified according to the shapes of their nuclei and the presence or absence of cytoplasmic granules. Leukocytes containing large granules are called **granular leukocytes,** or **granulocytes.** Granulocytes also have a nucleus consisting of several lobes, or segments, and therefore are called **polymorphonuclear leukocytes.** The types of granulocytes, named according to the staining properties of their granules, are the **neutrophil,** the **eosinophil,** and the **basophil.**

Leukocytes without a granular cytoplasm are called **agranulocytes.** They have a simple nucleus without lobes and thus are designated as **monomorphonuclear leukocytes.** The two basic types of agranulocytes are the **monocyte** and the **lymphocyte.**

On average, whole blood contains 7000 to 10,000 leukocytes/μL; this figure represents the **combined white blood cell (WBC) count.** Because leukocytes have different functions, however, it is diagnostically useful to count the white blood cells according to type; this count is called the **differential WBC count.** When a differential count is performed, the number of each type of leukocyte is usually expressed as a percentage of the total or combined WBC count. A neutrophil count of 65%, for example, means that neutrophils make up 65% of all leukocytes present in a blood sample. Normal differential counts for the leukocytes are given in Figure 17–14.

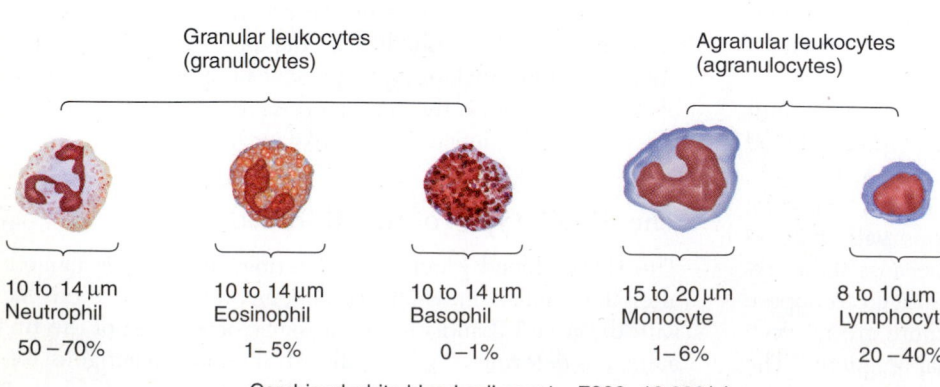

Granular leukocytes (granulocytes) | Agranular leukocytes (agranulocytes)

10 to 14 μm	10 to 14 μm	10 to 14 μm	15 to 20 μm	8 to 10 μm
Neutrophil	Eosinophil	Basophil	Monocyte	Lymphocyte
50–70%	1–5%	0–1%	1–6%	20–40%

Combined white blood cell count = 7000–10,000/μl

Figure 17–14

Types of leukocytes.

Leukocytes function collectively to combat foreign substances that enter the body. They are part of a system of defensive cells that **phagocytize** (engulf and destroy) material, detoxify poisons, produce antibodies, and release chemical messengers, enzymes, and other substances. Each type of leukocyte performs different functions in body defense, but each function is necessary for an integrated and effective defense.

Leukocytes are not confined to the blood or lymph; they also may be found in other tissues, particularly in loose connective tissue. They are motile cells, traveling via **ameboid movement** (movement similar to that of an amoeba, in which the cell projects protoplasmic extensions ahead and then follows them). They are capable of passing through the capillary wall in a process called **diapedesis** to enter the spaces around the blood vessels whenever necessary. Leukocytes are attracted to sites of injury, inflammation, or bacterial invasion when chemical substances are liberated by damaged cells or bacteria and when immune complexes form during immune responses. The attraction and subsequent movement of leukocytes toward these chemical substances is known as **chemotaxis** (Fig. 17–15).

Granulocytes: Granular Leukocytes

Neutrophils

The neutrophil functions as the body's first line of defense against invasion by **pyogenic** (pus-forming) **bacteria.** Neutrophils are the most numerous white blood cells. They are very motile and actively phagocytic. They can ingest and destroy several kinds of microbes, small particulate matter, and fibrin from blood clots. Most granules in the cytoplasm of neutrophils are lysosomes containing **hydrolytic enzymes,** which digest ingested matter. Other cytoplasmic granules contain **phagocytins,** a group of antibacterial proteins. Neutrophils are usually the first white cells to arrive at the site of a bacterial invasion, and they arrive in large numbers. In the process of defending the body against microbial invasion, many neutrophils die (Fig. 17–16), forming what is called **pus** at the invasion site. Dying neutrophils also release digestive enzymes into surrounding connective tissue. To a limited extent, these enzymes damage tissue and contribute to localized pain and swelling at a site of inflammation. After the invading or traumatizing factor has been controlled, many neutrophils remain to assist in healing.

Eosinophils

The eosinophil is less motile than the neutrophil and, although capable of phagocytosis, it is not as active in this regard. The eosinophil does not, for example, normally phagocytize bacteria. The lysosomes of the eosinophils contain many enzymes, such as **oxidase, peroxidases,** and **phosphatases,** the presence of which indicates that the primary function of eosinophils is detoxification of foreign proteins and other substances. As do neutrophils, eosinophils exhibit chemotaxis; however, their attraction to traumatized tissues appears to depend on the presence of antibodies specific to foreign proteins. Antigen–antibody complexes have been shown to cause eosinophils to travel from the

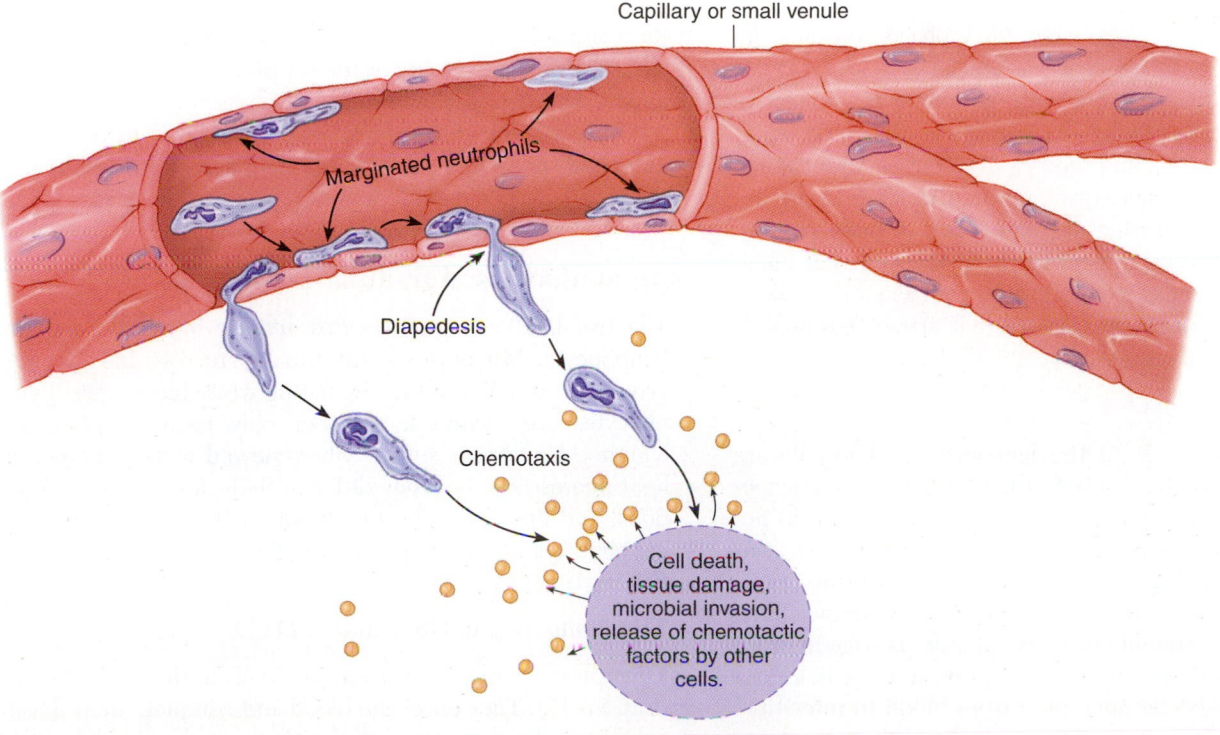

Capillary or small venule

Marginated neutrophils

Diapedesis

Chemotaxis

Cell death, tissue damage, microbial invasion, release of chemotactic factors by other cells.

Figure 17–15

Leukocyte diapedesis and chemotaxis.

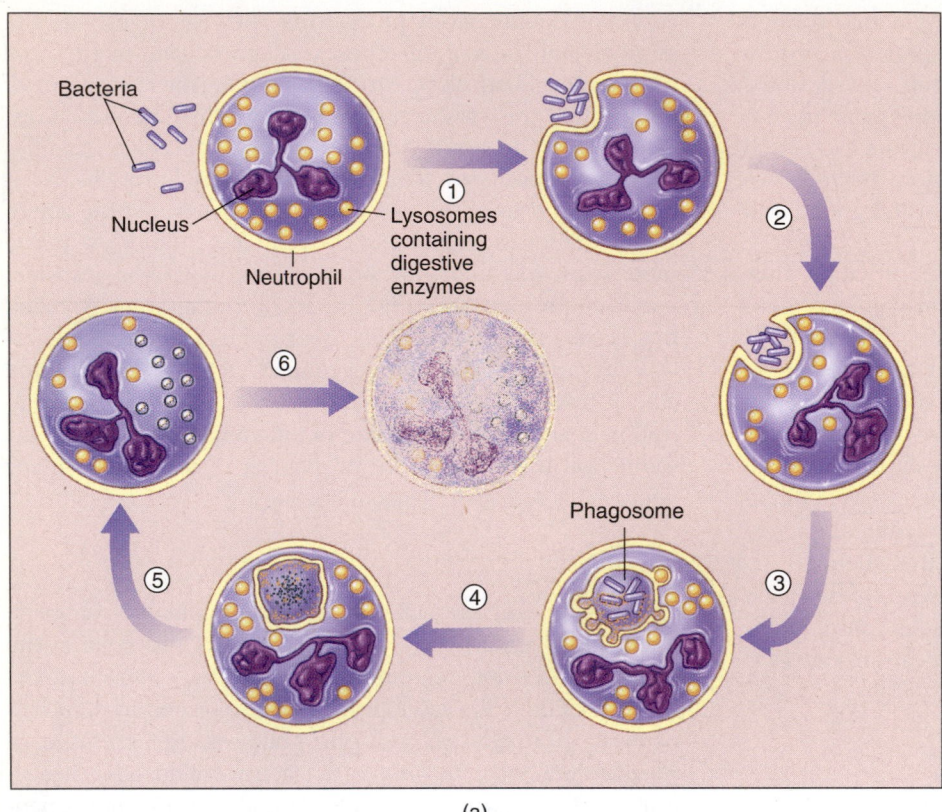

1. Neutrophil encounters and engulfs bacteria.

2. Phagosome forms around bacteria.

3. Degranulation of lysosomes to form digestive vacuole.

4. Bacterial lysis by digestive enzymes.

5. Dispersement of phagosome in cytosol.

6. Neutrophil lysis.

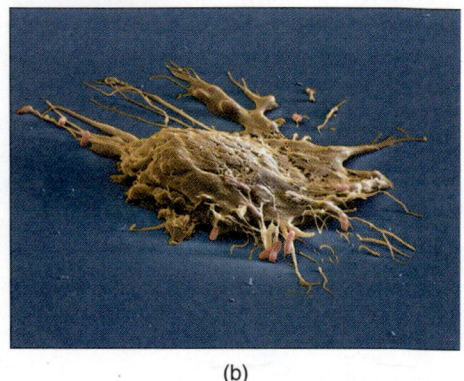

(a)   (b)

Figure 17–16

(a) The processes of phagocytosis, bacterial lysis, and neutrophil death.
(b) Scanning electron micrograph of an angry macrophage engulfing a yeast cell.
(© *Biology Media/Science Source/Photo Researchers, Inc.*)

blood into connective tissues, where they phagocytize and destroy the complexes.

Normally, eosinophils are more abundant in connective tissues other than the blood—in particular, the lung, mammary glands, omentum, and inner wall of the small intestine. They are observed to increase in number in allergic and autoimmune reactions, during decomposition of body protein, and in certain parasitic infections. It is believed that blood basophils and mast cells (connective tissue basophils) release **eosinophilic chemotactic factors** that attract eosinophils to an area of tissue damage.

Basophils

Basophils are the rarest of the leukocytes and usually are found in greater numbers outside the blood, in loose connective tissue. They show little ameboid movement and do not contain lysosomes. Many of their cytoplasmic granules contain either **heparin**, an anticoagulant, or **histamine**, a vasodilator substance that also increases capillary permeability. Basophils release histamine in areas of tissue damage in order to increase blood flow, attract neutrophils, and facilitate the movement of leukocyte emigration from blood to interstitial fluids in the area of damage. The effects of histamine release have been experienced by everyone who has suffered from the common cold: invasion of the nasal tissue by viruses induces basophils to gather at the invasion site and release histamine, which causes swelling and congestion of the tissue. Antihistamines and vasoconstrictors contained in decongestant nasal sprays may provide relief but actually interfere with the body's normal defense and may prolong infection.

Agranulocytes: Agranular Leukocytes

The two kinds of agranulocytes are the monocyte and the lymphocyte. Monocytes constitute 1% to 6%, and lymphocytes, 20% to 40%, of the circulating white blood cells. Lymphocytes are second in number only to neutrophils. All lymphocytes appear similar when viewed with a compound light microscope, but they differ in their development, functions, and life span. The two functionally distinct classes of lymphocytes are **T-lymphocytes (CD8$^+$)** and **(CD4$^+$),** and **B-lymphocytes.**

T-Lymphocytes (CD8$^+$) and (CD4$^+$)

T-lymphocytes originate from stem cells in the bone marrow (Fig. 17–17). They enter the blood and complete their development as they pass through the **thymus gland,** a lymphoid organ in the thorax. T-lymphocytes are the most numerous

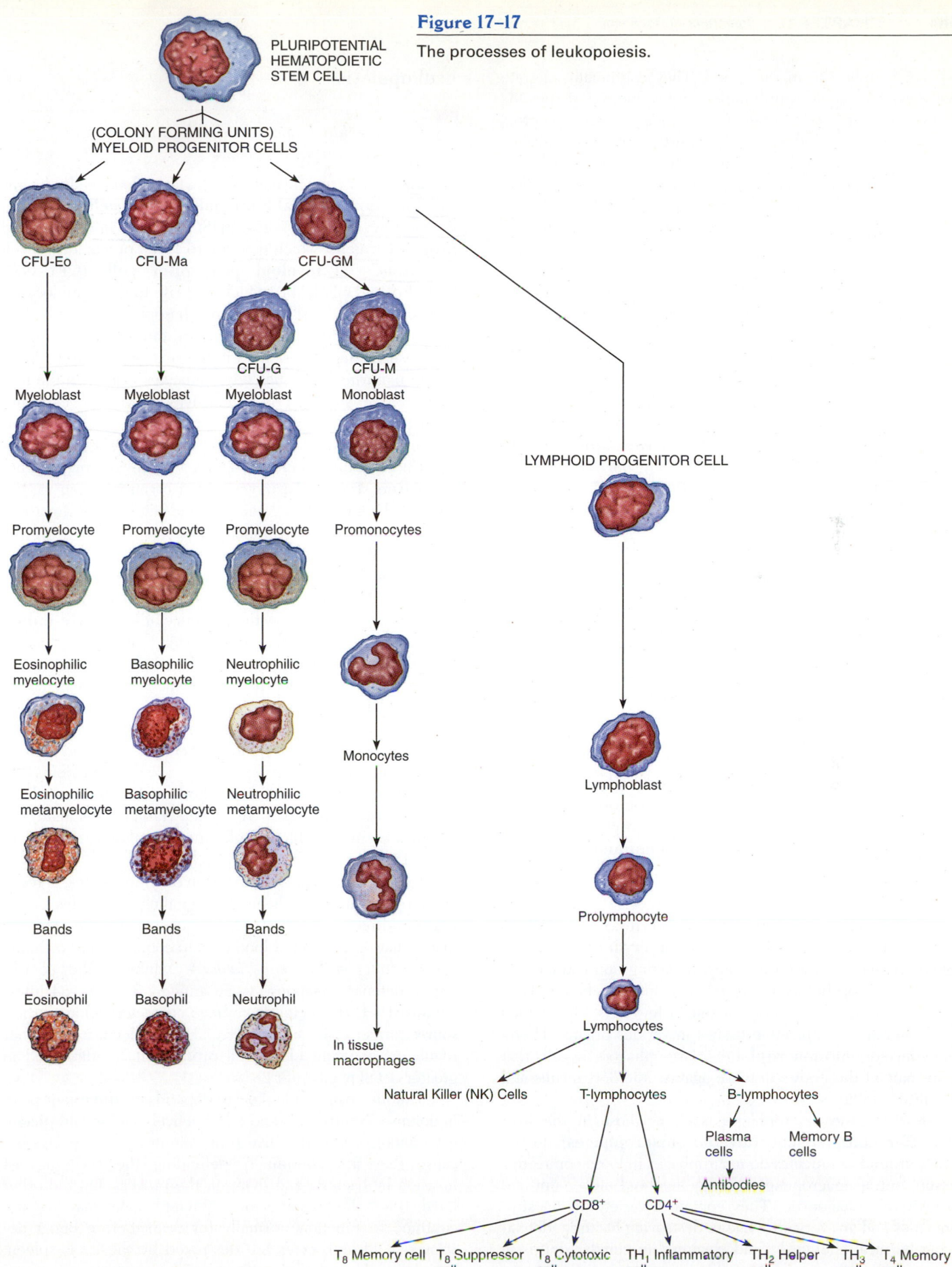

PLURIPOTENTIAL HEMATOPOIETIC STEM CELL

(COLONY FORMING UNITS) MYELOID PROGENITOR CELLS

CFU-Eo   CFU-Ma   CFU-GM

CFU-G   CFU-M

Myeloblast   Myeloblast   Myeloblast   Monoblast

Promyelocyte   Promyelocyte   Promyelocyte   Promonocytes

Eosinophilic myelocyte   Basophilic myelocyte   Neutrophilic myelocyte   Monocytes

Eosinophilic metamyelocyte   Basophilic metamyelocyte   Neutrophilic metamyelocyte

Bands   Bands   Bands

Eosinophil   Basophil   Neutrophil   In tissue macrophages

Figure 17–17

The processes of leukopoiesis.

LYMPHOID PROGENITOR CELL

Lymphoblast

Prolymphocyte

Lymphocytes

Natural Killer (NK) Cells   T-lymphocytes   B-lymphocytes

Plasma cells   Memory B cells

Antibodies

CD8$^+$ CD4$^+$

T$_8$ Memory cell   T$_8$ Suppressor cell   T$_8$ Cytotoxic cell   TH$_1$ Inflammatory cell   TH$_2$ Helper cell   TH$_3$ cell   T$_4$ Memory cell

type of lymphocyte in the blood. They continually migrate between the spleen, lymph nodes, and connective tissues. As T-lymphocytes pass through the thymus, surface receptors designated CD4 and CD8 are inserted into their plasma membrane, allowing the T-lymphocytes to recognize peptide antigens. T-lymphocytes bearing the CD4 receptors are designated as CD4$^+$. T-lymphocytes bearing the CD8 receptors are designated as CD8$^+$. The roles of CD4$^+$ and CD8$^+$ are discussed in greater detail in Chapter 28.

These lymphocytes are said to be "preconditioned," or "competent." Competent T-lymphocytes respond to antigens either by attacking them directly or by releasing chemicals called **lymphokines,** which attract granulocytes to the area and stimulate B-lymphocytes and other T-lymphocytes. Activated T-lymphocytes also may release nonspecific toxins that neutralize the antigens. T-lymphocytes become activated when antigens contact them and bind to specific membrane receptors. Activated T-lymphocytes give rise, through cell division, to clones of lymphocytes that are responsive to the activating antigen, thus conveying **cell-mediated immunity.** T-lymphocytes also play important roles in helping B-lymphocytes function normally.

B-lymphocytes

B-lymphocytes develop in the bone marrow and migrate by way of the blood to lymphoid tissues in the spleen, tonsils, lymph nodes, and walls of the small intestine, where they remain for variable periods of time before recirculating in the blood and lymph. When stimulated by an antigen, some B-lymphocytes become specialized to make and secrete **antibodies,** proteins that react with the antigens and help destroy them. Other B-lymphocytes become specialized to recognize the antigen should it re-enter the body at a later date; these B-lymphocytes are called **memory cells.** B-lymphocytes can live for many years and are the basis of the body's humoral or **antibody-mediated immunity.**

Monocytes

Monocytes are the largest of the leukocytes, some of them being nearly three times the size of an erythrocyte. Monocytes originate in the bone marrow, taking approximately 3 days to develop before being released into the blood. They remain in circulating blood for only a few days, after which they migrate to connective tissues in various organs. There, they develop into **macrophages,** large phagocytic cells that form part of the body's defense against microorganisms and harmful chemicals.

Some tissue macrophages, such as those in the liver **(Kupffer cells)** or lung **(alveolar macrophages)** do not move around as much as do macrophages in loose connective tissue, but all macrophages display ameboid movement and are actively phagocytic. They contain a variety of digestive enzymes and are very effective in destroying bacteria, detoxifying harmful chemicals, and cleaning up cellular debris in areas of tissue damage.

Leukopoiesis

 How do white blood cells develop?

Leukopoiesis, the formation of white blood cells, begins with the **pluripotential hemopoietic stem cell (PHSC)** in the marrow (see Fig. 17–17). PHSCs develop into **lymphoid progenitor cells,** which give rise to T-cell precursors, B-cell precursors, and **myeloid progenitor cells** (CFU-GM, CFU-Eo, and CFU-Ma), which give rise to the granulocytes, monocytes, and mast cells (tissue basophils). Granulocytes and monocytes develop to maturity in the marrow. Lymphocytes usually begin their lives in the marrow, but most leave before maturing. Potential T-lymphocytes mature in the thymus and the spleen. Potential B-lymphocytes develop and mature in lymphoid tissue of the intestines, the spleen, and the bone marrow.

Uncommitted stem cells can leave the marrow and become **free stem cells,** moving from organ to organ via the blood and lymph. Thus, hematopoietic centers in the marrow may serve to maintain lymphoid tissue via "seeding" with free stem cells. In turn, lymphoid organs may become sites of blood cell formation when the marrow fails to function normally.

Granulopoiesis (formation of granulocytes) has been shown to be stimulated by the presence of **leukocyte-inducing factor,** a specific factor in plasma that can be demonstrated when leukocytes are depleted. Various other chemicals released into the blood when tissue is damaged also stimulate leukopoiesis.

Granulocyte, monocyte, and megakaryocyte formation has been shown to be regulated by chemicals called colony-stimulating factors (CSFs), which are glycoproteins secreted by a variety of cell types both within and outside of the bone marrow. Six CSFs, their target cells, and the colonies of mature blood cells they regulate are listed in Table 17–6.

Lymphopoiesis, the formation of lymphocytes, is regulated by an array of diffusible signaling molecules called **lymphokines** or **interleukins,** chemicals secreted by macrophages and white blood cells to communicate with one another (*inter* = "between," *leuko* = "white") as they coordinate a defensive response to an antigen. To date, 24 interleukins (IL-1, IL-2, etc.) have been characterized as to their source, target cells, and effects. These molecules and other regulators of white blood cell production are discussed in greater detail in Chapter 28.

The life span of a leukocyte depends on the role it plays in defense. Neutrophils and other actively motile and phagocytic leukocytes tend to live from minutes to a few days because they are continually defending the body against invasion by bacteria and dying in the process. On the other hand, B-lymphocytes and some T-lymphocytes may live and continue to function normally for years before being destroyed. Old leukocytes are destroyed by the liver, spleen, marrow, and lymph nodes.

TABLE 17–6

Colony Stimulating Factors

Factor	Principal Target Cell	Regulated Colony
Monocyte colony stimulating factor (M-CSF)	Colony forming unit-monocyte (CFU-M)	Monocyte
Granulocyte colony stimulating factor (G-CSF)	Colony forming unit-granulocyte (CFU-G)	Neutrophil
Granulocyte–monocyte colony stimulating factor (GM-CSF)	Colony forming unit-granulocyte–monocyte (CFU-GM)	Neutrophil, monocyte
Multicolony stimulating factor (Multi-CSF) (IL-3)	Colony forming unit-granulocyte, eosinophil, monocyte, megakaryocyte (CFU-GEMMe)	Neutrophil, eosinophil, monocyte, megakarocyte
Megakaryocyte colony stimulating factor (Meg-CSF)	Colony forming unit-megakaryocyte (CFU-Me)	Megakaryocyte
Erythropoietin	Colony forming unit-erythrocyte (CFU-E)	Erythrocyte

Leukocytosis

Leukocytosis is an abnormal increase in the total number of circulating leukocytes—sometimes as many as 400,000 cells/μL. Leukocytosis may have physiological or pathological causes. All combined white blood cell counts above 50,000 cells/μL indicate the latter. **Physiological leukocytosis** (WBC count, 10,000–20,000 cells/μL) is nonpathological and occurs in newborn infants and in pregnancy, emotional disturbances, menstruation, and intense exercise. **Pathological leukocytosis** occurs in bacterial and viral infections, metabolic and hormonal disturbances, allergies, and malignancies.

Leukemia is a malignant disease characterized by the proliferation of hematopoietic cells that are immature and therefore functionally impaired. Environmental factors (e.g., radiation or benzene), genetic factors, and RNA viruses have been implicated. Leukemias are classified by the course and duration of the illness and the abnormal type of cells and tissue involved. Acute leukemias, such as acute lymphocytic leukemia and acute myelogenous leukemia, are characterized by rapid onset, massive numbers of immature leukocytes, severe anemia, and other signs. Chronic leukemias, such as chronic granulocytic leukemia and chronic lymphocytic leukemia, are characterized by gradual onset and increased numbers of mature leukocytes. Chronic leukemias may lead to acute leukemias. Although symptoms and complications vary with the type of leukemia, they generally include anemia, excessive bleeding due to impairment of thrombocytes, and local and systemic infection due to impaired granulocyte function.

Leukopenia

Leukopenia is an abnormal reduction in the total number of circulating leukocytes. It may result from bone marrow defects, arrest of cell development and maturation, or excessive destruction of leukocytes. **Agranulocytosis** is a type of leukopenia characterized by a marked reduction in neutrophils. Often, it is caused by slowed functioning of the bone marrow due to irradiation or cytotoxic drugs. As a result of agranulocytosis, the body's defenses against bacterial invasion are considerably weakened, and the risk for dying from a bacterial infection is greatly increased.

THROMBOCYTES (PLATELETS)

 What are platelets; what do they do?

A **platelet** or **thrombocyte** (*thrombus* = "clot") is a cell fragment split from a large cell called a **megakaryocyte** (Fig. 17–18). One megakaryocyte gives rise to approximately 6000 thrombocytes. Because thrombocytes are not actually cells produced by a mitotic division of a parent cell, the term **platelet** is more commonly used.

Megakaryocytes derive from pluripotential hemopoietic stem cells in the bone marrow. PHSCs become committed progenitor cells called **megakaryocyte colony-forming units** (CFU-Me), which in turn develop into colonies of mature megakaryocytes (see Fig. 17–18). The development of megakaryocytes (**megakaryocytopoiesis**) is regulated by several interleukins (IL-3, IL-6, and IL-11) and thrombopoietin, a hormone that stimulates both megakaryocyte development and the release of platelets from megakaryocytes.

Most megakaryocytes remain in the marrow, releasing platelets into the circulating blood. Some megakaryocytes enter the blood and travel to other organs (particularly the lung), where they remain and produce platelets.

Platelets are small, membrane-bounded bodies without nuclei, approximately 2 to 4 μm in diameter. They are packed with granules, vesicles, microfilaments, microtubules, and occasionally mitochondria. They are normally present

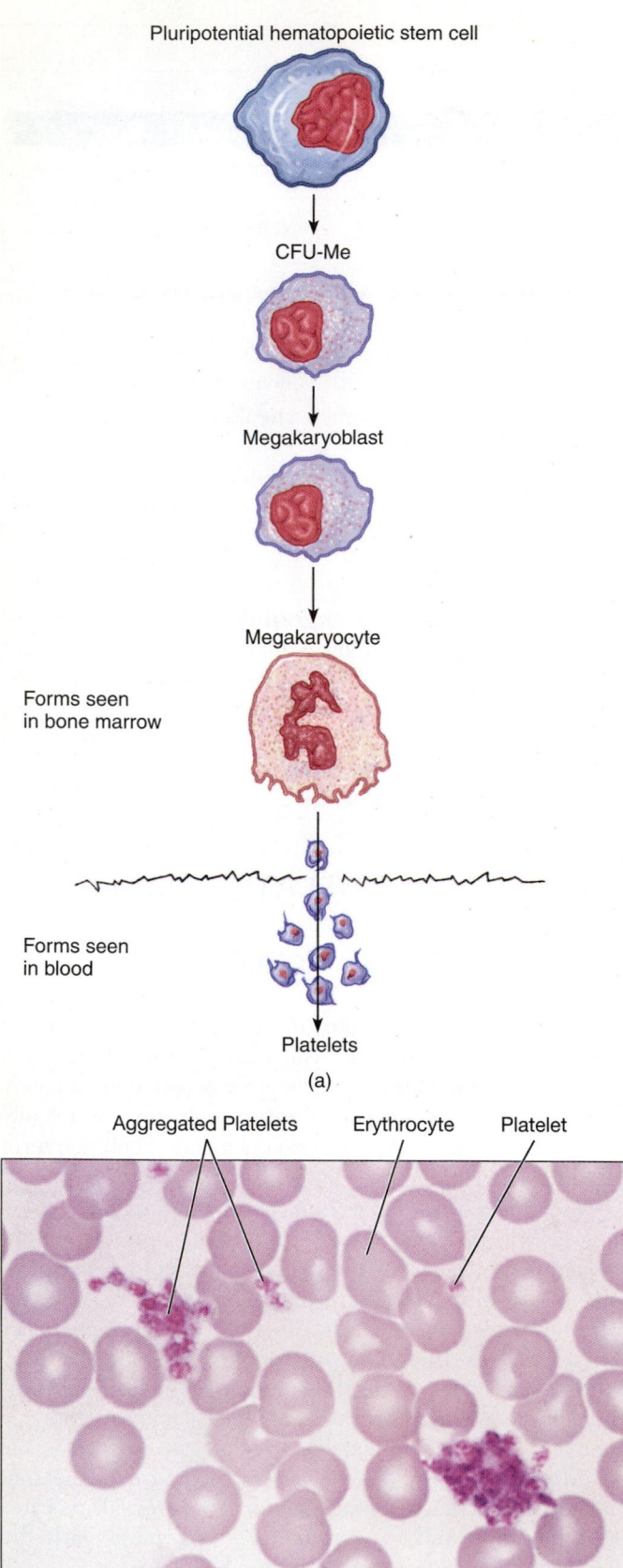

Pluripotential hematopoietic stem cell

CFU-Me

Megakaryoblast

Megakaryocyte

Forms seen in bone marrow

Forms seen in blood

Platelets

(a)

Aggregated Platelets   Erythrocyte   Platelet

(b)

Figure 17–18

(a) The development of platelets. *(b)* Photomicrograph of thrombocytes (platelets). *(© Manfred Kage/Peter Arnold, Inc.)*

in numbers ranging from 150,000 to 350,000 cells/μL of blood and play several important roles in hemostasis. Following injury, chemicals contained within or absorbed in platelets are released to the lining of the blood vessels, where they stimulate contraction of the injured vessels to minimize blood loss. Because of their adhesive properties, they clump together at the site of vascular damage, effectively plugging the defective end. In addition, they participate in the formation of factors that initiate coagulation of the blood, and they release **platelet-derived growth factor (PDGF),** a protein that promotes repair after vascular injury. The role of platelets in hemostasis will be discussed later.

Circulating platelets have a life span of approximately 1 to 2 weeks. If not consumed in the process of blood coagulation, platelets eventually are destroyed by macrophages in the liver and spleen. The spleen is also an important storage organ for platelets. In certain kinds of stress, such as a hemorrhage or burns, sympathetic stimulation of the spleen results in the release of large numbers of stored platelets into circulating blood.

HEMOSTASIS: MECHANISMS TO PREVENT BLOOD LOSS

 How do blood clots form?

The term **hemostasis** (*hemo* = "blood"; *sta* = "remain") refers to mechanisms that minimize or prevent the loss of blood when a blood vessel is opened. Hemostasis is vital because unchecked hemorrhage (blood loss) eventually leads to cardiovascular collapse and death. Four interrelated events constitute hemostasis: (1) local vasoconstriction, (2) formation of a platelet aggregate (clump), (3) formation of a blood clot, and (4) clot retraction and dissolution.

Local Vasoconstriction

When a blood vessel is injured, its immediate response is to constrict and thereby reduce blood flow. This initial vasoconstriction is due to local spasm of the smooth muscle in the wall of the blood vessel and to sympathetic reflexes. In smaller vessels, vasoconstriction can be maintained by the release of vasoconstrictor chemicals from platelets that begin to accumulate at the damaged site.

Formation of a Platelet Aggregate

Damaged cells of the injured blood vessel release **adenosine diphosphate (ADP),** which attracts platelets and causes them to clump together at the damaged site. Platelets coming in contact with exposed collagen of the vascular wall **degranulate** (release stored chemicals), releasing ADP, **serotonin** (a vasoconstrictor), and **platelet**

factors necessary for coagulation of blood. The release of ADP attracts more platelets and causes them to swell and become sticky; they adhere in increasing numbers to the damaged site and form a plug called a **platelet aggregate.** Activated platelets also produce **thromboxane A$_2$,** a powerful vasoconstrictor and platelet aggregator derived from platelet **prostaglandin H$_2$.** In addition, the spherical platelets extend **pseudopodia** (footlike extensions of their cytoplasm) to nearby exposed collagen, anchoring the platelet plug and establishing a framework on which coagulation can proceed.

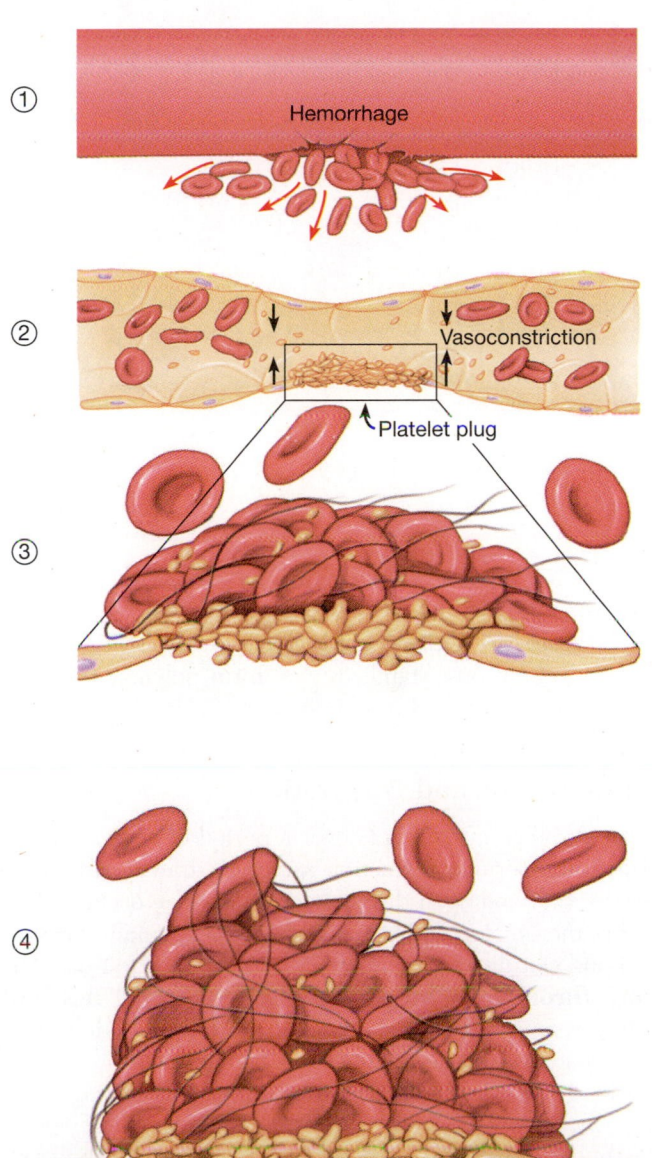

Figure 17–19

Formation of a clot. Fibrin forms long threads in which blood cells, platelets, and plasma become trapped.

Formation of a Blood Clot

Coagulation is the process by which some of the blood loses its fluid consistency and becomes a clot (a semisolid mass similar in consistency to gelatin). In the formation of a clot, an enzyme called **thrombin** converts **fibrinogen,** a soluble plasma protein, into an insoluble plasma protein, **fibrin.** Fibrin is a threadlike protein. Fibrin aggregates to form a mesh-like network at the site of vascular damage, trap red blood cells and plasma, and form a clot (Fig. 17–19). The complex sequence of chemical events that produce fibrin are divided into three phases or stages, designated as **stage I, stage II,** and **stage III** (Fig. 17–20). Common synonyms for blood clotting factors involved in each stage are given in Table 17–7.

Stage I: Formation of a Prothrombin Converting Factor

Stage I may begin when blood comes in contact with injured tissue (the **extrinsic pathway**), or it may be initiated in the absence of tissue damage (the **intrinsic pathway**).

When blood comes in contact with injured tissue, clotting is initiated via the extrinsic pathway, owing to the release of the **tissue lipoprotein thromboplastin** (factor III). Tissue thromboplastin interacts with a plasma protein, **proconvertin** (factor VII), and **calcium ions** to form an agent that activates the **Stuart factor** (factor X). The activated Stuart factor, in the presence of **calcium ions,** forms complexes with **accelerin** (factor V) on phospholipid micelles provided by tissue thromboplastin to form the **prothrombin-converting factor.**

Intrinsic coagulation may occur inside or outside of the body. In either case, the first step is the activation of the **Hageman factor** (factor XII). In the body, activation of the Hageman factor may occur from collagen, fibrin, or platelet membrane during platelet aggregation. In addition, it can apparently be activated under conditions of stress, anxiety, fear, and other states. Outside the body (e.g., in a test tube), activation of the Hageman factor occurs when blood comes in contact with foreign substances whose common property appears to be a negative surface charge.

The Hageman factor activates the plasma enzyme, **plasma thromboplastin antecedent (PTA;** factor XI), which, in the presence of calcium ions, activates a plasma protein, the **Christmas factor** (factor IX). Activated Christmas factor interacts with another protein, **antihemophilic factor** (factor VIII), on the surface of phospholipids and in the presence of calcium, to form a complex that activates the Stuart factor. The succeeding steps in the formation of prothrombin-converting factor are the same as for the extrinsic mechanism.

Stage II: Conversion of Prothrombin to Thrombin

Prothrombin is a plasma globulin manufactured by the liver and normally present in circulating plasma. It is the inactive precursor of an active enzyme called **thrombin.**

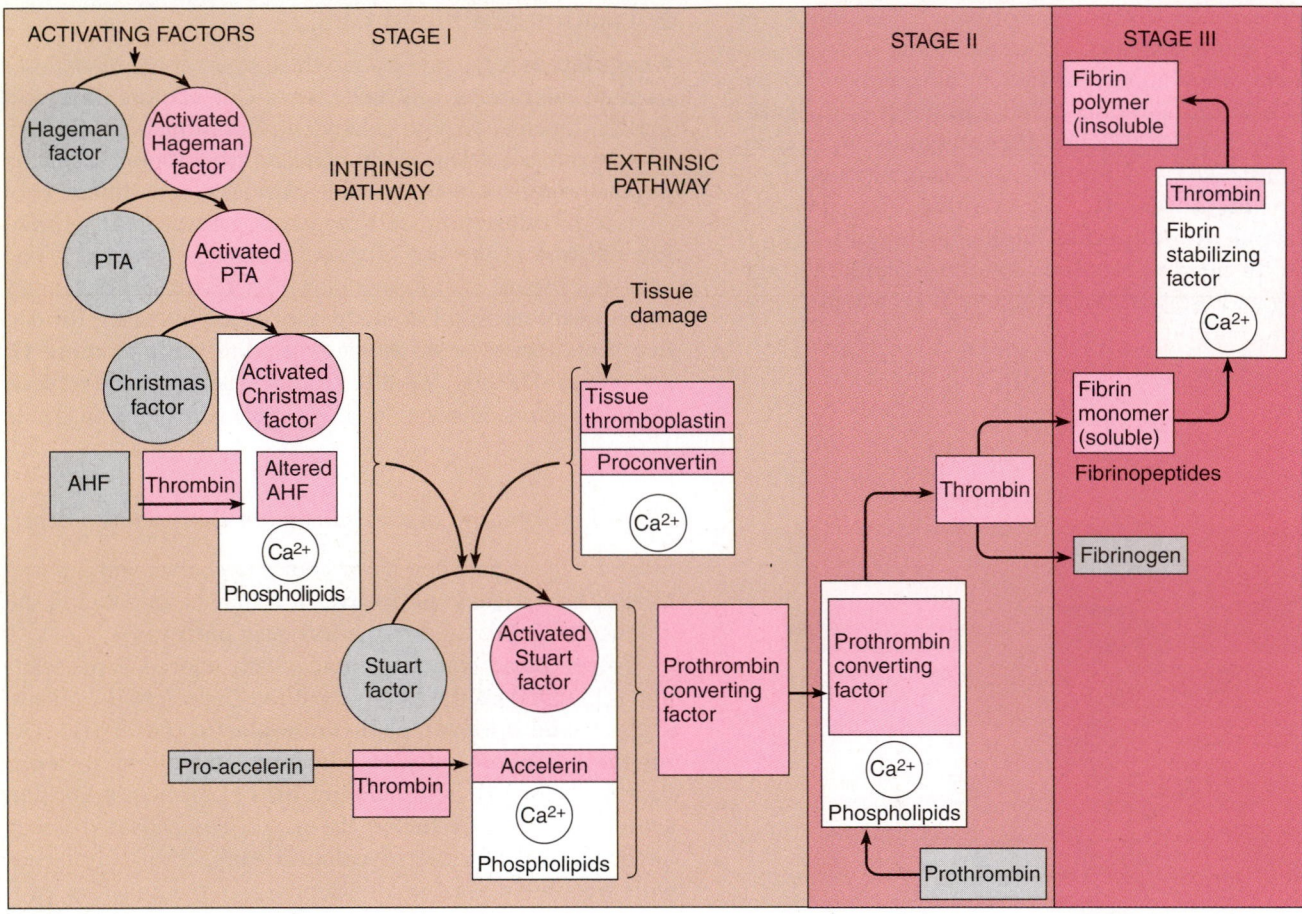

Figure 17–20

A schema of blood coagulation.

Thrombin is not normally present in plasma unless blood is clotting. In the presence of calcium ions and the prothrombin-converting factor, prothrombin is enzymatically split into two fragments—one inert and the other possessing the properties of thrombin. Initially, the conversion of prothrombin proceeds too slowly to produce significant amounts of thrombin needed for coagulation. Thrombin itself, however, increases its own rate of formation by converting an unstable plasma protein, **proaccelerin** (factor V), into **accelerin,** which then accelerates the formation of thrombin. Thrombin also activates the antihemophilic factor and is needed to activate a fibrin-stabilizing factor in stage III.

Stage III: Conversion of Fibrinogen to Fibrin

Fibrinogen is a soluble plasma protein produced by the liver and normally circulating in the plasma. Thrombin converts fibrinogen to **fibrin** by cleaving two pairs of small polypeptides (fibrinopeptides) from each fibrin molecule, leaving a **fibrin monomer** (a single link in a chain of molecules that make up fibrin). Fibrin monomers spontaneously polymerize, forming the insoluble, threadlike protein, fibrin.

In addition, thrombin activates another plasma enzyme, **fibrin-stabilizing factor** (factor XIII), which, in the presence of calcium ions, stabilizes the fibrin polymer through covalent bonding of the fibrin monomers.

Clot Retraction and Dissolution

After a clot forms, it retracts over a period of several hours and extrudes serum. Retraction serves to draw wound surfaces together and open the vessel if it has been occluded by the clot, thereby increasing blood flow and promoting tissue repair and clot dissolution. Retraction is caused by a platelet factor, **thrombosthenin,** a contractile protein that, to shorten, requires the presence of thrombin and adenosine triphosphate (ATP).

Blood clots are not permanent. After hemorrhage has been checked and tissue repair is well underway, the clot is gradually dissolved by breaking down fibrin (fibrinolysis) into soluble fragments by the enzyme **plasmin.** Plasmin is derived from **plasminogen,** an inactive β-globulin normally present in plasma. The formation of plasmin from plasminogen requires an enzyme called **activator.** Activa-

TABLE 17–7

Clotting Factors

International Committee Designation	Synonyms	Location
Factor 1	Fibrinogen	Plasma
Factor 2	Prothrombin	Plasma
Factor 3	Tissue thromboplastin	Tissue cells
Factor 4	Calcium ion	Plasma
Factor 5	Proaccelerin Prothrombin accelerator Accelerator globulin Labile factor	Plasma
Factor 6	Obsolete	
Factor 7	Serum prothrombin conversion accelerator (SPCA) Proconvertin Autoprothrombin 1 Stable factor	Plasma
Factor 8	Antihemophilic factor (AHF) Platelet cofactor 1 Thromboplastinogen Antihemophilic factor A	Plasma
Factor 9	Plasma thromboplastin component (PTC) Christmas factor Platelet cofactor 2 Antihemophilic factor B Autoprothrombin 2	Plasma
Factor 10	Stuart-Prower factor Stuart factor Autoprothrombin 3	Plasma
Factor 11	Plasma thromboplastin antecedent (PTA)	Plasma
Factor 12	Hageman factor Contact factor	Plasma
Factor 13	Fibrin-stabilizing factor Plasma transglutaminase Laki-Lorand factor	Plasma
Platelet factor	Platelet factor 3	Platelets

tor is normally absent from plasma, but its precursor, **proactivator**, is present and may be converted into activator by **cytofibrokinase** (a tissue enzyme); **staphylokinase** and **streptokinase** (bacterial enzymes); **plasma kinase** (Hageman factor); and other enzymes in the body excreted in urine (**urokinase**), tears, and saliva. **Tissue plasminogen activator (TPA)** formed by endothelial cells appears to be responsible for most of the physiological **fibrinolysis.** TPA has been synthesized by recombinant DNA techniques and is being used to treat coronary thrombosis (see Applications of Physiology).

Normally, a balance between deposition of fibrin and fibrinolysis limits coagulation to the area of vascular injury. Ad-

ditional limits on excessive clotting include plasma enzymes, which inactivate each clotting factor shortly after each is activated, and anticoagulants.

Anticoagulants

 Why don't blood clots form spontaneously in blood vessels?

An **anticoagulant** is a chemical that blocks or inhibits the coagulation process. Some anticoagulants normally are produced in the body and are referred to as **physiological**

APPLICATIONS OF PHYSIOLOGY

Clot Busters

Heart muscle (myocardium) contracts and relaxes more than 100,000 times per day and requires an abundant blood supply to meet metabolic requirements. Blood is supplied to heart muscle by way of branches of the left and right coronary arteries. Partial or complete blockage of a coronary artery interrupts the normal blood supply to the heart muscle, producing ischemia (a below-normal reduction in the blood supply of a tissue or organ), which may lead to death (infarction) of the affected muscle cells. Ischemic heart disease, commonly called *coronary heart disease* (CHD), frequently manifests with chest pain (angina pectoris) that worsens with exertion, difficulty in breathing, sweating, nausea, vomiting, pallor, restlessness, cold extremities, and a slowing of the heart rate with a fall in blood pressure. These manifestations characterize myocardial infarction, commonly referred to as a "heart attack." If the infarct is large, the heart may stop, resulting in sudden death.

Typically the underlying cause of myocardial infarction is a coronary blood vessel narrowed by atherosclerosis and occluded by a thrombus (clot). Damage to the heart muscle distal to the occlusion begins within minutes, but the muscle deteriorates gradually instead of dying immediately; therefore, damage can be minimized if blood flow can be restored quickly (ideally within 1 hour after symptoms of a heart attack first appear).

One way to restore blood flow is by means of angioplasty, a surgical procedure in which a balloon is introduced by catheter into the blocked coronary artery and inflated at the site of obstruction, flattening the obstruction against the arterial wall (see Chapter 19). Another way to improve coronary blood flow is to surgically bypass the blockage by grafting a vein around it.

A nonsurgical approach to the problem of quickly restoring blood flow through a vessel blocked by a thrombus uses a thrombolytic agent, a "clot buster," to dissolve the clot. One of the most recent thrombolytic agents is recombinant tissue plasminogen activator (TPA), synthesized using recombinant DNA technology. Although effective, it is very expensive to use (more than $2000 per dose). Another older, but effective and less costly (approximately $200 to $300 per dose), thrombolytic agent is streptokinase, manufactured using more conventional technology.

Heart-attack patients who are candidates for thrombolytic therapy should be treated with "clot busters" within 60 minutes of the onset of symptoms and within 30 minutes of arriving at emergency rooms in order to derive maximum benefit from the drugs. Studies have shown that "clot busters" administered after 12 hours or more offer the patient little or no benefit. Clearly, when heart attack symptoms appear, time is of the essence.

Thrombolytic therapy, coronary angioplasty, and coronary bypass surgery, administered separately or in combination, are used to treat coronary heart disease after symptoms of ischemia appear.

---

anticoagulants. Others, called **therapeutic anticoagulants,** are manufactured and can be administered to block coagulation.

Unless there is trauma to the blood vessels or to the blood itself, coagulation does not normally occur. Although tissue breakdown and platelet destruction are normal occurrences in the absence of trauma, intravascular clotting does not usually occur because (1) the amounts of tissue and platelet procoagulants released are very small and (2) natural coagulation inhibitors, such as **antithromboplastin, antithrombin III,** and **heparin,** are present.

Antithrombin III inhibits activated Stuart factor and, to a lesser extent, inhibits other enzymes required for production of prothrombin-converting factor. Antithrombin III also inhibits thrombin, thereby blocking the conversion of fibrinogen to fibrin. Heparin, produced by basophils and mast cells

of connective tissues, acts in anticoagulation by combining with antithrombin III and greatly potentiating its effects. Acting alone, heparin is not a significant anticoagulant. Although heparin is produced in the body, it is normally absent from blood.

Therapeutic anticoagulants, on the basis of action, may be classified into three general categories: (1) agents that depress the action of procoagulants, (2) agents that remove calcium from the blood by precipitation or chelation (binding), and (3) agents that inhibit the synthesis of prothrombin in the liver.

In cases of excessive coagulation, heparin is used widely as a therapeutic anticoagulant. It inhibits the conversion of fibrinogen to fibrin and also inhibits the production of thrombin. Its effects can be countered by administering protamine sulfate.

Agents that remove calcium ions from the blood are used only to prevent coagulation in blood outside the body because removal of adequate amounts of calcium to prevent clotting would interfere with the normal role of calcium in the heart, skeletal muscle, and nerves. **Sodium citrate** binds and makes unavailable (chelates) calcium ions, whereas **sodium oxalate** precipitates calcium ions out of solution.

Liver cells synthesize four clotting factors: prothrombin, (factor II), factor VII, the Christmas factor (factor IX), and the Stuart factor (factor X). Vitamin K is required and acts as an enzyme cofactor during synthesis of the clotting factors. **Bishydroxycoumarin** (Dicumarol), an agent originally isolated from spoiled sweet clover, inhibits coagulation by interfering with the liver's use of vitamin K and, hence, clotting-factor production.

Abnormalities of Hemostasis

Thrombocytopenia is an abnormally low number of circulating platelets in blood. If the platelet count falls below 50,000 cells/μL, there is an increased risk for internal hemorrhage associated with accidental trauma or surgery. Platelet counts of approximately 20,000 cells/μL are associated with multiple small bruises **(purpura),** hemorrhagic spots **(petechiae)** in the skin, and sometimes spontaneous bleeding from mucosal surfaces. Potentially fatal hemorrhage in the intestines or brain can occur with platelet counts of less than 10,000 cells/μL.

Thrombocytopenia can result from decreased production of platelets by the bone marrow due to toxins, radiation, or infection. It also can result from sequestration of platelets, such as in congestive **splenomegaly,** or from increased destruction of platelets by autoimmune process, as in **idiopathic thrombocytopenic purpura.** Disorders characterized by increased coagulation of blood and thus increased platelet consumption, such as disseminated intravascular coagulation (discussed in the following paragraph), also produce thrombocytopenia.

Disseminated intravascular coagulation (DIC) involves widespread coagulation that produces thrombosis in small blood vessels, increased fibrinolysis, and depletion of coagulation factors, which in turn collectively result in generalized bleeding. Causes of DIC include bacterial infections that cause widespread endothelial damage; disseminated cancers that cause increased release of procoagulants; and complications of pregnancy, such as the release of amniotic fluid (procoagulant material) into the circulation.

Hemophilia is an inability of the blood to properly coagulate due to a lack of a coagulation factor. **Hemophilia A** involves a deficiency in factor VIII and **hemophilia B** (Christmas disease) involves a deficiency in factor IX. Hemophilia A or B are transmitted genetically and affect only males; females carry the gene but do not show symptoms. **Hemophilia C** is not sex-linked and results from a deficiency in factor XI. Approximately 85% of hemophilia is type A. Hemophilia is characterized by spontaneous or traumatic subcutaneous hemorrhage; blood in the urine; and bleeding in the mouth, lips, tongue, and joints.

Absence or lack of adequate amounts of other factors also interferes with normal coagulation. **Afibrinogenemia,** a lack of circulating fibrinogen, may be hereditary or may result from liver disease. **Hypoprothrombinemia,** a lack of prothrombin, may be congenital, a result of liver disease, or a complication of vitamin K deficiency. Vitamin K is required for liver manufacture of prothrombin. Newborn infants, for example, lack a store of vitamin K at birth. Vitamin K normally is synthesized by bacteria in the gastrointestinal tract, but infants are born with sterile gastrointestinal tracts. Therefore, hypoprothrombinemia may develop, and life-threatening hemorrhage may occur. The disorder is called **hemorrhagic disease of the newborn.**

Hemorrhagic disease also may occur because of an increase in heparin release into the blood **(hyperheparinemia).** Allergies, collagen diseases, and disorders of the liver or bone marrow are sometimes accompanied by hyperheparinemia and a reduction in the ability of blood to coagulate effectively.

CHAPTER REVIEW

Summary

- Blood transports molecules and heat from one area of the body to another, defends against foreign agents, and assists in the maintenance of extracellular fluid pH and osmolarity.
- Blood is a connective tissue containing erythrocytes, leukocytes, thrombocytes, and plasma.
- Average adult blood volume is 5 L.

- Plasma is the intercellular fluid of the blood. Plasma is approximately 93% water and 7% solutes (inorganic and organic).
- Plasma proteins are divided into the following fractions: albumin, globulins, and fibrinogen. Plasma proteins serve as amino acid reserves, carriers, and buffers and also participate in blood coagulation and the regulation of fluid distribution within the

body. The immune globulins play important roles in the body's defense.

- Hematopoiesis is the generation of blood cells from pluripotential hemopoietic stem cells (PHSCs).
- PHSCs develop into committed progenitor cells called **colony-forming units** (CFUs), which generate colonies of specific blood cell types.
- The erythrocyte is a biconcave, anucleated cell that numbers approximately 5 million cells/μL of whole blood.
- Erythrocytes contain hemoglobin, which transports oxygen. One gram of hemoglobin can transport 1.34 mL of O_2. The amount of hemoglobin in blood is approximately 15 g/dL; therefore, the oxygen-carrying capacity of blood is approximately 20 mL of O_2 per deciliter.
- Erythropoiesis is the formation of erythrocytes. Normally, approximately 230 billion red blood cells are produced each day. Substances required for erythropoiesis include vitamin B_{12}, folic acid, iron, and protein. Erythropoiesis is controlled by a renal hormone called **erythropoietin.**
- Approximately 230 billion red blood cells are destroyed each day. When an erythrocyte is destroyed, most parts of the hemoglobin are conserved. The remainder is excreted as bilirubin in bile.
- Anemia is an abnormal reduction in the oxygen-carrying capacity of the blood. Anemias are characterized by changes in one or more of the following red blood cell indices: MCV, MCH, and MCHC.
- Polycythemia is a relative or absolute increase in the number of circulating red blood cells above normal. Two main types are polycythemia vera and secondary polycythemia.
- The ABO blood group contains blood types based on the presence or absence of antigens A and B on the red blood cell membrane.
- Major agglutination occurs when donor red blood cells containing a strong antigen possessed by the recipient are infused into the recipient's body.
- Minor agglutination occurs when donor plasma containing antibodies against the recipient's red cell antigens is infused into the recipient's body.

- The Rh blood group contains blood types based on the presence or absence of antigen D. If antigen D is present on the red blood cell membrane, the blood type is Rh-positive. If antigen D is absent, the blood type is Rh-negative. The plasma of an Rh-negative person may or may not contain anti-D.
- Leukocytes are blood cells that do not contain hemoglobin. Two principal categories are granulocytes and agranulocytes. On the average, whole blood contains approximately 7000 to 10,000 leukocytes/μL. Their principal function is body defense.
- T-lymphocytes convey cell-mediated immunity, and B-lymphocytes convey antibody-mediated immunity.
- Leukopoiesis, the formation of leukocytes, occurs primarily in the bone marrow, although some leukocytes mature outside the marrow in lymphoid organs.
- Thrombocytes, or platelets, are produced from megakaryocytes. One megakaryocyte gives rise to approximately 6000 platelets.
- Platelets play several roles in hemostasis.
- Local vasoconstriction helps minimize blood loss when a blood vessel is opened.
- Platelets clump together, forming a plug at the site of vascular damage. Platelets also release procoagulants.
- Coagulation of blood, or formation of a blood clot, at the site of vascular injury helps seal the opened vessel. Coagulation involves the formation of fibrin thread, which is the foundation of the clot.
- Fibrin formation occurs in three states: stage I—formation of prothrombin-converting factor, stage II—conversion of prothrombin to thrombin, and stage III—conversion of fibrinogen to fibrin.
- After formation of a clot, clot retraction helps draw wound surfaces together.
- Clots gradually are dissolved by enzymes that break down fibrin.
- Abnormalities of hemostasis can result from platelet defects, coagulation defects, and/or excessive production of anticoagulants. Principal abnormalities include thrombocytopenia and hemophilia.

Review Questions

Choose the Correct Answer

1. In a 20-mL sample of whole blood, packed red cell volume is 9 mL. The hematocrit is therefore:
   a. 1190.
   b. 4.5 mL.
   c. 11 mL.
   d. 45%.
   e. 45 mL.
2. If hemoglobin content is 13 g/dL and the combining power of hemoglobin for oxygen is 1⅓ mL of oxygen per gram of hemoglobin, then the oxygen-carrying capacity of the blood:
   a. cannot be computed without more data.
   b. is approximately 20 vol. percent.
   c. is approximately 15 mL of O_2 per liter.

   d. is 17 to 18 mL of O_2 per deciliter of blood.
   e. is 17.4 g/mL.
3. Which of the following can safely be infused into a B− recipient without agglutination and without sensitizing the recipient to the Rh antigen?
   a. Plasma of whole blood type AB+
   b. Whole blood type B+
   c. Whole blood type A−
   d. Serum of whole blood type O+
   e. Two of the preceding
4. Upon death of the erythrocyte, the hemoglobin contained within the cell:
   a. is completely destroyed and excreted into the intestine.
   b. remains intact and is incorporated into new erythrocytes.

c. is broken down into heme and globin from which iron and amino acids are extracted and conserved.
d. remains intact and is excreted into bile.
e. is released into the plasma and returned to the bone marrow for reuse.

5. The fraction of the total blood volume occupied by erythrocytes is:
   a. elevated in anemia.
   b. normally approximately 12% to 16% in females.
   c. depressed in polycythemia.
   d. termed the *hematocrit*.
   e. higher in females than in males.

6. If a person's combined WBC count were 7×10^3 cells/μL, a normal monocyte count would be (in cells per microliter):
   a. 3500.
   b. 350.
   c. 35.
   d. 5.
   e. 1×10^3.

7. Which of the following contains antibodies against antigens A, B, and D?
   a. Serum of whole blood type O+
   b. Whole blood type A−
   c. Whole blood type B+
   d. Plasma of whole blood type AB+
   e. None of the preceding

8. Which of the following values for cell count (per μL) would indicate polycythemia vera?
   a. Neutrophils, 7500
   b. Lymphocytes, 3000
   c. Erythrocytes, 9.5 million
   d. Monocytes, 500
   e. None of the preceding

9. Which of the following combinations suggests hypochromic anemia?
   a. Hemoglobin, 12 g/dL; RBC count, 4.8 million cells/μL
   b. Hemoglobin, 18 g/dL; RBC count, 5.4 million cells/μL
   c. Hemoglobin, 9 g/dL; RBC count, 4.0 million cells/μL
   d. Hemoglobin, 14 g/dL; RBC count, 5.0 million cells/μL
   e. Two of the preceding

10. When bacteria invade the body, which one of the following cells is most likely to suffer the highest mortality rate?
    a. B lymphocytes
    b. Monocytes
    c. Eosinophils
    d. Basophils
    e. Neutrophils

11. Which of the following bloods could safely be the recipient for a donation of 1 pint of AB+ serum?
    a. O+
    b. A−
    c. AB−
    d. O−
    e. All of the preceding

12. The largest white blood cell, the blood cell that is part of the body's macrophage defense system, is the:
    a. neutrophil.
    b. eosinophil.
    c. monocyte.
    d. basophil.
    e. lymphocyte.

13. The most common white blood cell, and the highly motile and phagocytic cell that forms the primary line of cell defense against bacterial invasion, is the:
    a. eosinophil
    b. neutrophil
    c. lymphocyte
    d. monocyte
    e. basophil

14. Numerically, which of the following is the largest?
    a. Monocyte count
    b. Eosinophil count
    c. Basophil count
    d. Lymphocyte count
    e. Neutrophil count

15. During stage I of the blood coagulation process:
    a. thrombin is formed.
    b. fibrin is formed.
    c. prothrombin is converted into thrombin.
    d. prothrombin converting factor is formed.
    e. fibrin is dissolved.

16. Which of the following whole blood types can be safely infused into a B− recipient?
    a. AB−
    b. A−
    c. B−
    d. B+
    e. Two of the preceding

17. Which one of the following is abnormally high?
    a. WBC count, 8000 cells/μL
    b. RBC count, 5.3×10^6 cells/μL
    c. ESR, 6.0 mm/h
    d. Hematocrit, 60%
    e. MCH, 30 pg

18. Which of the following values is abnormal?
    a. Hematocrit, 43%
    b. RBC count, 5 million cells/μL
    c. WBC count, 8000 cells/μL
    d. O_2 carrying capacity, 20 mL O_2/dL
    e. MCHC, 26%

19. Assume that 230 billion red blood cells are destroyed and replaced each day, that the RBC count is 5 million cells/μL, and that the total blood is 5 L. How many days would the body require to replace all erythrocytes currently present?
    a. 14
    b. 30
    c. 109
    d. 130
    e. 150

20. Given a hematocrit of 45% and an RBC count of 5 million cells/μL, calculate the MCV.
    a. 90 pL
    b. 85 pL
    c. 70 pL
    d. 87 pL
    e. 75 pL

21. Which of the following can be safely infused into an AB− recipient, even if the recipient has been previously sensitized to the Rh antigen, and no agglutination will occur?
    a. Whole blood type A−
    b. Whole blood type B+

c. Serum of whole blood type O+
d. Plasma of whole blood type AB+
e. Two of the preceding

22. Which of the following, if infused into an A− recipient, would result in minor but not major agglutination?
   a. Plasma from B+ whole blood
   b. Plasma from A+ whole blood
   c. Serum from A+ whole blood
   d. AB− whole blood
   e. More than one of the preceding

23. When is prothrombin activated?
   a. During agglutination of blood
   b. During stage I of blood coagulation
   c. During stage II of blood coagulation
   d. During stage III of blood coagulation
   e. During clot retraction in hemostasis

24. Which of the following contains anti-B but not anti-D antibodies?
   a. Whole blood AB−
   b. Serum from whole blood O+
   c. Plasma from whole blood B−
   d. Whole blood A+
   e. Two of the preceding

25. On destruction of the red blood cells and metabolism of hemoglobin:
   a. all components of Hb are excreted except iron, which is conserved.
   b. all components of the molecule are conserved.
   c. all components of the molecule are excreted.
   d. iron and protein (amino acids) are conserved and the remainder is excreted as waste in the urine or feces.
   e. the globin is conserved and the heme is excreted.

Answers to Case History Questions

1. Anemia.
2. A reduction in oxygen-carrying capacity of the blood and thus a reduction in the delivery of oxygen to various body tissues.
3. An iron-deficiency anemia.
4. Most cases of iron-deficiency anemia result from internal blood loss. Dark, tarry, loose stools suggest bleeding from the gastrointestinal tract and warrant further tests to determine the exact cause.

Key Terms

agglutination (p. 535)
albumin (p. 526)
anemia (p. 533)
antibody (p. 537)
antibody-mediated immunity (p. 544)
anticoagulant (p. 549)
antigen (p. 536)
atherosclerosis (p. 550)
B cell (p. 543)
basophil (p. 528)
cell-mediated immunity (p. 544)

clot (p. 547)
colony-forming unit (p. 545)
colony-stimulating factors (p. 545)
eosinophil (p. 540)
erythropoiesis (p. 528)
extrinsic clotting pathway (p. 547)
fibrinolysis (p. 549)
globulin (p. 526)
hematopoiesis (p. 528)
hemoglobin (p. 530)

immunoglobulin (p. 528)
interleukin (p. 544)
intrinsic clotting pathway (p. 547)
intrinsic factor (p. 535)
leukocyte (p. 540)
leukocytosis (p. 545)
leukopenia (p. 545)
lymphocyte (p. 540)
macrophage (p. 532)
monocyte (p. 528)
neutrophil (p. 540)

oxygen-carrying capacity (p. 530)
phagocytosis (p. 541)
plasma (p. 522)
pluripotential hemopoietic stem cell (p. 531)
polycythemia (p. 535)
red cell indices (p. 524)
reticulocyte (p. 532)
Rh factor (p. 540)
T cell (p. 543)
thrombocyte (p. 545)
thrombus (p. 545)

Suggested Readings

Baugh, R. F., "Platelets and whole blood coagulation." *Perfusion,* 15:41–50, 2000.

Bloom, W., and Fawcett, D. W. *A Textbook of Histology,* ed 12. London, Chapman and Hall, 1994.

Esmon, C. T., "Regulation of blood coagulation." *Biochem Biophys Acta,* 1477:349–60, 2000.

Gleich, G. J., Adolphson, C. R., and Leiferman, K. M. "The biology of the eosinophilic leukocyte," *Annual Review of Medicine,* 44:85–101, 1993.

Golpe, D. W., and Gasson, J. D. "Hormones that stimulate the growth of blood cells." *Scientific American,* 259:1, 1988

Johansen, K. M., Skorpe, S. Olsen, J. O., and Osterlid, B. "The effect of red wine on the fibrinolytic system and the cellular activation reactions before and after exercise." *Thrombosis Research,* 96:355–63, 1999.

Mavrommatis, A. C., Theodoridis, T., Orfanidou, A., Roussos, C., Christopoulou-Kokkinou, V., and Zakynthinos, S. "Coagulation system and platelets are fully activated in uncomplicated sepsis." *Critical Care Medicine,* 28:451–7, 2000.

Metcalf, D. "Thrombopoietin." *Nature*, 369:519–20, 1994.

Priessner, K. T. "Vascular protease receptors: Integrating haemostasis and endothelial cell functions." *Journal of Pathology*, 190:360–72, 2000.

Spangrude, G. J. "Biologic and clinical aspects of hematopoietic stem cells." *Annual Review of Medicine*, 45:93–104, 1994.

Spivak, J. L. "Recombinant erythropoietin." *Annual Review of Medicine*, 44:243–53, 1993.

Stamler, J., and Neaton, J.D. "Benefits of lower cholesterol." *Scientific American Science and Medicine*, 1:28–37, 1994.

Wan, H., Liu, Z., Xia, X., Gu, J., Wang, B., Liu, X., Zhu, M., Li, P., and Ruan, C., "A recombinant antibody-targeted plasminogen activator with high affinity for activated platelets increases thrombolytic potency in vitro and in vivo." *Thrombosis Research*, 97 (3):133–41, 2000.

Answers to Review Questions

1. d	**2.** d	**3.** a	**4.** c	**5.** d	**6.** b	**7.** e	**8.** e
9. c	**10.** e	**11.** e	**12.** c	**13.** b	**14.** e	**15.** d	
16. c	**17.** d	**18.** e	**19.** c	**20.** a	**21.** d	**22.** a	
23. c	**24.** e	**25.** d					

Chapter 18

THE HEART

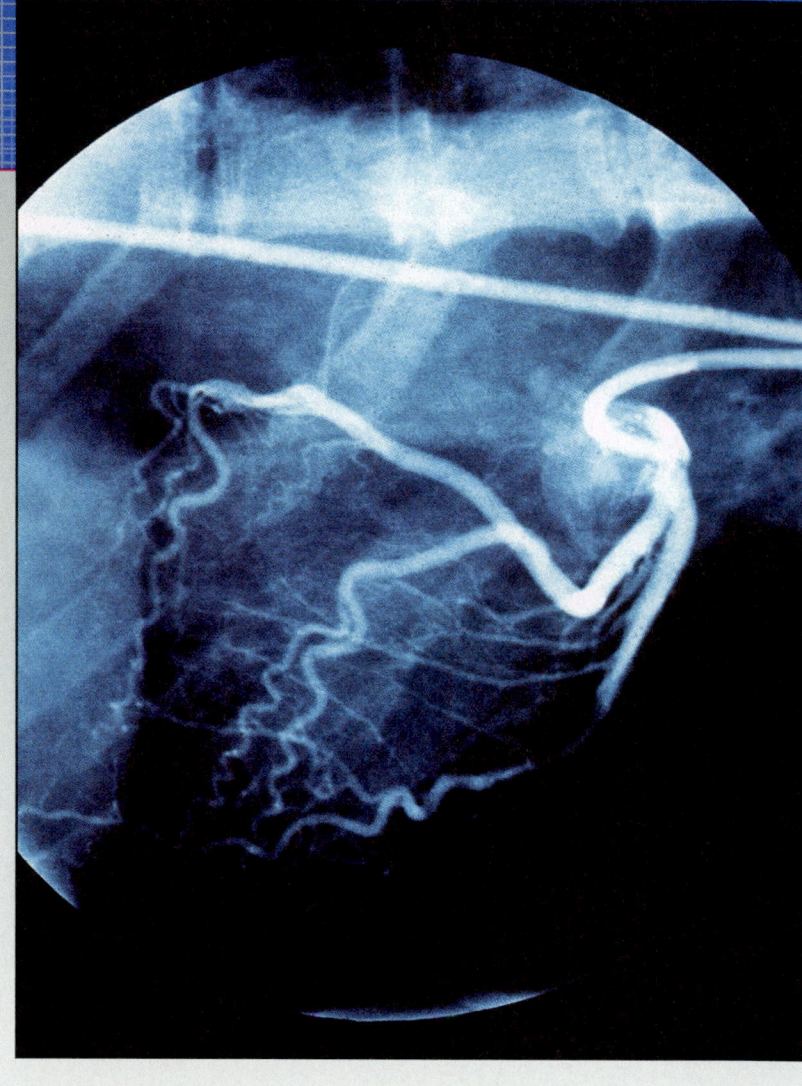

- *Arteriogram showing a catheter inserted into the right coronary artery.*
(© CNR/Phototake)

KEY CONCEPTS

- The heart consists of two separate pumps that generate the pressure that drives the unidirectional flow of blood through the pulmonary circulation, where gas exchange occurs, and through the systemic circulation, where exchange of nutrients, metabolites (including heat), and hormones occurs.

- Each beat of the heart, or cardiac cycle, involves a number of interrelated and highly coordinated events, including electrical activation of the atria and ventricles, contraction and relaxation of the atria and ventricles, closing and opening of the cardiac valves, and filling and emptying of the atria and ventricles.

- The cardiovascular system consists of a pump (the **heart**), which propels blood throughout the body by way of a series of tubes (**blood vessels**). The **circulating blood** provides a vehicle for the rapid movement of cells and molecules from one part of the body to another.

CASE HISTORY

A 55-year-old male patient has experienced periodic bouts of lightheadedness and dizziness. He tells his physician that he is aware of his heart beating forcefully but slowly during these episodes. His electrocardiogram (ECG) is normal at the time of his examination; however, his physician outfits him with a Holter monitor that continuously monitors the patient's ECG and records it on tape with a portable tape recorder attached to the patient's belt. Examination of the Holter recording shows periodic episodes of second-degree atrioventricular block, with every other atrial depolarization failing to conduct through the AV node to the ventricles (2:1 AV block). During these episodes, which typically lasted for 3 to 4 minutes, the patient's atrial rate was 66 beats per minute (bpm) and the ventricular rate was 33 bpm. The physician prescribes atropine, which reduced the incidence of AV block. Six years later, the patient begins to experience sudden, and unpredictable episodes of syncope (fainting). The patient reports that he quickly regains consciousness but that he frequently continues to feel lightheaded for some time afterward and that his heart rate is very slow. Again, he is outfitted with a Holter monitor, and the recordings show that the patient is now experiencing third-degree (complete) AV block. Typically, the onset of AV block was followed by a period of 3 to 4 seconds of normal P-waves without associated QRS complexes. When the QRS complexes reappeared, they were abnormal (wider) in appearance, had no temporal relationship to the P-waves, and occurred at a regular rate of approximately 40/min. After a variable period of time, a normal rhythm, in which each P-wave was followed by a normal-looking QRS complex, was reestablished. A demand-pacemaker, which would electrically stimulate the ventricles during these episodes, was surgically implanted. The pacemaker would pace the ventricles only when it failed to detect ventricular depolarization following atrial depolarization. The pacemaker would continue to monitor atrial depolarization and pace the ventricles after a delay that was slightly longer than this patient's normal P-R interval. Pacing would continue only until a normally conducted QRS complex was detected by the pacemaker.

Questions

1. Why would the patient experience more forceful contractions of the heart during the episodes of second-degree block?

2. Why would atropine prevent episodes of second-degree heart block?

3. Why is the patient's heart rate only 40 bpm during episodes of complete heart block; what is the significance of the abnormal (wide) QRS complexes?

4. Why is a demand-pacemaker better than one that simply paced the ventricles at a fixed rate all of the time?

5. Why was the demand-pacemaker set to give a slightly longer than normal P-R interval?

OVERVIEW OF THE CARDIOVASCULAR SYSTEM

> *How is the "plumbing" of the cardiovascular system arranged?*

William Harvey demonstrated in the seventeenth century that the cardiovascular system forms a closed loop through which blood is pumped by the heart. As shown in Figure 18–1, this loop consists of two pumps, the **left heart** and the **right heart,** and two vascular systems, the **pulmonary circulation** and the **systemic circulation.** From an anatomical perspective, there is only one heart; however, from a physiological perspective, there are two independent pumps, as shown in Figure 18–1. Although in practice you will see that they are both mechanically and electrically interdependent, it is useful to visualize them independently when considering the path of blood flow through the body.

In Figure 18–1, the left and right hearts *(brown)* and the pulmonary *(dark gray)* and systemic circulations *(light gray)* are arranged in **series** *(pink arrows)*, which means that blood must flow through these four components in sequence, one after the other. Blood flows, via the pulmonary circulation, through the lungs, where oxygen is added and carbon dioxide is removed. After passing through the left heart, this oxygenated blood then flows, via the systemic circulation, to all the cells of the body, which extract oxygen and other nutrients from the blood and add carbon dioxide and other waste products to it. The blood then flows through the right heart, and the cycle repeats.

Notice in Figure 18–1 that the systemic circulation consists of many separate pathways that supply blood to various organ systems within the body *(red arrows)*. Notice also that these individual **vascular beds,** which make up the systemic circulation, are arranged in **parallel** *(red arrows)* with one another. This means that the total volume of blood flowing through the systemic circulation is partitioned, or divided, among the different vascular beds. As we will see later, our bodies are able to dynamically adjust the distribution of blood flow among the various tissue beds. For example, during exercise, the blood flow to skeletal muscle and to the skin can be increased several-fold, while the flow to other organ systems, such as the digestive tract, is simultaneously reduced. We will discuss in the next chapter the mechanisms that allow for this optimization of blood flow in response to the needs of different parts of the body. The relative distribution of the **total blood flow** among the various organs of the body during rest is shown in Figure 18–2.

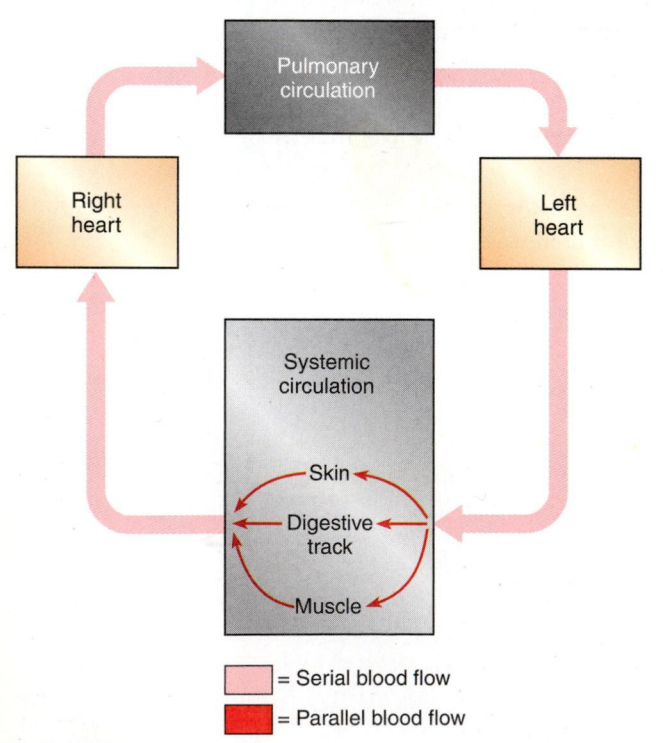

Figure 18–1

Arrangement of the circulatory system.

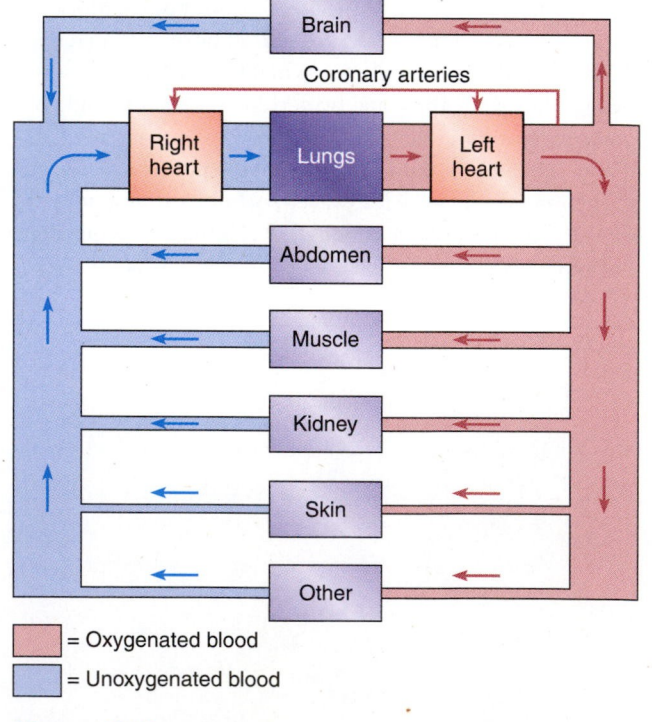

Figure 18–2

Relative distribution of blood flow in the cardiovascular system at rest. Vessel diameter is proportional to blood flow.

The Systemic Circulation Carries Oxygenated Blood to the Tissues of the Body

The vessels that carry blood *from* the heart make up the **arterial system;** the vessels that carry blood *back to* the heart make up the **venous system** (Fig. 18–3). The systemic circulation begins with a single artery, the **aorta,** which receives all the blood that the heart pumps out. The aorta branches into a number of smaller **arteries,** which partition and direct blood flow to the various organs of the body. Once it enters an organ, an artery soon branches into progressively smaller arteries, which in turn branch into smaller **arterioles.** This arteriolar tree distributes blood flow within individual organs. The arterioles then branch into an extensive network of **capillaries,** the smallest blood vessels in the body. It is here, in the capillaries, that exchange occurs between the blood and the interstitial fluid. Blood then flows from the capillaries into the venous system, which channels the blood back to the heart.

The venous system begins at the point where the capillaries reunite to form larger vessels called **venules,** which in turn unite to form larger **veins.** All the veins from the upper portion of the body drain into a single large vein called the **superior vena cava,** and the veins from the lower portion of the body drain into the **inferior vena cava.** The two venae cavae unite to return all of the blood to the right heart.

The Pulmonary Circulation Carries Unoxygenated Blood Through the Lungs for Gas Exchange

A similar pattern of branching occurs in the **pulmonary circulation** (see Fig. 18–3). The blood leaves the right heart by way of the **pulmonary trunk,** which soon divides into the **pulmonary arteries.** The pulmonary arteries distribute the blood to the two lungs. Notice that the entire **cardiac output** passes through the lungs before being distributed throughout the systemic circulation.

As in the systemic circulation, the pulmonary arteries progressively branch into smaller and smaller vessels. The **pulmonary capillaries** are the site of gas exchange between the blood and the air in the lungs. As the blood flows through the pulmonary capillaries, its oxygen content increases and its carbon dioxide content decreases as a result of diffusional exchange with the gas in the lungs. This oxygenated blood then flows from the pulmonary capillaries through progressively larger venules and veins to the **pulmonary veins** and then into the left heart.

Not all arterial blood has a high oxygen content. Both the **pulmonary venous blood** and the **systemic arterial blood** are oxygenated. The distinguishing factor between arteries and veins is that arteries carry blood *from* the heart *to* capillaries, and veins carry blood back *to* the heart *from* capillaries.

Blood Flow Is Determined by the Blood Pressure and the Resistance to Blood Flow

 How do we define blood flow?

We now know that the role of the cardiovascular system is to provide for a continuous flow of blood through the tissues. Let's consider for a moment what the term *flow* means and how it might be quantified.

Blood flow can be defined as the volume of blood that moves past a particular point in the cardiovascular system during a given period of time — for example, the amount that flows through the left heart or through the aorta in 1 minute. A practical example of fluid flow with which we are all familiar is the flow of water through a garden hose. If we were to collect the water coming out of a hose over a period of 1 minute, we could calculate the flow of water through the hose. If, after 1 minute, we had 5 liters of water in our bucket, then the calculated flow would be 5 liters in 1 minute, or 5 L/min. We have measured the volume of water that moved past a particular point in the garden hose (in this case, the nozzle) during a period of time (1 minute).

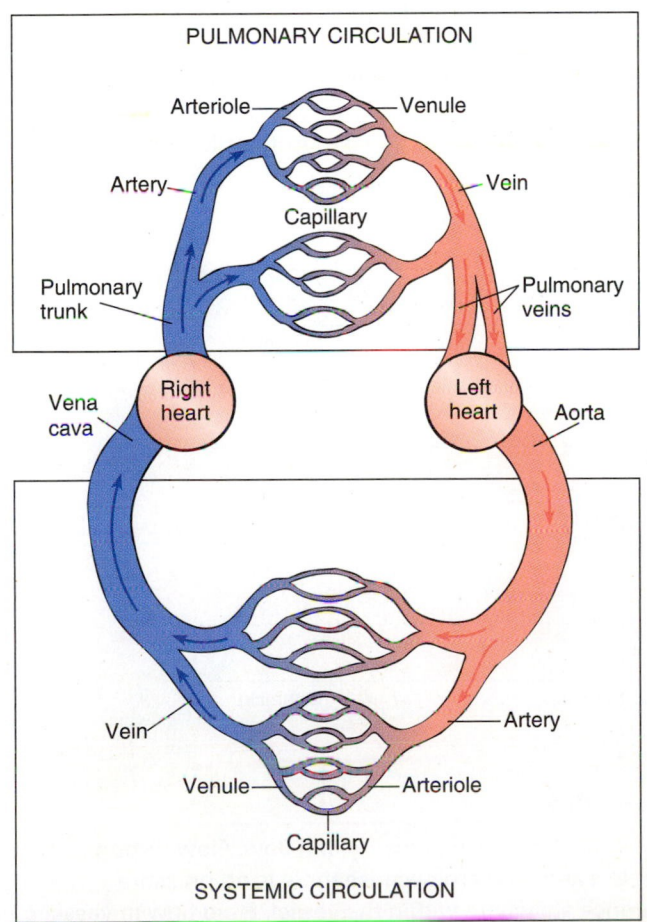

Figure 18–3

Branching pattern of blood vessels in the cardiovascular system.

Does this number tell us anything about the flow of water elsewhere in the hose? The answer is yes, because the flow of water will be the same at all points in the hose. We can generally state that the flow in a closed system will be the same through all portions that are arranged in **series.** In terms of the cardiovascular system, this means that we could measure the mean blood flow from the heart into the aorta and know that it is the same as the mean blood flow through the lungs, systemic circulation, left heart, and right heart, because these portions of the cardiovascular system are all connected in series and constitute a closed system.

Could we also say what the blood flow was through the kidneys on the basis of this one measurement? The answer is no, because the individual vascular beds that make up the systemic circulation are arranged in parallel, and therefore only a portion of the total systemic blood flow goes through each of these vascular beds (see Fig. 18–2). As illustrated in Figure 18–4, the combined flow through the vascular beds arranged in parallel with one another must be equal to the flow through portions of the system that are in series with these vascular beds. Furthermore, as mentioned previously, the proportion of the total systemic blood flow channeled through a particular vascular bed can change in response to the changing needs of the body, even though the total blood flow through the systemic circulation remains constant.

The force that causes blood to flow through the vessels can be measured as **blood pressure.** Blood pressure measurements are usually expressed in units of millimeters of mercury (**mm Hg**). For example, average **aortic blood pressure** is 90 mm Hg. This means that the pressure inside the aorta (relative to atmospheric pressure outside the aorta)

is equal to the downward pressure exerted by a column of mercury 90 mm in height. It is important to realize that the pressure exerted by a column of mercury, or any other fluid, is determined by the height of the column and not the diameter of the column.

A more familiar unit of pressure is pounds per square inch (psi). Automobile tires, for example, have an internal air pressure of approximately 35 psi. A blood pressure of 100 mm Hg is equivalent to 2 psi. Blood pressure is a measure of the driving force propelling blood through the blood vessels. Just as the high air pressure within a tire will cause air to rush out of the tire through an open valve, the differences in blood pressure within the circulatory system will cause blood to flow from one point to another. Flow does not occur in response to any particular pressure (*P*), but rather to a pressure difference or **pressure gradient** ($P_1 - P_2$, or ΔP) between two points. This is analogous to net diffusion between two points; the rate of diffusion is determined by the magnitude of the concentration gradient and not by the absolute concentration of a diffusing substance. Thus, the blood flow between two points in the circulatory system is proportional to the *difference* in blood pressure between these same two points and not the *absolute* pressures at either point. This concept is illustrated in Figure 18–5.

The reason that a physical force or pressure is required to maintain blood flow through the circulatory system is that frictional forces tend to impede blood flow. Friction exists among the molecules that make up the blood and between

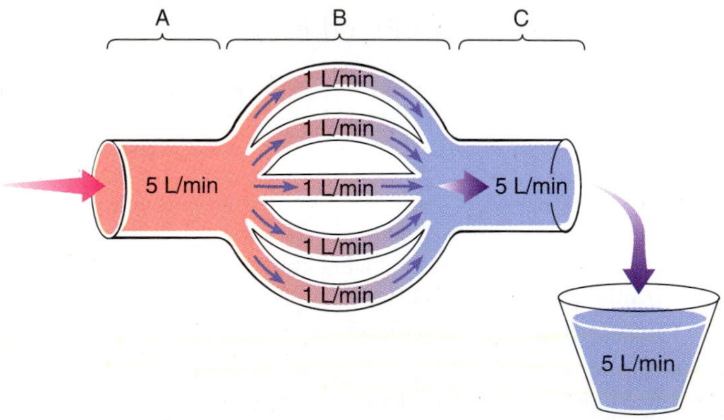

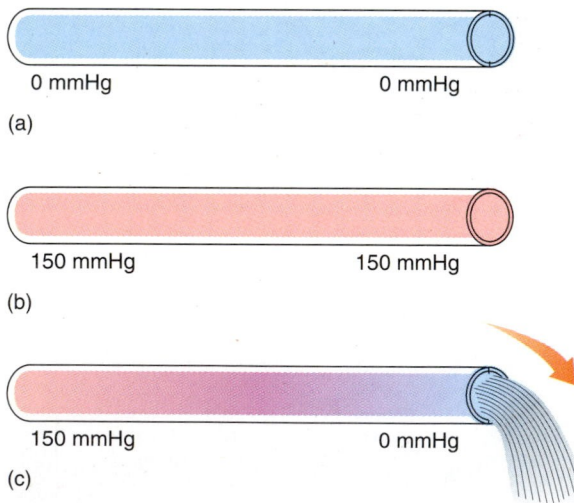

Figure 18–4

A comparison of flow through tubes arranged in series and in parallel. Blood flow through all portions of a closed system that are arranged in series (*A, B,* and *C*) must be the same (5 L/min). Blood flow through the individual vessels arranged in parallel (small vessels in section *B*) must add up to the total blood flow through the system (5 × 1 L/min = 5 L/min).

Figure 18–5

Relationship between pressure and flow. Flow through vessels *a* and *b* is zero because there is no pressure difference anywhere within the vessel. Fluid flow in vessel *c* is driven by the pressure difference (150 mm Hg − 0 mm Hg = 150 mm Hg) between the ends of the vessel. The arrow indicates the direction of blood flow.

the blood and the walls of the vessel. Resistance is a quantity that summarizes the net effect of all the frictional components that oppose blood flow in a particular situation. Thus, the greater the resistance, the lower the blood flow for any given pressure difference. The equation that describes the relationship between blood flow *(F)*, driving pressure *($P_1 - P_2$ or ΔP)*, and resistance *(R)* is:

$$F = (P_1 - P_2)/R$$

As we will see in the next chapter, the resistance to blood flow through a vessel is determined by a number of factors, including the length and diameter of the vessel, as well as the consistency or viscosity of the blood itself.

THE HEART: A MUSCULAR PUMP

The major role of the heart is to maintain an adequate level of blood flow throughout the cardiovascular system by pumping blood under pressure into the vascular system. As we will see in the next chapter, our bodies actually monitor and regulate arterial pressure, not blood flow, in order to ensure adequate blood flow to all the tissues in the body. To under-

stand how this system functions, we must first consider the anatomy of the heart and the manner in which cardiac muscle contraction is coordinated within the heart to generate pressure and flow.

The Heart Consists of Two Separate Pumps that Maintain the Unidirectional Flow of Blood

What is the anatomy of the heart?

As previously mentioned, the heart consists of two functionally distinct pumps: the left heart and the right heart. The left and right hearts, in turn, each contain two chambers: an **atrium** (pl., *atria*) and a ventricle. The atria are thin-walled muscular chambers that receive blood returning to the heart from either the venae cavae (in the case of the right heart) or the pulmonary veins (in the case of the left heart). Blood then is pumped from each atrium into the associated **ventricle.** The ventricles (particularly the left ventricle) have thicker muscular walls capable of pumping blood out of the heart at higher pressures. Figure 18–6 shows schematically how the chambers of the heart are organized and the direction of blood flow through the heart.

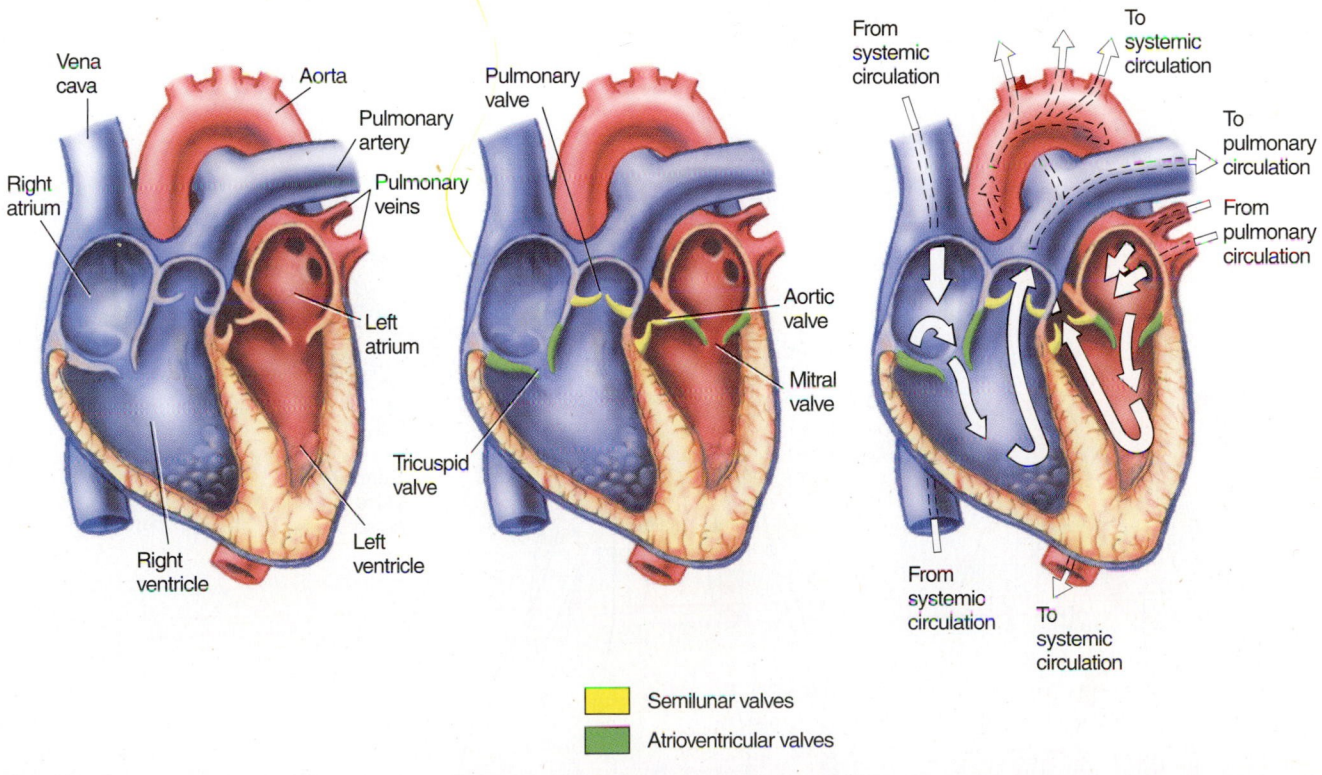

Semilunar valves

Atrioventricular valves

Figure 18–6

Overview of the heart. The direction of blood flow through the chambers of the heart and the major vessels leading into and out of the heart is indicated by the arrows.

Valves in the Heart

Notice that there is a valve at the entrance and exit of each ventricle (Fig. 18–7). These valves allow blood to flow in only one direction. We will see later how these one-way valves operate to ensure unidirectional flow of blood through the heart. The **atrioventricular (AV) valves** are located between the atria and the ventricles. The AV valve associated with the right heart is called the **tricuspid valve,** and the AV valve associated with the left heart is called the **bicuspid** or, more commonly, the **mitral valve.** The valves located at the junction of the ventricles with the pulmonary and systemic arteries are called **semilunar valves.** The semilunar valve between the right ventricle and the pulmonary trunk is called the **pul-** **monary valve,** and the valve between the left ventricle and the aorta is called the **aortic valve.** The opening and closing of the valves are passive processes that occur in response to the **blood pressure gradient** across the valve. For example, when the pressure in either of the two atria is greater than that in the adjoining ventricles, the AV valves open and allow blood to flow into the ventricle (Fig. 18–8). If the pressure gradient is reversed — with ventricular pressure being greater than atrial pressure — the AV valves will close in response to the momentary reversal of blood flow and then be held closed by the reversed pressure gradient; this prevents the flow of blood from the ventricle back into the atrium during ventricular systole.

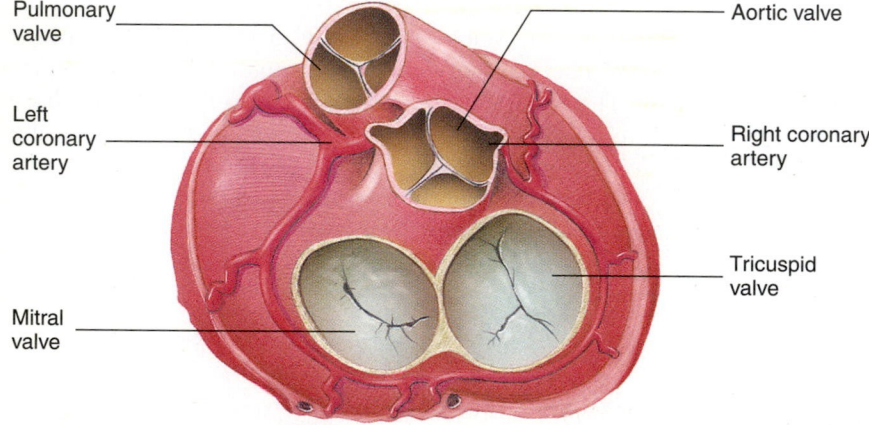

Pulmonary valve

Left coronary artery

Mitral valve

Aortic valve

Right coronary artery

Tricuspid valve

Figure 18–7

Cardiac valves. Top view of the heart with the major vessels cut away to show the location of the four cardiac valves. Both the aortic and pulmonary valves consist of three crescent-shaped (semilunar) cusps. The left atrioventricular valve consists of two valve cusps (bicuspid or mitral valve), while the right atrioventricular valve consists of three valve cusps (tricuspid valve). All the heart valves are connected to a tough connective tissue framework or cardiac skeleton that lies between the atria and the ventricles.

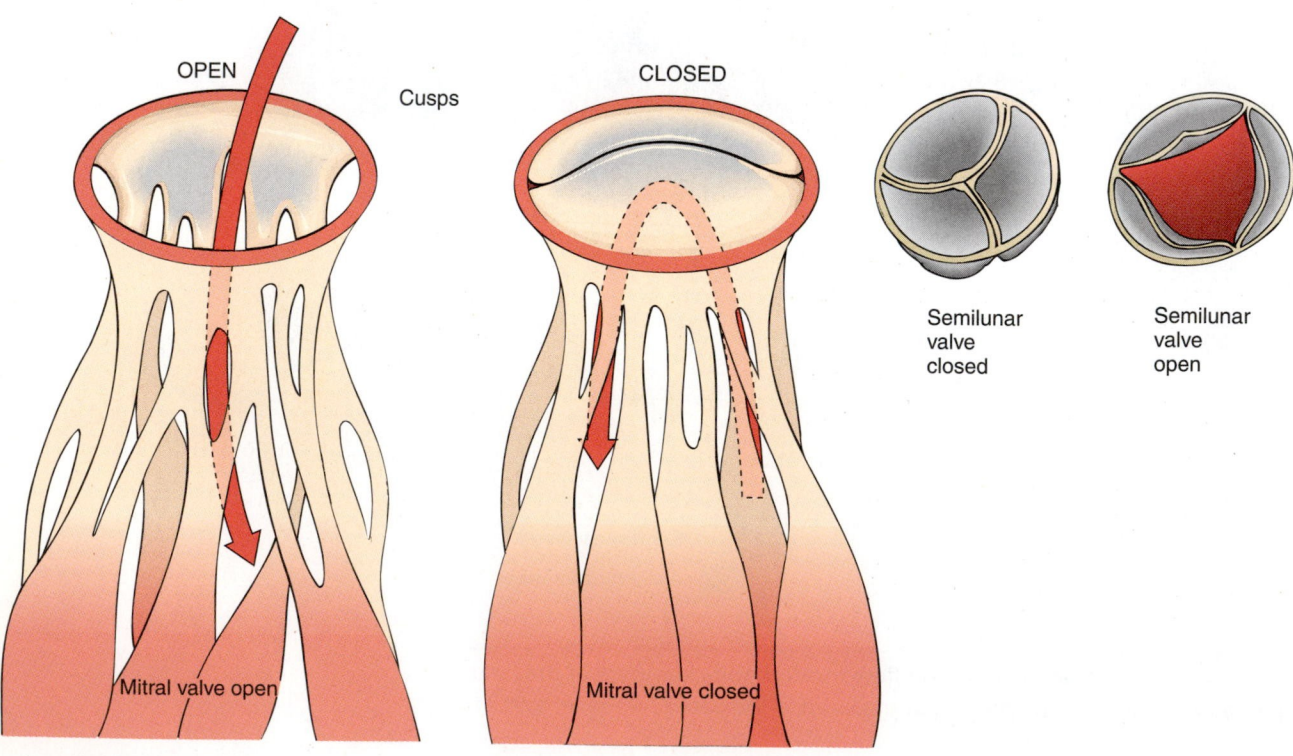

OPEN

Cusps

CLOSED

Mitral valve open

Mitral valve closed

Semilunar valve closed

Semilunar valve open

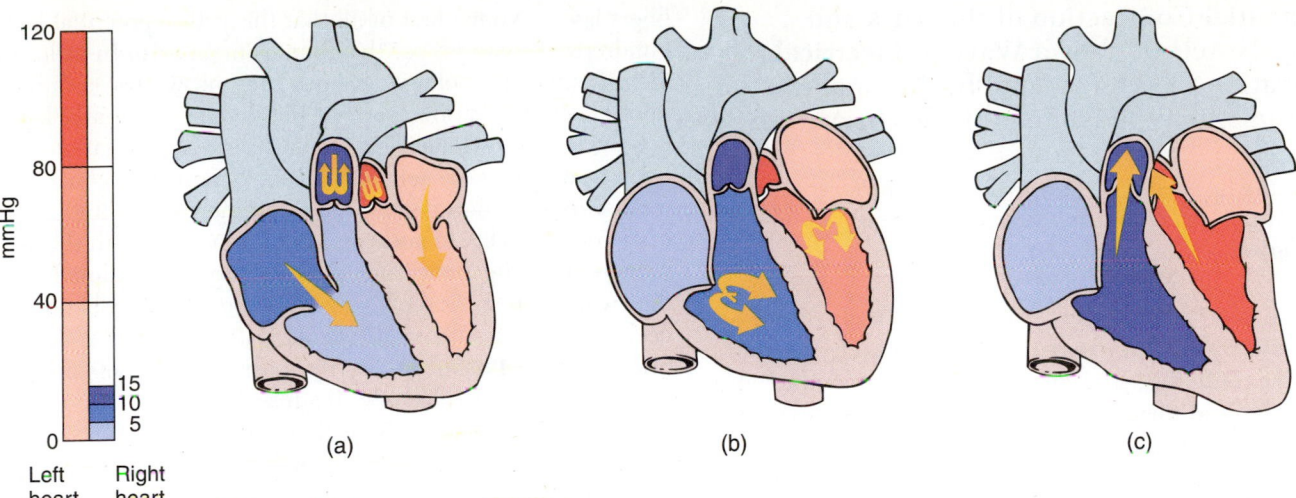

(a)   (b)   (c)

Figure 18–8

Cardiac valve operation. *(a)* Valves in the heart normally open when the pressure gradient across the valve is increasing in the direction that blood normally flows through the heart (forward pressure gradient). In contrast, a reverse pressure gradient *(b)* forces the valve closed. This prevents reverse blood flow in response to reverse pressure gradients that occur as a result of the pumping action of the heart. *(c)* A forward pressure gradient forces the semilunar valve to open. Darker colors correspond to higher pressure.

The Walls of the Heart

The walls of the heart are composed of three anatomically distinct tissue layers (Fig. 18–9). The middle, muscular layer, the **myocardium,** is composed primarily of cardiac muscle cells. (The special properties of this muscle type were discussed in Chapter 16.) The inner surface of the myocardium, which forms the inside wall of the atria and ventricles, is lined with a thin layer of endothelial cells, the **endocardium.** The endothelial layer forms a continuous lining that covers not only the inner surfaces of the heart and the heart valves but also the entire vascular network. The outer surface of the myocardium is covered by the **epicardium,** which consists primarily of connective tissue and fat. The main **coronary vessels** that supply blood to the myocardium (see Fig. 18–7) are located within this tissue layer.

The terms *left heart* and *right heart* refer to separate parts of the one anatomical heart as shown in Figure 18–6. While the left and right hearts are distinctly separate pumps, they are hydraulically, mechanically, and electrically coupled to each other, and, in fact, must remain coupled in this fashion if they are to function effectively. The anatomical placement of the left and right hearts within a single organ facilitates both the mechanical and the electrical coupling. This will become apparent when we discuss these aspects of cardiac physiology. While both of the pumps have unique aspects, many properties are common to both. The term *heart* will subsequently be used to refer to aspects that are common to both, whereas *left heart* and *right heart* will be used to refer to specific aspects of these two pumps that are not necessarily common to both.

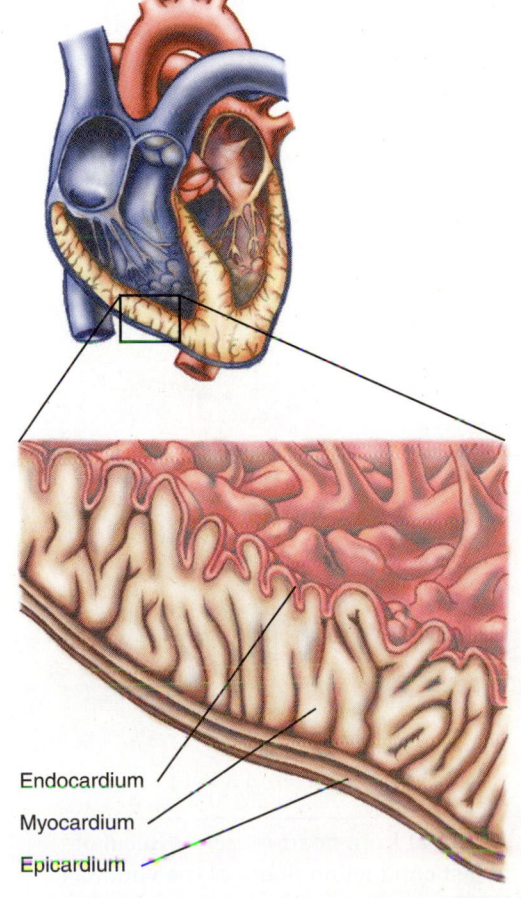

Endocardium
Myocardium
Epicardium

Figure 18–9

Anatomy of the heart wall.

The Sequential Contraction of the Atria and Ventricles Is Activated by a Wave of Electrical Depolarization that Is Initiated by Pacemaker Cells in the Sinoatrial Node in the Right Atrium

How is the heart stimulated, and how do different heart muscle cells differ with respect to their action potentials?

Although the heart is composed primarily of muscle, it must function quite differently from the way in which skeletal muscle works. Different parts of the heart must contract at different times, and they must do it in a proper sequence if the heart is to pump blood efficiently. The key to understanding much of the special function of the heart lies in a knowledge of the electrical properties of its muscle.

Phases of the Cardiac Action Potential

As do most other muscles, the heart muscle cells have plasma membranes that can generate action potentials. It is difficult to speak of a "typical" heart muscle action potential because each of the specialized regions of the heart has special action potentials. While they are similar in many ways to the relatively simple action potentials found in nerves and skeletal muscles, action potentials in the heart involve additional membrane processes. Cells found in the ventricular region have most of the special features that we need to discuss, so a cell of this sort can serve as our typical example.

Figure 18–10 shows such a complex action potential. This is the type of electrical activity found in the **Purkinje fibers** of the ventricles (we will discuss the function of these

fibers later). Notice first of all that this action potential lasts hundreds of times longer than one from a nerve. In order to identify the extra voltage changes present, we use numbers to designate the various parts of this action potential. **Phase 0** is the rapid upstroke. This is a very fast depolarization that **overshoots** the zero millivolt level. As in nerve tissue, this rapid depolarization is due to a rapid increase in the membrane permeability to **sodium ions.** Associated with this is a reduction of the potassium permeability to a value lower than when at rest. As the high sodium permeability begins to decline, a small amount of repolarization takes place. This brief event is **Phase 1.** Shortly thereafter, the membrane potential becomes relatively steady at a value near zero millivolts. This is **Phase 2,** also called the **plateau** of the action potential. Here a slow inward membrane current due to the flow of **calcium ions** is balanced against the efflux (outward flow) of **potassium ions,** even though the potassium permeability is reduced at this time. The membrane channels through which calcium passes also allow a small amount of sodium to pass inward. Late in the plateau, the calcium permeability declines, and with the falling membrane potential, the potassium permeability begins to increase. These factors produce a **repolarization,** which is called **Phase 3.** During this time, the membrane potential falls to its original value and remains steady until the next beat. This steady portion is **Phase 4,** which is equivalent to the resting potential of other muscle cells. This sequence of membrane potential changes lasts for approximately 300 msec, depending on the heart rate.

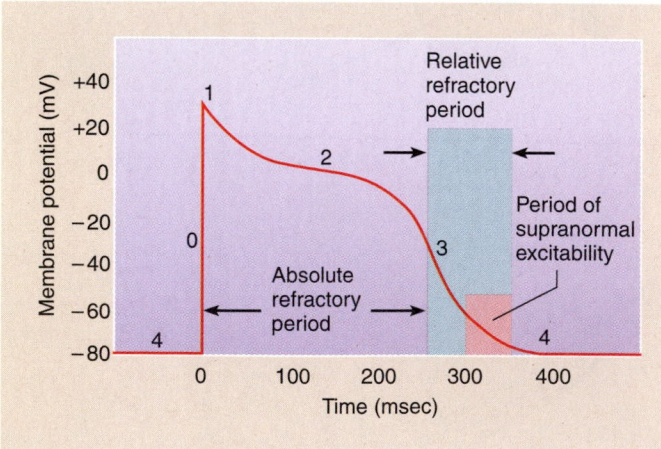

Figure 18–10

A sample action potential from heart muscle, typical of those found in the fast conduction fibers of the ventricle, such as the Purkinje fibers. The numbers above the curve designate the phases described in the text, and the approximate durations of the refractory periods are shown.

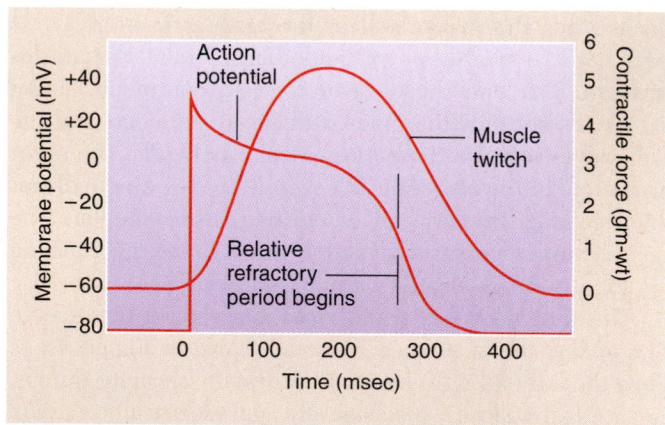

Figure 18–11

The relationship between the action potential and the twitch. The isometric twitch of a sample of heart muscle is shown in relation to the action potential that caused it. To some extent the duration of the twitch is determined by the duration of the action potential. Agents that prolong the plateau of the action potential make the twitch last longer, and the force of contraction may be increased because of the greater opportunity for calcium to enter the cell during the action potential.

As in nerve and other types of muscles, these phases of the action potential are associated with refractory periods. These are diagrammed in Figure 18–11. During the **absolute refractory period,** the muscle cannot be stimulated again. The **relative refractory period** then follows, and latest of all is the period of **supernormal excitability.** Here the muscle is actually easier to stimulate because the potassium permeability is still a bit lower than normal. The action potential produced is smaller than normal because not all of the sodium channels from Phase 0 have had time to "reset." Normally, the muscle is never stimulated during this time period. If such stimulation were to occur, the result could be very dangerous because the activity could spread to other tissue normally at rest at that time.

The Cardiac Action Potential and Contraction of Cardiac Muscle

 How does stimulation of cardiac muscle differ from that of skeletal muscle?

Notice in Figure 18–11 that the absolute refractory period lasts almost as long as the isometric twitch. This relationship prevents the muscle from contracting again before it has relaxed because the muscle will not respond to a second stimulus so soon after the first. This means that a tetanus is not possible in heart muscle; such a long-lasting contraction would prevent the heart from filling with blood for the next beat, making it a poor pump indeed.

Action potentials are necessary to make heart muscle contract. In skeletal muscle, action potentials are produced as a result of stimulation by a motor nerve. Each cell in a skeletal muscle is individually supplied by its own motor nerve terminal, and coordinated activity is due to the timing of impulses from the central nervous system (CNS).

The heart, in contrast, has no motor nerve supply, and its cells are very small. A separate nerve supply to each of the millions of muscle cells in the heart would make it a huge and complex organ. Instead, the cells of heart muscle communicate with one another directly, and muscle takes over the function of nerve. This means that the heart can be stimulated to contract at a single location, and the message will be passed along to the rest of the heart muscle. Within the individual cells of the heart muscle, the process of activation is similar to that in skeletal muscle. The t-tubule system of the ventricular muscle cells aids in conducting the surface action potential inward, where it causes the release of stored calcium for the activation of the troponin-tropomyosin regulation system. Significant amounts of calcium enter the cell during the plateau of the action potential, and this calcium is added to the internal stores. The entering calcium is also directly responsible for causing some release of internal stored calcium from the sarcoplasmic reticulum. Because of this continual entry of calcium, cardiac muscle cells must (and do) have a way of getting rid of excess internal calcium by exchanging it for incoming sodium.

Initiation of the Action Potential by Pacemaker Cells

 What determines heart rate?

The single location from which the stimulus for the heart action arises is called the **pacemaker.** Normally, the pacemaker region is the **sinoatrial node** (also called the **SA node**), a specialized region of muscle cells in the right atrium. A pacemaker action potential is shown in Figure 18–12. These cells do not have a steady resting potential. After each action potential, the cells repolarize to a value called the **diastolic potential.** Very soon after repolarization, the membrane begins a slow depolarization. This depolarization quickly brings the membrane potential to its **threshold,** and another action potential takes place. This action is repeated again and again, as many as 3 billion times in a 70-year lifetime. At resting heart rates, it is repeated approximately 70 times per minute, and it can reach as high as 200 times per minute during heavy exercise. The rate of the heart is controlled by the rate at which the diastolic potential reaches the threshold.

Normally, only one pacemaker is active at one time in the heart. This is because whichever pacemaker cell "fires"

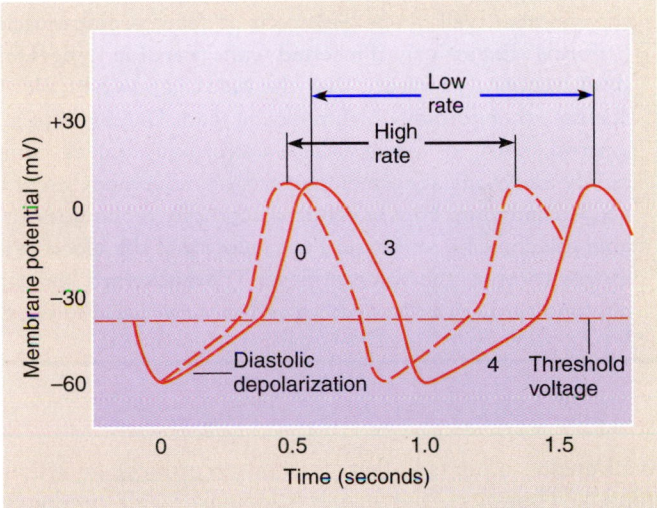

Figure 18–12

An action potential from a pacemaker cell. The slow diastolic depolarization (Phase 4) lets the membrane potential reach the threshold voltage shortly after the cell has repolarized, resulting in another action potential. The solid line shows a pacemaker potential at a low heart rate, which is determined by the spacing between successive action potentials. In this case the low rate corresponds to a heart rate of 65 beats per minute. The dotted line shows the rate that would result from a more rapid diastolic depolarization. The action potentials occur closer together, and the new spacing corresponds to a rate of 72 beats per minute.

APPLICATIONS OF PHYSIOLOGY

Cardiac Imaging

A number of techniques are available for imaging the heart chambers, heart valves, and coronary arteries. A frequently used noninvasive (not requiring surgery) technique is **echocardiography.** High-frequency sound waves are used to image the chamber walls and valves in the heart. The basic principle is the same as that used for radar imaging of airplanes. An echocardiogram (Figure a) provides a two-dimensional image of a cross-section through the heart and shows the same structural organization that you would see if you cut a heart into two pieces. A small, hand-held device both emits and receives reflected sound waves. This device (a **transducer**) is placed against the chest wall; different cross-sectional views are generated, depending on the position of the transducer. This imaging method provides information on the thickness and motion of the muscular walls, the size of the heart chambers, and the functioning of the heart valves. Absence of normal motion in a particular area of the chamber wall can be diagnostic of dead cardiac muscle (**infarct**). Excessively thickened walls (**cardiac hypertrophy**) frequently indicate that the heart is pumping blood against an abnormally high pressure load. Damaged (e.g., scarred, torn, or congenitally malformed) valves commonly show an abnormal pattern of movement on the echocardiogram. Newer echocardiography techniques enable the clinician to measure the velocity of the blood as it moves through the valve orifices. The backward leakage of blood through a damaged valve (**valvular insufficiency**) or the high-velocity jet of blood moving through an obstructed valve (**valvular stenosis**) often is detected by echocardiography.

A relatively new technique for imaging the heart is **magnetic resonance imaging (MRI)**, an imaging technique that uses powerful magnets to look inside the body (Figure b). MRI is based on the principles of nuclear magnetic resonance (NMR). The original name for the technique, *nuclear magnetic resonance imaging* was shortened to *magnetic resonance imaging* because of the negative connotation of the word *nuclear* in the late 1970s. This noninvasive technique can provide high-resolution images of the heart and larger blood vessels (Figure c). Because the patient is positioned in the center of a very powerful magnet, it is not an appropriate technique for patients with pacemakers or implanted defibrillators. An alternative technique that provides similar information is **X-ray computed tomography (CT)**. It also is a noninvasive technique but does involve the use of ionizing radiation in the form of X-rays.

Angiography is an invasive technique that requires the insertion of a catheter into the heart (**cardiac catheterization**), followed by an injection of a radiopaque dye. The movement of the dye-bolus through the heart sequentially delineates the chambers. If the catheter is inserted into one of the coronary arteries (**coronary angiography**), the size of the artery lumen can be visualized (Figure d). Obstructions of the coronary artery (e.g., in atherosclerosis and coronary artery spasms) can be visualized easily with this method. Recent developments in computer technology and digital image-enhancement techniques have significantly enhanced the image quality of coronary angiograms.

---

first stimulates other cells that are about to produce an action potential. Therefore, the fastest pacemaker cell predominates. The heart actually contains many cells capable of becoming pacemakers, but their activity is not usually expressed because they are inherently slower. Under some conditions, if these "hidden" pacemakers do not receive a stimulus from other cells, their activity will become apparent. As we will see later, this can be an important safety device.

The rate of the SA node pacemaker is subject to control by the **autonomic nervous system.** The **parasympathetic** nervous system decreases the rate of diastolic depolarization and makes the diastolic potentials more negative (by making the cell membranes more permeable to potassium ions). This means that it takes longer for the pacemaker cells to reach threshold, and so the heart rate is decreased. The **sympathetic** nervous system speeds up the diastolic depolarization

of the pacemaker cells (and the repolarization as well) and allows the cells to reach threshold sooner; this increases the heart rate. Normally, the resting rate of the heartbeat is somewhat slowed by the parasympathetic nervous system, and if parasympathetic activity is blocked, the heart speeds up. The pacemakers of heart transplant patient's (whose hearts are not connected to the autonomic nervous system) can still respond to circulating hormones (e.g., epinephrine) to produce an increase in heart rate.

The rate of the heart varies in response to physiological needs, increasing during exercise and decreasing at rest. If the resting rate is very low (< 60 bpm or so), the condition is termed **brachycardia;** an elevated resting heart rate (> 100 bpm) is called **tachycardia.** Normally, these rates are under the control of the SA node pacemaker; in other cases, blockages of conduction elsewhere in the heart may lead to a very

(*a*) Echocardiogram of a normal heart. (© Phototake/ Phototake, NYC) (*b*) MRI scan showing a four-chamber view of the heart. (*Dr. Brett Cowan, University of Auckland, Auckland, New Zealand*) (*c*) Arteriogram showing a catheter inserted into the right coronary artery. (© CNRI/Phototake) (*d*) A schematic of the arteriogram.

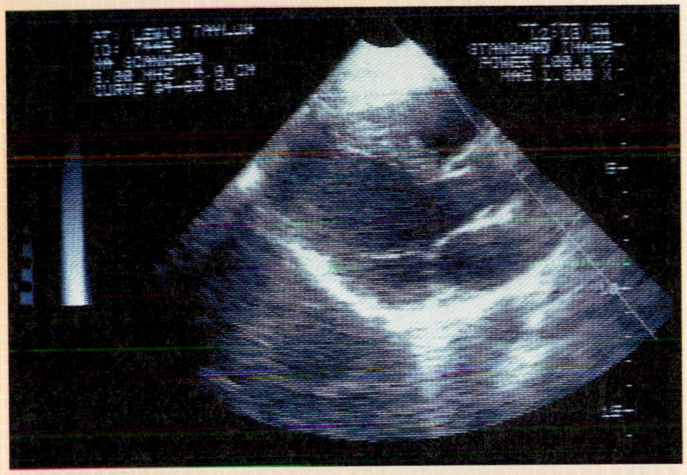

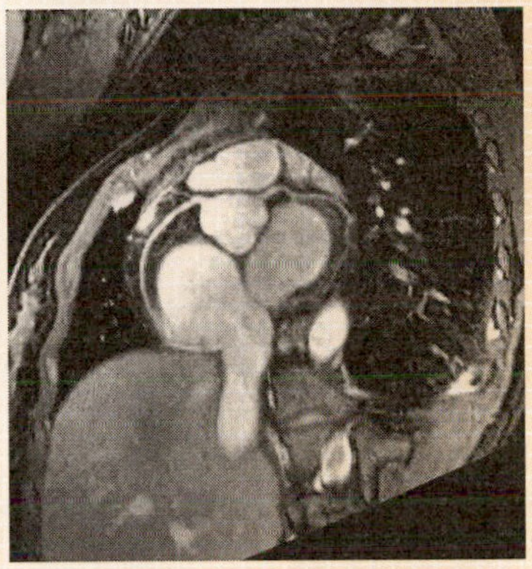

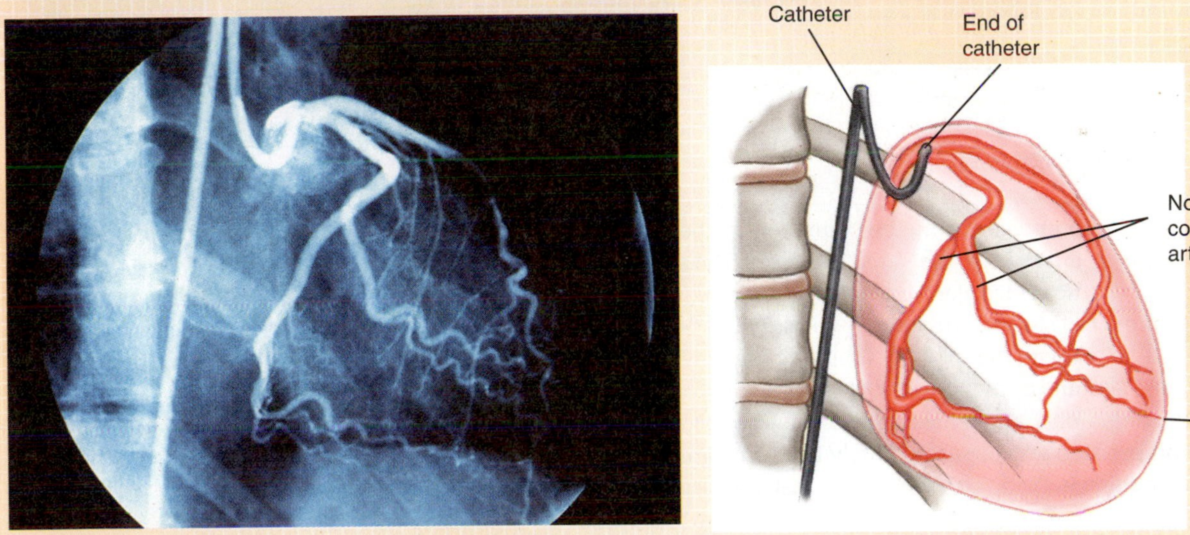

Catheter End of catheter

Normal coronary arteries

Approximate outline of heart

severe brachycardia, and abnormal pacemaker activity at locations other than the SA node can lead to tachycardias with rates as high as 300 bpm.

Drugs and other influences that affect the rate of the heart are called **chronotropic** influences, while those agents that affect the strength of contractions are called **inotropic**.

The Sequence of Cardiac Activation

 What is the pattern of excitation in the heart?

Activation of the Atria. The rhythmic activity of the SA node pacemaker is conducted first into the musculature of the right atrium and then to the left atrium (Fig. 18–13). This impulse conduction occurs when the action potential in one cell causes current to flow in an adjacent resting cell. This local current, which flows between the cells in the region of their intercalated disks (see Chapter 16), brings the membrane potential of the resting cell to its threshold. It then produces its own action potential, which in turn stimulates the next adjacent cell. This process occurs (with varying degrees of efficiency) in all regions of the heart. Conduction of action potentials between separate cells in heart muscle is similar in many ways to conduction along a nerve axon (a single cell), in which active regions stimulate the resting regions next to them. Cell-to-cell conduction in heart muscle is made possible by the low intercellular resistance associated with the intercalated disks.

Conduction through the atria is quite rapid. Atrial action potentials are conducted quickly because the muscle cells are relatively large and have a very negative resting potential. In addition, some specialized atrial conducting pathways conduct very rapidly. The rapid conduction and fast action

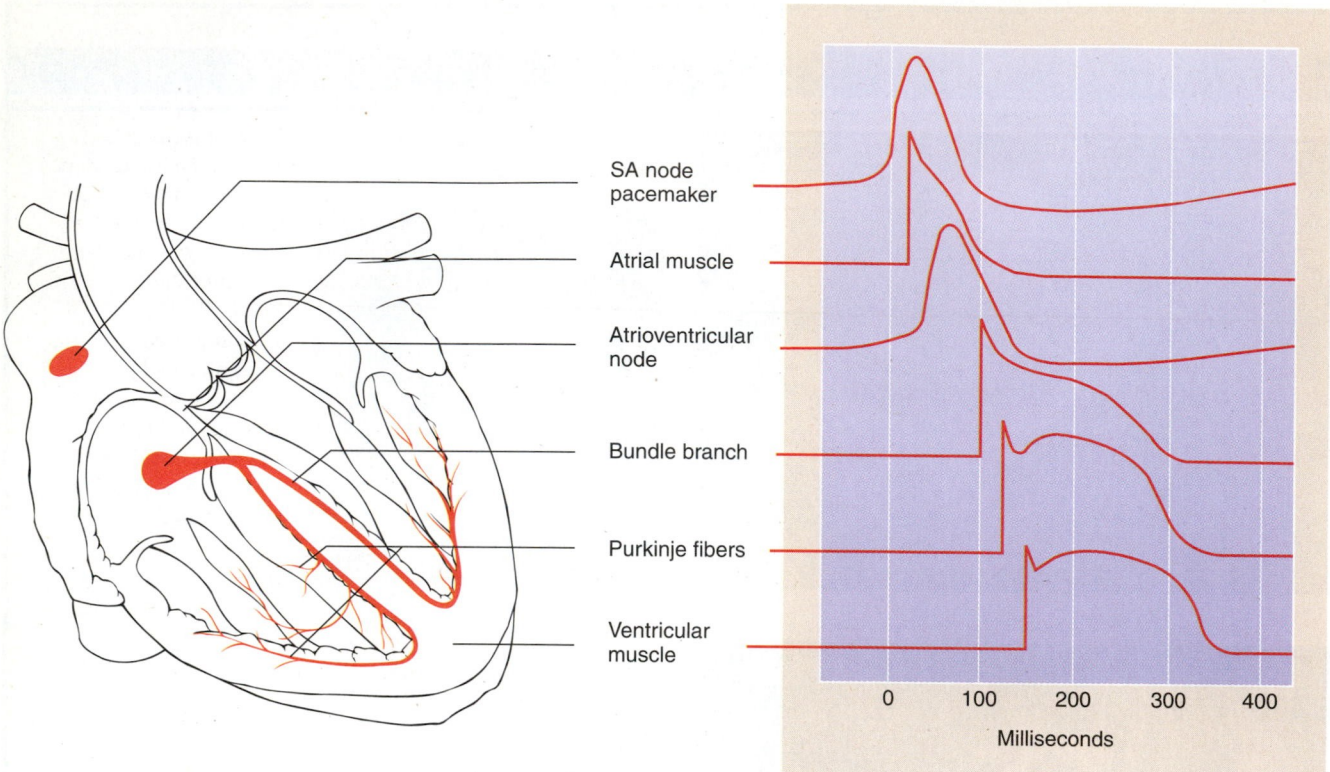

SA node pacemaker

Atrial muscle

Atrioventricular node

Bundle branch

Purkinje fibers

Ventricular muscle

0   100   200   300   400
Milliseconds

Figure 18–13

The succession of action potentials during the course of one heartbeat. The sequence starts with the pacemaker potential at the sinoatrial node. The activity spreads rapidly to the atrial muscle. The atrioventricular node action potential, similar in form to the SA node potential, is conducted slowly, and there is a longer space between this and the first ventricular potential shown, that of the left bundle branch. (This is similar to the slightly earlier action potential of the bundle of His, which is not shown.) Action potentials of the Purkinje fibers and the ventricular muscle follow quickly in succession. The entire activation process takes approximately two tenths of a second from the time the action potential begins in the SA node until the last ventricular muscle is activated.

potentials result in a very rapid Phase 0 depolarization with a large overshoot in the cells, which produces a large local current flow. These conditions favor the rapid conduction of the impulse throughout the tissue.

Conduction of the Action Potential from the Atria to the Ventricles. At the lower portion of the right atrium is another region of specialized cells called the **atrioventricular (AV) node.** These cells are quite small and have resting potentials that are not as negative as those of the atrial cells. As action potentials pass through this region, they become weaker and travel more slowly. This slowed impulse travel is called **decremental conduction.** If the distance is great enough, or the impulse becomes weak enough, it may die out altogether. Under normal conditions, the impulse is slowed but not stopped at the AV node. The delay this causes ensures that the atria finish their contraction before the ventricles begin theirs. The AV node is the only normal connection between the atria and the ventricles, and its delay serves as a

guard against very high atrial heart rates ($\approx$ 200 bpm) being conducted to the ventricles.

Activation of the Ventricles. After the impulse crosses the AV node, it enters the **bundle of His** (also called the **common bundle**). This conducting structure is composed of muscle cells specialized for very rapid conduction, and here the impulse travels at its greatest speed. The common bundle soon divides into two **bundle branches,** one for each of the ventricles. Conduction is very rapid here as well.

The bundle branches divide further in each ventricle, eventually becoming an array of Purkinje fibers, which form a distribution system over the endocardial surface and even penetrate slightly into the main muscle mass of the ventricles. The Purkinje fibers are also specialized muscle fibers with a high speed of conduction.

From the Purkinje fibers, the impulse passes into the muscle of the ventricle wall, stimulating contraction. The ordinary ventricular muscle fibers are somewhat slower con-

APPLICATIONS OF PHYSIOLOGY

Cardiac Pacemakers

The **artificial cardiac pacemaker** is an electronic device that delivers an electrical stimulus to the heart to initiate contraction of the ventricles. Pacemakers frequently are used to treat patients who suffer from permanent or recurrent bradycardias (slow heart rates) with undesirable symptoms (e.g., fainting). The first totally implantable pacemakers were reported in the 1950s; today, millions of patients worldwide have them. Implantable pacemakers are powered by lithium batteries, which may last for six to ten years. Replacement of the battery requires surgery with a local anesthetic to open the skin over the pacemaker.

Early pacemakers were unable to sense the electrical activity of the heart and stimulated it (either the atrium or ventricle) at a fixed rate. Modern pacemakers can sense the heart's normal electrical activity and are activated only when they fail to detect appropriate spontaneous activity. Such **demand-pacemakers** are commonly programmable to respond in a specified manner when a particular arrhythmia is detected. For example, to treat **complete heart block**, a pacemaker is used that senses the depolarization of the atria and then delivers a stimulus to the ventricles after a delay equal to the normal AV conduction time. This method of pacing retains the normal temporal relationship between atrial and ventricular contraction and therefore preserves the normal contribution of atrial contraction to ventricular filling. In addition, by sensing the normal rate of atrial depolarization, the pacemaker can respond appropriately to increases in heart rate associated with exercise and stress. Programmable pacemakers can be reprogrammed without being removed, which allows their operating parameters to be revised to fit the patient's changing needs.

In the past, pacemakers could not correct an arrhythmia called **ventricular fibrillation** in which the pathway of electrical depolarization of the heart becomes completely chaotic. This arrhythmia, if untreated, is fatal because the heart is no longer able to pump any blood. Until recently, the only treatment was to deliver a large shock to the chest of the patient with a machine called a **defibrillator**. Typically these devices are found only in hospitals and ambulances; however, some airlines now put defibrillators on their airplanes. While a patient could be maintained by **cardiopulmonary resuscitation (CPR)** until they reached a hospital or until an ambulance arrived, more often than not, ventricular fibrillation is fatal. Recently, however, a new pacemaker device has become available called an **implanted cardioverter defibrillator (ICD)** or *AICD* for *automatic ICD*. An ICD is a small electric generator that is implanted under the skin of the abdomen. One wire or lead from the generator is attached to the lower surface of the heart. The other two wires are inserted through a neck vein. One lead goes into the right atrium, and the other goes into the right ventricle. An ICD is a type of demand-pacemaker that delivers a large shock to the heart whenever it fails to detect normal electrical activity in the heart. The amplitude of the shock is sufficient to electrically reset all of the excitable cells in the heart, allowing the heart to reestablish normal electrical activity.

---

ductors than the specialized conducting tissues, but by this time the activity is well distributed throughout the heart, and the distance remaining is relatively short. The arrangement of the specialized conducting system ensures a nearly simultaneous activation of all of the ventricular muscle.

The overall effect of the distribution of the stimulating impulse by the rapid conduction system is a wave of depolarization that sweeps over the heart from its base to its apex and from the endocardial to the epicardial surface. Because of the rapid conduction of the impulse and the fast rise of the action potential, the boundary between active and resting tissue is very sharp and moves very quickly. During repolarization, there is also an apparent movement of the active areas.

As active cells return to rest, a less-pronounced wave of repolarization sweeps over the heart in the opposite direction. However, the directionality of the repolarization is determined not by the conduction system but by the fact that the cells activated last have the shortest action potentials and begin to repolarize first. This produces an apparent movement in the opposite direction from the wave of depolarization. Because this is not a conducted phenomenon, and because the repolarization is not nearly as rapid as the depolarization, the speed of the repolarization wave is lower and the boundary between recovering and still-active cells is not very sharp.

Abnormal Activation of the Heart. The proper function of the heart as a pump depends on the correct sequence of activation, and this in turn depends on the normal function of the conducting system. For example, failure of the AV node to conduct the impulse causes **heart block,** in which the ventricles are not stimulated by the atrial impulse. At such times, cells other than those of the SA node can adopt a pacemaker function and provide a source of stimulation, although the rate from such auxiliary pacemakers is usually lower than normal. These "emergency" pacemakers usually lie somewhere in the conduction system, and their activity is normally suppressed by the faster pacemaker rhythm of the SA node. A typical pacemaker from the AV node might have an inherent rate of around 50 bpm, while a ventricular pacemaker could have a rate of 30 bpm.

Regular muscle tissue rarely shows any spontaneous activity. If there is such a pacemaker in the ventricular region — and if for some reason it becomes active from time to time — its impulse is conducted both in the normal direction and "backward" toward the atria. When it stimulates the AV node, it produces refractoriness in the nodal tissue. The next atrial impulse arriving may not be able to be conducted through the AV node, and the ventricles "drop" a beat. This can be very apparent to the person in whom it happens, because the delay gives the ventricles a chance to become filled with extra blood, and the next heartbeat is larger than usual. Such "skipping a beat" is not rare and is usually of no concern.

However, a dangerous condition may arise if the conducting system of the ventricle is damaged so that conduction along parts of it is slowed. Such damage is not uncommon following a heart attack (a **myocardial infarction**) because areas of the heart muscle may have been deprived of oxygen for some time. This slowed conduction can cause cells that are in the process of normal repolarization to be stimulated by a late-arriving action potential, which may arrive during the period of supranormal excitability. The resulting stimulation can give rise to even more action potentials out of their proper time sequence, which are then conducted to normally resting cells. The final result may be an uncoordinated stimulation of the ventricular muscle called **ventricular fibrillation.** Such uncoordinated contractile activity is useless for pumping blood, and death follows shortly. These conduction and rhythm disturbances, and many others related to them, are called **arrhythmias** and are often fatal if not treated immediately.

> *Clinically, how can we determine if the conduction system is working properly?*

The Electrocardiogram. The action potentials and their conduction sequence as just described can be observed only by painstaking study of exposed or isolated hearts. While the knowledge gained is important in understanding the operation of the heart, it is obviously not a realistic approach to the study or treatment of human (or veterinary) patients. Fortunately, the electrical activity of the heart may be harmlessly detected at the body surface by a technique called **electrocardiography.**

The heart may be thought of as an electrical generator lying in a conducting medium made up of the body tissue and fluids. During the phases of the cardiac cycle in which depolarization or repolarization is sweeping over the heart muscle mass, some portions are positively charged and others negatively charged. This potential difference, or **voltage gradient** (as much as 100 mV), causes current to flow in the external medium between these regions of the heart. These currents are strongest close to the heart, and they become weaker at greater distances from the heart. The largest voltage measured at the body surface is only approximately 1 mV; however, this is sufficient to be detected by a sensitive instrument called an **electrocardiograph.** This device converts the minute surface currents to movements of a pen on paper or a spot of light on a cathode ray tube (CRT), producing the familiar **electrocardiogram (ECG or EKG).**

Figure 18–14 is a diagram of a typical ECG recording. For convenience, the principal features of the recording have been designated by the letters P, Q, R, S, and T. The so-called **P-wave** is caused by the depolarization of the atrial muscle, and the **QRS complex** is caused by the depolarization of the ventricles. The **T-wave** is the result of the repolarization of the ventricles. Repolarization of the atria takes place during the ventricular depolarization, and its weak signal is lost. When the entire heart is either depolarized (active) or is resting, there is no potential difference to set up a current flow, and no deflection is seen on the ECG. The line of no deflection is called the **isoelectric line.** Figure 18–14 also shows the relationship between cellular action potentials and the surface ECG.

It is important to realize what the ECG can and cannot show. First, only electrical events are shown, and one can

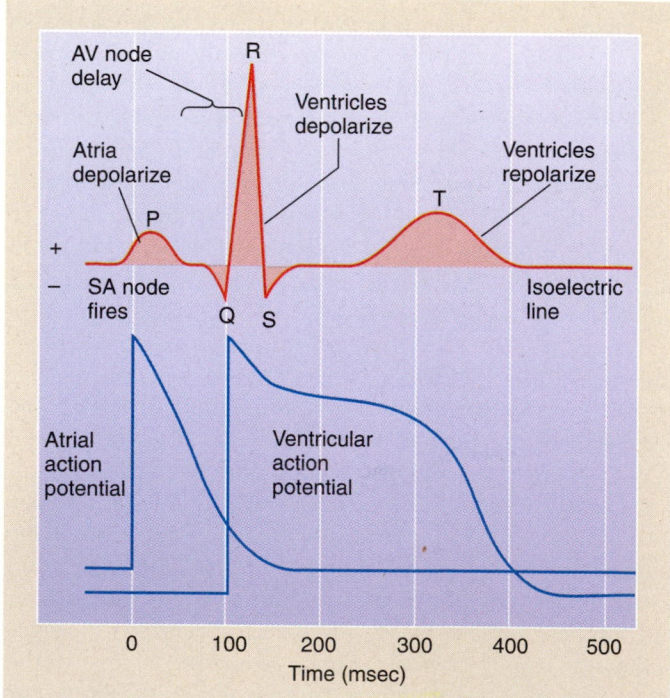

Figure 18–14

The time relationship between the action potentials and the electrocardiogram (ECG). All three traces share the same time axis, and the ECG recording may be used to infer when the actual electrical changes took place in the heart muscle.

only *infer* that a contraction of the muscle has taken place (sometimes it has not). The ECG is a record of **voltage** and **time** only. Second, only the electrical events that involve large amounts of muscle show up in the ECG record. Because of the small tissue mass involved, activity of the conducting system and nodal tissues can only be inferred, but these inferences can be very revealing. For example, if a P-wave is not followed by a QRS complex, then one can assume that conduction was blocked somewhere between the atria and the ventricular muscle. If the QRS complex is very broad or abnormally shaped, one can assume that some areas of the muscle of the ventricle were activated later than they should have been. This implies a defect in the ventricular conducting system. A very long delay between the P-wave and the QRS complex most probably means that the conduction in the AV node was more decremental than usual.

Further information can be gained about the direction of the spread of electrical activity by recording from electrodes located on different areas of the body. The portion of the electrical activity that the recording electrodes intercept depends on their location and orientation. Clinical recording of the ECG assumes that the heart lies at the center of an equilateral triangle and that the recording connections are made at the vertices of this triangle (Fig. 18–15). In practice, the electrical connections to the right and left arms and to the left leg are considered to connect to these vertices, and the standard **limb leads** are recordings be-

tween any two of the connections. A recording made between the right and left arms is designated as **Lead I** and is most sensitive to activity spreading laterally across the heart. **Leads II** and **III** are recordings made between the left leg and the right and left arms, respectively. Records from these leads are most sensitive to activity proceeding from the top to the bottom (base to apex) of the heart. By combining the timing and voltage information recorded from a number of leads, someone skilled in the interpretation of the ECG can build up a very complete picture of the functioning of the heart.

Each Beat of the Heart Involves Integrated Electrical, Biochemical, and Mechanical Events Referred to as the Cardiac Cycle

How does the correlation between electrical and mechanical events in the heart contribute to the heart's capacity to pump blood?

As discussed in the previous section, the electrical activation of the left and right hearts occurs almost simultaneously. As a result, the sequence of mechanical events occurring during a single beat or cycle of the heart is very similar for both hearts. Because of this similarity, portions of the following discussion of the cardiac cycle focus on either the left or the

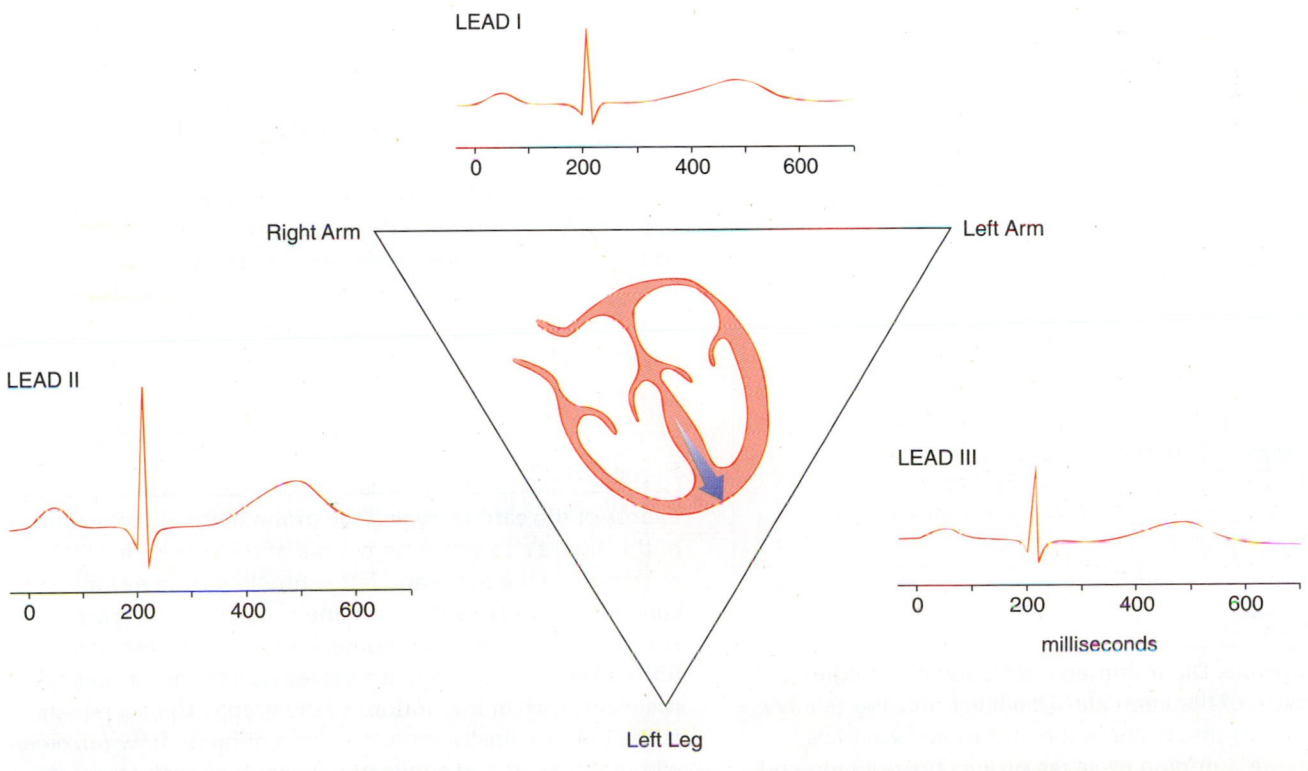

Figure 18–15

Standard electrocardiographic limb leads.

right heart. Keep in mind that similar events are occurring at the same time in both sides of the heart.

The Ventricular Pump

Basically, the ventricles are two-phase pumps; they have a filling phase (Fig. 18–16a and b) and an ejection phase (Fig. 18–16c). The filling phase is a two part process. Initially (see Fig. 18–16a), the ventricle fills passively as venous blood re-

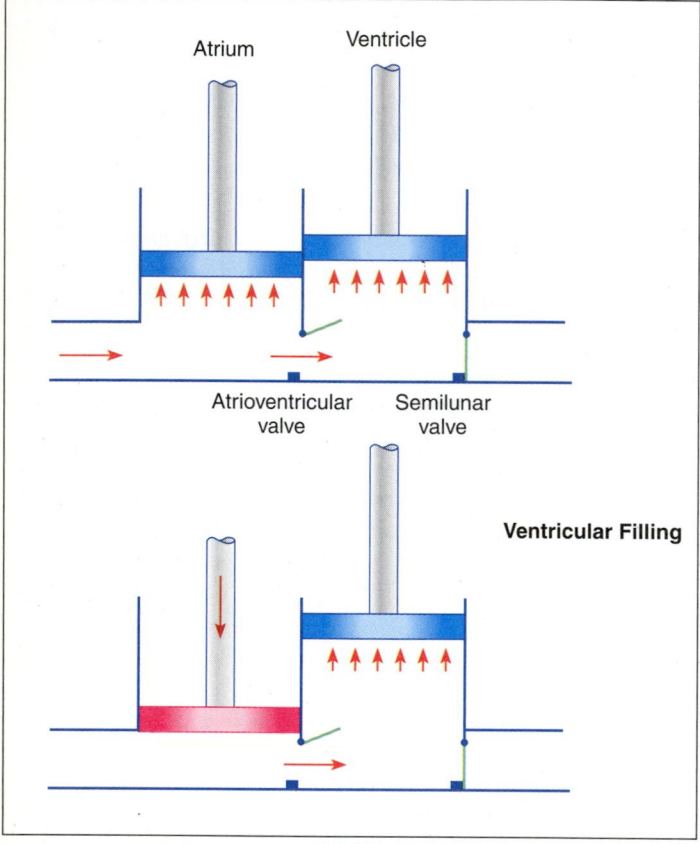

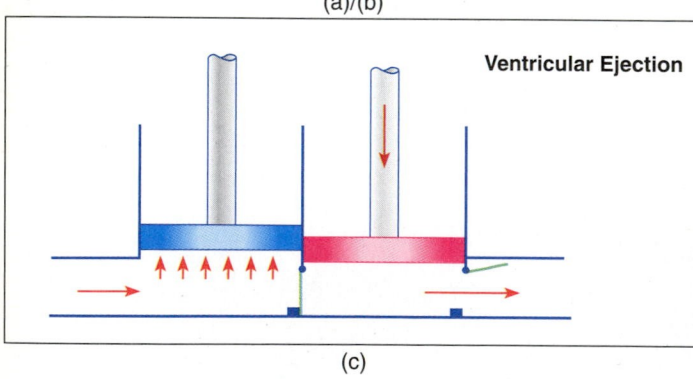

(c)

Figure 18–16

The cardiac cycle. Diagrammatic depiction of a single pumping cycle of the ventricle. This illustrates the relative volume changes that occur in the atrium and ventricle during a single pumping cycle, as well as the opening and closing sequence of the inflow (atrioventricular) and outflow (semilunar) valves.

turns to the heart. At the same time, the atria also fill with blood. Just prior to ventricular contraction, the atria contract and add additional blood to the ventricle (see Fig. 18–16b). As we will see in more detail later, the atrial contraction normally adds little to the filling of the ventricles because there is sufficient time between beats for the relaxed ventricles to fill almost to capacity with blood flowing back to the heart. At very high heart rates, however, the atrial component of filling takes on a much greater importance because of the reduced time available for filling the ventricles. Critical for the unidirectional pumping of blood by the ventricles are the one way atrioventricular and semilunar valves, which prevent the reverse flow of blood. Notice that, during the ventricular filling phase (see Fig. 18–16a and b), the atrioventricular valves open in response to the flow of blood into the ventricles. The semilunar valves, however, are held closed by the higher pressure in the arteries relative to that in the ventricles. This prevents the reverse flow of blood from the artery back into the heart. The second phase (Fig. 18–16c) is the ejection of blood from the ventricles. During this phase the atrioventricular valves are pushed closed by the increased pressure in the contracting ventricle, and this prevents the reverse flow of blood into the atria. In contrast, the increased pressure in the ventricles now pushes the semilunar valves open as blood is ejected from the ventricles. After ejection of blood, the ventricles relax and once again begin to fill with blood (*top diagram*). This completes one cardiac cycle.

The Cardiac Cycle

Now let's look in greater detail at the sequence of events occurring during a single cardiac cycle as they occur in the heart. The sequence of events is depicted in Figure 18–17. Notice that the overall sequence of events occurs almost simultaneously in the left heart and the right heart.

Ventricular Filling. The term **diastole** refers to the period of the cycle during which the ventricles are relaxed and filling with blood, and **systole** refers to the portion of the cycle during which the ventricles are actively contracting and pumping blood out of the heart. Let's begin our examination of the cardiac cycle with the **ventricular filling** phase of diastole

Figure 18–17

Events of the cardiac cycle. The graph in the upper portion of this figure shows a time-course of the pressure measured in the aorta and left ventricle during a single cardiac cycle. Also shown are the occurrence of heart sounds, left ventricular volume, and an electrocardiogram (ECG). The state of the heart valves during the cardiac cycle is shown at the bottom of the graph. The six panels at the bottom illustrate the direction of blood flow (arrows) within the heart and connecting vessels at each stage of the cardiac cycle. Higher pressures are indicated by darker colors.

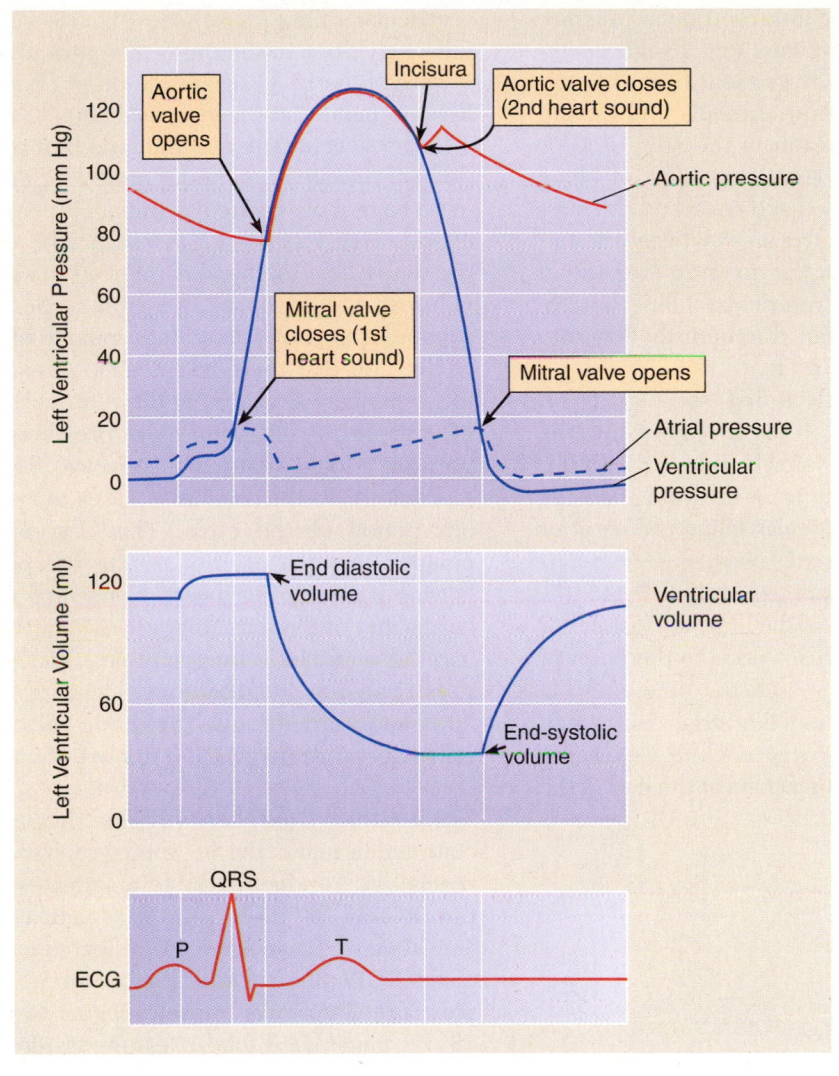

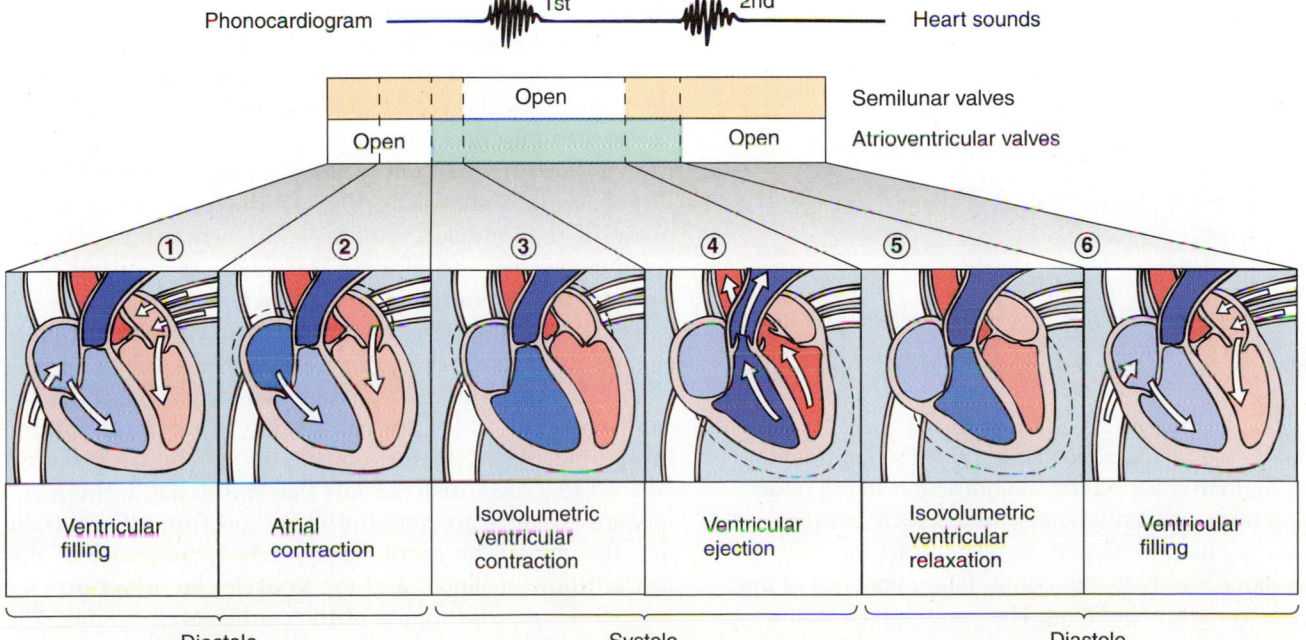

(see Fig. 18–17, *panel 6*). When the intraventricular pressure falls below the pressure in the atria (see Fig. 18–17, *beginning of panel 6*), the forward pressure gradient from the atria to the ventricles forces open the atrioventricular valves, and ventricular filling begins. The pressure in the atria exceeds that in the ventricles because the continuous flow of blood into the atria from the veins has filled and stretched the elastic walls of the atria. Because the atria are relatively compliant, as compared to the ventricles, the pressure increase in the atria is sufficient to facilitate ventricular filling but remains low enough so that blood can continuously flow into the atria during most of the cardiac cycle.

Ventricular filling occurs rapidly at first (see Fig. 18–17, *panel 6*), as the accumulated blood in the atria flows into the relaxed ventricles and then more slowly (see Fig. 18–17, *panel 1*) as blood continues to flow from the veins into the atria. Notice that most of the ventricular filling occurs prior to atrial contraction. Normally, atrial contraction increases the filling of the ventricle by only approximately 10% to 30%. This is because adequate time is available during diastole for nearly complete filling of the ventricles prior to the onset of atrial contraction. As heart rate increases, the time available for ventricular filling is reduced, and the atrial contraction becomes an important contributor to complete ventricular filling (Fig. 18–18). Atrial contraction occurs at the end of the

ventricular filling phase (see Fig. 18–17, *panel 2*) and immediately prior to the onset of ventricular contraction. The P-wave of the ECG signals the onset of atrial depolarization. Keep in mind that the electrical depolarization precedes the mechanical contraction of the atria by a few milliseconds because of the latency period required for muscle activation.

The end of ventricular filling corresponds to the end of diastole (see Fig. 18–17, *panel 2*). The volume of blood in the ventricle at this time is referred to as the **end-diastolic volume.** As we will see in a later section, the end-diastolic volume is an important determinant of cardiac function. Notice also that there is a reverse pressure gradient across the semilunar valves during the ventricular filling phase (aortic pressure > left ventricular pressure; pulmonary artery pressure > right ventricular pressure). This reverse pressure gradient forces the semilunar valves to remain closed during this period of the cycle. This prevents blood that was pumped into the arteries during the previous cycle from flowing back into the heart. Congenital or pathological abnormalities of the semilunar valves allow blood to squirt back into the ventricles during diastole. This condition is referred to as a **valvular insufficiency.** The condition in which blood flows backward through the aortic valve during diastole is called **aortic regurgitation** or **aortic insufficiency.**

Ventricular Contraction. The initiation of ventricular contraction marks the beginning of systole. Recall that the specialized conduction system activates the ventricles approximately 100 msec after atrial activation. This electrical activation of the ventricles is reflected by the QRS complex of the ECG. Shortly after the occurrence of the QRS complex, the ventricular muscle begins to contract (see Fig. 18–17, *panel 3*). Blood pressure inside the ventricles increases rapidly as the ventricular muscle contracts. Ventricular pressure quickly increases to a value that is greater than atrial pressure. This reversal of the pressure gradient causes a momentary reversal of the flow of blood through the atrioventricular valves. The valves fill like boat sails and are pushed into their closed position and held there by the reverse pressure gradient, preventing any additional blood flow back into the atria. During the remainder of this phase of the cardiac cycle, both the atrioventricular and semilunar valves are closed. This phase is referred to as **isovolumetric ventricular contraction** (see Fig. 18–17, *panel 3*) because the volume of the ventricles cannot change while both the inflow and outflow valves remain closed. If the ventricular volume cannot change, then the cardiac muscle fibers are contracting isometrically during this period. When pressure inside the left ventricle has increased to a level that exceeds that in the aorta, there is a forward pressure gradient for blood flow from the ventricle into the aorta. The aortic valve opens in response to this forward flow of blood, and the **ventricular ejection** (see Fig. 18–17, *panel 4*) phase of the cardiac cycle begins. The ejection of blood from the left ventricle occurs quite rapidly at first and then slows as pressure builds up in the

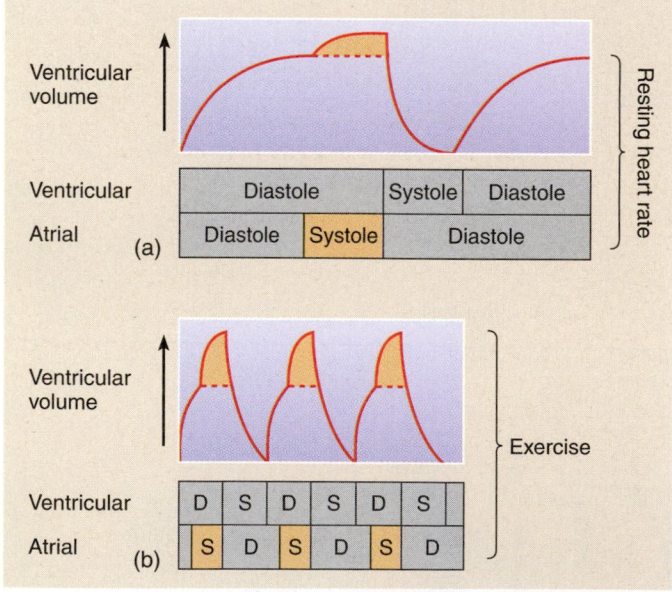

Figure 18–18

Effect of heart rate on diastolic filling. *(a)* Atrial contraction makes only a small contribution to ventricular filling under normal conditions. *(b)* At high heart rates, such as those occurring during heavy exercise, atrial contraction contributes significantly to ventricular filling because of the reduced time available for filling. The ventricular volume added during atrial contraction is indicated by the orange area.

aorta. Just as pressure increases inside a balloon as it is inflated with air, pressure increases inside the aorta as it is inflated with blood. Keep in mind that even as blood is being pumped from the heart into the aorta, blood is flowing out of the aorta into the systemic vasculature. The increase in aortic pressure during ventricular ejection indicates that the flow of blood into the aorta from the heart is greater than the flow of blood out of the aorta into the systemic circulation.

Ventricular Relaxation. The onset of ventricular relaxation immediately follows ventricular repolarization. Recall that the T-wave of the ECG is a measure of the electrical repolarization of the myocardium. Notice in Figure 18–17 that ventricular pressure begins to decrease rapidly after the occurrence of the T-wave and is soon less than the blood pressure in the aorta. This reversal of the pressure gradient (aortic > ventricular) reverses the direction of blood flow in the region of the aortic valve. This reverse blood flow closes the aortic valve, preventing any further flow of blood back into the heart from the aorta. This momentary reversal in blood flow results in a decrease in the aortic pressure that appears as a dip or notch on the aortic pressure tracing that is referred to as the **incisura** (see Fig. 18–17). The closure of the semilunar valve in response to ventricular relaxation marks the end of systole and the beginning of diastole.

The first phase of diastole is referred to as **isovolumetric relaxation** (see Fig. 18–17, *panel 5*); relaxation of the ventricles occurs at constant volume because both the atrioventricular and semilunar valves are closed. As soon as ventricular pressure falls below atrial pressure, the forward pressure gradient from the atria to the ventricles initiates blood flow into the ventricles through the one-way atrioventricular valves (see Fig. 18–17, *panel 6*).

Heart Sounds

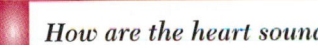

> *How are the heart sounds generated?*

One of the physical manifestations of the intense mechanical activity occurring with each contractile cycle of the heart is a series of distinct heart sounds that correlate with specific events of the cardiac cycle. These sounds can normally be heard through a stethoscope placed against the chest, directly over the heart. Heart sounds can also be recorded electronically, as shown by the **phonocardiogram** in Figure 18–17.

Although four heart sounds can be observed on a phonocardiogram, typically only the first two can be heard with a stethoscope and have a sound similar to a low-pitch "lub-dub." The **first heart sound** occurs at the time of closure of the AV valves, and, as previously mentioned, marks the beginning of systole. The **second heart sound** occurs at the time of closure of the semilunar valves and defines the end of systole. These sounds are the result of vibrations in the heart

and chest initiated by the closure of the heart valves and the sudden cessation of blood flow. A common example of how vibrations of this sort can occur is seen in the plumbing of older buildings. The rumbling or clanging noise that occurs when a faucet is turned off is the result of vibrations set up in the pipes as the moving column of water is suddenly brought to a halt.

A number of additional heart sounds associated with abnormal conditions of the heart can occur. For example, valvular obstructions can result in a high velocity jet of blood flowing through the narrowed opening of an obstructed valve, which produces higher pitch sounds called **murmurs.** Likewise, valve abnormalities that allow blood to leak backward (e.g., from the aorta to the ventricle) often are associated with audible murmurs that can be heard during the normal periods of silence. The interpretation of abnormal heart sounds is based on the quality and intensity of the sound as well as on the timing of the sound relative to other normal heart sounds.

The Heart Is Controlled by Changing both the Strength and Rate of Contractions to Produce Whatever Cardiac Output Is Required

> *What factors contribute to determining the cardiac output?*

As we have already stated, the major purpose of the heart is to maintain an adequate flow of blood through the pulmonary and systemic circulations. The amount of blood pumped out of the left ventricle and into the aorta during a period of one minute is defined as the **cardiac output.**

Recall that the left and right hearts are in series and must pump exactly the same volume of blood over any extended period of time. Therefore, if the average output of the left heart is 5 L/min, then the average output of the right heart must also be 5 L/min. Even though the right heart generates significantly lower pressures during systole than the left heart, the volume of blood pumped by the two hearts must be the same.

The average cardiac output for a resting male is approximately 5 L/min; however, this value can increase or decrease significantly in response to the changing needs of the body or in response to disease. In order to understand how the cardiac output can be altered, we must understand the many factors that determine its magnitude.

We know from our previous discussion that the heart is a pulsatile pump that ejects a volume of blood with each beat or cycle. The volume of blood ejected in a single beat is called the **stroke volume** and is simply the difference between the volume of blood in the heart at the end of diastole (end-diastolic volume, or EDV) and the volume of blood remaining in the heart at the end of systole (end-systolic volume, or ESV), as shown in Figure 18–19. The cardiac output is simply the sum of all of the stroke volumes

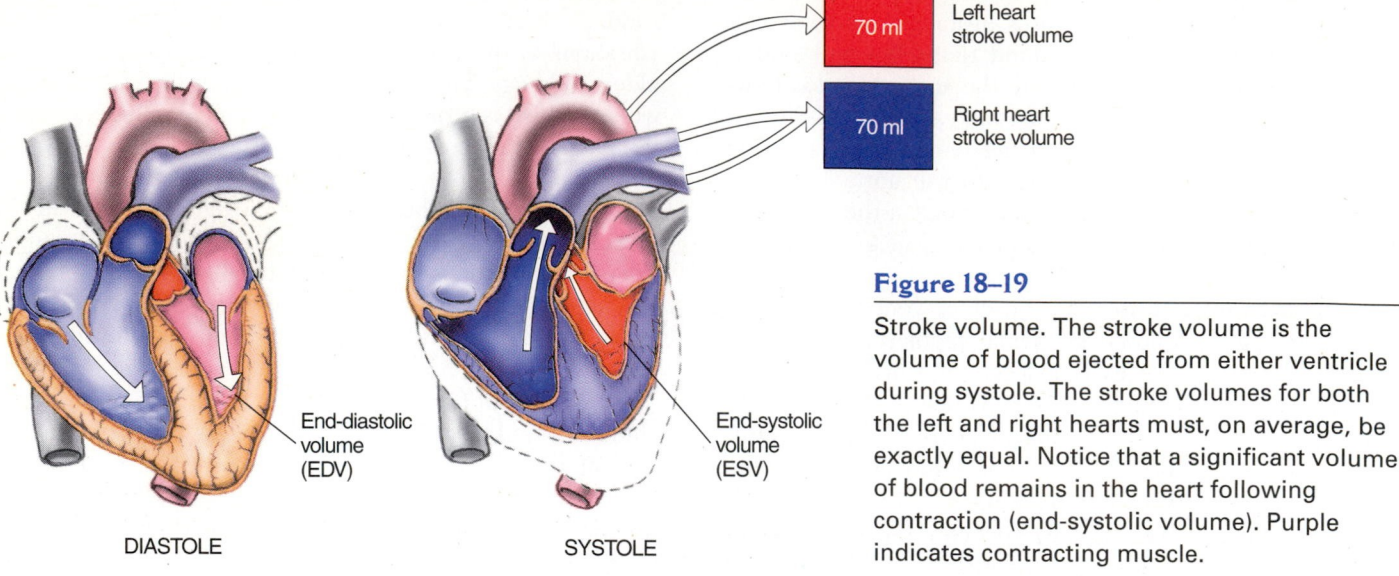

Figure 18–19

Stroke volume. The stroke volume is the volume of blood ejected from either ventricle during systole. The stroke volumes for both the left and right hearts must, on average, be exactly equal. Notice that a significant volume of blood remains in the heart following contraction (end-systolic volume). Purple indicates contracting muscle.

ejected from the heart over a period of 1 minute. If we know the stroke volume (*SV*) and the heart rate (*HR*; number of beats in 1 minute), then we can calculate cardiac output (*CO*) as follows:

$$CO = HR \times SV$$

An average adult at rest might have a heart rate of 72 bpm and a stroke volume of 70 ml/min (70 ml = 0.070 L = 2.4 oz). On this basis, the cardiac output would be:

$$CO = 72 \text{ bpm} \times 0.070 \text{ L/beat} = 5.0 \text{ L/min}$$

The total volume of blood contained in all the blood vessels and the heart of a 70-kg male is approximately 5 liters, or slightly more than 1 gallon. Therefore, a volume equal to the total blood volume is pumped through the heart each minute. This means that, on average, a blood cell will require 1 minute to travel from the left heart, through the systemic circulation, through the right heart, through the pulmonary circulation, and back to the left heart. It is interesting to consider that, over a 70-year life span the heart will pump about 200 million L of blood. This is approximately 53 million gallons — enough to fill 564 million 12-oz cans, which, if placed end to end, would circle the earth twice.

During strenuous exercise, cardiac output may increase four to five times in a physically fit individual and by as much as seven times in a well-trained marathon runner. From the equation for calculating cardiac output, we can see that changes in either stroke volume or heart rate can alter cardiac output. During strenuous exercise, for example, heart rate may increase to 180 to 200 bpm. Let's look at each of these factors in turn and the mechanisms by which they are controlled.

Control of Heart Rate

You will recall from the previous discussion of the electrical activation of the heart that heart rate is determined by the rate of spontaneous depolarization of the SA node. The rate of SA nodal depolarization can be altered by a number of influences, including stimulation of autonomic nerve fibers that innervate the SA node; circulating hormones, such as epinephrine; plasma electrolyte concentrations; and body temperature.

Let's consider the effects of autonomic nerve activity on heart rate. Figure 18–20 shows how an increase in the firing rate of the sympathetic and parasympathetic cardiac nerve fibers alter the rate of spontaneous depolarization of the SA node. Norepinephrine, the sympathetic neurotransmitter, increases the rate of spontaneous depolarization of the SA nodal cells so that threshold is reached more quickly and the heart rate is increased. Acetylcholine, the parasympathetic neurotransmitter, in contrast, decreases the rate of spontaneous de-

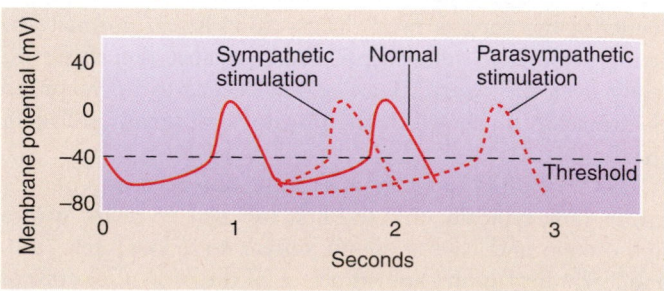

Figure 18–20

Effect of autonomic nerve activity on spontaneous depolarization of the SA node.

polarization and, in addition, hyperpolarizes the pacemaker cells. These two effects of parasympathetic stimulation work together to increase the time required for the membrane potential to reach threshold, with a resultant decrease in heart rate. Under resting conditions, both the parasympathetic and sympathetic cardiac fibers are active. If the heart is denervated, either surgically or pharmacologically, the heart rate increases to approximately 100 bpm. The fact that the usual resting rate for an innervated heart is approximately 70 bpm indicates that, under resting conditions, the parasympathetic nerves dominate in the control of the SA node.

There are definite limits on how much cardiac output can be increased by increasing heart rate alone. First, the upper limit for conduction of impulses through the AV node limits heart rate to approximately 250 bpm. Secondly, with rapid heart rates (**tachycardia**) of more than approximately 150 to 170 bpm, cardiac output begins to decrease because of inadequate time for complete filling of the ventricles during diastole (see Fig. 18–24). If the amount of blood returning to the ventricle during diastole is decreased, we might suspect that the amount of blood pumped by the heart (stroke volume) also would be decreased. As we will see in the next section, stroke volume is *very* dependent on ventricular filling. Recall that a normal mechanism for increasing heart rate is to increase sympathetic nerve activity to the heart. In the next section you will see how an increase in sympathetic nerve activity increases the available filling time.

Changes in body temperature also produce changes in heart rate. This is of little use for normal control of heart rate because body temperature remains relatively constant; however, cardiac surgeons often decrease the body temperature to reduce the heart rate and decrease the motion of the heart during surgery.

Control of Stroke Volume

The volume of blood pumped by the heart with each beat is highly regulated both by mechanisms that are **intrinsic** to cardiac muscle itself and by **extrinsic** factors such as neural stimulation and circulating hormones. Let's begin by looking at the intrinsic mechanisms for regulation of stroke volume.

Intrinsic Control of Stroke Volume. The heart has the intrinsic ability to adjust its output (stroke volume) in response to changes in the input (venous return). This property of the heart is called the **Frank-Starling law of the heart** in honor of two physiologists, Otto Frank and Ernest Starling, who first described this property of the heart in the early 1900s. As shown in Figure 18–21, a proportional relationship exists between the diastolic volume of the heart and the stroke volume over a relatively wide range of end-diastolic volumes. If the rate of blood flow to the heart increases, the end-diastolic volume of the heart also increases, causing an increase in the stroke volume. Simply stated, the Frank-Starling law says that, within defined limits, the heart pumps whatever volume of blood it receives.

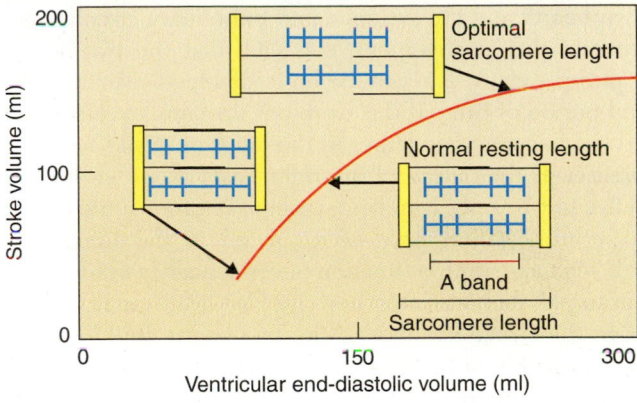

Figure 18–21

Frank-Starling law of the heart. This graph illustrates the relationship between stroke volume and changes in ventricular end-diastolic volume. The insets, showing diagrammatic sarcomeres, illustrate the relationship between end-diastolic volume and myofilament overlap. At normal resting ventricular volumes, sarcomere length is less than the optimal length for contraction.

The mechanism underlying the Frank-Starling relationship is partially due to the same mechanism that was discussed in Chapter 16 for the muscle length-tension curve. When the end-diastolic volume of the heart is increased, the length of the cardiac muscle fibers in the wall of the ventricle is also increased (see Fig. 18–21). Recall from Chapter 16 that cardiac muscle, like skeletal muscle, has a well-defined length-tension relationship, which indicates that there is an optimal length for muscle contraction (see Fig. 16–13). Although skeletal muscle generally is constrained by the skeletal attachments so that it operates near the optimal length for contraction, this is not true of cardiac muscle. In fact, in a normal resting heart, the muscle fibers are *shorter* than their optimal length. Therefore, an increase in end-diastolic volume increases the muscle fiber length toward the optimal length for contraction. This increases the vigor of ventricular contraction, resulting in the expulsion of a greater stroke volume from the heart. Conversely, a decrease in end-diastolic volume results in an immediate reduction in the vigor of contraction and a decrease in the stroke volume. Other factors that contribute to the Frank-Starling relationship are changes in the calcium sensitivity of the contractile proteins with sarcomere length, and effects of muscle length on the rate of influx of extracellular calcium.

The Frank-Starling mechanism affects each ventricle independently and adjusts the vigor of contraction on a beat-by-beat basis. This ensures that each ventricle pumps out the same volume that it receives. Although the full significance of this mechanism for regulation of cardiac output will become more apparent as we discuss the overall regulation of the cardiovascular system, one aspect can be appreciated now, based on our previous discussion of the serial arrangement of the two hearts and the systemic and pulmonary circulations. Because of this arrangement, it is vital that the two hearts each pump exactly the same volume of blood over any extended period of time. If this were not the case, excess blood would soon accumulate either in the lungs or in the systemic circulation. If the output of the right heart exceeded that of the left heart by only 1 ml/beat, within 90 minutes the entire blood volume would have accumulated in the lungs. Although such an extreme situation never actually would occur, this example demonstrates how rapidly even small imbalances in the stroke volumes of the two hearts could lead to the accumulation of blood in one portion of the cardiovascular system. An increase in the amount of blood in the pulmonary vasculature can compromise gas exchange in the lungs, with severe and often fatal results. The ability of the two hearts to independently adjust their output to match their input ensures that the average cardiac output of the two

hearts remains exactly equal. Later, we will discuss how the Frank-Starling mechanism contributes to the maintenance of normal arterial blood pressure.

The Frank-Starling mechanism also provides intrinsic regulation of cardiac output in response to changes in aortic pressure. Consider the effect on the heart of a sudden increase in mean aortic pressure from 90 mm Hg to 120 mm Hg. In order for the heart to eject blood, the ventricular pressure must now increase to 120 mm Hg before the aortic valve will open, and blood must be ejected against a greater pressure. As a result of this increased pressure load, more time is required for the ventricles to develop sufficient pressure to open the semilunar valves, and in addition, the muscle fibers shorten more slowly against this increased afterload. As a result, less blood is ejected. If less blood is ejected (reduced stroke volume), more blood remains in the heart at the end of systole (increased end-systolic volume).

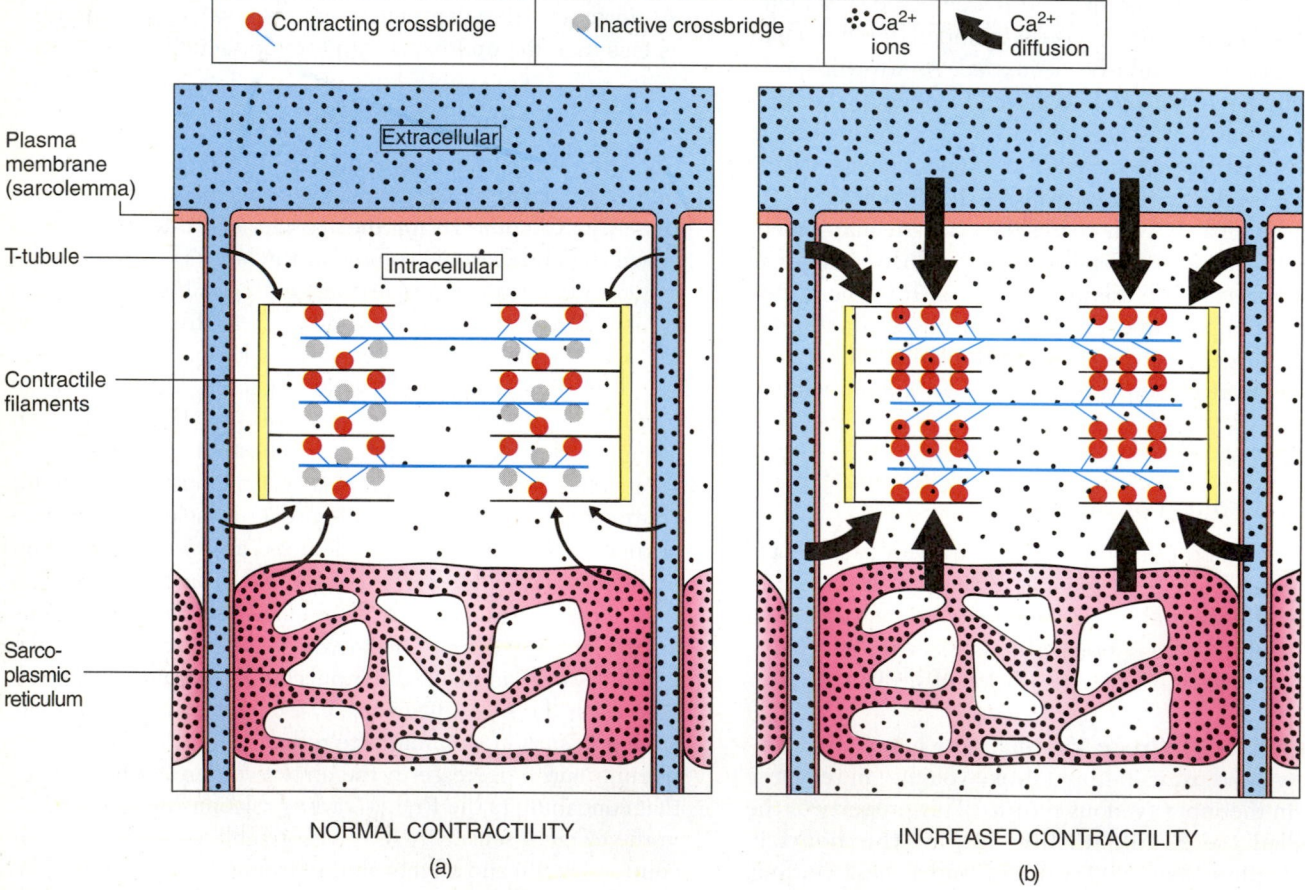

NORMAL CONTRACTILITY
(a)

INCREASED CONTRACTILITY
(b)

Figure 18–22

Relationship between contractility and intracellular calcium. *(a)* Normal contractility. *(b)* An increase in contractility is the result of an increase in the cytoplasmic calcium concentration. This is the result of both an increased release of calcium from the sarcoplasmic reticulum and an increased influx of calcium into the cell from the extracellular space. The increased concentration of intracellular calcium results in the activation of additional crossbridges, with a resultant increase in the vigor of cardiac muscle contraction. Active crossbridges are indicated in red.

This extra blood in the heart, when added to the normal diastolic filling from the pulmonary vein, results in an increase in the end-diastolic volume of the heart. The Frank-Starling mechanism provides for a more forceful contraction on the next beat of the heart and results in an increase in the stroke volume back toward normal. In this manner, the Frank-Starling mechanism helps to maintain a constant cardiac output in the face of changes in aortic blood pressure.

If aortic blood pressure remains chronically elevated, the heart can adjust its capacity to pump blood in a more permanent fashion. Just as skeletal muscle **hypertrophies** in response to exercise, cardiac muscle also responds to chronically elevated pressure or demands for increased cardiac output by increasing the size and contractile protein content of the individual muscle cells of the heart.

Cardiac hypertrophy of both ventricles is readily apparent in athletes who perform intensive aerobic exercise (e.g., marathon runners). Cardiac hypertrophy can also be a response to an abnormal load placed on the heart because of disease, such as hypertension. In many cases, this results in selective hypertrophy of a single chamber of the heart. For example, blockage of the pulmonic valve (**pulmonic stenosis**) results not only in a systolic murmur but also in hypertrophy of the right ventricle because the right heart must generate higher than normal pressures to force blood through the partially occluded valve. The mechanism by which an increased load on the heart ultimately leads to increased production of contractile proteins by the myocardial cells is not currently understood.

Extrinsic Control of Stroke Volume. A number of factors extrinsic to the heart can also influence the vigor of ventricular contraction without changing end-diastolic volume. Any changes in the vigor of cardiac contraction that occur independently of changes in end-diastolic volume are referred to as changes in **contractility.** A change in contractility is mechanistically different from the altered vigor of contraction seen with changes in muscle length. Changes in the contractility of the heart are the direct result of changes in the rate and extent of calcium movement into the cytoplasm, as indicated diagrammatically in Figure 18–22. Recall from Chapter 16 that the concentration of calcium ions in the cytoplasm determines the degree of muscle activation. Increased firing of cardiac sympathetic nerve fibers results in an increase in both the rate and extent of calcium movement into the cytoplasm, from both the sarcoplasmic reticulum and from outside the cell. This results in a more rapid and more forceful contraction of the ventricles and an increase in the stroke volume ejected from the heart with each beat. As shown in Figure 18–23, an increase in stroke volume due to an increase in contractility results in a reduced end-systolic volume. Cardiac drugs are available that increase the contractility of the heart, primarily by increasing the intracellular concentration of calcium. One of the oldest and most widely used drugs is **digoxin,** a plant extract.

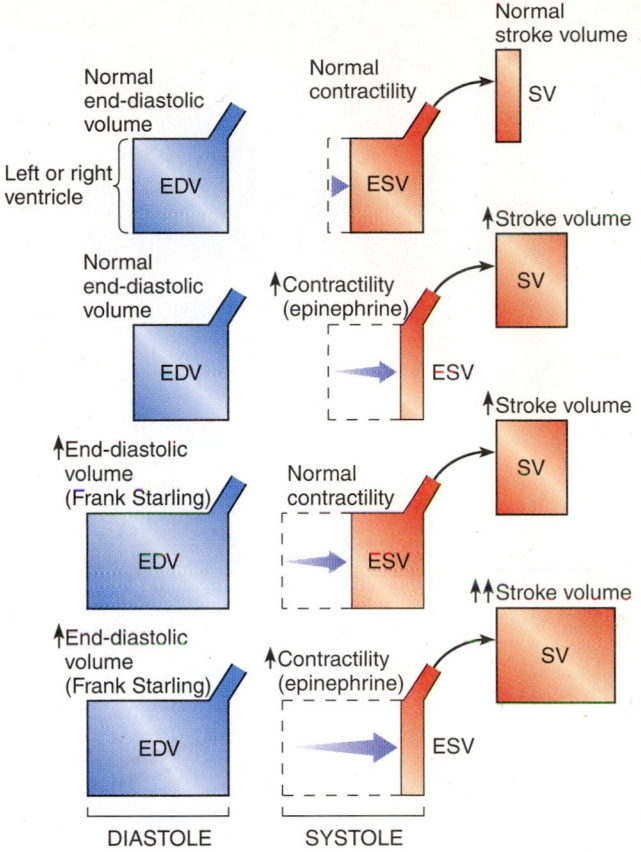

Figure 18–23

Changes in stroke volume due to changes in contractility are mechanistically different from those occurring as a result of increased end-diastolic volume. Therefore, the two mechanisms can operate simultaneously to increase stroke volume, as shown in the lower panel. EDV = end-diastolic volume; ESV = end-systolic volume; SV = stroke volume.

Sympathetic stimulation also increases the rate of relaxation by increasing the rate at which the sarcoplasmic reticulum removes calcium from the cytoplasm. Because sympathetic stimulation increases both the rate of contraction and the rate of relaxation, the duration of systole is decreased. This effect can indirectly lead to a further increase in cardiac output. Sympathetic stimulation of the heart generally results in an increase in heart rate. Recall from the previous section that increases in cardiac output due to increases in heart rate were limited at higher rates by inadequate filling time during diastole (see Fig. 18–24a and b). The effect of shortening the duration of systole, at any given heart rate, is to increase the duration of diastole, and consequently, increase ventricular filling time (Fig. 18–24c).

Increased firing of the cardiac parasympathetic nerve fibers decreases the contractility of the heart but not by a direct action of the parasympathetic neurotransmitter (acetylcholine) on cytoplasmic calcium concentrations. Acetylcholine blocks or antagonizes the effects of sympathetic stimulation on the heart and, in addition, inhibits release of norepinephrine from nearby

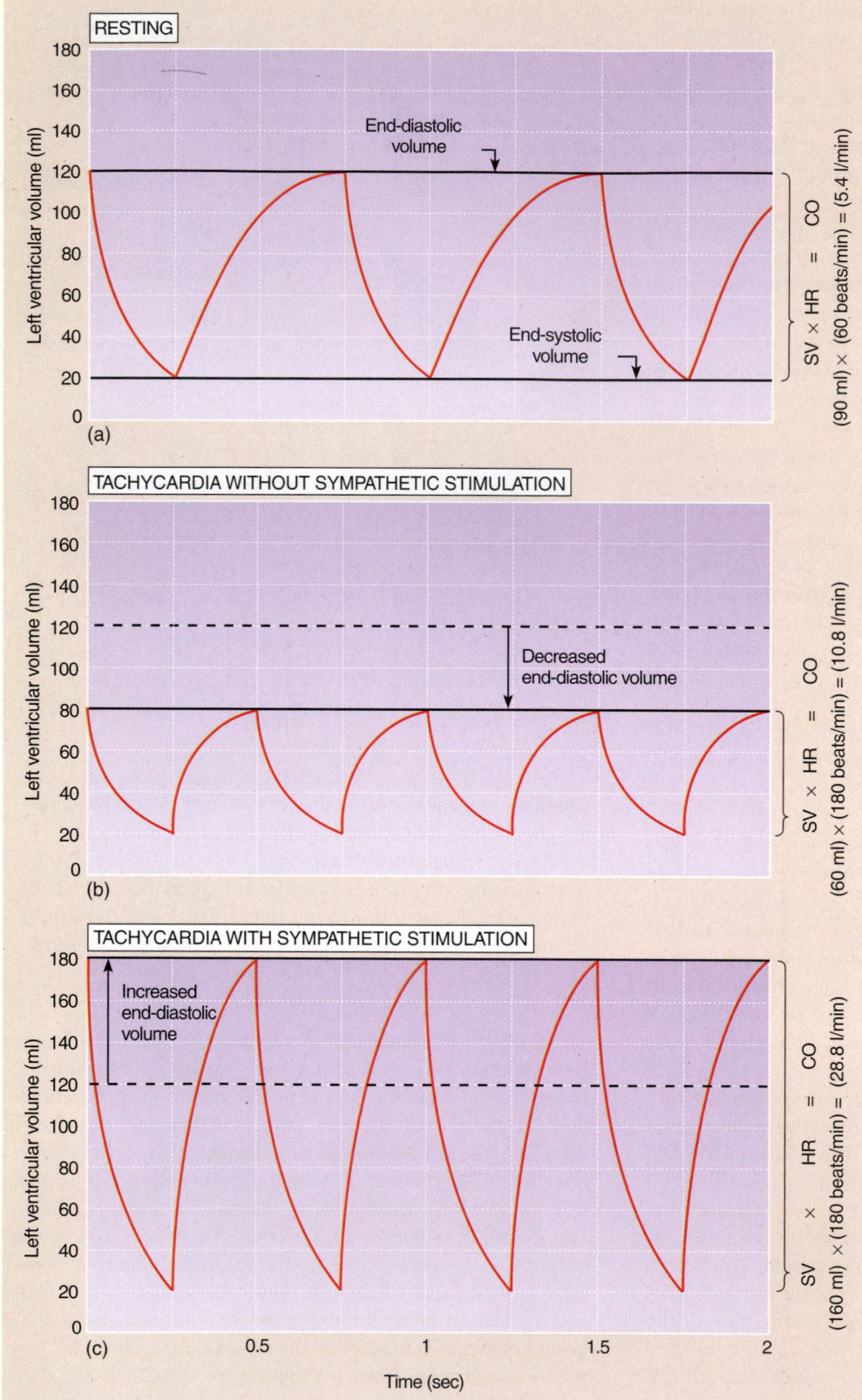

(a) RESTING

End-diastolic volume

End-systolic volume

SV × HR = CO
(90 ml) × (60 beats/min) = (5.4 l/min)

(b) TACHYCARDIA WITHOUT SYMPATHETIC STIMULATION

Decreased end-diastolic volume

SV × HR = CO
(60 ml) × (180 beats/min) = (10.8 l/min)

(c) TACHYCARDIA WITH SYMPATHETIC STIMULATION

Increased end-diastolic volume

SV × HR = CO
(160 ml) × (180 beats/min) = (28.8 l/min)

Time (sec)

Figure 18–24

Increased cardiac output with sympathetic stimulation (effect of increasing the rate on contraction and relaxation of the heart). *(a)* Changes in left ventricular volume in the resting heart of a marathon runner and calculated cardiac output *(right)*. *(b)* Effect of tachycardia without sympathetic stimulation on ventricular volume and cardiac output. Tachycardia of this nature results from cardiac arrhythmias, such as atrial tachycardia. *(c)* Tachycardia induced by sympathetic stimulation and associated cardiac output. The increased rate of contraction and relaxation increases the time available for diastolic filling of the ventricle, leading to increased stroke volume and cardiac output compared with tachycardia *(b)*.

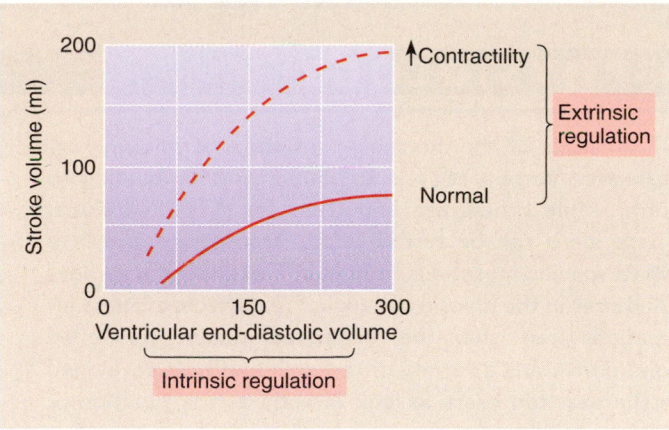

Figure 18–25

Summary of factors that affect stroke volume.

sympathetic nerve fibers. In the absence of sympathetic nerve activity, increased firing of ventricular parasympathetic nerve fibers has only a small depressant effect on contractility.

In summary, stroke volume can be altered by two basic mechanisms. Changes in end-diastolic volume result in changes in stroke volume (Frank-Starling law). Changes in contractility, on the other hand, are the result of changes in intracellular calcium and do not require a change in end-diastolic volume. Because these two mechanisms for changing stroke volume are based on independent mechanisms, they can add together to produce even greater changes in stroke volume. The interaction of these two mechanisms is shown graphically in Figures 18–23 and 18–25. Notice that changes in stroke volume due to changes in contractility can occur at any end-diastolic volume. Likewise, changes in stroke volume due to changes in end-diastolic volume can occur at any level of contractility. Figure

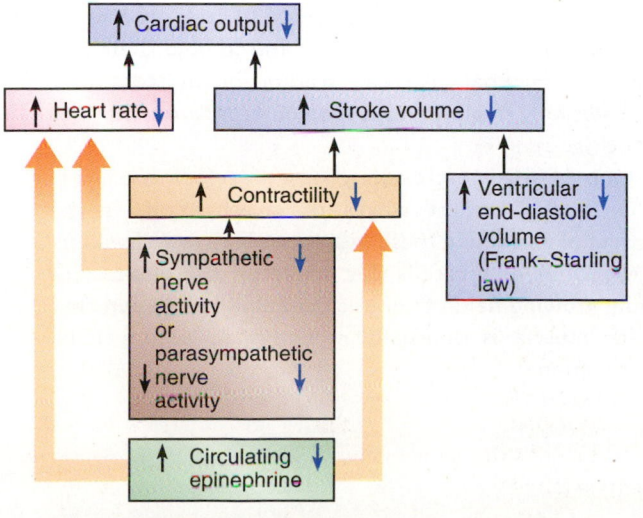

Figure 18–26

Summary of mechanisms that affect cardiac output.

18–26 summarizes the different mechanisms that we have talked about by which the cardiac output can be altered. Cardiac output can be increased by increases in stroke volume or heart rate. Stroke volume can be increased by increases in end-diastolic volume (Frank-Starling law) or by increases in contractility. Both contractility and heart rate can be increased by increases in the level of sympathetic nerve activity to the heart or by increases in the release of epinephrine from the adrenal medulla. A reduction in aortic pressure increases stroke volume by reducing the load on the heart. Although an increase in either heart rate or stroke volume results in a proportional increase in cardiac output, the metabolic costs of increasing cardiac output by these two mechanisms are not equivalent, as we will see in the next section.

Energy Consumption by the Heart

 How does cardiac muscle get energy for contraction?

Contraction of all muscle types is associated with the hydrolysis of ATP. Cardiac muscle derives ATP primarily from the metabolism of circulating glucose and fatty acids. An important difference between cardiac muscle and other muscle types is that cardiac muscle must contract continuously, without interruption, throughout the life of an individual. Recall from Chapter 16 that skeletal muscle is able temporarily to sustain high levels of contractile activity by anaerobic metabolism of circulating glucose and stored glycogen. Because skeletal muscle can metabolize glycogen stored within the muscle cells, the supply of oxygen and glucose to the muscle by the circulating blood does not limit the production of ATP. Anaerobic metabolism results in the accumulation of lactate and the depletion of muscle glycogen. Upon cessation of muscular activity, lactate levels are reduced and glycogen stores are replenished.

Cardiac muscle, in contrast, cannot stop contracting to accumulate stored glycogen for support of anaerobic metabolism. The heart must instead rely upon continuous, oxidative metabolism of circulating glucose and fatty acids to produce ATP. This means that the ability of the heart to contract is strictly limited by the availability of oxygen supplied by the coronary circulation. As we will see in the next chapter, a number of diseases can prevent adequate cardiac blood flow. In conditions in which the heart is receiving inadequate oxygen, we might anticipate that cardiac function would be impaired. This is indeed the case, and one goal in the treatment of heart failure is to maintain adequate cardiac output while minimizing the oxygen requirements of the heart. Although at first glance this might appear to be an unresolvable situation, it turns out that the relationship between oxygen supply and cardiac output is not the same under all conditions. For example, at high heart rates and low stroke volumes, the heart uses more nutrients, and therefore more oxygen, to maintain a given cardiac output than at lower heart rates and larger stroke volumes. Blood pressure also affects the energy cost of maintaining cardiac

APPLICATIONS OF PHYSIOLOGY

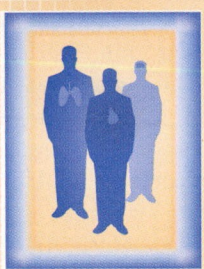

New Tools for Investigating Congenital Heart Disease

Less than three years before Boston Celtics star Reggie Lewis collapsed on the basketball court from a heart attack in the summer of 1993, scientists made an important discovery that would ultimately reveal the link between genetics and heart disease. The first major clue came when the mutation of a single amino acid in myosin, one of the major contractile proteins in the heart, was discovered in 1990. Since that time, more than 100 mutations that result in heart disease have been discovered. Some mutations cause anatomical defects in the heart. Other mutations disrupt normal function of ion channels that govern the electrical activity of the heart. Mutations of myosin and other proteins that are directly involved in muscle contraction lead to **cardiomyopathies**. Hypertrophic cardiomyopathy is a disease that is associated with abnormal growth (i.e., hypertrophy) of the ventricles of the heart. This devastating disease appears usually in late adolescence with no, or very little, advance warning. Approximately 3% of these individuals die each year from sudden cardiac death. Roughly half of all hypertrophic patients have an inherited form of the disease known as *familial hypertrophic cardiomyopathy*.

The mystery of how a single mutation in one of the millions of amino acids that make up the proteins of the heart could lead to such devastating disease has been most clearly unraveled for mutations of myosin. This is largely the result of a number of recent discoveries about the molecular structure and function of myosin molecules. One of the most recent, and most exciting advances was the publication of the crystallographic structure of myosin. By analyzing the diffraction pattern produced by directing a beam of X-rays through a crystal of purified myosin, scientists have deduced the atomic structure of the myosin molecule. Using sophisticated modeling programs, three-dimensional representations of the myosin molecule can be generated (Fig. a). It is now possible to begin to visualize exactly how myosin binds to actin and how this interaction is coupled to the breakdown of ATP to generate molecular motion and force (Fig. b).

Analysis of the molecular structure of myosin suggests that certain regions are important for binding to actin, while others are important for ATP hydrolysis. These ideas can be tested using "transgenic mice," in which specific mutations have been introduced at predefined sites in the myosin molecule. The effects of these alterations can then be examined under controlled conditions using a variety of research methods developed in the past ten years to look directly at the function of single myosin molecules. It was first discovered that a myosin mutation could alter mechanical function using a "sliding-filament" motility assay. In this assay, purified mutant myosin molecules are attached to the surface of a glass coverslip, as shown in the diagram (Fig. c). To this are added chemically modified actin filaments that fluoresce (i.e., give off red light) when exposed to green light. The light emanating from the actin filaments can be visualized (Fig. d) with a light microscope, appropriate color filters, and a very sensitive video camera of the sort first developed for nighttime recognizance by the military (i.e., night-vision devices). In the presence of ATP, the actin filaments move over the surface with a velocity that is determined primarily by the source of the myosin. For example, cardiac myosin moves actin slightly slower than fast skeletal muscle myosin and approximately three times faster than smooth muscle myosin. Many of the mutations in myosin cause a reduction in the velocity of filament motion.

It is also now possible to measure the force generated by a single myosin molecule pulling on a single actin filament. The measured force is in the range of 1 to 10 pN; this is roughly equal to the gravitational attraction between your body and the textbook you are now reading. This technique uses a laser beam to grab onto single actin filaments floating in solution and to hold them stationary (somewhat like the tractor beam from Star Trek) while a single myosin molecule pulls on the filament (Fig. e).

One of the most exciting new developments, currently in the earliest stages of investigation, is the prospect of correcting these mutations in myosin by altering the DNA of the cells that make up the heart. This rapidly growing field of genetic medicine has shown that cardiac muscle is unusually receptive to such genetic manipulation.

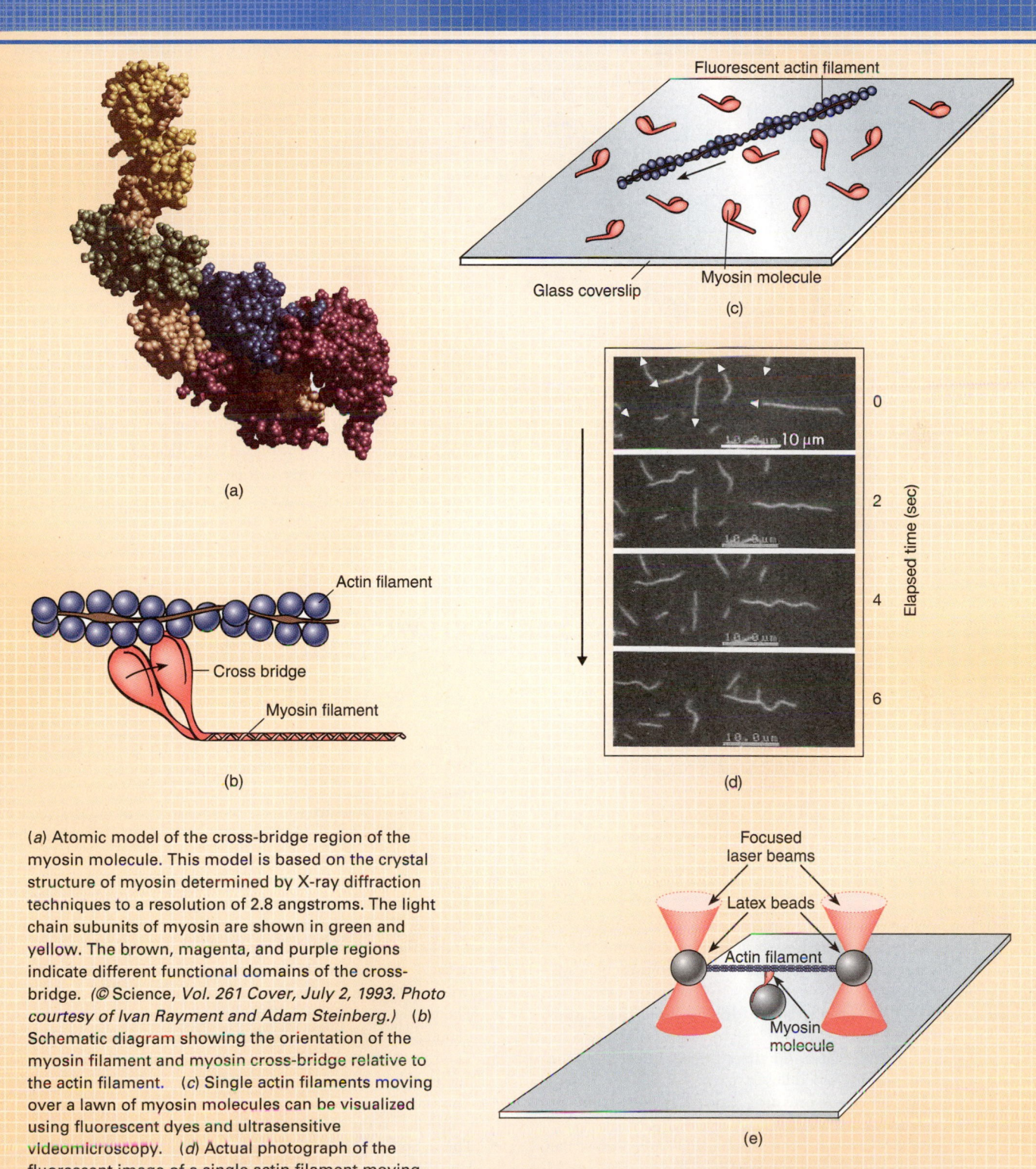

(a)

(b)

Actin filament

Cross bridge

Myosin filament

Fluorescent actin filament

Glass coverslip

Myosin molecule

(c)

10 µm

Elapsed time (sec)

0

2

4

6

(d)

Focused
laser beams

Latex beads

Actin filament

Myosin
molecule

(e)

(*a*) Atomic model of the cross-bridge region of the myosin molecule. This model is based on the crystal structure of myosin determined by X-ray diffraction techniques to a resolution of 2.8 angstroms. The light chain subunits of myosin are shown in green and yellow. The brown, magenta, and purple regions indicate different functional domains of the cross-bridge. (© *Science, Vol. 261 Cover, July 2, 1993. Photo courtesy of Ivan Rayment and Adam Steinberg.*) (*b*) Schematic diagram showing the orientation of the myosin filament and myosin cross-bridge relative to the actin filament. (*c*) Single actin filaments moving over a lawn of myosin molecules can be visualized using fluorescent dyes and ultrasensitive videomicroscopy. (*d*) Actual photograph of the fluorescent image of a single actin filament moving over myosin, as shown schematically in Figure c, at a velocity of 2 µm/sec in the direction indicated by the arrow. (© *Dr. Joe Haeberle, University of Vermont*) (*e*) Cartoon depicting the laser light-trap apparatus, in which focused laser beams are used to trap 50-nm-diameter latex beads attached to the ends of an actin filament. This device can be used to measure the movement and force produced by a single myosin molecule interacting with the trapped actin filament.

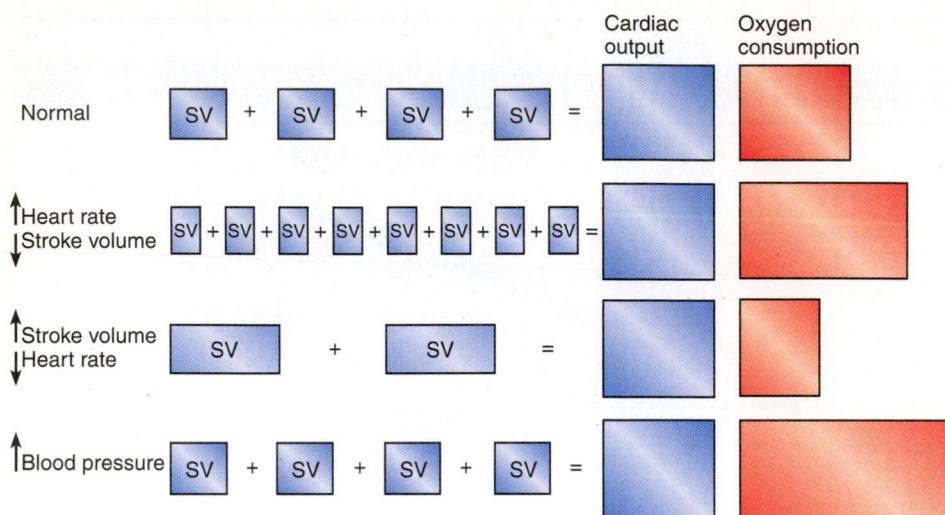

Figure 18–27

Effect of heart rate, stroke volume (SV), and arterial pressure on oxygen consumption by the heart.

output. The energy cost of producing a given cardiac output increases as blood pressure increases. The most economical method for increasing cardiac output is to increase the end-diastolic volume (Frank-Starling law) without increasing either heart rate or blood pressure. Figure 18–27 illustrates the relationship between increases in cardiac output and increases in oxygen consumption. In the next chapter, we will see how this information can be useful in the design of effective therapies for the clinical treatment of heart failure.

CHAPTER REVIEW

Summary

- The function of the cardiovascular system is to rapidly transport blood throughout the body.
- The cardiovascular system can be viewed as a closed loop consisting of the left and right hearts and the pulmonary and systemic circulations.
- Exchange between the blood and the cells occurs at the level of the capillaries.
- Blood flow is the volume of blood that moves past a particular point in the cardiovascular system during a period of time.
- Blood flow is identical in all portions of the cardiovascular system arranged in series.
- Blood pressure is a measure of the driving force that causes blood to flow.
- Resistance is a quantity that summarizes the net effect of all the frictional forces opposing blood flow.
- The relationship between pressure, flow, and resistance is: Flow = $\triangle$ Pressure/Resistance.
- The left and right hearts both consist of a thin-walled muscular atrium and a thick-walled muscular ventricle.
- A series of one-way cardiac valves ensures that blood flows in only one direction through the heart.
- Blood flows from the atria to the ventricles through the atrioventricular valves (right heart: tricuspid valve; left heart: bicuspid or mitral valve).

- Blood flows out of the ventricle through the semilunar valves (right heart: pulmonic valve; left heart: aortic valve).
- The wall of the heart consists of a middle, muscular layer (myocardium), an inner layer (endocardium), and an outer layer (epicardium).
- The coronary vessels supply blood flow to the heart.
- The sequential activation of different parts of the heart is governed by the conduction of a cardiac action potential from the atria to the ventricles.
- Because the absolute refractory period is as long as the muscle twitch (300 msec), tetanus is not possible in cardiac muscle.
- The action potential is initiated by pacemaker cells located in a region of the right atrium called the *sinoatrial node*.
- The rate of diastolic depolarization determines the rate at which the heart beats.
- The action potential is conducted from the right atrium to the ventricles via the atrioventricular node (AV node).
- The action potential is then conducted through the bundle of His, the bundle branches, and the Purkinje fibers to the ventricular muscle cells.
- This specialized conduction system ensures the nearly simultaneous activation of all of the ventricular muscle.
- Electrocardiography is a method for measuring the electrical activity of the heart.

- The cardiac cycle is the sequence of mechanical events that occur during a single contraction of the heart.
- The sequence of events is the same for both the left and right hearts, which beat simultaneously.
- The ventricles are the primary pumping chambers of the heart.
- Ventricular contraction and ventricular ejection occur during systole.
- The systolic period begins with the occurrence of the first heart sound and ends with the occurrence of the second heart sound.
- Ventricular relaxation and ventricular filling occur during diastole.
- Diastole begins with the second heart sound and ends with the first heart sound.
- The volume of blood ejected from the heart with each beat is called the *stroke volume*.
- The sum of all the stroke volumes ejected from the heart over a period of 1 minute is equal to the cardiac output.
- Heart rate is determined by both sympathetic and parasympathetic innervation of the SA node.

- The intrinsic ability of the heart to respond to an increase in end-diastolic volume with an increase in stroke volume is referred to as the Frank-Starling law of the heart, which ensures that the volume of blood pumped by the left and right hearts is equal.
- A change in the vigor of contraction that is not dependent on a change in end-diastolic volume is referred to as a *change in contractility*.
- Contractility changes are the result of changes in the concentration of intracellular calcium ions.
- Cardiac muscle derives ATP from the metabolism of circulating glucose and fatty acids.
- The ability of the heart to contract is strictly limited by the supply of nutrients carried to the cardiac muscle by the coronary circulation.
- The energy cost of maintaining a given level of cardiac output is related to a number of factors, including heart rate, stroke volume, and arterial pressure.

Review Questions

Choose the Correct Answer

1. Spontaneous depolarization of cells within the sinoatrial node:
   a. prevents the electrical activation of the heart.
   b. results in a propagated action potential when the membrane potential reaches threshold.
   c. occurs more slowly with increased sympathetic stimulation.
   d. occurs more quickly with increased parasympathetic stimulation.
   e. does not occur in a healthy heart.
2. Stroke volume is 70 ml, and heart rate is 80 bpm. What is the cardiac output?
   a. 70 ml/min
   b. 80 ml/min
   c. 5600 ml/min
   d. 5600 ml
   e. 6400 ml
3. The normal location of the pacemaker in the human heart is in the:
   a. atrioventricular node.
   b. sinoatrial node.
   c. right ventricle.
   d. Purkinje fibers.
   e. left ventricle.
4. If two successive QRS complexes are 0.8 sec apart, heart rate is:
   a. 65 bpm.
   b. 70 bpm.
   c. 75 bpm.
   d. 80 bpm.
   e. 85 bpm.
5. Failure of just the right heart would not cause:
   a. decreased blood flow to the lungs.
   b. decreased systemic blood flow.
   c. increased end-diastolic volume of the right ventricle.
   d. pulmonary edema.
   e. systemic edema.

6. Which of the following contributes to the increased cardiac output that occurs with increased sympathetic stimulation of the heart?
   a. Decreased heart rate
   b. Decreased diastolic filling time
   c. Increased contractility
   d. Decreased rate of ventricular contraction
   e. Decreased rate of ventricular relaxation
7. Which of the following could not be detected by an ECG?
   a. Change in heart rate
   b. Atrioventricular block
   c. Ventricular fibrillation
   d. Change in contractility
   e. Atrial fibrillation
8. Atrial contractions are more important at high heart rates (e.g., 150 bpm) than at normal resting heart rates because, at high heart rates:
   a. systole is shorter.
   b. diastole is longer.
   c. arterial blood pressure is higher.
   d. ventricular filling time is decreased.
   e. cardiac output is lower.
9. If mean arterial blood pressure is 90 mm Hg and total peripheral resistance is 18 mm Hg/liter/min, what is total blood flow?
   a. 2 L/min
   b. 3 L/min
   c. 4 L/min
   d. 5 L/min
   e. 6 L/min
10. Total flow through which of the following is not equal?
   a. Right heart and left heart
   b. Left heart and pulmonary circulation
   c. Pulmonary circulation and systemic circulation
   d. Pulmonary circulation and the brain
   e. Systemic circulation and the right heart

11. The rate of oxygen and glucose consumption by the heart would be decreased by:
    a. increased arterial pressure.
    b. increased heart rate.
    c. increased rate of ATP hydrolysis.
    d. increasing stroke volume and decreasing heart rate with no change in cardiac output.
    e. increased stroke volume.

12. Which event occurs after the first heart sound and before the second heart sound?
    a. Ventricular filling
    b. P-wave of the ECG
    c. Closure of the mitral valve
    d. Atrial contraction
    e. Isovolumetric ventricular contraction

13. Which of the following represent parallel blood flows?
    a. Total systemic blood flow and total pulmonary blood flow
    b. Blood flow to muscle and blood flow to the skin
    c. Total capillary blood flow and total arterial blood flow
    d. Total capillary blood flow and total venous blood flow
    e. Right heart output and left heart output

14. An increase in the concentration of circulating epinephrine would:
    a. decrease arterial blood pressure.
    b. decrease heart rate.
    c. increase cardiac output.
    d. decrease contractility of the heart.
    e. decrease stroke volume.

15. In a normal, healthy heart, stroke volume would be increased by:
    a. increased sympathetic stimulation of the heart.
    b. increased parasympathetic stimulation of the heart.
    c. decreased contractility.
    d. decreased circulating epinephrine.
    e. decreased end-diastolic volume.

16. Contractility of the heart is *not* altered by which of the following?
    a. A change in end-diastolic volume
    b. A change in cytoplasmic calcium concentration
    c. A change in extracellular calcium concentration
    d. A change in the level of sympathetic stimulation to the heart
    e. The drug digoxin

17. The exchange of nutrients and gases between the blood and the tissues occurs in the:
    a. aorta.
    b. arteries.
    c. arterioles.
    d. capillaries.
    e. large veins.

18. The rate of spontaneous depolarization of sinoatrial nodal cells:
    a. is increased by acetylcholine.
    b. is decreased by epinephrine.
    c. is decreased by norepinephrine.
    d. determines heart rate.
    e. is relatively constant.

19. Cardiac muscle cannot contract in a tetanic fashion because:
    a. the long absolute refractory period prevents the muscle from becoming restimulated while still contracting.
    b. the action potentials travel too slowly along the conducting tissue to restimulate the muscle.
    c. contraction is possible only when the heart is filled with blood.
    d. the autonomic nervous system blocks rapid action potentials.
    e. calcium uptake by the sarcoplasmic reticulum is too fast.

20. In the electrocardiogram, the T-wave is associated with:
    a. atrial depolarization.
    b. atrial repolarization.
    c. ventricular depolarization.
    d. ventricular repolarization.
    e. repolarization of AV node.

Answers to Case History Questions

1. The slower rate (33 bpm) allows more time for diastolic filling of the ventricles and therefore increase stroke volume as described by the Frank-Starling relationship. The decrease in arterial pressure at this slow heart rate would increase sympathetic stimulation of the heart and increase contractility, which would further increase stroke volume, primarily by decreasing end-systolic volume.

2. Atropine is a cholinergic antagonist (blocks acetylcholine receptors) and would decrease the effect of parasympathetic stimulation on the AV node. This would facilitate conduction of the impulse through the AV node.

3. In the absence of depolarization from the atrium, pacemakers in the AV node (40–50 bpm), Purkinje fibers (30–40 bpm), or ventricular muscle (30 bpm) take over pacing the ventricles. The slow rate suggests that the impulse is originating below the bundle of His, either in the Purkinje fibers or ventricular muscle. The wide QRS complex confirms this because it indicates that ventricular depolarization is following an abnormal pathway (spreading from one ventricle to the other). The QRS complex would be normal looking if the impulse originated in the AV node or the bundle of His.

4. If the pacemaker stimulated the ventricles at a fixed rate of, for example, 70 bpm all of the time, several potential problems might develop. Because atrial depolarization is occurring independently of ventricular depolarization, there is a real risk that occasionally an impulse will be conducted through the AV node during the period of supernormal excitability of the ventricles (during the T-wave), and this can abruptly initiate ventricular fibrillation. A second problem would be that the patient's ventricular rate would not change in response to changes in autonomic stimulation of the heart, and this would limit the patient's ability to engage in physically demanding activities. Because a demand-pacemaker fires only when it detects failure of conduction following a detected P-wave, the possibility of a conducted impulse arriving during the period of supernormal excitability is reduced and requires a premature atrial contraction at just the right time in the cycle. By monitoring the atrial rate of depolarization and pacing the ventricles at the same rate, the patient's response to stress and exercise is normal.

5. If the demand-pacemaker was set to deliver a stimulus with a P-R interval shorter than or equal to normal, then the pacemaker would not be able to detect (and stop pacing) when the AV nodal block terminated.

Key Terms

aorta (p. 559)
aortic valve (p. 562)
arrhythmia (p. 570)
arterial system (p. 559)
arterioles (p. 559)
atrioventricular node (AV node) (p. 568)
atrioventricular valve (p. 562)
blood pressure (p. 560)
brachycardia (p. 566)
bundle branches (p. 568)
bundle of His (p. 568)

cardiac cycle (p. 571)
cardiac output (p. 575)
diastole, diastolic pressure (p. 572)
electrocardiogram (ECG or EKG) (p. 570)
endocardium (p. 563)
epicardium (p. 563)
first heart sound (p. 575)
Frank-Starling law of the heart (p. 577)
heart sounds (p. 575)

limb leads (I, II, III) (p. 571)
mitral valve (p. 562)
myocardium (p. 563)
P-wave (p. 570)
pacemaker (p. 565)
pulmonary circulation (p. 559)
pulmonary valve (p. 562)
Purkinje fiber (p. 564)
QRS complex (p. 570)
second heart sound (p. 575)
semilunar valve (p. 562)
sinoatrial (SA) node (p. 565)

stroke volume (p. 575)
systemic circulation (p. 559)
systole, systolic pressure (p. 572)
T-wave (p. 570)
tachycardia (p. 566)
tricuspid valve (p. 562)
venous system (p. 559)
ventricle (p. 561)
ventricular fibrillation (p. 570)

Suggested Readings

Berne, R. M., and Levy, M. N. *Physiology,* ed 4. St. Louis, C.V. Mosby, 1998.

Dickman, S. "Mysteries of the heart." *Discover,* 8:116–119, 1997.

Gibbs, W. W. "Helping heartache." *Scientific American,* 277:34–36, 1997.

Harken, A. H. "Surgical treatment of cardiac arrhythmias." *Scientific American,* 269:68–74, 1993.

Jarvik, R. K. "The total artificial heart." *Scientific American,* 244:74–80, 1981.

Levy, M. N., and Berne, R. M. *Cardiovascular Physiology,* ed 8. St. Louis, C.V. Mosby, 2001.

Mohrman, D. E., and Heller, L. J. *Cardiovascular Physiology,* ed 4. New York, McGraw-Hill, 1997.

Rhoades, R. A., and Tanner, G. A. *Medical Physiology,* ed 1. Boston, Little, Brown & Co., 1995.

Robinson, T. F., Factor, S. M., and Sonnenblick, E. H. "The heart as a suction pump." *Scientific American,* 254:84–91, 1986.

Smith, O. "Contortions of the heart." *Science,* 288:453, 2000.

Solaro, J. R., "Myosin and why hearts fail." *Cell,* 85:1945–1946, 1992.

Teitz, C. C. *Scientific Foundations of Sports Medicine.* Philadelphia, B. C. Decker, 1989.

Answers to Review Questions

1. b   2. c   3. b   4. c   5. d   6. c   7. d   8. d
9. d   10. d   11. d   12. a   13. b   14. c   15. a
16. a   17. d   18. d   19. a   20. d

Chapter 19

CIRCULATION

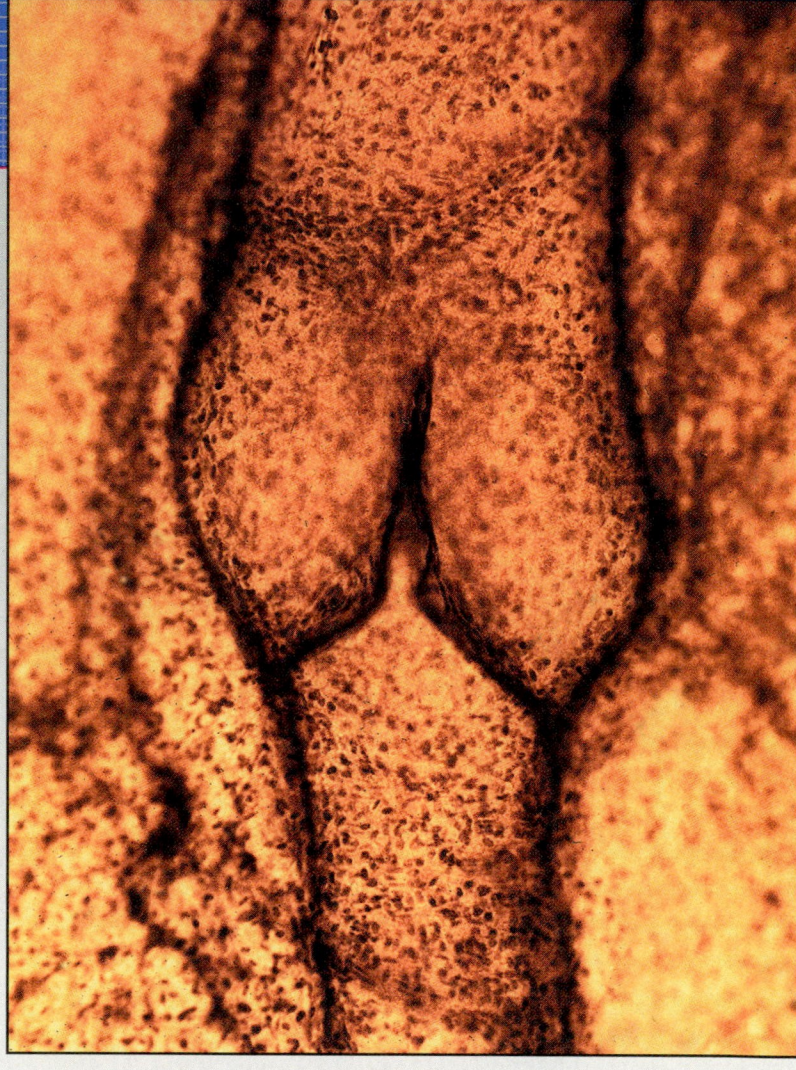

- *Photomicrograph of lymphatic valve.*
(© John D. Cunningham/ Visuals Unlimited)

KEY CONCEPTS

- *The vascular system is both a conduit for the flowing blood and a dynamic system that controls the distribution of blood in the body.*

- *The elastic arteries dampen the pulsatile outflow of blood from the heart to provide a more continuous flow of blood to the tissues. The arterioles, because of their smaller internal diameter, are the major site of resistance to blood flow. Changes in the diameter of the arterioles determine the amount of blood flowing through a particular tissue bed, as well as the distribution of blood flow among different tissue beds.*

- *Capillaries are the major site of molecular exchange between the blood and the extracellular fluid. The balance of hydrostatic and oncotic forces (i.e., Starling's forces) determines whether fluid leaks out of the capillaries are retained by the capillaries, or are absorbed into the capillaries.*

- *The veins contain the major portion of the circulating blood volume and, because of the surrounding smooth muscle and venous one-way valves, play a major role in regulating the distribution of the blood volume within the body.*

- *The lymphatic system returns fluid from the extracellular space to the general circulation, but only after it has been filtered through the lymph nodes.*

- *Arterial pressure is regulated by the medullary cardiovascular control center that integrates sensory input from many sources and subsequently modulates the autonomic nerve activity to the heart and blood vessels to maintain a relatively constant arterial blood pressure.*

- *Any number of factors can cause an inability of the heart to maintain normal cardiac output, resulting in a variety of compensatory responses by the body to adapt to this less than ideal situation.*

CASE HISTORY

A 56-year-old man has come to the emergency room complaining of fatigue, weakness, shortness of breath (dyspnea), and exercise-induced chest pain that radiates to his left arm and neck (angina). His history indicates that, in the past couple of years, he has had several minor episodes of angina during periods of sudden exertion. Recently, he has awakened several times at night with the feeling that he cannot breathe; however, this sensation has gone away after a few minutes of sitting up in bed. A physical examination shows a relatively low blood pressure of 110/80 mm Hg, a weak pulse, and a fast heart rate of 100 beats per minute (tachycardia). The patient has swollen ankles (edema), and the jugular veins in his neck are visibly distended. An electrocardiographic (ECG) analysis indicates that the electrical activation of his heart is proceeding normally from the atria to the ventricles (normal sinus rhythm). Normally, an ECG records little or no electrical potential during the time the ventricular muscle is actually contracting (ST segment), however, this patient's ECG indicates a positive potential during this period (elevated ST segment), which can be indicative of ventricular muscle that is receiving insufficient blood flow and oxygen (ischemia). By beaming high-frequency sound waves into the patients heart from outside of the chest (echocardiography), the physician can look at the anatomy and motion of the heart in much the same way that radar is used to locate and identify aircraft. The echocardiographic analysis indicates that the patient's heart is dilated and contains more blood than does a normal heart. It also shows that portions of the walls of both the right and left ventricles are not showing the normal or expected amount of motion during contraction. This reduced wall motion, along with the elevated ST segment on the ECG, suggests that there may be partial or complete blockage of one or more of the coronary arteries that supplies blood to the heart. To visualize the coronary arteries, a dye, which is visible on a portable X-ray screen, is injected into the coronary arteries.

The dye is introduced through a catheter that is inserted into a femoral artery and, while being visualized with the X-ray machine, is advanced into one of the main coronary arteries. The arterial tree that appears after the injection of a bolus of dye indicates significant narrowing of several segments of the coronary arteries. The patient was treated with bed rest to decrease activity of the heart and, thereby, decrease its need for oxygen and other nutrients supplied by the blood. The drug digitalis was prescribed to increase the heart's capacity to pump blood. In addition, a diuretic was prescribed to increase the rate of urine formation by the kidneys. After three months, the patient was admitted for coronary artery bypass surgery to replace the narrowed segments of coronary artery with pieces of blood vessel (vein) removed from the patient's leg.

Questions

1. Why does the patient experience shortness of breath while sleeping; why do the symptoms disappear after he sits up in bed?

2. What is the cause of the peripheral edema and the distended neck veins?

3. Why is the patient's blood pressure low even though his heart rate is relatively high, and what relationship does this have to the fact that he has a dilated heart and partially obstructed coronary arteries?

4. How will the prescribed medications help with the patient's symptoms?

OVERVIEW OF THE VASCULAR SYSTEM

Recall from our previous discussion that the major subdivisions of both the systemic and pulmonary vascular systems are arteries, arterioles, capillaries, venules, and veins. As we will see, the exchange of substances between the blood and the interstitial fluid occurs as blood passes through the capillaries. In large part, the role of the arteries, arterioles, venules, and veins is to direct the flow of blood to and from the capillaries. Far from being passive conduits, however, these vessels also play an active role in determining what portion of the total cardiac output reaches each of the various tissue beds in the body. In addition, the venules and veins help to ensure the efficient operation of the heart as a pump by maintaining an optimal distribution of blood volume between the venous and arterial systems.

THE ARTERIAL SYSTEM: CARRYING BLOOD TO THE TISSUE

The aorta and the arteries are conduits that direct the flow of blood from the heart to the various tissue beds within the body. The aorta, a large vessel with an average internal diameter of 2.5 cm (1 inch), must be of sufficient size to accommodate the combined blood flow to all of the tissue beds. The arteries that branch off of the aorta to distribute blood to the various tissues have internal diameters on the order of 0.4 cm. Relative to the diameter of the arterioles and capillaries, these are large vessels that offer comparatively little resistance to the flow of blood (see Fig. 19–5).

The arterial system plays a second role in the operation of the cardiovascular system. The walls of the arteries are made of smooth muscle and elastic tissue. When the ventricle contracts and suddenly ejects a volume of blood, the aorta and arteries stretch to accommodate the ejected blood. During diastole, when the heart is no longer ejecting blood, the elastic arteries recoil and force blood through the downstream blood vessels.

Blood Pressure in the Arteries Propels the Blood Through the Capillaries and Back to the Heart

 What causes blood to flow through the blood vessels?

Most people are aware that blood is under pressure within the arterial system. Cutting a large artery results in a serious medical emergency because blood is rapidly lost from such a wound. The high pressure inside the artery forces blood out through the cut. This high arterial blood pressure is necessary for the normal operation of the cardiovascular system. As mentioned previously, blood pressure is a measure of the driving force that causes blood to flow through the blood vessels. The magnitude of the driving pressure must be ade-

quate to overcome the resistance to blood flow, which is a result of the friction between the flowing blood and the walls of the blood vessels and the friction between the individual molecules and cells that make up the blood. We all deal with frictional forces every day. It is necessary to apply pressure to one side of a heavy box to slide it across the floor because of the friction between the bottom of the box and the floor. Likewise, pressure is necessary to move blood through the vascular system.

We can estimate how much pressure must be generated by the heart to overcome this resistance by using the equation presented in Chapter 18 that describes the relationship between flow (F), pressure (P), and resistance (R).

$$F = (P_1 - P_2)/R$$

Because we are looking at flow from the aorta to the vena cava, it is the pressure difference between the aorta (P_1) and vena cava (P_2) that determines flow (i.e., aortic pressure − vena cava pressure = pressure difference). We can simplify our analysis by recognizing two points. First, the pressure in the vena cava (P_2) is small compared with that in the aorta and can be considered to be zero for our purposes. Second, pressure within the aorta (P_1) is approximately equal to the arterial pressure measured in any of the major systemic arteries; this equality indicates that the resistance to flow through the arteries is relatively small. Knowing that blood pressure in the vena cava is approximately zero and that arterial pressure equals aortic pressure, we can modify the equation for blood flow as follows:

$$F = (P_1 - P_2)/R$$
$$F = (aortic\ pressure - vena\ cava\ pressure)/R$$
$$F = (arterial\ pressure - 0)/R$$
$$F = arterial\ pressure/R$$

Under normal resting conditions, the arterial pressure required to provide adequate tissue flow is approximately 100 mm Hg. Figure 19–1 illustrates the values for blood pressure at various points in the circulatory system. Notice first that the systemic arterial pressure starts out approximately five times higher than the pulmonary arterial pressure, and yet the pressures in both systems return to very nearly the same value by the time the blood reaches the veins. The decrease in blood pressure reflects the conversion of kinetic energy to heat as a result of the resistance to blood flow. The greater the resistance to flow, the greater the pressure drop as blood flows through the vessels. On this basis, we can reason that the resistance to blood flow must be approximately five times greater in the systemic circulation than in the pulmonary circulation.

Notice that the greatest pressure drop, in the systemic circulation, occurs at the level of the arterioles. This tells us that a major portion of the overall resistance to blood flow occurs at this point in the circulation. As we will see, this has

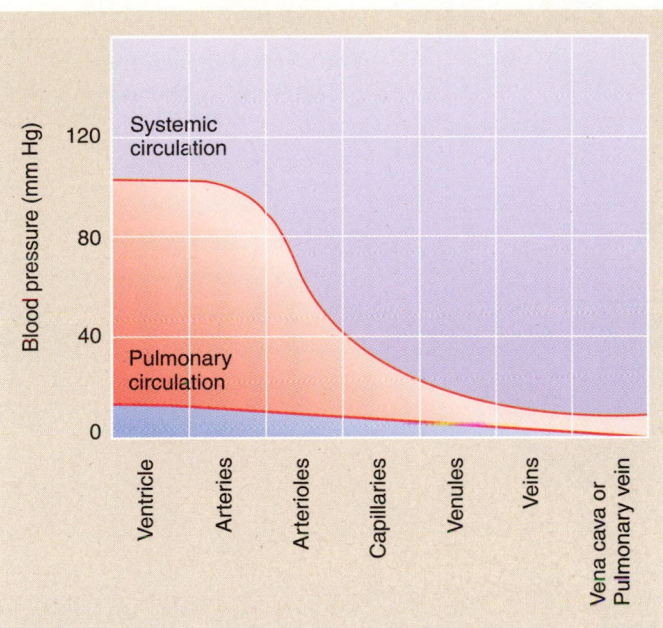

Figure 19–1

The range of blood pressures throughout the systemic and pulmonary circulations.

important implications for the distribution of blood flow within the body.

Pulse Pressure

What is the pulse pressure; what does it tell us about the arterial system?

If we were to directly measure arterial blood pressure, the first thing we would notice is that it is pulsatile. Recall from Chapter 18 that arterial pressure is lowest at the end of ventricular diastole and rapidly increases to a peak during ventricular systole. When the ventricle contracts, a volume of blood (the stroke volume) is ejected rapidly from the heart into the arterial vessels. As shown in Figure 19–2, the flow of blood into the arteries during systole is greater than the flow of blood out of the arteries into the capillaries. Blood flows out of the arteries more slowly because of the high resistance to flow through the arterioles. The elastic tissue that makes up the walls of the aorta and arteries allows them to stretch and increase in diameter in response to the blood ejected from the ventricle. Just as the pressure increases inside an elastic balloon as it is inflated with air, the pressure in the arterial system increases as the vessels inflate with blood. During diastole, when blood is no longer flowing into the arteries from the heart, the elastic recoil of the arteries forces blood to flow out of the arteries into the arterioles. In other words, a portion of the pressure generated by the heart during systole is stored in the stretched walls of the arteries and is then slowly dissipated during diastole as blood flows out of the arterial system. The elastic properties of the arteries help to convert the pulsatile flow of blood from the heart into a more continuous flow of blood through the rest of the circulation.

As blood flows out of the arteries, the pressure progressively decreases until it reaches a minimum called **diastolic pressure** (Fig. 19–3). The peak arterial pressure, which occurs during contraction of the ventricle, is called **systolic pressure.** Your nurse or physician may have measured your blood pressure as "120 mm Hg over 80 mm Hg." These two numbers refer to the systolic and diastolic pressures, respectively, and are normally written as 120/80 mm Hg. The difference between systolic and diastolic pressures $(120 - 80 = 40$ mm Hg) is referred to as the **pulse pressure.**

As you might guess, the amplitude of the pulse pressure indicates the vigor of contraction by the ventricle; it also tells us something about the elasticity of the arteries. As shown in Figure 19–3, the magnitude of the pulse pressure is determined by two factors: (1) the stroke volume and (2) the distensibility of the arteries. As we discussed in Chapter 18, the stroke volume of the heart is dynamically

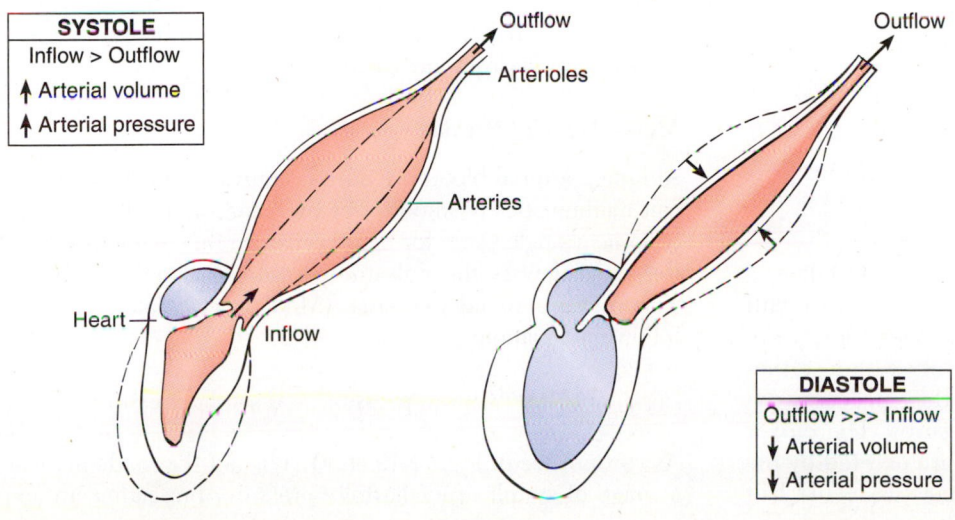

Figure 19–2

Role of the arteries as a high-pressure storage reservoir. The flow of blood into the arteries (from the heart) during systole exceeds the flow out of the arteries (through the arterioles), leading to an increase in arterial volume and pressure. During diastole, the elastic recoil of the arterial walls provides the driving force to propel blood out of the arteries.

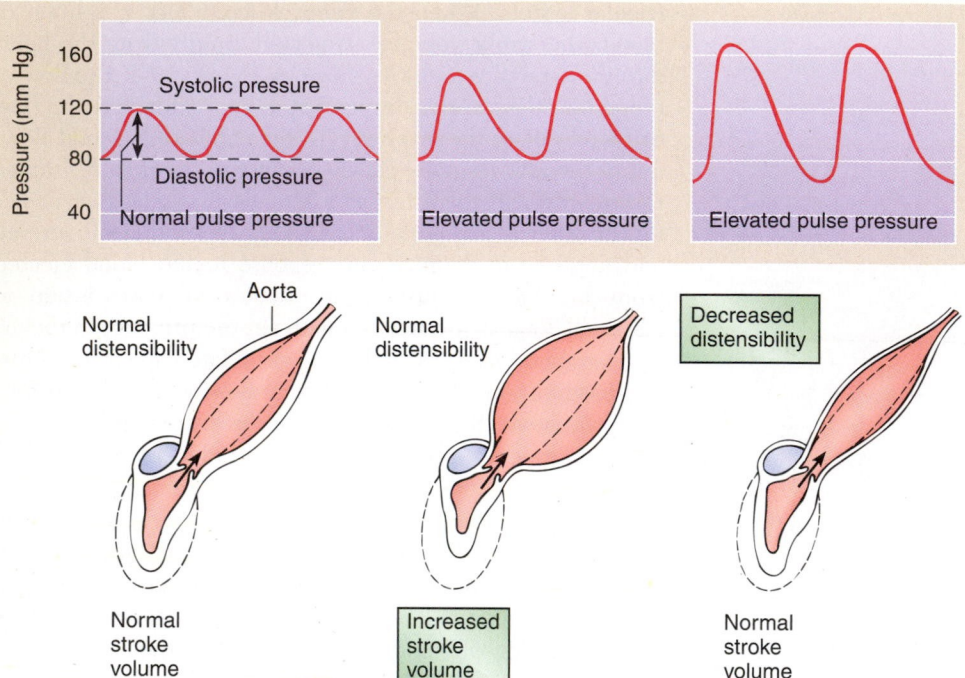

Figure 19–3

The dependence of pulse pressure on stroke volume and aortic distensibility.

regulated and can change significantly from beat to beat. In order for the heart to inject an increased stroke volume into the arteries, it must generate greater pressure to further stretch the elastic walls of the arteries. A sudden increase in pulse pressure, therefore, indicates an increase in the stroke volume ejected by the heart. Although the distensibility of the arteries can also affect the magnitude of the pulse pressure, arterial distensibility is relatively constant from day to day. However, progressive arterial disease, such as atherosclerosis, results in a decrease in the distensibility of the arteries and therefore an increase in pulse pressure. Analogously, greater pressure is required to blow up a thick-walled, less distensible balloon than a thin-walled, easily distensible one. A progressive increase in pulse pressure as an individual ages can be diagnostic of atherosclerotic arterial disease commonly referred to as "hardening" of the arteries (see Applications of Physiology: Edema, Lymphatic Pathology, and Varicose Veins).

Measurement of Arterial Pressure

 How can we measure arterial pressure?

Everyone has had blood pressure measurements taken at one time or another. The physician or nurse places a cuff around the arm, inflates it with air, and then slowly deflates it while listening to the brachial artery with a stethoscope. Figure 19–4 illustrates how this process can measure both systolic and diastolic pressures. The cuff is first inflated (Fig. 19–4a) until the pressure exerted by the cuff is sufficient to collapse the underlying artery and halt

blood flow. The pressure in the cuff is then gradually decreased. When the cuff pressure falls below the peak systolic pressure (Fig. 19–4b), blood is able to momentarily flow through the partially collapsed artery. This momentary, high-velocity spurt of blood through the artery can be heard through a stethoscope placed over the artery (downstream from the cuff) as a soft tapping sound. When this sound first occurs, the cuff pressure is recorded as systolic pressure. As the cuff pressure is progressively decreased (Fig. 19–4c), the duration of blood flow during each cardiac cycle gets longer and the nature of the sounds heard with the stethoscope changes in a characteristic manner. When the cuff pressure has decreased below the diastolic pressure (Fig. 19–4d), the vessel remains open at all times and flow is again continuous. Just after the cuff pressure falls below diastolic pressure, the sounds disappear entirely. The cuff pressure at which the sounds disappear is recorded as diastolic pressure.

Mean Arterial Pressure

Although arterial blood pressure is constantly fluctuating, we can mathematically smooth out the peaks and valleys to determine a single value for blood pressure that would produce the same flow as the pulsatile pressure that actually exists. This **mean arterial pressure (MAP)** is calculated with the following equation:

$$\text{MAP} = \tfrac{1}{3}\text{ pulse pressure} + \text{diastolic pressure}$$

As you can see in Figure 19–4, MAP is not the mathematical average of systolic and diastolic pressure, but rather an ap-

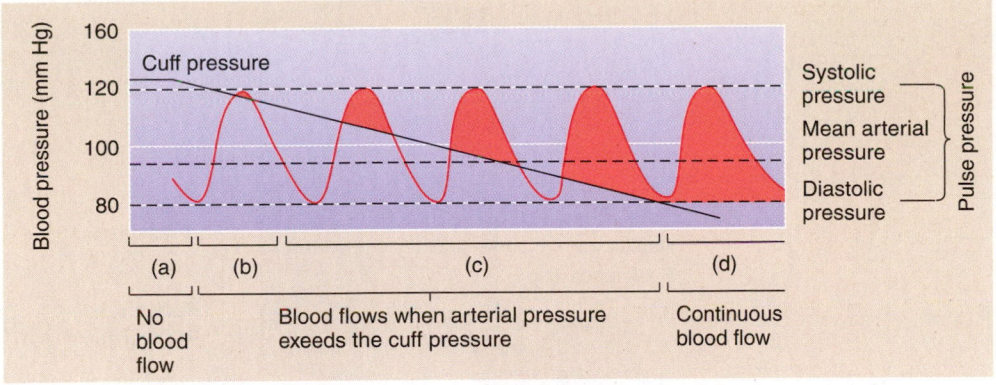

Figure 19–4

Measurement of systolic and diastolic blood pressures by sphygmomanometry. The shaded areas indicate when blood is able to flow through the brachial artery as the cuff pressure is lowered.

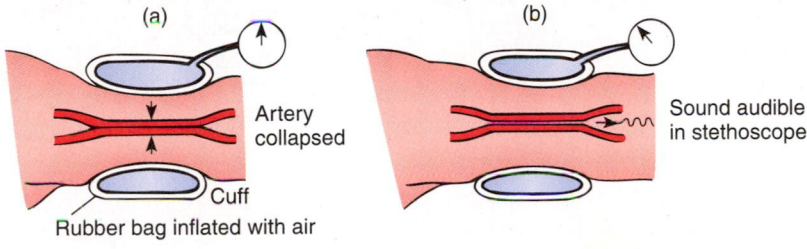

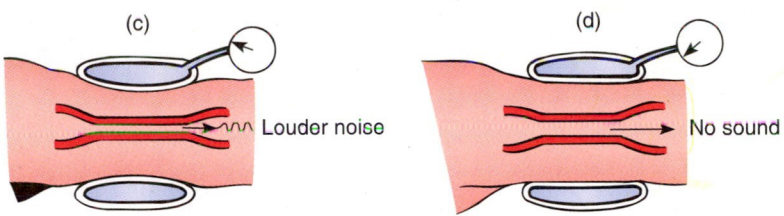

proximation of the geometric mean. By calculating the mean blood pressure, we can significantly simplify the calculation of blood flow because it is the MAP that is related to blood flow. Hence, we can further modify the equation for flow as follows:

$$F = \text{arterial pressure}/R = \text{MAP}/R$$

Without calculating the MAP, it would be rather difficult, for example, to evaluate how a change in blood pressure from 120/90 to 160/70 would alter blood flow. If we calculate the MAP in each case:

$$\text{MAP} = \tfrac{1}{3}(120 - 90) + 90 = 100$$
$$\text{MAP} = \tfrac{1}{3}(160 - 70) + 70 = 100$$

we quickly realize that flow remains the same. As we will see in a later section, our bodies have homeostatic systems that sense MAP and then change the functional state of the heart or the vasculature, or both, to maintain MAP constant. If we recall that the purpose of the cardiovascular system is to maintain a continuous flow of blood to the tissues, it should not be too surprising to see that our bodies regulate MAP and not pulsatile pressure.

The Major Site of Resistance to Blood Flow is the Arteriole

How is the distribution of blood flow to various organs determined?

Recall that soon after the arteries reach the various tissue beds within the body, they divide into smaller arterioles. As shown in Figure 19–5, the diameter of the arteriole is considerably smaller than that of the artery. This sudden decrease in diameter results in a relatively large increase in the resistance to blood flow and, consequently, a relatively rapid decrease in pressure as illustrated in Figure 19–1.

The Diameter of Arterioles

Although a number of factors contribute to the resistance of flow through blood vessels, the diameter of a blood vessel is one of the more important contributing factors, for two reasons. First, small changes in diameter cause relatively large changes in resistance. The resistance to flow through a blood vessel is inversely proportional to the radius raised to the fourth power:

$$R \propto 1/\text{radius}^4$$

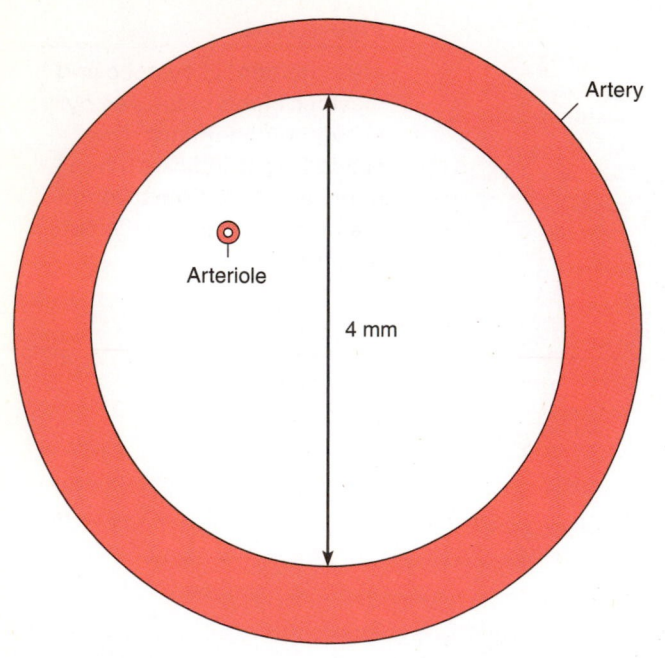

Figure 19–5

Comparison of the relative diameters and wall thicknesses of arteries and arterioles.

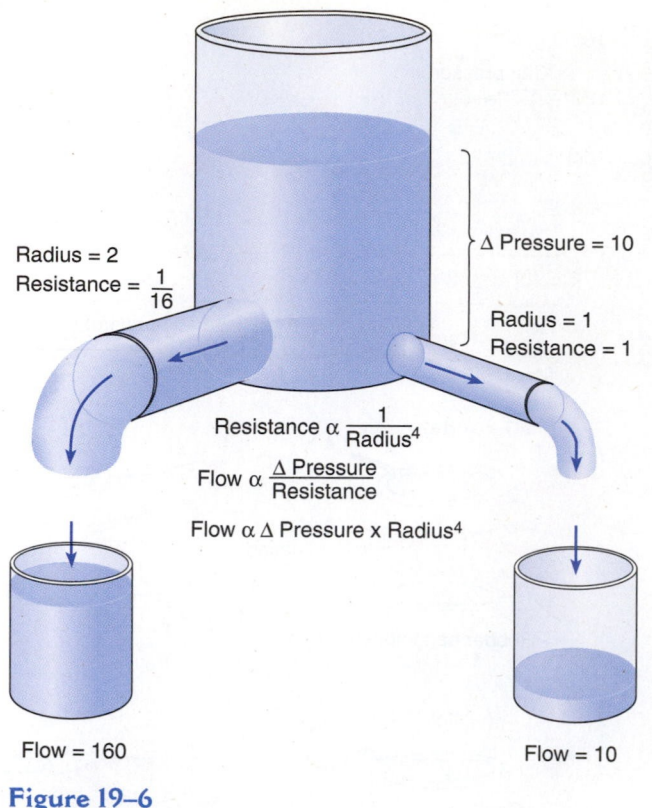

Radius = 2
Resistance = $\frac{1}{16}$

Δ Pressure = 10

Radius = 1
Resistance = 1

Resistance $\alpha \ \dfrac{1}{\text{Radius}^4}$

Flow $\alpha \ \dfrac{\Delta \text{ Pressure}}{\text{Resistance}}$

Flow $\alpha \ \Delta$ Pressure x Radius^4

Flow = 160

Flow = 10

Figure 19–6

The effect of vessel radius on resistance and flow. Note that doubling the radius increases flow 16-fold.

Therefore, halving the diameter of an arteriole increases the resistance to flow by a factor of 16 (Fig. 19–6). Second, arterioles are able to dynamically adjust their diameters. The walls of arterioles contain relatively more smooth muscle cells than do other blood vessels in the body. Moreover, these smooth muscle cells are arranged so that when they contract, they cause the diameter of the vessel to decrease, as illustrated in Figure 19–7 (see also Fig. 19–13). These two properties of the arterioles (high resistance and variable diameter) provide a mechanism by which the body can adjust the blood flow through different organs or tissues.

Recall from Chapter 18 that the blood vessels supplying different tissues in the body are arranged into parallel circuits (see Fig. 18–2). Because of this arrangement, the total cardiac output is distributed among the various tissues, with the sum of all the individual flows equal to the cardiac output. The amount of flow through the kidneys, for example, is determined by the resistance to blood flow through the kidneys relative to the resistance to flow through the other systemic beds. Because the resistance to flow through the arterioles is so much greater than that through the other vessels, the resistance to blood flow through a tissue is determined almost entirely by the diameter of the arterioles. Therefore, by adjusting arteriolar diameter, individual tissue beds can control the local flow of blood. Figure 19–8 shows how relative changes in arteriolar diameter can produce shifts in blood flow. The overall resistance to the flow of blood from the aorta back to the heart (previously designated as *R*) is referred to as the **total peripheral resistance** **(TPR)**. This term reflects the fact that essentially all this re-

sistance is located in the peripheral regions of the circulatory system at the level of the resistance arteriole. Substituting *TPR* for *R*, our flow equation becomes: F = MAP/TPR.

The Water System Analogy for the Regulation of Blood Flow

A good analogy for this aspect of the vascular system is a public water system. One large pipe (the aorta) carries water (the blood) from the water company's pump (the heart) to a series of distribution lines (arteries) that direct water to individual houses (organs). Within each house are a number of spigots (arterioles) that can be adjusted (variable resistance) to control the flow of water to different appliances (tissue beds). In this system, the distribution of water flow among the various houses and appliances is determined by local demand. The water company (the CNS) has only to ensure that the pressure in the main line (aorta) remains constant, and the flow can then be distributed among the various appliances (tissue beds) according to local demand. If you want to wash the dishes, your clothes, the dog, and the car all at the same time, you don't have to notify the water company of your intentions; you open the appropriate spigots (decrease the local resistance to flow).

Such a system makes the water company's job of providing adequate flow to all the users rather easy because it doesn't have to monitor the usage by individuals to set the

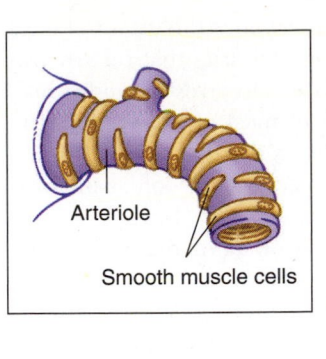

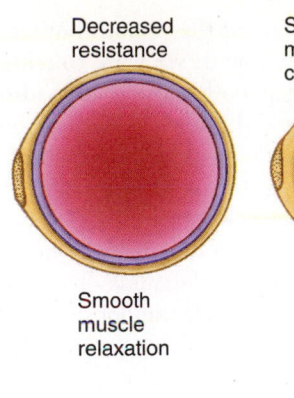

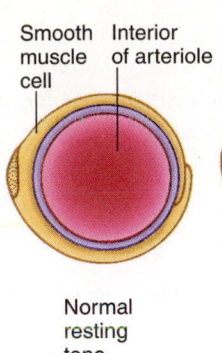

(a)

(b)

Figure 19–7

(a) Smooth muscle cells wrap around the wall of an arteriole. *(b)* Changes in resistance of an arteriole due to contraction or relaxation of the smooth muscle cells in the vessel wall.

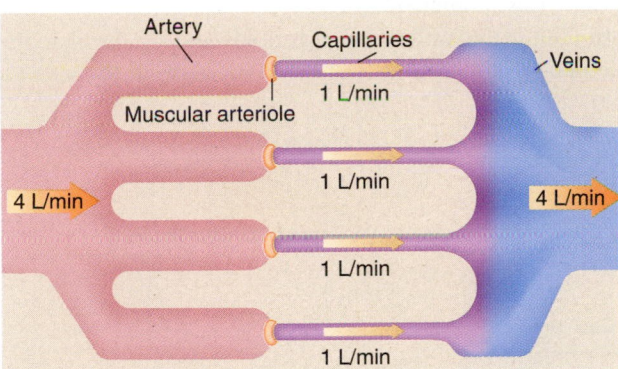

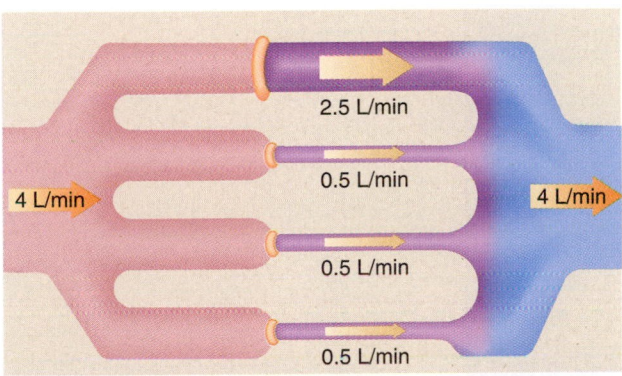

Figure 19–8

The effects of arteriolar resistance on distribution of blood flow among capillaries arranged in parallel. Note that total flow in arteries, arterioles, capillaries, and veins must all remain equal because they are arranged in series within a closed system. The arteriolar smooth muscle has been represented by a single band of smooth muscle in these operational diagrams (Figs. 19–8 to 19–12). See Figures 19–7 and 19–13 for more realistic depictions of the continuous distribution of smooth muscle cells along the entire length of the arteriole.

pumping rate (cardiac output). It simply monitors the water pressure (MAP) and adjusts the output of the pump (cardiac output) to keep the pressure constant. As long as the demand for water does not exceed the capacity of the water company's pump (the heart), all of the consumers (the organs) will receive whatever flow they demand. If the demand on the system does exceed the capacity of the pump, then pressure will fall, and flow to the consumers will decrease. We have all experienced such a drop in water pressure during a morning shower. Because many other people are also taking showers at the same time, the demand may temporarily exceed the capacity of the pump, and both pressure and flow will decrease. If you wanted to remedy this loss of pressure and flow, you could call your neighbors (other organs) and ask them to turn off their spigots (increase local resistance to flow), while at the same time opening your spigot further (decreasing local resistance) to direct a greater portion of the limited flow to your house. This would be a reasonable thing to do if you had an urgent need (your house was on fire) compared with that of your neighbor's (watering the grass). Notice that this cooperative approach to distribution of limited flow requires communication (the ANS) among the various users (the organs). However, as long as the capacity of the pump (the heart) exceeds the demand of the users (the organs), local regulation (by the arterioles) is adequate to adjust flow according to local needs.

Let's summarize the points made by this analogy. First, the individual organs and tissue beds can adjust local blood flow to meet demand by changing the resistance to flow through changes in the diameter of the arterioles. Second, the body can correctly adjust cardiac output to match the demand for flow simply by monitoring mean arterial pressure and adjusting the cardiac output to keep pressure constant. Third, as long as the demand for flow does not exceed the pumping capacity of the heart, individual tissue beds can function autonomously. Fourth, if demand does exceed flow, then extrinsic regulatory mechanisms must be used to redistribute flow to prevent a loss of pressure and flow to tissues with crucial needs.

In order to fully understand how blood flow is distributed in the body, we must understand (1) how the resistance of the arterioles changes in response to local demand, (2) how the blood flow through the different tissue beds in the body is coordinated by extrinsic regulatory mechanisms, and (3) how the body monitors and adjusts cardiac output to maintain mean arterial pressure. Let's first look at the mechanisms by which local factors change arteriolar resistance.

Local Factors Such as Blood Pressure and Metabolites Autoregulate Blood Flow to Some Vascular Beds

 How is blood flow to various tissue beds controlled?

The smooth muscle content of the arteriolar wall provides the means for dynamic control of arteriolar diameter and consequently of arteriolar resistance. Therefore, the contrac-

tile state of the smooth muscle determines the extent of local blood flow. If we were to administer a drug that relaxes vascular smooth muscle (a **vasodilator**) to a resting individual, we would find that the diameter of most of the arterioles in the body would increase. This shows that the normal resting state of the arteriolar smooth muscle is one of maintained contraction, referred to as a resting vascular **"tone."**

Resting tone is important because an increase in activity of an organ demands greater blood flow and thus relaxation of the arteriolar smooth muscle to produce dilation of the arteriole. Several factors contribute to this level of resting tone. Like the sinoatrial node of the heart, the resting membrane potential of vascular smooth muscle undergoes spontaneous depolarization. As discussed in Chapter 16, such depolarization is characteristic of single-unit smooth muscle and results in the initiation of a burst of action potentials, which in turn lead to the activation of muscle contraction. As we will see in the next few sections, both mechanical and chemical factors can modify this spontaneous activity.

Stretching the wall of an arteriole, and of the smooth muscle cells included in it, causes an increase in the rate of

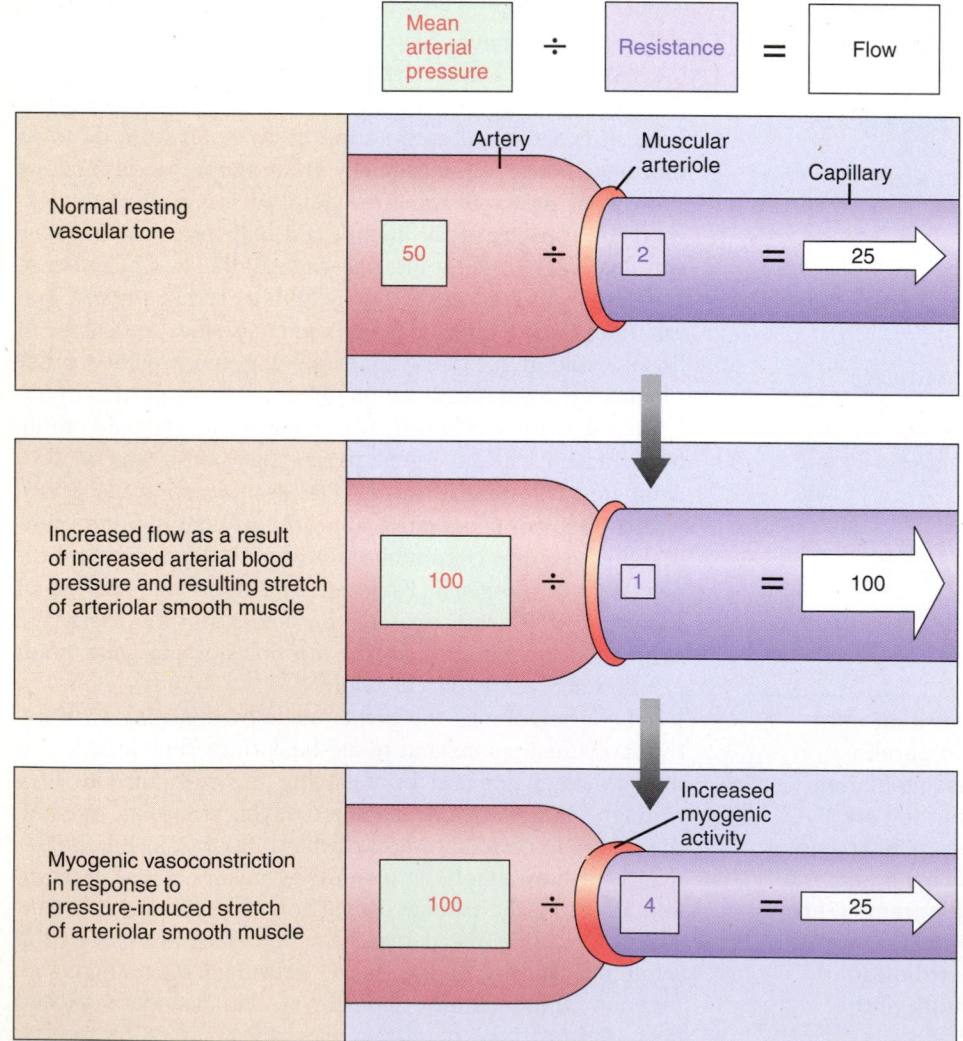

Figure 19–9

Autoregulation of capillary blood flow by myogenic activity of arteriolar and precapillary sphincter smooth muscle. Flow remains constant as perfusion pressure changes because of changes in vessel resistance due to myogenic activity.

spontaneous membrane depolarization and leads to an increased level of contraction. The increase in contractile activity in response to stretch is often referred to as **myogenic activity.** Let's examine how this might contribute to the local regulation of blood flow. An increase in arterial blood pressure would initially cause all arteries and arterioles to increase their diameters by stretching the elastic elements (including smooth muscle cells) in the vessel walls. If we recall the equation for blood flow (F = MAP/TPR), we can see that an increase in pressure increases flow not only by increasing the driving force (MAP) but also by increasing the vessel radius and decreasing the resistance to flow. Both the physical stretching of the smooth muscle and the increased blood flow stimulate myogenic contraction of the smooth muscle. This myogenic contraction decreases the diameter of the vessel and decreases blood flow back toward the initial level. Therefore, the increased myogenic activity of the smooth muscle in response to increased arterial pressure tends to minimize increases in both vessel diameter and flow in response to increased pressure. The sequence of events associated with an increase in blood pressure is shown in Figure 19–9. Conversely, a decrease in blood pressure initially decreases flow and unstretches the vascular smooth muscle. Again, both effects lead to decreased myogenic activity. The ensuing vasodilation minimizes the overall reduction in flow in response to a reduction in arterial pressure.

Experiments have shown that the blood flow to many organs remains relatively constant despite large changes in blood pressure. This ability of individual vascular beds to maintain constant blood flow is referred to as **autoregulation.** Although the mechanisms underlying autoregulation remain to be fully described, considerable evidence exists to

suggest that myogenic activity is a contributing factor. As we will see in the next section, a number of other factors also contribute to autoregulation of blood flow.

Blood flow to most tissues increases in proportion to the metabolic demand of the tissue **(active hyperemia).** This is dramatically illustrated in skeletal muscle, in which blood flow may increase as much as tenfold in response to intense exercise. Although some of this increase is mediated by factors external to the muscle, most of it is mediated by the change in the chemical composition of the interstitial fluid, which in turn, causes the arterioles to dilate.

Consider the chemical changes that occur as a result of cellular metabolism. The oxygen concentration of interstitial fluid decreases as a result of cellular use of oxygen for energy production. A wide range of metabolic end-products are generated, including carbon dioxide, adenosine, hydrogen ions, and heat. In addition, the potassium ion concentration of the interstitial fluid increases with increased cellular metabolism because of potassium loss from inside of the cells. All of these changes (increased carbon dioxide, decreased oxygen, increased adenosine, increased hydrogen ion, increased potassium ion concentration, and increased temperature) have been shown to promote vasodilation of arterioles (Fig. 19–10). Vasodilation results in increased local blood flow, which tends to increase the oxygen concentration and reduce the metabolite concentrations in the interstitial fluid back toward resting values. Realize that local vasodilation in response to increased metabolism can be an effective method for regulating local blood flow only if the blood vessels are partially contracted at rest (resting tone).

The autoregulatory response to changes in blood pressure that were discussed previously are in part due to

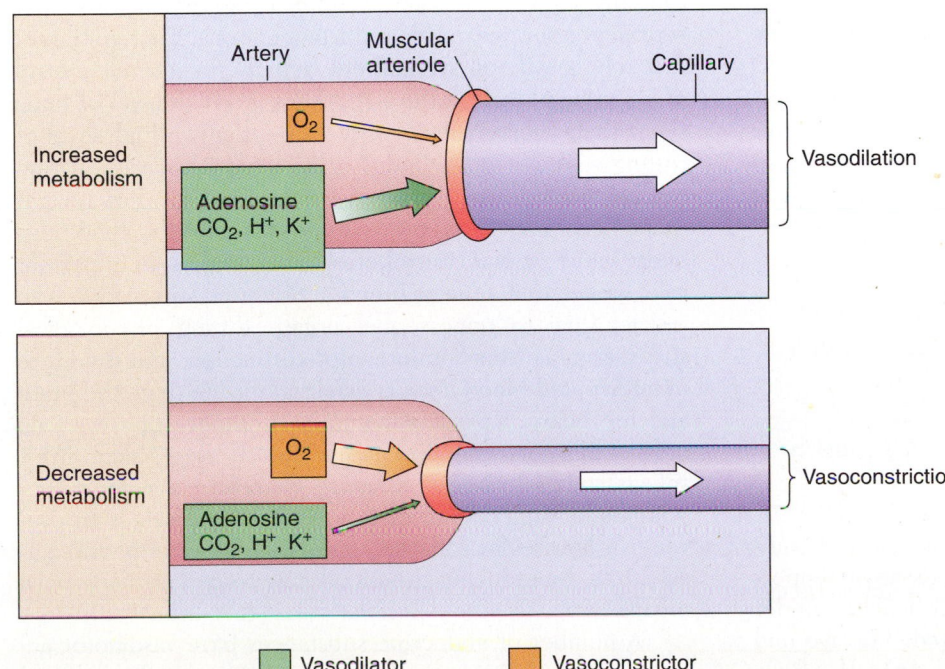

Figure 19–10

Autoregulation of capillary blood flow by local tissue factors. At high metabolic levels, the increased amount of vasodilator substances and decreased amounts of vasoconstrictors shift the balance such that there is vasodilation. At low metabolic levels, the balance is shifted so as to cause vasoconstriction. Relative concentrations are indicated by the size of the labels.

alterations of the composition of the interstitial fluid. A decrease in arterial pressure tends to decrease tissue blood flow, promoting the accumulation of chemical substances associated with metabolism and the depletion of oxygen. These chemical changes tend to promote vasodilation and increased blood flow. Conversely, an increase in arterial pressure increases flow and washes out accumulated metabolic byproducts, resulting in vasoconstriction.

One extreme form of chemically mediated autoregulation is seen following the complete occlusion (blockage) of blood flow to a tissue. When the occlusion is removed, an extreme reactive vasodilation occurs (**reactive hyperemia).** This is apparently due to the depletion of oxygen and the accumulation of metabolites that occur during the absence of blood flow. Next time you have your blood pressure measured, watch the changes in the blood flow to the skin in your forearm during and after inflation of the pressure cuff. Upon inflation of the blood pressure cuff, blood flow to the forearm is diminished. If the cuff is left inflated for several seconds and then deflated, the skin of the forearm flushes as a result of the reactive vasodilation of the arteriolar smooth muscle associated with the subcutaneous vasculature. As the increased blood flow restores the normal chemical balance of the interstitial fluid, vascular tone increases and blood flow to the skin returns to normal.

A number of other vasoactive substances produced locally by cells may also modify the level of vascular tone. For example, **histamine** released from mast cells during an inflammatory response (see Chapter 17) stimulates vasodilation. **Prostaglandins** may also be released from cells to cause either vasodilation or vasoconstriction, depending on the specific prostaglandin released. Vasoactive substances are also released from the endothelial cells that line the blood vessels. The first of these to be described was **endothelium-derived relaxing factor (EDRF),** which has been shown in many cases to be **nitric oxide (NO).** In addition to being a very potent vasodilator, NO also appears to be a neurotransmitter in the central nervous system. An endothelium-derived vasoconstrictor called **endothelin** has been described. The specific roles of many such locally released vasoactive agents remain to be clarified.

Nerves and Circulating Hormones Coordinate the Blood Flow Through All the Vascular Beds with the Pumping Activities of the Heart

> *How does the nervous system affect the distribution of blood flow?*

Earlier in this discussion, we noted that some form of communication between individual organs and tissues is important in the distribution of blood flow, particularly under conditions in which cardiac output is limited. Massive and complete vasodilation of all the vascular beds in the body would result in a drastic decrease in blood pressure because the heart would not be able to pump blood as fast as it flowed out of the arterial system.

This is precisely what happens with heat exhaustion. The vasculature to the skin is so extensively vasodilated (to dissipate excessive heat from the body) that blood pressure decreases and blood flow to other crucial organs (e.g., the brain, heart, and kidneys) is seriously compromised. A less extreme example of vasodilation occurs during strenuous exercise where blood flow to the muscles may increase to greater than 15 L/min. In both of these situations, the demand for blood flow exceeds the pumping capacity of the heart, and therefore redistribution of flow must occur. As pointed out in the water system analogy, redistribution of flow requires some form of communication between the various users. The communication link in our bodies involves the CNS and ANS as well as circulating hormones.

Most arterioles in the body are richly innervated by postganglionic sympathetic nerve fibers. Exceptions include the coronary vessels, which supply the heart, and the cerebral vessels in the brain; these two vascular beds are predominantly regulated by local metabolic factors. The sympathetic neurotransmitter **norepinephrine** binds to **alpha (α) receptors** on the vascular smooth muscle cells and stimulates contraction (Fig. 19–11). Under resting conditions, the systemic arterioles are constantly being stimulated by sympathetic nerve impulses originating from **vasoactive centers** in the medulla. This continuous neural activity contributes to the basal level of resting tone, which we previously discussed.

Although the parasympathetic neurotransmitter **acetylcholine** is a vasodilator, parasympathetic nerve fibers are not generally involved in the neural control of blood flow. This presents us with a rather one-sided control system that can only stimulate vasoconstriction. However, this does not represent a serious problem for two reasons. First, in tissues that rely heavily on sympathetic activity for normal control of blood flow, such as the skin, there is a high level of tonic sympathetic nerve activity and consequently a high level of resting vascular tone. Under these conditions, vasodilation can be accomplished by a reduction of the level of sympathetic nerve activity. Second, if one examines the conditions under which neural control of blood flow plays an important role, we see that a major function of neural stimulation is to restrict flow to nonessential organs during physiological stress. Organs whose uninterrupted function, and therefore uninterrupted blood flow, is essential for life (e.g., the brain and the heart) have less sympathetic innervation than do organs that can function intermittently (e.g., skin, liver, spleen, and gastrointestinal tract). Therefore, a major role of sympathetic innervation of the resistance vessels is to prevent cardiovascular collapse, under conditions of physiological stress, by reducing the blood flow to certain noncrucial organs.

A number of endocrine substances have vasomotor activity. Epinephrine, which is released from the adrenal

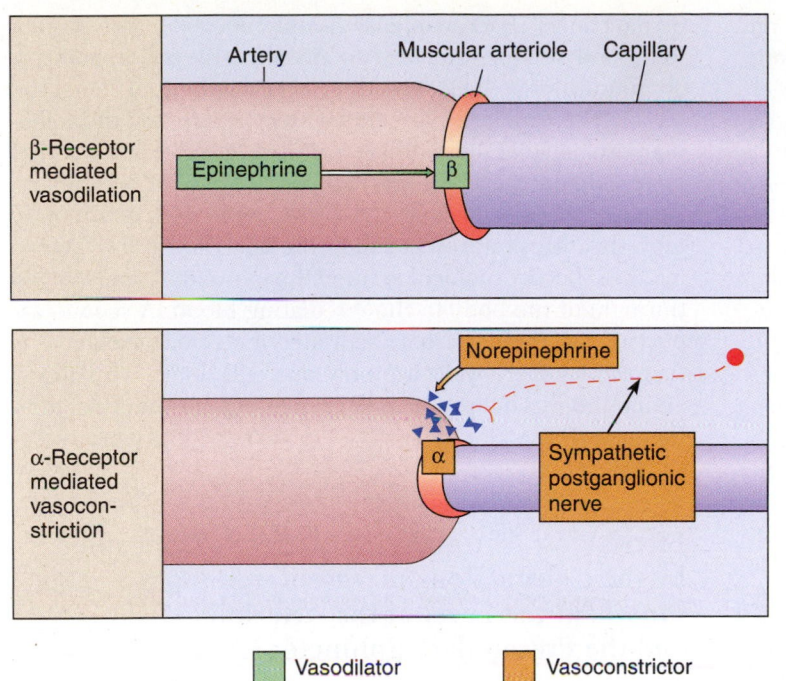

Figure 19–11

Opposing actions of the neurotransmitter norepinephrine and the hormone epinephrine on the contractile state of vascular smooth muscle. While some tissues, like skeletal muscle, contain both receptor types, most tissues are dominated by one receptor type. For example, the blood vessels of the skin contain predominantly α-receptors, whereas the blood vessels of the heart (coronary arteries) contain predominantly β-receptors.

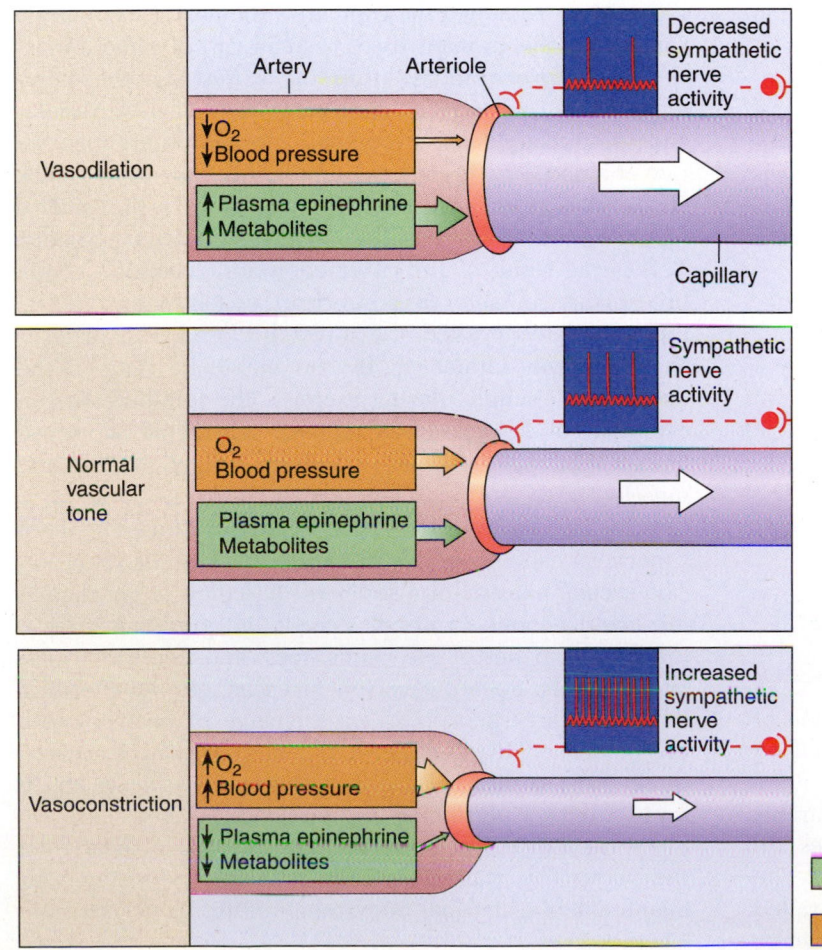

Figure 19–12

Summary of local tissue factors that produce either vasodilation or vasoconstriction of vascular smooth muscle. The resting level of vascular tone represents a balance between vasoconstrictor and vasodilator substances. An increase in the amount of vasoconstrictors and/or a decrease in the amount of vasodilators shifts the balance to produce vasoconstriction *(lower panel)* relative to resting tone *(middle panel)*. A decrease in the amount of vasoconstrictors and/or an increase in the amount of vasodilators shifts the balance to produce vasodilation *(upper panel)* relative to resting tone *(middle panel)*.

medulla, can bind to both **α** and **beta (β) receptors** on smooth muscle cells. The binding of epinephrine to smooth muscle β-receptors produces relaxation and vasodilation of the arterioles. Although epinephrine can also bind to α-receptors and produce vasoconstriction, this occurs only at relatively high concentrations of the hormone. The concentrations of epinephrine found in the circulating blood due to medullary stimulation are generally assumed to be too low to allow for significant binding to α-receptors. Therefore, the effect of circulating epinephrine is to antagonize the vasoconstrictor activity of sympathetic stimulation (see Fig. 19–11). The antagonism of norepinephrine-mediated vasoconstriction is most pronounced in the heart, skeletal muscle, and liver which have more β-receptors associated with the resistance vessels than do other tissues which have primarily α-receptors. During periods of elevated sympathetic stimulation, when both norepinephrine and epinephrine are being released, blood flow is shifted away from vascular beds containing predominantly α-receptors and to vascular beds containing both α and β-receptors. Figure 19–12 summarizes the effects of these vasoactive agents on arteriolar resistance.

One of the most potent vasoconstrictor substances is **angiotensin II** (see Chapter 23). Angiotensin II may play a role in the regulation of the renal circulation and may be a contributing factor in hypertension. The major role of angiotensin II is probably the control of aldosterone production (see Chapter 23).

Atrial natriuretic factor is a potent vasodilator secreted from cells in a variety of tissues, including the atria of the heart. This peptide causes both vasodilation and increased excretion of salt and water by the kidneys and may play a role in the maintenance of normal blood volume. The function of this peptide in the overall control of vascular tone, however, remains to be determined.

THE CAPILLARY SYSTEM: THE EXCHANGE OF MOLECULES

The Exchange of Molecules Between the Blood and the Extracellular Space Occurs Primarily in the Capillary Beds

 How do substances move from capillaries to tissue cells?

We have reached an important point in our discussion of the cardiovascular system. Thus far, we have focused on how the cardiovascular system functions to generate blood flow and to distribute this flow appropriately to the various tissues. We are now ready to consider how molecules are exchanged between the blood and the interstitial fluid. The **capillaries** are an extensive network of very thin-walled blood vessels that allow for the rapid diffusional exchange

of molecules. Each capillary is only approximately 1 mm long, and yet together they represent the major point of communication between the interstitial fluid and the blood. The capillaries are well suited to this task of molecular exchange. The capillary network is so extensive that most cells in the body are within 0.02 mm of a capillary. Because the distances are so small, molecules can rapidly move by diffusion from a cell to the nearest capillary. Once in the blood, molecules are then rapidly transported throughout the body in the circulating blood. A second aspect of the capillary that facilitates molecular exchange is that the capillary walls are only one cell thick. We will examine the dynamics of capillary exchange in some detail in the next few sections.

Blood Flow Through Capillaries Is Controlled by the Contractions of Vascular Smooth Muscle at the Level of the Arteriole and the Precapillary Sphincter

The anatomy of the **microcirculation** is shown in Figure 19–13. The capillary beds are at the center of the microcirculatory system, which includes the **arterioles, capillaries,** and **venules.** The capillaries (5–10 μm in diameter) branch out from the arterioles and then empty into the venules. Although the capillaries themselves contain no smooth muscle, in many cases, a small cuff of smooth muscle called a **precapillary sphincter** is present at the beginning of the capillary. The contractile state of the arteriolar and precapillary sphincter smooth muscle determines the rate of blood flow through the capillaries. If we were to look at the flow of blood through the capillaries with a microscope, we would see that the blood flow is often sporadic. This is the result of the contraction and relaxation of the precapillary sphincter (**vasomotion**), which tends to be either completely opened or completely closed. This vasomotion is heavily influenced by the metabolic state of the tissue. For example, during exercise, the number of open capillaries in skeletal muscle increase severalfold as precapillary sphincters vasodilate to increase blood flow to the muscle.

The skin contains two distinct capillary beds. The more superficial capillary bed is similar to that discussed earlier. The second consists of a series of muscular, larger diameter, arteriovenous shunts that are found predominantly in the fingers, palms of the hands, toes, and the face. These latter vessels are not involved in exchange of molecules, but rather are specialized for exchange of heat between the blood and the surface of the body. Reduced sympathetic tone to these vessels causes vasodilation, which, in turn, shunts a portion of the circulating blood closer to the surface of the body to facilitate the loss of heat from the body. This represents one of the major systems for maintenance of normal body temperature as discussed in Chapter 27.

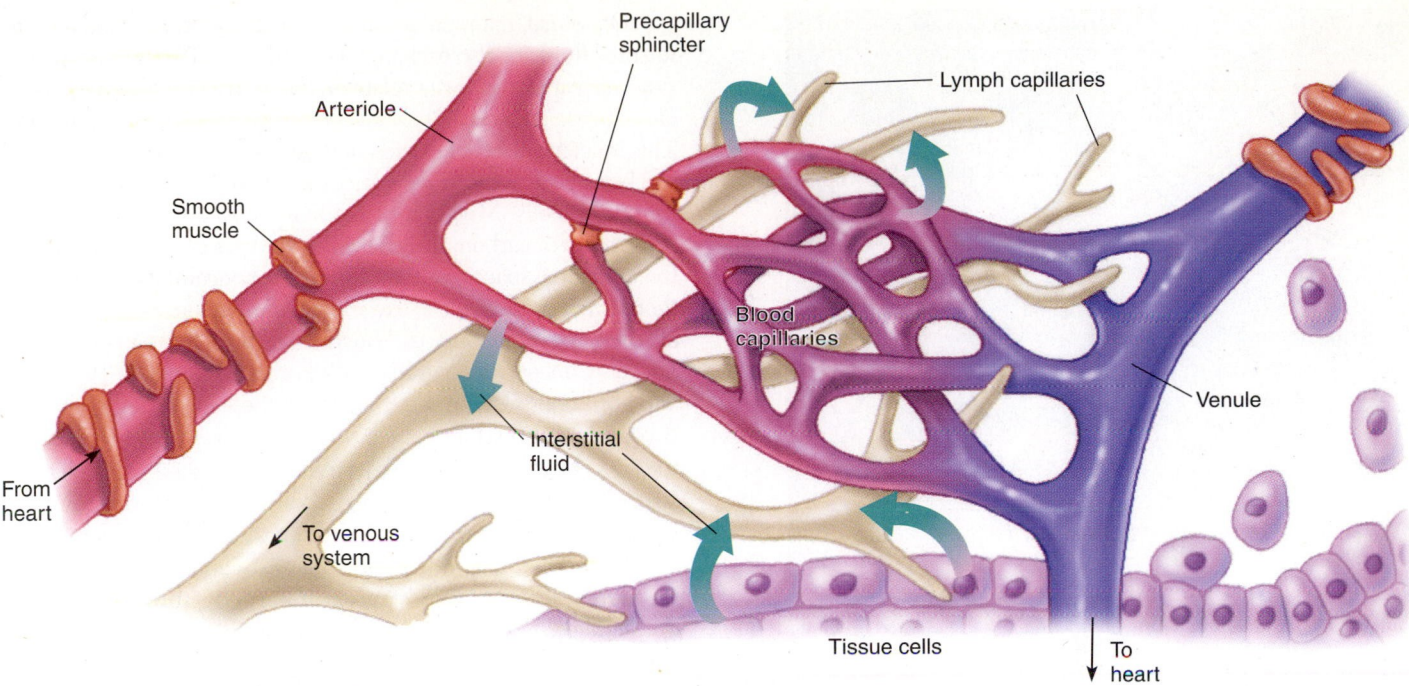

Figure 19–13

Representation of the microcirculation. The diameter of the capillaries in humans ranges from approximately 4 to 12 μm.

The Absence of Smooth Muscle Cells Distinguishes the Capillary from Other Small Vessels

As shown in Figure 19–14, the capillary wall is made up entirely of endothelial cells and contains no smooth muscle. The flat, thin, endothelial cells are joined at their edges like patches in a patchwork quilt. Small pores (**clefts**) between the endothelial cells allow for the passage of molecules. The size and number of these **endothelial pores** vary greatly from tissue to tissue. The capillaries in the brain, for example, may not contain any pores, which accounts for the impermeability of the blood-brain barrier to a variety of substances. The capillaries in the kidney, on the other hand, contain relatively large pores that allow the passage of large molecules. Although the endothelial pores represent an obvious point of exchange, in most capillaries they represent a very small percentage of the capillary surface area. As we will see in a later section, direct diffusion of molecules across the endothelial cell membranes accounts for the major portion of molecular exchange.

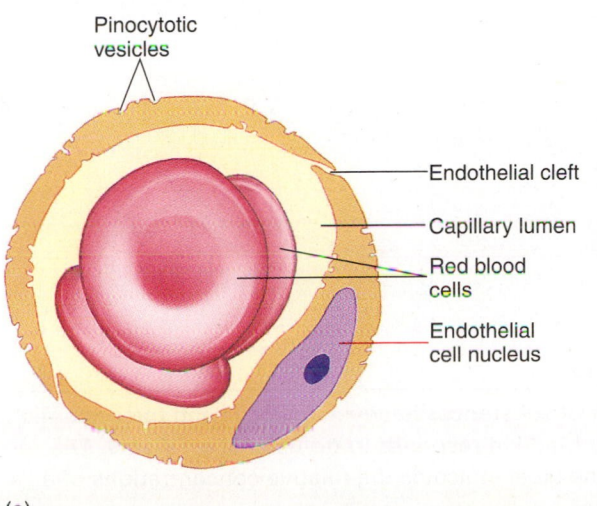

(a)

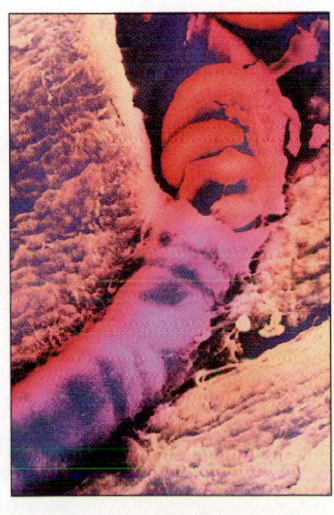

(b)

Figure 19–14

(a) Cross-sectional view of a single large capillary. The capillary wall is made up of a single layer of endothelial cells. **(b)** SEM of a single capillary surrounded by cells and interstitial space. The cells in the lumen of the capillary are red blood cells (erythrocytes). *(© Don W. Fawcett/Visuals Unlimited.)*

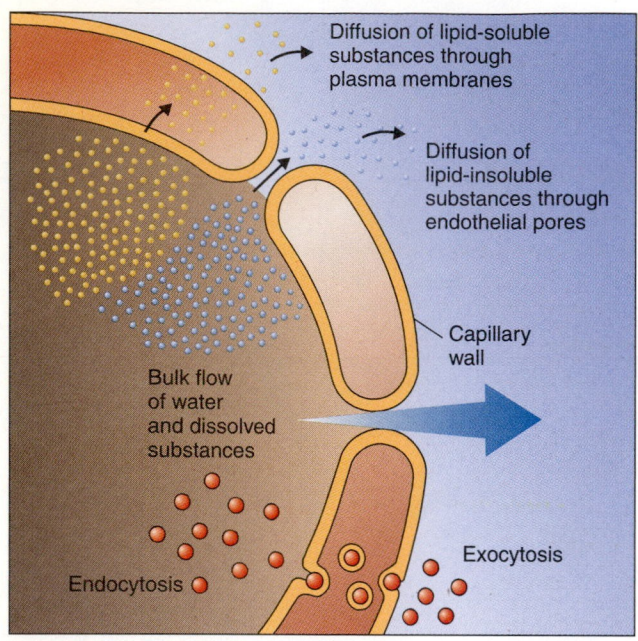

Figure 19–15

Summary of transport pathways across the capillary wall. Movement of substances is in response to either a hydrostatic pressure gradient (bulk flow) or a diffusional gradient.

Molecules Exchange Across the Capillary Wall by Transcytosis, Free Diffusion, and Bulk Flow

The three fundamentally different mechanisms by which substances can cross the capillary wall are transcytosis, diffusion, and bulk flow (Fig. 19–15).

Transcytosis and Endocytosis

Relatively large molecules can cross the capillary wall by an active process called **transcytosis,** in which molecules cross the epithelial cells either in discrete shuttling vesicles or through transient vesicle-derived channels. True **endocytosis** (e.g., phagocytosis and pinocytosis) is a less common occurrence. The extent of transcytosis and endocytosis varies greatly among different vascular endothelia. Pinocytosis is prominent in the postcapillary venules of lymph nodes, whereas transcytosis is the dominant event in most other capillaries.

Diffusion

By far, the most important mechanism for exchange of water and dissolved substances is diffusion. We can get some idea of this exchange by comparing the rate of blood flow through a capillary with the rate of water exchange by diffusion across the capillary wall. The rate of exchange is as much as 100 times greater than the rate of blood flow down the length of the capillary. Because water is moving across the capillary wall in both directions, there is little net gain or loss of water from the circulation by diffusion; however, the exchange rate is very high.

Of course, many solute molecules dissolved in blood and interstitial fluid are also exchanged by diffusion. Blood entering the systemic capillaries has a relatively high concentration of oxygen and glucose and a relatively low concentration of carbon dioxide. The concentrations of oxygen and glucose in the interstitial fluid are less because these substances are being continuously taken up by the cells. Metabolism of these substances results in the production of carbon dioxide, which diffuses out of the cells into the interstitial space. As a result of these concentration gradients, there is net movement of oxygen and glucose out of the capillary and net movement of carbon dioxide into the capillary (Fig. 19–16). Concentration gradients for other nutrients, metabolites, and hormones result in net transfer of these substances between the interstitial fluid and the blood. In all cases, the direction of the concentration gradient determines the direction of exchange. Likewise, heat generated by metabolism is transferred by convection down its thermal gradient from the cell to the blood.

Molecules can diffuse across the capillary walls either through the water-filled pores or directly through the endothelial cells. The path a molecule takes is determined by its relative solubility in water and lipids. Molecules that are only slightly soluble in lipid (e.g., Na^+, K^+, Cl^+, and glucose) are more likely to diffuse through pores. Molecules such as oxygen, carbon dioxide, and urea, which are more lipid-soluble, can diffuse directly through the plasma membranes of the endothelial cells. Water molecules are able to cross either through pores or by diffusion through the endothelial cells; recent studies have demonstrated the existence of specific water channels in some tissues.

Because pores occupy less than 1% of the capillary wall area, the surface area available for diffusion of lipid-soluble substances is more than 100 times greater than that for lipid-insoluble substances. We can begin to appreciate how well suited this system is for the very rapid and continuous exchange of oxygen and carbon dioxide, which are both lipid-soluble.

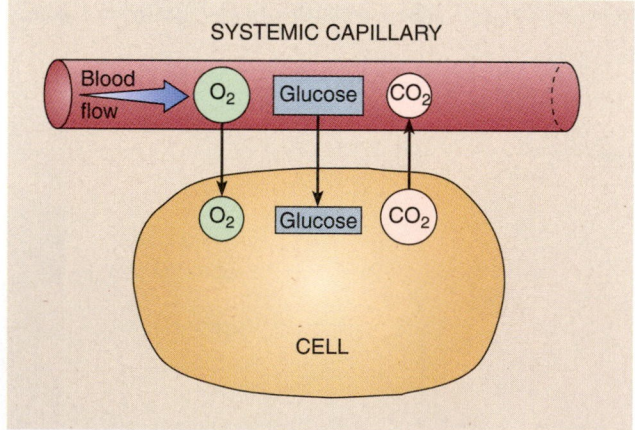

Figure 19–16

Transfer of substances between the cells and the systemic capillary blood in response to diffusional gradients. The size of the label indicates the relative concentrations of a substance.

A number of additional factors contribute to the rapid diffusional exchange of substances across the capillary wall through either pores or the endothelial cell. First, the distances that molecules must move are very small. Most metabolically active cells are within 20 μm of a capillary. Second, the total surface area of the capillary walls is extensive. Both diffusion distance and surface area influence the rate of net diffusion. The total surface area of all the capillary walls may be as high as 7000 ft². Imagine spreading the entire blood volume of the capillaries ($\approx$ 250 ml, or 1cup) over an area 1.5 times larger than a basketball court; this gives you some idea of the ideal conditions for diffusion that exist in the capillaries.

A third factor is the relatively slow rate of blood flow that exists within the capillaries. The velocity of blood flow through the capillaries is greatly reduced because of the large total cross-sectional area of all the capillary beds (Fig. 19–17). The velocity of blood flow is inversely proportional to the cross-sectional area of the vessel(s) through which it is flowing. We see this principle in action whenever we observe the flow of water down a river. The velocity is greatest where the stream narrows (as in rapids) and slowest where the stream widens; total flow, however, is the same at both points. When a stream empties into a lake, the velocity of water movement in the lake is imperceptible because of the dramatic increase in cross-sectional area. For precisely the same reason, the average velocity of blood flow in the aorta may be more than 1000 times greater than that in the capillaries. The velocity of blood flow in a capillary ranges from 0.1 to 1.0 mm/sec compared with 40 cm/sec in the aorta. The dramatic reduction in flow velocity in the capillaries means that the time for exchange is substantially increased.

Bulk Flow and Oncotic Pressure

Exchange of water (and dissolved solutes) also occurs through endothelial pores by **bulk flow** in response to the pressure gradient between the inside and outside of the capillary (Fig. 19–18). As we will see, the pressure gradient is always from inside the capillary to outside, which causes water to flow out of the capillary. Although the total exchange of fluid by this means is relatively small (except in the kidney), it represents an extremely important mechanism for the maintenance of the circulating blood volume, because it occurs in response to a pressure gradient rather than to a diffusional gradient.

The **hydrostatic pressure** gradient that drives water through the endothelial pores is simply the difference between the hydrostatic pressure inside the capillary (capillary blood pressure) and the hydrostatic pressure outside the capillary (interstitial hydrostatic pressure). The hydrostatic pressure outside the capillary is extremely difficult to measure, and there is some uncertainty about the actual value. Although it may in fact be slightly subatmospheric (negative), for purposes of our discussion, we will assume that it is zero. Given this assumption, the hydrostatic pressure gradient (capillary hydrostatic-interstitial hydrostatic pressure) is equal to the capillary hydrostatic pressure. When a blood vessel (usually a superficial vein) is cut, blood flows out of the vessel in response

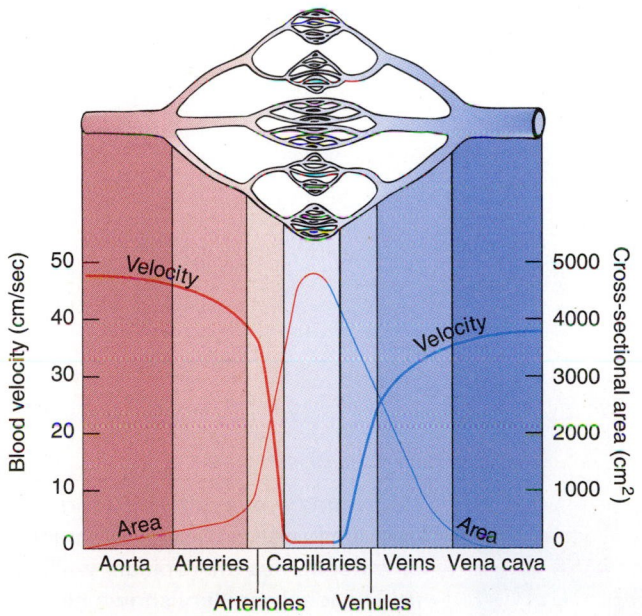

Figure 19–17

Vascular cross-sectional area and velocity of blood flow. Cross-sectional area is the total area of all of the vessels added together at each level in the circulatory system. Blood velocity is that which would be measured in a single vessel at each level in the circulatory system.

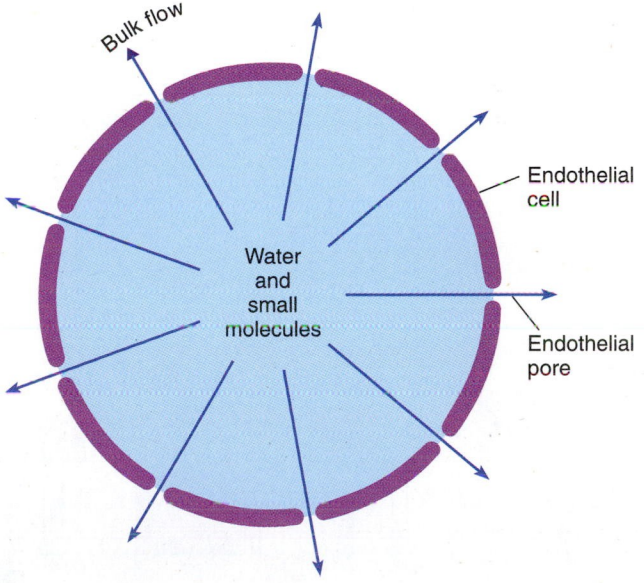

Figure 19–18

Bulk flow of water and dissolved substances out of the capillary through endothelial pores in response to a hydrostatic pressure gradient. Hydrostatic pressure inside of the capillary is always greater than the pressure of the interstitial fluid surrounding the capillary.

to the pressure difference between the inside and the outside of the vessel. Because water can move through the capillary pores by bulk flow, we can reason that water and dissolved solutes should be forced out of the capillaries. Clearly, if this were the whole story, we would have a great deal of trouble maintaining a normal circulating blood volume. There is in fact an opposing force that minimizes the loss of fluid by bulk flow.

The opposing force that prevents this loss of vascular volume is the **osmotic pressure** created by the presence of proteins in the blood (plasma proteins). Although all dissolved solutes contribute to the **total osmotic pressure** of blood, we can ignore the contribution of all solutes that are freely exchanged across the capillary wall because these solutes are at diffusional equilibrium. Only the plasma proteins, which do not freely cross the capillary wall, contribute

to the osmotic pressure gradient between the inside and outside of the capillary. The **oncotic pressure** or **colloid osmotic pressure** (π) is that portion of the total osmotic pressure of blood that is due to the presence of plasma proteins. Recall from Chapter 4 that net diffusion of water across a selectively permeable membrane occurs in response to a concentration gradient for water. The capillary wall functions as a selectively permeable membrane because of its limited permeability to plasma proteins. Recall that the major plasma protein is albumin. Most plasma proteins cannot leave the capillary and therefore establish an oncotic pressure gradient that favors the net diffusion of water from the interstitial space to the intravascular space. You have probably guessed by now that the oncotic pressure balances the hydrostatic pressure so that there is normally only a very

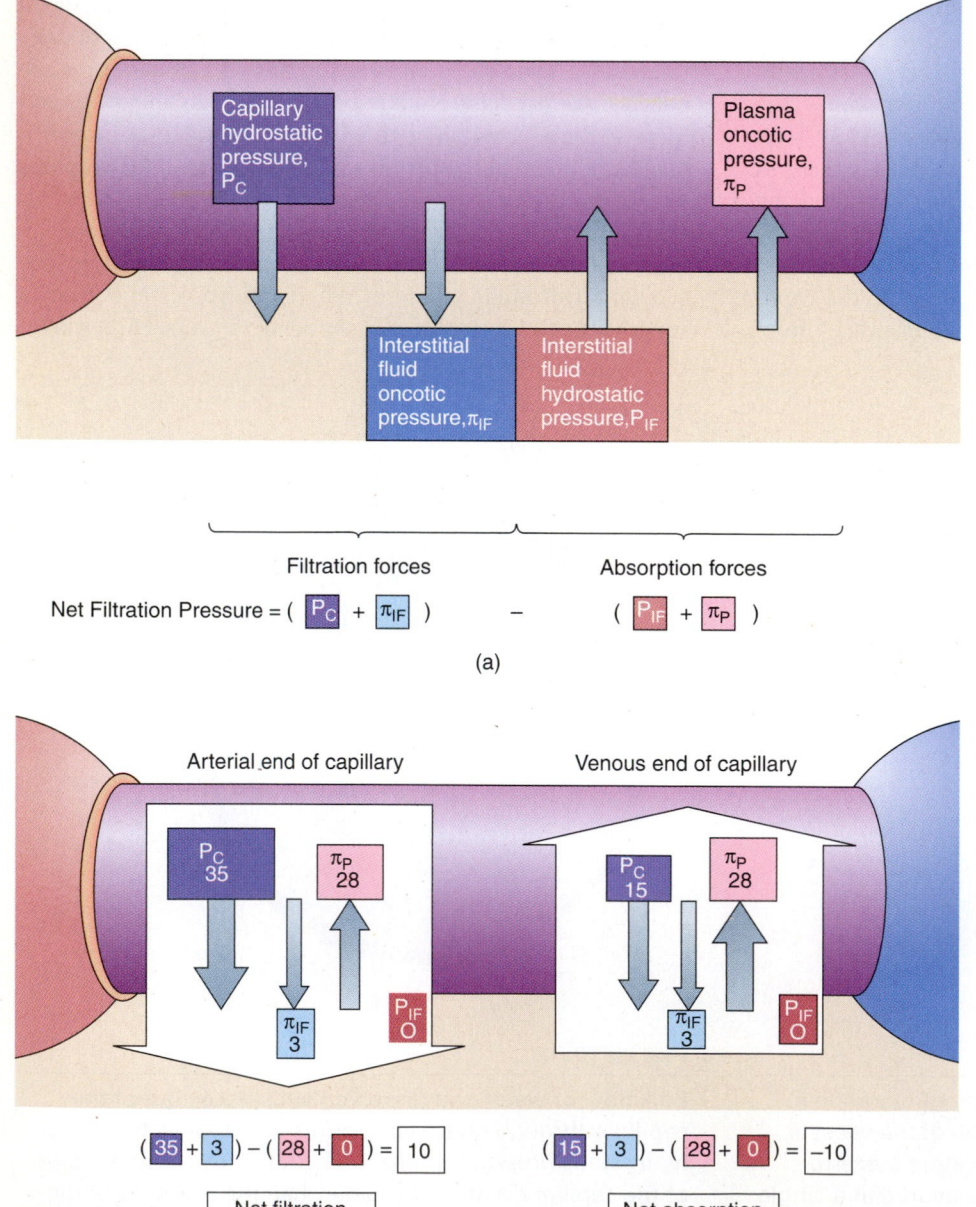

(a)

Net Filtration Pressure = (P_C + π_{IF}) − (P_{IF} + π_P)

Filtration forces — Absorption forces

(b)

(35 + 3) − (28 + 0) = 10 Net filtration

(15 + 3) − (28 + 0) = -10 Net absorption

Figure 19–19

(a) Summary of hydrostatic and osmotic forces causing movement of water across the capillary wall. *(b)* The ideal capillary shown here represents the average of all capillaries in the body. Pressures within individual capillaries vary considerably. *(Modified from Vander, A. J., Sherman, J. H., Luciano, D. S. Human Physiology, ed 4. New York, McGraw-Hill Publishing Company, 1985, Fig. 11–53, p 346.)*

slow leakage of water from the circulating blood. Water is lost from the capillary system at a rate of approximately 3 to 4 ml/min (5 L/day) and is returned to the general circulation via the **lymphatic system,** as discussed in a later section.

Filtration and Absorption: Starling's Hypothesis

This balance of hydrostatic and oncotic pressures across the capillary endothelium was originally described by Starling in what is now called **Starling's hypothesis** (Fig. 19–19).

Let's begin by looking at an idealized average capillary. (The situation in any particular capillary might differ significantly from this model, but the principles remain the same.) The oncotic pressure gradient is determined by the difference between the oncotic pressure of the capillary blood and that of the interstitial fluid. The oncotic pressure of blood is normally about 28 mm Hg. Although the capillaries are relatively impermeable to plasma proteins, some proteins do leak out of the capillaries. This small amount of protein in the interstitial fluid results in a tissue oncotic pressure of approximately 3 mm Hg. The oncotic pressure gradient (28 − 3 = 25 mm Hg) causes a net flow of water into the capillary. Remember that the osmotic flow of water is in the direction of increasing osmotic pressure. This is easy to remember if you recall that osmosis is the diffu-

sion of water down its concentration gradient. Water is less concentrated in blood because of the presence of plasma proteins.

If we examine a typical capillary, we will measure a hydrostatic pressure of approximately 35 mm Hg at the arterial end of the capillary and approximately 15 mm Hg at the venous end. The pressure drop along the length of the capillary reflects the resistance to flow of blood through the capillary. The hydrostatic pressure causes fluid to move out of the capillary; this is opposed by the oncotic pressure that moves water back into the capillary (see Fig. 19–19). At the arterial end of the capillary, hydrostatic pressure (35mm Hg) exceeds the oncotic pressure gradient (25mm Hg), resulting in net filtration (filtration pressure = 10mm Hg). On the other hand, at the venous end of the capillary, hydrostatic pressure (15mm Hg) is less than the oncotic pressure gradient (25mm Hg), resulting in net absorption (filtration pressure = −10mm Hg). In this example, filtration exactly equals absorption, which means that no change in the overall capillary volume occurs.

We stated earlier that capillary flow often changes dramatically, following the contraction of smooth muscle in the terminal arterioles and precapillary sphincters. These changes in flow are associated with similarly large changes in intracapillary pressure. As shown in Figure 19–20, when

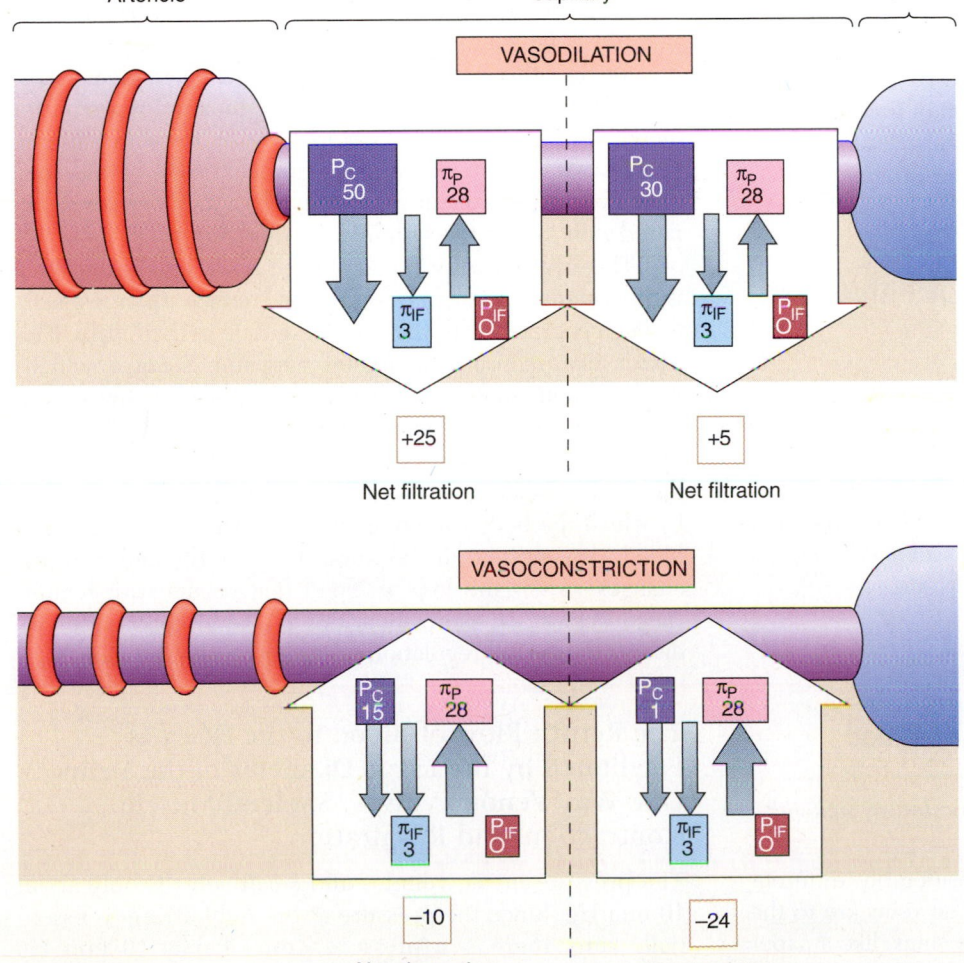

Figure 19–20

Effect of changes in arteriolar smooth muscle contraction on the balance of hydrostatic and osmotic forces in the capillary. Vasodilation promotes filtration, and vasoconstriction promotes absorption.

precapillary smooth muscles are fully dilated, the associated capillaries show filtration along their entire length. Fully contracted precapillary muscles decrease the capillary pressure enough to cause absorption of fluid along the entire length of the capillary. Changes in arterial pressure can also lead to changes in capillary pressure to favor either filtration or absorption. We can get some idea of what the "average" state of the capillaries is by looking at the net loss or gain of fluid from the cardiovascular system. Under normal conditions, approximately 2 to 5 liters of fluid are returned from the interstitial space to the circulation by the lymphatic system during a single 24-hour period. Thus, the average situation slightly favors filtration over absorption.

Filtration and absorption provide an important mechanism for the dynamic maintenance of the circulating blood volume. Although the cardiovascular system is a closed system at the "macro" level, it is an open system at the level of the microcirculation. The constancy of the circulating blood volume in fact represents a dynamic balance of water moving into and out of the capillaries. Recall that cardiac output is determined in part by filling of the heart (end-diastolic volume). Cardiac filling, in turn, is greatly influenced by the total volume of circulating blood. Any significant reduction in blood volume leads to a reduction in cardiac output. As we will see later, transfer of fluid from the extravascular space into the capillary represents an important and automatic compensatory response to abnormally low blood pressure as occurs, for example, with blood loss (**hemorrhage**). The distribution of fluid can also be affected by disease. Heart failure can lead to an increase in capillary pressure, causing the accumulation of fluid in the interstitial space (edema). Other diseases may also lead to edema because of a decrease in plasma protein concentration.

THE VENOUS SYSTEM: RETURNING BLOOD TO THE HEART

The capillaries empty into thin-walled vessels called *venules*, which represent the beginning of the venous system. The venules contain endothelial clefts like those in the capillaries. Although fewer in number, these allow for additional exchange of substances. The capillaries and venules together are often referred to as the **exchange vessels.** The venules then converge to form progressively larger veins.

The Veins Provide a Conduit for the Return Flow of Blood to the Heart and Function as a Variable Capacity Reservoir for Blood

 What role do the veins play in the circulatory system?

The physical nature of the veins is considerably different from that of the arteries and arterioles. The veins are to the arteries as a paper bag is to a balloon. The veins, like a paper

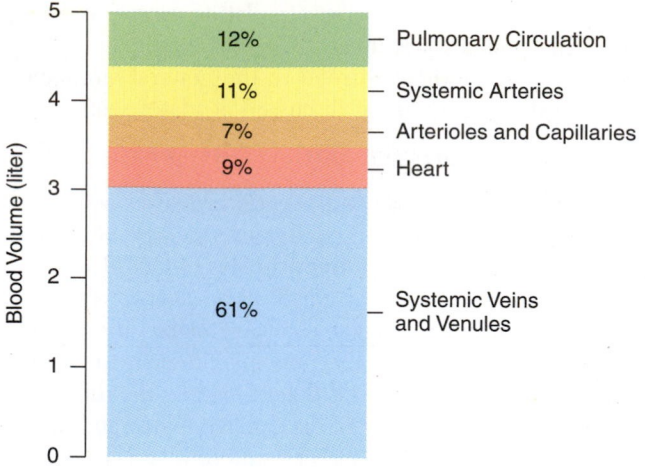

Figure 19–21

Distribution of the total blood volume among the various regions of the circulatory system. Recall from Figure 18–1 that the total blood flow through each region is the same because these regions are arranged in series with one another.

bag, can accommodate large changes in volume with little change in internal pressure up to a point after which pressure increases rapidly with further increases in volume. In addition, the total volume of the venous system is several times greater than that of the arteriolar system. These characteristics allow the veins to function as a low-pressure reservoir for blood; the veins are often referred to as the **capacitance vessels.** At any one time, approximately 61% of the total blood volume is contained within the systemic veins, compared with 11% in the systemic arteries (Fig. 19–21).

This large capacity in and of itself is not significant. What is important is that the veins can reduce their capacity through contraction of the smooth muscle in their walls. This occurs in response to increased sympathetic nerve activity. As the venous smooth muscle contracts, blood is transferred from the veins to the heart and the arterial system. This ability of the veins to dramatically shift blood from one part of the cardiovascular system to another provides a mechanism by which the body can compensate for the redistribution of blood that occurs in response to exercise and postural changes, or for the loss of blood that occurs with hemorrhage. We will say more about this in a later section when we discuss the overall regulation of cardiovascular function.

The Return Flow of Blood to the Heart Is Facilitated by the Large Diameter of the Veins, One-Way Venous Valves, Skeletal Muscle Contraction, and Respiration

The pressure in the venules and small veins is only about 10 mm Hg. Since the pressure at the right atrium is essentially zero, there is a pressure drop of only 10 mm Hg

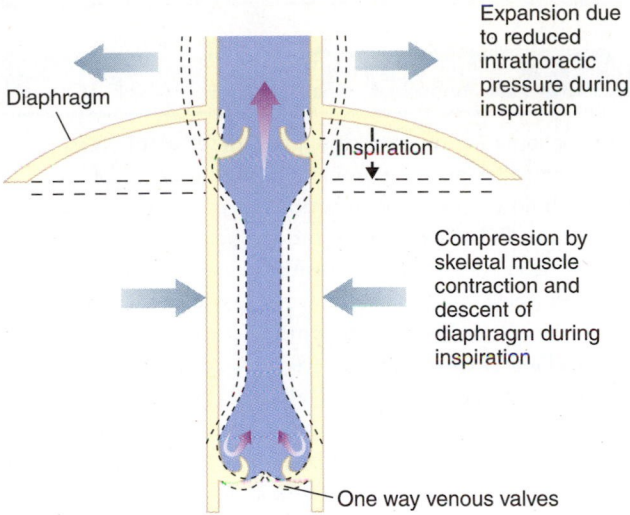

Diaphragm

Expansion due to reduced intrathoracic pressure during inspiration

 Inspiration

Compression by skeletal muscle contraction and descent of diaphragm during inspiration

One way venous valves

Figure 19–22

Factors promoting venous return to the heart. The combination of one-way venous valves and the compression/decompression of veins by contraction/relaxation of skeletal and respiratory muscles enhances the flow of blood from the veins back to the heart.

between the venules and the heart, indicating that the resistance to venous blood flow is low. This low resistance is due to the large diameter of the veins. Recall that the resistance to blood flow decreases as the fourth power of the radius.

Several other factors also contribute to the flow of blood back to the heart. Some veins contain one-way valves that function much like the valves in the heart (Fig. 19–22). The venous valves serve two major functions. First, by preventing backflow of blood away from the heart, they tend to minimize the pooling of blood in the feet and legs upon standing. Second, they allow the contractile activity of the muscles surrounding the veins to contribute to the flow of blood back to the heart. Contraction of the surrounding muscle tends to squeeze the veins. Blood is forced out of the locally constricted vein and, because of the valves, is constrained to flow toward the heart. So you can see that the combination of venous valves and skeletal muscle contraction constitutes a pump that assists the heart in maintaining the unidirectional flow of blood through the circulatory system. The muscular activity associated with normal respiration also acts through the venous system to pump blood back to the heart. When the diaphragm descends during inspiration to reduce intrathoracic pressure and draw air into the lungs, the contents of the abdomen are compressed along with the abdominal veins. As before, this compression of the veins

forces blood back toward the heart. The reduced intrathoracic pressure also draws blood into the thoracic veins and the right heart.

THE LYMPHATIC SYSTEM

> *What role do lymph vessels play in cardiovascular function?*

The lymphatic system is a second system of vessels that differ anatomically from the blood vessels (Figs. 19–13 and 19–23). The function of the lymphatic vessels is to return fluid and plasma protein that has leaked out of the capillaries into the interstitial space, back to the circulating blood. The extent of this loss increases significantly during exercise and with certain diseases. The normal operation of the lymphatics prevents the edema that would occur otherwise.

Lymphatic Vessels Begin as Blind Sacs that Empty into Lymphatic Vessels

The lymphatic vessels, like the capillaries, are made up of a single layer of epithelial cells. As shown in Figure 19–23a, the lymphatic system begins as a series of sacs with low internal hydrostatic pressure. These terminal sacs have relatively large endothelial pores, allowing for the movement of both fluid and protein into the lymphatic system. The **lymph** capillaries drain into larger vessels, which themselves drain into the subclavian veins.

Lymph Flow Is Facilitated by One-Way Lymphatic Valves and the Pumping Effect of Skeletal Muscle Contraction

The lymphatic vessels have an extensive series of one-way valves like those in the veins (Fig. 19–23b). Phasic contractions of the smooth muscle associated with lymphatic vessels, as well as compression of the vessels by skeletal muscle activity, pumps lymph through the one-way valves. In 24 hours, the lymphatics may return a volume of fluid equal to the total circulating blood volume (5L). The lymphatic system represents the only means by which plasma proteins can be returned to the blood. Approximately one fourth to half of the total plasma protein is returned to the blood during this period. The lymphatics also transport substances absorbed from the gastrointestinal tract — primarily fat and fat soluble vitamins — to the circulating blood. Periodic swellings of the lymphatic vessels, called **lymph nodes,** are packed with lymphocytes and phagocytes. As the lymph percolates through this meshwork of reticular cells, microorganisms and other foreign material are removed from the lymph by phagocytosis.

APPLICATIONS OF PHYSIOLOGY

Edema, Lymphatic Pathology, and Varicose Veins

Edema

Any condition that causes blockage or interruption of normal lymph flow leads to peripheral edema. It results from an accumulation of protein (which has leaked out of the capillaries) in the interstitial spaces. The osmotic pressure caused by the presence of this extra protein promotes the retention of extravascular water, causing swelling, or edema. Obstruction of lymph flow may result from injury, inflammation, parasitic infection, or surgery. Lymph nodes are surgically removed during a radical mastectomy in an attempt to arrest the spread of cancer cells from the breast to the rest of the body via the lymphatics. This may lead to temporary edema in the patient's arms; new lymph vessel growth usually alleviates the problem. Another cause of impaired lymph flow is parasitic infection, as shown in the figure on filariasis. Filariasis is the result of a parasitic infection of the lymphatic system. The parasitic larvae are transmitted by mosquitoes, and the adult worms reside in the lymph nodes. The resultant blockage of lymph flow results in progressive swelling of tissues in the affected region. As shown in the figure, affected limbs may swell to several times their normal size. The appearance of such swollen limbs has led to the more descriptive name for this disease — *elephantiasis*.

The other major cause of edema is accelerated leakage of fluid from the blood to the interstitial space. If the rate of leakage exceeds the rate of fluid removal by the lymphatics, edema will develop. A frequent cause of excessive lymph formation is congestive heart failure (see Fig. 19–34). When the heart fails, blood backs up in the veins and capillaries due to the reduced rate of blood transfer by the heart to the arterial system. The distension of the veins causes increased venous (and therefore increased intracapillary) hydrostatic pressure, followed by increased filtration of water (see Figs. 19–19 and 19–20). Increased fluid leakage may also occur as a result of reduced concentrations of plasma proteins secondary to either renal failure or severe protein malnutrition (see Figs. 19–19 and 19–20). With a decrease in plasma protein concentration, the edema is not localized and fluid accumulates in the most readily expandable areas of the body, such as the peritoneal cavity (see figure on kwashiorkor).

Filariasis causes the formation of edema and swelling of the afflicted areas as illustrated here.

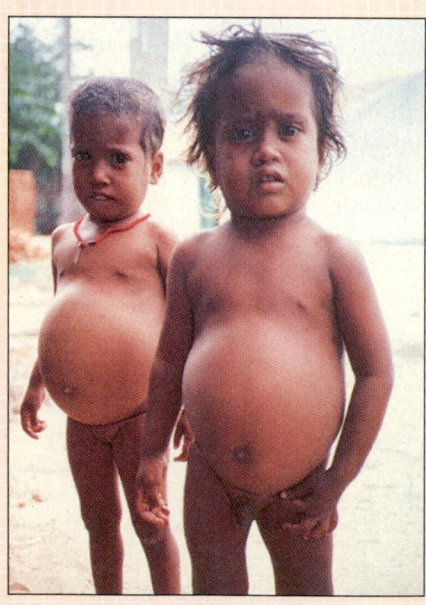

Children suffering from kwashiorkor, San Salvador, Brazil.

Varicose Veins

Varicose veins are usually superficial veins that have become excessively dilated, creating a distended and tortuous appearance. The primary cause for this abnormal distension is the failure of the venous valves (valvular incompetence) to prevent pooling of blood in the peripheral veins. Leaky venous valves can occur at any site in the leg, but the great majority of varicose veins are caused by faulty valves in the groin or behind the knee. When standing, the weight of the vertical column of blood in the veins creates pressure that tends to stretch the veins, particularly the more superficial veins that are not supported by surrounding tissues. To see this effect, add some water to a balloon and then try lifting it by one end; the weight of the water and the compliance of the balloon allow all of the water to remain near the bottom. The degree of distension will be greatest at the very bottom of the balloon because the outward pressure at any point in the balloon is directly proportional to the height of the column of water above that point. Recall that skeletal muscle activity associated with walking and running normally compresses the veins and pushes blood back to the heart. This is equivalent to squeezing the bottom of the water-filled balloon with your hands to force the water back up into the elevated end. But because the balloon does not have one-way valves, the water will fall back as soon as you stop compressing the balloon. You can also observe this effect if you have prominent superficial veins in your arms and backs of your hands. Hang your hands to your side for a few minutes and you will see the veins become distended from the pressure created by the weight of the column of venous blood. Next, elevate your hands above your head, and the veins will collapse as blood drains to the heart. To locate a venous valve, lower your hands as before and let the veins become distended. Using the index finger of your left hand, compress a vein on the back of your right hand or arm. Next, while continuing to compress the vein with your left index finger, use your left thumb to compress the vein just above your index finger, and then move your thumb up and away from your index finger while still compressing the vein. This will squeeze the blood out of the segment of vein between your thumb and index finger. If you now lift your index finger, the segment of vein will immediately refill with blood flowing back towards the heart. But if, instead, you lift your thumb while leaving your index finger in place, you would see that blood flows backward into the vein only until it reaches a one-way valve. The portion of the vein between the valve and your index finger cannot refill with blood. You may have to try this at several different places until you locate a venous valve. This is the precise method used at the beginning of the seventeenth century by Sir William Harvey to deduce that blood flowed back to the heart by way of the veins.

There is evidence that weak valves may be inherited in some people. Venous valves can also be stretched and caused to leak by obesity and pregnancy. Valvular incompetence is precipitated by factors that increase the volume of blood in the veins. During pregnancy, the growing fetus causes compression of all the internal organs, including the inferior vena cava. The compression increases the resistance to venous return and therefore increases the pressure inside the capillaries, and inside the veins between the capillaries and the compression point; the increased internal pressure leads to venous distension. Prolonged distension of the veins eventually causes permanent morphologic changes in the vessel wall, with a resultant increase in diameter and a loss of elasticity. Hemorrhoids are varicosities of the veins in the anal region. Formation of hemorrhoids is promoted by pregnancy for the same reasons discussed previously, and also by chronic constipation, which leads to excessive straining (attempted expiration against a closed glottis) and hence to elevated venous pressure during defecation. While people who spend a great deal of time on their feet are certainly more likely to notice varicose veins in their legs and any symptoms from them, it is unlikely that prolonged standing actually causes varicose veins.

Varicose veins can be removed surgically, but usually this is not necessary. Elastic stockings often alleviate the majority of symptoms. However, in some cases, the veins can be painful and can even lead to ulceration or thrombophlebitis (clotting and acute tender inflammation of the varicose veins). Patients are commonly concerned about the effect of tying and removal of veins on the circulation of their legs. In fact, the veins that are removed in varicose vein surgery are superficial veins collecting blood only from the skin and contributing very little overall to the major blood drainage from the leg, which occurs through quite separate, deep veins within the leg. Fortunately, the leg contains a complex, interconnected network of both superficial and deep veins, with considerable spare capacity, so that blood can easily find another route out of the leg after varicose veins are tied or removed. If surgical removal is necessary, a sophisticated ultrasound scanner can be used prior to surgery to produce a detailed "roadmap" of superficial and deep veins in the leg.

It is sometimes possible to obliterate varicose veins by injecting an irritant substance in a segment of the vein and then applying a small pressure pad over the vein. The injected irritant produces damage and inflammation of the lining of the vein. Opposite walls of the vein will then adhere together if the vein is kept empty and compressed. This method enjoyed great popularity in the 1970s, particularly because it avoided hospital admission and surgery. This method of treatment is commonly reserved for cosmetic treatment of smaller "spider veins."

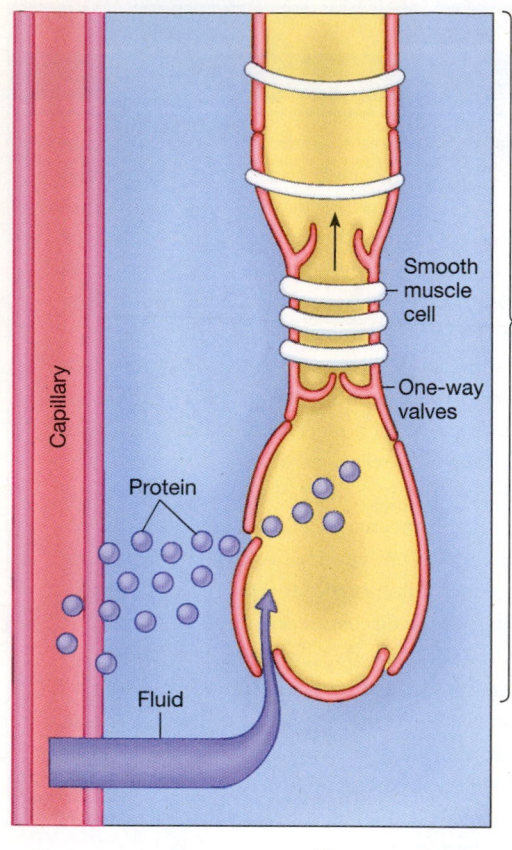

Capillary

Smooth
muscle
cell

Lymphatic
capillary

One-way
valves

Protein

Fluid

(a)

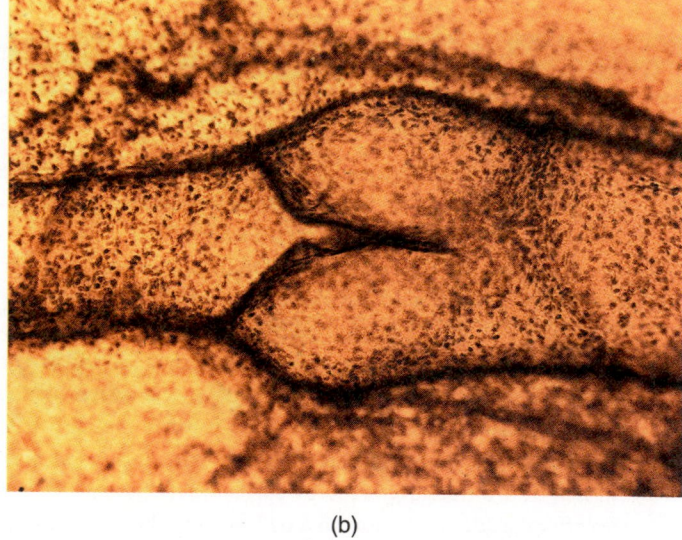

(b)

Figure 19–23

(a) Lymphatic capillary, showing entry of fluid and proteins and the pumping action of surrounding smooth muscle and lymphatic valves. *(b)* Photomicrograph of lymphatic valve. (© *John D. Cunningham/Visuals Unlimited.*) See Figure 19–13 for review of the relationship of the lymphatic capillary to that of surrounding blood capillaries and interstitial cells.

REGULATION OF SYSTEMIC ARTERIAL PRESSURE

  How is arterial blood pressure regulated?

Recall from our previous discussions that the body maintains normal tissue flow by monitoring and maintaining mean arterial blood pressure (MAP). As pointed out in the water system analogy, this approach simplifies the regulatory process and at the same time allows for a great deal of local regulation of blood flow by individual tissues. Regulating arterial pressure provides a simple method for adjusting cardiac output to meet the blood flow requirements of the tissues. You will also recall that under conditions in which the demand for flow exceeded the capacity of the pump, it was necessary to selectively decrease the flow to certain users in order to maintain flow to others. Based on our accumulated knowledge of how the heart and vasculature function, we are now in a position to investigate the mechanisms by which the body achieves the necessary coordination of the heart and vasculature to maintain arterial pressure.

As previously noted, the relationship between cardiac output (CO), mean arterial pressure (MAP), and total peripheral resistance (TPR) is described by the equation:

$$CO = MAP/TPR$$

Algebraic rearrangement gives us the equation:

$$MAP = CO \times TPR$$

We can see from this equation that both CO and TPR determine the magnitude of MAP. CO is determined by the heart, and TPR reflects the operational status of all the resistance arterioles in the body. Figure 19–24 summarizes all the factors that we have discussed that can change either CO or TPR. This figure indicates only where relationships exist, without delineating how a change in one component alters related components. The ways in which these various parameters are integrated to achieve normal cardiovascular function are the subject of the next several sections.

Before proceeding, let's look briefly at the meaning of *total peripheral resistance*. TPR is the total resistance to the flow of blood from the aorta back to the right atrium. Keep in mind, however, that the major portion of the resistance to blood flow is posed by the arterioles and precapillary sphincters. More importantly, changes in peripheral resistance are due entirely to changes in the diameters of these vessels. Therefore, when we talk about changes in TPR, we will be referring to changes in the overall resistance posed by all of the resistance arterioles in the body. Although other factors, such as hematocrit, can alter the viscosity of the blood — and, consequently, the resistance to flow — these changes do not constitute a mechanism for the homeostatic maintenance

Figure 19–24

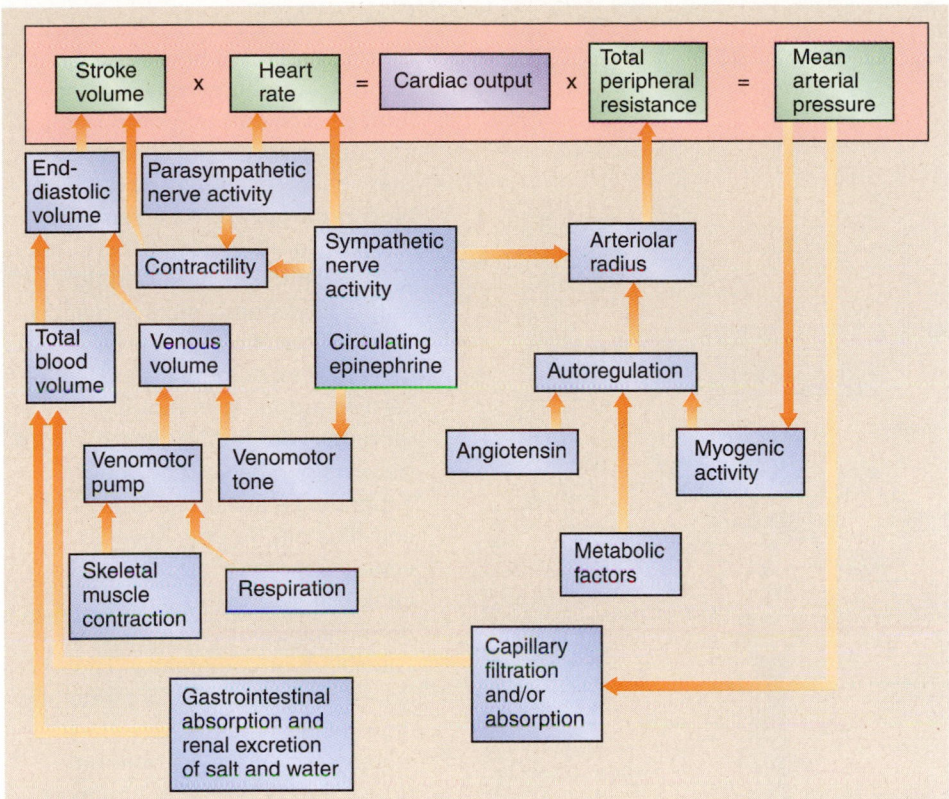

Summary of the factors that can alter cardiac output and arterial pressure.

of arterial pressure. We will therefore consider the viscosity of the blood to be constant under most conditions. It is also important to understand that while small veins and venules do vasoconstrict, this causes very little increase in TPR. This is because the overall cross-sectional area of all the venules and veins is much greater than that of the arterioles (see Fig. 19–17).

Cardiac Output Is Determined by the End-Diastolic Volume of the Heart, Which in Turn Is Determined by the Venous Return

These terms must be defined clearly if we are to understand how the factors interact. The terms *cardiac output* and *venous return* obviously refer to blood flow from the heart to the capillaries and from the capillaries to the heart, respectively. What is not obvious is whether they refer to the average flow over a period of time or to the instantaneous flow at any given point in time.

If we use these terms to describe average flow over a period of many minutes, then cardiac output and venous return must be equal because this is a closed system. Frequently, however, we will consider transient changes in flow that occur in selected portions of the system. Consider for a moment the changes following an increase in venomotor tone. Prior to this event, cardiac output and venous return were equal. As the veins contract and shift blood volume from the peripheral veins to the heart, venous return is, for a short

time, just slightly higher than cardiac output because of the extra volume of blood flowing back to the heart from the peripheral veins. As long as venous return to the heart exceeds cardiac output, the volume of blood in the heart increases. The increase in end-diastolic volume then leads to an increase in stroke volume and cardiac output (according to the Frank-Starling law). After the venoconstriction, cardiac output and venous return are again equal (but greater than before the venoconstriction occurred because of the increased cardiac end-diastolic volume). We might then state that an increase in venous return (temporarily) produced an increase in cardiac output, leading to an increase in venous return (average).

This sounds like a circular argument if it is not made clear that, in the first case, the term *venous return* refers to instantaneous flow, whereas in the second case it refers to average flow. A clearer description of the same sequence of events is that the redistribution of blood volume from the peripheral veins to the heart produces an increase in end-diastolic volume, leading to an increase in cardiac output and venous return (average). The important point is that cardiac output is very sensitive to changes in end-diastolic volume, which in turn depends on the total blood volume and the distribution of blood within the vascular system.

This point is illustrated by the system diagrammed in Figure 19–25. When the pump is set to 5, water is transferred from the large container on the left to the narrow container on the right. As the water rises in the narrow

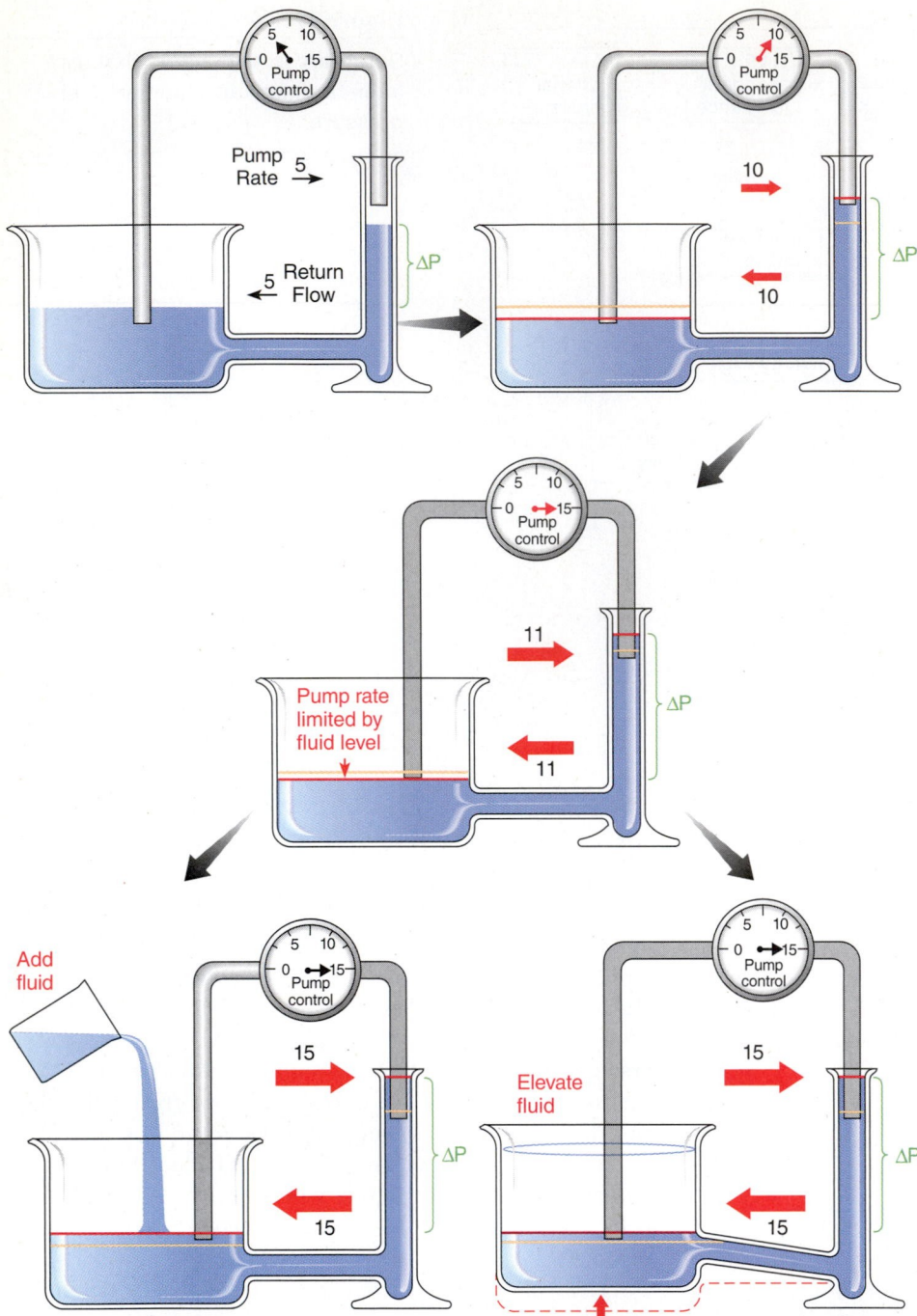

Figure 19–25

Demonstration of how venous return (return flow) can limit cardiac output (pump rate). The upper two panels demonstrate that an increase in pump rate shifts blood volume from the venous (large reservoir on left) to arterial vessels (small reservoir on right) until a new steady-state is reached where return flow = pump rate. The center panel demonstrates that cardiac output (pump rate) is ultimately limited by venous return (return flow) because of the dependence of cardiac output on end-diastolic volume (level in venous reservoir is below pump intake line) even though the potential cardiac output (pump control) might be greater. The lower two panels demonstrate how this can be remedied by either increasing total blood volume *(lower left)* or redistributing existing blood volume by venoconstriction *(lower right)*.

container, water begins to flow back through the connecting tube at the bottom. If the rate of the pump is increased to 10, the water level in the narrow chamber again increases while the level in the larger container goes down. This continues until the hydrostatic pressure gradient (ΔP) is sufficient to drive water back through the connecting tube at exactly the same rate (10) as the pump rate. Notice, however, that turning up the pump rate to 15 has little additional effect because the pumping rate becomes limited by the water level in the large container on the left. The pumping rate (11) is limited

by the filling rate (11). Two methods by which the actual pumping rate could be increased are shown in the bottom two panels. Adding more water to the left container would raise the water level and increase the filling rate of the pump *(bottom left)*; this could also be accomplished by raising the container as illustrated on the right. By analogy to the cardiovascular system, adding more water would be analogous to increasing the circulating blood volume (e.g., by ingesting liquid), and raising the container would be analogous to redistribution of available blood volume (e.g., by venous con-

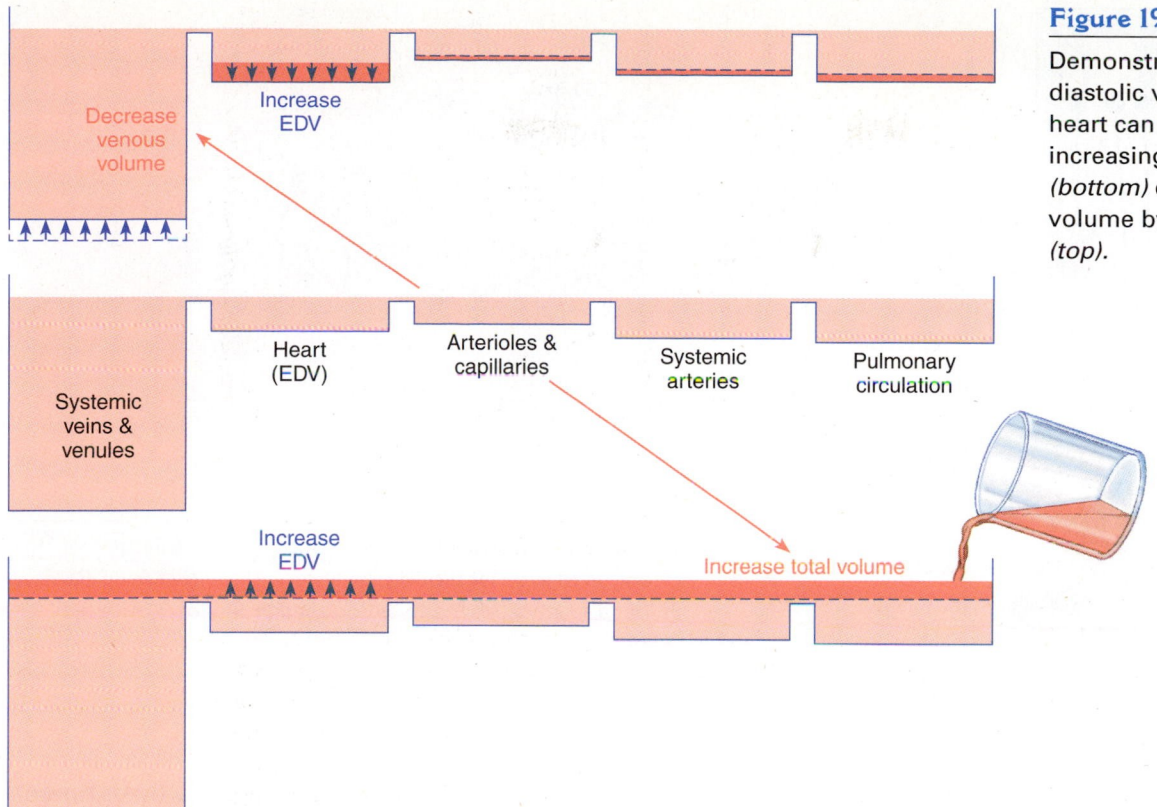

Figure 19–26

Demonstration of how end-diastolic volume (EDV) of the heart can be increased either by increasing total blood volume *(bottom)* or by shifting blood volume by venoconstriction *(top).*

striction). The bottom line is that it does not matter whether you increase end-diastolic volume by increasing total blood volume by shifting blood volume to the heart by venoconstriction (Fig. 19–26); the effect on cardiac output is the same. So you can see that the pumping capacity of the heart can be influenced equally by changes in total blood volume and by changes in the distribution of blood volume in the circulatory system.

Arterial Baroreceptors Are the Sensing Element of a Negative Feedback System That Regulates Mean Arterial Pressure

You know from previous discussion about reflexive control mechanisms that an essential part of any such system is the presence of a receptor that can monitor the variable to be held constant — in this case, mean arterial pressure. The receptors that perform this operation in the cardiovascular system are called **baroreceptors.**

Location of Baroreceptors

Arterial baroreceptors are located in the wall of the arch of the aorta and in the carotid sinus (Fig. 19–27). The **carotid sinus** is a thin-walled, highly innervated region of the carotid artery located high in the neck at the point where the carotid artery branches into two smaller arteries.

Activation of Baroreceptors

The baroreceptor nerve endings function as stretch receptors and are distributed within the walls of the elastic arteries. You have already encountered stretch receptors in other systems in the body, and the characteristics of all stretch receptors are similar. These receptors respond to stretch with an increase in the rate of action potential firing. Because of the elastic nature of the arterial wall, an increase in arterial pressure causes the arterial wall and the associated receptors to stretch. Stretching the nerve endings initiates the firing of action potentials, which are transmitted by the associated afferent (sensory) neurons to the cardiovascular control center in the brain.

Patterns of Baroreceptor Activation by Changes in Blood Pressure

Figure 19–28 shows the pattern of action potential firing, with changes in both mean arterial pressure and pulsatile arterial pressure. Notice that action potential firing increases in response to the pulsatile pressure changes during a single heartbeat. This tells us that these receptors respond very quickly to changes in pressure. Therefore, the baroreceptors are effective in monitoring acute changes in blood pressure. As mean arterial pressure increases, the overall firing rate of the receptor increases.

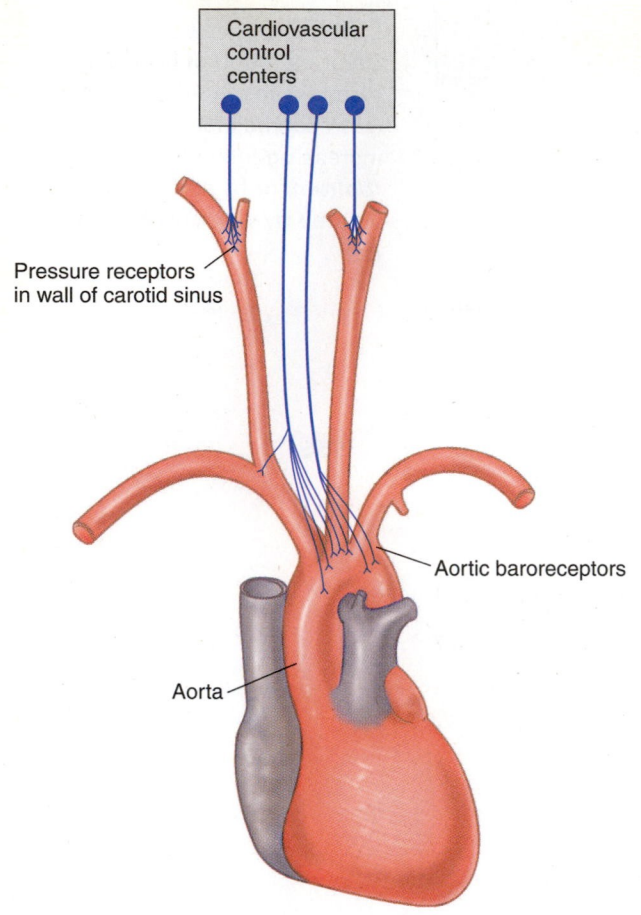

Figure 19–27

Location of the arterial baroreceptor nerve endings in the systemic arterial system. Afferent nerves from the baroreceptors travel to the medullary cardiovascular control centers.

The Cardiovascular Control Center in the Brain Determines the Level of Autonomic Nerve Activity to Both the Heart and the Blood Vessels

The afferent nerves from the baroreceptors travel to the brain, where they synapse with neurons that make up the **cardiovascular control centers.** The primary control center for the baroreceptor reflex is located in the medulla. The medullary cardiovascular control center is a complex and diffuse network of neurons that integrates neural inputs from the baroreceptors with inputs from other control centers in the brain. For now, we will concentrate on how changes in the level of input from the baroreceptors lead to changes in autonomic nerve activity to the heart and vasculature.

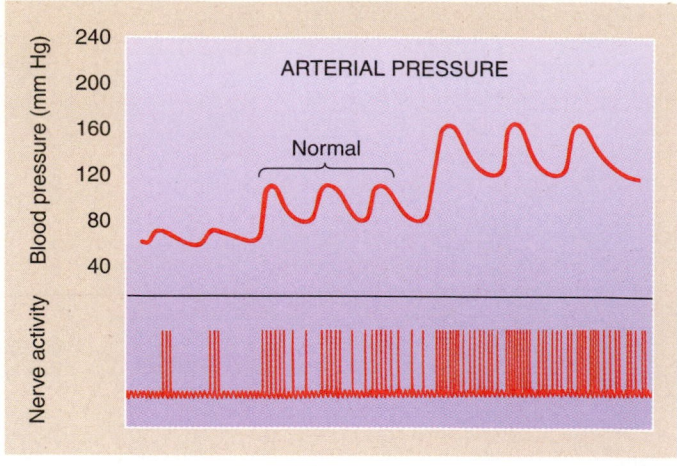

Figure 19–28

Changes in carotid sinus nerve activity as a function of changes in arterial blood pressure. *(Modified from Berne, R. M., and Levy, M. N. Physiology, ed 2. St. Louis, Mosby, 1988, Fig. 33–10, p 519.)*

Cardiac Output and Total Peripheral Resistance Are Controlled by the Cardiovascular Control Center so That Mean Arterial Pressure Remains Relatively Constant

You will recall from our previous discussions that the heart is innervated by both parasympathetic and sympathetic nerves and that the blood vessels are innervated primarily by sympathetic nerves. Preganglionic efferent nerve fibers from the cardiovascular centers synapse with the cell bodies of both parasympathetic and sympathetic neurons that innervate the heart and blood vessels (Fig. 19–29). Changes in baroreceptor activity produce reciprocal changes in parasympathetic and sympathetic activity to the heart and blood vessels. An increase in the rate of nerve activity from the baroreceptors leads to a decrease in the rate of stimulation of sympathetic nerves and an increase in the rate of stimulation of parasympathetic nerves. Conversely, a decrease in baroreceptor nerve activity leads to increased sympathetic and decreased parasympathetic activity.

Cardiac Output and Total Peripheral Resistance

Let's now examine the overall response of the system to a change in blood pressure (Fig. 19–30). If arterial pressure increases, the baroreceptor firing rate increases. This leads to a decrease in sympathetic nerve activity to the heart and blood vessels and an increase in parasympathetic nerve activity to the heart. This, in turn, causes the heart rate to decrease and contractility to decrease, leading to a reduction in cardiac output. The decreased sympathetic nerve activity to the blood vessels promotes dilation of the heavily innervated

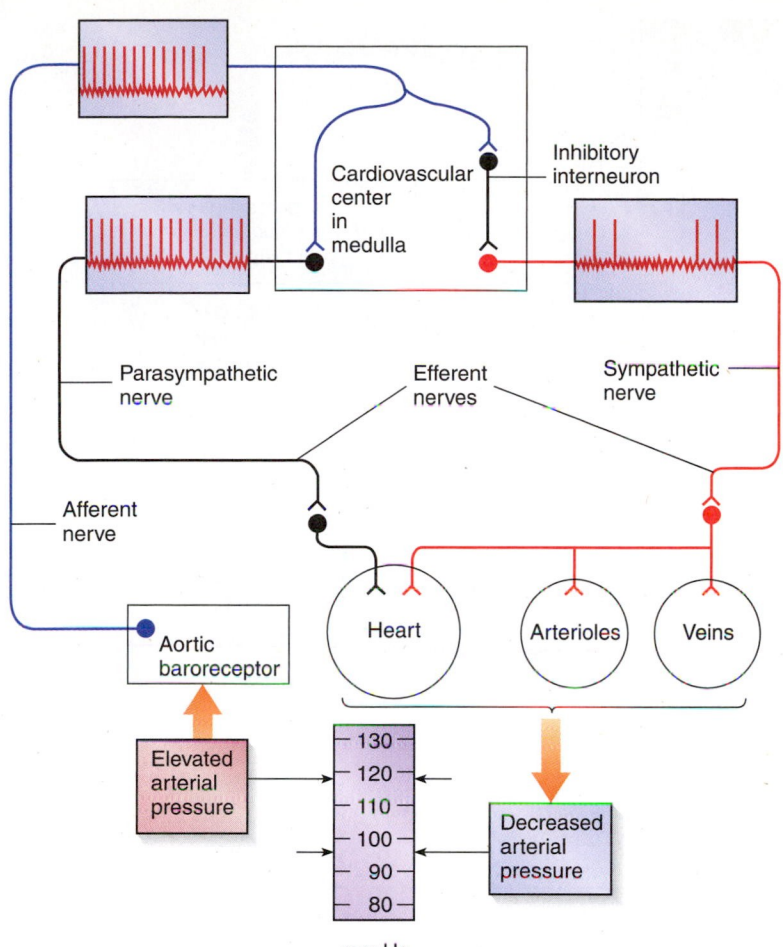

Figure 19–29

Increased aortic blood pressure leads to increased afferent nerve activity from the baroreceptors, which results in reciprocal changes in efferent sympathetic and parasympathetic nerve activity to the heart and blood vessels with a resultant decrease in aortic blood pressure back to normal.

vascular beds, such as the skin, muscle, and splanchnic circulations, leading to a reduction of total peripheral resistance. The reduction in both cardiac output and total peripheral resistance causes a reduction of arterial pressure (MAP = CO × TPR). Clearly, the response of the system is to minimize changes in arterial pressure by changing both cardiac output and total peripheral resistance in the appropriate directions. If arterial pressure were to decrease, decreased baroreceptor firing would lead to increased cardiac output and increased total peripheral resistance. As before, this compensatory response would tend to restore arterial pressure back to normal.

Venomotor Tone

In our previous discussion of the venous system, we pointed out that contraction of venous smooth muscle (venomotor tone) occurs in response to sympathetic stimulation. Changes in venomotor tone also play a role in the baroreceptor reflex to maintain arterial blood pressure. In response to increased arterial pressure and increased baroreceptor firing, there is a decrease in sympathetic activity to the veins, resulting in venodilation. Venodilation results in a shift of blood volume from

the heart and arteries to the peripheral veins. This leads to a decrease in end-diastolic volume, which decreases cardiac output, according to the Frank-Starling law. The decrease in cardiac output then leads to a decrease in mean blood pressure back toward normal.

Long-Term Regulation of Blood Pressure

Many physiologists contend that regulation of blood volume is one of the most important aspects of blood pressure regulation, particularly on a long-term basis. Whereas reflexive mechanisms, such as those just discussed, play a dominant role in short-term pressure regulation, the maintenance of adequate circulating blood volume is primarily under the control of the kidneys (which mediate the excretion of salt and water from the body) and the gastrointestinal system (which mediates absorption of salt and water). You will see more clearly how important the maintenance of blood volume is when we discuss the ways in which hemorrhage affects cardiovascular performance. Also, we will discuss the ways in which venomotor tone and alteration of blood volume by changes in filtration and absorption in the capillaries can temporarily compensate for inadequate blood volume. If

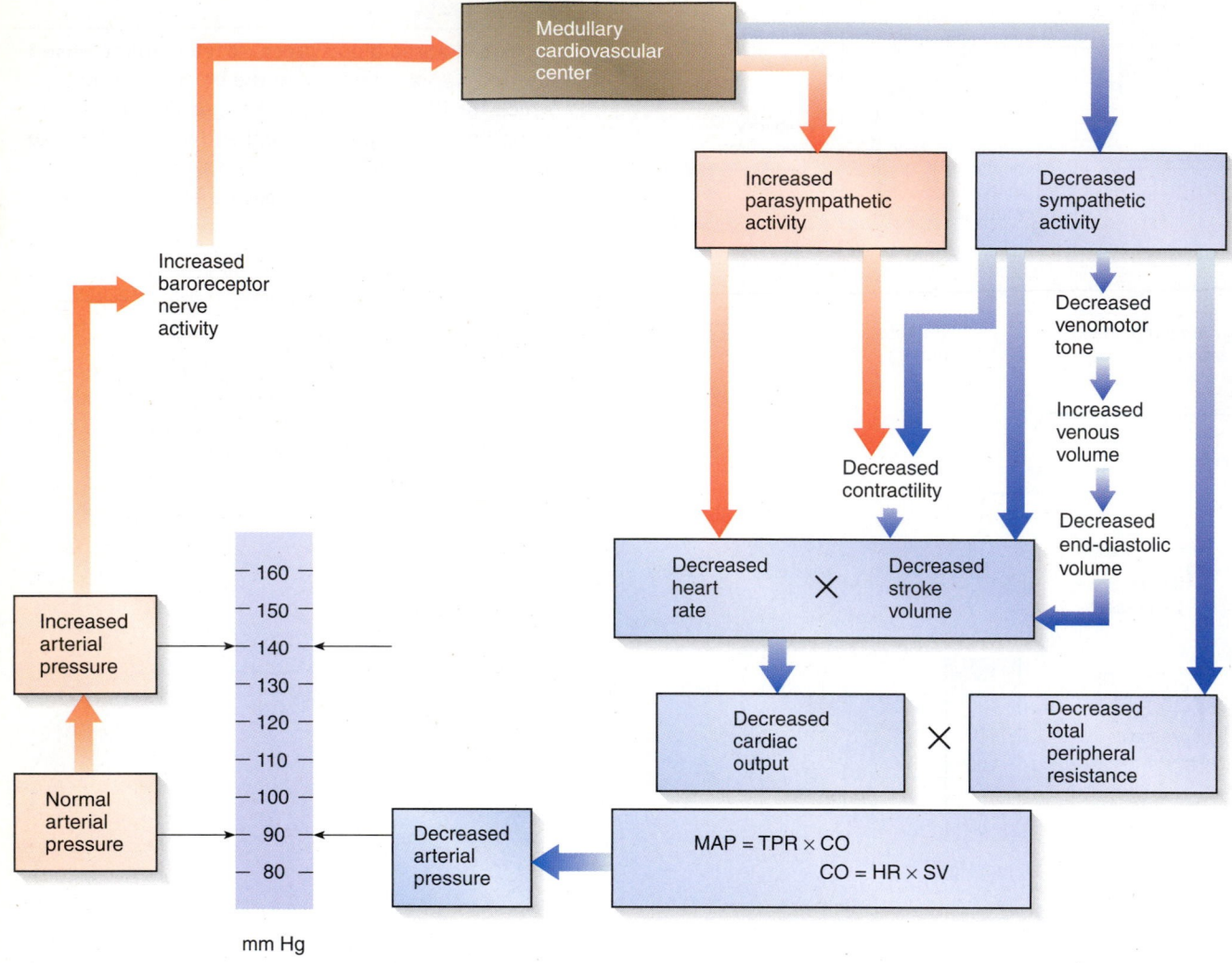

Figure 19–30

Summary of cardiovascular changes in response to increased arterial blood pressure and increased baroreceptor nerve activity. This is an example of negative feedback. An increase in blood pressure leads to compensatory changes that tend to restore blood pressure back toward normal.

cardiac output is compromised by inadequate end-diastolic volume, this can be corrected by (1) increasing the total circulating blood volume, (2) shifting of blood volume from peripheral vascular beds to the heart by increasing venomotor tone, or (3) increasing the reabsorption of interstitial fluid by the capillaries (see Fig. 19–20). These mechanisms are equivalent in terms of their effect on the heart. Remember, however, that they are not equivalent in terms of overall physiological homeostasis. For example, you will see that, during hemorrhage, the body compensates for loss of blood by severely decreasing flow to organs such as the kidneys. While this may temporarily serve to maintain blood pressure and blood flow to the heart and brain, the kidney can be damaged severely as a result of prolonged compensation of this nature.

Hemorrhage, Hypotension, and Exercise All Cause Primary Changes in Either Cardiac Output or Total Peripheral Resistance That Must Be Matched by Compensatory Changes Elsewhere

We have now examined the structural and functional characteristics of the individual components of the cardiovascular system. We have discussed how the baroreceptor reflex functions to maintain arterial blood pressure by initiating changes in cardiac output and total peripheral resistance. As we have stated on several occasions, the purpose of the cardiovascular system is to maintain adequate blood flow to the tissues to support cellular metabolism. What we have begun to see is that not all tissues are equal in terms of their need for blood flow and that, under some conditions, blood flow is directed from some tissue beds to others to sustain temporarily elevated levels of metabolism (e.g., exercising muscle). Moreover, blood flow to certain tissues, such as the brain and heart, must be maintained, even at the expense of other tissues, because of the body's need for their uninterrupted function. Perhaps the best way to see how all of these factors operate together is to look at the response of the cardiovascular system to changes from its normal resting state. To illustrate these points, we will examine three cardiovascular disturbances: (1) hemorrhage, (2) postural (orthostatic) hypotension, and (3) exercise.

Hemorrhage: Loss of Blood

The immediate effect of blood loss (hemorrhage) is a decrease in arterial blood pressure. This occurs as a result of an associated decrease in cardiac output. If blood volume is removed from the vascular system, the end-diastolic volume of the heart decreases and cardiac output is reduced, as described by the Frank-Starling law. We know from the equation $MAP = CO \times TPR$ that restoration of arterial pressure requires an increase in either cardiac output or total peripheral resistance. The decrease in arterial pressure following hemorrhage immediately causes a reduction in the rate of baroreceptor nerve activity. This leads to increased sympathetic nerve activity to the heart and resistance arterioles. This increases both heart rate and cardiac output, which in turn, increases arterial pressure back toward normal. Also contributing to the increase in arterial pressure is the vasoconstriction of the resistance arterioles in response to the increase in sympathetic nerve activity.

As we know from previous discussions, not all vascular beds receive the same degree of sympathetic innervation. The vascular beds in the skin, skeletal muscle, gastrointestinal tract, and kidneys are all innervated by sympathetic nerves. Arteriolar vasoconstriction of these vascular beds minimizes the fall in blood pressure by increasing total peripheral resistance. The vascular resistance to the heart and brain does not increase because of the sparse sympathetic innervation of these vascular beds and because of the high degree of local control.

We can now begin to appreciate the consequences of the selective innervation of different vasculature beds in the body. Blood flow to those tissues whose continuous function is not essential for life can be temporarily reduced so that a limited blood flow can be distributed to those organs, such as the brain and heart, that must function continuously. Loss of blood flow to the heart would weaken the heart and prevent it from participating in the restoration of blood pressure. Loss of blood flow to the brain would disrupt the function of the cardiovascular centers responsible for the mediation of the compensatory response. The kidneys can sustain a short-term reduction in blood flow, but long-term reductions lead to renal failure and death. As you will see in more detail later, the temporary reduction of blood flow to the kidney has the added benefit of maintaining blood volume by reducing the rate of urine formation. Blood flow to the skin, muscles, and

the gastrointestinal system can be substantially reduced with little ill effect. The net result is to shift most of the available blood flow to the heart and brain.

Figure 19–31 provides an overview of the compensatory responses to hemorrhage. You can see that the compensatory response consists of a rapid component that includes increased heart rate, increased cardiac contractility, increased total peripheral resistance, and venoconstriction. This is clearly a temporary response that sacrifices the immediate needs of some tissues to maintain those of the heart and brain. A second, more prolonged response is a consequence of the decreased capillary pressure that results in fluid reabsorption from the interstitial compartment, which helps to restore total blood volume. While this tends to alleviate the hemodynamic crisis, complete recovery involves several additional processes. Complete restoration of total body water involves increased fluid ingestion, decreased urine production, stimulation of erythropoiesis to increase the hematocrit, and synthesis of plasma proteins.

One of the more pronounced changes that occurs with hemorrhage is a reduction in stroke volume; this results in a weak or shallow pulse. Although an increase in heart rate and contractility can partially compensate for this reduction, both of these compensatory mechanisms are limited by the diastolic filling rate of the heart. Recall from earlier discussions that, even under normal conditions, increasing cardiac output by increasing heart rate is limited by the progressively diminished diastolic filling time. As blood is transferred to the arterial system to maintain blood pressure following hemorrhage, less blood is available for filling of the heart. So you can see that the extent of compensation by increased heart rate, increased contractility, and increased peripheral resistance becomes self-limiting. The only way to overcome this limitation is to increase the diastolic filling rate and the end-diastolic volume of the heart. As was previously demonstrated in Figures 19–25 and 19–26, this can be accomplished by increasing the total blood volume or by redistributing blood volume.

Obviously, a blood transfusion is an effective way of returning the system to normal and is the treatment of choice for hemorrhage. There are, however, compensatory mechanisms that allow the body to recruit blood volume from other sources. One response to hemorrhage is generalized venoconstriction. Although this does not increase the total circulating blood volume, it does increase the volume of blood available to the heart. Consider what happens as the veins constrict. Because of the reduction in vessel diameter, blood shifts from the veins to the heart and arterial system. From the standpoint of the heart, this is as effective as an increase in total blood volume.

There is also a mechanism by which the body is able to create a real increase in total blood volume. Consider what is happening to capillary pressure, particularly in vascular beds, where arteriolar vasoconstriction has taken place (see Fig. 19–20). Mean arterial blood pressure and venous pressure have been reduced because of the hemorrhage, and

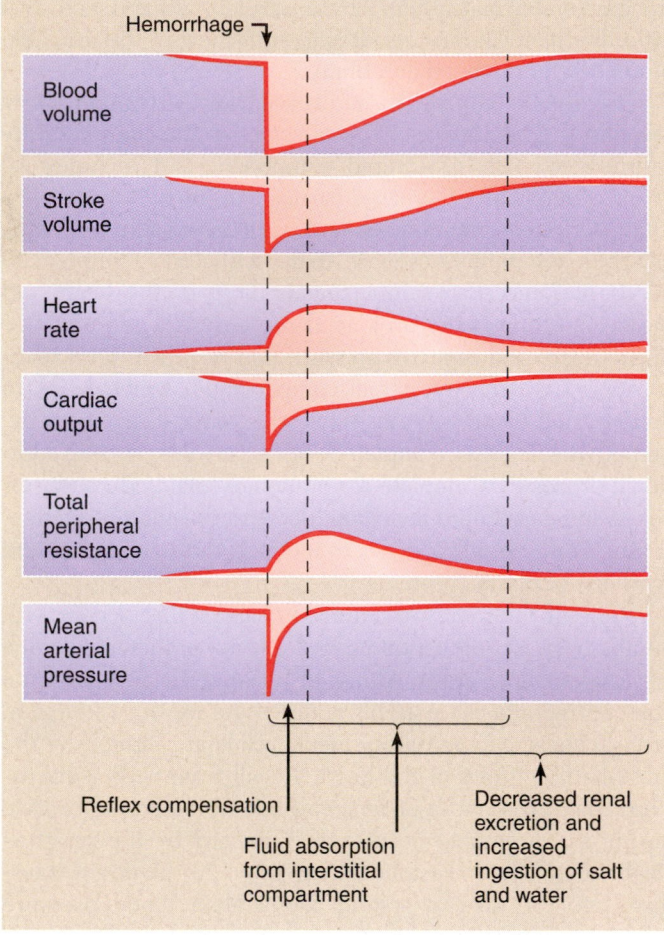

Figure 19–31

Summary of the hemodynamic responses of the cardiovascular system to mild hemorrhage. A rapid, reflexive compensation is followed by a slower response to return blood volume to normal by increasing fluid reabsorption and decreasing fluid excretion.

capillary pressure therefore has also been reduced. The plasma protein concentration, however, has not changed. According to Starling's hypothesis, reabsorption of fluid is favored, and interstitial fluid volume will be transferred into the capillaries. As much as 1 L/h of fluid can be transferred to the vascular compartment by this mechanism following severe hemorrhage.

This reabsorption of fluid does not occur immediately and may take as long as 12 to 24 hours to reach completion, depending on the extent of hemorrhage. For example, if you donate a pint of blood, within 12 hours, your circulating blood volume will be restored almost to normal. The fluid transferred from the interstitial compartment is free of red blood cells, and what little plasma protein this fluid contains is filtered during reabsorption. Remember that the capillary wall is relatively impermeable to protein. As you might expect, hematocrit and plasma protein concentration are re-

duced as a result of fluid reabsorption. The dilutional effect of fluid reabsorption limits the extent of this compensatory mechanism. As interstitial fluid is transferred into the blood, the protein concentration of the interstitial fluid increases while that of the blood decreases. This progressively reduces the osmotic gradient, which is the driving force for fluid reabsorption.

The mechanisms just described can adequately compensate for blood losses of as much as 1.0 to 1.5 liters (2–3 pints), with few ill effects. Greater losses of blood, however, can lead to severe tissue damage, irreversible circulatory collapse, and death. This progressive deterioration of cardiovascular function is referred to as **circulatory shock.** Depending on the extent of blood loss and the elapsed time, a point is reached after which no known therapy, including massive transfusions, can prevent death. Although the details of this process are complex, it is clear that contributing factors include irreversible damage to many organ systems, including the kidneys, heart, and brain, as a result of inadequate blood flow. A phenomenon that has gained general recognition recently is that of "reperfusion injury." This refers to the events that occur following restoration of blood flow to any tissue that has experienced a prolonged episode of ischemia. At least a part of this process involves the release of molecular free radicals from immune cells after reperfusion. These free radicals (e.g., oxygen free radicals) are extremely reactive molecules that cause cellular damage by oxidation of proteins, lipids, and nucleic acids. These effects can be partially protected against by the use of free radical scavengers and antioxidants. Natural antioxidants present in the body include vitamins C and E.

Orthostatic Hypotension

From our previous discussion, we know that hemorrhage produces a form of **hypotension** (low blood pressure) that results from a loss of blood volume. Recall that one of the compensatory processes for hemorrhage involves the translocation of blood volume from the veins to the heart and arterial system by venoconstriction. From the point of view of the heart, this is entirely equivalent to a transfusion of an equal volume of blood. As you can see, many of the regulatory and compensatory mechanisms that we have been discussing involve blood volume redistribution.

An everyday example of the effects of hypotension are fainting spells that might occur when someone stands abruptly after waking or stands for a long time in the hot sun while wearing warm, heavy clothing. The problems here are gravity and inadequate peripheral vascular tone, which result in inadequate filling of the heart. From our earlier analogy in Figure 19–25, this would be the equivalent of lowering the large reservoir on the left until the water was below the level of the input pipe going to the pump.

In the first example given, the sleeper probably has a heart rate of less than 60 beats per minute (bpm), and the veins in the legs are dilated. When this person stands up, blood pools in the veins of the legs and feet, and suddenly no blood is available to fill the heart. Cardiac output decreases precipitously, leading to inadequate blood flow to the brain, causing unconsciousness. In the second instance, the person is standing quietly, and the lack of venomotor pumping activity results in the transfer of blood volume away from the heart. This, in turn, leads to reduced end-diastolic volume, reduced cardiac output, and a further reduction of arterial blood pressure. Because the person is also overheated, the subcutaneous blood vessels are dilated to bring blood close to the surface of the body for heat dissipation. This reduction in peripheral resistance also contributes to the reduction of arterial pressure. The fall in arterial blood pressure ultimately results in insufficient blood flow to the brain. The body involuntarily compensates by fainting, which adjusts the posture to promote greater blood flow to the heart and brain. Such problems are normally avoided by a combination of compensatory mechanisms. Normal levels of physical activity result in periodic contraction of the skeletal muscles in the legs, which prevents pooling of blood in the veins of the lower extremities. In addition, normal levels of both mental and physical activity, are associated with slight increases in sympathetic nerve activity and therefore with increases in heart rate and venomotor tone.

Exercise and Cardiovascular Homeostasis

> ### How does the cardiovascular system respond to exercise?

The cardiovascular changes occurring during exercise are in many ways the opposite of those occurring during hypotension and hemorrhage. With hypotension, the major disturbance is reduced arterial blood pressure resulting from reduced cardiac output. The compensatory response (increased cardiac output and increased vascular contraction) is mediated primarily by decreased activation of the arterial baroreceptors in response to reduced arterial blood pressure. During both mild and intense exercise, there is an increase in arterial pressure. Under resting conditions, an increase in arterial pressure tends to decrease cardiac output and decrease constriction of systemic vascular beds via the baroreceptor reflex. Yet we know that cardiac output always increases during exercise; well-trained athletes may experience an increase in cardiac output from 5 L/min to as much as 35 L/min during maximal exercise.

It is now well accepted that the cardiovascular changes observed during exercise are not in fact mediated by the baroreceptor reflex. A number of local and humoral factors contribute to vasodilation of the skeletal muscle vasculature. In addition, reflexes originating from within the contracting muscles lead to activation of sympathetic nerves to the heart and peripheral vasculature to further adapt the cardiovascular system to the demands of exercise.

Let's first consider what demands exercise places on the cardiovascular system. Exercise, by definition, is an increase

in the rate and extent of contraction of skeletal muscle. We can correctly assume that this is associated with a substantial increase in the metabolic rate of muscle. A good measure of the level of muscle metabolism is oxygen consumption. During maximal exertion, muscle oxygen consumption for the well-trained athlete may increase by as much as 50 to 60 times resting levels. This means that a significant increase in blood flow must occur (up to 20 times resting flow) to deliver oxygen to the muscle and to remove metabolic by-products. If we consider that resting muscle blood flow is approximately 1.0 to 1.5 L/min, then the blood flow to skeletal muscle alone during exercise can be as high as 15 to 20 L/min. It is clear that cardiac output has to increase substantially to support this increased demand by exercising muscle (Fig. 19–32).

The necessary cardiovascular changes required for exercise actually begin even before the onset of muscular activity. The mental anticipation of exercise leads directly to

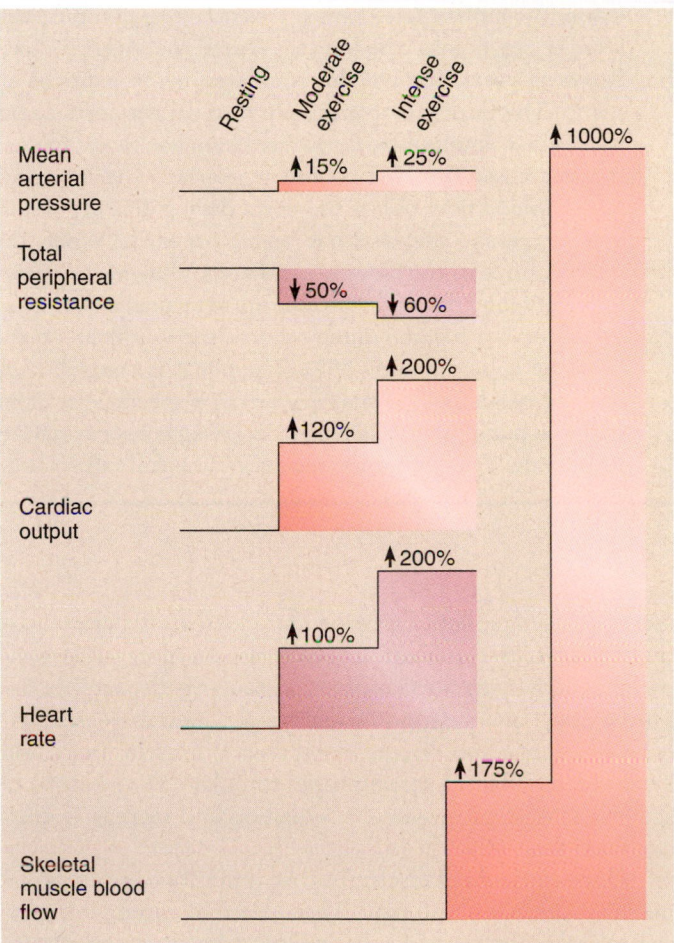

Figure 19–32

Summary of the hemodynamic changes in the cardiovascular system in response to mild and intense exercise expressed as the percentage of change from resting.

APPLICATIONS OF PHYSIOLOGY

Aortic Balloon Pump

One of the newest methods, currently in clinical trials, for treating left ventricular failure is the permanent implantation of a balloon pump in the aorta. This device, called the Kantrowitz CardioVad System (LVAD Technology, Detroit, Michigan) was developed by Adrian Kantrowitz, MD, who was the first US surgeon to perform a heart transplant. He has spent over 30 years finding the right combination of materials and technology required for a permanently implantable balloon pump. One of the early spinoffs of this work was a temporary assist device that is now used 100,000 times each year worldwide to support patients with acute heart failure. With the permanently implantable pumps, a 6-inch-long pumping balloon is sewn into the aorta. When the heart relaxes between beats, the bladder is inflated by an external air pump. This pushes blood forward through the aorta to the tissues and simultaneously pushes blood backward toward the heart and into the coronary arteries. The increased rate of blood flow out of the aorta during diastole results in a decreased end-diastolic aortic pressure. When the heart contracts, the bladder is deflated, allowing the heart to inject blood normally into the aorta against a decreased pressure. The balloon pump reduces the workload on the heart by approximately 50%. The pump is controlled by an external computer that is worn in a special vest along with the pump, which, all together, weighs approximately 10 pounds. The computer receives electrical information

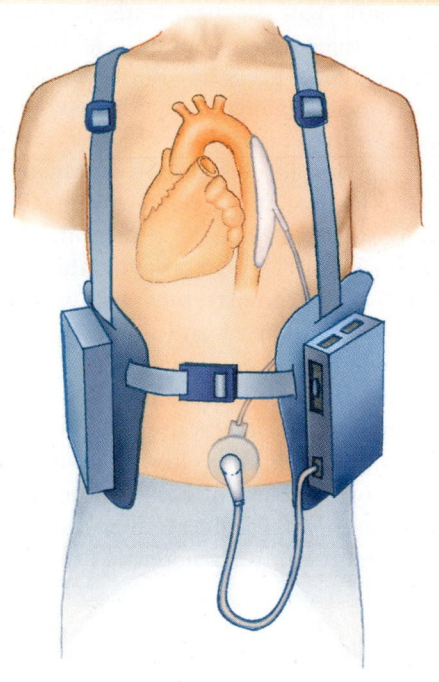

The Kantrowitz Cardio Vad™ System

about the timing of the heartbeat via a permanently implanted electrode. Because this device only assists the heart, the patient can unplug themselves for short periods of time. Researchers are hoping that some patients will only have to use the device while they sleep, allowing them to be free from the device during the day.

increased sympathetic activity and decreased parasympathetic activity to the heart. There is also an increase in sympathetic activity to the peripheral vascular beds, such as the skin, kidney, and gastrointestinal tract, which increases vascular resistance and directs blood flow to muscle. Increased levels of circulating epinephrine stimulate β-receptors in skeletal muscle to produce vasodilation and increased muscle blood flow.

Once muscular exercise begins, a number of additional mechanisms come into play. The release of vasodilator substances from the muscle cells, such as potassium and adenosine, leads to vasodilation of the precapillary resistance vessels. As exercise progresses and body temperature begins to increase, there is a reflexive vasodilation of the subcutaneous vessels to allow for dissipation of excess body heat. Chemoreceptors within the muscle respond to the re-

lease of metabolites and initiate a reflexive increase in sympathetic nerve activity to the heart and resistance vessels of the gastrointestinal tract and kidneys to increase cardiac output and to shift the available blood flow to exercising muscle. The level of afferent nerve activity is roughly proportional to the level of muscle work being performed and therefore provides a means for matching cardiac performance and blood flow with the requirements of the active muscle.

Consider for a moment how the situation during exercise differs from that during compensation for hemorrhage. In both cases, there is increased stimulation of heart rate and contractility. In both cases, blood is being shunted from inactive to active tissues. With hemorrhage, cardiac output is limited by inadequate cardiac filling; this is clearly not the case with exercise, because cardiac output may reach 35

L/min. The difference with exercise is that there has been no reduction of blood volume. Compensation for blood loss due to hemorrhage requires a net transfer of blood from the venous system to the heart and arterial side of the circulatory system in order to maintain adequate perfusion pressures for the heart and brain. This process is self-limiting because depletion of venous blood volume decreases the rate of cardiac filling and end-diastolic volume. We can deduce that, during exercise, there is little increase in arterial blood volume, even as cardiac output increases, because mean blood pressure increases only slightly. No transfer of blood volume from the venous to arterial systems occurs because total peripheral resistance has decreased in proportion to the increase in cardiac output; blood flow from the arterial side to the venous side increases in direct proportion to the increase in cardiac output. If we think back to the analogy in Figure 19–25, the decreased total peripheral resistance with exercise would be analogous to increasing the diameter of the connecting return tube. As long as water returns to the bucket as fast as it is removed, the only limitation to the pumping rate is the maximum rate at which the pump can operate. For the heart, this is determined by the minimum time required for contraction and relaxation of the ventricle muscle. The effective upper limit for heart rate is approximately 180 bpm.

CARDIOVASCULAR DISEASE

Heart Failure Refers to a Variety of Primary Cardiac Pathologies in Which the Heart Is Unable To Maintain Normal Cardiac Output

As the name implies, heart failure is the result of the inability of the heart to maintain adequate cardiac output. Although many factors may cause failure of the heart, the resulting changes in cardiovascular performance are similar.

Cardiovascular Effects of Heart Failure

As before, let's first specify what cardiovascular parameter undergoes a change. With heart failure, the problem is a reduction in cardiac contractility. Recall that this means that the ability of the heart to pump blood at any end-diastolic volume is decreased (Fig. 19–33, *bottom curve*). Let's start with a normal cardiovascular system and then impose a reduction in contractility on the heart. We know that the decrease in contractility decreases stroke volume and cardiac output. Furthermore, we can reason that a decrease in cardiac output will lead to a decrease in mean arterial pressure.

Compensation for Heart Failure

As the end-diastolic volume of the heart increases, the Frank-Sterling mechanism tends to increase cardiac output. Recall that in hemorrhage, the kidneys respond to the reduc-

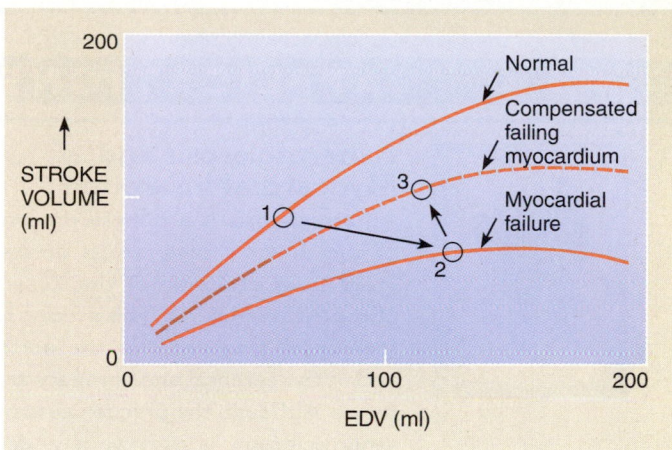

Figure 19–33

Change in the relationship between end-diastolic volume and stroke volume as a normal heart (1) progresses into heart failure (2). The decrease in stroke volume with myocardial failure is lessened by the increase in EDV that occurs as a result of the reduced stroke volume. Increased sympathetic nerve activity to the heart can compensate for mild failure by increasing the contractility *(dotted curve)*, which in turn decreases the end-diastolic volume back toward normal (3). At this point (3), stroke volume has returned to normal because of the combined effects of increased EDV and increased contractility.

tion in blood pressure by increasing the retention of water and sodium. Exactly the same process of fluid retention occurs with heart failure. The increase in circulating blood volume further increases the end-diastolic volume of the heart to increase cardiac output (Fig. 19–33). Unfortunately, this compensatory mechanism is limited. Recall that increases in end-diastolic volume lead to increased vigor of contraction because, in a normal resting heart, the cardiac muscle is shorter than its optimal length for contraction (see Fig. 18–12). With progressive heart failure, additional increases in volume stretch the muscle beyond its optimal length. Once this point is reached, further increases in heart volume actually reduce the ability of the heart to pump blood. This in turn leads to further increases in venous and cardiac volume, and a vicious cycle ensues. This progressive congestion of the veins and heart with blood is described as **congestive heart failure.**

Clearly, the reduced blood pressure will stimulate the baroreceptor reflex, which results in increased sympathetic and decreased parasympathetic stimulation of the heart as well as sympathetic stimulation of the vasculature. The result is increased contractility, increased heart rate, increased arteriolar constriction, and increased venomotor tone. Notice that these changes are very similar to those seen with hemorrhage. The major difference concerns the

APPLICATIONS OF PHYSIOLOGY

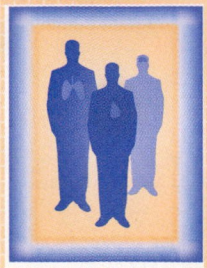

Atherosclerosis and Myocardial Infarction

Atherosclerosis is a disease process of the arteries that results in the progressive occlusion of the lumen of the vessel (Fig. A). This process is seen most frequently in the aorta and in the cerebral and coronary arteries. Although the precise cause of atherosclerosis is still widely debated, the progression of events is well documented. The initial event probably involves damage to the endothelial lining of the vessel wall. This is followed by proliferation of the underlying smooth muscle cells that make up the muscular wall of the vessel. A progressive reduction in the size of the vessel lumen results from the thickening of the vessel wall. As the disease progresses, lipid from the blood accumulates along with fibrous tissue (atherosclerotic plaque), resulting in the progressive occlusion of the vessel lumen.

Occlusion of the coronary arteries in this manner (coronary artery disease) is the most frequent cause of cardiac dysfunction. The occlusion of the vessel lumen gradually reduces blood flow to the heart. As the dis-

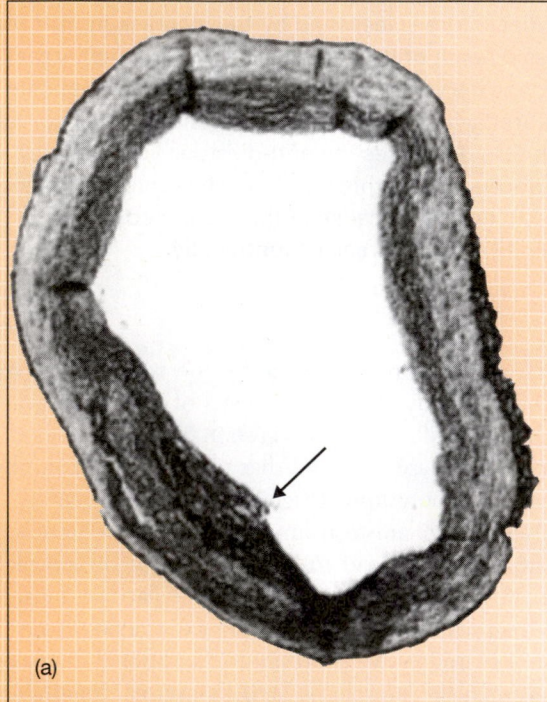

(a)

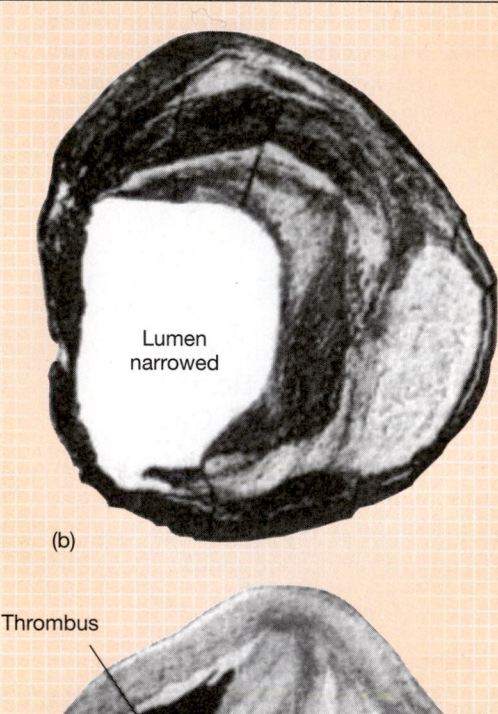

Lumen narrowed

(b)

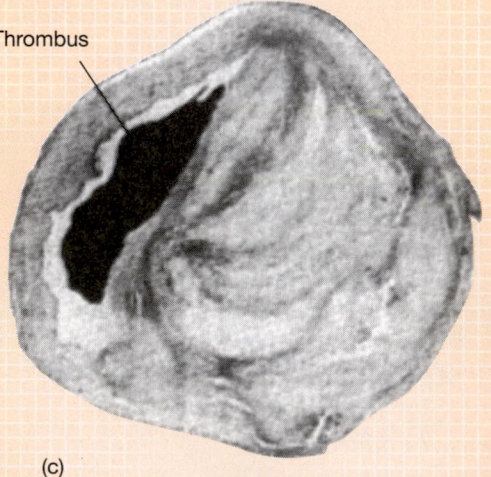

Thrombus

(c)

Figure A   Tissue sections from an atherosclerotic vessel. (*a*) Early stage of atherosclerosis with slight thickening of the arterial wall indicated by the arrow. (*b*) Progressive accumulation of smooth muscle cells, lipid, and calcium results in plaque formation. (*c*) The progressive accumulation of plaque results in almost complete occlusion of the vessel lumen and a substantial increase in the resistance to blood flow.

ease progresses, individuals may begin to experience periodic episodes of inadequate coronary blood flow, often associated with chest pain referred to as *angina pectoris*. Episodes of angina are frequently associated with increased physical exertion or emotional stress. A heart attack is the result of an acute coronary occlusion, which may occur because of either clot formation at the site of the plaque (thrombus) or intense spasm of the coronary smooth muscle near the plaque. Since the heart receives its blood supply from the coronary vessels, occlusion can lead to injury and death of that portion of the myocardium supplied by the occluded vessel. The area of damaged myocardium is referred to as an *infarct*.

An infarct can reduce the ability of the heart to pump blood in a number of ways. Depending on the extent and location of the infarct, the pumping capability of the heart may be compromised because of the loss of functional muscle. The infarct may also initiate cardiac arrhythmias. Either or both of these situations can lead to decreased cardiac output. This, in turn, may lead to reduced arterial blood pressure, depending on the ability of the body to compensate by increasing peripheral resistance. Autoregulation of coronary blood flow compensates for small drops in arterial pressure; larger decreases in arterial pressure result in reduced coronary blood flow. This, in turn, further diminishes the pumping ability of the heart, leading to further reductions in cardiac output and arterial pressure. Unless halted, this downward spiral leads to collapse of the cardiovascular system and death.

The compensatory responses of the body are limited, and therefore the consequences of advanced coronary artery disease can be severe. Arterial pressure can be increased following a heart attack only by increasing cardiac contractility, heart rate, or peripheral vascular resistance. Unfortunately, these compensatory responses all require that the heart do more physical work, which in turn increases the utilization of oxygen by the heart. Because the primary problem is inadequate blood flow and insufficient oxygen delivery to the heart, these compensatory responses frequently only advance the extent of myocardial damage.

As with all cardiac disorders, reversal of the primary disturbance is desirable. Several therapeutic methods are available to increase coronary blood flow. Nitroglycerin provides relief from angina by dilating the coronary arteries and increasing coronary blood flow. A class of drugs known as *calcium antagonists*, or *calcium channel*

blockers, which block the entry of calcium into the smooth muscle cells, also provide relief by dilating the coronary arteries. Agents that dissolve blood clots in the coronary arteries are being used to unblock arteries occluded by thrombus formation. If progressive occlusion of a coronary vessel is detected prior to the onset of a heart attack, measures are available to alleviate the occlusion. Coronary bypass surgery involves the grafting of a vein (removed from elsewhere in the patient's body) around the occluded segment of the coronary artery so that blood flow can bypass the diseased area (Fig. B). Another approach is to insert a catheter with a small balloon at the tip into the occluded coronary artery (Fig. C). When the catheter tip is positioned within the occluded portion of the vessel, the balloon is inflated and the atherosclerotic coronary plaque is fractured, allowing the vessel to dilate. This procedure is referred to as *coronary angioplasty*. To keep the vessel open following angioplasty, a wire mesh tube, called a *stent*, can be inserted via a catheter (Fig. D). The stent remains permanently in the coronary artery. After a while, tissue will grow over the stent keeping it in place. Unfortunately, this can sometimes result in reblockage of the vessel (restenosis). In recent years, stents have been impregnated with drugs to help minimize restenosis.

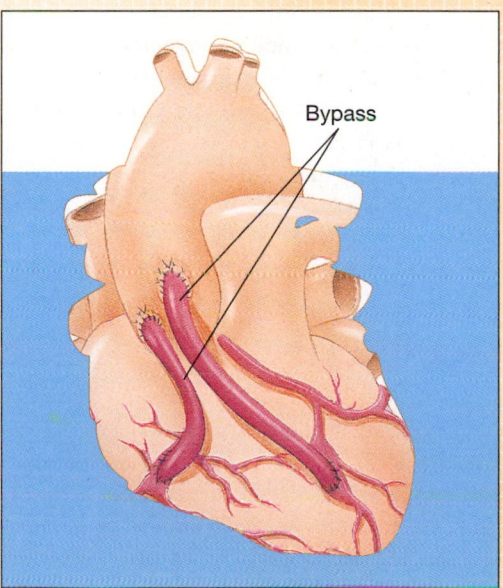

Figure B Coronary Bypass.

Atherosclerosis and Myocardial Infarction *(continued)*

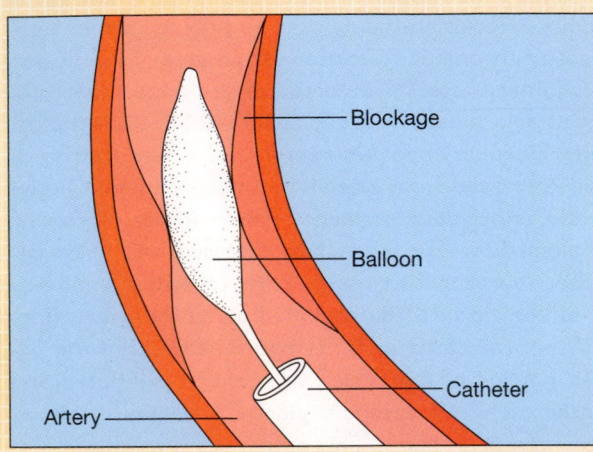

Figure C A small catheter is introduced into the blocked coronary artery. A guide wire containing a small balloon at its tip is pushed through the blockage. When the balloon is in place, it is inflated to flatten the blockage against the arterial wall.

Prevention is clearly the best treatment. Although the cause of atherosclerosis is unknown, a number of risk factors have been associated with an increased incidence of the disease. These include smoking, hypertension, elevated plasma cholesterol, lack of exercise, and diabetes. Current evidence suggests that a reduction of cholesterol in the diet, as well as regular exercise, may offer some protection against the progression of the disease.

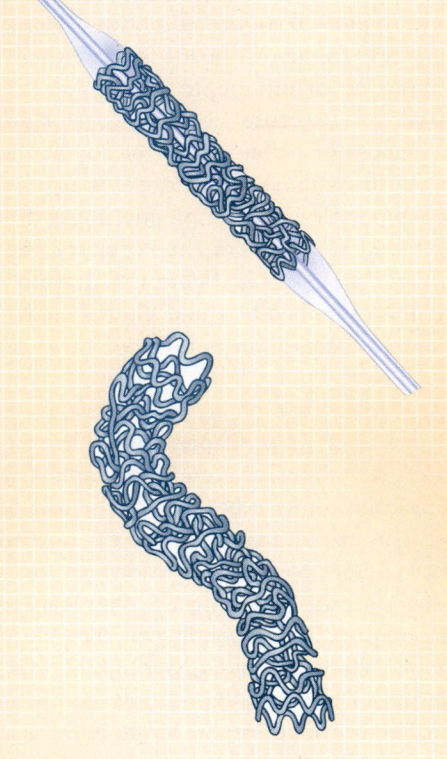

Figure D After angioplasty, a wire mesh tube, called a stent, can be inserted via a catheter and left to help keep open the diseased artery.

distribution of blood volume. Following hemorrhage, heart rate and vigor of contraction increase to transfer venous volume to the arterial system in order to maintain arterial pressure. Recall that full compensation is limited by inadequate filling of the heart. With cardiac failure, just the opposite occurs. Because of the reduced pumping capacity of the heart, blood accumulates in the heart and veins. The heart fails to maintain the flow of blood to the arterial system. This, of course, results in a fall in arterial pressure and an increase in the volume of blood in the venous system and in the heart.

Left Heart Failure Versus Right Heart Failure

Until now, we have considered the heart as a single unit. In reality, either or both ventricles can undergo failure. Failure of the right ventricle leads to an accumulation of blood in the right heart and the venous system. As blood accumulates in the veins and ultimately in the capillaries, capillary filtration increases because of the increase in capillary pressure. This leads to the accumulation of excess fluid in the tissues in the form of edema (Fig. 19–34). The formation of edema is especially pronounced in the lower ex-

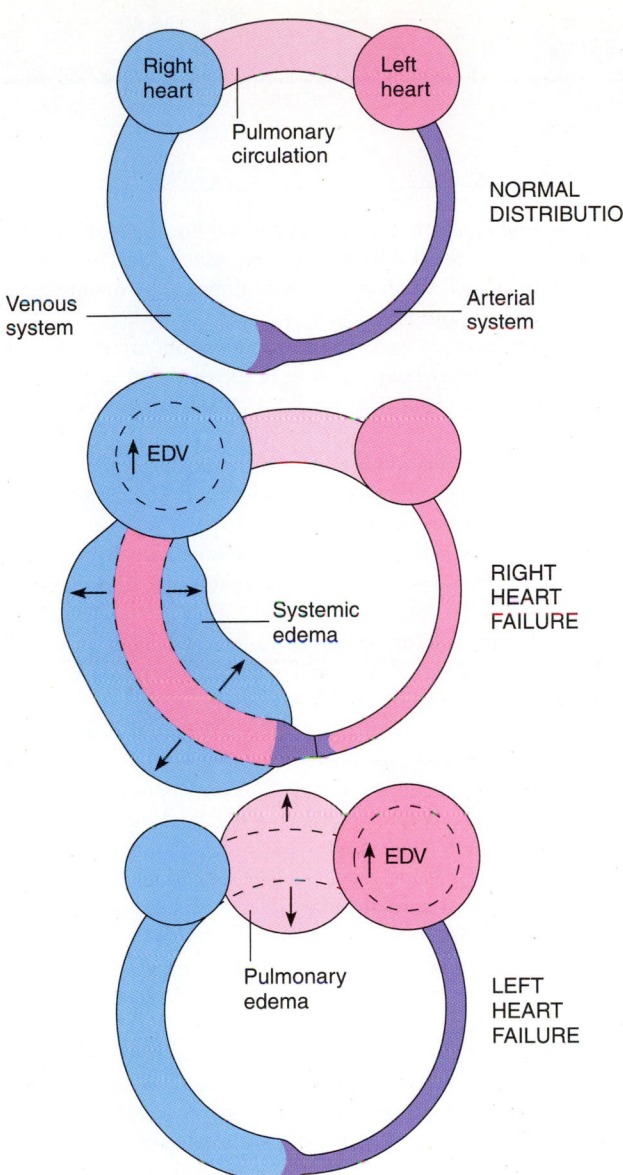

Right heart

Left heart

Pulmonary circulation

NORMAL DISTRIBUTION

Venous system

Arterial system

↑EDV

RIGHT HEART FAILURE

Systemic edema

↑EDV

Pulmonary edema

LEFT HEART FAILURE

Figure 19–34

Blood volume redistribution in the cardiovascular system due to failure of either the left or right heart. The increased volume in either the pulmonary or systemic veins leads to increased capillary pressure and loss of fluid to the interstitial space (edema).

tremities because gravity also contributes to the high capillary pressure.

If, on the other hand, the left heart fails, blood accumulates in the pulmonary circulation. The formation of pulmonary edema can be particularly serious because this accumulation of fluid impairs gas exchange. This can lead to a vicious cycle; poor gas exchange further exaggerates the loss of ventricular contractility, which in turn leads to further edema.

Treatment of Heart Failure

How can cardiac failure be treated medically? Obviously, if contractility could be returned to normal, the remaining symptoms would be alleviated. Administration of the drug

digitalis tends to improve contractility by increasing intracellular calcium concentrations. The excess fluid accumulation can be reversed by the administration of drugs that enhance the loss of sodium and water by the kidneys (i.e., diuretics). This latter treatment must be used with caution because excessive reduction in volume would compromise cardiac output as a result of decreased cardiac filling and reduced end-diastolic volume. A more recent therapy involves the use of angiotensin-converting enzyme (ACE) inhibitors, which inhibit the enzyme that converts angiotensin I to angiotensin II (see Chapter 14). At present, the mechanism for the beneficial effects of these drugs is unclear.

CHAPTER REVIEW

Summary

- The purpose of the vascular system is to distribute blood flow throughout the body.
- The arteries distribute blood to the various tissues in the body.
- Blood flow = mean arterial pressure/total peripheral resistance.
- Mean arterial pressure = $\frac{1}{3}$ pulse pressure + diastolic pressure.
- Arterioles are the major resistance vessels.
- Contraction of smooth muscle changes the diameter and, therefore, the resistance of arterioles.
- In many tissues, blood flow is autoregulated according to demand.
- In other tissues, blood flow is regulated by the autonomic nervous system.
- Epinephrine binds to β-receptors to produce vasodilation of the arterioles in the heart, skeletal muscle, and the liver.
- Capillaries are thin-walled vessels that are optimized for diffusional exchange.
- The precapillary sphincter is a small cusp of smooth muscle at the beginning of some capillaries.
- Contraction of the precapillary sphincter modulates capillary flow in an all-or-none fashion.
- Substances move between the capillary blood and the extracellular fluid by transcytosis, diffusional exchange, or bulk flow.
- Starling's hypothesis describes how the balance of hydrostatic and oncotic pressures in the capillaries determines whether water is filtered or absorbed.
- The veins serve as a conduit for the return flow of blood from the capillaries to the heart.
- Veins also provide a variable-capacitance storage reservoir for blood.
- The unidirectional flow of blood in the veins is facilitated by one-way valves.
- The pumping action of skeletal muscle contraction and respiration facilitate the flow of blood to the heart.
- The lymphatic vessels return fluid and protein that leaks from the capillaries back to the circulating blood.
- Absorbed fat digestion products are also transported by the lymphatics from the gut to the blood.

- Cardiac output is matched with tissue blood flow by maintaining constant mean arterial blood pressure, much like municipal water utilities match water supply with demand by maintaining constant water line pressure.
- Baroreceptors provide sensory information on arterial blood pressure to the cardiovascular center in the medulla.
- Autonomic outflow from the cardiovascular centers maintains blood pressure constant.
- Average cardiac output and venous return must be equal.
- Total blood volume affects cardiac output by changing end-diastolic volume of the ventricles.
- Hemorrhage leads to reduced blood pressure due to reduced cardiac output as a result of reduced ventricular end-diastolic volume.
- The normal compensatory response to hemorrhage is vasoconstriction of resistance vessels and capacitance veins and increased cardiac contractility and heart rate.
- Hypotension can also result from sudden postural changes or prolonged quiet standing (orthostatic hypotension).
- The compensatory response to hypotension is vasoconstriction of resistance vessels and capacitance veins, and increased cardiac contractility and heart rate.
- During exercise, the blood flow to active skeletal muscle may increase 10 to 15 times normal and cardiac output may increase to as much as 35 L/min.
- Failure of the left heart to maintain normal cardiac output leads to an accumulation of blood in the lungs, which inhibits gas exchange.
- Failure of the right heart leads to accumulation of blood in the systemic veins, increased capillary filtration, and edema.
- Compensation for decreased arterial pressure caused by heart failure, includes increases in heart rate, total peripheral resistance, vasoconstriction of capacitance veins, and retention of salt and water by the kidneys.

Review Questions

Choose the Correct Answer

1. Which of the following would not occur in response to vigorous exercise?
   a. Dilation of resistance arterioles in skeletal muscle
   b. Dilation of resistance arterioles in the small intestine
   c. Increased cardiac output
   d. Increased stroke volume
   e. Vasoconstriction of small veins
2. Starling's law of the heart predicts that:
   a. decreased filling of the ventricles would increase stroke volume.
   b. increased stroke volume from the right heart is balanced by decreased stroke volume from the left heart.
   c. increased ventricular end-diastolic volume would lead to increased stroke volume.

   d. the stroke volumes of the left and right hearts will never be the same.
   e. decreased arterial pressure will increase stroke volume.
3. Factors that have a direct impact on arterial pulse pressure do not include:
   a. arterial stiffness.
   b. stroke volume.
   c. hematocrit.
   d. cardiac contractility.
   e. total peripheral resistance.
4. Which of the following changes from normal resting levels would you expect to see with both hemorrhage and vigorous exercise?
   a. Increased total peripheral resistance
   b. Increased cardiac output

c. Increased arterial blood pressure

d. Increased heart rate

e. Decreased contractility

5. Left heart failure would not cause:

a. increased heart rate.

b. increased sympathetic stimulation of the heart.

c. increased circulating epinephrine.

d. vasoconstriction of resistance arterioles.

e. vasodilation of capacitance veins.

6. In a standing individual, arterial blood pressure at the level of the aortic arch is 90 mm Hg and 170 mm Hg at the level of the feet. Pressure in the right atrium is approximately 0 mm Hg. The pressure gradient determining systemic blood flow is:

a. 80 mm Hg.

b. 90 mm Hg.

c. 140 mm Hg.

d. 170 mm Hg.

e. 260 mm Hg.

7. Pulse pressure is 50 mm Hg, and systolic pressure is 120 mm Hg. What is the diastolic pressure?

a. 50 mm Hg

b. 60 mm Hg

c. 70 mm Hg

d. 90 mm Hg

e. 170 mm Hg

8. Capillary hydrostatic pressure is 35 mm Hg, capillary colloidal osmotic pressure is 28 mm Hg, and interstitial fluid colloidal osmotic pressure is 3 mm Hg. The net filtration or absorption pressure is:

a. 4 mm Hg (absorption).

b. 4 mm Hg (filtration).

c. 10 mm Hg (absorption).

d. 10 mm Hg (filtration).

e. 7 mm Hg (filtration).

9. Mean arterial blood pressure is 100 mm Hg, and total peripheral resistance is 20 mm Hg/min/L. What is the total blood flow?

a. 1 L/min

b. 2 L/min

c. 5 L/min

d. 10 L/min

e. 50 L/min

10. An individual's blood pressure is measured as 120/90 mm Hg. What is the mean arterial pressure?

a. 30 mm Hg

b. 90 mm Hg

c. 100 mm Hg

d. 120 mm Hg

e. 210 mm Hg

11. Diastolic blood pressure is:

a. equal to the cuff pressure at which flow through the brachial artery begins.

b. less than mean capillary blood pressure.

c. equal to mean blood pressure minus systolic pressure.

d. equal to systolic pressure minus mean blood pressure.

e. equal to the cuff pressure at which flow through the brachial artery is continuous.

12. Which of the following causes vasoconstriction of the arterioles in the skin?

a. Increased heart rate

b. Increased tissue metabolism

c. Increased concentration of carbon dioxide in the blood

d. Increased circulating epinephrine

e. Increased body temperature

13. In which of the following organs are the resistance arterioles not regulated by sympathetic innervation?

a. Skin

b. Liver

c. Brain

d. Gastrointestinal tract

e. Kidney

14. A decrease in vessel radius from 10 mm to 5 mm increases the resistance to blood flow by how much?

a. 4 times

b. 6 times

c. 8 times

d. 16 times

e. 32 times

15. Which of the following blood vessels is not surrounded by smooth muscle cells?

a. Artery

b. Arteriole

c. Capillary

d. Venule

e. Vein

16. The overall resistance to blood flow is greatest at the level of the:

a. aorta.

b. arterioles.

c. capillaries.

d. veins.

e. venules.

17. Which one of the following promotes reabsorption of water into the capillary?

a. Plasma proteins

b. Capillary hydrostatic pressure

c. Interstitial fluid oncotic pressure

d. Soluble electrolytes

e. Red blood cells

18. In a healthy individual, the overall balance of filtration and absorption for all of the capillaries in the body is such that over a 24-hour period:

a. filtration of water exactly equals absorption of water.

b. there is neither net filtration or net absorption of water across the capillary wall.

c. there is net absorption of water.

d. there is net filtration of water.

e. the concentration of plasma proteins decreases.

19. On average:

a. left heart output is slightly greater than right heart output.

b. total systemic blood flow exceeds total pulmonary blood flow.

c. cardiac output equals venous return.

d. total capillary blood flow exceeds total arterial blood flow.

e. resistance to blood flow is greater in the pulmonary circulation than in the systemic circulation.

20. An increase in arterial blood pressure causes which of the following?

a. A decrease in baroreceptor afferent nerve activity

b. An increase in heart rate

c. A decrease in efferent parasympathetic nerve activity from the medullary cardiovascular control center

d. An increase in total peripheral resistance

e. A decrease in sympathetic nerve activity to the resistance arterioles

Answers to Case History Questions

1. The patient has trouble breathing at night because fluid accumulates in the lungs as a result of elevated pulmonary capillary pressure. This elevation of capillary pressure is partially due to the fluid retention by the kidneys in response to the decreased systemic arterial pressure and partially due to the fact that blood backs up behind the failing left heart. When the patient sits up, gravity redistributes blood volume away from the lungs to the peripheral veins. This decreases pulmonary capillary pressure and reduces the filtration of fluid into the lungs.

2. Failure of the right heart allows blood to backup in the veins, and, as discussed in the previous answer, fluid is retained by the kidneys. The increased volume of blood in the venous system causes distension of the neck veins. This also increases systemic capillary pressure. The increased systemic capillary pressure results in increased fluid filtration and formation of peripheral edema, particularly in the lower periphery, where capillary hydrostatic pressures are highest due to the effects of gravity.

3. Because the heart is receiving inadequate coronary blood flow, contractility is reduced. This results in decreased stroke volume, decreased cardiac output, and decreased arterial pressure. The increased heart rate is a compensatory response to the decreased baroreceptor activity. The dilated heart and reduced wall motion both indicate the myocardium is not able to contract normally as a result of the reduced coronary blood flow through the partially occluded coronary arteries. Once the ventricles dilate beyond the peak of the length-tension curve, further dilation results in further reduction in stroke volume as described by the Frank-Starling relationship.

4. Digitalis increases the contractility of the heart, however, the effectiveness of this drug may be limited by the fact that the reduced coronary flow places a limit on the amount of work the heart can do. Diuretics decrease circulating blood volume, and this decreases capillary pressures (both systemic and pulmonary) and reduces both systemic and pulmonary edema. The reduced blood volume also decreases arterial blood pressure, reducing the load on the heart and decreasing the rate of oxygen consumption by the heart. This minimizes further ischemic damage to the myocardium. The reduced circulating blood volume and reduced arterial pressure would predispose the patient to orthostatic hypotension and fainting, and therefore, bed rest is appropriate. Bed rest also further reduces the load on the heart by reducing blood flow to tissues such as skeletal muscle. Coronary bypass surgery can restore blood flow to the heart and alleviate most of the symptoms by increasing cardiac contractility and decreasing dilation of the heart.

Key Terms

angina pectoris (p. 622)
arterial pressure (p. 592)
arterioles (p. 593)
atherosclerotic plaque (p. 622)
atrial natriuretic factor (p. 600)
autoregulation of blood flow
  (p. 597)
baroreceptor (p. 613)
cardiovascular control center
  (p. 614)

carotid sinus (p. 613)
circulatory shock (p. 618)
congestive heart failure
  (p. 621)
diastolic pressure (p. 591)
edema (p. 608)
endothelium-derived relaxing
  factor (EDRF) (p. 598)
filtration pressure (p. 605)
hemorrhage (p. 617)

hypotension (p. 618)
infarct (p. 622)
mean arterial pressure (MAP)
  (p. 592)
microcirculation (p. 600)
precapillary sphincter
  (p. 600)
pulse pressure (p. 591)
Starling's hypothesis (p. 605)

systolic pressure (p. 591)
tone (p. 596)
total peripheral resistance
  (TPR) (p. 594)
transcytosis (p. 602)
vasodilator (p. 596)
vasomotion (p. 600)
venous return (p. 611)
venules (p. 606)

Suggested Readings

Berne, R. M., and Levy, M. N. *Physiology,* ed 4. St. Louis, C.V. Mosby, 1998.

Cabe, D. K. "Saving hearts that grow old." *Scientific American* Special Issue, June, 2000.

Eisenberg, M. S., Bergner, L., Hallstrom, A. P., and Cummins, R. O. "Sudden cardiac death." *Scientific American,* 254:37–43, 1986.

Goldstein, G. W., and Betz, A. L. "The blood-brain barrier." *Scientific American,* 255:74–83, 1986.

Grady, P. "Can heart disease be reversed?" *Discover,* 8:54–68, 1987.

Hall, E. D., and Hooper, C. "Free radicals: Research on biochemical bad boys comes of age." *Journal of the National Institutes of Health Research,* 1:101–106, 1989.

Keating, M. T., and Sanguinetti, M. C. "Molecular genetic insights into cardiovascular disease." *Science,* 272:681–685, 1996.

Levy, M. N., and Berne, R. M. *Cardiovascular Physiology,* ed 8. St. Louis, C. V. Mosby, 2001.

Little, R. C., and Little, W. C. *Physiology of the Heart and Circulation,* ed 4. Chicago, Year-Book Medical Publishers, 1988.

Michael, D., and McGoon, M. D. *Mayo Clinic Heart Book.* New York, William Morrow & Co., 1993.

Mohrman, D. E., and Heller, L. J. *Cardiovascular Physiology,* ed 4. New York, McGraw-Hill, 1997.

Mukerji, B., Alpert M. A., and Mukerji, V. "Cardiovascular changes in athletes." *American Family Physician,* 40:169, 1989.

Rhoades, R. A., and Tanner, G. A. *Medical Physiology,* ed 1. Boston, Little Brown & Co., 1995.

Ross, R. "The pathogenesis of atherosclerosis: A perspective for the 1990s." *Nature,* 362:801–809, 1993.

Taubes, G. "Nutrition: The soft science of dietary fat." *Science,* 291:2536–2545, 2001.

Answers to Review Questions

1. b   **2.** c   **3.** c   **4.** d   **5.** e   **6.** b   **7.** c   **8.** d
9. c   **10.** c   **11.** e   **12.** d   **13.** c   **14.** d   **15.** c
16. b   **17.** a   **18.** d   **19.** c   **20.** e

Chapter 20

RESPIRATION

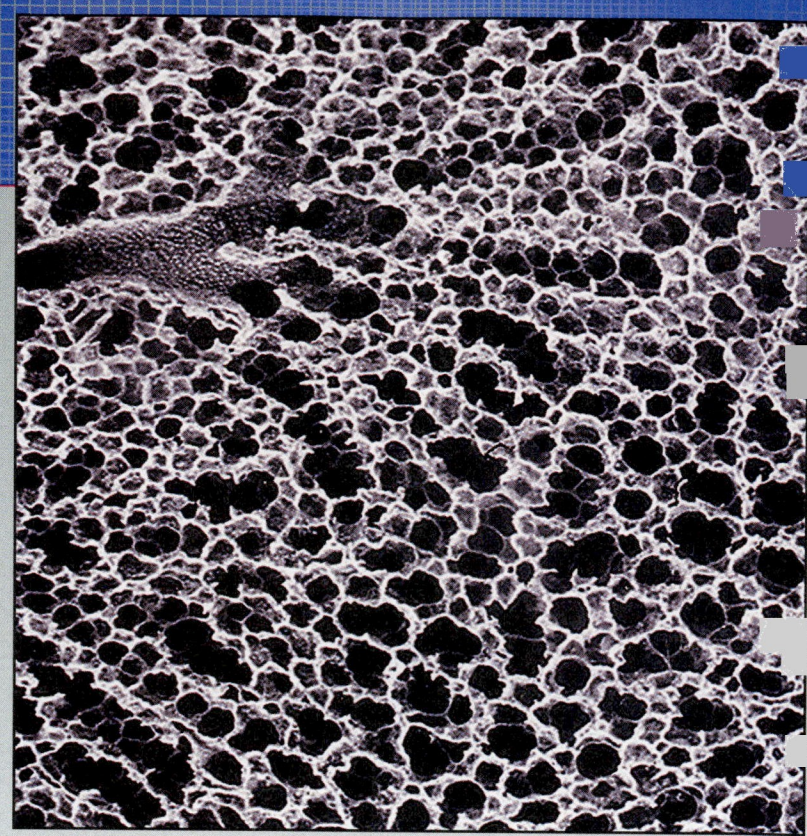

• *Scanning electron microscope (SEM) of a normal human lung with numerous alveoli arising from two alveolar ducts. Note tha alveoli are not all the same size.*
(© Fawcett/Gehr/Science Source/ Photo Researchers, Inc.)

KEY CONCEPTS

• *The main function of the lung is for gas exchange, that is, taking up oxygen and removing carbon dioxide from the blood.*

• *Groups of alveolar ducts and their alveoli merge with pulmonary capillaries to form terminal respiratory units for gas exchange.*

• *Contraction of the diaphragm, the main muscle of breathing, creates a small negative pleural pressure.*

• *Changes in pleural pressure move air into and out of the lungs.*

• *Spirometry is used to measure lung function.*

• *Alveolar ventilation regulates carbon dioxide levels in the blood.*

• *Dead space ventilation is inspired air that does not participate in gas exchange.*

• *Compliance is a measure of lung distensibility.*

• *Surfactant is important in maintaining alveolar stability.*

• *Airway resistance increases during expiration.*

• *Lung compliance has a marked effect on airway resistance during forced ventilation.*

• *During inspiration, work is required to expand the lung and chest wall.*

CASE HISTORY

Mr. James Miller, a 65-year-old man, enters the university hospital emergency room because of a five-day history of shortness of breath on exertion. He also complains of a cough producing green sputum. He said he felt feverish at home but denies any shaking chills, sore throat, nausea, vomiting, or diarrhea. He appears pale and diaphoretic, and is breathing through pursed lips. You walk him into one of the examination rooms, sit him on the stretcher, and begin a quick evaluation as you question him. Mr. Miller slowly explains that he has trouble like this all the time. It has become a lot worse over the past year and acutely worse over the past week. He is impatient with your questioning. You sense that he expects you to know what is wrong and to do something right away to relieve his obvious distress. Mr. Miller states that he has been smoking two packs of cigarettes a day for the past 30 years. However, he recently decreased his habit to one pack a day. Mr. Miller then tells you that he is a longtime smoker, and for the past 10 years he has had a cough productive of thick, white sputum. His condition is especially bad in the morning but improves after his first cigarette of the day. His breathing becomes worse whenever he gets an upper respiratory infection or when he goes out on hot, "muggy" days. Approximately one week ago, he began to have increased sputum production and the color of his sputum changed from whitish-gray to greenish-yellow. He has not been hospitalized previously. He is a retired truck driver and lives with his wife. They have no pets. Although he has had dyspnea upon exertion for the past two years, he continues to maintain an active lifestyle. He still mows his own lawn without much difficulty and can walk one to two miles on a flat surface at a moderate pace. Mr. Miller says he rarely drinks alcohol. He denies any other significant past medical history, such as a history of heart disease, hypertension, edema, childhood asthma, or any allergic diseases. He does state that his father, also a heavy smoker, died of emphysema at age 55.

Mr. Miller slowly explains that he has had trouble like this for the past two years but that his situation has become worse over the past year and acutely worse over the past month. Two to three days ago, he was not able to perform his usual daily activities without getting severely short of breath. He has not been able to eat his normal diet. After his dyspnea worsened over the past 24 hours, he called an ambulance to bring him to the emergency room.

Mr. Miller's physical examination shows that he is a somewhat thin, but well-developed, man in moderate respiratory distress. You have blood drawn for blood gas analysis. His blood gases were: Po_2 = 40 mm Hg (normal values = 85–95 mm Hg); pH = 7.32 (normal values = 7.35–7.45). He is hemodynamically stable with vital signs of BP = 130/80, respiratory rate of 28 to 32, and heart rate of 92 bpm. His temperature is 37.9°C orally. He has decreased but audible breath sounds in both lung fields, with expiratory wheezing and a prolonged expiratory phase. Head, eye, ear, nose, and throat findings are unremarkable.

Pulmonary function tests are ordered. The spirometry test results reveal altered pulmonary function, epecially severe limitation of expiratory airflow. Mr. Miller has been diagnosed with pulmonary emphysema.

Questions

1. What are the common spirometry measurements associated with emphysema?

2. What are the mechanisms of airflow limitation with emphysema?

3. What is the most commonly held theory explaining the development of emphysema?

INTRODUCTION

The process of exchanging oxygen and carbon dioxide between the body and the environment is known as **respiration,** sometimes referred to as the "breath of life." A breath in and a breath out, 12 to 15 times every minute, seems like a simple and unimpressive process in which all of the oxygen is supplied and all of the carbon dioxide is removed from the trillions of cells in the human body. However, this simplicity is deceptive because breathing is extremely responsive to small changes in the blood chemistry, mood, level of alertness, and body activity. The human lungs are so efficient in gas exchange that they have played a key role in the success and extraordinary adaptability of the human species. For example, a marathon runner who staggers across the 26-mile finish line in less than 3 hours, or a swimmer who crosses the English Channel in record time, is not limited by gas exchange in the lungs. The reason is that the lungs can increase gas exchange more than 20-fold to meet the body's energy demands.

Respiration takes place in two stages. The first, known as **gas exchange,** is the process of transferring oxygen (O_2) and carbon dioxide (CO_2) between the atmosphere to the cells. More specifically, gas exchange occurs in two phases: (1) the exchange of O_2 and CO_2 between the external environment and the lungs and (2) the exchange of O_2 and CO_2 between peripheral capillaries and the tissues. The first phase of gas exchange (i.e., the transfer of O_2 and CO_2 between the external environment and the lungs) involves three steps. The first step involves moving air into and out of the lungs and is termed **ventilation.** The second step involves the transfer of gases from the lung to the pulmonary capillary blood and is termed **gas diffusion.** The third step is carrying oxygen and carbon dioxide by the blood and is termed **gas transport.**

The second stage of respiration is known as **cellular respiration,** which consists of a series of complex metabolic reactions that break down the food we eat into metabolic energy. Driving the body's chemical reactions, especially those of muscular contraction, requires large amounts of energy, and little energy is available until the food we eat is digested and **oxidized** (burned with oxygen). Oxidation occurs when oxygen is taken up by the cells and is used in cellular respiration. Recall from Chapter 6 that oxygen is required in the final step of cellular respiration to serve as an electron acceptor in the process by which cells obtain energy. The human respiratory system is so efficiently designed to supply oxygen and remove carbon dioxide that gas exchange and transport rarely limit our activity.

The function of the human respiratory system can be divided into three main sections: (1) ventilation and the mechanics of breathing, (2) gas transfer and transport, and (3) the control of breathing. This chapter discusses lung gas exchange, mechanical properties, and the work of breathing. Chapter 21 discusses the circulatory design, gas diffusion between the lung and blood, transport of respiratory gases by the blood to the cells, and how the control of breathing is connected to our metabolic as well as our voluntary activities.

STRUCTURAL AND FUNCTIONAL RELATIONSHIPS OF THE LUNG

Movement of gases into and out of cells occurs by simple diffusion. In unicellular organisms, the process by which oxygen and carbon dioxide diffuse between the cell and the surrounding environment is very simple, and no special respiratory structures are required. In more complex organisms, however, cells deep within the body cannot exchange enough oxygen and carbon dioxide between the cells and environment because the diffusion distance is too great. As a result, special gas exchange organs have developed. These include tracheal tubes in insects; gills in fish; and lungs in higher animals, such as humans. All of the respiratory organs mentioned, from the simplest to the most complex, have three features in common. First, they all have very thin walls so that gas diffusion can readily occur. Second, they are always moist so that oxygen and carbon dioxide can dissolve in fluids. Third, they are richly supplied with blood vessels to ensure efficient gas exchange and transport of the respiratory gases to the cells.

The Airway Tree Divides Repeatedly To Increase Total Cross-Sectional Area

 What is the functional anatomy of the lungs?

The human gas exchange organ consists of two lungs and five lobes — two in the left lung and three in the right (Fig. 20–1). The lungs are composed of two treelike structures, the **vascular tree** and the **airway tree,** which are embedded in the lung tissue. The vascular tree consists of arteries and veins connected by capillaries. The airway tree consists of a series of hollow branching tubes that decrease in diameter at each branching (Fig. 20–2). The main airway **(trachea)** branches into two **bronchi,** each entering a lung. Within each lung, these bronchi branch many times into progressively smaller bronchi, which in turn form **bronchioles.**

A functional model of the airway tree is presented in Figure 20–3. The first 17 generations (i.e., trachea and 16 airway branches) make up the **conducting zone.** The conducting units (trachea, bronchi, and bronchioles) have three important functions: (1) to warm and humidify inspired air, (2) to distribute air evenly to the deeper parts of the lungs, and (3) to serve as a part of the body's defense system (removal of dust, bacteria, and noxious gases from the lung). The latter function is linked to the **mucociliary transport system,** which keeps microorganisms, dust particles, and noxious gases from entering the alveoli (Fig. 20–4). The conducting units are lined with cilia that beat synchronously.

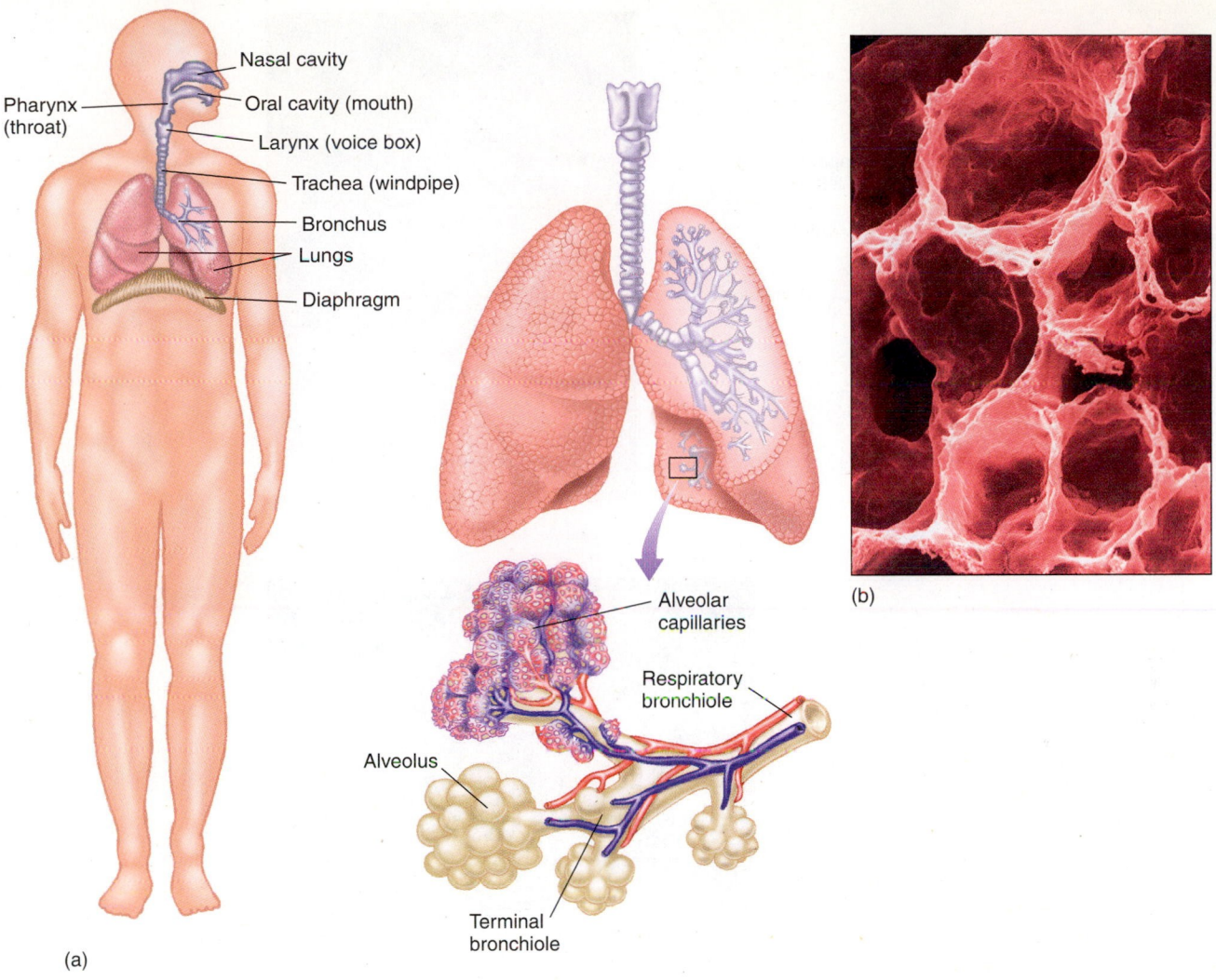

(a)

(b)

Nasal cavity
Pharynx (throat)
Oral cavity (mouth)
Larynx (voice box)
Trachea (windpipe)
Bronchus
Lungs
Diaphragm

Alveolar capillaries
Respiratory bronchiole
Alveolus
Terminal bronchiole

Figure 20–1

Organization of the human lung. The right lung consists of three lobes (upper, middle, and lower), and the left lung consists of two lobes (upper and lower). **(a)** Human lung with each alveolus enmeshed in a network of capillaries. **(b)** Scanning electron microscope (SEM) of normal lung tissue showing alveolar-capillary membrane, where gas exchange takes place at the blood-gas interface. (original magnification × 2200.) *(© SECCHI-LECAQUE-ROUSSEL-UCLAF/CNRI/SPL/Photo Researchers, Inc.)*

These cilia are buried in a carpet of sticky mucus that traps foreign material. The mucociliary carpet "floats" on top of the beating cilia. For example, particles more than 10 μm in diameter, which are visible, are trapped on the hairs and moist mucus lining of the nasal passage. However, small particles (< 10 μm), which are not visible, get inhaled into the lungs. At every bend in the airway, these small particles fail to make the turn and stick on the mucous film. The mucus is "streamed" toward the throat by the beating cilia and can be swallowed or coughed up when its presence is noticed as it stimulates receptors in the throat.

The cilia are greatly affected by irritant gases, such as tobacco smoke. When they come in contact with irritants, they become partially paralyzed, slowing the mucociliary transport system. As a result, microorganisms and excess mucus accumulate in the lower airways (e.g., the bronchioles). Often, the excess mucus plugs the lower airways, and a cough is required to dislodge and remove it. If excess mucus accumulates over a long period of time, bacterial infection may develop. When this occurs, the mucus is no longer a light yellow but instead is green as a result of the infiltration and rupturing of white blood cells.

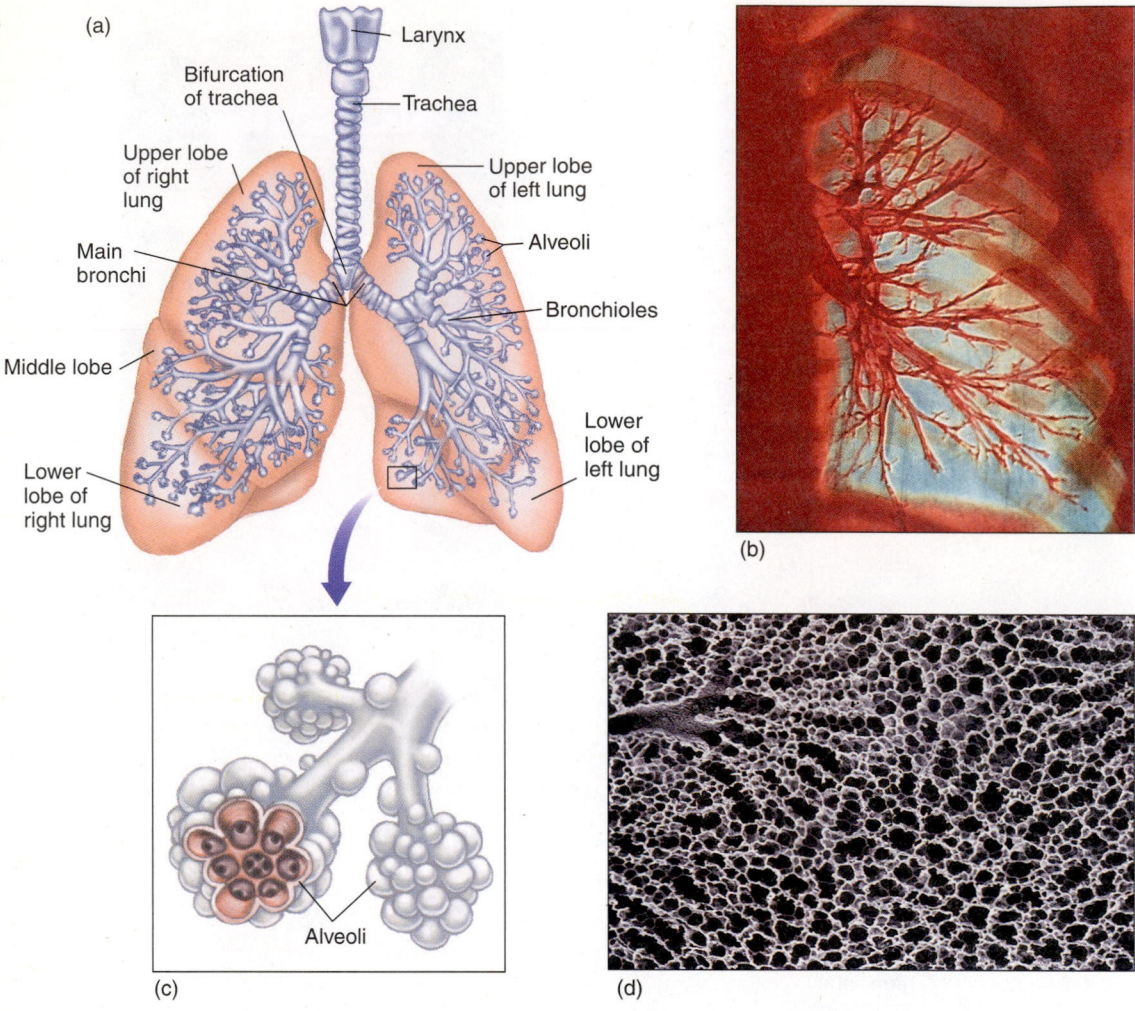

(a)

Larynx

Bifurcation of trachea

Trachea

Upper lobe of right lung

Upper lobe of left lung

Main bronchi

Alveoli

Bronchioles

Middle lobe

Lower lobe of left lung

Lower lobe of right lung

(b)

Alveoli

(c)

(d)

Figure 20–2

Organization of the airway tree. **(a)** The trachea subdivides to form bronchi, which branch repeatedly, leading to bronchioles and then to alveoli. **(b)** False-color arteriograph of a normal right lung showing the pattern of bronchial and bronchiolar branching. *(© CNRI/SPL/Photo Researchers, Inc.)* **(c)** The respiratory unit, which consists of respiratory bronchioles, alveolar ducts, and alveoli. **(d)** Scanning electron microscope (SEM) of a normal human lung with numerous alveoli arising from two alveolar ducts. Note that alveoli are not all the same size. *(© Fawcett/Gehr/Science Source/Photo Researchers, Inc.)*

Another striking feature of the conducting zone is the cartilage arrangement. The first four generations of the conducting zone are subjected to lung pressures during forced expiration and contain a considerable amount of cartilage to prevent airway collapse during forced expiration. In the trachea and main bronchi, the cartilage consists of U-shaped rings. Further down, in the lobar and segmental bronchi, the cartilaginous rings give way to small plates of cartilage. In the bronchioles, the cartilage disappears altogether. The smallest airways in the conducting zone are the **terminal bronchioles.** Both the bronchioles and terminal bronchioles are suspended by elastic tissue in the lung parenchyma. The

conducting zone has its own separate circulation, the **bronchial circulation,** which originates from the descending aorta and drains into the pulmonary veins. An important feature to recognize is that no gas exchange occurs in the conducting zone.

The last seven generations make up the **respiratory zone,** the site of gas exchange. The respiratory zone is composed almost entirely of alveolar ducts and their alveoli. Like the conducting zone, the respiratory zone has its own separate and distinct circulation, the **pulmonary circulation.** An important feature to recognize is that one pulmonary branch accompanies each airway and branches with it. The lung has

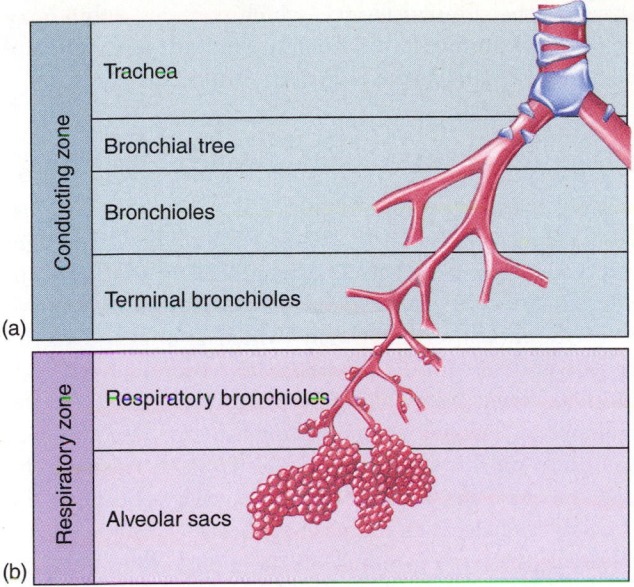

Figure 20–3

The airways are divided into the conducting zone and the respiratory zone. *(a)* The conducting zone is illustrated from the trachea to the terminal bronchioles. The conducting zone has its own separate circulation called the bronchial circulation. *(b)* The respiratory zone is the site of gas exchange and is illustrated from the respiratory bronchioles to the alveolar sacs.

the most extensive capillary network of any organ in the body; pulmonary capillaries occupy 70% to 80% of the alveolar surface area.

The increase in internal surface area is accomplished by formation of outpockets from the small airways to form **alveoli** (see Fig. 20–3). As mentioned earlier, a network of

capillaries surrounds each alveolus and brings blood into close proximity with the gas inside of the alveolus. Oxygen diffuses across the lung alveoli into the blood and carbon dioxide diffuses from the blood into the alveoli. The adult lung contains 300 to 500 million alveoli, with a combined internal surface area of approximately 75 m², roughly the size of a tennis court. This represents one of the largest biologic membranes in the body. Alveolar surface area increases as the number and size of alveoli increase from birth to adolescence (Table 20–1). However, after adolescence, alveoli increase only in size and, if damaged, have limited ability to repair themselves.

The Vascular and Airway Trees Merge To Form a Blood-Gas Interface

In the respiratory zone, a group of alveolar ducts and their alveoli merge with pulmonary capillaries to form a **terminal respiratory unit,** which is a thin interface where air and blood are brought into close contact. Approximately 60,000 of these terminal respiratory units are present in the human lung, and the **alveolar-capillary membrane** of these units separates the blood in the pulmonary capillaries from the gas in the alveoli. This interface is referred to as the **blood-gas interface** (Fig. 20–5) and is exceedingly thin (in some places, < 0.5 μm; approximately half of the thickness of a small strand of human hair). The blood-gas interface is composed of alveolar epithelium, an interstitial fluid layer, and capillary endothelium. Gas is brought to one side of the interface by ventilation (i.e., movement of air to and from the alveoli). Blood is brought to the other side of the blood-gas interface by pumping blood through the pulmonary circulation. Oxygen and carbon dioxide cross the blood-gas interface during gas exchange by diffusion.

Figure 20–4

Mucociliary transport system. Mucus layer traps small inhaled particles (< 10 μm). The particles stick to the thick carpet layer of mucus, which is propelled by cilia toward the trachea to remove bacteria and particulate matter.

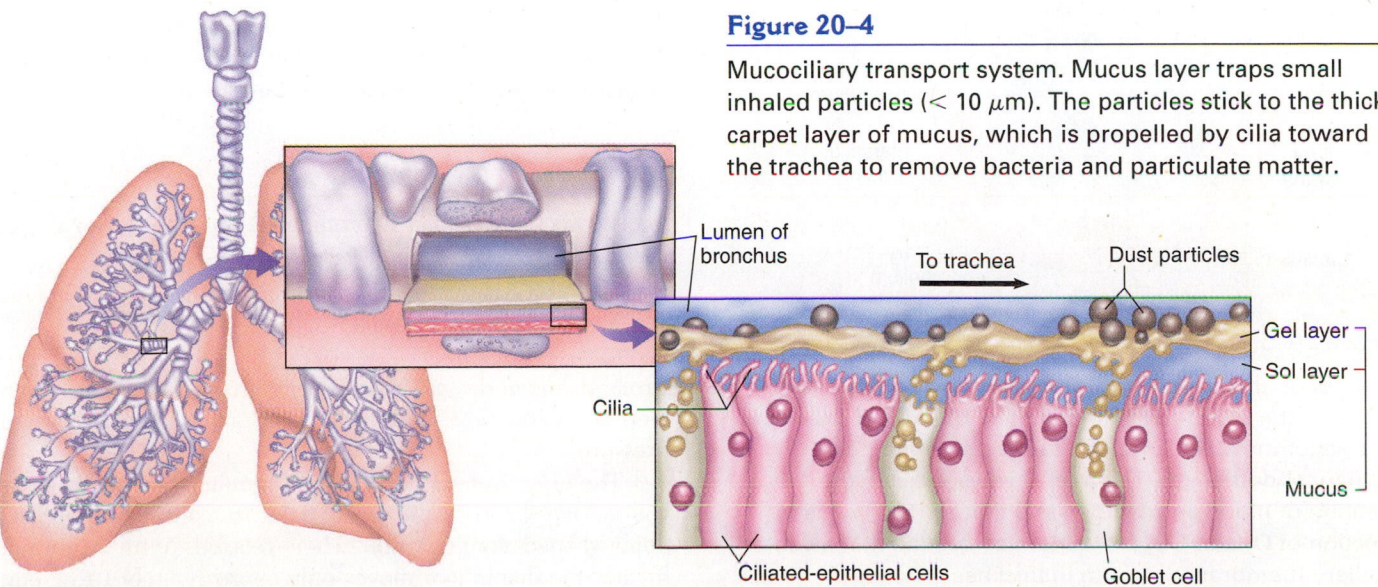

TABLE 20–1

Alveolar Number and Surface Area Changes with Age in the Human Lung

Age	No. Alveoli (10^6)	Alveolar Surface Area (m^2)	Skin Surface Area (m^2)
Birth	24	2.8	0.2
8 years	300	32.0	0.9
Adult	300	75.0	1.8
Fold increase	12	26	8.5

PRESSURE AND AIRFLOW CHANGES DURING BREATHING

Before examining how air gets into the lung, a brief discussion of the pressure changes and the muscles that are involved in breathing is helpful. The lungs are housed in an airtight chest cavity, the **thoracic cavity,** and are separated from the abdomen by a large dome-shaped muscle, the **diaphragm.** The thoracic cage is made up of 12 pairs of **ribs,** a **sternum,** and a set of internal and external intercostal muscles that lie between the ribs. The ribcage is hinged to the vertebral column, allowing it to be raised and lowered during breathing (Fig. 20–6). The space between the lungs and

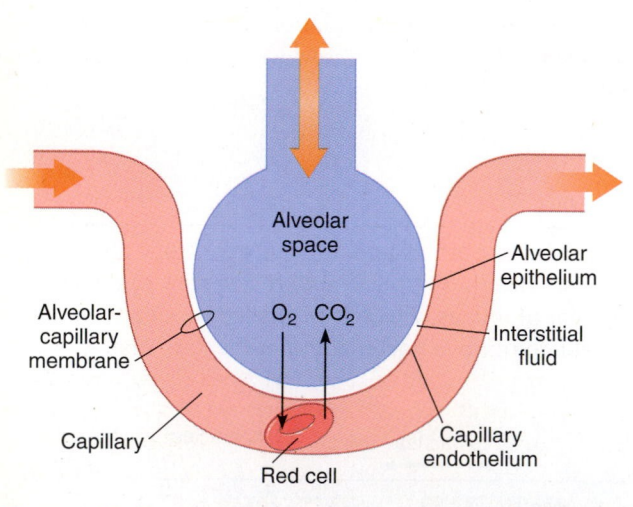

Figure 20–5

A model of an alveolus and a pulmonary capillary depicting the blood-gas interface. The blood-gas interface consists of the alveolar epithelium, interstitium, and capillary endothelium. The thick arrows indicate the direction of blood and airflow, and thin arrows indicate direction of O_2 and CO_2 movement across the thin alveolar-capillary membrane ($\approx 5\ \mu m$ in thickness).

chest wall is the **pleural cavity,** which contains a thin layer of fluid ($\approx 10\ \mu m$ thick) that function, in part, as a lubricant so the lungs can slide against the chest wall.

The Diaphragm Is the Main Muscle of Breathing

Inflation of the lungs is due to contraction of a skeletal muscle, the diaphragm (see Fig. 20–6). When the diaphragm contracts, the thoracic cavity is expanded, and the lungs inflate automatically. This enlargement is accomplished in two ways. First, when the diaphragm (which is attached to the lower ribs and sternum) contracts, the abdominal contents are pushed down, thus enlarging the thoracic cavity in the vertical plane. Second, when the diaphragm descends and pushes down on the abdominal contents, it also pushes the ribcage outward, further enlarging the cavity. Because the chest cavity is airtight, an increase in thoracic volume causes the **pleural pressure** (pressure in the pleural fluid between lung and chest wall) to decrease. The decrease in pleural pressure causes the lung to expand and fill with air. This key pressure-volume relationship in breathing is based on two gas laws. **Boyle's law,** which states that at a constant temperature the pressure (P) of the gas varies inversely with the volume (V) of gas:

$$P \approx 1/V.$$

As a result of Boyle's law, if either pressure or volume changes and if temperature remains constant, the product of pressure and volume remain constant:

$$P_1V_1 = P_2V_2.$$

Charles's law states that if pressure is constant, volume and temperature vary proportionately:

$$V \approx T$$

As a result of Charles's law, if either temperature or volume changes and pressure remains constant, then:

$$V_1/T_1 = V_2/T_2.$$

These two gas laws can be combined into the **general gas law:**

$$P_1V_1/T_1 = P_2V_2/T_2.$$

From the general gas law, at a constant temperature, an increase in thoracic volume leads to a decrease in pleural pressure.

The effectiveness of the diaphragm in changing thoracic volume is related to the strength of its contraction and its dome-shaped configuration when relaxed. With a normal breath, the diaphragm moves only approximately 1 to 2 cm,

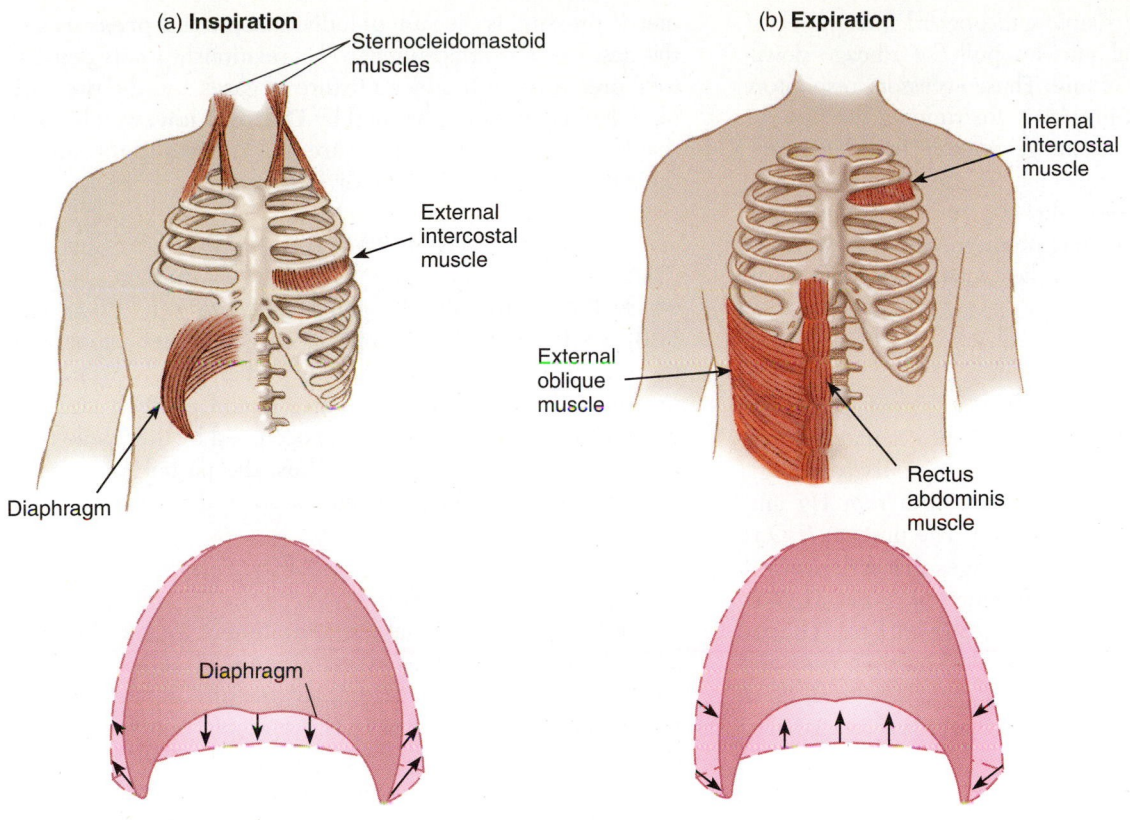

(a) **Inspiration**

Sternocleidomastoid muscles

External intercostal muscle

Diaphragm

Diaphragm

(b) **Expiration**

Internal intercostal muscle

External oblique muscle

Rectus abdominis muscle

Figure 20–6

Movement of the diaphragm and ribcage changes thoracic volume, which allows the lungs to inflate *(a)* and deflate *(b)*. At rest, during inspiration, the diaphragm contracts and pushes the abdominal contents downward. The lever action of the diaphragm on abdominal contents pushes the ribcage up and outward during inspiration *(a, lower panel)*. With deep and heavy breathing, the accessory muscles (the external intercostals and sternocleidomastoid) contract and pull the ribcage upward and outward. During resting conditions, expiration is passive. The diaphragm relaxes and returns to the dome shaped position, and the ribcage is lowered *(b, lower panel)*. During forced expiration, however, the intercostal muscles contract and pull the ribcage downward and inward. The abdominal muscles also contract and help pull the ribcage downward, compressing the thoracic volume.

but with forced inspiration, a total excursion of 10 to 12 cm can occur. Obesity, pregnancy, and tight clothing around the abdominal wall can impede the effectiveness of the diaphragm in enlarging the thoracic cavity. Two phrenic nerves, one to each lateral half, innervate the diaphragm, and damage to the phrenic nerves can also lead to paralysis of the diaphragm. When one of the phrenic nerves is damaged, that portion of the diaphragm does not move down during inspiration.

During forced inspiration, **accessory muscles** are used to bring in large volumes of air into the lung. These include the **external intercostal** muscles that are used in addition to the diaphragm (Fig. 20–6a). The intercostal nerves innervate them, and their contraction raises the anterior end of the ribcage, causing the ribcage to be pulled upward and outward. To see how this works, stand sideways in front of a mirror and

take a deep breath. The chest wall moves upward while the sternum moves outward, thereby enlarging the thoracic cavity. The last group of accessory muscles, which also become active with forced inspiration, are the **sternocleidomastoids**, which are inserted into the top of the sternum. These accessory muscles are brought into play during deep and heavy breathing, such as during exhaustive exercise, and are used to elevate the upper ribcage to further increase the thorax volume.

At rest, breathing out is a much simpler process than breathing in. At the end of a normal inspiration, the diaphragm relaxes and the ribcage drops. The lowering of the ribcage causes the lungs to deflate. Thus, during normal breathing at rest, expiration is purely passive. However, with exercise or forced expiration, the expiratory muscles become active. These muscles include those of the **abdominal wall** and the **internal intercostal muscles** (Fig. 20–6b). Contraction of the abdominal wall pushes the diaphragm upward into the chest and the internal intercostal muscles pull the ribcage down, thereby reducing thoracic volume. These accessory respiratory muscles are important and necessary for running but also for such functions as coughing, straining, vomiting, and defecating. The expiratory muscles are extremely important in endurance races and are one of the reasons that competitive long-distance runners, as a part of their training program, often do exercises to strengthen their abdominal and chest muscles.

Relative Pressures Are Used To Express Pressure Changes in the Lungs

In Chapter 18, which discusses the cardiovascular system, the units for pressure were millimeters of mercury (mm Hg). When discussing pressure of breathing, both mm Hg and centimeters of water (cm H_2O) are used. The unit *cm H_2O* is used because of the small pressure changes that are involved in breathing. Remember that a pressure of 1 cm H_2O is equal to 0.74 mm Hg (or 1 mm Hg = 1.36 cm H_2O). The atmosphere, the air we breathe and live in, exerts a pressure *(P)* known as **barometric pressure** (P_B). At sea level, P_B is equal to 760 mm Hg (Fig. 20–7). The total pressure or baro-

metric pressure is the sum of individual **partial pressures** of the gases in the atmosphere. The relationship between the total pressure exerted by a mixture of gases and the pressure of individual gases is governed by **Dalton's law,** which states that the total barometric pressure (P_B) is equal to the sum of the partial pressures of the individual gases:

$$P_B = PN_2 + PO_2 + PH_2O + PCO_2,$$

where PN_2 represents nitrogen and where PO_2, PH_2O, and PCO_2 are the partial pressures of oxygen, water vapor, and carbon dioxide, respectively. The partial pressures are the pressures that the individual gases would exert if each gas were present alone in the volume occupied by the whole mixture at the same temperature. Thus, the partial pressure of oxygen (PO_2), according to Dalton's law, is determined as:

$$PO_2 = P_B \times FO_2$$

where FO_2 is the fractional concentration of oxygen. Because 21% of air is made up of oxygen, then 0.21 is the fractional concentration of 760 mm Hg of air pressure at sea level that is accounted for by oxygen. In this example, the partial pressure (PO_2) exerted by oxygen is 160 mm Hg (760 × 0.21). If all of the other gases in a container of air were removed, the remaining oxygen by itself would still exert a pressure of 160 mm Hg at sea level. Partial pressure of a gas is often referred to as **gas tension,** and *partial pressure* and *gas tension* are used synonymously. When air is inspired, it is warmed and humidified. The inspired air becomes saturated with water vapor at 37°C. The water vapor also exerts a partial pressure and is a function of body temperature and not barometric pressure. At 37°C, water vapor exerts a partial pressure (PH_2O) of 47 mm Hg. Water vapor pressure does not change the percent of oxygen or nitrogen in the dry gas mixture; however, water vapor does decrease the partial pressure of oxygen inside of the lung. The partial pressures of gases in the lung are calculated based on a dry gas pressure. Thus, water vapor pressure is subtracted when the partial pressure of a gas is determined. The dry gas pressure in the trachea is 760 − 47 = 713 mm Hg, and the individual partial pressures of O_2 and N_2 are:

$$PO_2 = 0.21 \times (760 - 47) = 150 \text{ mm Hg}$$

and

$$PN_2 = 0.79 \times (760 - 47) = 563 \text{ mm Hg}$$

Changes in respiratory pressures during breathing are often expressed as a **relative pressure,** which is defined as a pressure relative to atmospheric pressure. For example, during inspiration, the pressure inside of the alveoli can be −2 cm H_2O. The negative sign indicates the pressure is 2 cm H_2O *less than* atmospheric pressure. Conversely, during expiration, the pressure inside of the alveoli can be 3 cm H_2O. This

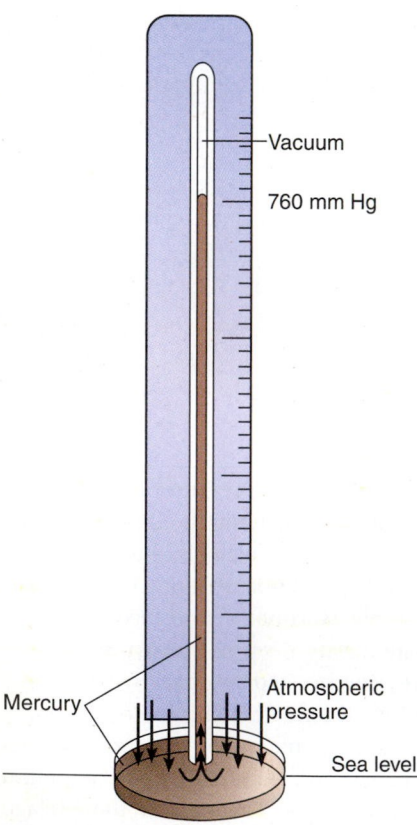

Vacuum

760 mm Hg

Mercury

Atmospheric pressure

Sea level

Figure 20–7

At sea level, atmospheric pressure can push a column of mercury (Hg) to a height of 76 cm. This is also known as 1 atmosphere (atm) of pressure (760 mm Hg).

TABLE 20–2

Pressure Changes During Breathing

	Pleural Pressure (cm H$_2$O)	Alveolar Pressure (cm H$_2$O)
Beginning of inspiration	−5	0
Mid-inspiration	−7	−1
End inspiration	−8	0
Mid-expiration	−6	+1
End expiration	−5	0

means the pressure is 3 cm H$_2$O *more than* atmospheric pressure. Remember a *positive* or *negative* pressure indicates what the pressure is relative to atmospheric pressure and is either more or less than P_B. When relative pressures are used in respiration, atmospheric pressure is set at zero. For example, if airway pressure is zero, pressure inside of the airway equals atmospheric pressure. Unless otherwise specified, the pressures of breathing are relative pressures and the units are cm H$_2$O. An example of the pressure changes during a breathing cycle at rest is shown in Table 20–2.

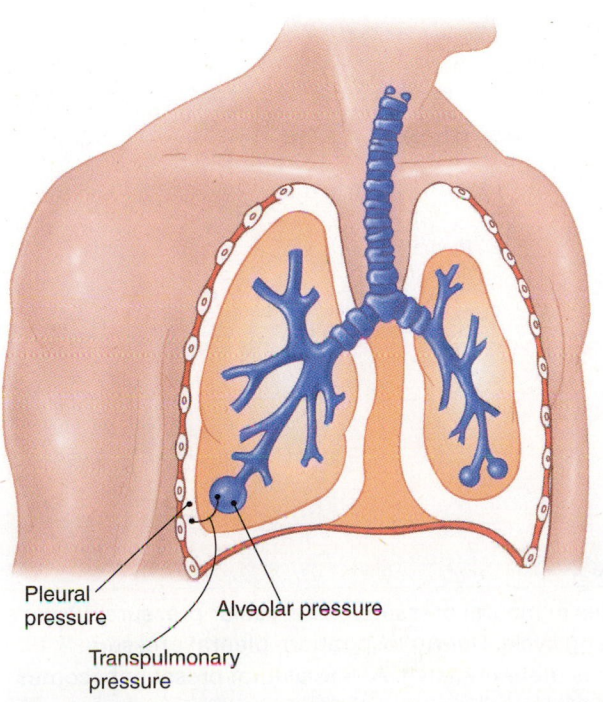

Pleural pressure

Alveolar pressure

Transpulmonary pressure

Figure 20–8

The basic pressures involved in breathing are transpulmonary pressure, pleural pressure, and alveolar pressure.

Several important pressures are associated with breathing and airflow (Fig. 20–8). The pleural pressure (P_{pl}) is the pressure in the pleural fluid and is the pressure between the lung and chest wall. The **alveolar pressure** (P_A) is the pressure inside of the alveoli; and **transmural pressure** (P_{tm}) is the pressure difference across an airway or across the lung. The transmural pressure is defined as the pressure *inside* minus the pressure *outside*. Two major types of transmural pressures are involved in breathing. First is **transpulmonary pressure** (P_{TP}), which is the pressure difference across the lung (see Fig. 20–8). Transpulmonary pressure is measured by subtracting pleural pressure from alveolar pressure:

$$P_T = P_A - P_{pl}$$

Transpulmonary pressure, seen at the beginning of inspiration and calculated from the values seen in Table 20–2, would be:

$$+5 \text{ cm H}_2\text{O } [0 - (-5) = +5 \text{ cm H}_2\text{O}]$$

Transpulmonary pressure is always positive and is the pressure that keeps the lungs inflated and also prevents the lungs from collapsing. Transpulmonary pressure is sometimes referred to as the **distending pressure** because it is this pressure that keeps the lungs inflated. The second transmural pressure important in the mechanics of breathing is **transairway pressure** (P_{ta}), which is the pressure difference across the airways and is the pressure difference inside of the airways (P_{aw}) minus the pressure in the pleural fluid:

$$P_{ta} = P_{aw} - P_{pl}$$

Transairway pressure is very important in keeping the airways open during forced expiration. One way to remember how to calculate transairway or transpulmonary pressure is "*In* minus *Out,*" where P_{pl} is the pressure outside of the lung or airway.

Why is pleural pressure negative or subatmospheric (see Table 20–2)? The reason is that the lung and chest wall are both **distensible** (i.e., capable of being stretched). At the end of expiration, the lung and chest wall are stretched in equal but opposite directions (Fig. 20–9, *left side*). The stretched lung has the potential to recoil inward, and the stretched chest wall has the potential to recoil outward. These equal but opposing forces cause the pleural pressure to decrease to less than atmospheric pressure. Pleural pressure is negative during quiet breathing and becomes more negative with deep inspiration. Only during forced expiration does pleural pressure become positive. The importance of pleural pressure is seen when the chest wall is punctured (Fig. 20–9, *right side*). The stretched lungs collapse immediately (recoil inward), and the ribcage simultaneously expands outward (recoil outward). Because the normal pleural pressure is subatmospheric, any time the chest wall or lung is punctured, air rushes into the pleural cavity because air moves from regions of high to low pressure. In this situation, transpulmonary pressure is zero ($P_{TP} = 0$) because the pressure difference across the lung is eliminated. As a result of this, the stretched lung collapses, a condition known as

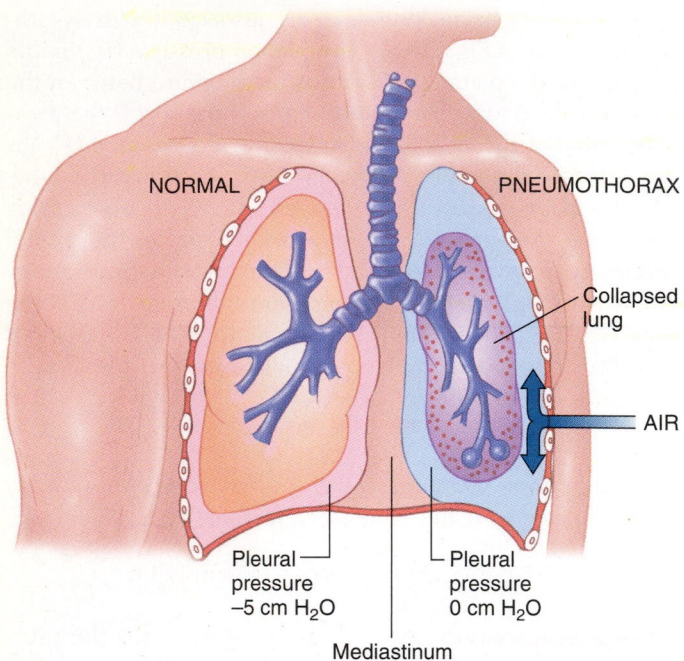

NORMAL   PNEUMOTHORAX

Collapsed lung

AIR

Pleural pressure −5 cm H₂O

Pleural pressure 0 cm H₂O

Mediastinum

Figure 20–9

Elastic recoil of the chest wall and lungs create a negative pleural pressure. Because of the elastic recoil property, the stretched lung (at the end of a normal expiration) tends to pull inward and chest wall tends to pull outward, but in *equal* and *opposite* directions *(depicted on left side)*. Consequently, pleural pressure (P_{pl}) becomes negative (i.e., less than atmospheric pressure). Rupture or a puncture of the lung or chest wall results in pneumothorax *(depicted on the right side)*. During a pneumothorax, transpulmonary pressure becomes zero and the lungs no longer stay distended because the lung elastic recoil causes them to collapse. Note that the mediastinal membrane prevents the other lung from collapsing.

a **pneumothorax** (see Fig. 20–9). A pneumothorax occurs with a knife or gunshot wound in which the chest wall is punctured, or when the lung ruptures from an abscess or severe coughing. In the treatment of some lung disorders (e.g., tuberculosis), a pneumothorax is performed surgically by inserting a sterile needle between the ribs and injecting nitrogen into the pleural fluid to rest the diseased lung.

Changes in Alveolar Pressure Move Air into and out of the Lungs

 How does the system cause gas to move in and out of the lungs?

Pressure changes during a normal breathing cycle are illustrated in Figure 20–10. At the end of expiration, respiratory muscles are relaxed and no airflow occurs. At this point,

alveolar pressure is zero (equal to atmospheric pressure or barometric pressure). Pleural pressure is −5 cm H₂O; transpulmonary pressure is therefore +5 cm H₂O. Remember, the more positive transpulmonary pressure becomes, the more the lung inflates.

Inflation of the lungs is initiated by contraction of the diaphragm. If inspiration is started from the end of a maximal expiration, the chest wall can be felt to expand as we inhale. At no time, as our lungs fill, do we feel the need to close our epiglottis to keep the air in. This is because air is held in our lungs by only a slight pressure difference in pleural pressure. In the example shown in Figure 20–10, pleural pressure goes from −5 to −8 cm H₂O. One of the basic characteristics of fluids is that the pressures between two regions tend to equilibrate. So, as the pressure in the pleural fluid decreases, transpulmonary pressure increases, and the lungs inflate. Inflation of the lungs causes the alveolar diameter to increase, and alveolar pressure (P_A), therefore, decreases to less than atmospheric pressure (see Fig. 20–10). This produces a pres-

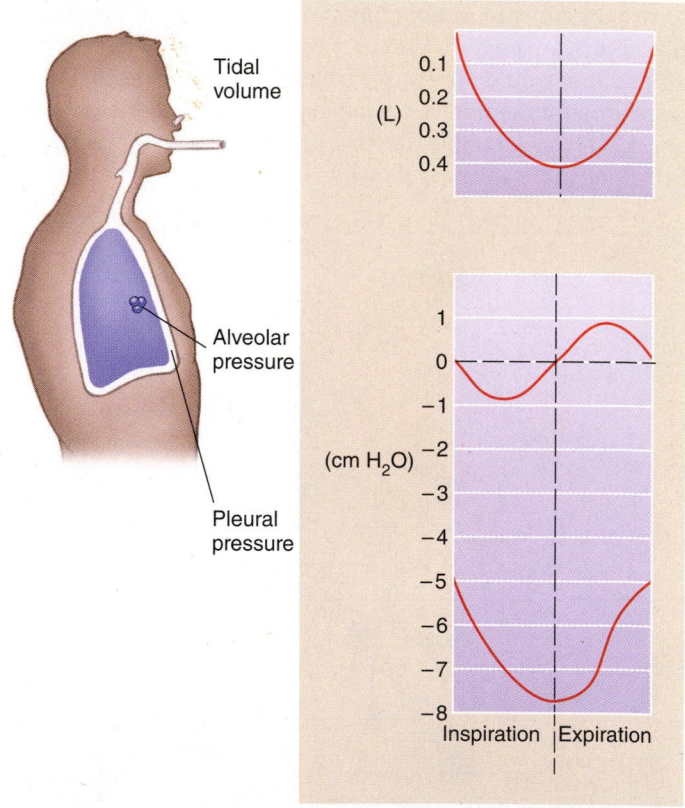

Tidal volume

(L)

0.1
0.2
0.3
0.4

Alveolar pressure

Pleural pressure

(cm H₂O)

1
0
−1
−2
−3
−4
−5
−6
−7
−8

Inspiration | Expiration

Figure 20–10

Changes in pleural pressure and alveolar pressure during a breathing cycle. During inspiration, pleural pressure becomes more negative. As the pleural pressure becomes more negative, alveoli are distended, causing a lower alveolar pressure. As a result of the pressure difference between the mouth and alveoli, air enters the lung. The amount of air entering the lung during inspiration is the tidal volume.

TABLE 20–3

Sequence of Events of Moving Air in and out of the Lungs

Inspiration	Expiration
1. Respiratory muscles contract	1. Respiratory muscles relax
2. Thoracic cavity expands	2. Thoracic cavity decreases
3. Decrease in pleural pressure	3. Pleural pressure become less negative
4. Transpulmonary pressure increases	4. Transpulmonary pressure decreases
5. Lungs inflate	5. Lungs deflate
6. Alveolar pressure < atmospheric pressure	6. Alveolar pressure > atmospheric pressure
7. Air flows into the lungs	7. Air flows out of the lungs

sure difference between the mouth and alveoli ($P_B - P_A$), which causes air to rush into the alveoli. Airflow stops at the end of inspiration because alveolar pressure again equals atmospheric pressure. The sequence of events in inspiration is shown in Table 20–3. A change in pressure between the alveoli and the mouth (ΔP) determines airflow; the greater the ΔP, the greater the airflow. When $\Delta P = 0$, then flow = 0.

During expiration, the inspiratory muscles relax, the ribcage drops, pleural pressure becomes less negative, transpulmonary pressure decreases, and the inflated lungs deflate. When alveolar diameter decreases during deflation, alveolar pressure becomes greater than the atmospheric pressure and pushes air out of the lung. Airflow out of the lungs occurs until alveolar pressure equals atmospheric pressure. The sequence of events in expiration during quiet breathing is shown Table 20–3.

Typical pleural and alveolar pressures during a normal breathing cycle are shown in Table 20–2. Note from the values presented in Table 20–2 and Figure 20–10 that only a small pressure gradient is required for a typical breath of air (500 ml).

SPIROMETRY AND LUNG VOLUMES

Spirometry Is Used To Measure Lung Volumes and Air flow

> *What vocabulary is used to refer to different lung volumes?*

The volume of air that is breathed into or out of the lungs is measured by using a **spirometer** (Fig. 20–11). This device is a simple volume recorder consisting of a double-walled cylin-

der in which an inverted bell is immersed in water to form a seal. A pulley attaches the bell to a pen that writes on a rotating drum. When air enters the spirometer from the lungs, the bell rises. Because of the pulley arrangement, the pen is lowered. Thus, a downward pen deflection represents expiration, and an upward pen deflection represents inspiration. The recording is known as a **spirogram.** The slope of the spirogram measures the rate of airflow and the amplitude of the pen deflection measures the volume of air. Volume is plotted on the vertical axis (ordinate, or y-axis) and time on the horizontal axis (abscissa, or x-axis).

The volume of air leaving the lungs during a single breath is called **tidal volume** (V_T). Under resting conditions, V_T is approximately 500 ml in the average adult and represents only a fraction of the air in the lungs. The maximum amount of air in the lungs is **total lung capacity (TLC)** and is approximately 6 liters in a young adult. Another important spirometry measurement is **functional residual capacity (FRC),** defined as the volume of air remaining in the lungs at the end of a normal expiration. The FRC is approximately 2.4 liters in a normal adult. Note the use of *volume* in the first term and *capacity* in the next two. When referring to the subdivisions of the lung, the terms *volume* and *capacity* are both used. **Capacity** is used when a volume can be broken down into two or more smaller volumes; for example, FRC is made up of two lung volumes: **expiratory reserve volume (ERV)** + **residual volume (RV).** The latter lung volume is the volume of air remaining in the lungs following a maximal expiration. The lung volumes and capacities are summarized in Table 20–4.

Forced Vital Capacity Is Used To Test Lung Function

The maximum amount of air that can be exhaled is **vital capacity (VC).** As seen from Figure 20–11, VC is a summation of expiratory reserve volume, tidal volume, and inspiratory reserve volume. However, when exhalation is performed forcefully and rapidly into a spirometer after maximal inspiration, the measurement is termed **forced vital capacity (FVC).** Thus, FVC is determined directly from spirometery during a maximal forced expiration (Fig. 20–12). Forced vital capacity is approximately 5 liters in the normal adult. However, a well-trained athlete can have a FVC of more than 7.0 liters, whereas an asthmatic or an emphysema patient may have a FVC no greater than 3 liters. The small airways of these patients collapse toward the end of forced expiration and results in trapped air in the lungs. When trapped air occurs in the lungs, this leads to a decrease in FVC and an increase in RV. A similar problem exists with individuals who smoke or have **bronchitis,** a condition in which the lining of the bronchioles becomes inflamed, and swollen, which reduces airway diameter.

One of the most useful tests to assess the overall lung function is FVC, which is determined primarily by four factors: (1) strength of chest and abdominal muscles, (2) airway

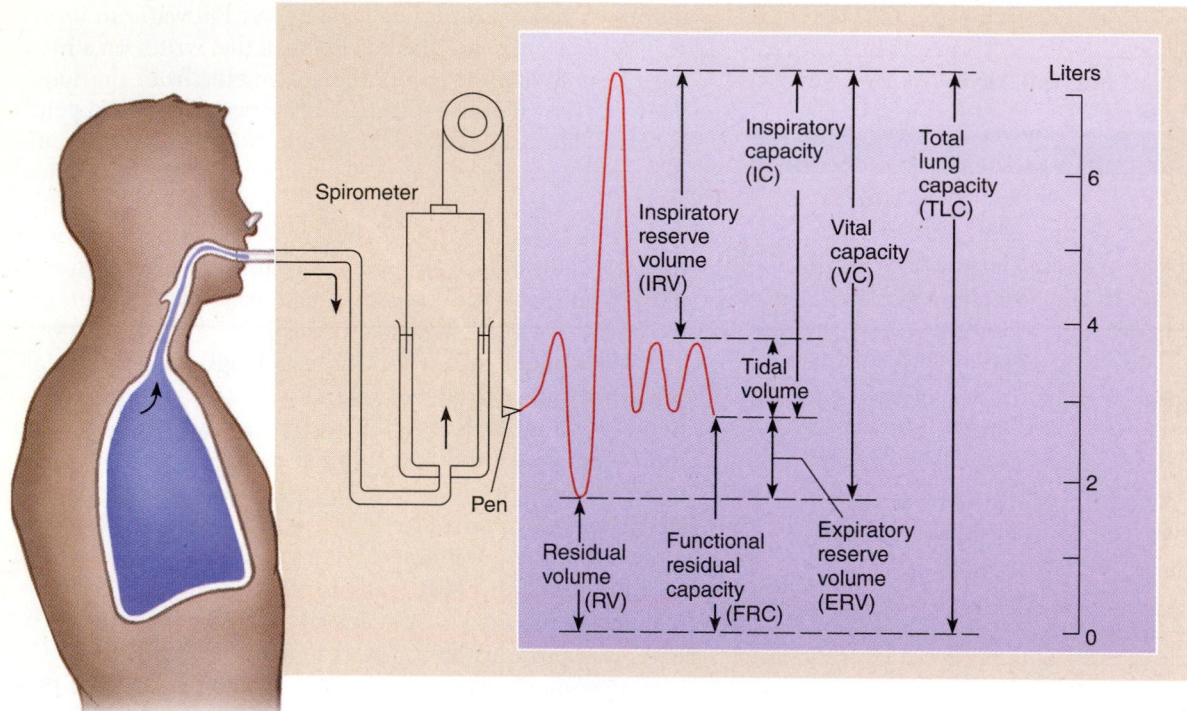

Figure 20–11

Lung volumes are measured by a spirometer. A cross-section of the spirometer is shown on the left. With inspiration, the pen shows an upward deflection on the spirogram, and with expiration, a downward deflection. On the right, a spirogram illustrating the different lung volumes is depicted. Note that the functional residual capacity (FRC), residual volume (RV), and total lung capacity (TLC) cannot be measured directly with a spirometer.

resistance, (3) lung size, and (4) elastic properties of the lung. Any condition that decreases the strength of the respiratory muscles (e.g., poliomyelitis), decreases lung volume (e.g., tuberculosis), increases airway resistance (e.g., asthma or bronchitis), or conditions that make the lung stiffer will lead to a significant decrease in FVC.

From FVC, another important determination that can be obtained is the forced volume exhaled in 1 second. This volume is termed **forced expired volume in 1 second (FEV$_1$),** has one of the least variability of the measurements obtained from a forced expiratory maneuver, and is considered one of the most reliable measurements (see Fig. 20–12). Another useful way of using FEV$_1$ is as a percentage of FVC (i.e., FEV$_1$/FVC $\times$ 100), which corrects for differences in lung size. Normally, FEV$_1$/FVC ratio is at least 0.8, which means that 80% of an individual's FVC can be exhaled in the first second. FVC and FEV$_1$ are important measurements in the diagnosis of certain types of lung diseases. Lung diseases are often classified as either obstructive or restrictive disorders. With an **obstructive** disorder, expiratory flow is obstructed. In obstructive lung disorders, FVC and FEV$_1$, as well as the ratio of FEV$_1$/FVC, are drastically reduced below normal. Therefore, FEV$_1$/FVC ratio < 0.8. Examples of ob-

structive disorders include asthma and emphysema. In **restrictive** disorders, lung inflation is restricted, and both FVC and FEV$_1$ are reduced. However, the ratio of FEV$_1$/FVC is 80% or slightly higher than normal. Thus, the ratio of FEV$_1$/FVC is used to differentiate between obstructive and restrictive lung disease. Examples of restrictive disorders include pulmonary edema and pulmonary fibrosis.

To summarize, the measurement of FVC is most commonly and easily obtained from a forced expiratory maneuver. From a FVC determination, we can readily measure FEV$_1$ and the FEV$_1$/FVC ratio. These measurements, by themselves, provide a considerable amount of information about ventilatory function during normal and abnormal conditions.

Not All Lung Volumes Can Be Measured Directly by Spirometry

Because the lungs cannot be emptied completely following forced expiration, neither residual volume nor functional residual capacity can be measured directly by spirometery. These two volumes are measured indirectly by a dilution technique shown in Figure 20–13. The technique used to measure RV and FRC involves the dilution of helium, an

TABLE 20–4

Definitions of Standard Lung Volumes, Capacities, and Pulmonary Function Tests for a 70-kg Adult

Term	Value	Definition
V_T (tidal volume)	0.5 liter	Volume of exhaled air with each breath
VC (vital capacity)	4.8 liters	Volume of air that can be exhaled after maximal inspiration (VC = IRV + V_T + ERV)
RV (residual volume)	1.2 liters	Volume of air remaining in the lungs after maximal expiration
ERV (expiratory reserve volume)	1.2 liters	Maximal volume of air that can be exhaled at the end of a tidal volume
IRV (inspiratory reserve volume)	3.1 liters	Volume of air that can be inhaled at the end of a normal inspiration
FVC (forced vital capacity)	4.8 liters	Volume of air that can be exhaled as forcibly and rapidly as possible from total lung capacity
IC (inspiratory capacity)	3.6 liters	Maximal volume of air that can be inhaled after a normal expiration (IC = IRV + V_T)
FRC (functional residual capacity)	2.4 liters	Volume of air remaining in the lungs at the end of a normal expiration (FRC = ERV + RV)
TLC (total lung capacity)	6.0 liters	Total volume of air in the lungs after maximal inspiration (TLC = VC + RV)
RV/TLC (residual volume/total lung capacity) × 100	20%	Percent of total lung capacity made up of residual volume
FEV_1 (Timed forced expiratory volume in 1 second)	3.8 liters	Volume of VC forcibly exhaled in 1 second
FEV_1/FVC (Timed forced expiratory volume/forced vital capacity) × 100	80%	Percent of VC that can be forcibly exhaled in 1 second

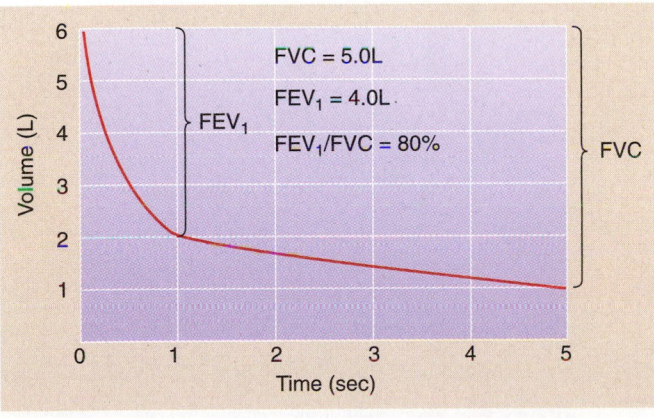

Figure 20–12

Spirogram of forced vital capacity (FVC). The subject inspires maximally to total lung capacity and then exhales forcefully and as completely as possible into a spirometer. From forced vital capacity, another measurement can be obtained, i.e., the forced expiratory volume in 1 second (FEV_1). Both FVC and FEV_1 are measured in liters. These are often expressed as a ratio (FEV_1/FVC).

inert and insoluble gas. The subject is connected to a spirometer filled with 10% helium in air (see Fig. 20–13). Because the lung initially contains no helium, the helium concentration in the lungs becomes the same as in the spirometer after the subject rebreathes the helium-air mixture. Because helium is insoluble and not taken up by the blood, the concentration of helium in the lung before and after equilibration can be represented by the following relationship:

$$C_1 \times V_1 = C_2 \times V_2$$

where C_1 = the initial 10% concentration of helium in the spirometer, V_1 = the initial volume of helium-oxygen in the spirometer, C_2 = helium concentration after equilibration, and V_2 = unknown volume in the lungs. By conservation of mass,

$$C_1 V_1 = C_2 (V_1 + V_2)$$

and by rearranging:

$$V_2 = \frac{V_1(C_1 - C_2)}{C_2}$$

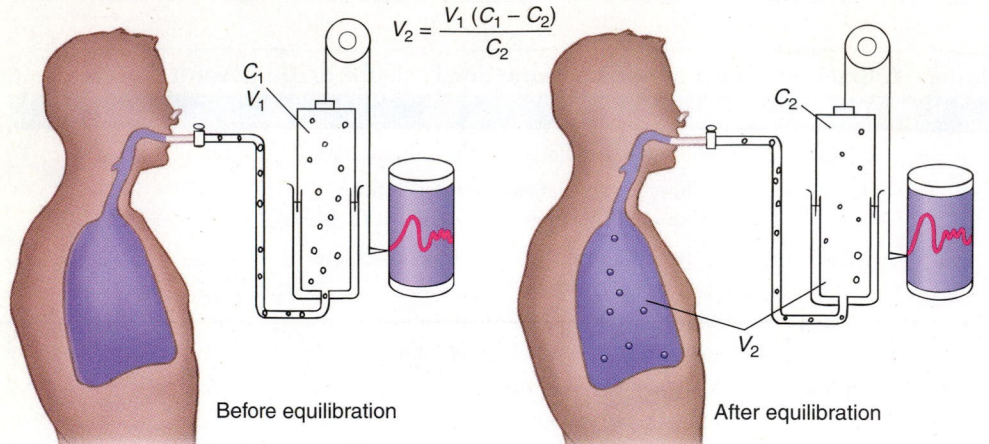

Starting the test at precisely the right time is important. If the test begins at the end of a normal tidal volume (end of expiration), the volume of air remaining in the lung represents FRC. If the test begins at the end of an FVC, then the test measures RV. Similarly, if the test starts after a maximal inspiration, then V_2 would equal total lung capacity. In practice, carbon dioxide is absorbed and oxygen is added to the spirometer to make up for the oxygen consumed by the subject during the test.

ventilation becomes wasted air. For each 500 ml tidal volume, approximately 150 ml remains in the conducting airways. This volume of wasted air is known as **dead space volume (V_D).** Picture, then, what occurs during a normal breathing cycle. A normal tidal volume of 500 ml is expired. During the next inspiration, another 500 ml is taken in, but the first 150 ml of air entering the alveoli is dead space (old alveolar gas left behind). Thus, only 350 ml of fresh air

ALVEOLAR AND MINUTE VENTILATION

Some Inspired Air Does Not Reach the Alveoli

 How is ventilation defined?

Ventilation is a dynamic process and involves the amount of air brought into and out of the lungs in a minute. If a person has a V_T of 500 ml and the breathing rate is 14 breaths/min, then the total volume of air that leaves the lungs each minute is $500 \times 14 = 7000$ ml/min, or 7 L/min.

This volume of air per minute is known as **minute ventilation ($\dot{V}$).** Minute ventilation is determined by measuring the amount of expired air that leaves the lungs per minute. Thus, minute ventilation is represented by:

$$\dot{V}_E = V_T \times f$$

where V_T = volume/breath and f = breaths/min.

Not all of the gas from minute ventilation participates in gas exchange. The tidal volume (V_T) is distributed between the conducting airways and alveoli (Fig. 20–14). Because gas exchange occurs only between the alveoli and the blood in the pulmonary capillaries and not in the conducting airways (trachea, bronchi, bronchioles), then a part of the minute

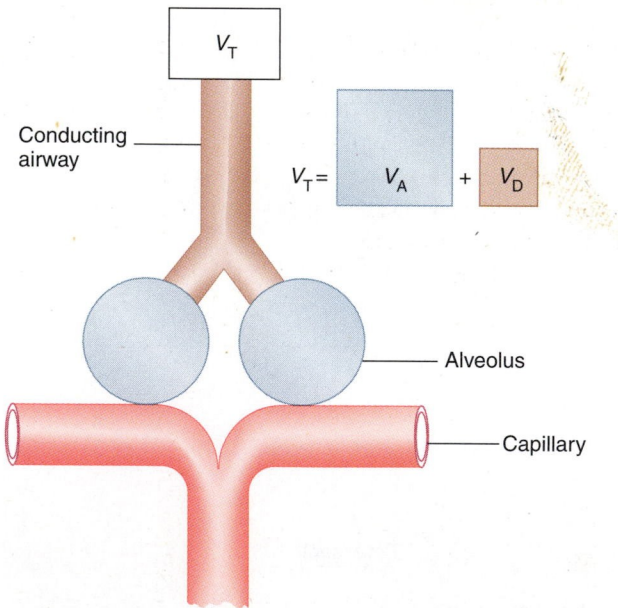

Figure 20–14

Tidal volume (V_T) is distributed between the conducting airways and alveoli. V_D is dead space volume in the conducting airways and constitutes anatomic dead space. V_A is the volume of alveolar air from a tidal volume, which participates in gas exchange. Note that V_A is not the total volume of air in the alveolar space.

APPLICATIONS OF PHYSIOLOGY

Hyperventilation Syndrome

Hyperventilation is a condition in which alveolar ventilation is in excess and results in a lowering of carbon dioxide in the blood. Hyperventilation causes too much carbon dioxide to be ventilated out of the lungs and results in a low arterial PCO_2 (hypocapnia), which leads to an increase in blood pH. Hyperventilation and hypocapnia are much more common than are hypoventilation and hypercapnia, even though much more attention has been devoted to the latter. A distinction must be made between the terms *hyperpnea* and *hyperventilation*. *Hyperpnea* indicates increased minute ventilation, but without a change in arterial PCO_2, whereas *hyperventilation* means increased alveolar ventilation with a concomitant decrease in arterial PCO_2. An example of hyperpnea is seen with the increased minute ventilation with exercise, in which arterial PCO_2 is not markedly changed.

Hyperventilation is always associated with a decreased arterial PCO_2. The far-reaching consequences of hyperventilation with the associated increased blood pH (alkalosis) are linked primarily to the pathophysiologic effects on the heart and brain. Hypocapnic alkalosis causes vasoconstriction in the coronary arteries, resulting in arrhythmias and cerebral ischemia as a result of vasoconstriction of the cerebral blood vessels. The transport of oxygen to all of the organs is also impaired because of the decreased cardiac output and the decreased availability of oxygen to the tissues. The latter is due to the increased alkalosis, which shifts the oxyhemoglobin equilibrium curve to the left and increases the oxygen binding affinity; resulting in less oxygen available to the tissues.

Voluntary hyperventilation is often followed by long periods of apnea (cessation of breathing). The apnea is due to the powerful inhibitory effect of the hypocapnic alkalosis on the respiratory control center and often results in severe consequences. Many deaths have occurred in underwater swimmers who hyperventilated before submerging in an attempt to prolong their time under water. Assisted ventilation (i.e., mechanical ventilation) often results in hyperventilation and is often a persistent problem with anesthesia and comatose patients. Maternal hyperventilation is also a persistent problem during pregnancy and usually starts three to four weeks after fertilization, lasts the duration of pregnancy, and continues approximately three weeks after delivery. The mechanism that mediates maternal hyperventilation is thought to be the effect of progestational hormones on the central nervous system.

Hypoventilation occurs with many diseases. In particular, these include cerebral disorders, such as strokes, meningitis, and brain tumors. Pulmonary diseases, such as asthma, pulmonary edema, and pulmonary embolism, often are complicated with hyperventilation and low arterial PCO_2 values. The common mechanism responsible in each of these conditions is stimulation of receptors, resulting in increased drive to breathe. Fever, sepsis, and severe liver disease also can cause hyperventilation. However, the underlying stimuli and responding respiratory centers are unknown.

Anxiety and other psychogenic influences can lead to acute or chronic hyperventilation. Unfortunately, there is a tendency to regard hyperventilation of unknown causes as psychogenic in origin. Consequently, some of the subtle diseases associated with hyperventilation are overlooked.

reaches the alveoli, and 150 ml is left in the conducting airways:

$$\text{tidal volume} = 500 \text{ ml}$$
$$\text{dead space volume} = -150 \text{ ml}$$
$$\text{fresh air entering alveoli} = 350 \text{ ml}$$

The normal ratio of dead space volume to tidal volume (V_D/V_T) is in the range of 0.20 to 0.35. In the example shown above, the ratio (150/500) is 0.30. This means that 30% of the air that is brought into the lungs does not participate in the gas exchange between alveoli and pulmonary capillary blood and represents dead space ventilation.

To determine how much fresh air reaches the alveoli per minute, dead space volume is subtracted from the tidal volume and the result is multiplied by breathing frequency. For example, if a subject has a breathing rate of 14 breaths per minute, a $V_T = 500$ ml, and a $V_D = 150$ ml, then the volume of air entering the alveoli 4900 ml/min [(500 − 150) × 14 = 4900 ml/min]. This represents the volume of "fresh air" entering the alveolar compartment for gas exchange each minute and is termed **alveolar ventilation ($\dot{V}_A$).**

Alveolar ventilation is the most important variable in gas exchange and can be represented as:

$$\dot{V}_A = (V_T - V_D) \times f$$

TABLE 20–5

Influence of Breathing Patterns on Alveolar Ventilation

Subject	Tidal Volume (ml)	×	Frequency (breaths/min)	=	Minute ventilation (ml/min)	−	Dead space ventilation (ml/min)	=	Alveolar ventilation (ml/min)
A	1000		6		6000		900 (150 × 6)	=	5100
B	500		12		6000		1800 (150 × 12)	=	4200
C	150		40		6000		6000 (150 × 40)	=	0

The reason that alveolar ventilation is so important is because the capillaries containing the blood that gas enters and leaves are from the alveoli. Minute ventilation does not represent the amount of fresh gas reaching the alveoli. For instance, a swimmer using a snorkel breathes through a tube that increases dead space volume. Similarly, a patient connected to a mechanical ventilator also has increased dead space volume. Indeed, if minute ventilation is held constant, then alveolar ventilation is decreased with snorkel breathing or with mechanical ventilation.

The significance of dead space volume, minute ventilation, and alveolar ventilation can readily be seen in Table 20–5. In this experiment, subject A's breathing is slow and deep, whereas subject B's breathing is normal and subject C's breathing is rapid and shallow. Note that each subject has the same minute ventilation (i.e., the total amount of expired gas per min is the same), but each has marked differences in alveolar ventilation. Subject C has no alveolar ventilation and would die in a matter of minutes, whereas subject A has alveolar ventilation greater than normal. The important message from the examples presented in Table 20–5 is that increasing the **depth** of breathing is far more effective in elevating alveolar ventilation than increasing the frequency or rate of breathing. This fact is important in exercise. In most exercise situations, increased ventilation is accomplished by increases in the depth of breathing more than in the rate. For example, a well-trained athlete can often increase alveolar ventilation during moderate exercise with little or no increase in breathing frequency.

ELASTIC AND MECHANICAL PROPERTIES OF THE LUNG

How are the mechanical properties of the lungs related to function?

The lung, airway, and vascular trees are embedded in elastic tissue. Thus, the degree of lung expansion at any given time in the breathing cycle is proportional to transpulmonary pressure. How well a lung inflates and deflates with a change in transpulmonary pressure is a measure of its **recoil property.**

An important feature of elastic material is that, once stretched, it recoils back to its unstretched position. Thus, the lung is like a spring; it recoils back when stretched.

Elastic Properties Affect Lung Inflation and Deflation

The ease with which the lung can be stretched or inflated is termed **distensibility.** Distensibility and elastic recoil are inversely related to each other. That is, the greater the distensibility, the less the elastic recoil. Thus, a lung that is easily inflated has elastic recoil. An analogy of this relationship is a coiled spring. Overstretching the lung also causes it to lose its elastic recoil. A simple analogy is that of nylon stockings, which can be easily stretched out but do not recoil to fit the contour of the legs and sags or become "baggy." In the same way, lungs that lose their elastic recoil also become "baggy." In other words, a "baggy lung" is easy to inflate but is very difficult to deflate because of its inability to recoil back.

Elastic properties of a lung are determined by measuring the changes in lung volume that occur with changes in pressure and plotting them as a **pressure-volume curve.** A simple analogy is the inflation of a balloon with a syringe (Fig. 20–15). For each change in pressure, the balloon inflates to a new volume. The slope of the line is known as **compliance** ($C_L = \Delta$volume/Δpressure). Compliance is a measure of distensibility. A similar pressure-volume curve can be generated for the human lung by measuring a simultaneous change in lung volume with a spirometer and a change in pleural pressure with a pressure gauge (Fig. 20–16). In practice, a pressure-volume curve is produced by first having the subject inspire maximally to total lung capacity and then expire slowly while lung volume and pleural pressure measurements are recorded. A steep slope (high compliance) indicates that the lung is very distensible and is easily inflated. In other words, with the same transpulmonary pressure, the lung with the higher compliance inflates to a higher volume because it is more distensible. Lung compliance indirectly measures another elastic property, lung elastic recoil. A lung with a low compliance (decreased slope) is not as distensible, but has a greater elastic recoil. Remember that distensibility and elas-

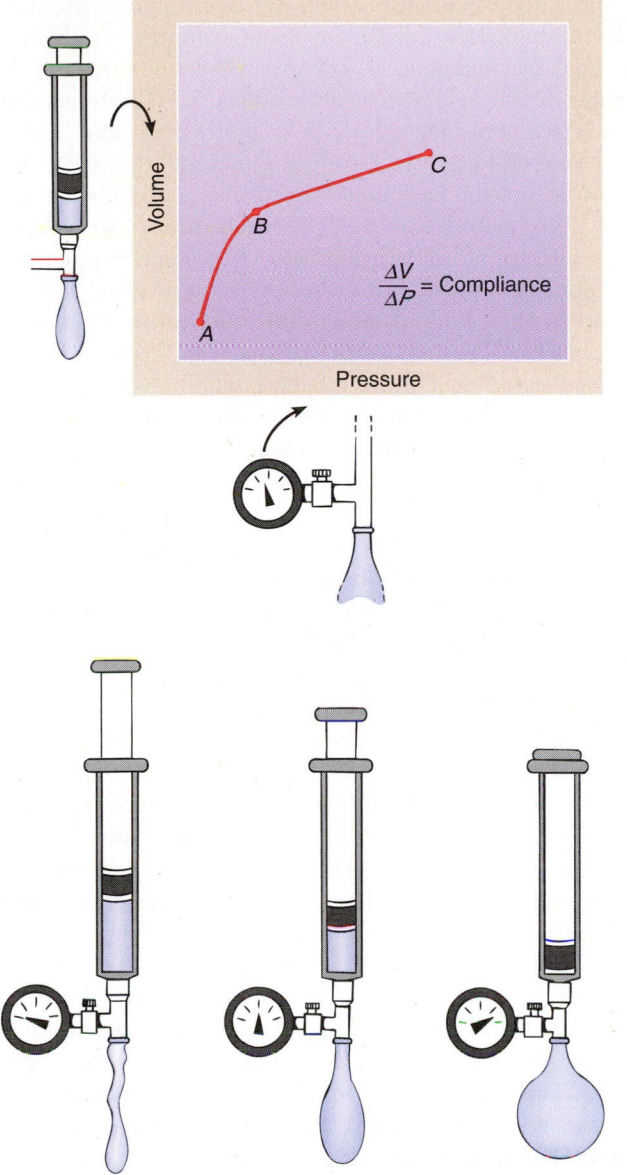

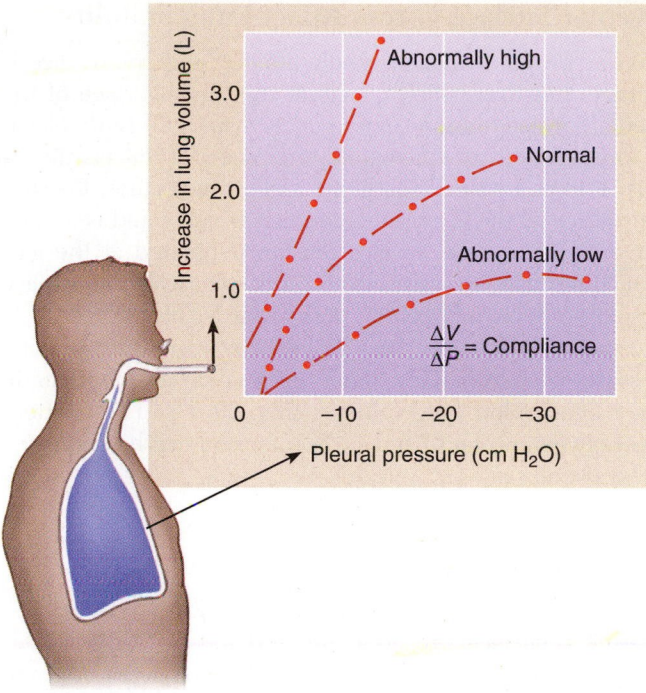

Figure 20–16

Lung compliance (C_L) is determined from a pressure-volume curve. The subject first inspires maximally to total lung capacity (TLC) and then expires slowly, while airflow is periodically stopped for the simultaneous measurement of pleural pressure and lung volume. The unit of measure for lung compliance is L/cm H_2O. A lung with an abnormally high compliance would have a slope that is greater than normal. Conversely, a lung with an abnormally low compliance would have a slope that is less than normal. Patients with emphysema exhibit abnormally high lung compliance, and patients with fibrosis (stiff lungs) exhibit abnormally low lung compliance.

Figure 20–15

A pressure-volume curve is used to assess the elastic properties of the lungs (distensibility, stiffness, and elastic recoil). A simple analogy is inflating a balloon and measuring the change in pressure and volume. For each change in pressure (shown by movement of the arrow in the manometer dial), the balloon inflates to a new volume. *Compliance* is defined as the slope of the line between any two points on a pressure-volume curve.

tic recoil are inversely related (i.e., a highly compliant lung has less recoil):

$$\text{Lung compliance} = 1/\text{lung recoil}$$

What is the significance of having an abnormally high or abnormally low compliance? Low compliance means that more work is required to inflate the lungs in order to bring in a normal tidal volume. Any time the lung becomes injured (by infection or by toxic environmental insults), the elastic properties are often altered, and the lung loses its distensibility (becomes stiffer). A very high compliance, however, is just as detrimental as a low compliance. Consider the disease **emphysema,** which is an obstructive disorder, with lungs being overstretched from chronic coughing and congested airways. Lungs of patients with emphysema (which is strongly linked with smoking) have a high compliance (high distensibility) and are extremely easy to inflate. However, getting air out again is another matter! Lungs with abnormally high compliance have low elastic recoil. Thus, a lung affected by emphysema is easily distended but does not recoil during expiration. Consequently, a lot of effort is required to get air out of the lungs. Thus, lungs with emphysema retain an abnormally high residual volume of air.

Alveolar Surface Forces Affect Lung Stability

Another property that markedly affects lung compliance is **surface tension,** which occurs at the inner surface of the alveoli. Surface tension (measured in dynes/cm) is a molecular force created at a gas-liquid interface and reflects the attractive force between the liquid and air molecules. Because the surface of the alveolar membrane is moist and is in contact with air, a large surface tension is created at the gas-liquid interface. Because alveoli are spherical, the surface tension produces a force that pulls inward. This tension tends to cause alveoli to become unstable by increasing alveolar pressure proportionately more in smaller alveoli than in larger ones at low lung volumes (Fig. 20–17). The relationship between pressure and surface tension can be expressed in the alveolus as:

$$P = \frac{2T}{r}$$

where T = surface tension (dynes/cm) and r = radius (mm). As seen in Figure 20–17, if surface tension were constant (e.g., 50 dynes/cm), the pressure in the smaller alveoli would be greater. Because alveoli are connected in parallel, alveoli

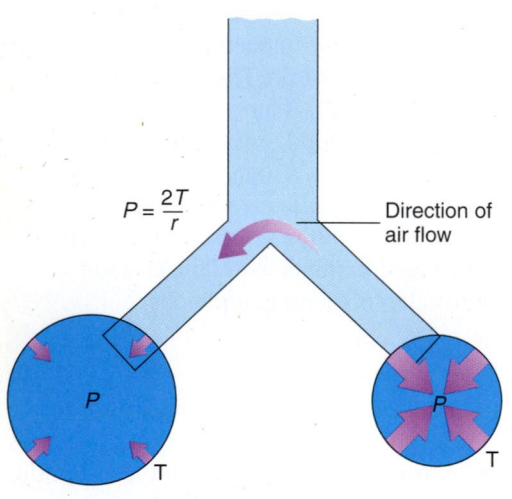

$$P = \frac{2T}{r}$$

Direction of air flow

Figure 20–17

Surface tension alters alveoli stability. Alveoli that are interconnected and vary in diameter cannot coexist if surface tension remains constant. If alveolar surface tension remained constant, smaller alveoli would generate a greater pressure and cause air to flow into larger units. At low lung volumes, these smaller alveoli become unstable and collapse because of the higher surface tension. Surfactant prevents this from occurring. Surfactant promotes alveolar stability by lowering surface tension proportionately more in the small alveoli. The relationship between alveolar pressure and surface tension is expressed as P = 2T/r, where *P* = pressure, *T* = surface tension, and *r* = radius.

with the larger diameter would be overinflated and the smaller alveoli, especially at low lung volumes, would become unstable and tend to collapse. The phenomenon of collapsing alveoli is known as **atelectasis.** To counteract the high surface tension, specialized epithelial cells lining the alveoli secrete a surface-reducing material that coats the inner surface of the alveoli and drastically decreases surface tension. This material is known as **pulmonary surfactant** and is a mixture of lipids and protein. The main component of pulmonary surfactant responsible for reducing surface tension is a phospholipid, **phosphatidylcholine.** With surfactant, surface tension varies with surface area and thus with alveolar radius. During inflation, when alveoli get bigger, surface tension increases. During deflation, pulmonary surfactant decreases surface tension, prevents atelectasis (alveolar collapse), and makes the lung more distensible (compliant).

Pulmonary surfactant is extremely important because the effects of surface tension on lung compliance are just as great as are the effects of elastic forces. A striking example of what happens when inadequate amounts of surfactant are present is seen with a disorder known as **respiratory distress syndrome.** This disorder frequently afflicts adults and premature newborns. During the gestation period, the lungs are one of the last organs to develop. Consequently, premature infants are often born with structurally intact alveoli, but with insufficient amounts of surfactant. In these newborns, breathing is extremely labored because high surface tension makes the lungs stiffer and results in breathing difficulties. As a result of high surface tension, these infants also have atelectasis. Each year in North America, more than 50,000 infants are afflicted with respiratory distress syndrome. These infants are at high risk until the lung becomes mature enough to secrete surfactant. Consequently, respiratory distress syndrome in newborns is termed **infant respiratory distress syndrome (IRDS)** and is still a major problem among newborns in North America. Respiratory distress syndrome also exists in the adult lung as well and often occurs with acute lung injury, in which surfactant is destroyed. Adult respiratory distress syndrome (ARDS) presents a similar pattern and also constitutes a major cause of death in adults.

AIRWAY RESISTANCE

When you take a breath of air into your lungs, you can hear the sound of air rushing in, especially if you breathe rapidly. The sound comes from gas molecules that literally collide with each other and the airway, resulting in a noise that can be heard on inhalation or exhalation. Thus, the faster you inhale or exhale gas, the greater the noise. The collision of gas molecules also leads to resistance of air movement in the lung.

CURRENT CONCEPTS IN PHYSIOLOGY

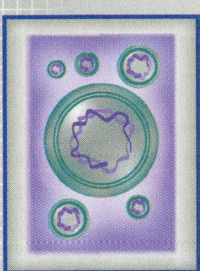

Pulmonary Fibrosis: Pathways for Pathogenesis Are Coming to Light

Jim is a 45-year-old professional piano player who complains of increasing shortness of breath and fatigue over the past three years. His shortness of breath is particularly noticeable during a performance or climbing stairs. At the same time, Jim began to feel unusually fatigued and experienced some weight loss. Upon examination, Jim had rapid, shallow breathing and was unable to make a large inspiration. A chest X-ray showed contracted lungs with a reticular pattern suggesting interstitial pulmonary fibrosis. This diagnosis was confirmed by a lung biopsy.

Pulmonary fibrosis is a condition that is associated with excessive collagen deposition (fibrosis) in lung tissue. In most cases, pulmonary fibrosis is characterized as diffuse fibrosis at interstitial sites that leads to progressive deposition of type I and type III collagen in the alveolar walls. As the alveoli become more fibrotic, there is a progressive decrease in lung compliance, with marked impair-

ment of gas exchange in the lungs caused by the thickened alveolar-capillary membrane, which results in hypoxemia and heart failure. The initiating mechanism seems to be an inflammatory response that results in an infiltration of cells into the alveoli not present in the normal lung. However, the etiologic factors are largely unknown. Patients with idiopathic fibrosis have a median survival of four to five years after onset of symptoms; idiopathic fibrosis represents a major cause of death in North America.

Some pathways that reveal the underlying mechanism of pulmonary fibrosis are beginning to emerge. Several laboratories have suggested that transforming growth factor (TGF)-β is important in pulmonary fibrosis, and treatment with interferon-γ decreases levels of this cytokine, which slows the progression of this disease. The specific source of TGF-β appears to be from three cell types in the lung, which include endothelial cells, epithelial cells, and alveolar macrophages. TGF-β is chemotactic for fibroblasts and neutrophils, which leads to pneumonitis and subsequent fibrosis. In addition, TGF-β is involved in programmed cell death (apoptosis) of alveolar epithelial cells. The potential pathways of the TGF-β induced pulmonary fibrosis are outlined below.

$$\text{Inflammation} \rightarrow \textbf{TGG-}\boldsymbol{\beta} \begin{cases} \rightarrow \textbf{Leukocyte influx} \rightarrow \textbf{Fibroblast proliferation} \rightarrow \textbf{Fibrosis} \\ \rightarrow \textbf{Epithelial cell apoptosis} \end{cases}$$

Airway Resistance Decreases Airflow in the Lung

Airway resistance is expressed as cm H_2O/L/sec and is defined as the ratio of driving pressure (ΔP) to airflow ($\dot{V}$). For total airway resistance (R_{aw}), the driving pressure is the pressure difference between the mouth (P_{mouth}) and the alveoli (P_A). The equation can be written as:

$$R_{aw} = \frac{\Delta P}{\dot{V}}$$

where ΔP = mouth pressure-alveolar pressure [$P_M - P_A$ (cm H_2O)], $\dot{V}$ = airflow (L/sec), and R_{aw} = airway resistance (cm H_2O/L/sec).

The major site of airway resistance is in the medium-sized airways (lobar and segmental bronchi) and in bronchi down to approximately the seventh generation (Fig. 20–18). Another important factor affecting airway resistance is airway diameter, especially narrowing of the bronchioles. Recall from Chapter 18 that resistance (R) to flow in small vessels is

inversely proportional to the radius raised to the fourth power:

$$R \approx \frac{1}{radius^4}$$

Therefore, if peripheral airway constricts such that the radius is halved, then airflow in that tube decreases 16-fold. Smaller airways are capable of compression or distension. Because bronchi and smaller airways are embedded in lung parenchyma, "guy wires" to surrounding tissue connect them. As the lung enlarges, airways expand and the diameter increases. As a result, airway resistance (R_{aw}) decreases during inspiration. Conversely, at low lung volumes, airways are compressed, and airway resistance increases during expiration, especially at lower lung volumes.

Bronchial smooth muscle tone also affects airway resistance. A change in smooth muscle tone changes airway diameter. The smooth muscle in the walls of the airway, from the trachea down to the terminal bronchioles, is under autonomic

APPLICATIONS OF PHYSIOLOGY

Adult Respiratory Distress Syndrome: The Case of Scott

Scott is a healthy, 20-year-old man who is starting his second year of college. He is in very good condition and works out regularly by lifting weights and swimming 1 mile each day. However, one fall afternoon, he began to feel ill. He went to the student health center with complaints of a severe headache and pain in the neck and lower abdomen. He was told that he was coming down with the flu and was advised to go back to the dorm and rest. The next day, his symptoms became worse and included fever, cold sweats, skin rash, and nausea. His roommate took him back to the health center, and Scott was placed in one of the student beds for closer observation. When his condition started deteriorating, he was rushed to the emergency room at the university medical center. Upon arrival, Scott was in shock, with a high fever and a breathing rate of 36 breaths/min (normal = 12 to 15 breaths/min). Because of the danger of respiratory failure, the attending physician immediately had Scott intubated and placed on assisted ventilation. The decision was made to transfer him to the intensive care unit (ICU). Upon arrival at the ICU, Scott's chest X-ray showed patchy infiltrate and pulmonary edema. He was diagnosed as having adult respiratory distress syndrome (ARDS), with a secondary widespread infection of unknown origin.

Scott's case is similar to the other 200,000 to 250,000 persons who develop ARDS each year in the United States. Despite Scott's youth and previous history, the anticipated mortality rate for ARDS is approximately 80%. Scott was lucky and was able to return to his classes after 12 weeks in the hospital and a bill of more than $200,000.

ARDS is the name given to diffuse lung injury, the causes of which vary widely. Few patients have histories of previous lung disorders. The causes include trauma from chest injury (e.g., car accidents), long bone injury, and pelvic injury. During the Vietnam era, the trauma that caused ARDS was coined *Da Nang lung*. Two other major causes include sepsis (presence of pathogen or toxin in the blood) and aspiration of gastric contents. The latter occurs with gastric reflux and usually occurs at night, during sleep. Other causes include cardiopulmonary bypass, smoke inhalation, high altitude, and exposure to irritant gases. Exposure to irritant gases was the cause of the widespread ARDS cases that occurred at the disaster in Bhophal, India.

Although the cause of ARDS varies, the pathophysiology is nearly identical in every case. ARDS is characterized by decreased lung compliance, pulmonary edema, focal atelectasis, and hypoxemia (low PO_2 in the arterial blood), with an inflammatory reaction that leads to an infiltration of neutrophils into the lung. The increase in lung stiffness (i.e., decreased compliance) is due to loss of surfactant, which is unrelated to edema.

Neutrophil aggregation is a key underlying mechanism of ARDS. Aggregation of neutrophils causes capillary endothelial damage by releasing a number of toxic products. These include oxygen free radicals, proteolytic enzymes, arachidonic acid metabolites (leukotrienes, thromboxane, prostaglandin), and platelet-activating factor.

New approaches to therapy involve ways to reduce neutrophil chemoattraction and aggregation in the pulmonary capillaries and ways to reduce the amount of toxic substances released by neutrophils.

---

control. Parasympathetic stimulation of the cholinergic postganglionic fibers causes bronchial constriction as well as increased mucus secretion. Sympathetic stimulation of adrenergic fibers causes dilation of bronchial and bronchiolar airways and inhibition of mucus secretion. Drugs such as isoproterenol and epinephrine, which stimulate β_2-adrenergic receptors in the airways, cause dilation. This relationship between airflow and radius in the small airways explains why a person who suffers from asthma (a condition caused by spasmodic contraction of the bronchiole smooth muscles) has great difficulty getting air into and out of the alveoli. During asthma attacks, many asthmatics take epinephrine to relax the smooth muscle and in turn dilate the airways.

Turbulent Airflow Increases Airway Resistance

Three types of airflow in the lung exist. One type, occurring in the trachea and large bronchi, is **turbulent flow** (Fig. 20–19a). Turbulent flow occurs at high flow rates and consists of completely disorganized patterns of airflow in which the molecules literally collide with each other, resulting in a noise that can be heard on inhalation or exhalation. This is the reason that the faster you inhale or exhale air, the more turbulent the noise.

An important consideration regarding turbulent flow is that turbulence in the large airways is the primary source of resistance to breathing, seen in Figure 20–18. Turbulent flow

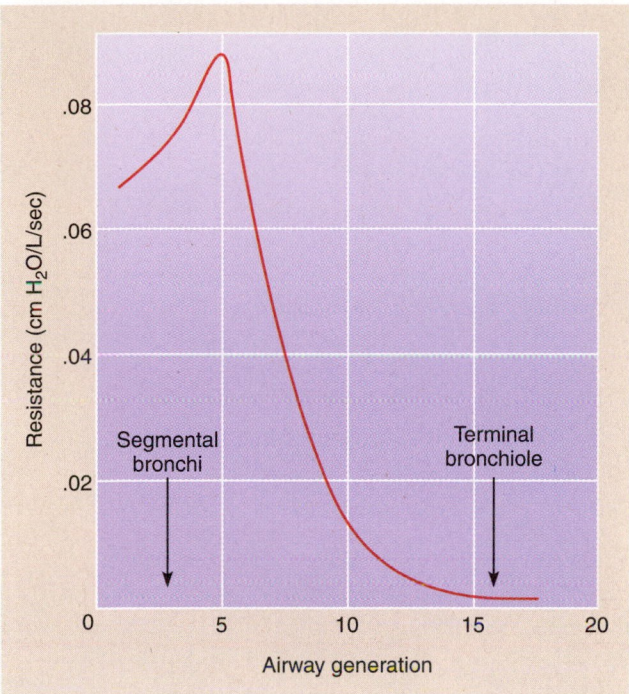

Figure 20–18

Site of airway resistance in the lung. Resistance to air flow is the greatest in the large and medium airways, namely the lobar and segmental bronchi.

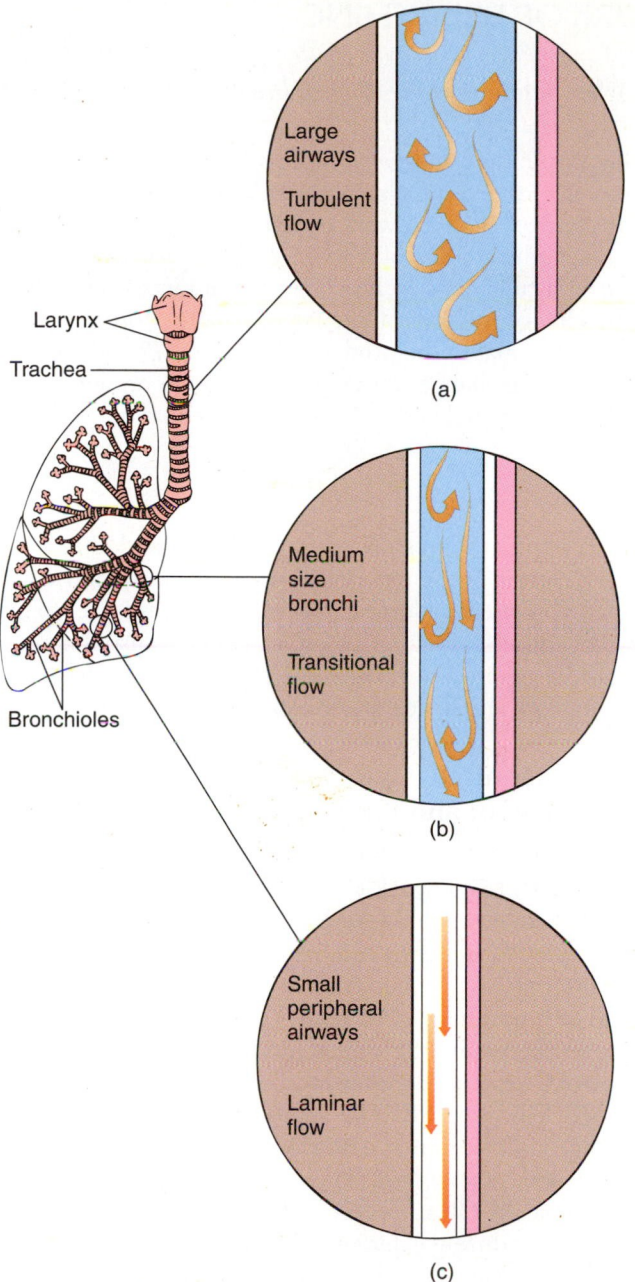

Figure 20–19

Patterns of air flow in the lung. Three types of airflow exist in the lung. Turbulent flow occurs at high flow rates in large airways *(a)*. Transitional flow is a combination of laminar and turbulent flow and occurs in medium-sized bronchi *(b)*. Laminar flow occurs at low flow rates in small peripheral airways *(c)*.

is also greatly affected by gas density. If a helium-oxygen mixture is breathed, which is much lighter than air, less turbulence is produced, meaning that less resistance and less work are involved in breathing. Remember, atmospheric pressure increases 1 atmosphere (1 atm) for every 33 feet below sea level. Thus, at 33 feet, the pressure is 2 atm, and at 66 feet the pressure is 3 atm. With the increased atmospheric pressure, the density of the gas becomes a major concern in breathing during a dive. Gas density also affects the vocal cords in the larynx by changing the pitch of our voices, making us sound like Donald Duck when we talk after breathing a helium-oxygen mixture.

Airflow in the Small Airways Is Silent

The second type of airflow in the lungs is **laminar airflow,** which is characterized by streamlined flow that parallels the side of the airways (Fig. 20–19c). Laminar flow is silent because air molecules slide over each other. This type of airflow occurs mainly in the small peripheral airways, where airflow is extremely slow. Laminar flow is greatly affected by airway diameter, and, as mentioned earlier, the resistance to flow is inversely proportional to the radius raised to the fourth power.

At lower flow rates during expiration — particularly at branches of the airway tree where flow from separate tubes comes together and empties into a single airway — the airflow pattern is a mix. This is referred to as **transitional flow,** a combination of turbulent and laminar flow (Fig. 20–19b). Transitional flow occurs most commonly at bifurcations, and high flow, in medium-sized airways.

WORK OF BREATHING

> *What determines the work of breathing?*

During inspiration, muscular work is involved in expanding the thoracic cavity, inflating the lungs, and overcoming airway resistance. Because work is defined as force multiplied by distance, the amount of work involved in breathing can be expressed as a change in lung volume (distance) multiplied by the change in transpulmonary pressure (force). Thus, work (*W*) is equal to the product of pressure (*P*) and volume (*V*). If pressure changes during work, then the product is replaced by an integral and is defined by the equation:

$$W = \int PdV$$

where *P* = pleural pressure, d = derivative, and *V* = lung volume. During work, energy is expended with muscular contraction to create a force (transpulmonary pressure) to inflate the lungs. Therefore, when a greater transpulmonary pressure is required to bring more air into the lungs, more muscular work and hence greater energy is required.

Work Is Required To Expand the Chest Wall and Lung

A pressure-volume curve can be used to determine the work required for breathing (Fig. 20–20). In the normal lung, point **A** represents pleural pressure at the end of expiration. At point A, airflow is zero and the lung and chest wall are at equilibrium at FRC. Point **C** represents pleural pressure at the end of inspiration, when flow is again zero, and the lung volume is 1 liter *above* functional residual capacity. The slope of line **AEC** would, therefore, represent the compliance of the lung. The area **ABCDA** represents the total work of breathing. The clear area, **ABCEA**, represents the work just to overcome airway resistance during inspiration. The hatched area, **AECFA**, represents the work to overcome airway resistance during expiration. No additional energy is required to force air out of the lungs under resting conditions because the kinetic energy in the stretched lung is used to overcome airway resistance during expiration.

What factors affect the amount of work involved in ventilating the lungs? First is lung compliance. As discussed earlier, lung inflation requires energy to overcome elastic forces. Thus, the stiffer the lung, the more difficult it is to inflate the lungs and consequently more muscular work is required to increase transpulmonary pressure in order to inflate the lungs to bring in the same volume of air (see Fig. 20–20b).

Work of Breathing Is Increased with High Airway Resistance

A second factor affecting the work of breathing is airway resistance, which can be increased in two ways: (1) by an increase in flow rate and (2) by abnormal restriction of the

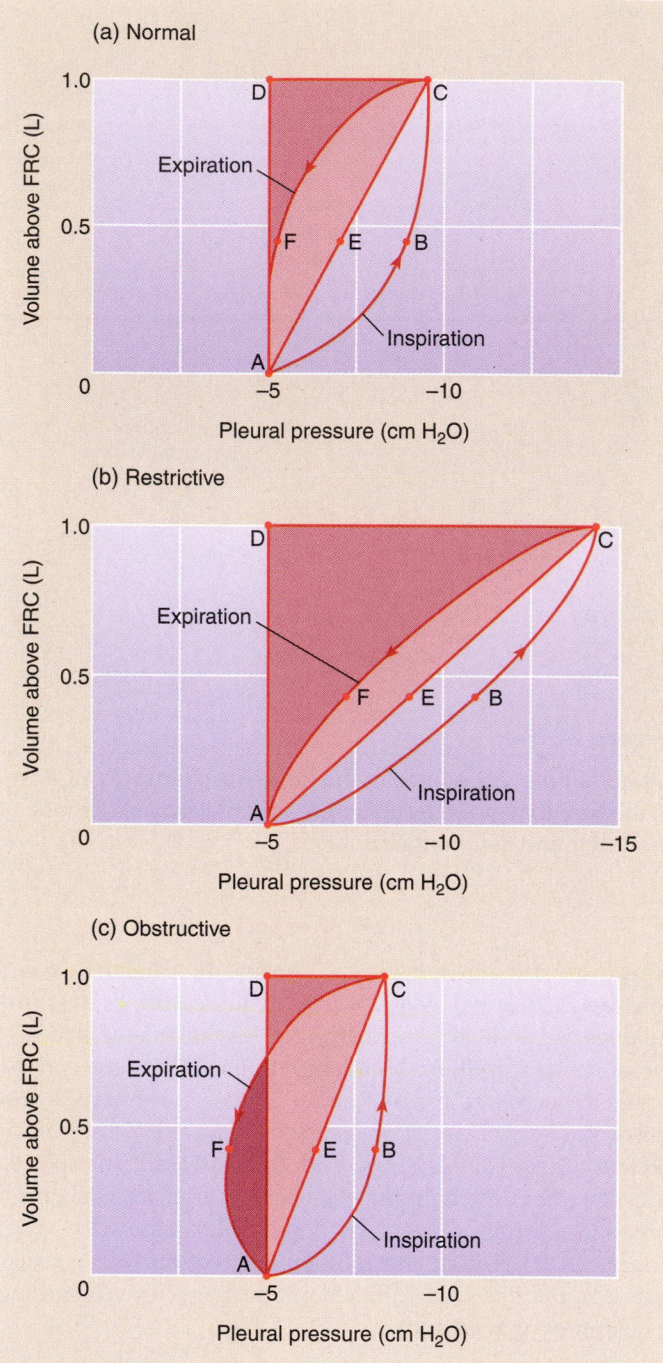

Figure 20–20

A pressure-volume curve is used to determine the work of breathing. Pleural pressure is plotted on the horizontal axis, and volume at any given pressure is plotted on the vertical axis. Work = force × distance, and the change in pressure represents force and the change in volume represents distance. Thus, the area under the pressure-volume curve represents work and is represented by ABCDA. Line AEC represents mean lung compliance.

airways. Both of these lead to greater energy expenditure because of the greater pressure difference (between alveoli and mouth) required to maintain adequate airflow during expiration. This is illustrated in Figure 20–20*c* with an obstructive disorder. The work to get air out of the lung (area AECDFA) is much greater than normal, seen in Figure 20–20*a*. The stippled area outside of the inverted triangle represents additional energy cost from accessory muscles to help force air out of the lungs.

How much energy is involved in breathing? In a normal person, the energy needed for breathing represents approximately 1% to 2% of the body's total energy expenditure. During heavy exercise, approximately 15% of the total energy expenditure is involved. Breathing is most economical when there is a balance of elastic and resistive forces. This balance requires the least amount of work. The faster the breathing rate, the greater the amount of work required to overcome the increase in airway resistance. In contrast, the greater the depth of breathing, the greater the amount of work required to overcome elastic forces. Thus, patients with stiff lungs, seen in Figure 20–20*b*, compensate rapidly by breathing smaller tidal volume and more rapidly, whereas patients with severe airway obstruction tend to take deeper breaths but breathe more slowly. Both of these breathing patterns help to minimize the amount of work required for breathing under abnormal conditions.

CHAPTER REVIEW

Summary

- Gas exchange is the process of transferring O_2 and CO_2 between the atmosphere and the capillary blood in the lungs and then between the blood in the systemic capillaries and the surrounding cells.
- Total lung capacity is made of several volumes and overlapping capacities. All can be measured directly with a spirometer, with the exception of residual volume (RV), functional reserve volume (FRC), and total lung capacity (TLC).
- One of the most useful spirometry measurements to assess lung function is forced vital capacity (FVC).
- Ventilation is the process of moving air into and out of the lungs. Minute ventilation ($\dot{V}_E$) is the total volume of air moved per minute out of the lungs and is measured using expired air ($\dot{V}_E$), whereas alveolar ventilation ($\dot{V}_A$) is volume of fresh air that reaches the alveoli per minute.
- Dead space volume represents the volume of air in the conducting zone (large airways and bronchioles) and does not participate in gas exchange.

- The V_D/V_T ratio represents the fraction of the total minute ventilation that does not participate in gas exchange.
- An increase in transpulmonary pressure is required to overcome the elastic forces of the lung and chest wall in order to inflate the lungs.
- A volume-pressure curve is used to assess the elastic properties of lungs. The slope of the volume-pressure curve represents compliance, which is a measure of lung distensibility.
- Surface tension occurs at the alveolar gas-liquid interface and is a major force affecting lung compliance.
- Surfactant is a lipoprotein that coats the inner surface of the alveoli and acts as a surface-reducing agent that lowers surface tension and stabilizes alveoli.
- Work is required to expand the chest wall, inflate lungs, and overcome airway resistance. Energy costs for the work of breathing represents 1% to 2% of total oxygen consumption.

Review Questions

Choose the Correct Answer

1. From a lung volume-pressure curve, a change in lung volume per unit of transpulmonary pressure change is a measure of:
   a. alveolar recruitment.
   b. compliance.
   c. alveolar pressure.
   d. elastance.
   e. hysteresis.
2. A patient has a V_T of 500 ml, a breathing rate of 16 breaths/min, a dead space volume of 150 ml, and a FRC of 3 liters. Minute ventilation for this patient would be:
   a. 2.4 L/min.
   b. 2.9 L/min.
   c. 4.8 L/min.
   d. 5.6 L/min.
   e. 8 L/min.
3. The airway tree is divided into two zones, the conducting zone and respiratory zone. Which of the following is not part of the conducting zone?
   a. Trachea
   b. Bronchi
   c. Bronchioles
   d. Alveoli
   e. Both c and d
4. Total lung capacity is the sum of:
   a. residual volume + functional residual capacity.
   b. residual volume + vital capacity.
   c. residual volume + expiratory reserve volume + tidal volume.

d. residual volume + functional residual capacity + tidal volume.

e. residual volume + inspiratory reserve volume.

5. Functional residual capacity is the volume of air remaining in the lungs:
   a. after at the end of a half maximal inspiration.
   b. after a full inspiration.
   c. at the end of a normal expiration.
   d. at the end of a maximal expiration.
   e. after a forced expired volume in 1 second (FEV_1).

6. Minute ventilation represents:
   a. total amount of expired air per minute.
   b. total amount of expired air minute minus dead space volume per minute.
   c. alveolar ventilation minus dead space volume per minute.
   d. alveolar ventilation minus residual volume per minute.
   e. tidal volume divided by frequency of breathing.

7. Which of the following pressures is/are subatmospheric at the end of a normal expired tidal volume?
   a. Alveolar pressure
   b. Transpulmonary pressure
   c. Pleural pressure
   d. Tracheal pressure
   e. Both b and d

8. A lung with an abnormally high compliance would:
   a. be easy to inflate.
   b. be easy to deflate.
   c. have greater elastic recoil.
   d. be less distensible.
   e. Both c and d

9. Pleural pressure is the most negative at:
   a. residual volume.
   b. functional residual capacity.
   c. tidal volume.
   d. total lung capacity.
   e. expiratory reserve volume.

10. A trained athlete has a forced vital capacity of 5.0 liters, a functional residual volume of 2.4 liters, and a residual volume of 1.2 liters. What is the person's total lung capacity?
    a. 3.8 liters
    b. 5.0 liters
    c. 6.2 liters
    d. 7.4 liters
    e. 9.6 liters

11. A patient has an alveolar ventilation of 5 L/min, a frequency of 10 breaths/min, and a tidal volume of 700 ml. What is the patient's dead space ventilation?
    a. 0.7 L/min
    b. 1.0 L/min
    c. 2.0 L/min
    d. 4.3 L/min

12. A person expires into a spirometer for 10 minutes. Her expired volume was 54 liters, her respiration rate was 12 breaths/min, and her alveolar ventilation was 4.2 L/min during the 10-minute period. What is this person's tidal volume?
    a. 1.2 liters
    b. 450 ml
    c. 420 ml
    d. 300 ml

Answers to Case History Questions

1. The chief pathophysiologic hallmark of emphysema is the limitation of airflow rates out of the lungs. In emphysema, expiratory flow rates (FVC, FEV_1, and FEV_1/FVC ratio) are significantly decreased. However, key lung volumes (TLC, FRC, and RV) are increased, and the increase in these volumes is due to the loss of lung elasticity (increased compliance).

2. The mechanisms that restrict expiratory airflow in emphysema include (1) hypersensitivity of airway smooth muscle, (2) mucus hypersecretion and bronchial wall inflammation, and (3) increased airway compression due to an increased compliance.

3. In emphysema, many of the physiologic changes are due to the loss of alveolar-capillary membrane and lung elastic recoil. Most authorities believe that these are due to an imbalance between the proteases and antiproteases (α_1-antitrypsin) in the lower respiratory tree. Normally, proteolytic enzyme activity is inactivated by antiproteases (e.g., α_1-antitrypsin). In emphysema, excess proteolytic activity destroys elastin and collagen, the major extracellular matrix proteins responsible for maintaining the integrity of the alveolar-capillary membrane and the elasticity of the lung. Cigarette smoke increases proteolytic activity, which may arise through an increase in protease levels, a decrease in antiprotease activity, or a combination of the two.

Key Terms

airways (p. 630)
alveolar pressure (p. 637)
alveolar ventilation (p. 643)
alveoli (p. 633)
atmospheric pressure (p. 636)
atelectasis (p. 646)
barometric pressure (p. 636)
bronchi (p. 630)
compliance (p. 644)

dead space (p. 642)
diaphragm (p. 634)
forced vital capacity (p. 639)
functional residual capacity (p. 639)
gas exchange (p. 630)
intercostal muscle (p. 635)
laminar flow (p. 649)
lung volumes (p. 639)

partial pressure (p. 636)
pleural pressure (p. 634)
pneumothorax (p. 638)
relative pressure (p. 636)
residual volume (p. 639)
respiration (p. 630)
spirogram (p. 639)
spirometry (p. 639)

surfactant (p. 646)
thoracic cavity (p. 634)
trachea (p. 630)
transpulmonary pressure (p. 637)
turbulent flow (p. 648)
ventilation (p. 630)

Suggested Readings

Cotes, J. E. *Lung Function: Assessment and Application in Medicine*, ed 5. Boston, Blackwell Scientific Publications, 1993.

Decramer, M. "Respiratory muscle interaction." *News in Physiological Sciences*, 8:121–124, 1993.

Demling, R. H. "Current concepts in the adult respiratory distress syndrome." *Circulatory Shock*, 30:297–309, 1990.

Massaro, G. D., Massaro, D. "Formation of pulmonary alveoli and gas exchange surface area: quantitation and regulation." *Annual Review of Physiology*, 58:73–92, 1996.

Piiper, J., and Scheid, P. "Gas exchange in vertebrates through lungs, gills and skin." *News in Physiological Sciences*, 7:199–203, 1992.

Raffin, T. A. "ARDS: Mechanisms and management." *Hospital Practice*, Nov 15:65–80, 1987.

Rhoades, R. A. *In* Rhoades, R. A., and Tanner, G. A. *Medical Physiology*. Boston, Little, Brown, 1984.

Sieck, G. "Neural control of the inspiratory pump." *News in Physiological Sciences*, 6:260–265, 1991.

Snider, G. L. "State of the art: Emphysema: The first two centuries and beyond." *American Review of Respiratory Diseases*, 146:1334–1344, 1992.

Staub, N. C. *In* Berne, R., and Levy, M. *Physiology*, ed 3. St. Louis, C. V. Mosby, 1992.

Van Golde, L. M. C., Batenburg, J. J., and Robertson, B. "The pulmonary surfactant system." *News in Physiological Sciences*, 9:13–20, 1994.

Answers to Review Questions

1. b   **2.** e   **3.** d   **4.** b   **5.** c   **6.** a   **7.** c   **8.** a
9. d   **10.** d   **11.** c   **12.** b

Chapter 21

PULMONARY CIRCULATION, GAS EXCHANGE, AND CONTROL OF BREATHING

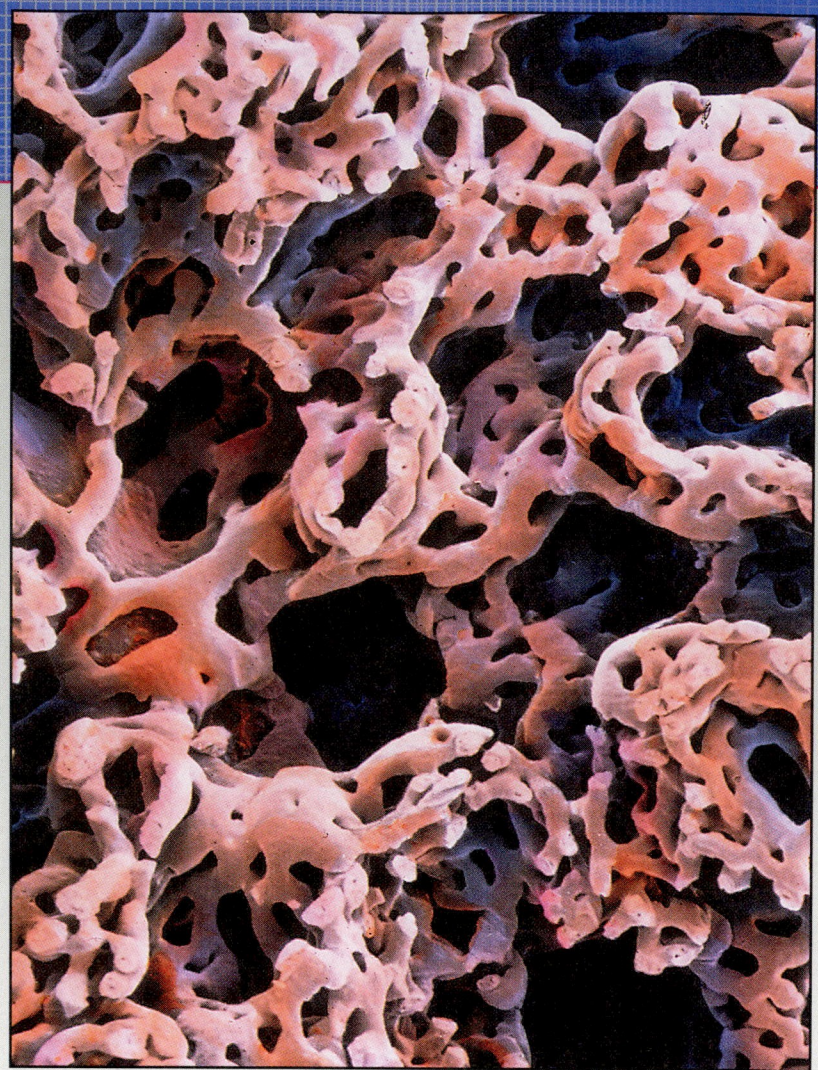

- SEM of the aveolar capillary bed shows very complex network that offers the red cells a very large array of paths.
(With permission from Guentheroth, et al. Journal of Applied Physiology, 53:510–515, 1982.)

KEY CONCEPTS

- The airway tree and the vascular tree merge to form an alveolar-capillary bed for gas exchange.

- Pulmonary circulation is a high flow, low-pressure, and a low-resistance system that accepts all of the cardiac output.

- The downward gravitational pull on the lungs causes more blood flow at the bottom part of the lungs.

- Hypoxic pulmonary vasoconstriction helps to match blood flow with ventilation.

- Capillary recruitment and increased capillary flow increase gas exchange.

- Respiratory gases move across the alveolar-capillary membrane by diffusion.

- Most of the O_2 in the blood is carried by hemoglobin, whereas most of the CO_2 is transported as bicarbonate ions in the plasma.

- Arterial oxygen saturation is a measure of the amount of O_2 present in the blood.

- Control of breathing is tightly linked to metabolic and physical activity.

- The basic breathing rhythm is generated by neurons in the medulla that can be modified by ventilatory reflexes.

- Chemoreceptors detect changes in the partial pressure of O_2 and CO_2 and to changes in pH.

- Breathing patterns change during sleep.

CASE HISTORY

Jeff is a 26-year-old patient who started suffering with extreme dyspnea (labored breathing) in the past six months that required constant oxygen therapy. Jeff is a fourth-year graduate student who is getting his PhD. Three years ago, he started working for his advisor in cardiology. Jeff was doing very well academically until he stopped his graduate program six months ago because he heard rumors that there were dangerous gases in the building's air supply. During the past six months, he has been temporarily employed but began to have shortness of breath upon exertion. His physician could not find any evidence from the physical examination or X-ray to explain his respiratory problem. Shortly after his visit to see his physician, he developed an extreme bout of dyspnea and was admitted to the emergency room at a local city hospital, where he was placed on oxygen therapy. His dyspnea promptly disappeared. Jeff's condition has progressed to such a point that he has to be maintained on constant oxygen to avoid these debilitating episodes of dyspnea. During his last visit, Jeff reported to his physician that he had experienced several attacks of convulsive seizures. At this point, his physician referred Jeff to the pulmonary function laboratory at the university medical center. A blood sample revealed a normal hematocrit and normal blood gases with an oxygen saturation of 98.5% while breathing room air. Pulmonary function tests were performed on the patient while breathing room air. The patient became extremely nervous, and his rate and depth of breathing were so irregular that the tests had to be performed while breathing oxygen. As soon as Jeff saw that he was breathing oxygen, he became quite calm and the pulmonary function tests (forced vital capacity, forced expired volume in 1 second, and maximum ventilatory volume) were within normal range for his height, weight, and age. The patient was referred back to his physician with a diagnosis of respiratory neurosis.

Questions

1. What important information do the normal blood gas data provide?

2. What explanation can be given for disappearance of the chaotic behavior of the respiratory pattern on the administration of oxygen?

3. What would account for his convulsive-like seizures?

FUNCTIONAL DESIGN OF THE PULMONARY CIRCULATION

> *How does the pulmonary circulation compare to the systemic circulation?*

The way in which fresh air travels from the atmosphere to the alveoli was presented in Chapter 20. The other components of gas exchange to be discussed include the diffusion of gases across the alveolar-capillary membrane, blood flow from the heart to alveoli, and the transport of oxygen (O_2) and carbon dioxide (CO_2) by the blood to tissues. Blood is oxygenated in the capillaries of the alveoli and then returned to the left side of the heart to be pumped, together with the oxygen it contains, to the rest of the body.

Pulmonary Vascular Tree Accepts All of the Cardiac Output

The circulation that carries blood through the lungs, the **pulmonary circulation,** is unique because it is the only circulation in the body that must accommodate all of the cardiac output. The pulmonary, in contrast to the systemic, circulation is characterized as a high-flow, low-pressure, and low-resistance system. Blood vessels are designed to operate at low pressure, and there are many more pulmonary capillaries than are needed to oxygenate the blood

during resting conditions. The extra capillaries become perfused with blood as cardiac output increases during exercise and thereby oxygenates more blood. The blood leaves the right ventricle via the pulmonary artery and follows the same kind of treelike branching pattern found in the airways (Fig. 21–1). In fact, each time that the airway tree branches, the arterial tree branches, so that the two shadow each other.

Once the arteries reach the alveoli, the vessels break up into a fine network of tiny vessels, the pulmonary capillaries, which form a dense double network embedded in the alveolar wall (Fig. 21–2). This complex system results from the way that the fetal lung develops. As the alveoli grow, each develops its own set of capillaries. During lung maturation, the capillaries between adjacent alveoli join, but the joining is incomplete (see Fig. 21–2). This leaves a complex capillary bed with a large number of pathways through which blood can travel in each alveolar wall.

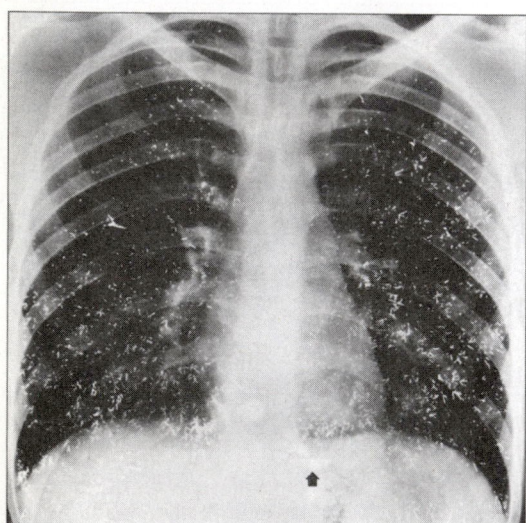

Figure 21–1

When a dye that is opaque to X-rays is injected into the pulmonary artery and an X-ray image is made, the dye shows the pathways followed by the blood through the pulmonary arteries. If another X-ray image is taken a few seconds later, the dye-filled veins show the pulmonary vein tree.

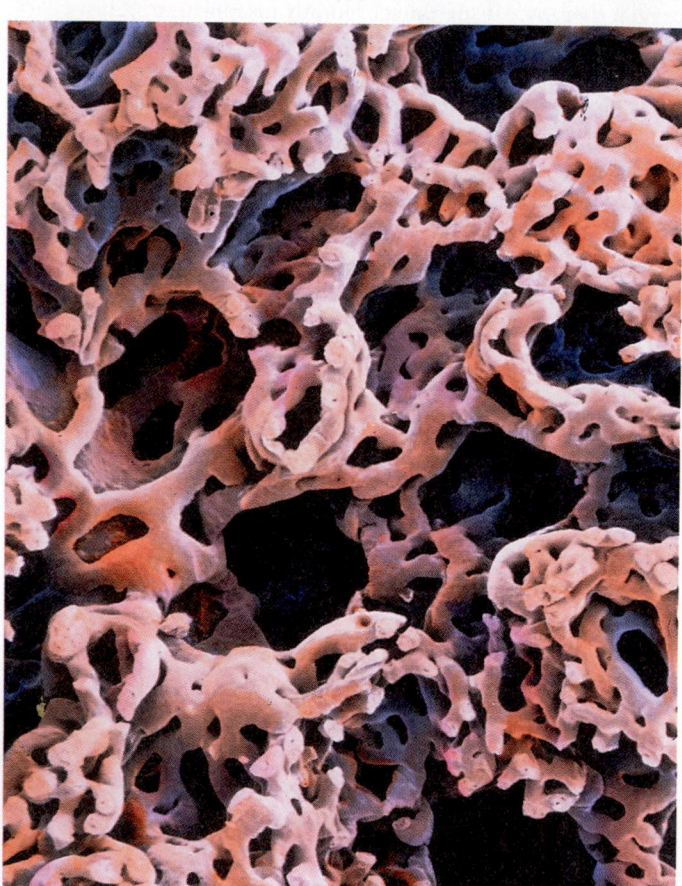

Figure 21–2

SEM of the alveolar capillary bed shows very complex network that offers the red cells a very large array of paths. *(With permission from Guentheroth, et al. Journal of Applied Physiology, 53:510–515, 1982.)*

APPLICATIONS OF PHYSIOLOGY

Pulmonary Edema

Karen is a 60-year-old attorney who was well until approximately one year ago, when she began to have central chest discomfort during physical exertion, such as climbing stairs. She did not seek medical advice. After nine holes of golf with a colleague, she developed severe central chest pain, which she described as crushing and radiated into her left shoulder. Shortly after being admitted to the hospital emergency department, she became very short of breath and started coughing up small amounts of frothy fluid. Karen had no previous history of shortness of breath or ankle swelling. She has smoked a pack of cigarettes a day for the past 30 years. Karen was diagnosed with pulmonary edema-caused left heart failure following a myocardial infarction.

Pulmonary edema is a common and important disorder of the pulmonary circulation. Pulmonary edema is an abnormal accumulation of fluid that leaks from the pulmonary capillaries and occurs when fluid accumulation exceeds the rate of removal by the lymphatics. The condition is severe, and patients with pulmonary edema have difficulty breathing (tachypnea: rapid and shallow breathing) and suffer from hypoxemia.

Pulmonary edema occurs in two stages. The first is interstitial edema, which causes flooding and engorgement in the interstitial space. The second is alveolar edema, which is flooding of the alveoli. In severe pulmonary edema, the fluid moves into the small and large airways and is coughed up as frothy sputum. Edematous lungs interfere with the gas exchange (i.e., increase the diffusion distance for O_2 and CO_2) and impede ventilation by interfering with surfactant (i.e., increase surface tension), which makes alveoli shrink (i.e., decrease surface area for gas diffusion) and makes lungs less compliant (stiffer).

The etiology of pulmonary edema includes (1) increased hydrostatic pressure, (2) increased capillary permeability, (3) lymphatic insufficiency, (4) decreased interstitial pressure, and (5) decreased colloid osmotic pressure. By far, the most common cause of pulmonary edema is an abnormal increase in hydrostatic pressure. Events that lead to an increase in hydrostatic pressure include heart or renal failure. When the left ventricle is unable to empty itself during systole, left atrial pressure increases, which causes hydrostatic pressure to increase. The second most common cause of pulmonary edema is increased permeability. Conditions that lead to increased permeability include inhaled toxins (e.g., chlorine, sulfur dioxide, or nitrogen oxides), circulatory toxins (e.g., endotoxin), oxygen toxicity, radiation, and adult respiratory distress syndrome.

Lymphatic insufficiency is due to carcinoma of the lymph glands and to silicosis (damage from silicon dust). Events that can cause decreased interstitial pressure include a pneumothorax and hyperinflation of the lung. Starvation (hypoproteinemia) and overtransfusion are the causes of decreased colloid osmotic pressure.

The pattern of edema formation is much different depending on the etiology. For example, in adult respiratory distress syndrome, a patchy pattern of edema occurs around the lung periphery. In heart failure or fluid overload (overhydration or renal failure), a central edema pattern occurs, principally at the base of the lung. These patterns of edema formation help the pulmonary physician to postulate the etiology of pulmonary edema.

The arterioles and venules are separated by four to eight alveoli. Thus, the blood can follow many pathways through a given alveolar wall, as well as travel over different alveolar walls; these combinations result in a huge number of possible routes by which the blood can travel across the gas exchange surface to pick up O_2 and unload CO_2.

The erythrocytes (i.e., red blood cells) are separated from the capillary wall by a thin layer of plasma that acts as a lubricant. The distance across the capillary wall and plasma layer is short, measuring approximately 1 μm (in more familiar terms, a typical human hair is approximately 100 μm in diameter, so that the distance between gas in the alveolus and the blood is 1/100th of a hair width). The alveolar gas reaches the blood quickly because the distance from the gas to the blood phase is so small.

The passage of blood through the capillaries takes approximately 0.75 sec. In that time, the blood unloads CO_2 and picks up all of the O_2 that it can carry. The blood then enters the smallest venules and returns through the pulmonary venous tree to the left side of the heart, from which the newly oxygenated blood is then pumped to the body.

Another tree-shaped conducting system in the pulmonary system is the lymphatic system. The lymph vessels are auxiliary drainage vessels. A small amount of some fluid is continually being forced from the capillaries, into the inter-

APPLICATIONS OF PHYSIOLOGY

Coach Class Syndrome

Larry is a 40-year-old executive whose business takes him overseas frequently. He never boards a flight without first donning a pair of knee-high compression stockings. He eschews alcohol on the flight and makes a point of getting up regularly to walk about the cabin. He also starts popping aspirin 24 hours before his flight. The reason that Larry takes these precautions is that 2 years ago, on a 12-hour flight from Heathrow to Hong Kong, he suddenly developed a sharp, stabbing pain over his left chest after debarking the Hong Kong airport. His pain was made worse by a sudden inspiration. At the same time, he had shortness of breath and developed a cough with a small amount of blood in the sputum. Larry was taken to the hospital and was diagnosed with a pulmonary thromboembolism. Larry was more fortunate than a healthy 28-year-old woman who died of a pulmonary thromboembolism that developed from a Sidney-to-London flight.

A growing number of travelers are emphasizing new preflight rituals thanks to a mounting concern about a medical condition known as *coach class syndrome*. This syndrome describes a fate that can befall passengers on long flights: combined conditions of immobility, cramped seating, and dehydration, which can cause a fatal blood clot.

The design of the pulmonary circulation, with its many branching arteries feeding into the capillary net, acts as a sieve for blood as it circulates through the body. This design has an important protective effect in that material in the circulation larger than red blood cells plugs small vessels and prevents the flow of blood. In the systemic circulation, such blockages can be life-threatening if they occur in the coronary vessels (causing heart attack) or in the brain (causing stroke). The most common abnormally large objects in the circulation are blood clots. These often form in the veins in the legs, especially when blood flow is slowed because of varicose veins or prolonged bed rest. A blood clot stuck in a vein is called a *thrombus*. If the clot breaks loose, it is called an *embolus*, or a thromboembolus. The embolus proceeds through the right side of the heart and out into the pulmonary arteries, moving farther and farther into the lung until it wedges in a branch of the pulmonary arterial tree. The resultant trapping blocks blood flow to a part of the lung. Because the gas-exchange surface area in the pulmonary circulation is so vast, oxygenation of

---

stitial space. The capillaries reabsorb much of the fluid, but part of it remains behind and must be drained away through the lymphatics. The lymphatic fluid drains into systemic veins. If a pathologic condition exists and the leak rate increases, the fluid accumulates, the diffusion distance between the alveolar gas and the capillary blood increases, and the exchange of gases decreases between the blood and alveoli. Accumulation of fluid in the lung is called **pulmonary edema** and is a severe condition that can result in death in extreme cases.

Pulmonary Circulation Has Metabolic Functions

The alveoli have a large surface area, as discussed in Chapter 20. Because the capillary bed takes up approximately 80% to 90% of the alveolar wall, the surface area of the capillaries is impressively large as well. The inside of the capillary wall is completely covered by metabolically active endothelial cells. Because of the central location of the pulmonary circulation, all of the blood is exposed to the capillary endothelium each time it is pumped through the lungs. The endothelial cells act on a number of substances. For example, **angiotensin II,** a potent vasoconstrictor, is an important hormone in the regulation of blood flow, as mentioned in Chapter 19. The hormone reaches the lungs in an inactive form, angiotensin I. During its passage through the pulmonary circulation, angiotensin I is converted to active angiotensin II by an enzyme located on the surface of the capillary endothelium. Angiotensin II is delivered by the blood to the systemic resistance vessels (arterioles) to cause constriction, which results in an increase in blood pressure. Many other vasoactive substances, such as bradykinin, serotonin, and the prostaglandins, are inactivated or removed from the blood by the endothelium. In this way, the pulmonary circulation serves important metabolic as well as gas-exchange functions.

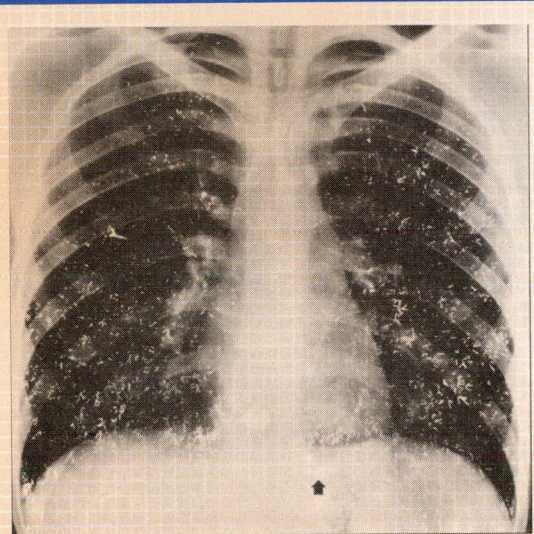

X-ray of an embolism created by an injection of liquid mercury. *(Courtesy of Dr. William Beamish, University Hospital, Edmonton, Alberta. Fraser and Paré: Diseases of the Chest, Saunders.)*

blood is not adversely affected unless the blockage is massive. Sometimes the thromboembolus is so large that it can block the main pulmonary artery and cause death. Smaller clots can resolve over a period of one month or so, allowing the pulmonary circulation to return to normal.

The pulmonary circulation filters any large material in the venous circulation. Bone marrow, from broken bones, for example, sometimes embolizes to the lungs. If air enters the venous circulation, bubbles form and cause air embolization of the lungs. The air diffuses across the air-blood barrier and is exhaled with alveolar gas, so air emboli do not last long. Much air would have to be injected to cause harm.

In the accompanying figure, the pulmonary circulation has been embolized by liquid mercury (the small white lines in the lung field). The X-ray was taken of a drug user who injected mercury intravenously for a special "kick." It is difficult to tell from an X-ray whether the mercury is in blood vessels or small airways. The differential diagnosis is made by the pool of mercury *(arrow)* in the bottom of the right ventricle.

PULMONARY PRESSURE AND BLOOD FLOW

Pulmonary Circulation Is a Low-Pressure System

 How is blood flow distributed in the lung?

One of the striking contrasts between pulmonary circulation and systemic circulation is the pressure profile. Pressure in the pulmonary artery is much lower than in the arteries of the rest of the body. The right ventricle generates a pressure of approximately 25 mm Hg, which becomes the pulmonary arterial systolic pressure as the ejected blood passes through the pulmonary valve. Once the pulmonary valve closes, the pressure decreases to diastolic pressure, just as it does in the systemic circulation. Pulmonary arterial diastolic pressure is approximately 10 mm Hg. Mean pressure for a normal adult

is 15 mm Hg. These numbers are stated the same way as systemic arterial pressure: 25/10, with a mean of 15 mm Hg (Table 21–1). Pulmonary arterial pressure is usually reported as a mean pressure. These pressures, shown in Figure 21–3, are from a pressure tracing as a cardiac catheter is passed through the right atrium and right ventricle and into the pulmonary artery. If there are abnormalities in the heart, such as a narrowed valve or a defect (hole) between the chambers, the pressure tracing is abnormal and permits the physician to make the diagnosis.

The low pulmonary arterial pressure is just enough to pump blood to the top of the lung. Although the entire lung receives some perfusion, even during resting conditions, the majority of the blood flows through the lower lung (Fig. 21–4). This creates a gradient of blood flow, with most of the flow going to the bottom of the lung and little going to the top. Almost all of the capillaries in the lower lung are open and have blood flowing rapidly through them; under these

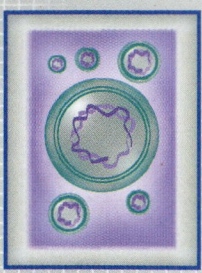

Transit of Leukocytes Through the Pulmonary Circulation

Of all of the white blood cells that circulate in the blood, neutrophils make up the largest proportion and are one of the body's chief defenders against invading microorganisms. For unknown reasons, neutrophils travel through the lung far more slowly than do red blood cells. This causes neutrophils to concentrate in the small pulmonary blood vessels. From that site, they can either defend the body against inhaled bacteria or move quickly through the circulation to any part of the body.

Sometimes, however, neutrophils misuse their powerful bacteria-killing ability and damage the blood vessels, which causes plasma to leak into the tissue. In the lung, edema in alveoli increases the thickness of the blood-gas barrier and impedes the exchange of oxygen and carbon dioxide. In its severe form, this condition is called the Adult Respiratory Distress Syndrome.

Because neutrophils are present in large numbers in the lungs of patients with Adult Respiratory Distress Syndrome, they are implicated in the etiology of this disorder. Therefore, considerable investigative attention has been focused on what causes the neutrophils to stop under normal conditions in the pulmonary circulation. Two hypotheses are currently being investigated. The first, the mechanical impediment hypothesis, suggests that the highly viscous nature of the neutrophils impedes their travel through the pulmonary capillary bed. When the neutrophils reach capillary segments with diameters smaller than their own, they require considerable time to deform sufficiently to fit through the narrow lumens of the obstructing capillary segments. The second hypothesis, the adhesion hypothesis, is based on the fact that the neutrophil surface membrane has many adhesion glycoproteins that recognize receptors on the cells lining the blood vessel walls. There, the binding of the neutrophils to the endothelial cells may slow them down. These adhesion sites are crucial in the cells' response to invading bacteria because they anchor the neutrophils to the vessel wall, allowing them to then move through it and attack the microorganisms. Their function under normal conditions, however, is unknown.

The mechanical impediment hypothesis is being investigated by treating neutrophils with chemicals that make them either more rigid or more flexible. After treatment, the transit times of the neutrophils through the pulmonary circulation are measured to determine whether the altered viscosity affects transit. To test the adhesion hypothesis, scientists are determining the effect on neutrophil transit times of antibodies that specifically block the functioning of the adhesion molecules.

TABLE 21–1

Pressures in the Pulmonary Artery During Various Conditions

Condition	Systolic/ Diastolic Pressure (mm Hg)	Mean Pressure (mm Hg)
Rest	25/10	15
Moderate exercise	30/15	20
Maximal exercise	50/25	33
At high altitude	50/25	33
At high altitude with maximal exercise	75/45	55
Severe pulmonary hypertension	175/100	125

From Groves, B. M., et al., *J. Appl. Physiol.* 63: 521–530, 1987.

circumstances, the capillary bed is said to be highly "recruited." In the upper lung, only a few capillary pathways are open (recruited), and the blood cells flowing through them travel slowly.

Hypoxic Vasoconstriction Balances Blood Flow with Ventilation

As long as inspired air is distributed evenly through the lung, venous blood in the pulmonary artery (the blue blood in Fig. 21–5a) can flow to any alveolar region and pick up oxygen from the alveolar gas; as the blood flows pass the alveoli, it becomes oxygenated. Inspired air, however, is not always evenly distributed to all alveoli. This causes **regional hypoxia** (low oxygen) to exist from time to time in groups of alveoli. Of course, blood passing through such areas is not completely oxygenated (see Fig. 21–5b). This blood mixes with the rest of the oxygenated blood and reduces the total

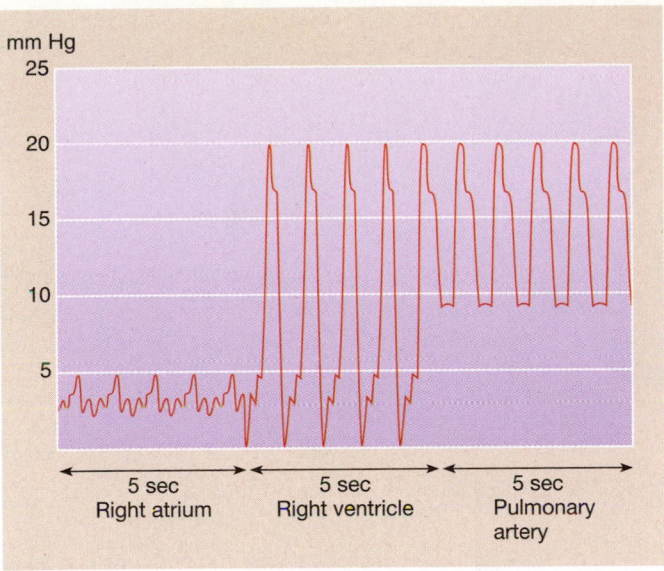

Figure 21–3

A pressure tracing from a cardiac catheter as it is passed from the right atrium into the right ventricle, past the pulmonic valve, and into the pulmonary artery.

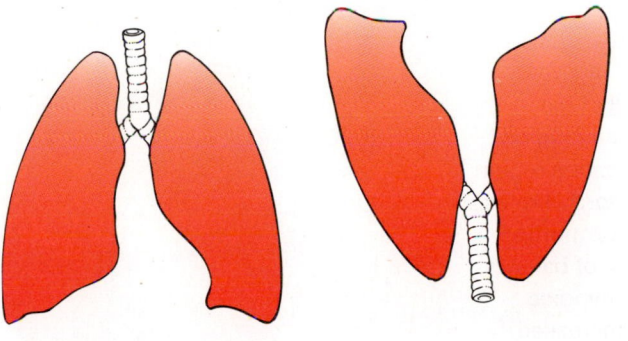

Figure 21–4

Because of the low values of both pulmonary arterial pressure and vascular resistance, most of the blood flows to the lower part of the lung during rest. If a person stands on his or her head, blood flows to the apex of the lung. To calculate the height to which the right ventricle can pump the blood, consider the following. Blood has a specific gravity near 1, whereas mercury has a specific gravity of 13.6—that is, mercury is 13.6-fold heavier than water, with its specific gravity of 1.0. Multiplying the mean pulmonary arterial pressure of 15 mm Hg by 13.6 gives 204 mm H_2O. Thus, mean pulmonary arterial pressure is sufficient to pump the blood 200 mm high, which is approximately the height of the top of the lung above the heart.

amount of oxygen that can be delivered to the body. The condition of low blood oxygen is called **hypoxemia** (*-emia* refers to blood; e.g., *anemia* refers to decreased red blood cell count).

The lung has a mechanism called **hypoxic vasoconstriction** for balancing the flow or perfusion of blood with the availability of regional ventilation. When a region of lung becomes hypoxic, the small arteries feeding that region sense the hypoxia in the alveolar gas, and the arteries constrict. Their constriction causes the resistance to local blood flow to increase and forces the blood to flow away from the hypoxic region to other parts of the lung, where resistance is low because oxygen is available (see Fig. 21–5c). In this way, the balance between ventilation and perfusion is maintained, and compensation for disturbances in delivery of inspired air to the alveoli can be made. When an imbalance between ventilation and perfusion occurs due to lung disease, hypoxemia, sometimes of a severe nature, can result.

When someone experiences hypoxia throughout the entire lung (e.g., at high altitude) all of the local hypoxic vasoconstrictor responses are triggered (see Fig. 21–5d). This situation causes the resistance to blood flow in the entire pulmonary circulation to increase, leading to an increase in pulmonary arterial pressure. The increase in pressure has an interesting effect. The alveoli in the upper part of the lungs have more oxygen than the alveoli in the lower part of the lungs because most of the blood flow goes to the lower lung (see Fig. 21–4) and extracts more oxygen from the lower-most alveoli. The elevated pulmonary arterial pressure caused by whole lung hypoxia is useful because it enables more blood to be pumped to the upper regions of lung, where capillaries are recruited and thus increase the surface area for gas exchange. Because recruitment occurs in the upper lungs, where alveolar oxygen is relatively abundant, the lungs become a more effective gas exchanger. If, however, the high pulmonary arterial pressure is sustained for long periods, there are detrimental effects because pulmonary hypertension can cause vessel damage and eventually failure of the right ventricle. Therefore, the redistribution of blood from the bottom to the top of the lungs can be useful only if the resulting pulmonary hypertension is either mild or transient (see Table 21–1).

Exercise Recruits Capillaries and Decreases Transit Time

As cardiac output increases during exercise, pulmonary arterial pressure increases due to the increased quantity of blood flowing through the lungs. The pressure increase is modest at low levels of exercise, such as walking (see Table 21–1). As the level of exercise increases, pulmonary arterial pressure continues to increase, resulting in an adequate driving force to propel the blood to the uppermost parts of the lung

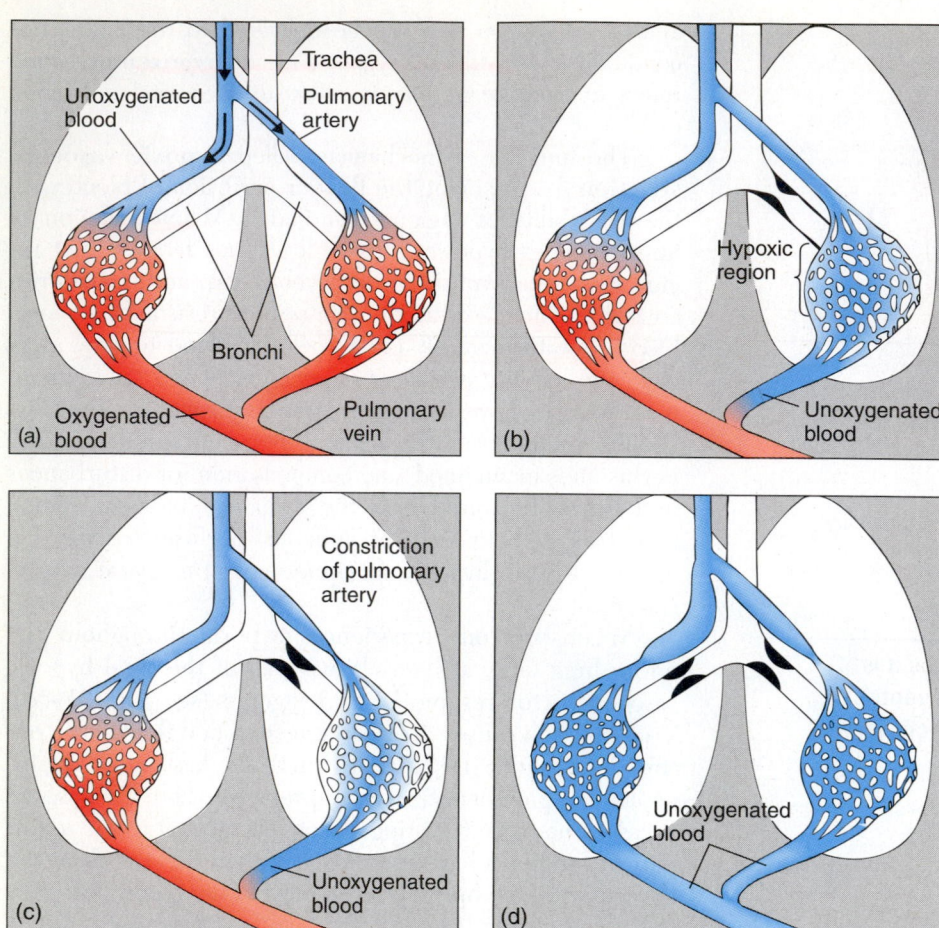

Figure 21–5

The two-alveolus model of the lung can be used to represent two alveoli or two regions of the lung with many alveoli in each region. **(a)** In this figure, all of the alveoli are well ventilated so the unoxygenated (blue) blood can flow to any region and pick up oxygen, shown here turning from blue to red. **(b)** In this case, one region of lung is hypoxic. If blood continues to flow past the hypoxic alveoli at normal rates, unoxygenated (blue) blood enters the pulmonary veins and reduces the proportion of oxygen in the blood. Ventilation (by air) and perfusion (by blood) is out of balance. **(c)** In some undiscovered way, the pulmonary artery accompanying the hypoxic airway detects the low oxygen and constricts. The narrowed vessel has increased resistance and causes local blood flow to be reduced and directed to other, better oxygenated parts of the lung. In this way, ventilation-perfusion balance is maintained and the amount of oxygen in the arterial blood is kept at high levels. **(d)** If all regions of the lung become hypoxic, as happens at high altitude or in certain diseases of the airways in which gas movement is impeded, the arteries detect hypoxia in the airways and constrict. Because all of the arteries constrict, the resistance to blood flow through the lung increases and pulmonary arterial pressure increases.

(Fig. 21–6). This results in more even perfusion of the lung from top to bottom and improves the ability of the lungs to exchange gas.

As blood flow increases, two important changes take place in the capillaries to improve oxygen uptake from the alveolar gas. The first change involves the number of capillar-ies with blood flowing through them. In the lower part of the lungs, blood flows through most of the capillary segments (see Fig. 21–6a), even during resting conditions, when cardiac output is low. In the upper part of the lungs, relatively few capillaries are perfused during rest, which leaves capillaries in reserve that can be recruited as the demand for

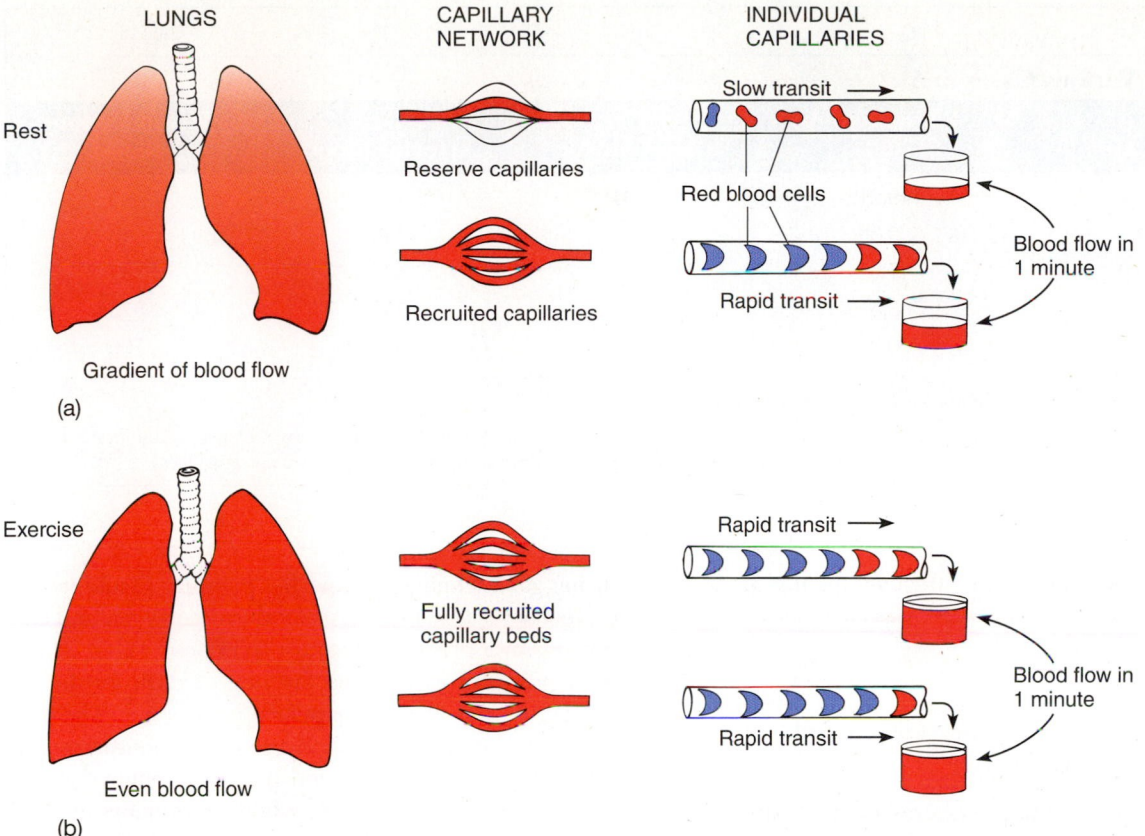

LUNGS	CAPILLARY NETWORK	INDIVIDUAL CAPILLARIES

Rest — Gradient of blood flow (a)

Reserve capillaries
Recruited capillaries

Slow transit
Red blood cells
Rapid transit
Blood flow in 1 minute

Exercise — Even blood flow (b)

Fully recruited capillary beds

Rapid transit
Blood flow in 1 minute
Rapid transit

Figure 21–6

During resting conditions, most of the capillaries in the lower lungs are open (recruited), with blood flowing briskly through them. In the upper lungs, few capillaries are recruited and what flow there is moves at a slow pace *(a)*. With the increased blood flow during exercise, more blood perfuses both the upper and lower lung *(b)*. In the lower lungs, the primary change is that the blood moves more rapidly through the capillaries. In the upper lungs, not only does the blood move more rapidly but also capillary recruitment occurs.

oxygen increases. As cardiac output increases to higher levels during exercise, pulmonary arterial pressure continues to increase as well, and more blood flow is directed to the upper part of the lungs, where reserve capillaries are recruited (see Fig. 21–6b). In this way, the surface area for gas exchange is increased.

The second change that takes place during exercise to improve oxygen uptake is more rapid movement of the blood through the capillaries. Even though the normal transit time through the capillaries of one second is rapid, the blood is completely oxygenated in only 0.25 sec. Thus, a large part of the **capillary transit time** is wasted. As cardiac output increases with exercise, the transit time decreases until it approaches a 0.25-sec minimum time. This change causes more blood to be oxygenated every second during exercise. Again, reserve is present in the upper lung, where the transit times are considerably longer than 1 sec at rest. Evidence suggests that, as blood flow is shifted upward during exercise, not only

are there capillaries to be recruited, but also the slowly flowing blood is speeded up.

GAS UPTAKE AND TRANSPORT

How does oxygen move from the alveolar gas to the tissues; how does carbon dioxide move from the tissues to the alveolar gas?

Once the inspired air approaches the alveoli, the gas flow slows nearly to zero. The predominant means of gas movement is no longer the bulk flow that moves along a pressure gradient, but diffusion, which causes individual gases to move along a concentration gradient. In the lung, the diffusion distance across an alveolus is small, in the range of 100 to 200 μm. Gas molecules diffuse very quickly over such

TABLE 21-2

Partial Pressures of Various Gases in Air*

Pressure	Dry Air	Moist Tracheal Air (37°C)	Alveolar Air	Arterial Blood	Mixed Venous Blood
P_{O_2}	159.1†	149.2†	104†	100	40
P_{CO_2}	0.3	0.3	40	40	46
P_{H_2O}	0.0	47.0	47	47	47
P_{N_2}	600.6	563.5	569	573	573
Total P_B	760.0	760.0	760	760	706

*Usual values in a resting, healthy individual at sea level (barometric pressure = 760 mm Hg).

†This is an approximate value and holds only for individuals breathing air at sea level. Because the total atmospheric pressure at Denver, Albuquerque, or Salt Lake City is approximately 640 mm Hg, the partial pressure of O_2 in inspired and alveolar gas is significantly below values at sea level.

short distances; hence, diffusion is an effective means of moving oxygen and carbon dioxide molecules rapidly across the alveolar-capillary membrane.

Partial Pressures Move Respiratory Gases Across Alveolar-Capillary Membrane

Recall from Chapter 20 that partial pressures of gases are used to tell us how much pressure is exerted by one particular gas in a mixture of gases. Atmospheric pressure (760 mm Hg), which is the sum of all the partial pressures in air, is shown in Table 21–2. Note that percentages for each gas remain constant regardless of the barometric pressure.

The gas in the lung is moisturized as it passes through the airways until it becomes saturated with water. The water vapor exerts a partial pressure, just as oxygen and nitrogen do. The water vapor in saturated gas is directly related to the temperature. In the case of a normal body temperature of 37°C, water vapor is 47 mmHg. Therefore, to compute the partial pressure of O_2 (P_{O_2}) in moist tracheal air, 47 mmHg must first be subtracted from the barometric pressure:

$$P_{O_2} = (760 - 47) \times 0.21$$

The partial pressures of gases are shown in Table 21–2.

As inspired gas reaches the respiratory zone of the lung, the incoming gas mixes with the alveolar gas (see Chapter 20). The final composition of alveolar gas has a partial pressure of oxygen of 104 mmHg. Even though that figure is reduced from inspired partial pressure of oxygen (150 mmHg), a sufficient partial pressure gradient still exists to drive oxygen from the alveolar gas into the capillary blood. The initial O_2 diffusion gradient is 64 mmHg. (Alveolar oxygen tension is 104 mmHg, whereas mixed venous blood **oxygen tension** is 40 mmHg—see Table 21–2). A diffusion gradient also exits for carbon dioxide and the initial gradient is 6 mmHg (mixed venous P_{CO_2} = 46 mmHg, whereas alveolar P_{CO_2} = 40 mmHg).

During gas exchange, the partial pressure gradients between alveolar gas and capillary blood cause oxygen and carbon dioxide to move in opposite directions (Table 21–2). Fick's law (Chapter 2) governs the rate of diffusion for O_2 and CO_2 across the alveolar-capillary membrane. The oxygen in the alveolar gas dissolves in the moist membrane just as it would in water. The oxygen passes through the thin membrane and into the plasma, where it remains in solution. Diffusion continues through the plasma and then through the very thin membrane of the red blood cell. It is inside the red cell that the majority of oxygen is transported to the tissues.

If alveolar ventilation is increased through a larger tidal volume, increased respiratory rate, or both, then more oxygen is delivered to the alveoli and more carbon dioxide is removed. This **hyperventilation** increases the partial pressure of oxygen in alveolar gas and decreases the alveolar carbon dioxide, which increases the gradients for both gases and improves gas exchange. Hyperventilation occurs during exercise. The opposite condition, **hypoventilation**, means that alveolar ventilation is reduced. When that occurs, alveolar oxygen decreases and carbon dioxide increases. Clinicians, for example, encounter this condition in patients with depressed respiratory centers. Blood gases deteriorate under these circumstances, and the physician may use supplemental oxygen or place the patient on a mechanical ventilator.

Oxygen Is Transported by Hemoglobin

Every 100 ml of blood that passes through tissue delivers approximately 5 ml of oxygen. A normal young adult at rest requires approximately 250 ml/min of oxygen. This means that the heart must pump 5 liters per minute [(250/5) × 100 = 5000 ml/min] at rest.

A schema of the circulatory delivery is shown in Figure 21–7. During each minute, 7.5 liters of gas (minute ventilation) is moved into and out of the lungs. From that, 250 ml/min of oxygen is taken up through the pulmonary capillaries and transported to the tissue capillaries, where the tissue takes up 250 ml/min of oxygen. The venous blood is still partially loaded with oxygen after passage through the tissue capillaries (see Fig. 21–7). After returning to the right heart, the blood is circulated once again through the lungs.

With exercise, the body's requirement for oxygen increases considerably. During heavy exercise, a normal, physically fit adult might require 12-fold the resting amount of oxygen, or 12 × 250 ml/min = 3000 ml/min. A cardiac output of 20 L/min is required to circulate enough blood to deliver 3000 ml/min of oxygen. In highly trained athletes, the oxygen demands may exceed 5000 ml/min O_2. Cardiac outputs may be more than 30 L/min in these endurance athletes.

Even though these numbers are large, they are much smaller than would be required if, in order to be transported, oxygen had to dissolve in blood. If such were the case, the cardiac output would have to be more than 1000 L/min. That flow rate would completely fill an automobile gasoline tank in 3 sec! Such a delivery rate would require a huge heart. Each beat of the heart would cause the blood to exceed the sound barrier.

A remarkable molecule, hemoglobin, solves the problems of transporting oxygen. Hemoglobin is a pigmented protein that permits large amounts of oxygen to be carried to the tissues in an efficient way. The structure of hemoglobin was presented in Chapter 17. Each hemoglobin molecule contains four heme molecules, each of which contains an iron atom. A single iron atom is capable of binding with a molecule of oxygen. When the partial pressure of oxygen is high, oxygen combines with the heme portion of hemoglobin in the lung to form **oxyhemoglobin.** When the oxyhemoglobin reaches the tissue where Po_2 is low, the hemoglobin releases the oxygen. Because of the loose way in which O_2 binds with iron atoms, the oxygen is delivered to the tissue fluids as dissolved molecular oxygen.

Because of hemoglobin's strong affinity for oxygen, the hemoglobin carries almost 99% of the oxygen in the blood. A small amount of oxygen is dissolved in blood, approximately 0.3 ml of O_2 for every 100 ml of whole blood. Thus, every 100 ml of whole blood carries some dissolved oxygen plus oxygen loosely bound to hemoglobin. The amount of oxygen carried by hemoglobin is calculated in the following way. Assuming that 1 g of hemoglobin (Hb) carries 1.34 ml O_2 and a normal hematocrit of 15 g per 100 ml of whole blood, then the amount of oxygen carried by hemoglobin in 100 ml of blood is 20.1 ml (1.34 ml O_2/g Hb × 15 g/100 ml blood = 20.1 ml O_2/100 ml). For whole blood that is fully saturated with oxygen, the maximum amount of O_2 that can be carried is 20.1 + 0.3 = 20.4 ml and is called the **oxygen carrying capacity** of whole blood and is frequently written as 20.4 "volumes percent" (vol.%). Arterial blood usually has less oxygen bound to hemoglobin than 20 ml O_2 per 100 ml of blood. The amount actually bound to hemoglobin is called **oxygen content** (where capacity is the amount that can potentially be bound). The **percentage of O_2 saturation of hemoglobin (So_2)** is calculated as:

$$\text{Percentage of HbO}_2 \text{ saturation} = \text{HbO}_2 \text{ content/HbO}_2 \text{ capacity} \times 100$$

For example, if the hemoglobin in 100 ml of whole blood is carrying 19 ml of O_2, then the percentage of saturation would be (19/20.1) × 100 = 94.5%.

Now we can see how O_2 loads with hemoglobin in the lung. Fig. 21–8a shows an important physiological curve that is S shaped and called the **oxyhemoglobin dislocation curve.** If the blood is 100% saturated, then all of the binding sites on the hemoglobin contain an oxygen molecule. If the blood is 50% saturated, then only half of the hemoglobin binding sites contain oxygen molecules. Saturation gives an indication of how completely the binding sites are filled and

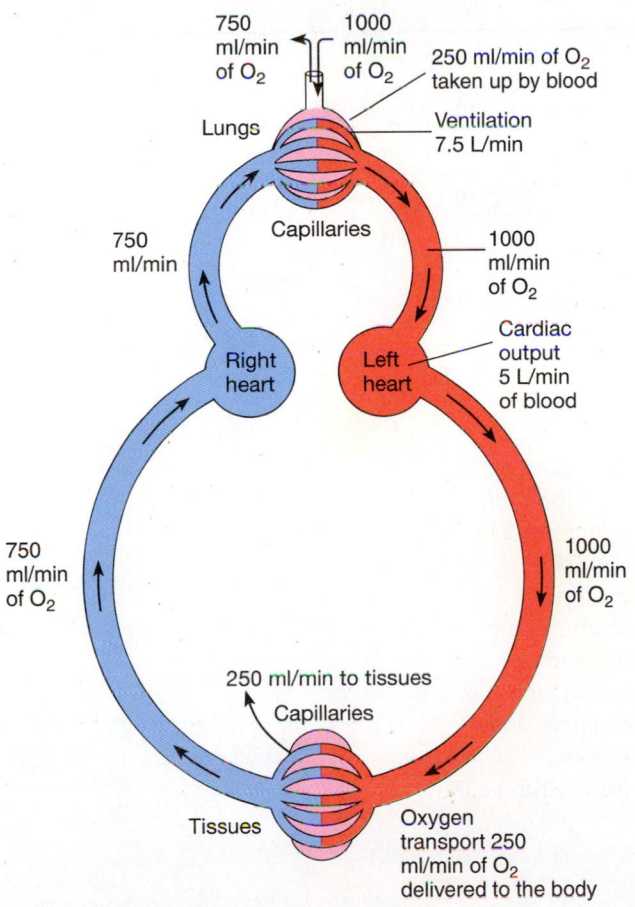

Figure 21–7

Schema of the pulmonary and systemic circulation showing oxygen delivery to the tissue. The heart pumps 5 L/min in this example. Because each 100 ml of blood carries 20 ml of oxygen, a total of 1000 ml/min of oxygen is delivered (5,000 ml ÷ 100 ml = 50 aliquots of blood, each carrying 20 ml of oxygen); thus 50 × 20 = 1000 ml/min oxygen delivery.

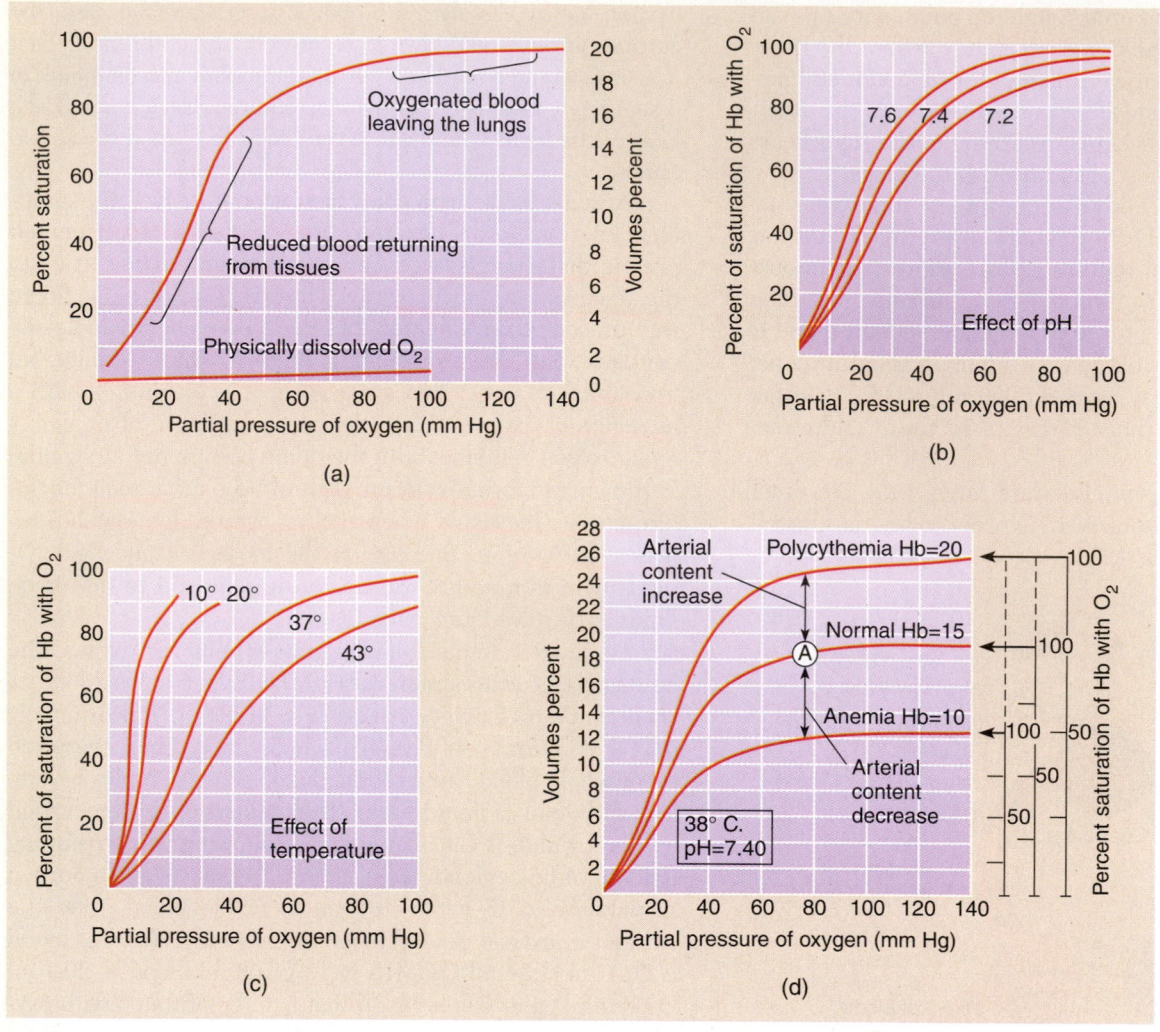

Figure 21–8

(a) The oxyhemoglobin-equilibrium curve. Oxygen is loaded onto the hemoglobin in the lung and is unloaded in the tissues. Note that as oxygen tension increases in the midrange, the hemoglobin loads rapidly with oxygen. As the hemoglobin molecules become saturated with oxygen, loading slows as oxygen tension increases. *(b)* and *(c)* A change in pH and temperature causes the oxyhemoglobin-equilibrium curve to shift. A shift in the curve causes the hemoglobin to load or unload more easily. In the lungs, where the temperatures are relatively cool and the pH is relatively alkaline, loading is facilitated by the leftward shift of the curve. In an exercising muscle, in which the environment is relatively warm and acidotic, unloading is facilitated. *(d)* Oxygen-equilibrium curve in anemia and polycythemia.

how many are left in reserve. Another way to measure the amount of oxygen in the blood is to measure oxygen content (i.e., milliliters of O_2 actually carried by each 100 ml of blood). As we have seen before, the unit for content is vol.% and is shown on the right-hand axis of Figure 21–8a. Under normal resting conditions, systemic arterial blood is 97% saturated, and the O_2 content is 19.4 vol.% (20 vol.% × 0.97). Venous blood has a partial pressure of oxygen of 40 mm Hg and a content of 15 vol.% (see Table 21–2). Under certain

conditions, the amount of hemoglobin in the blood can vary. The effect is shown in Figure 21–8d. The normal curve in the middle has 15 g of hemoglobin in every 100 ml of blood and when it is 100% saturated it carries approximately 20 ml of oxygen. In cases of anemia, in which the number of red blood cells is reduced, there is less hemoglobin in each 100 ml of blood. In the lower curve with only 10 g of hemoglobin per 100 ml of blood, less oxygen can be carried (1.34 × 10.0 = 13.4 ml O_2). However, the blood is still 100% saturated when this amount of oxygen is carried because all of the hemoglobin binding sites have oxygen attached (see saturation axis in Figure 21–8d). In the upper curve in Figure 21–8d, abnormally large numbers of red blood cells are present, a condition known as **polycythemia.** In this case, more oxygen can be carried ($20.0 \times 1.34 = 26.8$ ml O_2) when the blood is 100% saturated.

As shown in Figure 21–8a, dissolved oxygen makes a small contribution to the total carrying capacity of the blood. For example, in a healthy person at rest, with a partial pressure of oxygen in the blood of 100 mm Hg, the resulting value for dissolved oxygen is 0.3 vol.% and the hemoglobin is 97% saturated. If the hemoglobin is 15 g per 100 ml of blood, then the oxygen combined with hemoglobin is $0.97 \times 15.0 \times 1.34 = 19.5$ vol.%. The total content is $19.5 + 0.3 = 19.8$ vol.%.

More Oxygen Is Extracted from the Blood During Exercise

Under normal conditions, in which the alveolar oxygen tension is 100 mm Hg, the hemoglobin saturation of systemic arterial blood is 97% and the oxygen content is 20 vol.%. Normal venous blood with its oxygen tension of 40 mmHg is 75% saturated (15 vol.%). Thus, 5 ml of oxygen per 100 ml of blood is removed as the blood passes through the systemic capillaries, which means that the arteriovenous oxygen difference is 5 vol.%. In other words, the tissues extract 5 ml of oxygen for every 100 ml of blood coming through them.

During exercise, the oxygen tension in the muscles decreases to lower values than when at rest because the working muscle is using oxygen so rapidly. Heavy exercise can reduce muscle oxygen tension to less than 15mm Hg. This increases the gradient that favors the movement of more oxygen from the capillary blood and into the tissue. Thus, the delivery of oxygen is increased because of greater extraction. For example, if the venous oxygen tension decreases from 40 mmHg to 15mmHg, then the content decreases to approximately 4 vol.%. Now much more oxygen is delivered by each 100 ml of blood: $20 - 4 = 16$ ml of oxygen, which triples the delivery simply by increasing the extraction.

Some interesting characteristics of hemoglobin facilitate oxygen transport. Blood acidity (see Fig. 21–8b) shifts the position of the oxyhemoglobin equilibrium curve to the right. For a given oxygen tension, the amount of oxygen (oxygen content) is lower. This means that for a given oxygen tension in the tissue, the hemoglobin unloads more oxygen. Exercising muscle has greater acidity owing to increased carbon dioxide and lactic acid production. The unloading of oxygen from hemoglobin is therefore facilitated. Similarly, increased temperature causes a rightward shift of the curve (see Fig. 21–8c). Blood temperature increases when it reaches a hot exercising muscle, and oxygen unloading is enhanced.

A leftward shift in the oxyhemoglobin equilibrium curve facilitates oxygen loading. At a given Po$_2$, the content and saturation are higher. As the blood enters the alveolar capillaries, carbon dioxide leaves the blood, the pH becomes more basic, and oxygen loading is enhanced. Similarly, cooler temperatures cause a leftward shift of the curve and, because the lungs are cooler than exercising muscle, a leftward shift is also promoted. Thus, the oxyhemoglobin curve shifts in a remarkable way from right to left and back to the right with each passage around the circulation in order to favor oxygen loading in the lungs and unloading in the tissues.

Carbon Dioxide Is Transported in Three Forms

How is carbon dioxide transported from the tissues to the alveoli?

Carbon dioxide is transported in the circulation more readily than is oxygen because carbon dioxide is a nonpolar molecule that is highly soluble in lipid and therefore moves easily across plasma membranes. Because the maintenance of pH in the body is crucial to so many chemical reactions, it is important that carbon dioxide remain normal. Carbon dioxide in the blood is intimately linked to acid-base balance, as described in Chapter 25.

Carbon dioxide diffuses from the tissue into systemic capillaries and is transported in the blood in three ways (Fig. 21–9). The largest fraction of carbon dioxide ($\sim 60\%$) is transported as bicarbonate ions (HCO_3^-). These ions are formed by an important series of reactions that take place inside the red blood cell.

First, carbon dioxide and water in the red blood cell combine to form carbonic acid (H_2CO_3), as shown in this equation:

$$CO_2 + H_2O \underset{\text{Carbonic anhydrase}}{\rightleftharpoons} H_2CO_3$$

Ordinarily, this is a slow reaction requiring seconds for completion, but when catalyzed by the enzyme **carbonic anhydrase,** the reaction is accelerated by a factor of 5000 and is complete in a fraction of a second. In the next chemical step, the carbonic acid dissociates into a bicarbonate ion (HCO_3^-) and a hydrogen ion (H^+):

$$H_2CO_3 \rightleftharpoons HCO_3^- + H^+$$

The bicarbonate ion moves from the red cell into the plasma, where it is readily dissolved. This leaves a hydrogen ion free in the red cell, where it rapidly combines with hemoglobin in the following way:

$$H^+ + Hb^- \rightleftharpoons HHb$$

As the bicarbonate ions diffuse into the plasma, chloride ions diffuse into the red blood cells to replace the bicarbonate. The bicarbonate-chloride exchange is carrier mediated

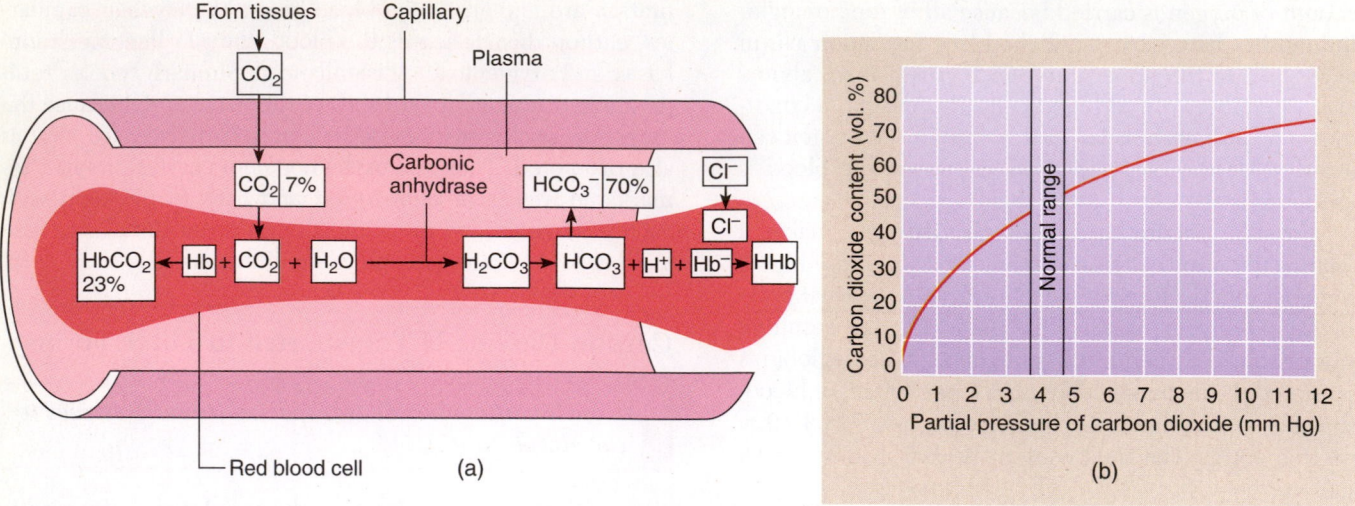

Figure 21–9

(a) Carbon dioxide is carried by the blood in three forms (dissolved, as bicarbonate, and bound to hemoglobin) from the tissues to the lungs. *(b)* The carbon dioxide-equilibrium curve.

(facilitated diffusion) and is known as the **chloride shift,** which causes red blood cells in venous blood to have a higher chloride content than red blood cells in arterial blood (see Fig. 21–9*a*). These reactions are summarized in Figure 21–9.

Carbon dioxide can combine directly with hemoglobin to form **carbaminohemoglobin.** The carbon dioxide does not bind at the same site on the iron molecules as oxygen but instead binds loosely by a direct reaction with some of the amine groups that form the hemoglobin molecule.

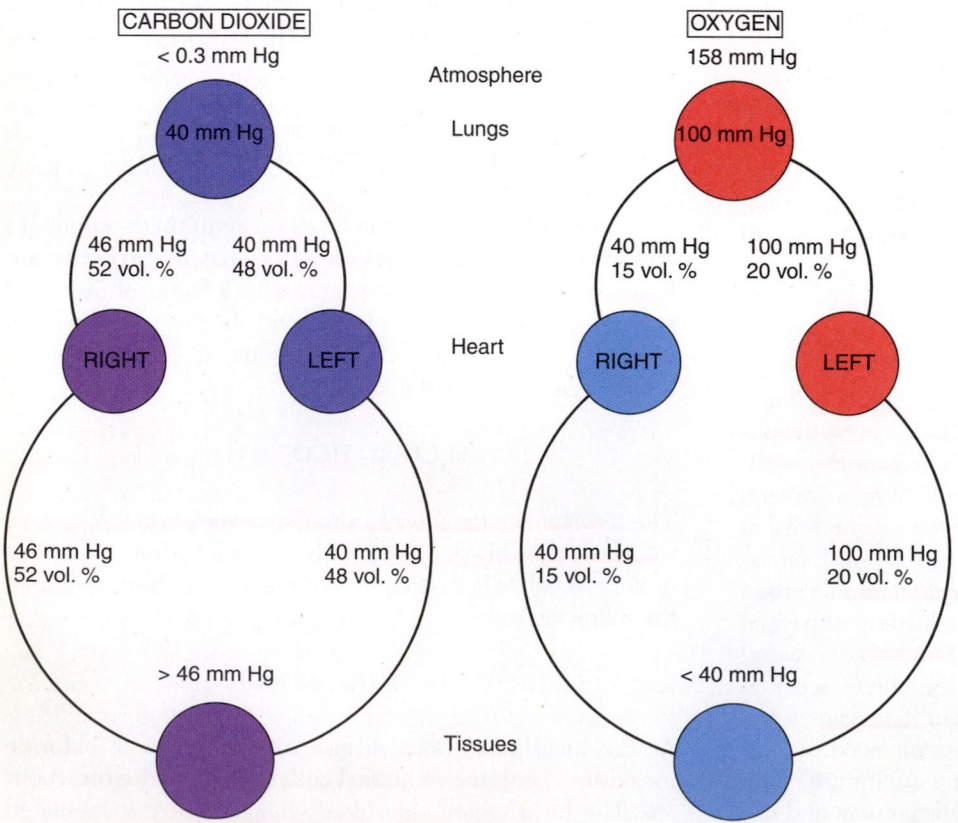

Figure 21–10

A comparison of the partial pressures of carbon dioxide and oxygen in the circulation during resting conditions. Arterial blood contains carbon dioxide, and venous blood contains oxygen. That is, the blood is not scrubbed clean of carbon dioxide in the lungs, and not all of the oxygen is removed from the blood by the tissues. Partial pressures are in mm Hg, and contents, in vol.%.

Approximately (30%) of carbon dioxide transport is accomplished in this form (Fig. 21–9a). The hemoglobin molecule can carry more carbon dioxide when it is oxygen-desaturated—exactly the condition that exists in venous blood, in which oxygen is low and carbon dioxide is high. The bond between hemoglobin and carbon dioxide is loose enough to be reversible in the lung. The carbaminohemoglobin reaction is much slower than the enzyme-catalyzed reaction that forms bicarbonate.

Approximately 10% of the carbon dioxide is transported in a dissolved state in the plasma and red blood cells. This accounts for only approximately 0.3 vol.% of the total carbon dioxide transport.

All three of these methods of transport are readily reversible in the pulmonary capillaries, where the concentration gradient favors the movement of carbon dioxide from the blood into the alveolar gas. Because the equations read in the reverse as well as the forward direction, the transport cycle is complete when carbon dioxide is given off in the lung.

The three ways in which carbon dioxide can be transported in the blood (dissolved in plasma, as bicarbonate, or as carbaminohemoglobin) are all dependent on the partial pressure of carbon dioxide (Pco_2), just as is the case with oxygen. The carbon dioxide dissociation curve, with its narrow normal range of operation of between 40 and 46 mmHg, is shown in Figure 21–9b. The total carrying capacity of blood for carbon dioxide is considerably higher than for oxygen (compare Figs. 21–9b and 21–8a). As blood passes through the systemic capillaries, the carbon dioxide content increases to approximately 52 vol.%. As the blood passes through the pulmonary capillaries, 4 vol.% of carbon dioxide are unloaded into alveolar gas and the carbon dioxide content decreases to 48 vol.%.

An important fact to remember is that the blood does not completely empty itself of carbon dioxide in the lung or of oxygen in the tissues. Even during the most strenuous exercise, when oxygen extraction is high in the tissues and ventilation is high in the lungs, oxygen and carbon dioxide always remain in the blood. The partial pressures (mmHg) and contents (vol.%) of these gases in various regions of the circulation are shown in Figure 21–10.

FETAL AND NEONATAL CIRCULATORY PHYSIOLOGY

Before birth, the lungs have no respiratory function because air is unavailable. The fetus must depend on the mother for gas exchange. This is accomplished through the placenta (Fig. 21–11), in which gas exchange vessels from fetus and mother pass close to each other. Such a system is less effective than lungs exposed directly to the atmosphere and results in a low fetal arterial Po_2, on the order of 25 to 30 mmHg. Blood travels to and from the placenta through the umbilical arteries and veins. The oxygenated blood flows into the right heart, but only 10% to 12% flows through the lungs. Approximately half of the blood flows from the right atrium to the left atrium through a passage called the **foramen ovale.** This oxygenated blood is pumped by the left ventricle out the aorta and perfuses mostly the head and upper body. The remainder of the blood follows the usual pathway through the right heart into the pulmonary artery. The bulk of pulmonary arterial blood, however, flows through the **ductus arteriosus** and into the descending aorta to perfuse the lower body. More than half of the cardiac output passes through the umbilical artery and into the gas-exchanging vessels of the placenta (see Fig. 21–11).

The changes that occur in the circulation at birth are almost as important as is the act of breathing itself. As soon as the umbilical cord is severed, the increased blood flow in the aorta immediately increases pressure. As the lungs expand, the pulmonary vessels are no longer compressed, and hypoxic vasoconstriction is relieved; these have the combined effect of decreasing pulmonary vascular resistance nearly tenfold. Because left atrial pressure is 2 to 4 mmHg higher than right atrial pressure, the flap that lies on the left side of the atrial septum closes the foramen ovale. A few hours after birth, the powerful muscular wall of the ductus arteriosus constricts and closes that vessel. Growth of fibrous tissue into the lumen of the ductus over the course of the next few months causes permanent closure.

Congenital Heart Disorders Interfere with Gas Exchange

In some newborns, the ductus does not close after birth. This condition, known as **patent ductus arteriosus** (*patent* means "open"), can lead to a left-to-right shunt of blood from the aorta into the pulmonary artery. The high aortic pressure causes pulmonary hypertension, which in severe cases leads to irreversible damage to the pulmonary arteries.

Sometimes, the pulmonary hypertension becomes so great that the shunt reverses and flows from right to left. Such a shunt forces venous blood into the aorta and causes hypoxemia. Patients with patent ductus arteriosus can experience right ventricular failure and die. The patent ductus can be readily closed by surgery, and if the diagnosis is made in time, the disorder can be repaired with few untoward effects. The diagnosis is made by listening to the chest with a stethoscope. Blood traveling through the patent ductus makes a sound called a *murmur* that alerts the physician that an abnormality is present.

Numerous other defects can occur during the development of the fetus. For example, the septum between the atria can be incompletely formed, causing an **atrial septal defect.** A similar defect in the ventricles is called a **ventricular septal defect.** Although there are a number of other congenital heart defects, fortunately these disorders as a group are rare and can often be repaired by modern surgical techniques.

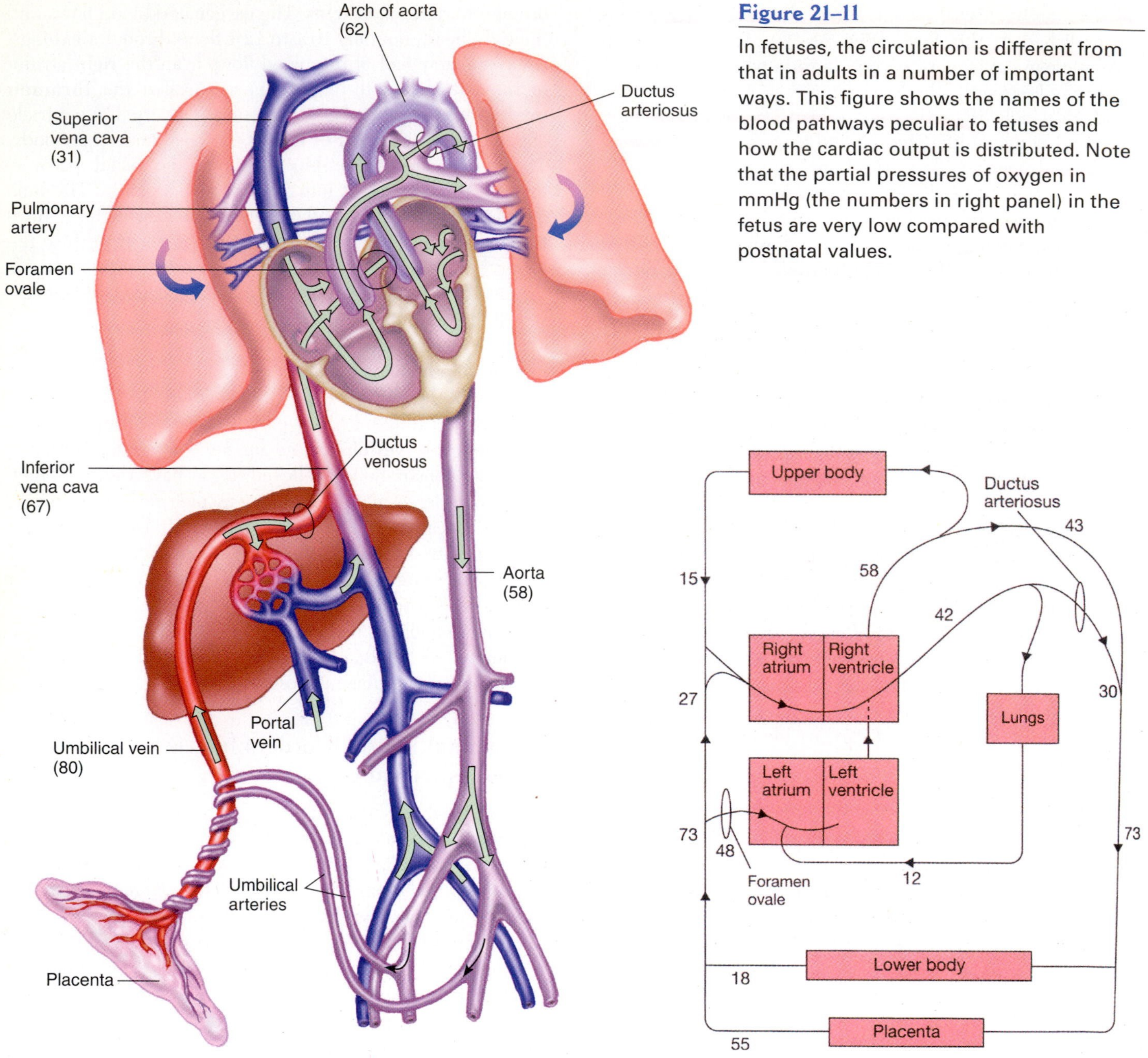

Arch of aorta
(62)

Ductus
arteriosus

Superior
vena cava
(31)

Pulmonary
artery

Foramen
ovale

Inferior
vena cava
(67)

Ductus
venosus

Aorta
(58)

Portal
vein

Umbilical vein
(80)

Umbilical
arteries

Placenta

Figure 21–11

In fetuses, the circulation is different from that in adults in a number of important ways. This figure shows the names of the blood pathways peculiar to fetuses and how the cardiac output is distributed. Note that the partial pressures of oxygen in mmHg (the numbers in right panel) in the fetus are very low compared with postnatal values.

Upper body		Ductus arteriosus

15

58

43

27

42

| Right atrium | Right ventricle | | Lungs | 30 |

| Left atrium | Left ventricle |

73

48

12

73

Foramen
ovale

| Lower body |

18

| Placenta |

55

CONTROL OF BREATHING

> *How does the nervous system control the depth and rate of breathing?*

Breathing is an automatic process and occurs without any conscious effort while we are awake and asleep and even while we are under anesthesia. We can, however, exert some conscious control over our own ventilation by voluntarily changing the rate and depth of breathing (Fig. 21–12). We can also voluntarily stop breathing for a short period of time until carbon dioxide builds up in the blood, which then stimulates breathing regardless of how hard we try to hold our breath. Clearly, the basic automatic rhythm of breathing can be modified by both neural and chemical stimuli (see Fig. 21–12).

During quiet breathing, inspiration is brought about by a progressive increase in activation of inspiratory muscles, most importantly the diaphragm. This increase in activity of the diaphragm causes the lungs to fill at a nearly constant rate until tidal volume has been reached. The end of inspiration is associated with a rapid decrease in excitation of inspi-

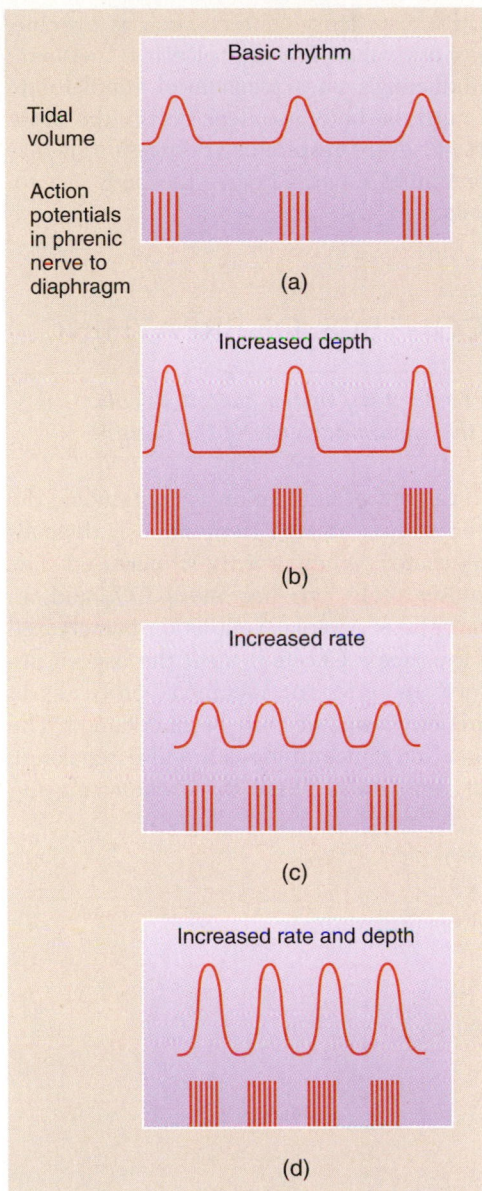

Figure 21–12

Recordings of tidal volume and phrenic nerve action potential to the diaphragm. *(a)* Normal breathing pattern with basic rhythm. *(b)* The rate of discharge determines depth of breathing (tidal volume), whereas *(c)* the time interval between action potentials determines breathing rate. *(d)* Both rate and depth of breathing are increased.

ratory muscles, after which expiration occurs, passively due to the elastic recoil of the lungs and chest wall. As more ventilation is demanded, such as during exercise, other inspiratory muscles (external intercostals, cervical muscles) are recruited. In addition, expiration becomes an active process through use of, most notably, muscles of the abdominal wall.

The Medulla Controls the Basic Breathing Rhythm

The neural basis of these breathing patterns depends on the generation and subsequent tailoring of cyclic changes in activity of cells primarily located in the **medulla oblongata.** Although breathing is automatic, it depends entirely on the cyclic excitation of the respiratory muscles. The diaphragm receives an electrical signal from the **phrenic nerve,** which leaves the spinal cord in the upper half of the neck and passes down through the chest to innervate the diaphragm. Breathing has a basic rhythm, and in a resting individual, the normal rate of breathing is approximately 14 breaths per minute. The central pattern for the basic breathing rhythm has been localized to fairly discrete areas in the medulla that discharge action potentials in a phasic pattern with respiration.

Two different aggregates of cells have been found in the medulla. Their anatomic locations are indicated in Figure 21–13. One, called the **dorsal respiratory group (DRG)** because of its dorsal location in the region of the nucleus tractus solitarius, predominantly contains cells that are active during inspiration. The other, the **ventral respiratory group (VRG),** is a column of cells in the general region of the nucleus ambiguus that extends caudally nearly to the bulbospinal border. The VRG contains both

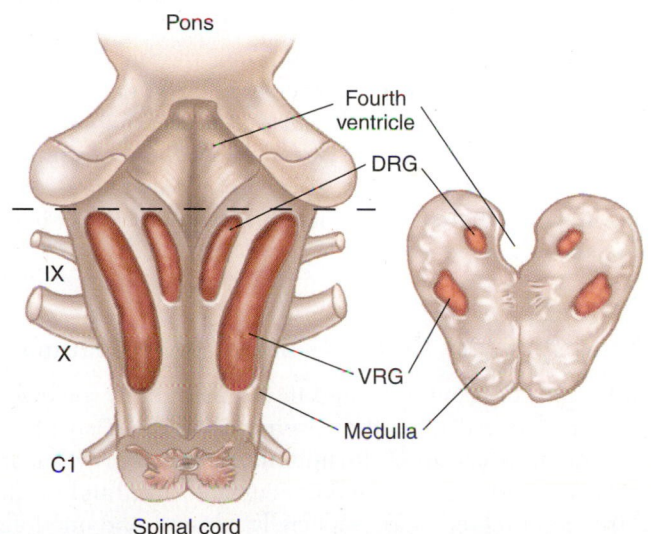

Figure 21–13

A schematic of the medulla illustrating the location of the respiratory center. The diagram represents the dorsal aspect of the medulla showing the general locations of two clusters of neurons: the dorsal (DRG) and ventral (VRG) respiratory groups. C1, first cervical nerve; X, vagus nerve; IX, glossopharyngeal nerve, located in the medulla. The center consists of inspiratory and expiratory neurons.

inspiration- and expiration-related neurons. Both groups contain cells projecting ultimately to the bulbospinal motor neuron pools. The DRG and VRG are bilaterally paired, but cross-communication occurs such that they behave in synchrony; as a consequence, respiratory movements are symmetric. Thus, the neural networks forming the **central pattern generator for breathing** are contained within the DRG-VRG aggregate, but the exact anatomic and functional description remains unclear. Central pattern generation probably does not arise from a single pacemaker or by reciprocal inhibition of two pools of cells, one having inspiratory-related activity and the other expiratory-related activity.

Inspiratory Activity Is Switched Off to Initiate Expiration

Two groups of neurons, located within the VRG, serve as an inspiratory off-switch (see Fig. 21–13). Switching occurs abruptly when the excitatory inputs to the off-switch reaches a threshold. Adjustment of the threshold level is one of the ways in which depth of breathing can be varied. Other potential "cutoff" signals come from **proprioreceptors.** These include **pulmonary stretch receptors** embedded in the smooth muscles of the airways. When activated, these stretch receptors send impulses (action potentials) to the medulla and inhibit the inspiratory neurons. This feedback mechanism from the airways helps to not only stop inspiration but also prevent overinflation, which can damage the lungs and alter the blood flow through the lungs. In newborns, the pulmonary stretch receptors play a very important role in regulating the basic rhythm and preventing overinflation. However, in adults, the threshold for these receptors is high and the stretch receptors play a role only under conditions in which there is a very large tidal volume (e.g., heavy exercise).

The Respiratory Center Is Sensitive to Trauma

The respiratory center located in the medulla is somewhat fragile and can disrupt the respiratory cycle. Two of the more common causes of disruption of the basic rhythm are a blow to the head (cerebral concussion) and fluid on the brain (cerebral edema), especially around the medulla. Both of these cause blood vessel collapse, which stops blood flow and causes the respiratory rhythm to cease functioning.

Another common cause of respiratory failure is barbiturate overdose. Barbiturates depress the inspiratory neurons and stop the rhythm. General anesthesia and narcotic pain relievers also act in a similar fashion, inhibiting the respiratory center. If too much anesthesia is given, breathing stops altogether. Once breathing stops, it is extremely difficult for the body to start the cycle up again. Very few drugs can be used to reactivate the inspiratory rhythm. In general, those

available to excite the respiratory centers, such as caffeine, are too weak to be of any value. The only effective treatment is to place these individuals on a mechanical ventilator to sustain ventilation until the body itself can rectify the problem. In some cases of severe respiratory depression, it may take weeks and sometimes months before the body can resume breathing on its own.

CHEMICAL CONTROL OF BREATHING

How does the respiratory system match alveolar ventilation to the metabolic needs of the body?

Although the basic rhythm of ventilation is continuous, the rate and depth of breathing can vary tremendously depending on metabolic demands. When activity is increased (i.e., during exercise) muscle cells produce more CO_2 and increase their oxygen uptake. Alveolar ventilation is increased to rid the body of the excess CO_2 and meet the oxygen demands of the tissues. In order for metabolic processes to occur, a constant pH environment must be maintained. The appropriate hydrogen ion concentrations are also regulated, in part, by alveolar ventilation. These three chemicals, O_2,

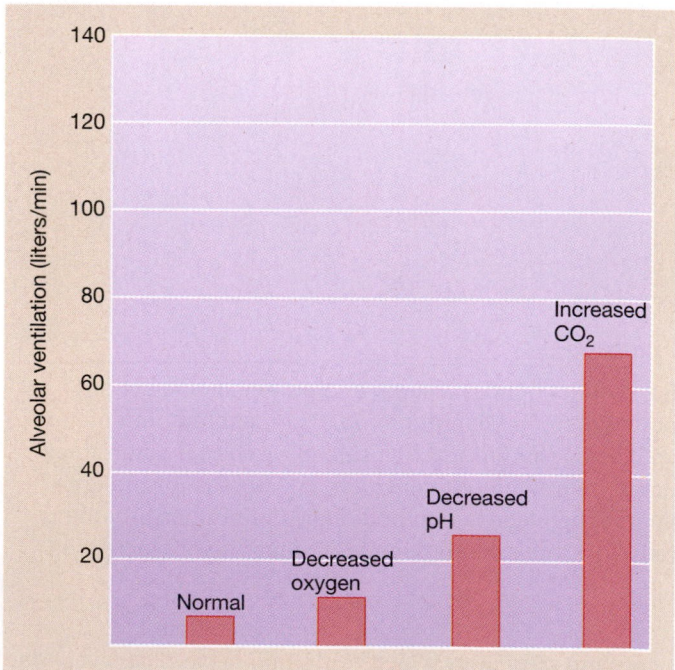

Figure 21–14

The effects of changes in arterial blood pH, P_{O_2}, and arterial P_{CO_2} on alveolar ventilation. Note that increased arterial CO_2 the most powerful stimulus can increase alveolar ventilation approximately 15-fold above normal.

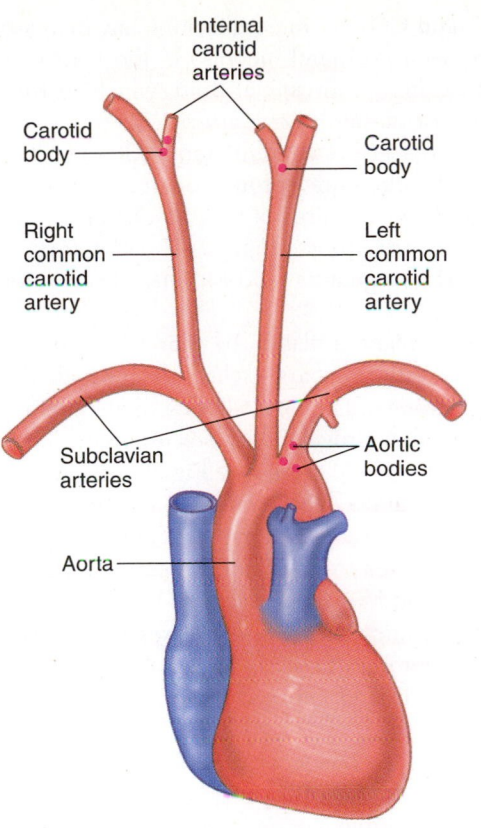

Figure 21-15

Peripheral chemoreceptors located in the aortic arch (aortic bodies) and carotid artery (carotid bodies). The latter are the most important in humans.

CO_2, and H^+, can profoundly change alveolar ventilation by affecting both the rate and the depth of breathing (Fig. 21-14).

Chemoreceptors Detect Changes in Blood Gases

These three chemicals (O_2, CO_2, and H^+) are detected by two sets of **chemoreceptors**, which are strategically located in the body to detect changes in P_{O_2}, P_{CO_2}, and hydrogen ion concentration ($[H^+]$). One set of receptors is located in the medulla and is referred to as **central chemoreceptors.** The other set of chemoreceptors, located outside of the central nervous system, is called **peripheral chemoreceptors.** The peripheral chemoreceptors (Fig. 21-15) are located in the arch of the aorta (**aortic bodies**) and in the neck at the bifurcation of the common carotid artery (**carotid bodies**). In humans, the carotid bodies are the most important peripheral chemoreceptors.

As shown in Figure 21-16, the central chemoreceptors only respond directly to changes in $[H^+]$ in the cerebral spinal fluid (and hence indirectly to arterial P_{CO_2}) while the

peripheral chemoreceptors are activated by changes in arterial oxygen tension (Pa_{O_2}), carbon dioxide tension (Pa_{CO_2}), and hydrogen ion concentration ($[H^+]$). The afferent (sensory) nerve fibers that arise from these cells extend to the respiratory center in the medulla.

Of the three blood chemicals that stimulate alveolar ventilation, carbon dioxide is the most powerful (see Fig. 21-14). When arterial carbon dioxide levels increase to above normal, all portions of the respiratory center become excited. A small increase of only 2 to 5 mmHg in arterial carbon dioxide can more than double alveolar ventilation. The second most powerful chemical stimulus to ventilation is a change in arterial $[H^+]$. Arterial oxygen has the smallest effect on ventilation. The reason that carbon dioxide is such a powerful stimulus to respiration is because the only effective way the body can regulate short-term changes in blood $[H^+]$ is by the removal of CO_2 through alveolar ventilation.

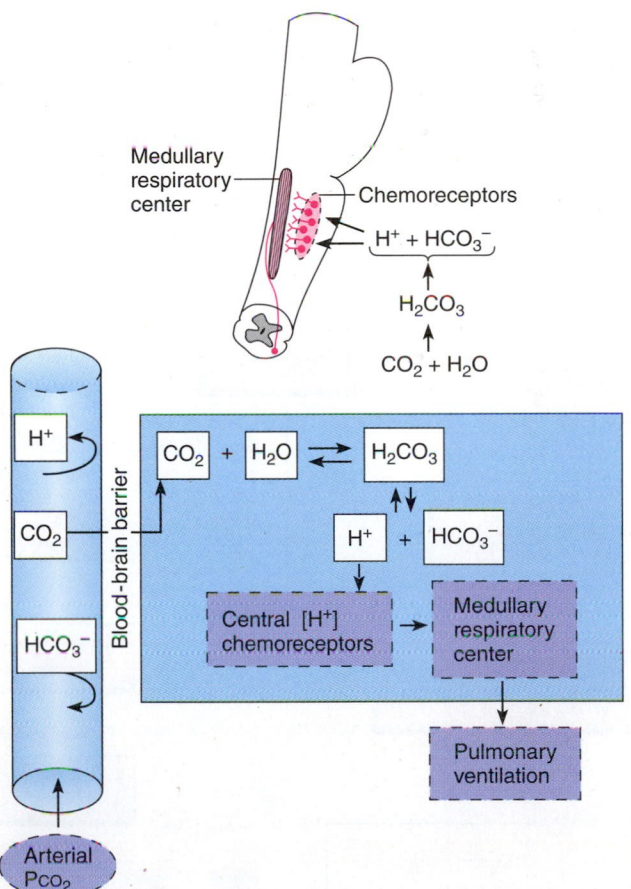

Figure 21-16

Central chemoreceptors are located in the medulla next to the respiratory center and are sensitive to changes in cerebral spinal fluid $[H^+]$. Note the blood-brain barrier is impermeable to hydrogen ions $[H^+]$ and to bicarbonate (HCO_3^-). The central chemoreceptors are more sensitive to changes in $[H^+]$ than the peripheral chemoreceptors.

If arterial CO_2 levels suddenly become too high, all chemical reactions essentially stop because most enzymatic reactions operate within a precise range of pH. To prevent this, the respiratory center regulates alveolar ventilation so that Pco_2 stays within a narrow range. Alveolar ventilation is increased, which results in "blowing off" excess CO_2 and thereby decreasing arterial Pco_2 back to normal. This increase in ventilation is known as **hyperpnea.** A distinction must be made between the terms *hyperpnea* and *hyperventilation*. *Hyperpnea* indicates increased minute ventilation but without a change in arterial Pco_2, whereas *hyperventilation* means increased alveolar ventilation with a concomitant decrease in arterial Pco_2. Conversely, if arterial Pco_2 becomes too low, the blood becomes too alkaline (basic), which depresses the respiratory center and causes a decreased alveolar ventilation. With decreased alveolar ventilation, CO_2 is retained, and arterial Pco_2 increases back to normal.

The control of ventilation by carbon dioxide is mediated by both the central and peripheral chemoreceptors. When CO_2 diffuses across the blood-brain barrier (see Fig. 21–16), it combines with water by the enzyme carbonic anhydrase, to form carbonic acid (H_2CO_3), which rapidly dissociates into HCO_3^- and H^+ in the following manner:

$$H_2O + CO_2 \underset{\text{Carbonic anhydrase}}{\rightleftharpoons} H_2CO_3 \rightleftharpoons HCO_3^- + H^+$$

Since H^+, HCO_3^-, and CO_2 are in equilibrium, any increase in arterial Pco_2 causes a profound increase in the hydrogen ion concentration in the cerebrospinal fluid, which in turn stimulates alveolar ventilation. Low arterial Pco_2 has an equally strong inhibitory effect on ventilation. Although the central and peripheral chemoreceptors work in concert to regulate CO_2, more than two thirds of the stimulation comes from the central chemoreceptors (Fig. 21–17). For further information about $[H^+]$ regulation and buffers, please refer to Chapter 25.

The control of alveolar ventilation by changes in arterial hydrogen ion concentration occurs exclusively through the peripheral chemoreceptors. The central chemoreceptors do not respond to arterial hydrogen ions because the blood-brain barrier is impermeable to them (see Fig. 21–16). An increase in systemic arterial hydrogen ion concentration activates the peripheral chemoreceptors, which in turn stimulate the respiratory center to increase alveolar ventilation. The increased ventilation blows off excess CO_2, thereby decreasing arterial Pco_2, and indirectly decreases the arterial hydrogen ion concentration (Fig. 21–18). The converse is true for low arterial hydrogen ion concentrations, which inhibit ventilation.

In many situations, a change in arterial hydrogen ion concentration is caused by factors other than altered blood carbon dioxide. For example, an increase in blood $[H^+]$ occurs with severe diabetes owing to excess fatty acids and

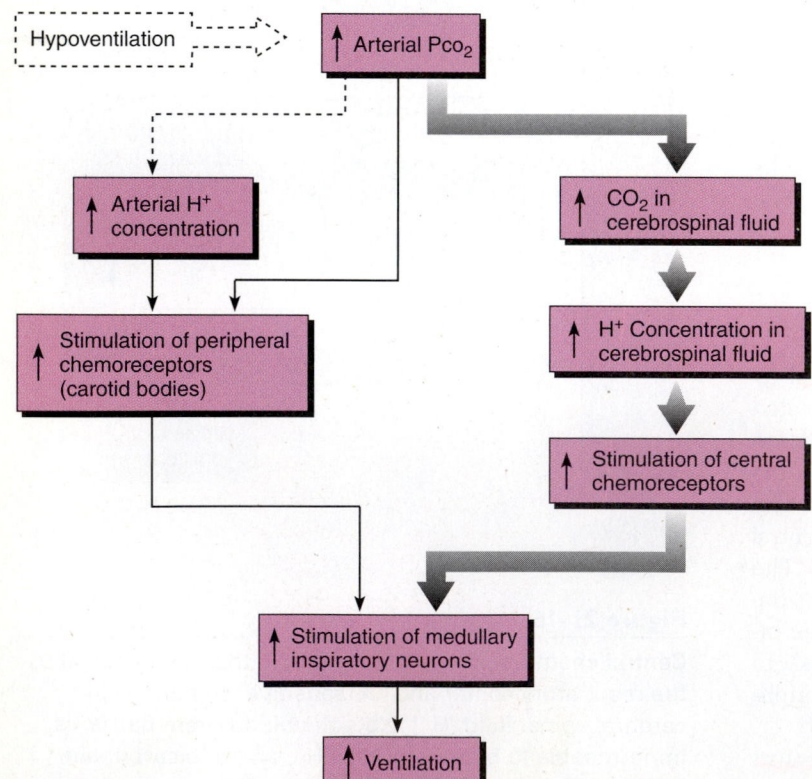

Figure 21–17

Stimulation of alveolar ventilation by carbon dioxide. The main stimulus is via the central chemoreceptors *(thick, solid line)*. Note that the direct effect of arterial Pco_2 on the peripheral chemoreceptors *(thin, solid line)* is secondary to the central chemoreceptors. The dashed line indicates stimulation of pH resulting from the increase in arterial Pco_2.

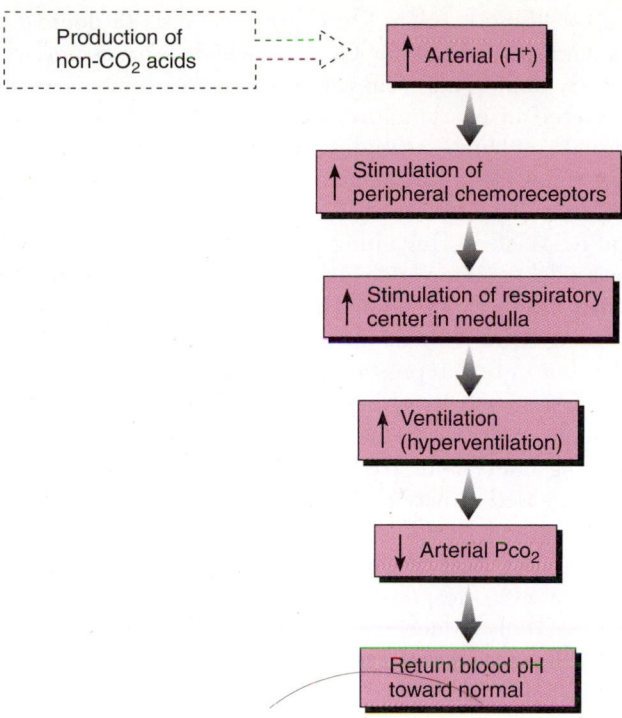

Figure 21–18

Increase in alveolar ventilation by arterial blood pH. Changes in hydrogen ion concentration can occur through the production of non-CO_2 acids, such as lactic acid from muscles, or from excess fatty acid in the blood with severe diabetes.

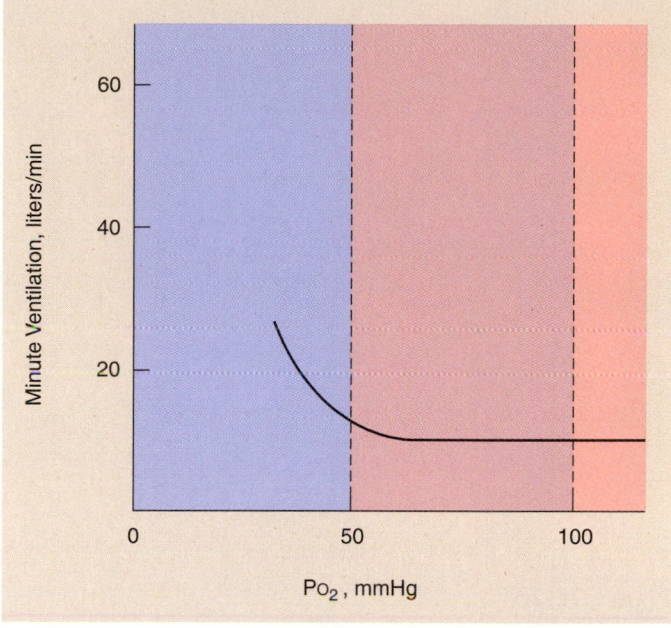

Figure 21–19

Hypoxia-induced increased ventilation occurs via a decrease in arterial P_{O_2} that is detected by the peripheral chemical receptors, mainly the carotid bodies.

ketone bodies in the blood. Acidosis stimulates ventilation through the peripheral chemoreceptors. Another example is during vomiting, in which hydrogen ions are lost. With the consequent decrease in arterial hydrogen ion concentration, ventilation is depressed because the alkalosis inhibits the peripheral chemoreceptors.

Hypoxia Stimulates Ventilation

 How does the respiratory system respond to changes in arterial oxygen level?

As mentioned earlier, of the three chemicals that stimulate ventilation, oxygen has the smallest effect. It may seem surprising that the control of ventilation is so insensitive to reduced oxygen tension because most of us think intuitively that breathing is controlled by the body's need for oxygen. The reason that arterial P_{O_2} has so little effect on the ventilation is that the peripheral chemoreceptors are not sensitive to arterial P_{O_2} of more than 60 mm Hg. When the arterial P_{O_2} becomes too low (< 85 mm Hg), the blood becomes **hypoxemic.** Thus, the **hypoxic response** is due to a change in

alveolar oxygen tension (Pa_{O_2}) and not to a change in the percentage of oxygen concentrations (F_{O_2}), or the % O_2. Hypoxemia occurs in many pulmonary disorders (e.g., asthma, emphysema, cystic fibrosis, and pulmonary edema) as well as at high altitude. When the Pa_{O_2} decreases, the body makes noticeable effort to deliver normal amounts to tissue. Chief among the hypoxic responses to altitude is hyperventilation. As seen in Figure 21–19, hypoxia-induced hyperventilation does not occur until the arterial P_{O_2} decreases to less than 60 mm Hg.

CONTROL OF BREATHING DURING SLEEP

 How do breathing patterns change during sleep?

People spend approximately one third of their lives in sleep, and disorders of sleep and of breathing during sleep are common and often of physiological consequence. Chapter 11 described the two different neurophysiologic sleep states: (1) rapid eye movement (REM) sleep and (2) non-rapid eye movement (NREM) sleep, or slow-wave sleep. Sleep is the result of withdrawal of a wakefulness stimulus that arises from the brain stem reticular

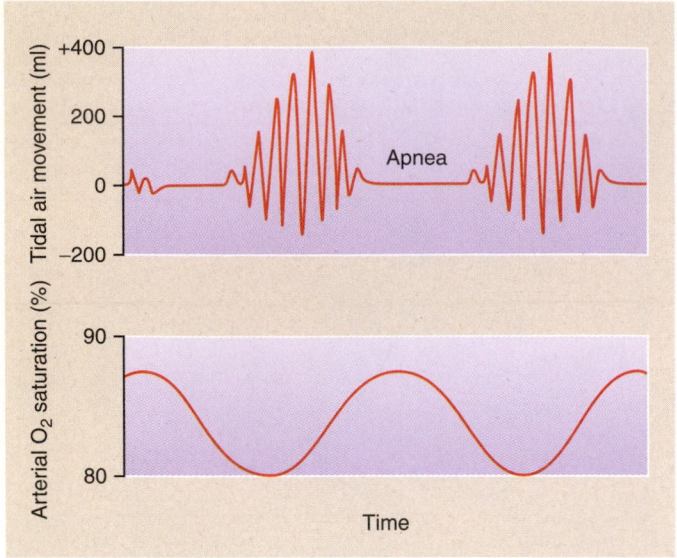

Figure 21–20

Cheyne-Stokes breathing and its effect on arterial O_2 saturation. Cheyne-Stokes breathing occurs frequently during breathing. A decrease in arterial Po_2 and an increase in arterial Pco_2 during apneic periods ultimately induce a response and breathing returns again.

formation. This wakefulness stimulus is one component of the tonic excitation of brain stem respiratory neurons, and sleep results in a general depression of breathing. There are, however, other changes, and the effects of REM and NREM sleep on breathing differ.

Sleep Changes the Responses to Respiratory Stimuli

During NREM sleep, breathing frequency and inspiratory flow rate are reduced and minute ventilation decreases. This is due, in part, to the reduced physical activity during sleep and a change in the set point of the chemoreceptors for carbon dioxide. During sleep, there is a small (≈ 3 mm Hg) increase in $Paco_2$, and the change in either sensitivity or the set point allows minute ventilation to decrease during sleep. In the deepest stage of NREM sleep, breathing is slow, deep, and very regular. However, during light sleep, the depth of breathing sometimes varies periodically. When returned, breathing is excited not only by the wakefulness stimulus but also by the carbon dioxide retained during the interval of sleep. This periodic pattern of breathing, illustrated in Figure 21–20, is known as **Cheyne-Stokes breathing,** named after a Scottish physician (Cheyne) and an Irish physician (Stokes) who first characterized the abnormality. In most cases, Cheyne-

Stokes breathing is related to central asphyxia (a decrease in Po_2 and an increase in CO_2) in which the respiratory center is depressed. Cheyne-Stokes breathing is also encountered in heart failure, brain injury, and in normal individuals at high altitudes. Cheyne-Stokes breathing is sometimes seen in newborns, especially those born prematurely.

In REM sleep, breathing frequency varies erratically, whereas tidal volume varies little. This results in a slight reduction in alveolar ventilation. Unlike NREM sleep, the variations during REM sleep do not reflect a changing wakefulness stimulus but instead represent responses to increased central nervous system activity of behavioral, rather than autonomic or metabolic, control systems.

During intervals of REM sleep in which there is little sign of increased activity in other domains, the breathing response to carbon dioxide is slightly reduced, resembling the response during NREM sleep. However, during intervals of increased activity, response to carbon dioxide during REM sleep is markedly reduced; at those times, breathing seems to be under the control of the behavioral control system. It is interesting that subservience of breathing to the behavioral control system during REM sleep, rather than to carbon dioxide, is similar to the way breathing is controlled during speech.

Ventilatory responses to hypoxia are often reduced during both NREM and REM sleep, especially in people who have high sensitivity to hypoxia during wakefulness. No difference seems to exist between the effects of NREM and REM sleep on hypoxic responsiveness, and the irregular breathing of REM sleep is unaffected by hypoxia.

Arousal Mechanisms Protect the Sleeper

In general, arousal from REM sleep is more difficult than from NREM sleep. Several stimuli cause arousal from sleep. In humans, an increase in arterial Pco_2 is a more potent arousal stimulus than is hypoxia, the former requiring a $Paco_2$ of approximately 55 mm Hg and the latter requiring a Pao_2 less than 40 mm Hg. Airway irritation and airway occlusion induce arousal readily in NREM sleep but much less readily during REM sleep. All of these arousal mechanisms probably are effective through activation of a reticular arousal mechanism similar to the wakefulness stimulus. They serve a very important role in protecting the sleeper from airway obstruction, alveolar hypoventilation of any cause, and entrance into the airways of irritating substances. Recall that cough depends on the aroused state and without arousal airway irritation leads to apnea. It should be obvious that wakefulness altered by other than natural sleep, such as during drug-induced sleep, brain injury, or anesthesia leaves the individual exposed to risk of a coma because arousal from those states is impaired or blocked.

Airway Patency Is Compromised During Sleep

Both NREM and REM sleep cause an important change in responses to airway irritation. Specifically, a stimulus that causes a cough, and airway constriction during wakefulness, causes apnea (cessation of breathing) and airway dilation during sleep unless the stimulus is sufficiently intense to cause arousal. The limited information available suggests that the lung stretch reflex is unchanged or somewhat enhanced during arousal from sleep, but the effect of stretch receptors on upper airways during sleep may be important.

A general reduction in skeletal muscle tone also occurs during sleep and is particularly prominent during REM sleep. Muscles of the larynx, pharynx, and tongue share in this relaxation, which often causes obstruction of the upper airway. A common consequence of this upper airway obstruction is snoring. However, in many individuals, most often men, the degree of obstruction may at times be sufficient to cause essentially complete occlusion. In such individuals, an intact arousal mechanism prevents disaster, and this sequence is not in itself unusual or abnormal. In some subjects, however, obstruction is more often complete and more frequent and the arousal threshold may be increased. Repeated obstruction leads to significant hypercapnia and hypoxemia, and repeated arousals cause sleep deprivation that leads to excessive daytime sleepiness, often interfering with daily activity.

Sleep Apnea Causes Hypoxemia

In some individuals, the degree of obstruction may be sufficient to cause complete occlusion of the airway that can lead to **sleep apnea,** a condition in which respiration stops for relatively long periods of time (30–60 sec). Obstruction of the upper airway during sleep is often accentuated with flabby, fatty tissue that causes additional narrowing. When breathing stops, arterial CO_2 and O_2 changes become exaggerated (high P_{CO_2} and low P_{O_2}). Eventually, it is the rising levels of arterial CO_2 that stimulate the respiratory center. The individual reacts with grunting and gasping breaths and immediately returns to a deeper sleep until the next apneic episode occurs. Sleep apnea victims may stop breathing 20 to 30 times per hour, although they are usually unaware of the sleep disruptions. During these episodes, the arterial P_{O_2} becomes dangerously low. The loss of sleep can lead to pronounced daytime drowsiness, a classic symptom of sleep apnea. Some of the victims are involved in automobile accidents because they fall asleep at the wheel. Sleep apnea occurs most frequently in obese people, especially men who suffer from the obstructive form. The real danger of sleep apnea is that death from cardiac arrhythmia can occur during an apneic episode.

In another type of sleep apnea, called *nonobstructive,* the respiratory muscles stop working, producing a similar cycle of repeated sleep disruptions. This form often causes insomnia rather than excessive daytime sleepiness. No one knows what causes the nonobstructive form of sleep apnea. Some respiratory physiologists think that the cause of the nonobstructive form may be reduced sensitivity of the chemoreceptors to carbon dioxide during certain phases of sleep. At present, sleep apnea is being studied extensively because it may be linked to **sudden infant death syndrome (SIDS),** the disease in which a seemingly normal, healthy infant is found dead in his or her crib. Many of these infants, upon autopsy, have been found to have poorly developed carotid bodies, which are located in carotid arteries.

UNUSUAL BREATHING PATTERNS

A respiratory movement that is very common and with which we are all familiar is the **yawn.** The yawn is associated with slow or shallow breathing and is thought to be involved in preventing alveoli from collapsing (atelectases). Recall from Chapter 20 that, at low lung volumes, alveoli tend to become very unstable. The yawn is a deep involuntary inspiration with the mouth open, often accompanied with stretching. The yawn is thought to aid the spreading of surfactant to stabilize alveoli.

Another respiratory movement is breath-holding, in which normal breathing patterns are voluntarily suspended. Breath-holding can occur until the arterial P_{CO_2} increases and overrides the consciousness voluntarily effort. Breath-holding is a common maneuver with diving and is usually harmless. However, breath-holding becomes dangerous if hyperventilation precedes the event. For example, if a diver hyperventilates before going under water, he or she can hold the breath longer because of the strong inhibitory effect of blowing off excess CO_2. The problem with hyperventilation is that arterial P_{CO_2} is drastically decreased, with no appreciable change in oxygen content. As a result of the low arterial CO_2 tension, the respiratory centers become depressed. At the same time, the exercising muscles rapidly take up oxygen, which decreases both the arterial oxygen content and P_{O_2}. The low arterial P_{O_2} stimulates the carotid chemoreceptors. However, the hypoxia-induced stimulation is overridden by the strong inhibitory effect of the low P_{CO_2}. As a result, the brain becomes hypoxic, causing the swimmer to black out under water and drown, commonly referred to as **shallow-water blackout.**

Another breathing pattern associated with water is the **diving reflex.** The diving reflex is most pronounced in newborns and young children and is elicited when the face is submerged in water. The diving reflex initiates a number of cardiopulmonary responses. The reflex causes the individual to gasp; breathing stops (apnea), heart rate decreases (bradycardia), and blood flow to the peripheral tissue also decreases (peripheral vasoconstriction). Diverting blood away from the peripheral tissues allows for a heart-brain perfusion. The diving reflex that causes these cardiopulmonary changes is a protective reflex. The gasp component of the diving reflex stops breathing and prevents water from entering the lung. The reflex component that induces bradycardia reduces energy requirements for cardiac tissue, and the peripheral vasoconstriction conserves oxygen to the brain. A number of cases have been reported in which young children have broken through ice on a frozen pond or have fallen into a lake and survived under water for 20 to 30 minutes. Often, the diving reflex by itself is not enough to save victims who remain under water for periods of time longer than 10 to 15 minutes. Decreased body temperature caused by the cold water is equally important. A lower body temperature means a lower metabolic rate, which means that less oxygen is required. The reason that children survive better than do adults is twofold. First, children have a more

pronounced diving reflex. Second, their body temperature decreases at a faster rate because of their high surface area-to-body weight ratio. For further details about the relationship between body temperature and surface area, see Chapter 27.

Another type of altered breathing pattern that affects ventilation is called the **Pickwickian syndrome.** This disorder occurs in severely obese people who often suffer from hypoventilation. The Pickwickian syndrome was named after "Joe," the boy in Charles Dickens' novel *The Pickwick Papers.* Joe was excessively sleepy during the day and snored loudly. Pickwickian patients suffer from hypoventilation because of their excess weight. These individuals usually have abnormalities in the control of breathing and often suffer from sleep apnea, which may account for the fact that they are always sleepy during the day.

CHAPTER REVIEW

Summary

- Pulmonary arteries connect to a fine network of capillaries embedded in the walls of alveoli.
- Gas exchange is extremely efficient and takes place in the alveolar surface area capillaries, where the blood loads with oxygen and unloads carbon dioxide.
- The lymphatic system is an auxiliary network of vessels that help prevent pulmonary edema by draining excess fluid leaked from the capillaries into the interstitial space.
- Blood pressure in the pulmonary circulation is low in comparison to arterial pressure in the rest of the body, usually with a mean pressure of 15 mmHg. This low-pressure, low-resistance circuit accommodates all of the cardiac output.
- In upright individuals, gravitational pull causes more blood to flow to the lower part of the lungs.
- Hypoxic pulmonary vasoconstriction is important in balancing blood flow with alveolar ventilation in the lungs. Hypoxic pulmonary vasoconstriction increases pulmonary vascular resistance in the hypoxic area, and routes blood to the better oxygenated regions of the lung.
- The body compensates for increased oxygen demand during exercise by enhancing the oxygen uptake from the alveolar gas in two ways: (1) by increasing the number of capillaries through which the blood flows, thus increasing gas-exchange surface area and (2) by shortening the capillary transit time.
- During exercise, the body increases cardiac output, which causes pulmonary arterial pressure to increase. Some of the increased blood flow perfuses the upper regions, resulting in a more even perfusion of the lungs.
- At rest, the blood delivers 5 ml of oxygen for every 100 ml of blood that circulates through the tissues. Resting cardiac output is 5 L/min; therefore, 250 ml/min of oxygen is delivered to the

tissues. With exercise, this delivery can be increased to 15- to 20-fold.

- Hemoglobin greatly facilitates transport of oxygen by loosely binding O_2 to its four binding sites in the alveolar capillaries and then releasing the oxygen in the tissues. The oxyhemoglobin equilibrium curve describes how hemoglobin is loaded and unloaded with oxygen at a given oxygen tension.
- During rest in normal adults, 4 vol.% of carbon dioxide are transported to the lungs to be exhaled each minute. Carbon dioxide is transported in the blood in three ways: (1) bicarbonate ions, (2) carbaminohemoglobin, and (3) dissolved carbon dioxide.
- Before birth, the lungs of the fetus do not serve any respiratory function. Gas exchange takes place between the mother and the fetus through the placenta. In the fetus, blood is shunted past the lungs by the foramen ovale and the ductus arteriosus.
- After birth, the blood flow through the lungs increases, systemic arterial pressure increases, and the foramen ovale and ductus arteriosus close. Examples of congenital heart disease are patent ductus arteriosus (ductus remains open) and ventricular septal defect (a hole in the ventricular septum that separates the right and left ventricles).
- Ventilation is controlled by negative and positive feedback systems, which maintain normal arterial blood gases and minimize the work of breathing in response to changes in the environment, activity, and lung function.
- The basic breathing rhythm is generated by neurons in the medulla, and ventilatory reflexes can modify the basic rhythm.
- Mechanical and chemical irritation of airways stimulate coughing, bronchoconstriction, shallow breathing, and excess mucous production. These responses are mediated by vagal

nerve endings, which are sensitive to mechanical and chemical irritation.

- Arterial P_{CO_2} is the most important determent in the ventilatory drive in resting individuals.
- The central chemoreceptors detect changes only in arterial P_{CO_2}, but peripheral chemoreceptors, especially the carotid body, sense changes in arterial P_{O_2}, P_{CO_2}, and pH.
- Exercise is the most common cause of increased ventilation in healthy normal individuals. However, the known reflexes cannot explain the ventilatory drive to increase minute ventilation.
- The hypoxic drive to increase minute ventilation is not very sensitive until the arterial P_{O_2} decreases to less than 60 mmHg.
- Sleep results in a general depression of breathing and is induced by the withdrawal of a wakefulness stimulus arising from the reticular formation in the brain stem.
- Sleep leads to disturbances in airway patency that can lead to sleep apnea.

Review Questions

Choose the Correct Answer

1. Which of the following causes the oxyhemoglobin-equilibrium curve to shift to the right?
   a. Low CO_2
   b. Low O_2
   c. Low pH
   d. Low temperature
   e. All of the above
2. Carbonic anhydrase catalyzes the reaction between:
   a. water and carbon dioxide.
   b. the dissociation of carbonic acid to bicarbonate ions and hydrogen ions.
   c. the combination of hydrogen ions and hemoglobin.
   d. carbon dioxide and hemoglobin to form carbaminohemoglobin.
   e. Both b and d
3. Blood flow in the lung:
   a. is equal to the total blood flow to the rest of the body.
   b. goes mostly to the upper lung.
   c. is less per minute than in the rest of the body.
   d. goes mostly to the lower lung.
   e. Both a and d
4. Oxygen removal from the alveolar gas can be facilitated by:
   a. increased alveolar P_{O_2}
   b. capillary recruitment.
   c. increased capillary transit times.
   d. increased hematocrit.
   e. All of the above
5. The pulmonary and systemic circulation have the same:
   a. systolic pressure.
   b. volume flow of blood per minute.
   c. mean pressure.
   d. diastolic pressure during rest.
   e. vascular resistance.
6. Hypoxic vasoconstriction:
   a. helps to maintain ventilation-perfusion balance.
   b. directs blood away from hypoxic alveoli.
   c. causes an increase in pulmonary artery pressure at high altitude.
   d. increases pulmonary capillary recruitment.
   e. All of the above
7. The average time that a red blood cell spends in the pulmonary capillaries during resting conditions is approximately:
   a. 0.25 sec.
   b. 0.50 sec.
   c. 0.75 sec.
   d. 1.5 sec.
   e. 3.0 sec.
8. The pulmonary circulation:
   a. converts angiotensin to its active form.
   b. has a large surface area.
   c. has a high systolic pressure.
   d. Both a and b
   e. a, b, and c
9. In fetal circulation:
   a. arterial oxygen tension closely approximates maternal oxygen tension.
   b. pulmonary circulation receives all of the cardiac output.
   c. blood flows from the right to the left side of the circulation through the patent ductus arteriosus.
   d. all blood flows through the placenta.
   e. a, c, and d
10. Most of the carbon dioxide is transported in the blood in the form of:
    a. dissolved CO_2 in plasma.
    b. carbonic acid in plasma.
    c. carbaminohemoglobin in the red cell.
    d. bicarbonate ions in plasma.
11. Hypoxia stimulates ventilation primarily through:
    a. central chemoreceptors in the pons.
    b. peripheral chemoreceptors.
    c. medullary chemoreceptors.
    d. direct stimulation of the chemoreceptors in the diaphragmatic muscle.
    e. All of the above
12. Control of basic rhythms of breathing is located in the:
    a. medullary center.
    b. pneumotaxic center.
    c. respiratory center located in the cortex.
    d. respiratory center of the pons.
    e. Both b and d
13. Which of the following causes the greatest stimulus to ventilation?
    a. Blood bicarbonate ions in blood
    b. P_{O_2} in arterial blood
    c. Hydrogen ions in blood
    d. P_{CO_2} in arterial blood
    e. P_{O_2} in alveoli
14. Which of the following is true about cerebrospinal fluid?
    a. Its protein content is equal to that of plasma.
    b. Its P_{CO_2} equals that of systemic arterial blood.

c. It is freely accessible to blood hydrogen ions.

d. Its composition is essentially that of a plasma ultrafiltrate.

e. Its pH is a function of $PaCO_2$.

15. Which of the following is *not true* during sleep?

    a. Airway irritation evokes apnea.

    b. Airway irritation evokes cough.

    c. Airway irritation evokes arousal.

    d. Airway occlusion evokes arousal.

    e. Hypercapnia evokes arousal.

16. With regard to the control of minute ventilation by carbon dioxide:

    a. changes in arterial P_{CO_2} are mediated through the peripheral chemoreceptors.

b. central effects are mediated by direct effects of H^+ on cells of the DRG-VRG complex.

c. sensitivity of the control system is directly related to the prevailing PaO_2.

d. this mechanism is more sensitive than is control in response to oxygen.

e. All of the above

Answers to Case History Questions

1. The issue is to determine whether the alleged history of exposure to a toxic atmosphere is the cause of the patient's problem. The fact that he has normal blood gases suggests that the patient is not suffering from cyanosis and argues against a diffusion block due to alveolar-capillary membrane injury.

2. The nature of this patient's problem is demonstrated in the spirogram. The bizarre respiratory pattern observed on air breathing, which became normal when switched to oxygen is very unusual. This type of breathing pattern is not in accord with medullary control of breathing. No known physiological response exists to change pulmonary function tests when going from air breathing to breathing oxygen. This patient who is suffering from a respiratory neurosis is using higher brain centers to drive the respiratory motor neurons, presumably via the corticospinal system. These patients bypass the medullary control centers normally responsive for the control of breathing. Anxiety and psychogenic influences can lead to respiratory neurosis. Respiratory neurosis is often overlooked, but these patients are just as severely crippled by this functional disturbance as by a severe organic lung disease.

3. Bouts of respiratory alkalosis associated with periods of hyperventilation can lead to tetany, which could have accounted for the report of "convulsive seizures."

Key Terms

alveolar-capillary (p. 664)
angiotensin (p. 658)
aortic bodies (p. 673)
apnea (p. 677)
breathing patterns (p. 677)
capillary transit time (p. 663)
carbon dioxide equilibrium
  curve (p. 668)

carbonic anhydrase (p. 667)
chemoreceptors (p. 673)
cartoid bodies (p. 673)
chloride shift (p. 668)
ductus arteriosus (p. 669)
foramen ovale (p. 669)
gas uptake (p. 663)
hypoxemia (p. 661)

hypoxia (p. 661)
hypoxic vasoconstriction (p. 661)
oxygen tension (p. 664)
oxyhemoglobin (p. 665)
oxyhemoglobin dissociation
  curve (p. 665)
partial pressure (p. 664)

pulmonary circulation
  (p. 656)
pulmonary edema (p. 657)
pulmonary hypertension
  (p. 660)
septal defect (p. 669)
ventilation-perfusion (p. 662)

Suggested Readings

Altman, L.K. *Who Goes First?* New York, Random House, 1987.

Cotes, J.E. *Lung Function: Assessment and Application in Medicine*, ed 5. Boston, Blackwell Scientific, 1993.

Haddad, G.G., and Jian, C. "O$_2$-sensing mechanisms in excitable cells: Role of plasma membrane K$^+$ channels." *Annual Review of Physiology*, 59:23–41, 1997.

Patterson, D. J. "Potassium and breathing in exercise." *Sports Medicine*, 23:149–163, 1997.

Schoene, R. B. "Control of breathing at high altitude." *Respiration*, 64:407–415, 1997.

Sieck, G. "Neural control of the inspiratory pump." *News in Physiological Sciences*, 6:260–265, 1991.

Simon, T. *The Heart Explorers.* New York, Basic Books, 1966.

Snapper, J. R. "Pulmonary edema." *Hospital Practice*, May: 87–101, 1986.

Staub, N. C. *Basic Respiratory Physiology*. Oxford, Churchill-Livingstone, 1991.

Thalhofer, S., Dorow, P. "Central sleep apnea." *Respiration*, 64:2–9, 1997.

Wagner, P. D. "Determinants of maximal oxygen transport and utilization." *Annual Review of Physiology*, 58:21–50, 1996.

West, J. B. *Ventilation/Blood Flow and Gas Exchange*, ed 4. Oxford, Blackwell, 1995.

Answers to Review Questions

1. a   **2.** a   **3.** e   **4.** e   **5.** b   **6.** e   **7.** c   **8.** d
9. c   **10.** d   **11.** b   **12.** a   **13.** d   **14.** e   **15.** a
16. e

Chapter 22

THE GASTROINTESTINAL SYSTEM

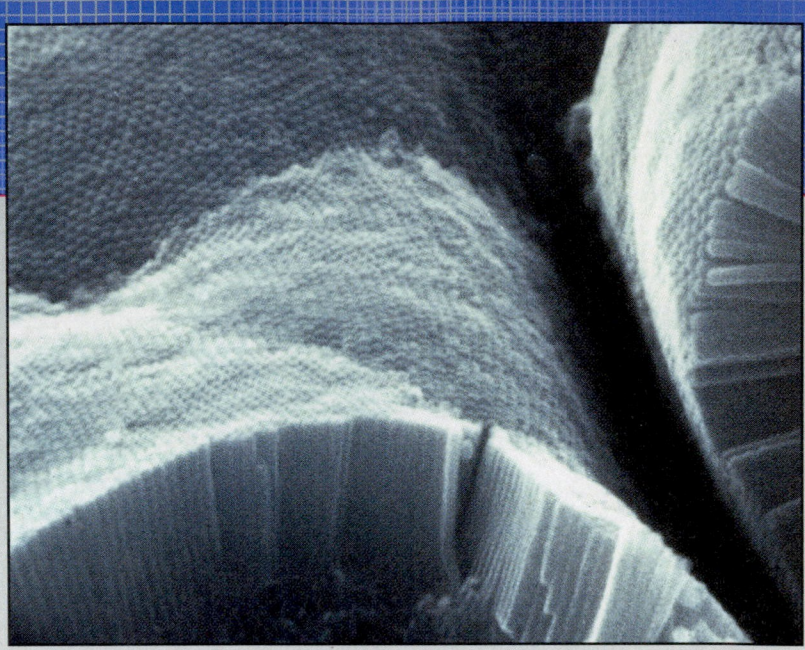

KEY CONCEPTS

- The gastrointestinal system comprises a group of glands and organs that work collectively to accomplish the digestion and absorption of food.

- Most gastrointestinal activities are coordinated and controlled by the autonomic nervous system, enteric nervous system, and numerous hormones.

- Propulsive movements, mixing movements, and tonic contractions of gastrointestinal smooth muscle keep materials moving in the right direction and promote the digestion and absorption of a meal.

- Chewing and salivary secretion prepare the food for swallowing. The salivary amylase initiates the digestion of carbohydrates.

- The stomach receives the food, mixes it, and breaks it up into small particles. It secretes hydrochloric acid, pepsinogen, gastric

lipase, and intrinsic factor. The stomach and duodenum have protective mechanisms to prevent damage by acid and pepsin.

- The pancreas is the major source of digestive enzymes and also secretes bicarbonate to neutralize gastric acid.

- Bile salts are produced by the liver, stored in the gallbladder, and promote the digestion and absorption of fat.

- Carbohydrates must be digested to monosaccharides, proteins to tripeptides, dipeptides, and amino acids, and fats (triglycerides) to monoglycerides and fatty acids before they can be absorbed.

- The small intestine is the major site of absorption of nutrients, vitamins, minerals, and water.

- The large intestine dehydrates the feces and discharges its contents at appropriate times.

CASE HISTORY

A 51-year-old man visited his doctor because of a persistent burning sensation behind his sternum (**heartburn**) that awakened him at night. The patient stated that he often felt bloated after supper and experienced heartburn that came in waves and sometimes was accompanied by a bitter or sour taste in his mouth. The patient had experienced heartburn for many years and had obtained relief with over-the-counter alkalinizing agents (antacids), such as sodium bicarbonate. Recently, the symptoms had become worse.

After a complete history and physical examination, the family physician concluded that the patient had **gastroesophageal reflux disease.** He recommended that the patient should stop smoking, restrict his consumption of alcohol, and sleep with the head of his bed elevated. He was advised to avoid meals before bedtime and fatty foods. Because the patient was obese, it was recommended that he lose some weight. Treatment with an H_2-receptor antagonist was started, and the patient felt considerably better. Gradually, however, the symptoms recurred, and he was advised to see a gastroenterologist.

The gastroenterologist passed an endoscope tube into the patient's esophagus and noted some inflammation in the mucosa of the lower part of the esophagus. He prescribed a proton pump (H^+/K^+-ATPase) inhibitor, which resulted in relief of symptoms, but eventually these returned.

A year later, 24-hour pH monitoring was arranged, with a fine pH probe positioned in the lower esophagus. The monitor recorded a decrease in esophageal pH when the patient had heartburn, demonstrating a linkage between reflux and the symptoms. The gastroenterologist adjusted the dosing of the proton pump inhibitor and recommended an H_2-receptor antagonist, to be taken at bedtime. The patient has had complete remission of his symptoms for three years but notices that the symptoms recur when he forgets to take his medication.

Questions

1. What are some possible causes of reflux of stomach acid into the lower esophagus?

2. Why does gastroesophageal reflux cause pain and damage to the esophagus?

3. Why did the doctor recommend avoidance of fatty meals?

4. Why were the symptoms especially severe at night?

5. How do H_2-receptor antagonists and proton pump inhibitors work?

INTRODUCTION

The gastrointestinal system provides the body with the nutrients that are essential for life. Dietary carbohydrates, proteins, and fats are broken down into simpler molecules before they are absorbed by the epithelial cells lining the intestine and distributed to the rest of the body via the blood and lymph. Water, minerals, and vitamins are also absorbed. The processing of food by the gastrointestinal system is controlled and coordinated by nerves and hormones.

The gastrointestinal system comprises many organs with distinct functions (Fig. 22–1). The **digestive tract** includes the mouth, pharynx, esophagus, stomach, small intestine (duodenum, jejunum, ileum), and large intestine (cecum; ascending, transverse, descending, and sigmoid colon; rectum and anal canal). The **accessory structures** include the salivary glands, exocrine pancreas, and biliary system (liver, bile ducts, and gallbladder).

Gastrointestinal function is classified as motility, secretion, digestion, and absorption. In this chapter, we will discuss these topics in turn, recognizing that all of these processes are coordinated with each other.

Motility refers to the mechanical movements in the gastrointestinal system. These movements are mainly accomplished by contraction and relaxation of smooth muscle. It is only at the beginning of the digestive tract (mouth, pharynx, and upper esophagus) and at the very end (external anal sphincter) where voluntary skeletal muscle is important. Ingested food is propelled from the mouth to the anus by **propulsive movements**. **Mixing movements** of the distal stomach serve to break up the food into small digestible particles; mixing movements of the intestines facilitate digestion and absorption. Sustained or **tonic contraction** and relaxation of sphincters keeps materials moving in the right direction. The timing of propulsive and mixing movements and relaxation of sphincters is precisely controlled to allow for proper digestion of food and absorption of nutrients.

The digestive tract and its accessory organs also add water, electrolytes, mucus, enzymes, and other substances to the lumen of the digestive tract. This **secretion** is important for proper digestion of food and for protection of the digestive tract. The stomach produces a strongly acidic secretion, whereas the salivary glands, small and large intestine, pancreas, and liver produce alkaline secretions. Acid and enzymes can be quite harmful to the organs that secrete them, and so the gastrointestinal system has various **protective mechanisms.** Secretion by the accessory glands (salivary glands, exocrine pancreas, and liver) is controlled so as to allow for efficient digestion and absorption of foodstuffs after a meal.

Digestion is the breakdown of food into molecules that can be absorbed by the intestine. Gastric (from the Greek *gaster,* "stomach") juice accounts for some digestion, but most digestion is a consequence of enzymes produced by the accessory glands (mainly the pancreas) and enzymes that are present on the luminal surface of absorptive cells of the small intestine. Before nutrients can be absorbed, polysaccharides must be broken down to monosaccharides; proteins, to

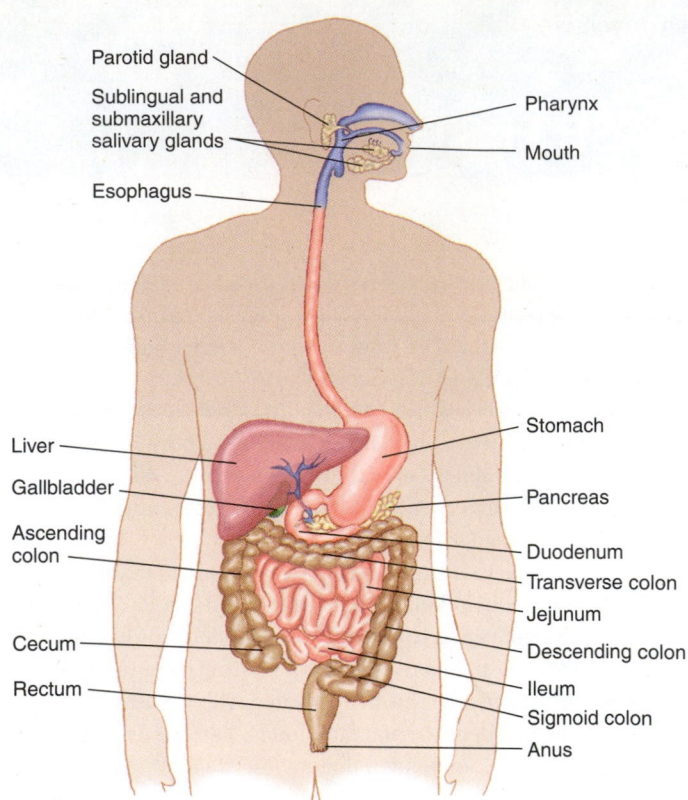

Figure 22–1

Gross anatomy of the gastrointestinal system.

tripeptides, dipeptides, and amino acids; and fat (triglycerides), to fatty acids and monoglycerides.

Absorption is the movement of materials across the epithelial cells lining the digestive tract into the blood or lymph. Different substances are absorbed by different mechanisms. Absorption of some substances (e.g., water) is passive, but in many cases, it is an active, energy-demanding process (e.g., glucose absorption). The small intestine is quantitatively the major site for absorption of organic nutrients, water, minerals, and vitamins.

Motility, secretion, digestion, and absorption are interrelated and are coordinated by nerves and hormones.

GASTROINTESTINAL MOTILITY

> *How does the arrangement of smooth muscle in the wall of the digestive tract reflect the types of movement seen in the gastrointestinal system?*

Smooth Muscle Contraction Is Responsible for Gastrointestinal Motility

Smooth muscle in the digestive tract is mainly responsible for motility. Figure 22–2 shows a schematic view of the layers of the digestive tract. The external muscle layer (**muscularis externa**) is organized into an outer **longitudinal layer** and an inner **circular layer.** In general, the circular layer is thicker; its contraction causes the lumen to become narrower and longer. Contraction of the longitudinal layer causes the intestine to become shorter and fatter. The **mucosa** is the innermost part of the intestinal wall and comprises epithelium, lamina propria (a layer of loose connective tissue that contains many blood vessels, lymphatics, and glands), and muscularis mucosa. The **muscularis mucosa,** a thin muscle layer that separates the lamina propria and **submucosa,** may function to change the absorptive and secretory area of the mucosa.

Most of the smooth muscle in the digestive tract is capable of spontaneous electrical and contractile activity. Adjacent smooth muscle cells are coupled by **gap junctions** that allow electrical currents to spread from cell to cell through these low-resistance pathways (Chapter 16). Gut smooth muscle behaves like a **functional syncytium** because hundreds of smooth muscle cells typically are depolarized and contract at the same time. Smooth muscle activity is controlled by nerves and hormones. The intrinsic nervous system of the gut controls the distance and direction of spread of electrical and mechanical activity.

In the stomach, small intestine, and large intestine, but not in the esophagus, spontaneous waves of electrical activity called **slow waves** occur (Fig. 22–3). These are due to a rhythmic pattern of membrane depolarization and repolar-ization of smooth muscles. Slow-wave electrical activity spreads from cell to cell via gap junctions. The slow waves determine the maximum frequency, direction, and velocity of rhythmic contractions. The pacemaker for the slow waves is not in nerve or muscle cells but comes from a specialized group of interstitial cells called **interstitial cells of Cajal.** These are found between the external muscle layers and in the submucosa. These musclelike, stellate cells form gap junctions with each other and with smooth muscle cells and determine the frequency of the slow waves.

Slow-wave potentials can be recorded from single muscle cells (see Fig. 22–3). If the slow-wave potential change is of sufficient magnitude to exceed the threshold potential, then one or more action potentials may occur at the peak of the membrane depolarization. The action potential is due to the rapid influx of calcium ions through voltage-gated calcium channels and is followed by a brief (phasic) muscle contraction. The amount of force developed by muscle contraction is directly related to the number of action potential firings. Generally speaking, stimuli do not affect slow wave frequency but do affect the number of action potentials fired and, in this way, the strength of contraction.

The frequency of slow waves determines the **basic electrical rhythm.** The basic electrical rhythm varies along the gastrointestinal tract. For example, in the stomach, it is three slow waves per minute; consequently, a stomach contraction

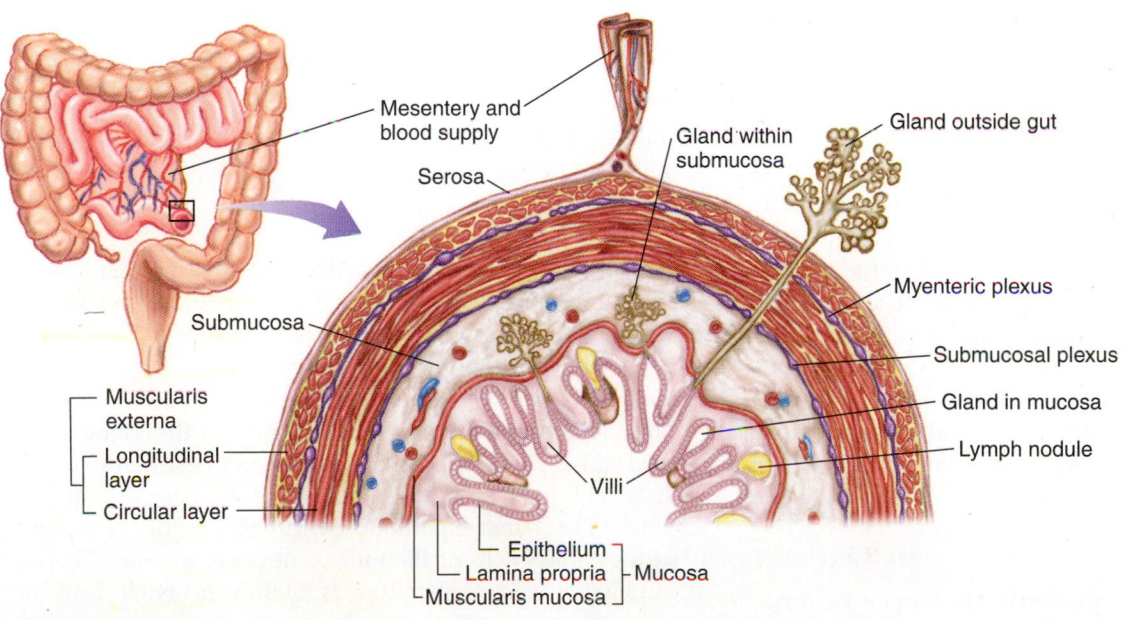

Figure 22–2

General plan of the digestive tract. Starting from the outside in, the layers include a serosal covering, muscularis externa (longitudinal and circular smooth muscle layers), submucosa, and mucosa (muscularis mucosa, lamina propria, and epithelium). Glands may be external to the digestive tract (e.g., exocrine pancreas), in the submucosa, or in the mucosa (e.g., crypts). Nerve plexuses are found between the two external muscle layers (myenteric plexus) and in the submucosa (submucosal plexus). *(Modified from Ham, A. B.* Histology, *ed 3. Philadelphia, J. B. Lippincott, 1957)*

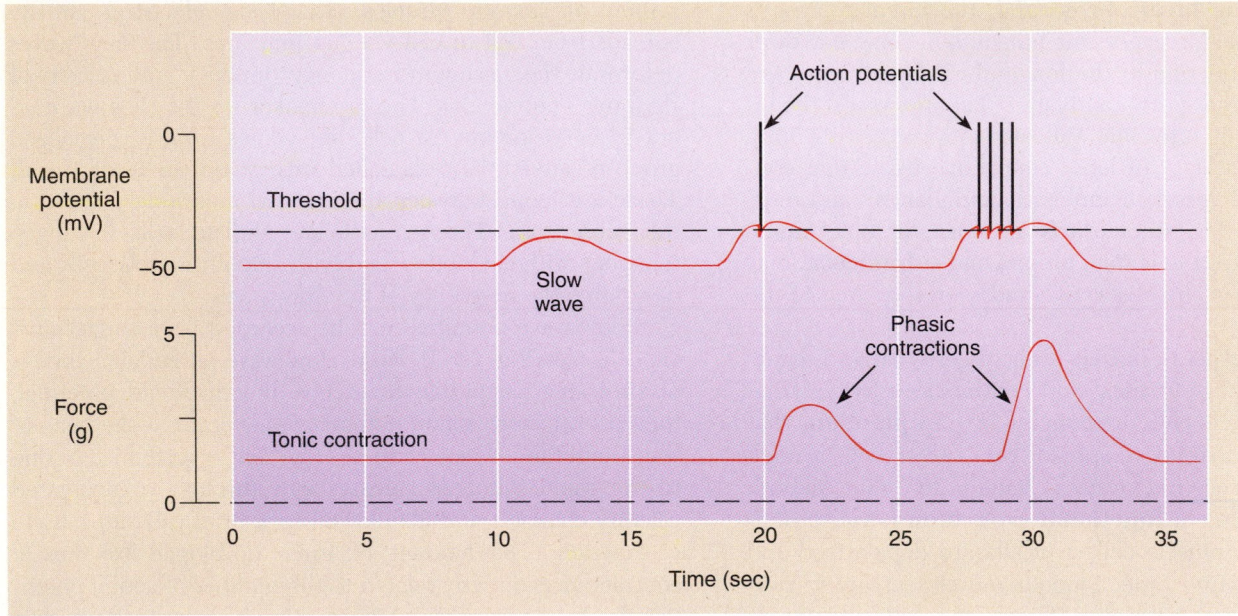

Figure 22–3

Relations between slow waves, action potentials, and muscle contraction in intestinal smooth muscle. The slow waves occur at regular intervals, last several seconds, and do not necessarily produce a phasic contraction *(bottom trace)*. If excitatory stimuli are present (e.g., acetylcholine released from nerve fibers), the slow wave depolarizes the muscle membrane sufficiently to exceed the threshold potential and brief action potentials are produced. The higher the number of action potentials, the greater the force produced by muscle contraction.

can occur at 20-second intervals but not more frequently. Not every slow wave results in a contraction. In the small intestine, the basic electrical rhythm decreases stepwise in the *aboral* (away from the mouth) direction. It is highest in the duodenum ($\approx$ 11–12 waves/min) and lowest in the terminal ileum ($\approx$ 8 waves/min). This gradient favors the aboral movement of the intestinal contents. In the large intestine, the slow-wave frequency is lowest in the proximal colon and cecum ($\approx$ 8 waves/min) and highest in the distal (sigmoid) colon ($\approx$ 16 waves/min). This reversed gradient of activity in the large intestine favors mixing of the luminal contents, allowing the absorptive surface to extract water and electrolytes efficiently.

Parasympathetic and Sympathetic Nerve Fibers Innervate the Digestive Tract

 How does the autonomic nervous system exert its influence on the digestive tract?

The activity of smooth muscle cells in the gastrointestinal tract is profoundly influenced by nerves and by hormones. The **extrinsic efferent innervation** of the digestive tract is composed of the parasympathetic and sympathetic divisions of the autonomic nervous system (Fig. 22–4). Most of the extrinsic efferent fibers project onto intrinsic (enteric) neurons in the gut wall.

The **parasympathetic division** has cranial (brain) and sacral outflows (Chapter 10). The **dorsal motor nucleus of the vagus** in the medulla supplies efferent vagus nerve fibers to the lower esophagus, stomach, small intestine, and proximal colon (including the transverse colon). The sacral outflow arises from neurons in the sacral spinal cord (S2–S4), which, via the **pelvic nerves,** innervate the distal colon (descending and sigmoid), rectum, and anal canal. The parasympathetic preganglionic fibers synapse with ganglion cells of the enteric nervous system. The preganglionic neurotransmitter is mainly acetylcholine. Increased parasympathetic nerve activity is mostly stimulatory; for example, vagus nerve stimulation increases gastric, pancreatic, and small intestinal secretion and increases blood flow and smooth muscle contraction. Some inhibitory influences are also exerted, such as relaxation of the lower esophageal sphincter and receptive relaxation of the stomach.

The **sympathetic division** arises from the thoracic and upper lumbar spinal cord. Preganglionic fibers synapse in

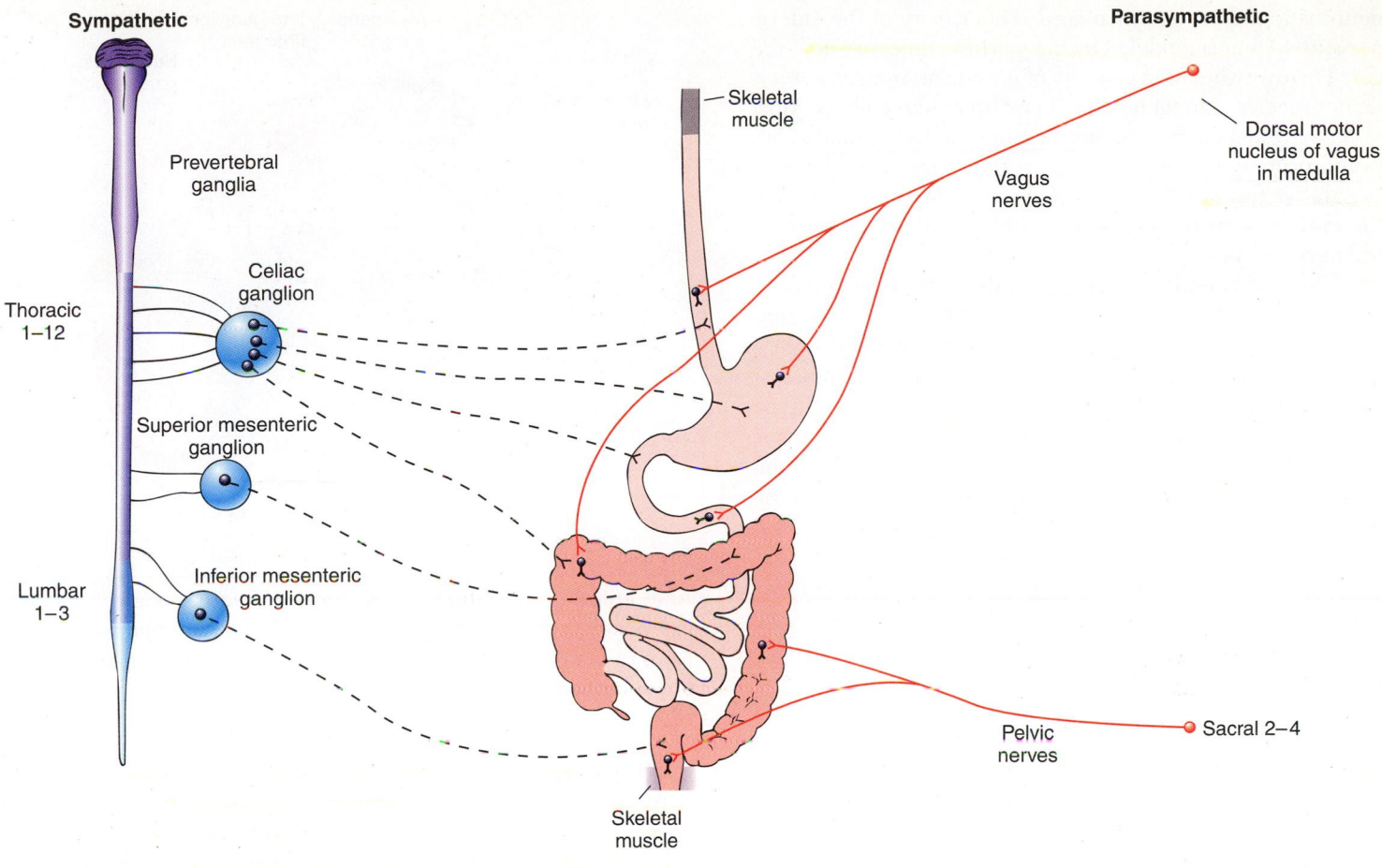

Sympathetic

Prevertebral ganglia

Thoracic 1–12

Celiac ganglion

Superior mesenteric ganglion

Lumbar 1–3

Inferior mesenteric ganglion

Skeletal muscle

Parasympathetic

Dorsal motor nucleus of vagus in medulla

Vagus nerves

Pelvic nerves

Sacral 2–4

Skeletal muscle

Figure 22–4

Innervation of the digestive tract. Sympathetic preganglionic nerve fibers arise from the thoracolumbar region of the spinal cord and synapse in prevertebral ganglia. The postganglionic sympathetic fibers are represented by dashed lines. Parasympathetic fibers in the vagus nerves supply most of the digestive tract (up to the transverse colon). More distal parts of the colon are innervated by parasympathetic fibers in the pelvic nerves. Skeletal muscle in the upper esophagus and external anal sphincter is innervated by the somatic motor system.

prevertebral ganglia: the celiac, superior mesenteric, and inferior mesenteric ganglia. The preganglionic neurotransmitter is, in general, acetylcholine; the postganglionic neurotransmitter is norepinephrine. Postganglionic fibers mainly innervate enteric neurons but can also directly innervate intramural blood vessels, intestinal crypts, and the smooth muscle of sphincters. Stimulation of sympathetic nerves generally has an inhibitory effect on the digestive tract and results in decreased motility, secretion, and blood flow. Sympathetics do not ordinarily play much of a role in gastrointestinal control. Removal of sympathetic nerve fibers (sympathectomy) produces no generalized disturbance of gastrointestinal function.

The digestive tract is also innervated by sensory neurons that are sensitive to mechanical, chemical, and thermal stimuli.

The Enteric Nervous System Plays a Major Role in Coordinating Gut Activities

What provides for local nervous control of the digestive tract?

The **enteric nervous system** is composed of neurons, the cell bodies of which are located in ganglia within the wall of the digestive tract. This intrinsic nervous system has been called the "little brain in the gut" because it is organized as an independent integrative system, with sensory neurons, interneurons, motor neurons, organized reflex pathways, and motor programs. This is similar to the way the rest of the nervous system is organized. The number of enteric neurons ($\approx$ 100 million) is approximately the same as the number of neurons found in the spinal cord. The activity of the enteric nervous system is modified by the extrinsic innervation of the gut. The overwhelming majority of enteric neurons, however, do not receive a direct input from extrinsic nerve fibers. Most of the enteric neurons are interneurons, neurons situated between primary sensory and final motor neurons that have an integrative function. Sensory information can be processed in the enteric nervous system quite independently of the central nervous system.

The vast majority of neurons in the enteric nervous system are *not* cholinergic (releasing acetylcholine) or adrenergic (releasing norepinephrine). There are many different neurotransmitters (at least 20) in the enteric nervous system (Table 22–1). These include **nonpeptide** transmitters, such as acetylcholine, serotonin, ATP, dopamine, and nitric oxide (NO). **Peptide** transmitters include cholecystokinin, opioid peptides (enkephalins), gastrin-releasing peptide (GRP), neuropeptide Y, substance P, somatostatin, and vasoactive intestinal peptide (VIP). Some of these same substances are also found in mucosal **endocrine cells** and may function in paracrine or endocrine control. This is not surprising because the peptide-secreting endocrine cells of the gut have the same embryological origin as do nerve cells. Most gut neurotransmitters are also present in the central nervous system. For example, VIP was first discovered in the intestine and then found in the brain. VIP is a leading candidate for the major inhibitory transmitter in the enteric nervous system. Some neurotransmitter substances were first found in the brain and then later discovered in the gut; an example is somatostatin, which is produced by the hypothalamus and inhibits growth hormone release from the pituitary (Chapter 13).

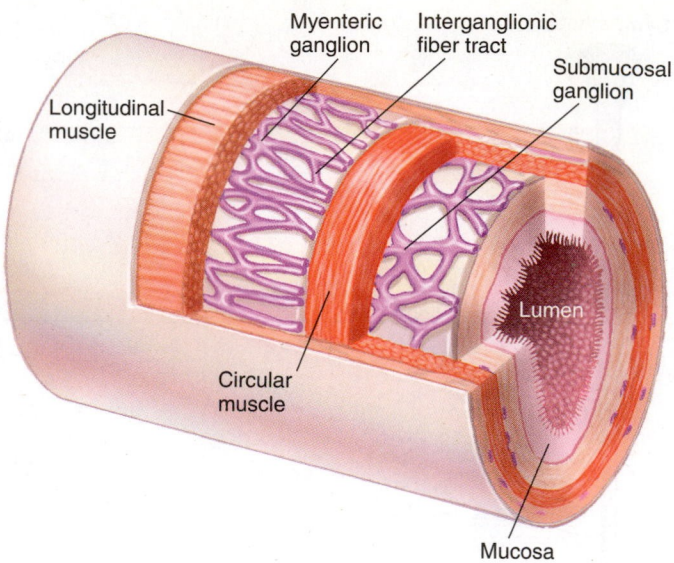

Figure 22–5

Anatomical organization of the enteric nervous system. Nerve cell bodies in ganglia and interganglionic fiber tracts form two major nerve plexuses: the myenteric plexus and the submucosal plexus.

The enteric nervous system is organized into two major plexuses (Fig. 22–5). The **myenteric plexus** is located between the circular and longitudinal muscle layers and provides most of the innervation for these muscle layers. The **submucosal plexus** is located between the circular muscle coat and the submucosa and innervates the mucosa. The nerve cell bodies are in the ganglia of the plexuses; the ganglia within each plexus are connected via **interganglionic fiber tracts.**

Figure 22–6 shows a conceptual model of the enteric nervous system. The enteric nervous system controls not only the contraction or relaxation of muscles and the diameters of blood vessels (hence blood flow) but also gut endocrine cells and the secretory activity of epithelial cells. The effector pathways are activated in a coordinated fashion, so that secretion of electrolytes, water, and mucus protects and lubricates the gut lumen, blood flow increases to support secretory activity, and muscle contraction propels the luminal contents. The enteric nervous system has motor programs for peristalsis, segmenting movements, vomiting, and defecation. It also produces different patterns of activity for the fed state (during a meal and for 2–3 hours thereafter) and fasting, or the **interdigestive state** (between digestion of meals).

Nerve Reflexes Control Motor and Secretory Activity

Nerve reflexes in the digestive tract may be classified as intrinsic (local) or extrinsic (short or long) (Fig. 22–7). **Extrinsic reflexes** occur via connections to the prevertebral ganglia, spinal cord, or brain. **Intrinsic reflexes** occur wholly within the enteric nervous system. In a **local** (or intrinsic) **reflex,**

TABLE 22–1	
Some Neurotransmitters of the Enteric Nervous System and Their Effects on Digestive Tract Motility	
Neurotransmitter	**Effect on Motor Activity**
Nonpeptides	
Acetylcholine	Usually excitatory
Serotonin	Excitatory
ATP	Inhibitory
Dopamine	Inhibitory
Nitric oxide	Inhibitory
Peptides	
Cholecystokinin	Excitatory
Enkephalins	Excitatory
Gastrin-releasing peptide	Excitatory
Neuropeptide Y	Excitatory
Substance P	Excitatory
Somatostatin	Inhibitory
Vasoactive intestinal peptide	Inhibitory

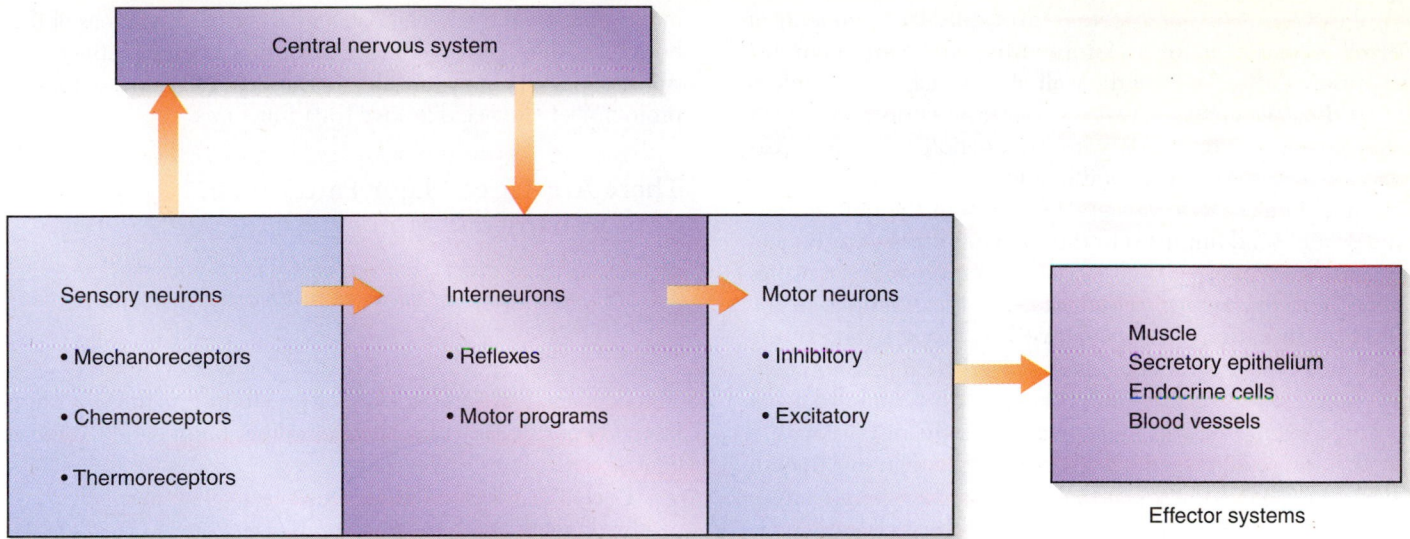

Figure 22–6

Conceptual model for the enteric nervous system. The enteric nervous system plays a key role in organizing gut activities. In addition to simple reflexes, it contains the neural circuitry for more complex motor programs of activity. *(From Wood, J. D. "Gastrointestinal neurophysiology." In Rhoades R. A., and Tanner, G. A., eds. Medical Physiology. Boston, Little, Brown and Co., 1995, pp 487–504)*

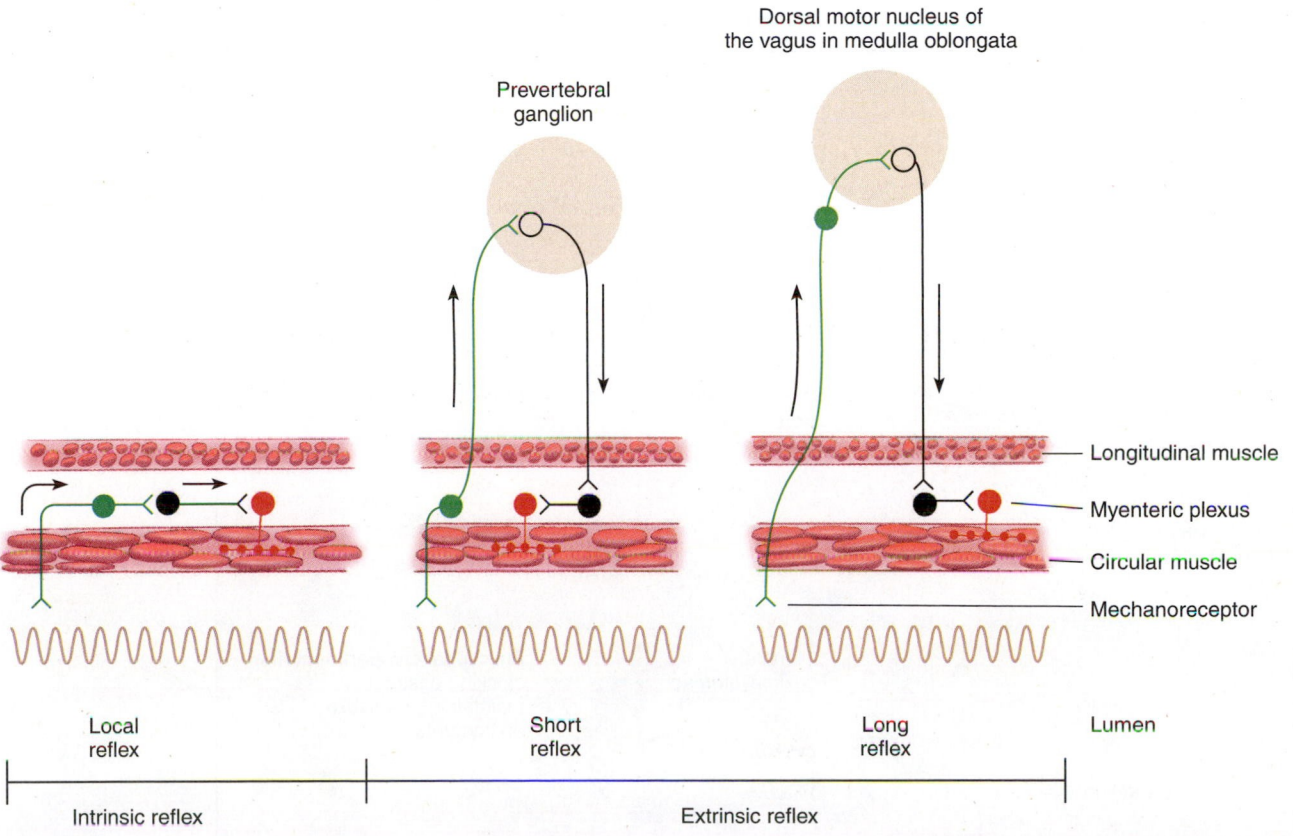

Figure 22–7

Types of reflexes in the gastrointestinal system. Sensory afferents (e.g., from a mechanoreceptor) are depicted in green. Enteric nervous system interneurons are colored black, and motor neurons are red.

only the enteric nervous system is involved. All components of the reflex (sensory neurons, interneurons, and motor neurons) are present entirely within the wall of the tract. An example is the **peristaltic reflex:** when the intestine is distended by a bolus of food, it responds with contraction in back of the bolus and relaxation ahead of the bolus.

In a **short reflex,** sensory receptors in the enteric nervous system send impulses to the **prevertebral ganglia** and synapse there with postganglionic noradrenergic neurons. These neurons, in turn, inhibit activity. For example, in the **intestino-intestinal inhibitory reflex,** if one region of the intestine is overdistended, this is sensed by mechanoreceptors, which reflexly cause inhibition of contraction in adjacent intestinal segments. This helps to alleviate the distension.

In a **long reflex,** sensory fibers from the digestive tract or other areas send impulses to the spinal cord or brain, and efferent impulses travel back to the digestive tract via autonomic nerves. An example is that distension of the stomach with food increases gastric acid secretion (but only in part) via afferent and efferent vagus nerve fibers. Another example is the **gastroileal reflex**: when a meal is swallowed, motor activity in the ileum increases and the ileocecal sphincter relaxes. This reflex helps to move the ileal contents into the large intestine

in anticipation of the new meal. Signals from other areas of the body (e.g., the sight, smell, or taste of food) can also affect gastrointestinal function. As will be discussed later, hormones also profoundly influence digestive tract functions.

There Are Three Major Patterns of Digestive Tract Motor Activity

 How are the three patterns of digestive tract motor activity generated?

Three major patterns of motility are found in the digestive tract: (1) peristalsis, (2) rhythmic segmentation, and (3) tonic contraction.

Peristalsis is a wave of contraction that normally proceeds along the gut in an aboral direction (Fig. 22–8). A major stimulus for initiation of peristalsis is distension of the gut wall (peristaltic reflex). Sensory neurons in the submucosal plexus act as mechanoreceptors and synapse with excitatory and inhibitory interneurons in the myenteric plexus. These interneurons synapse with motor neurons and coordinate the contraction of the muscle layers. A complete ring of

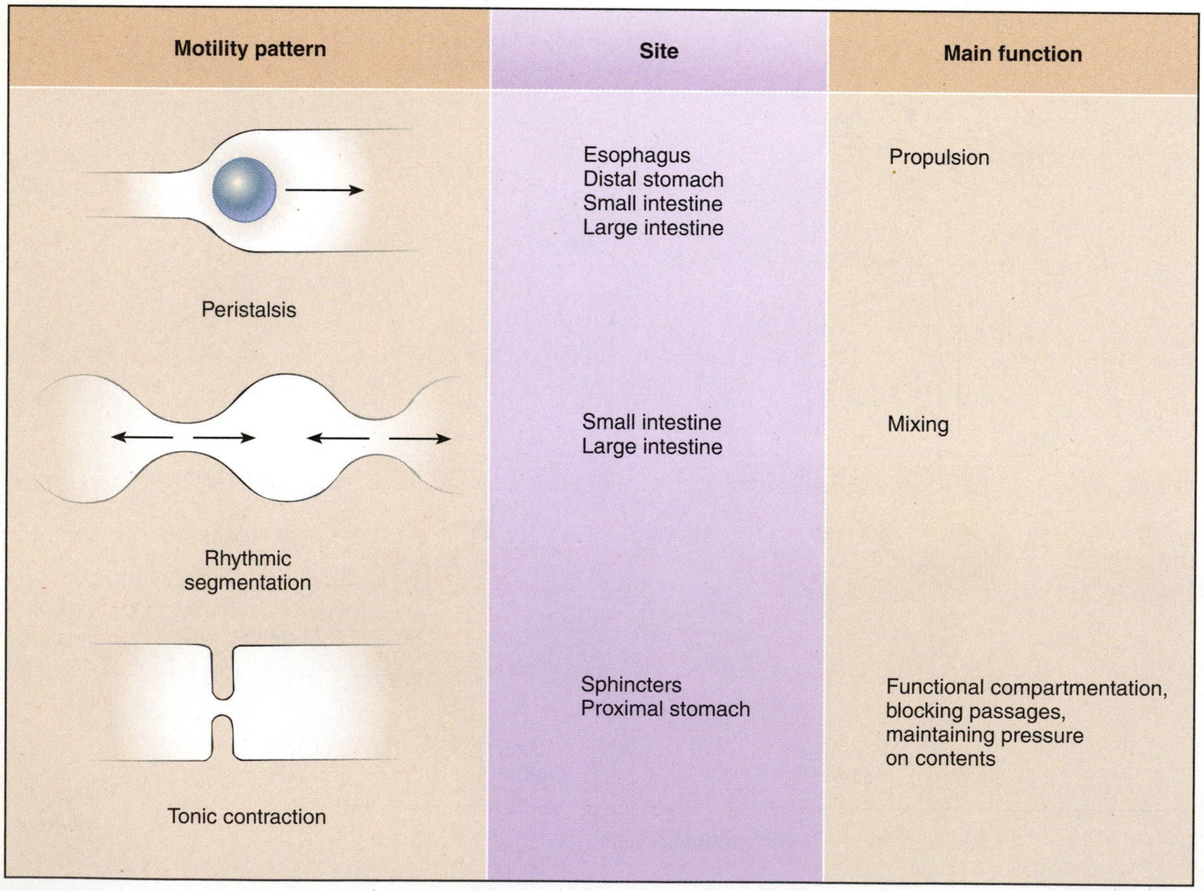

Motility pattern	Site	Main function
Peristalsis	Esophagus Distal stomach Small intestine Large intestine	Propulsion
Rhythmic segmentation	Small intestine Large intestine	Mixing
Tonic contraction	Sphincters Proximal stomach	Functional compartmentation, blocking passages, maintaining pressure on contents

Figure 22–8

Motility patterns of the gastrointestinal system.

contraction develops in the circular muscle layer behind the point of stimulation. The longitudinal muscle in this **propulsive segment** relaxes. At the same time, segments in front of the wave of contraction undergo **receptive relaxation.** This relaxation is due to neural inhibition of circular muscle activity and excitation of the longitudinal muscle layer. This widens the gut lumen and reduces the force necessary to propel the luminal contents into the **receiving segment.** Subsequently, receiving segments are converted to propulsive segments, so that a bolus of food is propelled aborally.

Segmentation or **mixing movements** are rhythmic contractions of the circular muscle layer that serve to mix and divide the intestinal contents. Ringlike contractions appear over short areas of the gut, giving it the appearance of a string of sausages. Segmentation in the small intestine is characteristic of the fed state and slowly moves the intestinal contents toward the large intestine because the frequency of segmentation is higher in the duodenum than in the ileum. Segmentation within the large intestine is called *haustration* and produces longer-lasting contractions than in the small intestine. No segmentation occurs in the esophagus or stomach.

Tonic contractions, that is, sustained contractions, are found in the proximal part of the stomach and in **sphincters.** Important smooth muscle sphincters include the lower esophageal sphincter, pyloric sphincter, sphincter of Oddi (at the entrance of bile and pancreatic ducts into the duodenum), ileocecal sphincter, and internal anal sphincter. The tonic type of contraction in these sphincters is a property of their smooth muscle cells. The magnitude of tone is controlled by nerves and hormones. The upper esophageal sphincter and external anal sphincter are also tonically contracted but consist of skeletal muscle under voluntary control. The function of tonic sphincter contraction is to block passages, separate compartments, and prevent the reflux of luminal contents. Figure 22–9 gives the location of various sphincters.

Swallowing Involves Coordinated Activity of Skeletal and Smooth Muscles

> *How is swallowing accomplished?*

Chewing is a complicated movement under voluntary control, mediated by elaborate neural control of two major reciprocal reflexes: the **masseteric** (jaw-closing) and **digastric** (jaw-opening) **reflexes.** Chewing serves to grind food into smaller particles, which are easier to swallow and digest. Chewing mixes food with saliva; saliva lubricates the food and makes it easier to swallow. Chewing food into small particles also decreases the wear and tear on the mucosal lining of the digestive tract.

Swallowing is the propulsion of a bolus of material through the mouth and down the pharynx and esophagus to the stomach (Fig. 22–10). Normally, liquids are propelled immediately from the mouth to the oropharynx by a backward movement of the tongue. When solid material is swallowed, the tip of the tongue is placed against the hard palate and separates a bolus of material from the rest of the mouth (Fig. 22–10a). The tongue then sweeps backward, forcing the bolus into the oropharynx. At the same time, the nasopharynx is closed by a downward movement of the soft palate and contraction of the superior (upper) constrictor muscles of the pharynx (see Fig. 22–10b). This prevents regurgitation into the nose. Simultaneously, respiration is inhibited and the laryngeal muscles contract to close the glottis (entrance to the larynx) and elevate the larynx. (Feel your larynx rise and move anteriorly during a swallow). These events prevent the bolus from entering the airway. A peristaltic contraction now begins in the superior constrictor muscles and proceeds through the middle and inferior constrictors, propelling the bolus through the pharynx into the esophagus (Fig. 22–10c, d). The **upper**

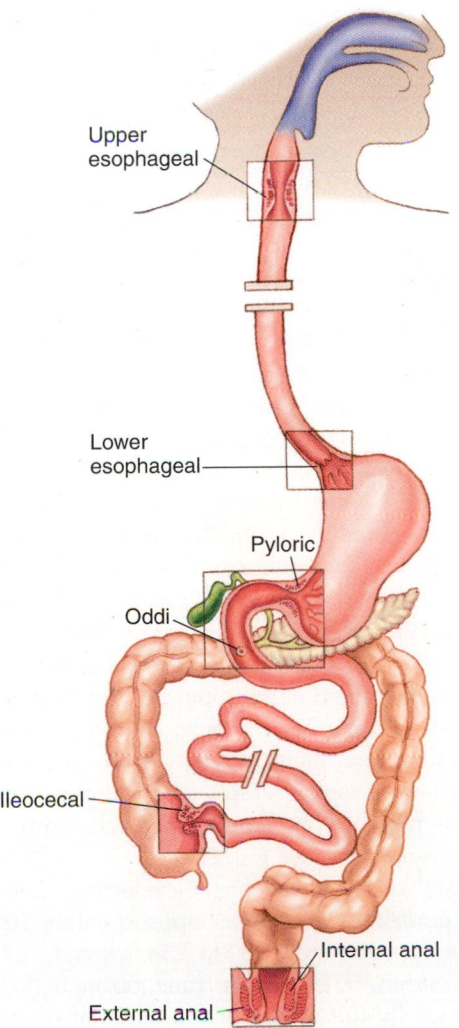

Upper esophageal

Lower esophageal

Pyloric

Oddi

Ileocecal

Internal anal

External anal

Figure 22–9

Sphincters are composed of tonically contracted circular muscle that relaxes in response to nerve or hormonal stimulation. Sphincters promote the one-way movement of luminal contents. The sphincter of Oddi controls flow from the common bile duct.

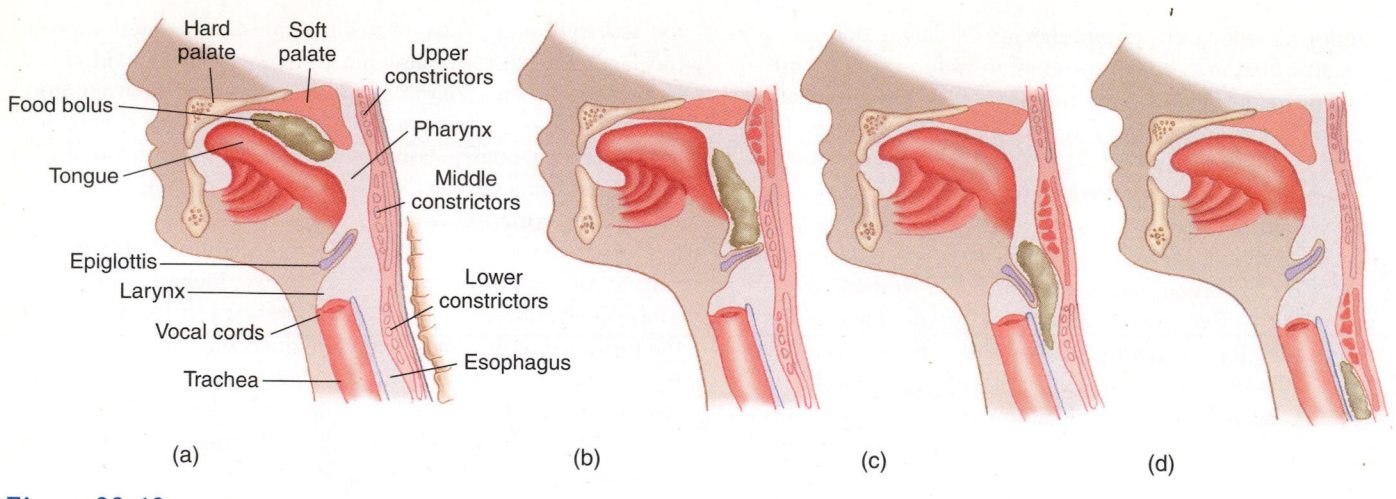

Figure 22–10

Passage of a bolus of food from the mouth through the pharynx into the upper esophagus during a swallow.

esophageal sphincter relaxes, allowing contractions of the pharyngeal muscles to move the material into the esophagus.

Swallowing is often initiated as a voluntary act but then is taken over by involuntary reflexes. Sensory information is transmitted to the brain by glossopharyngeal and vagal afferents via mechanoreceptors in the pharynx and esophagus. The information is coordinated by a **swallowing center** located in the medulla, which produces the sequential excitation or inhibition of different muscle groups. Clearly, breathing must be coordinated with swallowing. Motor fibers to the various skeletal muscles involved in swallowing travel in the Vth (trigeminal), VIIth (facial), IXth (glossopharyngeal), Xth (vagus), and XIIth (hypoglossal) cranial nerves. Relaxation of the upper esophageal sphincter is brought about by decreased motor nerve activity.

The final, or esophageal, phase of swallowing includes the movement of a bolus of food all the way to the stomach. The esophagus is a muscular tube that contains skeletal muscle in its upper one third, both skeletal and smooth muscle in its middle one third, and smooth muscle only in its lower one third. Contractile activity along the esophagus accompanying a swallow can be recorded by using a **manometric catheter** passed into the esophagus; this catheter incorporates several small tubes connected to individual pressure transducers that allow recordings to be made simultaneously at several different levels (Fig. 22–11). The pressure in the upper esophageal sphincter, before a swallow, is approximately 60 cm H_2O and prevents a person from swallowing air. Pressure in the body of the esophagus, as the esophagus traverses the thorax, is −5 to −10 cm H_2O because of the subatmospheric pleural pressure (Chapter 20), and decreases and increases with inspiration and expiration, respectively. The manometric catheter in the region of the **lower esophageal sphincter** records a positive pressure of approximately 30 cm H_2O. This zone of increased pressure is important because it prevents the reflux of acidic stomach contents into the esophagus, which would

have damaging effects (see Case History). Intragastric pressure is approximately 10 cm H_2O.

At the beginning of a swallow, pressure in the upper esophageal sphincter decreases due to relaxation of the sphincter (see Fig. 22–11). The bolus is swept into the esophagus by pharyngeal muscle contraction; then the sphincter contracts to prevent reflux and swallowing of air. A **primary peristaltic contraction** passes down the body of the esophagus. This is seen as a positive pressure wave behind the advancing bolus (see Fig. 22–11). In the upper esophagus, adjacent motor units are sequentially activated in an aboral direction by the swallowing center. Peristalsis in the upper esophagus is abolished if its skeletal muscles are denervated by cutting the vagus nerves. In the lower esophagus, peristalsis is due to smooth muscle contraction and is not abolished by cutting the vagus nerves because it is organized by the enteric nervous system. When the bolus approaches the lower esophageal sphincter, it encounters a sphincter that is relaxed by vagus nerve impulses. The wave of contraction behind the bolus pushes the bolus into the stomach; then the lower esophageal sphincter contracts to prevent reflux. Notice that the lower esophageal sphincter and proximal stomach relax (fall in pressure) before the peristaltic wave reaches the end of the esophagus (see Fig. 22–11).

Esophageal peristalsis is slow, moving at a velocity of 2 to 6 cm/s, so that the peristaltic wave takes approximately 10 seconds to cover the approximately 18- to 25-cm length of the esophagus in an adult. The passage time through the esophagus is influenced by the nature of the material swallowed and by gravity. Liquids, when swallowed in an upright position, reach the stomach before the peristaltic contraction wave because of gravity. For most materials, however, peristalsis is essential for transport to the stomach.

If the original (or primary) peristaltic contraction does not clear the esophagus of food, then distension of the esophagus initiates a **secondary peristaltic contraction.** This is a

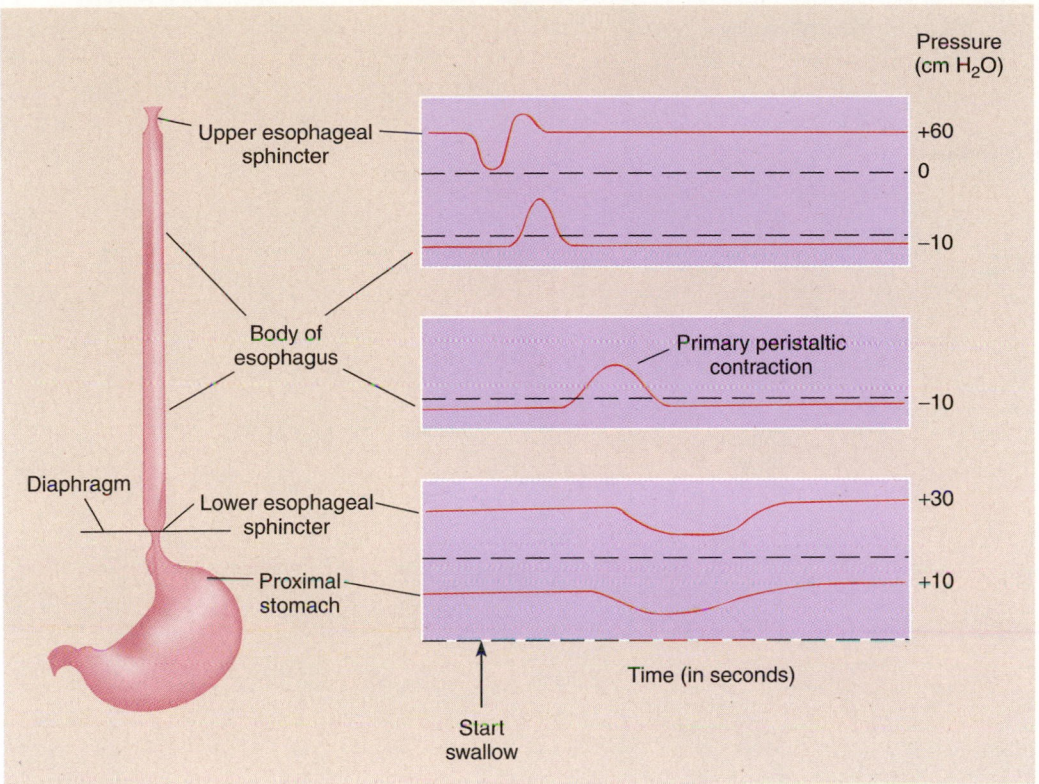

Pressure (cm H₂O)

+60

0

−10

Upper esophageal sphincter

Body of esophagus

Primary peristaltic contraction

−10

Diaphragm

Lower esophageal sphincter

Proximal stomach

+30

+10

Time (in seconds)

Start swallow

Figure 22–11

Recording of esophageal pressures with manometric catheters during the esophageal phase of swallowing. The dashed lines indicate the "zero" level of pressure (i.e., same as atmospheric). The position of the recording is indicated in the anatomical drawing on the left. Before the swallow, pressures in the upper and lower esophageal sphincters are approximately +60 and +30 cm H₂O, respectively; the elevated pressures mean that both sphincters are closed. With the start of a swallow, the upper esophageal sphincter relaxes temporarily, and pressure decreases toward atmospheric. A wave of peristaltic contraction then passes down the esophagus and sweeps the food bolus ahead of it. The lower esophageal sphincter and proximal stomach relax; the latter response allows the stomach to receive the food without much of an increase in intragastric pressure.

reflex response that serves to clear the esophagus of material left from a swallow or refluxed from the stomach. Several secondary peristaltic waves may be required to remove all material from the esophagus.

The Proximal Stomach Is a Reservoir and the Distal Stomach Grinds the Food into Small Particles

 How do the motor functions in the proximal and distal stomach aid in digestion?

The stomach is divided anatomically into several regions: cardia, fundus, corpus (body), antrum, and pylorus (pyloric sphincter) (Fig. 22–12). From a motility standpoint, the stomach is divided into proximal and distal stomach. The **proximal stomach** is important as a reservoir for ingested food. The **distal stomach** is important in mixing and grinding the food and in controlling the delivery of chyme to the duodenum. **Chyme** is the semifluid, homogenous, creamy or gruel-like material produced by gastric digestion of food. The proximal stomach includes the cardia, fundus, and proximal one third of the corpus. The distal stomach includes the distal two thirds of the corpus, the antrum, and the pylorus.

The proximal stomach is less muscular and is more distensible than is the distal stomach. Its smooth muscle cells show slow, sustained (tonic) contractions and infrequently display phasic contractions (Chapter 16). The proximal stomach relaxes to accommodate food during a meal. It relaxes in response to swallowing; this is called **receptive relaxation.** This is a reflex mediated by mechanoreceptors in the pharynx,

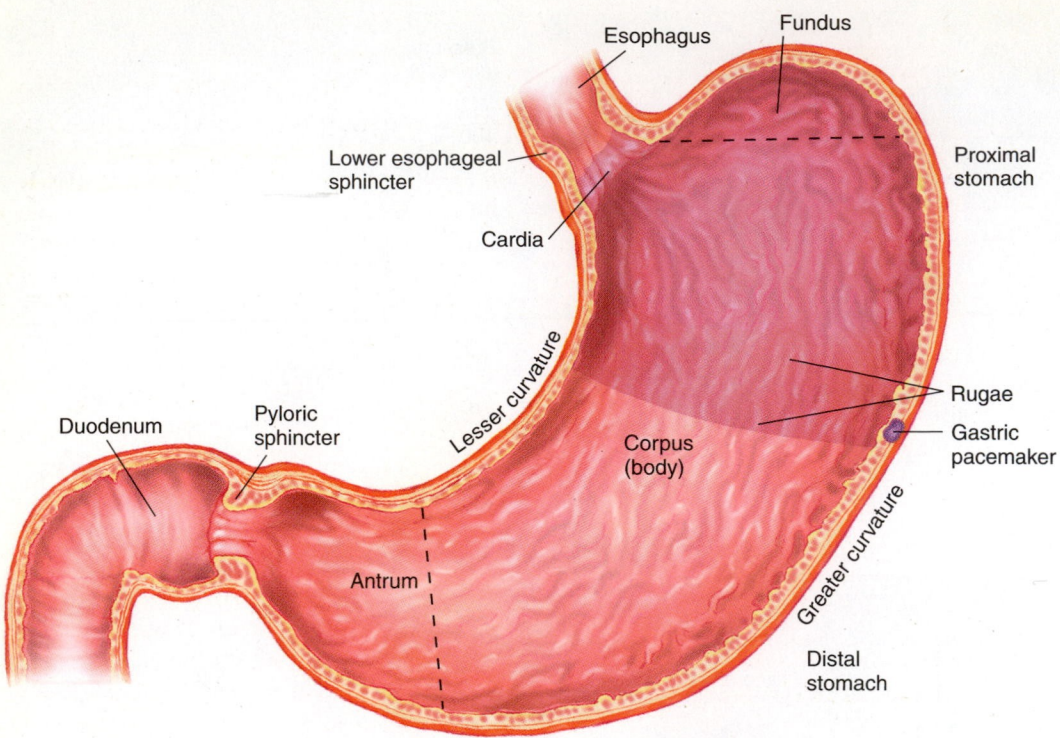

Figure 22–12

Parts of the stomach. The proximal stomach (shaded area) displays slow tonic contractions and has a high distensibility. It functions as a reservoir. No slow waves are seen here. By contrast, the distal stomach shows slow waves and peristaltic phasic contractions that sweep toward the pylorus. Distensibility here is low. The distal stomach is responsible for mixing and grinding food into small particles and for the controlled delivery of chyme to the duodenum.

afferent nerve fibers to the medulla, and efferent vagal nerve fibers to inhibitory motor neurons of the enteric nervous system. The stomach also relaxes in response to distension; this is a **vagovagal reflex** in which both afferent and efferent nerve fibers are in the vagus nerves. Tonic contraction in the proximal stomach controls intragastric pressure, which in turn influences the rate of emptying of the stomach.

The distal stomach is more muscular and less distensible. Between meals, the stomach is quiescent, except for periods (≈ every 90 minutes) when contractions occur in the proximal and distal stomach that clear the stomach of any food residues (the pylorus remains open). These periodic contractions are called **migrating motor complexes** (**MMCs**) and will be discussed later.

After a meal, the distal stomach shows phasic contractions at a rate of approximately three per minute. These peristaltic contractions are initiated by action potentials that are superimposed on slow waves. The slow waves originate from a **gastric pacemaker** located along the greater curvature of the stomach near the boundary between proximal and distal stomach (see Fig. 22–12). The slow waves are propagated distally and are accompanied by action potentials and *two successive rings of contraction* of the stomach's circular muscle layer. The slow wave moves down the body of the stomach toward the pylorus, picking up speed as it approaches the pylorus; the two contractile waves are approximately 2 or 3 seconds apart. When the leading contractile wave arrives at the pylorus, it squeezes a thin stream of chyme through the open sphincter into the duodenum. Then the second wave hits, but the sphincter is now closed, so the rest of the bolus is forced back into the stomach, a process called **retropulsion.** Retropulsion serves to mix the food with the gastric juice and to grind food into smaller particles.

Vagal stimulation increases the force and frequency of antral contractions, whereas sympathetic stimulation has the opposite effects. If the vagus nerves to the antrum are cut (vagotomy) or degenerate, the strength of antral peristalsis decreases, and this leads to delayed gastric emptying.

The Rate of Gastric Emptying Depends on Several Factors

 How is gastric emptying controlled?

The rate of **gastric emptying** is adjusted to provide a reasonable rate of chyme delivery for digestive and absorptive events in the small intestine. The duodenum should not be

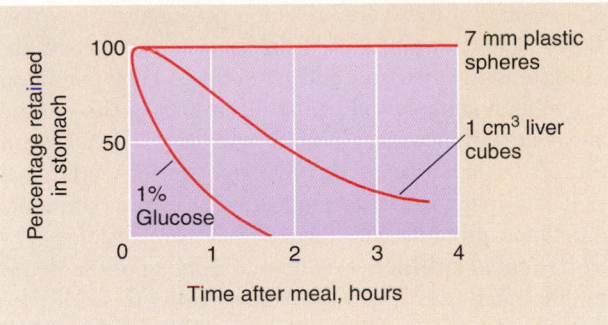

Figure 22–13

Gastric emptying after a meal. The liquid meal (1% glucose) is emptied much faster than the digestible solid meal (liver cubes). Plastic spheres of 7 mm diameter (large indigestible solids) were not emptied at all during the 4 hours of the experiment that was done in a dog. *(From Hinder, R. A., and Kelly, K. A. "Canine gastric emptying of solids and liquids."* American Journal of Physiology, *233:E335–E340, 1977)*

overwhelmed with too much stomach acid or by too much hypotonic or hypertonic chyme that could produce large fluid shifts across the small intestinal wall. Gastric emptying is restricted by the pyloric sphincter.

The rate of gastric emptying into the duodenum is determined by the nature of the meal and conditions in the stomach and duodenum. Figure 22–13 shows that a liquid meal (1% glucose solution) is emptied without delay and more rapidly than a solid meal (liver cubes). The rate of emptying of liquid is directly proportional to the volume of liquid in the stomach and so the volume left in the stomach falls exponentially with time. The greater the volume in the stomach, the greater the intragastric pressure, which is the pressure head that pushes liquid through the pyloric sphincter. Also, gastric distension reflexly increases the force and frequency of antral contractions, thereby causing more rapid emptying at larger volumes. Emptying of a solid meal is delayed and slower than is emptying of a liquid meal because large particles must be broken down to smaller particles before they can leave the stomach, a process that takes time.

The rate of emptying of the solid meal depends on the particle size. Smaller particles are emptied more rapidly than larger particles because the smaller particles are more easily squirted through the pyloric sphincter into the duodenum. Plastic spheres 7 mm or greater in diameter do not pass through the pyloric sphincter as long as food is in the stomach (see Fig. 22–13). Normally, the chyme emptied by the stomach contains almost no particles greater than 2 mm in diameter, and more than 90% of the particles are smaller than 0.25 mm.

Conditions in the duodenum also influence the rate of stomach emptying. A decrease in duodenal pH inhibits gastric motility and slows the rate of gastric emptying. This effect appears to be mediated by chemoreceptors that sense the intraluminal hydrogen ion concentration and by in-

hibitory vagovagal reflexes. Hypotonic or hypertonic fluids are emptied more slowly than isotonic fluids; an osmoreceptor appears to be present in the duodenum. Lipid digestion products (fatty acids, monoglycerides, and diglycerides) in the lumen of the duodenum inhibit gastric emptying by stimulating the release of the hormone cholecystokinin (CCK) from endocrine cells located in the epithelium of the duodenum. CCK stimulates contraction of the pyloric sphincter. Because of the slower emptying, fatty foods stay in the stomach longer than do protein- or carbohydrate-rich foods. The slower gastric emptying allows more time for fat to mix with pancreatic lipase and bile salts in the duodenum.

Delayed gastric emptying or too-rapid gastric emptying is found in a number of disorders. The most common cause of delayed gastric emptying in adults is obstruction of the gastric outlet by edema and scarring caused by peptic ulcer disease. Delayed gastric emptying also occurs in diabetes mellitus and is probably due to the nerve damage caused by this disease. Too-rapid emptying results in the **dumping syndrome.** In this case, entry of hypertonic fluid (chyme is usually hypertonic) into the duodenum causes a shift of extracellular water into the intestinal lumen, and blood volume decreases. There is also an exaggerated release of enteric hormones that have significant vascular effects. In people with this problem, symptoms of pallor, sweating, tachycardia, dizziness, and fainting may be observed soon after meals are eaten.

Peristaltic and Mixing Movements Occur in the Small Intestine

 How are movements in the small intestine controlled?

The main functions of the small intestine are the digestion and absorption of nutrients and the absorption of water and electrolytes. The motility pattern of the small intestine is organized to optimize these functions. Movements here include (1) segmental contractions that mix the food with digestive secretions and expose fresh materials to the absorptive cells and their surface enzymes and (2) propulsive movements that drive the intestinal contents in the aboral direction.

In subjects in the interdigestive state (2–3 hours after a meal or fasting), there are usually only a few contractions that occupy small lengths (1–5 cm) of small intestine. Approximately every 90 to 120 minutes, however, a strong peristaltic contraction wave passes down the length of the intestine, from stomach to the ileocecal valve. These are the migrating motor complexes (MMCs). They differ from the usual peristaltic activity in that the contraction wave travels down the entire length of the small intestine; most peristaltic waves usually die out after a few centimeters. MMCs clear food debris, mucus, and sloughed epithelial cells from the intestine between meals. If they are pathologically absent, then bacterial overgrowth results; this in turn produces impaired absorption. MMCs keep the stomach and small intestine clean and are often described as having a "housekeeping" function. MMCs empty large particles from the stomach between meals.

MMCs require an intact enteric nervous system. They are inhibited by feeding and by increased vagal nerve efferent activity, which is associated with feeding. The hormone motilin stimulates MMC-like patterns of motility, whereas the hormones gastrin and CCK, which are released after ingestion of a meal, inhibit MMCs.

Feeding converts the small intestine from intermittent to continuous contractile activity. The contractile activity is initiated by smooth muscle action potentials, which are superimposed upon intestinal slow waves. Normal peristaltic movements propel the intestinal contents aborally over only short distances, before dying out, but occur repetitively over the entire length of the small intestine, so that the intestinal contents are moved downstream. Segmenting contractions move materials toward and away from the mouth.

Intestinal motility is influenced by hormones and nerves. The physiological action of various hormones (e.g., gastrin and CCK) on small intestinal movements is controversial. Parasympathetic stimulation increases intestinal motility, whereas sympathetic activity depresses it. Noxious substances in the intestinal lumen can induce reversed peristalsis (as is seen with vomiting) or strong power propulsive movements in the aboral direction. Most people appreciate the fact that emotions can influence bowel functions.

In some circumstances, intestinal movements may be strongly inhibited. This occurs with a condition called **paralytic ileus.** *Ileus* means "obstruction of the intestine." In paralytic ileus, there is no real mechanical obstruction of the lumen, but *contractile activity in the small intestine is abolished.* Slow waves still occur, but there are no action potentials or contractions. Bowel sounds are absent. Patients may experience abdominal distension, nausea, and vomiting. This condition reflects a general suppression of motor circuits in the enteric nervous system and increased inhibitory neuron activity. It can be produced by handling of the intestine during abdominal surgery. It is a reflex response mediated via visceral afferents and sympathetic efferents. This form of ileus is reversible.

The **ileocecal sphincter** controls emptying of the ileum into the colon and acts as a valve that permits forward flow from ileum to cecum and prevents retrograde flow (Fig. 22–14). The sphincter is normally kept closed by an enteric reflex from the cecum. Increased fluid content, pressure, or chemical irritation of the ileum promotes relaxation of the sphincter and emptying of the ileum, whereas pressure or chemical irritation of the cecum causes further sphincter contraction and prevents reflux. Material moves intermittently from ileum to colon.

Motility Patterns in the Large Intestine Delay Transit or Permit Defecation

How does motility in the large intestine affect its contents?

The large intestine has two major functions: (1) absorption of fluid and electrolytes and (2) formation, storage, and periodic elimination of the feces. Its circular smooth muscle layer is con-

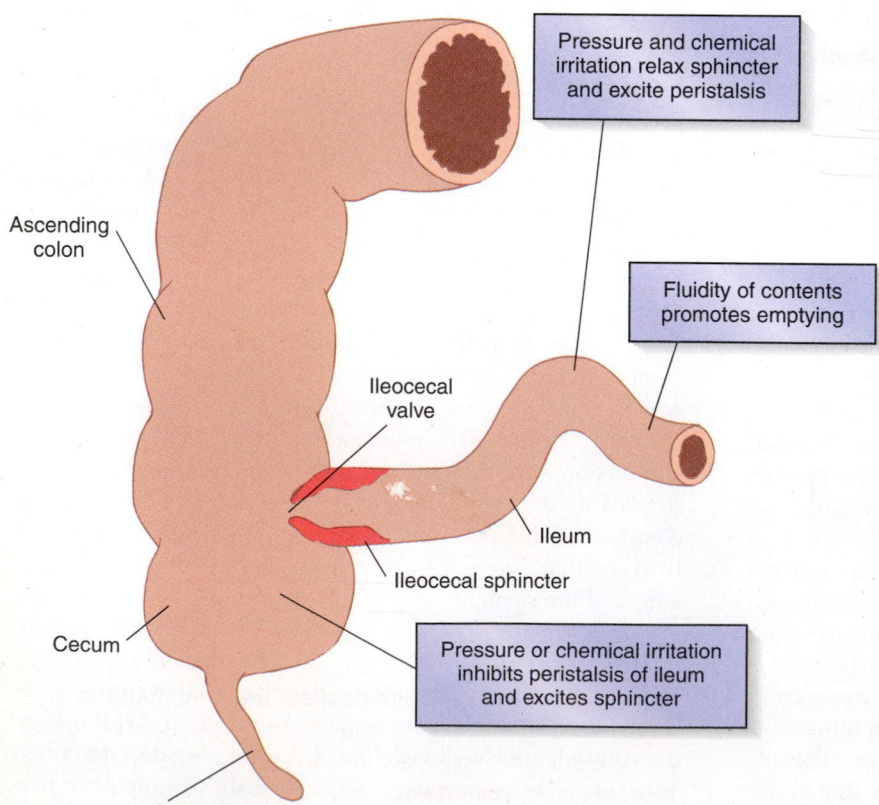

Pressure and chemical irritation relax sphincter and excite peristalsis

Fluidity of contents promotes emptying

Pressure or chemical irritation inhibits peristalsis of ileum and excites sphincter

Ascending colon

Ileocecal valve

Ileum

Ileocecal sphincter

Cecum

Appendix

Figure 22–14

Emptying at the ileocecal sphincter.
(From Guyton, A. C., and Hall, J. E.
Textbook of Medical Physiology, *ed 10.*
Philadelphia, W. B. Saunders, 2000)

tinuous from the cecum to the anal canal and thickens around the anal canal to form the **internal anal sphincter.** The longitudinal muscle layer is organized into three bands, the **teniae coli,** which merge at the junction of sigmoid colon and rectum to form a thick outer longitudinal layer around the rectum. Layers of skeletal (voluntary) muscle distal to and overlapping the internal anal sphincter form the **external anal sphincter.**

In contrast to the small intestine, where individual meals do not mix with each other because of clearing by MMCs, the large intestine contains a mixture of the remnants of several meals eaten over a day or two. Material commonly resides in the colon for 16 to 48 hours. The longest time is spent in the transverse colon, the segment chiefly responsible for storing and dehydrating the feces.

Motor activity in the large intestine comprises four main types: (1) haustration (or mixing movements), (2) peristaltic propulsive movements, (3) mass movements, and (4) defecation. **Haustration** produces saccules ("haustra") of the colon, mainly as a result of contraction of circular smooth muscle (Fig. 22–15). It is similar to segmentation and has a mixing function. It also serves to compact the feces. The sites of contraction are not fixed but often recur at the same points along the colon. The colonic contractions and relaxations of haustration last many minutes, in contrast to segmenting movements in the small intestine, which last only seconds. **Peristaltic propulsive movements** consist of a progressive wave of relaxation followed by a progressive wave of contraction. These normally occur in both oral and aboral directions in the large intestine, in contrast to what happens in the small intestine, where peristalsis normally occurs only in an aboral direction. Movements toward the mouth occur mainly in the ascending and transverse colon and contribute to the long transit time of materials in the large intestine.

Mass movements ("mass peristalsis") occur one to three times a day. During a mass movement, the circular muscle layer in a relatively long stretch (e.g., 20 cm) of colon contracts powerfully and propels material distally over long distances (Fig. 22–16). The haustra disappear in the contracting segment but reappear after the mass movement. Most of the net movement of feces through the colon occurs by mass movements.

When feces enter the rectum, an awareness of the need for **defecation** arises. Stretch receptors in the rectum are activated, causing the myenteric plexus of the enteric nervous system to initiate a mass movement in the sigmoid colon and rectum. Feces are forced toward the anus; the internal anal sphincter relaxes reflexly in response to distension of the rectum **(rectosphincteric reflex).** Internal sphincter relaxation

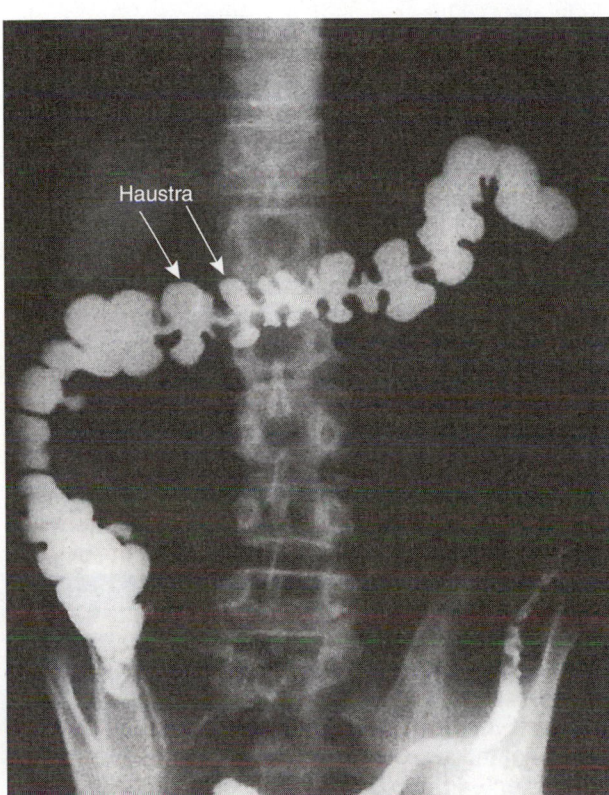

Figure 22–15

Radiograph showing haustra in the ascending and transverse colon. The lumen of the large intestine was rendered opaque to X-rays by ingesting barium sulfate. Rings of circular muscle contraction produce the haustra. (*From* Wood, J. D. "Gastrointestinal motility." *In Rhoades, R. A., and Tanner, G. A., eds.* Medical Physiology. Boston, Little, Brown and Co., 1995, pp 505–529)

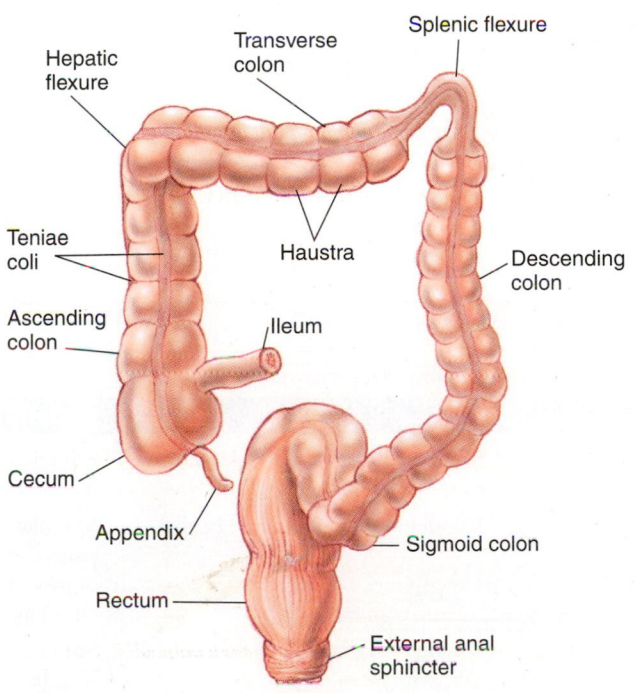

Figure 22–16

Mass movement. A peristaltic mass movement at the splenic flexure pushes material from the transverse colon into the descending colon. In the region of the mass movement, the circular smooth muscle is strongly contracted and the haustra temporarily disappear.

allows the rectal contents to come into contact with the upper anal canal, with its rich sensory innervation, thus providing information as to whether the material distending the rectum is solid, liquid, or gas. The external anal sphincter (skeletal muscle) is tonically active and under voluntary control; the sphincter closes more tightly if conditions are not appropriate. Muscles of the distal colon and rectum then relax. Stretch receptors in the rectum decrease their firing rate, internal anal sphincter tone increases, and defecation is postponed until the arrival of more feces in the rectum reactivates the stretch receptors and the external anal sphincter is voluntarily relaxed. The **defecation reflex,** which involves coordinated activities of both smooth and skeletal muscles, was described in Chapter 10.

Disorders of motility in the large intestine are common. **Constipation** may be caused by many factors: improper diet with lack of dietary fiber or bulk, drugs, metabolic disorders, tumors or other obstruction of the colon, or disorders of muscle or neural function. In **Hirschsprung's disease,** a developmental disorder, ganglia in the colon are reduced in number or absent; this leads to continuous contractile activity of the circular muscle because of loss of inhibitory enteric neurons. The colon upstream to the functional obstruction becomes greatly expanded. Surgical removal of the segment with missing ganglia usually restores normal function. **Incontinence** may reflect incompetence or weakness of internal or external sphincters. **Diarrhea** may be due to increased colonic motility, but it has many other causes.

GASTROINTESTINAL HORMONES

Gastrointestinal hormones play an important role in controlling gastrointestinal processes. They have important actions on motility, secretion, and growth. The main gastrointestinal hormones are gastrin, CCK, secretin, glucose-dependent insulinotropic polypeptide (GIP), and motilin (Table 22–2).

Gastrin and CCK are members of the same polypeptide family. Gastrin is produced by **G cells** located mainly in the gastric antrum and to a lesser extent in the duodenum. CCK is produced by **I cells** in the upper small intestine. These endocrine cells are found in the epithelial cell layer of the mucosa. The hormones are released into the blood, which carries them to distant target cells or tissues that can respond because they have the appropriate receptors. Gastrin stimulates acid secretion by stomach parietal cells and pepsinogen secretion by stomach chief cells. It also has an important long-term effect in that it stimulates growth of the gastric oxyntic gland mucosa. Cholecystokinin is so-named from the Greek *chole* ("bile"), *kystis* ("bladder"), and *kinein* ("to move"), because it was first identified as a hormone that stimulates contraction of the gallbladder. It also relaxes the sphincter of Oddi, which allows bile to flow into the duodenum. As will be discussed later, bile salts are important in lipid digestion. It is not surprising then that fat (or fatty acids) in the lumen of the upper small intestine causes the release of CCK. CCK also stimulates pancreatic enzyme secretion and inhibits gastric emptying. Overall, CCK promotes the right conditions for digestion and absorption of a meal in the upper small intestine. CCK induces satiety and reduces food intake in humans. Long-term effects include a stimulatory effect on the growth of the exocrine pancreas.

Secretin is a member of another family of polypeptides that also includes GIP, vasoactive intestinal peptide (VIP), and glucagon. It was the first hormone identified as a blood-borne messenger. Secretin is produced by **S cells** in the duodenum. Its release is stimulated by the presence of hydrogen ions in the duodenum; it counteracts acidity by stimulating pancreatic and biliary bicarbonate secretion and by inhibiting gastric acid secretion.

TABLE 22–2

The Gastrointestinal Hormones

Hormone	Source	Action	Stimuli for Release
Gastrin	Gastric antrum, and to a lesser extent the duodenum (G cells)	Stimulates gastric secretion and growth of gastric oxyntic gland mucosa	Peptides, amino acids, distension, vagal stimulation
CCK	Duodenum and jejunum (I cells)	Stimulates gallbladder contraction, pancreatic enzyme secretion and growth of exocrine pancreas; inhibits gastric emptying	Peptides, amino acids, fatty acids
Secretin	Duodenum (S cells)	Stimulates pancreatic and biliary bicarbonate secretion; inhibits gastric acid secretion and trophic effect of gastrin	Acid
GIP	Duodenum and jejunum (K cells)	Stimulates insulin secretion	Glucose, amino acids, fatty acids
Motilin	Duodenum and jejunum (M cells)	Stimulates migrating motor complex	Interdigestive state

TABLE 22–3

Magnitude of Gastrointestinal Secretions (L/day)*

Saliva	1.5
Gastric juice	1–3
Pancreatic juice	2
Bile	0.6–1.2
Intestinal secretions	2
Total	7–10

*The water lost in the feces is only approximately 100 ml/day.

GIP or **glucose-dependent insulinotropic polypeptide** is produced in the upper small intestine. It was originally called *gastric inhibitory polypeptide* because it inhibits gastric acid secretion, but it is now believed that this is not an important physiological function of this hormone in humans. The physiological function of GIP is controversial, but it appears to stimulate the secretion of insulin in response to the presence of glucose in the intestinal lumen.

Motilin is a polypeptide produced in the upper small intestine by **M cells.** It increases gastric and intestinal motility between meals. It may initiate the interdigestive migrating motor complex.

GASTROINTESTINAL SECRETIONS: SALIVARY, GASTRIC, PANCREATIC, AND BILIARY

In an adult, approximately 7 to 10 liters of fluid are added to the digestive tract every day in various secretions (Table 22–3). These secretions are essential to the proper digestion of food and absorption of nutrients and may also play a role in protecting the gastrointestinal system. We consider next the following secretions: saliva, gastric juice, pancreatic juice, and bile. In each case, we consider where and how these secretions are produced, what functions they serve, and what factors affect their production.

Salivary Gland Secretion Has Roles in Digestion, Lubrication, and Protection

The three major pairs of salivary glands are (1) the **parotid,** (2) **submaxillary** (submandibular), and (3) **sublingual glands,** all of which empty saliva via ducts into the mouth. These glands are listed in Table 22–4, along with their parasympathetic innervation, cell types, and contribution to total secretion of saliva. Numerous small salivary glands also are present in the oral cavity. The largest of the major salivary glands are the parotids; the smallest are the sublinguals. The submaxillary glands produce the largest fraction (70%) of total salivary flow in a day. All of the salivary glands receive a dual innervation from the autonomic nervous system. Parasympathetic nerve fibers go to the parotid gland via the glossopharyngeal nerve (cranial nerve IX) and the otic ganglion. The submaxillary and sublingual glands receive their parasympathetic innervation via the chorda tympani branch of the facial nerve (cranial nerve VII). Preganglionic sympathetic nerve fibers to the salivary glands come from thoracic spinal cord segments, and these neurons synapse in the superior cervical ganglia to send postganglionic nerve fibers to all three pairs of major salivary glands.

Different salivary glands contain differing proportions of **serous cells** and **mucous cells** (Table 22–4). These cells are organized as a single layer surrounding a lumen in a saclike structure called an **acinus.** The lumen of the acinus is connected to a branching duct system. The acinus and its associated ducts (intercalated and striated ducts) form the fundamental secretory unit of the salivary gland, the **salivon** (Fig. 22–17). Serous cells can be identified by their content of stored enzyme precursor granules called **zymogen granules.** They produce a watery secretion containing electrolytes and the enzyme **salivary amylase.** The mucous cells secrete **mucin,** which, when mixed with the watery secretion of the serous cells, produces a solution of high viscosity called **mucus.**

Saliva serves a number of functions. The saliva moistens and lubricates the mouth and food; this is important for swallowing. Lubrication is also needed for speech. By dissolving various food constituents, the saliva is important for taste.

TABLE 22–4

The Principal Salivary Glands

Gland	Parasympathetic Innervation	Histologic Type	Percentage of Total Salivary Secretion
Parotid	Glossopharyngeal	Serous	25
Submaxillary	Facial	Mixed*	70
Sublingual	Facial	Mixed†	5

*The ratio of serous to mucous acini is 4:1.
†The ratio of serous to mucous acini is 1:4.

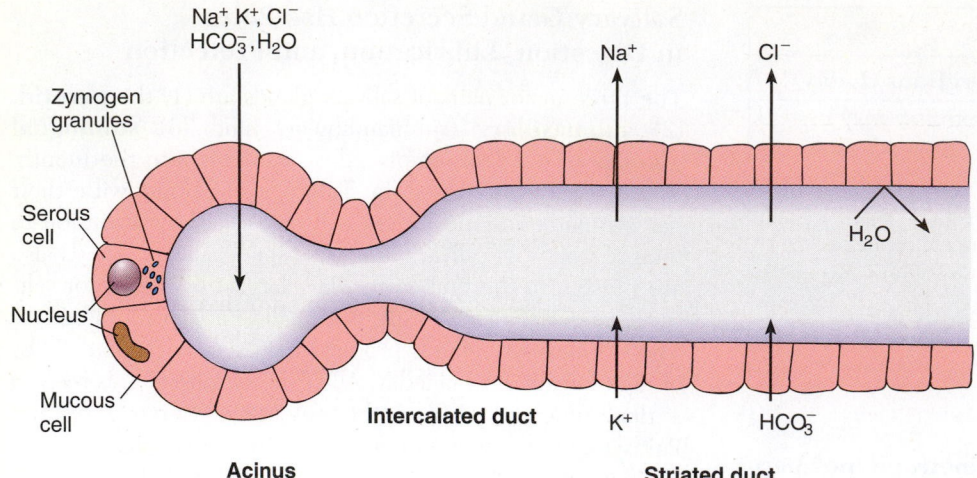

Figure 22–17

Schematic view of a salivon, the basic unit of salivary gland function. The acinar cells secrete electrolytes and water. The duct cells reabsorb sodium and chloride ions and secrete lesser amounts of potassium and bicarbonate ions. Because the duct epithelium is quite water-impermeable, the final saliva is more dilute than plasma.

The salivary amylase initiates the digestion of complex carbohydrates, such as starch. **Starch** is a polysaccharide consisting of many branched and straight chains of the monosaccharide glucose. The amylase catalyzes the hydrolysis of starch to form maltose (two glucose units), **maltotriose** (three glucose units), and **oligosaccharides** (fragments consisting of several molecules of glucose) (Fig. 22–18). The optimum pH of salivary amylase is 7, and so it is inactivated by stomach acid. Because a large portion of a meal may not be immediately mixed with acid in the stomach, this enzyme can hydrolyze up to 75% of ingested starch.

Saliva protects the teeth against the development of dental caries. It has bacteriocidal, buffering, and cleansing actions. It contains **lysozyme,** an enzyme that attacks the bacterial cell wall; **lactoferrin,** a protein that binds iron and makes it unavailable for the growth of microorganisms; immunoglobulins that defend against bacteria; **haptocorrin,** a protein that binds vitamin B_{12}; growth factors; and steroid hormones. Saliva dilutes and buffers harmful substances. It neutralizes acids from mouth bacteria and acid refluxed from the stomach. The decreased salivary secretion that occurs with dehydration is important in the conscious definition of thirst.

Salivary flow can vary over a wide range, from approximately 0.1 ml/min in a resting gland to 4 ml/min in a maximally stimulated gland. The ionic composition of saliva varies with flow rate. As flow rate increases, sodium and chloride concentrations increase. Saliva is usually slightly alkaline because of its bicarbonate content. At all flow rates, saliva is more dilute than the blood plasma; osmolalities range from approximately 10 to 50 mosm/kg H_2O compared with a plasma osmolality of approximately 300 mosm/kg H_2O.

The fluid leaving an acinus is isosmotic to plasma and is plasmalike in terms of its concentrations of Na^+, K^+, Cl^-, and bicarbonate. As fluid flows through the striated ducts, sodium and chloride are reabsorbed, and potassium and bicarbonate are secreted. Because the duct epithelium is relatively water impermeable, and because more sodium and chloride ions are absorbed than potassium and bicarbonate ions are secreted, the saliva becomes hypo-osmotic (see Fig. 22–17).

Salivary gland secretion, unlike secretion by other glands associated with the gastrointestinal system, is not stimulated or inhibited by hormones and is almost exclusively controlled by autonomic nerves. Centers in the medulla oblongata, the **salivatory nuclei,** control salivary secretion. Stimuli that cause an increase in salivary flow include chewing, and the smell or taste of food. Salivation may also be increased by the thought of food and as a conditioned reflex. Salivary flow is inhibited by sleep, fatigue, dehydration, and fear.

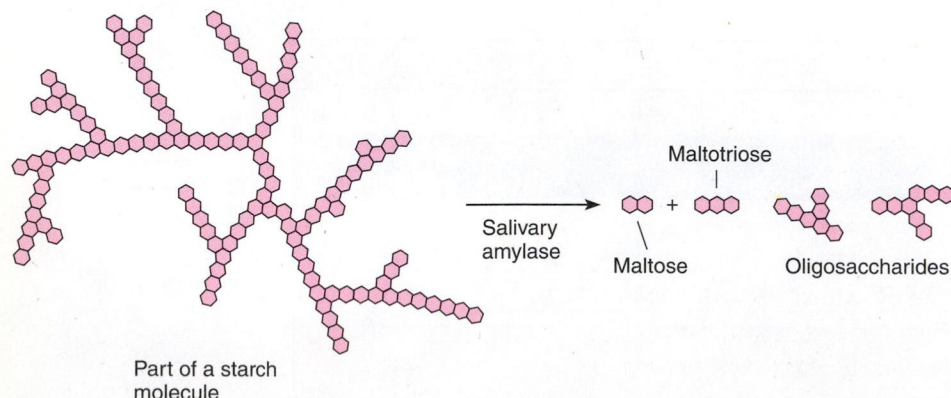

Figure 22–18

Starch is a branched polymer of glucose units that is digested into maltose, maltotriose, and branched oligosaccharides by salivary amylase.

Parasympathetic nerve fiber stimulation results in an increased secretion of saliva that is rich in electrolytes and salivary amylase. Blood flow is also increased as much as tenfold that in a resting gland. The effects on blood flow may be mediated in part by specific vasodilator fibers that release vasoactive intestinal polypeptide (VIP), along with acetylcholine, and in part by release of kallikrein from acinar cells. **Kallikrein** is a proteolytic enzyme produced by acinar cells that results in the production of vasodilator kinins, such as lysyl-bradykinin and bradykinin.

The salivary glands atrophy when their parasympathetic nerves are cut. By contrast, cutting the sympathetic nerves has no major effect on gland size.

Sympathetic nerve fiber stimulation increases salivary secretion, but less than with parasympathetic stimulation. The saliva is richer in mucins and is more viscous. Blood flow tends to be acutely decreased with sympathetic stimulation but then increases as vasodilator metabolites build up.

Xerostomia ("dry mouth") is a symptom caused by a lack of salivary secretion. Medications, such as antidepressants, through their anticholinergic effect, are the most common cause of xerostomia. Xerostomia leads to difficulty speaking and swallowing and impaired taste. Dental caries are common, and the oral mucosa becomes inflamed.

Parietal Cells of the Stomach Secrete Hydrochloric Acid

 How is gastric acid secretion controlled?

The stomach secretes 1 to 3 L/day of **gastric juice.** Gastric juice is a mixture composed mainly of the secretions of gastric glands but also secretions of surface epithelial cells. The gastric juice has a high concentration of **hydrochloric acid** (as high as 150 mmol/L, pH = ~1) and contains pepsin, mucus, and intrinsic factor. The acid and pepsin (a proteolytic enzyme) and grinding action of the stomach aid in the digestion of food, but the stomach is actually not absolutely essential for digestion. The acidity of the gastric juice provides a hostile environment for most ingested microorganisms; if stomach acid secretion is greatly reduced, the incidence of intestinal infections is increased. It is remarkable that, despite the acidity and pepsin in its lumen, the stomach does not digest itself.

Most of the gastric juice is produced by the **oxyntic glands** located in the body and fundus of the stomach (Fig. 22–19). These glands contain (*1*) **parietal (or oxyntic) cells,** which secrete hydrochloric acid (HCl) and

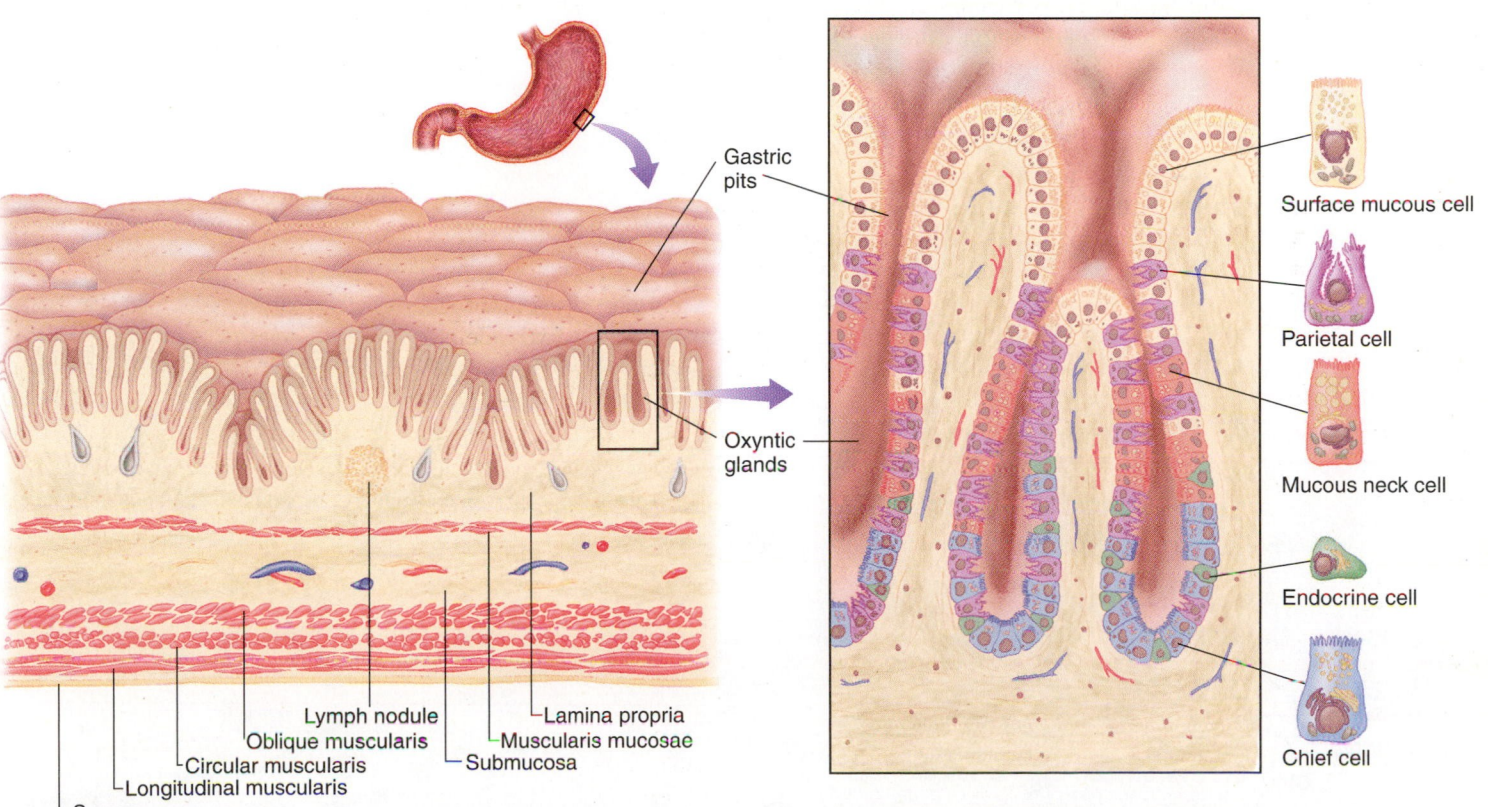

Gastric pits

Oxyntic glands

Surface mucous cell

Parietal cell

Mucous neck cell

Endocrine cell

Chief cell

Lymph nodule
Oblique muscularis
Circular muscularis
Longitudinal muscularis
Serosa
Lamina propria
Muscularis mucosae
Submucosa

Figure 22–19

The mucosa of the body of the stomach with an oxyntic gland. The oxyntic glands are at the bottom of gastric pits and contain mucous neck cells, parietal cells, endocrine cells, and chief cells.

intrinsic factor (a protein important in vitamin B_{12} absorption); (2) the more numerous **chief** (or **peptic**) **cells,** which produce **pepsinogen** (the precursor of pepsin) and **gastric lipase;** (3) endocrine cells; and (4) **mucous neck cells.** The most important endocrine cells in the oxyntic gland are the histamine-producing **enterochromaffin-like (ECL) cells** and the somatostatin-producing **(D) cells.** The mucous neck cells are the germinal cells (stem cells) for most other cell types and, by mitotic activity, replace cells in the glands and on the surface. **Cardiac glands** are located in the cardia of the stomach and secrete mainly mucus. **Pyloric glands** are found in the antral mucosa and contain cells that produce mucus and pepsinogen; **G cells,** which produce gastrin; and D cells, which produce somatostatin. The surface of the stomach is lined by a simple, columnar epithelium that secretes mucus and an alkaline fluid rich in bicarbonate.

The parietal cell of the gastric oxyntic gland is responsible for secretion of hydrochloric acid (Fig. 22–20). This cell can produce a secretion with a hydrogen ion concentration more than 1 million-fold higher than in the blood. The active secretion of H^+ is mediated by an **H^+/K^+-ATPase** in the luminal cell membrane. This proton pump has been recently cloned and sequenced and is highly homologous to Na^+/K^+-ATPase. It is inhibited by a drug called *omeprazole.* The source of the secreted hydrogen ions may be water or carbonic acid. The K^+ ion taken up by the cell during secretion of H^+ is recycled by way of a potassium ion channel in the luminal cell membrane. Bicarbonate leaves the cell, down its electrochemical potential gradient, via a chloride-bicarbonate exchanger in the **basolateral cell membrane** (the cell membrane not exposed to the lumen) and powers the active uptake of chloride ions. Once chloride is in the cell, it is above electrochemical equilibrium because the cell interior is negative compared with extracellular fluid. The negative membrane potential drives diffusion of negatively charged chloride ions into the lumen through a chloride channel. Water follows the secreted HCl passively. Gastric juice is essentially isosmotic to plasma. When HCl is secreted into the lumen of the gastric oxyntic gland, $NaHCO_3$ is added to the blood. This produces the so-called **alkaline tide,** an increase in blood pH that may occur after a heavy meal.

The three most powerful and important stimuli for gastric acid secretion by parietal cells are (1) histamine, (2) acetylcholine, and (3) gastrin (Fig. 22–21). Histamine is produced by enterochromaffin-like (ECL) cells, which are found in the epithelium of the bottom part of the gastric glands. Histamine acts as a *paracrine* signal, i.e., it reaches neighboring parietal cells by diffusion. The histamine receptors on the parietal cells are of the H_2 type and are blocked by drugs such as cimetidine. Acetylcholine is a *neurocrine* signal (neurotransmitter) released at vagal postganglionic nerve terminals. Gastrin is an *endocrine* signal that is produced mainly by

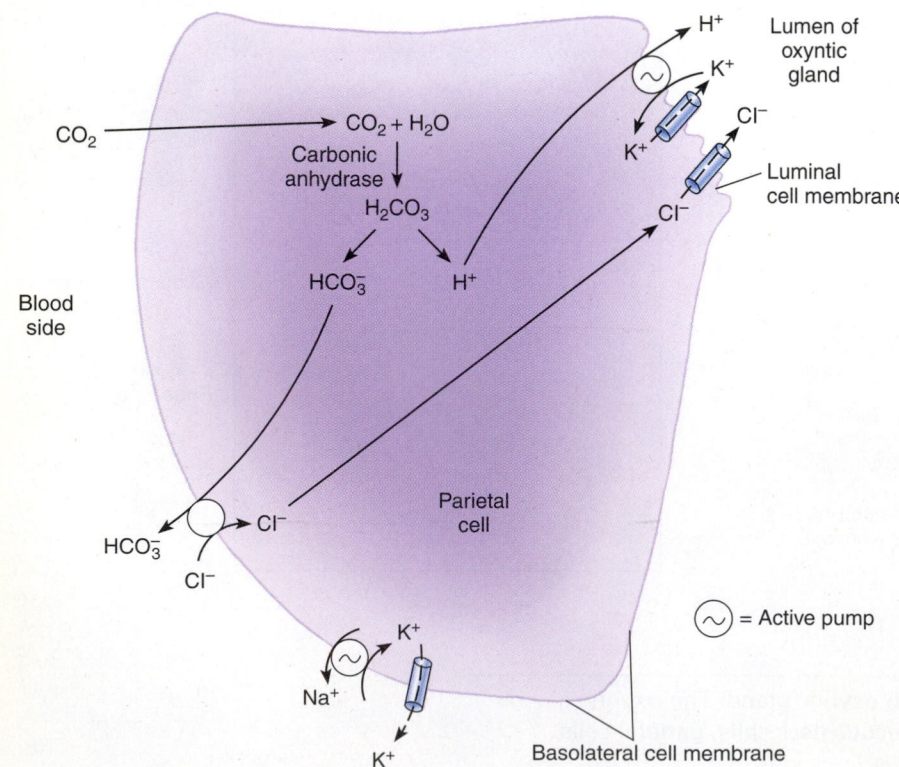

= Active pump

Figure 22–20

Model for secretion of HCl by the parietal cell. An active H^+/K^+-ATPase pump (proton pump) secretes hydrogen ions into the lumen of the oxyntic gland.

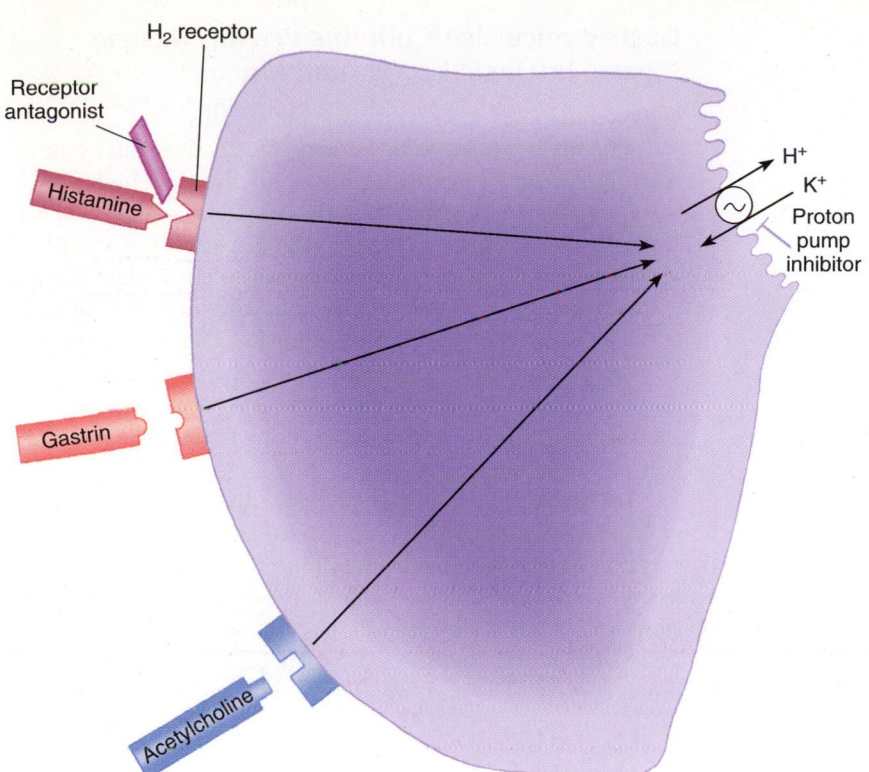

Figure 22–21

Major stimuli that promote secretion by gastric parietal cells. The chemicals bind to membrane receptors and activate various cell-signaling pathways.

antral G cells. Gastrin travels in the bloodstream to stimulate parietal cell secretion. All three of these stimuli increase gastric secretion, and they appear to potentiate each other (i.e., the response to any two agents exceeds the summed response of each agent tested alone). This concept of **potentiation** is important because it helps to explain why histamine H_2-receptor antagonists are so effective; in effect, they reduce acid secretion induced by all three secretory stimulants.

Gastric Secretion in Response to a Meal Can Be Divided into Cephalic, Gastric, and Intestinal Phases

Gastric secretion and other digestive functions in response to a meal have classically been divided into three phases: (1) cephalic, (2) gastric, and (3) intestinal. The phases are named according to where the stimuli originate, and they overlap in time.

During the **cephalic phase** of gastric secretion, the stomach is prepared to receive food (Fig. 22–22). The sight, smell, or taste of food; the thought of food or some conditioned stimulus; or chewing and swallowing results in efferent vagal impulses that stimulate acid secretion. The cephalic phase is abolished by severing the vagus nerves. The stimulation of gastric acid secretion occurs (1) directly via acetylcholine acting on the parietal cells and (2) by release of gastrin. Gastrin release is stimulated by (1) **gastrin-releasing peptide (GRP)** liberated by vagal postganglionic efferents

and (2) cholinergic inhibition of antral somatostatin release (somatostatin inhibits gastrin release). The cephalic phase accounts for approximately 40% of acid secretion in response to a meal.

The **gastric phase** occurs when food has entered the stomach and accounts for approximately 50% of the acid secretion in response to a meal (Fig. 22–23). Acid secretion is stimulated by several mechanisms. First, *distension* of the stomach stimulates gastric acid secretion by local and long (vagovagal) reflexes, which stimulate gastrin release and also directly stimulate parietal cells. Second, *protein digestion products* (amino acids and peptides) in the stomach lumen are detected by G cells in the antrum. The G cells then release gastrin, which stimulates parietal cells in the fundus and body of the stomach. Third, when food enters the stomach, it buffers H^+, *pH increases*, and this removes inhibition of gastrin release; this pathway is illustrated in Figure 22–24 and is explained next.

Before a meal, the $[H^+]$ in the stomach lumen is very high (pH <2), resulting in stimulated release of somatostatin by D cells. The somatostatin inhibits gastrin secretion and results in decreased HCl secretion. When food enters the stomach, the pH of the stomach lumen increases (e.g., to pH 5 or 6) due to the buffering action of the food. The lower $[H^+]$ results in decreased somatostatin release, increased gastrin secretion, and an increased rate of HCl secretion. As the stomach empties, the amount of food in the stomach and, hence, its buffering capacity decrease. Consequently, the gastric lumen $[H^+]$

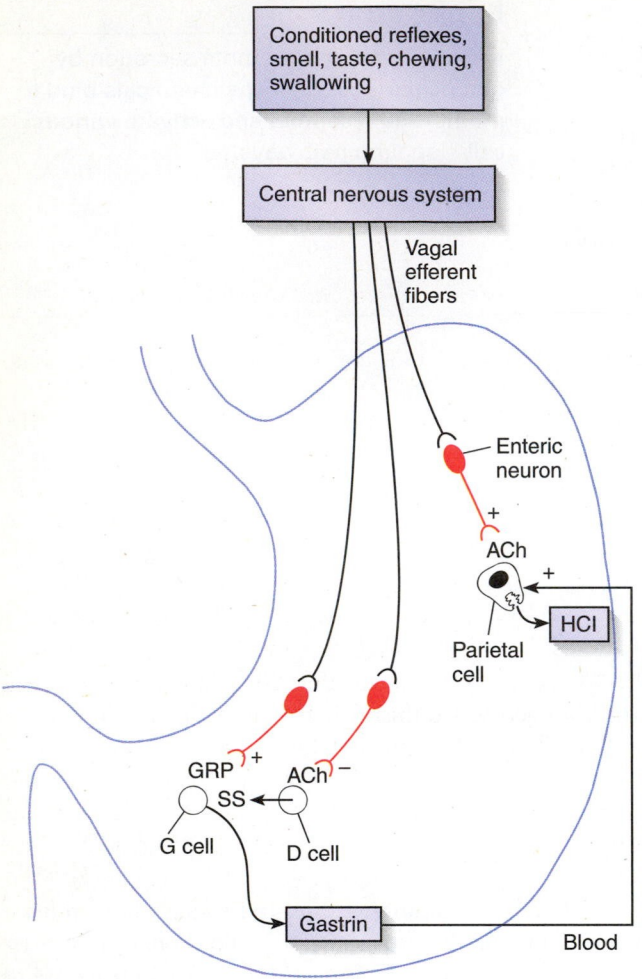

Figure 22–22

The cephalic phase of gastric secretion.
ACh = acetylcholine; GRP = gastrin-releasing peptide;
SS = somatostatin; + = stimulation; − = inhibition.

Gastric Juice also Contains Pepsin, Gastric Lipase, Intrinsic Factor, and Mucus

Gastric juice contains several organic constituents.

Pepsin is a proteolytic enzyme that is secreted as an inactive proenzyme, **pepsinogen,** mainly by the chief cells in the base of oxyntic glands in the fundus and body of the stomach. Pepsinogen is converted to pepsin when the pH is less than 5, and this enzyme is maximally active when the pH is approximately 2. Pepsin can catalyze the formation of additional pepsin from pepsinogen. Pepsin is an endopeptidase (cleaves proteins from the inside), and the products of its digestive activity consist of a mixture of polypeptides of diverse size. **Gastric lipase** is also produced by chief cells and initiates the digestion of fats.

Intrinsic factor is a glycoprotein produced by parietal cells. It binds vitamin B_{12} (cobalamin) and is essential for absorption of this vitamin in the ileum. The production of intrinsic factor by the stomach is the only reason that the stomach is essential for life.

Mucins are synthesized by surface mucous cells and mucous neck cells and are the main constituents of gastric

increases, somatostatin release starts to increase, gastrin release is inhibited, and the rate of HCl secretion decreases. These responses allow for secretion of acid when food is in the stomach and reduced acid secretion when the stomach is empty.

The **intestinal phase** occurs when chyme enters the duodenum. It accounts for approximately 10% of acid secretion in response to a meal. Absorbed protein digestion products (amino acids) stimulate acid secretion by way of the circulation. Distension of the duodenum releases a poorly characterized hormone that stimulates acid secretion. On the other hand, solutions with a high osmotic pressure, acid in the duodenum, or fatty acids in the duodenum inhibit gastric acid secretion. These inhibitory effects on acid secretion may be mediated by neural reflexes or by secretin or other hormones (called **enterogastrones**) produced by the duodenum.

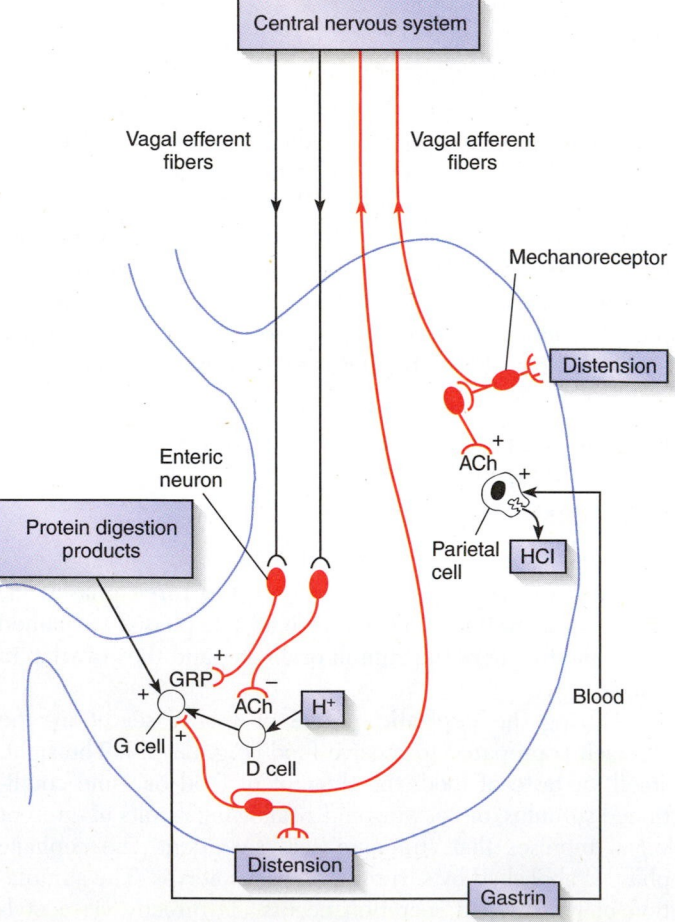

Figure 22–23

The gastric phase of gastric secretion.

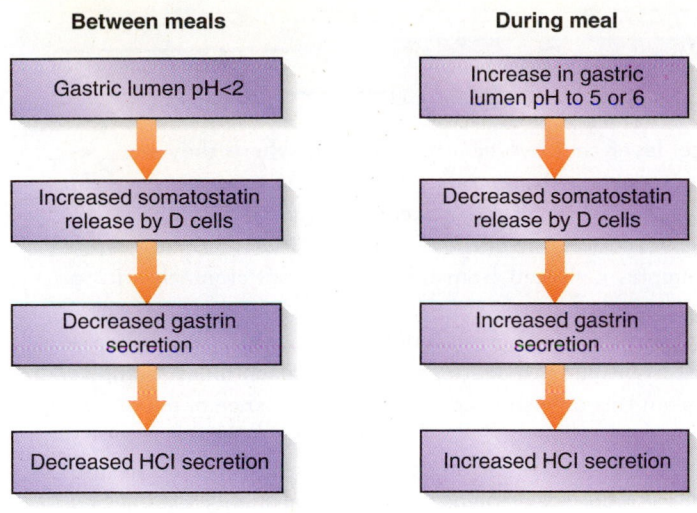

Between meals

- Gastric lumen pH<2
- Increased somatostatin release by D cells
- Decreased gastrin secretion
- Decreased HCl secretion

During meal

- Increase in gastric lumen pH to 5 or 6
- Decreased somatostatin release by D cells
- Increased gastrin secretion
- Increased HCl secretion

Figure 22–24

The pH level of the stomach contents is lowest between meals. The addition of food to the stomach increases the pH and leads to an increase in gastric acid secretion.

mucus. Mucins are large glycoproteins rich in carbohydrates. Mucus also contains phospholipids and nucleic acids. Mucus coats and protects the mucosal surface of the stomach and duodenum. Mucus secretion is stimulated by mechanical and chemical stimuli (e.g., acetylcholine and prostaglandins). Mucus secretion is inhibited by corticosteroids and by aspirin.

The Stomach and Duodenum Have Protective Mechanisms Against the Harmful Effects of Acid and Pepsin

> *How are the walls of the stomach protected from the acid and gastric enzymes that are in the lumen?*

The acidity of gastric juice and its content of the proteolytic enzyme, pepsin, present a rather hostile environment for cells. It has been known for more than two centuries that after death, the stomach digests itself. But why does the gastric juice not destroy the stomach in life? The answer is the presence of a protective **gastric mucosal barrier** that can be attributed to a number of factors (Fig. 22–25). A similar barrier is found in the duodenum.

Overlying the surface epithelial cells is a **mucus gel layer** that is rich in **bicarbonate.** In the stomach, the adherent mucus layer averages approximately 200 μm in thickness. The mucus layer acts as a diffusion barrier to hydrogen ions and is neutral or alkaline. The surface mucus-secreting cells produce a bicarbonate-rich secretion that is trapped in the mucus gel. The pH in the mucus layer is close to 7 at the surface of the epithelium, even when the pH in the stomach lumen is 2. The mucus layer provides physical protection for the underlying

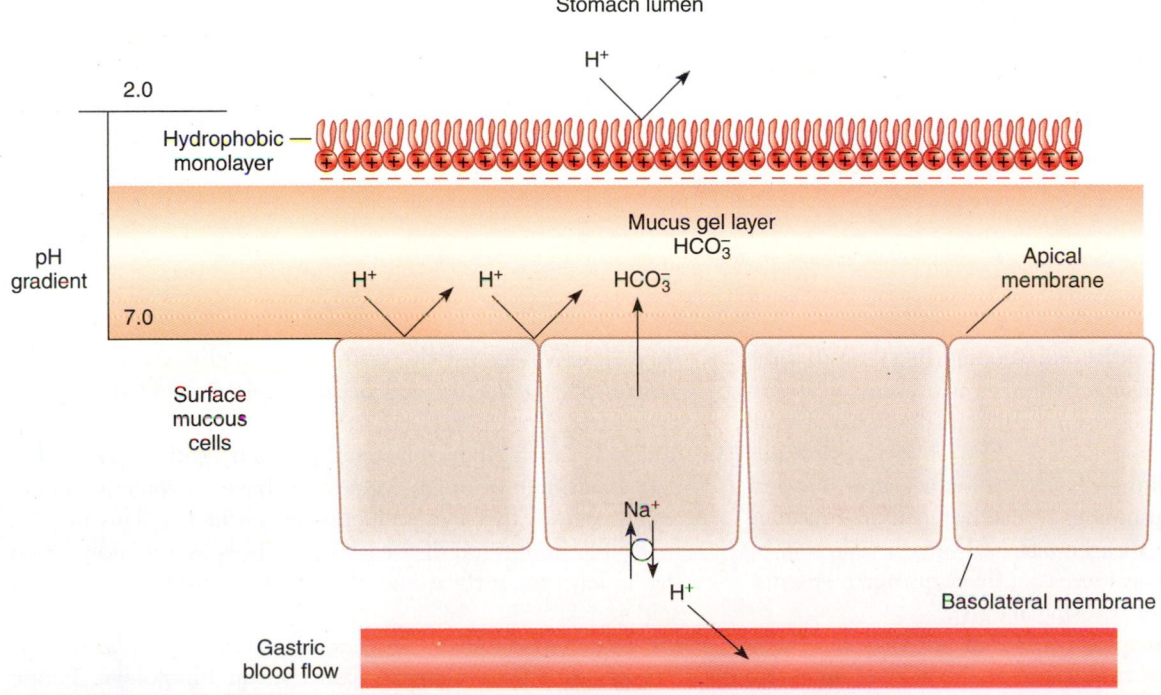

Stomach lumen

Figure 22–25

The gastric mucosal barrier. The alkaline mucus gel layer and other factors protect the stomach mucosa against the injurious effects of stomach acid and pepsin.

CURRENT CONCEPTS IN PHYSIOLOGY

Peptic Ulcers

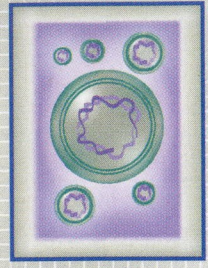

A **peptic ulcer** is a mucosal lesion of the stomach or duodenum caused by acid and pepsin. The injury can extend to the muscularis mucosa and beyond and may damage blood vessels, leading to severe bleeding. Ulcers usually require weeks or months to heal. A **perforated ulcer** extends entirely though the wall of the digestive tract; the resulting leakage of food, gastrointestinal secretions, and bacteria into the peritoneal cavity can have dire consequences.

The corrosive effects of gastric acid and the digestive actions of pepsin play a key role in the development of peptic ulcers. Ulcer formation, however, is not necessarily due to overactivity of these "aggressive" factors. Ulcers can develop if mucosal defense mechanisms are impaired. **Duodenal ulcers** are the most common; patients with this problem sometimes have excessive gastric acid secretion, too-rapid stomach emptying, or inadequate bicarbonate addition to the duodenum. **Gastric ulcers** are less common; patients with this problem most often do not secrete acid at an excessive rate, suggesting that mucosal defense mechanisms are inadequate. If the lower esophageal sphincter is not competent and allows the reflux of gastric juice, this can lead to ulceration of the esophagus.

It is now generally accepted that infection with ***Helicobacter pylori*** plays an important role in production of peptic ulcers. These spiral-shaped bacteria were first found in stomach mucosa biopsy samples from patients with chronic active gastritis (inflammation of the stomach) and peptic ulcer disease by two Australian researchers in 1983. The bacteria were present in the mucus gel layer that overlies the stomach, where they are protected from the extreme acidity of the gastric juice. In order to prove that the bacteria actually cause the stomach disorder, one of the researchers, Marshall, had biopsy samples collected from his own normal stomach with a gastroscope tube. No *H. pylori* were present. Subsequently, he swallowed a culture of *H. pylori,* and ten days later was ill with acute gastritis. Stomach biopsy samples taken this time showed histological evidence of inflammation and colonies of *H. pylori.* The gastritis resolved, and at 14 days, *H. pylori* could no longer be detected in mucosal biopsy specimens. These experiments established that the *H. pylori* previously detected in patients with gastritis were not secondary invaders, but the primary cause of disease. Marshall proposed that some people are chronically infected with *H. pylori* and that this can predispose a person to ulcers.

The damaging effects of *H. pylori* infection may be due to disruption of the mucosal barrier or stimulated acid secretion in some patients. Virtually all patients with duodenal ulcers and 75% to 85% of patients with gastric ulcers are infected with *H. pylori.* Treatment with antibiotics that eradicate *H. pylori* results in a much lower recurrence of peptic ulcer disease.

In addition to antibiotics, treatment of peptic ulcers is usually combined with drugs that reduce acid secretion, such as histamine (H_2) receptor blockers (e.g., cimetidine) and inhibitors of the H^+/K^+-ATPase (e.g., omeprazole). Bismuth salts are helpful. Patients are advised to stop smoking and avoid alcohol and caffeine. The importance of psychological factors in the development of peptic ulcer disease is controversial. Chronic anxiety or stress may aggravate ulcer disease.

---

cells by providing a lubricating surface and by separating the cells from the digestive juices. It also provides chemical protection (buffering action); 100 ml of mucus neutralizes 40 ml of 0.1 mol/L HCl. The mucus layer also physically prevents pepsin in the stomach lumen from gaining access to the surface epithelial cells. Pepsin is also inactivated by the alkaline pH in the mucus layer. Bicarbonate and mucus secretion by the surface mucous cells is increased by vagus nerve stimulation and by hormones (e.g., prostaglandins).

Surface-active phospholipids that are associated with the mucus gel layer are important in the barrier function. These lipids accumulate at the luminal surface of the gel. The gel is negatively charged and attracts the positive heads of the phospholipid molecules; the tails form a **hydrophobic monolayer** on the mucus gel surface. This kind of barrier is very effective against the diffusion of hydrogen ions. In industry, during the etching of metal plates, similar nonwettable layers (e.g., paraffin or other waxes) are used to prevent attack by acid. Many phospholipid-rich foods, such as bananas and milk or cream, appear to have a protective effect against agents that may induce stomach ulcers. This may be related to deposition of the phospholipids as a monolayer on the mucus gel surface and the resulting protection against acid attack.

Many peptic ulcers are associated with *Helicobacter pylori* infection (see Current Concepts in Physiology: Peptic Ulcers). These bacteria may compromise the hydrophobic phospholipid barrier in two ways. First, they produce phospholipases that catalyze the hydrolysis of surface-active phospholipids. Second, they have high activities of the enzyme urease, which results in the generation of high concentrations of ammonium ions within the mucus gel layer. The ammonium ions compete with the phospholipids for binding sites on the surface of the mucus gel layer and destroy the integrity of the phospholipid monolayer.

The apical cell membranes and tight junctions of surface epithelial cells are highly impermeable to hydrogen ions (see Fig. 22–25). The surface mucous cells are able to regulate their intracellular pH closely and possess basolateral cell membrane Na^+/H^+ exchangers that can dispose of excess hydrogen ions.

The interstitium below the epithelial cells is usually alkaline because of the addition of bicarbonate during gastric acid secretion. Gastric blood flow serves to remove any hydrogen ions that may have leaked into the mucosa. Irritation of the gastric mucosa increases production of nitric oxide and prostaglandins, and these increase gastric blood flow, which sweeps away hydrogen ions.

Finally, the ability of undifferentiated mucous cells to reconstitute the epithelium by migrating rapidly and covering the surface of a denuded area is an important protective response. Within minutes to hours after injury to the surface epithelium, viable cells from adjacent gastric pits and glands flatten, extend lamellipodia, and migrate over the denuded basement membrane to reseal the epithelial lining. This response occurs in a time frame too short for stimulated cell division. After this rapid restitution, the epithelium is renewed by cell division over the course of a few days. The turnover of

epithelial cells in the stomach (as in other parts of the intestine) is naturally quite high, and the cells are replaced every few days.

The Exocrine Pancreas Secretes Bicarbonate Ions and Digestive Enzymes

> *How does the pancreas contribute to digestion?*

The pancreas is both an endocrine and exocrine gland. Only approximately 1% of its mass consists of the endocrine component, the islets of Langerhans, which produce insulin and glucagon; the bulk of the pancreas is devoted to its exocrine functions. The exocrine secretory cells are arranged in saclike acini connected to a system of branching ducts that ultimately empty into the duodenum. The acini secrete a small volume of juice containing digestive enzymes. The small ducts (ductules) secrete a relatively large volume of juice with a high concentration of the alkaline salt sodium bicarbonate. The pancreatic juice is the major source of digestive enzymes. The bicarbonate component of pancreatic juice is important in neutralizing stomach acid in the duodenum and in providing an optimal pH for digestive enzyme activity.

The mechanism of production of a bicarbonate-rich secretion by pancreatic ductule cells is illustrated in Figure 22–26. Sodium ions are pumped out of the cell by a basolateral cell membrane Na^+/K^+-ATPase; this creates an inwardly

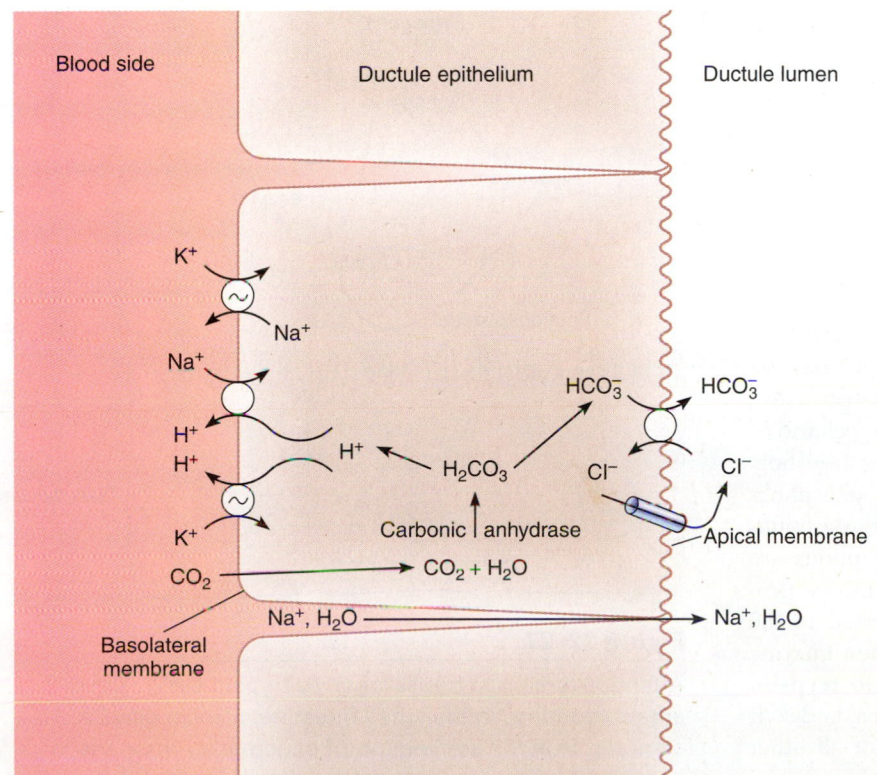

Figure 22–26

Cell model for secretion by pancreatic ductule cells. These cells produce a bicarbonate-rich fluid. ~ = active ion pump; cylinder symbol = CFTR (chloride channel).

TABLE 22–5

Digestive Enzymes of Pancreatic Juice

Enzyme	Substrate	Action	Products of Digestion
Trypsin, chymotrypsin, elastase	Proteins and polypeptides	Hydrolysis of interior peptide bonds	Small peptides
Carboxypeptidases	Proteins and polypeptides	Hydrolysis of carboxy-terminal peptide bonds	Peptides and amino acids
Amylase	Polysaccharides	Splitting of internal glucose bonds	Oligosaccharides, maltotriose, and maltose
Lipase	Triglycerides	Release of two fatty acids	Free fatty acids and monoglyceride
Phospholipase A_2	Lecithin (and other phospholipids)	Splitting off of one fatty acid	Lysolecithin and fatty acid
Cholesterol esterase	Cholesterol esters	Splitting of ester bond	Cholesterol and fatty acid
Ribonuclease, deoxyribonuclease	Nucleic acids	Hydrolysis of phosphate ester linkages	Oligonucleotides and mononucleotides

directed sodium gradient. CO_2 is hydrated with water to form carbonic acid, a reaction catalyzed by **carbonic anhydrase.** Carbonic acid dissociates immediately into hydrogen and bicarbonate ions. The hydrogen ion is extruded by a basolateral cell membrane Na^+/H^+ exchanger and H^+/K^+-ATPase. Bicarbonate accumulates in the cytoplasm and is secreted into the lumen via a **chloride/bicarbonate exchanger.** The chloride needed by the chloride-bicarbonate exchanger enters the lumen via an apical cell membrane chloride channel called the **cystic fibrosis transmembrane conductance regulator (CFTR).** Sodium ions diffuse passively into the lumen through cation-selective tight junctions to accompany the secreted bicarbonate ions, and water follows along to maintain isotonicity. A prominent defect in patients with cystic fibrosis, in addition to their pulmonary problems (Chapter 4), is a decrease in the watery secretion of the pancreas. This is due to deficient CFTR and, hence, deficient chloride secretion by the pancreatic ductule cells; this leads to plugging of the pancreatic ductules and eventual destruction of the pancreas.

The pancreatic juice contains almost all of the enzymes needed for nutrient digestion, including proteolytic enzymes (e.g., trypsin, chymotrypsin, elastase, and carboxypeptidases), amylase (which breaks down starches to oligosaccharides, maltotriose, and maltose), lipase (which produces free fatty acids and 2-monoglycerides from triglycerides), phospholipases, cholesterol esterase, ribonuclease, and deoxyribonuclease (Table 22–5). Most of the enzymes (exceptions are pancreatic lipase and amylase) are secreted as inactive proenzymes **(zymogens).** The zymogens are converted to the active enzymes when an intestinal **brush border** enzyme called **enteropeptidase** converts **trypsinogen** to **trypsin.** Trypsin, in turn, activates the pancreatic zymogens, and so it is the central enzyme that controls the activity of all other pancreatic zymogens (Fig. 22–27).

Why does the pancreas not digest itself? The pancreas protects itself against the potentially harmful effects of its own digestive enzymes in several ways. First, those enzymes that can digest membranes are synthesized and secreted as inactive zymogens; the activation of these zymogens normally occurs only after they have reached the intestine. Second, all digestive enzymes are confined in vesicles (zymogen granules) within the acinar cells. Third, acinar cells also synthesize and

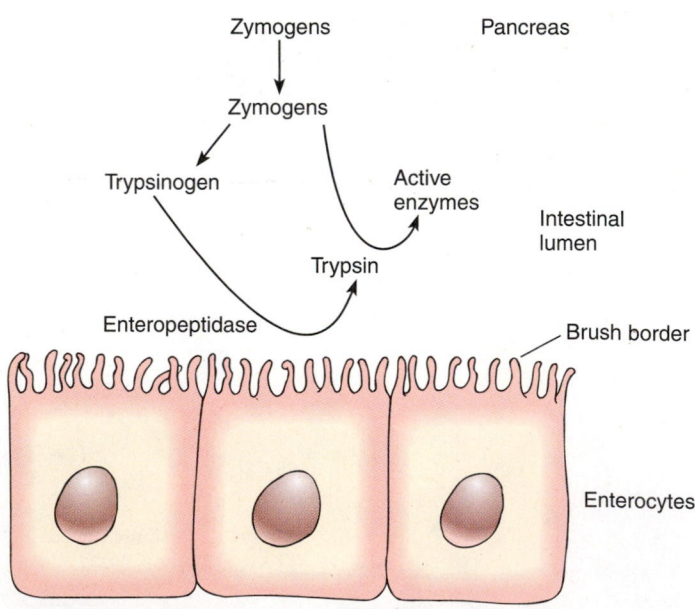

Figure 22–27

After conversion of trypsinogen to trypsin by enteropeptidase in the small intestine, trypsin plays a central role in the conversion of pancreatic zymogens to active enzymes.

secrete a **trypsin inhibitor,** a protein that is protective against early activation of trypsinogen. The inhibitor is packaged along with trypsinogen in the zymogen granules. If pancreatic enzymes are activated within the pancreas, the pancreas may destroy itself.

Pancreatic Secretion After a Meal Can Be Divided into Cephalic, Gastric, and Intestinal Phases

 How is pancreatic secretion controlled?

Pancreatic secretion averages approximately 2 L/day of fluid, but most of this secretion occurs on demand (with eating) and over relatively short periods. Pancreatic secretion is divided into cephalic, gastric, and intestinal phases.

During the **cephalic phase,** the sight, smell, taste, or thought of food, chewing, or swallowing stimulates pancreatic secretion. This is mediated via vagal cholinergic nerve fibers mainly to the pancreatic acinar cells. The secreted pancreatic juice is rich in enzymes. The cephalic phase accounts for approximately 20% of the pancreatic response to a meal.

During the **gastric phase,** distension of the stomach initiates long vagovagal reflexes. This results in increased pancreatic secretion via release of acetylcholine by nerve fibers to acini and ducts. Enzyme output is increased more than bicarbonate secretion. The percentage of pancreatic secretion due to the gastric phase is small, amounting to only 5% to 10% of the total.

The **intestinal phase** of pancreatic secretion is most important and accounts for 70% to 80% of the response to a meal. Three mechanisms are involved. First, an increase in $[H^+]$ in the duodenum stimulates S cells to release the hormone secretin, which in turn stimulates the duct cells to increase bicarbonate secretion. Second, fatty acids and protein digestion products (amino acids and peptides) stimulate duodenal and jejunal I cells to release the hormone CCK (see Table 22–2). CCK is the major stimulator of the secretion of enzymes by pancreatic acinar cells and also potentiates the action of secretin on the duct cells, leading to increased bicarbonate secretion. Third, hydrogen ions, fatty acids, and peptides and amino acids in the intestinal lumen also stimulate pancreatic secretion, especially the enzymatic component, via vagovagal reflexes.

Exocrine pancreatic secretion is decreased in patients with cystic fibrosis, pancreatic tumors, or pancreatitis. Chronic **pancreatitis** (inflammation of the pancreas) may lead to autodigestion of the pancreas and typically results in progressive loss first of the exocrine and then of the endocrine functions of the pancreas. Clinically evident pancreatic insufficiency requires loss of more than 90% of pancreatic function because the pancreas normally secretes a huge excess of digestive enzymes. Loss of exocrine pancreas function leads to inadequate absorption and weight loss.

Bile Is Produced by the Liver

 What is bile and what does it do?

Bile is a yellow fluid produced by the liver. In an average adult, 0.6 to 1.2 liters of bile are produced in a day. Most of the bile is produced by **hepatocytes** (liver cells) and is secreted into minute channels called **bile canaliculi.** These channels drain into **biliary ductules.** The epithelium lining the biliary ductules produces a small volume of bicarbonate-rich fluid. Secretion by the ductule cells is stimulated by secretin, similar to what is seen in the pancreatic ductules.

The branched biliary duct system forms the **hepatic ducts** (Fig. 22–28). When food is not being digested, bile is diverted by way of the **cystic duct** into the **gallbladder,** which stores bile between meals. During a meal, contractions of the gallbladder are stimulated by CCK and vagal impulses. The bile spurts through the **common bile duct** and the relaxed **sphincter of Oddi** into the small intestine, where it plays a role in the digestion and absorption of fat.

The main organic constituents of bile are the **bile salts** (Table 22–6). These are synthesized by hepatocytes from cholesterol and are amphipathic molecules, that is, they have both hydrophilic and lipophilic aspects. They play a key role in solubilizing cholesterol and phospholipids in the bile and in emulsifying dietary lipids and solubilizing lipid digestion products in the small intestine. The bile contains **bile pigments,** such as **bilirubin** (Chapter 17). These molecules are breakdown products of hemoglobin and are, for the most part, excreted in the

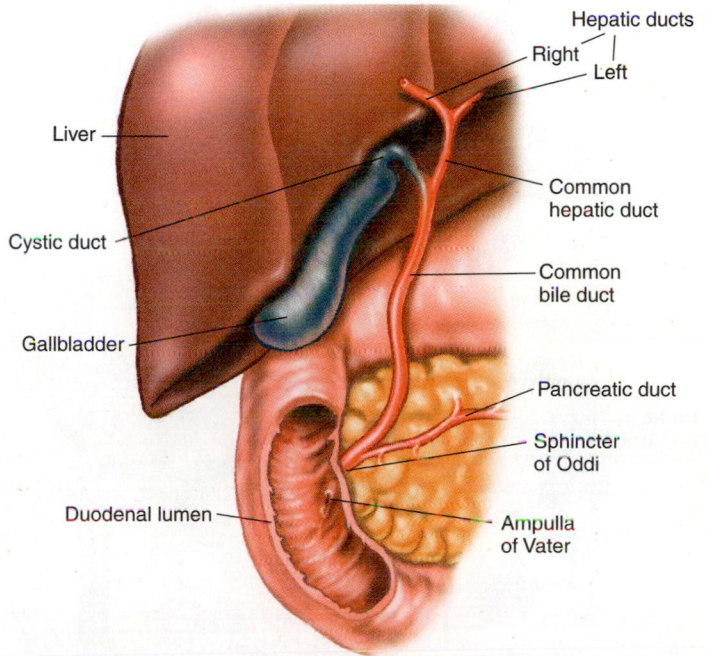

Figure 22–28

Gross anatomy of the biliary system.

TABLE 22–6

Composition of Human Hepatic and Gallbladder Bile*

Component	Hepatic Bile	Gallbladder Bile
Na^+	140–160	230–240
K^+	4–5	6–14
Ca^{2+}	1.0–2.5	2.5–16
Cl^-	62–112	1–10
HCO_3^-	20–50	8–10
Bile salts	20–40	200–300
Bilirubin	1–3	5–30
Cholesterol	2–4	10–25
Phospholipids	3–7	18–40

*Concentrations are in mmol/L.

feces. The bile also contains cholesterol and phospholipids, such as lecithin. Approximately two thirds of the cholesterol added to the intestine via bile is usually excreted in the feces. Cholesterol is virtually insoluble in water but is kept in solution in the bile by the formation of aggregates of bile salts, lecithin, and cholesterol. These aggregates can form vesicles or **micelles** (see subsequent discussion). If too much cholesterol is present in the bile in relation to the amounts of bile salts and lecithin, cholesterol may precipitate out of solution, and cholesterol-containing **gallstones** may result.

The Gallbladder Concentrates the Bile

The composition of bile collected directly from the liver (hepatic bile) is distinctly different from bile collected from the gallbladder (see Table 22–6). This is because the gallbladder epithelium absorbs water (secondary to NaCl absorption) and concentrates the bile.

Both hepatic and gallbladder bile are isosmotic to plasma. At first glance (see Table 22–6), it seems that the total solute concentration in gallbladder bile might be greater than in plasma. However, the bile salts, cholesterol, and phospholipids and associated counter-ions (e.g., Na^+) form aggregates (micelles and vesicles), so the osmolality is not higher than that of plasma. Recall that osmolality depends on the number of dissolved particles in solution (Chapter 4). If particles form aggregates or are attracted to each other and do not behave as independent particles, then the osmolality is considerably less than one might predict from just adding up the number of millimoles per liter.

The Bile Salts Are Reused Because of the Enterohepatic Circulation

> *How are bile salts recycled?*

The total amount of bile salts in the body is much smaller than the amounts needed to promote daily lipid digestion and absorption. Even with an average meal, the amount of bile salts entering the intestine is approximately twofold the total amount of bile salts in the body. The **enterohepatic circulation** of bile salts, by recycling bile salts between the intestine and liver, allows the body to use the same bile salts over and over again (Fig. 22–29). Bile salts are absorbed by the intestine and enter hepatic portal venous blood, are efficiently taken up by hepatocytes, are actively secreted into the bile, and return to the intestine.

Approximately 90% of bile salts entering the duodenum are absorbed by a sodium-dependent active transport process in the **terminal ileum.** The location of the active absorptive mechanism at the end of the small intestine

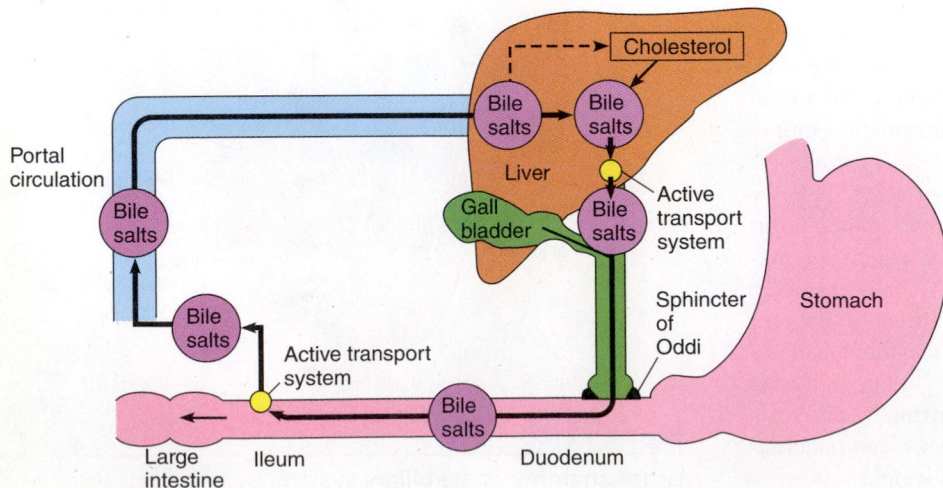

Figure 22–29

The enterohepatic circulation of bile salts as shown by the heavy, solid arrows. The dashed arrow indicates that bile salts inhibit bile salt synthesis from cholesterol.

ensures the presence of adequate amounts of bile salts for micelle formation until all of the fat digestion products have been absorbed. The absorption of fat is usually completed by the end of the jejunum. Small quantities of bile salts are passively absorbed in the small and large intestines. Fecal excretion of bile salts is normally approximately 5% of the quantity that entered the small intestine in the bile. In the steady state, the fecal loss is matched by new synthesis of bile salts by the liver.

The rate of *synthesis* of new bile salts depends on the hepatic portal blood concentration of bile salts. The rate of synthesis is high when the portal blood (and hence hepatocyte) concentration is low. This makes sense because this relation would serve to replace bile salts if there has been excessive loss from the body. If the bile salt concentration in hepatic portal blood (or hepatocyte) is too high, then bile salt synthesis is inhibited. The bile salts inhibit a rate-limiting step in their synthesis from cholesterol.

The rate of bile salt *secretion* increases when the hepatic portal bile salt concentration increases, as occurs during a meal. The increase in bile salt secretion results in an increase in bile flow from the liver.

Parasympathetic (vagal) stimulation increases bile secretion during a meal. During the intestinal phase of digestion, secretin is released from the duodenum after entry of acidic chyme from the stomach. Secretin stimulates secretion of bicarbonate by the bile ductule cells. This secretion, together with the pancreatic juice, helps to neutralize acid in the duodenum.

DIGESTION AND ABSORPTION

> *How does the small intestine absorb the digestion products of carbohydrates, proteins, and lipids?*

Absorption of nutrients is the principal purpose of the gastrointestinal system. **Absorption** is the net transfer of materials from the intestinal lumen to the circulatory system (blood or lymph). Absorption of nutrients is almost entirely accomplished by the small intestine.

The *small intestine* has a very large surface area for absorption (Fig. 22–30). The human small intestine in an adult is approximately 280 cm (or 9 ft) long. The total surface area is increased by circular folds of mucosa and submucosa, villi, and microvilli on the absorptive cells. **Villi** are minute, fingerlike processes approximately 0.5 to 1.5 mm in length. They are covered with two major types of epithelial cells: (1) columnar **absorptive cells** (also called **enterocytes**) and (2) mucus-secreting **goblet cells.** The **microvilli,** which collectively form the brush border of the absorptive cells, possess many digestive enzymes (e.g., oligosaccharidases, disaccharidases, and peptidases) and transport proteins on their surface. Because of the amplification, the membrane surface area of the small intestine is approximately the size of a doubles tennis court.

The core of a villus contains a **central lacteal** (a lymph vessel), **arteriole, capillaries,** and **venules** (Fig. 22–31). The villi have a high blood flow, especially after a meal. Venous

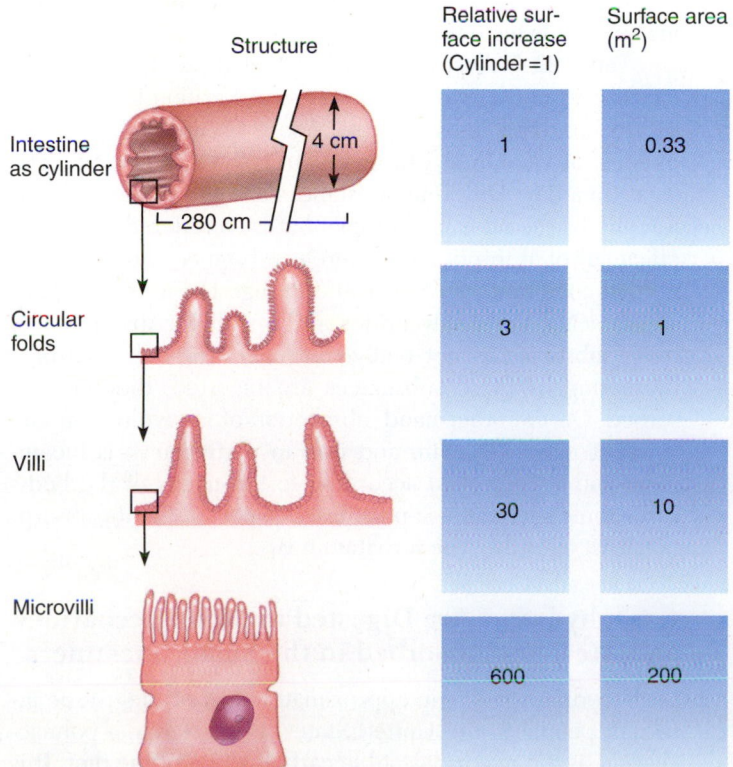

Figure 22–30

Three ways by which the surface area of the small intestine is increased from that of a simple cylinder of the same gross dimensions as those of the small intestine (280 cm in length and 4 cm in diameter).

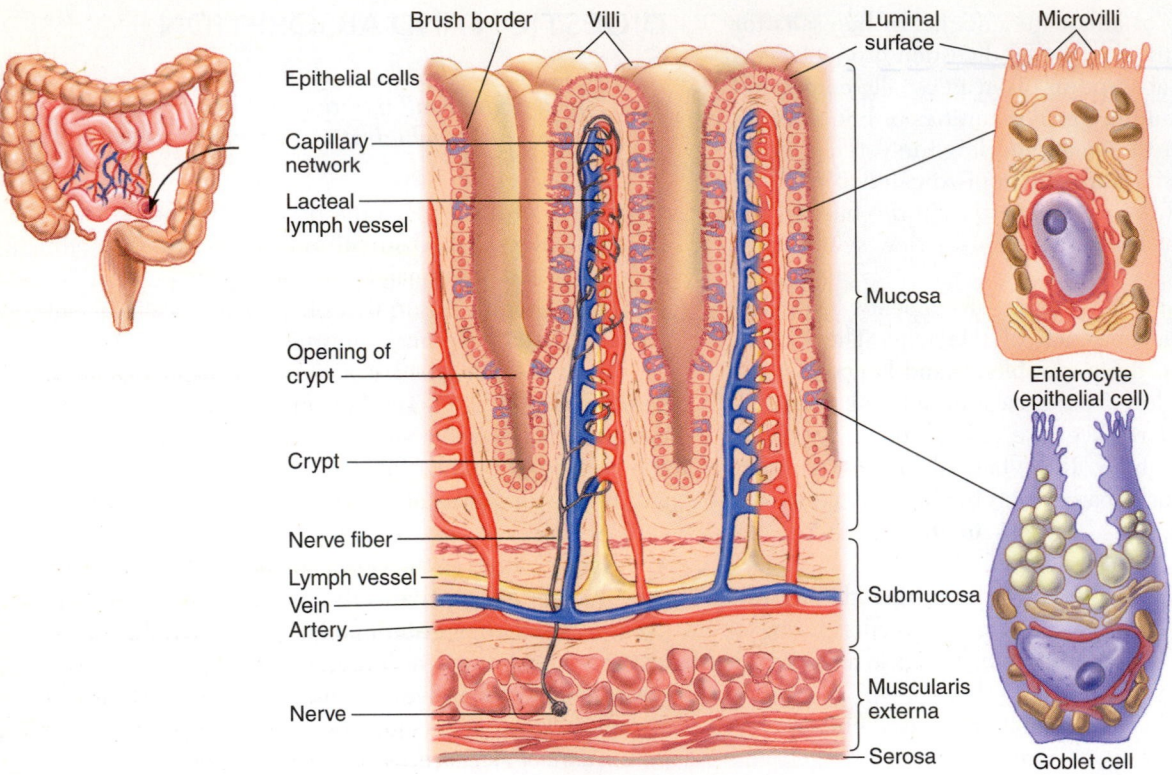

Figure 22–31

Structure of the small intestine. The villi project into the lumen, and tubular intestinal glands (crypts) extend down into the mucosa. The luminal surface of intestinal epithelial cells is covered by microvilli, which form a brush border.

blood draining the intestine flows first to the liver via the **hepatic portal vein.** Lymph ultimately joins the venous system in the thorax. The villus epithelial cells are constantly replaced by new cells produced from a dividing stem cell population that resides in tubular glands called **crypts.** The cells migrate up the surface of the villus and are shed from the villus tip after living only a few days.

Table 22–7 illustrates the fact that there are enormous differences in the maximum rates of absorption for different substances. The intestine can absorb as little as 0.000001 mmoles of vitamin B_{12} to as much as 1,000,000 mmoles of water in a day. Different substances are clearly handled differently by the intestine, a topic that is considered later. The efficiency of absorption of different substances varies. Under normal conditions, 95% to 100% of ingested water, glucose, amino acids, and triglycerides are absorbed. Absorption of these substances is not really controlled; the only control is by dieting. If these substances are ingested, they are absorbed. On the other hand, absorption of many divalent and trivalent ions — calcium and iron in particular — is incomplete and is controlled according to the needs of the body. Also, some incredibly sophisticated mechanisms for absorption exist, as is the case for vitamin B_{12}.

Carbohydrates Are Digested to Monosaccharides and Are Then Absorbed in the Small Intestine

Carbohydrates make up approximately half of the caloric intake of people in the United States. Plant starch, a polysaccharide, is the major digestible carbohydrate of the diet. It is

TABLE 22–7

Intestinal Transport Capacity for Different Nutrients in an Adult

Substance	mmoles/day
Water	1 million (18 L/day)
Glucose	20,000 (3.6 kg/day)
Amino acids	5,000
Triglycerides	900
Cholesterol	10
Iron	0.2
Vitamin B_{12}	0.000001

From Wilson, T. H. *Intestinal Absorption.* Philadelphia, W. B. Saunders, 1962.

APPLICATIONS OF PHYSIOLOGY

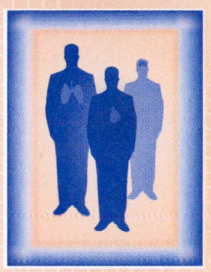

Celiac Disease

Celiac disease (gluten-dependent enteropathy) is a chronic disorder in which there is damage to the mucosa of the small intestine that is caused by ingestion of certain cereal grains. The first report of a condition resembling celiac disease was made in the 2nd century by Aretaeus of Cappadochia, and in the 19th century, Gee first described the clinical features of celiac disease in children. The relation between this disease and the ingestion of cereals was recognized only after World War II. An astute Dutch pediatrician, Dicke, noted the increased incidence of celiac disease after the war years and associated this with increased consumption of wheat and rye grains. Dicke and associates identified the alcohol-soluble component (gliadin) of the water-insoluble protein **gluten** as the injurious agent. Gluten is the protein fraction of wheat, rye, and barley that conveys stickiness and thus allows the baking of bread. Celiac disease in children and nontropical sprue in adults were recognized to be the same disease in the early 1960s.

Mucosal damage in celiac disease is most severe in the upper small intestine (i.e., duodenum and jejunum) because the offending proteins are usually digested to harmless products before reaching the distal small intestine. Villi are lost, crypts hypertrophy, and increased numbers of plasma cells, lymphocytes, and eosinophils infiltrate the lamina propria. The loss of brush border digestive enzymes and epithelial surface area leads to inadequate intestinal absorption. Symptoms that ensue include diarrhea, weight loss, abdominal distension, bloating, and weakness. Anemia, bone disease, and peripheral neuropathy can result because of deficient absorption of vitamins and minerals.

Celiac disease appears to be the result of a complex interplay of genetic and environmental factors. There is a definite genetic susceptibility to the disease; it is much more common in twins and close relatives than in the general population. Celiac disease is more common in persons who have certain histocompatibility markers. Dietary cereals clearly provoke the disease. Immunological factors are key in the pathogenesis of the disease and result in the mucosal injury. Gluten peptides drive the immune response that occurs mainly in the connective tissue of the lamina propria. Certain antibodies (antiendomysial antibody) are found in the plasma of patients with celiac disease and can be used to screen for this disease in asymptomatic people. Patients with celiac disease are more likely to have other autoimmune diseases.

Most, but not all, patients with celiac disease who adhere to a gluten-free diet have a relief of their symptoms. Patients on a gluten-free diet should avoid the cereals wheat, barley, rye, and oats (and foods made with these, including many processed foods) but can substitute corn, millet, buckwheat, sorghum, and rice.

composed of many straight and branched chains of the monosaccharide glucose (see Fig. 22–18). Starch is rapidly hydrolyzed to maltose, maltotriose, and oligosaccharides by salivary and pancreatic amylases. The disaccharides **sucrose,** or table sugar (made up of one glucose and one fructose unit), and **lactose,** or milk sugar (made up of one glucose and one galactose unit), are also consumed in the diet.

The diet contains nondigestible carbohydrates (dietary fiber), such as **cellulose.** Cellulose is a glucose polymer, but the linkages between glucose units is different from the kind found in starch, and human tissues do not produce an enzyme capable of breaking this linkage. Cellulose is a dietary fiber and is excreted in the feces.

Only monosaccharides can be absorbed, so oligosaccharides and disaccharides must be further hydrolyzed by enzymes that are found in the brush border of intestinal epithelial cells. These enzymes include (1) **isomaltase,** which hydrolyzes oligosaccharides to glucose; (2) **maltase,** which hydrolyzes maltose and maltotriose to glucose; (3) **sucrase,** which hydrolyzes sucrose to glucose and fructose; and (4) **lactase,** which hydrolyzes lactose to glucose and galactose.

Deficiency of the lactase enzyme is extremely common. The enzyme is usually present at birth but commonly disappears from the intestinal epithelial cells when a person is two to six years old, especially in certain racial and ethnic groups (African Americans, Ashkenazic Jews, Arabs, Greek Cypriots, Japanese, Formosans, and Filipinos) and leads to **lactose intolerance.** If lactose is not broken down, it remains in the intestinal lumen. Because of its osmotic activity, lactose causes an accumulation of fluid in the intestine. Metabolism of lactose by bacteria in the large intestine produces lactic acid, carbon dioxide, and hydrogen gas. The extra gas and fluid in the intestine produce symptoms of fullness and distension. Distension of the intestine reflexly produces muscle contractions (cramps). Diarrhea may result from increased intestinal motility and the osmotic effect of the lactic acid and undigested lactose. The distressing symptoms produced by lactose intolerance can be eliminated by avoiding milk and milk products, by taking lactase pills, or by treating milk with lactase before drinking it.

Usually the digestion and absorption of sugars are so efficient that all monosaccharides in the intestinal lumen are completely absorbed by the end of the jejunum. Fructose is absorbed passively by the intestinal epithelial cells, but the absorption of glucose and galactose requires secondary active transport. Glucose and two sodium ions are transported into the absorptive cell via a cotransporter known as **sodium-glucose transporter 1 (SGLT1).** SGLT1 was the first transport protein cloned and sequenced. The downhill movement of sodium into the cell energizes the uphill movement of glucose into the cell. Glucose accumulates in the cell at a higher concentration than in the blood and leaves the cell via a sodium-independent, facilitated diffusion mechanism in the basolateral cell membrane. Galactose is transported by the same carriers and competes with glucose for absorption. A similar transport pathway is seen in the kidney tubule.

Proteins Are Digested to Small Peptides and Amino Acids That Are Absorbed by Specific Transporters

Dietary proteins are essential for growth of the young and for the maintenance of health in adults. The minimum daily requirement for protein is approximately 2 g/kg body weight in infants and 0.8 g/kg body weight in an adult (56 g for a 70-kg man). An average adult intake in the United States is approximately 60 to 120 g/day. The proteins of the body consist of approximately 20 different amino acids, of which only approximately half can be synthesized in the human body. The rest, the **essential amino acids,** must be derived from the diet. Animal proteins (e.g., fish, meat, eggs, and milk) generally contain all of the amino acids needed, but vegetable proteins may be lacking in one or more of the essential amino acids. For example, corn protein is lacking in lysine and tryptophan. For this reason, vegetarians should eat a variety of vegetables (e.g., corn, beans, and squash) to avoid amino acid deficiencies.

Protein added to the intestine not only is due to oral intake but also comes from the various digestive secretions (e.g., saliva, gastric juice, pancreatic juice, and bile), mucin, and shed cells. The amount of protein from these endogenous sources is almost as high as the usual dietary protein intake. The small intestine absorbs more than 90% of the total protein load. Fecal protein loss is mainly in the form of bacteria and shed cells, and mucoproteins from the distal ileum and colon.

Protein digestion starts in the stomach with pepsin. The amount of protein digested in the stomach is small and is not crucial. People who lack pepsin show no impairment in overall protein digestion. The pancreatic enzymes (see Table 22–5) are most important in the digestion of proteins. The microvilli of enterocytes contain peptidases that cleave oligopeptides to amino acids, dipeptides, and tripeptides.

The end product of protein digestion is a mixture of free amino acids (30%) and small peptides (70%).

It used to be thought that proteins had to be completely broken down to amino acids before absorption, but this is no longer considered to be true. **Dipeptides** and **tripeptides** are efficiently absorbed by intestinal epithelial cells and are now considered to be the major pathway for absorption of the products of protein digestion. A specific carrier in the luminal cell membrane actively transports peptides into the cell. The energy for uphill movement of peptide is supplied by an inwardly directed H^+ gradient, not the more usual Na^+ gradient. This carrier transports a broad range of dipeptides and tripeptides and also certain antibiotics (e.g., aminocephalosporins). Tetrapeptides are only poorly transported. The dipeptides and tripeptides that enter the cells are converted to amino acids by intracellular peptidases, so the cells release free amino acids into portal blood.

Amino acids are absorbed more slowly than are dipeptides and tripeptides. Six different amino acid transporters that transport different groups of related amino acids have been identified in the brush border. Most, but not all, of these are capable of active transport and are sodium dependent. Other carriers that mediate exit of amino acids from the intestinal absorptive cell, usually by facilitated diffusion, are located in the basolateral cell membrane. Most amino acids are absorbed in the duodenum and jejunum.

Only minute quantities of undigested proteins are absorbed by the intestine in adults. The intestinal epithelium is normally too impermeable for diffusion of macromolecules through tight junctions, and pinocytosis is insignificant. In some newborn mammals, significant passive immunity is conveyed by absorbing immunoglobulins in the milk by pinocytosis, but this does not appear to be important in humans. Absorption of antigenic quantities of proteins appears to be greater in infants less than three months of age than in older children. As a result, food allergies are more common in newborns and infants.

The Digestion and Absorption of Lipids Present Special Problems

Dietary intake of lipids is approximately 60 to 100 g/day. Triglycerides account for approximately 90% of the lipids, with the rest composed of cholesterol esters, phospholipids, and minute quantities of the fat-soluble vitamins. The digestion of triglycerides is emphasized here because these are quantitatively the most important lipids. **Triglycerides** (also called *triacylglycerols*) consist of a glycerol backbone linked by ester bonds to three fatty acids, mostly of the long-chain type, such as palmitic, stearic, oleic, and linoleic acids (Fig. 22–32). Most lipids are quite insoluble in water, so a major problem in lipid absorption is to make fats soluble in the fluid in the intestinal lumen.

Efficient digestion of triglycerides requires that they be thoroughly emulsified as small (approximately 1 μm diameter) droplets of oil with a large surface area. **Emulsification** is the process of producing a suspension of small globules of one liquid within another liquid. This is accomplished by the churning action of the gastric antrum and by forcing the chyme through the narrow pylorus. The emulsion droplets are stabilized by fatty acids, proteins, and other amphipathic molecules, such as the bile salts.

Fatty acids are produced in the stomach by the action of **gastric lipase,** an enzyme produced by the fundus of the stomach. This enzyme normally accounts for approximately 10% to 30% of triglyceride digestion in adults, has an acidic pH optimum (pH, 4–6), and has a preference for the 3-ester bond of triglycerides, resulting in the release of a fatty acid and diglyceride (see Fig. 22–32). The fatty acids produced by gastric lipase activity stimulate the release of CCK from the duodenum (see Table 22–2), which in turn stimulates secretion of pancreatic lipase. The fatty acids also help stabilize the fat emulsion that is produced by muscular activity of the stomach.

When the emulsified, partially digested lipid reaches the duodenum it is acted upon by **pancreatic lipase.** This is the main enzyme responsible for digestion of triglycerides in the normal adult and is usually secreted in great excess. This enzyme requires a neutral pH for optimal activity. It specifically cleaves the 1- and 3-ester bonds of triglycerides to produce a *2-monoglyceride* and two fatty acids (see Fig. 22–32). Pancreatic lipase is secreted in an active form, not as a zymogen. It is an unusual enzyme in that it is not effective except when acting at an oil-water interface. Bile salts in the intestinal lumen promote

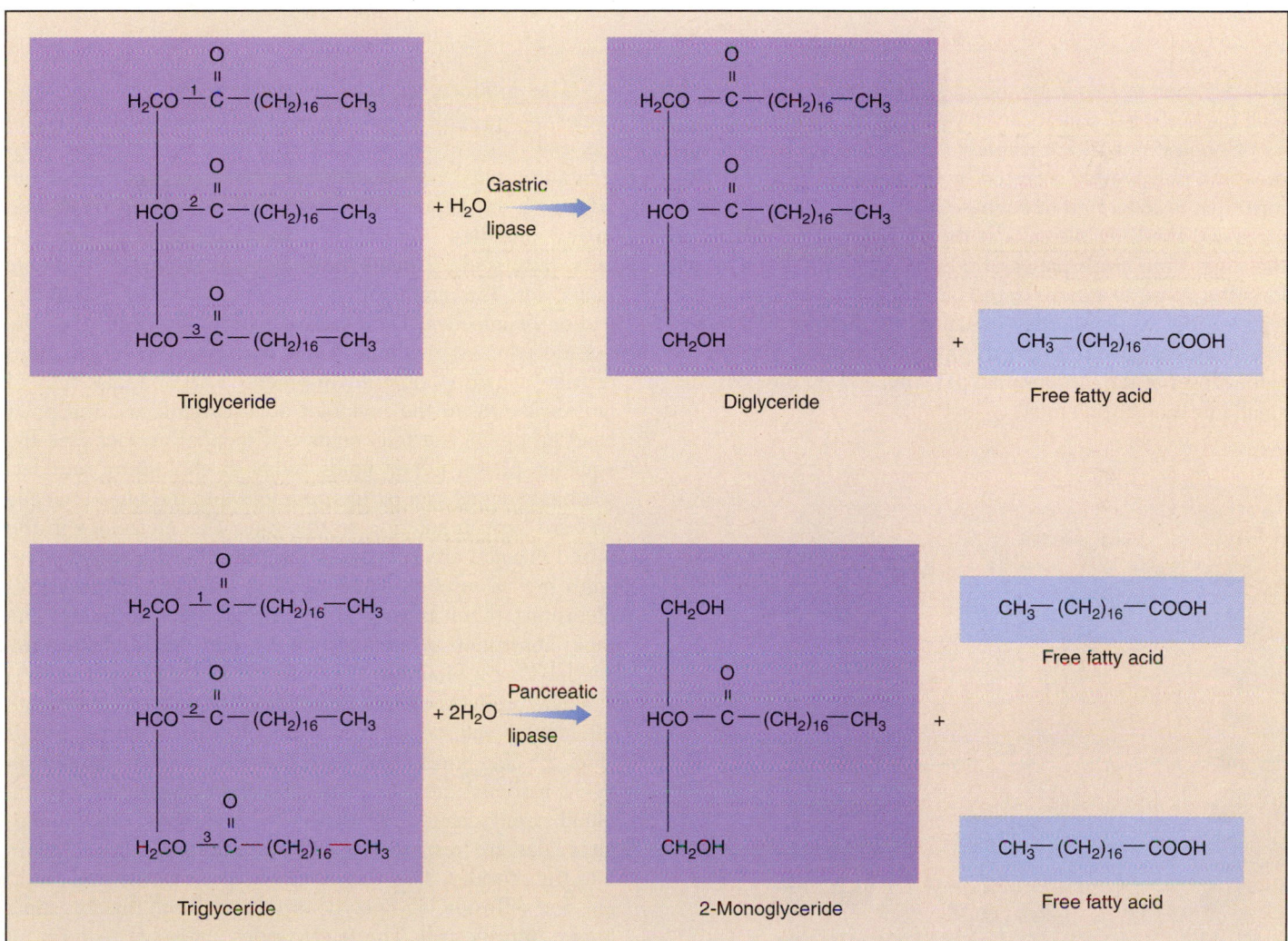

Figure 22–32

Hydrolysis of triglyceride by gastric and pancreatic lipases. The positions of the ester bonds in the triglyceride (triacylglycerol) molecule are labeled 1, 2, and 3. Triglycerides are hydrolyzed to diglyceride (diacylglycerol) and one fatty acid by gastric lipase, and to a 2-monoglyceride (2-monoacylglycerol) and two fatty acids by pancreatic lipase.

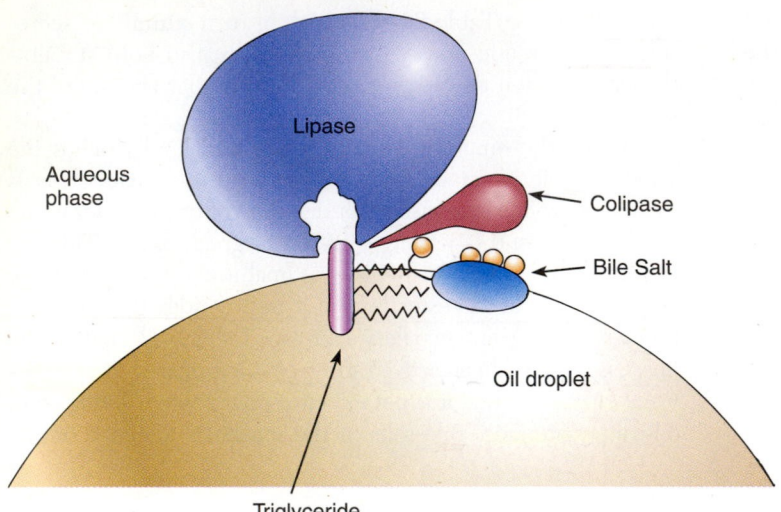

Figure 22–33

Role of colipase. Colipase anchors the pancreatic lipase to the surface of oil droplets, thereby facilitating hydrolysis of triglycerides.   *(From Turnberg, L. A., and Riley, S. A. "Digestion and Absorption of Nutrients and Vitamins". In Sleisenger, M. H., and Fordtran, J. S., eds.* Gastrointestinal Disease, *ed 5. Philadelphia, W. B. Saunders, 1993, pp 977–1008.)*

lipid digestion by stabilizing the fat emulsion. However, they inhibit the binding of pancreatic lipase to the oil (fat) droplets.

Colipase provides a remedy for this last problem. **Colipase** is a polypeptide secreted by the pancreas in an inactive form that is converted to colipase by trypsin. Colipase recognizes and binds to bile salts at the oil-water interface and at the same time binds pancreatic lipase (Fig. 22–33). This allows the lipase to adhere to the oil droplet and hydrolyze the triglycerides. Colipase also enhances the activity of pancreatic lipase by lowering its pH optimum from 8.5 to 6.5, a value closer to the intraluminal pH (pH, 6–7) in the proximal small intestine after a meal.

The products of lipolysis (fatty acids and monoglycerides) are present at an extremely low concentration as single molecules in aqueous solution (because of their poor solubility), but large concentrations of these substances can effectively be kept in solution by incorporating them into **mixed micelles** (Fig. 22–34). These are aggregates of bile salts, fatty acids, 2-monoglycerides, and other lipid-soluble molecules. The amphipathic bile salts play a key role in formation of micelles. Their polar (hydrophilic) sides face the outside aqueous medium. Their nonpolar (lipophilic) sides provide a hydrophobic environment. The nonpolar tails of fatty acids are in the nonpolar environment, and the polar head groups of the fatty acids and monoglycerides face the aqueous phase. Other lipids, such as cholesterol and fat-soluble vitamins, can be incorporated into the mixed micelle and are kept in solution in this way. The molecules in the mixed micelles are in dynamic equilibrium with single molecules free in solution and serve as a reservoir for these, replenishing them as they are absorbed. This facilitates the rapid absorption of monoglycerides, fatty acids, cholesterol, and fat-soluble vitamins.

The products of lipid digestion diffuse passively through the luminal cell membranes of the intestinal absorptive cells (Fig. 22–35). Short-chain (<6 carbons) and medium-chain (8–12 carbons) fatty acids are absorbed directly into portal blood. Long-chain fatty acids (>12 carbons) and monoglycerides are resynthesized in the endoplasmic reticulum into triglycerides. This resynthesis helps to maintain a gradient for diffusion of lipolytic products from the intestinal lumen into the cell. The triglycerides aggregate to form oil droplets. Re-esterified cholesterol and newly synthesized phospholipids are added to the oil droplets. Proteins called **apolipoproteins** are added to the surface. The result is a large (100- to 500-nm diameter) lipoprotein particle called a **chylomicron.** Chylomicrons are packaged into secretory vesicles and are expelled from the cell by exocytosis through the basolateral side of the cells.

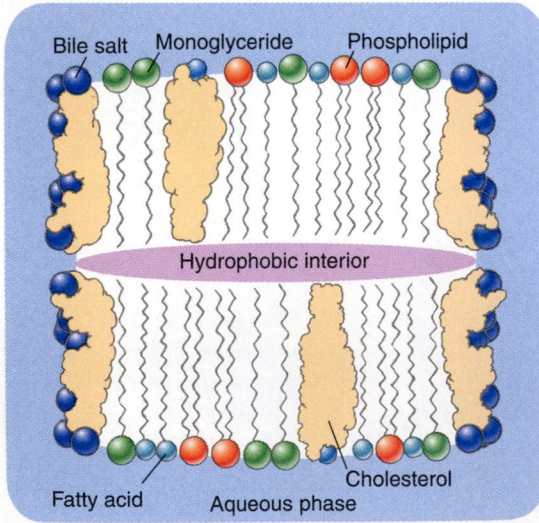

Figure 22–34

Longitudinal section of a mixed micelle. The circles represent polar (hydrophilic) portions of the molecules. Micelle formation allows the products of lipid digestion to stay in solution in the intestinal lumen.

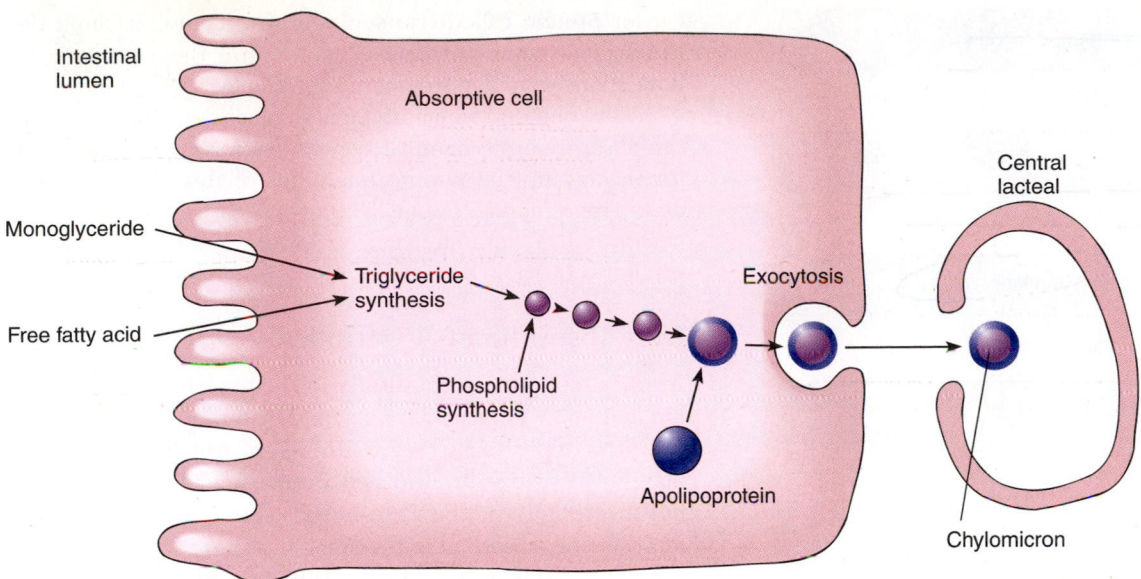

Figure 22–35

Pathway for lipid absorption in the small intestine. Free fatty acids and monoglycerides diffuse through the luminal cell membrane into the absorptive cell and are resynthesized into triglycerides. Phospholipids are also resynthesized. Droplets containing triglycerides, phospholipids, and cholesterol are produced and apolipoproteins are added to the surface to form chylomicrons. Chylomicrons are extruded from the cell by exocytosis and are removed by intestinal lymphatic vessels.

The chylomicrons, because of their size, cannot get into intestinal capillaries, but they can enter the lacteals of the intestinal villi. During absorption of a fatty meal, the lymphatics have a milky-white appearance due to the presence of numerous chylomicrons. Chylomicrons bypass the liver and enter the systemic circulation via the thoracic duct, which empties into the subclavian vein. Chylomicron triglyceride is hydrolyzed in the blood by **lipoprotein lipase,** an enzyme found on the surface of endothelial cells. The fatty acids liberated are bound to albumin and are taken up and metabolized by the tissues. The remainder of the chylomicron, now enriched in cholesterol and phospholipid, is termed a **chylomicron remnant.** It is sufficiently small to pass through fenestrations of the endothelial cells of the hepatic sinusoids. The remnants are taken up by the hepatocytes and degraded.

Intestinal epithelial cells also produce small lipoprotein particles called **very-low-density lipoproteins (VLDLs).** The VLDLs contain more cholesterol and protein and less triglyceride than do chylomicrons and are smaller (30–80 nm in diameter) than are chylomicrons. The processes of VLDL formation and release are much like those described for chylomicrons.

Normally ingested fat is absorbed completely. If fat digestion or absorption is impaired, the supply of calories and absorption of fat-soluble vitamins may be inadequate. **Steatorrhea,** the presence of excessive amounts (>6 g/day) of fat in the feces, may result in especially foul-smelling stools and diarrhea. Impaired fat absorption occurs with (1) removal of the stomach (i.e., inadequate emulsification), (2) pancreatic disease (i.e., inadequate digestion by pancreatic lipase), (3) hepatic disease (i.e., inadequate bile salts to solubilize lipid digestion products), (4) small intestine disease (i.e., inadequate absorptive area and impaired production of chylomicrons), and (5) intestinal lymphatic disease (i.e., inadequate transport of chylomicrons).

 Where and how are other substances absorbed?

Vitamins Are Absorbed by Many Different Mechanisms

The absorption of the **fat-soluble vitamins (A, D, E, and K)** is very similar to that of other lipids. They are incorporated into mixed micelles, are usually passively transported into absorptive cells, are incorporated into chylomicrons or VLDLs, and are carried away by intestinal lymphatics.

The **water-soluble vitamins** include the B vitamins, folic acid, and vitamin C. These vitamins are absorbed by diffusion or by active transport mechanisms.

The absorption of **vitamin B₁₂ (cobalamin)** is an especially intricate and sophisticated process (Fig. 22–36). This cobalt-containing vitamin serves as a coenzyme in amino acid metabolism. Deficiency of this vitamin produces **pernicious anemia,** a disease that can be fatal. This disorder is associated

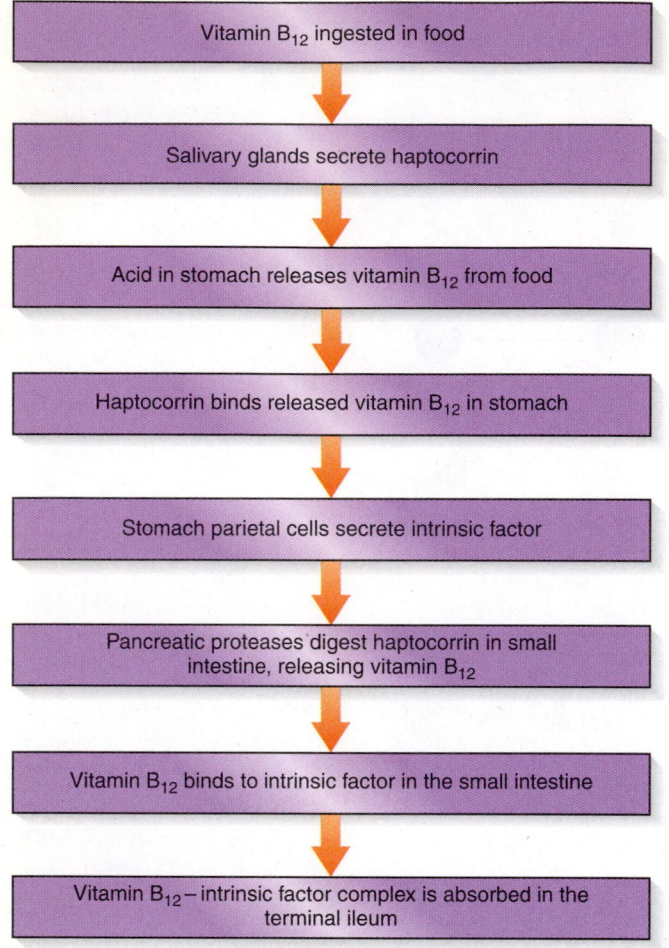

Vitamin B_{12} ingested in food

↓

Salivary glands secrete haptocorrin

↓

Acid in stomach releases vitamin B_{12} from food

↓

Haptocorrin binds released vitamin B_{12} in stomach

↓

Stomach parietal cells secrete intrinsic factor

↓

Pancreatic proteases digest haptocorrin in small intestine, releasing vitamin B_{12}

↓

Vitamin B_{12} binds to intrinsic factor in the small intestine

↓

Vitamin B_{12}–intrinsic factor complex is absorbed in the terminal ileum

Figure 22–36

Steps in the absorption of vitamin B_{12}.

with a macrocytic anemia (low hematocrit, but large red cells) and neurological disease and usually takes one to three years to develop because there are large reserves of the vitamin in the liver. The adult daily requirement is approximately 6 µg/day, and the average Western diet contains 5 to 15 µg/day. Vitamin B_{12} can be synthesized only by bacteria; it is found in animal food products (liver is an excellent source) but not in vegetables.

In the stomach, vitamin B_{12} is first released from dietary proteins by acid and pepsin. It then binds to glycoproteins, called **haptocorrins,** found in the saliva swallowed with the ingested food. The **intrinsic factor** is a glycoprotein produced by stomach parietal cells; it binds to vitamin B_{12} after the haptocorrins are digested by pancreatic proteases (e.g., trypsin) in the duodenum. The alkaline environment of the duodenum facilitates vitamin B_{12}-intrinsic factor binding. The vitamin B_{12}-intrinsic factor complex is resistant to digestion by pancreatic proteases. It binds to receptors in the brush border membrane of absorptive cells in the *terminal ileum* and is absorbed by receptor-mediated endocytosis (Chapter 12). Vitamin B_{12} is absorbed into villus capillary blood and binds to a

carrier protein called **transcobalamin.** Transcobalamin delivers the vitamin to tissues that require the vitamin (e.g., bone marrow) and to storage sites in the liver and kidneys.

This cumbersome but highly specific arrangement for the absorption of vitamin B_{12} provides a recognition system for picking up the minute quantities of this vitamin in the diet. The remarkable specificity of this process prevents the absorption of unwanted analogues of vitamin B_{12} that are produced by some bacteria.

Iron Absorption Is Controlled

Iron is an important constituent of hemoglobin and myoglobin and many intracellular enzymes. The average daily intake of iron is approximately 12 to 15 mg, of which only approximately 10% is usually absorbed. Stomach acidity is important in converting ferric (Fe^{3+}) to ferrous (Fe^{2+}) iron; the latter is more readily absorbed in the small intestine. Iron can be absorbed along the whole length of the intestine, but efficiency of absorption is greatest in the duodenum and upper jejunum.

The iron in meat is present mainly as **heme iron,** a bound form also present in hemoglobin. Heme iron is transported into the cell by a facilitated transport mechanism, and its ferrous ion is released by the enzyme heme oxygenase (Fig. 22–37). **Free iron** is absorbed by a different carrier-mediated transport mechanism in the brush border. Once inside the cell, iron may be transported by an intracellular transport (carrier) protein to the basolateral cell membrane, where iron is actively extruded, or it can combine with a cell protein, **apoferritin,** to form **ferritin.** Ferritin is an intracellular storage form of iron; one apoferritin molecule can bind as many as 4500 Fe^{3+} atoms. Iron is transported in the blood bound to a plasma protein called **transferrin** that specifically binds 2 Fe^{3+} atoms.

The absorption of iron is controlled according to the body's need for iron. In an iron-deficient subject, such as a person with chronic blood loss, the rate of intestinal absorption of iron is increased. The number of transporters present in the luminal cell membrane is increased and iron in the cell is also more rapidly transferred *out* of the cell and into the plasma. The plasma iron concentration and saturation of plasma transferrin are reduced, which would favor absorption. The cell contents of apoferritin and ferritin are decreased, and so less iron is lost when enterocytes are shed into the lumen.

On the other hand, if there is excess iron in the body, plasma transferrin becomes saturated and can accept no more iron from the absorptive cells. Synthesis of apoferritin and ferritin in the absorptive cell is increased, and most of the iron entering the cell is sequestered in ferritin. Eventually, the iron-rich cells are shed at the villus tip, releasing iron into the lumen, and the iron is excreted in the feces.

Absorption of Calcium Is Controlled by 1,25-Dihydroxy-Vitamin D₃

With a high dietary intake of calcium, calcium is absorbed throughout most of the small intestine via passive diffusion through the tight junctions between epithelial cells. With low

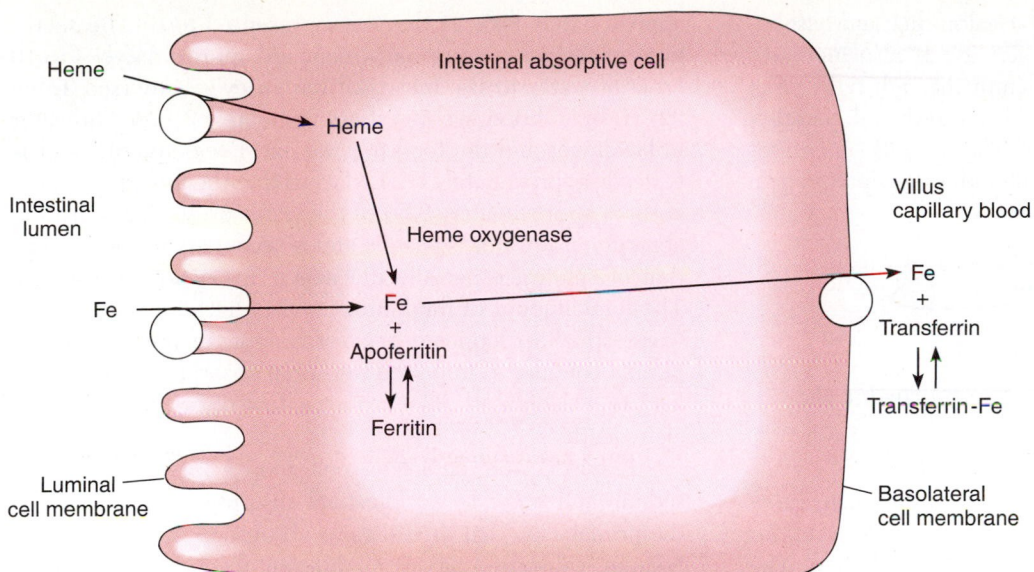

Figure 22–37

Model for intestinal iron (Fe) absorption. Heme iron and free iron enter the absorptive cells separately. Iron transport mechanisms are poorly understood. Iron is bound to apoferritin for storage in the cell and is bound to a plasma protein, transferrin, in the blood.

intake, absorption is mainly via an active, transcellular route (Fig. 22–38). Transcellular calcium absorption takes place mainly in the duodenum and upper jejunum, with very little transport in the ileum. Calcium uptake occurs via specific calcium channels located in the brush border; Ca^{2+} enters the cell down an electrochemical gradient. A calcium-binding protein **(calbindin)** found in the cytosol minimizes an increase in free cytosolic Ca^{2+} and transports calcium to the basolateral side of the cell. Active extrusion out the basolateral cell membrane occurs via a Ca^{2+}-ATPase. The hormone **1,25-dihydroxy-vitamin D_3,** also known as **calcitriol,** increases transcellular transport by increasing calcium entry

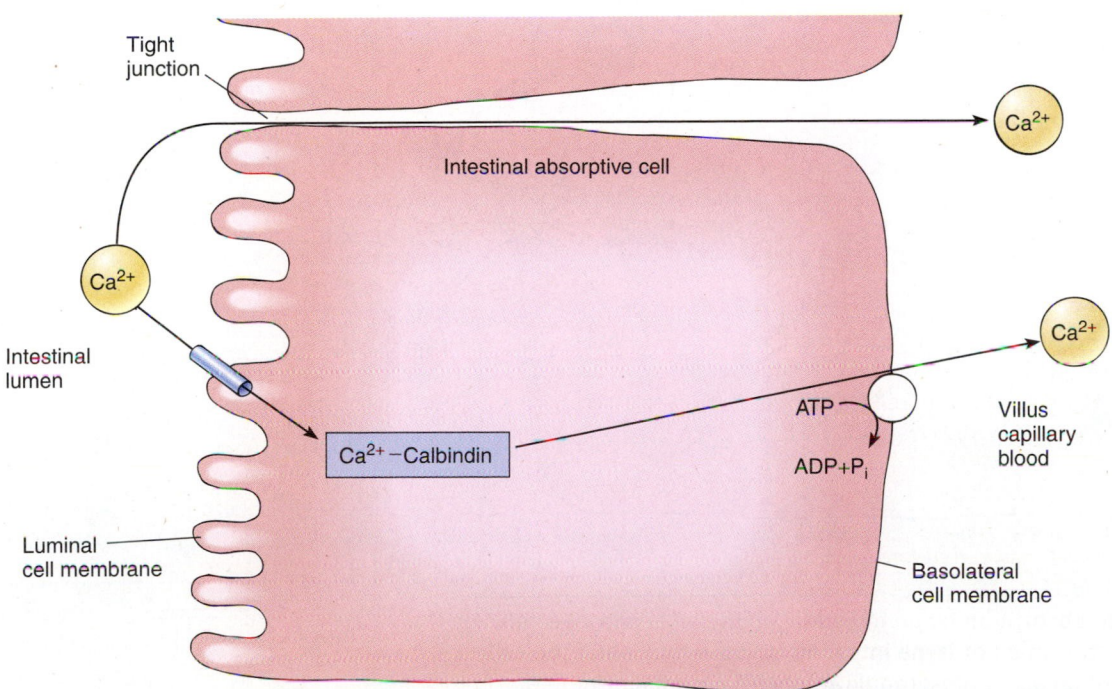

Figure 22–38

Model for intestinal calcium absorption. Calcium is absorbed via a transcellular route (luminal membrane calcium channel, intracellular binding by calbindin, and basolateral cell membrane Ca^{2+}-ATPase) and via a paracellular route through tight junctions.

into the cell, synthesis of intracellular calbindin, and basolateral cell membrane Ca^{2+}-ATPase activity. It also increases some paracellular absorption of calcium through tight junctions. When plasma free calcium levels are reduced, parathyroid hormone secretion is stimulated and increases $1,25(OH)_2$-vitamin D_3 synthesis in the kidneys, which in turn stimulates intestinal calcium absorption.

The Small and Large Intestine Absorb Salt and Water

> *How and where are salt and water absorbed in the digestive tract?*

Absorption of salt (NaCl) and water is one of the major functions of the intestine. The various gastrointestinal secretions add roughly 30 g/day of NaCl to the intestinal lumen. The diet typically adds an additional 5 to 15 g/day NaCl. Fecal loss of NaCl is typically 100 mg/day, so the intestine absorbs

approximately 99% of the NaCl presented to it. The secretions of the gastrointestinal system add approximately 7 to 10 liters of water to the intestinal lumen every day (see Table 22–3). In addition, we ingest approximately 2 L/day of water in beverages and the food that we eat. Fecal loss of water is typically approximately 0.1 L/day, so that the intestine overall absorbs approximately 99% of the water presented to it. The absorption of water is passive and is secondary to the absorption of solutes, especially the active absorption of sodium. The total amount of fluid entering the digestive tract in gastrointestinal secretions every day is approximately twice the plasma volume, so impaired intestinal absorption can rapidly lead to a decreased blood volume, low blood pressure, and even death.

Figure 22–39 depicts a cell model for absorption of sodium and water by an intestinal absorptive cell. Sodium is the primary ion that drives water absorption in the intestine. Sodium is absorbed via several different mechanisms in the luminal cell membranes. This varies from one region of the intestine to another. In the basolateral membrane of the cells,

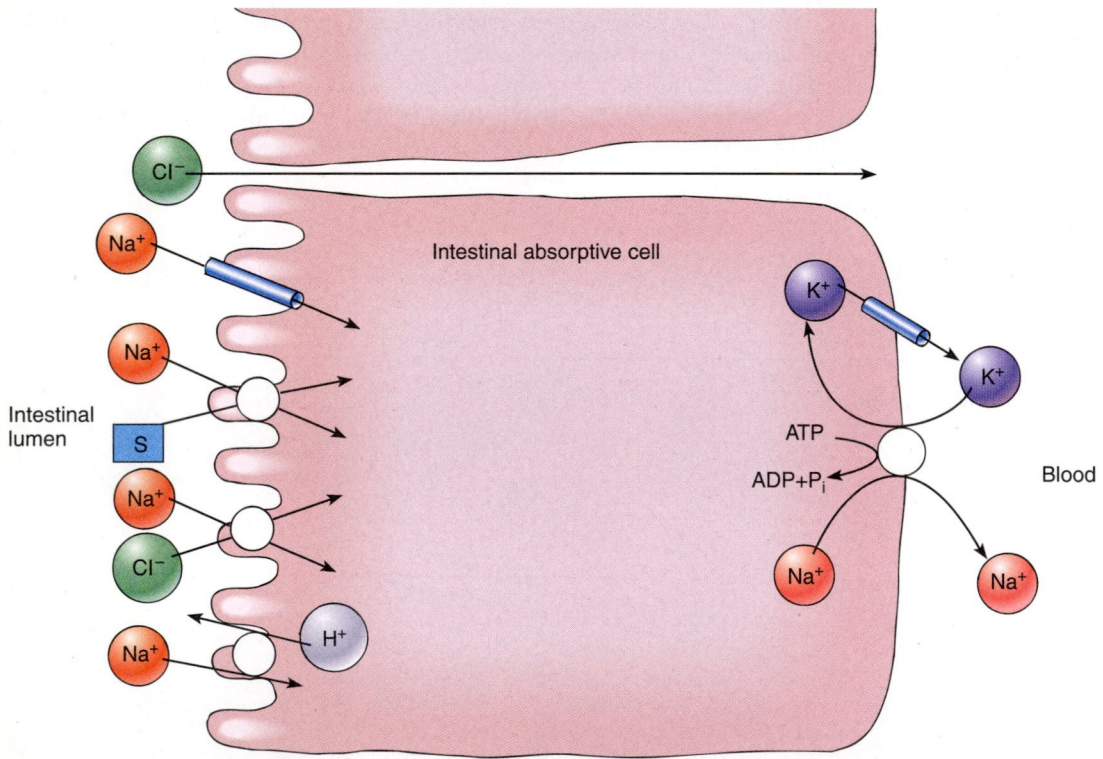

Figure 22–39

Model for sodium and water absorption by an intestinal epithelial absorptive cell. Depending on the region of the small or large intestine, sodium may enter the cell via (1) a sodium ion channel, (2) a sodium-organic solute (S) cotransporter (e.g., SGLT1), (3) a sodium-chloride cotransporter, or (4) a sodium ion/hydrogen ion exchanger. The sodium is pumped out of the cell by the basolateral membrane Na^+/K^+-ATPase. Water absorption is passive, accompanying the absorbed solutes (mainly sodium), and occurs either between the cells (paracellular path through the tight junctions) or through the cells (transcellular path).

there is always a Na^+/K^+-ATPase, which maintains a low intracellular sodium concentration and pumps sodium out of the cell toward the blood side. Water follows the absorbed solutes passively, along an osmotic gradient created by removal of solutes from the lumen. The water permeability of the small intestine is high and, hence, only a very small transepithelial osmotic gradient suffices to bring about water absorption. The large intestine has less water permeability. Fecal water may actually be hypo-osmotic (due to solute absorption without water) or hyperosmotic (due to bacterial degradation of organic molecules) compared with plasma. Intestinal epithelial cells also absorb or secrete potassium ions and secrete bicarbonate ions.

The stomach is *not* an important site of fluid or electrolyte absorption. The duodenum also is not: its length is rather short, and its main function appears to be osmotic equilibration of the luminal contents with the blood. Water readily moves across the duodenal epithelium in either direction, depending on the transepithelial osmotic gradient. The jejunum is the major site of sodium chloride and water absorption.

The small intestine has a much larger capacity to absorb sodium and water than does the large intestine. Its efficiency is, however, less. Thus the small intestine ordinarily absorbs 75% of the sodium presented to it, but the large intestine absorbs 95%. The small intestine can establish only a modest transepithelial sodium gradient; for example, in the jejunum, the lowest luminal sodium concentration that can be achieved is 130 mEq/L compared with a luminal sodium concentration of 30 mEq/L achieved in the large intestine. This is because the small intestine has leaky tight junctions that permit the back-diffusion of sodium from the blood side (where the sodium concentration is 140 mEq/L) into the lumen. In the large intestine, tight junctions are moderately tight, so there is little back-diffusion of sodium, and a lower luminal sodium concentration can therefore be established. Because the large intestine is at the end of the system, it has a major impact on sodium and water loss in the feces.

The Intestine Also Secretes Fluid

The intestine secretes approximately 2 L of fluid in a day (see Table 22–3), although the exact volume is hard to measure because of simultaneous absorption of fluid. The crypts are the main source of secreted fluid. The intestinal secretions also include mucus produced by goblet cells and shed mucosal cells.

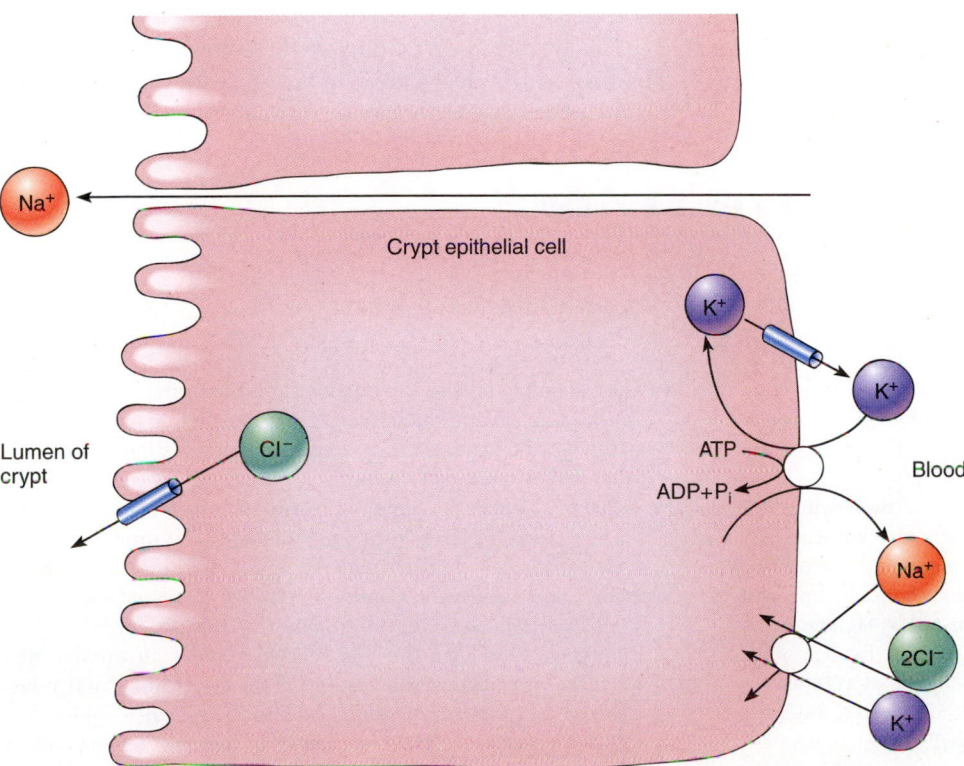

Figure 22–40

Mechanism of intestinal fluid secretion. Crypt epithelial cells have a Na/K/2Cl symporter in their basolateral cell membrane that can accumulate chloride in the cell. The chloride then diffuses into the lumen via an apical cell membrane chloride channel. The secreted chloride ions are accompanied by sodium ions and water.

The crypt cells secrete an isosmotic watery fluid that is richer in bicarbonate than in chloride and so is somewhat alkaline. This fluid is thought to be important in maintaining the liquidity of the chyme. A cell model for secretion is illustrated in Figure 22–40. Chloride accumulates in the cell via a Na/K/2Cl symporter in the basolateral cell membrane. Chloride then diffuses downhill into the lumen via a chloride channel (CFTR). Sodium probably moves between the cells, and water follows to maintain isotonicity.

Secretion of fluid is stimulated by mechanical irritation of the mucosa and distension of the intestine. These responses are mediated by local nerve reflexes in the gut. Parasympathetic stimulation augments secretion. Secretion is also stimulated by several gastrointestinal hormones (e.g., gastrin, CCK, secretin, and GIP). Secretion is inhibited by the sympathetic nervous system and by epinephrine. The absorptive and secretory functions of the intestine are independent and have different control mechanisms. Therefore, the absorptive function may remain intact when the secretory function is excessively stimulated, such as by bacterial toxins.

Diarrhea May Have Many Different Causes

Diarrhea is defined as a fecal water output of greater than 500 ml/day in an adult. Diarrhea is the main cause of death or disability in the world today, afflicting individuals of all ages. Death is due to decreased blood volume and circulatory collapse, compounded by metabolic acidosis and low plasma potassium. The causes of diarrhea can be divided into four categories.

First, diarrhea may be due to impaired absorption of NaCl by the small or large intestine. This can result from a congenital defect, infection, inflammation, or anything that causes a reduction of total effective surface area of the intestine. Some bacterial enterotoxins, such as from certain strains of *Escherichia coli*, inhibit intestinal sodium absorption. Excessive amounts of bile salts in the large intestine also inhibit sodium absorption here.

Second, if nonabsorbable or poorly absorbed solutes accumulate in the intestinal lumen, then this causes increased loss of sodium and water in the feces. This effect contributes to the diarrhea that people with lactose intolerance sometimes experience. Epsom salts (magnesium sulfate) produce diarrhea because magnesium ions are only slowly absorbed.

Third, excessive stimulation of secretory cells present in the crypts of the small and large intestine results in a **secretory diarrhea**. **Cholera** results from infection with *Vibrio cholerae* and can cause the intestine to secrete as much as 20 L/day of fluid, resulting in massive diarrhea and fatal consequences. The cholera bacteria produce a polypeptide enterotoxin (**cholera toxin**) that turns on a G protein in crypt cells. This leads to persistently high levels of cAMP in crypt cells and excessive chloride secretion. Administration of an oral rehydration solution containing glucose is useful in treating cholera because intestinal glucose absorption is not impaired and glucose absorption is accompanied by sodium and water absorption.

Finally, increased motility of the colon can lead to diarrhea because the feces are eliminated before the maximum amount of salt and water can be extracted.

CHAPTER REVIEW

Summary

- The major function of the gastrointestinal system, which comprises the digestive tract and its accessory organs, is to process food, and, by so doing, the gastrointestinal system provides nutrients to the cells of the body.
- The processing of food begins with chewing and swallowing. The salivary glands produce saliva, a fluid that acts as a lubricant and solvent; has bacteriocidal, buffering, and cleansing actions; and contains an amylase that initiates the digestion of starch, the major dietary carbohydrate. Swallowing is coordinated by centers in the brain and accomplishes the rapid movement of food into the stomach by a peristaltic wave that travels along the esophagus.
- The proximal stomach relaxes to receive the ingested food; the distal stomach grinds the food into small particles. Stomach parietal cells secrete hydrochloric acid; stomach chief cells produce pepsinogen, a proenzyme that is converted to the proteolytic enzyme pepsin at acidic pH levels. Gastric lipase initiates some digestion of triglycerides (fat).
- The secretion of gastric juice is stimulated by vagal efferents and by gastrin, a hormone produced mainly by G cells in the gastric antrum. The stomach has a gastric mucosal barrier that protects it against damage by acid and pepsin.

- Gastric emptying into the duodenum is controlled by conditions in the stomach and duodenum via various neural and hormonal pathways and is adjusted so that the small intestine is not overwhelmed by too much chyme.
- A low pH level in the lumen of the duodenum stimulates the release of secretin. This hormone stimulates the production of a bicarbonate-rich fluid by pancreatic and bile duct cells. Fatty acids, amino acids, and peptides stimulate duodenal and jejunal I cells to produce CCK, a hormone that inhibits further gastric emptying, stimulates the pancreas to secrete an enzyme-rich juice, and stimulates contraction of the gallbladder and relaxation of the sphincter of Oddi. The bile and pancreatic juice flow into the duodenum. The bicarbonate in these secretions buffers the gastric acid in the lumen of the duodenum. Pancreatic amylase digests polysaccharides to oligosaccharides, maltotriose, and maltose; proteolytic enzymes (i.e., trypsin, chymotrypsin, elastase, and carboxypeptidases) digest proteins and polypeptides to tripeptides, dipeptides, and amino acids; and pancreatic lipase digests triglycerides to monoglycerides and fatty acids.
- Peristaltic movements and mixing movements, coordinated by the enteric nervous system, are reflexly stimulated by distension

of the small intestine. These movements mix the chyme with the various secretions and expose the surface epithelial cells to new materials for absorption.

- The small intestine is the major site of absorption of nutrients, minerals, water, and vitamins. Its surface absorptive cells (enterocytes) have a brush border (microvilli) rich in enzymes that can break down oligosaccharides and disaccharides to monosaccharides and peptides to smaller peptides and amino acids. These cells also possess a variety of transporters. The products of triglyceride digestion (monoglycerides and free fatty acids) are incorporated into mixed micelles together with bile salts. Long chain fatty acids and monoglycerides enter the absorptive cells by simple diffusion and are resynthesized into triglycerides, incorporated into chylomicrons, and transported away from the intestine via lymphatics. The bile salts are absorbed in the termi-

nal ileum and undergo an enterohepatic circulation so that they can be used more than once during a meal.

- In most cases, absorption is an uncontrolled process; exceptions are the absorption of iron, calcium, and vitamin B_{12}.
- Between meals, a migrating motor complex passes down the intestinal tract, from the stomach to the end of the small intestine, and cleanses these areas of food residues.
- The large intestine contains the residues of several meals and many bacteria. Segmenting contractions (haustrations) promote the absorption of water and result in the formation of a firm fecal mass. This dehydrated material is moved into the distal large intestine by mass movements and is stored there for variable times until the defecation reflex is initiated by distension of the rectum, and defecation occurs.

Review Questions

Choose the Correct Answer

1. Intestinal peristalsis:
   a. always occurs in an aboral direction.
   b. is decreased by parasympathetic nerve stimulation.
   c. is dependent on an intact enteric nervous system.
   d. is increased by sympathetic nerve stimulation.
   e. is purely myogenic.

2. A hormone that stimulates secretion of an enzyme-rich fluid by pancreatic acinar cells is:
   a. CCK.
   b. gastrin.
   c. GIP.
   d. motilin.
   e. secretin.

3. Which pancreatic enzyme plays a key role in the conversion of zymogens to active enzymes in the lumen of the small intestine?
   a. Chymotrypsin
   b. Enteropeptidase
   c. Lipase
   d. Pepsin
   e. Trypsin

4. During fasting or between meals:
   a. blood gastrin and secretin levels are highest.
   b. blood motilin levels are lowest.
   c. migrating motor complexes pass down the small intestine.
   d. release of somatostatin by stomach D cells is lowest.
   e. stomach luminal pH level is highest.

5. A gastrointestinal process that is not controlled is:
   a. gastric emptying.
   b. gastric secretion of HCl.
   c. secretion of pancreatic enzymes.
   d. small intestinal absorption of calcium.
   e. small intestinal absorption of glucose.

6. The organ of greatest importance in providing the enzymes required for digestion of food is the:
   a. liver.
   b. parotid salivary gland.
   c. pancreas.
   d. stomach.

7. Intrinsic factor:
   a. is produced by chief cells in the stomach.
   b. binds vitamin B_{12} in the stomach lumen.

   c. binds vitamin B_{12} after it is released from haptocorrin.
   d. is rapidly hydrolyzed by pancreatic enzymes in the small intestine.
   e. is stored in the liver.

8. In which of the following regions is intraluminal pressure lowest when there is no swallowing activity?
   a. Mouth
   b. Upper esophageal sphincter
   c. Body of the esophagus
   d. Lower esophageal sphincter
   e. Stomach

9. A blood-borne hormone that stimulates secretion of HCl by gastric parietal cells is:
   a. CCK.
   b. gastrin.
   c. motilin.
   d. secretin.
   e. somatostatin.

10. The cephalic phase of gastric HCl secretion is abolished by:
    a. surgical removal of the antrum.
    b. vagal denervation of the fundus and body of the stomach.
    c. vagal denervation of the antrum of the stomach.
    d. None of the above

11. The pancreatic ductule cells are primarily responsible for which component of the pancreatic juice?
    a. Bicarbonate
    b. Bilirubin
    c. Chymotrypsinogen
    d. Pancreatic lipase
    e. Trypsinogen

12. Which hormone causes contraction of the gallbladder and relaxation of the sphincter of Oddi?
    a. Gastrin
    b. CCK
    c. GIP
    d. Motilin
    e. Secretin

13. The intestinal digestion of triglycerides is facilitated by the incorporation of these molecules into:
    a. large oil droplets.
    b. small emulsified oil droplets.
    c. micelles.
    d. chylomicrons.

14. The proximal stomach is characterized by:
    a. phasic contractions.
    b. production of gastrin.
    c. receptive contraction.
    d. slow waves.
    e. tonic contraction.

15. The enterohepatic circulation refers to the:
    a. hepatic artery blood flow.
    b. hepatic portal venous blood flow.
    c. lymphatic drainage of the intestines and liver.
    d. microcirculation of the gut and liver.
    e. recycling of bile salts between intestine and liver.

16. Which of the following factors in the intestinal lumen stimulates secretin release from duodenal S cells?
    a. A high hydrogen ion concentration
    b. Amino acids and peptides
    c. Fatty acids
    d. Glucose

17. The migrating motor complex:
    a. is inhibited by feeding.
    b. is inhibited by motilin.
    c. is responsible for mass movements in the large intestine.
    d. promotes the mixing of several meals in the small intestine.
    e. starts in the esophagus and travels to the rectum.

18. It is necessary to remove the terminal ileum of an individual. Which of the following is not a likely consequence of this surgical procedure?
    a. Increased hepatic synthesis of bile salts
    b. Inadequate absorption of glucose
    c. Pernicious anemia
    d. Steatorrhea

19. Micelle formation in the small intestine is important in the absorption of all of the following vitamins *except* vitamin:
    a. A.
    b. C.
    c. D.
    d. E.
    e. K.

20. The portion of the digestive tract that normally absorbs the largest amount of sodium and water is the:
    a. esophagus.
    b. stomach.
    c. jejunum.
    d. proximal colon.
    e. distal colon.

21. The intestinal absorption of lactose:
    a. depends on the enzymatic hydrolysis of lactose by pancreatic amylase.
    b. involves the binding of lactose to a carrier in the intestinal brush border membrane.
    c. is controlled by hormonal mechanisms.
    d. requires the presence of sodium ions in the intestinal lumen for optimal absorption of its digestion products.

22. Which of the following secretions is almost entirely under neural control?
    a. Gastric secretion
    b. Intestinal secretion
    c. Pancreatic secretion
    d. Salivary secretion

23. Chief cells of the stomach secrete:
    a. gastrin.
    b. hydrochloric acid.
    c. intrinsic factor.
    d. pepsin.
    e. pepsinogen.

24. When the pH level of the stomach lumen decreases to less than 3, the antrum of the stomach releases a peptide that acts in a paracrine fashion to inhibit gastrin release. This peptide is:
    a. acetylcholine.
    b. gastrin-releasing peptide (GRP).
    c. GIP.
    d. somatostatin.
    e. vasoactive intestinal peptide (VIP).

25. The volume of fluid added to the lumen of the digestive tract every day in saliva, gastric juice, pancreatic juice, bile, and intestinal secretions is equal to:
    a. half of the plasma volume.
    b. plasma volume.
    c. twofold the plasma volume.
    d. tenfold the plasma volume.

Answers to Case History Questions

1. Gastroesophageal reflux occurs when the lower esophageal sphincter fails as a barrier between the stomach and esophagus. This may be due to inappropriate transient relaxations of the lower esophageal sphincter, chronic low pressure in the sphincter zone, anatomical abnormalities, or other causes. Increased intragastric pressure favors reflux.

2. The acidity and pepsin in gastric juice damage the esophageal mucosa and cause pain. Complications of chronic gastroesophageal reflux include esophageal ulcers; scarring and narrowing of the esophagus; and, in some patients, esophageal cancer.

3. Fat delays gastric emptying via the hormone cholecystokinin. This results in maintenance of an elevated intragastric pressure, and so favors reflux into the esophagus.

4. The symptoms are worse at night because, in the recumbent position, gravity does not help to drain refluxed acid from the esophagus. Increased intragastric pressure associated with lying flat also promotes **gastroesophageal reflux.** Salivary secretion is inhibited during sleep, so there is less swallowed saliva to neutralize refluxed acid.

5. H$_2$-receptor antagonists block the stimulation of stomach parietal cells by histamine. The proton pump inhibitors block the H$^+$/K$^+$-ATPase in the luminal cell membrane of parietal cells, reducing active secretion of hydrogen ions.

Key Terms

absorption (p. 684)
amylase (p. 699)
basic electrical rhythm (p. 685)
bile (p. 709)
bile salts (p. 709)
CCK (cholecystokinin) (p. 698)
celiac disease (p. 713)
cephalic phase (p. 703)
chylomicron (p. 716)
chyme (p. 693)
colipase (p. 716)
crypt (p. 712)
defecation (p. 697)
diarrhea (p. 722)
digestion (p. 684)
dipeptide (p. 714)
distal stomach (p. 693)
emulsification (p. 714)
enteric nervous system
    (p. 687)

enterocyte (p. 711)
enterohepatic circulation
    (p. 710)
enteropeptidase (p. 708)
fat-soluble vitamins (p. 717)
gallbladder (p. 710)
gastric emptying (p. 694)
gastric juice (p. 701)
gastric lipase (p. 702)
gastric mucosal barrier (p. 705)
gastric phase (p. 703)
gastrin (p. 698)
gastroesophageal reflux (p. 724)
GIP (glucose-dependent
    insulinotropic polypeptide)
    (p. 699)
haustration (p. 697)
hydrochloric acid (p. 701)
ileocecal sphincter (p. 696)
interstitial cell of Cajal (p. 685)

intestinal phase (p. 709)
intrinsic factor (p. 704)
lactose intolerance (p. 713)
mass movement (p. 697)
micelle (p. 710)
migrating motor complex
    (p. 694)
mixing movements (p. 684)
motilin (p. 699)
motility (p. 684)
mucosa (p. 685)
oxyntic gland (p. 701)
pancreatic lipase (p. 715)
paralytic ileus (p. 696)
pepsin (p. 704)
peptic ulcer (p. 706)
peristalsis (p. 690)
propulsive movements
    (p. 684)
proximal stomach (p. 693)

receptive relaxation (p. 691)
retropulsion (p. 694)
salivon (p. 699)
secretin (p. 698)
secretion (p. 684)
secretory diarrhea (p. 722)
segmentation (p. 691)
slow wave (p. 685)
somatostatin (p. 688)
sphincter (p. 691)
steatorrhea (p. 717)
swallowing (p. 691)
trypsin (p. 708)
vagovagal reflex (p. 694)
very-low-density lipoprotein
    (p. 717)
vitamin B_{12} (p. 717)
water-soluble vitamins (p. 717)
xerostomia (p. 701)
zymogen (p. 699)

Suggested Readings

Allen, A., Flemström, G., Garner, A., and Kivilaasko, E. "Gastro-duodenal mucosal protection." *Physiological Reviews,* 73:823–857, 1993.

Castell, D. O. "A practical approach to heartburn." *Hospital Practice,* 34:89–98, 1999.

Cave, D. R., and Hoffman, J. S. "Management of *Helicobacter pylori* infection in ulcer disease." *Hospital Practice,* 31:63–75, 1999.

Chang, E. B., Sitrin, M. D., and Black, D. B. *Gastrointestinal, Hepatobiliary, and Nutritional Physiology*. Philadelphia, Lippincott-Raven, 1996.

Grelot, L., and Miller, A. D. "Vomiting: Its ins and outs." *News in Physiological Sciences,* 9:142–147, 1994.

Henderson, J. M., ed. *Gastrointestinal Pathophysiology*. Philadelphia, Lippincott-Raven, 1996.

Hofer, D., Asan, E., and Drenckhahn, D. "Chemosensory perception in the gut." *News in Physiological Sciences,* 14:18–24, 1999.

Johnson, L. R., ed. *Gastrointestinal Physiology*, ed 6. St. Louis, Mosby, 2001.

Lichtenberger, L. M. "The hydrophobic barrier properties of gastrointestinal mucus." *Annual Review of Physiology,* 57:565–583, 1995.

Liddle, R. A. "Cholecystokinin cells." *Annual Review of Physiology,* 59:221–242, 1997.

Mäki, M., and Collin, P. "Coeliac disease." *Lancet.* 349:1755–1759, 1997.

Mawe, G. M. "Nerves and hormones interact to control gallbladder function." *News in Physiological Sciences,* 13:84–90, 1998.

Mittal, R. K., and Balaban, D. H. "The esophagastric junction." *New England Journal of Medicine,* 336:924–932, 1997.

Phillips, S. F., and Wingate, D. L., eds. *Functional Disorders of the Gut*. London, Churchill Livingstone, 1998.

Sawada, M., and Dickinson, C. J. "The G cell." *Annual Review of Physiology,* 59:273–298, 1997.

Answers to Review Questions

1. c	**2.** a	**3.** e	**4.** c	**5.** e	**6.** c	**7.** c	**8.** c
9. b	**10.** d	**11.** a	**12.** b	**13.** b	**14.** e	**15.** e
16. a	**17.** a	**18.** b	**19.** b	**20.** c	**21.** d	**22.** d
23. e	**24.** d	**25.** c				

Chapter 23

THE KIDNEY

KEY CONCEPTS

- The kidneys regulate the composition and volume of the extracellular fluid.

- The kidney glomeruli filter the blood plasma at a high rate, and the kidney tubules reabsorb and secrete substances, so that appropriate amounts of various substances are excreted in the urine.

- The kidneys regulate the amount of sodium in the body by excreting just the correct amount to maintain homeostatic balance.

- Human kidneys can form urine that has a higher or lower total solute concentration (osmolality) than plasma.

- Micturition (urination) is a complex reflex act controlled by higher brain centers.

CASE HISTORY

A 12-year-old boy has a headache, malaise, and poor appetite. He complains of back pain. His mother notices that his face is puffy, especially around his eyes. She becomes alarmed when her son passes a smoky-colored urine and immediately brings him to the doctor. Ten days ago, the patient had a bad sore throat, fever, and upper respiratory infection and missed several days of school. Physical examination reveals a minimally inflamed pharynx, elevated blood pressure (145/100 mm Hg), abdominal distension, and swollen ankles. His height is 4 feet, 8 inches (1.42 m), and his weight is 91 lb (41 kg), with a calculated body surface area of 1.26 m^2. Blood urea nitrogen (BUN) concentration is 30 mg/dl (normal, 7–18 mg/dl) and plasma creatinine concentration is 2.0 mg/dl (normal, 0.8–1.3 mg/dl). A 24-hour urine sample was collected and had a volume of 900 ml and contained 1.30 g creatinine and 0.90 g protein. Microscopic examination of the urine sediment revealed odd-shaped red cells, red cell casts, and leukocytes. A tentative diagnosis of **poststreptococcal glomerulonephritis** was made and was later supported by the finding of antibodies against streptococcal exoenzymes in the patient's serum. Certain strains of streptococci provoke an immunological response that damages the kidney glomeruli. Treatment with penicillin, diuretic drug, a reduced salt intake, and rest at home were prescribed. The diuretic led to a brisk increase in urine output, and the puffiness in the face and hypertension promptly disappeared. The urine color became normal in a few days, but the urine still tested positive for red cells and protein for several weeks. Laboratory values indicated normal renal function one year after this episode.

Questions

1. What is the evidence that the bleeding occurred in the kidneys?

2. What is the patient's glomerular filtration rate? Is it normal?

3. Why are the BUN and plasma creatinine concentration elevated?

4. Why is urinary protein excretion high?

5. What might explain the puffy face and elevated blood pressure?

INTRODUCTION

What do the kidneys do?

Most of us take our kidneys for granted. We know that if we drink a lot, urine output increases, and if we limit our water intake, then urine output decreases and the urine looks very concentrated. It is common knowledge that the kidneys eliminate wastes from the body. But they really do much more than this. The kidneys play a dominant role in regulating the volume and composition of our extracellular fluids, and thereby maintain an ideal "internal environment" in our body. When kidney function is impaired, as it was in the boy described at the beginning of this chapter, many body systems are adversely affected. Fortunately, renal injury in this boy was self-limited, and complete recovery of renal function occurred. This is not always the case, however, and many kinds of renal disease can lead to irreversible, complete failure of the kidneys. When this happens, life can be maintained only by some form of dialysis or, better, a kidney transplant. In this chapter, we consider the basics of kidney function: the functional anatomy of the kidneys, renal blood flow, the processes involved in urine formation, and the ways in which the kidneys handle a variety of solutes and water. We end with a brief discussion of the functions of the lower urinary tract (ureters, bladder, and urethra).

Before considering kidney function in detail, we list some of the many functions of the kidneys.

1. The kidneys regulate the total solute concentration (osmolality) of the body fluids.
2. By controlling the excretion of sodium and water, the kidneys regulate the volume of the extracellular fluid.
3. The kidneys regulate the individual concentrations of numerous electrolytes in the extracellular fluid, including sodium, potassium, calcium, magnesium, chloride, sulfate, and phosphate ions.
4. The kidneys regulate the plasma pH and so play a crucial role in acid-base balance. This topic will be discussed in Chapter 25.
5. The kidneys eliminate metabolic waste products, such as urea (an end-product of protein metabolism), uric acid (an end-product of purine metabolism), and creatinine (an end-product of muscle metabolism). They also eliminate many foreign compounds from the body, including many drugs and toxins.
6. The kidneys produce a number of special substances. These include **erythropoietin,** a hormone that stimulates the rate of production, maturation, and release of red blood cells from bone marrow; **renin,** a proteolytic enzyme important in the regulation of extracellular fluid volume and blood pressure; **kallikrein,** a proteolytic enzyme that leads to the formation of kinins, which are vasodilators; and various **prostaglandins** and **thromboxane,** fatty acid derivatives that act as local hormones. Prostaglandins

E_2 and I_2 have several actions in the kidneys, including vasodilation, enhancement of renal excretion of salt and water, and stimulation of renin release. Thromboxane is a vasoconstrictor that may be responsible for reduced renal blood flow in a number of renal diseases.

7. The kidneys have several special metabolic functions. They are responsible for converting the inactive form of vitamin D to its active form, **1,25-dihydroxy vitamin D₃.** The latter is a hormone that stimulates intestinal calcium absorption. The kidneys synthesize ammonia from amino acids. This is important in acid-base regulation. The kidneys can synthesize glucose from noncarbohydrate sources (e.g., amino acids), a process called **gluconeogenesis.** During a prolonged fast, the glucose added to the blood by the kidneys helps to maintain the blood sugar concentration. The kidneys are an important site of degradation (hence inactivation) of several polypeptide hormones, including angiotensin II, glucagon, insulin, and parathyroid hormone.

FUNCTIONAL ANATOMY OF THE KIDNEY

How does the anatomy of the kidney reflect its function?

The functions of the kidneys can be understood only in terms of their unique anatomy. Figure 23–1 presents an overview of the urinary system and the human kidney. The kidneys are bean-shaped organs that lie in back of the abdominal cavity on either side of the vertebral column. They are drained by the **ureters,** which carry the urine to the **urinary bladder.** The tube draining the bladder is called the **urethra.** If a kidney is cut through, two parts are easily distinguished: an outer part, called the **cortex,** and an inner part, called the **medulla** (see Fig. 23–1b). The cortex is reddish-brown and has a granular appearance. It contains all of the **glomeruli** and **convoluted tubules** (proximal and distal). The medulla is a lighter color and is striated (striped). Striations result from the parallel arrangement of **loops of Henle, medullary collecting ducts,** and blood vessels. The medulla is usually divided into an **outer medulla** (closer to the cortex) and an **inner medulla** (farther from the cortex).

The human kidney is divided into approximately a dozen **lobes.** Each consists of a **pyramid** of medullary tissue plus the cortical tissue overlying its base and covering its sides. The apex of a medullary pyramid forms a renal **papilla,** which drains the urine into a **minor calyx.** Minor calyces unite to form a **major calyx;** major calyces lead to the expanded **renal pelvis,** which is drained by the ureter. The ureter, renal artery, renal vein, nerves, and lymphatic vessels enter or leave the kidney at the **renal hilum,** a depression of its medial surface.

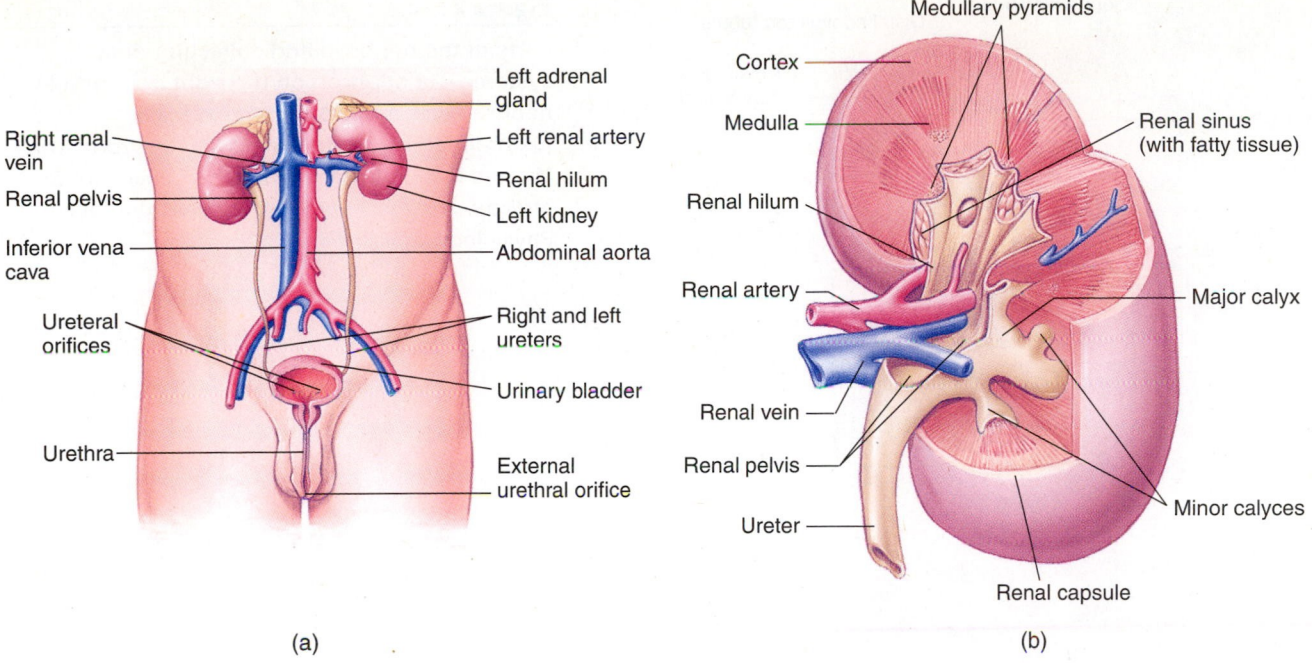

(a)   (b)

Figure 23–1

(a) The urinary tract. The urine formed in the kidneys flows into the renal pelvis. It is carried by the ureters to the urinary bladder, which is drained by the urethra. *(b)* The human kidney.

The Nephron Is the Basic Unit of Kidney Structure and Function

Each human kidney contains approximately 1 million **nephrons.** Each nephron consists of a **renal corpuscle** and its attached **renal tubule** (Fig. 23–2).

The renal corpuscle consists of a tuft of capillaries, the **glomerulus,** surrounded by an expanded, double-walled cup, **Bowman's capsule.** The **urinary space** of Bowman's capsule is continuous with the lumen of the renal tubule. Traditionally, the renal tubule has been divided into several anatomically distinct segments. The first segment is called the **proximal tubule.** This is divided into two parts: the **proximal convoluted tubule,** which coils and twists in the neighborhood of its renal corpuscle, and the **proximal straight tubule,** which plunges toward the renal medulla. The **loop of Henle** is the portion of the nephron between the proximal convoluted and distal convoluted tubules. Its first part is the proximal straight tubule. Next is a **thin limb** (descending and ascending portions), and finally a **thick ascending limb.** This is followed by a short **distal convoluted tubule.** Distal convoluted tubules join **connecting tubules,** which lead to **cortical collecting ducts.** These lead to the **outer medullary collecting ducts,** which pass straight through the outer medulla. In the inner medulla, **inner medullary collecting ducts** unite to form larger and larger ducts, which convey the urine into the minor calyces.

Strictly speaking, the collecting ducts are not part of the nephron. These structures have different embryological origins, and there are many more nephrons than collecting ducts. Functionally, the collecting ducts are not just conduits; they are responsible for determining the final composition and volume of the urine. Because the collecting ducts continue the tasks performed by the nephrons, they are often considered part of the nephron unit.

Not all nephrons are alike (see Fig. 23–2). Nephrons with renal corpuscles in the outer cortex, called **cortical nephrons,** have relatively small glomeruli, short or absent thin limbs, and loops that bend wholly within the cortex or outer medulla. Nephrons with renal corpuscles in the inner cortex (i.e., near the medulla), called **juxtamedullary nephrons,** have large glomeruli, long thin limbs, and long loops, some of which may reach the papillary tips. Functional differences between these two nephron populations include differences in filtration rate, tubular transport properties, and renin content. In the human kidney, the majority of nephrons (85%) are of the cortical type, and the rest (15%) are of the juxtamedullary type.

The renal tubule and collecting duct are composed of a single layer of epithelial cells surrounding a tubule lumen. Cell structure and function vary considerably from one segment to another. As an example of the specialized architecture of the renal tubule epithelium, we will consider a cell from the proximal tubule (Fig. 23–3). The cell surface that contacts the fluid in the tubule lumen, the

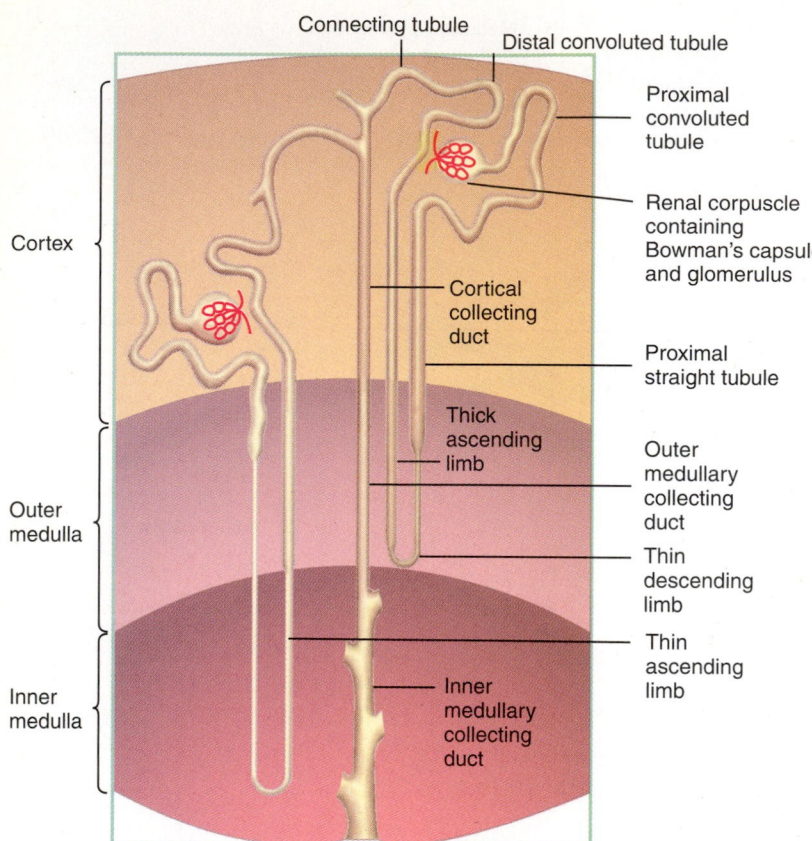

Connecting tubule

Distal convoluted tubule

Proximal convoluted tubule

Renal corpuscle containing Bowman's capsule and glomerulus

Cortical collecting duct

Proximal straight tubule

Thick ascending limb

Outer medullary collecting duct

Thin descending limb

Thin ascending limb

Inner medullary collecting duct

Cortex

Outer medulla

Inner medulla

Figure 23–2

Parts of the nephron and collecting duct system. The nephron on the right is a cortical nephron, the nephron on the left is a juxtamedullary nephron.   *(Modified from: Kriz, W., and Bankir, L. "A standard nomenclature for structures of the kidney." American Journal of Physiology, 254:F1-F8, 1988.)*

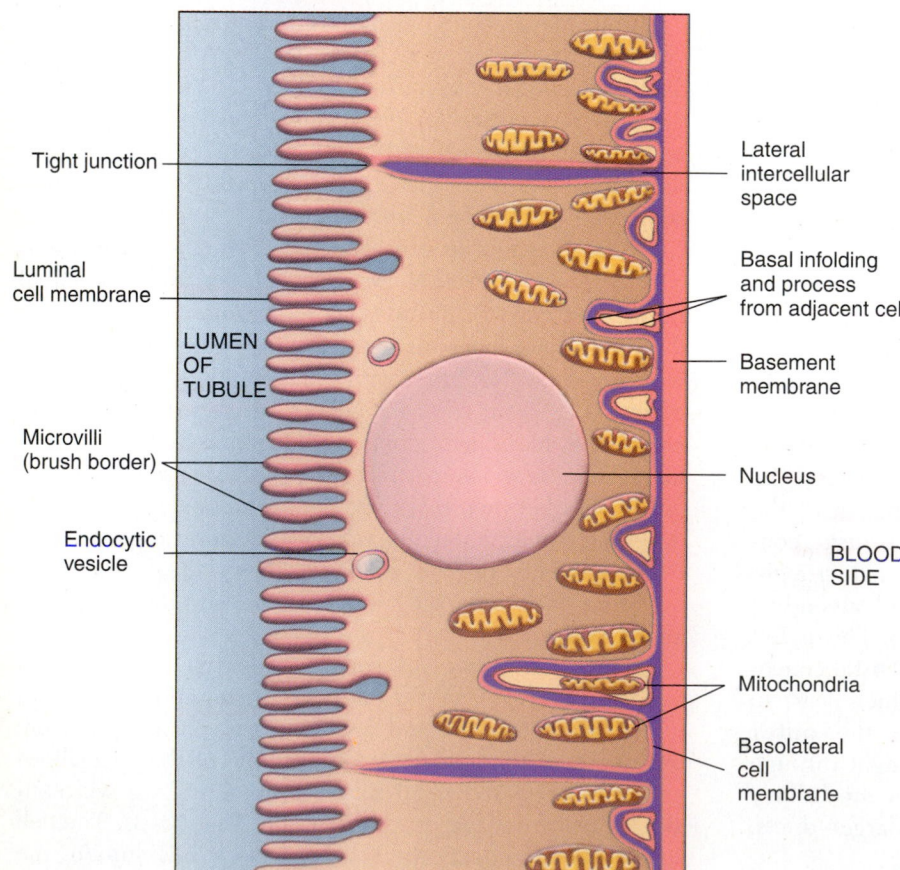

Tight junction

Luminal cell membrane

LUMEN OF TUBULE

Microvilli (brush border)

Endocytic vesicle

Lateral intercellular space

Basal infolding and process from adjacent cell

Basement membrane

Nucleus

BLOOD SIDE

Mitochondria

Basolateral cell membrane

Figure 23–3

Schematic view of a proximal tubular cell.

luminal or **apical cell membrane,** has numerous microvilli, collectively called the *brush border*. The microvilli greatly increase the membrane surface area, thereby facilitating the transfer of materials between cells and tubular fluid. At the apical side of the cell are numerous **endocytic vesicles,** thought to be involved with the reabsorption of proteins. Adjacent cells are linked to each other near the apical end of the cell by a **tight junction.** In the proximal tubule, this junction is leaky and permits some movement of small ions and water. The surface area of the **basolateral cell membrane** is considerably increased by numerous ridges and processes that are formed as adjacent cells interdigitate with each other. The space between adjacent cells, the **lateral intercellular space,** is thought to be an important route for salt and water movement across the epithelial cell layer. Within the cell, numerous mitochondria lie close to the basolateral cell membrane and furnish the energy for an important sodium-potassium pump (Na^+/K^+-ATPase) located in these membranes. The tubule cells rest on a **basement membrane,** which acts as a supportive structure. If the cells are damaged, the basement membrane provides a framework on which regenerating cells can reline the tubule. The overall appearance of the proximal tubule cell is that of a metabolically active cell, and we will see later that it is heavily engaged in transport activities.

Blood Vessels and Nerves Enter or Leave the Kidney at Its Hilum

Each **renal artery** arises from the abdominal aorta and divides into progressively smaller and more numerous branches: segmental, interlobar, arcuate, and cortical radial arteries (Fig. 23–4). The **cortical radial arteries** course toward the surface of the cortex and supply the short, muscular **afferent arterioles.** Each afferent arteriole leads to a **glomerulus.** The glomerular capillaries re-form the **efferent arteriole,** which usually has a thin muscular coat. This vessel breaks up into the **peritubular capillaries,** which form an extensive network surrounding the tubules in the cortex. The peritubular capillaries lead to veins that generally parallel the course of the arterial vessels and ultimately empty into the **renal vein** (see Fig. 23–4).

The blood supply to the kidney medulla is derived from the efferent arterioles associated with juxtamedullary nephrons. These give rise to long, straight capillaries called **vasa recta.** The vasa recta descend into and ascend from the medulla in parallel with the tubular structures.

The renal circulation is unusual in that there are two capillary beds in series (see Fig. 23–4). The blood pressure is high in the glomeruli and low in the peritubular capillaries. The glomeruli are specialized for filtration, whereas the peritubular capillaries are specialized for uptake of fluid reabsorbed by the tubules.

The kidneys are richly innervated. Sympathetic nerve fibers travel to the kidneys mainly in thoracic nerves X, XI, and XII and lumbar spinal nerve I. The sympathetic efferent

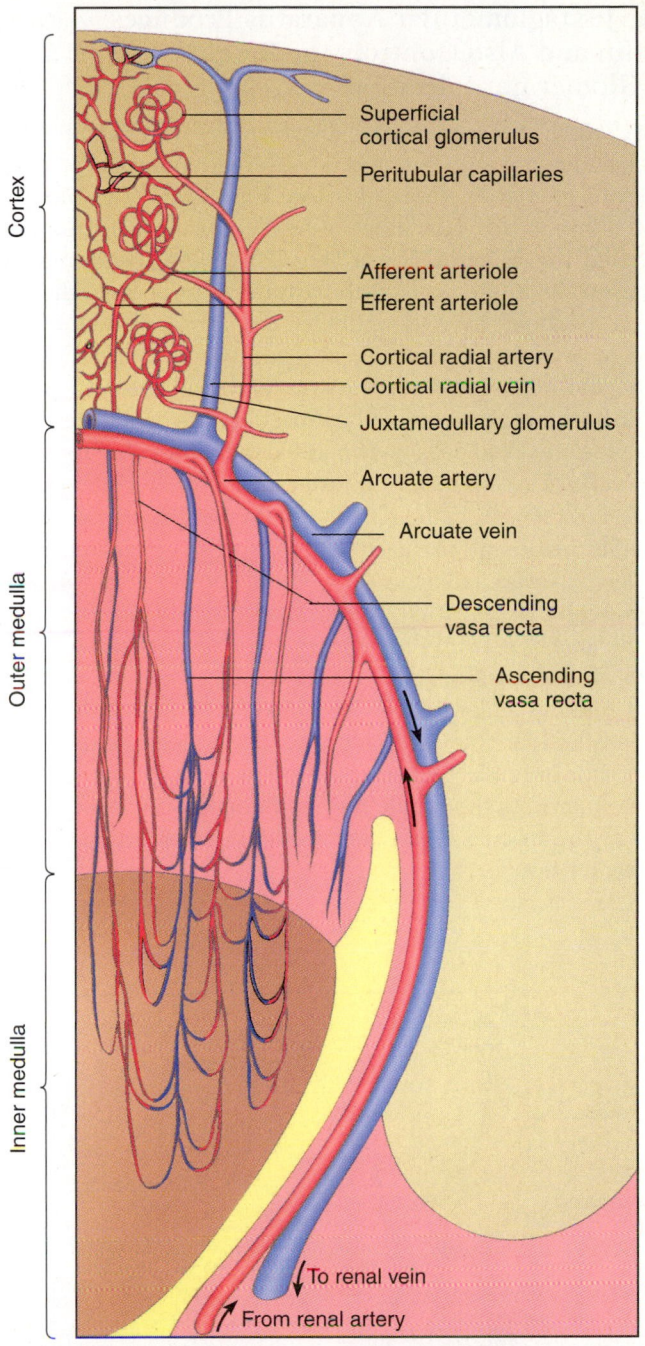

Figure 23–4

Blood vessels within the kidney. *(Modified from: Koushanpour, E., and Kriz, W.* Renal Physiology. Principles, Structure, and Function, *ed 2. New York, Springer, 1986.)*

fibers predominantly innervate blood vessels and cause vasoconstriction and release of renin. Sympathetic fibers are also found adjacent to tubular cells and may increase sodium reabsorption. Afferent (sensory) nerve fibers from the renal pelvis and ureter mediate pain of renal origin.

Lymphatic vessels are found in the cortex, but not in the medulla.

The Juxtaglomerular Apparatus Produces Renin and Also Controls the Rate of Glomerular Filtration

Just before it becomes the distal convoluted tubule, the thick ascending limb contacts the afferent and efferent arterioles of its parent glomerulus (see Fig. 23–2). The tubular epithelium at this spot shows a dense crowding of cells and is called the **macula densa** (Fig. 23–5). Adjacent smooth muscle cells in the walls of the arterioles (especially the afferent arteriole) have an epithelial cell-like appearance and contain granules of renin. These **granular cells** synthesize, store, and release this enzyme into the blood and interstitial fluid. In the region bounded by macula densa and arterioles are closely packed cells known as **extraglomerular mesangial cells.** These cells may transfer information from macula densa cells to cells in the blood vessel walls. The entire group of cells made up of macula densa cells, extraglomerular mesangial cells, and granular cells is known as the **juxtaglomerular apparatus.** The intimate relationship between tubule and blood vessels at this site provides the anatomical basis for **tubuloglomerular feedback;** changes in tubule fluid sodium chloride concentration produce changes in glomerular blood flow and filtration rate at a single nephron level. For example, an increase in flow through the loop of Henle increases the tubule fluid sodium chloride concentration at the macula densa and results in constriction of the afferent arteriole.

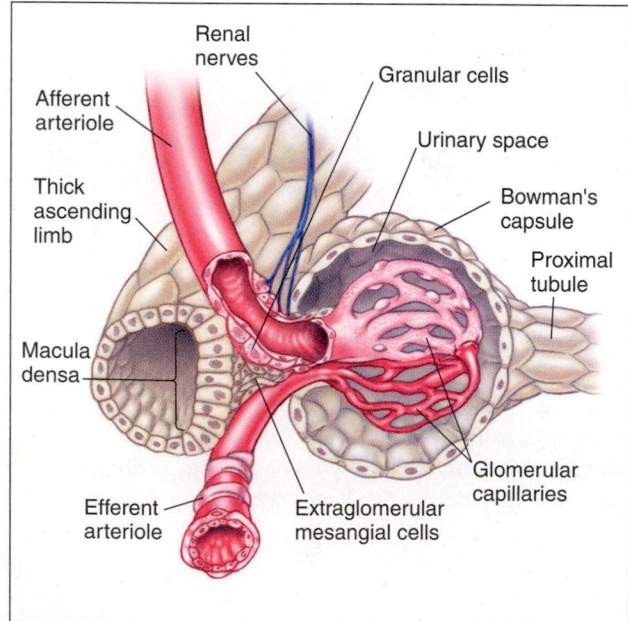

Figure 23–5

The juxtaglomerular apparatus consists of macula densa, extraglomerular mesangial cells, and granular cells. This structure may play a role in controlling glomerular blood flow and filtration rate at the level of a single nephron.

The Kidneys Have a High Blood Flow

Total renal blood flow in a resting, young adult, 70-kg male averages approximately 1200 ml/min or 20% to 25% of the cardiac output (5 L/min), even though both kidneys weigh less than 0.5% of body weight (300 g/70 kg). Blood flow per gram of kidney tissue averages 4 ml/min ([1200 ml/min]/ 300 g), which is much higher than in other organs considered to be well perfused (e.g., brain, heart, and liver).

Why is blood flow to the kidneys so high? A high blood flow is needed to sustain a high rate of filtration of plasma in the glomeruli. Essentially all of the blood perfusing the kidneys flows into the glomeruli, where a filtrate is formed at a rate normally equal to approximately 10% of the blood flow rate. The high renal blood flow does not reflect excessive metabolic demands. Although the kidneys use approximately 8% of the body's oxygen consumption in a resting adult, renal blood flow is more than adequate to deliver the needed oxygen. In fact, renal venous blood typically has a bright red color owing to its high oxygen content, a result of low oxygen extraction.

Blood flow is not distributed uniformly within the kidney. It is highest in the renal cortex, where all of the glomeruli are located and where the blood is filtered. Less blood flows to medullary structures, and flow decreases with depth in the medulla. The relatively low blood flow helps to preserve the high solute concentration in the kidney medulla.

Renal blood flow and glomerular filtration rate change relatively little if arterial blood pressure is varied between 80 and 180 mm Hg in an isolated, perfused, denervated kidney. This phenomenon is called **renal autoregulation;** it is a self-regulating mechanism in the kidney. Changes in the caliber of blood vessels upstream to the glomerulus (cortical radial arteries and afferent arterioles) are thought to be primarily involved. If blood pressure is increased, these vessels constrict, thereby increasing vascular resistance and minimizing an increase in blood flow. If blood pressure is decreased, these vessels dilate, thereby decreasing vascular resistance and minimizing a decrease in blood flow. The importance of renal autoregulation is that it prevents changes of arterial blood pressure from producing large changes in glomerular filtration rate. Such changes could lead to dramatic changes in sodium excretion, which could produce abnormalities in extracellular fluid volume.

The autoregulatory response has a limited range, and so if blood pressure is decreased below 80 mm Hg, renal blood flow and glomerular filtration rate may decrease markedly. Also, in an intact animal or person, a low blood pressure (for example, due to blood loss) reflexly activates the sympathetic nervous system. This then overrides renal autoregulation, and renal blood vessels constrict in this situation.

Total renal blood flow is decreased in a number of stressful situations, including cold, deep anesthesia, fright, hemorrhage, pain, and strenuous exercise. This decrease is a reflex response primarily due to activation of renal sympathetic nerve fibers, which cause vasoconstriction. The increase in

APPLICATIONS OF PHYSIOLOGY

Polycystic Kidney Disease

Polycystic kidney disease (PKD) is a disorder, usually inherited, in which numerous cysts develop in both kidneys. The cysts are fluid-filled epithelial sacs that arise from nephrons or collecting ducts. The growth of cysts can produce massive enlargement of the kidneys and ultimately complete renal failure. PKD is the most common of all life-threatening genetic diseases, and affects 600,000 Americans and millions more worldwide.

PKD in people may be produced by several genes. The most common form (PKD-1) is inherited in an autosomal dominant fashion and is due to a defective gene on the short arm of chromosome 16. The gene was isolated in 1994 and the gene product is called **polycystin-1**, a very large cell membrane protein that may be a receptor for cell-cell or cell-extracellular matrix interactions. A defect in this gene accounts for 85% to 90% of cases of autosomal dominant PKD (ADPKD). Approximately 10% to 15% of people with ADPKD have PKD-2, with a defective gene located on chromosome 4. This gene was isolated in 1996 and the gene product, **polycystin-2**, may be a calcium channel. Polycystins 1 and 2 interact with each other and with other cell proteins, such as cadherins. A third gene, which has not yet been identified, may account for 1% of ADPKD patients. PKD-1 is usually more severe than is PKD-2. There is also an autosomal recessive form of PKD (ARPKD), which occurs in approximately 1 in 10,000 to 40,000 live births and results in a high infant mortality rate. This defective gene is on chromosome 6.

The phenotypic expression of ADPKD is quite variable. Some people show symptoms in childhood, whereas others may lead a long and healthy life with cystic kidneys detected only upon autopsy. The usual pattern is for patients to develop symptoms (e.g., hypertension or pain in the back and sides) in their 30s and 40s and end-stage renal failure occurs in approximately 50% of patients by the age of 60 years. The specific gene affected, the nature of the mutation in a particular gene, the genetic background of an individual, and environmental factors may all play a role in determining the development of the disease.

It is generally believed that only a minority ($\approx$ 1%) of nephrons produce cysts in ADPKD, even though every kidney cell has a mutant copy of a dominant gene. A provocative idea that might help explain why the disease is so variable and why only relatively few nephrons produce cysts is the "**two-hit hypothesis.**" According to this idea, the production of a cyst requires a second (somatic) mutation, so that the gene (allele) accompanying the inherited defective gene is also abnormal. Only when this happens does a cyst develop. Researchers have demonstrated that in many cysts in human ADPKD, two abnormal genes are indeed present.

Because PKD is basically a genetic disease, it seems that it is an ideal candidate for some form of gene therapy. At the present time, however, this approach is remote. Clinicians and laboratory scientists are examining several possible treatments to slow the progression of the disease.

An elevated blood pressure is a common finding in patients with ADPKD and is associated with a faster progression of renal disease and increased cardiovascular mortality. Recent studies of patients suggest that early treatment of hypertension with an angiotensin-converting enzyme inhibitor or calcium channel blocker may slow disease progression.

It is also possible that dietary manipulations might affect the course of the disease. Studies on rats with PKD had demonstrated that a low protein intake results in improved kidney function. In a large-scale, multicenter, randomized clinical trial (the Modification of Diet in Renal Disease Study), patients with ADPKD were provided with a reduced protein intake, but no beneficial effect could be demonstrated. The relatively short duration of this study (patients were studied for an average of 2.2 years) and the fact that treatment was started at a relatively late stage of the disease may have resulted in these disappointing results. In rats with PKD, a diet containing soy protein (instead of animal protein), flaxseed oil, or citrate (an alkalinizing diet) leads to greatly improved kidney function; whether these results are applicable to people remains to be seen.

Epidermal growth factor (EGF) and other growth factors may be important in PKD. EGF is found in high concentrations in cyst fluid, and functional EGF receptors are mislocated to the apical cell membrane of cyst cells. EGF stimulates the proliferation of cyst cells and hence the growth of cysts. Researchers have demonstrated, in an experimental mouse model of ARPKD, that inhibition of the EGF receptor with a new anticancer drug dramatically reduces cyst growth and prolongs the life of newborn mice with the disease. Whether this therapy will be successful in human ARPKD is not known at this time.

Further information on PKD can be obtained from the Polycystic Kidney Disease Foundation and their Web site, www.pkdcure.org.

renal vascular resistance causes blood flow to shift to the brain and heart, which are more vital to immediate (short-term) survival.

A number of substances also affect renal blood flow. For example, angiotensin II, endothelin, epinephrine, norepinephrine, and thromboxane cause renal vasoconstriction. Bacterial pyrogens, kinins (e.g., bradykinin), nitric oxide, and many prostaglandins (e.g., prostaglandins E_2 and I_2) have vasodilator effects. Some of these substances are synthesized and act locally within the kidneys. Increased amounts of vasodilator prostaglandins are produced in circumstances in which kidney blood flow is threatened — for example, by intense stimulation of renal sympathetic nerves. This production of prostaglandins opposes excessive vasoconstriction, which might otherwise damage the kidneys.

PROCESSES INVOLVED IN FORMATION OF URINE

 What happens to fluid that is filtered at the glomerulus?

Three basic processes are involved in the formation of urine: glomerular filtration, tubular reabsorption, and tubular secretion (Fig. 23–6).

Glomerular filtration is the starting point for urine formation. Because of the high hydrostatic blood pressure in the glomerular capillaries, a filtrate of plasma is pushed out of the capillaries into the urinary space of Bowman's capsule. This process of filtration is often referred to as **ultrafiltration,** and the filtrate is referred to as an **ultrafiltrate** of plasma because the glomerular capillary wall, the **glomerular filtration barrier,** behaves like an ultrafine filter. It allows for free passage of small molecules (molecules with a molecular weight of less than 10,000) but restricts the passage of large molecules. In particular, the large plasma proteins (serum albumin, globulins, and fibrinogen) are normally virtually excluded from the glomerular filtrate. Glomerular filtration is rather nonselective, and both useful substances and waste products are filtered. The final urine differs radically from the glomerular filtrate in both volume and composition. Most of the valuable constituents of the filtrate, such as salts, water, and metabolites, are returned to the blood by the tubules.

Tubular reabsorption is the process in which substances are transferred out of the tubular fluid and returned to the capillaries surrounding the kidney tubules. In this way, substances that have been temporarily lost from the plasma in the process of glomerular filtration are returned to the circulation. Reabsorption is a selective process, and the tubules handle different substances differently.

Tubular secretion is movement of substances across the tubule epithelium in a direction opposite to that of reabsorption. Certain substances — for example, the organic anion *p*-aminohippurate (PAH) — are taken up by the tubule cells from the blood surrounding the tubules and deposited into the tubular fluid. Other substances, such as hydrogen ions and ammonia, are generated within the kidney tubule cells and then secreted into the tubular urine. The terms *reabsorption* and *secretion* simply indicate the direction of transport across the tubular epithelium, either out of or into the tubular urine, respectively.

Excretion refers to what comes out in the urine that leaves the body. In most cases, excretion depends on both glomerular filtration and tubular transport. If a substance is reabsorbed, then less is excreted than is filtered. If a substance is secreted, then more is excreted than is filtered (Fig. 23–7).

Table 23–1 presents data on the daily (24-hour) filtration, reabsorption, and excretion of some normal plasma constituents. Because filtration rate is so high (180 L/day), enormous quantities of water, salts, and organic compounds are filtered. More than 99% of the filtered water, sodium, chloride, and bicarbonate is usually reabsorbed. The small percentage of these substances that is excreted is of crucial importance because in order for a person to stay in balance — that is, to show no net change in the amount of these substances in the body — exactly the right amount needs to be excreted.

A small change in the percentage of reabsorption of a filtered substance may lead to a large change in the absolute excretion of the substance. For example, if the reabsorbed percentage of filtered sodium decreases from 99.6% (23,900 mmoles/day divided by 24,000 mmoles/day) to 99.2% (23,800 mmoles/day divided by 24,000 mmoles/day), the amount of sodium excreted will double from 100 to 200 mmoles/day.

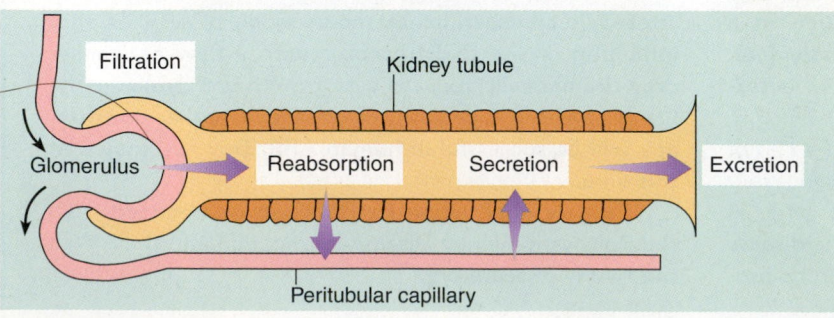

Figure 23–6

Processes involved in urine formation.

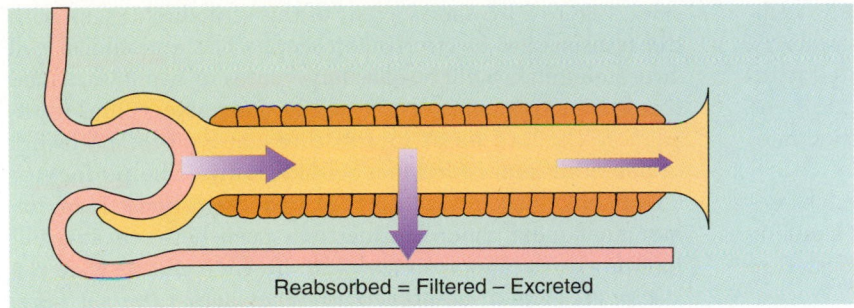
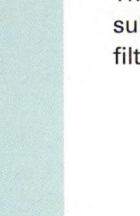

Figure 23–7

The net rate of reabsorption or secretion of a substance can be calculated from the amounts filtered and excreted per unit of time.

Reabsorbed = Filtered – Excreted

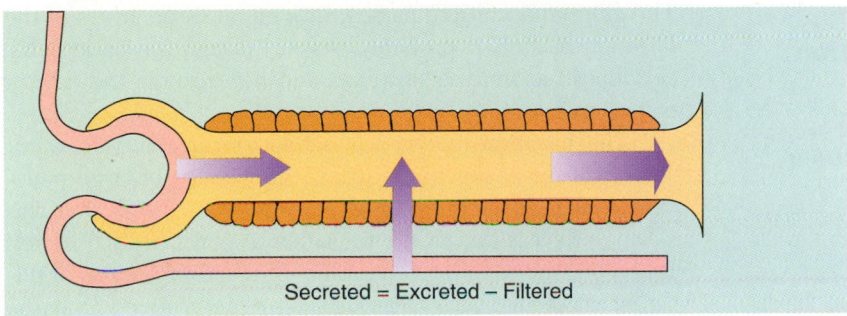

Secreted = Excreted – Filtered

Clearly, tubular reabsorption of electrolytes and water must be carefully controlled to avoid excessive losses or abnormal retention.

Do not memorize the numbers in Table 23–1, but try to appreciate the magnitude of filtration, reabsorption, and excretion of different substances and recognize that different substances are handled differently by the tubules. Notice that potassium is listed as "reabsorbed" because the excreted amount is less than the filtered amount. Actually, nearly all of the filtered potassium is reabsorbed by proximal portions of the nephron, and most of the excreted potassium is secreted by the cortical collecting ducts. Filtered calcium is mostly reabsorbed. Phosphate is incompletely reabsorbed; the phosphate in the urine keeps us in balance and is also an important

TABLE 23–1

Daily Filtration, Reabsorption, and Excretion of Some Normal Plasma Constituents*

Constituent	Filtered	Reabsorbed	Excreted
Water	167.5 liters	166 liters	1.5 liters
Sodium	24,000 mmoles	23,900 mmoles	100 mmoles
Potassium	720 mmoles	630 mmoles	90 mmoles
Calcium	270 mmoles	266 mmoles	4 mmoles
Magnesium	135 mmoles	120 mmoles	15 mmoles
Chloride	19,500 mmoles	19,400 mmoles	100 mmoles
Bicarbonate	4,500 mmoles	4,498 mmoles	2 mmoles
Phosphate	6 g	5 g	1 g
Glucose	150 g	150 g	0 g
Urea	50 g	25 g	25 g
Uric acid	8 g	7.2 g	0.8 g
Creatinine[†]	1.5 g	0 g	1.8 g

*Data are for a normal, young, 70-kg man.

[†]The amount of creatinine excreted exceeds the amount filtered because of tubular secretion of creatinine.

pH buffer. Filtered glucose is normally completely reabsorbed; it would make little sense to lose this valuable metabolite in the urine. Filtered creatinine is excreted without reabsorption. Urea and uric acid are reabsorbed to some extent. For all three of these waste products, production rates in the body and urinary excretion are normally equal.

The rates of excretion in Table 23–1 can vary widely, depending on the diet and other circumstances. Renal handling of each of the substances listed in this table is discussed in more detail later.

Glomerular Filtration Involves Ultrafiltration of Plasma at a High Rate

> *What factors play a role in determining glomerular filtration rate?*

The Glomerular Filtration Barrier

The glomerular filtration barrier consists of three layers: (1) capillary endothelium, (2) basement membrane, and (3) podocyte cell layer (visceral epithelium of Bowman's cap-

sule). Figure 23–8 shows a part of this structure as seen with the transmission electron microscope. The endothelial cell layer lining the capillaries has large pores or windows, called **fenestrae,** approximately 50 to 100 nanometers (nm) in diameter. The **basement membrane** consists of a meshwork of fine fibrils embedded in a gel-like matrix. The **podocytes** ("foot cells") have numerous **foot processes** that rest on the basement membrane. The narrow space between adjacent foot processes is approximately 20 nm wide and is called the **slit pore.** A thin membrane stretches across the slit pore; this **filtration slit diaphragm** has a fine porous structure. The glomerular filtrate passes through all three layers of the filtration barrier. It probably passes through the endothelial cell fenestrae and the slit pores and not through the thicker cell cytoplasm.

The filtration barrier has a high permeability to water and small molecules but restricts the passage of large molecules (macromolecules). It is convenient to think that this sieving effect is due to the presence of pores in this barrier. Based on studies of the filterability of macromolecules of different molecular size, scientists have determined that the glomerular filtration barrier behaves as though it were penetrated by cylindrical pores 7.5 to 10.0 nm in diameter. But no

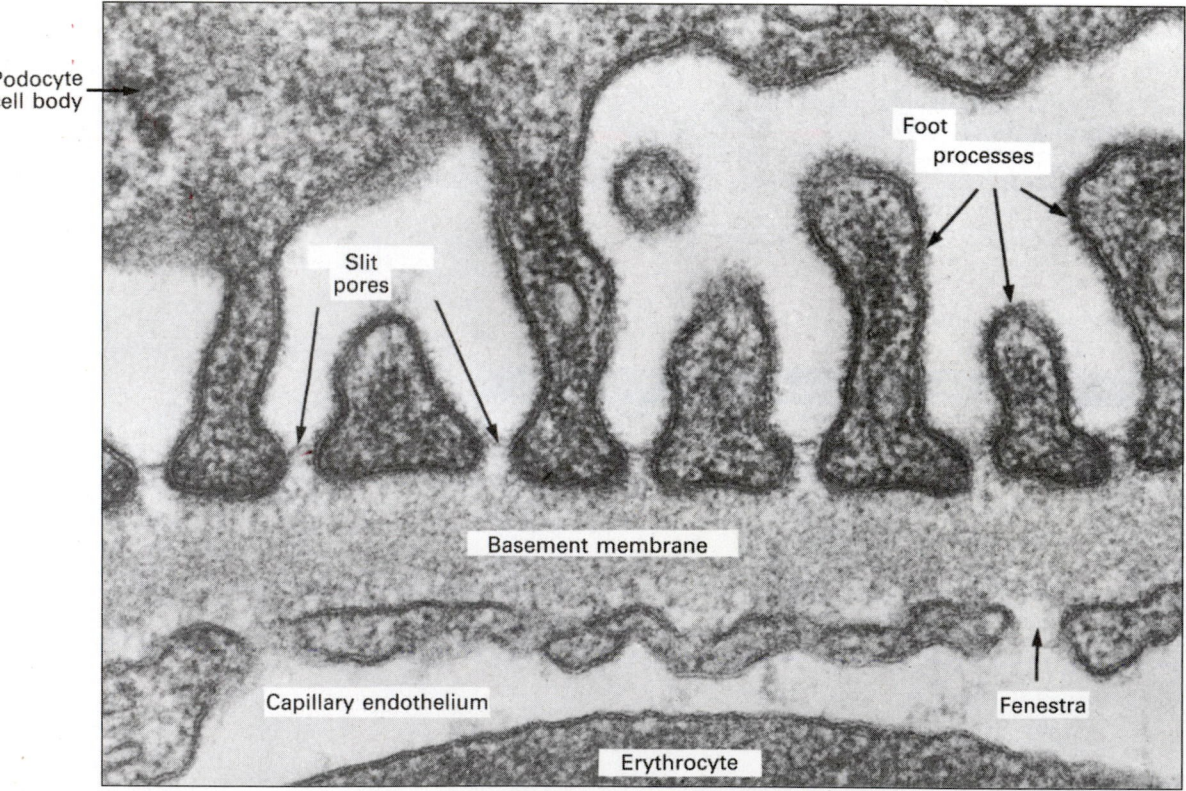

Figure 23–8

SEM (x 70,000) of the glomerular filtration barrier. Note the pore (fenestra) in the thin capillary endothelium and the slit pores between the foot processes of the podocytes. *(From: Fawcett, D. W. A Textbook of Histology, ed 11. Philadelphia, Saunders, 1986, p. 765; Courtesy of D. Friend.)*

one has ever seen pores of these dimensions; it is possible that the pores may be lost during the preparation of tissue for electron microscopy. In any case, most scientists believe that the basement membrane is probably the major size-selective barrier. The filtration slit diaphragm may also be a significant second barrier.

Electrical charge also affects the passage of macromolecules through the filtration barrier. The endothelial pores, basement membrane, and surface coat of the podocytes contain negatively charged glycoproteins. These negatively charged structural elements impede the filtration of negatively charged plasma proteins, such as serum albumin.

One of the hallmarks of glomerular disease is the abnormal urinary excretion of plasma proteins, as in the boy described at the very beginning of this chapter. The proteinuria may be due to (1) an increase in size of glomerular pores or (2) a loss of fixed negative charges from the glomerular filtration barrier.

Factors That Affect Glomerular Filtration Rate

The factors that affect **glomerular filtration rate (GFR)** are basically the same as those that influence fluid movement across capillary walls anywhere in the body (Chapter 19). The driving force for filtration is the glomerular capillary hydrostatic pressure (P_{GC}). This pressure ultimately depends on the energy imparted to the blood by the beating of the heart. Filtration is opposed by the hydrostatic pressure in the urinary space of Bowman's capsule (P_{BS}) and by the intracapillary colloid osmotic pressure (oncotic pressure) due to the plasma proteins (π_{GC}) (Fig. 23–9). Because the glomerular filtrate is essentially protein-free, the colloid osmotic pressure of fluid in the urinary space is zero and can be disregarded. The net filtration pressure gradient is equal to the difference between pressures favoring and opposing filtration. In other words,

$$\text{Net filtration pressure} = P_{GC} - P_{BS} - \pi_{GC}$$

The glomerular filtration rate depends not only on the pressure gradient but also on the glomerular filtration coefficient (K_f). We can write:

$$\text{GFR} = K_f \times \text{Net filtration pressure}$$

This equation is analogous to the equation that describes fluid flow rate along a tube, $F = \Delta P/R$ (Chapter 19), except that here we are considering fluid movement across a capillary wall. K_f can be thought of as the conductance (1/resistance) of the glomerular filtration barrier; its magnitude depends on the fluid permeability and surface area of the barrier.

Figure 23–9 summarizes the average pressures thought to exist in the normal human kidney. Glomerular hydrostatic pressure is probably approximately 55 mm Hg, about twice as high as in most other capillaries in the body. The high pressure in the glomerulus results from the relatively low resistance of wide upstream blood vessels (cortical radial artery, afferent arteriole) and high resistance of narrower downstream blood vessels (efferent arteriole). Glomerular hydrostatic pressure probably changes very little (perhaps 2 mm Hg) along the length of the glomerular capillaries. This is because there are many (30–50) parallel capillary loops in a glomerulus, making the total resistance to blood flow very low. Hydrostatic pressure in the urinary space of Bowman's capsule is approximately 15 mm Hg. This is the pressure left after filtration, and it opposes further filtration. It provides the driving force for fluid flow down the entire length of the nephron. The colloid osmotic pressure of glomerular capillary blood plasma averages approximately 30 mm Hg. This pressure increases along the length of the glomerular capillary because a protein-free filtrate is forced out of the glomerular plasma and the plasma proteins are left behind. Consequently, filtration rate decreases from the beginning (afferent arteriole) to the end (efferent arteriole) of the glomerular capillaries. In the normal human kidney, outward filtration probably occurs along the entire length of the

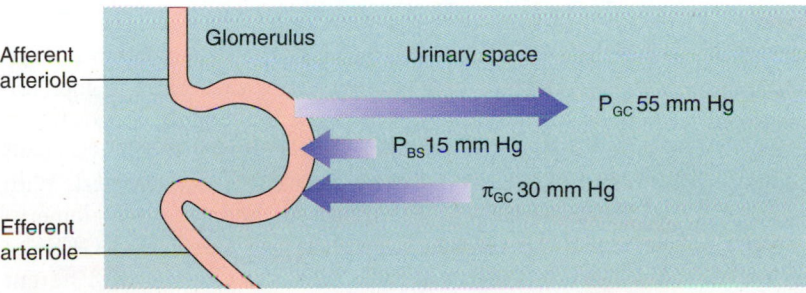

Figure 23–9

Pressures involved in glomerular filtration. The glomerular capillary hydrostatic pressure (P_{GC}) favors filtration, and the hydrostatic pressure in the urinary space of Bowman's capsule (P_{BS}) and the colloid osmotic pressure of proteins in the glomerular capillary plasma (π_{GC}) both oppose filtration. The pressures indicated are for the human kidney and are estimates based on measurements in other mammals.

glomerular capillaries. The net filtration pressure gradient, averaged over the entire glomerulus, is approximately 10 mm Hg (55 mm Hg − 15 mm Hg − 30 mm Hg = 10 mm Hg).

The rate of filtrate formation in the glomeruli of the kidney greatly exceeds that in all other capillary beds in the human body. The high rate of glomerular filtration in the kidneys, approximately 180 L/day, is due to several factors. First, the glomerular filtration coefficient is very high, which reflects, in part, a high fluid permeability of the glomerular filtration barrier. This barrier appears to be very "porous"; it behaves as though it contains many more pores than are present in most other capillaries. In part, the high glomerular filtration coefficient also reflects a large surface area. If all of the glomerular capillaries were spread out as a flat sheet, the total surface area occupied would be approximately 2 m² (roughly equal to body surface area for an adult). Second, glomerular filtration rate is high because glomerular hydrostatic pressure is unusually high for a capillary bed. Third, the high rate of renal blood flow permits a rapid rate of filtration. This effect on filtration rate is subtle. Consider what would happen, however, if glomerular blood flow were low. In this case, forcing a small quantity of protein-free filtrate out of the capillary would lead to a sharp increase in intracapillary colloid osmotic pressure early along the length of a glomerular capillary. Filtration would soon be brought to a halt; the rest of the capillary would no longer function in filtration. Therefore, a high renal blood flow is necessary for a high GFR.

GFR may be altered in a variety of physiological and pathophysiological conditions. The conditions that affect filtration rate do so by affecting the factors already mentioned — namely, filtration coefficient, glomerular hydrostatic pressure, hydrostatic pressure in Bowman's space, and plasma colloid osmotic pressure.

Filtration coefficient (K_f) — and with it GFR — is decreased if there is a decrease in either filtration barrier fluid permeability or surface area. K_f may be altered physiologically by a number of hormones (e.g., angiotensin II). In chronic renal failure, glomeruli are destroyed, leading to a reduction in the filtering surface and a diminished GFR.

Glomerular hydrostatic pressure (P_{GC}) depends mainly on the arterial blood pressure and the resistances of upstream blood vessels (cortical radial arteries, afferent arterioles) and downstream blood vessels (efferent arterioles). Within the range of operation of renal autoregulation (an arterial pressure of 80–180 mm Hg), glomerular hydrostatic pressure changes very little. If, however, blood pressure is reduced to less than this range, glomerular hydrostatic pressure and GFR may decrease substantially. At a mean arterial blood pressure of 40 to 50 mm Hg (as in shock or hypotension), GFR and urine output are practically zero because the glomerular hydrostatic pressure is inadequate to form a filtrate.

Glomerular hydrostatic pressure may be altered by changing the diameter of afferent and/or efferent arterioles (Fig. 23–10); these blood vessels are affected by sympathetic

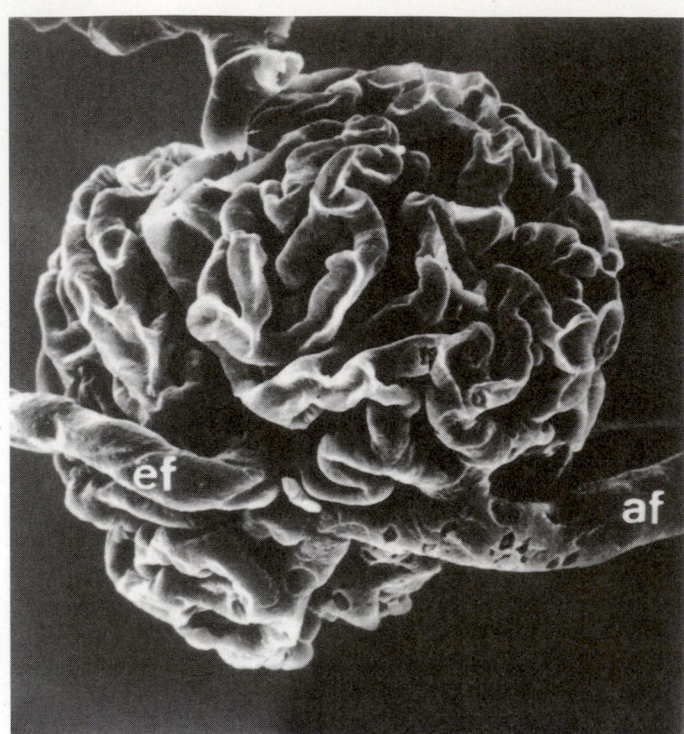

Figure 23–10

SEM of a cast of a glomerulus with afferent and efferent arterioles. Constriction or dilation of the arterioles changes glomerular blood flow and pressure. *(From: Kimura, K., et al. "Effects of atrial natriuretic peptide on renal arterioles: morphometric analysis using microvascular casts."* American Journal of Physiology, *259:F936-F944, 1990.)*

nerve fibers and by many vasoconstrictor and vasodilator substances. Constriction of afferent arterioles decreases the downstream glomerular pressure. At the same time, blood flow is decreased. For both of these reasons, GFR decreases. Dilation of afferent arterioles increases glomerular hydrostatic pressure and blood flow, which leads to an increase in GFR. Constriction of efferent arterioles leads to an increase in the upstream glomerular pressure; blood flow decreases. The effect on GFR of efferent arteriolar constriction is complex. An increase in glomerular pressure tends to increase GFR; on the other hand, a decrease in blood flow tends to decrease GFR. So, depending on the degree of efferent constriction, GFR may be either increased or decreased. With modest constriction, GFR increases (the increase in glomerular pressure predominates); with severe constriction, GFR decreases (the decrease in blood flow predominates). Efferent arteriolar dilation leads to a decrease in glomerular pressure and an increase in glomerular blood flow; GFR decreases.

If the urinary tract is obstructed (e.g., by stones or prostate enlargement), **hydrostatic pressure in Bowman's space (P_{BS})** increases, and consequently GFR decreases. A very high rate of urine output also may be accompanied by an increase in hydrostatic pressure in Bowman's space and a decrease in GFR because an increased pressure head is required to force a large volume flow down the tubules and collecting ducts.

A decrease in plasma protein concentration (e.g., due to intravenous infusion of a large volume of isotonic saline) decreases **plasma colloid osmotic pressure (π_{GC})** and leads to an increase in GFR.

The Clearance Concept and the Measurement of Glomerular Filtration Rate

The determination of GFR is one of the most important measurements of kidney function, both clinically and in experimental studies of renal function. This is done by applying the rather ingenious clearance concept. First, we will explain what is meant by *clearance*.

When the kidneys excrete substances in the urine, they, in effect, free (or clear) the blood plasma of these same substances. Different substances are cleared from plasma at different rates. The **renal plasma clearance** of a substance is defined as the minimum volume of plasma per unit of time that supplies the quantity of the substance excreted in the urine per unit of time. It is calculated from the rate of excretion of the substance divided by its plasma concentration. By convention, the symbol C_x indicates the clearance of a substance x; U_x, the **urine concentration** of the substance x; P_x, the **plasma concentration** of the substance x; and $\dot{V}$, the **urine flow rate.** The rate of excretion of a substance is equal to the product of its urine concentration and the urine flow rate, or in symbols, $U_x\dot{V}$. From the definition of clearance, we can write the **clearance equation:**

$$C_x = \frac{U_x \dot{V}}{P_x}$$

Plasma and urine concentrations of substances are often expressed in terms of mg/ml, and urine flow is given in ml/min. Substituting these units into the clearance formula, we see that the units of clearance are ml plasma/min.

$$C_x = \frac{(\text{mg } x/\text{ml urine}) \times (\text{ml urine/min})}{\text{mg } x/\text{ml plasma}}$$

$$= \text{ml plasma/min}$$

To measure GFR, we need a substance that is cleared from the plasma solely by glomerular filtration. This substance, therefore, should not be reabsorbed or secreted by the kidney tubules. It should not be metabolized, synthesized, or stored by the kidneys. It should pass through the glomerular filtration barrier unhindered; in other words, it should not be a molecule that is too large or that is bound to plasma proteins. It should be nontoxic. Finally, we should be able to measure this substance in plasma and urine using simple analytical methods.

Such an ideal substance is the polysaccharide **inulin**, a fructose polymer found in the roots of certain plants; it has an

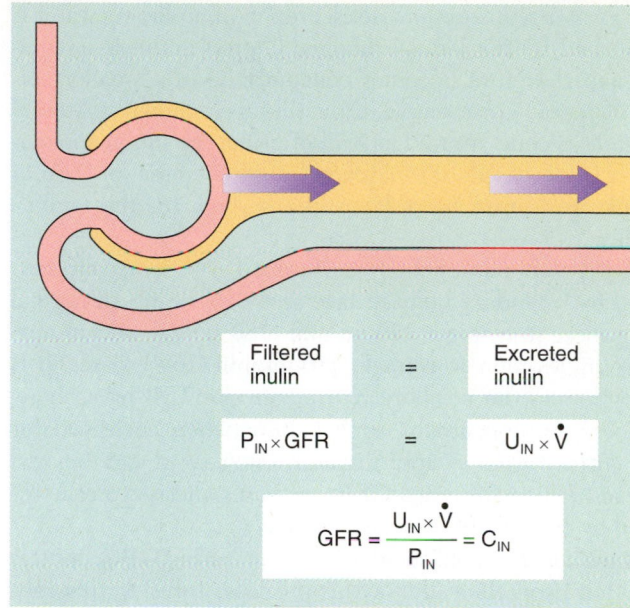

Figure 23–11

Principle behind the measurement of glomerular filtration rate (GFR). Inulin is freely filterable, so the amount filtered per unit of time equals its plasma concentration (P_{IN}) multiplied by the GFR. Inulin is not reabsorbed, secreted, synthesized, metabolized, or stored by the kidney tubules, so filtered and excreted amounts are equal. Rearranging the equation, we find that GFR is equal to the inulin clearance (C_{IN}).

average molecular weight of approximately 5000. Figure 23–11 shows that the amount of inulin (IN) filtered per unit of time, $P_{IN} \times$ GFR, equals the amount of inulin excreted per unit of time, $U_{IN}\dot{V}$. The inulin clearance, C_{IN}, is defined by the expression $U_{IN}\dot{V}/P_{IN}$ and is therefore equal to the GFR.

The way in which inulin is used to measure GFR can be illustrated by an example. An inulin solution is infused intravenously to achieve a constant plasma inulin concentration, and the person is given water to drink to improve the urine output. A timed urine sample is collected, and average urine flow rate is calculated by dividing the urine volume by the duration of collection. In the middle of the urine collection period, a blood sample is taken. Subsequently, plasma and urine samples are analyzed for inulin. Suppose that the following values are found: P_{IN} = 0.30 mg/ml, U_{IN} = 30 mg/ml, and $\dot{V}$ = 1.25 ml/min. Substituting in the inulin clearance equation, $C_{IN} = U_{IN}\dot{V}/P_{IN}$ = GFR, we get:

$$GFR = \frac{30 \text{ mg/ml urine} \times 1.25 \text{ ml urine/min}}{0.30 \text{ mg/ml plasma}}$$

$$= 125 \text{ ml plasma/min}$$

You may have noticed that, in this example, the inulin concentration in the urine is 100-fold that in plasma. This is due to water reabsorption, not secretion of inulin. As water is reabsorbed by the kidney tubules, filtered inulin is left behind and therefore becomes concentrated in a smaller volume of water. For example, if the inulin contained in 100 ml of filtrate is concentrated in 1 ml of urine, the inulin concentration is increased 100-fold, and 99 ml of water (or 99% of the filtrate) must have been reabsorbed by the kidney tubules.

GFR depends on body surface area; it is usually corrected to a standard body surface area of 1.73 m². In normal young men, it averages 125 ml/min (180 L/day), and in normal young women, it averages 110 ml/min (160 L/day). GFR is very low in the newborn, 20 ml/min per 1.73 m² of body surface area, and attains adult values (when corrected for body surface area) by approximately one year of age. Beyond the age of 45 to 50 years, GFR decreases and is typically reduced by 30% to 40% by age 80 years.

Inulin is the "gold standard" for measuring GFR, but it is not often used clinically for this purpose. Infusing this substance is somewhat inconvenient. Instead, the **endogenous creatinine clearance** is used. Creatinine does not have to be infused because it is normally produced by muscles in the body. Its plasma levels are normally quite constant, and it can be easily measured in plasma and urine. In people, creatinine is actually secreted in addition to being filtered, but the error introduced by tubular secretion is usually small, and the endogenous creatinine clearance does provide a clinically valuable estimate of GFR.

If GFR decreases, substances that depend primarily on glomerular filtration for their elimination from the body, such as creatinine and urea, accumulate, and their plasma concentrations increase. The extent to which the plasma concentration of creatinine is elevated may serve as an index of the degree of impairment of filtration rate. BUN is a poorer indicator of GFR than is plasma creatinine concentration because BUN is profoundly affected by other factors, such as the rate of protein catabolism and the urine flow rate.

We have devoted considerable attention to the measurement of GFR because of its importance in assessing kidney function. Glomerular filtration is a major process by which the kidneys regulate the volume and composition of the body fluids. A reduction in GFR to far below normal impairs a person's ability to excrete various waste products and the proper amounts of water and mineral electrolytes and leads to an abnormal internal environment. In patients, the measurement of GFR provides the most valuable indicator of the severity of renal failure. Also, in studying tubule transport functions, a topic we turn to next, the measurement of GFR is important. In order to tell how much of a substance is reabsorbed, we must first know how much was filtered. To tell whether a substance is secreted by the kidney tubules, we cannot simply measure the amount excreted; we must also know how much was filtered. Measurements of GFR are thus important to the study of transport by the kidney tubules.

Tubular Reabsorption Returns Filtered Substances Back to the Blood

 What membrane transport mechanisms govern solute and water reabsorption in the proximal tubule?

As a result of glomerular filtration, large quantities of mineral electrolytes, water, and organic compounds are presented to the tubules. Varying amounts of some of these substances are excreted (see Table 23–1). Most of the glomerular filtrate is not excreted because the tubules selectively reabsorb the constituents of the glomerular filtrate.

Reabsorption may be either active or passive. **Active reabsorption** (active transport) is a process requiring the local expenditure of metabolic energy by the tubular epithelium, and it can effect the net movement of a substance against concentration or electrical gradients, or both. Examples of actively reabsorbed substances are sodium, glucose, and phosphate. **Passive reabsorption** (passive transport) does not depend directly on the expenditure of metabolic energy, and movement occurs down concentration, electrical, or osmotic gradients. Passively reabsorbed substances include urea, chloride, and water. In this section, we discuss the mechanisms for reabsorbing various organic compounds. Later we will consider the reabsorption of various mineral electrolytes and water.

Reabsorption of Glucose

Glucose is filtered by the glomeruli and actively reabsorbed by the tubules. The glucose reabsorptive mechanism, which appears to be confined to the proximal tubule, is so effective that normally all of the filtered glucose is reabsorbed, with none excreted in the urine.

The cellular mechanism for glucose reabsorption by the proximal tubule is depicted in Figure 23–12. Glucose is cotransported with sodium across the luminal (brush border) cell membrane. This movement of glucose is energetically "uphill" and leads to accumulation of glucose in the cell. The energy required is derived from the sodium gradient across the luminal cell membrane. The cell sodium concentration is less than that in the luminal fluid and is maintained at a low level by the Na^+/K^+-ATPase in the basolateral cell membrane. Also, the cell interior is electrically negative (by approximately 70 mV) compared with the fluid in the tubule lumen. As sodium moves into the cell down its concentration and electrical gradients, the cotransporter moves glucose against its concentration gradient. Glucose leaves the cell by moving down a concentration gradient (cell-to-peritubular capillary blood); this movement is via a carrier-mediated, sodium-independent mechanism in the basolateral cell membrane. Overall, glucose reabsorption is an example of "secondary" active transport coupled to "primary" active transport of sodium.

The ability of proximal tubules to reabsorb filtered glucose is limited. At normal plasma glucose levels (70–110

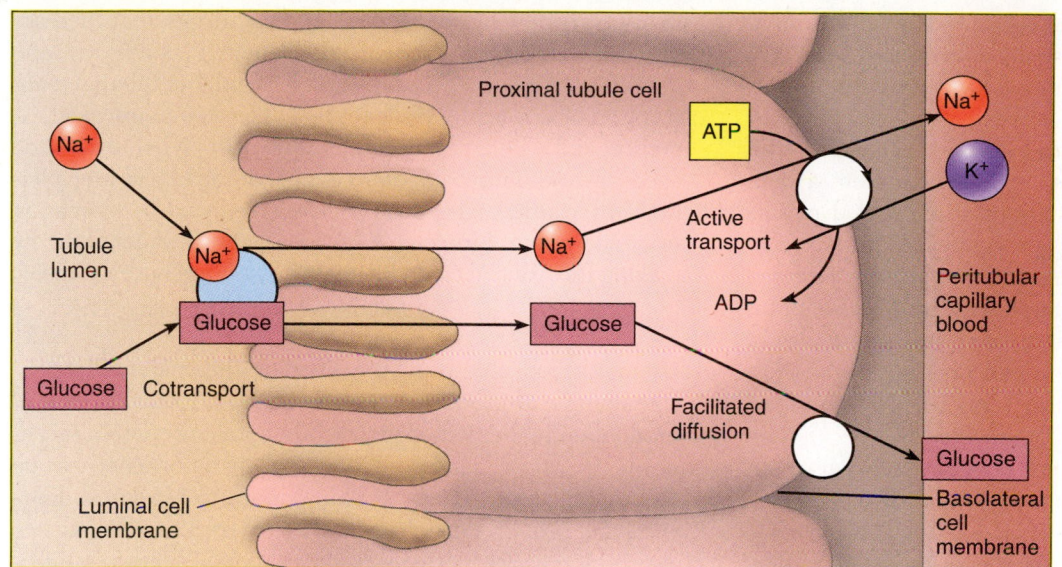

Figure 23–12

A model of glucose reabsorption by the proximal tubule cell. Glucose is cotransported with sodium across the luminal cell membrane, resulting in a high intracellular glucose concentration. Glucose passes through the basolateral cell membrane via facilitated diffusion and then diffuses into the peritubular capillary blood.

mg/dl, fasting values), the filtered glucose load can be completely reabsorbed by the proximal tubules, and so no glucose is excreted (Fig. 23–13*a*). If the plasma level is increased sufficiently — for example, to 180 to 200 mg/dl — the filtered glucose load can no longer be completely reabsorbed. The plasma level at which glucose first appears in the urine in appreciable amounts is called the **glucose threshold** (Fig. 23–13*b*). Glucose appears in the urine because the proximal tubules have a limited reabsorptive capacity. At sufficiently high filtered glucose loads, all of the glucose carriers become saturated, and any extra filtered glucose is excreted (Fig. 23–13*c*). The maximal rate of

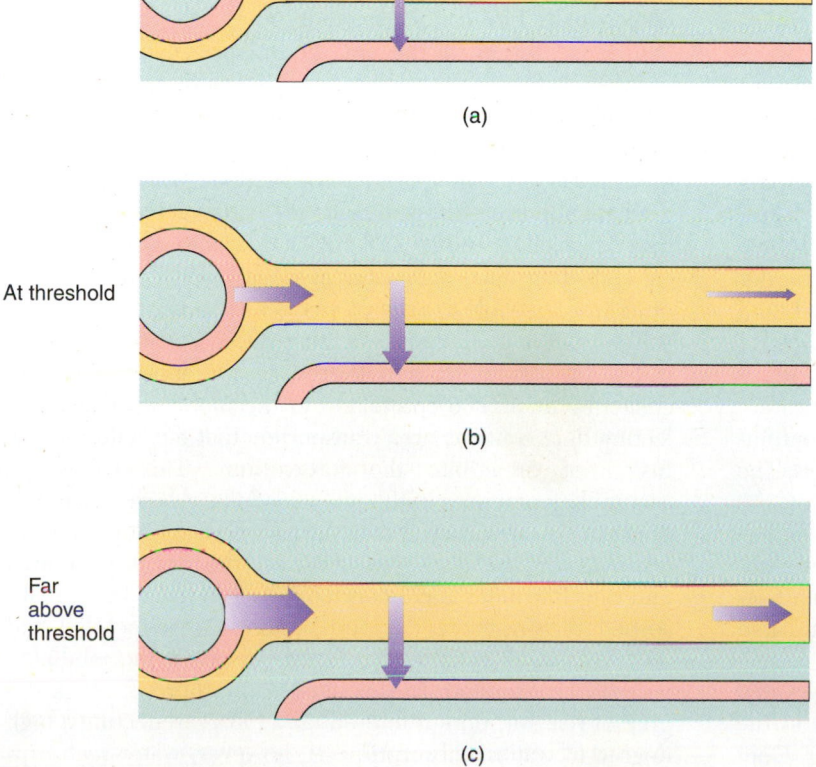

Figure 23–13

Effect of increasing the filtered amount of glucose on glucose excretion. In the normal condition *(a)*, all of the filtered glucose is reabsorbed, and none is excreted. At threshold *(b)*, the amount of glucose filtered is increased to a sufficiently high level (by increasing the plasma glucose concentration) so that not all of the filtered glucose is completely reabsorbed. *(c)* Far above threshold, further increases in the amount of glucose filtered result in increased glucose excretion, because the reabsorptive capacity of the tubules is operating at a maximal rate (glucose Tm). Note that the arrows for reabsorbed glucose in *(b)* and *(c)* are identical in size.

glucose reabsorption is called the **tubular transport maximum (Tm) for glucose (G),** usually abbreviated Tm_G. Tm is analogous to the V_{max} (maximum velocity) of enzyme-catalyzed reactions (Chapter 2).

Glucose in the urine, or **glucosuria,** is often a sign of uncontrolled **diabetes mellitus.** In this condition, a deficiency in the insulin system results in an increased plasma glucose concentration. Filtered glucose exceeds the reabsorptive capabilities of the tubules, and so glucose is excreted in the urine. Many of the early symptoms of diabetes mellitus result from the glucosuria. Excretion of glucose causes loss of water and salt in the urine. The result is a high urine flow rate (increased frequency of urination), dehydration, and thirst.

But glucose in the urine is not necessarily a sign of uncontrolled diabetes mellitus. Glucosuria may also occur because of a high blood glucose level after eating a heavy meal or in conjunction with stress that results in increased sympathetic nervous system activity. There are also certain kidney tubule disorders (e.g., renal glucosuria and Fanconi syndrome) in which the ability of the tubules to reabsorb glucose is diminished.

The glucose reabsorptive mechanism reabsorbs other simple sugars, such as galactose, xylose, and possibly fructose. Because the capacity of the transporting mechanism is limited, two sugars may depress each other's reabsorption if they are simultaneously presented to the tubules. One sugar may displace another from their common carrier, depending on their relative concentrations and affinities for the carrier. This phenomenon is called **competition** for transport.

Reabsorption of Amino Acids

Plasma amino acids are filtered by the glomeruli and extensively reabsorbed by the renal tubules. Normally, only traces appear in the urine. There are several distinct transport mechanisms for amino acids, each reabsorbing a group of structurally similar compounds. The amino acid transport mechanisms, although separate from the glucose reabsorbing mechanism, show several similarities to the latter. Thus, amino acids are actively reabsorbed in the proximal tubule via sodium-dependent cotransporters in the brush border membrane. Amino acid reabsorption is Tm-limited and shows within-group competition.

Reabsorption of Uric Acid

Uric acid, an end-product of purine metabolism, is continuously produced in the body and excreted by the kidneys. The amount excreted is normally approximately 10% of the amount filtered. Hence, uric acid is reabsorbed by the kidney tubules. The kidney tubules also secrete uric acid, but reabsorption normally predominates. Both reabsorption and secretion occur in the proximal tubule by carrier-mediated processes.

Gout is a condition in which plasma uric acid is elevated and uric acid crystals precipitate in the joints (often in the great toe), causing inflammation and painful swelling. Gout

may be treated by **uricosuric agents,** drugs that increase uric acid excretion. Probenecid (Benemid), sulfinpyrazone (Anturane), and high doses of aspirin inhibit tubular reabsorption of uric acid. This leads to an increase in uric acid excretion and a decrease in plasma uric acid. When uricosuric agents are administered, there is a risk for uric acid precipitation in the urine. This danger can be reduced by increasing urine flow rate (drinking large amounts of water) and by making the urine alkaline by ingesting sodium bicarbonate.

Reabsorption of Urea

Urea, formed chiefly in the liver, is the major nitrogen-containing end-product of metabolism in humans. Its rate of production in the body depends on the rate of protein breakdown. Under ordinary conditions, this is determined by the dietary intake of protein. It is important to recognize, however, that protein catabolism, and hence urea production, can also be increased by tissue trauma, internal bleeding, infection, fever, and other conditions. Urea elimination from the body depends on an adequate rate of glomerular filtration.

Figure 23–14 shows that the renal handling of urea along the tubules is fairly complex. The numbers in this figure refer to relative amounts, not concentrations, present at different sites. The amount of urea filtered per unit of time is taken as "100." Fifty percent of the filtered urea is left at the end of the proximal convoluted tubule. The explanation is as follows: as water is reabsorbed, filtered urea becomes more concentrated in the tubular fluid. Because the proximal tubule epithelium is somewhat permeable to urea, urea diffuses out of the tubular fluid back into the blood surrounding the tubules. Some 50% of the filtered urea is reabsorbed in this portion of the nephron. If we examine fluid from the distal convoluted tubule, we see that more urea is present here than was present at the end of the proximal convoluted tubule (see Fig. 23–14). Hence, urea was added (or secreted) into tubular fluid in the loop of Henle; again, this is thought to be a passive movement. A very high concentration of urea exists in the interstitial fluid of the medulla, and so there is a gradient for urea diffusion into the loop of Henle. Notice that the thick ascending limb, distal convoluted tubule, cortical collecting duct, and outer medullary collecting duct are quite impermeable to urea. Water is reabsorbed by the cortical and outer medullary collecting ducts, and so a concentrated urea solution is delivered to the inner medullary collecting duct. The epithelium of the inner medullary collecting duct contains urea transporters that facilitate the diffusion of urea into the interstitium. This results in accumulation of urea in the inner medulla at high concentrations; this is important for the urinary concentrating mechanism, as will be discussed later. Of the quantity of urea entering the inner medullary collecting duct, 80% is reabsorbed (50% re-enters the loops of Henle to be recycled, and 30% is carried away by the blood vessels of the medulla), and 20% is excreted. In this illustration, there is a high degree of reabsorption of filtered urea (80%), reflecting a high degree of water reabsorption. If, however, water reabsorp-

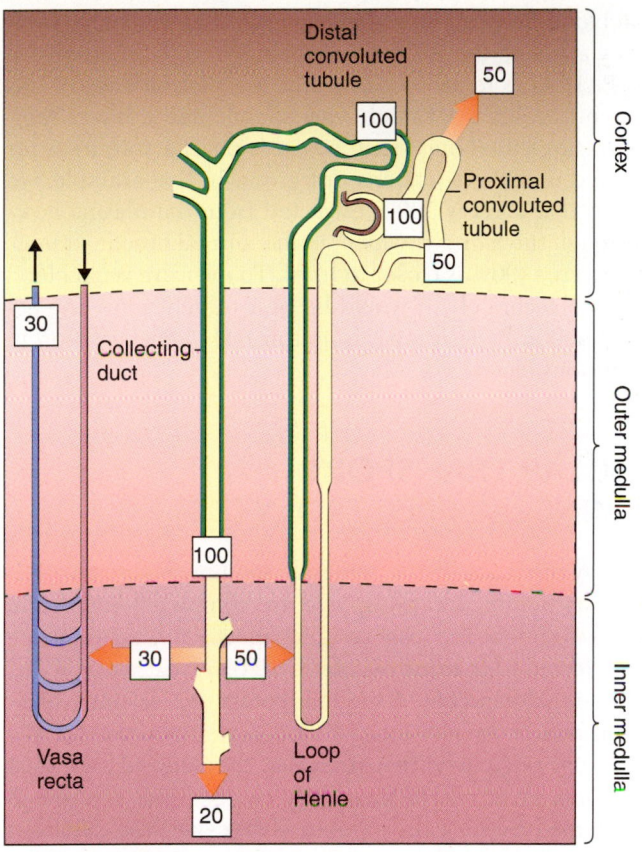

Figure 23–14

Movements of urea along the nephron. The numbers indicate relative amounts (100 = filtered urea). The heavy green outline indicates relative impermeability of these nephron segments to urea. Urea is reabsorbed by the inner medullary collecting duct; most of this urea re-enters the loop of Henle, and some is removed by the medullary blood vessels (vasa recta).

tion is diminished, then urea reabsorption is also less because a reduction in tubular water reabsorption results in less of a concentration gradient for diffusion of urea from tubule fluid back into the blood. Also, less time is available for back-diffusion when the urine flow rate is high. With urine flow rates greater than 2 ml/min, approximately 40% of filtered urea is reabsorbed and 60% is excreted.

Reabsorption of Proteins and Other Substances

The glomerular filtrate is nearly, but not completely, protein-free. Quantitatively speaking, serum albumin is the major protein in plasma, and so we consider this substance. Serum albumin concentration in the glomerular filtrate is probably less than 1 mg/dl; compare this to a plasma concentration of 4.5 g/dl. Clearly, the glomerular barrier is very effective in hindering the passage of this macromolecule. If we consider the high rate of filtrate formation per day, the amount of serum albumin filtered per day (approximately 1 mg/dl ×

180 L/day = 1.8 g/day) is not trivial. Normally, less than 0.03 g/day of serum albumin is excreted. This comparison suggests that the kidney tubules reabsorb filtered serum albumin. Indeed, evidence shows that the proximal tubules reabsorb a variety of proteins by endocytosis. The urine normally contains less than 0.15 g of protein per 24-hour urine sample. **Proteinuria** is usually due to an abnormal leakiness of the glomerular filtration barrier. Heavy proteinuria (more than 4 g/day) is almost always the result of glomerular disease.

Many other organic compounds are reabsorbed by the kidney tubules. Specific tubular reabsorptive mechanisms are known for the Krebs cycle acids (citric, α-ketoglutaric, and malic acids), lactic acid, ascorbic acid (vitamin C), and the ketone body acids (acetoacetic and β-hydroxybutyric acids). Most of these mechanisms have properties similar to the glucose-reabsorbing mechanism.

Proximal Tubules Actively Secrete Many Organic Anions and Organic Cations

Many organic compounds are actively transported from the blood surrounding the kidney tubules into the tubular urine by specific secretory mechanisms. The amount excreted is the sum of filtered and secreted amounts.

A few secretory mechanisms (carriers) transport a variety of organic anions; these compounds are mainly carboxylic and sulfonic acids, which at blood pH (7.4) are mostly ionized and bear a negative charge. Other secretory mechanisms transport a variety of organic cations (organic bases); these are mainly amine and ammonium compounds, which at blood pH bear a positive charge. These active secretory mechanisms are present only in the proximal tubule. The secretory mechanisms show a maximal transport rate (Tm) at high plasma levels and show within-group competition for transport. Some of the more lipid-soluble organic acids and bases may also be passively reabsorbed or secreted.

Table 23–2 lists a few of the many secreted organic compounds. Many of these substances are foreign to the body and can be of medical importance. By supplementing filtration, tubular secretion results in rapid excretion. Urinary excretion is an important consideration in adjusting the dosage of some drugs. In patients with renal failure, drug dosage may have to be reduced to avoid excessively high plasma levels.

PAH (*p*-aminohippuric acid) is so avidly secreted by the proximal tubules that it is almost completely cleared (removed) from all of the plasma in one passage of blood through the kidneys. Therefore, the clearance of PAH can be used to estimate the rate of plasma flow through the kidneys. The explanation for this follows: the amount of PAH per unit of time delivered to the kidneys in the renal blood flow is equal to the product of the renal plasma flow (RPF) and the plasma PAH concentration (P_{PAH}). (The amount of PAH in red blood cells is negligible.) The amount of PAH per unit of time leaving the kidneys, assuming that all is extracted from the renal plasma and excreted in the urine, is equal to $U_{PAH}\dot{V}$ (the PAH excretion rate). From the principle of conservation

TABLE 23–2

Some Organic Compounds Secreted by the Kidney Tubules

Compound	Use
Organic Anions	
Phenol red (phenol-sulfonphthalein, PSP)	pH indicator dye
p-Aminohippuric acid (PAH)	Measurement of renal plasma flow and tubular secretory activity
Penicillin	Antibiotic
Probenecid (Benemid)	Inhibitor of penicillin secretion and uric acid reabsorption
Chlorothiazide (Diuril)	Diuretic drug
Acetazolamide (Diamox)	Carbonic anhydrase inhibitor
Creatinine°	Normal end product of muscle metabolism
Organic Cations	
Histamine	Vasodilator, stimulator of gastric aid secretion
Norepinephrine	Neurotransmitter
Quinine	Antimalarial drug
Quinidine	Antiarrhythmic drug
Tetraethylammonium (TEA)	Ganglionic blocking agent
Creatinine°	Normal end product of muscle metabolism

°Creatinine is an unusual compound because it is secreted by both the organic anion and cation mechanisms. This is probably because of negatively and positively charged groups in the molecule at physiological pH.

of matter, we can write:

$$input = output$$
$$RPF \times P_{PAH} = U_{PAH} \times \dot{V}$$

Rearranging,

$$RPF = \frac{U_{PAH} \times \dot{V}}{P_{PAH}}$$

By definition, the expression on the right is the PAH clearance *(C_PAH)*. Therefore, the renal plasma flow equals the PAH clearance. To determine the renal blood flow (RBF), we need to know the blood hematocrit (HCT) and use the following equation:

$$RBF = RPF/(1 - HCT)$$

Renal blood flow can be readily measured in humans or animals by applying the PAH clearance method. One simply infuses PAH so as to achieve a steady, low plasma level of this compound; collects a timed urine sample and a blood sample; measures the hematocrit; analyzes plasma and urine for PAH; and uses the equations just given. In this example, we assumed that PAH is 100% extracted from the plasma flowing through the kidneys, which, in the normal human kidney, is almost true (90% is closer to true). To measure renal blood flow more accurately, it would be best to determine the actual extraction by also measuring the PAH concentration in renal venous blood.

TUBULAR TRANSPORT OF MINERAL ELECTROLYTES

The inorganic salts of the minerals sodium, potassium, calcium, magnesium, and phosphate are completely ionized in aqueous solution and can conduct electricity, so they are often referred to as **mineral electrolytes.** Next, we will discuss the renal handling of sodium, potassium, calcium, magnesium, and phosphate ions. Although we will consider the various ions separately, recognize that these ions do not exist in isolation. Positively or negatively charged ions are always accompanied by ions of opposite charge because solutions are electrically neutral.

Sodium Reabsorption Is the Major Activity of the Kidney Tubules

 How does sodium reabsorption affect the reabsorption of water and of other substances?

The renal handling of sodium is important for several reasons. First, the kidneys are major regulators of the amount of sodium in the body. Because we take in variable amounts of sodium in our diets, the kidneys must adjust the excretion of sodium to maintain balance. We consider this topic in detail in Chapter 24. Second, sodium reabsorption by the kidney tubules is the major driving force for water reabsorption. Sodium is, for the most part, actively reabsorbed; water follows passively. Third, sodium reabsorption is coupled to the reabsorption of both organic substances (e.g., glucose and amino acids) and inorganic substances (e.g., chloride and phosphate). Fourth, sodium reabsorption is intimately related to the tubular secretion of hydrogen and potassium ions. Thus, it affects acid-base balance and potassium balance. Finally, in terms of energy demands, reabsorption of sodium is the major operation in the kidneys. Sodium and its major accompanying anions, chloride and bicarbonate, constitute the solutes reabsorbed in greatest amounts. Normally, 80% of the oxygen consumed by the kidneys is devoted to powering active sodium reabsorption. Later, we will consider

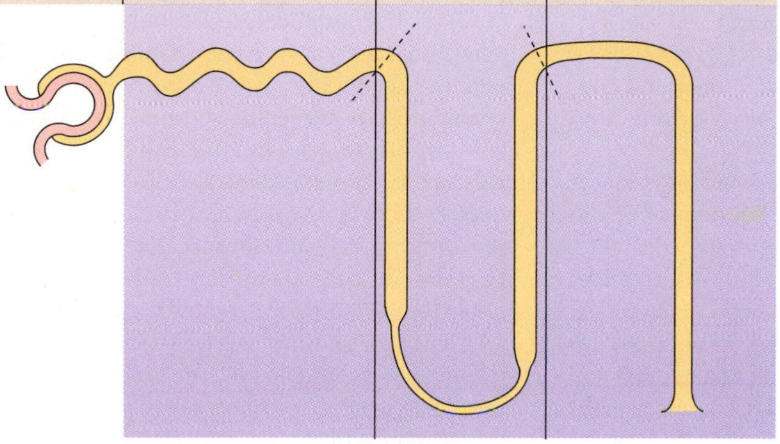

Figure 23–15

Percentages of filtered sodium and water reabsorbed by different nephron segments under normal conditions. Approximately 1% of filtered sodium and water is excreted.

	Proximal convoluted tubule	Loop of Henle	Distal convoluted tubule and collecting duct
Percent of filtered sodium reabsorbed	70	20	9
Percent of filtered water reabsorbed	70	10	19

how sodium reabsorption is affected by various factors and how it is controlled in order to keep the body in sodium balance. For now, we discuss how sodium is reabsorbed in various parts of the nephron. Figure 23–15 presents a summary of the percentages of filtered sodium and water that are reabsorbed under normal conditions by various nephron segments.

Sodium Transport in the Proximal Convoluted Tubule

The logical place to start is the proximal convoluted tubule. This segment reabsorbs 70% of the filtered sodium and water. The cellular mechanisms involved are depicted in Figure 23–16. Sodium enters the cell across the luminal membrane (brush border) and moves down chemical and electrical

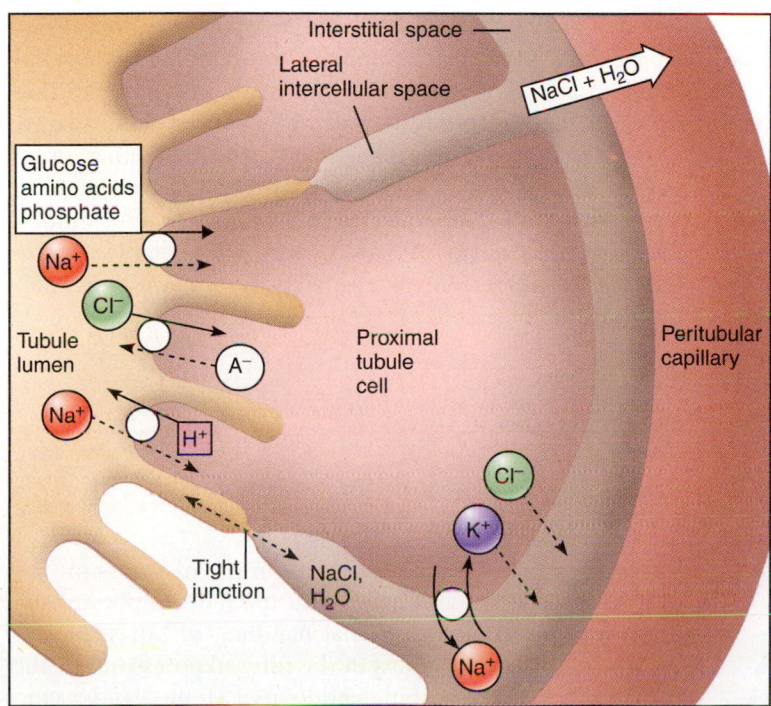

Figure 23–16

Model of ion and water transport in the proximal convoluted tubule. Solid arrows indicate active transport; dashed arrows indicate passive transport. A luminal membrane Cl⁻/anion (A⁻) exchanger participates in transcellular chloride reabsorption.

gradients. The intracellular sodium concentration is kept low ($\approx$ 30–40 mEq/L) by the basolateral membrane Na$^+$/K$^+$-ATPase. Inside negativity of the cell also favors sodium entry. The entry of sodium into the cell is coupled to the entry of other reabsorbed solutes, such as glucose, amino acids, and phosphate. Also, luminal sodium ions can exchange for cellular hydrogen ions via a luminal cell membrane Na$^+$/H$^+$ exchanger. Sodium is actively pumped out the basolateral cell membrane by its Na$^+$/K$^+$-ATPase. This is an example of primary active transport. The energy needed to move sodium out of the basolateral cell membrane, against combined concentration and electrical gradients, is directly derived from the hydrolysis of ATP.

The reabsorption of sodium and accompanying solutes results in the reabsorption of water. The epithelium of the proximal convoluted tubule has a high water permeability, and so only a very small osmotic gradient is necessary to account for the observed rate of water reabsorption. Removal of solute is thought to lead to an effective osmotic pressure difference of approximately 2 to 5 mosm/kg H$_2$O across the tubular wall; this is considered sufficient to drive water reabsorption by osmosis. Because the gradient is so small, proximal tubular fluid is essentially isosmotic to plasma. Sodium and its accompanying anions comprise the major osmotically active solutes in the glomerular filtrate and proximal tubule fluid. Because 70% of the filtered sodium is reabsorbed, 70% of the filtered water is also reabsorbed in the proximal convoluted tubule.

The final step in the reabsorption of sodium is the uptake of the reabsorbed fluid by the peritubular capillaries. The forces involved here are usually referred to as **peritubular capillary Starling forces** and involve the hydrostatic and colloid osmotic pressure differences that affect fluid movement across any capillary wall (see Chapter 19). The hydrostatic pressure in these capillaries is low because the blood has passed through several upstream resistance vessels (i.e., the cortical radial arteries, afferent arterioles, and efferent arterioles). Also, the colloid osmotic pressure of the peritubular capillary plasma is high because the plasma proteins were concentrated by glomerular filtration. Consequently, a net pressure gradient favors uptake of reabsorbed fluid. Under some circumstances (e.g., loading a person with isotonic saline), peritubular capillary pressures are altered in a direction that diminishes the uptake force. The reabsorbed fluid then accumulates in the renal interstitium. The lateral intercellular spaces widen, and the tight junctions become expanded and very leaky. This results in back-leak of sodium and water into the tubule lumen and an overall reduction in net sodium reabsorption.

Sodium Transport in More Distal Nephron Segments

The loop of Henle reabsorbs approximately 20% of the filtered sodium and 10% of the filtered water (see Fig. 23–15). Note that more sodium than water is reabsorbed here because the ascending limb is impermeable to water but reabsorbs salt. The transcellular transport of sodium in the thick ascending limb is accomplished by a coupled Na/K/2Cl cotransporter in the luminal cell membrane (which is inhibited by "loop" diuretic drugs, e.g., furosemide and bumetanide) and a Na$^+$/K$^+$-ATPase in the basolateral cell membrane. As will be discussed later, the removal of salt along the ascending limb and its deposition in the medulla are essential steps in the mechanism in which the kidneys produce osmotically dilute or concentrated urine.

The distal convoluted tubule and collecting duct accomplish the reabsorption of most, but not all, of the remaining salt and water (see Fig. 23–15). In distal convoluted tubule cells, sodium is reabsorbed via a Na/Cl cotransporter in the luminal cell membrane and a Na$^+$/K$^+$-ATPase in the basolateral cell membrane. In collecting duct cells, sodium is reabsorbed via a sodium channel in the luminal cell membrane and a Na$^+$/K$^+$-ATPase in the basolateral cell membrane. Tight junctions in these nephron segments are really tight, so steep concentration gradients for sodium can be established. For example, if a person is depleted of salt, the collecting ducts can decrease the final urine sodium concentration to 1 mEq/L (compared with a plasma concentration of 140 mEq/L). Aldosterone, a hormone secreted by the adrenal cortex, stimulates sodium reabsorption by the distal convoluted tubule and collecting ducts. These nephron segments have an intrinsically low permeability to water. The antidiuretic hormone (ADH), secreted from the posterior pituitary, increases collecting duct water permeability. The collecting ducts are the final site of control of sodium and water excretion, so it is not surprising that reabsorption here is under hormonal control.

To summarize and compare reabsorption in the early (proximal convoluted tubule) and late (collecting duct) portions of the nephron, we note that in the proximal tubule, large quantities of salt and water are reabsorbed along small gradients. The proximal tubule has a leaky epithelium and cannot sustain large gradients for small ions or water. Reabsorption here can be considered as a large, but coarse, operation. By contrast, the collecting ducts have a smaller capacity for reabsorption and can establish steep gradients. The collecting ducts have a tight epithelium and are the site of fine control; the hormones aldosterone and ADH control, respectively, sodium and water reabsorption here.

> ■ *How are other ions handled by the nephron?*

Potassium Is Filtered, Reabsorbed, and Secreted in the Kidney

The kidneys are normally the major route of potassium excretion from the body and serve as the most important site of potassium regulation. Renal handling of potassium involves reabsorption of most of the filtered potassium by the proximal convoluted tubule and loop of Henle. Under con-

CURRENT CONCEPTS IN PHYSIOLOGY

Genetic Disorders of Tubular Transport

More than 30 genetic disorders of kidney tubule transport are known to occur in people. With recent advances in molecular biology and genetics, the defective proteins have now been isolated in many instances. This information has helped to explain many puzzling diseases, and, at the same time, has improved our understanding of normal and abnormal physiology.

Renal glucosuria is a rare condition in which glucose appears in the urine even at normal plasma glucose levels. It may be due to a defective sodium-glucose transporter 2. This transporter, located in the brush border of the proximal convoluted tubule, reabsorbs one sodium ion together with one glucose molecule.

Cystinuria is a condition in which there is abnormal excretion of the sulfur-containing amino acid cystine in the urine. Cystine is rather insoluble in water and so it precipitates in the urine, forming kidney stones. In fact, cystine was first isolated from a glistening yellow urinary bladder ("cyst") stone by Wollaston in 1810; this is how it got its name. Patients with cystinuria typically show elevated urinary excretion of lysine, arginine, and ornithine, which agrees with other evidence that these three amino acids and cystine share a common apical membrane transporter in the proximal tubule.

Bartter's syndrome was for many years a puzzling disorder in which affected patients showed extreme salt wasting in the urine; hypokalemia (i.e., low plasma potassium); metabolic alkalosis; and other disturbances, such as short stature. Plasma renin and aldosterone levels are high, reflecting the volume depletion. We now know that the problem is caused by defective sodium, potassium, and chloride transport in the thick ascending limb. In many of these patients, the loop-diuretic (bumetanide)-sensitive Na/K/2Cl cotransporter is defective. In others, the apical membrane potassium channel or basolateral membrane chloride channel in the thick ascending limb cells is abnormal. The disorder is transmitted in an autosomal recessive mode.

Gitelman's syndrome is rather similar to Bartter's syndrome. The defect here is due to diminished functioning of a Na/Cl cotransporter in the luminal cell membrane of distal convoluted tubule cells.

Liddle's syndrome is a rare autosomal dominant condition in which severe hypertension develops. It is caused by overactivity of the epithelial sodium channel (ENaC), which is located in the luminal cell membrane of collecting duct cells (see Fig. 23–17). This leads to excessive reabsorption of sodium, volume expansion, inhibition of renin and aldosterone secretion, and hypertension. The association between sodium retention by the kidneys and hypertension in these patients is strong evidence for a link between salt and high blood pressure.

Nephrogenic diabetes insipidus is a disorder in which the collecting ducts of the kidneys do not show the normal increase in water permeability in response to the antidiuretic hormone (ADH). This leads to excessive excretion of a dilute urine, dehydration, and thirst. There is an X-linked form in which the vasopressin receptor, located in the basolateral cell membrane of collecting duct cells, is defective. There are also autosomal recessive and autosomal dominant forms in which the water channel (aquaporin-2), which needs to be inserted and function in the apical cell membrane of collecting duct cells, appears to be deficient.

Treatment of these various inherited disorders is greatly aided by understanding the underlying pathophysiology. Whether it will be possible, in the future, to treat these disorders by some form of gene therapy, i.e., by correcting or replacing the defective gene, remains unclear. Further information on these and other human genetic diseases may be obtained by consulting www.ncbi.nlm.nih.gov/Omim.

---

ditions of potassium depletion, potassium continues to be reabsorbed along the collecting ducts. Under normal conditions or with potassium excess, potassium is secreted by the collecting ducts, especially by the cortical collecting ducts. The major cell type of the collecting ducts, the **principal cells,** appear to be responsible for this secretion. Most of the excreted potassium usually represents secreted potassium.

A number of factors affect potassium secretion, most of which can be understood by examining the cell model for a cortical collecting duct principal cell in Figure 23–17. According to this model, potassium is taken up by the cell via a Na^+/K^+-ATPase located in the basolateral membrane. This leads to an increase in intracellular potassium ion concentration. Potassium can then diffuse out of the cell into the lumen and in this way be secreted. The intracellular potassium

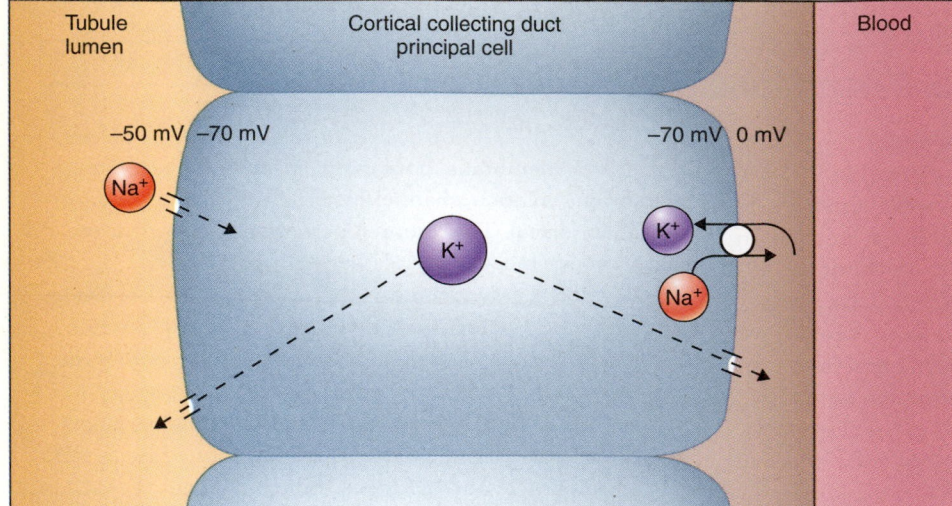

Figure 23–17

A model of potassium and sodium transport by a principal cell in the cortical collecting duct. The two steps in potassium secretion are (1) uptake into the cell via the basolateral membrane Na^+/K^+-ATPase and (2) passive diffusion across the luminal cell membrane. Sodium reabsorption involves passive diffusion into the cell through a sodium-selective ion channel in the luminal cell membrane (called ENaC, for epithelial sodium channel), and active pumping of sodium across the basolateral cell membrane.

concentration is a key element in determining the rate of potassium secretion. An increase in extracellular potassium concentration or increased plasma levels of aldosterone enhances cell uptake of potassium and overall secretion. Aldosterone, in addition, increases the luminal membrane permeability to both potassium and sodium, which also favors potassium secretion and sodium reabsorption.

A negative tubular lumen also favors potassium secretion. There is a **transtubular potential difference** across many nephron segments. If we advance a potential-measuring microelectrode through cortical collecting duct epithelium, we first record a negative deflection of 70 mV when the microelectrode enters the cell. (The zero reference potential is always on the blood side of the tubules.) A negative deflection of approximately 50 mV is recorded when the microelectrode penetrates the lumen. Therefore, the electrical potential opposing potassium movement out of the cell is much smaller at the luminal side, $-70\,mV - (-50\,mV) = -20\,mV$, than at the basolateral side of the cell, $-70\,mV$. Consequently, there is less force opposing movement into the lumen, and potassium secretion is favored. Increased amounts of poorly reabsorbed anions in the tubule lumen increase the transtubular potential difference (make the lumen more negative) and so enhance potassium secretion.

Increased amounts of sodium in the cortical collecting duct lumen result in increased potassium secretion. More sodium in the lumen results in increased diffusion of sodium into cortical collecting duct principal cells, further depolarization of the luminal cell membrane, and consequently less of an electrical force opposing potassium diffusion from cell to lumen. Also, more sodium in the cell results in increased activity of the basolateral membrane Na^+/K^+-ATPase, which then brings more potassium into the cell from the blood side, increases the intracellular potassium concentration, and thereby enhances potassium secretion. An increased rate of

fluid flow through the collecting duct lumen maintains the cell to lumen potassium gradient and so promotes potassium secretion. These effects explain why most diuretics increase potassium excretion.

Calcium, Magnesium, and Phosphate Are Reabsorbed by the Kidney Tubules

Renal excretion of **calcium** represents only approximately 10% of the total daily output of calcium from our bodies on a normal diet. The remainder (90%) of the ingested calcium is excreted in the feces and represents mainly ingested calcium that was not absorbed by the small intestine. The regulation of plasma calcium involves control of intestinal calcium absorption, renal excretion, and exchanges with bone. The parathyroid hormone and vitamin D play important roles in this regulation. Calcium balance is discussed in Chapter 26.

Calcium is filtered and reabsorbed by the kidneys. Approximately 50% to 60% of the filtered calcium is reabsorbed in the proximal convoluted tubule, and most of the rest in the thick ascending limb, distal convoluted tubule, and collecting ducts. The quantity excreted is approximately 1% to 2% of the filtered load. Control of urinary excretion of calcium occurs primarily in the connecting tubule and initial part of the cortical collecting duct, where parathyroid hormone stimulates calcium reabsorption.

Approximately two thirds of the **magnesium** in our diet is normally excreted in the feces; like calcium, this ion is incompletely absorbed by the intestine. Approximately one third of the dietary magnesium is excreted in the urine. The proximal convoluted tubules reabsorb approximately 20% to 30% of filtered magnesium, and 50% to 60% is reabsorbed in the loop of Henle, mainly in the thick ascending limb. Only 3% to 5% of the filtered magnesium is excreted under normal conditions. The kidneys play a major role in regulating the plasma magnesium concentration. Excesses of magnesium are rapidly excreted in the urine; in magnesium-deficient states, magnesium is avidly reabsorbed by the kidney tubules and virtually disappears from the urine.

The kidneys play an important role in regulating the plasma concentration of **inorganic phosphate.** We usually excrete in the urine most of the phosphate ingested in our diet. Filtered phosphate is reabsorbed by sodium-dependent, secondary active transport in the proximal tubule. The reabsorptive mechanism has a limited rate, or Tm. The Tm is normally exceeded by the quantities of phosphate filtered, and so phosphate is excreted in the urine. The situation here differs from that of glucose, where the Tm is reached only when the plasma glucose concentration is increased dramatically. Because the filtered phosphate load and Tm are normally so close to one another, the kidneys can participate in regulating the plasma phosphate concentration.

If phosphate intake is increased, plasma phosphate levels increase, and the filtered phosphate load will be increased. The Tm is exceeded more than usual, and so phosphate excretion increases, thereby ridding the body of the extra phosphate. Conversely, if phosphate intake is decreased, plasma phosphate and filtered phosphate amounts decrease, and all of the filtered phosphate is reabsorbed, thus conserving phosphate for the body. Phosphate excretion is also controlled by changing the Tm. For example, the parathyroid hormone inhibits proximal tubule phosphate reabsorption, thus promoting phosphate loss in the urine.

TUBULAR REABSORPTION OF WATER

The rate of water excretion in the urine depends on the rates of filtration of water (GFR) and tubular water reabsorption. Because the urine volume is mostly water, we can assume that the renal water excretion rate is simply equal to the urine flow rate. Urine flow rate can vary widely, depending on conditions. If a person drinks a large quantity of water, urine flow rate may be as high as 20 ml/min. The osmolality[1] of this urine may be as low as 30 to 40 mosm/kg H_2O, much more dilute than plasma ($\approx$ 300 mosm/kg H_2O). On the other hand, in a dehydrated person, urine flow rate may be as low as 0.3 ml/min. The osmolality of this urine may be as high as 1200 to 1400 mosm/kg H_2O, or approximately four- to fivefold the normal plasma osmolality. A "normal" urine flow rate is approximately 1 ml/min (1.5 L/day), and the urine is usually modestly concentrated (600–800 mosm/kg H_2O).

---

[1]The terms *osmolality* and *osmolarity* (Chapter 4) are often confused. Both are measures of solute concentration. Osmolality is expressed as osmoles per kilogram of H_2O, and osmolarity as osmoles per liter of solution. Biological fluids are mostly water, so the two terms are practically identical.

The Kidneys Save Water When They Produce a Hyperosmotic Urine

 How is water reabsorption controlled in the kidney?

We will consider now the mechanisms for producing an osmotically concentrated (hyperosmotic) or dilute (hypoosmotic) urine — that is, urine with a total solute concentration greater than or less than that of plasma. If the body contains excess water (e.g., due to drinking a lot of liquid), it makes sense for the kidneys to get rid of the extra water and excrete the urinary solutes in a large volume of osmotically dilute urine. The importance of excreting osmotically concentrated urine is more subtle. Recall that the kidneys are always called upon to excrete solutes. These solutes include waste products of metabolism and various mineral electrolytes consumed in the diet. The excretion of these solutes requires the excretion of water because this is the vehicle in which these substances are dissolved. Suppose someone had to excrete 600 mosm/day of solutes. Suppose also that the person could form only an isosmotic urine — that is, a urine with an osmolality of 300 mosm/kg H_2O. How much water would have to be excreted? It would take 2 kg (or 2 L) of water to excrete 600 mosm. Suppose now that the same amount of solutes could be concentrated in a urine with an osmolality of 1200 mosm/kg H_2O. How much water would have to be excreted? One half-kilogram (or 0.5 L) would contain 600 mosm solute as a 1200 mosm/kg H_2O solution. We can see that by excreting the urinary solutes in osmotically concentrated (hyperosmotic) urine, the kidneys save water for the body. In the example given, by excreting the same amount of solutes in urine with an osmolality of 1200 mosm/kg H_2O instead of 300 mosm/kg H_2O, the kidneys, in effect, saved 1.5 L of pure water for the body.

Because water is often scarce, the ability to form osmotically concentrated urine is an important adaptation to a terrestrial environment. By getting rid of the urinary solutes in hyperosmotic urine, the need for water intake is diminished. In the animal kingdom, only birds and mammals can produce urine that is osmotically more concentrated than is the blood. These animals are the only classes that have loops of Henle. Furthermore, in general, those mammals with relatively long loops of Henle are able to produce the most concentrated urine. For example, the kangaroo rat, a small rodent that inhabits the southwestern deserts of the United States, has an exceptionally long loop of Henle and can produce urine with an osmolality of 5500 mosm/kg H_2O, some 18-fold that of plasma. This mammal's high metabolic rate (due to its small size) results in especially vigorous active sodium transport by the thick ascending limb of Henle's loop, a key step in concentrating the urine. These facts from comparative anatomy and physiology suggest an association between loops of Henle and the ability to produce concentrated urine.

The Countercurrent Hypothesis Explains the Formation of Osmotically Concentrated Urine

The **countercurrent hypothesis** is now universally accepted as the explanation for how osmotically concentrated urine is formed. This hypothesis proposes that the loops of Henle and the vasa recta form a countercurrent mechanism that produces and maintains the osmotic gradient in the kidney medulla. The term *countercurrent* indicates that fluid flows in opposite directions in adjacent limbs of tubules and blood vessels in the kidney medulla. Scientists are still not sure about certain aspects of this hypothesis—in particular, how the osmotic gradient is established in the inner medulla.

There are actually two countercurrent mechanisms. The loops of Henle act as **countercurrent multipliers;** they set up a gradient of osmolality that increases from the junction of the cortex and medulla to the tips of the renal papillae. The blood vessels of the medulla (vasa recta) act as passive **countercurrent exchangers;** they help to preserve the osmotic gradient in the medulla. The collecting ducts act as **osmotic equilibrating devices;** it is here that the urine finally becomes osmotically concentrated. Depending on the plasma level of ADH, the fluid in the collecting ducts tends to equilibrate osmotically with the surrounding interstitium.

If kidney tissue from a water-deprived animal is taken from various levels of the medulla, a progressively higher osmolality is found with increasing depth of the medulla. The highest osmolalities are found at the papillary tip. All of the structures in the medulla participate in this increasing gradient. Also, in a dehydrated animal, with high plasma levels of ADH, the final urine has the same osmolality as tissue fluid at the tip of the kidney papilla.

This gradient is established by countercurrent multiplication in the loops of Henle. Figure 23–18 shows a simplified model for the countercurrent multiplication process. Consider a loop filled with fluid iso-osmotic to plasma (300 mosm/kg H_2O). Assume that the membrane separating the two loops is impermeable to water. Also assume that the loop is able to establish an osmotic gradient of 200 mosm/kg H_2O between the two limbs of the loop at any level (see Fig. 23–18b). This step is often called the **single effect.** This gradient could be produced by transport of salt out of the ascending limb and its deposition in the descending limb. Water is left behind (because the intervening membrane is water-impermeable), and so the osmolality of the ascending limb fluid decreases and that of the descending limb increases by the same amount. Next, we add new fluid to the loop (see Fig. 23–18c), as fresh fluid flows in from the proximal convoluted tubule. Again allow the 200 mosm/kg H_2O gradient to be established (see Fig. 23–18d). Repeat the process several times (see Fig. 23–18e to h). What is the final result? Notice that at any level of the loop a gradient of 200 mosm/kg H_2O exists. Along the length of the loop, however, a larger osmotic gradient ($\approx$ 400 mosm/kg H_2O) has been established. This is what is meant by countercurrent multiplication; namely, a gradient existing at any level of the loop is multiplied by countercurrent flow so as to produce a larger gradient along the length of the loop.

The model just discussed should not be taken too literally. For example, flow through the loop of Henle is not a discontinuous process. Also, ascending and descending limbs do not share a common membrane; rather, there is an intervening interstitial space. The model is realistic, however, in that salt is transported out of the ascending limb across a water-impermeable membrane. The descending limb is water-permeable, and its osmolality is increased primarily by water removal, not by active deposition of salt into its lumen. Notice that fluid at the end of the ascending limb is osmotically dilute, even in a kidney that will be forming concentrated urine. This is so because salt has been removed along the ascending limb.

Another important feature of countercurrent multiplication is that it is an energy-consuming process. In other words, in order to establish the gradient in the medulla, there must

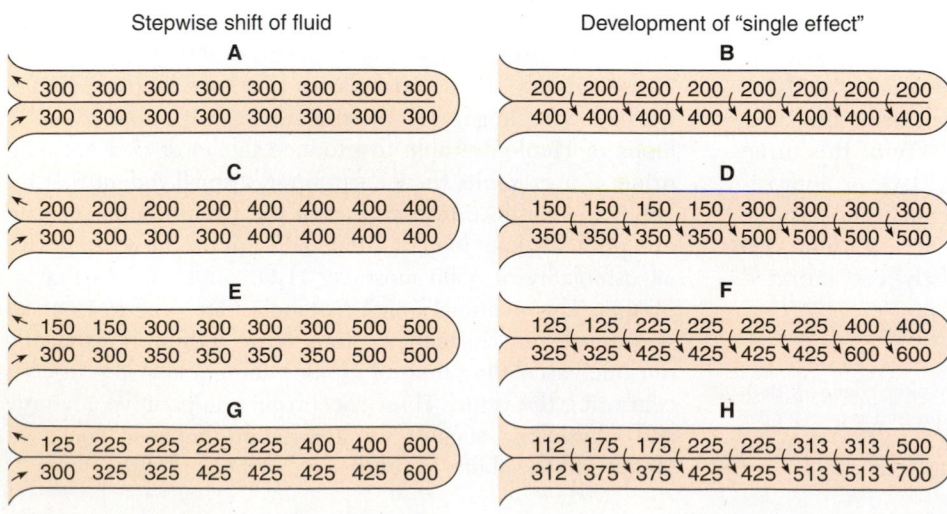

Figure 23–18

The principle of countercurrent multiplication, based on the assumption that, at any level along the loop, an osmotic gradient of 200 mosm/kg H_2O can be established by salt transport between ascending and descending limbs. *(Modified from: Pitts, R. F. Physiology of the Kidney and Body Fluids, ed 3. Chicago, Year Book, 1974.)*

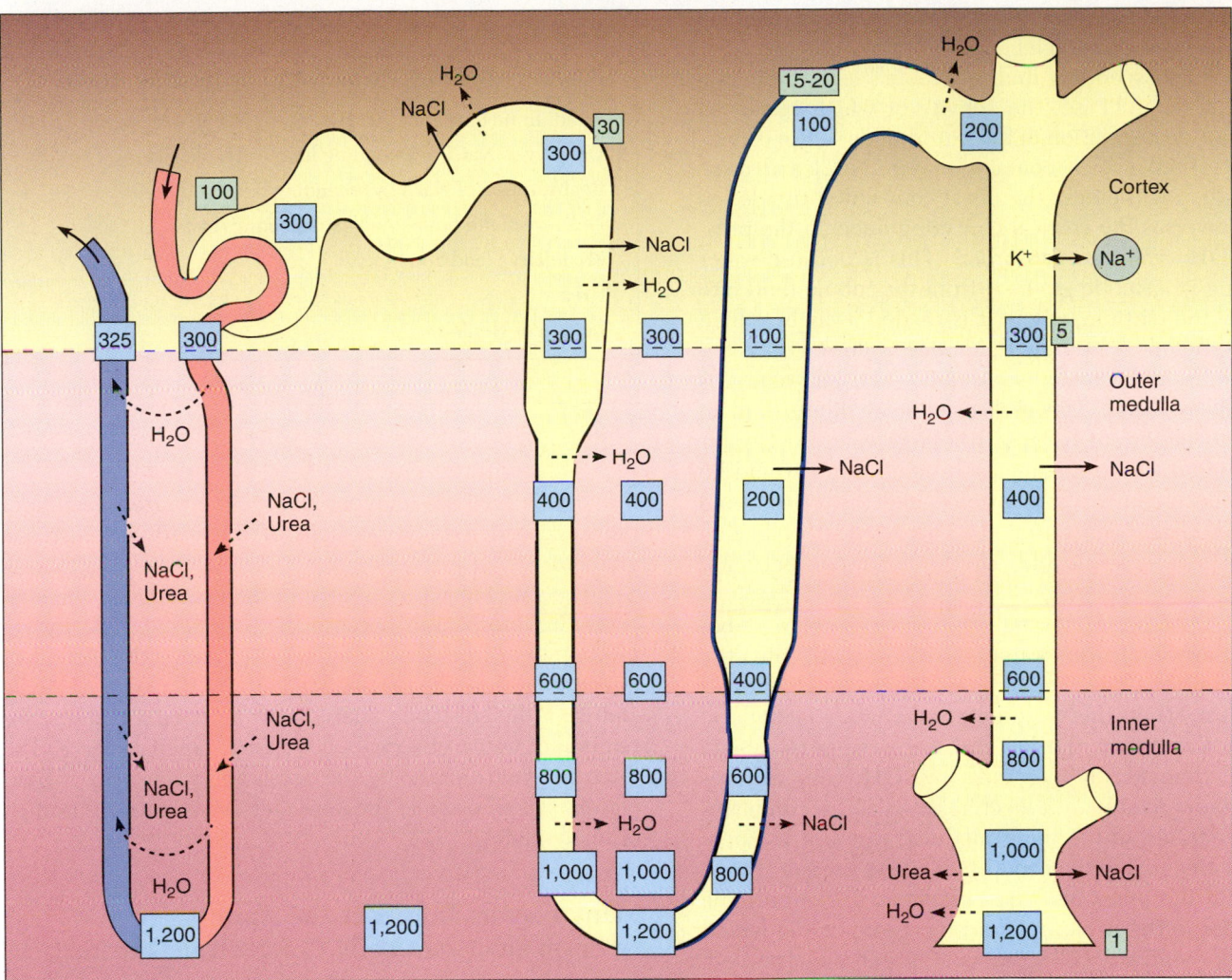

Figure 23–19

Summary of movements of water, ions, and urea in the kidney during elaboration of a maximally concentrated urine (1200 mosm/kg H_2O). Numbers boxed in blue give osmolality in mosm/kg H_2O. Numbers boxed in green give the relative amount of water present at each level of the nephron. Solid arrows indicate active transport; dashed arrows passive transport. The heavy blue outline along the thin ascending and thick ascending limbs of Henle's loop indicates that these segments are relatively water-impermeable.

be an energy source. The energy source for operating the countercurrent multiplier is ultimately active sodium transport, which is linked to ATP hydrolysis. Note that the magnitude of the gradient established depends on the size of the single effect and the length of Henle's loop. If the single effect is reduced, or if the loops are short, then a large gradient along the length of the loop cannot be established.

Figure 23–19 shows a model for the operation of the countercurrent mechanism in the kidney. In this model, we are assuming that maximally concentrated urine is formed. The numbers in blue give the osmotic concentration in mosm/kg H_2O. The boxed numbers in green give the relative amounts of filtered water present at various points along the nephron. The heavy blue shading along the ascending limb indicates low water permeability. A juxtamedullary nephron, with a long loop of Henle extending to the tip of the papilla, is depicted.

Seventy percent of the filtered sodium and water is reabsorbed along the proximal convoluted tubule. This reabsorption is essentially iso-osmotic. Fluid entering the descending limb of Henle's loop has an osmolality of approximately 300 mosm/kg H_2O. The descending limb is water-permeable, and because the interstitium surrounding the tubules is hyperosmotic, water leaves the descending limb and the osmolality increases. Because NaCl is the main solute in the descending limb fluid, its concentration increases.

In the thick ascending limb of Henle's loop, there is a vigorous Na^+/K^+-ATPase, the activity of which results in sodium chloride deposition in the outer medulla. An osmotically dilute fluid, which contains mainly NaCl and urea, leaves the loop and enters the distal convoluted tubule. As the fluid traverses the cortical collecting ducts in the presence of ADH, water is reabsorbed. This is because water moves along its osmotic gradient from the tubule fluid into the cortical interstitium, where it is carried away by a high blood flow. Before the tubule fluid once again re-enters the outer medulla, it becomes iso-osmotic, and its volume is substantially reduced. The reabsorption of dilute fluid in the cortex is important because if a large volume of water were presented to the medulla, too much water would be added to the medullary interstitium, and the urine could not be maximally concentrated. The cortical collecting ducts are urea-impermeable, so the urea concentration increases.

As the fluid passes through the outer medulla, it becomes hyperosmotic as water is reabsorbed into the medullary interstitium. The urea concentration increases further. Next, when the urine enters the inner medulla, it encounters a segment, the low intrinsic urea permeability of which is dramatically increased by ADH. Urea diffuses out of the inner medullary collecting ducts and accumulates in the interstitium of the inner medulla, where it balances osmotically the urea in the urine. This allows the salt (NaCl) deposited in the inner medulla to balance osmotically the other solutes in the urine. As fluid moves down the length of the inner medullary collecting duct, water is reabsorbed. In the presence of high levels of ADH, fluid in the collecting ducts achieves the same osmolality as in the surrounding interstitium. Urine of high osmolality and low volume is finally excreted.

The Vasa Recta Are Countercurrent Exchangers

 How does blood flow to the medulla affect the medullary concentration gradient?

The blood vessels of the medulla, the vasa recta, supply its nutritional needs. They also act as countercurrent exchangers. Countercurrent exchange is a passive process that helps to maintain a gradient established by some other means. For example, countercurrent exchange of heat between blood flowing into and out of the limbs allows the core body temperature to be maintained higher than that of fingers and toes. Blood flowing into and out of the medulla exchanges water and solutes between ascending and descending vasa recta (see Fig. 23–19). This exchange reduces the extent to which blood flowing into the medulla tends to dissipate the osmotic gradient. A relatively low blood flow to

TABLE 23–3
Factors Affecting Urinary Concentrating Ability
Antidiuretic hormone (ADH)
Delivery of NaCl to ascending limb of Henle's loop
Reabsorption of NaCl by ascending limb
Delivery of fluid to medullary collecting ducts
Medullary blood flow
Urea
Length of Henle's loop

the kidney medulla (compared with the cortex) facilitates exchange and reduces washout of the medullary osmotic gradient.

The vasa recta are essential to the operation of the concentrating mechanism because they are responsible for removing water from the medulla. Water enters the medullary interstitium from the descending limbs of the loops of Henle and from the collecting ducts; there must be a pathway for removal of water; otherwise, it would accumulate there. The force for fluid uptake by the ascending vasa recta is their plasma colloid osmotic pressure and elevated concentration of small solutes.

Many Factors Influence the Ability to Form an Osmotically Concentrated Urine

Antidiuretic hormone is necessary for the production of osmotically concentrated urine (Table 23–3). In the absence of ADH, the urine is not iso-osmotic to plasma but is actually markedly hypo-osmotic. The range of urine osmotic concentration, from very dilute to very concentrated urine, is normally due to varying plasma levels of ADH. **Diabetes insipidus** is a condition in which there is an ADH deficiency, and as much as 20 L of dilute urine may be excreted in a day. To survive, a person with a disorder of this magnitude would have to drink 80 glasses of water a day and would spend a large part of the day and night urinating and drinking. In **neurogenic diabetes insipidus,** synthesis or secretion of ADH is inadequate. In **nephrogenic diabetes insipidus,** ADH is present, but the collecting ducts are unresponsive, usually due to a deficiency of ADH receptors or water channels.

An adequate delivery of salt to the ascending limb is necessary for maximal urine concentration. If glomerular filtration rate is abnormally low, such as following severe hemorrhage, then concentrating ability is impaired.

Not only must delivery of salt be adequate, but also the rate of reabsorption of salt along the ascending limb must be maintained. After all, this is the single effect essential for countercurrent multiplication. The "loop" diuretic drugs inhibit salt reabsorption in the thick ascending limb; they decrease concentrating ability and cause a marked increase in salt and water excretion.

The urine can be maximally concentrated only if fluid flow through the collecting duct system is low. If fluid reabsorption is impaired in early portions of the nephron, as occurs during an osmotic diuresis, then a maximal urine osmolality cannot be achieved.

Medullary blood flow can affect the osmotic gradient in the medulla and thus affect concentrating ability as well. An excessive blood flow to this region tends to wash out the gradient and thereby reduces the maximum urine osmolality that can be attained.

Urea plays a crucial role in the concentrating mechanism, especially in the establishment of an osmotic gradient in the inner medulla. With a low-protein diet and the resulting decrease in urea production, concentrating ability is impaired.

Finally, the length of Henle's loops affects concentrating ability. Some desert-living mammals have very long loops for their body sizes and a correspondingly enhanced ability to concentrate the urine and save water.

Excretion of an Osmotically Dilute Urine Rids the Body of Excess Water

In principle, the production of osmotically dilute urine is quite straightforward. All that is needed is the reabsorption of solute across a relatively water-impermeable membrane, with water being left behind. In the thick ascending limb, active reabsorption of sodium (accompanied by chloride) results in considerable dilution of the tubular fluid, as we have already seen. This segment is often called the **diluting segment.** In the absence of ADH, the collecting ducts are very water-impermeable. Even though the medullary interstitium is hyperosmotic (but not as much as in a concentrating kidney), collecting duct water mostly stays in the duct lumen. Continued reabsorption of salt results in further dilution of the urine along the collecting ducts. The end result is the excretion of a large volume of hypoosmotic urine.

FUNCTIONS OF URETERS AND URINARY BLADDER

> **What happens to urine after it leaves the collecting ducts?**

The kidneys form urine constantly. Urine is conveyed to the urinary bladder by the ureters and stored there until the bladder is emptied by way of the urethra.

The Ureters Transmit the Urine to the Urinary Bladder

The two ureters are muscular tubes that carry the urine from the pelvis of each kidney to the urinary bladder. Peristaltic movements originate in the region of the calyces, which contain specialized smooth muscle pacemaker cells. Waves of electrical activity (action potentials) and contractions pass over the pelvis and down the ureter. The muscular contractions force the urine toward the bladder and help overcome the gradually increasing pressure as urine accumulates in the bladder. The ureters enter the base of the bladder obliquely, thus forming a valvular flap that passively prevents reflux of urine during bladder contraction. The ureters are innervated by sympathetic and parasympathetic nerve fibers. Afferent sensory nerve fibers are present, as evidenced by the excruciating pain felt when a stone is in the ureter.

The Urinary Bladder Stores the Urine and Is Periodically Emptied

The urinary bladder is a distensible, hollow organ containing smooth muscle in its wall (see Fig. 10–10). The muscle is called the **detrusor,** which means in Latin "that which pushes down." The neck of the bladder, or internal sphincter, also contains smooth muscle. The body of the bladder and bladder neck are innervated by parasympathetic **pelvic nerves** and sympathetic **hypogastric nerves.** The external sphincter is composed of skeletal muscle and is innervated via somatic nerves, the **pudendal nerves.** Pelvic, hypogastric, and pudendal nerves contain both motor and sensory fibers.

The bladder functions in two ways: (1) it serves as a distensible reservoir for storage of the urine and (2) it empties its contents at suitable intervals. As the bladder fills, the wall tension is adjusted to its volume, so that minimal increases in bladder pressure occur. The external sphincter is kept closed via impulses along the pudendal nerves. The first awareness of bladder filling occurs at a volume of approximately 100 to 150 ml, and the desire to void is usually experienced when the bladder contains 150 to 250 ml of urine. A person becomes uncomfortably aware of a full bladder when the volume is 350 to 400 ml; at this volume, pressure in the bladder is approximately 10 cm of water. With further increases in volume, bladder pressure increases steeply, in part due to reflex contractions of the detrusor. An increase in volume to 700 ml creates pain and often loss of control. The sensations of bladder filling, of conscious desire to void, and of painful distension are mediated by afferent fibers in the pelvic nerves.

Micturition Is a Complex Act Controlled by Autonomic and Somatic Nerves

The periodic complete emptying of the urinary bladder is called **micturition,** or **urination.** It involves coordinated activity over both autonomic and somatic nerve pathways and several reflexes that can either be facilitated or inhibited by higher centers in the brain. The basic reflexes occur at the level of the sacral spinal cord and are modified by centers in the midbrain and cerebral cortex. Distension is sensed by stretch receptors in the bladder wall; these induce reflex contraction of the bladder and relaxation of internal and external sphincters. This reflex is released by removing inhibitory impulses from the cerebral cortex. Fluid flow through the urethra reflexly causes further contraction of the detrusor and relaxation of the external sphincter. Increased parasympathetic nerve activity is responsible for contraction of the detrusor and relaxation of the internal sphincter. The perineal and levator ani muscles relax during micturition; this shortens the urethra and decreases its resistance. Descent of the diaphragm and contraction of abdominal muscles increase intra-abdominal pressure, thus aiding in the expulsion of urine from the bladder.

Fortunately, micturition is under voluntary control. In young children, however, it is purely reflex and occurs whenever the bladder is sufficiently distended. At approximately 2.5 years of age, it begins to come under cortical control, and complete control is usually achieved by 3 years of age.

CHAPTER REVIEW

Summary

- The kidneys play a dominant role in regulating the volume and composition of the extracellular fluid.
- The basic unit of kidney structure and function is the nephron.
- The juxtaglomerular apparatus consists of specialized cells thought to play a role in regulating glomerular filtration rate and renin release.
- The kidneys have a huge blood flow because of the need to sustain a high rate of plasma filtration.
- Urine formation starts with the filtration of plasma in the kidney glomeruli. Glomerular filtration is favored by the high hydrostatic pressure of the blood in the glomerular capillaries and is opposed by the hydrostatic pressure in the urinary space of Bowman's capsule and by the glomerular capillary colloid osmotic pressure. Glomerular filtration is rather nonselective; proteins are mostly retained in the plasma by the glomerular filtration barrier, but all low-molecular-weight substances (provided they are not bound to plasma proteins) are freely filtered. Glomerular filtration rate is most accurately measured by determining the inulin clearance.
- Glucose is reabsorbed by the proximal tubule. Reabsorption is active and sodium-dependent and shows a maximal rate (Tm). Normally, all of the filtered glucose is reabsorbed. Urea is filtered by the kidney glomeruli and is passively reabsorbed by the kidney tubules. Active reabsorption of sodium provides the driving force for tubular reabsorption of water, glucose, amino acids, chloride, and phosphate.
- Many medically important organic anions and cations are secreted by the kidney proximal tubules.
- Most of the filtered sodium is reabsorbed by the kidney tubules; the quantity of sodium excreted plays an important role in body sodium balance. The proximal convoluted tubule reabsorbs the greatest percentage (70%) of the filtered sodium and water; the tubule fluid stays essentially iso-osmotic to plasma in this nephron segment. The loop of Henle reabsorbs approximately 20% of the filtered sodium and 10% of the filtered water. The distal convoluted tubule and collecting ducts reabsorb approximately 9% of the filtered sodium and 19% of the filtered water. The collecting ducts are the site of final regulation of sodium and water excretion; aldosterone and ADH increase sodium and water reabsorption, respectively, by the collecting ducts.
- Potassium is filtered, reabsorbed, and secreted by the kidney tubules. The cortical collecting duct normally secretes potassium and is an important site that determines potassium excretion.
- The kidneys play a major role in regulating the concentrations of calcium, magnesium, and inorganic phosphate in the plasma.
- The mammalian kidney can form urine that is osmotically more dilute or concentrated than plasma. The formation of osmotically concentrated urine depends on the establishment of an osmotic gradient in the medulla by the loops of Henle. These structures are countercurrent multipliers. The vasa recta are countercurrent exchangers and remove water from the kidney medulla. The addition of urea to the inner medulla allows for efficient operation of the urinary concentrating mechanism.
- Factors affecting urinary concentrating ability are ADH, salt reabsorption by the ascending limb of Henle's loop, tubule fluid and blood flows, urea, and the length of Henle's loops.
- Production of osmotically dilute urine involves the reabsorption of solute across a membrane, with water left behind.
- The two ureters are muscular tubes that carry the urine from the kidneys to the bladder. The urinary bladder functions as a reservoir for urine and is periodically emptied (micturition).
- Micturition is a complex act involving autonomic and somatic nerves, spinal reflexes, and higher brain centers.

Review Questions

Choose the Correct Answer

1. Which of the following produces renin?
   a. Granular cells
   b. Intercalated cells
   c. Macula densa
   d. Podocytes

2. Which of the following produces an increase in renal blood flow?
   a. Angiotensin II
   b. Norepinephrine
   c. Prostaglandin I_2
   d. Thromboxane

3. Approximately what percentage of filtered water is usually reabsorbed by the kidney tubules?
   a. 1%
   b. 20%
   c. 70%
   d. 99%

4. Which nephron segment secretes PAH and penicillin?
   a. Collecting duct
   b. Distal convoluted tubule
   c. Proximal tubule
   d. Thick ascending limb
   e. All of the above

5. Which of the following substances normally has the highest renal plasma clearance?
   a. Glucose
   b. Inulin
   c. PAH
   d. Urea

6. Active reabsorption of glucose across the luminal membrane of proximal tubule cells is accomplished by:
   a. A glucose pump
   b. Facilitated diffusion
   c. Glucose-sodium cotransport
   d. Simple diffusion

7. The most accurate determination of glomerular filtration rate is obtained from measurement of:
   a. blood urea nitrogen (BUN) concentration.
   b. endogenous creatinine clearance.
   c. inulin clearance.
   d. PAH clearance.
   e. plasma creatinine concentration.

8. Dilation of efferent arterioles in the kidneys results in:
   a. an increase in glomerular blood flow.
   b. an increase in glomerular capillary hydrostatic pressure.
   c. an increase in glomerular filtration rate.
   d. All of the above
   e. None of the above

9. Antidiuretic hormone (ADH) increases the epithelial water permeability of the:
   a. proximal tubule.
   b. thin descending limb.
   c. thick ascending limb.
   d. distal convoluted tubule.
   e. collecting duct.

10. Approximately 70% of filtered sodium is reabsorbed in the:
    a. collecting duct.
    b. distal convoluted tubule.
    c. loop of Henle.
    d. proximal convoluted tubule.

11. Which of the following is associated with an increased ability to concentrate the urine osmotically?
    a. Administration of a "loop" diuretic drug
    b. A low-protein diet
    c. An increased medullary blood flow
    d. Long loops of Henle

12. If the glomerular filtration rate is 125 ml/min, glomerular capillary hydrostatic pressure is 55 mm Hg, hydrostatic pressure in Bowman's space is 15 mm Hg, and mean glomerular capillary colloid osmotic pressure is 30 mm Hg, what is the glomerular filtration coefficient?
    a. 0.08 mm Hg per ml/min
    b. 1.25 ml/min per mm Hg
    c. 3.13 ml/min per mm Hg
    d. 12.5 ml/min per mm Hg

13. A renal clearance study was done on a normal, 70-kg man. Plasma PAH concentration was 0.02 mg/ml, urine PAH concentration was 2.64 mg/ml, urine flow rate was 5.00 ml/min, and blood hematocrit ratio was 0.45. If we assume that PAH was essentially cleared from all of the plasma flowing through the kidneys (i.e., 100% extraction), what is his renal blood flow?
    a. 297 ml/min
    b. 363 ml/min
    c. 660 ml/min
    d. 1200 ml/min
    e. 1467 ml/min

14. In a renal clearance study on a young woman, the plasma glucose concentration was increased to 5 mg/ml by intravenous infusion of a concentrated glucose solution. If the glomerular filtration rate is 110 ml/min, urine glucose concentration is 25 mg/ml, and urine flow rate is 10 ml/min, what is the rate of tubular glucose reabsorption?
    a. 50 mg/min
    b. 300 mg/min
    c. 360 mg/min
    d. 660 mg/min

15. What percentage of the cardiac output goes directly to the kidneys in a normal resting subject?
    a. 5%
    b. 10%
    c. 25%
    d. 50%

16. In a kidney producing a maximally concentrated urine, the osmolality of tubule fluid at the beginning of the distal convoluted tubule is approximately:
    a. 100 mosm/kg H_2O.
    b. 300 mosm/kg H_2O.
    c. 600 mosm/kg H_2O.
    d. 900 mosm/kg H_2O.
    e. 1200 mosm/kg H_2O.

17. A decrease in glomerular filtration rate results in an increased plasma concentration of:
    a. glucose.
    b. sodium.
    c. creatinine.

d. serum albumin.

e. All of the above

18. Most of the energy consumed by the kidneys powers:

a. reabsorption of filtered sodium.

b. reabsorption of filtered glucose, amino acids, and other important metabolites.

c. glomerular filtration.

d. secretion of hydrogen ions.

e. secretion of organic solutes.

19. Which segment has the leakiest tight junctions?

a. Proximal convoluted tubule

b. Thick ascending limb

c. Distal convoluted tubule

d. Collecting duct

20. Loss of negatively charged macromolecules from the glomerular filtration barrier results in increased urinary excretion of:

a. chloride.

b. glucose.

c. inulin.

d. serum albumin.

e. water.

Answers to Case History Questions

1. The presence of red cell casts in the urine, an abnormal finding, indicates that bleeding occurred in the kidneys. A red cell cast is a mold of the kidney tubule, typically cylindrical in form, which results when red cells and precipitated proteins clump together in the tubule lumen. Red cell casts usually result from glomerular bleeding. The presence of red cells in the urine and a dark (smoky) color (due to oxidized hemoglobin) are signs of bleeding but do not necessarily indicate that bleeding occurred in the kidneys because, in some circumstances, blood may come from the lower urinary tract (i.e., the ureters, urinary bladder, and urethra).

2. The glomerular filtration rate can be calculated from the endogenous creatinine clearance or the formula $C_{CR} = U_{CR}\dot{V}/P_{CR}$. In this case, C_{CR} = 1300 mg/1440 min ÷ 0.02 mg/ml = 45 ml/min. Creatinine clearance is usually normalized to a standard body surface area (BSA) of 1.73 m^2. The patient had a body surface area of 1.26 m^2, hence GFR = 62 ml/min per 1.73 m^2 BSA,

which is below the normal range (95–155 ml/min per 1.73 m^2 BSA).

3. The clearances of both creatinine and urea depend on filtration of the plasma in the kidney glomeruli. When glomerular filtration rate decreases, these substances accumulate in the plasma.

4. Immunologic damage to the glomeruli results in abnormal leakiness of the glomerular filtration barrier to plasma proteins. The proteins are incompletely reabsorbed by proximal tubules and appear in the urine (proteinuria). Urine protein excretion in this patient was abnormally high (normal, < 150 mg/day).

5. Decreased renal function may lead to abnormal retention of salt and water. This leads to accumulation of fluid in the interstitial fluid spaces (edema), as evidenced by the facial puffiness, abdominal distension, and ankle swelling. The salt and water retention, and release of renin by the injured kidneys, may contribute to the elevated blood pressure.

Key Terms

apical (luminal) cell membrane (p. 731)

basolateral cell membrane (p. 731)

cortical nephron (p. 729)

countercurrent exchange (p. 750)

countercurrent multiplication (p. 750)

diabetes insipidus (p. 747)

endogenous creatinine clearance (p. 740)

excretion (p. 734)

glomerular filtration (p. 734)

glomerular filtration coefficient (p. 738)

glucose threshold (p. 741)

glucosuria (p. 742)

juxtaglomerular apparatus (p. 732)

juxtamedullary nephron (p. 729)

micturition (p. 753)

nephron (p. 729)

peritubular capillary Starling forces (p. 746)

proteinuria (p. 743)

renal autoregulation (p. 732)

renal plasma clearance (p. 739)

tubular reabsorption (p. 734)

tubular secretion (p. 734)

tubular transport maximum (p. 742)

tubuloglomerular feedback (p. 732)

ultrafiltration (p. 734)

vasa recta (p. 731)

Suggested Readings

Agre, P. "Aquaporin water channels in kidney." *Journal of the American Society of Nephrology,* 11:764–777, 2000.

Brooks, V. L., and Vander, A. J. "Refresher course for teaching renal physiology." *Advances in Physiology Education,* 20(suppl): 114–245, 1998.

Koeppen, B. M., and Stanton, B. A. *Renal Physiology,* ed 3. St. Louis, Mosby, 2001.
Kriz, W., and Bankir, L. "A standard nomenclature for structures of the kidney." *American Journal of Physiology,* 254:F1-F8, 1988.

Rose, B. D., and Post, T. W. *Clinical Physiology of Acid-Base and Electrolyte Disorders,* ed 5. New York, McGraw-Hill, 2001.

Scheinman, S. J., Guay-Woodford, L. M., Thakker, R. J., and Warnock, D. G. "Genetic disorders of renal electrolyte transport." *New England Journal of Medicine,* 340:1177–1187, 1999.

Seldin, D. W., and Giebisch, G. *The Kidney: Physiology and Pathophysiology,* ed 3. Philadelphia, Lippincott Williams & Wilkins, 2000.
Smith, H. W. *From Fish to Philosopher.* Boston, Little, Brown and Co., 1953.

Valtin, H., and Schafer, J. A. *Renal Function,* ed 3. Boston, Little, Brown and Co., 1995.
Vander, A. J. *Renal Physiology,* ed 5. New York, McGraw-Hill, 1995.

Answers to Review Questions

1. a **2.** c **3.** d **4.** c **5.** c **6.** c **7.** c **8.** a
9. e **10.** d **11.** d **12.** d **13.** d **14.** b **15.** c
16. a **17.** c **18.** a **19.** a **20.** d

Chapter 24

FLUID AND ELECTROLYTE BALANCE

KEY CONCEPTS

- Approximately two thirds of the body water is in intracellular fluid and one third is in extracellular fluid; intracellular and extracellular fluids are in osmotic equilibrium.

- The kidneys help to keep us in water, sodium ion, and potassium ion balance, by excreting the appropriate amounts of these substances in the urine.

- Renal failure results in an abnormal internal environment. Dialysis and renal transplantation are used to treat patients with terminal (end-stage) renal disease.

CASE HISTORY

A 45-year-old marathon runner became disoriented and confused 30 minutes after completing a 50-mile (80-km) ultramarathon race in 10 hours and 36 minutes. The race took place in Chicago, and the weather was unseasonably hot for an October day, with a peak temperature of 32°C (89°F). There was a brisk wind on the lakefront. During the race, the man drank water or electrolyte-replacement glucose solution (14 mmol NaCl/L, 11 mmol KCl/L, 5% glucose, osmolality 344 mosm/kg H_2O) alternately at each aid station, at 1-mile intervals, for the first 30 miles. After this, he drank only water at each station plus a minimal amount of cola. It was estimated that he consumed approximately 24 liters of fluid during the race and 110 mEq of sodium.

The man was rushed to a hospital emergency room. He was described as conscious but restless, with slurred speech, unpurposeful movements, and hyperventilation. Blood pressure was 130/80 mm Hg; respirations, 28 breaths/min; heart rate, 100 beats per min (bpm); and rectal temperature, 38°C. He did not appear to be dehydrated. Blood plasma values on admission were: sodium, 118 mEq/L (normal, 135–148 mEq/L); potassium, 3.8 mEq/L (normal, 3.5–5.0 mEq/L); BUN, 10 mg/dl (normal, 8–15 mg/dl); glucose, 153 mg/dl (normal fasting, 70–110 mg/dl); and osmolality, 248 mosm/kg H_2O (normal, 281–297 mosm/kg H_2O). A slow intravenous infusion of 3% (513 mmol/L) NaCl solution was started. His urine output was extremely high (4.1 liters in 8 hours), and the urine was very dilute. Three hours after admission, he was noted to be fully alert and oriented. He was discharged after 8 hours in the emergency room. He has returned to his usual running but has not run in additional ultramarathons. (*Case reported in JAMA 255:772–774, 1986; used with permission.*)

Questions

1. Why is the electrolyte-replacement glucose solution a hypo-osmotic solution in the body even though it has an osmolality higher than that of plasma?

2. What would be the effect of sweating on plasma volume and osmolality if no fluid loss is replaced by drinking?

3. What is the likely explanation for the low plasma sodium (hyponatremia) in this patient?

4. What is a possible reason for the neurological symptoms (e.g., disorientation, slurred speech, and abnormal movements)?

5. Why was the intravenous hypertonic sodium chloride solution given slowly?

INTRODUCTION

In Chapter 23, we discussed the processes involved in urine formation. In this chapter, we first consider the volume and composition of the various body fluids. Then we discuss how we keep in water, sodium ion, and potassium ion balance. The kidneys clearly play a key role in keeping us in balance for these substances. We will indicate some of the body disturbances that occur when our kidneys fail and discuss some of the methods used for treating patients with renal failure.

BODY FLUIDS AND COMPARTMENTS

 How is water distributed within the body?

The fluids of the body can be divided into two categories: (1) **intracellular fluid** (fluid within cells) and (2) **extracellular fluid** (fluid outside of cells). These fluids differ strikingly in composition and in volume. Because most cell membranes are water permeable, however, the total solute concentration inside and outside of our cells is the same.

Body Water Is Distributed in Several Fluid Compartments

The body fluids contain both solutes and water, but in terms of amount or volume occupied, water is the major constituent. Therefore, we will consider the water content of the body first and later cover the various solutes dissolved in the body fluids.

The percentage of total body weight that is water varies from 45% to 75% in different individuals. This range mostly reflects differences in the amount of body fat. The water content of adipose (fat) tissue is approximately 10% by weight; most other tissues contain 70% to 75% water by weight. An obese individual has a low percentage of body weight made up of water, and a lean individual has a high percentage. Young men average approximately 60% water by weight; young women average approximately 50% water. The difference is due to females' smaller muscle mass and greater amount of subcutaneous fat. With aging, the percentage of body weight that is water decreases because water-rich muscle tissue tends to be replaced by water-poor adipose tissue.

Total body water is distributed in several divisions, or compartments (Fig. 24–1). Approximately two thirds of body water is contained in the **intracellular fluid compartment,** and one third, in the **extracellular fluid compartment.**

The intracellular fluid compartment is not one continuous space but is made up of trillions of cells. The intracellular space is separated from the extracellular space by cell plasma membranes. If we assume an idealized, young adult, 70-kg male, in whom total body water is 60% of body weight, we can calculate that total body water is 42 kg (or 42 liters), intracellular water is 28 liters, and extracellular water is 14 liters.

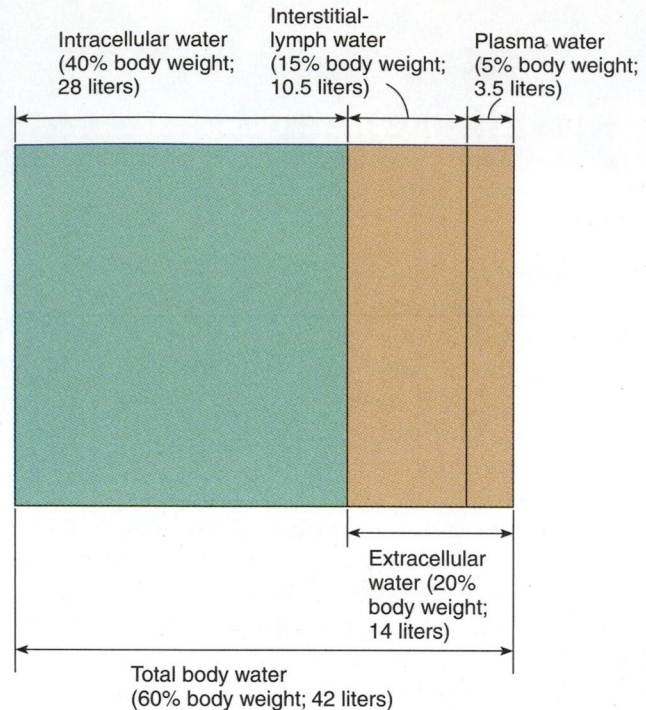

Figure 24–1

Distribution of water in the body of an average, young, 70-kg man. In an average, young, 55-kg woman, total body water is 50% of body weight, intracellular water is 30% of body weight, and extracellular water is 20% of body weight.

The extracellular fluid can be further divided into two major subcompartments, separated from each other by the endothelium of the blood vessels. Blood vessels contain **blood plasma,** the part of the blood exclusive of blood cells and platelets. The plasma is approximately 93% water (by weight or volume). Plasma water accounts for approximately one quarter of the extracellular water (3.5 liters in a 70-kg, young man). Outside of the blood vessels are the **interstitial fluid** and **lymph.** Interstitial fluid is the fluid directly bathing most body cells, and lymph is the fluid in lymphatic vessels. The interstitial fluid and lymph together contain three quarters of the extracellular water (10.5 liters in a 70-kg man). Blood plasma, interstitial fluid, and lymph are nearly identical in chemical composition, except for a higher protein concentration in plasma.

Figure 24–1 does not show an additional, small, extracellular fluid compartment, the **transcellular fluid.** This fluid compartment includes specialized fluids, such as **cerebrospinal fluid,** the **aqueous humor** of the eye, secretions of the digestive glands, sweat, synovial fluid, and renal tubular fluid and urine. These fluids are separated from the plasma by an endothelium as well as by a continuous epithelial cell layer. This epithelium modifies the chemical composition of the transcellular fluid, so it is not a simple ultrafiltrate of plasma, as is interstitial fluid or lymph. Transcellular fluid amounts to only 1% to 3% of body weight but is extremely important, physiologically speaking. The transcellular fluids, in general, are continuously formed, and abnormal loss of these fluids or blockage of fluid drainage may have serious consequences for the organism.

Body Fluids Differ in Electrolyte Composition

 How are electrolytes distributed within the body?

The body fluids contain many uncharged organic substances (e.g., glucose and urea), but in terms of concentration, electrolytes are quantitatively more important. An **electrolyte** is a substance (e.g., NaCl) that dissociates into ions in solution and thus becomes capable of conducting electricity. Electrolytes contribute most to the total solute concentration (osmolality) of the body fluids and consequently are crucial in the distribution of body water.

The concentrations of various electrolytes in plasma, interstitial fluid, and intracellular fluid are summarized in Table 24–1. The intracellular fluid values are based on measurements in skeletal muscle cells. These cells were chosen as a representative cell type because they account for approximately two thirds of the cell mass in the human body. Concentrations are expressed in terms of milliequivalents per liter of solution or per kilogram of H_2O. An **equivalent** is equal to the product of the moles and valence of an ion; one **milliequivalent (mEq)** is equal to 1/1000 Eq. For singly charged (**univalent**) ions, 1 mEq is the same as 1 millimole (mmole). Thus, we can express plasma sodium concentration as 142 mEq/L or 142 mmol/L. For doubly charged (**divalent**) ions, 2 mEq is the same as 1 mmole. For example, a

plasma Ca^{2+} concentration of 5 mEq/L is the same as 2.5 mmol/L. Some electrolytes, such as proteins, are **polyvalent,** so there are several milliequivalents per millimole. The virtue of expressing concentration in terms of mEq/L is that in any fluid compartment, the sum of the positive ions (cations) must equal the sum of the negative ions (anions) because of the requirement of electroneutrality in solutions. This allows us to calculate the concentration of unmeasured anions ("Others") in Table 24–1.

Plasma concentrations are listed in the first column of Table 24–1. Sodium is the major cation in plasma, and chloride and bicarbonate are the major anions. The plasma proteins (mostly serum albumin) bear a net negative charge at physiological pH. The electrolytes are actually dissolved in the watery phase of plasma, so concentrations in plasma water are presented in the second column of Table 24–1. These values were calculated by assuming a plasma water content of 93%. For example, for Na^+, 142 mEq/L plasma divided by 0.93 liter of water per 1 liter of plasma equals 153 mEq/L water. Because 1 liter of water weighs 1 kg, the concentrations could also be expressed per kilogram of water.

The interstitial fluid (third column of Table 24–1) is an ultrafiltrate of plasma. It contains essentially the same concentrations as plasma of all low-molecular-weight substances, but little protein. The differences in concentrations between the small ions in plasma water and interstitial fluid (compare columns 2 and 3 in Table 24–1) arise because of differences in protein concentration between these two fluids. Two factors are involved. The first is an electrostatic effect: because the large plasma proteins are negatively charged and kept within the circulation, there is a lesser concentration of small diffusible cations and a higher concentration of small

TABLE 24–1

Electrolyte Composition of the Body Fluids

Electrolytes	(1) Plasma (mEq/L)	(2) Plasma Water (mEq/kg H_2O)	(3) Interstitial Fluid (mEq/kg H_2O)	(4) Intracellular Fluid (Skeletal Muscle) (mEq/kg H_2O)
Cations				
Na^+	142	153	145	10
K^+	4	4.3	4	159
Ca^{2+}	5	5.4	3	1
Mg^{2+}	2	2.2	2	40
Total	153	165	154	210
Anions				
Cl^-	103	111	117	3
HCO_3^-	25	27	28	7
Protein	17	18	—	45
Others	8	9	9	155
Total	153	165	154	210

diffusible anions in a plasma ultrafiltrate. Second, Ca^{2+} and Mg^{2+} are also bound to some extent (approximately 40% and 30%, respectively) to plasma proteins; consequently, the total concentrations of these ions in interstitial fluid are less than in plasma.

If we examine intracellular fluid composition (column 4 of Table 24–1), major differences between this fluid and the extracellular fluids are apparent. The concentrations of potassium, magnesium, and protein in the cell are much higher than in the surrounding interstitial fluid. The concentrations of sodium, calcium, chloride, and bicarbonate are much less in the cell. The anions in muscle cells labeled "Others" are mainly organic phosphate compounds (e.g., creatine phosphate and ATP) and various other anions to which the cell membrane is mostly impermeable.

The high internal potassium concentration and low internal sodium concentration are maintained by activity of the cell membrane Na^+/K^+-ATPase, which extrudes sodium ions and takes up potassium ions (Chapter 4). The lower intracellular concentrations of chloride and bicarbonate can be mostly accounted for by the membrane potential difference, approximately –90 mV, which favors the outward movement of these negatively charged ions.

The distribution of ions within the cell is not uniform; different organelles may have different ion concentrations. For example, the cell nucleus has a higher sodium concentration than does the cell as a whole. In skeletal muscle, most of the intracellular calcium is in the sarcoplasmic reticulum and mitochondria, and the cytoplasmic calcium concentration is very low.

Do not be misled by the higher number of milliequivalents per kilogram of H_2O in the cell than in the interstitial fluid (compare totals in columns 3 and 4 of Table 24–1). A single protein molecule or phosphate compound may bear several negative charges, so the number of milliequivalents in the cell is greater than the number of milliosmoles. Also, magnesium is divalent (40 mEq/kg H_2O corresponds to 20 mmol/kg H_2O) and is largely protein-bound within the cell, so it is not effective osmotically. It is generally accepted that the total solute concentrations (osmolalities) in the cell and surrounding extracellular fluid are identical. Water moves freely through most plasma membranes, so there is a condition of osmotic equilibrium between cellular and extracellular fluids.

The Osmolalities of Intracellular and Extracellular Fluids Are Normally Identical

Osmotic pressure is of prime importance in determining the distribution of water between cells and extracellular fluid. Osmotic pressure is directly related to total solute concentration (osmolality). An increased solute concentration decreases the activity (or free energy) of water; water moves spontaneously from a region of lower osmolality (higher free energy) to one of higher osmolality (lower free energy). This water movement could be prevented by applying a hydrosta-

tic pressure to the more concentrated solution. Animal cells cannot employ such a mechanism because a hydrostatic pressure difference of a few centimeters of H_2O leads to rupture of the plasma membrane. In multicellular animals, osmotic pressures inside and outside of cells are equal. If solute or water is added to or lost from the extracellular fluid, osmotic equilibrium will be temporarily upset. Water will move into or out of the cells until a new osmotic equilibrium is achieved.

Most of the volume in any body fluid compartment is occupied by water. At a given osmolality, the total volume of (or amount of water in) a compartment is directly related to the amount of solute in the compartment. This important relationship follows from the equation that defines the term *concentration* — namely, concentration = amount/volume. If solute concentration (osmolality) is maintained constant, then the volume must change directly with the amount of solute.

In the extracellular fluid, the amount of sodium present is a major determinant of the volume of water present. This relationship results because sodium and its accompanying anions, chloride and bicarbonate, are the major osmotically active solutes present in extracellular fluid. If we add up the concentrations of these ions in interstitial fluid (see Table 24–1), we get 145 + 117 + 28 = 290 mmol/kg H_2O. The corresponding osmolality is approximately 270 mosm/kg H_2O. (This figure is less than 290 because of interactions among ions in solution; it is only in infinitely dilute solutions that ions behave as completely independent particles.) Because interstitial fluid (or plasma) osmolality averages 287 mosm/kg H_2O ($\approx$ 300 mosm/kg H_2O), it is apparent that sodium salts account for more than 90% of the osmolality. The osmolality of the extracellular fluid is closely regulated by the thirst sensation and control of renal excretion of water by ADH. Consequently, if osmolality (sodium concentration) is kept constant, it follows that the amount of extracellular sodium will determine the amount of extracellular water. To express it another way, if a person takes in extra salt (NaCl), thirst results, water is ingested, and extracellular fluid volume is increased. If a person loses salt, extracellular fluid volume is decreased.

Cell volume, also mostly water, is similarly influenced by the amount of contained solute. In the cell compartment, potassium is the major osmotically active solute. Hence, a loss of potassium from the cell results in a decrease in cell volume, and a gain in potassium results in an increase in cell volume. The large amount of impermeable anions (e.g., proteins and organic phosphates) within cells could affect cell volume. In theory, these solutes would cause a redistribution of small, permeable ions so that an excess of solute would be present in the cell, and, therefore, water would tend to enter the cell. This tendency is, however, counteracted by the activity of the Na^+/K^+-ATPase, which effectively excludes sodium and thereby decreases the total amount of solute in the cell. In this way, the Na^+/K^+-ATPase plays a key role in the regulation of cell volume. If Na^+/K^+-ATPase activity is decreased (e.g., by low temperature, oxygen lack, metabolic poisons, or cardiac glycosides), cells gain sodium and water, and they swell.

The normal distribution of water between cellular and extracellular compartments changes in a variety of circumstances. Figure 24–2 provides some examples. Note that the height of the boxes corresponds to solute concentration. The width of the boxes corresponds to volume. The area of each box (height × width) corresponds to the total amount of solute in a compartment (concentration × volume = amount). The dotted lines represent the normal condition and are the same in all four parts of this figure. The solid lines represent conditions after a new osmotic equilibrium has been achieved. Note that the height of the boxes is always identical in intracellular and extracellular fluids because osmotic equilibrium is attained.

Figure 24–2*a* shows the normal condition. Osmolalities are identical in cellular and extracellular compartments (300 mosm/kg H_2O). The volume of the intracellular compartment (40% of body weight) is twice the volume of the extracellular compartment (20% of body weight).

In Figure 24–2*b*, pure water is added to the extracellular fluid; ingestion of water would have the same effect. Because the extracellular fluid osmolality is decreased, water moves into the cells until the osmolality in both compartments is decreased to the same level. The entry of water into the cells increases their volume. Because the cells contain two thirds of

the body fluid solutes, two thirds of the added water enters the cells, and one third remains in the extracellular space.

In Figure 24–2*c*, isotonic saline is added to the extracellular fluid. This can be done by intravenous infusion or ingestion of a 0.9% NaCl solution. Note that the final osmolality is not different from normal. This is not surprising because isotonic saline is also iso-osmotic. Note also that cell volume does not change because, by definition, cells do not shrink or swell when exposed to an isotonic solution. All of the isotonic saline is retained in the extracellular fluid and accordingly increases only this volume.

Although it is not illustrated, a loss of isotonic fluid from the extracellular compartment has predictable consequences: no change in body fluid osmolality or in cell volume and a decrease in extracellular fluid volume. Hemorrhage and diarrhea are two cases in which an isotonic fluid is lost from the body.

Figure 24–2*d* shows the effect of adding sodium chloride without water to the extracellular compartment. Extra salt intake (e.g., eating salted potato chips) would be a corresponding circumstance. The added salt increases the osmolality of the extracellular fluid, which results in water movement out of the cells. Extracellular fluid volume is increased at the expense of cell volume. At equilibrium, osmolality is uniformly increased. When pure NaCl (or a hypertonic NaCl solution) is added to the extracellular fluid, the NaCl is diluted by the total body water, even though all of the NaCl stays in the extracellular fluid.

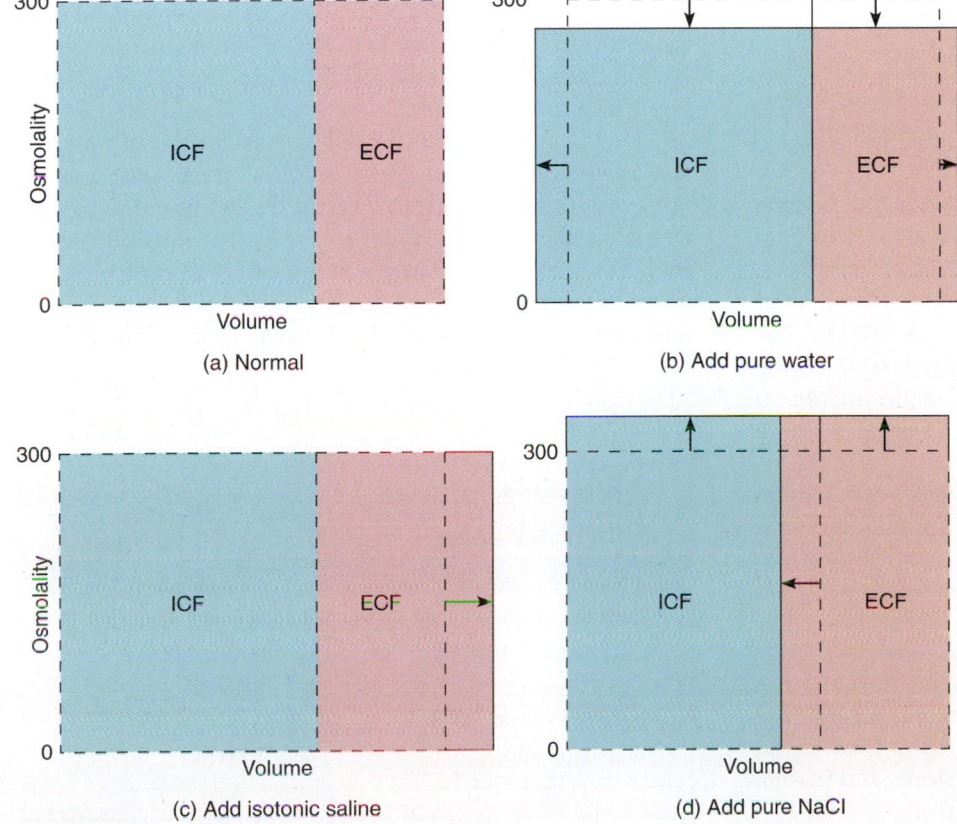

Figure 24–2

The effects on osmolalities and volumes of the intracellular fluid (ICF) and extracellular fluid (ECF) of adding pure water *(b)*, isotonic saline *(c)*, and pure NaCl *(d)* to the extracellular fluid. *(a)* The normal condition.

The Distribution of Water Between Blood Plasma and Interstitial Fluid Is Determined by Starling Forces

The relative volumes of plasma and interstitial fluid are primarily determined by the Starling forces that operate across capillary walls. As discussed in Chapter 19, movement of fluid across capillary walls is determined by the balance between capillary and tissue hydrostatic and colloid osmotic pressures. The hydrostatic pressure in the capillaries tends to push fluid out of the capillaries; the colloid osmotic pressure due to the plasma proteins acts to retain fluid within the vascular system. An increase in capillary hydrostatic pressure, a decrease in plasma colloid osmotic pressure, or an increase in capillary membrane permeability to proteins would favor a net movement of fluid out of the vascular system and result in a decreased plasma volume and an increased interstitial fluid volume (edema). On the other hand, a decrease in capillary hydrostatic pressure or an increase in plasma colloid osmotic pressure would favor movement of interstitial fluid into the circulation. In most vascular beds, the net force favoring outward movement of the fluid at the arterial end of the capillaries is roughly the same as the net force favoring inward movement of fluid at the venous end of the capillaries. Therefore, little fluid leaves the circulation. Extra fluid that has left the circulation is returned by the lymphatics. Inadequate lymphatic drainage can lead to an appreciable increase in the size of the interstitial space.

THE CONCEPT OF FLUID AND ELECTROLYTE BALANCE

 How are fluid and electrolyte balance maintained?

Despite varying daily intake of food and water, the volume and composition of body fluids remain constant. This suggests that we stay in balance with respect to many substances. A person in a **stable balance** maintains the same amount of a specified substance in the body over a period of time. The rates of gain or input (due to intake or production in the body) and loss or output (due to excretion or destruction in the body) of a particular substance are exactly equal, so there is no net accumulation or loss.

If fluids and electrolytes are not in balance, two alternative conditions occur. If input exceeds output, then accumulation of a substance in the body results, and a **positive balance** exists. For example, if the intake of salt exceeds the rate at which it is excreted, then salt and water will accumulate in the body and extracellular fluid volume and body weight will increase. If output exceeds input, then the net loss of a substance from the body results, and a **negative balance** exists. For example, if urinary excretion exceeds the daily intake of potassium, body potassium stores and plasma potassium concentration decrease.

The kidneys play a major role in maintaining balance with respect to water, mineral electrolytes, hydrogen ions, and various organic compounds. They accomplish this by adjusting the output of these substances in the urine to match the rate of addition to the body. Renal excretion of water is under the continuous control of the antidiuretic hormone and is normally adjusted to maintain water balance. The kidneys are the primary site of control of electrolyte output. The mineral electrolytes, such as sodium, potassium, and phosphate, are ingested in variable amounts. They are not produced or destroyed in the body, and so, in a balanced system, the amounts taken in are excreted, mostly in the urine. Some substances, such as hydrogen ions or urea, are produced by metabolic reactions in the body. Because these substances do not normally accumulate, they are excreted at a rate matching their production rate.

Water Input and Output Are Normally Equal

Table 24–2 presents a balance chart for water for an average 70-kg adult. We assume that the person is in water balance — that is, total input and output are equal (2500 ml/day). The input of water is normally derived from the diet. In a hospital setting, intravenous infusions may also be a source of water (and electrolytes). The various beverages we drink contain water. This intake of water is largely conditioned by habit but can also be stimulated by the thirst sensation. Water is also contained in solid food. For example, most fruits and vegetables are 80% to 90% water, and a hamburger is approximately 55% water. Water is also produced by oxidation of foodstuffs in the body. For example, for each mole of glucose oxidized, 6 moles of water are produced. (Recall from Chapter 6 the following reaction: $C_6H_{12}O_6 + 6 O_2 \rightarrow 6 CO_2 + 6 H_2O$.)

On the output side, we need to consider loss of water by way of the skin, lungs, gastrointestinal tract, and kidneys. Some water is always lost by way of the lungs and skin. This loss is usually not sensed and so is called **insensible water loss.** On a cold day, however, it is possible to perceive this water loss because exhaled air is saturated with water vapor, and the water condenses to form visible droplets in cold air.

TABLE 24–2	
Daily Water Balance in an Average 70-kg Man	
Input	**Output**
Water in beverages, 1000 ml	Skin and lungs, 900 ml
Water in food, 1200 ml	Gastrointestinal tract (feces), 100 ml
Water of oxidation, 300 ml	Kidneys (urine), 1500 ml
Total, 2500 ml	Total, 2500 ml

Large quantities of water also can be lost from the skin through sweating. In a hot environment or during heavy exercise, as much as 4 L/h water can be lost in perspiration. When we perspire, we lose not only water but also electrolytes, especially sodium chloride.

Gastrointestinal losses of water are normally small. The salivary glands, stomach, liver, pancreas, and intestine add approximately 7 liters of fluid to the gastrointestinal tract every day, but almost all of this is absorbed, and only 100 ml of water is normally lost in the feces (Chapter 22). With vomiting or diarrhea, large losses of water (and electrolytes) can result.

Finally, the kidneys are a site of water loss. Normally, renal water loss is regulated to maintain water balance. If there is a water deficit, renal excretion of water is decreased. If there is an excess of water in the body, renal excretion of water is increased. Renal excretion of water is controlled by the antidiuretic hormone (ADH).

Table 24–2 presents approximate daily water input and output for a normal 70-kg man. You should recognize, however, that water intake and output vary widely, even in normal people, depending on conditions such as habit, climate, and activity. Children have a greater need for water and are more susceptible to dehydration than are adults because children have more surface area per unit volume and a greater metabolic rate for their size.

The regulation of body water balance involves two basic mechanisms: (1) the thirst sensation and (2) control of renal water excretion by the antidiuretic hormone.

Thirst and the Control of Fluid Intake

Thirst is a conscious sensation associated with a craving for drink and is ordinarily interpreted as a desire for water. The function of the thirst mechanism is to repair a fluid deficit.

People drink largely by habit, and such drinking normally covers water needs. Thirst is an emergency mechanism not ordinarily experienced. In other words, the thirst threshold is usually not reached. Constant thirst in temperate climates and with moderate activity may indicate an underlying disease or disorder.

The thirst sensation is interesting because — unlike many sensations — it reaches the level of consciousness. With thirst, we can sense how body mechanisms generally act to bring about normalization of body conditions. The greater the need for water, the greater the intensity of thirst and the tendency for corrective action.

In humans, thirst is primarily associated with dryness of the mouth and throat, due to a reflex decrease in secretion by salivary and buccal (cheek) glands. A dry mouth plays a role in the conscious definition of thirst. It is also important in monitoring water intake and so helps us to know when to stop drinking. A dry mouth is not essential in regulating water intake.

Thirst is primarily controlled by centers in the hypothalamus. The hypothalamic centers relay information to the cerebral cortex, where thirst becomes a conscious sensation. Interestingly, the hypothalamus also controls the release of ADH and is therefore concerned with controlling both water intake and output. The hypothalamus is concerned with several other functions related to water balance: temperature regulation, food intake, and cardiovascular function.

The two major stimuli for thirst are (1) cellular dehydration and (2) extracellular dehydration. Cellular dehydration means a decrease in water within cells, or cell shrinkage. If the effective osmotic pressure of the plasma is increased, water moves out of all body cells. In the anterior hypothalamus, there are **osmoreceptor cells;** when these neurons shrink, they signal the cerebral cortex to give rise to the thirst sensation. Not all solutes are effective stimuli for the osmoreceptor cells. For example, urea and ethanol are ineffective because they readily penetrate the osmoreceptor cells and therefore do not cause them to shrink. Sodium chloride is an effective stimulus. An increase in plasma osmolality of 1% to 2% is needed to reach the thirst threshold.

Extracellular dehydration, which is synonymous with **hypovolemia,** or a decrease in blood volume, stimulates thirst. To be more exact, a decrease in **effective arterial blood volume** or **effective circulating blood volume** is the stimulus here. The effective arterial blood volume is the volume of blood pumped by the heart into the arterial system that perfuses the tissues. People with severe congestive heart failure may experience intense thirst, even though their blood volume is typically greater than normal, because, due to inadequate cardiac function, their effective arterial blood volume is decreased. Thirst is stimulated by loss of iso-osmotic fluid, such as occurs with hemorrhage, diarrhea, or vomiting.

There appear to be several receptors for hypovolemia. The various baroreceptor areas in the cardiovascular system (Chapter 19), such as the arterial baroreceptors in the carotid sinus and aortic arch, and the stretch (volume) receptors in the cardiac atria and great veins in the thorax, are involved. These receptors transmit impulses to the thirst control centers in the hypothalamus. The kidneys may also act as volume receptors. When blood volume is decreased, the kidneys release renin into the circulation. This results in the production of angiotensin II, which acts on neurons near the third ventricle of the brain to stimulate thirst. The kidneys not only participate in water balance by adjusting water output but also affect water intake.

Extracellular dehydration has to be considerable before thirst is stimulated. Blood donors usually do not experience thirst after donating 500 ml of blood (10% of blood volume). If a person loses a lot of blood, however, intense thirst results.

The two stimuli, cellular dehydration and extracellular dehydration, usually work together and reinforce each other. For example, if you have been active in hot weather and have lost a lot of water by sweating, plasma osmolality is increased and blood volume is decreased; thirst is stimulated by both pathways. In some situations, however, the two signals may oppose each other. For example, as stated before, people with congestive heart failure experience thirst because of a decrease in effective arterial blood volume. When they drink water, their kidneys tend to retain it in the body because ADH secretion is also stimulated. The result is that the body fluids become hypo-osmotic. Because the plasma sodium is diluted by excess water, the condition that develops is often called **hyponatremia** (low sodium concentration in the blood). In this situation, plasma osmolality is less than normal, yet thirst persists. Apparently, the body's attempt to restore an effective arterial blood volume is a more powerful drive, and so plasma osmolality is sacrificed.

Antidiuretic Hormone and the Control of Water Output

 How does antidiuretic hormone act to influence water reabsorption in the kidney?

Antidiuretic hormone (ADH) plays a key role in controlling water excretion by the kidneys by facilitating the reabsorption of water by the collecting ducts (Chapter 23). This peptide hormone is produced in the hypothalamus and released from the posterior pituitary into the blood (Chapter 13). The two major factors affecting ADH release are the same as those affecting thirst: (1) cellular dehydration and (2) extracellular dehydration.

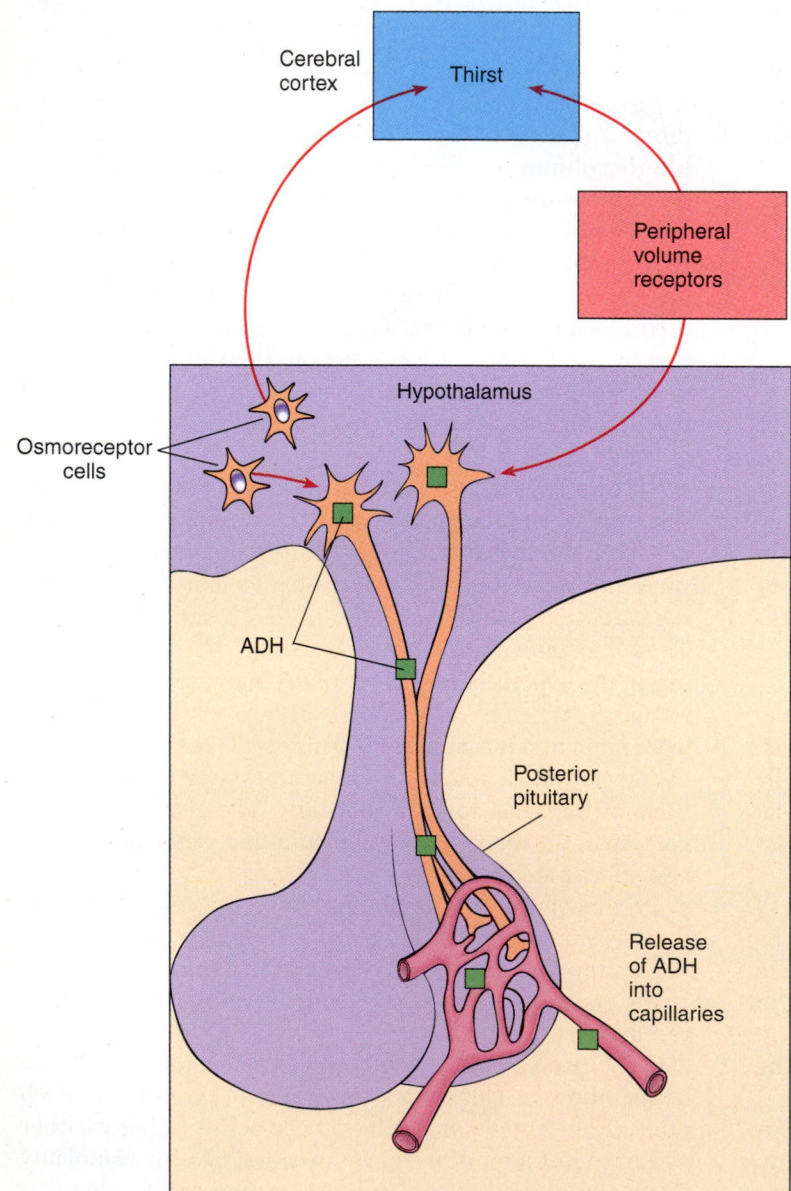

Figure 24–3

Control of thirst and ADH release. ADH-producing neurons are found in the supraoptic and paraventricular nuclei of the anterior hypothalamus. ADH is synthesized in the nerve cell bodies and is transported by axoplasmic transport to nerve terminals in the posterior pituitary, where it is stored in vesicles. When these cells are brought to threshold, they fire action potentials, the ADH–containing vesicles fuse with the nerve cell membrane, and ADH is released and transported by the blood to the kidneys. The hypothalamus also contains osmoreceptor cells that, via synaptic contacts, affect ADH release and thirst.

The main stimulus under ordinary conditions is cellular dehydration, caused by an increase in effective plasma osmolality. When plasma osmolality is increased, osmoreceptor cells in the anterior hypothalamus (distinct from those that cause thirst) shrink and signal the nearby ADH-producing cells to release ADH from their nerve terminals in the posterior pituitary.

The threshold for ADH release is normally exceeded, and so the plasma contains ADH at a measurable level ($\approx$ 2.5 picograms/ml). This plasma level is associated with a modestly concentrated urine (urine osmolality approximately 600 mosm/kg H_2O). This means that plasma ADH can be induced to decrease (by decreasing the plasma osmolality) or increase (by increasing the plasma osmolality). Thus, changes in plasma osmolality are continuously translated into changes in plasma ADH levels and, in turn, into changes in renal water excretion. The fact that the osmotic stimulus threshold for ADH release is considerably less than that for thirst is consistent with the view that thirst is mainly an emergency mechanism.

Extracellular dehydration (hypovolemia or a decrease in effective arterial blood volume) stimulates ADH release by acting at three sites. First, there are stretch receptors in the left atrium of the heart and in the pulmonary veins within the pericardium; less stretch results in fewer impulses transmitted to the brain via vagal afferents. An increase in intrathoracic blood volume inhibits ADH release and produces an increase in urine flow. Second, a decrease in pressure in the carotid sinuses and aortic arch reflexly stimulates ADH release. Third, granular cells in the afferent arterioles of the kidneys are sensitive to stretch. A decrease in pressure here stimulates renin release, renin increases angiotensin I and angiotensin II formation, and angiotensin II stimulates ADH release from the posterior pituitary.

Relatively large changes in blood volume ($\approx$ 10%) are required to affect ADH release. A large blood loss (e.g., 15–20% of blood volume), however, produces very large increases in plasma ADH. With severe hemorrhage, the high circulating levels of ADH may cause a significant increase in blood pressure due to the vasoconstrictor properties of this peptide (which is also called **vasopressin**).

The two stimuli, cellular and extracellular dehydration, often work synergistically in promoting ADH release. But (as discussed previously) in certain conditions (e.g., congestive heart failure) they may act in opposite directions. Other stimuli, primarily unrelated to water balance control, may also affect ADH release and were discussed in Chapter 13. Figure 24–3 is a schema for the control of thirst and ADH release by osmo– and volume receptors.

The primary target for ADH action in the kidney is the collecting duct epithelium (Fig. 24–4). In the absence of ADH, the luminal membrane is relatively water impermeable, and fluid in the lumen is hypo-osmotic to the cell or blood. ADH combines with a receptor in the basolateral cell membrane. By way of a guanine nucleotide stimulatory

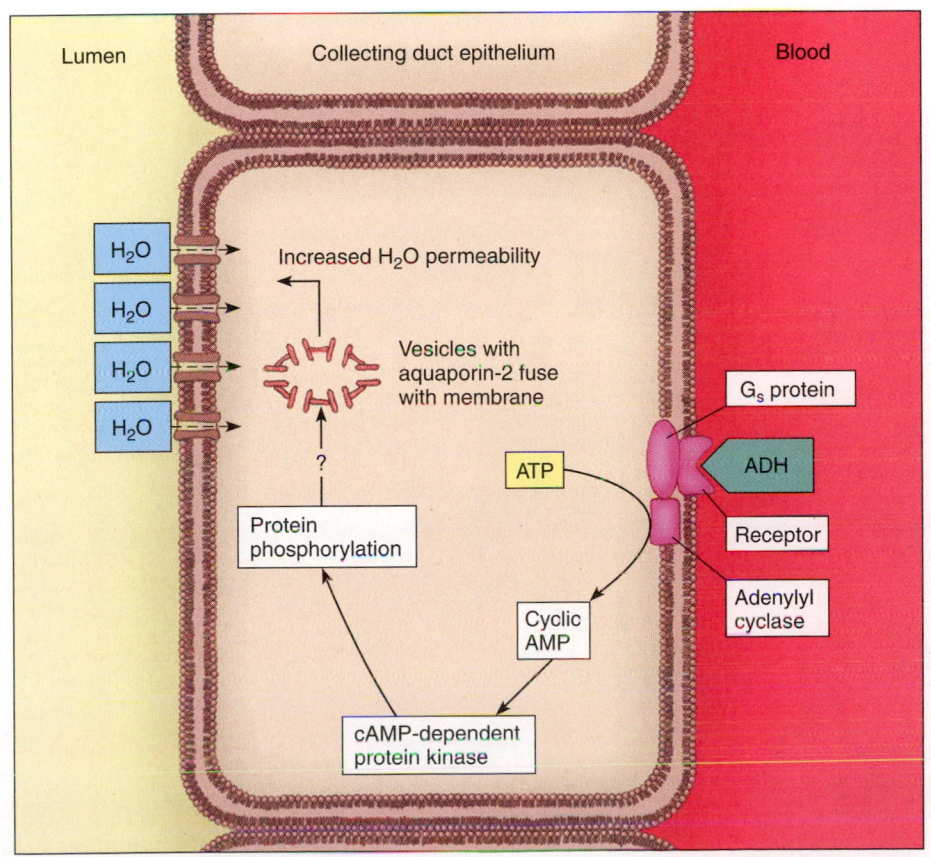

Figure 24–4

Mechanism of action of ADH on the collecting duct epithelium.

protein (G$_s$), the membrane–bound enzyme adenylyl cyclase is activated. This enzyme catalyzes the formation of cyclic AMP from ATP. In turn, cyclic AMP activates a protein kinase that phosphorylates other proteins. The subsequent steps are not entirely clear, but the final result is an increase in water permeability of the luminal cell membrane. This appears to be due to fusion of intracellular vesicles, which contain water channel proteins, called **aquaporin-2,** with the luminal cell membrane. When the collecting duct epithelium becomes water permeable, fluid in the duct lumen tends to become concentrated by osmotic withdrawal of water.

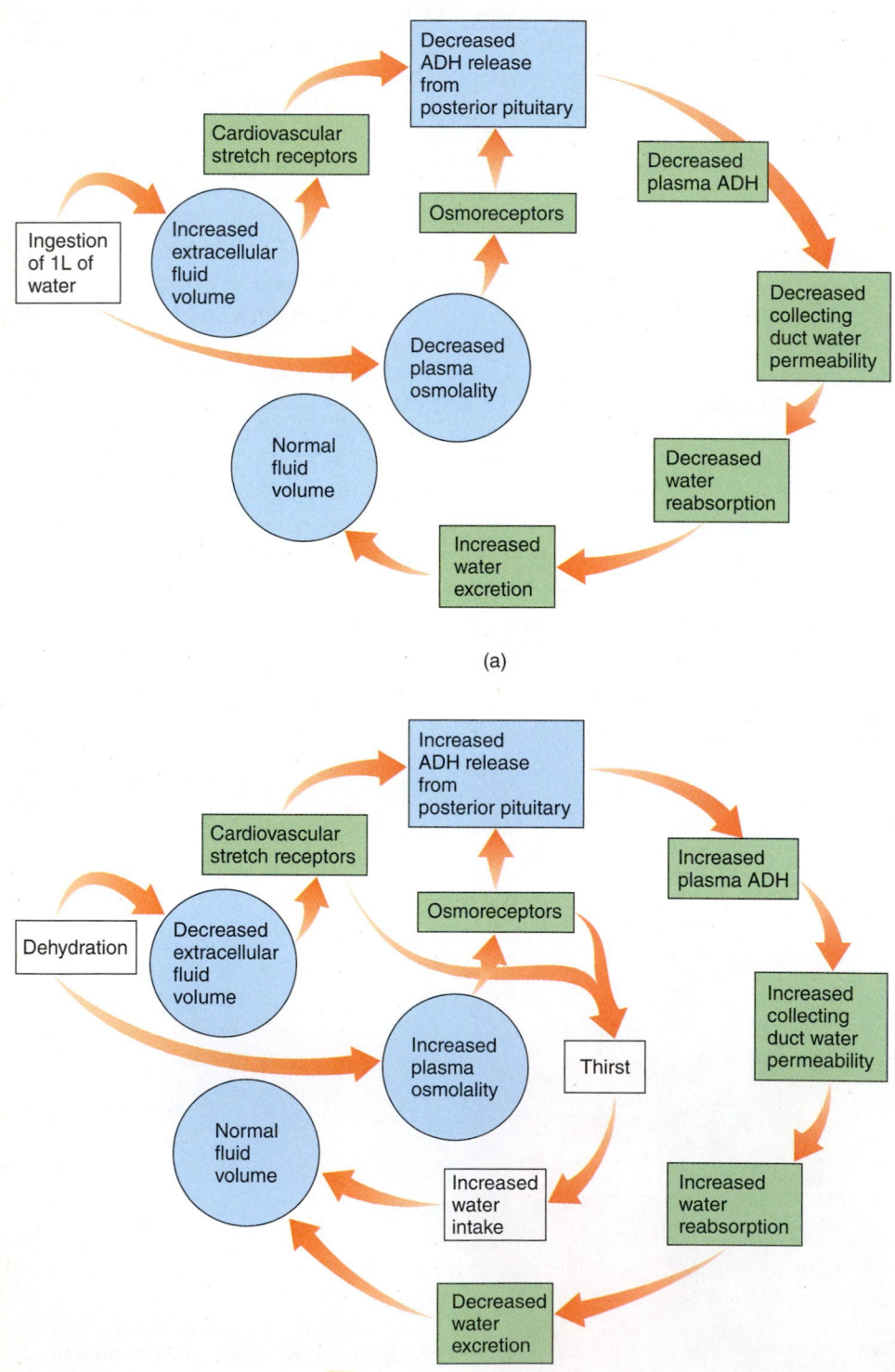

(a)

(b)

Figure 24–5

(a) Response to excess hydration.
(b) Response to dehydration.

CURRENT CONCEPTS IN PHYSIOLOGY

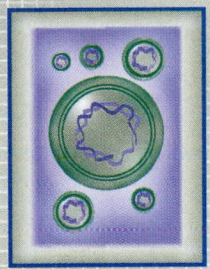

Aquaporin Water Channels in the Kidney

The aquaporins are a family of proteins that facilitate water movement through cell membranes (Chapter 4). Of the ten aquaporins currently known to be present in mammals, six are expressed in the kidneys, and four have been most extensively studied: AQP1, AQP2, AQP3, and AQP4. The aquaporins permit rapid movement of large quantities of water across the tubular epithelium.

AQP1, the first water channel protein, was discovered in the early 1990s. This aquaporin is expressed in the kidney in the luminal and basolateral cell membranes of proximal tubules and thin descending limbs, and on the endothelial cells of descending vasa recta. The proximal tubule contains approximately 20 million copies of this molecule per cell, so AQP1 is highly abundant.

AQP1 is not essential for survival in humans or mice. People who have an inherited deficiency in both copies of the *AQP1* gene, an extremely rare condition, have no obvious related clinical problems. If, however, they are subjected to water deprivation, they can only concentrate their urine osmotically to less than half of normal. The concentrating defect may be related to a lack of AQP1 in the thin descending limbs and descending vasa recta. Absence of AQP1 along the descending thin limb diminishes the loss of water along this segment and would lead to impaired countercurrent multiplication. Loss of AQP1 from the descending vasa recta would result in an increased blood flow to the inner medulla and washout of the osmotic gradient necessary to concentrate the urine.

Mice in which the *AQP1* genes have been knocked out (disrupted) exhibit an extreme degree of polyuria (increased urine output) and consequent polydipsia (increased fluid intake). In studies of isolated proximal tubules from *AQP1* knockout mice, it was found that the transepithelial water permeability was decreased to only 20% of normal. Micropuncture studies on these mice demonstrated that *AQP1* deletion resulted in a 50% reduction in water reabsorption in the proximal tubule. Furthermore, unlike the normal mouse, in which the late proximal tubule fluid/plasma osmolality ratio averages 0.98 (close to unity), this ratio was significantly reduced to 0.89 in the *AQP1*-deficient mice. This change indicates an abnormally high transepithelial osmotic gradient developed in the knockout mice. It appears from these experiments that most of the transepithelial water movement in the normal proximal tubule occurs through the cells (transcellular route) and not between the cells (paracellular route) and takes place through AQP1 water channels. Also, AQP1 normally functions to allow reabsorbed water to follow reabsorbed solute across the proximal tubule epithelium so rapidly that the fluid absorption is nearly an iso-osmotic process. Besides defective water reabsorption in the proximal tubule, impaired water movement in the thin descending limb and vasa recta contributes to the excessive urine output in these mice.

AQP2 is found in the apical cell membrane of collecting duct principal cells and in intracellular vesicles just below the apical cell membrane (see Fig. 24–4). The apical cell membrane is the rate-limiting barrier for transepithelial water reabsorption in the collecting duct. As described in this chapter, insertion of AQP2 into the apical cell membrane is stimulated by antidiuretic hormone. Inherited defects in the *AQP2* gene lead to nephrogenic diabetes insipidus in mice and people. Several drugs (e.g., lithium, used for the treatment of bipolar disorder) decrease the expression of AQP2 in the kidneys and lead to an impaired concentrating ability. AQP2 expression is increased in some conditions associated with water retention, such as pregnancy and congestive heart failure.

AQP3 and AQP4 are found in the basolateral cell membrane of collecting duct principal cells and provide the pathway for water to leave these cells. These water channels are constitutively expressed. Mice lacking these aquaporins show defects in concentrating ability.

Studies of the aquaporins have provided new insights into how the kidneys normally reabsorb water and a better understanding of what may go wrong in disease. In principle, blockers of these water channels might be developed in the future to increase water output by the kidneys and treat abnormal water retention.

Renal Responses to Extra Water Intake or to Dehydration

Renal excretion of water is continuously adjusted by changes in plasma ADH levels, so that plasma osmolality and extracellular fluid volume are kept at normal levels. To illustrate this adjustment, we will consider the response to two extreme states: (1) ingestion of a liter of tap water and (2) dehydration due to sweating.

Figure 24–5a summarizes the response to ingestion of a liter of tap water. In this case, the body is threatened by dilution and excess volume, and the appropriate renal response is excretion of a large volume of osmotically dilute urine, called a **water diuresis.** The ingested water is absorbed chiefly in the small intestine, which leads to a decrease in plasma osmolality. The decrease in plasma osmolality causes osmoreceptors in the hypothalamus to swell, resulting in

APPLICATIONS OF PHYSIOLOGY

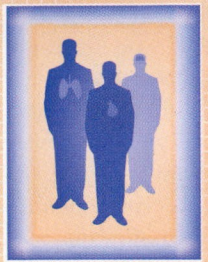

Polyuria

Polyuria is usually defined as a urine output of more than 3 L/day in adults. Such a urine output requires that the urinary bladder be emptied more frequently than the usual four to six times in a day. Polyuria can be distressing because of frequent bladder emptying and disturbed sleep. It may also be an important sign of some underlying pathology.

Frequency of urination may occur without polyuria; in other words, a person may frequently void small volumes. This is a symptom of decreased urinary bladder filling capacity and can be caused by urinary tract infection, stones, tumors, prostatic hyperplasia, or other conditions. Polyuria is best substantiated by making a 24-hour urine collection.

Causes of polyuria include neurogenic diabetes insipidus, nephrogenic diabetes insipidus, primary polydipsia, solute diuresis, and salt-losing syndromes. In neurogenic diabetes insipidus, polyuria arises because of deficient production or secretion of antidiuretic hormone (ADH), so that filtered water is poorly reabsorbed in the collecting ducts of the kidneys. It is usually caused by damage to the hypothalamus or pituitary, the normal sites of synthesis and secretion, respectively, of ADH. In nephrogenic diabetes insipidus, the response of the collecting duct to ADH is impaired. This disorder may be caused by various drugs (e.g., lithium carbonate, methoxyflurane anesthetic, and the antibiotic amphotericin), diseases of the renal medulla, inherited defects, or electrolyte disturbances (hypercalcemia and hypokalemia). In psychogenic diabetes insipidus, often called **primary polydipsia**, the polyuria is a consequence of increased water intake, as a result of habit, a psychiatric disorder, a lesion in the brain, or certain medications. A common form of polyuria is due to a solute diuresis, for example, as occurs in uncontrolled diabetes mellitus. Abnormal excretion of glucose in the urine leads to enhanced urinary loss of salt and water, dehydration, and thirst. In various salt-losing conditions — for example, diuretic drug abuse — the polyuria simply reflects decreased sodium reabsorption. Recall that tubular water reabsorption depends on sodium reabsorption, and when the latter is inhibited, increased excretion of water naturally follows.

To distinguish among the various causes of polyuria, a patient history and physical examination, analysis of the urine for glucose, and measurements of plasma creatinine and osmolality are helpful. Polyuria due to uncontrolled diabetes mellitus is easily identified by measuring glucose in the urine. Knowledge of the medications that a patient may be taking may supply important information. A person with primary polydipsia tends to have a reduced plasma osmolality because of excessive water intake. On the other hand, persons with neurogenic or nephrogenic diabetes insipidus tend to have elevated plasma osmolalities due to excretion of dilute urine.

A common test used to distinguish the various forms of diabetes insipidus is the **water deprivation test.** This test is easily interpretable if the lesions are severe but is less conclusive with partial (incomplete) lesions. Under constant observation, a patient is deprived of water for 8 to 14 hours, and urine osmolality is measured. The person is then given 5 units of vasopressin (ADH) subcutaneously, and 2 hours later, another urine sample is collected. If the patient has complete neurogenic diabetes insipidus, the urine osmolality is persistently dilute (< 200 mosm/kg H_2O) during dehydration but increases substantially in response to injected vasopressin. If the person has complete nephrogenic diabetes, the urine osmolality stays dilute in the presence of both dehydration and injected vasopressin. With primary polydipsia, the urine should become hyperosmotic (> 400 mosm/kg H_2O) during simple dehydration, and injected vasopressin produces little additional increase in urine osmolality because plasma levels of ADH are already high due to the dehydration protocol. Once the underlying cause of polyuria has been identified, a proper mode of treatment can be instituted.

---

decreased release of ADH from the posterior pituitary. The increased volume of the extracellular fluid, resulting from water ingestion, also acts (by way of stretch receptors in the cardiovascular system) to inhibit ADH release. (This pathway appears to be less important under normal conditions.) When ADH release is decreased, plasma ADH decreases because this hormone is continuously destroyed in the body. With a low plasma ADH level, the collecting duct expresses its intrinsically low water permeability. Less filtered water is then reabsorbed at this site. The result is the excretion of a large volume (up to 20 ml/min) of osmotically dilute urine. Urine osmolality may be as low as 30 to 40 mosm/kg H_2O. The increase in urine flow starts a few minutes after water ingestion, and the entire extra load of water is excreted in a few hours. By getting rid of the extra water, the kidneys are able to restore plasma osmolality and volume back to normal.

Consider now what happens in the opposite case, dehydration (see Fig. 24–5b). Suppose you exercise in hot weather; you sweat a lot and don't replace the water losses. The sweat you lose is a hypo-osmotic fluid, and so the body fluids become hyperosmotic. In addition to water lost in sweat, there is also insensible water loss from the lungs and skin. These losses lead to an increase in plasma osmolality, stimulation of hypothalamic osmoreceptors, and increased release of ADH. Also, a decrease in extracellular volume causes the cardiovascular stretch receptors to increase ADH release reflexly. The increase in plasma ADH levels causes increased water reabsorption from the collecting ducts, with resulting excretion of the urinary solutes in a hyperosmotic urine. By excreting the urinary solutes in a minimum volume, the body saves water, which can delay or minimize the effects of dehydration. To correct the dehydration, the thirst mechanism, stimulated by the same changes that affect ADH release, is engaged. Thirst stimulates water intake, and so the water deficit is repaired.

The Kidneys Are the Major Site of Control of Sodium Balance

 How do the kidneys participate in the control of extracellular fluid sodium?

Sodium Input and Output

In a normal person on a normal diet, the amount of sodium ingested is closely matched by the amount of sodium lost from the body. Figure 24–6 summarizes sodium balance relationships. On the input side, sodium intake in the diet may be quite variable, but in an average American diet, sodium amounts to approximately 100 to 300 mEq/day (6–18 g NaCl/day). In a hospital setting, intravenous fluids can also be a source of sodium.

On the output side, sodium may be lost from the skin, gastrointestinal tract, or kidneys. Loss of sodium from the skin may occur with sweating, burns, or hemorrhage. Sweat contains 10 to 100 mEq sodium per liter, and so with heavy exercise or in warm environments, an appreciable amount of sodium may be lost in sweat. Gastrointestinal losses of sodium are normally small, but with diarrhea or vomiting, large losses may result. The urine normally accounts for approximately 95% of the total output of sodium from the body. The kidneys constitute the major site of control of sodium output and normally adjust excretion to maintain balance.

A positive sodium balance results if sodium input exceeds output. Because water is retained along with sodium, and because sodium is primarily an extracellular ion, extracellular fluid volume is increased. If sufficient salt and water are retained, expansion of the interstitial fluid space will lead to generalized edema. This condition is associated with weight gain (due to the extra water) and skin puffiness, espe-

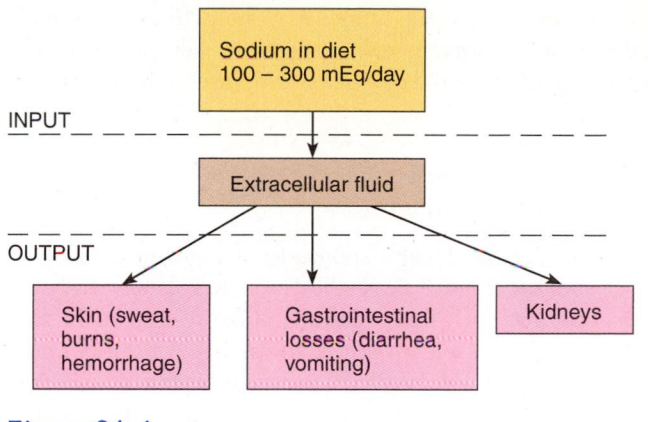

Figure 24–6

Sodium balance. The kidneys are usually responsible for 95% of the sodium output.

cially in the feet and ankles (due to the effects of gravity). A positive sodium balance and generalized edema occur in a number of clinically important conditions, including congestive heart failure, hepatic cirrhosis (a liver disease), and nephrotic syndrome (a kidney disease). In order to reduce the edema, it is common practice to (1) prescribe a low-salt diet and (2) use diuretic drugs to promote sodium excretion by the kidneys. In this way, the physician tries to relieve the edema and bring the patient back into sodium balance.

A negative sodium balance results if sodium output exceeds input. This may occur if there is excessive loss of sodium via the skin, gastrointestinal tract, or kidneys. A negative sodium balance leads to a reduction of extracellular fluid volume. In turn, this produces a decrease in plasma volume and consequently may lead to an inadequate circulation, hypotension, and even death.

It should be clear that sodium balance is closely related to regulation of extracellular fluid volume. This is so because sodium salts are the major solutes of the extracellular fluid, as discussed earlier. Because the osmolality of the extracellular fluid is closely regulated by ADH, the kidneys, and thirst, it follows that retention or loss of sodium from the body will be accompanied by parallel changes in the amount of water in, and hence volume of, the extracellular fluid. The control of sodium balance depends mainly on the kidneys, although there is some evidence for a **sodium appetite** (which stimulates sodium intake) when a sodium deficit exists. Because the kidneys are intimately concerned with control of sodium balance and regulation of extracellular fluid volume, it should not be surprising that extracellular fluid volume is a major determinant of renal sodium excretion. We will elaborate on this later.

There is compelling evidence that hypertension (high blood pressure) may often be due to a disturbance in sodium (salt) balance. Excessive dietary intake of salt or inadequate renal excretion of salt tends to increase intravascular fluid volume; this is translated into an increase in blood pressure.

Avoidance of a high salt intake may help to prevent high blood pressure in some people. A reduced salt intake and diuretic drugs are used to keep blood pressure under control in people with hypertension.

The Renal Responses to Changes in Dietary Intake of Sodium

Let us consider the renal response to changes in dietary intake of sodium. Figure 24–7 illustrates a balance study. A person was put on a fixed sodium intake of 100 mEq/day. During the first three days of the study (the control period), the person was in a stable balance, as indicated by the fact that urinary excretion matched the sodium intake (in this study, we will ignore extrarenal losses of sodium, which are usually small). On day 4, the sodium intake was increased to 300 mEq/day and was maintained at that level until the end of day 10. What happened? Sodium excretion rose, but for the first few days (days 4 to 7) was less than the sodium input. Consequently, there was a phase of positive sodium balance. During this time, the person retained salt, and because the subject was allowed free access to water, water intake increased, and salt and water were retained as an iso-osmotic solution. If we measured plasma sodium concentration, we would find that it was essentially constant throughout this study. The body's sodium content (that is, the amount of sodium), the extracellular fluid volume, and body weight, however, increased during days 4 to 7. During days 8 to 10, the subject was back in a stable balance; that is, input and output of sodium were once again equal. When the person switched to a low-sodium diet (10 mEq/day), sodium excre-

tion decreased. During days 11 to 14, sodium excretion was higher than sodium intake, and so a negative sodium balance resulted. During this time, extracellular fluid volume and body weight decreased. Finally, a new stable balance on the low-salt diet was established (days 15 and 16).

This study demonstrates that changing the dietary sodium intake leads to appropriate changes in sodium excretion by the kidneys. An increase in sodium intake leads to increased urinary sodium excretion, and a decrease in sodium intake results in decreased sodium excretion. The kidneys can adjust sodium excretion over a wide range. Note, however, that the renal response is sluggish, and so it takes a few days to reach the appropriate rate of renal sodium excretion. During the time that renal sodium excretion is less than sodium intake (days 4 to 7), extracellular fluid volume is increasing. Extracellular fluid volume is stable, but higher than the control volume, during days 8 to 10. The increased extracellular fluid volume may be thought of as the stimulus that promotes renal sodium loss. When the sodium intake is decreased, a decrease in extracellular fluid volume may be thought of as the cause of the reduced sodium excretion.

Effector Mechanisms Activated by Altered Extracellular Fluid Volume

What mechanisms allow the kidneys to adjust their output of sodium to maintain balance and relative stability of extracellular fluid volume? Figure 24–8 summarizes how sodium excretion may be controlled by a negative feedback system. According to this scheme, the regulated variable is extracellular fluid volume. Changes in extracellular fluid volume are

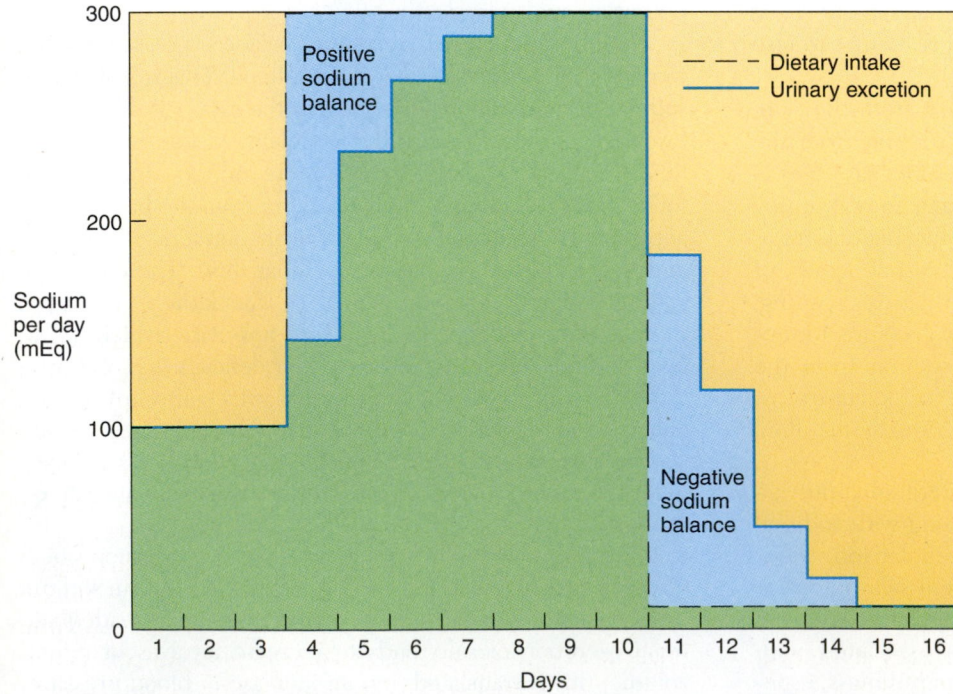

Figure 24–7

A sodium balance study. Dietary intake of sodium was initially 100 mEq/day, was then increased to 300 mEq/day (days 4 to 10), and was finally reduced to 10 mEq/day (days 11 to 16). Daily urinary excretion of sodium (*solid blue lines*) was measured.

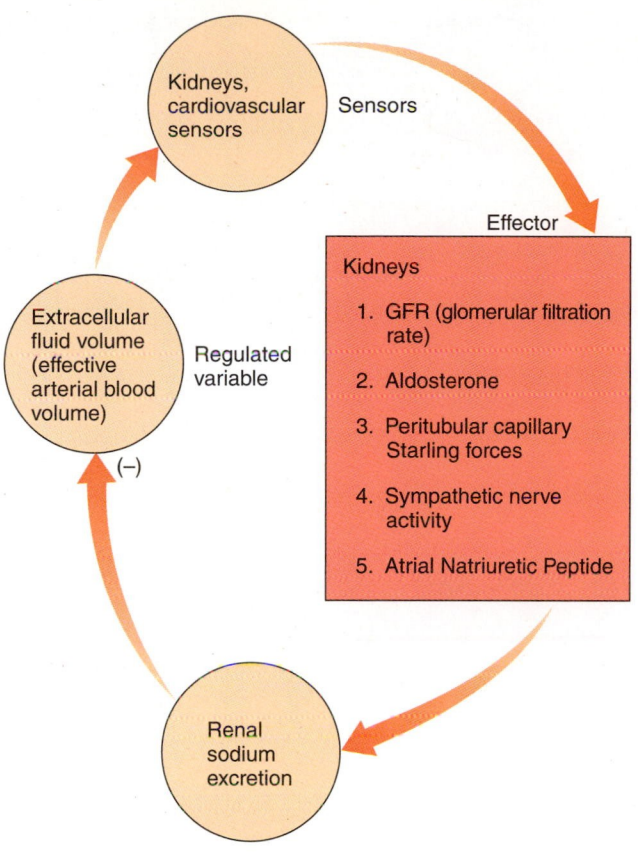

Figure 24–8

Regulation of extracellular fluid volume (or effective arterial blood volume) by a negative feedback control system. Changes in effective arterial blood volume are sensed by cardiovascular and renal sensors and are translated into changes in renal effector pathways, leading to changes in sodium excretion that bring effective arterial blood volume back to normal.

sensed by cardiovascular volume receptors and by the kidneys. The effector mechanisms include changes in (1) glomerular filtration rate, (2) plasma aldosterone levels, (3) peritubular capillary Starling forces, (4) renal sympathetic nerve activity, and (5) plasma atrial natriuretic peptide. Additional effector mechanisms may exist. Changes in these mechanisms lead to changes in sodium excretion and bring extracellular fluid volume back to normal.

We will first discuss the five effector mechanisms just listed and then the nature of the sensors and what they may actually be sensing. Sodium excretion represents the difference between filtered and reabsorbed amounts, and so factors that affect sodium excretion may act on the glomeruli, kidney tubules, or both.

Glomerular Filtration Rate. Glomerular filtration, as discussed in Chapter 23, delivers an enormous amount of sodium to the tubules. Most of this is reabsorbed. In-

creases or decreases in glomerular filtration rate (GFR) lead to directionally similar changes in urinary sodium excretion. Expansion of the extracellular fluid volume with isotonic saline tends to increase GFR and so may contribute to increased sodium excretion. An increase in GFR actually leads to only a slight increase in sodium excretion, due to a phenomenon called **glomerulotubular balance.** In response to an increase in GFR, the proximal convoluted tubule and loop of Henle increase their rates of sodium reabsorption, thereby preventing excessive loss of sodium in the urine.

Severe hemorrhage is accompanied by a decrease in blood pressure and renal vasoconstriction and, hence, a decrease in GFR. The kidney tubules can then reabsorb the reduced filtered sodium load more completely, and sodium excretion therefore decreases. This is a valuable adjustment because, with hemorrhage, the kidneys ideally should retain sodium to aid restoration of the depleted extracellular volume.

Aldosterone and the Renin-Angiotensin System. Aldosterone is the most important mineralocorticoid hormone in humans. It is a salt-retaining steroid produced by the zona glomerulosa region of the adrenal cortex (Chapter 14). It increases the reabsorption of sodium (and secretion of potassium) by a variety of epithelial tissues, including sweat glands, salivary glands, and intestine. In the kidney, it acts chiefly on the distal convoluted tubule and collecting ducts. In the absence of aldosterone, the final urine sodium concentration is high, and abnormal quantities of sodium are excreted in the urine; instead of reabsorbing 99.6% of the filtered sodium load, 98.0% of the filtered sodium load may be reabsorbed. This may seem like a small difference, but remember that if 24,000 mmoles of sodium are filtered in a day (see Table 23–1), a difference of 1.6% amounts to 384 mmoles sodium, roughly the quantity of sodium in 2.5 liters of extracellular fluid. Maintained excretion of this amount of sodium over a period of days would result in severe (and even fatal) depletion of the extracellular fluid. To some extent, the effects of such salt loss can be reversed by a high sodium chloride intake. Patients with **Addison's disease** (adrenocortical insufficiency) may show a heightened sodium appetite, which acts to compensate for the excessive salt loss in the urine.

The most important pathway controlling aldosterone secretion in response to volume changes is the renin-angiotensin system (Chapter 14). Briefly, a decrease in extracellular fluid volume (or, to be more exact, a decrease in effective arterial blood volume) results in increased renin release from the granular cells of the juxtaglomerular apparatus (see Fig. 23–5). Increased renin release may be due to three mechanisms: (1) a decrease in blood pressure at the level of the afferent arteriole (renal baroreceptor mechanism), (2) an increase in activity of renal sympathetic nerves that innervate the granular cells, and (3) a decrease in sodium chloride delivery to and transport by the macula densa cells. All three mechanisms are activated by volume depletion, such as that caused by hemorrhage. **Renin,** a proteolytic enzyme, acts on

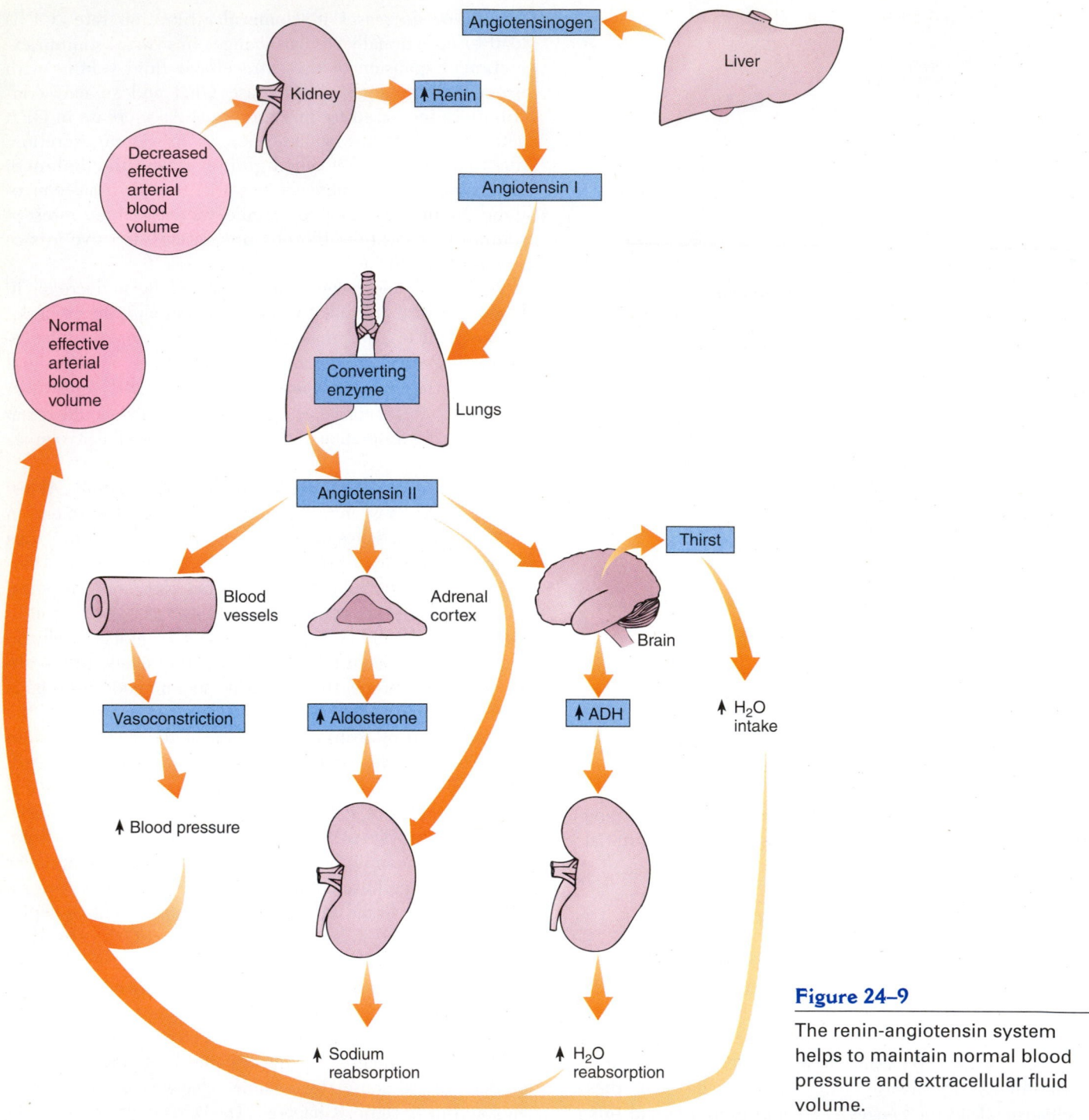

Figure 24–9

The renin-angiotensin system helps to maintain normal blood pressure and extracellular fluid volume.

a plasma protein, **angiotensinogen,** to produce **angiotensin I** (Fig. 24–9). Angiotensin I is converted to the active peptide, **angiotensin II,** by an enzyme called **angiotensin-converting enzyme,** mainly as blood passes through the lungs. Angiotensin II has several important actions. First, it is a powerful vasoconstrictor and so increases blood pressure. Second, it increases the synthesis and re-

lease of aldosterone by a direct action on the adrenal cortex. Aldosterone causes enhanced sodium reabsorption by distal parts of the nephron. Third, angiotensin II directly stimulates sodium reabsorption by the kidney tubules. Fourth, angiotensin II acts on the thirst center in the hypothalamus, stimulating increased water intake. Fifth, angiotensin II may stimulate ADH release from the posterior pituitary, resulting in renal water retention. In all of these actions, angiotensin II helps to restore a normal blood pressure and extracellular fluid volume. In the day-to-day control of sodium balance, changes in aldosterone secretion, mediated by changes in renin release, are probably most important in bringing about changes in sodium excretion. Aldosterone secretion is also stimulated directly by a decrease in plasma sodium concentration of the blood perfusing the adrenal cortex.

Peritubular Capillary Starling Forces. In Chapter 23, we learned that changes in hydrostatic and colloid osmotic pressures in the peritubular capillaries affect the net reabsorption of salt and water by the proximal convoluted tubule. Specifically, an increase in hydrostatic pressure or a decrease in colloid osmotic pressure reduces fluid uptake. Such a change occurs, for example, when a large volume of isotonic saline is added to the body. The result is an increase in sodium excretion, an appropriate response.

Renal Sympathetic Nerve Activity. Stimulation of renal sympathetic nerves results in a decrease in renal sodium excretion. If stimulation is intense, the decrease in sodium excretion may be explained largely by hemodynamic changes — that is, by a decrease in glomerular filtration rate and renal blood flow. Low levels of renal sympathetic nerve stimulation, at intensities less than those needed to produce measurable hemodynamic changes, also result in decreased sodium excretion. This latter response may be due to direct stimulation of tubular cells, so that they reabsorb more sodium. There is anatomical evidence that renal sympathetic nerve fibers lie in close proximity to tubular cells. Finally, as was mentioned earlier, stimulation of renal sympathetic nerves increases renin release and results in a decrease in renal sodium excretion via increased angiotensin II and aldosterone. Inhibition of renal sympathetic nerve activity — for example, by volume expansion— promotes sodium excretion.

Atrial Natriuretic Peptide. The atria of the heart produce a peptide called **atrial natriuretic peptide (ANP).** ANP is released by the atria in response to mechanical stretch — for example, by an increase in blood volume — and it increases sodium excretion. The **natriuresis** (increased sodium excretion) results from several actions: inhibition of sodium reabsorption by collecting ducts in the inner medulla, an increase in glomerular filtration rate, and inhibition of renin and aldosterone secretion. Other hormonal factors promote sodium excretion — urodilatin (a kidney natriuretic hormone similar to ANP), brain natriuretic peptide, prostaglandins (PGE_2 and PGI_2) and bradykinin (produced in the kidneys), and guanylin and uroguanylin (natriuretic hormones produced by the intestine). Additional natriuretic substances are being investigated.

Clearly, multiple effector mechanisms influence sodium excretion.

The Nature of the Stimuli and the Location of Receptors Affecting Sodium Excretion

What exactly is the sensed variable that leads to changes in renal sodium excretion? The idea that it is the extracellular fluid volume seems a satisfactory explanation in a normal person. However, in a number of disease states, generalized edema develops because the kidneys retain salt and water; sodium is conserved despite an expanded extracellular fluid volume. The argument has been presented that it is not extracellular fluid volume per se, but a component thereof, the plasma volume, that is regulated. But this argument is not convincing, because blood volume is commonly increased in patients with congestive heart failure whose kidneys avidly retain sodium.

To explain these difficulties, researchers have postulated that the effective arterial blood volume is the regulated variable that affects sodium excretion. The effective arterial blood volume is an index of the degree of filling of the systemic arterial tree. It determines whether blood flow will be adequate to meet the metabolic demands of the tissues. In congestive heart failure, the heart muscle is weak and does not adequately transfer blood from the venous to the arterial side of the circulation. Blood tends to "dam up" in back of a failing ventricle, which decreases effective arterial blood volume. Effective arterial blood volume is also decreased in other conditions in which generalized edema occurs, such as liver cirrhosis, malnutrition, arteriovenous fistula, and nephrotic syndrome. Effective arterial blood volume or extracellular fluid volume is decreased with hemorrhage, diarrhea, severe sweating, and low salt intake. Therefore, all of these conditions are accompanied by renal salt retention.

Where are the volume receptors? Convincing evidence shows that various cardiovascular stretch receptors can affect sodium excretion. These include stretch receptors in the wall of the left atrium of the heart and in intrathoracic veins. An increase in volume of blood in the thorax is a well-known stimulus that increases sodium excretion. The arterial baroreceptors also function as stretch receptors affecting sodium excretion.

The kidneys act as volume sensors. When effective arterial blood volume is decreased, renal blood flow is threatened, and the kidneys retain salt and water, a response that improves their blood flow. How might changes in effective arterial blood volume be sensed by the kidneys and translated into the appropriate response? Three possible pathways may be involved. In the first one, increases or decreases in effective arterial blood volume lead to increases or decreases, respectively, in glomerular filtration rate, due to changes in glomerular blood flow and capillary pressure. Directionally similar changes in sodium excretion ensue. In the second, there is an intrarenal baroreceptor in the afferent arteriole. A decrease in pressure at this site increases renin release and ultimately sodium reabsorption. Finally, in the third pathway, changes in peritubular capillary Starling forces could result directly from changes in extracellular fluid volume and once again would lead to appropriate changes in sodium excretion.

Re-examination of Figure 24–8 will help to review some of the foregoing points. When effective arterial blood volume is decreased (e.g., after hemorrhage), renal sodium excretion is decreased. This makes sense because renal conservation of sodium aids in restoring effective arterial blood volume to normal. The reduction in renal sodium excretion may be due to a reduced glomerular filtration rate, increased plasma angiotensin II and aldosterone levels, decreased peritubular capillary hydrostatic pressure, increased renal sympathetic nerve activity, and decreased plasma levels of the atrial natriuretic peptide. When effective arterial blood volume is increased (e.g., because of increased dietary sodium intake), sodium excretion is enhanced because of an increase in glomerular filtration rate (GFR), decreased plasma angiotensin II and aldosterone levels, increased hydrostatic pressure and decreased colloid osmotic pressure in the peritubular capillaries, reduced sympathetic nerve activity, and increased plasma levels of the atrial natriuretic peptide. The relative importance of the various effector mechanisms varies with the circumstances and is not always known. Because regulation of extracellular fluid volume or effective arterial blood volume is so important, it is not surprising that multiple mechanisms, which interact and overlap, control renal sodium excretion.

Other Factors That Influence Sodium Excretion

A number of other factors are known to affect sodium excretion. These include (1) glucocorticoids, (2) estrogens, (3) osmotic diuretics, (4) poorly reabsorbed anions, and (5) diuretic drugs.

Glucocorticoids. The glucocorticoids (e.g., cortisol) are steroid hormones produced by the adrenal cortex (Chapter 14). They increase tubular sodium reabsorption but also cause an increase in GFR, which may mask the tubular effect. In general, glucocorticoids cause a decrease in sodium excretion.

Estrogens. Estrogens may increase sodium reabsorption by the renal tubules. High plasma levels of estrogens, together with numerous other hormonal and hemodynamic changes, contribute to sodium retention during pregnancy.

Osmotic Diuretics. **Osmotic diuretics** are solutes excreted in the urine that obligate the excretion of water. Urine flow rate, sodium excretion, and potassium excretion are all increased.

Clinically important osmotic diuretics include mannitol, glucose, and urea. Mannitol is a 6-carbon sugar alcohol filtered by the glomerulus but not reabsorbed. Increased excretion of mannitol leads to increased excretion of salt and water. A concentrated mannitol solution is sometimes given intravenously to promote urine output. In uncontrolled diabetes mellitus, more glucose is filtered than can be reabsorbed by the tubules. Again, increased excretion of solute (glucose) leads to enhanced excretion of salt and water. The loss of salt leads to a decrease in extracellular fluid volume and so stimulates thirst. Urea excretion also promotes renal excretion of salt; this may be important in sustaining sodium excretion in chronic kidney disease.

Poorly Reabsorbed Anions. Poorly reabsorbed anions result in increased sodium excretion. Because of the requirement for electroneutrality in solutions, whenever more anion is excreted, more cation must also be excreted. Increased excretion of phosphate, bicarbonate, ketone body acids (as occurs in uncontrolled diabetes mellitus), or sulfate results in increased sodium excretion. To some extent, sodium can be replaced by other cations, such as potassium, ammonium, and hydrogen ions.

Diuretic Drugs. Diuretic drugs inhibit tubular reabsorption of sodium. The increase in urine volume that ensues is basically due to impaired sodium reabsorption. Remember that tubular reabsorption of water depends on sodium reabsorption.

The Kidneys Are the Major Site of Control of Potassium Balance

 How do the kidneys participate in the control of extracellular potassium concentration?

Most of the body's potassium is found within cells. The concentration of potassium in cells is typically approximately 150 mEq/L cell water. In extracellular fluid, the potassium concentration averages approximately 4.5 mEq/L.

Potassium plays a number of important roles in the body. First, it influences the electrical excitability of cells. As discussed in Chapter 7, the resting membrane potential of cells is largely determined by the unequal distribution of potassium inside and outside of cells and by a high membrane permeability to potassium ions. Disturbances in potassium balance lead to altered excitability of nerves and muscles.

Second, potassium is the major osmotically active solute in cells, and so the amount of potassium in cells affects cell volume. Loss of potassium from cells leads to a decrease in cell volume.

Third, potassium balance is intimately related to acid-base balance. The relationships are complex, but there is

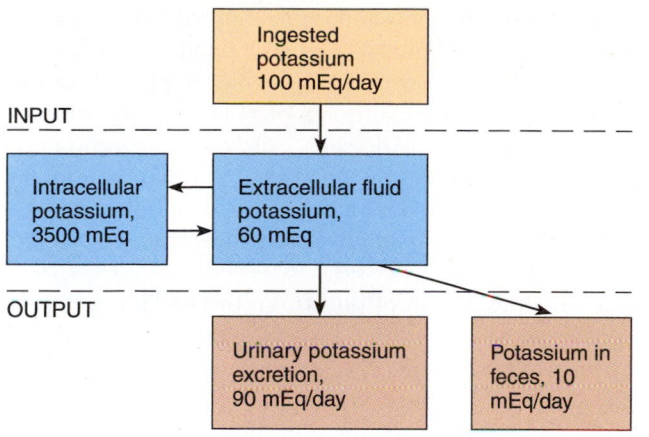

Figure 24–10

Potassium balance (input and output) and the distribution of potassium between the two major body fluid compartments.

abundant evidence to suggest that disturbances in acid-base status can affect potassium balance, and vice versa.

Finally, intracellular potassium affects cell metabolism. Tissue growth and repair require potassium. Conversely, tissue breakdown or increased protein catabolism leads to a release of cell potassium into the extracellular fluid.

The normal range for plasma potassium concentration is 3.5 to 5.0 mEq/L. A plasma potassium concentration of less than 3.5 mEq/L is called **hypokalemia.** This condition may cause skeletal muscle weakness and even paralysis. A plasma potassium concentration of more than 5.0 mEq/L is called **hyperkalemia.** Plasma potassium concentrations of more than 7 mEq/L demand immediate attention because of possible cardiac irregularities. A plasma potassium concentration of more than 10 to 12 mEq/L is usually fatal; the cause of death is cardiac arrhythmia or arrest. The changes in skeletal and cardiac muscle function that occur with abnormal plasma potassium levels result from the influence of potassium on cell membrane potentials.

Figure 24–10 summarizes potassium balance. Normally, we ingest approximately 100 mEq of potassium per day. The amount we take in can be quite variable, depending on what we eat and drink. Approximately 10% of the ingested potassium is excreted in the feces; the rest (90%) is absorbed by the intestine and is excreted in the urine. Loss of potassium in sweat (which contains 5–10 mEq K^+/L) is usually negligible. The kidneys are the major site of potassium loss from the body, and they are the major site of control of potassium balance.

Figure 24–10 also reemphasizes that most of the body's potassium is inside of cells. The amount of potassium in extracellular fluid can be calculated from its concentration (4.5 mEq/L) and the extracellular fluid volume ($\approx$14 liters) and is

equal to approximately 60 mEq. This amounts to only 2% of the body's potassium.

Several factors affect the distribution of potassium between cells and extracellular fluid. First, activity of the cell membrane Na^+/K^+-ATPase, which pumps potassium into cells, is of key importance. Impaired cell metabolism or cardiac glycosides (e.g., digitalis) inhibit the pump and tend to produce lower cell and higher extracellular potassium levels. Second, acid-base status affects the distribution of potassium between intracellular and extracellular compartments. A decrease in plasma pH causes a shift of hydrogen ions into cells in exchange for cell potassium ions. The consequence is an increase of plasma potassium concentration. Third, the availability of insulin is an important factor affecting potassium distribution. Insulin promotes the movement of potassium into skeletal muscle and liver cells; a lack of insulin causes more of the body's potassium to stay outside of cells. Fourth, hyperosmolality tends to increase the extracellular fluid potassium concentration by causing cell shrinkage and an increased concentration gradient for potassium to diffuse out of cells. Finally, shifts of potassium from cells to extracellular fluid occur with cell breakdown, due to tissue trauma, infection, ischemia (inadequate blood flow), or heavy exercise. Because so much of the body's potassium is within cells, a small loss from or gain by this compartment could profoundly affect the plasma potassium concentration.

Plasma potassium must be closely regulated. A part of this regulation involves the hormones insulin, epinephrine, and aldosterone, which stimulate the uptake of potassium by cells. The kidneys also play an important part in regulating plasma potassium. Normally, when increased potassium loads are presented to the body, the kidneys can rapidly excrete excess potassium.

In Chapter 23, we discussed some of the factors that affect potassium excretion by the kidneys. Recall that most of the filtered potassium is reabsorbed by early portions of the nephron and that the collecting ducts secrete potassium. Some of the factors that affect potassium excretion include intracellular potassium concentration, aldosterone, excretion of anions, and urine flow rate.

Figure 24–11 shows the mechanisms whereby an increase in potassium intake in the diet leads to increased renal potassium excretion. Increased potassium intake tends to increase the plasma potassium concentration. By a direct effect on the adrenal cortex, an elevated extracellular potassium concentration stimulates the release of aldosterone. Aldosterone then travels in the blood to the kidneys, where it acts on the principal cells of the collecting duct. It increases the sodium permeability of the luminal cell membrane, the activity of the Na^+/K^+-AT-Pase in the basolateral cell membrane, and potassium permeability of the luminal membrane of these cells, thereby favoring potassium secretion (see Figure 23–17). An increased potassium intake increases the potassium

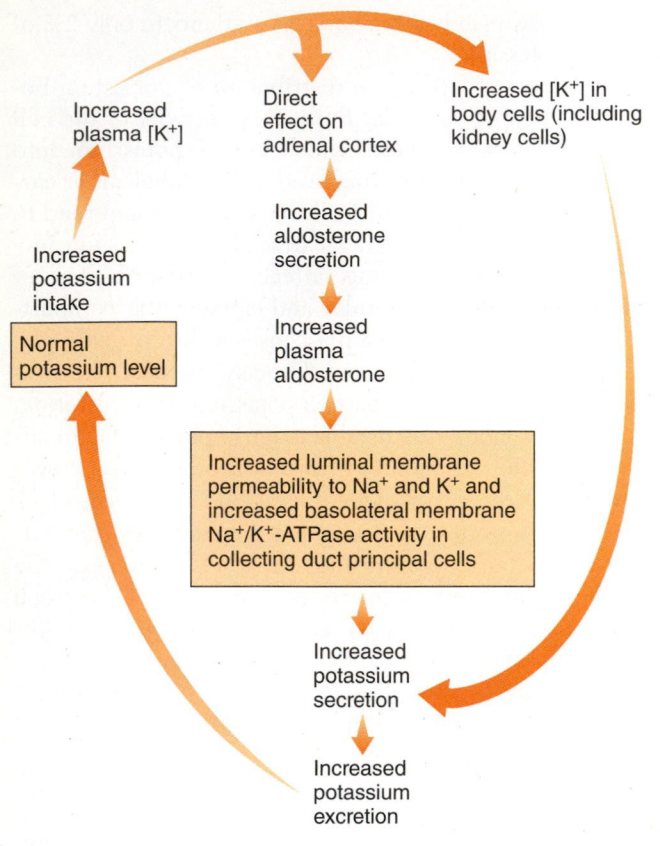

Figure 24–11

Pathways by which an increase in dietary potassium intake leads to increased renal potassium excretion.

concentration in most body cells, including the principal cells, and so potassium secretion is directly increased in this way.

The opposite changes occur if potassium intake is reduced. Plasma and cell potassium levels tend to decrease, aldosterone secretion is inhibited, and secretion of potassium by the collecting duct is diminished. The result is a decreased rate of potassium excretion.

The kidneys normally do a remarkable job of maintaining potassium balance. But abnormal renal excretion of potassium (either too much or too little) is a common cause of disturbed potassium balance.

Too little renal excretion of potassium causes hyperkalemia. Most foods are rich in potassium, so continued food intake with inadequate renal excretion leads to a positive potassium balance. In acute renal failure — that is, with sudden renal shutdown — life-threatening hyperkalemia may develop. Often compounding inadequate renal excretion in patients with acute renal failure are tissue trauma, infection, and acidosis, all of which tend to increase the plasma potassium level. Chronic renal disease leads to hy-

perkalemia when the glomerular filtration rate decreases to less than 15 to 20 ml/min. Up to that point, hyperkalemia does not develop because each surviving nephron can excrete a greater-than-normal quota of potassium. In adrenocortical insufficiency (Addison's disease), secretion of aldosterone is inadequate. A decreased plasma aldosterone level results in decreased stimulation of renal potassium secretion and excretion, so that ingested potassium tends to be retained in the body.

Too much excretion of potassium by the kidneys leads to hypokalemia. With adrenocortical hyperfunction, aldosterone levels are abnormally high, and potassium secretion is stimulated. Treatment with glucocorticoids can also result in excessive potassium excretion. In a number of kidney diseases, the ability of the tubules to reabsorb filtered potassium is impaired, and excessive potassium excretion results. Excessive renal potassium loss occurs in uncontrolled diabetes mellitus for several reasons: (1) an osmotic diuretic effect of glucose (increased urine output produces potassium loss), (2) increased production of ketone body acids (excretion of anions results in increased excretion of cations), and (3) increased plasma aldosterone levels (secondary to volume contraction caused by excessive loss of salt and water in the urine). Diuretic drugs (used in the treatment of hypertension and edema) are the most common cause of abnormally high renal potassium loss. This effect is probably due mostly to the increased sodium and rate of fluid flow in the collecting ducts that these agents produce (Chapter 23). Patients taking diuretics are often advised to eat bananas, a rich source of potassium.

Excessive amounts of potassium can also be lost from the body through the gastrointestinal tract. Diarrheal fluid may have a potassium concentration of 50 to 60 mEq/L, so a disturbance of the lower gastrointestinal tract may rapidly lead to potassium depletion. Potassium depletion also develops during vomiting.

KIDNEY FAILURE AND ITS TREATMENT

Kidney failure can occur at any age. It may develop suddenly (acute renal failure) and lead to death in a matter of days, or it may be due to a loss of functional nephrons that occurs over decades (chronic renal disease). Kidney disease has a variety of causes, among which are bacterial infections, congenital defects, immunological injury, impaired renal blood flow, toxins, tumors, and urinary tract obstruction. The specific type of kidney disease can usually be diagnosed from the patient's history and clinical tests. In all forms of renal failure, however, the signs and symptoms are very similar. This symptom complex has been called **uremia** (urine in the blood). Basically, uremia indicates a disturbed internal environment, due to renal failure, that impairs the function of all body cells.

Renal Failure Results in Disturbances in Many Body Systems

Following are some of the disturbances that occur in uremia. This alphabetical list is not comprehensive but should serve to reemphasize many of the important functions of normal, healthy kidneys.

Anemia, with a decrease in blood hematocrit to 20% to 35%, may develop in chronic kidney failure. The anemia is mainly due to deficient production of **erythropoietin,** a hormone produced by the kidneys. Erythropoietin stimulates the bone marrow to produce red blood cells (erythropoiesis). Recombinant human erythropoietin is administered to patients with chronic renal failure to correct the anemia.

Azotemia, the accumulation of nitrogenous waste products in the blood, develops as a consequence of a markedly decreased GFR. Plasma urea, creatinine, and uric acid levels are elevated.

Calcium and phosphate homeostasis are altered with renal failure. In uremia, plasma calcium is typically decreased, and plasma phosphate is increased. Secondary hyperparathyroidism, deficient synthesis of the active form of vitamin D by the kidneys, and acidosis all lead to bone disease.

Hypertension (high blood pressure) is often present in renal failure. Its origin is not always known, but there is evidence for several possible pathways. The diseased kidney may produce excessive amounts of renin. Renin causes increased production of angiotensin II (a potent vasoconstrictor) and aldosterone (which favors salt retention). The inability of the kidneys to excrete a normal amount of sodium could also lead to hypertension, owing to excess volume. Finally, the diseased kidneys may fail to produce vasodilator substances.

Metabolic acidosis typically accompanies severe renal failure. Acid end-products of metabolism are not excreted at a normal rate and hence accumulate in the body. Impaired renal synthesis of ammonia contributes to the decreased ability to excrete hydrogen ions.

Potassium balance is usually well maintained until GFR decreases to less than 15 to 20 ml/min. Retention of potassium can cause plasma levels to increase to fatal levels.

Sodium balance is typically impaired in patients with severe renal failure. On a normal sodium diet, excessive salt and water may be retained, leading to generalized edema. On a low-salt diet, patients may become salt-depleted.

Urinary concentrating ability is impaired with renal failure. The osmotic gradient in the medulla is reduced or abolished. Urine with an osmolality close to that of plasma is excreted.

Numerous other cardiovascular, gastrointestinal, muscular, and nervous abnormalities occur in the uremic state. In many instances, it is not understood why renal failure causes these disturbances.

Dialysis and Transplantation Are Used to Treat Patients with Terminal Renal Failure

Most of the signs and symptoms of uremia can be relieved by dialysis. **Dialysis** is the separation of smaller from larger molecules in solution by diffusion of the small molecules through a selectively permeable membrane. In treating uremia, two methods of dialysis may be used.

In **peritoneal dialysis,** approximately 2 liters of hyperosmotic glucose-salt solution are introduced into the abdominal cavity. The peritoneum acts as a dialyzing membrane. Small molecules and ions (for example, water, urea, and potassium) diffuse into the injected solution, which is then drained and discarded. This procedure is used by some patients with chronic renal failure on an outpatient basis and is called **continuous ambulatory peritoneal dialysis.** The patient can perform self-dialysis and usually does this several times in a day.

Hemodialysis (Fig. 24–12) is a more efficient process involving an **artificial kidney.** The patient's blood is pumped through cellophane-like tubing immersed in a bath of balanced salt solution. Urea, potassium, and other substances in excess in the body diffuse through the tubing membrane out of the blood and into the bath. Patients are usually dialyzed three times a week, and they can often undergo this procedure at home for 4 to 6 hours while watching television or reading.

Dialysis can allow patients with otherwise fatal chronic renal failure to live useful and productive lives. It has some drawbacks, however. It does not correct the anemia. It controls hypertension only with difficulty. Uremia develops between the periods of dialysis. There is a constant risk for infection and, with hemodialysis, hemorrhage. Dialysis does not maintain normal growth and development in children. Finally, it is costly. Dialysis is, however, invaluable in keeping patients alive and functioning until suitable kidney transplants become available.

Renal transplantation is the only "cure" for patients with terminal chronic renal disease. It may restore complete health and function. In 1999, approximately 12,500 renal transplant operations were performed in the United States. At present, approximately 94% of kidneys grafted from a living related donor function for at least one year; grafts derived from an unrelated person who has just died (cadaver donors) have a one-year survival rate of approximately 90%.

A major problem in transplantation today is an inadequate number of kidney donors; the organ transplant waiting list has reached all-time highs. The immunological rejection of the transplanted kidney by the host is a major biological problem. Finally, transplantation (or dialysis) remains quite costly.

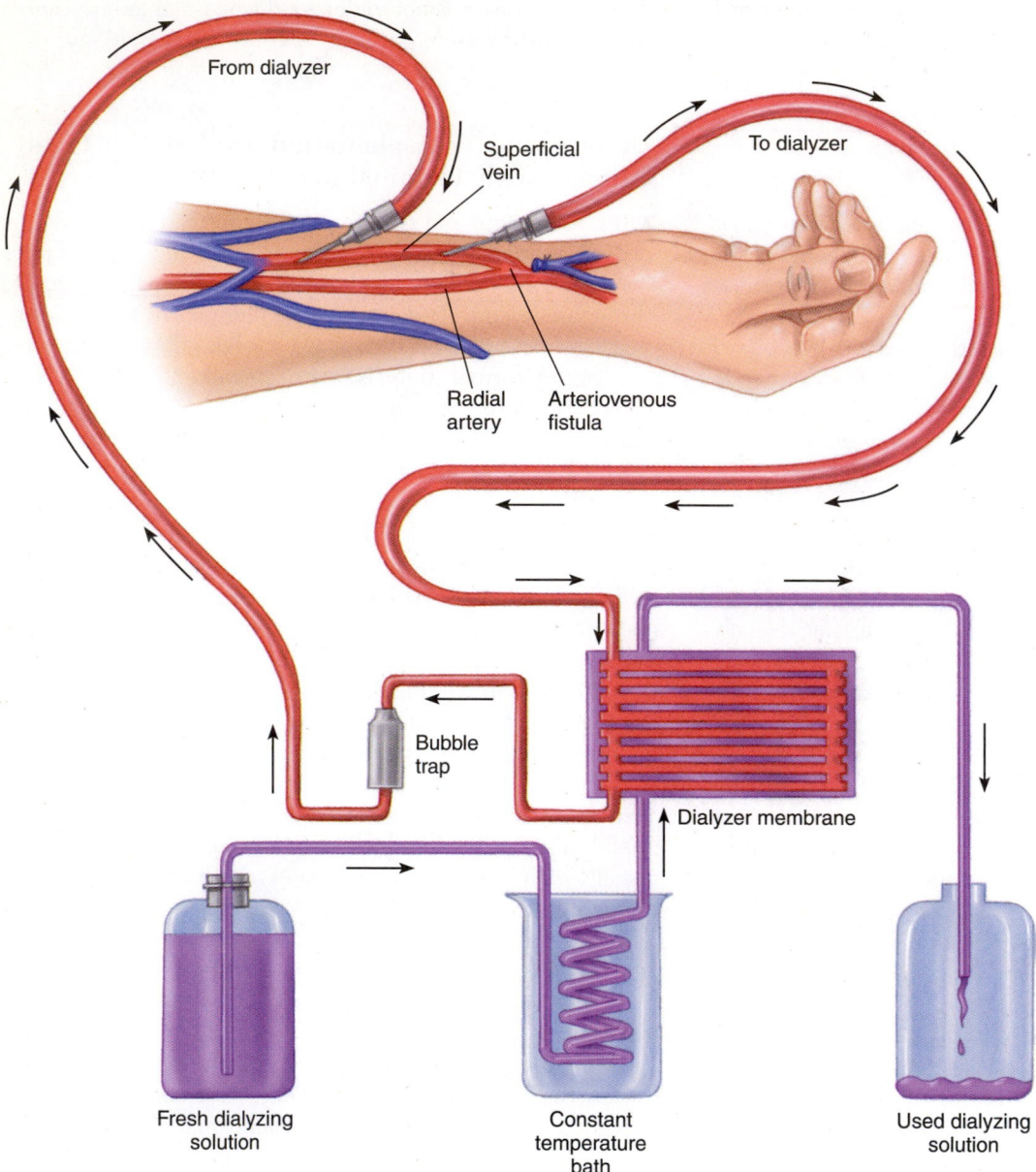

Figure 24–12

Arrangement for hemodialysis. An arteriovenous fistula is created between the radial artery and a nearby superficial wrist vein. The vein is punctured with needles, which allows blood to be carried to and from the dialyzer ("artificial kidney"). Alternatively, a shunt made of silicone rubber tubing can be permanently implanted and can be opened and connected to the blood lines of the dialyzer. In the dialyzer, the blood is separated from dialysis fluid by a thin membrane. Urea, other waste products, and potassium ions diffuse out of the blood and into the dialysis fluid. The blood may also be ultrafiltered, to remove excess fluid, by applying a hydrostatic pressure gradient across the membrane.

CHAPTER REVIEW

Summary

- In an average young woman, total body water is 50% of body weight, intracellular water is 30% of body weight, and extracellular water is 20% of body weight. In an average young man, total body water is 60% of body weight, intracellular water is 40% of body weight, and extracellular water is 20% of body weight. The differences between the sexes are due to the relatively higher proportion of adipose tissue and lower proportion of muscle tissue in women compared with men.

- Potassium ions are the major osmotically active solutes in cells; potassium influences cell volume, excitability, and metabolism.
- Sodium and accompanying anions, chloride and bicarbonate, are the major osmotically active solutes in extracellular fluid. The amount of water in, and hence volume of, the extracellular fluid space is determined primarily by the amount of sodium in this compartment.

- In general, cell membranes are highly permeable to water, and so osmotic equilibrium between intracellular and extracellular fluids results.
- In the vascular system, the colloid osmotic pressure exerted by the plasma proteins keeps fluid within the circulation.
- The kidneys play a major role in keeping us in a stable balance with respect to water, sodium and potassium ions, and numerous other substances. Abnormal renal excretion is a common cause of imbalances (positive or negative balances).
- Plasma osmolality is regulated by renal excretion of water, which is controlled by ADH, and the thirst mechanism. ADH release and thirst are stimulated by cellular dehydration and by extracellular dehydration. Thirst is an emergency mechanism. Under ordinary circumstances, the main stimulus for ADH release is an increase in plasma osmolality detected by osmoreceptor cells in the hypothalamus.
- The kidneys are the major site of sodium output and regulation of extracellular fluid volume. An increase in extracellular fluid volume (or effective arterial blood volume) leads to increased renal sodium excretion and vice versa. Renal excretion of sodium represents the difference between filtered and reabsorbed amounts. The following factors influence sodium excretion: (1) glomerular filtration rate, (2) mineralocorticoids (aldosterone), (3) peritubular capillary Starling forces, (4) renal sympathetic nerve activity, (5) atrial natriuretic peptide, (6) glucocorticoids, (7) estrogens, (8) osmotic diuretics, (9) poorly reabsorbed anions, and (10) diuretic drugs.
- Most of the body's potassium is within cells. The kidneys normally maintain potassium balance by excreting most of the ingested potassium. Changes in dietary potassium intake cause changes in plasma aldosterone levels and kidney cell potassium concentration, and in this way change secretion and excretion of potassium by the kidneys.
- The symptom complex (uremia) of renal failure is much the same no matter what the cause, and reflects an abnormal internal environment.
- Dialysis can correct some of the abnormalities that occur in uremia (e.g., acidosis, hyperkalemia, and high plasma urea and creatinine concentrations), but it does not effectively correct anemia and hypertension, so it is not a cure for renal failure.
- Successful renal transplantation is the best hope for patients with chronic renal failure.

Review Questions

Choose the Correct Answer

1. The greatest fraction of the body's water is contained in:
   a. blood plasma.
   b. cells.
   c. extracellular fluid.
   d. transcellular fluid.
2. Intravenous infusion of 1 liter of isotonic saline will cause:
   a. a 1-liter increase in intracellular fluid volume.
   b. a 1-liter increase in extracellular fluid volume.
   c. a 0.5-liter increase in intracellular fluid volume and a 0.5-liter increase in extracellular fluid volume.
3. Intravenous infusion of 1 liter of hypertonic saline will cause:
   a. a decrease in intracellular fluid volume.
   b. an increase in extracellular fluid volume.
   c. an increase in plasma osmolality.
   d. All of the above
4. Release of antidiuretic hormone from the posterior pituitary is stimulated by:
   a. a decrease in plasma osmolality.
   b. severe hemorrhage.
   c. stimulation of arterial baroreceptors.
   d. stretch of left atrial receptors.
5. Which of the following produces a decrease in renal sodium excretion?
   a. Decreased plasma aldosterone level
   b. Increased plasma level of atrial natriuretic peptide
   c. Increased GFR
   d. Increased renal sympathetic nerve activity
6. Which of the following produces an increase in renal sodium excretion?
   a. Administration of glucocorticoids
   b. Decreased peritubular capillary hydrostatic pressure

   c. Increased plasma estrogen levels
   d. Uncontrolled diabetes mellitus
7. Renin release is stimulated by:
   a. increased blood pressure in afferent arterioles.
   b. increased effective arterial blood volume.
   c. increased sodium chloride transport by macula densa cells.
   d. stimulation of renal sympathetic nerves.
8. Which of the following are commonly seen in patients with severe congestive heart failure?
   a. Elevated plasma ADH levels
   b. Generalized edema
   c. Hyponatremia
   d. Thirst
   e. All of the above
9. The most abundant intracellular cation is:
   a. calcium.
   b. chloride.
   c. potassium.
   d. sodium.
10. Which of the following promotes a shift of potassium from cells to extracellular fluid?
    a. A decrease in plasma pH
    b. An overdose of digitalis
    c. Inadequate blood flow
    d. Lack of insulin
    e. All of the above
11. Which of the following produces excessive urinary excretion of potassium?
    a. Acute renal failure
    b. Inadequate aldosterone secretion
    c. Severe chronic renal failure (for example, GFR < 10 ml/min)
    d. Uncontrolled diabetes mellitus

12. The largest fraction of cell mass in the human body is comprised of:
    a. blood cells.
    b. connective tissue cells.
    c. epithelial cells.
    d. nerve cells.
    e. skeletal muscle cells.
13. What is the magnesium concentration in mEq/L and the approximate osmolality of a 2 mmol/L $MgCl_2$ solution?
    a. 2 mEq/L, 2 mosm/kg H_2O
    b. 2 mEq/L, 3 mosm/kg H_2O
    c. 2 mEq/L, 6 mosm/kg H_2O
    d. 4 mEq/L, 3 mosm/kg H_2O
    e. 4 mEq/L, 6 mosm/kg H_2O
14. The stimulus for excessive renal salt and water retention in patients with congestive heart failure is most likely:
    a. a decrease in effective arterial blood volume.
    b. a decrease in extracellular fluid volume.
    c. a decrease in total blood volume.
    d. an increase in interstitial fluid volume.
    e. an increase in total blood volume.
15. What solute is the major determinant of the amount of water in the extracellular fluid compartment?
    a. Glucose
    b. Potassium
    c. Serum albumin
    d. Sodium
    e. Urea
16. Which tissue has the lowest water content?
    a. Adipose tissue
    b. Blood
    c. Bone
    d. Skeletal muscle
17. A hypertensive patient was given an angiotensin-converting enzyme (ACE) inhibitor. Which of the following is expected?
    a. A decrease in renal sodium excretion
    b. An increase in blood pressure
    c. An increase in plasma aldosterone concentration
    d. An increase in plasma angiotensin I concentration
    e. An increase in plasma angiotensin II concentration
18. An 80-kg man with congestive heart failure consumes 250 mmoles of NaCl every day for 5 days but excretes only 100 mmoles of NaCl/day. How much will he weigh at the end of this time? Assume that he has free access to water and that 1 liter of extracellular fluid contains 150 mmoles of NaCl and weighs 1 kg.
    a. 75 kg
    b. 80 kg
    c. 85 kg
    d. 88 kg
19. Which of the following is a transcellular fluid?
    a. Cerebrospinal fluid
    b. Glomerular filtrate
    c. Interstitial fluid
    d. Lymph
    e. All of the above
20. Which of the following is a common problem in patients with end-stage, or terminal, renal disease?
    a. Elevated plasma phosphate levels (hyperphosphatemia)
    b. Hypokalemia
    c. Hypotension
    d. Increased hematocrit (polycythemia)
    e. Metabolic alkalosis

Answers to Case History Questions

1. Most (approximately 85%) of the osmolality of the electrolyte replacement glucose solution is due to its glucose. The glucose is metabolized in the body and leaves behind pure water.
2. Sweat is a hypo-osmotic fluid compared to plasma, with an osmolality of 30 to 230 mosm/kg H_2O and sodium concentration of 10 to 100 mEq/L. If sweat is lost and no fluid is replaced, then plasma osmolality (and sodium concentration) increases and plasma volume decreases.
3. The low plasma sodium (hyponatremia) and hypo-osmolality in the runner are likely due to excessive intake and retention of water and failure to replace the salt lost in sweat. Renal excretion of water did not keep up with water intake during the race, possibly because of a decreased glomerular filtration rate and high plasma ADH level associated with the stress of running.
4. When plasma osmolality decreases, cells swell. The brain is particularly sensitive to hypo-osmolality because it is enclosed in the rigid cranium. Intracranial pressure may increase considerably when brain cells swell. This compromises brain blood flow, which may contribute to the neurological symptoms.
5. The hypertonic NaCl solution was given to increase the plasma sodium concentration level and reduce cell swelling but was given slowly and cautiously, with repeated measurements of plasma sodium concentration, to avoid possible severe brain damage (due particularly to demyelination of nerve fibers in the pons).

Key Terms

antidiuretic hormone (p. 766)
aquaporin (p. 768)
atrial natriuretic peptide (p. 775)

azotemia (p. 779)
dialysis (p. 779)
effective arterial blood volume (p. 765)

electrolyte (p. 761)
extracellular fluid (p. 760)
glomerulotubular balance (p. 773)

hyperkalemia (p. 777)
hypokalemia (p. 777)
hyponatremia (p. 766)
hypovolemia (p. 765)

Suggested Readings

Agre, P. "Aquaporin water channels in kidney." *J Am Soc Nephrol* 11:764–777, 2000.

Brenner, B. M., ed. *Brenner and Rector's The Kidney,* ed 6. Philadelphia, W. B. Saunders, 2000.

Narins, R. G., ed. *Maxwell and Kleeman's Clinical Disorders of Fluid and Electrolyte Metabolism,* ed 5. New York, McGraw-Hill, 1994.

Rose, B. D., and Post, T. W. *Clinical Physiology of Acid–Base and Electrolyte Disorders,* ed 5. New York: McGraw-Hill, 2001.

Seldin, D. W., and Giebisch, G., eds. *The Kidney: Physiology and Pathophysiology,* ed 3. Philadelphia, Lippincott Williams & Wilkins, 2000.

Answers to Review Questions

1. b **2.** b **3.** d **4.** b **5.** d **6.** d **7.** d **8.** e **9.** c **10.** e **11.** d
12. e **13.** e **14.** a **15.** d **16.** a **17.** d **18.** c **19.** a **20.** a

Chapter 25

REGULATION OF ACID-BASE PHYSIOLOGY

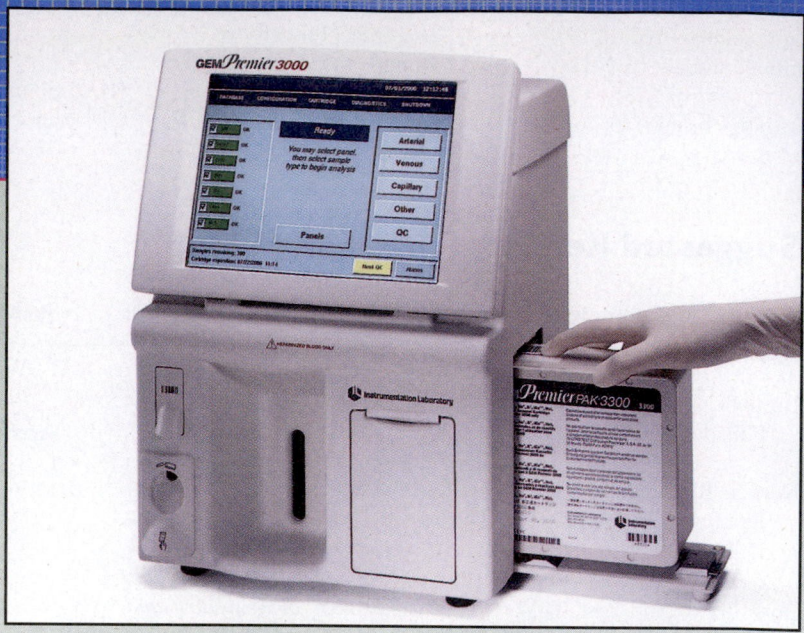

- An automated blood analyzer that measures blood gases, electrolytes, and metabolites important in acid-base physiology.

KEY CONCEPTS

- Stability of pH (or H^+ concentration) is achieved by buffers.

- The HCO_3^-/CO_2 buffer pair is especially important in the body. The lungs influence plasma pH by controlling the arterial blood P_{CO_2}, and the kidneys influence plasma pH by controlling the plasma bicarbonate (HCO_3^-) concentration.

- As a result of metabolism of foodstuffs (especially proteins), the body is usually threatened by the net production of acids. Normal kidneys maintain acid-base balance by excreting the excess hydrogen ions, which are mainly combined with urinary buffers, such as phosphate (titratable acid) and ammonia.

- There are four simple types of acid-base disturbance.

CASE HISTORY

A 20-year-old college junior came to the hospital emergency room after taking an overdose of aspirin (acetylsalicylic acid) in a suicide attempt. She confessed that she had been despondent about the recent breakup with her boyfriend and had consumed approximately a half-bottle of adult aspirin tablets (325 mg/tablet) approximately 2 hours before. She complained of ringing in her ears (tinnitus), nausea, dizziness, and shortness of breath.

The physical examination, electrocardiogram, and a chest X-ray were all normal. Blood pressure was 115/75 mm Hg, and heart rate, 80 bpm. Respiratory rate (30 breaths/min) and depth of breathing were abnormally high.

Vomiting was induced by administering ipecac syrup. Several undissolved tablet fragments were recovered. This was followed by administration by mouth of a slurry of activated charcoal to reduce absorption of the aspirin from the stomach and small intestine.

Supplemental oxygen was given, and an arterial blood sample showed a pH of 7.48, Po_2 of 120 mm Hg, and Pco_2 of 18 mm Hg. Plasma sodium concentration was 142 mEq/L; potassium, 3.8 mEq/L; chloride, 105 mEq/L; BUN, 11 mg/dl; and glucose, 90 mg/dl. The plasma salicylate level on admission was 2.8 mmol/L. The patient was hospitalized, and an intravenous infusion containing sodium bicarbonate was started. This produced a large increase in urine output and an increase in urine pH to 8.0. Arterial blood gases, pH, and plasma electrolytes were closely monitored over the next 24 hours.

The following day, the patient's serum salicylate level was 0.6 mmol/L, her symptoms were gone, and she felt considerably better. She was discharged in the company of her parents. Psychiatric counseling was arranged.

Questions

1. What is the plasma bicarbonate concentration in this patient?

2. What is the evidence that she has a respiratory alkalosis?

3. Why does she also have a metabolic acidosis?

4. What is the rationale for the intravenous infusion of the sodium bicarbonate solution?

INTRODUCTION

In this chapter, we will explain why pH is important, how pH changes are minimized by chemical buffers, how the lungs and kidneys defend a normal blood pH, and what types of acid-base disturbances occur in people.

This chapter will focus on the regulation of extracellular fluid pH or hydrogen ion concentration ($[H^+]$). A normal extracellular fluid $[H^+]$ is an important part of the ideal internal environment that our cells need (Chapter 1). In turn, constancy of extracellular $[H^+]$ promotes stability of intracellular (cytoplasmic) $[H^+]$. Intracellular $[H^+]$ is important because it influences the activity of enzymes that control metabolic reactions and energy production. It also affects many physiological activities, including (1) muscle contraction, (2) the opening and closing of cell membrane ion channels, (3) secretion by endocrine gland cells, and (4) cell division and proliferation. Disturbances in $[H^+]$ profoundly affect function and can eventually become fatal.

PRINCIPLES OF ACID-BASE PHYSIOLOGY

An Acid Is a Proton (H^+) Donor and a Base Is a Proton Acceptor

 What are the chemical properties of acids and bases?

As first presented in Chapter 2, an **acid** is a substance that can donate hydrogen ions. A hydrogen ion (H^+) consists of a bare proton, a hydrogen atom without its orbiting electron. Examples of acids include **hydrochloric acid (HCl), sulfuric acid (H_2SO_4), nitric acid (HNO_3), phosphoric acid (H_3PO_4), ammonium ion (NH_4^+), lactic acid, acetic acid,** and **carbonic acid (H_2CO_3).** An acid donates its H^+ to a base.

Accordingly, a **base** is a substance that can accept or bind hydrogen ions. Examples of bases include **sodium hydroxide (NaOH), potassium hydroxide (KOH), ammonia (NH_3),** and **lactate, acetate,** and **bicarbonate (HCO_3^-)** ions.

Not all acids have the word *acid* in their names. For example, the ammonium ion (NH_4^+) is an acid. If we mix aqueous solutions of the salts ammonium chloride (NH_4Cl) and sodium bicarbonate ($NaHCO_3$), the following reversible reaction takes place:

$$NH_4^+Cl^- + Na^+HCO_3^- \rightleftharpoons NH_3 + H_2CO_3 + Na^+Cl^-$$

In this reaction (left to right), the ammonium ion is an acid; it donates a H^+ to the base bicarbonate.

In a **neutralization reaction,** an acid reacts with a base to form a salt and water. For example, if we add HCl to a solution of NaOH, the following reaction occurs:

$$H^+Cl^- + Na^+OH^- \rightarrow Na^+Cl^- + H_2O$$

The strong acid HCl donates a H^+ to the hydroxide ion of NaOH to form an ionized salt (NaCl) and water.

The amount of an acid or base is often expressed in terms of **equivalents** or **milliequivalents (mEq)** (1 mEq is 1/1000th of an equivalent). One equivalent of acid neutralizes one equivalent of base. One mole of HCl contains one equivalent of acid. One mole of H_2SO_4 contains two equivalents of acid because sulfuric acid can donate two H^+s. Therefore, 2 moles (two equivalents) of NaOH are required to neutralize 1 mole of sulfuric acid, according to the following two steps:

$$(1)\ H_2SO_4 + NaOH \rightarrow NaHSO_4 + H_2O$$
$$(2)\ NaHSO_4 + NaOH \rightarrow Na_2SO_4 + H_2O$$

Amphoteric Substances Can Function as Acids and Bases

Some chemical substances can function as both an acid and a base and are referred to as **amphoteric substances** (*amphoteros* in Greek means "pertaining to both"). These include amino acids and proteins. For example, the amino acid **glycine,** $^+H_3N\text{---}CH_2\text{---}COO^-$, acts as an acid in the following reaction:

$$NaOH + {}^+H_3N\text{---}CH_2\text{---}COO^- \rightarrow$$
$$H_2N\text{---}CH_2\text{---}COO^-Na^+ + H_2O$$

The ammonium group of glycine donates H^+ to the hydroxide ion of NaOH. Glycine acts as a base in the following reaction:

$$HCl + {}^+H_3N\text{---}CH_2\text{---}COO^- \rightarrow Cl^-{}^+H_3N\text{---}CH_2\text{---}COOH$$

The carboxylate group of glycine accepts H^+ from HCl. Proteins contain many different side groups that can act as acids or bases.

The Acid Dissociation Constant Is Directly Related to the Strength of an Acid

When an acid (generically indicated as **HA**) is added to water, the following reversible reaction takes place:

$$HA + H_2O \rightleftharpoons H_3O^+ + A^-$$

For simplicity, it is common practice to ignore water in this reaction and to write:

$$HA \rightleftharpoons H^+ + A^-$$

The reaction going from left to right is called the **dissociation reaction,** and the reaction going from right to left is the **association reaction.** The rate of the dissociation reaction is equal to the product of the concentration of HA and the **dis-**

sociation rate constant k_1, a specific value for this reaction. The rate of the association reaction is equal to the product of the concentrations of H^+ and A^- and the **association rate constant k_2,** a specific value for this reaction. At equilibrium, the rates of dissociation and association reactions are equal. Therefore, we can write:

$$k_1 \times [HA] = k_2 \times [H^+] \times [A^-]$$

Rearranging, we get $[H^+] \times [A^-]/[HA] = k_1/k_2$. We define a new constant K_a as equal to the ratio k_1/k_2. K_a is the **equilibrium constant** for this reaction and is called the **dissociation constant,** or **ionization constant,** for the acid.

The higher the acid dissociation constant, the more completely an acid is dissociated or ionized and, consequently, the more acidic the solution because more free (unbound) H^+s are present. Acids with high dissociation constants are **strong acids.** These include hydrochloric, sulfuric, nitric, and phosphoric acids. Hydrochloric acid, for example, is essentially completely dissociated into H^+s and chloride ions in dilute aqueous solution. There is very little undissociated HCl. In a 0.1 mol/L HCl solution, the free $[H^+]$ is nearly 0.1 mol/L.

A low acid dissociation constant is characteristic of **weak acids.** In other words, a weak acid is incompletely dissociated in solution. For example, the dissociation constant of acetic acid is equal to 1.8×10^{-5}. If we have a 0.100 mol/L solution of acetic acid in water, the concentration of undissociated acid (CH_3COOH) is 0.0987 mol/L, and the concentrations of free hydrogen and acetate (CH_3COO^-) ions are both equal to 0.0013 mol/L. In other words, nearly 99% of the acetic acid molecules are not dissociated in this solution. Note especially that the concentration of free H^+s (or acidity) is low for a solution of a weak acid.

Acid dissociation constants are usually small and awkward to manipulate mathematically, so they are often presented in a logarithmic form. The **pK_a** is defined as the logarithm, to the base 10, of the *inverse* of K_a, or:

$$pK_a = \log (1/K_a).$$

For example, for acetic acid, $pK_a = \log (1/[1.8 \times 10^{-5}]) = \log (1/0.000018) = \log 55556 = 4.7$. A low pK_a corresponds to a high dissociation constant (strong acid), and a high pK_a corresponds to a low dissociation constant (weak acid).

Water can act as either an acid or base, as described by the following reaction:

$$H_2O + H_2O \rightleftharpoons H_3O^+ + OH^-$$

In this reaction, one water molecule is a proton acceptor (and forms a **hydronium ion,** a hydrated hydrogen ion) and the other is a proton donor (and forms the hydroxide ion, OH^-). This reaction is usually simplified as:

$$H_2O \rightleftharpoons H^+ + OH^-$$

and describes the dissociation of water, which is actually very small, because 1 liter (or 1 kg) of pure water contains approximately 56 moles ([1000 g]/[18 g/mole]) of undissociated water and 10^{-7} moles of dissociated water. Because the dissociation of water is so low, water is an extremely weak acid or base.

The pH of Aqueous Solutions Ranges Widely

What does pH mean?

pH is defined as the logarithm, to the base 10, of the *inverse* of the molar activity (concentration) of free H^+s. In equation form:

$$pH = \log (1/[H^+])$$

Because of the inverse relationship between pH and $[H^+]$, a low pH indicates a high $[H^+]$, and a high pH indicates a low $[H^+]$. The pH scale was devised by chemists to compress the very wide range of $[H^+]$s possible in aqueous solutions into a manageable scale. The pH scale usually goes from 0 to 14, with a pH of 0 indicating a very acidic solution (e.g., 1 mol/L HCl which has a $[H^+]$ of 1 mol/L), and a pH of 14 indicating a very alkaline solution (e.g., 1 mol/L NaOH, which has a $[H^+]$ of 10^{-14} mol/L).

The Henderson-Hasselbalch Equation

The equilibrium equation for dissociation of an acid is often expressed in logarithmic form. We can rearrange the equation:

$$[H^+] \times [A^-]/[HA] = K_a$$

and solve for $[H^+]$:

$$[H^+] = K_a \times [HA]/[A^-]$$

Taking the logarithms of this equation, we get:

$$\log [H^+] = \log K_a + \log ([HA]/[A^-])$$

Multiplying both sides of the equation by -1, we get:

$$-\log [H^+] = -\log K_a + \log ([A^-]/[HA])$$

Because $pH = \log (1/[H^+]) = -\log [H^+]$ and $pK_a = \log (1/K_a) = -\log K_a$, we can write the equation:

$$pH = pK_a + \log ([A^-]/[HA])$$

This logarithmic form of the acid dissociation equation is known as the **Henderson-Hasselbalch equation** and is very useful, as we will see later. From this equation, we can see that when $[A^-] = [HA]$, the solution pH equals the pK_a (because the logarithm of 1 is 0).

pH Values for Solutions of Acids and Bases, Body Fluids, and Beverages

Table 25–1 presents pH values for some aqueous solutions of chemicals, body fluids, and beverages. Notice that the pH of various solutions of acids depends not only on the concentration of acid but also on the dissociation constant of the acid. Strong acids produce solutions with a lower pH than do the same concentrations of weak acids because the stronger acid is dissociated (ionized) more and so yields a greater concentration of free H^+s in solution.

TABLE 25–1

pH Values of Aqueous Solutions of Chemicals, Body Fluids, and Beverages

Chemicals	pH
0.100 mol/L hydrochloric acid	1.1
0.010 mol/L hydrochloric acid	2.0
0.001 mol/L hydrochloric acid	3.0
0.100 mol/L lactic acid (pK_a 3.9)	2.5
0.100 mol/L acetic acid (pK_a 4.7)	2.9
Pure water	7.0
0.100 mol/L sodium bicarbonate	8.4
0.100 mol/L ammonia	11.1
0.100 mol/L sodium hydroxide	13.0
Body fluids	
Arterial blood	7.40
Venous blood	7.35
Cerebrospinal fluid	7.35
Skeletal muscle cell cytoplasm	6.9
Saliva	5.8–7.1
Gastric juice	0.7–3.8
Gallbladder bile	5.6–8.0
Pancreatic juice	7.5–8.8
Intestinal juice	7.0–8.0
Feces	5.9–8.5
Urine	4.5–8.0
Beverages	
Lemon juice	2.2–2.4
Carbonated soft drinks	2.8–3.7
Orange juice	3.7
Beer	4.4
Coffee	5.0
Milk	6.7
Tea	6.9

The pH of arterial blood normally averages 7.40, which corresponds to a $[H^+]$ of 4×10^{-8} mol/L or 40 nmol/L. (A nmol, or nanomole, is 10^{-9} mole.) The pH of venous blood is slightly lower (average 7.35) owing to its higher content of carbon dioxide (hence, H_2CO_3). Most body fluids, with the exception of gastric juice, have pH values in the range 5 to 8. Cell cytoplasm is more acidic than is extracellular fluid; the pH of skeletal muscle cell cytoplasm averages 6.9. Within a cell, pH differs in different intracellular organelles; for example, pH is approximately 4.5 in lysosomes, which contain enzymes that operate best in an acidic environment. Gastric juice is acidic owing to its high concentration of hydrochloric acid. Pancreatic juice is slightly more alkaline than is blood because of its high bicarbonate concentration. The pH of urine can vary between 4.5 and 8.0, depending on acid-base conditions in the rest of the body.

Many foods are acidic or alkaline. When we consume food, the load of acid or base depends not only on the free $[H^+]$ but also on how much undissociated acid (or base) is present. The effect of a chemical substance on acid-base balance also depends on how the substance is metabolized in the body, a subject we will discuss later.

Buffers Promote Stability of pH

> *What is a buffer, and how does a chemical buffer work?*

A buffer, by definition, is something that minimizes a shock and thereby promotes relative stability. A **pH buffer** *minimizes* the change in $[H^+]$ when either acid or base is added to a solution; it *does not prevent* a pH change. Whenever an acid is added, pH decreases. If a base is added, pH increases. The extent of change in pH depends on the amount and nature of added acid or base and on the amount and nature of the pH buffer.

Previously, we wrote the equation for dissociation of an acid: $HA \rightleftharpoons H^+ + A^-$. In this reaction, HA is the acid and A^- is the **conjugate base.** *Conjugate* means "joined in a pair." A **chemical buffer** consists of a weak acid and its conjugate base (or a weak base and its conjugate acid). Following are some examples of conjugate acid-base buffer pairs:

$$\underset{\substack{\text{dihydrogen}\\\text{phosphate}}}{\underline{\text{acid}}\atop H_2PO_4^-} \rightleftharpoons \underset{\substack{\text{monohydrogen}\\\text{phosphate}}}{\underline{\text{conjugate base}}\atop HPO_4^{2-}} + H^+$$

$$\underset{\substack{\text{carbonic acid}}}{H_2CO_3} \rightleftharpoons \underset{\substack{\text{bicarbonate}}}{HCO_3^-} + H^+$$

$$\underset{\substack{\text{ammonium}}}{NH_4^+} \rightleftharpoons \underset{\substack{\text{ammonia}}}{NH_3} + H^+$$

The equilibrium expression for dissociation of an acid can be written in the Henderson-Hasselbalch equation form:

$$pH = pK_a + \log([\text{conjugate base}]/[\text{acid}])$$

For example, for a mixture of monohydrogen and dihydrogen phosphate,

$$pH = 6.8 + \log ([HPO_4^{2-}]/[H_2PO_4^{-}])$$

This equation demonstrates that if the ratio of conjugate base to acid is defined for a buffer of known pK_a, then the pH is automatically determined. This principle is useful in making up pH buffer solutions.

What, for example, is the pH of an aqueous solution containing a mixture of 0.1 mol/L Na_2HPO_4 (sodium monohydrogen phosphate) and 0.1 mol/L NaH_2PO_4 (sodium dihydrogen phosphate)? Because the ratio $HPO_4^{2-}/H_2PO_4^{-}$ is 1 and the logarithm of 1 is 0, the pH is 6.8 + 0 = 6.8. In another example, if the ratio of $HPO_4^{2-}/H_2PO_4^{-}$ is 4, then the pH of this solution is 6.8 + log 4 = 6.8 + 0.6 = 7.4.

Chemical buffers stabilize the $[H^+]$ of aqueous solutions. For example, if we add strong acid (HCl) to a phosphate buffer solution, the following reaction takes place:

$$H^+Cl^- + Na_2^+HPO_4^{2-} \rightarrow Na^+H_2PO_4^- + Na^+Cl^-$$

In this reaction, the base component of the buffer pair (HPO_4^{2-}) binds H^+s. The strong acid (HCl) is, in effect, converted into a weak acid ($H_2PO_4^-$), thereby minimizing the increase in free $[H^+]$ (decrease in pH). If we add strong base (NaOH) to the phosphate buffer solution, the following reaction takes place:

$$Na^+OH^- + Na^+H_2PO_4^- \rightarrow Na_2^+HPO_4^{2-} + H_2O$$

The acid component of the buffer pair ($H_2PO_4^-$) liberates H^+s, thereby diminishing the increase in pH caused by addition of strong base. The strong base (NaOH) is, in effect, converted into a weak base (HPO_4^{2-}).

Figure 25–1 shows a titration curve for a phosphate buffer solution. The x axis gives the amount of strong acid or base added to the solution. The y axis gives the pH. Going from left to right along the curve, $H_2PO_4^-$ is converted to HPO_4^{2-} by the addition of a strong base. Going from right to left along the curve, HPO_4^{2-} is converted to $H_2PO_4^-$ by the addition of a strong acid. The slope of the titration curve is an index of the effectiveness of the buffer in resisting a change in pH. The slope of the curve is flattest when the pH is equal to the pK_a of the phosphate buffer. This means that for a given amount of added acid or base, the pH change is least near the pK_a of the buffer. As we move away from the pK_a, the effectiveness of the buffer declines. In choosing a buffer to stabilize a certain pH, it is always best to select a buffer pair with a pK_a close to the desired pH.

The effectiveness of a buffer in minimizing pH changes depends on two factors: the pK_a of the buffer in relation to the desired pH and the concentration (or amount) of buffer. The pK_a of the imidazole group of histidine (an amino acid found in hemoglobin) is close to 7.4, and hence this is an ideal buffer

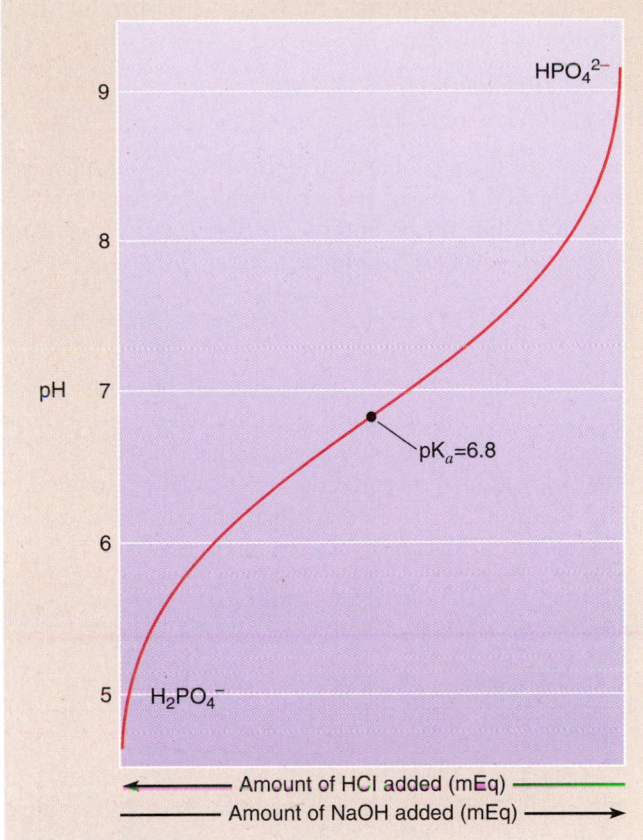

Figure 25–1

Titration curve for inorganic phosphate buffer. HPO_4^{2-} predominates at alkaline pHs. When strong acid (HCl) is added, the reversible reaction $H^+ + HPO_4^{2-} \rightarrow H_2PO_4^-$ takes place. When strong base (NaOH) is added, $H_2PO_4^-$ is converted to HPO_4^{2-}: $OH^- + H_2PO_4^- \rightarrow HPO_4^{2-} + H_2O$. The change in pH produced by a given amount of acid or base is least when the solution pH is at the pK_a (6.8) of the buffer pair $HPO_4^{2-}/H_2PO_4^{-}$.

in blood. The greater the concentration of buffer, the greater the ability to bind (when acid is added) or release (when base is added) H^+s. If no buffers are present, as in pure water or in an aqueous solution of pure NaCl, large pH changes are produced by the addition of very small amounts of acid or base.

Hydrogen Ions Are Produced in Metabolic Reactions

 How are hydrogen ions added to the body?

The metabolism of food consumed in our diet is usually the major source of H^+s. For most people eating a mixed diet of meat and vegetables, metabolism results in net addition of acid to the body. Vegetarians can be faced with the opposite situation, net addition of base. Because the usual problem is one of eliminating excess acid, we will emphasize this situation for most of this chapter.

Metabolic Production of CO_2 as a Source of Hydrogen Ions

In the course of metabolism, a normal adult produces approximately 300 liters of carbon dioxide per day. Carbon dioxide from the tissues enters the blood and reacts with water to produce an acid, carbonic acid:

$$CO_2 + H_2O \rightleftharpoons H_2CO_3 \rightleftharpoons H^+ + HCO_3^-$$

In the lungs, these reactions are reversed because carbon dioxide is expired. No acid burden is imposed on the body, as long as carbon dioxide is not allowed to accumulate. Carbon dioxide production and expiration rates (usually measured in ml/min) are usually matched so that arterial blood carbon dioxide tension and pH normally stay quite constant.

Metabolism of Carbohydrates and Fatty Acids as a Source of Hydrogen Ions

Most carbohydrates (e.g., glucose) and fatty acids are normally completely oxidized to carbon dioxide and water. No net addition of acid to the body results because the carbon dioxide produced in the tissues is blown off by the lungs.

Incomplete oxidation of carbohydrates and fats produces nonvolatile organic acids. If glucose is not oxidized completely, then lactic acid is produced. With severe exercise or an inadequate circulation, insufficient delivery of oxygen to the tissues results in lactic acid production and a blood pH below normal. If fatty acids are incompletely oxidized, as occurs in uncontrolled diabetes mellitus, starvation, and alcoholism, ketone body acids (acetoacetic and β-hydroxybutyric acids) are produced. This situation also results in an abnormally low blood pH.

Metabolism of Proteins as a Source of Hydrogen Ions

The metabolism of dietary proteins results in the production of strong acids. Sulfuric acid is produced by the oxidation of sulfur-containing amino acids (cysteine, cystine, and methionine). Hydrochloric acid is produced by the oxidation of cationic amino acids (e.g., lysine and arginine). Phosphoric acid is produced by the oxidation of phosphorus-containing proteins.

To some extent, these acid-producing reactions are neutralized by acid-consuming reactions. Thus, the oxidation of dietary organic anions (e.g., citrate, lactate, and acetate) uses up H^+s. For example, citrate, which is contained in fruits and vegetables, is oxidized in the body according to the following simplified reaction:

$$citrate^- + O_2 + H^+ \rightarrow CO_2 + H_2O$$

In this reaction, H^+ is consumed. The net result of acid-producing and acid-consuming reactions on the usual American diet is such that an excess of approximately 50 to 100 mEq of H^+s is produced in a day.

REGULATION OF EXTRACELLULAR HYDROGEN ION CONCENTRATION: BUFFERING MECHANISMS IN THE BODY

 How is hydrogen ion concentration in the extracellular fluid regulated?

Despite the net metabolic production of acids in the body, a normal $[H^+]$ in extracellular fluid is preserved in a healthy person. The $[H^+]$ is stabilized by buffers in the body. The first line of defense of $[H^+]$ consists of chemical buffers in extracellular and intracellular fluids and in bone. The second is the lungs, which buffer blood $[H^+]$ by disposing of carbon dioxide as rapidly as it is formed. When there is excessive production of acids (or bases) in the body, lung ventilation changes in order to reduce the deviation from a normal blood $[H^+]$. The third is the kidneys, which buffer blood $[H^+]$ by excreting buffered H^+s and the anions (e.g., sulfate and chloride) liberated from acids. Figure 25–2 sketches the pathways for maintaining a stable blood pH despite the acid threat posed by cell metabolism of foodstuffs.

Many Chemical Buffers Stabilize Extracellular Fluid Hydrogen Ion Concentration

 How do chemical buffer systems in the body interact to stabilize hydrogen ion concentration?

We will consider chemical pH buffers in the body. These include (1) phosphates, (2) proteins, and (3) the bicarbonate-CO_2 system.

Phosphate as a Chemical Buffer

Phosphate, by virtue of its pK of 6.8, is a good buffer for stabilizing the pH of the blood close to its normal value of 7.4. The concentration of inorganic phosphate in extracellular fluid, however, is rather low (about 1.0 mmol/L). Because of its low concentration, inorganic phosphate has a limited capacity to buffer added acid or base. There are, however, large quantities of inorganic phosphate salts in bone and organic phosphates in cells, and these can participate in the buffering of extracellular fluid pH.

Proteins as Chemical Buffers

Proteins compose the largest amount of buffer in the body and are excellent pH buffers. Proteins contain many different amino acids with ionizable groups, which can donate or accept H^+s.

Serum albumin and plasma globulins act as pH buffers in blood plasma. The large amounts of proteins in cells also participate in buffering H^+s. An important example of an intracellular protein buffer is hemoglobin, found in red blood cells. When carbon dioxide is added to the blood in the tissues, H^+ liberated from carbonic acid is mainly bound by hemoglobin (Chapter 21).

The Bicarbonate-CO_2 System as a Chemical Buffer

The bicarbonate-CO_2 system is of special importance in pH buffering in extracellular fluid. The components of this buffer system are present in large quantities. A practically limitless supply of CO_2 is available from metabolism, and the concentration of bicarbonate in extracellular fluid, approximately 24 mEq/L, is substantial. Despite a pK of 6.10, which is far from the desired pH of 7.40, this buffer pair is quite effective because the lungs and kidneys can adjust its components.

We will now examine the bicarbonate-CO_2 relationship in the light of acid-base chemistry. We can write the following reversible reactions:

$$CO_2(g) \rightleftharpoons CO_2(d)$$
$$+$$
$$H_2O \underset{\text{Carbonic anhydrase}}{\rightleftharpoons} H_2CO_3 \rightleftharpoons H^+ + HCO_3^-$$
$$\updownarrow$$
$$H^+ + CO_3^{2-}$$

in which $CO_2(g)$ represents gaseous CO_2 in the lung alveoli, and $CO_2(d)$ represents CO_2 dissolved in pulmonary capillary blood. CO_2 in the lung alveoli equilibrates with CO_2 dissolved in pulmonary capillary blood. $CO_2(d)$ reacts with water in a **hydration reaction** to form carbonic acid. The reverse reaction, the formation of $CO_2(d)$ and water from H_2CO_3, is called the **dehydration reaction.**

Both of these reactions are slow if uncatalyzed but are speeded up by the zinc-containing enzyme **carbonic anhydrase.** This enzyme is present in red blood cells and in cells of the kidneys, pancreas, stomach, ciliary body of the eye, and choroid plexuses of the brain. The dissociation of carbonic acid to H^+ and bicarbonate or the association of H^+ and bicarbonate to form carbonic acid occur essentially instantaneously. The dissociation of bicarbonate into hydrogen and carbonate (CO_3^{2-}) ions occurs to an appreciable extent only at pH values much more alkaline than those encountered in the body fluids, and so we will not discuss this last reaction. It should be noted, however, that bone contains considerable amounts of calcium carbonate, which can buffer acids.

For the bicarbonate-CO_2 system, we can write the Henderson-Hasselbalch equation as:

$$pH = 6.10 + \log\,([HCO_3^-]/0.03\,P_{CO_2})$$

Carbonic acid does not appear explicitly in this expression, but is implicit. The denominator of the log term is the concentration of dissolved CO_2. In blood plasma at a temperature of 37°C, the concentration of dissolved CO_2 is related to the P_{CO_2} by the equation:

$$[CO_2(d)] = 0.03 \times P_{CO_2}$$

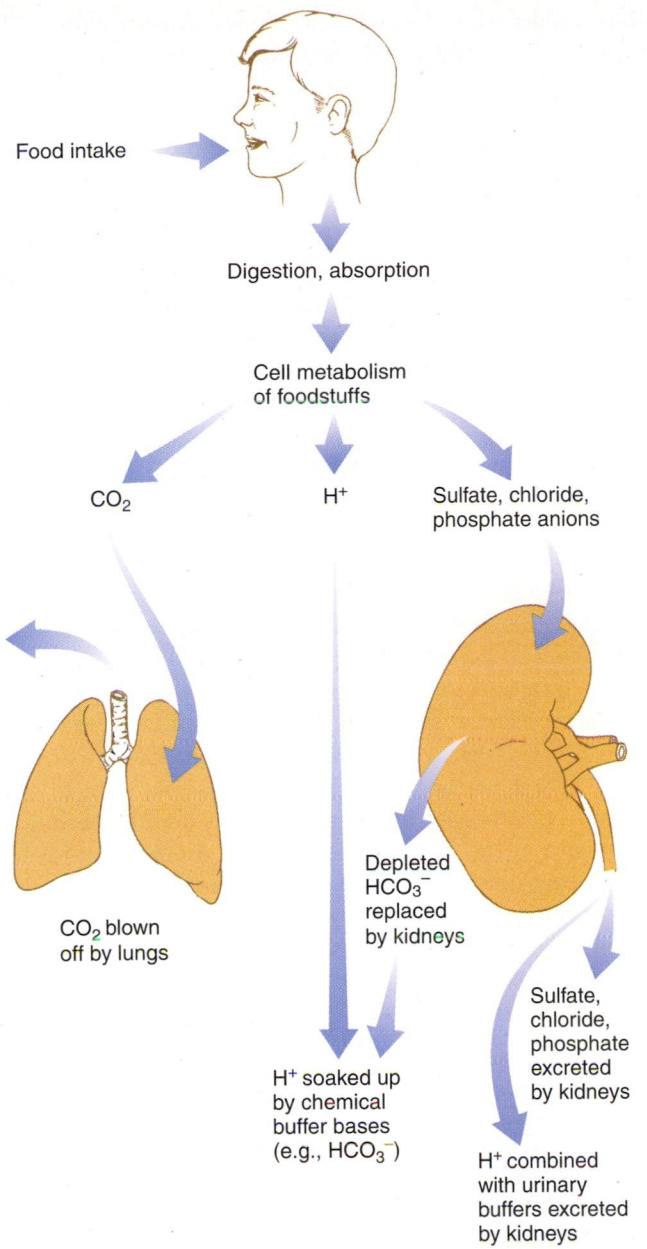

Food intake

Digestion, absorption

Cell metabolism of foodstuffs

CO_2 H^+ Sulfate, chloride, phosphate anions

CO_2 blown off by lungs

Depleted HCO_3^- replaced by kidneys

H^+ soaked up by chemical buffer bases (e.g., HCO_3^-)

Sulfate, chloride, phosphate excreted by kidneys

H^+ combined with urinary buffers excreted by kidneys

Figure 25–2

Overall schema for maintenance of acid-base balance. On the usual mixed diet, pH is threatened by production of strong acids (e.g., sulfuric, hydrochloric, and phosphoric), which result mainly from protein metabolism. These strong acids are buffered by chemical buffers in the body. Removal of extra H^+s and the accompanying anions from the body is accomplished by renal excretion. When the kidneys excrete H^+s, they add new bicarbonate to the blood, thereby restoring depleted body buffer bases. The respiratory system eliminates CO_2 produced by metabolism. CO_2 is not a threat to acid-base balance, provided its partial pressure in arterial blood is kept at a normal value.

In this equation, $CO_2(d)$ is in mmol/L, Pco_2 is in mm Hg, and 0.03 is the solubility coefficient in mmol/L per mm Hg. $CO_2(d)$ and H_2CO_3 are usually in equilibrium (remember carbonic anhydrase); in blood plasma at 37°C, the ratio of these two chemical species is approximately 400 to 1 (1.2 mmol/L versus 3 μmol/L). Because the H_2CO_3 concentration is so low and hard to measure, $CO_2(d)$ is substituted for H_2CO_3 in the Henderson-Hasselbalch equation just given. The constant (6.10) incorporates both the acid dissociation constant for H_2CO_3 and the constant for the H_2CO_3-$CO_2(d)$ equilibrium. The real source of H^+ is H_2CO_3, not CO_2; the H_2CO_3 concentration is directly proportional to the Pco_2 by way of dissolved CO_2.

As stated previously, the pK of the bicarbonate-CO_2 buffer pair, 6.10, is remote from the pH of 7.40 that we want to maintain in blood. From this alone, we might conclude that this buffer pair is not effective. This is not true, however, because it operates in an "open system." In an **open system,** a component can be removed or added. An example of how this works is presented in Figure 25–3. Suppose we had extracellular fluid containing 24 mmol/L bicarbonate and a Pco_2 of 40 mm Hg ($CO_2(d)$ = 1.2 mmol/L). The pH of the extracellular fluid can be calculated as follows:

$$pH = 6.10 + \log([HCO_3^-]/0.03\, Pco_2)$$
$$= 6.10 + \log(24/[0.03 \times 40])$$
$$= 6.10 + \log 20 = 6.10 + 1.30 = 7.40.$$

Suppose we now add 10 mmoles of strong acid to each liter of extracellular fluid. For simplicity, we will ignore the fact that bicarbonate is not the only buffer base that combines with the added H^+s. According to the following reactions, 10 mmoles of dissolved CO_2 will be formed:

$$H^+ + HCO_3^- \rightarrow H_2CO_3 \rightarrow H_2O + CO_2$$

In a closed container (Fig. 25–3b), from which none of the CO_2 formed can escape, the resulting pH is equal to 6.10 + log ([24 − 10]/[1.2 + 10]) = 6.20. In an open system (see Fig. 25–3c), the CO_2 produced can be removed as it is formed, and Pco_2 can be maintained constant at 40 mm Hg ($CO_2(d)$ = 1.2 mmol/L). In this case, the pH is equal to 6.10 + log ([14]/[1.2]) = 7.17, which is much better than a pH of 6.20 (a fatal pH level). In the body, a reduced extracellular fluid pH (elevated [H^+]) causes hyperventilation so that the Pco_2 is not maintained constant but actually decreases, such as to 30 mm Hg ($CO_2[d]$ = 0.90 mmol/L) (see Fig. 25–3d). In this case, the pH is equal to 6.10 + log ([14]/[0.90]) = 7.29. We can see, therefore, that the ability to remove CO_2 and to decrease its level by hyperventilation diminishes the decrease in pH produced by adding strong acid to the system.

The bicarbonate-CO_2 buffer system is "open" in other respects, too. There is a continuous source of CO_2 from metabolism, which can replace H_2CO_3 consumed when strong base is added to the body. The kidneys can change the amount of bicarbonate in the extracellular fluid by excreting bicarbonate in the urine (when there is excess base in the body) or by synthesizing new bicarbonate and adding it to the blood (when there is excess acid in the body). The ability of the body to change the amounts of components of this particular buffer pair makes the bicarbonate-CO_2 system a remarkably effective buffer.

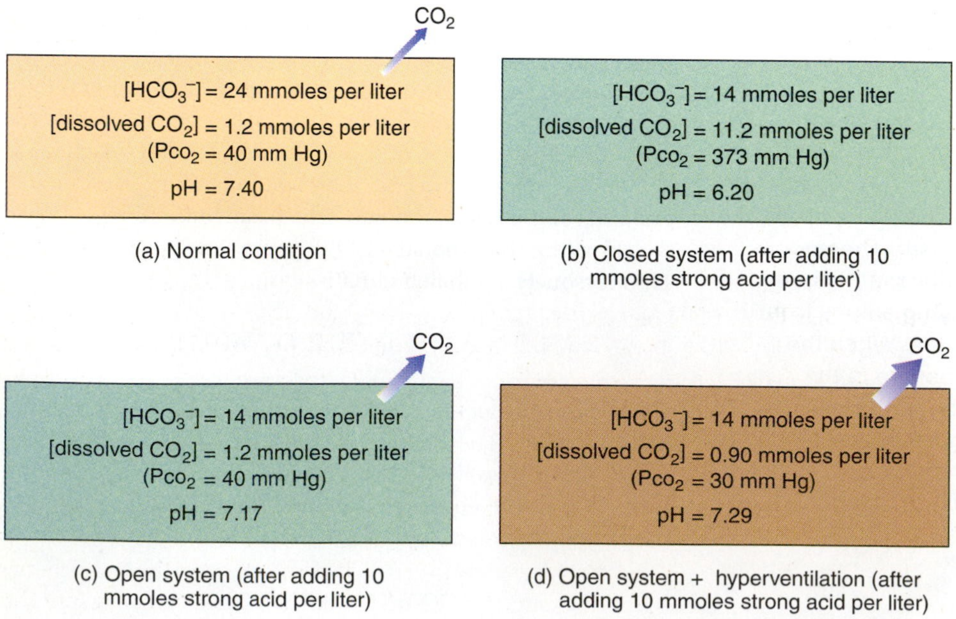

CO_2

[HCO_3^-] = 24 mmoles per liter
[dissolved CO_2] = 1.2 mmoles per liter
(Pco_2 = 40 mm Hg)
pH = 7.40

(a) Normal condition

CO_2

[HCO_3^-] = 14 mmoles per liter
[dissolved CO_2] = 11.2 mmoles per liter
(Pco_2 = 373 mm Hg)
pH = 6.20

(b) Closed system (after adding 10 mmoles strong acid per liter)

CO_2

[HCO_3^-] = 14 mmoles per liter
[dissolved CO_2] = 1.2 mmoles per liter
(Pco_2 = 40 mm Hg)
pH = 7.17

(c) Open system (after adding 10 mmoles strong acid per liter)

CO_2

[HCO_3^-] = 14 mmoles per liter
[dissolved CO_2] = 0.90 mmoles per liter
(Pco_2 = 30 mm Hg)
pH = 7.29

(d) Open system + hyperventilation (after adding 10 mmoles strong acid per liter)

Figure 25–3

Schema to illustrate the effectiveness of the bicarbonate-CO_2 system in buffering added strong acid. *(a)* The normal condition. The decrease in pH is much greater in the closed system *(b)* than in an open system *(c and d)* that allows removal of CO_2.

The Isohydric Principle

We have discussed the various buffer systems separately, but in reality they all work together. In a solution containing multiple buffers, all buffers are in equilibrium with the same $[H^+]$. This idea is called the **isohydric principle.** The term *isohydric* means "same H^+." We can therefore write for blood plasma:

$$pH = 6.8 + \log([HPO_4^{2-}]/[H_2PO_4^-])$$
$$= pK_{protein} + \log([proteinate^-]/[H\text{-proteinate}])$$
$$= 6.1 + \log([HCO_3^-]/0.03Pco_2)$$

When an acid or base is added to the body, all buffer pairs participate in buffering. The importance of each buffer depends on its pK, amount, and accessibility. The expression just given reemphasizes the point that the ratio of conjugate base to acid for any buffer pair of known pK defines the pH. For the remainder of this chapter, we will stress the bicarbonate-CO_2 system and its effects on blood pH. Recognize, however, that other buffers are also important and interact with the bicarbonate-CO_2 buffer pair. Changes in the ratio HCO_3^-/CO_2 indicate that other body buffers are changing too. The reason for emphasizing the bicarbonate-CO_2 system is mainly that the lungs and kidneys set the levels of its components and, in this way, affect blood pH. The respiratory system controls the Pco_2, and the kidneys influence the plasma bicarbonate concentration.

The Lungs and Kidneys Normally Stabilize Extracellular Fluid pH

The Respiratory System as a Buffer System: Control of the Pco_2

> *How does the respiratory system contribute to the regulation of extracellular fluid hydrogen ion concentration?*

The respiratory system regulates the partial pressure of CO_2 in arterial blood. Recall from Chapter 20 that the level of alveolar ventilation is a determinant of the Pco_2 in the alveolar spaces of the lungs. CO_2 tensions in alveoli and arterial blood are equal because of equilibration of CO_2 between alveolar gas and pulmonary capillary blood. For a constant rate of CO_2 production, alveolar ventilation and alveolar Pco_2 are inversely related. The higher the rate of alveolar ventilation, the lower is the alveolar Pco_2, and vice versa. Normally, CO_2 is expired at the same rate that it is produced from metabolism, and the systemic arterial blood Pco_2 remains near 40 mm Hg.

In a state of hyperventilation, the Pco_2 in the alveoli and arterial blood decreases, and the reactions:

$$CO_2 + H_2O \rightleftharpoons H_2CO_3 \rightleftharpoons H^+ + HCO_3^-$$

are pulled to the left. The $[H^+]$ of the blood decreases; in other words, the blood becomes more alkaline. The above reactions have a more profound effect on the $[H^+]$ than on the $[HCO_3^-]$ because $[H^+]$ is in the nmol/L range, whereas $[HCO_3^-]$ is in the mmol/L range.

During hypoventilation, Pco_2 in the alveoli and arterial blood increases. The above reactions are pushed to the right, so that the blood $[H^+]$ increases (blood pH decreases).

The respiratory system normally acts to minimize pH changes in the blood. Such changes might be produced by adding a "fixed" acid or a base to the blood or by adding or removing too much CO_2.

A **fixed acid** is an acid other than carbonic acid. In contrast to carbonic acid, the concentration of a fixed acid in blood is not affected by lung function. If, for example, hydrochloric acid is added to or produced in the body, it is neutralized by chemical buffers, and then the H^+s (combined with urinary buffers) and chloride ions are excreted by the kidneys. A fixed acid is effectively nonvolatile and cannot be blown off by the lungs.

If the blood is made more acidic by addition of fixed acid, pulmonary ventilation is increased. The receptors that sense the increased $[H^+]$ and reflexly increase ventilation are mainly the **peripheral chemoreceptors,** the carotid and aortic bodies (Chapter 10). Increased ventilation serves to decrease the arterial blood Pco_2 and carbonic acid concentration and so reduces the increase in blood $[H^+]$. Figure 25–3*d* illustrates how hyperventilation in response to added acid minimizes a decrease in extracellular fluid pH.

If the blood is made more alkaline by infusing base, the opposite changes occur. The blood $[H^+]$ decreases, ventilation is depressed, and this leads to retention of CO_2. With a higher carbonic acid concentration in the blood, there is less of an alkaline shift in blood pH. In normal individuals, hypoventilation is limited because the ensuing accumulation of CO_2 and a decrease in arterial oxygen tension stimulate ventilation.

If CO_2 accumulates in the body because of inadequate alveolar ventilation or breathing of CO_2-enriched air, the increase in Pco_2 stimulates ventilation. CO_2 is a very powerful stimulus; an increase in arterial blood Pco_2 of 2 to 3 mm Hg may cause the minute ventilation to double. The receptors mediating this increase in ventilation are chiefly **central chemoreceptors** in the medulla of the brain (Chapter 21). Carbon dioxide diffuses from the blood into brain interstitial and cerebrospinal fluids, where it forms carbonic acid, which liberates H^+s. It is generally believed that chemoreceptors respond to $[H^+]$ and not to molecular CO_2 directly. The peripheral chemoreceptors are also stimulated by an increase in arterial blood Pco_2 but are less important than are the central chemoreceptors in the ventilatory response to CO_2. The increase in ventilation caused by CO_2 is useful in bringing about removal of CO_2, thereby minimizing carbonic acid accumulation in the blood and an acidic shift in blood pH.

Buffering of acid-base disturbances by the respiratory system is rapid but relatively coarse. Respiratory responses begin within seconds and are maximal in approximately 12 to 24 hours. The respiratory system brings blood pH back closer to normal but cannot eliminate a fixed acid or base from the body and so cannot restore a normal pH. By contrast, renal responses to acid-base disturbances are slower and more complete. If an excess of fixed acid or base is added to the body, normal kidneys slowly (over many hours or days) eliminate it from the body and return the blood pH to normal. The kidneys exert a fine control over acid-base balance.

The Kidneys as a Buffer System: Acidification of the Urine

> *How do the kidneys contribute to the regulation of extracellular fluid hydrogen ion concentration?*

The kidneys play a major role in acid-base regulation by excreting H^+s (mainly bound to urinary buffers) when there is an excess of H^+s in the body or by excreting bicarbonate when there is excess base in the body. As mentioned previously, the usual condition is one of excess acid because metabolism of food produces approximately 50 to 100 mEq of strong acid in a day. This acid is first buffered by bicarbonate (and other chemical buffer bases) in the body. Then the kidneys excrete the H^+s and at the same time add new bicarbonate to the blood, bringing the blood pH back to normal.

Most of the H^+s excreted in the urine are combined with urinary buffers. The limits of urine pH are from a maximum of 8.0 to a minimum of 4.5. Suppose we excrete 1.5 L of pH 4.5 urine in a day. How many *free* H^+s are excreted? This can be calculated by multiplying the urine volume (1.5 L) by the $[H^+]$ ($10^{-4.5}$ Eq /L or $10^{-1.5}$ mEq/L) and amounts to approximately 0.05 mEq. This amount is clearly inadequate to take care of the amount of acid that needs to be excreted (50–100 mEq/day). Most of the H^+s excreted in the urine are not free but are combined with buffers and are in the form NH_4^+ and titratable acid (e.g., $H_2PO_4^-$). Daily net acid excretion is calculated from the sum of urinary titratable acid excretion and ammonium excretion, minus excreted bicarbonate (the latter is usually small—see Table 23–1).

Titratable acid is measured in the clinical laboratory by determining how many milliequivalents of strong base (NaOH) are needed to bring a urine sample back to the pH of arterial blood (usually 7.4). Most of the titratable acid is normally dihydrogen phosphate ($H_2PO_4^-$). Depending on the circumstances, other substances in the urine (e.g., creatinine and various organic acids) may also be included. In the test tube, the following reaction takes place:

$$H_2PO_4^- + OH^- \rightarrow HPO_4^{2-} + H_2O$$

This reverses the reaction that took place along the kidney tubules when H^+s were added to the tubular urine:

$$H^+ + HPO_4^{2-} \rightarrow H_2PO_4^-$$

Urinary **ammonia** is usually measured by a chemical method and includes the free base NH_3 and the ammonium ion NH_4^+. These two forms are in equilibrium; because the pK_a of NH_4^+ is so high (approximately 9.0), most of the ammonia in the urine is NH_4^+. The high pK_a for NH_4^+ also means that it is not significantly titrated when titratable acid is measured, and so it is not a part of titratable acid. NH_3 buffers secreted H^+s in the urine, according to the following reaction:

$$H^+ + NH_3 \rightleftharpoons NH_4^+$$

Of the H^+s excreted in urine, normally nearly one third are in the form of titratable acid, close to two thirds are present as NH_4^+, and a very small amount is free.

The urine is acidified by secretion of H^+s by the tubular epithelium (Fig. 25–4). The nature of the secretory processes, the amounts of H^+s secreted, and the pH changes produced differ in different segments of the nephron. For example, most of the H^+s are secreted by the proximal convoluted tubule. This secretion is primarily via a Na^+/H^+-exchanger in the brush border membrane. Because the glomerular filtrate contains large amounts of buffer (especially bicarbonate) and because the proximal convoluted tubule cannot develop steep $[H^+]$ gradients, the decrease in tubular fluid pH along the proximal convoluted tubule is modest (e.g., to pH 6.8). More distal portions of the nephron (i.e., distal convoluted tubule and collecting duct) secrete smaller amounts of H^+s. Distal H^+ secretion involves primary active transport by a H^+-ATPase and H^+/K^+-ATPase in the luminal cell membrane. The distal portions of the nephron can establish steep transepithelial pH gradients but cannot decrease the pH to less than 4.5. This is probably because of back-leak of H^+ from urine to blood and an inability of the collecting duct cell H^+-secretory pumps to pump against an extremely high luminal $[H^+]$. A urine pH of 4.5, compared with a blood pH of 7.4, means that the urine $[H^+]$ is 800-fold ($10^{(7.4-4.5)} = 10^{2.9}$) higher than in plasma. For simplicity, we treat the tubular epithelium as a whole in the cell model in Figure 25–4 and do not consider differences in different nephron segments.

Three processes are involved in urinary acidification: (1) reabsorption of filtered bicarbonate, (2) excretion of titratable acid, and (3) excretion of ammonia (see Fig. 25–4). All three involve H^+ secretion; furthermore, all three processes are associated with the addition of bicarbonate to the peritubular capillary blood by the kidney tubules. H^+ secretion into the lumen and addition of bicarbonate to the blood by the tubules are best viewed as opposite sides of the same coin. Reabsorption of filtered bicarbonate conserves the normal amount of bicarbonate in the extracellular fluid. In the process of glomerular filtration, large quantities (4500 mEq/day) of bicarbonate are filtered, and only traces (e.g., 2 mEq/day) are usually excreted. Reabsorption of filtered bicarbonate reclaims bicarbonate for the body. Most of the H^+s secreted are used to bring about reabsorption of filtered

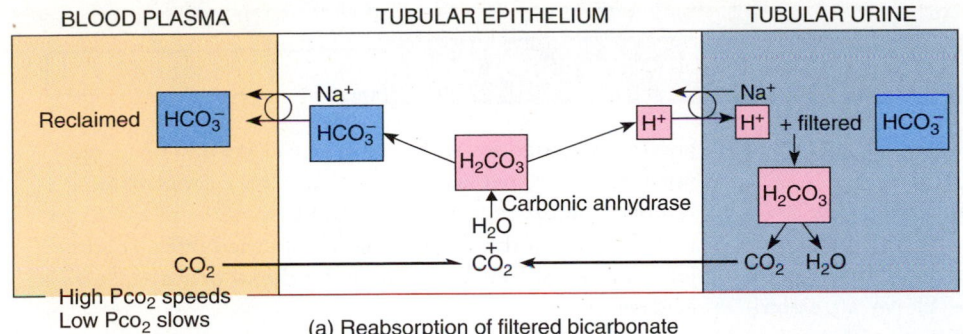

(a) Reabsorption of filtered bicarbonate

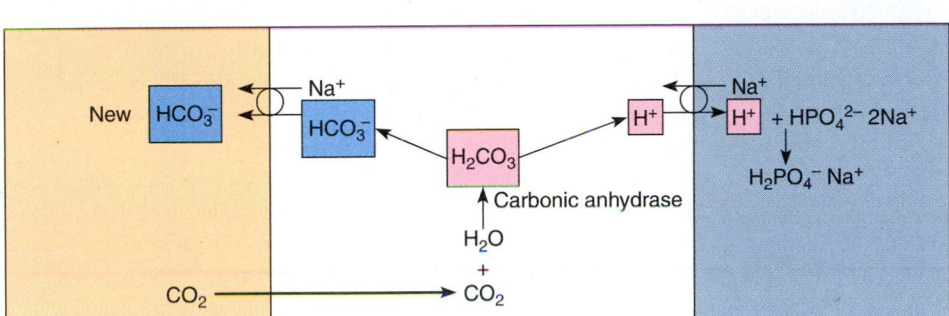

(b) Formation of titratable acid

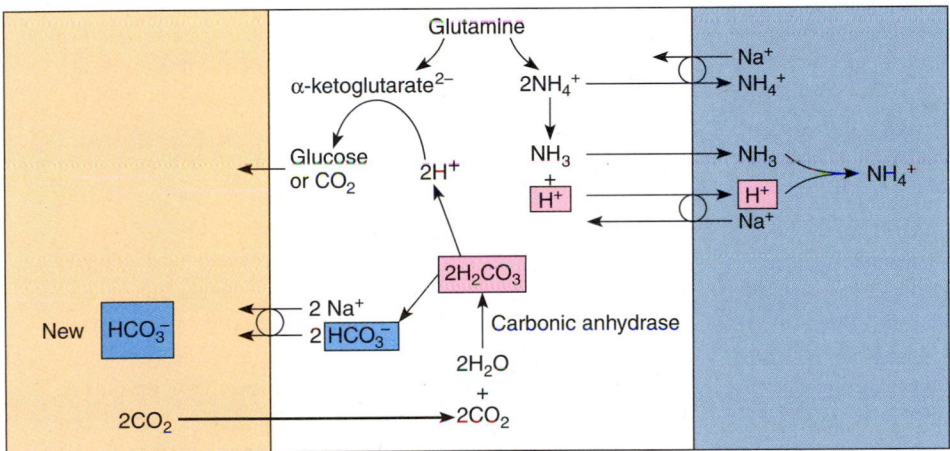

(c) Excretion of ammonia

Figure 25–4

Renal cell models of the three processes involved in urinary acidification: *(a)* reabsorption of filtered bicarbonate, *(b)* formation of titratable acid, and *(c)* excretion of ammonia. In the distal nephron, H^+ secretion occurs via a H^+-ATPase or H^+/K^+-ATPase and not via the Na^+/H^+-exchanger illustrated here.

bicarbonate and are not excreted in the final urine. Those that are excreted are combined with phosphate (titratable acid) and ammonia. In these latter processes, *new* bicarbonate is generated in the kidney tubule cells. This new bicarbonate is added to the blood and replaces bicarbonate consumed in the buffering of strong acids produced in the body.

Reabsorption of Filtered Bicarbonate. Figure 25–4a depicts the reabsorption of filtered bicarbonate. This occurs mainly in the proximal convoluted tubule. H^+s are secreted into the tubular urine in exchange for sodium ions (Na^+/H^+-exchanger) and combine with bicarbonate that has been filtered by the glomeruli. This leads to formation of carbonic acid in the tubular urine. The dehydration of this acid to CO_2

and water is catalyzed in the proximal convoluted tubule lumen by carbonic anhydrase in the brush border membrane. PCO_2 values in tubular urine, cells, and peritubular capillary blood are essentially identical because CO_2 diffuses rapidly through cell membranes. In the cell, CO_2 combines with water to form carbonic acid, a reaction catalyzed by cytoplasmic carbonic anhydrase. Carbonic acid dissociates into H^+s and bicarbonate. The H^+s are secreted into the lumen of the kidney tubule. Bicarbonate ions exit from the basolateral side of the cell along with sodium ions. In the proximal tubule, this occurs via a Na/HCO_3 cotransporter in the basolateral cell membrane. In this scheme, it should be apparent that filtered bicarbonate is not reabsorbed by direct transport out of the tubular urine; rather, filtered bicarbonate is reabsorbed indirectly by H^+ secretion.

CURRENT CONCEPTS IN PHYSIOLOGY

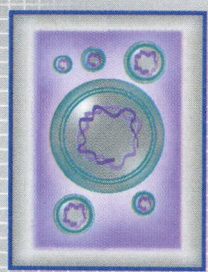

Sodium-Hydrogen Ion Exchangers

Sodium-hydrogen ion (Na$^+$/H$^+$) exchangers are integral membrane proteins that transport one Na$^+$ in exchange for one H$^+$ in an electrically neutral fashion. These exchangers are involved in the regulation of intracellular [H$^+$] and cell volume and in epithelial cell H$^+$ secretion and Na$^+$ reabsorption.

Six different isoforms of sodium-hydrogen ion exchangers (abbreviated NHE1–NHE6) have been identified. The first of these, NHE1, was cloned in 1989 and is present in essentially all cells. It has a "housekeeping" function; it helps to maintain a normal cell pH and volume. The other isoforms are distributed differently in various tissues, where they have special functions. All NHE isoforms isolated so far are large proteins that are believed to have 12 membrane spanning segments. The different isoforms are inhibited to varying degrees by the diuretic drug amiloride and its derivatives. Numerous growth factors and hormones control the expression and activity of the various NHE isoforms. In so doing, they change cell pH, which may affect complex processes, such as cell proliferation and differentiation.

Most body cells are constantly threatened by excess H$^+$s. Acids are produced by cell metabolism. Also, because of the inside negativity in body cells (e.g., a membrane potential of −90 mV in skeletal muscle cells), H$^+$ ions tend to leak into cells from the extracellular fluid down the electrical gradient. (The transmembrane chemical concentration gradient for H$^+$ actually opposes H$^+$ entry because the cell [H$^+$] is higher than in extracellular fluid.) The cells extrude H$^+$ via NHEs, a process that keeps cell [H$^+$] at a value lower (or cell pH higher) than would be expected from simple passive distribution of H$^+$s across the cell membrane. The NHEs do not consume ATP directly but use the sodium ion gradient (outside versus inside of cells) established by the Na$^+$/K$^+$-ATPase to energize the extrusion of H$^+$ ions. Activity of NHEs is greatly increased by an increase in cell [H$^+$]. This is due not only to an increase in substrate (H$^+$) concentration but also the exchanger is activated by binding of H$^+$ to a modifier site on the cytoplasmic side of the NHE. This allows the cell to cope better with the threat of intracellular acidosis.

NHE1 is activated by hyperosmotic stress and plays a role in recovery of cell volume. If cells are exposed to a hyperosmotic medium, they shrink. Activation of NHE leads to an influx of sodium ions into the cells, which is accompanied by chloride ions and water, leading to a return of cell volume back to a more normal value.

In the kidneys, NHE1 is present in the basolateral cell membrane of tubule epithelial cells in multiple nephron segments. NHE2 is found in the apical membranes of thick ascending limb and distal convoluted tubule cells. NHE3 is found in the apical cell membrane of proximal tubule and thick ascending limb cells. NHE4 is found in the basolateral membrane of thick ascending limb, distal convoluted tubule, and collecting duct cells. The localization of NHE5 and NHE6 needs further study. Abnormally increased activity of NHE3, a major pathway for sodium reabsorption in the proximal tubule, may lead to hypertension.

In the heart, increased activity of NHE1 may contribute to cardiac damage during states of low blood flow (ischemia) and reperfusion. Reduced blood flow causes intracellular acidosis and increased expression and activity of NHE1 by myocardial cells. This leads to an increase in intracellular [Na$^+$], which in turn results in an increased intracellular [Ca^{2+}] because of diminished Na$^+$/Ca^{2+} exchange. Ca^{2+}, in turn, activates various cell enzymes. During reperfusion, excessive activity of the NHE1 exchanger results in further ionic imbalances and leads to enhanced activity of cell proteases and other enzymes that damage cell membranes. Numerous studies using NHE inhibitors have shown protective effects against ischemic and reperfusion injury in animal hearts. Clinical trials are currently underway to test whether these drugs will reduce cardiac damage in patients with coronary heart disease.

Excretion of Titratable Acid. Normally, we ingest phosphate in our diet. Filtered phosphate is incompletely reabsorbed by the kidney tubules, and the unreabsorbed phosphate acts as a pH buffer in the urine. The cell scheme for formation of titratable acid is shown in Figure 25–4*b*. H$^+$s are secreted into the tubular urine, and they titrate buffers in the urine in the acidic direction. For example, the basic form of phosphate (HPO$_4^{2-}$) is converted to the acidic form (H$_2$PO$_4^-$). One of the sodium ions accompanying HPO$_4^{2-}$ is reabsorbed. The intracellular and membrane events are similar to those described in Figure 25–4*a*. For each mEq of H$^+$ excreted as titratable acid, a new bicarbonate is added to the blood. Excretion of titratable acid, therefore, eliminates H$^+$s from the body and, at the same time, restores depleted plasma bicarbonate reserves.

Excretion of Ammonia. Figure 25–4*c* summarizes the role of ammonia in urinary acidification. Ammonia is synthesized by kidney tubule cells, mainly in the proximal tubule. The metabolism of the amino acid glutamine results in formation of two ammonium ions and one α-ketoglutarate molecule. Each ammonium ion yields one molecule of free base ammonia (NH_3) and one H^+. Ammonia is secreted into the urine by two mechanisms. In the proximal tubule, ammonia is actively secreted by the Na^+/H^+-exchanger (operating in a Na^+/NH_4^+ exchange mode). NH_3 also diffuses readily through cell membranes into the urine and combines with a secreted H^+ to form an ammonium ion. Recall from Chapter 4 that, mainly because of its charge, the ammonium ion does not readily diffuse through the lipid phase of plasma membranes and so is, in effect, "trapped" in an acidic urine, a process called **diffusion trapping.** Diffusion trapping of ammonia is especially important in the medullary collecting ducts because they have the lowest intraluminal pH. The ammonium ion is accompanied by an anion (e.g., chloride) in the urine and allows the body to conserve a sodium ion, which it replaces. In the cell, hydration of CO_2 produces H^+s and bicarbonate ions, as described previously. These H^+s are consumed when α-ketoglutarate is metabolized to form glucose or CO_2. The bicarbonate ions are reabsorbed together with sodium ions. Note that for each H^+ excreted in the urine as ammonium ion, an equivalent quantity of new bicarbonate is added to the blood. The kidneys' ability to synthesize ammonia from glutamine is increased under conditions of excess acid in the body. This adaptive increase in ammonia synthesis takes several days to develop fully but allows the body to dispose of unusually large acid loads.

Effects of CO_2 on Urinary Acidification. The effects of CO_2 on urinary acidification are also indicated in Figure 25–4. A high blood P_{CO_2}, by mass action, favors generation of H^+s in the cell. This results in more complete reabsorption of filtered bicarbonate and increased excretion of H^+s combined with titratable acids and ammonia. Conversely, a low blood P_{CO_2} results in a lower rate of H^+ secretion; filtered bicarbonate is less completely reabsorbed, and excretion of a bicarbonate-rich urine is favored. Fewer H^+s are excreted in combination with titratable acids and ammonia. These effects of CO_2 on kidney tubule H^+ secretion are key to understanding the renal compensations for respiratory disturbances of acid-base balance.

If we reconsider the Henderson-Hasselbalch equation, $pH = 6.10 + \log([HCO_3^-]/0.03P_{CO_2})$, we can see that blood pH is affected by changes in plasma bicarbonate concentration. In cases of acid excess, the kidneys increase acid excretion in the urine and increase the plasma $[HCO_3^-]$. This helps to make the blood less acidic; that is, it increases the pH toward a normal value. In cases of base excess, the kidneys increase the output of bicarbonate in the urine, thereby decreasing the plasma bicarbonate concentration toward normal. This helps to make the blood less alkaline; that is, it decreases the pH toward a normal value. From these relations, we can see that it is useful to think of the kidneys as regulating plasma $[H^+]$ by changing the plasma bicarbonate concentration in the appropriate direction. We will discuss later the renal compensation for various acid-base disturbances.

DISTURBANCES OF ACID-BASE BALANCE

 How are acid-base disturbances classified?

In numerous pathophysiological conditions, the pH of systemic arterial blood may not be within the normal range, 7.35 to 7.45. If the arterial blood pH is less than 7.35, **acidemia** exists. If the blood pH is greater than 7.45, **alkalemia** exists. The range of blood pH values compatible with life is approximately 6.8 to 7.8, corresponding to $[H^+]$s of 160 and 16 nmol/L, respectively.

There are four fundamental (or "simple") types of acid-base disturbance: (1) **respiratory acidosis,** (2) **respiratory alkalosis,** (3) **metabolic acidosis,** and (4) **metabolic alkalosis.** This classification is based on the nature of the process leading to the disturbed pH and on the direction of pH change. If the problem is too much or too little CO_2, then a **respiratory disturbance** of acid-base balance is present. If the problem is an excess or deficit of a fixed acid or base, then a **metabolic (nonrespiratory) disturbance** of acid-base balance is present. Acidosis leads to acidemia, and alkalosis leads to alkalemia. These four disturbances of acid-base balance are each accompanied by characteristic changes in arterial blood pH, P_{CO_2}, and $[HCO_3^-]$ (Table 25–2). We will describe for each disturbance the nature of the abnormal process, characteristic changes in blood chemistry, some common causes, and how buffering mechanisms in the body minimize the deviation from normal pH.

Accumulation of CO_2 Results in Respiratory Acidosis

 What is the response to abnormal accumulation or loss of CO_2?

Respiratory acidosis is an abnormal process characterized by an accumulation of CO_2. Most often, it is due to failure to expire (blow off) metabolically produced CO_2 at an adequate rate. In other words, alveolar ventilation is inadequate, and CO_2 accumulates in the blood. The problem may be caused by insufficient neural drive for ventilation, inadequate

TABLE 25–2

Summary of Directional Changes in Blood Values in the Four Simple Types of Acid-Base Disturbance

Disturbance	Arterial Blood			Compensatory Response
	pH	Plasma [HCO₃⁻]	PCO₂	
Respiratory acidosis	↓	↑	⇑	Kidneys increase H⁺ excretion (increase plasma [HCO₃⁻])
Respiratory alkalosis	↑	↓	⇓	Kidneys increase HCO₃⁻ excretion (decrease plasma [HCO₃⁻])
Metabolic acidosis	↓	⇓	↓	Alveolar hyperventilation; normal kidneys increase net acid excretion
Metabolic alkalosis	↑	⇑	↑	Alveolar hypoventilation; kidneys increase HCO₃⁻ excretion

Heavy arrows indicate the main change.

movements of the respiratory muscles or thoracic cage, airway obstruction, pulmonary disease, or inadequate ventilation on an artificial respirator. Respiratory acidosis is also produced if a person breathes CO_2-enriched air. If CO_2 is allowed to build up in the arterial blood, the following reactions are pushed to the right:

$$CO_2 + H_2O \rightleftharpoons H_2CO_3 \rightleftharpoons H^+ + HCO_3^-$$

Clearly, retention of CO_2 (the primary chemical change) leads to an increase in $[H^+]$ and so to a decrease in pH. The plasma $[HCO_3^-]$ also is elevated because the other body buffer bases (e.g., proteins and phosphates) combine with H^+s generated in the above reactions, leaving behind bicarbonate. The increase in $[HCO_3^-]$ is proportionately less than the increase in P_{CO_2}. (Otherwise, the pH would not decrease. See the Henderson-Hasselbalch equation.)

Buffering of H^+s produced by excess CO_2 is first accomplished by chemical buffers in the body. These buffers include the large amounts of protein and organic phosphate buffers in cells and inorganic phosphate in bone. For example, CO_2 diffuses into red cells, combines with water to form carbonic acid (a reaction catalyzed by intracellular carbonic anhydrase), and the H^+s liberated when carbonic acid dissociates are then buffered by hemoglobin (Chapter 21).

Respiratory acidosis is most often due to inadequate ventilation or imbalances in the matching of ventilation and perfusion in the lungs (Chapter 21). The retention of CO_2 and increase in blood $[H^+]$ that ensue stimulate ventilation. In addition, hypoxia is often also present, and a low arterial P_{O_2} acts on the peripheral chemoreceptors to stimulate ventilation. Despite these stimuli, however, the respiratory system is the problem, so it cannot return the blood $[H^+]$ to normal.

The kidneys compensate for respiratory acidosis by increasing the output of H^+s in the urine. As was mentioned, a high P_{CO_2} stimulates the renal tubular epithelium to secrete H^+s into the tubular urine. The result is that all of the filtered bicarbonate is reabsorbed, and increased amounts of H^+s, combined with urinary buffers, are excreted in the urine. By getting rid of H^+s in the urine, the kidneys increase

the plasma bicarbonate concentration. In other words, they add buffer base (bicarbonate) to the blood, diminishing the severity of the acidemia. The renal response takes a few days, so with chronic respiratory acidosis, higher plasma levels of bicarbonate are achieved than those seen with respiratory acidosis of short duration (acute respiratory acidosis).

Excessive Loss of CO_2 Results in Respiratory Alkalosis

Respiratory alkalosis is easily understood as the opposite of respiratory acidosis. In this case, we have an abnormal process leading to an excessive loss of CO_2. This causes a decrease in arterial blood P_{CO_2}, a decrease in plasma $[H^+]$ (an increase in pH), and a decrease in plasma $[HCO_3^-]$. Respiratory alkalosis is produced by alveolar hyperventilation, which results in excessive removal of CO_2 from the blood. Hyperventilation may be caused by voluntary effort, anxiety, or stimulation of respiratory centers owing to some abnormal influence acting directly on the brain (e.g., aspirin intoxication, fever, and meningitis). Hyperventilation may also be produced reflexly. For example, hyperventilation and respiratory alkalosis are observed at high altitudes because the low ambient P_{O_2} results in a reflex stimulation of ventilation.

The alkaline shift in pH produced during respiratory alkalosis is diminished by the action of buffers in the body. Chemical buffers (e.g., hemoglobin) are titrated in the alkaline direction and liberate H^+s, thereby tending to diminish the decrease in free $[H^+]$. As is the case for respiratory acidosis, most (95%) of the chemical buffering occurs within cells. Although the origin of respiratory alkalosis is hyperventilation, the resulting loss of CO_2 and increase in pH limit the extent of hyperventilation. CO_2 and H^+ are two important stimuli of ventilation, so a decrease in these stimuli leads to diminished ventilatory drive. Excessive ventilation leads to a decreased P_{CO_2} in the brain, cerebral vasoconstriction, and even a loss of consciousness.

The kidneys compensate for respiratory alkalosis by excreting bicarbonate in the urine. A low P_{CO_2} leads to decreased H^+ secretion by the tubular epithelium. The result is that filtered bicarbonate is incompletely reabsorbed and so is excreted. The loss of bicarbonate in the urine results in a further decrease in plasma bicarbonate. In other words, the kidneys reduce the amount of base (bicarbonate) in the blood and so diminish the severity of the alkalemia. The kidneys work relatively slowly, so it takes them a few days to compensate fully for a persistent respiratory alkalosis.

Metabolic (Nonrespiratory) Acidosis Results from a Gain of Fixed Acid or a Loss of Bicarbonate

> *What is the response to nonrespiratory addition or loss of hydrogen ions?*

Metabolic acidosis is an abnormal process characterized by a gain of fixed acid (i.e., an acid other than H_2CO_3) or a loss of bicarbonate. Many conditions can produce metabolic acidosis. In renal failure, the acid end products of metabolism are not adequately excreted and accumulate in the body. In uncontrolled diabetes mellitus, large quantities of ketone body acids are produced. If the circulation is inadequate—for example, due to cardiogenic or hemorrhagic shock—glucose is converted anaerobically to lactic acid. In heavy exercise too, lactic acid production is increased. Ingestion of certain acidifying salts (e.g., ammonium chloride) or toxins (e.g., methanol, ethylene glycol, or aspirin overdose) produce metabolic acidosis. Loss of bicarbonate from the body, due to diarrhea (loss of alkaline intestinal fluids) or abnormal excretion of bicarbonate in the urine, also results in metabolic acidosis.

If a strong acid is added to the body, the reactions:

$$H^+ + HCO_3^- \rightleftharpoons H_2CO_3 \rightleftharpoons H_2O + CO_2$$

are pushed to the right. We can then correctly predict that the $[H^+]$ should be increased, and the plasma bicarbonate concentration, decreased. Further, we might predict that Pco_2 should be elevated (but recall that CO_2 operates in an open system). Pco_2 is increased only transiently during acute infusion of large amounts of strong acid; in established metabolic acidosis, Pco_2 is actually less than normal due to respiratory compensation (see Table 25–2).

Buffering during metabolic acidosis is accomplished by chemical buffers in extracellular fluid, cells, and bone. Roughly half of the buffering occurs in extracellular fluid, where bicarbonate is the principal buffer. Due to the elevated blood $[H^+]$, the respiratory system is stimulated. Ventilation increases, decreasing the Pco_2 (or $[H_2CO_3]$) in the blood and making the blood less acidic (more alkaline). Respiratory compensation for a metabolic acidosis is prompt (seconds to hours) but usually does not restore the blood pH to normal. Figure 25–3 illustrates how the bicarbonate-CO_2 system participates in buffering strong acid.

Compensation by normal kidneys is slower (takes days) but more complete than is respiratory compensation. The kidneys increase the output of H^+s in the urine. Because the plasma bicarbonate is mainly decreased, the filtered bicarbonate load is reduced, and all of the filtered bicarbonate is reabsorbed. Increased amounts of H^+s, together with urinary buffers (e.g., phosphate and ammonia), are excreted in the urine, so new bicarbonate is added to the blood by the kidneys to replenish depleted bicarbonate reserves. If the underlying cause of the metabolic acidosis is corrected, then normal kidneys bring the pH of the blood back to normal.

Metabolic (Nonrespiratory) Alkalosis Results from a Gain of Strong Base or Bicarbonate or a Loss of Acid

Metabolic alkalosis is an abnormal process characterized by a gain of strong base or bicarbonate or a loss of acid (other than H_2CO_3). Ingestion of excess sodium bicarbonate or oxidation of organic anions (e.g., citrate, lactate, and acetate) to bicarbonate produces metabolic alkalosis. A common cause of metabolic alkalosis is vomiting of acidic gastric juice. Every time a H^+ is secreted into the lumen of the stomach, a bicarbonate ion is added to the blood (Chapter 22). Loss of the stomach contents through vomiting results in net loss of HCl and the addition of bicarbonate to the blood. The characteristic arterial blood changes include a primary increase in plasma $[HCO_3^-]$, an increase in pH, and a compensatory increase in Pco_2 (see Table 25–2).

Chemical buffers in the body act to limit the alkaline shift in pH by liberating H^+s as the buffers are titrated in the alkaline direction. Respiratory compensation for a metabolic alkalosis is alveolar hypoventilation, brought about by the decreased blood $[H^+]$. Respiratory compensation is limited, however, because hypoventilation leads to an increase in Pco_2 and a decrease in Po_2, both of which stimulate ventilation.

Renal compensation involves increased excretion of bicarbonate in the urine. Because the plasma bicarbonate is primarily elevated, more bicarbonate is filtered than can be reabsorbed by the tubules. Increased excretion of bicarbonate in the urine leads to a decrease in plasma bicarbonate concentration and a return of blood pH back toward normal.

Clinical Evaluation of Acid-Base Status Involves a History, Physical Examination, and Arterial Blood Chemistry Values

We have described individually each of the four simple disturbances of acid-base balance. Proper identification of an acid-base disturbance in a patient requires clinical evidence (medical history and examination) and supporting laboratory findings. From blood acid-base data (pH, Pco_2, and plasma $[HCO_3^-]$), it is often possible to suggest what type of disturbance might be present.

APPLICATIONS OF PHYSIOLOGY

Metabolic Acidosis in Diabetes Mellitus

Diabetes mellitus ("diabetes") is a common disorder in which there is insufficient secretion of insulin or resistance of target tissues to the actions of this hormone (Chapter 15). In uncontrolled diabetes mellitus, severe metabolic acidosis may develop. The most common factors that cause diabetic patients to develop metabolic acidosis are infections of any kind and failure to take insulin.

Acidosis occurs because insulin deficiency leads to disturbed carbohydrate metabolism, diversion of metabolism toward utilization of fatty acids, and overproduction of the ketone body acids (acetoacetic and β-hydroxybutyric acids). The ketone body acids are fairly strong acids, with pK_as of approximately 4. Chemical buffer bases, especially bicarbonate, are consumed in buffering these acids, and so the plasma bicarbonate concentration decreases. Interestingly, the production of ketone body acids in the body is actually inhibited by a more acidic pH; this mechanism helps to oppose a further decrease in blood pH.

The elevated blood $[H^+]$ reflexly stimulates pulmonary ventilation via peripheral chemoreceptors in the carotid and aortic bodies. This produces a lower alveolar and arterial Pco_2 and blood H_2CO_3 concentration, which helps to compensate for the acidic blood pH. The labored, deep breathing that accompanies severe, uncontrolled diabetes mellitus is called **Kussmaul's respiration**.

The kidneys compensate for the metabolic acidosis in several ways. First, they reabsorb essentially all of the filtered bicarbonate so as to prevent loss of this valuable buffer base in the urine. Second, they increase excretion of titratable acids. A part of the titratable acid is composed of the ketone body acids. These acids, however, can be only partially titrated to their acidic form because the urine pH cannot go below 4.5. The result is that ketone body acids are excreted largely in their anionic form, and because of the requirement of electroneutrality in solutions, increased amounts of sodium and potassium are also lost in the urine. Third, the kidneys synthesize and excrete more ammonia (as the ammonium ion). This most important adaptive change takes several days to develop fully and allows the patient to dispose of large amounts of acid. For each NH_4^+ excreted in the urine, a new bicarbonate is added to the blood to replace depleted bicarbonate reserves. The ammonium in the urine substitutes for sodium and potassium ions, thereby resulting in renal conservation of these valuable cations.

Severe acidosis, electrolyte disturbances, and volume depletion can occur in uncontrolled diabetes mellitus and may lead to coma and death. Correction of the acidosis is best achieved by correcting the *underlying cause* rather than just treating the symptoms. Administration of a suitable dose of insulin is usually the key element of therapy. In some patients with severe acidemia, sodium bicarbonate solutions may be infused to speed recovery. Losses of sodium, potassium, and water should be replaced.

In physiology and clinical medicine, systemic arterial blood is used to evaluate acid-base status. The pH of whole blood, measured with a pH meter, is actually the pH of plasma (not the cells) and so is a measurement of extracellular fluid pH. Blood samples need to be handled anaerobically (without exposure to air) so as to prevent CO_2 loss and the resulting alkaline shift in pH.

The **pH-bicarbonate diagram** (Fig. 25-5) is a useful way to look at arterial blood data and determine what type of acid-base disturbance may be present in a person. In the center of the diagram are normal values for arterial blood pH (7.35–7.45), Pco_2 (35–45 mm Hg), and plasma $[HCO_3^-]$ (22–26 mEq/L). On the left side of the diagram are the processes that produce acidemia, namely, respiratory or metabolic acidosis. On the right side of the diagram are the processes that produce alkalemia, namely, respiratory and metabolic alkalosis. The lines that curve upward and to the

right are Pco_2 isobars; the partial pressure of CO_2 is constant along each isobar (*iso* = "same"; *bar* = "pressure"). Note that any point on the diagram satisfies the Henderson-Hasselbalch equation, so if, for example, pH and Pco_2 are known, then the $[HCO_3^-]$ is automatically defined. The shaded areas include 95% of people with the designated simple acid-base disturbance. Notice that a distinction is made between acute and chronic *respiratory* disturbances of acid-base balance. This is because the renal compensation for respiratory acid-base disturbances (namely, a change in plasma bicarbonate concentration) takes a few days to develop fully. Notice in Figure 25-5 that for the same Pco_2, the blood pH is closer to normal in a chronic, as opposed to an acute, respiratory disturbance of acid-base balance. No distinction is made between acute and chronic *metabolic* disturbances of acid-base balance on this diagram because the respiratory compensation (an appropriate change in Pco_2) is prompt.

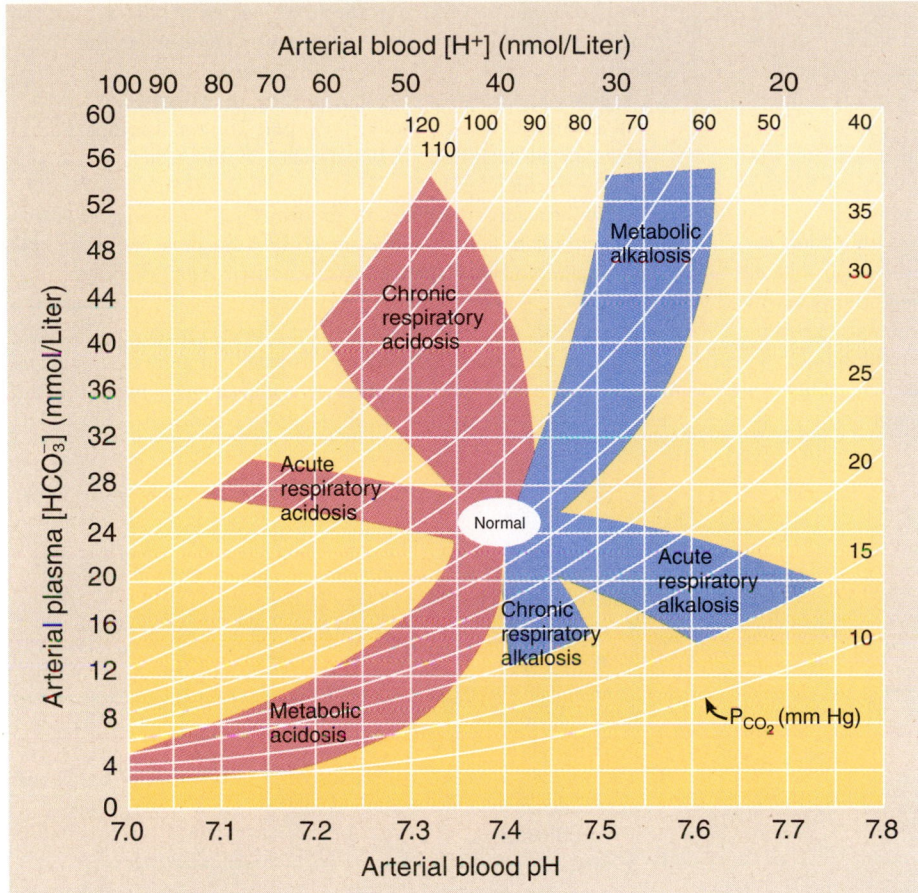

Figure 25–5

The pH-bicarbonate diagram. The shaded areas show values for the simple acid-base disturbances.

Sometimes, more than one independent disturbance of acid-base balance is present in the same person. This is called a **mixed acid-base disturbance.** The case of aspirin overdose (salicylate intoxication) at the beginning of this chapter illustrates such a situation. Mixed disturbances often, but not always, fall outside of the shaded areas (simple acid-base disturbances) on the pH-bicarbonate diagram. Note that a person can have a completely normal blood pH but still have a mixed acid-base disturbance, for example, respiratory acidosis and metabolic alkalosis that completely offset each other.

CHAPTER REVIEW

Summary

- An acid is a substance that can donate hydrogen ions. A base is a substance that can accept or bind hydrogen ions.
- An acid in aqueous solution reversibly dissociates into a H^+ and conjugate base. A high acid dissociation constant (low pK_a) is characteristic of a strong acid; a low acid dissociation constant (high pK_a) is characteristic of a weak acid.
- The pH of a solution is equal to the logarithm, to the base 10, of the inverse of the molar $[H^+]$. The pH of systemic arterial blood normally averages 7.40.
- pH buffers minimize changes in pH produced when acid or base is added to a solution.

- The body obtains H^+ ions from the metabolism of foods and from carbonic acid formed when CO_2 reacts with water.
- Chemical buffers consist of a weak acid and its conjugate base (or a weak base and its conjugate acid). Chemical buffers in the extracellular fluid include the bicarbonate-CO_2 system, plasma proteins, and inorganic phosphate.
- The Henderson-Hasselbalch equation for the bicarbonate-CO_2 system is pH = 6.10 + log ($[HCO_3^-]$/0.03Pco_2).
- The bicarbonate-CO_2 buffer system has special importance in acid-base regulation because the lungs and kidneys act on the components of this buffer pair. The respiratory system

influences plasma pH by controlling the partial pressure of CO_2 (P_{CO_2}) in arterial blood by varying the level of alveolar ventilation. The kidneys influence plasma pH by getting rid of acid or base in the urine—in other words, by adding new bicarbonate to the blood or excreting bicarbonate.

- On the usual mixed diet, people are challenged by excess production of nonvolatile (fixed) acids. Although chemical buffers can minimize a decrease in pH, the burden of eliminating H^+s from the body falls upon the kidneys. Urinary acidification involves three processes: (1) reabsorption of filtered bicarbonate, (2) excretion of titratable acid, and (3) excretion of ammonia. All involve H^+ secretion by the kidney tubule cells. Renal reabsorption of filtered bicarbonate returns bicarbonate to the blood.

- Respiratory acidosis is characterized by an accumulation of CO_2 and a decrease in blood pH. The renal compensation is increased excretion of H^+s in the urine, which results in addition of bicarbonate to the blood and thereby counteracts the acidic shift in blood pH.

- Respiratory alkalosis is characterized by an excessive loss of CO_2 and a consequent increase in blood pH. The renal compensation is increased excretion of bicarbonate in the urine, which decreases the blood pH.

- Metabolic acidosis involves an abnormal gain of fixed acid or loss of bicarbonate and is characterized by a decrease in plasma bicarbonate and pH. Respiratory compensation decreases the blood P_{CO_2}, making the blood less acidic.

- Metabolic alkalosis is associated with an abnormal gain of base or loss of fixed acid and is characterized by an increase in plasma bicarbonate and pH. Respiratory compensation increases the blood P_{CO_2}, bringing the blood pH closer to normal.

Review Questions

Choose the Correct Answer

1. Which of the following is a base?
   a. NH_3
   b. Ca^{2+}
   c. H_3PO_4
   d. K^+
   e. Na^+

2. The formula for the hydronium ion is:
   a. H^+.
   b. H_3O^+.
   c. H_2O^-.
   d. OH^-.

3. Solution A has a pH of 7.0 and solution B has a pH of 4.0. The $[H^+]$ of solution B is:
   a. one third that of solution A.
   b. threefold that of solution A.
   c. 1/1000th that of solution A.
   d. 1000-fold that of solution of A.

4. Which of the following is an amphoteric substance?
   a. HCO_3^-
   b. Glycine
   c. $H_2PO_4^-$
   d. Water
   e. All of the above

5. Physiologically speaking, which of the following is a fixed acid?
   a. HCl
   b. H_2CO_3
   c. CO_2
   d. All of the above

6. What is the approximate pH of a 0.01 mol/L HCl solution?
   a. 1
   b. 2
   c. 3
   d. 5
   e. 9

7. Which acid has the lowest pK_a?
   a. Acetic acid
   b. NH_4^+
   c. β-hydroxybutyric acid
   d. Hydrochloric acid
   e. Lactic acid

8. The reaction $H_2CO_3 \rightarrow CO_2 + H_2O$ is:
   a. catalyzed by carbonic anhydrase.
   b. reversible.
   c. slow if uncatalyzed.
   d. All of the above
   e. None of the above

9. The major source of fixed acids on a mixed diet of meat and vegetables is from the oxidation of:
   a. carbohydrates.
   b. fats.
   c. organic anions.
   d. proteins.

10. If the pK_a of NH_4^+ is 9.0 and if urine pH is 6.0, what fraction (or %) of ammonia in the urine is present as NH_3?
    a. 0.001 (0.1%)
    b. 0.01 (1%)
    c. 0.03 (3%)
    d. 0.10 (10%)
    e. 1.00 (100%)

11. Most of the H^+s secreted by the kidney tubules are:
    a. consumed by filtered bicarbonate.
    b. excreted as ammonium ions.
    c. excreted as free H^+s.
    d. excreted as titratable acid.

12. Which process in the kidneys leads to generation of *new* bicarbonate to replenish depleted bicarbonate reserves?
    a. Excretion of ammonium ions
    b. Excretion of titratable acid
    c. Reabsorption of filtered bicarbonate
    d. Both a and b

13. The nephron segment that secretes the greatest number of H^+s is the:
    a. proximal convoluted tubule.
    b. thick ascending limb.
    c. distal convoluted tubule.
    d. collecting duct.

14. What is the net acid excretion by the kidneys if a person excretes 24 mEq titratable acid, 48 mEq NH_4^+, and 2 mEq HCO_3^- in a day?
    a. 24
    b. 48

c. 70

d. 72

e. 74

15. In a respiratory acidosis, most of the chemical buffering in the body occurs in:

 a. blood plasma.

 b. bone.

 c. extracellular fluid.

 d. intracellular fluid.

16. A person on a mountain-climbing expedition has been living at high altitude for several days. Arterial blood values are pH = 7.48, P_{CO_2} = 25 mm Hg, and plasma $[HCO_3^-]$ = 18 mEq/L. What type of acid-base disturbance is present?

 a. Acute respiratory acidosis

 b. Acute respiratory alkalosis

 c. Chronic respiratory acidosis

d. Chronic respiratory alkalosis

e. Metabolic acidosis

f. Metabolic alkalosis

17. The renal compensation for chronic respiratory alkalosis is:

 a. increased excretion of bicarbonate.

 b. increased excretion of titratable acid.

 c. increased synthesis and excretion of ammonia.

 d. Both b and c

18. A person with uncontrolled diabetes mellitus also has emphysema (pulmonary disease). Arterial blood values are pH = 7.21, P_{CO_2} = 60 mm Hg, and plasma $[HCO_3^-]$ = 23 mEq/L. This patient has a:

 a. metabolic acidosis.

 b. mixed metabolic acidosis and respiratory acidosis.

 c. mixed metabolic acidosis and respiratory alkalosis.

 d. respiratory acidosis.

Answers to Case History Questions

1. From the Henderson-Hasselbalch equation, 7.48 = 6.10 + log $([HCO_3^-]/[0.03 \times 18])$, hence $[HCO_3^-]$ = 0.54 × $10^{1.38}$ = 13 mEq/L. This value could also be determined by plotting pH and P_{CO_2} on the pH-bicarbonate diagram (see Fig. 25–5).

2. The low P_{CO_2} and elevated pH, together with increased pulmonary ventilation, indicate respiratory alkalosis. Salicylate stimulates the respiratory centers in the brain medulla.

3. Salicylate, at toxic concentrations, interferes with cell metabolism and results in increased production of acids, such as lactic acid and ketone body acids. These acids are buffered by body buffer bases, including bicarbonate, so the plasma $[HCO_3^-]$ is abnormally low. If the patient only had an acute respiratory alka-

losis, then the plasma bicarbonate would not be as low as it was. Renal compensation is too slow to explain the very low plasma bicarbonate concentration. Notice where the patient's data fall on the pH-bicarbonate diagram (see Fig. 25–5).

4. The sodium bicarbonate infusion was given to counteract the metabolic acidosis and to promote urinary excretion of salicylate. Salicylate is extensively reabsorbed if the urine is acidic (because more is present as the uncharged, undissociated acid). It is less completely reabsorbed and is excreted if the urine is made alkaline (because more salicylate is present in the charged, anionic form, which diffuses with difficulty from tubule lumen to blood).

Key Terms

acid (p. 786)

acid dissociation constant (p. 787)

acidemia (p. 797)

alkalemia (p. 797)

ammonia (p. 794)

amphoteric substance (p. 786)

base (p. 786)

carbonic anhydrase (p. 791)

chemical pH buffer (p. 788)

diffusion trapping (p. 797)

fixed acid (p. 793)

Henderson-Hasselbalch equation (p. 787)

hydronium ion (p. 787)

isohydric principle (p. 793)

Kussmaul's respiration (p. 800)

metabolic acidosis (p. 797)

metabolic alkalosis (p. 797)

neutralization (p. 786)

open system (p. 792)

pH-bicarbonate diagram (p. 800)

respiratory acidosis (p. 797)

respiratory alkalosis (p. 797)

sodium-hydrogen ion exchanger (p. 796)

strong acid (p. 787)

titratable acid (p. 794)

weak acid (p. 787)

Suggested Readings

DuBose, T. D., Jr. "Acid-base disorders." *In* Brenner, B. A., ed. *Brenner & Rector's The Kidney,* ed 6. Philadelphia, W. B. Saunders, 2000.

Knepper, M. A., Packer, R., and Good, D. W. "Ammonium transport in the kidney." *Physiological Reviews,* 69:179–249, 1989.

Lowenstein, J. *Acid and Basics, A Guide to Understanding Acid-Base Disorders.* New York, Oxford, 1993.

Rose, B. D., and Post, T. W. *Clinical Physiology of Acid-Base and Electrolyte Disorders,* ed 5. New York, McGraw-Hill, 2001.

Valtin, H., and Gennari, F. J. *Acid-Base Disorders: Basic Concepts and Clinical Management.* Boston, Little, Brown and Co., 1987.

Answers to Review Questions

1. a **2.** b **3.** d **4.** e **5.** a **6.** b **7.** d **8.** d

9. d **10.** a **11.** a **12.** d **13.** a **14.** c **15.** d

16. d **17.** a **18.** b

Chapter 26

CALCIUM, PHOSPHATE, AND BONE HOMEOSTASIS

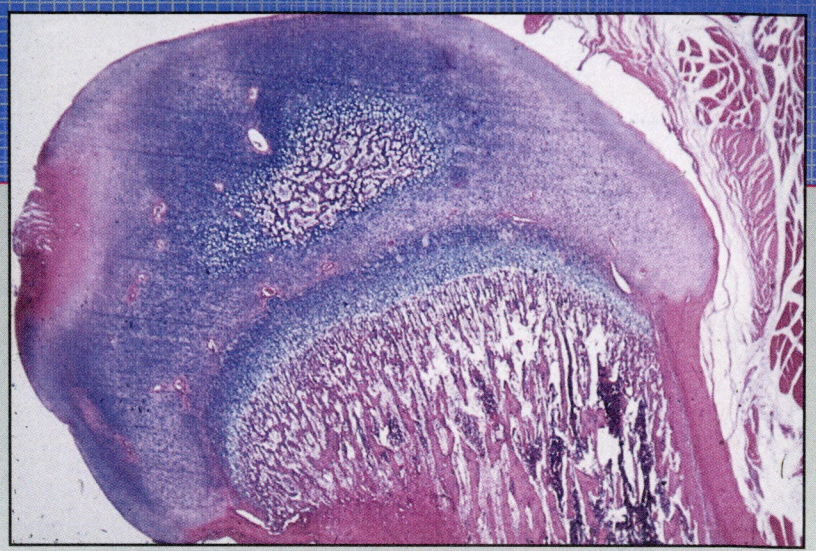

KEY CONCEPTS

- Calcium and phosphate ions serve important roles in the body. Each ion exists in several forms and is regulated by mechanisms involving several tissues and organs.

- The majority of ingested calcium is excreted in the feces, whereas the majority of ingested phosphate is absorbed into the blood.

- Calcium and phosphate metabolism in the gastrointestinal tract, kidneys, and bone is coordinated by hormonal signals. The activity of cells involved in bone formation and bone resorption are regulated by hormones. Not only does bone serve an important structural role in the body, but also it serves as an important reservoir of body calcium.

- Three hormones are involved in the regulation of plasma calcium levels: parathyroid hormone, calcitonin, and metabolites of vitamin D.

- The most common abnormalities of bone mineral metabolism are grouped together under the term metabolic bone disease, of which there are two principal types. Osteoporosis results from an equivalent loss of bone mineral and organic matrix. Osteomalacia and rickets result from inadequate mineralization of new bone.

CASE HISTORY

George is a 38-year-old Caucasian man who up until recently was in relatively good health. He is not a smoker and uses alcohol only moderately, but his only form of exercise is cutting the lawn on weekends during the summer months. He has not required any major surgeries over his lifetime and had only minor bouts with standard childhood illnesses. However, at age eight he was diagnosed with asthma after he suffered severe respiratory problems during a baseball game on a hot summer day. He has been treated ever since with a daily tablet of a synthetic glucocorticoid and the occasional use of an inhaler when needed to relieve acute symptoms of the disease.

George recently came to the attention of his physician when he suffered the second of two bone fractures in the past year and a half. One was to the left wrist and the other to the right forearm. In both cases, the trauma that caused the fracture was relatively minor. Suspecting that there may be an underlying problem, his physician ordered a series of bone density scans. Results of these studies showed that George had a considerable reduction in bone mass compared to other men of the same age.

Questions

1. What is the most probable diagnosis that the doctor would make?

2. What is the most probable underlying cause for George's problem?

3. What risk factors for osteoporosis are present (or absent) in George's case?

INTRODUCTION

As described in several earlier chapters, numerous homeostatic mechanisms are at work in the body. The primary focus of these is maintenance of a constant environment in which cells can develop and function. If we were to monitor a number of different physiological parameters in the body—including, for example, blood sugar concentration, core body temperature, or blood carbon dioxide concentration—results would show that one of the most closely regulated of all parameters is the plasma calcium concentration. Such strict regulation usually suggests that the factor in question plays a vital role in one or more processes in the body. This chapter will describe calcium homeostasis in the body. Although not as closely regulated, phosphate also plays several important roles in the body, and these will be described as well.

OVERVIEW: FUNCTIONS OF CALCIUM AND PHOSPHATE

 How do calcium and phosphate contribute to the physiology of the body?

Calcium, or more correctly the **Ca²⁺ ion,** is involved in a number of different physiological processes, several of which are listed in Table 26–1. The first of these, the effect of calcium on nerve function, is the most important with regard to an acute dependency on normal calcium ion concentrations. External calcium ions dramatically influence the permeability of nerves to sodium ions. Therefore, maintenance of extracellular calcium concentration is crucial if the resting membrane potential of nerves is to be maintained and nerve impulse transmission is to occur normally. When extracellular calcium ion concentrations are significantly decreased, sodium permeability is increased and spontaneous action potentials can be generated in nerves. If

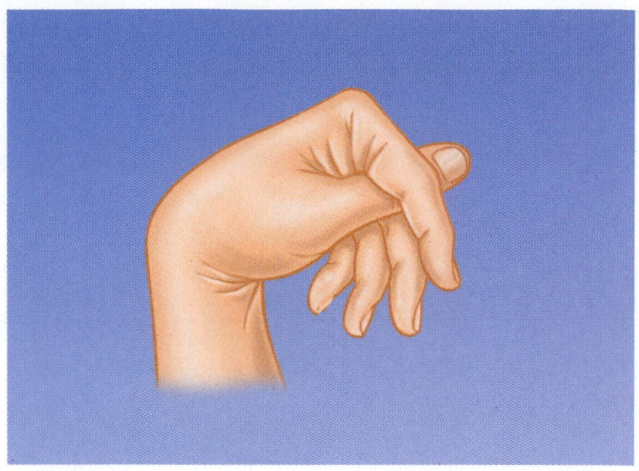

Figure 26–1

Typical position of the hand in hypocalcemic tetany. The spastic contractions of muscles in the forearm and hand are the result of hyperexcitability in nerves, which occurs when the plasma calcium concentration decreases to below normal.

a motor neuron is involved, the result can be **tetany** of the muscles in that motor unit. Figure 26–1 illustrates a classic clinical sign indicating the first stages of **hypocalcemic (low-calcium) tetany.** Muscles in the hand and forearm contract to produce the characteristic appearance shown. With prolonged hypocalcemia, muscles of the larynx become tetanized to the extent that the individual dies from suffocation. Other functions of calcium ion in the body are also listed in Table 26–1. At neuromuscular junctions, an influx of calcium into nerve endings serves to trigger release of acetylcholine. As described in Chapter 16, calcium fluxes in muscle provide the signal for excitation-contraction coupling. In other cell types, calcium influx provides a signal analogous to a second messenger; thus, calcium ion is important in the mechanism of action of some hormones. In addition, a number of enzymes and proteins are known to require calcium to function properly, and the process of protein secretion is also known to depend on the presence of calcium ion.

A specialized case in which calcium ion is needed for enzymatic activity involves the formation of enzyme complexes responsible for blood clotting (see Table 26–1). Some of the compounds used clinically to prevent clotting in blood samples taken from patients (e.g., citrate or EDTA) produce their effect by binding calcium and preventing its participation in the clotting reactions.

Finally, calcium is a constituent of bone, and thus in this regard it serves an important structural role in the body. As later sections will describe, the bones provide an important reservoir of calcium and therefore play an important role in the regulation of extracellular calcium ion homeostasis.

TABLE 26–1
Some of the Physiological Actions of Calcium
1. Required for the maintenance of normal sodium permeability in nerves
2. Involved in triggering the release of acetylcholine from nerve endings at the neuromuscular junction
3. Involved in excitation-contraction coupling in muscle cells
4. Serves as an intracellular signal for some hormones
5. Required by some enzymes for normal activity
6. Required for protein secretion
7. Required for blood clotting to occur normally
8. Constituent of bone

TABLE 26–2

Some of the Physiological Actions of Phosphate

1. Functions as part of the intracellular buffer system
2. Important constituent of a variety of macromolecules, such as nucleic acids, phospholipids, metabolic intermediates, and phosphoproteins
3. Constituent of bone

TABLE 26–4

Major Inorganic Constituents of Bone

Constituent	Total Body Content Present in Bone Percent
Calcium	99
Phosphate	85
Carbonate	80
Magnesium	50
Sodium	35
Water	9

Phosphorus is also present in considerable amounts in bone. As will be described later, the majority of phosphorus in the body exists in the form of **phosphate ion (PO₄).** Extracellular phosphate is not as strictly regulated as is calcium, however. As indicated in Table 26–2, phosphate serves the important role of providing a major portion of the intracellular buffering capacity. In addition, when coupled in organic form, phosphate is a component of a number of important cellular constituents and macromolecules. Nucleic acids, phospholipids, and various metabolic intermediates, to name a few, all contain phosphate. Chapter 5 introduced another important aspect of phosphate metabolism involving the regulation of certain metabolic processes by phosphorylation or dephosphorylation of key regulatory enzymes.

Calcium and Phosphorus Are Distributed Differently in Tissues

Relative distributions of calcium and phosphate among bone and other tissues in the body are presented in Table 26–3. As this table indicates, approximately 99% of the total calcium in the body is located in bone. Even in muscle, where calcium plays such a vital role in excitation-contraction coupling, only approximately 0.3% of the total is present. The remaining 0.7% is distributed between the extracellular fluid and the rest of the cells in the body. Phosphate is also located pre-

dominantly in bone, although not nearly to the extent as is calcium; approximately 85% of the total body phosphate is present in bone.

Although the majority of body calcium and phosphate is present in bone, these are not the only constituents found in relatively high abundance there. As Table 26–4 indicates, considerable amounts of total body carbonate, magnesium, and sodium are also located in bone. The large amount of carbonate present is of particular importance in that this can be passively added to the extracellular fluid to combat acidosis, and thus bone provides a fourth line of defense in acid-base regulation along with buffers, lungs, and kidneys. Long-standing uncorrected acidosis can lead to a loss of bone mineral for this very reason. Perhaps surprisingly, bone also contains a considerable amount of water. Approximately 20% of the weight of bone is actually due to water, and thus approximately 9% of all the water present in the body is located in bone.

Calcium and Phosphorus Exist in Several Forms in the Blood

 How are calcium and phosphate carried in the blood?

Forms of Calcium in the Blood

As indicated previously, calcium content of the extracellular fluid is maintained within fairly narrow limits. Total calcium content of plasma in humans is approximately 10 mg/100 ml. Of this total, calcium can exist in one of several different forms. Functionally, there are two major forms of calcium in plasma. As indicated in Figure 26–2, these are **filterable calcium** and the **protein-bound** calcium. Each of these forms behaves differently in physiological systems.

Filterable Calcium: Ionized or Complexed. **Filterable calcium** represents approximately 60% of total calcium in the plasma (see Fig. 26–2 and Table 26–5) and includes

TABLE 26–3

Typical Body Content and Tissue Distribution of Calcium and Phosphorus

	Calcium	Phosphorus
Total body content*	≈ 1000 g	≈ 500 g
Tissue distribution	Relative distribution (% of total body content)	
Bone	99.0	85.0
Muscle	0.3	6.0
Other tissues	0.7	9.0

*Based on a body weight of 70kg (150 lb) and an average body composition.

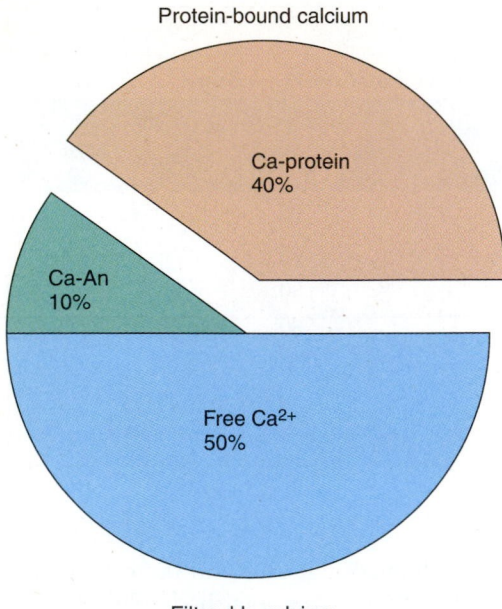

Protein-bound calcium

Ca-protein
40%

Ca-An
10%

Free Ca²⁺
50%

Filterable calcium

Figure 26–2

Relative distribution of different forms of calcium in the plasma. Calcium bound to plasma proteins is represented by "Ca-protein." Calcium bound to an anion, such as bicarbonate or citrate, is represented by "Ca-An," and the free ionized calcium in the plasma is represented by "Free Ca^{2+}."

those forms of calcium that are readily filtered in the glomerulus of the kidney. This calcium must therefore be reabsorbed in later segments of the renal tubules if it is not to be lost in the urine. Filterable calcium can be further subdivided into the **ionized** and the **complexed** forms.

TABLE 26–5		
Distribution of Plasma Calcium		
Form*	Amount (mg/ 100 ml)	Percentage of Total (approximate)
Filterable		
Ionized calcium	4.8	50
Complexed, ultra-filterable, bicarbonate (0.5), citrate (0.3), phosphate (0.4), (and others)	1.2	10
Protein-bound	4.0	40
Albumin (3.0)		
Globulins (1.0)		
Total	10.0	100

Numbers in parentheses represent the contribution (in mg/100 ml) of each of several forms within a category.

Free Ca^{2+} ion constitutes roughly half of the total calcium present in plasma (see Table 26–5). This is the **physiologically active** form of calcium. As the free ion, it is able to diffuse across plasma membranes and participate in processes such as those listed in Table 26–1. This is the form of calcium that is strictly regulated in plasma.

A smaller portion of filterable calcium is complexed with an anion (see Table 26–5). **Complexed calcium** normally represents approximately 10% of the total calcium present in plasma. As indicated in Table 26–5, bicarbonate is the major anion to which calcium is bound, although significant amounts are also found in association with phosphate and citrate. A variety of other anions make up the remainder of this category.

Nonfilterable Calcium: Protein-Bound Calcium. The second major category is that of **protein-bound calcium** (see Fig. 26–2 and Table 26–5). Calcium bound to relatively large plasma proteins cannot freely pass through plasma membranes, nor is it filtered in the kidney glomeruli. As indicated in Table 26–5, the majority of protein-bound calcium is bound to **albumin.** Albumin is a relatively large protein produced by the liver and is normally the most abundant protein present in the blood and interstitial fluid. A smaller proportion of protein-bound calcium is associated with various globulins in the blood.

Forms of Phosphorus in the Blood

In contrast to the plasma calcium concentration, which is normally regulated within very narrow limits, the phosphate concentration in blood may fluctuate on a daily basis from as low as half of normal up to 1.5-fold normal. A plasma concentration ranging from 3.0 to 4.5 mg/100 ml (expressed as milligrams of **phosphorus**) is considered to be normal in humans.

In plasma, phosphorus exists primarily as **orthophosphate,** or **PO₄.** At a normal blood pH, approximately 80% of the phosphate is present as HPO_4^{2-}, and approximately 20%, $H_2PO_4^-$. When present as the phosphate ion, the majority of the phosphorus readily diffuses across membranes and is also filterable in kidney glomeruli. In addition to the phosphate ion, small amounts of phosphorus are also present in blood in organic form, such as in hexose or triose phosphates, but these represent a relatively small proportion of the total.

PATHWAYS OF CALCIUM AND PHOSPHATE HOMEOSTASIS

Diet is, of course, the ultimate source of both calcium and phosphate in the body. Once these ions are ingested, **intestines, kidneys,** and **bone** are primarily involved in determining their plasma concentrations. As we will see, however, the overall patterns of movement in the body of the two ions differ considerably.

Understanding Pathways of Calcium Homeostasis Furthers Understanding of How Plasma Calcium Is Regulated

> **By what routes does calcium enter and leave the body?**

Figure 26–3 shows both tissue distribution and average daily flow of calcium among different tissues. In this figure, the quantity of calcium present is indicated by blue numerals, and daily flow of calcium among tissues is indicated by red numerals. Average calcium intake is approximately 1 g/day (1000 mg/day) (see Fig. 26–3). Most normal people can consume up to twice that amount without any deleterious effects. However, excessive calcium intake can result in problems with deposition of calcium in tissues (soft tissue **calcification**) or kidney stone formation. Of the 1 g of calcium that is ingested, normally approximately one third is absorbed from the intestines (300 mg/day, see Fig. 26–3). The percentage can change with differing calcium intake and with differing plasma calcium status. For example, when the requirements increase, such as in growing children and pregnant or nursing women, this percentage tends to be

somewhat higher, whereas in people of advanced age, it tends to be lower.

Note also in Figure 26–3 that approximately 150 mg of calcium enters the intestines from the body each day. This occurs in part as a result of sloughing of cells that line the digestive tract and also secretion of fluids and various digestive enzymes into the intestines. This component of calcium flux into and out of the intestines tends to change relatively little with varying calcium intake or plasma calcium status. Therefore, as Figure 26–3 indicates, the major route of excretion of *ingested* calcium is via the feces, simply because the majority of ingested calcium is not absorbed. However, as we will see later, regulation of calcium absorption from the intestines is an important aspect of overall calcium homeostasis.

Figure 26–3 also indicates that the kidneys excrete approximately 150 mg of calcium into the urine each day. This represents approximately 1% of the total amount of calcium filtered through kidney glomeruli, as the remaining 99% is reabsorbed and returned to the blood. A small change in calcium reabsorption by the kidneys can therefore dramatically affect calcium homeostasis.

Although we may think of bone as being relatively inert, bone mineral is constantly being turned over and remodeled to meet differing mechanical needs, as well as contributing to

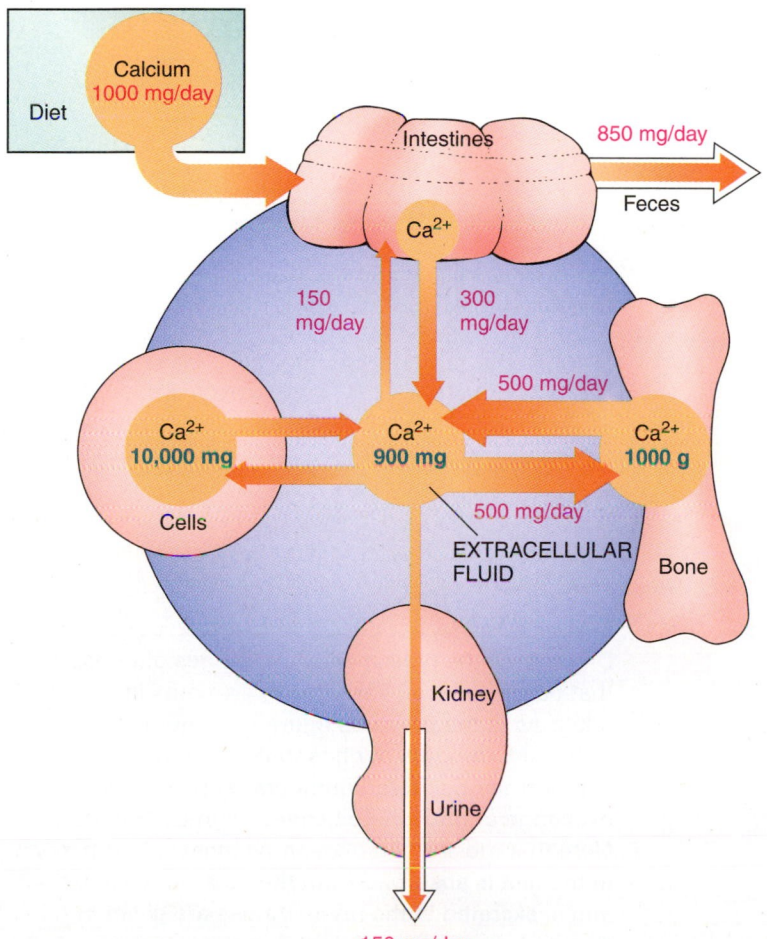

Figure 26–3

Diagram of the typical daily exchanges of calcium that occur between tissue compartments of a normal adult. Numbers shown in blue represent the amount of calcium in each compartment in grams or milligrams. Numbers in red represent the exchange of calcium between compartments in milligrams per day. Note that most calcium ingested each day in the diet is excreted in the feces; also note the high daily exchange of calcium between bone and the extracellular fluid. [1000 mg = 1 g]

calcium homeostasis on both a short-term and long-term basis. Roughly 500 mg of calcium both enter and leave bone each day (see Fig. 26–3), indicating that bone mineral turnover is rather extensive. In a nongrowing adult, the fluxes of calcium into and out of bone will of course be equal, as indicated in Figure 26–3. In growing children, the rate of calcium deposition is greater than the rate of calcium loss, and as bone mineral is lost with advancing age, the converse becomes true.

Each of these processes—calcium absorption from the intestines, calcium reabsorption by the kidneys, and bone mineral formation and resorption—is precisely controlled. The mechanisms involved in this control will be discussed in later sections.

Understanding Pathways of Phosphate Homeostasis also Provides Important Basic Concepts

Figure 26–4 shows the various pathways of phosphate metabolism in the body. As previously indicated, the extracellular phosphate concentration is not regulated as tightly as is the calcium concentration. Normally, approximately 1.4 g (1400 mg) of phosphorus is ingested each day. Because phosphate is relatively abundant in most foods, a nutritional phosphorus deficiency is generally not likely as long as sufficient food is taken in to meet caloric and protein needs. As Figure 26–4 shows, the majority of ingested phosphate is absorbed from the intestines. This is in direct contrast to calcium metabolism because, as shown in the previous figure, the majority of ingested calcium is not absorbed. As with calcium, however, a small amount of phosphate is also lost into the intestinal lumen each day. Thus, there is a *net* uptake of approximately 1100 mg of phosphorus each day.

The primary route for excretion of this daily phosphate load is the kidneys (see Fig. 26–4). Remember that most of the phosphorus present in the plasma exists as the phosphate ions, which are readily filtered in the glomeruli of the kidneys. Phosphate would thus be readily lost from the body if not for mechanisms that reabsorb and return it to the plasma. As will be described later, the reabsorption of phosphate by kidney tubule cells is a major mechanism of the hormonal regulation of phosphate homeostasis.

As with calcium, considerable amounts of phosphate flow both into and out of bone each day. Again, this reflects the very active turnover and remodeling of bone mineral that normally occurs.

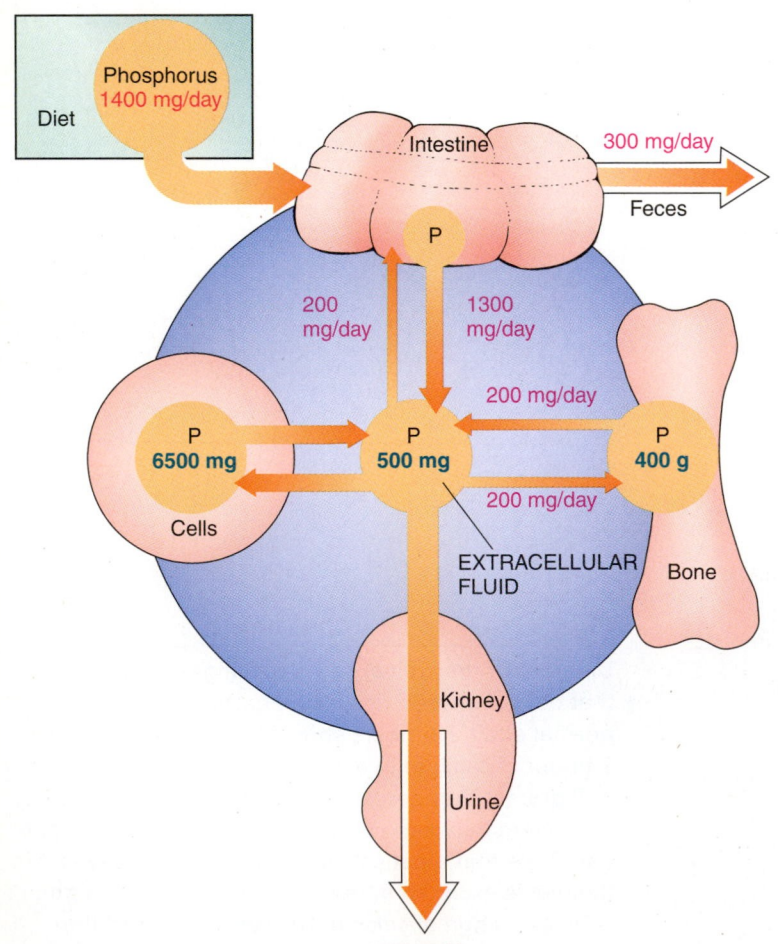

Figure 26–4

Diagram of the typical daily exchanges of phosphate that occur between tissue compartments in a normal adult. As in the previous figure, blue numbers represent amounts of phosphate in each compartment, and red numbers represent the exchange of phosphate between compartments. Note that most of the phosphate ingested each day in the diet is absorbed from the gastrointestinal tract and is excreted in the urine. Values are given in terms of grams or milligrams of phosphorus.

MECHANISMS INVOLVED IN CALCIUM AND PHOSPHATE HOMEOSTASIS

Previous sections referred to processes of absorption, reabsorption, and resorption of calcium and phosphate as they occur in the gastrointestinal tract, kidney, and bone, respectively. We will now describe in more detail specific mechanisms involved in handling calcium and phosphate in each of these tissues.

Calcium and Phosphate Absorption in the Gastrointestinal Tract Differ in Magnitude

 How are calcium and phosphate absorbed in the gastrointestinal tract?

Calcium Absorption in the Gastrointestinal Tract

In the gastrointestinal tract, calcium is absorbed predominantly from the small intestine. Calcium is taken up from the small intestine by both active transport and simple passive diffusion. Relative contributions of each of these two processes depend on the region of small intestine. Active transport predominates in the duodenum and jejunum, whereas simple diffusion predominates in the ileum. Because it is longer than the other two regions of the intestine (and therefore has a larger surface area), the ileum is where the most calcium is absorbed.

As with other active transport processes, calcium transport exhibits characteristics of saturability. It requires energy and moves calcium against a concentration gradient. Calcium uptake by passive diffusion is directly related to the concentration gradient of calcium between the lumen of the small intestine and the plasma.

The relative importance of active transport and passive diffusion depends on a number of factors. For example, with low calcium intake, active transport predominates because uptake by diffusion is low. With ingestion of large amounts of calcium, the active transport system becomes saturated, and uptake by diffusion predominates.

Calcium uptake from intestines is controlled through alterations in active calcium transport mechanisms. As we will see later, metabolites of vitamin D that function as hormones act to control calcium absorption by intestinal mucosal cells. Specific calcium-binding proteins involved in calcium transport are present in intestinal mucosal cells. Under the influence of vitamin D metabolites, these mucosal cells produce more calcium-binding proteins, thereby increasing their capacity for calcium transport into the body.

Phosphate Absorption in the Gastrointestinal Tract

Phosphate absorption also occurs by both active transport processes and passive diffusion, with active transport predominating. The small intestine is the primary site of phosphate absorption from the gastrointestinal tract. As Figure 26–4 indicates, phosphate is very efficiently absorbed from the intestines, as usually 70% to 90% of ingested phosphate is eventually absorbed. Active transport of phosphate in the small intestine is in part coupled to calcium transport. Therefore, in situations in which active transport of calcium is low (e.g., with vitamin D deficiency) phosphate absorption is also low. The coupling of phosphate transport to calcium transport offers certain physiological advantages that will be discussed later in this chapter. However, phosphate absorption from the small intestine differs from calcium absorption in that there is only minor control. As will be described subsequently, the primary site of regulation of phosphate metabolism is the kidney.

Calcium and Phosphate Reabsorption in Kidneys Differ by Tubular Location and Extent

 How does the kidney contribute to calcium and phosphate homeostasis?

Calcium Reabsorption in the Kidney

Recall from Table 26–5 and Figure 26–2 that filterable calcium consists of free calcium ion and calcium bound to filterable anions. Together, these make up approximately 60% of the total calcium in plasma; the remaining 40% circulates bound to protein and is not filtered in kidney glomeruli. Ordinarily, approximately 99% of filtered calcium is eventually reabsorbed by tubule cells and does not appear in the urine. Reabsorption occurs mainly by active transport processes in proximal and distal tubules. The nature of calcium reabsorption (active or passive) in the loop of Henle is not well established. Approximately 60% of the filtered calcium is reabsorbed from the proximal tubule, approximately 30% from the loop of Henle, and another 9% from the distal tubule; the remaining 1% is excreted in the urine. Although calcium reabsorption in the distal tubule is a small percentage of the total, it is a major site of regulation. Major adjustments in calcium reabsorption occur at this site as a result of an action of parathyroid hormone, and thus parathyroid hormone serves as a major factor in determining how much calcium is retained in the body and how much leaves in the urine. Parathyroid hormone is one of three hormones involved in regulating calcium and phosphate homeostasis, as later sections of this chapter will describe. The effect of parathyroid hormone to increase transfer of calcium from tubular fluid to plasma is mediated by an increase in cAMP in distal tubule cells (refer back to Table 12–3).

Phosphate Reabsorption in the Kidney

As was indicated in Figure 26–4, a major route by which ingested phosphate leaves the body is via the urine. The kidney is therefore an important site of regulation of extracellular phosphate concentrations. Ordinarily, about 85% of filtered phosphate is reabsorbed, and the remaining 15% is excreted in the urine. Reabsorption of phosphate is due to active transport processes, with the principal site of phosphate reabsorption being the proximal tubules, but some reabsorption occurs in distal tubules as well. Parathyroid hormone exerts a major regulatory effect on phosphate excretion by inhibiting phosphate reabsorption in proximal tubules. As a result, parathyroid hormone causes an increase in urinary phosphate excretion with a resulting decrease in plasma phosphate.

Calcium and Phosphate Homeostasis in Bone Is a Highly Regulated Process

 How is bone formed and resorbed?

The Composition of Bone

Simply described, mature bone consists of inorganic bone mineral deposited on a framework of organic support material. Imagine what would happen if you dipped a sponge into wet plaster and then allowed the plaster to harden. The sponge is analogous to the organic support material, and the hardened plaster is analogous to the bone mineral.

The mineral fraction of bone is composed largely of calcium phosphate, in the form of what are known as **hydroxyapatite crystals.** The hydroxyapatite crystals in bone have the general formula

$$Ca_{10}(PO_4)_6(OH)_2$$

Normally, the mineral portion of bone makes up approximately one quarter of the bone volume, but because of its high density, the mineral fraction makes up approximately half of bone weight. Recall from Table 26–4 that, in addition to calcium phosphate, inorganic matter of bone also contains a considerable amount of carbonate, magnesium, and sodium.

The organic matrix of bone is known as **osteoid,** 95% of which is **collagen.** The type of collagen in bone is similar to that present in tendons and in skin, although collagen in bone does have biochemical properties that give it greater mechanical strength than other types of collagen. The remaining 5%, or noncollagen portion of organic matter, is known as **ground substance,** which consists of a mixture of various **proteoglycans,** which are high-molecular-weight compounds made up of carbohydrate and protein.

If one closely examines bone structure using an electron microscope, one sees needlelike hydroxyapatite crystals lying alongside the collagen fibers and surrounded by the somewhat amorphous ground substance. Association of hydroxyapatite with the collagen fibers is responsible for the strength and hardness characteristic of bones. If a bone is **demineralized**—that is, if all the inorganic material is removed—the shape of the bone will be preserved because of the collagen that remains, but it will be as flexible as a tendon. Conversely, removal of the organic matter will also leave the bone with its original shape, but in this case the mineral that remains will be very brittle and easily broken. Thus, it is the combination of organic and inorganic material that gives bone its high degree of mechanical strength.

Bone Cells

There are three primary cell types involved in formation and resorption of mature bone. These are osteoblasts, osteocytes, and osteoclasts (Fig. 26–5).

Osteoblasts. Located side by side on the surfaces of bone, **osteoblasts** are the cells responsible for synthesis of osteoid. Osteoblasts that are actively producing osteoid have a cuboidal shape and exhibit an abundant rough endoplasmic reticulum and Golgi complex, two characteristics typical of cells actively synthesizing proteins for export. Osteoblasts not actively engaged in bone formation have a considerably flattened appearance. Osteoblasts also have numerous cytoplasmic processes that serve to bring them into contact with neighboring osteoblasts.

As Figure 26–5 indicates, osteoid secreted from osteoblasts is deposited on the surface of bone. Bone mineral is then deposited on the organic osteoid matrix, and as this process occurs, the osteoblasts eventually become surrounded by mineralized bone.

Osteocytes. At this stage, osteoblasts progressively lose their bone-forming capability and are termed **osteocytes.** Cell-to-cell connections that were present among osteoblasts are maintained through the osteocyte stage of development. These interconnections are channels known as **canaliculi** in the bone and represent direct contacts between cells. This coupling among cells provides a "bucket brigade" mechanism whereby nutrients and ions can travel from blood vessels to distant osteocytes deeply embedded in bone. Hormones that control bone growth and development are also believed to pass from one osteocyte to another in this way.

Osteoclasts. The process of bone resorption is carried out by **osteoclasts.** These are large, multinucleated cells, each containing approximately four to six nuclei. The surface of osteoclasts adjacent to bone has a very ruffled appearance (see Fig. 26–5). This ruffling increases surface area and allows cells to perform the task of bone resorption more effectively. When stimulated, osteoclasts secrete both acids and proteolytic enzymes into the extracellular space adjacent to bone. The acidic environment increases the solubility of bone mineral, while the proteolytic enzymes attack the organic matrix of bone. Together, these two factors promote bone resorption.

Formation of Bone

During early fetal development, the skeleton consists solely of cartilage, which is laid down in the shape and form of the bony skeleton. This cartilage "model" serves as the template

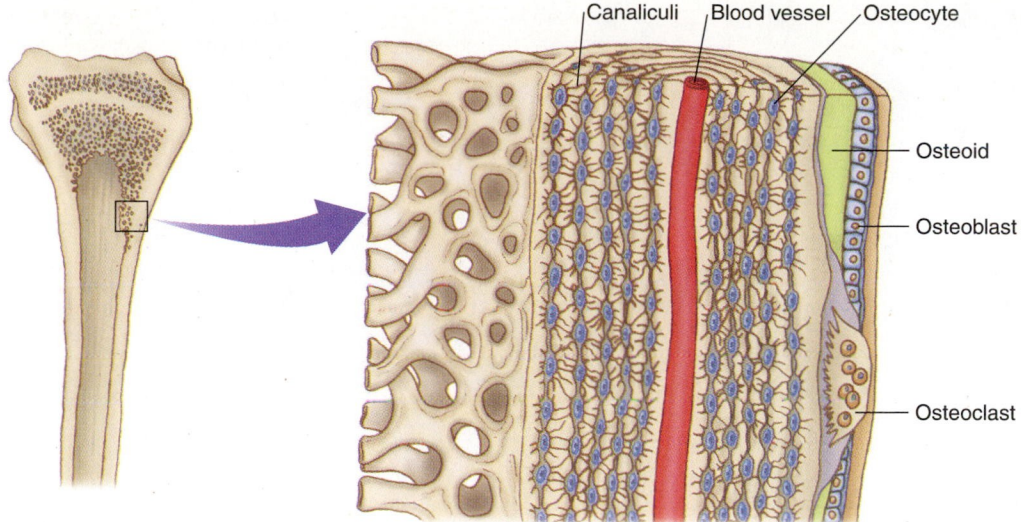

Figure 26–5

Diagram illustrating the relationships among different types of bone cells. Osteoblasts line the surface of bone and secrete osteoid into the space adjacent to the bone. An osteoclast actively engaged in bone resorption is represented by the large cell on the right. Note the ruffled appearance of the osteoclast membrane on the side next to the bone surface.

for development of the bony skeleton and is eventually replaced by bone.

The Epiphyseal Plate. The process of cartilage replacement begins in the middle or center and progresses out toward each end of what will later form the bone (Fig. 26–6*a* to *c*). As this progression occurs, the bone increases in both length and thickness. Figure 26–6*d* illustrates the typical structure of a long bone in a growing adolescent. The histological appearance of bone is the result of an extension of a series of events shown in parts *a* to *c* of Figure 26–6. The area of bone designated as the **epiphyseal plate** is of particular interest because it is here that elongation or growth of bone occurs after birth.

Chondrocytes. As indicated in Figure 26–7, the epiphyseal plate shows considerable differences between its leading and trailing edges. On the leading edge are cells known as **chondrocytes,** which are actively engaged in synthesis of cartilage of the epiphyseal plate. As cartilage is formed, it surrounds the chondrocytes and gradually becomes calcified. The trapped chondrocytes are replaced by new cells on the surface of the cartilage, so that active synthesis of cartilage continues. The calcified cartilage that is formed *does not* itself become bone but is later replaced by bone. As the cartilage becomes increasingly calcified, chondrocytes embedded in it begin to die off, and the calcified material begins to erode. Osteoblasts then move into the area, and the actual process of bone formation begins. As a result of continuing cycles of

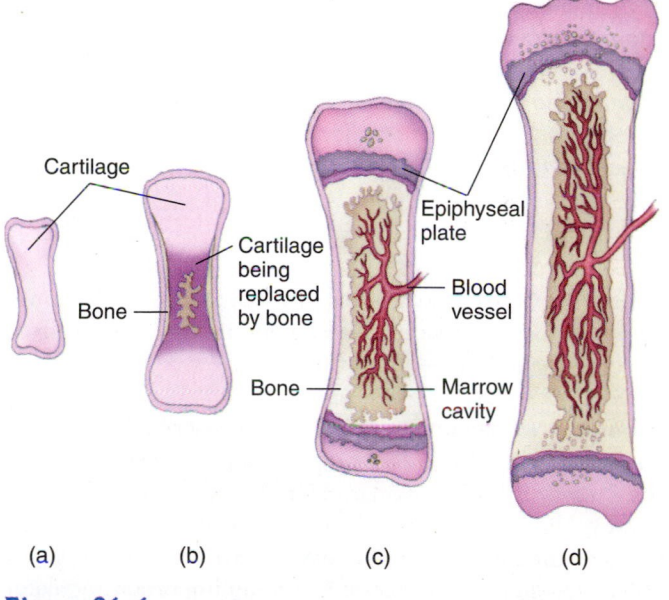

Figure 26–6

The process of growth and elongation of a typical long bone. **(a)** During fetal development, cartilage "models" of the bone are formed. **(b** and **c)** The cartilage is then replaced by mineralized bone, beginning in the middle of the bone and progressing toward each end. In the process, the bone becomes thicker and longer. **(d)** The center of the bone shaft becomes hollowed to form the marrow cavity. The epiphyseal plate is the zone where elongation of bone occurs.

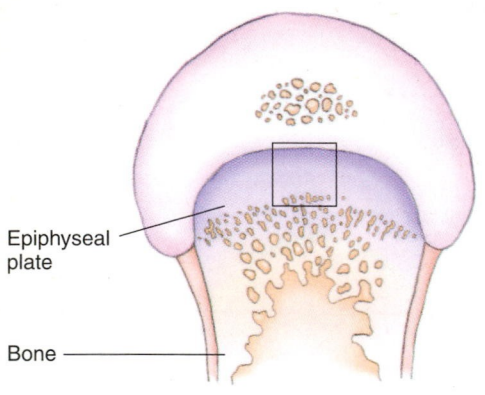

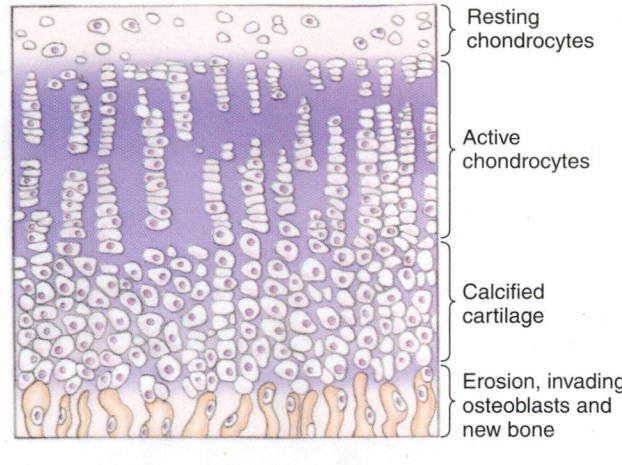

(a)

Figure 26–7

(a) The appearance of an epiphyseal plate. Zones of different cellular activities within the epiphyseal plate are illustrated. Active chondrocytes secrete collagen, which eventually becomes calcified. The calcified cartilage erodes and is replaced by osteoblasts, which then form new bone. The entire sequence is repeated many times as the zone of new bone formation gradually moves upward. *(b)* Histological photo of cellular activities of an epiphyseal plate showing chondrocytes, calcified cartilage, and osteoblasts forming new bone.   (© Biophoto Associates/Photo Researchers)

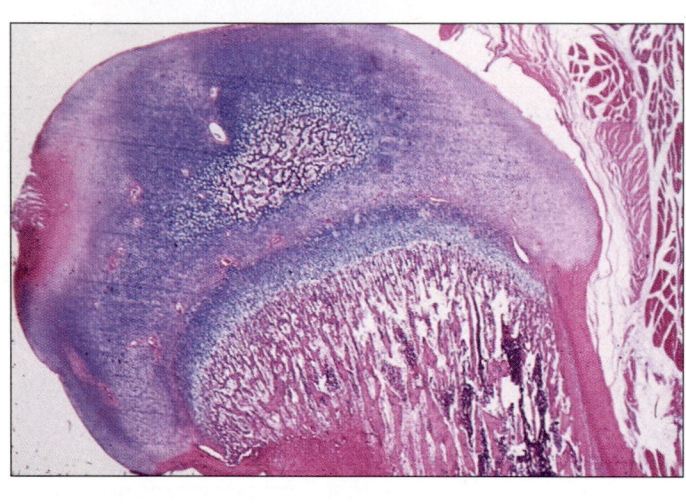

(b)

cartilage synthesis, calcification, erosion, and osteoblast invasion, the zone of active bone formation moves outward from the center toward the ends of the bones.

Hormones in Bone Formation. As described earlier in Chapter 13, chondrocytes of epiphyseal plates are under hormonal control. In particular, **IGF-I**, a somatomedin produced by liver in response to growth hormone (GH), is the primary stimulator of chondrocyte activity. This effect of IGF-I accounts for the role of GH as an important regulator of growth in young children. Other hormones, such as insulin and thyroid hormones, also tend to promote chondrocyte activity and therefore promote bone growth as well.

Beginning around the time of puberty, the epiphyseal plates of the long bones, such as those in the thighs or arms, begin to "close," or become unresponsive to hormonal stimuli. In most individuals, this process is complete by approximately age 20, so that after this time linear growth is no longer possible. Bones of the fingers, feet, and skull remain hormonally responsive, however, which ac-

counts for the skeletal changes that occur with GH oversecretion in the adult, a condition known as *acromegaly* (see Chapter 13).

Bone Remodeling. As indicated earlier, bone is continually undergoing synthesis and resorption, even after linear growth has ceased. This is reflected by the large amounts of both calcium and phosphate that flow into and out of bone each day, as shown in Figures 26–3 and 26–4. This process of continued bone turnover is termed **remodeling.** Synthesis or resorption of bone can occur along most of the outer surface of bones, making them thicker or thinner, respectively. In the long bones, synthesis and resorption can also occur along inner surfaces of the bone shaft, adjacent to the bone marrow cavity. Remodeling reshapes and re-forms bone to meet the changing mechanical needs that occur throughout life. A good illustration of such a case is a cowboy's bowed legs, in which the curvature of the legs results partly from the continued pressure on the legs as he rides his horse. In addition, because of the role of bone as a reservoir for both calcium and phosphate, it must be able to store and rapidly mobilize these two ions as required. The factors involved in regulating these processes will be described shortly.

REGULATION OF CALCIUM AND PHOSPHATE IN PLASMA

 How is plasma calcium concentration held constant?

Because calcium ions are so important in physiological processes, and because disturbances in the plasma concentration of calcium have such grave consequences, most of the regulatory processes to be discussed are directed toward controlling the plasma concentration of this ion. Calcium ions can move between extracellular fluid and bone as a result of either rapid, nonhormonal processes or slower, hormonally regulated mechanisms. Both processes are important in maintaining a constant extracellular fluid calcium ion concentration.

Nonhormonal Mechanisms for Calcium and Phosphate Homeostasis Provide a Rapid, but Low-Capacity, System

Recall from Figure 26–3 and Table 26–3 that approximately 1000 g, or 99%, of the total body calcium is present in bone. The vast majority of this calcium exists as hardened bone mineral, but approximately 1% of the calcium present in bone exists in a readily exchangeable form that can rapidly move between bone and extracellular fluid. This rapidly exchangeable source of calcium is primarily located on the surface of bone and in newly formed bone. Calcium can move rapidly by this mechanism because it involves purely physical events. Free ionized calcium in the extracellular fluid exists in chemical equilibrium with calcium present in hydroxyapatite crystals of bone. Any decrease in free calcium in the extracellular fluid therefore directly results in movement of calcium out of hydroxyapatite and into extracellular fluid until a new chemical equilibrium is attained.

This mechanism, while permitting very rapid movement of calcium both into and out of bone, has a very limited capacity because of the small proportion of calcium that exists in this readily exchangeable pool.

Hormonal Mechanisms for Calcium and Phosphate Homeostasis Provide for Long-Term Control

 How does the control system that regulates the extracellular calcium ion work?

The second mechanism involves hormones, and although these are somewhat slower than are nonhormonal processes, they can cause large shifts in calcium between bone and the

extracellular fluid over a prolonged period of time. Three hormones are primarily involved in regulating calcium ion concentrations in the extracellular fluid: (1) parathyroid hormone, (2) calcitonin, and (3) metabolites of vitamin D. As will be described, the actions of these hormones are intimately interrelated by a series of feedback loops, with the concentration of ionized calcium (Ca^{2+}) in the plasma as the factor being regulated.

Chemistry and Regulation of Hormone Production and Secretion

Production and Secretion of Parathyroid Hormone. **Parathyroid hormone (PTH, parathormone)** is a polypeptide hormone produced by chief cells of the parathyroid glands. Humans have four parathyroid glands, found in pairs on the dorsal surface of each of the two lobes of the thyroid gland (Fig. 26–8). As is true of other endocrine glands, the parathyroid glands have a very rich blood supply.

The primary stimulus for parathyroid hormone secretion is a *decrease* in the plasma calcium concentration. As Figure 26–9 illustrates, as the calcium concentration in the plasma decreases to below its normal value of approximately 10 mg/100 ml, the parathyroid gland is stimulated to secrete more parathyroid hormone. Although Figure 26–9 shows the relationship between the *total* plasma calcium concentration and the rate of parathyroid hormone secretion, the parathyroid glands actually respond to changes in free *ionized* calcium. Note also in Figure 26–9 that as the plasma calcium

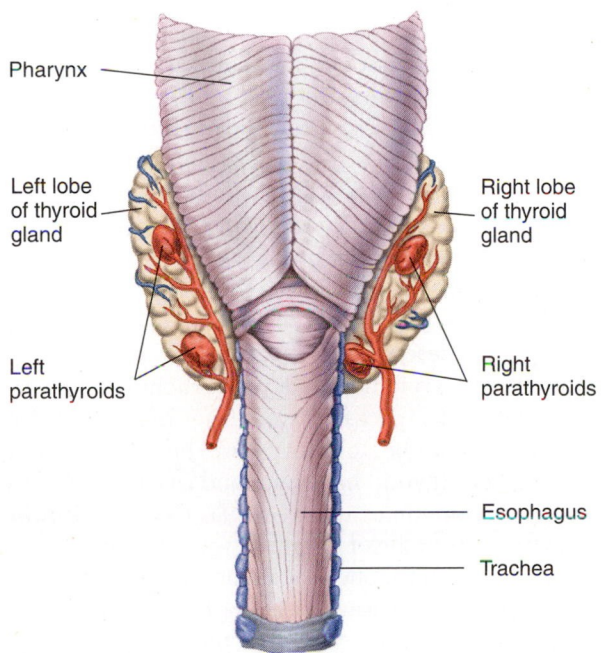

Figure 26–8

The anatomical location of the four parathyroid glands in humans. The view depicts the dorsal surface of the two lobes of the thyroid gland.

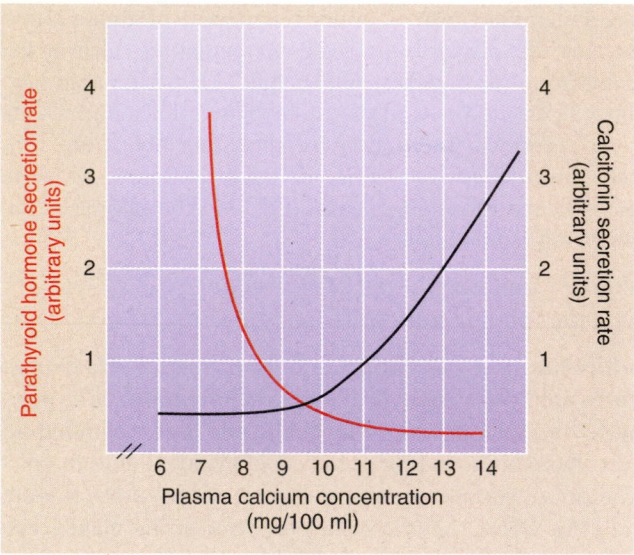

Figure 26–9

Relationship between the total plasma calcium concentration, parathyroid hormone secretion, and calcitonin secretion. The normal plasma calcium concentration is approximately 10 mg/100 ml.

first begins to decrease to below normal, the secretion of parathyroid hormone increases rather gradually. As the plasma calcium concentration decreases to below approximately 8.5 mg/100 ml, however, the curve becomes much steeper. As will be described, the net effect of parathyroid hormone action is to increase the plasma calcium concentration. Thus, dramatic decreases in plasma calcium produce an equally dramatic increase in parathyroid hormone secretion in an attempt to maintain plasma calcium at normal values.

Production and Secretion of Calcitonin. Calcitonin (sometimes abbreviated as CT) is a polypeptide hormone (32 amino acids) that in humans is produced by cells known as **parafollicular cells** or **C-cells** of the thyroid gland (the "C" stands for *clear,* which is how the cytoplasm appears under the microscope). As described in Chapter 13, thyroid follicle cells are the site of thyroid hormone production. Parafollicular cells are located in spaces between the thyroid follicles and are distinct from the follicle cells. Thus, the thyroid gland produces both thyroid hormones and calcitonin. Calcitonin is therefore sometimes referred to as **thyrocalcitonin (TCT),** to indicate its tissue of origin.

Like parathyroid hormone, calcitonin is secreted in response to changes in the plasma concentration of ionized calcium, except that an *increase* rather than a decrease in plasma calcium stimulates calcitonin secretion (see Fig. 26–9). Calcitonin secretion also increases in response to several gastrointestinal hormones, particularly gastrin. The net effect of calcitonin actions, which will be described in detail shortly, is to cause an increased deposition of calcium in bones, with an accompanying decrease in the plasma calcium concentration. Stimulation of calcitonin secretion by gastrointestinal hormones coordinates increased calcium uptake into bone with the anticipated absorption of calcium associated with a meal.

Production of Vitamin D Metabolites. The third hormone involved in regulating plasma calcium is **vitamin D,** or, more precisely, a metabolite of the vitamin that functions as the active hormone. D vitamins are a group of lipid-soluble vitamins long known to prevent the childhood disease of **rickets.** Only in the past several decades has a **hormonal** role of vitamin D metabolites in regulating plasma calcium been determined.

D vitamins have steroidlike structures and are derived from cholesterol, as are other steroids. The natural form is **vitamin D_3, or cholecalciferol,** which is shown in Figure 26–10. Vitamin D_3 can either be obtained from the diet or formed in the skin by the action of **ultraviolet light** on a precursor, **7-dehydro-cholesterol,** which is derived from cholesterol (Fig. 26–11). With adequate exposure to sunlight, sufficient vitamin D_3 can be formed in the skin to supply the body's needs without any dietary supplementation.

Another related compound of interest is **vitamin D_2,** or ergocalciferol (see Fig. 26–10). Ergocalciferol is a form of the vitamin found in plants and yeasts, and thus it can be a significant dietary source of D vitamin. Vitamin D_2 is metabolized and has actions essentially identical to those of vitamin D_3. Because it can be obtained relatively inexpensively from plant sources, it is often used to fortify foods, such as milk, with vitamin D. The following sections will discuss the metabolism and actions of vitamin D_3 because this is the natural form of the vitamin. But keep in mind that vitamin D_2 undergoes the same modifications and has the same actions as does D_3.

Vitamins D_2 and D_3 by themselves have relatively little biological activity. The following paragraphs describe steps by which these substances are transformed into potent hormones involved in regulating plasma calcium concentrations.

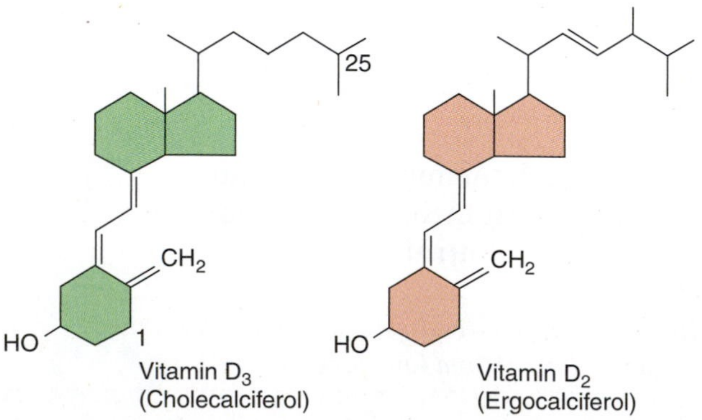

Figure 26–10

Structures of vitamin D_3 and vitamin D_2.

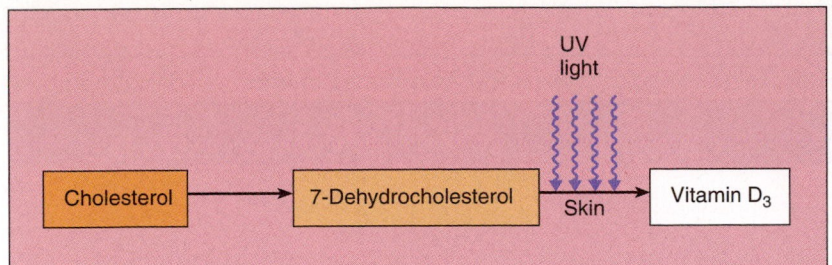

Figure 26–11

Illustration of the pathway of formation of vitamin D$_3$, beginning with cholesterol.

The hormonally active form of vitamin D is produced when inactive molecules are modified by liver and kidney. The modifications involve addition of a hydroxyl (OH) group first to carbon number 25 and then to carbon number 1 of vitamin D$_3$. As indicated in Figure 26–12, the first step involved in activation of vitamin D occurs in liver. Vitamin D$_3$, derived either from the diet or formed in the skin, is carried to the liver via the blood. In liver, a specific enzyme catalyzes the addition of a hydroxyl group to carbon number 25 to form 25-hydroxy-vitamin D$_3$. 25-hydroxy-vitamin D$_3$ (sometimes referred to as *calcidiol*) then leaves the liver and travels via the blood to the kidneys. In proximal tubule cells of the kidney, a specific enzyme known as **1α-hydroxylase** catalyzes the conversion of 25-hydroxy-vitamin D$_3$ to 1,25-dihydroxy-vitamin D$_3$ (1,25-OH-vitamin D$_3$) (sometimes referred to as *calcitriol*). It is this form of the vitamin that functions as a hormone and is involved in regulating plasma calcium concentrations.

Formation of 1,25-OH-vitamin D$_3$ is a highly controlled process. The primary site of control is the kidney. Activity of 1α-hydroxylase is markedly increased by parathyroid hormone (see Fig. 26–12). Thus, activity of the enzyme changes indirectly as a result of changes in the plasma calcium concentration. (As mentioned, if the plasma calcium concentration decreases to below normal, parathyroid hormone secretion is stimulated. This results in an increase in 1α-hydroxylase activity, leading to increased formation of 1,25-OH-vitamin D$_3$ (see Fig. 26–12).

In addition, a decrease in plasma phosphate causes an increase in activity of 1α-hydroxylase, which also leads to increased formation of 1,25-OH-vitamin D$_3$. In this case, phosphate appears to directly modulate activity of the enzyme. Therefore, either a decrease in plasma calcium, with its associated increase in parathyroid hormone, or a decrease in plasma phosphate results in increased production of the hormonally active form of vitamin D (see Fig. 26–12).

Actions of Hormones Involved in Regulating Calcium and Phosphate Homeostasis

Actions of parathyroid hormone, calcitonin, and 1,25-OH-vitamin D$_3$ are summarized in Table 26–6. The specific effects of these hormones will be described in the following sections.

Actions of Parathyroid Hormone in Calcium and Phosphate Homeostasis. As a result of its actions in regulating plasma calcium concentrations, parathyroid hormone is *acutely* required for life. Removal of the parathyroid glands without any subsequent intervention usually leads to death from hypocalcemic tetany within 48 to 72 hours. Figure 26–1 shows the initial signs of hypocalcemic tetany, involving spasm of muscles of the hand and forearm.

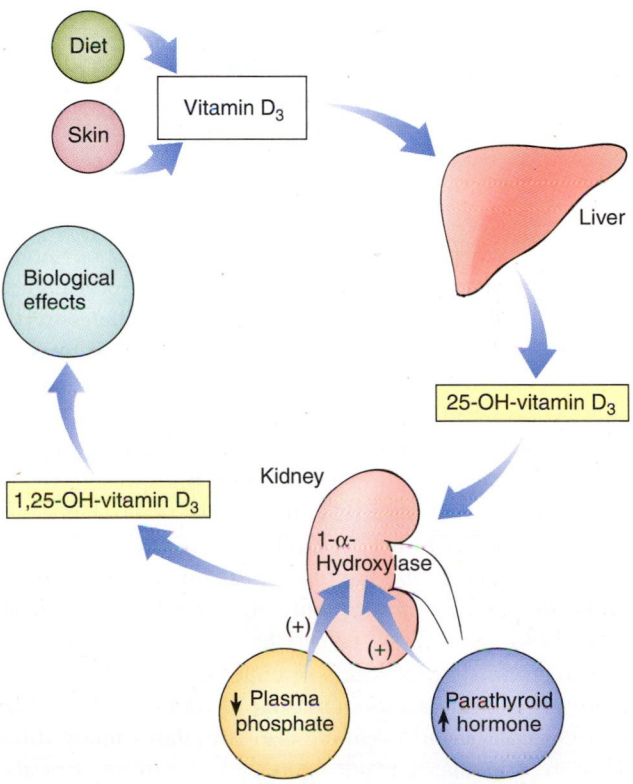

Figure 26–12

Diagram of the pathway for conversion of vitamin D$_3$ into 1,25-dihydroxy-vitamin D$_3$. Hydroxylation of vitamin D$_3$ on the 25 position occurs first in the liver, followed by hydroxylation on the 1 position in the kidney. Parathyroid hormone stimulates the kidney enzyme, which catalyzes the addition of a hydroxyl group to carbon number 1.

TABLE 26–6

Summary of the Effects of Hormones Involved in Regulating Calcium and Phosphate Balance

Parameter or Target Tissue	Parathyroid Hormone	Calcitonin	1,25-OH-vitamin D$_3$
Plasma calcium concentration	↑	↓	↑
Plasma phosphate concentration	↓	↓	↑
Kidney	↑ Reabsorption of calcium ↓ Reabsorption of phosphate ↑ Activity of 1α-hydroxylase	↓ Reabsorption of calcium ↓ Reabsorption of phosphate	↑ Reabsorption of calcium ↑ Reabsorption of phosphate
Bone	↑ Resorption of bone	↓ Resorption of bone	Promotes PTH actions
Small intestine	Indirect effects by increasing 1,25-OH-vitamin D$_3$ formation	No effect	↑ Absorption of calcium and phosphate

As indicated in Table 26–6, the overall effect of parathyroid hormone is to increase plasma calcium and decrease plasma phosphate. With these effects in mind, what specific responses does the hormone elicit that lead to these changes?

One important effect of parathyroid hormone in kidneys is to increase the activity of 1α-hydroxylase, which catalyzes the formation of 1,25-OH-vitamin D$_3$, the active metabolite of vitamin D. Parathyroid hormone also stimulates active transport mechanisms for calcium reabsorption in distal tubules, leading to an increase in calcium retention and a decrease in urinary excretion of calcium. Parathyroid hormone also decreases proximal tubular reabsorption of phosphate, resulting in **phosphaturia** (i.e., increased phosphate excretion in the urine). This phosphaturic effect of parathyroid hormone is an important one because other effects of parathyroid hormone tend to increase the flow of *both* calcium and phosphate into the blood. Without decreased phosphate reabsorption, both calcium and phosphate would accumulate in plasma and then simply recrystallize in bone mineral or in soft tissues. This would counteract the desired effect, which is to increase the plasma calcium concentration.

In bone, parathyroid hormone has several effects, all of which lead to increased net resorption of bone. The primary effect is to stimulate dissolution of bone by increasing osteoclast activity. In addition to increasing the activity of existing osteoclasts, parathyroid hormone also stimulates maturation of immature osteoclasts, resulting in a greater number of active osteoclasts. A third effect of parathyroid hormone in bone is to decrease collagen synthesis in osteoblasts, resulting in a decrease in bone matrix formation. Thus, parathyroid hormone impairs new bone formation while simultaneously stimulating bone resorption. Together, these effects of parathyroid hormone result in an increased net transfer of calcium from bone into plasma.

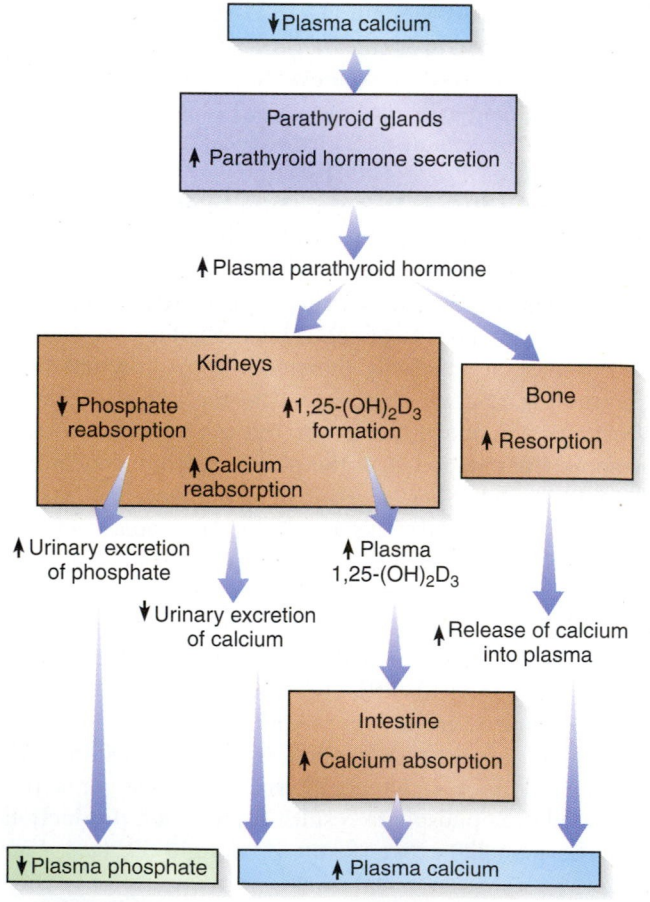

Figure 26–13

Diagram showing the changes that occur in response to a decrease in plasma calcium. The end result is that plasma calcium concentration is restored to its normal value.

Parathyroid hormone does not appear to have any direct effects on the gastrointestinal tract but indirectly results in increased absorption of both calcium and phosphate. These effects are due to the fact that it stimulates the formation of 1,25-OH-vitamin D_3, which in turn acts on the gastrointestinal tract. A summary of the effects of parathyroid hormone is shown in Figure 26–13. Note that the net effect of PTH is to restore plasma calcium to normal.

Actions of Calcitonin in Calcium and Phosphate Homeostasis.

The overall effect of calcitonin is to decrease both plasma calcium and phosphate (see Table 26–6). These changes occur as the result of an effect of the hormone primarily in bone, although it has some weak effects in the kidney as well.

In bone, calcitonin decreases resorption by inhibiting the activity of osteoclasts. Therefore, this effect on osteoclasts is opposite to that of parathyroid hormone and results in increased bone formation. In kidney, calcitonin decreases tubular reabsorption of both calcium and phosphate, which results in an increased urinary excretion of both ions. These effects of calcitonin on bone and kidney each contribute to the overall effect of the hormone to decrease the concentrations of these two ions in plasma.

Despite the distinct and demonstrated biological effects of calcitonin and its usefulness in certain clinical situations, this hormone plays only a minor role in the regulation of plasma calcium and phosphate in humans. This general conclusion results from observations made on persons having marked abnormalities of calcitonin secretion. Neither individuals with a complete deficiency of calcitonin secretion nor those with a marked oversecretion of the hormone display any overt abnormalities of calcium or phosphate homeostasis. Thus, it is believed that the role of calcitonin may only be to provide a "fine tuning" for the regulation of calcium and phosphate homeostasis. However, because calcitonin is often used clinically, it is worth bearing in mind its known physiological actions.

Actions of 1,25-OH-Vitamin D_3 (Calcitriol) in Calcium and Phosphate Homeostasis.

As indicated in Table 26–6, the net effect of 1,25-OH-vitamin D_3 action is to increase plasma concentrations of both calcium and phosphate. Its primary target tissue is the small intestine, although it also has some actions in kidney and bone.

In kidneys, 1,25-OH-vitamin D_3 increases tubular reabsorption of both calcium and phosphate. This leads to a decreased urinary excretion of both ions. However, this appears to be only a minor effect of the hormone. In bone, the primary effect of 1,25-OH-vitamin D_3 is to promote the actions of parathyroid hormone. Parathyroid hormone has a greater effect in the presence of 1,25-OH-vitamin D_3 than it does alone. Therefore, the net effect of 1,25-OH-vitamin D_3 in bone is to promote bone resorption. This in turn contributes to an increase in plasma calcium and phosphate concentrations.

As indicated, 1,25-OH-vitamin D_3 has its primary effect in the gastrointestinal tract, where it stimulates both calcium and phosphate absorption from the small intestine. Again, this contributes to an increased concentration of these two ions in plasma. The effect of this hormone on calcium absorption is seen only after a delay of several hours, which is associated with increased production of calcium transport proteins in intestinal mucosal cells as a result of the hormone's actions via the gene expression model of hormone action (Chapter 12). Increased numbers of calcium transport proteins in turn lead to greater calcium absorption from the small intestine.

ABNORMALITIES OF BONE MINERAL HOMEOSTASIS

 What characterizes abnormalities of bone mineral homeostasis?

The most common abnormality of bone metabolism is not a single disorder but a group of disorders termed **metabolic bone disease,** in which there is a generalized abnormality of the ongoing processes of bone formation and bone resorption. The two major categories of metabolic bone disease are (1) **osteoporosis** and (2) **osteomalacia and rickets.**

Osteoporosis Results from an Equivalent Loss of Bone Mineral and Organic Matrix

Osteoporosis is a condition in which there is a reduction in overall bone mass, with equal losses of both mineral and organic matrix. The ratio of mineral relative to the amount of organic matrix therefore does not change. Because there is a net decrease in the amount of bone, it follows that bone resorption must be occurring at a faster rate than is bone formation.

Although a number of factors can contribute to osteoporosis, in many cases the exact cause of this imbalance is not known. Long-term calcium deficiency can lead to osteoporosis owing to mobilization of bone mineral in an attempt to maintain normal plasma calcium concentrations. In addition, pronounced vitamin C deficiency can also result in a net loss of bone. Vitamin C is required for normal synthesis of bone collagen matrix and, thus, without synthesis of new matrix bone formation, cannot occur. Immobilization, such as might occur after injury of a limb, also can lead to osteoporosis.

Most commonly, osteoporosis is associated with advancing age in both men and women. However, women tend to suffer from osteoporosis far more frequently than do men.

APPLICATIONS OF PHYSIOLOGY

Anabolic Effects of PTH on Bone

As described in the text, peak bone mass occurs around age 30 and then begins to decrease. As bone mass decreases, the likelihood of fractures increases (see figure, noting the reversed left and right axes). For each individual, there is a hypothetical "fracture threshold," beyond which fractures are highly likely with the ensuing diagnosis of osteoporosis.

Currently available treatments for osteoporosis all act as antiresorptives in that they slow the rate of loss of bone mineral (see figure). Treatments currently available include calcium supplements, estrogen, and selective estrogen receptor modulators (SERMs), such as raloxifene. While they are helpful in slowing the approach to the fracture threshold, they do not either totally prevent bone loss nor promote the accumulation of bone. In other words, once the "fracture threshold" has been passed, there is no going back.

The ideal treatment for osteoporosis would be an agent with **anabolic** action on bone that could promote the accumulation of bone mineral, even after the fracture threshold has been passed (see figure). A surprising candidate for such a role is parathyroid hormone (PTH). As described in the text, PTH is classically considered to be a bone **catabolic** agent. However, recent studies have shown that when delivered intermittently at low doses, PTH stimulates bone growth in both animal and clinical human studies.

PTH is a moderate-sized peptide of 84 amino acids, which would necessitate that it be injected if it were to be used for the treatment of osteoporosis. However, it has been known for quite some time that the full or nearly full biological activity of PTH resides in the first 31 to 34 amino acids at the amino terminal end of the molecule. At this size PTH (1–31) or PTH (1–34) can be absorbed by the nasal mucosa, and nasal sprays of these smaller PTH peptides have been shown to be effective in building bone mass. This exciting combination of physiology and technology will likely lead to a novel form of treatment for osteoporosis in the very near future.

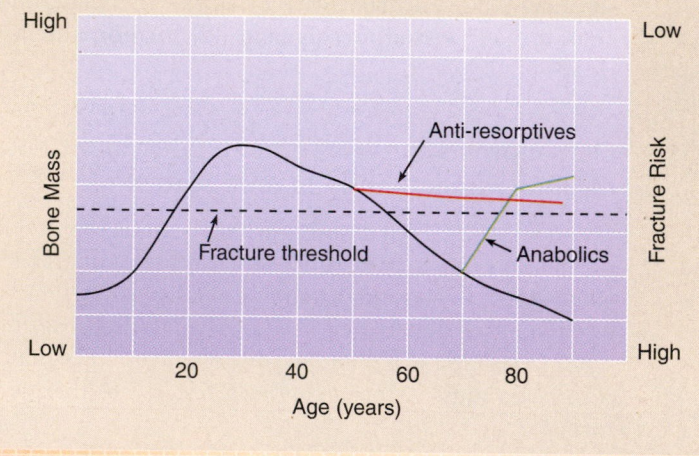

Age-related changes in bone mass and fracture risk. Fracture threshold is arbitrarily set.

Figure 26–14 shows the amount of bone present in males and females according to age. Note that up until approximately 10 to 12 years of age, boys and girls have roughly equivalent amounts of bone. At about this time, which coincides with the onset of puberty, boys begin to acquire bone at a more rapid rate than do girls, presumably as a result of the male hormone testosterone. At the point between ages 30 and 40 when maximal bone mass is reached, men have 20% to 30% more bone than do women, on average. After approximately age 40, both men and women begin to lose

bone mass and do so at roughly equivalent rates up until approximately age 50 (see Fig. 26–14). At menopause (at approximately age 50), women begin to lose bone more rapidly as a result of decreasing estrogen levels (see Chapter 31). This more rapid loss of bone can result in **postmenopausal osteoporosis.**

Osteoporosis is a major public health problem owing to the increased risk for fractures occurring with only minimal stress or trauma placed on the bone and the resulting disability that occurs.

CURRENT CONCEPTS IN PHYSIOLOGY

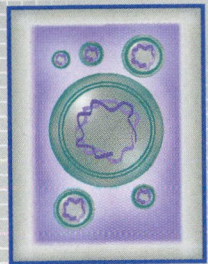

The Economic Impact of Osteoporosis

Osteoporosis is often called the "silent disease" because bone loss initially occurs without any noticeable symptoms. Osteoporosis is a major public health threat in the United States, as it affects some 28 million individuals. Approximately 10 million people have been diagnosed with the disease in the United States and another 18 million have low bone mass, placing them at increased risk for osteoporosis. Approximately 80% of those affected by osteoporosis are women. Although osteoporosis is often thought of as a disease mainly striking older persons, it can strike at any age. The seeds of osteoporosis are often sown in childhood due to low calcium intake or other complicating (risk) factors. A large number of children do not get the exercise, vitamin D, and calcium needed to ensure that they are safe from the disease later in life. Osteoporosis is responsible for more than 1.5 million fractures annually, including 300,000 hip fractures, 700,000 vertebral fractures, 250,000 wrist fractures, and 300,000 fractures at other sites. It is estimated that osteoporosis costs some $10 billion to $15 billion a year for hospitalization and nursing homes alone. Nearly one third of people who have hip fractures go into nursing homes within a year, and nearly 20% die within a year.

What causes osteoporosis, and what can be done to prevent or treat the disease? While it is known that a diet low in calcium or vitamin D puts one at risk and that certain medications, such as glucocorticoids, anticonvulsants, and aluminum-containing antacids, can cause osteoporosis, in the majority of cases, the exact cause is unknown. However, a number of risk factors associated with the disease have been identified:

Being female, and especially postmenopausal
Being Caucasian or Asian
Advanced age
Having a family history of the disease
Low testosterone levels in men
Inactive lifestyle
Cigarette smoking
Excessive use of alcohol

A comprehensive program to help prevent osteoporosis includes a balanced diet rich in calcium and vitamin D, weight-bearing exercise, a healthy lifestyle with no smoking or excessive alcohol use, and bone density testing and medication when appropriate.

There is no cure for osteoporosis at present; however, the US Food and Drug Administration (FDA) has approved medications that slow the progression or severity of the disease in women. Although 20% of all cases of osteoporosis occur in men, only two medications are currently FDA approved for use in men, and only for cases of corticosteroid-induced osteoporosis. Testosterone replacement therapy may be beneficial for men with low testosterone levels.

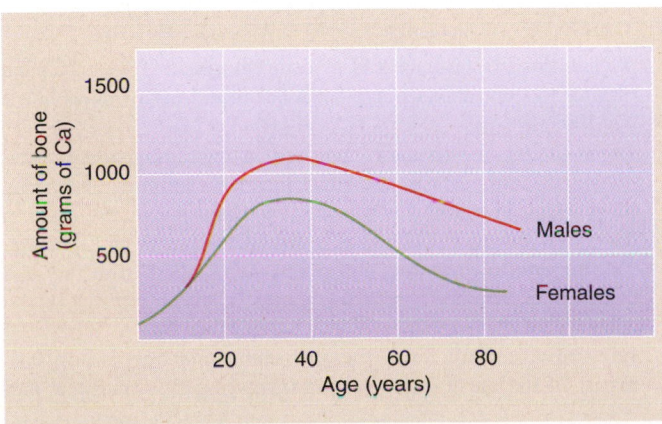

Figure 26–14

Diagram showing the amount of bone present in males and females of different ages. Note that maximal bone mass is higher in males than in females. Also illustrated is the more rapid loss of bone experienced by women compared with men, beginning around age 50.

Osteomalacia and Rickets Result from Inadequate Mineralization of New Bone

The second major category of metabolic bone disease is that of osteomalacia and rickets. Both diseases are characterized by **inadequate mineralization** of new bone matrix, such that the ratio of mineral to organic matrix is lower than normal. When this disorder occurs in children, it is referred to as **rickets** (Fig. 26–15). In adults, the same disorder is termed **osteomalacia.** In children with rickets, bones have decreased mechanical strength and are subject to distortion, which may lead to a bowed appearance, particularly in the long bones of the legs. In adults, osteomalacia can produce severe bone pain.

Rickets and osteomalacia are most commonly caused by a deficiency in vitamin D *activity* as a result of one of three general causes. The first is a combined dietary deficiency of vitamin D and insufficient exposure to sunlight. Because, in developed nations, foods are often fortified with vitamin D, severe nutritional deficiencies of vitamin D are relatively rare in these countries, although this can be a health problem in many other parts of the world. A second cause of rickets or osteomalacia involves an impairment in converting vitamin D in liver and kidney to its hormonally active form. Numerous factors can contribute to this type of deficiency, including liver disease; kidney disease; or an absence of parathyroid hormone, which normally stimulates activity of 1α-hydroxylase in the kidney. The third cause of impaired vitamin D activity involves an inability of 1,25-OH-vitamin D_3 to act on its target tissues. Some anticonvulsant drugs are known to interfere with vitamin D action, but the mechanisms involved are poorly understood. However, long-term treatment with these drugs, which are used to treat conditions such as epilepsy, may lead to osteomalacia or rickets.

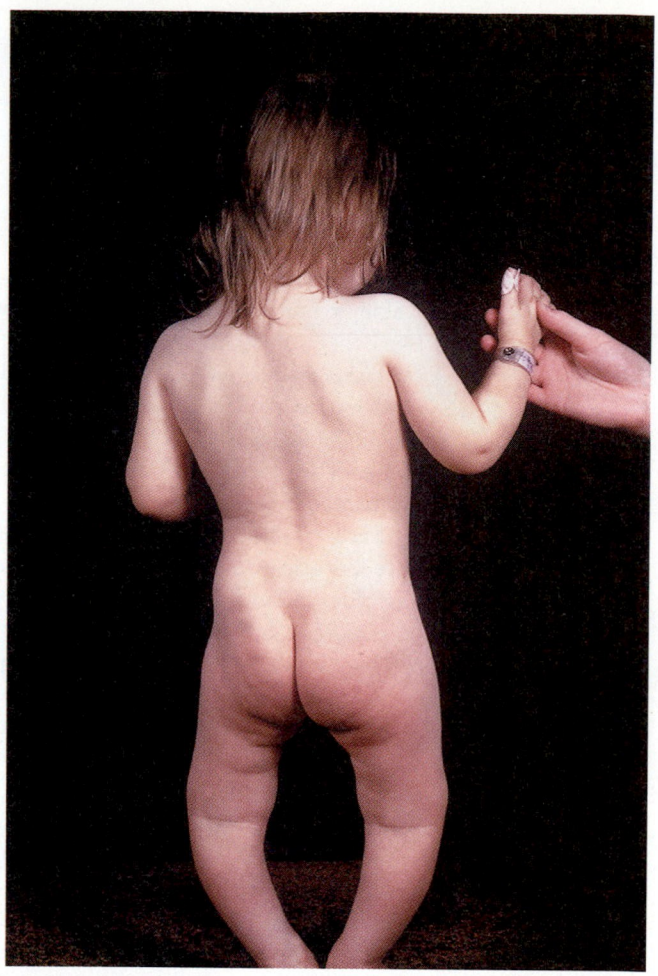

Figure 26–15

Classic appearance of rickets in a child. (© Biophoto Associates/Photo Researchers, Inc.)

CHAPTER REVIEW

Summary

- Because calcium is involved in a number of important physiological processes in the body, its concentration in extracellular fluid must be maintained within fairly narrow limits. Phosphate also serves several important functions, but regulation of plasma phosphate concentrations is not as crucial as that of calcium. Approximately 99% of the total body calcium is located in bone. The majority (85%) of phosphate is also present in bone.
- Total plasma calcium is normally about 10 mg/100 ml. Two primary forms of calcium in plasma are filterable calcium and protein-bound calcium. Filterable calcium consists of free calcium ion and calcium complexed with an anion, such as bicarbonate. Free calcium is the physiologically active form of calcium. Most

phosphorus in plasma exists as orthophosphate (PO_4). Normal plasma concentrations range from 3.0 to 4.5 mg of phosphorus per 100 ml.
- Three tissues are primarily involved in regulating calcium and phosphate homeostasis: the small intestine, kidneys, and bone. Most ingested calcium is not absorbed and leaves the body in the feces; absorbed calcium is eventually excreted in the urine. The majority of ingested phosphate is absorbed from the gastrointestinal tract; its primary route for excretion is via the urine.
- Calcium absorption from the gastrointestinal tract occurs primarily in the small intestine and involves active transport and

passive diffusion; regulation occurs via changes in active transport. Phosphate absorption from the small intestine occurs by active transport and passive diffusion and does not appear to be regulated.

- Approximately 99% of calcium filtered by the kidneys is reabsorbed. Reabsorption involves both active transport and passive diffusion. Active transport in distal tubules is stimulated by parathyroid hormone. In the case of phosphate, approximately 85% of that filtered is absorbed. Reabsorption occurs primarily by active transport in proximal tubules and is inhibited by parathyroid hormone.

- Bone consists of inorganic bone mineral deposited on an organic support matrix. The mineral fraction is composed primarily of hydroxyapatite crystals; the organic matrix consists primarily of collagen. Three primary cell types in mature bone are osteoblasts, osteocytes, and osteoclasts. Osteoblasts and osteocytes are involved in bone formation and bone maintenance; osteoclasts are responsible for bone resorption. In growing bone, chondrocytes of the epiphyseal plate are under hormonal control. Stimulation of chondrocytes leads to bone growth.

- Three hormones are involved in regulating plasma calcium and phosphate concentrations: (1) parathyroid hormone, (2) calcitonin, and (3) vitamin D metabolites. Parathyroid hormone is secreted in response to decreased plasma calcium. Its target tissues are kidney; bone; and, indirectly, the small intestine, with a net effect of increasing calcium and decreasing phosphate in plasma.

- Calcitonin plays only a minor role in regulation. It is secreted in response to increased plasma calcium. Acting primarily on bone, its net effect is to decrease both calcium and phosphate in plasma.

- Vitamin D_3 is converted into 1,25-OH-vitamin D_3 by sequential conversion in liver and kidney. The final step of vitamin D activation in kidney is tightly regulated and is stimulated by parathyroid hormone. The net effect of 1,25-OH-vitamin D_3 action is to increase plasma calcium and phosphate, acting primarily on the intestines to increase calcium absorption.

- The most common abnormality of bone metabolism is a group of disorders known as *metabolic bone disease*. One category of metabolic bone disease is osteoporosis, which is characterized by an overall loss of bone mass with equal loss of bone mineral and organic matrix. Osteomalacia and rickets make up the second major category of metabolic bone diseases. These are characterized by inadequate mineralization of bone, most commonly caused by a deficiency of vitamin D activity.

Review Questions

Choose the Correct Answer

1. The physiologically active form of calcium in the body is:
   a. calcium bound to albumin.
   b. calcium citrate.
   c. calcium ion (Ca^{2+}).
   d. unable to cross membranes.
   e. calcium phosphate.

2. At pH of 7.4, most of the phosphate in the blood is present as:
   a. hydroxyapatite.
   b. HPO_4^{2-}.
   c. hexose phosphate.
   d. $H_2PO_4^-$.
   e. ribose phosphate.

3. The majority of ingested calcium leaves the body via the:
   a. feces.
   b. sweat.
   c. urine.
   d. saliva.
   e. expired air.

4. Parathyroid hormone acts to _____ the reabsorption of calcium in the _____ of the kidney.
   a. increase, proximal convoluted tubules
   b. decrease, proximal convoluted tubules
   c. increase, distal tubules
   d. increase, loops of Henle
   e. decrease, distal tubules

5. The conversion of 7-dehydrocholesterol to vitamin D_3 occurs in the:
   a. GI tract.
   b. skin.
   c. liver.

   d. epiphyseal plate.
   e. kidney.

6. The kidney enzyme 1α-hydroxylase catalyzes the conversion of:
   a. calcium to hydroxyapatite.
   b. 25-OH-vitamin D_3 to 1,25-OH-vitamin D_3.
   c. vitamin D_2 to vitamin D_3.
   d. cholecalciferol to dehydrocholesterol.
   e. 1,25-OH-vitamin D_3 to 25-OH-vitamin D_3.

7. A deficiency in vitamin D action can lead to:
   a. rickets.
   b. osteoporosis.
   c. osteomalacia.
   d. postmenopausal osteoporosis.
   e. a or c.

8. Hypocalcemic tetany is the result of:
   a. inadequate calcium at the motor neuron terminal, resulting in a reduction in acetylcholine release.
   b. hyperexcitability of nerves resulting from an effect on sodium permeability.
   c. reduced calcium release from skeletal muscle sarcoplasmic recticulum.
   d. a direct effect of calcium to stimulate the gamma motor neurons.
   e. a stimulation of acetylcholine release from postganglionic sympathetic motor neurons.

9. The majority of phosphorus in the body exists:
   a. in muscle.
   b. in bone.
   c. in adipose tissue.
   d. as phosphate ion (PO_4).
   e. Both b and d

10. Filterable calcium comprises approximately _____ % of the total plasma calcium.
   a. 3.4
   b. 10
   c. 40
   d. 50
   e. 60

11. On an average day, the kidneys excrete approximately _____ mg of calcium into the urine.
   a. 2
   b. 150
   c. 300
   d. 500
   e. 1000

12. The primary route by which ingested phosphate leaves the body is via the:
   a. urine.
   b. feces.
   c. sweat.
   d. nasal discharges.
   e. phosphate canal.

13. Which type of bone cell is responsible for the synthesis of osteoid?
   a. Osteoblast
   b. Chondrocyte
   c. Osteoclast
   d. Canalicular
   e. Osteocyte

14. The area of long bones where growth and elongation occurs is known as the:
   a. zone of chondrogenesis.
   b. epiphyseal plate.
   c. zona twilighta.
   d. techtonic plate.
   e. red zone.

15. Parathyroid hormone (PTH) is produced by:
   a. thyroid follicle cells.
   b. parafollicular cells.
   c. parathyroid gland cells.
   d. osteoblasts.
   e. the pharynx.

16. One effect of parathyroid hormone (PTH) is to produce:
   a. hypocalcemia.
   b. hyponatremia.
   c. hypoglycemia.
   d. hypervitaminosis D.
   e. phosphaturia.

17. Most commonly, osteoporosis results from:
   a. equivalent loss of bone mineral and bone matrix.
   b. predominantly a loss of bone mineral.
   c. predominantly a loss of bone organic matrix.
   d. ingestion of excess calcium.
   e. lack of dietary phosphate.

18. The net effect of 1,25-OH-vitamin D_3 is to:
   a. decrease plasma calcium.
   b. increase plasma phosphate.
   c. decrease plasma phosphate.
   d. increase plasma calcium.
   e. Both b and d

19. The primary stimulus for an increase in parathyroid hormone secretion is:
   a. increased plasma calcitonin.
   b. decreased plasma calcitonin.
   c. increased plasma phosphate.
   d. decreased plasma calcium.
   e. decreased plasma phosphate.

20. The primary mechanisms by which long-term regulation of plasma calcium and phosphate occurs is via _____ mechanisms.
   a. hormonal
   b. intrinsic
   c. extrinsic
   d. neural
   e. homeopathic

Answers to Case History Questions

1. Osteoporosis (and perhaps glucocorticoid-induced osteoporosis)

2. Because George is only 38 years of age and lives a relatively healthy lifestyle, it seems unlikely that his osteoporosis is age dependent or risk factor dependent. The most probable cause is his 30-year history of treatment with glucocorticoids for his asthma.

3. George lacks the risk factors of smoking, excessive alcohol abuse, and being female. However, he does appear to have the risk factor of a somewhat sedentary lifestyle.

Key Terms

calcitonin (p. 819)
calcidiol (p. 817)
calcitriol (p. 819)
calcium (p. 806)
canaliculi (p. 812)
chondrocytes (p. 813)
epiphyseal plate (p. 813)

ground substance (p. 812)
hydroxyapatite (p. 812)
hypocalcemia (p. 806)
metabolic bone disease (p. 819)
osteoblast (p. 812)
osteoclast (p. 812)
osteocyte (p. 812)

osteoid (p. 812)
osteomalacia (p. 819)
osteoporosis (p. 819)
parathyroid gland (p. 815)
parathyroid hormone (PTH) (p. 815)

phosphate (p. 807)
phosphaturia (p. 818)
proteoglycan (p. 812)
rickets (p. 816)
tetany (p. 806)
vitamin D_3 (p. 816)

Suggested Readings

Bilezikian, J. P., Kurland, E. S., and Rosen, C. J. "Male skeletal health and osteoporosis." *Trends in Endocrinology and Metabolism,* 10:244–250, 1999.

Cizza, G., Ravn, P., Chrousos, G. P., and Gold, P. W. "Depression: A major unrecognized risk factor for osteoporosis?" *Trends in Endocrinology and Metabolism,* 12:198–203, 2001.

Ganong, W. F. *Review of Medical Physiology,* ed 19. Los Altos, Lange Medical Publications, 1999.

Goodman, H. M. *Basic Medical Endocrinology,* ed 2. New York, Raven Press, 1994.

Griffin, J. E., and Ojeda, S. R. *Textbook of Endocrine Physiology,* ed 4. New York, Oxford University Press, 2000.

Kacsoh, B. *Endocrine Physiology.* New York, McGraw Hill, 2000.

Norman, A. W., and Litwack, G. *Hormones,* ed 2. Orlando, Academic Press, 1997.

Walters, M. R. "Newly identified actions of the vitamin D endocrine system." *Endocrine Reviews,* 13:719–764, 1992.

Wilson, J. D., and Foster, D. W., eds. *Williams' Textbook of Endocrinology,* ed 7. Philadelphia, W.B. Saunders Co., 1985.

Web sites

http://www.nof.org
Home page for the National Osteoporosis Foundation.

http://merckusa.com.pro/osteoporosis/inde21.htm
A series of educational modules concerning bone physiology.

Answers to Review Questions

1. c **2.** b **3.** a **4.** c **5.** b **6.** b **7.** e **8.** b
9. e **10.** e **11.** b **12.** a **13.** a **14.** b **15.** c
16. e **17.** a **18.** e **19.** d **20.** a

Chapter 27

REGULATION OF BODY TEMPERATURE

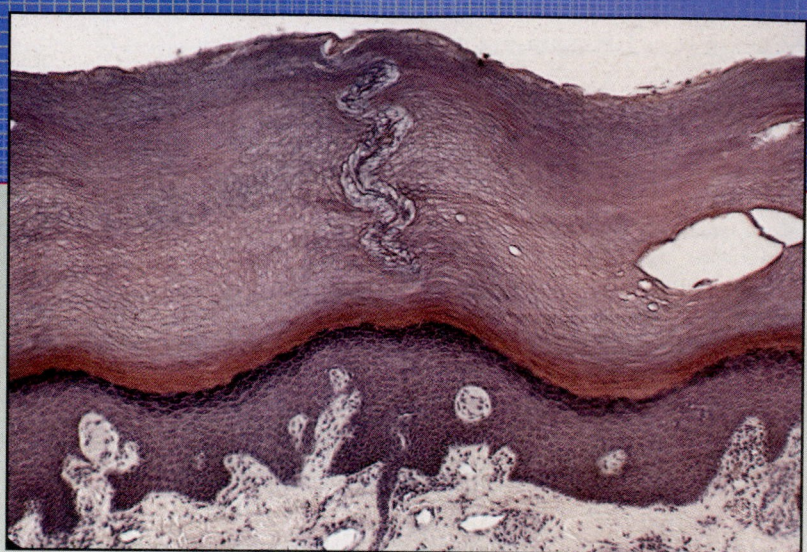

KEY CONCEPTS

- Temperature deep within the body is carefully controlled within a narrow normal range.

- Heat gain and heat loss change body temperature.

- The temperature control system includes skin and core temperature receptors.

- Information from skin and core temperature receptors is integrated in the hypothalamus.

- Humans maintain constant core temperature by balancing heat gain and heat loss.

- Heat gain and loss occur by the processes of radiation, conduction, convection, and evaporation.

- Vasoconstriction of the skin reduces heat transfer from core to skin.

- Evaporation of sweat from the skin causes heat loss from the skin.

- Fever is a regulated increase in the set point for core temperature.

- Exercise increases energy expenditure, heat production, and core temperature.

CASE HISTORY

A college distance runner dresses for a summer workout in an air-conditioned room maintained at 21°C, 30% humidity. Because she felt slightly ill the previous day, she checks her oral temperature before beginning her run; it is 36.8°C, and today she is feeling well. Stepping outside, where the air temperature is 33°C with 60% humidity, she immediately notices that her skin feels warm and has reddened. She begins running at a constant pace on level terrain and notices within a very short time that she is sweating on her face, trunk, arms, and legs. She continues to sweat steadily throughout an hour of running, but this exercise is something she performs daily, and she completes it without difficulty. Immediately after she finishes her run, feeling tired but excellent overall, she checks her oral temperature again; it is 38.4°C.

Questions

1. Why did her skin warm and redden immediately after going outside?

2. Why did her sweating increase during exercise?

3. Why did she sweat on her face, trunk, arms, and legs?

4. Why did her oral temperature increase during exercise?

5. How is the temperature increase in exercise different from a fever?

INTRODUCTION

Like all mammals and birds, humans control temperature deep inside the body (**"core" temperature**) within a relatively narrow range, even during exposure to wide-ranging environmental conditions. In other animals—for example, fish and reptiles—body temperatures fluctuate widely with changes in ambient temperature. Unlike a fish or reptile, a nude human can be exposed to temperatures as low as 12°C or as high as 60°C in dry air and still maintain core temperature near 37°C. Although higher core temperatures of 40°C or even 41°C can be associated with strenuous exercise or fever, core temperatures higher than 42°C are associated with a breakdown of cellular proteins and death. Humans, therefore, regulate core temperature only a few degrees below the point of thermal death.

The regulation of deep body temperature in humans at or near 37°C ensures optimal cellular function. All cellular enzymatic reaction rates are temperature dependent. At normal body temperature, cellular enzymatic reactions proceed optimally. If temperature were to decrease by 10°C, however, these reaction rates would decrease dramatically. Human energy expenditure, in fact, changes approximately 12% for every 1°C change in body temperature. Animals that keep core temperature constant (homeotherms) are ensured stable enzymatic activity that makes high levels of physical activity possible, independent of the ambient temperature. In contrast, animals whose core temperature varies with the environment (poikilotherms) become sluggish when core temperature decreases.

A NEGATIVE FEEDBACK SYSTEM CONTROLS CORE TEMPERATURE

 How is the body temperature regulated?

The addition or subtraction of heat from the human body predictably will increase or decrease the body temperature. The heat capacity of human tissue is similar to that of water: If 1 kcal of heat is added to one kg of body tissue, the temperature of that tissue increases nearly 1°C. Consequently, mean body temperature and total body heat content are linked, both depending upon the relative rates of heat gain or loss. In humans, core temperature is carefully regulated near 37°C by a negative feedback control system. The control system includes temperature sensors, a central controller, and several effectors that can be modified to alter heat gain or heat loss (Fig. 27–1). Temperature sensors are found in the skin and deep within the body, and these provide information to a central controller located in the **hypothalamus** (Fig. 27–1). In response to these temperature inputs, the hypothalamus initiates physiological processes that modify heat loss or heat gain, in a fashion that returns core temperature to the normal level (see Fig. 27–1).

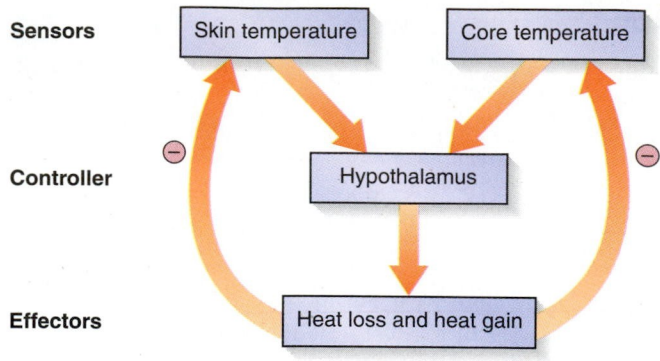

Figure 27–1

Overview of temperature regulation. Temperature changes sensed in the body core and skin are integrated in the hypothalamus. The hypothalamus then initiates physiological responses that alter heat loss and heat gain, restoring core and skin temperature to the regulated level.

Core Thermoreceptors

Because the circulatory system efficiently distributes heat within an area that includes the heart, lungs, liver, kidneys, and brain, temperature within this region is nearly uniform, providing the basis for the concept of core temperature. Temperature-sensitive neurons within the body core are found in the hypothalamus, spinal cord, abdominal viscera, and great veins. Specific thermoreceptors are sensitive to decreases or increases in core temperature. Nerve impulses from core receptors are integrated with peripheral thermal information at the hypothalamus.

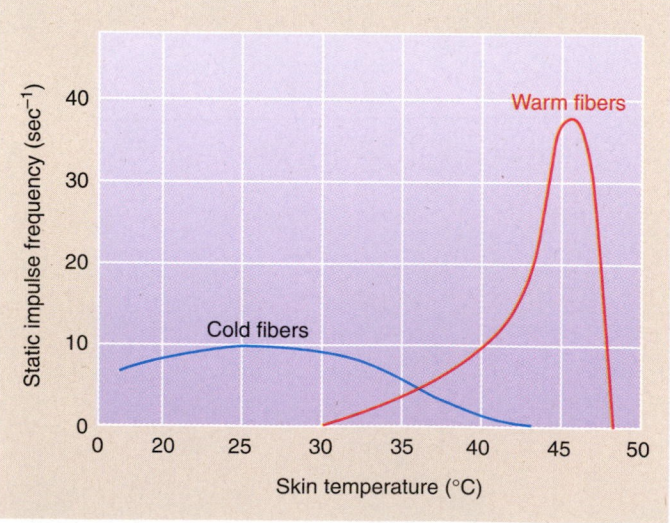

Figure 27–2

Static discharge frequency of cold and warm nerve fibers as a function of skin temperature.

Peripheral Thermoreceptors

Peripheral receptors measure temperature in the skin. These receptors—naked, temperature-sensitive nerve endings—selectively respond to cold or warm stimuli (Fig. 27–2). While both types of temperature receptors are found throughout the body surface, cold receptors are ten times more numerous. Nerve impulses from peripheral receptors enter the spinal cord and ascend to the brain, to be integrated in the hypothalamus with temperature information from the body core.

Central Integration of Thermal Information

Core and skin temperatures are integrated in the hypothalamus. Because the hypothalamus itself is maintained at core temperature, changes in hypothalamic temperature are the single most important input determining thermoregulatory responses. When skin and core temperatures deviate from a regulated value, the hypothalamus initiates a number of physiological responses that modify heat loss or heat gain (Fig. 27–3). These responses include regulation of sympathetic neural outflow to arterioles in the skin, and sympathetic neural control of sweat glands (see Fig. 27–3). The hypothalamus also can directly stimulate motor nerves that control skeletal muscle **shivering** (see Fig. 27–3). This pathway is largely involuntary. In contrast, the hypothalamus also can modify voluntary skeletal muscle activity through an influence on the cerebral cortex (see Fig. 27–3).

HOW PHYSIOLOGICAL EFFECTOR MECHANISMS MODIFY HEAT GAIN AND HEAT LOSS

 How does the body gain or lose heat?

The ability of physiological effector mechanisms to modify heat gain and heat loss is determined in part by the physical processes involved in heat exchange. These processes include **radiation, conduction, convection,** and **evaporation.** Heat is also produced within the body as energy is expended; physiological processes can also modify the amount of heat generated in this fashion.

Heat Gain or Loss by Radiation

All objects emit heat from their surfaces as waves in the infrared portion of the electromagnetic spectrum. This heat emission, which depends upon the surface temperature of the object, transfers heat between objects that are not in contact (Fig. 27–4). Important examples of radiant heat exchange include radiant heat loss from warm skin to cooler surfaces in a cold environment, and radiant heat gain from exposure of skin to direct sun (see Fig. 27–4). Net radiation of heat between our surface (the skin) and the environment depends upon the temperature difference between the skin

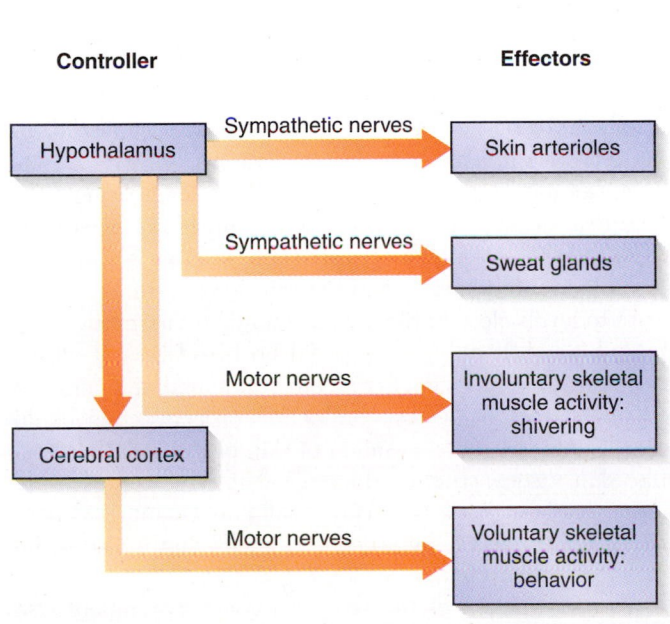

Figure 27–3

Hypothalamically controlled physiological mechanisms for heat loss or heat gain.

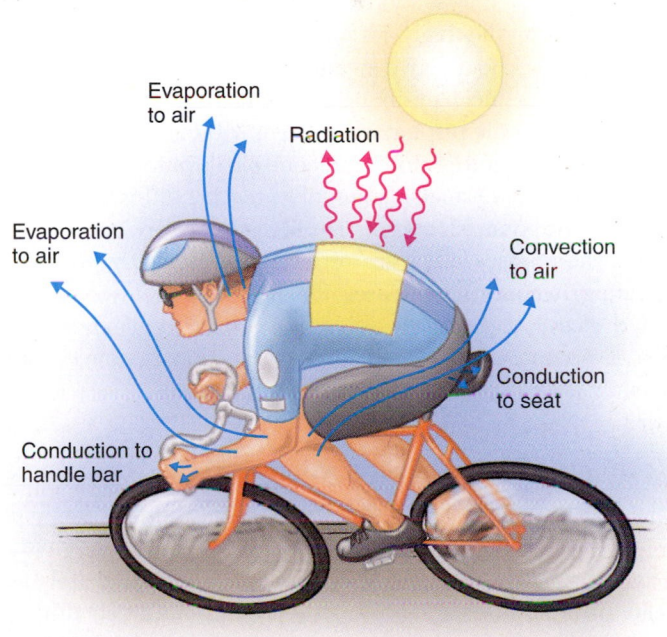

Figure 27–4

Mechanisms of heat exchange between the body and the environment.

and the average temperature of all of the objects in the environment.

Heat Gain or Loss by Conduction

Conduction is heat exchange in the form of kinetic energy between molecules of objects in direct contact. The greater the temperature difference between the two objects, the larger the heat transfer by conduction from the warmer to the cooler object. In humans, only relatively small amounts of heat are usually gained or lost from the body by direct contact with other objects (see Fig. 27–4).

Heat Gain or Loss by Convection

Conduction of heat into or out of the skin from surrounding air or water may be increased when the air or water is moving. This process, called *convection,* enhances heat loss from the body when cooler air replaces air that has been warmed during contact with the skin. When wind, fans, or movement of the body through the air (as during cycling) increase the rate of air movement ("forced convection"), the rate of heat loss can increase dramatically (see Fig. 27–4). In cold air, increasing wind velocity accounts for the **wind chill factor.** Heat loss by convection at a given cool air temperature is roughly proportional to the square root of wind velocity.

Heat Loss by Evaporation

Thermal energy is required for the conversion of water from a liquid to a gaseous state. When water evaporates from the skin or respiratory passages, approximately 0.58 Calories of heat are removed from the body for each gram of water (see Fig. 27–4). Even without active **sweating,** evaporation is a continuous heat loss mechanism because small amounts of water evaporate from the respiratory passages and the mucous membranes of the mouth. Another uncontrolled route for evaporative heat loss is the small amount of water that diffuses through and then evaporates from the skin. Because we are unaware of these water losses, they are termed *insensible perspiration.* Sweating, in contrast with insensible perspiration, involves regulated, active secretion of water by specialized glands in the skin. Evaporation, unlike radiation, conduction, or convection, can only remove heat from the body surface.

Heat Gain from Energy Expenditure

All animals convert chemical energy stored in foods into energy for cellular processes. Approximately 60% of the energy released during breakdown of foods appears immediately as heat. In addition, nearly all of the energy liberated during food breakdown and used for diverse processes (including active transport processes and muscle contraction) eventually will be converted to heat. Only energy used to perform external work, and energy stored in newly created tissues during growth, escapes ultimate conversion to heat. Heat produced from expenditure of energy always is a source of heat gain for the body.

PHYSIOLOGICAL CONTROL OF HEAT GAIN AND HEAT LOSS

When temperature information relayed to the hypothalamus triggers effector responses to return core temperature to normal levels, these responses modify heat gain and heat loss until temperature is normalized. The specific processes invoked depend upon the direction, magnitude, and rate of change of the deviations in core and skin temperature.

Physiological Responses to Decreased Core and Skin Temperature

Vasoconstriction in the Skin

How do body mechanisms affect heat exchange with the environment?

Because the arterial blood temperature is virtually identical to the body core temperature (37°C), and because blood efficiently transfers heat by convection, highly perfused tissues typically have temperatures close to core temperature. This is also true for the skin: high levels of blood flow in the skin can cause skin temperature to nearly match core temperature. However, because skin is the major tissue in contact with the environment, changing the amount of blood flow in the skin also changes the temperature gradient between the body surface and the environment. Changing the temperature gradient between the skin and the environment alters conductive, convective, and radiant heat exchange. The control of skin blood flow is consequently a major regulated response for maintenance of core temperature.

As sensors in the skin and core report decreasing temperatures, the hypothalamus initiates sympathetic **vasoconstriction** of the skin. Vasoconstriction reduces the flow of warm blood into the skin, and the skin temperature may decrease to levels close to the environmental temperature (Fig. 27–5). Indeed, if the skin received no blood flow at all, its temperature would match the environmental temperature and no heat exchange by conduction or convection could occur. While complete cessation of skin perfusion is not possible, skin vasoconstriction does result in large temperature differences between a relatively small core maintained near 37°C, and a relatively large, cooler "shell" that includes the skin and neighboring tissues (see Fig. 27–5).

Vasoconstriction of the skin in a cold environment also limits heat loss from core to skin because convection of heat from core to skin is minimized and conduction of heat from core to skin is a relatively inefficient process (Fig. 27–6). The

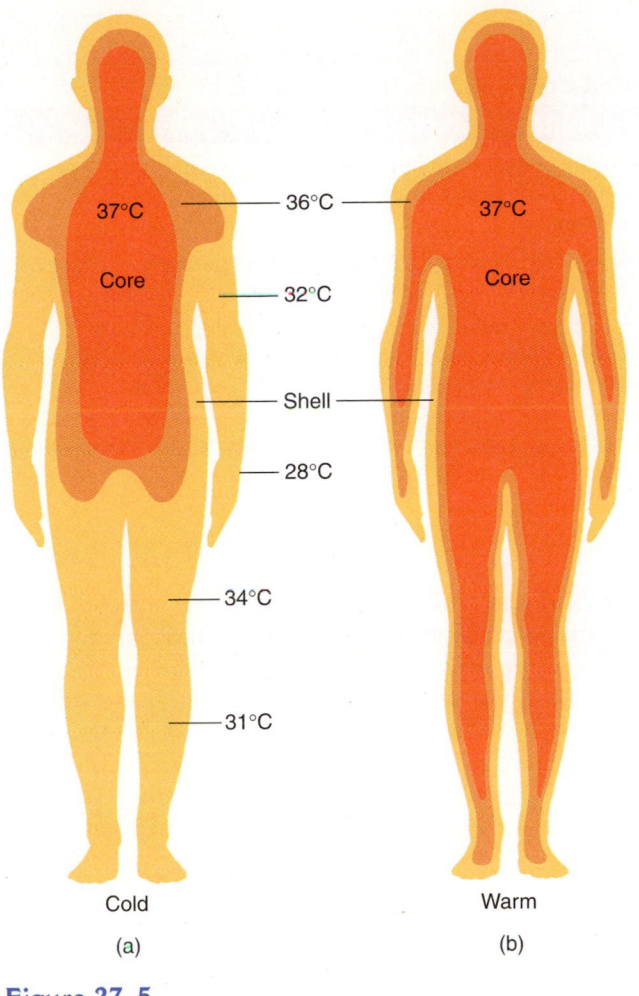

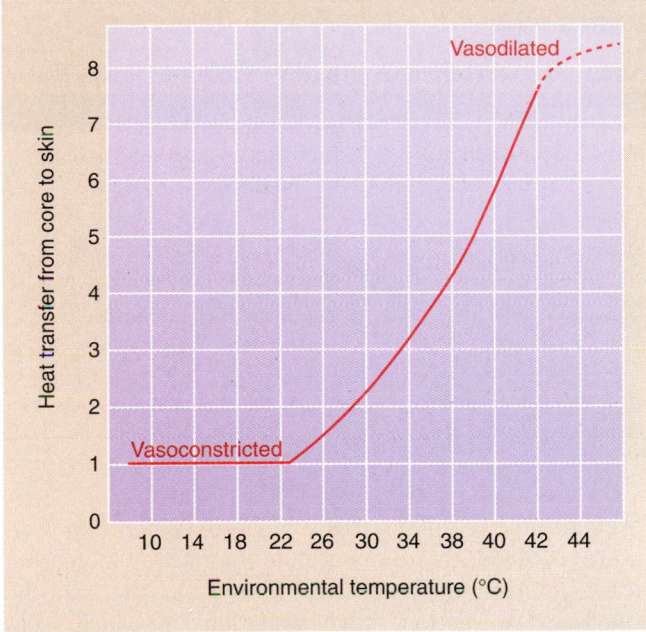

Figure 27–6

Effect of changes in environmental temperature on the heat transferred from the body core to the skin surface.

Figure 27–5

Compared with a warm environment *(b)*, skin vasoconstriction in the cold *(a)*, reduces the portion of the body maintained at core temperature (37°C).

differences in core and shell volume and temperature in cold and warm environments show that the primary controlled variable in human thermoregulation is core temperature (which is identical in the two conditions shown in Fig. 27–5), not skin temperature, total body heat content, or mean body temperature (which are greatly reduced in a cold environment).

Involuntary Muscle Activity

When skin and core temperature decrease sufficiently, the hypothalamus directly activates skeletal muscle. Core and skin temperature act together to initiate shivering, and the intensity of shivering increases as these temperatures decrease. The involuntary muscle contractions of shivering, which vary in frequency, amplitude, duration, and intensity, increase energy expenditure and heat production. However, the upper limit of shivering-induced heat production is perhaps only a threefold increase above resting levels, and shivering is inefficient for two reasons. First, shivering increases

blood flow into the activated muscles, increasing the local skin surface temperature. The greater skin temperature increases heat loss to the environment by conduction, convection, and radiation. Second, the shaking motions of shivering increase heat loss by convection.

Voluntary Muscle Activity

Decreased core and skin temperature also provoke behavioral responses that minimize heat loss. These responses, mediated through the hypothalamus to the cerebral cortex and then to voluntary skeletal muscle, may include postural changes that reduce the surface area exposed to the cold environment, movement into a warmer environment, and addition of clothing. Each of these responses limits heat loss by conduction, convection, and radiation. A summary of the effector responses to decreased core and skin temperature is provided in Table 27–1.

Physiological Responses to Increased Core and Skin Temperature

Vasodilation in the Skin

When thermal receptors in the core and skin report temperature elevations above the regulated level, the hypothalamus responds by reducing sympathetic neural outflow to resistance vessels in the skin, increasing skin blood flow. Warm blood perfusing the skin increases skin temperature by convectively transferring heat from the

TABLE 27–1		
Summary of Thermoregulatory Effector Responses to Decreased Core and Skin Temperature		
Response	**Mechanism**	**Effect on Heat Gain or Loss**
Skin vasoconstriction	Increased sympathetic outflow to skin resistance vessels	Cooler skin reduces heat lost to environment; reduced convective heat transfer from core to skin maintains core temperature
Shivering	Involuntary skeletal muscle contraction	Increased energy expenditure increases heat production
Behavior	Voluntary skeletal muscle contraction	Posture changes reduce surface exposed to cold; movement to warmer environment reduces heat loss; increased clothing traps air near the skin, reducing heat lost by convection

body core to the periphery (see Fig. 27–6). When skin temperature exceeds environmental temperature, increasing skin temperature increases heat loss from the skin by conduction, convection, and radiation. Maximal skin **vasodilation** causes skin temperature to approach core temperature, reducing the volume of the body "shell" and increasing the volume of body core (see Fig. 27–5). The process of skin vasodilation and vasoconstriction takes place without conscious awareness.

Sweating

As core and skin temperatures increase to a threshold, the hypothalamus increases sympathetic outflow to sweat glands, causing active secretion of water called *sweating* (Fig. 27–7). There are 2 to 3 million sweat glands distributed over most of the body surface. Core and skin temperatures act synergistically on the hypothalamus to initiate the sweating response. At any given skin temperature, sweating

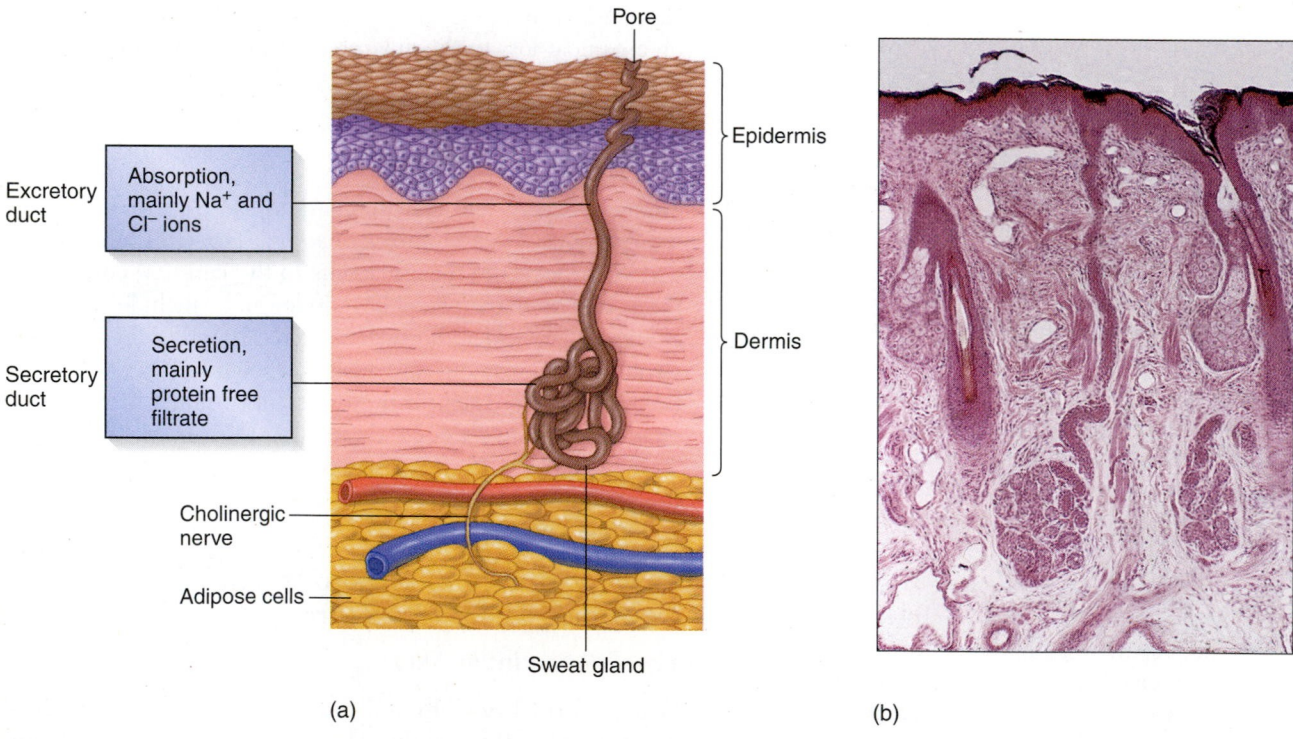

(a)   (b)

Figure 27–7

(a) A human sweat gland innervated by a sympathetic nerve. A protein-free filtrate is formed by the secretory coil, and most of the electrolytes are reabsorbed along the excretory duct, producing a dilute, watery sweat. *(b)* Photomicrograph of a human sweat gland (magnification × 35). *(© John D. Cunningham/Visuals Unlimited.)*

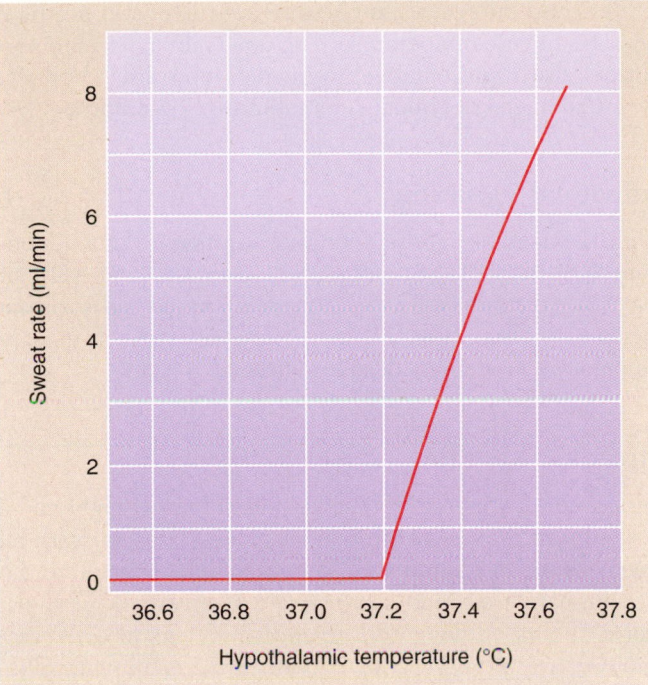

Figure 27–8

Effect of changes in hypothalamic temperature on sweat rate. Sweating begins at a critical temperature and increases linearly with increasing hypothalamic temperature.

increases linearly with increases in core temperature (Fig. 27–8). Heat loss from sweating, like heat loss from insensible perspiration, can occur only if sweat evaporates from the surface of the skin. The rate of sweat evaporation depends upon environmental humidity and the rate of air movement at the skin surface. Under optimal conditions for evaporation, up to two liters of sweat can be evaporated from the skin surface in 1 hour, providing an avenue for

heat loss of sufficient magnitude to balance the very high heat production of exercise.

Sweating increases water and electrolyte loss from the body. Because Na^+ and Cl^- are substantially reabsorbed along the excretory duct of the sweat gland, normal sweat osmolarity is much lower than is plasma osmolarity. High rates of sweating can be maintained indefinitely if the lost water is replaced; the loss of electrolytes is usually adequately balanced by dietary intake in all but the most extreme conditions.

Voluntary Muscle Activity

The behavioral responses provoked by increasing core and skin temperature increase heat loss and reduce heat gain by modifying radiation, conduction, and convection. Examples of these behavioral responses include (1) seeking shelter from the sun (to reduce radiant heat gain), (2) decreasing clothing (to increase heat loss by conduction and convection), and (3) the use of fans (to increase heat loss by conduction and convection). The physiological effector responses to increased core and skin temperature are summarized in Table 27–2.

VARIATION IN CORE TEMPERATURE

 What factors contribute to the core temperature?

Although a regulatory system maintains core temperature within a narrow range, several different mechanisms cause normal core temperature to vary under certain conditions. These normal fluctuations differ from **fever, hyperthermia,** and **hypothermia,** all of which involve changes in core temperature outside this normal range. Hypothalamic temperature can range approximately 0.4°C, from approximately 36.8°C to 37.2°C, without either sweating or shivering being initiated. This temperature range is termed the *hypothalamic thermoneutral zone.*

TABLE 27–2

Summary of Thermoregulatory Effector Responses to Increased Core and Skin Temperature

Response	Mechanism	Effect on Heat Gain or Loss
Skin vasodilation	Decreased sympathetic outflow to skin resistance vessels	Warmer skin increases heat lost to environment; increased convective heat transfer from core to skin reduces core temperature
Sweating	Sympathetic stimulation of sweat glands	Increased evaporation of water from the skin increases heat loss
Behavior	Voluntary skeletal muscle contraction	Shelter from the sun reduces radiant heat gain; reduced clothing increases conduction and convection of heat from the skin; use of fans increases conduction and convection of heat from the skin

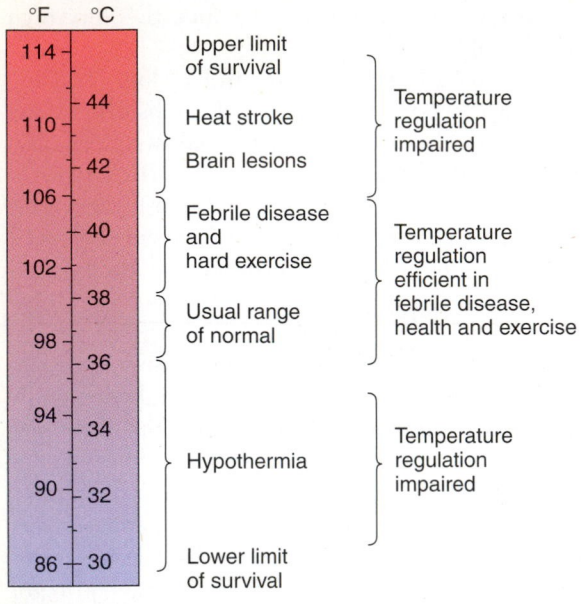

Figure 27–9

Survival range of body temperature in humans.

Measurement of Core Temperature

Because fever, hyperthermia, and hypothermia may require intervention to normalize temperature and diagnose underlying disease, the determination and interpretation of body core temperature has medical significance (Fig. 27–9). In humans, it is not possible to directly measure the hypothalamic temperature. However, both the tympanic membrane tem-

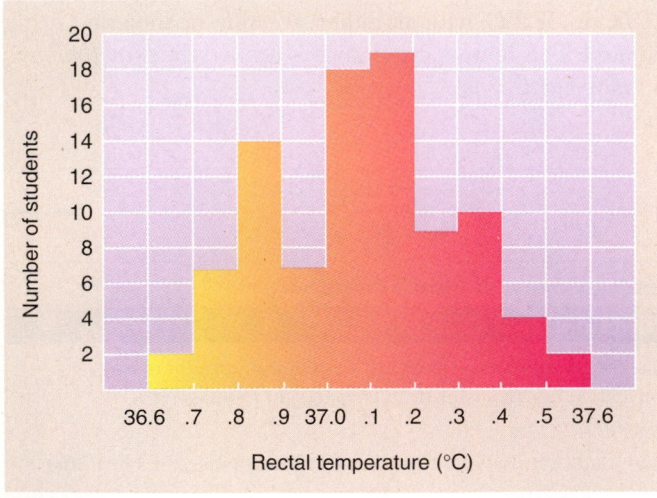

Figure 27–10

Range of rectal temperatures measured under well-controlled conditions in a group of healthy young students. The mean rectal temperature is 37.1°C.

perature and the rectal temperature are reasonably accurate estimates of core temperature. The oral (sublingual) temperature is also representative, averaging approximately 0.6°C lower than rectal or tympanic membrane temperature.

Individual Variation

While the average core temperature (as measured in a moderate environment, without recent exercise, and without fever) is very close to 37.0°C, core temperature differs from person to person (Fig. 27–10). These differences in core temperature have no medical significance.

Circadian Rhythm

Human core temperature reaches a minimum at night and a maximum in the mid-afternoon (Fig. 27–11). Even with all time cues removed, and with sleep schedules disrupted, the core temperature continues to cycle with a periodicity of just over 24 hours. Although circadian fluctuations in core temperature are a marker of an inherent, brain-controlled rhythm, the significance of these daily temperature changes remains unclear.

The Menstrual Cycle and Body Temperature

The temperature changes of the menstrual cycle are superimposed on diurnal temperature changes (see Fig. 27–11). Several days before menstruation, the core temperature usually decreases approximately 0.6°C and is maintained at the lower level until just before ovulation, when it may decrease an additional 0.2°C. After ovulation, core temperature increases approximately 1°C and remains at this level until the onset of the next cycle (see Fig. 27–11).

Menstrual cycle temperature oscillations are variable from cycle to cycle and from person to person, making temperature monitoring an inexact method for contraception. In fact, core temperature measurements show only a 30% correlation with ovulation. In pregnancy, the premenstrual decrease in core temperature is inhibited, and core temperature increases (see Fig. 27–11). While the mechanism for core temperature changes during the menstrual cycle and pregnancy are not known, they are closely associated with the ratio of progesterone to estradiol in the blood, suggesting that both of these hormones influence hypothalamic control of core temperature.

Menopause

Many women after menopause experience sudden episodes of skin flushing and sweating (**"hot flashes"**). Hot flashes occur because estrogen deficiency narrows the hypothalamic thermoneutral zone from approximately 0.4°C to less than 0.1°C (Fig. 27–12). Persons with hot

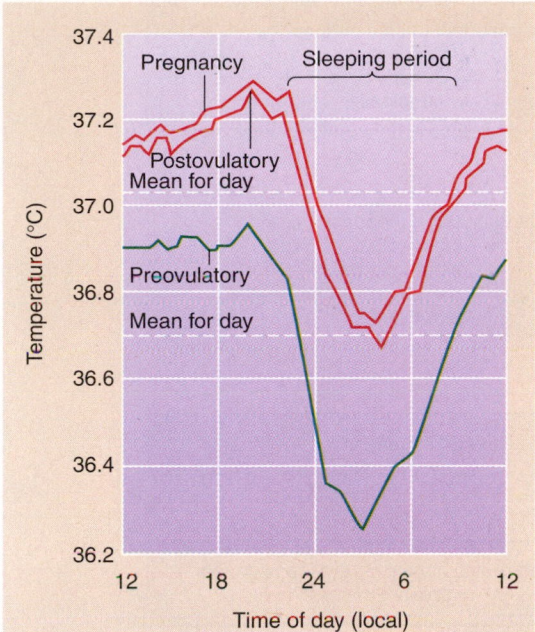

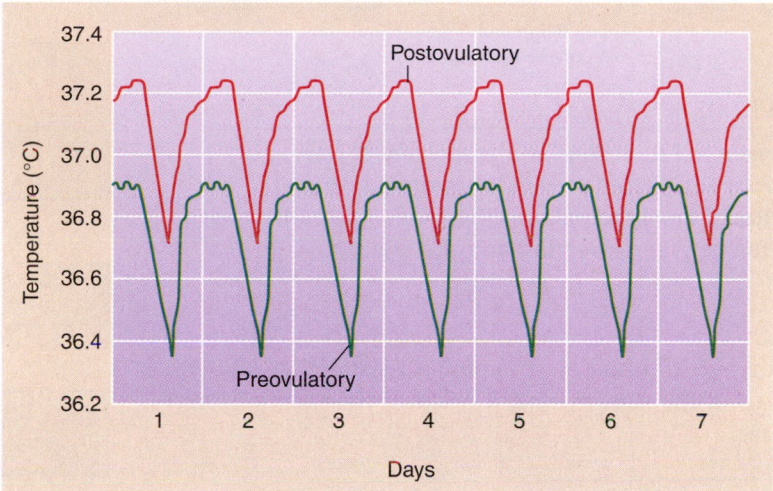

Figure 27–11

Circadian rhythm in core temperature, with the influence of ovulation and pregnancy superimposed.

flashes find that flushing and sweating (in response to elevated temperature) may be followed immediately by shivering (in response to decreased temperature) because the hypothalamic thermoneutral zone is so limited (see Figs. 27–12 and 27–13). Increases in hypothalamic norepinephrine release, secondary to estrogen reduction, mediate these changes. Estrogen replacement therapy restores the normal hypothalamic thermoneutral zone and eliminates hot flashes.

Age and Body Temperature

Infants have a high ratio of surface area to volume. Since heat loss to a cool environment is proportional to surface area, while heat production is roughly proportional to volume, infants are at risk for rapid heat loss and hypothermia. Furthermore, premature infants may have poorly developed body temperature control systems, further increasing the risk for hypothermia. In older age, the risk

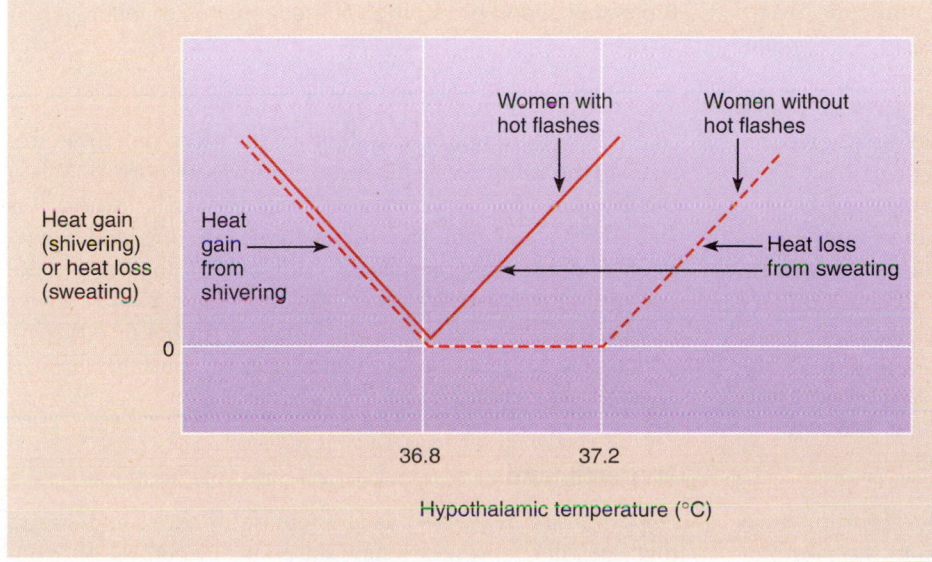

Figure 27–12

Activation of sweating and shivering in women with and without hot flashes. Sweating is initiated at lower hypothalamic temperature in women with hot flashes, and there is no detectable hypothalamic thermoneutral zone.

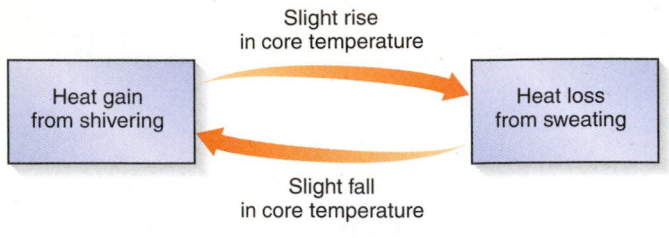

Figure 27–13

Mechanism of hot flashes and associated shivering in postmenopausal women with low estrogen levels. Reduction of the hypothalamic thermoneutral zone causes oscillation between sweating and shivering.

for both hyperthermia and hypothermia increase due to decreases in the sensitivity and onset of thermoregulatory responses.

Fever: A Regulated Increase in Core Temperature

 What is fever and how does it happen?

When matter recognized as alien enters the body, the immune system generates a complex, stereotyped defense response. The invaders may be foreign organisms (infectious disease) or may be self-generated (neoplastic or autoimmune disease). Fever represents one portion of this defensive response to perceived foreign invasion.

Fever begins when activated immune response cells (including leukocytes, mesangial cells, vascular endothelial cells, and astrocytes) produce interleukins 1 and 6. The interleukins, sometimes referred to as *pyrogens*, elevate prostaglandin E_2 biosynthesis in the hypothalamus, changing the temperature set point (Fig. 27–14). Antipyretic drugs block prostaglandin E_2 production, decreasing set point temperature.

During a fever, the hypothalamus compares skin and core temperature with an elevated set point temperature, activating heat production and inhibiting heat loss (Figs. 27–14 and 27–15). Vasoconstriction, shivering, and behavioral responses are initiated and maintained until core temperature has increased to the new reference temperature. Fever represents a regulated increase in core temperature because thermoregulatory effector responses operate normally at the higher hypothalamic set point. For example, during a fever, attempts to decrease body temperature by increasing convective or evaporative heat loss will be actively opposed by increased shivering and vasoconstriction. Circadian rhythms in core temperature are also maintained during fever, so that typically febrile core temperature is higher in the afternoon than in the morning (see Fig. 27–15).

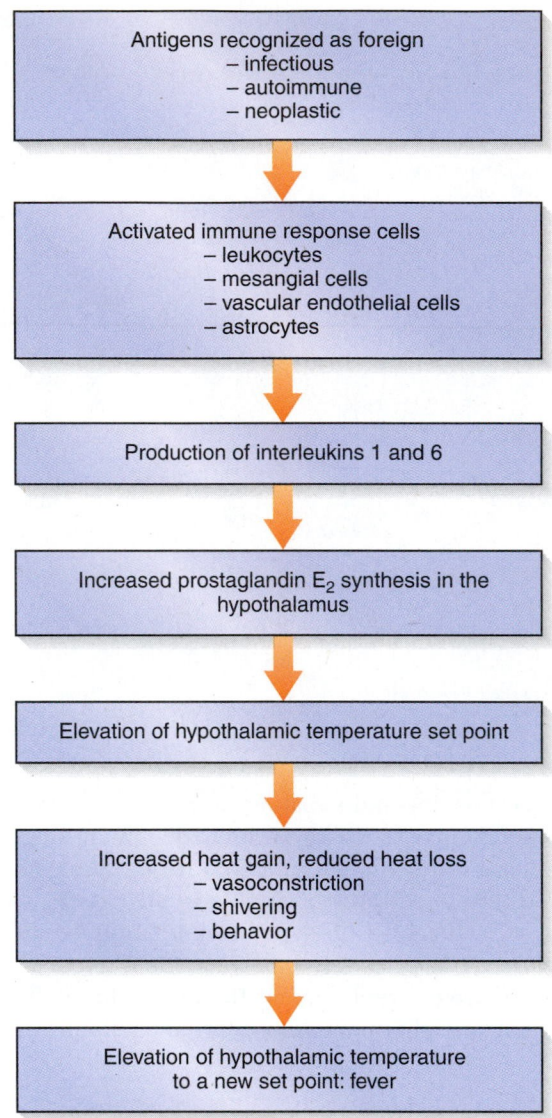

Figure 27–14

Biochemical and physiological mechanisms of fever.

When the immune system ceases the defensive response, the reference temperature returns to normal. When this happens, the hypothalamic integrator activates heat loss and inhibits heat gain, and vasodilation, sweating, and behavioral responses are sustained until core temperature decreases to the normal reference level (see Fig. 27–15). Fever is suppressed in pregnant women near term because the hypothalamic set point fails to increase in response to elevated levels of prostaglandin E_2.

Hyperthermia

Hyperthermia is defined as an increase in core temperature that does not represent a new set point for control of hypothalamic temperature. Hyperthermia may occur despite nor-

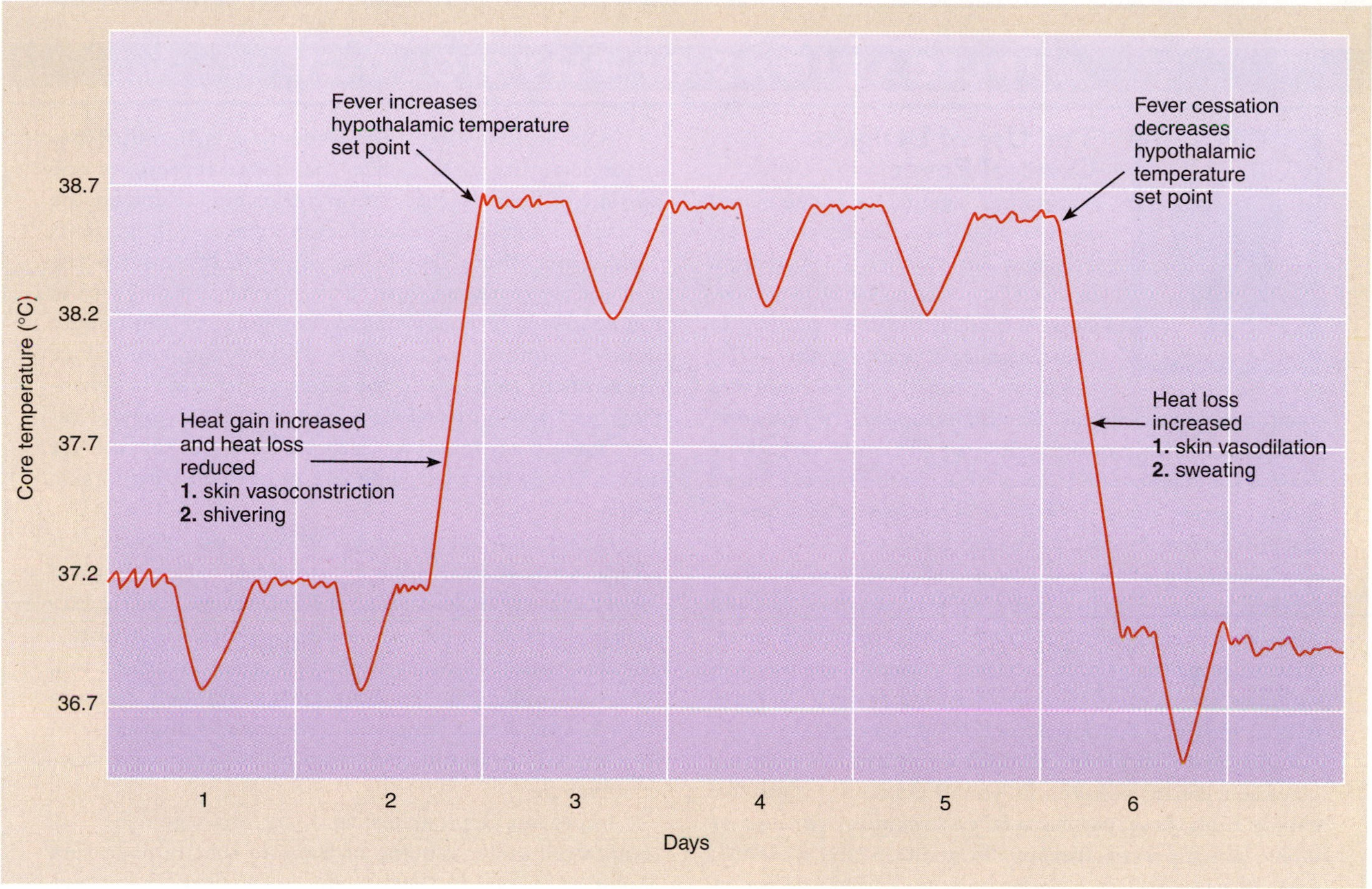

Figure 27–15

Time course of a typical fever. Temperature changes during fever are superimposed on the normal circadian temperature rhythm. Fever onset increases the hypothalamic temperature set point, triggering skin vasoconstriction and shivering that contribute to elevation in core temperature. Fever cessation returns hypothalamic set point temperature to normal; skin vasodilation and sweating contribute to reduction of core temperature.

mal hypothalamic activation of effector responses (skin vasodilation and sweating; behavioral responses that minimize heat gain and increase heat loss), or it can occur due to defects in the temperature regulatory system that prevent adequate responses to increasing core temperature. Perhaps the most familiar example of hyperthermia occurs during exercise. Exercise increases energy expenditure and heat production, and despite activation of heat loss mechanisms, core temperature increases.

Factors Affecting Heat Production

> *What factors contribute to heat production?*

Heat Production at Rest. Because all of the energy liberated from the combustion of foods—except energy used for external work or for the creation of new body tissue—

eventually appears as heat, the rate of energy expenditure has important consequences for temperature regulation. The greatest increases in heat production occur during skeletal muscle contraction, either involuntary (shivering) or voluntary (exercise). In addition, several factors alter energy expenditure at rest, and these factors, as well as skeletal muscle contraction, may alter thermoregulatory responses.

Diet-induced Thermogenesis. The energy expenditure of a resting person increases 10% to 15% following a meal. This increase, which is termed "diet-induced thermogenesis," arises from insulin-induced synthesis of glycogen and fat. Perhaps because insulin levels increase more when carbohydrate and protein are ingested, compared with fat, diet-induced thermogenesis is slightly higher when persons eat high-carbohydrate and -protein meals.

APPLICATIONS OF PHYSIOLOGY

The Use of Drugs to Control Fever

Antipyretic agents have been used for over 2000 years to control fever. The first antipyretics, derived from extracts from the bark of the willow, were shown in the 19th century to contain salicylic acid. By the 1890s, the Bayer company in Germany was manufacturing acetylsalicylic acid as aspirin, the first commercially available antipyretic drug. The introduction of antipyretic medications and clinical thermometers forced the competing view—that fever is one aspect of a beneficial host response—into the background.

Treatment of fever—whether of infectious or noninfectious origin—remains today an unquestioned goal for patients and caregivers. Scientific evidence that fever reduction benefits patients, however, remains elusive. The rationale for treating fever is that temperature elevation is inherently noxious and potentially dangerous.

Advocates of fever reduction argue that temperature elevation causes seizures in young children. Controlled clinical trials, however, have failed to confirm a protective effect of antipyretic therapy. In addition, the metabolic costs of fever, which include shivering and higher energy expenditure at higher body temperature, presumably increase risks for persons with heart or lung disease. Careful analysis, however, fails to find benefits of temperature reduction in such patients. Furthermore, the common experience of greater well-being after antipyretic treatment may stem more from the analgesic, or pain-reducing, effects of these drugs than from decreased temperature.

Is it possible that fever reduction is harmful? Because scientists now realize that fever represents one portion of host defense, antipyretics could worsen the course of infectious or noninfectious diseases. In patients with sepsis, those unable to generate fever have the poorest prognosis, suggesting that fever is a measure of the adequacy of host defense. In studies of adults with rhinovirus infections, aspirin and acetaminophen reduce the antibody response to the disease, increase viral shedding, and worsen nasal signs and symptoms. Similar results have been found in children with chickenpox, for whom antipyretics lengthen the time to crusting of skin lesions.

Finally, antipyretic drugs do carry side effects. Aspirin can increase the risk for gastrointestinal bleeding and renal dysfunction. Acetaminophen can cause liver toxicity at high doses. In an effort to enhance antipyretic therapy, physical methods for fever reduction are often attempted, using skin cooling by convection or evaporation. Because fever is a regulated temperature increase, however, heat loss will be vigorously countered by shivering and vasoconstriction.

Improved approaches to fever management will require full understanding of the role that temperature elevation plays in host defense against infectious, autoimmune, and cancerous disease. Ideal treatment would be targeted, blocking unsafe or uncomfortable aspects of host defense while enhancing useful portions of that response. Future studies could prove that fever is merely a secondary phenomenon, or instead that temperature elevation is indispensable for optimal host defense.

Hormones. The most important hormones that influence energy expenditure are the catecholamines and thyroid hormone. Catecholamine release during stimulation of the sympathetic nervous system increases the rates of a number of enzymatic reactions, increasing energy expenditure. Increased sympathetic activity contributes to increased energy expenditure and heat production during emotional stress. Thyroid hormone levels also influence resting heat production. Hypothyroid individuals have reduced resting energy expenditure and heat production and often experience cold extremities and easy chilling. In contrast, hyperthyroidism increases the resting expenditure of energy and

TABLE 27–3

Primary Hormonal Influences on Resting Energy Expenditure

Hormone	Effect on Resting Energy Expenditure	Speed of Response
Epinephrine, norepinephrine	Increase	Immediate
Thyroid hormone	Increase	Slow

CURRENT CONCEPTS IN PHYSIOLOGY

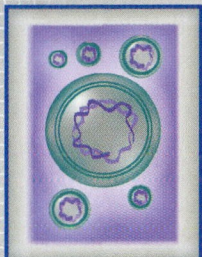

Energy Expenditure and Heat Production Without Work?

We assume that the food we eat generates predictable amounts of energy. This energy, stored temporarily in ATP, is used for ion pumping, protein anabolism, and muscle contraction. Calories burned thus equal the demands of the living organism. Only by regularly increasing those demands, primarily by exercise, can energy usage and caloric expenditure be chronically increased.

In undeveloped societies, where the physical demands of manual labor are great and food is scarce, the primary problem is finding adequate calories to supply the human machine. To partially uncouple metabolism from energy production (i.e., to make less ATP from each molecule of metabolized carbohydrate, fat, and protein) would be counterproductive, because ATP is vitally needed and food supplies are restricted. In affluent Western societies, however, a reverse problem of excess energy storage (obesity) has emerged, in response to abundant food and reduced exercise demands. Calls for voluntary reductions in food intake (dieting) or self-imposed increases in energy demand (exercise) have been remarkably ineffective. Under these conditions, uncoupling ATP production from foodstuff use would usefully increase caloric expenditure without requiring increased cellular work.

Can ATP production and caloric expenditure be uncoupled? Researchers have found that hibernating rodents face a related problem: In these animals, ATP requirements are very low, yet heat production must remain high to prevent hypothermia. The animals solve the problem by activating brown fat, a tissue rich in an uncoupling protein (UCP) that increases heat production without generating significant ATP. Unfortunately, brown adipose tissue and its specific uncoupling protein (UCP-1) are found at very low levels in primates, so the activation of UCP-1 may have little relevance in human physiology. Recently, however, two other uncoupling proteins have been discovered: UCP-2 has widespread distribution in the human body, while UCP-3 is found in skeletal muscle. All of these molecules decrease the efficiency of ATP production and dissipate energy as heat without the organism having to perform cellular work. Although the role that any of the uncoupling proteins may play in human physiology or disease remains unclear, researchers are exploring at least three possibilities.

First, a flurry of interest has developed around sympathetic activation of the β_3-receptor, which stimulates fat breakdown and activates uncoupling proteins. Selectively stimulating β_3-receptors increases basal heat production, potentially benefiting elderly persons at risk for hypothermia and helping obese individuals lose weight. Second, studies of populations at risk for obesity suggest that UCP activity contributes to genetic risk for weight gain, suggesting that direct manipulation of the UCPs could be a future obesity treatment. Third, cytokines involved in the onset of fever stimulate the uncoupling proteins, suggesting that UCP expression may be essential for adequate defense against foreign invasion.

resting heat production, contributing to a reduced ability to tolerate hot environments. Unlike the rapid effects of catecholamines, thyroid hormone has a slow onset of action (Table 27–3).

Heat Production in Exercise. Voluntary skeletal muscle contraction (exercise) is easily the most powerful stimulus for increasing energy expenditure and heat production. In an active young adult, the increased energy expended during exercise can increase heat production up to 15-fold above resting levels for several minutes. Table 27–4 lists the energy expended during different activities. The enormous heat gain during heavy exercise would quickly cause severe hyperthermia unless heat loss mechanisms were activated in matching amounts.

TABLE 27–4

Estimated Energy Expenditure and Heat Production of Various Activities for a 70-kg Person

Activity	Energy Expenditure (kcal/h)
Sleeping	65
Sitting at rest	100
Walking	300
Jogging	700
Swimming	900
Stair climbing	1100

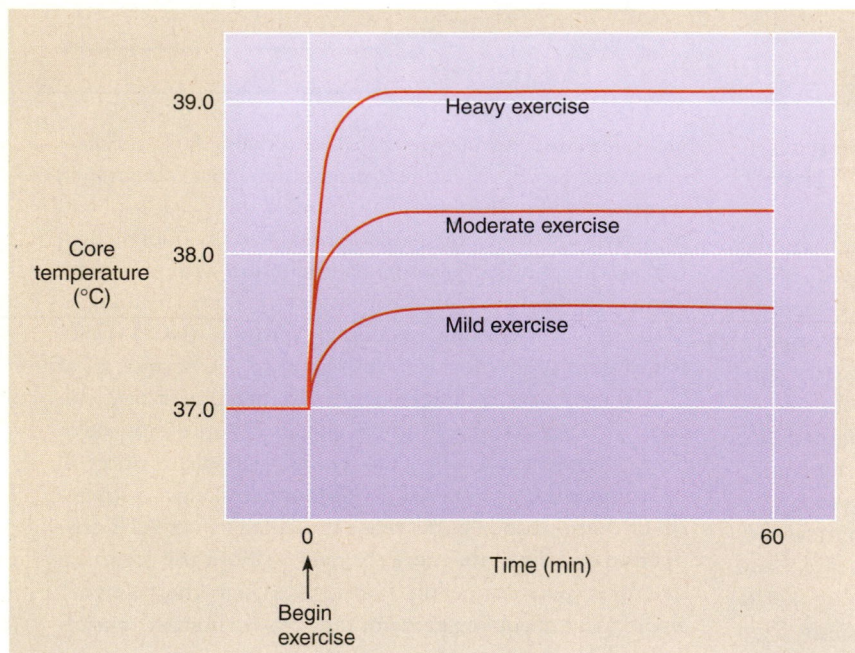

Figure 27–16

Core temperature increase during exercise of varying severity in a moderate environment. Core temperature increases in proportion to exercise intensity and is stable over time as exercise continues.

Exercise Hyperthermia

During exercise, heat is produced primarily in active muscles. Temperature deep within these active muscles typically exceeds core temperature, and convective heat transfer from active skeletal muscles to the core contributes to a rapid increase in hypothalamic temperature during the first minutes of exercise. Increased skeletal muscle heat production in exercise can increase core temperature to levels that approach 40°C. Increasing core temperature in turn activates heat loss mechanisms, primarily skin vasodilation and increased sweating. If work continues at a constant rate in moderate or cool environmental conditions, the core temperature increase is predictable, proportional to exercise intensity, and stable over time (Fig. 27–16). Increased evaporation of sweat is the major mechanism for heat loss during exercise. The ability to evaporate up to two liters of sweat per hour depends upon the activation of sweat glands over nearly the entire body surface. Evaporation of two liters of sweat per hour from the skin surface causes the loss of roughly 1200 kcal, a rate similar to heat production during intense exercise (see Table 27–4). Heat loss from skin vasodilation and sweating is compromised in proportion to increasing environmental temperature and humidity, explaining in part the reduced ability to perform exercise in warm, humid conditions. Under more moderate conditions, the core temperature increase during exercise represents a form of regulated increase in core temperature that differs from fever. During exercise compared with a fever at the same hypothalamic temperature, energy expenditure and heat production are

TABLE 27–5

Comparison of Thermoregulatory Responses to Fever and Exercise at Matched Core Temperature

	Healthy, at Rest	Fever	Exercise
Core temperature (°C)	37	39	39
Energy expenditure (kcal/h)	100	120	900
Skin blood flow	Intermediate	Vasoconstricted	Vasodilated
Sweat rate (ml/h)	0	0	1500

much greater, and heat loss mechanisms are strongly activated (Table 27–5).

Heat Exhaustion and Heat Stroke

 What are heat exhaustion and heat stroke?

Vasodilation of the skin in response to increasing core temperature increases cardiovascular demands in exercise. As exercise continues, loss of plasma volume (due to prolonged sweating) and reduced total systemic resistance (due to vasodilation in active muscles and in the skin) combine to reduce blood pressure. The decrease in blood pressure during prolonged exercise depends on the exercise intensity, the fitness of the individual at work, and environmental conditions. Elevated core temperature itself, to levels at or below 40°C, contributes only indirectly to decreasing blood pressure and the onset of symptoms of dizziness and nausea. Development of these symptoms, which are usually posture dependent, is termed **heat exhaustion.** Oral fluid replacement during demanding, long-term exercise is essential to prevent or mitigate blood pressure decreases during work.

If heat and exercise exposure is continued, however, until body temperature increases beyond a critical temperature in the range of 41°C to 42°C, the person may develop **heat stroke.** Symptoms include dizziness, abdominal distress, absence of sweating, delirium, and eventually loss of consciousness and death if core temperature is not decreased quickly. The high core temperature may damage the brain and lead to sweating cessation, which further elevates core temperature toward fatal levels.

Heat exhaustion and heat stroke can also occur in the absence of exercise, most often during hot, humid weather. As air temperature increases, conduction and convection of heat from the skin to the environment decreases, then is reversed when air temperature exceeds skin temperature. Furthermore, as environmental humidity increases, the rate of evaporation of sweat decreases. If the relative humidity is 100% (representing saturation of air with water vapor) at an air temperature higher than skin temperature, no sweat can evaporate. Consequently, in extreme hot and humid conditions, continued skin vasodilation and sweat secretion can provide little cooling while still contributing to decreasing blood pressure.

Other Causes of Hyperthermia

Uncontrolled, potentially fatal hyperthermia (malignant hyperthermia) can also develop after drug treatment or during anesthesia. Drug-induced malignant hyperthermia is most common in patients receiving antipsychotic medications. These drugs alter brain neurotransmitter levels and occasionally can provoke inappropriate catecholamine-induced vasoconstriction and increased heat production.

These responses are unrelated to hypothalamic temperature and cause an unregulated increase in core temperature. Malignant hyperthermia during anesthesia is a rare genetic disorder. Muscle rigidity and increased energy expenditure develop from interaction of specific anesthetics with skeletal muscle proteins involved with calcium release and storage.

Hypothermia

 What clinical conditions can result from hypothermia?

Hypothermia is defined as a decrease in body core temperature that does not represent a new set point for control of hypothalamic temperature. Instead, decreasing hypothalamic temperature activates heat gain and heat conservation mechanisms (vasoconstriction, shivering, and behavioral changes). Depending upon environmental conditions, duration of exposure, and other factors, these responses may be adequate to maintain core temperature at or above 35°C. However, when hypothalamic temperature decreases below approximately 35°C, thermoregulatory responses are progressively inactivated, leading to further core temperature reductions in cold environments. Hypothermia due to long-term exposure to cold air or cold water can lead to death at a body core temperature of approximately 26°C to 28°C, usually due to myocardial fibrillation.

Cold environments can also cause peripheral cold injuries, often without substantially decreasing core temperature. In response to decreasing skin and core temperature, the hypothalamus appropriately activates vasoconstriction of the skin, which reduces heat loss to the environment and reduces heat transfer from core to skin. However, skin vasoconstriction, by reducing blood-borne convective heat transfer into the skin, can lead to hazardous reductions in skin temperature when heat loss from the skin to a cold environment is very large. In cold environments, skin temperature decreases to the greatest extent in distal portions of the anatomy (fingers, toes, nose, earlobes) for which skin represents a sizable fraction of the tissue mass (see Fig. 27–5). Consequently, exposed skin and these distal portions of the anatomy are at greatest risk for tissue freezing (frostbite) in severely cold environments. Increasing wind chill accelerates skin temperature decreases in the cold by increasing convective heat loss from the skin. However, tissue cannot freeze unless the environmental temperature is below freezing. Treatment of frostbite can be difficult because freezing and thawing of tissue generates a complex cascade of cellular events, including cellular dehydration, shrinkage, and collapse, extreme vasoconstriction and subsequent vasodilation, platelet activation, and eventual thrombosis.

Summary

- Humans regulate core temperature within a narrow range.
- A negative feedback control system regulates core temperature.
- The control system that regulates core temperature includes temperature sensors in the skin and core, a central controller in the hypothalamus, and effector responses that modulate heat gain and heat loss.
- Heat gain and heat loss from the body take place by the processes of convection, conduction, radiation, and evaporation.
- Energy expenditure by the body results in the production of heat.
- The physiological responses to decreasing skin and core temperature include vasoconstriction of the skin, shivering, and behavioral changes that reduce heat loss.
- Skin vasoconstriction reduces the transfer of heat from the core to the skin, preserving core temperature in a cool environment.
- The physiological responses to increasing skin and core temperature include vasodilation of the skin, sweating, and behavioral changes that reduce heat gain and increase heat loss.
- Hypothalamic temperature can vary slightly, perhaps approximately 0.4°C, without provoking either sweating or shivering; this range represents the hypothalamic thermoneutral zone.
- The core temperature of a healthy person varies with the time of day (circadian rhythm) and with the menstrual cycle in young women.

- Estrogen deficiency after menopause narrows the hypothalamic thermoneutral zone, leading to hot flashes.
- The high surface-to-volume ratio of infants increases the risk for hypothermia.
- Fever is a regulated increase in the set point for core temperature regulation.
- Hyperthermia is defined as an increase in core temperature that does not represent a new set point for temperature regulation.
- The catecholamines and thyroid hormone influence resting levels of energy expenditure and heat production.
- Exercise increases energy expenditure and heat production, causing core temperature to increase.
- The core temperature increase during exercise in a moderate environment is predictable given the intensity of the work.
- Core temperature elevation in exercise causes proportional increases in skin vasodilation and sweating.
- Exercise and fever represent different forms of regulated core temperature increase; the set point for thermoregulation is not changed during exercise.
- Hypothermia is defined as a decrease in core temperature that does not represent a new set point for temperature regulation.

Review Questions

Choose the Correct Answer

1. The hypothalamic thermoneutral zone ranges from approximately:
   a. 30 to 40°C.
   b. 38 to 42°C.
   c. 36.8 to 37.2°C.
   d. 37 to 43°C.
   e. 12 to 60°C.
2. Humans are called homeotherms because we maintain constant:
   a. rates of heat loss in any environment.
   b. core temperature.
   c. skin temperature.
   d. rates of heat production in any environment.
   e. skeletal muscle tension.
3. The sensors for temperature regulation are found:
   a. in the skin and body core.
   b. evenly distributed throughout the body.
   c. only in the hypothalamus.
   d. primarily in the skin.
   e. only in the central nervous system.

4. The integrative center for human temperature regulation is located in the:
   a. spinal cord.
   b. heart and lungs.
   c. hypothalamus.
   d. viscera and great veins.
   e. autonomic nervous system.
5. Regulated responses for maintenance of constant core temperature involve the:
   a. parasympathetic nerves.
   b. control of renal blood flow.
   c. loss of water from the lungs.
   d. diffusion of water through the skin.
   e. sympathetic nerves.
6. Conduction removes heat from the body whenever the body is:
   a. in contact with an object that is cooler than core temperature.
   b. placed away from bright sunlight.
   c. sweating.
   d. in contact with an object that is cooler than skin temperature.
   e. in contact with an object that is warmer than skin temperature.

7. The thermoregulatory responses to decreasing core and skin temperature include:
   a. vasoconstriction in the skin.
   b. sweating.
   c. increased thyroid hormone production.
   d. behaviors that increase heat loss.
   e. behaviors that decrease heat gain.

8. When the skin is maximally vasoconstricted, heat transfer from the core to the skin is:
   a. maximal.
   b. determined primarily by radiant heat transfer.
   c. minimal.
   d. determined primarily by convective heat transfer.
   e. determined primarily by wind chill.

9. In response to increasing skin and core temperature, the hypothalamus stimulates sweat production by increasing:
   a. skeletal muscle tension.
   b. parasympathetic outflow to sweat glands.
   c. behaviors that increase heat loss.
   d. behaviors that increase energy expenditure and heat production.
   e. sympathetic outflow to sweat glands.

10. Compared with the osmolarity of plasma, the osmolarity of sweat is:
    a. similar.
    b. lower.
    c. higher.
    d. highly variable from place to place on the body surface.
    e. determined by the environmental temperature.

11. The circadian rhythm in body temperature ensures that for most persons core temperature is higher:
    a. during sleep than during wakefulness.
    b. when awake than when asleep.
    c. in the afternoon than in the morning.
    d. during episodes of jet lag or sleep deprivation.
    e. during dreaming than during deep, nondreaming sleep.

12. Core temperature varies with the menstrual cycle, suggesting that:
    a. estrogen causes hot flashes.
    b. progesterone and estrogen influence temperature regulation.
    c. progesterone causes hot flashes.
    d. fever cannot occur during ovulation.
    e. fever cannot occur during menstruation.

13. Hot flashes occur in some estrogen-deficient women after menopause because:
    a. skin temperature decreases too readily during cold exposure.
    b. a lack of estrogen prevents vasodilation and sweating.
    c. estrogen blocks normal shivering and sweating.
    d. exercise can too readily increase energy expenditure and heat production.
    e. the hypothalamic thermoneutral zone becomes quite narrow.

14. Fever is initiated by biochemical processes that cause increased hypothalamic levels of:
    a. leukocytes.
    b. prostaglandin E_2.
    c. bacteria.
    d. antigens recognized as foreign.
    e. viruses.

15. One line of evidence that fever represents a regulated increase in the set point for core temperature regulation is that:
    a. attempts to cool the body during fever will be opposed by vasoconstriction and shivering.
    b. core temperature during fever does not show diurnal variation.
    c. persons with fever cannot exercise.
    d. the hypothalamus is no longer an effective control center for thermoregulation during fever.
    e. during fever, sweating and vasodilation are maintained at maximal rates.

16. The primary hormone(s) that influence energy expenditure and heat production at rest are:
    a. parathyroid hormone and calcitonin.
    b. aldosterone.
    c. growth hormone.
    d. the catecholamines and thyroid hormone.
    e. cortisol, glucagon, and insulin.

17. At matched core temperature elevation, exercise and fever differ in that:
    a. hypothalamic effector responses are inactivated during fever.
    b. hypothalamic effector responses are inactivated during exercise.
    c. energy expenditure and heat production are much greater during exercise.
    d. sweating or shivering cannot occur during fever.
    e. skin vasodilation cannot occur during exercise.

18. The extent of core temperature increase in exercise in a moderate environment is determined primarily by:
    a. exercise intensity.
    b. thyroid hormone levels.
    c. the time of day chosen for exercise.
    d. the age and sex of the person performing the exercise.
    e. behavioral responses during exercise that decrease heat loss.

19. During exposure of unclothed skin to 2°C air, an increase in wind velocity from 0 to 50 miles/hour can cause skin temperature to decrease to:
    a. 0°C.
    b. 2°C.
    c. slightly below 0°C, causing frostbite.
    d. values far below 0°C, depending on skin blood flow.
    e. near 10°C to 15°C because increased wind chill increases skin blood flow.

20. Heat exhaustion during exercise occurs as a consequence of:
    a. inability to elevate core temperature during exercise.
    b. excessively cool skin due to inappropriate vasodilation.
    c. excessively warm skin due to inappropriate vasoconstriction.
    d. plasma volume loss due to sweating, and reduced total systemic resistance.
    e. inability to generate adequate sweating and vasodilation during exercise.

21. Moving from sun to shade on a 38°C summer afternoon decreases heat gain from:
    a. conduction.
    b. convection.
    c. evaporation.
    d. radiation.
    e. energy expenditure.

Answers to Case History Questions

1. Her skin warms and reddens immediately after going outside because the much warmer air temperature quickly increases skin temperature. Temperature receptors in the skin report the increasing temperature to the hypothalamus, which responds by increasing the flow of blood to the skin, which quickly warms and reddens the skin.

2. Her sweating increases during exercise because increased energy expenditure in active skeletal muscles increases the heat produced by the body. Increased heat production increases the temperature of the hypothalamus, which responds by both increasing skin blood flow and sweat rate.

3. She sweats on her face, arms, trunk, and legs because sweat glands are distributed over nearly the entire skin surface. For a given environmental temperature and humidity, distributing sweat over a large surface increases the amount of water that can be evaporated per unit time.

4. Her oral temperature increases during exercise because the temperature of all of the tissue deep within her body increases during exercise. Her sweating response is proportional to this temperature increase. Oral temperature is a rough approximation of the hypothalamic temperature.

5. The temperature increase in exercise differs from a fever in several ways. In exercise, core temperature increases because energy expenditure and heat production increase, and responses that increase heat loss (e.g., sweating) are activated proportionally. Fever, in contrast, occurs despite little or no change in energy expenditure or heat production. Instead, the immune response to foreign material changes the set point of the hypothalamic control system to a higher level. The body reaches this higher temperature set point by conserving heat (by reducing the flow of warm blood to the skin) and by involuntary increases in heat production (by shivering).

Key Terms

conduction (p. 829)
convection (p. 829)
core temperature (p. 828)
evaporation (p. 829)
fever (p. 833)

heat exhaustion (p. 841)
heat stroke (p. 841)
hot flashes (p. 834)
hyperthermia (p. 833)

hypothalamus (p. 828)
hypothermia (p. 841)
radiation (p. 829)
shivering (p. 829)

sweating (p. 830)
vasoconstriction (p. 830)
vasodilation (p. 832)
wind chill factor (p. 830)

Suggested Readings

Boulant, J. A. "Role of the preoptic-anterior hypothalamus in thermoregulation and fever." *Clinical Infectious Diseases,* 31(Suppl 5):157–161, 2000.

Enrich, D. "An 'unacceptable' death: An NFL player, Korey Stringer, has died of heat stroke at training camp." *US News and World Report,* August 13, 2001.

Freedman, R. R. "Physiology of hot flashes." *American Journal of Human Biology,* 13:453–464, 2001.

Gordon, C. J. "The therapeutic potential of regulated hypothermia." *Emergency Medicine Journal,* 18:81–89, 2001.

Hasday, J. D., Fairchild, K. D., and Shanholtz, C. "The role of fever in the infected host." *Microbes and Infection,* 2:1891–1904, 2000.

Lyon, A. J., and Oxley, C. "HeatBalance: A computer program to determine optimum incubator air temperature and humidity. A comparison against nurse settings for infants less than 29 weeks gestation." *Early Human Development,* 62:33–41, 2001.

Pecqueur, C., Couplan, E., Bouillaud, F., and Ricquier, D. "Genetic and physiological analysis of the role of uncoupling proteins in human energy homeostasis." *Journal of Molecular Medicine,* 79:48–56, 2001.

Plaisance, K. I., and Mackowiak, P. A. "Antipyretic therapy: Physiologic rationale, diagnostic implications, and clinical consequences." *Archives of Internal Medicine,* 160:449–456, 2000.

Porter, A. M. "The death of a British officer-cadet from heat illness." *Lancet,* 355:569–571, 2000.

Rehm, K. P. "Fever in infants and children." *Current Opinion in Pediatrics,* 13:83–88, 2001.

Sawka, M. N., and Montain, S. J. "Fluid and electrolyte supplementation for exercise heat stress." *American Journal of Clinical Nutrition,* 72(suppl):564–572, 2000.

Sehdev, P. S., and Mackowiak, P. A. "Fever." *In* Rakel, R. E. (ed): *Conn's Current Therapy.* Philadelphia, W. B. Saunders, 2001, pp 20–23.

Smith, I. K. "The trouble with fat-burner pills." *Time Magazine,* Vol 158, No. 8, August 27, 2001.

Answers to Review Questions

1. c	**2.** b	**3.** a	**4.** c	**5.** e	**6.** d	**7.** a	**8.** c
9. e	**10.** b	**11.** c	**12.** b	**13.** e	**14.** b	**15.** a	
16. d	**17.** c	**18.** a	**19.** b	**20.** d	**21.** d		

Chapter 28

BODY DEFENSE AND THE IMMUNE RESPONSE

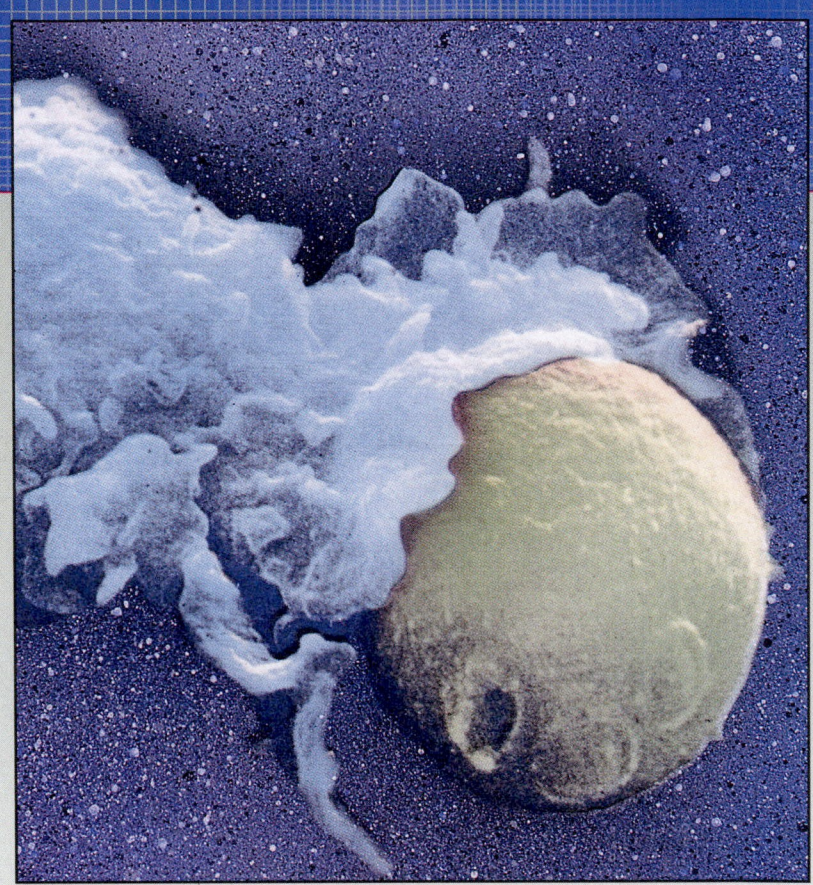

KEY CONCEPTS

- Lymph is chemically similar to plasma and interstitial fluid and is drained into systemic venous blood.

- Primary lymphoid tissues and organs are sites where lymphocytes develop and mature.

- Secondary lymphoid tissue and organs are where mature lymphocytes reside and encounter antigen.

- Internal defenses comprise mechanisms that deter pathogen invasion of the body and mechanisms that destroy pathogens after invasion.

- Internal defense mechanisms must allow for recognition of self versus nonself.

- Nonspecific defenses, such as complement and phagocytosis, deter a variety of pathogens regardless of type.

- Specific defenses, such as the cell-mediated immune response and the antibody-mediated immune response, engage and destroy specific kinds of pathogens.

- Active and passive immunity against pathogens can be induced.

- Autoimmunity occurs when lymphocytes and antibodies react with normal components of the body.

CASE HISTORY

Ellen's pregnancy at age 39 years was unexpected. Three previous pregnancies and recoveries, all before she was 30, were normal and uneventful. Although the delivery was routine and the baby was healthy, this pregnancy had been more stressful and the recovery, prolonged.

Several weeks after the birth, Ellen complained about muscle tremors and weakness. She was more easily fatigued, restless, and unable to sleep peacefully. She attributed her nervousness and increased irritability to her lack of rest. She began to experience heart palpitations, tachycardia, heat intolerance and increased sweating, and blurred vision. Although her appetite had increased, she was losing weight. Easily distracted and unable to concentrate, she blamed her erratic behavior on lack of sleep. She sought the advice of her family physician, hoping he would prescribe a tranquilizer to enable her to sleep and thereby restore her mental and physical fitness. A thorough physical examination followed by thyroid function tests suggested a preliminary diagnosis of hyperthyroidism. Immunological tests confirmed a diagnosis of Graves' disease.

Hyperthyroidism is a condition defined by an excessive secretion of the calorigenic thyroid hormones triiodothyronine (T_3) and tetraiodothyronine (thyroxine) (T_4). When T_3 and T_4 levels increase in target tissues, tissue metabolic rate increases as reflected by an increase in tissue activities. In the gut, an increase in motility and secretory activity may lead to increased frequency of stools and diarrhea. In the nervous system, the result may be hyperactivity, manifested by sleeplessness, irritability, inability to concentrate, and abnormal behavior. An increase in the body's basal metabolic rate produces excessive heat, leading to increased sweating and, in general, heat intolerance. Other systems, such as the circulatory system, are similarly affected. The leading cause of hyperthyroidism is Graves' disease.

Secretion of T_3 and T_4 is controlled by TSH (thyroid-stimulating hormone) produced by the adenohypophysis. TSH engages surface receptors on thyroid follicle cells and, by way of a G-protein-adenylate cyclase-cyclic AMP second-messenger system, increases T_3 and T_4 secretion when it is needed. In Graves' disease, the body's immune system produces immunoglobulins (antibodies) that recognize and bind to the TSH receptors on the follicle cells, abnormally increasing T_3 and T_4 secretion when it is not necessary, leading to hyperthyroidism.

Graves' disease is one of many autoimmune disorders in which the body's internal defenses turn on the body itself, causing disease instead of preventing it. Although there is a genetic predisposition to autoimmune diseases, the cause of Graves' disease is unknown. Infections and physical and emotional stress may play a role in the onset of the disease. There is no cure for Graves' disease, but it is completely treatable.

Questions

1. Why, despite increased appetite, do people with hyperthyroidism tend to lose body weight?

2. What is the fundamental cause of hyperthyroidism in people with Graves' disease?

3. Suggest two ways of treating Graves' disease.

INTRODUCTION

The term *immunity* is derived from the Latin *immunis*, meaning "to make safe." An immune response is the body's defensive reaction to an invasion attempt by a disease-producing organism, called a **pathogen,** or the presence in the body of a chemical the body deems to be foreign or harmful. Usually, the chemical is a protein, and because it illicits an immune response, it is called an **antigen.** The response may include activation of chemical defensive mechanisms, such as the production of an **antibody** that can react with the antigen to reduce its harmful effect, and activation of a variety of defensive cells that comprise the body's immune system.

Many cells of the immune system originate, mature, and function defensively in the lymphatic system. Therefore, it is appropriate to begin with the lymphatic system and then proceed to more specific aspects of body defense.

TABLE 28–1		
Chemical Composition of Lymph and Plasma		
Constituent	**Lymph**	**Plasma**
Calcium	4.2 mEq/L	5.0 mEq/L
Chloride	98 mEq/L	98 mEq/L
Sodium	141 mEq/L	141 mEq/L
Potassium	4.7 mEq/L	5.0 mEq/L
Phosphorus	2.5 mEq/L	2.5 mEq/L
Protein	4 gm/100 ml	7 gm/100 ml
Albumin	2 gm/100 ml	4 gm/100 ml
Globulin	2 gm/100 ml	3 gm/100 ml
Lipid	7 gm/100 ml (intestinal)	0.5 gm/100 ml

THE LYMPHATIC SYSTEM

> *What is the functional anatomy of the lymphatic system and what kind of cells are found in the lymph system?*

The lymphatic system consists of (1) lymphatic vessels, which transport lymphocytes; (2) lymph, a transparent, amber-tinted fluid derived from serum and interstitial fluid; (3) lymphoid tissue; and (4) lymphoid organs, such as the lymph nodes and spleen, which modify lymph. The lymphatic system has three principal functions: (1) it returns filtered plasma proteins and excess interstitial fluid to the blood, (2) it transports lipids absorbed into intestinal lymphatic vessels to the blood, and (3) it helps defend the body against pathogens. This chapter focuses on the third function.

Lymph

Usually, **lymph** is a transparent, amber-tinted fluid and has a chemical composition similar to plasma and interstitial fluid (Table 28–1). Lymph has a lower protein content than does plasma, with an albumin-globulin ratio near 1.0, and generally contains less carbohydrate and more lipid. Lymph in intestinal lymphatic vessels may be opalescent due to high fat content. It is unusual to find erythrocytes in lymph unless blood vessels have been damaged; however, lymph contains all kinds of leukocytes. Some leukocytes emigrate from blood into the lymphatic vessels, and others enter from the lymph nodes.

Lymphatic Vessels

Beginning with the vessel in which lymph forms and in order of increasing diameter, the three principal kinds of lymphatic vessels are **lymphatic capillaries, collecting vessels,** and **lymphatic ducts.**

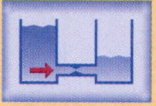

MODELS IN PHYSIOLOGY:
Mass and Heat Flow

The lymphatic system begins as a network of tiny, porous, closed-end channels called **lymphatic capillaries** (Fig. 28–1). Lymphatic capillaries are microscopic endothelial tissues that resemble blood capillaries, except for the blind end of the terminal lymphatic capillary, and they are more porous, enabling large molecules, such as proteins, to enter the lymph. The flattened cells that form the wall of the terminal lymphatic capillary loosely overlap. The free end of the overlapping cell acts as a flapper valve, allowing interstitial fluid to enter the lymphatic capillary when the pressure is greater outside of the capillary, and preventing fluid from leaving by the same route when pressure is greater inside. Lymphatic capillaries are interconnected, forming extensive networks within connective tissue before emptying into collecting vessels.

Collecting vessels are of two kinds: **afferent lymphatic vessels** and **efferent lymphatic vessels.** Afferent vessels collect lymph from lymphatic capillaries and carry it to lymph nodes. Efferent vessels carry lymph away from nodes and toward collecting trunks or lymphatic ducts. Collecting vessels have a connective tissue covering containing scattered smooth muscle cells and elastic fibers that surround the endothelium. These components enable the vessels to contract and relax, helping to move lymph toward the lymphatic ducts. The collecting lymphatic vessels also are equipped with valves that prevent the backward flow of lymph in the same way that venous valves prevent the backward flow of blood.

Two main lymphatic ducts, the **thoracic duct** and the **right lymphatic duct,** collect all of the body's lymph from efferent lymphatic vessels and empty it into the subclavian

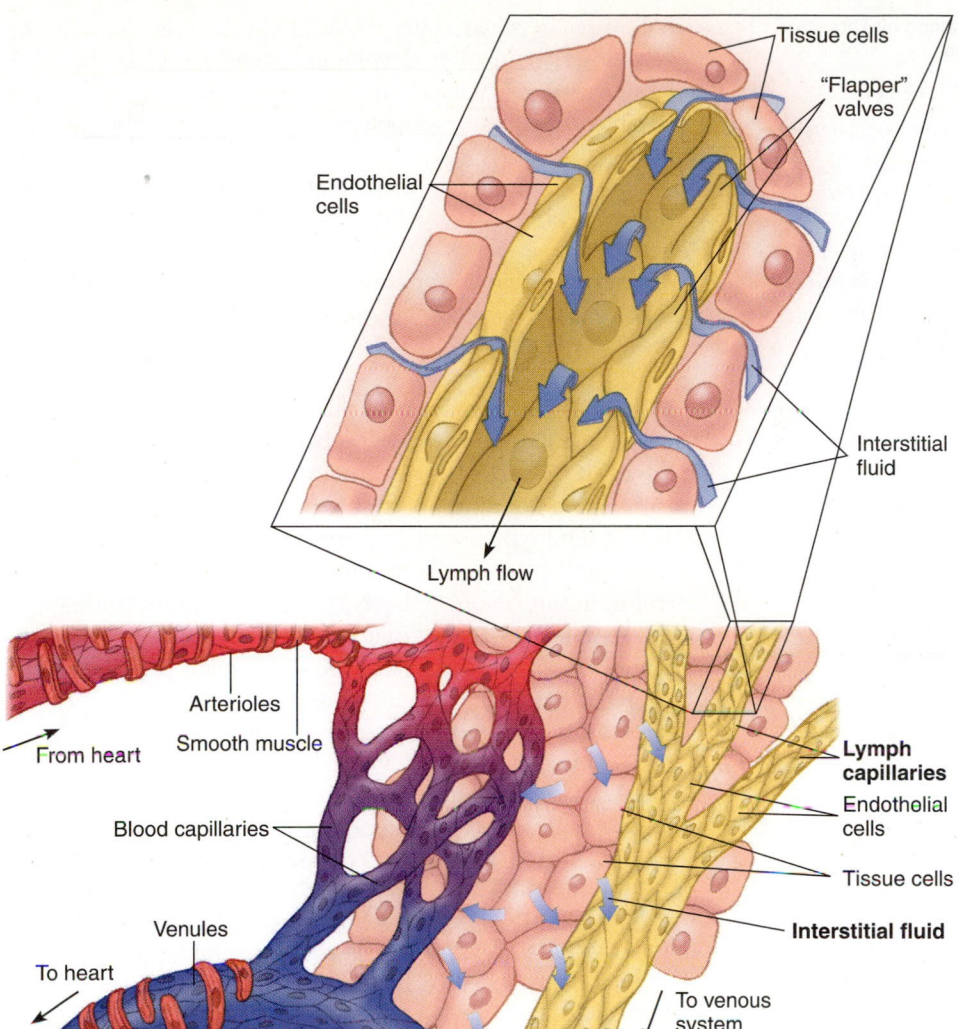

Figure 28–1

Diagram of terminal lymphatic capillaries and the formation and flow of lymph.

veins near the base of the neck. The right lymphatic duct drains lymph from the right half of the head and neck, the right half of the upper thorax, and the right upper extremity, and empties into the right subclavian vein (Fig. 28–2). The remainder of the body is drained by the thoracic duct, which empties into the left subclavian vein. Each lymphatic duct is equipped with a valve at its termination so that blood cannot enter the duct. Forces causing lymph to be formed and to move through lymphatic vessels toward the subclavian veins were outlined in Chapter 19.

Lymphoid Tissue

Lymphoid tissue consists of fixed, interconnected reticular cells and reticular fibers that form a loose network (reticulum) called **stroma.** The interstices of the stroma contain free (unfixed) cells, such as T-lymphocytes, B-lymphocytes, other **leukocytes** (e.g., eosinophils), and macrophages.

Five distinct kinds of lymphoid tissue are present in the body: (1) *diffuse unencapsulated lymphoid tissue,* found just beneath the epithelia of organs; (2) *lymphoid nodules,* which

are dense, discrete, unencapsulated solitary aggregates of lymphocytes and other cells (nodules are found in the loose connective tissue beneath the epithelium of the respiratory passageways and in the walls of the alimentary canal and genitourinary tract; in general, diffused lymphoid tissue and lymphoid nodules are located in areas of the body most likely to be invaded by pathogens, and thus they form a primary line of defense); (3) small encapsulated collections of nodules that form a *lymph node* (lymph nodes always are located along the course of a lymphatic vessel); (4) large encapsulated collections of lymphoid cells, which form the *lobules of each lobe of the thymus gland*; (5) nodules that form the *white pulp of the spleen.*

Primary Lymphoid Tissues

Primary lymphoid tissues are sites where lymphocytes develop and mature. During fetal development, lymphocytes are produced in the yolk sac and liver, but after birth, the red marrow of bone becomes the major source of lymphocytes. Other primary lymphoid tissues make up the thymus gland in the neck and Peyer's patches in the wall of the intestine.

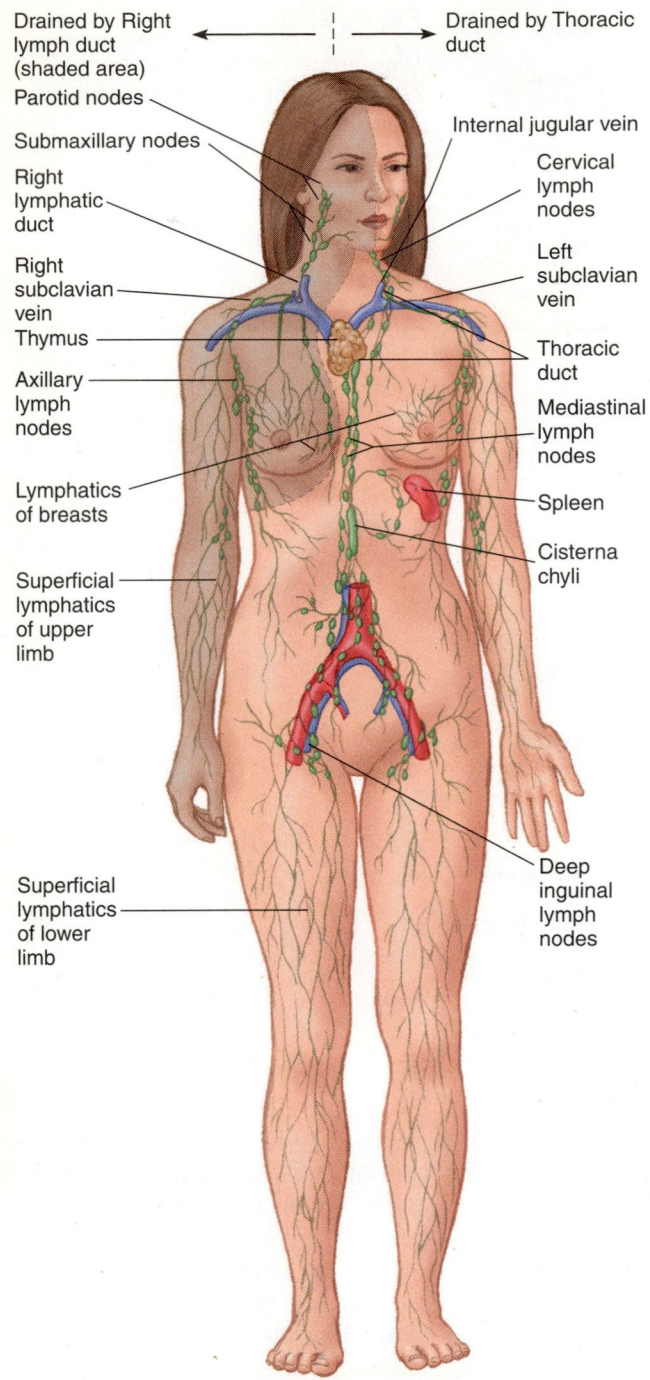

Drained by Right
lymph duct
(shaded area)

Parotid nodes

Submaxillary nodes

Right
lymphatic
duct

Right
subclavian
vein

Thymus

Axillary
lymph
nodes

Lymphatics
of breasts

Superficial
lymphatics
of upper
limb

Superficial
lymphatics
of lower
limb

Drained by Thoracic
duct

Internal jugular vein

Cervical
lymph
nodes

Left
subclavian
vein

Thoracic
duct

Mediastinal
lymph
nodes

Spleen

Cisterna
chyli

Deep
inguinal
lymph
nodes

Figure 28–2

Major lymphatic drainage routes. All lymph formed
eventually is emptied into the left and right subclavian veins
by the thoracic duct and right lymphatic duct, respectively.

Bone Marrow. The bone marrow (Fig. 28–3) is the source
of both red and white blood cells in adults. More than 1 billion white blood cells are produced and released each day,
replacing those that die in defense of the body and those that
die of old age. Within the marrow, some of the developing
lymphocytes mature and become competent in body defense

before being released into blood. Other lymphocytes are released earlier in their development and travel by blood to
other lymphoid tissues, where they complete their development and become competent to defend the body against
pathogens.

In addition to lymphocyte and other blood cell production, the bone marrow serves as an important site for B-
lymphocyte antibody production. The blood sinuses of the
marrow also contain numerous resident phagocytic cells that
remove foreign materials from the blood as it circulates
through the marrow.

Thymus. The thymus is a soft, pale, bilobed gland located
in the mediastinum (middle chest cavity) near the base of
the heart (Fig. 28–4). It is situated anterior to the aortic arch
and behind the upper part of the sternum (breastbone). The
size of the thymus varies inversely with age. Relative to surrounding structures, the thymus is large in infants and young
children, but before puberty, the gland begins gradually to
decrease in size (involute) and is partially replaced with fatty
tissue. In adults, the gland is very small, and in elderly
people it has been so reduced as to be nearly absent, except
for scattered bits of thymic remnants among the adipose
replacement.

The internal structure of the thymus (Fig. 28–5) consists
of lymphoid tissue loosely arranged into irregularly shaped,
interconnected lobules by connective-tissue partitions that
extend inward from the outer connective-tissue capsule.
Each lobule has a peripheral cortex and a central medulla.
The medulla consists of densely packed epithelial cells;
smaller numbers of lymphocytes; and, occasionally, granulocytes. The cortex contains scattered but interconnected reticular cells that form a network within which are large
numbers of rapidly dividing, immature lymphocytes called
thymocytes. The cortex also contains macrophages and small
numbers of plasma cells.

Thymocytes. **Thymocytes** develop from immature lymphocytes released from bone marrow into circulating blood.
The precursors, called **pre-T cells,** are attracted from
thymic blood by **thymotaxin,** a chemotactic peptide secreted by reticular cells in the thymic capsule. Within the
thymic lobules, the thymocytes mature into **T lymphocytes
(T cells)** and reenter the blood prepared to defend the body
against foreign material. Maturation of T cells is discussed in
greater detail later in this chapter.

Peyer's Patches. Great numbers of lymphocytes are found in
the mucosa of the intestinal tract, where they help protect the
body against bacterial invasion through the wall of the intestine. Some lymphocytes are scattered individually along the
entire length of the intestinal tract, but others aggregate to
form lymphoid follicles or nodules. Many of the lymphoid
nodules are small and individual, but some coalesce to form
large aggregates of the nodules known as **Peyer's patches**
(Fig. 28–6). Peyer's patches are numerous in the mucosa of

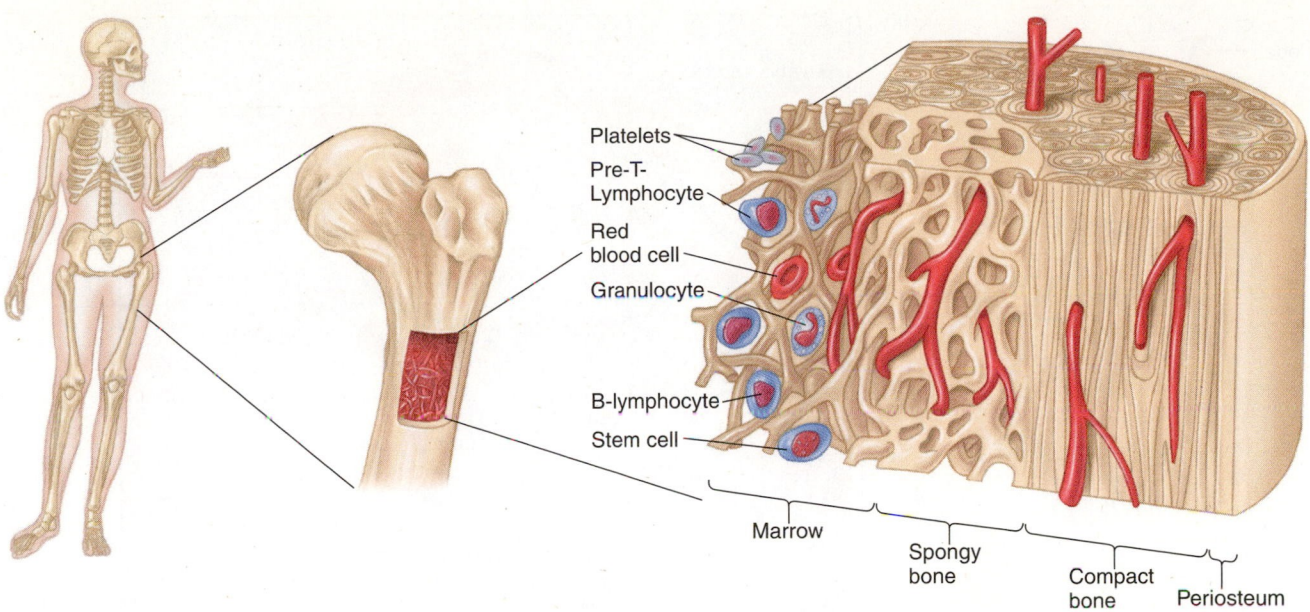

Platelets

Pre-T-Lymphocyte

Red blood cell

Granulocyte

B-lymphocyte

Stem cell

Marrow

Spongy bone

Compact bone

Periosteum

Figure 28–3

The bone marrow is the principal source of red blood cells, white blood cells, and platelets. Within the marrow, precursors develop into B-lymphocytes and pre-T-lymphocytes. Some of the B-lymphocytes remain in the marrow and develop into antibody-secreting plasma cells. Others enter the blood of the sinusoidal capillaries and travel to lymphoid organs, such as the spleen and lymph nodes, where they colonize and clone plasma cells in response to antigen. Pre-T-lymphocytes enter the blood and travel to the thymus to complete their development before recirculating in blood and lymphoid tissue.

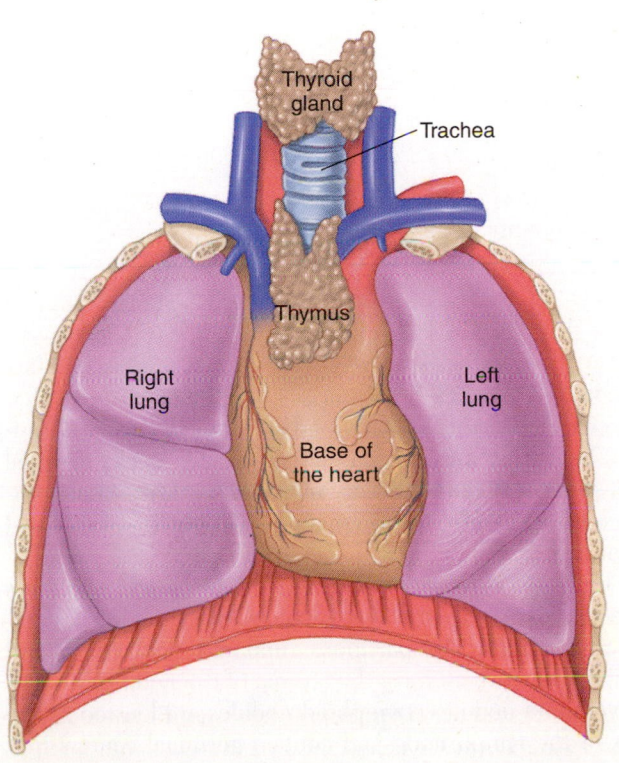

Thyroid gland

Trachea

Thymus

Right lung

Left lung

Base of the heart

the ileum but also are located to a lesser extent in other parts of the intestine, such as the jejunum. The lymphoid nodules contain a germinal center of proliferating **B-lymphocytes (plasma cells)** specialized to secrete antibodies, macrophages, dendritic cells, and different kinds of T-lymphocytes that help coordinate defense against antigens contacting the intestinal epithelium.

The intestinal epithelium covering Peyer's patches contains specialized antigen-binding cells called **M cells.** These cells express receptors on their luminal surface to which antigens bind. Bound antigens are internalized by endocytosis, transported across the M cell's cytoplasm, and released by exocytosis into the interstitium, where antigen processing is completed by local macrophages, T-lymphocytes, and B-lymphocytes.

◄ **Figure 28–4**

The thymus gland has two lobes and is located at the base of the heart anterior to the great blood vessels and immediately behind the sternum.

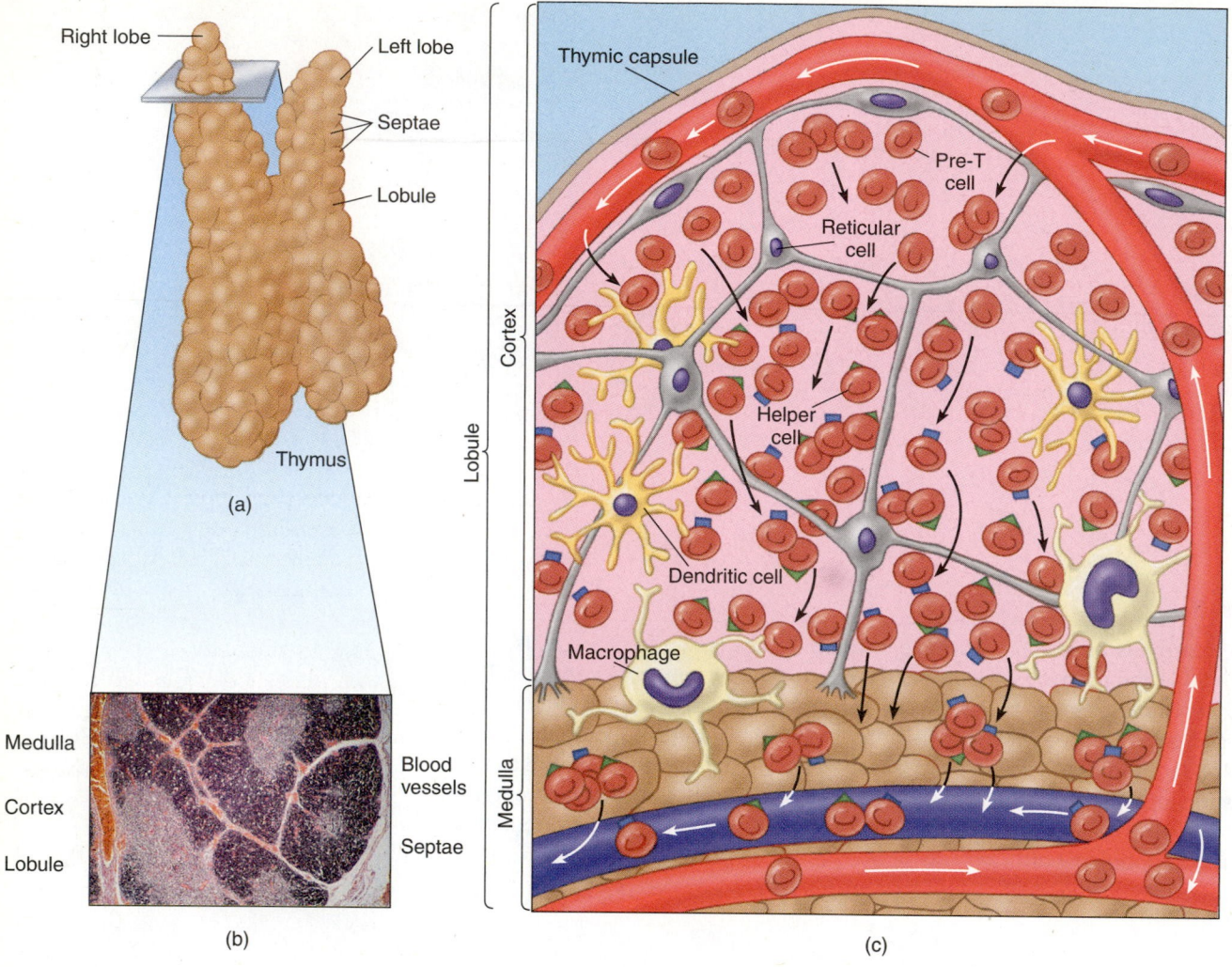

Figure 28–5

(a) The thymus consists of irregularly shaped lobules separated from one another by connective-tissue septa. *(b)* Each lobule has an outer cortex and inner medulla. Pre-T-lymphocytes begin their development in the outer cortex and complete it in the medulla. *(c)* Developing T cells test their immunological competence against dendritic cells, macrophages, and reticular cells. T cells that fail to develop into competent defenders of the body self-destruct (undergo apoptosis) and remnants are removed by macrophages.   *(b, © Vu/Fred E. Hossler)*

Secondary Lymphoid Tissues

Secondary lymphoid tissues and organs develop late in fetal life and continue to develop and function through adulthood. They include lymph nodes and the spleen, as well as previously mentioned bone marrow and Peyer's patches. Also, at some sites, lymphoid cells reside and take position for body defense.

Lymph Nodes. Lymph nodes, sometimes erroneously called "glands," are small (1–2 cm), round or slightly flattened, oval, encapsulated bodies located in series with lym-

phatic vessels (Fig. 28–7). Groups of lymph nodes are located in specific regions of the body and are named after those regions. Examples are cervical nodes of the neck, inguinal nodes of the groin, and axillary nodes of the arm. Many others occur irregularly in the course of lymphatic vessels and are not named.

The lymph node is surrounded by a dense, white, fibrous connective-tissue capsule that projects into the node as trabecular partitions. The partitions support reticular cells and reticular fibers, which form the stroma of the node, and separate lymphoid nodules. Lymphoid nodules are located in the cortex of the lymph node and contain germinal centers that

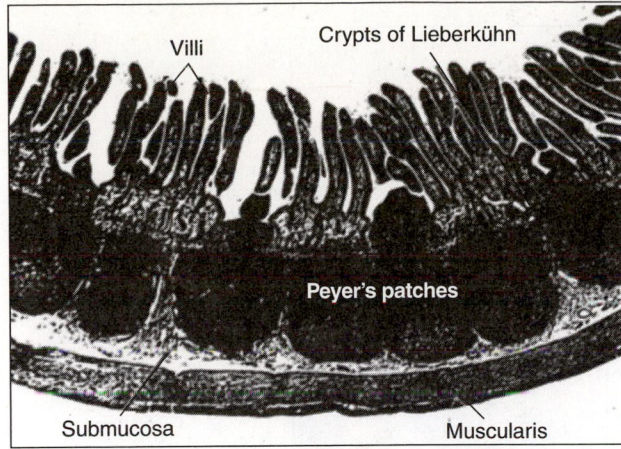

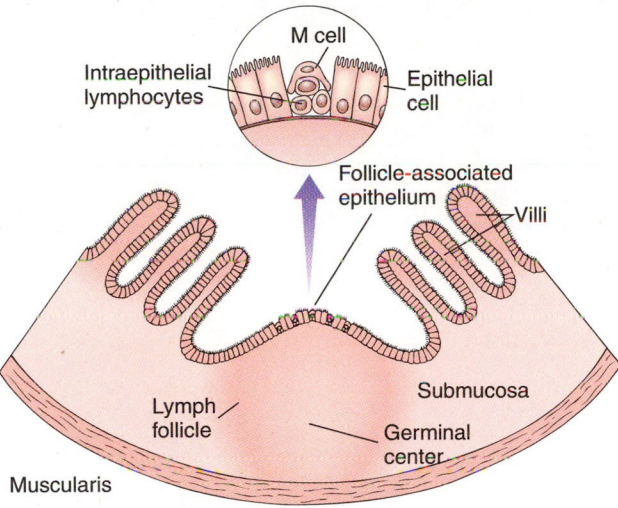

Figure 28–6

Peyer's patches in the cat ileum. *(© Biophoto Associates/Photo Researchers, Inc.)*

are active sites of B-lymphocyte proliferation and macrophage phagocytosis. The central medulla is made up of medullary sinuses and cords of lymphocytes, macrophages, and reticular cells. The area between the cortex and the medulla, called the **paracortical region** or **thymic-dependent zone,** contains densely packed T-lymphocytes, whereas the area immediately beneath the capsule that forms the subcapsular sinus is relatively cell-free.

Lymph is carried from drainage sites in the body to the periphery of the lymph node by valve-equipped afferent lymphatic vessels (see Fig. 28–7). Lymph enters the subcapsular sinus, then flows through cortical and medullary sinuses before exiting the node by way of valve-equipped efferent lymphatic vessels. In its one-way passage through the lymph node, lymph is modified. Antibodies produced by B-lymphocytes (plasma cells) may be added to the lymph. Foreign materials, such as bacteria and particulate matter, are removed by phagocytosis. B-lymphocytes and T-lymphocytes enter the lymph and are carried to the blood to perform defensively at

other sites in the body. Basically, the lymph node performs three distinct yet related functions in body defense: (1) it is a center for lymphocyte proliferation, (2) it acts as a phagocytic organ for purification of lymph, and (3) it is an important site for antibody production and release.

Spleen. The spleen is located in the upper left quadrant of the abdominal cavity, below the dome of the diaphragm, and is protected by the lower ribs. The spleen is the largest lymphoid organ in the body, and its construction is similar to a large lymph node; however, the spleen is designed to filter blood, not lymph. No afferent lymphatic vessels bring lymph to the spleen.

The splenic capsule (Fig. 28–8) contains elastic connective tissue and some smooth muscle, suggesting the possibility of slight distension and contraction, as when storing and releasing blood into the systemic circulation. Contraction of the splenic capsule is probably unimportant in humans because of the small amount of smooth muscle present. Internally, trabeculae incompletely partition the spleen into lobules, each containing red pulp and white pulp.

Red pulp is the major component and consists of large thin-walled sinuses filled with blood and separated by cords of lymphoid tissue containing numerous macrophages. As blood enters the spleen and passes through the red pulp, aged erythrocytes are destroyed by resident macrophages. During fetal development, the red pulp is a temporary site of hematopoiesis.

White pulp consists of clusters of lymphocytes anchored by reticular cells and fibers around a small central artery. Numerous T-lymphocytes are located immediately adjacent to the central artery. In response to a blood-borne antigen, they can enter the blood quickly and mount a cell-mediated defense. Most of the white pulp consists of germinal centers (splenic nodules) containing densely packed B-lymphocytes. Because B-lymphocytes develop into antibody-secreting plasma cells in response to an antigen, the spleen is important for antibody-mediated defense as well as cell-mediated defense.

INTERNAL DEFENSE

 How does the body defend itself against pathogens, toxins, and other harmful chemicals?

Disease-producing microorganisms are always present in our environment. We are exposed to these organisms through the food we eat and drink, the air we breathe, and the animate and inanimate objects that we contact in daily life. Internal defenses comprise mechanisms that deter pathogen invasion of the body and mechanisms that destroy pathogens after invasion. Bodily defense against these pathogens, their toxins, and other harmful chemicals comprise both nonspecific and specific responses.

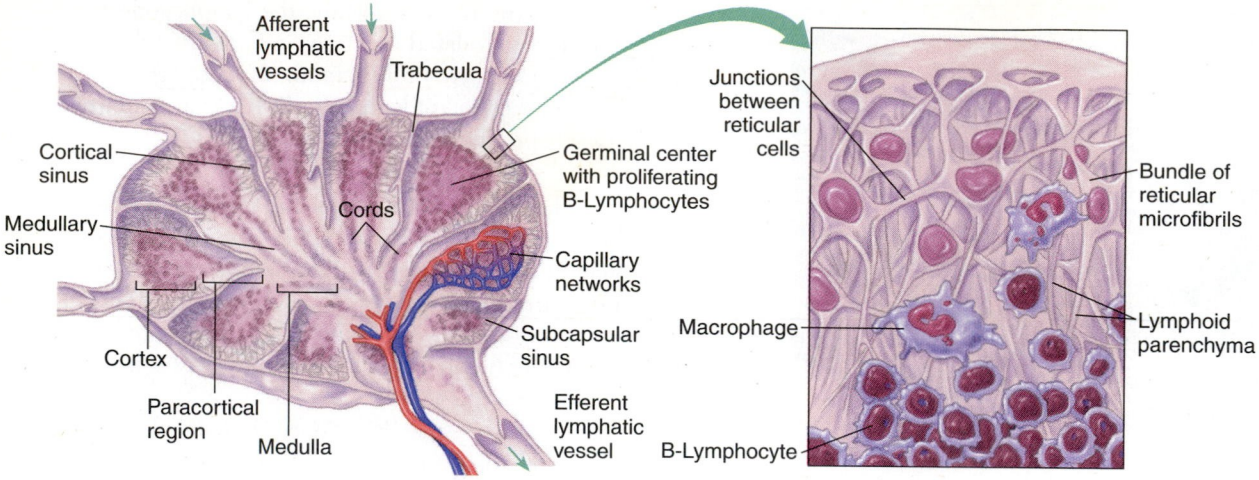

Figure 28–7

Flow through a lymph node. Lymph is brought to the node by afferent lymphatic vessels and enters the subcapsular sinus. Lymph flows through cortical sinuses to medullary sinuses and exits the node by way of efferent lymphatic vessels. As lymph passes through the node, B cells, T cells, and macrophages remove antigens. T cells and antibodies produced by plasma cells also enter blood capillaries within the node and leave by way of veins.

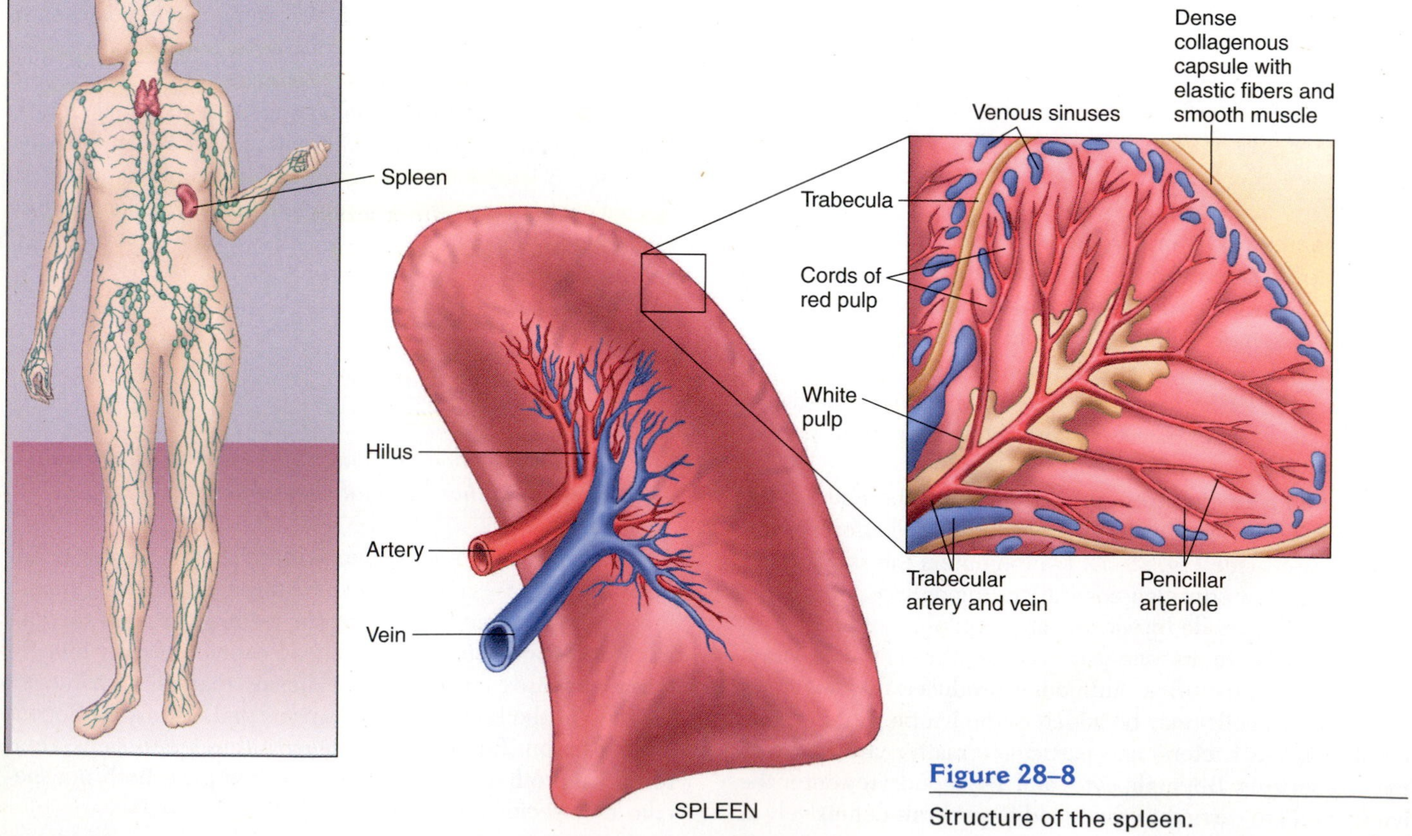

SPLEEN

Figure 28–8

Structure of the spleen.

Nonspecific Defense

Nonspecific defenses deter or destroy a variety of pathogens regardless of pathogen type. Nonspecific defense does not depend on specific antigenic determinant recognition and includes mechanisms that deter a variety of pathogens by preventing their entrance into the body as well as mechanisms that quickly destroy an invader if it gets in. A neutrophil encountering, phagocytizing, and destroying a bacterium is an example of general, nonspecific defense. Nonspecific defenses include surface-protective mechanisms; groups of defensive proteins, such as interferons and complement; and defensive processes, such as inflammation and phagocytosis.

Surface-Protective Mechanisms

The human skin is the first line of defense against pathogens. Most pathogens, as well as many harmful chemicals, cannot penetrate the densely packed, keratinized cells of the epidermis. Sweat and sebum (a secretion of the oil glands of the skin) contain chemicals that inhibit the growth of certain pathogenic bacteria. Many nonpathogenic bacteria called **commensal bacteria** normally inhabit the skin, and their presence, because of competition for nutrients and environment, inhibits the growth of pathogenic bacteria. Also, the pH of the skin is normally acidic because of the chemicals in sebum and sweat and the chemical products of commensal bacterial metabolism; an acidic environment inhibits the growth of many pathogenic bacteria.

Mucous membranes line body cavities that open to the exterior and, like the skin, are potential sites for invasion by pathogens. The linings of the alimentary canal, respiratory passageways, urinary tract, and reproductive tract are examples of frequently invaded sites. Mucus is a sticky, viscous glycoprotein secretion of the epithelial cells that form part of the mucous membrane. It traps microorganisms and prevents the membrane from drying out. The mucus and trapped microbes in the upper respiratory passageways are moved by ciliated epithelial cells toward the nasal and oral cavities (via the aptly named *ciliary escalator*) from which they may be discharged or swallowed. Swallowed microbes are destroyed by the highly acidic secretions of the stomach. Microbes trapped in the mucus of the lower respiratory tract and the genitourinary tracts usually are destroyed by macrophages. In addition, the continual secretion and swallowing of saliva and the periodic flow of urine and feces help prevent pathogens from colonizing on associated mucous membranes. Millions of commensal microbes also inhabit mucous membranes and, as on the skin, they effectively reduce the risk that pathogens will multiply and invade the body. Commensal bacteria also help to inhibit pathogen invasion by producing acids and peroxides that form an unfavorable environment for potential pathogens.

Finally, most of the body's secretions contain antimicrobial chemicals. Sweat, saliva, tears, and urine, for example, contain **lysozyme,** an enzyme that effectively digests the cell wall of many pathogenic bacteria. Figure 28–9 summarizes primary physical and chemical defense mechanisms.

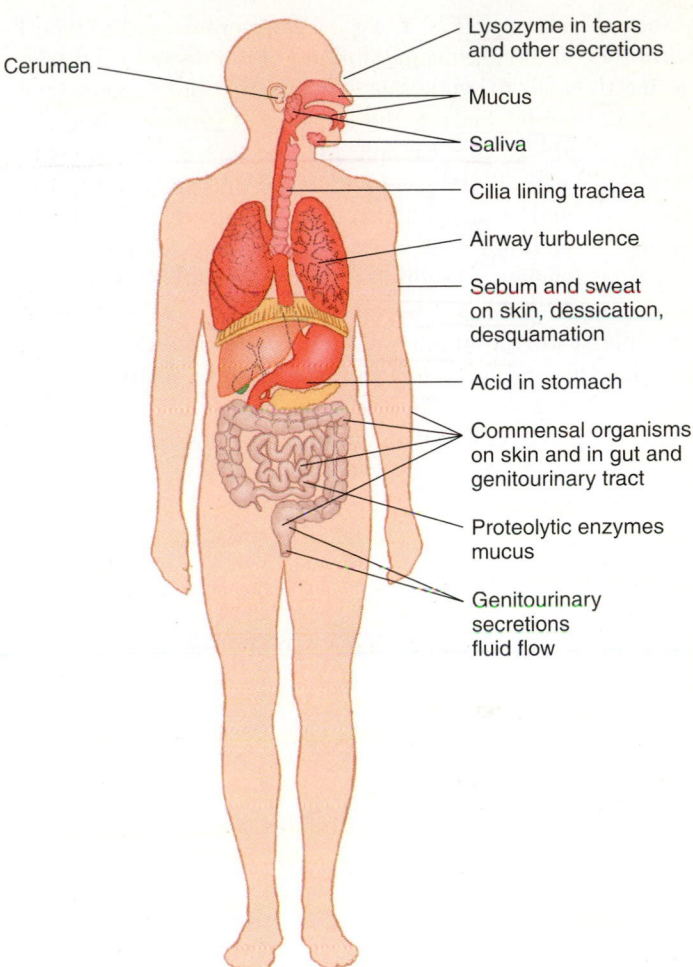

Figure 28–9

Primary physical and chemical defenses of the body.

Interferons

Interferons (IFNs) are a group of small proteins that have antiviral activities. Interferons are released by many types of cells as part of a secondary defense system that is activated when a virus invades the body. They were discovered in 1957 and named for their ability to "interfere" with viral replication. The two main types of interferons are **type I** interferons (alpha [α], beta [β], tau [τ], and omega [ω]), each comprising a single chain of amino acids, and **type II** (gamma [γ]) interferon, a dimer of two identical proteins.

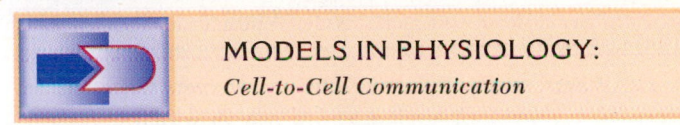

MODELS IN PHYSIOLOGY:
Cell-to-Cell Communication

Interferon alpha (IFN-α) and **interferon beta** (IFN-β) are the two principal type I interferons. IFN-α is made by almost any type of cell infected with a virus; fibroblasts (fiber-forming cells found in connective tissues) are the principal source of IFN-α. Type I interferons bind to type I receptors on the plasma membranes of uninfected cells, triggering the cells to synthesize enzymes that break down viral mRNA, thereby making the cells virus resistant. Because viruses need host cells to replicate, interferons interfere with viral replication, thereby helping to prevent viral spread. The U.S. Food and Drug Administration recently licensed IFN-α for use in treating genital warts, hairy cell leukemia, acquired immune deficiency syndrome (AIDS)-related Kaposi's sarcoma, hepatitis B, and hepatitis C. IFN-β is used to treat multiple sclerosis, an autoimmune disease in which various lymphocytes and macrophages collaborate to destroy myelin in the CNS. Although the exact mechanism of action is not clear, IFN-β is believed to suppress the involved immune cells, thereby alleviating symptoms of the disease.

Interferon gamma (IFN-γ) is produced mainly by activated T-lymphocytes and natural killer cells. These cells need not be virally infected, only alerted to the presence of bacteria, viruses, parasites, or cancer cells to secrete IFN-γ. IFN-γ binds to type II receptors on plasma membranes, inducing within the target cell the production of proteins that inhibit viral replication, inhibit the growth of tumors, and orchestrate specific immune defense. IFN-γ induces macrophages to kill cells infected by viruses, bacteria, or parasites, and tumor cells. It also stimulates production of class I or II major histocompatibility complex (MHC) proteins (recognition proteins expressed on the cell surface) and subsequent antigen presentation, key steps in cell-mediated and antibody-mediated immune responses. IFN-γ currently is used to treat chronic granulomatous disease, a rare hereditary disease of blood granulocytes in which neutrophils ingest bacteria but fail to kill them, resulting in severe, chronic infections. IFN-γ stimulates macrophage destruction of bacteria, thereby boosting defense against bacteria.

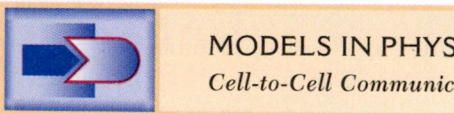

MODELS IN PHYSIOLOGY:
Cell-to-Cell Communication

Complement

Complement is a set of approximately 20 proteins, found in normal plasma or serum, that interact in a defined sequence to promote the lysis of certain microbes and the release of chemical signals. The system is called "complement" because its functions complete—that is, are complementary to—other internal defense mechanisms. When an antibody reacts with an antigen, for example, an immune complex is formed that initiates sequential interactions between complement proteins in a manner similar to the way in which chemicals initiate the reaction sequences in Stage I of blood coagulation. Activated complement proteins (Fig. 28–10) exhibit four biological actions: (1) some lyse cell walls of certain microbes, (2) some increase capillary permeability by stimulating histamine release from tissue cells and leukocytes,

(3) some coat the surface of the microbe and signal the neutrophils and macrophages to ingest and destroy the microbe (this process is called *opsonization,* and the complement proteins that function in this matter are called *opsonins*), and (4) some attract leukocytes (by chemotaxis) to the site of the reactions.

Inflammation

Inflammation is usually a localized body response to tissue damage or injury, such as occurs when pathogens invade tissues. It signifies the body's attempt to defend itself against the pathogen and prevent the pathogen from spreading and damaging other tissues (Fig. 28–11). The clinical signs of inflammation are redness, heat, edema, and pain. They are caused by the following inflammatory responses:

1. Injured cells, basophils, mast cells, and others release *histamine* and other chemicals that dilate blood vessels and increase the blood supply to the injured area. This process makes Caucasian skin look red and feel warm.
2. Histamine and other chemicals increase the permeability of local capillaries, which allows antibody proteins and other chemicals that mediate an immune response to pass out of the capillary and enter the tissues. The increased flow of materials out of the blood and into the tissues causes local swelling or edema (fluid accumulation), which, in turn with chemicals released from damaged cells, causes pain.
3. Neutrophils and, to a lesser extent, other leukocytes migrate out of the capillaries (*diapedesis*) and into the surrounding tissues (Fig. 28–12). Neutrophils and tissue macrophages are attracted (*chemotaxis*) to the area of tissue damage by chemicals released from damaged cells, activated complement proteins, lymphokines (from lymphocytes), and other substances.

Phagocytosis

After a pathogen penetrates the body's epithelium, it encounters *phagocytes,* cells that ingest and destroy microorganisms by the processes of phagocytosis. Although there are several types of phagocytes, they are all derived from a common stem cell in the bone marrow. Phagocytes are divided into two groups: (1) the neutrophils, or *polymorphonuclear leukocytes,* and (2) the monocytes/macrophages.

Neutrophils are the most numerous of the leukocytes, making up between 50% and 70% of the total in normal peripheral blood. They are highly phagocytic, engulfing all kinds of particles (including pathogens) and destroying them. They are short-lived, existing for just 1 to 5 days. They are continually being replaced and are the first phagocytes to arrive at an invasion site.

Monocytes migrate out of the blood and into the tissues, where they develop into **macrophages.** Tissue macrophages include the *Kupffer cells* of the liver, *microglia* in the brain, *alveolar macrophages* in the lung, and others. They are strategically located (near the blood, near the

Functions of Complement Proteins

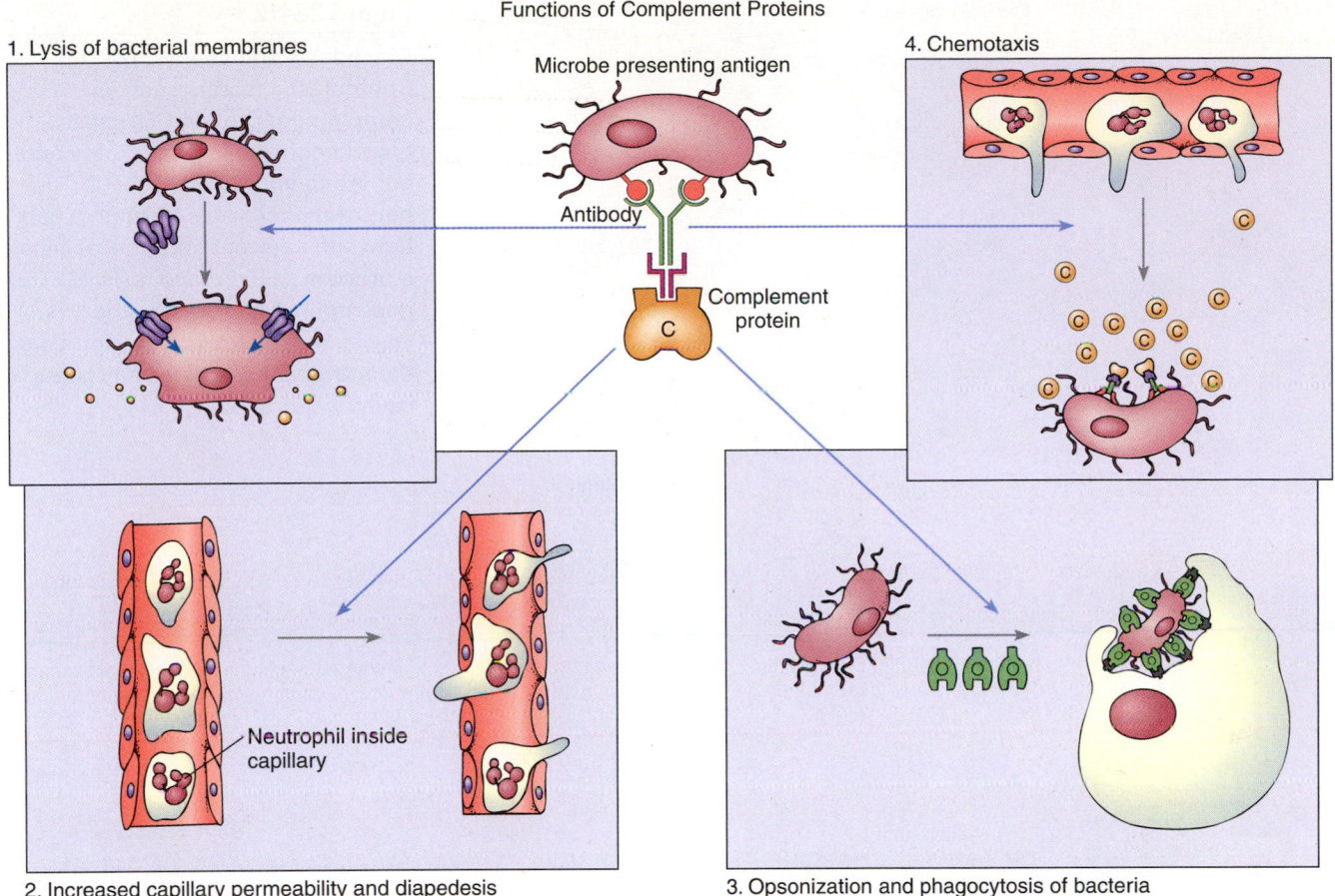

1. Lysis of bacterial membranes

Microbe presenting antigen

Antibody

Complement protein

C

4. Chemotaxis

Neutrophil inside capillary

2. Increased capillary permeability and diapedesis

3. Opsonization and phagocytosis of bacteria

Figure 28–10

Four different ways in which complement proteins function.

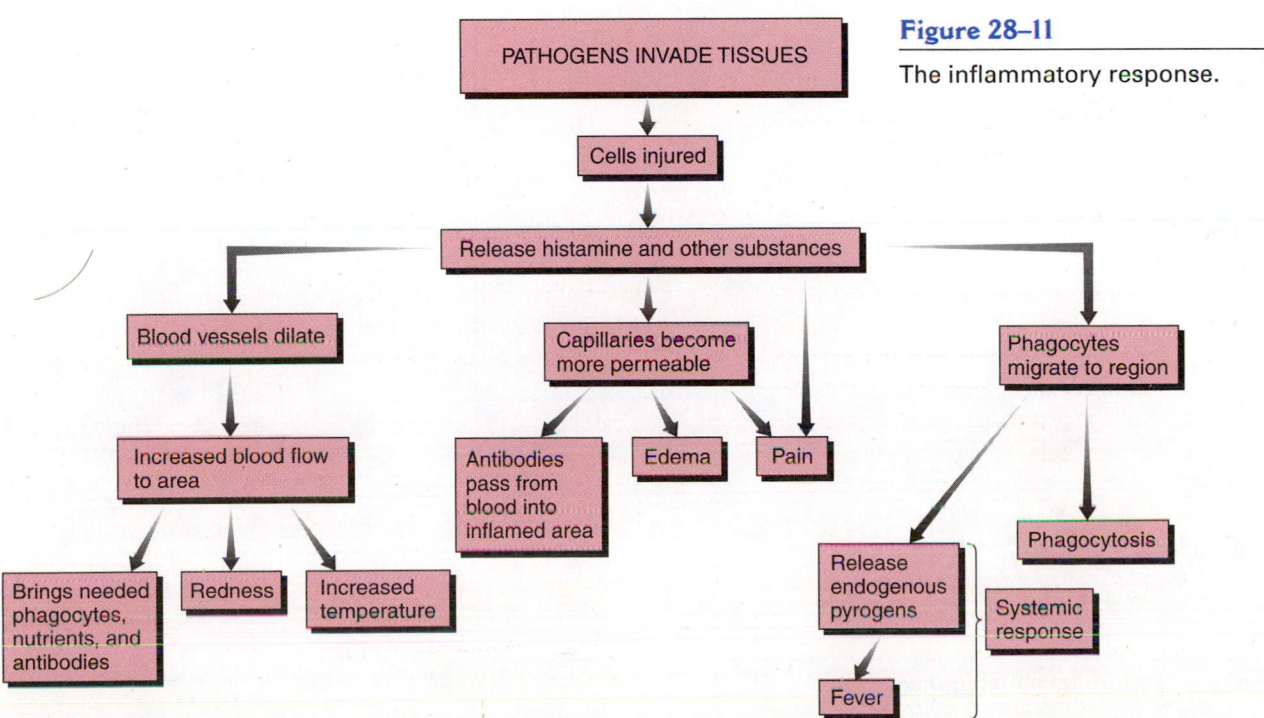

Figure 28–11

The inflammatory response.

PATHOGENS INVADE TISSUES

Cells injured

Release histamine and other substances

Blood vessels dilate

Capillaries become more permeable

Phagocytes migrate to region

Increased blood flow to area

Antibodies pass from blood into inflamed area

Edema

Pain

Brings needed phagocytes, nutrients, and antibodies

Redness

Increased temperature

Release endogenous pyrogens

Systemic response

Phagocytosis

Fever

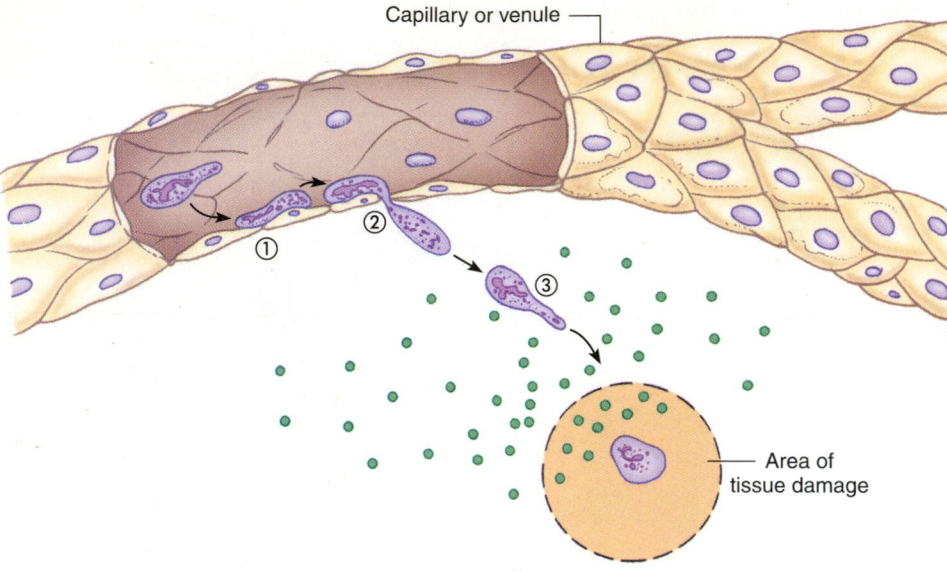

Figure 28–12

Diapedesis and chemotaxis. (1) Chemotactic factors such as complement protein C5a cause blood neutrophil to marginate and adhere to endothelial cell. (2) The neutrophil migrates through a gap between adjacent endothelial cells, a process called *diapedesis*. (3) The neutrophil, once outside the blood vessel, moves toward the source of the attractant chemical, a process called *chemotaxis*.

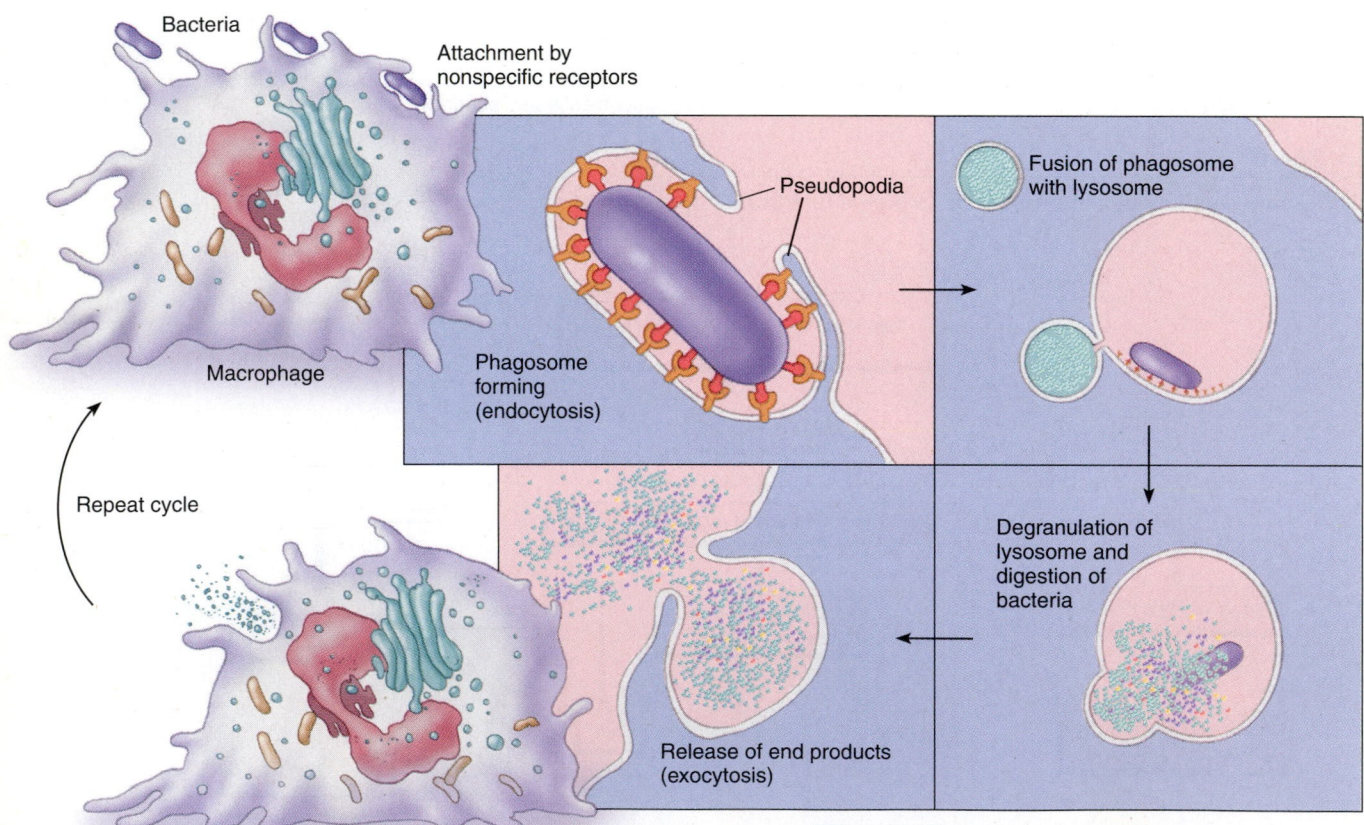

Figure 28–13

Phagocytosis of a bacterium by a macrophage.

alveoli of the lung, near the lumen of the intestine, and so on) where they are most likely to encounter foreign substances, including infectious agents. Some, such as Kupffer cells, remain relatively stationary (fixed macrophages) in the organs where they settle; others (wandering macrophages) roam about through the connective tissues.

The monocyte/macrophage differs from the neutrophil in that it is not as phagocytic unless it is "activated" by complement proteins, antigen-antibody complexes, interferon, or other chemicals. Once activated, the macrophage becomes highly phagocytic and is known as an "activated" macrophage. Macrophages are usually slower in arriving at an invasion site but live longer than do neutrophils (1–3 weeks) and can ingest and destroy more material.

Phagocytosis of a bacterium by a macrophage is illustrated in Figure 28–13. The macrophage attaches to the bacterium by way of nonspecific receptors and extends pseudopodia around the attached bacterium, engulfing it and forming a membrane-bound intracellular vesicle called a *phagosome*. Lysosomes bind to the phagosome and release digestive enzymes into it, a process known as *degranulation*. The bacterium is digested, and the end-products of the digestion are released or, in some cases, used by the macrophage. For example, antigenic parts of the bacterium may be displayed with MHC proteins on the surface of the macrophage so as to activate specific immune defense mechanisms.

Specific Defense

Specific defenses engage and destroy specific kinds of pathogens. As defined earlier, the term *immunity* is derived from the Latin *immunis,* meaning "to make safe." The two kinds of protective immunity are (1) *natural,* or *innate, immunity* and (2) *acquired,* or *adaptive, immunity*. The nonspecific defense mechanisms previously discussed, such as physical and chemical barriers, provide us with natural immunity. Specific defense mechanisms, geared toward the identification and destruction of specific pathogens, give us acquired immunity. The two main types of specific defense responses are (1) cell-mediated immune responses and (2) antibody-mediated immune responses.

MODELS IN PHYSIOLOGY:
Cell-to-Cell Communication

Cell-Mediated Immune Response

> *How do the various cells of the immune system develop and interact to defend against pathogens?*

The cell-mediated immune response is the primary responsibility of the macrophages and the T-lymphocytes (T cells): both types of cells collaborate in the direct destruction of viruses and

foreign cells that enter the body. Collaboration is required because most T cells cannot by themselves recognize free antigens in tissue fluids; the antigen must be presented to the T cell by another cell called an *antigen-presenting cell*. Usually, the antigen-presenting cell is a macrophage, but B cells and dendritic cells also can present antigen. Macrophages and lymphocytes coordinate their defensive functions and communicate by sending and receiving molecular signals by way of *cytokines*, a family of small proteins that act as intercellular messengers. Cytokines include *interferons, monokines,* and *lymphokines*. Macrophages secrete monokines that stimulate T-cell development or help destroy a pathogen, and various T cells secrete lymphokines that regulate T- and B-lymphocyte function, attract macrophages, and in other ways help with specific defense. Before examining cell-to-cell interactions of the immune response, we must understand how a T cell becomes competent to function and learn the various types of T cells.

Maturation of T-lymphocytes takes place in an orderly manner. Pre-T cells migrate from the blood into the cortex of the thymus glands and descend through a web of *reticular cells, dendritic cells,* and *macrophages* (Fig. 28–14) toward the medulla. As they descend, they divide repeatedly and acquire distinctive surface proteins that contain receptor sites. As we shall see, some receptors allow the T cell to be identified, and other receptors allow the T cell to identify and distinguish between the body's own cells and antigens.

T-lymphocyte precursor cells are carried in the blood from the yolk sac, embryonic liver, and bone marrow to the thymus, where they are induced by *thymotaxin* to leave the blood and enter the outer cortex of the thymic lobule. In the outer cortex, the developing T-lymphocytes (*thymocytes*) begin to express surface proteins on their plasma membranes, which act as receptors or cell markers. Developing T-lymphocytes possessing a surface marker designated as CD4 are called **CD4⁺** lymphocytes, or simply **T4 cells.** Developing T-lymphocytes possessing a surface marker designated as CD8 are called **CD8⁺** lymphocytes, or simply **T8 cells**. The "+" means the CD4 or CD8 can be detected when it binds an appropriate monoclonal antibody. The binding means the presence ("+") of the CD4 or CD8 marker. T-lymphocytes also acquire surface proteins called **major histocompatibility complex** (MHC), two classes of which we will consider here. **Class I MHC** molecules are present on the surface of all nucleated cells in the body. **Class II MHC** molecules are present on the surface of some cells of the immune system. CD8⁺ cells express class I MHC, whereas CD4⁺ cells express class II MHC.

As CD4⁺ and CD8⁺ cells move through the inner cortex, they begin to synthesize and insert into their plasma membranes T-cell receptors (TCRs) for recognition of foreign antigens. TCRs consist of two peptide chains, usually alpha and beta chains, which determine the antigen and MHC specificity of the receptor. Extensive rearrangement of the genes coding for the alpha and beta peptides during T-cell development results in the expression of a great variety of TCRs in the lymphocyte population of the thymic cortex.

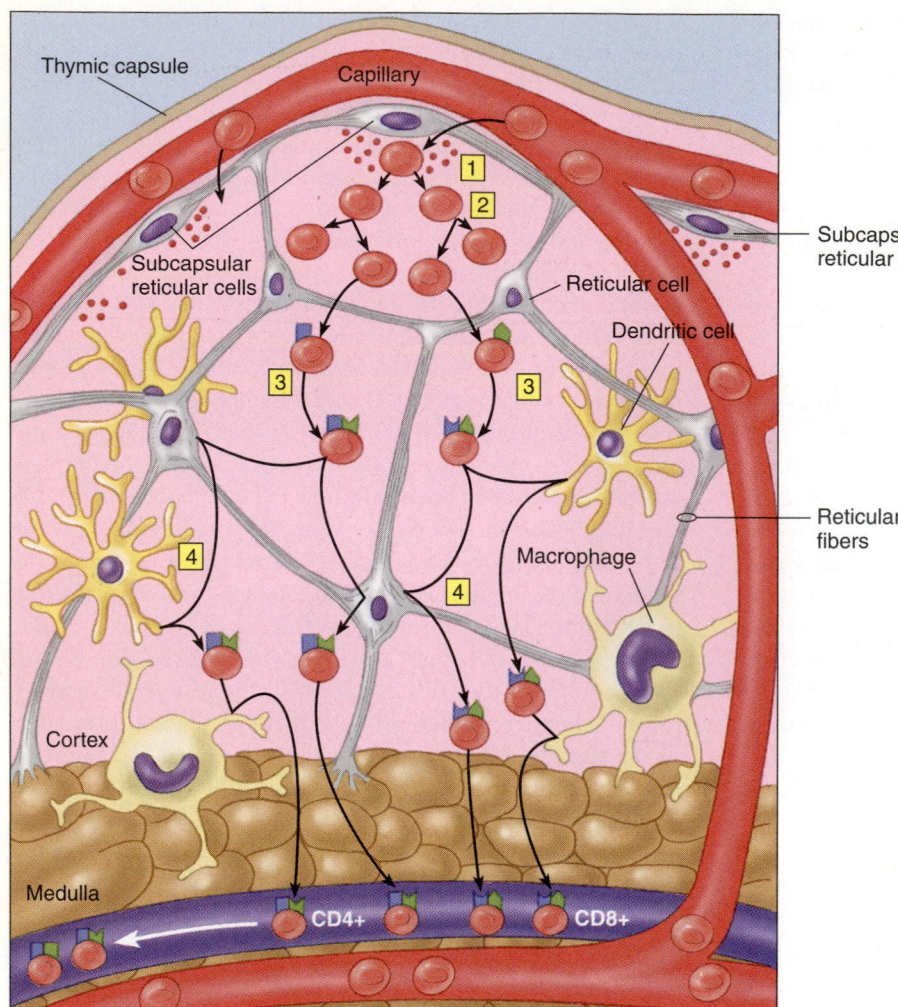

Figure 28–14

Maturation of T-lymphocytes in the thymus. (1) Thymosin secreted by subcapsular reticular cells attracts pre-T cells from blood to the outer cortex of the thymic lobule. (2) Thymocytes divide repeatedly as they descend toward the medulla. (3) Thymocytes synthesize and insert surface receptor proteins that allow them to be identified (CD4, CD8, and others) and allow them to identify MHC antigen. (4) Developing thymocytes test themselves against surface proteins of reticular cells, macrophages, and dendritic cells. Thymocytes that react strongly to self antigens undergo apoptosis (programmed cell death) and are removed by macrophages. (This process is not shown.) Other thymocytes become mature, multiply, and enter venous blood.

T cells that by chance express receptors that react with self MHC in combination with other self-proteins undergo programmed self-destruction **(apoptosis),** and their remains are removed by macrophages and other cells in the thymic cortex. This is a process of **negative selection.** It has been estimated that approximately 96% to 99% of the developing T cells are negatively selected and destroyed or rendered unresponsive (anergic) in the thymus, leaving 1% to 4% to mature into immunocompetent lymphocytes. The latter move from the inner cortex through the medulla, reenter the blood, and leave the thymus, ready to colonize other lymphoid tissue and battle antigens.

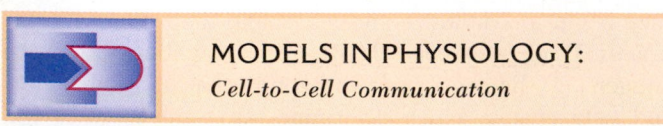

MODELS IN PHYSIOLOGY:
Cell-to-Cell Communication

An **immunocompetent T cell** possesses surface receptors that allow the cell to recognize, bind, and react to an antigen presented in combination with either class I MHC or

class II MHC. Thousands of different kinds of competent T cells exist, each capable of responding to a different antigen. Many of these T-cell clones also are capable of rapidly expanding their numbers (clonal expansion) when stimulated by an appropriate antigen.

Functionally, the two types of CD8$^+$ cells are (1) **cytotoxic T cells** (T$_C$) and (2) **suppressor T cells** (T$_S$). Cytotoxic/supressor cells are CD8$^+$ class I MHC restricted cells. *Restriction* means the cell can recognize an antigen only if the antigen is presented in combination with an appropriate MHC protein on the surface of another cell. Functionally, the two kinds of CD4$^+$ cells are (1) **helper cells** (T$_H$) and (2) **memory cells** (T$_M$). Helper/memory cells are CD4$^+$ class II MHC restricted cells.

The cell-mediated immune response depends on cooperation among the various types of T cells and macrophages. We can gain a basic understanding of the interplay between cellular elements of the cell-mediated immune response by looking at the responses to bacterial and viral infection.

When a pathogen, such as a bacterium, invades the body tissues, it is ingested by a macrophage, an **antigen-presenting cell.** Within the macrophage, bacterial proteins are bro-

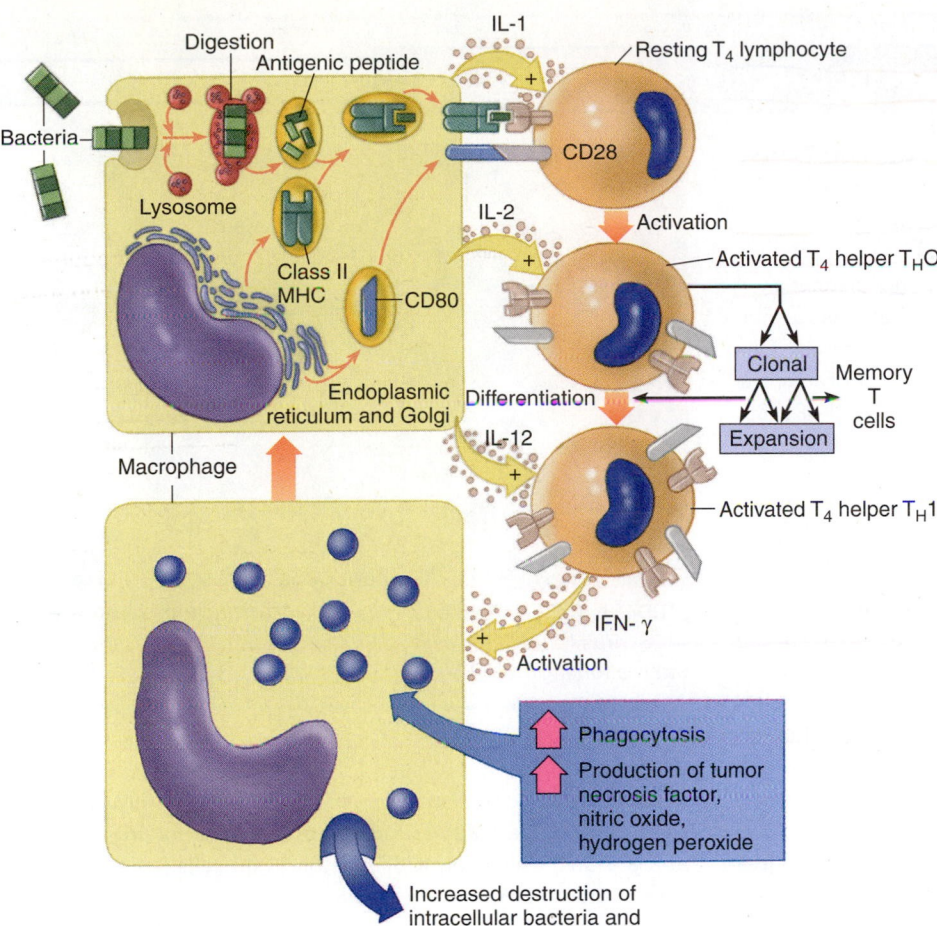

Figure 28–15

Cell-mediated immune response to bacterial infection.

ken down into peptides or epitopes (antigenic determinants), which then are combined with newly synthesized class II MHC molecules and transported to the plasma membrane for display on the surface of the membrane as MHC II-associated antigen. The macrophage presents the MHC II antigen to a competent resting CD4$^+$ cell (Fig. 28–15), which binds to the MHC II antigen. To activate the bound CD4$^+$ cell, the macrophage synthesizes a protein called CD80 and inserts it into its plasma membrane. CD80 binds to a CD4$^+$ cell surface protein (designated CD28), and together these interactions activate the bound CD4$^+$ cell. The binding also stimulates the macrophage to secrete **interleukin 1** (IL-1), a cytokine that stimulates the activated T4 cell to become a T4 helper cell designated T$_H$0. Stimulated by another lymphokine produced during the immune response, **interleukin 2** (IL-2), T$_H$0 clonally expands to produce some memory T cells and cells that are stimulated by **interleukin 12** (IL-12) to become T$_H$1 cells. CD4$^+$ T$_H$1 cells secrete IFN-γ, which activates macrophages, stimulating them to destroy ingested antigens and to present the digested antigens more efficiently on the macrophage plasma membrane.

When a virus invades the body tissues, it enters cells and moves to the infected cell's nucleus. Using the host cell's

DNA, RNA, and endoplasmic reticulum, viral proteins are produced. Some of the viral proteins are broken down into peptides (Fig. 28–16), returned to the endoplasmic reticulum, combined with class I MHC, and transported to the plasma membrane for incorporation as MHC I-associated antigen. A competent resting CD8$^+$ cell binds to the presented MHC I antigen and becomes activated when the antigen-presenting cell synthesizes CD80 protein, inserts it into its plasma membrane, and binds to CD28 of the CD8$^+$ cell. An activated T$_H$1 cell secretes IL-2, which stimulates the CD8$^+$ cell to differentiate into CD8$^+$ suppressor cells and CD8$^+$ cytotoxic cells. The cytotoxic CD8$^+$ lymphocytes bind to virally infected cells, which present class I MHC antigen and secrete cytotoxic chemicals, which disrupt the plasma membrane and kill the infected cells, thereby limiting the spread of the virus.

Suppressor T cells develop more slowly from activated CD8$^+$ cells and, with the aid of helper T cells, eventually suppress or turn off cytotoxic cells when the pathogen is no longer a threat. Suppressor T cells also act to turn off helper T cells, thereby preventing a "runaway" immune response.

(Text continued on page 864)

CURRENT CONCEPTS IN PHYSIOLOGY

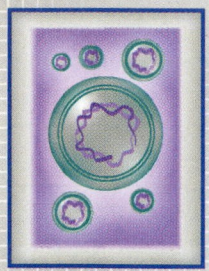

Human Immunodeficiency Virus (HIV)/Acquired Immune Deficiency Syndrome (AIDS)

Acquired immune deficiency syndrome (AIDS) is a viral disease first recognized by the Centers for Disease Control and in Atlanta in the summer of 1981. By the end of 1985, approximately 14,000 cases of HIV infection had been reported; by June 1987, the number had risen to approximately 37,000. According to recent (January 2002) estimates by the Centers for Disease Control and Prevention (CDC), approximately 800,000 to 900,000 people are currently living with HIV in the United States, with approximately 40,000 new HIV infections occurring in the United States every year. HIV infection is a worldwide epidemic, with approximately 34 million people infected with the virus. No cure has been discovered, no human vaccine has been developed that prevents infection, and the disease is fatal.

The cause of AIDS is an RNA retrovirus called the *human immunodeficiency virus* (HIV). When a retrovirus infects a host cell, the viral RNA is used as a template to create a DNA copy—in a reverse ("retro") sequence of that ordinarily used by cells. Using the host cells' enzymes, the viral DNA directs the synthesis of new viral RNA and proteins, which are assembled into new viruses. While doing so, the virus disrupts normal cell functions, ultimately causing the cell to die and release the new viruses, which then infect more cells.

As the number of viruses and infected cells increases, the rate of cell death increases, and overt symptoms of the disease appear. In the case of AIDS, the incubation period or the time between HIV infection and the appearance of symptoms may vary from several months to decades, but typically it is approximately 7 to 8 years, with death occurring 9 to 10 years after initial infection.

The principal (but not exclusive) target cells of HIV are the CD4$^+$ T cells, which play central roles in orchestrating both cell-mediated and antibody-mediated immune responses (see text). CD4$^+$ T cells bear a surface receptor, designated CD4, which HIV uses to attach to and infect the cells. A healthy adult has approximately 1000 CD4$^+$ T cells per microliter of blood. HIV infection causes the number of CD4$^+$ T cells to decline to approxi-

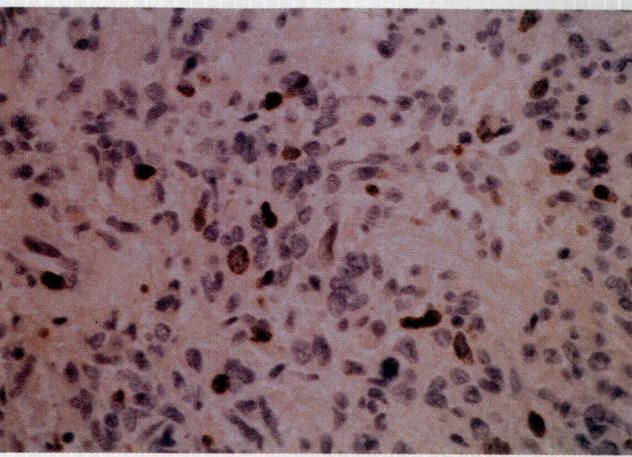

A CD4$^+$ lymphocyte called a *helper T cell* being attacked by the AIDS virus (*blue particles*). (© Boehringer Ingelheim International GmbH. Photo by L. Nilsson)

mately 800 after the first year, to approximately 400 after the seventh year. AIDS is defined as either having HIV and a CD4$^+$ T-cell count of less than 200 cells per microliter of blood, or as having HIV and an opportunistic infection normally associated with AIDS. As the CD4$^+$ T-cell population decreases, immune defenses gradually are compromised, allowing opportunistic infections or malignancies to develop and cause death.

Although HIV infects and kills CD4$^+$ T cells, this does not appear to be the sole reason for the decrease in CD4$^+$ T-cell population in patients with AIDS. It is possible that the viral infection triggers an autoimmune response in which uninfected killer T cells destroy uninfected and infected CD4$^+$ T cells. Other suggested mechanisms for the decline of the CD4$^+$ T-cell population include inhibition of CD4$^+$ T-cell clonal expansion and a triggering of CD4$^+$ T-cell apoptosis (genetically programmed cell death). Many scientists believed that HIV depleted CD4$^+$ T cells by blocking new T-cell production, but recent research (December 2001) indicates that HIV depletes CD4$^+$ T cells by accelerating the rate of CD4$^+$ T-cell death.

When the CD4$^+$ helper cell count falls to approximately 300 to 200 cells per microliter of blood, minor skin and mucous-membrane infections develop, including fungal infections of the mouth (thrush) and feet (athletes foot); whitish patches on the tongue (leukoplakia); and shingles, a viral infection of nerves and skin caused by

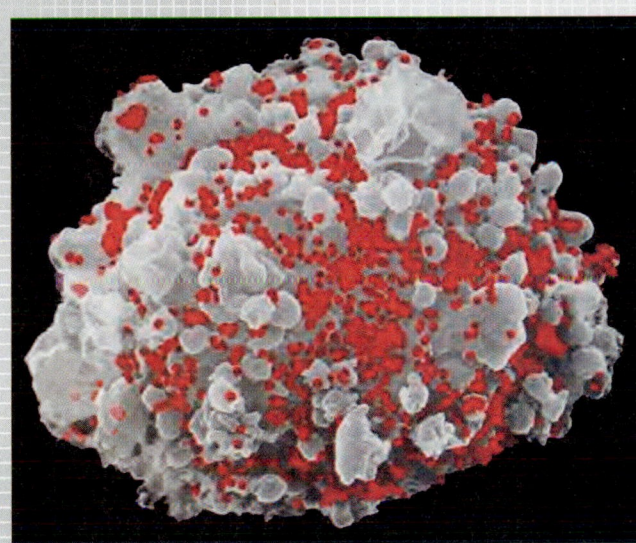

T-lymphocyte infected with the human immunodeficiency virus (HIV). The green, granular-like structures in this electron photomicrograph are the virus particles in the process of budding. *(© NIBSC/Science Photo Library)*

varicella-zoster virus (VZV). Associated symptoms include chronic, low-grade fever; fatigue; diarrhea; muscle aches; night sweats; and unexplained weight loss. Collectively, these early infections and associated symptoms are known as *AIDS-related complex* (ARC). ARC signifies the interval between initial HIV infection and the development of more severe AIDS-defining infections.

When the $CD4^+$ helper cell count falls to approximately 200 cells per microliter of blood (≈8 years after initial HIV infection), fatal opportunistic infections develop, including *Pneumocystis carinii* pneumonia (a protozoan-caused pneumonia), toxoplasmosis (a parasitic infection of the brain), cryptococcal meningitis (a fungal inflammation of the brain), and a relatively rare form of skin cancer called *Kaposi's sarcoma*. Shortness of breath; dry cough; sore throat; sharp chest pains; enlarged lymph nodes in the neck, axilla, and groin; fever; and the appearance of painless purple or brownish spots on the skin or mucous membranes of the mouth and rectum may accompany the infections.

The AIDS virus does not appear to penetrate intact skin or membranes. No epidemiological evidence suggests that insects, air, water, food, or casual contact spreads AIDS. The common modes of transmission are sexual contact (homosexual or heterosexual) and blood contact, such as by infusion of infected blood, use of infected syringes, spattering of infected blood on skin cuts or abrasions, or transmission from maternal to fetal blood.

Current AIDS research is being concentrated in three major endeavors: (1) the development of a vaccine that will prevent HIV infections; (2) the development of drugs that will suppress or cure HIV infection; and (3) the development of drugs and therapies that will suppress or cure AIDS-related infections, thereby improving survival time and the quality of the life of patients with AIDS.

The three types of drugs that fight the HIV are (1) *nucleoside reverse transcriptase inhibitors*, such as azidothymidine (AZT) and dideoxycytidine (ddC); (2) *nonnucleoside reverse transcriptase inhibitors*, such as dideoxyinosine (ddI); and (3) *protease inhibitors*. AZT, ddC, and ddI block reverse transcription by hindering reverse transcriptase, the enzyme required to produce DNA from the viral RNA. Unfortunately, if these drugs are used alone, their effectiveness is short-lived because HIV mutates, producing variant forms of reverse transcriptase that are able to bypass the blocking drugs. Most people with HIV infection are on combination therapy called *highly active antiretroviral therapy* (HAART), which involves two or more drugs. Current U.S. guidelines advise that HIV-infected patients begin treatment when $CD4^+$ T-cell counts drop below 350 cells per microliter of blood, or plasma HIV RNA levels exceed 30,000 copies per mL. Recent studies (December 2001) have shown that HIV infection leads to a growing subpopulation of rapidly proliferating $CD4^+$ T cells that quickly die out. Treatment with HAART decreases the size of this subpopulation, thereby slowing $CD4^+$ T-cell loss. HAART does not cure HIV/AIDS, but it does help to keep the $CD4^+$ T-cell count higher and the HIV count lower, improving the person's condition.

References

Kovacs, J. A., et al. "Identification of dynamically distinct subpopulations of T lymphocytes that are differentially affected by HIV." *Journal of Experimental Medicine*, 194:1731–1741, 2001.

Mohri, H. et al. "Increased turnover of T lymphocytes in HIV-1 infection and its reduction by antiretroviral therapy." *Journal of Experimental Medicine*, 194:1277–1287, 2001.

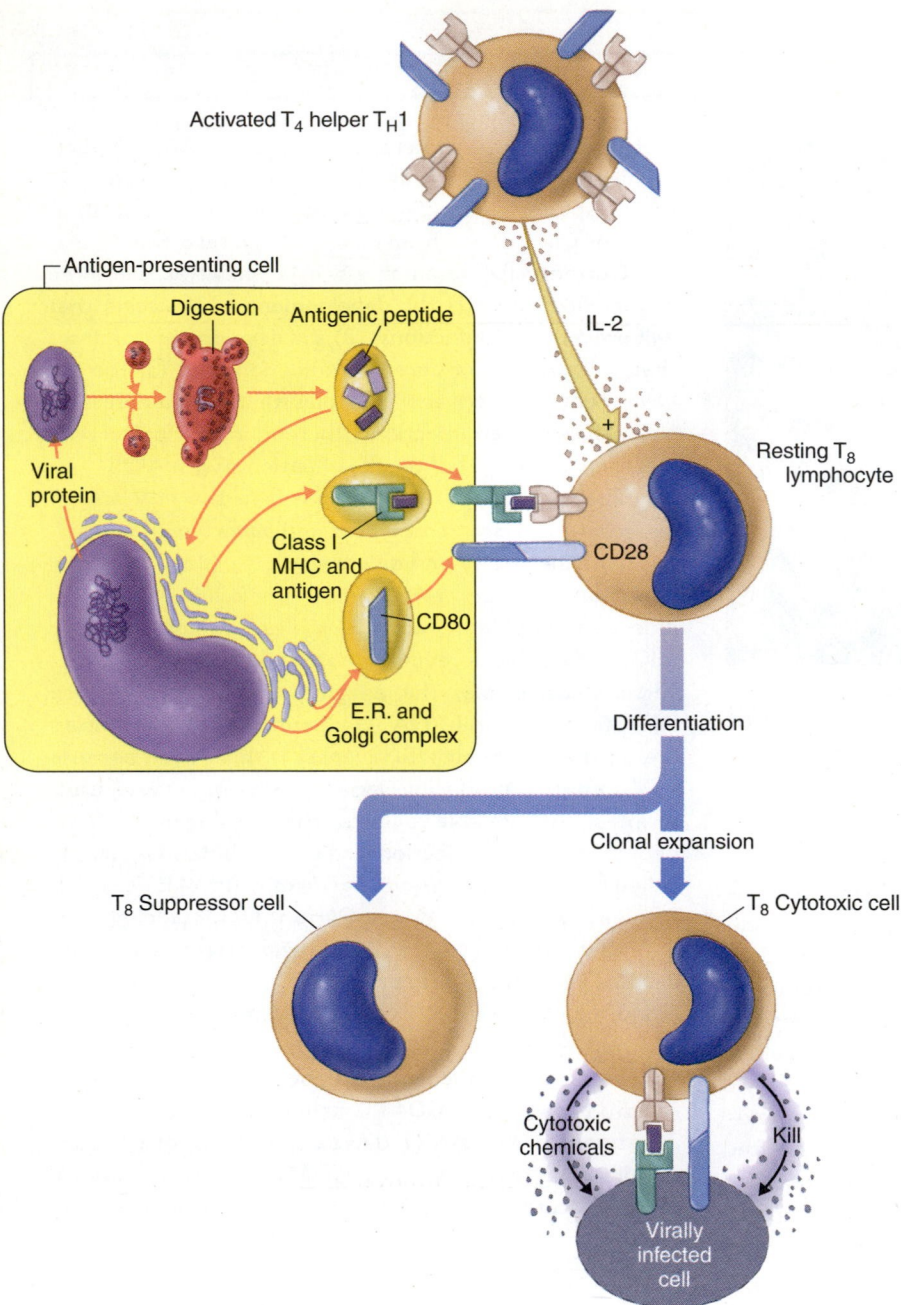

Figure 28–16
Cell-mediated immune response to viral infection.

Activated T$_4$ helper T$_H$1

IL-2

Antigen-presenting cell

Digestion   Antigenic peptide

Viral protein

Class I MHC and antigen

CD80

E.R. and Golgi complex

+

Resting T$_8$ lymphocyte

CD28

Differentiation

Clonal expansion

T$_8$ Suppressor cell

T$_8$ Cytotoxic cell

Cytotoxic chemicals

Kill

Virally infected cell

Following suppression of the cell-mediated immune response, some of the CD4$^+$ cells and some of the CD8$^+$ cells remain and function as memory T cells, which are capable of initiating an accelerated response to subsequent encounters with the antigen.

We have seen that the entire cell-mediated immune response is orchestrated by the CD4$^+$ cells; in fact, a deficiency in CD4$^+$ cells caused by viral infection is ultimately responsible for death in patients with AIDS. CD4$^+$ cells also are required to initiate antibody-mediated immune responses.

Antibodies

What are antibodies and how do they contribute to the body's defense?

Antibodies are glycoproteins called **immunoglobulins** that are produced and secreted by specialized B-lymphocytes in response to specific antigens. The basic unit of all immunoglobulins is a structure consisting of two identical **light polypeptide chains** and two identical **heavy polypeptide chains,** stabilized and linked by disulfide (—s—s—) bonds

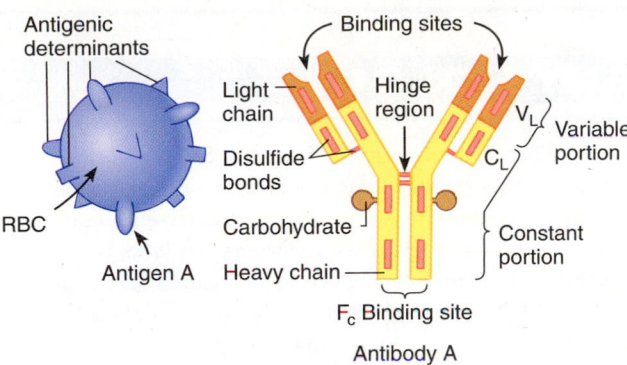

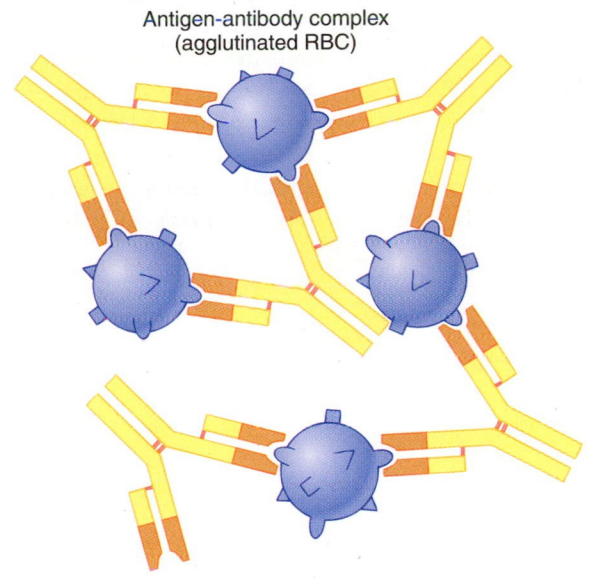

Figure 28–17

General structure of an antibody and agglutination of red cells.

between cysteines to form a Y-shaped molecule (Fig. 28–17). The light (smaller) chains exist in two forms, called kappa and lambda. The heavy (larger) chains exist in five major forms (alpha, gamma, delta, epsilon, and mu) and are used to designate the class of antibody (e.g., IgA contains two alpha heavy chains, IgD contains two delta heavy chains, and so on). The heavy chains also contribute significantly to the functional character of that antibody class. Both heavy chains and light chains have a constant portion (C_H and C_L, respectively) and a variable portion of (V_H and V_L, respectively).

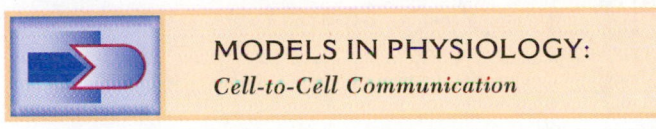

MODELS IN PHYSIOLOGY:
Cell-to-Cell Communication

Various constant regions of the heavy chains bind complement proteins and thus can initiate complement reactions. They also bind to receptors on macrophages, monocytes, neutrophils, and natural killer cells.

The variable portions of the heavy and light chains also contain the antigen-specific binding sites, one on each arm of the Y-shaped molecule. A typical antibody molecule therefore can combine with two of the same kind of antigens, forming an antigen-antibody complex (see Fig. 28–17), as well as bind to receptors on cells such as macrophages, which in turn ingest and destroy the immune complex.

Immunoglobulins are grouped into five classes (and several subclasses) according to the structure of the heavy chains. Using **Ig** as an abbreviation for immunoglobulin, the five major classes are designated IgG, IgA, IgM, IgD, and IgE.

IgG is the major immunoglobulin in normal blood and accounts for 70% to 75% of all immunoglobulins present. It contributes immunity against many kinds of pathogens, including bacteria, viruses, and fungi, as well as chemical defense against toxins. Anti-Rh antibodies are IgG immunoglobulins. Usually, IgG molecules are distributed evenly between the blood and extravascular fluids.

IgA immunoglobulins represent about 15% to 20% of the total immunoglobulin pool. They are the principal antibodies found in various secretions, such as mucus (respiratory and digestive tract), saliva, milk, tears, and fluids secreted into the genitourinary and digestive tracts. IgA antibodies provide defense against pathogens that contact the body surface or are ingested or inhaled.

IgM immunoglobulins account for approximately 10% of the body's immunoglobulins. Almost all IgM molecules are in the blood. Their primary importance is in bacterial defense, working in conjunction with complement and blood phagocytes. IgM molecules are also the antibodies that are used to characterize ABO blood type (e.g., anti-A, anti-B, and so on).

IgD immunoglobulins make up less than 1% of the blood immunoglobulins but are present on the plasma membranes of many circulating B-lymphocytes. IgD antibodies are believed to be involved in antigen-triggered lymphocyte differentiation—that is, the development of plasma cells and memory cells from B-lymphocytes.

IgE immunoglobulins rarely are found as free circulating antibodies but commonly are found on the surface of the plasma membrane of basophils and mast cells of connective tissue. When engaged by an antigen, IgE molecules stimulate basophils and mast cells to release histamine and other chemicals that mediate the local, often allergic, response to the presence of an antigen. Thus, IgE molecules are involved directly in diseases characterized by hypersensitivity, an exaggerated immune response to certain antigens; asthma and hay fever are well-known examples.

IgE molecules also play an important role in defense against helminthic parasites (worms). For example, IgE binds to mast cells, stimulating the mast cell to secrete eosinophilic chemotactic factor A, which attracts eosinophils to the infested area to destroy the parasites.

(Text continued on page 868)

APPLICATIONS OF PHYSIOLOGY

Hypersensitivity

Generally, the word *hypersensitive* means overreactive, that is, having an excessive or exaggerated response to a stimulus to which most persons react in a more subdued manner. The immunological definition of *hypersensitivity* is similar: an exaggerated adaptive immune response to an antigen, resulting in tissue damage.

Hypersensitivity begins with initial exposure to an antigen and stimulation of the body's normal adaptive immune response to the antigen, including activation of cell-mediated and antibody-mediated defense mechanisms. This initial "gearing-up" of defenses against the particular antigen is called "sensitization." In a susceptible person, hypersensitivity reactions occur when a second or subsequent exposure to the antigen evokes an excessive immune response, which often includes inflammation and tissue damage. Hypersensitivity reactions are classified into four types (types I-IV). These reactions may occur singly or in multiple and are characteristic of the individual.

Type I hypersensitivity (*immediate hypersensitivity*) occurs after an initial exposure to an antigen (often an innocuous antigen, such as pollen) stimulates IgE production. Large numbers of IgE antibodies bind to plasma membrane receptors on basophils and mast cells during the sensitization process. Exposure to the same antigen at a later date cross-links the bound IgE molecules, thereby stimulating these cells to degranulate, releasing histamine (a potent

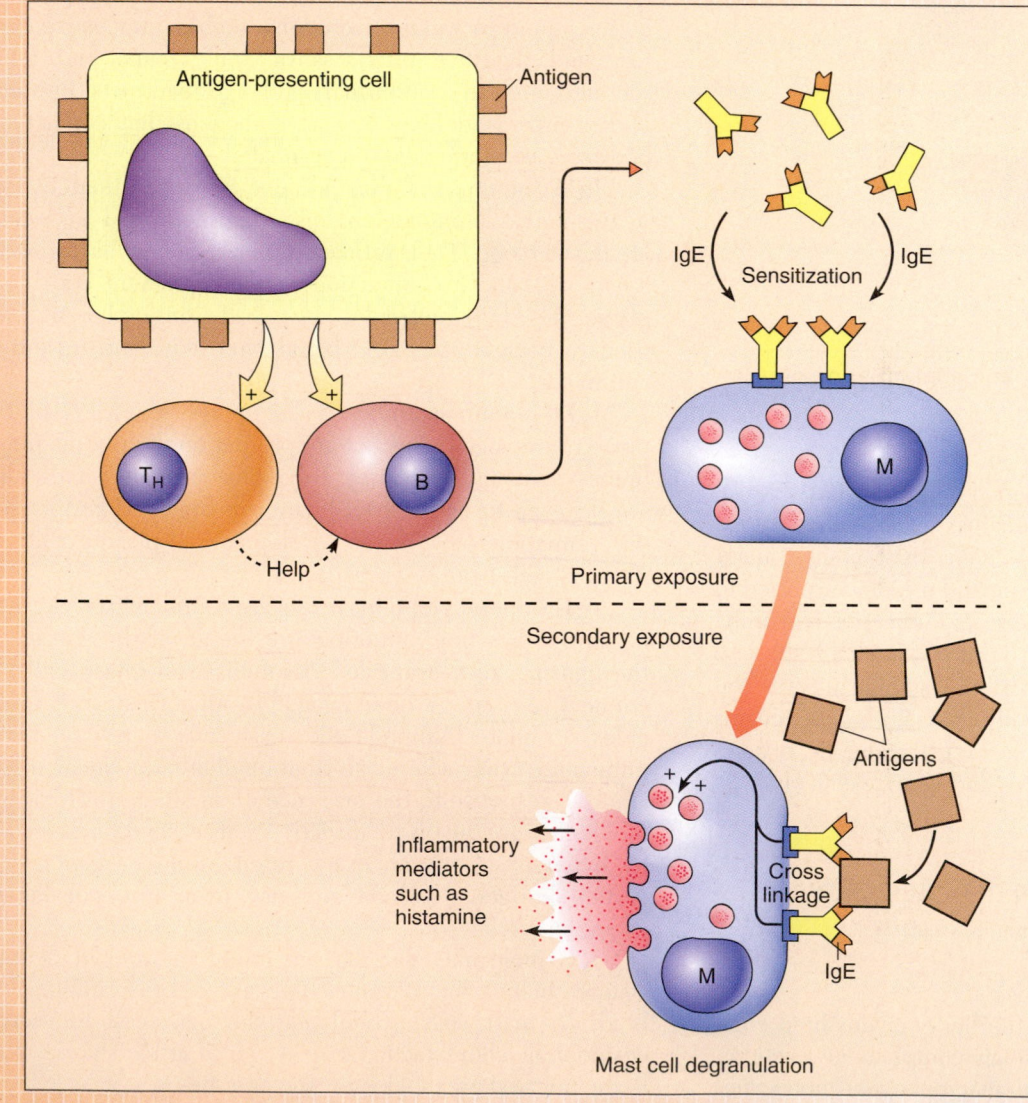

Figure 1. Type I (immediate) hypersensitivity.

local vasodilator), leukocyte chemotactic factors, and other chemicals which promote local acute inflammation (Fig. 1). Symptoms of local acute inflammation include swelling in the affected area (raised welts or hives on the skin, congestion in nasal passages and airways, and so on), redness and increased local temperature of the skin, itching, watery eyes, and sneezing (see text Fig. 28–11). Because the inflammatory response to the particular antigen occurs within minutes of the secondary exposure, the response is termed *immediate hypersensitivity*. Type I hypersensitivity includes several kinds of allergies (type I responses to pollen, dust mite feces, animal dander, and so on), asthma, and rhinitis.

If exposure to the antigen occurs at the body surface (skin), respiratory epithelium, gastrointestinal mucosa, and so on, the response may remain localized, but if the antigen enters the circulatory system, the response is widespread and could be dangerous. In *anaphylaxis* (*ana* = "throughout," *phylax* = "guard") circulating antigens of the particular type bind to mast cells and basophils throughout the body, producing systemic effects within a matter of minutes. Hives and intense itching of the skin all over the body, accompanied by swelling and increased skin temperature, are cardinal signs of anaphylaxis. In severe cases, increased mucus secretion and contraction of smooth muscles along

the airways restrict airflow and make breathing more difficult. Extensive peripheral vasodilation accelerates heat loss and causes blood pressure to fall, leading toward circulatory collapse and death. These more severe effects of the immune response characterize anaphylactic shock.

Type II hypersensitivity (*cytotoxic hypersensitivity*) is antibody dependent and occurs when IgG and IgM antibody bound to a cytotoxic cell also binds to antigen on the surface of a target cell, thereby initiating cytotoxic cell destruction of the target cell (Fig. 2). The cytotoxic cell is usually a $CD8^+$ lymphocyte but may be a monocyte, eosinophil, neutrophil, B-lymphocyte, or natural killer cell. Type II reaction also occurs when soluble antibody binds to cell surface antigen and initiates complement-mediated lysis of the target cell. Inflammation occurs when the destroyed target cells release inflammatory chemicals. Examples of type II hypersensitivity include immune reactions against blood cells (transfusion reactions, hemolytic disease of the newborn, and so on) and graft rejection by a transplant recipient.

Type III hypersensitivity (*immune-complex hypersensitivity*) develops when large numbers of immune complexes form faster than they can be cleared by phagocytes. The excess immune complexes are deposited in tissues and activate complement (Fig. 3). Complement proteins C3a and C5a attract neutrophils that attempt to phagocytize the immune complexes, fail, and release digestive enzymes by exocytosis, causing tissue damage. Activated complement

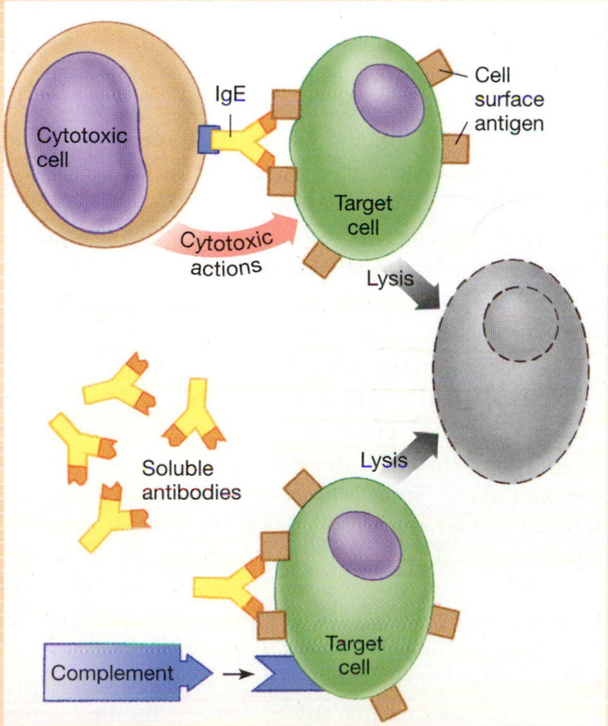

Figure 2. Type II (cytotoxic) hypersensitivity.

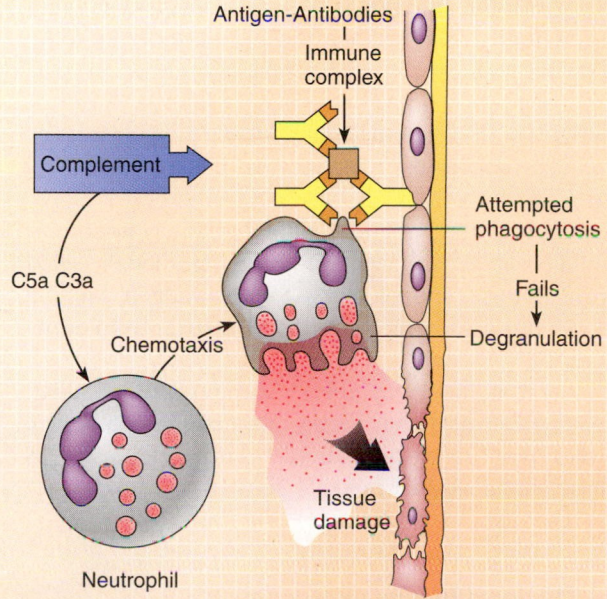

Figure 3. Type III (immune-complex) hypersensitivity.

(continued)

Hypersensitivity *(continued)*

also stimulates histamine release from mast cells and basophils, increasing blood flow, capillary permeability, and inflammation. Systemic lupus erythematosus, an autoimmune disease, and farmers lung disease, characterized by repetitive exposure and an allergic response to fungal spores in hay, are typical of type III hypersensitivity.

Type IV hypersensitivity (*delayed hypersensitivity*) occurs when antigen-sensitized T cells release vasoactive and chemotactic lymphokines following secondary contact with the same antigen. The lymphokines induce local inflammatory reactions and attract and activate macrophages that amplify the inflammatory response by releasing additional inflammatory mediators (Fig. 4). The response takes at least 24 hours to reach maximal intensity and therefore is called *delayed hypersensitivity*. Type IV reactions are seen in response to graft rejection by a transplant recipient and in several bacterial (e.g., tuberculosis) and viral diseases.

In many instances, the symptoms of hypersensitivity reactions can be prevented or at least minimized. For example, a person allergic to a certain food can avoid the food and thus the allergic response.

Suggest another method of preventing many of the symptoms of type I hypersensitivity.

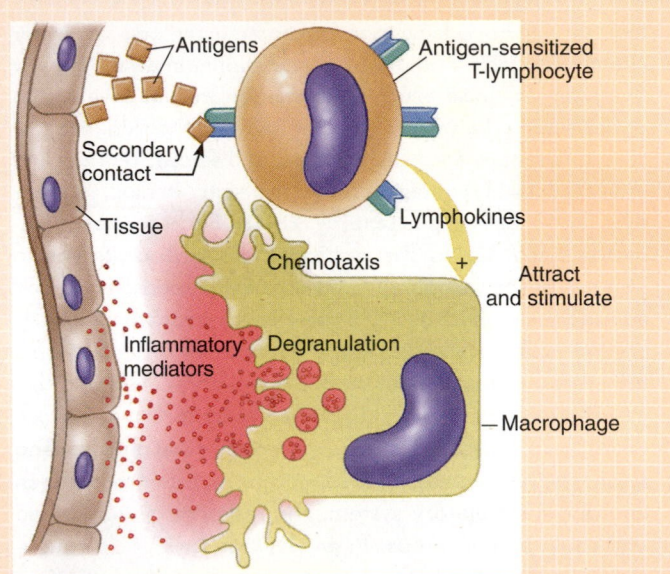

Figure 4. Type IV (delayed) hypersensitivity.

All antibodies work by first combining with the specific antigen for which the antibody is designed. The formation of an antigen-antibody complex then may result in one or more of the following:

1. The antigen-antibody complex stimulates phagocytosis by neutrophils, monocytes, lymphocytes, macrophages, and other phagocytic cells.
2. In reacting with the antigen, the antibody may neutralize the antigen's ability to harm. Some IgG antibodies called *antitoxins*, for example, block the toxin's (antigen) binding to cell receptors, thereby preventing the toxin from exerting its effect.
3. The antigen-antibody complex activates complement, which in turn attracts phagocytes, opsonizes the pathogen, perforates membranes, and stimulates other defense cells to release chemicals that mediate general and specific defensive responses to the antigen.

Antibody-Mediated Immune Response

The B-lymphocytes, with assistance from T4 helper cells, are responsible for the antibody-mediated immune response. As with the T-lymphocytes, B-lymphocytes are able to recognize and bind to a specific antigen once they become competent,

and, as with T-lymphocytes, thousands of different kinds of B-lymphocytes exist. In contrast to T-lymphocytes, most B-lymphocytes develop competency in the bone marrow before entering the blood and lymph. As an example of the antibody-mediated immune response, let us look again at the invasion of the body by a pathogenic bacterium (Fig. 28–18). A macrophage recognizes the bacterial antigen, binds to it, phagocytizes the bacterium, chemically removes the antigen, and displays the bacterial antigen along with class II MHC protein on its own plasma membrane. A competent CD4+ cell binds to the antigen-MHC II complex on the antigen-presenting cell and is activated when its CD28 protein binds to CD80 of the macrophage. IL-1, secreted by the macrophage, stimulates the CD4+ cell to develop into a helper cell (T_H0). **Interleukin 4** causes further differentiation of the helper cell into T_H2, a CD4+ helper cell that mediates the antibody-related immune response.

A competent B-lymphocyte binds to an antigen, processes it internally, and displays it with MHC II on its surface. A T_H2 cell binds to the MHC II antigen and synthesizes a protein called CD40 ligand, which binds to CD40 protein of the antigen-presenting B-lymphocyte, thereby activating the B cell. Interleukins 4, 5, and 10 stimulate the activated B-lymphocyte to differentiate into and clone plasma cells and memory B cells.

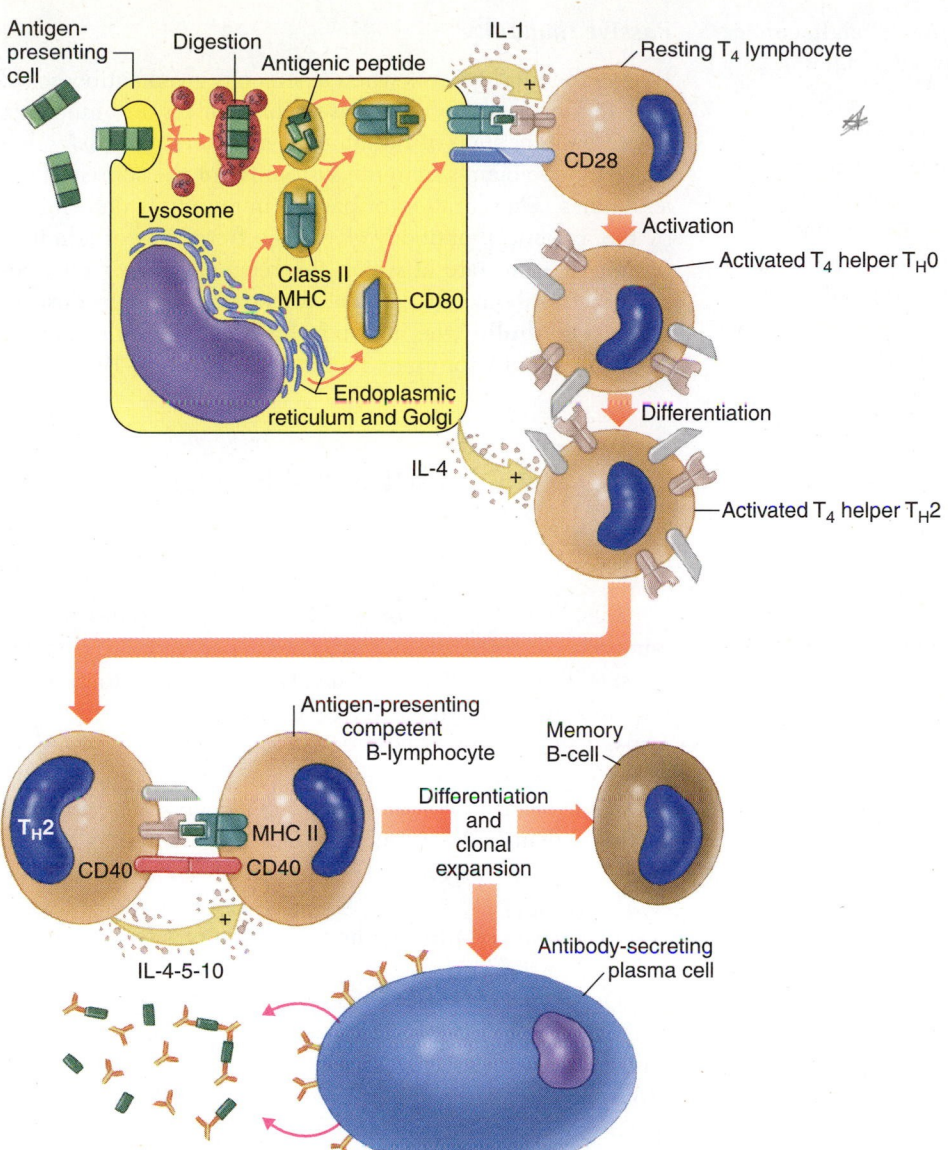

Figure 28–18

Antibody-mediated immune
response to bacterial infection.

Plasma cells are enlarged B-lymphocytes with a highly developed and extensive endoplasmic reticulum. Plasma cells remain in the lymphatic system and secrete antibody molecules that are chemically identical to the antigen receptors and thus specific for the antigen. The antibodies travel via the lymph and blood to the invasion site, where they combine with the antigen, thereby marking it for destruction or reducing its capacity to cause disease.

Memory B cells function as do memory T cells; they permit a rapid mobilization of the immune response if subsequent exposure to the same antigen occurs. Memory B cells secrete small amounts of antibodies for several years after the initial infection is overcome. These antibodies circulate as part of the gamma globulin proteins in plasma and are available for immediate defense should infection occur again. Recurrence of infection also activates memory T cells, which in turn help memory B cells form a new clone of antibody-

secreting plasma cells. Memory B and memory T cells form the basis of lasting immunity and often prevent us from suffering twice from the effects of the same pathogen: We may contract measles (a viral disease) as children, for example, but it is uncommon for us to suffer from it again as adults.

Acquired Immunity

 How is immunity acquired, and how does active immunity differ from passive immunity?

The immune system can be manipulated to prevent a person from becoming ill with disease or to lessen the severity of the disease and speed recovery. *Immunization,* a process of conferring immunity against disease, is an example. *Passive immunity* confers temporary, short-term protection. *Active*

immunity confers long-lasting, continuing protection. In either case, the protection conferred is called *acquired immunity*.

Active Immunity

Active immunization procedures involve giving an antigen to a recipient to stimulate the person's immune system so that an immune response is provoked and resistance to the antigen builds up. Usually, the antigen is administered in the form of a vaccine by a process called *vaccination*. Pathogenic bacteria and viruses that have been treated in such a way as to be no longer virulent (able to cause disease) yet retain their antigenic properties often are used in vaccines. Such pathogens are said to be *attenuated*. The vaccines for tuberculosis, measles, and mumps are examples of vaccines containing attenuated organisms. Recombinant DNA technology has been used to mass-produce antigenic proteins that can be incorporated into a vaccine. One of the vaccines for hepatitis B (an inflammatory, sometimes fatal liver disease caused by a virus) is a recombinant type of vaccine.

Antibody titer, a measure of the amount of a specific antibody present in a sample of serum, can be determined periodically in order to follow the development of active immunity after antigen exposure. After first exposure to an antigen, the antibody titer rises and then falls within the first six weeks (Fig. 28–19). This is termed the **primary response.** If antigen is given again during the decline of the primary response, a **secondary response** may occur, in which the antibody titer rises more rapidly to a level higher than that before. In general, the decline of the secondary response is much slower than for the primary response, conferring a longer-lasting immunity. A third, "booster" vaccination during decline of the secondary response elevates the antibody titer higher than achieved with the second "booster" and usually confers prolonged immunity against the disease caused by the antigen-bearing organism.

Passive Immunity

Passive immunity occurs when a person is given antibodies to help him or her combat disease. This confers an immediate but temporary state of immunity that is termed *passive* because the recipient's immune system is not the source of the antibodies. The duration of passive immunity is determined by the amount of antibody given, the frequency of administration, and the rate at which the recipient's body metabolizes the antibody. In general, for adults, a single dose of **gamma globulin** (the antibody fraction of serum) confers passive immunity for up to 3 months (see Fig. 28–19).

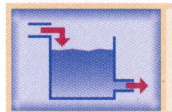

MODELS IN PHYSIOLOGY:
Conservation of Mass

Usually, the source of antibodies used to confer passive immunity is another human who has active immunity against the specific disease. Occasionally, horses are used to produce immunoglobulins, which can be harvested and given to humans. Examples of equine antisera include those used to treat people exposed to rabies, tetanus, and botulism. Because horse serum contains proteins recognized as antigens by the human body, a recipient of equine immunoglobulin may develop an immune response against the serum and become ill, especially after a repeated dose of the serum is given a few weeks after the first. The illness is termed *serum sickness*.

Development of techniques for harvesting human monoclonal antibodies could eliminate the use of equine immunoglobulins for conferring passive immunity in humans. **Monoclonal antibodies** are identical antibodies against one specific antigen, and they are produced by plasma cells (hy-

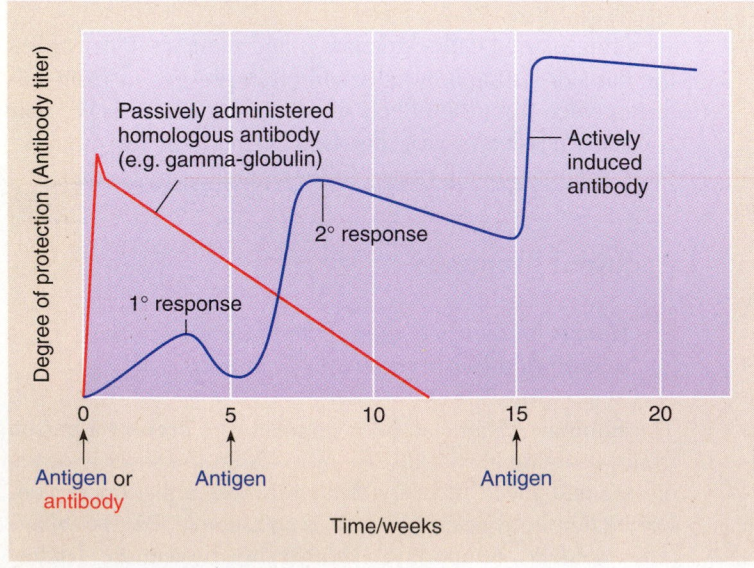

Figure 28–19

Passive versus active immunity.

bridomas) cloned from the same B lymphocyte. At present, monoclonal antibodies are available and in use for diagnostic purposes, such as pregnancy testing, but they usually are derived from mice and therefore could cause serum sickness if used to confer passive immunity.

An important kind of passive immunity naturally occurs in newborn infants as a result of antibodies that have crossed the placenta from the mother's blood. Although this passive immunity lasts for just a few months, it is important in helping infants resist infection until their immune systems become more capable. Breast-feeding prolongs passive immunity because the infant receives antibodies in the mother's milk.

Recognition of Self Versus Nonself

All specific defenses and many nonspecific defenses require the body to be able to recognize substances that are foreign—that is, chemicals not usually present in or a part of the body. The ability to recognize "self" and "nonself" is possible because each organism is chemically unique: No two organisms, even though they may be individuals of the same species, are chemically identical. Plasma membranes of human and other animal cells, as well as cell walls of bacteria and other forms of life, contain surface molecules of protein, glycolipid, and glycoprotein that allow the cell to be recognized as "foreign" (nonself) or "self." A molecule, or part of a

---

TABLE 28–2

Antibody Targets in Autoimmune Diseases

Organ-Specific Diseases

Organ	Disease	Antibody Target
Skeletal muscle	Myasthenia gravis	Acetylcholine receptors
Thyroid	Graves' disease	TSH receptor
	Thyroiditis	Thyroglobulin, thyroid peroxidase
Pancreas	Type I diabetes	Islet cells, insulin, GAD, ICA69, GM2-1
	Insulin-resistant diabetes with acanthosis nigricans	Insulin receptor
Adrenal	Addison's disease	Adrenal cell 17α hydroxylase
Parathyroid	Idiopathic hypoparathyroidism	Parathyroid cells
Gastrointestinal tract	Pernicious anemia	Parietal cell K^+, H^+-ATPase, intrinsic factor
Biliary tract	Primary biliary cirrhosis	Mitochondrial dihydrolipoyl acetyltransferase complex
Liver	Chronic active hepatitis	Cytochrome P450
Skin	Pemphigus	Keratinocyte surface antigens
	Vitiligo	Melanocytes
Blood	Autoimmune hemolytic anemia	Erythrocyte antigens
	Idiopathic thrombocytopenic purpura	Platelet membrane glycoproteins
Collagen-vascular system	Ménière's disease	Type II collagen

Systemic Diseases

Organ	Disease	Antibody Target
	Goodpasture's syndrome	Basement membrane, type IV collagen
	Rheumatoid arthritis	Gamma globulin
	Sjögren's syndrome	Ribonuclear proteins, gamma globulin
	Systemic lupus erythematosus	Nucleosome, spliceosomal complex, Ro La particle
	Polymyositis	Aminoacyl-tRNA synthases, tRNA, EF-1, nuclear MI-2
	Rheumatic fever	Myocardium, heart valves, choroid plexus

Source: From George Eisenbarth and Donald Bellgrau, "Autoimmunity," in *Science & Medicine*, 1(2):May/June 1994.

molecule, that is recognized by the body as foreign is called an *antigen* because it activates specific defense mechanisms geared toward its destruction or removal.

Autoimmunity occurs when lymphocytes and antibodies called *autoantibodies* react with normal components of the body. *Autoimmune* literally means "to protect the body against itself." An autoimmune response arises from, and is directed against, an individual's own tissues. Most often, autoimmunity is thought of solely in terms of disease, in which the immune system attacks normal tissues as though they were foreign. But normal or physiological autoimmune responses are also part of the body's daily maintenance of homeostasis. A good example of physiological autoimmunity is the mechanism whereby senescent erythrocytes are removed from circulating blood and destroyed.

Red blood cells normally live approximately 125 days before being removed from circulation and destroyed, usually within the spleen. When a red blood cell nears the end of its life span, changes occur in some of the proteins on its plasma membrane. The changes expose a binding site for an IgG molecule that binds to the red blood cell and induces its phagocytosis by a macrophage in the spleen. The macrophage destroys the aged red blood cell and recycles some of its constituents. This kind of autoimmune response to aged red blood cells causes the destruction of approximately 2.5 billion red blood cells each day.

Autoimmune disease occurs when there is a breakdown in self tolerance—that is, when self proteins that normally are not recognized as being foreign (one could say they are "tolerated") begin to be treated as though they were foreign. The immune system mounts an attack, resulting in inflammation and destruction of affected tissues. The disease may be widespread and involve many organs, such as in systemic lupus erythematosus, or it may be specific for a single organ, such as type 1 diabetes mellitus (Table 28–2). The specific molecular events that trigger autoimmune disease remain unclear but may include events such as exposure to toxic chemicals, infection by certain pathogens, and chronic physical and emotional stress. There appears to be a genetic predisposition for developing some of the autoimmune disorders.

As discussed earlier, maturation of B-lymphocytes and T-lymphocytes involves a process of negative selection, whereby many autoreactive lymphocytes are killed prior to release, and a positive selection process in which lymphocytes that do not react to self antigens are permitted to mature and enter circulation. In fact, some autoreactive cells (a very small number) do escape and enter circulation; however, they do not respond to self antigens unless a triggering event occurs. Possible inducers of autoimmune disease include loss of suppressor T cells, which suppress autoreactive lymphocytes, and viral infections that activate autoreactive T cells or directly infect B cells, stimulating B-cell proliferation and polyclonal antibody production—the antibodies reacting with self antigens to cause disease.

The development and the severity of autoimmune disease is known to be determined in part by the type of MHC molecules present in the affected individual. Class II MHC molecules exist in many forms; the type present in an individual is determined genetically. The type of class II MHC molecules present determines which peptides (part of an antigenic protein) can be bound and presented to TCRs. Autoimmune disease may occur when an individual inherits genes that code for a class II MHC protein that allows presentation of self antigens to activated T cells.

CHAPTER REVIEW

Summary

- The lymphatic system functions to return filtered proteins and interstitial fluid to the blood, to transport absorbed lipids from the intestine to the blood, and to defend the body against pathogens.
- Normal lymph is usually a clear, light-amber fluid similar in composition to plasma and interstitial fluid.
- Lymphoid tissue contains fixed reticular cells and fibers, and free cells, such as T cells, B cells, granulocytes, and macrophages.
- Primary lymphoid organs, where lymphocytes initially develop and mature, include red bone marrow, the thymus gland, and Peyer's patches.
- Secondary lymphoid tissue and organs, where lymphocytes mature, reside, and encounter antigens, include lymph nodes and the spleen.

- *Internal defense* refers to physiological mechanisms that defend the body against toxins and pathogens.
- Nonspecific defenses are general in nature and deter a wide variety of pathogens. These defenses include surface-protective mechanisms, interferons, complement, and processes of inflammation and phagocytosis.
- Surface-protective mechanisms include physical and chemical barriers against pathogen invasion, such as the skin; sweat; sebum; saliva; and the linings of the gastrointestinal, urinary, respiratory, and reproductive tracts.
- Interferons are antiviral proteins that help prevent the spread of viruses and stimulate leukocytes that destroy virally infected cells. The two major types are type 1 (alpha, beta) and type 2 (gamma) interferons.

- Complement is a set of proteins in the blood that interact to promote lysis of microbes and signal leukocytes for defense.
- Inflammation is a localized response to an invading microbe that signals active defense; it is characterized by redness, heat, edema, and pain.
- Phagocytosis is a process whereby a phagocyte ingests and destroys toxins, invading microbes, foreign matter, and immune complexes. Neutrophils, monocytes, and macrophages are phagocytes.
- Specific defenses are geared toward identification and destruction of specific pathogens. Two types of specific defense responses are cell-mediated and antibody-mediated responses.
- Cell-mediated immune responses involve coordinated efforts of antigen-presenting cells (macrophages and B-lymphocytes) and two kinds of T-lymphocytes: CD4$^+$ (T4 cells) and CD8$^+$ (T8 cells). Coordination between these cells involves cytokines, such as interleukins and interferons.
- Antibodies are proteins produced and released by B-lymphocytes called *plasma cells* in response to the presence of an antigen. Antibodies are usually specific for a given antigen and help to destroy the antigen's harmful effect.

- Antibody-mediated immune responses involve coordinated efforts of T4 cells, antigen presenting cells, and B-lymphocytes. Coordination involves cytokines and results in the clonal expansion of plasma cells that secrete antibody specific to the antigen present.
- Immunization is a process of conferring immunity against disease. Active immunity occurs when an individual is given an antigen that stimulates specific antibody production by plasma cells. Repeat doses elevate antibody titer and confer long-lasting immunity against the disease. Passive immunity occurs when a person is given antibodies to help him or her combat disease or minimize infection. This form of protection is short-lived.
- Internal defense mechanisms allow for recognition of self versus nonself.
- *Autoimmunity* refers to physiological and pathological processes in which self proteins are recognized by the immune system and attacked. An example of physiological autoimmunity is the recognition and destruction of aged red blood cells. An example of pathological autoimmunity (autoimmune disease) is myasthenia gravis, a disease of the neuromuscular junction in which cholinergic receptors are destroyed.

Review Questions

Choose the Correct Answer

1. The largest leukocyte cell is part of the body's macrophage defense system. It is the:
   a. eosinophil.
   b. basophil.
   c. neutrophil.
   d. monocyte.
   e. lymphocyte.
2. CD4$^+$ helper cells:
   a. act as memory cells for antibody production.
   b. suppress the immune response.
   c. are involved in both cell-mediated and antibody-mediated immune responses.
   d. recognize antigen-MHC I.
   e. Two of the preceding
3. The following cell recognizes an antigen presented in combination with MHC II:
   a. cytotoxic cell.
   b. suppressor cell.
   c. neutrophil cell.
   d. competent CD4$^+$ cell.
   e. basophil.
4. The lymphocyte most responsible for orchestrating the body's immune response is the:
   a. CD8$^+$ lymphocyte.
   b. plasma cell.
   c. CD4$^+$ lymphocyte.
   d. memory cell.
   e. suppressor cell.
5. Which cell secretes interleukins that stimulate B-lymphocyte development?
   a. T$_H$ cell
   b. CD8$^+$ cytotoxic cell
   c. Liver macrophage

   d. Neutrophil
   e. Basophil
6. The principal antibody in circulation, it attacks microorganisms and their toxins. It is class:
   a. IgE.
   b. IgD.
   c. IgG.
   d. IgM.
   e. IgA.
7. The type of immunity conferred by intravenous infusion of gamma globulins (antibodies) is:
   a. active.
   b. passive.
   c. cell-mediated.
   d. autoimmunity.
   e. secondary.
8. The term describing the movement of a leukocyte through a thin-walled blood vessel is:
   a. chemotaxis.
   b. inflammation.
   c. diapedesis.
   d. cytokinesis.
   e. leukopoiesis.
9. Antibodies against the red cell antigens A and B are ____ immunoglobulins.
   a. IgA
   b. IgG
   c. IgD
   d. IgE
   e. IgM
10. Complement proteins that coat the surface of a microbe, thereby marking it for phagocytosis, are called:
    a. interleukins.
    b. cytokines.

c. monokines.
d. agglutinins.
e. opsonins.

11. In Graves' disease, autoantibodies bind to:
    a. T_3 receptors.
    b. acetylcholine receptors.
    c. TSH receptors.
    d. T_4 receptors.
    e. Two of the preceding

12. The effectiveness of both antibody-mediated immune responses and cell-mediated immune responses to the presence of an antigen is most dependent on:
    a. eosinophils.
    b. $CD4^+$ lymphocytes (T4 cells).
    c. $CD8^+$ lymphocytes (T8 cells).
    d. plasma cells.
    e. basophils.

13. Cytotoxic chemical production is the primary function of which cell?
    a. Erythrocyte
    b. Eosinophil
    c. T8 cell ($CD8^+$)
    d. T4 cell ($CD4^+$)
    e. B-lymphocyte

14. The lymphocyte most responsible for initiating the body's secondary immune response is the:
    a. T8 lymphocyte.
    b. plasma cell.
    c. T4 lymphocyte.
    d. memory cell.
    e. suppressor cell.

15. When bacteria invade the body, which one of the following cells is most likely to initially suffer the highest mortality rate?
    a. B-lymphocytes
    b. Monocytes
    c. Eosinophils
    d. Basophils
    e. Neutrophils

16. All of the following characterize the cell-mediated immune response except:
    a. macrophages and the T-lymphocytes are the dominant defensive cells.
    b. T-lymphocytes recognize free antigens in tissue fluid.
    c. helper T4 cells regulate cytotoxic T8 and suppressor T8 cells.
    d. T4 cells induce T8 lymphocytes to differentiate into cytotoxic T8 and suppressor T8 cells.

e. suppressor T8 cells function to turn off cytotoxic T8 cells and helper T4 cells.

17. All of the following characterize the antibody-mediated immune response except:
    a. B-lymphocytes and helper T4 cells are involved in this type of response.
    b. B-lymphocytes divide and differentiate into plasma cells and memory cells.
    c. plasma cells secrete antibodies.
    d. re-infection activates the memory cells to develop into antibody-secreting plasma cells.
    e. suppressor T8 cells function to turn off the plasma cell activity.

18. Which of the following produces histamine?
    a. Plasma cell
    b. T8 cell
    c. T4 cell
    d. Erythrocyte
    e. Basophil or mast cell

19. T4 lymphocytes:
    a. become cytotoxic lymphocytes.
    b. engulf and destroy immune complexes.
    c. suppress other types of lymphocytes.
    d. induce T8 lymphocytes to clone.
    e. Two of the preceding

20. T4 helper lymphocytes:
    a. suppress the immune response by inhibiting inducer cells.
    b. are also known as natural killer cells.
    c. act as memory cells for antibody production and secretion.
    d. secrete interleukin 1 to activate neutrophils.
    e. assist cytotoxic and suppressor cells.

21. Antigen presentation is a primary function of which cell?
    a. Macrophage
    b. Eosinophil
    c. T8 cell
    d. T4 cell
    e. Basophil

22. The kind of immunity acquired by receiving an antigen for the purpose of stimulating the production of specific antibodies is called:
    a. adaptive immunity.
    b. passive immunity.
    c. active immunity.
    d. autoimmunity.
    e. secondary immunity.

Answers to Case History Questions

1. Hypersecretion of T_3 and T_4 mobilizes body fat for use as an energy substrate, increases the degradation of body proteins to constituent amino acids for use as an energy substrate, and accelerates basal metabolic rate.

2. The fundamental cause of hyperthyroidism in Graves' disease is autoantibody activation of TSH receptors on thyroid follicle cells, thereby increasing production and secretion of T_3 and T_4.

3. (a) Surgical removal of some or most of the thyroid tissue to reduce the gland's capacity to secrete. (b) Administration of radioactive iodine to destroy part or all of the thyroid follicle cells. (c) Administration of antithyroid drugs that block T3 and T4 synthesis.

Key Terms

active immunity (p. 870)
antibody-mediated immunity
(p. 868)
antigen (p. 848)
antigen-presenting cell
(p. 860)
autoimmunity (p. 872)
autoantibody (p. 872)
B cell (p. 869)
bone marrow (p. 850)
CD4+ cell (p. 859)

CD8+ cell (p. 859)
cell-mediated immunity
(p. 859)
class I MHC (p. 859)
class II MHC (p. 859)
complement (p. 856)
dendritic cell (p. 859)
histamine (p. 856)
immunity (p. 848)
immunoglobulins (p. 864)
inflammation (p. 856)

interferons (p. 855)
interleukins (p. 861)
leukocyte (p. 849)
lymph (p. 848)
lymph node (p. 852)
lymphokines (p. 859)
lysozyme (p. 855)
monoclonal antibody
(p. 870)
monokine (p. 859)
nonspecific defense (p. 855)

passive immunity (p. 870)
pathogen (p. 848)
Peyer's patches (p. 850)
phagocytosis (p. 856)
plasma cell (p. 851)
specific defense (p. 859)
T cell (p. 850)
thymotaxin (p. 850)
thymocyte (p. 850)
thymus (p. 850)

Suggested Readings

Akbar, A., and Salmon, M. "Selection and survival of activated lymphocytes." *Science & Medicine,* 2:48–57, 1995.

Eisenbarth, G. S., and Bellgrau, D. "Autoimmunity." *Science & Medicine,* 1:38–47, 1994.

Engelhard, V. "How cells present antigens." *Scientific American,* 271:54–61, 1994.

Heumann, D., and Glauser, M. "Pathogenesis of sepsis." *Science & Medicine,* 1:28–37, 1994.

Hogg, R. S., et al. "Rates of disease progression by baseline CD4 cell count and viral load after initiating triple-drug therapy." *Journal of the American Medical Association,* 286:2568–2577, 2001.

Janeway, C. A., Travers, P., Walport, M., and Capra, J. D. *Immunobiology,* ed 4. New York, Garland Publishing, 1999.

Johnson, H., et al. "How interferons fight disease." *Scientific American,* 270:68–75, 1994.

Mills, J., and Masur, H. "AIDS-related infections." *Scientific American,* 263:50–59, 1990.

Monney, L., et al. "Th1-specific cell surface protein Tim-3 regulates macrophage activiation and severity of an autoimmune disease." *Nature,* 415: 536–541, January 2002.

Nizet, V., et al. "Innate antimicrobial peptide protects the skin from invasive bacterial infection." *Nature,* 414:454–457, 2001.

Nossal, G., et al. "Life, death, and the immune system." *Scientific American,* 269:52–144, 1993.

Phillips, A. N., et al. "HIV viral load response to antiretroviral therapy according to the baseline CD4 cell count and viral load." *Journal of the American Medical Association,* 286:2560–2567, 2001.

Romagani, S. "T$_{h1}$ and T$_{h2}$ subsets of CD4$^+$ T lymphocytes." *Science & Medicine,* 1:68–77, 1994.

Rosen, F. S., and Geha, R. S. *Case Studies in Immunology,* ed 2. New York, Garland Publishing, 1999.

Telford, I. R., and Bridgman, C. F. "Lymphatic system." In: *Introduction to Functional Histology,* ed 2. New York, Harper Collins, 1995, pp. 235–255.

Tizard, I. R. *Immunology: An introduction,* ed 4. Philadelphia, Saunders College Publishing, 1995.

Answers to Review Questions

1. d **2.** c **3.** d **4.** c **5.** a **6.** c **7.** b **8.** c **9.** e **10.** e **11.** c
12. b **13.** c **14.** d **15.** e **16.** b **17.** e **18.** e **19.** d **20.** e
21. a **22.** c

Chapter 29

ENVIRONMENTAL PHYSIOLOGY

KEY CONCEPTS

- *Acute environmental stress can lead to acclimatization, in which functional changes benefit the individual but are not passed on to the next generation.*

- *Long-term environmental stress can lead to adaptation, in which functional changes benefit the individual and are transmitted to the next generation through a genetic change.*

- *Intrinsic clocks, via the neuroendocrine system, control biological rhythms.*

- *High altitude and deep-sea diving lead to different types of acclimatization changes.*

- *Breathing artificial gases can cause toxic effects.*

- *Acclimation to space flights is a future challenge.*

- *Pollutants are classified as either respiratory irritants or systemic poisons.*

- *Pollutants have different modes of action that can lead to specific diseases.*

CASE HISTORY

A 56-year-old Caucasian man is in the emergency room with complaints of fever, fatigue, and malaise. He reports having intermittent mouth sores and a rash over his lower extremities during the past two weeks. He has experienced mid-chest pain for the past four days upon swallowing only. He denies smoking, alcohol, or drug usage. Physical examination revealed a slight fever of 38.6°C, with a mild tachycardia of 108 bpm. Head, eyes, ears, nose, and throat findings consist of a few petechiae (round, purple-red spots due to submucosal bleeding), which were present over the soft tissue. Multiple white plaques were seen on the oral mucosa of his mouth with some swelling of the gums. During the examination, multiple petechiae were present over the distal lower extremities. Other examination findings were normal. Specifically, no swelling of the lymph glands or enlarged liver was apparent. Chest x-ray was also normal.

Laboratory findings showed a low hemoglobin concentration and a low red cell count. White blood cell count was abnormally low, with a significant increase in blastocytes. His platelet and neutrophil counts were abnormally low. A peripheral blood smear showed nucleated red cells, few platelets, and many larger cells containing finely reticulated nuclei. Other laboratory findings showed his electrolyte, blood urea nitrogen (BUN), creatinine, and transaminase levels to be normal. His uric acid level was slightly increased at 9.4 mg/dl (normal range, 3.5–8.0 mg/dl), as was his LDH level at 373 IU/L (normal, 30–220 IU/L).

A brief history revealed that the 56-year-old patient was employed as plant foreman at a nearby chemical plant, a job that he has held for the past eight years. The patient was diagnosed with acute leukemia.

Questions

1. What is the pathophysiology of acute leukemia?

2. What are the primary classifications of acute leukemia?

3. What are the predisposing factors associated with acute leukemia?

RESPONSES TO THE ENVIRONMENT

Environmental physiology studies functional mechanisms of healthy individuals who are exposed to different types of environmental stress. Environmental **stress** refers to the body's ability to respond to external stimuli (stressors) in the environment to such conditions as extreme cold or high altitudes, which are potential disruptions to life and health. The body's control system coordinates and integrates the nervous and endocrine systems in response to an environmental stress. Environmental physiologists are interested in both the immediate physiological adjustments induced by the environment, as well as the long-term effects that occur with environmental challenges. Environmental physiology focuses more on responses of individuals than on members of populations or communities.

Environmentally Induced Stress Leads to Acclimatization or to Adaptation

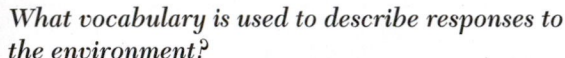

What vocabulary is used to describe responses to the environment?

In the study of environmental physiology, two key concepts are important: **acclimation** and **adaptation**. Adaptation involves a genetic change that is induced from a long-term exposure to an environmental stress. An adaptive change arises from mutation or recombination, and individuals who possess it may be more likely to survive. Sweat glands and sweating are examples of adaptations because the genes that encoded them are passed on from one generation to the other. The genes were formed as a result of a mutation that occurred by chance to heat stress, and then were selected.

Acclimation refers to a functional change in response to a *single environmental stress* that is artificially induced in the laboratory. Acclimation may last months, years, or a lifetime but does not involve genetic alteration and is not passed on to the next generation. Acclimation is illustrated by a study using identical twins exposed to heat stress in the laboratory. In this experiment, both girls are exposed for 30 minutes to a 95°F temperature and are asked to complete a standardized exercise test. In the hot environment, both will produce sweat in very similar amounts during the heat stress testing. Then, a second experiment is performed on the identical twins. For two weeks twin A enters the heat chamber for 2 hours/day for three weeks and exercises. Twin B maintains her normal activity and does not participate in the second experiment. At the end of the three-week period, both twins will again be brought into the heat chamber and tested for their sweating rates. Twin B's sweating rate is about the same as before. In contrast, twin A begins to sweat earlier and more profusely than twin B. Twin A has acclimated to the heat stress. That is, she has undergone functional changes induced by the environment that allow her to adjust better, but with no permanent change in her genome. Thus, acclimation can be viewed as an environment-induced improvement in

the functioning of an already existing genetically based homeostatic mechanism. In most cases, acclimation is reversible. For example, if the heat stress is discontinued, the sweating rate of twin A will eventually return to pre-acclimated values.

Acclimatization is another term used by environmental physiologists and is often confused with acclimation. Acclimatization refers to physiological changes that occur in response to natural environmental stresses. Acclimatization usually involves multiple environmental stresses in the natural environment, whereas acclimation involves a single environmental factor artificially induced in the laboratory. Acclimatization occurs in mountain climbing, a situation in which multiple environmental factors are present (cold stress, altitude, and physical stress), whereas the twin study is an example of acclimation because only a single environmental factor is involved (heat stress). Acclimatization, like acclimation, can last months to years but with no permanent change in the genome.

Organs during early development are most vulnerable to environmental stresses. Often acclimatization has a most pronounced effect during a crucial period of organ development. For example, early malnutrition will cause permanent stunting of growth. Another example is a newborn exposed to high altitude, an environmental stress leading to permanent changes in some lung functions, whereas an adult exposed to the same altitude will experience changes in lung function, which can return to pre-acclimatized values upon return to sea level.

Not all types of acclimatization are beneficial, and in some cases they can be harmful. An example is exposure to low level of air pollutants (dust, pollen, chemicals) that irritate the lining of the airways. These pollutant-induced irritations cause hypersecretion in the airways, leading to excessive production of mucus, which protects the airways against the irritant pollutants. Chronic mucus production can lead to airway plugging, infection, and a persistent cough that can develop into bronchitis and emphysema—a disease that leads to permanent lung damage.

BIOLOGICAL CLOCKS

To what extent are rhythmic events found in normal physiological phenomena?

There are countless examples of rhythmic activities in humans. These rhythms occur at the cellular as well as the whole-body level. Examples at the cellular level include the "pacemaker cells" of the heart and the neuronal activity controlling breathing rhythms. At the whole-body level, examples include metabolic rate, temperature, and daily activity. These rhythms are referred to as *biological rhythms,* and they vary in time from seconds to minutes to days. Biological rhythms do not simply track light-dark, seasonal, or lunar cycles but often occur in the absence of these environmental cycles. The mechanisms of biological rhythms are poorly understood, but in many instances they are neuronal in origin and often are located in some component of the nervous system.

Circadian Rhythms Are Linked to the Light-Dark Cycles

One of the most common rhythms seen in nature is the circadian rhythm, which can be detected in physiological responses of humans and other animals. The word *circadian* means "nearly 24 hours" and was introduced because daily activity does not always encompass exactly 24 hours.

Many common circadian rhythms are tied to the light-dark cycle. The light-dark cycle has a significant effect on mammals because it is the most error free and is the sole environmental factor common to all species. For example, metabolic rate, temperature, and heart rate are highest during the day, or light (L), cycle and lowest during the night, or dark (D), cycle (Fig. 29–1). In nocturnal animals, these rhythmic activities are reversed. Many hormones are secreted in a circadian rhythm. In humans, for example, **melatonin,** a hormone produced by the pineal gland, is secreted only during the dark period. The concentration of melatonin in the urine is highest from 11:00 PM to 7:00 AM. The concentration of cortisol, one of the principal hormones produced by the adrenal cortex, also follows a circadian rhythm (see Fig. 29–1).

In some mammals, many of the daily rhythms persist even if the light-dark cycles are altered, either in constant light (LL) or constant darkness (DD), as seen in Figure 29–1. Examples in nature are the Arctic and Antarctic regions, where each year there are 82 days of continuous light. Studies in which radiotelemetry was used in a number of animals (foxes, wolves, and ground squirrels) show that heart rate, metabolism, and activity follow a circadian rhythm despite the continued light. Also, continuous darkness has been suggested to have an adverse effect, which is most apparent in the Arctic and Antarctic regions as well as in some of the Scandinavian countries. One

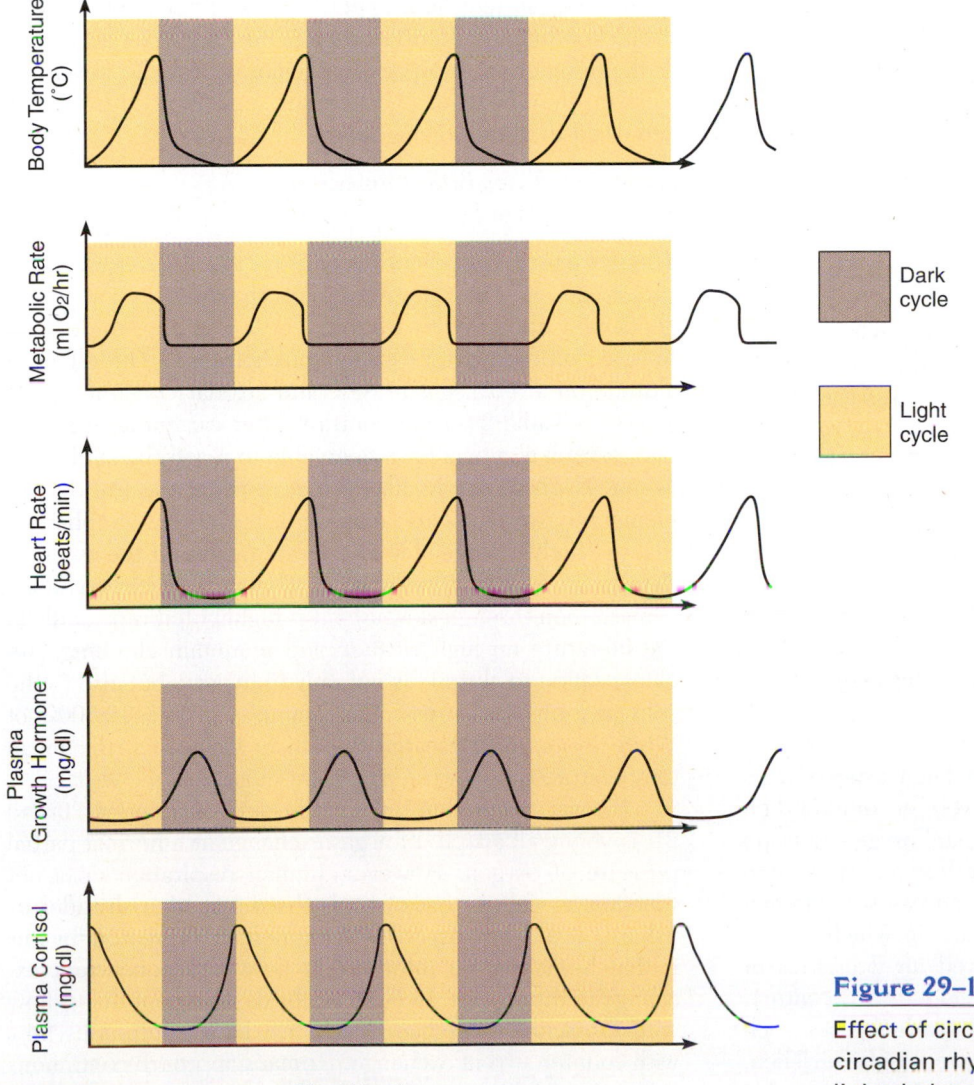

Time (days)

■ Dark cycle

□ Light cycle

Figure 29–1

Effect of circadian rhythm on body functions. Most circadian rhythms continue in the absence of the light-dark cycle.

example is **seasonal affective disorder** (SAD), which is linked to the lack of adequate daylight and is a problem in the winter months with extended periods of darkness. This disorder has a negative effect on personality and performance. Milk production in dairy cattle also is affected and can be a problem in the Scandinavian countries during prolonged periods of darkness. Recent evidence shows that SAD may be opioid mediated in the CNS and that opioid antagonists, such as naloxone, may help people with SAD.

The Lunar Cycle Has a Periodicity of Approximately 29.5 Days

The phase of the moon affects the light intensity at night and has some effect on the circadian rhythm of nocturnal animals. In humans, the lunar cycle has little effect on biological rhythms. Interestingly, the 28-day menstrual cycle follows the lunar cycle, as does the length of human gestation—10 lunar months, or 40 weeks. Most of the evolved life occurred before the invention of artificial light. Babies conceived during a full moon would clearly stand a better chance of survival if they were born on a full moon, when their parents could see 24 hr/day to protect them against predators during the vulnerable time. Mating can now occur at all times of the month without regard to the phase of the moon.

The cellular mechanisms underlying circadian rhythms are not well understood. In mammals, the central biological clock is thought to be located in the suprachiasmatic nucleus of the hypothalamus. Transplantation of the suprachiasmatic nucleus in animals transfers the biological rhythm of the donor to the recipient. The nervous connection of the eye to the suprachiasmatic nucleus provides the light-dark signal for the circadian rhythm. The suprachiasmatic-eye connection is an independent nervous pathway of the normal visual pathway. The pineal gland is also involved with circadian rhythmicity. In mammals, neural connections can be observed between the pineal gland and the suprachiasmatic nucleus.

RESPONSES TO THE PHYSICAL ENVIRONMENT

> *How does the body respond to various types of stresses?*

For much of our history, humans have been exposed to a range of environmental stresses. These were encountered on ascents to high altitude where alveolar oxygen tension dropped as barometric pressure fell as well as on exposure to temperature extremes. There are reported cases, beginning in Grecian times, of underwater activities in which divers, using primitive diving bells, encountered air pressures of greater than 1 atmosphere (atm). In the eighteenth century, the chemistry of gases blossomed, and individual gases, such as oxygen, nitrogen, and carbon dioxide, were isolated. Then, the study of unusual gases began in earnest. Joseph Priestley

said of pure oxygen, "Only two mice and myself have had the privilege of breathing it." Today we breathe high-oxygen, low-oxygen, polluted air, and special gases mixtures such as those in space capsules and underwater devices. We have contrived all sorts of artificial breathing devices to keep us alive in hostile environments and during times when our natural ability to breathe is inadequate.

High Altitude Causes Hypoxia

The condition of low oxygen (**hypoxia**) is most commonly encountered during travel to high altitudes. Remember from Chapter 20 that barometric pressure decreases with altitude. The percentage of oxygen in the air, however, remains constant at near 21%. Thus, the decreasing barometric pressure causes the partial pressure of oxygen to fall. This fact was not appreciated until the last century, when the French physiologist Paul Bert was shocked by the deaths of balloonists who ascended to altitudes of more than 15,000 feet. Bert's lengthy scientific work promoted by this tragedy was published in the classic book *La Pression Barometricque*, in which he emphasized, "Oxygen tension is everything; barometric pressure in itself is nothing or almost nothing." Bert has been acclaimed as the founder of aerospace medicine; he also did important work on the effects of high pressure experienced by divers, and discovered the "diving reflex."

Hypoxia Causes Both Immediate- and Long-Term Effects

The detrimental effects of hypoxia are amplified if the exposure is sudden. For example, a pilot who loses his or her oxygen supply at 30,000 feet has just over 1 minute to descend to lower altitude before losing consciousness. The effect of altitude on alveolar gas tension and arterial O_2 saturation is shown in Table 29–1. In contrast, after careful acclimatization, several climbers have been able to reach the summit of Mount Everest, nearly 30,000 feet, without the use of supplemental oxygen. This amazing achievement emphasizes that the body is more able to tolerate hypoxia if the ascent is slow so the body can acclimatize. The ideal of successful acclimatization through slow ascent is highlighted repeatedly in the literature on high altitude and mountain climbing. Acclimatization to altitude is not only to hypoxia but also to the cold and physical stress. For example, at the 12,500-foot White Mountain Research Station in California, the mean temperature is below freezing eight months of the year.

Human respiration, as pointed out in Chapters 20 and 21, is very well adapted for gas exchange at a normal partial pressure of oxygen. However, human respiration does not function as well as that of birds that fly at high altitudes or fish that live in low-oxygen environments. The reason for this is that humans have intermittent flow with concurrent exchange in their lungs. In contrast, birds have continuous flow with crosscurrent exchange and fish have continuous flow with countercurrent exchange. Humans approach continuous flow only with high ventilatory rates.

APPLICATIONS OF PHYSIOLOGY

Altitude Sickness

Altitude sickness, sometimes referred to as *acute mountain sickness*, is a condition caused by a lack of oxygen. The hypoxia-induced sickness is due to a decrease in partial pressure of oxygen (P_{O_2}) and not to a decrease in the percentage of oxygen. With increasing altitude, the drop in barometric pressure makes the air thinner and lowers the P_{O_2}.

Altitude can catch the unwary by surprise. For example, college students who take an extended weekend to go to a ski resort at an elevation of 8000 to 10,000 feet and start pounding down the slopes as soon as they arrive may wake up the next morning with what feels like a terrible hangover and a strong desire to vomit. These students are experiencing the early onset of acute mountain sickness. The tourist industry estimates that one fourth of the 25 million people who visit the Rockies each year experience mild symptoms of altitude sickness, which costs the tourist industry $40 million a year because these tourists feel too ill to ski, eat at a restaurant, or go to the bar and drink.

How does hypoxia affect the body? How do natives who live at high altitude become acclimatized and immune to altitude sickness? Symptoms of altitude sickness date back to 30 BCE, when Chinese documents describe the "Great Headache Mountain." The early symptoms of altitude sickness, which include headaches, dizziness, and nausea, are due to increased pressure on the brain. Hypoxia causes cerebral vessels to dilate and also causes cerebral vessels to leak fluid that seeps into the brain, both of which increase cerebral pressure. How fluid leaks into the brain is still open for debate. Some physiologists think that hypoxia causes the capillary endothelial cells simply to "run out of gas" and the energy-dependent transport mechanisms for salt and water falter. Others believe that, with hypoxia, cerebral vessels overdilate and in the process endothelial cells are stretched and start leaking. Both mechanisms may occur, and the fluid that leaks out of the cerebral vessels puts added pressure on the brain that triggers headaches, dizziness, nausea, and muscle spasms. Cerebral edema also can cause a victim to lose motor skills, resulting in staggering and slurring of words as with intoxication.

Pulmonary edema is even more severe and can strike at 9000 feet. Pulmonary endothelial cells become leaky, and recent studies have shown that severe hypoxia constricts the pulmonary vein, which adds further insult by increasing capillary hydrostatic pressure. The problem of pumping blood through water-soaked lungs as well as trying to oxygenate the blood in the edematous lung becomes life-threatening. As lungs fill up with fluid, victims become increasingly short of breath and often spit up foamy blood. Severe edema causes death from heart failure due to hypoxemia and the added strain of pumping blood through edematous lungs.

Altitude sickness can be avoided with a bit of care. One precaution is to ascend the mountain slowly. Most climbers go up the mountain in plateaus and work patiently to acclimatize. Acclimatization includes step-up breathing to capture more O_2, a faster heart rate to pump more oxygenated blood to the tissues, renal compensation for respiratory alkalosis, and increased production of red blood cells to carry more O_2. In addition, there are other acclimatization processes that take place at the cellular level. One is the increase in the number of mitochondria, the "power plants" of cells that transform glucose into usable ATP, which energizes all cellular activity. The second is glucose utilization. The body's two biggest energy consumers are the brain and heart. Unlike the brain, which uses only glucose, the heart typically burns fat and some glucose as well. However, at higher altitudes the heart switches substrates and has a higher preference for glucose because fat is a far less efficient fuel to burn when oxygen levels are low. The brain also develops a lower utilization of glucose. Recent studies have shown that, with acclimatization, the brain down-regulates its demand for glucose as well and therefore uses less O_2. Third, there is also mitochondrial acclimatization to altitude. The cell burns glucose two ways. One is anaerobically (without O_2) and the other is aerobically. The anaerobic step produces lactate and is not very efficient. The aerobic step produces 18 times more energy with O_2. Recent studies have shown that individuals who have acclimatized to altitude produce less lactate than do nonacclimatized individuals. This means that the mitochondrial enzymes that control aerobic metabolism have developed a way of keeping this pathway open during prolonged hypoxia.

QUESTION: Why do caffeine and alcohol exacerbate altitude sickness?

TABLE 29–1

Effect of Altitude on Atmospheric Pressure (P_B) and Alveolar Gas Tensions

Altitude (ft)	P_B (mm Hg)	Percent O_2	P_{AO_2} (mm Hg)	P_{ACO_2} (mm Hg)	Hb Saturation (%)
0	760	21	104	40	97
10,000	523	21	67	36	90
20,000	349	21	40	24	20
30,000	226	21	21	24	20
40,000	141	21	8	24	5

P_{AO_2} = alveolar partial pressure of oxygen; P_{ACO_2} = alveolar partial pressure of carbon dioxide; Hb = hemoglobin.

As altitude increases, however, the body makes noticeable efforts to deliver more oxygen to the tissues. Chief among these responses to altitude is **hyperventilation**—that is, deeper breaths taken more rapidly in which alveolar ventilation is increased. The sequence of events for hyperventilation is as follows: (1) a fall in inspired oxygen tension (P_{IO_2}), (2) decreased alveolar oxygen tension (P_{AO_2}), (3) decreased arterial oxygen tension (P_{aO_2}), (4) increased firing of peripheral chemoreceptors (the carotid bodies), (5) hyperventilation, and (6) increased alveolar and arterial oxygen tension. The loop is regulated in part by a negative feedback limb: (1) hyperventilation, (2) elimination of alveolar carbon dioxide, (3) decreased arterial carbon dioxide tension, (4) increased arterial and cerebrospinal fluid (CSF) pH, (5) decreased firing of both central and peripheral chemoreceptors, and (6) decreased ventilation. The increased ventilatory response to hypoxia is mediated directly through the peripheral chemoreceptors, namely the carotid bodies (see Chapter 21 for control of breathing). The stimulus for ventilation during hypoxia is a decrease in arterial P_{aO_2} rather than O_2 content or percent O_2 saturation. Although the ventilatory response to hypoxia is immediate, the arterial P_{aO_2} must drop below 50 mm Hg before ventilation is stimulated (Fig. 29–2). Ventilation is essentially unchanged at values above 50 mm Hg. It is important to realize that there is also an immediate rise in cardiac output to match the increase in ventilation.

Hypoxic-induced hyperventilation appears in two stages. In the first stage there is an immediate increase in ventilation. However, the increase in ventilation is small compared with the second stage, in which ventilation continues to rise slowly over 8 to 12 hours, with sustained ventilation after 12 hours (Fig. 29–3). The reason for the two phases is the acid-base disturbance seen with hyperventilation (Fig. 29–4). In the first phase, the increase in ventilation is rapid but is slowed because the increased ventilation results in increased CO_2 elimination, which causes a decreased arterial P_{CO_2}. A decreased arterial P_{CO_2} also causes an increased pH (respiratory alkalosis), but the decreased P_{aCO_2} and respiratory alkalosis inhibits respiration. Thus, low arterial P_{CO_2} and respiratory alkalosis become antagonistic to the hypoxic drive. During the second phase, the kidneys begin to return the hydrogen ion concentration back

toward normal. Recall from Chapter 26 that the kidneys regulate HCO^-_3 and the lungs regulate arterial P_{CO_2}. From the Henderson–Hasselbach equation, it is the ratio of HCO_3^-/P_{aCO_2} in the blood, and not the absolute amounts, that is a key component in regulating pH. This is illustrated in Figure 29–5, in which the normal ratio of HCO_3^-/P_{aCO_2} is approximately 20:1. With hypoxia, the ratio is increased during hyperventilation and the blood pH increases. The kidney compensates by excreting more bicarbonate, which brings the ratio back toward normal. The antagonistic effect of low arterial P_{CO_2} and increased pH is minimized, which allows the hypoxic drive to further increase ventilation. Renal compensation occurs within 10 to 12 hours and corresponds to the second phase of hypoxic hyperventilation.

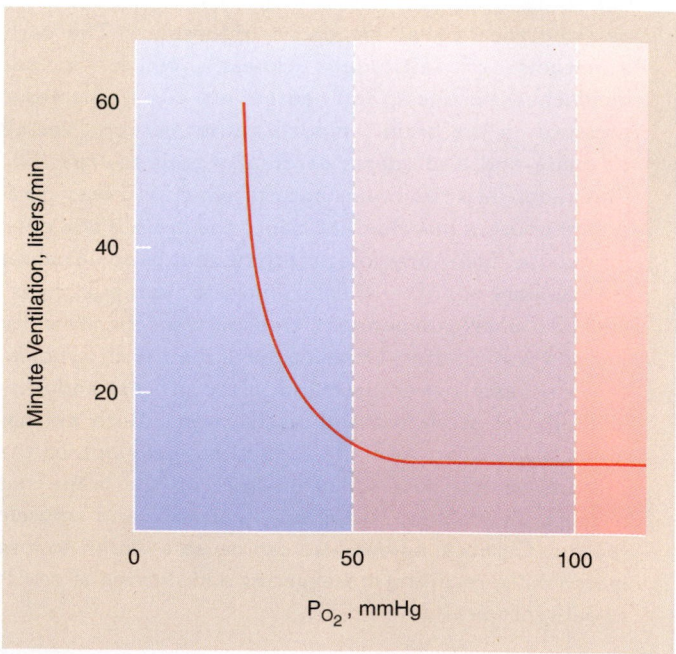

Figure 29–2

Ventilatory response to hypoxia. Note that minute ventilation is unchanged until the arterial P_{O_2} drops below 50 mm Hg.

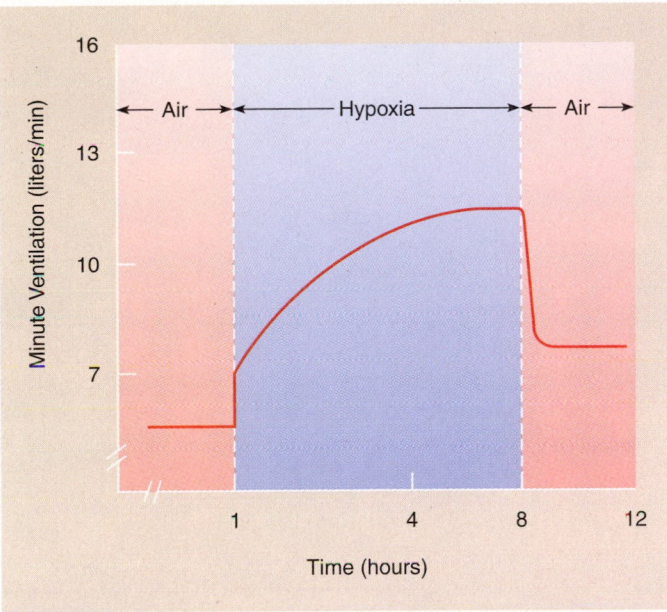

Figure 29–3

Hypoxic drive occurs in two stages. The first stage is a rapid increase in minute ventilation, and the second stage is a slow, sustained increase in ventilation.

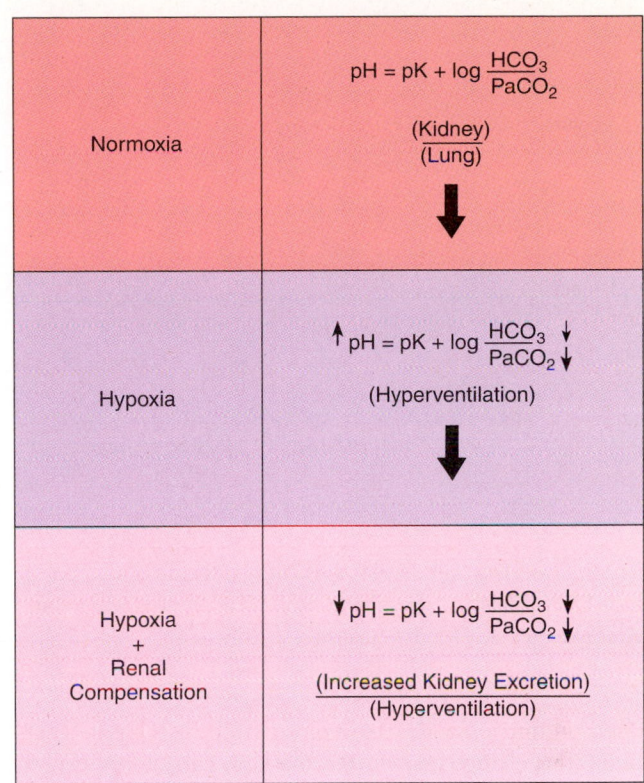

Figure 29–5

Renal compensation to high altitude respiratory alkalosis.

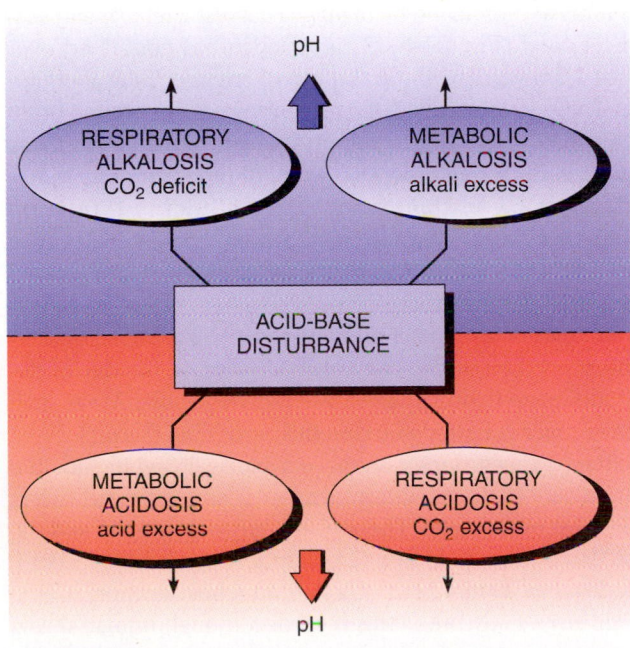

Figure 29–4

Various pathways of acid-base disturbance. High altitude takes the pathway of respiratory alkalosis because hyperventilation creates a carbon dioxide deficit.

Long-Term Effects of Hypoxia

Because millions of people travel to high altitudes annually for recreation, the acute effects of hypoxia are of both physiological and of clinical interest. Some effects are present at even moderate elevations of 5000 feet, an altitude at which major cities are found throughout the world. There are noticeable changes in ventilation, especially during exercise, even at these relatively low elevations. Acclimatization occurs over a matter of days to weeks (Fig. 29–6). Most residents of elevations of 5000 feet soon do not notice any effect of the altitude. The highest settlement in the world is a mining camp at 17,500 feet in Peru. The residents work a mine that is at 19,000 feet, with a daily climb of 1500 feet. Interestingly, when the camp was moved to 18,500 feet, the miners complained that they had no appetite, could not sleep, and lost weight. From these observations, it appears that 17,500 feet is the highest altitude at which even acclimatized humans can live permanently.

In addition to ventilatory adjustments, the body undergoes other physiological changes to adjust to low environmental oxygen. One important step in acclimatization is that more red blood cells are manufactured in the red bone marrow, which causes an elevated hematocrit (hypoxia→stimulates erythropoietin by kidney→stimulates bone marrow→which leads to an

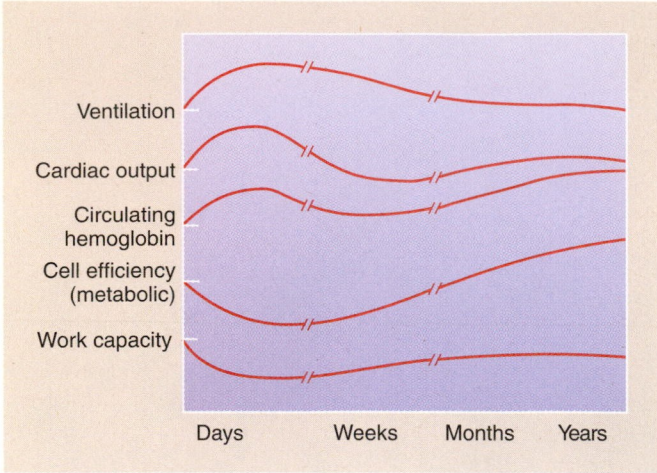

Figure 29–6

Acclimatization that takes place at high altitude (14,000–16,000 feet).   *(From Houston, C.S.* Hypoxia: Man at Altitude. *New York, Thieme-Stratton, 1982.)*

Figure 29–7

Sherpas are mountain-dwelling people who have highly adapted to the low oxygen environment of high altitudes (Nepal).   *(© Steve Covello/Mountain Stock/1990.)*

increased production and release of erythrocytes). Up to a certain point, this change enables the blood to carry more oxygen to the tissues. However, the higher the hematocrit, the more viscous (thicker) the blood becomes, a change that increases the workload of the heart. Normal hematocrit is near 45%. An increase in hematocrit is called **polycythemia,** and, in some cases, high-altitude residents have hematocrit levels of over 70%. These individuals must have blood withdrawn periodically to reduce their hematocrit to a range that permits the heart to pump effectively.

The body increases ventilation, cardiac output, and circulating hemoglobin (see Fig. 29–6) during the first few weeks to months at high altitude. Over a long period, cardiac output and ventilation tend to return to low-altitude values, whereas hematocrit continues to climb. The cells also change to become metabolically more efficient. For example, the mitochondria increase in number, and there are alterations of metabolic pathways. Even with these changes, the maximum work capacity never reaches sea-level values. Nevertheless, the acclimatization of long-term residents can be impressive. In the Andes, the natives return from the high mines in the evenings to their "low-" altitude homes for a rousing soccer game at 15,000 feet.

The best acclimatized of all high-altitude residents are the Sherpas in Nepal, whose climbing feats are legendary (Fig. 29–7). Other people—for example, the residents of high altitudes (10,000 feet) in Colorado—are often not as well adapted to low oxygen. The percentage of Coloradans, aged 60 years and older, that live at high altitude is much lower than the percentage of older people who live at low altitude. High-altitude residents tend to leave for health reasons, usually because of heart and lung disease. The former residents report that their health is improved after they settle at low altitude.

Korean Women Are Champion Divers

Off the shores of Korea and southern Japan, over 30,000 women make their living by diving to harvest shellfish and seaweed for food (Fig. 29–8). This practice has continued over the past 1500 years, without the use of any mechanical equipment. These divers, called *Ama,* can descend to depths of 80 feet and hold their breath for up to two minutes. The Ama dive repeatedly to these depths, often 30 times per hour, coming up only for a few breaths of air and a brief rest. In warm weather they work 4 hr/day, with 1-hour rest intervals. The Ama work in the winter in a water temperature of 50°F. Diving is a lifelong profession for these women, often starting at 10 to 12 years of age and continuing to age 65. Pregnancy does not interrupt their work. They often dive up to the day of delivery and nurse their babies between diving shifts. As we shall see, the female is better suited to this work than the male.

Other diving groups who make a similar living include the sponge divers in the Mediterranean area and the pearl divers off the coast of Australia. These divers are mostly men, and the record descent was held by a Greek sponge diver who, in 1913, dived down to 200 feet to put a line on a boat anchor. A U.S. Navy diver recently broke this record. Although human divers do not stay submerged as long as other diving mammals (Table 29–2), they have similar physiological responses.

Preparing for the dive, divers hyperventilate for 5 to 10 seconds before they make the plunge. Hyperventilation does two things: it blows off considerable amounts of CO_2 and inual lung volume), they are subjected to a painful "lung squeeze." More importantly, if the hydrostatic pressure in the pulmonary vessels exceeds the air pressure in the lungs, small pulmonary blood vessels can rupture. Surprisingly, the seal also inhales to 85% of its total lung capacity before a dive.

The Ama divers have evolved another technique. During hyperventilation they purse their lips and emit a loud whistle with each forced expiration. The whistling is a trademark of the Ama and can be heard for long distances. The practice of pursing their lips helps to protect their lungs by preventing excessive hyperventilation, which can lead to underwater blackout (unconsciousness) during the dive (see Chapter 21). Not all divers whistle when they hyperventilate. Nevertheless, this is another example of how humans, without any

Figure 29–8

The *Ama* are Korean and Japanese women who make a living descending unassisted to a depth of 60 to 80 feet to collect shellfish. At the end of each dive, a helper in a boat at the surface pulls the Ama up by means of a long rope attached to her waist. (The other rope belongs to another diver.) In warmer weather, the Ama wear only loincloths during their dives. In the winter, they wear only light cotton bathing suits. *(From Hong S.K., and Rahn H. "The diving women of Korea and Japan." Scientific American, May 1967. Photo by Fosco Maraini.)*

creases the concentration of O_2 in the lungs (Fig. 29–9). The final breath, however, is approximately 85% of their total lung capacity. This is done to avoid the uncomfortable pressure on the lungs during the dive and to restrict the body's buoyancy. If divers go deeper than the level of lung compression (resid-

TABLE 29–2	
Breath-Holding Capacities of Some Mammals	
Species	**Time (Min)**
Polar bear	1.5
Pearl divers (human)	2.5
Sea otter	5
Hippopotomus	15
Beaver	20
Seals	30
Sperm whale	90

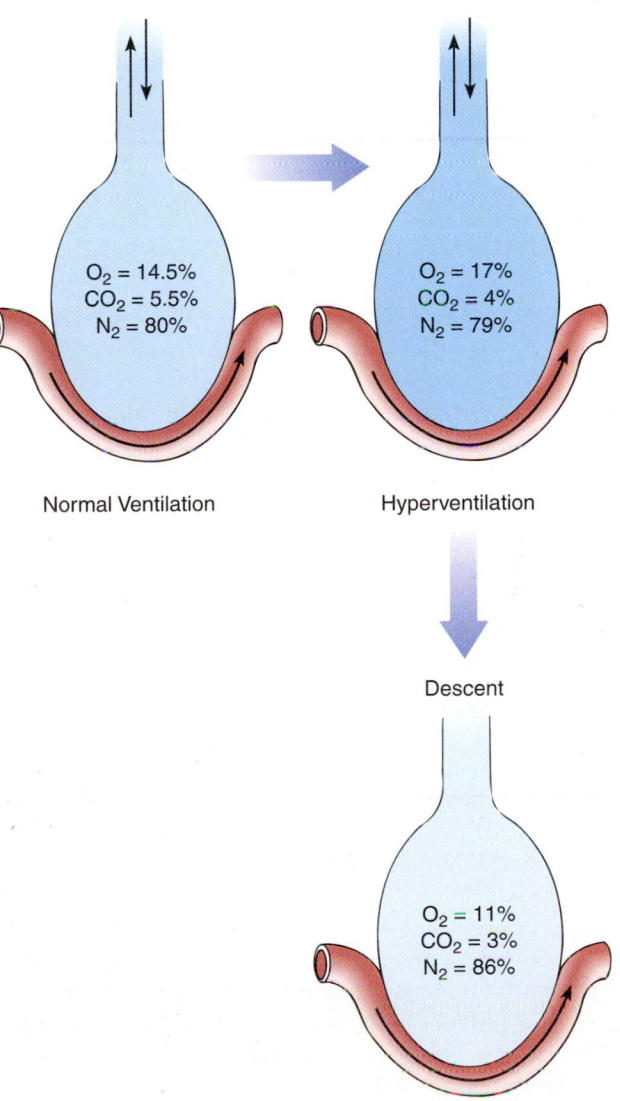

Normal Ventilation — $O_2 = 14.5\%$, $CO_2 = 5.5\%$, $N_2 = 80\%$

Hyperventilation — $O_2 = 17\%$, $CO_2 = 4\%$, $N_2 = 79\%$

Descent — $O_2 = 11\%$, $CO_2 = 3\%$, $N_2 = 86\%$

Figure 29–9

Changes in gas concentration of O_2, N_2, and CO_2 during diving. During descent, pressure on the lungs causes all gases to enter the lungs; during ascent, the situation is reversed.

knowledge of respiratory physiology, learned that (1) hyperventilation blows off excess CO_2, which extends their breath-holding; (2) whistling during hyperventilation prevents too much CO_2 from being blown off; and (3) not taking a full breath before diving can extend the depth of the dive.

The effect of diving on blood gases is seen in Figure 29–9. When divers reach the bottom at a depth of 40 feet, the O_2 concentration is reduced to approximately 11% because of breath holding and the continual uptake of O_2 by the blood. However, at this depth the pressure has compressed the lungs to approximately half of their predive volume, but the alveolar partial pressure of O_2 (PAO_2) has increased from 100 mm Hg before the dive to 149 mm Hg at the bottom of the dive. Another paradox occurs in CO_2 uptake during the dive. Normally, when someone holds his breath, the carbon dioxide in the lungs increases. However, during a dive, CO_2 decreases in the lungs. As seen in Figure 29–9, alveolar carbon dioxide drops from 4% after hyperventilation to 3% at the end of the dive. The reason for this is that at a depth of 40 feet or greater the compression of lung raises the partial pressure of CO_2 in the alveoli that is greater than the partial pressure of CO_2 in the pulmonary capillary blood. As a result, CO_2 is actually taken up from the lungs and retained in the tissues.

Upon ascent from the bottom, the diver's lungs expand, drastically reversing the situation of compression. With the reduction in pressure, CO_2 leaves the blood very rapidly. However, the situation with O_2 transport is quite different. With the precipitous drop in pressure, partial pressure of O_2 in the alveoli goes from 149 to 40 mm Hg in a matter of 30 seconds. The PAO_2 of 40 mm Hg is no greater than the partial pressure of O_2 in the pulmonary capillary blood. Hence, O_2 cannot readily bind to hemoglobin, and the O_2 content of the arterial blood drops. In some cases the blood actually loses O_2 to the lungs. Thus, the cumulative O_2 deficiency in tissues, especially the brain, is sharply accentuated during ascent and can be the cause of death among divers (especially sport divers) returning to the surface after a deep and lengthy dive.

Another important physiologic event that occurs is the **"diving reflex,"** in which blood vessels to all parts of the body constrict except those to the heart, brain, and exercising muscles that require undiminished O_2 supply to be maintained. The heart rate also decreases, thereby decreasing myocardial O_2 consumption (Fig. 29–10). Most diving mammals have the diving reflex that assists in the conservation of O_2 and extends their diving time. At a depth of 300 feet, a seal's heart rate is only 4 bpm, whereas on the surface it is 120 bpm. Interestingly, breath-holding alone in humans will not initiate the diving reflex. The reflex is initiated by submerging of the face in water and is a part of the gasp reflex seen in the newborn human and other mammals.

Acclimatization to Diving

Many divers, especially the Ama, have acclimatized to diving. The Ama women have a larger vital capacity than do nondiving Korean women—due, in part, to development of the diver's inspiratory muscles, which also helps to resist compression of the

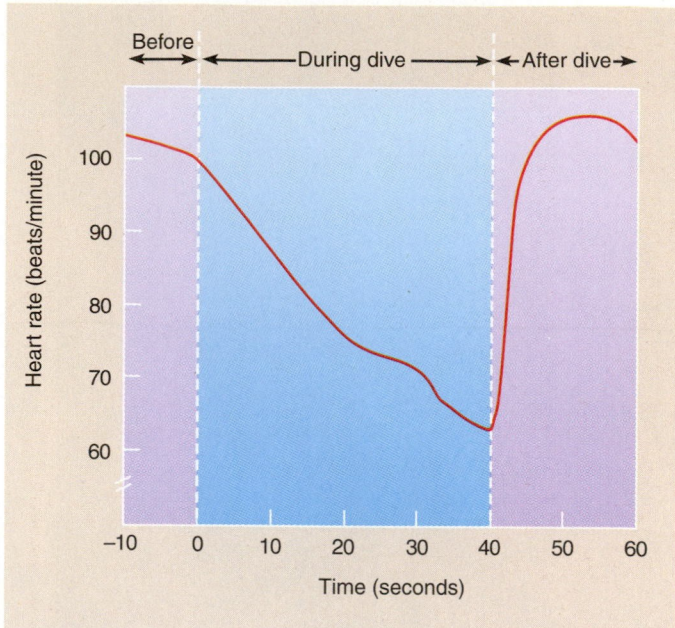

Figure 29–10

Diving reflex causes a slowing of the heart rate during descent.

chest and lungs under water. The acclimatization is acquired through training. Daughters of Korean farmers can become just as competent divers as the daughters of divers. The other physiologic change that occurs in the acclimatization process is the elevated basal metabolic rate to compensate for the heat loss during diving. The Ama on the average lose about 1000 cal/day from cold exposure. They compensate by eating more than their nondiving sisters and by having a significantly higher basal metabolic rate. Normally, humans all over the world—in cold climates and in warm climates—have the same basal metabolic rate. In the winter months, the Ama basal metabolic rate is approximately 25% higher than that of nondiving women in the same community with the same economic background.

Heat production is one defense against cold. The other is prevention of heat loss. Women have better protection against heat loss because they have more subcutaneous fat, which explains why women dominate the diving profession of Korea and Japan. However, the diving women have the same thickness of subcutaneous fat as do nondiving women, but have less heat loss. This means that, as a part of the acclimatization to diving, vasoconstriction restricts the heat loss from the blood vessels to the skin.

Increased Gas Pressures Can Be Lethal

> *What steps have been taken to operate in an underwater environment?*

As divers submerge under water, the pressure surrounding their bodies rises because of the weight of the water. To visualize how the pressure rise takes place, imagine diving with a balloon that is filled with air at a pressure of 1 atm. The deeper the balloon goes, the greater the water pressure pushing in on it. This increased pressure causes the volume of the balloon to decrease. Every 33 feet below the surface, the pressure increases by 1 atm. As the balloon gets smaller and smaller, the air molecules are compressed closer and closer together, increasing the pressure in the balloon. Similarly, as a diver submerges, the lungs become compressed and the gas inside the lungs also becomes compressed (Fig. 29–11). This means if a diver goes 33 feet under water, the alveolar pressure is 2 atm instead of 1 atm. If he goes 330 feet below sea level, the pressure is 11 atm.

For thousands of years, people have used equipment to help them remain under water for prolonged periods. The earliest devices were probably snorkels (Fig. 29–12). These devices provide a practical way to breathe to a depth of approximately 18 inches. Below that depth, two problems occur. First, the water pressure makes it difficult to expand the chest to draw in fresh air. Second, the snorkel adds dead space through which the diver must draw air. If the dead space becomes too large, then the diver is simply moving air back and forth in the snorkel and does not get any fresh air into the alveoli.

The next apparatus to be invented was most likely the diving bell. These devices have been used for centuries. The first ones were simple, inverted chambers, open at the bottom, in which the diver stood and breathed the air in the chamber as it submerged (Fig. 29–13). These diving bells had many limitations, including air that becomes rapidly fouled by carbon dioxide. Also, the air volume was com-

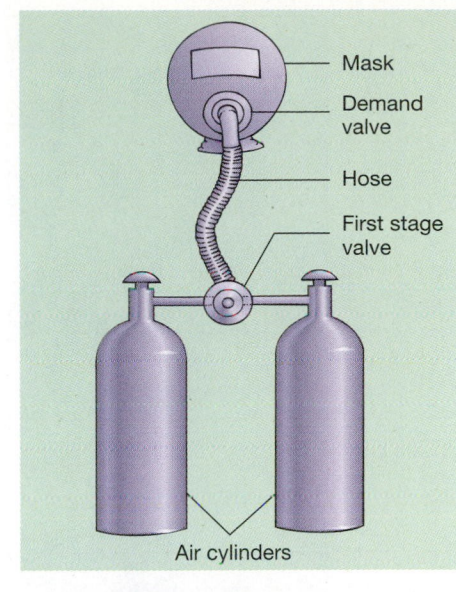

Mask
Demand valve
Hose
First stage valve
Air cylinders

(a)

(b)

Figure 29–12

(a) The development of the self-contained underwater breathing apparatus (SCUBA) has enabled many people to pursue underwater activities. It is critically important, however, to be cautious during deep or prolonged dives because gases, such as nitrogen, can be absorbed into the body and cause problems during both the dive and the ascent. *(b)* The diver in the photo makes a decompression safety stop during ascent from a dive of 120 feet. This stop facilitates the gradual release of nitrogen from the tissues, thereby preventing the bends. *(Courtesy of C. Petree and J. Holtmeier.)*

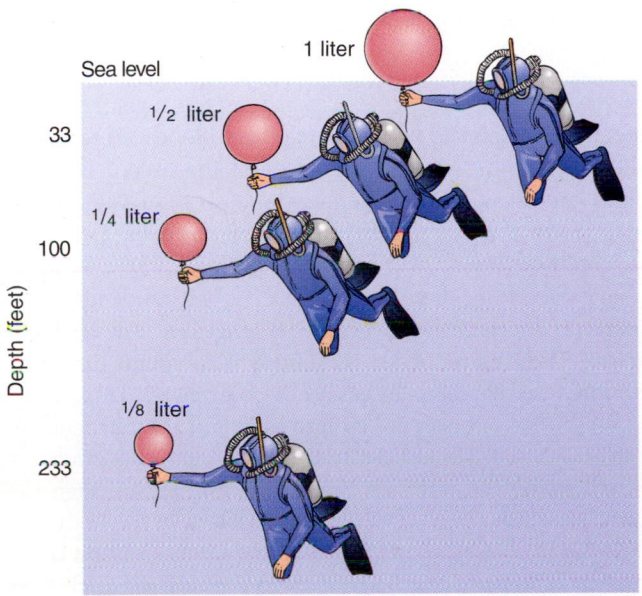

1 liter
Sea level
1/2 liter
33
1/4 liter
100
1/8 liter
233
Depth (feet)

Figure 29–11

Air in a container is compressed progressively as the container is lowered into water, where the surrounding pressure increases.

pressed to a smaller and smaller volume as the pressure increased during descent. Alexander the Great is said to have used a diving bell in the fourth century BC. Modern diving bells are completely enclosed, so compressed air volume is no longer a problem. By the early nineteenth century, the enclosed diving suit had been invented (see Fig. 29–13).

(a)

(b)

Figure 29–13

(a) Diving bells have been used since the time of the ancient Greeks. *(b)* During the eighteenth century, the advent of pumps allowed for the development of diving suits, which gave the diver greater mobility under water. *(© The Bettmann Archive.)*

Figure 29–14

Although a swimmer can remain submerged for long periods with a snorkel, the depth of a dive is limited because the snorkel increases respiratory dead space and because the increasing pressure on the chest makes it difficult to draw in fresh air.   *(© Maria Paraskevas)*

Fresh air was fed from the surface into the suit by a pump to sustain the diver for long periods under water. However, pressure plays an important role in delivering compressed air under water. If, for example, 1 cubic foot of air is pumped down to a depth equivalent to 4 atm (approximately 100 feet), then this 1 cubic foot, when it reaches the diver's helmet, will be only 1/4 cubic foot. Therefore, the fixed amount of air that must be pumped to the helmet per minute to wash out the carbon dioxide is directly proportional to the depth to which the diver descends.

During World War II, work began in earnest on the self-contained underwater breathing apparatus **(SCUBA)** (Fig. 29–14). Much of the pioneering work has been credited to Jacque Cousteau. This kind of gear has become very popular for recreation, exploration, and work on underwater oil rigs and pipelines.

Effects of Nitrogen during Deep-Sea Diving

A characteristic of all these devices is that air enters the lungs at higher-than-normal pressure. Unfortunately, breathing such high pressures has some potentially serious consequences. The problems come not during the descent, but when the diver ascends. When the pressure becomes high enough, particularly at depths below 100 feet, significant amounts of nitrogen in the compressed air are dissolved into the tissue. The longer the time spent under water, the more nitrogen is dissolved. As the diver rises to the surface, the nitrogen is released from the tissues. If the ascent is slow, the

nitrogen can be picked up by the capillary blood and returned to the lungs to be exhaled. However, if the ascent is rapid, bubbles can form, both in the blood and in the tissue. When bubbles form near the joints, they cause agonizing pain. Because the diver bends over to relieve the pain, this syndrome has been called the "bends." The risk for the bends increases as both the depth of the dive and the length of time at depth increase. The treatment for the bends is to spend time in a decompression chamber in which the barometric pressure is raised until the bubbles are compressed and the nitrogen redissolves into the tissue. Pressure in the chamber is gradually lowered, giving the nitrogen time to be exhaled through the respiratory system instead of forming bubbles in the body.

At depths below about 200 feet, the concentration of nitrogen molecules in the body rises to a point at which they have a narcotic effect, causing "raptures of the deep." Under these conditions, divers can often make unintelligent, life-threatening decisions. One way they can avoid this is to switch from a nitrogen-oxygen mixture to a helium-oxygen mixture. Helium is less soluble in tissues than nitrogen, so

Figure 29–15

Wilhelm Bauer's submarine sank in 60 feet of water, but the captain and crew escaped by swimming easily to the surface.   *(Photo courtesy of Submarine Force Library and Museum, Groton, CT.)*

fewer helium molecules dissolve in the body. This reduces the narcotizing effects as well as the number of molecules that must be removed from the tissue during ascent. Also, helium has a smaller molecular size than nitrogen, so it diffuses more rapidly, which aids in elimination of the gas.

Respiration in Submarines

The first practical submarines were built in the 1800s, although crude, hand-propelled models were fabricated in the 1700s. By the start of World War I, Britain had 77 submarines; France, 45; the United States, 35; Germany, 29; and Russia, 28. The submarines in the early 1900s could stay submerged only for several hours and operated exclusively in shallow coastal waters. The effectiveness of submarines in warfare made them popular with every major power. To date, thousands of these vessels have been built. During World War II, a big problem was the building up of CO_2 either from "CO_2 scrubbers" or in battle, when submersion for long periods of time was required. It was not uncommon for CO_2 to build up to 4% inside the submarine (sea-level concentration is 0.03%).

One of the unpleasant aspects of submarines is that, when they sink, they usually take all hands aboard and trap them below. That grim reality has prompted much work on methods of escape. All sorts of SCUBA-type gear have been developed, but usually to no avail. During World War II, mechanical lungs saved only five men. Ironically, a simple, excellent escape technique effective to depths of 300 feet had been known for nearly a century. Wilhelm Bauer used it in 1851, when his hand-propelled submarine sank in 60 feet of water and the hull began to collapse from the pressure (Fig. 29–15). Bauer waited until the air pressure in the vessel was equalized with the water pressure and then opened the hatch. The captain and his crew members by then were breathing air at 3 atm, but not for long enough to build up nitrogen pressure in the tissue. Thus, they could ascend very rapidly without experiencing the bends. As they rose to the surface, the compressed air in their lungs expanded. This caused them to exhale steadily during the ascent. They never ran out of air even though they exhaled the whole time. As Captain Bauer explained, "We came to the surface like bubbles in a glass of champagne."

This technique was ignored and forgotten. It was accidentally repeated in several instances during World War II, but was never developed. Instead, with technology gone awry, navies continued to develop impractical mechanical contrivances that did not work, while all the time the body was superbly adapted to escape. In the 1950s, the British and U.S. navies recognized that the best technology for escape was no technology. Water-filled escape training towers have been built, and unaided escapes at depths to 300 feet are now a routine part of submarine training.

One step has been added to the natural escape method: Sailors now inflate small life vests at the moment of departure. This extra buoyancy causes them to ascend very rapidly to the surface. The decompression is so rapid under these circumstances that the sailors exhale as fast as they can. The air they breath while at 300 feet is so highly compressed that it expands rapidly in their lungs as they ascend and, even with continual maximal forced expiration, they never run out of air on the way up. In navy jargon, this technique is known as "blow and go."

"Homo Aquaticus"

Two methods currently in the research stage offer promise for underwater use. One is the actual breathing of liquid. If saline (salt solution) or some other fluid, such as fluorocarbon, is put under several atmospheres of pressure by 100% oxygen, then the oxygen tension of the saline rises to thousands of millimeters of mercury. If an experimental animal, such as a rat, is held under the surface, it will eventually be forced to breathe the hyperoxygenated saline. Rats can breathe such fluid and swim around for periods of up to 18 hours.

There are two major problems with liquid breathing. The first is that the removal of carbon dioxide is difficult because the respiration depends on the bellows action of the lungs to wash carbon dioxide out. If the lungs are filled with fluid, it is much more difficult for the bellows to work effectively, and carbon dioxide builds up in the blood. The second problem is that the fluid washes the surfactant out of the lungs, leading to alveolar collapse when the animal attempts to return to air. This makes the transition from fluid back to air difficult. It has been done successfully, however, in a number of rats, mice, and even some dogs.

Another possibility for prolonged undersea activities is the development of artificial gills. This is not a small engineering feat, but it seems possible. If a gill could be developed that could trail behind a diver while being held in the mouth, its large surface area could extract oxygen from the water and excrete carbon dioxide just as the gills of fish do. Animals such as the mudpuppy have external gills as well as lungs. They use the gills to stay successfully under water for prolonged periods. Liquid breathing techniques, artificial gills, or some similar method might lead to the development of what Jacques Cousteau describes as *Homo aquaticus,* an underwater man capable of living indefinitely beneath the surface of the sea.

Respiration in High-Oxygen Environments

> **How does a high-oxygen environment affect the lungs?**

The usual problem in breathing is getting enough oxygen into the body. However, with modern chemistry and technology it is possible to breathe 100% oxygen and actually get too much. High-oxygen environments are often found in hospitals, where oxygen is administered to patients with oxygen deficiency, and in aviation and space flights, in which the pilot or astronaut breathes a gas composition other than air.

Alveolar Collapse in Oxygen-Rich Environments

One of the physical dangers of breathing pure oxygen is **atelectasis** (collapsed alveoli), which can occur in pilots, astronauts, and hospital patients. This respiratory complication occurs when small airways become blocked by mucus; the oxygen in the plugged alveoli continues to be absorbed, and the alveoli rapidly collapse (Fig. 29–16). The total pressure of the trapped gas is close to 760 mm Hg, but O_2 pressure in the venous blood is only 40 to 45 mm Hg. Consequently, the oxygen is continuously absorbed, and the alveoli collapse (see Fig. 29–16a and b). Normally, when gas is trapped in the

course of breathing air, nitrogen acts to deter collapse because it is not utilized and therefore is not absorbed into the blood. However, when one is breathing pure oxygen, there is no nitrogen.

Postoperative atelectasis occurs commonly in patients who are ventilated with high-oxygen mixtures. Atelectasis is also prevalent in fighter pilots, who breathe high oxygen and experience large gravitational forces (see the section on space travel). The added gravitational effects augment alveolar collapse. Atelectasis is most likely to occur at the bottom of the lung, where alveoli are poorly expanded and airways are partially closed. Reopening atelectatic areas is difficult because the surface tension effects are more pronounced on small units. Atelectic portions of the lung can, however, be reopened by deep breaths.

Toxic Effect of Breathing High Oxygen

The second problem associated with breathing high oxygen, much more serious than atelectasis, is its toxicity. This occurs when a person breathes high concentrations of oxygen for more than one day. The primary target organ for oxygen toxicity is the lung, where tissue injury can occur within 24 hours and can cause pulmonary edema.

The first site of injury is the endothelial cell of the pulmonary capillary. The mechanism for tissue damage is the formation of **free radicals.** These highly reactive molecules are formed as by-products from molecular oxygen. Before molecular oxygen (O_2) can be used in the oxidation of fuels for energy, it first must be converted to "active forms." These active forms are called **oxidizing free radicals,** and the most important one in the cell is the **superoxide anion radical** (O_2^-). Free radicals are continuously being produced from dissolved molecular oxygen under normal oxygen tensions (tissue $P_{O_2} = 40$ mm Hg). However, under these normoxic conditions, protective cellular enzymes—namely, superoxide dismutase, catalase, and peroxidase, rapidly remove the free radicals. As long as P_{O_2} is normal, the oxidizing free radicals are rapidly removed and no cellular damage occurs. When tissue P_{O_2} is greatly increased—such as occurs when an individual breathes a high concentration of oxygen—excess free radicals are formed that overwhelm the protective enzymes. As a result, these free radicals cause destructive and lethal cellular effects by (1) oxidizing membrane lipids, thus altering essential properties of membranous structures of the cells; (2) oxidizing cellular proteins and DNA; and (3) oxidizing enzymes, thus shutting down cellular respiration.

High oxygen is not the only source of oxidant damage from free radicals. Breathing ozone (O_3) or nitrogen dioxide (NO_2) from polluted air will also cause free radical injury to the lung. Some herbicides—such as Paraquat, are extremely toxic because of the free radical—induced injury. If someone smokes tobacco or marijuana that has been treated with Paraquat, the Paraquat in the smoke penetrates the lungs and forms free radicals. Crop dusters and migrant workers are also at risk because of the exposure to Paraquat through the lungs or skin.

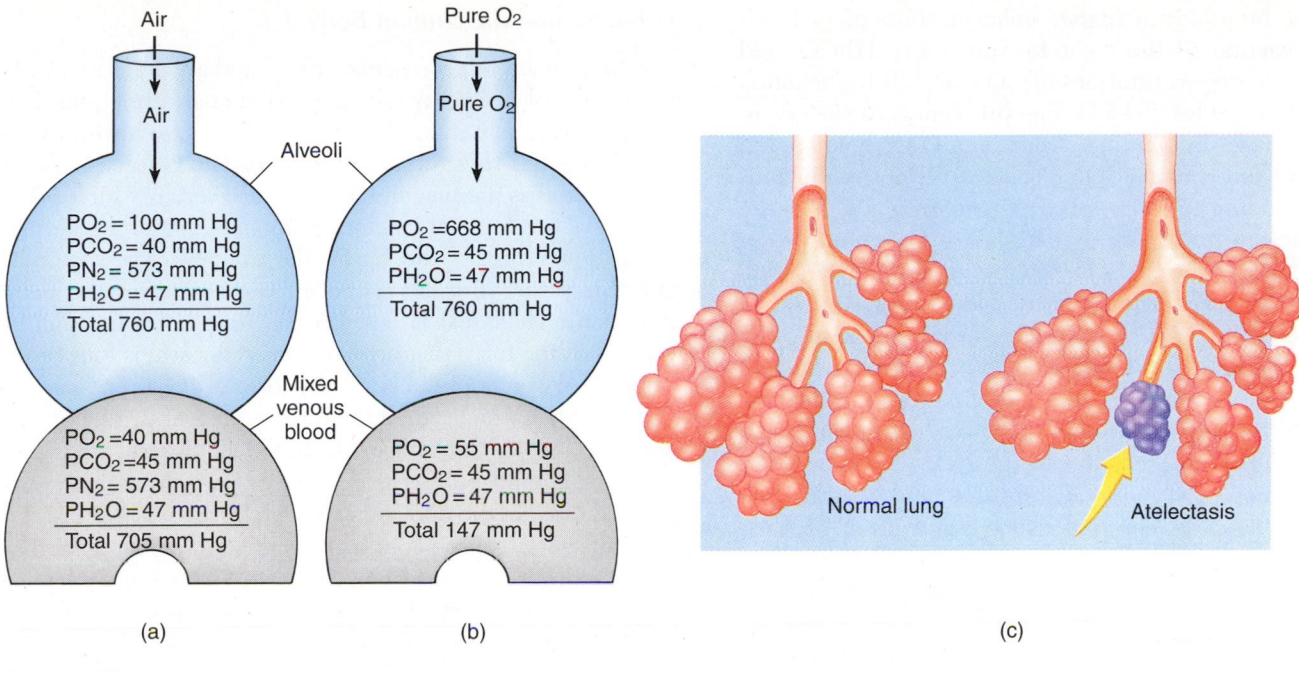

Figure 29–16

Cause of atelectasis as a result of a plugged airway. When *(a)* air is breathed and *(b)* oxygen is breathed, alveolar pressure in both cases is the same, but the sum of the gas pressures in the mixed venous blood is smaller. *(c)* Atelectasis is caused by airway obstruction. Trapped air is absorbed into the pulmonary capillaries, causing alveoli to collapse.

Another hazard of breathing high concentrations of oxygen is seen in intensive care units in which premature infants develop blindness from **retrolental fibrosis.** The mechanism for the blindness is deterioration of the retina and inadequate development of retinal blood vessels. The poor vascularization leads to the formation of fibrosis tissue behind the lens (retrolental fibrosis). If the PO_2 is kept below 140 mm Hg in the incubator, retrolental fibrosis can be avoided.

A third type of oxygen toxicity occurs when oxygen is breathed at pressures exceeding 760 mm Hg. Oxygen is breathed at pressures greater than 1 atm in deep-sea diving or in hyperbaric (high-pressure) chambers. When oxygen is breathed at 2 to 4 atm, it stimulates the central nervous system (CNS), leading to convulsions. The exact cause for the toxic effect on brain tissue is not fully known, but the high oxygen pressure is thought to inactivate dehydrogenase enzymes. Twitching of the face, ringing in the ears, and nausea often precede the convulsions. Convulsions can occur when an amateur SCUBA diver inadvertently uses a tank filled with O_2.

Respiration in Unusual Gaseous Environments

Breathing high carbon dioxide levels usually does not occur because the normal Pco_2 in the air we breathe has a partial pressure of 0.3 mm Hg (0.03%), which is negligible. However, in certain types of diving gear and closed environments (e.g., submarines and spacecraft), carbon dioxide can build up and be rebreathed because of defects in equipment. Rebreathing higher-than-normal levels of carbon dioxide stimulates ventilation (see Chapter 21). An alveolar Pco_2 of 80 mm Hg, which is about twice normal, is the maximum that an individual can tolerate. With these high concentrations, hypoventilation occurs, and the individual becomes severely acidotic. When Pco_2 levels are greater than 80 mm Hg, the situation becomes intolerable, and the respiratory centers are no longer stimulated but instead are inhibited. Respiration begins to fail, and acidosis becomes more severe. Eventually, the CO_2 acts like a narcotic, and the individual becomes lethargic and subsequently becomes unconscious.

Carbon Monoxide Poisoning and Oxygen Transport

Carbon monoxide (CO), unlike carbon dioxide, does not stimulate respiration. The problem with carbon monoxide, which is an odorless, colorless, and poisonous gas, comes from incomplete combustion of fuels. Carbon monoxide cannot be detected by the body and becomes lethal because it competitively binds at the same site as oxygen on the hemoglobin molecule to form **carboxyhemoglobin (COHb).** The formation of COHb interferes with the transport of oxygen to the cells. The reaction (Hb+CO⇌HbCO) is reversible and is a function of the partial pressure of CO. This means that breathing a higher concentration of CO will shift the reaction to the right to form more HbCO, and breathing lower concentrations of CO will shift the reaction to the left to form less HbCO. The problem is further complicated because the binding affinity of CO for hemoglobin is about 210 times that of O_2. Because of this greater tendency for binding to Hb, even trace amounts of CO in the respired air can be extremely dangerous. For example, an alveolar PCO of 0.5 mm Hg (1/210 that of alveolar PO_2) will cause half of the hemoglobin in the blood to bind with CO and half to bind with O_2. Under these conditions, the amount of oxygen carried by the blood to the cells is halved. Thus, a PCO of 0.7 mm Hg (0.1% of air) can be lethal because the cells become **anoxic** (deprived of oxygen). The first organ to become sensitive to the lack of oxygen is the brain. Alteration of vision, reaction times, and coordination are the first signs of cerebral hypoxia and often explains why there are more accidents on the freeway when CO levels rise. The real danger of CO poisoning is that it cannot be detected in the body. With hypoxemia, there are physical signs; the individual will turn blue around fingertips and lips. However, the formation of HbCO makes the blood crimson red and masks the bluish tint of hypoxemia. There are no chemoreceptors to detect CO, the problem is further compounded because arterial PO_2 is still normal (PaO_2 95 mm Hg), and the chemoreceptors that detect changes in PO_2 are not stimulated.

An individual severely poisoned with CO can best be treated by administration of 100% O_2. Administering pure oxygen elevates the blood PO_2, which displaces CO from the hemoglobin. Also beneficial is the simultaneous administration of 5% to 7% carbon dioxide. Breathing the high concentration of carbon dioxide stimulates ventilation and reduces alveolar CO, which concomitantly releases CO from the blood. When the combination of high oxygen and carbon dioxide is breathed, CO can be removed 10 to 15 times faster than when just fresh air is breathed.

Space Travel Imposes Unusual Stresses

> *How does the body respond to changes in gravitational force?*

With the development of supersonic planes and spacecrafts, the human body can now be subjected to acceleratory forces resulting from sudden changes in velocity or direction. For example, a spacecraft must attain a speed of 25,000 miles/hr in order to sufficiently escape the Earth's gravitational field and attain a permanent orbit. High acceleration rates are required to attain these speeds. When an aircraft or spacecraft takes off, the crew are in a lying position and are subjected to **positive acceleration.** (The inertial forces move in a direction from front to back.) When an aircraft or spacecraft slows down at the end of a flight, **negative acceleration** occurs (inertial forces move in a foot-to-head direction).

G Forces and the Human Body

Any time an aircraft turns, centrifugal and acceleratory forces come into play as a result of gravitational attraction. The human body's sensation of weight is due to **gravitational attraction** and is referred to as **G force.** The G force is a unit used to express the magnitude of the acceleratory force. One G is equal to a force of magnitude equal to that of the normal gravitational force of the Earth. For example, when an individual is sitting in a chair, the force that presses the body against the chair seat is equal to 1G (one times the pull of gravity). If the force that presses against the seat is four times the normal weight of the body, the force is 4G. If force is less than the body weight, then the force is a negative G. Positive and negative G forces are illustrated in Figure 29–17, in which an aircraft pulls out of a dive or goes through an outside loop so that the pilot is held down by the seat belt. In the dive, the pilot experiences a positive G force, with inertial forces acting in a head-to-foot direction. In the loop, the pilot experiences a negative G force, in which the inertial forces are directed from foot to head.

The physiological effects of these acceleratory forces are due to the fact that some tissues (blood in particular) are "loose" with respect to body. When a positive G force is exerted in a head-to-foot direction, blood is pulled toward the lower parts of the body. In Figure 29–17, when the pilot "pulls 4Gs," venous pressure in the leg veins is four times the normal pressure. The heart cannot pump unless venous blood is returned, and consequently there is considerable pooling of blood in the lower extremities. At 4G, a pilot in the seated position experiences a fall in arterial systemic pressure, at the level of the heart, of approximately 40 mm Hg. Under these conditions, blood flow to the brain almost stops. Consequently, loss of vision, referred to as **blackout,** occurs. Other symptoms are also prevalent. At 3G to 4G, a pilot or astronaut experiences difficulty in the use of muscles. Facial tissue sags and eyelids droop. At 5G, movement of the body is almost impossible, and respiration stops because the respiratory muscles cease to function. Between 5G and 9G, legs become congested and calf muscle cramps are prevalent. Pulmonary edema occurs because of venous engorgement in the base of the lung, and vision as well as hearing are lost. Finally, consciousness is lost due to cerebral hypoxia. The effect of G forces are minimized in pilots of high performance aircraft, because they wear G-suits that provide compression of the legs and abdomen, and prevent the blood from pooling.

Most of these symptoms quickly vanish when the positive G force is removed. To minimize the G force, astronauts wear special antigravity suits and often lie in an reclining position during takeoff or reentry. On acceleration, these suits automatically pump up, and increased pressure exerted on the body prevents pooling of the blood. With negative G force (foot-to-head), the motion of blood is toward the head. At up to 3G, there is pressure in the eye sockets. Neck veins are overdistended and the face becomes puffy. Ringing in

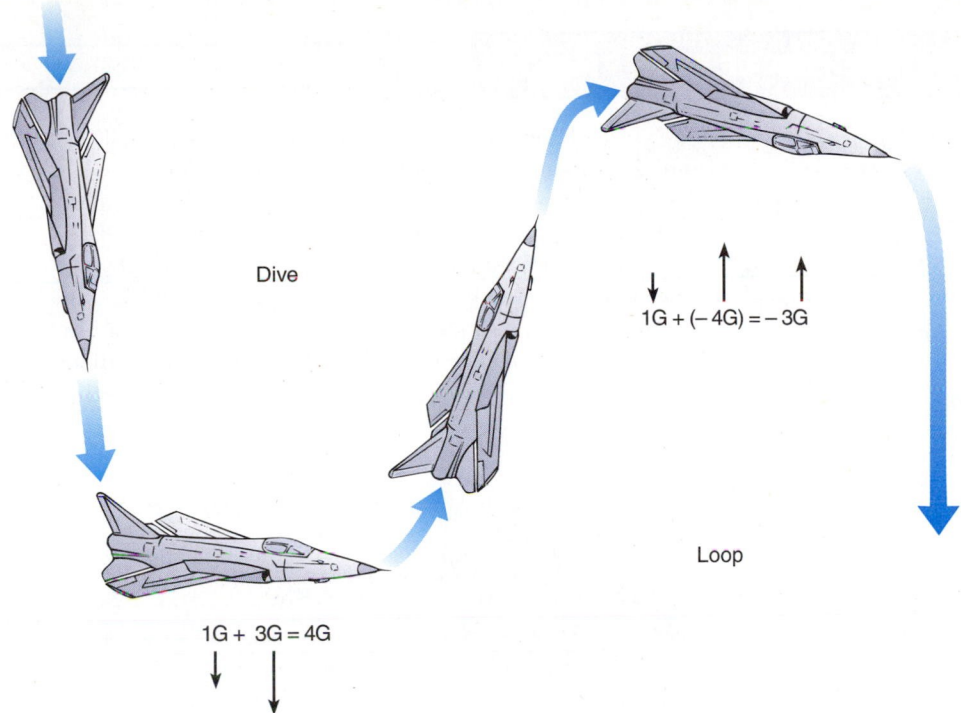

Figure 29–17

Illustration of positive and negative G forces on an airplane.

Dive

$1G + (-4G) = -3G$

Loop

$1G + 3G = 4G$

the ears occurs, and headaches may ensue. Retinal vessels and vision become engorged and vision is impaired at 3 to 4G. The light filtering through the blood gives a sensation of redness, a phenomenon known as **redout.** Often, at high negative G (greater than 3G), retinal and cerebral hemorrhages can occur. The physiological effects of negative G are much more lasting than those of positive G, and problems with vision and headache may persist for hours following a negative G exposure.

Weightlessness and Body Function

Once a spacecraft leaves the Earth's gravitational forces, astronauts are free of the problem of acceleratory forces but are faced with the problem of weightlessness (Fig. 29–18). The physiological problems associated with weightlessness do not appear to be severe. One concern is the lack of gravity, which results in decreased capillary pressure in the lower

(Text continued on page 896)

Figure 29–18

Astronaut James Bagian floats through the Spacelab Life Sciences module aboard the Earth-orbiting *Columbia* (SLS-40 mission). Bagian and six other crew members spent over nine days in space conducting life science research. *(Courtesy of NASA, Lyndon B. Johnson Space Center, Houston, TX.)*

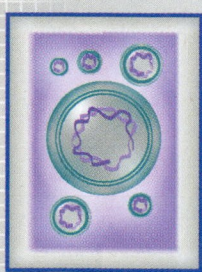

CURRENT CONCEPTS IN PHYSIOLOGY

Space Shuttle *Columbia:* First Space Laboratory Dedicated to Life Science Research

When the United States made the decision to explore space, many Americans were very skeptical about whether humans could live in outer space and many raised questions. Could cosmic particles affect the brain? Did space harbor foreign organisms that could be lethal to humans back on Earth? Could humans live for long periods of time in weightlessness? Since the National Aeronautics and Space Administration (NASA) was formed over 30 years ago, we have learned that people can indeed survive—and, more importantly, can live and work productively—in space.

Successful exploration of space depends on the health and well-being of the astronauts who travel and work there. For this reason, NASA has dedicated several shuttle missions to examine how space flight affects the human body. Spacelab Life Sciences-1 (SLS-1) was the first of these dedicated missions. It was launched as the 11th flight of the Space shuttle *Columbia* and was completed in June 1991. The SLS-1 was the first space laboratory dedicated to life science research. Its goal was to study how the human body adapted to microgravity.

In previous space flights, body fluid shifts, fluid losses, baroreflex deconditioning, deconditioning of muscles, calcium loss from bones, and reduction of red blood cell production were just some of the effects observed in astronauts. Missing from the previous studies were data on the interrelationships among different body systems in microgravity. Space shuttle *Columbia*'s crew of seven used themselves as human guinea pigs to enable scientists to understand how the body acclimates to weightlessness. Twenty different studies were carried out on five body systems during the SLS-1 mission. These systems included (1) cardiopulmonary, (2) renal/endocrine, (3) blood-immune, (4) musculoskeletal, and (5) neurovestibular (brain, nerves, eyes, and inner ear). Specific experiments were designed to determine how much blood was delivered to the various organs, how pulmonary function changes, and what reflex mechanisms regulate blood pressure. The problem of motion sickness in space was investigated by subjecting the crew to conflicting visual, vestibular, and tactile information to determine the effect of sense on orientation. Studies on the other systems included muscle metabolism, inhibition of bone formation, and blood and lymphocyte production.

In addition to the human studies, over 2000 jellyfish and 29 laboratory rats were carried on the mission and were used for complementary studies. In the jellyfish experiments, thousands of jellyfish polyps were encased in flasks containing artificial seawater. Early in the mission, thyroxine or iodine was injected into the flasks to induce the polyps to metamorphose into free-swimming ephyrae. The tiny jellyfish were filmed to observe their swimming motions compared with a control on Earth to determine whether

Columbia shuttle liftoff, June 5, 1991. The Spacelab Life Sciences-1 Laboratory was the first Spacelab focusing solely on life sciences research. During the trip, the crew explored physiological changes that occur during space flight, as well as the consequences of the body's adaptation to microgravity and the readjustment to gravity upon return to Earth. *(Courtesy of NASA, John F. Kennedy Space Center.)*

SLS-40 mission insignia. The SLS-40 patch, like the space mission, focused on human beings living and working in space. Against a background of the universe, seven silver stars pictured around the orbital path of *Columbia* represent the seven crew members. The orbiter's flight path forms the double-helix of the DNA molecule that is common to all living creatures. In the words of a crew spokesman, "[The helix] affirms the ceaseless expansion of human life and American involvement in space while simultaneously emphasizing the medical and biological studies to which this flight is dedicated." Da Vinci's *Vitruvian Man,* silhouetted against the blue darkness of the heavens, also serves as a powerful embodiment of the extension of human inquiry from the boundaries of Earth to the limitless laboratory of space. Drawn by artist Sean Collins, the SLS-40 Space Shuttle patch was designed by the crewmembers for the flight. *(Courtesy of NASA, Lyndon B. Johnson Space Center, Houston, TX.)*

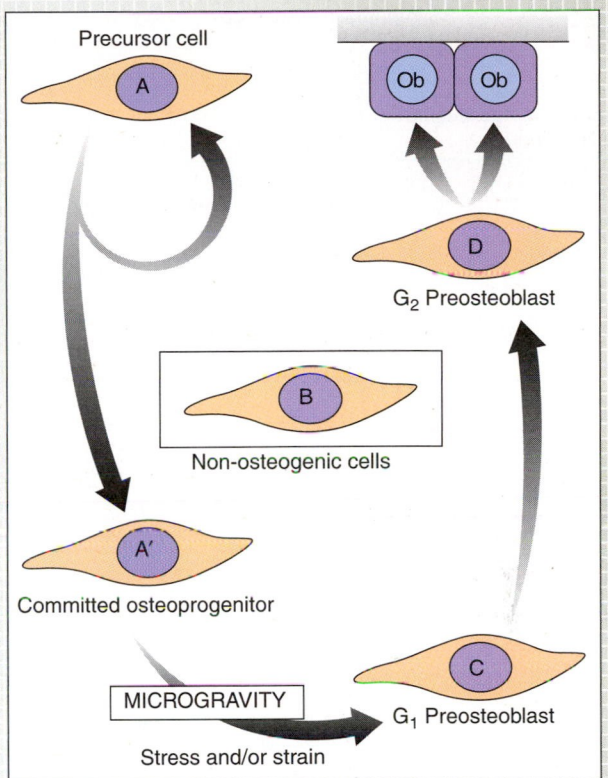

Microgravity inhibits conversion of the committed osteoprogenitor cell to a G_1 preosteoblast. The result is decreased bone formation. The mechanisms for this inhibition in microgravity conditions are under investigation.

there is any difference between jellyfish gravity receptors that develop in space and those that develop on Earth.

The rodents were used for research on muscle, bone, and inner ear function. In one particular study, young rats in rapid growth stage were studied for alterations in bone physiology. Two scientists from the Indiana University School of Dentistry, Drs. W.E. Roberts and L. Garetto, in conjunction with Dr. E.M. Holton at NASA Ames Research Center, are studying microgravity-induced osteopenia (loss of bone) in the jawbones. Under normal conditions, the conversion of the committed osteoprogenitor cell to a G_1 preosteoblast requires a certain amount of mechanical loading (stress/strain). In space, however, this

conversion step is inhibited, and bone formation is decreased (see figure). Drs. Roberts and Garetto are investigating two possible mechanisms that could inhibit this process in microgravity. First, is there a general decrease in mechanical loading, which fails to induce committed osteoprogenitor cells to differentiate into preosteoblast cells? Second, is the change in tissue geometry due to the loss of extracellular water? Differentiation of preosteoblasts is inhibited by excessive cell contact. Data from the SLS-1 mission will provide crucial information on how microgravity affects bone physiology.

QUESTION: Would calcium supplementation help space-induced osteopenia?

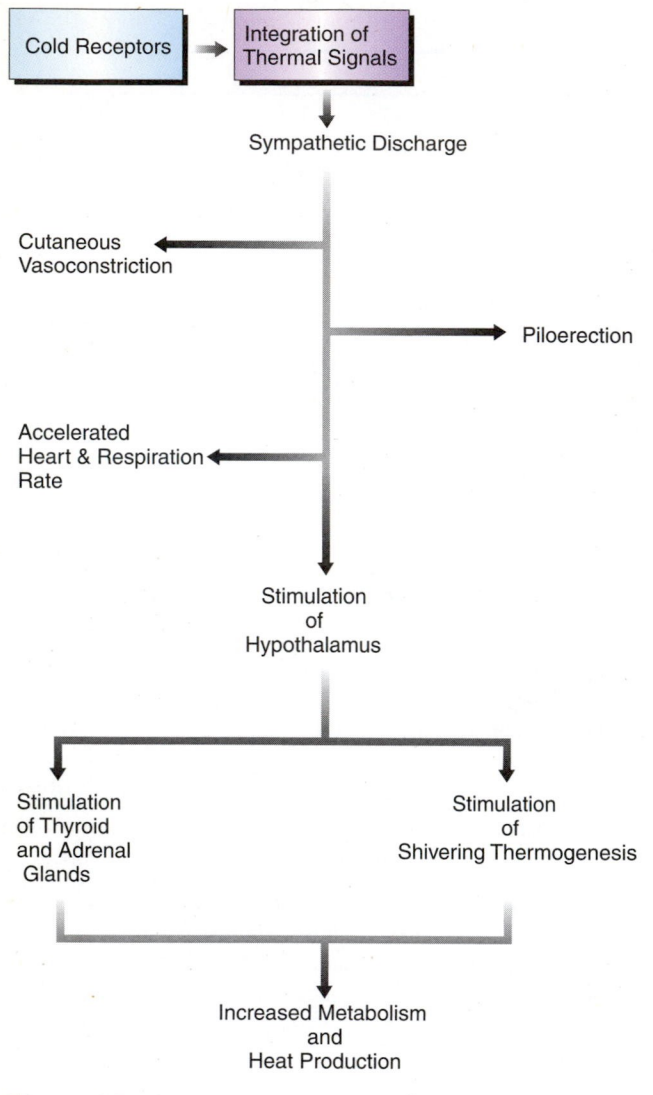

Figure 29–19

Integration of thermal signals in response to cold stress.

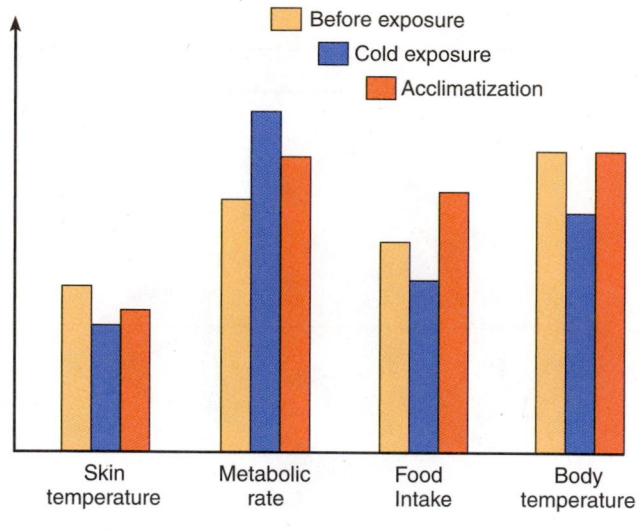

Figure 29–20

Acclimatization to cold.

extremities, decreased blood volume, red cell mass, and decreased cardiac output. The second physiological problem related to weightlessness is diminished physical activity, because very little muscle contraction is required for movement. The decrease in physical activity leads to a decrease in work capacity and loss of calcium from bone.

Artificial Atmospheres in Spacecraft

There is no atmosphere in space; consequently astronauts must be provided with an artificial atmosphere that is pressurized and equipped with life-sustaining oxygen. The types of atmospheric conditions selected depend, in part, on the sophistication of the engineering. For example, the Russian space program used 21% oxygen in nitrogen at a pressure equivalent to sea level, while the early Apollo missions used 100% oxygen at one third of atmospheric pressure, which gives a Po_2 equivalent of 1 atm. The reason for these two different gas mixtures was based on weight and space. In the

earlier Apollo missions, NASA did not have booster rockets sufficient to launch a heavy spacecraft. Using 100% oxygen at 1/3 atm gives a lighter-weight load than a thick-shelled spacecraft using air at 1 atm. One of the dangers of 100% oxygen is the possibility of an explosion or a fire. Such an accident did occur aboard one of the early Apollo crafts on the launch pad, resulting in a fatal fire and forcing NASA to change gas mixtures. Now all U.S. space flights use air as a gas mixture.

Temperature Extremes Cause Injury

 How does the body respond to extreme temperature environments?

Modern heating and air conditioning allow cars, homes, and offices to maintain temperatures at 72°F. The average North American is not aware of the physiological responses involved in temperature extremes. Crossing the desert at temperatures of 115°F or the plains at temperatures of −30°F can now be done without much discomfort.

Even our society somewhat influences the degree of cold or heat exposure that we experience. For example, back in the 1950s, before the energy crisis, a study showed that the average temperature for dressing children for school in the winter for British children of university-educated parents was 50°F, while the room temperature averaged 75°F for U.S. children with similar backgrounds. Also, office temperatures in the United States are maintained at much lower temperatures in the summer and at higher temperatures in the winter compared with many other countries.

Response to Cold

The mechanisms of heat loss have been discussed previously, in Chapter 27. Environmental physiology deals with these mechanisms but focuses primarily on acute and chronic effects of cold stress. When a lightly clad human subject is placed in a cold room at −20°F, a number of physiological events take place very quickly (Fig. 29–19).

1. **Cutaneous vasoconstriction** shifts blood from the body surface to the core and decreases heat conductivity from the interior by lowering heat loss from skin to the environment.
2. **Piloerection** (erection of hair) in the skin, often called "goose bumps" or "goose flesh," increases insulation by trapping air on the skin. In mammals such as dogs and cats, piloerection is very effective in trapping air in fur, but it is not very effective in humans, who have little body hair.
3. Neurohormonal activation of the hypothalamus triggers the release of hormones from the anterior pituitary, especially those that stimulate the thyroid and adrenal cortex.
4. Increased activity of skeletal muscle results in shivering and often occurs as soon as a lightly clad subject is exposed to cold. Shivering increases metabolic rate and can lead to three- to fourfold increase in heat production as a result of the energy expended by this type of muscular activity. Because all of this energy goes into producing heat and none into work, shivering thermogenesis is a very economical way of elevating heat production. Interestingly, **hypothermia** cannot be induced in experimental animals unless shivering is blocked, for example, by a light anesthetic. Women have a higher shivering threshold than do men; i.e., men shiver before women at the same environmental temperature because women's extra subcutaneous fat layer insulates against the cold. Similarly, subjects who are heavily built, with extracutaneous fat, do not shiver as quickly as those who are lean.
5. Body position by both humans and animals is changed in order to reduce heat loss. This is accomplished by tucking in arms and legs and curling up in a ball position.
6. Heart rate and ventilation increase.

These responses are somewhat paradoxical because increased ventilation accelerates heat loss from the lungs, and increased heart rate places an added stress on the cardiovascular system because of the massive peripheral vasoconstriction that occurs. This means that there is not only increased blood flow to the internal organs but also increased systolic and diastolic blood pressures.

Most of these responses to cold exposure are initiated through the sympathetic stimulation, which tends to conserve heat. Cold receptors located on the skin trigger a sympathetic discharge that causes vasoconstriction, piloerection, increased heart/ventilation rate, and increased shivering thermogenesis (see Fig. 29–19).

Cold Acclimatization

Acclimatization to cold occurs over a two- to three-week period. One of the first criteria of cold acclimatization is the switch from shivering thermogenesis to nonshivering thermogenesis. During continuous exposure to cold, nonshivering thermogenesis is optimal at the end of two to three weeks. At the end of this period, shivering disappears and the sustained heat production is now maintained solely by the elevated metabolism of visceral organs. Metabolic rate is elevated approximately 20% above normal during cold exposure due to the level of thyroid hormones (Fig. 29–20). If acclimatized individuals are removed from a cold environment and placed into a warm environment, metabolism and O_2 consumption remain elevated for several weeks. The next step in cold acclimatization is increased food intake to maintain the elevated metabolim. Other features of acclimatization are a rise in skin temperature, warmer hands and feet, and a higher core temperature (see Fig. 29–20).

Response to Heat Stress

Temperature extremes have been recorded all over the world. A temperature of 127°F was reported in Queensland, Australia, and a temperature of 136°F was recorded in Libya. Death Valley, California, has similar temperatures. Unfortunately, these high temperatures are also in areas where water is not available for evaporative cooling and for osmotic regulation. In a dramatic account of a man lost in the Mexican desert for eight days with only one day's supply of water, he survived by drinking his own urine, chewing the end of cactus spines for water, and drinking blood from scorpions.

Humans suffer more stress from heat than from cold exposure because of our high internal heat production and the limited avenues for unloading heat. Essentially, the only physiological mechanisms for cooling the body are sweating, vasodilation, and panting (see Chapter 27 for mechanisms of heat loss). Heat stress usually is classified as hot-dry or hot-humid. For hot-dry, the temperature is greater than 95°F with a low humidity. For hot-humid, the upper limit for ambient temperature is 90°F to 95°F with a relative humidity of greater than 75%. Abrupt functional changes occur in the first two days of heat stress. There is an immediate increase in sweating, followed by cutaneous vasodilation. Second, there is a transitory increase in plasma volume as fluid shifts from the interstitial space to the vascular bed. When water lost by sweat is not replaced, the blood volume decreases. Third, there is an increase in oxygen consumption due to the effect of elevated body temperature on metabolic rate. Fourth, there is a slight increase in food consumption, even though body weight decreases with heat stress. Fifth, there is dehydration because the thirst mechanism is not sufficient to replenish the body's loss of water. Dehydration progresses with heat stress. Surprisingly, sweating continues at the same rate regardless of the dehydration. At 2% weight loss, dehydration is very severe. At 4% weight loss, the throat and mouth become extremely dry, and at 8% weight loss salivary function stops and speech is slurred. Humans are unable to function if dehydration continues and there is a 10% weight loss. Much of the pioneering work in this area was done in the United States during World War II on conscientious objectors.

Acclimatization to heat takes 7 to 14 days. Specific changes include (1) increased peripheral thermal (heat) conductance, (2) increased sweating capacity with a fall in salt concentration in the sweat, and (3) a rise in skin temperature, a fall in core temperature, and more even distribution of sweat over the body. Women do not acclimatize to heat as well as men. This is due, in part, to a lower thermal gradient to remove metabolic heat, lower capacity to increase blood flow, and a lower rate of sweating.

Heat Exhaustion

Failure to tolerate heat stress leads to **heat exhaustion,** a condition marked by a subnormal body temperature, dizziness, headache, nausea, and sometimes collapse. Each year several hundred deaths occur in the United States from heat exhaustion. Mild heat exhaustion is a common occurrence in hot work environments and exposure to direct sunlight. Heat exhaustion is prevalent in the military, especially during training or troop mobilization in hot environments, and physical fitness alone does not protect against it. There are many instances of highly trained combat troops who suffered heat exhaustion even though they were in top physical condition. Interestingly, increases in core temperature that result from high external environmental temperatures are not tolerated as well as similar core temperature increases that result from internal causes. For example, heat exhaustion does not occur when core temperatures reach 42°C (normal, 37°C) with a fever. During strenuous exercise, rectal temperatures can reach 40°C without harm. Yet, if a rectal temperature reaches 39°C when heated from the external environment, a person can lose consciousness.

The first stage of heat exhaustion, termed *heat shock*, occurs when thermal regulation fails and there is a critical rise in body temperature. Heat shock can also occur through dehydration without a rise in body temperature. One of the early symptoms of heat shock is heat-induced tetany (muscle cramps and uncontrolled muscle spasms). This appears to be induced by cellular changes in CO_2 tension and pH rather than a rise in temperature or increased intracellular calcium.

The second stage of heat exhaustion is **heat stroke** (sometimes referred to as *sunstroke*) and is marked by convulsions, coma, and brain damage. Death occurs from cardiovascular failure. Individuals usually go into circulatory shock with internal massive vasodilation and a concomitant decrease in central blood volume.

RESPONSES TO THE CHEMICAL ENVIRONMENT

 How does the body respond to chemical pollutants?

The human body is continually subjected to a wide variety of chemicals in the environment. Some toxic poisons, such as lead, have been known for 2000 years. However, with the Industrial Revolution, the release of new pollutants into the environment has been staggering. There are over 10,000 new chemicals produced annually that are "foreign" and not normally found in nature. These and other environmental chemicals are in the air, water, and food that we use.

What are the effects of these environmental chemicals on the body? How does the body handle them? How successfully does the body acclimatize to chemical stress? These questions and other closely related questions will be the focus of this section.

Three important issues with chemical pollutants are mode of entry, removal, and tissue storage. Chemicals gain entry into the body through the lungs, gastrointestinal tract, and skin. Many organic compounds enter through the lining of the lung and gut by either simple diffusion or carrier-mediated transport. Chemicals are removed by excretion in the urine and feces. The liver helps to remove these molecules by transforming them to the less lipid-soluble form, resulting in enhanced urinary excretion. The liver is one of the most important organs in the body's defense mechanisms against toxic agents. The liver is capable of detoxifying many chemicals or transforming them so that they can be readily eliminated from the body. However, transformation by the liver is not always beneficial. In some cases, foreign chemicals are nontoxic until liver enzymes transform them. Many of these transformed chemicals are carcinogenic. A classic example is DDT. This pesticide is essentially harmless until the liver enzymes transform DDT into toxic molecules that cause most of the adverse biological effects of DDT.

Respiratory Pollutants Exhibit Different Modes of Action

Pollutants are classified as respiratory or systemic (Table 29–3). A number of pollutants irritate and damage the lungs. Each exhibits a different mode of action. The gaseous pollutants (e.g., sulfur dioxide, nitrogen dioxide, ozone, chlorine, and ammonia) damage the epithelial and mucous lining of the respiratory tract. Dust particles (e.g., silica, quartz, carbon, asbestos, cobalt, and iron oxides) damage the pulmonary interstitial tissue, and other irritants, such as beryllium, nickel, and chromium, are carcinogenic and affect many different types of lung cells.

Gaseous Pollutants

Sulfur oxide is one of the oldest air pollutants. Long before coal was used as a fuel, hydrogen sulfide (H_2S) was added to the atmosphere from volcanoes, swamps (swamp gas), and shallow lakes. However, sulfur oxides became a major pollutant with the advent of the industrial revolution and the burning of coal and oil, both of which contain sulfur as an impurity. Sulfur content of oil and coal is approximately 3%, with soft coal having a higher content than hard coal. The major sulfur pollutants in the atmosphere are hydrogen sulfide (H_2S), sulfur dioxide (SO_2), sulfuric acid (H_2SO_4), and sulfates (SO_4). Hydrogen sulfide readily combines with O_2 to form sulfur dioxide ($H_2S + O_2 \rightarrow SO_2 + H_2O$). Sulfur dioxide can undergo a series of reactions to form sulfuric acid ($SO_2 + O_2 \rightarrow SO_3$, which immediately reacts with water vapor: $SO_3 + H_2O \rightarrow H_2SO_4$. These reactions are extremely fast in the sunlight. Sulfur dioxide is a colorless gas with a pungent irritating odor that can be detected at 0.5 to 0.8 parts per million (ppm). Sulfur dioxide is very soluble, and more than 95% of the inhaled SO_2 is absorbed in the moist lining of the upper airways and never reaches the small airways or alveoli. SO_2 penetrates the alveoli only when it is absorbed onto the surface of a small

TABLE 29–3

Respiratory and Systemic Pollutants and Their Principal Effects

Classification	Pollutants	Effects
Respiratory pollutants		
Gaseous irritants		
	Sulfur dioxide	Respiratory lining
	Nitrogen dioxide	Respiratory lining; bronchitis and emphysema
	Ozone	Respiratory lining; bronchitis and emphysema
	Choline	Pulmonary edema
	Ammonia	Pulmonary edema
Granuloma producing		
	Beryllium	Lung parenchyma cells
	Nickel	Lung parenchyma cells
	Chromium	Lung parenchyma cells
Respiratory dusts		
	Silica	Pulmonary interstitial inflammation and fibrosis
	Quartz	Pulmonary interstitial inflammation and fibrosis
	Carbon	Pulmonary interstitial inflammation and fibrosis
	Asbestos	Pulmonary interstitial inflammation and fibrosis
	Cobalt	Pulmonary interstitial inflammation and fibrosis
	Iron oxides	Pulmonary interstitial inflammation and fibrosis
Systemic pollutants		
	Lead	Nerve tissue
	Mercury	Brain and gut
	Fluoride	Bones and teeth
	Cadmium	Kidneys and blood vessels
	Chlorinated	
	Hydrocarbons	Liver and fatty tissues

dust particle (0.5 μm in diameter) or is converted to H_2SO_4. SO_2 at 1 ppm causes irritation of eyes, nose, and throat and causes constriction of the bronchi at higher concentrations. During the two-day London air pollution crisis in 1952, the average concentration of SO_2 was 1.7 ppm. The most serious of the sulfur oxides is H_2SO_4. In the atmosphere this exists as sulfuric acid mist, and in the presence of small dust particles H_2SO_4 can be absorbed and transported to the alveoli. H_2SO_4 causes swelling of mucous membranes, increased mucous secretion, narrowing of airway passages, and alveolar thickening.

Both ozone and the oxides of nitrogen are oxidants and are capable of attacking biological membranes. The principal industrial source of nitrogen oxides is from the burning of oil or natural gas, and motor vehicle fuel combustion. Over 20 million tons of nitrogen oxides are produced annually in the United States. When nitrogen is exposed to high temperatures (e.g., furnace, combustion engine), nitrogen is converted to nitric oxide (NO; the new nomenclature is *nitrogen monoxide*). When NO is vented and cooled, a fraction is converted to nitrogen dioxide (NO_2). Further complex reactions in the presence of sunlight result in the formation of ozone (O_3).

Nitrogen dioxide (NO_2) is a reddish-brown gas that is highly toxic and irritating, with a pungent odor. Nitrogen dioxide poisoning was first recognized in farmers who filled silos with silage (chopped green corn). NO_2 builds up shortly after

filling, and if a farmer ventured into the top of the silo after filling he could be exposed to lethal concentrations of NO_2. Similar accidents have been reported by firefighters fighting burning chemical plants, welders, and workers inside of boilers. The symptoms of NO_2 poisoning include headache and a dry, hacking cough shortly after exposure, followed by pulmonary edema one to two days later. Patients with pulmonary edema experience a shortness of breath, fever, and wheezing. The increased burden of oxygenating the blood as well as pushing blood through water-soaked lungs leads to hypoxemia and a life-threatening strain on the right heart. If pulmonary edema is not reversed, these individuals can become comatose and die from severe hypoxemia and right heart failure. One of the most tragic incidents of NO_2 poisoning occurred in a fire at the Cleveland Clinic in 1929. The burning of X-ray films killed 97 persons within 2 hours, and 26 other victims died within three weeks.

Nitrogen dioxide can be detected at concentrations of 1 to 2 ppm, which are known to occur in polluted air. Low-level exposure (of 1–3 ppm) leads to irritation of the lung lining. A high incidence of bronchitis and emphysema has been reported in cities with high NO_2 levels. Nitrogen dioxide in cities is often part of photochemical **smog** that makes epidemiologic studies difficult. Nitrogen dioxide is also an ingredient of tobacco smoke. Cigarette, pipe, and cigar smoke contain approximately 300, 900, and 1200 ppm NO_2, respectively.

Tobacco smoke also contains other harmful compounds, which again makes it difficult to separate the NO_2 effects from other long-term hazards of smoking. However, smokers working in an environment already polluted by auto exhaust are at high risk for the harmful effects of NO_2.

Ozone is an unstable blue gas with an oxidizing power that is surpassed only by fluorine. The name comes from its pungent odor, which is derived from the Greek *Ozein*: the smell that occurs from lightning during a severe thunderstorm. Ozone constitutes over 90% of the oxidants in urban smog.

Ozone, like NO_2, acts on the lung, where it can harm the respiratory lining. Ozone can be detected at 0.05 to 0.1 ppm and at 0.1 ppm can cause irritation of the eyes, nose, and throat. Higher concentrations can cause bronchitis and emphysema. In Los Angeles, a concentration of 0.5 ppm constitutes a "first alert." A second effect of long-term, low-level exposure to ozone is that it is involved in premature aging. A third effect is its leading to a higher incidence of lung tumors. Both the premature aging and high incidence of tumors are due to the formation of free radicals from ozone.

Respiratory Dusts

When a small glass mirror covered with petroleum jelly is placed in the backyard on the picnic table of any urban home, a number of particles of all sizes and shapes accumulate on the glass. These include an abundance of carbon, quartz, and silica particles ranging from 500 μm down to 0.5 μm. The glass can also show rubber particles and trace amounts of cobalt and iron oxide. These are examples of the kinds of particulate matter in our city parks, shopping centers, and factories that become a part of our respired air. Dust particles produce a disease known as *pneumoconiosis*, a condition characterized by fibrosis (scarring of lung tissue). Pneumoconiosis caused by a certain type of particulate is given a specific name. For example, silicosis is a specific type of pneumoconiosis caused by silica.

Inhaled particles in a range from 0.5 to 5.0 μm constitute **respiratory dust.** These are the particles suspended in air and often are seen in the light beam from a projector shining onto a screen. Particles measuring 15 μm or more are subject to gravitational settling. With air-pollution control, the large particles (>15 μm) have decreased. Removal of these large particles has caused a shift in the particle size distribution in the atmosphere and has actually increased the number of respiratory dust particles (0.5–5.0 μm).

The hairs in the nose and moist membranes in the nasopharyngeal passage stop particles from 5 to 10 μm. Particles ranging in diameter from 2 to 5 μm are caught by a sheet of mucous lining the tracheobronchial tree (Fig. 29–21). These particles are removed by this mucociliary transport system and are either coughed up or swallowed. Particles measuring less than 2 μm are capable of reaching the alveolar/capillary bed. Alveoli have neither mucus nor cilia, and can remove these small particles only by alveolar macrophages. These are phagocytic cells that creep along the floor of the alveoli and scavenge foreign particulate, including bacteria. The foreign matter is ingested and then destroyed by proteolytic enzymes. Because of their mobility, macrophages move up the respiratory tract and are also removed by the mucociliary transport system. Many of the particles that penetrate the alveolar/capillary bed are not removed and eventually get into the lung interstitial tissue, causing injury.

Silica and Other Dusts

The disease caused by silicon dioxide (usually called *silica*) is **silicosis.** At one time, silica was thought to be responsible for the specific disease that coal miners get from breathing coal dust ("black lung disease"). Coal miners' pneumoconiosis is now recognized as a distinct disease due to coal dust alone and not to silica.

Asbestos is present in nature, and, if handled improperly, becomes a respiratory pollutant. Inhalation of asbestos

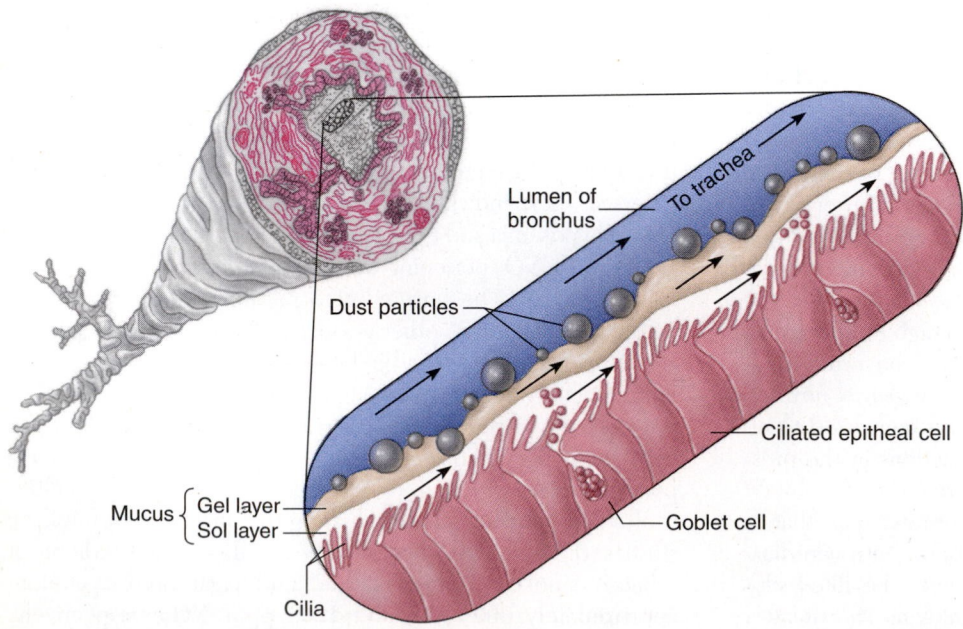

Lumen of bronchus

To trachea

Dust particles

Ciliated epitheal cell

Mucus { Gel layer / Sol layer

Goblet cell

Cilia

Figure 29–21

The mucociliary transport system acts as a biological scrubber to trap and remove inhaled dust particles and chemical irritants.

dust has long been recognized as the cause of **asbestosis**. Asbestos is a general name given to a variety of minerals that can unravel into fibers and are found in nearly every country in the world. The most important deposits in the United States are California, Arizona, and Vermont. The value of asbestos comes from its indestructibility and heat-resistance properties. Before the ban on asbestos, it was manufactured into more than 3000 different products. One microgram of asbestos suspended in air can unravel into more than 200 million tiny fibers, and, because of the geometry, asbestos fibers 200 μm long and 3.5 μm in diameter can penetrate the lung. Asbestos causes an inflammatory response that leads to excess collagen synthesis, which in turn causes fibrosis.

Iron oxide is another particulate that will cause pneumoconiosis. At autopsy, lungs show a brick-red discoloration. Usually atmospheric iron binds with other particulate, namely silica, to cause lung injury. Iron oxide also has been shown to be carcinogenic.

Cobalt is another inhaled metal that causes pneumoconiosis. Cobalt exists in nature, and many species of bacteria, protozoa, fungi, and insects require cobalt to sustain life (it is an essential part of vitamin B_{12}). Only during the twentieth century has cobalt become an important air pollutant, particularly since World War II.

Pneumoconiosis, regardless of its cause (dust from silica, coal, or asbestos) has similar pathophysiology. The first stage leads to scarring of lung tissue (fibrosis), increased lung elastic recoil, shortness of breath, and a decline in pulmonary function (difficulty inflating the lungs). The latter stage leads to alveolar hypoxia and to low oxygen in the blood (hypoxemia). Alveolar hypoxia also triggers vasoconstriction of the pulmonary arterial vessels, causing pulmonary hypertension and eventually right heart failure.

Some Pollutants Act as Systemic Poisons

In contrast to pollutants that mainly involve the lung, there is another class of pollutants that affect numerous organs and act like systemic poisons (see Table 29-3). These include such compounds as lead, mercury, fluoride, and cadmium. They are emitted from both moving and stationary sources and constitute a great threat to our environment because of their toxicity and their ability to accumulate in the body.

Lead is a heavy soft metal that is used in a number of utensils, paints, and glazes in pottery. The toxic properties have been known for more than 2000 years. The Greeks knew about lead poisoning, but the Romans apparently did not. Major sources of lead in the environment and its path into the body are shown in Figure 29-22. The sources of entry of lead into the body are the lungs and digestive tract. Only 5% to 10% of the ingested lead is actually absorbed, with the remaining amount excreted in the feces. However, over 40% of the lead inhaled by the lungs enters the blood and is deposited in tissue. Lead has four sites of action (Fig. 29-23): blood; gastrointestinal tract; and, in the advanced stages, the nervous system (brain and spinal cord) and kidneys. Early symptoms are anemia, headaches, and muscle

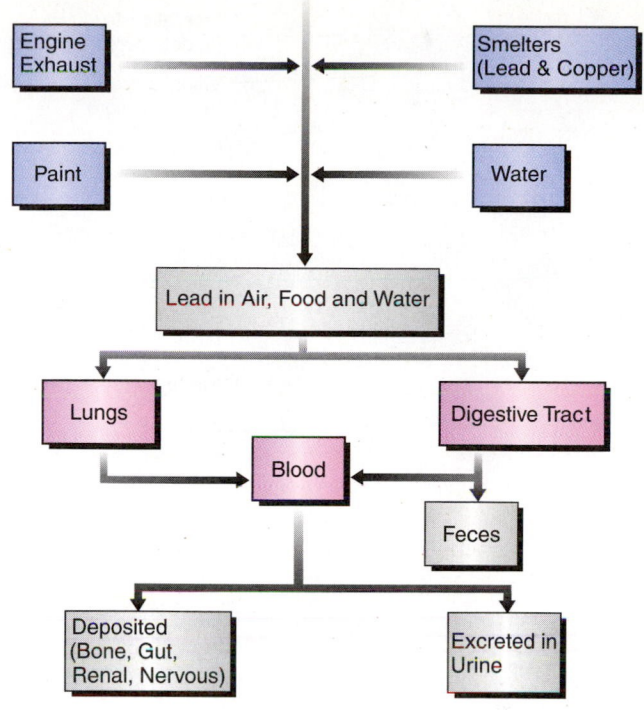

Figure 29-22

Sources of lead and its path into the body.

pain. Persons affected also become clumsy and irritable. In the later stages, symptoms include vomiting, abdominal pain, and black stools from bleeding of the colon. Advanced stages include brain damage, convulsions, and renal failure.

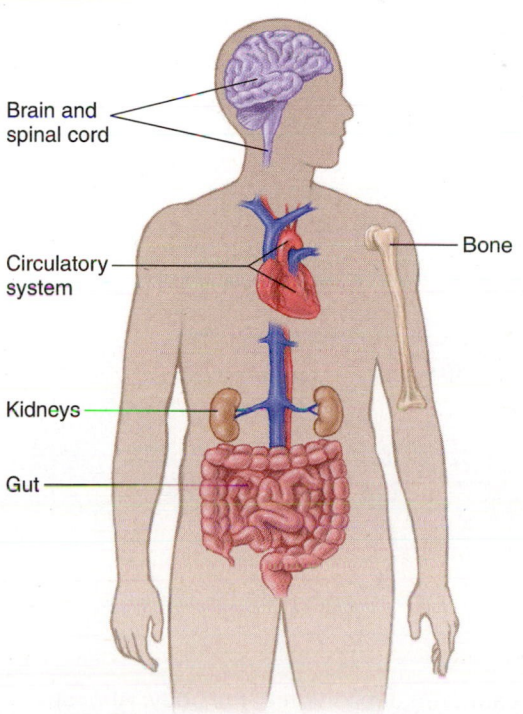

Figure 29-23

Principal sites of action of ingested and inhaled lead.

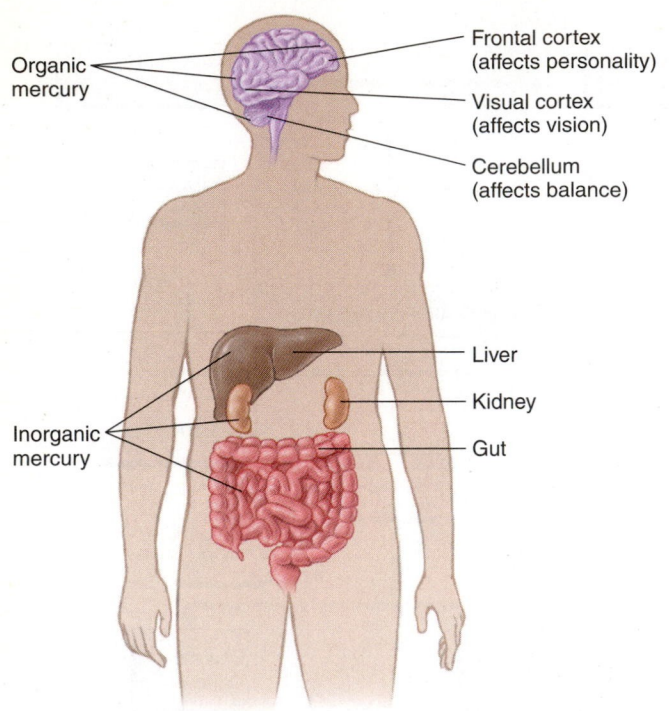

Figure 29–24

Sites of action of mercury.

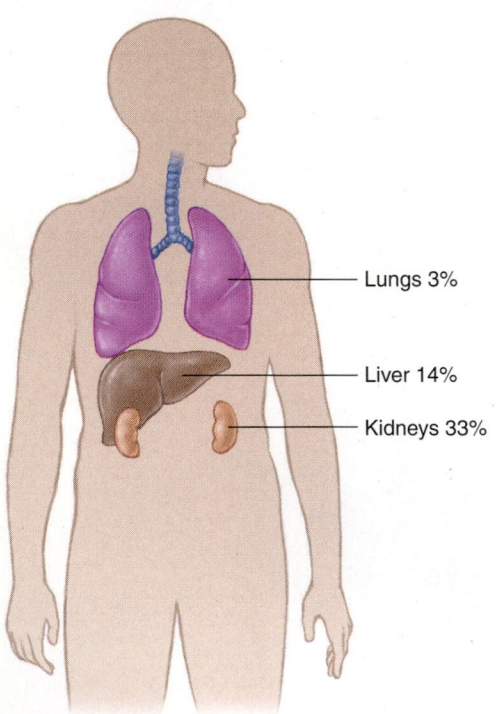

Figure 29–25

Major sites of cadmium deposition in the body. Almost half of the cadmium in the body is stored in kidney and liver.

Removing it from paints and gasoline has dramatically reduced lead contamination in the environment. Removing lead from gasoline has been extremely beneficial because atmospheric lead was a major way that it entered the body. However, lead poisoning is still a major problem in metropolitan areas around the world.

Mercury, or "quicksilver," has long been regarded as a special metal because of its unique property of being a liquid at room temperature. The Greeks were aware of its toxic properties, and mercury ore (cinnabar) was mined in Spain for 27 centuries. As early as the year 1500 ACE, mercury was used in the treatment of syphilis and undoubtedly contributed to a number of deaths. Mercurous chloride (calomel) was used as a cathartic (causing evacuation of the bowels) until 50 years ago, when physicians became aware of its toxic action. Sources of mercury include batteries, photographic materials, paints, steel, and paper mills. Mercury enters the body in two forms: either as organic or inorganic mercury, primarily through the food chain, but it can also be taken in by drinking or by breathing. Inorganic mercury settles in and damages the liver, kidney, and gastrointestinal tract, and organic mercury settles in the brain and destroys neuronal cells that control coordination, vision, and personality (Fig. 29–24).

Cadmium is a soft, silver-white metal that resists corrosion and is used in batteries, electrical wires, engine parts, and radio/TV parts, as well as in nuclear reactors to control the rate of nuclear fission. In the early 1900s, cadmium was used in the treatment of syphilis and malaria, until its toxic properties became recognized. Cadmium enters the body primarily through

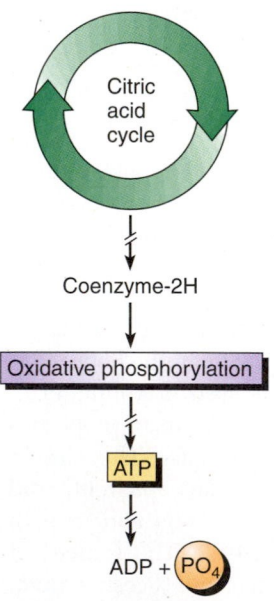

Figure 29–26

Sites of action of cadmium on cellular ATP. Cadmium (1) blocks transfer of electrons from citric acid cycle, (2) blocks ATP synthesis, and (3) inhibits ATPase (i.e., the breakdown of ATP to ADP).

food and water and is absorbed through the intestinal wall from the gut lumen. Cadmium also enters the lungs from polluted air and from smoking cigarettes. Cadmium is deposited in a variety of tissues but is concentrated in three organs: kidney, liver, and lungs (Fig. 29–25). Long-term cadmium poisoning leads to pneumonia, pulmonary emphysema, and cirrhosis of the liver. Cadmium is also linked to hypertension (high blood pressure). Humans with a high concentration of cadmium in the kidney have a high incidence of arterial hypertension and excrete 40 times more cadmium in their urine than do individuals with normal blood pressures.

The cellular action of cadmium has been clearly pinpointed to the membrane of the mitochondria. Recall from Chapter 6 that two major energy-producing pathways exist in the mitochondria: the citric acid cycle and the electron transport (respiratory) chain. The citric acid cycle generates nicotinamide and flavin, which contain electrons and hydrogen atoms. The electron transport chain takes these electrons and hydrogen atoms and converts them to usable ATP. Cadmium acts on mitochondrial enzymes and blocks ATP synthesis (Fig. 29–26). Cadmium also binds to the enzyme adenosine triphosphatase (ATPase), which is required to split a phosphate from ATP. Inhibition of ATPase prevents the body from utilizing ATP as a source of cellular energy.

Is Environmental Toxicology a Modern Issue?

Systemic poisons vividly illustrate how difficult it is to answer the central question of modern environmental toxicology: what are the effects of long-term exposure to small doses of a chemical pollutant? It is relatively easy to identify symptoms of acute poisoning (e.g., lead) but extremely difficult to detect long-term low-level effects. For most environmental pollutants, a major question that still has not been addressed is, Does every potential toxic agent have a threshold below which there is no damage?

CHAPTER REVIEW

Summary

- Environmental physiology is the study of functional mechanisms of healthy individuals that occur in response to their external environment.
- Environmental stress can lead to (1) adaptation, a genetically induced change that is passed on to the next generation and (2) acclimatization, which is environmentally induced and leads to improvement of an existing physiological mechanism, but with no change in the genome.
- Biological clocks lead to intrinsic rhythms that vary from minutes to days.
- Circadian rhythms follow a 24-hour cycle and are often linked to a light-dark cycle.
- Ascent to high altitude exposes the body to hypoxia, and rapid ascent to high altitude can lead to altitude sickness.
- Acclimatization to altitude allows the body to adjust to hypoxia and includes improved regulation of pH as well as increased cardiac output, ventilation, and hematocrit.
- Acclimatization to underwater diving leads to improved diving reflexes, improved respiration, increased metabolism, and increased body insulation.
- Deep-sea diving leads to an increase in gas pressures and causes nitrogen to be absorbed by blood and tissues. Rapid ascent leads to "the bends."

- Breathing pure oxygen leads to atelectasis and pulmonary edema. Cellular damage is due to the formation of free radicals, which oxidize membrane lipids, proteins, and enzymes. High oxygen also causes blindness in newborns (retrolental fibrosis); oxygen under high pressure causes convulsions.
- Space travel leads to unusual stresses. Astronauts are subjected to G forces, artificial environments, and weightlessness.
- Temperature extremes lead to a number of physiologic changes. Response to cold stress leads to improved neuroendocrine function, cardiovascular changes, and heat production. Humans have lower options in response to heat stress, which often leads to heat exhaustion.
- Lungs and liver are the most important organs in the body's defense mechanism against environmental pollutants.
- Gaseous pollutants (e.g., nitrogen dioxide, sulfur dioxide, and ozone) damage epithelial and mucous linings of the respiratory tract, whereas dust particles (e.g., silica, carbon, and quartz) damage the interstitium to cause fibrosis and cancer.
- Other pollutants, such as lead, mercury, and cadmium, are classified as systemic pollutants because they affect numerous organs.

Review Questions

Choose the Correct Answer

1. Which of the following will facilitate the lowering of body temperature with warm stress?
   a. Curling up
   b. Increased food intake
   c. Increased circulating epinephrine
   d. Increased peripheral dilation to a cold environment

2. A person who is acclimatized, compared with an unacclimatized person, has:
   a. a higher metabolic rate in the cold to produce more heat.
   b. a lower peripheral blood flow in the cold to retain the heat.
   c. higher blood flow in the hands and feet in the cold to preserve their function.

d. increased food intake to produce more heat.
e. All of the above
3. Dust particles that are inhaled have a size range of:
   a. 10 to 15μm.
   b. 5 to 10μm.
   c. 2 to 5μm.
   d. 0.5 to 5.0μm.
4. The main detoxifying organ in the body is the:
   a. spleen.
   b. kidney.
   c. lung.
   d. liver.
   e. stomach and small intestine.
5. For every 33 feet below sea level, barometric pressure will increase by:
   a. 1/4 atm.
   b. 1/3 atm.
   c. 1 atm.
   d. 2 atm.
6. The stimulus for hyperventilation at altitude is low arterial:
   a. pH.
   b. oxygen tension.
   c. bicarbonate.
   d. [H+].
   e. a and b
7. Bends associated with deep-sea diving are a consequence of:
   a. excess oxygen dissolved in brain tissue.
   b. excess carbon dioxide dissolved in brain tissue.
   c. excess gas in gut tissue.
   d. excess nitrogen dissolved in fat tissue.
8. Which of the following does not occur upon acute exposure to cold?
   a. Shivering thermogenesis
   b. Vasoconstriction to skeletal muscles
   c. Sympathetic stimulation to heart
   d. Release of epinephrine
9. As one ascends to high altitude, which of the following does not decrease?
   a. Partial pressure of O_2
   b. Barometric pressure
   c. Water vapor pressure in the trachea
   d. Percent O_2
   e. c and d
10. During acclimatization to altitude, which of the following will increase?
    a. Vital capacity
    b. Minute ventilation

c. Cardiac output
d. Alveolar ventilation
e. All of the above
11. A student climbs a mountain, looks at his barometric gauge on his watch, and discovers that the barometric pressure is half what it was at sea level (P_B = 760 mm Hg). Which of the following partial pressures would best represent the partial pressure of oxygen at the mountain site where she is standing?
    a. 38 mm Hg
    b. 70 mm Hg
    c. 76 mm Hg
    d. 80 mm Hg
12. Estimate how much oxygen is delivered to the tissues for a cardiac output of 5 L/min in a high mountain dweller (15,000 feet) if his arterial oxygen tension is 40 mm Hg and his venous oxygen tension is 20 mm Hg.
    a. 0.5 L/min
    b. 1 L/min
    c. 1.5 L/min
    d. 2.0 L/min
    e. 2.5 L/min
13. A diver goes 99 feet under water. What is the barometric pressure (in mm Hg) at this depth?
    a. 760 mm Hg
    b. 1520 mm Hg
    c. 2280 mm Hg
    d. 3040 mm Hg
    e. 3800 mm Hg
14. Calculate the total G force of a pilot who pulls a negative G force of 3.
    a. −2
    b. −3
    c. −4
    d. −5
    e. +4
15. What is the partial pressure of O_2 in a space cabin filled with 60% O_2 at 1/2 atm?
    a. 152 mm Hg
    b. 190 mm Hg
    c. 228 mm Hg
    d. 633 mm Hg
    e. 684 mm Hg

Answers to Case History Questions

1. Acute leukemia is the abnormal production of blood cell precursors. The abnormality occurs at different stages of maturation of the cell and explains the different types of leukemia. Leukemia is a rapid progressive disease, although there are occasional patients whose disease remains stable for several months. However, the latter situation is rare. A lack of normal blood cells is the cause of the morbidity and mortality of this disease, and not the leukemic cells per se. The lack of normal blood cells leads to anemia, thrombocytopenia, and leukopenia and is brought about by the leukemic cells "crowding out" the normal cells in the bone marrow. Recent studies suggest that leukemic cells may have an inhibitory effect on the production of normal bone marrow cells, which leads to the life-threatening hemorrhage and infection.

2. Acute leukemia is classified into two primary types: acute lymphocytic leukemia (ALL), and acute nonlymphocytic leukemia

(ANLL). Each type has unique cytology that distinguishes them, and the distinction is important because the therapy differs for each type. ANLL occurs most commonly in adults, whereas ALL occurs more commonly in children.

3. Certain environmental and genetic factors are linked to acute leukemia. With respect to the latter, many chromosome alterations are known to be important predisposing factors in causing leukemia. For example, the incidence of leukemia is significantly increased with congenital disorders, such as Down's syndrome, Fanconi's syndrome, celiac disease, as well as with other specific types of congenital defects. Environmental factors are the most important etiological links in the development of acute leukemia. These include exposure to specific chemicals and to ionizing radiation. Radiation exposure significantly increases the risk for acquiring acute leukemia. People with occupational exposure to radiation and patients who receive radiation treatments are at high risk. Chemicals, particularly benzene in industrial use and several therapeutic drugs (chloramphenicol, phenylbutazone, melphalan, chlorambucil, as well as others) are linked to leukemia.

Key Terms

acclimatization (p. 878)
adaptation (p. 878)
asbestosis (p. 901)
atelectasis (p. 890)
biological rhythms (p. 878)
blackout (p. 892)

carboxyhemoglobin (p. 891)
circadian rhythms (p. 879)
diving reflex (p. 886)
free radicals (p. 890)
G force (p. 892)

heat stroke (p. 898)
hypoxia (p. 880)
hypothermia (p. 897)
oxygen radicals (p. 890)
oxygen toxicity (p. 890)

pneumoconioses (p. 900)
polycythemia (p. 884)
SCUBA (p. 888)
silicosis (p. 900)
weightlessness (p. 893)

Suggested Readings

Edmunds, L. N. *Cellular and Molecular Bases of Biological Clocks: Models and Mechanisms for Circadian Timekeeping.* Berlin, Springer-Verlag, 1988.

Fenn, W. O., and Rahn, H., eds. *Handbook of Physiology: Respiration,* vol II, sect 3. Bethesda, MD, American Physiological Society, 1965.

Gantenbein, D. "The high life." *Discover,* October 1993: 115–122.

Gelfond, J. A., Dinarello, C. A., and Sheldon, M. W. "Fever." *In* Iseelbacher, K. J., Braunwald, E., Wilson, J., et al, eds. *Harrison's Principles of Internal Medicine,* ed 13. New York, McGraw-Hill, 1994, pp 81–90.

Gordon, C. J., and Heath, J. E. "Integration and central processing in temperature regulation." *Annual Review of Physiology,* 48:595–612, 1986.

Hong, S. K., and Rahn, H. "The diving women of Korea and Japan." *In Human Physiology and the Environment in Health and Disease.* New York, Scientific American/W.H. Freeman & Co., 1976, pp. 92–102.

Hook, G. E., ed. *Environmental Health Perspectives.* Volumes 55 and 56, 1984.

Hubbard, R. W., Gaffin, S. L., and Squire, D. "Heat-related illnesses." *In* Auerbach, P. S., ed. *Wilderness Medicine: Management of Wilderness and Environmental Emergencies,* ed 3. St. Louis, Mosby-Year Book, 1995.

Lambertson, C. J. *Underwater Physiology.* Baltimore, Williams & Wilkins, 1967.

Mangum, C. P., and Towle, D. W. "Physiological adaptation to unstable environments." *American Scientist,* 65:67–75, 1977.

Marrow, P. E. "Animals in toxic environments: Mammals in polluted air." *In* Dill, D. B., Adolph, E. F., and Wilbur, C. G., eds. *Handbook of Physiology: Adaptation to the Environment.* Bethesda, MD, American Physiological Society, 1964, pp 773–794.

Pandolf, K. B., Sawka, M. N., and Gonzalez, R. R., eds. *Human Performance Physiology and Environmental Medicine at Terrestrial Extremes.* Indianapolis, Benchmark, 1988.

Rowell, L. B. "Cardiovascular aspects of human thermoregulation." *Circulation Research,* 52:367–379, 1983.

Web sites

www.osteo.org/
The official Web site of the NIH Osteoporosis and Related Bone Disease.

www.nof.org/
The official Web site of the National Osteoporosis Foundation

www.nasa.gov
The official Web site of NASA.

www.niehs.nih.gov
The official Web site of NIH Environmental Health Sciences.

Answers to Review Questions

1. d 2. e 3. d 4. d 5. c 6. b 7. d 8. b 9. e 10. e 11. d 12. c 13. d 14. a 15. c

Chapter 30

EXERCISE PHYSIOLOGY

KEY CONCEPTS

- Exercise is defined as increased activity of skeletal muscle.

- ATP is the immediate source of energy for all skeletal muscle contraction.

- Oxygen consumption increases during dynamic exercise in proportion to exercise intensity.

- The maximal oxygen consumption is the single best predictor of the capacity to perform dynamic exercise.

- Muscle requires carbohydrate as fuel during intense exercise.

- Chronic activity increases the capacity for fat metabolism in skeletal muscle.

- Control of food intake is linked to daily caloric expenditure.

- Acute and chronic exercise increase insulin sensitivity and glucose transport into muscle.

- Dynamic exercise increases cardiac output, and skeletal muscle and coronary blood flow.

- Ventilation increases in exercise help to maintain acid-base balance.

CASE HISTORY

Todd, a sedentary 19-year-old college student, enters a one-week study of his daily energy intake, as conducted by a classmate who is examining how exercise affects nutrition. Todd's weight is recorded as 78 kg, and his daily energy intake is measured as 2100 kcal. The classmate then persuades Todd to join the dormitory's bicycle racing team. On Todd's first day with the team, the warm-up is a 30-minute ride at 24 km/h on a level course. This is mild exercise for all of the experienced cyclists. Todd finds it very difficult. In the 10th minute of the ride, all of the cyclists measure their heart rates. For veteran riders, 130 bpm is average, while Todd's heart rate is 180. While the others casually comment about the light exercise involved in this warm-up, Todd's leg muscles ache. Todd also notices that while the experienced riders are able to easily talk during this exercise, he is breathing much harder than anyone else and is unable to join the conversations around him. Exhausted, Todd retires for the day. For the next few days, the muscles in his legs are extremely sore.

Despite his initial fatigue, Todd persists in daily workouts. Within a few days, the soreness in his muscles has disappeared despite daily exercise that matches or exceeds what he carried out on his first day with the team. Todd notices gradual improvement in his performance, and within a week he begins to steadily increase both his daily distance and speed. After two months of daily exercise, he is riding more than an hour every day. His roommate comments that Todd is eating much more than before he joined the team. Todd notices that despite increasingly long and strenuous rides each day, he is never sore. After six months of training, a novice rider joins the group, and Todd begins casual conversation with the newcomer during the usual 30-minute warm-up at 24 km/h on the level terrain. The new rider is breathing so hard that he cannot reply to Todd's questions. Todd realizes that he is no longer out of breath during the warm-up. He then checks his heart rate. It is 130 bpm, much reduced from his first day of exercise and matching that of other highly conditioned riders. Re-entering his classmate's nutrition project, Todd's energy intake is measured as 3100 kcal/day, an increase of nearly 50% above the pretraining value. Despite the increase, Todd has lost 5 kg during the six months of training.

Questions

1. Why did he experience muscle soreness only during the first few days of exercise?

2. Why, after training, was his heart rate reduced during the standardized "warm-up" activity?

3. Why was he breathing much harder than the others during the "warm-up" on the first day?

4. Why was he able to steadily increase his daily distance and speed?

5. Why did he lose weight during training despite eating more?

INTRODUCTION

Exercise evokes images of elite athletes, trained to peak fitness and competing in high-profile events of widespread spectator interest. However, exercise is much more than the limits of human performance and includes mundane activities, such as sitting, walking, and doing routine chores. The physiological responses to acute exercise maintain homeostasis while enabling coordinated muscle contraction to proceed. These responses, in broad outline, are identical in individuals across the entire range of fitness. While the acute responses to exercise are compelling, the body's adjustments to chronic activity are equally important. These adaptations include changes in multiple organ systems that allow future exercise to be completed with less fatigue. Such alterations, commonly called *training,* enable the untutored young athlete to eventually manifest his or her maximal abilities. The physiological adaptations to chronic activity, however, also include several responses more significant for health than for human performance. These modifications, which are initiated by even the most modest physical activity, play a major role in prevention of the obesity, adult-onset diabetes, and cardiovascular disease associated with sedentary living.

EXERCISE IS DEFINED AS INCREASED ACTIVITY OF SKELETAL MUSCLE

> *How do muscle cells obtain the energy necessary to perform exercise?*

The acute and chronic responses to exercise depend upon the specific muscles involved, the total amount of muscle mass used, the duration and intensity of the work, and the type of muscle contraction. For discussion of physiological responses, exercise is divided into two general categories. **Isometric exercise** refers to muscle contractions at constant length and with constant force development. **Dynamic exercise** refers to muscle activation involving rhythmic cycles of contraction and relaxation. While many daily activities mix isometric and dynamic exercise, the physiological responses to physical activity are most easily understood by considering isometric and dynamic exercise separately.

ATP is the Immediate Source of Energy for all Skeletal Muscle Contraction

Exercising muscle greatly increases its need for energy. The primary energy source for skeletal muscle contraction, as for other types of cellular work, is the high-energy phosphate compound **adenosine triphosphate (ATP).** Like all cells, skeletal muscle fibers store little ATP, so immediate and continued replenishment of ATP must occur if exercise is to continue. Exercise can increase rates of ATP formation and breakdown in skeletal muscle more than tenfold; during

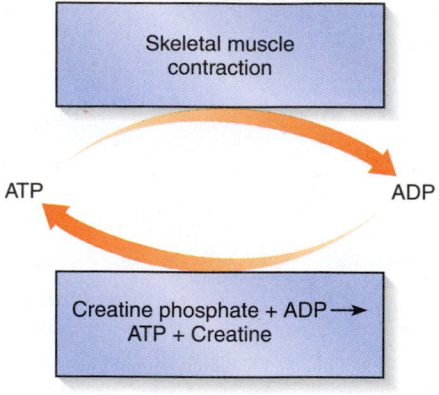

Figure 30–1

Muscle contraction requires energy released when ATP is broken down to ADP and inorganic phosphate. Creatine phosphate stored in the muscle is an immediate source for creation of new ATP.

strenuous exercise, the body can produce and consume ATP at a rate exceeding 0.5 kg/min. In the first few seconds of exercise, ATP is replenished from creatine phosphate, a stored high-energy phosphate that transfers its phosphoryl group to adenosine diphosphate (ADP) (Fig. 30–1). Creatine phosphate stores in muscle are sufficient, by themselves, for very brief periods of exercise. For example, during a 100-m sprint, creatine phosphate is the primary source of newly formed ATP for the first 4 seconds of running. Levels of creatine

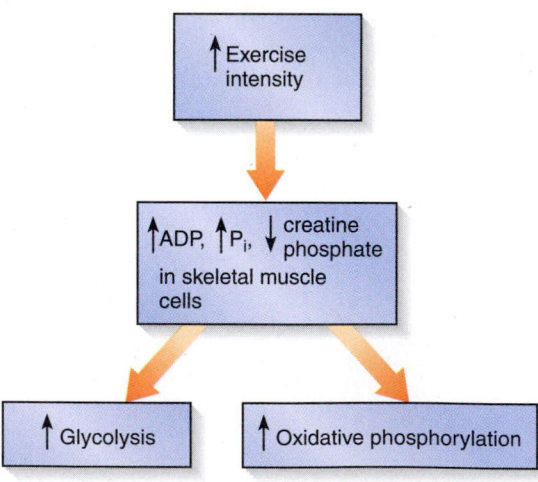

Figure 30–2

Increasing intensity of either isometric or dynamic exercise increases ATP breakdown, leading to increases in ADP and inorganic phosphate (P_i), and decreases in creatine phosphate. Changes in ADP, P_i, and creatine phosphate stimulate glycolysis and oxidative phosphorylation.

phosphate in the muscle, and the ability to perform short-term, high-intensity exercise, can be enhanced slightly by dietary creatine supplementation.

In contracting skeletal muscle cells, ATP breakdown increases the concentration of ADP and inorganic phosphate (P_i), and reduces the concentration of creatine phosphate, in proportion to the intensity of exercise. Increasing levels of ADP and P_i, and decreasing levels of creatine phosphate stimulate ATP formation from glycolysis and from oxidative phosphorylation (Fig. 30–2). In mild and moderate exercise, sufficient oxygen and metabolic machinery are available within the muscle fiber to allow all of the pyruvic acid formed by glycolysis to be converted to acetyl CoA. Acetyl CoA then enters the citric acid cycle prior to oxidative phosphorylation. In such exercise, all of the energy after the first few seconds is provided by the oxidation of carbohydrate, fat, and protein.

Oxygen Consumption During Exercise

The primary dependence on oxidative mechanisms for ATP production after the first few seconds of mild and moderate exercise allows steady-state exercise intensity to be measured in terms of oxygen consumption. During dynamic exercise, oxygen consumption increases with increasing intensity of the exercise (Fig. 30–3). For a given muscle mass performing a given level of external work (e.g., pedaling a bicycle at a work rate of 100 watts), oxygen consumption is similar in all persons. As the intensity of exercise is increased, a plateau in O_2 consumption is reached. This plateau is termed the peak oxygen consumption and is influenced by the age, sex, and training level of the person performing the exercise. Higher levels of peak oxygen consumption are associated with higher levels of peak dynamic exercise performance and with decreased levels of fatigue

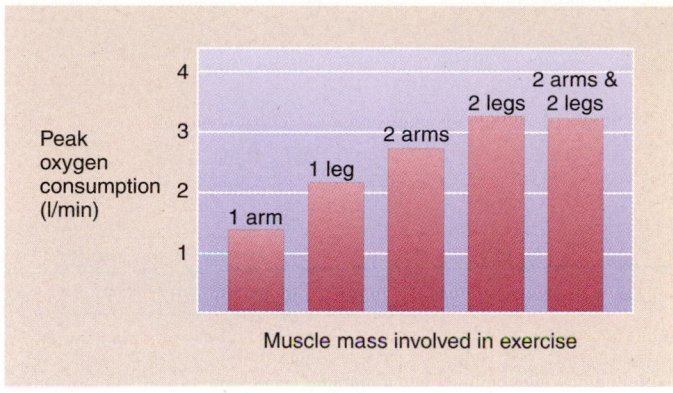

Figure 30–4

Peak oxygen consumption during dynamic exercise involving different muscle masses. The maximal oxygen consumption is reached during exercise with two legs and is not exceeded when arm exercise is added to leg exercise.

during submaximal exercise. As more muscle mass is added to dynamic exercise, the peak oxygen consumption increases before reaching a plateau (Fig. 30–4). The plateau in peak oxygen consumption, reached during dynamic exercise involving a sufficiently large muscle mass, represents the **maximal oxygen consumption** attainable by the body ($\dot{V}O_2$max; see Fig. 30–4). Much as peak oxygen consumption reached using a smaller muscle mass, the maximal oxygen consumption is influenced by age, sex, and training status.

Oxygen consumption during exercise is limited by the ability of the cardiovascular system to deliver oxygen to skeletal muscle and by the ability of working skeletal muscle to utilize the provided oxygen. Peak oxygen consumption of a relatively small active muscle mass is limited by the peak blood flow into that muscle and by the capacity within the mitochondria of those fibers to generate ATP from the delivered oxygen. In contrast, exercise that demands the maximal oxygen consumption is limited by the maximal cardiac output as well as by muscle **oxidative capacity.** Several of the steps in the delivery of oxygen from atmosphere to working muscle, including the maximal cardiac output and maximal O_2 use in the muscle fibers, have some adaptive capacity given appropriate training (Fig. 30–5).

In addition to serving as the best single measure of dynamic exercise capacity, the peak and maximal oxygen consumption also provide a scale for calibrating exercise intensity. Exercise can be measured in absolute terms, which refers to an external standard (e.g., cycling at 24 km/h). In a trained and untrained individual, a given external work rate requires the same, constant requirement for oxygen uptake. While this external standard will determine some of the physiological responses to the work, many other responses will be scaled to the relative demands on the

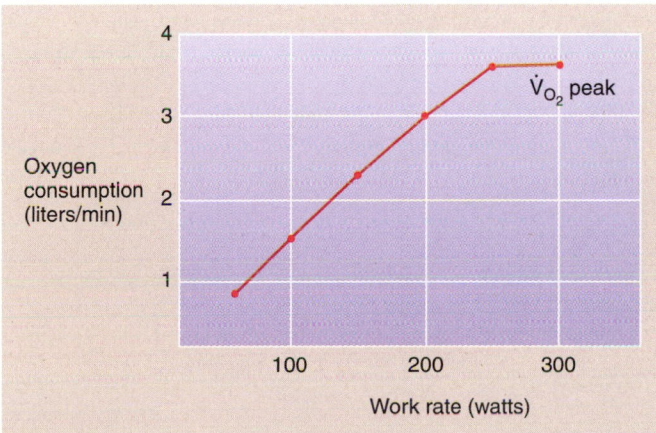

Figure 30–3

Oxygen consumption increases during graded dynamic exercise to a plateau termed the $\dot{V}O_2$ peak.

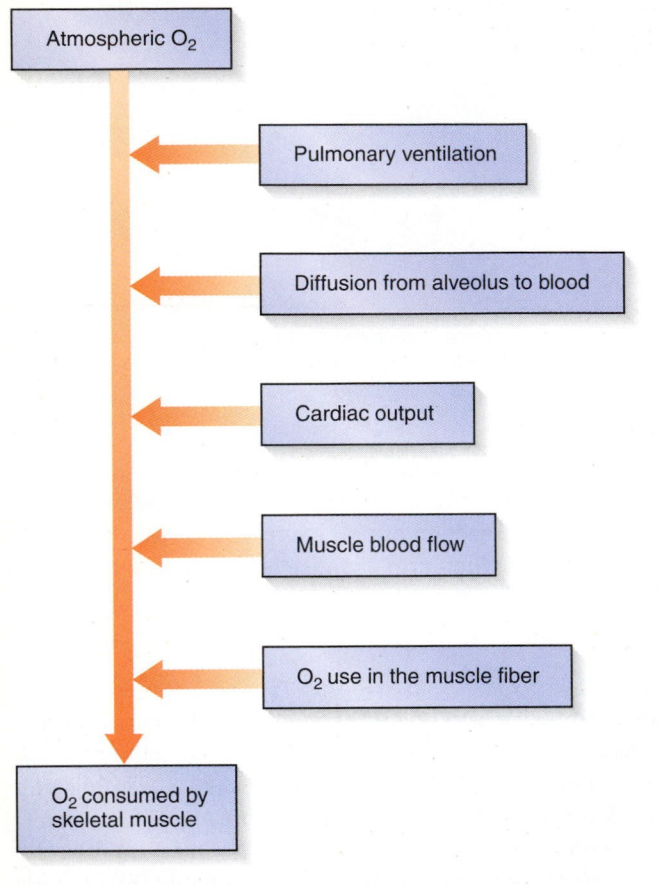
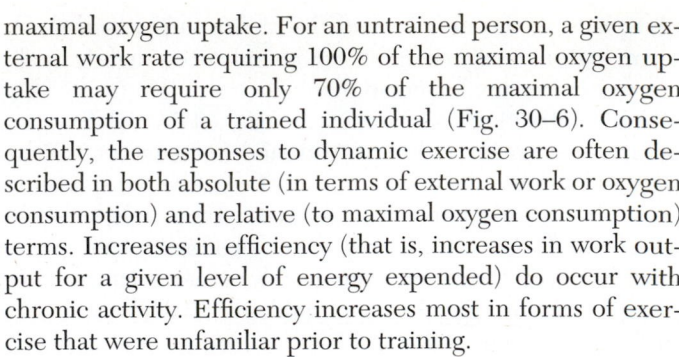

Figure 30–5

Maximal oxygen uptake demands maximal capacity of one or more stages of oxygen delivery and use from the atmosphere to the active muscle.

maximal oxygen uptake. For an untrained person, a given external work rate requiring 100% of the maximal oxygen uptake may require only 70% of the maximal oxygen consumption of a trained individual (Fig. 30–6). Consequently, the responses to dynamic exercise are often described in both absolute (in terms of external work or oxygen consumption) and relative (to maximal oxygen consumption) terms. Increases in efficiency (that is, increases in work output for a given level of energy expended) do occur with chronic activity. Efficiency increases most in forms of exercise that were unfamiliar prior to training.

Lactic Acid Production During Exercise

During intense exertion, when ADP and P_i concentrations are high, and creatine phosphate levels are low, stimulation of glycolysis is intense. In this situation, the production of pyruvate may exceed the capacity of the muscle to process pyruvate through oxidative pathways (aerobically). However, glycolysis can continue to generate ATP without oxygen (anaerobically) if pyruvic acid is converted to lactic acid. Consequently, as exercise intensity increases, ATP demands in some muscle fibers begin to exceed the capacity to process pyruvate aerobically, and lactic acid is formed in these fibers. Lactic acid then diffuses out of muscle into the blood. Further increases in exercise intensity cause more fibers to release lactate, so that blood lactate levels continue to increase (Fig. 30–7). Some tissues, especially the heart, can utilize lactate for energy, so the blood lactate level always represents a balance between active muscle lactate production and lactate metabolism in other tissues. Exercise that requires the greatest ATP production simultaneously demands both the

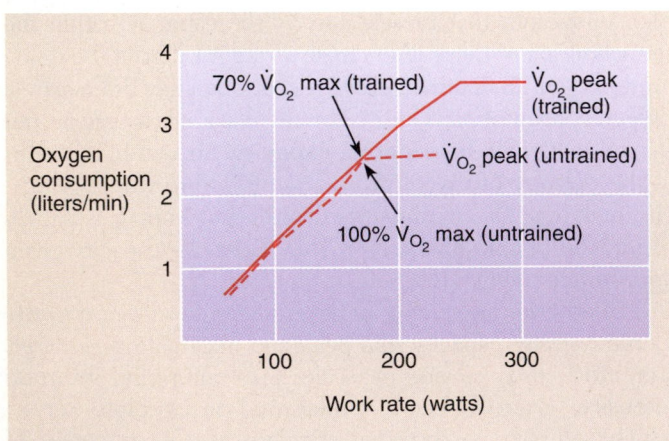

Figure 30–6

Exercise at a given external work rate is less demanding, in relative terms, in trained persons with a higher $\dot{V}O_2$ peak.

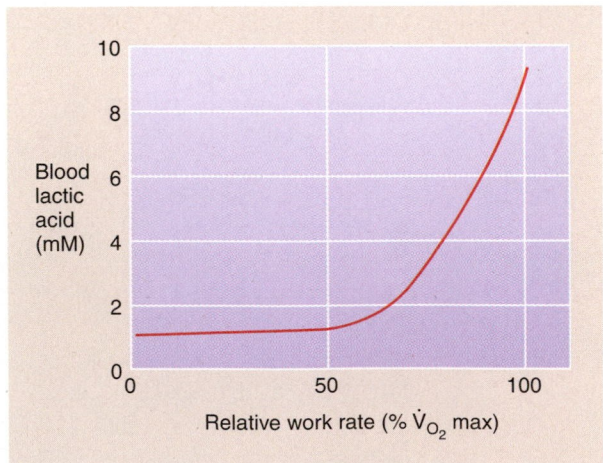

Figure 30–7

The increase in blood lactic acid during dynamic exercise is a function of the relative work rate.

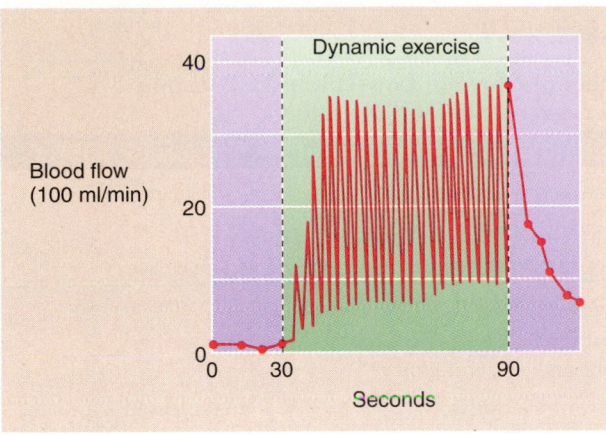

Figure 30–8

Effect of dynamic exercise on human calf muscle blood flow.

greatest rate of oxidative phosphorylation and the greatest rate of glycolysis. As a consequence, dynamic exercise requiring the maximal oxygen uptake also generates the highest levels of blood lactic acid (Fig. 30–7).

The cellular processes of energy expenditure in skeletal muscle are identical in dynamic and isometric exercise. However, during isometric exercise, fatigue occurs much more readily than during dynamic exercise. Fatigue is rapid in isometric exercise because high intramuscular pressure throughout sustained contraction blocks blood flow—and oxygen delivery—into the muscle. It is only during the relaxation phase of rhythmic, dynamic exercise that **muscle blood flow** reaches high levels (Fig. 30–8). Force production during isometric exercise is usually graded in terms of an individual's maximal voluntary contraction (MVC). An isometric contraction can be sustained indefinitely only if the force developed is less than approximately 15% of the MVC.

Skeletal Muscle Fatigue is Determined by Intramuscular Levels of ADP, P_i, and Creatine Phosphate

Fatigue in exercise does not occur either in motor nerves or at the level of the neuromuscular junction. Rather, fatigue in skeletal muscle contracting either dynamically or isometrically is caused by accumulation of ADP and P_i, and reduction of creatine phosphate. These changes reduce the ability of membrane pumps to sequester calcium ions within, and then release calcium ions from, the terminal cisternae of the sarcoplasmic reticulum. By blocking movement of calcium ions in both directions, fatigue inhibits both muscle contraction and muscle relaxation. Muscle fatigue is normal, not pathological. By reducing the demand for ATP, fatigue represents the muscle's defense against excessive reductions in the abil-

ity to form new ATP. If calcium ion pumping continued unabated in the face of reduced energy to form new ATP, the end-point of exercise would be rigor mortis, not fatigue. Although pH inside working muscles can decrease to 6.80 during intense exercise as a consequence of lactic acid formation, acidosis within the muscle does not contribute to fatigue.

Substrate Usage During Exercise

Carbohydrate, fat, and protein may be oxidatively broken down to CO_2 and H_2O with a high yield of ATP. Intramuscular glycogen and circulating free fatty acids are the major exercise energy substrates; amino acid usage is very low at any work intensity. The relative usage of carbohydrate and fat depends upon exercise intensity. During mild or moderate work (exercise intensities roughly corresponding to, for dynamic exercise, < 50% of the peak oxygen uptake achievable for that exercise, or < 15% of the MVC for isometric exercise), fat is the primary energy source for contracting muscle, in part due to relatively weak stimulation of glycolysis. Increased sympathetic stimulation of adipose tissue during exercise accelerates lipolysis, which increases circulating levels of fatty acids and glycerol. Body fat stores are very large relative to the energy demands of even the most prolonged mild exercise. Consequently, mild exercise is not limited by depletion of energy substrate and can be continued almost indefinitely.

At higher dynamic exercise work rates above 50% of the peak oxygen consumption, changing ADP, P_i, and creatine phosphate levels in working muscle strongly stimulate glycolysis, and breakdown of glycogen stores in muscle becomes essential to maintain adequate rates of ATP production. When intramuscular **glycogen** stores are exhausted,

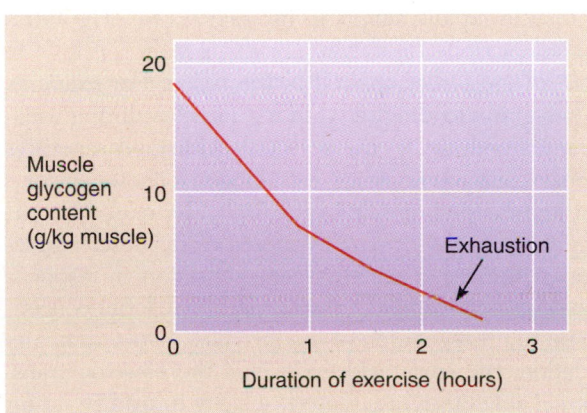

Figure 30–9

Muscle glycogen content during dynamic exercise requiring 70% of the maximal oxygen consumption. Exhaustion occurs when glycogen in active muscle is depleted.

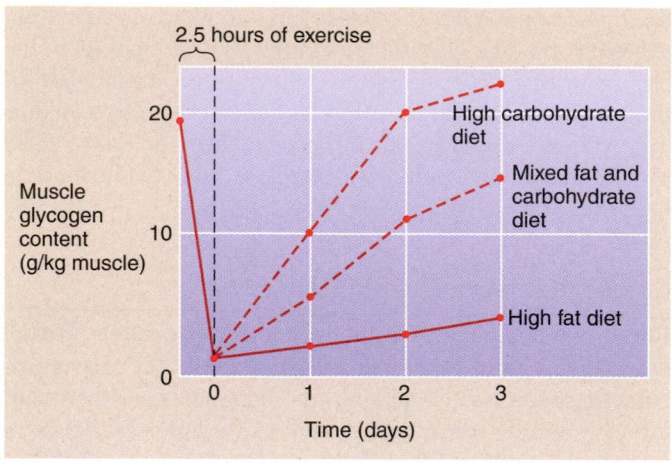

Figure 30–10

Rate of replenishment of muscle glycogen with different dietary carbohydrate following exhausting exercise.

TABLE 30–1	
Effect of Chronic Low-Intensity Dynamic Exercise on Skeletal Muscle	
Skeletal Muscle Constituent	**Exercise Effect**
Muscle cross-sectional area	Unchanged
Capillary density	Increased
Myoglobin concentration	Increased
Fatty acid oxidative capacity	Increased
Citric acid cycle enzyme concentration	Increased
Cytochrome oxidase	Increased

exercise ceases (Fig. 30–9). Because endurance of exercise at these intensities depends upon stored muscle glycogen, exercise duration is maximized when pre-exercise muscle glycogen levels are at their highest. Muscle glycogen levels are altered by prior activity and diet. Consequently, optimizing the ability to perform heavy dynamic exercise requires several days of prior rest, while ingesting sufficient carbohydrates (Fig. 30–10).

Muscle Responses to Chronic Activity Depend Upon Exercise Load and Duration

 How do muscles respond to exercise?

Skeletal muscle rapidly adapts to use and disuse. The intramuscular adaptations to activity are exquisitely sensitive to the nature of the forces applied to the tissue. For example, chronic low-grade loading (e.g., in leg muscles during walking) increases endurance, the ability to withstand repetitive submaximal contractions with less fatigue. These changes occur without causing muscle hypertrophy. Increased fatigue resistance is mediated in part by increases in muscle oxidative capacity. Increased oxidative capacity occurs in response to increased muscle capillary density and myoglobin concentration, and to up-regulation of the enzymes involved in fatty acid oxidation, the citric acid cycle, and the electron transport chain (Table 30–1). Increased capacity to oxidize fat for energy shifts the energy source from carbohydrate to fat. This adaptation preserves limited muscle glycogen stores and increases fatigue resistance in moderate and heavy exercise. Taken together, these adaptations increase the maximal oxygen consumption. Changes with training also allow a higher fraction of the maximal oxygen consumption to be utilized

without measurable accumulation of lactic acid within working muscles or in the blood (Fig. 30–11).

Within the skeletal muscle cell, increased oxidative capacity is linked to increased numbers and concentration of mitochondria. Because exercise at a fixed external work rate requires the same level of ATP production and utilization before and after training, the level of oxidative phosphorylation required from each mitochondrion is reduced after training. Consequently, local changes in the levels of ADP, P_i, and creatine phosphate—changes that cause fatigue—are reduced at a fixed external workload and are scaled to the increase in maximal oxygen consumption. This adaptation accounts for the similarity in many physiological responses, among them cardiovascular adjustments, lactic acid production, and muscle fatigue, when dynamic exercise is expressed as a percentage of the maximal oxygen uptake.

In contrast to the responses seen after chronic, low-grade loading, brief, high-intensity muscle contractions cause

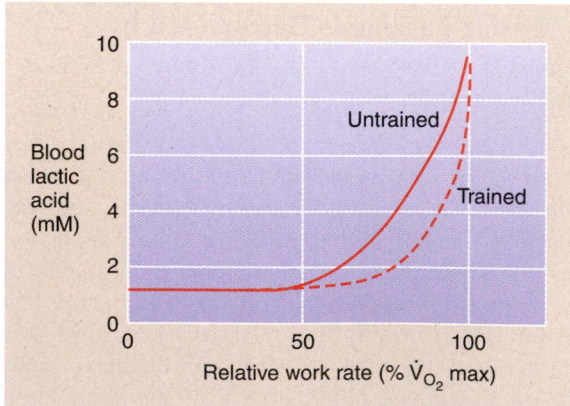

Figure 30–11

Training reduces blood lactic acid levels at work rates between approximately 50% and 100% of $\dot{V}O_2$ max.

TABLE 30–2

Responses 24–72 Hours After Eccentric Exercise in Untrained Muscle

Muscle or Systemic Response	Eccentric Exercise Effect
Maximal force generation	Decreased
Muscle soreness	Increased
Muscle edema	Increased
Myofibrillar inflammation	Increased
Intramuscular neutrophil accumulation	Increased
Plasma creatine kinase	Increased
Serum myoglobin	Increased

as it contracts (concentric contraction) or is lengthened by external forces as it contracts (**eccentric contraction**) profoundly influences muscle adaptations. The demand for energy is less when external forces lengthen the muscle during contraction, making such work (e.g., resisting acceleration while walking downhill) relatively light compared to a force-matched concentric comparison (e.g., walking downhill is less demanding than walking uphill). However, perhaps because the force per active motor unit is greater in eccentric work, such exercise in untrained muscle causes a range of delayed responses that signify muscle damage. These responses include weakness (seen within the first day), soreness and edema (typically delayed 1–3 days), and elevated plasma levels of intramuscular enzymes and myoglobin (delayed 2–6 days; Table 30–2). Histological evidence of damage may persist for two weeks. Such muscle damage may be the essential element in muscle hypertrophy. Experiments comparing concentrically and eccentrically contracting limbs, at matched force, find hypertrophy only in the eccentrically contracted limb. Repeated exposure to identical eccentric exercise leads to adaptations that abolish subsequent soreness and muscle damage.

hypertrophy and increase strength but do not alter oxidative capacity. As in low-intensity exercise, the muscle response to heavy, short-term loading is precisely matched to the demands placed on the contractile elements. In fact, for the same force development, whether or not a muscle shortens

---

CURRENT CONCEPTS IN PHYSIOLOGY

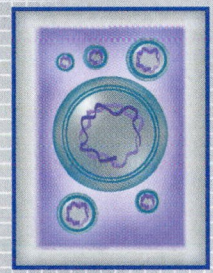

Exercise has, in general, favorable effects on the skeleton, and current interventions to prevent or reverse osteoporosis include regimens that load bone with high peak force and strain. However, the response of patients with established osteoporosis to exercise interventions is often marginal, and postmenopausal women, in whom osteoporosis is most prevalent, have been little studied. Greater understanding of the precise mechanisms linking strain to alterations in bone density, strength, or mass remains an important goal for future research. These studies will help to explore the interface between genetic factors (which explain perhaps 70% of the variance among healthy persons in bone phenotype) and environmental influences on the skeleton.

To be sure, there is no question that the reverse of exercise, microgravity or bed rest, in which even postural strains on bone are minimized, leads to rapid bone loss. Under these conditions of unloading, accelerated bone resorption increases urinary calcium excretion, and indices of bone formation are reduced. Additionally, resistance to insulin-like growth factor I may develop as a consequence of reduced force applied to the skeleton. The microgravity of space flight is associated with a 33% decrease in trabecular bone mass within six months.

Experiments in cell culture use computer-generated microstrains to examine in detail the responses of osteoblasts and osteoclasts, the cells responsible for bone formation, resorption, and remodeling. Applying strain reduces recruitment and activation of osteoclasts (bone-resorbing cells). In contrast, when strain is applied to osteoblasts (bone-forming cells) in culture, these cells proliferate and increase bone matrix production. However, measurements in cell culture cannot integrate every factor affecting bone formation and resorption. The levels of complexity include the relevance of microstrains to real-world exercise; the time dependence of the response; and the role of paracrine, hormonal, and nutritional influences.

Studies of healthy, postmenopausal women suggest that six months of daily, site-specific exercise, designed to maximize strain on a selected joint, have no impact upon total bone mass at that location. However, the cross-sectional area and the density of cortical bone (the more dense form of bone, with greater strength) increases at the site in response to training, in part by converting weaker, trabecular bone to cortical bone. This change increases the strength of the affected bone. More studies are required to determine what role exercise may play in the treatment of persons with established osteoporosis, in whom the ability of weakened bone to respond to strain may be compromised.

PHYSICAL ACTIVITY PLAYS A MAJOR ROLE IN THE CONTROL OF APPETITE

 How does physical activity affect appetite?

Activities that increase energy expenditure above a resting baseline can be quantified in terms of oxygen consumption. Units of oxygen consumption are readily converted to kilocalories (kcal). Using one liter of oxygen to oxidize carbohydrate yields 5.0 kcal of energy. (Similarly, oxidation of fat releases 4.7 kcal per liter of oxygen.) While the energy costs of activities are usually expressed as kilocalories utilized per unit time (Table 30–3), for appetite control the more important measure is the total kilocalories expended per day. Daily energy expenditure is determined by considering both the intensity and the duration of an activity. For example, walking or light housework elevates oxygen consumption one-third as much as swimming or running, but the energy expenditure of an hour of housework equals the energy expenditure of 20 minutes of running (see Table 30–3).

Even the most modest elevations in daily energy expenditure alter the control of food intake. Appetite increases in proportion to caloric expenditure, and energy intake and expenditure are precisely balanced across a broad range of high daily activity (Fig. 30–12). When activity levels increase, several factors increase caloric intake to match caloric expenditure. Peripherally, these inputs include more rapid gastric emptying and small bowel absorption, leading to earlier gastrointestinal signals for feeding. Activity also causes blood levels of glucose and insulin to decrease more quickly, intensifying these appetite stimuli. In the brain, exercise may stimulate appetite by activating hypothalamic neural pathways involving neuropeptide Y.

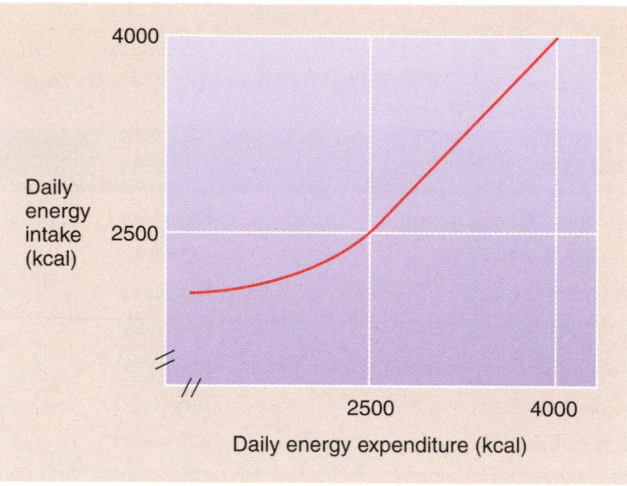

Figure 30–12

Appetite and daily energy intake is precisely matched to daily energy expenditure in active persons. Energy intake often exceeds energy expenditure in less active individuals.

The tight correlation between caloric expenditure and appetite is lost in more sedentary persons (see Fig. 30–12). The factors that uncouple caloric intake from caloric expenditure when daily caloric expenditure is low remain unclear. Leptin, an appetite-suppressant hormone secreted by adipocytes, may play a role in this response, but leptin levels decrease, rather than increase, in persons who lose weight by increasing daily activity.

ACUTE AND CHRONIC EXERCISE INCREASE INSULIN SENSITIVITY AND GLUCOSE TRANSPORT INTO MUSCLE

 How is glucose transport into skeletal muscle affected by exercise?

In addition to increasing sympathetic activity in adipose tissue, exercise also elevates sympathetic activity at the pancreatic islets, reducing insulin release (Fig. 30–13). However, despite increased circulating fatty acids and reduced circulating insulin, insulin-dependent and **insulin-independent muscle glucose uptake** increase in exercise. Contraction of skeletal muscle recruits insulin-independent glucose transporters from intracellular storage sites to the plasma membrane. These effects are lost within 3 hours of the cessation of exercise. Exercise also increases the sensitivity of skeletal muscle to insulin. For a fixed level of circulating insulin, exercise increases recruitment of insulin-dependent glucose receptors to the skeletal muscle cell membrane. This effect is maintained for approximately 18 hours after exercise (Fig. 30–14). Dynamic and isometric

TABLE 30–3

Energy Requirements of Various Activities

Activity	Energy Cost (kcal/h)
Sleeping	65
Sitting	100
Standing	180
Dressing, undressing	180
Walking (3 mph)	300
Making a bed	300
Dancing	420
Gardening/shoveling	680
Jogging	700
Swimming	900
Climbing stairs	1100

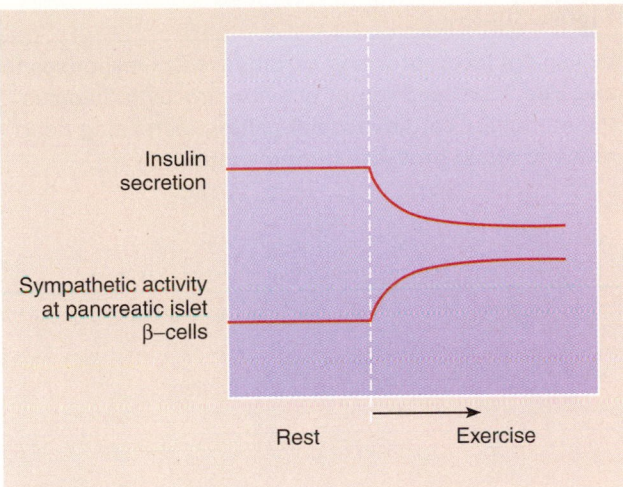

Figure 30–13

Acute exercise suppresses insulin secretion by increasing sympathetic tone at the pancreatic islet beta cells.

exercise both increase insulin sensitivity and glucose uptake in active skeletal muscle.

THE CARDIOVASCULAR RESPONSE TO EXERCISE

 How does the cardiovascular system respond to exercise?

Cardiovascular Control During Acute Dynamic Exercise

The cardiovascular responses to dynamic exercise begin as neural traffic flows from the motor cortex, brain stem, and spinal cord to skeletal muscle. Neural traffic from motor areas also flows to the cardiovascular center in the medulla, in rough proportion to the number of active skeletal muscle motor units. As motor activity and exercise intensity increases, the cardiovascular center reduces the parasympathetic neural influence on the heart and increases the blood pressure set point of the arterial baroreceptors. These changes increase heart rate and mean arterial pressure in exercise above resting levels (Fig. 30–15).

When dynamic exercise becomes more demanding, an additional cardiovascular control response is added. As lactic acid is formed in working muscle, muscle afferent nerves begin sending information to the medullary cardiovascular center. This response, termed the **muscle chemoreflex,** increases sympathetic neural outflow to the heart and systemic arterioles. Increased sympathetic activity elevates heart rate and cardiac contractility, and vasoconstricts arterioles in skeletal muscle, skin, and viscera. The major com-

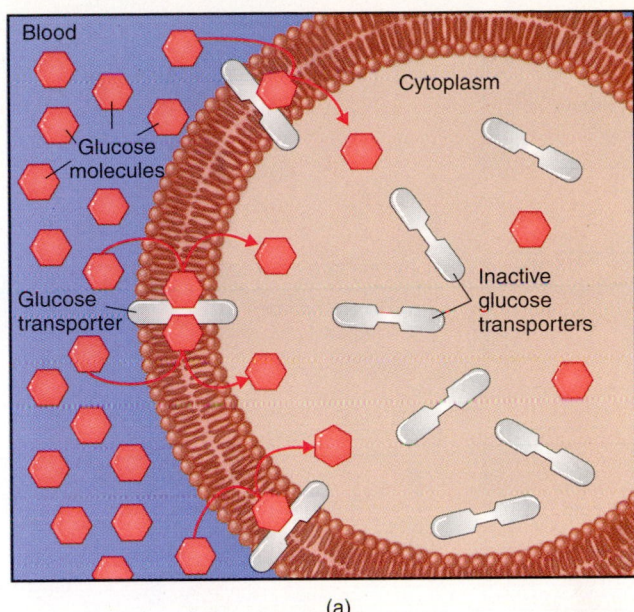

(a)

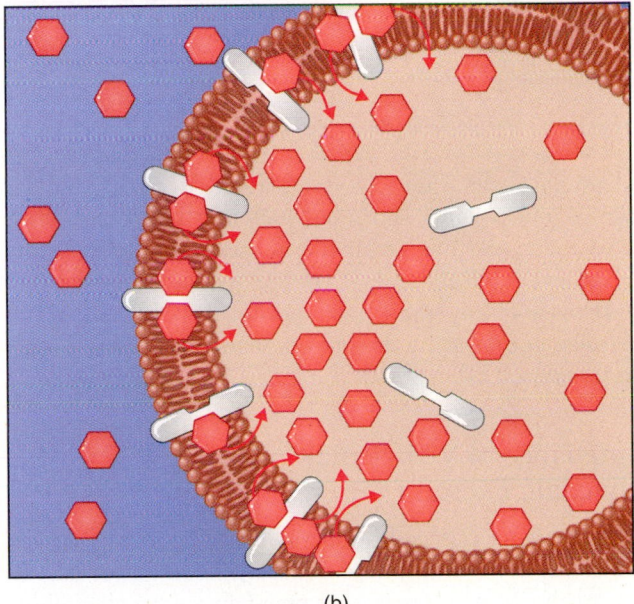

(b)

Figure 30–14

Illustration of the effect of exercise on glucose uptake into active skeletal muscle cells. This figure depicts the situations that might exist before **(a)**, compared with during and after **(b)** exercise. Exercise increases glucose uptake by increasing the number of glucose transporters in the cell membrane, through both insulin-dependent and insulin-independent effects.

ponents of cardiovascular control in heavy exercise are shown in Figure 30–16.

In active skeletal muscle, increased vasoconstrictor sympathetic drive is superceded by local factors, linked to high energy expenditure. These local factors cause profound

Locomotor Response   **Cardiovascular Response**

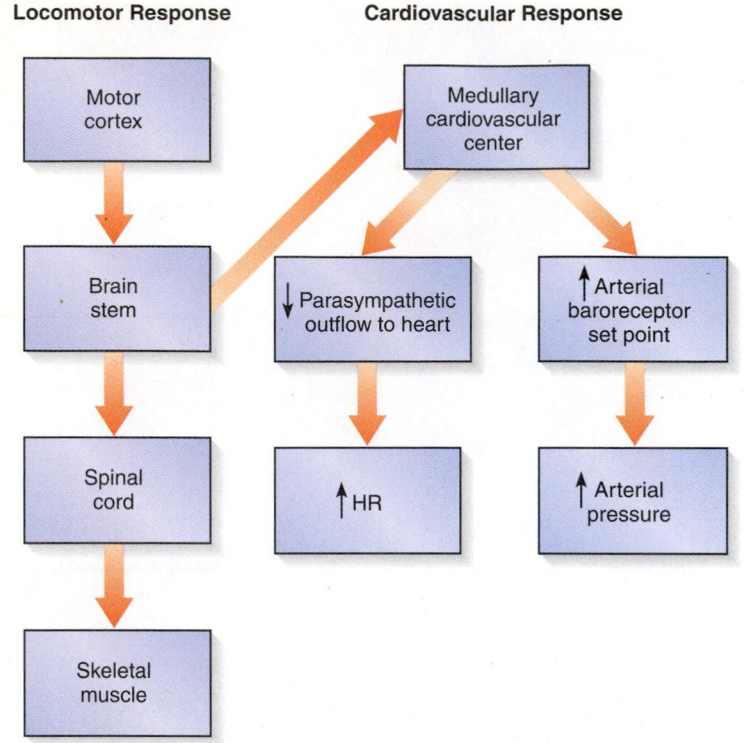

Figure 30–15

Linkage of locomotor and cardiovascular responses to exercise. Increased motor activity directly influences the medullary cardiovascular center, increasing heart rate and blood pressure during exercise.

vasodilation. Skeletal muscle vasodilation decreases vascular resistance and permits enormous increases in muscle blood flow. At the same time, sympathetic vasoconstriction in the skin, kidneys, and abdominal viscera reduce blood flow to these tissues. These adjustments permit the vast majority of the cardiac output to be directed to working skeletal muscle during exercise (Table 30–4).

Maintenance of Ventricular Filling in Dynamic Exercise

Increases in heart rate and cardiac contractility in exercise can only increase cardiac output and muscle blood flow if venous return to the heart increases to maintain ventricular filling.

During dynamic exercise, several factors aid the return of blood to the heart (Fig. 30–17). First, the rhythmic action of skeletal muscle allows high flow into and out of this tissue during the relaxation-contraction cycle (see Fig. 30–8). This "muscle pump," in concert with one-way venous valves, is a primary factor aiding venous return during dynamic exercise. Second, exercise augments venous return by changing the distribution of blood volume in the circulatory system. Sympathetic vasoconstriction in the skin and splanchnic circulations moves blood out of these compliant vascular beds, making it available to the active muscles and heart. The muscle vascular bed is relatively noncompliant, and despite high blood flow, little blood is stored in skeletal muscle during exercise (see Fig. 30–17). Because ventricular filling is maintained and car-

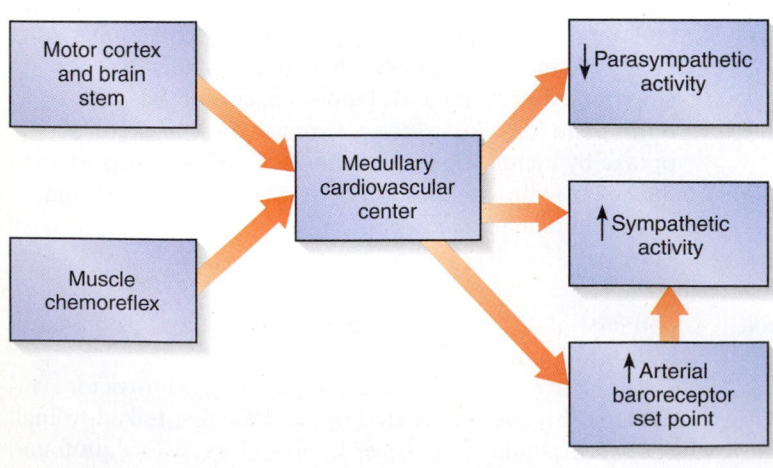

Figure 30–16

The cardiovascular response to heavy exercise includes inputs from motor areas in the brain and from the muscle chemoreflex response to increased lactic acid.

TABLE 30–4

Blood Flow Distribution During Rest and Heavy Dynamic Exercise

	Rest		Heavy Exercise	
Area	ml/min	%	ml/min	%
Splanchnic	1,400	24	300	1
Renal	1,100	19	300	1
Brain	750	13	750	3
Coronary	250	4	1,250	6
Skeletal muscle	1,200	21	19,900	86
Skin	500	9	300	1
Others	600	10	450	2
Total cardiac output	5,800	100	23,250	100

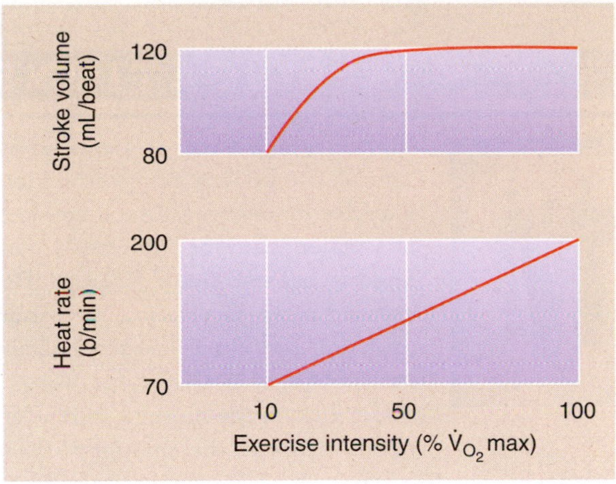

Figure 30–18

Heart rate and stroke volume increase with increasing dynamic exercise intensity.

diac contractility is augmented by increased sympathetic activity, stroke volume and heart rate both increase with increasing dynamic exercise intensity (Fig. 30–18).

Cardiovascular Response to Isometric Exercise

The different contraction/relaxation cycle in dynamic and isometric exercise generates different cardiovascular responses. While muscle blood flow in dynamic exercise can increase 20-fold over resting levels and can represent the majority of the cardiac output (see Table 30–4), compression of intramuscular arteries and veins during isometric exercise prevents muscle vasodilation and increased blood flow. Decreased oxygen

delivery in isometric exercise causes rapid accumulation of lactic acid and stimulation of muscle chemoreceptors, resulting in elevation of baroreceptor set point and sympathetic drive. As a result, blood pressure is higher during isometric compared with dynamic exercise (Fig. 30–19).

Cardiovascular Adaptations to Dynamic Exercise Training

Dynamic exercise training increases the maximal oxygen consumption in part by increasing the maximal cardiac output. This adaptation arises from increases in maximal stroke

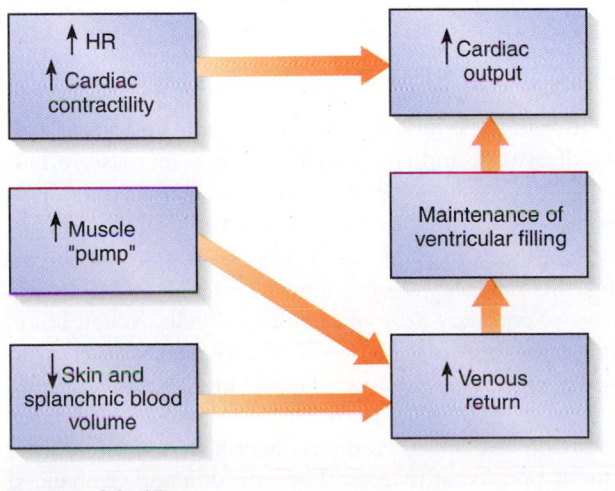

Figure 30–17

Factors that increase cardiac output and venous return in dynamic exercise.

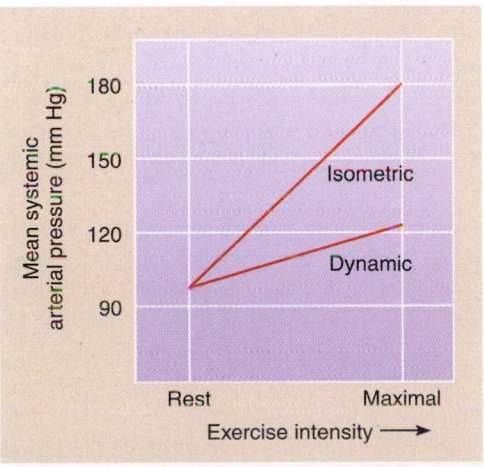

Figure 30–19

During exercise using a large muscle mass, blood pressure is greater during isometric than during dynamic exercise.

APPLICATIONS OF PHYSIOLOGY

Efficient, high-quality sleep is an essential component of health. How exercise influences subsequent sleep is a major issue for the elderly confined to assisted-living facilities. Decreased sleep quality is a persistent problem for these persons. Daily mild physical activity, for 1 to 2 hours in both the morning and afternoon, increases the amount of slow-wave (deep, nondreaming) sleep without altering the circadian rhythm or the body temperature. The improved sleep enhances memory during the daytime.

A second sleep-related issue concerns adjustments to new time zones (jet lag) or adaptation to shift work. The circadian sleep-wake cycle is linked to bright light exposure, and judicious use of light in the morning or evening can facilitate adaptation to new schedules. In contrast, controlled studies show that exercise has no circadian effect. These findings caution that physical activity cannot be selectively scheduled in the morning or evening to expedite circadian phase shifting. A positive view of the results is that exercise can be scheduled at any time of the day, without fear that timing will shift sleep onset or offset.

Of course, because exercise in a competitive atmosphere has potent arousal effects beyond the exercise per se, competitive activity near bedtime will often cause sleep-onset insomnia. In noncompetitive settings, a typical daily workout can be scheduled within 30 minutes of bedtime without affecting sleep onset or sleep quality.

Exercise is also often used to combat daytime sleepiness. Because the act of exercising itself increases arousal, during exercise feelings of sleepiness are reduced. However, studies of sleepy drivers show that stopping driving for a brief period of mild exercise is an ineffective intervention to prevent dozing at the wheel. A central problem here is that the act of driving itself is sedentary, and the resumption of the activity quickly allows relaxation and sleep to reassert their dominance. Instead, even a very brief nap (15 minutes), followed by caffeine intake, prevents sleep onset during driving.

The routine use of physical activity to resist daytime sleepiness can have deleterious effects. Repeated nights of inadequate sleep lead to accumulation of sleep debt, with steady increases in daytime sleepiness and deterioration of mood and cognitive function. While the sleep-deprived body and mind do have the ability to respond normally to enforced exercise, the steep decline in motivation that accompanies long-term sleep loss ensures that self-selected exercise will decrease in quality and quantity.

---

volume because maximal heart rate is unchanged or even reduced by training. Increases in pericardial volume, in response to repeated high preload, mediate the increase in ventricular volume with dynamic exercise training. These same volume changes also help to reduce resting heart rate after training. All of the increase in maximal exercise cardiac output is distributed to working skeletal muscles, as brain, splanchnic, renal, and skin blood flow remain unchanged.

Many of the cardiovascular responses to dynamic exercise are predictable on the basis of relative usage of the maximal oxygen consumption. That is, for the trained and untrained, heart rate and blood pressure are similar if these persons are compared at the same relative, but not at the same absolute, workload (see Fig. 30–18). The basis for this originates in the active muscles themselves. Because intramuscular lactic acid accumulation stimulates muscle chemoreceptors, adaptations that increase muscle oxidative capacity and delay lactate production also reduce muscle chemoreflex influence on the cardiovascular system. Because chemoreflex input elevates baroreceptor set point and increases sympathetic nerve activity, training lowers blood pressure and heart rate during identical external work.

Coronary Blood Flow During Dynamic Exercise

Because dynamic exercise increases heart rate and stroke volume, cardiac work and coronary blood flow increase in tandem with exercise intensity (coronary oxygen extraction is already high at rest). Coronary flow during exercise can exceed resting levels fivefold (see Table 30–4). Chronically increasing the coronary blood flow alters the physiological responses of coronary arterial endothelial cells. When blood flow increases, **endothelial cell shear stress** increases, causing these cells to generate vasodilator nitric oxide and prostacyclin (Fig. 30–20). Increased endothelial production of vasodilators enhances coronary arterial relaxation during subsequent periods of increased oxygen demand. Enhanced endothelium-dependent vasodilation is also seen in the vascular beds of active skeletal muscle after chronic activity.

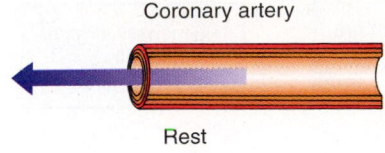

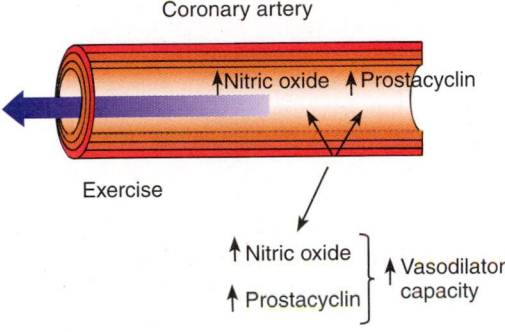

Figure 30–20

Increased shear stress in coronary arteries during exercise increases vasodilator nitric oxide and prostacyclin production, which in turn increase coronary vasodilator capacity.

THE RESPIRATORY RESPONSE TO EXERCISE

 How does the respiratory system respond to exercise?

Ventilation and Arterial Blood P_{O_2}, P_{CO_2}, and pH During Dynamic Exercise

As dynamic exercise intensity and muscle oxygen usage increase, oxygen is depleted from venous blood returning to the lung, even as cardiac output and pulmonary blood flow increase. Ventilation increases linearly with work intensity during mild and moderate work, then more steeply in intense exercise (Fig. 30–21). Increased ventilation maintains oxygen partial pressure and hemoglobin saturation unchanged in arterial blood in even the most intense exercise (Fig. 30–22). During dynamic exercise, increases in both tidal volume and breathing frequency contribute to increasing ventilation.

Increased exercise energy expenditure elevates carbon dioxide production in active skeletal muscle. Easily diffused across the alveolus, CO_2 excretion is intimately tied to blood acid-base balance. The linear increase in ventilation with work intensity, seen in mild and moderate work, maintains acid-base homeostasis in arterial blood. CO_2 is removed in exact proportion to CO_2 produced, leaving arterial P_{CO_2} unchanged from rest through moderate work levels (see Fig. 30–22).

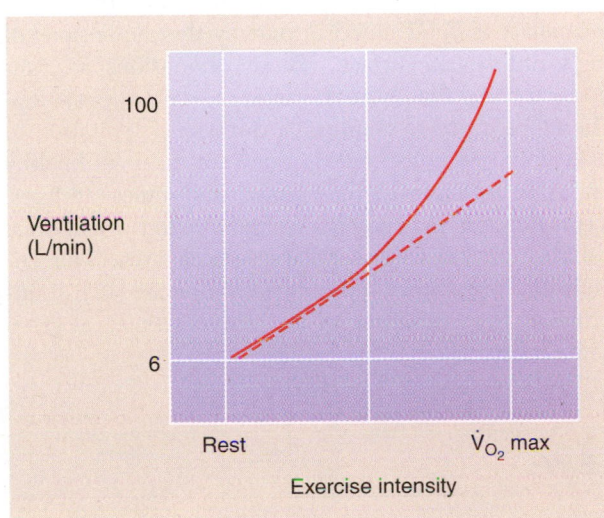

Figure 30–21

During dynamic exercise of increasing intensity, ventilation increases linearly over the mild to moderate range, then more rapidly in intense exercise.

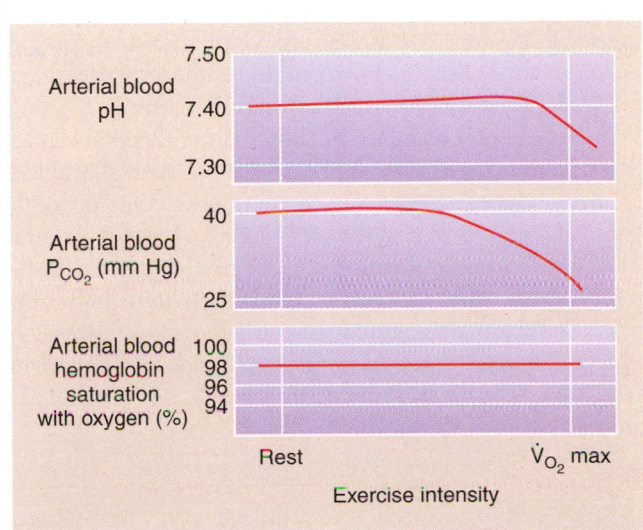

Figure 30–22

During dynamic exercise of increasing intensity, arterial blood hemoglobin saturation remains identical to resting levels. Arterial P_{CO_2} decreases, preventing a decrease in arterial pH until the most intense levels of exercise are reached.

As work intensity increases further, increasing amounts of lactic acid are added to the circulation. As lactic acid enters the blood, the acid-base demands of exercise intensify because both CO_2 and lactic acid contribute to potential acidosis. Because the kidney cannot rapidly respond to the acid-base challenges of acute dynamic exercise, increasing CO_2 excretion through the lungs represents the primary pathway to acid-base homeostasis in intense exercise. The lungs meet this demand by further increasing ventilation, lowering P_{CO_2} below resting levels in intense exercise. Consequently, despite CO_2 production rates that may exceed resting rates by 10- or 15-fold, arterial P_{CO_2} is reduced compared with rest in intense work (see Fig. 30–22). In very intense dynamic exercise, the lungs cannot excrete sufficient CO_2 to maintain constant arterial pH, and **lactic acidosis** prevails in arterial blood despite concomitant arterial P_{CO_2} reductions (see Fig. 30–22).

Control of Ventilation in Exercise

In a fashion analogous to the cardiovascular responses to exercise, the primary signal for increased exercise ventilation arises from motor areas in the brain. Neural traffic from these areas flows to the respiratory center in the medulla, causing tidal volume and breathing frequency to increase. This ventilatory response is rapid and roughly proportional to the number of activated motor units. In isometric exercise, this "central command" pathway appears to be the only determinant of the ventilatory response.

In dynamic exercise, feedback from peripheral receptors modulates the respiratory response. The peripheral chemoreceptors, located in the carotid body, sense deviations in arterial P_{CO_2} and P_{O_2} and adjust ventilation to maintain these variables constant during mild exercise. Pulmonary stretch receptors modify the respiratory pattern, limiting tidal volume in heavy exercise to about half of the vital capacity. The factors that cause ventilation to increase sharply in more intense dynamic exercise remain unclear. One possibility is that decreased blood pH, and increased catecholamine and K^+ concentration, further stimulate the carotid body chemoreceptors. Alternatively, increased activity from motor areas in the hypothalamus may, possibly in response to muscle chemoreceptor stimulation, provoke hyperventilation in intense work. The factors regulating ventilation in dynamic exercise are summarized in Figure 30–23.

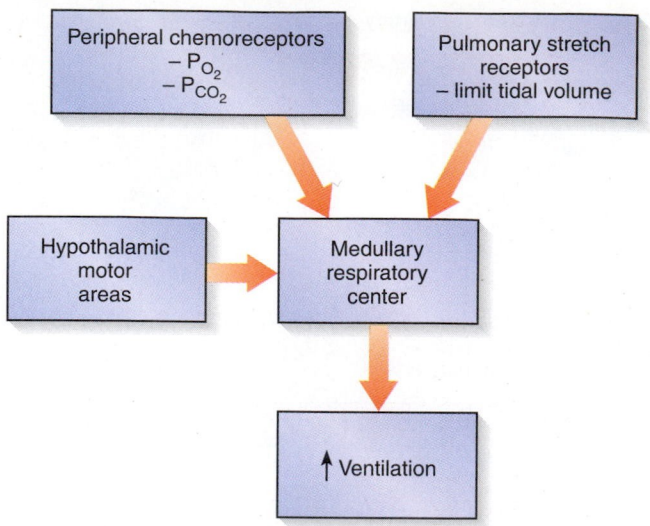

Figure 30–23

Factors controlling ventilation in dynamic exercise. Peripheral chemoreceptors and pulmonary stretch receptors play a minor role in control of ventilation in isometric exercise.

Ventilatory Adaptations to Dynamic Exercise Training

In contrast to maximal cardiac output and skeletal muscle oxidative capacity, both of which increase with training, much of the pulmonary system, including the airways, alveoli, and diffusional exchange barrier, is unaffected by chronic dynamic exercise. The capacity for oxygen to diffuse from alveolus to capillary is determined in part by the dimensions of the lung's diffusional surface. These dimensions are unchanged by training that increases maximal oxygen consumption. Therefore, prior to training, a diffusional limitation to the maximal oxygen consumption does not occur in healthy persons, while in some highly trained endurance athletes, oxygen consumption may reach a ceiling due to the anatomic limits of the lung alveolar-capillary surface area. The endurance of the respiratory muscles does increase with training, reducing the sensation of breathlessness at a given exercise intensity.

CHAPTER REVIEW

Summary

- The acute and chronic responses to exercise depend upon the muscle mass involved, the type of contraction (dynamic or isometric), and the duration and intensity of the work.

- ATP, the immediate source of energy for all skeletal muscle contraction, can be quickly and briefly replenished from creatine phosphate, or produced long-term by oxidative mechanisms.

- Maximal oxygen consumption, the single best index of dynamic exercise capacity, is limited by the cardiovascular system (maximal cardiac output) and by the oxidative capacity of exercising muscle.
- Many physiological responses to exercise are best predicted by the level of exercise oxygen consumption, expressed as a percentage of the maximal oxygen consumption.
- Training increases skeletal muscle oxidative capacity and reduces lactic acid production and dependence on carbohydrate at a fixed external workload.
- Acute exercise decreases insulin secretion and increases both insulin-dependent and insulin-independent glucose transport into muscle.
- During exercise, neural input from motor areas of the brain to the cardiovascular center reduces parasympathetic out-

flow to the heart and increases the arterial baroreceptor set point.
- Lactic acid stimulation of muscle chemoreceptors increases sympathetic drive in heavy exercise.
- Local factors dilate active skeletal muscle despite sympathetic vasoconstrictor drive, leading to vastly increased skeletal muscle blood flow in dynamic exercise.
- Isometric exercise quickly leads to fatigue and sharply increases blood pressure because blood vessel compression blocks nutrient delivery, leading to lactic acid accumulation and chemoreceptor stimulation.
- Input from motor areas of the brain stimulate the respiratory center to increase tidal volume and breathing frequency in exercise.

Review Questions

Choose the Correct Answer

1. Which of the following is not a major determinant of the acute and chronic long-term physiological responses to exercise?
   a. Isometric versus concentric contractions
   b. Exercise duration
   c. Exercise intensity
   d. Time of day of exercise
   e. Amount of muscle mass involved

2. Intramuscular creatine phosphate is:
   a. not used when exercise is mild.
   b. not used when exercise is intense.
   c. the only energy source needed during sustained, mild exercise.
   d. not used during isometric exercise.
   e. depleted when fatigue occurs.

3. Dynamic exercise intensity is often measured in terms of oxygen consumption because:
   a. creatine phosphate levels have not decreased from rest.
   b. ATP is not the energy source during mild, dynamic exercise.
   c. oxidative mechanisms provide all or almost all of the energy for ATP production.
   d. glycolysis is not activated during mild, dynamic exercise.
   e. muscle blood flow increases during dynamic exercise.

4. The maximal oxygen consumption is usually limited by:
   a. pulmonary ventilation.
   b. cardiac output and muscle oxidative capacity.
   c. lactic acid accumulation in active skeletal muscle.
   d. decreasing arterial oxygen content.
   e. carbohydrate stored in skeletal muscle.

5. In a regularly active person compared with a sedentary person, the peak blood glucose level seen after a standard meal would be:
   a. reduced because gastric emptying is slower.
   b. unchanged because chronic activity does not change insulin sensitivity.
   c. unchanged because decreased insulin secretion is balanced by increased insulin receptor density.
   d. increased because insulin receptor density is decreased.
   e. reduced because insulin sensitivity is increased.

6. During dynamic exercise, the vascular resistance within active muscles is, as compared with rest:
   a. unchanged.
   b. greatly reduced.
   c. increased.
   d. enormously increased.
   e. variable, depending primarily upon the degree of lactic acid production.

7. Fatigue occurs much more readily during isometric than during dynamic exercise because isometric exercise limits:
   a. blood flow to working muscle.
   b. utilization of anaerobic energy sources.
   c. muscle contractile force.
   d. consumption of stored phosphagen energy sources.
   e. muscle temperature.

8. Skeletal muscle made chronically active shifts its substrate preference toward:
   a. nucleic acids.
   b. protein.
   c. amino acids.
   d. fat.
   e. carbohydrate.

9. The relative use of carbohydrate and fat for energy in dynamic exercise is best predicted from the:
   a. oxygen consumption, expressed as a fraction of the maximal oxygen consumption.
   b. body fat stores.
   c. body temperature.
   d. oxygen consumption, expressed in absolute terms (L/min).
   e. absolute amount of muscle blood flow.

10. Within working skeletal muscle, fatigue is linked to:
    a. loss of acetylcholine receptors at the neuromuscular junction.
    b. increased activity of calcium pumps in the sarcoplasmic reticulum.
    c. increased ADP and P_i, and reduced creatine phosphate.
    d. loss of calcium ions to the extracellular fluid.
    e. neurotransmitter depletion.

11. A sedentary lifestyle is typically associated with weight gain during adulthood because:
    a. energy intake exceeds energy expenditure.
    b. appetite decreases compared with appetite when extremely active.
    c. resting skeletal muscle is unable to oxidize fat.
    d. hypothalamic appetite centers are directly linked to the motor cortex.
    e. sympathetic outflow is reduced.

12. The increased heart rate seen in mild dynamic exercise compared with rest occurs in response to:
    a. increased circulating lactic acid.
    b. parasympathetic withdrawal.
    c. decreasing blood pressure.
    d. increased sympathetic activity.
    e. muscle chemoreceptor stimulation.

13. While walking uphill is much more energetically demanding than walking downhill, paradoxically downhill walking typically results in greater muscle soreness, indicating that downhill walking involves considerable:
    a. lactic acidosis.
    b. eccentric contractions.
    c. sympathetic activation.
    d. skeletal muscle vasodilation.
    e. creatine phosphate depletion and accumulation of ADP and P_i.

14. Chronic dynamic exercise increases fatigue resistance in skeletal muscle, in part due to decreases within the muscle in:
    a. resting lactic acid concentration.
    b. tendon tensile strength.
    c. capillary diffusion distance.
    d. myoglobin concentration.
    e. mitochondrial density.

15. Isometric exercise can substantially increase blood pressure by strongly stimulating:
    a. cardiac output.
    b. carotid body chemoreceptors.
    c. muscle vasodilation.
    d. peripheral chemoreceptors.
    e. muscle chemoreceptors.

16. In healthy persons, intense dynamic exercise leads to:
    a. homeostasis of arterial blood PO_2, PCO_2, and pH.
    b. arterial acidosis despite reduced arterial PCO_2.
    c. exercise cessation due to dyspnea.
    d. a plateau in minute ventilation.
    e. peripheral chemoreceptor dominance of respiratory control.

17. Venous return is maintained in dynamic exercise in part by a reduction in:
    a. skeletal muscle blood flow.
    b. arterial blood pressure.
    c. splanchnic blood volume.
    d. pulmonary blood flow.
    e. muscle pump activity.

18. Blood pressure increases during mild dynamic exercise because:
    a. the baroreceptors are reset to a higher level.
    b. total systemic resistance decreases.
    c. active muscles vasodilate.
    d. arterial oxygen saturation decreases.
    e. carotid body chemoreceptors are activated.

19. Dynamic exercise training increases:
    a. the number of alveoli.
    b. the caliber of conducting airways.
    c. ventilatory muscle endurance.
    d. resting pulmonary diffusing capacity.
    e. arterial oxygen content at rest and in exercise.

20. The primary factor increasing ventilation in exercise is information provided to the respiratory center in the medulla from:
    a. carotid body chemoreceptors.
    b. muscle chemoreceptors
    c. peripheral sensors of reduced arterial PO_2.
    d. motor areas in the brain.
    e. peripheral sensors of increased arterial PCO_2.

21. Increased coronary arterial shear stress during exercise:
    a. occurs only during isometric exercise.
    b. increases oxygen extraction within the coronary vasculature.
    c. occurs only during dynamic exercise.
    d. stimulates vasodilator prostacyclin and nitric oxide release.
    e. blocks further increases in coronary flow.

Answers to Case History Questions

1. Muscle soreness occurs in response to eccentric muscle contractions and is followed by adaptations within the muscle that prevent soreness during similar exercise.

2. Training increases the oxidative capacity of skeletal muscle and reduces lactic acid production at a fixed external workload. Muscle chemoreceptors are activated by intramuscular lactic acid, so training reduces chemoreceptor stimulation. Because chemoreceptor activation increases heart rate, training reduces exercise heart rate at a fixed external workload.

3. Dynamic exercise training increases the endurance of the respiratory muscles, decreasing the sensation of breathlessness during exercise at a fixed intensity.

4. Chronic dynamic exercise increases both cardiac output and muscle oxidative capacity. These changes elevate maximal oxygen consumption and increase exercise capacity.

5. Appetite (energy intake) is closely regulated to match energy expenditure in active persons. In contrast, in sedentary individuals, energy intake often exceeds energy expenditure. Consequently, previously sedentary persons often lose weight after increasing daily energy expenditure.

Key Terms

adenosine triphosphate (ATP) (p. 908)
dynamic exercise (p. 908)
eccentric contraction (p. 913)

endothelial cell shear stress (p. 918)
glycogen (p. 911)
insulin-independent muscle glucose uptake (p. 914)

isometric exercise (p. 908)
lactic acidosis (p. 920)
maximal oxygen consumption (p. 909)

muscle blood flow (p. 911)
muscle chemoreflex (p. 915)
oxidative capacity (p. 909)

Suggested Readings

Adami, S., Gatti, D., Braga, V., Bianchini, D., and Rossini, M. "Site-specific effects of strength training on bone structure and geometry of ultradistal radius in postmenopausal women." *Journal of Bone Mineral Research,* 14:120–124, 1999.

Clapp, J. F. "Exercise during pregnancy: A clinical update." *Clinics in Sports Medicine,* 19: 273–286, 2000.

Dempsey, J. A., Adams, L., Ainsworth, D. M., Fregosi, R. F., Gallagher, C. G., Guz, A., Johnson, B. D., and Powers, S. K. "Airway, lung, and respiratory muscle function during exercise." In Rowell, L. B., and Shepherd, J. T., *Handbook of Physiology, Section 12: Exercise: Regulation and Integration of Multiple Systems.* New York, Oxford University Press, 1996.

Higashi, Y., Sasaki, S., Kurisu, S., Yoshimizu, A., Sasaki, N., Matsuura, H., Kajiyama, G., and Oshima, T. "Regular aerobic exercise augments endothelium-dependent vascular relaxation in normotensive as well as hypertensive subjects: Role of endothelium-derived nitric oxide." *Circulation,* 100:1194–1202, 1999.

Inoue, M., Tanaka, H., Moriwake, T., Oka, M., Sekiguchi, C., and Seino, Y. "Altered biochemical markers of bone turnover in humans during 120 days of bed rest." *Bone,* 26:281–286, 2000.

Jiang, Y., Zhao, J., Rosen, C., Geusens, P., and Genant, H. K. "Perspectives on bone mechanical properties and adaptive response to mechanical challenge." *Journal of Clinical Densitometry,* 2:423–433, 1999.

Johnson, B. D., Weisman, I. M., Zeballos, R. J., and Beck, K. C. "Emerging concepts in the evaluation of ventilatory limitation during exercise: The exercise tidal flow-volume loop." *Chest,* 116:488–503, 1999.

Kaufman, M. P., and Forster, H. V. "Reflexes controlling circulatory, ventilatory and airway responses to exercise." In Rowell, L. B., and Shepherd, J. T., *Handbook of Physiology, Section 12: Exercise: Regulation and Integration of Multiple Systems.* New York, Oxford University Press, 1996.

Meyer, R. A., and Foley, J. M. "Cellular processes integrating the metabolic response to exercise." In Rowell, L. B., and Shepherd, J. T., *Handbook of Physiology, Section 12: Exercise: Regulation and Integration of Multiple Systems.* New York, Oxford University Press, 1996.

Rowell, L. B., O'Leary, D. S., and Kellogg, D. L. Jr. "Integration of cardiovascular control systems in dynamic exercise." In Rowell, L. B., and Shepherd, J. T., *Handbook of Physiology, Section 12: Exercise: Regulation and Integration of Multiple Systems.* New York, Oxford University Press, 1996.

Vita, J. A., and Keaney, J. F. "Exercise: Toning up the endothelium?" *New England Journal of Medicine,* 342:503–504, 2000.

Votruba, S. B., Horvitz, M. A., and Schoeller, D. A. "The role of exercise in the treatment of obesity." *Nutrition,* 16:179–188, 2000.

Answers to Review Questions

1. d 2. e 3. c 4. b 5. e 6. b 7. a 8. d
9. a 10. c 11. a 12. b 13. b 14. c 15. e
16. b 17. c 18. a 19. c 20. d 21. d

Chapter 31

REPRODUCTIVE PHYSIOLOGY

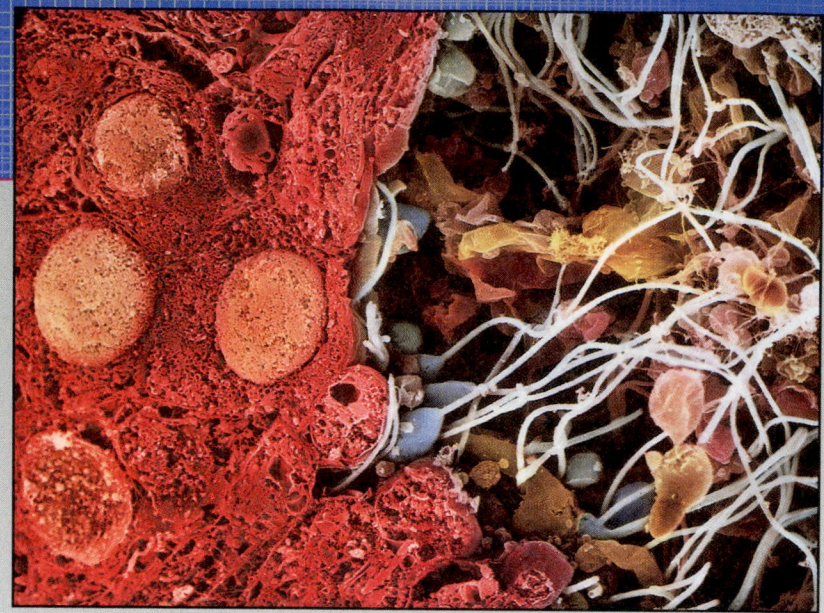

KEY CONCEPTS

- *Reproduction in both males and females is controlled by hormonal interactions among the hypothalamus, pituitary, and gonads.*

- *GnRH influences the production and release of the pituitary gonadotropins, FSH, and LH in both males and females.*

- *The pituitary hormones FSH and LH mediate germ cell maturation and sex steroid hormone production in both male and females.*

- *Production of germ cells as well as synthesis and secretion of sex hormones are common major functions of the ovaries and testes.*

- *Sex steroid hormones are necessary for fertility and secondary sexual physical characteristics in males and females.*

- *Female reproductive capacity involves cyclic changes in sex steroid hormone levels and uterine histology.*

CASE HISTORY

A 27-year-old white man, accompanied by his wife, presents with a history of two years of unprotected intercourse and the inability to conceive. Neither has had a child. He has no history of testicular surgery, injury, or chemical exposure at work or through his hobbies. He denies symptoms consistent with thyroidal disease, delay of puberty, or diabetes. There is no family history of endocrine disorders. He denies smoking; drinking; or the use of prescription, herbal, or illegal drugs. His physical examination reveals a blood pressure of 140/90. He is 5′ 10″ and 99 kg (220 lb) and extremely well muscled. He has early male-pattern baldness, modest acne, and a normal male hair distribution. His thyroid appears normal. The abdomen shows no unusual signs, and the penis and scrotum are normal. Both testes are withered and measure no more that 2 cm in length. He is azoospermic (i.e., no sperm in his semen), and has a normal glucose level. His serum testosterone level is 64 ng/dl (normal, 241–860 ng/dl). After further discussion, he does admit to the use of "anabolic steroids and testosterone shots" while playing college football. He was assured by his locker-room supplier that the combination of medicines he was taking included a "blocker" to protect the testicles and later sperm production. He has occasionally used the testosterone shots since college to boost a somewhat failing erection. His wife was unaware of these practices. Two years later, his sperm count remains zero and he now has to use testosterone shots twice monthly to maintain a normal level of male hormone.

Questions

1. Is there a link between the patient's well-muscled physique, his acne, and his male-pattern baldness?

2. What is the cause of the patient's infertility and reduced testicular size?

INTRODUCTION

In preceding chapters, we have seen that cell specialization produces a complex interdependence among the different cell types. These interrelationships help to maintain a stable internal environment, or homeostasis. In contrast, the physiological processes we are about to discuss involve cells from two separate individuals that interact to form a third individual. The sperm cell is structurally and chemically designed to leave the male and join with the female ovum. The ovum, while remaining inside the female, is designed to react with the sperm. These are the only two human cell types that are so adapted.

In this chapter, we will examine the male and female anatomy involved in reproduction and discuss how the nervous and the endocrine systems act together to control reproductive functions. In Chapter 32, we will examine how fertilization (the union of sperm and ovum) occurs in the female and how pregnancy alters female physiology to allow for the development of the fetus. The special physiological features of the developing fetus also will be examined. In Chapter 33, we will examine sexual physiology as separate from reproductive physiology and look at the factors that determine an individual's sex.

REPRODUCTIVE PHYSIOLOGY IN THE MALE

The major male reproductive functions are (1) secretion of sex hormones, (2) production of sperm, and (3) transport of sperm from the male to the female.

The Reproductive Anatomical Structures in the Male Include Both the Internal and External Genitalia

 What is the functional anatomy of the male reproductive system?

The principal reproductive organs in the male are external. They include the **penis** and the **scrotum,** which contains the testes (Fig. 31–1). The penis is covered by thin, loose skin and contains the **urethra, erectile tissue,** nerves, and many blood vessels. The scrotum is a pouch of skin made of connective tissue and muscle. This pouch contains the **testes,** the **epididymis,** and the **spermatic cord,** which contains the vas deferens. The location of the testes in the scrotal compartment outside the body allows for a testicular temperature approximately 2°C below body temperature, which is essential for normal sperm production. In response to cold, the cremaster muscle of the scrotum contracts involuntarily and lifts the testes closer to the body. The muscle of the scrotum relaxes when the testes are exposed to heat, allowing them to move away from the body. These muscular reflexes ensure that the testes are maintained at a temperature conducive to the formation of spermatozoa (spermatogenesis).

Anatomy of the Testes

The testes of the adult male have two major functions: (1) production of sperm and (2) production of male sex hormones. Each testis contains a large number of tightly coiled tubules called **seminiferous tubules** (Fig. 31–2). The semi-

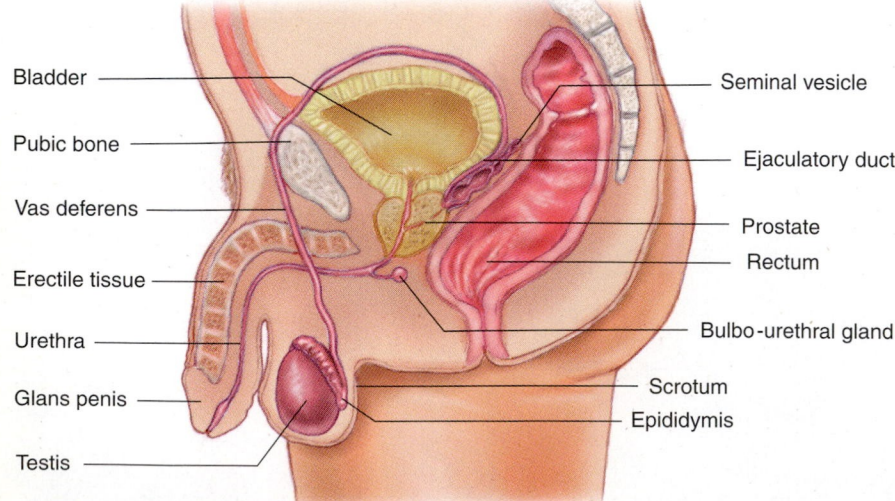

Bladder
Pubic bone
Vas deferens
Erectile tissue
Urethra
Glans penis
Testis

Seminal vesicle
Ejaculatory duct
Prostate
Rectum
Bulbo-urethral gland
Scrotum
Epididymis

Figure 31–1

Side view of the male reproductive tract. *(Redrawn from Masters and Johnson, Human Sexuality, ed 2.)*

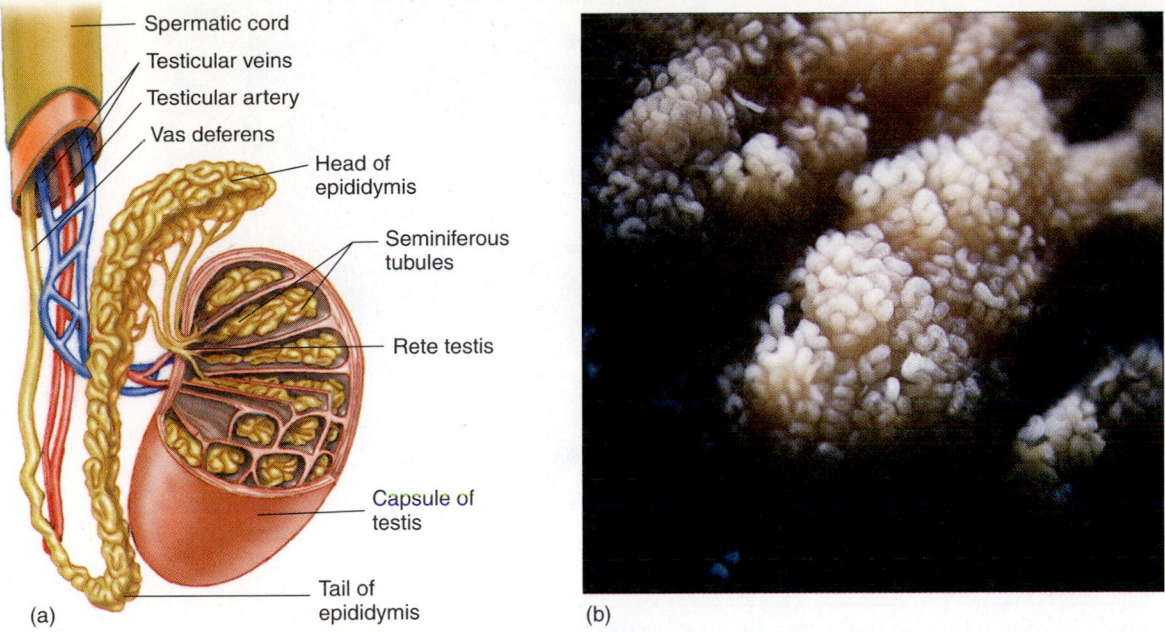

Spermatic cord
Testicular veins
Testicular artery
Vas deferens
Head of epididymis
Seminiferous tubules
Rete testis
Capsule of testis
Tail of epididymis

(a)

(b)

Figure 31–2

(a) A cross-section of the internal structure of the testis, illustrating compartmentalization of the seminiferous tubules. *(b)* The convolutions of a single seminiferous tubule. *(© Nilsson, L., A Child Is Born, Dell Publishing Company, 1977.)*

niferous tubules contain **Sertoli cells,** which surround developing germ, or sperm-producing cells. The cells that synthesize male sex hormones, such as **testosterone,** are called **Leydig cells** and are located in the interstitial spaces between the seminiferous tubules in the testes. In addition to the Leydig cells, the interstitial space also contains lymphatics, blood vessels, nerve fibers, connective tissue, and extracellular fluid. The nerve fibers and blood vessels enter and leave the testes via the spermatic cord.

The Formation of Sperm Takes Place in the Seminiferous Tubules of the Testes

 How does sperm form?

The process of sperm production, called **spermatogenesis,** takes place within the coiled seminiferous tubules of the testes. Primitive germ cells, called **spermatogonia,** are present in the testes during embryonic development and childhood. Prior to puberty, the spermatogonia divide by mitosis to produce more spermatogonia. However, spermatogenesis, or the formation of mature sperm, does not begin until puberty. Once begun, the process of sperm production never ceases. Although sperm production diminishes with age, the number of sperm produced is still usually sufficient to maintain fertility.

During spermatogenesis, germ cells undergo a series of divisions and differentiations to become mature sperm (Figs. 31–3 and 31–4). Some of the spermatogonia enlarge and differentiate to become **primary spermatocytes.** Each primary spermatocyte then undergoes meiotic cell division to produce two **secondary spermatocytes.** The secondary spermatocytes undergo a second meiotic division to produce **spermatids.** During the final stage of spermatogenesis, referred to as **spermiogenesis,** the spermatids differentiate into mature sperm, or **spermatozoa.** This process of spermatogenesis takes approximately 70 to 80 days. Spermatogenesis is constantly being repeated, and all stages of sperm cell development are present in the seminiferous tubules at any given time.

As many as several hundred million sperm may be produced daily. Two decades ago, the average college age male produced approximately 90 to 100 million sperm per milliliter of semen. Today, the average has dropped to approximately 60 million/ml, but whether this decline represents a decrease in the number of viable sperm is not known. It has been speculated that some environmental pollutants may inhibit spermatogenesis. If sperm counts drop below 35 million/ml, fertility is markedly impaired.

Differentiation of Germ Cells

Two types of cells are involved in spermatogenesis: (1) the **spermatogenic cells,** from which mature sperm arise, and (2) the Sertoli cells, which provide mechanical and

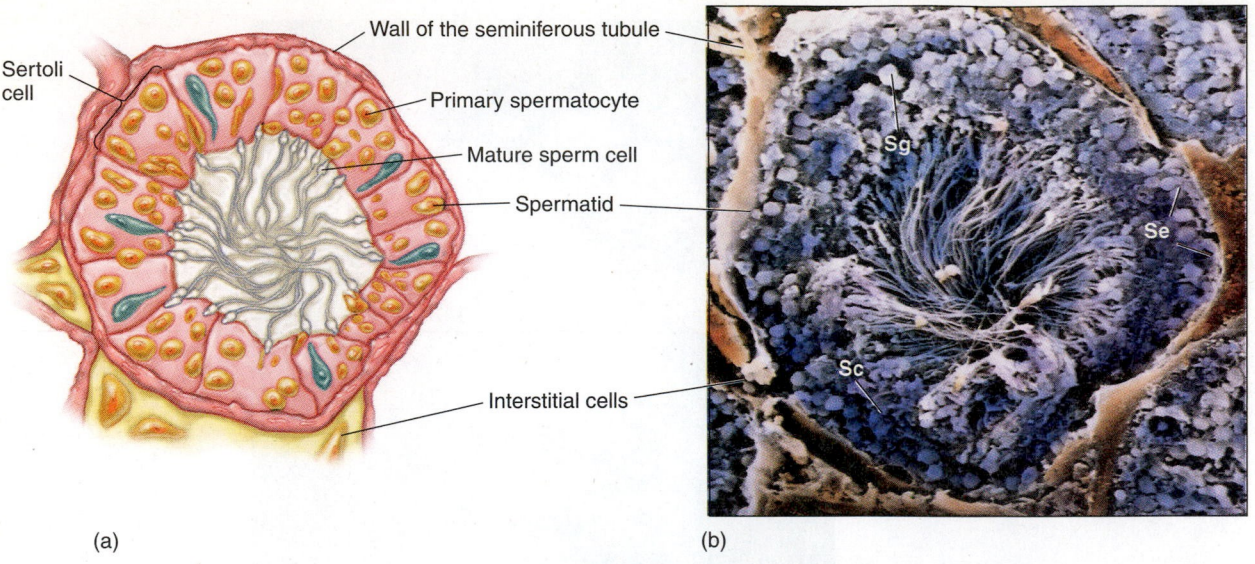

(a)

(b)

Figure 31-3

(a) A cross-section of a seminiferous tubule in the testes, illustrating the anatomical relationships among different cell types. *(b)* Scanning electron micrograph of a transverse section through a seminiferous tubule.   *(© Professor P. Motts/Department of Anatomy/University "La Sapienza," Rome/Science Photo Library/Custom Medical Stock Photo.)*

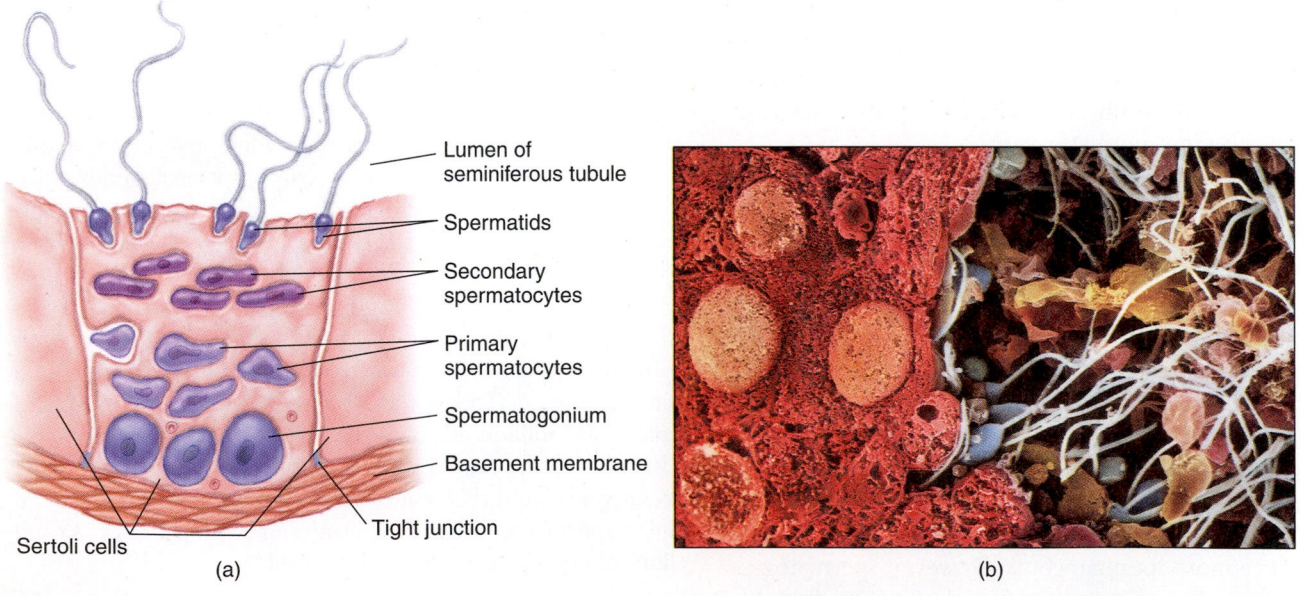

(a)

(b)

Figure 31-4

(a) Cross-section of the seminiferous tubule, illustrating the ring formed by Sertoli cells and the location of the spermatogenic cells. *(b)* Mature sperm that have been pushed into the central duct of a seminiferous tubule.   *(© Nilsson, L., A Child Is Born, Dell Publishing Company, 1977.)*

nutritional support for the spermatogenic cells. Both spermatogenic and Sertoli cells are located inside the seminiferous tubules. Each seminiferous tubule is surrounded by a basement membrane that separates the tubule from the surrounding Leydig cells and connective tissue. The anatomical relationship among the spermatogenic, Sertoli, and Leydig cells is illustrated in Figure 31–3.

The immature or primitive germ cells are found along the outer border of each seminiferous tubule in contact with the basement membrane. The primitive germ cells undergo a mitotic division and differentiate into spermatogonia (sing., *spermatogonium*). In the adult, spermatogonia represent the first cell stage of spermatogenesis. Spermatogenesis begins when some of the spermatogonia differentiate into primary spermatocytes. As the spermatogonia differentiate, they increase in size, undergo a second mitotic division, and move away from the basement membrane and toward the center of the seminiferous tubule (see Fig. 31–4). The remaining spermatogonia continue to divide mitotically and provide a constant source of new germ cells and additional generations of primary spermatocytes.

Mitotic and Meiotic Divisions During Spermatogenesis. Spermatogonia, in common with all other cells of the body, are diploid and contain 46 chromosomes, 23 contributed by the mother and 23 by the father. During mitotic division, the chromosomes are duplicated before cell division. When the cell divides, each of the two new cells receives one copy of each of the 46 chromosomes present in the parental cell. Therefore, primary spermatocytes, formed during mitotic division of spermatogonia, have chromosomes identical to those of the spermatogonia that produced them (23 maternal and 23 paternal chromosomes).

Primary spermatocytes undergo two more divisions through a special form of cell division known as **meiosis.** The first meiotic division (meiosis I) reduces the number of chromosomes in the cell by half. During meiotic division, there is a random assortment and recombination of maternal and paternal chromosomes. In addition, homologous chromosomes (similar chromosomes, one of maternal origin and the other of paternal origin) exchange chromosomal segments in a process known as **crossing over** (Fig. 31–5a). Crossing over can occur at random along the length of the

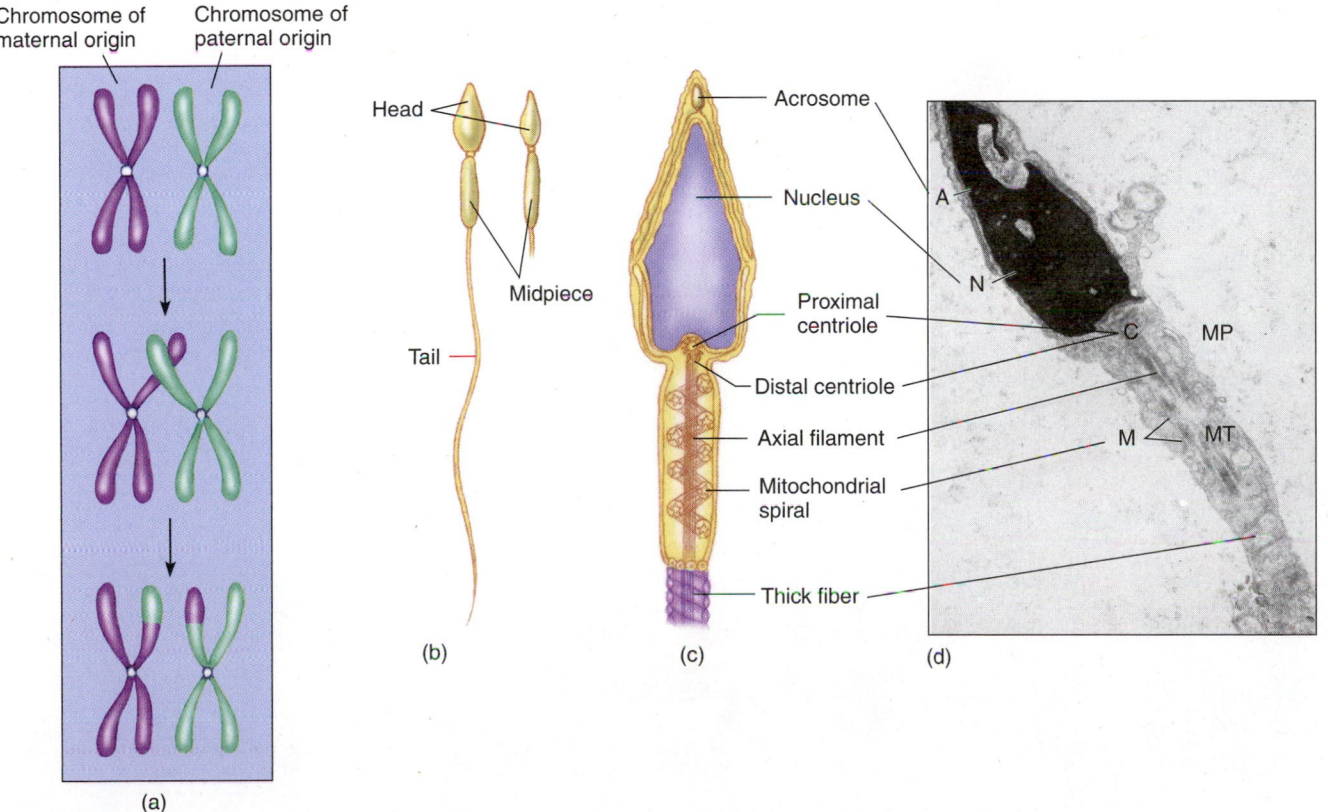

(a)   (b)   (c)   (d)

Figure 31–5

(a) Diagrammatic representation of crossing over, or the exchange of chromosomal segments between homologous chromosomes, during meiotic division of the spermatocyte. *(b)* Top and side view of a sperm. *(c)* The head and midpiece of a sperm. *(d)* Electron micrograph of a human sperm. A, acrosome; N, nucleus; MP, midpiece; M, mitochondria; MT, microtubules; C, centrioles.   (© Dr. Lyle C. Dearden.)

paired homologous chromosomes, and several exchanges can occur at different physical points during a single meiotic division. The term *recombination* refers to any process that leads to new gene combinations. Crossing over produces a recombination of genetic information or the formation of new gene combinations not present in the parental generation. Crossing over creates the possibility that genetic material can recombine in an infinite number of ways, leading to infinite individual variation.

Before the first meiotic division, the 23 maternal and 23 paternal chromosomes in the cell are duplicated through DNA synthesis (Fig. 31–6). The homologous maternal and paternal chromosomes pair with each other to form 23 pairs of chromosomes, with each chromosomal pair known as a *bivalent*. Each chromosome has two chromatids, and therefore each pair of chromosomes has four chromatids. The primary spermatocyte, which contains 23 pairs of chromosomes, then undergoes two very rapid meiotic divisions that do not

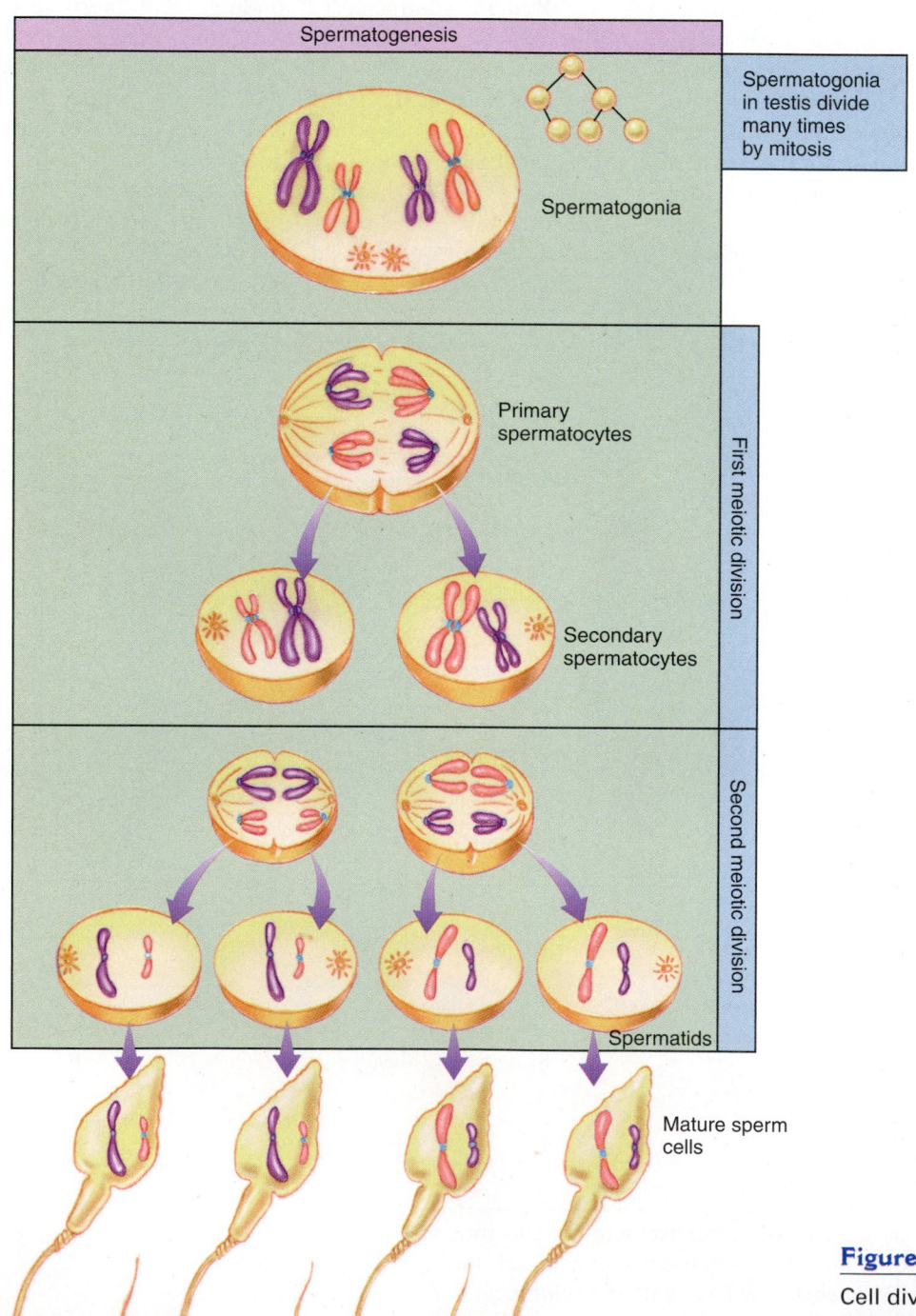

Figure 31–6

Cell division during spermatogenesis. The primary spermatocyte undergoes meiosis and gives rise to four spermatids.

involve chromosomal duplication. During the first meiotic division, each primary spermatocyte gives rise to two secondary spermatocytes, which receive a random mixture of maternal and paternal chromosomes. Therefore, individual secondary spermatocytes differ from each other and from primary spermatocytes in terms of their genetic composition.

Each secondary spermatocyte then undergoes a second meiotic division and gives rise to two spermatids, each of which contains 23 chromosomes. Through this process, four spermatids are formed from the original primary spermatocyte. The spermatids then differentiate into mature sperm. This process of meiotic cell division results in great diversity in the genetic composition of sperm.

Physiology of the Mature Sperm

As primary spermatocytes differentiate into spermatids, they migrate away from the basement membrane of the seminiferous tubule toward the fluid-filled lumen of the tubule (see Fig. 31–4). Near the lumen, the spermatids undergo the final phase of spermatogenesis and differentiate into mature **spermatozoa** or **sperm,** which enter the lumen of the seminiferous tubule. From the lumen, the sperm leave the seminiferous tubule and enter the epididymis.

During the final phase of spermatogenesis, the spermatids do not undergo cell division but rather go through a complex series of cytological changes. The **chromatin,** or genetic material of the spermatid, is contained within the nucleus in the head of the sperm and is surrounded by a membrane. As the spermatid changes into a mature sperm, it increases in size and becomes elongated (Fig. 31–5b). The anterior half of the head of the sperm is covered by a small, caplike structure called the **acrosome** (Fig. 31–5c), which contains enzymes that aid the sperm during fertilization of the ovum. The midpiece of the sperm contains a spiral arrangement of mitochondria that generate the energy necessary for sperm movement. The tail of the sperm, or flagellum, contains filaments that slide across one another to produce movement of the flagellum, which propels the sperm forward (see Fig. 31–5c).

Sertoli Cells Serve Several Functions That Are Important for Spermatogenesis

 How do Sertoli cells support the process of spermatogenesis?

The Sertoli cells are located inside the seminiferous tubules. Each Sertoli cell extends from the basement membrane to the fluid-filled lumen of the seminiferous tubule (see Fig. 31–4). The Sertoli cells are joined to one another by specialized attachments called **tight junctions.** As a result, the Sertoli cells form a ring around the inner diameter of the seminiferous tubule (Fig. 31–7). All the spermatogenic cells, in different stages of differentiation, are interspersed between and surrounded by the Sertoli cells (see Fig. 31–4).

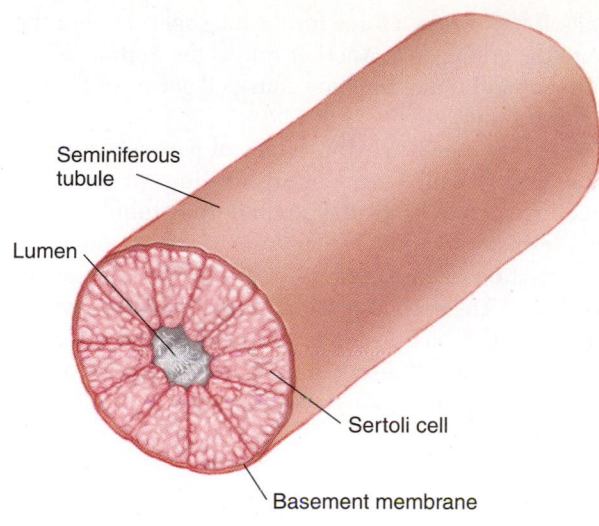

Figure 31–7

Cross-section of a seminiferous tubule, illustrating the ring formed by Sertoli cells.

The Sertoli cells have many important functions (Table 31–1).

The Sertoli cells form a **blood–testes barrier,** which develops at puberty just before spermatogenesis begins and prevents many substances from entering or leaving the seminiferous tubules. For instance, the Sertoli cell barrier excludes from the tubule a variety of proteins, hormones, ions, and drugs that could adversely affect the developing spermatogenic cells. Conversely, this barrier keeps the sperm in the tubules from diffusing into the blood. Mature sperm are very immunogenic. If mature sperm entered the blood, they would be recognized as foreign antigens, resulting in the production of sperm-specific antibodies that could destroy new sperm as they were produced.

Sertoli cells secrete most of the fluid in the lumen of the seminiferous tubules. This fluid is rich in proteins; enzymes; nutrients; androgens (e.g., testosterone); and certain ions, such as potassium and bicarbonate. These chemicals stimulate and nourish the developing sperm. The fluid produced by the

TABLE 31–1
Functions of Sertoli Cells
Maintenance of the blood–testes barrier
Nourishment of developing germ cells
Production of seminiferous tubular fluid
Removal of damaged germ cells
Synthesis of androgen-binding protein
Synthesis of inhibin

Sertoli cells also provides a force for flushing sperm out of the tubules into the epididymis. Another role of the Sertoli cells is the engulfment and removal of any damaged germ cells in the seminiferous tubules.

Sertoli cells synthesize two proteins of particular importance. The first is called **androgen-binding protein,** or **ABP,** which is secreted into the seminiferous tubular fluid. ABP binds testosterone, which allows for a greatly increased local concentration of testosterone to exist within the seminiferous tubules. The second protein is called **inhibin,** which is secreted into the blood and transported to the pituitary gland, where it inhibits the secretion of **follicle-stimulating hormone (FSH).** Both of these proteins are important for spermatogenesis.

Sperm Are Transported Through the Internal Reproductive Structures

 How are sperm transported to the outside of the body?

The seminiferous tubules meet at the top of the testes in a network of ducts called the **rete testis** (see Fig. 31–2). Sperm move from the tubules through the rete testis and into the epididymis as a result of pressure created by the continuous formation of fluid in the tubules. The epididymis lies along the back portion of each testis and is lined with smooth muscle. The muscle undergoes rhythmic peristalsis, which transports the sperm through the epididymis and toward the vas deferens (see Fig. 31–2). The duct system involved in sperm transport is illustrated in Figure 31–8.

When sperm enter the epididymis, they are immature, nonmotile, and incapable of fertilization. The epididymis secretes a fluid containing nutrients, enzymes, and hormones that aid in sperm maturation. Sperm take approximately two weeks to pass through the epididymis, during which time they become mature and mobile. From the epididymis, they enter the vas deferens. As the sperm pass through the epididymis, approximately 90% of the fluid surrounding them is reabsorbed, hence concentrating the sperm entering the vas deferens.

Storage of Mature Sperm

Sperm can be stored in the caudal portion of the epididymis for several months in the absence of ejaculation. Sperm travel from the epididymis to the **vas deferens (ductus deferens),** which are components of the spermatic cords that arise from the tail of the epididymis. The vas deferens are composed of long, slender tubes, or ducts, which run from the scrotum up, into the abdominal cavity, and behind the bladder, where they empty into the **ejaculatory duct** (see Fig. 31–8). The ejaculatory duct runs through the center of the **prostate gland,** located at the base of the bladder surrounding the urethra (see Fig. 31–1). The thick wall of the vas deferens is composed of smooth muscle and innervated by sympathetic nerves. As the vas deferens enter the ejaculatory duct, they are joined by the secretory ducts of the seminal vesicles.

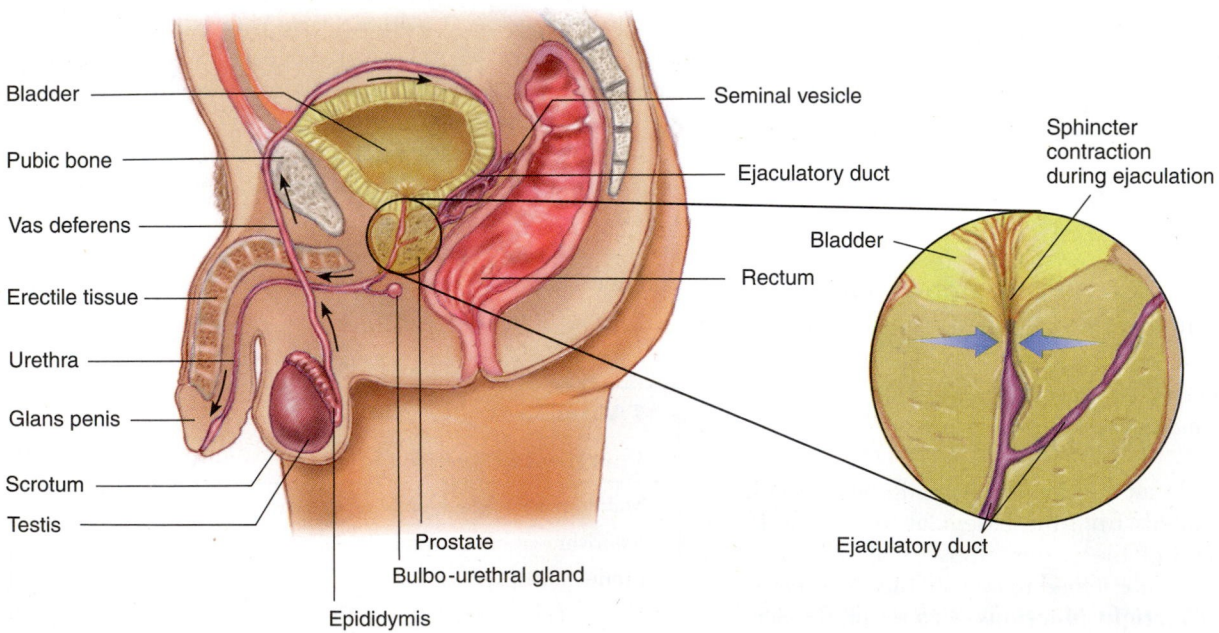

Figure 31–8

Reproductive duct system of the male, illustrating sperm pathway and sphincter contraction in the prostate during ejaculation.

Addition of Nutrient Fluid by the Seminal Vesicles

The **seminal vesicles** are composed of two sacs, one on either side of the bladder. The seminal vesicles secrete a viscous material rich in prostaglandins, fructose, and other nutrients. Prostaglandins, which derive their name from the fact that they were thought to be secreted by the prostate gland, stimulate contractions of the uterus in the female and play a role in moving the sperm through the female reproductive tract following ejaculation. Nutrients in seminal vesicle fluid serve as an energy source for the sperm following ejaculation. Seminal vesicle fluid accounts for approximately 60% of the total volume of ejaculated semen. During ejaculation, the contents of the vas deferens are emptied into the ejaculatory duct immediately before the contents of the seminal vesicles. Seminal vesicle fluid washes sperm out of the ejaculatory duct.

Addition of Alkaline Fluid by the Prostate Gland

The **prostate gland** secretes a liquid that contains citric acid, calcium, and various enzymes. Fluid from the prostate gland adds to the volume of the semen. Prostatic fluid has a pH of approximately 6.5, which may be very important in helping to maintain sperm viability and motility in the acidic environment of the vagina.

The prostate gland contains an elaborate system of valves. During ejaculation, a valve in the prostate opens, and the semen is ejected into the urethra. A sphincter in the prostate contracts during ejaculation and seals off the exit from the bladder, which prevents the passage of urine from the bladder into the urethra (see Fig. 31–8). During ejaculation, semen passes from the ejaculatory duct into the urethra, where it is joined by secretions from the bulbourethral glands, also called Cowper's glands (see Fig. 31–8).

Addition of Lubricant Fluid by the Bulbourethral Glands

The **bulbourethral glands** are located on each side of the urethra. They secrete a clear, viscous liquid that adds to the volume of the semen. During sexual arousal, a small amount of fluid from the bulbourethral glands may appear at the tip of the penis before ejaculation. This fluid can contain sufficient sperm to fertilize a female ovum. The **urethra** transports the semen through the shaft of the penis to the outside of the body during ejaculation (see Fig. 31–8).

The average volume of semen ejaculated is 3 to 5 ml (5 ml ≈ 1 teaspoon). The normal concentration of sperm in the ejaculate is highly variable (40–120 million/ml) and depends in part on frequency of ejaculation. Once sperm are ejaculated, their maximum life span is only 24 to 72 hours at body temperature. The major organs involved in sperm production, maturation, and transport are summarized in Table 31–2.

TABLE 31–2
Major Organs Involved in Sperm Production, Maturation, and Transport
Testes: Sperm production—mitosis, meiosis, and differentiation
Epididymis: Sperm transport and maturation—motility and fertility
Vas deferens: Sperm storage
Seminal vesicles: Production of seminal fluid containing nutrients, fructose, and prostaglandins
Prostate: Production of prostatic fluid that is alkaline pH ≈ 6.5 and contains calcium and citric acid
Bulbo-urethral gland: Production of "pre-ejaculatory" fluid
Penis: Erection and ejaculation

Neuroendocrine Control of Reproduction in the Male Involves an Interaction Between the Hypothalamus, Pituitary, and Testes

 How are reproductive functions in the male controlled?

The **pituitary gland** acts as a relay station, receiving neural and hormonal input from the brain and relaying this information, in the form of hormonal messages, to other organs of the body.

Neuroendocrine Control of Spermatogenesis

The hypothalamus secretes a releasing factor called **gonadotropin-releasing hormone,** or **GnRH** (Fig. 31–9). GnRH is transported to the anterior pituitary gland via a complex set of blood vessels known as the **hypothalamic-pituitary portal vessels.** When GnRH reaches the anterior pituitary in the male, it stimulates the release of two major reproductive hormones, termed the **gonadotropins.** One of the gonadotropins, **luteinizing hormone (LH),** is secreted by the anterior pituitary into the blood and is transported to the testes. There it stimulates the Leydig cells to synthesize and secrete testosterone. The other gonadotropin, **follicle-stimulating hormone (FSH),** is secreted by the anterior pituitary into the blood and is also transported to the testes, where it binds to specific receptors on the plasma membranes of Sertoli cells. FSH stimulates the conversion of spermatogonia into spermatocytes in the seminiferous tubules of the testes in a process that also requires testosterone. Both LH and FSH are required for spermatogenesis (see Fig. 31–9).

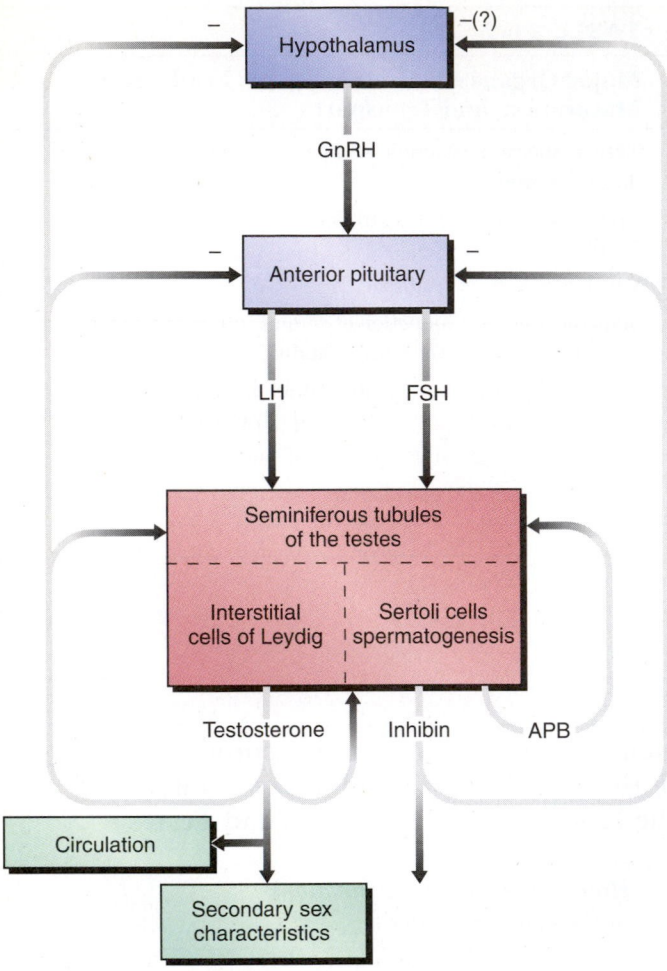

Figure 31–9

Hormonal feedback loops within the hypothalamic–pituitary–gonadal system. The minus sign indicates inhibition.

Control of Neuroendocrine Function by Sensory Stimuli and Feedback Regulation

The hypothalamus receives neural input from a variety of brain areas involved in processing sensory stimuli. The amount of GnRH released by the hypothalamus is controlled by the excitatory and inhibitory neural signals that the hypothalamus receives. Visual, olfactory, or tactile stimuli are capable of altering GnRH release and hence the release of LH, FSH, and testosterone. In addition, thoughts and moods can alter the secretion of these sex hormones. Stress, such as protracted aerobic exercise, can inhibit gonadotropin secretion in both males and females, which may result in transient inhibition of sperm production in the male and disruption of menstruation in the female.

Nevertheless, the secretion of GnRH, LH, and FSH is fairly constant from day to day in the male, in marked contrast to the cyclic secretion of these hormones in the female. The constancy of hormone secretion in the male is due to the presence of a continuous inhibitory **feedback loop** between the hypothalamus, pituitary gland, and testes (see Fig. 31–9). For example, the hypothalamus is sensitive to changes in the circulating levels of sex hormones. It monitors the level of these hormones and responds by increasing or decreasing the production of its releasing factor, GnRH. If the level of circulating testosterone in the blood is too high, testosterone inhibits further production of GnRH in the hypothalamus, leading to a drop in the secretion of LH and FSH from the anterior pituitary. A decrease in LH in the blood results in reduced production and secretion of testosterone by the Leydig cells of the testes; hence blood levels of testosterone fall. Conversely, if the circulating level of testosterone is too low, the hypothalamus increases synthesis and release of GnRH, which stimulates the anterior pituitary to secrete additional LH. The increase in circulating LH stimulates the Leydig cells of the testes to increase the synthesis and release of testosterone into the blood. In addition to this feedback loop between testosterone and the hypothalamus, testosterone also can act directly on the anterior pituitary to inhibit the release of LH in response to GnRH secreted by the hypothalamus by decreasing the sensitivity of the pituitary to GnRH.

Inhibin is a peptide hormone secreted by the testes that exerts an important inhibitory feedback effect on the release of FSH from the pituitary. Inhibin is secreted by the Sertoli cells in the seminiferous tubules of the testes and is known to directly inhibit the release of FSH from the anterior pituitary in a negative feedback loop.

Testosterone: Biological Effects

 What are the actions of testosterone in the male?

Testosterone is the primary **androgen** (male sex steroid) produced in the testes. Testosterone is a steroid formed from the steroid precursor cholesterol. The general features of testosterone biosynthesis were discussed in Chapter 12. The Leydig cells do not store testosterone, but they do store large quantities of the cholesterol precursor in the form of lipid droplets. Following synthesis, testosterone is secreted into the extracellular fluid surrounding the seminiferous tubules, where it diffuses into the tubules to produce local effects on sperm production. Testosterone also is transported from the extracellular fluid into the blood vessels of the testes and from there to the general circulation, where it produces a variety of systemic effects.

When testosterone enters the blood, it is bound to albumin and to another plasma protein, **sex steroid-binding globulin (SSBG)** or **testosterone-estradiol binding globulin (TeBG).** A certain proportion of all steroid hormones are "stored" in the circulation in their bound form, which increases the duration of time that they remain in the plasma. However, only free, unbound hormone acts on target tissues.

Testosterone has some biological effect on almost every tissue of the body. Unaltered testosterone has direct effects on target tissues, such as muscle, kidney, and bone. In other tissues, testosterone is transformed into another more potent steroid, **dihydrotestosterone (DHT),** by the target tissue. This steroid then mediates effects on target tissues. DHT mediates maturation of the male reproductive tract and is an important stimulant of bone growth.

Effects of Testosterone on Puberty

Before puberty, the hypothalamic-gonadal system is relatively dormant. Puberty begins with the secretion of increasing amounts of GnRH from the hypothalamus. This stimulates the release of LH and FSH from the pituitary, which becomes more sensitive to GnRH during puberty. FSH produces enlargement of the testes, and LH stimulates the secretion of testosterone from the Leydig cells. This increase in testosterone secretion at puberty results in the enlargement of the penis, scrotum, and testes and stimulates the development of the male secondary sex characteristics (Table 31–3). Testosterone stimulates muscular development, bone growth, thickening of the skin, and growth of facial and body hair. Testosterone also produces growth and thickening of the vocal cords and enlargement of the larynx, thereby lowering the pitch of the voice. These anatomical changes, stimulated by testosterone, mark the onset of puberty. If testosterone secretion is interrupted prior to puberty, as a result of destruction or removal of the testes, a male will not undergo the anatomical changes characteristic of puberty.

Effects of Testosterone on Spermatogenesis

Spermatogenesis requires the presence of high levels of testosterone in the seminiferous tubules. A decrease in testosterone production can result in sterility. Testosterone

is thought to stimulate both the formation of spermatogonia and the second meiotic division that results in the formation of spermatids in the seminiferous tubules. An interesting interdependence exists between testosterone and the Sertoli cells (see Fig. 31–9). Testosterone stimulates protein synthesis and fluid secretion from the Sertoli cells, and the Sertoli cells in turn secrete androgen-binding protein (ABP), which binds testosterone and keeps testosterone levels high in the seminiferous tubules (see Fig. 31–9).

REPRODUCTIVE PHYSIOLOGY IN THE FEMALE

The major reproductive functions in the female are (1) secretion of sex hormones, (2) production of ova, and (3) gestation and delivery of the fetus.

The Principal Reproductive Anatomical Structures in the Female

> *What is the functional anatomy of the female reproductive system?*

The external genitalia of the female, which are collectively called the **vulva,** include the **labia majora, labia minora, clitoris,** and **vaginal opening** (Fig. 31–10). The labia majora consist of folds of skin that cover fat and smooth muscle. The labia majora meet on the midline and protect the urinary opening and the entrance to the vagina. The labia minora are composed of skinfolds rich in small blood vessels, that meet above the clitoris and provide mechanical protection for the clitoris. The only visible part of the clitoris is termed the **glans clitoris.** This structure, which has an exclusively sexual function, contains a dense supply of nerve endings, which makes it highly sensitive to touch, pressure, and temperature.

Anatomy of the Female Internal Genitalia

The internal reproductive organs of the female include the **vagina, uterus, uterine tubes,** and **ovaries** (Fig. 31–11). In contrast to those in the male, the urinary and reproductive tracts in the female are anatomically separate.

The Vagina. The vagina is a muscular organ that extends from the external opening in the vulva to the uterus. The muscular vaginal walls have the ability to contract and expand, allowing for the passage of a baby during childbirth. The vaginal lining, called the **vaginal mucosa,** secretes a fluid during sexual arousal that acts as a lubricant. The vagina contains few sensory nerves and is relatively insensitive to touch and pressure.

TABLE 31–3

Testosterone-Dependent Changes at Puberty

1. Penis and testes enlarge and become pigmented; the scrotal skin becomes wrinkled.

2. Growth of facial hair results in the development of mustache and beard, and scalp line recedes.

3. Pubic hair grows in an upward pattern on the abdomen; hair appears on the chest, armpits, and extremities.

4. Growth of the long bones increases to a rate of about 3 inches per year. Bones increase in size and strength due to an increase in the quantity of bone matrix and increased calcium retention.

5. Vocal cords and larynx enlarge, thereby lowering the pitch of the voice.

6. Muscle mass and skin thickness increase, and the skin deepens in hue due to melanin deposition. Secretions from the sebaceous glands of the face increase, resulting in the development of acne.

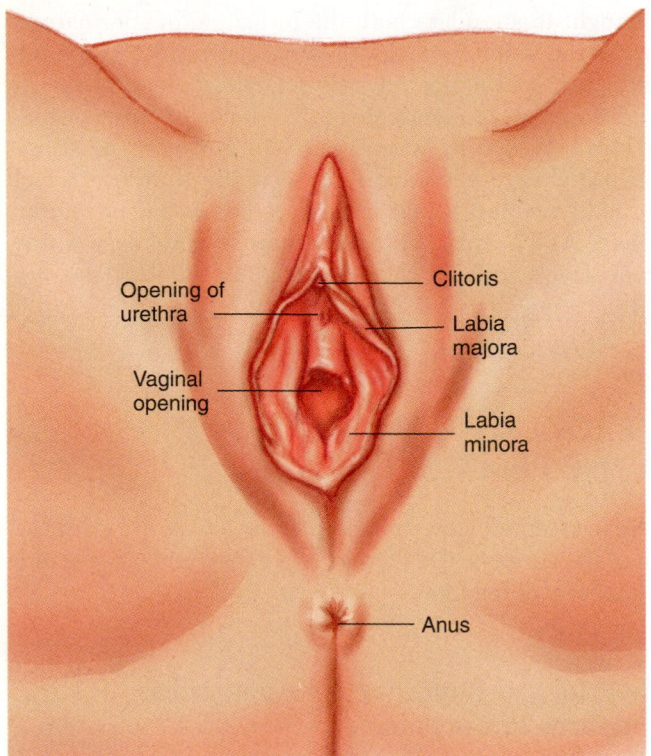

Figure 31–10

External genitalia of the female. *(Redrawn from Masters and Johnson,* Human Sexuality, *ed 2.)*

The Uterus. The uterus, which is located in the lower abdomen, is a hollow, fist-sized organ with walls composed of smooth muscle. The term **cervix** applies to the portion of the uterus that extends into the vagina. The lining of the uterus, termed the **endometrium,** changes in thickness during the menstrual cycle. At the beginning of pregnancy, the fertilized ovum implants in the endometrium of the uterus. The mus-

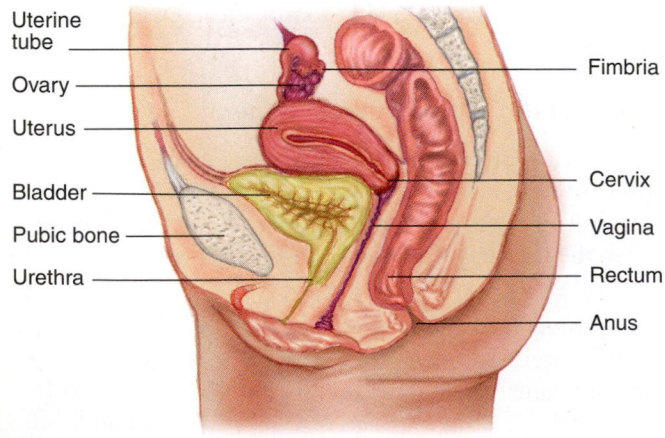

Figure 31–11

Side view of the female reproductive tract. *(Redrawn from Geer, J., Heiman, J., and Leitenberg, H.,* Human Sexuality.)*

cular walls of the uterus, called the **myometrium,** expand and contract during labor and delivery and during sexual stimulation.

The Uterine Tubes. The uterine tubes, also called **fallopian tubes,** or **oviducts,** arise from the uterus and extend into the abdominal cavity (Fig. 31–12). These muscular, tubelike structures terminate near the ovaries. The end of each uterine tube is surrounded by long, fingerlike extensions called the **fimbriae.** The uterine tubes are lined with cilia, which propel the ovum through the uterine tube to the uterus.

The Ovaries. The ovaries are oval-shaped organs located in the pelvic cavity below the open ends of the uterine tubes (see Fig. 31–12). The ovaries are the site of the production of germ cells and hormones.

Oogenesis, or the Formation of Ova, Takes Place in the Ovaries

The primary germ cells in the female, termed **oogonia,** are analogous to spermatogonia in the male. During the second and third months of embryonic development, meiosis begins in some oogonia, converting them to primary oocytes. The **primary oocytes** increase in size and are surrounded by a layer of flattened follicular cells called *granulosa cells.* The entire structure is then called a **primary follicle,** or *primordial follicle.* Primary oocytes that are not converted to primary follicles degenerate.

During the fourth and fifth months of fetal development, the number of primary follicles in the ovary increases, and the number reaches a peak during the sixth month of fetal development (Fig. 31–13). The primary follicles increase in size and are surrounded by a layer of cuboidal granulosa cells called the *stratum granulosum.* The granulosa cells secrete mucopolysaccharides that form a coating, or membrane, called the **zona pellucida,** which lies between the follicular cells and the oocyte (Fig. 31–14a, b). The granulosa cells provide nutrients and chemical signals that stimulate further maturation of the follicle (see Fig. 31–14 *a* and *b*).

The **granulosa cells** form a barrier around the developing follicle that prevents the entrance of substances such as proteins, ions, and drugs, which could destroy the developing germ cell. During the fifth and sixth months of gestation, the connective tissue around the follicle differentiates to form a highly vascular cell layer, the **theca interna,** which surrounds the follicle. The primary follicles remain in this first stage of meiotic division until puberty, when ovulation begins. The first meiotic division of the oocyte is completed just before ovulation; the second meiotic division does not occur unless and until the germ cell is fertilized by a spermatozoon.

The growth patterns of female germ cells differ from that seen in the male. In the female, meiotic division of the

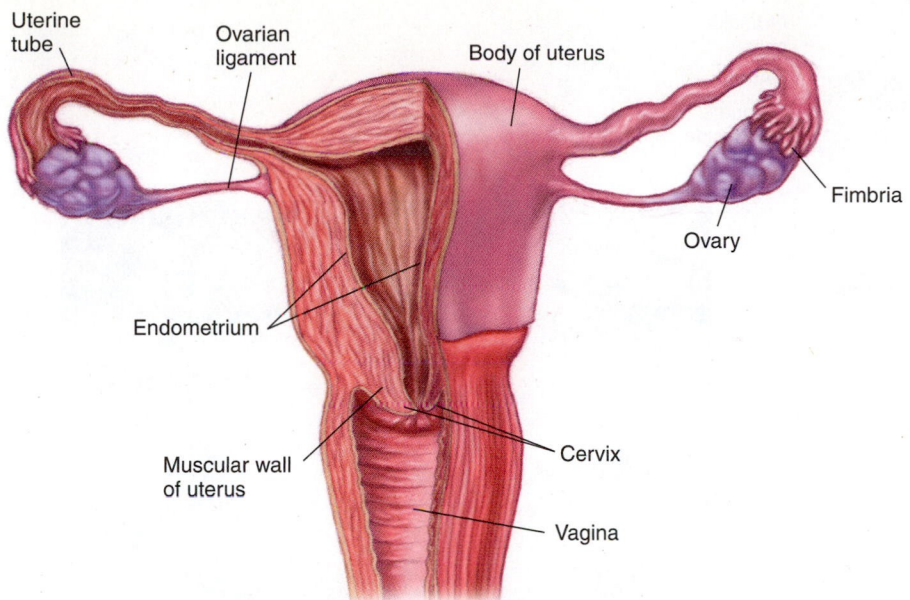

Figure 31–12

Front view of the female reproductive tract. *(Redrawn from Masters and Johnson, Human Sexuality, ed 2.)*

germ cells begins during embryonic development but then stops until puberty, when ovulation begins. While millions of sperm may be produced daily in the testes during the entire reproductive life span of the male, all of the primary follicles that a female will ever produce, a maximum of approximately 7 million, are formed in the ovaries during the first six months of embryonic life and most will disappear. At birth, only approximately 2 million primary follicles remain in the ovaries, and at the onset of puberty this number has fallen to approximately 400,000 (see Fig. 31–13). When the female reaches **menopause,** at approximately 50 years of age, very

few primary follicles are left in the ovaries, and reproduction ceases.

Ova (female germ cells; singular **ovum**) are formed from the primary follicles in the ovary. During the middle of each ovulatory cycle, a single ovum is expelled by one of the two ovaries into the abdominal cavity. The ovum enters one of the uterine tubes and passes through into the uterus. If the ovum has been fertilized by a sperm, it implants in the uterus. If it has not been fertilized, it disintegrates and leaves the uterus along with the menstrual flow. Usually only one ovum is released from one of the two ovaries each month during the reproductive life span of the female, which begins at the onset of puberty and ends with menopause. Thus, only approximately 400 to 500 follicles will ever develop into mature ova and be released from the ovaries during a woman's reproductive life span.

Ovulation Occurs Once Each Month from Puberty Until Menopause

> *What happens during the various stages of ovulation?*

Once each month, from puberty until menopause, 6 to 20 primary follicles enter the next stage of follicular development, which usually culminates in the release of a single ovum from one of the two ovaries.

Growth of the Follicle

Each month, the spindle cells, surrounding 6 to 20 of the primary follicles, proliferate to form the **theca externa.** These mature follicles are called **Graafian follicles** (see Fig. 31–14*a*). Although 6 to 20 follicles may begin to

Figure 31–13

Number and type of follicles in the ovary during development.

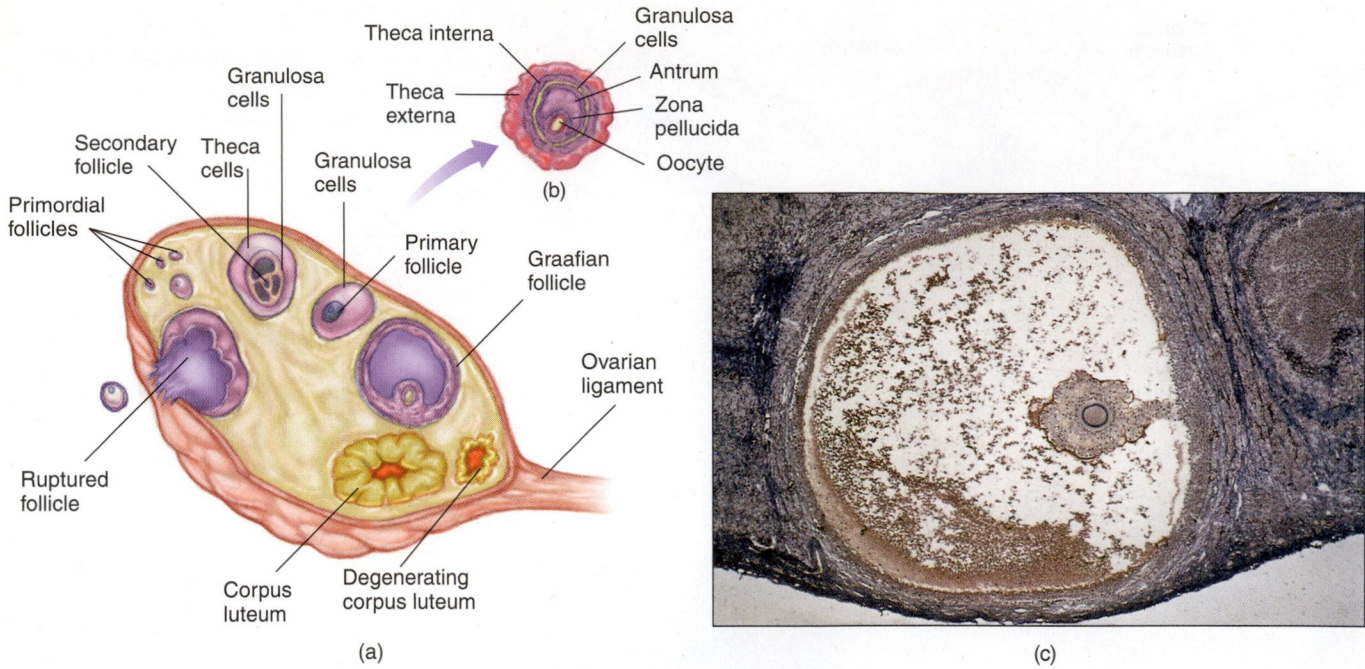

Figure 31–14

(a) Cross-section of the ovary, containing follicles in different stages of development. *(b)* Cross-section of a Graafian follicle. *(c)* Cross-section of a developing follicle in the ovary.   *(© Biophoto Associates.)*

develop in the ovary each month, only one reaches maturation. As one follicle increases in size faster than the others, the remaining follicles in the ovary degenerate and undergo **atresia** (collapse and withering). Degeneration of the remaining follicles can occur at any stage of follicular development, and atretic follicles can always be found in the ovary.

Several days before ovulation, a single Graafian follicle increases dramatically in size to reach 10 to 20 mm in diameter within 48 hours. The granulosa cells surrounding this follicle begin to secrete increasing amounts of fluid that contain proteins, electrolytes, polysaccharides, and hormones. The follicular fluid in the Graafian follicle collects in a single central area called the **antrum** (see Fig. 31–14*b* and *c*). The fluid-filled antrum displaces the ovum to the side of the follicle, and the follicle itself forms a bulge on the surface of the ovary. The follicle secretes proteolytic enzymes that break down the surface of the ovary, and the antral fluid and the ovum are released from the follicle into the peritoneal cavity in a process known as **ovulation** (see Fig. 31–14*a*). Just before ovulation, the first meiotic division of the ovum is resumed and, following ovulation, the ovum enters the uterine tube.

Growth of the Corpus Luteum

During the first few hours after ovulation, the remaining granulosa and theca cells of the follicle undergo complex physical changes, and the follicle becomes the **corpus lu-**

teum (see Fig. 31–14*a*). The cells of the corpus luteum increase in size and develop a rich vascular supply for 7 to 8 days following ovulation. The corpus luteum secretes large amounts of **progesterone** and smaller amounts of **estrogen,** which feed back to inhibit pituitary hormones and thus retard the growth of new follicles. If fertilization and implantation of the ovum do not occur, the corpus luteum degenerates between days 8 and 21 following ovulation. When the corpus luteum degenerates, new follicles begin to grow again in the ovary, marking a new ovarian cycle.

Neuroendocrine Control of Reproduction in the Female Involves an Interaction Between the Hypothalamus, Pituitary, Ovaries, and Uterus

 How are the ovarian and uterine cycles controlled?

Neuroendocrine control of reproduction in the female is dependent on a continuous interaction between the hypothalamus, the pituitary gland, and the ovaries. Hormone secretion and reproductive processes in the female are cyclic rather than continuous.

In the female, the hypothalamus secretes gonadotropin-releasing hormone, or GnRH, which is transported to the anterior pituitary, where it stimulates the release of LH and FSH. LH and FSH stimulate germ cell development and hormone secretion in the ovaries (Fig. 31–15).

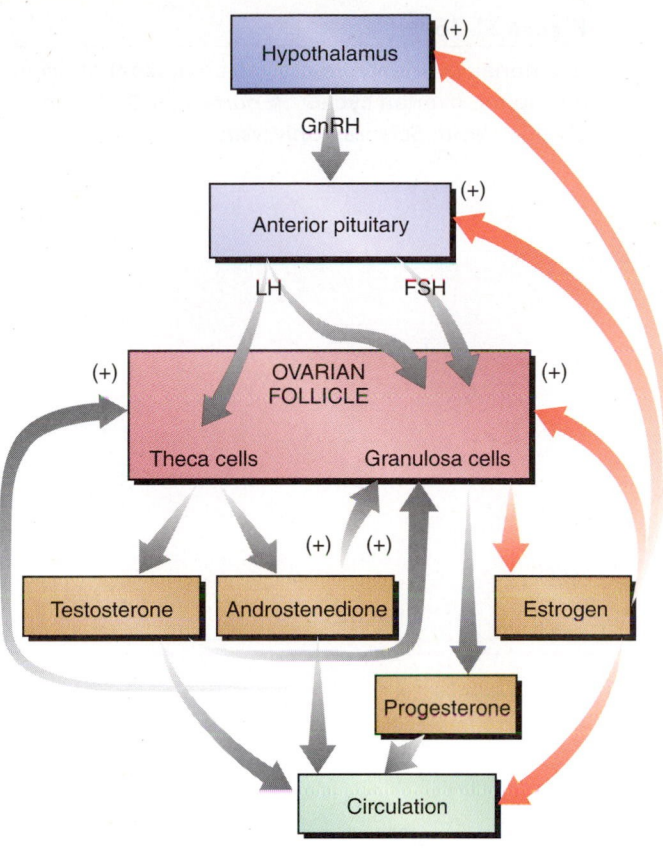

Figure 31–15

Hormonal interactions between the pituitary gland and ovary during the follicular phase of the ovarian cycle. The plus sign indicates stimulation.

In the female, the effects of LH and FSH on the ovary vary during different phases of the ovarian cycle. The **ovarian cycle** is approximately 28 days long and can be divided into three phases: (1) the follicular phase, which lasts for approximately 12 days; (2) the luteal phase, which lasts for approximately 11 days; and (3) the menstrual phase, which lasts for approximately 5 days (Fig. 31–16).

During the **follicular phase,** a single follicle matures and is released during ovulation. During the **luteal phase,** which follows ovulation, the postovulatory follicular cells in the ovary change into the corpus luteum. During the **menstrual phase,** the uterine endometrium is shed in a process called **menstruation.**

Neuroendocrine Control of the Follicular (Preovulatory) Phase

Once each month, just after menstruation ceases, growth and development of follicles begin in the ovary. Follicular growth is stimulated by the pituitary hormone FSH. FSH stimulates both hypertrophy (increase in cell size) and **hyperplasia** (increase in cell number) of the granulosa cells that surround the primary follicle. In response to FSH, the granulosa cells synthesize estrogen from androgen precursors. Estrogen secretion from the granulosa cells results in rising levels of circulating estrogen (see Fig. 31–15). Estrogen stimulates further GnRH secretion from the hypothalamus and sensitizes the pituitary to GnRH, thereby increasing LH and FSH secretion from the pituitary (see Fig. 31–15). **Estrogen,** which is secreted by the granulosa cells, acts locally on the granulosa cells to increase the number of receptors in the follicle for both estrogen and FSH. As the number of estrogen receptors in the follicle rises, more estrogen is retained in the follicle, and follicular estrogen content increases. This autocrine action induces an increase in the number of FSH receptors on the granulosa cells in the follicle and serves to augment the growth-promoting effect of FSH on the follicular granulosa cells. As the follicle grows, it secretes more estrogen, which continues to augment the action of FSH in the follicle. FSH secreted by the pituitary and estrogen secreted by the follicle together increase the number of LH receptors on the granulosa cells of the follicle.

During the follicular phase (see Figs. 31–15 and 31–16), the increasing estrogen, secreted by the granulosa cells of the follicle, stimulates the pituitary to synthesize and secrete additional LH. This increase of LH in the blood stimulates the theca cells of the follicle to synthesize and secrete two androgens: testosterone and **androstenedione.** These androgens diffuse into the follicular granulosa cells and serve as precursors for testosterone production in the granulosa cells. This diffusion of an agent from one cell type to influence function in a neighboring cell is called a **paracrine** function.

Gradual elevation of blood levels of LH, together with the increased number of LH receptors in the granulosa cells, stimulate the granulosa cells to secrete progesterone. Progesterone enters the circulation, and blood levels of progesterone begin to increase (see Fig. 31–16). In addition, some of the progesterone from the granulosa cells diffuses back into the theca cells and acts as a substrate for further androgen synthesis by the theca cells (Figs. 31–16 and 31–17).

The mechanisms that control the selection of a single follicle for full maturation are not understood. Atresia may result from an insufficient number of FSH receptors in some follicles, or from an inability of the granulosa cells of certain follicles to produce adequate amounts of estrogen from androgen precursors. Fluid from atretic follicles has been found to contain lower levels of both FSH and estrogen than fluid from mature follicles.

Neuroendocrine Control of Ovulation

During the early follicular phase of the ovarian cycle, increased release of estrogen from the ovary stimulates LH and FSH release from the anterior pituitary, which results in continued follicular growth for approximately 10 days (see Fig. 31–15). On day 14, just prior to ovulation, blood levels of LH and FSH are higher than at any other time in the ovarian

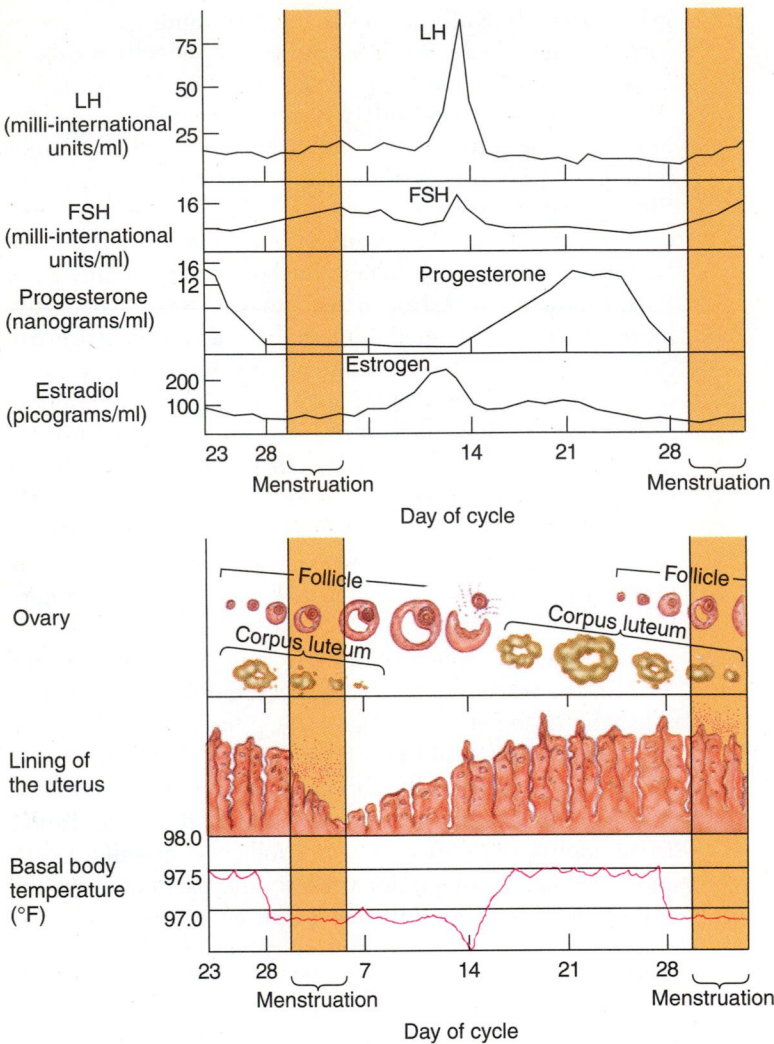

Figure 31–16

Hormonal, ovarian, and uterine changes that occur during the ovarian cycle.   *(Courtesy of Dr. Resco, Oregon Health Sciences University.)*

cycle. These peaks in blood levels of LH and FSH are referred to as the **preovulatory surges** of LH and FSH. The surge of LH induces changes in the structure of the follicle that result in ovulation (see Fig. 31–16) and include an increase in the production of antral fluid; compression of the ovum toward the wall of the follicle; protrusion of the follicle against the wall of the ovary; and, finally, enzymatic degeneration of the ovarian wall and release of the antral fluid and ovum into the abdominal cavity during ovulation. One or two days before ovulation, the follicle stops secreting estrogen, and blood levels of estrogen begin to fall (see Fig. 31–16). Ovulation marks the end of the follicular phase of the ovarian cycle in the female.

In many women, ovulation may be accompanied by a slight decrease, followed by an upward shift of approximately 0.5° to 1.0° F, in basal body temperature, which persists throughout the luteal phase as a result of progesterone secretion (see Fig. 31–16). However, because ovulation can occur in the absence of a detectable change in body temperature, measurements of body temperature are not reliable indicators of ovulation.

Neuroendocrine Control of the Luteal (Postovulatory) Phase

Immediately following the release of the ovum from the follicle on day 14 of the ovarian cycle, the remaining granulosa cells of the follicle increase in size and number and undergo chemical changes to form the corpus luteum (see Fig. 31–16). This process is called **luteinization.** Change and growth of the corpus luteum are stimulated by LH from the anterior pituitary.

During the luteal phase, the corpus luteum secretes a large amount of **progesterone** and a smaller amount of **estrogen,** increasing blood levels of these two hormones (see Fig. 31–16). The estrogen secreted by the corpus luteum is structurally identical to the estrogen secreted by the follicle during the follicular phase. However, while estrogen stimulates release of GnRH, LH, and FSH during

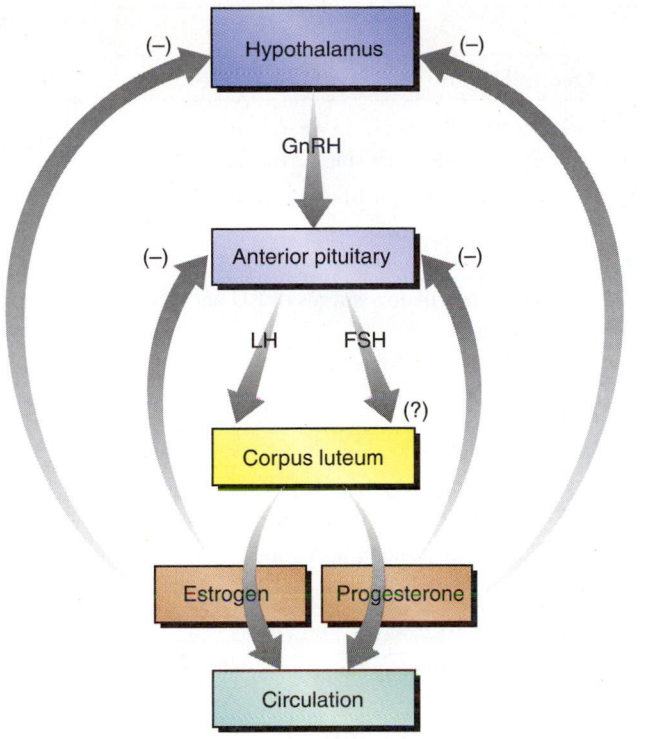

Figure 31–17

Hormonal interactions between the pituitary gland and ovary during the luteal phase of the ovarian cycle. The minus sign indicates inhibition.

the follicular phase prior to ovulation, the estrogen and progesterone secreted together during the luteal phase *inhibit* both GnRH secretion from the hypothalamus and LH and FSH secretion from the pituitary (see Fig. 31–17). Consequently, during the luteal phase, blood levels of LH and FSH fall (see Fig. 31–16). The fact that estrogen and progesterone have different effects on GnRH, LH, and FSH release during the follicular versus the luteal phase of the ovarian cycle is due to a change in hypothalamic sensitivity to estrogen and progesterone during the follicular and luteal phases.

As LH and FSH blood levels fall during the luteal phase, the signal for follicular growth diminishes, and no new follicles begin to grow in the ovary. The corpus luteum, following ovulation, begins to degenerate, perhaps in response to falling blood levels of LH (see Fig. 31–16). The corpus luteum itself is thought to secrete prostaglandins, which induce further degeneration.

As the corpus luteum degenerates, it secretes less progesterone and estrogen (see Fig. 31–16). The feedback inhibition of GnRH, LH, and FSH release is now removed and blood levels of LH and FSH begin to rise again. This stimulates growth of another set of follicles in the ovary and a new follicular cycle (see Fig. 31–16).

Neuroendocrine Control of the Menstrual Phase

Dramatic changes in the structure and secretory capacity of the endometrium of the uterus occur during the ovarian cycle in the female in response to estrogen and progesterone. Estrogen stimulates growth and proliferation of the endometrium of the uterus during the follicular phase of the menstrual cycle (see Fig. 31–16). The endometrium thickens as cells enlarge and new cells are produced. Blood vessels invade the thickened tissue, and tubular endometrial glands are formed. Estrogen also increases the number of progesterone receptors in the uterus.

During the luteal phase of the ovarian cycle, estrogen and progesterone (secreted by the corpus luteum) stimulate continued growth and proliferation of the coiled blood vessels of the endometrium (see Fig. 31–16). Progesterone stimulates the endometrial glands to secrete a nutrient-rich mucus. The tubular endometrial glands fill with glycogen, and enzymes accumulate in the connective tissue of the endometrium. During the luteal phase, the endometrium swells and becomes approximately 4 to 6 mm thick. This is called the **secretory phase** of the uterine cycle.

The function of the proliferation of endometrial cells is to provide a continuing source of nutrients for the ovum if it is fertilized and implants in the uterus. If the ovum is not fertilized, the corpus luteum begins to degenerate on day 8 following ovulation and stops secreting progesterone and estrogen. This drop in circulating levels of estrogen and progesterone removes the hormonal support for the uterine lining and causes **hemorrhagic changes** in the uterine endometrium that result in the onset of menstruation (see Fig. 31–16). In the absence of stimulation from estrogen and progesterone, the endometrium becomes **necrotic** (i.e., dead). Arterioles of the endometrium constrict, which results in a slowing of the circulation. Blood enters the vascular layer of the endometrium, resulting in blood pooling. Gradually, the outer layers of the endometrium separate from the uterus at the site of hemorrhage.

Prostaglandins, produced by the uterus, stimulate the uterine smooth muscles to contract rhythmically. These uterine contractions result in the discharge of blood and endometrial cells that constitutes the **menstrual flow.** The menstrual flow lasts for approximately 5 to 7 days (see Fig. 31–16), during which time the uterus expels an average of 20 to 200 ml of blood. An overproduction of prostaglandins in the uterus during the menstrual phase of the ovarian cycle can lead to very strong and protracted uterine contractions in many women, causing menstrual cramps, a condition called **dysmenorrhea.** Prostaglandin action on smooth muscles in other areas of the body can also result in the nausea, vomiting, and headache that sometimes accompany the menstrual phase. A summary of the hormonal, ovarian, and uterine changes that occur during the ovarian cycle is contained in Table 31–4.

TABLE 31–4
Summary of the Hormonal, Ovarian, and Uterine Changes That Occur During the Ovarian Cycle
1. During the early part of the follicular phase, blood levels of LH and FSH begin to rise.
2. This rise in LH and FSH stimulates growth of 6 to 12 primary follicles and the full maturation of a single Graafian follicle.
3. The growing follicles secrete large amounts of estrogen, lesser amounts of progesterone, and small amounts of testosterone and androstenedione.
4. Rising blood estrogen stimulates growth of the endometrium of the uterus.
5. Rising blood estrogen also stimulates further secretion of GnRH, which results in the preovulatory surges of LH and FSH.
6. Just prior to ovulation the follicle stops secreting estrogen and blood estrogen levels fall.
7. The surges of LH and FSH at midcycle induce ovulation.
8. Following ovulation, the remaining follicular cells in the ovary become the corpus luteum.
9. The corpus luteum secretes large amounts of progesterone and lesser amounts of estrogen. Blood levels of these hormones begin to rise.
10. Progesterone and estrogen from the corpus luteum stimulate further growth of the endometrium of the uterus.
11. Progesterone and estrogen also inhibit secretion of LH and FSH, and blood levels of these pituitary hormones begin to fall.
12. As blood levels of LH and FSH fall, the signal for follicular growth diminishes and no new follicles begin to grow.
13. On day 8 following ovulation, the corpus luteum begins to degenerate and secretes less progesterone and estrogen.
14. Falling blood levels of progesterone and estrogen result in hemorrhagic changes in the uterine endometrium, which culminate with the onset of menstruation.
15. As blood levels of progesterone and estrogen fall at the end of the luteal phase, the inhibition of GnRH, LH, and FSH, which was exerted by progesterone and estrogen, is removed. Blood levels of LH and FSH begin to rise, thus marking the beginning of another follicular phase (see point 1).

Estrogen, the Primary Sex Hormone in the Female, Has a Biological Effect on Many Tissues of the Body

> *What other effects do estrogen and progesterone have on the body?*

The major sex hormones in the female are the steroids estrogen and progesterone. The term *estrogen* refers to a group of steroids that includes estradiol, estrone, and estriol. In addition to their roles in regulating the ovarian cycle, estrogen and progesterone have widespread effects on many other physiological and metabolic processes (Table 31–5).

Effects of Estrogen at Puberty

Estrogen stimulates the development of the sex organs and the appearance of the secondary sex characteristics in the female. At puberty, estrogen stimulates growth of the vagina, uterus, and fallopian tubes, as well as the external

TABLE 31–5
Effects of Estrogen
Puberty
Stimulates growth of internal reproductive organs, external genitalia, and breasts
Stimulates growth of body hair
Stimulates growth of long bones and early closure of the epiphyses
Stimulates "female" pattern of fat distribution
Hormonal
Stimulates secretion of GnRH, LH, and FSH during the follicular phase and inhibits secretion of these same hormones during the luteal phase
Decreases sensitivity of peripheral tissues to insulin
Increases circulating levels of renin and angiotensin II
Metabolic
Protects against bone loss

CURRENT CONCEPTS IN PHYSIOLOGY

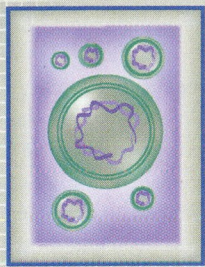

Estrogen and the Cardiovascular System

Recently, physiologists have become aware that men and women differ in their normal physiology and in their response to diseases in areas beyond those related to the obvious differences in their reproductive biology. In particular, it has been shown that the incidence and severity of cardiovascular disease between men and women is markedly different. This is potentially important because more people in the United States die from cardiovascular diseases than from all other causes of death combined. Men, however, are over four times more likely to suffer from some form of cardiovascular disease compared with women and are more than 40 times as likely to suffer a heart attack. This suggests that something protects women from cardiovascular disease. However, it is also well known that this so-called "cardiovascular protection" does not exist for women throughout their entire lifetime. Once women reach their mid-40s and beyond, they begin to suffer from cardiovascular diseases at an accelerated rate. By the time they are in their 70s, they experience cardiovascular disease and heart attacks at a rate similar to that of men of the same age.

Approximately age 45 is typically when most women experience menopause. Menopause is characterized by a significant drop in the production of female sex steroid hormones, especially estrogen. Clinical studies have shown that, at similar ages, premenopausal women have a lower incidence of cardiovascular diseases than do postmenopausal women. This has led many physiologists to examine whether estrogen may in some way protect women from cardiovascular disease.

High blood cholesterol levels, high blood pressure, and formation of blood clots in the coronary arteries are major contributors to heart attacks. Alterations of these risk factors by estrogen could result in marked differences in the incidence of heart attacks between men and women. However, estrogen actually may promote blood clot formation and has negligible effects on blood pressure. Although estrogen favorably affects the ratio of HDL (good) to LDL (bad) cholesterol in women, it has been determined that this effect is too small to account for the large difference in the incidence of cardiovascular disease between men and premenopausal women. Recent physiological research has shown that estrogen, at concentrations typically seen in premenopausal women, stimulates the production of nitric oxide in the endothelial cells of arteries, including those of the heart. Nitric oxide is one of the most important vascular regulatory molecules in the body. It is a potent relaxation agent of arterial smooth muscle and helps prevent high blood pressure. Many drugs, as well as naturally occurring hormones and neurotransmitters in our bodies, relax arteries by stimulating the production of nitric oxide. Scientists have reported that coronary arteries exposed to physiological concentrations of estrogen relax more to drugs and agents that stimulate the arterial endothelial nitric oxide system. It has been proposed that this effect of estrogen may be related to the lower incidence of cardiovascular disease in premenopausal women.

Nevertheless, many unanswered questions remain before one can definitively characterize why premenopausal women differ so markedly compared with men regarding cardiovascular disease. Although it is now widely accepted that estrogen enhances endothelium dependent, nitric oxide-mediated relaxation of coronary arteries, it is difficult to ascribe cardioprotection in women solely to this effect. Heart attacks can be caused by severe contraction (spasm) of coronary arteries; therefore, any factor that helps to relax arteries could help to prevent this form of heart attack. However, it is known that this particular form of heart attack is rare. This has led investigators to look for other means by which estrogen's enhancement of the arterial nitric oxide system might benefit women. The recognition that nitric oxide prevents blood clot formation and inhibits the development of atherosclerosis has led investigators to examine whether this may be the link between estrogen, nitric oxide, and cardioprotection.

---

genitalia, breasts, and body hair. Estrogen stimulates linear growth and also accelerates closure of the epiphyses of the long bones, which results in cessation of bone growth—which is why many women are shorter than men. Stimulation by estrogen results in the particular pattern of distribution of body fat around the hips, buttocks, and abdomen that characterizes the female form. When the ovary is absent, or when it secretes inadequate amounts of estrogen, puberty either fails to occur or progresses very slowly.

APPLICATIONS OF PHYSIOLOGY

Detection of Gonadotropins and Their Use as Markers of Perimenopause and Menopause

Menopause is the cessation of menstruation and is characterized by low plasma levels of estrogen. It is often stated that this occurs naturally in sexually mature women at approximately age 45 to 50. However, unlike the onset of menstruation, which marks the start of regular monthly menstrual cycles within a few months or less from the start of the first menses, the transition to menopause is often preceded for many years by irregular periods, missed periods, and unstable hormone levels. In addition, the progression from plasma estrogen states in mature premenopausal women to the general sustained low levels of plasma estrogen seen in menopause does not occur suddenly but, instead, progresses over several years prior to menopause. This syndrome of unstable menstrual cycles and progressively declining levels of plasma estrogen is called perimenopause. Physicians can treat women in peri-

menopause and menopause more effectively if they can effectively identify markers of these conditions in a woman's body. Current physiological research indicates that such tools may be available to physicians in the near future. LH and FSH levels are high in menopausal women due to low estrogen levels in the plasma, with resultant loss of negative feedback on LH and FSH production. However, gonadotropin patterns become markedly altered in the perimenopausal transition to menopause. The exact nature of these alterations and how they differ in perimenopausal women from those in pre- or postmenopausal women is currently under investigation. Current research is being directed at creating simple, reliable tests for the detection of these patterns. It has been shown that stable protein fragments of LH and FSH appear in the urine of women, and sensitive assays for the detection of these fragments are now being developed. With the development of reliable assays for LH and FSH using urine samples, physicians may some day be able to easily and accurately characterize perimenopausal and menopausal states in women and use such information to provide better diagnosis and treatment for women during these stages of life.

Metabolic Effects of Estrogen

Estrogen decreases the sensitivity of peripheral tissues to insulin, which can result in a decreased ability of the body to reduce plasma levels of glucose after ingestion of food (i.e., decreased glucose tolerance). Estrogen increases the levels of plasma proteins that bind thyroxine, cortisol, and progesterone. In addition, estrogen increases circulating levels of renin and thus angiotensin II, which can result in sodium and water retention.

Estrogen inhibits breakdown of bone in women. Removal of the ovaries, or inadequate ovarian secretion of estrogen, results in the acceleration of bone loss known as **osteoporosis.** In such cases, replacement of estrogen can dramatically slow the rate of bone loss. Postmenopausal women, in whom ovarian secretion of estrogen is very low, exhibit osteoporosis most commonly.

Effects of Progesterone

Progesterone is the most active naturally occurring antagonist of androgens. Progesterone antagonizes the effects of both testosterone and dihydrotestosterone by competing for the androgen-binding sites present in androgen-dependent tissues. Progesterone prevents target tissues from responding

to the normally low levels of circulating androgens in the female. Progesterone and its derivatives are very useful in treating abnormally heavy hair growth in the female, a condition termed **hirsutism,** which is thought to be due to an overproduction of androgens from the adrenal cortex.

Progesterone, secreted by the corpus luteum, also may promote the salt and fluid retention that occurs in many women prior to the onset of menstruation. Progesterone is thought to increase circulating levels of renin and angiotensin, which stimulates aldosterone secretion and results in salt and fluid retention.

Menopause: The Cessation of Menstruation in the Female

 What happens during menopause?

Menopause, the cessation of menstruation in the female, occurs on the average at age 50. However, there is considerable variation in the age of onset of menopause. For several years before the onset of menopause, menstruation occurs less frequently and at variable intervals. The number of follicles in the ovary declines with age; hence estrogen secretion declines as well. When most of the follicles have disappeared from the ovary, the ovary stops secreting estrogen. The low circulating levels of estrogen found in postmenopausal women mainly are due to a reduction in estrogen production by the ovaries, despite the fact that androgen precursors produced by the adrenal cortex are converted to estrogen in the periphery.

During menopause, the remaining follicles in the ovary become less sensitive to LH and FSH, and, as a consequence, blood levels of these pituitary hormones begin to rise. The decline in circulating estrogen during menopause results in a loss of vaginal epithelium, a decrease in breast mass, and an increase in bone loss. Other symptoms that accompany menopause include vascular flushing and sweating ("hot flashes"), rapid shifts in mood and emotion, and an increased risk for cardiovascular disease. All these symptoms appear to be related to the loss of estrogen (**estrogen withdrawal**). Estrogen withdrawal influences neural transmitters in such a way that the body's temperature control center in the hypothalamus gets reset to a higher level, which then stimulates heat-generating mechanisms in the body. Hot flashes often interrupt nighttime sleep patterns, resulting in fatigue and irritability.

CHAPTER REVIEW

Summary

- Spermatogenesis takes place in seminiferous tubules of the testes.
- The Sertoli cells form a protective blood-testes barrier around the inside of the seminiferous tubules.
- Sperm mature in the epididymis and are stored in the vas deferens.
- The seminal vesicles, prostate gland, and bulbourethral gland contribute important fluids that nourish sperm.
- LH stimulates the Leydig cells of the testes to secrete testosterone. FSH stimulates spermatogenesis in the testes.
- Release of GnRH, LH, and FSH is controlled through feedback loops involving both testosterone and inhibin.
- Testosterone stimulates the onset of puberty and promotes spermatogenesis in the testes, and secondary male sexual characteristics.
- Once each month in females, beginning at puberty, one of the primary follicles in the ovaries grows at a rapid rate and differentiates into a mature Graafian follicle containing an ovum.

- At midcycle, the mature follicle releases its ovum in a process termed *ovulation*. Following ovulation, the ovarian follicle becomes the corpus luteum.
- During the follicular phase of the ovarian cycle, FSH stimulates the growth of ovarian follicles, which in turn secrete estrogen.
- The luteal phase of the ovarian cycle is marked by growth of the corpus luteum in response to LH stimulation.
- The corpus luteum secretes both estrogen and progesterone.
- Degeneration of the corpus luteum results in necrotic changes in the endometrium that culminate in the onset of menstruation.
- Estrogen has a wide variety of effects on other endocrine systems and on many tissues of the body.
- Menopause, the cessation of menstruation, marks the end of the reproductive years in the female.

Review Questions

Choose The Correct Answer

1. Full function of the seminiferous tubules requires:
   a. prostaglandins.
   b. LH.
   c. GnRH.
   d. FSH only.
   e. androgens and FSH.

2. Puberty does not normally occur in humans aged less than 8 years. Which of the following would you propose is most likely responsible for this delay in sexual development?
   a. Reproductive tissues cannot respond to gonadal steroids in humans less than 8 years of age.
   b. Ovaries and testes are unresponsive to gonadotropins until after age 8 years.
   c. The pituitary is incapable of manufacturing gonadotropins before age 8 years.
   d. The brain secretes steroid hormone receptor antagonists prior to puberty.
   e. Prior to this age, GnRH secretion by the hypothalamus is extremely sensitive to feedback inhibition by gonadal steroids.

3. Progesterone:
   a. decreases motility of the uterus.
   b. stimulates follicular development.
   c. promotes degeneration of the endometrium.

d. blocks implantation of a fertilized ovum.

e. reaches peak levels on day 7 of the menstrual cycle.

4. Menopause occurs primarily because:

   a. the uterus is unable to accept fertilized ovum.

   b. the ovarian follicles are depleted.

   c. plasma levels of FSH and LH become too high.

   d. inhibin production produces suppression of FSH secretion.

   e. testosterone production is increased.

5. Testosterone:

   a. is synthesized from estrogen.

   b. is less potent than DHT in male reproductive tissues.

   c. is not required for full spermatogenesis.

   d. does not induce feedback inhibition on pituitary gonadotropins.

   e. has anabolic effects on metabolism.

6. The function of inhibin is to:

   a. bind testosterone thereby increasing its plasma half-life.

   b. potentiate spermatogenesis.

   c. suppress FSH production.

   d. suppress effects of testosterone on male hair growth.

   e. inhibit GnRH release.

7. Spermatogenesis:

   a. is unaffected by estrogen.

   b. occurs in Leydig cells of the testes.

   c. requires testosterone and progesterone.

   d. yields male gametes with diploid chromosome number.

   e. involves two meiotic divisions to produce four spermatids for every primary spermatocyte.

8. All of the following are true of FSH except:

   a. it stimulates spermatogenesis.

   b. its secretion is inhibited by inhibin.

   c. it synergizes with testosterone to maximize sperm production.

   d. it inhibits synthesis of androgen-binding protein by Sertoli cells.

   e. its plasma concentration would increase following removal of the testes.

9. In the ovarian cycle peak progesterone levels:

   a. occur during the luteal phase.

   b. occur during ovulation.

   c. are responsible for ovulation.

   d. inhibit testosterone secretion.

   e. None of the above

10. The Sertoli cells mediate negative feedback inhibition of FSH secretion by:

    a. production of inhibin.

    b. synthesis of testosterone.

    c. synthesis of DHT.

    d. maintaining the blood-testes barrier.

    e. releasing estrogen.

11. A high school junior football player has been given testosterone illicitly by his coach in an attempt to improve his performance. Which of the following will not occur as a result of this testosterone administration?

    a. An increase in basal metabolic rate

    b. Weight loss due to protein catabolism

    c. Inhibition of LH secretion

    d. Development of acne

    e. Inhibition of GnRH release

12. Which of the following cells contains a half-set of doubled chromosomes?

    a. Spermatogonium

    b. Primary spermatocyte

    c. Spermatid

    d. Secondary spermatocyte

    e. Leydig cells

13. Sertoli cells perform all of the following functions except:

    a. binding LH.

    b. contributing to seminiferous tubule fluid production.

    c. secreting inhibin.

    d. producing androgen binding protein.

    e. nourishing developing sperm.

14. Which of the following is false concerning Sertoli cells?

    a. They form part of the seminiferous tubules.

    b. They form the blood-testes barrier.

    c. They produce sperm precursor cells.

    d. They stimulate testosterone secretion by production of androgen binding protein.

    e. Their products feed back to inhibit FSH release.

15. Semen has all of the following properties except that:

    a. it contains fructose.

    b. it contains testosterone.

    c. it neutralizes vaginal acid.

    d. it contains fluid produced by the seminal vesicles.

    e. it contains fluid produced by the prostate gland.

16. The plasma LH concentration of a 49-year-old postmenopausal woman:

    a. is greater than that in the woman prior to menopause.

    b. is the same as that in the same woman prior to menopause.

    c. is lower than that in the same woman prior to menopause.

    d. is lower than that in a 49-year-old man.

    e. cannot be measured.

17. Which of the following is not true concerning the menstrual cycle?

    a. LH secretion is inhibited by estrogen and progesterone in the luteal phase of the cycle.

    b. The corpus luteum secretes estradiol and progesterone.

    c. FSH release is partly responsible for follicular maturation.

    d. The corpus luteum does not regress toward the end of the menstrual cycle if human chorionic gonadotropin is produced.

    e. Ovulation is prevented by follicular estrogen secretion.

18. Which of the following is primarily responsible for the basal body temperature increase in females following ovulation?

    a. FSH

    b. Progesterone

    c. Estradiol

    d. Testosterone

    e. LH

19. Which of the following is primarily responsible for follicular recruitment during the menstrual cycle?

    a. Progesterone

    b. LH

    c. FSH

    d. Estrogen

    e. Prolactin

20. Ovulation can be prevented by:

    a. high levels of estrogen and progesterone.

    b. FSH.

    c. LH.

    d. GnRH.

    e. None of the above

Answers to Case History Questions

1. Increased muscle mass, acne, and male-pattern baldness all can be caused by overproduction or ingestion of androgens. Illicit anabolic steroids are testosterone-like drugs that can create secondary physical changes in an individual similar to those seen from testosterone.

2. Long-term use of anabolic steroids, which mimic testosterone, result in sustained high plasma concentrations that produce feedback inhibition of pituitary gonadotropin secretion, especially that of FSH. This results in testicular atrophy inhibition of spermatogenesis, which over time will lead to infertility.

Key Terms

androgen-binding protein (ABP) (p. 934)
corpus luteum (p. 940)
dysmenorrhea (p. 943)
endometrium (p. 938)
estrogen (p. 940)

fallopian tube (p. 938)
gonadotropin-releasing hormone (GnRH) (p. 935)
inhibin (p. 934)
luteinization (p. 942)
meiotic division (p. 931)

menopause (p. 939)
oogenesis (p. 938)
ovary (p. 938)
ovum (p. 939)
progesterone (p. 940)
puberty (p. 944)

sperm (p. 933)
spermatogenesis (p. 929)
testis (p. 928)
testosterone (p. 936)
vas deferens (p. 934)

Suggested Readings

Bulletti, C., et al. "The uterus: Endometrium and myometrium." *Annals of the New York Academy of Sciences*, 1997.

Chard, T., and Grudzinskas, J. G. *The Uterus.* New York, Cambridge University Press, 1994.

Jones, R. E. *Human Reproductive Biology,* ed 2. San Diego, Academic Press, 1997.

Kirby, R., Kirby, M. G., and Farah, R.N. *Men's Health.* Oxford, Isis Medical Media, 1999.

Knobil, E., and Neill, J. D. *Encyclopedia of Reproduction.* San Diego, Academic Press, 1998.

Knobil, E., and Neill, J. D. *The Physiology of Reproduction,* ed 2. New York, Raven Press, 1994.

Lol, R., Kelsey, J., and Marcus, R. *Menopause: Biology and Pathobiology.* San Diego, Academic Press, 2000.

Yen, S. S. C., Jaffe, R.B., and Barbieri, R. L. *Reproductive Endocrinology: Physiology, Pathophysiology, and Clinical Management,* ed 4. Philadelphia, WB Saunders, 1999.

Web sites

www.merck.com/pubs/mmanual_home/contents.htm
Chapters 228, 231, and 232. The Merck Manual of Medical Information-Home Edition. Hormones: Male and Female Reproduction.

www.discoveryhealth.com
His Health and Her Health sections. General Health Topics.

Answers to Review Questions

1. e **2.** e **3.** a **4.** b **5.** b **6.** c **7.** e **8.** d
9. a **10.** a **11.** b **12.** d **13.** a **14.** d **15.** b
16. a **17.** e **18.** b **19.** c **20.** a

Chapter 32

PREGNANCY, FETAL DEVELOPMENT, AND LACTATION

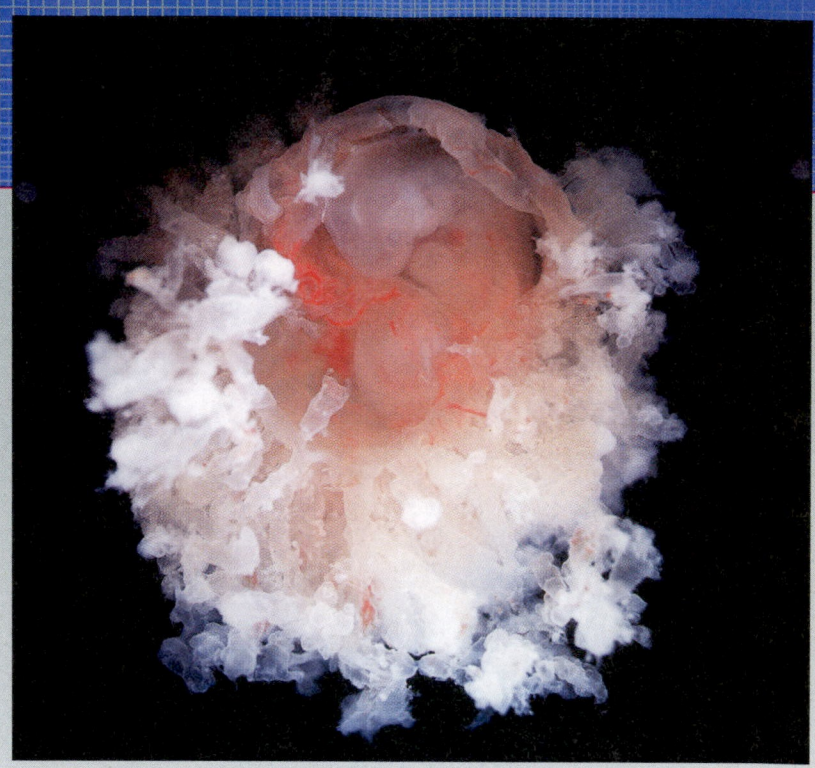

KEY CONCEPTS

- Because both sperm and ovum are viable for a limited period of time, fertilization will only occur when sperm are deposited in the vagina close to the time of ovulation.

- Cell division, which begins prior to implantation, marks the beginning of embryonic development.

- The placenta delivers oxygen and nourishment to the developing fetus, absorbs and removes waste products, provides immunity to various diseases, and secretes hormones necessary for both the maintenance of pregnancy and delivery of the fetus.

- The plasma hormonal profile of pregnant women is very different from the hormonal profile seen in nonpregnant women during the ovarian cycle.

- Parturition and lactation are under hormonal control.

CASE HISTORY

Pregnancy is the normal outcome of fertilization of the female ovum by the male sperm. However, pregnancy can often become complicated and life-threatening. A 23-year-old woman who is 32 weeks pregnant arrives at her physician's office for a routine scheduled prenatal examination. She has received normal prenatal care for the past 26 weeks and has no complaints today. However, her weight is 73.8 kg (164 lb), which is 6.3 kg (14 lb) more than it was at her last visit, two weeks ago. Her blood pressure is 150/96 mm Hg. It was 92/62 mm Hg at her last visit. Her urinary protein, which has always been negative in the past, is now 4+, which is very high. The height of her uterus is 28 cm, which is unchanged from 2 weeks ago. Fetal heart tones are heard at 154 bpm in the left upper quadrant of the uterus. You admit her to the hospital, and on rounds 6 hours later, her blood pressure is 148/98 and urinary protein is still 4+.

In the hospital, the following laboratory tests are performed: complete blood count (CBC) with platelets, 24-hour urinary quantitative protein, creatinine clearance, liver function tests, fetal ultrasound, and a nonstress test (NST), which is a measure of fetal stress. You order 12 mg of betamethasone intramuscularly now and repeat in 24 hours. Twenty-four hours later, her hemoglobin is 12.1 g/dl and platelets were 210,000/ul of blood. Urinary protein is 1200 mg per 24 hours. Creatinine clearance is 110 ml/min. Liver function tests are all at the upper end of the normal range. Fetal ultrasound shows a fetus that is the size of a 28-week fetus, although by your early ultrasound, she is 32 weeks pregnant. The estimated fetal weight is 900 g,

and the female fetus is in a breech position. The amniotic fluid index (AFI) is 9 cm. The NST is reactive.

Forty-eight hours later, her blood pressure is now 170/116 and she is highly agitated. Her hemoglobin is now 16 and platelets are 53,000. She is complaining of having a severe left-sided temporal headache, which is throbbing in nature and has not been relieved by acetaminophen. She also is seeing "shooting stars," or streaks of bright light. The right upper quadrant of her abdomen is now severely tender. Her urinary output has been 400 ml for the past 24 hours. AFI is now 3 cm and the NST is nonreactive.

Questions

1. What medical condition characterizes this patient upon admission to the hospital?

2. Explain why these tests were ordered upon admission to the hospital.

3. What assessment can be made based on the results of these tests?

4. Explain the physiology of the changes in this individual 48 hours after admission to the hospital.

INTRODUCTION

Creation of new life involves the joining of sperm and ova in the process of fertilization, or conception. This is followed by a 40-week period of fetal development in the uterus culminating in birth.

THE PROCESS OF FERTILIZATION

 What happens during fertilization?

Following ejaculation, sperm normally retain their capacity to fertilize an ovum in the female reproductive tract for only 24 to 72 hours, whereas the ovum is receptive to fertilization for approximately 10 to 15 hours after ovulation. Therefore, in order for the ovum to be fertilized, sperm must be present in the female reproductive tract during the 72 hours before or the 15 hours after ovulation. This restricted time interval makes the odds against fertilization rather high. Only one in four women becomes pregnant after one month of repeated intercourse without contraception.

Following Ovulation, the Ovum Is Transported Through the Uterine Tube to the Uterus

During ovulation, a single ovum usually is released from an ovary into the abdominal cavity at a point directly below the open end of the uterine tube (Fig. 32–1*a*). The open end of the uterine tube is composed of long, fingerlike projections called **fimbriae.** The fimbriae and uterine tubes are lined with **cilia,** which move or sway in a rhythmic pattern toward the uterus. Following ovulation, the ovum is retained in the ampulla of the uterine tube for several days. This retention is caused by estrogen-mediated contraction of the isthmus of the tube and allows time for the uterine lining to be adequately prepared by progesterone secreted in the luteal phase of the ovarian cycle. Eventually the ovum is drawn from the abdominal cavity and into the uterine tube by the current induced by the sweeping motion of the cilia (see Fig. 32–1*a* and *b*).

After entering the uterine tube, the ovum is transported to the uterus by the wavelike motion of the cilia (see Fig. 32–1*a*). The ovum takes approximately three to five days to travel from the ovary through the uterine tube to the uterus. Fertilization normally takes place in the uterine tube before the ovum has reached the uterus.

Following Ejaculation into the Vagina, Sperm Are Transported by Way of the Vagina, Cervix, and Uterus into the Uterine Tubes

Following ejaculation into the vagina, sperm are transported by way of the vagina, cervix, and uterus into the uterine tubes (Fig. 32–1*c*). Some sperm can reach the uterine tube in min-utes, although substantial numbers may take a few hours to reach this location.

The transport time of sperm through the uterus and the uterine tubes is much too short to be attributed solely to the movement in the tail of the sperm. Sperm movement through the uterus and uterine tubes is assisted by increases in the muscle motility of these structures resulting from exposure to prostaglandins in semen as well as **oxytocin** and prostaglandins released in the female during intercourse.

Cervical Mucus and Sperm Transport

The cervix contains glands that secrete a mucus composed of glycoproteins, salts, and water. The volume and consistency of the cervical mucus change during the ovarian cycle in response to changes in the circulating levels of estrogen and progesterone.

Estrogen secretion by the ovary increases at midcycle just prior to ovulation and stimulates the cervix to secrete large amounts of a clear, watery, nonviscous mucus. Upon estrogen stimulation, the glycoproteins in the cervical mucus assemble to form elongated fibers arranged in channels that allow sperm to penetrate the mucus and pass upward through the cervix at a time that coincides with ovulation.

Following ovulation, progesterone stimulates the cervix to secrete a thick, viscous, sticky mucus that lacks glycoprotein channels. This mucus plugs the cervical opening into the uterus and acts as a barrier that impedes sperm migration into the uterus following ovulation.

An average of 100 to 500 million sperm are deposited in the vagina during intercourse. Because only one sperm can fertilize the ovum, this number may seem much larger than is necessary to ensure fertilization. However, of this number, only a few thousand sperm reach the uterine tubes, and only a few hundred reach the vicinity of the ovum. **Sperm loss** is high for many reasons. Up to 50% of the sperm in a normal ejaculate may be incapable of fertilization because of abnormal shape or reduced motility. Some sperm also are destroyed by the acidic secretions of the vagina. Upon reaching the cervix, sperm entry into the uterus may be blocked by the cervical mucus at certain times in the ovarian cycle, as we have previously seen. Of the sperm that manage to reach the uterus, about half will enter the uterine tube that does not contain an ovum. Although these factors result in a large reduction in the number of sperm that reach the ovum, only one sperm is required for fertilization of the ovum.

Sperm Must Undergo the Acrosomal Reaction Before They Are Capable of Fertilizing an Ovum

In the epididymis, sperm mature and acquire the ability to move. However, mature sperm from the epididymis are not capable of fertilizing an ovum. Sperm must first attain a **capacitated state,** which occurs in the female reproductive

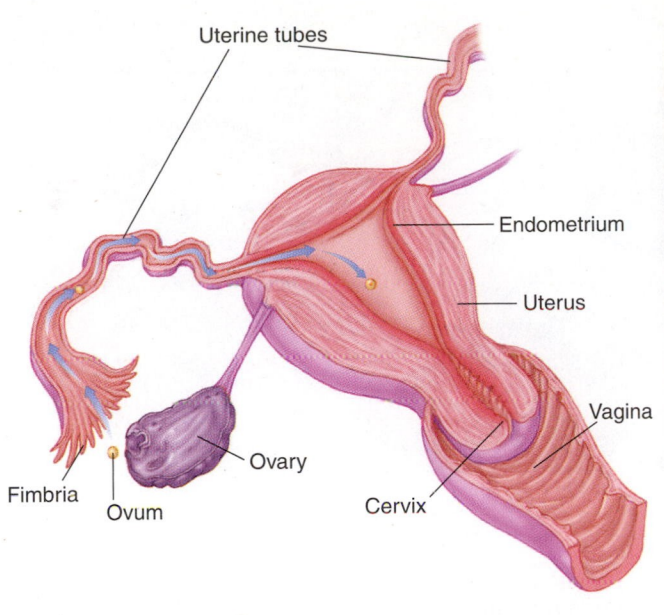

(a)

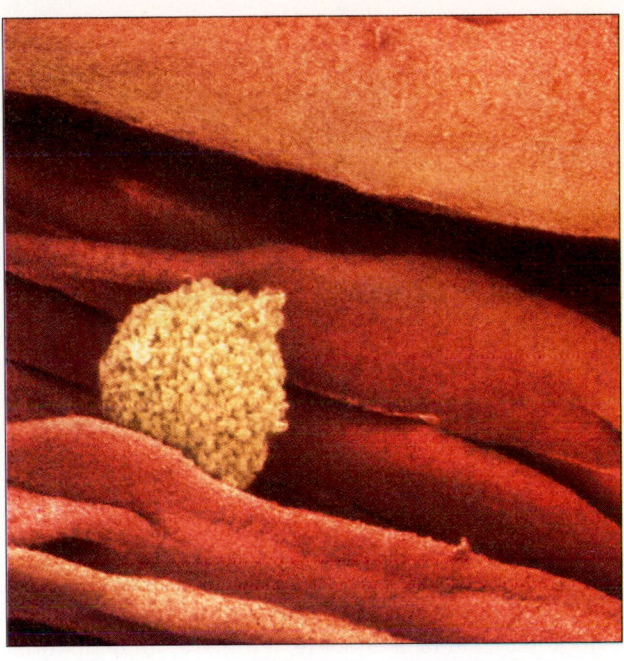

(b)

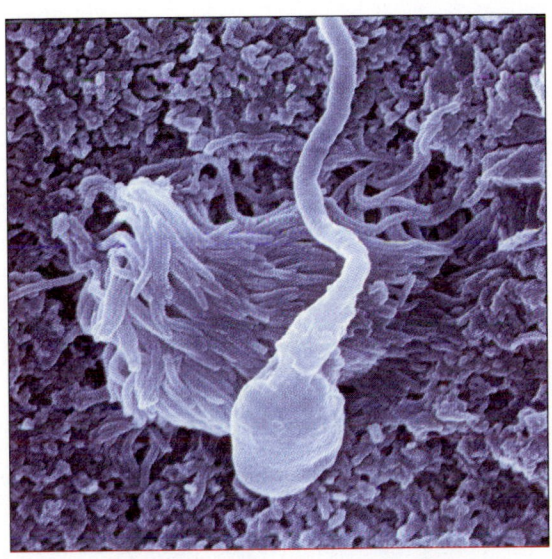

(c)

Figure 32–1

(a) Path of the ovum through the uterine tube and into the uterus following ovulation. *(b)* An unfertilized ovum in the folds of the uterine tube. Photograph taken with an electron microscope (original magnification × 100). *(c)* A sperm cell migrates through the uterine tube toward the ovum. *(b, © Nilsson, L. A Child Is Born, Dell Publishing Co., 1977; c, Nilsson, L. Behold Man, Little, Brown & Co.)*

tract prior to fusion of the sperm and ovum. During **capacitation,** the plasma membrane of the sperm is altered. It loses its acrosomal cap and increases its motility. Capacitation enables the sperm to undergo the **acrosome reaction,** which enables the sperm to penetrate and fertilize the ovum.

Fertilization Refers to the Fusion of Sperm and Ovum

Figure 32–2a and b illustrate the general structure of a mature sperm. The head of the sperm is composed of a dense, compact nucleus, an **acrosome,** which contains enzymes

that play an important role in the process of fertilization and a surface membrane. The midpiece of the sperm contains spirals of mitochondria, while the tail of the sperm is composed of microtubules surrounded by a fibrous sheath (see Fig. 32–2b).

When the sperm reaches the ovum, glycoproteins that make up the ovum's **zona pellucida** induce the sperm to undergo the acrosome reaction. The acrosome reaction (Fig. 32–3) involves a series of fusions between the outer membranes of the acrosome, resulting in the formation of pores or channels in the outer acrosomal membrane. During the acrosome reaction, these pores expand to allow macromolecules

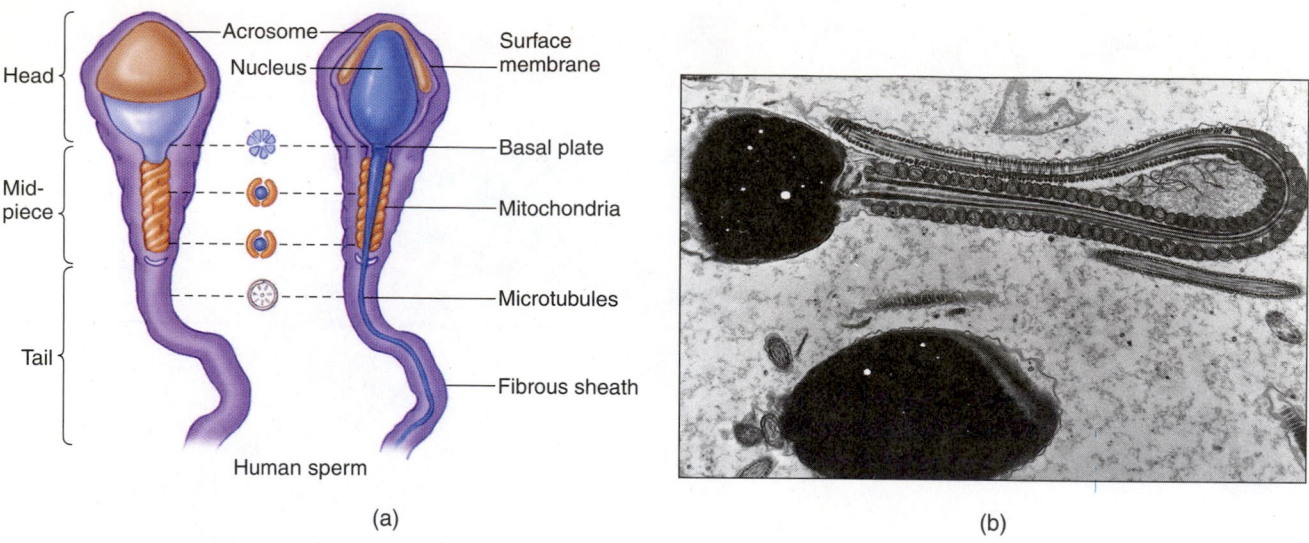

Figure 32–2

(a) Structure of a human sperm. *(b)* Transmission electron micrograph of a human sperm, showing the nucleus, acrosome, midpiece, mitochondria, and flagella. *(a, Redrawn from Warsaw, J.B., ed.,* The Biological Basis of Reproductive and Developmental Medicine, *1983; b, © David M. Phillips/Visuals Unlimited.)*

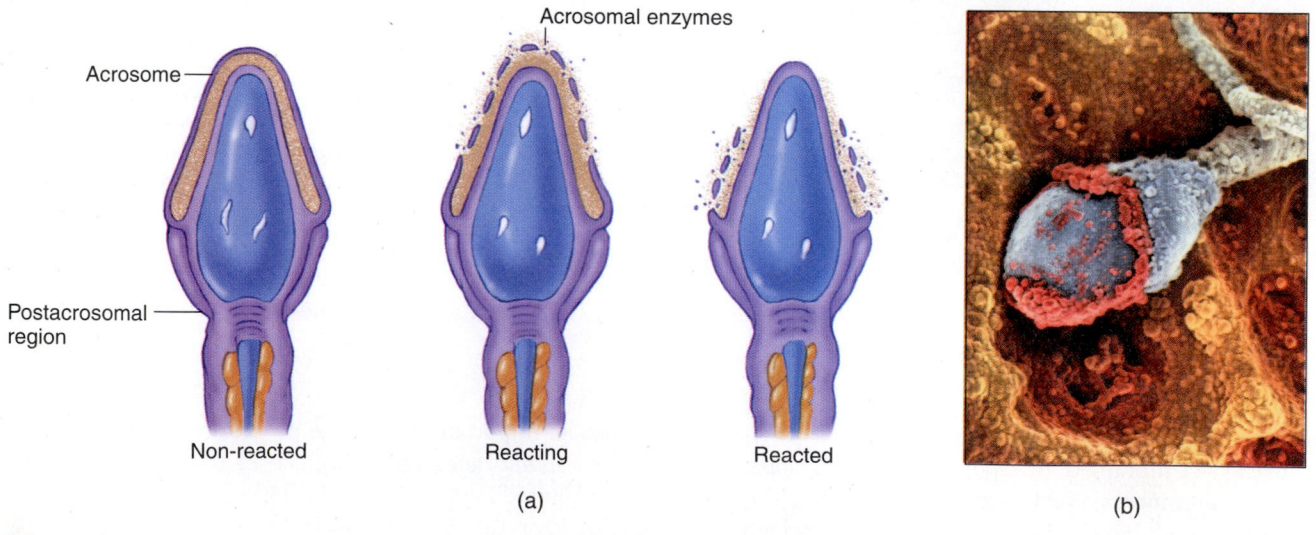

Figure 32–3

(a) Successive stages of the acrosome reaction of the human sperm. *(b)* The cap of the sperm gradually dissolves during the acrosome reaction, allowing enzymes to be released, which help the sperm to penetrate the ovum. *(a, Redrawn from Sathananthan et al.,* Atlas of Fine Structure of Human Sperm Penetration, Eggs, and Embryos Cultured in Vitro, *1986; b, © Nilsson, L.* A Child Is Born, *Dell Publishing Co.)*

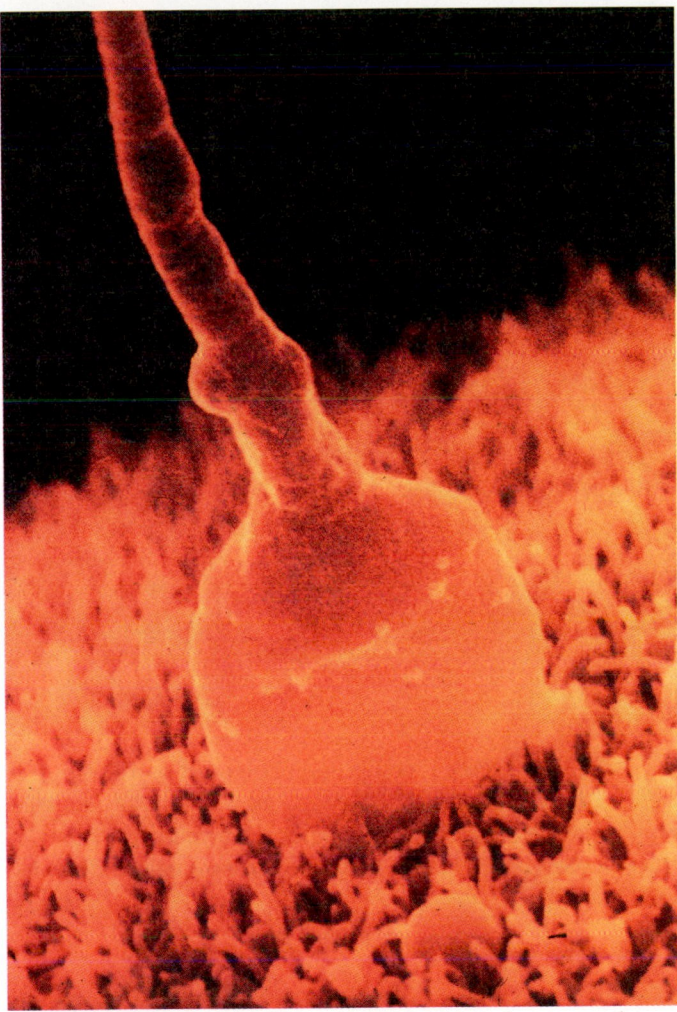

Figure 32–4

Photograph of a sperm beginning to penetrate an ovum. (© *David M. Phillips/Visuals Unlimited.*)

to pass in and out of the pores. Extracellular calcium has access to the acrosomal matrix, which then swells and begins to disperse (Fig. 32–3*a* and *b*). Next, proteolytic acrosomal enzymes, which disperse the cells surrounding the ovum and aid the sperm in penetrating the zona pellucida, are released from the acrosomal matrix. Following penetration of the zona pellucida, the sperm binds to the plasma membrane of the ovum. The acrosome-reacted sperm is now capable of fusing with the ovum plasma membrane.

Once fusion is complete, the sperm enters the ovum and fertilization is underway. The term *fertilization* refers to the fusion of the sperm and ovum. Figure 32–4 illustrates the initial meeting of sperm and ovum. The fertilized ovum, or zygote, is the first cell of a new organism that will divide and eventually change to become an embryo.

Polyspermy Is Lethal for the Ovum

 What prevents more than one sperm from fertilizing the ovum?

Fertilization of the ovum by more than one sperm, a condition called **polyspermy,** is lethal for the fertilized ovum. The prevention of polyspermy is therefore very important for reproduction. Although many sperm may attach themselves to the surface of the ovum, normally only one sperm penetrates to the inner portion of the zona pellucida surrounding the ovum.

The ovum has efficient ways of blocking the entry of more than one sperm. First, attachment of a sperm to the surface of the ovum induces membrane depolarization and increased permeability to calcium, which triggers an electrical block on the surface of the ovum, preventing fusion of other sperm. In addition, fusion of the first sperm with the ovum results in the exocytosis of cortical granules from the periphery of the ovum into the space between the plasma membrane and the zona pellucida. These granules alter the surface membrane of the ovum, making penetration by additional sperm more difficult. In addition, the cortical granules released from the ovum during fertilization alter the binding sites for sperm on the zona pellucida and the plasma membrane of the ovum. Normally, after one sperm enters the inner portion of the zona pellucida, all other sperm are virtually stopped before they reach the inner zona pellucida (Fig. 32–5).

Multiple Births

Occasionally, two or more ova, instead of the usual one, may be released during ovulation. Fertilization of each of these ova by individual sperm results in the formation of **dizygotic** (or **fraternal) twins,** triplets, or quadruplets that are as genetically different from each other as are any other siblings. **Monozygotic** (or **identical) twins,** triplets, or quadruplets, on the other hand, result from the fertilization of a single ovum by a single sperm followed by cell division that results in a single duplication or multiple duplications of the fertilized ovum prior to implantation. Monozygotic siblings are genetically identical.

CELL DIVISION AND IMPLANTATION FOLLOWING FERTILIZATION

 What happens after fertilization?

When the ovum is released from the follicle at midcycle, it has completed the first meiotic division. Fertilization stimulates the ovum to complete the second meiotic division, which results in the formation of the **haploid ovum,** containing 23

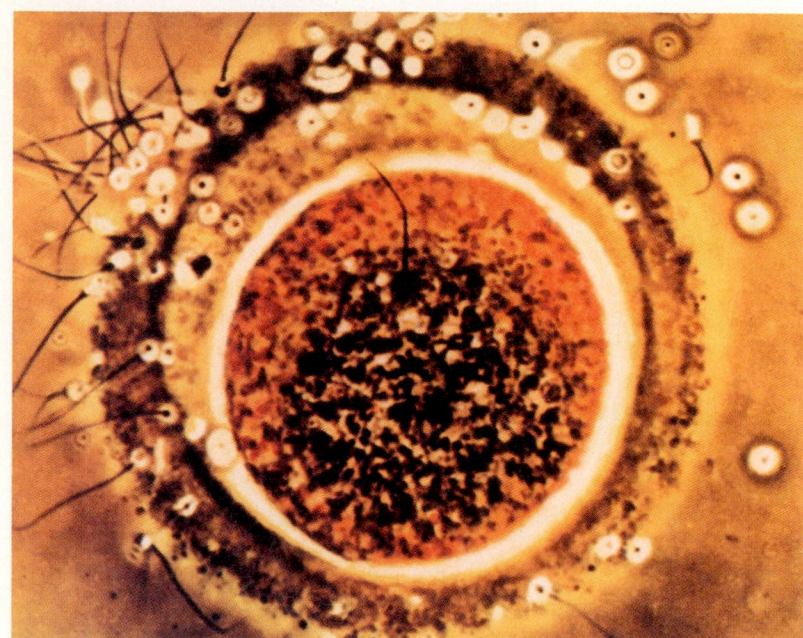

Figure 32–5

Fertilization of the human ovum. Note that as one sperm enters the ovum, the entrance of all other sperm is blocked, which prevents polyspermy. Phase contrast microscopy (original magnification × 400).
(© Dr. Landrum B. Shettles.)

chromosomes. Within hours of fertilization, metabolic changes occur that initiate protein synthesis and the transformation of the haploid set of chromosomes to form the **female pronucleus.** At the same time, the sperm head swells to form a **male pronucleus.**

Once the sperm has entered the ovum, the nucleus of the sperm migrates toward the nucleus of the ovum. The two sets of unpaired chromosomes, 23 from the ovum and 23 from the sperm, align themselves to form a complete set of 46 chromosomes in the **diploid** ovum. Thus each parent contributes half of the genetic material that will determine the characteristics of the embryo.

The Fertilized Ovum Undergoes Cell Division During Transport Through the Uterine Tube

The Zygote

Initially, the fertilized ovum is a single cell containing 46 chromosomes and is called the **zygote.** The zygote begins to divide mitotically approximately 30 hours after fertilization. Cell division, which begins prior to implantation, marks the beginning of embryonic development. Embryonic cells are the only cells, other than cancer cells, that divide on their own initiative. All other cells divide only in response to a need for tissue replacement. The signal for division of embryonic cells is contained in the genetic material of the zygote.

In the uterine tube, the zygote undergoes cleavage and first divides into two cells; then these two cells divide into four cells, and so on, as illustrated in Figure 32–6. During mitotic division, each of the 46 chromosomes in the dividing cell duplicates itself before division, which is why all the cells of the body contain the same 46 chromosomes.

When the zygote is composed of eight or more cells, it is called a **morula** (see Fig. 32–6). As cells of the morula continue to divide, a hollow area begins to form in the center of the cell mass and fills with fluid. This is now called a **blastocyst** (see Fig. 32–6).

The Blastocyst

The blastocyst is the precursor of the embryo. Blastocysts are composed of two major cell types. An outer layer of tightly connected cells forms the **trophoblast,** which protects and maintains the fluid-filled cavity of the blastocyst. This layer of trophoblastic cells has a tendency to stick to the uterine endometrium and thus play an important role in the initiation of implantation. The trophoblastic cells will eventually change into the placenta. The inner cell mass of the blastocyst is composed of cells called **embryoblasts** because they give rise to the embryo itself. Hence, both the placenta and the embryo arise from cells of the fertilized ovum (Fig. 32–7).

The Blastocyst Implants in the Endometrium of the Uterus

The fertilized ovum, while undergoing cell division, is transported through the uterine tube to the uterus over the course of approximately three to five days. The blastocyst undergoes continuous cell divisions for an additional two to four days before becoming implanted in the endometrium. Once the blastocyst has fully implanted in the endometrium, at approximately 2 weeks after fertilization, it is called an **embryo.** After approximately 10 weeks of embryonic development, the embryo is called a **fetus.**

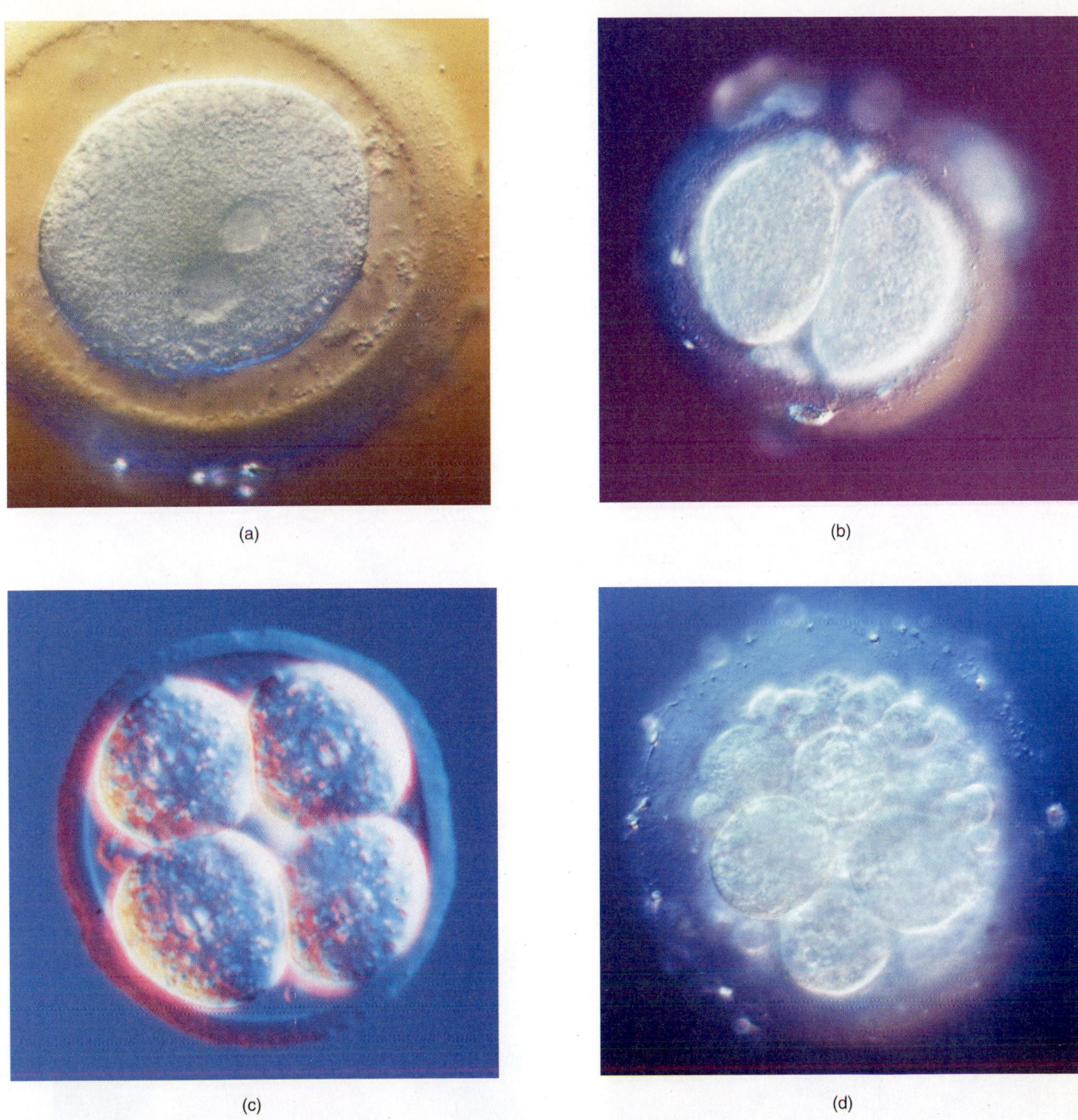

Figure 32–6

Stages in the early development of the fertilized ovum. *(a)* The single cell contains all the genetic information necessary to produce a complete human being. *(b)* The two-cell stage of development. *(c)* The eight-cell stage of development. *(d)* Cell division continues and gives rise to a cluster of cells called the morula.
(© *Nilsson, L. Being Born, pp 14, 15, 17.*)

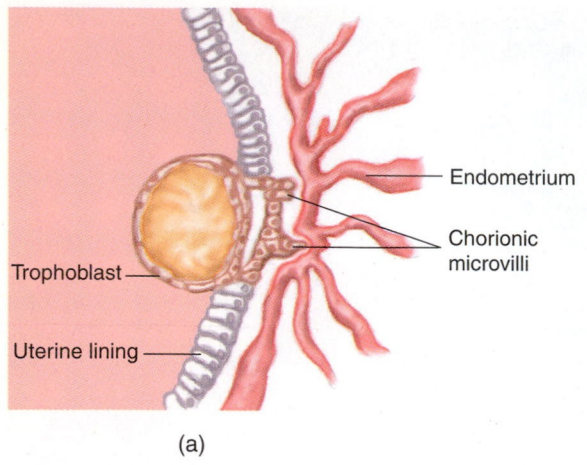

Endometrium

Chorionic microvilli

Trophoblast

Uterine lining

(a)

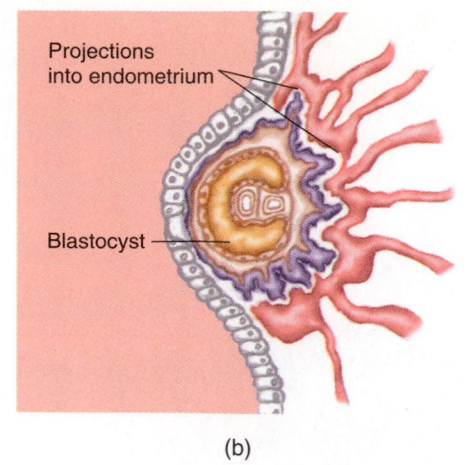

Projections into endometrium

Blastocyst

(b)

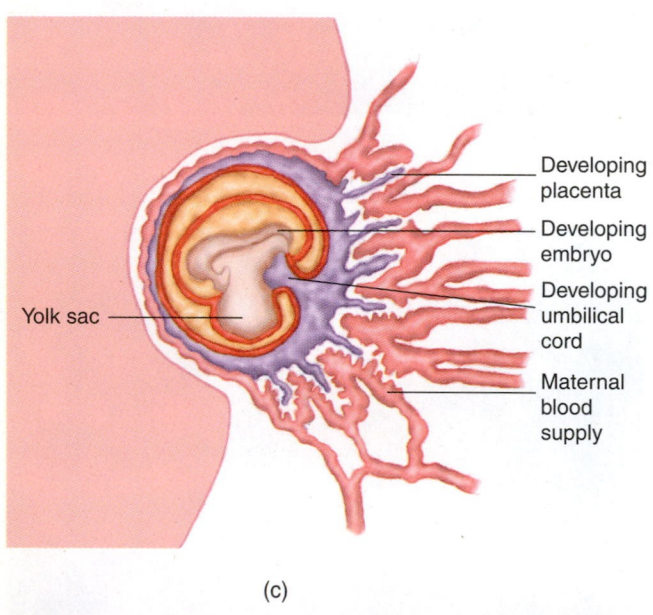

Developing placenta

Developing embryo

Developing umbilical cord

Maternal blood supply

Yolk sac

(c)

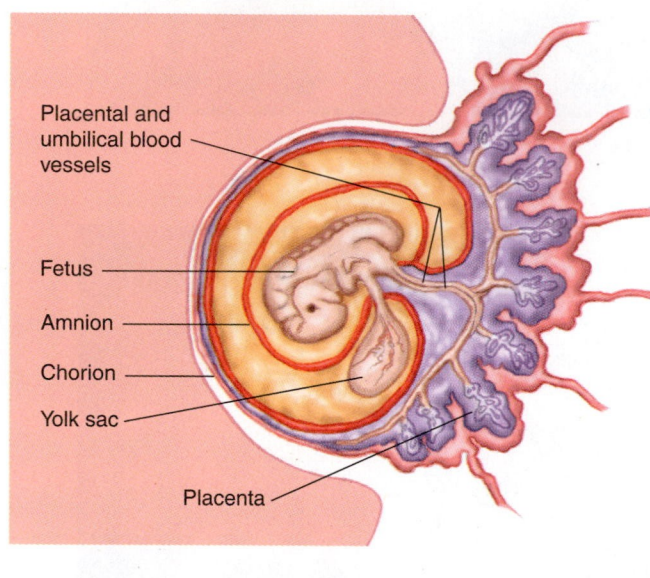

Placental and umbilical blood vessels

Fetus

Amnion

Chorion

Yolk sac

Placenta

(d)

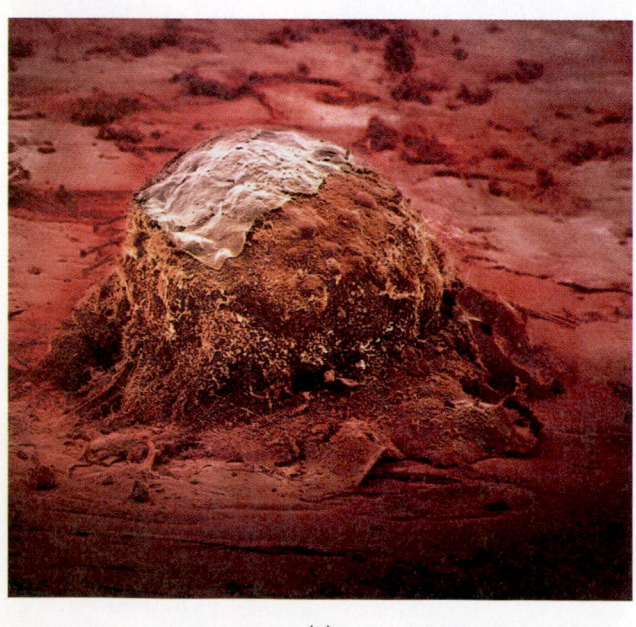

(e)

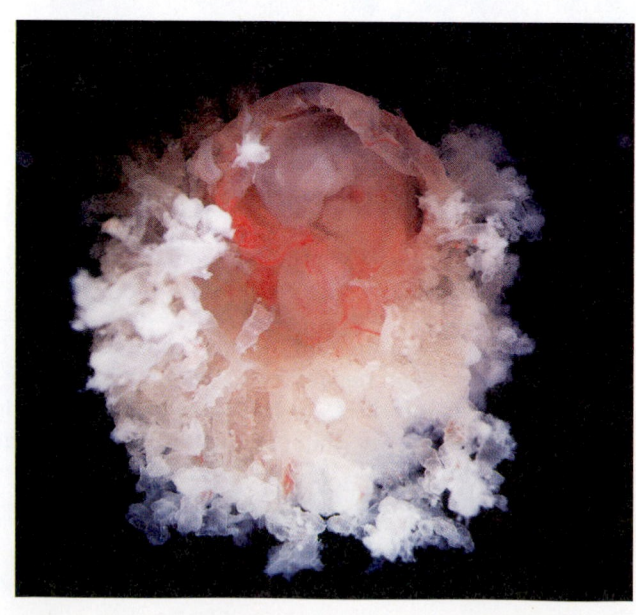

(f)

◀ **Figure 32-7**

Sequential development of the human placenta. *(a)* Initial implantation of the blastocyst in the wall of the uterus. *(b)* Blastocystic invasion of the endometrium. *(c)* Development of the placenta and the umbilical cord, which connects the fetus and the endometrium. *(d)* Formation of the blood vessels of the placenta and the umbilical cord. *(e)* On the twelfth day following fertilization, the blastocyst is securely anchored in the endometrium. *(f)* The embryo at four to five weeks of development. The chorion, which is shaggy in appearance, and the amniotic sac have been opened to visualize the embryo.   *(d, Adapted from Beaconsfield, Birdwood, and Beaconsfield. "The placenta." Scientific American, 243:94, 1980; e and f, © Nilsson, L. A Child Is Born, Dell Publishing Co.)*

Secretion of Enzymes and Hormones by the Blastocyst

Prior to implantation, while the blastocyst is floating free in the uterus, it begins to secrete hormones and proteolytic enzymes. These enzymes cause some of the neighboring endometrial cells to degenerate, forming small cavities in the endometrium, a process referred to as the **decidual reaction.** The cavities in the endometrium become filled with blood from the arteries of the uterus. Within nine days following fertilization, the trophoblast cells begin to secrete **human chorionic gonadotropin (hCG)** a luteinizing hormone (LH)-like hormone that can be detected in maternal plasma and urine. Maternal plasma levels of hCG peak at 7 to 12 weeks' gestation

then decline to a stable plateau for the remainder of pregnancy. Detection of hCG is the most commonly used and most specific test for pregnancy. hCG binds to LH receptors on luteal cells, preventing degeneration of the corpus luteum and thus maintaining secretion of estrogen and progesterone. This estrogen stimulates the secretion of prostaglandins from the endometrium of the uterus, which then increase vascular permeability and swelling of the endometrium.

Formation of Embryonic Membranes

During implantation, the trophoblast cells form microvilli on the outer surface of the blastocyst. These microvilli begin to invade the endometrium of the uterus. The trophoblasts penetrate to the level of the first muscle layer, and the blastocyst becomes buried in the endometrium. The trophoblasts differentiate into two cell layers that will form the **amnion** and the **chorion.** The amnion is a tough extraembryonic membrane that surrounds and protects the developing embryo. The chorion, which is the outermost extraembryonic membrane, gives rise to the placenta (see Fig. 32–7).

Secretion of Progesterone by the Corpus Luteum

Implantation is completed around day 9 following ovulation, which corresponds to day 23 of the ovarian cycle. On day 23, the corpus luteum is secreting large amounts of progesterone, so named because it has "pro-gestational" properties that help to prepare the uterus for implantation. Progesterone stimulates proliferation of the uterine endometrium, making it thick and vascular in preparation for receiving and nourishing the blastocyst (Fig. 32–8). It also suppresses uterine muscle motility that helps to prevent potential spontaneous abortion after implantation.

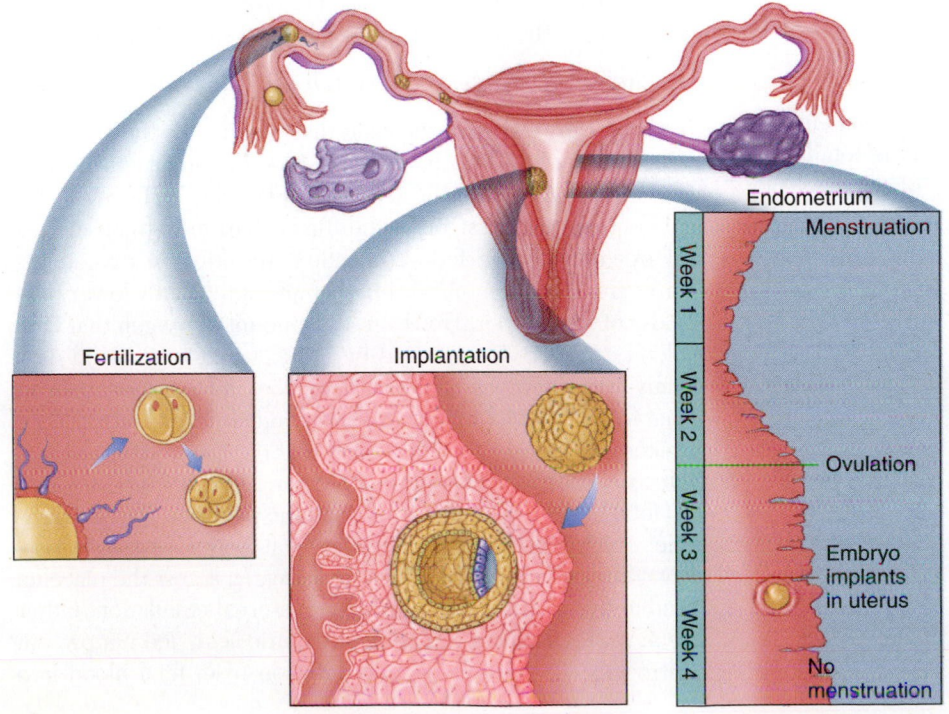

Figure 32–8

The menstrual cycle is interrupted when pregnancy occurs. The corpus luteum does not degenerate, and menstruation does not take place. Instead, the wall of the uterus remains thickened so that the embryo can develop within it.

Abnormal Implantation

Occasionally, the presence of scar tissue in the uterine tube following infections or surgery, or tumors located in the uterine tube, slow or block transport of the fertilized ovum from the uterine tube to the uterus. In this case, the fertilized ovum becomes implanted in the uterine tube, or, in rare cases, it may be expelled from the uterine tube back into the abdominal cavity. In both of these cases, implantation occurs outside of the uterus and is called an **ectopic pregnancy.**

Ectopic pregnancy occurs in approximately 1 in 100 pregnancies. Approximately 97% of ectopic pregnancies occur in the uterine tubes, called **tubal pregnancies,** and do not continue to term because there is not enough space in the uterine tube for fetal growth. Most tubal pregnancies abort at a relatively early stage. When they do not, surgical removal may be necessary in order to reduce the risk for tubal rupture and bleeding.

THE PLACENTA

The needs of the embryo, and later of the fetus, are met by the **placenta,** which delivers oxygen and nourishment, absorbs and removes waste products, provides for transfer of immunity to various diseases, and secretes hormones necessary for both the maintenance of pregnancy and the expulsion of the fetus at term (Table 32–1).

The Embryo Is Dependent on the Maternal Circulation for Delivery of Nutrients and Oxygen and for Removal of Waste Products

 What are the functions of the placenta?

Following implantation, the embryo is no longer independent. All of its physiological needs and removal of its waste products are dependent on its link with the maternal circulation through the placenta. Although the fetus is totally dependent on the placenta for survival, the placenta is not reciprocally dependent on the fetus and can function in the uterus even after a fetus has died.

The **placenta** is a flat organ that is connected to the uterus on one side and to the embryo, via the umbilical cord, on the other side. The **umbilical cord** and amniotic cavity are formed from trophoblastic cells. The cells that line the inner surface of the amniotic cavity form the amnion, which forms an expandable, fluid-filled cavity surrounding the entire embryo. The clear amniotic fluid keeps the embryo moist and serves as a shock absorber to protect it from physical injury.

The placenta is derived from the embryonic chorion and the maternal endometrium. When the blastocyst implants in the endometrium, the tissue and blood vessels of the endometrium break down at the site of implantation to form small spaces in the endometrium that fill with maternal blood. The developing embryo sends out rootlike villi into the pools of maternal blood. In the first few weeks following implantation, the villi are composed of thick columns of cells that contain capillaries. These capillaries serve as the route for exchange of nutrients from the mother to the fetus and of waste products from the fetus to the mother. These small vessels become the arteries and veins of the umbilical cord and placenta. Within 5 weeks after implantation, the vasculature of the umbilical cord and placenta is well established. Figure 32–7 illustrates the different stages in the development of the placenta.

The placenta anatomically separates the circulatory systems of the fetus and mother and acts as an interface between these two circulatory systems; these circulations normally do not mix. Maternal blood enters the placenta through the uterine artery, perfuses the placenta, and exits through the uterine vein (Fig. 32–9). Fetal blood enters the placenta through the umbilical artery, perfuses the placenta, and exits through the umbilical vein. The maternal side of the placenta is divided into thick, highly vascularized lobes that provide a very large surface area for exchange of materials between the fetal and maternal blood supplies.

Exchange of Materials Through the Placenta

Oxygen diffuses from the maternal blood into the placenta, where it is picked up by the fetal blood and carried through the umbilical cord to the fetus (see Fig. 32–9). However, the placenta presents a significant diffusion barrier to the transport of oxygen into the fetal circulation, resulting in oxygen gas pressures in the umbilical vein that are significantly lower than those of the mother. However, the amount of oxygen that fetal blood can carry is improved by the fact that fetal blood contains 50% more hemoglobin than does maternal blood and that fetal hemoglobin has a higher affinity for oxygen than does maternal hemoglobin. Substances that the fetus needs, such as glucose, sodium, potassium, and chloride, pass from maternal to fetal blood by simple diffusion in the placenta. Carbon dioxide, which is continually formed in fetal tissues, crosses by passive diffusion more easily than does oxygen across the placenta and in this manner enters into the maternal circulation. Other fetal waste products, including urea, uric acid, and nonprotein nitrogens, also pass by simple diffusion from fetal blood into

TABLE 32–1
Functions of the Placenta
Exchange of gases between fetus and mother
Delivery of nutrients from mother to fetus
Delivery of antibodies from mother to fetus
Removal of fetal waste
Secretion of hormones including human chorionic gonadotropin (hCG), progesterone, estrogen, and human chorionic somatomammotropin (hCS)

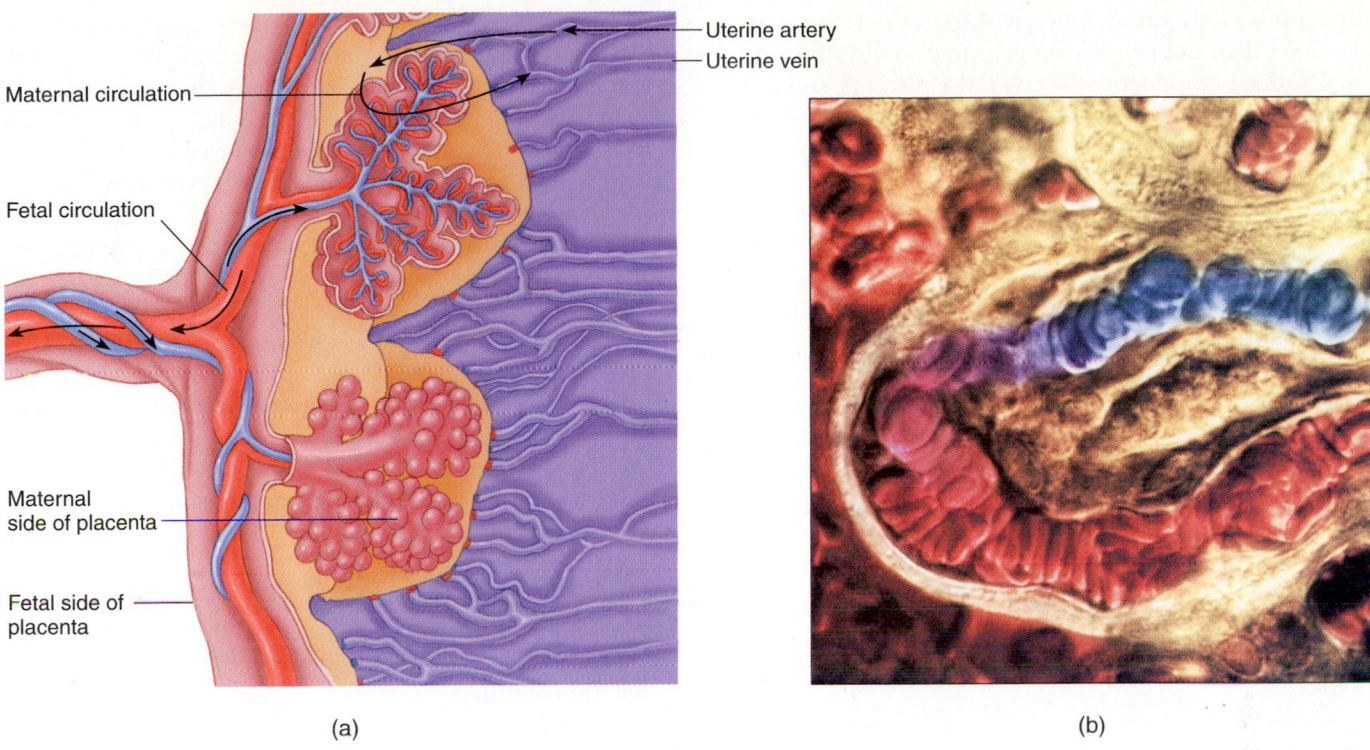

Maternal circulation

Fetal circulation

Maternal side of placenta

Fetal side of placenta

Uterine artery
Uterine vein

(a)

(b)

Figure 32–9

(a) Cross-section of placental tissue illustrating the directions of fetal and maternal blood flow. *(b)* Fetal blood cells in a U-shaped capillary protruding from the placenta. The fetal blood cells *(blue)* are separated from the maternal blood cells *(red)* by a thin membrane that transfers oxygen and nutrients from the maternal blood to the fetal blood. As fetal blood cells are oxygenated they turn from blue to red.   (a, Adapted from Beaconsfield, Birdwood, and Beaconsfield. "The placenta," Scientific American, *243:100, 1980; b,* © Nilsson, L. A Child Is Born, *Dell Publishing Co.)*

maternal blood in the placenta and are then excreted by the mother along with her own wastes.

The fetus grows at a more rapid rate than do tissues in the adult body. To sustain this growth, the fetus needs a variety of structural proteins. The placenta extracts amino acids from the maternal blood and synthesizes proteins that are then transported into the fetal circulation and delivered to the developing fetus. The placenta synthesizes proteins at a higher rate than does any other organ of the body. On average, the placenta uses approximately one third of all of the oxygen and glucose that the maternal blood supplies in order to meet the metabolic needs of the fetus and the placenta itself.

Placental Transport of Drugs

The same efficient placental transport system necessary for fetal development also can expose the fetus to many drugs and pollutants to which the pregnant woman is exposed. Such exposure is dangerous during the first three months of pregnancy, when fetal tissue is differentiating and especially susceptible to damage. Alcohol, aspirin, the chemicals contained in tobacco and illicit drugs can reach the fetus via the

placenta and influence its growth and development. For this reason, fetal heart rate changes can be observed in women who smoke cigarettes, and babies born to women who use heroin during pregnancy experience severe heroin withdrawal symptoms after birth.

Placental Transport of Antibodies

The placenta also provides the fetus with antibodies formed by the mother during her contact with a variety of antigens. These antibodies are too large to cross the placenta by simple diffusion, so selective active transport systems convey antibodies to the fetus. This protects the newborn infant from a variety of infections.

The Placenta and Embryo Are Protected from Immunological Rejection

Embryonic and placental tissue arising from the embryo are immunologically foreign to the mother and hence might be expected to stimulate the synthesis of maternal antibodies that would cause tissue rejection of the embryo. However, this does not occur. During the early stages of development, the blastocyst does not express antigens that would stimulate maternal antibody production. Second, the placenta secretes many hormones, such as hCG, that suppress lymphocytes that induce tissue rejection. Third, the embryo is physically protected from the maternal system by membranes that surround it, and the fetal and maternal blood supplies do not mix.

HORMONAL AND PHYSICAL CHANGES DURING PREGNANCY

During pregnancy, the placenta secretes four important hormones that help to continue pregnancy and control physiological changes associated with early pregnancy. These are hCG, progesterone, estrogen, and human chorionic somatomammotropin (hCS), also called *human placental lactogen* (hPL).

 How does the placenta function as an endocrine organ?

Human Chorionic Gonadotropin

The fall in progesterone seen seven days after ovulation, the degeneration of the corpus luteum, necrosis of the endometrium, and resultant menstruation seen in the absence of fertilization would expel a fertilized ovum if they were allowed to occur following fertilization. These phenomena are prevented, however, by the secretion of hCG (Table 32–2), a glycoprotein similar in structure to LH. hCG is secreted first by trophoblast cells of the blastocyst before implantation and later by the placenta. hCG diffuses from fetal blood into maternal circulation and stimulates the corpus luteum of the pregnant woman and prevents its degeneration and involution. This ensures continued secretion of progesterone. In response to hCG, the corpus luteum doubles in size during the first two months of pregnancy. As pregnancy continues, hCG

TABLE 32–2

Physiological Functions of Human Chorionic Gonadotropin

Prevents degeneration of the corpus luteum

Stimulates the corpus luteum to secrete estrogen and progesterone, which in turn stimulate continued growth of endometrium

Stimulates steroid synthesis in the developing fetal adrenals

Stimulates fetal gonads, especially testosterone production by the fetal testes

Suppresses maternal lymphocytes and reduces the possibility of immunorejection of the fetus

---

stimulates the corpus luteum to secrete large amounts of both estrogen and progesterone. These two hormones are necessary for the continued growth of the endometrium, which is essential for adequate development of placental and fetal tissue. If the corpus luteum is removed during the first 7 to 10 weeks of pregnancy, spontaneous abortion will occur. During the second month of pregnancy, the placenta becomes capable of secreting estrogen and progesterone on its own and maintains high levels of estrogen and progesterone in the mother. At this time, the placenta stops secreting hCG, maternal plasma levels of hCG fall, and the corpus luteum becomes less functional.

Human chorionic gonadotropin also stimulates steroid synthesis in the developing fetal adrenal gland (see Table 32–2). In the male fetus, hCG stimulates the synthesis of testosterone by the fetal testes, which stimulates the development of the internal genitalia of the male fetus.

Placental Secretion of Progesterone

The placenta extracts cholesterol from the maternal circulation and uses this cholesterol to form progesterone, which then diffuses back into maternal circulation. The placenta begins to secrete progesterone at about the sixth week of pregnancy, and by weeks 12 to 14, the placenta produces enough progesterone to replace that from the corpus luteum.

Progesterone, secreted by the placenta, is essential for the maintenance of pregnancy. Progesterone stimulates growth of the endometrium and secretion of nutrients from glands in the endometrium, which nourish the developing embryo. It also inhibits contractions of the uterus, which prevents premature expulsion of the fetus. Progesterone and prolactin, together with oxytocin from the posterior pituitary, stimulate the breast in preparation for milk production and release following delivery. Placental progesterone also serves as a precursor for the synthesis of estrogen by the fetoplacental unit (see subsequent discussion).

Placental Secretion of Estrogen

During the first month of pregnancy, the corpus luteum secretes estrogen in response to hCG stimulation. After the first month, hCG begins to stimulate estrogen secretion from the placenta (see Table 32–2). Circulating levels of estrogen in maternal plasma continue to rise up to the time of delivery and reach levels several hundred times higher than the highest concentration seen during the ovarian cycle.

Estrogen, secreted by the placenta, stimulates enlargement of the uterus in the pregnant woman to accommodate the growing fetus. Estrogen also stimulates breast growth in the pregnant woman and development of the duct system within the breasts. Estrogen stimulates enlargement of the external genitalia and relaxation of the pelvic ligaments, which facilitates the passage of the fetus through the canal at birth.

Estrogen production in the placenta is unique. The synthesis of estrogen requires a complex interaction between the placenta and the adrenal gland of the fetus. This system is often referred to as the **fetoplacental unit.** The placenta can produce pregnenolone and progesterone from cholesterol but lacks the enzymes to convert pregnenolone to **dehydroepiandrosterone (DHEA),** a precursor for testosterone and estrogens, without involving the fetus. The fetal adrenal cortex contains the enzymes needed to form DHEA-sulfate and synthesizes the androgen DHEA-sulfate from pregnenolone produced by the placenta. This androgen is then transported to the placenta, where it is converted to **estriol,** a form of estrogen. Estriol is then released from the placenta into maternal circulation.

Human Chorionic Somatomammotropin

The placenta secretes hCS starting at approximately four weeks of pregnancy. hCS levels rise steadily during pregnancy. hCS is a protein with a molecular structure similar to that of human growth hormone. hCS has weak growth-promoting effects that may play a role in fetal development and maternal breast development. hCS is partly responsible for the condition later in pregnancy called "accelerated starvation." hCS has anti-insulin actions and therefore decreases maternal utilization of glucose. This makes more glucose available to the developing fetus, which is the prime fetal energy substrate, but also may create a gestational diabetes mellitus (discussed later) in the mother. The anti-insulin action of hCS also promotes the release of free fatty acids from maternal fat stores. This allows the mother to use fatty acids as an energy source, further sparing glucose for use by the fetus.

Other Hormonal Changes Occur in the Pregnant Woman During Gestation

Inhibition of Pituitary Gonadotropins

High circulating levels of estrogen and progesterone in maternal plasma during pregnancy inhibit the secretion of gonadotropin-releasing hormone (GnRH) from the maternal hypothalamus and hence inhibit the release of LH and follicle-stimulating hormone (FSH) from the maternal pituitary. This results in low levels of LH and FSH that are not adequate to permit follicular development, ovulation, or menstruation, all of which are then absent during pregnancy.

Stimulation of Prolactin and Relaxin

Prolactin secreted by the maternal pituitary rises throughout pregnancy. Prolactin augments estrogen and progesterone stimulation of maternal breast development and milk synthesis.

During the first two months of pregnancy, **relaxin** is secreted by the corpus luteum in response to hCG stimulation. Later in pregnancy, relaxin is secreted by the endometrium of the uterus. Relaxin relaxes pelvic ligaments in preparation for parturition and, together with progesterone, inhibits uterine contractions during pregnancy.

Stimulation of Insulin

Secretion of **insulin** from the maternal pancreas increases after the third month of pregnancy in response to decreased maternal sensitivity to insulin. Insulin insensitivity during pregnancy may become so severe that it produces a temporary form of diabetes, called **gestational diabetes mellitus.** After delivery, maternal sensitivity to insulin usually returns to normal.

Stimulation of Aldosterone

Estrogen and progesterone directly increase the secretion of renin, which increases angiotensin II and **aldosterone** secretion from the maternal adrenal cortex. Angiotensin II and aldosterone promote sodium and water retention in the mother, which causes the maternal volume expansion seen in pregnancy. These hormones also help to provide sodium to the developing fetus. Estrogen appears to diminish the vascular actions of the high plasma angiotensin II levels.

Stimulation of Cortisol

Levels of both free and protein-bound **cortisol** increase in the maternal plasma during pregnancy. These changes are due to increased secretion of cortisol by the maternal adrenal gland and increased synthesis of **cortisol binding globulin (CBG)** in maternal plasma. The liver increases the synthesis of CBG during pregnancy in response to estrogen stimulation. Increased plasma cortisol contributes to an increase in maternal body fat and the development of the mammary glands. The increase in body fat results from the action of cortisol on the CNS to increase appetite and, from its effect, to increase blood glucose levels. The increased blood glucose level, in turn, increases insulin secretion, which results in stimulation of lipid storage in adipose tissue.

Stimulation of Thyroxine (T4) and Triiodothyronine (T3)

During pregnancy, high circulating levels of estrogen stimulate thyrotropin releasing hormone (TRH) release from the hypothalamus which, in turn, stimulates TSH release from the anterior pituitary. TSH stimulation results in increases in thyroid gland size, thyroid secretion of thyroxine (T4) and triiodothyronine (T3), basal metabolic rate, and cardiopulmonary function during pregnancy. T3 and T4 increase resting heart rate in the mother, which results in an increase in the rate of delivery of oxygen and nutrients to the developing fetus.

Pregnancy Is Divided into Three-Month Periods Called Trimesters, Each Marked by Dramatic Physical Changes in Many Tissues of the Body

How are the stages of pregnancy and development classified clinically?

Pregnancy begins when a fertilized ovum becomes implanted in the endometrium of the uterus. Pregnancy, which usually lasts for approximately 280 days, is divided into three-month

CURRENT CONCEPTS IN PHYSIOLOGY

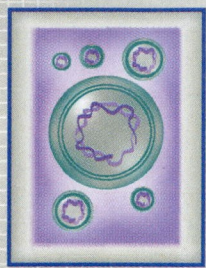

Fetal Alcohol Syndrome

Women who drink alcohol during pregnancy expose themselves to the same risks that alcohol has in non-pregnant women or men. However, a fetus is much more susceptible to the dangerous effects of alcohol than its mother. Alcohol easily crosses the placental barrier and enters the fetal circulation. Importantly, the fetus cannot metabolize alcohol as well as an adult, so alcohol levels in fetal blood are higher than in the mother's blood and the alcohol remains in the fetal blood stream longer. **Fetal alcohol syndrome (FAS)**, sometimes called *fetal alcohol abuse syndrome*, constitutes a wide range of growth, mental and physical birth defects, as well as later childhood learning and behavior difficulties, that result from alcohol use by the mother during pregnancy. The key physical characteristics of this syndrome are intrauterine growth retardation resulting in small birth weight; a small head, upper jaw and thin smooth upper lip; small narrow set eyes with drooping lids; flattened cheeks; and **atrial** or **ventricular septal defects** (holes in the walls of the heart between the atria or ventricles). The newborn with FAS often exhibits "failure to thrive" and abnormally slow growth. The heart defects in this syndrome often require surgical correction. Children with FAS are mildly to moderately mentally retarded, show delayed development of gross and fine motor skills, and impaired memory and learning difficulties. They are often hyperactive, display poor judgment and are easily distracted. It appears that the severity of a child's physi-cal abnormalities in FAS correlate closely with its mental deficits. It is not surprising that children affected with FAS exhibit attention deficit disorder (ADD), conduct disorders, depression and even psychotic episodes. FAS children have higher rates of school suspension and expulsion, are more likely than their peers to experience trouble with the law and show a high incidence of drug and alcohol abuse as adults.

FAS is seen in about 1 in every 1000 live births and there is no medical treatment to prevent the syndrome. In addition, there is currently no medical treatment available for mental retardation. However, if detected early, professional educational, health and psychological services can prevent some of the secondary problems associated with FAS. Early detection can include measurement of fetal body and head size in the uterus with ultrasound, conducting an electroencephalogram (EEG) and electro-cardiogram (ECG) on the newborn, and then monitoring the newborn's development. If a pregnant woman is suspected of alcohol abuse, her blood alcohol level should be measured as a screen for alcoholism. It is important to note that there is *no recognized safe level of alcohol ingestion during pregnancy;* any alcohol consumption places the fetus at risk, especially in the early stages of pregnancy. FAS is one of the leading causes of birth defects and is believed to be the most common cause of preventable mental retardation in the United States. The only certain way to avoid FAS is to abstain from alcohol consumption throughout pregnancy. Pregnant women who suffer from alcoholism should be directed to alcohol-abuse rehabilitation programs and monitored closely by medical professionals during their pregnancy.

---

periods, called **trimesters.** The first trimester is composed of the first three months after fertilization of the ovum; the second trimester includes months 4 to 6 of pregnancy; and the third trimester refers to the time between month 7 of pregnancy and parturition, or delivery.

Physical Changes During the First Trimester

Tiredness, nausea, and vomiting often accompany the first trimester of pregnancy. These symptoms, referred to as **morning sickness,** usually disappear within the first few months. Other physical changes that occur during this time are frequent urination due to compression of the bladder by the enlarging uterus, breast swelling and tenderness, and increased vaginal secretions. A pregnant woman cannot feel the changes occurring in her uterus during the first trimester.

Physical Changes During the Second Trimester

During the second trimester of pregnancy, a woman becomes very aware of the changes in her uterus because the fetus begins to move. Toward the end of the fourth month, the fetus begins to twist and kick, and these movements intensify during the following weeks. In response to aldosterone and estrogens, most women begin to retain fluid during the second trimester. Increased fluid retention results in swelling, or edema, of the hands and feet. Increased production of red blood cells by maternal bone marrow, together with fluid retention, results in a large increase in maternal blood volume. The increased nutritional demands of the fetus during this time increase the mother's appetite. An increased demand for oxygen by the fetus is accompanied by an increase in maternal blood flow to the placenta and an increase of 30% to 40% in maternal cardiac output in response to increased maternal thyroid function.

Growth of the Uterus. Expansion of the uterus during the second trimester results in protrusion of the abdomen and widening of the waist. Rapid growth of the abdomen can cause stretch marks to form in the abdominal skin. As the uterus grows, it compresses other internal organs, such as the stomach and intestines, which can lead to indigestion, constipation, and hemorrhoids.

Development of the Breasts. In response to estrogen, progesterone, and prolactin, the breasts increase dramatically in size during the second trimester to the extent that stretch marks may occur. During the second trimester of pregnancy, the breasts begin to secrete a thin, yellowish fluid called **colostrum,** a precursor of breast milk that contains antibodies which, when ingested and absorbed, give the newborn a measure of immune protection.

Physical Changes During the Third Trimester

The largest weight gain of pregnancy usually occurs during the third trimester. The average woman gains approximately 9.0 to 11.3 kg (20–25 lb), with the fetus, placenta, and amniotic fluid accounting for approximately 4.5 to 6.8 kg (10–15 lb) of this added weight. During the third trimester, the uterus reaches its largest size and protrudes forward in the standing woman, altering her center of gravity. To compensate for this, a pregnant woman often walks with head and shoulders held back, which can result in backache.

The increased size of the uterus puts constant pressure on the bladder, which results in the need to urinate four to five times during the night, disrupting sleep patterns and increasing fatigue. Increased pressure on the blood vessels of the lower part of the body results in muscle cramps in the legs, and compression of the diaphragm can cause shortness of breath.

During the third trimester, the muscles of the uterus undergo short periods of spasmodic contractions, called **Braxton-Hicks contractions.** The fetus changes position in the uterus during the last few weeks of pregnancy, and the head of the fetus commonly becomes oriented toward the pelvis in preparation for parturition.

Development of the Fetus Is Divided into Three Stages: The Germinal, Embryonic, and Fetal Stages

Prenatal development, which occurs throughout the nine months of pregnancy, can be divided into three stages: (1) the **germinal stage,** the first two weeks of development following fertilization of the ovum; (2) the **embryonic stage,** weeks 2 to 8 following fertilization; and (3) the **fetal stage,** the time between week 8 and birth.

The Germinal Stage of Development (Weeks 1 and 2)

During the first two weeks of pregnancy, the fertilized ovum, although microscopic in size, has reached the blastocyst stage of development. The amnion from the blastocyst forms a thin sac of tissue that surrounds the fetus and is filled with a liquid called **amniotic fluid.** Amniotic fluid is constantly absorbed by the placenta and replaced with new fluid every 2 to 3 hours throughout pregnancy.

The inner cells of the blastocyst change into three cell layers. The outer layer of cells becomes the skin, sense organs, and nervous system of the developing fetus. The middle layer forms the skeleton, muscles, heart, and blood vessels. The inner cell layer develops into the liver, pancreas, and digestive system of the fetus.

The Embryonic Stage of Development (Weeks 2 to 8)

The embryonic stage is a period of remarkable growth for the developing embryo even though it is less than 2.5 cm (1 inch) long by the end of the eighth week of development (Table 32–3). During this time, all of the embryonic organs become partially formed, although cellular development is far from complete. By the end of the first month, the primitive heart has begun to beat; the digestive system, brain, spinal cord, and nervous system are developing; and buds of tissue that will become the arms and legs begin to form (Fig. 32–10). By the sixth week, the head of the embryo is quite large, and the outlines of eyes can be seen. During the sixth and seventh weeks, the eyes and ears develop more fully, and the teeth, bones, and facial muscles begin to form.

The Fetal Stage of Development (Week 8 to Birth)

At approximately nine weeks of development, the embryo is now called a **fetus.** At the beginning of the ninth week of development, the head makes up almost half of the total length of the fetus. All primitive organs and structures become more differentiated.

By the second month, distinct hands and feet have formed, and major blood vessels are forming (see Fig. 32–10). By the third month, the fetal stomach is producing digestive

TABLE 32–3

Average Fetal Growth Patterns During Gestation

Age of Fetus	Crown to Rump Length	Weight
(in weeks)	(in inches)	(in pounds)
5–8	0.2–1.0	0.01–0.03
13–16	3.5–5.5	0.12–0.43
21–24	7.9–9.1	1.06–1.81
29–32	11.0–11.8	3.10–4.60
37–40	13.8–14.2	6.60–7.50

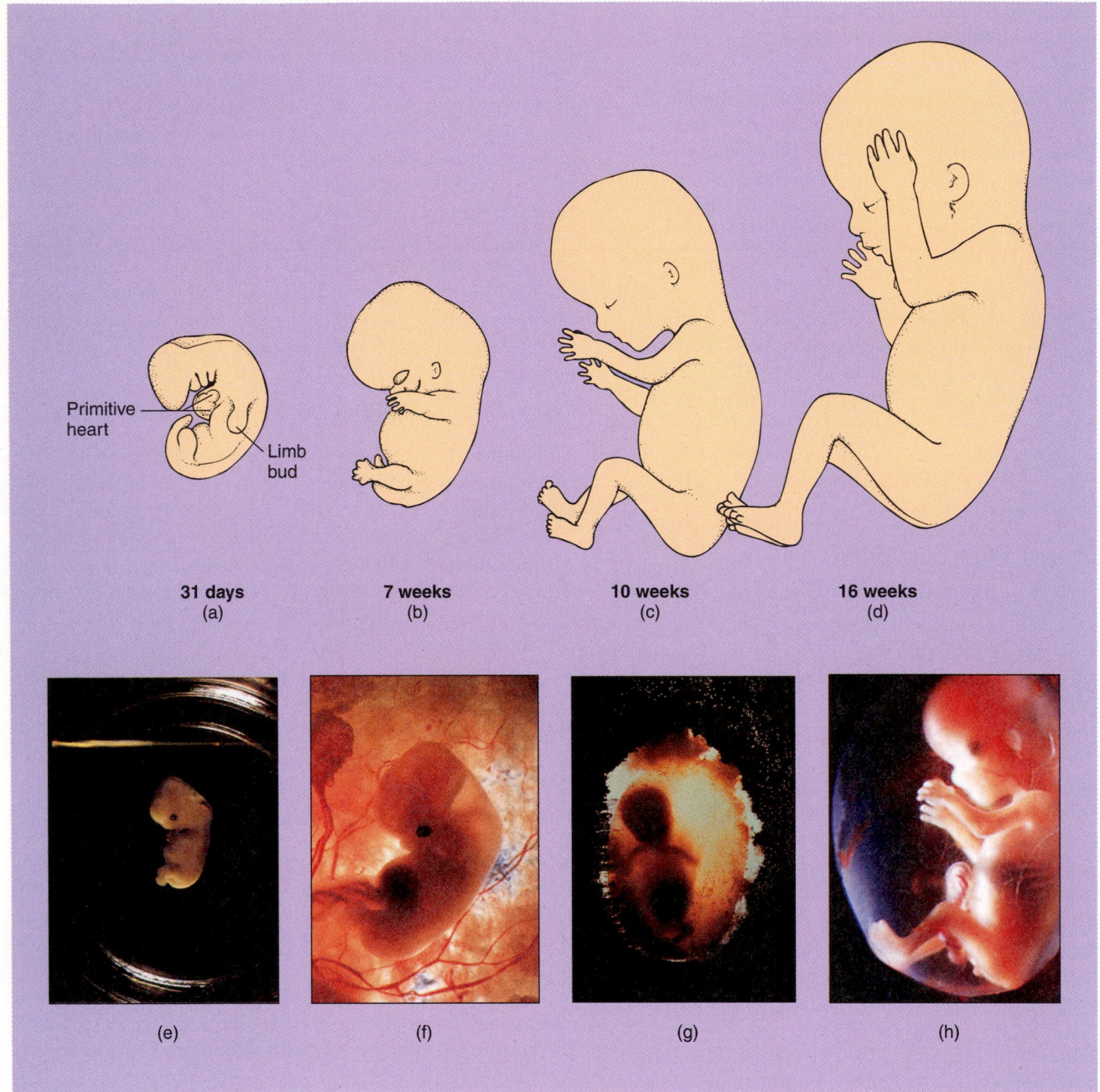

Primitive
heart

Limb
bud

| 31 days | 7 weeks | 10 weeks | 16 weeks |
| (a) | (b) | (c) | (d) |

(e)   (f)   (g)   (h)

Figure 32–10

Appearance of the embryo (**a** and **b**) and the fetus (**c** and **d**) at various times during gestation. (**e**) The embryo at 31 days. (**f**) The embryo at seven weeks. (**g**) The fetus at 10 weeks. (**h**) The fetus at 16 weeks. *(e, © Dr. Landrum B. Shettles; f, © Petit Format and Guigoz; g, © CNRI/Phototake, N.Y.; h, © Petit Format/Science Source.)*

fluids. The heart is pumping blood through the fetal arteries and veins, the kidneys have begun to function, and the bone marrow is manufacturing red blood cells. Also by the third month, the external genitalia of the fetus appear, and the sex of the fetus becomes anatomically distinguishable.

By the fourth month, the fetus has developed vocal cords, lips, fingernails, toenails, and hair on its head (see Fig. 32–10). Sucking motions and respiratory movements begin during the fourth month, and the fetus swallows and absorbs amniotic fluid. Many peripheral reflexes are well developed by this time, although the central nervous system is still very underdeveloped and incapable of higher functions.

During the fifth month, the fetal heartbeat can be heard through the maternal abdomen. The body of the fetus be-

APPLICATIONS OF PHYSIOLOGY

Infant Respiratory Distress Syndrome

The lungs are designed for exchange of oxygen and carbon dioxide between air and the circulation. In the aqueous environment of the fetus, this function is not necessary and the fetal lungs are essentially collapsed.

At birth, the lungs must adapt to an air-breathing environment. This ordinarily would be difficult because of the high surface tension at the fluid-air interface of the alveoli. This high surface tension makes it difficult to expand the alveoli and creates a tendency for the alveoli to collapse. As discussed earlier in this text, the alveolar epithelium produces surfactant, which greatly reduces the surface tension at the fluid-air interface in the alveoli. This greatly reduces the work required to inflate the lungs and helps to prevent alveolar collapse.

Perhaps because the gas-exchange function of the lungs is not needed until birth, surfactant is not produced in the lungs of the fetus until the sixth or seventh month of gestation and may not reach sufficient levels until the eighth month of pregnancy. If an infant is born prematurely, before sufficient surfactant is produced in its lungs, it develops what is called *infant respiratory distress syndrome* (IRDS) and its survival is at great risk. Lack of adequate surfactant production leads to the multiple symptoms that characterize this syndrome. Without adequate surfactant, many alveoli collapse, greatly reducing the surface area for exchange of oxygen and carbon dioxide in the lungs. The greatly increased surface tension at the lungs requires that the newborn expend tremendous amounts of energy just to breathe, a task made more difficult by the reduced alveolar surface area available for the exchange of oxygen between the alveoli and pulmonary circulation. Increased carbon dioxide levels in the newborn form carbonic acid, which impairs many physiological functions, especially those of the brain. If this syndrome persists, the respiratory membrane becomes inflamed, further increasing the diffusion barrier for gas exchange. This inflammation led to the old name for this syndrome, *hyaline membrane disease*. In short order, a newborn with IRDS becomes hypoxic, acidotic, and exhausted and, without intervention, the newborn will die.

It was not long ago that clinical medicine offered little in the way of help for premature infants with this syndrome other than instituting positive pressure, or forced, ventilation of the lungs. The physician and parents could do little else but hope that this would buy enough time for the premature infant's lungs to develop enough surfactant, before hypoxia, acidosis, and physical exhaustion resulted in death.

Today, many more tools are at the physician's disposal that can help to reduce the mortality associated with premature birth. A woman in premature labor can be evaluated and treated so that IRDS can be avoided. Through the process of amniocentesis, in which a sample of the amniotic fluid can be obtained without harm to the fetus, the physician can determine whether sufficient amounts of surfactant are being produced. If this is the case, the physician can allow the delivery of the infant to occur. If this test shows that surfactant levels are not adequate, the physician can take steps to delay labor and delivery. Uterine smooth muscle contains β-adrenergic receptors that mediate muscle relaxation. Several drugs are now available that bind these receptors and induce uterine relaxation. This helps to prevent uterine contractions that might result in premature delivery, thereby buying time for the fetal lungs to develop further in utero. In addition, surfactant production can be stimulated by administering cortisol and cortisone like drugs to the mother. These drugs accelerate fetal surfactant production, thus increasing the possibility that a prematurely born infant will have sufficient surfactant at birth to breathe easily on its own.

Nevertheless, there are still cases in which efforts to delay delivery and increase fetal surfactant production are insufficient to prevent IRDS. Furthermore, these measures occasionally reduce mortality at the expense of creating survivors with chronic lung disease. Consequently, additional therapies are being tested in clinical trials to improve the survivability and long-term health of infants born with IRDS. One of the more promising therapies under consideration is the introduction of surfactant directly into the trachea of newborns with IRDS within the first 15 minutes of life. Several natural and synthetic surfactants are under study, with clinical trials indicating that animal-derived surfactants improve gas exchange, lung function, and infant survivability.

comes covered with fine, downy hair called **lanugo.** At the end of the sixth month, the fetus is only approximately 30 cm (12 inches) long and weighs approximately 0.68 kg (1.5 lb) (see Table 32–3). If the fetus is born prematurely at six months of development, its chances for survival are very poor.

During the seventh month, the brain and nervous system become more developed, although many important higher functions of the central nervous system will undergo further development throughout fetal life and during the first six months following birth. During the seventh month, fatty tissue forms under the skin of the fetus, and the downy hair covering its body begins to disappear. During the eighth month, many fetal organs become mature, and the fetal lungs produce a substance called **surfactant** (see Chapter 21), which is important for survival of the fetus after birth.

During the ninth month of pregnancy, the fetus migrates to a lower part of the uterus and normally positions itself head downward in preparation for parturition. The average fetus is 50 cm (20 inches) long and weighs 3.15 kg (7 lb) at birth (see Table 32–3).

PARTURITION

As parturition approaches, uterine contractions become stronger, more regular, and more frequent. Delivery of the fetus and expulsion of the placenta result from strong contractions of the uterus that push the fetus and the placenta through the cervix and out of the vagina (Fig. 32–11). **Labor** is said to begin when uterine contractions start to occur once every 10 to 15 minutes. In some women, the amniotic membrane surrounding the fetus ruptures, and amniotic fluid flows out of the vagina before delivery begins. But in most women, the amniotic membrane ruptures during delivery.

Parturition, or Delivery, Involves Both Cervical Dilation and Uterine Contractions, Which Are Under Hormonal Control

Uterine Contractions During Parturition

Uterine contractions at delivery are facilitated by a number of hormones. Muscular contractions begin in the upper portion of the uterus and move down toward the cervix. These contractions force the fetus against the cervix. Under hormonal stimulation, and with pressure from the fetus, the cervix widens or dilates to a maximum diameter of approximately 10 cm (4 inches). **Cervical dilation,** combined with pressure generated by the mother, forces the fetus and placenta through the cervix and out of the vagina canal (see Fig. 32–11). At birth, the infant is still connected by the umbilical cord to the placenta. Within minutes following delivery, the maternal and infant blood vessels in the placenta and umbilical cord constrict as a result of their exposure to oxygen and catecholamines, and the infant is separated from the maternal blood supply. The placenta becomes separated from the wall of the uterus and is expelled through the vagina (see Fig. 32–11). Because the placenta leaves the vagina after the infant is delivered, it is sometimes referred to as the **afterbirth.**

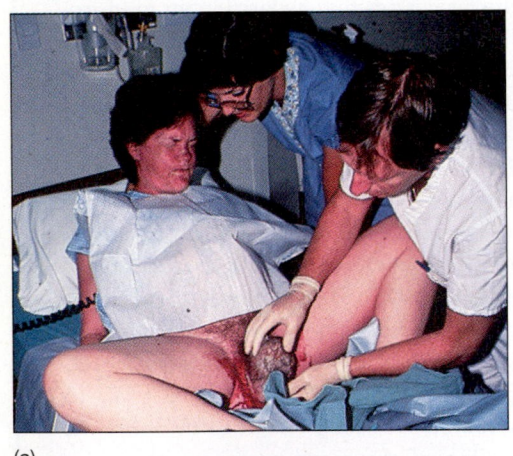
(a)

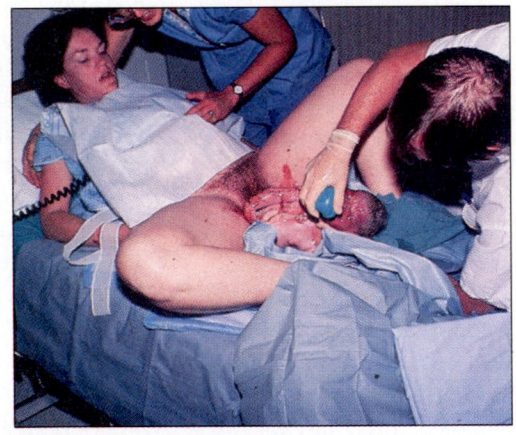
(b)

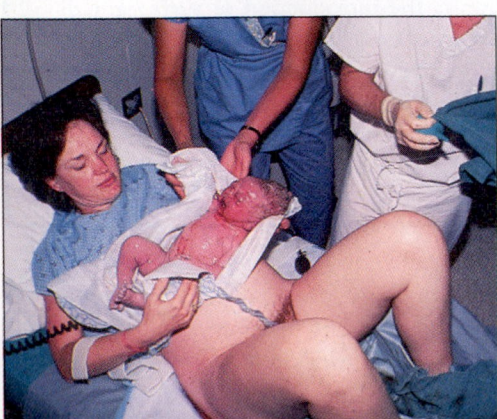
(c)

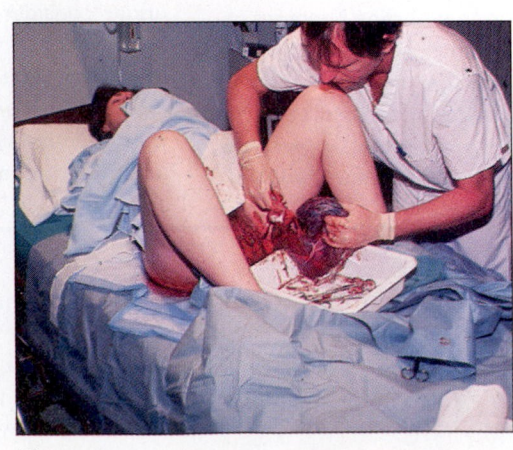
(d)

Figure 32–11

Birth of a baby. In approximately 95% of all human births, the baby descends through the vagina head downward. *(a)* The abdominal muscles of the mother are used to push the baby out through the vagina. *(b* and *c)* When the baby emerges, the mouth and pharynx are gently aspirated to remove any amniotic fluid, mucus, or blood that could impede respiration. *(d)* The placenta emerges following delivery of the baby.

When the fetus and placenta are expelled from the uterus, all of the hormones secreted by the fetoplacental unit begin to decrease markedly in maternal plasma. Within three days after delivery, the concentration of steroid and protein hormones in maternal plasma returns to nonpregnant levels.

What hormones are involved in parturition?

Many hormones, including estrogen, progesterone, relaxin, oxytocin, and prostaglandins, have been found to play important roles in the initiation of parturition (Fig. 32–12). In response to estrogen and relaxin stimulation, the number of receptors for oxytocin in the uterus increases progressively during the later part of pregnancy. This increases uterine sensitivity to oxytocin at the end of pregnancy. Dilation of the cervix before delivery provides afferent neural input to the hypothalamus, which results in the release of oxytocin from the posterior pituitary gland. Oxytocin stimulates strong rhythmic contractions of uterine muscles.

Prostaglandins also stimulate contractions of uterine smooth muscle. In response to oxytocin, the smooth muscle cells of the uterus secrete prostaglandins, which, through a paracrine action, stimulate further uterine contractions. The prostaglandins are so powerful in stimulating uterine contractions that exogenous administration of prostaglandins can induce labor and terminate pregnancy at almost any stage of gestation.

Progesterone strongly inhibits uterine contractions. In some women, circulating levels of progesterone decline just prior to delivery, which removes progesterone inhibition of uterine contractions and facilitates the onset of labor. However, a decrease in total circulating progesterone prior to delivery has not been found in all women. Recent evidence indicates that the placenta may secrete a progesterone-binding protein before delivery, which reduces the amount of free progesterone in maternal circulation and facilitates delivery.

Relaxin, secreted first by the corpus luteum and later by the endometrium, may facilitate delivery. Relaxin increases the number of oxytocin receptors in the uterus. It also softens the cervix, making it more pliable and facilitating passage of the fetus through the cervix during delivery. Finally, relaxin relaxes the pelvic ligaments, easing passage of the fetus through the birth canal.

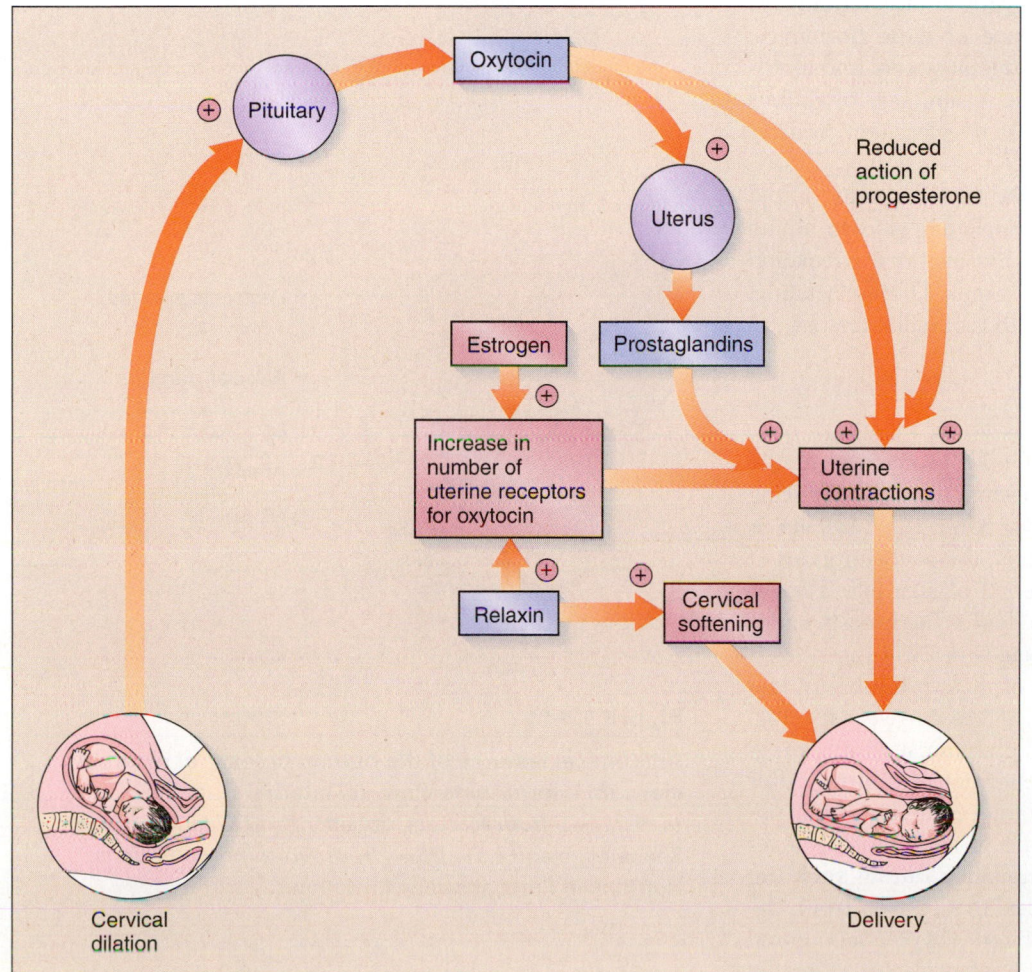

Figure 32–12

Hormonal factors that contribute to the onset of parturition. The plus sign indicates stimulation.

Lactation, the Synthesis and Release of Breast Milk, Is Under Hormonal Control

 How is lactation controlled?

Anatomy of the Breast

The outer surface of the female breast is illustrated in Figure 32–13a. In the center of the breast is a pigmented area, called the **areola,** which surrounds the nipple. The internal structure of the breast is illustrated in Figure 32–13b and c. The mammary, or breast, tissue lies directly over the pectoralis major muscle and is separated from this muscle by a layer of fat (see Fig. 32–13c). The breast is composed of **glandular epithelium** and a **ductal system** embedded in interstitial tissue and fat. The ducts of the breast, which terminate in the nipple, branch through the breast tissue and are connected to glands called the alveoli (see Fig. 32–13b). The **alveoli,** which secrete breast milk, have been described as resembling clusters of grapes on stems. The alveoli are surrounded by contractile cells called **myoepithelial cells** that are involved in the expulsion of milk from the breasts.

Development of the Breast

Estrogen, progesterone, hCS, and prolactin stimulate breast development during pregnancy. Estrogen also stimulates prolactin release. Under the influence of these hormones, the breasts increase in size, and the ductal system and alveolar structures become more complex. A summary of various hormones and their effects on breast development prior to parturition is shown in Table 32–4.

Under hormonal stimulation, the alveolar cells of the breast extract a variety of substances, including glucose, amino acids, fatty acids, and glycerol, from the maternal circulation and convert these materials into breast milk. During lactation, the breasts may produce as much as 1.5 L of milk each day.

Effects of Estrogen, Progesterone, and Prolactin on Lactation

Estrogen and progesterone inhibit milk production in the breast prior to parturition by blocking the milk-inducing effects of prolactin on the breasts. After the placenta is expelled from the uterus at parturition, the concentrations of estrogen and progesterone in maternal plasma fall. The removal of estrogen and progesterone leaves the breasts under the influence of prolactin and oxytocin (Fig. 32–14).

Prolactin stimulates synthesis of milk in the breast and secretion of milk into the alveoli of the breasts (see Fig. 32–14). Prolactin levels in maternal plasma rise during pregnancy and reach very high concentrations at birth. After birth, plasma prolactin levels begin to decline, but the hypothalamus then becomes very sensitive to neural signals it receives from the breasts. Breast stimulation during **suckling** stimulates afferent neural impulses that travel from the breast, through the spinal cord, and to the hypothalamus and stimulate release of a large amount of prolactin from the

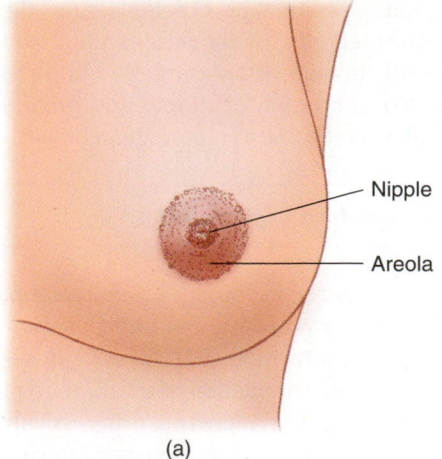

(a)

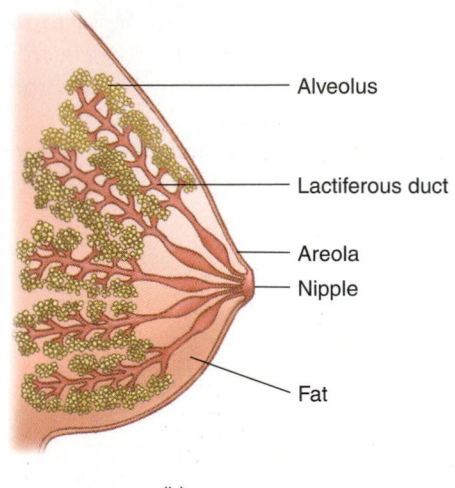

(b)

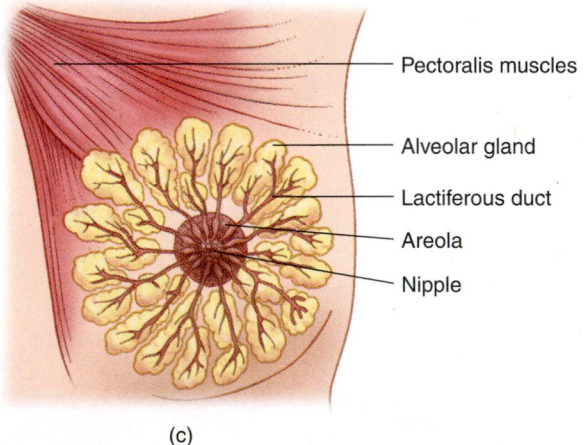

(c)

Figure 32–13

Anatomical features of the human breast. **(a)** External view. **(b)** Internal side view. **(c)** Internal front view. *(a, Redrawn from Masters, Johnson, and Kolodny,* Human Sexuality; *b and c, Redrawn from Worthington-Roberts,* Nutrition in Pregnancy and Lactation.)

TABLE 32–4

Hormones Involved in Breast Development During Pregnancy

Hormone	Origin	Function During Pregnancy
Prolactin	Anterior pituitary	Stimulates mammary growth
Oxytocin	Posterior pituitary	Generally no effect on mammary growth, but sensitivity of myoepithelial cell to oxytocin increases during pregnancy
Estrogen	Ovary and placenta	Stimulates proliferation of glandular tissue and ducts of the breast; stimulates prolactin release but blocks action of prolactin on the breast
Progesterone	Ovary and placenta	Stimulates proliferation of glandular tissue and ducts of the breast; blocks action of prolactin on the breast
Human chorionic somatomammotropin (hCS)	Placenta	Stimulates mammary growth
Human chorionic gonadotropin (hCG)	Placenta	Stimulates mammary growth

maternal pituitary. Suckling results in a tenfold rise in plasma prolactin that lasts for approximately 1 hour, after which plasma prolactin levels return to normal. Therefore, prolactin concentrations in maternal plasma rise and fall with every episode of nursing (Fig. 32–15). Prolactin, released in response to a single nursing episode, stimulates **milk production** in the breast, which makes additional milk available for the next nursing episode. If an infant is not breast-fed or when nursing is terminated, there is no stimulus for prolactin release, milk synthesis is not stimulated, and the breast stops secreting milk within a few days. With prolonged nursing, milk production in the breast begins to decline within seven to nine months after birth, but the breast can still produce significant quantities of milk for several years.

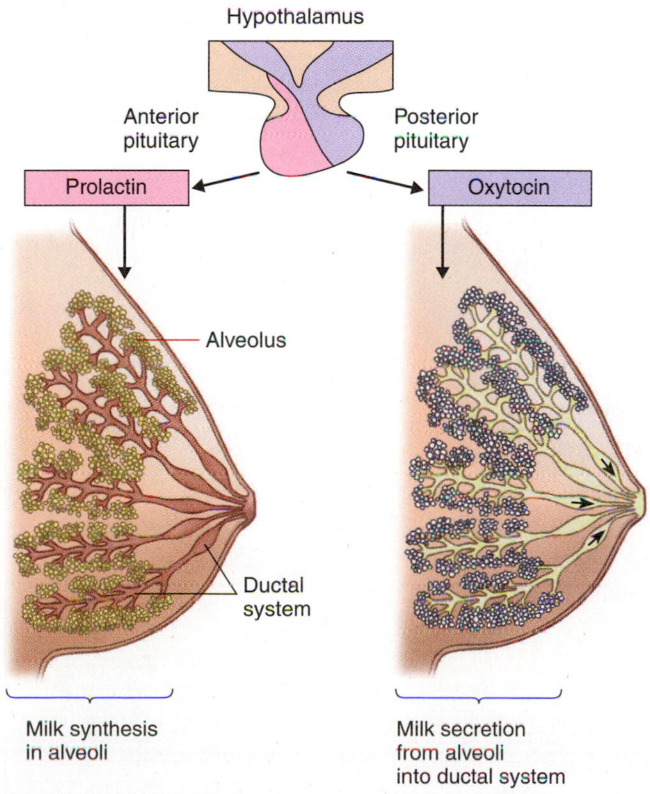

Prolactin inhibits the release of GnRH from the hypothalamus and hence inhibits LH and FSH release from the pituitary. This results in the absence of follicular growth and ovulation. Therefore, breast-feeding tends to inhibit ovulation and reduce fertility. However, due to the large amount of individual variation in the time during which prolactin inhibits GnRH release with prolonged nursing, breast-feeding is not a reliable method of birth control.

Effects of Oxytocin on Lactation

Oxytocin is released in response to breast stimulation during nursing. Suckling stimulates sensory nerves in the breast that transmit afferent neural impulses to the spinal cord and from there to the hypothalamus, resulting in the release of oxytocin from the pituitary gland. Oxytocin is released episodically in response to nursing episodes (see Fig. 32–15). Oxytocin stimulates the release of milk from the breast, a process termed **milk let-down.** In addition to breast stimulation, certain auditory stimuli, such as the sound of a baby crying, also can stimulate the release of oxytocin and hence milk ejection from the breast. Oxytocin stimulates contraction of the myoepithelial cells that surround the outer walls of the alveoli and forces milk from the alveoli into the ductal system of the breast (see Fig. 32–14). Once milk reaches the ductal system, it can be pulled from the breast by the negative pressure that results from suckling.

The Composition of Breast Milk

Breast milk contains water, protein, **lactose** (milk sugar), fat, and a variety of vitamins and minerals. Table 32–5 compares the nutrient content of human milk and cow's milk. The

Figure 32–14

Hormonal control of milk production and expulsion in the breast.

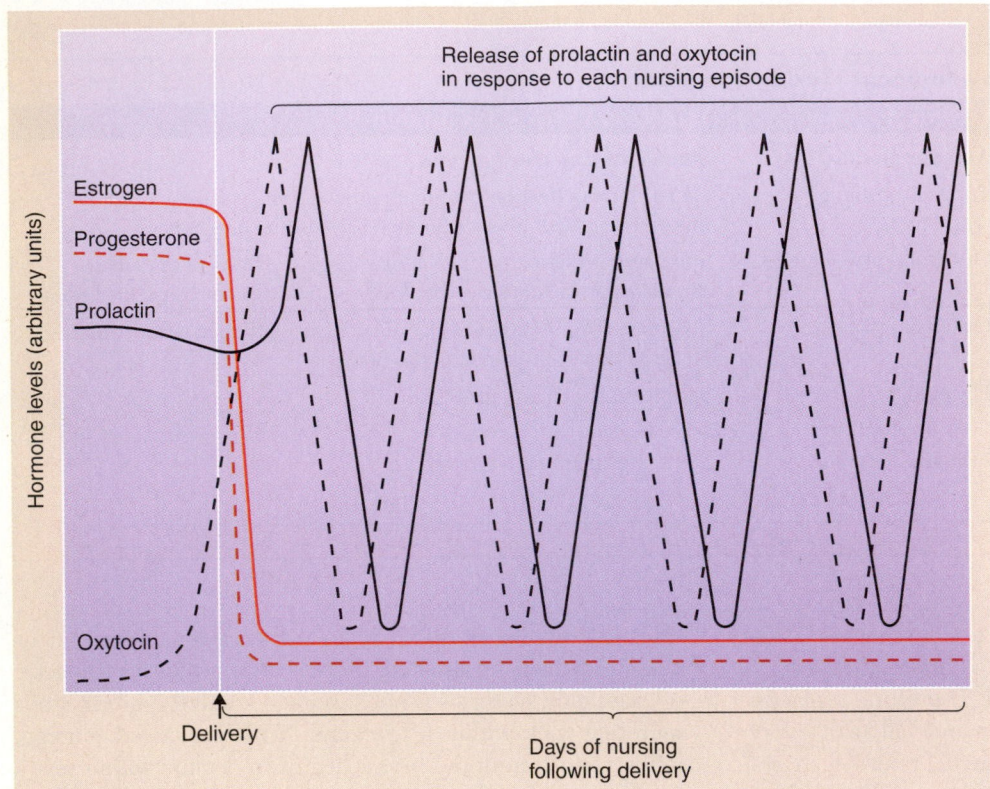

Figure 32–15

Changes in maternal hormones following delivery and during episodes of nursing.

TABLE 32–5

Nutrient Content of Human Milk and Cow's Milk

Constituent (per liter)	Human Milk	Cow's Milk	Constituent (per liter)	Human Milk	Cow's Milk
Energy (kcal)	690	660	Sodium (mg)	160	506
Protein (g)	9	35	Potassium (mg)	530	1570
Fat (g)	40	38	Chlorine (mg)	400	1028
Lactose (g)	68	49	Magnesium (mg)	38–41	120
Vitamin A (IU)	1898	102	Sulfur (mg)	140	300
Vitamin D (activity)	40	14	Iron (mg)	0.3–0.56	0.5
Vitamin E (IU)	3.2	0.4	Iodine (mg)	200	80
Vitamin K (μg)	34	170	Manganese (μg)	4.0–5.9	20–40
Thiamin (μg)	150	370	Copper (μg)	60	110
Riboflavin (μg)	380	1700	Zinc (mg)	0.5–4	3–5
Niacin (mg)	1.7	0.9	Selenium (μg)	20	5–50
Pyridoxine (μg)	130	460	Fluoride (mg)	0.05	0.03–0.1
Folic acid (μg)	41–84.6	2.9–68	Chromium (μg)	4	2
Cobalamine (μg)	0.5	4			
Ascorbic acid (μg)	44	17			
Minerals					
Calcium (mg)	241–340	1200			
Phosphorus (mg)	150	920			

Adapted from Worthington-Roberts, B.S., Vermeersch, J., and Williams, S.R. *Nutrition in Pregnancy and Lactation.* St. Louis, Times Mirror/Mosby, 1985, p. 257.

protein in human milk is largely in the form of **α-lactalbumin,** whereas that of cow's milk is in the form of **casein.** The caseins are milk proteins that form curds when exposed to digestive enzymes, pepsin, or changes in pH. The lower casein content of human milk makes it easier for the infant to digest than cow's milk.

CHAPTER REVIEW

Summary

- Ovulation occurs approximately 14 days after the beginning of the last menstrual cycle in response to stimulation from high circulating levels of LH and FSH. The ovum is transported through the uterine tubes to the uterus.
- Following ejaculation into the vagina, sperm are transported through the vagina, cervix, and uterus into the uterine tubes.
- Capacitation and the acrosome reaction prepare the sperm to fertilize the ovum in the uterine tube.
- Initially, the fertilized ovum is a single cell, called a zygote, which contains 46 chromosomes. The zygote begins to divide mitotically approximately 30 hours after fertilization, while it is still in the uterine tube.
- The blastocyst sends out tissue projections that invade the endometrium, resulting in the implantation of the blastocyst in the endometrium.
- The placenta acts as a transport interface between the fetal and maternal blood supplies.
- hCG stimulates growth of the corpus luteum during early pregnancy and ensures continued secretion of estrogen and progesterone from the corpus luteum, which stimulates continued growth of the endometrium.
- During the second month of pregnancy, the placenta secretes estrogen and progesterone on its own, and the corpus luteum is no longer needed.
- Progesterone secreted by the placenta is essential for the maintenance of pregnancy and breast development.
- Estrogen secreted by the placenta stimulates enlargement of the uterus to accommodate the growing fetus, development of the ducts in the breasts in preparation for nursing, and relaxation of pelvic ligaments at birth.

- High circulating levels of estrogen and progesterone in maternal plasma during pregnancy inhibit the secretion of GnRH, inhibiting follicular development and menstruation during pregnancy.
- Maternal plasma prolactin, insulin, aldosterone, cortisol, and thyroxine increase during pregnancy.
- Major physical and physiological changes occur in the pregnant woman during the second and third trimesters of pregnancy.
- The germinal stage of fetal development is marked by the formation of the blastocyst and implantation of the blastocyst in the endometrium.
- During the embryonic stage of development, internal organs, including the heart, digestive system, and nervous system, as well as external organs, including the eyes and ears, begin to form in the developing embryo.
- During the fetal stage of development, the fetus develops from a partially formed organism, completely dependent on the maternal blood supply, into an organism capable of sustaining life outside of the uterus.
- As delivery approaches, oxytocin released by the pituitary and prostaglandins secreted by the uterus stimulate uterine contractions and cervical dilation. Relaxin, secreted by the endometrium of the uterus, softens the cervix and relaxes pelvic ligaments.
- A variety of hormones, including prolactin, estrogen, progesterone, hCG, and hCS, stimulate breast development during pregnancy in preparation for nursing.
- After delivery, production and release of milk from the breasts are largely controlled by prolactin, which is secreted by the anterior pituitary, and oxytocin, which is secreted by the posterior pituitary.

Review Questions

Choose the Correct Answer

1. The normal location for fertilization of the ovum occurs in the:
   a. vagina.
   b. cervix.
   c. uterine cavity.
   d. fallopian tube.
   e. abdominal cavity.
2. Movement of sperm through the female reproductive tract:
   a. is due to the movement of the flagella on sperm.
   b. is aided by contractions in the uterus.
   c. is facilitated by the properties of cervical mucus during the luteal phase of the ovarian cycle.
   d. is inhibited by prostaglandins produced in the female during intercourse.
   e. All of the above are correct.

3. Which of the following is true concerning the ovum?
   a. It undergoes the second meiotic division by the time of ovulation.
   b. It is normally fertilized by sperm while in the uterine cavity.
   c. It has completed the first meiotic division at the time of ovulation.
   d. It has a haploid number of chromosomes and DNA at ovulation.
   e. It is transported within 2 hours of ovulation into the uterine cavity.
4. Implantation of a fertilized ovum:
   a. normally occurs in the ampulla of the fallopian tube.
   b. follows deciduation of the endometrium.
   c. is inhibited by progesterone.
   d. occurs when it is in the blastocyst stage.
   e. is most likely to occur during the follicular phase of the ovarian cycle.

5. Which of the following is correct?
   a. Trophoblast cells form the placental membranes involved in exchange of materials between the fetus and mother.
   b. Fetal blood is less saturated with oxygen than maternal blood because fetal blood contains less hemoglobin.
   c. Glucose crosses the placental surface by simple passive diffusion.
   d. Fetal oxygen carrying capacity is enhanced by high fetal blood carbon dioxide concentration.
   e. Alcohol in the maternal circulation cannot enter the fetus in utero.

6. Concerning steroid hormones in pregnancy, which of the following is true?
   a. Progestins are converted to androgens by the placenta.
   b. Progesterone production by the corpus luteum early in pregnancy is replaced by placental production at approximately 7 to 10 weeks of gestation.
   c. Hormones are regulated by hCS.
   d. Hormones do not require fetal endocrine glands.
   e. Hormones require production of DHEA by fetal gonads.

7. All of the following are true except:
   a. Fetal hormones initiate parturition (birth).
   b. The corpus luteum is supported early in pregnancy by hCG.
   c. Estrogen increases myometrial sensitivity to oxytocin.
   d. Cervical dilation promotes uterine contraction.
   e. Inhibition of effects of progesterone on the myometrium facilitates birth.

8. Suckling:
   a. inhibits oxytocin secretion.
   b. stimulates secretion of estradiol.
   c. stimulates development of the mammary ducts.
   d. stimulates secretion of prolactin.
   e. None of the above

9. Lactation is suppressed in the third trimester of pregnancy by:
   a. low levels of prolactin.
   b. high prolactin levels.
   c. a lack of fetal adrenal DHEA production.
   d. high levels of blood estrogen and progesterone.
   e. suppression of the maternal anterior pituitary.

10. Oxytocin secretion is increased by all of the following except:
    a. crying of an infant.
    b. suckling.
    c. dilation of the cervix.
    d. increased estrogen.
    e. tactile stimulation of the nipple.

11. During the second and third trimesters, estrogen levels are elevated by:
    a. the corpus luteum.
    b. maternal ovaries and fetal adrenal.
    c. maternal adrenal gland and fetal liver.
    d. the placenta and fetal adrenal.
    e. None of the above

12. Consumption of alcohol in the mother can adversely affect fetal development because:
    a. this drug can enter the fetal circulation via the placenta.
    b. alcohol reduces uterine blood flow.
    c. alcohol suppresses estrogen secretion by the mother.
    d. alcohol cause diabetes in the mother.
    e. alcohol increases maternal heart rate.

13. Failure of which of the following would likely result in the condition of polyspermia?
    a. Suppression of exocytosis of cortical granules from the ovum during sperm contact

b. Increased calcium permeability of the ovum upon contact with sperm
   c. Suppression of sperm binding sites
   d. Ovum membrane depolarization
   e. Increased number of nonmotile sperm in the ejaculate

14. Spontaneous abortion of the developing zygote in the first two weeks of pregnancy is most likely related to:
    a. suppression of uterine motility.
    b. overproduction of estrogen during the follicular phase of the previous month's ovarian cycle.
    c. insufficient progesterone production by the corpus luteum.
    d. insufficient progesterone production by the placenta.
    e. overproduction of LH.

15. *Abruptio placentae* is a condition in which the placenta detaches prematurely from the wall of the uterus and can result in fetal death. The most likely reason for fetal death in this condition would be:
    a. a lack of placental steroid production.
    b. vaginal blood loss.
    c. inability to get oxygen and nutrients from the mother to the fetus.
    d. hypersecretion of pituitary gonadotropins in response to loss of fetal steroid production.
    e. release of epinephrine in response to maternal stress.

16. Which of the following pair of conditions is most related to physical growth changes in the mother in the third trimester of pregnancy?
    a. Increased urination and leg edema
    b. Increased metabolism and water retention
    c. Salt retention and hypertension
    d. Hyperinsulinemia and hyperprolactinemia
    e. Increased GnRH and FSH levels

17. A woman with no previous history of diabetes suddenly develops diabetes mellitus in the sixth month of pregnancy. This diabetes is likely related to:
    a. increased lipolysis in the accelerated starvation phase of pregnancy.
    b. a shift from a catabolic to anabolic state in the mother.
    c. anti-insulin actions of prolactin, estrogen and progesterone.
    d. enhanced sensitivity to insulin caused by exposure to increasing levels of estrogen and progesterone.
    e. increased GI absorption of glucose.

18. Spinal anesthesia is often given to women to relieve pain in childbirth. This procedure can prolong labor. The most likely explanation for this problem is:
    a. spinal anesthesia stimulates uterine contractions.
    b. the anesthesia interferes with sensory fibers detecting cervical stretch and intrauterine pressure.
    c. spinal anesthesia removes inhibition of relaxin release.
    d. anesthesia inhibits estrogen release.
    e. anesthesia relaxes the cervix.

19. During birth, uterine contractions are facilitated by all of the following except:
    a. prostaglandins.
    b. release of oxytocin in response to cervical dilation.
    c. inhibition of release of progesterone.
    d. increased oxytocin receptor production in response to estrogen.
    e. release of relaxin.

20. Breast stimulation during nursing:
    a. is not related to milk production.
    b. stimulates milk production by stimulating release of oxytocin.

c. stimulates filling of the ductal system with milk by causing myoepithelial contraction in response to prolactin.

d. inhibits milk let-down caused by psychological stimuli.

e. stimulates milk production by stimulating release of prolactin.

21. A kit used for at home pregnancy testing would best be designed to test:

a. urine FSH levels.

b. urine estriol levels.

c. urine progesterone levels.

d. plasma estrogen levels.

e. urine hCG levels.

Answers to Case History Questions

1. She has hypertension (sustained for 6 hours), proteinuria (protein in the urine), and edema (accumulation of water in the body as evidenced by a 6.3-kg [14-lb] weight gain in just two weeks). These symptoms are all consistent with the potentially dangerous condition called preeclampsia, formerly known as *toxemia of pregnancy*. In severe cases, this condition produces the HELLP syndrome characterized by (1) hemolysis (breakdown of red blood cells), (2) elevated liver enzymes, indicative of liver damage, and (3) low platelet count, indicating impaired blood clotting ability (potentially a major problem during labor and delivery). This condition also damages the maternal kidneys.

2. CBC with platelets. As preeclampsia becomes more advanced, the hemoglobin level will go up due to hemoconcentration and a drop in platelets to less than 100,000; this is a part of the HELLP syndrome. The 24-hour quantitative protein helps to document the severity of renal damage that has occurred. The greater the damage, the greater the proteinuria. Creatinine clearance is a measure of renal function. Liver function tests are elevated as a part of the HELLP syndrome; this is caused by

liver edema. A fetal ultrasound determines whether or not intrauterine growth retardation (IUGR) has occurred. The nonstress test (NST) is an indicator of fetal well-being. Because there is a good likelihood that she will need to be delivered soon, betamethasone is given to enhance fetal lung maturity (surfactant production) and to strengthen intracranial blood vessels.

3. At this time, she has mild preeclampsia. But many of the tests are on the borderline for severe preeclampsia. Some authorities would say that due to the IUGR, the condition is severe.

4. Renal damage leads to leakage of her protein into the urine, causing her to become hypoproteinemic. Her capillaries are also damaged by the preeclampsia, which in turn leads to extravasation of intravascular fluid. Edema of the brain can lead to acute personality changes and headaches. Edema of the retina causes scintillating scotomata (shooting stars). Extravasation of fluid (leakage of fluid out of the capillaries of the circulation) while maintaining red blood cells within the vascular system causes hemoconcentration. Edema of the liver will cause swelling of the liver with stretching of the capsule. This causes severe pain in the right upper quadrant.

Key Terms

acrosome reaction (p. 953)
amniotic fluid (p. 965)
blastocyst (p. 956)
capacitation (p. 953)
colostrum (p. 965)
embryo (p. 956)

fertilization (p. 955)
fetus (p. 956)
human chorionic gonadotropin (hCG) (p. 959)
lactation (p. 970)

myoepithelial cells (p. 970)
oxytocin (p. 952)
parturition (p. 968)
placenta (p. 960)
polyspermy (p. 955)

pregnancy (p. 962)
prolactin (p. 970)
relaxin (p. 963)
umbilical cord (p. 960)
zygote (p. 956)

Suggested Readings

Coustan, D.R., ed.; Haning, R.V., Jr., and Singer, D.B., assoc. eds. *Human Reproduction: Growth and Development*, Boston, Little, Brown & Co., 1995.

Jones, R.E. *Human Reproductive Biology*, ed 2. San Diego, Academic Press, 1997.

Knobil, E., and Neill, J.D., eds. *Encyclopedia of Reproduction*. San Diego, Academic Press, 1998.

Knobil, E., and Neill, J.D., eds; Greenwald, G.S., Markert, C.L., and Pfaff, D.W., assoc. eds. *The Physiology of Reproduction*, ed 2. New York, Raven Press, 1994.

Riordan, J., and Auerbach, K.G. *Breastfeeding and Human Lactation*, ed 2. Boston, Jones and Barlett Publishers, 1999.

Web Sites

http://www.merck.com/pubs/mmanual_home/contents.htm
The Merck Manual of Medical Information–Home Edition. Pregnancy, chapters 241–57.

http://www.pregnancyguideonline.com
StorkNet's Week-by-Week Guide to Your Pregnancy. Parental information and advice.

http://tlc.discovery.com/tlcpages/maternity/maternity.htm
Birth.

Answers to Review Questions

1. d	**2.** b	**3.** c	**4.** d	**5.** a	**6.** b	**7.** a	**8.** d
9. d	**10.** d	**11.** d	**12.** a	**13.** a	**14.** c	**15.** c	
16. a	**17.** c	**18.** b	**19.** e	**20.** e	**21.** e		

Chapter 33

SEXUAL PHYSIOLOGY

- *Scanning electron micrograph of a human four cell embryo, day 2, on the tip of a pin.*

KEY CONCEPTS

- Male and female embryos are identical during the first six weeks of embryonic life.

- The characteristics that define a male and female are established through sexual determination and differentiation.

- Many of the reproductive organs in both sexes serve both sexual and reproductive functions.

- Four stages of sexual arousal occur in both males and females.

- Fertility in males and females depends on the integrity of the hypothalamus, pituitary, and gonads. A disruption of normal function at any of these levels can result in infertility.

- Contraception, or methods to prevent pregnancy, is available for both men and women.

CASE HISTORY

A 46-year-old man is referred for impotency, otherwise known as *erectile dysfunction* (ED). He relates a slow loss of the abilities to both obtain and maintain an erection and cannot maintain this for more than a few moments. Vaginal intromission is not possible. He does not have nocturnal erections. He is married and has been faithful. All forms of sexual stimulation, with and without a partner, meet with failure. There is no previous history of failure. He quit smoking 10 years ago but had a history of smoking approximately a pack a day for 18 years prior to quitting. He is a nondrinker. He denies the use of herbal or illegal drugs. He does take a multivitamin and runs 10 miles weekly. His father died at age 56 years of myocardial infarction. His mother and two brothers are well. He has an eight-year history of insulin-dependent diabetes and has taken the β-adrenergic blocker propranolol (Inderal) for hypertension for four years. He denies visual, gastrointestinal, or urologic complaints. There are no reported neurological changes. His physical examination is remarkably benign. Retinal examination is normal, his blood pressure is 118/68, and all pulses are full. A random glucose, testosterone, and lipid profile are normal. By changing his antihypertensive medication to prazosin, an α_1-adrenergic blocker, his erections were restored.

Questions

1. What is the significance of a lack of nocturnal erections in this individual?

2. Why was the physician concerned whether the patient smoked?

3. What is the potential role of diabetes in this patient's ED?

4. What is the potential role of antihypertensive drug therapy in this patient's ED?

INTRODUCTION

Sexual physiology involves physiological functions in the mature male and female that physically allow the introduction of sperm into the female reproductive tract in the act of **sexual intercourse.** This requires the proper development of anatomical structures in the male and female as well as proper functioning of those structures. Although sexual intercourse is involved in the process of impregnation, not all acts of sexual intercourse in humans can result in pregnancy, and humans have devised means of preventing the possibility of pregnancy resulting from sexual intercourse.

SEXUAL DETERMINATION AND DIFFERENTIATION

 What factors determine how the gonads and genitalia will develop?

The Sex Chromosomes

Every cell in the body, except the sperm and ovum, contains 23 pairs of chromosomes. Twenty-two of these chromosomal pairs are termed **autosomes;** the twenty-third pair are the **sex chromosomes.** Autosomes provide the genetic information for the development and differentiation of most of the cells in the body. The sex chromosomes provide the genetic information that directs the differentiation of gonadal cells into ovaries or testes and hence, the **gonadal sex** of the individual.

Meiosis and the Sex Chromosomes

The two identical sex chromosomes in the female are designated **XX;** the two nonidentical sex chromosomes in the male are designated **XY.** Prior to fertilization, the male germ cells (sperm) and the female germ cells (ova) undergo **meiosis.** During meiotic division, each daughter cell receives only 23 chromosomes, one from each of the 23 chromosomal pairs. Given that the twenty-third pair (the sex chromosomes) in the male are XY, and in the female, XX, half of the sperm formed during meiosis receive 22 autosomes and an X sex chromosome, and half receive 22 autosomes and a Y sex chromosome. All the ova formed during meiosis receive 22 autosomes and an X sex chromosome.

The Sex Chromosomes Provide the Genetic Information That Directs Differentiation of the Gonads into Ovaries or Testes

During fertilization, when the sperm and ovum fuse, the 23 chromosomes from the sperm and the 23 chromosomes from the ovum combine to produce a single-celled zygote that contains a total of 46 chromosomes. When a sperm containing an X chromosome fertilizes the X-bearing ovum, the **chromosomal sex** of the zygote is XX, or **female.** When a sperm containing a Y chromosome fertilizes the X-bearing ovum, the chromosomal sex of the zygote is XY, or **male.** During the first six weeks of pregnancy, the fetus contains undifferentiated gonads as well as internal and external genitalia that can develop into male or female organs. The Y chromosome carries genes, most notably the *sex-determining region Y (SRY)* gene, that direct the gonads to differentiate as testes. If an ovum is not fertilized by a Y-bearing sperm, the testes, male genital ducts, and male external genitalia do not develop. Therefore, the Y chromosome determines the male phenotype, or *maleness* (Fig. 33–1). This is why a person with Klinefelter's syndrome, a chromosomal abnormality in which a genetic male receives an extra X chromosome (XXY), appears as a nearly normal male.

It appears that the X chromosome may carry genes for femaleness. A gene, preliminarily termed the *DSS* gene, may direct the gonads to differentiate as ovaries. In the female, both of the X chromosomes are essential for normal differentiation of the ovaries. A person with Turner's syndrome, or a single X chromosome, has the appearance of a female but the development of the ovaries and secondary sex characteristics is impaired and menstruation does not occur (Table 33–1).

The X chromosome contains many genes that are required by both sexes, and some traits are determined by genes that are located only on the X chromosome. These are termed *sex-linked traits,* and they follow the pattern of transmission of the X chromosome. Colorblindness and hemophilia are two such sex-linked traits. For most X-linked

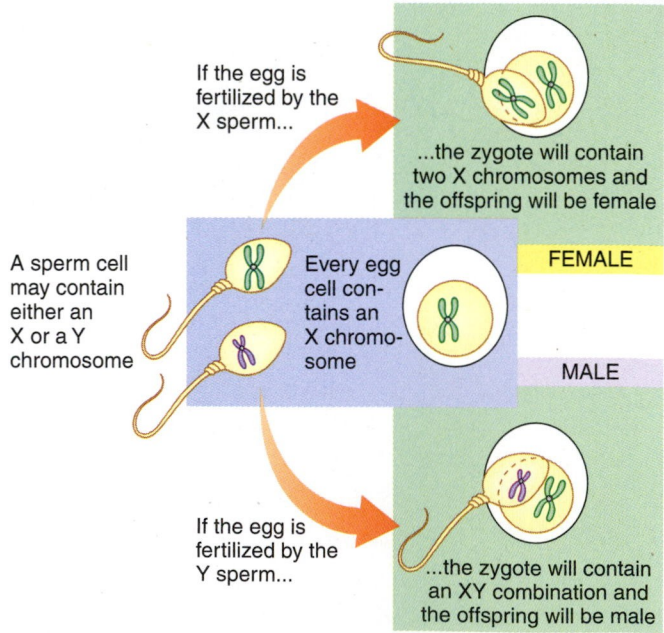

If the egg is fertilized by the X sperm...

...the zygote will contain two X chromosomes and the offspring will be female

A sperm cell may contain either an X or a Y chromosome

Every egg cell contains an X chromosome

FEMALE

MALE

If the egg is fertilized by the Y sperm...

...the zygote will contain an XY combination and the offspring will be male

Figure 33–1

Sex of the zygote is determined at the time of fertilization by the genetic material contributed by the sperm. An X-bearing sperm produces a female; a Y-bearing sperm produces a male.

TABLE 33–1

Abnormalities of Sexual Differentiation

Chromosomal Complement	Enzyme or Receptor Defects	Gonads	Internal Genitalia	External Genitalia
XX Normal female	—	Ovaries	Female	Female
XO Turner's syndrome	—	Ovaries with impaired development—"streak gonads"	Female	Female
XXY Klinefelter's syndrome	—	Testes with impared development of seminiferous tubules and deficient spermatogenesis	Male	Male
XX Adrenogenital syndrome	Adrenal 11- or 21-hydroxylase deficiency	Ovaries	Female	Male/Female
XY Testicular feminization	Lacks androgen receptors	Testes	None	Female
XY	Lacks ability to synthesize testosterone	Testes	Male with impaired development	Female/Male
XY	5 α-reductase deficiency	Testes	Male	Female/Male

chromosomal loci, the uncommon allele is recessive and the common allele is dominant in the female. Therefore, the uncommon phenotype usually is not expressed in the female. In order for a female to express an uncommon phenotype, such as colorblindness, she must inherit two recessive X-linked alleles, from her mother and her father. This is a rare combination. In contrast, the male receives only one X chromosome from his mother and a Y chromosome from his father. In the male, every X chromosome allele present is expressed whether the allele was dominant or recessive in the mother. Therefore, an uncommon recessive allele that may have been carried by the mother, but not expressed in her, is expressed in her male offspring. Thus, X-linked recessive traits are much more common in males than in females.

The Role of the Y Chromosome

Although this process is not completely understood, it appears that the Y chromosome contains genetic loci, such as the *SRY* gene that determines testicular development. A protein also is produced by a gene on the Y chromosome to produce the **histocompatibility-Y (H-Y) antigen,** which is antigenic, and induces immunological rejection in females. The H-Y antigen was once thought to be responsible for inducing testicular differentiation in the primitive gonad.

However, the *SRY* gene appears to be most important in regulating the development of the testis. Lack of H-Y antigen does, however, impair spermatogenesis and fertility.

Differentiation of the Internal Genitalia

During the third to the seventh week of embryonic development, two sexually undifferentiated gonads (one on each side) develop, which differentiate into ovaries in the female or testes in the male (Fig. 33–2).

The Embryonic Ductal Systems

Two distinct ductal systems develop in the embryo regardless of genetic sex (XX or XY). One ductal system, termed the **Mullerian duct system,** is the precursor of the internal genitalia of the female. The other duct system, called the **Wolffian duct system,** is the precursor of the internal genitalia of the male (see Fig. 33–2). In the genetically female embryo, the Mullerian duct system differentiates into uterine tubes, uterus, cervix, and vagina, and the Wolffian duct system regresses. In the genetically male embryo, the Wolffian duct system differentiates into the epididymis, vas deferens, seminal vesicles, and ejaculatory duct, and the Mullerian duct system regresses. Differentiation and development of one or the other of the duct systems is due to the presence or absence of hormones secreted by the developing gonads.

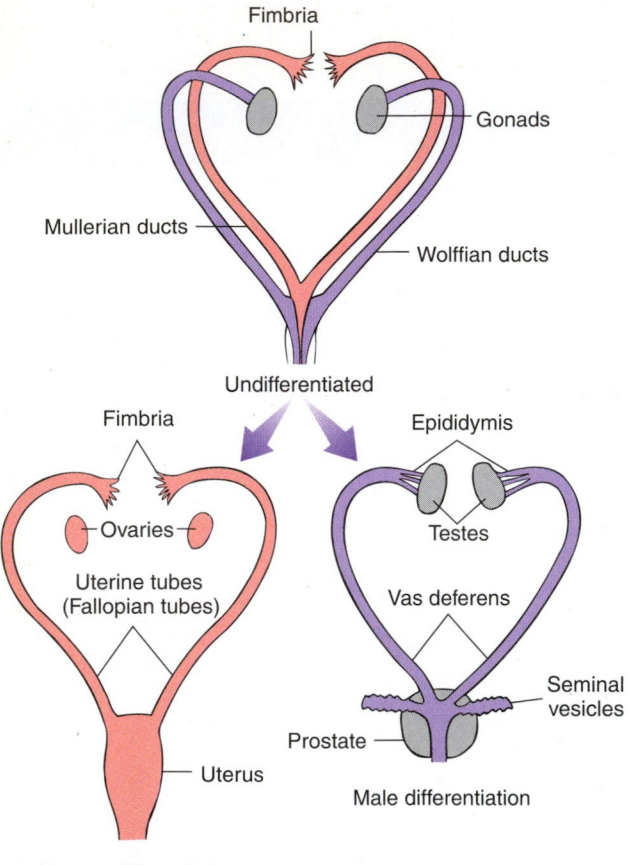

Figure 33–2

Embryonic differentiation of the male and female internal genitalia.

Differentiation of the Internal and External Genitalia Is Under Hormonal Control

Anti-Mullerian Hormone and Gonadal Differentiation

Differentiation of the testes occurs between the sixth and seventh week of development. As the testes develop in the male embryo, the Sertoli cells of the testes begin to secrete a substance called **anti-Mullerian hormone (AMH).** AMH is a glycoprotein that acts in a paracrine fashion on cells surrounding the gonads to induce regression, or involution, of the Mullerian duct system. That AMH acts locally is evidenced by the fact that if one of the two testes is removed at an early stage of development, Mullerian regression occurs only on the side of the body that has an intact testis.

Testosterone and Gonadal Differentiation

Although AMH is responsible for regression of the Mullerian ducts, differentiation of the male internal genitalia is under the control of **testosterone.** Leydig cells of the developing testes begin to secrete testosterone at approximately seven to nine weeks of gestation (Fig. 33–3). Testosterone, which diffuses from the testes, acts locally to stimulate differentiation of the Wolffian duct system during weeks 9 to 16 of embry-

onic development. If testosterone biosynthesis is deficient in the developing testes, or if the Wolffian ducts do not respond to testosterone, the internal genitalia of the male embryo do not develop, resulting in a condition called **testicular feminization.**

Differentiation of Internal Genitalia in the Female

Ovarian differentiation in the female occurs later than it does in the male. At approximately 12 weeks of gestation, the germ cells (oogonia) in the ovary begin to divide meiotically to form oocytes, marking the beginning of ovarian differentiation. By week 20 to 25 of gestation, the female gonads can be structurally characterized as ovaries. During the third month of gestation, the Mullerian duct system in the female embryo undergoes further development, while the Wolffian duct system degenerates, owing to the absence of testosterone stimulation (see Fig. 33–2). The presence of the ovaries is not necessary for Mullerian duct development in the female embryo; differentiation of the uterus and uterine tubes occurs even when no ovaries are present. It appears that the ovaries do not contribute hormonal factors that govern differentiation of the female genital tract.

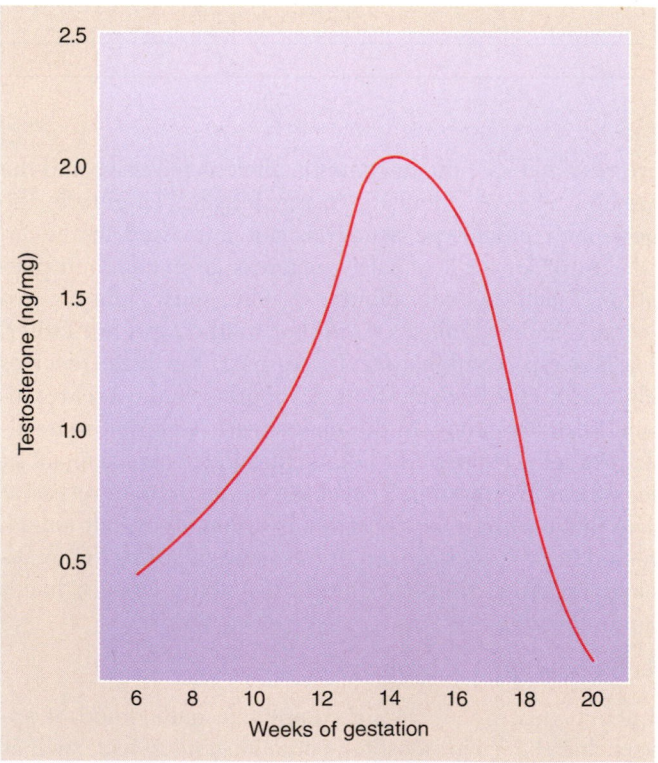

Figure 33–3

Testicular content of testosterone in the male fetus during gestation.   *(Adapted from Shearman.* Clinical Reproductive Endocrinology, *1985, p. 349, originally reproduced from Tapanainen, et al., 1981.)*

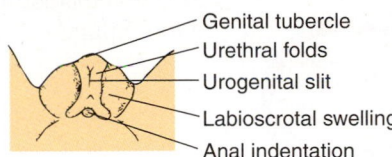

Undifferentiated

- Genital tubercle
- Urethral folds
- Urogenital slit
- Labioscrotal swelling
- Anal indentation

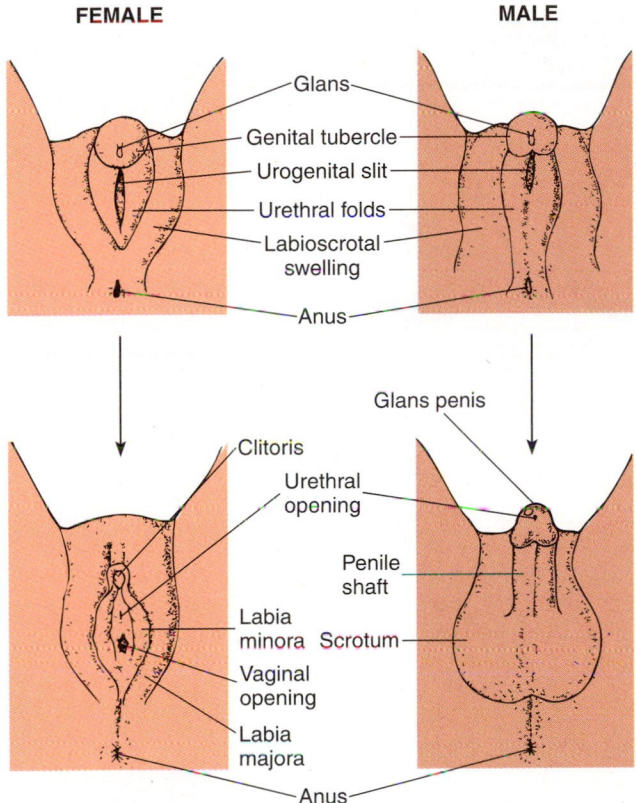

FEMALE   **MALE**

- Glans
- Genital tubercle
- Urogenital slit
- Urethral folds
- Labioscrotal swelling
- Anus

Glans penis

- Clitoris
- Urethral opening
- Penile shaft
- Labia minora
- Scrotum
- Vaginal opening
- Labia majora
- Anus

Figure 33–4

Embryonic differentiation of the male and female external genitalia.

Differentiation of External Genitalia in the Female

The external genitalia of both the male and female embryo are identical during the first six weeks of gestation (Fig. 33–4). The external genitalia consist of a urogenital slit surrounded by urethral folds, which are in turn surrounded by the labioscrotal swellings. A genital tubercle is located above the urogenital slit, and the anal indentation is located below. Sexual differentiation of these embryonic structures into the external genitalia of the male and female is summarized in Table 33–2.

Differentiation of the external genitalia in the female occurs because of the **absence of testosterone.** In the female embryo, the urethral folds and the labioscrotal swellings remain separate and differentiate into the labia minora and labia majora, respectively (see Fig. 33–4). The genital tubercle differentiates first into the glans and then into the clitoris. The urogenital slit differentiates into the vaginal opening and the urethra.

Differentiation of External Genitalia in the Male

Dihydrotestosterone (DHT) derived from testosterone by the developing testes stimulates differentiation of the external genitalia in the male embryo. The external genitalia are androgen-sensitive tissues that contain the enzyme **5α-reductase,** which converts testosterone to DHT. Deficiencies of testosterone secretion, 5α-reductase activity, or receptors for DHT in external genital tissues can result in impaired masculinization of the male external genitalia.

In response to stimulation from DHT, the urethral folds fuse and surround the urethra. Fusion of the urethral folds forms the shaft of the penis (see Fig. 33–4). DHT also stimulates fusion of the labioscrotal swellings on the midline, forming the scrotum and the skin that covers the ventral portion of the penis. It also induces differentiation of the prostate and inhibits the development of a vagina. The factors controlling sexual determination and differentiation are schematically illustrated in Figure 33–5.

TABLE 33–2		
Structural Origin of Male and Female External Genitalia		
Undifferentiated Embryonic Structure	**Differentiation in the Female**	**Differentiation in the Male**
Genital tubercle	Clitoris	Glans of penis
Urethral folds	Labia minora	Shaft of penis
Labioscrotal swellings	Labia majora	Scrotum
Urogenital slit	Urethra and vaginal opening	Urethra
Anal indentation	Anus	Anus

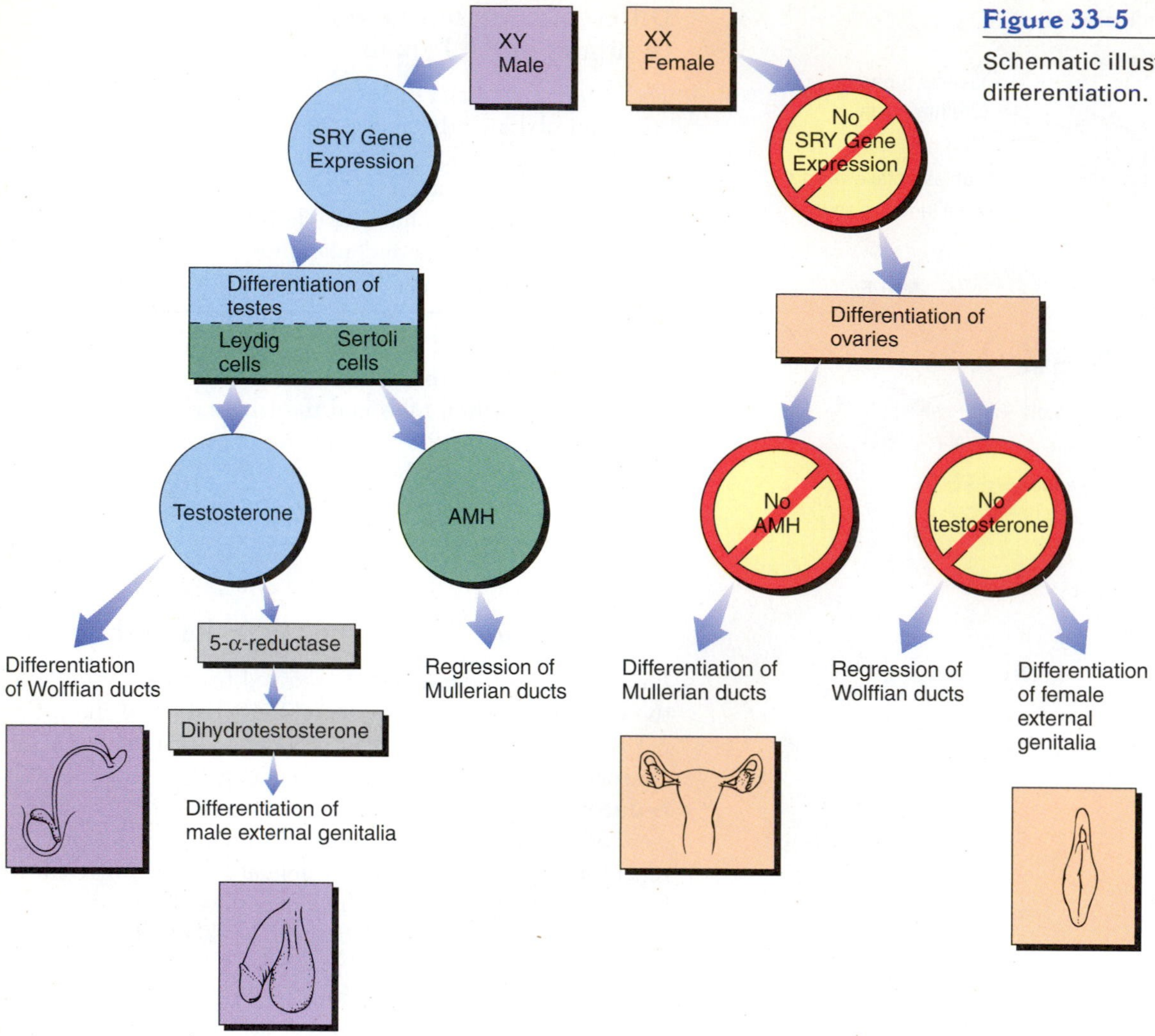

Figure 33–5

Schematic illustration of sexual differentiation.

Abnormalities of Sexual Differentiation Can Occur at Several Levels

> *What factors can contribute to abnormalities in sexual differentiation?*

Chromosomal Abnormalities in the Female and Male

In females, both of the X chromosomes are essential for normal differentiation of the ovaries, but only one X is necessary for differentiation of the Mullerian ducts and the female external genitalia. **Turner's syndrome** is a chromosomal abnormality in which only one X sex chromosome, designated as XO, exists in the female (see Table 33–1). The uterus, uterine tubes, and external genitalia develop normally but the female gonads do not fully develop, resulting in the appearance of an incompletely formed gonad, termed a **vestigial gonadal streak.**

The normal male embryo has 22 autosomal pairs and one pair of sex chromosomes, XY (Fig. 33–6). Some genetically male individuals receive an extra X chromosome

(XXY), which results in **Klinefelter's syndrome** (see Table 33–2). In these cases, the presence of the Y chromosome directs differentiation of the testes, but they are reduced in size. Testosterone secretion from the testes in XXY males stimulates the development of the Wolffian duct system and the development of the male external genitalia, while AMH secretion from the testes induces Mullerian regression. However, the presence of the extra X chromosome leads to testicular dysfunction, owing to the impaired development of the seminiferous tubules and a deficiency in spermatogenesis.

Enzyme Deficiencies in the Female

The presence of androgens (testosterone and dihydrotestosterone) during early embryonic development results in the differentiation of male external genitalia. Exposure of a genetically female embryo (XX) to androgens during embryonic development can result in **masculinization,** or **virilization,** of the female embryo's external genitalia. Some genetic females (XX) lack the enzymes **11-hydroxylase** or **21-hydroxylase** in their adrenal glands. These enzymes are necessary for the syn-

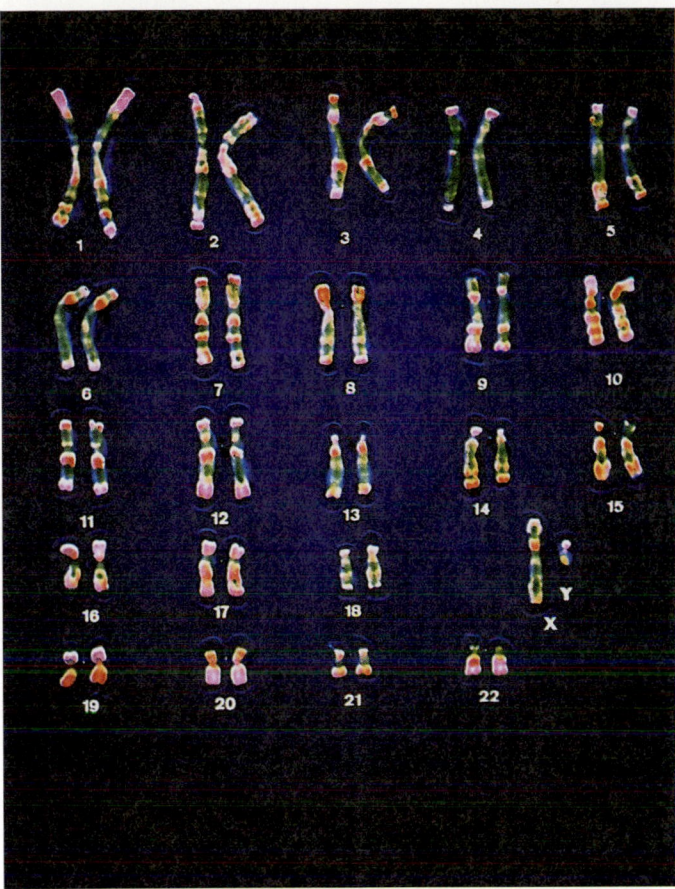

Figure 33–6

Photograph of normal human karyotype. *(© Leonard Lessin/Photo Researchers, Inc.)*

thesis of gluco- and mineralocorticoids. In the absence of these enzymes, the adrenal glands synthesize androgens from cholesterol. Hypersecretion of adrenal androgens results in the development of the **adrenogenital syndrome,** also called **congenital adrenal hyperplasia** (see Table 33–1). These individuals have normal ovaries, owing to a full chromosomal XX complement. The absence of a Y chromosome precludes the development of testes and ensures normal development of the Mullerian duct system (including the uterus and uterine tubes). However, hypersecretion of adrenal androgens can result in masculinization of the external genitalia that can range from mild hypertrophy of the clitoris to fusion of the urethral folds and the labioscrotal swellings which then resemble a penis and scrotum. Severe masculinization of the female external genitalia, a condition termed **female pseudohermaphroditism,** can lead to mistakes in identifying the gender of the infant.

Enzyme and Receptor Deficiencies in the Male

Genetic males (XY) who lack androgen receptors cannot respond to testosterone or DHT, which results in a condition known as **testicular feminization syndrome** (see Table 33–1). In these individuals, the presence of the Y chromosome results in development of the testes and regression of the Mullerian ducts. However, these individuals have no internal genitalia at all because AMH induces regression of the Mullerian ducts while the lack of androgen receptors precludes development of the Wolffian ducts.

A deficiency of the enzymes necessary for testosterone biosynthesis also results in varying degrees of Wolffian duct underdevelopment and reduced masculinization of external genitalia (see Table 33–1). In such cases, the presence of the Y chromosome ensures the development of the testes and regression of the Mullerian ducts. A deficiency of the enzyme 5α-reductase, necessary for converting testosterone to dihydrotestosterone, results in reduced masculinization of the external genitalia without interfering with the formation of normal testes (due to the presence of the Y chromosome). Mullerian regression occurs because of the presence of AMH, and the Wolffian duct system develops in response to testosterone stimulation (see Table 33–1).

Events that control sexual determination and differentiation are summarized in Figure 33–5 and Table 33–3.

TABLE 33–3
Summary of Events Controlling Sexual Determination and Differentiation
Female
1. Two X chromosomes are needed for normal differentiation of the ovaries.
2. Only one X chromosome is required for differentiation of the Mullerian ducts and the female external genitalia.
3. The absence of ovaries will not preclude development of the Mullerian ducts or the female external genitalia.
4. Exposure of the genetically female embryo to androgens during early development will result in varying degrees of masculinization of the external genitalia, depending on when during development the exposure occurs. Exposure to androgens will not alter development of the ovaries or the Mullerian ducts.
Male
1. The Y chromosome, which codes for or regulates the expression of the H-Y antigen, is required for differentiation of the testes.
2. Mullerian inhibiting substance, secreted by the Sertoli cells of the testes, induces regression of the Mullerian ducts.
3. Testosterone, secreted by Leydig cells of the testes, stimulates Wolffian duct development but has no effect on the Mullerian ducts.
4. Testosterone is converted to dihydrotestosterone (DHT) in tissues of the external genitalia, and DHT stimulates differentiation of the male external genitalia. An absence of DHT formation results in reduced masculinization of the external genitalia but does not affect normal development of the Wolffian ducts, which respond directly to testosterone.

HUMAN SEXUAL RESPONSE

 What happens during the human sexual response?

With the widespread use of contraceptives, scientists have been able to view sexual physiology as independent from reproduction. Nevertheless, sexual and reproductive physiologies are inextricably intertwined. Most of the reproductive organs in both sexes serve both sexual and reproductive functions. In addition, sexual arousal in the male is necessary for reproduction because ejaculation, which results in sperm transmission, does not occur in the absence of sexual arousal, and such arousal in the female is now believed to be important in determining her fertility.

The Physiology of Sexual Responses in Males and Females: The Four-Stage Model of Sexual Arousal

Sexual arousal can result from either **physical stimulation,** such as touching or kissing or **mental stimulation,** such as movies or books. Although the human sexual re-sponse can be significantly altered by thoughts, feelings, and learned values, the physiological responses to sexual arousal are the same whether arousal is due to physical or mental stimulation. Sexual arousal, which occurs throughout life from infancy into old age, involves activation of the nervous system, the endocrine system, and the sex organs.

Masters and Johnson developed a standard four-stage model of sexual arousal. These stages are defined as **excitement, plateau, orgasm,** and **resolution.** While distinctions among the four stages are not clear-cut and individual variations are many, these stages occur in both males and females and in both heterosexuals and homosexuals.

Excitement Phase in the Male

Sexual arousal during the excitement phase increases parasympathetic nerve activity, which causes the arteries in the penis to dilate, resulting in increased blood flow to the penis. Acetylcholine, vasoactive intestinal peptide (VIP), and nitric oxide have been identified as vasodilating neurotransmitters involved. As blood becomes concentrated in the tissues of the penis, a process called **vasocongestion,** the penis

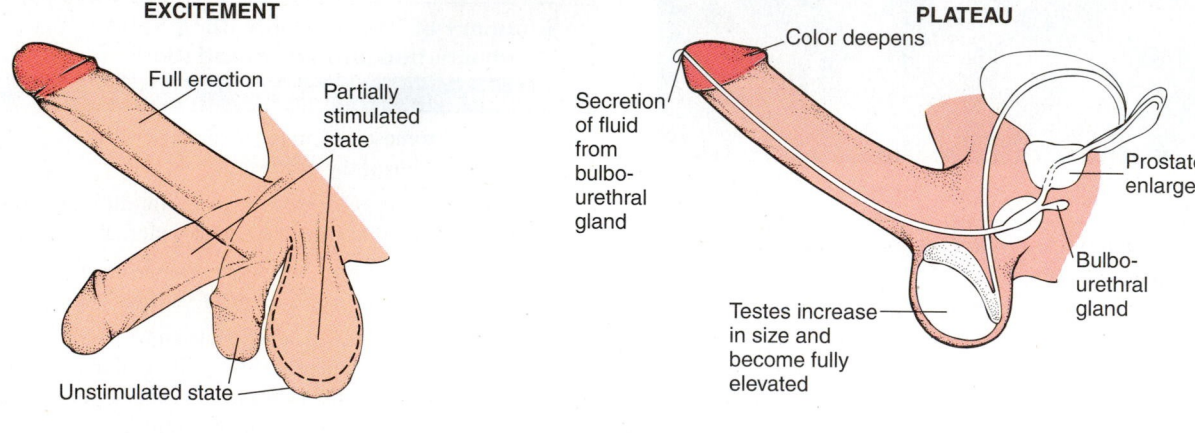

EXCITEMENT

Full erection

Partially stimulated state

Unstimulated state

PLATEAU

Color deepens

Secretion of fluid from bulbo-urethral gland

Prostate enlarges

Bulbo-urethral gland

Testes increase in size and become fully elevated

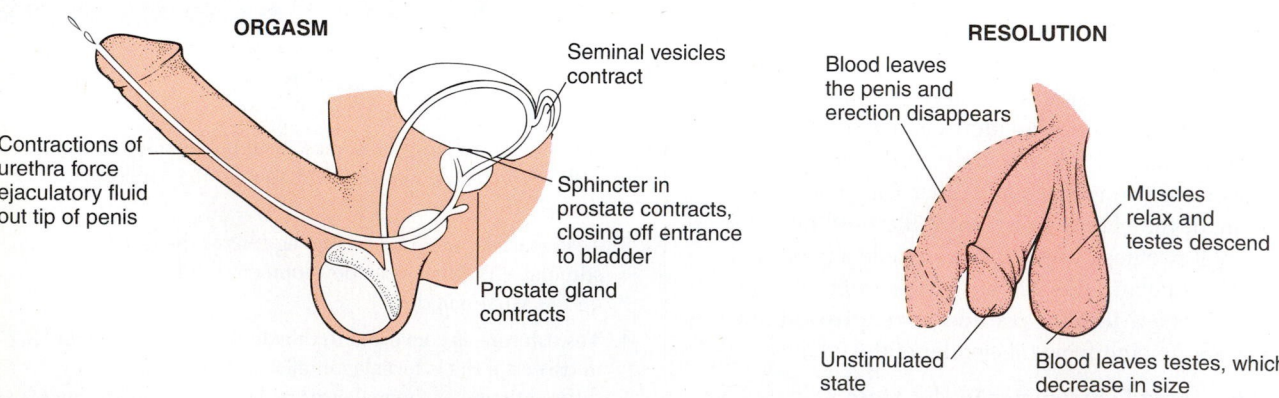

ORGASM

Seminal vesicles contract

Contractions of urethra force ejaculatory fluid out tip of penis

Sphincter in prostate contracts, closing off entrance to bladder

Prostate gland contracts

RESOLUTION

Blood leaves the penis and erection disappears

Muscles relax and testes descend

Unstimulated state

Blood leaves testes, which decrease in size

Figure 33–7

Changes in the external and internal genitalia of the male during the sexual response cycle.   *(Redrawn from Masters, Johnson, and Kolodny,* Human Sexuality. *Little, Brown & Co., 1985.)*

increases in diameter, length, and firmness, which defines **penile erection** (Fig. 33–7). Unlike lower mammals, male humans do not have a penile bone. Instead, penile erection depends on increased blood flow to the penis and increased intrapenile pressure. Penile vasoconstriction during sexual arousal in the male is not constant and can result in short-lived reductions in penile size and firmness even though sexual stimulation and neuromuscular tension have not diminished.

Sexual arousal also results in increased blood flow to the testes, causing an increase in testicular size (see Fig. 33–7). During the excitement phase, the muscles surrounding the vas deferens contract, lifting the testes closer to the body and smoothing the scrotal skin (see Fig. 33–7). An increase in general muscle tone or tension, termed **myotonia,** also accompanies sexual arousal. Erection of the nipples of the breast occurs in some (but not all) men.

Excitement Phase in the Female

One of the physiological responses to sexual arousal during the excitement phase in the female is **vaginal lubrication** (Fig. 33–8). Vaginal lubrication in the female occurs in response to vasocongestion. Sexual arousal results in increased blood flow to the vagina, vaginal congestion, and a release of

moisture from the vaginal lining, a process termed **transudation.** As vaginal secretion of moisture increases, some moisture may reach the vaginal opening and the surrounding labia. Vaginal secretion of moisture lubricates the vagina and facilitates insertion of the penis without discomfort. There are great individual variations in the amount of moisture secreted by the vagina during sexual arousal and therefore the amount of moisture present cannot be used as an index of sexual arousal.

During the excitement phase, the vagina expands, and the cervix and uterus are lifted upward away from the vagina (see Fig. 33–8). Increased blood flow to the external genitalia results in increased size of the **clitoris**, labia minora, and labia majora (Fig. 33–9a). Muscle contractions in the breast tissue often result in nipple erection, and increased blood flow to the breast causes breast enlargement (see Fig. 33–9b). There is an increase in general muscle tension, or myotonia, in the female during the excitement phase.

Plateau Phase in the Male

The plateau phase of sexual arousal precedes orgasm. Time spent in the plateau phase varies widely within and among individuals. Men who ejaculate quickly spend little time in this

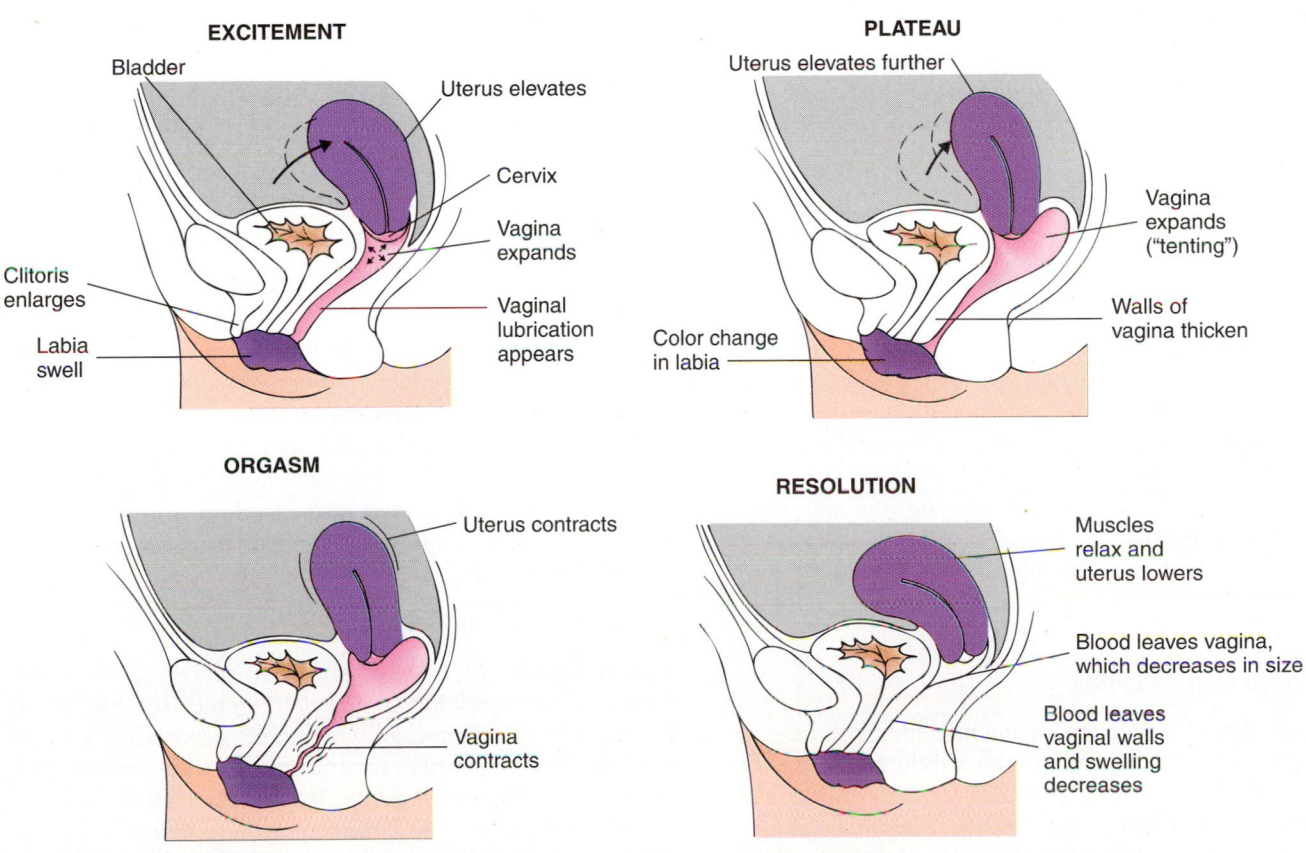

Figure 33–8

Changes in the external and internal genitalia of the female during the sexual response cycle. *(Redrawn from Masters, Johnson, and Kolodny.* Human Sexuality. *Little, Brown & Co., 1985.)*

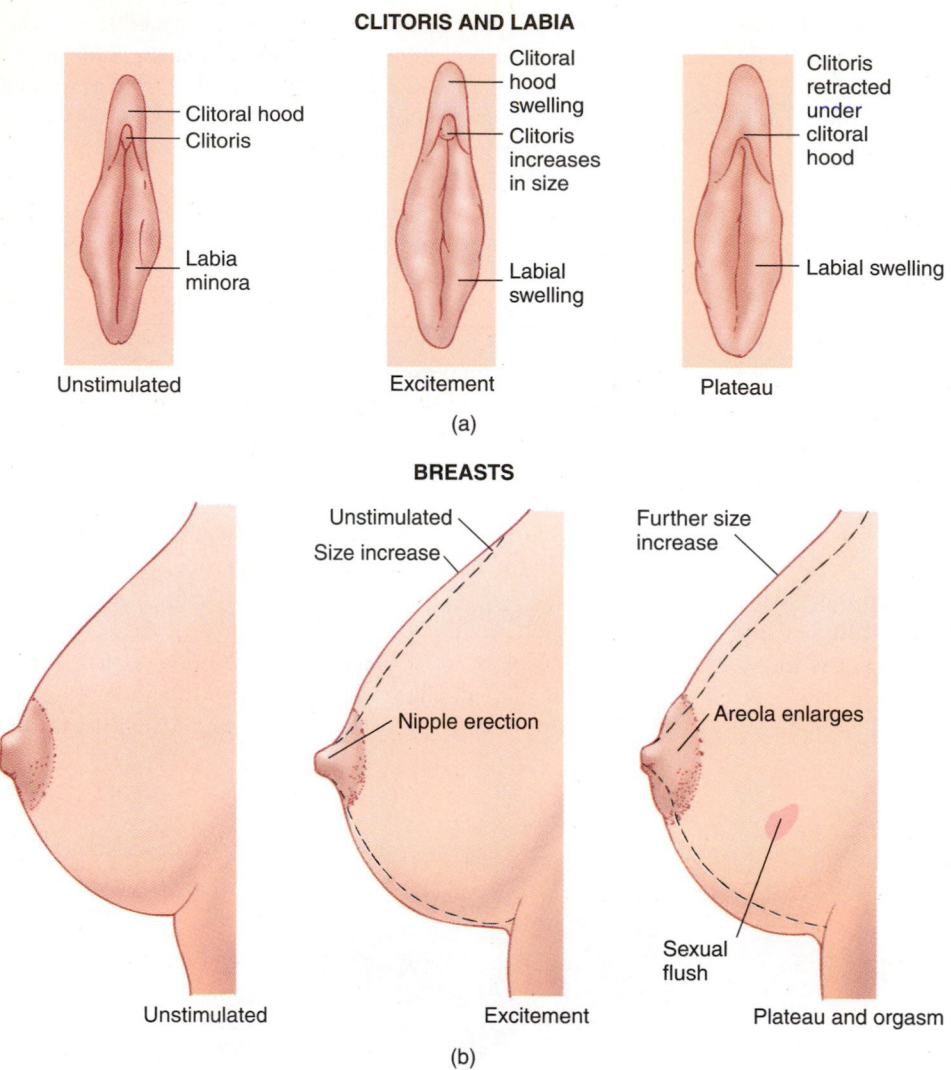

CLITORIS AND LABIA

Unstimulated
- Clitoral hood
- Clitoris
- Labia minora

Excitement
- Clitoral hood swelling
- Clitoris increases in size
- Labial swelling

Plateau
- Clitoris retracted under clitoral hood
- Labial swelling

(a)

BREASTS

Unstimulated

Excitement
- Unstimulated
- Size increase
- Nipple erection

Plateau and orgasm
- Further size increase
- Areola enlarges
- Sexual flush

(b)

Figure 33–9

Changes in the *(a)* clitoris, labia, and *(b)* breasts during the female sexual response cycle. *(Redrawn from Masters, Johnson, and Kolodny. Human Sexuality. Little, Brown & Co., 1985.)*

phase. During the plateau phase, blood pooling in the head of the penis causes this area to enlarge and become darker in color (see Fig. 33–7). The testes continue to increase in size because of increased blood flow, and muscle contractions in the scrotum lift the testes up and back toward the anus (see Fig. 33–7). Pre-ejaculatory fluid from the bulbourethral glands, which may contain viable sperm, often is secreted at the tip of the penis during this phase (see Fig. 33–7), and heart rate, blood pressure, respiration rate, and muscle tension all increase.

Plateau Phase in the Female

In the female, blood flow to the vagina increases, resulting in swelling of the walls of the vagina, which reduces the size of the vaginal opening (see Fig. 33–8). This increases vaginal contact with the penis during intercourse. The uterus is elevated away from the vagina in a process termed *tenting* (see Fig. 33–8). The clitoris, which is highly sensitive to touch during this phase, is retracted against the pubic bone and becomes partially covered by the skin of the clitoral hood, which reduces direct physical contact of the clitoris (see Fig.

33–9*a*). Increased blood flow to the labia produces a deepening in the color of the labia minora and labia majora. Blood flow to the breasts results in increased size of the areolae and reddening or darkening of the skin color in areas on the breasts and other areas of the body, referred to as the *sexual flush* (see Fig. 33–9*b*). Heart rate, blood pressure, respiration rate, and muscle tension all increase during the plateau phase in the female.

Orgasmic Phase in the Male

Ejaculation in the male involves two processes: the **emission** and the **expulsion** of ejaculatory fluid (see Fig. 33–7). During emission, increased sympathetic nerve activity stimulates rhythmic muscular contractions of the vas deferens, seminal vesicles, and prostate, forcing ejaculatory fluid into the urethra. Once these contractions have begun, they cannot be stopped until ejaculation is complete. Ejaculation can occur in the absence of penile erection. During the expulsion phase, rhythmic muscular contractions of the urethra force ejaculatory fluid out the tip of the penis. These two phases of ejaculation usually last only a few seconds. A sphincter in the prostate contracts during ejaculation, sealing off the entrance to the bladder and preventing urine from mixing with ejaculatory fluid (see Fig. 33–7).

Orgasmic Phase in the Female

Orgasm in the female is characterized by rhythmic muscular contractions of the uterus and vagina that may last for as long as 15 seconds (see Fig. 33–8). As in the male, muscles in various parts of the body involuntarily contract during female orgasm, resulting in muscular rigidity and spasms.

It appears that most women, while capable of achieving orgasm in response to vaginal stimulation, experience more intense orgasms when clitoral and vaginal stimulation are combined. Although most females do not ejaculate fluid from the urethra during orgasm, ejaculation-like responses have been reported by some females.

Recent evidence obtained on film with fiber optics in the vagina during intercourse has revealed a potential role for female orgasm in fertility. It has been observed that uterine contractions during female orgasm actually cause repetitive "dipping" of the cervix into the vaginal semen pool, thereby enhancing contact of the entrance of the uterus with sperm in the semen.

Resolution Phase in the Male

Following ejaculation, most males enter a refractory period (lasting for a few minutes to many hours) during which no further orgasm or ejaculation is possible. The refractory period is prolonged by fatigue, time elapsed since the last orgasm, and age. During the refractory period, increased sympathetic nerve activity constricts the arteries in the penis and blood leaves the penis through the veins, resulting in a loss of penile erection (see Fig. 33–7). Decreased blood flow to the testes also results in a decrease in testicular size. As muscles relax, the testes are lowered away from the body (see Fig. 33–7). Heart rate, blood pressure, respiration, and muscle tension all return to normal during the resolution phase. When intense sexual arousal is not followed by orgasm, the length of the resolution phase is prolonged and can be accompanied by aching sensations in the testes due to continued vasoconstriction.

Resolution Phase in the Female

Many females are **multiorgasmic** (capable of experiencing more than one orgasm before entering the resolution phase) if intense sexual arousal is maintained. During the resolution phase, muscular relaxation results in a lowering of the uterus and cervix toward the vagina (see Fig. 33–8). Blood flow away from the breasts and genitals results in a reduction in the size of the breasts, clitoris, labia, and vagina. Heart rate, blood pressure, respiration, muscle tension, and skin flushing all return to normal during the resolution phase. Breast and genital tissues are extremely sensitive to touch following orgasm in both sexes, and continued physical stimulation of these tissues during the resolution phase can be irritating or painful.

Sexual Response During Infancy

Penile erection, in response to physical stimulation, occurs even before birth. Penile erections have been detected by ultrasound in the male fetus during the last few months of gestation. At birth, many male infants exhibit penile erections, while female infants manifest vaginal lubrication and clitoral enlargement. Following birth, certain forms of physical contact, including breastfeeding, can result in penile erection in male infants and vaginal lubrication and enlargement of the clitoris in female infants. Although both males and females are capable of achieving orgasm from birth, the ability of the male to ejaculate does not occur until the internal and external genitalia have matured at puberty.

 How does the aging process affect the sexual response?

Although physiological changes in the sexual response occur in both males and females during aging, the ability to become sexually aroused does not normally disappear.

Aging and the Female Sexual Response

In the female, the breasts, clitoris, and vagina remain sensitive to stimulation in old age. However, after menopause, low levels of circulating estrogen results in diminished blood flow to the vagina during sexual arousal, which in turn leads to decreased vaginal lubrication, a condition termed **dyspareunia.** Use of an artificial lubricant can circumvent the pain that may result from intercourse in the absence of sufficient vaginal lubrication. Some evidence suggests that regular sexual activity following menopause slows the decline in vaginal lubrication that occurs during aging. A decrease in the flexibility of the vaginal walls in aging women leads to reduced vaginal expansion during sexual arousal. In addition, enlargement of the breast, development of the sexual flush, and increased muscle tension during arousal are significantly reduced in older women. Changes in the female response during aging are summarized in Table 33–4.

Aging and the Male Sexual Response

Unlike the rapid decline in circulating estrogen levels in the female following menopause, circulating testosterone levels decline gradually after age 55 to 60, without a dramatic drop. Healthy males can maintain the capacity for penile erection throughout their entire life spans, and sperm production, although slowed after age 40, is still present until age 80 or 90.

However, physiological changes in the male sexual response do occur with aging. The sensitivity of penile sensory nerves to tactile stimulation decreases with age, and hence more prolonged and direct penile stimulation may be needed to induce erection. Blood flow to the penis during sexual arousal also may be reduced with age, and so the penis may be less firm when fully erect. Testicular elevation during sexual arousal is slower and less complete in the older male. The amount of fluid ejaculated at orgasm is often reduced, the refractory period following orgasm is prolonged, and muscle

TABLE 33–4

Changes in the Sexual Response During Aging

Female

1. A decrease in circulating estrogen following menopause results in decreased blood flow to the vagina, which in turn results in diminished production of vaginal lubrication. Decreased vaginal lubrication can result in painful intercourse, or dyspareunia.
2. The vaginal walls lose elasticity, and sexual arousal produces less vaginal expansion.
3. Many responses to sexual arousal—including enlargement of the breast, reddening of skin color (sexual flush), and muscle tension—are significantly reduced.

Male

1. Penile erection occurs more slowly and may require more prolonged stimulation.
2. The penis is less firm when fully erect.
3. Time spent in the refractory period following orgasm, during which no erection or ejaculation is possible, is prolonged.
4. The volume of ejaculatory fluid expelled at orgasm is decreased.
5. Testicular size and elevation during sexual arousal are reduced.
6. Many responses to sexual arousal—including reddening of skin color (sexual flush) and muscle tension—are reduced.

tension during sexual arousal also is reduced. A summary of the changes in the male sexual response during aging is contained in Table 33–4.

SEXUAL DYSFUNCTION

 What types of pathophysiology can affect sexual function?

The term **sexual dysfunction** refers to any condition in which the normal physical responses to sexual stimulation are impaired. Sexual dysfunction can result from hormonal or neural disorders, certain diseases, or drugs. It can be a source of anxiety, frustration, and distress, disrupting personal relationships and family life.

Estrogen Deficiencies in the Female

Estrogen stimulates proliferation of the vascular bed beneath the vaginal epithelium and increases blood flow to the vagina, which results in secretion of moisture from the vaginal walls and is therefore important in maintaining vaginal lubrication.

Inadequate estrogen stimulation can result in dryness and thinning of the vaginal mucosal membrane and decreased flexibility and muscle tone of the vagina, called **atrophic vaginitis.** Following menopause, when circulating estrogen levels are decreased, vaginal distensibility and vascularization are reduced, and blood flow to the vagina can be decreased by as much as 80%. Estrogen replacement therapy following menopause can increase blood flow to the vagina by fourfold.

Premenopausal women with very low levels of circulating estrogen (for instance, following ovarian disease or surgical removal of the ovaries) also can experience atrophic vaginitis with accompanying loss of vaginal lubrication and distensibility. Inadequate vaginal lubrication can produce discomfort or pain during sexual intercourse (dyspareunia). Women with very low levels of circulating estrogen also can experience symptoms of peripheral neural dysfunction, which can include slowed response to tactile stimulation or loss of clitoral sensation. Estrogen replacement therapy can reduce the severity of these symptoms.

Testosterone Deficiencies in the Male

Some evidence suggests that testosterone is necessary for normal sexual activity in the male. Testosterone facilitates ejaculation or sexual desire by unknown mechanisms. It has been proposed that androgens are involved in sensory perception and the capacity for development of muscle tension within the pelvis. Although males with low circulating levels of testosterone can attain penile erection, the capacity to ejaculate during orgasm may be delayed, diminished, or absent. In addition, low levels of testosterone often are associated with reduced sexual desire or sexual libido. Androgen replacement therapy often increases both **ejaculatory capacity** and **sexual desire.**

Neural Disorders and Sexual Dysfunction

The excitement phase of sexual arousal in both males and females involves increased afferent neural input from sensory nerves in the genitals and, through the brain and spinal cord, efferent neural induction of vascular changes in the internal and external genitalia. Orgasm involves increased efferent motor activity, which induces muscular contractions in the internal and external genitalia. Therefore, both mechanical and disease-related injuries to the spinal cord or brain can result in impairment of sexual arousal and orgasm in both males and females. Interruption of afferent sensory pathways can interfere with sexual arousal. Interruption of efferent pathways may disrupt increased blood flow to the genitals during sexual arousal and hence may reduce penile erection in the male and vaginal swelling and lubrication in the female, as well as muscle contractions during orgasm in both sexes.

Multiple sclerosis, a disease characterized by patches of demyelinization in the spinal cord and brain, can be accompanied by a reduction or elimination of sexual arousal and orgasmic capacity in both males and females. Severe **diabetes mellitus** can be accompanied by dysfunction of the sensory fibers that innervate the penis and clitoris, resulting in a reduction or absence of the genital sensations necessary for sexual arousal and orgasm. Diabetic neural dysfunction also can result in decreased blood flow to the genitals, which can result in insufficient penile rigidity to achieve vaginal penetration.

Drugs and Sexual Dysfunction

Although small amounts of alcohol are often used to reduce sexual inhibitions, large amounts have a depressant effect on the central nervous system (CNS) and can interfere with sexual arousal. Chronic and excessive consumption of alcohol, as seen in alcoholics, can lead to damage of the somatic and autonomic nerves in the periphery, spinal cord, or brain. This damage can result in a reduction or absence of sexual arousal and orgasmic capacity.

Sedatives (e.g., barbiturates, chloral hydrate), narcotics (e.g., codeine, morphine), and tranquilizers (e.g., diazepam, meprobamate), when taken in high doses, can retard or inhibit sexual arousal and orgasm in both males and females because of a depression of nerve conductance in the CNS. Some drugs used to treat hypertension can alter neural control of blood flow to the internal and external genitalia in both males and females, resulting in decreased vaginal swelling and lubrication in the female and difficulty in achieving erection in the male. Some antihistamines that contain ephedrine decrease blood flow to the vagina and vaginal lubrication, resulting in dyspareunia.

Anorgasmia, Infection, and Vaginismus

Difficulty in reaching orgasm, which is referred to as **anorgasmia**, represents the largest category of sexual dysfunction in the female. However, organic disorders that can interfere with orgasm, such as hormone deficiencies, alcoholism, diabetes, or neurological disorders, account for fewer than 5% of the cases of anorgasmia. Anorgasmia appears to be largely due to psychological factors, which may include sexual inhibitions acquired during early training, unrealistic performance expectations for oneself or one's partner, difficulty with sexual communication, and negative self-image. For further information on these issues, which are beyond the scope of this chapter, consult the volumes about human sexuality listed at the end of this chapter.

Vaginal infections, allergic reactions, or chemical irritants can produce a condition termed **nonatrophic vaginitis.** This condition, which is quite common and usually of short duration, can result in vaginal dryness or tenderness.

Endometriosis (see section about infertility) and **pelvic inflammatory disease** both can cause pain during sexual arousal and intercourse as well as infertility. Vasocongestion and pelvic muscle contractions during sexual arousal place increased pressure on diseased tissues. Further pressure due to intercourse can result in deep pelvic pain. Pain associated with sexual arousal and intercourse can result in decreased vaginal lubrication, loss of orgasmic capacity, loss of sexual desire, or sexual aversion.

Vaginismus is defined as involuntary contractions and spasms of the vaginal (or pubococcygeal) muscles, which make vaginal penetration very difficult or impossible (Fig. 33–10). Vaginismus may stem from emotional problems or traumatic or painful sexual experiences; or it may arise as a protective reflex to avoid pain in the case of pelvic disease. Prolonged dyspareunia, or painful intercourse, can sometimes also lead to vaginismus. Note that dyspareunia and/or vaginismus in the female can cause sexual problems in male partners when they feel responsible for unsuccessful attempts at intercourse. Feelings of inadequacy and guilt can result in erectile difficulty and loss of sexual desire.

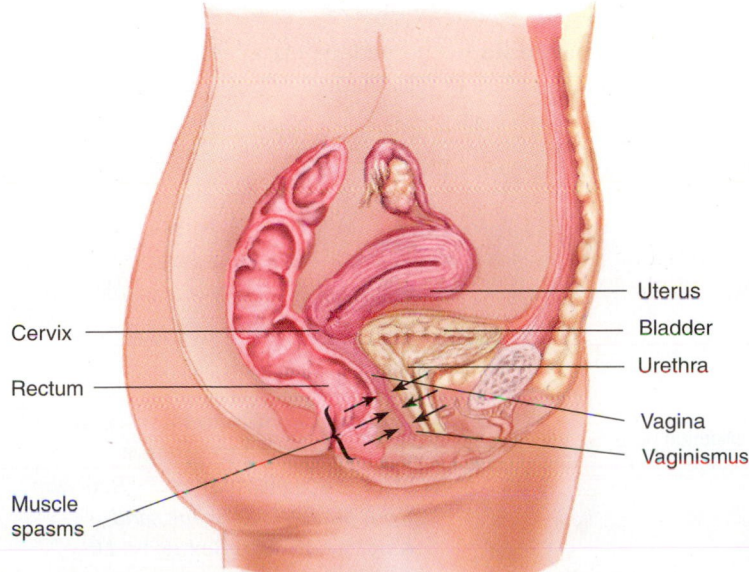

Cervix
Rectum
Muscle spasms

Uterus
Bladder
Urethra
Vagina
Vaginismus

Figure 33–10

The involuntary muscle spasms of vaginismus. *(Redrawn from Masters, Johnson, and Kolodny.* Human Sexuality. *Little, Brown & Co., 1985.)*

APPLICATIONS OF PHYSIOLOGY

The Pharmacological Treatment of Impotence

Impotence, or male penile erectile dysfunction (ED), can be caused by a variety of physical or emotional problems in the male. It has been known for some time that the penis is innervated by a class of autonomic nerves responsible for erection that use neither norepinephrine nor acetylcholine as their neurotransmitter. These nerves are known as nonadrenergic, noncholinergic **(NANC) nerves**. A few years ago, it was discovered that NANC nerves use the potent vasodilating chemical nitric oxide as their neurotransmitter. It is this nitric oxide that is primarily responsible for penile arteriole dilation, increased penile blood flow, and the resultant penile vasoconstriction that causes erection.

The discovery that nitric oxide is involved in penile erection has led to investigations into ways this system might be manipulated to treat male impotence. The vasodilating activity of nitric oxide is due to its ability to bind to and activate the smooth muscle enzyme guanylate cyclase, which converts GTP to cyclic GMP (cGMP). cGMP is a potent intracellular second messenger that mediates changes in smooth muscle, primarily intracellular calcium reduction, that cause the muscle to relax. It is now known that cGMP levels can be altered, not only by alterations in nitric oxide production but also by altering the activity of another smooth muscle enzyme, phosphodiesterase, which causes the breakdown of cGMP into inactive products. This observation has led to the development of the phosphodiesterase inhibitor sildenafil sulfate **(Viagra)**, which is now widely used for the successful treatment of many forms of ED.

Erectile Dysfunction, Premature Ejaculation, and Dyspareunia

Erectile dysfunction or **impotence,** is defined as the inability to achieve penile erection sufficient for vaginal penetration. ED is classified as either primary or secondary. **Primary ED,** which is very rare, defines the condition in which a male has never been able to achieve erection sufficient for vaginal penetration. **Secondary ED** is defined as a period of inability to achieve or maintain erection. Secondary ED can occur in response to anxiety, stress, illness, fatigue, lack of privacy, alcohol consumption, or change of sexual partner. Episodes of secondary ED are usually short-lived, and isolated episodes occur in most men.

Premature ejaculation can be defined as consistent, unintentional ejaculation prior to vaginal penetration.

When defined in this way, very few men experience premature ejaculation. More typically, loss of ejaculatory control results in ejaculation soon after vaginal penetration. The probability of premature ejaculation is higher in young men than in older men and increases with time elapsed since last orgasm, novelty of the sexual partner, and prolonged stimulation prior to vaginal intromission. Manual application of pressure to the penis for a few seconds, from front to back and just below the coronal ridge or at the base of the penis, can reduce the urge to ejaculate and help to control premature ejaculation (Fig. 33–11). Application of pressure to the penis also may cause a partial loss of erection that is only temporary.

Dyspareunia in the male is classified as painful erection, intromission, or ejaculation. Painful erection and/or

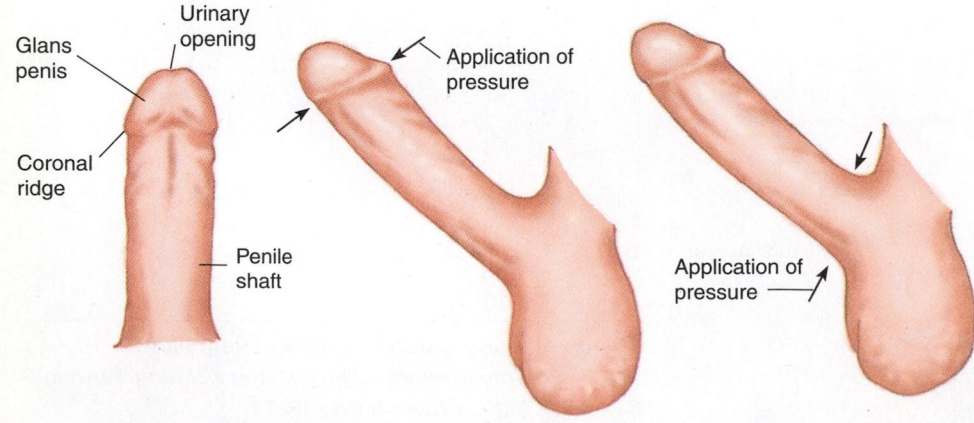

Glans penis

Urinary opening

Coronal ridge

Penile shaft

Application of pressure

Application of pressure

Application of pressure

Figure 33–11

Application of manual pressure to the penis, used to prevent premature ejaculation. *(Redrawn from Masters, Johnson, and Kolodny.* Human Sexuality. *Little, Brown & Co., 1985.)*

TABLE 33–5

Origins of Sexual Dysfunction

Male

1. Neurological disorders resulting from physical or disease-related injury (multiple sclerosis, diabetes mellitus) to the brain, spinal cord, or peripheral nerves.
2. Testosterone deficiency characterized by decreased penile erection, capacity to ejaculate, or sexual desire.
3. Excessive use of drugs including alcohol, sedatives, narcotics, tranquilizers, antihistamines, and antihypertensive agents.
4. Primary or secondary erectile dysfunction resulting from physiological or psychological causes.
5. Premature ejaculation.
6. Dyspareunia due to physical, chemical, or infectious damage to the penis or urethra.

Female

1. Neurological disorders resulting from physical or disease-related injury (multiple sclerosis, diabetes mellitus) to the brain, spinal cord, or peripheral nerves.
2. Chronic estrogen deficiency that results in atrophic vaginitis characterized by vaginal dryness, thinning, and decreased elasticity.
3. Acute estrogen deficiency that results in nonatrophic vaginitis characterized by vaginal dryness.
4. Excessive use of drugs including alcohol, sedatives, narcotics, tranquilizers, antihistamines, and antihypertensive agents.
5. Dyspareunia due to physical, chemical, or infectious damage to the vagina, clitoris, or labia.
6. Endometriosis and pelvic inflammatory disease.
7. Involuntary contractions of the vaginal muscles (vaginismus).

pain on intromission can result from many factors, including inflammation of the foreskin of the penis, termed **balanitis,** reduced flexibility of the skin covering the penis, and penile scarring. These conditions can result from physical trauma to, or infection of, penile tissue. Painful ejaculation can result from physical, infectious, or chemical irritation of the urethra, termed **urethritis.** Sources of sexual dysfunction in the male and female are summarized in Table 33–5.

Infertility

Fertility in the male depends on several factors: adequate secretion of the gonadotropins that act to stimulate spermatogenesis, testes capable of responding to gonadotropins by producing sperm, glandular production of seminal fluid, an intact ductal system for sperm delivery, and an intact nervous system for the control of penile erection and ejaculation. A disruption of normal function at any of these levels can result in infertility.

In addition, three aspects of the sperm themselves are important for sperm penetration and fertilization of the ovum: sperm number, sperm morphology (shape and structure), and sperm motility. Although it is always difficult to establish normal ranges for physiological processes, it is generally accepted that 15 to 250 million sperm per milliliter of semen represents a normal sperm count. A further definition of sperm normality requires that at least 60% of the sperm have a normal appearance and that at least 60% are active or motile.

A low sperm count and/or decreased sperm motility can result from gonadotropin deficiencies or from damage to the testes. Testicular damage can result from infections; inflammation; impaired circulation; and certain diseases, such as mumps. In addition, a variety of environmental factors, such as fever, prolonged heat exposure, chemical agents, and irradiation can result in testicular damage and impaired spermatogenesis. The extent of testicular damage produced by these agents is determined by the intensity and duration of testicular exposure. Abnormal sperm shape or structure, which includes a nonoval head section, a broken midpiece, or a bent or coiled tail, can result in infertility by preventing sperm migration through the cervical mucus or penetration of the ovum. Infection and inflammation of the epididymis, vas deferens, or ejaculatory ducts can reduce or prevent sperm transport, which also can result in infertility.

Infertility in the female can be due to a number of factors. One of the most common causes is failure to ovulate. Ovulatory failure can result from inadequate secretion of gonadotropins, failure of the ovary to respond to gonadotropins, variations in the anatomical structure of the ovaries, and ovarian damage resulting from infection or inflammation. Other factors that can cause female infertility include overproduction of thick cervical mucus, which can block sperm entry into the uterus, and infection or disease of the uterus or uterine tubes, with accompanying scarring that can block ovum transport to, or implantation in, the uterus. For instance, endometriosis, a disease characterized by the presence of endometrial-like tissue in locations other than the uterus, such as the uterine tubes or ovaries, can produce such scarring. Tissue formation, which may appear as cysts or nodules, can reduce the diameter of the uterine tubes and block ovum transport to the uterus (Fig. 33–12).

Immunological factors also may contribute to infertility. Antisperm antibodies have been found in both males and females. Antibodies, produced by the male, coat the surface of the sperm and reduce its ability to penetrate the cervical mucus. Antibodies produced by the female can induce **agglutination,** or aggregation of sperm into clumps, which also reduces their ability to penetrate the cervical mucus.

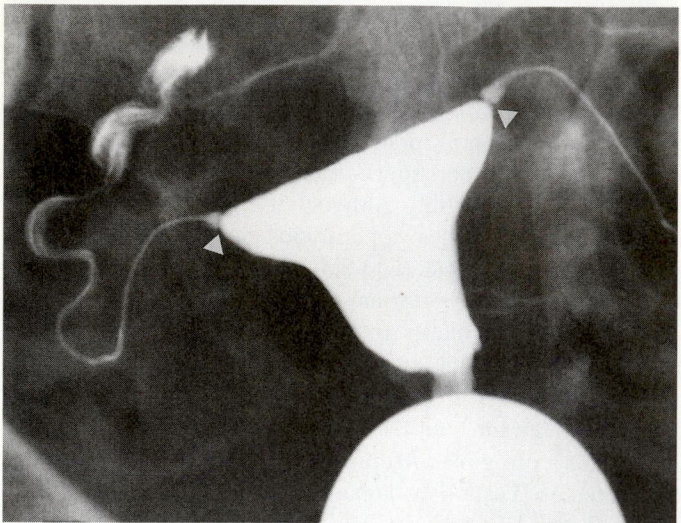
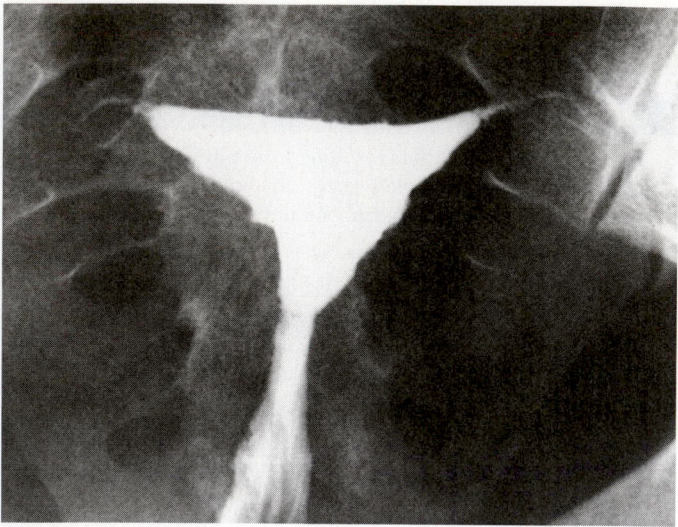

Figure 33–12

Radiographic image of the uterus and *(a)* normal bilateral uterine tubes and *(b)* bilateral uterine tube blockage.   (© *Ott and Fayez.* Hysterosalpingography: A Text and Atlas. *Urban & Schwarzenberg, 1991.)*

CONTRACEPTIVE METHODS

The ability to block conception (i.e., contraception) has resulted in increased independence of human sexuality from reproduction. Currently, there are four basic categories of contraception: **barrier methods, chemical methods, surgical methods,** and **abstinence.** All methods of contraception work by disrupting one of the three major steps in the reproductive process: ovulation, sperm and ovum transport, or implantation of the fertilized ovum in the uterus. The sites of actions of various methods of contraception are illustrated in Figure 33–13.

Barrier Methods

All barrier methods of contraception prevent sperm transport. Barrier methods include condoms, diaphragms, and cervical caps. Condoms are thin sheaths, usually made of latex, that fit over the penis and prevent sperm from entering the vagina. Use of condoms has increased in recent years because they not only prevent sperm transport but also provide some degree of protection against sexually transmitted diseases (STDs). Condoms are used once and then must be replaced prior to each sexual encounter. Diaphragms and cervical caps are rubber or polyurethane devices that cover

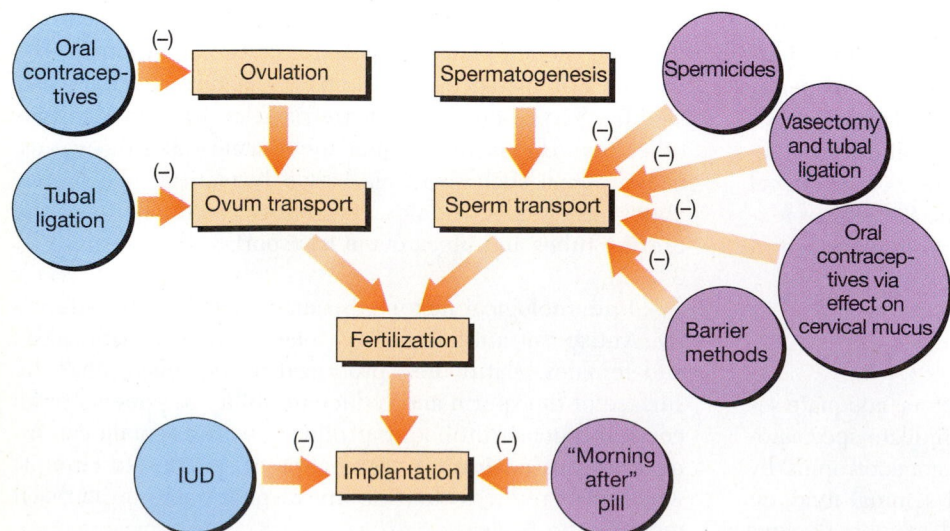

Figure 33–13

Sites of action of various methods of contraception. Minus sign indicates inhibition.

the entrance to the cervix and prevent sperm from entering the uterus. Diaphragms can be removed and reinserted. Cervical caps, which work similarly to the diaphragm, can be worn for longer periods prior to removal; they are held in place by suction.

The effectiveness of barrier methods of contraception varies, and they often are combined with chemical methods to increase their effectiveness. For instance, **spermicides**, contained in cream or jelly, are applied to the inside of the diaphragm to kill sperm that may pass the rim of the diaphragm.

Barrier methods of contraception have minimal side effects, which include the possibility of an allergic reaction to the spermicide or to the material used in construction, and a chance of introducing infection into the vagina.

Chemical Methods of Contraception: Birth Control Pills and Synthetic Variations of Gonadal Hormones

The most popular method of contraception, the **birth control pill,** is taken orally and acts to prevent pregnancy by blocking ovulation (see Fig. 33–13). There are many types of birth control pills, but they all contain synthetic variations of the gonadal hormones estrogen and/or progesterone, and they act to inhibit the release of follicle-stimulating hormone (FSH) and luteinizing hormone (LH) from the pituitary. When FSH and LH release are suppressed by oral contraceptives (Fig. 33–14), follicular maturation does not occur, and ovulation is inhibited. Continuous use of oral contraceptives inhibits follicular maturation, ovulation, and the onset of menstruation. Discontinuation of oral contraceptives for one week each month results in a drop in circulating levels of estrogen and progesterone, which stimulates the onset of menstruation. The contraceptive efficacy of birth control pills does not require monthly discontinuation of use.

Progesterone, contained in oral contraceptives, also interferes with conception by increasing the viscosity of cervical mucus, which blocks the entry of sperm into the uterus and by inhibiting estrogen-induced preparation of the endometrial lining of the uterus, which is required for implantation.

Birth control pills can produce a variety of side effects, including nausea, constipation, elevated blood pressure, skin rashes, and salt retention with accompanying weight gain. However, the most severe side effect occurs in women who smoke and use the pill. These women develop a great risk for blood clot formation and heart attack. The pill does not induce such risks in women who do not smoke.

Another type of oral contraceptive, termed the **morning-after pill,** contains a very high level of **estrogen,** which alters the endometrial lining of the uterus in such a way to prevent implantation of the fertilized ovum. However, the high levels of estrogen contained in the morning-after pill often induce severe nausea and vomiting, which have limited

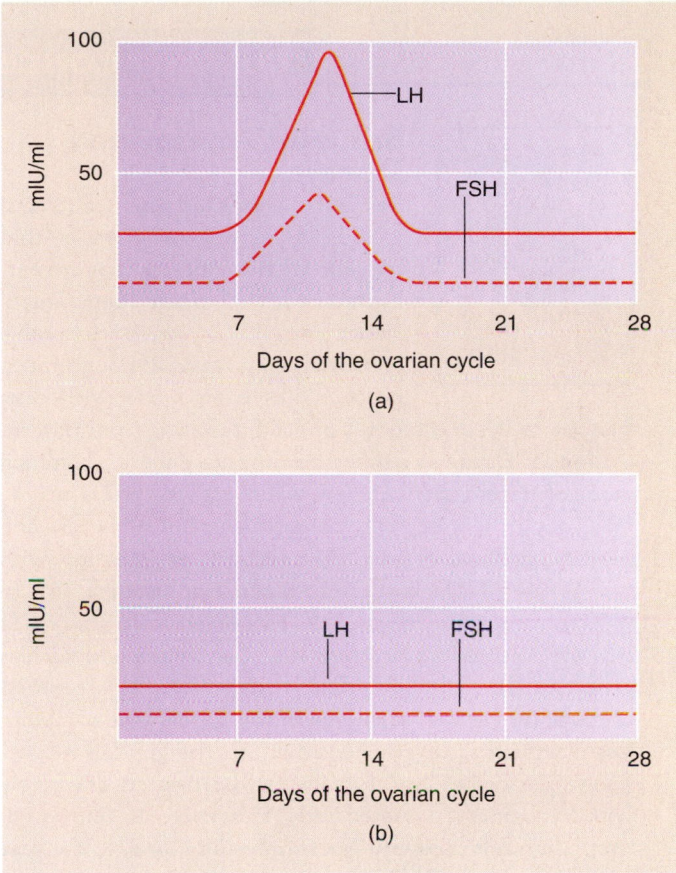

Figure 33–14

Plasma levels of gonadotropins during the ovarian cycle in women *(a)* not using oral contraceptives and *(b)* using oral contraceptives.

the use of this type of oral contraceptive. Recently, progestin- and estrogen-containing pills or levonorgestrel alone have been used as morning-after pill contraception and have been shown to produce less nausea than do pills containing only estrogen. Morning-after pills also have been used as emergency contraceptives (see subsequent discussion).

Recently, time-release forms of hormonal contraception have become available that remove the need to take a daily hormonal pill. Progestin-implant contraception (Norplant, NorplantII, Jadelle, Implanon) involves placement of small, plastic rods impregnated with progestins, such as levonorgestrel or ketodesogestrel, under the skin. These rods then release small amounts of these steroids into the bloodstream and can prevent pregnancy continuously for up to five years after implantation. This form of contraception has proven to be safe and effective. Fertility in women with these implants is restored after their removal. Depo-Provera is an injectable, time-release form of progestin that can be injected intramuscularly at three- to four-month intervals to prevent pregnancy. Progestin implants and injections have proven to be 99% effective in preventing pregnancy.

CURRENT CONCEPTS IN PHYSIOLOGY

Sexually Transmitted Diseases

Sexually transmitted diseases (STDs) include a variety of diseases that range from mild illnesses that are easily treated to human immunodeficiency virus (HIV), which is incurable and ultimately fatal, although much progress has been made with drug therapy to delay the onset of AIDS resulting from HIV infection. STDs are so defined because they share a common mode of transmission, namely sexual activity. STDs include HIV, gonorrhea, syphilis, chlamydia, genital herpes, and human papillomaviruses. Co-infection, or infection with two or more STDs, is not uncommon. For instance, 15% to 20% of men and 35% to 40% of women with gonorrhea also are infected with chlamydia. Gonorrhea and syphilis generally are easily identified, respectively, as a pustular urethral (or vaginal) discharge or the appearance of a painless ulcerations, called *chancroids*, on the genitals respectively. Gonorrhea and syphilis can be treated effectively with antibiotics if detected early. However, left untreated, they can result in severe illness; infertility; and, in the case of syphilis, severe CNS disease and death.

Three of the most common STDs are described below.

Chlamydia. Chlamydia is caused by the *Chlamydia trachomatis* bacteria. It is one of the most widespread STDs in the United States, producing approximately 4 million cases a year. It is transmitted during sexual intercourse but also can be passed to a newborn during delivery from an infected mother or into the eye from the hand or other skin areas containing infected secretions. Initial infection is often asymptomatic (i.e., without symptoms). When symptoms do occur, they most often include vaginal discharge and dysuria (i.e., painful urination). In women, both chlamydia and gonorrhea can progress asymptomatically from a lower genital tract infection to an ascending tubal infection. Chlamydial infections are a major cause of pelvic inflammatory disease (PID), which is characterized by inflammation of the uterine tubes (termed *salpingitis*), and is associated with an increased risk for both ectopic pregnancy and infertility. The risk for infertility increases with each episode of tubal inflammation. The cervix is the most common female organ infected by *Chlamydia trachomatis*, although the urethra and rectum also may be infected. Use of a barrier contraceptive is effective in preventing cervical infection. Secondary prevention involves screening for, and treating, the cervical infection before it ascends to the uterine tubes. Tertiary prevention consists of treating the inflammation of the uterine tubes. Asymptomatic chlamydial infections are detectable only by screening pelvic examinations.

Chlamydial infections in men can be asymptomatic as well, while producing inflammation of the urethra (termed *urethritis*) and increased risk of pain and swelling of the epididymis (termed *epididymitis*). Because chlamydial infections are accompanied by few, if any, symptoms, the rate of sexual transmission is high. Antibiotics are effective in treating chlamydial infections.

Genital Herpes. Genital herpes, caused by **herpes simplex virus type 2 (HSV-2)**, is the most common cause of genital ulcers in the United States. Lesions can occur on the skin or epithelium of the vulva, vagina, cervix, urethra, or rectum and can be spread, by touch, to other skin areas or to the eyes. HSV-2 is a very difficult virus to control. Initial infection with this virus is characterized by clusters of vesicles that rupture to form painful, shallow ulcers. These ulcers dry up, form a crust, and disappear in approximately three weeks. The virus then establishes a permanent latent infection in dorsal root ganglia, which results in recurrent episodes that are usually milder and of shorter duration than the first. Recurrent episodes often are preceded by pain, burning, itching, or tingling of the skin in the area of subsequent outbreak. Risk for infection is highest when active lesions are present, but asymptomatic shedding of the virus occurs between episodes and serves as a common mode of transmission. Many adult men and women have serologic evidence of past infection, are capable of periodically shedding the virus and thereby transmitting genital herpes to a sex partner, and yet have no clinical history of outbreaks. While there is no known cure for genital herpes, drugs are often effective in suppressing outbreaks, and some patients inexplicably cease having recurrences.

Current research is being directed at elucidating the genetic code of the virus to understand better how the virus works. Recently, a protein called HveC, which promotes HSV-2 infection, was discovered. This protein is being targeted as a potential site of prevention or therapy for HSV-2 infections.

Human Papillomaviruses. There are approximately 70 types of human papillomaviruses (HPVs). HPVs are one of the most common causes of STD in the world. Some HPVs cause genital warts, which usually are found on the skin in the genital and rectal areas or on the mucosa of the urethra, vagina, or rectum and occasionally in the oropharyngeal region. These warts are self-limiting, hyperproliferative lesions. Several treatments are available for eliminating genital warts, but none can eliminate the virus responsible for the infection. Small warts usually are removed by cryotherapy, using liquid nitrogen or a cryoprobe. Intravaginal or perianal warts are surgically removed. Various chemical creams and solutions are FDA-approved for use in removing or treating the warts.

Some HPVs are associated with malignant disease. For instance, certain strains of HPVs are a major cause of cervical cancer in women, although not all women infected with an oncogenic HPV develop malignant cervical cancer. In men, some strains of HPVs cause malignant disease of the penis, or penile cancer, which is rare in the United States but is much more common in Central and South America and parts of Africa. Some strains of HPVs are associated with anal cancer in homosexual men.

Many STDs are accompanied by symptoms that alert the patient to the presence of the disease, which allows for medical diagnosis and treatment. The presence of symptoms also provides the patient with the opportunity to identify potentially infected sexual partners. All states require reporting of HIV, syphilis, and gonorrhea, and many states require reporting of chlamydia. Public health authorities usually can assist in notification, diagnosis, and treatment of infected individuals, which serves to limit further disease transmission. However, some STDs can be initially asymptomatic (see earlier section about chlamydia) and hence may go undetected. Therefore, screening for asymptomatic infections is suggested for all sexually active populations and is particularly important for individuals engaging in high-risk behaviors.

Given that STDs are transmitted via sexual activity, one of the most effective approaches for the prevention and control of STDs is to change behaviors that increase the risk for STD transmission. Factors that influence the risk for acquiring STDs include number of sexual partners, rate of acquiring new partners, incidence and diversity of sexual activity of the partners, sexual preference, and type of sexual practice. Behaviors that can contribute to increased transmissions of STDs include failure to use condoms, delay in seeking medical care following onset of infection, noncompliance with suggested medical treatment, and failure to refer infected partners.

Prevention of Implantation

Another type of contraceptive method is the **intrauterine device (IUD),** which can be worn for extended periods. The IUD is made of plastic or metal and is inserted by a physician into the uterus through the vagina and cervix. Although the mechanism of action of the IUD is not completely understood, it is thought to induce a local inflammatory reaction in the endometrium of the uterus, which prevents implantation of the ovum. Side effects of the IUD include uterine bleeding, spasms and cramping of uterine muscles, pelvic infection, and perforation of the uterine wall.

Emergency Contraception

Pregnancy can be prevented on an emergency basis using hormonal or IUD mechanisms. Such prevention can be used after sexual assault, after failure of another form of contraception (i.e., if a condom breaks or a diaphragm dislodges during intercourse), or after consensual unprotected sex.

The Copper T-IUD, inserted within five days of intercourse, is 99.9% effective in preventing pregnancy and can be removed after the woman's next period, if desired. This form of emergency contraception is unsuitable, however, for women who have been sexually assaulted or face possible STDs.

Recently, the drug **mifepristone**, otherwise known as **RU 486,** has been approved for use as an emergency contraceptive. Mifepristone is a progesterone blocker. As such, it antagonizes the action of progesterone on the uterine lining. A single dose of this drug, taken within 72 hours of unprotected sex, can prevent pregnancy.

Surgical Methods of Contraception

The most effective form of birth control in sexually active individuals is surgical sterilization.

Vasectomy is a simple, 15- to 20-minute surgical sterilization procedure performed under local anesthesia in males (Fig. 33–15). A vasectomy consists of a small surgical incision on each side of the scrotum to expose a small section of the vas deferens. The vas deferens are severed, and the ends are tied, clamped, or cauterized to prevent sperm transport.

Vasectomy does not interfere with the synthesis or release of testosterone from the testes and does not affect erection, ejaculation, orgasm, or sperm production. Following vasectomy, sperm accumulate in the epididymis of the testes, where are engulfed and destroyed by phagocytosis. Vasectomy is a very effective method of contraception, with failure rates below 0.15%. Postsurgical complications, which can include bleeding, infection, and swelling of scrotal tissue due to sperm leakage, are rare. Few long-term health risks appear to be associated with vasectomy.

Tubal ligation in females consists of an abdominal incision to permit insertion of an instrument called a **laparoscope,** which is used to locate and sever the uterine tubes on each side of the body (Fig. 33–16). The cut ends of the uterine tubes are occluded with rings, clips, or cautery, blocking both ovum transport to the uterus and sperm transport through the uterine tubes. Tubal ligation is a very effective method of preventing pregnancy, and postsurgical complications, which can include bleeding and infection, are rare. As with vasectomy, few long-term health risks appear to be associated with tubal ligation.

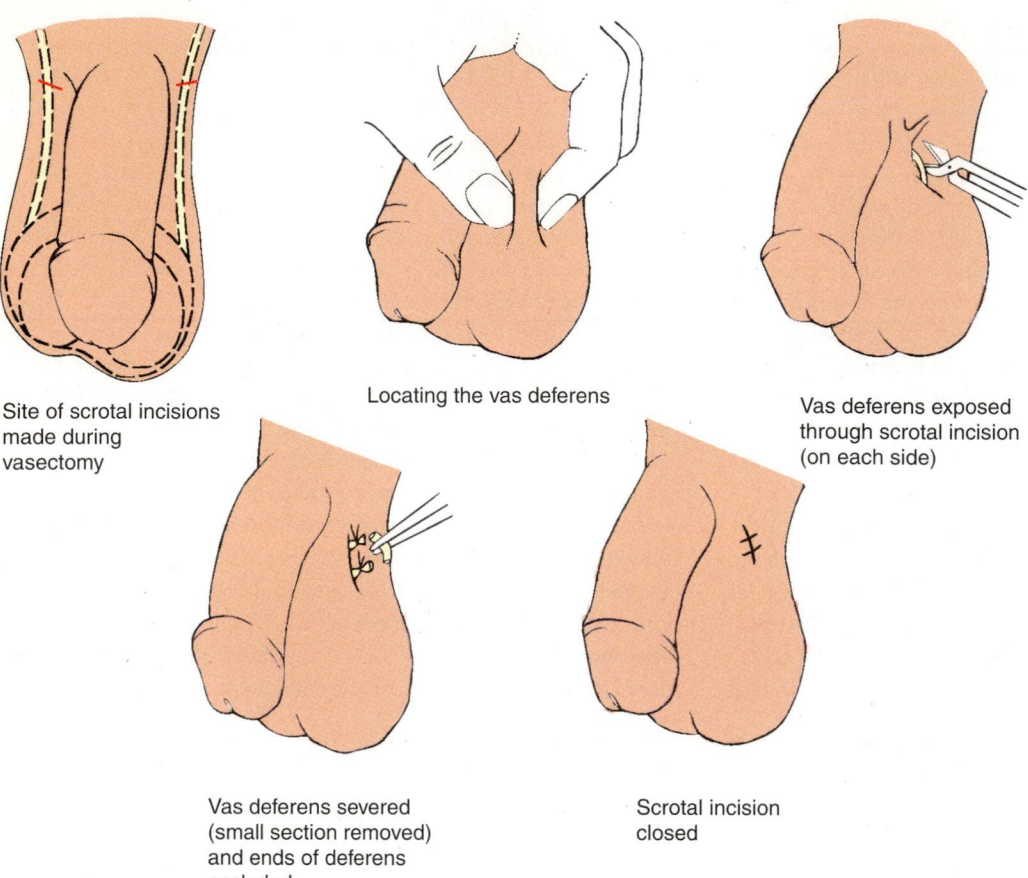

Figure 33–15

Steps involved in a vasectomy.

Site of scrotal incisions made during vasectomy

Locating the vas deferens

Vas deferens exposed through scrotal incision (on each side)

Vas deferens severed (small section removed) and ends of deferens occluded

Scrotal incision closed

Abstinence

Complete abstinence is the only completely effective form of birth control. It is unpopular for obvious reasons. However, many people practice **periodic abstinence** (also termed the **rhythm method**) during the time of ovulation when fertility is the greatest. The efficacy of this method depends on the accuracy of predicting the time of ovulation.

Three methods are used to predict ovulation. The first involves monitoring the ovarian cycle and length of menstrua-tion for at least six months and avoiding intercourse on the days preceding and following ovulation (at midcycle). This method has a high failure rate because many women exhibit a large variation in length of monthly cycles. The second method involves daily monitoring of basal body temperature and avoiding intercourse on the days preceding and following ovulation. Body temperature decreases prior to ovulation and rises again following ovulation. This method also has a high failure rate because ovulation occurs before the rise in body temperature is observed and the temperature shifts that occur during the ovarian cycle are small (usually less than 1°F) and difficult to identify accurately. The third method of identifying the time of ovulation requires monitoring the consistency of cervical mucus. Cervical mucus becomes clear, thin, and stringy around the time of ovulation. Predicting ovulation on the basis of consistency of cervical mucus has a high failure rate because of the difficulty of accurately identifying these changes and because illness or variations in the hormonal state of the woman may alter cervical mucus properties.

New Approaches to Contraception

Several new approaches to contraception are currently being developed and clinically tested. The important variables that should be identified for all new methods of contraception are efficacy, duration of action, reversibility, ease of use, and inci-dence of both short- and long-term side effects. In the fe-

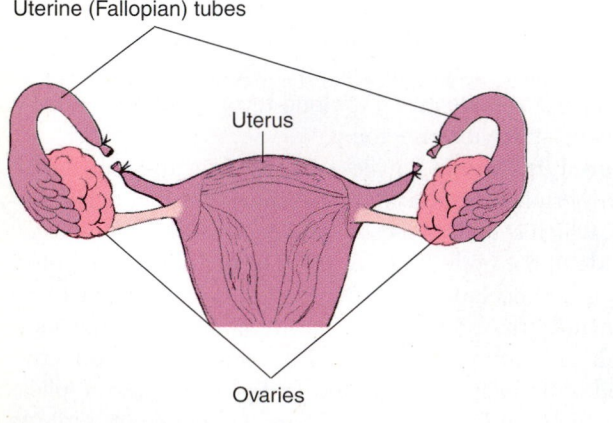

Uterine (Fallopian) tubes

Uterus

Ovaries

Figure 33–16

A tubal ligation.

male, research efforts are being directed toward blocking ovulation and implantation. In the male, the focus is on blocking production, maturation, motility, and/or transport of sperm.

Hormonal analogues that suppress LH and FSH release are being investigated, as are substances that antagonize the actions of progesterone and human chorionic gonadotropin (hCG) on the endometrium and hence interfere with implantation of the ovum. Synthetic analogues of the prostaglandins, which induce uterine contractions, are being investigated for their efficacy in inducing the onset of menstruation, which would evacuate the contents of the uterus whether or not a fertilized ovum is present.

Combinations of estrogen, progesterone, and testosterone currently are being examined for their ability to transiently block sperm production. Other naturally occurring substances, such as an extract from cottonseed oil, termed *gossypol*, which are capable of immobilizing or killing sperm, also are being investigated. This agent destroys the lining of tubules in the testes. Efforts are focused as well on identifying and testing substances, such as inhibin, that inhibit FSH release from the pituitary and hence remove the hormonal signal for sperm production in the testes.

CHAPTER REVIEW

Summary

- The presence or absence of a Y chromosome determines the sex of the zygote.
- During the first six weeks of embryonic development, the gonads in all embryos are anatomically and morphologically identical.
- Hormones secreted by the developing gonads direct gonadal differentiation and the development of the internal genitalia.
- The external genitalia in the male and female differentiate in response to the presence and absence of testosterone, respectively.
- Abnormal patterns of sexual differentiation can occur at four levels: chromosomal sex (genotype), gonadal sex, internal genital sex, and external genital sex.
- Sexual arousal in adult males and females is divided into four phases: excitement, plateau, orgasm, and resolution.
- Endocrine deficiencies can result in reduced sexual arousal and decreased capacity for orgasm in both males and females.

- Mechanical or disease-related damage to the peripheral nerves, spinal cord, or brain can result in impaired sexual arousal and/or impaired capacity for orgasm in both sexes.
- Drugs such as alcohol, antihistamines, sedatives, narcotics, and tranquilizers can retard or inhibit sexual arousal.
- Infertility can result from hormonal deficiencies and destruction of gonadal tissue resulting from inflammation or disease.
- Barrier methods of contraception interfere with sperm transport.
- Birth control pills contain synthetic hormones that inhibit the secretion of pituitary hormones that control ovulation.
- Surgical methods of contraception include vasectomy in the male and tubal ligation in the female.
- Rhythm methods of contraception are not very effective.
- Future trends in contraception include the development of long-acting natural and synthetic chemicals that inhibit ovulation, spermatogenesis, sperm transport, ovum transport, and/or implantation of the ovum in the uterus.

Review Questions

Choose the Correct Answer

1. The rhythm method of birth control:
   a. relies on timing intercourse to coincide with ovulation.
   b. depends on the ability to reliably predict the time of ovulation, every month, for the entire duration of time during which a woman does not wish to become pregnant.
   c. is more effective than the pill in preventing pregnancy.
   d. involves examining changes in cervical mucus composition following ovulation.
   e. is based on cyclic sperm production in the male.
2. The external genitalia of a woman with congenital 11-hydroxylase deficiency would:
   a. appear female.
   b. appear male.
   c. have male and female characteristics.
   d. have indeterminate gender characteristics.
   e. not be affected by this enzyme deficiency.

3. Testicular feminization results from deficient androgen receptors in an individual with a 46XY genotype. Which of the following would be expected to be seen in an individual with this condition?
   a. Accelerated development of Wolffian ducts
   b. Development of Mullerian ducts
   c. No functional internal genitalia
   d. Male external genitalia
   e. Overexpression of pubic and axillary hair
4. 5α-reductase deficiency in the male:
   a. suppresses DHT formation.
   b. results in normal internal genitalia formation.
   c. results in incomplete masculinized external genitalia.
   d. All of the above are correct.
   e. Only b and c are correct.
5. Development of the testes in the embryo is due to:
   a. production of placental estrogen.
   b. presence of Mullerian inhibiting hormone.

   c. presence of a single Y chromosome.
   d. increased fetal FSH secretion.
   e. effects of biochemical conversion of steroid precursors by the fetal adrenal.

6. Male external genitalia do not develop:
   a. in the presence of an X chromosome.
   b. without androgen secretion by fetal testes.
   c. without prior regression of the Mullerian ducts.
   d. if Wolffian ducts develop.
   e. in the absence of estrogen in the fetus.

7. Mullerian inhibiting hormone:
   a. is synthesized and secreted by Leydig cells.
   b. stimulates development of female internal genitalia.
   c. prevents the development of female internal reproductive organs in an XY male.
   d. is produced in response to hCG.
   e. is required for differentiation of the Wolffian ducts.

8. During the second trimester of pregnancy, fetal FSH and LH levels:
   a. increase because the maternal hypothalamus secretes increasing amounts of GnRH.
   b. decrease because the maternal hypothalamus secretes decreasing amounts of GnRH.
   c. increase because the fetal hypothalamus and pituitary are relatively insensitive to inhibition by steroids.
   d. decrease because the fetal hypothalamus and pituitary are hypersensitive to inhibition by steroids.
   e. increase because the placenta begins to synthesize FSH and LH.

9. All of the following are true concerning the excitement phase of sexual arousal except:
   a. increase in ventilatory rate.
   b. increase blood flow to the external genitals.
   c. increase in heart rate.
   d. this phase is dependent on physical stimuli.
   e. this phase can occur in the absence of physical contact with the external genitals.

10. Nursing may reduce the possibility of conception because:
   a. release of prolactin during nursing suppresses GnRH release.
   b. release of oxytocin during nursing suppresses GnRH release.
   c. nursing increases feedback sensitivity of the pituitary to estrogen.
   d. nursing increases LH and FSH levels in response to release of prolactin.
   e. prolactin directly inhibits follicular development.

11. Women with the condition called polycystic ovarian syndrome have high circulating levels of androgens, estrone, and LH. They also have enlarged ovaries with a beaded-pearl appearance. Which of the following would be expected in a woman with this syndrome?
   a. She would have facial hair and acne.
   b. She would be hyperovulatory due to circulating estrone levels.
   c. She would be anovulatory due to suppression of FSH secretion.
   d. a and b are correct.
   e. a and c are correct.

12. Adrenogenital syndrome is a condition in which there is hypersecretion of adrenal androgens in an XX female. Which of the following accurately describes the consequences of this syndrome?
   a. Development of testes in the female
   b. Suppression of Mullerian development
   c. Masculinization of the external genitalia

   d. Secretion of AMH
   e. Atrophied ovaries

13. External genital development in the fetus:
   a. is directed by estrogen in the female.
   b. is expressed as male or female within the first month of pregnancy.
   c. is determined by the Y chromosome in the male independent of any steroid influence.
   d. develops as female in an XX female due to lack of testosterone production by testes.
   e. is directed by LH and FSH in both males and females.

14. The excitement phase of sexual arousal in both men and women is characterized by all of the following except:
   a. lubricating secretions prior to orgasm.
   b. vasocongestion in the genital area.
   c. reflex orgasm.
   d. increased genital size.
   e. myotonia.

15. Inability to maintain an erection could be caused by all of the following except:
   a. venodilation in response to nitroglycerin taken for angina.
   b. penile nerve damage caused by uncontrolled diabetes mellitus.
   c. lower spinal cord injury.
   d. venoconstriction in the penis.
   e. atherosclerosis of the arteries to the penis.

16. Orgasm:
   a. has no reproductive value in the female.
   b. can result in ejaculation in males in the absence of penile erection.
   c. is essential for fertility in the female.
   d. results in rhythmic genital contractions in males only.
   e. is a reflex consequence of all forms of sexual stimulation.

17. After menopause:
   a. vaginal muscle tone increases, making intercourse difficult.
   b. women can no longer achieve orgasms.
   c. vaginal lubrication is absent.
   d. vaginal lubrication during sexual stimulation is reduced due to low estrogen production.
   e. vaginal lubrication during sexual stimulation is reduced due to low progesterone production.

18. As men age:
   a. sexual drive diminishes due to absence of testosterone production.
   b. the refractory period of sexual arousal is shortened.
   c. sperm production diminishes and is absent after age 60.
   d. impotence is inevitable.
   e. penile erections are diminished.

19. All of the following can result in infertility in a sexually active couple except:
   a. endometriosis in the woman.
   b. pelvic inflammatory disease in the woman.
   c. anabolic steroid abuse in the man.
   d. frequent intercourse.
   e. chronic elevated testicular temperature.

20. The efficacy of birth control pills in preventing conception are related to all of the following except:
   a. feedback inhibition of FSH secretion by synthetic steroids in the pill.
   b. blockade of ovulation.
   c. absence of menstruation.
   d. increase in cervical mucus viscosity.
   e. inhibition of estrogen-mediated endometrial proliferation.

Answers to Case History Questions

1. Inability to achieve or maintain an erection can be caused by physical, physiological, or psychological problems in the male. In the latter case, a healthy individual is able to have erections during sleep, when psychological interference with erection is absent. The fact that this individual cannot achieve nocturnal erections suggests an underlying physical or physiological problem causing erectile dysfunction.

2. Tobacco smoking damages arterial endothelial cells and nitric oxide-mediated vasorelaxation. Because this type of relaxation is involved in penile erection, tobacco smoking can impair erectile function. In fact, tobacco smoking is now recognized as a leading cause of impotence among men.

3. Diabetes mellitus leads to nerve and arterial endothelial cell damage with resultant loss in arterial dilating capacity. Impotence is a major pathological consequence of diabetes. However, laboratory tests and the history of this patient suggest that his diabetes is under control and therefore may not be the cause of his erectile dysfunction.

4. Many antihypertensive drugs can cause impotence. In particular, β-adrenergic blockade often results in unopposed α-mediated sympathetic nerve vasoconstriction and reduction in blood flow. This factor, along with a decrease in blood pressure, could likely be the cause of this patient's impotence. The fact the an α-adrenergic blocking agent, which may actually dilate blood vessels to the penis, corrected his erectile dysfunction suggests that his condition was a side effect of his initial antihypertensive drug therapy.

Key Terms

autosomes (p. 978)
chromosomal sex (p. 978)
clitoris (p. 985)
contraception (p. 993)

dyspareunia (p. 987)
ejaculation (p. 986)
endometriosis (p. 989)
gonadal sex (p. 978)

impotence (p. 990)
infertility (p. 991)
Mullerian duct system (p. 979)
orgasm (p. 984)

spermicides (p. 993)
transudation (p. 985)
vasectomy (p. 995)
Wolffian duct system (p. 979)

Suggested Readings

Glover, T.G., and Barratt, C.L.R., eds. *Male Fertility & Infertility*. New York, Cambridge University Press, 1999.

Kirby, R., Kirby, M.G., and Farah, R.N. *Men's Health*. Oxford, Isis Medical Media, 1999.

Knobil, E., and Neill, J.D., eds. *Encyclopedia of Reproduction*. San Diego, Academic Press, 1998.

Knobil, E., and Neill, J.D., eds., Greenwald, G.S., Markert, C.L., and Pfaff, D.W., assoc. eds. *The Physiology of Reproduction*, ed 2. New York, Raven Press, 1994.

Masters, W.H., Johnson, V.E., and Kolodny, R.C., eds. *Human Sexuality*, ed 4. New York, Harper Collins, 1991.

Morris, D. *The Human Animal*. New York, Crown Publishers, 1994.

Reinisch, J.M. *The Kinsey Institute New Report on Sex*. New York, St. Martin's Press, 1990.

Stanberry, L., and Bernstein, D., eds. *Sexually Transmitted Diseases: Vaccines, Prevention and Control*. San Diego, Academic Press, 2000.

Trounson, A.O., and Gardner, D.K. *Handbook of In Vitro Fertilization*, ed 2. Boca Raton, CRC Press, 2000.

Web Sites

http://www.merck.com/pubs/mmanual_home/contents.htm
The Merck Manual of Medical Information–Home Edition. STD's, Gynecology, Infertility, Family Planning and HIV. Chapters 187, 189, 229–230, 233–242, and 261.

http://www.mayoclinic.com/mayo/common/htm/menspg2.htm
Menopause, Contraception, and Infertility.

http://www.urologychannel.com/erectiledysfunction/index.shtml
Urology Channel and Erectile Dysfunction.

http://www.ashastd.org/
American Social Health Organization. Sexually Transmitted Diseases.

http://www.womenfirst.com/pageControl.asp?SubjectID=72
Women First Healthcare.

Answers to Review Questions

1. b 2. b 3. c 4. a 5. c 6. b 7. c 8. c
9. d 10. a 11. e 12. c 13. d 14. c 15. d
16. b 17. d 18. e 19. d 20. c

Glossary

A-band A transverse band of thick filaments of cardiac and skeletal muscle located in the center portion of the sarcomere. (Ch. 16)

absolute refractory period The period of time in which an axon is unable to generate an action potential regardless of the applied stimulus strength. (Ch. 7)

absorption (ab-**sorp**-shun) The movement of digestion products and other nutrients across the wall of the small intestine to the blood or lymph. (Ch. 22)

acclimatization The process of becoming accustomed to a new environment. (Ch. 29)

accumbens (a-**kum**-binz) **nucleus** A region of the basal ganglia that is considered as a reward center of the brain. (Ch. 11)

A-cell A glucagon-secreting cell of islets of Langerhans; also called *alpha cell*. (Ch. 14)

acetone (**as**-ih-tone) A volatile, sweet-smelling ketone body produced from acetyl coenzyme A during prolonged fasting or untreated severe diabetes mellitus. (Ch. 25)

acetyl coenzyme A (uh-**see**-til koh-**ehn**-zime) An important intermediate in the Krebs cycle and the main precursor to lipids; it is formed in the mitochondrial matrix when all of the acetyl groups produced from food molecules are coupled to the terminal—SH group of coenzymes. (Ch. 6)

acetylcholine (uh-see-til-**koe**-leen) (Often abbreviated as ACh.) A low-molecular-weight neurotransmitter; found in the central and peripheral nervous systems. (Ch. 10)

acetylcholinesterase (uh-*see*-til-koe-leen-**ess**-tur-aze) Enzyme essential for breaking down acetylcholine to an inactive form once it has engaged a receptor. (Ch. 16)

acid (**a**-sid) Any substance whose dissociation in water releases hydrogen ions. (Ch. 2)

acid dissociation constant The ionization constant for an acid. (Ch. 25)

acidemia (*as*-i-**dehm**-ee-uh) An arterial blood pH less than 7.35. (Ch. 25)

acidosis (*as*-i-**doe**-sis) An abnormal process that tends to produce acidemia. (Ch. 25)

acinus (**a**-sih-nuhs) Saclike clusters of exocrine secretory cells in the pancreas that secrete digestive enzymes. (Ch. 22)

acromegaly (*ak*-roe-**meh**-gah-lee) A disfiguring condition resulting from the overproduction of GH in adults whose epiphyses have closed; stimulates excessive growth of soft tissues and transverse diameter of bones leading to bony thickening and deformity. (Ch. 13)

acrosome (**ak**-roe-sohm) **reaction** A controlled disruption of the head of a sperm that allows it to penetrate the cumulus oophorus and the zona pellucida of the ovum. (Ch. 32)

ACTH Adrenocorticotropic hormone, secreted by the anterior lobe of the pituitary. Also called *corticotropin*, the hormone regulates the synthesis and secretion of cortisol by the adrenal cortex. (Ch. 14)

actin (**ak**-tin) A globular protein to which myosin cross bridges bind; located in muscle thin filaments and microfilaments of cytoskeleton. (Ch. 16)

actin-linked regulation Muscle regulation that is accomplished through some action on the thin filaments. (Ch. 16)

action potential A rapid change in the voltage of the membrane potential that occurs when adequate stimulus is applied. The response is comprised of a depolarization phase followed immediately by a repolarization phase. Propagation of the action potential is the basis of a nerve or muscle impulse. (Ch.7)

activation energy The initial energy required to break the existing bonds in a molecule and form new ones. (Ch. 2)

activation state One of three states of the sodium channel; if the nerve cell is stimulated and the membrane potential is depolarized to the threshold value, the sodium channel switches from its resting state to its activating state; sodium ions will then flow until the inactivating gate closes the channel. (Ch. 7)

active immunity Long-lasting continuing protection against disease conferred through immunization with an antigen. (Ch. 28)

active site Areas on the surfaces of enzymes that permit reactant molecules to bind. (Ch. 2)

active tension The force (tension) in a muscle produced in response to stimulation. (Ch. 16)

active transport A situation in which a solute must move from a less concentrated region outside the cell to a more concentrated region inside, against the prevailing gradient and thus requiring energy. When the movement of solute is coupled directly to an energy-yielding reaction, the process is termed **primary active transport**; when the active transport of a solute is not coupled directly to the energy-yielding reaction, it is described as **secondary active transport**. (Ch. 4)

active zone Restricted areas of the presynaptic terminal where neurotransmitters are released. (Ch. 7)

acute mountain sickness An illness typically seen in high-altitude mountain climbers, in which too much carbon dioxide is blown off during hyperventilation. (Ch. 29)

adaptation A genetic change that enhances the ability of an organism to adjust to the environment. (Ch. 29)

Addison's disease A condition caused by an overall deficiency in steroid production by the adrenal cortex. It is a progressive disease with symptoms such as diarrhea, hypotension, increased skin pigmentation, fatigue, and digestive difficulties. (Ch. 14)

adenohypophysis (a-*den*-oh-high-**pof**-i-sis) The anterior lobe of the pituitary gland. (Ch. 13)

adenosine (a-**den**-oh-seen) A nucleoside that has been shown to promote vasodilation of arteriolar smooth muscle. (Ch. 19)

adenosine diphosphate (a-**den**-oh-seen die-**fos**-fate) **(ADP)** An important molecule that is formed from adenosine triphosphate (ATP) when the terminal phosphate is removed. This reaction produces the energy that drives other energy-requiring reactions. (Ch. 6)

adenosine monophosphate (a-**den**-oh-seen mah-noe-**fos**-fate) **(AMP)** A molecule that is formed from ATP when pyrophosphate is removed (also known as *adenosine phosphate*). (Ch. 6)

adenosine triphosphate (a-**den**-oh-seen try-**fos**-fate) **(ATP)** The universal energy compound that is produced in all cells and is the source of energy in most of the energy-requiring reactions of metabolism. ATP is comprised of adenosine — plus three phosphate groups attached. (Ch. 6)

adenylate cyclase (a-**den**-ih-late **sye**-klase) An enzyme located in the inner surfaces of cell membranes that catalyzes the conversion of adenosine triphosphate (ATP) to cyclic adenosine monophosphate (cAMP) following the binding of a signal molecule to a membrane-bound receptor. (Ch. 12)

adequate stimulus The type of stimulus to which a sensory receptor is most sensitive. (Ch. 8)

adipose tissue (**ad**-i-poz) Tissue composed mainly of fat-storing cells. (Ch. 11)

adrenal cortex (ad-**ree**-nal **kore**-teks) The outer two-thirds shell of the adrenal gland that produces the steroid hormones — cortisol, aldosterone, and androgens. (Ch. 12)

adrenal glands A pair of endocrine glands, one located above each kidney: Each consists of outer adrenal cortex and inner adrenal medulla. (Ch. 12)

adrenal medulla (ad-**ree**-nal meh-**dul**-ah) The inner core of the adrenal gland that secretes the catecholamines, epinephrine, and norepinephrine. (Ch. 12)

adrenergic (*ad*-ren-**er**-jik) Pertaining to epinephrine or norepinephrine. (Ch. 10)

adrenocorticotropic (ad-*ree*-no-kore-tih-koe-**trope**-ik) **hormone (ACTH)** A polypeptide hormone that is synthesized, stored, and released from the anterior pituitary gland. It stimulates the production and secretion of cortisol by the zona fasciculata of the adrenal cortex. (Ch. 11)

adrenogenital (ad-*ree*-no-**jen**-ih-tahl) **syndrome** A genetic disorder in females in which hypersecretion of adrenal androgens results in the development of female internal genitalia but male/female external genitalia. (Ch. 33)

adult respiratory distress syndrome A form of pulmonary edema caused by increased permeability of the capillaries; damage is caused by leakage of oxygen radicals. (Ch. 21)

aerobic (air-oh-bik) Any oxygen-requiring process. (Ch. 6)

afferent (af-er-ent) Carrying toward. (Ch. 9)

afferent arteriole (af-er-ent ar-**tee**-ree-ole) The arteriole that carries blood toward the glomerulus. (Ch. 23)

afferent nerve fiber Nerve fiber that transmits stimuli to the CNS. (Ch. 9)

affinity chromatography (ah-**fih**-nih-tee kro-muh-**tog**-ruff-ee) A technique that uses the specificity of an antigen-antibody reaction to study antibodies. (Ch. 3)

afterload The force a weight on a lever system will provide during an isotonic contraction. (Ch. 16)

agglutination (a-*glue*-ti-**nay**-shun) The process that is described by the aggregation, or clumping, of red blood cells, which is caused by the attachment of substances such as antibodies to cell surfaces that, in turn, bind the cells together. (Ch. 17)

agglutinin (a-**glue**-ti-nin) Specific antibodies in the blood that cause clumping, or agglutination. (Ch. 17)

agglutinogen (a-**glue**-ti-nin-oh-jen) Specific antigens on red blood cell membrane surfaces that causes the production of an agglutinin-type of immunoglobulin. (Ch. 17)

agonist (**ah**-goh-nist) Any ligand capable of both binding to a plasma membrane receptor and evoking a physiologic response. (Ch. 5)

agranulocyte (a-**grahn**-you-low-site) Leukocyte that does not have a granular cytoplasm; monocyte and lymphocyte. (Ch. 17)

airway A series of highly branched hollow tubes within the lung that carry gas to and from the alveoli, trachea, bronchi, and bronchioles. (Ch. 20)

alactacid oxygen debt The first stage of repayment of the oxygen debt during exercise; refers to the oxygen needed to replenish the oxygen stores of the body as well as the phosphagen system. (Ch. 30)

albumin (al-**byou**-min) The most common plasma protein; produced in the liver. It regulates plasma osmotic pressure. (Ch. 12)

aldosterone (al-**dos**-ter-own) A mineralocorticoid (steroid) hormone produced by the zona glomerulosa of the adrenal cortex that helps to regulate body fluid volume by controlling sodium chloride and potassium excretion from the kidney tubular cells. (Ch. 12)

alkalemia (al-kuh-**lee**-me-uh) An arterial blood pH greater than 7.45. (Ch. 25)

alkaline (**al**-kuh-line) Basic; a pH greater than 7. (Ch. 25)

alkalosis (al-kuh-**low**-sis) An abnormal process that tends to produce alkalemia. (Ch. 25)

allosteric enzyme (al-o-**steer**-ik **en**-zime) A catalytic protein that has two reactive sites. One site activates the enzyme. The other one, the allosteric site, can enable a substance other than the normal substrate to bind to the enzyme molecule, thereby changing the shape of the molecule and activity of the enzyme. (Ch. 6)

alpha adrenergic (**al**-fah *ad*-ree-**ner**-jik) **receptor** A type of receptor site for norepinephrine and epinephrine that, when activated, opens a calcium channel in the plasma membrane where it is located. (Ch. 7)

alpha cells of the pancreas One cell type found in the islets of Langerhans; produces glucagon and is responsible for regulation of blood glucose concentrations and fuel homeostasis. (Ch. 15)

alpha motor neuron An efferent neuron originating in the CNS that synapses with extrafusal muscle fibers. (Ch. 9)

alpha-helix A type of conformation that results when a single polypeptide chain coils about itself to form a uniform cylindrical structure. (Ch. 2)

alpha receptor A receptor to which norepinephrine binds. (Ch. 10)

alveolar (al-**vee**-oh-lar) **capillary membrane** The membrane through which gas exchange takes place between the alveolus and the blood in the pulmonary capillaries. (Ch. 21)

alveolar pressure The pressure of air inside the alveoli. (Ch. 20)

alveolar surface area The site of gas exchange in the lung; roughly 80 times greater than the external body surface. (Ch. 20)

alveolar ventilation The volume of atmospheric air entering alveoli each minute. (Ch. 20)

alveolus (al-**vee**-oh-lus) A microscopic air sac in the lung; effects gas exchange (plural: *alveoli*). (Ch. 20)

ameboid (a-**mee**-boyd) **movement** Movement similar to that of an ameba, in which the cell projects protoplasmic extensions ahead and then follows them. (Ch. 17)

amenorrhea (a-men-oh-**ree**-ah) The absence of menstruation. (Ch. 33)

amino acid The basic unit of protein molecules whose chemical substrate can be written as R—CHNH$_2$COOH. There are 20 naturally occurring amino acids that appear in biological proteins. (Ch. 2)

amino group —NH$_2$. (Ch. 2)

aminoacyl attachment site of tRNA A feature of transfer RNA in which an amino acid destined to join a growing polypeptide chain attaches for transport to a ribosome. (Ch. 5)

ammonia NH$_3$. (Ch. 25)

amniotic (am-**nee**-ot-ik) **fluid** The liquid that fills the amnion. (Ch. 32)

amphetamine (am-**fet**-uh-meen) A sympathomimetic drug that stimulates the CNS, increases blood pressure, and reduces the appetite. (Ch. 11)

amphipathic (am-fee-**path**-ik) **molecule** Molecule that has a polar region at one location and a nonpolar region at another location, making it both hydrophobic and hydrophilic. Examples of amphipathic molecules are the phospholipids in cell membranes. (Ch. 4)

amphoteric (am-fo-**tear**-ik) Capable of acting as either an acid or base. (Ch. 25)

amylase (**am**-uh-laze) A digestive enzyme that helps break down starch. (Ch. 22)

amylin (**am**-uh-lyn) Amylopectin; the insoluble constituent of starch. (Ch. 15)

anabolic (an-a-**ball**-ik) **hormone** A hormone that promotes tissue growth. (Ch. 13)

anabolic steroid A steroid hormone having anabolic effects. (Ch. 13)

anabolism (a-**nab**-o-lizm) **(biosynthesis)** A series of steps in which the energy extracted in catabolism is used to assemble building block molecules into proteins, nucleic acids, lipids, and other macromolecules for the cell. (Ch. 6)

anaerobic (an-air-**oh**-bik) Refers to processes that can occur in the absence of molecular oxygen. (Ch. 6)

analgesia (an-al-**jee**-zee-ah) The first stage of anesthesia; a reduced response to pain without loss of consciousness. (Ch. 29)

anatomical planes Imaginary lines used to divide the body into parts. (Ch. 1)

anatomy Structure and organization of the body. (Ch. 1)

androgen (**an**-drow-jen) Any chemical with testosteronelike action. (Ch. 11)

androgen-binding protein (ABP) A protein synthesized by Sertoli cells that binds testosterone and increases the concentration of testosterone in seminiferous tubules of the testes. (Ch. 31)

anemia (ah-**nee**-mee-uh) A condition characterized by an abnormally low oxygen-carrying capacity of the blood, resulting from marked deficiencies in the number of circulating red blood cells, the amount of hemoglobin, or both. (Ch. 17)

anesthesia (*an*-es-**thee**-zee-ah) Loss of feeling or sensation. (Ch. 29)

angina pectoris (an-**jie**-nah **pek**-toe-ris) Chest pain due to hypoxia of the myocardium that may occur because of overexertion, stress, or excitement. (Ch. 19)

angiogram (**an**-jee-oh-gram) A roentgenogram of blood vessels filled with a contrast medium. (Ch. 21)

angiotensin I (*an*-jee-o-**ten**-sin) A compound in the blood formed by the action of renin, an enzyme secreted by the kidney. Angiotensin I becomes angiotensin II by means of a converting enzyme of the pulmonary capillaries. (Ch. 10)

angiotensin II A hormone that causes the hypothalamus to release ADH into the circulatory system; stimulates aldosterone secretion from adrenal cortex, vascular smooth muscle contraction, and thirst. (Ch. 10)

angiotensin-converting enzyme An enzyme that converts angiotensin I into angiotensin II through the removal of two amino acids. (Ch. 14)

angiotensinogen A plasma protein synthesized by the liver and secreted into the blood; precursor for angiotensin I. (Ch. 14)

angular gyrus The association cortex thought to be involved with the association of visual, auditory, and tactile sensations. (Ch. 11)

anion (*an*-eye-on) An atom that gains one or more electrons, thus acquiring a negative charge. (Ch. 2)

anorgasmia (*an*-ore-**gaz**-me-uh) Difficulty in reaching orgasm. (Ch. 33)

anoxia (ah-**nok**-see-uh) Reduced oxygen supply. (Ch. 29)

antagonist (an-**tag**-o-nist) A ligand that binds with high affinity to a receptor but evokes no response. (Ch. 5)

antagonist muscle A muscle that opposes an agonist muscle at a joint where two bones meet; contraction of the agonist muscle results in the extension of the antagonist and vice versa. (Ch. 9)

anterior hypothalamus The anterior portion of the hypothalamus; appears to be the center for sexual drive in both males and females. (Ch. 11)

anterior pituitary gland Anterior portion of the pituitary gland. (Also called *adenohypophysis*.) (Ch. 13)

anterograde fast transport The movement of organelles and materials from the soma to the axon terminal of a nerve cell. (Ch. 7)

anterolateral pathway Pathways in the anterior and lateral white matter of the spinal cord that transmit somatic sensory information from the surface of the body to the neocortex. (Ch. 8)

antibiotic (*an*-ti-bye-**ot**-ik) Molecules that kill or inhibit the growth of microorganisms. (Ch. 5)

antibody (**an**-ti-*bod*-ee) A highly specific protein produced by plasma cells in response to the introduction of foreign matter (antigens). The antibody reacts against the antigen to remove or destroy it. The new term for antibody is *immunoglobulin*. (Ch. 28)

antibody-mediated immunity Humoral immunity; protection against bacteria and viruses in extracellular fluid as a result of stimulating antibody production. (Ch. 28)

anticoagulant (*an*-ti-koh-**ag**-you-lant) A chemical that blocks or inhibits the coagulation process. (Ch. 17)

anticodon A set of three nucleotides in tRNA complementary to a codon on an mRNA molecule. (Ch. 5)

antidiuretic (an-ti-die-your-**et**-ik) **hormone (ADH)** A hormone that directs kidneys to increase the absorption of water; *see also* vasopressin. (Ch. 10)

antidrop response Vasoconstriction and shivering due to stimulation of the hypothalamus by excessive cooling. (Ch. 27)

antigen (**an**-tih-jen) Any molecule that stimulates the immune system to respond. (Ch. 28)

antigen-presenting cell A cell, usually a macrophage or B-lymphocyte, that displays an antigen in combination with major histocompatibility complex (MHC) proteins so as to activate specific immune defense mechanisms. (Ch. 28)

anti-Mullerian hormone (AMH) A glycoprotein, secreted by the Sertoli cells of the testes, which induces regression of the Mullerian duct system; earlier known as *müllerian inhibiting substance* (MIS). (Ch. 33)

antiparallel DNA strands When the 3′ ends of the two complementary nucleic acid strands of the DNA molecule are at opposite ends of the double helix. (Ch. 5)

antiport The membrane transport of different solutes by the same transporter in opposite directions, such as sodium ions out of the cell and potassium ions into the cell by the Na^+/K^+ pump. (Ch. 4)

antipyretics Agents that prevent or reduce the fever response. (Ch. 27)

antirise response Vasodilation and sweating due to stimulation of the hypothalamus by overheating. (Ch. 27)

antithrombin (an-tee-**throm**-bin) A natural coagulation inhibitor. Prevents thrombin from converting fibrinogen to fibrin. (Ch. 17)

antithromboplastin (an-tee-**throm**-bo-plas-tin) A natural coagulation inhibitor. Inactivates tissue thromboplastin. (Ch. 17)

antrum (**an**-trum) The distal portion of the stomach, closest to the pyloric sphincter. (Ch. 22)

aorta (ay-**or**-tah) The artery that receives all of the blood the left heart pumps out; branches into a number of smaller arteries that direct blood flow to organs. (Ch. 18)

aortic arch The part of the aorta in which baroreceptors that sense blood pressure are located. (Ch. 10)

aortic bodies Chemoreceptors of the peripheral nervous system within the wall of the aorta; sensitive to levels of O_2 and CO_2 in blood. (Ch. 10)

aortic valve The valve between the left ventricle and the aorta. (Ch. 18)

aphasia (ah-**fay**-zee-ah) Disorders of language resulting from damage to the brain. (Ch. 11)

apoptosis An organized series of steps which, once triggered, leads to cell death. (Ch. 3)

appendicitis An inflammation of the appendix; usually preceded by obstruction of the lumen of the appendix. (Ch. 22)

appendix A small, fingerlike blind tube that extends from the cecum. (Ch. 22)

aprosodia (a-pro-**so**-dea) Damage to the areas of the nondominant hemisphere of the brain; results in an inability to understand or express intonation. (Ch. 11)

aqueous (**ak**-wee-us) Prepared with water. (Ch. 2)

areola (ah-**ree**-o-lah) The pigmented area of skin that surrounds the nipple of the breast. (Ch. 32)

arm projection neuron A neuron of the posterior parietal cortex that generates action potentials when the arm reaches for an object. (Ch. 9)

arrhythmia (ah-**rith**-mee-ah) Conduction and rhythm disturbances of the heart. (Ch. 18)

arterial (ar-**teer**-ee-al) **baroreceptors** Nerve endings sensitive to stretch or distortion produced by arterial blood pressure changes; located in carotid sinus or aortic arch arteries. (Ch. 19)

arterial pressure The pressure within arteries. (Ch. 19)

arterial system The system of blood vessels that carries blood from the heart to the capillaries. (Ch. 18)

arteriole (ar-**tee**-ree-ole) The smaller blood vessels that branch off an artery once the artery has entered an organ. (Ch. 18)

arteriosclerosis (ar-**teer**-ee-o-skleh-roe-sis) A group of diseases characterized by thickening and loss of elasticity of arterial walls. (Ch. 17)

artery (**ar**-ter-ee) A blood vessel that transports blood away from the heart to the various organs of the body. (Ch. 1)

asbestosis A lung disease caused by breathing asbestos fibers. (Ch. 29)

associative learning A type of learning in which a subject acquires an understanding of the relationships between stimuli or between a behavior and subsequent reward or punishment. (Ch. 11)

asthma (**az**-muh) A breathing disorder brought on by narrowing of the bronchi. (Ch. 20)

astrocytes Star-shaped glial cells of the CNS that form the blood-brain barrier and provide nutrients, support, and insulation for CNS neurons (also called *astroglia*). (Ch. 7)

atelectasis (at-ee-**lek**-tah-sis) Alveolar collapse due to high surface tension in the lungs. (Ch. 20)

atherosclerosis (*ath*-er-or-skleh-**roe**-sis) A disease process of arteries that results in the progressive occlusion of the lumen of the vessel. (Ch. 17)

atherosclerotic plaque Accumulation of lipid and fibrous tissue inside blood vessels resulting in narrowing of the vessel lumen. (Ch. 19)

atmospheric pressure Air pressure surrounding the body; equal to 760 mm Hg at sea level. (Ch. 20)

atom (**a**-tom) The smallest unit of an element that retains the properties of the element. (Ch. 2)

atomic number (a-**tom**-ic **num**-ber) A specific number of protons in the nucleus of an atom, by which it can be identified. (Ch. 2)

ATP synthetase An enzyme that transforms the energy of the proton electrochemical gradient into chemical energy in the form of ATP. (Ch. 6)

atrial (**ay**-tree-al) **natriuretic peptide** A peptide secreted from the cells in the walls of the atria; causes both vasodilation and increased secretion of salt and water by the kidneys. (Ch. 19)

atrial septal defect A congenital birth defect in which the septum between the atria is not completely formed. (Ch. 21)

atrioventricular (*ay*-tree-o-ven-**trik**-you-lar) **(AV) node** A region of specialized cells located in the right atrial wall near the interventricular septum. It relays the contractile stimulus to the bundle of His, the Purkinje fibers, and the ventricular myocardium. (Ch. 18)

atrioventricular valve Valve located between an atrium and a ventricle. The left AV valve is called the *bicuspid* or *mitral valve*. The right AV valve is called the *tricuspid valve*. (Ch. 18)

atrium (**ay**-tree-um) A thin-walled chamber of the heart that receives blood returning to the heart from either the vena cava or the pulmonary vein. (Ch. 18)

atrophic (a-tro-fik) **vaginitis** Drying and thinning of vaginal mucosal membrane and decreased elasticity and muscle tone of the vagina due to inadequate estrogen stimulation. (Ch. 33)

atrophy (**at**-roe-fee) Wasting away. (Ch. 16)

auditory cortex (**aw**-di-*toe*-ree **kor**-teks) That part of the neocortex that detects sounds via the ear. (Ch. 7)

auditory sensory aphasia A condition produced by lesions in the auditory association cortex or Wernicke's area, in which voices can be heard but the words appear to have no meaning. (Ch. 8)

auditory system The system of the body that pertains to the sense of hearing. (Ch. 8)

autocrine (*aw*-**toe**-krin) Mechanism by which a hormonelike substance is released into the interstitial fluid surrounding the cells in which it is produced, and engages receptors on the same cells. (Ch. 12)

autoimmunity (*aw*-toe-ih-**mew**-ni-tee) An immune response arising from and directed against a person's own tissues. (Ch. 28)

autonomic (aw-toe-**nom**-ik) **nervous system** The portion of the nervous system that controls the visceral functions of the body such as regulation of the cardiac and smooth muscles and glands; comprises the sympathetic and parasympathetic divisions. (Ch. 10)

autophagy (**aw**-toe-*fay*-jee) A process in which lysosomes break down the cell's own organelles and other matter; can be a response to cell injury or can occur during remodelling of the cell or during adverse nutritional conditions. (Ch. 3)

autoradiography (*aw*-toe-ray-**dee**-og-ra-fee) A technique that locates the radioactive molecules within cells or tissues; has been useful in discovering the secretory pathways in cells and in determining the role of the Golgi complex in this process. (Ch. 3)

autoreceptor Alpha-2 and beta-2 receptors located on the membrane of the presynaptic terminal that appear to regulate the amount of neuroepinephrine released. (Ch. 7)

autoregulation of blood flow The ability of individual vascular beds to maintain constant blood flow despite changes in arterial blood pressure; the intrinsic ability of the heart to regulate the strength of ventricular contraction. (Ch. 19)

autosome (**aw**-toe-sowm) Chromosomes that provide genetic information for the development and differentiation of most of the cells of the body. (Ch. 33)

auxotonic (**ox**-o-tohn-ik) **contraction** Muscle movement in which the force continually increases as the muscle shortens. (Ch. 16)

axial (**ak**-si-al) **muscles** Muscles that produce movement of the skull, backbone, and ribs. (Ch. 9)

axoaxonic (**ak**-so-ak-son-ik) **synapse** Synaptic contact made between two axons. (Ch. 7)

axodendritic synapse Synaptic contact made between axons and dendrites. (Ch. 7)

axon (**ak**-son) A fiberlike extension of the neuron soma that transmits nerve impulses to effector cells. (Ch. 7)

axon terminal The ends of axons; the point of contact with effector cells at synapses to transmit information. (Ch. 7)

axonal transport The movement of cellular components from the cell body to the terminals. (Ch. 7)

axosomatic synapse Synaptic contacts made between axons and somata. (Ch. 7)

Babinski (bah-**bin**-skee) **sign** Extension of the big toe and fanning of the other toes in response to stroking the bottom of the foot; due to upper-motor-neuron lesions. (Ch. 9)

bacteria (bak-**teer**-ee-ah) Unicellular organisms that have an outer cell wall and a plasma membrane. (Ch. 17)

barometric pressure Atmospheric pressure. The pressure on the surface of the Earth due to the weight of the atmosphere. At sea level, atmospheric pressure is approximately 760 mm Hg. (Ch. 20)

baroreceptor (*bar*-oh-re-**sept**-tor) A stretch receptor sensitive to pressure and to rate of change in pressure. (Ch. 10)

basal (**bay**-sal) Refers to resting level. (Ch. 3)

basal ganglia (**bay**-sal **gang**-lee-ah) A group of nuclei in the brain involved in the programming of motor patterns. (Ch. 9)

basal metabolic rate (BMR) Metabolic rate measured under basal conditions. (Ch. 27)

base Any substance that can accept or bind hydrogen ions. (Ch. 2)

basement membrane The thin proteinaceous layer of extracellular material upon which epithelial and endothelial cells fit. (Ch. 3)

basic electrical rhythm Cyclic spontaneous depolarization and repolarization of pacemaker cells in the stomach causing gastric peristalsis at a frequency of three contractions per minute. (Ch. 22)

basilar (**bas**-ih-lar) **membrane** Thin membrane that divides the scala tympani from the cochlear duct in the inner ear. It supports the organ of Corti and helps to record the frequency of sound waves. (Ch. 8)

basolateral membrane The side of the epithelial cell facing away from the lumen. (Ch. 22)

basophil (**bay**-so-fil) The rarest type of leukocyte; found in greater numbers outside the blood in loose connective tissue. (Ch. 17)

B cell A lymphocyte that can become an antibody-secreting cell. (Ch. 17)

belly The wide or thick portion of a muscle. (Ch. 16)

beta cells of the pancreas One cell type found in the islets of Langerhans; produces insulin and is responsible for regulation of blood glucose concentrations and fuel homeostasis. (Ch. 15)

beta (**bay**-tah) **receptor** One of two types of plasma membrane receptor sites for norepinephrine and epinephrine. When stimulated, the effects include accelerated heart rate, dilation of bronchioles, and glycogenolysis. (Ch. 10)

beta sheet A secondary structural conformation of a protein molecule that results when the polypeptide chain runs back and forth upon itself. (Ch. 2)

beta-lipotropin (**bay**-tah li-**po**-tro-pin) A large polyprotein that breaks down into a number of opiates that can inhibit the sensation of pain. (Ch. 7)

bicuspid (bye-**kus**-pid) **valve** The atrioventricular valve located between the atrium and the ventricle of the left heart; also called the *mitral valve*. (Ch. 18)

bile The fluid secreted by the liver; contains bicarbonate, bile salts, cholesterol, lecithin, and bile pigments. (Ch. 22)

bile canaliculi (kan-ah-**lik**-u-li) Small ducts next to liver cells into which bile is secreted. (Ch. 22)

bile salt An organic component secreted by the liver; promotes solubilization and digestion of fat in the small intestine. (Ch. 22)

bilirubin (bil-ee-**roo**-bin) A red-colored product of the breakdown of hemoglobin by the liver cells. It is then excreted as waste in the bile. (Ch. 17)

biliverdin (*bil*-ee-**vur**-din) A green-colored product of the breakdown of hemoglobin by the liver cells. It is the first product of catabolism and is then converted to bilirubin. It is excreted as waste. (Ch. 17)

biological rhythms Biological processes that occur in a rhythmic or regular periodic manner; for example, the menstrual cycle (once a month) or the circadian (24-hour) cycle of core body temperature. (Ch. 29)

bipolar cell A type of nerve cell found in the retina. (Ch. 8)

bipolar neuron A type of neuron that has one axon and one dendrite, each extending from opposite sides of the soma. (Ch. 7)

birth control pill An oral contraceptive containing synthetic estrogen and/or progesterone that prevents pregnancy by blocking ovulation. (Ch. 33)

blackout Loss of vision; often due to reduced blood flow to the brain. (Ch. 29)

blastocyst (**blas**-toe-sist) The precursor of the embryo; composed of a ball of developing cells surrounding a central cavity. (Ch. 32)

blood group A system of related blood types, such as the ABO blood group or the Rh blood group. (Ch. 17)

blood pressure The force that causes the blood to flow through the vessels; usually expressed in units of millimeters of mercury (mm Hg). (Ch. 18)

blood type Method of classifying blood by determining the presence or absence of the specific antigens on the red blood cells. (Ch. 17)

blood vessels A series of tubes that convey blood from the heart throughout the body and back to the heart. (Ch. 18)

blood-brain barrier A group of anatomical barriers and transport systems that controls kinds of substances entering the brain extracellular space from blood and their rates of entry. (Ch. 7)

B-lymphocyte A blood cell that develops in the bone marrow and that is the basis of humoral or antibody-mediated immunity. (Ch. 28)

bolus A rounded mass of food. (Ch. 22)

Bowman's capsule A double-walled cup that surrounds the glomerulus and collects the glomerular filtrate. (Ch. 23)

bradycardia An abnormally low heart rate. (Ch. 18)

bradykinin (brad-i-kin-in) A hormone produced by the kidneys, whose primary biological effect is vasodilation. (Ch. 10)

brain The organ of the nervous system encased in the skull. (Ch. 7)

brain stem The section of the brain consisting of the medulla oblongata, pons, and midbrain; located between the spinal cord and the forebrain. (Ch. 7)

Braxton-Hicks contractions Spasmodic muscle contractions of the uterus that occur during the third trimester of pregnancy. (Ch. 32)

breast alveolus A gland located in the breast that stores and secretes breast milk. (Ch. 32)

Broca's (broe-kaz) **area** Region of the neocortex responsible for the motor control of speech; usually located in the left frontal lobe. (Ch. 7)

Brodmann area 3 The location of cells in the somatosensory cortex that exhibit the annular receptive field. (Ch. 8)

bronchiole (brong-kee-ol) A small division of a bronchus. (Ch. 20)

bronchitis (brong-kye-tis) A condition in which the lining of the bronchioles becomes inflamed and swollen, thereby reducing airway diameter. (Ch. 20)

bronchus A large-diameter airway branching off the trachea and entering the lung. (Ch. 20)

brush border membrane The microvilli of all the epithelial cells lining the small intestine. (Ch. 22)

buffer An agent that minimizes the change in pH when either acid or base is added to a solution. (Ch. 2)

bulk Material that cannot be digested. (Ch. 22)

bulk flow The movement of fluids or gases from a region of higher pressure to one of lower pressure. (Ch. 19)

bundle branches Right and left branches of the AV bundle going to each ventricle. (Ch. 18)

bundle of His Group of modified muscle cells between the right atrium and the interventricular septum specialized for very rapid conduction of impulses between the AV node and the Purkinje network; also called the *atrioventricular* (AV) *bundle*. (Ch. 18)

calcitonin (*kal-*sih-**toe-**nin) A peptide hormone produced by c-cells of the thyroid gland; secreted in response to increases in plasma calcium concentration causing increased deposition of calcium in bones. (Ch. 26)

calcium (kal-see-um) A chemical element essential for nerve function, muscle contraction, blood coagulation, bone formation, and many other physiological processes. (Ch. 26)

calcium antagonist A class of drugs that block the entry of calcium into the smooth muscle cells and dilate coronary arteries. (Ch. 19)

calcium ion A chemical that acts as a second messenger to initiate the intracellular reactions that modify the behavior of the target cell. (Ch. 5)

calmodulin (kal-**mod-**you-lin) An intracellular calcium-binding protein that mediates many of calcium's second-messenger functions. (Ch. 11)

caloric (kal-**or-**ik) **density** The amount of calories per gram of basic foodstuff. (Ch. 15)

calorigenesis (kal-o-ri-**jen-**i-sis) The production of heat as a result of chemical reactions of cell metabolism. (Ch. 13)

calyx (kayl-ikz) An extension of the renal pelvis; collects urine from the renal papillae. (Ch. 23)

cAMP *See* cyclic AMP.

canaliculus (*kan-*ah-**lik-**you-lus) A narrow tubular passage or channel between cells. (Ch. 26)

canaliculi (*kan-*ah-**lik-**you-lye) Plural of canaliculus. (Ch. 26)

capacitation A process involving physiological change in the plasma membrane of the sperm that increases motility and permits penetration of the surface of an ovum. (Ch. 32)

capillary (kap-ih-lar-ee) The smallest type of blood vessel between arteriole and venule. (Ch. 1)

capillary recruitment An increase in the number of capillaries that have blood flowing through them. (Ch. 21)

capillary transit time The amount of time necessary for blood to move through the capillaries; normal transit time is 1 second. (Ch. 21)

carbaminoglobin (*kar-*bam-**in-**oh-globe-in) The binding of carbon dioxide and hemoglobin to facilitate the transport of carbon dioxide in venous blood. (Ch. 21)

carbohydrate (*kar-*boe-**hye-**drayt) A class of organic compounds containing carbon, hydrogen, and oxygen in a ratio of 1:2:1. Sugars, starches, and cellulose are carbohydrates. This class is divided into monosaccharides, disaccharides, and polysaccharides. (Ch. 2)

carbon dioxide dissociation curve A graph of the relationship between blood P_{CO_2} and CO_2 content. (Ch. 21)

carbon monoxide (CO) A colorless, odorless gas that readily binds to hemoglobin, thereby decreasing blood-oxygen capacity. (Ch. 22)

carbonic anhydrase (*kar-***bon-**ik an-**hi-**draz) An enzyme that catalyzes the reversible reaction $CO_2 + H_2O \Leftrightarrow H_2CO_3$. (Ch. 7)

carboxyhemoglobin (COHb) (*kar-*bok-see-hee-moh-**glo-**bin) The binding of carbon monoxide, rather than oxygen, to hemoglobin, interfering with transport of oxygen to the cells. (Ch. 29)

carboxyl group COOH. (Ch. 2)

carboxypeptidase (car-*bok-*see-**pep-**tih-daze) An enzyme, the inactive precursor of which is procarboxypeptidase, secreted by the exocrine pancreas into the small intestine; breaks the peptide bond at the carboxyl end of the peptide chain. (Ch. 22)

carcinogen (car-**sin-**oh-jin) Any agent that induces cancerous transformation of cells. (Ch. 3)

cardiac cycle The sequence of electrical and mechanical events comprising one heartbeat. (Ch. 18)

cardiac muscles Striated, involuntary muscles that form the myocardium of the heart. (Ch. 9)

cardiac output The amount of blood pumped out of one ventricle in 1 minute. (Ch. 18)

cardiovascular control center A region of the medulla that integrates sensory input from many sources and subsequently modulates the autonomic nerve activity to the heart and blood vessels to maintain a relatively constant arterial blood pressure. (Ch. 19)

carotid (kar-**rot-**id) **body** A chemoreceptor located at the bifurcation of the carotid artery. It is sensitive to changes in levels of oxygen, carbon dioxide, and pH in the blood. (Ch. 10)

carotid sinus The dilated portion of the internal carotid artery located just above the branching of the common carotid artery. Pressure receptors that help regulate blood pressure are located here. (Ch. 10)

carrier proteins Membrane proteins that allow ions and large water-soluble molecules to pass through the lipid bilayer via facilitated diffusion. (Ch. 4)

catabolic hormone A hormone that promotes tissue breakdown. (Ch. 13)

catabolism (kah-**tab-**o-lizm) The breakdown of organic molecules into simpler compounds accompanied by the release of energy. (Ch. 6)

catalase (**kah-**tuh-laz) A cellular enzyme that catalyzes the decomposition of hydrogen peroxide. (Ch. 23)

catecholamine (kat-e-**kole-**ah-mean) A hormone that acts on the heart to increase both heart rate and the volume of blood pumped, thereby increasing blood pressure; dopamine, epinephrine, and norepinephrine. (Ch. 10)

cathartic A purgative; facilitates defecation by increasing bulk. (Ch. 22)

catheter (**kath-**eh-ter) A tube that is inserted into a body cavity or along a blood vessel for the introduction or removal of body fluids. (Ch. 23)

cation (**cat-**eye-on) An atom that loses one or more electrons, thus acquiring a positive charge. (Ch. 2)

caudal Anatomical term of position that means nearer to the tail, or further from the head; also called *inferior*. (Ch. 1)

caudate (**kaw-**dayt) One type of basal ganglia; receives inputs from the neocortex, thalamus, and substantia nigra of the brain stem. The **caudate nucleus** is one pair of basal ganglia associated with the reward system in the brain. (Ch. 9, 11)

C-cells Parafollicular cells in the thyroid gland, cells that produce the peptide hormone calcitonin. (Ch. 26)

CD4$^+$ cell A T-lymphocyte possessing a surface marker (receptor) designated as CD4; also called T_4 cell. (Ch. 28)

CD8$^+$ cell A T-lymphocyte possessing a surface marker (receptor) designated as CD8; also called T_8 cell. (Ch. 28)

cecum (see-come) The blind sac that marks the first part of the large intestine. (Ch. 22)

cell adhesion molecule (CAM) A glycoprotein important in nervous system development. (Ch. 7)

cell cycle The period of time from the beginning of one cell division to the beginning of the next division. (Ch. 3)

cell differentiation *See* differentiation.

cell fractionation The separation of different organelles of the cell in order to study each on its own in a strictly controlled system. (Ch. 3)

cell-mediated immunity A type of specific defense response wherein several kinds of T-cells and macrophages work together to destroy an antigen. (Ch. 28)

cell theory The broad basis for all modern biology and physiology; key concepts are (1) all organisms are made up of cells; (2) new cells arise only from preexisting cells; (3) all cells have the same fundamental chemical makeup and metabolic processes; and (4) activities and processes of an organism result from interdependent and cooperative workings of groups of cells. (Ch. 1)

cellular respiration The process by which organic molecules are catabolized by reactions that require atmospheric O_2, producing CO_2 and H_2O; the reverse of photosynthesis. (Ch. 6)

cellulose A polysaccharide composed of glucose subunits; found in plant cells. (Ch. 22)

central nervous system (CNS) The brain and spinal cord. (Ch. 7)

centrifuge A machine that allows materials to be spun at high speeds, thus separating them by density. (Ch. 3)

cerebellar (*ser*-eh-**bel**-ar) **anterior lobe** The portion of the cerebellum involved with planning, initiation, and control of limb movement. (Ch. 9)

cerebellar flocculonodular (*flock*-u-low-**nod**-you-lar) **lobe** The portion of the cerebellum involved with the maintenance of equilibrium and posture. (Ch. 9)

cerebellar granule cell Fibers in the cerebellum that synapse with the dendrites of the Purkinje cells, which in turn generate simple patterns of action potentials. (Ch. 9)

cerebellar posterior lobe The portion of the cerebellum involved with planning, initiation, and control of limb movement. (Ch. 9)

cerebellum (*ser*-eh-**bel**-um) The area of the brain responsible for locomotor coordination. (Ch. 7)

cerebrospinal (se-*ree*-broe-**spy**-nal) **fluid (CSF)** The fluid that fills the cerebral ventricles and the subarachnoid space surrounding the brain and the spinal cord; provides an extracellular environment for brain cells. (Ch. 10)

cervical mucosa (**ser**-vih-kul mew-**ko**-sah) Secretions of the cervix composed of glycoproteins, salt, and water. (Ch. 32)

cervix (**ser**-viks) The lower portion of the uterus; connects uterus and vagina. (Ch. 32)

channel gating Mechanism by which the opening and closing of ion channels is regulated. Channel gating is regulated by channel proteins. (Ch. 4)

channel protein Membrane proteins that allow ions and large, water-soluble molecules to pass through the lipid bilayer via facilitated diffusion. (Ch. 4)

characteristic frequency, f_0 The frequency that most effectively activates a particular auditory nerve fiber. (Ch. 8)

chemical energy The energy of the cell used to do work; released by reactions that take place within the cell. (Ch. 2)

chemical equation A representation using chemical symbols to show a chemical reaction. (Ch. 2)

chemical pH buffer A weak acid and its conjugate base. (Ch. 25)

chemical reaction Process in which chemical bonds are broken down and recombine into new compounds. (Ch. 2)

chemical stimuli Humoral or fluid-related stimuli. (Ch. 29)

chemical symbol An abbreviation for an element. (Ch. 2)

chemiosmotic (*kem*-ee-oz-**moh**-tik) **mechanism** Mitochondrial process by which movement of protons down an electrochemical gradient is used to convert adenosine diphosphate (ADP) into adenosine triphosphate (ATP). (Ch. 6)

chemiosmotic synthesis of DNA The overall process by which oxidative phosphorylation generates ATP. (Ch. 6)

chemoreceptor An afferent nerve ending (or cells) that is sensitive to concentrations of certain chemicals. (Ch. 10)

chemotaxis (*kee*-moe-**tak**-sis) The attraction by and subsequent movement toward a chemical by a leukocyte or macrophage. (Ch. 28)

chloride shift The exchange of chloride ions for bicarbonate ions by the erythrocyte as carbon dioxide enters and leaves the erythrocyte. (Ch. 21)

chloroform An inhalation anesthetic, first used for women in childbirth. (Ch. 29)

cholecalciferol (*koe*-lee-kal-**sif**-er-al) Natural vitamin D. (Ch. 26)

cholecystectomy (*koe*-lee-sis-**tek**-toe-mee) The surgical removal of the gall bladder. (Ch. 22)

cholecystokinin (CCK) A polypeptide hormone produced by the gastrointestinal tract; stimulates enzyme secretion and growth in the pancreas, stimulates gallbladder contraction, and relaxes the sphincter of Oddi. (Ch. 22)

cholesterol (*koe*-**les**-ter-ol) The most abundantly found steroid located in cell membranes; it is actually classified as a lipid. Cholesterol is a component of gallstones and is a precursor in the production of many steroid hormones. (Ch. 4)

cholinergic (*koe*-lin-**er**-jik) **receptor** A receptor of acetylcholine. (Ch. 7)

chondrocyte (**kon**-droe-site) A cell involved in cartilage formation. (Ch. 13)

chromaffin (*kroe*-**maf**-in) **cell** The functional unit of the adrenal medulla; functions as a neuroendocrine cell. A cell that is readily stained with solutions containing chromium salts. (Ch. 14)

chromaffin granules Dense, membrane-bound vesicles in chromaffin cells that contain epinephrine. (Ch. 14)

chromatid (**kroe**-mah-tyd) One of two identical strands of chromatin that result from DNA replication during meiosis and mitosis. (Ch. 3)

chromatin (**kroe**-mah-tin) A combination of DNA and nuclear proteins that is the principal component of chromosomes. (Ch. 3)

chromosomal sex The genotypic sex of an individual. (Ch. 33)

chromosome (**kroe**-moe-sowm) A highly coiled, condensed form of chromatin formed in the cell nucleus during mitosis and meiosis. (Ch. 3)

chylomicron (*kie*-loe-**my**-kron) An aggregate of triglycerides, cholesterol, and phospholipids surrounded by protein that is absorbed from the intestinal cells into the lacteal of a villus. (Ch. 22)

chymotrypsin (*kie*-moe-**trip**-sin) An enzyme secreted by the exocrine pancreas; breaks down certain peptide bonds in proteins and polypeptides. (Ch. 22)

circadian (ser-**kay**-dee-an) **rhythms** Rhythms of the body that vary on a daily basis. (Ch. 11)

circulatory convection A method of transfer of heat within the body via the movement of body fluids. (Ch. 27)

circulatory shock The progressive deterioration of cardiovascular function; generally caused by great losses of blood, severe tissue damage, and irreversible circulatory collapse. (Ch. 19)

citric acid cycle A series of reactions occurring in the mitochondrion that represent the final common pathway for the breakdown of all the fuel molecules in the cell. Also called the *Krebs cycle* or the *tricarboxylic acid cycle*. (Ch. 6)

classical conditioning One type of associative learning in which an animal acquires an understanding about the relationship between one stimulus and another. (Ch. 11)

clearance The volume of plasma per unit time needed to account for the rate of removal of a particular substance. (Ch. 23)

clitoris (**kli**-tor-iss) A small body of erectile tissue in the female external genitalia; homologous to the penis in males. (Ch. 33)

cloning vector Plasmids used in DNA cloning studies. (Ch. 3)

clot A semisolid mass similar in consistency to gelatin. (Ch. 17)

clot retraction The decrease in the size of a clot; caused by thrombosethenin. (Ch. 17)

CNS Central nervous system. (Ch. 7)

CO_2 fixation The series of steps in which CO_2 from the atmosphere is used in the creation of carbohydrates and other high-energy organic compounds. (Ch. 6)

coagulation (koh-*ag*-you-**lay**-shun) The process of clot formation. (Ch. 17)

cocaine (coh-**cane**) A drug that enhances release of dopamine and augments the rate of self-stimulation. (Ch. 11)

cochlea (**koke**-lee-ah) A spirally coiled, fluid-filled inner tube of the inner ear; contains the organ of Corti, where nerve impulses are generated in response to sound waves. (Ch. 8)

cochlear hair cells Cells in the cochlea that transduce the vibrations of the basilar membranes into action potentials. (Ch. 8)

codon (**koh**-don) Three-base sequence in a polynucleotide of DNA or mRNA that specifies the location of a single amino acid in a polypeptide chain. (Ch. 5)

coenzyme (koe-**en**-zime) Complex organic co-factors that serve to transfer atoms and small groups from one enzymatic substrate to another acceptor. Vitamins function as coenzymes. (Ch. 2)

coenzyme A (CoA) A coenzyme also known as *pantothenic acid*; involved in many reactions, such as the oxidation of pyruvate or fatty acids to acetyl CoA in the citric acid cycle. (Ch. 6)

cofactor (**koe**-fak-tor) The nonprotein component that some enzymes need to function as catalysts. (Ch. 2)

cold receptors Peripheral temperature receptors that are naked nerve endings sensitive to temperature; characterized by increasing steady-state discharge rates as the skin is cooled. (Ch. 27)

collagen (**kol**-a-jen) A fibrous protein found in many connective tissues. (Ch. 26)

colloid (**kah**-loyd) A mixture in which there is suspended material of large molecular size. The protein-containing material found in the interior of the follicles of the thyroid gland. (Ch. 13)

colloidal osmotic pressure (also called **oncotic pressure**) Portion of the total osmotic pressure of blood that is due to the presence of plasma proteins. (Ch. 19)

colon (**koe**-lon) Part of the large intestine between the cecum and the rectum; subdivided into the ascending colon, transverse colon, descending colon, and sigmoid colon. (Ch. 22)

colony-forming unit A group of progenitor cells committed to forming a colony of one kind of blood cell. (Ch. 17)

color-blind Refers to visual defect in which an individual is unable to distinguish between certain colors. (Ch. 8)

colostrum (koe-**las**-strum) Thin yellowish fluid excreted by the breast; precursor to breast milk. (Ch. 32)

combined white blood cell count The number of leukocytes per microliter of whole blood. (Ch. 28)

common bundle *See* bundle of His.

complement (**kom**-plih-ment) A system of proteins that are normally present in plasma and that are necessary for the lysis of foreign cells in immunological defenses. (Ch. 28)

compliance The change in volume of a hollow organ that results from a change in pressure within the organ. (CL = Δ volume/Δ pressure.) (Ch. 20)

compound A combination of two or more elements. (Ch. 2)

concentration gradient The gradation in concentration that occurs between two regions having different concentrations of the same chemical. Expressed as a difference divided by the distance that separates the two regions. (Ch. 7)

conditioned response The behavior that results from repeated use of conditioned stimulus. (Ch. 11)

conditioned stimulus The stimulus used to create an association with another stimulus to acquire the desired response. (Ch. 11)

conductance The ease with which ions can flow through an ion channel; measured in Siemens. (Ch. 7)

conduction Heat exchange in the form of kinetic energy between molecules of objects that are in direct contact. (Ch. 27)

conductive hearing loss Hearing loss resulting from damage to the middle ear. (Ch. 8)

cone photoreceptor A light-sensitive cell located in the retina that is cone-shaped; responsible for color vision. (Ch. 8)

congestive heart failure Heart failure that results from the inability of the heart to maintain adequate cardiac output due to progressive congestion of the veins and heart. (Ch. 19)

conjugation (kon-ja-**gay**-shun) To join together. For example, a process by which a steroid is coupled to a polar sulfate or glucuronide group to make it more water-soluble to facilitate elimination. (Ch. 12)

connective tissue A loosely arranged group of cells, separated by an intercellular matrix, that support, bind, protect, and partition all bodily components. (Ch. 1)

connexin A protein channel that spans the gap between presynaptic and postsynaptic cells. (Ch. 7)

connexon A protein subunit of connexin; there are six, arranged in hexagonal assembly. (Ch. 7)

constipation (kon-stih-**pay**-shun) The accumulation of feces beyond what is normal in the large intestine. (Ch. 22)

continence (**kon**-tih-nence) The ability to keep urine in the bladder and to retain fecal material within the colon and rectum. (Ch. 10)

contraception (kon-trah-**sep**-shun) The ability to block or prevent conception. (Ch. 33)

contractility (kon-trak-**til**-ih-tee) The ability of muscle to generate force. (Ch. 16)

contraction (kon-**trak**-shun) Any form of muscle activity in response to stimulation. (Ch. 16)

control systems Mechanisms by which the body maintains homeostasis. (Ch. 1)

convection (kon-**vek**-shun) The transfer of heat between the body and a gas or liquid moving over the body surface. (Ch. 27)

coronary angioplasty A procedure in which a catheter is inserted into the occluded portion of a coronary artery and inflated, compressing the plaque and expanding the vessel's diameter. (Ch. 19)

coronary bypass surgery Surgery in which a vein from elsewhere in the body is grafted around the occluded segment of the coronary artery so that blood flow can bypass the diseased area. (Ch. 19)

coronary circulation The arrangement of blood vessels and their distribution through the myocardium. (Ch. 18)

coronary disease Disease that compromises blood flow through arteries which supply heart muscle. (Ch. 30)

corpus luteum (**kor**-pus **loo**-tee-um) An ovarian structure formed from the follicle after ovulation; secretes estrogen and progesterone. (Ch. 31)

cortex (**kor**-teks) The outer part of an organ. (Ch. 23)

cortical efferent zone (CEZ) Functional columns of cells in the motor cortex; all the cells in a given zone are involved in the contraction of a given muscle. (Ch. 9)

cortical nephron A nephron whose renal corpuscle is in the outer cortex. (Ch. 23)

corticobulbar (kor-ti-koe-**bul**-bar) **pathway** The avenue by which information from the motor cortex is transmitted to the brain stem. (Ch. 9)

corticospinal (kor-ti-koe-**spy**-nal) **pathway** The major motor pathway within the brain and spinal cord, used for voluntarily activating skeletal muscle. (Ch. 9)

corticospinal tract lesion A damaged area of the brain or spinal cord that disrupts inputs to lower motor neurons; characterized by loss of strength and movement in muscle groups, loss of strength in voluntary muscles, and the Babinski sign. (Ch. 9)

corticosteroid-binding (kor-tih-koe-**stear**-oid) **globulin (CBG)** A protein that binds with cortisol to facilitate transport in blood. (Ch. 14)

corticotropin-releasing hormone (CRH) A hormone produced by the hypothalamus that regulates the secretion of adrenocorticotrophic hormone (ACTH). (Ch. 10)

cortisol (**kor**-ti-sol) A glucocorticoid secreted by the adrenal glands. (Ch. 12)

cortisol-binding globulin (CBG) A carrier protein that binds cortisol for transport in the blood. (Ch. 12)

cortisone (**kor**-tih-zone) A hormone regulated by the hypothalamus. (Ch. 11)

countercurrent exchange The passive exchange of solutes, water, or heat between two adjacent streams flowing in opposite directions, as in the vasa recta of the kidney medulla. (Ch. 23)

countercurrent multiplication An energy-demanding process that sets up a solute (osmotic) gradient along the length of two streams flowing in opposite directions, as in Henle's loop in the kidney medulla. (Ch. 23)

coupled reactions Endergonic reactions that are paired with exergonic reactions to obtain the free energy necessary to proceed. (Ch. 6)

covalent (koe-**vay**-lint) **bond** The chemical bond formed when two or more atoms share electrons. (Ch. 2)

c-peptide A peptide involved in insulin synthesis. (Ch. 15)

cranial An anatomical term of position that means nearer to the head; also called *superior* or *caudal*. (Ch. 1)

creatine phosphate (CP) A high-energy compound capable of transferring its phosphate and energy to adenosine diphosphate (ADP) to create adenosine triphosphate (ATP). (Ch. 6)

creatinine (kree-**at**-ih-nine) An end-product of muscle metabolism excreted in the urine; its renal clearance can be used to estimate glomerular filtration rate. (Ch. 23)

crenation The shriveling of an erythrocyte as it loses an excessive amount of water. (Ch. 17)

cretinism (**kree**-tin-izm) A mental handicap that results from an untreated thyroid hormone deficiency; linear growth is also impaired, resulting in dwarfism. (Ch. 13)

CRH Corticotropin-releasing hormone, produced by the hypothalamus; controls the release of adrenocorticotrophic hormone (ACTH) from the anterior pituitary. (Ch. 14)

cross bridge Myosin molecule head that projects from the ends of the thick myofilaments within a muscle cell; can bind to an active site on a thin myofilament in the presence of calcium, causing the filaments to slide past one another. (Ch. 16)

crossing over A process in mitosis in which homologous chromosomes exchange chromosomal segments, ensuring the infinite recombination of genetic material. (Ch. 31)

crossmatching A process by which agglutinin in the serum of blood is tested to confirm blood type; the presence or absence of red cell agglutination indicates the type of antibody present, thus confirming the type of antigen present on unknown red cells. (Ch. 17)

current flow The movement of electric charge; in biologic systems the current flow is the movement of ions. (Ch. 7)

Cushing's syndrome A condition resulting from excessive secretion of the adrenal cortex. General appearance of the patient is thick arms and legs, increased fat in the face, trunk, shoulder blades, and base of the neck, as well as masculinization of the female. (Ch. 14)

cutaneous (cue-**tay**-nee-us) Pertaining to the skin. (Ch. 8)

cyclic AMP (cAMP) Cyclic 3′,5′-adenosine monophosphate; the compound derived from nucleotide that mediates the action of many hormones in cells. (Ch. 2)

cyclic GMP Cyclic 3′,5′ guanosine monophosphate, an intracellular phosphate that opens sodium channels in photoreceptor plasma membranes. (Ch. 12)

cyclins Specialized activating proteins that undergo a cycle of synthesis and degradation during the cell cycle. They help control progression from one stage of the cell cycle to the next. (Ch. 3)

cyclosporine (sye-cloe-**spore**-in) An immunosuppressive agent that reduces the possibilities of rejection of foreign substances by the immune system. (Ch. 18)

cystic fibrosis (**sis**-tik fi-**broe**-sis) An inherited genetic disorder in which secretions of body glands are abnormally thick and sticky, clogging glands and preventing them from working properly. (Ch. 4)

cytochemistry (sye-to-**khem**-is-tree) A research technique used with microscopy to produce color or enhanced contrast at the sites of specific molecules. (Ch. 3)

cytochrome (**sye**-toe-krome) A protein that serves as an electron carrier in oxidative phosphorylation. (Ch. 6)

cytokine (**sye**-toe-keen) A general term for a protein messenger secreted by macrophages and monocytes or by lymphocytes. (Ch. 30)

cytokinesis (sye-toe-kih-**nee**-sis) The process by which the cytoplasm divides, each half taking with it one of the two new nuclei to form a new cell. (Ch. 3)

cytoplasm (**sye**-toe-plazm) One of two major compartments of the cell; composed of the cytosol and the organelles. (Ch. 3)

cytoskeleton (sye-toe-**skel**-eh-ton) The internal structure of the cell responsible for maintaining cell shape. (Ch. 3)

cytosol (**sye**-toe-sol) The fluid surrounding the nucleus. (Ch. 3)

data interpretation The comparison of amounts, rates, times, and other measurable evidence of the body processes that reveals relationships between function and structure. (Ch. 1)

dead space air The amount of air in conducting airways, approximately 150 ml for each 500 ml inhaled, that does not participate in gas exchange. (Ch. 20)

deamination (dee-*am*-in-**ay**-shun) **of amino acids** Removal of an amino acid group (NH_2) from a molecule; a process by which amino acids are broken down by acetyl compounds so they can be used to generate adenosine triphosphate (ATP). (Ch. 6)

decremental conduction The slowed travel of a pacemaker signal as it passes through the atrioventricular node. (Ch. 18)

defecation reflex-defecation (*deaf*-ih-**kay**-shun) A sacral spinal reflex elicited by distension of the rectum and, unless voluntarily inhibited, resulting in propulsion of feces through the anus. (Ch. 22)

dehydration (*dee*-hi-**dray**-shun) A deficiency of water in the body. (Ch. 15)

7-dehydrocholesterol A precursor to cholecalciferol, which itself is derived from cholesterol; changes to cholecalciferol through the action of ultraviolet light. (Ch. 26)

dehydroepiandrosterone (dee-*hy*-dro-ep-i-an-**dro**-stir-own) **(DHEA)** A steroid hormone secreted by the adrenal cortex. (Ch. 12)

dehydrogenation reaction The removal of an electron during oxidation through the loss of an entire atom, almost always hydrogen. (Ch. 6)

deiodinase A compound involved in the breakdown of thyroglobulin through the removal of iodide from MIT and DIT to produce iodine and tyrosine. (Ch. 13)

delta cells of the pancreas Cells found in the islet of Langerhans responsible for the production of somatostatin. (Ch. 15)

dendrite (**den**-drite) A fiberlike structure that branches out from the cell body. (Ch. 7)

dentate nucleus A deep cerebellar nucleus that forms synaptic connections with the axons of Purkinje cells. (Ch. 9)

deoxyribonucleic (dee-ok-see-*ri*-boe-nyoo-**klee**-ik) **acid (DNA)** A double-helical molecule consisting of two strands of four different nucleotides. Contains the genetic information that is duplicated in cell division. (Ch. 3)

deoxyribonucleic ligase An enzyme that joins together broken strands of different DNA molecules, allowing any DNA fragment to be grafted into the DNA of a bacterium. (Ch. 3)

dependent variable The variable that changes as a result of changes in the independent variable; usually represented on the vertical axis. (Ch. 1)

depolarization (dee-*poe*-lar-i-**zay**-shun) The gradual movement of the membrane potential toward a more positive value. (Ch. 8)

dermatome A particular area of the skin whose receptors send sensory information to specific spinal cord segments. (Ch. 8)

dermis (**der**-mis) The thick layer of skin located beneath the epidermis; composed of irregular, dense connective tissue. (Ch. 27)

desmosome (**dez**-moe-zome) Type of cell-to-cell junction where membrane differentiation takes place. (Ch. 16)

DHEA Dehydroepiandrosterone, a steroid hormone secreted by the adrenal cortex and converted to testosterone in peripheral tissues. (Ch. 14)

diabetes insipidus (*dye*-ah-**bee**-teez in-**sip**-ih-des) A disease due to a lack of ADH; marked by great thirst and excretion of a large volume of dilute urine. (Ch. 13)

diabetes mellitus (mel-**ih**-tus) A disease in which plasma glucose control is defective because of insulin deficiency or decreased target cell response to insulin. (Ch. 15)

diabetic neuropathy A secondary complication of diabetes involving the impairment of nerve function, resulting in abnormalities in the bladder or gastrointestinal tract function, impotence, or loss of feeling in the lower limbs. (Ch. 15)

diabetic retinopathy (reh-tin-**op**-uh-thee) A secondary complication of diabetes that results in the deterioration of blood vessels that nourish the retina, eventually resulting in blindness. (Ch. 15)

diacylglycerol (dye-*ace*-ill-**glis**-er-all) **(DG)** The second messenger that activates the protein kinase C, which then phosphorylates a large number of other proteins. (Ch. 5)

dialysis (dye-**al**-ih-sis) The separation of smaller from larger molecules in solution by diffusion of the small molecules through a selectively permeable membrane. (Ch. 24)

diapedesis (*dye*-ah-peh-**dee**-sis) The process by which leukocytes pass through the capillary wall to enter spaces around blood vessels. (Ch. 17)

diaphragm (**dye**-ah-fram) A large, dome-shaped skeletal muscle that separates the chest cavity from the abdomen. (Ch. 20)

diarrhea (*dye*-ah-**ree**-ah) Frequent defecation of highly fluid material. (Ch. 22)

diastole (dye-**as**-toe-lee) The period of cardiac cycle during which the ventricles are filling with blood. (Ch. 18)

dichromat (dye-**kroh**-mat) Males who are color-blind due to possessing only two types of cones; a sex-linked trait. (Ch. 8)

dichromatic color vision Color vision that relies on two types of cones. (Ch. 8)

diencephalon (die-en-**sef**-ah-lon) The core of the anterior part of the brain; lies beneath the cerebral hemispheres and contains the thalamus and the hypothalamus. (Ch. 9)

dietary fiber Cellulose and other indigestible polysaccharides found in plant food sources. (Ch. 22)

differential white blood cell count The number of each kind of white blood cell, usually expressed as a percentage of the total or combined white blood cell count. (Ch. 17)

differentiation (dif-er-en-she-**aye**-shun) The process by which a single cell type develops into many different cell types. (Ch. 1)

diffuse synapse A chemical synapse in which the neurotransmitter release is not limited to specific active zones. (Ch. 7)

diffusing capacity of the lung The milliliters of a gas diffusing across the pulmonary membrane per minute per millimeter of mercury pressure difference between the alveolar air and the pulmonary blood. (Ch. 30)

diffusion (dif-**you**-zhun) A process that involves the random movement of particles, such as ions and molecules in solution, such that the substances become evenly mixed. (Ch. 2)

diffusion coefficient $D = 1.0310^{25}$ $cm^2/sec.$ (Ch. 4)

diffusion trapping A phenomenon in which the neutral form of a weak base will diffuse across a cell membrane, whereas the charged form will not. (Ch. 4)

digestion (dye-**jes**-chun) The conversion of large organic molecules in the diet to smaller molecules. (Ch. 23)

digestive enzymes Organic compounds that break down large organic molecules into simpler, smaller components. (Ch. 22)

digestive tract A tube about thirty feet long that traverses the body from lips to anus and is composed of the mouth, pharynx, esophagus, stomach, and small and large intestines; also called *alimentary canal.* (Ch. 22)

digitalis A drug that tends to increase the contractility of cardiac muscle by increasing intracellular calcium concentrations. (Ch. 19)

dihydrotestosterone (dye-*hi*-droe-tes-**tos**-ter-own) **(DHT)** A steroid that stimulates sexual differentiation of the external genitalia in the male. (Ch. 33)

1,25 dihydroxyvitamin Vitamin D_3. (Ch. 12)

diiodotyrosine (DIT) An iodinated compound found primarily within the thyroid follicle cell; used to make T_3 and T_4. (Ch. 13)

direct calorimetry A method of measuring metabolic rate that uses a large insulated chamber to determine the heat loss from a person's body. (Ch. 27)

disaccharidase (dye-**sak**-er-ide-aze) An enzyme that can hydrolyze disaccharides. (Ch. 22)

disaccharide (dye **sak** er-ide) A carbohydrate molecule composed of two monosaccharides. (For example, maltose = glucose-glucose.) (Ch. 2)

discrete synapse A chemical synapse in which the chemical neurotransmitter is released from restricted areas of the presynaptic terminal into a small synaptic cleft. (Ch. 7)

distal (**dis**-tal) Refers to the insertion, or less stationary point of attachment, of a muscle. (Ch. 16)

distal convoluted tubule A short nephron segment between the thick ascending limb of Henle's loop and the connecting segment. (Ch. 23)

distal muscles Muscles involved with the movement of the skeleton, primarily the limbs. (Ch. 9)

distal tubule The distal convoluted tubule, connecting segment, and first part of the cortical collecting duct. (Ch. 23)

diuresis (dye-you-**ree**-sis) An increased urine flow rate. (Ch. 23)

diuretic (dye-you-**ret**-ik) An agent that increases urine output. (Ch. 24)

diurnal (die-**urn**-al) Occurring on a 24-hour cycle. (Ch. 11)

diving bell A chamber, either open at the bottom or enclosed, which holds air for a diver to breathe. (Ch. 29)

diving reflex A protective reflex, elicited when the face is submerged in water, that shunts blood away from peripheral tissues and allows for heart-brain perfusion. (Ch. 20)

dizygotic twins Siblings resulting from fertilization of two ova by individual sperm. (Ch. 32)

DNA polymerase An enzyme that during DNA replication forms a new DNA strand by joining together nucleotides already base-paired with an existing DNA strand. (Ch. 5)

DNA replication The copying of the chromatin in the nucleus. (Ch. 5)

DNA template An existing strand of DNA that guides synthesis of a new strand. During DNA synthesis the template strand is incorporated into the new DNA molecule. During RNA synthesis the DNA template is not incorporated. (Ch. 5)

DNA transcription The formation of mRNA containing, in linear sequence of its nucleotides, genetic information in a specific gene; the first stage of protein synthesis. (Ch. 5)

dopamine (**dope**-ah-mean) A catecholamine synthesized from tyrosine; a precursor of epinephrine and norepinephrine. (Ch. 7)

dopamine receptor A receptor that causes the hyperpolarization of the postsynaptic potential through the activation of chemically sensitive potassium ion channels. (Ch. 7)

dorsal column nuclei A collection of second-order sensory neurons located in the medulla oblongata. (Ch. 8)

dorsal column pathway A major pathway located in the dorsal (posterior) white matter of the spinal cord for the transmission of somatic sensory information from the surface of the body to the neocortex. (Ch. 8)

dorsal root ganglion A cluster of first-order sensory cell bodies located along the posterior root of a spinal nerve. (Ch. 8)

double helix A configuration in which two nucleotide strands of DNA are coiled around each other. (Ch. 2)

down regulation of receptor number A sustained elevation of the concentration of an agonist circulating in the blood, leading to a decrease in the number of receptors present in target cells. (Ch. 5)

ductus arteriosus (**duk**-tus ar-*tee*-ree-o-sus) A fetal bypass vessel that connects the pulmonary artery to the aorta. (Ch. 21)

duodenum (doo-o-dee-num) The initial 30-cm segment of the small intestine. (Ch. 22)

dwarfism The state of being an abnormally undersized person. (Ch. 13)

dynamic exercise Activities such as bicycling or running in which the length of muscles changes and which can be quantified in physical units such as watts. (Ch. 30)

dynamic neurons Neurons that change their activity when there is an increase in the level of force applied. (Ch. 9)

dynein A protein from the microtubules of cilia and flagella; functions as an ATP-splitting enzyme. (Ch. 7)

dyslexia (dis-**lek**-see-ah) A congenital disorder that affects an individual's ability to read, write, or spell. (Ch. 11)

dysmenorrhea (*dis*-men-o-**ree**-ah) Excessive uterine contractions or menstrual cramps. (Ch. 31)

dyspareunia (*dis*-**par**-ee-oo-nee-ah) The absence of sufficient vaginal lubrication. (Ch. 33)

dysphagia Difficulty in swallowing. (Ch. 22)

E_m The difference in electrical potential across a membrane. (Ch. 7)

ear drum The membrane located between outer and middle ear that vibrates in response to sound waves; also called *tympanic membrane.* (Ch. 8)

ECG, EKG Electrocardiogram (electro + *kardia* ["heart"] + *gamma* ["a drawing"]). *See* electrocardiogram. (Ch. 18)

ectopic (ek-**top**-ik) **pregnancy** The implantation and development of a fetus at a site other than the uterus. (Ch. 32)

edema (e-**dee**-mah) An abnormal accumulation of fluid in the interstitial fluid spaces. (Ch. 23)

EEG *See* electroencephalogram.

efferent (**ef**-er-ent) Carrying away from. (Ch. 9)

efferent arteriole The arteriole that carries blood away from the glomerulus. (Ch. 23)

efferent neuron Nerve cell that carries signals away from the CNS. (Ch. 7)

efficiency The amount of work done for a certain amount of energy. (Ch. 16)

ejaculation (e-*jak*-yoo-**lay**-shun) The emission and expulsion of ejaculatory fluid. (Ch. 33)

elastic recoil The property of stretched tissue that allows it to return to a prestretched state. (Ch. 20)

elasticity The ability of deformed material to return to its original size and shape. (Ch. 20)

electrical potential (E) The difference in net charge between the inside and the outside of a cell. (Ch. 7)

electrocardiogram (ECG or EKG) (e-*lek*-troe-**kar**-dee-o-gram) A record of the electrical activity of the heart. (Ch. 18)

electrochemical gradient The combination of chemical and electrical gradients of a solution that determines the flux of positive and negative ions. (Ch. 4)

electroencephalogram (e-*lek*-troe-en-**sef**-ah-loe-gram) **(EEG)** A record of the frequency and amplitude of the brain's electrical activity. (Ch. 11)

electrolyte (e-**lek**-troe-lite) An ionized substance in solution; capable of conducting electricity. (Ch. 23)

electromagnetic radiation A continuum of electromagnetic energy of varying wavelengths, ranging from radio waves to X-rays. (Ch. 8)

electron Negatively charged particle of an atom. (Ch. 2)

electron carrier A molecule that donates electrons to proteins. (Ch. 6)

electron transport chain (respiratory chain) The transfer of electrons from reduced coenzymes to O_2, a process that is coupled with phosphorylation of ADP to ATP. (Ch. 6)

element A substance that cannot be broken down into simpler substances by chemical reactions. (Ch. 2)

embolus (**em**-boe-lus) A foreign object floating in the blood; for example, a clot (thromboembolus) or an air bubble (air embolus). (Ch. 17)

embryo (**em**-bree-o) A blastocyst after it has fully implanted in the endometrium of the uterus. (Ch. 32)

embryonic stage Weeks 2 to 8 following fertilization. (Ch. 32)

emetic (**em**-et-ik) Chemical agent that causes vomiting by stimulating chemoreceptors in the stomach and small intestine or brain. (Ch. 22)

emphysema (em-fih-**see**-mah) A lung disease that is characterized by destruction of the alveolar wall and consequent impairment of gas exchange. (Ch. 20)

emulsion (ee-**mul**-shun) A preparation of one liquid distributed in globules throughout a second liquid. (Ch. 22)

en passant An overlapping system of synapses in passage. (Ch. 7)

end-diastolic (dye-ah-**stol**-ik) **volume (EDV)** The volume of blood in the ventricle at the end of ventricular diastole. (Ch. 18)

end-plate current A small local current that flows across the postsynaptic membrane at the neuromuscular junction. (Ch. 16)

end-plate potential The depolarization of motor end-plate of skeletal muscle fiber in response to acetylcholine; initiates action potential in muscle plasma membrane. (Ch. 16)

end-systolic volume The volume of blood in the ventricle at the end of ventricular systole. (Ch. 18)

endergonic reaction A reaction that cannot proceed without the input of free energy. (Ch. 6)

endocardium (en-doe-**kar**-dee-um) A thin layer of endothelial cells that lines the inner surface of the myocardium, which forms the inside walls of the atria and ventricles. (Ch. 18)

endocrine (**en**-doe-krin) **gland** Any of the several ductless-type glands that secrete hormones into the blood stream, thereby influencing the activity of cells located away from the gland. (Ch. 12)

endocrine pancreas Small portion of the pancreas that secretes insulin and glucagon and other hormones (also called *islets of Langerhans*). (Ch. 15)

endocrine system All of the body's hormone-secreting glands. (Ch. 12)

endocytosis (en-doe-sye-**toe**-sis) The process that allows the entrance into a cell of substances that are otherwise unable to penetrate the cell membrane; includes pinocytosis and phagocytosis. (Ch. 3) **Receptor-mediated endocytosis** involves larger peptide hormones that are internalized by their receptor tissue cells. (Ch. 12)

endogenous (en-**doj**-e-nus) Produced within the body or due to internal causes. (Ch. 23)

endogenous creatinine clearance The renal clearance of endogenous (i.e., not administered) creatinine; it is used clinically to estimate glomerular filtration rate. (Ch. 23)

endogenous protein Digestive enzymes, epithelial cells, mucoproteins, and plasma proteins that leak into the digestive tract and are absorbed. (Ch. 23)

endometriosis (en-doe-mee-tree-**oh**-sis) A disease characterized by the presence of endometrial-like tissue in locations other than the uterus, such as the fallopian tubes or ovaries. (Ch. 33)

endometrium (en-doe-**mee**-tree-um) The lining of the uterus. (Ch. 30)

endomysium (en-doe-**meez**-ee-um) The connective tissue sheath surrounding individual muscle fibers. (Ch. 16)

endoplasmic reticulum (en-doe-**plaz**-mik re-**tik**-you-lum) **(ER)** A network of membranous tubules, flattened sacs, and vesicles located in the cytoplasm that function in intracellular transport, synthesis, storage, packaging, and secretion. (Ch. 3)

endorphin (en-**dor**-fin) Any of the "endogenous-morphine" family of peptides; they function as a neurotransmitter at synapses activated by opiate drugs and as a paracrine hormone. (Ch. 12)

endothelial (en-doe-**thee**-lee-al) **clefts (or pores)** Small openings between endothelial cells that allow for the passage of molecules. (Ch. 19)

endothelium-derived relaxing factor (EDRF) Vasodilator released from the endothelial cells that line the blood vessels; currently thought to be nitric oxide (NO). (Ch. 19)

energy transformation Conversion of energy from one form to another. Plants convert radiant energy into chemical energy stored in carbohydrates. Animals convert this chemical energy into a form (ATP) that can be used for biological work. (Ch. 6)

enkephalin (en-**kef**-ah-lin) A peptide of the nervous system that affects the perception of pain. (Ch. 8)

enterohepatic (en-teh-roe-he-**pat**-ik) **circulation of bile salts** Recycling of bile salts between liver and small intestine. (Ch. 22)

enterokinase (en-teh-roe-**kin**-aze) An enzyme present in the luminal membrane of the epithelial lining of the small intestine; converts inactive trypsinogen to active trypsin. (Ch. 22)

enzyme (**en**-zime) A natural catalyst that increases the rate of chemical reactions in the body. (Ch. 2)

eosinophil (ee-o-**sin**-o-fil) A leukocyte whose primary function is detoxification of foreign proteins and other substances. (Ch. 17)

epicardium (ep-i-**kard**-ee-um) The outer surface of the myocardium, which consists primarily of connective tissue and fat. (Ch. 18)

epididymis (ep-i-**did**-i-mis) The portion of the male reproductive duct system in which sperm mature between the seminiferous tubules and the vas deferens. (Ch. 31)

epiglottis (ep-ih-**glot**-is) Cartilage guarding the superior opening into the larynx. (Ch. 22)

epimysium (ep-ih-**meez**-ee-um) A connective tissue sheath that covers the entire surface of a muscle. (Ch. 16)

epinephrine (ep-ih-**nef**-rin) One of the two catecholamines that is produced by the adrenal medulla; stimulates the sympathetic nervous systems, increasing blood pressure, stimulating the cardiac muscle, and causing glycogenolysis. Also known as *adrenaline*, it acts together with norepinephrine to prepare the body for "fight-or-flight" response. (Ch. 14)

epiphyseal (ep-ih-**fizz**-e-al) **plate** The region near the end of a bone where linear growth occurs. (Ch. 13)

epithelial tissue Closely packed cells arranged in flat sheets ranging from one to several layers in thickness that cover exposed body surfaces and line body cavities, tubes, and organs. (Ch. 1)

equilibrium A state at which the rate of formation of products of a chemical reaction is exactly balanced by the rate of reformation of the original reactants. (Ch. 2)

equilibrium potential The voltage of electrical force required to draw ions across a cell membrane at the same rate as they pass because of the concentration gradient. (Ch. 7)

ergocalciferol Vitamin D_2, the form of vitamin D found in plants and yeasts. (Ch. 26)

erythrocytosis (eh-rith-roe-sigh-**toe**-sis) An increase in the number of red blood cells in response to low oxygen environments and other states in which oxygenation of blood is reduced. (Ch. 17)

erythropoiesis (eh-rith-roe-**poy**-ee-sis) Erythrocyte formation. (Ch. 17)

erythropoietin (eh-rith-roe-**poy**-ih-tin) A glycoprotein hormone, produced chiefly by the kidneys, that stimulates the rate of production, maturation, and release of red blood cells from bone marrow. (Ch. 23)

esophagus (eh-**sof**-ah-gus) The muscular tube located mostly in the thorax that conducts material from the pharynx to the stomach. (Ch. 22)

essential amino acids Amino acids that cannot be synthesized and must be ingested as food. (Ch. 6)

essential hypertension Chronically elevated arterial blood pressure due to primary disturbance of the cardiovascular system. (Ch. 19)

estradiol (es-trah-**die**-ol) The principal sex steroid in females. (Ch. 13)

estrogen (es-**trow**-jen) A hormone secreted by the ovaries and the placenta that stimulates the growth and development of tissues in the reproductive tract. (Ch. 12, 32)

ether An anesthetic gas. (Ch. 22)

eukaryotic cell A cell that has a well-defined nucleus contained within a nuclear membrane. (Ch. 3)

evaporation The process of converting a liquid to a gas. (Ch. 27)

excision repair of DNA A process under enzyme control by which a section of damaged DNA is selectively removed and replaced. (Ch. 5)

excitatory postsynaptic potentials (EPSP) Potentials that depolarize the postsynaptic membrane and tend to excite nerve cells to discharge action potentials. (Ch. 7)

excretion (eks-**kree**-shun) The elimination of a substance from the body in the urine or feces. (Ch. 23)

exergonic reaction A chemical reaction that releases free energy and is likely to proceed spontaneously. (Ch. 6)

exocrine (ek-so-krin) **gland** A type of gland that secretes its products by way of ducts that open onto lining or covering epithelium. (Ch. 22)

exocrine pancreas A portion of the pancreas responsible for the secretion of fluid and various enzymes involved in the process of digestion. (Ch. 15, 22)

exocytosis (*eks*-oh-sih-**toe**-sis) A process in which a secretory vesicle fuses with the plasma membrane and materials are expelled into the extracellular fluid. (Ch. 3)

exogenous (eks-**oj**-en-us) Foreign to the body. (Ch. 23)

exon (**eks**-on) Segment of the eukaryotic gene that consists of DNA coding for a sequence of nucleotides in a specific molecule of messenger RNA. (Ch. 5)

experimental design Construction of experiments so that specific conditions are created deliberately to produce the observations in question, ideally so that only one explanation exists for the observation. (Ch. 1)

expiration (*eks*-pih-**ray**-shun) Movement of air out of the lungs. (Ch. 10)

expiratory neurons Neurons in the medulla that stimulate expiratory muscles during exercise. (Ch. 20)

extension The movement of one body part away from another; for example, movement of the forearm away from the upper arm when extension occurs at the elbow. (Ch. 16)

extensor A muscle that, by contracting, causes extension; for example, the triceps brachius is an extensor of the forearm. (Ch. 9)

external Of two sites, the one farther from the center of the body. (Ch. 1)

external anal sphincter Skeletal muscles at the end of the large intestine that are relaxed during defecation. (Ch. 22)

external intercostal muscles A set of inspiratory muscles that pull the ribcage upward and outward. (Ch. 20)

external sphincter A group of striated muscles that close off the exit from the bladder. (Ch. 10)

extracellular fluid The fluid outside of cells. (Ch. 1, 23)

extrafusal fiber Skeletal muscle fibers that form the bulk of skeletal muscles. Contraction is controlled by alpha motor neurons. (Ch. 9)

extrapolation The extension of a relationship beyond the range of available experimental data. (Ch. 1)

extrinsic (eks-**trin**-sik) Having an external origin, arising from outside the body. (Ch. 1)

extrinsic (peripheral) proteins Membrane proteins that do not penetrate the lipid bilayer. (Ch. 4)

extrinsic clotting pathway The process by which clotting begins after blood comes in contact with injured tissue. (Ch. 17)

F cells of the pancreas Cells that produce pancreatic polypeptide. (Ch. 15)

facilitated (fah-**sil**-ih-tay-ted) **diffusion** Downhill movement of substances, in or out of the cell, that is enabled by the presence of a carrier protein. (Ch. 4)

fallopian (fal-**low**-pee-an) **tubes** Muscular, tubelike structures that arise from the uterus and extend into the abdominal cavity to terminate near the ovaries. Also known as the *uterine tubes*. (Ch. 31)

fasciculi Bundles of individual muscle fibers. (Ch. 16)

fast axonal transport The movement of newly produced membranous organelles down the axon. (Ch. 7)

fast skeletal muscles Muscles that contract rapidly but fatigue quickly because they are poorly vascularized. (Ch. 9)

fastigial nucleus One of three deep cerebellar nuclei involved in the maintenance of balance and posture. (Ch. 9)

fat cells Adipose tissue that stores triglycerides, a major source of energy for the body. (Ch. 11)

fats Lipids that are poorly soluble in water and highly soluble in organic liquids; *see also* triacylglycerol. (Ch. 2)

fatty acid A long chain of covalently linked carbon atoms to which only hydrogen atoms are attached, except for a carboxylic acid (—COOH) group at one end. (Ch. 2)

fatty acid oxidation The breakdown of fatty acids in the mitochondrial matrix. (Ch. 5)

feedback excitation A process by which the activation of beta-2 receptors increases the release of norepinephrine from the presynaptic terminal. (Ch. 7)

feedback inhibition A process by which the activation of alpha-2 autoreceptors inhibits the release of norepinephrine from the presynaptic terminal. (Ch. 7)

female pseudohermaphroditism Severe mascularization of the female external genitalia. (Ch. 33)

ferritin (**fer**-ih-tin) An iron-protein complex that regulates iron storage and transport. (Ch. 17)

fertilization The union of sperm and ovum. (Ch. 32)

fetal stage The period between week 8 after fertilization and birth. (Ch. 32)

fetoplacental (*fee*-toe-plah-**sen**-tahl) **unit** The fetal adrenal gland and placenta, which together stimulate estrogen synthesis. (Ch. 31)

fetus (**fee**-tus) An embryo after second month of development. (Ch. 32)

fever (**fee**-ver) A regulated increase in body temperature beyond the normal range. (Ch. 27)

fibrin (**fye**-brin) An insoluble protein derived from fibrinogen, which forms the foundation of a blood clot. (Ch. 17)

fibrinogen (fye-**brin**-o-jin) A soluble plasma protein manufactured in the liver. (Ch. 17)

fibrinolysis The enzymatic dissolution of a blood clot. (Ch. 17)

"fight-or-flight" response The physiological changes that prepare the body for the stress of threatening situations. (Ch. 27)

filterable calcium One of two major forms of calcium in extracellular fluid. (Ch. 26)

filtered load The quantity of a substance, per unit of time, that is filtered by the glomeruli and is therefore presented to the tubules; for a freely filterable substance, this is equal to the product of the plasma concentration and the glomerular filtration rate. (Ch. 23)

filtration (fil-**tray**-shun) The loss of capillary water to the extravascular space; *see* filtration pressure. (Ch. 19)

filtration pressure A net pressure leading to loss of capillary water; occurs when the hydrostatic pressure in the capillaries exceeds the oncotic pressure of the blood. (Ch. 19)

first heart sound The closure of the atrioventricular valves shortly after the beginning of ventricular systole. (Ch. 18)

fixed acid Physiologically speaking, a nonvolatile acid (i.e., an acid other than carbonic acid). (Ch. 25)

flatus (**flay**-tus) Gas present in the large intestine. (Ch. 22)

flavin adenine dinucleotide (FAD) A coenzyme derived from riboflavin. (Ch. 5)

flexion (**flek**-shun) A muscle action that causes a joint to close and a limb to assume a more withdrawn position. (Ch. 16)

flexor (**flek**-sor) A muscle that by contracting causes a joint to close up. (Ch. 9)

fluid mosaic model A generalized model of the structure of cell plasma membranes. (Ch. 4)

follicle-stimulating hormone (FSH) A glycoprotein hormone, created by the anterior pituitary, that stimulates growth and development of the gonads. (Ch. 12)

foramen ovale (foe-**ray**-men o-va-lee) A passage that allows blood to flow between the right and left atria in the fetal heart (Ch. 21)

force-velocity curve A plot of the velocity of muscle shortening over a range of forces ranging from zero to full isometric. (Ch. 16)

forced vital capacity The maximum amount of air that can be forcefully and rapidly exhaled after a deep breath. (Ch. 20)

fovea (**foe**-vee-ah) The most sensitive part of the retina. (Ch. 8)

Frank-Starling law of the heart The ability of the heart to adjust its output (stroke volume) in response to the input (venous return). (Ch. 18)

free energy The energy available to be put to work. (Ch. 2)

free radicals Highly reactive molecules formed as byproducts from molecular oxygen. (Ch. 29)

frequency code of contractile force The control of muscle tension as determined by the frequency of action potentials generated by upper motor neurons. (Ch. 9)

frequency code of stimulus intensity The number of action potentials generated per unit of time as a function of intensity of a stimulus. (Ch. 8)

frontal plane The plane that divides the body into front and back portions (also called the *coronal plane*). (Ch. 1)

functional residual capacity The volume of air remaining in the lungs after a normal expiration; the sum of expiratory reserve volume and residual volume. (Ch. 20)

gallbladder A small sac that stores bile between meals; propels bile to the small intestine where it participates in the emulsification and absorption of fat. (Ch. 22)

gallstones Small crystals of cholesterol that form in the gallbladder, this formation occurs when the capacity of bile salts and lecithin to dissolve cholesterol is exceeded. (Ch. 22)

gamma amino butyric acid (GABA) A neurotransmitter found throughout the CNS; a potent inhibitory neurotransmitter synthesized from glutamate by glutamic acid decarboxylase. (Ch. 7)

gamma motor neuron Motor neurons that synapse with intrafusal muscle fibers. (Ch. 9)

ganglion (gang-lee-on) **cells** Nerve cells in the retina that generate action potentials even in the absence of input from the bipolar cells. (Ch. 8)

gap junction The intercellular junction that allows ions and small molecules to flow between the cytoplasm of adjacent cells; the two adjacent plasma membranes are joined by small tubes. (Ch. 5)

gas diffusion The net movement of gas molecules from an area of higher to lower concentration. (Ch. 21)

gas exchange Process involved with delivering oxygen to and removing carbon dioxide from the body's cells. (Ch. 20)

gas uptake The process whereby the blood absorbs oxygen from alveolar air. (Ch. 21)

gastric (gas-trik) **chief cells** Cells, located in the basal region of exocrine secretory glands, that secrete pepsinogen, the precursor to pepsin, the enzyme that initiates the digestion of protein in the digestive tract. (Ch. 22)

gastric inhibitory peptide (GIP) A hormone of gastrointestinal origin that is important in the control of motility and secretion; released in response to intraluminal stimuli acting directly on the endocrine cells in the luminal lining. (Ch. 22)

gastric mucosal barrier The ability of the cell membranes of the gastric mucosa to restrict entrance of hydrogen ions to the interior of these cells because of the low permeability of the cell membrane to hydrogen ions. (Ch. 22)

gastric parietal cells Cells that secrete an isosmotic solution of HCl. (Ch. 22)

gastrin A hormone released by the gastric antrum in response to either impulses over vagal nerves or the stretching of the gastric wall and the contacting of chemical substances with the gastric antral mucosa; excites secretion of gastric juice. (Ch. 22)

gastrointestinal hormone Gastrin, cholecystokinin, secretin, and gastric inhibitory peptide. (Ch. 22)

gastrointestinal (*gas*-tro-in-**tes**-tin-al) **system** A collection of organs specialized to handle food through the processes of motility, secretion, digestion, and absorption. (Ch. 22)

G-cells Cells in the epithelial lining of the antral mucosa, which secrete the hormone gastrin into the portal blood. (Ch. 22)

gene Each section of chromosomal DNA that carries the information necessary for synthesis of a single polypeptide. (Ch. 3)

generator potential A change in the membrane potential of a sensory receptor cell that generates action potentials. (Ch. 8)

gene amplification An increase in the number of copies of a specific gene on a chromosome. (Ch. 5)

genetic code A reference to the way in which information about protein synthesis is stored in DNA, translated into RNA, and then used to make proteins. (Ch. 5)

genetic expression The phenomenon of the fact that only certain genes are tapped for information about how to make specific proteins they describe. (Ch. 5)

genome The sum of all the chromosomal genes of a cell. (Ch. 3)

germinal stage In prenatal development, the first two weeks of development following fertilization of the ovum. (Ch. 32)

gestation The period of development of the young from the time of fertilization of the ovum until birth. (Ch. 11)

gestational diabetes A transient form of diabetes that occurs in about 3% of pregnant women. (Ch. 15)

G-force The human body's sensation of weight due to gravitational attraction. (Ch. 29)

Gigantism *See* acromegaly.

glia (glee-al) Nonneuronal cellular elements of nervous tissue; provide physical and functional support to adjacent neurons. (Ch. 7)

glial scar Formed by reactive astrocytes in conjunction with fibroblasts after a brain injury; the scar blocks the regeneration of new axons. (Ch. 7)

globulin (glob-you-lin) A type of plasma; classified into alpha, beta, and gamma groups. (Ch. 17)

glomerular (glo-**mer**-yoo-lar) **filtration** The process in the kidney glomeruli in which an ultrafiltration of plasma is pushed out of the capillaries into the urinary space of Bowman's capsule. (Ch. 23)

glomerular filtration coefficient A coefficient which depends on the surface area and fluid permeability of the glomerular filtration barrier; it relates glomerular filtration rate to the net filtration pressure. (Ch. 23)

glomerular filtration rate (GFR) The rate at which the plasma is filtered by the kidney glomeruli; usually approximately 125 ml/min. (Ch. 23)

glomerulotubular balance The proportionate change in sodium reabsorption by the proximal convoluted tubule and Henle's loop when glomerular filtration rate is changed. (Ch. 23)

glomerulus (glo-**mer**-you-lus) In the kidney, the tuft of capillaries, surrounded by Bowman's capsule, where the blood is filtered. (Ch. 23)

glottis (glot-iss) The true vocal cords and the opening between them. (Ch. 22)

glucagon (gloo-kuh-gon) Hormone released by the islets of Langerhans of the pancreas; elevates the level of blood sugar. (Ch. 15)

glucagonlike peptide Peptides that are synthesized in the small intestine and hypothalamus. (Ch. 15)

glucocorticoid (*gloo*-koe-**kor**-tih-koyd) A steroid hormone produced by the adrenal cortex and having major effects on glucose metabolism. (Ch. 10)

gluconeogenesis (*gloo*-koe-*nee*-oh-**jen**-ih-sis) The formation of glucose by the liver and kidney from noncarbohydrate precursors. (Ch. 6)

glucose threshold The plasma level at which glucose first appears in the urine. (Ch. 23)

glucosidase An enzyme that splits a glucoside. α-Glucosidase (maltase) occurs in intestinal juice and β-glucosidase (cellobiase) in the kidney, liver, and intestinal mucosa. (Ch. 2)

glucosuria (*gloo*-koe-**soo**-ree-uh) A condition in which glucose is excreted in the urine. (Ch. 15)

glutamate (gloo-tah-mate) One of the most potent excitatory neurotransmitters in the nervous system. (Ch. 7)

glycerol-phosphate A three-carbon molecule that is combined with free fatty acids in triacylglycerol synthesis. (Ch. 15)

glycogen (gly-ko-jen) A polysaccharide carbohydrate resembling starch that occurs in many tissues, particularly the liver and skeletal muscles; a polymer of glucose. (Ch. 6)

glycogenolysis (*gly*-koe-jeh-**nol**-ih-sis) The process by which glycogen is broken down into units of glucose-1-phosphate, which enter gluconeogenesis after conversion of glucose-6-phosphate. (Ch. 6)

glycolipid (*gly*-koe-**lip**-id) Lipids, occurring in cell membranes, that are derived from sphinogosine and contain one or more monosaccharide units. (Ch. 2)

glycolysis (gly-**kol**-ih-sis) A process by which glucose obtained from food is converted to pyruvic acid. (Ch. 6)

glycophorin An intrinsic glycoprotein in the red blood cell membrane. (Ch. 4)

glycoprotein (*gly*-koe-**pro**-teen) A member of a class of conjugated proteins composed of protein with an attached carbohydrate group. (Ch. 2)

glycoprotein hormones Hormones that contain peptide and carbohydrate parts. Examples: thyroid-stimulating hormone (TSH), follicle-stimulating hormone (FSH). (Ch. 12)

goiter Any abnormal enlargement of the thyroid gland. (Ch. 13)

goitrogen An agent that inhibits iodide uptake or organification reactions, resulting in a blockage of thyroid hormone formation and the formation of a thyroid goiter. (Ch. 13)

Goldman equation The relationship between the intra- and extracellular concentrations of Na^+, K^+, and Cl^- ions, as well as their membrane permeabilities, P_{Na}, P_K, and P_{Cl}, and membrane potential. (Ch. 7)

Golgi (**goal**-jee) **complex** Part of the cell responsible for modifying macromolecules synthesized in the ER; processes and packages the macromolecules inside vesicles to be transported to the cell surface or cytosol. (Ch. 3)

Golgi tendon organ A type of muscle receptor located in series with the extrafusal muscle fiber and that transmits information about the force or tension produced by the contraction of a muscle. (Ch. 9)

gonadal sex The presence of testes or ovaries that determines the gender of the embryo. (Ch. 33)

gonadotropin (*gon*-ad-oh-**trow**-pin) Luteinizing hormone and follicle-stimulating hormone; hormones whose target tissue is the gonads. (Ch. 13)

gonadotropin-releasing hormone (GnRH) A hormone secreted by the hypothalamus that stimulates the release of luteinizing hormone and follicle-stimulating hormone. (Ch. 31)

gout A condition in which plasma uric acid is elevated and uric acid crystals precipitate in the joints, causing inflammation and painful swelling of the affected joints (often in the big toe). (Ch. 23)

G-proteins A group of integral membrane proteins that bind and require guanosine triphosphate to function. (Ch. 12)

gradient (**gray**-dee-ent) The continuous decrease or increase of a variable over a distance. (Ch. 2)

granulocyte (**gran**-yoo-low-site) A leukocyte that contains large granules and has a complex lobulated nucleus; also called *polymorphonuclear granulocyte*. (Ch. 17)

graph A diagram that expresses a relationship between two or more quantities. (Ch. 1)

graphic interpretation Learning and understanding the relationship between the variable quantities depicted in a graph. (Ch. 1)

Graves' disease Hyperthyroidism that occurs due to abnormality in the immune system that leads to the production of antibodies that recognize and bind to sites on the surface of thyroid follicle cells. (Ch. 13)

ground substance The noncollagenic portion of organic matter in bone; consists of a mixture of various proteoglycans. (Ch. 26)

growth cone A specialized enlargement at the top of the growing neuron. (Ch. 7)

growth hormone A large peptide hormone that stimulates body growth; also called *somatotropin*. (Ch. 12)

gustatory receptors Taste receptors; a type of epithelial cell clustered with other cells to compose a taste bud. (Ch. 8)

gustatory system The peripheral and central parts of the nervous system involved in the processing of taste information. (Ch. 8)

habituation (hah-*bit*-you-**a**-shun) A type of behavior in which the individual learns to decrease a response to a repeated stimulus. (Ch. 11)

hair receptor A mechanoreceptor that is a hair follicle on the surface of the skin and that has a nerve fiber wrapped around its base. (Ch. 8)

Haldane effect Describes the amount of carbon dioxide the blood can load, which is greater the more the hemoglobin is deoxygenated. (Ch. 21)

hand manipulation neurons Neurons in the posterior parietal cortex that increase their firing rate when the hand explores an object. (Ch. 9)

hand-eye coordination neurons Neurons in the posterior parietal cortex that are most active when the eye fixates on an object that is being touched. (Ch. 9)

haptoglobin A protein that binds with hemoglobin; prevents hemoglobin from being excreted in the urine. (Ch. 17)

heart block A situation in which the ventricles are not stimulated by the atrial impulse; due to failure of the AV node or AV bundle to conduct the impulse. (Ch. 18)

heart failure The inability of the heart to maintain adequate cardiac output (Ch. 19)

heart sounds A series of distinct sounds that correlate with specific events of the cardiac cycle; four heart sounds are distinguishable in most healthy individuals. (Ch. 18)

heartburn An unpleasant sensation due to irritation of the esophagus caused by reflux of gastric acid. (Ch. 22)

heat exhaustion An illness caused by circulatory insufficiency and hypotension; takes the form of fainting due to decreased blood pressure. (Ch. 27)

heat stroke A form of unregulated hyperthermia or heat illness associated with a breakdown of the body's temperature-regulating system. (Ch. 27)

hematocrit (hee-**mat**-o-krit) The volume occupied by packed red blood cells in a centrifuged 100-ml sample of whole blood. (Ch. 17)

hematopoeisis (hem-*ah*-toe-poy-**ee**-sis) The formation of blood cells in the red bone marrow. (Ch. 17)

hemocytoblast (hee-moh-**sigh**-tow-blast) A colony-forming unit or stem cell of the bone marrow that gives rise to several types of committed stem cells. (Ch. 17)

hemoglobin (hee-moh-*glo*-bin) A pigment molecule that gives the red blood cells its color; can take up oxygen or release it. (Ch. 17)

hemolysis (he-**mol**-ih-sis) The destruction of erythrocytes and the resultant escape of hemoglobin. (Ch. 17)

hemolytic disease of the newborn Hypoprothrombinemia causing possible life-threatening hemorrhage in newborns. (Ch. 17)

hemophilia (*hee*-moh-**fil**-ee-ah) A disease in which one of the clotting factors is absent due to a genetic mutation. (Ch. 17)

hemorrhage (**hem**-or-ij) Loss of blood. (Ch. 19)

hemostasis (hee-moh-**stay**-sis) Mechanisms that minimize or prevent the loss of blood when a blood vessel is opened. (Ch. 17)

Henderson-Hasselbalch equation The logarithmic form of the acid dissociation equation: $pH = pK_a \log([A^2]/[HA])$. (Ch. 25)

heparin (**hep**-ah-rin) A mucopolysaccharide acid, abundant in the lungs and liver; inhibits the clotting process. (Ch. 17)

hierarchical motor organization Organization of the components of the somatic motor system into serial levels of complexity beginning with the simplest. (Ch. 9)

hierarchical organization The range of complexity from lowest level to highest; also called *serial organization*. (Ch. 9)

high-altitude cerebral edema Cerebral edema that occurs at altitudes of 10,000 to 12,000 feet; characterized by severe headache, hallucinations, ataxia, weakness, impaired mental ability, stupor, and death. (Ch. 29)

high-altitude pulmonary edema Edema that occurs at altitudes of 9000 to 10,000 feet; characterized by dyspnea, cough, weakness, headache, stupor, and rarely death. (Ch. 21)

high-energy phosphate compound A compound that has a strong tendency to transfer its phosphate group to an acceptor molecule; the transfer is accompanied by the release of a large amount of free energy. (Ch. 6)

histamine A vasoactive substance released from mast cells during an inflammatory response that stimulates vasodilation. (Ch. 19)

histocompatibility-Y antigen (H-Y antigen) An antigenic protein that induces an immunologic rejection in females; thought to be responsible for inducing testicular differentiation in the primitive gonad. (Ch. 33)

histone A specialized protein that binds to DNA and organizes its structure in the nucleus. (Ch. 3)

homeostasis (*ho*-me-o-**sta**-sis) The stable state of the internal environment, which is maintained by regulatory physiologic processes, despite changes that may occur in the external environment. (Ch. 1)

homeothermy (*ho*-mee-oh-**thur**-me) The maintenance of a constant body temperature despite changes in environmental temperature. (Ch. 20)

"homo aquaticus" Jacques Cousteau's description of an underwater human capable of living indefinitely beneath the surface of the sea due to liquid breathing techniques or artificial gills. (Ch. 29)

hormone (**hoar**-moan) A chemical messenger that helps regulate the activity of other tissues and organs; secreted by endocrine or ductless glands. (Ch. 3)

hormone receptor A hormone that allows the signaling molecule to recognize and bind to the target cell. (Ch. 12)

hormone sensitive lipase An enzyme that catalyzes the breakdown of triglycerides into their component fatty acids and glycerol. (Ch. 15)

human calorimeter A large, insulated chamber used for measuring metabolic rate. (Ch. 27)

human chorionic gonadotropin (*ko*-ree-**on**-ik *gon*-ah-do-**trow**-pin) **(hCG)** A glycoprotein that is secreted by cells of the trophoblast, that signals the corpus luteum that pregnancy has begun. (Ch. 12)

human chorionic somatomammotropin (hCS) A peptide hormone secreted by the placenta during pregnancy. (Ch. 13)

humoral stimuli Chemical or fluid-related stimuli. (Ch. 30)

hydration The addition of water molecules to an ion. (Ch. 7)

hydrochloric acid (HCl) A highly dissociated acid (dissociates to H and chloride ions) produced by parietal cells in the gastric gland. HCl provides an optimal pH for the activity of the pepsin enzyme and is, therefore, an aid in digestion. HCl kills bacteria that enter the digestive system through the mouth. It contributes to the breakdown of muscle fiber and connective tissue and is a major factor in the formation of ulcers in the upper digestive tract. (Ch. 22)

hydrogen bond Similar to van der Waals bond, but slightly stronger; forms when the negatively charged electron cloud formed by an asymmetrical orbit of electrons in one atom is attracted to the positively charged nucleus of a neighboring atom. (Ch. 2)

hydrogen ion A hydrogen atom without its orbiting electron; a proton. (Ch. 2)

hydrogenation reaction The reduction process in which a hydrogen atom is gained. (Ch. 6)

hydrophilic (*high*-droe-**fil**-ik) **compound** (literally, "water loving") A compound that readily absorbs water or readily dissolves in water. (Ch. 2)

hydrophobic (*high*-droe-**foe**-bik) **compound** (literally, "water fearing") A compound that is insoluble in water; repels water molecules. (Ch. 2)

hydrostatic (*high*-droe-**stat**-ik) **pressure** Pressure created by a fluid that is either moving or stationary. (Ch. 19)

hydroxyapatite Calcium phosphate deposits usually found in crystalline form in the bone. (Ch. 26)

α-hydroxylase A specific enzyme in the proximal tubule cells of the kidneys that catalyzes the conversation of 25-hydroxy-vitamin D_3 to 1,25-dihydroxy-vitamin D_3. (Ch. 26)

5-hydroxytryptamine (5-HT) *See* serotonin.

hyperalgesic (*high*-pur-al-**jee**-zik) **zones** Painful areas on the surface of the body that are associated with inflammations of the internal organs. (Ch. 10)

hyperbilirubinemia An elevation of blood bilirubin above normally low levels. (Ch. 17)

hypercalcemia (*high*-pur-kal-**see**-mee-ah) An excessive amount of calcium in the blood. (Ch. 26)

hyperchromic A morphologic description of red cells; indicates an erythrocyte that is deeper in color than normal. (Ch. 17)

hyperglycemia (*high*-pur-gly-**see**-me-ah) An above-normal increase in the blood glucose concentration due to the absence of either insulin secretion or insulin action. (Ch. 15)

hyperkalemia (*high*-pur-kah-**lee**-me-ah) A plasma potassium concentration greater than 5.5 mEq/L. (Ch. 24)

hypernatremia (*high*-pur-nah-**tree**-me-ah) A plasma sodium ion concentration greater than 145 mEq/L. (Ch. 24)

hyperpnea Increased ventilation. (Ch. 30)

hyperpolarization (*high*-per-*pohl*-er-ih-**zay**-shun) When a membrane becomes more negative on its inner surface than it was in the resting state. (Ch. 7)

hypertension (*high*-pur-**ten**-shun) A chronically elevated blood pressure above 140 systolic or 90 diastolic. (Ch. 24)

hyperthermia (*high*-pur-**ther**-mee-ah) An elevation above normal body temperature. (Ch. 27)

hyperthyroidism The excess secretion of thyroid hormones that results in markedly increased heart rate, considerable weight loss, and high responsiveness to external stimuli. (Ch. 13)

hypertonic (*high*-per-**ton**-ik) Having an osmotic pressure or solute concentration greater than that of some other solution, which is taken as standard. (Ch. 17)

hypertrophy (*high*-**per**-trow-fee) An increase in the mass of a muscle without an increase in the actual number of cells. (Ch. 16)

hyperventilation (*high*-pur-ven-tih-**lay**-shun) Increased alveolar ventilation. (Ch. 20)

hypervolemia (*high*-pur-vo-**lee**-mee-ah) An abnormal increase in the volume of circulating blood in the body. (Ch. 19)

hypocalcemia (*high*-poe-kal-**see**-me-ah) Abnormally low calcium concentration in blood. (Ch. 26)

hypochromic A morphologic description of red cells; indicates an erythrocyte that is paler in color than normal. (Ch. 17)

hypoglycemia (*high*-poe-gly-**see**-me-ah) A decrease in the blood glucose concentration to below normal. (Ch. 15)

hypokalemia (*high*-poe-kahl-**ee**-me-ah) A plasma potassium concentration level less than 3.5 mEq/L. (Ch. 24)

hyponatremia (*high*-poe-nay-**tree**-me-ah) A plasma sodium concentration less than 137 mEq/L. (Ch. 24)

hypoosmotic Having an osmolality less than a reference solution (usually plasma). (Ch. 24)

hypophysis (*high*-**pof**-ih-sis) The pituitary gland; located at the base of the brain. (Ch. 13)

hypotension Low blood pressure; less than 100 systolic, 60 diastolic. (Ch. 19)

hypothalamic-pituitary portal system A complex of blood vessels that consists of the primary capillary network, the secondary capillary network, and the vessels that connect them; carries factors that regulate the secretion of anterior pituitary hormones from the hypothalamus to the anterior pituitary. (Ch. 13)

hypothalamic-pituitary-adrenal axis The relationship between CRH, ACTH, and the adrenal cortex. (Ch. 13)

hypothalamus (*high*-poe-**thal**-ah-mus) A portion of the brain that regulates the pituitary gland, the autonomic nervous system, emotional response, body temperature, water balance, and appetite. (Ch. 10)

hypothermia (*high*-poe-**ther**-mee-ah) The lowering of body temperature below normal. (Ch. 27)

hypothesis (*high*-**poth**-e-sis) A statement that tentatively explains an observation. (Ch. 2)

hypothyroidism A condition that exists when there is a deficiency in thyroid hormone production; affected individuals cannot tolerate cold temperatures and exhibit lethargy, sometimes accompanied by myxedema. (Ch. 13)

hypotonic (*high*-poe-**ton**-ik) Refers to a solution whose osmotic pressure or solute content is less than that of some standard of comparison. (Ch. 17)

hypoventilation (*high*-poe-ven-tih-**lay**-shun) Decreased alveolar ventilation below normal. (Ch. 20)

hypovolemia (*high*-poe-vo-**lee**-mee-ah) An abnormally decreased volume of circulating blood in the body. (Ch. 19)

hypoxemia (*high*-pock-**see**-me-ah) The condition of low blood oxygen. (Ch. 21)

hypoxia (*high*-**pock**-see-ah) Oxygen deficiency. (Ch. 21)

hypoxic pulmonary vasoconstriction A mechanism in the lung that balances the flow or perfusion of blood with the availability of regional alveolar oxygen. (Ch. 21)

H-zone A somewhat lighter region in the sarcomere; found in the center region of the A-band where the thin filaments of the I-band have not penetrated. (Ch. 16)

I-band A region of a muscle where only thin filaments are present. (Ch. 16)

IGF-I, IGF-II Insulin-like growth factor one and two; peptides produced by the liver that exhibit many of the growth-promoting activities of insulin; part of a group of growth-promoting factors called *somatomedins*. (Ch. 13)

ileocecal (ill-ee-oh-**see**-kal) **sphincter** The last 2 to 3 cm of the musculature of the ileum; controls passage of material to the large intestine. (Ch. 22)

ileum (il-lee-um) The distal portion of the small intestine; extends from the jejunum to the cecum. (Ch. 22)

immune (ih-**mewn**) **response** Any reaction that is designed to defend the body against pathogens or any other foreign substances. (Ch. 28)

immunity (ih-**mew**-nih-tee) Either nonspecific (natural) or specific (adaptive) physical and chemical barriers or defense mechanisms geared toward the identification and destruction of specific pathogens. (Ch. 28)

immunoglobulin (*im*-yoo-no-**glob**-yoo-lin) **(Ig)** Any glycoprotein secreted by B-lymphocytes in response to specific antigens. (Ch. 28)

immunosuppression The inhibition of the body's ability to mount an effective immune response; may be due to certain drugs or exposure to ionizing radiation. (Ch. 28)

implantation (*im*-plan-**tay**-shun) The penetration of the epithelial lining of the uterus by a blastocyst to embed in the endometrium and form an embryo. (Ch. 32)

impotence (**im**-poe-tence) The inability to achieve penile erection sufficient for vaginal penetration. (Ch. 33)

in utero (**yoo**-ter-oh) Within the uterus. (Ch. 32)

in vitro (**vee**-tro) In an artificial environment, such as of cells grown in culture dishes. (Ch. 2)

in vivo (**vee**-vo) Within the body. (Ch. 2)

inactivating state One of three states of operation of the sodium channel of membranes; the time-dependent gate is closed. (Ch. 7)

incontinence (in-**kon**-tih-nence) The inability to retain urine or feces due to loss of sphincter control. (Ch. 22)

incus The middle of the three ossicles of the ear; serves to conduct vibrations from the tympanic membrane to the inner ear; also known as the *anvil*. (Ch. 8)

independent variable A datum plotted on the horizontal axis of a line graph. (Ch. 1)

indirect calorimetry An estimate of metabolic rate based on oxygen consumption. (Ch. 27)

infarct (**in**-farkt) An area of tissue death resulting from circulatory obstruction, usually a thrombus or embolus. (Ch. 19)

inferior (in-**fe**-ree-or) Nearer to the lower end of the body. (Ch. 1)

inferior vena cava The large vein that returns blood from the lower part of the body to the heart. (Ch. 18)

infertility A diminished or absent ability to reproduce. (Ch. 33)

inflammation (in-flah-**may**-shun) A localized body response to tissue damage or injury; the clinical signs are redness, heat, edema, and pain. (Ch. 28)

inhibin (in-**hib**-in) A protein secreted by Sertoli cells of the seminiferous tubules that regulates the secretion of follicle-stimulating hormone (FSH). (Ch. 31)

inhibitory postsynaptic potential (IPSP) A potential that hyperpolarizes the membrane potential and inhibits the nerve cell from generating an action potential. (Ch. 7)

initial segment The first portion of an axon plus part of the cell body where the axon joins. (Ch. 7)

inner ear The part of the ear containing receptors for hearing and equilibrium. (Ch. 8)

inner mitochondrial membrane A substrate for proteins in the electron transport chain of oxidative phosphorylation. (Ch. 6)

inositol triphosphate (IP$_3$) One of a pair (with diacylglycerol) of intracellular second-messenger molecules. (Ch. 5)

inotropic agent A compound that affects the contractility of heart muscle. (Ch. 16)

insensible perspiration Water loss by evaporation from the lungs and skin. (Ch. 27)

insertion (in-**sir**-shun) The distal, or less stationary, point of attachment of a skeletal muscle. (Ch. 16)

insomnia (in-**som**-nee-uh) The inability to sleep. (Ch. 11)

inspiration (in-spih-**ray**-shun) The act of drawing air into the lungs. (Ch. 10)

inspiratory neurons Neurons within the medulla oblongata responsible for rhythmic breathing when the body is at rest. (Ch. 20)

insulin (**in**-suh-lin) A double-chain protein hormone produced by the beta cells of the islets of Langerhans in the pancreas that allows cells to take in glucose in response to an increase in blood sugar levels. (Ch. 15)

insulin/glucagon ratio (I/G) The ratio of the blood concentrations of insulin and glucagon, pancreatic hormones that have opposing actions on cell metabolism. The ratio is used to assess metabolic status in the body. A high ratio suggests an anabolic state; a low ratio suggests a catabolic state. (Ch. 15)

insulin resistance A lack of response by tissues to insulin in the blood. (Ch. 15)

insulin-dependent diabetes mellitus (IDDM) A disease caused by inadequate insulin secretion; type 1 diabetes. (Ch. 15)

insulin-like growth factor Somatomedins produced by the liver; IGF-I stimulates skeletal growth; IGF-II stimulates tissue growth and repair. (Ch. 13)

integument (in-**teg**-yoo-ment) Any covering, especially the skin. (Ch. 28)

intercalated (in-**ter**-kah-lay-ted) **disc** A specialized structure that provides mechanical and electrical connection between cells of cardiac muscle. (Ch. 16)

intercalated duct System of small ducts that delivers the exocrine secretion of the pancreas to larger ducts that conduct pancreatic juice into the small intestine; also contributes a water-bicarbonate secretion to pancreatic juice. (Ch. 22)

intercellular Between cells. (Ch. 23)

intercostal muscles Skeletal muscles between the ribs. Internal intercostals depress the ribcage and contract during forced expiration. External intercostals elevate the ribcage and contract during inspiration. (Ch. 20)

interferon (in-ter-**feer**-on) A group of proteins secreted by cells that have been infected by a virus, inducing noninfected cells to form antiviral proteins that inhibit viral multiplication. (Ch. 28)

interleukins (in-ter-**loo**-kins) Hormonelike messengers that stimulate T lymphocytes; **IL-1** stimulates activated T$_4$ lymphocytes to differentiate and reproduce, forming a clone of inducer T cells and helper T cells; **IL-2** induces T$_8$ lymphocytes to differentiate and reproduce, forming cytotoxic T cells and suppressor T cells. (Ch. 28)

internal Nearer the center of an organ, cavity, or part of the body. (Ch. 1)

internal anal sphincter The smooth muscle that is relaxed during defecation. (Ch. 22)

internal environment The extracellular fluid that surrounds the cell, outside the cell but inside the organism. (Ch. 1)

internal intercostal muscle The expiratory muscle that contracts to pull the ribcage down. (Ch. 20)

internal sphincter A group of smooth muscles that close off the exit from the bladder. (Ch. 10)

interneuron Any neuron in a chain of neurons situated between the sensory and the final motor neuron. (Ch. 7)

interoceptor (in-ter-o-**sep**-ter) A sensory receptor that transmits information from viscera to CNS. (Ch. 8)

interomedial lateral nuclei (ILN) The location in the spinal cord of the preganglionic nerve cells of the sympathetic nervous system. (Ch. 10)

interphase (**in**-ter-faze) The period between the end of one cell division and the beginning of the next. (Ch. 3)

interpolation Prediction of data values that fall between existing data points. (Ch. 1)

interstitial (in-ter-**stish**-al) **fluid** The fluid outside blood vessels that directly bathes most body cells. (Ch. 24)

interstitium (in-ter-**stish**-ee-um) The fluid-filled space between tissue cells. (Ch. 23)

intracellular (in-tra-**sell**-yoo-lar) Within a cell. (Ch. 23)

intracellular fluid The volume of fluid within cells. (Ch. 24)

intracellular pH The pH level within the cell; generally lower than extracellular pH. (Ch. 25)

intrafusal muscle fibers Fibers in the inside of skeletal muscles; arranged parallel to extrafusal fibers. The fibers are part of the neuromuscular spindle and supplied by gamma motor neurons. (Ch. 9)

intrauterine (in-trah-**yoo**-ter-in) **device (IUD)** A contraceptive device inserted into the uterus to prevent implantation of the ovum. (Ch. 33)

intrinsic clotting pathway The initiation of clotting in the absence of tissue damage; can occur inside or outside the body. (Ch. 17)

intrinsic factor A glycoprotein, secreted by the lining of the stomach, that is essential for absorption of vitamin B$_{12}$ in the small intestine. (Ch. 17)

intrinsic (in-**trin**-sik) **protein** A protein that partially or completely penetrates the lipid bilayer of a plasma membrane. (Ch. 4)

intron (**in**-trahn) Noncoding region of a eucaryotic gene that is transcribed into RNA but removed by splicing during production of messenger RNA. (Ch. 5)

inulin A fructose polymer whose renal clearance is used to measure glomerular filtration rate. (Ch. 23)

inverse myotatic reflex A reflex that inhibits skeletal muscle contraction when the Golgi tendon organ detects excessive force that might damage the muscle. (Ch. 9)

inverse relationship The relationship between data such that *y*-values get smaller as *x*-values get larger. (Ch. 1)

involuntary muscle The smooth or cardiac muscle whose actions are controlled by the autonomic nervous system in response to some internal reflex. (Ch. 16)

iodide trap The transport process that pumps iodide into cells of the thyroid. (Ch. 13)

ion (**eye**-on) **channel** The pores in a nerve cell membrane that allow passage of certain ions. A **passive ion channel** allows a specific ion to pass through; a **chemically activated ion channel** can be opened or closed by a chemical transmitter; a **voltage-activated ion channel** is opened when it detects a certain voltage. (Ch. 7)

ion product of water The product of [H$^+$] and [OH$^-$] in aqueous solutions; it has a value of 1310^{214} M^2 at 25°C. (Ch. 25)

ionic bond The equal and opposite electrical charge that holds two atoms together. (Ch. 2)

ionic current   The flow of ions. (Ch. 7)

ionization (*eye*-on-eye-**zay**-shun)   The dissociation of a substance in solution into ions. (Ch. 26)

ionized calcium   A free Ca^{2+} ion, the physiologically active form of calcium. (Ch. 26)

ischemia (is-**kee**-me-ah)   Reduced blood flow to an organ due to functional constrictions or obstruction of a blood vessel. (Ch. 17)

islets of Langerhans (**eye**-lits of **Lahng**-er-hanz)   Small groups of cells interspersed within the cells of the exocrine pancreas; secrete insulin, glucagon, and somatostatin. (Ch. 15)

isoelectric line   The baseline of an electrocardiogram with no positive or negative deflection. (Ch. 18)

isohydric principle   The idea that all buffer systems are in equilibrium with the same hydrogen ion concentration. (Ch. 25)

isometric contraction   A contraction in which a muscle does not shorten but develops force or tension while pulling against immovable attachments. (Ch. 16)

iso-osmotic (*eye*-so-oz-**mat**-ik)   Having the same osmotic pressure. (Ch. 4)

isotonic (*eye*-so-**tahn**-ik)   Having the same osmolality, equal to that of a reference solution. (Ch. 16)

isotonic (*eye*-so-**tahn**-ik) **contraction**   A contraction in which the force in a muscle remains constant while the muscle shortens. (Ch. 16)

isotope (*eye*-so-tope)   An atom of an element that has a differing number of neutrons. (Ch. 2)

isovolumetric ventricular contraction   Early phase of the cardiac cycle in which both the atrioventricular valves and the semilunar valves are closed and ventricular volume does change. (Ch. 18)

isovolumetric ventricular relaxation   Early phase ventricular relaxation when atrioventricular and aortic valves are closed. (Ch. 18)

jaundice (**jawn**-dis)   A condition in which the skin is yellowish due to hyperbilirubinemia and deposition of bile pigment in the skin. (Ch. 17)

jejunum (jeh-**joo**-num)   The middle portion of the small intestine; extends from the duodenum to the ileum. (Ch. 22)

joint   The site of junction or union between two or more bones of the skeleton. (Ch. 16)

junctional complex   The cellular structure that prevents the lateral diffusion of proteins in the plasma membrane. (Ch. 4)

junctional folds   A series of deep grooves of the muscle plasma membrane. (Ch. 4)

juxtaglomerular (*juks*-tah-glo-**mer**-you-lar) **apparatus**   A structure in the kidney consisting of macula densa, extraglomerular mesangial cells, and granular cells, located in the region where the distal straight tubule touches afferent and efferent arterioles of its parent glomerulus. (Ch. 23)

juxtamedullary nephron   Nephrons with renal corpuscles in the inner cortex; have large glomeruli, long thin limbs, and long loops; 15% of all nephrons are of this type. (Ch. 23)

kainate   One of the three subtypes of glutamate receptors. Its activation results in excitatory postsynaptic potentials by opening ion channels that increase in Na^+ and K^+ conductance. (Ch. 7)

kallikrein (kal-li-**kre**-in)   A proteolytic enzyme that leads to the formation of kinins. (Ch. 23)

keloid (**key**-loyd)   An elevated, enlarging scar due to excessive amounts of collagen in the dermis during connective tissue repair. (Ch. 8)

keto (**key**-toe) **acid**   Acid formed by the removal of an amino group from amino acid. (Ch. 15)

ketoacidosis   An acid/base imbalance caused by an increase in the concentration of ketones in the blood. (Ch. 15)

ketogenesis   The formation of ketones in the liver; stimulated by glucagon. (Ch. 15)

ketone (**key**-tone) **bodies**   Four-carbon compounds that are released by the liver accumulate in the blood during starvation and untreated diabetes mellitus; acetoacetate, acetone, or ß-hydroxybutyrate acid. (Ch. 15)

ketosis   A condition characterized by an abnormally elevated concentration of ketone bodies in body tissue and fluids; a complication of diabetes mellitus and starvation. (Ch. 15)

kinesthesia (*kin*-es-**the**-ze-ah)   The sense of movement. (Ch. 8)

kinetic energy   Energy in use that is producing motion. (Ch. 2)

kinin (**kye**-nin)   An endogenous peptide derived from kallikrein that causes vasodilation. (Ch. 23)

knee-jerk reflex   A stretch reflex activated by tapping the patellar tendon below the knee. (Ch. 9)

Krebs cycle   Aerobic cycle of respiratory chemical reactions that occur in the mitochondria and produce cellular ATP (also called the *citric acid cycle*). (Ch. 3)

labor   Strong contractions of the uterus that occur with increasing frequency and intensity during the latter part of pregnancy; results in expulsion of the fetus. (Ch. 32)

lactase (**lak**-tase)   ß-Galactosidase. An enzyme that digests lactose into glucose and galactose. (Ch. 22)

lactation (lak-**tay**-shun)   The production and secretion of milk by the breast. (Ch. 32)

lacteal (**lak**-tee-al)   Pertaining to milk; any of the intestinal lymphatics that transport chyle. (Ch. 22)

lactic (**lak**-tik) **acid**   A three-carbon molecule formed as a byproduct of anaerobic metabolism of the energy pathways of cells. (Ch. 16)

lactic acid oxygen debt   The second stage of repayment of the oxygen debt that involves removal of lactic acid from the body. (Ch. 30)

lactose (**lak**-tose)   A disaccharide composed of one molecule of glucose and one molecule of galactose. (Ch. 22)

laminar air flow   The streamlined flow of air in the lungs that parallels the sides of the airways. (Ch. 20)

larynx (**lar**-inks)   The upper part of the airway between the pharynx and the trachea that contains the vocal cords. (Ch. 22)

latent period   Period of time (several milliseconds) between stimulus delivery and muscle contraction. (Ch. 16)

lateral (**lat**-er-al)   Position farther from the midline or midsagittal plane. (Ch. 1)

lateral reticulospinal tract   A descending pathway from the brain stem to the spinal cord; used to control flexor muscles involved in the maintenance of posture. (Ch. 9)

laxative   A cathartic such as mineral oil that hydrates or lubricates material in the large intestine, promoting evacuation of the bowel. (Ch. 22)

L-dopa   Dopamine precursor that is often administered to patients with Parkinson's disease. (Ch. 9)

lead I   An electrocardiograph recording made with electrical connections between the right and left arms; **lead II** with electrical connections between the left leg and the right arm; **lead III** with electrical connections between the left leg and left arm. (Ch. 18)

length-tension curve   The relationship between the force capability of a muscle and its length. (Ch. 16)

leukemia (loo-**key**-mee-ah)   A malignant disease characterized by proliferation of hematopoietic cells that are immature and therefore functionally impaired. (Ch. 17)

leukocyte (**loo**-ko-site)   A white blood cell (WBC). (Ch. 17)

leukocyte inducing factor   A specific factor in plasma that can be demonstrated when leukocytes are depleted; stimulates leukopoiesis. (Ch. 17)

leukocytosis (*loo*-ko-sigh-**toe**-sis)   An abnormal increase in the total number of circulating leukocytes. (Ch. 17)

leukopenia (*loo*-ko-**pee**-nee-ah)   An abnormal reduction in the total number of circulating leukocytes: <5000 cells/ul[1]. (Ch. 17)

leukopoiesis (*loo*-ko-**poe**-ee-sis)   The development of white blood cells. (Ch. 17)

ligand (**lihg**-and)   Any compound that binds with a high degree of specificity to a receptor. (Ch. 5)

limb leads   An electrocardiograph recording made between any two of the standard connections: the right and left arms and the left leg. (Ch. 18)

limbic system   Parts of the prefrontal cortex, hippocampus, amygdala, and hypothalamus that collectively function in the expression of emotional behavior. (Ch. 8)

linear relationship   A relationship that can be described by a straight line. (Ch. 1)

lipase (**lye**-pase)   An enzyme produced by the pancreas that digests triacylglycerols to monoglycerides and free fatty acids. (Ch. 22)

lipids   A group of fat and fatlike substances that are water-insoluble and highly soluble in organic liquids. (Ch. 2)

lipogenesis (*lipe*-oh-**jen**-eh-sis)   The production of lipids from precursors such as glucose and amino acids. (Ch. 15)

lipolysis   The breakdown of triglycerides to fatty acids and glycerol. (Ch. 14)

lipoprotein lipase   An enzyme that promotes the uptake of lipoproteins from the blood. (Ch. 15)

lipoproteins   A complex of lipids and proteins, synthesized in the endoplasmic reticulum lumen, that has a critical role in the transport and distribution of lipids. (Ch. 3)

local chemical mediator A chemical signal released by a cell that affects only nearby cells and that is rapidly destroyed if not used. (Ch. 5)

long-term memory Memory that is stable, less volatile, and long-lasting. (Ch. 11)

loop of Henle The hairpin portion of the nephron situated between proximal convoluted and distal convoluted tubules. (Ch. 23)

lordosis A receptive posture adopted by female animals during mating. (Ch. 11)

low-density lipoprotein A lipoprotein aggregate that transports most of the cholesterol of the cell. (Ch. 17)

lower esophageal sphincter A ring of smooth muscle that acts to restrict entrance of gastric contents into the esophagus; also called the *gastroesophageal sphincter*. (Ch. 22)

lower-motor-neuron lesion Occurs with spinal cord or brain stem injuries that affect motor neurons directly; characterized by ipsilateral hypoactive reflexes, paralysis, and limp, atrophic muscles. (Ch. 9)

lumen (**loo**-men) The cavity within a tubelike structure such as a blood vessel, lymphatic vessel, or intestine. (Ch. 23)

luminal membrane The cell surface that contacts fluid in the tubule lumen of the kidney. (Ch. 23)

lung volumes Four primary subdivisions of air in the lungs: inspiratory reserve volume, tidal volume, expiratory reserve volume, and residual volume. (Ch. 20)

luteinization The formation of the corpus luteum from the remnants of the ovarian follicle after ovulation. (Ch. 31)

luteinizing (**loot**-eh-*nigh*-zing) **hormone (LH)** Gonadotropin hormone released by the anterior pituitary that initiates ovulation. (Ch. 11)

lymph (**limf**) A transparent, slightly yellow fluid of alkaline reaction that is found in lymphatic vessels and derived from tissue fluids. (Ch. 19)

lymph node Small organelle located along lymph vessels that contain lymphocytes; filters microorganisms and other foreign material. (Ch. 11)

lymphatic (lim-**fat**-ik) **system** A system of vessels that directs fluid and plasma protein from interstitial spaces to the blood. (Ch. 19)

lymphocyte (**lim**-foe-site) An agranular leukocyte, the second most common type of white blood cell. (Ch. 17)

lymphokines (**lim**-foe-kines) Chemicals secreted by lymphocytes that regulate T- and B-leukocyte function, attract macrophages, activate complement proteins, and help with specific defense. (Ch. 28)

lysis (**lye**-sis) The destruction of cells by a specific antibody. (Ch. 4)

lysosome (**lye**-so-sowm) A cytoplasmic, membrane-bound organelle that contains digestive enzymes used for intracellular digestion. (Ch. 3)

lysozyme (**lye**-so-zime) An enzyme that digests the cell wall of many pathogenic bacteria. Present in sweat, saliva, tears, and urine. (Ch. 28)

macrocytosis (*mak*-roe-sye-**toe**-sis) The production of red blood cells that are larger than normal; also called *macrocythemia*. (Ch. 17)

macromolecule A very large molecule having a polymeric chain structure; all contain carbon and assume huge proportions and a great variety of shapes. (Ch. 2)

macrophage (**mak**-roe-faje) A large phagocytic cell that forms part of the body's defense against microorganisms and harmful chemicals. (Ch. 17)

macula densa (**mak**-you-la **den**-sa) Portion of the straight distal tubular epithelium that helps make up the juxtaglomerular apparatus of the kidney. (Ch. 23)

major histocompatibility complex II (MCHII) A protein present on the plasma membrane of antigen-presenting cells; involved in cell-mediated immune response. (Ch. 28)

malignant hyperthermia An inherited defect in skeletal muscle metabolism. (Ch. 27)

malleus (**mal**-lee-us) The largest of the auditory ossicles, the one attached to the tympanic membrane; also known as the *hammer*. (Ch. 8)

maltase (**mawl**-tase) α-Glucosidase. A digestive enzyme that breaks down maltose into glucose. *See also* glucosidase. (Ch. 22)

mass movement A propulsive contraction of the colon, moving material into the distal large intestine. (Ch. 22)

mass number The total number of protons and neutrons in the nucleus of an element. (Ch. 2)

mast cell A specialized type of cell in connective tissue; contains histamine and is important in allergic reactions. (Ch. 5)

mastocytosis (*mas*-toe-sigh-**toe**-sis) A condition in which greatly elevated numbers of mast cells accumulate in body tissues. (Ch. 5)

matrix (**ma**-triks) The intercellular substance of a tissue or the tissue from which a structure develops. (Ch. 1)

maximal aerobic power A means of determining an individual's ability to sustain high-intensity exercise; the maximal oxygen consumption is an indicator of this state. (Ch. 30)

maximal oxygen consumption (VO$_2$max) The state at which the maximum oxygen transport and uptake in the tissues are reached and a plateau is attained. (Ch. 30)

mean arterial pressure A calculation of a single value for blood pressure that factors both the pulse and diastolic pressures. (Ch. 19)

mean body temperature An estimate of the fractional sum of core and mean skin body temperatures. (Ch. 27)

mean corpuscular hemoglobin (MCH) An index used to determine the concentration of hemoglobin in red blood cells. (Ch. 17)

mean corpuscular volume (MCV) A quantitative assessment of the volume of red blood cells; reported in picoliters (pl). (Ch. 17)

mean skin temperature The temperature of the body shell, calculated as the mean of the temperatures at several representative parts of the body. (Ch. 27)

mechanoreceptor A sensory receptor that is excited by mechanical pressures or distortions, such as bending, twisting, or compressing. (Ch. 8)

medial (**mee**-dee-al) A term of position; literally means nearer to the midsagittal plane. (Ch. 1)

medial lemniscal pathway A pathway in the brain stem through which sensory fibers of the dorsal column ascend toward the thalamus. (Ch. 8)

medial reticulospinal tract A component of the ventromedial pathway; originates in the reticular formation and is involved in maintaining posture by activation of extensor muscles. (Ch. 9)

median eminence A specialized region in the hypothalamus where arterial blood vessels branch into a network of capillaries. (Ch. 13)

medulla (mah-**dull**-ah) The innermost portion of an organ or structure. (Ch. 10)

megakaryocyte (*meg*-ah-**kare**-ee-oh-site) A type of blood cell from which thrombocytes, or platelets, are derived. (Ch. 17)

meiotic (my-**ah**-tik) **division** A division of germ cells that reduces the number of chromosomes in the cell by half; random assortment and recombination of maternal and paternal chromosomes occur as well as crossing over. (Ch. 31)

meiotonic (*my*-oh-**tahn**-ik) **contraction** A muscle contraction in which the force becomes progressively less as the contraction goes on. (Ch. 16)

melatonin (*mel*-ah-**toe**-nin) A hormone derived from tryptophan; suspected to have a role in puberty onset and control of body rhythms. (Ch. 12)

membrane conductance The ease with which ions can flow through channels in the plasma membrane. (Ch. 7)

membrane potential Voltage difference between inside and outside of cell due to an ionic concentration gradient. (Ch. 7)

memory cell A B-lymphocyte cell that permits rapid mobilization of immune response upon subsequent exposure to an antigen. (Ch. 28)

menopause (**men**-oh-pawz) The cessation of menstruation in the human female; occurs on the average at age 50. (Ch. 31)

menstrual (**men**-stroo-al) **cycle** The approximately 28-day ovarian cycle; consists of the follicular phase, the luteal phase, and the menstrual phase. (Ch. 31)

menstruation (*men*-streh-**way**-shun or men-**stray**-shun) Normal flow of menstrual fluid from the uterus that occurs approximately once a month in most nonpregnant females. (Ch. 31)

mesentery (**mez**-en-*ter*-ee) A membrane attaching various organs to the body wall, especially the peritoneum connecting the intestine to the posterior abdominal wall. (Ch. 10)

messenger RNA The transcribed segment of DNA containing the information specifying the amino acid sequence of a polypeptide chain. (Ch. 5)

metabolic acidosis An abnormal process characterized by a gain of acid which leads to a fall in blood pH; due to any cause other than elevated carbon dioxide. (Ch. 22, 24, 25)

metabolic alkalosis An abnormal process characterized by a gain of strong base, which leads to a rise in blood pH; due to any cause other than a decrease in blood carbon dioxide levels. (Ch. 22, 24, 25)

metabolic bone disease A group of disorders in which there is generalized abnormality of the ongoing process of bone formation and bone resorption. (Ch. 26)

metabolic pathway The sequence of enzyme-mediated chemical reactions by which molecules are synthesized and broken down in cells. (Ch. 6)

metabolic rate Total energy production of an individual per unit time. (Ch. 3)

metabolism (meh-**tab**-oh-lizm) The sum of all physical and chemical processes by which living organized substance is produced and maintained; also the transformation by which energy is made available for the organism's use. (Ch. 3)

metarteriole Blood vessels that are slightly larger than capillaries. (Ch. 19)

MHC Major histocompatibility complex; a group of proteins on the surface of nucleated cells. Class I MHC molecules are present on the surface of all nucleated cells in the body. Class II MHC molecules are present on the surface of some cells (e.g., CD4+ cells) of the immune system. MHC proteins play important roles in cell recognition and antigen recognition. (Ch. 28)

micelle (my-**sel**) Small aggregates of the bile salts, monoglycerides, and free fatty acids that diffuse in aqueous solution of intestinal contents. (Ch. 22)

microcirculation The circulation of blood through the arterioles, metarterioles, capillaries, and venules. (Ch. 19)

microcytosis (*my*-kroe-si-**toe**-sis) The production of red blood cells that are smaller than normal; also called *microcythemia*. (Ch. 17)

microfilament A component of cytoskeletal structure; a rodlike cytoplasmic protein believed to have a supportive and cytoskeletal function and/or mediate movement of the cell and the organelles within it. (Ch. 3)

microglia (my-**krog**-lee-ah) Small, nonneural, interstitial cells of mesodermal origin that form part of the supporting structure of the CNS. (Ch. 7)

microtubule (*my*-krow-**too**-bewl) A tubular cytoplasmic filament that provides internal support for cells; allows change in cell shape and organelle movement in cell. (Ch. 7)

microtubule-associated protein (MAPS) Accessory proteins that are thought to be responsible for the distribution of material to dendrites and axons. Dendrites have high-molecular-weight MAPS, while axons have low-molecular-weight MAPS. (Ch. 7)

microvilli (mi-krow-**vill**-ee) Small projections that cover epithelial cells of the digestive tract. (Ch. 22)

micturition (*mik*-tu-**rish**-un) Emptying of the bladder. (Ch. 10)

middle ear The portion of the ear between the tympanic membrane and the cochlea; contains the ossicles and internal auditory meatus (eustachian tube). (Ch. 8)

midsagittal plane An imaginary plane that passes directly along the midline of the body, dividing it into right and left halves. Also called *median sagittal plane*. (Ch. 1)

milliequivalent (*mil*-ee-**kwiv**-a-lent) For an ion, an amount equal to the product of millimoles and valence (e.g., 2.5 mmol Ca²⁺ is the same as 5 milliequivalents [mEq] of Ca²⁺). (Ch. 24)

mineral A nonorganic homogeneous solid substance, usually a constituent of the Earth's crust. (Ch. 2)

mineralocorticoid (*min*-er-al-oh-**kor**-ti-koyd) A steroid hormone produced by the adrenal cortex, involved in regulation of electrolyte and water balance. (Ch. 12)

minute ventilation The total volume of inspired air that enters the lungs per minute; equals tidal volume times frequency. (Ch. 20)

mitochondrion (*my*-tow-**kon**-dree-un) An organelle of the cell enclosed in a double membrane; the site of ATP production through the process of aerobic respiration. (Ch. 3)

mitosis (my-**tow**-sis) A process of cell division whereby each of the two daughter cells receives a complete set of chromosomes. (Ch. 3)

mitral (**my**-tral) **cell** Cells located in the olfactory bulbs that form the olfactory tract. (Ch. 8)

mitral valve The atrioventricular valve on the left side of the heart; also called the *bicuspid*. (Ch. 18)

M-line A structure in muscle cells that holds the thick filaments in side-by-side alignment. (Ch. 16)

molality A unit of concentration; the number of moles of solute per kilogram of solvent in a solution. (Ch. 2)

molarity A unit of concentration; the number of moles of solute per liter of solution. (Ch. 2)

molecular formula A symbol that shows the number of each type of atom in a molecule or group of atoms. (Ch. 2)

molecular mass The sum of the atomic masses of the elements in a compound. (Ch. 2)

monochromatic color vision Color vision that relies on one type of cone. (Ch. 8)

monoclonal antibody A specific antibody produced in vitro by hybridoma cell lines; the hybridoma cell lines are produced through fusion of activated β-cells with cancer cells. (Ch. 3)

monocyte (**mon**-oh-site) One of two types of agranulocyte white blood cells; they migrate out of blood and into tissues to form macrophages. (Ch. 17)

monoiodotyrosine (MIT) A compound found within the thyroid follicle cell; used in the synthesis of T₃ and T₄. (Ch. 13)

monokine A chemical secreted by macrophages that stimulate T-lymphocyte development. (Ch. 28)

monosaccharide (*mon*-oh-**sak**-ah-ride) A simple sugar; cannot be degraded by hydrolysis to a simpler sugar. (Ch. 2)

monozygotic twins Twins that result from the fertilization of a single ovum by a single sperm, followed by cell division that results in a single duplication or multiple duplications of the fertilized ovum prior to implantation; also called *identical twins*. (Ch. 32)

mossy fiber Neurons that originate in the neocortex and spinal cord; sensory stimuli and voluntary movements enhance activity. (Ch. 9)

motility (mow-**til**-i-tee) Mixing and propulsive movements of the digestive tract. (Ch. 22)

motor cortex The area of the brain that controls movement. (Ch. 7)

motor end-plate The connection between a muscle fiber and the myelinated axon of a motor neuron; also called *myoneural junction*. (Ch. 9)

motor neuron A neuron that carries instructions about muscle movements from the CNS in response to sensory stimuli; also called an *efferent neuron*. (Ch. 7)

motor neuron pool A functional grouping in the spinal cord of motor neurons that innervate a single muscle. (Ch. 9)

motor system Collectively, the areas of the brain and spinal cord that control and coordinate skeletal muscle contractions. (Ch. 7)

motor unit A motor neuron plus the muscle fibers it innervates. (Ch. 9)

mRNA Messenger ribonucleic acid.

mucin (**mew**-sin) A nonenzymatic protein found in mucus; lubricates and protects the lining of the digestive tract. (Ch. 22)

mucosa (mew-**ko**-sah) A mucous membrane. (Ch. 22)

mucus (**mew**-kus) A viscous mixture of mucin and watery secretions from mucus and serous cells of salivary glands. (Ch. 22)

müllerian duct system An embryonic ductal system that is the precursor of the internal genitalia of the female. (Ch. 33)

multipolar neuron A type of neuron with several short dendrites and one long axon. (Ch. 7)

multiunit smooth muscle A class of muscles that are closely controlled by the nervous system. (Ch. 16)

muscarinic acetylcholine receptor A category of cholinergic receptor, a receptor on target cells such as muscle. (Ch. 10)

muscle (**mus**-ul) A number of muscle fibers bound together by connective tissue. (Ch. 9)

muscle contraction The shortening of muscle tissue. (Ch. 9)

muscle fiber A type of contractile cell that is elongated in shape; also called *muscle cell*. (Ch. 9)

muscle relaxation The lengthening of muscle tissue. (Ch. 9)

muscle spindle A receptor complex attached to somatic sensory neurons that detect the length of the muscle and its velocity of contraction. (Ch. 9)

muscle spindle receptor A capsule-enclosed arrangement of afferent nerve fiber endings in skeletal muscle; sensitive to stretch. (Ch. 8)

muscle tissues Tissues whose cells are elongated and contain many myofibrils; specialized for contraction; functions to accomplish the organism's movement. (Ch. 1)

muscle tone The amount of resistance of muscle to passive stretch. (Ch. 9)

myasthenia gravis (*my*-as-**then**-ee-ya **gra**-vis) A disorder of neuromuscular function characterized by muscle weakness; may affect any muscle of the body, but especially those of the eye, face, lips, tongue, throat, and neck. (Ch. 16)

myelin (**my**-eh-lin) An insulating lipid elaborated by oligodendrocytes (centrally) or Schwann cells (peripherally) that surrounds the fiber extension of neurons in order to speed the transmission of information. (Ch. 7)

myenteric plexus A network of nerve cells that lie between the two layers of the muscularis externa of the digestive tract. (Ch. 22)

myocardial infarction (my-oh-**kar**-dee-al in-**fark**-shun) A heart attack; necrosis of the myocardium when a coronary blood vessel is occluded. (Ch. 18)

myocardium (my-oh-**kar**-dee-um) The middle and thickest layer of the heart; composed of cardiac muscle. (Ch. 18)

myoepithelial (my-oh-ep-ah-**thee**-lee-yel) **cells** Contractile cells in the breast that help the flow of milk. (Ch. 32)

myofibril (my-oh-**fye**-bril) A functional unit of muscle cells formed by hundreds of sarcomeres connected in series. (Ch. 16)

myofilament (my-oh-**fil**-ah-ment) A protein filament that provides for contraction of skeletal, cardiac, or smooth muscle. (Ch. 16)

myogenic activity The increase in the contractile activity of smooth muscle cells in response to stretch. (Ch. 19)

myoglobin (**my**-oh-glow-bin) A respiratory pigment that transports oxygen and imparts a red color to muscle tissue. (Ch. 16)

myoneural (my-oh-**new**-ral) **junction** The connection between a muscle fiber and the myelinated axon of a motor neuron; also called *motor end-plate*. (Ch. 9)

myosin (**my**-oh-sin) A very large protein that composes the thick filaments of muscle cells; 68% of protein in muscle is myosin. (Ch. 16)

myosin light-chain kinase (MLCK) An enzyme involved in phosphorylation and the regulation of muscle contraction. (Ch. 16)

myosin-linked regulation The regulation of muscle contraction that takes place via thick filaments. (Ch. 16)

myotonia (my-oh-**toe**-nee-ah) A condition that is characterized by tonic spasms or temporary rigidity of muscle. (Ch. 33)

myxedema (mik-seh-**dee**-mah) The accumulation of water in the skin, leading to a thickened and puffy appearance. (Ch. 13)

narcolepsy A severe disorder of sleep in which an individual suddenly lapses into sleep during the middle of the day; thought to be the result from activation of the REM-sleep generator. (Ch. 11)

natriuretic An agent that promotes the excretion of sodium in urine. (Ch. 23)

negative balance The condition in which there is net loss of a substance from the body. (Ch. 23)

negative feedback A class of control systems active in the body; works to restore normal values of a variable and thus to exert a stabilizing influence. (Ch. 1)

neocortex The outer gray matter of the cerebrum. (Ch. 7)

nephron (**nef**-ron) The basic unit of kidney structure and function; consists of a renal corpuscle and its attached tubule. (Ch. 23)

Nernst equation A chemical equation that allows the equilibrium potential for any ion to be calculated. (Ch. 7)

nerve A group of many nerve fibers traveling together in the peripheral nervous system. (Ch. 7)

nerve cell A cell specialized to initiate, integrate, and conduct electric signals; also called *neuron*. (Ch. 7)

nervous tissue One of four primary types of tissue; made up of neurons and glial cells. (Ch. 1)

neural fat *See* triacylglycerol.

neuroendocrinology An area of study that covers the overlap between the nervous and endocrine systems. (Ch. 12)

neurofilament A rodlike cytoplasmic protein that forms a major component of the cytoskeleton of neurons. (Ch. 7)

neurohormone Chemical messenger released by a neuron and carried by the blood to its target cell. (Ch. 12)

neurohypophyseal (new-row-high-pawf-**is**-ee-al) **peptide** A neuropeptide released from the posterior pituitary that functions to regulate plasma osmolarity and lactation. (Ch. 7)

neuromuscular junction The location of synaptic connections between alpha motor neurons and skeletal muscles where acetylcholine is released from the nerve fiber terminal. (Ch. 9)

neuron (**new**-ron) *See* nerve cell.

neuropeptide transmitter A family of at least 50 neurotransmitters composed of two or more amino acids; often function as chemical messengers in nonneural tissues. (Ch. 7)

neurotransmitter A chemical messenger used by neurons to communicate with each other or with effectors. (Ch. 5)

neutralization The reaction of an acid and base to produce a salt and water. (Ch. 25)

neutron (**new**-tron) The electrically neutral particle found in the nucleus of the atom. (Ch. 2)

neutrophil (**new**-tro-fil) A polymorphonuclear granulocytic leukocyte whose granules show preference for neither eosin nor basic dyes; functions as a phagocyte and releases chemicals involved in inflammation. (Ch. 17)

nexus (**nek**-sus) A structure that connects numerous portions of smooth muscle cells; involves a fusion of the outer layers of the membrane of each cell so that there is no extracellular space at the region between cells; also called *gap junction*. (Ch. 16)

NH$_3$ Ammonia; a base. (Ch. 23)

NH$_4^+$ Ammonium ion; an acid. (Ch. 23)

nicotinamide adenine dinucleotide (NAD) A coenzyme that is most frequently employed as a hydrogen acceptor in metabolic reactions. (Ch. 6)

nicotinic A cholinergic receptor that also responds to nicotine; located on postganglionic autonomic neurons. (Ch. 10)

nitroglycerin (nigh-troh-**glis**-eh-rin) A therapeutic drug that provides relief from angina by dilating the coronary arteries to increase blood flow. (Ch. 19)

nitrous oxide A colorless, sweetish gas used as an anesthetic; commonly called *laughing gas*. (Ch. 29)

NMDA (*N*-methyl-D-aspartate) One of three subtypes of glutamate receptors; its activation results in an increase in Ca^{2+} conductance. It is both a ligand and a voltage-gated channel. (Ch. 7)

nociceptor (**no**-sih-sep-tor) A receptor that detects painful stimuli; a **mechanical nociceptor** is activated by intense mechanical stimulation such as a slap to the head; a **heat nociceptor** responds to temperatures above 45°C. (Ch. 8)

nodes of Ranvier A space between adjacent myelin-forming cells along a myelinated axon where axonal plasma membrane is exposed to extracellular fluid. (Ch. 7)

nonassociative learning A type of learning in which we adapt to a single type of stimulus according to its relevance to our survival. (Ch. 11)

nonatrophic vaginitis Vaginal dryness or tenderness due to allergic reactions or chemical irritations. (Ch. 33)

nonessential amino acids Amino acids that can be synthesized within the body by relatively simple pathways. (Ch. 6)

non-insulin-dependent diabetes mellitus (NIDDM) A form of diabetes in which the target tissues of insulin lose their responsiveness to the hormone; also known as *type II diabetes*. (Ch. 15)

nonspecific defense Defense mechanisms that deter or destroy a variety of pathogens regardless of pathogen type (e.g., phagocytosis). (Ch. 28)

norepinephrine (nor-*ep*-ih-**nef**-rin) A catecholamine neurotransmitter released from the adrenal medulla and at most sympathetic postganglionic endings and in many CNS regions. (Ch. 7)

normovolemia The normal blood volume. (Ch. 17)

nuclear bag fibers Intrafusal fibers that help make up muscle spindle; innervated by type Ia nerve fibers that transmit information about muscle length and velocity of contraction to the CNS. (Ch. 9)

nuclear chain fibers Intrafusal fibers that help make up muscle spindle; innervated by type II nerve fibers that transmit information about muscle length to the CNS. (Ch. 9)

nuclear envelope The double membrane that surrounds the cell nucleus. (Ch. 3)

nucleic (new-**klee**-ik) **acid** A nucleotide polymer in which phosphate of one nucleotide is linked to the sugar of the adjacent one; stores and transmits genetic information; includes DNA and RNA. (Ch. 2)

nucleolus (new-**klee**-oh-lus) Densely stained structure within the nucleus that contains the DNA that codes for rRNA; site of ribosome assembly. (Ch. 3)

nucleotide A compound consisting of a pentose sugar, one of a set of special groups called *bases*, and one to three phosphate groups. (Ch. 2)

nucleus (**new**-klee-us) A large membrane-bound organelle that contains the cell's DNA; (neural) cluster of neuron cell bodies in the CNS. (Ch. 3)

nucleus accumbens The lower forebrain structure that is part of the reward system. (Ch. 11)

nucleus of the vagus A region of the brain stem that, when activated, directly reduces the heart rate. (Ch. 10)

ocular (ah-**kew**-lar) **dominance column** The organization within the visual cortex in which columns of cells respond to light impinging on either the right or left eye. (Ch. 8)

"off" center and "on" surround cells Ganglion cells for which the annulus is inhibitory and periphery is excitatory. (Ch. 8)

olfactory (ohl-**fak**-toe-ree) **bulb** The site of the mitral cells of the olfactory system. (Ch. 8)

olfactory receptor cells Cells within the nasal cavity that are activated by odors; enervate mitral cells in the olfactory bulb. (Ch. 8)

olfactory system The physiological system that provides the ability to smell odors. (Ch. 8)

olfactory tract A bundle of fibers that lead from mitral cells in the olfactory bulb and innervate areas of the cortex. (Ch. 8)

oligodendroglia (ohl-ih-go-den-**dro**-glee-ah) One of two types of glial cells that wrap around the fiber extensions of neurons to form myelin. (Ch. 7)

oligosaccharide A carbohydrate that on hydrolysis yields a small number of monosaccharides. (Ch. 22)

oliguria (ohl-ih-**goo**-ree-ah) The excretion of a small amount of urine in relation to fluid intake. (Ch. 24)

"on" center and "off" surround cells Ganglion cells for which the center of the annulus is excitatory and the periphery is inhibitory. (Ch. 8)

oncotic (on-**kot**-ik) **pressure** The osmotic pressure due to the presence of colloids in a solution; in the case of plasma interstitial fluid interaction, it is the force that tends to counterbalance the capillary blood pressure. (Ch. 17)

oogenesis (oh-ah-**gen**-eh-sis) The formation of ova. (Ch. 31)

operant conditioning A type of associative learning that involves the relationship between a behavior and a subsequent reward or punishment. (Ch. 11)

opioid (oh-**pea**-oid) One of a family of neuropeptides that are structurally similar to opiates and that modulate the perception of pain. (Ch. 7)

opsonin (**op**-so-nin) A complement protein that is part of the internal defense mechanism; coats bacteria and other cells so that they are susceptible to phagocytosis. (Ch. 28)

ordinate The vertical axis of a graph; used to plot a dependent variable. (Ch. 1)

organ (**or**-gan) A group of tissues that have been combined for the performance of a specific function or a series of related functions. (Ch. 1)

organelle An organized, reproducible, permanent subcellular structure, membranous or nonmembranous, that provides for specific functions of the cell. (Ch. 3)

orgasm (**or**-gazm) The apex and culmination of sexual excitement; characterized in the male by ejaculation and in the female by rhythmic muscular contractions in the uterus and vagina. (Ch. 31)

orientation column Nerve cells of the primary visual cortex that are sensitive to bars of light oriented at different angles. (Ch. 8)

origin The more stationary and proximal point of the attachment of a muscle. (Ch. 16)

orthophosphate PO_4; the primary form of phosphorus in plasma. (Ch. 26)

osmolality Total solute concentration, expressed as osmoles per kg of H_2O. (Ch. 23)

osmolarity The concentration of osmotically active particles in solution, expressed as osmoles of a solute per liter of solution. (Ch. 2)

osmoreceptor Specialized neurons in the anterior hypothalamus that are stimulated by increased osmolality of the extracellular fluid, their excitation promotes the output of antidiuretic hormone and stimulates the thirst sensation. (Ch. 9)

osmosis (oz-**mow**-sis) The net movement of solvent (usually water) from a solution of lesser to one of greater solute concentration when the two solutions are separated by a membrane that selectively prevents passage of solute molecules, but is permeable to the solvent. (Ch. 2)

osmotic (oz-**mot**-ik) **concentration** Total concentration of all osmotically active solutes present in the aqueous solution. (Ch. 4)

osmotic fragility The ability of a cell to withstand an osmotic stress. (Ch. 4)

osmotic pressure A property of a solution that is proportional to the non-diffusible solute concentration; a cause of water diffusion across membranes. (Ch. 2)

ossicle (**os**-ih-kul) A small bone within the inner ear. (Ch. 8)

osteoblast (**os**-tee-oh-*blast*) A cell that arises from a fibroblast and which at maturity is associated with the production of bone. (Ch. 26)

osteoclast (**os**-tee-oh-*klast*) A cell that carries out the process of bone resorption. (Ch. 26)

osteocyte (**os**-tee-oh-*site*) An osteoblast that has progressively lost its bone-forming capability. (Ch. 26)

osteoid (**os**-tee-oyd) The organic matrix of bone. (Ch. 26)

osteomalacia (*os*-tee-oh-mal-**ay**-see-ah) A metabolic bone disease in which there is softening of the bone due to inadequate mineralization, with excess accumulation of osteoid. (Ch. 26)

osteoporosis (*os*-tee-oh-poe-**row**-sis) A metabolic bone disease in which there is reduction of overall bone mass. (Ch. 26)

otolithic (*oh*-toe-**lith**-ik) **membrane** A membrane in which the stereocilia of hair cells in the utricle of the inner ear are embedded; contains calcium carbonate crystals called *statoconia*. (Ch. 9)

otolithic organ The utricle and saccule of the inner ear; detects linear acceleration. (Ch. 8, 9)

otosclerosis (*oh*-toe-skler-**oh**-sis) Abnormal growth of bones in the middle ear, causing conductive hearing loss. (Ch. 8)

outer ear The portion of the ear that channels sound waves to the ear drum. (Ch. 8)

oval window A membranous structure of the inner ear that vibrates with movement of the ossicles. (Ch. 8)

ovarian (oh-**var**-ee-an) **cycle** The approximately 28-day cycle of the female reproductive system; also called the *menstrual cycle*. (Ch. 31)

ovary (**oh**-var-ee) The female gonad; an oval-shaped organ located in the pelvic cavity below the open ends of the fallopian tubes; produces germ cells and hormones. (Ch. 31)

Overton's rules Relationships that characterize the permeability of cell membranes to polar and nonpolar solutes. (Ch. 4)

ovulation (ov-vu-**lay**-shun) The release of a single ovum from one of the two ovaries. (Ch. 31)

ovum (**oh**-vum) The female reproductive cell, which after fertilization develops into an embryo. (Ch. 31)

oxidation (*oks*-ih-**day**-shun) **reaction** A chemical reaction in which electrons are lost by a reactant. (Ch. 6)

oxidation-reduction reaction A class of chemical reactions in which one reactant accepts electrons that are removed from another. (Ch. 2)

oxidative phosphorylation The chemical pathway that provides ATP to the cell by phosphorylation of ADP in the mitochondrion. (Ch. 6)

oxygen An element with a high affinity for electrons; the primary receptor of electrons in the mitochondrial respiratory chain. (Ch. 6)

oxygen debt The amount of oxygen required to metabolize excess lactic acid formed in skeletal muscle during exercise. (Ch. 16)

oxygen radicals Highly reactive byproducts of molecular oxygen that are toxic to cells. (Ch. 29)

oxygen tension The partial pressure of oxygen in a liquid or a gas. (Ch. 21)

oxygen toxicity Damage to the lung and other body parts that occurs after breathing high concentrations of oxygen for more than 1 day. (Ch. 29)

oxygenated blood Blood that has passed through the pulmonary capillaries and carries about 20 ml of O_2 per deciliter of whole blood. (Ch. 21)

oxygen-carrying capacity The amount of oxygen that the hemoglobin in 100 ml of whole blood is capable of transporting. (Ch. 17)

oxyhemoglobin (*ok*-see-**he**-mow-glow-bin) Hemoglobin that is saturated with oxygen after passing through the lungs. (Ch. 17)

oxyhemoglobin dissociation curve An S-shaped curve that illustrates the way that hemoglobin binds to oxygen in the lung and unloads oxygen in the tissues. (Ch. 21)

oxytocin (*ok*-see-**toe**-sin) A hormone secreted by the posterior pituitary; stimulates contraction of uterine muscles during labor and milk ejection actions. (Ch. 13)

pacemaker Neurons that set the rhythm of biological clocks independent of external cues; any nerve or muscle cell that has an inherent autorhythmicity and determines activity patterns of other cells. (Ch. 18)

pacemaker cell A cardiac cell that regulates heart rate. (Ch. 10)

Pacinian corpuscle A somatic sensory mechanoreceptor, located beneath the skin and in walls of viscera, that responds to pressure. (Ch. 9)

pain receptor A mechanical or heat nociceptor. (Ch. 9)

palmitic acid A fatty acid; one of the most common types found in phospholipids. (Ch. 4)

pancreas (**pan**-kree-as) An organ of the digestive system; located in the abdomen near the stomach; connected by ducts to the small intestine; contains endocrine gland cells, which secrete insulin, glucagon, and somatostatin into the bloodstream, and exocrine gland cells, which secrete digestive enzymes and bicarbonate into the intestine. (Ch. 15, 22)

pancreatic polypeptide A polypeptide produced by the F-cells of the pancreas; its physiological function is not fully characterized. (Ch. 15)

pancreatitis (*pan*-kree-ah-**tie**-tis) An acute or chronic inflammation of the pancreas; due to autodigestion of pancreatic tissue by its own enzymes. (Ch. 22)

paracellular transport pathway Junctions that are relatively permeable and allow water and ions to move between the epithelial cells. (Ch. 4)

paracrine (**pear**-ah-krin) The mechanism whereby a hormone secreted by one cell affects an adjacent or nearby target cell. (Ch. 12)

parafollicular cells Thyroid cells that produce calcitonin; also called *C-cells*. (Ch. 26)

parallel organization The arrangement of the motor system whereby information from multiple sources can simultaneously and independently affect the same target. (Ch. 9)

parasympathetic nervous system The portion of the autonomic nervous system that coordinates the body's vegetative activities. (Ch. 10)

parathormone *See* parathyroid hormone.

parathyroid (*par*-ah-**thy**-roid) **gland** Two pairs of glands located on the dorsal surface of each of the two lobes of the thyroid glands. (Ch. 26)

parathyroid hormone (PTH) A polypeptide protein produced by the parathyroid glands. (Ch. 26)

paraventricular nucleus The location of osmoreceptors in the hypothalamus. (Ch. 11)

partial pressure That part of the total gas pressure that is due to molecules of one gas species; a measure of concentration of a gas in a gas mixture. (Ch. 2)

parturition (*par*-too-**rish**-in) Birth; the delivery of an infant and placenta. (Ch. 32)

passive immunity Short-term protection against disease as a result of receiving antibodies. (Ch. 28)

passive tension The force in a muscle at rest that is due to the elastic nature of its anatomical structure. (Ch. 16)

patellar (pah-**tell**-ar) **tendon** A tendon of the anterior thigh muscles that passes through the knee, contains the patella (knee cap), and is inserted into the anterior tibia of the leg. (Ch. 9)

pathogen (**path**-oh-jen) Any disease-producing organism or agent. (Ch. 28)

PCO$_2$ Carbon dioxide partial pressure or tension. (Ch. 30)

penile (**pee**-nile) **erection** The increase in diameter, length, and firmness in the penis; due to vasocongestion. (Ch. 33)

pentose phosphate pathway The pathway for oxidation of glucose in the cytosol; also called the *hexose monophosphate shunt*. (Ch. 6)

pepsin (**pep**-sin) An enzyme, secreted by the stomach, that initiates digestion of protein. (Ch. 22)

peptidase Any enzyme that hydrolyzes specific peptides to their constituent amino acids. (Ch. 22)

peptide (**pep**-tide) A compound that is composed of two or more amino acids; forms the constituent parts of proteins. (Ch. 7)

peptide bond A polar covalent chemical bond joining two amino acids; forms the protein backbone. (Ch. 2)

perfusion Capillary blood flow. (Ch. 19)

perimysium (*per*-ih-**mis**-ee-um) A type of connective tissue that groups bundles of muscle fibers. (Ch. 16)

peripheral nervous system (PNS) Nerve fibers that emerge from the brain stem and spinal cord; composed of somatic and autonomic portions. (Ch. 7)

peripheral proprioreceptors Neuroreceptors in joints and muscle spindles that detect movement of limbs. (Ch. 30)

peripheral thermoreceptors Temperature receptors that detect changes in skin and shell temperature; located just beneath the skin. (Ch. 27)

peristalsis (*per*-ih-**stal**-sis) The propulsive movement of semisolid food mass through the digestive system by a wave of muscle contractions. (Ch. 22)

peritoneum (*per*-ih-tow-**nee**-um) A membrane that lines the abdominal cavity and covers some of the abdominal viscera. (Ch. 22)

peritubular capillary The capillary network surrounding the tubules of the kidney. (Ch. 23)

peritubular capillary starling forces The hydrostatic and colloid osmotic pressures that influence fluid movement across the capillaries that surround the kidney tubules. (Ch. 23)

permeability coefficient A number expressing the ease with which a solute can penetrate a plasma membrane; governed by the thickness of the membrane and the lipid solubility of the solute; can be expressed as P. (Ch. 4)

permissive action The amplifying effect of one hormone on another. (Ch. 14)

peroxidase (per-**ox**-ih-dase) A cellular enzyme that readily reacts with oxidizing free radicals. (Ch. 29)

peroxide An oxide of any element that contains more oxygen than any other. (Ch. 29)

peroxisome (per-**ox**-ih-zome) A small, membrane-bounded bag, found in a cell, that is involved in digestion and food waste; also important in detoxifying a number of molecules. (Ch. 3)

peyer's patches Aggregates of lymphoid nodules in the mucosa of the intestine. (Ch. 28)

pH A measure of the acidity or alkalinity of a solution; it is equal to the negative logarithm (to base 10) of the hydrogen ion concentration (the latter measured in moles per liter). (Ch. 2)

phagocytosis (*fag*-oh-**sigh**-toe-sis) A special type of enocytosis involving internalization of particulate matter. (Ch. 3)

pharynx (**fare**-inks) The portion of the digestive tract between the mouth and the esophagus. (Ch. 22)

phase contrast microscope A microscope that enhances the contrast between the internal organelles so that structures can be seen clearly in living, unstained, and nonfixed cells. (Ch. 3)

phasic (**faze**-ik) **contraction** A contraction that occurs intermittently. (Ch. 16)

phasic discharge A very brief burst of action potentials by a proprioceptor. (Ch. 9)

phenotype (**fee**-no-type) The visible expression of the genotype; physical or other characteristics, which are produced by genes and influenced by the interaction between genes and environmental factors. (Ch. 11)

pheromone A substance secreted to the outside of the body by an individual and perceived (as by smell) by a second individual. (Ch. 10)

phonocardiogram The electronic recording of heart sounds. (Ch. 18)

phosphagen system The energy system of muscles based on ATP and phosphocreatine; used for short bouts of maximal exercise. (Ch. 30)

phosphate PO$_4$. (Ch. 26)

phosphatidylinositol (fass-fah-**tid**-el-in-**os**-i-tahl) A minor phospholipid constituent of the plasma membrane. (Ch. 12)

phosphatidylinositol 4,5-bisphosphate (PIP2) The immediate precursor for the generation of two intracellular second messengers, derived from phosphatidylinositol. (Ch. 12)

phosphaturia (*fos*-fa-**tur**-ee-ah) Increased phosphate excretion in the urine. (Ch. 26)

phosphocreatine A high-energy phosphate molecule used as an energy resource in the metabolism of muscle cells. (Ch. 30) *See also* phosphate.

phosphodiesterase (*fass*-foe-die-**es**-ter-aze) An enzyme present in the disc membrane of photoreceptors; catalyzes the conversion of cyclic AMP to AMP. (Ch. 9)

phospholipase (*fos*-fo-**lie**-pase) **C** A receptor-controlled plasma-membrane enzyme that catalyzes phosphatidyl bisphosphate breakdown to inositol trisphosphate and diacylglycerol. (Ch. 5)

phospholipid (fos-fo-**li**-pid) A lipid subclass similar to tracylglycerol except that a phosphate group and a small nitrogen-containing molecule are attached to the third hydroxyl group of glycerol; a major component of the cell membrane. (Ch. 2)

phosphoprotein phosphatase An enzyme that removes phosphate from protein, restoring the protein's original shape. (Ch. 5)

phosphorylation The addition of a phosphate group to an organic molecule. (Ch. 6)

photon A particle of light; photons have wavelengths and energies associated with different colors. (Ch. 9)

photoreceptor (**fo**-toe-re-*sep*-tor) **cones** Conically shaped photoreceptors that are the basis of color perception and day vision. (Ch. 9)

photoreceptor rods Cylindrically shaped photoreceptors that are very sensitive to low levels of illumination. (Ch. 9)

photoreceptors   Receptor cells of the retina that absorb photons. (Ch. 9)

photosynthesis   The use of radiant energy to manufacture ATP, which drives biosynthesis in plant cells. (Ch. 6)

phrenic (**free**-nik) **nerve**   The nerve that sends respiratory messages to the diaphragm. (Ch. 20)

physiology (*fiz*-ee-**ol**-oh-jee)   A branch of biology dealing with the mechanisms by which the body functions. (Ch. 1)

Pickwickian syndrome   A type of altered breathing pattern that affects ventilation; occurs in severely obese individuals who often suffer hypoventilation. (Ch. 20)

pituitary (pit-**too**-ih-*terr*-ee) **dwarfism**   Dwarfism that is due to a deficiency in GH secretion, which occurs as a result of a defect in pituitary function. (Ch. 13)

pituitary gland   The endocrine gland that lies in a bony pocket below the hypothalamus; includes the anterior, middle, and posterior pituitary. (Ch. 13)

pituitary stalk   A thin segment of tissue that connects the pituitary gland to the hypothalamus. (Ch. 13)

placenta (plah-**sen**-tah)   An interlocking fetal and maternal organ of molecular exchange between fetal and maternal circulation. (Ch. 32)

plasma (**plas**-muh)   The liquid portion of blood; a component of extracellular fluid. (Ch. 17)

plasma cell   A cell that differentiates from activated β-lymphocytes and secretes antibodies. (Ch. 28)

plasma membrane   The membrane that forms the outer surface of the cell and separates the cell's contents from extracellular fluid. (Ch. 1)

plasmid (**plas**-mid)   A small accessory DNA molecule used for DNA cloning. (Ch. 3)

plateau   The phase of sexual excitement that precedes orgasm. (Ch. 33)

platelet (**plate**-let)   Thrombocyte; a cell fragment, suspended in plasma, that functions in blood clotting. (Ch. 17)

platelet plug   An aggregation of platelets at the site of vascular damage; effectively plugs the defect end. (Ch. 17)

pleural (**ploor**-ahl) **pressure**   The pressure in the pleural fluid between the lung and the chest wall. (Ch. 20)

pluripotential hematopoietic stem cell   An undifferentiated cell that is capable of developing into progenitor cells of the erythrocyte, leukocyte, or megakaryocyte. (Ch. 17)

pneumatology   An early theory of physiology put forward by Galen; involved the belief that the blood took on natural spirits and carried them through the veins to the bodily organs. (Ch. 1)

pneumoconiosis   (new-mo-koe-ni-oh-seis)   A disease that causes scarring of the lung from breathing dust particles. (Ch. 29)

pneumothorax   The condition when air enters the pleural space and the lungs collapse. (Ch. 20)

Po₂   Partial pressure of oxygen. (Ch. 30)

poikilocytosis (poy-*kee*-lo-sigh-**to**-sis)   A variation in the shape of an erythrocyte. (Ch. 17)

poikilotherm (poy-**kee**-lo-therm)   An animal whose body temperature may fluctuate with ambient temperature. (Ch. 27)

polycythemia (*pol*-ee-sie-**thee**-me-ah)   An increase in the number of red blood cells in the blood. (Ch. 17)

polypeptide   A polymer consisting of amimo acid subunits joined by peptide bonds; also called *peptide* and *protein*. (Ch. 2)

polysome   A structure composed of ribosomes attached to a single strand of mRNA, playing a role in peptide synthesis; also called a *polyribosome*. (Ch. 5)

polyspermy   Fertilization of the ovum by more than one sperm; lethal for the fertilized ovum. (Ch. 32)

polyuria (*pah*-ly-ee-**yoo**-ree-ah)   The excessive excretion of urine; characteristic of diabetes. (Ch. 15)

population code   A method of informing the brain about the intensity of a stimulus; the strength of a stimulus is related to the number of receptor cells that are activated. (Ch. 9)

population code of contractile force   A means of control of the amount of force that a muscle generates during a contraction; an increase in the number of motor neurons and motor units can increase a muscle's force of contraction. (Ch. 9)

portal circulation   Blood flow from the stomach, intestines, and pancreas through the portal vein and to the liver. (Ch. 22)

positive balance   The condition in which there is net gain of a substance in the body. (Ch. 24)

positive feedback   A class of control system that destabilizes normal values of a variable and thus has limited value in maintaining homeostasis. (Ch. 1)

posterior (pos-**teer**-ee-or)   A descriptive term of position that means nearer to the back of the body. (Ch. 1)

posterior parietal cortex   An area of the brain that processes sensory stimuli, leading to purposeful movement. (Ch. 9)

posterior pituitary gland   The posterior lobe of the pituitary gland from which ADH and oxytocin are released into blood. (Ch. 13)

postganglionic (post-gang-glee-**on**-ik) **cell**   The autonomic neuron that directly innervates the effector cell. (Ch. 10)

postganglionic fibers   Axons of postganglionic autonomic neurons. (Ch. 14)

postmenopausal osteoporosis   The rapid loss of bone due to decreasing estrogen levels in postmenopausal women. (Ch. 26)

post-translational modification of proteins   A series of reactions in which newly synthesized RNA molecules are covalently modified; also known as *RNA processing*. (Ch. 5)

potential energy   Stored energy. (Ch. 2)

power output   The rate of doing work. (Ch. 16)

precapillary sphincter (**sfink**-ter)   A smooth muscle ring around the arteriole that regulates the flow of blood into the capillaries supplied by that arteriole. (Ch. 19)

preganglionic cells   Efferent sympathetic and parasympathetic neurons whose cell bodies are located within the CNS but whose fibers leave the CNS and synapse within peripheral autonomic ganglia. (Ch. 10)

pregnancy   The condition of having a developing embryo or fetus within the body. (Ch. 32)

pregnenolone   A steroid hormone that is an early constituent of the synthesis of all steroids; its precursor is cholesterol. (Ch. 12)

premature ejaculation   Consistent, unintentional ejaculation prior to vaginal penetration; more commonly defined as loss of ejaculatory control resulting in ejaculation soon after vaginal penetration. (Ch. 33)

premise   Given information stated in the form of an absolute rule using a word such as "all" or "always." (Ch. 1)

premotor cortex   A component of the frontal, cerebral cortex where movements are planned and programmed. (Ch. 9)

preoptic area   An area of the hypothalamus in which electrical stimulation induces non-REM sleep; destruction of this area of the brain causes insomnia in laboratory rats. (Ch. 11)

pressure gradient P_1–P_2 or ΔP; the difference in pressure between two points. (Ch. 18)

presynaptic facilitation   An underlying mechanism that enhances an animal's response to a particular stimulus; causes sensitization to the stimuli. (Ch. 11)

presynaptic (*pre*-sin-**ap**-tik) **inhibition**   The action of an interneuron to inhibit the release of substance P by the release of the peptide enkephalin. (Ch. 9)

primary somatosensory area   A narrow strip of cortical tissue, the location of cells in the somatosensory cortex that exhibit the annular receptive field; also called *Brodmann area 3*. (Ch. 9)

primitive gonads   Gonads found in the zygote before they have differentiated into testes or ovaries. (Ch. 33)

procarboxypeptidase (*pro*-car-box-y-**pep**-tid-aze)   An inactive precursor of carboxypeptidase, which is a proteolytic enzyme secreted by the pancreas. (Ch. 22)

progenitor cell   A committed stem cell that is capable of developing into precursor cells of a specific blood cell. (Ch. 17)

progesterone (pro-**jes**-ter-own)   A steroid hormone secreted by the corpus luteum and placenta; stimulates uterine-gland secretions, inhibits uterine smooth-muscle contractions, and stimulates breast growth. (Ch. 12)

progestin (pro-**jes**-tin)   One of the major categories of steroid hormones; involved in the maintenance of pregnancy as well as being produced by the ovaries and the placenta. (Ch. 12)

proglucagon (pro-**glue**-ka-gon)   The immediate precursor of glucagon; synthesized in the pancreas, small intestine, and hypothalamus, but only converted to glucagon in the pancreas. (Ch. 15)

proinsulin   A large precursor molecule from which insulin is derived; synthesized in the rough endoplasmic reticulum of the beta cells in the pancreas. (Ch. 15)

prokaryotic cells   Cells in which the DNA is not in a separate compartment from the rest of the cell. (Ch. 3)

prolactin (pro-**lak**-tin) A peptide hormone secreted by the anterior pituitary; stimulates milk secretion in the mammary glands. (Ch. 13)

promoter site The place where transcription of a given gene begins when the RNA polymerase binds on a DNA template. (Ch. 5)

pro-opiomelanocortin (**pro**-*op*-e-o-mel-**uh**-no-**cort**-in) (**POMC**) A large protein precursor for ACTH, endorphins, and several other hormones. (Ch. 13)

proprioception (*pro*-pre-oh-**sep**-shun) The sensation of a change in the position of the limbs. (Ch. 9)

proprioceptor (*pro*-pre-oh-**sep**-tor) A somatic sensory receptor activated by movement of the limbs. (Ch. 9)

prosomatostatin (pro-so-**ma**-to-stat-in) The immediate precursor to somatostatin. (Ch. 15)

prostaglandins (*pros*-tah-**glan**-dins) A group of naturally occurring hydroxy fatty acids derived from arachidonic acid; affect a wide variety of physiological processes. (Ch. 23)

protease (**pro**-tee-ase) A general term for a proteolytic enzyme; *see* peptidase. (Ch. 22)

protein (**pro**-teen) A large polymer consisting of one or more sequences of amino acid subunits joined by peptide bonds. (Ch. 2)

protein asymmetry in membranes The condition when extrinsic and intrinsic membrane proteins are distributed unequally between the two halves of the bilayer. (Ch. 4)

protein conformation The distinctive three-dimensional shape of a protein molecule formed when the linear polypeptide chain folds up in a precise way. (Ch. 1)

protein kinase One of a family of enzymes that phosphorylate certain other proteins by transferring them to a phosphate group from ATP. (Ch. 5)

protein-bound calcium One of two major forms of calcium in plasma; calcium bound to relatively large plasma proteins that cannot pass through plasma membranes. (Ch. 26)

proteinuria (*pro*-tee-**nyoo**-re-ah) The presence of an excess of proteins in the urine. (Ch. 23)

proteoglycan High-molecular-weight compounds that form the noncollagen portion of bone. (Ch. 26)

prothrombin (pro-**throm**-bin) An inactive precursor of thrombin; produced by the liver and normally present in plasma. (Ch. 17)

prothrombin converting factor The activated Stuart factor, in the presence of calcium ions; forms complexes with accelerin on phospholipid micelles provided by tissue thromboplastin; converts prothrombin to thrombin. (Ch. 17)

proton A positive charged subatomic particle. (Ch. 2)

protoplasm A complex mixture of chemicals that displays the attributes of life; that is, is organized into specific kinds of structural units, has the ability to enter into chemical activities that include the transformation of energy and the maintenance or synthesis of protoplasm, the ability to respond to changes in the environment, and the ability to grow and reproduce. (Ch. 1)

proximal Nearer, closer to a reference point; the opposite of distal. (Ch. 16)

proximal convoluted tubule The first part of the renal tubule. (Ch. 23)

pseudopodia Footlike extensions of the cytoplasm. (Ch. 17)

puberty The attainment of sexual maturity, when conception becomes possible; as commonly used, refers to 3 to 5 years of sexual development that culminates in sexual maturity. (Ch. 31)

pulmonary (**pul**-mow-*ner*-ee) **artery** The large vessel leading from the right ventricle that carries systemic venous blood to the lungs. (Ch. 18)

pulmonary circulation The circulation through the lungs; the portion of the cardiovascular system between the pulmonary trunk, as it leaves the right ventricle, and the pulmonary veins, as they enter the left atrium. (Ch. 18)

pulmonary edema An abnormal accumulation of fluid in the lung interstitium and alveoli. (Ch. 21)

pulmonary embolism Blockage in the pulmonary circulation due to emboli which are often blood clots (thrombi) that have formed in the systemic venous system. (Ch. 21)

pulmonary hypertension Abnormally high pulmonary arterial pressure; can cause vessel damage and eventual failure of the right ventricle. (Ch. 21)

pulmonary stretch receptors Afferent nerve endings lying between the smooth muscle cells of the airway and activated by lung inflation. (Ch. 20)

pulmonary trunk A large artery that carries blood from the right ventricle of the heart to pulmonary arteries. (Ch. 18)

pulmonary valve The valve between the right ventricle of the heart and the pulmonary trunk. (Ch. 18)

pulmonary veins Veins that take oxygenated blood into the left heart. (Ch. 18)

pulmonary ventilation Tidal volume times respiratory rate; normally 5 to 6 liters per minute at rest. (Ch. 30)

pulmonic stenosis Narrowing of the right atrioventricular orifice or pulmonary trunk resulting in hypertrophy of the right ventricle. (Ch. 18)

pulse pressure The difference between systolic and diastolic arterial blood pressures. (Ch. 19)

Purkinje (per-**kin**-jee) **cell** The most prominent nerve cell in the cerebellum. (Ch. 9)

Purkinje fiber A specialized myocardial cell that constitutes part of the conducting system of the heart; conveys excitation from bundle branches to ventricular muscle. (Ch. 18)

putamen One of the basal ganglia involved with involuntary control of skeletal muscle. (Ch. 9)

P-wave A component of an electrocardiogram that reflects atrial depolarization. (Ch. 18)

pyloric (pie-**lor**-ik) **sphincter** A thick bundle of circular smooth muscle and connective tissue that controls the passage of chyme from the stomach to the small intestine. (Ch. 22)

pylorus (pie-**lor**-us) The distal funnel-shaped end of the stomach; stomach contents are emptied through it to the duodenum. (Ch. 22)

pyramidal tract Nerve fibers that descend from the motor cortex to the brain stem and down the lateral corticospinal tract. (Ch. 9)

pyrogens Chemical agents that cause fever. (Ch. 27)

pyruvic acid A three-carbon intermediate in the glycolytic pathway that, in the absence of oxygen, forms lactic acid or, in the presence of oxygen, enters the Krebs cycle. (Ch. 30)

QRS complex The component of an electrocardiogram corresponding to ventricular depolarization. (Ch. 18)

qualitative Pertaining to quality. (Ch. 1)

quantal release Refers to the release of a fixed number of neurotransmitter molecules from synaptic vesicles. (Ch. 7)

quantitative Denoting or expressible as a quantity; relating to the proportionate quantities or the amount of the constituents of a compound. (Ch. 1)

quisqualate One of three subtypes of glutamate receptors that, when activated, results in excitatory postsynaptic potentials by opening ion channels that increase in Na^+ and K^+ conductance. (Ch. 7)

radiation Heat transfer by electromagnetic radiation between objects that are not in contact. (Ch. 27)

radioimmunoassay (**ray**-dee-o-im-mew-no-**as**-ay) A procedure used to measure very small amounts of molecules with great accuracy; uses radioisotopes to label cyclic AMP in a sample. (Ch. 3)

radioisotope An unstable isotope of an element that undergoes radioactive decay. (Ch. 3)

raphe nucleus A portion of the brain stem that is responsible for the generation of REM sleep. (Ch. 11)

rapid eye movement (REM) sleep A phase of sleep during which eye movement becomes rapid, the heart rate and blood pressure increase, and gastrointestinal motility decreases. (Ch. 11)

rate-limiting enzyme An enzyme in the metabolic pathway that is most easily saturated with substrate; determines the rate of the entire metabolic pathway. (Ch. 6)

rate-limiting reaction The slowest reaction in a metabolic pathway. (Ch. 6)

Rathke's pouch A pocket of cells from the primitive oral cavity that are involved in the embryological development of the pituitary. (Ch. 13)

ratio An expression that compares two numbers or quantities by division. (Ch. 1)

reactive hyperemia (*hi*-per-**ee**-me-ah) Extreme reactive vasodilation. (Ch. 19)

receptive field The small area of the skin that can activate the receptor of a nerve cell in the dorsal root ganglion. (Ch. 8)

receptor A macromolecule, usually a protein, that binds a signal molecule with high specificity and triggers intracellular events leading to a change in cell function. (Ch. 1)

receptor-mediated endocytosis A process of endocytosis whereby a hormone and its plasma membrane receptor are internalized. The hormone is degraded and the receptor is recycled to the plasma membrane. (Ch. 12)

recombinant (re-**kom**-bih-nent) **DNA (rDNA)** A DNA molecule created by combining pieces of DNA from markedly different organisms. (Ch. 3)

recovery oxygen Net amount of oxygen consumed during recovery from exercise. (Ch. 30)

rectum (**rek**-tum) The portion of the large intestine between the sigmoid colon and the anus. (Ch. 22)

red cell indices Quantitative assessments of erythrocyte size, shape, and color using the hematocrit, hemoglobin content, and RBC count of blood. (Ch. 17)

redout A loss of vision due to high negative G-forces. (Ch. 29)

redox potential The oxidation-reduction potential; a measure of the affinity of molecules to give up or accept electrons. (Ch. 6)

reduction reaction A chemical reaction in which a molecule is reduced by gaining electrons. (Ch. 6)

referred pain Pain that is attributed to the surface of the body rather than its internal source. (Ch. 10)

reflex (**ree**-flex) An automatic, involuntary action. (Ch. 9)

refractory (re-**frak**-tor-ee) **period** The period between the peak of an action potential and the return of a membrane to the resting state. (Ch. 16)

regenerative process The basis for the generation of an action potential; the depolarization of a membrane after a threshold is reached. (Ch. 7)

relative pressure Pressure relative to the atmospheric pressure. (Ch. 20)

relative refractory period The period near the end of an action potential during which the threshold voltage needed to initiate a second action potential is large. (Ch. 7)

relaxin (re-**lax**-in) A hormone that relaxes pelvic ligaments in preparation for parturition. (Ch. 32)

REM generator *See* raphe nucleus.

REM sleep *See* rapid eye movement sleep.

renal (**ree**-nul) Pertaining to the kidneys. (Ch. 23)

renal hilum The indentation of the medial aspect of the kidneys. (Ch. 23)

renal pelvis The expanded upper portion of the ureter that collects the urine from the renal calyces. (Ch. 23)

renal plasma clearance The volume of plasma per unit time that must be completely freed (cleared) of a substance to supply the quantity of the substance excreted per unit time in the urine. (Ch. 23)

renin (**reh**-nin) A proteolytic enzyme, produced by granular cells in the kidney, that is involved in the regulation of thirst, osmolarity, and water volume. (Ch. 10)

renin-angiotensin system Sometimes called the *renin-angiotensin-aldosterone system*; a hormonal system that plays an important role in blood pressure and extracellular fluid volume regulation. (Ch. 14, 24)

replication fork A structure that is produced by the separation of the two original DNA strands at the site of active synthesis. Many replication forks can exist on a DNA molecule at once. (Ch. 5)

repolarization The period during an action potential in which the membrane potential is restored to its original value. (Ch. 18)

residual volume The amount of air remaining in the lungs after a person expires as forcefully as possible. (Ch. 20)

respiration (res-per-**ay**-shun) The act or function of breathing; the act by which air is drawn in and expelled from the lungs; also called *external respiration*. (Ch. 20)

respiratory acidosis An abnormal process characterized by CO_2 accumulation, which leads to a fall in blood pH. (Ch. 25)

respiratory alkalosis An abnormal process characterized by the loss of too much CO_2, which leads to a rise in blood pH. (Ch. 25)

respiratory center The portion of the medulla that is responsible for the control of breathing. (Ch. 20)

respiratory distress syndrome Labored breathing that afflicts premature neonates born with insufficient amounts of surfactants in the lungs. (Ch. 20)

respirometer An instrument for the accurate measurement of oxygen consumption. (Ch. 27)

restriction endonuclease An enzyme that can break a strand of DNA at a specific site. (Ch. 3)

resting tension *See* passive tension. (Ch. 16)

retching A process by which the stomach contents are forced into the esophagus but not expelled. (Ch. 22)

reticular (reh-**tik**--yoo-lur) **formation** A diffuse network of neurons in the brain stem; involved in maintaining posture by the activation of extensor muscles. (Ch. 9)

reticulocytes Stem cells that have lost their nuclei but are not yet mature erythrocytes. (Ch. 17)

retina (**ret**-ih-na) The interior surface of the eye; contains photoreceptors and other visual system neurons. (Ch. 8)

retinotopic map The organization of the visual cortex in the form of a map of the external visual field. (Ch. 8)

retrograde (**ret**-row-grade) **transport** Transport against the main direction of flow. (Ch. 7)

retrolental fibrosis Blindness due to deterioration of the retina and inadequate development of the retinal blood vessels; the formation of fibrosis tissue behind the lens. (Ch. 29)

retroperitoneal (*reh*-trow-*per*-ih-tow-**nee**-al) External to the peritoneal lining of the abdominal cavity. (Ch. 22)

retropulsion The closure of the pyloric sphincter and forcible squirting of stomach contents back into the body of the stomach. (Ch. 22)

re-uptake A mechanism by which transmitter molecules are returned into the presynaptic terminal. (Ch. 7)

reward system Brain structures that provide pleasurable inner emotions in response to certain behaviors and drugs. (Ch. 11)

Rh factor Antigen D, which is genetically determined to be present or absent on red blood cells. (Ch. 17)

rhodopsin (row-**dop**-sin) A visual pigment; a photosensitive purple-red chromoprotein in

the retinal rods that is bleached to visual yellow by light. (Ch. 8)

ribonucleic (*rye*-bo-new-**clay**-ik) **acid (RNA)** A single strand of nucleotides that represent a copy of a limited region of the nucleotide sequence of a strand of DNA. (Ch. 3)

ribosomal (rye-bo-**zome**-al) **RNA (rRNA)** One of three kinds of RNA synthesized; the nucleotide sequence on rRNA molecules provides information for protein synthesis. (Ch. 5)

ribosome (**rye**-bo-sown) Cellular organelles that make proteins according to the instructions carried by RNA. (Ch. 3)

rickets (**rik**-ets) A metabolic bone disease of children, characterized by inadequate mineralization of new bone matrix; caused by deficiency of vitamin D. (Ch. 26)

right heart The chambers of the heart that receive systemic venous blood and pump it to the lungs. (Ch. 18)

rigor mortis After death, a stiffness of skeletal muscles resulting from loss of ATP. (Ch. 16)

RNA polymerase Enzyme that synthesizes RNA during the transcription of DNA. (Ch. 5)

RNA processing (post-transcriptional modification) A series of reactions that modify newly synthesized RNA before it leaves the nucleus. (Ch. 5)

RNA splicing A type of RNA processing in which axon segments are joined together to form functional mRNA. (Ch. 5)

RNA translation The translation of codons on mRNA in the synthesis of proteins. (Ch. 5)

rod photoreceptors Cylindrically shaped photoreceptors that contain one type of pigment capable of absorbing photons representing a broad range of wavelengths. (Ch. 8)

Ruffini capsule A slowly adapting mechanoreceptor sensitive to the stretching of skin. (Ch. 8)

saccule (**sak**-yool) One of two sacs in the membranous labyrinth of the inner ear. (Ch. 9)

saltatory (**sal**-tah-*tow*-ree) **conduction** Conduction of a nerve impulse (action potential) by a myelinated fiber in which the impulse jumps from one node of Ranvier to the next. (Ch. 7)

sanguine Abounding in blood. (Ch. 1)

sarcolemma (*sar*-koe-**lem**-ma) The plasma membrane of a muscle cell, especially a skeletal muscle cell. (Ch. 16)

sarcomere (**sar**-koe-mere) A repeating structural unit of myofibril; composed of thick and thin filaments and extending between two adjacent Z-lines. (Ch. 16)

satiety (sah-**tie**-ih-tee) A feeling of fullness or of not being hungry. (Ch. 11)

saturation The degree to which protein-binding sites are occupied by ligands. (Ch. 2)

scanning electron microscope A microscope that images the deflection of electrons by a specimen. (Ch. 3)

Schwann (shvon) **cell** A type of cell that forms the myelin sheath around a motor axon. (Ch. 7)

scientific measurements Quantitation of functions and processes, using the metric system of units. (Ch. 1)

scientific method A form of inquiry that involves posing questions and then searching for answers to those questions through experimentation and testing to validate answers. (Ch. 1)

scientific notation Quantities expressed as products of numbers plus a power of ten. (Ch. 1)

SCUBA An acronym for self-contained underwater breathing apparatus. (Ch. 29)

second heart sound A heart sound that occurs at the time of closure of the semilunar valves and defines the end of systole; the sound is caused by the vibrations in the heart and chest that are initiated by the closure of the heart valves and the sudden cessation of blood flow. (Ch. 18)

second messenger An extracellular signal that travels from a cell to a target cell, where it reacts with receptor molecules. (Ch. 5)

secondary somatosensory area The area of the somatosensory cortex that does not exhibit an annular receptive field; also called *Brodmann areas 1 and 2*. (Ch. 8)

secretin (**seek**-reh-tin) A strongly basic polypeptide hormone secreted by the upper small intestine; stimulates the pancreas to secrete bicarbonate into the small intestine. (Ch. 12, 22)

secretion (se-**kree**-shun) The elaboration and release of organic molecules, ions, and water by cells in response to specific stimuli. (Ch. 22)

sedimentation rate The rate at which red cells settle; the normal rate is between 2 and 20 mm/h. (Ch. 17)

segmenting contraction The most frequently occurring type of muscular movement in the small intestine; moves the contents of different regions of the small intestine back and forth over short distances. (Ch. 22)

semicircular canal The three sensory organs of the vestibular apparatus of the ear that detect position and motion of the head in space; the **superior, inferior,** and **horizontal semicircular canals** are each situated perpendicular to the other two. (Ch. 9)

semiconservative replication A process in DNA replication in which the structure of the parent DNA molecule is disrupted but the structure of the nucleotide sequence of each strand is conserved. (Ch. 5)

semilunar (*sem*-ee-**loo**-nar) **valve** A valve located at the junction of the ventricle with the pulmonary and systemic arteries; consists of flaps shaped like half-moons. (Ch. 18)

sense organ Any organ of the sensory system, such as the eyes or ears. (Ch. 8)

sensitization An increase in the response to a stimulus. (Ch. 11)

sensor A mechanism that detects changes to bodily or environmental conditions. (Ch. 1)

sensorineural hearing loss Hearing loss due to the damage to hair cells located between the basilar and tectorial membranes of the ear. (Ch. 8)

sensory adaptation The gradual loss of perception of a stimulus that is continuously applied. (Ch. 8)

sensory neuron A neuron that carries information about external or internal stimuli from sense receptors to the central nervous system; also called an *afferent neuron*. (Ch. 7)

sensory system Parts of the peripheral and central nervous systems involved with the reception and processing of sensory information; sometimes subdivided into special sensory systems—vision, audition, equilibrium, gustation, olfaction, and somatic sensory systems—pain, touch, temperature, pressure, proprioception. (Ch. 8)

sensory transduction The conversion of energy (light, heat, sound, etc.) into nerve impulses by a sensory receptor. For example, the process by which the sensory receptors of the eye convert various wavelengths of light into electrical impulses that nerve cells transmit to the thalamus. (Ch. 8)

septal defect An opening in the wall (septum) that separates blood in the right heart from blood in the left heart. (Ch. 21)

serotonin (*sair*-oh-**toe**-nin) A vasoconstrictor synthesized in the brain stem; inhibits gastric secretion, stimulates smooth muscle, and serves as a central neurotransmitter. (Ch. 7)

sex chromosomes The X and Y chromosomes; females have two X chromosomes and males have one X and one Y; provide the genetic information for the development and differentiation of most of the cells in the body. (Ch. 33)

sex steroid binding globulin (SSBG) A plasma protein that binds with testosterone in the blood; aids in storing hormones in the blood. (Ch. 31)

sexual dysfunction Any condition in which the normal physical responses to sexual stimulation are impaired. (Ch. 33)

sexual flush A darkening of the body's skin color during sexual intercourse. (Ch. 33)

shivering Involuntary trembling or quivering of the body caused by contraction or twitching of the muscles; a physiologic method of heat production. (Ch. 27)

short portal system A vascular system that connects the posterior and anterior lobes of the pituitary; may carry substances secreted by the posterior lobe to the anterior lobe where they regulate hormone secretion from the anterior pituitary. (Ch. 13)

short-reflex A regulatory reflex that is mediated over sensory neurons, interneurons, and motor and secretory neurons that lie entirely within the wall of the digestive tract. (Ch. 22)

short-term memory A section of the brain's memory bank that is easily disrupted and short-lived. (Ch. 11)

Siemens Dimensional units of conductance. (Ch. 7)

silicosis (*sil*-ih-**koe**-sis) A lung disease caused by breathing silicon (sand) dust. (Ch. 29)

single channel current The number of ions that flow through a channel per unit of time. (Ch. 7)

sino-atrial (*sigh*-no-**ay**-tree-al) **node (SA node)** The region, in the right atrium of the heart, containing specialized cardiac cells that depolarize spontaneously faster than other heart cells; determines heart rate; also called *pacemaker of the heart*. (Ch. 18)

skeletal muscle The muscle that produces voluntary movement of body parts. (Ch. 9)

sleep apnea A condition in which respiration can stop for long periods of time (often 30–60 seconds) during sleep. (Ch. 20)

sliding-filament hypothesis The hypothesis that explains muscle contraction on the basis of overlapping, interdigitating, thick (myosin) and thin (actin) filaments that are drawn past each other by the action of myosin cross bridges, with ATP providing the energy. (Ch. 16)

slow axonal transport The movement of proteins along the axon from the soma to the terminal region. (Ch. 7)

slow skeletal muscle The muscle that contracts slowly and is more resistant to fatigue. (Ch. 9)

slowly adapting tactile mechanoreceptor A type of receptor that continues to generate action potentials as long as a stimulus is applied. (Ch. 8)

small intestine A portion of the gastrointestinal system consisting of the duodenum, jejunum, and ileum; digestion is completed here by enzymes in the epithelial cells that line the inner surface, and the products of digestion are absorbed into the blood and lymph. (Ch. 22)

smooth muscle A muscle type associated with blood vessels and other organ systems. (Ch. 9)

sodium appetite A craving for salt. (Ch. 24)

sodium pump The Na/K-ATPase that uses energy from ATP hydrolysis to drive active transport of Na^+ ions out of cells and K^+ ions in. (Ch. 4)

soft tissue calcification The deposition of calcium in tissues. (Ch. 26)

solute (**sol**-yoot) A substance that is uniformly distributed in a solvent to form a solution. (Ch. 2)

solution A type of mixture in which one substance is uniformly distributed throughout another substance, which is usually a liquid. (Ch. 2)

solvent (**sol**-vent) A substance in which a solute is uniformly distributed to form a solution. (Ch. 2)

soma The body and metabolic center of a neuron. (Ch. 7)

somatic Pertaining to the skin and skeletal muscles of the body. (Ch. 7)

somatic (so-**mat**-ik) **nervous system** Sensory nerves that transmit information from sense receptors to the central nervous system and motor nerves that transmit instructions about appropriate responses of skeletal muscles. (Ch. 7)

somatic sensory cortex The portion of the neocortex that responds to sensations felt on the surface of the skin and the position of limbs relative to the body. (Ch. 7)

somatomammotropic hormones A group of related peptide hormones that stimulate body growth and breast development; human growth hormone (hGH) and human chorionic somatomammotropin (hCS) are examples. (Ch. 13)

somatomedin (so-*mah*-tow-**me**-din) A growth factor found in many tissues, including the liver, that mediate the effect of growth hormone on cartilage. (Ch. 13)

somatostatin (so-*mat*-oh-**stat**-in) A hypothalamic hormone that inhibits growth hormone and TSH secretion by the anterior pituitary; a possible neurotransmitter; also in the stomach and pancreatic islets. (Ch. 13)

somatotropin *See* growth hormone.

spatial summation The additive effects of two simultaneously active synaptic inputs. (Ch. 7)

specific defense Defense mechanisms that engage and destroy specific kinds of pathogens (e.g., antibody production). (Ch. 28)

specific dynamic action The increased heat production that results from processing food. (Ch. 26)

specific gravity The ratio between the weight of a given volume of a liquid and the weight of an equal volume of pure water. (Ch. 17)

specificity The ability of a binding site to react with only one (or a limited number of) type of molecule. (Ch. 5)

spectrin An extrinsic protein found only on the cytoplasmic side of the red blood cell membrane; thought to be important in determination of red blood cell shape. (Ch. 4)

sperm The male germ cell; also called *spermatozoa*. (Ch. 31)

sperm loss Nonviable sperm. (Ch. 32)

sperm viability The ability of a sperm to fertilize an ovum. (Ch. 33)

spermatogenesis (*spur*-mah-tow-**jen**-eh-sis) The formation of sperm. (Ch. 31)

spermicide (**spur**-muh-side) A chemical that kills sperm; one form of contraception. (Ch. 33)

spherical micelle A small sphere formed by phospholipids in solution. (Ch. 4)

sphincter (**sfink**-ter) **of Oddi** A smooth muscle ring surrounding the bile duct at its entrance into the duodenum. (Ch. 22)

spinal (**spy**-nal) **autonomic reflex arc** A loop of nerve cell activity; begins with the detection of visceral information by autonomic sensory receptors in an organ that in turn activates interneurons within the spinal cord. These neurons synapse with the cells of the IML; the IML cells integrate incoming information and discharge to activate postganglionic neurons. These neurons convey information to the target organ, causing it to make the appropriate response. (Ch. 10)

spinal cord The part of the CNS that transmits information to and receives information from most of the body's internal organs, muscles, and skin; located caudal to the brain stem. Consists of 31 segments, each with a pair of spinal nerves. (Ch. 7)

spirogram The record of lung volumes measured with a spirometer. (Ch. 20)

spirometry The measurement of pulmonary volumes and capacities. (Ch. 20)

splanchnic circulation The circulation in the stomach, small and large intestines, pancreas, and liver. (Ch. 22)

stable balance The condition in which there is no net gain or loss of a substance. (Ch. 3)

stance phase The phase of the walk cycle in which the foot is on the ground and supporting the weight of the body. (Ch. 9)

standard free energy change $\Delta G°$, or the difference in total free energy between reactants and products of a chemical reaction under a set of standard conditions: temperature of 25°C, pressure of 1 atmosphere, and a concentration of 1.0 M for all reactants and products. (Ch. 6)

stapes (**stay**-peas) The innermost of the auditory ossicles, shaped somewhat like a stirrup; also called *stirrup*. (Ch. 8)

Starling's hypothesis Description of the principle that net filtration pressure is proportional to the capillary hydrostatic pressure minus the oncotic pressure of the blood; not to be confused with the Frank-Starling law of the heart. (Ch. 19)

start codon A codon that initiates translation of nucleotide triplets. (Ch. 5)

stasis (**stay**-sis) A slowing down or stoppage of flow, such as of blood or other body fluid. (Ch. 17)

static exercise Exercise that involves isometric muscle contractions. (Ch. 30)

statistics The science that deals with the collection, tabulation, and analysis of numerical facts. (Ch. 1)

statoconia Calcium carbonate crystals contained within the otolithic membrane. (Ch. 9)

steady state A condition unvarying with respect to time. (Ch. 5)

steatorrhea The appearance of excessive amounts of undigested fat in the feces; usually due to the absence of lipase in the gastrointestinal system. (Ch. 22)

stereocilia Fiberlike elements that protrude from hair cells located between the basilar and tectorial membranes of the ear. (Ch. 8)

steroid (**steer**-oid) A lipid subclass; the molecule consists of four interconnected carbon rings to which polar groups may be attached. (Ch. 11)

steroid hormones Hormones derived from cholesterol. They are the glucocorticoids, mineralocorticoids, androgens, estrogens, and progestins. (Ch. 12)

stomach (**stum**-ak) The portion of the gastrointestinal system located between the esophagus and the small intestine; functions as a storage organ for food. (Ch. 22)

stop codon A codon that terminates translation of nucleotide triplets. (Ch. 5)

stretch receptors Afferent nerve endings that are depolarized by stretching; *see also* muscle spindle receptor. (Ch. 30)

stretch reflex Reflex contraction of skeletal muscle in response to its stretch. (Ch. 9)

stria A narrow bandlike structure; a general term for such longitudinal collections of nerve fibers in the brain. (Ch. 14)

striate (**stri**-ate) **muscle** *See* skeletal muscle.

striatum An area beneath the neocortex that is involved in the initiation and coordination of movement. (Ch. 7)

stroke volume The blood volume ejected by a ventricle during one heartbeat. (Ch. 18)

strong acid An acid with a high dissociation constant, which consequently is highly ionized in solution. (Ch. 25)

structural formula A formulaic illustration of the arrangement in space of the atoms of a molecule. (Ch. 2)

subcutaneous (*sub*-kyoo-**tay**-nee-us) Beneath the skin. (Ch. 27)

sublingual (sub-**ling**-gwal) Below the tongue. (Ch. 27)

submucosa (*sub*-myoo-**koe**-sah) A connective tissue layer under the mucosa in the gastrointestinal tract. (Ch. 22)

submucosal plexus A nerve-cell network in the submucosa of the esophageal, stomach, and intestinal walls. (Ch. 22)

substantia gelatinosa A group of nerve cells located in the dorsal horn of the spinal cord. (Ch. 8)

substantia nigra An area of the brain associated with the reward system. (Ch. 11)

substrate (**sub**-straight)**-level phosphorylation** The direct formation of high-energy phosphate compounds such as ATP by a chemical reaction that incorporates a phosphate group. (Ch. 6)

suckle To derive or to provide nourishment by feeding. (Ch. 33)

sucrase An enzyme that digests sucrose to glucose and fructose. (Ch. 22)

sudden infant death syndrome (SIDS) A disease in which a seemingly normal, healthy infant is found dead in its crib; sleep apnea is being studied extensively as a causative link. (Ch. 20)

sulfonyl ureas Oral hypoglycemic agents that act by stimulating insulin secretion from pancreatic beta cells. Used to treat type II diabetes mellitus. (Ch. 15)

summation The cumulative effects of a number of stimuli applied to a muscle nerve or reflex arc. (Ch. 16)

superior Literally, nearer to the head. (Ch. 1)

superior vena cava (**vee**-nah **kay**-vah) A large vein that carries blood from the upper half of the body to the right atrium of the heart. (Ch. 18)

superoxide anion radical An active form of free radical. (Ch. 29)

superoxide dismutase A protective cellular enzyme that removes free radicals. (Ch. 29)

supplemental motor cortex An area of the brain that designs the necessary movements required to accomplish a task; also called *Brodmann area 6*. (Ch. 9)

suprachiasmatic nucleus The nucleus of the hypothalamus, thought to be responsible for coordinating various biological rhythms. (Ch. 11)

supraoptic nucleus Cells in the hypothalamus that secrete vasopressin into the blood in response to increases in osmolarity. (Ch. 10)

surface tension A molecular force created at the gas-liquid interface; specifically, the attractive forces between the liquid and air molecules. (Ch. 20)

surfactant (ser-**fak**-tant) A lipoprotein produced by pulmonary alveolar cells; a detergent-like material that reduces surface tension of fluid film lining alveoli. (Ch. 20)

swallowing The propulsion of a food bolus through the pharynx and esophagus into the stomach; the bolus moves in response to skeletal and smooth muscle contractions that occur sequentially from above to below. (Ch. 22)

sweating The active secretion of water and electrolytes by specialized sweat glands in the skin for the purpose of increasing evaporative heat loss; activated by sympathetic nerves in response to overheating; increases linearly with abnormal increases in body temperature. (Ch. 27)

swing phase The phase of the walk cycle in which the foot is off the ground and swinging forward. (Ch. 9)

sympathetic (*sim*-pah-**thet**-ik) **nervous system** One portion of the autonomic nervous system; preganglionic fibers leave the CNS at the thoracic and lumbar portions of the spinal cord. (Ch. 10)

sympathoadrenal (*sim*-pah-thow-ah-**dree**-nal) **system** The sympathetic nervous system and the adrenal medulla together. (Ch. 14)

sympathomimetic (*sim*-pah-thow-my-**met**-ik) Adrenergic; producing effects resembling those produced by the sympathetic nervous system. (Ch. 10)

symport The transport of two different solutes by the same carrier in the same direction. (Ch. 4)

synapse (**sin**-aps) The junction between nerve cells; **electrical synapses** consist of gap junctions that permit direct passage of ions, and **chemical synapses** prevent the direct passage of ions. (Ch. 7)

synaptic (sin-**ap**-tik) **cleft** In a chemical synapse, the space, between one cell and another, that prevents the direct passage of ions. (Ch. 7)

synaptic transmission The process by which there is communication between nerve cells. (Ch. 7)

synaptic vesicles Small structures filled with neurotransmitting chemicals that fuse with the presynaptic membrane. (Ch. 7)

syncytium (**sin**-sish-e-um) A multinucleate mass of protoplasm produced by the merging of cells. (Ch. 16)

synergistic (sin-er-**jis**-tik) **muscle** A muscle that works together with another to produce a given movement. (Ch. 9)

systemic (sis-**tem**-ik) **circulation** The circulation from the left ventricle through all the organs except the lungs and back to the heart. (Ch. 18)

systemic hypertension A complex disease state resulting in chronically elevated arterial blood pressure. (Ch. 19)

systole (**sis**-toe-lee) Contraction of cardiac muscle. Clinically, a period of ventricular contraction. (Ch. 18)

systolic pressure The peak arterial pressure, which occurs during contraction of the ventricle. (Ch. 19)

T cell A lymphocyte derived from a precursor that differentiated in thymus; *see also* cytotoxic T cell, helper T cell, suppressor T cell. (Ch. 28)

tachycardia (*tak*-ih-**kar**-dee-ah) An abnormally rapid heart rate, usually defined as more than 100 beats per minute. (Ch. 18)

tactile (**tak**-tile) **receptor** A mechanoreceptor that detects touch, pressure, and vibrations applied to the skin. (Ch. 8)

target cell A cell whose activity is affected by a specific hormone. (Ch. 5)

target tissue Tissue containing cells that have receptors for a particular neurotransmitter or hormone. (Ch. 12)

taste bud A cluster of gustatory receptors, basal cells, and supporting cells. (Ch. 8)

tectorial (tek-**toe**-ree-al) **membrane** A membranous structure in which the hair cells of the cochlea are embedded. (Ch. 8)

tectospinal tract A component of the ventromedial pathway that originates in the tectum. (Ch. 9)

tectum (**tek**-tum) A structure involved in the coordinated control of heat and eye movements. (Ch. 9)

temperature The metabolic heat produced within the body. (Ch. 27)

temperature regulating system A system that works to maintain thermal homeostasis in the body. (Ch. 30)

temporal summation The addition of postsynaptic potentials that arise from activation at one synaptic input. (Ch. 7)

tendon A collagen fiber bundle that connects muscle to bone and transmits muscle contractile force to the bone. (Ch. 16)

tenting The elevation of the uterus away from the vagina as the blood flow to the vagina increases during the plateau phase of sexual arousal. (Ch. 33)

terminal cisterna Enlarged end portions of the sarcoplasmic reticulum; also called *lateral sacs*. (Ch. 16)

terminator site on DNA A sequence of nucleotides immediately following the 5' end of the gene; acts as a recognition site for the ending of the RNA strand. (Ch. 5)

testicular feminization syndrome A condition in which the internal genitalia of the male do not develop; usually due to a deficiency of testosterone biosynthesis or when Wolffian ducts do not respond to testosterone. (Ch. 33)

testis The male gonad. (Ch. 11)

testosterone (tes-**tos**-teh-rone) A steroid hormone produced in interstitial cells of the testes; essential for spermatogenesis and maintaining growth and development of reproductive organs and secondary sexual characteristics of males. (Ch. 12)

tetanic fusion frequency The lowest frequency that will produce a fused tetanus; around 20 to 60 times per second for most skeletal muscles. (Ch. 16)

tetanus (**tet**-ah-nus) A maintained mechanical response of muscle to high-frequency stimulation; the disease lockjaw. (Ch. 16)

tetany (**tet**-ah-nee) A syndrome manifested by sharp flexion of the wrist and ankle joints; occurs in response to low concentrations of extracellular calcium ion. (Ch. 26)

thalamus (**thal**-uh-mus) An area of the brain that receives all types of sensory information. (Ch. 8)

thalassemia (*thal*-ah-**see**-me-ah) A group of inherited hemolytic anemias characterized by decreased synthesis of one or more hemoglobin polypeptide chains. (Ch. 17)

theory A formulated hypothesis. (Ch. 1)

thermal homeostasis Physiological regulatory mechanisms continuously operating to keep heat production and heat loss approximately equal. (Ch. 27)

thermal receptor A sensory receptor that responds to temperature. (Ch. 8)

thermography A technique for measuring mean skin temperature in humans. (Ch. 27)

thermoreceptor (*ther*-mow-ree-**sep**-tor) A receptor that detects changes in temperature. (Ch. 1)

thermostat A controller to prevent excessive body heating or cooling. (Ch. 27)

thick myofilament A 12- to 18-nm myosin filament in a muscle cell. (Ch. 16)

thin myofilament A 5- to 8-nm myosin filament in a muscle cell; consists of actin, troponin, and tropomyosin. (Ch. 16)

thirst A conscious sensation associated with a craving for drink; ordinarily interpreted as a desire for water. (Ch. 24)

thoracic (thow-**ras**-ik) **cavity** The chest cavity. (Ch. 20)

threshold The point when a membrane potential is depolarized sufficiently to produce an action potential. (Ch. 18)

thrombin An enzyme that catalyzes conversion of fibrinogen to fibrin. (Ch. 17)

thrombocyte (**throm**-bow-site) A fragment of cytoplasm, enclosed in plasma membrane, that plays a role in blood clotting; found in the circulation. Also called a *platelet*. (Ch. 17)

thrombocytopenia (*throm*-boe-sigh-toe-**pee**-nee-ah) A deficiency in circulating platelets; results in excessive bleeding. (Ch. 17)

thromboplastin (**throm**-bow-plas-tin) A complex of several phospholipids and a proteolytic enzyme released from damaged tissue; important in the clotting process. (Ch. 17)

thrombosis (throm-**bow**-sis) The formation or presence of a clot in an unbroken blood vessel or heart chamber. (Ch. 17)

thrombus A blood clot. (Ch. 17)

thymus A primary lymphoid gland located in the lower anterior neck. T-lymphocytes mature in the thymus. (Ch. 28)

thyrocalcitonin A peptide hormone, secreted by the parafollicular cells of the thyroid, that causes increased deposition of calcium in bones; also called *calcitonin*. (Ch. 26)

thyroglobulin (*thy*-row-**glob**-you-lin) A large protein to which thyroid hormones bind in the thyroid gland; a storage form of thyroid hormones. (Ch. 13)

thyroid (**thy**-roid) **follicle** The internal structure of the thyroid gland as a collection of closely packed follicles; normal thyroid contains about 3 million follicles. (Ch. 13)

thyroid hormone (TH) A collective term for amine hormones released from the thyroid gland; that is, thyroxine (T4) and triiodothyronine (T3). (Ch. 13)

thyroid-stimulating hormone (TSH) A glycoprotein hormone secreted by the anterior pituitary; induces secretion of thyroid hormone. Also called *thyrotropin*. (Ch. 13)

thyrotropin-releasing hormone (TRH) A hypothalamic hormone that stimulates thyrotropin and prolactin secretion by the anterior pituitary. (Ch. 13)

thyroxin (thy-**rok**-sin) A hormone, secreted by the thyroid gland, that accelerates the rate at which cells consume oxygen, metabolize glucose, and produce heat. (Ch. 11)

thyroxine-binding globulin (TBG) A carrier protein that binds and carries mainly thyroxine in the blood. (Ch. 12)

tidal volume The volume of air inhaled or exhaled during a single breath. (Ch. 20)

"tight" epithelia Epithelia that do not permit leakage between the cells. (Ch. 4)

tight junction A specialized attachment that connects Sertoli cells to one another inside the seminiferous tubules; the area where plasma membranes are joined. (Ch. 31)

time period The time between the peaks of a sound wave of a tone; the inverse of the frequency of a sound wave. (Ch. 8)

tinnitus (tie-**nie**-tis) A constant ringing in the ear due to the inappropriate discharge of action potentials by particular nerve fibers associated with specific frequencies. (Ch. 8)

tissue An aggregate of differentiated cells of similar type united in the performance of a particular function; also denotes the general cellular fabric of a given organ. (Ch. 1)

titratable acid A solution whose acid concentration can be determined by measuring the milliequivalents of strong base needed to bring the pH to a certain level; for example, the amount of base needed to bring a urine sample back to the pH of arterial blood. (Ch. 25)

tone *See* muscle tone.

tonic contraction A long, sustained contraction in response to single or continued stimulation by nerves, drugs, or hormones. (Ch. 16)

tonic discharge An action potential generated at one specific frequency in response to the position of a joint. (Ch. 8)

tonotopic map The organization of auditory nerve fibers such that fibers with a high characteristic frequency innervate hair cells at the base of the cochlea, while those with a low characteristic frequency innervate hair cells near the apex of the cochlea. (Ch. 8)

tonus The constant steady force of contraction of most smooth muscles; also present in skeletal muscles. (Ch. 16)

total peripheral resistance (TPR) The total resistance to flow in systemic blood vessels from the beginning of the aorta to the end of the venae cavae. (Ch. 19)

trachea (**tray**-kee-ah) The main airway of the respiratory system, which branches into two bronchi. (Ch. 20)

transcellular fluid Specialized extracellular fluid that is separated from the blood plasma not only by a capillary endothelium but by a continuous layer of epithelial cells; it includes aqueous humor, cerebrospinal fluid, digestive secretions, sweat, and renal tubular and bladder urine. (Ch. 24)

transcellular transport pathway The movement of water and solutes through epithelial cells; entry and exit pathways are on opposite sides of the cell. (Ch. 4)

transcription (tran-**skrip**-shun) The synthesis of mRNA using a DNA template; the first step in the transfer of genetic information from DNA to proteins. (Ch. 2)

transcytosis (tran-si-**toe**-sis) An active process by which relatively large molecules can cross the capillary wall by moving through the epithelial cells either in discrete shuttling vesicles or through transient vesicle-derived channels. (Ch. 19)

transducin A protein present in the disc membrane of photoreceptors. (Ch. 8)

transfer RNA (tRNA) A type of RNA; different tRNAs combine with different amino acids and with codon on mRNA specific for that amino acid, thus arranging amino acids in sequence to form specified proteins. (Ch. 5)

transferrin An iron-binding protein carrier for iron in plasma. (Ch. 17)

translation The synthesis of a polypeptide using mRNA as a template. (Ch. 2)

transmission electron microscope A microscope that uses magnetic lenses to focus a beam of electrons in much the same way that glass lenses in ordinary microscopes focus a beam of light; specimens must be fixed and stained to be seen under a transmission electron microscope. (Ch. 3)

transmitter *See* neurotransmitter.

transpulmonic pressure The pressure across the lung; measured by subtracting the pleural pressure from the alveolar pressure. (Ch. 20)

transthyretin (trans-**thigh**-re-tin) A plasma protein that acts as a transporter for thyroxine. (Ch. 12)

transudation The vasoconstriction of the vagina, leading to the release of moisture from the vaginal lining as a result of vasocongestion in the vaginal wall; lubricates the vagina to facilitate insertion of the penis without discomfort. (Ch. 33)

transverse plane An anatomical position that describes a plane dividing the body into upper and lower portions. (Ch. 1)

transverse tubule (t-tubule) A tubule extending from striated muscle plasma membrane into the fiber, passing between opposed sarcoplasmic-reticulum segments; conducts muscle action potential into muscle fiber. (Ch. 16)

triacylglycerol (tri-ace-il-**gliss**-er-al) A subclass of lipids composed of glycerol and three fatty acids; also called *fat, neural fat,* or *triglyceride*. (Ch. 6)

triad The region of contact in skeletal muscles of one t-tubule and the terminal cisternae from two adjacent sarcomeres. (Ch. 16)

trichromatic color vision The ability to discriminate between different colors on the basis of three types of photoreceptors: blue, red, and green cones. (Ch. 8)

tricuspid valve The valve between the right atrium and the right ventricle of the heart. (Ch. 18)

triglyceride *See* triacylglycerol.

triiodothyronine (*try*-eye-oh-dow-**thy**-row-neen) **(T3)** An iodine-containing amine hormone secreted by the thyroid gland. (Ch. 13)

trimesters The three 3-month periods of pregnancy. (Ch. 32)

trophic Of or pertaining to nutrition. (Ch. 7)

tropomyosin A regulatory protein of the I-band that inhibits contraction unless its position is modified by troponin so that myosin molecules can make contact with actin. (Ch. 16)

troponin A regulatory protein bound to actin and tropomyosin of striate-muscle thin filaments; a site of calcium binding that initiates contractile activity. (Ch. 16)

trypsin (**trip**-sin) A protein-digesting enzyme secreted by the pancreas. (Ch. 2, 22)

tryptophan (**trip**-toe-fan) An essential amino acid; a precursor to serotonin. (Ch. 7)

tubal ligation (lie-**gay**-shun) The cutting and tying off of both fallopian tubes, resulting in sterilization. (Ch. 33)

tubular reabsorption The transport of materials by the kidney tubule epithelium out of the tubular urine. (Ch. 23)

tubular secretion The transport of materials by the kidney tubule epithelium into the tubular urine. (Ch. 23)

tubular transport maximum (T_m) The maximum rate at which a particular substance is reabsorbed (e.g., glucose) or secreted (e.g., PAH) by the kidney tubules. (Ch. 23)

turbulent flow Completely disorganized patterns of air flow, which occur at high flow rates. (Ch. 20)

Turner's syndrome A chromosomal abnormality in females in which there is only one X sex chromosome; characterized by impaired development of female gonads, resulting in the appearance of an incompletely formed gonad. (Ch. 33)

T-wave A component of an electrocardiogram, corresponding to ventricular repolarization. (Ch. 18)

twitch The mechanical response of muscle to a single maximal stimulus. (Ch. 16)

tympanic (tim-**pan**-ik) **membrane** A thin, semitransparent membrane that stretches across the ear canal; separates outer and middle ear. (Ch. 8)

type I diabetes Diabetes in which insulin is completely or almost completely absent; also called *insulin-dependent diabetes*. (Ch. 15)

type II diabetes Diabetes in which insulin is present at near-normal or even above-normal levels, but the tissues do not respond optimally to it; also called *insulin-independent diabetes*. (Ch. 15)

tyrosine An amino acid; the precursor of catecholamines and thyroid hormones. (Ch. 12)

tyrosine hydroxylase An enzyme that converts tyrosine to dihydroxyphenylalanine. (Ch. 14)

ulcer (**ul**-ser) An erosion of the mucosa of certain regions of the digestive tract. (Ch. 22)

ultrafiltrate (ul-tra-**fill**-trate) An essentially protein-free fluid formed from plasma as it is forced through capillary walls by a pressure gradient. (Ch. 23)

ultrafiltration Filtration through a membrane with exceedingly fine pores. (Ch. 23)

umbilical (um-**bil**-ih-kal) **cord** An artery or vein that transports blood between fetus and placenta. (Ch. 32)

unconditioned stimulus A stimulus that is paired with a conditioned stimulus to evoke a conditioned response. (Ch. 11)

unitary postsynaptic potential A change in the membrane potential due to the flow of synaptic current resulting from the release of transmitters. (Ch. 7)

unitary smooth muscle The smooth muscle that makes up the bulk of the visceral organs; less strictly controlled by the nervous system; a large number of its cells function together as a single unit. (Ch. 16)

upper esophageal sphincter A band of skeletal muscle between the pharynx and the body of the esophagus. (Ch. 22)

urea (you-**ree**-ah) The major nitrogen-containing end-product of metabolism; formed chiefly by the liver. (Ch. 23)

urea cycle A cyclic series of reactions that produce urine; a major route for removal of ammonia. (Ch. 14)

uremia (you-**ree**-mee-ah) The symptom that results from renal failure. (Ch. 23)

ureter (you-**ree**-ter) The tube that carries urine from the kidney to the urinary bladder. (Ch. 23)

urethra (yoo-**ree**-thra) The tube that carries urine from the urinary bladder to the outside. (Ch. 23)

uric acid An end-product of purine metabolism, secreted by the kidneys. (Ch. 23)

uricosuric An agent that promotes excretion of uric acid in the urine. (Ch. 23)

urinary bladder A hollow, muscular organ that collects the urine and is periodically emptied. (Ch. 23)

utricle (**yoo**-trih-kul) The largest of the two divisions of the membranous labyrinth of the inner ear. (Ch. 9)

vacuole (**vac**-you-ohl) A cavity enclosed by a membrane and located in cytoplasm. (Ch. 3)

vaginismus Involuntary contractions and spasms of the vaginal or pubococcygeal muscles; makes vaginal penetration difficult or impossible. (Ch. 33)

vagus (**vay**-gus) The tenth cranial nerve. Motor functions include control of laryngeal and pharyngeal muscles and muscles of the esophagus, stomach, heart, lungs, and intestines (swallowing, control of heart rate, gastric secretion, etc.). Sensory functions include input from viscera or thorax and abdomen (blood pressure from aorta, pain from stomach, etc.). (Ch. 10)

Van Allen belt Particles of cosmic radiation trapped in the Earth's magnetic field. (Ch. 29)

van der Waals bond Weak attractive forces between nonpolar regions of molecules. (Ch. 2)

variables The related quantities displayed on a graph. (Ch. 1)

vas deferens (**def**-ur-enz) Components of the spermatic cords that arise from the tail of the epididymis; also called *ductus deferens*. (Ch. 31)

vasa recta The blood vessels of the medulla that supply the nutritional needs of the kidney medulla as well as act as countercurrent exchangers. (Ch. 23)

vascular (**vas**-kyoo-lar) **bed** Many separate circulatory pathways that supply blood to various organ systems within the body. (Ch. 18)

vascular reactivity The degree of responsiveness or sensitivity of vascular smooth muscle to agents which stimulate contraction or elevate blood pressure. Glucocorticoids, for example, enhance vascular reactivity. (Ch. 14)

vascular resistance The total frictional force opposing blood flow through a particular vascular system. *See* total peripheral resistance (TPR). (Ch. 18)

vasectomy (vah-**sec**-tow-mee) The cutting and tying-off of both ductus deferens, which results in sterilization of the male without loss of testosterone. (Ch. 33)

vasocongestion The concentration of blood in the tissues of the penis, which causes an increase in diameter, length, and firmness. (Ch. 33)

vasoconstrictor (vas-oh-kon-**strik**-tor) A substance that causes contraction of the smooth muscle associated with blood vessels leading to a decrease in vessel diameter. (Ch. 19)

vasodilator (vas-oh-die-**lay**-tor) A substance that produces relaxation of the smooth muscle associated with blood vessels leading to an increase in vessel diameter. (Ch. 19)

vasomotion (vas-oh-**moe**-shun) Contraction and relaxation of the precapillary sphincter smooth muscle which typically results in intermittent blood flow through capillaries. (Ch. 19)

vasopressin (vas-oh-**press**-in) A hormone, secreted by the paraventricular nucleus of the hypothalamus, that causes the kidneys to reabsorb more water. (Ch. 10)

vein (vane) Any vessel that returns blood to the heart. (Ch. 1)

venomotor tone The contraction of venous smooth muscle. (Ch. 19)

venous capacitance The total blood volume in the veins. (Ch. 19)

venous return The rate (ml/min) of blood return to the right atrium; usually equal to cardiac output. (Ch. 19)

venous system The vessels that carry the blood back to the heart; begins at the point where the capillaries reunite to form venules. (Ch. 18)

ventilation The process of moving air in and out of the lungs. (Ch. 20)

ventilation-perfusion balance A mechanism for balancing regional blood flow with regional alveolar ventilation. (Ch. 21)

ventral An anatomical position that describes the area of the body; specifically, toward, or at the front of, the body. (Ch. 1)

ventral corticospinal tract Fibers from the motor cortex that descend to the spinal cord without crossing the midline of the body; primarily innervate motor neurons in the medial region of the ventral horn associated with axial muscles of the body. (Ch. 9)

ventral posterior lateral (VPL) thalamus The region of the thalamus that receives somatic sensory information. (Ch. 8)

ventral posterior medial (VPM) thalamus The region of the thalamus that receives information from taste receptors. (Ch. 8)

ventricle (**ven**-trih-kuhl) Any cavity. (Ch. 18)

ventricular ejection Part of the cardiac cycle when the semilunar valve is forced open and blood flows out of the ventricle; this occurs when ventricular pressure exceeds the pressure in the outflow tract. (Ch. 18)

ventricular fibrillation The uncoordinated stimulation of the ventricular muscle; useless for pumping blood. (Ch. 18)

ventricular filling Part of the cardiac cycle when the ventricle is being filled with blood from the atrium. (Ch. 18)

ventricular relaxation Part of the cardiac cycle when the ventricle relaxes and pressure inside the ventricle falls; the semilunar valve is forced closed as blood attempts to flow from the outflow track back into the ventricle. (Ch. 18)

ventromedial pathway A pathway that descends along the ventral and medial region of the spinal cord. (Ch. 9)

venule (**ven**-yule) A small vessel that carries blood from the capillary network to a vein. (Ch. 18)

vestibular (ves-**tib**-you-lar) **nerve fiber** A nerve fiber that innervates the hair cells of the vestibular apparatus of the ear. (Ch. 9)

vestibular system Parts of the peripheral and central nervous systems involved with the reception and processing of sensory information regarding body balance, orientation in space, linear acceleration and rotational (angular) acceleration. (Ch. 9)

vestibulospinal tract A component of the ventromedial tract that originates in the vestibular nucleus and carries information for the reflex control of equilibrium. (Ch. 9)

villus A small, fingerlike projection that creates the convoluted surface of the small intestine. (Ch. 22)

viscera (**vis**-ur-uh) Organs of the thoracic and abdominal cavities. (Ch. 10)

visceral (**vis**-er-al) **afferent fiber** Nerve fibers that carry information from an organ to the spinal cord. (Ch. 10)

visceral efferent fiber Nerve fibers that carry information from the spinal cord to an organ. (Ch. 10)

visceral muscles Muscles that line the walls of the internal organs. (Ch. 16)

viscosity (vis-**kos**-ih-tee) A property of fluid that makes it resist flow. (Ch. 17)

visual agnosia Difficulty recognizing familiar objects or colors due to lesions in the secondary visual cortices. (Ch. 8)

visual cortex The area of the neocortex that responds to visual images seen by the eye. (Ch. 7)

visual field The area that can be seen through either the right or the left eye. (Ch. 8)

visual orientation columns Columns of nerve cells in the visual cortex that are sensitive to bars of light oriented at different angles. (Ch. 8)

visual pigment The photoreceptor pigment that absorbs photons. (Ch. 8)

visual system The physiological system that perceives light to determine the shape, color, and grouping of objects in the environment. (Ch. 8)

visuotopic map A precise mapping of the visual field onto the visual cortex. (Ch. 8)

vitamin A general term for a number of unrelated organic molecules that occur in many foods in trace amounts and that are necessary for normal growth and development. They may be water-soluble or fat-soluble. (Ch. 26)

vitamin D₃ Cholecalciferol, the natural form of vitamin D, obtained from the diet or formed in the skin by the action of ultraviolet light on 7-dehydrocholesterol. (Ch. 26)

volt A unit of electric potential. (Ch. 7)

volume regulation Mechanism by which most cells are able to maintain normal volume during periods of osmotic stress. (Ch. 4)

voluntary muscles Muscles that move in response to conscious effort. (Ch. 16)

vomiting The rapid expulsion of gastric contents through the mouth. (Ch. 22)

warm receptors Peripheral temperature receptors sensitive to temperature; their steady-state discharge rates increase in response to warming of the skin. (Ch. 27)

warm-blooded *See* homeothermy.

water diuresis Excretion of a large volume of osmotically dilute urine. (Ch. 24)

watt A unit measurement of power. (Ch. 30)

wavelength A characteristic of electromagnetic radiation that is unique for each part of the continuum of radio waves to X-rays. (Ch. 8)

weak acid An acid that does not dissociate completely when dissolved in water. (Ch. 4)

weak base A base that does not dissociate completely when dissolved in water. (Ch. 4)

"wedge" pressure A pressure measurement made during cardiac catheterization; near-left atrial pressure. (Ch. 21)

weightlessness Lacking gravitational pull. (Ch. 29)

Wernicke's area An area of the neocortex responsible for the interpretation of language. (Ch. 7)

Wolffian duct system A duct system in the embryo that is the precursor to the internal genitalia of the male. (Ch. 33)

x-axis The abscissa or horizontal axis of a graph. (Ch. 1)

y-axis The ordinate or vertical axis of a graph. (Ch. 1)

Z-disc Another term for Z-line to emphasize its two-dimensional nature. (Ch. 16)

Z-line A structure running across a myofibril at each end of a striated muscle sarcomere; anchors one end of the thin filaments. (Ch. 16)

zona fasciculata (**zow**-nah fah-*sik*-you-**lay**-tah) The thick middle layer of the adrenal cortex that secretes glucocorticoid hormones. (Ch. 14)

zona glomerulosa (glow-*mer*-you-**low**-sah) The outer layer of the adrenal cortex that secretes mineralocorticoid hormones. (Ch. 14)

zona pellucida (peh-**loo**-sih-da) The thick, clear layer separating the ovum from surrounding granulosa cells. (Ch. 31)

zona reticularis (reh-**tik**-you-lar-is) The inner layer of the adrenal cortex that secretes sex hormones, mainly androgens. (Ch. 14)

zygote (**zye**-goat) A fertilized ovum; the cell resulting from the fusion of male and female gametes. (Ch. 31)

zymogen (**zye**-mow-jen) **granules** Vesicles of the enzyme salivary amylase that are stored in the serous cells. (Ch. 22)

Appendix A

Measurement, Computation, and Graphic Analysis Provide a Quantitative Approach

In the study of physiology, **quantitative methods** are invaluable for understanding bodily functions and processes. Often in physiology it is the comparison of amounts, rates, times, and other measurable evidence of the body processes that reveals relationships between function and structure.

UNITS OF MEASUREMENT

The metric system is almost exclusively the system used for physiological measurements. Length is expressed in **meters;** mass, in **grams;** and volume, in **liters.** A length of 1 meter is equivalent to 39.37 inches. A mass of 1 gram is equivalent to 0.035 ounces, and a volume of 1 liter is the same as 0.26 gallons. These metric terms usually are abbreviated for convenience to **m** for meters, **g** for grams, and **L** for liters. A length of 1000 meters would be written as 1000 m.

Prefixes are used to indicate multiples or fractions of these basic units, and the prefixes are abbreviated also. The prefix **kilo,** abbreviated as **k,** indicates a multiple of 1000 of the basic unit. A length of 2000 meters (m), for example, is equivalent to 2×1000 m, or 2 kilometers (km). The combinations of prefixes and units that are most commonly used in physiological measurements are summarized and defined in Table 1.

These prefixes avoid cumbersome notations across the enormous range of sizes encountered in living organisms.

SCIENTIFIC NOTATION

In physiology, as in many of the sciences, it is often awkward or inconvenient to express measurements using standard digits (e.g., the diameter of a blood vessel as 10,000 μm, a blood glucose concentration of 0.005 moles per liter). Instead of writing out all the "zeroes," scientists use a system of **scientific notation,** in which quantities are expressed as products of a number and a power of ten. Powers of ten are expressed with **exponents,** which are small numbers written above and to the right of the base number, ten. The exponent indicates the number of times the base should be used as a factor in the expression. For example, $10^2 = 10 \times 10 = 100$. Thus,

$$10,000 \ \mu m = 1 \times 10^4 \ \mu m$$
$$0.005 \text{ moles/liter} = 5 \times 10^{-3} \text{ moles/liter}$$
$$0.001 \text{ grams} = 1 \times 10^{-3} \text{ grams}$$

(Note that negative exponents indicate the number of times 10 is used as a *divisor* for quantities less than one. For example, $10^{-3} = 1/10 \times 1/10 \times 1/10$.) Numbers without zeroes can also be expressed in scientific notation (e.g., 46,782 = 4.6782×10^4).

Following are some simple rules for converting a number expressed in scientific notation back to the original number:

1. For positive exponents: Shift the decimal point to the right by the number of places indicated by the exponent.
2. For negative exponents: Move the decimal point to the left by the number of places indicated by the exponent.
3. When multiplying exponential quantities that have the same base, add the exponents ($5^3 \times 5^2 = 5^5$).

TABLE 1
Commonly Used Units of Measurement
The meter (m) is the unit of length.
1 kilometer (km) = 1×10^3 = 1000 m
1 decimeter (dm) = 1×10^{-1} = 0.1 m
1 centimeter (cm) = 1×10^{-2} = 0.01 m
1 millimeter (mm) = 1×10^{-3} = 0.001 m
1 micrometer (μm) = 1×10^{-6} = 0.000001 m
1 nanometer (nm) = 1×10^{-9} = 1 Angstrom
The gram (g) is the unit of mass.
1 kilogram (kg) = 1×10^3 = 1000 g
1 milligram (mg) = 1×10^{-3} = 0.001 g
1 microgram (μg) = 1×10^{-6} = 0.000001 g
1 nanogram (ng) = 1×10^{-9} = 0.000000001 g
1 picogram (pg) = 1×10^{-12} = 0.000000000001 g
1 femtogram (fg) = 1×10^{-15} = 0.000000000000001 g
The liter (L) is the unit of volume.
1 deciliter (dL) = 1×10^{-1} = 0.1 L
1 milliliter (mL) = 1×10^{-3} = 0.001 L
1 microliter (μL) = 1×10^{-6} = 0.000001 L
1 nanoliter (nL) = 1×10^{-9} = 0.000000001 L

RATIOS AND PROPORTIONS

A **ratio** is an expression that compares two numbers or quantities by division. For example, to compare the numbers 100 and 10, we divide 100 by 10 to get 10/1. Ratios can be expressed in several different ways with the same meaning:

1:250 means 1 part to 250 parts.

1/5 means 1 part out of 5 parts.

2 cats to 6 dogs equals a ratio of 1 cat to every 3 dogs.

A **proportion** is a mathematical statement of the equality of two ratios. By arbitrarily using the letters *A, B, C,* and *D* to express quantities, we can state a proportion in the following way:

$$A \text{ is to } B \text{ as } C \text{ is to } D \text{ or } A/B = C/D$$

This is equivalent to saying: *A* times *D* equals *B* times *C*. For example, if $A = 15$, $B = 25$, $C = 3$, and $D = 5$, then

$$A/B = C/D = 15/25 = 3/5$$
$$A \times D = B \times C; 15 \times 5 = 25 \times 3$$

Therefore, it follows that if three of the quantities are known, the value of the fourth can be determined.

For example, assume that an electrocardiogram is being recorded on a moving strip of paper (Fig. 1). The speed of the moving paper is 25 mm/sec. If each repeating cycle of the electrocardiogram represents one heartbeat, how many heartbeats are occurring each minute? The problem can be solved as follows:

1. The distance between cycles is 20 mm (as measured from the record).
2. The time interval between cycles (beat to beat) is unknown; it equals *x* sec.
3. The ratios relating distance to time can be expressed as a proportion:

$$25 \text{ mm/sec} = 20 \text{ mm}/x \text{ sec}$$

4. $25x = 20$; $x = 0.8$. The interval between cycles is 0.8 seconds.
5. The number of cycles (beats) occurring each minute = *y* cycles (beats):

$$1 \text{ beat}/0.8 \text{ sec} = y \text{ beats}/60 \text{ sec} \quad 0.8y = 60$$
$$y = 75 \text{ beats/minute}$$

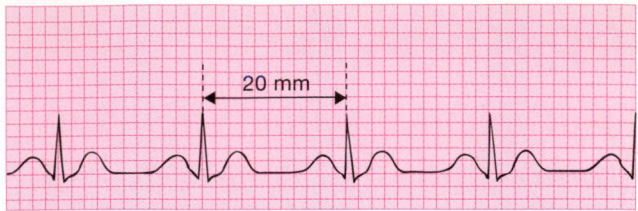

Paper speed: 25 mm/sec

Figure 1

A proportion based on the speed of the paper and the distance between points on this electrocardiogram can be used to determine the number of heartbeats per minute.

INTERPRETATION OF GRAPHIC DATA

A **graph** is a diagram that expresses a relationship between two or more quantities. In some cases there is a definite cause-and-effect relationship, whereas in others the association is not as direct but may be due to a third factor. Graphic presentation of data may not explain the reason for the relationship, but the shape of it can provide clues. A graph puts into visual form abstract ideas or experimental data, so that their relationships become apparent.

The related quantities displayed on a graph are called **variables.** The simplest sort of graph uses a system of coordinates or axes to represent the values of the variables. Usually the relative size of the variable is represented by its position along the axis, and numbers along the axis allow the reader to estimate the values. If the relationship being plotted is one of cause and effect, the variable that expresses the cause is called the **independent variable.** Usually this is represented by the horizontal axis (also sometimes called the *x-axis,* or the **abscissa**). The variable that changes as a result of changes in the independent variable is the **dependent variable.** It is usually represented on the vertical axis (also called *y-axis,* or **ordinate**). The two axes are arranged at right angles to each other and cross at a point called the **origin** (Fig. 2).

To show the relationship between two variables that are directly related (at some specific value, such as point A in Fig. 2), the value on the *x*-axis (X_1) is extended vertically, and the corresponding value on the *y*-axis (Y_1) is extended horizontally. The point *A* at which these lines cross is determined by their relationship. If another pair of points (X_2 and Y_2) is chosen, their point of intersection on the graph also can be plotted; this is point *B*. A line drawn between points *A* and *B* can then give information about how all other *x* and *y* values on this graph should relate to each other.

Parts of a graph

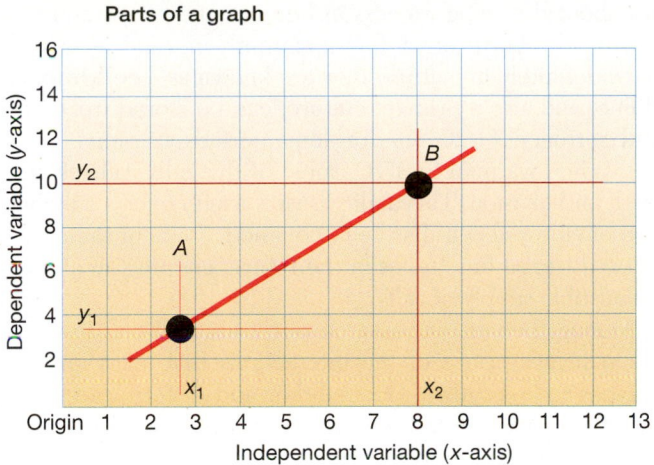

Figure 2

In a typical line graph, the vertical axis (the *y*-axis, or ordinate) shows the values for the dependent variable, whereas the horizontal axis (the *x*-axis, or abscissa) shows values for the independent variable. The two axes meet at a point called the origin.

Types of Relationships

One of the simplest kinds of relationship that a graph can represent is called a **direct relationship:** the *y*-values get larger as the *x*-values get larger. An example of this type of graph is shown in Figure 3. The data plotted here could have come from an experiment in which various concentrations of an enzyme were used to study how fast a particular chemical reaction happened at each concentration.

When there is no direct relationship and the data show "scatter." (Sometimes this sort of graph is called a **scatter diagram.**) In many cases a mathematical procedure may be carried out to determine the "best-fitting" line to describe the relationship. This is called the **ideal line** in Figure 3. A relationship that can be described by a straight line is called a **linear relationship.**

Inverse relationships are also common. In such relationships, the *y*-values get *smaller* as the *x*-values get larger. Such relationships may still be linear if straight lines describe them. Other inverse relationships may be curvilinear, as in Figure 4. The graph shown here summarizes the experimental finding that a muscle exerting a large force must contract more slowly than one exerting only a small force. A graph of this sort is adequate for the process of interpolation within the range of data, but it is poor for extrapolation outside that range.

Interpolation and Extrapolation

If an experimenter had enough confidence in the reliability of the data, two kinds of predictions could be made from such a graph. Predicting data values that fall "between the

Direct relationship

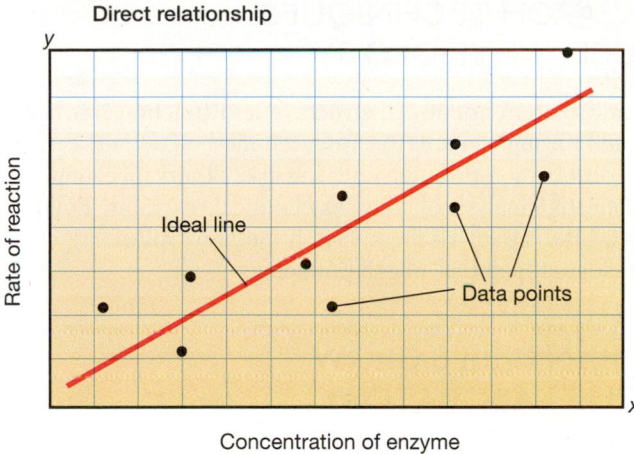

Concentration of enzyme

Figure 3

In a direct relationship, *y*-values increase as *x*-values increase. Data points frequently do not fall exactly on a straight line but are "scattered" about an ideal line, which is drawn to show the general shape of the relationship. A linear relationship can be shown by a straight line. The equation for a straight line is $y = mx + b$, where mx is the slope and b is the *y*-intercept.

points" is called **interpolation.** This process is useful if the curve is to be used as a guide for interpreting or testing the reliability of newly obtained data. A more risky procedure is **extrapolation,** which involves extending the ideal line into ranges where experimental data are not present. If there is good reason to believe that the same relationship should hold outside this range, then this prediction could be valid and might permit useful information to be gained.

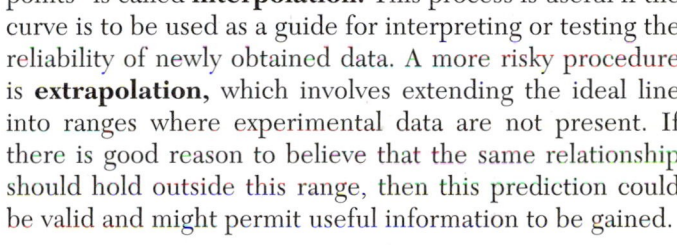

Inverse relationship

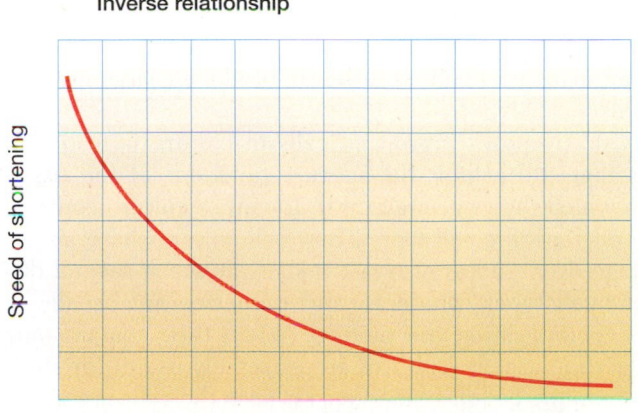

Force exerted by muscle

Figure 4

In an inverse relationship, *y*-values decrease as *x*-values increase, thus producing a downward-sloping line or curve.

RESEARCH TECHNIQUES
IN CELL PHYSIOLOGY

The information about cell structure and function has been obtained through a variety of research methods, some as simple as observation through a microscope, others involving the sophisticated analysis of complex chemical reactions. Following is a brief survey of the research methods still being used to elucidate the details of cell physiology.

HUMAN CELLS GROW
IN THE LABORATORY

Cells taken out of mammalian tissues will grow and multiply in the laboratory if provided with appropriate nutrients and other factors. Human cells have been used almost routinely for **cell culture** since 1952. Many types of cells can be grown as a single layer attached to glass or plastic surfaces and covered with a liquid **growth medium,** or mixture of nutrients. Under these conditions, the cells are ideally suited for direct study by various techniques of microscopy.

Alternatively, cells can be grown in suspension in liquid media for the purpose of producing large numbers. Special conditions are used to avoid bacterial growth in the liquid medium, and antibiotics are frequently added as an extra precaution. The growth medium contains glucose, various salts, amino acids, and vitamins and is often supplemented with serum derived from the blood of a horse or calf. (**Serum** is the fluid that remains in the blood after the blood has clotted.) The precise composition of serum is always uncertain, but it is an adequate supplement to the growth medium for routine cell culture.

Cells that are transferred directly from the tissues of an animal to a plastic dish containing artificial growth medium are known as *primary* cultures. As they grow and multiply, by dividing to produce daughter cells, they can be **subcultured**—that is, the cells can be removed and dispersed among many new culture dishes to produce a large number of cells.

One purpose of growing cells in culture is to obtain cells that still exhibit the specific functions of the tissue from which they were removed. Ideally, a culture contains only one specific cell type. Thus, cell cultures have an advantage over studies of intact tissues, in which several different cell types, products, and processes are usually at work, complicating the picture. When they contain only the pertinent cells, cultures are extremely useful for studies of unique cellular functions because the cells are growing under carefully controlled conditions and are far more accessible than when part of an intact tissue inside a living animal.

Most normal cells will divide only a limited number of times in culture before dying. Some researchers believe that studies of this built-in "senescence" will provide important clues about the aging process in humans and other animals. Occasionally, however, a few apparently normal cells will grow indefinitely in culture: they are known as a **cell line.** A cell line, and also a primary culture, can be stored frozen in liquid nitrogen (−200°C) for long periods of time, even years. When warmed to 37°C, some of the cells will resume growth and division. This ability is very useful once a suitable group of cultured cells has been obtained. Some of them can be stored frozen for studies in the future, creating an almost inexhaustible supply of cells.

Among their many research purposes, cell cultures are used in investigations of the interactions that allow similar cell types to "recognize" one another and to form the close associations that lead to the development of a specific tissue. When cells are cultured on a plastic surface, they grow and multiply only to the point at which most of the surface is covered by a monolayer of cells. Further growth is inhibited by close contact with other cells (so-called **contact inhibition**), and the monolayer is said to be **confluent.** Interestingly, cancer cells that grow in an uncontrolled way in the body exhibit the same uncontrolled growth pattern in culture. They are much less sensitive to contact inhibition.

Finally, the study in culture of **mutant** cells, abnormal cells that cannot synthesize a specific protein, often reveals a great deal of information about the role of that protein in normal cells. The specific functions that protein normally direct are absent from the cultured mutant cell. The altered behavior of the mutant cell is a strong clue to the function of the protein in normal cells of the same type.

A word of caution: Cells in culture—especially cell lines that have been cultured for a number of years—are not growing in their normal physiological environment. Therefore, despite the usefulness of the technique, the behavior of cells in culture may be different from those in an intact tissue. The observations made about cultured cells must be checked eventually against cells in their normal situation.

MICROSCOPES ALLOW VISUALIZATION
OF CELLS AND ORGANELLES

First used to study cells and tissues in the seventeenth century, the microscope in its modern form is essential in physiological research. A significant advantage of microscopic studies is that the researcher can see the complete cell and its organelles in a relatively undisrupted state.

The Light Microscope

The **light microscope** (Fig. 5) is familiar to all students of science. It is extremely useful for examining basic sizes and shapes of cells, and it can be used to detect the presence of nuclei and mitochondria when the cells are stained with dyes that change the color of only these organelles. The tissue of

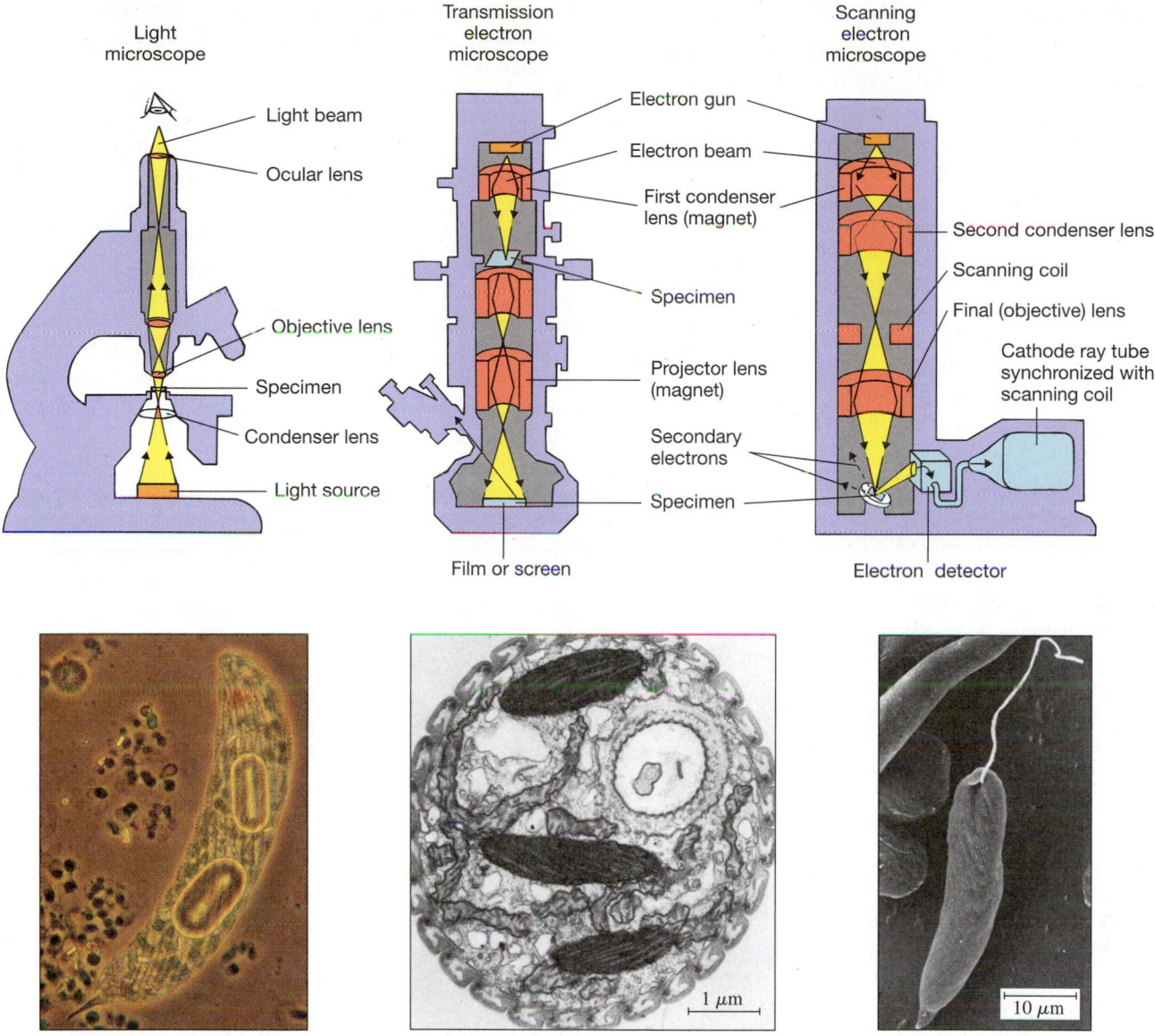

Figure 5

A light microscope focuses a beam of light through curved glass to produce an image, whereas a transmission electron microscope focuses a beam of electrons using an electromagnet. The electrons pass through the specimen to produce an image that can be made visible to the human eye on a screen or photographic film. A scanning electron microscope images the deflection of electrons by the specimen. The photographs show views of the unicellular organism Euglena. The light microscope shows the whole cell; the transmission electron microscope shows a thin cross-section; the scanning electron microscope shows the cell's outer surface with its pattern of fine ridges.

interest is usually **fixed** with a solution that preserves the cell structure and **embedded** in a hard wax to provide support so that thin slices can be prepared for viewing.

Students and researchers are often concerned that these treatments may distort the appearance of the cells. This problem can be avoided by using a different optical system, such as that used in a **phase contrast microscope.** This in-

strument enhances the contrast between the internal organelles so that mitochondria, lysosomes, chromosomes, nucleoli, and other structures can be seen quite clearly in living, unstained, and nonfixed cells. Light microscopes are rarely used for magnifications greater than 1000 to 1500 times normal size because beyond this point the **resolution,** or clarity of the image, is poor.

The Electron Microscope

The development of the **transmission electron microscope** (see Fig. 5) in the 1940s and 1950s made greater magnifications (more than 100,000 times normal size) possible. The use of this type of microscope has provided a wealth of information about the organization of cells and the structure of their organelles.

In an electron microscope, magnetic lenses focus a beam of electrons in much the same way that glass lenses in ordinary microscopes focus a beam of light (see Fig. 5). To be seen by an electron microscope, specimens must be fixed and then stained with solutions that introduce atoms such as lead or uranium into the various structures of the cell. The presence of these large atoms permits fewer electrons to pass through the specimen and gives rise to a dark image of the specimen on a bright fluorescent screen like a television screen. This image can then be recorded on a photographic plate to provide a permanent record.

The **scanning electron microscope** (see Fig. 5) is especially useful for examining the surface of a specimen. These microscopes image the deflection of electrons by the specimen. The **high-voltage electron microscope** uses an energy source of one million volts and stands about 32 feet high. It allows much thicker tissue sections to be viewed with high resolution and in considerable depth, producing an almost three-dimensional view of the cell interior. The use of this technique led to the concept of the microtrabecular lattice in the cytosol.

Cytochemical Studies

Cytochemistry is used in combination with microscopy to produce color or enhanced contrast at the sites of specific molecules. Specific stains for DNA and RNA have been used to demonstrate the location of these molecules within the cell. If the molecule being studied is an enzyme, chemicals can be added to sections of a tissue to cause the enzyme to generate an insoluble reaction product. This material will be deposited wherever the enzyme is located. Researchers can identify these sites precisely by looking at the section through a light microscope or electron microscope. They can see the reaction product readily because it is too dense to allow the passage of light or electrons.

AUTORADIOGRAPHY DETECTS THE LOCATION OF SPECIFIC MOLECULES

Some isotopes are unstable and tend to break down to a more stable form. Tritium, an isotope of hydrogen, is a specific example. This process is termed **radioactive decay** because it releases high-energy electrons or other particles known as **radiation.** Unstable isotopes that undergo radioactive decay are **radioactive isotopes,** or **radioisotopes.** Although radioisotopes are rare in nature, radioisotopes of phosphorus, iodine, sulfur, carbon, calcium, and hydrogen are readily manufactured for use in cell research.

Because radiation can be detected with extreme sensitivity, the amounts required in the laboratory are very small and well within the accepted safety limits. The use of radioactive molecules (molecules to which a radioisotope has been attached) is a powerful technique that allows researchers to follow the course of almost any process in a cell. A radioactive molecule performs identically to the naturally occurring molecule. Its biochemical properties are unaltered, and the cell reacts as if it were the natural nonradioactive form. However, because it is "labeled" with a radioisotope, any changes in its chemical form or its location can be followed as the molecule is processed within the cell. Researchers worked out the complex metabolic pathways of cells largely with the aid of this technique.

The location of radioactive molecules within cells or tissues is best determined by the technique of **autoradiography.** After incubating the cells with an appropriate radioactive compound to allow its incorporation into the macromolecules of the cells, the researcher washes the cells or slices of tissue to remove unincorporated radioactivity, fixes and embeds them, and cuts them into thin sections for either light or electron microscopy. Each preparation is then overlaid with a film of photographic emulsion and left in the dark. The radioactivity in the preparations reacts with the emulsion in much the same way that photographic film reacts to light. After developing the film, the researcher can determine the position of the radioactivity in the cells from the position of the dark spots on the emulsion when viewed in the microscope. Autoradiography has been very useful in discovering the secretory pathways in cells and in determining the role of the Golgi complex in this process.

ANTIBODIES ARE USED TO DETECT, ISOLATE, OR MEASURE SPECIFIC MOLECULES

The mammalian **immune system** is an important defense mechanism that recognizes and destroys invading microorganisms and cells not normally present in blood. An important phase of this reaction is the synthesis and release of special proteins, known as **antibodies,** into the bloodstream. A molecule on the surface of the foreign cell triggers their production. A foreign molecule, such as a protein, polysaccharide, or nucleic acid, can elicit a similar response. Any molecule that stimulates the production of antibodies is termed an **antigen.** The antibodies that are released are highly specific molecules that recognize and bind only to the antigen. This reaction may inactivate the cell bearing the antigen, or, alternatively, the antibody-antigen complex on the surface of the cell may stimulate removal of the cell from the bloodstream by phagocytic cells specifically designed for this purpose.

Immunocytochemistry

Mammals can produce an enormous variety of antibodies. The specificity of these molecules makes them powerful research tools because they can be used to pick out specific molecules within a cell. For example, antibodies to tubulin can be used to study the organization of microtubules in a cell. A researcher labels the antitubulin antibodies by attaching a fluorescent molecule. The antibodies then are allowed to react with thin sections of cells or with intact cells that have been treated to allow them to enter. The antibodies will bind only to microtubules because only the microtubules contain tubulin. Because the antibodies are fluorescent, they can be traced in a microscope fitted with an ultraviolet light source. Only the fluorescent label will be seen, but it suggests the presence of the antitubulin antibodies, which in turn indicate the microtubules to which they are bound (Fig. 6*a* and *b*). Thus, the fluorescent sites within the cell indicate the location of microtubules. If the antibodies to an antigenic molecule are labeled with **ferritin,** an iron-protein complex that does not permit the passage of electrons, the location of the antigen within cells can be seen in great detail in the electron microscope.

Radiolabeled Antibodies

In a method similar to using fluorescent markers, a radioactive isotope can be attached to an antibody to give a **radiolabeled** antibody. The radiolabeled antibody can be used in combination with autoradiography to pinpoint the location of specific molecules in the cell. Researchers have used this approach to follow the movement and fate of the plasma membrane that is internalized by cells during endocytosis. In this case, the antibody used is one that binds to molecules that are components of the plasma membrane.

Radioimmunoassays

Another use of radioisotopes and immunochemistry provides the means to measure very small amounts (1×10^{12} mole) of certain molecules with great accuracy. The measuring procedure is known as a **radioimmunoassay.** A specific example of its use is the measurement of cyclic AMP in various tissues, in blood, and in urine.

The basic principle of the method is to extract the cyclic AMP and allow it to bind to an anticyclic AMP antibody. A trace amount of radiolabeled cyclic AMP is added to the mixture, which is then set aside for several hours. The unlabeled and labeled cyclic AMP are identical molecules that compete for binding to the antibody. The amount of radiolabeled cyclic AMP present is always the same. The only variable is the amount of unlabeled cyclic AMP in the sample; this determines how much labeled cyclic AMP can be bound to the antibody. When there is very little cyclic AMP in the sample, a lot of **labeled** cyclic AMP can bind to the antibody because only a few unla-

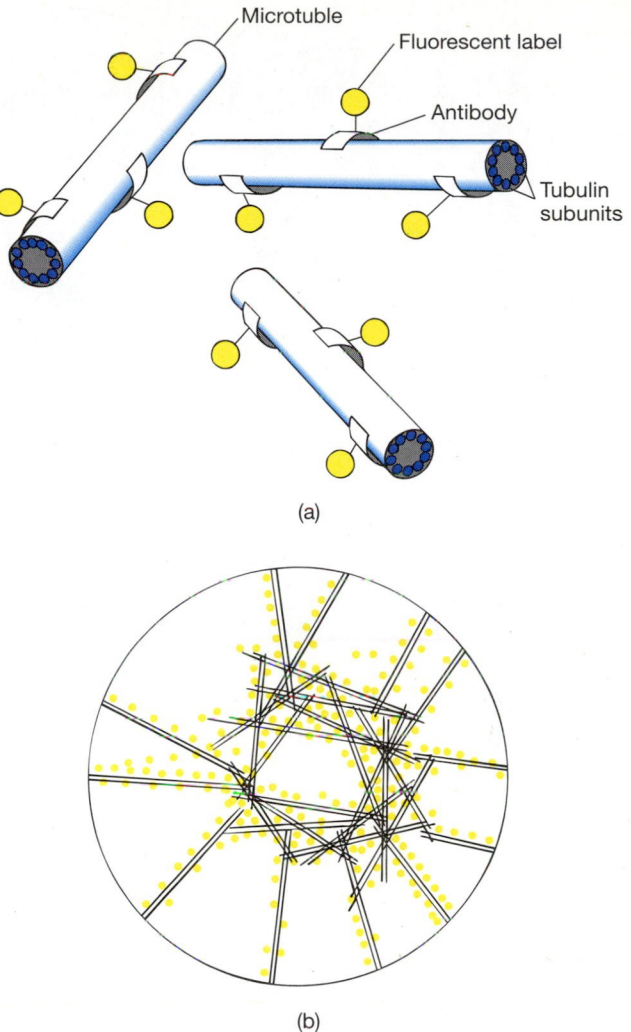

(a)

(b)

Figure 6

In the fluorescent-labeled antibody technique, an antibody to tubulin, which is the protein that makes up microtubules, is injected into a cell. The antibody has been labeled with a fluorescent marker. **(a)** The antibody attaches to the tubulin molecules as they form microtubules throughout the cell. A special microscope allows a view of the fluorescent markers, which indicate the presence of the antibodies, in turn showing the presence of microtubules **(b)**.

beled cyclic AMP molecules will compete for binding. If there is a lot of cyclic AMP in the sample, however, much more will bind to the antibody and only a small amount of labeled cyclic AMP will be able to bind (Fig. 7). A researcher can determine the amount of radiolabeled cyclic AMP bound by the antibody by isolating the antibody-antigen complex and measuring the radioactivity present. The amount of radiolabeled cyclic AMP bound is inversely related to the amount of cyclic AMP present in the sample. The latter is easily calculated.

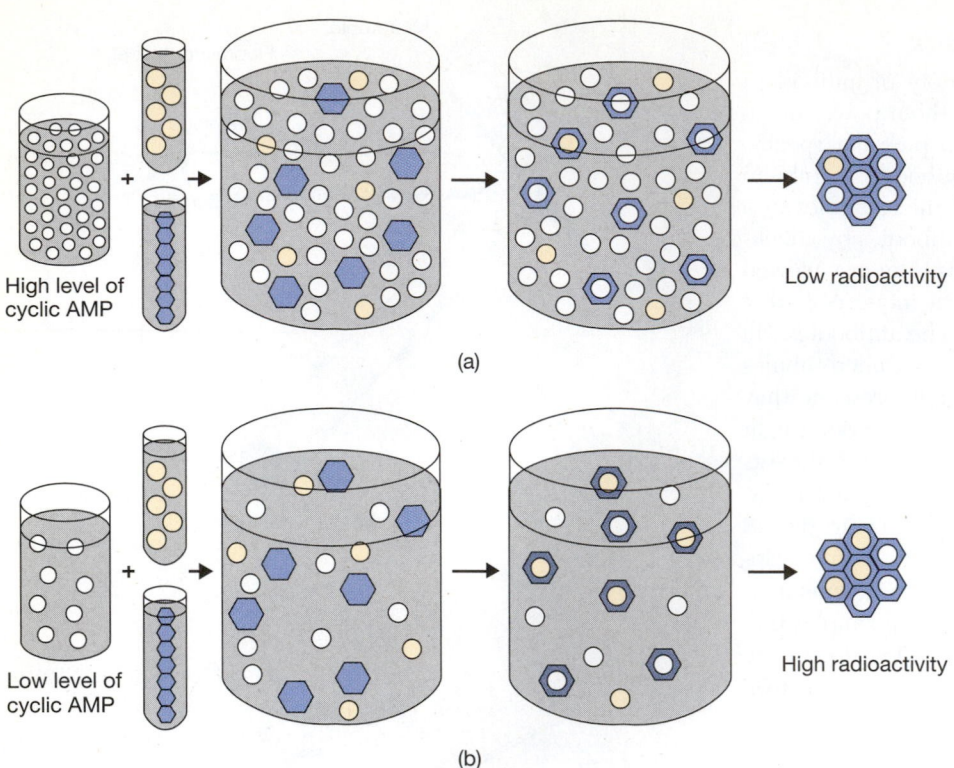

(a)

(b)

Figure 7

A radioimmunoassay is used to determine the relative quantity of cyclic AMP present in a solution. *(a)* Radioactive (*yellow*) cyclic AMP molecules and antibodies (*blue*) specific to cyclic AMP are added to a solution containing an unknown amount of unlabeled cyclic AMP (*white*). When the antibody–antigen complex is isolated, the presence of radioactive cyclic AMP is relatively low, revealing that a large amount of cyclic AMP was present in the original solution. *(b)* The same process is followed in another solution, in which the amount of cyclic AMP is unknown. The same numbers of radioactive molecules and antibodies are used, but the isolated antigen–antibody complex has a higher proportion of radioactive cyclic AMP. This indicates that the original solution had a low level of cyclic AMP.

High level of cyclic AMP

Low radioactivity

Low level of cyclic AMP

High radioactivity

Affinity Chromatography

Another technique that uses the specificity of antigen-antibody reactions is **affinity chromatography.** Following is an example of how this technique might be used to study a particular antibody. A researcher produces the antibody to be studied by injecting a purified sample of an antigen several times into an animal (such as a goat or a rabbit) in which the antigen is a foreign molecule. When the researcher draws blood from the animal, it now contains the antibodies the researcher wants to use as well as proteins and other antibodies normally present in the blood. The antibody in question must be separated out. The researcher allows the blood sample to clot in order to remove red blood cells. The remaining serum is passed through a column of an insoluble carbohydrate matrix to which an antigen to the antibody is attached. The antibody will bind to the antigen and will be retained on the column while the other serum proteins pass straight through (Fig. 8). The antibody can then be washed off the column and recovered separately in a relatively pure form.

A disadvantage of this technique is that antibodies to a single injected antigen are produced by a variety of antibody-producing cells in the animal. Each of these cells produces its own antibody that recognizes and binds to a part of the structure of the antigen, but it is usually not the same part recognized by the antibodies produced by the other cells. Thus, the antibodies specific for a given antigen are not a singular molecular species, and the different antibody molecules vary widely in the efficiency with which they bind to the antigen. The antibody preparation is described as **heterogeneous,** and it has limited use.

Monoclonal Antibodies

The problem of heterogeneity was solved in 1976 by the use of **hybridoma cell lines** to produce large quantities of identical antibody molecules known as **monoclonal antibodies.** The cells that produce antibodies have a limited lifetime outside the body. But researchers can make them grow and multiply indefinitely in a culture dish by joining them to cancer cells, which have an unlimited ability to grow under these conditions. The two joined cells become a single cell that still produces antibodies but now has an unlimited lifetime; the cell line is known as a **hybridoma.** The hybrid cells that produce the specific needed antibody can be selected, and a single hybrid cell is used to start the growth of large numbers of new cells, a procedure known as **cloning.** Because all the new cells were originally derived from the same hybrid cell, the antibody molecules that the new cells produce will be identical.

An additional advantage is that the technique allows a selection step to identify the cell producing the required anti-

Serum sample
poured
through column

Antibody to
antigen A

Antigen A

Other serum
proteins

Figure 8

Affinity chromatography. Antibodies are produced when
an antigen is injected into the blood of an animal such as a
goat or rabbit. To isolate these antibodies, serum is
collected and poured through a column that contains an
antigen specific to the antibody being studied. The
antibodies bind to the antigens in the column, whereas
other proteins in the serum pass through the column. The
antibodies being studied can then be washed from the
column and measured.

body. This means that researchers do not need to purify the
antigen used at the start of the process because cells producing antibodies against the impurities can be eliminated at the
selection step. Initial purification of the antigen is a critical
step in the conventional procedure for antibody production,
as described previously. The high specificity of monoclonal
antibodies makes them tools that are as sensitive as radioisotopes for detecting specific molecules in a complex mixture
or in a section of tissue.

HPLC Is Important for Rapid Amino Acid Sequencing

When a protein has been isolated, scientists often wish to
determine the exact sequence of amino acids that make it
up. Various methods exist for **amino acid sequencing.** Researchers usually begin by breaking the protein into shorter
peptide chains, using enzymes that act to split peptide
bonds only between specific amino acids. The peptide fragments can then be analyzed by various chromatographic
means. One recent advance in this area is the development
of **high-performance liquid chromatography (HPLC).**
Solutions of amino acids are poured into columns containing densely packed separation substances, which separate
amino acids based on size or other factors. In contrast with
other column separation techniques, however, HPLC involves the use of great pressures to force the solution
through the column in a very short time. HPLC also provides a greater degree of resolution than do other chromatographic techniques.

Cell Organelles Are Isolated by Cell Fractionation

The purpose of **cell fractionation** is to separate the different organelles of the cell in order to study each organelle on
its own in a strictly controlled, "cell-free" system. This isolation permits direct, detailed analysis of the structure and the
function of the organelle.

Fractionation is begun by a physical disruption of the
tissue and its cells, a process that can be controlled to minimize damage to the internal organelles. Before being broken up, the tissue usually is placed in a solution of sucrose,
which helps to protect organelle structure. A common
method of tissue disruption is **homogenization,** in which
the tissue and its cells are broken by being forced between
a rotating pestle and the walls of a glass vessel (Fig. 9). The
clearance between the two surfaces is small enough to
break cells but too large to harm most of the internal organelles. Mitochondria, lysosomes, and nuclei remain intact, while large structures such as the endoplasmic
reticulum are broken into small pieces, as is the plasma
membrane. The resultant thick suspension of broken cells
in sucrose is called a **homogenate.**

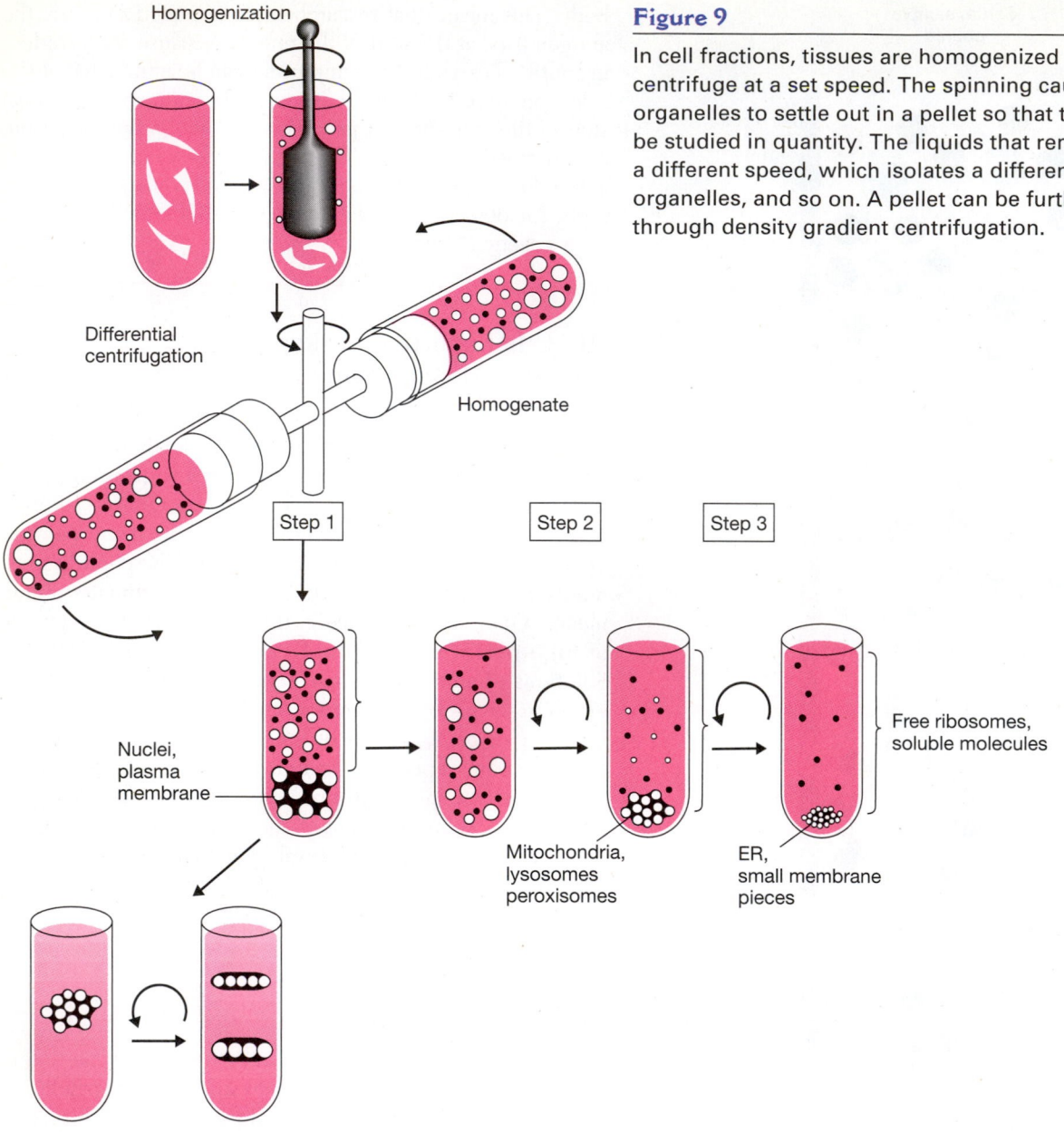

Homogenization

Differential centrifugation

Homogenate

Step 1   Step 2   Step 3

Nuclei, plasma membrane

Mitochondria, lysosomes peroxisomes

ER, small membrane pieces

Free ribosomes, soluble molecules

Density gradient centrifugation

Figure 9

In cell fractions, tissues are homogenized and then spun by a centrifuge at a set speed. The spinning causes these organelles to settle out in a pellet so that those organelles can be studied in quantity. The liquids that remain can be spun at a different speed, which isolates a different sample of organelles, and so on. A pellet can be further analyzed through density gradient centrifugation.

Once the tissue and its cells have been broken up, centrifugal force is used to separate the organelles. The most widely used technique for separating the cell organelles present in a mixture such as a homogenate is **differential centrifugation.** The homogenate is placed in a special tube that can be spun at high speed in a machine called a **centrifuge.** The centrifugal force generated by these machines can exceed 100,000 times the force of gravity (g). Under the influence of a centrifugal force, the structures present in the homogenate begin to move toward the bottom of the centrifuge tube (see Fig. 9). The large and dense structures (e.g., the nuclei) move at the fastest rate and collect as a "pellet" at the bottom of the tube. The centrifugal force can be adjusted so that only the nuclei collect during one centrifugation, whereas the other organelles remain suspended in the solution or "supernatant."

The supernatant is then poured into a new tube, which is spun at a higher centrifugal force. A different group of smaller organelles is collected as a separate pellet. In this way, the homogenate can be divided by centrifugal force into several "fractions," each containing a specific organelle. The total centrifugal force applied to the homogenate depends not only on the force used but also on the time for which it is applied. The sequence of different forces and times that is commonly used to fractionate tissue homogenates is as follows:

Step	Centrifugation	Pellet Contents
1. Homogenate	1000 × g for 10 min	Nuclei, large pieces of plasma membrane
2. Supernatant from 1	10,000 × g for 20 min	Mitochondria, lysosomes, peroxisomes
3. Supernatant from 2	100,000 × g for 1 h	Endoplasmic reticulum, small pieces of plasma membrane
4. Supernatant from 3 contains free ribosomes and soluble molecules.		

Although the pellet fractions are enriched in certain organelles, they are not pure because more than one type of organelle is usually present. This is one of the limitations of differential centrifugation. Different organelles in each pellet can be separated by resuspending the pellet material and placing it as a thin layer on top of a sucrose solution that increases in density from the top to the bottom of the tube (see Fig. 9). This technique, **density gradient centrifugation,** allows the separation of organelles that have the same size but different densities. Upon application of a centrifugal force, the organelles with different densities will separate from one another because they will move at different rates through the gradient. At the end of the centrifugation the gradient usually contains two to three separate "zones" or bands of material, each one representing a relatively pure fraction of a specific organelle (see Fig. 9). The purity and integrity of a mitochondrial fraction, for example, can be determined by electron microscopy and by checking to see whether enzymes characteristic of lysosomes, peroxisomes, endoplasmic reticulum, and other structures are present.

Isolated fractions of cell organelles are used widely to study the processes that occur within cells, because the process of interest can be isolated and studied without interference from the complex side reactions that occur in the intact cell. The use of a cell fraction that could synthesize polypeptides when provided with mRNA was a crucial step in understanding the mechanism of protein synthesis.

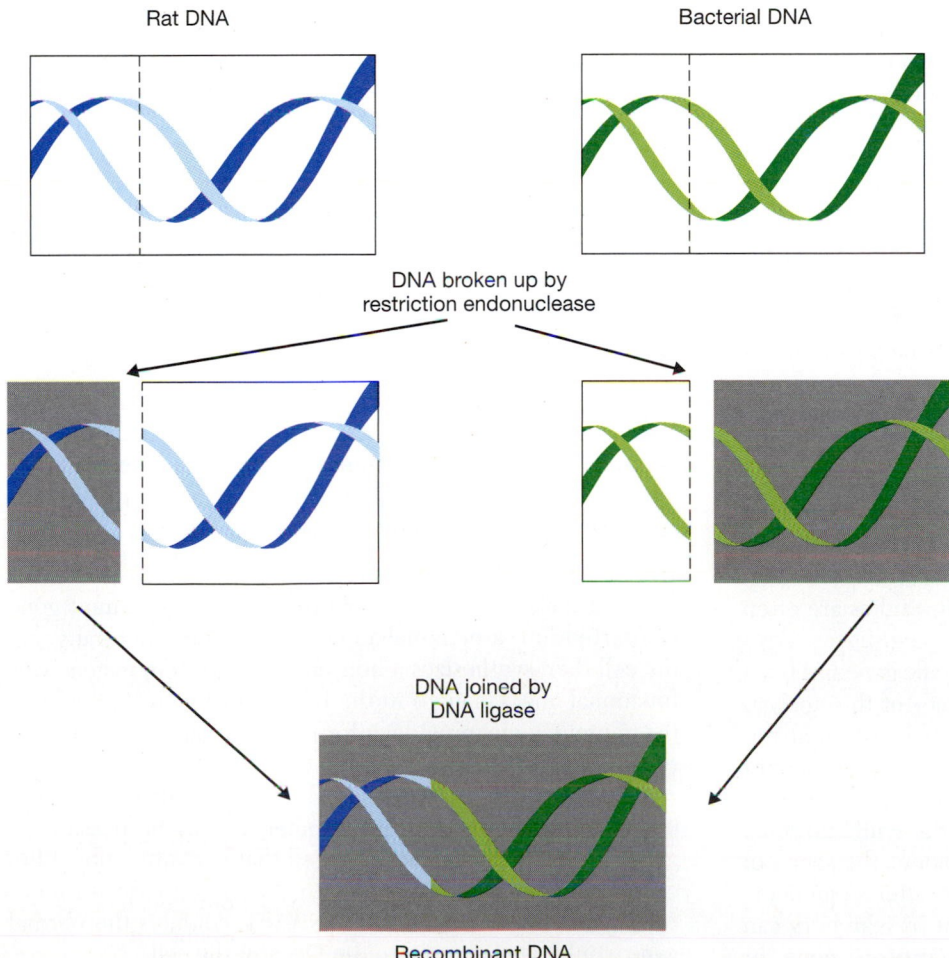

Rat DNA

Bacterial DNA

DNA broken up by restriction endonuclease

DNA joined by DNA ligase

Recombinant DNA

Figure 10

Recombinant DNA is prepared by splicing together DNA strands from two different organisms.

NEW DNA MOLECULES ARE FORMED BY GENETIC ENGINEERING

Recall that a **gene** is a section of a DNA molecule that contains all the information needed for the synthesis of a single polypeptide chain. The techniques of genetic engineering allow scientists to form new DNA molecules by combining pieces of DNA from markedly different organisms. The new DNA is known as **recombinant DNA.** The cell that contains this new DNA will use it to synthesize a new protein, one that the cell does not normally produce.

One of the early commercial successes of genetic engineering was the insertion of the gene for human insulin, a polypeptide, into the DNA of the bacterium *Escherichia coli* (*E. coli*) so that the bacterium synthesized human insulin in addition to its own proteins. Large amounts of insulin now can be produced at relatively low cost when the *E. coli* are grown on a large scale. The method is one of the most powerful tools for understanding how the information stored in chromosomal genes is used and how it is regulated. Here we briefly review some of the basic techniques, terminology, and potential applications of genetic engineering.

The first step in the preparation of recombinant DNA is to break both strands of the native DNA at specific sites, using enzymes known as **restriction endonucleases.** Next, the broken strands of two different DNA molecules are joined together by a different enzyme called **DNA ligase** (Fig. 10). The combined use of these enzymes allows any DNA fragment to be grafted into the DNA of a bacterium.

If the inserted DNA fragment represents a single gene, these recombination experiments can produce large amounts of the gene because it is copied and reproduced as the bacteria rapidly divide and multiply. This process is known as **DNA cloning.** Researchers recover the gene from the bacterial DNA using the same restriction endonuclease used to break open the original DNA. The amount of recombinant DNA that is now available is sufficient for detailed studies of the precise sequence of nucleotides.

The bacterial DNA used for DNA cloning is not the chromosomal DNA; rather, it is a small accessory DNA molecule known as a **plasmid.** Researchers use these molecules because they are easily separated from the chromosomal DNA and, once the gene of interest has been attached to them, they can be put back inside living bacterial cells (Fig. 11). Plasmids that are used in DNA cloning studies are often referred to as **cloning vectors.**

Once the nucleotide sequence of a specific gene has been identified, it is possible to check the accuracy of this analysis by chemically synthesizing the gene from its constituent nucleotides. The gene can be incorporated into a cloning vector and inserted into bacteria for cloning and to see whether the bacteria produce a functional protein from the synthetic gene.

The ability to use synthetic genes broadens the scope of these techniques enormously. For instance, after sequencing a gene that produces a functional protein, researchers can synthesize one that is slightly different, a **mutant** gene, or

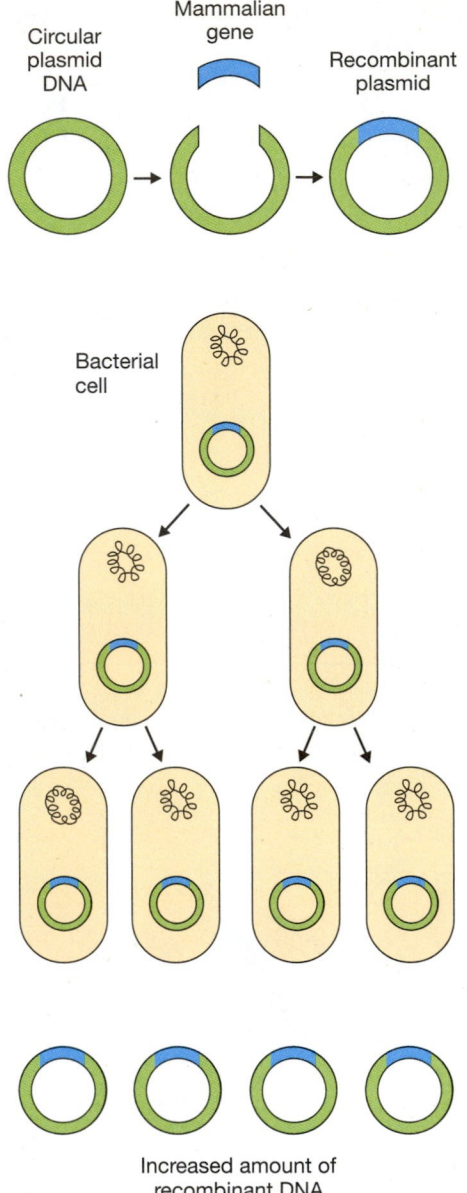

Figure 11

Large quantities of recombinant DNA can be produced by placing them in bacterial cells, which duplicates the recombinant molecule each time they divide.

one that makes a nonfunctional protein. If this mutant gene is inserted into a mammalian cell to replace a normal gene, the cell then synthesizes a nonfunctional protein instead of a functional one. Changes in the behavior of cells that contain the mutant gene provide information about the biological role of the normal protein within the cell.

At present it is extremely difficult to introduce genes into mammalian DNA, but ultimately it may be possible to genetically repair a mammalian cell that is nonfunctional because of the presence of a mutant gene. In this case, a chemically synthesized fragment of DNA that contains the normal gene would be inserted into the DNA of the cell.

Appendix B

Normal Physiological Values for Adults

Total Body Water (% body weight)

Female	50
Male	60

Core Temperature

	°C	°F
	36–38	96.8–100.4

Blood

Hematocrit (volume % RBC in blood)	37–52
Hemoglobin content (g/dl)	12–18
Erythrocytes (million/ul^3)	4.6–6.2
Leukocytes (thousand/ul^3)	7–10
Plasma glucose (mg/dl)	80–120
Plasma calcium (ionized—mEq/L)	2.0–2.5
Plasma iron (mg/dl)	50–150
Plasma sodium (mEq/L)	136–145
Plasma osmolarity (mOsm/kg H_2O)	290–300

Cardiovascular

Resting heart rate (beats/min)	60–100
Systolic pressure (mm Hg)	100–140
Diastolic pressure (mm Hg)	60–95
Mean systemic arterial pressure (mm Hg)	80–100
Mean pulmonary arterial pressure (mm Hg)	10–15
Cardiac output (L/min)	5–8

Renal

Glomerular filtration rate (ml/min)	115–130
Urine output (ml/day)	600–6000
Urine pH	4.5–8.0
Urine osmolarity (mOsm/kg H_2O)	500–800

Respiratory

Alveolar ventilation (L/min)	4.2
Respiratory rate (breaths/min)	12–15
Tidal volume (ml)	500
Total lung capacity (L)	6
Vital capacity (L)	4.8
Oxygen uptake (ml/min)	240
Carbon dioxide production (ml/min)	192
Respiratory exchange ratio (CO_2 output/O_2 uptake)	0.8
Alveolar gas tension	
O_2 (mm Hg)	104
CO_2 (mm Hg)	40
Arterial blood gases	
pH	7.35–7.45
O_2 saturation (% saturation of HB with O_2)	95–98
O_2 tension (mm Hg)	85–100
CO_2 tension (mm Hg)	35–45
Oxygen content (ml/dl)	20
Carbon dioxide content (ml/dl)	50
Bicarbonate (mEq/L)	23–28

Appendix C
Medical Terminology

Prefixes

Prefix	Meaning (and Example)
a-, ab-	away from (abnormal)
a-, an-	negative, not, without (anaerobic)
abdomin-	pertaining to abdomen (abdominal)
ad-	toward, adhere to (adduction)
aer-	pertaining to air (aerobic)
af-	movement toward a point (afferent nerve)
algi-	pain (algesic)
all-	other, different (allele)
amb-	both, on both sides (ambidextrous)
amph-	on both sides (amphiarthrosis)
ana-	up (anaphylaxis)
angio-	pertaining to blood or lymph vessels (angiogram)
ante-	before (antebrachium)
anti-	against (antibiotic)
apo-	from, opposed to (apocrine)
aqua-	water (aqueous)
arche(i)-	first (archetype)
arthr-	joint (arthritis)
aud-	pertaining to hearing (audiology)
auto-	pertaining to self (autoimmunity)
bi-	double (binocular)
bio-	life, living (biofeedback)
blast-	germinal, bud (blastocyst)
brachi-	arm (brachium)
brachy-	short (brachycephalic)
brady-	slow (bradycardia)
bucc-	cheek (buccinator muscle)
cac-	bad, evil (cachexia)
calc-	pertaining to calcium, lime, or stone (calcification)
carcin-	cancer (carcinogen)
cardi-	heart (cardiac)
cata-	down, lower (catatonia)
caud-	tail (caudal)
centri(o)-	center, middle (centrilobular)
cephal-	pertaining to head (cephalomeningitis)
cerebro-	brain (cerebrovascular accident)
chol-	bile (cholangiography)
chondr-	cartilage (chondrocranium)
chromo-	color (chromosome)
circum-	around (circumcision)
co-	together (coarctation)
coel-	hollow cavity (coelom)
com-	together (communicable)
con-	together, with (congested)
contra-	against (contraceptive)
corn-	hardened, horny (cornea)
corp-	body (corpus cavernosum)
cryo-	pertaining to cold (cryopreservation)
crypt-	hidden (cryptocephalus)
cyan-	blue in color (cyanotic)
cyst-	bladder, baglike (cystocarcinoma)
cyto-	cell (cytoplasm)

Prefix	Meaning (and Example)
dacry-	tears, lacrimal (dacryostenosis)
dactyl-	pertaining to fingers (dactylospasm)
de-	from, not, down (decongestant)
deci-	tenth (decibel)
dent-	pertaining to teeth (dentition)
derm-	skin (dermatitis)
di-	two, double (digastric)
dia-	through, between, separate (diastema)
dipl-	double (diplococcus)
dis-	apart, away from, negative (dislocation)
duct-	to lead or conduct (ductus arteriosus)
dys-	difficult, bad, painful (dysmenorrhea)
ecto-	external, outside (ectoderm)
ef-	out (effusion)
endo-	internal, inside (endoderm)
entero-	intestine (enteroscope)
ento-	within (entocyte)
epi-	upon, over, in addition to (epicardium)
erythro-	red (erythrocyte)
eu-	well, healthy, normal (euphoria)
ex-	from, out of (excrement)
exo-	outside (exoskeleton)
extra-	beyond, outside of (extraembryonic)
fasci-	band (fasciculus)
febr-	pertaining to fever (febricide)
feto-	pertaining to a fetus (fetoscope)
fibr-	fibrous (fibroma)
for-	opening (foramen rotundum)
fore-	in front of (forebrain)
galact-	pertaining to milk (galactosemia)
gastr-	pertaining to the stomach (gastroenteritis)
gloss-	tongue (glossectomy)
gluco-	pertaining to sugar (glucogenesis)
glyco-	pertaining to sugar (glycosemia)
gonado-	pertaining to the gonads (gonadotropin)
gran-	particle, grain (granulocyte)
gravi-	heavy (gravimetric)
gyn-	female sex (gynecologist)
haplo-	simple, single (haploid)
hema(o)-	blood (hemapoiesis)
hemato-	blood (hematologist)
hemi-	half (hemisphere)
hepat-	liver (hepatitis)
hetero-	different, other (heterosexual)
histo-	tissue (histocompatibility)
holo-	whole (holocrine)
homo-	same (homosexual)
hydro-	water (hydrocephalic)
hyper-	excessive, above (hypertension)
hypo-	under, below (hypoglycemia)

Prefix	Meaning (and Example)
idio-	individual, distinct (idiopathic)
ilei(o)-	pertaining to the ileum (ileitis)
ilio-	pertaining to the ilium (iliosacral)
in-	in, within, negative (incontinence)
infra-	beneath (infraorbital)
inter-	between, among (intercostal)
intra-	within, inside (intramuscular)
irid-	pertaining to the iris (iridectomy)
iso-	like, equal (isometric)
karyo-	nucleus (karyotype)
kerato-	pertaining to tough substances, cornea (keratosis)
keto-	pertaining to ketone bodies (ketosis)
kine-	movement (kinesiology)
kypho-	humped (kyphosis)
labi-	lip (labia majora)
lacri-	tears (lacrimal duct)
lacto-	pertaining to milk (lactogenic)
laparo-	pertaining to loin or abdomen (laparoscopy)
laryngi(o)-	pertaining to larynx (laryngoscope)
later-	side (lateral)
leuk-	white (leukocyte)
lipo-	pertaining to fat (lipocyte)
lith-	stone (lithodialysis)
lymph-	pertaining to lymphatic system (lymphoma)
macro-	large (macrophage)
mal-	abnormal, bad, ill (malnutrition)
mamm-	pertaining to breast (mammography)
mast-	pertaining to breast (mastectomy)
medi-	middle (mediastinum)
mega-	large (megakaryocyte)
mela-	black (melanoma)
meningo-	meninges (meningopathy)
meso-	middle (mesoderm)
meta-	beyond, after (metacarpal)
metra(o)-	pertaining to uterus (metrocystosis)
micro-	small (microcephaly)
mio-	less, smaller (mioplasmia)
mito-	thread (mitosome)
mono-	one, alone (mononucleosis)
morph-	shape (morphogenesis)
muco-	pertaining to mucous (mucoenteritis)
multi-	many (multicellular)
myelo-	pertaining to bone marrow (myeloma)
myo-	muscle (myocardium)
narc-	numbness, stupor (narcolepsy)
naso-	nose (nasoscope)
necro-	dead, corpse (necrosis)
neo-	young, new (neonatal)
nephro-	kidney (nephropathy)
neuro-	nerve (neurosurgery)
nucleo-	nucleus (nucleotide)
ob-	against, in front of (obstruction)
oc-	against (occlusion)
oculo-	eye (oculomotor)
odont-	tooth (odontoblast)
oligo-	small, few (oligospermia)
omo-	shoulder (omoclavicular)
oo-	egg (oocyte)
oophoro-	ovary (oophorotomy)
opisth-	backward, behind (opisthotic)

Prefix	Meaning (and Example)
orchi-	testis (orchidectomy)
ortho-	straight, normal (orthopedics)
osteo-	bone (osteocyte)
oto-	ear (otopathy)
ovi(o)-	egg (oviduct)
pachy-	thick, heavy (pachyderma)
pan-	all (panarthritis)
para-	beside, near (paranasal)
path-	disease (pathology)
pedia-	children (pediatrics)
per-	through (perfusion)
peri-	surrounding, near (periosteum)
phag-	pertaining to eating (phagocyte)
pharmaco-	drugs, medicine (pharmacotherapy)
phlebi(o)-	vein (phlebitis)
photo-	pertaining to light (photoreceptor)
platy-	flat, broad (platycephaly)
pleur-	pertaining to pleura (pleurisy)
pneumo-	air, lung (pneumonia)
pod-	foot (podiatry)
poly-	much, many (polydactyly)
post-	after, behind (postmortem)
pre-	before (prenatal)
pro-	before (prochondral)
proct-	anus (proctology)
proto-	first (protoplasm)
pseudo-	false (pseudogout)
psycho-	mental (psychogenesis)
pyo-	pus (pyodermatitis)
quad-	pertaining to four (quadriplegia)
re-	again, back (reabsorb)
reno-	pertaining to the kidney (renography)
rete(i)-	network (reticulocyte)
retro-	behind, backward (retroperitoneal)
rhino-	nose (rhinopathy)
saccharo-	sugar (saccharuria)
sarco-	flesh (sarcolemma)
sclero-	hard (scleroderma)
semi-	half (semilunar)
steno-	narrow (stenosis)
sub-	under, below, beneath (subcutaneous)
super-	beyond, above, upper (supervirulent)
supra-	over, above (supramandibular)
sym-	joined together (symphysis)
syn-	joined together (synarthrosis)
tachy-	fast (tachycardia)
tele-	far (telencephalon)
tetra-	four (tetraploid)
therm-	pertaining to heat (thermogenesis)
thorac-	chest (thoracic cavity)
thrombo-	clot, lump (thrombosis)
tox-	poison (toxicology)
trans-	over, across (transfuse)
tri-	three (triceps brachii)
tropho-	pertaining to nourishment (trophoblast)
ultra-	excess, beyond (ultrasonogram)
uni-	one (unicellular)
uro-	urinary organs, urine (urogenital)
vas-	vessel (vasodilation)

Prefix	Meaning (and Example)
viscer-	internal organs (visceroreceptor)
vit-	life (vitality)
xanth-	pertaining to yellow (xanthoderma)
zoo-	animal (zoology)
zygo-	union, join (zygote)

Suffixes

Suffix	Meaning (and Example)
-able	capable of (digestable)
-a(e)sthesia	sensation (anesthesia)
-algia	pain (gastralgia)
-ase	enzyme (lactase)
-asis	condition (homeostasis)
-cele	tumor, cyst (hydrocele)
-cide	destroy (fungicide)
-coele	enlarge cavity, swelling (blastocoele)
-cyte	cell (erythrocyte)
-dynia	pain (ophthalmodynia)
-ectomy	surgical removal (appendectomy)
-emesis	vomiting (hyperemesis)
-emia	pertaining to the blood (glycemia)
-ferent	moving, carrying (afferent)
-form	shape (penniform)
-fuge	drive away (centrifuge)
-gen	causative agent (pathogen)
-genic	causing (carcinogenic)
-gram	recording, record (electrocardiogram)
-ia	condition (uremia)
-iatrics	medical specialties (geriatrics)
-ism	condition (rheumatism)
-itis	inflammation (arthritis)
-kinesis	movement (orthokinesis)
-lith	stone (otolith)
-logy	study of (osteology)
-lysis	dissolve, break apart (hemolysis)
-megaly	enlarged (acromegaly)
-oid	resembling (nephroid)
-oma	tumor (melanoma)
-ory	pertaining to (sensory)
-ose	full of (adipose)
-osis	condition (scoliosis)
-pathy	disease, abnormality (colonopathy)
-penia	deficiency (leukopenia)
-phil	attraction to (neutrophil)
-phobia	abnormal fear (hematophobia)
-phylaxis	protection (prophylaxis)
-plasty	to reconstruct (rhinoplasty)
-plegia	stroke, paralysis (paraplegia)
-pnea	pertaining to breathing (apnea)
-poiesis	formation of (hematopoiesis)
-rrhage	overflow (hemorrhage)
-rrhea	discharge, flow (menorrhea)

Suffix	Meaning (and Example)
-sclerosis	hardness, dryness (arteriosclerosis)
-sis	action, process (dialysis)
-tomy	cut, incision (tracheostomy)
-trophy	pertaining to nutrition (hypertrophy)
-tropic	changing, influencing (gonadotropic)
-uria	pertaining to urine (glycosuria)

Latin and Greek Terms for Body Parts

ankle: (L): tars-, tarsi-, -tarsus
anus: (L): ano-, -anus; (G): procto-, -proctus
arm: (G): brachi-, -brachium
back: (L): dors-, dorsi-, -dorsum; (G): noto-, -notus; (L): terg-, -tergum
bag: See bladder.
beak: (G): rhyncho-, -rhychus; (L): rostr-, -rostrum
belly: (L): -venter, ventr-, ventro-
bladder: (G): asco-, -ascus; (G): cysto-, -cystis; (G): -physa, physo-; (L): -vesica, vesico-
blood: (G): -haema, haemato-, haemo-; (L): sanguini-, -sanguis
body: (L): corporo-, -corpus; (G): -soma, somato-
bone: (L): -os, ossi-; (G): osteo-, -osteum
border: (G): chilo-, -chilus; (G): craspedo-, -craspedum
brain: (L): cereb-, cerebr-, -cerebrum; (G): encephalo-, -encephalus
breast: (L): pector-, -pectus; (L): stern-, -sternum; see also chest.
bristle: (G): -chaeta, chaeto-; (L): -seta, seti-
cartilage: (G): chondro-, -chondrus
cell: (L): -cella, celli-; (G): cyto-, -cytus
cheek: (L): bucc-, -bucca; (L): gen-, -gena, geno-
chest: (G): stetho-, -stethus; see also breast and thorax.
claw: (G): chel-, -chela, chelo-; (G): onycho-, -onyx; (L): ungui-, -unguis
crest: (L): crist-, -crista; (G): lopho-, -lophus
crown: (L): coron-, -corona; (G): stephano-, -stephanus
digit: (G): dactylo-, -dactylus; (L): digiti-, -digitus
ear: (L): auri-, -auris; (G): otido-, oto-, -ous
egg: (G): oo-, -oum; (L): ovi-, -ovum
eye: (L): oculi-, -oculus; (G): -omma, ommato-; (G): ophthalmo-, -ophthalmus; (G): opo-, -ops, opto-
eyelash: (G): -blepharis, blepharo-
eyelid: (L): cili-, -cilium
face: (L): faci-, -facies; (G): -ops
feather: (L): -pinna, pinni-; (L): plum-, -pluma, plumi-; (G): -pteryla, pterylo-; (G): ptilo-, -ptilum
finger: See digit.
flesh: (L): carni-, -caro; (G): sacro-, -sarx
foot: (L): pedi-, -pes; (G): podo-, -pus
forehead: -frons, front-; (G): metopo-, -metopus, -metopius
gill: (G): branch-, -branchium, brancho-
gland: (G): -aden, adeno-; (L): glandi-, -glans
groin: -inguen, inguini-
hair: (L): capill-, -capillus; (L): crini-, -crinis; (L): pil-, pili-, -pilus; (G): -thrix, tricho-
hand: (G): -chir, chiro-; (L): mani, manu-, -manus
head: (L): capit-, capiti-, -caput; (G): -cephala, cephalo-
heart: (G): cardi-, -cardia; (L): -cor, cordi-
heel: (L): calcan-, calcane-, -calcaneum; (L): talari-, tali-, -talus
horn: (G): -cera, cerato-; (L): corn-, -cornus
jaw: (G): genyo-, -genys; (G): gnatho-, -gnathus; (L): maxill-, -maxilla
joint: (G): arthro-, -arthrum; (L): articuli-, articulus, -artus
kidney: (G): nephro-, -nephrus; (L): ren-, -ren, reni-
knee: (L): genu-, -genu; (G): gony-, gonyo-, -gonys; (G): -gonatium, gonato-
knuckle: (G): condylo-, -condylus
leg: (G): cnemi-, -cnemis; (L): crur-, -crus; (G): -scelis, scelo-, scelido-
lip: (G): chilo-, -chilus; (L): labi-, labio-, -labium, labr-, -labrum

liver: (G): -hepar, hepato-; (L): jecori-, -jecur

lung: (G): -pneuma, pneumo-; (L): pulmo-, -pulmo, pulmono-

membrane: (G): chorio, -chorium; (G): -hymen, hymeno-; (L): membran-, -membrana; (G): meningo-, -meninx

mouth: (L): ora-, ori-, -os; (G): -stoma, stomato-

mucus: (G): blenno-, -blennus

muscle: (G): myo-, -mys; see also flesh.

neck: (G): -auchen, aucheno-; (L): cervic-, -cervix; (L): coll-, -collum; (G): -dera, dero-; (G): trachelo-, -trachelus

nose: (L): nasi-, -nasus; (G): rhino-, -rhis

rib: (L): cost-, -costa, costi-; (G): scelido-, -scelis

rump: (G): gluteo-, -gluteus; (G): -pyga, pygo-

scale: (G): lepido-, -lepis; (L): squam-, -squama, squami-

shell: (G): -concha, concho-; (G): ostraco-, -ostracum

shoulder: (G): omo-, -omus

skin: (L): -byrsa, byrso-; (G): chorio-, -chorium; (L): cutan-, cuti-, -cutis; (G): derm-, -derma, dermo-, dermato-; (G): scyto-, -scytus

skull: (G): cranio-, -cranium

snout: See beak.

sperm: (L): -semen, semin-; (G): -sperma, spermato-

spine: (G): -acantha, acantho-; (G): rhachi-, -rhachis; (L): -spina, spini-

stomach: (G): -gaster, gastro-; (L): -venter, ventr-

suture: (G): -rhapha, rhapho-

tail: (L): caud-, -cauda; (G): cerco-, -cercus; (G): -ura, uro-

thigh: (L): femor-, -femur; (G): mero-, -merus

thorax: (G): thoraco-, -thorax

throat: (L): gula-, -gula; (L): guttur-, -guttur; (G): laemo-, -laemus; (G): pharyngo-, -pharynx; (G): trachelo-, -trachelus

tissue: (G): histo-, -histus; (L): tel-, -tela, teli-

toe: See digit.

tongue: (G): -glossa, glosso-, -glotta, glotto-; (L): -lingu, -lingua

tooth: (L): -dens, dent-, denti-; (G): odonto-, -odous, -odus

vein: (G): phlebo-, -phleps; (L): ven-, -vena, veni-

windpipe: (G): broncho-, -bronchus; (G): trache-, -trachea

wing: (L): ala-, -ala, ali-; (G): ptero-, -pterum; (G): pterygo-, -pteryxs

wrist: (L): carpo-, -carpus

Index

Clinical Abbreviations

a.c.	before eating
abs. feb.	without fever
ACTH	adrenocorticotrophic hormone
ad effect.	until effective
ad lib.	as desired
ad us.	as customary
ad us. ext.	for external use
ADH	antidiuretic hormone
ADP	adenosine diphosphate
adst. feb.	when fever is present
ag. feb.	when fever increases
AIDS	acquired immunodeficiency syndrome
alt. dieb.	every other day
alt. hor.	every other hour
ANS	autonomic nervous system
AQ	aqueous
ARC	AIDS-related complex
ARDS	adult respiratory distress syndrome
ARF	acute respiratory failure; acute renal failure
AS	aortic stenosis
atm	atmosphere
ATP	adenosine triphosphate
AV	atrioventricular
BBB	bundle branch block; blood-brain barrier
bid	twice daily
bis in 7d	twice a week
BMR	basal metabolic rate
BP	blood pressure
BSA	body surface area
BUN	blood urea nitrogen
C	Celsius; complement; gallon
$\bar{c}$	with
C_{cr}	renal creatinine clearance
C1, etc.	first cervical vertebra, etc.
ca	about
CAD	coronary artery disease
cal	calorie
Cal	kilocalorie
cAMP	cyclic adenosine monophosphate
CAPD	continuous ambulatory peritoneal dialysis
CAT	computerized axial tomography
CBC	complete blood count
cc	cubic centimeter (cm^3)
CCK	cholecystokinin
CCK-PZ	cholecystokinin-pancreozymin
CDC	Centers for Disease Control and Prevention
CF	complement fixation; cystic fibrosis
CHD	congenital heart disease
CHF	congestive heart failure
cm	centimeter
CNS	central nervous system
CO	cardiac output
CoA	coenzyme A
COAD	chronic obstructive airway disease
comp	compound
contra	against
COPD	chronic obstructive pulmonary disease
CPAP	continuous positive airway pressure
CPR	cardiopulmonary resuscitation
Cr	creatinine
CRF	chronic renal failure
CRH	corticotrophin-releasing hormone
CSF	cerebrospinal fluid
CT	computed tomography
CV	cardiovascular
CVA	cerebrovascular accident
CVP	central venous pressure
CVS	chorionic villus sampling
d	dexto (right); day
D & C	dilation and curettage
da	day
DBP	diastolic blood pressure
def	defecation
DH	delayed hypersensitivity
DHAP	dihydroxyacetone
dl	deciliter
DNA	deoxyribonucleic acid
dr	dram
DTP	diphtheria-tetanus-pertussis (vaccine)
ECF	extracellular fluid
ECG	electrocardiogram
ECT	electroconvulsive therapy
EDV	end-diastolic volume
EEG	electroencephalogram
EM	electron micrograph
EMG	electromyogram
ENS	enteric nervous system
ENT	ear, nose, and throat
EPSP	excitatory postsynaptic potential
ER	endoplasmic reticulum
ERV	expiratory reserve volume
ESR	erythrocyte sedimentation rate
ESRD	end-stage renal disease
ESV	end-systolic volume
F	Fahrenheit
F1, F2	first, second filial generation
FDA	US Food and Drug Administration
FEV	forced expiratory volume
fl oz	fluid ounce
FSH	follicle-stimulating hormone
FUO	fever, unknown origin
G6PD	glucose 6-phosphate dehydrogenase
GABA	gamma-aminobutyric acid
GFR	glomerular filtration rate
GI	gastrointestinal
GnRH	gonadotrophin-releasing hormone
gr	grain
GTP	guanosine triphosphate
gtt	drops
GU	genitourinary
Hb	hemoglobin
Hct	hematocrit
HD	Huntington's disease
HDL	high-density lipoprotein
Hg	mercury
hgb	hemoglobin
HGH	human growth hormone
HIV	human immunodeficiency virus
HLA	human leukocyte antigen
HTLV-III	human T-lymphotrophic virus type III
hypo	hypodermic
Hz	hertz (cycles per second)
IBS	irritable bowel syndrome
ICF	intracellular fluid
IDL	intermediate density lipoprotein
Ig	immunoglobulin
IM	intramuscular
inf	infusion
inhal	inhalation
inj	injection
IPPB	inspiratory positive pressure breathing
IPSP	inhibitory postsynaptic potential
IRV	inspiratory reserve volume
IU	international unit
IUD	intrauterine device
IV	intravenous